handbook of HUMAN TOXICOLOGY

edited by

Edward J. Massaro, Ph.D.
National Health and Environmental Effects
Research Laboratory
Research Triangle Park, North Carolina

CRC Press
Boca Raton New York

Acquiring Editor:	Bob Stern
Project Editor:	Sarah Fortener
Assistant Managing Editor	Gerry Jaffe
Marketing Manager:	Susie Carlisle
Direct Marketing Manager:	Becky McEldowney
Cover design:	Dawn Boyd
PrePress:	Carlos Esser

Library of Congress Cataloging-in-Publication Data

Handbook of human toxicology / edited by Edward J. Massaro.
p. cm.
Includes bibliographical references and index.
ISBN 0-8493-4493-X
1. Toxicology. I. Massaro, Edward J.
[DNLM: 1. Poisoning. 2. Poisons. QV 600 H2345 1997]
RA1211.H286 1997
615.9--dc21
DNLM/DLC
for Library of Congress

97-721
CIP

International Standard Book Number 0-8493-4493-X
Library of Congress Card Number 97-721
Printed in the United States of America 1 2 3 4 5 6 7 8 9 0
Printed on acid-free paper

Preface

Toxicology is a vast and growing polymorphic field, and the sheer quantity and variety of toxicological and toxicology-related information that has been and currently is being generated preclude comprehensive coverage of the field in any single volume. Our goal, therefore, was to develop a compendium of toxicologic information on selected topics that, if successful, would be supplemented with additional volumes to increase the scope of coverage of the field. Even within the areas selected for inclusion within this volume, it was possible to cover only a limited number of topics. Furthermore, aspects of the individual topics included within each section of this volume could not be covered completely, and some were not covered at all.

Modern toxicology is an experimental science — it has to be. As in any scientific field in which nonphenomenological information is of primary importance, understanding toxicologic effects requires carefully controlled experimental investigation. Furthermore, meaningful analysis of the mechanisms through which toxicants operate requires accurate data from experimental manipulation of model systems. Because one biologic system may not respond to toxic insult precisely like another, comparative data are a necessity. If the differences, as well as the similarities, among diverse systems can be understood and appreciated, it may be possible to reliably extrapolate data obtained from one system to another. Although the search for methods that would allow reliable extrapolation of *in vivo* animal data and/or *in vitro* data to the human condition continues, no methods, except those relating the most fundamental toxicologic effects, are yet of acceptable reliability. It may not be possible to develop useful extrapolation methodology from the state of current knowledge, vast though it may be. Major impediments include identifying and locating important information. Just what is out there? Useful handbooks can aid in alleviating such problems. Our ultimate goal is the development of useful handbooks. Your input is solicited and will be appreciated.

The Editor

Edward J. Massaro, Ph. D., is a cell physiologist/biochemist/experimental pathologist and an established international authority on mechanisms of cytotoxicology of drugs, toxicants, and teratogens. He is Senior Research Scientist, Developmental Biology Branch, National Health and Environmental Effects Research Laboratory (NHEERL), Research Triangle Park, NC; Research Professor of Biophysical Sciences, State University of New York at Buffalo (SUNYAB) School of Medicine and Biomedical Sciences; and Adjunct Professor of Toxicology at Duke University, Durham, NC, and The University of North Carolina-Chapel Hill. Dr. Massaro has served as the Director of the Inhalation Toxicology Division of NHEERL; Director of the Center for Air Environment Studies; Professor of Cytotoxicoloy, Department of Veterinary Science, and Senior Research Associate, Institute for the Study of Human Development, at The Pennsylvania State University. Dr. Massaro also has served as the Director of Toxicology and Chemical Carcinogenesis at the Mason Research Institute, Worcester, MA, and as Professor of Biochemistry at SUNYAB.

Dr. Massaro was graduated from Rutgers University in 1955 with an A.B. degree in biology and chemistry. He obtained his M.A. degree in biochemistry and cell physiology from the University of Texas-Austin in 1958 and his Ph.D. degree in anatomy, biochemistry, and physiology from the University of Texas-Austin and the University of Texas Medical Branch-Galveston in 1962. Dr. Massaro held both USPHS pre- and postdoctoral fellowships.

Dr. Massaro is a member of the American Association for the Advancement of Science, American Association for Cancer Research, American Chemical Society, American Society for Biological and Molecular Chemistry, American Society for Cell Biology, American Society for Investigative Pathology, American Society for Pharmacology and Experimental Therapeutics, Biophysical Society, International Association of Bioinorganic Scientists, International Society for the Study of Xenobiotics, International Society for Trace Element Research in Humans, International Union of Pharmacology — Section on Toxicology, Metals Specialty Section of the Society of Toxicology, New York Academy of Sciences, Sigma Xi, Society of Toxicology, Teratology Society, and The Johns Hopkins Medical and Surgical Association.

Dr. Massaro's research accomplishments have been recognized both nationally and internationally. He was elected a Fellow of the American Association for the Advancement of Science in 1986, was the recipient of the Achievement Award of the international journal, *Neurotoxicology,* in 1989, and received the Scientific and Technological Achievement Award of the U.S. Environmental Protection Agency in 1992.

Currently, Dr. Massaro is editor of the journal, *Cell Biochemistry and Biophysics,* and the multivolume series, *Methods in Toxicology.* He served as editor (with J.A. Crapo and D.E. Gardner) of *Toxicology of the Lung,* is a member of the editorial board of the *Modern Nutrition* series, and is a member of the editorial advisory board of the series, *Current Topics in Toxicology and Biological Effects of Heavy Metals.* Dr. Massaro has been a member of the editorial board of the journal, *Biological Trace Element Research,* for many years and served as associate editor of the journal, *Neurotoxicology,* and on the editorial boards of the journals *Neurotoxicology, Environmental Research, Environmental Health Perspectives, Toxicology, Toxicology and Environmental Health, Toxicology and Applied Pharmacology, Toxicology and Industrial Health,*

Drug and Chemical Toxicology, *Environmental Pathology and Toxicology*, *Environmental Science and Health*, *Gerontology*, and *Copeia*. Dr. Massaro has served as a reviewer for *Chemical Research in Toxicology* since its inception and *Teratology*, *Toxicology*, *Applied Pharmacology*, and the *American Journal of Physiology* for many years.

Dr. Massaro has published extensively in the field of metals toxicology. In addition, he has presented numerous papers at national and international meetings and has been actively involved in national, state, and local health (human and veterinary), educational, environmental, and scientific affairs.

Contributors

Yasunobu Aoki, Ph.D.
Research Scientist
Environmental Chemistry Division
National Institute for Environmental Studies
Ibaraki, Japan

Cheng-Long Bai, Ph.D.
Senior Toxicologist
Therapeutic Goods Administration
Commonwealth Department of Health and Family Services
Woden, ACT
Australia

John B. Barnett, Ph.D.
Professor and Chair
Microbiology and Immunology Department
West Virginia University
Morgantown, West Virginia

Leila M. Barraj, D.Sc.
Senior Statistician
TAS, Inc.
Washington, D.C.

David A. Beckman, Ph.D.
Associate Professor
Department of Pediatrics
Thomas Jefferson University and duPont Hospital for Children
Wilmington, Delaware

Abdellaziz Ben-Jebria, Ph.D.
Associate Professor
Department of Chemical Engineering
Penn State University
University Park, Pennsylvania

Robert L. Brent, M.D., Ph.D., D.Sc. (Hon)
Distinguished Professor
Department of Pediatrics
Thomas Jefferson University and duPont Hospital for Children
Wilmington, Delaware

Edward J. Carson
Southern College of Optometry
Memphis, Tennessee

Mitchell D. Cohen, Ph.D.
Research Assistant Professor
Nelson Institute of Environmental Medicine
New York University School of Medicine
Tuxedo, New York

Christine F. Colie, M.D.
Assistant Professor
Department of Obstetrics and Gynecology
Georgetown University Medical Center
Washington, D. C.

Stephanie D. Collier, M.S.
Doctoral Student
Department of Biological Sciences
Mississippi State University
Mississippi State, Mississippi

Alan R. Dahl, Ph.D.
Scientist
ITRI
Lovelace Research Institute
Albuquerque, New Mexico

Robert Devlin, Ph.D.
Chief, Clinical Research Branch
Human Studies Division
National Health and Environmental Effects Research Laboratory
U.S. Environmental Protection Agency
Research Triangle Park, North Carolina

Judith S. Douglass, M.S., R.D.
Senior Nutritionist
TAS, Inc.
Washington, D.C.

M. Duane Enger, Ph.D.
Professor and Chair
Department of Zoology and Genetics
Iowa State University
Ames, Iowa

Lynda B. Fawcett, Ph.D.
Instructor
Department of Pediatrics
Thomas Jefferson University and duPont Hospital for Children
Wilmington, Delaware

Kathryn H. Fleming, Ph.D.
Senior Nutritionist
TAS, Inc.
Washington, D.C.

Lawrence J. Folinsbee, Ph.D.
Chief, Environmental Media Assessment Group
National Center for Environmental Assessment
U.S. Environmental Protection Agency
Research Triangle Park, North Carolina

Susan Groziak, Ph.D., R.D.
Manager, Nutrition Research and Scientific Publications
Department of Nutrition Research
National Dairy Council
Rosemont, Illinois

Helen G. Haggerty, Ph.D.
Senior Research Investigator
Department of Biologics Evaluation
Bristol-Myers Squibb
Syracuse, New York

Jack R. Harkema, Ph.D., D.V.M.
Department of Pathology
Michigan State University
East Lansing, Michigan

James T. Heimbach, Ph.D.
Chief Operating Officer
TAS, Inc.
Washington, D.C.

Kimiko Hirayama, Ph.D
Professor
College of Medical Science
Kumamoto University
Kumamoto, Japan

Robert V. House, Ph.D.
Senior Immunologist
Life Sciences Department
IIT Research Institute
Chicago, Illinois

Claude L. Hughes, Jr., M.D., Ph.D.
Associate Professor of Comparative Medicine and Obstetrics and Gynecology
Comparative Medicine Clinical Research Center
Bowman Gray School of Medicine
Wake Forest University
Winston-Salem, North Carolina

Dallas M. Hyde, Ph.D.
Department of Anatomy, Physiology, and Cell Biology
School of Veterinary Medicine
University of California
Davis, California

Hiroshi Imai, Ph.D.
Laboratory Chief
Laboratory of Reproductive Biotechnology
National Institute of Animal Industry
Ibaraki, Japan

Y. James Kang, Ph.D., D.V.M.
Associate Professor
Departments of Medicine, and Pharmacology and Toxicology
University of Louisville
Louisville, Kentucky

Howard R. Kehrl, M.D.
Medical Officer
Human Studies Division
National Health and Environmental Effects Research Laboratory
U.S. Environmental Protection Agency
Research Triangle Park, North Carolina

Deborah Keil, Ph.D.
Assistant Professor
Department of Medical Laboratory Sciences
Medical University of South Carolina
Charleston, South Carolina

Chong S. Kim, Ph.D.
Senior Research Scientist
Human Studies Division
National Health and Environmental Effects Research Laboratory
U.S. Environmental Protection Agency
Research Triangle Park, North Carolina

Jane Q. Koenig, Ph.D.
Professor
Department of Environmental Health
University of Washington
Seattle, Washington

Shinji Koizumi, Ph.D.
Senior Researcher
Department of Experimental Toxicology
National Institute of Industrial Health
Kawasaki, Japan

Rebecca Liebes, Ph.D.
Graduate Fellow
Department of Human Nutrition and Food Management
Ohio State University
Columbus, OH

Daniel L. Luchtel, Ph.D.
Professor
Department of Environmental Health
University of Washington
Seattle, Washington

Andrew T. Mariassy, Ph.D.
Department of Anatomy
Nova Southeastern University
Coral Gables, Florida

Edward J. Massaro, Ph.D.
Senior Research Scientist
Developmental Biology Branch
National Health and Environmental Effects Research Laboratory
Research Triangle Park, North Carolina

Richard Mattes, Ph.D.
Member
Monell Chemical Senses Center
Philadelphia, Pennsylvania

John T. McBride, M.D.
Department of Pediatrics
University of Rochester Medical Center
Rochester, New York

Denis M. Medeiros, Ph.D.
Professor of Human Nutrition, and Associate Dean for Research
Department of Human Nutrition and Food Management
The Ohio State University
Columbus, Ohio

Robert R. Mercer, Ph.D.
Department of Medicine
Duke University Medical Center
Durham, North Carolina

Gregory D. Miller, Ph.D., F.A.C.N.
Vice President, Nutrition Research and Technology Transfer
National Dairy Council
Rosemont, Illinois

Daniel L. Morgan, Ph.D.
Head, Respiratory Toxicology
Laboratory of Toxicology
National Institute of Environmental Health Sciences
Research Triangle Park, North Carolina

Kevin T. Morgan, B.V.Sc., Ph.D.
Department of Experimental Pathology and Toxicology
Chemical Industry Institute of Toxicology
Research Triangle Park, North Carolina

Iyabo O. Obasanjo, D.V.M., Ph.D.
Fellow
Comparative Medicine Clinical Research Center
Bowman Gray School of Medicine
Wake Forest University
Winston-Salem, North Carolina

Kent E. Pinkerton, Ph.D.
Department of Anatomy, Physiology, and Cell Biology
School of Veterinary Medicine
University of California
Davis, California

Charles G. Plopper, Ph.D.
Department of Anatomy, Physiology, and Cell Biology
School of Veterinary Medicine
University of California
Davis, California

James D. Prah, Ph.D.
Research Psychologist
Human Studies Division
National Center for Environmental Assessment
U.S. Environmental Protection Agency
Research Triangle Park, North Carolina

Stephen B. Pruett, Ph.D.
Professor
Department of Biological Sciences
Mississippi State University
Mississippi State, Mississippi

Daniel J. Raiten, Ph.D.
Senior Staff Scientist
Life Sciences Research Office
Federation of American Sciences for Experimental Biology
Bethesda, Maryland

Krystyna M. Rankin, Ph.D.
Postdoctoral Fellow
Monell Chemical Senses Center
Philadelphia, Pennsylvania

Kathleen E. Rodgers, Ph.D.
Associate Professor
Livingston Research Center
School of Medicine
University of Southern California
Los Angeles, California

James L. Schardein, M.S., A.T.S.
Senior Vice President, Director of Research
Will Research Laboratories
Ashland, Ohio

Richard B. Schlesinger, Ph.D.
Professor
Department of Environmental Medicine
New York University Medical Center
Tuxedo, New York

Steven Schrader, Ph.D.
Chief, Functional Toxicology Section
Robert A. Taft Laboratory
Centers for Disease Control
National Institute for Occupational Safety and Health
Cincinnati, Ohio

Anthony R. Scialli, M.D.
Associate Professor
Department of Obstetrics and Gynecology
Georgetown University Medical Center
Washington, D.C.

Joseph Scimeca, Ph.D.
Associate Technology Principal
Nutrition Department
Kraft Foods Technology Center
Glenview, Illinois

Elizabeth E. Sikorski, Ph.D.
Scientist
Human Safety Department
Miami Valley Laboratories
The Procter & Gamble Company
Cincinnati, Ohio

Morris B. Snipes, Ph.D.
Scientist
Inhalation Toxicology Research Institute
Albuquerque, New Mexico

Neill H. Stacey, Ph.D.
Head, Research Science and Statistics
Worksafe Australia
Sydney, New South Wales
Australia

Judith A. St. George, Ph.D.
Genzyme Corporation
Farmingham, Massachusetts

James Ultman, Ph.D.
Professor
Department of Engineering
Penn State University
University Park, Pennsylvania

Paul A. Weiss, M.S.
Research Assistant
College of Pharmacy
Washington State University
Pullman, Washington

Robert E. C. Wildman, Ph.D.
Assistant Professor of Human Nutrition
Department of Nutrition and Dietetics
The University of Delaware
Wilmington, Delaware

Chris Winder, Ph.D.
Associate Professor
Department of Safety Science
University of New South Wales
Sydney, New South Wales
Australia

Wen-Jun Wu, D.V.M.
Pre-doctoral Fellow
Department of Biological Sciences
Mississippi State University
Mississippi State, Mississippi

Akira Yasutake, Ph.D.
Chief, Biochemistry Section
National Institute for Minamata Disease
Kumamoto, Japan

Jun Yoshinaga, Ph.D.
Research Scientist
Environmental Chemistry Division
National Institute for Environmental Studies
Ibaraki, Japan

Judith T. Zelikoff, Ph.D.
Associate Professor
Nelson Institute of Environmental Medicine
New York University School of Medicine
Tuxedo, New York

Table of Contents

Part 5: Reproductive and Developmental Toxicology

James L. Schardein, Editor

PART 1

Metal Toxicology

Y. James Kang, Editor

CHAPTER 1

METHODS OF METALS TOXICOLOGY

Jun Yoshinaga, Yasunobu Aoki, Akira Yasuake, Kimiko Hirayama, Hiroshi Imai, and Shinji Koizumi

CONTENTS

0-8493-4493-X/97/$0.00+$.50

1 ANALYSIS OF METALS IN HUMAN TISSUES

1.1 Atomic Absorption Spectrometry

Jun Yoshinaga

1.1.1 *Electrothermal Atomic Absorption Spectrometry*

Atomic absorption spectrometry (AAS) is an established method for analyzing trace elements, including heavy metals, in a variety of matrices. Although the sensitivity of flame AAS is not adequate for many biomedical applications, electrothermal AAS (ETAAS) provides much higher sensitivity and low cost, simple operation. Table 1 shows general analytical information about ETAAS.[1]

A small volume of sample, typically 5 to 50 μl, is injected into a graphite furnace, where it is electrically heated to be dried, charred, and atomized according to the program. Liquid samples, such as serum, plasma, and urine, can be introduced directly into the furnace, but solid samples should be decomposed to liquid form.

ETAAS is prone to interference from the matrix, but it has overcome the problem by the introduction of the stabilized temperature platform furnace (STPF) concept.[2] Table 2 shows typical operational parameters of ETAAS. "Interference-free" analysis by the STPF concept was further developed by inclusion of a more effective matrix modifier. A matrix modifier is defined as that which "... facilitates analysis through *in situ* changing (modifying) of the thermochemical behaviour of both the analyte and the matrix".[3] Appropriate choice of matrix modifier composition for a desired analyte is essential for successful analysis by ETAAS. The palladium and Mg $(NO_3)_2$ mixed modifier is called a "universal modifier"[4] and is applicable to many analytes.[5] Table 3 lists furnace heating conditions for 21 elements with or without Pd/Mg $(NO_3)_2$ modifier[5] under the STPF concept. It is clear that a higher char temperature is applicable when matrix modifier is used. Inorganic and organic matrix can be removed efficiently by applying a higher char temperature, thus interference becomes less significant.

Correction of nonspecific absorbance arising from the sample matrix is another critical point for accurate and precise analysis by ETAAS. Background correction based on deuterium arc emission is the traditional means, but correction on a Zeeman-effect system is preferred because the latter can cope with structured backgrounds. Table 4 shows recent applications of ETAAS to biomedical samples.

1.1.2 *Hydride Generation Atomic Absorption Spectrometry*

Hydride generating elements, such as As, Bi, Ge, Pb, Sb, Se, Sn, and Te, can be analyzed more sensitively (by 2 to 3 orders of magnitude) by hydride generation AAS (HGAAS) than by conventional nebulization (Table 5)[52] or even by ETAAS. In particular, HGAAS is used extensively for analysis of As and Se in biomedical samples.

Sample digestion for HGAAS is critical; all of the analytes in the sample must be brought to inorganic ions. For instance, certain organic forms of As in biological samples cannot be brought to the hydride-generating form by simple digestion with HNO_3.[53] Digestion with $HNO_3/H_2SO_4/HClO_4$ or $HNO_3/HClO_4$ at the boiling temperature of $HClO_4$ is essential to decompose organic species.

Either Zn and $NaBH_4$ is used for reduction of elements to generate gaseous hydride, but the latter is preferred currently because of its more effective reducing capacity. The oxidation state of the element in an acidic sample solution must be unified prior to hydride generation even when $NaBH_4$ is used as reductant; for example, As(V) must be reduced to As(III), which forms covalent hydride AsH_3. Prior reduction of analyte by KI or HCl, therefore, is essential to

TABLE 1

General Analytical Information of ETAAS

Element	Wavelength (nm)	Sensitivity (pg)[a]
Al	328.1	1.4
Ag	309.3	10
As	193.7	17
Au	242.8	11
B	249.7	700
Ba	553.6	6.5
Be	234.9	0.5
Bi	223.0	15
Ca	422.7	0.8
Cd	228.8	0.35
Co	240.7	6.0
Cr	357.9	3.0
Cu	324.8	4.0
Fe	248.3	5.0
Hg	253.7	150
In	303.9	11
K	766.5	0.8
Li	670.8	1.4
Mg	285.2	0.3
Mn	279.5	2.0
Mo	313.3	9.0
Na	589.0	1.0
Ni	232.0	13
Pb	283.3	10
Pt	265.9	90
Rb	780.0	2.3
Sb	217.6	38
Se	196.0	25
Si	251.6	40
Sn	224.6	10
Sr	460.7	1.4
Te	214.3	15
Ti	364.3	43
Tl	276.8	7.0
U	351.5	12,000
V	318.4	30
Zn	213.9	0.1

[a] Mass of element, in pg, which gives 0.0044 absorbance (characteristic mass).

Source: Adapted from *Graphite Furnace Owner's Manual*, Perkin-Elmer, Norwalk, CN.

TABLE 2

Recommended Operational Parameters of ETAAS Under the STPF Concept

Sample volume	20 μl
Drying temperature	250°C for 60 sec (no ramp)
Char (ashing) time	45 sec after 1 sec ramp
Atomization and integration time (gas stop)	4 sec (volatile metals)
	6 sec (less volatile metals)
Cleanout temperature	2600°C for 6 sec (1 sec ramp)
Cool-down time	20 sec

TABLE 3

Recommended Char (Ashing) Temperature When Pd/Mg $(NO_3)_2$ Modifier Is Used

	Maximum Char (Ashing) Temperature (°C)	
Element	Without Modifier	$Pd/Mg(NO_3)_2$ Modifier[a]
Ag	650	1000
Al	1400	1700
As	300	1400
Au	700	1000
Bi	600	1200
Cd	300	900
Cu	1100	1100
Ga	800	1300
Ge	800	1500
Hg	—	250
In	700	1500
Mn	1100	1400
P	200	1350
Pb	600	1200
Sb	900	1200
Se	200	900
Si	1100	1200
Sn	800	1200
Te	500	1200
Tl	600	1000
Zn	600	1000

[a] For details, see Reference 5.

TABLE 4

ETAAS Applications to Biomedical Samples (1990–1995)

Analyte	Sample Matrix	Matrix Modifier	Detection Limit[a]	Background Correction[b]	Remarks	Ref.
Al	Bone, soft tissues	Not used	9 pg CM in bone; 8 pg CM in soft tissues	Z		6
Al	Serum, urine	Not used	2.6 μg/l serum, 1.3 μg/l urine	Z		7
Al	Milk	$Mg(NO_3)_2$	0.32 μg/l	D_2		8
As	Urine	$Pd/Mg(NO_3)_2/K_2S_2O_8$	30 pg for 0.01 A	Z	Modifier composition studied	9
Bi	Soft tissues, bone	$PtCl_4$	25 ng/g wet	Z		10
Cd	Seminal fluid		0.05 μg/l	Z	Pb also studied	11
Cd	Urine, whole blood	$NH_4H_2PO_4/(NH_4)_2MoO_4$	0.4 μg/l in urine, 0.1 μg/l in blood	D_2	Pb also studied	12
Cd	Biological CRMs	Not used		D_2	W furnace	13
Cd	Urine	Pd/NH_4NO_3	0.1 μg/l	Z		14
Cd	Urine	Not used	0.3 μg/l	D_2	Probe atomization	15
Cd	Serum	$Pd/Mg(NO_3)_2$	0.45 pg (CM)	D_2	*In situ* removal of modifier contamination	16

TABLE 4 (continued)

ETAAS Applications to Biomedical Samples (1990–1995)

Analyte	Sample Matrix	Matrix Modifier	Detection Limit[a]	Background Correction[b]	Remarks	Ref.
Cd	Whole blood, urine	$Pd/Mg(NO_3)_2$	0.22 μg/l	Z	Modifier composition studied	17
Co	Liver	$Pd/Mg(NO_3)_2$		D_2	Chelate extraction; Se, Mo also studied	18
Co	Serum, urine	Not used		Z	Solvent extraction	19
Co	Whole blood	$Mg(NO_3)_2$	0.3 μg/l		On-line sample mineralization-flow injection	20
Cr	Urine, serum, others	$Mg(NO_3)_2$/ $Ca(NO_3)_2$	0.04 μg/l	D_2 or Z		21
Cr	Serum, others	NH_4VO_3, $NH_4VO_3/NaMoO_4$	0.2 μg/l		Modifier composition studied	22
Cr	Whole blood, blood component, bone, urine	Not necessary	2.7 pg (CM) in D_2, 5.0 pg (CM) in Zeeman	Z and D_2	Superior detection limit in D_2	23
Cu	Serum, urine	Not used			Fast furnace heating program	24
Hg	Whole blood	Not used	0.6 μg/l	Z	Solvent extraction	25
In	Whole blood, bone, soft tissues	$PdCl_3$	5–10 ng/g wet		Ion-pair extraction	26
Li	Serum, urine, rat kidney	Ta	0.98 pg CM		Furnace treated with Ta	27
Mn	Serum	EDTA	1.45–2.90 nmol/l	Z and D_2	Z and D_2 compared	28
Mn	Whole blood	$NH_4H_2PO_4$/EDTA	0.2 μg/l	Z	Pb also studied	29
Mo	Liver	$Pd/Mg(NO_3)_2$		D_2	Chelate extraction; Co, Se also studied	18
Mo	Milk	BaF_2	12 pg CM	D_2		30
Ni	Serum	Not used	14 pg CM or 0.15 μg/l	D_2		31
Pb	Whole blood	$(NH_4)_2HPO_4$		Z	Extraction of Pb from dried blood-spot of newborns	32
Pb	Whole blood	$NH_4H_2PO_4$		Z		33
Pb	Biological CRMs	Pd	0.8 pg		Flow injection on-line digestion	34
Pb	Urine, whole blood	$NH_4H_2PO_4$/ $(NH_4)_2MoO_4$	4 μg/l	D_2	Cd also studied	12
Pb	Whole blood	$NH_4H_2PO_4$/EDTA	12 μg/l	Z	Mn also studied	29
Pb	Urine	Not used	2 μg/l (217.0 nm); 4 μg/l (283.3 nm)	Z	Probe atomization	35
Pb	Urine	Pd	0.3 μg/l	D_2	Probe atomization	36
Pt	Cell suspension	Not used	1 ng/10^6 cell	Z	HNO_3-H_2SO_4-$HClO_2$ digestion	37
Sb	Urine	Not used	0.69 μg/l	D_2	Chelate extraction	38
Se	Se compounds, others	$Pd/Mg(NO_3)_2$	18 pg (CM)	Z	Modifier composition studied	39
Se	Biological CRMs, others	Pd	0.002 μg/g	D_2	Co-extraction with Pd	40

TABLE 4 (continued)

ETAAS Applications to Biomedical Samples (1990–1995)

Analyte	Sample Matrix	Matrix Modifier	Detection Limit[a]	Background Correction[b]	Remarks	Ref.
Se	Liver	$Pd/Mg(NO_3)_2$		D_2	Chelate extraction; Co, Mo also studied	18
Se	Serum, urine	$Pd/Mg(NO_3)_2$		Z	Modifier composition studied	41
Se	Serum	Ag/Cu/Mg		Z	Total and concanavalin A-bound Se in serum	42
Se	Urine	$Pd/Mg(NO_3)_2/Ba(NO_3)_2$	6 μg/l			43
Se	Hair, nail	Pd	0.02 μg/l			44
Si	Bone, soft tissue	$La(NO_3)_3/CaCl_3/NH_4H_2PO_4$, $La(NO_3)_3$/tartaric acid	0.90 μg/g in bone, 0.14 μg/g in soft tissues	Z		45
Si	Plasma, urine	$EDTA/KH_2PO_4/Ca/NaCl$	30 pg (CM)		Platform treated with Mo	46
Si	Serum, urine	W-coated tube	3.5 μg/l	Z	Furnace treated with various modifiers	47
Sn	Whole blood, soft tissues	Ascorbic acid	0.02 μg/g tissue	D_2		48
Te	Urine	Pt/Mg		Z	Comparison with chelation-GC/MS	49
Zn	Milk		0.052 μmol/l	D_2	Diluent studied	50
Zn	Milk	Not used	26.4 nmol/l	D_2	Ultrafiltrable Zn	51

[a] CM = characteristic mass (see footnote of Table 1).
[b] Mode of background correction: Z = Zeeman effect; D_2 = deuterium lamp.

TABLE 5

Comparison of Detection Limit by Solution-Nebulization and Hydride-Generation AAS (ng/ml)

Element	Nebulization	Hydride Generation
As	630	0.8
Bi	44	0.2
Ge	20	3.8
Pb	17	0.6
Sb	60	0.5
Se	230	1.8
Sn	150	0.5
Te	44	1.5

diminish kinetic interference and to provide reliable analysis by HGAAS. Interference in the hydride formation/liberation step resulting from the presence of transition metals such as Ni, Cu, or Fe[54-56] must also be eliminated or reduced by increasing $NaBH_4$ concentration, increasing sample acidity, or by the use of masking agents, such as EDTA and KCN.[52]

For atomization of evolved hydride, conventional flame can be used, but a flame-heated or electrically heated quartz tube is widely used because of the higher sensitivity obtainable.[52] Another advantage of using HGAAS is its capability for speciation analysis.[57,58] It is based on the difference in reaction rates of different valences of an element or the difference in boiling points of different hydrides (e.g., AsH_3, CH_3AsH_2, and $(CH_3)_2AsH$).

1.1.3 Cold Vapor Atomic Absorption Spectrometry

Mercury is detected by AAS after Hg vapor generation (cold vapor AAS, or CVAAS) preceded by wet oxidation with various acids. The oxidation step is critical in CVASS analysis because (1) loss of Hg during this step is liable to occur due to volatilization, and (2) complete destruction of organic matter to produce ionic Hg is essential for the subsequent reduction step. In order to avoid loss by volatilization, sample digestion must be done with reflax apparatus or by using a sealed digestion apparatus. For the complete destruction of organic matter, appropriate use should be made of various mixtures of acids, such as HNO_3, H_2SO_4, $HClO_4$, H_2O_2, $KMnO_4$, and/or other oxidizing agents such as V_2O_5. The recommended digestion procedure for total mercury analysis by CVAAS has been described in detail.[59] Tin chloride is commonly used for reduction of ionic Hg to vapor Hg.

Direct introduction of solid or liquid sample followed by combustion to release Hg vapor is the alternate method of choice,[60,61] and such an instrument is available commercially. This allows rapid sample throughput because it does not require sample pretreatment. Amalgamation of evolved Hg vapor with Au or Ag foil/granule is effective for removing interfering gases in both modes of the Hg analysis.

Particular attention during Hg analysis should be paid to its instability in solution at low concentration levels, i.e., loss by adsorption to the container wall and by volatilization.[62-64] A mixture of 1% H_2SO_4 and 0.05% $K_2Cr_2O_7$ medium or 2% HCl and 2% H_2O_2 medium is reported to be effective for long-term storage of part-per-billion-level inorganic Hg. Although not generally agreed upon, glass usually is preferred to polyethylene or polypropylene containers.

REFERENCES

1. *Graphite Furnace Operator's Manual,* Perkin-Elmer, Norwalk, CN.
2. Slavin, W., Manning, D.C., and Carnick, G. R., The stabilized temperature platform furnace, *At. Spectrosc.,* 2, 137–145, 1981.
3. Tsalev, D. L., Slaveykova, V. I., and Mandjukov, P. B., Chemical modification in graphite-furnace atomic absorption spectrometry, *Spectrochim. Acta Rev.,* 13, 225–274, 1990.
4. Schlemmer, G. and Weltz, B., Palladium and magnesium nitrates, a more universal modifier for graphite furnace atomic absorption spectrometry, *Spectrochim. Acta,* 41B, 1157–1165, 1986.
5. Welz, B., Schlemmer, G., and Mudakavi, J. R., Palladium nitrate-magnesium nitrate modifier for electrothermal atomic absorption spectrometry. Part 5. Performance for the determination of 21 elements, *J. Anal. At. Spectrom.,* 7, 1257–1271, 1992.
6. Liang, L., D'Haese, P. C., Lamberts, L. V., and De Broe, M. E., Direct calibration for determining aluminum in bone and tissues by graphite furnace atomic absorption spectrometry, *Clin. Chem.,* 37, 461–466, 1991.
7. Wang, S. T., Pizzolato, S., and Demshar, H. P., Aluminum levels in normal human serum and urine as determined by Zeeman atomic absorption spectrometry, *J. Anal. Toxicol.,* 15, 66–70, 1991.
8. Arruda, M. A. Z., Quintela, M. J., Gallego, M., and Valcarcel, M., Direct analysis of milk for aluminum using electrothermal atomic absorption spectrometry, *Analyst,* 119, 1695–1699, 1994.
9. Nixon, D. E., Mussmann, G. V., Eckdahl, S. J., and Moyer, T. P., Total arsenic in urine: palladium-persulfate vs. nickel as a matrix modifier for graphite furnace atomic absorption spectrophotometry, *Clin. Chem.,* 37, 1575–1579, 1991.
10. Slikkerveer, A., Helmich, R. B., and de Wolff, F. A., Analysis of bismuth in tissue by electrothermal atomic absorption spectrometry, *Clin. Chem.,* 39, 800–803, 1993.
11. Jurasovic, J. and Telisman, S., Determination of lead and cadmium in human seminal fluid by electrothermal atomic absorption spectrometry, *J. Anal. At. Spectrom.,* 8, 419–425, 1993.

12. D'Haese, P. C., Lamberts, L. V., Liang, L., Van de Vyver, F. L., and De Broe, M.C., Elimination of matrix and spectral interferences in the measurements of lead and cadmium in urine and blood by electrothermal atomic absorption spectrometry with deuterium background correction, *Clin. Chem.,* 37, 1583–1588, 1991.
13. Gine, M. F., Krug, F. J., Sass, V. A., Reis, B. F., Noabrega, J. A., and Berndt, H., Determination of cadmium in biological materials by tungsten coil atomic absorption spectrometry, *J. Anal. At. Spectrom.,* 8, 243–245, 1993.
14. Smeyers-Verbeke, J., Yang, Q., Penninckx, W., and Vandervoort, F., Effectiveness of palladium chemical modification for the determination of cadmium by graphite furnace atomic absorption spectrometry, *J. Anal. Atom. Spectrom.,* 5, 393–398, 1990.
15. Marchante-Gayon, J. M., Sanz-Medel, A., Fellows, C., and Rock, P., Determination of cadmium in human urine using electrothermal atomic absorption spectrometry with probe atomization and deuterium background correction, *J. Anal. At. Spectrom.,* 7, 1079–1083, 1993.
16. Bulska, E., Grobenski, Z., and Schlemmer, G., *In situ* removal of contamination from a palladium-magnesium chemical modifier within an electrothermal atomizer for the determination of cadmium in serum, *J. Anal. At. Spectrom.,* 5, 203–204, 1990.
17. Moreira, M. R., Curtius, A. J., and Campos, R. C., Determination cadmium in whole blood and urine by electrothermal atomic absorption spectrometry using palladium-based modifiers and *in situ* decontamination, *Analyst,* 120, 947–950, 1995.
18. Parsely, D. H., Determination of copper, cobalt, selenium and molybdenum in liver by flame and electrothermal atomic absorption spectrometry, *J. Anal. At. Spectrom.,* 6, 289–293, 1991.
19. Baruthio, F. and Pierre, F., Cobalt determination in serum and urine by electrothermal atomic absorption spectrometry, *Biol. Trace Elem. Res.,* 39, 21–31, 1993.
20. Burguera, M., Burguera, J. L., Rondon, C., Rivas, C., Carreno, P., Gallignani, M., and Brunetto, M. R., *In vivo* sample uptake and on-line measurements of cobalt in whole blood by microwave-assisted mineralization and flow injection electrothermal atomic absorption spectrometry, *J. Anal. At. Spectrom.,* 10, 343–347, 1995.
21. Cimadevilla, E. A-C., Wrobel, K., Gayon, J. M. M., and Sanz-Medel, A., Determination of chromium in biological fluids by electrothermal atomic absorption spectrometry using wall, platform and probe atomization from different graphite surfaces, *J. Anal. At. Spectrom.,* 9, 117–123, 1994.
22. Manzoori, J. L. and Saleemi, A., Determination of chromium in serum and lake water by electrothermal atomic absorption spectrometry using vanadium and molybdenum modifier, *J. Anal. At. Spectrom.,* 9, 337–339, 1994.
23. Wang, S. T. and Demshar, H. P., Rapid Zeeman atomic absorption determination of copper in serum and urine, *Clin. Chem.,* 39, 1907–1910, 1993.
24. Granadillo, V. A., Parra de Machado, L., and Romero, R. A., Determination of total chromium in whole blood, bone, and urine by fast furnace program electrothermal atomization AAS and using neither analyte isoformation nor background correction, *Anal. Chem.,* 66, 3624–3631, 1994.
25. Emteborg, H., Bulska, E., Frech, W., and Baxter, D. C., Determination of total mercury in human whole blood by electrothermal atomic absorption spectrometry following extraction, *J. Anal. At. Spectrom.,* 7, 405–408, 1992.
26. Zheng, W., Sipes, I. G., and Carter, D. E., Determination of parts-per-billion concentrations of indium in biological material by electrothermal atomic absorption spectrometry following ion pair extraction, *Anal. Chem.,* 65, 2174–2176, 1993.
27. Sampson, B., Determination of low concentrations of lithium in biological samples using electrothermal atomic absorption spectrometry, *J. Anal. At. Spectrom.,* 6, 115–118, 1991.
28. Neve, J. and Leclercq, N., Factors affecting determinations of manganese in serum by atomic absorption spectrometry, *Clin. Chem.,* 37, 723–728, 1991.
29. Christensen, J. M., Poulsen, O. M., and Anglov, T., Protocol for design and interpretation of method evaluation in atomic absorption spectrometric analysis. Application to the determination of lead and manganese in blood, *J. Anal. At. Spectrom.,* 7, 329–334, 1992.
30. Bermejo-Barrera, P., Calvo, C. P., and Bermejo-Martinez, F., Comparative study of chemical modifiers for the determination of molybdenum in milk by electrothermal atomisation atomic absorption spectrometry, *Analyst,* 115, 549–551, 1990.
31. Patriarca, M. and Fell, G. S., Determination of nickel in serum of haemodialysed patients by means of electrothermal atomic absorption spectrometry with deuterium background correction, *J. Anal. At. Spectrom.,* 9, 457–461, 1994.
32. Wang, S. T. and Demshar, H. P., Determination of blood lead in dried-spot specimens by Zeeman-effect background-corrected atomic absorption spectrometry, *Analyst,* 117, 959–961, 1992.
33. Jacobson, B. E., Lockitch, G., and Quigley, G., Improved sample preparation for accurate determination of low concentrations of lead in whole blood by graphite furnace analysis, *Clin. Chem.,* 37, 515–519, 1991.
34. Burguera, J. L. and Burguera, M., Determination of lead in biological materials by microwave-assisted mineralization and flow injection electrothermal atomic absorption spectrometry, *J. Anal. At. Spectrom.,* 8, 235–241, 1993.

35. Chen, T. and Littlejohn, D., Determination of lead in urine by electrothermal atomic absorption spectrometry with probe atomization, *Analyst,* 118, 541–543, 1993.
36. Marchante-Gayon, J. M., Enrique Sanchez Uria, J., and Sanz-Medel, A., Determination of lead in human urine using electrothermal atomic absorption spectrometry with probe atomization and deuterium background correction, *J. Anal. At. Spectrom.,* 8, 731–736, 1993.
37. Tabary, T. H. and Millart, H., Flameless atomic absorption spectrophotometry analysis of cell-linked platinum after wet ashing, *Anticancer Res.,* 11, 343–346, 1991.
38. Smith, M. M., White, M. A., and Wilson, H. K., Determination of antimony in urine by solvent extraction and electrothermal atomization atomic absorption spectrometry for the biological monitoring of occupational exposure, *J. Anal. At. Spectrom.,* 10, 349–352, 1995.
39. Deaker, M. and Maher, W., Determination of selenium in seleno compounds and marine biological tissues using electrothermal atomization atomic absorption spectrometry, *J. Anal. At. Spectrom.,* 10, 423–431, 1995.
40. Hocquellet, P. and Candillier, M-P., Evaluation of microwave digestion and solvent extraction for the determination of trace amounts of selenium in feeds and plant and animal tissues by electrothermal atomic absorption spectrometry, *Analyst,* 116, 505–509, 1991.
41. Johannessen, J. K., Gammelgaard, B., Jons, O., and Hansen, S. H., Comparison of chemical modifiers for simultaneous determination of different selenium compounds in serum and urine by Zeeman-effect electrothermal atomic absorption spectrometry, *J. Anal. At. Spectrom.,* 8, 999–1004, 1993.
42. Haher, R. and Van Lente, F., Concanavalin A-bound selenoprotein in human serum analyzed by graphite furnace atomic absorption spectrometry, *Clin. Chem.,* 40, 62–70, 1994.
43. Drake, E. N. and Hain, T. D., Palladium (II), magnesium (II), and barium (II) nitrate combinations for matrix modification in electrothermal atomic absorption measurement of total selenium in human urine, *Anal. Biochem.,* 220, 336–339, 1994.
44. Harrison, I., Littlejohn, D., and Fell, G. S., Determination of selenium in human hair and nail by electrothermal atomic absorption spectrometry, *J. Anal. At. Spectrom.,* 10, 215–219, 1995.
45. Zhuoer, H., Silicon measurement in bone and other tissues by electrothermal atomic absorption spectrometry, *J. Anal. At. Spectrom.,* 9, 11–15, 1994.
46. Perez Parajon, J. M. and Sanz-Medel, A., Determination of silicon in biological fluids using metal carbide-coated graphite tubes, *J. Anal. At. Spectrom.,* 9, 111–116, 1994.
47. Gitelman, H. J. and Alderman, F. R., Determination of silicon in biological samples using electrothermal atomic absorption spectrometry, *J. Anal. At. Spectrom.,* 5, 687–689, 1990.
48. Itami, T., Ema, M., Amano, H., and Kawasaki, H., Simple determination of tin in biological materials by atomic absorption spectrometry with a graphite furnace, *J. Anal. Toxicol.,* 15, 119–122, 1991.
49. Aggaewal, S. K., Kinter, M., Nicholson, J., and Herold, D. A., Determination of tellurium in urine by isotope dilution gas chromatography/mass spectrometry using (4-fluorophenyl) magnesium brimide as a derivatizing agent and a comparison with electrothermal atomic absorption spectrometry, *Anal. Chem.,* 66, 1316–1322, 1994.
50. Arnaud, J., Favier, A., and Alary, J., Determination of zinc in human milk by electrothermal atomic absorption spectrometry, *J. Anal. At. Spectrom.,* 6, 647–652, 1991.
51. Arnaud, J. and Favier, A., Determination of ultrafiltrable zinc in human milk by electrothermal atomic absorption spectrometry, *Analyst,* 117, 1593–1598, 1992.
52. Nakahara, T., Application of hydride generation techniques in atomic absorption, atomic fluorescence and plasma atomic emission spectroscopy, *Prog. Anal. At. Spectrosc.,* 6, 163–223, 1983.
53. Jin, K., Ogawa, H., and Taga, M., Study on wet digestion method of total arsenic in marine organisms by continuous flow arsine generation and atomic absorption spectrometry using some model compounds, *Bunseki Kagaku,* 32, E171–E176, 1983.
54. Smith, A. E., Interferences in the determination of elements that form volatile hydrides with sodium borohydride using atomic-absorption spectrophotometry and argon-hydrogen flame, *Analyst,* 100, 300–306, 1975.
55. Pierce, F. D. and Brown, H. R., Inorganic interference study of automated arsenic and selenium determination with atomic absorption spectrometry, *Anal. Chem.,* 48, 693–695, 1976.
56. Pierce, F. D. and Brown, H. R., Comparison of inorganic interferences in atomic absorption spectrometric determination of arsenic and selenium, *Anal. Chem.,* 49, 1417–1422, 1977.
57. Braman, R. S., Johnson, D. L., Foreback, C. C., Ammons, J. M., and Bricker, J. L., Separation and determination of nanogram amounts of inorganic arsenic and methylarsenic compounds, *Anal. Chem.,* 49, 621–625, 1977.
58. Andreae, M. O., Determination of arsenic species in natural waters, *Anal. Chem.,* 49, 820–823, 1977.
59. Analytical Methods Committee, Determination of mercury and methylmercury in fish, *Analyst,* 102, 769–776, 1977.
60. Anderson, D. H., Evans, J. H., Murphy, J. J., and White, W. W., Determination of mercury by a combustion technique using gold as a collector, *Anal. Chem.,* 43, 1511–1512, 1971.
61. Wittmann, Z., Determination of mercury by atomic-absorption spectrophotometry, *Talanta,* 28, 271–273, 1981.

62. Carron, J. and Agemian, H., Preservation of sub-p.p.b. levels of mercury in distilled and natural fresh waters, *Anal. Chim. Acta,* 92, 61–70, 1977.
63. Stoeppler, M. and Matthes, W., Storage behaviour of inorganic mercury and methylmercury chloride in sea water, *Anal. Chim. Acta,* 98, 389–392, 1978.
64. Krivan, V. and Haas, H. F., Prevention of loss of mercury (II) during storage of dilute solutions in various containers, *Fresenius. Z. Anal. Chem.,* 332, 1–6, 1988.

1.2 Inductively Coupled Plasma Atomic Emission Spectrometry

Jun Yoshinaga

Inductively coupled plasma atomic emission spectrometry (ICP-AES) utilizes ICP as an excitation source of elements. The advantages of ICP-AES over AAS include: (1) simultaneous or high-speed sequential multi-element analysis, (2) dynamic range (linearity of calibration curve) as wide as 10^5, and (3) freedom from chemical interferences. Sensitivity is intermediate, between flame AAS and ETAAS; detection limits for many elements are in the range of 5 to 50 ng/ml (Table 6).

Because of the multi-element capability and wide dynamic range, a multi-element standard solution is usually used for calibration. Tables 7 and 8 show examples of standard solutions used for biomedical samples. The standard solutions in Table 7 are prepared for the quantification of major and essential trace elements in digested, tenfold diluted plasma or serum samples. The solutions in Table 8 are for digested, 600-fold diluted bone samples. Since bone contains Ca and P at high levels, it is recommended that Ca and P be added to all of the standard solutions to match the physical and spectral properties of standard solutions and the samples (matrix matching). In this case, simultaneous determination of Ca and P is not possible; they must be determined from another more diluted (e.g., an additional 20-fold) sample solution. Acid concentration should also be matched between samples and standards. Another factor to be considered is that elements in the same group should not spectrally interfere with each other. In Table 7, for instance, Cu and P were in separate groups because of interference. The chemical property of the elements in the same standard solution should also be considered; for example, Ba will precipitate as $BaSO_4$ when Ba ion is added to H_2SO_4 medium.

Spectral interference from major constituents of the sample (organic matter and alkali and alkaline earth elements in the case of biomedical samples) may sometimes be problematic, particularly for an analyte at a low concentration level. Preconcentration and separation from major elements is often necessary to analyze elements at lower levels in biological materials. In order to take advantage of the multi-element capability of ICP-AES, less selective separation schemes have been preferred: (1) extraction by using carbamates[1,2] and 8-quinolinol[2,3] as chelate and (2) liquid-solid extraction by chelating resin.[4-6]

Gaseous hydride introduction of hydride-forming elements (HG-ICP-AES) is advantageous, as is the case with AAS, in terms of interference elimination and increased sample introduction efficiency when compared to solution nebulization ICP. The advantage of using ICP-AES as a detector of hydride-forming elements over AAS lies in the fact that simultaneous multi-element analysis is possible.[7,8] The necessity of unifying the valences of analytes in a sample solution and of masking interference from transition metals in the hydride-generating step also holds true for HG-ICP-AES.[9,10] Table 9 lists recent applications of ICP-AES, including HG-ICP-AES, to biomedical samples.

Microwave-induced helium plasma (He-MIP) is a more efficient excitating source than Ar-ICP for emission spectrometry. Since He-MIP is suited for the introduction of gaseous samples, it is used extensively for Hg analysis after the Hg vapor generation technique. It also allows considerably higher sensitivity (detection limit, 0.004 ng/ml) for Hg.[11]

TABLE 6
ICP-AES Analysis General Information

Element	Wavelength (nm)[a]	Detection Limit (ng/ml)
Ag	328.1 I	7.0
Al	308.2 I	45
As	193.8 I	53
Au	242.8 I	17
B	249.8 I	4.8
Ba	455.4 II	1.3
Be	313.0 II	0.27
Bi	223.1 I	34
Ca	393.4 II	0.19
	317.9 II	19
Cd	228.8 I	2.7
	226.5 II	3.4
Co	228.6 II	7.0
Cr	267.7 II	7.1
Cu	324.8 I	5.4
Fe	259.9 II	6.2
Hg	253.7 I	61
K	766.5 I	60
Li	670.8 I	1.8
Mg	279.6 II	0.15
Mn	257.6 II	1.4
Mo	202.0 II	7.9
Na	589.0 I	29
Ni	231.6 II	15
P	214.9 I	76
Pb	220.4 II	42
Pt	265.9 I	81
Rb	780.0 I	150
S	182.0 I	30
Se	196.1 I	75
Si	288.2 I	27
Sn	190.0 II	25
Sr	421.6 II	0.77
Ti	334.9 II	3.8
Tl	190.9 II	40
	377.6 I	230
U	386.0 II	250
V	292.4 II	7.5
Zn	213.9 I	1.8

[a] I = atomic line; II = ionic line.

TABLE 7

Example of Standard Solutions Used for Biomedical Samples (Plasma or Serum)

Solution	Element	Concentration (ppm)	Expected Conc. in 10-fold Sample (ppm)
Standard 1		1 *M* HNO_3 (blank)	
Standard 2	Na	300	300
	Mg	10	1
	Ca	10	8
	K	20	20
	Fe	10	0.1
	Cu	10	0.1
	Zn	10	0.08
Standard 3	P	10	10
	S	10	100

TABLE 8

Example of Standard Solutions Used for Biomedical Samples (Bone)

Solution	Element [a]	Concentration (ppm)	Expected conc. in 600-fold Sample (ppm)
Standard 1 (Blank)	(Ca)	400	
	(P)	200	
Standard 2	Na	10	10
	Mg	10	5
	K	10	<0.1
	Fe	10	0.1
	Zn	10	0.2
	Sr	1	0.1
	Ba	10	<0.01
	(Ca)	400	
	(P)	200	
Standard 3	Cr	10	<0.01
	Mo	10	<0.01
	(Ca)	400	
	(P)	200	
Standard 4	Ti	10	<0.01
	V	10	<0.01
	(Ca)	400	
	(P)	200	

[a] Ca and P are not for calibration but rather for matrix matching.

TABLE 9

Application of ICP-AES to Biological Samples (1990–1995)

Analyte	Sample Matrix	Mode of Analysis	Detection Limit	Remarks	Ref.
Se	Whole blood, tissues	Continuous flow-HG	4 ng/g	HNO_3-$HClO_4$-H_2SO_4 digestion; HCl reduction	12
Na, K, Mg, Ca, P, Fe, Cu, Zn	Serum	Solution nebulization		Protein precipitation, Fe reduction; Y as internal standard	13
Sn	Tissues	Continuous flow-HG	30 pg/ml in sample solution	HNO_3-$HClO_4$ digestion; standard addition	14
Fe	Liver	Solution nebulization		Non-heme Fe; liver homogenate-hydrolysis-centrifugation-supernatant; Y as internal standard	15
As	Whole blood, tissues	Continuous flow-HG	4 ng/g	HNO_3-$HClO_4$-H_2SO_4 digestion; KI reduction	16
Hg	Whole blood, plasma, erythrocyte, urine	Continuous flow-HG		Oxidation is not necessary	17
Co, Ni, Cu, Cd, Pb	Urine	Solution nebulization	<10 ng/ml	APDC-IBMK extraction	18
Cd	Biological reference materials	Solution nebulization	0.1 ng/ml	Extraction with 1,5-bis(di-2-pyridylmethylene) tricarbonohydrazide	19
Ni	Urine, biological reference materials	Solution nebulization	0.3 ng/ml	Extraction with 1,5-bis(phenyl-2-pyridylmethylene) thiocarbonohydrazide	20
C, Ca, Cu, Cd, Fe, Mg, Mn, Zn	Biological reference materials	Solution nebulization		Coupled with microsample (50–165 mg) digestion method	21
Ca, Cu, Fe, Mg, Zn	Human autopsy sample	Solution nebulization		Effect of formalin fixation studied	22

Note: APDC-IBMK = ammonium pyrrolidine dithiocarbamate-isobuthyl methyl ketone.

References

1. McLeod, C. W., Otsuki, A., Okamoto, K., Haraguchi, H., and Fuwa, K., Simultaneous determination of trace metals in sea water using dithiocarbamate pre-concentration and inductively coupled plasma emission spectrometry, *Analyst,* 106, 419–428, 1981.
2. Miyazaki, A., Kimura, A., Bansho, K., and Umezaki, Y., Simultaneous determination of heavy metals in waters by inductively-coupled plasma atomic emission spectrometry after extraction into diisobutyl ketone, *Anal. Chim. Acta,* 144, 213–221, 1982.
3. Nojiri, Y., Kawai, T., Otsuki, A., and Fuwa, K., Simultaneous multi-element determination of trace metals in lake waters by ICP emission spectrometry with preconcentration and their background levels in Japan, *Water Res.,* 4, 503–509, 1985.
4. Kingston, H. M., Barnes, I. L., Brady, T. J., Rains, T. C., and Champ, M. A., Separation of eight transition elements from alkali and alkaline earth elements in estuarine and seawater with chelating resin and their determination by graphite furnace atomic absorption spectrometry, *Anal. Chem.,* 50, 2064–2070, 1978.
5. Paulsen, A. J., Effects of flow rate and pretreatment on the extraction of trace metals from estuarine coastal seawater by Chelex-100, *Anal. Chem.,* 183–187, 1986.
6. Barnes, R. M., Determination of trace elements in biological materials by inductively coupled plasma spectroscopy with novel chelating resins, *Biol. Trace Elem. Res.,* 6, 93–103, 1984.
7. Thompson, M., Pahalavanpour, B., Walton, S. J., and Kirkbright, G. F., Simultaneous determination of trace concentrations of arsenic, antimony, bismuth, selenium and tellurium in aqueous solution by introduction of the gaseous hydrides into an inductively coupled plasma source for emission spectrometry. Part 1. Preliminary studies, *Analyst,* 103, 568–579, 1978.
8. Pahlavanpour, B., Thompson, M., and Thorne, L., Simultaneous determination of trace amounts of arsenic, antimony and bismuth in herbage by hydride generation and inductively coupled plasma atomic-emission spectrometry, *Analyst,* 106, 467–471, 1981.
9. Thompson, M., Pahalavanpour, B., Walton, S. J., and Kirkbright, G. F., Simultaneous determination of trace concentrations of arsenic, antimony, bismuth, selenium and tellurium in aqueous solution by introduction of the gaseous hydrides into an inductively coupled plasma source for emission spectrometry. Part 2. Interference studies, *Analyst,* 103, 705–713, 1978.
10. Nakahara, T., Hydride generation techniques and their applications in inductively coupled plasma-atomic emission spectrometry, *Spectrochim. Acta Rev.,* 14, 95–109, 1991.
11. Tanabe, K., Chiba, K., Haraguchi, H., and Fuwa, K., Determination of mercury at the ultratrace level by atmospheric pressure helium microwave-induced plasma emission spectrometry, *Anal. Chem.,* 53, 1450–1453, 1981.
12. Tracy, M. T. and Moller, G., Continuous flow vapor generation for inductively coupled argon plasma spectrometric analysis. Part 1. Selenium, *J. Assoc. Off. Anal. Chem.,* 73, 404–410, 1990.
13. Melton, L. A., Tracy, M. L., and Moller, G., Screening trace elements and electrolytes in serum by inductively-coupled plasma emission spectrometry, *Clin. Chem.,* 36, 247–250, 1990.
14. Yokoi, K., Kimura, M., and Itokawa, Y., Determination of tin in biological samples using gaseous hydride generation-inductively coupled plasma-atomic emission spectrometry, *Anal. Biochem.,* 190, 71–77, 1990.
15. Yokoi, K., Kimura, M., and Itokawa, Y., Determination of nonheme iron using inductively coupled plasma-atomic emission spectrometry, *Biol. Trace Elem. Res.,* 31, 265–279, 1991.
16. Tracy, M. L., Littlefield, E. S., and Moller, G., Continuous flow vapor generation for inductively coupled argon plasma spectrometric analysis. Part 2. Arsenic, *J. Assoc. Off. Anal. Chem.,* 74, 516–521, 1991.
17. Buneaux, F., Buisine, A., Bourdon, S., and Bourdon, R., Continuous flow quantification of total mercury in whole blood, plasma, erythrocytes and urine by inductively coupled plasma atomic emission spectroscopy, *J. Anal. Toxicol.,* 16, 99–101, 1992.
18. Lopez-Artiguez, M., Camean, A., and Repetto, M., Preconcentration of heavy metals in urine and quantification by inductively coupled plasma atomic emission spectrometry, *J. Anal. Toxicol.,* 17, 18–22, 1993.
19. Espinosa-Almendro, J. M., Bosch-Ojeda, C., Garcia de Torres, A., and Cano Pavon, J. M., Determination of cadmium in biological samples by inductively coupled plasma atomic emission spectrometry after extraction with 1,5-bis(di-2-pyridylmethylene)thiocarbonohydrazide, *Analyst,* 117, 1749–1751, 1992.
20. Vereda-Alonso, E., Garcia de Torres, A., and Cano Pavon, J. M., Determination of nickel in biological samples by inductively coupled plasma atomic emission spectrometry after extraction with 1,5-bis[phenyl-(2-pyridyl)methylene]thiocarbonohydrazide, *J. Anal. At. Spectrom.,* 8, 843–846, 1993.
21. Amarasiriwardena, D., Krushevska, A., Argentine, M., and Barnes, R. M., Vapour-phase acid digestion of microsamples of biological material in a high-temperature, high-pressure asher for inductively coupled plasma atomic emission spectrometry, *Analyst,* 119, 1017–1021, 1994.
22. Bush, V. J., Moyer, T. P., Batts, K. P., and Parisi, J. E., Essential and toxic element concentrations in fresh and formalin-fixed human autopsy tissues, *Clin. Chem.,* 41, 284–294, 1995.

1.3 Inductively Coupled Plasma Mass Spectrometry

Jun Yoshinaga

Inductively coupled plasma mass spectrometry (ICP-MS) utilizes ICP as an ion source. Since many elements are ionized nearly 100%, ICP is a suitable ion source for inorganic mass spectrometry. ICP-MS possesses the multi-element analysis capability of ICP-AES and the high sensitivity of ETAAS; in fact, the sensitivity is superior to that of ETAAS for most of the elements. Detection limits are typically in the range of parts-per-trillion to sub-parts-per-billion (Table 10).[1] At present, ICP-MS is the most sensitive analytical method for inorganic analysis.

TABLE 10

ICP-MS General Information

Element	Isotope Commonly Used (abundance, %)	Detection Limit (pg/ml)	Spectral Interference
Ag	^{107}Ag (51.8)	1.0	
Al	^{27}Al (100)	1.0	$^{13}C^{14}N$, $^{12}C^{14}NH$
As	^{75}As (100)	14	$^{40}Ar^{35}Cl$
Au	^{197}Au (100)	2.7	
B	^{10}B (19.8), ^{11}B (80.2)	—	^{12}C
Ba	^{138}Ba (71.7)	1.2	
Be	^{9}Be (100)	2.2	
Bi	^{209}Bi (100)	0.1	
Cd	^{111}Cd (12.8)	0.6	
Ce	^{140}Ce (88.5)	0.2	
Co	^{59}Co (100)	0.5	
Cr	^{52}Cr (83.8)	41	$^{12}C^{40}Ar$
Cs	^{133}Cs (100)	0.3	
Cu	^{63}Cu (69.2)	2.4	
Dy	^{163}Dy (24.9)	0.3	
Er	^{166}Er (33.4)	0.2	
Eu	^{153}Eu (52.1)	0.2	
Fe	^{57}Fe (2.15)	77	$^{40}Ar^{16}OH$
Ga	^{69}Ga (60.1)	1.5	$^{37}Cl^{16}O_2$
Gd	^{157}Gd (15.7)	0.6	
Ge	^{74}Ge (36.5)	94	$^{37}Cl_2$
Hf	^{180}Hf (35.2)	1.2	
Hg	^{202}Hg (29.8)	4.4	
Ho	^{165}Ho (100)	0.06	
In	^{115}In (95.7)	0.2	
La	^{139}La (99.911)	0.09	
Li	^{7}Li (92.5)	2.7	
Lu	^{175}Lu (97.39)	0.06	
Mg	^{24}Mg (78.99)	11	$^{12}C_2$
Mn	^{55}Mn (100)	2.6	
Mo	^{98}Mo (24.1)	1.3	
Na	^{23}Na (100)	2.6	
Nb	^{93}Nb (100)	0.4	
Nd	^{146}Nd (17.2)	0.3	
Ni	^{61}Ni (1.13)	10	
Pb	^{208}Pb (52.3)	0.8	
Pd	^{105}Pd (22.2)	1.6	
Pr	^{141}Pr (100)	0.1	
Pt	^{195}Pt (33.8)	0.8	
Rb	^{85}Rb (72.17)	0.2	
Re	^{185}Re (37.4)	0.1	

TABLE 10 (continued)

ICP-MS General Information

Element	Isotope Commonly Used (abundance, %)	Detection Limit (pg/ml)	Spectral Interference
Rh	^{103}Rh (100)	0.1	
Sb	^{121}Sb (57.3)	0.6	
Sc	^{45}Sc (100)	3.6	$^{12}C^{16}O_2H$
Se	^{77}Se (7.6), ^{82}Se (9.2)	30 (^{82}Se)	$^{40}Ar^{37}Cl$
Sm	^{149}Sm (13.9)	1.2	
Sn	^{118}Sn (24.3)	0.8	
Sr	^{88}Sr (82.6)	0.3	
Ta	^{181}Ta (99.9877)	1.0	
Tb	^{159}Tb (100)	0.07	
Te	^{130}Te (34.5)	4.1	
Th	^{232}Th (100)	0.3	
Ti	^{47}Ti (7.4)	4.8	$^{31}P^{16}O$
Tl	^{205}Tl (70.5)	2.5	
Tm	^{169}Tm (100)	0.08	
U	^{238}U (99.275)	0.07	
V	^{51}V (99.750)	23	$^{35}Cl^{16}O$
W	^{184}W (30.7)	3.1	
Y	^{89}Y (100)	0.2	
Yb	^{172}Yb (21.9)	0.3	
Zn	^{66}Zn (27.9)	5.6	
Zr	^{90}Zr (51.5)	1.7	

There are two major analytical problems with ICP-MS: spectral interference and difficulty of analyzing samples with high matrix content.[2] Spectral interference is an overlap of analyte and polyatomic molecular mass/charge (m/z). Typical examples are $^{40}Ar^{12}C$ on ^{52}Cr, $^{40}Ar^{16}O$ on ^{56}Fe, $^{40}Ar^{35}Cl$ on ^{75}As, or $^{40}Ar_2$ on ^{80}Se. This problem is inherent to ICP-MS because most of the currently available instruments are equipped with low-resolution quadrupole mass filters. Table 11 lists major molecular interferences to biologically relevant elements in biological matrixes.[3,4] To eliminate these interferences, the choice of other isotopes (e.g., ^{82}Se instead of ^{80}Se), interferent separation (e.g., elimination of Cl), addition of other gases into nebulizer gas (H_2 or N_2), or instrumental manipulation (e.g., use of a special torch) have been investigated. Use of a high-resolution sector type of mass spectrometer solves most of the molecular interferences.[5,6]

Introduction of a high salt content sample to ICP-MS results in clogging of pinholes through which ions in ICP are introduced to MS. It is generally recommended that total salt content of a sample should be less than 100 ppm. The presence of a concomitant element also induces suppression (sometimes enhancement) of analyte signal, a process known as the matrix effect.[7,8] In order to eliminate or correct the matrix effect, dilution of the sample, the use of an internal standard, or the use of the standard addition technique are recommended. The most commonly used internal standards for biomedical samples are ^{9}Be for low-mass analytes, ^{89}Y or ^{115}In for mid-mass analytes, and ^{205}Tl or ^{209}Bi for high-mass analytes. Table 12 shows recent applications of ICP-MS to biomedical samples.

One of the important applications of ICP-MS in the biomedical field is stable isotope tracer studies. In earlier metabolic and nutritional research, radio-isotopes were commonly used. Stable isotopes are preferred to radio-isotopes because of the absence of potential risk and they can be safely administered to humans, even to pregnant women. A number of studies have been devoted to investigating the applicability of ICP-MS to these research areas,[28-31] and the accuracy and precision obtainable have been found to be satisfactory. Applications of ICP-MS to the study of human metabolic trace elements such as Fe, Ni, or Zn have been reported.[32-34]

TABLE 11

Major Molecular Ion Interferences in Biological Matrices

Isotope (abundance, %)	Interfering Molecular Ions
^{50}V (0.2), ^{50}Cr (4.3)	$^{36}Ar^{14}N$
^{51}V (99.8)	$^{35}Cl^{16}O$, $^{34}S^{16}OH$, $^{37}Cl^{14}N$
^{52}Cr (83.8)	$^{40}Ar^{12}C$, $^{35}Cl^{16}OH$, $^{38}Ar^{14}N$, $^{36}S^{16}O$
^{53}Cr (9.6)	$^{37}Cl^{16}O$
^{54}Fe (5.8), ^{54}Cr (2.4)	$^{37}Cl^{16}OH$
^{55}Mn (100)	$^{39}K^{16}O$
^{56}Fe (91.7)	$^{40}Ar^{16}O$, $^{40}Ca^{16}O$
^{57}Fe (2.2)	$^{40}Ar^{16}OH$, $^{40}Ca^{16}OH$
^{58}Fe (0.3), ^{58}Ni (68.3)	$^{42}Ca^{16}O$, $^{40}Ca^{18}O$, $^{40}Ar^{18}O$
^{59}Co (100)	$^{43}Ca^{16}O$, $^{42}Ca^{16}OH$
^{60}Ni (26.2)	$^{44}Ca^{16}O$, $^{43}Ca^{16}OH$, $^{12}C^{16}O_3$
^{61}Ni (1.3)	$^{44}Ca^{16}OH$
^{62}Ni (3.7)	$^{23}Na_2{}^{16}O$
^{63}Cu (69.1)	$^{40}Ar^{23}Na$, $^{31}P^{16}O_2$
^{64}Zn (48.9), ^{64}Ni (1.2)	$^{32}S^{16}O_2$, $^{31}P^{16}O_2H$
^{65}Cu (30.9)	$^{32}S^{16}O_2H$, $^{31}P^{16}O^{18}O$
^{66}Zn (27.8)	$^{34}S^{16}O_2$
^{67}Zn (4.1)	$^{34}S^{16}O_2H$, $^{32}S^{16}O^{18}OH$, $^{35}Cl^{16}O_2$
^{68}Zn (18.6)	$^{36}S^{16}O_2$, $^{34}S^{16}O^{18}O$
^{70}Zn (0.6)	$^{35}Cl_2$
^{75}As (100)	$^{40}Ar^{35}Cl$
^{76}Se (9.4)	$^{40}Ar^{36}Ar$
^{77}Se (7.6)	$^{40}Ar^{37}Cl$
^{78}Se (23.5)	$^{40}Ar^{38}Ar$
^{80}Se (49.8)	$^{40}Ar_2$
^{82}Se (9.2)	^{81}BrH, $^{34}S^{16}O_3$

TABLE 12

ICP-MS Applications to Biomedical Samples (1990–1995)

Analyte	Sample Matrix	Detection Limit	Remarks	Ref.
Fe, Co, Cu, Zn, Br, Rb, Sr, Mo, Cs	Serum		Result compared with neutron activation analysis and PIXE	9
24 Elements	Biological reference materials	0.0007–11 μg/g sample	Microwave dissolution, good agreement with certified values	10
19 Elements	Urine	0.006–6 ng/ml	Sb, Cd, Hg results compared with other methods	11
Pt	Tissue	0.1 μg/l sample solution	Performance compared with ETAAS, detection limit 100-fold superior to ETAAS	12
La, Ce, Gd, Tb, Yb	Plasma, urine	0.003–0.02 μg/l	6-fold dilution with 1% HNO_3; mean of 28 men: <0.3 μg/l for all elements; Ce detected in one urine sample (1.5 μg/l)	13
Multi-elements	Serum, whole blood		Semiquantitative scan	14
As	Urine, others		N_2 in carrier reduced $^{40}Ar^{35}Cl$ formation	15, 16
11 Elements	Biological reference materials, human tissues			
As	Urine	0.46 ng/ml	Hydride generation speciation	17

TABLE 12 (continued)

ICP-MS Applications to Biomedical Samples (1990–1995)

Analyte	Sample Matrix	Detection Limit	Remarks	Ref.
Pt	Plasma	0.05 μg/l	Eu as internal standard, total Pt vs. free Pt	18
As	Urine		Cl interference correction based on $^{35}Cl^{16}O$ studied	19
As, V, Ni	Urine, others	3–7 μg As per liter, 1 μg V per liter, 10–20 μg Ni per liter	Molecular interference reduced by H_2 in carrier and cryogenic desolvation	20
Ni	Serum, urine		Rh as internal standard, interference corrected based on principal component analysis	21
11 Elements	Serum	0.007–0.5 ng Bi per milliliter	5-fold dilution with 0.14 *M* HNO_3; Be, In, Tl as internal standard	22
As, Cr, Se, V	Biological reference materials	1.2 (V), 6 (Cr), 9 (As), 65 (Se), in ng/ml	Interference eliminated with flow-injection and on-line ion exchange	23
V, Fe, Cu, Zn, Ag	Serum	10 pg/ml order	Interference eliminated with use of high-resolution MS	24
Pb	Plasma	16 fg (10–15)	Isotope dilution, electrothermal vaporization	25
Pb	Whole blood		Comparison with TIMS	26
As	Feces, others	2 ng/g dried feces	Digestion methods studied	27

Note: PIXE = particle-induced X-ray emission spectrometry.

References

1. Suzuki, C., Yoshinaga, J., and Morita, M., Determination of trace elements in pure water by inductively coupled plasma mass spectrometry, *Anal. Sci.,* 7(Suppl.), 997–999, 1991.
2. Houk, R. S. and Thompson, J. J., Inductively coupled plasma mass spectrometry, *Mass Spectrom. Rev.,* 7, 425–462, 1988.
3. Morita, M., ICP-MS — application to biological samples, in Tomita, H., Ed., *Trace Elements in Clinical Medicine,* Springer-Verlag, Tokyo, 1990, pp. 427–436.
4. Vanhoe, H., Goossens, J., Moens, L., and Dams, R., Spectral interferences encountered in the analysis of biological materials by inductively coupled plasma mass spectrometry, *J. Anal. At. Spectrom.,* 9, 177–185, 1994.
5. Bradshaw, N., Hall, E. F. H., and Sanderson, N. E., Inductively coupled plasma as an ion source for high-resolution mass spectrometry, *J. Anal. At. Spectrom.,* 4, 801–803, 1989.
6. Morita, M., Ito, H., Uehiro, T., and Ohtsuka, K., High resolution mass spectrometry with inductively coupled argon plasma ionization source, *Anal. Sci.,* 5, 609–610, 1989.
7. Olivares, J. A. and Houk, R. S., Suppression of analyte signal by various concomitant salts in inductively coupled plasma mass spectrometry, *Anal. Chem.,* 58, 20–25, 1986.
8. Tan, S. H. and Horlick, G., Matrix-effect observations in inductively coupled plasma mass spectrometry, *J. Anal. At. Spectrom.,* 2, 745–763, 1987.
9. Vandecasteele, C., Vanhoe, H., Dams, R., and Versieck, J., Determination of trace elements in human serum by inductively coupled plasma-mass spectrometry. Comparison with nuclear analytical technique, *Biol. Trace Elem. Res.,* 30, 553–560, 1990.
10. Friel, J. K., Skinner, C. S., Jackson, S. E., and Longerich, H. P., Analysis of biological reference materials, prepared by microwave dissolution, using inductively coupled plasma mass spectrometry, *Analyst,* 115, 269–273, 1990.
11. Mulligan, K. J., Davidson, T. M., and Caruso, J. A., Feasibility of the direct analysis of urine by inductively coupled argon plasma mass spectrometry for biological monitoring of exposure to metals, *J. Anal. At. Spectrom.,* 5, 301–306, 1990.
12. Tothill, R., Matheson, L. M., Smyth, J. F., and McKay, K., Inductively coupled plasma mass spectrometry for the determination of platinum in animal tissues and a comparison with atomic absorption spectrometry, *J. Anal. At. Spectrom.,* 5, 619–622, 1990.

13. Allain, P., Berre, S., Premel-Cabic, A., Mauras, Y., and Delaporte, T., Concentrations of rare earth elements in plasma and urine of healthy subjects determined by inductively coupled plasma mass spectrometry, *Clin. Chem.*, 2011–2012, 1990.
14. Vaughan, M.-A., Baines, A. D., and Templeton, D. M., Multi-element analysis of biological samples by inductively coupled plasma-mass spectrometry. II. Rapid survey method for profiling trace elements in body fluids, *Clin. Chem.*, 37, 210–215, 1991.
15. Branch, S., Ebdon, L., Ford, M., Foulkes, M., and O'Neill, P., Determination of arsenic in samples with high chloride content by inductively coupled plasma mass spectrometry, *J. Anal. At. Spectrom.*, 6, 151–154, 1991.
16. Lyon, T. D. B., Fell, G. S., McKay, K., and Scott, R. D., Accuracy of multi-element analysis of human tissue obtained at autopsy using inductively coupled plasma mass spectrometry, *J. Anal. At. Spectrom.*, 6, 559–564, 1991.
17. Story, W. C., Caruso, J. A., Heitkemper, D. T., and Perkins, L., Elimination of the chloride interference on the determination of arsenic using hydride generation inductively coupled plasma mass spectrometry, *J. Chrom. Sci.*, 30, 427–432, 1992.
18. Allain, P., Berre, S., Mauras, Y., and Le Bouil, A., Evaluation of inductively coupled plasma mass spectrometry for the determination of platinum in plasma, *Biol. Mass Spectrom.*, 21, 141–143, 1992.
19. Kershisnik, M. M., Kalamegham, R., Owen-Ash, K., Nixon, D. E., and Ashwood, E. R., Using $^{16}O^{35}Cl$ to correct for chloride interference improves accuracy of urine arsenic determinations by inductively coupled plasma mass spectrometry, *Clin. Chem.*, 38, 2197–2202, 1992.
20. Alves, L. C., Allen, L. A., and Houk, H. S., Measurement of vanadium, nickel, and arsenic in seawater and urine reference materials by inductively coupled plasma mass spectrometry with cryogenic desolvation, *Anal. Chem.*, 65, 2468–2471, 1993.
21. Xu, S. X., Stuhne-Sekalec, L., and Templeton, D. M., Determination of nickel in serum and urine by inductively coupled plasma mass spectrometry, *J. Anal. At. Spectrom.*, 8, 445–448, 1993.
22. Vanhoe, H., Dams, R., and Versieck, J., Use of inductively coupled plasma mass spectrometry for the determination of ultra-trace elements in human serum, *J. Anal. At. Spectrom.*, 9, 23–31, 1994.
23. Ebdon, L., Fisher, A. S., and Worsfold, P. J., Determination of arsenic, chromium, selenium and vanadium in biological samples by inductively coupled plasma mass spectrometry using on-line elimination of interference and preconcentration by flow injection, *J. Anal. At. Spectrom.*, 9, 611–614, 1994.
24. Moens, L., Verrept, P., Dams, R., Greb, U., Jung, G., and Laser, B., New high-resolution inductively coupled plasma mass spectrometry technology applied for the determination of V, Fe, Cu, Zn and Ag in human serum, *J. Anal. At. Spectrom.*, 9, 1075–1078, 1994.
25. Bowins, R. J. and McNutt, R. H., Electrothermal isotope dilution inductively coupled plasma mass spectrometry method for the determination of sub-ng ml^{-1} levels of lead in human plasma, *J. Anal. At. Spectrom.*, 9, 1233–1236, 1994.
26. Paschal, D. C., Caldwell, K. L., and Ting, B. G., Determination of lead in whole blood using inductively coupled argon plasma mass spectrometry with isotope dilution, *J. Anal. At. Spectrom.*, 10, 367–370, 1995.
27. Lasztity, A., Krushevska, A., Kotrebai, M., Barnes, R., and Amarasiriwardena, D., Arsenic determination in environmental, biological and food samples by inductively coupled plasma mass spectrometry, *J. Anal. At. Spectrom.*, 10, 505–510, 1995.
28. Sun, X. F., Ting, B. T. G., Zeisel, S. H., and Janghorbani, M., Accurate measurement of stable isotopes of lithium by inductively coupled plasma mass spectrometry, *Analyst,* 112, 1223–1228, 1987.
29. Schuette, S., Vereault, D., Ting, B. T. G., and Janghorbani, M., Accurate measurement of stable isotopes of magnesium in biological materials with inductively coupled plasma mass spectrometry, *Analyst,* 113, 1837–1842, 1988.
30. Ting, B. T. G., Mooers, C. S., and Janghorbani, M., Isotopic determination of selenium in biological materials with inductively coupled plasma mass spectrometry, *Analyst,* 114, 667–674, 1989.
31. Friel, J. K., Longerich, H. P., and Jackson, S. E., Determination of isotope ratios in human tissues enriched with zinc stable isotope tracers using inductively coupled plasma-mass spectrometry (ICP-MS), *Biol. Trace Elem. Res.*, 37, 123–136, 1993.
32. Ting, B. T. G. and Janghorbani, M., Inductively coupled plasma mass spectrometry applied to isotopic analysis of iron in human fecal matter, *Anal. Chem.*, 58, 1334–1340, 1986.
33. Serfass, R. E., Thompson, J. J., and Houk, R. S., Isotope ratio determinations by inductively coupled plasma/mass spectrometry for zinc bioavailability studies, *Anal. Chim. Acta,* 188, 73–84, 1986.
34. Templeton, D. T., Xu, S. X., and Stuhne-Sekalec, L., Isotope-specific analysis of Ni by ICP-MS: application of stable isotope tracers to biokinetic studies, *Sci. Total Environ.*, 148, 253–262, 1994.

1.4 Neutron Activation Analysis

Jun Yoshinaga

Neutron activation analysis (NAA) is an established method for trace element analysis of a wide variety of matrices. Elements in a sample that is commonly in dried form and sealed in a polyethylene bag or in quartz ampoule is activated by a neutron flux. The generated radionuclide decays with a characteristic half-life by β- or γ-emission of energy to the element. Quantitative analysis is possible by detecting the γ-emission intensity which is proportional to the generated radionuclides and also to the existing parent stable element. The detectors used frequently are Ge(Li) or pure Ge(HPGe) because of the higher resolution (~2 keV) obtainable.

The major advantages of NAA are high sensitivity, accuracy, and precision and the capability for simultaneous multi-element analysis. NAA is considered to be a reference method for several heavy metals such as As, Se, and Hg. The capability for nondestructive analysis may be another advantage in some circumstances.

The most significant drawback is that accessibility to NAA is restricted because a nuclear reactor is essential for the analysis. Relatively longer analysis time required for irradiation and decay time together with high cost may be other drawbacks.

The analytical merits of NAA are obtainable only after several precautions are heeded: correction of neutron flux fluctuation during irradiation and correction or elimination of interferences in the activation step (nuclear interference) and in the emission detection step (spectral interference). The use of a high resolution detector is essential for the elimination of spectral interferences. Chemical separation of analytes from an interfering matrix prior to or after irradiation (radiochemical neutron activation analysis, RNAA) is an effective means of interference elimination and of obtaining higher sensitivity, accuracy, and precision, although the cost and time required for analysis significantly increases. Various separation techniques, including solvent extraction, chelating resin, and chromatography have been used. Table 13 shows nuclear data essential in NAA analysis.[1]

NAA has been applied to the analysis of so-called "difficult elements", such as Cr, Co, As, Se, Sb, or Hg, which are difficult to analyze with adequate sensitivity and accuracy in biomedical matrices by other analytical techniques. Recent remarkable applications of NAA to biomedical samples are found in the RNAA analysis of heavy metals such as V, Cr, Co, Sn, and Hg in human serum at their normal levels (less than a part-per-billion level in the sample).[2-5] From these results, normal values of these elements in human serum have been determined. Analysis of Al in tissues by NAA has been of particular interest due to the toxic effect of Al.[6,7]

TABLE 13

Nuclear Data for Some Toxicologically Relevant Elements

Element	Target Isotope	Abundance (%)	Product Isotope	Cross-Section (Barn)	Half-Life	γ-Energy (Absolute Intensity) (MeV)
Al	^{27}Al	100	^{28}Al	0.230 ± 0.003	2.25 m	1.7789 (1.00)
As	^{75}As	100	^{76}As	4.3 ± 0.1	26.4 hr	0.5591 (0.446)
						0.6570 (0.064)
Cd	^{114}Cd	28.80	^{115}Cd	0.300 ± 0.015	2.22 d	0.4923 (0.081)
						0.5239 (0.275)
			^{115m}In		4.5 hr	0.3363 (0.45)
Co	^{59}Co	100	$^{60m+g}$Co	37.2 ± 0.2	5.272 y	1.1732 (0.999)
						1.3325 (0.999)
Cr	^{50}Cr	4.35	^{51}Cr	15.9 ± 0.2	27.7 d	0.3201 (0.202)
Cu	^{63}Cu	69.09	^{64}Cu	4.5 ± 0.1	12.74 hr	0.511 (0.37)
						1.3458 (0.005)
	^{65}Cu	30.91	^{66}Cu	2.17 ± 0.03	5.10 m	1.0390 (0.09)
Fe	^{58}Fe	0.33	^{59}Fe	1.15 ± 0.02	45 d	1.0993 (0.565)
						1.2919 (0.432)
Hg	^{196}Hg	—	^{197m}Hg	120 ± 13	24 hr	0.1340 (0.34)
						0.2793 (0.051)
	—	—	^{197}Hg	3080 ± 200	65 hr	0.0775 (0.18)
						0.1916 (0.006)
	^{202}Hg	29.80	^{203}Hg	5.04 ± 0.38	47 d	0.2791 (0.815)
Mn	^{55}Mn	100	^{56}Mn	13.3 ± 0.2	2.58 hr	0.8466 (0.99)
						1.8112 (0.300)
						2.115 (0.155)
Ni	^{64}Ni	0.95	^{65}Ni	1.49 ± 0.03	2.52 hr	1.1154 (0.152)
						1.4817 (0.254)
Sb	^{121}Sb	57.25	^{122}Sb	6.25 ± 0.20	2.72 d	0.5639 (0.66)
						0.6928 (0.063)
	^{123}Sb	42.75	^{124}Sb	4.33 ± 0.16	60.3 d	0.6027 (0.98)
						0.7228 (0.119)
						1.691 (0.49)
Se	^{74}Se	0.9	^{75}Se	51.8 ± 1.2	120 d	0.1360 (0.54)
						0.2646 (0.58)
						0.2795 (0.25)
						0.4006 (0.116)
	^{82}Se	9.19	$^{83m+g}$Se	0.045 ± 0.003	22.4 m	0.226 (0.35)
						0.356 (0.75)
						0.512 (0.4)
						0.720 (0.224)
V	^{51}V	99.76	^{52}V	4.89	3.75 m	1.4342 (1.00)
Zn	^{64}Zn	48.89	^{65}Zn	0.78 ± 0.02	244 d	1.1154 (0.508)
	^{68}Zn	18.57	^{69m}Zn	0.072 ± 0.004	13.9 hr	0.4387 (0.95)

Source: Adapted from Kim, J. I., *J. Radioanal. Chem.*, 63, 121–144, 1981.

REFERENCES

1. Kim, J. I., Monostandard activation analysis: evaluation of the method and its accuracy, *J. Radioanal. Chem.*, 63, 121–144, 1981.
2. Versieck, J., Hoste, J., Barbier, F., Steyaert, H., De Rudder, J., and Michaels, H., Determination of chromium and cobalt in human serum by neutron activation analysis, *Clin. Chem.*, 24, 303–308, 1978.
3. Byrne, A. R. and Versieck, J., Vanadium determination at the ultratrace level in biological reference materials and serum by radiochemical neutron activation analysis, *Biol. Trace Elem. Res.*, 26–27, 529–540, 1990.

4. Versieck, J., Vanballenberghe, L., Wittoek, A., Vermeir, G., and Vandecasteele, C., Determination of mercury in human serum and packed blood cells by neutron activation analysis, *Biol. Trace Elem. Res.*, 26–27, 683–689, 1990.
5. Versieck, J. and Vanballenberghe, L., Determination of tin in human blood serum by radiochemical neutron activation analysis, *Anal. Chem.*, 63, 1143–1146, 1991.
6. Blotcky, A. J., Classen, J. P., Roman, F. R., Rack, E. P., and Badakhsh, S., Determination of aluminum by chemical and instrumental neutron activation analysis in biological standard reference material and human urine, *Anal. Chem.*, 64, 2910–2913, 1992.
7. Alfassi, Z. B. and Rietz, B., Determination of aluminum by instrumental neutron activation analysis in biological samples with special reference to NBS SRM 1577 bovine liver, *Analyst*, 119, 2407–2410, 1994.

1.5 Speciation

Jun Yoshinaga

Speciation analysis of chemical forms of metals in biological systems (Table 14) is of particular importance because it is recognized that toxicity and metabolism vary to a considerable extent among chemical forms of metals, e.g., inorganic and organic As and Hg. There are two major methods of speciation analysis: off-line and on-line. In off-line analysis, chemical species are separated by an appropriate method, and the resulting fractions are analyzed one by one. There are no special requirements for the particular analytical method used for detection. In on-line analysis, the separation and detection systems are connected; the detection system must be specifically matched for continuous monitoring of the eluent. Combining modern chromatographic techniques and element-specific detectors is a recent trend in metal speciation analysis. Gas chromatography (GC) is used for the separation of volatile and heat-resistant species. Use of capillary GC provides high-resolution separation. Liquid chromatography (LC) is applicable to less volatile and less heat-resistant species, and various modes of separation (e.g., ion exchange, size exclusion, or reversed phase) are available. AAS, AES, and ICP-MS are commonly used as detectors because these are highly element specific and are suited for continuous monitoring of eluent from chromatography.

1.5.1 Mercury

The significance of Hg speciation has long been recognized. Westöö's method[1] has been the standard for organic Hg analysis. It is based on extraction of organic Hg species with HCl (or with HBr or HI) into organic solvent (benzene or toluene). Clean-up with back extraction and re-extraction is followed by separation by GC and detection by electron capture detector (GC-ECD). The formation of emulsion between the aqueous and solvent layers significantly decreases extraction efficiency, particularly in fat-rich sample. Alkali digestion is recommended for effective extraction of Hg species.[2] The detector has been replaced with an Hg-selective detector, such as AAS,[3,4] ICP-AES,[5] microwave-induced helium plasma AES (MIP-AES),[6,7] and ICP-MS.[8] Use of the Hg-selective detector allows less extensive sample pretreatment.

Selective reduction of inorganic Hg followed by AAS detection is another speciation method.[9,10] The difference between total and inorganic Hg is assumed to be organic Hg, but it does not allow speciation among organic species.

Recent developments in Hg speciation can be found in aqueous phase ethylation-GC-AFS or AAS detection.[11,12] It is based on formation of volatile ethyl derivatives of Hg species by reaction with NaB $(C_2H_5)_4$ in aqueous phase. It does not require Hg species extraction from the sample, and simultaneous analysis of both inorganic and organic Hg species is possible.

LC is an alternative separation method for Hg. Organic Hg separation by reversed-phase, high-performance liquid chromatography (HPLC) using AAS,[13,14] ICP-AES,[15] and ICP-MS[16]

TABLE 14
Applications of Speciation Analysis to Biomedical Samples (1990–1995)

Element	Species	Sample matrix	Separation	Detection	Remarks	Ref.
As	IAs, MMA, DMA	Urine	HPLC	HG-AAS		35
As	IAs, MMA, AB, AC	Urine	HPLC	ICP-AES		36
As	IAs, MMA, DMA	Urine	IC	ICP-MS		37
As	IAs, MMA, DMA	Urine	HPLC	ICP-MS	AB, AC, TMAO cannot be separated	38
As	IAs, MMA, DMA	Urine	HPLC	HG-ICP-AES	Column treated with didodecyl-dimethyl-ammonium bromide	39
As	Roxarsone	Chicken tissue	HPLC	ICP-MS		40
As	AB and other As compounds	Erythrocyte, plasma	HPLC	ICP-MS	Only AB was detected among 17 As compounds separable by the system	41
As	IAs, MMA, DMA, TMAO, TMA^+, AB, AC	Urine	HPLC	ICP-MS		42
As	IAs, MMA, DMA	Urine	HPLC	ICP-MS	Micellar HPLC	43
Au	Anti-arthritis drugs and metabolites	Urine	HPLC	ICP-MS	Reversed phase	44
Fe	Ferritin, transferrin, other hemoprotein	Liver, heart, hepatocyte	HPLC	ETAAS, ICP-MS	Off-line ETAAS; on-line ICP-MS	45
Hg	IHg, MeHg	Hair, fish, tissue	GC	CV-AFS	KOH digestion; aqueous phase ethylation	46
Hg	IHg, MeHg	Urine	HPLC	CV-AAS		47
I	I^-, iodo amino acids	Thyroid	HPLC	ICP-MS	Reversed phase	48
Pt	Cisplatin metabolites		HPLC	ICP-MS	Separation condition studied	49

Se	Selenonio choline, TMSe	Urine	HPLC	HG-AAS	Thermochemical HG	50
Se	Protein	Rat tissue	Gel electrophoresis	NAA	HPLC also used for protein separation	51
Se	ISe, TMSe	Urine	HPLC	ICP-MS		52
Zn	Proteins	Milk	HPLC	DCP-AES	Off-line gel permeation	53
Zn	Proteins	Protein mixture chicken meat	HPLC	ICP-MS	Stable isotope labelling study	54
Al, Si	Serum proteins	Serum	HPLC	ETAAS, gel electrophoresis	Off-line anion exchange separation of serum protein	55
Al, Fe	Serum proteins	Serum	HPLC	ETAAS	Off-line anion exchange separation of serum proteins	56
Cd, Cu, S, Zn	Metallothionein	Serum	HPLC	ICP-AES	Gel permeation	57
Cd, Cu, Zn	Metalloprotein	Cytoplasm	HPLC	ICP-MS	Gel permeation	58
Co, Fe, Zn	Metalloporphyrin	Whole blood	HPLC	ICP-MS		59
Cu, Zn, P, Fe, S	Whey proteins	Milk, whey	HPLC	ICP-AES	Gel permeation	60
I, Br, Cl	Inorganic halogens	Urine	HPLC	ICP-MS		61
Pb, Fe, Zn	Serum proteins	Serum	HPLC	ICP-MS	Gel permeation	62
V, Cr	Inorganic species	Urine, reference material	IC	ICP-MS		63
SHs	SHs in chicken ovalbumin	Chicken ovalbumin	HPLC	ICP-MS	Indirect SHs determination using mercaptides of organo Hg	64

Abbreviations: IAs, inorganic As; MMA, monomethyl As; DMA, dimethyl As; TMAO, trimethyl As; TMA^+, tetramethylarsonium ion; AB, arsenobetaine; AC, arsenocholine; IHg, inorganic Hg; MeHg, methyl Hg; ISe, inorganic Se; TMSe, trimethylselenonium ion.

as the Hg-selective detector has been reported. This technique also allows simultaneous analysis of inorganic and organic forms of Hg in aqueous sample.

1.5.2 Arsenic

Toxicity of As differs markedly between organic and inorganic forms. Although various forms of organic As species are identified in nature, speciation of inorganic and methylated forms of As — such as monomethyl (MMA), dimethyl (DMA), and trimethyl (TMA) As compounds — is important because methylation is considered to be the detoxification of inorganic As. Arsenic speciation in biological systems has developed in two ways: hydride generation and HPLC.

The hydride generation technique utilizes the different boiling points of various hydrides of As species evolved by reaction with $NaBH_4$ in acidic medium and tripped cryogenically.[17-19] Use of GC for separation of evolved hydrides is also applicable.[20,21] As(III), As(V), MMA, DMA, and TMA are the target species. Arsine species thus separated are detected by AAS and ICP-AES.

HPLC separation of As species has been studied extensively. Either ion exchange[22-24] or ion-pair, reversed-phase separation modes[25-27] are adopted. AAS, ICP-AES, and ICP-MS are the commonly used detectors.

1.5.3 Selenium

Speciation of Se has been studied in relation to the essentiality and toxicity of this element. Speciation of inorganic Se —Se(VI) and Se(IV) — is carried out by GC-based and fluorimetric methods preceded by the formation of a volatile piazoselenol complex with diamine reagent. Only Se(IV) reacts with such a reagent; the difference between total and Se(IV) is assumed to be Se(VI). However, this technique can be applied to relatively simple matrices such as water samples.

Analysis of trimethylselenonium ion (TMSe) is important in toxicology because TMSe is considered to be a detoxicification metabolite of Se. Off-line and on-line ion exchange separation from Se(IV) or other organoselenium species followed by various detectors including γ-counting, AAS,[28] and ICP-MS[29] has been reported. Use of NAA as an off-line detector of Se is also reported.[30]

1.5.4 Metalloprotein and Other Compounds

Gel permeation HPLC (GPC) coupled with AAS, ICP-AES, and ICP-MS detection systems has been used for the characterization of plasma or tissue metalloproteins.[31-34] The capability for simultaneous detection of C, P, and S, as well as other elements, in chromatographic eluent of ICP-AES is advantageous to protein identification and characterization, in addition to the molecular weight information provided by GPC. An anion exchange is also used for protein separation. The main target molecular species in biomedical samples so far are serum (plasma) metalloproteins and metallothionein.

References

1. Westöö, G., Determination of methylmercury salts in various kinds of biological material, *Acta Chem. Scand.*, 22, 2277–2280, 1968.
2. Analytical Methods Committee, Determination of mercury and methyl-mercury in fish, *Analyst,* 102, 769–776, 1977.
3. Bye, R. and Paus, P. E., Determination of alkylmercury compounds in fish tissue with an atomic absorption spectrometer used as a specific gas chromatographic detector, *Anal. Chim. Acta,* 107, 169–175, 1979.

4. Gui-bin, J., Zhe-ming, N., Shunronhg, W., and Heng-bin, H., Organic mercury speciation in fish by capillary gas chromatography interfaced with atomic absorption spectrometry, *Fresenius Z. Anal. Chem.*, 334, 27–30, 1989.
5. Kato, T., Uehiro, T., Yasuhara, A., and Morita, M., Determination of methylmercury species by capillary column gas chromatography with axially viewed inductively coupled plasma atomic emission spectrometric detection, *J. Anal. At. Spectrom.*, 7, 15–18, 1992.
6. Bache, C. A. and Lisk, D. J., Gas chromatographic determination of organic mercury compounds by emission spectrometry in a helium plasma. Application to the analysis of methylmercuric salts in fish, *Anal. Chem.*, 43, 950–952, 1971.
7. Quimby, B. D., Uden, P. C., and Barnes, R. M., Atmospheric pressure helium microwave detection system for gas chromatography, *Anal. Chem.*, 50, 2112–2118, 1978.
8. Prange, A. and Jantzen, E., Determination of organometallic species by gas chromatography inductively coupled plasma mass spectrometry, *J. Anal. At. Spectrom.*, 10, 105–109, 1995.
9. Magos, L., Selective atomic-absorption determination of inorganic mercury and methylmercury in undigested biological samples, *Analyst*, 96, 847–853, 1971.
10. Oda, C. E. and Ingle, Jr., J. D., Speciation of mercury by cold vapor atomic absorption spectrometry with selective reduction, *Anal. Chem.*, 53, 2305–2309, 1981.
11. Bloom, N., Determination of picogram levels of methylmercury by aqueous phase ethylation, followed by cryogenic gas chromatography with cold vapor atomic fluorescence detection, *Can. J. Fish. Aquat. Sci.*, 46, 1131–1140, 1989.
12. Rapsomanikis, S. and Craig, P. J., Speciation of mercury and methylmercury compounds in aqueous samples by chromatography-atomic absorption spectrometry after ethylation with sodium tetraethylborate, *Anal. Chim. Acta*, 248, 563–567, 1991.
13. Holak, W., Determination of methylmercury in fish by high-performance liquid chromatography, *Analyst*, 107, 1457–1461, 1982.
14. Fujita, M. and Takabatake, E., Continuous flow reducing vessel in determination of mercuric compounds by liquid chromatography/cold vapor atomic absorption spectrometry, *Anal. Chem.*, 55, 454–457, 1983.
15. Krull, I. S., Bushee, D. S., Schleicher, R. G., and Smith, Jr., S. B., Determination of inorganic and organomercury compounds by high-performance liquid chromatography-inductively coupled plasma emission spectrometry with cold vapor generation, *Analyst*, 111, 345–349, 1986.
16. Bushee, D. S., Speciation of mercury using liquid chromatography with detection by inductively coupled plasma mass spectrometry, *Analyst*, 113, 1167–1170, 1988.
17. Braman, R. S., Johnson, D. L., Foreback, C. C., Ammons, J. M., and Bricker, J. L., Separation and determination of nanogram amounts of inorganic arsenic and methylarsenic compounds, *Anal. Chem.*, 49, 621–625, 1977.
18. Andreae, M. O., Determination of arsenic species in natural waters, *Anal. Chem.*, 49, 820–823, 1977.
19. Howard, A. G. and Arbab-Zabar, M. H., Determination of "inorganic" arsenic(III) and arsenic(V), "methylarsenic" and "dimethylarsenic" species by selective hydride evolution atomic-absorption spectrometry, *Analyst*, 106, 213–220, 1981.
20. Talmi, Y. and Bostick, D. T., Determination of alkylarsenic acids in pesticide and environmental samples by gas chromatography with a microwave emission spectrometric detection system, *Anal. Chem.*, 47, 2145–2150, 1975.
21. Odanaka, Y., Tsuchiya, N., Matano, O., and Goto, S., Determination of inorganic arsenic and methylarsenic compounds by gas chromatography and multiple ion detection mass spectrometry after hydride generation-heptane cold trap, *Anal. Chem.*, 55, 929–932, 1983.
22. Morita, M., Uehiro, T., and Fuwa, K., Determination of arsenic compounds in biological samples by liquid chromatography with inductively coupled plasma-atomic emission spectrometric detection, *Anal. Chem.*, 53, 1806–1808, 1981.
23. Spall, W. D., Lynn, J. G., Andersen, J. L., Valdez, J. G., and Gurley, L. R., High-performance liquid chromatographic separation of biologically important arsenic species utilizing on-line inductively coupled argon plasma atomic emission spectrometric detection, *Anal. Chem.*, 58, 1340–1344, 1986.
24. Heitkemper, D., Creed, J., Caruso, J., and Fricke, F. L., Speciation of arsenic in urine using high-performance liquid chromatography with inductively coupled plasma mass spectrometric detection, *J. Anal. At. Spectrom.*, 4, 279–284, 1989.
25. Nisamaneepong, W., Ibrahim, M., Gilbert, T. W., and Caruso, J. A., Speciation of arsenic and cadmium compounds by reversed-phase ion-pair LC with single-wavelength inductively coupled plasma detection, *J. Chromatgr. Sci.*, 22, 473–477, 1984.
26. Shibata, Y. and Mirita, M., Speciation of arsenic by reversed-phase high performance liquid chromatography-inductively coupled plasma mass spectrometry, *Anal. Sci.*, 5, 107–109, 1989.
27. Thomas, P. and Sniatecki, K., Inductively coupled plasma mass spectrometry: application to the determination of arsenic species, *Fresenius J. Anal. Chem.*, 351, 410–414, 1995.

28. Blais, J.-S., Huyghues-Despointes, A., Momplaisir, G. M., and Marshall, W. D., High-performance liquid chromatography-atomic absorption spectrometry interface for the determination of selenonicholine and trimethylselenonium cations: application to human urine, *J. Anal. At. Spectrom.*, 6, 225–232, 1991.
29. Yang, K.-L. and Jiang, S.-J., Determination of selenium compounds in urine samples by liquid chromatography-inductively coupled plasma mass spectrometry with an ultrasonic nebulizer, *Anal. Chim. Acta,* 307, 109–115, 1995.
30. Blotcky, A. J., Hansen, G. T., Opelanio-Buencamino, L. R., and Rack, E. P., Determination of trimethylselenonium ion in urine by ion-exchange chromatography and molecular neutron analysis, *Anal. Chem.*, 57, 1937–1941, 1985.
31. Suzuki, K. T., Direct connection of high-speed liquid chromatography (equipped with gel permeation column) to atomic absorption spectrophotometer for metalloprotein analysis: metallothionein, *Anal. Biochem.*, 102, 31–34, 1980.
32. Morita, M., Uehiro, T., and Fuwa, K., Speciation and elemental analysis of mixtures by high performance liquid chromatography with inductively coupled argon plasma emission spectrometric detection, *Anal. Chem.*, 52, 349–351, 1980.
33. Thompson, J. J. and Houk, R. S., Inductively coupled plasma mass spectrometric detection for multi-element flow injection analysis and elemental speciation by reversed-phase liquid chromatography, *Anal. Chem.*, 58, 2541–2548, 1986.
34. Sunaga, H., Kobayashi, E., Shimojo, N., and Suzuki, K. T., Detection of sulfur-containing compounds in control and cadmium-exposed rat organs by high-performance liquid chromatography-vacuum-ultraviolet inductively coupled plasma-atomic emission spectrometry (HPLC-ICP), *Anal. Biochem.*, 160, 160–168, 1987.
35. Hakala, E. and Pyy, L., Selective determination of toxicologically important arsenic species in urine by high-performance liquid chromatography-hydride generation atomic absorption spectrometry, *J. Anal. At. Spectrom.,* 7, 191–196, 1992.
36. Mürer, A. J. L., Abildtrup, A., Poulsen, O. M., and Christensen, J. M., Effect of seafood consumption on the urinary level of total hydride-generating arsenic compounds. Instability of arsenobetaine and arsenocholine, *Analyst,* 117, 677–680, 1992.
37. Sheppard, B. S., Caruso, J. A., Heitkemper, D. T., and Wolnik, K. A., Arsenic speciation by ion chromatography with inductively coupled plasma mass spectrometric detection, *Analyst,* 117, 971–975, 1992.
38. Larsen, E. H., Pritzl, G., and Hansen, S. H., Speciation of eight arsenic compounds in human urine by high-performance liquid chromatography with inductively coupled plasma mass spectrometric detection using antimonate for internal chromatographic standardization, *J. Anal. At. Spectrom.,* 8, 557–563, 1993.
39. Liu, Y. M., Fernádez-Sánchez, M. L., González, E. B., and Sanz-Medel, A., Vesicle-mediated high performance liquid chromatography coupled to hydride generation inductively coupled plasma atomic emission spectrometry for speciation of toxicologically important arsenic species, *J. Anal. At. Spectrom.,* 8, 815–820, 1993.
40. Dean, J. R., Ebdon, L., Foulkes, M. E., Crews, H. M., and Massey, R. C., Determination of the growth promotor, 4-hydroxy-3-nitrophenyl-arsenic acid, in chicken tissue by coupled high-performance liquid chromatography-inductively coupled plasma mass spectrometry, *J. Anal. At. Spectrom.,* 9, 615–618, 1994.
41. Shibata, Y., Yoshinaga, J., and Morita, M., Detection of arsenobetaine in human blood, *Appl. Organomet. Chem.,* 8, 249–251, 1994.
42. Le, X.-C., Cullen, W. R., and Reimer, K. J., Human urinary arsenic excretion after one-time ingestion of seaweed, crab, and shrimp, *Clin. Chem.*, 40, 617–624, 1994.
43. Ding, H., Wang, J., Dorsey, J. G., and Caruso, J. A., Arsenic speciation by micellar liquid chromatography with inductively coupled plasma mass spectrometric detection, *J. Chromatogr. A,* 694, 425–431, 1995.
44. Zhao, Z., Jones, W. B., Tepperman, K., Dorsey, J. G., and Elder, T. C., Determination of gold-based antiarthritis drugs and their metabolites in urine by reversed-phase ion-pair chromatography with ICP-MS detection, *J. Pharm. Biomed. Anal.,* 4, 279–287, 1992.
45. Stuhne-Sekalec, L., Xu, S. X., Parkes, J. G., Olivieri, N. F., and Templeton, D. M., Speciation of tissue and cellular iron with on-line detection by inductively coupled plasma-mass spectrometry, *Anal. Biochem.,* 205, 278–284, 1992.
46. Liang, L., Bloom, N. S., and Horvat, M., Simultaneous determination of mercury speciation in biological materials by GC/CVAFS after ethylation and room-temperature precollection, *Clin. Chem.,* 40, 602–607, 1994.
47. Aizpún, B., Fernández, M. L., Blanco, E., and Sanz-Medel, A., Speciation of inorganic mercury(II) and methylmercury by vesicle-mediated high-performance liquid chromatography coupled to cold vapor atomic absorption spectrometry, *J. Anal. At. Spectrom.,* 9, 1279–1284, 1994.
48. Takatera, K. and Watanabe, T., Speciation of iodo amino acids by high-performance liquid chromatography with inductively coupled plasma mass spectrometric detection, *Anal. Chem.,* 65, 759–762, 1993.
49. Zhao, Z., Tepperton, K., Dorsey, J. G., and Elder, R. C., Determination of cisplatin and some possible metabolites by ion-pairing chromatography with inductively coupled plasma mass spectrometric detection, *J. Chromatgr.,* 615, 83–89, 1993.
50. Blais, J.-S., Huyghues-Despointes, A., Momplaisir, G. M., and Marshall, W. D., High-performance liquid chromatography-atomic absorption spectrometry interface for the determination of selenoniocholine and trimethylselenonium cations: application to human urine, *J. Anal. At. Specrom.,* 6, 225–231, 1991.

51. Behen, D., Weiss-Nowak, C., Kalcklösch, M., Westphal, C., Gessner, H., and Kyriakopoulos, A., Application of nuclear analytical methods in the investigation of new selenoproteins, *Biol. Trace Elem. Res.*, 31, 287–297, 1994.
52. Yang, K.-L. and Jiang, S.-J., Determination of selenium compounds in urine samples by liquid chromatography-inductively coupled plasma mass spectrometry with an ultrasonic nebulizer, *Anal. Chim. Acta,* 307, 109–115, 1995.
53. Michalke, B., Münch, D. C., and Schramel, P., Contribution to Zn-speciation in human breast milk: fractionation of organic compounds by HPLC and subsequent Zn-determination by DCP-AES, *J. Trace Elem. Electrolytes Health Dis.,* 5, 251–259, 1991.
54. Owen, L. M. W., Crews, H. M., Hutton, R. C., and Walsh, A., Preliminary study of metals in proteins by high-performance liquid chromatography-inductively coupled plasma mass spectrometry using multi-element time resolved analysis, *Analyst,* 117, 649–655, 1992.
55. Wróbel, K., González, E. B., Wróbel, K., and Sanz-Medel, A., Aluminum and silicon speciation in human serum by ion-exchange high-performance liquid chromatography-electrothermal atomic absorption spectrometry and gel electrophoresis, *Analyst,* 120, 809–815, 1995.
56. Van Landeghem, G. F., D'Haese, P. C., Lamberts, L. V., and De Broe, M. E., Quantitative HPLC/ETAAS hybrid method with an on-line metal scavenger for studying the protein binding and speciation of aluminum and iron, *Anal. Chem.,* 66, 216–222, 1994.
57. Brätter, P., Brunetto, R., Gramm, H.-J., Recknagel, S., and Siemes, H., Detection by HPLC-ICP of metallothionein in serum of epileptic child with valproate-associated hepatotoxicity, *J. Trace Elem. Electrolytes Health Dis.,* 6, 251–255, 1992.
58. Mason, A. Z., Storms, S. D., and Jenkins, K. D., Metalloprotein separation and analysis by directly coupled size exclusion high-performance liquid chromatography inductively coupled plasma mass spectrometry, *Anal. Biochem.,* 186, 187–201, 1990.
59. Kumar, U., Dorsey, J. G., Caruso, J. A., and Evans, E. H., Metalloporphyrin speciation by liquid chromatography and inductively coupled plasma-mass spectrometry, *J. Chromatogr. Sci.,* 32, 282–285, 1994.
60. Suzuki, K. T., Tamagawa, H., Hirano, S., Kobayashi, E., Takahashi, K., and Shimojo, N., Changes in element concentration and distribution in breast-milk fractions of a healthy lactating mother, *Biol. Trace Elem. Res.,* 28, 109–121, 1991.
61. Salov, V. V., Yoshinaga, J., Shibata, Y., and Morita, M., Determination of inorganic halogen species by liquid chromatography with inductively coupled argon plasma mass spectrometry, *Anal. Chem.,* 64, 2425–2428, 1992.
62. Gercken, B. and Barnes, R. M., Determination of lead and other trace element species in blood by size exclusion chromatography and inductively coupled plasma/mass spectrometry, *Anal. Chem.,* 63, 283–287, 1991.
63. Tomlinson, M. J., Wang, J., and Caruso, J. A., Speciation of toxicologically important transition metals using ion chromatography with inductively coupled plasma mass spectrometric detection, *J. Anal. At. Spectrom.,* 9, 957–964, 1994.
64. Takatera, K. and Watanabe, T., Determination of sulfhydryl groups in ovalbumin by high-performance liquid chromatography with inductively coupled plasma mass spectrometric detection, *Anal. Chem.,* 65, 3644–3646, 1993.

1.6 A Note About Reference Materials for Metal Analysis

The importance of analytical quality control is well recognized not only in the biomedical field but also in other fields in which chemical analysis is carried out. There are two aspects to the quality of chemical analysis: accuracy and precision. The accuracy denotes how close the obtained result is to the "true" value. The precision stands for reproducibility of analytical results. The accuracy and precision of metal analysis used in any research must be reported properly to other researchers. Use of reference material (RM) is the most practical way of testing the accuracy of metal analysis, though there are other means such as interlaboratory comparison, recovery tests, etc. A reference material is a homogeneous sample for which the metal content is known (a certified value or tentative "true" value). The accuracy and precision of chemical analysis employed can be evaluated by repeated analyses of the same RM; the mean value compared with the certified value and the variability of the results reflect the precision.

A variety of RMs for a variety of uses, such as clinical, biological, environmental, or metallurgical, are currently available from several institutions worldwide. Since it is well recognized that any analytical methods may suffer from interference from the matrix of the

TABLE 15

Reference Materials for Metal Analysis of Biomedical Samples Available in 1995

Supplier	Name	Form	Composition Certified
BCR	CMR 184 bovine muscle	Powder	Element composition
BCR	CRM 185 bovine liver	Powder	Element composition
BCR	CRM 186 pig kidney	Powder	Element composition
BCR	CRM 397 human hair	Powder	Element composition
IAEA	A-13 animal blood	Powder	Element composition
NIES	CRM No. 13 human hair	Powder	Element composition, including MeHg
NIST	SRM 909a human serum	Powder (to be reconstituted)	Element composition and clinical parameters
NIST	SRM 955a lead in blood	Powder (to be reconstituted)	Pb; set of four different Pb levels
NIST	SRM 1400 bone ash	Powder	Element composition
NIST	SRM 1486 bone meal	Powder	Element composition
NIST	SRM 1577b bovine liver	Powder	Element composition
NIST	SRM 1598 bovine serum	Liquid (frozen)	Element composition
NIST	SRM 2670 freeze-dried human Urine	Powder (to be reconstituted)	Element composition; set of low and elevated levels of toxic elements
NIST	SRM 2671a freeze-dried urine	Powder (to be reconstituted)	F; set of low and elevated levels
NIST	SRM 2672a freeze-dried urine	Powder (to be reconstituted)	Hg; set of low and elevated levels

Abbreviations: BCR = Community Bureau of Reference, Brussels, EU; IAEA = International Atomic Energy Agency, Vienna, Austria; NIES = National Institute for Environmental Studies, Tsukuba, Japan; NSIT = National Institute of Standards and Technology, Gaithersburg, MD.

sample, it is essential to use a RM of similar chemical composition to the real sample being analyzed. Table 15 lists currently available biomedical RMs. Careful choice of RM and proper description of the analytical results of the material are requisites of any scientific work.

1.7 Western Blotting Method

Yasunobu Aoki

The western blotting technique was developed to identify an antigen protein corresponding to a specific antibody. This technique was applied to identify a metal-binding protein using the radio-isotope of a divalent metal ion instead of an antibody. The sample proteins in Table 16 are separated by SDS/polyacrylamide gel electrophoresis.[1] Separated proteins are electrophoretically transferred to the membrane made of nitrocellulose or polyvinyliden fluoride (PVDF) and are immobilized on it.[2] After the membrane immobilizing the sample proteins is equilibrated in a buffer solution, it is incubated in a buffer containing the radio-isotope of a metal ion (a binding buffer) and then washed with a washing solution. The radio-isotope-binding proteins on the membrane are visualized by autoradiography. A divalent metal ion binds to proteins nonspecifically. In order to reduce nonspecific binding of the radio-isotope to proteins, an appropriate competitor ion (a divalent metal ion) is often added to a binding buffer. Concentration of the competitor ion, ion strength, and pH of the buffers are important factors in detecting specific metal-binding proteins.

TABLE 16

Detection of Metal-Binding Proteins by Western Blotting Technique

Metal	Equilibrium Buffer	Binding Buffer	Washing Buffer	Identified Proteins	Membrane[a]
Cadmium	10 m*M* Tris/HCl (pH 7.4)	1 μCi/ml $^{109}CdCl_2$, 0.1 *M* KCl, 1[3] or 0.1[3-7] m*M* zinc acetate, and 10 m*M* Tris/HCl (pH 7.4)	10 m*M* Tris/HCl buffer (pH 8.0) and 0.1 *M* KCl	Metallothionein,[3-5] Cd-binding protein,[6] ornithine carbomyltransferase[7]	NC[3] or PVDF[4-7]
Calcium	60 m*M* KCl, 5 m*M* $MgCl_2$, and 10 m*M* imidazole/HCl (pH 6.8)	1[8-10] or 1.2[11] μCi/ml ^{45}Ca, 60 m*M* KCl, 5 m*M* $MgCl_2$, and 10 m*M* imidazole/HCl (pH 6.8)	Distilled water[8,9] or 50% ethanol[8-11]	Troponin C,[8] myosin DTNB light chain,[8] calmodulin,[8] 55 K Ca-binding protein,[8] parvalbumin,[8] 21,000-dalton Ca binding protein,[9] calbindin D-28K,[10] phosphatidylinositol transfer protein[11]	NC
Copper	50 m*M* KCl and 100 m*M* Tris/HCl (pH 7.5)	0.5–1 μCi/ml $^{67}CuCl_2$, 50 m*M* KCl, and 100 m*M* Tris/HCl (pH 7.4)	50 m*M* KCl and 100 m*M* Tris/HCl (pH 7.5)	Manganese superoxide dismutase[12]	NC
Nickel	50 m*M* KCl, 5 m*M* $CaCl_2$, and 100 m*M* Tris/HCl (pH 7.4)[13-16] (Buffer F)	3.3 μCi/ml $^{63}NiCl_2$-containing Buffer F[13-16]	Buffer F[13-16]	Hepatic serpin,[14] lipovitellin 2β,[15] fructose-1, 6-bisphosphate aldolase A[16]	NC
	Buffer F and Buffer F with 0.5 m*M* dithiothreitol (DTT)[17]	3.3 μCi/ml $^{63}NiCl_2$-containing Buffer F[17]	Buffer F and Buffer F with 0.5 m*M* DTT[17]	Transcription factor IIIA,[17 b] p43[17 b]	
Zinc	10 m*M* Tris/HCl (pH 7.5)	0.2 μCi/ml $^{65}ZnCl_2$, 0.1 *M* KCl, and 10 m*M* Tris/HCl (pH 7.5)	0.1 *M* KCl and 10 m*M* Tris/HCl (pH 7.5)	Alcohol dehydrogenase,[18] thermolysin,[18] pancreas ribonuclease A,[18] transcription factor IIIA,[18] protein of RNP particle,[18] T4 gene 32 protein,[18] poly(ADP-ribose)-polymerase,[18,19] nuclear protein associated with nucleosomes[20]	NC
	50 m*M* NaCl and 100 m*M* Tris/HCl (pH 6.8)	0.25–2 μCi/ml $^{65}ZnCl_2$, 50 m*M* NaCl, and 100 m*M* Tris/HCl (pH 6.8)	50 m*M* NaCl and 100 m*M* Tris/HCl (pH 6.8)	Retroviral gag protein[21]	NC
			1 m*M* DTT, 50 m*M* NaCl, and 100 m*M* Tris/HCl (pH 6.8)	Papillovirus polypeptides E6 and E7[22]	PVDF
	10 m*M* DTT, 50 m*M* NaCl, and 100 m*M* Tris/HCl (pH 6.8)	10–15 μ*M* $^{65}ZnCl_2$, 50 m*M* NaCl, and 100 m*M* Tris/HCl (pH 6.8)	0.05% Tween 20 and phosphate-buffered saline (pH 7.0)	Tubulin,[23] alcohol dehydrogenase,[23] leucine aminopeptidase,[23] bovine serum albumin,[23] carboxypeptidase A,[23] alkaline phosphatase,[23] carbonic anhydrase[23]	NC

TABLE 16 (continued)

Detection of Metal-Binding Proteins by Western Blotting Technique

Metal	Equilibrium Buffer	Binding Buffer	Washing Buffer	Identified Proteins	Membrane[a]
Zinc (cont.)	10 m*M* β-mercaptoethanol, 1 m*M* $MnCl_2$, and 10 m*M* Tris/HCl (pH 7.5)	1 μCi/ml $^{65}ZnCl_2$, 10 m*M* β-mercaptoethanol, 1 m*M* $MnCl_2$, and 10 m*M* Tris/HCl (pH 7.5)	10 m*M* β-mercaptoethanol and 10 m*M* Tris/HCl (pH 7.5);[24] 10 m*M* Tris/HCl (pH 7.5)[25,26]	Cystein-rich intestinal protein,[24,25] metallothionein,[25 c] Zn-binding protein in mammary gland[26]	NC

[a] NC = nitrocellulose; PVDF = PVDF membrane.

[b] Binding of Zn and Cd to these proteins was also detected.

[c] Binding of Cd to these proteins was also detected.

REFERENCES

1. Laemmli, U. K., Cleavage of structural proteins during the assembly of the head of bacteriophage T4, *Nature*, 227, 680–685, 1970.
2. Towbin, H., Ataehelin, T., and Gordon, J., Electrophoretic transfer of proteins from polyacrylamide gels to nitrocellulose sheets: procedure and some applications, *Proc. Natl. Acad. Sci. USA*, 76, 4350–4354, 1979.
3. Aoki, Y., Kunimoto, M., Shibata, Y., and Suzuki, K. T., Detection of metallothionein on nitrocellulose membrane using western blotting technique and its application to identification of cadmium-binding proteins, *Anal. Biochem.*, 157, 117–122, 1986.
4. Aoki, Y., Tohyama, C., and Suzuki, K. T., A western blotting procedure for detection of metallothionein, *J. Biochem. Biophys. Method.*, 23, 207–216, 1991.
5. Aoki, Y. and Suzuki, K. T., Detection of metallothionein by western blotting, in *Methods in Enzymology*, Vol. 205, Riordan, J. F. and Vallee, B. L., Eds., Academic Press, San Diego, 1991, p. 108.
6. Aoki, Y., Hatakeyama, S., Kobayashi, N., Sumi, Y., Suzuki, T., and Suzuki, K. T., Comparison of cadmium-binding protein induction among mayfly larvae of heavy metal resistant (*Baetis thermicus*) and susceptible species (*B. yoshinensis* and *B. sahoensis*), *Comp. Biochem. Physiol.*, 93C, 345–347, 1989.
7. Aoki, Y., Sunaga, H., and Suzuki, K. T., A cadmium-binding protein in rat liver identified as ornithine carbomyl transferase, *Biochem. J.*, 250, 735–742, 1988.
8. Maruyama, K., Mikawa, T., and Ebashi, S., Detection of calcium binding protein by ^{45}Ca autoradiography on nitrocellulose membrane after sodium dodecyl sulfate del electrophoresis, *J. Biochem.*, 95, 511–519, 1984.
9. McDonald, J. R., Groschel-Stewart, U., and Walsh, M. P., Isolation of two isoforms of a 21,000-dalton Ca^{2+}-binding protein of bovine brain, *Biochem. Int.*, 15, 587–597, 1987.
10. Dunn, M. A., Johnson, N. E., Liew, M. Y. B., and Ross, E., Dietary aluminum chloride reduces the amount of intestinal calbindin D-28K in chicks fed low calcium or low phosphorus diets, *J. Nutr.*, 123, 1786–1793, 1993.
11. Vihtelic, T. S., Goebl, M., Milligan, S., O'Tousa, J. E., and Hyde, D. R., Localization of Drosophila retinal degeneration B, a membrane-associated phosphatidylinositol transfer protein, *J. Cell Biol.*, 122, 1013–1022, 1993.
12. Kaler, S. G., Maraia, R. J., and Gahl, W. A., Human manganese superoxide dismutase is readily detectable by a copper blotting technique, *Biochem. Med. Metab. Biol.*, 46, 406–415, 1991.
13. Lin, S. M., Hopfer, S. M., Brennan, S. M., and Sunderman, Jr., F. W., Protein blotting for detection of nickel-binding proteins, *Res. Comm. Chem. Pathol. Pharmacol.*, 65, 275–288, 1989.
14. Beck, B. L., Henjum, D. C., Antonijczuk, K., Zaharia, O., Korza, G., Ozols, J., Hopfer, S. M., Barber, A. M., and Sunderman, Jr., W. F., pNiXa, a Ni^{2+}-binding protein in *Xenopus* oocytes and embryos, shows identity to Ep45, an estrogen-regulated hepatic serpin, *Res. Comm. Chem. Pathol. Pharmacol.*, 77, 3–16, 1992.
15. Grbac-Ivankovic, S., Antonijczuk, K., Varghese, A. H., Plowman, M. C., Antonijczuk, A., Korza, G., Ozols, J., and Sunderman, Jr., W. F., Lipovitellin 2β is the 31 kD Ni^{2+}-binding protein (pNiXb) in *Xenopus* oocytes and embryos, *Mol. Reproduct. Dev.*, 38, 256–263, 1992.
16. Antonijczuk, K., Kroftova, O. S., Varghese, A. H., Antonijczuk, A., Henjum, D. C., Korza, G., Ozols, J., and Sunderman, Jr., W. F., The 40 kDa $^{63}Ni^{2+}$-binding protein (pNiXc) on western blots of *Xenopus laevis* oocytes and embryos in the monomer of fructose-1,6-bisphosphate aldolase A, *Biochim. Biophys. Acta*, 1247, 81–179, 1995.
17. Makowski, G. S., Lin, S. M., Brennan, S. M., Smilowitz, H. M., Hopfer, S. M., and Sunderman, Jr., W. F., Detection of two Zn-finger proteins of *Xenopus laevis*, TFIIIA, and p43, by probing western blots of ovary cytosol with $^{65}Zn^{2+}$, $^{63}Ni^{2+}$, $^{109}Cd^{2+}$, *Biol. Trace Elem. Res.*, 29, 93–109, 1991.
18. Mazen, A., Gerard, G., and de Murcia, G., Zinc-binding proteins detected by protein blotting, *Anal. Biochem.*, 172, 39–42, 1988.
19. Mazen, A., Menissier-de Murcia, J., Molinete, M., Simonin, F., Gradwohl, G., Poirier, G., and de Murcia, G., Poly(ADP-ribose)polymerase: a novel finger protein, *Nucl. Acids Res.*, 17, 4689–4698, 1989.
20. Liew, C. C. and Chen, H. Y., A nuclear protein associated with actively transcribed nucleosomes exhibits Zn^{2+}-binding activity, *FEBS Lett.*, 258, 116–118, 1989.
21. Schiff, L. A., Nibert, M. L., and Fields, B. N., Characterization of a zinc blotting technique: evidence that a retroviral gag protein binds zinc, *Proc. Natl. Acad. Sci. USA*, 85, 4195–4199, 1988.
22. Barbosa, M. S., Lowy, D. R., and Schiller, J. T., Papillomavirus polypeptides E6 and E7 are zinc-binding proteins, *J. Virol.*, 63, 1404–1407, 1989.
23. Serrano, L., Dominguez, J. E., and Avila, J., Identification of zinc-binding site of proteins: zinc binds to the amino-terminal region of tubulin, *Anal. Biochem.*, 172, 210–218, 1988.
24. Hempe, J. M. and Cousins, R. J., Cysteine-rich intestinal protein binds zinc during transmucosal zinc transport, *Proc. Natl. Acad. Sci. USA*, 88, 9671–9674, 1991.
25. Hempe, J. M. and Cousins, R. J., Cysteine-rich intestinal protein and intestinal metallothionein: an inverse relationship as a conceptual model for zinc absorption in rats, 122, 89–95, 1992.
26. Lee, D. Y., Shay, N. F., and Cousins, R. J., Altered zinc metabolism occurs in murine lethal milk syndrome, 122, 2233–2238, 1992.

2 ANIMAL MODELS

2.1 Toxicological Study Using Animals

Akira Yasutake and Kimiko Hirayama

The various toxic effects caused by a number of metals have been documented in experimental animals. Specific organ toxicity closely related to the tissue distribution has been documented for each metal. Because susceptibility to metal toxicity sometimes varies significantly between sex or strain, this information has also been cited. Here, we focus on recent rodent data. Since the amount of data cited here may not be sufficient for some colleagues, previous publications[1,2] are recommended for reference.

2.1.1 Arsenic (As)

Arsenic occurs in the environment in its inorganic and organic forms, the toxicity of the latter being very low. Among the number of inorganic arsenic compounds, arsenite (trivalent), arsenate (pentavalent), arsenic oxide, gallium arsenide, and arsine are mentioned in the literature. Toxic effects of arsenite and arsenate are shown in Table 17. The LD_{50} value for sodium arsenite ($NaAsO_2$) was shown to be 6.7 mg As per kg in mice after s.c. injection.[6] Hyperglycemia and glucose intolerance were documented as acute effects of sodium arsenite in rats.[3] Oral exposure to arsenite using drinking water (100 ppm) for 4 to 11 days caused decreased acetylcholine esterase and sorbitol dehydrogenase activities in axons and increased leucine amino peptidase activity in glial cells.[4] Giving 0.5- to 10-ppm As-containing water for 3 weeks caused immunosuppressive effects in mice.[7] Fowler and Woods[8] observed mitochondria swelling and decreased monoamine oxidase (MAO) activity in the liver of mice given water containing 40 or 85 ppm As in the form of sodium arsenate for 6 weeks. Chronic exposure of rats to arsenite or arsenate in food (up to 400 ppm) caused a reduction in body weight and the survival period and an enlargement of the common bile duct.[5] When $NaAsO_2$ was given to pregnant mice at dose levels of 40 (for p.o.) or 12 (for i.p.) mg As per kg, fetal death or resorption resulted.[9,10] Injection of arsenate (45 mg As per kg, i.p.) also caused fetal resorption.[11]

Gallium arsenide (GaAs) is an excellent semiconductor material used in microcircuits. Intratracheal injection of GaAs up to 200 mg/kg or oral administration of up to 2000 mg/kg caused pathological changes in lung and kidney,[14] inhibition of tissue δ-aminolevulinic acid dehydratase (ALAD) activities,[12,14] and increased urinary elimination of porphyrin and δ-aminolevulinic acid (ALA) (Table 18).[12-14] Acute fibrogenic responses of the lung were also documented in rats intratracheally injected with GaAs (100 mg/kg).[15] Sikorski et al.[16] found that female mice intratracheally treated with 50 to 200 mg GaAs per kg showed reduced IgM antibody response to sheep erythrocytes.

The LD_{50} value for arsenic oxide (As_2O_3) was shown to be 11.3 mg/kg after s.c. injection in mice.[17] Webb et al.[15] reported that As_2O_3 caused effects similar to GaAs in rat lung. Serial oral administration of As_2O_3 (3 mg/kg) for 10 days resulted in increased motor activity, whereas higher dose level (10 mg/kg × 10 days) brought about the reverse result.[18] Inhalation of As_2O_3 aerosol (270 to 940 μg As/m^3) for 3 hr caused decreased pulmonary bacterial activity and increased infectious mortality in mice.[19] Arsine gas was documented to cause various hematological alterations in mice, such as decreased hematocrit, erythropoiesis, and hemolytic anemia.[20,21]

TABLE 17

Toxic Effects of Arsenate and Arsenite

Animal Strain, Sex, (Weight)[a]	Chemical	Dose (mg As per kg or ppm As)	Route	Effects	Ref.
Rats					
CD, M	$NaAsO_2$	5 or 10	i.p.	Hyperglycemia, glucose intolerance (1.5–3 hr)	3
Wistar, M	$NaAsO_2$	100 ppm/water × 4–11 d		Axons: acetylcholine esterase, sorbitol dehydrogenase ↓ Glial cells: leucine amino peptidase ↑	4
Osborne, M + F	$NaAsO_2$	125–400 ppm/food × 2 y		Body weight, survival ↓; enlargement of common bile duct	5
	Na_2HAsO_4	250–400 ppm/food × 2 y			
Mice					
CD-1, M	$NaAsO_2$	6.7	s.c.	50% mortality	6
Swiss cross, M	$NaAsO_2$	0.5–10 ppm/water × 3 wk		Immunosuppressive effect	7
C57BL/6, ?, (10–15 g)	Na_2HAsO_4	40, 85 ppm/water × 6 wk		Liver: mitochondrial swelling, MAO ↑	8
CD-1, pregnant	$NaAsO_2$	40 or 45	p.o.	Fetal death (max: G.D. 13 injection)[b]	9
CD-1, pregnant	$NaAsO_2$	12 (G.D. 9 or 13)[b]	i.p.	Fetal death, resorption (G.D. 13, 44–51%; G.D. 9, 73–95%)[b]	10
Swiss, pregnant	Na_2HAsO_4	45	i.p.	Fetal abnormality, resorption (G.D. 8 injection)[b]	11

[a] Data from adult animals, unless otherwise specified.
[b] G.D. = gestation day; MAO = monoamine oxidase.

2.1.2 Beryllium (Be)

Toxic effects of beryllium have been well investigated by the inhalation route, because most cases in humans have occurred via this route. The lung is the major target in beryllium inhalation (Table 19). Consecutive exposure for 14 days to 2.59 mg Be per m^3 for 2 hr as $BeSO_4$ caused 80% mortality in male rats.[22] Single (1-hr) exposure to the same salt in a relatively higher dose (3.3 to 13 mg Be per m^3) resulted in pathological changes[23] or an increase in lavage lactic dehydrogenase (LDH) and alkaline phosphatase activities.[24] Inhalation of beryllium oxide (0.45 mg Be per m^3 for 1 hr) also caused similar effects.[25] Chronic exposure to a much lower dose caused lung tumors.[26] Intratracheal injection of beryllium was also effective to examine pulmonary toxicity. Groth et al.[27] showed that beryllium (metal or passivated) and its alloy with a high beryllium content could induce lung neoplasms in rats 16 to 19 months after intratracheal injection (>0.3 mg Be), but alloys containing a low amount (<4%) of beryllium failed. If rats were intravenously injected with $BeSO_4$, inhibition of hepatic enzyme inductions by the specific drugs was observed.[28] Vacher et al.[29] demonstrated that the toxic effect of the soluble salt of beryllium was much higher than that of insoluble salt after i.v. injection in mice. Mathur et al.[30] found that if nitrate salt (14.6 μg Be per kg, i.v.) was injected into pregnant rats on gestation day 11, all embryos were resorbed. However, no resorption occurred from its injection on other days.

TABLE 18

Toxic Effects of GaAs and As_2O_3

Animal Strain, Sex, (Age)[a]	Chemical	Dose (mg GaAs or As_2O_3/kg)	Route	Effects	Ref.
Rats					
?, M	GaAs	500–2000	p.o.	Blood: ALAD, Zn-protoporphyrin ↑ Brain, liver: ALAD ↑	12
F344, M	GaAs	10, 30, 100 1000	i.t. p.o.	Lung weight, urinary porphyrin ↑ Urinary porphyrin ↑	13
CD, M	GaAs	100, 200	i.t.	Blood, liver, kidney: ALAD ↓ Urine: ALA ↑ Lung, kidney: pathological change	14
F344, M	GaAs	100	i.t.	Lung: tissue weight, protein, DNA, 4-HyPro ↑; acute fibrogenic response	15
	As_2O_3	17	i.t.	Lung: tissue weight protein, DNA, 4-HyPro ↑; acute fibrogenic response	
Mice					
B6C3F1, F	GaAs	50, 100, 200	i.t.	Splenic accessory cell function ↓ IgM antibody response to sheep RBC ↓	16
NMRI, M	As_2O_3	11.3	s.c.	50% mortality (30 d)	17
ddY, ?, (4 wk)	As_2O_3	3 × 10 d 10 × 10 d	p.o. p.o.	Motor activity ↑ Motor activity ↓ Brain: alteration in monoamine metabolite	18
CD-1, F, (4–5 wk)	As_2O_3 (aerosol)	270–940 μg As per m³ × 3 hr		Infectious mortality ↑ Bactericidal activity ↓	19

[a] Data from adult animals, unless otherwise specified.

TABLE 19

Pulmonary Toxicity of Beryllium Aerosol Inhalation or Intratracheal Injection

Animal Strain, Sex	Chemicals	Dose (mg Be per m³ × hr)	Duration of Exposure	Effect	Ref.
Rats					
F344, M	$BeSO_4$	2.59 × 2	14 d	80% mortality (15 d)	22
F344, M	$BeSO_4$	13 × 1		Pathological change (8 d)	23
F344, M	$BeSO_4$	3.3 or 7.2 × 1		Lavage LDH, alkaline phosphatase (21 d) ↑	24
F344, M	BeO	0.45 × 1		Lavage LDH, acid, and alkaline phosphatase ↑	25
?, M + F	$BeSO_4$	0.035 × 7	9 months (5 d/wk)	Tumor	26
Mice					
BALB/c, M	$BeSO_4$	13 × 1		Pathological change (5 d)	23
BALB/c, M	$BeSO_4$	7.2 × 1		Lavage LDH, alkaline phosphatase (21 d) ↑	24

2.1.3 Cadmium (Cd)

Cadmium damages various organs, such as liver, kidney, testis, bone, and lung, of experimental animals. The organ toxicity varies widely depending on its chemical forms, administration routes, and dosing schedule.

TABLE 20

Acute Effects of Cadmium Toxicity

Animal Strain, Sex, (Age)[a]	Dose[b] (mg Cd per kg)	Route	Effects	Ref.
Rats				
Wistar, F	1.8	i.v.	LD_{50}	31
SD, M	3.9	i.v.	Liver lesions (1 hr)	32
SD, M	2	i.v.	Liver: stress protein synthesis (2 hr) ↑	33
SD, F	7	i.v.	55% mortality (96 hr); ovary, liver: lesions	34
SD, M	6.6	s.c.	Testis: lesions	
SD, M	1.65	s.c.	Testis, epididymis: *in situ* pH (24 hr) ↑	35
F344, M	2.24	s.c.	Testis: lesions (24 hr)	36
SD, M	2.24 (acetate)	s.c.	Sertoli and Leydig cells: GPx ↑; GSR, GSH ↓	37
F344, M	3.4	s.c.	>50% mortality (7 d)	38
Wistar, M	3.4	s.c.	Testis: hemorrhage (12 hr)	39
LE, M + F, (5 d)	4	s.c.	Decreased motor activity (8–11 d)	40
SD, M, (4 d)	4	s.c.	Brain lesions, hyperactivity (4–18 d)	41
SD, M + F, (7 d)	6	s.c.	30% mortality (48 hr)	42
SD, M + F, (28 d)	6	s.c.	4% mortality (48 hr)	
SD, M	4	i.v.	Liver: lesions (10 hr)	43
SD, M + F, (10 d)	4–6		No damage	
SD, M	8.3	s.c.	10% mortality	44
	0.18	s.c.	hGC-induced serum testosterone ↓	
?, (3 d)	2 × 3 (on 3, 10, and 17 d)	i.p.	Brain: hypomyelination (5 d)	45
Wistar (d 19 fetus)	11.2 μg/fetus	i.p.	Hydrocephalus, brain necrosis	46
Wistar, M	2.6	i.p.	Heart: GSH, LPO, MT ↑	47
Wistar, F	64	p.o.	22% mortality (96 hr)	48
SD, M	75	p.o.	Liver: MT induction; GSH (4 hr) ↓	49
SD, M	3.35	i.v.	LD_{50} (14 d)	50
	3.55	i.p.	LD_{50} (14 d)	
	9.3	s.c.	LD_{50} (14 d)	
	225	p.o.	LD_{50} (14 d)	
Mice				
Swiss-Webster, M	4	i.v.	60% mortality (14 d)	51
Swiss-Webster, M	0.2	i.p.	Liver: MT induction	52
CFLP, M	1	i.p.	Testis: pathological change (3 d)	53
ICR, F	1.2	i.p.	Blood: ALAD (1 d) ↑	54
ICR, M	5	i.p.	40% mortality (10 d)	55
	112	p.o.	60% mortality (10 d)	
C57BL/6, M, (50 d)	4.08	i.p.	LD_{50} (7 d)	56
C57BL/6, M, (7 d)	1.65	i.p.	LD_{50} (7 d)	
C3H, F	2.8	s.c.	Sheep RBC-induced immune response ↓	57
5 Strains, M	3.36	s.c.	Liver, testis: alterations in essential metal levels	58
CBA, M	53	p.o.	11% mortality (10 d)	59

[a] Data from young adults were shown, unless otherwise specified.
[b] $CdCl_2$ was used, unless otherwise specified.

2.1.3.1 Acute Toxicity

When injected into animals as chloride or acetate, cadmium damages various tissues of the liver, kidney, heart, brain, and immune and reproductive systems in the acute phase, and sometimes even has lethal effects (Table 20). Numerous investigators reported the lethal effect of a single injection of Cd in mice and rats. LD_{50} values showed wide variation depending upon the administration route. Oral administration yielded the highest LD_{50} value, followed by s.c., i.p., and i.v. injections.[50] The lethal response is higher in weanling animals than in

TABLE 21
Effects of Cadmium on Pregnant Animals

Animal Strain	Dose (mg Cd per kg or ppm Cd)	Injection Time (G.D)[a]	Route	Effects	Ref.
Rats					
SD	2.1 × 4	8, 10, 12, 14	i.p.	Offspring: 75% mortality (12 d)	62
Wistar	4.5	18	s.c.	Placental necrosis (24 hr) Fetal death	63
Wistar	1.25	12	i.v.	Fetus: DNA and protein synthesis ↓	64
Wistar	1.25	12	i.p. or i.v.	Fetus: skeletal malformation	65
Wistar	0.49 × 20	1–20	s.c.	Neonatal: thymus weight and liver Zn ↓	66
LE	4.9 × 4	12–15	s.c.	Fetal: lung DNA and protein ↓	67
SD	50, 100 ppm water × 15 d	6–20		Fetus: body weight and liver Zn ↓ Mother: body weight, serum ALAD, and alkaline phosphatase ↓	68
Wistar	60 ppm water × 20 d	1–20		Offspring: impaired movement	69
Mice					
CD-1	5.6	10	s.c.	Embryo: malformation	70

[a] G.D. = gestation day.

adults.[42,56] In adult animals, it should be noted that susceptibility sometimes varies widely among strains. For example, more than 50% of male F344 rats died within 7 days following s.c. injection of 3.4 mg Cd per kg,[38] whereas male Wistar rats showed only chronic nephropathy or tumors, with a slight shortening of life span after the same treatment.[81] Of course, similar variations should be considered between sexes or species. At an acute phase of cadmium injection, liver and testis damage was most frequently observed in adult animals, while damage to neural tissue was often observed in young animals.[40,41,45,46] Cd-induced liver damage is recognized by conventional markers, such as increased blood levels of hepatic enzymes[32,43,49] and morphological changes.[32,33,43] Less hepatic damage was apt to be caused in weanling rats than adults.[43]

Like the liver, reproductive tissues show a very high susceptibility to acute cadmium toxicity. Testicular damage could be detected by hemorrhage,[39] decreased glutathione (GSH) levels,[37] and pathological changes.[39,43] The testicular lipid peroxidation (LPO) level may not be a good indicator of Cd exposure because it tends to decrease after low-dose injection.[60] Laskey et al.[44] suggested that a decrease in serum testosterone levels induced by human chorionic gonadotropin (h-GC) was the most sensitive marker of cadmium toxicity. Maitani and Suzuki[58] reported alterations in essential metal levels of liver, testis, and kidney in five $CdCl_2$-treated strains of mice. Testis of young animals showed a higher susceptibility to Cd toxicity than that of adults.[61]

Since the placenta is one of the target organs of Cd toxicity, Cd injection into pregnant animals might cause adverse effects on the fetus via damage to the placenta (Table 21). High dosing-induced placental necrosis may result in fetal death.[63] Lower dosing leads to embryo malformation[65,70] and decreases in body weight,[68] DNA, and protein.[64,67] Early death of offsprings was also documented.[62]

2.1.3.2 Subacute to Chronic Toxicities

Continuous exposure to Cd was achieved by repeated injection or providing Cd-contaminated food or drinking water. Long-term treatment at low levels increased the damage to other organs, including the kidney (Table 22). Chronic and subchronic renal damage

TABLE 22

Subacute to Chronic Effects of Cadmium

Animal Strain, Sex, (Weight)[a]	Dose × Time (mg/kg or ppm Cd)	Route	Effects	Ref.
Rats				
SD, M	1.8 × 14 d	i.v.	Liver, kidney: pathological change	71
SD, M	0.6 × 2–6 wk (5 d/wk)	s.c.	Kidney: membrane degeneration	72
SD, M	0.025 × 6 wk (5 d/wk)	p.o.	Liver: cytochrome c-oxidase ↓	73
Wistar, M	1.5 × 26 d	s.c.	Urinary protein, AST, amino acids ↑	74
BN, F	0.49 × 3 wk (5 d/wk)	s.c.	S-phase thymocytes ↓	75
Lewis, F	0.49 × 3 wk (5 d/wk)	s.c.	G_2-phase thymocytes ↑	75
?, M, (40–50 g)	0.4 × 30 d	i.p.	Brain: SOD ↓; LPO ↑	76
SD, M	0.5 × 26 wk (6 d/wk)	s.c.	Liver damage (4 wk); renal damage (8 wk)	77
Wistar, F	0.4 × 13 wk (5 d/wk)	s.c.	Urinary protein, HyPro, HyLys ↑	78
Wistar, F	6.12 × 1–3 months	p.o.	Duodenum: Ca^{2+} transport ↓	79
Wistar, M	1.49 × 11 wk (5 d/wk)	s.c.	Intestinal mucosa: alkaline phosphatase ↓	80
Wistar, M	3.4 × 1		Chronic nephropathy, testicular tumors (90 wk)	81
SD, F	100 ppm/water × 7 months		Urine: transferrin, IgG, β_2-M, albumin ↑	82
Wistar, M	50 ppm/water × 10 months		Alteration in skeletal muscle ultrastructure	83
Wistar, M	50–100 ppm/food × 6–8 wk		Bone: pathological change; lysil oxidase ↓	84
	50 ppm/food × 52 wk		Bone: collagen cross-linking ↓	
LE, F	1 ppm/water × 18 months		Hypertension (2 months)	85
			Heart, liver, kidney: ATP ↓; ADP ↑	
SD, M	50 ppm/water × 30 d		Small intestine: hemoxygenase activity ↑	86
Mice				
C57BL/6, M	50–200 ppm/water × 3 wk		Proliferative response of spleen cell ↑	87
CF1, F	50 ppm/food × 252 d		Femur Ca levels ↓	88
QS, F	10, 100 ppm/water × 22 wk		Brain: degenerative damage in choroid plexus	89

[a] Data from adult animals, unless otherwise specified.

induced by cadmium salt was recognized by the pathological changes[71,72,77] or the appearance of abnormal urinary components.[74,78,82] Urinary β_2-microglobulin (β_2-M) was reported to be an earlier marker than albumin.[82] Abnormalities of bone tissue were also documented as a chronic effect of cadmium.[84,88] It should be noted that, particularly in a repeated injection experiment, injection of even a sublethal dose of Cd induced metallothionein (MT) synthesis in the target organs that led to increased resistance to Cd (and other metal) toxicities.[52,90,91] Nishimura et al.[92] suggested urinary trehalase activity was a sensitive indicator for cadmium-induced chronic renal failure in rabbit.

When animals were exposed to aerosol of CdO or $CdCl_2$ or the metal was instilled intratracheally, the lung was primarily damaged (Table 23). An increase in tissue weight[93,96,99] and various cytosolic enzyme activities[93,97] and a decrease in mitochondrial enzymes[94,98] have been documented. Long-term treatment caused an increase in the connective-tissue components.[96]

Damage to the kidney in the acute phase of cadmium salt injection is very rare. However, when cadmium was injected as Cd-metallothionein or co-injected with an SH compound such as cysteine, even via an alternative route, the kidney became a primary target organ (Table 24). Suppression of *p*-aminohippuric acid (PAH) uptake by the renal slice prepared from the intoxicated rats was suggested to be a sensitive marker for Cd-plus-cysteine-induced nephrotoxicity.[109] Goering et al.[106] found induction of renal stress proteins at the initial phase of renal damage in rats treated by Cd-cysteine.

TABLE 23

Pulmonary Toxicity of Aerosol Inhalation or Intratracheal Injection of Cadmium

Animal Strain, Sex	Chemicals	Cd Level (mg/m^3)	Exposure Time (hr × days)	Effects on Lung	Ref.
Rats					
SD, M	CdO, $CdCl_2$	0.45–4.5	2 × 1	Tissue weight, GSR, GST, and G6PD ↑	93
SD, M	CdO	4.5	0.5 × 1	Monooxygenase and cytochrome P-450 ↓	94
Wistar, M	CdO	5	3 × 1	Nonprotein SH ↑	95
F344, F	$CdCl_2$	1	6 × 62	Tissue weight, elastin, collagen ↑	96
Lewis, M	CdO	1.6	3 × 20	G6PD, GSR, Cat, and GPx ↑	97
		8.4	3 × 1	Alkaline and acid phosphatase, LDH, and protein ↑	
CD-1, M	$CdCl_2$	0.85	2 × 1	Mitochondrial enzymes ↓	98
Wistar, M	CdO, $CdCl_2$	0.5–10 μg Cd per rat (i.t.)		Tissue weight, lavage fluid cell number ↑	99
F344, F	$CdCl_2$	0.1, 0.4 mg Cd per kg (i.t.)		HyPro ↑	100
Mice					
BALB/c, F	$CdCl_2$	4.9	1 × 1	Pathological change	101
BALB/c, F	$CdCl_2$	4.9	1 × 1	Cell proliferation ↑	102

TABLE 24

Renal Toxicities by Cadmium-Metallothionein or Cadmium Plus Thiol Co-Injection

Animal Strain, Sex	Chemicals	Dose (mg Cd/kg)	Routes	Effects	Ref.
Rats					
SD, M	Cd-MT	0.3	i.p.	Pathological changes; PAH uptake by kidney slice ↓	103
Wistar, M	Cd-MT	0.4	s.c.	Ca^{2+} uptake by luminal and batholateral membrane vesicles ↓	104
Wistar, M	Cd + Cys	1.3–1.7	i.v.	Urinary protein, glucose, amino acids ↑	105
	Cd + Cys-peptide	0.51–0.64			
	Cd-MT	0.16–0.23			
SD, M	Cd + Cys	2	i.v.	Stress protein synthesis ↑	106
SD, M	CD + β-ME	1.68	i.p.	Urinary protein, amino acids ↑	107
Mice					
4 strains, M	Cd-MT	0.4–1.6	s.c.	Urinary glucose ↑	108
ICR, M	Cd + Cys	1.5	i.v.	Urinary glucose, protein ↑ PAH uptake by kidney slice ↓	109

2.1.4 Chromium (Cr)

Trivalent chromium is essential in animals (including humans), while hexavalent chromium, which is easily absorbed from the gastrointestinal tract, is very toxic. Hepatic and renal damages have been reported in the acute phase of chromium parenteral injection at dose levels of 7.9 to 15.8 mg Cr per kg in rats (Table 25).[110-113] Tsapakos et al.[110] detected DNA-protein cross-links in liver and kidney of Cr-injected rats, suggesting a relationship to the carcinogenicity and toxicity of Cr(VI). In the subacute to chronic phase, reproductive tissues were also affected. Serial i.p. injection of 1 to 4 mg Cr per kg as dichromate for 5 to 90 days caused pathological changes in testis cells or altered enzyme activities in rats.[114-116] Vyskočil et al.[117] showed that female rats manifested renal dysfunction after 6-month exposure to 25-ppm

TABLE 25
Toxic Effect of Hexavalent Chromium

Animal Strain, Sex, (Age)[a]	Chemical	Dose (mg Cr per kg or ppm Cr)	Route	Effects	Ref.
Rats					
SD, M	$Na_2Cr_2O_7$	7.9, 19.8	i.p.	Liver, kidney: DNA-protein cross-links	110
SD, M	$Na_2Cr_2O_7$	10.5	i.p.	Liver: GSH ↑	111
Wistar, M	Na_2CrO_4	4.7 × 3	i.p.	Liver microsome: Cr(VI) reductase, cytochrome P-450, cytochrome b_5 ↓	112
SD, M	$Na_2Cr_2O_7$	7.9, 15.8	s.c.	Serum: BUN, lactate, glucose ↑; insulin ↓	113
Wistar, M	Na_2CrO_4	1, 2 or 4 × 5	i.p.	Testis: tissue weight, epididymal sperm number ↓; pathological change	114
ITRC, M, (weanling)	$K_2Cr_2O_7$	1–3 × 90	i.p.	Testis: pathological change; γ-GTP, LDH ↑; sorbitol dehydrogenase, G6PD ↓	115
Druckrey, M	$K_2Cr_2O_7$	2 × 15	i.p.	Testis: pathological change in epithelial cells	116
Wistar, M + F	Na_2CrO_4	25 ppm/water × 6 months		Female: urinary albumin, β_2-M ↑ Male: no change	117
F344, M	K_2CrO_4	100–200 ppm/water × 3–6 wk		Liver: DNA-protein cross-link	118

[a] Data from adult animals, unless otherwise specified.

Cr (as chromate) in drinking water, but males did not. Increased DNA-protein cross-links were detected in lymphocytes and liver of rats given drinking water contaminated by 100 to 200 ppm Cr as chromate.[118] Chronic inhalation experiments of Cr aerosol were reported to cause adverse effects in the lung. Glaser and co-workers[119] found that, although exposure to low Cr level caused activation of alveolar macrophages, high-level exposure inactivated it. They also showed that Cr_5O_{12} aerosol was much more toxic to lung and blood cells than $Na_2Cr_2O_7$ aerosol.[120] If pregnant mice were exposed to $K_2Cr_2O_7$ in drinking water containing 250 to 1000 ppm Cr throughout the gestation period, embryonic death and malformation of offsprings occurred.[121]

2.1.5 *Cisplatin (CDDP)*

Cisplatin (*cis*-diaminodichloroplatinum, CDDP) is an effective anticancer drug widely used in cancer chemotherapy; however, it also acts as a nephrotoxin. The nephrotoxic actions of cisplatin have been well documented in laboratory animals (Table 26); 50% mortality was documented, for example, in rats (7.7 mg/kg, i.p.)[123] and mice (9.5 to 13.4 mg/kg, s.c. and i.v.).[142,143] Cisplatin-induced renal damage was detected in the form of increased blood urea nitrogen (BUN) and serum creatinine levels, pathological changes or decreased enzyme activities in the kidney, and alterations in urine constituents. Urinalysis was suggested as a more sensitive method.[124] De Witt et al.[126] demonstrated that an increase in the Ca^{2+} pump activity of the renal endoplasmic reticulum (ER) was the earlier marker, followed by renal failure in the cisplatin-treated rat. Litterst[125] showed that a high salt concentration of vehicle markedly lowered the toxic effects of cisplatin in rat.

In ways other than its nephrotoxic action, cisplatin also affects neural tissue, glucose metabolism, testis, blood cells, and embryo. Neurotoxic action could be detected by pathological change,[132,133] electrophysiological methods,[134] and abnormal behavior.[147] Goldstein et al.[135,136] demonstrated the increased plasma glucagon half-life and impaired glucose tolerance that possibly led to renal failure. Effects on rat testes were observed to be decreased plasma

TABLE 26

Toxic Effects of CDDP

Animal Strain, Sex, (Age)[a]	Dose (mg CDDP per kg)	Route	Effects	Ref.
Rats				
F344, M	6, 25	i.p.	GFR ↓; jejunal crypt cell survival (3–5 d) ↓	122
F344, M	0.5–12	i.p.	LD_{50}: 7.7 mg/kg; renal damage: BUN ↑; histology	123
F344, M	2.5–15	i.v.	Renal damage: urine analysis as sensitive method	124
SD, M	9	i.p.	Decreased toxicity (lethality) by high salt vehicle	125
SD, M	5, 7.5	i.p.	Renal endoplasmic reticulum: Ca^{2+} pump activity, Ca^{2+} content (4–24 hr) ↑	126
SD, M	7	i.v.	Renal cytochrome P-450, b_5, γ-GCS, γ-GTP (7 d) ↓	127
SD, ?	5	i.p.	Defect in papillary hypertonicity; damage at S_3 segment	128
Wistar, M	5	i.v.	Creatinine clearance ↓	129
Wistar, M	5	i.p.	Kidney, liver: LPO ↑ GPx, GST, Cat ↓; kidney SOD ↓	130
BN, M	3 × 3 (21-d interval)	i.p.	Kidney: cytochrome P-450, GSR, GST, GPx, GSH ↓; N-glucuronyl transferase, GSSG, LPO ↑	131
Wistar, (10 d)	5	s.c.	Brain: abnormal shape of the dendritic tree (24 hr)	132
Wistar, F	2 × 9 (1 or 2/wk)	i.p.	Pathological change in spinal ganglia neuron and sciatic and peroneal nerve	133
Wistar, F	1 × 15 or 34 (2/wk)	i.p.	Sensory nerve condition velocities (48 hr) ↓	134
	15	i.p.	Cisplatin-DNA binding in DRG satellite cells (6 hr)	
F344, M	5	i.v.	Plasma glucagon half-life ↑; renal failure (96 hr)	135
F344, M	5	i.v.	Impaired glucose tolerance (48 hr)	136
Wistar, M	9	i.v.	Plasma testosterone, testis cytochrome P-450 (72 hr) ↓	137
SD, M	2 × 5	i.p.	Leakage of the Sertoli cell tight junction (24 hr); abnormal Sertoli cell secretory function	138
F344, M	5	i.v.	Intestinal epithelium (ileum): pathological change (24 hr)	139
Wistar, pregnant		i.p.	Embryonic LD_{50}: G.D. 6, 2.88 mg/kg; G.D. 8, 1.28 mg/kg; G.D. 11, 1.0 mg/kg[b]	140
Wistar, pregnant	4, 7 (on G.D. 6)[b]	i.p.	Serum: LH, progesterone, 20 α-hydroxysteroid dehydrogenase ↓; embryonal resorption	141
Mice				
BDF1, ?	9.5	s.c.	Plasma: BUN, CRT ↑; 50% mortality (30 d)	142
Swiss, M + F		i.v.	LD_{50}: M, 13.4 mg/kg; F, 12.32 mg/kg	143
Swiss, M	18, 20	i.p.	Decreased nephrotoxicity in high NaCl vehicle	125
B6D2F1, M	5 × 2 *M* (weekly)	i.v.	Lesion in renal cortical tubules and bone marrow; circadian rhythm affected	144
B6C3F1, M + F	15.5	i.v.	Reticulocytes ↓	145
B6D2F1, M	6.5	i.v.	Immature WBC, PMN ↑; immature RBC ↓	146
CD1, M	10	i.p.	Tail flick temperature and distal sensory latency ↑	147
Swiss, pregnant	5.24	i.p.	Embryonic LD_{50} (day 8)	140

[a] Data from adult animals, unless otherwise specified.
[b] G.D. = gestation day.

testosterone levels[137] and functional and morphological alterations of Sertoli cells.[138] Reduction in reticulocytes was documented in cisplatin-treated mice.[145,146] Aggarwal and coworkers[140,141] found that embryonic resorption or death occurred in cisplatin-treated rats and mice. They suggested that the cisplatin-induced decreases of sex hormone levels are responsible for the embryonic toxicity.

TABLE 27
Toxic Effects of Cobalt in Experimental Animals

Animal Strain, Sex[a]	Compound	Dose (mg/kg or ppm Co)	Route	Effects	Ref.
Rats					
SD, F	Co (metal dust)	10	i.t.	Lung: moderate inflammatory response, weight ↑	148
SD, M	$CoCl_2$	10 × 42	i.p.	Glomus cell hypertrophy; hematocrit ↑	149
SD, M	$CoCl_2$	20 × 69	Food	Testicular atrophy; slower lever press	150
SD, M	$CoCl_2$	10–20 × 2	s.c.	Liver: cytochrome P-450 (24 hr) ↓	151
SD, pregnant	$CoCl_2$	100 × 10 (G.D. 6–15)[b]	p.o.	No fetotoxicity	152
SD, M	Co-protoporphyrin	0.88–3.5	s.c.	Liver: cytochrome P-450, cytochrome b_5, NADPH-P-450 reductase ↓	153
CD-1, M	$CoCl_2$	400 ppm/water × 13 wk		Pathological change in testis cells	154
Hamsters					
Syrian, M	Co-protoporphyrin	5.3	s.c.	Liver: cytochrome P-450 ↓	155

[a] Data from adult animals, unless otherwise specified.
[b] G.D. = gestation day.

2.1.6 *Cobalt (Co)*

Cobalt is an essential component of vitamin B_{12}. Its toxicity is summarized in Table 27. The increase of lysozyme level in lavage fluid was documented in rabbits exposed to $CoCl_2$ aerosol (0.5 mg Co per m^3 × 6 hr/day) for a month.[156] Increased tissue weight and inflammatory response of the lung were documented in rat after intratracheal injection (10 mg Co per kg) of metallic Co dust.[148] Chronic exposure experiments using Co-contaminated food or water revealed testicular atrophy and reduced behavioral activity in rats[150] and abnormal testis morphology in mice.[154] Di Giulio et al.[149] indicated that consecutive i.p. injection of $CoCl_2$ (10 mg Co per kg) for 42 days caused increased hematocrit and hypertrophy of glomus cells in the rat carotid body. $CoCl_2$ treatment (10 to 20 mg Co per mg × 2, s.c.) caused a significant reduction in the hepatic P-450 levels of rats in the acute phase,[151] an effect much more prominent in Co-protoporphyrin-treated rats[153] and hamsters.[155] Since serial injection of 100 mg Co per kg as $CoCl_2$ in pregnant rats for 10 days caused no damage to the fetus,[152] embryo toxicity of $CoCl_2$ was considered to be very low.

2.1.7 *Copper (Cu)*

Although copper is an essential metal, excessive intake has been shown to have a variety of toxic effects. Metabolism and pulmonary toxicity of intratracheally instilled $CuSO_4$ or CuO have been reported in rats at dose levels of 2.5 to 50 μg Cu per rat.[157,158] The biochemical and elemental inflammatory indices in bronchoalveolar lavage fluid reached maximum values at 12 to 72 hr after instillation of 5 μg Cu per rat. Toxic effects on liver or kidney have been reported in rats fed a diet containing 1500 ppm Cu as $CuSO_4$ for 15 or 16 weeks, respectively.[159,160] Hemolytic anemia has also been documented in rats after consecutive intraperitoneal injection of Cu-nitrilotriacetate at a dose level of 4 to 7 mg Cu per kg.[161] Mascular mutant mouse, which has been proposed as an animal model of Menkes' kinky-hair disease, was reported to be sensitive to the acute hepatotoxic effects of Cu as compared to normal

TABLE 28

Toxic Effect of Iron Overload in Experimental Animals

Animal Strain, Sex, (Age)[a]	Fe-compound	Dose (mg/kg or ppm Fe)	Route	Effects	Ref.
Rats					
Wistar, M	Dextran	500 mg/kg	i.p.	Liver: chemiluminescence, LPO ↑; cytochrome P-450, SOD, Cat ↓	164
Wistar, M	Dextran	116 mg Fe per kg × 3 d	i.p.	Liver: dimethylhydrazine demethylase, UDP glucronyltransferase ↓	165
Wistar, F	Carbonyl	25,000 ppm/food × 10 wk		Proliferative activity after a mitogenic stimulus ↓	166
Wistar, F, (4 wk)	Saccharate	5 mg Fe per kg × 12 wk	i.p.	Mesothelioma	167
SD, M	Carbonyl	20,000 ppm/food × 4–15 months		Hemochromatosis	168
SD, M	Carbonyl	25,000 ppm/food × 12 months		Hemochromatosis, hepatic fibrosis	169
SD, M	NTA	2 mg Fe per kg × 14 wk	i.p.	Liver mitochondria: LPO ↑	170
SD, M	Carbonyl	25,000 ppm Fe per food × 28–44 d		Liver mitochondria and microsome: LPO ↑	
SD, M	Sulfate	305 ppm Fe per food × 10 wk		Liver: LPO ↑; non-Se-GPx ↑	171
Mice					
SWR, M	Dextran	600 mg Fe per kg	s.c.	Porphyria (25 wk)	172
A/J, M + F	NTA	1.8–2.7 mg Fe per kg × 12 wk (6 d/wk)	i.p.	Nephrotoxicity, renal carcinoma	173
Gerbil					
Mongolian	Dextran	1000 mg Fe per kg × 7 wk (1 d/wk)	s.c.	Liver, heart: hemosiderosis, hemochromatosis (12 wk)	174

[a] Data from adult animals, unless otherwise specified.

mice.[162] Clastogenic effects of Cu on the bone marrow chromosomes were shown in mice intraperitoneally injected with $CuSO_4$ (1.1 to 6.6 mg/kg).[163]

2.1.8 Iron (Fe)

Despite its abundance and necessity in almost all living organisms, excess iron causes various toxic effects if accumulated in human and animal tissues. The effects of iron overload in experimental animals vary among species, strains, and sex (Table 28). Since the liver is the major storage organ of the excess iron, hepatotoxicity is the most common finding in animals undergoing iron overload experiments. Hepatic fibrosis was induced in rats by feeding them an iron carbonyl-contaminated diet for as long as 8 months.[168,169] On the other hand, single dosing of iron dextran caused hepatic fibrosis in gerbils,[175] which were suggested to be sensitive models for induction of hepatic fibrosis.

Cardiotoxicity is the other major effect of iron overload observed in humans, but no experimental model has been reported in mice or rats. Recently, Carthew et al.[174] demonstrated that repeated s.c. injection of gerbils with iron dextran resulted in hemochromatosis with a heart pathology similar to human cases.

Since iron has a catalytic action in reactive oxygen generation *in vivo*, a considerable part of its toxicity may be the oxidative damage caused. Several studies demonstrated elevated lipid peroxidation in the tissue of iron-overloaded animals.[164,165,170,171] The catalytic action of iron was also suggested to be involved in the carcinogenic action of polyhalogenated aromatic hydrocarbons.[176]

TABLE 29

The LD_{50} of Lead Salts for Rats and Mice After Intraperitoneal (i.p.) Dose

Animal Strain	Sex	Age or Weight	Salt	LD_{50} (mg/kg)	Days	Ref.
Rats						
?	M	3 wk	Acetate	225	8	178
	F	3 wk		231	8	
	M	18 wk		170	8	
	F	18 wk		258	8	
SD	M	20 wk	Acetate	172	8	179
	F	20 wk		280	8	
Mice						
DBA/2	M	60–70 d	Nitrate	74	10	180
C57BL/6	M			102	10	
Swiss-Webster	M			148	10	
ICR	M	26 g	Acetate	278	8	181
	F	23 g		280	8	

2.1.9 *Lead (Pb)*

Lead may cause various adverse effects in experimental animals in both acute and chronic phases. Toxic effects of inorganic lead related to the hematopoietic, nervous, gastrointestinal, and renal systems, while those of organic lead largely related to the nervous system. Although rats are the most frequently used animals in studies of metal toxicity, adult rats are relatively insensitive to lead toxicity,[177] whereas perinatal animals are very sensitive. Accordingly, numerous experimental studies have been carried out using young animals.

2.1.9.1 *Inorganic Lead Toxicity*

There are marked differences in LD_{50} values for lead toxicity depending upon species, sex, and age (Table 29). Acute lead encephalopathy occurs easily in young animals but only rarely in adults.[184,188] Kumar and Desiraju[184] reported that about 20% of rat pups orally administered lead acetate (0.4 g Pb per pup) for 10 to 11 days developed hind limb paralysis and died within 24 hr.

Characteristic disturbances in the central nervous system (CNS) functions during chronic lead exposure (lead encephalopathy) have stimulated numerous behavioral, pathological, and neurochemical investigations (Table 30). CNS disturbance seems more apt to occur during development,[186,189] possibly due to high lead absorbability[193,194] and high susceptibility.[195] In the last decade, an increasing amount of evidence has accumulated to show that lead exposure, particularly in the perinatal period, disrupts the development of opioid peptide systems in the rat brain.[196] Disturbance of the peripheral nerve has been documented by histopathological[183] and electrophysiological methods.[197]

In the chronic phase, disturbances in the hematologic, nervous, and renal systems have been documented as results of lead toxicity in animals. Anemia is a common chronic systemic effect of lead, which is considered to be caused by a combined effect of the inhibition of hemoglobin (Hb) synthesis and shortened life-span of circulating erythrocytes (RBC). Lead has been shown to interfere with heme biosynthesis even at a low level of exposure.[209] Inhibition of the heme biosynthetic enzyme ALAD and elevation of free RBC and zinc protoporphyrin (ZPP) are the earliest effects, followed by an increase in urinary ALA and coprotoporphyrin and a fall in Hb and hematocrit (Ht) levels (Table 31). ALAD activity in RBC is shown to be the most sensitive indicator of lead exposure. Maes and Gerber[210] reported that when rats were severely intoxicated, marked shortening of RBC survival led to increased ALAD activity in circulating erythrocytes.

TABLE 30

Neurobehavioral Effects of Lead

Animal Strain, Sex	Age or Weight	Salt	Dose	Route	Effects	Ref.
Rabbits						
NZ, M + F	Newborn	Nitrate	4.5–18 mg/pup × 30 d	p.o.	Pathological changes in CNS	182
Rats						
Wistar, M	400 g	Carbonate	1000 mg/kg × 600 d	p.o.	Histopathological changes in CNS	183
Wistar, M + F	Newborn	Acetate	400 mg Pb per kg × 60 d	p.o.	Hind limb paralysis, brain edema and hemorrhages, biogenic amines and GABA/glutamate system changes	184
Wistar	Newborn	Acetate	45–180 mg Pb per kg × 19 d	p.o.	Behavioral changes	185
Wistar, M	Perinatal	Acetate	750 ppm/food × 17 d	Maternal	Behavioral changes	186
SD, M	Newborn	Nitrate	10 mg/kg × 15 d	i.p.	Histopathological changes in CNS	187
LE, M + F	Newborn	Acetate	600 mg/kg × 10 or 30 d	p.o.	Alteration in cerebellum development	188
LE, M + F	Newborn	Acetate	10–90 mg/kg × 19 d	p.o.	Behavioral changes	189
ITRC, M	220–240 g	Acetate	5–12 mg/kg × 14 d	i.p.	Behavioral changes and alteration in biogenic amine levels	190
Mice						
HET, M	60 d, 1 year	Acetate	5000 ppm/water × 7 wk		Behavioral changes	191
BK/W, M + F	Perinatal	Acetate	2500 ppm/water	Maternal	Behavioral changes	192

Kidney is also a target organ of chronic lead toxicity. Several transient effects on renal function in experimental animals are consistent with pathological findings of reversible lesions.[211-214] Irreversible lesions such as interstitial fibrosis have also been documented in animals following long-term lead exposure.[214,215]

Studies during the last decade have shown that chronic and low-level lead exposure might induce subtle alterations in the immune systems of experimental animals. Enhanced host susceptibility to bacteria and viral infections[216-218] and increased growth and metastasis of implanted tumors[219,220] have also been reported. Other studies demonstrated the ability of lead to reduce the number of antibody-forming cells,[221] to suppress antibody synthesis,[222,223] and to diminish the phagocytic function of the reticuloendothelial systems.[223] Lead has also been shown to induce perivascular edema,[183] teratogenicity,[224] and testicular toxicity[225] in experimental animals.

2.1.9.2 Organic Lead Toxicity

Triethyllead (TEL) and other organic lead compounds have been shown to affect preferentially the nervous system.[226] Degenerative changes have been observed in the cerebral cortex, cerebellum, and hippocampus of rabbits receiving lethal or near-lethal doses of tetraethyl lead.[227,228] Rats subjected to both acute and short-term repeated TEL exposures experienced alterations in reactivity, locomotor activity, and avoidance learning.[229,230] Moreover, organic lead has been shown to have several biochemical effects, including alteration of enkephalin levels,[231] dopaminergic processes,[232] and enhanced lipid peroxidation[233] in brain.

TABLE 31

Hematopoietic Effects of Lead

Animal Strain, Sex	Age or Weight	Salt	Dose	Route	Effects	Ref.
Rabbits						
NZ, F	2.5–2.8 kg	Acetate	0.2 mg/kg × 48 d (3 d/wk)	s.c.	RBC: ALAD ↓; urinary ZPP ↑	198
			0.8–1.2 mg/kg × 97–181 d (3 d/wk)		RBC: ALAD ↓; urinary ZPP and ALA ↑	
NZ, M	90 d	Subacetate	5000 ppm/food × 90–120 d		Hb ↓	199
NZ, M + F	Newborn	Nitrate	4.5 mg/pup × 20 d		Ht ↓	
Rats						
SD, M + F	Perinatal	Acetate	2500 ppm Pb/water × 7 wk	Maternal	RBC, kidney: ALAD ↓; Ht ↓; urinary ALA ↑	200
Wistar, M + F	Newborn	Acetate	20,000 ppm/food × 20–22 d	Maternal	Brain, liver: ALAD ↓	201
Wistar, F	Pregnant	Acetate	500 ppm Pb/water × 3 wk		RBC: ALAD ↓	202
Wistar, M + F	Prenatal			Maternal	RBC: ALAD ↓; Hb ↓; Ht ↓	
Wistar, M	180–250 g	Acetate	500 ppm Pb/water × 30 d		Urinary ALA and CP ↑	203
CF, M	180–200 g	Acetate	8 mg PB per kg × 7 d	i.p.	RBC: ALAD ↓	204
?, M	160–180 g	Acetate	550 ppm Pb/water × 1–4 months		RBC: ALAD ↓; ZPP ↑; urinary ALA ↑	205
?, ?	Newborn	Acetate	10,000 ppm/water × 40–60 d	Maternal	ALAD: RBC, liver, kidney spleen ↓ ALAS: spleen ↑; liver ↓ ALA: brain, spleen, kidney, urine ↑	206
	40–60 d	Acetate	2 mg Pb per kg × 3	i.p.	ALAD: RBC, liver, kidney, spleen ↓ ALAS: liver, spleen ↑; kidney ↓ ALA: spleen, kidney, urine ↑	
?, M	150–170 g	Acetate	10 mg Pb per kg × 4 wk	p.o.	RBC: ALAD ↓; ZPP ↑; urinary ALA ↑	207
Mice						
ddy, M + F	30–40 g	Acetate	500 ppm Pb/water × 30 d		Urinary ALA ↑	203
NMRI, M	20–25 g	Acetate	500–5000 ppm/water × 30 d or 500–2500 ppm/water × 90 d		RBC, liver, brain, bone marrow: ALAD ↓	208
		Acetate	0.1 mg/kg	i.v.	RBC, liver, brain, bone marrow: ALAD ↓	
DBA/2, M	60–70 d	Carbonate	4000 ppm/food × 12 d		Liver: ALAD ↓	180
C57BL/6, M						
Swiss-Webster, M					Liver: ALAD ↓; Ht ↓	

2.1.10 Manganese (Mn)

Manganese is an essential metal, considered to have low toxicity. The chronic and subchronic effects on the neural tissues have been widely investigated (Table 32). Serial i.p. injection of 4 mg Mn per kg for 30 days caused reduced LPO[236] or increased norepinephrine (NE) levels[237] in rats. Exposure via drinking water containing 1000 ppm Mn for 14 days effected increases in brain dopamine (DA) and NE levels and activated behavior.[243] Exposure for 30 days to the same level increased the turnover rates of DA and NE.[242] Bonilla and Prasad[244] have shown significant decreases of the biogenic amines in several regions of the brains of rats given water containing 100 or 1000 ppm Mn for 8 months. Komura and Sakamoto[250] have reported that the effects of several Mn compounds (chloride, acetate, carbonate, and dioxide) on the brain biogenic amine levels in mice differed, the highest

TABLE 32

Toxic Effects of Manganese Chloride

Animal Strain, Sex, (Age)[a]	Dose (mg/kg or ppm Mn)	Route	Effects	Ref.
Rats				
SD, (3 d)	150 × 44		Striatum and hypothalamus: homovanillic acid (15–22 d) ↓	234
Wistar, M	10	i.p.	Brain, heart: cholin esterase, carboxylesterase ↓	235
ITRC, M, (30 d)	4 × 30	i.p.	Brain: LPO ↓	236
MRC, M	4 × 30	i.p.	Brain: NE ↑; serum: Tyr, Trp ↓	237
Wistar, M	2, 8 × 48 (2/wk)	i.p.	Motor nerve conduction velocity ↑	238
?, M, (3 wk)	50 μg/rat × 60	p.o.	Brain: MAO activity ↑; morphological change (30 d)	239
SD, M, (20 d)	10,000 ppm/water × 60 d		Dorsal caudate putamen: DA ↑	240
SD, M	1000 ppm/water × 65 wk		High locomotor activity (5–6 wk)	241
SD, M, (50 g)	1000 ppm/water × 30 d		Brain: turnover rate of DA and NE ↑	242
ITRC, M	1000 ppm/water × 14 d		Brain: DA, NE ↑; SMA ↑; learning activity ↓	243
SD, M	100, 1000 ppm/water × 8 months		Brain: catecholamines, 5-HT, 5-HIAA ↓	244
SD, M	3 × 30	i.p.	Pancreatitis-like reaction	245
SD, M	100, 200	i.v.	Liver: cellular necrosis, cholestasis (12 hr)	246
?, M	200 ppm/water × 10 wk		Liver: change in ultrastructure	247
Mice				
CBA, A, C57BL/6, M	40, 80	i.m.	NK-cell activity ↑; plasma interferon ↑ Susceptibility: CBA, A > C57BL/6	248
CD-1, M	1, 3, 10 × 28	i.p.	Antibody production ↓	249
ddy, M	130 ppm/food × 1 year		Alteration in brain biogenic amines Toxicity: MnO_2 > $MnCl_2$	250

[a] Data from adult animals, unless otherwise specified.

toxicity being shown by MnO_2. Besides its neurotoxic effects, $MnCl_2$ was also documented to cause a pancreatitis-like reaction,[245] hepatic damage,[247] and reduced antibody production[249] in mice or rats. Rogers et al.[248] observed increases in the splenic natural killer cell activity and plasma interferon levels in three strains of mice.

Methylcyclopentadienyl manganese tricarbonyl (MMT) is used as an octane enhancer in unleaded gasoline. LD_{50} values of MMT after p.o. and i.p. injections were determined in rats to be 50 mg/kg[254] and 12.1 to 23 mg/kg,[251,254] respectively (Table 33). Fishman et al.[256] showed that the lethal effect of MMT was more potent in propyrene glycol vehicle than in corn oil vehicle. MMT-affected lung tissue of experimental animals was detected as a pathological change[253,254] and increased lavage protein levels.[252,255]

2.1.11 Mercury (Hg)

In the terms of its toxicological properties, mercury can be divided into metallic mercury, inorganic mercuric salt, and organic mercury. Among various inorganic and organic Hg compounds, we focus here on the toxic actions of mercuric chloride ($HgCl_2$) and methylmercury (MeHg).

Exposure to metallic mercury can take place by inhalation of Hg vapor. Since metallic mercury easily penetrates the blood-brain barrier, exposure to low-dose levels (<3 mg Hg per m^3) disturbs the neural tissues and causes behavioral changes in young and adult rats.[257,258] Prenatal exposure to Hg vapor was found to cause similar effects in offsprings after growing up.[259] An extremely high dose (30 mg Hg per m^3) for 2 hr damaged lung tissue, resulting in a significant mortality.[260]

TABLE 33

Toxic Effects of Methylcyclopentadienyl Manganese Tricarbonyl (MMT)

Animals Strain, Sex	Dose (mg MMT/kg)	Route	Effects	Ref.
Rats				
SD, M	6–37.4	i.p.	LD_{50}: 12.1 mg/kg (24 hr)	251
SD, M	4	s.c.	Lung: lavage protein (24 hr) ↑	252
S/A, F	5	i.p.	Lung cell damage	253
SD, M		p.o.	LD_{50} (14 d): 50 mg/kg; pulmonary hemorrhagic edema	254
		i.p.	LD_{50} (14 d): 23 mg/kg	
SD, M	0.5–2.5	s.c.	Bronchoalveolar lavage protein ↑	255
Mice				
CD-1, M		i.p.	LD_{50} (2 hr): 152 mg/kg (propyrene glycol vehicle), 999 mg/kg (corn oil vehicle)	256
BALB/c, F	120	i.p.	Lung cell damage	253
Hamster Syrian, F	180	i.p.	Lung cell damage	253

TABLE 34

Acute Effects of Mercuric Chloride

Animal Strain, Sex, (Age)[a]	Dose (mg Hg/kg)	Route	Effects	Ref.
Rats				
SD, M + F, (1, 29 d)	3.7	s.c.	29 d: 20% mortality; kidney damage 1 d: no effect	261
SD, M	0.37	i.p.	Urine: glucose, maltase (24 hr) ↑	262
SD, M	3	s.c.	Kidney damage: pathological (15 min), functional (6 hr)	263
SD, F	11	s.c.	Renal cortex amino acids ↓	264
SD, M	1.1	i.p.	Kidney mitochondria: GSH ↓; H_2O_2 formation, LPO ↑	265
Wistar, M	1 × 2	i.p.	Urine: Ca^{2+}, Mg^{2+}, MT ↑	266
Wistar, M	1.5, 3	s.c.	Urine: alkaline phosphatase, LDH ↑	267
Wistar, M	3	s.c.	Kidney: LPO ↑; vitamin C, E ↓	268
Wistar, M	0.5, 1	i.p.	Urine: alkaline phosphatase ↑; pathological change	269
Wistar, M	4.4	s.c.	100% mortality (48 hr)	270
Wistar, pregnant	0.79 (G.D. 8 or 16)[b]	i.v.	Placental transport activity ↓ LD_{50} (G.D. 8–19)[b]: 1.00–1.18 mg/kg	271
SD, pregnant	1	s.c.	Dam and newborn: urinary β_2-M, albumin ↑	272
Mice				
?, M	0.5	i.p.	Kidney: ribosome disaggregation	273
ICR, M	3.6	i.v.	Urine: NAG, LDH ↑; 100% mortality (7 d)	274
NMRI, F	5–40	p.o.	Renal GSH, GPx, protein: dose-dependent alteration; necrosis (>20 mg/kg)	275
C57BL/6, M + F	2, 4	i.v.	Urinary PSP excretion ↓; pathological change	276

[a] Data from adult animals, unless otherwise specified.
[b] G.D. = gestation day.

Inorganic mercuric mercury is well documented as a potent nephrotoxin. Numerous investigators have reported the acute and subacute toxic actions of $HgCl_2$ using various indicators (Table 34). The toxic effects could be detected by pathological change, enzyme activities, and LPO in the kidney and alterations in urinary components. However, because

TABLE 35

Subchronic to Chronic Effects of Mercuric Chloride

Animal Strain, Sex, (Age)[a]	Dose (mg/kg or ppm Hg)	Route	Effects	Ref.
Rat				
Charles Foster, M	0.037, 0.074 × 30	i.p.	Morphological change at epididymal epithelium; sperm count ↓	277
Wistar, (2 d)	3 × 59	p.o.	Brain: NE, DA ↑; acetylcholine esterase ↓	278
SD, M	50 ppm/water × 320 d		Alteration in cardiovascular response to epinephrine and NE	279
Mice				
Swiss, F	6 × 10	p.o.	Renal UDP-glucuronyltransferase ↑	280
STL/N, F	1.8 or 3.7 ppm/water × 10 wk		Autoimmunity, immune-complex disease	281
B6C3F1, M	3, 15, or 75 ppm/water × 7 wk		Bone marrow, thymus, and spleen: sugar metabolizing enzymes ↓	282

[a] Data from adult animals, unless otherwise specified.

of its poor absorbability by the gastrointestinal tract, the effective dose of this Hg species by oral administration is rather higher than by parenteral injection. Mortality within 7 days was documented in rats and mice with dose levels of 3.6 mg/kg or above, via s.c. or i.v. route.[261,274] It should be noticed that young rats hardly show nephrotoxic symptoms even after injection of toxic dose levels for adult rats.[261] Interestingly, $HgCl_2$, particularly at a low dose, increases the renal GSH level.[275] Bernard et al.[272] documented an increase in urinary albumin and β_2-M in newborn rats due to daily injection of dams during pregnancy with $HgCl_2$ (0.74 mg Hg per kg). Holt and Webb[271] found that variations of LD_{50} values for pregnant rats were rather small (1.00 to 1.18 mg/kg) despite a drastic increase of body weight. In animals (sub)chronically treated by $HgCl_2$, the toxic effects were observed in other tissues (Table 35). Abnormalities of the epididymis,[277] heart,[279] and, in the case of young animals, the brain[278] have been documented in rats. In mice, disturbance of the immune system was reported.[281,282]

Among various forms of organic Hg, toxicity of MeHg has been well investigated due to its natural occurrence and the history of Minamata disease. Acute and subacute toxic effects of MeHg after single and multiple injections are summarized in Tables 36 and 37. Although LD_{50} values were reported to be around 10 mg Hg per kg (i.p.) in rats[283] and mice[291] after single injection, its variation among sexes and strains should be considered. Female mice showed higher resistance to MeHg acute toxicity than did males;[276,290] however, after serial injection of a sublethal dose, male C57BL/6-strain mice survived much longer than females.[306] Acute and subacute effects were observed not only in brain but also in liver and kidney. In the liver, induction of protein synthesis,[284] mild glycogen accumulation,[304] and morphological changes[295] were observed. Nephrotoxic actions of MeHg in the acute phase were also documented,[276,288,301] although histochemical abnormality was very slight.[276] A significant decrease in serum albumin in mice was documented 24 hr after dosing of 16 mg Hg per kg.[307] It should be noted that, like $HgCl_2$, MeHg could also induce GSH synthesis in the kidney of rats and mice in the acute and subacute phases.[292,311]

MeHg exposure using contaminated food or water is a useful method, particularly for a long-term exposure experiment. Table 38 summarizes the toxic effects of MeHg from food and drinking water. Similar effects in the form of abnormalities in neural tissue and kidney have often been observed, as in repeated injection studies. Woods et al.[310] suggested urinary porphyrin might be a biomarker for renal damage induced by MeHg. Upon life-long exposure to MeHg, male mice and rats manifested neurotoxic symptoms earlier than did females.[312,316]

TABLE 36

Toxic Effects by Single Methylmercury Injection

Animal Strain, Sex, (Age)[a]	Dose (mg Hg per kg)	Route	Effects	Ref.
Rat				
SD, M	9.5	i.p.	LD_{50} (24 hr)	283
SD, M	8–40	s.c.	Protein synthesis, RNA polymerase: liver ↑; brain ↓	284
Wistar, M	8	i.p.	Cerebellum: granule cell swelling (1–3 d)	285
SD, ?, (10–20 d)	8	i.p.	Brain: t-RNA amino acylation ↓	286
CD, M	4.65	i.p.	Cerebellum: reactive oxygen species ↑	287
Mice				
C57BL/6, M	0.93	i.p.	Cerebellum: reactive oxygen species ↑	287
Swiss OF1, M	74	p.o.	Damage at renal proximal tubules	288
BALB/c, M + F, (2 d)	4	p.o.	Cerebellum: mitotic arrest; cell number ↓	289
RF, M + F	30	p.o.	Mortality (10 d): M > 70%; F 60%	290
ICR, M + F			Mortality (10 d): M > 70%; F 3%	
Swiss Webster, M	10.8	i.p.	LD_{50} (7 d)	291
C57BL/6, F	32	p.o.	Kidney: γ-GCS activity, GSH ↑	292
C57BL/6, F	40	p.o.	33% mortality (7 d)	276
C57BL/6, M	16	p.o.	67% mortality (7 d)	
C57BL/6, M	8	p.o.	Serum albumin (24 hr) ↓	307

[a] Data from adult animals, unless otherwise specified.

TABLE 37

Toxic Effects of Multiple Methylmercury Injection

Animal Strain, Sex, (Age)[a]	Dose (mg Hg per kg)	Route	Effects	Ref.
Rats				
Wistar, M	8 × 3	s.c.	MT induction in liver and kidney	293
Wistar, M	8 × 3	s.c.	Inflammation in kidney; urinary NE, DA (90 d) ↓	294
SD, M	8 × 4	s.c.	Ultrastructure change in liver	295
Wistar, M + F	8 × 5	p.o.	Damage at cerebellar glandular layer and dorsal root ganglion (M < F, 10–12 d)	296
LE, M	6.4 × 5	s.c.	Impaired auditory function (6 –7 wk)	297
SD, M	6.4 × 6	p.o.	Brain: general blood flow (at silent phase) ↓	298
Wistar, F	8 × 7	s.c.	Sciatic nerve: phosphorylation of specific proteins (15 d) ↓	299
SD, F	9.3 × 7	p.o.	^{14}Leu incorporation into cerebellar slice ↓	300
SD, M	0.8 × 20	i.p.	Kidney: lysosome and mitochondria dysfunction	301
Wistar, M	8 × 20	s.c.	Kidney, serum: LPO (2 d) ↑	302
SD, ?, (5 d)	5 × 10–27	s.c.	Neurological symptoms (23 d)	303
SD, M	0.8 × 8	s.c.	Liver: glycogen accumulation, SER proliferation (1–2 wk); pathological change (11 wk)	304
Mice				
ICR, F	10 × 5	s.c.	Brain: protein kinase C ↓	305
C57BL/6, M + F	4 × 49	p.o.	50% mortality: C57BL/6 M, 45 d; F, 21 d	306
BALB/c, M + F			50% mortality: BALB/c M, 15 d; F, 17 d	

[a] Data from adult animals, unless otherwise specified.

TABLE 38

Effects of Methylmercury Exposure From Food or Drinking Water

Animal Strain, Sex, (Age)[a]	Hg Level × Duration (ppm Hg)	Effect	Ref.
Rats			
SD (offspring)	3.9/food before mating, → gestation, → lactation, → 50 d postpartum	Cerebellar NE (50 d) ↑	308
Wistar, M	16/water × 95 d (2-d intervals)	Ataxia, Hg staining at CNS	309
F344, M	4.3/water × 2 wk, or 8.6/water × 1 wk	Urinary porphyrin (biomarker for renal damage) ↑	310
F344, M	4.3, 8.6/water × 4 wk	Kidney: GSH, γ-GCS ↑	311
SD, M + F	8/food × 130 wk	Ataxia (M, F); renal failure (M)	312
SD, M + F	1.6, 8/food × 130 wk	Pathological changes in spinal ganglion, spinal dorsal root, proximal tubules	313
Mice			
CBA, M	8–32/water × 2 wk	LPO in liver, kidney, brain ↑	314
BALB/c, F	10/water × 60–71 d 20/water × 20–75 d 40/water × 7 d	Ataxia	315
Swiss cross, M	0.5–10/water × 3 wk	Immunosuppressive effects	7
B6C3F1, M + F	8/food × 104 wk	Neurotoxic signs, chronic nephropathy (M > F)	316

[a] Data from adult animals, unless otherwise specified.

MeHg induced congenital abnormalities through prenatal exposure (Table 39). Effects were documented as fetal death,[317,323] renal failure,[319] and neural disorders detected by behavioral[320] and pathological methods.[318,321,322] It should be noted that the susceptibility of the fetus to MeHg toxicity varied with the time of exposure during the gestation period. For example, the induction rate of hydrocephalus in offsprings of B10D2 strain mice was highest by injection on gestation day 15.[322] Similar variations were seen with other metals.[9,10,140,326] Inouye and Kajiwara[324] observed abnormal morphology in the fetal brain of the guinea pig (whose gestation period is much longer than for rats or mice) after a single injection of MeHg (7.5 mg Hg per animal) during pregnancy.

2.1.12 Nickel (Ni)

Nickel is considered to be an essential element in several animal species. Inhalation of nickel compounds causes lesions in the lung. Dunnick et al.[340] demonstrated that in rats and mice the lung toxicity and lethality of aerosols containing nickel sulfate and subsulfide depended on the solubilities of the salts. Parenteral injection of nickel salts affected various tissues in experimental animals (Table 40). Intraperitoneal injection of 6.75 mg Ni per kg as acetate in adult rats caused 57% mortality during the 14 days after injection.[329] Hogan[338] found LD_{50} values were higher in weanling mice than in adults. The acute effects of nickel toxicity proved to be increased tissue hemoxygenase activities,[325] hyperglycemia,[329,333] hepatic dysfunction,[331,335] and decreased natural killer cell activity.[328,337] The incidence of sarcoma was documented in nickel subsulfide-treated rats.[327] Nickel also caused adverse effects in embryos. LD_{50} values in rat embryos varied with the time of injection during the gestation period.[326] Smith et al.[334] reported that the lowest observed adverse effect level for pups was 10 ppm Ni in drinking water, which caused a significant number of embryo deaths in the second of two successive gestations.

TABLE 39

Effect of Prenatal Exposure to Methylmercury

Animal Strain	Dose (mg/kg or ppm Hg)	Injection time (G.D.)[a]	Route	Effect	Ref.
Rats					
Wistar	8	18	i.p.	Fetal death (4 hr)	317
SD	6.4	15	p.o.	DA receptor density (14 d) ↑	318
SD	2.4 or 4.8 × 3	8, 10, 12	i.p.	Urine: γ-GTP, alkaline phosphatase, NAG (3 or 6 d) ↑	319
Wistar	0.01 or 0.05 × 4	6–9	p.o.	Behavioral performance deficits (4 months)	320
Mice					
C57BL/6	3.2 × 3	14–16	i.p.	Abnormal neuronal migration at cortical layers II and III in newborn	321
B10D2, C57BL/10, DBA/2	8	15	p.o.	Hydrocephalus in newborn: B10D2, 88%; C57BL/10, 54%; DBA/2, 0%	322
IVCS	3.2 or 6.4 ppm/food	–30–18		Litter size ↓; resorption, dead embryo ↑	323
Guinea pigs					
Hartley	7.5/animal	21, 28, 35, 42, or 49	p.o.	Abnormal morphology in fetal brain	324

[a] G.D. = gestational day.

2.1.13 *Selenium (Se)*

Selenium is one of the essential trace elements, and its deficiency has been documented widely in human and experimental animals; however, excess selenium also causes various adverse effects (Table 41). Mortality from selenium overload has been documented in mice and rats. Jacobs and Forst[343] showed that giving water containing 16 ppm Se (as Na_2SeO_3) to rats for 35 days caused up to 80% mortality. They also found that susceptibility varied with sex and age. Mortalities in mice were documented from an experiment in which they were given water containing 64 ppm for 46 days.[346] Since selenium functions as a potent SH oxidizing reagent *in vivo*, decreased GSH and increased glutathione disulfide (GSSG) were documented in tissues of Na_2SeO_3-treated rats. David and Shearer[341] found that a decreased lens GSH level was accompanied by cataract formation and increased insoluble protein levels in selenite-treated rats. LeBoeuf and Hoekstra[342] observed increased GSSG, GSR, and γ-glutamyl transpeptidase (γ-GTP) in the liver of Na_2SeO_3-treated rats. Watanabe and Suzuki[345] reported that selenite injection caused transient hypothermia and cold-seeking behavior in mice. Inhalation of dimethylselenide gas up to 8000 ppm for 1 hr caused alterations of DNA, RNA, and protein levels in lung, liver, and spleen of rats, but no pathological change was observed.[347]

2.1.14 *Thallium (Tl)*

Soluble thallium salts are easily absorbed from the gastrointestinal tract and widely distributed among various tissues to cause adverse effects (Table 42). The LD_{50} value (4 days) was documented as 24.8 mg Tl per kg in TlOAc-injected (i.p.) male rats.[348] Around this dose level, increases in the serotonine turnover rate and MAO activity in the brain were observed within 24 hr.[351] Peele et al.[349] showed that oral administration of Tl_2SO_4 caused a flavor

TABLE 40

Toxic Effects of Nickel

Animal Strain, Sex, (Age)[a]	Compound	Dose (mg/kg or ppm Ni)	Route	Effects	Ref.
Rats					
F344, M	$NiCl_2$	15	s.c.	Kidney, liver, lung, brain: hemeoxygenase (17 hr) ↑	325
F344, pregnant	$NiCl_2$	16	i.m.	Fetal death: 18.5–20% (G.D. 8 and 18)[b] LD_{50} (dam, 14 d): 22 mg Ni per kg (G.D. 8);[b] 16 mg Ni per kg (G.D. 18)[b]	326
F344, M	Ni_3S_2	0.88	i.m.	Sarcoma in 77% of rats (2 years)	327
F344, M + F	$NiCl_2$	10–20	i.m.	NK-cell activity (24 hr) ↓	328
F344, M	$Ni(OAc)_2$	6.75	i.p.	57% mortality (14 d)	329
		5.6	i.p.	Hyperglycemia, renal cytochrome P-450 ↓	
F344, M	$NiCl_2$	3.6–29	s.c.	Alveolar macrophage: cAMP ↑; 5′-nucleochidase (1–4 hr) ↓; LPO (72 hr) ↑	330
F344, M	$Ni(OAc)_2$	6.28	i.p.	Liver, kidney: LPO ↑; Cat, GPx, GSH, GSR (3 hr) ↓ Serum: ALT, AST ↑	331
F344, M	$Ni(His)_2$	1.2	i.v.	Oxidative damage at DNA	332
Wistar, F	$NiCl_2$	4, 6	i.p.	Plasma: glucagon ↑; insulin ↓; hyperglycemia (1–4 hr)	333
LE, pregnant	$NiCl_2$	10–250 ppm/water		Fetal death	334
Mice					
C57BL/6, M	$NiCl_2$	15	s.c.	Kidney: hemeoxygenase (17 hr) ↑	325
C57BL/6, M C3H, M BALB/c, M B6C3F1, M	Ni $(OAc)_2$	10	i.p.	Liver: Cat, GSH, SOD, GST ↓; GSR, LPO ↑	335
B6C3F1, M	Ni_3S_2	2.85	i.t.	50% mortality (14 d)	336
C57BL/6, M CBA, M	$NiCl_2$	8.3	i.m.	NK-cell activity ↓	337
ICR, M + F, (3, 9, 14 wk)	Ni $(OAc)_2$		i.p.	LD_{50} (5 d): 3 wk M, 29.6; F, 32.2 (mg Ni per kg) 9 wk M, 16.6; F, 17.9 (mg Ni per kg) 14 wk M, 13.0; F, 15.9 (mg Ni per kg)	338
B6C3F1, F	$NiSO_4$	380–3800 ppm water × 180 d		Bone marrow cellularity ↓	339

[a] Data from adult animals, unless otherwise specified.
[b] G.D. = gestation day.

aversion effect in rats, whereas i.p. injection at the same dose level was much less effective. Decreased brain GSH levels[350] or the increased spontaneous discharge rate of Purkinje neurons[352] were documented after serial injections of lower doses.

Woods and Fowler[353] observed dose-related ultrastructural changes in the liver with concomitant increases of mitochondrial membranous enzyme activities and decreases of microsomal enzyme activities after i.p. injection of $TlCl_3$. Effects on the kidney included reduced glomerular filtration rate (GFR), proteinuria, and pathological changes in the loop of Henle in $TlSO_4$-treated female rats.[354] Effects on testes were reported in rats given drinking water containing 10 ppm Tl for 60 days.[355]

2.1.15 Tin (Sn)

Oral toxicities of inorganic tin, including metallic Sn, are thought to be rather low because of poor gastrointestinal absorption. High-dose exposure to inorganic Sn caused a decrease

TABLE 41

Toxic Effects of Sodium Selenite

Animal Strain, Sex, (Age)[a]	Dose (mg/kg or ppm Se)	Route	Effects	Ref.
Rats				
SD, M	1.6	s.c.	Lens: GSH, insoluble protein ↑; cataract formation (4 d)	341
SD, M	1.2 or 6 ppm/food × 6 wk	i.p.	Liver: GSSG, GSR, γ-GTP ↑	342
SD, M + F (5, 12 wk)	16 ppm/water × 35 d		Mortality: 5 wk M, 60%; F, 80%; 12 wk M, 0%; F, 20% Liver damage	343
Mice				
ICR, F	0.9 × 1–3 (2-d interval)	i.p.	WBC number (16 d) ↓	344
ICR, M	1.6–4.7	s.c.	Transient hypothermia, cold-seeking behavior (1 hr)	345
Swiss, M + F (7, 18 wk)	64 ppm/water × 46 d		Mortality: 20% (7 wk, M, F); 80% (18 wk, M); 40% (18 wk, F) Liver necrosis	346

[a] Data from adult animals, unless otherwise specified.

TABLE 42

Toxic Effects of Thallium on Rats

Strain, Sex	Chemicals	Dose (mg/kg or ppm Tl)	Route	Effect	Ref.
Wistar, M	TlOAc	16–54	i.p.	LD_{50} (4 d): 24.8 mg/kg	348
LE, M	Tl_2SO_4	1.7–13.6	p.o.	Flavor aversion effect; i.p. less effective	349
Charles Foster, M	TlOAc	5 × 6	i.p.	Brain: GSH, R-SH ↓	350
Wistar, M	TlOAc	23 or 39	i.p.	Brain: serotonin turnover, MAO activity (24 hr) ↑	351
SD, M	TlOAc	3.1 × 7	i.p.	Spontaneous discharge rate of Purkinje neurons ↑	352
SD, M	$TlCl_3$	50–200	i.p.	Liver: change in ultrastructure (16 hr) Mitochondrial membrane enzyme ↑; microsome enzyme ↓	353
Wistar, F	Tl_2SO_4	3.4–13.6	i.p.	Kidney: GFR ↓; protein level ↑; pathological change in loop of Henle	354
Wistar, M	Tl_2SO_4	10 ppm/water × 60 d		Pathological change in Sertoli cells; spermatozoa motility, testicular β-glucuronidase ↓	355

in weight gain and food intake due to damage to the gastrointestinal tract. de Groot et al.[356] reported that hemoglobin levels proved to be the most sensitive parameter in rats fed $SnCl_2$-contaminated food above 1000 ppm. Macroscopic and microscopic examinations were reported in $SnCl_2$-fed rats.[357,358] Zaręba and Chmielnicka[359] showed decreased ALAD activity after seven s.c. or i.p. injections of 2 mg Sn per kg as $SnCl_2$ in rats.

Toxicities of organic Sn compounds are much higher than those of inorganic compounds. Particularly, trialkylated Sn compounds are known to penetrate and damage the brain tissue owing to their high lipophilicity. Sublethal effects reported include abnormal behavior, such as hyperactivity, tremor, increase in hot plate or tail flick latency, disrupted learning, and flavor aversion (Table 43). Pathologically, lesions in the hippocampus were documented in trimethyltin-treated animals.[361,371] Triethyltin showed the highest toxicity, followed by trimethyl-, tripropyl-, and tributyltin.[371] Chang et al.[361] demonstrated that mice had higher susceptibility to trimethyltin toxicity than rats. In addition to neural tissue, inhibitions of protein phosphorylation and Ca^{2+}-ATPase activity in the heart[367] and natural killer cell activity[374] were documented.

TABLE 43

Toxic Effects of Trialkyltin Compounds

Animal Strain, Sex, (Age)[a]	Compound[b]	Dose (mg/kg or ppm Sn)	Route	Effects	Ref.
Rats					
LE, M	TMT	2.6 × 4	p.o.	Hippocampus: Synapsin I ↓	360
SD and LE, M	TMT	4.5	p.o.	Lesion in hippocampus, susceptibility: LE > SD	361
LE, M	TMT	3.6–5.1	p.o.	Length of pyramidal cell line (30 d) ↓	362
LE, M	TMT	3.6–5.1	p.o.	Disrupted learning and memory (21 d)	363
LE, M	TMT	2.9–5.1	p.o.	Visual system dysfunction	364
LE, M	TMT	2.2 (EC_{50})	i.p.	Flavor aversion	365
	TET	1.0 (EC_{50})	i.p.		
F344, M	TMT	2.5	s.c.	Hot plate latency (21–28 d) ↑	366
F344, M	TET	1.12	s.c.	Hot plate latency ↑	
		1.68	s.c.	80% mortality	
SD, M	TMT	1.8 × 6	p.o.	Heart: ^{45}Ca uptake by sarcoplasmic	367
	TET	0.87 × 6	p.o.	reticulum, Ca^{2+}-ATPase ↓; phosphorylation	
	TBT	1.2 × 6	p.o.	of specific protein ↓	
LE, M	TET	0.62 × 6	i.p.	Visual evoked potential ↓; CNS depression	368
F344, M	TET	0.21 × 14	s.c.	Latency in tail flick and hot plate ↑	369
LE, (3 d)	TMT	0.22 × 27	p.o.	Hyperactivity; learning memory function	370
	TET	0.17 × 27	p.o.	(180–200 d) ↓	
LE, (newborn)	TMT	2	p.o.	100% mortality (5–6 d)	371
	TET	1.3	p.o.	100% mortality (6–10 d)	
	TPT	4.2	p.o.	100% mortality (15–20 d)	
	TBT	10	p.o.	100% mortality (5–8 d)	
	TMT	0.66 × 24	p.o.	67% mortality	
Mice					
C57BL/6, M	TMT	2.2	i.p.	Hippocampus, front cortex; O_2-reactive species (48 hr) ↑	372
C57BL/6, M BALB/c, M	TMT	1.8	p.o.	Hippocampus: lesions (48 hr)	361
BALB/c, M	TMT	1.8	i.p.	SMA (24 hr) ↓	373
ICR, M	TET	0.58 or 2.8	p.o.	Anticonvulsant effect; interaction with adrenergic and GABAergic transmitter systems	374
C3H, M	TBT	3.6 or 36 ppm/water × 1 wk		NK-cell activity ↓	375

[a] Data from adult animals, unless otherwise specified.
[b] TMT = trimethyltin; TET = triethyltin; TPT = tripropyltin; TBT = tributyltin.

Different from trialkylated tin compounds, triphenyltin (TPT) was shown to damage hepatic functions in adult rats after three consecutive i.p. injections of 0.34 mg Sn per kg.[376] Lehotzky et al.[377] found that when pregnant rats were treated with TPT (1.74 mg/kg/day) during gestation days 7 to 15, the surviving offspring showed hyperactivity. Bis(tributyltin)oxide (0.5 ml/kg, intramuscular injection) brought about liver damage[378] and corneal edema[379] in rats. Feeding dioctyltin chloride (DOTC)-contaminated food (50 to 150 ppm) for several weeks caused thymus atrophy in rats.[380,381] Oral administration of DOTC (500 mg/kg) once a week for 8 weeks resulted in a suppressed anti-self RBC antibody response in mice.[382] Seinen et al.[381] compared the effectiveness of various dialkylated tin compounds and found that dioctyltin and dibutyltin compounds were most effective, whereas dimethyltin had no effect on the thymus.

2.1.16 Zinc (Zn)

Zinc is an essential metal and functions as a co-factor of various enzymes and insulin. Although zinc deficiency is well documented in humans and animals, its toxic effects have also been reported in experimental animals. Young mice showed twofold higher LD_{50} values (115.2 mg Zn per kg) after i.p. injection of zinc acetate than did adults (44.4 to 50.4 mg Zn per kg).[383] In the case of subchronic oral toxicity, the most severe histological lesions were observed in the kidney.[384] Inhalation exposure (5 mg Zn per m^3 × 3 hr)[385,387] or intratracheal instillation (20 μg/rat) of ZnO[386] has been demonstrated to cause functional, morphological, or biochemical changes in lungs of rats[385,386] and guinea pigs.[387]

References

1. Venugopal, B. and Luckey, T. D., Eds., *Metal Toxicity in Mammals,* Vol. 2, *Chemical Toxicity of Metals and Metalloids,* Plenum Press, New York, 1978.
2. Friberg, L., Nordberg, G. F., and Vouk, V. B., Eds., *Handbook of the Toxicology of Metals,* Vol. II, Elsevier Science, New York, 1986.
3. Ghafghazi, T., Ridlington, J. W., and Fowler, B. A., The effects of acute and subacute sodium arsenite administration on carbohydrate metabolism, *Toxicol. Appl. Pharmacol.,* 55, 126–130, 1980.
4. Valkonen, S., Savolainen, H., and Järvisalo, J., Arsenic distribution and neurochemical effects in peroral sodium arsenite exposure of rats, *Bull. Environ. Contam. Toxicol.,* 30, 303–308, 1983.
5. Byron, W. R., Bierbower, G. W., Brouwer, J. B., and Hansen, W. H., Pathologic changes in rats and dogs from two-year feeding of sodium arsenite or sodium arsenate, *Toxicol. Appl. Pharmacol.,* 10, 132–147, 1967.
6. Aposhian, H. V., Tadlock, C. H., and Moon, T. E., Protection of mice against the lethal effects of sodium arsenite: a quantitative comparison of a number of chelating agents, *Toxicol. Appl. Pharmacol.,* 61, 385–392, 1981.
7. Blakley, B. R., Sisodia, C. S., and Mukkur, T. K., The effects of methylmercury, tetraethyl lead, and sodium arsenite on the humoral immune response in mice, *Toxicol. Appl. Pharmacol.,* 52, 245–254, 1980.
8. Fowler, B. A. and Woods, J. S., The effects of prolonged oral arsenate exposure on liver mitochondria of mice: morphometric and biochemical studies, *Toxicol. Appl. Pharmacol.,* 50, 177–187, 1979.
9. Baxley, M. N., Hood, R. D., Vedel, G. C., Harrison, W. P., and Szczech, G. M., Prenatal toxicity of orally administered sodium arsenite in mice, *Bull. Environ. Contam. Toxicol.,* 26, 749–756, 1981.
10. Hood, R. D. and Vedel-Macrander, G. C., Evaluation of the effect of BAL (2,3-dimercaptopropanol) on arsenite-induced teratogenesis in mice, *Toxicol. Appl. Pharmacol.,* 73, 1–7, 1984.
11. Bosque, M. A., Domingo, J. L., Llobet, J. M., and Corbella, J., Effects of meso-2,3-dimercaptosuccinic acid (DMSA) on the teratogenicity of sodium arsenate in mice, *Bull. Environ. Contam. Toxicol.,* 47, 682–688, 1991.
12. Flora, S. J. S. and Gupta, S. D., Effect of single gallium arsenide exposure on some biochemical variables in porphyrin metabolism in rats, *J. Appl. Toxicol.,* 12, 333–334, 1992.
13. Webb, D. R., Sipes, I. G., and Carter, D. E., *In vitro* solubility and *in vivo* toxicity of gallium arsenide, *Toxicol. Appl. Pharmacol.,* 76, 96–104, 1984.
14. Goering, P. L., Maronpot, R. R., and Fowler, B. A., Effect of intratracheal gallium arsenide administration on δ-aminolevulinic acid dehydratase in rats: relationship to urinary excretion of aminolevulinic acid, *Toxicol. Appl. Pharmacol.,* 92, 179–193, 1988.
15. Webb, D. R., Wilson, S. E., and Carter, D. E., Comparative pulmonary toxicity of gallium arsenide, gallium (III) oxide, or arsenic (III) oxide intratracheally instilled into rats, *Toxicol. Appl. Pharmacol.,* 82, 405–416, 1986.
16. Sikorski, E. E., Burns, L. A., McCoy, K. L., Stern, M., and Munson, A. E., Suppression of splenic accessory cell function in mice exposed to gallium arsenide, *Toxicol. Appl. Pharmacol.,* 110, 143–156, 1991.
17. Kreppel, H., Reichl, F.-X., Szinicz, L., Fichtl, B., and Forth, W., Efficacy of various dithiol compounds in acute As_2O_3 poisoning in mice, *Arch. Toxicol.,* 64, 387–392, 1990.
18. Itoh, T., Zhang, Y. F., Murai, S., Saito, H., Nagahama, H., Miyate, H., Saito, Y., and Abe, E., The effect of arsenic trioxide on brain monoamine metabolism and locomotor activity of mice, *Toxicol. Lett.,* 54, 345–353, 1990.
19. Aranyi, C., Bradof, J. N., O'Shea, W. J., Graham, J. A., and Miller, F. J., Effects of arsenic trioxide inhalation exposure on pulmonary antibacterial defenses in mice, *J. Toxicol. Environ. Health,* 15, 163–172, 1985.
20. Hong, H. L., Fowler, B. A., and Boorman, G. A., Hematopoietic effects in mice exposed to arsine gas, *Toxicol. Appl. Pharmacol.,* 97, 173–182, 1989.
21. Blair, P. C., Thompson, M. B., Bechtold, M., Wilson, R. E., Moorman, M. P., and Fowler, B. A., Evidence for oxidative damage to red blood cells in mice induced by arsine gas, *Toxicology,* 63, 25–34, 1990.
22. Sendelbach, L. E. and Witschi, H. P., Protection by parenteral iron administration against the inhalation toxicity of beryllium sulfate, *Toxicol. Lett.,* 35, 321–325, 1987.

23. Sendelbach, L. E., Witschi, H. P., and Tryka, A. F., Acute pulmonary toxicity of beryllium sulfate inhalation in rats and mice: cell kinetics and histopathology, *Toxicol. Appl. Pharmacol.*, 85, 248–256, 1986.
24. Sendelbach, L. E. and Witschi, H. P., Bronchoalveolar lavage in rats and mice following beryllium sulfate inhalation, *Toxicol. Appl. Pharmacol.*, 90, 322–329, 1987.
25. Hart, B. A., Harmsen, A. G., Low, R. B., and Emerson, R., Biochemical, cytological, and histological alterations in rat lung following acute beryllium aerosol exposure, *Toxicol. Appl. Pharmacol.*, 75, 454–465, 1984.
26. Groth, D. H., Carcinogenicity of beryllium: review of the literature, *Environ. Res.*, 21, 56–62, 1980.
27. Groth, D. H., Kommineni, C., and Mackay, G. R., Carcinogenicity of beryllium hydroxide and alloys, *Environ. Res.*, 21, 63–84, 1980.
28. Witschi, H. P. and Marchand, P., Interference of beryllium with enzyme induction in rat liver, *Toxicol. Appl. Pharmacol.*, 20, 565–572, 1971.
29. Vacher, J., Deraedt, R., and Benzoni, J., Compared effects of two beryllium salts (soluble and insoluble): toxicity and blockade of the reticuloendothelial system, *Toxicol. Appl. Pharmacol.*, 24, 497–506, 1973.
30. Mathur, R., Sharma, S., Mathur, S., and Prakash, A. O., Effect of beryllium nitrate on early and late pregnancy in rats, *Bull. Environ. Contam. Toxicol.*, 38, 73–77, 1987.
31. Samarawickrama, G. P. and Webb, M., The acute toxicity and teratogenicity of cadmium in the pregnant rat, *J. Appl. Toxicol.*, 1, 264–269, 1981.
32. Dudley, R. E., Svoboda, D. J., and Klaassen, C. D., Acute exposure to cadmium causes severe liver injury in rats, *Toxicol. Appl. Pharmacol.*, 65, 302–313, 1982.
33. Goering, P. L., Fisher, B. R., and Kish, C. L., Stress protein synthesis induced in rat liver by cadmium precedes hepatotoxicity, *Toxicol. Appl. Pharmacol.*, 122, 139–148, 1993.
34. Lázár, G., Serra, D., and Tuchweber, B., Effect on cadmium toxicity of substances influencing reticuloendothelial activity, *Toxicol. Appl. Pharmacol.*, 29, 367–376, 1974.
35. Caflisch, C. R. and DuBose, Jr., T. D., Cadmium-induced changes in luminal fluid pH in testis and epididymis of the rat *in vivo*, *J. Toxicol. Environ. Health*, 32, 49–57, 1991.
36. Shiraishi, N., Barter, R. A., Uno, H., and Waalkes, M. P., Effect of progesterone pretreatment on cadmium toxicity in the male Fischer (F344/NCr) rat, *Toxicol. Appl. Pharmacol.*, 118, 113–118, 1993.
37. Chung, A.-S. and Maines, M. D., Differential effect of cadmium on GSH-peroxidase activity in the Leydig and the Sertori cells of rat testis: suppression by selenium and the possible relationship to heme concentration, *Biochem. Pharmacol.*, 36, 1367–1372, 1987.
38. Konishi, N., Ward, J. M., and Waalkes, M. P., Pancreatic hepatocytes in Fischer and Wistar rats induced by repeated injections of cadmium chloride, *Toxicol. Appl. Pharmacol.*, 104, 149–156, 1990.
39. Koizumi, T. and Li, Z. G., Role of oxidative stress in single-dose, cadmium-induced testicular cancer, *J. Toxicol. Environ. Health*, 37, 25–36, 1992.
40. Ruppert, P. H., Dean, K. F., and Reiter, L. W., Development of locomotor activity of rat pups exposed to heavy metals, *Toxicol. Appl. Pharmacol.*, 78, 69–77, 1985.
41. Wong, K.-L. and Klaassen, C. D., Neurotoxic effects of cadmium in young rats, *Toxicol. Appl. Pharmacol.*, 63, 330–337, 1982.
42. Bell, J. U., Induction of hepatic metallothionein in the immature rat following administration of cadmium, *Toxicol. Appl. Pharmacol.*, 54, 148–155, 1980.
43. Goering, P. L. and Klaassen, C. D., Resistance to cadmium-induced hepatotoxicity in immature rats, *Toxicol. Appl. Pharmacol.*, 74, 321–329, 1984.
44. Laskey, J. W., Rehnberg, G. L., Laws, S. C., and Hein, J. F., Reproductive effects of low acute doses of cadmium chloride in adult male rats, *Toxicol. Appl. Pharmacol.*, 73, 250–255, 1984.
45. Gulati, S., Gill, K. D., and Nath, R., Effect of cadmium on lipid composition of the weanling rat brain, *Acta Pharmacol. Toxicol.*, 59, 89–93, 1986.
46. White, T. E. K., Baggs, R. B., and Miller, R. K., Central nervous system lesions in the Wistar rat fetus following direct fetal injections of cadmium, *Teratology*, 42, 7–13, 1990.
47. Yáñez, L., Carrizales, L., Zanatta, M. T., Mejía, J. D. J., Batres, L., and Díaz-Barriga, F., Arsenic-cadmium interaction in rats: toxic effects in the heart and tissue metal shifts, *Toxicology*, 67, 227–234, 1991.
48. Scharpf, Jr., L. G., Ramos, F. J., and Hill, I. D., Influence of nitrilotriacetate (NTA) on the toxicity, excretion and distribution of cadmium in female rats, *Toxicol. Appl. Pharmacol.*, 22, 186–192, 1972.
49. Shimizu, M. and Morita, S., Effects of fasting on cadmium toxicity, glutathione metabolism, and metallothionein synthesis in rats, *Toxicol. Appl. Pharmacol.*, 103, 28–39, 1990.
50. Kotsonis, F. N. and Klaassen, C. D., Toxicity and distribution of cadmium administered to rats at sublethal doses, *Toxicol. Appl. Pharmacol.*, 41, 667–680, 1977.
51. Cantilena, Jr., L. R. and Klaassen, C. D., Comparison of the effectiveness of several chelators after single administration on the toxicity, excretion, and distribution of cadmium, *Toxicol. Appl. Pharmacol.*, 58, 452–460, 1981.
52. Probst, G. S., Bousquet, W. F., and Miya, T. S., Correlation of hepatic metallothionein concentrations with acute cadmium toxicity in the mouse, *Toxicol. Appl. Pharmacol.*, 39, 61–69, 1977.

53. Selypes, A., Serényi, P., Boldog, I., Bokros, F., and Takács, S., Acute and "long-term" genotoxic effects of $CdCl_2$ on testes of mice, *J. Toxicol. Environ. Health,* 36, 401–409, 1992.
54. Hogan, G. R. and Razniak, S. L., Split dose studies on the erythropoietic effects of cadmium, *Bull. Environ. Contam. Toxicol.,* 48, 857–864, 1992.
55. Jones, M. M., Basinger, M. A., Topping, R. J., Gale, G. R., Jones, S. G., and Holscher, M. A., Meso-2,3-dimercaptosuccinic acid and sodium N-benzyl-N-dithiocarboxy-D-glucamine as antagonists for cadmium intoxication, *Arch. Toxicol.,* 62, 29–36, 1988.
56. Thomas, D. J., Winchurch, R. A., and Huang, P. C., Ontogenic variation in acute lethality of cadmium in C57BL/6J mice, *Toxicology,* 47, 317–323, 1987.
57. Shippee, R. L., Burgess, D. H., Ciavarra, R. P., DiCapua, R. A., and Stake, P. E., Cadmium-induced suppression of the primary immune response and acute toxicity in mice: differential interaction of zinc, *Toxicol. Appl. Pharmacol.,* 71, 303–306, 1983.
58. Maitani, T. and Suzuki, K. T., Effect of cadmium on essential metal concentrations in testis, liver and kidney of five inbred strains of mice, *Toxicology,* 42, 121–130, 1986.
59. Andersen, O., Nielsen, J. B., and Svendsen, P., Oral cadmium chloride intoxication in mice: effects of chelation, *Toxicology,* 52, 65–79, 1988.
60. Manca, D., Ricard, A. C., Trottier, B., and Chevalier, G., Studies on lipid peroxidation in rat tissues following administration of low and moderate doses of cadmium chloride, *Toxicology,* 67, 303–323, 1991.
61. Laskey, J. W., Rehnberg, G. L., Laws, S. C., and Hein, J. F., Age-related dose response of selected reproductive parameters to acute cadmium chloride exposure in the male Long-Evance rat, *J. Toxicol. Environ. Health,* 19, 393–401, 1986.
62. Saillenfait, A. M., Payan, J. P., Brondeau, M. T., Zissu, D., and de Ceaurriz, J., Changes in urinary proximal tubule parameters in neonatal rats exposed to cadmium chloride during pregnancy, *J. Appl. Toxicol.,* 11, 23–27, 1991.
63. Levin, A. A. and Miller, R. K., Fetal toxicity of cadmium in the rat: decreased utero-placental blood flow, *Toxicol. Appl. Pharmacol.,* 58, 297–306, 1981.
64. Holt, D. and Webb, M., Comparison of some biochemical effects of teratogenic doses of mercuric mercury and cadmium in the pregnant rat, *Arch. Toxicol.,* 58, 249–254, 1986.
65. Holt, D. and Webb, M., Teratogenicity of ionic cadmium in the Wistar rat, *Arch. Toxicol.,* 59, 443–447, 1987.
66. Roelfzema, W. H., Roelofsen, A. M., Leene, W., and Copius Peereboom-Stegeman, J. H. J., Effects of cadmium exposure during pregnancy on cadmium and zinc concentrations in neonatal liver and consequences for the offspring, *Arch. Toxicol.,* 63, 38–42, 1989.
67. Daston, G. P., Toxic effects of cadmium on the developing rat lung. II. Glycogen and phospholipid metabolism, *J. Toxicol. Environ. Health,* 9, 51–61, 1982.
68. Sorell, T. L. and Graziano, J. H., Effect of oral cadmium exposure during pregnancy on maternal and fetal zinc metabolism in the rat, *Toxicol. Appl. Pharmacol.,* 102, 537–545, 1990.
69. Barański, B., Effect of maternal cadmium exposure on postnatal development and tissue cadmium, copper and zinc concentrations in rats, *Arch. Toxicol.,* 58, 255–260, 1986.
70. De, S. K., Dey, S. K., and Andrews, G. K., Cadmium teratogenicity and its relationship with metallothionein gene expression in midgestation mouse embryos, *Toxicology,* 64, 89–104, 1990.
71. Katsuta, O., Hiratsuka, H., Matsumoto, J., Tsuchitani, M., Umemura, T., and Marumo, F., Ovariectomy enhances cadmium-induced nephrotoxicity and hepatotoxicity in rats, *Toxicol. Appl. Pharmacol.,* 119, 267–274, 1993.
72. Goyer, R. A., Miller, C. R., Zhu, S.-Y., and Victery, W., Non-metallothionein-bound cadmium in the pathogenesis of cadmium nephrotoxicity in the rat, *Toxicol. Appl. Pharmacol.,* 101, 232–244, 1989.
73. Müller, L. and Stacey, N. H., Subcellular toxicity of low level cadmium in rats: effect on cytochrome C oxidase, *Toxicology,* 51, 25–34, 1988.
74. Kojima, S., Ono, H., Kiyozumi, M., Honda, T., and Takadate, A., Effect of *N*-benzyl-D-glucamine dithiocarbamate on the renal toxicity produced by subacute exposure to cadmium in rats, *Toxicol. Appl. Pharmacol.,* 98, 39–48, 1989.
75. Morsert, A. F. W., Leene, W., De Groot, C., Kipp, J. B. A., Evers, M., Roelofsen, A. M., and Bosch, K. S., Differences in immunological susceptibility to cadmium toxicity between two rat strains as demonstrated with cell biological methods. Effect of cadmium on DNA synthesis of thymus lymphocytes, *Toxicology,* 48, 127–139, 1988.
76. Shukla, G. S., Hussain, T., and Chandra, S. V., Possible role of regional superoxide dismutase activity and lipid peroxide levels in cadmium neurotoxicity: *in vivo* and *in vitro* studies in growing rats, *Life Sci.,* 41, 2215–2221, 1987.
77. Dudley, R. E., Gammal, L. M., and Klaassen, C. D., Cadmium-induced hepatic and renal injury in chronically exposed rats: likely role of hepatic cadmium-metallothionein in nephrotoxicity, *Toxicol. Appl. Pharmacol.,* 77, 414–426, 1985.
78. Nagai, Y., Sato, M., and Sasaki, M., Effect of cadmium administration upon urinary excretion of hydroxylysine and hydroxyproline in the rat, *Toxicol. Appl. Pharmacol.,* 63, 188–193, 1982.

79. Ando, M. and Matsui, S., Effect of cadmium on vitamin D-nonstimulated intestinal calcium absorption in rats, *Toxicology*, 45, 1–11, 1987.
80. O'Brien, I. G. and King, L. J., The effect of chronic parenteral administration of cadmium on isoenzyme levels of alkaline phosphatase in intestinal mucosa, *Toxicology*, 56, 87–94, 1989.
81. Waalkes, M. P., Kovatch, R., and Rehm, S., Effect of chronic dietary zinc deficiency on cadmium toxicity and carcinogenesis in the male Wistar (Hsd: (WI)BR) rat, *Toxicol. Appl. Pharmacol.*, 108, 448–456, 1991.
82. Cárdenas, A., Bernard, A., and Lauwerys, R., Incorporation of [^{35}S]sulfate into glomerular membranes of rats chronically exposed to cadmium and its relation with urinary glycosaminoglycans and proteinuria, *Toxicology*, 76, 219–231, 1992.
83. Toury, R., Stelly, N., Boissonneau, E., and Dupuis, Y., Degenerative processes in skeletal muscle of Cd^{2+}-treated rats and Cd^{2+} inhibition of mitochondrial Ca^{2+} transport, *Toxicol. Appl. Pharmacol.*, 77, 19–35, 1985.
84. Iguchi, H. and Sano, S., Effect of cadmium on the bone collagen metabolism of rat, *Toxicol. Appl. Pharmacol.*, 62, 126–136, 1982.
85. Kopp, S. J., Perry, Jr., H. M., Perry, E. F., and Erlanger, M., Cardiac physiologic and tissue metabolic changes following chronic low-level cadmium and cadmium plus lead ingestion in the rat, *Toxicol. Appl. Pharmacol.*, 69, 149–160, 1983.
86. Rosenberg, D. W. and Kappas, A., Induction of heme oxygenase in the small intestinal epithelium: a response to oral cadmium exposure, *Toxicology*, 67, 199–210, 1991.
87. Chowdhury, B. A., Friel, J. K., and Chandra, R. K., Cadmium-induced immunopathology is prevented by zinc administration in mice, *J. Nutr.*, 117, 1788–1794, 1987.
88. Bhattacharyya, M. H., Whelton, B. D., Peterson, D. P., Carnes, B. A., Moretti, E. S., Toomey, J. M., and Williams, L. L., Skeletal changes in multiparous mice fed a nutrient-sufficient diet containing cadmium, *Toxicology*, 50, 193–204, 1988.
89. Valois, A. A. and Webster, W. S., The choroid plexus as a target site for cadmium toxicity following chronic exposure in the adult mouse: an ultrastructural study, *Toxicology*, 55, 193–205, 1989.
90. Goering, P. L. and Klaassen, C. D., Altered subcellular distribution of cadmium following cadmium pretreatment: possible mechanism of tolerance to cadmium-induced lethality, *Toxicol. Appl. Pharmacol.*, 70, 195–203, 1983.
91. Wormser, U. and Nir, I., Effect of age on cadmium-induced metallothionein synthesis in the rat, *Arch. Toxicol.*, 62, 392–394, 1988.
92. Nishimura, N., Oshima, H., and Nakano, M., Urinary trehalase as an early indicator of cadmium-induced renal tubular damage in rabbit, *Arch. Toxicol.*, 59, 255–260, 1986.
93. Grose, E. C., Richards, J. H., Jaskot, R. H., Ménache, M. G., Graham, J. A., and Dauterman, W. C., A comparative study of the effects of inhaled cadmium chloride and cadmium oxide: pulmonary response, *J. Toxicol. Environ. Health*, 21, 219–232, 1987.
94. Boisset, M. and Boudene, C., Effect of a single exposure to cadmium oxide fumes on rat lung microsomal enzymes, *Toxicol. Appl. Pharmacol.*, 57, 335–345, 1981.
95. Buckley, B. J. and Bassett, D. J. P., Glutathione redox status of control and cadmium oxide-exposed rat lungs during oxidant stress, *J. Toxicol. Environ. Health*, 22, 287–299, 1987.
96. Kutzman, R. S., Drew, R. T., Shiotsuka, R. N., and Cockrell, B. Y., Pulmonary changes resulting from subchronic exposure to cadmium chloride aerosol, *J. Toxicol. Environ. Health*, 17, 175–189, 1986.
97. Hart, B. A., Voss, G. W., and Willean, C. L., Pulmonary tolerance to cadmium following cadmium aerosol pretreatment, *Toxicol. Appl. Pharmacol.*, 101, 447–460, 1989.
98. Prasada-Rao, P. V. V. and Gardner, D. E., Effects of cadmium inhalation on mitochondrial enzymes in rat tissues, *J. Toxicol. Environ. Health*, 17, 191–199, 1986.
99. Hirano, S., Tsukamoto, N., Higo, S., and Suzuki, K. T., Toxicity of cadmium oxide instilled into the rat lung. II. Inflammatory responses in broncho-alveolar lavage fluid, *Toxicology*, 55, 25–35, 1989.
100. Driscoll, K. E., Maurer, J. K., Poynter, J., Higgins, J., Asquith, T., and Miller, N. S., Stimulation of rat alveolar macrophage fibronectin release in a cadmium chloride model of lung injury and fibrosis, *Toxicol. Appl. Pharmacol.*, 116, 30–37, 1992.
101. Hakkinen, P. J., Morse, C. C., Martin, F. M., Dalbey, W. E., Haschek, W. M., and Witschi, H. R., Potentiating effects of oxygen in lungs damaged by methylcyclopentadienyl manganese tricarbonyl, cadmium chloride, oleic acid, and antitumor drugs, *Toxicol. Appl. Pharmacol.*, 67, 55–69, 1983.
102. Martin, F. M. and Witschi, H. P., Cadmium-induced lung injury: cell kinetics and long-term effects, *Toxicol. Appl. Pharmacol.*, 80, 215–227, 1985.
103. Suzuki, C. A. M. and Cherian, M. G., Renal glutathione depletion and nephrotoxicity of cadmium-metallothionein in rats, *Toxicol. Appl. Pharmacol.*, 98, 544–552, 1989.
104. Jin, T., Leffler, P., and Nordberg, G. F., Cadmium-metallothionein nephrotoxicity in the rat: transient calcuria and proteinuria, *Toxicology*, 45, 307–317, 1987.
105. Min, K.-S., Kobayashi, K., Onosaka, S., Ohta, N., Okada, Y., and Tanaka, K., Tissue distribution of cadmium and nephropathy after administration of cadmium in several chemical forms, *Toxicol. Appl. Pharmacol.*, 86, 262–270, 1986.

106. Goering, P. L., Kish, C. L., and Fisher, B. R., Stress protein synthesis induced by cadmium-cysteine in rat kidney, *Toxicology,* 85, 25–39, 1993.
107. Zhao, J. Y., Foulkes, E. C., and Jones, M., Delayed nephrotoxic effects of cadmium and their reversibility by chelation, *Toxicology,* 64, 235–243, 1990.
108. Sendelbach, L. E., Kershaw, W. C., Cuppage, F., and Klaassen, C. D., Cd-metallothionein nephrotoxicity in inbred strains of mice, *J. Toxicol. Environ. Health,* 35, 115–126, 1992.
109. Maitani, T., Watahiki, A., and Suzuki, K. T., Acute renal dysfunction by cadmium injected with cysteine in relation to renal critical concentration of cadmium, *Arch. Toxicol.,* 58, 136–140, 1986.
110. Tsapakos, M. J., Hampton, T. H., and Jennette, K. W., The carcinogen chromate induces DNA cross-links in rat liver and kidney, *J. Biol. Chem.,* 256, 3623–3626, 1981.
111. Standeven, A. M. and Wetterhahn, K. E., Possible role of glutathione in chromium (VI) metabolism and toxicity in rats, *Pharmacol. Toxicol.,* 69, 469–476, 1991.
112. Mikalsen, A., Alexander, J., Andersen, R. A., and Ingelman-Sundberg, M., Effect of *in vivo* chromate, acetone and combined treatment on rat liver *in vitro* microsomal chromium (VI) reductive activity and on cytochrome P-450 expression, *Pharmacol. Toxicol.,* 68, 456–463, 1991.
113. Kim, E. and Na, K. J., Effect of sodium dichromate on carbohydrate metabolism, *Toxicol. Appl. Pharmacol.,* 110, 251–258, 1991.
114. Ernst, E., Testicular toxicity following short-term exposure to tri- and hexavalent chromium: an experimental study in the rat, *Toxicol. Lett.,* 51, 269–275, 1990.
115. Saxena, D. K., Murthy, R. C., Lal, B., Srivastava, R. S., and Chandra, S. V., Effect of hexavalent chromium on testicular maturation in the rat, *Reproduc. Toxicol.,* 4, 223–228, 1990.
116. Murthy, R. C., Saxena, D. K., Gupta, S. K., and Chandra, S. V., Ultrastructural observations in testicular tissue of chromium-treated rats, *Reproduc. Toxicol.,* 5, 443–447, 1991.
117. Vyskočil, A., Viau, C., Čížková, M., and Truchon, G., Kidney function in male and female rats chronically exposed to potassium dichromate, *J. Appl. Toxicol.,* 13, 375–376, 1993.
118. Coogan, T. P., Motz, J., Snyder, C. A., Squibb, K. S., and Costa, M., Differential DNA-protein crosslinking in lymphocytes and liver following chronic drinking water exposure of rats to potassium chromate, *Toxicol. Appl. Pharmacol.,* 109, 60–72, 1991.
119. Glaser, U., Hochrainer, D., Klöppel, H., and Kuhnen, H., Low level chromium (VI) inhalation effects on alveolar macrophages and immune functions in Wistar rats, *Arch. Toxicol.,* 57, 250–256, 1985.
120. Glaser, U., Hochrainer, D., Klöppel, H., and Oldiges, H., Carcinogenicity of sodium dichromate and chromium (VI/III) oxide aerosols inhaled by male Wistar rats, *Toxicology,* 42, 219–232, 1986.
121. Trivedi, B., Saxena, D. K., Murthy, R. C., and Chandra, S. V., Embryotoxicity and fetotoxicity of orally administered hexavalent chromium in mice, *Reproduc. Toxicol.,* 3, 275–278, 1989.
122. Newman, R. A., Khokhar, A. R., Sunderland, B. A., Travis, E. L., and Bulger, R. E., A comparison in rodents of renal and intestinal toxicity of cisplatin and a new water-soluble antitumor platinum complex: *N*-methyl-iminodiacetato-diaminocyclohexane platinum (II), *Toxicol. Appl. Pharmacol.,* 84, 454–463, 1986.
123. Ward, J. M. and Fauvie, K. A., The nephrotoxic effects of *cis*-diammine-dichloroplatinum (II) (NSC-119875) in male F344 rats, *Toxicol. Appl. Pharmacol.,* 38, 535–547, 1976.
124. Goldstein, R. S., Noordewier, B., Bond, J. T., Hook, J. B., and Mayor, G. H., *cis*-Dichlorodiammineplatinum nephrotoxicity: time course and dose response of renal functional impairment, *Toxicol. Appl. Pharmacol.,* 60, 163–175, 1981.
125. Litterst, C. L., Alterations in the toxicity of *cis*-dichlorodiammineplatinum-II and in tissue localization of platinum as a function of NaCl concentration in the vehicle of administration, *Toxicol. Appl. Pharmacol.,* 61, 99–108, 1981.
126. De Witt, L. M., Jones, T. W., and Moore, L., Stimulation of the renal endoplasmic reticulum calcium pump: a possible biomarker for platinate toxicity, *Toxicol. Appl. Pharmacol.,* 92, 157–169, 1988.
127. Mayer, R. D. and Maines, M. D., Promotion of *trans*-platinum *in vivo* effects on renal heme and hemoprotein metabolism by D,L-buthionine-*S*,*R*-sulfoximine: possible role of glutathione, *Biochem. Pharmacol.,* 39, 1565–1571, 1990.
128. Safirstein, R., Miller, P., Dikman, S., Lyman, N., and Shapiro, C., Cisplatin nephrotoxicity in rats: defect in papillary hypertonicity, *Am. J. Physiol.,* 241, F175–F185, 1981.
129. Heidemann, H. Th., Müller, St., Mertins, L., Stepan, G., Hoffmann, K., and Ohnhaus, E. E., Effect of aminophylline on cisplatin nephrotoxicity in the rat, *Br. J. Pharmacol.,* 97, 313–318, 1989.
130. Sadzuka, Y., Shoji, T., and Takino, Y., Mechanism of the increase in lipid peroxide induced by cisplatin in the kidneys of rats, *Toxicol. Lett.,* 62, 293–300, 1992.
131. Bompart, G. and Orfila, C., Cisplatin nephrotoxicity in lead-pretreated rats: enzymatic and morphological studies, *Toxicol. Lett.,* 50, 237–247, 1990.
132. Scherini, E., Permanent alterations of the dendritic tree of cerebellar Purkinje neurons in the rat following postnatal exposure to *cis*-dichlorodiammineplatinum, *Acta Neuropathol.,* 81, 324–327, 1991.
133. Cavaletti, G., Tredici, G., Marmiroli, P., Petruccioli, M. G., Barajon, I., and Fabbrica, D., Morphometric study of the sensory neuron and peripheral nerve changes induced by chronic cisplatin (DDP) administration in rats, *Acta Neuropathol.,* 84, 364–371, 1992.

134. Terheggen, P. M. A. B., Van Der Hoop, R. G., Floot, B. G. J., and Gispen, W. H., Cellular distribution of *cis*-diamminedichloroplatinum (II)-DNA binding in rat dorsal root spinal ganglia: effect of the neuroprotecting peptide ORG.2766, *Toxicol. Appl. Pharmacol.*, 99, 334–343, 1989.
135. Goldstein, R. S., Mayor, G. H., Gingerich, R. L., Hook, J. B., Robinson, B., and Bond, J. T., Hyperglucagonemia following cisplatin treatment, *Toxicol. Appl. Pharmacol.*, 68, 250–259, 1983.
136. Goldstein, R. S., Mayor, G. H., Gingerich, R. L., Hook, J. B., Rosenbaum, R. W., and Bond, J. T., The effects of cisplatin and other divalent platinum compounds on glucose metabolism and pancreatic endocrine function, *Toxicol. Appl. Pharmacol.*, 69, 432–441, 1983.
137. Azouri, H., Bidart, J.-M., and Bohuon, C., *In vivo* toxicity of cisplatin and carboplatin on the Leydig cell function and effect of the human choriogonadotropin, *Biochem. Pharmacol.*, 38, 567–571, 1989.
138. Pogach, L. M., Lee, Y., Gould, S., Giglio, W., Meyenhofer, M., and Huang, H. F. S., Characterization of *cis*-platinum-induced sertoli cell dysfunction in rodents, *Toxicol. Appl. Pharmacol.*, 98, 350–361, 1989.
139. Choie, D. D., Longnecker, D. S., and Copley, M. P., Cytotoxicity of cisplatin in rat intestine, *Toxicol. Appl. Pharmacol.*, 60, 354–359, 1981.
140. Keller, K. A. and Aggarwal, S. K., Embryotoxicity of cisplatin in rats and mice, *Toxicol. Appl. Pharmacol.*, 69, 245–256, 1983.
141. Bajt, M. L. and Aggarwal, S. K., An analysis of factors responsible for resorption of embryos in cisplatin-treated rats, *Toxicol. Appl. Pharmacol.*, 80, 97–107, 1985.
142. Satoh, M., Naganuma, A., and Imura, N., Deficiency of selenium intake enhances manifestation of renal toxicity of *cis*-diamminedichloroplatinum in mice, *Toxicol. Lett.*, 38, 155–160, 1987.
143. Schaeppi, U., Heyman, I. A., Fleischman, R. W., Rosenkrantz, H., Ilievski, V., Phelan, R., Cooney, D. A., and Davis, R. D., *cis*-Dichlorodiammineplatinum (II) (NSC-119 875): preclinical toxicologic evaluation of intravenous injection in dogs, monkeys and mice, *Toxicol. Appl. Pharmacol.*, 25, 230–241, 1973.
144. Boughattas, N. A., Lévi, F., Fournier, C., Hecquet, B., Lemaigre, G., Roulon, A., Mathé, G., and Reinberg, A., Stable circadian mechanisms of toxicity of two platinum analogs (cisplatin and carboplatin) despite repeated dosages in mice, *J. Pharmacol. Exp. Ther.*, 255, 672–679, 1990.
145. Lerza, R., Bogliolo, G., Muzzulini, C., and Pannacciulli, I., Failure of *N*-acetylcysteine to protect against *cis*-dichlorodiammine-platinum (II)-induced hematopoietic toxicity in mice, *Life Sci.*, 38, 1795–1800, 1986.
146. Wierda, D. and Matamoros, M., Partial characterization of bone marrow hemopoiesis in mice after cisplatin administration, *Toxicol. Appl. Pharmacol.*, 75, 25–34, 1984.
147. Apfel, S. C., Arezzo, J. C., Lipson, L. A., and Kessler, J. A., Nerve growth factor prevents experimental cisplatin neuropathy, *Ann. Neurol.*, 31, 76–80, 1992.
148. Lasfargues, G., Lison, D., Maldague, P., and Lauwerys, R., Comparative study of the acute lung toxicity of pure cobalt powder and cobalt-tungsten carbide mixture in rat, *Toxicol. Appl. Pharmacol.*, 112, 41–50, 1992.
149. Di Giulio, C., Data, P. G., and Lahiri, S., Chronic cobalt causes hypertrophy of glomus cells in the rat carotid body, *Am. J. Physiol.*, 261, C102–C105, 1991.
150. Nation, J. R., Bourgeois, A. E., Clark, D. E., and Hare, M. F., The effects of chronic cobalt exposure on behavior and metallothionein levels in the adult rat, *Neurobehav. Toxicol. Teratol.*, 5, 9–15, 1983.
151. Suarez, K. A. and Bhonsle, P., The relationship of cobaltous chloride-induced alterations of hepatic microsomal enzymes to altered carbon tetrachloride hepatotoxicity, *Toxicol. Appl. Pharmacol.*, 37, 23–27, 1976.
152. Paternain, J. L., Domingo, J. L., and Corbella, J., Developmental toxicity of cobalt in the rat, *J. Toxicol. Environ. Health*, 24, 193–200, 1988.
153. Muhoberac, B. B., Hanew, T., Halter, S., and Schenker, S., A model of cytochrome P-450-centered hepatic dysfunction in drug metabolism induced by cobalt-protoporphyrin administration, *Biochem. Pharmacol.*, 38, 4103–4113, 1989.
154. Anderson, M. B., Pedigo, N. G., Katz, R. P., and George, W. J., Histopathology of tests from mice chronically treated with cobalt, *Reproduc. Toxicol.*, 6, 41–50, 1992.
155. Spaethe, S. M. and Jollow, D. J., Effect of cobalt protoporphyrin on hepatic drug-metabolizing enzymes, *Biochem. Pharmacol.*, 38, 2027–2038, 1989.
156. Lundborg, M. and Camner, P., Lysozyme levels in rabbit after inhalation of nickel, cadmium, cobalt, and copper chlorides, *Environ. Res.*, 34, 335–342, 1984.
157. Hirano, S., Ebihara, H., Sakai, S., Kodama, N., and Suzuki, K. T., Pulmonary clearance and toxicity of intratracheally instilled cupric oxide in rats, *Arch. Toxicol.*, 67, 312–317, 1993.
158. Hirano, S., Sakai, S., Ebihara, H., Kodama, N., and Suzuki, K. T., Metabolism and pulmonary toxicity of intratracheally instilled cupric sulfate in rats, *Toxicology*, 64, 223–233, 1990.
159. Fuentealba, I. C., Davis, R. W., Elmes, M. E., Jasani, B., and Haywood, S., Mechanisms of tolerance in the copper-loaded rat liver, *Exp. Molec. Pathol.*, 59, 71–84, 1993.
160. Fuentealba, I. C., Haywood, S., and Foster, J., Cellular mechanisms of toxicity and tolerance in the copper-loaded rat. III. Ultrastructural changes and copper localization in the kidney, *Br. J. Exp. Pathol.*, 70, 543–556, 1989.
161. Toyokuni, S., Okada, S., Hamazaki, S., Fujioka, M., Li, J.-L., and Midorikawa, O., Cirrhosis of the liver induced by cupric nitrolotriacetate in Wistar rats, *Am. J. Pathol.*, 134, 1263–1274, 1989.

162. Shiraishi, N., Taguchi, T., and Kinebuchi, H., Copper-induced toxicity in macular mutant mouse: an animal model for Menkes' kidney-hair disease, *Toxicol. Appl. Pharmacol.*, 110, 89–96, 1991.
163. Agarwal, K., Sharma, A., and Talukder, G., Clastogenic effects of copper sulfate on the bone marrow chromosomes of mice *in vivo*, *Mutation Res.*, 243, 1–6, 1990.
164. Galleano, M. and Puntarulo, S., Hepatic chemiluminescence and lipid peroxidation in mild iron overload, *Toxicology*, 76, 27–38, 1992.
165. Younes, M., Eberhardt, I., and Lemoine, R., Effect of iron overload on spontaneous and xenobiotic-induced lipid peroxidation *in vivo*, *J. Appl. Toxicol.*, 9, 103–108, 1989.
166. Pietrangelo, A., Cossarizza, A., Monti, D., Ventura, E., and Franceschi, C., DNA repair in lymphocytes from humans and rats with chronic iron overload, *Biochem. Biophys. Res. Commun.*, 154, 698–704, 1988.
167. Okada, S., Hamazaki, S., Toyokuni, S., and Midorikawa, O., Induction of mesothelioma by intraperitoneal injections of ferric sacchrate in male Wistar rats, *Br. J. Cancer*, 60, 708–711, 1989.
168. Iancu, T. C., Ward, R. J., and Peters, T. J., Ultrastructural observations in the carbonyl iron-fed rat, an animal model for hemochromatosis, *Virchows Arch. B*, 53, 208–217, 1987.
169. Park, C. H., Bacon, B. R., Brittenham, G. M., and Tavill, A. S., Pathology of dietary carbonyl iron overload in rats, *Lab. Invest.*, 57, 555–563, 1987.
170. Bacon, B. R., Tabill, A. S., Brittenham, G. M., Park, C. H., and Recknagel, R. O., Hepatic lipid peroxidation *in vivo* in rats with chronic iron overload, *J. Clin. Invest.*, 71, 429–439, 1983.
171. Lee, Y. H., Layman, D. K., Bell, R. R., and Norton, H. W., Response of glutathione peroxidase and catalase to excess dietary iron in rats, *J. Nutr.*, 111, 2195–2202, 1981.
172. Smith, A. G. and Francis, J. E., Genetic variation of iron-induced uroporphyria in mice, *Biochem. J.*, 291, 29–35, 1993.
173. Li, J.-L., Okada, S., Hamazaki, S., Ebina, Y., and Midorikawa, O., Subacute nephrotoxicity and induction of renal cell carcinoma in mice treated with ferric nitrilotriacetate, *Cancer Res.*, 47, 1867–1869, 1987.
174. Carthew, P., Dorman, B. M., Edwards, R. E., Francis, J. E., and Smith, A. G., A unique rodent model for both the cardiotoxic and hepatotoxic effects of prolonged iron overload, *Lab. Invest.*, 69, 217–222, 1993.
175. Carthew, P., Edwars, R. E., Dorman, B. M., and Francis, J. E., Rapid induction of hepatic fibrosis in the gerbil after the parenteral administration of iron-dextran complex, *Hepatology*, 13, 534–539, 1991.
176. Smith, A. G., Cabral, J. R. P., Carthew, P., Francis, J. E., and Manson, M. M., Carcinogenicity of iron in conjunction with a chlorinated environmental chemical, hexachlorobenzene, in C57BL/10ScSn mice, *Int. J. Cancer*, 43, 492–496, 1989.
177. Scharding, N. N. and Oehme, F. W., The use of animal models for comparative studies of lead poisoning, *Clin. Toxicol.*, 6, 419–424, 1973.
178. Kostial, K., Maljkovic, T., and Jugo, S., Lead acetate toxicity in rats in relation to age and sex, *Arch. Toxicol.*, 31, 265–269, 1974.
179. Hogan, G. R., Effects of ovariectomy and orchiectomy on lead-induced mortality in rats, *Environ. Res.*, 21, 314–316, 1980.
180. Garbar, B. and Wei, E., Lead toxicity in mice with genetically different levels of δ-aminolevulic acid, *Bull. Environ. Contam. Toxicol.*, 9, 80–83, 1973.
181. Hogan, G. R., Variation of lead-induced lethality in estradiol-treated mice, *J. Toxicol. Environ. Health*, 9, 353–357, 1982.
182. Lorenzo, A. V., Gewirtz, M., and Averill, D., CNS lead toxicity in rabbit offspring, *Environ. Res.*, 17, 131–150, 1978.
183. Nagatoshi, K., Experimental chronic lead poisoning, *Folia Psychiatrica Neurologica Japonica*, 33, 123–131, 1979.
184. Kumar, M. V. S. and Desiraju, T., Regional alterations of brain biogenic amines and GABA/glutamate levels in rats following chronic lead exposure during neonatal development, *Arch. Toxicol.*, 64, 305–314, 1990.
185. Kishi, R. and Uchino, E., Effects of low lead exposure on neuro-behavioral function in the rat, *Arch. Environ. Health*, 38, 25–33, 1983.
186. Altman, L., Weinsberg, F., Sveinsson, K., Lilienthal, H., Wiegand, H., and Winneke, G., Impairment of long-term potentiation and learning following chronic lead exposure, *Toxicol. Lett.*, 66, 105–112, 1993.
187. Sundström, R., Müntzing, K., Kalimo, H., and Sourander, P., Changes in the integrity of the blood-brain barrier in suckling rats with low dose lead encephalopathy, *Acta Neuropathol.*, 68, 1–9, 1985.
188. Lorton, D. and Anderson, W. J., The effects of postnatal lead toxicity on the development of cerebellum in rats, *Neurobehav. Toxicol. Teratol.*, 8, 51–59, 1986.
189. Overmann, S. R., Behavioral effects of symptomatic lead exposure during neonatal development in rats, *Toxicol. Appl. Pharmacol.*, 41, 459–471, 1977.
190. Chandra, S. V., Ali, M. M., Saxena, D. K., and Murthy, R. C., Behavioral neurochemical changes in rats simultaneously exposed to manganese and lead, *Arch. Toxicol.*, 49, 49–56, 1981.
191. Deluca, J., Donovick, P. J., and Burright, R. G., Lead exposure, environmental temperature, nesting and consummatory behavior of adult mice of two ages, *Neurotoxicol. Teratol.*, 11, 7–11, 1989.
192. Donald, J. M., Cutler, M. G., and Moore, M. R., Effects of lead in the laboratory mouse: development and social behavior after life long exposure to 12 μM lead in drinking fluid, *Neuropharmacology*, 26, 391–399, 1987.

193. Aiegler, E. E., Edwards, B. B., Jensen, R. L., Mahaffer, K. R., and Fomon, S. J., Adsorption and retention of lead by infants, *Pediatr. Res.*, 12, 29–34, 1978.
194. Mykkanen, H. M., Dickerson, J. W. T., and Lancaster, M. C., Effect of age on the tissue distribution of lead in the rat, *Toxicol. Appl. Pharmacol.*, 51, 447–454, 1979.
195. Brown, D. R., Neonatal lead exposure in the rat: decreased learning as a function of age and blood concentrations, *Toxicol. Appl. Pharmacol.*, 32, 628–637, 1975.
196. Kitchen, I., Lead toxicity and alterations in opioid systems, *Neurotoxicology*, 14, 115–124, 1993.
197. Yokoyama, K. and Araki, S., Alterations in peripheral nerve conduction velocity in low and high lead exposure: an animal study, *Ind. Health*, 24, 67–74, 1986.
198. Falke, H. E. and Zwennis, W. C. M., Toxicity of lead acetate to female rabbits after chronic subcutaneous administration. I. Biochemical and clinical effects, *Arch. Toxicol.*, 64, 522–529, 1990.
199. Hass, G. M., Brown, D. V. L., Eisenstein, R., and Hemmens, A., Relations between lead poisoning in rabbit and man, *Am. J. Pathol.*, 45, 691–715, 1964.
200. Oskarsson, A., Effects of perinatal treatment with lead and disulfiram on ALAD activity in blood, liver and kidney and urinary ALA excretion in rats, *Pharmacol. Toxicol.*, 64, 344–348, 1989.
201. Barlow, J. J., Baruah, J. K., and Davison, A., δ-Aminolevulinic acid dehydratase activity and focal brain hemorrhages in lead-treated rats, *Acta. Neuropathol.*, 39, 219–223, 1977.
202. Hayashi, M., Lead toxicity in the pregnant rat, *Environ. Res.*, 30, 152–160, 1983.
203. Tomokuni, K., Ichiba, M., and Hirai, Y., Species difference of urinary excretion of δ-aminolevulinic acid and coproporphyrin in mice and rats exposed to lead, *Toxicol. Lett.*, 41, 255–259, 1988.
204. Rehman, S. U., Effects of zinc, copper, and lead toxicity on δ-aminolevulinic acid dehydratase activity, *Bull. Environ. Contam. Toxicol.*, 33, 92–98, 1984.
205. Tandon, S. K. and Flora, S. J. S., Dose and time effects of combined exposure to lead and ethanol on lead body burden and some neuronal, hepatic and haematopoietic biochemical indices in the rat, *J. Appl. Toxicol.*, 9, 347–352, 1989.
206. Silbergeld, E. K., Hruska, R. E., Bradly, D., Lamon, J. M., and Frykholm, B. C., Neurotoxic aspects of porphyrinopathies: lead and succinylacetone, *Environ. Res.*, 29, 459–471, 1982.
207. Flora, S. J. S., Singh, S., and Tandon, S. K., Chelation in metal intoxication. XVIII. Combined effects of thiamine and calcium disodium versenate on lead toxicity, *Life Sci.*, 38, 67–71, 1986.
208. Schlick, E., Mengel, K., and Friedberg, K. D., The effect of low lead doses *in vitro* and *in vivo* on the d-ALA-d activity of erythrocytes, bone marrow cells, liver and brain of the mouse, *Arch. Toxicol.*, 53, 193–205, 1983.
209. Hernberg, S. J., Nikkanen, G., and Lilius, H., δ-Aminolevulinic acid dehydratase as a measure of lead exposure, *Arch. Environ. Health*, 21, 140–145, 1970.
210. Maes, J. and Gerber, G. B., Increased ALA dehydratase activity and spleen weight in lead-intoxicated rats. A consequence of increased blood cell destruction, *Experimentia*, 34, 381–382, 1978.
211. Karmakar, N., Saxena, R., and Ananda, S., Histopathological changes induced in rats tissues by oral intake of lead acetate, *Environ. Res.*, 41, 23–28, 1986.
212. Fowler, B. A., Kimmel, C. A., Woods, J. S., McConnell, E. E., and Grant, L. D., Chronic low level lead toxicity in the rat. III. An integrated assessment of long-term toxicity with special reference to the kidney, *Toxicol. Appl. Pharmacol.*, 56, 59–77, 1980.
213. Spit, B. J., Wibowo, A. A. E., Feron, V. J., and Zielhuis, R. L., Ultrastructural changes in the kidneys of rabbits treated with lead acetate, *Arch. Toxicol.*, 49, 85–91, 1981.
214. Khalil-Manesh, F., Gonic, H. C., Cohen, A., Bergamaschi, E., and Mutti, A., Experimental model of lead nephropathy. II. Effect of removal from lead exposure and chelation treatment with dimercaptosuccinic acid (DMSA), *Environ. Res.*, 58, 35–54, 1992.
215. Goyer, R. A., Lead toxicity: a problem in environmental pathology, *Am. J. Pathol.*, 64, 167–181, 1971.
216. Hemphill, F. E., Kaerberle, M. L., and Buck, W. B., Lead suppression of mouse resistance to *Salmonella typhimurium*, *Science*, 172, 1031–1033, 1971.
217. Blakley, B. R. and Archer, D. L., The effect of lead acetate on the immune response in mice, *Toxicol. Appl. Pharmacol.*, 61, 18–26, 1981.
218. Koller, L. D., The immunotoxic effects of lead in lead-exposed laboratory animals, *Ann. N.Y. Acad. Sci.*, 587, 160–167, 1990.
219. Kobayashi, N. and Okamoto, T., Effect of lead oxide on the induction of lung tumors in Syrian hamsters, *J. Natl. Cancer Inst.*, 52, 1605–1610, 1974.
220. Kerkvliet, N. I. and Baecher-Steppan, L., Immunotoxicology studies on lead: effect of exposure on tumor growth and cell-mediated tumor immunity after synergic or allogenic stimulation, *Immunopharmacology*, 4, 213–224, 1982.
221. Koller, L. D. and Kovacic, S., Decreased antibody response in mice exposed to lead, *Nature*, 250, 148–150, 1974.
222. Koller, L. D., Exon, J. H., and Roan, J. G., Humoral antibody response in mice after single dose exposure to lead or cadmium, *Proc. Soc. Exp. Biol. Med.*, 151, 339–342, 1976.
223. Trejo, R. A., DiLuzio, N. R., Loose, L. D., and Hoffman, E., Reticuloendothelial and hepatic alterations following lead acetate administration, *Exp. Mol. Pathol.*, 17, 145–158, 1972.

224. McClain, R. M. and Becker, B. A., Teratogenicity, fetal toxicity, and placental transfer of lead nitrate in rats, *Toxicol. Appl. Pharmacol.*, 31, 72–78, 1975.
225. Sokol, R. Z. and Berman, N., The effect on age of exposure on lead-induced testicular toxicity, *Toxicology*, 69, 269–278, 1991.
226. Walsh, T. J. and Tilson, H. A., Neurobehavioral toxicology of the organoleads, *Neurotoxicology*, 5, 67–86, 1984.
227. Niklowitz, W. J., Ultrastructural effects of acute teraethyl-lead poisoning on nerve cells of the rabbit brain, *Environ. Res.*, 8, 17–36, 1974.
228. Niklowitz, W., Neurofibrillary changes after acute experimental lead poisoning, *Neurology*, 25, 927–934, 1975.
229. Tilson, H. A., Mactutus, C. F., McLamb, R., and Burne, T. A., Characterization of triethyl lead chloride neurotoxicity in adult rats, *Neurobehav. Toxicol. Teratol.*, 4, 671–681, 1982.
230. Walsh, T. J., McLamb, R. L., and Tilson, H. A., Organometal-induced antinociception: a time and dose-response comparison of triethyl and trimethyl lead and tin, *Toxicol. Appl. Pharmacol.*, 73, 295–299, 1984.
231. Hong, J. S., Tilson, H. A., Hudson, P., Ali, S. F., Wilson, W. E., and Hunter, V., Correlation of neurochemical and behavioral effects of triethyl lead chloride in rats, *Toxicol. Appl. Pharmacol.*, 69, 471–479, 1983.
232. Walsh, T. J., Schulz, D. W., Tilson, H. A., and Dehaven, D., Acute exposure to triethyl lead enhances the behavioral effects of dopaminergic agonists: involvement of brain dopamine in organolead neurotoxicity, *Brain Res.*, 363, 222–229, 1986.
233. Ali, S. F. and Bondy, S. C., Triethyl lead-induced peroxidative damage in various regions on the rat brain, *J. Toxicol. Environ. Health*, 26, 235–242, 1989.
234. Kristensson, K., Eriksson, H., Lundh, B., Plantin, L.-O., Wachtmeister, L., Azazi, M. L., Morath, C., and Heilbronn, E., Effects of manganese chloride on the rat developing nervous system, *Acta Pharmacol. Toxicol.*, 59, 345–348, 1986.
235. Malik, J. K. and Srivastava, A. K., Studies on the interaction between manganese and fenitrothion in rats, *Toxicol. Lett.*, 36, 221–226, 1987.
236. Shukla, G. S. and Chandra, S. V., Manganese toxicity: lipid peroxidation in rat brain, *Acta Pharmacol. Toxicol.*, 48, 95–100, 1981.
237. Chandra, S. V., Shukla, G. S., and Murthy, R. C., Effect of stress on the response of rat brain to manganese, *Toxicol. Appl. Pharmacol.*, 47, 603–608, 1979.
238. Teramoto, K., Wakitani, F., Horiguchi, S., Jo, T., Yamamoto, T., Mitsutake, H., and Nakaseko, H., Comparison of the neurotoxicity of several chemicals estimated by the peripheral nerve conduction velocity in rats, *Environ. Res.*, 62, 148–154, 1993.
239. Chandra, S. V. and Shukla, G. S., Manganese encephalopathy in growing rats, *Environ. Res.*, 15, 28–37, 1978.
240. Eriksson, H., Lenngren, S., and Heilbronn, E., Effect of long-term administration of manganese on biogenic amine levels in discrete striatal regions of rat brain, *Arch. Toxicol.*, 59, 426–431, 1987.
241. Nachtman, J. P., Tubben, R. E., and Commissaris, R. L., Behavioral effects of chronic manganese administration in rats: locomotor activity studies, *Neurobehav. Toxicol. Teratol.*, 8, 711–715, 1986.
242. Chandra, S. V. and Shukla, G. S., Effect of manganese on synthesis of brain catecholamines in growing rats, *Acta Pharmacol. Toxicol.*, 48, 349–354, 1981.
243. Chandra, S. V., Ali, M. M., Saxena, D. K., and Murthy, R. C., Behavioral and neurochemical changes in rats simultaneously exposed to manganese and lead, *Arch. Toxicol.*, 49, 49–56, 1981.
244. Bonilla, E. and Prasad, A. L. N., Effects of chronic manganese intake on the levels of biogenic amines in rat brain regions, *Neurobehav. Toxicol. Teratol.*, 6, 341–344, 1984.
245. Scheuhammer, A. M., Chronic manganese exposure in rats: histological changes in the pancreas, *J. Toxicol. Environ. Health*, 12, 353–360, 1983.
246. Witzleben, C. L., Pitlick, P., Bergmeyer, J., and Benoit, R., A new experimental model of intrahepatic cholestasis, *Am. J. Pathol.*, 53, 409–423, 1968.
247. Wassermann, D. and Wassermann, M., The ultrastructure of the liver cell in subacute manganese administration, *Environ. Res.*, 14, 379–390, 1977.
248. Rogers, R. R., Garner, R. J., Riddle, M. M., Luebke, R. W., and Smialowicz, R. J., Augmentation of murine natural killer cell activity by manganese chloride, *Toxicol. Appl. Pharmacol.*, 70, 7–17, 1983.
249. Srisuchart, B., Taylor, M. J., and Sharma, R. P., Alteration of humoral and cellular immunity in manganese chloride-treated mice, *J. Toxicol. Environ. Health*, 22, 91–99, 1987.
250. Komura, J. and Sakamoto, M., Effects of manganese forms on biogenic amines in the brain and behavioral alterations in the mouse: long-term oral administration of several manganese compounds, *Environ. Res.*, 57, 34–44, 1992.
251. Cox, D. N., Traiger, G. J., Jacober, S. P., and Hanzlik, R. P., Comparison of the toxicity of methylcyclopentadienyl manganese tricarbonyl with that of its two major metabolites, *Toxicol. Lett.*, 39, 1–5, 1987.
252. McGinley, P. A., Morris, J. B., Clay, R. J., and Gianutsos, G., Disposition and toxicity of methylcyclopentadienyl manganese tricarbonyl in the rat, *Toxicol. Lett.*, 36, 137–145, 1987.
253. Hakkinen, P. J. and Haschek, W. M., Pulmonary toxicity of methylcyclopentadienyl manganese tricarbonyl: nonciliated bronchiolar epithelial (Clara) cell necrosis and alveolar damage in the mouse, rat, and hamster, *Toxicol. Appl. Pharmacol.*, 65, 11–22, 1982.

254. Hanzlik, R. P., Stitt, R., and Traiger, G. J., Toxic effects of methylcyclopentadienyl manganese tricarbonyl (MMT) in rats: role of metabolism, *Toxicol. Appl. Pharmacol.,* 56, 353–360, 1980.
255. Clay, R. J. and Morris, J. B., Comparative pneumotoxicity of cyclopentadienyl manganese tricarbonyl and methylcyclopentadienyl manganese tricarbonyl, *Toxicol. Appl. Pharmacol.,* 98, 434–443, 1989.
256. Fishman, B. E., McGinley, P. A., and Gianutsos, G., Neurotoxic effects of methylcyclopentadienyl manganese tricarbonyl (MMT) in the mouse: basis of MMT-induced seizure activity, *Toxicology,* 45, 193–201, 1987.
257. Kishi, R., Hashimoto, K., Shimizu, S., and Kobayashi, M., Behavioral changes and mercury concentrations in tissues of rats exposed to mercury vapor, *Toxicol. Appl. Pharmacol.,* 46, 555–566, 1978.
258. Fredriksson, A., Dahlgren, L., Danielsson, B., Eriksson, P., Dencker, L., and Archer, T., Behavioural effects of neonatal metallic mercury exposure in rats, *Toxicology,* 74, 151–160, 1992.
259. Danielsson, B. R. G., Fredriksson, A., Dahlgren, L., Gårdlund, A. T., Olsson, L., Dencker, L., and Archer, T., Behavioural effects of prenatal metallic mercury inhalation exposure in rats, *Neurotoxicol. Teratol.,* 15, 391–396, 1993.
260. Livardjani, F., Ledig, M., Kopp, P., Dahlet, M., Leroy, M., and Jaeger, A., Lung and blood superoxide dismutase activity in mercury vapor exposed rats: effect of *N*-acetylcysteine treatment, *Toxicology,* 66, 289–295, 1991.
261. Daston, G. P., Kavlock, R. J., Rogers, E. H., and Carver, B., Toxicity of mercuric chloride to the developing rat kidney. I. Postnatal ontogeny of renal sensitivity, *Toxicol. Appl. Pharmacol.,* 71, 24–41, 1983.
262. Kyle, G. M., Luthra, R., Bruckner, J. V., MacKenzie, W. F., and Acosta, D., Assessment of functional, morphological, and enzymatic tests for acute nephrotoxicity induced by mercuric chloride, *J. Toxicol. Environ. Health,* 12, 99–117, 1983.
263. McDowell, E. M., Nagle, R. B., Zalme, R. C., McNeil, J. S., Flamenbaum, W., and Trump, B. F., Studies on the pathophysiology of acute renal failure. I. Correlation of ultrastructure and function in the proximal tubule of the rat following administration of mercuric chloride, *Virchows Arch. B Cell Path.,* 22, 173–196, 1976.
264. Duran, M.-A., Spencer, D., Weise, M., Kronfol, N. O., Spencer, R. F., and Oken, D. E., Renal epithelial amino acid concentrations in mercury-induced and postischemic acute renal failure, *Toxicol. Appl. Pharmacol.,* 105, 183–194, 1990.
265. Lund, B. O., Miller, D. M., and Woods, J. S., Studies on Hg(II)-induced H_2O_2 formation and oxidative stress *in vivo* and *in vitro* in rat kidney mitochondria, *Biochem. Pharmacol.,* 45, 2017–2024, 1993.
266. Liu, X., Jin, T., and Nordberg, G. F., Increased urinary calcium and magnesium excretion in rats injected with mercuric chloride, *Pharmacol. Toxicol.,* 68, 254–259, 1991.
267. Fukino, H., Hirai, M., Hsueh, Y. M., Moriyasu, S., and Yamane, Y., Mechanism of protection by zinc against mercuric chloride toxicity in rats: effects of zinc and mercury on glutathione metabolism, *J. Toxicol. Environ. Health,* 19, 75–89, 1986.
268. Fukino, H., Hirai, M., Hsueh, Y. M. and Yamane, Y., Effect of zinc pretreatment on mercuric chloride-induced lipid peroxidation in the rat kidney, *Toxicol. Appl. Pharmacol.,* 73, 395–401, 1984.
269. Magos, L., Sparrow, S., and Snowden, R., The comparative renotoxicology of phenylmercury and mercuric chloride, *Arch. Toxicol.,* 50, 133–139, 1982.
270. Yamane, Y. and Koizumi, T., Protective effect of molybdenum on the acute toxicity of mercuric chloride, *Toxicol. Appl. Pharmacol.,* 65, 214–221, 1982.
271. Holt, D. and Webb, M., The toxicity and teratogenicity of mercuric mercury in the pregnant rat, *Arch. Toxicol.,* 58, 243–248, 1986.
272. Bernard, A. M., Collette, C., and Lauwerys, R., Renal effects of *in utero* exposure to mercuric chloride in rats, *Arch. Toxicol.,* 66, 508–513, 1992.
273. Pezerovic, D., Narancsik, P., and Gamulin, S., Effects of mercury bichloride on mouse kidney polyribosome structure and function, *Arch. Toxicol.,* 48, 167–172, 1981.
274. Tanaka, T., Naganuma, A., and Imura, N., Role of γ-glutamyltranspeptidase in renal uptake and toxicity of inorganic mercury in mice, *Toxicology,* 60, 187–198, 1990.
275. Nielsen, J. B., Andersen, H. R., Andersen, O., and Starklint, H., Mercuric chloride-induced kidney damage in mice: time course and effect of dose, *J. Toxicol. Environ. Health,* 34, 469–483, 1991.
276. Yasutake, A., Hirayama, K., and Inouye, M., Sex difference in acute renal dysfunction induced by methylmercury in mice, *Renal Failure,* 12, 233–240, 1990.
277. Chowdhury, A. R., Makhija, S., Vachhrajani, K. D., and Gautam, A. K., Methylmercury- and mercuric chloride-induced alterations in rat epididymal sperm, *Toxicol. Lett.,* 47, 125–134, 1989.
278. Lakshmana, M. K., Desiraju, T., and Raju, T. R., Mercuric chloride-induced alterations of levels of noradrenaline, dopamine, serotonin and acetylcholine esterase activity in different regions of rat brain during postnatal development, *Arch. Toxicol.,* 67, 422–427, 1993.
279. Carmignani, M., Finelli, V. N., and Boscolo, P., Mechanisms in cardiovascular regulation following chronic exposure of male rats to inorganic mercury, *Toxicol. Appl. Pharmacol.,* 69, 442–450, 1983.
280. Tan, T. M. C., Sin, Y. M., and Wong, K. P., Mercury-induced UDP glucuronyltransferase (UDPGT) activity in mouse kidney, *Toxicology,* 64, 81–87, 1990.
281. Hultman, P. and Eneström, S., Dose-response studies in murine mercury-induced autoimmunity and immune-complex disease, *Toxicol. Appl. Pharmacol.,* 113, 199–208, 1992.

282. Dieter, M. P., Luster, M. I., Boorman, G. A., Jameson, C. W., Dean, J. H., and Cox, J. W., Immunological and biochemical responses in mice treated with mercuric chloride, *Toxicol. Appl. Pharmacol.*, 68, 218–228, 1983.

283. Hoskins, B. B. and Hupp, E. W., Methylmercury effects in rat, hamster, and squirrel monkey: lethality, symptoms, brain mercury, and amino acids, *Environ. Res.*, 15, 5–19, 1978.

284. Omata, S., Tsubaki, H., Sakimura, K., Sato, M., Yoshimura, R., Hirakawa, E., and Sugano, H., Stimulation of protein and RNA synthesis by methylmercury chloride in the liver of intact and adrenalectomized rats, *Arch. Toxicol.*, 47, 113–123, 1981.

285. Syversen, T. L. M., Totland, G., and Flood, P. R., Early morphological changes in rat cerebellum caused by a single dose of methylmercury, *Arch. Toxicol.*, 47, 101–111, 1981.

286. Cheung, M. K. and Verity, M. A., Experimental methylmercury neurotoxicity: locus of mercurial inhibition of brain protein synthesis *in vivo* and *in vitro*, *J. Neurochem.*, 44, 1799–1808, 1985.

287. Ali, S. F., Lebel, C. P., and Bondy, S. C., Reactive oxygen species, formation as a biomarker of methylmercury and trimethyltin neurotoxicity, *Neurotoxicology*, 13, 637–648, 1992.

288. de Ceaurriz, J. and Ban, M., Role of γ-glutamyltranspeptidase and -lyase in the nephrotoxicity of hexachloro-1,3-butadiene and methylmercury in mice, *Toxicol. Lett.*, 50, 249–256, 1990.

289. Sager, P. R., Aschner, M., and Rodier, P. M., Persistent, differential alterations in developing cerebellar cortex of male and female mice after methylmercury exposure, *Develop. Brain Res.*, 12, 1–11, 1984.

290. Nomiyama, K., Matsui, K., and Nomiyama, H., Effects of temperature and other factors on the toxicity of methylmercury in mice, *Toxicol. Appl. Pharmacol.*, 56, 392–398, 1980.

291. Salvaterra, P., Massaro, E. J., Morganti, J. B., and Lown, B. A., Time-dependent tissue/organ uptake and distribution of ^{203}Hg in mice exposed to multiple sublethal doses of methylmercury, *Toxicol. Appl. Pharmacol.*, 32, 432–442, 1975.

292. Yasutake, A. and Hirayama, K., Acute effects of methylmercury on hepatic and renal glutathione metabolisms in mice, *Arch. Toxicol.*, 68, 512–516, 1994.

293. Sato, M., Sugano, H., and Takizawa, Y., Effects of methylmercury on zinc-thionein levels of rat liver, *Arch. Toxicol.*, 47, 125–133, 1981.

294. Kabuto, M., Chronic effects of methylmercury on the urinary excretion of catecholamines and their responses to hypoglycemic stress, *Arch. Toxicol.*, 65, 164–167, 1991.

295. Desnoyers, P. A. and Chang, L. W., Ultrastructural changes in rat hepatocytes following acute methyl mercury intoxication, *Environ. Res.*, 9, 224–239, 1975.

296. Magos, L., Peristianis, G. C., Clarkson, T. W., Brown, A., Preston, S., and Snowden, R. T., Comparative study of the sensitivity of male and female rats to methylmercury, *Arch. Toxicol.*, 48, 11–20, 1981.

297. Wu, M.-F., Ison, J. R., Wecker, J. R., and Lapham, L. W., Cutaneous and auditory function in rats following methylmercury poisoning, *Toxicol. Appl. Pharmacol.*, 79, 377–388, 1985.

298. Hargreaves, R. J., Eley, B. P., Moorhouse, S. R., and Pelling, D., Regional cerebral glucose metabolism and blood flow during the silent phase of methylmercury neurotoxicity in rats, *J. Neurochem.*, 51, 1350–1355, 1988.

299. Kawamata, O., Kasama, H., Omata, S., and Sugano, H., Decrease in protein phosphorylation in central and peripheral nervous tissues of methylmercury-treated rat, *Arch. Toxicol.*, 59, 346–352, 1987.

300. Verity, M. A., Brown, W. J., Cheung, M., and Czer, G., Methylmercury inhibition of synaptosome and brain slice protein synthesis: *in vivo* and *in vitro* studies, *J. Neurochem.*, 29, 673–679, 1977.

301. Stroo, W. E. and Hook, J. B., Renal functional correlates of methylmercury intoxication: interaction with acute mercuric chloride toxicity, *Toxicol. Appl. Pharmacol.*, 42, 399–410, 1977.

302. Yonaha, M., Saito, M., and Sagai, M., Stimulation of lipid peroxidation by methyl mercury in rats, *Life Sci.*, 32, 1507–1514, 1983.

303. O'Kusky, J. R. and McGeer, E. G., Methylmercury-induced movement and postural disorders in developing rat: high-affinity uptake of choline, glutamate, and γ-aminobutyric acid in the cerebral cortex and caudate-putamen, *J. Neurochem.*, 53, 999–1006, 1989.

304. Desnoyers, P. A. and Chang, L. W., Ultrastructural changes in the liver after chronic exposure to methylmercury, *Environ. Res.*, 10, 59–75, 1975.

305. Saijoh, K., Fukunaga, T., Katsuyama, H., Lee, M. J., and Sumino, K., Effects of methylmercury on protein kinase A and protein kinase C in the mouse brain, *Environ. Res.*, 63, 264–273, 1993.

306. Yasutake, A. and Hirayama, K., Sex and strain differences of susceptibility to methylmercury toxicity in mice, *Toxicology*, 51, 47–55, 1988.

307. Yasutake, A., Adachi, T., Hirayama, K., and Inouye, M., Integrity of the blood-brain barrier system against methylmercury acute toxicity, *Jpn. J. Toxicol. Environ. Health*, 37, 355–362, 1991.

308. Lindström, H., Luthman, J., Oskarsson, A., Sundberg, J., and Olson, L., Effects of long-term treatment with methylmercury on the developing rat brain, *Environ. Res.*, 56, 158–169, 1991.

309. Moller-Madsen, B. and Danscher, G., Localization of mercury in CNS of the rat. IV. The effect of selenium on orally administered organic and inorganic mercury, *Toxicol. Appl. Pharmacol.*, 108, 457–473, 1991.

310. Woods, J. S., Bowers, M. A., and Davis, H. A., Urinary porphyrin profiles as biomarkers of trace metal exposure and toxicity: studies on urinary porphyrin excretion patterns in rats during prolonged exposure to methylmercury, *Toxicol. Appl. Pharmacol.*, 110, 464–476, 1991.

311. Woods, J. S., Davis, H. A., and Baer, R. P., Enhancement of γ-glutamylcysteine synthetase mRNA in rat kidney by methylmercury, *Arch. Biochem. Biophys.*, 296, 350–353, 1992.
312. Mitsumori, K., Takahashi, K., Matano, O., Goto, S., and Shirasu, Y., Chronic toxicity of methylmercury chloride in rats: clinical study and chemical analysis, *Jpn. J. Vet. Sci.*, 45, 747–757, 1983.
313. Mitsumori, K., Maita, K., and Shirasu, Y., Chronic toxicity of methylmercury chloride in rats: pathological study, *Jpn. J. Vet. Sci.*, 46, 549–557, 1984.
314. Andersen, H. R. and Andersen, O., Effects of dietary α-tocopherol and β-carotene on lipid peroxidation induced by methyl mercuric chloride in mice, *Pharmacol. Toxicol.*, 73, 192–201, 1993.
315. Gilbert, S. G. and Maurissen, J. P. J., Assessment of the effects of acrylamide, methylmercury, and 2,5-hexanedione on motor functions in mice, *J. Toxicol. Environ. Health*, 10, 31–41, 1982.
316. Mitsumori, K., Hirano, M., Ueda, H., Maita, K., and Shirasu, Y., Chronic toxicity and carcinogenicity of methylmercury chloride in B6C3F1 mice, *Fundam. Appl. Toxicol.*, 14, 179–190, 1990.
317. Geelen, J. A. G., Dormans, J. A. M. A., and Verhoef, A., The early effects of methylmercury on the developing rat brain, *Acta Neuropathol.*, 80, 432–438, 1990.
318. Cagiano, R., De Salvia, M. A., Renna, G., Tortella, E., Braghiroli, D., Parenti, C., Zanoli, P., Baraldi, M., Annau, Z., and Cuomo, V., Evidence that exposure to methylmercury during gestation induces behavioral and neurochemical changes in offspring of rats, *Neurotoxicol. Teratol.*, 12, 23–28, 1990.
319. Saillenfait, A. M., Brondeau, M. T., Zissu, D., and De Ceaurriz, J., Effects of prenatal methylmercury exposure on urinary proximal tubular enzyme excretion in neonatal rats, *Toxicology*, 55, 153–160, 1989.
320. Bornhausen, M., Müsch, H. R., and Greim, H., Operant behavior performance changes in rats after prenatal methylmercury exposure, *Toxicol. Appl. Pharmacol.*, 56, 305–310, 1980.
321. Peckham, N. H. and Choi, B. H., Abnormal neuronal distribution within the cerebral cortex after prenatal methylmercury intoxication, *Acta Neuropathol.*, 76, 222–226, 1988.
322. Inouye, M. and Kajiwara, Y., Strain difference of the mouse in manifestation of hydrocephalus following prenatal methylmercury exposure, *Teratology*, 41, 205–210, 1990.
323. Nobunaga, T., Satoh, H., and Suzuki, T., Effects of sodium selenite on methylmercury embryotoxicity and teratogenicity in mice, *Toxicol. Appl. Pharmacol.*, 47, 79–88, 1979.
324. Inouye, M. and Kajiwara, Y., Developmental disturbances of the fetal brain in guinea-pigs caused by methylmercury, *Arch. Toxicol.*, 62, 15–21, 1988.
325. Sunderman, Jr., F. W., Reid, M. C., Bibeau, L. M., and Linden, J. V., Nickel induction of microsomal heme oxygenase activity in rodents, *Toxicol. Appl. Pharmacol.*, 68, 87–95, 1983.
326. Sunderman, Jr., F. W., Shen, S. K., Mitchell, J. M., Allpass, P. R., and Damjanov, I., Embryotoxicity and fetal toxicity of nickel in rats, *Toxicol. Appl. Pharmacol.*, 43, 381–390, 1978.
327. Sunderman, F. W., Jr., Kasprzak, K. S., Lau, T. J., Minghetti, P. P., Maenza, R. M., Becker, N., Onkelinx, C., and Goldblatt, P. J., Effects of manganese on carcinogenicity and metabolism of nickel subsulfide, *Cancer Res.*, 36, 1790–1800, 1976.
328. Smialowicz, R. J., Rogers, R. R., Rowe, D. G., Riddle, M. M., and Luebke, R. W., The effects of nickel on immune function in the rat, *Toxicology*, 44, 271–281, 1987.
329. Kasprzak, K. S., Waalkes, M. P., and Poirier, L. A., Effects of magnesium acetate on the toxicity of nickelous acetate in rats, *Toxicology*, 42, 57–68, 1986.
330. Sunderman, Jr., F. W., Hopfer, S. M., Lin, S.-M., Plowman, M. C., Stojanovic, T., Wong, S. H.-Y., Zaharia, O., and Ziebka, L., Toxicity to alveolar macrophages in rats following parenteral injection of nickel chloride, *Toxicol. Appl. Pharmacol.*, 100, 107–118, 1989.
331. Misra, M., Rodriguez, R. E., and Kasprzak, K. S., Nickel-induced lipid peroxidation in the rat: correlation with nickel effect on antioxidant defense systems, *Toxicology*, 64, 1–17, 1990.
332. Misra, M., Olinski, R., Dizdaroglu, M., and Kasprzak, K. S., Enhancement by L-histidine of nickel (II)-induced DNA-protein cross-linking and oxidative DNA base damage in the rat kidney, *Chem. Res. Toxicol.*, 6, 33–37, 1993.
333. Cartana, J. and Arola, L., Nickel-induced hyperglycaemia: the role of insulin and glucagon, *Toxicology*, 71, 181–192, 1992.
334. Smith, M. K., George, E. L., Stober, J. A., Feng, H. A., and Kimmel, G. L., Perinatal toxicity associated with nickel chloride exposure, *Environ. Res.*, 61, 200–211, 1993.
335. Rodriguez, R. E., Misra, M., North, S. L., and Kasprzak, K. S., Nickel-induced lipid peroxidation in the liver of different strains of mice and its relation to nickel effects on antioxidant systems, *Toxicol. Lett.*, 57, 269–281, 1991.
336. Fisher, G. L., Chrisp, C. E., and McNeill, D. A., Lifetime effects of intratracheally instilled nickel subsulfide on $B6C3F_1$ mice, *Environ. Res.*, 40, 313–320, 1986.
337. Smialowicz, R. J., Rogers, R. R., Riddle, M. M., Garner, R. J., Rowe, D. G., and Luebke, R. W., Immunologic effects of nickel. II. Suppression of natural killer cell activity, *Environ. Res.*, 36, 56–66, 1985.
338. Hogan, G. R., Nickel acetate-induced mortality in mice of different ages, *Bull. Environ. Contam. Toxicol.*, 34, 446–450, 1985.
339. Dieter, M. P., Jameson, C. W., Tucker, A. N., Luster, M. I., French, J. E., Hong, H. L., and Boorman, G. A., Evaluation of tissue disposition, myelopoietic, and immunologic responses in mice after long-term exposure to nickel sulfate in the drinking water, *J. Toxicol. Environ. Health*, 24, 357–372, 1988.

340. Dunnick, J. K., Benson, J. M., Hobbs, C. H., Hahn, F. F., Cheng, Y. S., and Edison, A. F., Comparative toxicity of nickel oxide, nickel sulfate hexahydrate, and nickel subsulfide after 12 days of inhalation exposure to F344/N rats and B6C3F$_1$ mice, *Toxicology,* 50, 145–156, 1988.
341. David, L. L. and Shearer, T. R., State of sulfhydryl in selenite cataract, *Toxicol. Appl. Pharmacol.,* 74, 109–115, 1984.
342. LeBoeuf, R. A. and Hoekstra, W. G., Adaptive changes in hepatic glutathione metabolism in response to excess selenium in rats, *J. Nutr.,* 113, 845–854, 1983.
343. Jacobs, M. and Forst, C., Toxicological effects of sodium selenite in Sprague-Dawley rats, *J. Toxicol. Environ. Health,* 8, 575–585, 1981.
344. Hogan, G. R., Decreased levels of peripheral leukocytes following sodium selenite treatment in female mice, *Bull. Environ. Contam. Toxicol.,* 37, 175–179, 1986.
345. Watanabe, C. and Suzuki, T., Sodium selenite-induced hypothermia in mice: indirect evidence for a neural effect, *Toxicol. Appl. Pharmacol.,* 86, 372–379, 1986.
346. Jacobs, M. and Forst, C., Toxicological effects of sodium selenite in Swiss mice, *J. Toxicol. Environ. Health,* 8, 587–598, 1981.
347. Al-Bayati, M. A., Raabe, O. G., and Teague, S. V., Effect of inhaled dimethylselenide in the Fischer 344 male rat, *J. Toxicol. Environ. Health,* 37, 549–557, 1992.
348. Rios, C. and Monroy-Noyola, A., D-penicillamine and Prussian blue as antidotes against thallium intoxication in rats, *Toxicology,* 74, 69–76, 1992.
349. Peele, D. B., MacPhail, R. C., and Farmer, J. D., Flavor aversions induced by thallium sulfate: importance of route of administration, *Neurobehav. Toxicol. Teratol.,* 8, 273–277, 1986.
350. Hasan, M. and Haider, S. S., Acetyl-homocysteine thiolactone protects against some neurotoxic effects of thallium, *Neurotoxicology,* 10, 257–262, 1989.
351. Osorio-Rico, L., Galván-Arzate, S., and Ríos, C., Thallium increases monoamine oxidase activity and serotonin turnover rate in rat brain regions, *Neurotoxicol. Teratol.,* 17, 1–5, 1995.
352. Marwaha, J., Freedman, R., and Hoffer, B., Electrophysiological changes at a central noradrenergic synapse during thallium toxicosis, *Toxicol. Appl. Pharmacol.,* 56, 345–352, 1980.
353. Woods, J. S. and Fowler, B. A., Alteration of hepatocellular structure and function by thallium chloride: ultrastructural, morphometric, and biochemical studies, *Toxicol. Appl. Pharmacol.,* 83, 218–229, 1986.
354. Appenroth, D., Gambaryan, S., Winnefeld, K., Leiterer, M., Fleck, C., and Bräunlich, H., Functional and morphological aspects of thallium-induced nephrotoxicity in rats, *Toxicology,* 96, 203–215, 1995.
355. Formigli, L., Scelsi, R., Poggi, P., Gregotti, C., Nucci, A. D., Sabbioni, E., Gottardi, L., and Manzo, L., Thallium-induced testicular toxicity in the rat, *Environ. Res.,* 40, 531–539, 1986.
356. de Groot, A. P., Feron, V. J., and Til, H. P., Short-term toxicity studies on some salts and oxides of tin in rats, *Food Cosmet. Toxicol.,* 11, 19–30, 1973.
357. Janssen, P. J. M., Bosland, M. C., Van Hees, J. P., Spit, B. J., Willems, M. I., and Kuper, C. F., Effects of feeding stannous chloride on different parts of the gastrointestinal tract of the rat, *Toxicol. Appl. Pharmacol.,* 78, 19–28, 1985.
358. Dreef-van der Meulen, H. C., Feron, V. J., and Til, H. P., Pancreatic atrophy and other pathological changes in rats following the feeding of stannous chloride, *Pathol. Eur.,* 9, 185–192, 1974.
359. Zaręba, G. and Chmielnicka, J., Aminolevulinic acid dehydratase activity in the blood of rats exposed to tin and zinc, *Ecotoxicol. Environ. Safety,* 9, 40–46, 1985.
360. Harry, G. J., Goodrum, J. F., Krigman, M. R., and Morell, P., The use of synapsin I as a biochemical marker for neuronal damage by trimethyltin, *Brain Res.,* 326, 9–18, 1985.
361. Chang, L. W., Wenger, G. R., McMillan, D. E., and Dyer, R. S., Species and strain comparison of acute neurotoxic effects of trimethyltin in mice and rats, *Neurobehav. Toxicol. Teratol.,* 5, 337–350, 1983.
362. Dyer, R. S., Deshields, T. L., and Wonderlin, W. F., Trimethyltin-induced changes in gross morphology of the hippocampus, *Neurobehav. Toxicol. Teratol.,* 4, 141–147, 1982.
363. Walsh, T. J., Gallagher, M., Bostock, E., and Dyer, R. S., Trimethyltin impairs retention of a passive avoidance task, *Neurobehav. Toxicol. Teratol.,* 4, 163–167, 1982.
364. Dyer, R. S., Howell, W. E., and Wonderlin, W. F., Visual system dysfunction following acute trimethyltin exposure in rats, *Neurobehav. Toxicol. Teratol.,* 4, 191–195, 1982.
365. MacPhail, R. C., Studies on the flavor aversions induced by trialkyltin compounds, *Neurobehav. Toxicol. Teratol.,* 4, 225–230, 1982.
366. Walsh, T. J., McLamb, R. L., and Tilson, H. A., Organometal-induced antiociception: a time- and dose-response comparison of triethyl and trimethyl lead and tin, *Toxicol. Appl. Pharmacol.,* 73, 295–299, 1984.
367. Kodavanti, P. R. S., Cameron, J. A., Yallapragada, P. R., Vig, P. J. S., and Desaiah, D., Inhibition of Ca^{2+} transport associated with cAMP-dependent protein phosphorylation in rat cardiac sarcoplasmic reticulum by triorganotins, *Arch. Toxicol.,* 65, 311–317, 1991.
368. Dyer, R. S. and Howell, W. E., Acute triethyltin exposure: effects on the visual evoked potential and hippocampal after discharge, *Neurobehav. Toxicol. Teratol.,* 4, 259–266, 1982.
369. Tilson, H. A. and Burne, T. A., Effects of triethyltin on pain reactivity and neuromotor function of rats, *J. Toxicol. Environ. Health,* 8, 317–324, 1981.

370. Miller, D. B., Eckerman, D. A., Krigman, M. R., and Grant, L. D., Chronic neonatal organotin exposure alters radial-arm maze performance in adult rats, *Neurobehav. Toxicol. Teratol.*, 4, 185–190, 1982.
371. Mushak, P., Krigman, M. R., and Mailman, R. B., Comparative organotin toxicity in the developing rat: somatic and morphological changes and relationship to accumulation of total tin, *Neurobehav. Toxicol. Teratol.*, 4, 209–215, 1982.
372. LeBel, C. P., Ali, S. F., McKee, M., and Bondy, S. C., Organometal-induced increases in oxygen reactive species: the potential of 2′,7′-dichlorofluorescin diacetate as an index of neurotoxic damage, *Toxicol. Appl. Pharmacol.*, 104, 17–24, 1990.
373. Wenger, G. R., McMillan, D. E., and Chang, L. W., Behavioral toxicology of acute trimethyltin exposure in the mouse, *Neurobehav. Toxicol. Teratol.*, 4, 157–161, 1982.
374. Fox, D. A., Pharmacological and biochemical evaluation of triethyltin's anticonvulsant effects, *Neurobehav. Toxicol. Teratol.*, 4, 273–278, 1982.
375. Ghoneum, M., Hussein, A. E., Gill, G., and Alfred, L. J., Suppression of murine natural killer cell activity by tributyltin: *in vivo* and *in vitro* assessment, *Environ. Res.*, 52, 178–186, 1990.
376. Nucci, A. D., Gregotti, C., and Manzo, L., Triphenyltin hepatotoxicity in rats, *Arch Toxicol.*, 9, 402–405, 1986.
377. Lehotzky, K., Szeberenyi, J. M., Gonda, Z., Horkay, F., and Kiss, A., Effects of prenatal triphenyltin exposure on the development of behavior and conditioned learning in rat pups, *Neurobehav. Toxicol. Teratol.*, 4, 247–250, 1982.
378. Yoshizuka, M., Hara, K., Haramaki, N., Yokoyama, M., Mori, N., Doi, Y., Kawahara, A., and Fujimoto, S., Studies on the hepatotoxicity induced by bis-(tributyltin) oxide, *Arch Toxicol.*, 66, 182–187, 1992.
379. Yoshizuka, M., Haramaki, N., Yokoyama, M., Hara, K., Kawahara, A., Umezu, Y., Araki, H., Mori, N., and Fujimoto, S., Corneal edema induced by bis-(tributyltin) oxide, *Arch. Toxicol.*, 65, 651–655, 1991.
380. Seinen, W. and Willems, M. I., Toxicity of organotin compounds. I. Atrophy of thymus and thymus-dependent lymphoid tissue in rats fed *di-n*-octyltindichloride, *Toxicol. Appl. Pharmacol.*, 35, 63–75, 1976.
381. Seinen, W., Vos, J. G., Spanje, I. V., Snoek, M., Brands, R., and Hooykaas, H., Toxicity of organotin compounds. II. Comparative *in vivo* and *in vitro* studies with various organotin and organolead compounds in different animal species with special emphasis on lymphocyte cytotoxicity, *Toxicol. Appl. Pharmacol.*, 42, 197–212, 1977.
382. Miller, K., Maisey, J., and Nicklin, S., Effect of orally administered dioctyltin dichloride on murine immunocompetence, *Environ. Res.*, 39, 434–441, 1986.
383. Hogan, G. R., Cole, B. S., and Lovelace, J. M., Sex and age mortality responses in zinc acetate-treated mice, *Bull. Environ. Contam. Toxicol.*, 39, 156–161, 1987.
384. Llobet, J. M., Domingo, J. L., Colomina, M. T., Mayayo, E., and Corbella, J., Subchronic oral toxicity of zinc in rats, *Bull Environ. Contam. Toxicol.*, 41, 36–43, 1988.
385. Cosma, G., Fulton, H., Defeo, T., and Gordon, T., Rat lung metallothionein and heme oxygenase gene expression following ozone and zinc oxide exposure, *Toxicol. Appl. Pharmacol.*, 117, 75–80, 1992.
386. Hirano, S., Higo, S., Tsukamoto, N., Kobayashi, E., and Suzuki, K. T., Pulmonary clearance and toxicity of zinc oxide instilled into the rat lung, *Arch. Toxicol.*, 63, 336–342, 1989.
387. Lam, H. F., Conner, M. W., Rogers, A. E., Fitzgerald, S., and Amdur, M. O., Functional and morphologic changes in the lung of guinea pigs exposed to freshly generated ultrafine zinc oxide, *Toxicol. Appl. Pharmacol.*, 78, 29–38, 1985.

2.2 Transgenic Animals

Hiroshi Imai

It is possible, through the use of biological technology, to transfer genetic information into mouse embryos to alter the genetic constitution of the mice. This approach to creating so-called "transgenic animals" has already been used widely for the study of gene function and regulation in many biological research fields.

Transgenic mice can be produced primarily by two methods (Figure 1): (1) microinjection of DNA constructs into a pronucleus of fertilized eggs, and (2) targeting embryonic stem (ES) cells to the endogenous genomic loci by a homologous mutated DNA sequence injected into host embryos to make transgenic chimera. Embryos prepared by these methods are then transferred into the oviducts of pseudopregnant females and allowed to develop to term. Some of the offspring have the foreign DNA sequences permanently integrated into the genome and thus become transgenic. By doing so, it is possible to study the effect of an

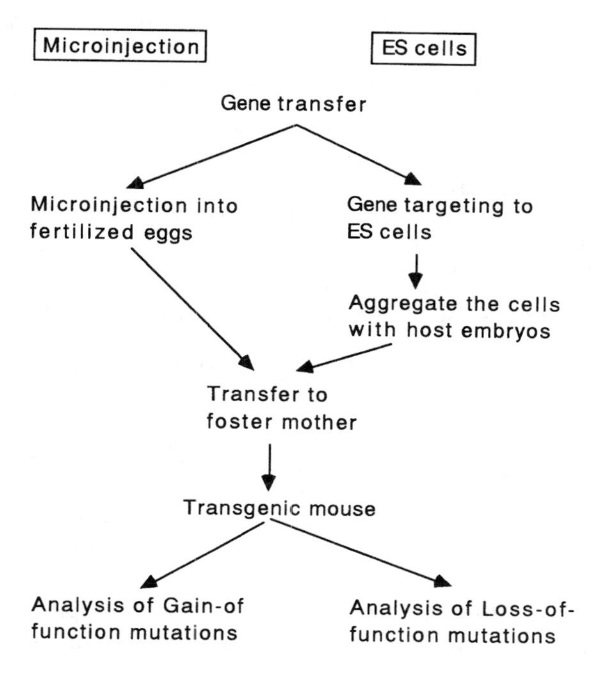

FIGURE 1
Two major methods of producing transgenic mice.

exogenous gene under the control of a tissue-specific promoter by microinjection and to analyze the functional mechanisms of endogenous genes by targeted disruption.

This review will focus on the experimental systems for the generation of transgenic mice and introduce recent research progress on the study of gene regulation of metal-response genes using transgenic mice. General descriptions of transgenic research[1-3] and methodological reviews[4-7] for production of transgenic animals are worth referring to.

2.2.1 Method for Producing Transgenic Animals

Most transgenic mice have been produced for two general purposes: (1) for studies of tissue-specific and developmental stage-specific gene regulation, and (2) for experiments involving the phenotypic effects of transgene expression in intact animals. Gene transfer into the mouse embryo has, therefore, provided the definitive experimental assay to define the *cis*-acting DNA sequences that dictate specific patterns of transcription in animals.

The first successful introduction of a genetically engineered material into mouse somatic tissues by pronuclear injection was reported by Gordon et al.[8] in 1980. Shortly thereafter, several groups were successful in introducing cloned genes into somatic tissues, as well as into the germ line, by this technique.[9-14]

While the DNA microinjection technique was being established, a remarkable cell line (ES cells) was isolated from pre-implantation mouse embryos by Martin[15] and Evans and Kaufman[16] in 1981. ES cells are of an undifferentiated and pluripotent state and thus have been shown to be capable of contributing to many different tissues in chimeras, including the germ cells, when injected into host blastocysts and returned to a foster mother.[17] When a fragment of genomic DNA is introduced into ES cells, it can locate and recombine with the endogenous homologous sequences. This type of homologous recombination is known as "gene targeting". Since the first reports of genetic manipulation of ES cells by Robertson et al.[18] and Gossler et al.,[19] this technique has been used widely for the functional analysis of genes in many different genomic loci and is the other subject of this chapter.

Female mouse
⇓
Induce superovulation by hormonal treatment
⇓
Mate mouse with male
⇓
Collect fertilized eggs
⇓
Inject DNA into male or female pronucleus
⇓
Transfer eggs to oviduct of pseudopregnant recipient female that previously had mated with vasectomized male
⇓
Use PCR and Southern blot analysis to identify DNA injected into resulting newborn animals
⇓
Analyze gene expression in transgenic animals

FIGURE 2
DNA microinjection into fertilized mouse eggs procedure.

2.2.1.1 Microinjection

The gene transfer method most extensively and successfully employed is the microinjection of DNA directly into a pronucleus of fertilized mouse eggs.[8] This method results in the stable chromosomal integration of the foreign DNA in 10 to 40% of the resulting mice.[9-11,13,14]

An alternative method for introducing genes into the mouse germ lines is to infect embryos with retroviral vectors. Pre- or post-implantation embryos can be infected by either wild-type or recombinant retroviruses, leading to the stable integration of viral genomes into the host chromosomes.[20-25] Although the use of retroviral vectors for transgenesis has some advantages,[2] the method has not been widely used for several reasons: the extra steps involved in producing high-titer recombinant retroviruses, a higher frequency of mosaicism and of multiple insertion events, and difficulties with the expression of genes introduced in retroviral vectors.

Whole procedures for DNA microinjection are shown in Figure 2. Artificially induced ovulation (superovulation) by hormonal treatment of female mice is necessary for efficient collection of the fertilized eggs from the oviduct. These procedures, including superovulation[26] and handling of fertilized eggs,[7,27] are described in detail elsewhere. Since fertilized mouse eggs are too small to handle by hand, being only approximately 80 μm in diameter, a special instrument fitted to an inverted or upright fixed-stage microscope with Normarski differential interference contrast optics — a micromanipulator (Figure 3) — should be used for the DNA injection procedure.[7,28]

Constructed DNA for microinjection should be prepared in a linearized form without prokaryotic vector sequences and dissolved in injection buffer (10 m*M* Tris-HCl, pH 7.6, and 0.1 m*M* EDTA). Linearized DNA has a fivefold higher integration efficiency compared with circular DNA.[29] Vector sequences derived from prokaryote can severely inhibit the expression of the introduced gene.[30-33] Purity of the DNA is an important factor for avoiding deleterious effects on embryo development after injection.[7,28] A DNA concentration of 1 to 2 μg/ml is usually used for injection into pronucleus at a volume of 1 to 2 pl of this solution by a fine-shaped glass micropipette (Figure 4), which corresponds to 200 to 400 copies of a 5-kb DNA fragment. After DNA injection, the eggs can be transferred immediately to the oviducts of pseudopregnant recipient mice.[7,28]

The length of the DNA construct used to produce transgenic mice has been limited only by cloning and handling considerations. A 50-kb bacteriophage λ clone[10] and a 60-kb cosmid insert[34] have been introduced intact. Recently, artificial yeast chromosomes carrying several hundred kilobases of mammalian genomic DNA have been also used to produce transgenic mice by pronuclear injection[35-36] and by ES cells.[37-38]

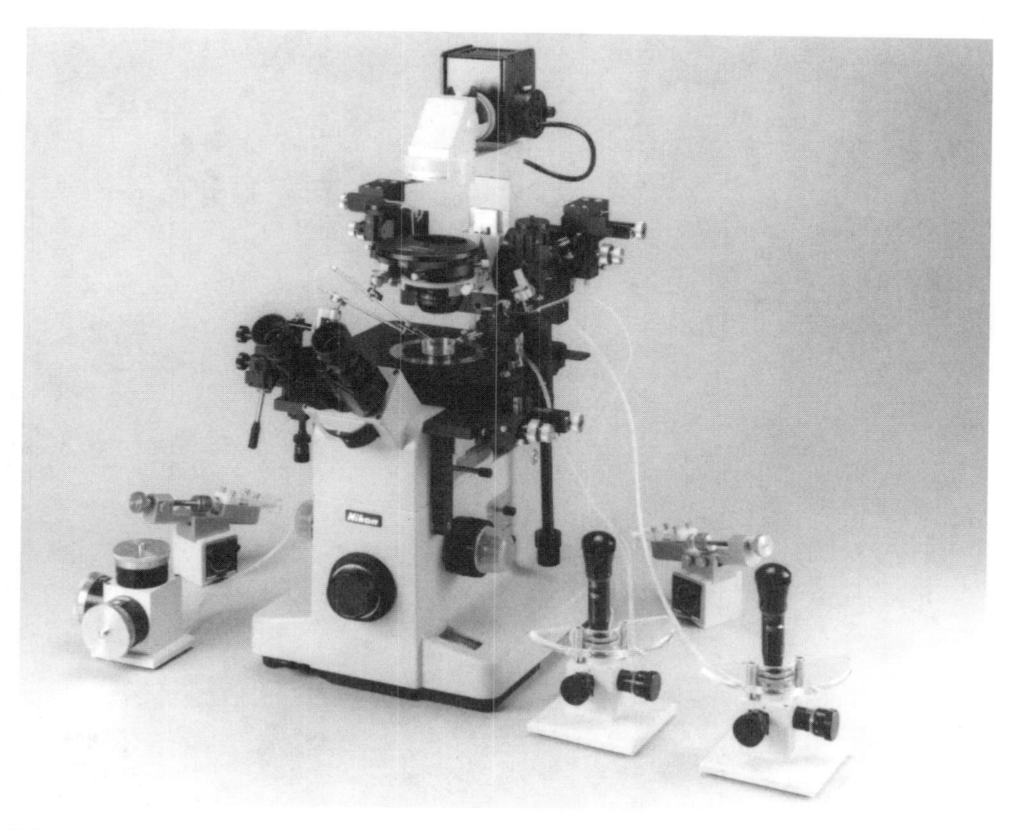

FIGURE 3
Micromanipulator fitted to the inverted microscope. (Courtesy of Narishige Co. Ltd., Japan.)

Constructed DNA for microinjection should be prepared in a linearized form without prokaryotic vector sequences and dissolved in injection buffer (10 m*M* Tris-HCl, pH 7.6, and 0.1 m*M* EDTA). Linearized DNA has a fivefold higher integration efficiency compared with circular DNA.[29] Vector sequences derived from prokaryote can severely inhibit the expression of the introduced gene.[30-33] Purity of the DNA is an important factor for avoiding deleterious effects on embryo development after injection.[7,28] A DNA concentration of 1 to 2 μg/ml is usually used for injection into pronucleus at a volume of 1 to 2 pl of this solution by a fine-shaped glass micropipette (Figure 4), which corresponds to 200 to 400 copies of a 5-kb DNA fragment. After DNA injection, the eggs can be transferred immediately to the oviducts of pseudopregnant recipient mice.[7,28]

The length of the DNA construct used to produce transgenic mice has been limited only by cloning and handling considerations. A 50-kb bacteriophage λ clone[10] and a 60-kb cosmid insert[34] have been introduced intact. Recently, artificial yeast chromosomes carrying several hundred kilobases of mammalian genomic DNA have been also used to produce transgenic mice by pronuclear injection[35-36] and by ES cells.[37-38]

It is often advantageous when constructing genes that a cDNA clone be used for expression in transgenic mice to provide the coding sequences for the desired gene product. However, many cases show that the levels of gene expression obtained with cDNA-based constructs are often lower than those obtained when genomic sequences, including introns and exons, are used.[39] In some cases, this may be due to the presence of enhancers in the introns,[39] but the addition of heterologous introns to the constructs can give significant increases in expression levels without altering the tissue specificity of expression.[40-41]

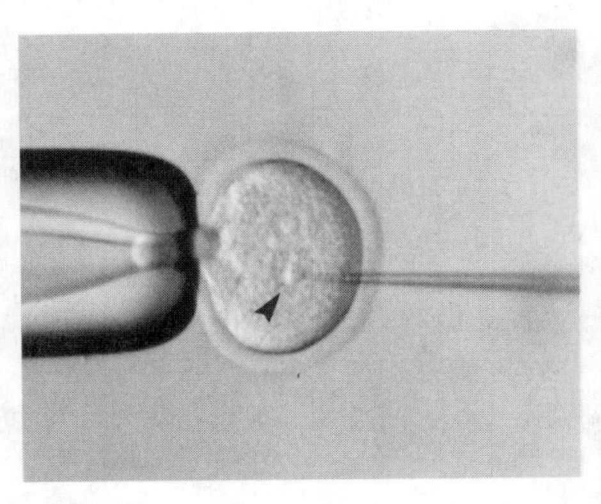

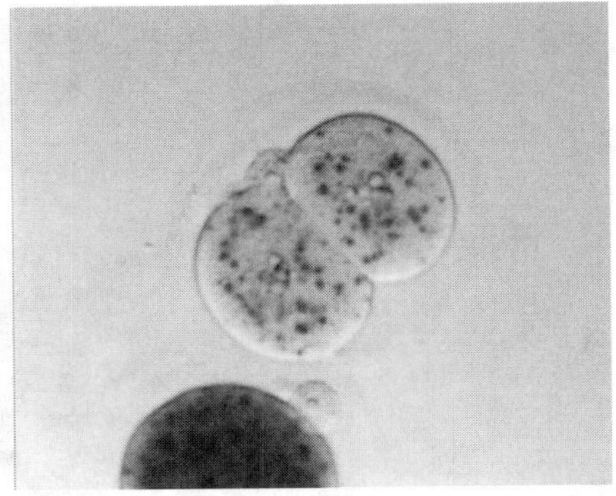

FIGURE 4
Microinjection of DNA into a fertilized mouse egg. (**left**) DNA injection by a glass pipette into a pronucleus (arrow) of the egg supported by a holding pipette. (**right**) Expression of bacterial marker gene (*lacZ*) in two-cell egg after DNA injection (dark spots).

FIGURE 5
Chimera mouse obtained from the embryo-injected ES cells.

In most cases, integration appears to occur at the one-cell stage because the foreign DNA is present in every cell of the transgenic animal, including germ cells. In approximately 20 to 30% of transgenic mice, however, the foreign DNA apparently integrates at a later stage, resulting in mice that are mosaic for the presence of foreign DNA.[42] The number of copies of the foreign DNA sequence integrated ranges from one to several hundred. When multiple copies are present, they are usually found at a single chromosomal locus; however, there occasionally may be separate integration sites on two different chromosomes.[43]

Mice obtained by embryo transfer to recipient mice may be initially identified by polymerase chain reaction (PCR) analysis.[44] Founder transgenic mice shown positive by PCR should always be retested by Southern blot analysis.[45] An estimate of the transgene copy number can be obtained by including standard amounts of the injected transgene. The apparent copy number measured by either Southern or dot-blot hybridization reflects both the actual copy number per diploid genome and the fraction of cells containing the transgene.[7,28] If a founder mouse is mosaic and some cells lack the injected gene, the copy number will be underestimated.

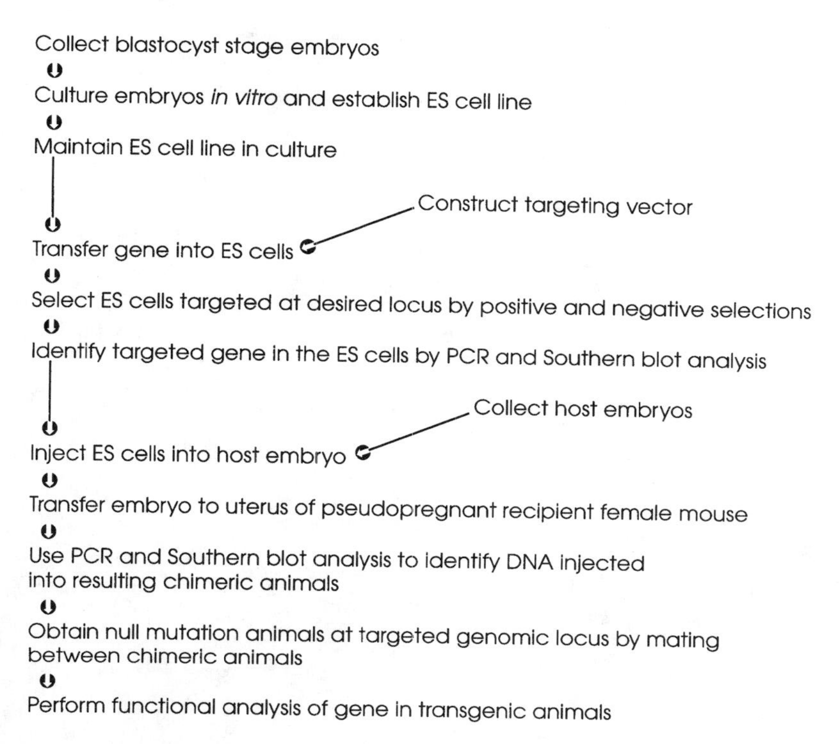

FIGURE 6
Genetic modification of animals using ES cells.

Embryonic stem cells are most useful when it is necessary to select for rare integration events into a genome; however, it may be desirable in some instances to introduce a transgene into ES cells and then produce chimeric mice, instead of microinjecting the gene into mouse eggs. Homologous recombination when applied to altering specific endogenous genes, referred to as "gene targeting", provides the highest possible level of control over producing mutations in cloned genes. Target mutations to specific genes by homologous recombination in ES cells was initially achieved using HPRT (hypoxanthine phosphoribosyl transferase).[53-57] To date, more than 100 genes have been disrupted by homologous recombination and transmitted through the germ line. Gene targeting in ES cells has been used primarily to make null mutations for testing where and when in development a gene is required. ES cells have also proven useful for other applications, such as enhancer and gene trap screens, where one can gain information on gene sequence, expression, and mutant phenotype all from a single insertion in ES cells.[58]

Technologies involved in ES cell research are shown in Figure 6 and are well covered in several reviews about the establishment and maintenance of mouse ES cell lines,[7,49-50] gene targeting procedures,[50,59] and handling of embryos and chimera production;[7,51,60] therefore, here, briefly, we add the following short notes about each procedure.

As shown in Table 44, several ES cell lines in mice have been established (Figure 7) and primarily derive from the 129 strain. Most researches have used blastocyst-stage embryos for the establishment of cell lines (Figure 8). Recently, another pluripotent stem cell line has been established from primordial germ cells (PGC),[68-69] designated as embryonic germ (EG) cells, which has the ability to contribute to the germ line in chimera.[70]

For establishment and maintenance of ES cells, Dulbecco's modified Eagle's medium (DMEM) supplemented with 20% serum, nonessential amino acids, nucleotides, and β-mercaptoethanol is usually used as the culture medium.[49] ES cells must be grown either on monolayers of mitotically inactivated fibroblast cells[15-16,61] or in the presence of leukemia

TABLE 44

Established Mouse ES Cell Lines

Cell Line	Strain	Ref.
D3	129/Sv	61
CCE	129/Sv	18
E14	129/Ola	54
AB1	129/Sv	62
J1	129/terSv	63
HM1	129/Sv	64
R1	129/Sv	65
BL/6-III	C57BL/6	66
TT2	F1 (C57BL/6 X CBA)	67

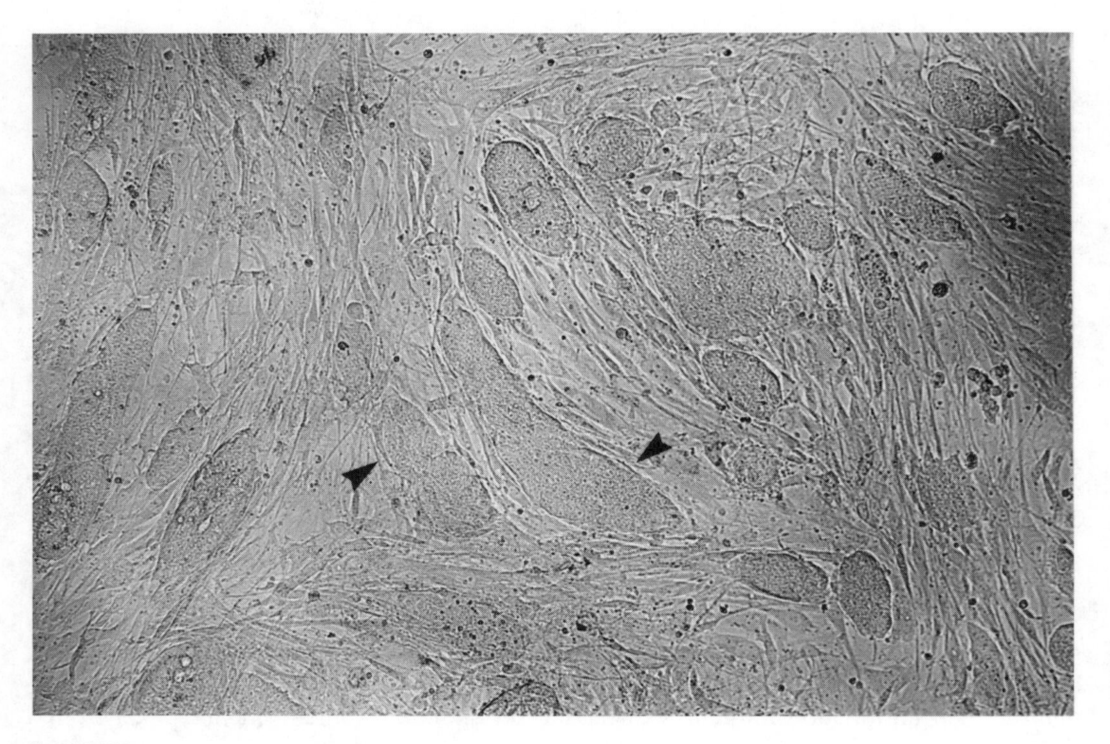

FIGURE 7
Established ES cell line (TT2). This cell line[67] derives from F1 embryo (C57BL/6 female mated with CBA male mouse). ES cells have a unique morphology: formation of colonies (arrows) containing hundreds of cells, cells packed tightly together without identifying individual cells, and a large nucleus in the cytoplasm.

inhibitory factor (LIF) to inhibit spontaneous differentiation of the cells.[71-72] During the maintenance of ES cells, it is important to dissociate the cultures into a single cell as much as possible after trypsinization, because the cell clumps tend to differentiate into endoderm.

All of the ES cell lines in common use are derived from male embryos and thus contain a Y chromosome. Male ES cells produce a higher proportion of phenotypic male chimeras. Female-derived XX cell lines, which have been shown to lose one X chromosome after several passages to become XO, are reputed to be unstable.[73] A sex bias among ES cell chimeras in favor of males is a common observation when male-derived XY ES cells are used. In combination with a female embryo, male cells will often produce a fertile, phenotypic male chimera.

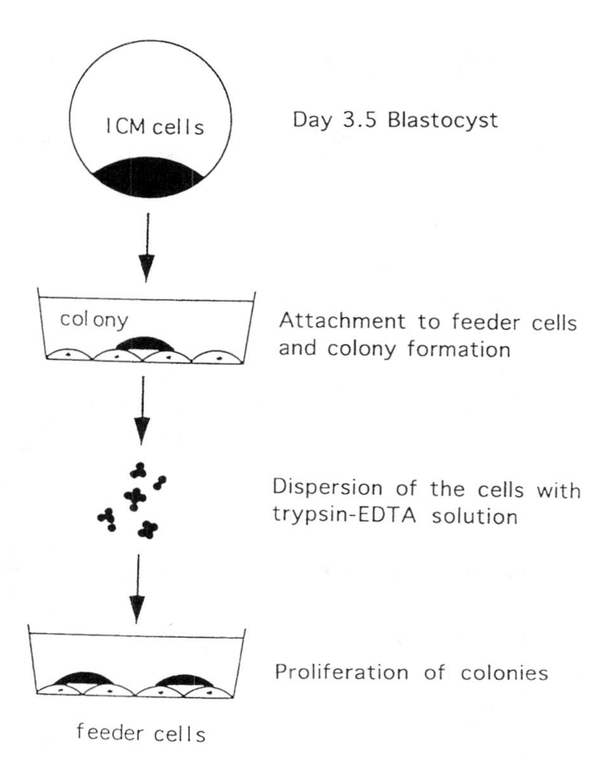

FIGURE 8
Procedures for establishing ES cell line.

Perhaps the most important part of the entire procedure of ES cell culture is to maintain both the starting cell population and the gene-engineered derivatives under optimal growth conditions in culture. The reason that culture conditions are so important is that the cells must retain a normal karyotype to form functional germ cells with high efficiency. Suboptimal conditions include those in which nutrients or growth factors are in limited supply and a culture regimen that involves leaving the cells too long at high density, which favors the differentiation of cells into endoderm on the surface of large clumps of ES cells. Some of these variants will have an obvious abnormal karyotype, and most are unlikely to be able to contribute to the germ line. Karyotype analysis of ES cells can be used to monitor gross chromosomal changes, but seemingly normal karyotype does not guarantee germ-line transmission.[7,49,52]

For introducing mutations into ES cells by gene targeting, the targeting vector is designed to recombine with and mutate a specific genomic locus. The minimal components of such a vector are a sequence homologous to the desired chromosomal integration site and bacterial plasmid sequences. The typical structure of targeting vector is shown in Figure 9. Following gene transfer to ES cells, this vector undergoes single reciprocal recombination with its homologous genomic target which is stimulated by a double-strand break or a gap in the vector.[59] Since both transfection and targeting frequency of such a vector can be low (Table 45), it is desirable to include other components in the vector, such as positive and negative selection markers which provide strong selections for the targeted recombination product (Table 46).

Gene transfer into ES cells has been performed by several methods including electroporation,[7,50] retroviral infection,[18] and lipofection,[75] but the more common and successfully used approach is electroporation. This has the advantage of being technically relatively simple. The major disadvantage is the low transformation efficiency that necessitates the use of selectable

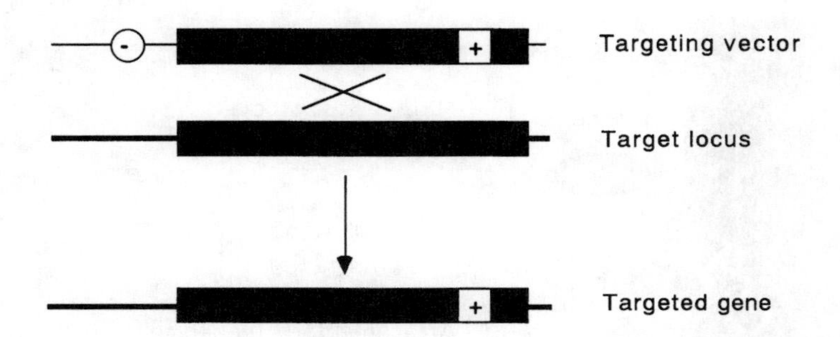

FIGURE 9
Structure of targeting vector and homologous recombination to target site. —— = plasmid vector sequences; ▬▬ = genomic sequences. (+) = positive selection marker gene, such as neomycine phosphoribocyl transferase. (–) = negative selection marker gene such as thymidine kinase. Random integration of vector in ES cells can be eliminated by positive selection (integration to transcription active site) or negative selection (integration to transcription nonactive site).

TABLE 45
Efficiency of Gene Targeting Into the Target Locus

Target Locus	No. of Cells After Electropolation	No. of Colonies After (+) Selection	No. of Colonies After (+) and (–) Selection	Fraction of Colonies Containing Targeting Events
hprt	1.3×10^7	1.5×10^5	48	19/24
int-2	1.5×10^7	1.6×10^6	81	4/81

Note: A targeting vector was introduced into ES cells (2.5×10^7) by electropolation to disrupt the *hprt* (hypoxanthin-guanine phosphoribosyl transferase) or proto-oncogene (*int*-2) locus.[74] Positive (+) and negative (–) selections were done with G418 and GANC (gancyclovir), respectively.

TABLE 46
Selection Marker for Efficient Isolation of Mutated ES Cells

Marker Gene	Positive Selection Drug	Negative Selection Drug
neo	G418	None
hph	Hygromycin B	None
gpt	Mycophenolic acid	6-Thioxanthine
HSVtk	None	GANC, FIAU
DT	None	Not required

Note: neo = neomycine phosphotransferase; hph = hygromycin B phosphotransferase; gpt = xanthine/guanine phosphoribosyl transferase; HSVtk = herpes simplex virus thymidine kinase; DT = diphtheria toxin; GANC = gancyclovir; FIAU = 1(1-2-deoxy-2-fluoro-β-darabinofuransyl)-5-iodouracil.

marker genes in the targeting vector as described previously. The best documented factor that affects the recombination frequency in mammalian cells is the length of homology to the target locus. A number of researchers have described a relationship between the length of homology and targeting frequency[56,76] and the degree of polymorphic variation between the vector and the chromosome.[77] As a general rule, greater length of homology will result

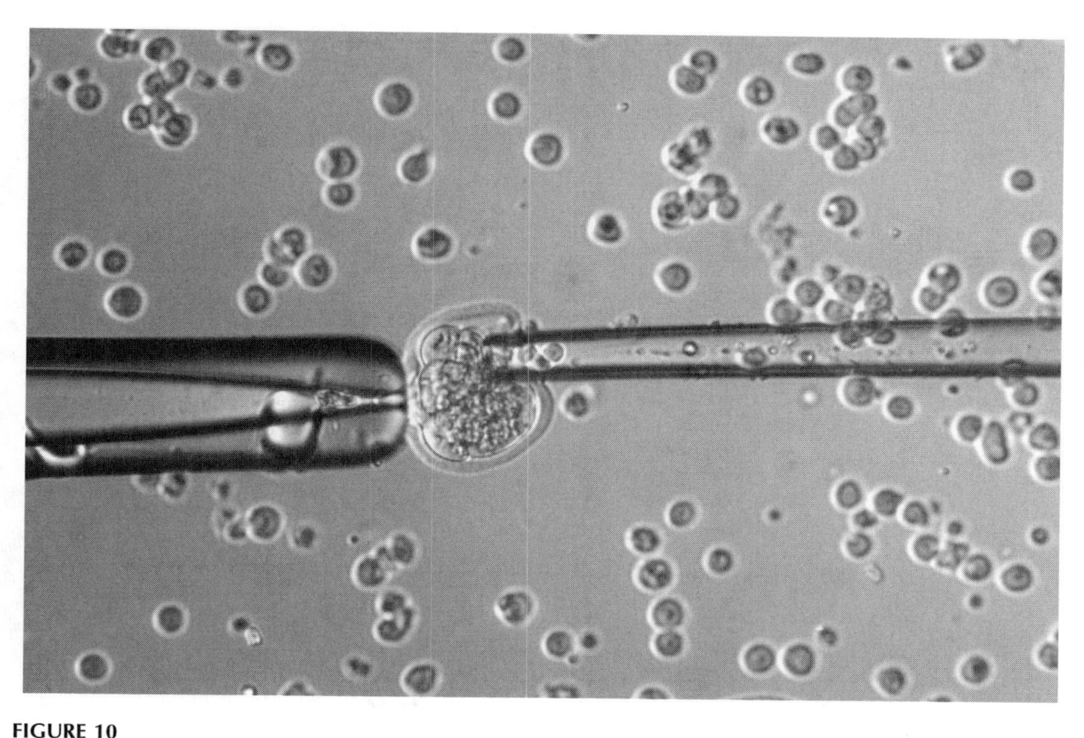

FIGURE 10
Injection of ES cells into host embryo for producing chimera mouse. This ES cell (TT-2) shows the ability for high contribution to the germ cell by injection into an eight-cell embryo, but not blastocyst.[67,78]

in higher targeting frequency. Most vectors are in the range of 5 to 10 kb, within the ideal length of the homologous sequences.

One of the most important aspects of any gene targeting experiment is to confirm that the desired genetic change has occurred. Following transfection of a vector, colonies that survive positive and negative selection should be clonally isolated by trypsinization and screened by either PCR or Southern blot analysis for a specific recombination allele.[50,59]

After screening transfected cells, these cells are used for producing the germ-line chimera to transmit the genetic change to the second generation. Although at present blastocyst injection is primarily used,[7,51,60] injection into embryos at the 8-cell stage is another means of producing chimeras, depending on the ES cell lines.[67,78] ES cells in an exponential growth phase are trypsinized to isolate a single cell. Following the injection of ES cells into the embryos by micromanipulator (Figure 10), the embryos can be transferred to pseudopregnant recipient females.[7,51,60]

When a chimera is obtained, the most convenient and readily apparent genetic marker of chimerism is coat color. Chimeric combinations of strains that differ at only one coat-color locus allow a simple visual appreciation of the degree of tissue contribution of each component in terms of the proportion of coat that expresses the ES cell allele. This evaluation of chimeric animals is necessarily subjective, but in general the degree of coat-color chimerism of a particular animal correlates with the degree of germ-line contribution.[51,60] Because most ES cell lines come from 129 mice, C57Bl/6J mice (which differ from 129 mice in coat color) are widely used as recipient females. Usually chimeras at the first generation have a single locus mutation at the targeted chromosomal site; therefore, two other matings — at first between chimera and normal female mice and then between chimeras — are necessary for obtaining null mutant mice. Identification and estimate of mutation can be done by PCR analysis and Southern blot analysis, as described in the previous section.

TABLE 47
Production of Transgenic Mice Expressing Various Genes Under the Control of Metallothionein Promoter

Introduced Gene	Source	Ref.
Growth hormone	Rat	83
Growth hormone	Ovine	96
SV-40, *v-myc* oncogene	Virus	84
Growth hormone releasing factor	Human	85, 88
Hepatitis B surface antigen	Virus	86
Interferon β mouse		87
Cholesteryl ester transfer protein	Human	89, 91
ret oncogene	Mouse	90
Apolipoprotein E	Rat	92
Transforming growth factor α	Human	93
Methyltransferase	*E. coli*	94
Polyoma large T-antigen	Virus	95
Kallikrein	Human	97

2.2.2 Expression of a Metal-Response Genes in Transgenic Mice

The regulation of gene expression *in vivo* can be studied by producing transgenic mice in which the transgene is composed of regulatory sequences of the gene of interest, including 5′- and 3′-flanking DNA. If the gene is expressed appropriately in transgenic mice, then a smaller DNA fragment with the 5′, 3′, or intragenic regions removed can be tested. If the gene is not expressed appropriately, then it may be regulated by more distant sequences that were absent from the initial construct. In defining the regulatory sequences of a gene, it is often advantageous to use gene fusions that contain a "reporter" gene. Reporter genes that have been used in transgenic mice include the *Escherichia coli* lacZ gene,[79] the *E. coli* chloramphenicol acetyltransferase (CAT) gene,[80] and the firefly luciferase gene.[81-82] A variety of tissue-specific genes has been used to produce constructs expressed in transgenic mice; in many cases, the correct pattern of gene expression has been obtained.

Frequently, it is desirable to introduce a transgene that will be silent until specifically activated by some experimental manipulation, such as administration of a drug. Regulatory elements of the metallothionein gene has been used exclusively for this purpose since the earliest studies of transgenic research (Table 47). Here, the regulatory element of two major metal-response genes, metallothionein and transferrin, using transgenic mice will be described.

2.2.2.1 Methallothionein

Two major isoforms of metallothionein (MT), designated MT-1 and MT-II, have been described in mammals. Both MTs are concordantly regulated in the animals, and the proteins are thought to be functionally equivalent.[98] The widespread inducible expression of MT-I and MT-II during development and in the adult has suggested that these proteins might play an essential role in zinc and copper metabolism during development and provide protection against various environmental stresses.[99-100]

DNA regions of 10 and 7 kb that flank the mouse MT-II and MT-I genes, respectively, were combined with a minimally marked MT-I (MT-I*) gene and tested in transgenic mice.[101] This construct resulted in (1) position-independent expression of MT-I* mRNA and copy number-dependent expression, (2) levels of hepatic MT-I mRNA per cell per transgene that were about half that derived from endogenous MT-I genes, (3) appropriate regulation by metals and hormones, and (4) tissue distribution of transgene mRNA that resembled that of endogenous MT-I mRNA. These MT-I flanking sequences also improved the expression of rat growth hormone reporter genes, with or without introns, that were under the control of the MT-I promoter. Deletion analysis indicated that regions known to have DNase I-hypersensitive sites were necessary but not sufficient for high-level expression.

2.2.2.2 *Transferrin*

Transferrin (TF) is a major plasma protein, mainly synthesized in the liver and regulated by iron at the level of translation,[102] that binds ferric iron and transports it to all target tissues of the body.[103] When chimeric genes composed of the human transferrin 5′ regulatory region fused to the CAT reporter gene were introduced into the genome of transgenic mice,[104] iron administration to the mice resulted in a significant decrease in human transferrin-directed CAT enzyme activity and CAT protein in liver, but no significant decrease in human transferrin-CAT mRNA levels. Binding of specific RNA iron regulatory elements by proteins in cytoplasmic extracts has been shown to regulate the transferrin system. A similar gene construct is also responsive to lead administration in the transgenic mice.[103] Transgene expression in liver was suppressed 31 to 50% by the lead treatment. Lead regulates human TF transgenes at the mRNA level, so that the regulation of human TF apparently differs from the regulation of mouse TF which is unresponsive to lead exposure.

The transferrin gene is transcribed at a high level in liver hepatocytes but is also active in several other cell types, including oligodendrocytes in the brain.[105] Enhancer elements between –560 and –44 bp of the transferrin gene promoter specifically activated transcription from a heterologous promoter in transgenic mouse liver and brain.[106] Within this region, a potent *cis*-acting element between –98 and –83 was found to be essential for gene activity in both cultured hepatocytes and transgenic mouse liver.[102,106] This element contains a CCAAT sequence and is specifically bound by a nuclear factor from mouse liver that is homologous to rat liver C/EBP (CAAT enhancer-binding protein). Point mutations within this binding site inhibit factor binding and abolish transcription in transfected hepatoma cells. The C/EBP binding site mutation causes a complete loss of transcription in transgenic mouse liver; however, transgene expression in the brain of the same animals was unaffected.

2.2.3 *Gene Targeting of the Metal-Response Gene*

Recently, gene targeting of MT-I and MT-II genes has been reported.[107,108] In the mouse, the genes encoding these two isoforms are 6 kb apart on chromosome 8.[98] A 2-kb Stu I restriction segment between the MT-I and MT-II genes was replaced with the neo gene. To disrupt the MT-I and MT-II genes, oligonucleotide pairs that encode in-frame stop codons were inserted into convenient restriction sites in the exons of the two genes (Figure 11).[108] The oligonucleotide insertions should produce transcripts that are longer than normal but nonfunctional due to in-frame stop codons. Then, the researchers disrupted the mouse MT-I and MT-II genes in embryonic stem cells and generated the mice homozygous for these mutant alleles. These mice were viable and reproduced normally but were more susceptible to hepatic poisoning by cadmium. This suggests that these widely expressed MTs are not essential for development but do protect against cadmium toxicity.

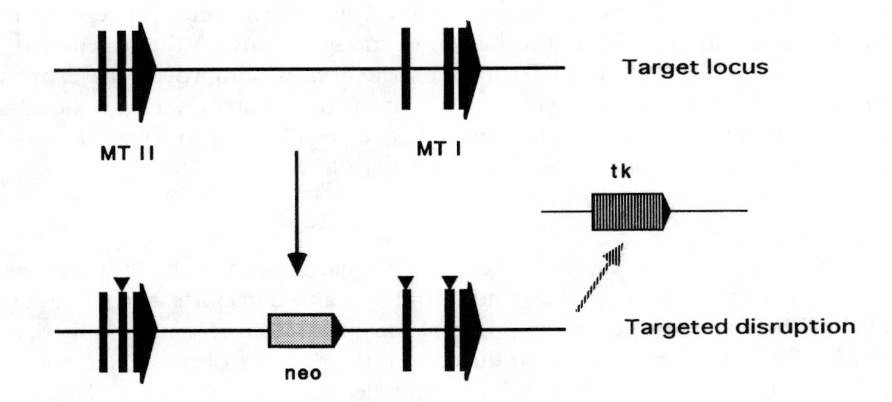

FIGURE 11
Targeted disruption of mouse metallothionein I and II genes. Mouse metallothionein I (MT-I) and II (MT-II) genes were disrupted by targeting vector containing homologous sequences of target locus but mutated by the oligonucleotide insertions (▼).[108] Neomycin (neo) and thymidine kinase (tk) genes were used as positive and negative selection markers, respectively. As a consequence of homologous recombination, the tk gene was removed from the target locus.

References

1. Palmiter, R. D. and Brinster, R. L., Germline transformation of mice, *Ann. Rev. Genet.,* 20, 3–60, 1986.
2. Jaenisch, R., Transgenic animals, *Science,* 240, 1468–1474, 1988.
3. Hanahan, D., Transgenic mice as probes into complex systems, *Science,* 246, 1265–1275, 1989.
4. Monk, M., *Mammalian Development: A Practical Approach,* Monk, M., Ed., IRL Press, Oxford, 1987.
5. Robertson, E. J., *Teratocarcinomas and Embryonic Stem Cells: A Practical Approach,* Robertson, E. J., Ed., IRL Press, Oxford, 1987.
6. Joyner, A. L., *Gene Targeting: A Practical Approach,* Joyner, A. L., Ed., IRL Press, Oxford, 1993.
7. Hogan, B., Beddington, R., Costantini, F., and Lacy, E., *Manipulating the Mouse Embryo: A Laboratory Manual,* 2nd ed., Hogan, B., Beddington, R., Costantini, F., and Lacy, E., Eds., Cold Spring Harbor Laboratory Press, New York, 1994.
8. Gordon, J. W., Scangos, G. A., Plotkin, D. J., Barbosa, J. A., and Ruddle, F. H., Genetic transformation of mouse embryos by microinjection of purified DNA, *Proc. Natl. Acad. Sci.,* 77, 7380–7384, 1980.
9. Brinster, R. L., Chen, H. Y., Trumbauer, M., Senear, A. W., Warren, R., and Palmiter, R. D., Somatic expression of herpes thymidine kinase in mice following injection of a fusion gene into eggs, *Cell,* 27, 223–231, 1981.
10. Costantini, F. and Lacy, E., Introduction of a rabbit beta-globin gene into the mouse germ line, *Nature,* 294, 92–94, 1981.
11. Gordon, J. W. and Ruddle, F. H., Integration and stable germ line transmission of genes injected into mouse pronuclei, *Science,* 214, 1244–1246, 1981.
12. Harbers, K., Jahner, D., and Jaenisch, R., Microinjection of cloned retroviral genomes into mouse zygotes: integration and expression in the animal, *Nature,* 293, 540–542, 1981.
13. Wagner, E. F., Stewart, T. A., and Mintz, B., The human beta-globin gene and a functional thymidine kinase gene in developing mice, *Proc. Natl. Acad. Sci.,* 78, 5061–5020, 1981.
14. Wagner, T. E., Hoppe, P. C., Jollick, J. D., Scholl, D. R., Hodinka, R. L., and Gault, J. B., Microinjection of a rabbit beta-globin gene into zygotes and its subsequent expression in adult mice and their offspring, *Proc. Natl. Acad. Sci.,* 78, 6376–6380, 1981.
15. Martin, G. R., Isolation of a pluripotent cell line from early mouse embryos cultured in medium conditioned by teratocarcinoma stem cells, *Proc. Natl. Acad. Sci.,* 78, 7634–7636, 1981.
16. Evans, M. J. and Kaufman, M. H., Establishment in culture of pluripotential cells from mouse embryos, *Nature,* 292, 154–156, 1981.
17. Bradley, A., Evans, M., Kaufman, M. H., and Robertson, E., Formation of germ-line chimeras from embryo-derived teratocarcinoma cell lines, *Nature,* 309, 255–256, 1984.
18. Robertson, E., Bradley, A., Kuehn, M., and Evans, M., Germ-line transmission of gene introduced into cultured pluripotential cells by retroviral vector, *Nature,* 323, 445–448, 1986.
19. Gossler, A., Doetschman, T., Korn, R., Serfling, E., and Kemler, R., Transgenesis by means of blastocyst-derived embryonic stem cell lines, *Proc. Natl. Acad. Sci.,* 83, 9060–9065, 1986.

20. Jaenisch, R., Germ line integration and Mendelian transmission of the exogenous Molony leukemia virus, *Proc. Natl. Acad. Sci.,* 73, 1260–1264, 1976.
21. Jaenisch, R., Jahner, D., Nobis, P., Simon, I., Lohler, J., Harbers, K., and Grotkopp, D., Chromosomal position and activation of retroviral genomes inserted into the germ line of mice, *Cell,* 24, 519–529, 1981.
22. Stuhlmann, H., Cone, R., Mulligan, R. C., and Jaenisch, R., Introduction of a selectable gene into different animal tissue by a retrovirus recombinant vector, *Proc. Natl. Acad. Sci.,* 81, 7151–7155, 1984.
23. Jahner, D., Haase, K., Mulligan, R., and Jaenisch, R., Insertion of the bacterial *gpt* gene into the germ line of mice by retroviral infection, *Proc. Natl. Acad. Sci.,* 82, 6927–6931, 1985.
24. van der Putten, H., Botteri, F. M., Miller, A. D., Rosenfeld, M. G., Fan, H., Evans, R. M., and Verma, I. M., Efficient insertion of genes into the mouse germ line via retroviral vectors, *Proc. Natl. Acad. Sci.,* 82, 6148–6152, 1985.
25. Soriano, P., Cone, R. D., Mulligan, R. C., and Jaenisch, R., Tissue-specific and ectopic expression of genes introduced into transgenic mice by retroviruses, *Science,* 234, 1409–1413, 1986.
26. Hetherington, C. M., Mouse husbandry, in *Mammalian Development: A Practical Approach,* Monk, M., Ed., IRL Press, Oxford, 1987, p. 1.
27. Pratt, H. P. M., Isolation, culture and manipulation of preimplantation mouse embryos, in *Mammalian Development: A Practical Approach,* Monk, M., Ed., IRL Press, Oxford, 1897, p. 13.
28. Allen, N. D., Barton, S. C., Surani, M. A. H., and Reik, W., Production of transgenic mice, in *Mammalian Development: A Practical Approach,* Monk, M., Ed., IRL Press, Oxford, 1987, p. 217.
29. Brinster, R. L., Chen, N. Y., Trumbauer, M. E., Yagle, M. K., and Palmiter, R. D., Factors affecting the efficiency of introducing foreign DNA into mice by microinjecting eggs, *Proc. Natl. Acad. Sci.,* 82, 4438–4442, 1985.
30. Chada, K. J., Magram, J., Raphael, K., Radice, G., Lacy, E., and Costantini, F., Specific expression of a foreign beta-globin gene in erythroid cells of transgenic mice, *Nature,* 314, 377–380, 1985.
31. Krumlauf, R., Chapman, V., Hammer, R., Brinster, R., and Tilghman, S. M., Differential expression of alpha-fetoprotein genes on the inactive X chromosome in extraembryonic and somatic tissues of a transgenic mouse line, *Nature,* 319, 224–226, 1986.
32. Townes, T. M., Lingrel, J. B., Brinster, R. L., and Palmiter, R. D., Erythroid specific expression of human beta-globin genes in transgenic mice, *EMBO J.,* 4, 1715–1723, 1985.
33. Hammer, R. E., Krumlauf, R., Camper, S. A., Brinster, R. L., and Tilghman, S. M., Diversity of alpha-fetoprotein gene expression in mice is generated by a combination of separate enhancer elements, *Science,* 235, 53–58, 1987.
34. Taylor, L. D., Carmack, C. E., Schramm, S. R., Mashayekh, R., Higgins, K. M., Kuo, C. C., Woodhouse, C., Kay, R. M., and Lonberg, N., A transgenic mouse that expresses a diversity of human sequence heavy and light chain immunoglobulins, *Nucleic Acid Res.,* 20, 6287–6295, 1992.
35. Peterson, K. R., Clegg, C. H., Huxley, C., Josephson, B. M., Haugen, H. S., Furukawa, T., and Stamatoyannopoulos, G., Transgenic mice containing a 248-kb yeast artificial chromosome carrying the human beta-globin locus display proper developmental control of human globin genes, *Proc. Natl. Acad. Sci.,* 90, 7593–7597, 1993.
36. Schedl, A., Montoliu, L., Kelsey, G., and Schutz, G., A yeast artificial chromosome covering the tyrosinase gene confers copy number-dependent expression in transgenic mice, *Nature,* 362, 258–261, 1993.
37. Choi, T. K., Hollenbach, P. W., Pearson, B. E., Ueda, R. M., Weddell, G. N., Kurahara, C. G., Woodhouse, C. S., Kay, R. M., and Loring, J. F., Transgenic mice containing a human heavy chain immunoglobulin gene fragment cloned in a yeast artificial chromosome, *Nature Genet.,* 4, 117–123, 1993.
38. Strauss, W. M., Dausman, J., Beard, C., Johnson, C., Lawrence, J. B., and Jaenisch, R., Germ line transmission of a yeast artificial chromosome spanning the murine alpha 1 (I) collagen locus, *Science,* 259, 1904–1907, 1993.
39. Brinster, R. L., Allen, J. M., Behringer, R. R., Gelinas, R. E., and Palmiter, R. D., Introns increase transcriptional efficiency in transgenic mice, *Proc. Natl. Acad. Sci.,* 85, 836–840, 1988.
40. Choi, T. K., Huang, M., Gorman, C., and Jaenisch, R., A generic intron increases gene expression in transgenic mice, *Mol. Cell. Biol.,* 11, 3070–3074, 1991.
41. Palmiter, R. D., Sandgreen, E. P., Avarbock, M. R., Allen, D. D., and Brinster, R. L., Heterogous introns can enhance expression of transgenes in mice, *Proc. Natl. Acad. Sci.,* 88, 478–482, 1991.
42. Wilkie, T. M., Brinster, R. L., and Palmiter, R. D., Germline and somatic mosaicism in transgenic mice, *Dev. Biol.,* 118, 9–18, 1986.
43. Lacy, E., Roberts, S., Evans, E. P., Burtenshaw, M. D., and Costantini, F., A foreign beta globin gene in transgenic mice: integration at abnormal chromosomal positions and expression in inappropriate tissues, *Cell,* 34, 343–358, 1983.
44. Saiki, R. K., Scharf, S., Faloona, F., Mullis, K. M., Horn, G. T., Erlich, H. A., and Arnheim, N., Enzymatic amplification of β-globin genomic sequences and restriction site analysis for diagnosis of sickle cell anemia, *Science,* 230, 1350–1354, 1985.
45. Southern, E., Detection of specific sequences among DNA fragments separated by gel electrophoresis, *J. Mol. Biol.,* 98, 503–517, 1975.

46. Magram, J. and Bishop, J. M., Dominant male sterility in mice caused by insertion of a transgene, *Proc. Natl. Acad. Sci.,* 88, 10327–10331, 1991.
47. Gridley, T., Insertional versus targeted mutagenesis in mice, *New Biol.,* 3, 1025–1034, 1991.
48. Meisler, M. H., Insertional mutation of "classical" and novel genes in transgenic mice, *Trends Genet.,* 8, 341–344, 1992.
49. Robertson, E. J., Embryo-derived stem cell lines, in *Teratocarcinomas and Embryonic Stem Cells: A Practical Approach,* Robertson, E. J., Ed., IRL Press, Oxford, 1987, p. 71.
50. Wurst, W. and Joyner, A. L., Production of targeted embryonic stem cell clones, in *Gene Targeting: A Practical Approach,* Joyner, A. L., Ed., IRL Press, New York, 1993, p. 33.
51. Papaioannou, V. and Johnson, R., Production of chimeras and genetically defined offspring from targeted ES cells, in *Gene Targeting: A Practical Approach,* Joyner, A. L., Ed., IRL Press, Oxford, 1993.
52. Nagy, A. and Rossant, J., Production of completely ES cell-derived fetuses, in *Gene Targeting: A Practical Approach,* Joyner, A. L., Ed., IRL Press, New York, 1993, p. 47.
53. Doetschman, T., Gregg, R. G., Maeda, N., Hooper, M. L., Melton, D. W., Thompson, S., and Smithies, O., Targeted correction of a mutant HPRT gene in mouse embryonic stem cells, *Nature,* 330, 576–578, 1987.
54. Hooper, M. I., Hardy, K., Handyside, A., Hunter, S., and Monk, M., HPRT-deficient (Lesch-Nyhan) mouse embryos derived from germ-line colonization by cultured cells, *Nature,* 326, 292–294, 1987.
55. Kuehn, M., Bradley, A., Robertson, E. J., and Evans, M. J., A potential animal model for Lesch-Nyhan syndrome through the introduction of HPRT mutations in mice, *Nature,* 326, 295–298, 1987.
56. Thomas, K. R. and Capecchi, M. R., Site-directed mutagenesis by gene targeting in mouse embryo-derived stem cells, *Cell,* 51, 503–512, 1987.
57. Koller, B. H., Marrack, P., Kappler, J. W., and Smithies, O., Normal development of mice deficient in β2M, MHC class 1 proteins, *Science,* 248, 1227–1230, 1990.
58. Gossler, A. and Zachgo, J., Gene and enhancer trap screens in ES cell chimeras, in *Gene Targeting: A Practical Approach,* Joyner, A. L., Ed., IRL Press, New York, 1993, p. 181.
59. Hasty, P. and Bradley, A., Gene targeting vectors for mammalian cells, in *Gene Targeting: A Practical Approach,* Joyner, A. L., Ed., IRL Press, New York, 1993, p. 1.
60. Bradley, A., Production and analysis of chimeric mice, in *Teratocarcinomas and Embryonic Stem Cells: A Practical Approach,* Robertson, E. J., Ed., IRL Press, Oxford, 1987, p. 113.
61. Doetschman, T. C., Eistetter, H., Kutz, M., Schmidt, W., and Kemler, R., The *in vitro* development of blastocyst-derived embryonic stem cell lines: formation of visceral yolk sac, blood islands and myocardium, *J. Embryol. Exp. Morphol.,* 87, 27–45, 1985.
62. McMahorn, A. P. and Bradley, A., The *wnt*-1 (*int*-1) proto-oncogene is required for development of a large region of the mouse brain, *Cell,* 62, 1073–1985, 1990.
63. Li, E., Bestor, T. H., and Jaenisch, R., Targeted mutation of the DNA methyltransferase gene results in embryonic lethality, *Cell,* 69, 915–926, 1992.
64. Magin, T. M., McEwan, C., Milne, M., Pow, A. M., Selfridge, J., and Melton, D. W., A position- and orientation-dependent element in the first intron is required for expression of the mouse *hprt* gene in embryonic stem cells, *Gene,* 122, 289–296, 1992.
65. Nagy, A., Rossant, J., Nagy, R., Abramov-Newerly, W., and Roder, J. C., Derivation of completely cell culture-derived mice from early-passage embryonic stem cells, *Proc. Natl. Acad. Sci.,* 90, 8424–8428, 1993.
66. Ledermann, B. and Burki, K., Establishment of a germ-line competent C57BL/6 embryonic stem cell line, *Exp. Cell Res.,* 197, 254–258, 1991.
67. Tokunaga, T. and Tsunoda, Y., Efficacious production of viable germline chimeras between embryonic stem (ES) cells and 8-cell stage embryos, *Dev. Growth. Differ.,* 34, 561–566, 1992.
68. Matsui, Y., Zsebo, K., and Hogan, B. L. M., Derivation of pluripotential embryonic stem cells from murine primordial germ cells in culture, *Cell,* 70, 667–669, 1992.
69. Resnick, J. L., Bixler, L. S., Cheng, L., and Donovan, P. J., Long-term proliferation of mouse primordial germ cells in culture, *Nature,* 359, 550–551, 1992.
70. Stewart, C. L., Gadi, I., and Bhatt, H., Stem cells from primordial germ cells can reenter the germ line, *Dev. Biol.,* 161, 626–628, 1994.
71. Smith, A. G., Heath, J. K., Donaldson, D. D., Wong, G. G., Moreau, J., Stahl, M., and Rogers, D., Inhibition of pluripotential embryonic stem cell differentiation by purified polypeptides, *Nature,* 336, 688–690, 1988.
72. Williams, R. L., Hilton, D. J., Pease, S., Wilson, T. A., Stewart, C. L., Gearing, D. P., Wagner, E. F., Metcalf, D., Nicola, N. A., and Gough, M. M., Myeloid leukemia inhibitory factor maintains the developmental potential of embryonic stem cells, *Nature,* 336, 684–687, 1988.
73. Stewart, T. A. and Mintz, B., Recurrent germ line transmission of the teratocarcinoma genome from the METT-1 culture line to progeny *in vivo, J. Exp. Zool.,* 224, 465–471, 1982.
74. Mansour, S. L., Thomas, K. R., and Capecchi, M. R., Disruption of the proto-oncogene *int*-2 in mouse embryo-derived stem cells: a general strategy for targeting mutations to non-selectable genes, *Nature,* 336, 348–352, 1988.
75. Strauss, W. M. and Jaenisch, R., Molecular complementation of a collagen mutation in mammalian cells using yeast artificial chromosomes, *EMBO J.,* 11, 417–422, 1992.

76. Hasty, P., Rivera-Perez, J., Chang, C., and Bradley, A., The length of homology required for gene targeting in embryonic stem cells, *Mol. Cell. Biol.,* 11, 5586–5591, 1991.
77. te Riele, H., Robanus-Maandag, E., and Berns, A., Highly efficient gene targeting in embryonic stem cells through homologous recombination with isogenic DNA constructs, *Proc. Natl. Acad. Sci.,* 89, 5128–5132, 1992.
78. Yagi, T., Tokunaga, T., Furuta, Y., Nada, S., Yoshida, M., Tsukada, T., Saga, Y., Takeda, N., Ikawa, Y., and Aizawa, S., A novel ES cell line, TT2, with high germline-differentiating potency, *Anal. Biochem.,* 214, 70–76, 1993.
79. Goring, D. R., Rossant, J., Clapoff, S., Breitman, M. L., and Tsui, L. C., *In situ* detection of beta-galactosidase in lenses of transgenic mice with a gamma-crystallin/*lacZ* gene, *Science,* 235, 456–458, 1987.
80. Overbeek, P. A., Chepelinsky, A., Khillan, J. S., Piatigorsky, J., and Westphal, H., Lens-specific expression and developmental regulation of the bacterial chloramphenicol acetyltransferase gene driven by the murine alpha-crystallin promoter in transgenic mice, *Proc. Natl. Acad. Sci.,* 82, 7815–7819, 1985.
81. Lira, S. A., Kinloch, R. A., Mortillo, S., and Wasserman, P. M., An upstream region of the mouse ZP3 gene directs expression of firefly luciferase specially to growing oocytes in transgenic mice, *Proc. Natl. Acad. Sci.,* 87, 7215–7219, 1990.
82. Lee, K. J., Ross, R. S., Rockman, H. A., Harris, A. N., Obrien, T. X., Van, B. M., Shubeita, H. E., Kondolf, R., Brem, G., Price, J., Evans, S. M., Zhu, H., Franz, W. M., and Chien, K. R., Myosin light chain-2 luciferase transgenic mice reveal distinct regulatory programs for cardiac and skeletal muscle-specific expression of a single contractile protein gene, *J. Biol. Chem.,* 267, 15875–15885, 1992.
83. Palmiter, R. D., Brinster, R. L., Hammer, R. E., Trumbauer, M. E., Rosenfeld, M. G., Birnberg, N. C., and Evans, R. M., Dramatic growth of mice that develop from eggs microinjected with metallothionein-growth hormone fusion genes, *Nature,* 300, 611–615, 1982.
84. Small, J. A., Blair, D. G., Showalter, S. D., and Scangos, G. A., Analysis of a transgenic mouse containing simian virus 40 and v-myc sequences, *Mol. Cell. Biol.,* 5, 624–628, 1985.
85. Hammer, R. E., Brinster, R. L., Rosenfeld, M. G., Evans, R. M., and Mayo, K. E., Expression of human growth hormone-releasing factor in transgenic mice results in increased somatic growth, *Nature,* 315, 413–416, 1985.
86. Chisari, F. V., Pinkert, C. A., Milich, D. R., Filippi, P., McLachlan, A., Palmiter, R. D., and Brinster, R. L., A transgenic mouse model of the chronic hepatitis B surface antigen carrier state, *Science,* 230, 1157–1160, 1985.
87. Iwakura, Y., Asano, M., Nishimune, Y., and Kawade, Y., Male sterility of transgenic mice carrying exogenous mouse interferon-beta gene under the control of the metallothionein enhancer-promoter, *EMBO J.,* 7, 3757–3762, 1988.
88. Frohman, L. A., Downs, T. R., Kashio, Y., and Brinster, R. L., Tissue distribution and molecular heterogeneity of human growth hormone-releasing factor in the transgenic mouse, *Endocrinology,* 127, 2149–2156, 1990.
89. Agellon, L. B., Walsh, A., Hayek, T., Moulin, P., Jiang, X. C., Shelanski, S. A., Breslow, J. L., and Tall, A. R., Reduced high density lipoprotein cholesterol in human cholesteryl ester transfer protein transgenic mice, *J. Biol. Chem.,* 266, 10796–10801, 1991.
90. Iwamoto, T., Takahashi, M., Ito, M., Hamatani, K., Ohbayashi, M., Wajjwalku, W., Isobe, K., and Nakashima, I., Aberrant melanogenesis and melanocytic tumour development in transgenic mice that carry a metallothionein/ret fusion gene, *EMBO J.,* 10, 3167–3175, 1991.
91. Hayek, T., Chajek-Shaul, T., Walsh, A., Agellon, L. B., Moulin, P., Tall, A. R., and Breslow, J. L., An interaction between the human cholesteryl ester transfer protein (CETP) and apolipoprotein A-I genes in transgenic mice results in a profound CETP-mediated depression of high density lipoprotein cholesterol levels, *J. Clin. Invest.,* 90, 505–510, 1992.
92. Shimano, H., Yamada, N., Katsuki, M., Shimada, M., Gotoda, T., Harada, K., Murase, T., Fukazawa, C., Takaku, F., and Yazaki, Y., Overexpression of apolipoprotein E in transgenic mice: marked reduction in plasma lipoproteins except high density lipoprotein and resistance against diet-induced hypercholesterolemia, *Proc. Natl. Acad. Sci.,* 89, 1750–1754, 1992.
93. Murakami, H., Sanderson, N. D., Nagy, P., Marino, P. A., Merlino, G., and Thorgeirsson, S. S., Transgenic mouse model for synergistic effects of nuclear oncogenes and growth factors in tumorigenesis: interaction of *c-myc* and transforming growth factor alpha in hepatic oncogenesis, *Cancer Res.,* 53, 1719–1723, 1993.
94. Nakatsuru, Y., Matsukuma, S., Nemoto, N., Sugano, H., Sekiguchi, M., and Ishikawa, T., O6-methylguanine-DNA methyltransferase products against nitrosamine-induced hepatocarcinogenesis, *Proc. Natl. Acad. Sci,* 90, 6468–6472, 1993.
95. Lebel, M. and Mes-Masson, A. M., Establishment and characterization of testicular cell lines from MT-PVLT-10 transgenic mice, *Exp. Cell Res.,* 213, 12–19, 1994.
96. Oberbauer, A. M., Pomp, D., and Murray, J. D., Dependence of increased linear bone growth on age at oMT1a-oGH transgene expression in mice, *Growth Dev. Aging,* 58, 83–93, 1994.
97. Simson, J. A., Wang, J., Chao, J., and Chao, L., Histopathology of lymphatic tissues in transgenic mice expressing human tissue kallikrein gene, *Lab. Invest.,* 71, 680–687, 1994.

98. Searle, P. F., Davison, B. L., Stuart, G. W., Wilkie, T. M., Norstedt, G., and Palmiter, R. D., Regulation, linkage, and sequence of the mouse metallothionein I and II genes, *Mol. Cell. Biol.,* 4, 1221–1230, 1984.
99. Kaji, J. H. R. and Kojima, Y., Metallothionein II, *Experientia (Suppl.),* 52, 25–80, 1987.
100. Suzuki, K., Imura, T., and Kimura, M., *Metallothionein III: Biological Roles and Medical Implications,* Suzuki, K., Imura, N., and Kimura, M., Eds., Birkhauser Verlag, Base, 1993, p. 417.
101. Palmiter, R. D., Sandgren, E. P., Koeller, D. M., and Brinster, R. L., Distal regulatory elements from the mouse methallothionein locus stimulate gene expression in transgenic mice, *Mol. Cell. Biol.,* 13, 5266–5275, 1993.
102. Schaeffer, E., Boissier, F., Py, M. C., Cohen, G. N., and Zakin, M. M., Cell type-specific expression of the human transferrin gene. Role of promoter, negative, and enhancer elements, *J. Biol. Chem.,* 264, 7153–7160, 1989.
103. Adrian, G. S., Rivera, E. V., Adrian, E. K., Lu, Y., Buchanan, J., Herbert, D. C., Weaker, F. J., Walter, C. A., and Bowman, B. H., Lead suppresses chimeric human transferrin gene expression in transgenic mouse liver, *Neurotoxicology,* 14, 273–282, 1993.
104. Cox, L. A. and Adrian, G. S., Posttranscriptional regulation of chimeric human transferrin genes by iron, *Biochemistry,* 32, 4738–4745, 1993.
105. Yang, F. M., Fridrichs, W. E., Buchanan, J. M., Herbert, D. C., Weaker, F. J., Brock, J. H., and Bowman, B. H., Tissue specific expression of mouse transferrin during development and aging, *Mech. Aging Dev.,* 56, 187–197, 1990.
106. Theisen, T., Behringer, R. R., Cadd, G. G., Brinster, R. L., and McKnight, G. S., A C/EBP-binding site in the transferrin promoter is essential for expression in the liver but not the brain of transgenic mice, *Mol. Cell. Biol.,* 13, 7666–7676, 1993.
107. Michalska, A. E. and Choo, K. H., Targeting and germ-line transmission of a null mutation at the methallothionein I and II loci in mouse, *Proc. Natl. Acad. Sci. USA,* 90, 8088–8092, 1993.
108. Masters, B. A., Kelly, E. J., Quaife, C. J., Brinster, R. L., and Palmiter, R. D., Targeted disruption of methallothionein I and II genes increases sensitivity to cadmium, *Proc. Natl. Acad. Sci. USA,* 91, 584–588, 1994.

3 CELLULAR AND MOLECULAR APPROACHES

3.1 Cell Cultures

Yasunobu Aoki

Cultured cells are strong tools for *in vitro* studies on metal toxicity. Three categories of cells are used for studying the mechanisms of toxicity or for testing the toxicity: (1) primary cultured cells and normal cells, (2) cell lines, and (3) blood cells. Not only soluble metallic compounds but also insoluble particles are subject to examination of their toxicity to cells. References for *in vitro* studies using the cells are summarized in Tables 48 to 50. The cells are usually cultured in a plastic culture dish with the defined medium,[170] but the procedures for culturing cells are very different among the various types of cells (See References 170 to 172). General problems involved in exposing metallic compounds to the cells follow.

The medium for culturing cells contains amino acids, and 10% (v/v) serum, such as a fetal bovine serum, is added to the medium. Some metallic compounds bind to amino acids and serum proteins in the medium, and the toxic effect of the metal is possibly altered. In order to avoid this problem, in some cases serum is omitted from the culture medium, or a balanced salt solution (BSS) not containing amino acids (e.g., Earle's or Hank's BSS) is used instead of a defined medium. The chemical form of a metal ion varies according to the pH of the culture medium, and its bioavailability of the metal ion is dependent on the chemical form. For example, most of the metal hydroxides (other than alkali metal) are insoluble in aquatic solution and are barely incorporated into the cells. For stabilizing the pH of the medium, Hepes buffer is often added to it.

TABLE 48

References for Studies on Metal Toxicology Using Primary Cultured Cells and Normal Cells

Elements	Adrenal Medulla	Arota	Bone/Bone Marrow	Brain/Neuron	Embryo/Fetus	Endothelium	Fibroblast	Kidney	Liver	Lymph node	Mucosa	Muscle	Skin	Testis	Trachea
Al			mo; 29	chick; 156 mo; 76 ra; 25, 88, 93, 156			catt; 104		ra;22						
As															
Cd		catt; 34	mo; 100	ra; 4,155	ch; 112	hu; 35, 90 mo; 140		dog; 38 hu; 63, 73 mo; 37 ra; 45, 121, 128, 151	ra; 101 mo; 140 ra; 26,39, 44,101 116, 132		hu; 55 ra; 55		hu; 64	ra; 26	
Co						hu; 75	catt; 104			gp; 135 mo; 135 ra; 135			hu; 13		
Cr							catt; 104			gp; 135 mo; 135 ra; 135	hu; 55 ra; 55		hu; 13		
Cu				ra; 4		hu; 131		mo; 37				hu; 23	hu; 64		
Fe				ra; 4, 30					ra; 53, 130			ra; 103			rabbit; 83
Hg				mo; 123									hu; 64		
Ni				ra; 4		hu; 75, 90					hu; 55 ra; 55		hu; 13		
Pb	catt; 86			ra; 4, 91, 142		hu; 162									
Zn			mo; 51	mo; 47 ra; 126											
Others			(Sr) chick; 48	(La) ra; 4 (Gd) ra; 4			(Ti) catt; 104							(Tl) ra; 166	

Note: catt = cattle; ch = Chinese hamster; gp = guinea pig; hu = human; mo = mouse; ra = rat.

TABLE 49A

References for Studies on Metal Toxicology Using Cell Lines Derived from Tissue

Elements	Connective Tissue	Embryo/ Fetus	Endothelium	Fibroblast	Ganglia	Kidney	Liver	Lung	Lymphoblast	Lymphocyte	Macrophage	Ovary	Osteoblast	Skin	Testis
Ag	mo; 148	mo; 79, 143													
Al		mo; 79	goat; 66		ra; 93			ch; 149							
Cd		ch; 114	catt; 5, 14, 117, 146, 169	ch; 114		dog; 160	hu; 69	ch; 110, 113, 114, 168		hu; 65		ch; 102	mo; 100	hu; 102	ra; 11
		mo; 10, 36, 79, 119, 143		mo; 36		hu; 145	mo; 50								
						pig; 115 ra; 45, 124 rabbit; 107	ra; 72								
Co				mo; 61 ra; 42				ch; 137							
Cr				ra; 42					hu; 71			ch; 80, 158			
Cu		mo; 79, 143	catt; 146	hu; 131		hu; 145		ch; 57							
Fe				mo; 6											
Ga		mo; 79, 143													
Hg				hu; 139								ch; 9			
Ni		mo; 79, 119, 143		hu; 129		hu; 21						ch; 19, 33, 56, 60, 70, 80	hu; 18		
Pb			catt; 5, 59					ch; 133							
V		mo; 79									mo; 106				
Zn		mo; 143		catt; 136		dog; 150 rabbit; 107		ch; 147						hu; 92	
Others	(Bi) mo; 148	(Ti) mo; 79											(Cs) hu; 84		

Note: catt = cattle; ch = Chinese hamster; gp = guinea pig; hu= human; mo = mouse; ra = rat.

TABLE 49B

References for Studies on Metal Toxicology Using Cell Lines Derived from Tumor

Elements	Adrenal Tumor	Burkitt Lymphoma	Carcinoma					Fibro-sarcoma	Glial Tumor	Glioma	Hepatoma	Leukemia and Lymphoma	Medullo-blastoma	Neuro-blastoma	Osteogenic Sarcoma	Pheochro-mocytoma
			Embryo	Kidney	Mouth	Ovary	Prostate									
Al												chick; 62		hu; 156 mo; 99, 141, 154, 156, 163		
Cd	mo; 85		mo; 87			hu; 17	hu; 24	mo; 108	mo; 94	ra; 153	ra; 32			mo; 94	hu; 96 ra; 96	ra; 74, 138
Co		hu; 20														
Cr		hu; 20														
Cu		hu; 15									ra; 43, 77	hu; 109				
Fe		hu; 20									ra; 53	hu; 7, 62				
Hg					hu; 9				mo; 94					mo; 94		
Ni		hu; 20														
Pb										ra; 153		hu; 152			hu; 96	
Zn										hu; 51 ra; 51	ra; 77, 122				ra; 96, 147	
Others		(Mn) hu; 15		(Sn) hu; 41			(Pt) hu; 24						(Ga) mo; 52			

Note: catt = cattle; ch = Chinese hamster; gp = guinea pig; hu = human; mo = mouse; ra = rat.

TABLE 50

References for Studies on Metal Toxicology Using Blood Cells

Element	Erythrocyte	Leukocyte	Lymphocyte	Macrophage/Monocyte	Alveolar Macrophage
Al					catt; 104
As			hu; 58 mo; 78		catt; 98 rabbit; 68
Cd		hu; 2, 82	hu; 1, 58, 81, 111, 114 mo; 127	hu; 1	goat; 118 gp; 27 rabbit; 68
Co				mo; 97, 167 ra; 31	catt; 104
Cr				mo; 161	catt; 98, 104
Cu	sheep; 125		hu; 1, 109	hu; 1	
Fe				mo; 97	catt; 98
Hg		ra; 144	hu; 1, 81	hu; 1 ra; 144	rabbit; 68
Mn					catt; 98 rabbit; 157
Ni			hu; 3, 12, 164, 165 mo; 127	ra; 95 rainbow trout; 67	goat; 118 gp; 28, 49 ra; 40, 134 rabbit; 68
Pb			hu; 1, 98 mo; 89 ra; 89	hu; 1	rabbit; 54
V			hu; 8		catt; 68, 98, 104
Zn			hu; 1	hu; 1	
Others			(Ga) mo; 78	(Sn) hu; 159 (W) mo; 167	(Ti) catt; 104

Note: catt = cattle; ch = Chinese hamster; gp = guinea pig; hu = human; mo = mouse; ra = rat.

3.1.1 Primary Cultured Cells and Normal Cells

The cells consisting of parenchyma are isolated after digestion with a protease (e.g., collagenase and trypsin) solution. The isolated cells are inoculated on a culture dish and cultured as monolayer cells, which are primary cultured cells. While several types of primary cultured cells are not divisible without a growth factor (e.g., liver parenchymal cells and nerve cells), many types of primary cultured cells are subcultured for several passages (e.g., kidney proximal tubule epithelial cells and fibroblast). These growing cells are used as normal cells for which the karyotype is diploid and stable. Primary cultures under adequate conditions maintain their own specific characteristics and functions expressed in original tissue or organ for several days. Normal cells also maintain their original characters and functions before being spontaneously transformed.

The cells derived from the target organs of metals and organo-metallic compounds (e.g., liver, kidney, and brain) are used for *in vitro* tests for the cytotoxicity and toxic effect on the organ or tissue specific function caused by the metals; however, it is difficult to remove the contamination of other cells (e.g., fibroblasts) completely. Procedures for preparing primary cultured rat cells used frequently for metal toxicology are briefly described below.

3.1.1.1 Preparation and Culture of Rat Liver Parenchymal Cells

The liver of a rat (150 to 200 g) was perfused *in situ* with Ca,Mg-free Hanks' BSS containing 0.5 m*M* EGTA and 10 m*M* Hepes/NaOH (pH 7.2) and then digested with 0.5

mg/ml collagenase solution in Mg and glucose-free Hanks' BSS containing 0.05 mg/ml soybean trypsin inhibitor, 5 m*M* $CaCl_2$, and 10 m*M* Hepes/NaOH (pH 7.5) according to the method originally developed by Seglen.[173] The cells were dispersed from digested liver in Hanks' BSS. The resulting cell suspension was filtrated through folded gauze and centrifuged at 600 rpm for 1 min. In order to remove nonparenchymal cells, precipitated cells were resuspended gently in Hanks' BSS and centrifuged at 600 rpm for 1 min three times. Viability of isolated cells should be over 90% to obtain well-conditioned primary cultures. The cells (0.85×10^6) were inoculated to plastic dishes (diameter, 35 mm; collagen-coated dish or Primaria®; Falcon, Franklin Lakes, NJ) with 1.5 ml of Williams' E medium containing 10% (v/v) fetal bovine serum, 100 n*M* dexamethasone, 1 n*M* insulin, 10 m*M* sodium pyruvate, 100 U/ml penicillin, 100 μg/ml streptomycin, 0.25 μg/ml fungizon, and 25 m*M* Hepes/NaOH (pH 7.4).[174] After incubation for 4 hr in a 5% CO_2/95% air incubator at 37°C, the medium was changed to remove dead cells. The cultures were able to continue for a week by changing the medium once every 2 days. Cultures were also maintainable in serum-free medium containing 0.1 μg/ml aprotinin for 2 to 3 days by changing the medium once every day.[174,175] Concentrations of dexamethasone and insulin in the medium are changeable within the range of 1 nM to 10 μ*M* and 1 n*M* to 1 μ*M*, respectively, depending on the condition of the cultures. These reagents were essential for maintaining primary cultures of rat liver parenchymal cells.[176] However, dexamethasone is an inducer of metallothionein, and this protein being induced in the cells possibly interferes with toxicity tests of metals. If it was necessary to omit dexamethasone from the medium, cultures were maintainable without dexamethasone in calcium-deficient Williams' E medium, which is not commercially available, containing 5% (v/v) fetal bovine serum, 1 μ*M* insulin, 10 m*M* sodium pyruvate, the same antibiotics, and 25 m*M* Hepes/NaOH (pH 7.4) for 4 days, changing the medium once every day.[177-179] However, if the cells were cultured without dexamethasone, protein secretion from the cells is reduced.[178] Mouse liver parenchymal cells were also isolated and cultured according to the same procedure as for the rat. Mouse cells were able to be attached to a culture dish not pre-coated with collagen, and the cells were maintainable in the medium without dexamethasone.

3.1.1.2 Preparation and Culture of Rat Kidney Tubule Epithelial Cells

Kidneys of a rat (150 to 200 g) were perfused via the aorta *in situ* with Ca,Mg-free Hanks' BSS containing 0.5 m*M* EGTA and 50 m*M* Hepes, for which the pH was adjusted to 7.3, and then a collagenase solution (100 U/ml) in Earle's BSS. Digested tissue containing proximal tubules were separated from connective tissue and minced into small pieces. Minced tissue were redigested in a collagenase solution at 37°C for 5 to 10 min with gentle shaking, and digestion was stopped by adding one tenth volume of the culture medium (Waymouth's medium containing 10% (v/v) fetal bovine serum, 100 U/ml penicillin, and 100 μg/ml streptomycin). After filtration through gauze, fragments of proximal tubule isolated from one kidney were collected by centrifugation at 600× *g* for 1 min and were inoculated in 12 to 18 collagen-coated plastic dishes (diameter, 35 mm) with 1.5 ml of the culture medium. After incubation overnight in a 5% CO_2/95% air incubator at 37°C, the medium was changed. The cultures were continued with the medium being changed once every 2 days. Proximal tubule epithelial cells grew out from inoculated fragments, and the cells became confluent after incubation for 5 days. Resulting confluent cells were used for toxicity testing of metals. The confluent cells were maintainable in serum-free medium (e.g., Earle's BSS containing 10 m*M* Hepes/NaOH, pH 7.2) for a day.[180]

3.1.1.3 Primary Cultures of Rat Cerebellar Cells

Cerebellar cells were prepared from neonatal rats (within 24 hr of birth) according to the same method as for the mouse.[181] The cerebellum was freed from the meninges and cut into small pieces. After these pieces were digested with Ca,Mg-free Hanks' BSS containing trypsin and DNaseI, the cells were dissociated from the digested tissue. Isolated cells were maintained in MEM medium containing 1 mg/ml bovine serum albumin, 10 μg/ml insulin, 30 n*M* sodium selenite, 0.25% glucose, 100 U/ml penicillin G, and 135 μg streptomycin sulfate in Primaria® 24-well plates.[182]

3.1.2 Established Cell Line

Cell lines are established from spontaneously transformed cells or oncogene-transfected cells and are also obtained from tumor cells. These cells are considered to continue to divide infinitely on a culture dish and are sometimes called immortalized cells, for which the karyotype is often unstable. Cell lines are generally used for testing cytotoxicity and genotoxicity of metals, but certain cell lines keep the specific character (e.g., specific gene expression) of the original tissue or organ in an adequate medium or express it in the presence of a factor (e.g., hormone and cytokine). Such cells provide a convenient way to examine the toxic effect of metals on the tissue or the organ-specific function. Recently, cell lines have been established from a mutant animal and used to examine the toxic mechanisms of metals.[10,46,50,102]

Well-characterized cell lines (e.g., 3T3 [mouse embryonic cells],[36,79,119] CHO [Chinese hamster ovary cells],[9,33,56,60,70,80,102,113,158,183] LLC-PK_1 [pig kidney tubule epithelial cells],[45,115,184] MDCK [dog kidney tubule epithelial cells],[150,160,185] PC12 [pheochromocytoma cells],[74,138] and V79 [Chinese hamster lung cells][57,133,137,149,168,186]) are often used for testing metal toxicity. Cell lines and normal cells are distributed from major cell banks as follows.[170]

United States

1. American Type Culture Collection; Rockville, Maryland
2. Human Genetic Mutant Cell Culture and Aging Cell Culture Repository, Coriell Institute for Medical Research; Camden, New Jersey

Europe

3. Cell Bank, Center for Applied Microbiology and Research; Porton, Salisbury, England
4. Department of Medical Virology, Pasteur Institute; Paris, France

Japan

5. Cell Bank, Institute for Physical and Chemical Research (RIKEN); Wako, Saitama, Japan

The cell lines are shipped frozen on dry ice. Basic protocols for culturing the cells are described in the cell bank's manual (a culture medium, a freeze medium, growth characteristics, etc.).

3.1.3 Blood Cells

Blood cells (e.g., lymphocyte and leukocyte) are important target cells of metal toxicity, because these cells act as immunocompetent cells and are sensitive to genotoxic metal compounds. Therefore, adverse effects on immunoresponse and genotoxicity caused by metals have been intensively examined using the blood cells. Metallothioneins are also inducible in the blood cells, as well as in liver and kidney cells.[187] Lymphocytes, leukocytes, and other types of cells are easily separated from blood of experimental animals or human.[188] Macrophages (monocytes), which are not blood cells precisely, are collected from a peritoneal cavity or the lungs.[188] Macrophages are often exposed to the particles of metallic compounds for testing their toxicity. Blood cells are also used for exposure assessment, as well as for *in vitro* experimentation.

Cytotoxicity is a simple but important index for examining the toxicity of metals. Many methods for testing cytotoxicity (Trypan blue exclusion,[172] protein content in the cells,[171] lactate dehydrogenase leakage,[180,189] tetrazolium assay (MTT assay),[190,191] natural red assay,[139] and ATP content in cells[27,28,41,193]) have been established. These methods are generally based on disturbances of the integrity of the plasma membrane and the depression of energy production caused by metal compounds. Several metal ions (e.g., nickel, chromium, iron, and copper) are known to produce active oxygen spices to cause this dysfunction of the cells and genotoxicity. Production of active oxygen spices by the metals is estimated by measuring thiobarbituric acid (TBA) reactants[49,92,130,193] or dichlorofluoroscein fluorescence.[10,19,70,194] The amount of reduced glutathione in the cells is also decreased by the effect of metals. Glutathione content in the cell is determined after the metal is exposed to the cells.[10,24,33,43,49,50,116,119,121,122,139,168,195,196]

Genotoxicity of metallic compounds is often assessed using cultured cells. Primarily, the effect on the cell growth is simply determined by counting the cell number or by measuring [^{3}H] thymidine incorporation into DNA,[171,172,197] and the duration of the cell cycle is also estimated.[12,33,159] Genotoxocity is estimated in terms of DNA damage or abnormality of the chromosomes. Active oxygen spices produced by metals (e.g., nickel and chromium) damage genomic DNA. In order to determine the DNA damage, single-strand[72,120,159,198] and double-strand[15,20,64,199] DNA breaks and unscheduled DNA synthesis[120,200-203] have been measured. DNA adduct produced by active oxygen spices is a potent marker for determining DNA damage caused by a metal ion and metallic compounds.[20,109,204] Mutation on a marker gene caused by the metal has been determined,[21,71,133,202,205] and recently the combination of a high-fidelity polymerase chain reaction and denaturing gel electrophoresis of DNA has been used to detect a point mutation on the marker gene (e.g., hypoxanthine guanine phosphoribosyl transferase).[71]

For testing the abnormality of chromosomes, chromosomal aberrations,[102,110,111,159,201,202,206] sister chromatid exchange (SCE)[33,58,159,164,201,202,207] (these methods are cited in the *OECD Guidelines on Genetic Toxicology* and *WHO/Environmental Health Criteria*), aneuploidy of chromosomes,[8,111-114] and the formation of micronuclei[81,109,110,114,164] have been determined.

References

1. Steffensen, I. L., Mesna, O. J., Andruchow, E., Namork, E., Hylland, K., and Andersen, R. A., Cytotoxicity and accumulation of Hg, Ag, Cd, Cu, Pb and Zn in human peripheral T and B lymphocytes and monocytes *in vitro*, *Gen. Pharmacol.*, 25, 1621–1633, 1944.
2. Theocharis, S. E., Souliotis, V. L., and Panayiotidis, P. G., Suppression of interleukin-1 beta and tumour necrosis factor-alpha biosynthesis by cadmium in *in vitro* activated human peripheral blood mononuclear cells, *Arch. Toxicol.*, 69, 132–136, 1994.
3. Zeromski, J., Jezewska, E., Sikora, J., and Kasprzak, K. S., The effect of nickel compounds on immunophenotype and natural killer cell function of normal human lymphocytes, *Toxicology*, 97, 39–48, 1995.
4. Talukder, G. and Harrison, N. L., On the mechanism of modulation of transient outward current in cultured rat hippocampal neurons by di- and trivalent cations, *J. Neurophysiol.*, 73, 73–79, 1995.
5. Kaji, T., Suzuki, M., Yamamoto, C., Mishima, A., Sakamoto, M., and Kozuka, H., Severe damage of cultured vascular endothelial cell monolayer after simultaneous exposure to cadmium and lead, *Arch. Environ. Contam. Toxicol.*, 28, 168–172, 1995.
6. Donohue, V. E., McDonald, F., and Evans, R., *In vitro* cytotoxicity testing of neodymium-iron-boron magnets, *J. Appl. Biomater.*, 6, 69–74, 1995.
7. Nagy, K., Pasti, G., Bene, L., and Nagy, I., Involvement of Fenton reaction products in differentiation induction of K562 human leukemia cells, *Leuk. Res.*, 19, 203–212, 1995.
8. Migliore, L., Scarpato, R., and Falco, P., The use of fluorescent *in situ* hybridization with a beta-satellite DNA probe for the detection of acrocentric chromosomes in vanadium-induced micronuclei, *Cytogenet. Cell. Genet.*, 69, 215–219, 1995.
9. Ariza, M. E., Holliday, J., and Williams, M. V., Mutagenic effect of mercury (II) in eukaryotic cells, *In-Vivo*, 8, 559–563, 1994.

10. Lazo, J. S., Kondo, Y., Dellapiazza, D., Michalska, A. E., Choo, K. H., and Pitt, B. R., Enhanced sensitivity to oxidative stress in cultured embryonic cells from transgenic mice deficient in metallothionein I and II genes, *J. Biol. Chem.*, 270, 5506–5510, 1995.
11. Shiraishi, N., Hochadel, J. F., Coogan, T. P., Koropatnick, J., and Waalkes, M. P., Sensitivity to cadmium-induced genotoxicity in rat testicular cells is associated with minimal expression of the metallothionein gene, *Toxicol. Appl. Pharmacol.*, 130, 229–236, 1995.
12. Sahu, R. K., Katsifis, S. P., Kinney, P. L., and Christie, N. T., Ni(II) induced changes in cell cycle duration and sister-chromatid exchanges in cultured human lymphocytes, *Mutat. Res.*, 327, 217–225, 1995.
13. Gueniche, A., Viac, J., Lizard, G., Charveron, M., and Schmitt, D., Effect of various metals on intercellular adhesion molecule-1 expression and tumour necrosis factor alpha production by normal human keratinocytes, *Arch. Dermatol. Res.*, 286, 466–470, 1994.
14. Mishima, A., Kaji, T., Yamamoto, C., Sakamoto, M., and Kozuka, H., Zinc-induced tolerance to cadmium cytotoxicity without metallothionein induction in cultured bovine aortic endothelial cells, *Toxicol. Lett.*, 75, 85–92, 1995.
15. Hiraku, Y., Inoue, S., Oikawa, S., Yamamoto, K., Tada, S., Nishino, K., and Kawanishi, S., Metal-mediated oxidative damage to cellular and isolated DNA by certain tryptophan metabolites, *Carcinogenesis,* 16, 349–356, 1995.
16. Bondemark, L., Kurol, J., and Wennberg, A., Orthodontic rare earth magnets — *in vitro* assessment of cytotoxicity, *Br. J. Orthod.*, 21, 335–341, 1994.
17. Lee, K. B., Parker, R. J., and Reed, E., Effect of cadmium on human ovarian cancer cells with acquired cisplatin resistance, *Cancer Lett.*, 88, 57–66, 1995.
18. Lin, X. and Costa, M., Transformation of human osteoblasts to anchorage-independent growth by insoluble nickel particles, *Environ. Health Perspect.*, 102 (Suppl. 3), 289–292, 1994.
19. Huang, X., Zhuang, Z., Frenkel, K., Klein, C. B., and Costa, M., The role of nickel and nickel-mediated reactive oxygen species in the mechanism of nickel carcinogenesis, *Environ. Health Perspect.*, 102 (Suppl. 3), 281–284, 1994.
20. Kawanishi, S., Inoue, S., and Yamamoto, K., Active oxygen species in DNA damage induced by carcinogenic metal compounds, *Environ. Health Perspect.*, 102 (Suppl. 3), 17–20, 1994.
21. Haugen, A., Maehle, L., Mollerup, S., Rivedal, E., and Ryberg, D., Nickel-induced alterations in human renal epithelial cells, *Environ. Health Perspect.*, 102 (Suppl. 3), 117–118, 1994.
22. Snyder, J. W., Serroni, A., Savory, J., and Farber, J. L., The absence of extracellular calcium potentiates the killing of cultured hepatocytes by aluminum maltolate, *Arch. Biochem. Biophys.*, 316, 434–442, 1995.
23. Benders, A. A., Li, J., Lock, R. A., Bindels, R. J., Bonga, S. E., and Veerkamp, J. H., Copper toxicity in cultured human skeletal muscle cells: the involvement of Na^+/K^+-ATPase and the Na^+/Ca^{2+}-exchanger, *Pflugers Arch.*, 428, 461–467, 1994.
24. Kondo, Y., Kuo, S. M., Watkins, S. C., and Lazo, J. S., Metallothionein localization and cisplatin resistance in human hormone-independent prostatic tumor cell lines, *Cancer Res.*, 55, 474–477, 1995.
25. Garrel, C., Lafond, J. L., Guiraud, P., Faure, P., and Favier, A., Induction of production of nitric oxide in microglial cells by insoluble form of aluminium, *Ann. N.Y. Acad. Sci.*, 738, 455–461, 1994.
26. Koizumi, T., Yokota, T., Ohmori, S., Kumagai, H., and Suzuki, K. T., Protective effect of metallothionein on intracellular pH changes induced by cadmium, *Toxicology,* 95, 11–17, 1995.
27. Leduc, D., Gressier, B., Gosset, P., Lheureux, P., de-Vuyst, P., Wallaert, B., and Yernault, J. C., Oxidant radical release by alveolar macrophages after cadmium chloride exposure *in vitro*, *J. Appl. Toxicol.*, 14, 381–385, 1994.
28. Arsalane, K., Gosset, P., Hildebrand, H. F., Voisin, C., Tonnel, A. B., and Wallaert, B., Nickel hydroxy carbonate increases tumour necrosis factor alpha and interleukin 6 secretion by alveolar macrophages, *J. Appl. Toxicol.*, 14, 375–379, 1994.
29. Garbossa, G., Gutnisky, A., and Nesse, A., The inhibitory action of aluminum on mouse bone marrow cell growth: evidence for an erythropoietin- and transferrin-mediated mechanism, *Miner. Electrolyte Metabl.*, 20, 141–146, 1994.
30. Prehn, J. H., Bindokas, V. P., Marcuccilli, C. J., Krajewski, S., Reed, J. C., and Miller, R. J., Regulation of neuronal Bcl2 protein expression and calcium homeostasis by transforming growth factor type beta confers wide-ranging protection on rat hippocampal neurons, *Proc. Natl. Acad. Sci. USA,* 91, 12599–12603, 1994.
31. Lison, D. and Lauwerys, R., Cobalt bioavailability from hard metal particles. Further evidence that cobalt alone is not responsible for the toxicity of hard metal particles, *Arch. Toxicol.*, 68, 528–531, 1994.
32. Wiegant, F. A., Souren, J. E., van-Rijn, J., and van-Wijk, R., Stressor-specific induction of heat shock proteins in rat hepatoma cells, *Toxicology,* 94, 143–159, 1994.
33. Lynn, S., Yew, F. H., Hwang, J. W., Tseng, M. J., and Jan, K. Y., Glutathione can rescue the inhibitory effects of nickel on DNA ligation and repair synthesis, *Carcinogenesis,* 15, 2811–2816, 1994.
34. Kaji, T., Ohkawara, S., Inada, M., Yamamoto, C., Sakamoto, M., and Kozuka, H., Alteration of glycosaminoglycans induced by cadmium in cultured vascular smooth muscle cells, *Arch. Toxicol.*, 68, 560–565, 1994.

35. Kishimoto, T., Oguri, T., Ohno, M., Matsubara, K., Yamamoto, K., and Tada, M., Effect of cadmium (CdCl2) on cell proliferation and production of EDRF (endothelium-derived relaxing factor) by cultured human umbilical arterial endothelial cells, *Arch. Toxicol.,* 68, 555–559, 1994.
36. Shopsis, C., Antagonism of cadmium cytotoxicity by differentiation inducers, *Cell. Biol. Toxicol.,* 10, 191–205, 1994.
37. Blumenthal, S., Lewand, D., Sochanik, A., Krezoski, S., and Petering, D. H., Inhibition of Na^+-glucose cotransport in kidney cortical cells by cadmium and copper: protection by zinc, *Toxicol. Appl. Pharmacol.,* 129, 177–187, 1994.
38. Hamada, T., Tanimoto, A., Iwai, S., Fujiwara, H., and Sasaguri, Y., Cytopathological changes induced by cadmium-exposure in canine proximal tubular cells: a cytochemical and ultrastructural study, *Nephron,* 68, 104–111, 1994.
39. Koizumi, T., Yokota, T., and Suzuki, K. T., Mechanism of cadmium-induced cytotoxicity in rat hepatocytes. Cd-induced acidification causes alkalinization accompanied by membrane damage, *Biol. Trace. Elem. Res.,* 42, 31–41, 1994.
40. Hirano, S., Asami, T., Kodama, N., and Suzuki, K. T., Correlation between inflammatory cellular responses and chemotactic activity in bronchoalveolar lavage fluid following intratracheal instillation of nickel sulfate in rats, *Arch. Toxicol.,* 68, 444–449, 1994.
41. Kurbacher, C. M., Nagel, W., Mallmann, P., Kurbacher, J. A., Sass, G., Hubner, H., Andreotti, P. E., and Krebs, D., *In vitro* activity of titanocenedichloride in human renal cell carcinoma compared to conventional antineoplastic agents, *Anticancer Res.,* 14, 1529–1533, 1994.
42. Evans, E. J., Cell damage *in vitro* following direct contact with fine particles of titanium, titanium alloy and cobalt-chrome-molybdenum alloy, *Biomaterials,* 15, 713–717, 1994.
43. Steinebach, O. M. and Wolterbeek, H. T., Role of cytosolic copper, metallothionein and glutathione in copper toxicity in rat hepatoma tissue culture cells, *Toxicology,* 92, 75–90, 1994.
44. Koizumi, T., Yokota, T., Shirakura, H., Tatsumoto, H., and Suzuki, K. T., Potential mechanism of cadmium-induced cytotoxicity in rat hepatocytes: inhibitory action of cadmium on mitochondrial respiratory activity, *Toxicology,* 92, 115–125, 1994.
45. Liu, J., Liu, Y., and Klaassen, C. D., Nephrotoxicity of $CdCl_2$ and Cd-metallothionein in cultured rat kidney proximal tubules and LLC-PK1 cells, *Toxicol. Appl. Pharmacol.,* 128, 264–270, 1994.
46. Kim, J. K., Yamada, T., and Matsumoto, K., Copper cytotoxicity impairs DNA synthesis but not protein phosphorylation upon growth stimulation in LEC mutant rat, *Res. Comm. Chem. Pathol. Pharmacol.,* 84, 363–366, 1994.
47. Koh, J. Y. and Choi, D. W., Zinc toxicity on cultured cortical neurons: involvement of *N*-methyl-D-aspartate receptors, *Neuroscience,* 60, 1049–1057, 1994.
48. Neufeld, E. B. and Boskey, A. L., Strontium alters the complexed acidic phospholipid content of mineralizing tissues, *Bone,* 15, 425–430, 1994.
49. Teissier, E., Shirali, P., Hannothiaux, M. H., Marez, T., and Haguenoer, J. M., Interactions of alpha Ni3S2 with guinea pig alveolar macrophages and liberation of inflammatory mediators, *J. Appl. Toxicol.,* 14, 167–171, 1994.
50. Shertzer, H. G., Bannenberg, G. L., Zhu, H., Liu, R. M., and Moldeus, P., The role of thiols in mitochondrial susceptibility to iron and tert-butyl hydroperoxide-mediated toxicity in cultured mouse hepatocytes, *Chem. Res. Toxicol.,* 7, 358–366, 1994.
51. Roosen, N., Doz, F., Yeomans, K. L., Dougherty, D. V., and Rosenblum, M. L., Effect of pharmacologic doses of zinc on the therapeutic index of brain tumor chemotherapy with carmustine, *Cancer Chemother. Pharmacol.,* 34, 385–392, 1994.
52. Whelan, H. T., Williams, M. B., Bajic, D. M., Segura, A. D., McAuliff, T. L., and Chitambar, C. R., Prevention of gallium toxicity by hyperhydration in treatment of medulloblastoma, *Pediatr. Neurol.,* 10, 217–220, 1994.
53. Itoh, H., Shioda, T., Matsura, T., Koyama, S., Nakanishi, T., Kajiyama, G., and Kawasaki, T., Iron ion induces mitochondrial DNA damage in HTC rat hepatoma cell culture — role of antioxidants in mitochondrial DNA protection from oxidative stresses, *Arch. Biochem. Biophys.,* 313, 120–125, 1994.
54. Cohen, M. D., Yang, Z., and Zelikoff, J. T., Immunotoxicity of particulate lead: *in vitro* exposure alters pulmonary macrophage tumor necrosis factor production and activity, *J. Toxicol. Environ. Health,* 42, 377–392, 1994.
55. Pool-Zobel, B. L., Lotzmann, N., Knoll, M., Kuchenmeister, F., Lambertz, R., Leucht, U., Schroder, H. G., and Schmezer, P., Detection of genotoxic effects in human gastric and nasal mucosa cells isolated from biopsy samples, *Environ. Mol. Mutagen,* 24, 23–45, 1994.
56. Costa, M., Zhuang, Z., Huang, X., Cosentino, S., Klein, C. B., and Salnikow, K., Molecular mechanisms of nickel carcinogenesis, *Sci. Total. Environ.,* 148, 191–199, 1994.
57. Held, K. D. and Biaglow, J. E., Mechanisms for the oxygen radical-mediated toxicity of various thiol-containing compounds in cultured mammalian cells, *Radiat. Res.,* 139, 15–23, 1994.
58. Hartmann, A. and Speit, G., Comparative investigations of the genotoxic effects of metals in the single cells gel (SCG) assay and the sister chromatid exchange (SCE) test, *Environ. Mol. Mutagen,* 23, 299–305, 1994.

59. Bressler, J., Forman, S., and Goldstein, G. W., Phospholipid metabolism in neural microvascular endothelial cells after exposure to lead *in vitro*, *Toxicol. Appl. Pharmacol.*, 126, 352–360, 1994.
60. Zhuang, Z., Huang, X., and Costa, M., Protein oxidation and amino acid-DNA crosslinking by nickel compounds in intact cultured cells, *Toxicol. Appl. Pharmacol.*, 126, 319–325, 1994.
61. Bondemark, L., Kurol, J., and Wennberg, A., Biocompatibility of new, clinically used, and recycled orthodontic samarium-cobalt magnets, *Am. J. Orthod. Dentofacial. Orthop.*, 105, 568–574, 1994.
62. Abreo, K., Glass, J., Jain, S., and Sella, M., Aluminum alters the compartmentalization of iron in Friend erythroleukemia cells, *Kidney Int.*, 45, 636–641, 1994.
63. Bylander, J. E., Li, S. L., Sens, M. A., Hazen-Martin, D., Re, G. G., and Sens, D. A., Induction of metallothionein mRNA and protein following exposure of cultured human proximal tubule cells to cadmium, *Toxicol. Lett.*, 71, 111–122, 1994.
64. Kappus, H. and Reinhold, C., Heavy metal-induced cytotoxicity to cultured human epidermal keratinocytes and effects of antioxidants, *Toxicol. Lett.*, 71, 105–109, 1994.
65. el-Azzouzi, B., Tsangaris, G. T., Pellegrini, O., Manuel, Y., Benveniste, J., and Thomas, Y., Cadmium induces apoptosis in a human T cell line, *Toxicology*, 88, 127–139, 1994.
66. Vorbrodt, A. W., Trowbridge, R. S., and Dobrogowska, D. H., Cytochemical study of the effect of aluminium on cultured brain microvascular endothelial cells, *Histochem. J.*, 26, 119–126, 1994.
67. Bowser, D. H., Frenkel, K., and Zelikoff, J. T., Effects of *in vitro* nickel exposure on the macrophage-mediated immune functions of rainbow trout (*Oncorhynchus mykiss*), *Bull. Environ. Contam. Toxicol.*, 52, 367–373, 1994.
68. Geertz, R., Gulyas, H., and Gercken, G., Cytotoxicity of dust constituents towards alveolar macrophages: interactions of heavy metal compounds, *Toxicology*, 86, 13–27, 1994.
69. Takeda, A., Norris, J. S., Iversen, P. L., and Ebadi, M., Antisense oligonucleotide of *c-myc* discriminates between zinc- and dexamethasone-induced synthesis of metallothionein, *Pharmacology*, 48, 119–126, 1994.
70. Huang, X., Klein, C. B., and Costa, M., Crystalline Ni3S2 specifically enhances the formation of oxidants in the nuclei of CHO cells as detected by dichlorofluorescein, *Carcinogenesis*, 15, 545–548, 1994.
71. Chen, J. and Thilly, W. G., Mutational spectrum of chromium(VI) in human cells, *Mutat. Res.*, 323, 21–27, 1994.
72. Coogan, T. P., Bare, R. M., Bjornson, E. J., and Waalkes, M. P., Enhanced metallothionein gene expression is associated with protection from cadmium-induced genotoxicity in cultured rat liver cells, *J. Toxicol. Environ. Health*, 41, 233–245, 1994.
73. Sens, M. A., Hazen-Martin, D. J., Bylander, J. E., and Sens, D. A., Heterogeneity in the amount of ionic cadmium necessary to elicit cell death in independent cultures of human proximal tubule cells, *Toxicol. Lett.*, 70, 185–191, 1994.
74. Hinkle, P. M. and Osborne, M. E., Cadmium toxicity in rat pheochromocytoma cells: studies on the mechanism of uptake, *Toxicol. Appl. Pharmacol.*, 124, 91–98, 1994.
75. Goebeler, M., Roth, J., Meinardus-Hager, G., and Sorg, C., The contact allergens nickel chloride and cobalt chloride directly induce expression of endothelial adhesion molecules, *Behring Inst. Mitt.*, Aug. (92), 191–201, 1993.
76. Sass, J. B., Ang, L. C., and Juurlink, B. H., Aluminum pretreatment impairs the ability of astrocytes to protect neurons from glutamate mediated toxicity, *Brain Res.*, 621, 207–214, 1993.
77. Toussaint, M. J. and Nederbragt, H., Copper and zinc toxicity in two rat hepatoma cell lines varying in differentiation, *Comp. Biochem. Physiol. C*, 104, 253–262, 1993.
78. Burns, L. A. and Munson, A. E., Gallium arsenide selectively inhibits T cell proliferation and alters expression of CD25 (IL-2R/p55), *J. Pharmacol. Exp. Ther.*, 265, 178–186, 1993.
79. Wataha, J. C., Hanks, C. T., and Craig, R. G., The effect of cell monolayer density on the cytotoxicity of metal ions which are released from dental alloys, *Dent. Mater.*, 9, 172–176, 1993.
80. Costa, M., Molecular targets of nickel and chromium in human and experimental systems, *Scand. J. Work Environ. Health*, 19 (Suppl. 1), 71–74, 1993.
81. Berces, J., Otos, M., Szirmai, S., Crane-Uruena, C., and Koteles, G. J., Using the micronucleus assay to detect genotoxic effects of metal ions, *Environ. Health Perspect.*, 101 (Suppl. 3), 11–13, 1993.
82. Horiguchi, H., Mukaida, N., Okamoto, S., Teranishi, H., Kasuya, M., and Matsushima, K., Cadmium induces interleukin-8 production in human peripheral blood mononuclear cells with the concomitant generation of superoxide radicals, *Lymphokine Cytokine Res.*, 12, 421–428, 1993.
83. Guilianelli, C., Baeza-Squiban, A., Boisvieux-Ulrich, E., Houcine, O., Zalma, R., Guennou, C., Pezerat, H., and Marano, F., Effect of mineral particles containing iron on primary cultures of rabbit tracheal epithelial cells: possible implication of oxidative stress, *Environ. Health Perspect.*, 101, 436–442, 1993.
84. Santini, M. T., Paradisi, S., Straface, E., and Malorni, W., Cesium ions influence cultured cell behavior by modifying specific subcellular components: the role of membranes and of the cytoskeleton, *Cell. Biol. Toxicol.*, 9, 295–306, 1993.
85. Mgbonyebi, O. P., Smothers, C. T., and Mrotek, J. J., Modulation of adrenal cell functions by cadmium salts. I. Cadmium chloride effects on basal and ACTH-stimulated steroidogenesis, *Cell. Biol. Toxicol.*, 9, 223–234, 1993.

86. Tomsig, J. L. and Suszkiw, J. B., Intracellular mechanism of Pb^{2+}-induced norepinephrine release from bovine chromaffin cells, *Am. J. Physiol.*, 265(6, Pt. 1), C1630–1636, 1993.
87. Piersma, A. H., Roelen, B., Roest, P., Haakmat-Hoesenie, A. S., van-Achterberg, T. A., and Mummery, C. L., Cadmium-induced inhibition of proliferation and differentiation of embryonal carcinoma cells and mechanistic aspects of protection by zinc, *Teratology*, 48, 335–341, 1993.
88. Kodavanti, P. R., Mundy, W. R., Tilson, H. A., and Harry, G. J., Effects of selected neuroactive chemicals on calcium transporting systems in rat cerebellum and on survival of cerebellar granule cells, *Fundam. Appl. Toxicol.*, 21, 308–316, 1993.
89. Lang, D. S., Meier, K. L., and Luster, M. I., Comparative effects of immunotoxic chemicals on *in vitro* proliferative responses of human and rodent lymphocytes, *Fundam. Appl. Toxicol.*, 21, 535–545, 1993.
90. Yamamoto, C., Kaji, T., Sakamoto, M., and Kozuka, H., Cadmium stimulation of plasminogen activator inhibitor-1 release from human vascular endothelial cells in culture, *Toxicology*, 83, 215–223, 1993.
91. Kern, M., Audesirk, T., and Audesirk, G., Effects of inorganic lead on the differentiation and growth of cortical neurons in culture, *Neurotoxicology*, 14, 319–327, 1993.
92. Fischbach, M., Sabbioni, E., and Bromley, P., Induction of the human growth hormone gene placed under human hsp70 promoter control in mouse cells: a quantitative indicator of metal toxicity, *Cell. Biol. Toxicol.*, 9, 177–188, 1993.
93. Busselberg, D., Platt, B., Haas, H. L., and Carpenter, D. O., Voltage gated calcium channel currents of rat dorsal root ganglion (DRG) cells are blocked by Al^{3+}, *Brain Res.*, 622, 163–168, 1993.
94. Huang, J., Tanii, H., Kato, K., and Hashimoto, K., Neuron and glial cell marker proteins as indicators of heavy metal-induced neurotoxicity in neuroblastoma and glioma cell lines, *Arch. Toxicol.*, 67, 491–496, 1993.
95. Jaramillo, A. and Sonnenfeld, G., Effect of nickel sulfide on induction of interleukin-1 and phagocytic activity, *Environ. Res.*, 63, 16–25, 1993.
96. Angle, C. R., Thomas, D. J., and Swanson, S. R., Osteotoxicity of cadmium and lead in HOS TE 85 and ROS 17/2.8 cells: relation to metallothionein induction and mitochondrial binding, *Biometals*, 6, 179–184, 1993.
97. Lison, D. and Lauwerys, R., Evaluation of the role of reactive oxygen species in the interactive toxicity of carbide-cobalt mixtures on macrophages in culture, *Arch. Toxicol.*, 67, 347–351, 1993.
98. Berg, I., Schluter, T., and Gercken, G., Increase of bovine alveolar macrophage superoxide anion and hydrogen peroxide release by dusts of different origin, *J. Toxicol. Environ. Health*, 39, 341–354, 1993.
99. Shi, B., Chou, K., and Haug, A., Aluminium impacts elements of the phosphoinositide signalling pathway in neuroblastoma cells, *Mol. Cell. Biochem.*, 121, 109–118, 1993.
100. Iwami, K. and Moriyama, T., Comparative effect of cadmium on osteoblastic cells and osteoclastic cells, *Arch. Toxicol.*, 67, 352–357, 1993.
101. Bauman, J. W., Liu, J., and Klaassen, C. D., Production of metallothionein and heat-shock proteins in response to metals, *Fundam. Appl. Toxicol.*, 21, 15–22, 1993.
102. Yamada, H., Miyahara, T., and Sasaki, Y. F., Inorganic cadmium increases the frequency of chemically induced chromosome aberrations in cultured mammalian cells, *Mutat. Res.*, 302, 137–145, 1993.
103. Parkes, J. G., Hussain, R. A., Olivieri, N. F., and Templeton, D. M., Effects of iron loading on uptake, speciation, and chelation of iron in cultured myocardial cells, *J. Lab. Clin. Med.*, 122, 36–47, 1993.
104. Maloney, W. J., Smith, R. L., Castro, F., and Schurman, D. J., Fibroblast response to metallic debris *in vitro*. Enzyme induction cell proliferation, and toxicity, *J. Bone. Joint. Surg. Am.*, 75, 835–844, 1993.
105. Haynes, D. R., Rogers, S. D., Hay, S., Pearcy, M. J., and Howie, D. W., The differences in toxicity and release of bone-resorbing mediators induced by titanium and cobalt-chromium-alloy wear particles, *J. Bone. Joint. Surg. Am.*, 75, 825–834, 1993.
106. Cohen, M. D., Parsons, E., Schlesinger, R. B., and Zelikoff, J. T., Immunotoxicity of *in vitro* vanadium exposures: effects on interleukin-1, tumor necrosis factor-alpha, and prostaglandin E2 production by WEHI-3 macrophages, *Int. J. Immunopharmacol.*, 15, 437–446, 1993.
107. Wan, M., Huuziker, P. E., and Kagi, J. H., Induction of metallothionein synthesis by cadmium and zinc in cultured rabbit kidney cells (RK-13), *Biochem. J.*, 292, 609–615, 1993.
108. Eichholtz-Wirth, H., Reidel, G., and Hietel, B., Radiation-induced transient cisplatin resistance in murine fibrosarcoma cells associated with elevated metallothionein content, *Br. J. Cancer*, 67, 1001–1006, 1993.
109. Zhang, L., Robertson, M. L., Kolachana, P., Davison, A. J., and Smith, M. T., Benzene metabolite, 1,2,4-benzenetriol, induces micronuclei and oxidative DNA damage in human lymphocytes and HL60 cells, *Environ. Mol. Mutagen.*, 21, 339–348, 1993.
110. Lynch, A. M. and Parry, J. M., The cytochalasin-B micronucleus/kinetochore assay *in vitro*: studies with 10 suspected aneugens, *Mutat. Res.*, 287, 71–86, 1993.
111. Sbrana, I., Di-Sibio, A., Lomi, A., and Scarcelli, V., C-mitosis and numerical chromosome aberration analyses in human lymphocytes: 10 known or suspected spindle poisons, *Mutat. Res.*, 287, 57–70, 1993.
112. Natarajan, A. T., Dulivenvoorden, W. C., Meijers, M., and Zwanenburg, T. S., Induction of mitotic aneuploidy using Chinese hamster primary embryonic cells. Test results of 10 chemicals, *Mutat. Res.*, 287, 47–56, 1993.
113. Warr, T. J., Parry, E. M., and Parry, J. M., A comparison of two *in vitro* mammalian cell cytogenetic assays for the detection of mitotic aneuploidy using 10 known or suspected aneugens, *Mutat. Res.*, 287, 29–46, 1993.

114. Natarajan, A. T., An overview of the results of testing of known or suspected aneugens using mammalian cells *in vitro*, *Mutat. Res.*, 287, 113–118, 1993.
115. Prozialeck, W. C. and Lamar, P. C., Surface binding and uptake of cadmium (Cd^{2+}) by LLC-PK1 cells on permeable membrane supports, *Arch. Toxicol.*, 67, 113–119, 1993.
116. Wan, X., Lachapelle, M., Marion, M., Fournier, M., and Denizeau, F., Recovery potential of hepatocytes from inhibition of albumin secretion by cadmium, *J. Toxicol. Environ. Health*, 38, 381–392, 1993.
117. Kaji, T., Mishima, A., Yamamoto, C., Sakamoto, M., and Kozuka, H., Zinc protection against cadmium-induced destruction of the monolayer of cultured vascular endothelial cells, *Toxicol. Lett.*, 66, 247–255, 1993.
118. Waseem, M., Bajpai, R., and Kaw, J. L., Reaction of pulmonary macrophages exposed to nickel and cadmium *in vitro*, *J. Environ. Pathol. Toxicol. Oncol.*, 12, 47–54, 1993.
119. Li, W., Zhao, Y., and Chou, I. N., Alterations in cytoskeletal protein sulfhydryls and cellular glutathione in cultured cells exposed to cadmium and nickel ions, *Toxicology*, 77, 65–79, 1993.
120. Gao, M., Binks, S. P., Chipman, J. K., and Levy, L. S., Hexavalent chromium produces DNA strand breakage but not unscheduled DNA synthesis at sub-cytotoxic concentrations in hepatocytes, *Toxicology*, 77, 171–180, 1993.
121. Chin, T. A. and Templeton, D. M., Protective elevations of glutathione and metallothionein in cadmium-exposed mesangial cells, *Toxicology*, 77, 145–156, 1993.
122. Steinebach, O. M. and Wolterbeek, H. T., Effects of zinc on rat hepatoma HTC cells and primary cultured rat hepatocytes, *Toxicol. Appl. Pharmacol.*, 118, 245–254, 1993.
123. Brookes, N., *In vitro* evidence for the role of glutamate in the CNS toxicity of mercury, *Toxicology*, 76, 245–256, 1992.
124. Kang, Y. J., Exogenous glutathione decreases cellular cadmium uptake and toxicity, *Drug. Metab. Dispos.*, 20, 714–718, 1992.
125. Langlois, C. J. and Calabrese, E. J., The interactive effect of chlorine, copper and nitrite on methaemoglobin formation in red blood cells of Dorset sheep, *Hum. Exp. Toxicol.*, 11, 223–228, 1992.
126. Swanson, R. A. and Sharp, F. R., Zinc toxicity and induction of the 72 kD heat shock protein in primary astrocyte culture, *Glia*, 6, 198–205, 1992.
127. Selgrade, M. K., Daniels, M. J., and Dean, J. H., Correlation between chemical suppression of natural killer cell activity in mice and susceptibility to cytomegalovirus: rationale for applying murine cytomegalovirus as a host resistance model and for interpreting immunotoxicity testing in terms of risk of disease, *J. Toxicol. Environ. Health*, 37, 123–137, 1992.
128. Fowler, B. A. and Akkerman, M., The role of Ca^{2+} in cadmium-induced renal tubular cell injury, *IARC Sci. Publ.*, 118, 271–277, 1992.
129. Putters, J. L., Kaulesar-Sukul, D. M., de Zeeuw, G. R., Bijma, A., and Besselink, P. A., Comparative cell culture effects of shape memory metal (Nitinol), nickel and titanium: a biocompatibility estimation, *Eur. Surg. Res.*, 24, 378–382, 1992.
130. Latour, I., Pregaldien, J. L., and Buc-Calderon, P., Cell death and lipid peroxidation in isolated hepatocytes incubated in the presence of hydrogen peroxide and iron salts, *Arch. Toxicol.*, 66, 743–749, 1992.
131. Kishimoto, T., Fukuzawa, Y., Abe, M., Hashimoto, M., Ohno, M., and Tada, M., Injury to cultured human vascular endothelial cells by copper ($CuSO_4$), *Nippon-Eiseigaku-Zasshi*, 47, 965–970, 1992.
132. Moffatt, P., Marion, M., and Denizeau, F., Cadmium-2-acetylaminofluorene interaction in isolated rat hepatocytes, *Cell Biol. Toxicol.*, 8, 277–290, 1992.
133. Roy, N. K. and Rossman, T. G., Mutagenesis and comutagenesis by lead compounds, *Mutat. Res.*, 298, 97–103, 1992.
134. Takahashi, S., Yamada, M., Kondo, T., Sato, H., Furuya, K., and Tanaka, I., Cytotoxicity of nickel oxide particles in rat alveolar macrophages cultured *in vitro*, *J. Toxicol. Sci.*, 17, 243–251, 1992.
135. Ikarashi, Y., Ohno, K., Tsuchiya, T., and Nakamura, A., Differences of draining lymph node cell proliferation among mice, rats and guinea pigs following exposure to metal allergens, *Toxicology*, 76, 283–292, 1992.
136. Kaji, T., Mishima, A., Koyanagi, E., Yamamoto, C., Sakamoto, M., and Kozuka, H., Possible mechanism for zinc protection against cadmium cytotoxicity in cultured vascular endothelial cells, *Toxicology*, 76, 257–270, 1992.
137. Kasten, U., Hartwig, A., and Beyersmann, D., Mechanisms of cobalt(II) uptake into V79 Chinese hamster cells, *Arch. Toxicol.*, 66, 592–597, 1992.
138. Giridhar, J., Rathinavelu, A., and Isom, G. E., Interaction of cadmium with atrial natriuretic peptide receptors: implications for toxicity, *Toxicology*, 75, 133–143, 1992.
139. Liu, Y., Cotgreave, I., Atzori, L., and Grafstrom, R. C., The mechanism of Hg^{2+} toxicity in cultured human oral fibroblasts: the involvement of cellular thiols, *Chem. Biol. Interact.*, 85, 69–78, 1992.
140. Liu, J., Kershaw, W. C., Liu, Y. P., and Klaassen, C. D., Cadmium-induced hepatic endothelial cell injury in inbred strains of mice, *Toxicology*, 75, 51–62, 1992.
141. Uemura, E., Minachi, M., and Lartius, R., Enhanced neurite growth in cultured neuroblastoma cells exposed to aluminum, *Neurosci. Lett.*, 142, 171–174, 1992.
142. Ujihara, H. and Albuquerque, E. X., Developmental change of the inhibition by lead of NMDA-activated currents in cultured hippocampal neurons, *J. Pharmacol. Exp. Ther.*, 263, 868–875, 1992.

143. Wataha, J. C., Hanks, C. T., and Craig, R. G., *In vitro* synergistic, antagonistic, and duration of exposure effects of metal cation on eukaryotic cells, *J. Biomed. Mater. Res.*, 26, 1297–1309, 1992.
144. Contrino, J., Kosuda, L. L., Marucha, P., Kreutzer, D. L., and Bigazzi, P. E., The *in vitro* effects of mercury on peritoneal leukocytes (PMN and macrophages) from inbred brown Norway and Lewis rats, *Int. J. Immunopharmacol.*, 14, 1051–1059, 1992.
145. Akman, S. A., Doroshow, J. H., and Kensler, T. W., Copper-dependent site-specific mutagenesis by benzoyl peroxide in the supF gene of the mutation reporter plasmid pS189, *Carcinogenesis,* 13, 1783–1787, 1992.
146. Kaji, T., Fujiwara, Y., Koyanagi, E., Yamamoto, C., Mishima, A., Sakamoto, M., and Kozuka, H., Protective effect of copper against cadmium cytotoxicity on cultured vascular endothelial cells, *Toxicol. Lett.*, 63, 13–20, 1992.
147. Sauk, J. J., Smith, T., Silbergeld, E.K., Fowler, B. A., and Somerman, M. J., Lead inhibits secretion of osteonectin/SPARC without significantly altering collagen or Hsp47 production in osteoblast-like ROS 17/2.8 cells, *Toxicol. Appl. Pharmacol.*, 116, 240–247, 1992.
148. McNamara, J. R., Heithersay, G. S., and Wiebkin, O. W., Cell responses to Hydron by a new *in vitro* method, *Int. Endod. J.*, 25, 205–212, 1992.
149. Brown, R. C., Sara, E. A., Hoskins, J. A., Evans, C. E., Young, J., Laskowski, J. J., Acheson, R., Forder, S. D., and Rood, A. P., The effects of heating and devitrification on the structure and biological activity of aluminosilicate refractory ceramic fibres, *Ann. Occup. Hyg.*, 36, 115–129, 1992.
150. Mills, J. W., Zhou, J. H., Cardoza, L., and Ferm, V. H., Zinc alters actin filaments in Madin-Darby canine kidney cells, *Toxicol. Appl. Pharmacol.*, 116, 92–100, 1992.
151. Chin, T. A. and Templeton, D. M., Effects of CdC12 and Cd-metallothionein on cultured mesangial cells, *Toxicol. Appl. Pharmacol.*, 116, 133–141, 1992.
152. Kafer, A., Zoltzer, H., and Krug, H. F., The stimulation of arachidonic acid metabolism by organic lead and tin compounds in human HL-60 leukemia cells, *Toxicol. Appl. Pharmacol.*, 116, 125–132, 1992.
153. Stark, M., Wolff, J. E., and Korbmacher, A., Modulation of glial cell differentiation by exposure to lead and cadmium, *Neurotoxicol. Teratol.*, 14, 247–252, 1992.
154. Zatta, P., Perazzolo, M., Facci, L., Skaper, S. D., Corain, B., and Favarato, M., Effects of aluminum speciation on murine neuroblastoma cells, *Mol. Chem. Neuropathol.*, 16, 11–22, 1992.
155. Winstel, C. and Callahan, P., Cadmium exposure inhibits the prolactin secretory response to thyrotrophin releasing hormone (TRH) *in vitro*, *Toxicology,* 74, 9–17, 1992.
156. Atterwill, C. K., Johnston, H., and Thomas, S. M., Models for the *in vitro* assessment of neurotoxicity in the nervous system in relation to xenobiotic and neurotrophic factor-mediated events, *Neurotoxicology,* 13, 39–53, 1992.
157. Thoren, S. A., Calorimetry: a new quantitative *in vitro* method in cell toxicology. A dose/effect study of alveolar macrophages exposed to particles, *J. Toxicol. Environ. Health,* 36, 307–318, 1992.
158. Xu, J., Wise, J. P., and Patierno, S. R., DNA damage induced by carcinogenic lead chromate particles in cultured mammalian cells, *Mutat. Res.*, 280, 129–135, 1992.
159. Ganguly, B. B., Talukdar, G., and Sharma, A., Cytotoxicity of tin on human peripheral lymphocytes *in vitro*, *Mutat. Res.*, 282, 61–67, 1992.
160. Quamme, G. A., Free cadmium activity in renal epithelial cells is enhanced by Mg^{2+} depletion, *Kidney Int.*, 41, 1237–1244, 1992.
161. Christensen, M. M., Ernst, E., and Ellermann-Eriksen, S., Cytotoxic effects of hexavalent chromium in cultured murine macrophages, *Arch. Toxicol.*, 66, 347–353, 1992.
162. Kaji, T., Yamamoto, C., Sakamoto, M., and Kozuka, H., Inhibitory effect of lead on the release of tissue plasminogen activator from human vascular endothelial cells in culture, *Toxicology,* 73, 219–227, 1992.
163. Sorek, N. and Meiri, H., Aluminium modifies the electrical response of neuroblastoma cells to a short hypertonic pulse, *Arch. Toxicol.*, 66, 90–94, 1992.
164. Arrouijal, F. Z., Marzin, D., Hildebrand, H. F., Pestel, J., and Haguenoer, J. M., Differences in genotoxic activity of alpha-Ni3S2 on human lymphocytes from nickel-hypersensitized and nickel-unsensitized donors, *Mutagenesis,* 7, 183–187, 1992.
165. Nordlind, K., Bondesson, L., Agerberth, B., and Mutt, V., Protecting effects of vasoactive intestinal polypeptide on lymphocytes against metal toxicity, *Immunopharmacol. Immunotoxicol.*, 14, 323–330, 1992.
166. Gregotti, C., Di-Nucci, A., Costa, L. G., Manzo, L., Scelsi, R., Berte, F., and Faustman, E. M., Effects of thallium on primary cultures of testicular cells, *J. Toxicol. Environ. Health,* 36, 59–69, 1992.
167. Lison, D. and Lauwerys, R., Study of the mechanism responsible for the elective toxicity of tungsten carbide-cobalt powder toward macrophages, *Toxicol. Lett.*, 60, 203–210, 1992.
168. Chubatsu, L. S., Gennari, M., and Meneghini, R., Glutathione is the antioxidant responsible for resistance to oxidative stress in V79 Chinese hamster fibroblasts rendered resistant to cadmium, *Chem. Biol. Interact.*, 82, 99–110, 1992.
169. Kaji, T., Mishima, A., Yamamoto, C., Sakamoto, M., and Koizumi, F., Effect of cadmium on the monolayer maintenance of vascular endothelial cells in culture, *Toxicology,* 71, 267–276, 1992.
170. Hay, R., Caputo, J., Chen, T. R., Macy, M., McClintock, P., and Reid, Y., Eds., *ATCC Cell Lines and Hybridomas,* 8th ed., American Type Culture Collection, Rochville, 1994.

171. Freshny, R. I., Ed., *Animal Cell Culture, A Practical Approach,* IRL Press, Oxford, 1986.
172. Jakoby, W. B. and Pastan, I. H., *Methods in Enzymology,* Vol. LVIII, *Cell Culture,* Academic Press, New York, 1979.
173. Seglen, P. O., Preparation of isolated rat liver cells, *Methods Cell Biol.,* 13, 29–83, 1976.
174. Aoki, Y., Satoh, K., Sato, K., and Suzuki, K. T., Induction of glutathione S-transferase in primary cultured rat liver parenchymal cells by co-planar polychlorinated biphenyl congeners, *Biochem. J.,* 281, 539–543, 1992.
175. Aoki, Y., Matsumoto, M., and Suzuki, K. T., Expression of glutathione S-transferase in primary cultured rat liver parenchymal cells by coplanar polychlorinated biphenyl congeners is suppressed by protein kinase inhibitors and dexamethason, *FEBS Lett.,* 333, 114–118, 1993.
176. Tanaka, K., Sato, M., Tomita, Y., and Ichihara, A., Biochemical studies on liver function in primary cultured hepatocyted of adult rats. I. Hormonal effects on cell viability and protein synthesis, *J. Biochem.,* 84, 937–946, 1978.
177. Aoki, Y. and Suzuki, K. T., Excretion process of copper from copper preloaded rat liver parenchymal cells, *Biochem. Pharmacol.,* 34, 1713–1716, 1985.
178. Mitane, Y., Aoki, Y., and Suzuki, K. T., Cadmium inhibits protein secretion from cultured rat liver parenchymal cells, *Biochem. Pharmacol.,* 36, 2647–2652, 1987.
179. Mitane, Y., Aoki, Y., and Suzuki, K. T., Accumulation of newly synthesized serum proteins by cadmium in cultured rat liver parenchymal cells, *Biochem. Pharmacol.,* 36, 3657–3661, 1987.
180. Aoki, Y., Lipsky, M. M., and Fowler, B. A., Alteration in protein synthesis in primary cultures of rat kidney proximal tubule epithelial cells by exposure to gallium, indium and arsenite, *Toxicol. Appl. Pharmacol.,* 106, 462–468, 1990.
181. Fischer, G., Cultivation of mouse cerebellar cells in serum free, hormonally defined media: survival of neurons, *Neurosci. Lett.,* 28, 325–329, 1982.
182. Kunimoto, M., Aoki, Y., Shibata, K., and Miura, T., Differential cytotoxic effects of methylmercury and organotin compounds on mature and immature neural cells and non-neuronal cells *in vitro, Toxicol. In Vitro,* 6, 349–355, 1992.
183. Kotkow, K. J., Roth, D. A., Porter, T. J., Furie, B. C., and Furie, B., Role of propeptide in vitamin K-dependent Ganma-carboxylation, in *Methods in Enzymology,* Vol. 222, Lorand, L. and Mann, K. G., Eds., Academic Press, San Diego, 1993, p. 439.
184. Napoli, J. L., Martin, C. A., and Horst, R. L., Induction, inhibition, and analysis of vitamin D metabolism in cultured cells, in *Methods in Enzymology,* Vol. 206, Waterman, M. R. and Johnson, E. F., Eds., Academic Press, San Diego, 1991, p. 493.
185. Podbiliewicz, B. and Mellman, I., Reconstitution of endocytosis and recycling using perforated madin-Darby canine kidney cells, in *Methods in Enzymology,* Vol. 219, Rothman, J. E., Ed., Academic Press, San Diego, 1992, p. 201.
186. Doehmer, J. and Oesch, F., V79 Chinese hamster cells genetically engineered for stable expression of cytochromes P-450, in *Methods in Enzymology,* Vol. 206, Waterman, M. R. and Johnson, E. F., Eds., Academic Press, San Diego, 1991, p. 120.
187. Kimura, M., Metallothioneins of monocytes and lymphocytes, in *Methods in Enzymology,* Vol. 205, Riordan, J. F. and Vallee, B. L., Eds., Academic Press, San Diego, 1991, p. 291.
188. Weir, D. M., Ed., *Handbook of Experimental Immunology,* Vol. 2, *Cellular Immunology,* 3rd ed., Blackwell Scientific Publications, London, 1978.
189. Wroblewski, F. and LaDue, J. S., Lactic dehydrogenase activity in blood, *Proc. Soc. Exp. Biol. Med.,* 90, 210–213, 1955.
190. Alley, M. C., Scudiero, D. M., Monks, A., Hursey, M. L., Czerwinski, M. J., Fine, D. L., Abbott, B. J., Mayo, J. G., Shoemaker, R. H., and Boyd, M. R., Feasibility of drug screening with panels of human tumor cell lines using a microculture tetrazolium assay, *Cancer Res.,* 48, 589–601, 1988.
191. Scudiero, D. M., Shoemaker, R. H., Paull, K. D., Monks, A., Tierney, S., Nofziger, T. H., Currens, M. J., Seniff, D., and Boyd, M. R., Evaluation of a soluble tetrazolium/formazan assay for cell growth and drug sensitivity in culture using human and other tumor cell cines, *Cancer Res.,* 48, 4827–4833, 1988.
192. McElroy, W. D. and Seliger, H. H., The chemistry of light emission, *Adv. Enzymol.,* 25, 119, 1963.
193. Wong, S. H. Y., Knight, J. A., Hopfer, S. M., Zaharia, O., Leach, Jr., C. N., and Sunderman, Jr., F. W., Lipoperoxides in plasma as measured by liquid-chromatographic separation of malondialdehyde-thiobarbituric acid adduct, *Clin. Chem.,* 33, 214–220, 1987.
194. Frenkel, K. and Gleichauf, C., Hydrogen peroxide formation by cells treated with a tumor promoter, *Free Radic. Res. Comm.,* 12–13, 783–794, 1991.
195. Hissin, P. J. and Hilf, R., A fluorometric method for determination of oxidized and reduced glutathione in tissues, *Anal. Biochem.,* 74, 214–226, 1976.
196. Issel, R. D. and Nagele, A., Influence of thiols on thermosensitivity of mammalian cells *in vitro,* in *Methods in Enzymology,* Vol. 186, Packer, L. and Glazer, A. N., Eds., Academic Press, San Diego. 1990, p. 696.
197. Krezoski, S. K., Show, III, C. F., and Petering, D. H., Role of metallothionein in essential, toxic, and therapeutic metal metabolism in Ehrlich cells, in *Methods in Enzymology,* Vol. 205, Riordan, J. F. and Vallee, B. L., Eds., Academic Press, San Diego, 1991, p. 311.

198. Patierno, S. R., Sugiyama, M., Basilion, J. P., and Costa, M., Preferential DNA-protein cross-linking by $NiCl_2$ in magnesium-insoluble regions of fractionated Chinese hamster ovarty cell chromatin, *Cancer Res.*, 45, 5787–5794, 1985.
199. Ito, K., Yamamoto, K., and Kawanishi, S., Manganese-mediated oxidative damage of cellular and isolated DNA by isoniazid and related hydrazines: non-febton-type hydroxyl radical formation, *Biochemistry*, 31, 11606–11613, 1992.
200. Chipman, J. K. and Davies, J. E., Reduction of 2-acetylaminofluorene-induced unscheduled DNA synthesis in human and rat hepatocytes by butylated hydroxytoluene, *Mut. Res.*, 207, 193–198, 1988.
201. WHO, *Guide to Short-Term Tests for Detecting Mutagenic and Carcinogenic Chemicals*, Environmental Health Criteria 51, World Health Organization, Geneva, 1985.
202. Venitt, S. and Parry, J. M., Eds., *Mutagenicity Testing, A Practical Approach*, IRL Press, Oxford, 1984.
203. OECD, *Guidelines for Testing of Chemicals*, No. 482, *DNA Damage and Repair, Unscheduled DNA Synthesis in Mammalian Cells in Vitro*, Organization for Economic Cooperation and Development, Paris, 1986.
204. Umemura, T., Sai, K., Takagi, A., Hasegawa, R., and Kurokawa, Y., Formation of 8-hydroxydeoxyguanosine (8-OH-dG) in rat kidney DNA after intraperitoneal administration of ferric nitrotriacetate, *Carcinogenesis*, 11, 345–347, 1990.
205. OECD, *Guidelines for Testing of Chemicals*, No. 476, In vitro *Mammalian Cell Gene Mutation Test*, Organization for Economic Cooperation and Development, Paris, 1984.
206. OECD, *Guidelines for Testing of Chemicals*, No. 473, In vitro *Mammalian Cytogenetic Test*, Organization for Economic Cooperation and Development, Paris, 1983.
207. OECD, *Guidelines for Testing of Chemicals*, No. 479, In vitro *Sister Chromatid Exchange Assay in Mammalian Cells*, Organization for Economic Cooperation and Development, Paris, 1986.

3.2 Analysis of Heavy Metal-Induced Gene Expression

Shinji Koizumi

It has been shown that transcription of certain genes is activated in response to several heavy metals. In this and following sections, major techniques used for the studies of heavy metal-regulated mammalian gene expression are described. The methods for detecting and quantifying specific gene expression are summarized in the present section, and those for analyzing regulatory DNA elements and the proteins interacting with them follow.

Before describing the techniques, a brief overview of the heavy metal-regulated genes analyzed so far is presented. The genes for which the transcripts have been observed to increase after heavy metal treatments are listed in Table 51. In some of these examples, activation of transcription has been confirmed by nuclear runoff analysis transcription (described later) to be responsible for the increased mRNA levels. The genes activated by heavy metals involve those encoding proteins considered to function directly in the detoxification of nonessential, toxic heavy metals such as cadmium and mercury and in the regulation of the levels of essential metals such as zinc and copper (note that even essential metals become toxic when their concentrations in the body deviate from the normal levels). The metallothionein gene family is a well-known example in this category.[1] Another type of metal-inducible genes consists of those encoding proteins assumed to be involved in the protection against and recovery from a number of detrimental effects caused by heavy metal toxicity. The 70-kD heat shock protein (hsp 70) gene is an example in this category;[2] these two types of genes are expected to be controlled by the heavy metal-specific regulatory mechanisms. However, for many of the genes listed in Table 51, it is not clear whether they are controlled in such a way, and the physiological significance in protecting against heavy metal toxicity is also ambiguous. It has been proposed that transcriptional activation of some of these genes is due to disturbance of the normal regulation caused by heavy metal toxicity.[3,4]

Although high concentrations of heavy metals usually suppress the entire cellular metabolism, repression of specific gene transcription caused by heavy metals has been described less often. It has been reported that transcription of the estrogen receptor gene is repressed by a concentration of cadmium that induces the expression of other genes.[5]

TABLE 51
Analysis of Heavy Metal-Inducible Gene Expression

Gene[a]	Inducer	Method[b]	Ref.
Metallothionein	Cd, Zn, Cu, Hg, etc.	T, N, D S, R, P, NR	6, 15, 62; 35, 36, 63; 25–27; 32–34; 51; 53–55; 33, 40, 58
hsp90	Cd	N	64
hsp70	Cd, Zn, Cu	N, S, NR	65–67, 39, 46, 47; 39
hsp60	Cd	N	68
hsp32	Cd	N	64
hsp27	Cd	N	69
31-kD stress protein	Cd	N	70
α1-acid glycoprotein	Hg, Cd, Pb, Cu, etc.	N	71
C-reactive protein	Hg, Cd, Zn, Ni, etc.	N	71
Heme oxygenase	Co, Cd	N, R, NR	52, 59, 60; 52; 52, 59, 60
c-*jun*	Cd	N	3
c-*fos*	Cd, Zn	N	4, 65
c-*myc*	Cd	N	3, 72
N-myc	Cd	N	64
TPA-inducible genes	Cd	N	4
MDR1	Cd	N, D, P	28, 64; 28; 28
IL-8	Cd	N	73
Erythropoietin	Co	NR	61
Progesterone receptor	Cd	R, NR	5; 5
pS2	Cd	R, NR	5; 5
αB-crystallin	Cd	N	69

[a] hsp = heat shock protein; TPA = tetradecanoyl phorbol acetate; MDR = multidrug resistance; IL = interleukin.
[b] T = *in vitro* translation; N = northern blotting; D = dot/slot blotting; S = S1 nuclease mapping; R = RNase protection; P = primer extension; NR = nuclear runoff transcription assay.

3.2.1 Cloning of Heavy Metal-Inducible Genes

At the first step of gene cloning, double-stranded complementary DNA (cDNA) is synthesized *in vitro* using a cellular mRNA mixture containing the gene transcript of interest as a template, and a library is constructed by inserting the cDNA into a plasmid or phage vector (phage vectors are commonly used nowadays). The colonies of bacteria transfected with the library plasmids or the bacterial plaques formed by infection of the library phages are then transferred onto a membrane filter, and the clones containing the DNA of interest are detected, usually by hybridization with an appropriate radio-labeled DNA or RNA probe. The clones thus obtained can be used for screening of the corresponding genomic sequences as well as the genes containing related sequences.

With regard to most of the heavy-metal inducible genes so far studied (see Table 51), the heavy metal-inducibility has been demonstrated using genes already cloned. In the case of the metallothionein (MT) genes, however, their metal-inducibility was known before gene cloning,[6,7] and cDNA coding for MT has been cloned by taking advantage of this property. In the cloning of the mouse MT-I cDNA, which is the first isolated MT-encoding DNA,[8] RNA was extracted from mouse livers enriched with MT-mRNA by injecting cadmium into the animals. MT-mRNA was further concentrated by fractionation of the liver RNA by sucrose gradient centrifugation and analysis of the *in vitro* translation products of each fraction and was used for the preparation of a cDNA library and hybridization probe. Similarly, an enriched mRNA preparation from cadmium-induced cultured cells was used for screening the Chinese hamster MT-cDNA sequences.[9] For the isolation of the human MT-II_A sequence,[10] cDNA synthesized on the mRNA templates from both cadmium-induced and uninduced HeLa cells were used as probes, and the colonies giving stronger hybridization signals with the "induced" probe than with the uninduced probe were selected. Such a strategy (differential hybridization screening) has also been used in the cloning of other inducible genes, for example, tumor promoter-induced genes.[11,12] For detailed technical information, see References 13 and 14.

3.2.2 *Analysis of Specific Gene Activation*

There are several techniques to analyze specific gene expression, and one should choose some of them depending on the situation and purpose. To estimate the level of a specific mRNA encoded by an uncloned gene, an *in vitro* translation system is useful. If an appropriate probe containing a cloned cDNA or genomic DNA sequence is available, mRNA levels can be determined by northern blotting or dot/slot blotting. S1 nuclease mapping, RNase protection assay, and primer extension analysis are suitable for the detection and quantification of a particular mRNA species starting from a definite transcription initiation site. All of these methods can be used for the determination of steady-state mRNA levels. Even if a fluctuation is observed in the level of a certain mRNA, it does not always reflect a change in gene transcription. To determine the rate of transcription, a nuclear runoff transcription assay is generally performed.

3.2.2.1 In Vitro *Translation*

In this technique, an mRNA-containing fraction is isolated from cells or tissues and translated in a reaction mixture containing the translation apparatus and a radio-labeled amino acid. For the translation apparatus, wheat germ extracts or rabbit reticulocyte lysates are generally used. Most generally, the labeled translation products are separated on a sodium dodecyl sulfate-polyacrylamide gel, detected by autoradiography, and quantified by densitometry. The amount of a specific translation product reflects the level of the translatable mRNA coding for it. Even in the studies of yet uncloned genes, this technique allows determination of specific translatable mRNA levels,[6,7,15,16] enrichment of selected mRNA species for cDNA cloning,[8,17] and analysis of cloned gene products,[8] only if the translation product of interest can be identified on a gel. In certain cases, this method also offers a convenient way for estimating the levels of several mRNA species at a time, without preparing a number of probes specific for the respective genes to be analyzed. Early studies of MTs often used *in vitro* translation.[6-8,15,16,18,19] Wheat germ extracts and reticulocyte lysates are commercially available from Promega (Madison, WI), Boehringer Mannheim (Mannheim, Germany), and others.

3.2.2.2 Northern Blotting

This technique is most widely used for determining specific mRNA levels. Total RNA or an mRNA fraction is isolated from cells or tissues, separated by agarose gel electrophoresis under denaturing conditions, and then transferred onto a nitrocellulose or nylon membrane. A specific mRNA is then hybridized with a radio-labeled DNA or RNA probe containing the sequence complementary to the mRNA, under conditions that minimize nonspecific binding of the probe to the membrane. After stringent washing, the signal of the hybridized probe is detected by autoradiography. This technique provides information on the size and amount of a specific transcript.

One of the most important points in this analysis is the quality control of mRNA. The methods for the preparation of mRNA from cells or tissues have been described in detail.[13,14] In the denaturing agarose gel electrophoresis, mRNA is denatured by formaldehyde,[20,21] glyoxal and dimethylsulfoxide,[22] or methylmercuric hydroxide.[23] Among these, the formaldehyde gel system is most widely used because of convenience and safety. Unless the mRNA of interest is one of a rare species, it can be detected using 10 to 20 μg of the total cellular RNA. Nitrocellulose and positively charged nylon membranes are usually used for blotting. Nylon membranes are physically more durable than nitrocellulose filters and therefore are often used when a blot is repeatedly hybridized with different probes, but they give relatively higher backgrounds. Membrane transfer is performed most conveniently by capillary blot.[13,14,23] More efficient and rapid blotting can be achieved by vacuum transfer using commercially available apparatus (Pharmacia Biotech, Uppsala, Sweden; Bio-Rad Laboratories, Hercules, CA).

There are several methods for preparing radioactive probes using *in vitro* reactions: synthesis of DNA on a nicked double-stranded DNA template with an α-^{32}P-deoxyribonucleotide triphosphate (dNTP) and the Klenow fragment of *Escherichia coli* DNA polymerase I (nick translation); synthesis of the complementary strand from a specific primer on a single-stranded DNA template with an α-^{32}P-dNTP and polymerase, 5′ end-labeling of an oligonucleotide with γ-^{32}P-ATP and T4 polynucleotide kinase; synthesis of RNA from a phage promoter with an α-^{32}P-ribonucleotide triphosphate (rNTP) and a phage RNA polymerase, etc. Convenient kits for labeling nucleic acids are commercially available from GIBCO BRL (Gaithersburg, MD), Pharmacia Biotech (Uppsala, Sweden), Boehringer Mannheim (Mannheim, Germany), and others. It is noteworthy that immunological detection systems recently have been developed and improved (for example, a digoxigenin-labeling and detection system available from Boehringer) and provide a safer alternative for probing.

When analyzing inducible gene expression by northern blotting, it is important to include an assay of a reference gene that is known not to be inducible by the inducer used. By normalizing to the reference gene expression, the induction of the gene of interest can be demonstrated unambiguously. Detailed technical information on northern blotting is given in laboratory manuals.[13,14] This technique has been used in most of the studies of metal-inducible genes, as shown in Table 51.

3.2.2.3 Dot/Slot Blotting

The dot blotting technique measures the level of a specific transcript with an RNA sample directly applied and fixed on a membrane,[24] without separating RNA on an agarose gel. The fixed RNA is then hybridized with an appropriate radioactive probe, detected by autoradiography, and quantified by densitometry. This method is more convenient and less time-consuming than northern blotting, and a number of samples can be assayed at a time. Slot blotting has been developed as an improved version of dot blotting. The use of a filtration manifold with a number of slots allows RNA samples to be loaded onto a membrane uniformly and the radioactive signals to be determined more accurately. Slot blotting apparatus are commercially available from GIBCO BRL (Gaithersburg, MD), Schleicher and Schuell (Keene, NH), and others. As in northern blotting, nonradioactive probes can also be used.

What should be noted when using these methods is the importance of ensuring that the detected signal reflects what one desires to observe. It is quite important to determine the conditions of hybridization and stringent washing that yield only the signals of interest, before adopting dot/slot blotting as a routine assay. To study metal-inducible gene expression, an assay of a reference gene should be included. Detailed technical information on these methods has been described.[13,14] Examples for dot blotting are found in the studies of MT gene expression.[25-27] Slot blotting was used for analyzing MDR1 gene expression.[28]

3.2.2.4 S1 Nuclease Mapping

Using the S1 nuclease mapping technique,[29,30] the level of an mRNA species initiated from a specific transcription initiation site can be determined. RNA extracted from cells is hybridized with a 5′ end-labeled probe that is complementary to the mRNA to be studied. The unlabeled end of the probe should be upstream of the transcription initiation site. The formed DNA-RNA hybrids are then digested with single-strand specific S1 nuclease to remove the unhybridized portion of the probe, and the resultant protected ^{32}P-DNA fragments are analyzed on a denaturing polyacrylamide/urea sequencing gel. The size of the detected fragment reflects the distance between the end-labeled 5′ terminus of the probe and the 5′ terminus of the mRNA. In earlier studies, a double-stranded DNA fragment was end-labeled and cut by a restriction enzyme to generate a probe labeled only at the 5′ end of the anti-sense strand. To avoid the interference of DNA-RNA hybrid formation by the sense strand of the double-stranded probe, a single-stranded probe was prepared by separating sense and anti-sense strands by electrophoresis. Currently, however, a single-stranded probe is generally prepared by enzymatic synthesis of DNA that is extended from a ^{32}P-labeled specific primer on a single-stranded DNA template or by end-labeling of a chemically synthesized oligonucleotide. To determine the accurate level of a specific transcript, the amount of a probe should be in large excess over RNA. As in the blotting experiments, an internal control should be included to verify that the assay works properly (this is also the case for RNase protection and primer extension described below). S1 nuclease mapping can also be used to map the 3′ end and splice junctions of an mRNA by choosing appropriate genome sequences for probes. Technical details of S1 nuclease mapping have been described.[13,14] This technique has been frequently used in the studies of heavy metal-induced gene expression (MT[31-38]; *hsp70*[39]) and also in the studies of metal-induced expression of transfected hybrid genes containing the metal-regulatory sequences (MT[40-45]; *hsp70*[46,47]). Mung bean nuclease is also used instead of S1 nuclease.[48]

3.2.2.5 *RNase Protection Assay*

RNase protection assay[49,50] is an alternative to S1 nuclease mapping. Before performing this assay, the probe sequence should be cloned downstream of a bacteriophage promoter. This DNA is cut with an appropriate restriction enzyme and used as the template in an *in vitro* transcription reaction with a phage RNA polymerase and an α-^{32}P-rNTP to generate a uniformly labeled single-stranded RNA of defined length. The template DNA is then removed by DNase digestion. The resultant RNA probe and RNases are used instead of a DNA probe and S1 nuclease. This technique allows convenient and efficient production of an RNA probe with a high specific activity and provides a high sensitivity of detection, but high background is sometimes a problem. (For technical details, see References 13 and 14.) RNase protection has been used in the studies of cadmium-inducible genes including MT,[51] heme oxygenase,[52] and estrogen-induced genes.[5]

3.2.2.6 *Primer Extension*

The primer extension analysis is another alternative technique for identifying and quantifying specific transcripts. The RNA to be analyzed is hybridized with a 5′ end-labeled, single-stranded DNA primer (usually a chemically synthesized oligonucleotide), and cDNA is synthesized from the primer with reverse transcriptase. The length of the resultant end-labeled DNA fragment reflects the distance between the 5′ terminus of the primer and the 5′ terminus of the mRNA. The level of the unique transcript is then determined by analyzing the cDNA product by denaturing gel electrophoresis. In contrast to the protection assays described above, primer extension does not detect splice junctions, but does detect the 5′ end of mRNA only. (For technical details, see References 13 and 14.) Primer extension was used in the studies of MT[53-55] and MDR1[28] gene expression.

3.2.2.7 *Nuclear Runoff Transcription Assay*

The nuclear runoff transcription assay[56,57] allows determination of the rate of transcription of a specific gene. This is based on the fact that transcription in the isolated nuclei system virtually reflects the elongation of transcripts that have been initiated prior to cell lysis. Nuclei are isolated from cells or tissues, and the ongoing mRNA synthesis is allowed to continue *in vitro* in the presence of an α-^{32}P-rNTP. The resultant labeled RNA is purified and hybridized with a specific DNA immobilized on a membrane. The signal of the hybridized RNA is detected by autoradiography. This analysis is often used for confirming that the change in the level of a specific transcript reflects the change in the specific gene transcription and does not result from the changes in mRNA stability, translation, etc. Technical details of this method have been described.[13] Transcriptional activation of MT,[33,40,58] estrogen-induced genes,[5] *hsp70*,[39] heme oxygenase,[52,59-61] and erythropoietin[61] genes has been demonstrated using this technique.

3.3 Analysis of Regulatory Mechanisms of Heavy Metal-Induced Transcription

Following the demonstration of gene activation using the techniques described in the previous section, the mechanism of activation has been studied enthusiastically. In almost all of the class II genes coding for proteins, the sequences that control gene expression are found upstream of the transcription initiation sites. These elements include those required for the basic process of transcription and those required for the specific regulation of respective genes. The DNA elements essential for the heavy metal-induced transcription belong to the latter. Recently, it has been shown that these regulatory elements are generally the recognition sites of specific regulatory proteins. As for the metal-inducible genes, the heavy metal-regulatory elements of MT,[44,74,75] *hsp70*,[39,46,47] and heme oxygenase[52] genes have been studied. In the case of the metal-regulatory element (MRE) of MT genes, there has been a number of detailed studies on the MRE-binding regulatory proteins (for review, see Reference 76). In this section, the techniques used to identify gene-specific regulatory DNA elements and to analyze proteins interacting with them are described.

3.3.1 *Identification of Transcriptional Regulatory Elements*

3.3.1.1 *Transfer of Hybrid Genes into Mammalian Cells*

The development of gene transfer techniques into mammalian cells[77,78] has enabled direct analysis of the *in vivo* function of any regulatory DNA sequence. In order to distinguish between the activities of an introduced gene and the corresponding endogenous gene, artificial genes created by fusing the regulatory sequence to be studied to a "reporter" structural gene are frequently used. As the reporter, non-mammalian genes whose products can easily be distinguished from the endogenous gene products are generally used: bacterial chloramphenicol acetyl transferase (CAT),[79] β-galactosidase,[80] firefly luciferase,[81-83] etc. The hybrid gene is introduced into cells by one of the gene transfer methods, including calcium phosphate transfection, DEAE dextran transfection, electroporation, liposome-mediated transfection, and microinjection. Most frequently used is calcium phosphate transfection,[77,78] in which DNA co-precipitated with calcium phosphate enters the cell by endocytosis. After gene transfer, the expression of the reporter gene is estimated within 1 to 4 days (transient expression) or after the gene is stably integrated in the chromosome (stable transformation). This is accomplished by measuring the enzyme activity in the cell lysate or the transcript level using a probe designed so that only the specific signal for the introduced gene is detected (see Section 3.2). As the intake of DNA often varies among experiments or among culture dishes in an experiment, it is necessary to co-transfect a reference gene that has regulatory and reporter sequences different from the gene to be studied and to use its expression for normalization. If the hybrid gene is observed to be regulated (e.g., induced by heavy metals) in the same way as the corresponding endogenous gene, the regulatory element of interest will lie within the sequence tested. (Technical details of the methods are given in References 13 and 14.) Studies that have demonstrated the presence of the heavy metal regulatory sequences using the gene transfer techniques are listed in Table 52.

3.3.1.2 *Analysis of Mutants of Regulatory Sequences*

Once a regulatory element is located within a limited length of DNA using the techniques described above, the position of the element can be mapped by transfecting a series of mutant gene constructs and estimating their expression. Deletion mutants in the regulatory region

TABLE 52

Identification of Heavy Metal-Regulatory Sequences by Gene Transfer Experiments

Gene[a]	Ref.	
	Stable Transformation	Transient Expression
Metallothionein		
mMT-I	40, 41, 111	44, 45, 55, 84, 85, 112
mMT-II	33	
hMT-I$_A$	113	
hMT-I$_B$	35	
hMT-I$_E$	34	
hMT-I$_F$	34	
hMT-I$_G$		37
hMT-I$_H$		114
hMT-I$_X$		114
hMT-II$_A$	42, 48, 53, 75	53, 54, 86, 115
rMT-1		36, 43
h Heme oxygenase		52, 116
h *hsp70*	46	47

[a] m = mouse; h = human; r = rat.

are conveniently generated by digestion of the DNA with appropriate restriction enzymes and re-ligation. For a more detailed analysis, the regulatory region is deleted from its 5′ or 3′ terminus by digesting with an exonuclease such as Bal31 to varied extents, generating a series of deletion mutants that have different lengths of the regulatory region. For example, a metal-regulatory element can be located between the 5′ and 3′ boundaries where heavy metal-inducibility is lost as a result of increasing deletion from the 5′ and 3′ termini, respectively. Using mutants generated by joining two appropriate 5′ and 3′ deletion mutants via a short oligonucleotide linker (linker-scanning mutants), the location of the regulatory element can be determined more directly (for example, Reference 74). By analyzing the expression of mutant genes, the heavy metal regulatory elements located in the upstream region of MT,[44,74,75] *hsp70*,[46,47] and heme oxygenase[52] genes have been mapped.

3.3.1.3 *Functional Analysis of Synthetic Regulatory Elements*

To demonstrate directly the regulatory function of DNA elements identified by deletion mapping, the activity of synthetic DNA elements is often examined. In the case of MT genes, a chemically synthesized oligonucleotide containing the MRE sequence was inserted into the upstream region of an uninducible gene[74,84,85] or directly fused to a promoter.[45,84-86] By measuring the expression of the transfected construct in the presence and absence of heavy metals, it was demonstrated unambiguously that the element is responsible for the metal inducibility. The function of a cadmium-responsive element located upstream of the heme oxygenase gene[52] has also been analyzed using this technique. Furthermore, the nucleotides required for the regulatory function can be determined by introducing point mutations into the regulatory element. The bases important for the function of MRE of the mouse MT-I gene have been identified by analyzing a series of point mutants.[55]

3.3.2 *Detection and Analysis of Proteins Interacting with Regulatory Elements*

In many cases, transcriptional regulatory elements have been proved to be the sites recognized and bound by specific regulatory proteins. To determine the properties of such proteins, several techniques for detecting their interaction with DNA elements have been developed. The most convenient and widely used method is the mobility shift assay (also called bandshift assay, gel retardation assay, etc.) To determine the protein binding site, DNase I footprinting or the methylation interference assay is used. For further information about the proteins, protein blotting and UV-crosslinking analysis are used; both of these methods give the molecular weight of the DNA-binding proteins. The most important point common to all the techniques described in this section is to verify that only specific interactions are detected.

3.3.2.1 Mobility Shift Assay

The mobility shift assay is based on the observation that the electrophoretic mobility of a ^{32}P-labeled DNA fragment is reduced by binding of proteins to it.[87,88] As a source of DNA-binding proteins, nuclear extracts are often prepared, most frequently using the method of Dignam et al.[89] Proteins and a ^{32}P-labeled double-stranded DNA probe are incubated under conditions that minimize nonspecific interactions (for this purpose, a synthetic copolymer is often added to the binding reaction mixture[90]). The protein-DNA complexes formed are electrophoresed through native polyacrylamide gel, and the ^{32}P-signals are detected by autoradiography of the dried gel. Since the protein-bound probe migrates much more slowly than the unbound probe, the protein-DNA interactions can easily be detected. In this kind of experiments, nonspecific interactions between proteins and the DNA probe could always occur, and care should be taken to obtain the conclusive evidence for a specific interaction. For this purpose, competitive binding experiments are usually performed. If an unlabeled DNA fragment containing the recognition site of the protein to be tested is added to the binding reaction in excess, it will compete with the labeled probe for the protein, resulting in a decrease of the signal intensity of the protein-^{32}P-probe complex. By contrast, if the competitor DNA does not have the protein binding site, it will have no effect on the protein-probe complex formation. Accordingly, both types of competitors should always be included in the competitive binding assay to verify the sequence-specific formation of the protein-probe complex. More strictly, oligonucleotide competitors carrying one or a few mutated bases within the binding sites are used as the nonfunctional competitors. This also provides a means to determine the bases important for the protein-DNA interaction. Moreover, such an analysis is also useful in determining whether the protein-DNA interaction directly correlates with transcriptional activation with a concomitant analysis of the effects of the mutations on transcription as described in Section 3.3.1. (Detailed technical information is available in Reference 13.) There are a number of examples for the mobility shift assay in the studies of MT,[45,86,91-93] *hsp70*,[39] and heme oxygenase[52] genes.

3.3.2.2 DNase I Footprinting

DNase I footprinting determines the protein binding sites on a DNA, taking advantage of the fact that the sites bound by proteins are protected from the attack of DNase I.[94] A double-stranded DNA fragment a few hundred bases long labeled at one end with ^{32}P is used as a probe. When the probe DNA is digested with DNase I under conditions that generate approximately one random break per DNA molecule, the products display a ladder of DNA bands on a denaturing polyacrylamide sequencing gel. The size of each band reflects the distance between the break and the ^{32}P-labeled terminal nucleotide. If a protein-bound DNA probe is digested in the same manner, the binding sites will be protected from DNase attack, resulting in missing bands on a sequencing gel. By referring to the chemical sequencing ladders[95] of the probe, the protein binding site can be accurately mapped. As it is for the mobility shift assay, it is quite important to confirm the sequence specificity of the protection by competition experiments. (For detailed technical information, see Reference 13). DNase I footprinting has been used in the studies of MRE of MT genes.[93,96,97] Based on a similar principle, exonuclease III footprinting is used to determine the boundary of the protein-binding sites.[97,98] For detecting protein-DNA interactions in living cells, *in vivo* footprinting is used.[43,51,99]

3.3.2.3 *Methylation Interference*

The methylation interference assay detects the purine nucleotides that are important for a protein-DNA interaction, based on the fact that methylated bases interfere with the contact of a protein.[100,101] A double-stranded DNA probe labeled at one end with ^{32}P is treated with dimethylsulfate to generate approximately one methylated base per molecule. Using this probe, a protein-DNA binding reaction is performed, and the formed complex and free DNA probe are separated on a native polyacrylamide gel, as for the mobility shift assay. The DNA molecules carrying methylated guanine (G) or adenine (A) residues in the protein-binding site cannot interact with the protein and are consequently recovered in the band of free probe, whereas the DNA molecules carrying unmethylated purines in the corresponding site can bind the protein and are recovered in the band of the protein-DNA complex. The DNAs extracted from both bands are then cleaved at the methylated bases by piperidine treatment, and electrophoresed side by side on a denaturing polyacrylamide sequencing gel. Among the series of DNA fragments labeled at one end, those having the other unlabeled termini within the protein binding site will be missing in the DNA derived from the complex band, as compared with the DNA from the protein-free band. This method thus allows one to determine the G and A residues that are in contact with the DNA-binding protein. (For detailed technical information, see Reference 13.) There are examples for methylation interference in the analysis of a factor binding to MRE of the mouse MT-I gene[45] and a factor binding to the heat-shock element of the *hsp70* gene.[39]

3.3.2.4 *Protein Blotting*

In this technique, often called southwestern blotting, proteins are electrophoresed and transferred onto a membrane, and specific DNA-binding proteins are detected with a ^{32}P-labeled DNA probe.[102] Proteins are separated by SDS-polyacrylamide gel electrophoresis and electrically transferred to a nitrocellulose membrane. Currently, a semi-dry type of blotting apparatus (available from Bio-Rad Laboratories, Hercules, CA; Pharmacia Biotech, Uppsala, Sweden; and others) is often used, because of convenience and even and efficient transfer. Prior to probing, the blot should be treated with a reagent that blocks nonspecific binding of the probe to the membrane (e.g., skim milk). The blot is then incubated with a small volume of binding reaction mixture containing a ^{32}P-labeled DNA probe. After washing away unbound probe, the ^{32}P signal is detected by autoradiography. This analysis gives the molecular weight of the DNA-binding protein. To verify a specific interaction, competition experiments are desirable. Proteins interacting with MRE of the mouse MT-I[97,103] and human MT-II_A[92,104,105] genes have been analyzed using this method.

3.3.2.5 *UV-Crosslinking*

DNA-binding proteins can also be identified by ultraviolet (UV)-crosslinking, which originally was used to analyze bacterial protein-DNA interactions[106-108] and was later applied to eukaryotic systems.[109] Probes for UV-crosslinking should be uniformly labeled at the specific protein binding site. This can be accomplished by *in vitro* synthesis of DNA in the presence of an α-^{32}P-dNTP from a short primer on a single-stranded DNA template, which has been prepared either by cloning of the protein-binding site sequence to an M13 vector or by chemical synthesis. The thymidine residues of the probe are usually substituted with bromodeoxyuridine to promote photoreaction. The ^{32}P-labeled DNA probe and proteins are incubated to form complexes as in the mobility shift assay. The binding reaction mixture is then irradiated by ultraviolet light, and the residues involved in the protein-DNA contacts are crosslinked. The complex is digested with DNase under conditions allowing only the crosslinked portion of the DNA to remain undigested. Since this short stretch of DNA has only a minor effect on the electrophoretic mobility of the protein, analysis of the ^{32}P-products on an SDS-polyacrylamide gel provides information about the amount and size of the DNA-binding protein. Prior to crosslinking, it should be confirmed that the probe can work properly by performing a mobility shift assay. The specificity of the protein-DNA interactions can be verified by including competitor DNAs in the binding reaction, as in the mobility shift assay. (Detailed technical information has been described in Reference 13.) This method has been used to identify proteins that interact with MRE of mouse[96,97,110] and human[92,93] MT genes.

REFERENCES

1. Suzuki, K. T., Imura, N., and Kimura, M., *Metallothionein III: Biological Roles and Medical Implications,* Birkhäuser Verlag, Basel, 1993.
2. Morimoto, R. I., Tissières, A., and Georgopoulos, C., *Stress Proteins in Biology and Medicine,* Cold Spring Harbor Laboratory Press, Cold Spring Harbor, NY, 1990.
3. Jin, P. and Ringertz, N. R., Cadmium induces transcription of proto-oncogenes c-*jun* and c-*myc* in rat L6 myoblasts, *J. Biol. Chem.,* 265, 14061–14064, 1990.
4. Epner, D. E. and Herschman, H. R., Heavy metals induce expression of the TPA-inducible sequence (TIS) genes, *J. Cell. Physiol.,* 148, 68–74, 1991.
5. Garcia-Morales, P., Saceda, M., Kenney, N., Kim, N., Salomon, D. S., Gottardis, M. M., Solomon, H. B., Sholler, P. F., Jordan, V. C., and Martin, M. B., Effect of cadmium on estrogen receptor levels and estrogen-induced responses in human breast cancer cells, *J. Biol. Chem.,* 269, 16896–16901, 1994.
6. Andersen, R. D. and Weser, U., Partial purification, characterization and translation *in vitro* of rat liver metallothionein messenger ribonucleic acid, *Biochem. J.,* 175, 841–852, 1978.
7. Shapiro, S. G., Squibb, K. S., Markowitz, L. A., and Cousins, R. J., Cell-free synthesis of metallothionein directed by rat liver polyadenylated messenger ribonucleic acid, *Biochem. J.,* 175, 833–840, 1978.
8. Durnam, D. M., Perrin, F., Gannon, F., Palmiter, R. D., Isolation and characterization of the mouse metallothionein-I gene, *Proc. Natl. Acad. Sci. USA,* 77, 6511–6515, 1980.
9. Griffith, B. B., Walters, R. A., Enger, M. D., Hildebrand, C. E., and Griffith, J. K., cDNA cloning and nucleotide sequence comparison of Chinese hamster metallothionein I and II mRNAs, *Nucleic Acids Res.,* 11, 901–910, 1983.
10. Karin, M. and Richards, R. I., Human metallothionein genes: molecular cloning and sequence analysis of the mRNA, *Nucleic Acids Res.,* 10, 3165–3173, 1982.
11. Angel, P., Pöting, A., Mallick, U., Rahmsdorf, H. J., Schorpp, M., and Herrlich, P., Induction of metallothionein and other mRNA species by carcinogens and tumor promoters in primary human skin fibroblasts, *Mol. Cell. Biol.,* 6, 1760–1766, 1986.
12. Lim, R. W., Varnum, B. C., and Herschman, H. R., Cloning of tetradecanoyl phorbol ester-induced "primary response" sequences and their expression in density-arrested Swiss 3T3 cells and a TPA non-proliferative variant, *Oncogene,* 1, 263–270, 1987.
13. Ausubel, F. M., Brent, R., Kingston, R. E., Moore, D. D., Seidman, J. G., Smith, J. A., and Struhl, K., *Current Protocols in Molecular Biology,* John Wiley & Sons, New York, 1995.
14. Sambrook, J., Fritsch, E. F., and Maniatis, T., *Molecular Cloning: A Laboratory Manual,* 2nd ed., Cold Spring Harbor Laboratory Press, Cold Spring Harbor, NY, 1989.
15. Enger, M. D., Rall, L. B., and Hildebrand, C. E., Thionein gene expression in Cd^{++}-variants of the CHO cell: correlation of thionein synthesis rates with translatable mRNA levels during induction, deinduction, and superinduction, *Nucleic Acids Res.,* 7, 271–288, 1979.
16. Karin, M., Andersen, R. D., and Herschman, H. R., Induction of metallothionein mRNA in HeLa cells by dexamethasone and by heavy metals, *Eur. J. Biochem.,* 118, 527–531, 1981.
17. Peterson, M. G., Lazdins, I., Danks, D. M., and Mercer, J. F. B., Cloning and sequencing of a sheep metallothionein cDNA, *Eur. J. Biochem.,* 143, 507–511, 1984.
18. Koizumi, S., Sone, T., Otaki, N., and Kimura, M., Cd^{2+}-induced synthesis of metallothionein in HeLa cells, *Biochem. J.,* 227, 879–886, 1985.
19. Koizumi, S. and Kimura, M., Characterization and measurement of metallothionein messenger RNA of C57BL mouse liver, *Chem.-Biol. Interact,* 54, 33–43, 1985.
20. Lehrach, H., Diamond, D., Wozney, J. M., and Boedtker, H., RNA molecular weight determinations by gel electrophoresis under denaturing conditions, a critical reexamination, *Biochemistry,* 16, 4743–4751, 1977.
21. Goldberg, D. A., Isolation and partial characterization of the *Drosophila* alcohol dehydrogenase gene, *Proc. Natl. Acad. Sci. USA,* 77, 5794–5798, 1980.
22. McMaster, G. K. and Carmichael, G. G., Analysis of single- and double-stranded nucleic acids on polyacrylamide and agarose gels by using glyoxal and acridine orange, *Proc. Natl. Acad. Sci. USA,* 74, 4835–4838, 1977.
23. Thomas, P. S., Hybridization of denatured RNA and small DNA fragments transferred to nitrocellulose, *Proc. Natl. Acad. Sci. USA,* 77, 5201–5205, 1980.

24. Kafatos, F. C., Jones, C. W., and Efstratiadis, A., Determination of nucleic acid sequence homologies and relative concentrations by a dot hybridization procedure, *Nucleic Acids Res.,* 7, 1541–1552, 1979.
25. Andersen, R. D., Birren, B. W., Ganz, T., Piletz, J. E., and Herschman, H. R., Molecular cloning of the rat metallothionein 1 (MT-1) mRNA sequence, *DNA,* 2, 15–22, 1983.
26. Peterson, M. G., Mercer, J. F. B., Structure and regulation of the sheep metallothionein-Ia gene, *Eur. J. Biochem.,* 160, 579–585, 1986.
27. Peterson, M. G. and Mercer, J. F. B., Differential expression of four linked sheep metallothionein genes, *Eur. J. Biochem.,* 174, 425–429, 1988.
28. Chin, K.-V., Tanaka, S., Darlington, G., Pastan, I., and Gottesman, M. M., Heat shock and arsenite increase expression of the multidrug resistance (*MDR1*) gene in human renal carcinoma cells, *J. Biol. Chem.,* 265, 221–226, 1990.
29. Berk, A. J. and Sharp, P. A., Sizing and mapping of early adenovirus mRNAs by gel electrophoresis of S1 endonuclease-digested hybrids, *Cell,* 12, 721–732, 1977.
30. Weaver, R. F. and Weissman, C., Mapping of RNA by a modification of the Berk-Sharp procedure: the 5′ termini of 15S β-globin mRNA precursor and mature 10S β-globin mRNA have identical map coordinates, *Nucleic Acids Res.,* 7, 1175–1193, 1979.
31. Glanville, N., Durnam, D. M., and Palmiter, R. D., Structure of mouse metallothionein-I gene and its mRNA, *Nature,* 292, 267–269, 1981.
32. Karin, M. and Richards, R. I., Human metallothionein genes — primary structure of the metallothionein-II gene and a related processed gene, *Nature,* 299, 797–802, 1982.
33. Searle, P. F., Davison, B. L., Stuart, G. W., Wilkie, T. M., Norstedt, G., and Palmiter, R. D., Regulation, linkage, and sequence of mouse metallothionein I and II genes, *Mol. Cell. Biol.,* 4, 1221–1230, 1984.
34. Schmidt, C. J., Jubier, M. F., and Hamer, D. H., Structure and expression of two human metallothionein-I isoform genes and ar related pseudogene, *J. Biol. Chem.,* 260, 7731–7737, 1985.
35. Heguy, A., West, A., Richards, R. I., and Karin, M., Structure and tissue-specific expression of the human metallothionein I_B gene, *Mol. Cell. Biol.,* 6, 2149–2157, 1986.
36. Andersen, R. D., Birren, B. W., Taplitz, S. J., and Herschman, H. R., Rat metallothionein-1 structural gene and three pseudogenes, one of which contains 5′-regulatory sequences, *Mol. Cell. Biol.,* 6, 302–314, 1986.
37. Foster, R., Jahroudi, N., Varshney, U., and Gedamu, L., Structure and expression of the human metallothionein-IG gene, *J. Biol. Chem.,* 263, 11528–11535, 1988.
38. Schmidt, C. J. and Hamer, D. H., Cell specificity and an effect of *ras* on human metallothionein gene expression, *Proc. Natl. Acad. Sci. USA,* 83, 3346–3350, 1986.
39. Mosser, D. D., Theodorakis, N. G., and Morimoto, R. I., Coordinate changes in heat shock element-binding activity and *hsp70* gene transcription rates in human cells, *Mol. Cell. Biol.,* 8, 4736–4744, 1988.
40. Mayo, K. E., Warren, R., and Palmiter, R. D., The mouse metallothionein-I gene is transcriptionally regulated by cadmium following transfection into human or mouse cells, *Cell,* 29, 99–108, 1982.
41. Pavlakis, G. N. and Hamer, D. H., Regulation of a metallothionein-growth hormone hybrid gene in bovine papilloma virus, *Proc. Natl. Acad. Sci. USA,* 80, 397–401, 1983.
42. Karin, M., Haslinger, A., Holtgreve, H., Cathala, G., Slater, E., and Baxter, J. D., Activation of a heterologous promoter in response to dexamethasone and cadmium by metallothionein gene 5′-flanking DNA, *Cell,* 36, 371–379, 1984.
43. Andersen, R. D., Taplitz, S. J., Wong, S., Bristol, G., Larkin, B., and Herschman, H. R., Metal-dependent binding of a factor *in vivo* to the metal-responsive elements of the metallothionein 1 gene promoter, *Mol. Cell. Biol.,* 7, 3574–3581, 1987.
44. Carter, A. D., Felber, B. K., Walling, M., Jubier, M.-F., Schmidt, C. J., and Hamer, D. H., Duplicated heavy metal control sequences of the mouse metallothionein-I gene, *Proc. Natl. Acad. Sci. USA,* 81, 7392–7396, 1984.
45. Westin, G. and Schaffner, W., A zinc-responsive factor interacts with a metal-regulated enhancer element (MRE) of the mouse metallothionein-I gene, *EMBO J.,* 7, 3763–3770, 1988.
46. Wu, B. J., Kingston, R. E., and Morimoto, R. I., Human *hsp70* promoter contains at least two distinct regulatory domains, *Proc. Natl. Acad. Sci. USA,* 83, 629–633, 1986.
47. Williams, G. T. and Morimoto, R. I., Maximal stress-induced transcription from the human *hsp70* promoter requires interactions with the basal promoter elements independent of rotational alignment, *Mol. Cell. Biol.,* 10, 3125–3136, 1990.
48. Karin, M., Cathala, G., and Nguyen-Huu, M. C., Expression and regulation of a human metallothionein gene carried on an autonomously replicating shuttle vector, *Proc. Natl. Acad. Sci. USA,* 80, 4040–4044, 1983.
49. Zinn, K., DiMaio, D., and Maniatis, T., Identification of two distinct regulatory regions adjacent to the human β-interferon gene, *Cell,* 34, 865–879, 1983.
50. Melton, D. A., Kreig, P. A., Rebagliati, M. R., Maniatis, T., Zinn, K., and Green, M. R., Efficient *in vitro* synthesis of biologically active RNA and RNA hybridization probes from plasmids containing a bacteriophage SP6 promoter, *Nucleic Acids Res.,* 12, 7035–7056, 1984.

51. Mueller, P. R., Salser, S. J., and Wold, B., Constitutive and metal-inducible protein: DNA interactions at the mouse metallothionein I promoter examined by *in vivo* and *in vitro* footprinting, *Genes Dev.,* 2, 412–427, 1988.
52. Takeda, K., Ishizawa, S., Sato, M., Yoshida, T., and Shibahara, S., Identification of a *cis*-acting element that is responsible for cadmium-mediated induction of the human heme oxygenase gene, *J. Biol. Chem.,* 269, 22858–22867, 1994.
53. Haslinger, A. and Karin, M., Upstream promoter element of the human metallothionein-II_A gene can act like an enhancer element, *Proc. Natl. Acad. Sci. USA,* 82, 8572–8576, 1985.
54. Karin, M., Haslinger, A., Heguy, A., Dietlin, T., and Cooke, T., Metal-responsive elements act as positive modulators of human metallothionein-II_A enhancer activity, *Mol. Cell. Biol.,* 7, 606–613, 1987.
55. Culotta, V. C. and Hamer, D. H., Fine mapping of a mouse metallothionein gene metal response element, *Mol. Cell. Biol.,* 9, 1376–1380, 1989.
56. Marzluff, W. F. and Huang, R. C., Transcription of RNA in isolated nuclei, in *Transcription and Translation: A Practical Approach,* Hames, B. D. and Higgins, S. J., Eds., IRL Press, Oxford, 1985, pp. 89–129.
57. Groudine, M., Peretz, M., and Weintraub, H., Transcriptional regulation of hemoglobin switching in chicken embryos, *Mol. Cell. Biol.,* 1, 281–288, 1981.
58. Durnam, D. M. and Palmiter, R. D., Transcriptional regulation of the mouse metallothionein-I gene by heavy metals, *J. Biol. Chem.,* 256, 5712–5716, 1981.
59. Alam, J., Shibahara, S., and Smith, A., Transcriptional activation of the heme oxygenase gene by heme and cadmium in mouse hepatoma cells, *J. Biol. Chem.,* 264, 6371–6375, 1989.
60. Lin, J. H.-C., Villalon, P., Martasek, P., and Abraham, N. G., Regulation of heme oxygenase gene expression by cobalt in rat liver and kidney, *Eur. J. Biochem.,* 192, 577–582, 1990.
61. Lutton, J. D., Griffin, M. O., Nishimura, M., Levere, R. D., Kappas, A., Abraham, N. G., and Shibahara, S., Co-expression of erythropoietin and heme oxygenase genes in Hep3B cells, *Hepatology,* 17, 861–868, 1993.
62. Karin, M., Andersen, R. D., Slater, E., Smith, K., and Herschman, H. R., Metallothionein mRNA induction in HeLa cells in response to zinc or dexamethasone is a primary induction response, *Nature,* 286, 295–297, 1980.
63. Varshney, U., Jahroudi, N., Foster, R., and Gedamu, L., Structure, organization, and regulation of human metallothionein I_F gene: differential and cell-type-specific expression in response to heavy metals and glucocorticoids, *Mol. Cell. Biol.,* 6, 26–37, 1986.
64. Murakami, T., Ohmori, H., Kato, T., Abe, T., and Higashi, K., Cadmium causes increases of N-myc and multidrug-resistance gene mRNA in neuroblastoma cells, *J. Univ. Occupation. Environ. Health,* 13, 271–278, 1991.
65. Andrews, G. K., Harding, M. A., Calvet, J. P., and Adamson, E. D., The heat shock response in HeLa cells is accompanied by elevated expression of the c-*fos* proto-oncogene, *Mol. Cell. Biol.,* 7, 3452–3458, 1987.
66. Mitani, K., Fujita, H., Sassa, S., and Kappas, A., Activation of heme oxygenase and heat shock protein 70 genes by stress in human hepatoma cells, *Biochem. Biophys. Res. Commun.,* 166, 1429–1434, 1990.
67. Hatayama, T., Asai, Y., Wakatsuki, T., Kitamura, T., and Imarhara, H., Regulation of hsp70 synthesis induced by cupric sulfate and zinc sulfate in thermotolerant HeLa cells, *J. Biochem.,* 114, 592–597, 1993.
68. Hiranuma, K., Hirata, K., Abe, T., Hirano, T., Matsuno, K., Hirano, H., Suzuki, K., and Higashi, K., Induction of mitochondrial chaperonin, hsp60, by cadmium in human hepatoma cells, *Biochem. Biophys. Res. Commun.,* 194, 531–536, 1993.
69. Head, M. W., Corbin, E., and Goldman, J. E., Coordinate and independent regulation of αB-crystallin and hsp27 expression in response to physiological stress, *J. Cell. Physiol.,* 159, 41–50, 1994.
70. Shuman, J. and Przybyla, A., Expression of the 31-kD stress protein in rat myoblasts and hepatocytes, *DNA,* 7, 475–482, 1988.
71. Yiangou, M., Ge, X., Carter, K. C., and Papaconstantinou, J., Induction of several acute-phase protein genes by heavy metals: a new class of metal-responsive genes, *Biochemistry,* 30, 3798–3806, 1991.
72. Tang, N. and Enger, M. D., Cadmium induces hypertrophy accompanied by increased *myc* mRNA accumulation in NRK-49F cells, *Cell Biol. Toxicol.,* 7, 401–411, 1991.
73. Horiguchi, H., Mukaida, N., Okamoto, S., Teranishi, H., Kasuya, M., and Matsushima, K., Cadmium induces interleukin-8 production in human peripheral blood mononuclear cells with the concomitant generation of superoxide radicals, *Lymphokine Cytokine Res.,* 12, 421–428, 1993.
74. Stuart, G. W., Searle, P. F., Chen, H. Y., Brinster, R. L., and Palmiter, R. D., A 12-base pair DNA motif that is repeated several times in metallothionein gene promoters confers metal regulation to a heterologous gene, *Proc. Natl. Acad. Sci. USA,* 81, 7318–7322, 1984.
75. Karin, M., Haslinger, A., Holtgreve, H., Richards, R. I., Krauter, P., Westphal, H. M., and Beato, M., Characterization of DNA sequences through which cadmium and glucocorticoid hormones induce human metallothionein-II_A gene, *Nature,* 308, 513–519, 1984.
76. Koizumi, S. and Otsuka, F., Factors involved in the transcriptional regulation of metallothionein genes, in *Metallothionein III,* Suzuki, K. T., Imura, N., and Kimura, M., Eds., Birkhäuser Verlag, Basel, 1993, pp. 457–474.

77. Graham, F. L. and van der Eb, A. J., A new technique for the assay of infectivity of human adenovirus 5 DNA, *Virology,* 52, 456–467, 1973.
78. Wigler, M., Pellicer, A., Silverstein, S., and Axel, R., Biochemical transfer of single-copy eucaryotic genes using total cellular DNA as donor, *Cell,* 14, 725–731, 1978.
79. Gorman, C. M., Moffat, L. F., and Howard, B. H., Recombinant genomes which express chloramphenicol acetyltransferase in mammalian cells, *Mol. Cell. Biol.,* 2, 1044–1051, 1982.
80. An, G., Hidaka, K., and Siminovitch, L., Expression of bacterial β-galactosidase in animal cells, *Mol. Cell. Biol.,* 2, 1628–1632, 1982.
81. Gould, S. J. and Subramani, S., Firefly luciferase as a tool in molecular and cell biology, *Anal. Biochem.,* 175, 5–13, 1988.
82. Brasier, A. R., Tate, J. E., and Habener, J. F., Optimized use of the firefly luciferase assay as a reporter gene in mammalian cell lines, *Bio-Techniques,* 7, 1116–1122, 1989.
83. Williams, T. M., Burlein, J. E., Ogden, S., Kricka, L. J., and Kant, J. A., Advantages of the firefly luciferase as a reporter gene: application to the interleukin-2 gene promoter, *Anal. Biochem.,* 176, 28–32, 1989.
84. Stuart, G. W., Searle, P. F., and Palmiter, R. D., Identification of multiple metal regulatory elements in mouse metallothionein-I promoter by assaying synthetic sequences, *Nature,* 317, 828–831, 1985.
85. Searle, P. F., Stuart, G. W., and Palmiter, R. D., Building a metal-responsive promoter with synthetic regulatory elements, *Mol. Cell. Biol.,* 5, 1480–1489, 1985.
86. Koizumi, S., Yamada, H., Suzuki, K., and Otsuka, F., Zinc-specific activation of a HeLa cell nuclear protein which interacts with a metal responsive element of the human metallothionein-II_A gene, *Eur. J. Biochem.,* 210, 555–560, 1992.
87. Garner, M. M. and Revzin, A., A gel electrophoresis method for quantifying the binding of proteins to specific DNA regions: application to components of the *Escherichia coli* lactose operon regulatory system, *Nucleic Acids Res.,* 9, 3047–3060, 1981.
88. Fried, M. and Crothers, D. M., Equilibria and kinetics of lac repressor-operator interactions by polyacrylamide gel electrophoresis, *Nucleic Acids Res.,* 9, 6505–6525, 1981.
89. Dignam, J. D., Lebovitz, R. M., and Roeder, R. G., Accurate transcription initiation by RNA polymerase II in a soluble extract from isolated mammalian nuclei, *Nucleic Acid Res.,* 11, 1475–1489, 1983.
90. Singh, H., Sen, R., Baltimore, D., and Sharp, P. A., A nuclear factor that binds to a conserved sequence motif in transcriptional control elements of immunoglobulin genes, *Nature,* 319, 154–158, 1986.
91. Searle, P. F., Zinc dependent binding of a liver nuclear factor to metal response element MRE-a of the mouse metallothionein-I gene and variant sequences, *Nucleic Acids Res.,* 18, 4683–4690, 1990.
92. Czupryn, M., Brown, W. E., and Vallee, B. L., Zinc rapidly induces a metal response element-binding factor, *Proc. Natl. Acad. Sci. USA,* 89, 10395–10399, 1992.
93. Otsuka, F., Iwamatsu, A., Suzuki, K., Ohsawa, M., Hamer, D. H., and Koizumi, S., Purification and characterization of a protein that binds to metal responsive elements of the human metallothionein II_A gene, *J. Biol. Chem.,* 269, 23700–23707, 1994.
94. Galas, D. J. and Schmitz, A., DNAase footprinting: a simple method for the detection of protein-DNA binding specificity, *Nucleic Acids Res.,* 5, 3157–3170, 1978.
95. Maxam, A. M. and Gilbert, W., A new method for sequencing DNA, *Proc. Natl. Acad. Sci. USA,* 74, 560–564, 1977.
96. Imbert, J. Zafarullah, M., Culotta, V. C., Gedamu, L., and Hamer, D., Transcription factor MBF-I interacts with metal regulatory elements of higher eucaryotic metallothionein genes, *Mol. Cell. Biol.,* 9, 5315–5323, 1989.
97. Labbé, S., Larouche, L., Mailhot, D., and Séguin, C., Purification of mouse MEP-1, a nuclear protein which binds to the metal regulatory elements of genes encoding metallothionein, *Nucleic Acids Res.,* 21, 1549–1554, 1993.
98. Séguin, C., A nuclear factor requires Zn^{2+} to bind a regulatory *MRE* element of the mouse gene encoding metallothionein-1, *Gene,* 97, 295–300, 1991.
99. Ephrussi, A., Church, G. M., Tonegawa, S., and Gilbert, W., B lineage-specific interactions of an immunoglobulin enhancer with cellular factors *in vivo, Science,* 227, 134–140, 1985.
100. Siebenlist, U. and Gilbert, W., Contacts between *Escherichia coli* RNA polymerase and an early promoter of phage T7, *Proc. Natl. Acad. Sci. USA,* 77, 122–126, 1980.
101. Hendrickson, W. and Schleif, R., A dimer of AraC protein contacts three adjacent major groove regions at the *ara I* DNA site, *Proc. Natl. Acad. Sci. USA,* 82, 3129–3133, 1985.
102. Miskimins, W. K., Roberts, M. P., McClelland, A., and Ruddle, F. H., Use of a protein-blotting procedure and a specific DNA probe to identify nuclear proteins that recognize the promoter region of the transferrin receptor gene, *Proc. Natl. Acad. Sci. USA,* 82, 6741–6744, 1985.
103. Séguin, C. and Prévost, J., Detection of a nuclear protein that interacts with a metal regulatory element of the mouse metallothionein 1 gene, *Nucleic Acids Res.,* 16, 10547–10560, 1988.
104. Koizumi, S., Suzuki, K., and Otsuka, F., A nuclear factor that recognizes the metal-responsive elements of human metallothionein II_A gene, *J. Biol. Chem.,* 267, 18659–18664, 1992.

105. Koizumi, S. and Otsuka, F., Nuclear proteins binding to the human metallothionein-II_A gene upstream sequences, *Ind. Health,* 32, 193–206, 1994.
106. Markovitz, A., Ultraviolet light-induced stable complexes of DNA and DNA polymerase, *Biochim. Biophys. Acta,* 281, 522–534, 1972.
107. Lin, S.-Y. and Riggs, A. D., Photochemical attachment of *lac* repressor to bromodeoxyuridine-substituted *lac* operator by ultraviolet radiation, *Proc. Natl. Acad. Sci. USA,* 71, 947–951, 1974.
108. Hillel, Z. and Wu, C.-W., Photochemical cross-linking studies on the interaction of *Escherichia coli* RNA polymerase with T7 DNA, *Biochemistry,* 17, 2954–2961, 1978.
109. Chodosh, L. A., Carthew, R. W., and Sharp, P. A., A single polypeptide possesses the binding and transcription activities of the adenovirus major late transcription factor, *Mol. Cell. Biol.,* 6, 4723–4733, 1986.
110. Labbé, S., Prévost, J., Remondelli, P., Leone, A., and Séguin, C., A nuclear factor binds to the metal regulatory elements of the mouse gene encoding metallothionein-I, *Nucleic Acids Res.,* 19, 4225-4231, 1991.
111. Hsiung, N., Fitts, R., Wilson, S., Milne, A., and Hamer, D., Efficient production of hepatitis B surface antigen using a bovine papilloma virus-metallothionein vector, *J. Mol. Appl. Gen.,* 2, 497–506, 1984.
112. Séguin, C., Felber, B. K., Carter, A. D., and Hamer, D. H., Competition for cellular factors that activate metallothionein gene transcription, *Nature,* 312, 781–785, 1984.
113. Richards, R. I., Heguy, A., and Karin, M., Structural and functional analysis of the human metallothionein-I_A gene: differential induction by metal ions and glucocorticoids, *Cell,* 37, 263–272, 1984.
114. Stennard, F. A., Holloway, A. F., Hamilton, J., and West, A. K., Characterization of six additional human metallothionein genes, *Biochim. Biophys. Acta,* 1218, 357–365, 1994.
115. Karin, M. and Holtgreve, H., Nucleotide sequence requirements for transient expression of human metallothionein-II_A-thymidine kinase fusion genes, *DNA,* 3, 319–326, 1984.
116. Tyrrell, R. M., Applegate, L. A., and Tromvoukis, Y., The proximal promoter region of the human heme oxygenase gene contains elements involved in stimulation of transcriptional activity by a variety of agents including oxidants, *Carcinogenesis,* 14, 761–765, 1993.

CHAPTER 2

Occupational and Environmental Exposures

Chris Winder, Cheng-Long Bai, and Neill H. Stacey

CONTENTS

0-8493-4493-X/97/$0.00+$.50

1 INTRODUCTION

The metals represent a different aspect of toxicology in that they do not undergo breakdown into other metals, and their absorption, disposition, and excretion are largely dependent on physical factors such as solubility, ionization, particle size, and chemical species (for metal salts).

About 70 to 80 elements in the Periodic Table (see Figure 1) are considered metals. Groups Ia and IIa, the "s block" metals, form monovalent and divalent cations, respectively. Groups IIIb to VIb constitute the "p block" elements and include metals that can have ions of different valencies. These are called the transition elements.

Of the metal elements, about 40 are considered to be "common" metals; however, less than 30 have compounds that have been reported to produce toxicity. The importance of some of the rarer metals may become apparent with emerging changes in technology, such as microelectronics and superconductors.

Metals are probably some of the oldest toxicants known to humans. Health effects, such as colic, were reported following exposure to lead, arsenic, and mercury over 2000 years ago. On the other hand, metals such as cadmium, chromium, and nickel belong to the modern era.

The toxicity of a metal is only partially related to its position in the Periodic Table. Toxicity decreases with the stability of the configuration of electrons in the atomic nuclei.[1] This produces a number of properties that can affect toxicity:

- The very light metals (Be and Li) have very small ionic radius and therefore higher charge-to-mass ratios; these are quite toxic metals.
- Increasing electropositivity increases toxicity (for example, Zn $<$ Cd $<$ Hg; Al $<$ Ga $<$In $<$ Tl). Electropositivity increases to the left and down the Periodic Table.
- Highly electropositive elements such as those in the first two columns (the alkali metals and alkali earth metals) appear in the biological environment primarily as free cations (that is, charged ions). This increases toxicity by virtue of increased bioavailability.
- Metals lower down the periodic table are potentially the most toxic; however, this toxicity only becomes apparent in those lead, mercury, and thallium salts that are relatively soluble.
- Various oxidation states (in the transition elements) are also important; for example, Mn(VII) is more toxic than Mn(II), and As(III) is more toxic than As(V). Again, electronic configuration is important in expression of toxicity.

One other factor important in the expression of the toxicity of metals is the relative amount of the metal produced. The annual production of a metal such as chromium is about 3 million tons. The potential for human or environmental exposure is much greater than, say, the platinum group metals, where annual production is in the range of 200 tons a year. While it is likely that chromium is generally more toxic than platinum, the huge amount in production and use increases the potential for its toxicity to be expressed.

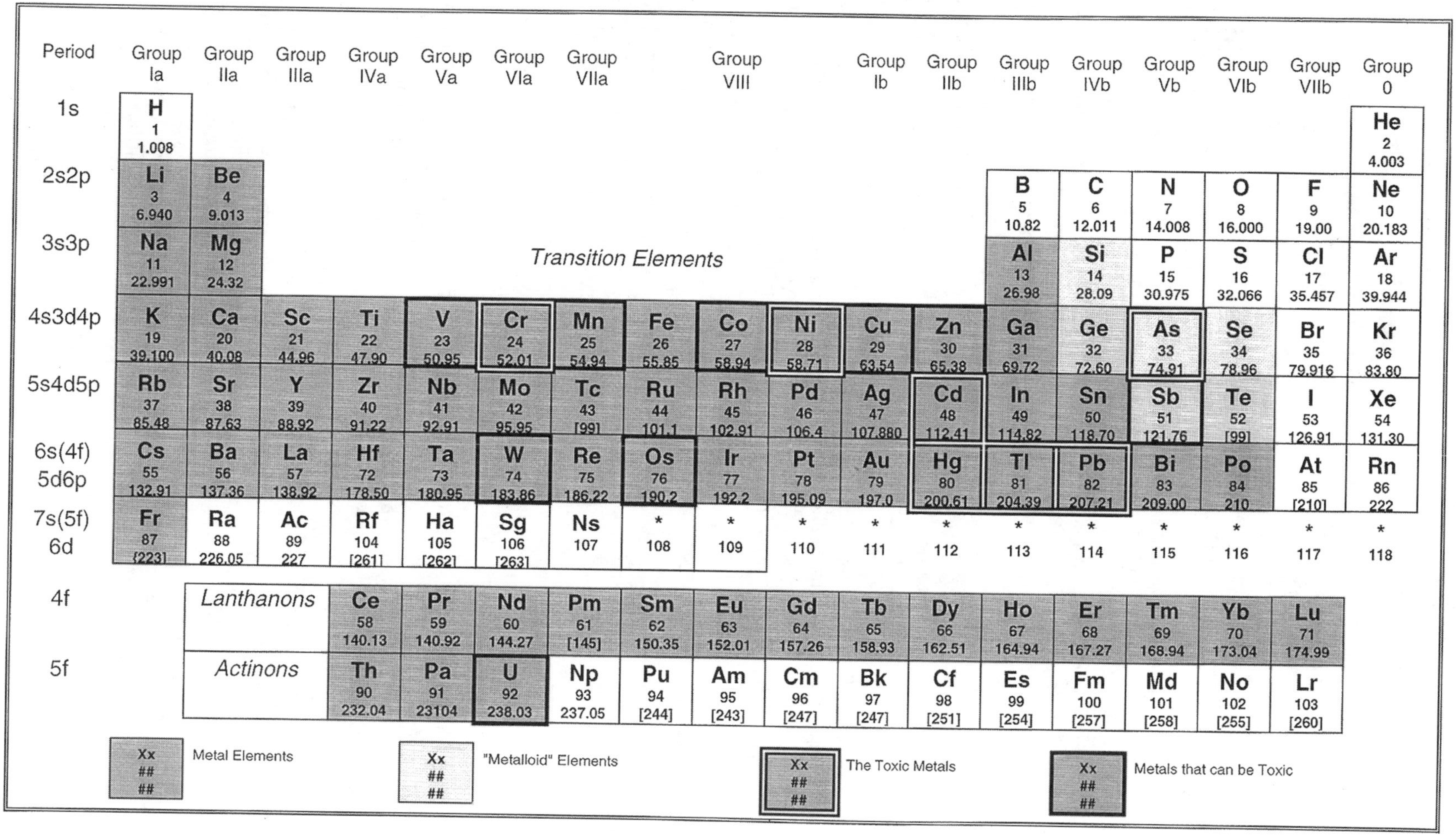

Period	Group Ia	Group IIa	Group IIIa	Group IVa	Group Va	Group VIa	Group VIIa	Group VIII			Group Ib	Group IIb	Group IIIb	Group IVb	Group Vb	Group VIb	Group VIIb	Group 0
1s	H 1 1.008																	He 2 4.003
2s2p	Li 3 6.940	Be 4 9.013											B 5 10.82	C 6 12.011	N 7 14.008	O 8 16.000	F 9 19.00	Ne 10 20.183
3s3p	Na 11 22.991	Mg 12 24.32					*Transition Elements*						Al 13 26.98	Si 14 28.09	P 15 30.975	S 16 32.066	Cl 17 35.457	Ar 18 39.944
4s3d4p	K 19 39.100	Ca 20 40.08	Sc 21 44.96	Ti 22 47.90	V 23 50.95	Cr 24 52.01	Mn 25 54.94	Fe 26 55.85	Co 27 58.94	Ni 28 58.71	Cu 29 63.54	Zn 30 65.38	Ga 31 69.72	Ge 32 72.60	As 33 74.91	Se 34 78.96	Br 35 79.916	Kr 36 83.80
5s4d5p	Rb 37 85.48	Sr 38 87.63	Y 39 88.92	Zr 40 91.22	Nb 41 92.91	Mo 42 95.95	Tc 43 [99]	Ru 44 101.1	Rh 45 102.91	Pd 46 106.4	Ag 47 107.880	Cd 48 112.41	In 49 114.82	Sn 50 118.70	Sb 51 121.76	Te 52 [99]	I 53 126.91	Xe 54 131.30
6s(4f) 5d6p	Cs 55 132.91	Ba 56 137.36	La 57 138.92	Hf 72 178.50	Ta 73 180.95	W 74 183.86	Re 75 186.22	Os 76 190.2	Ir 77 192.2	Pt 78 195.09	Au 79 197.0	Hg 80 200.61	Tl 81 204.39	Pb 82 207.21	Bi 83 209.00	Po 84 210	At 85 [210]	Rn 86 222
7s(5f) 6d	Fr 87 [223]	Ra 88 226.05	Ac 89 227	Rf 104 [261]	Ha 105 [262]	Sg 106 [263]	Ns 107	* 108	* 109	* 110	* 111	* 112	* 113	* 114	* 115	* 116	* 117	* 118
4f		*Lanthanons*		Ce 58 140.13	Pr 59 140.92	Nd 60 144.27	Pm 61 [145]	Sm 62 150.35	Eu 63 152.01	Gd 64 157.26	Tb 65 158.93	Dy 66 162.51	Ho 67 164.94	Er 68 167.27	Tm 69 168.94	Yb 70 173.04	Lu 71 174.99	
5f		*Actinons*		Th 90 232.04	Pa 91 23104	U 92 238.03	Np 93 237.05	Pu 94 [244]	Am 95 [243]	Cm 96 [247]	Bk 97 [247]	Cf 98 [251]	Es 99 [254]	Fm 100 [257]	Md 101 [258]	No 102 [255]	Lr 103 [260]	

Xx ## ## Metal Elements

Xx ## ## "Metalloid" Elements

Xx ## ## The Toxic Metals

Xx ## ## Metals that can be Toxic

FIGURE 1
The Periodic Table: toxic metals.

2 CONCEPTS OF EXPOSURE

The most fundamental principles of chemical safety relate to the dose-response, or exposure-effect, relationship. This relationship indicates a number of important toxicological and chemical safety concepts:

- Increase in exposure is directly proportional to an increase in effect. This means that the higher the exposure, the greater the likelihood of adverse outcomes or the greater the intensity of effects.
- The concept of a no-observable-adverse-effect level (NOAEL), where a small amount of exposure may not produce any effects: In chemical safety, the NOAEL is often used to establish a "safe" level of exposure.
- The concept of a "threshold" below which a safe level of exposure exists and above which effects are seen: Thresholds are used extensively in chemical safety and may be used in the occupational and environmental health context to establish exposure standards or permissible limits.

These three principles form the basic philosophy behind the prevention of exposure to chemical hazards. Preventive measures cover a range of options but their main aim is to reduce exposures to at least below the threshold, into the safe region of the exposure response relationship.

Humans are not laboratory animals in a toxicity test, where exposures can be carefully defined, administered, and measured. Therefore, the individual variability of humans in actual exposure and in their ability to respond and cope with exposure is much greater. There is also a large range in the types of exposure that can occur.

2.1 Questions To Ask in Assessing Exposure to Chemicals

In establishing levels of risk to chemical exposure, the most important task (after identification of the hazard) is assessment of the exposure. The following questions would need to be answered in an ideal exposure assessment, though not all would be required for estimating, say, worker exposure to workplace contaminants (see Table 1).

2.2 Concepts of Exposure

By the very nature of human activities, the concept of exposure is not immediately amenable to representation by a single number, as there is usually variable exposure over any given activity. There are two ways in which this variable exposure can be made presentable: the time-weighted average and the peak exposure.

2.2.1 *The Concept of Time-Weighted Average*

For some exposures, it is not the variability in exposure that is important; it is the actual amount of exposure. In such cases, it is possible to average the exposure, taking into account the amount of time spent at each concentration on a *pro rata* basis and calculating an average exposure over the duration of exposure. This is called a time weighted average, or TWA. For most chemical exposures, the TWA estimation is sufficient to quantify exposure. This concept is also known as the area under the curve (AUC), sometimes used to quantify exposure in pharmaco- or toxico-kinetics.

TABLE 1

Questions To Be Answered in an Exposure Assessment

Question	Issues To Be Addressed
Who is exposed?	Health status
	Age
	Gender
	Ethnicity
How many are exposed?	A single individual
	A small group of people
	Large groups
How are they exposed?	At work
	As consumers
	In the community
	Environmentally
By what route?	Oral
	Inhalational
	Dermal
	Multiple routes
In what pattern?	Amount per year
	Amount per case or incident
	As part of ambient environment
	Accumulatively
How frequently?	Continuously
	Regularly (8 hours/day, 3 pills/day)
	Periodically (once a month)
	Irregularly, but repeatedly
	In single incidents
Details of exposure	Manufacture
	Production or extraction
	Storage
	Transport
	Chemical use
	Discharge of waste
	At home
	Environmental

2.2.2 *The Concept of Peak Exposure*

For other exposures, it is not the amount of exposure that is important; it is the intensity. In such cases, the peak exposures can be used as a measure of exposure. Irritant chemicals, such as chlorine or ammonia welding fumes, produce many of their effects at peaks of exposure, rather than at lower and more constant exposures, and TWA estimations do not apply well in such circumstances. Depending on the situation or chemical hazard, either TWA or peak values can be used to represent exposure.

2.2.3 *Acute and Chronic Exposures*

"Acute" and "chronic" are words used to describe the types and intensities of exposures that can occur. Acute generally means single or short intense exposures, and chronic exposure usually means low-level exposure for long periods of time; however, the use of these terms to denote exposure is imprecise, can confuse, and should be discouraged. These words are best left to describe the health effects that may be caused.

2.2.4 *Single, Short-Term, and Long-Term Exposures*

The preferred terms to use when describing exposure are

- *Single exposure*, for one (only) exposure of any duration
- *Short-term repeated exposure*, for a few exposures, normally of short duration
- *Long-term repeated exposure*, for prolonged or regular exposure that has been occurring over a period of many months or years

It is important to note that the use of these terms does not imply anything about the intensity of exposure.

2.2.5 *Intensity of Exposure*

The final component in establishing the risk of adverse health effects is the amount of exposure that occurred. Chemicals have a range of toxicities from slight to severe. This means that the more toxic the chemical, the less exposure would be required to cause harm. Some chemicals, therefore, will cause effects with single or short-term exposures (for example, mercuric isocyanate), while others require long periods of exposure before they produce any effects (lead, cadmium). While exposures can be qualitatively described as low level, moderate, and high level, if it is at all possible, they should be quantified.

3 EXPOSURE TO METALS

Metals can be widely distributed in the environment by geological, meteorological, biological, environmental, and anthropogenic activities. For most individuals, the greatest cause of metal exposure is due to metal content in food, with a smaller additional component coming from air. Other potential exposures to metals in the nonoccupational environment include:

- Use of metals as therapeutic materials (barium as an X-ray contrast agent; lithium used in treatment of depression; platinum used as a chemotherapeutic agent; accumulation of some surgical implants that may be metal; aluminum accumulating in dialysis patients; gold colloids used in the treatment of arthritis)
- Pediatric mishaps (such as children swallowing mercury batteries)
- Consumer products such as deodorants (zirconium), vitamin and mineral supplements (selenium), hair dyes (silver and lead), and cosmetics (lead, antimony, and copper)
- Naturally occurring areas of high levels of some minerals, such as selenium
- Exposure to industrial wastes and pollution

For workers, occupational exposure to metals occurs in a wide variety of occupations.

3.1 General Principles

Exposure to metals and metal-containing compounds is common to many industrial, nonindustrial, and environmental situations. Absorption of metals can have may effects on the body, not all of them adverse. It must be remembered that some metals are essential for the normal function of the body. Examples include cobalt, copper, iron, magnesium, manganese, selenium, zinc.

Before a discussion of the exposure of metals, it is necessary to consider some general properties which have a bearing on the expression of toxicity:[2]

- Metals seldom interact with biological systems in the elemental form and are usually active in the ionic form.
- Availability of metal ions to biological processes is often dependent on solubility. Soluble salts of metals readily dissociate in the aqueous environment of biological membranes, making transport into the body easy; whereas, insoluble salts are poorly absorbed (for example, reduction of chromium(VI) to the less soluble chromium(III) will decrease its absorption).
- Absorption of soluble salts may be modified by formation of insoluble compounds in biological materials (for example, high dietary levels of phosphate will reduce absorption of lead because of the formation of insoluble lead phosphate).
- Some metals are produced as alkyl compounds. These often are very lipid soluble and pass readily across the lipid phase of biological membranes (examples include methylmercury and organotin compounds).
- Strong attractions between metals ions and organic compounds will influence the disposition of metals and their rate of excretion. Most of the toxicologically important metals bind strongly to tissues, are only slowly excreted, and therefore tend to accumulate on continuing exposure. Affinities for different tissues vary — elements such as lead are bound in bone, whereas mercury and cadmium localize in the kidney.

Some metals have only minor occupational uses, with an associated low risk of production of adverse effects; however, these metals may cause adverse effects in other applications (for example, in therapeutic use or pediatric poisoning).

3.2 Uses of Metals

Humanity has been using metals for millennia. The terms "bronze age" and "iron age" delineate early technological development of the human race. Further developments in alchemy, mining, engineering, industry, and modern chemistry have enabled human beings to identify, extract, refine, and ultimately exploit the mineral resources available on the Earth.

The uses of the metals are enormous, from the use of pure metal to alloys (mixtures of metals) to inorganic and organic compounds of metals. Table 2 shows some applications of some of the commercially useful metals.

The mining, extraction, industrial application, disposal, and environmental dispersion of metals are not without risks. Hazards to workers, the public, and the environment are possible in what is essentially a process of increasing the purity of the metal content of minerals to commercially useful concentrations.

4 SPECIFIC METAL EXPOSURES

Not all metals are of concern with regard to toxic effects, and not all toxic metals are important from the perspective of occupational and environmental health. Some metals, such as arsenic, lead, and mercury, have a long history, mainly because of their use as poisoning agents, but also for their now classic causes of disease. Other metals have come to the fore as new technologies and processes require their use (for example, beryllium, chromium, or uranium). The amount of information on health effects varies from metal to metal, and of course is dependent on a number of factors, including inherent toxicity, availability to workers, the volume of production, and the types of processes in which the metal is employed. Some of the more important occupational and environmental metals are described below.

TABLE 2
Commercial Uses of Metals

Metal	Uses
Aluminum	Packaging; building and transport; water treatment; medical (deodorants, antacids)
Antimony	Alloy manufacture
Arsenic and arsine	Pesticide (now declining); arsine is a contaminant
Barium	Radiographic agent
Beryllium	Nuclear industry
Bismuth	Low melting alloys; silvering of mirrors; dentistry; superconductors; cosmetics; therapeutic agents
Cadmium	Alloys; electroplating
Chromium	Electroplating; tanning (dichromates); safety match manufacture; pigments
Cobalt	Alloy manufacture (jet engines, turbines); radiation source
Copper	Plumbing; algicide; electrical industry; electroplating
Gallium	Integrated circuit boards; electronics; therapeutic agent
Germanium	Electronics; semiconductors
Gold	Jewelry; electroplating
Indium	Solder alloys; semiconductors
Iron	Iron and steel products
Lanthanides	Steel alloys; carbon arc electrodes; lens production; glass and ceramics pigments; cigarette lighter flints
Lead	Battery manufacture; pigment manufacture
Lithium	Nuclear industry; alkaline batteries; treatment of depression
Magnesium	Aluminum alloys; antacid/laxative
Manganese	Mining and refining; alloys (steel industry); dry cell batteries; MMT petrol additive
Mercury	Chlorine production; gold extraction; scientific instruments; dentistry; battery manufacture; fungicide
Molybdenum	Steel alloys; paint pigments
Nickel	Mining and refining; coinage; stainless steel production; electroplating
Niobium	Steel alloys; arc welding rods
Platinum	Jewelry; catalyst (chemical manufacturing); cytotoxic drugs
Selenium	Electronics; glass and ceramics production; pigments; topical therapeutics
Silver	Photography; electrical equipment; coinage; jewelry
Tantalum	Medical applications; electronics
Tellurium	Copper alloys; catalysts
Thallium	Semiconductors; electronics; lenses; contaminant in other metals; rodenticide (now discontinued)
Thorium	Incandescent mantles; nuclear industry
Tin	Tin plate; food containers; solder; marine antifouling paint
Titanium	Alloys (aeronautics); paint pigment; surgical prosthetics
Tungsten	Tool and drill manufacture; electric bulb filaments; pigment
Uranium	Nuclear industry
Vanadium	Alloys (steel industry); catalyst (chemical industry)
Yttrium	Television phosphor; superconductors
Zinc	Battery manufacture; galvanizing and electroplating; vulcanization of rubber; topical therapeutic
Zirconium	Alloys; nuclear industry; deodorants

4.1 ALUMINUM (AL)

A major use of aluminum metal is in the manufacture of aluminum cans — perhaps one third of all production is comprised of such packaging. The metal is also used in the construction industry, in transportation, in the electrical industry, and in consumer products. Alloys of alluminum have a range of uses, including powder metallurgical products, coatings, and reducing agents. Important inorganic aluminum compounds include aluminum sulphate (for water treatment) and potassium alum (for tanning and mordants), and synthetic zeolites are found in the paper industry and are used as a concrete accelerator. All soils contain aluminum compounds. However, not all minerals are economically viable. The term bauxite

TABLE 3

Human Exposure to Aluminum

Industrial/Occupational
Refining and smelting processes Production of aluminum alloys and compounds Production of fine aluminum dusts
Environmental
Air Mainly as aluminosilicates associated with dust particles. In rural areas air aluminum levels are normally less than 0.5 μg/m³, although in urban areas they are higher and can reach levels above 10 μg/m³ near point sources of pollution, such as cement plants.[4]
Water Some lakes, rivers, groundwaters, and domestic tapwater supplies can contain aluminum in high concentrations either naturally or because aluminum salts have been added as a flocculent in the purification process. Salinity, pH, and biological processes affect the concentration of dissolved aluminum, and concentrations can vary substantially.
Soil Exists almost exclusively in the form of silicate, hydroxides, and oxides. Release is possible from silicate in soil acidification (such as from acid rain). High concentrations of soil aluminum may cause root die back.
Domestic
Food Concentrations of aluminum in food are not high. Cereals, root vegetables, and meat and liver contain aluminum in the range of about 5 mg/kg. About 2.5 mg of aluminum is absorbed daily with food and undergoes minimal absorption through the gastrointestinal tract (0.1%), with total excretion of the absorbed dose through the kidneys.
Cooking in aluminum pots The contribution to total dietary intake is minimal.
Medical
Antacids A single dose normally represents about 50 times the average daily intake from other sources.
Dialysis[5] Implicated as a neurotoxic agent in the pathogenesis of Alzheimer's disease[6]

is used for sedimentary rocks that contain commercially extractable aluminum. Generally, bauxite is digested in the Bayer process at high temperatures and pressures with caustic soda. This produces aluminum hydroxide, which is reduced to alumina. Alumina is then reduced to metal aluminum in primary aluminum smelters using the Hall-Heroult method.[3] Occupational exposures to aluminum are considered less hazardous, however, than some of the medical exposures discussed in Table 3.

The concentrations of aluminum in blood of nonexposed subjects with normal renal function are extremely low (1 to 3 μg/l). Aluminum in plasma values of 5 to 15 mg/l have been reported in exposed workers, which is about ten times lower than the values observed in dialysis patients with encephalopathic symptoms.[7] Concentrations of aluminum in serum and urine reflect current exposure and the amount in the body. Care with respect to contamination is required with sample preparation, as the levels are low and the metal is widespread.[8,9] Urinary levels are regarded as more sensitive indicators of exposure in people with normal renal function, as concentrations in urine may be elevated while blood levels are barely altered. For workers chronically exposed to aluminum, samples collected one or two days after no exposure probably indicate body burden. Samples collected at the end of a shift are more likely to reflect the very recent exposure.

4.2 Arsenic (As)

Arsenic is widely distributed in the natural environment, especially in a large number of minerals. It is found most abundantly in sulfide ores as trivalent and pentavalent compounds. The Earth's crust and igneous rocks contain about 3 mg/kg arsenic, coal between 0.5 and 93 mg/kg with a mean value of 17.7 mg/kg, and brown coal up to 1500 mg/kg.[10] As a byproduct, arsenic trioxide is obtained during the production of copper and lead from sulfide ores. It is recovered from the flue dust in a reasonably pure form.[11] Occupational exposure to arsenic compounds takes place mainly among workers, especially those involved in the processing of copper, gold, and lead ores (see Table 4). In agriculture, the most frequent application of arsenic is in the preparation of insecticides, mainly as lead arsenate and less frequently as calcium arsenate and arsenite, sodium arsenite, cupric arsenite, cupric acetoarsenate. Other frequent applications of arsenicals include herbicides (weed killers for railroad and telephone posts), desiccants to facilitate mechanical cotton harvesting, fungicides, rodenticides, insecticides, algicides, and wood preservatives. Because of occupational and environmental risks, these applications are declining.[12,13] It has been noticed that the most significant human exposure to arsenic compounds originates from marine fish and shellfish. There may be more than 1 mg/kg of arsenic compounds in fish (often in the less toxic organic form), but normal is around 10 μg, which increases when contaminated water or food is consumed. Arsenic (As) in food (both fish arsenic and other compounds) is absorbed effectively from the gastrointestinal tract.[11,14,15] Although arsenic has almost exclusively been associated with criminal poisoning for many centuries, the matter of concern today is its contribution to occupational and environmental pollution through man's use of pesticides, nonferrous smelters, and coal-fired and geothermal power plants.[16]

The primary method used for biological monitoring for arsenic exposure determines inorganic arsenic, monomethylarsonic acid, and cacodylic acid in urine. This overcomes the problems of contamination from ingestion of seafood or water with high arsenic levels experienced when total arsenic is measured. The urinary concentration mainly reflects recent exposure. Sampling is best undertaken when the worker is into a normal routine for a day or two because it takes this much time for equilibrium to be reached. Arsenic has also been measured in the blood, and it also seems primarily to reflect more recent exposures. Hair and nails provide a good indication of inorganic arsenic entering the body during the growth period. External contamination of the hair with arsenic is a contamination problem for measurements involving those occupationally exposed.[19]

4.3 Beryllium (Be)

Beryllium is the 35th most abundant element in the Earth's crust, with an average content of about 6 mg/kg. Apart from the gemstones, emerald (chromium-containing beryl) and aquamarine (iron-containing beryl), only two beryllium minerals are of economic significance.[20,21] The annual global production of beryllium minerals in the period 1980 to 1984 was estimated to be about 10,000 tons, which corresponds to approximately 400 tons of beryllium.[20] In general, beryllium emissions during production and use are of minor importance compared with emissions that occur during the combustion of coal and fuel oil, which have natural average contents of 1.8 to 2.2 mg Be per kg dry weight and up to 100 μg Be per liter, respectively. Beryllium emission from the combustion of fossil fuels amounted to approximately 93% of the total beryllium emission in the U.S., one of the main producer countries.[1,21,22] Approximately 72% of the world production of beryllium is used in the form of beryllium-copper and other alloys in the aerospace, electronics, and mechanical industries. About 20% is used as the free metal, mainly in aerospace, weapons, and nuclear industries. The remainder is used as beryllium oxide for ceramics applications, principally in electronics

TABLE 4

Human Exposures to Arsenic

Industrial/Occupational
Alloying (accounts for about 90% of the production)
Metallic mining
Pest control with arsenical pesticides
Making gallium arsenide for dipoles and other electronic devices
Doping agent in germanium and silicon solid-state products
Solders
Cutting and sawing on wood pretreated with arsenic preservation
As catalyst in the manufacture of ethylene oxide
Semiconductor devices
Glass industry (As_2O_3, As_2Se, As_2O_5, metallic arsenic)
Colors for digital watches
Textile and tanning industries
Manufacture of pigments
Antifouling paints
Light filters (thin sheets of As_2O_5)
Ceramics industry (AsO_5)
Manufacture of fireworks (As_4S_4)[17]
Environmental
Naturally occurring sources
Volcanoes, sulfide ores
Artificial sources
Metallic mining, pesticide application, veterinary medicine, airborne emission from the smelting of metals (mainly nickel-copper smelters)
Air
Present mainly in particulate form as arsenic trioxide, with background levels of 1 to 10 ng/m^3 in rural areas and 20 ng/m^3 in urban areas[18]
Soil
Level in soil is about 7 mg/kg, but can be as high as 1000 mg/kg in the vicinity of metal smelters and in agricultural soil where extensive use was made of pesticides, herbicides, and defoliants
Water
Groundwater normally contaminated by metallic mining, metal smelters, pesticide application, and in appropriated industrial waste disposal[13,15]
Domestic
Criminal poison
In cosmetics as depilatory agents
Consumption of seafood (fish and shellfish), sprayed fruit, vegetables
Artwork (painting, photography, sculpture)
Ceramics
Tobacco smoking
Drugs[10,15,16]

and microelectronics (see Table 5).[20,23] In the environment, water contains very little beryllium because the small amount that escapes capture by clay minerals during rock weathering and soil formation is largely adsorbed by the surfaces of mineral grains. In soil, samples collected from about 1300 localities throughout the U.S. contained from less than 1 to 15 ppm of beryllium, averaging about 1 ppm.[23]

The usefulness of blood and urine beryllium levels as quantitative indicators of exposure remains to be established. At the moment, detection of this metal in blood and/or urine only indicates that exposure has occurred. There is some evidence that urinary beryllium reflects current exposure but it also seems that beryllium can remain detectable in the urine even years after exposure has ceased.[28]

TABLE 5
Human Exposures to Beryllium

Industrial/Occupational
Coal and fossil fuel combustion
Nuclear industry
Power plants
Hardening of copper
Space optics, missile fuel, and space vehicles
X-ray windows
Alloy component
Navigational systems, aircraft/satellite and missile parts[24]
Environmental
Naturally occurring sources
Abundance in Earth's crust, silicate minerals, and certain fossil fuels
Artificial sources
Coal and fuel combustion and beryllium-extraction plants, ceramics artists, and others
Air
Atmospheric beryllium concentrations at rural sites in the U.S. range from 0.03 to 0.06 ng/m^3; annual average beryllium concentrations in urban air in the U.S. found to range from <0.1 to 6.7 ng/m^3 [20,23]
Soil
Soil concentration generally range from 0.1 to 40 ppm, with the average around 6 ppm[25]
Water
Analysis of surface and rain waters in the U.S. have shown that beryllium concentrations are well below 1.0 µg/l[26]
Food chain
Food is not a significant source of human exposures; so far there is no evidence that beryllium is moving from soils into food or feed plants in amounts considered detrimental[27]
Domestic
Ceramics artists
Dental casting alloy
Jewelry
Food[23,24]

4.4 Chromium (Cr)

Chromium is an important commercial element with many uses. The metal exists in all oxidation states from (–II) to (VI), but only the trivalent (III) and hexavalent (VI) compounds and the metal are of practical importance. Of these, the hexavalent form is 100 to 1000 times more toxic than most common trivalent compounds. Biologically active chromium is an essential mineral, and a chromium deficiency has been identified in animals. Chromium metal is mostly used in the production of special (stainless) steels and is also used to electroplate other metals (see Table 6). Chromium is present in high concentrations in cement, where it may cause contact dermatitis. Chromium compounds are used as paint pigments and dyes, as a catalyst, to make magnetic tape, in tanning, in wood impregnation as a wood preservative, in safety match production, and so on.[29] The most important mineral of chromium is chromite. Virtually all (95%) of the world's economically viable resources of chromite are located in the southern part of Africa.

Due to the difference in the toxic effects between hexavalent and trivalent chromium, it is important that the contribution of each to levels measured in body fluids can be assessed. It seems that urinary and plasma levels mainly relate to hexavalent chromium exposures, but trivalent exposure may be responsible for some of the chromium measured. Sampling at the end of a work shift provides an indication of the recent exposure, but previous exposure may

TABLE 6

Human Exposures to Chromium

Industrial/Occupational
Extraction; most of the ore is reduced to ferrochrome, an iron-chromium alloy containing about 60% chromium, which is usually the main source of chromium, as the pure metal is not required for production
Production of ferrochrome
Pigment manufacture
Welding[30]
Electroplating
Timber preservation
Tanning
Environmental[a]
Air
Fly ash from incineration
Water
Disposal of industrial wastes, for example
Domestic
Biologically active chromium (high levels found in liver and cheese)
Cutlery (more than half of dietary chromium comes from sources other than the food itself, such as cutlery and in food preparation)

[a] Generally, chromium is found in nonindustrial environments in air at 10 ng/m³, in soil at concentrations of 10 to 90 ppm, and in fresh water at 1 to 10 μg/l. Under such conditions, chromium is found as the chromate. Chromium is found at concentrations in the air of industrial cities up to 70 ng/m³, and in water up to 25 μg/l.

contribute.[31] In regard to determinations in blood, it is relevant to note that exposure to hexavalent chromium results in its uptake by the red blood cells with subsequent reduction in the trivalent species. It has been suggested that differential estimation of plasma and red cell levels of chromium could provide a sensitive internal indication of exposure to the hexavalent form, but this requires further development. There is an indication that chromium levels are raised in hair after exposure.

4.5 Cadmium (Cd)

This metal, while being relatively rare, has a wide distribution in the Earth's crust and is often found in association with deposits of copper, zinc, and lead. Emissions from volcanoes are major sources of natural releases of cadmium to the environment. Cadmium has achieved prominence only in recent times.[32-34] Exposures associated with toxic effects in humans have occurred both environmentally and from work-related activities and are mainly related to production, consumption, and disposal of cadmium and other nonferrous metals (see Table 7). Primary uses of cadmium which result directly or indirectly in human exposures include production of batteries, as a protective plating for metals, and in plastics as pigments and stabilizers. It is also used as a component of some metal alloys, as well as in other products such as television phosphors, photographic equipment, lasers, and lithography. Disposal of materials containing cadmium presents a further source of potential exposure. Combustion of wood and fossil fuels releases cadmium, as does incineration of sewage sludge and refuse. Its presence in phosphate fertilizer may result in soil contamination and uptake by plants and, subsequently, grazing animals. Use of sewage sludge as a fertilizer can also increase soil levels of cadmium.

TABLE 7

Human Exposures to Cadmium

Industrial/Occupational
Plating of metals
Soldering
Battery production
Alloy production
Pigment production
Plastics production
Smelting of nonferrous metals
Welding
Environmental
Volcano emissions
Black shale deposits
Mining operations involving cadmium
Smelting of nonferrous metals
Production of cadmium products
Disposal of cadmium-containing materials
Phosphate fertilizers
Iron and steel production
Fossil fuel and wood combustion
Garbage and sewage sludge incineration
Cement manufacture
Domestic
Tobacco smoke
Foodstuffs

Cadmium may be measured in whole blood, this being of greater use than serum because of cadmium binding to red blood cells. Cadmium in blood reflects more recent exposure, as opposed to longer-term exposure. Cadmium can also be detected in urine, this being more representative of body burden than short-term exposure.[35,36] This is due in part to the very long half-life of cadmium which is between 10 and 30 years.[37] Great care is required when interpreting biological monitoring data for cadmium because a recent high exposure can increase urinary levels without a corresponding increase in body burden. Furthermore, when kidney damage is considerable (as a result of long-term exposure to cadmium), greater amounts of cadmium are released into the urine due to decreased reabsorption of metallothionein which has cadmium bound to it. This can be in the absence of any recent high exposures. Thus, an integrated picture of body levels and likely exposure patterns are needed to best interpret the data. Detection of low-molecular-weight proteins, such as β-microglobulin and retinol-binding protein, have been used to indicate early kidney damage as a result of long-term cadmium exposure. As they are reflective of an adverse effect, they are better regarded as a health surveillance technique. Assay for metallothionein in urine has been reported to be a specific and sensitive index of increased levels of cadmium in exposed workers. Organ concentrations of cadmium have been measured using neutron activation, but this is not a technique that can be used routinely. Hair and feces determinations are also possibilities that have been addressed but have not found regular application. Levels of 5 μg cadmium per gram of creatinine and 0.5 μg cadmium per 100 ml have been suggested as appropriate urinary and blood levels, respectively, which would not be associated with the adverse effects of this metal.

TABLE 8

Human Exposures to Cobalt

Industrial/Occupational
Metallic mining and metal refining
Cobalt powder handling and grinding
Blue and green ceramic glazes
Colored glass superalloys (jet engines, gas turbines)
Pigments
Acetic acid markers
Magnet steel workers
As catalyst in the synthesis of heating fuels and alcohol
Nuclear technology[39,40]
Environmental
Natural occurring sources
Natural mineral cobaltite
Artificial sources
Production of cobalt-containing metal and cobalt salts
Air
Cobalt can be detected during production of hard metals and cobalt salt
Soil
Exceptionally low levels of cobalt were observed in soil of some areas of Australia, New Zealand, and the U.S.; cobalt sulfate can be used for land treatment
Water
In uncontaminated sample of fresh water, cobalt concentrations are generally low, ranging from 0.1 to 10 μg/l; however, in waters of polluted rivers high cobalt levels have been observed, e.g., 4500 μg/l in Mineral Creek, U.S.
Food chain
Only a few plant species accumulate cobalt above 100 ppm, which causes severe phytotoxicity.[38,39]
Domestic
Pigments
Artwork (painting, ceramic art, and sculpture)
Ceramics
Alloy as dental materials
Tobacco smoking
Medication (used in gamma-ray therapy of cancer, e.g., ^{60}Co[41])
Drug

4.6 Cobalt (Co)

Cobalt is a nutritionally essential metal, and deficiency results in severe health consequences. It is a relatively rare metal produced primarily as a byproduct of the metals, chiefly copper, and there is a certain amount of cobalt in soil. It is used in high-temperature alloys and in permanent magnets (see Table 8). Its salts are useful in paint driers, as catalysts, and in the production of numerous pigments.[38]

Its deficiency can cause health problems in humans. Although cobalt deficiency is not yet widespread, it is likely to become a problem in the future, as the natural cobalt content of soil is low, and cobalt is depleted as rivers transport it to the oceans. Because of mining activities and its widespread industrial uses, however, cobalt also belongs to the class of metals posing potential dangers due to excessive exposures. At risk primarily are metal workers.[39] Traces of cobalt are found in all rocks, minerals, and soils. The average cobalt content in the Earth's crust is 18 ppm. Cobalt usually occurs together with nickel and iron. Annually, approximately 21,000 tons of cobalt are transported by rivers to the oceans and about the same amount is deposited in deep-sea sediments, the content of which is about 74 ppm.[39]

This metal can be measured in blood, urine, nails, and hair. Available data suggest that measurement of pre- and post-shift urinary levels allows estimation of exposure during the work shift by calculation of the difference. As the work week progresses, levels increase such that sampling at the end of the week represents accumulation over that period. A sample taken at the very beginning of the work week is thought to be more representative of long-term exposure.[42]

4.7 Lead (Pb)

Lead is a metal of antiquity and has been used for many purposes for thousands of years. Some of the uses for lead in ancient times (such as lead sheet for lining roofs) remain today. Other uses, such as white lead in paint, have been discontinued for obvious toxicological reasons, and still others have been developed for modern applications. About 40 to 50% of all lead production is used to make lead-acid batteries (both lead metal and lead oxides are used). A further 20% of production is used as the metal, for example, lead sheet, cable sheathing, solder, ammunition, alloys, weights, ballast, and low-melting alloys (see Table 9). Lead-based pigments have a long tradition of being used in paints, although they have virtually all been substituted by other pigments in the last 20 years. Lead compounds are used in glassware and ceramics and as a stabilizer in plastics. Red lead is used to make television tubes. About 10% is converted into alkyl-lead compounds and used as antiknock additives in gasoline, although this use is in decline as more and more countries move to limit the concentration of such additives in gasoline. The most important minerals of lead are galena (lead sulphide), cerussite (lead carbonate), and anglesite (lead sulphate). All are extracted through processing of the crude ore, roasting, sintering, reduction, and refining. The refining process also extracts zinc, copper, gold, silver, antimony, and arsenic.[43] While not strictly a lead exposure, physiological conditions such as pregnancy, infection, or menopause may mobilize lead sequestered in body bone stores. The total daily intake of lead varies considerably from area to area and country to country. Most studies report a daily intake between 20 and 200 μg/day in adults, with the recommendation that lead intake in children be much less, owing to a significantly higher lead intake on a body weight basis.

Background levels of lead in human blood are of the order of 5 to 15 μg/100 ml. While blood lead levels continue to fall, recent initiatives recommend that lead blood levels in children, particularly preschool infants, should not rise above 10 μg/100 ml.[47] Since about the late 1960s, biological monitoring of lead blood levels in workers has been employed to assess lead exposures, with removal of workers whose blood levels of lead exceed a mandated amount. While this practice is not preventive, it establishes the upper limit of exposure in workers; this value was 70 to 80 μg Pb per 100 ml in the early 1970s but has fallen to 40 to 60 μg Pb per 100 ml in the mid-1990s. The measurement of exposure of humans to lead can be carried out in virtually any biological media. Lead has been measured in blood, urine, sweat, hair, nails, saliva, milk, and so on. The most used methods are lead in capillary or venous blood and lead in urine (preferably 24-hour collection). Lead in blood represents recent exposure (over the past 3 to 8 weeks) and can range up to 10 to 15 μg Pb per 100 ml in nonexposed populations and 40 to 60 μg Pb per 100 ml in workers in lead processes (intervention and medical removal usually occurs in workers with blood leads above these values). Occupationally, lead in blood remains the method of choice for biological monitoring.

In addition to biological monitoring of absorption, it is also possible to carry out biological measurement of effect. The several enzymes in the haem and porphyrin synthetic pathway are inhibited by lead, and the rate-limiting enzyme in the synthetic pathway — δ-amino-levulinic acid dehydratase — is one of the most sensitive systems affected by lead, with inhibition beginning at a blood lead level of 10 to 20 μg Pb per 100 ml, and almost total inhibition at 70 to 80 μg Pb per 100 ml. A number of properties of this system have

TABLE 9

Human Exposures to Lead

Occupational[a]
Extraction process
Lead acid battery manufacture
Shipmaking
Car radiator repair
Welding
Paint manufacture or application
Plumbing
Gasoline manufacture
Plastics manufacture
Environmental
Mining, smelting
Processing
Use
Recycling
Disposal[44]
Air
The major part of lead found in the atmosphere results from the combustion of leaded gasoline (atmospheric transport of minute, particulate, airborne lead may range over hundreds and thousands of kilometers). Air lead concentrations vary from below 0.01 μg/m³ in remote areas, below 0.02 μg/m³ in rural areas, 0.02 to 2.0 μg/m³ in urban areas, and 1 to 2 μg/m³ near lead smelters. The World Health Organization (WHO) recommended limit is 0.5 to 1.0 μg/m³ as a long-term average. The stepwise decremental decreases of lead in air (and lead in blood in the population) from those countries removing lead from gasoline establishes car exhaust as a major source of environmental lead pollution.[45] Lead smelting and refining (including secondary refining/recycling) are also known to give rise to substantial lead emissions. Decreases in emissions have occurred, due to improvements in air pollution control technology and better air quality regulation.
Water
The major contributor to lead in water is from fallout of lead in the air. Lead in water comprises dissolved lead and suspended matter. Water lead concentrations vary from below 0.02 to 2.0 μg/l in rural areas, 1 to 40 μg/l in urban areas, and 10 to 1000 μg/l near lead smelters. However, concentrations in water remote from point sources of pollution can increase by two orders of magnitude (to about 100 to 1000 μg/l) where water is plumbosolvent (for example, areas of acidic or "soft" water).
Soil
Lead levels in unpolluted areas range from 10 to 40 ppm, with most values below 20 ppm. Except in areas where the underlying rock contains appreciable amounts of lead, the top layers of soils tend to have higher lead levels than do deeper levels, due to atmospheric deposition. Significantly increased lead levels have been reported in surface soils in inner city areas, near busy highways, and near lead processing industries.
Sediments
In areas where significant discharges have occurred, concentrations of lead in sediments can be substantial. For example, 50 ppm lead has been reported in sediments in the Rhône River.[46]
Domestic
Plants growing in high lead soils or from surface deposition
Inadvertent addition of lead from food processing
Leaching of lead from cans with soldered seams
Improperly glazed crockery
Lead in cooking water
Alcoholic drinks; notably, wine (from bottle sealing) and "moonshine" spirits (from lead solder in distilling equipment) may contain substantial amounts of lead.
Lead in paint (by far the biggest source of domestic lead exposure where lead in paint is found)
Lead plumbing
Hair dyes and color restorers
Asian cosmetics
Health tonics
Tobacco smoking

[a] In the past, significant domestic contamination has also occurred from workers taking home their overalls. This has decreased with better hygiene and laundry arrangements.

TABLE 10
Human Exposures to Lithium

Industrial/Occupational
Advanced technologies Tritium production Battery manufacture
Environmental
Lithium concentrations in the environment are generally low.
Others/Domestic
Medical Treatment of depression[48] Increase tolerance to the side effects of cancer therapy Topical treatment of genital herpes

been used in biological monitoring, including blood or erythrocyte levels of δ-amino-levulinic acid dehydratase, the level of its substrate d-δ-levulinic acid in blood or urine, levels of intermediates in the porphyrin synthetic pathway such as uroporphyrinilinogen and coproporphyrinilogen intermediates, zinc protoporphyrin (ZPP), and free erythrocyte protoporphyrin (FEP). Of these, ZPP and FEP are most closely correlated with blood lead and are used to measure lead exposure in workers.

4.8 Lithium (Li)

Lithium is the smallest of the alkali metal ions. It is highly electropositive and forms compounds with a range of metals. Lithium has found practical applications in medicine and battery production, is used for cooling in nuclear reactors, is a necessary raw material in the production of tritium, and is used in the form of its isotope, ^{6}Li, as a thermonuclear fuel (see Table 10). Lithium is obtained by electrolysis of volcanic brine from lithium-containing materials.

Lithium can be measured in urine and plasma, but routine biological monitoring for environmental or occupational exposures has not been undertaken.[49]

4.9 Manganese (Mn)

Manganese and its compounds are used in making steel alloys, dry-cell batteries, electrical coils, ceramics, matches, glass dyes; in fertilizers and welding rods; as oxidizing agents; and as animal food additives (see Table 11). Heavy exposure may occur in mines and during the production of metals, during which workers are mainly exposed to manganese dioxide by inhalation. Manganese is also used in the chemical industry. Organic manganese compounds such as methylcyclopentadienyl manganese tricarbonyl (MMT) have been used to increase the octane level of gasoline.[50,51] Manganese is widely distributed in soil as an abundant element. Its average concentration in soil is probably about 500 to 900 mg/kg. Manganese pollution of water does not appear to be a problem, except possibly in isolated cases of seepage or leaking from waste disposal activities. In fresh water, both soluble and suspended forms exist. Atmospheric concentrations of manganese observed in urban areas can be attributed primarily to man-made sources. A principal source of atmospheric emissions is metallurgical processing.[52] Manganese is an essential metal and is present in all living organisms. While it is present in urban air and in most water supplies, the principal portion of intake is derived from food. Human daily intake ranges from 2 to 9 mg.

TABLE 11

Human Exposures to Manganese

Industrial/Occupational
Mining and raw material transport
Metallurgical processing and reprocessing (90% used in the making of steel)
Boilermaking
Shipyards
Welding and cutting torches (the welding rod)
Power plant
Environmental
Metallurgical processing
Power plant
Welding and cutting torches
Waste disposal (waste water and solid waste)
Incinerators
Domestic
Food intake
Dye
Ceramics
Incinerators

While manganese can be determined in urine, blood, feces, and hair, it has been difficult to establish a clear relationship between levels in the body and chronic toxicity. Detection in urine would appear to be satisfactory only to confirm qualitatively that exposure has occurred.[53,54]

4.10 Mercury (Hg)

Toxicological effects and relevant human exposures of mercury have been illustrated over the past few centuries. On the basis of toxicologic characteristics, there are three forms of mercury: elemental, inorganic, and organic compounds. Mining, smelting, and industrial discharge of mercury (especially in the paper pulp industry) have been major factors in occupational and environmental contamination in the past (see Table 12). Fossil fuel may contain as much as 1 ppm of mercury, and it is estimated that about 5000 tons of mercury per year may be emitted from burning coal, natural gas, and the refining of petroleum products.[55] The major source of mercury is the natural degassing of the Earth's crust, including land areas, rivers, and oceans, which is estimated to be on order of 25,000 to 150,000 tons per year. Metallic mercury in the atmosphere represents the major pathway of global transport of mercury. As much as one third of atmospheric mercury may be due to industrial release of organic or inorganic forms. Regardless of source, both organic and inorganic forms of mercury may undergo environmental transformation. Metallic mercury may be oxidized to inorganic divalent mercury, particularly in the presence of organic material formed in the aquatic environment. Methylmercury, an important form of organic mercury, can be taken up by fish and eventually consumed by humans; however, the concentration of inorganic mercury in food is generally very low, with daily intake below 1 μg/day.[55-57]

Due to the different forms and sources of mercury to which humans are exposed, biological monitoring of the substance and data interpretation are somewhat complicated. Urinary levels relate mainly to exposure to mercury vapor or to inorganic mercury. The number of amalgam fillings in an individual may contribute to the overall levels in the bodily fluids. Coincident exposure to mercury-containing disinfectants could also result in higher

TABLE 12
Human Exposures to Mercury

Industrial/Occupational
Mining, smelting, and industrial discharge of mercury
Metal refining
Paper pulp mill
Fossil fuel (coal, natural gas) burning
Production of chlorine and caustic soda
Refining of petroleum products
Catalysts
Pesticide application
Military applications (such as detonators)
Production of steel
Cement production
Phosphate and smelting of metals from their sulfide ores
Dental applications
Laboratory uses
Environmental
Metallic industry
Burning of fossil fuel (such as power stations)
Production of steel
Cement production
Phosphate and smelting of metals from their sulfide ores
Incinerators
Waste disposal
Dental amalgam filling
Domestic
Dental amalgam
Painting and ceramics
Measurement and control systems (such as thermometers)
Foodstuffs (especially for organic mercury)
Contaminated water and plants
Paints
Battery
Drugs

levels. Considerable work has been done which indicates that health effects may be experienced at levels above about 50 μg/g creatinine. The duration of exposure must be considered because concentrations in people only recently exposed will not have reached equilibrium. Care also seems to be required for consistency of sampling time, correction for specific gravity, and ensuring that the person has not been absent from the exposure source for some days. The contribution to body levels of mercury from fish is of lesser importance for urinary levels than it is for blood because it is in the form of methylmercury, which is not excreted in the urine. Blood levels primarily reflect more recent exposures and are useful in confirming occasional high exposures, especially in comparison to urinary values. Determination of mercury in saliva has been reported to be consistent with blood and urine levels, but this method has not developed as a widely used one.[58,59]

4.11 Nickel (Ni)

The exposure and toxicity of nickel is based on the various classes of nickel compounds and the human activities associated with them (see Table 13):

TABLE 13

Human Exposures to Nickel

Occupational[a]
Extracting
Refining — either (1) electrolytic refining to yield nickel cathodes; or (2) the Mond process, whereby the oxide is reduced with hydrogen, reacted with carbon monoxide to produce nickel carbonyl, and thermally decomposed to produce pure nickel. Both processes produce exposures that are hazardous to workers, although the Mond process is relatively more toxic
Alloy production
Catalysts in the chemical industry and in petroleum refining
Electroplating processes
Environmental
Weathering
Dissolution of rocks and soil
Atmospheric fallout
Industrial processes
Waste disposal
Air
Volcanic emissions and windblown dusts from weathering, from combustion of fossil fuels (both coal and oil), from mining, from industrial processes, and from incineration of wastes. Atmospheric levels average 6 ng/m³ in non-urban areas in the U.S., and in urban areas 17 ng/m³ in summer and 25 ng/m³ in winter, indicating a source from energy demand for heating. Close to industrial areas, atmospheric nickel concentrations as high as 170 ng/m³ have been recorded; close to nickel refineries, air nickel levels can average as much a 1200 ng/m³.[62]
Water
Usually contains less than 20 μg/l, although this can increase owing to pollution of supply, and acid rain has a tendency to mobilize nickel from soil and to increase nickel concentrations in groundwaters. This can produce increased uptake in soil microorganisms and plants, and, in turn, animals. Nickel leached from dump sites can contribute to contamination of the aquifer, with potential ecotoxicity and risk to humans.
Soil
Depends on soil type and pH, with high mobility in acid soils. Insoluble nickel may be deposited by precipitation, complexation, adsorption onto clay or silica, and uptake by biota.[63] With variability in microbial activity, ionic strength, and particle concentration, these processes may be reversed. Anthropogenic sources of soil nickel include emissions and wastes from mining and refining, atmospheric deposition from other industrial activities, and disposal of sewage sludge.
Domestic
Food
Nickel levels in nuts (5 ppm) and cocoa (10 ppm) are high, compared with a nickel concentration in most foods of below 0.5 ppm.[64] Daily human intakes are in the range 100 to 800 μg/day, depending on dietary habits.
Jewelry
The incidence of allergic contact dermatitis through exposure to nickel-containing jewelry is relatively high and possibly occurs in about 10% of exposed individuals. The sensitizing exposure usually comes from the piercing of ears or other body parts.

[a] Representative exposure data are difficult to obtain. Warner[61] reported time-weighted average concentrations in air in different workplaces where nickel may be found. High concentrations (above 1 mg/m³) were found in roasting and smelting operations, in electrolytic refining, and in foundry operations. Moderate levels (0.05 to 1.0 mg/m³) were found in stainless-steel welding, electroplating, and nickel-cadmium battery manufacture.

- Divalent nickel compounds have a low toxicity at concentrations found in the environment.
- Exposure to nickel metal (for example, from its use in coinage or stainless steel) is not considered harmful.
- In humans, adverse effects of water-soluble nickel compounds occur after skin contact (causing contact dermatitis in perhaps 10% of exposed individuals) and after inhalation (which causes respiratory tract irritation and asthma) in workers such as electroplaters.

- Human exposure to inorganic, water-insoluble nickel compounds usually occurs through inhalation of fumes or dusts, which are associated with cancers of the respiratory tract among workers in nickel refineries.
- The organic compound nickel carbonyl ($Ni(CO)_4$) is produced in the Mond refining process and, because of its volatility (BP of 43°C) and lipid solubility, is highly toxic and carcinogenic.

Nickel is a constituent of over 3000 metal alloys and is used for a huge range of purposes, such as coinage (some coins may be 99.8% nickel), stainless steel (which contains about 10% nickel), cooking utensils, corrosion-resistant equipment, aircraft parts, magnetic equipment, jewelry, rechargeable batteries, medical applications, ceramics, and so on.[60] Most of the world's nickel is processed from sulfide ores and, to a lesser extent, oxides. The largest producers are Canada, the former U.S.S.R., and New Caledonia.

Wide variations have been reported in body nickel levels. Further, because of decreased emissions and better analytical methods, exposures are decreasing, suggesting that earlier data are not representative of contemporary conditions. Background levels of nickel in human blood are less than 0.5 μg/l and about 2 to 3 μg/l in urine. Both blood and urine are used to monitor workers, although more data are available from urinary measurements. Elevated urine nickel levels are shown in some nickel workers, with the highest being in nickel refinery workers (with a mean of over 200 μg/l).[65] All other occupational groups had urinary levels below 20 μg/l. Nickel concentrations in bile suggest that biliary excretion may be quantitatively significant. Workers exposed to nickel compounds have elevated levels in both urine and plasma, with plasma being suggested as the more reliable measure.[66,67] At the present time, there is no clear relationship between health risks and levels in bodily fluids, except for overexposure to nickel carbonyl. A potentially complicating factor is the exposure to nickel compounds of low solubility. Slow clearance of these compounds may be responsible for elevated levels in workers for some years after cessation of exposure. It is also these low-solubility nickel compounds (nickel subsulfide) that are linked more to lung cancer. Nickel has also been determined in nasal mucosa and in hair, although these methods do not seem to have gained wide use.

4.12 Platinum (Pt)

The use of platinum has followed a number of stepwise increases due to technological developments. Initially, the metal was used for jewelry and coinage in the nineteenth century. Then, the invention of the process of ammonia oxidation in which platinum served as a catalyst allowed large-scale fertilizer production in the earliest part of the twentieth century; this process was also used to make ammunition for two world wars. From the time of the second world war, platinum has been used extensively by the petroleum industry in the catalytic "cracking" of high boiling crude oil fractions. More recently, due to the advent of tighter emission controls on vehicles beginning in the 1970s, platinum (with palladium and rhodium) is now used for exhaust catalysts that oxidize harmful combustion byproducts, such as nitrogen oxides, partially oxidized hydrocarbons, and carbon monoxide. Modern day uses of platinum derive from its exceptional catalytic properties, and the major use (probably over two thirds of production) is in catalytic converters in vehicles. Further industrial applications also depend upon its catalytic properties and to other characteristics, including resistance to chemical corrosion, a high melting point, high mechanical strength, and good ductility.[68] Platinum is used as a catalyst in chemical manufacturing, in electrical engineering, in instruments for high temperature measurement; in glass technology, jewelry, and dentistry; and for surgical implants. Specific complexes of platinum (most notably cisplatin) are used therapeutically as cytotoxic agents to treat cancer. Possible emerging uses include cathodic protection of steel and as a catalyst in hydrogen fuel cells (see Table 14). Platinum is found both as the metal and in a number of minerals, mainly ores of nickel and copper. Extraction of platinum

TABLE 14
Human Exposures to Platinum

Industrial/Occupational
Mining and processing of platinum from copper and nickel sulfide ores
Soluble platinum salts in the manufacture of metal catalysts and alloys
The cytotoxic drug cisplatin and its analogues: In a study of exposure in hospital personnel, urinary platinum levels ranged from 0.6 to 23 μg/l, were at the limit of the analytical method, and were not different from controls (2.6 to 15 μg/l). By comparison, an average of 7 mg/l was measured in the urine of cisplatin-treated patients.[69,70]
Environmental
Loss of platinum from catalysts in normal use can occur, although major amounts are emitted only by platinum employed for ammonia oxidation and the catalytic conversion of nitric oxide to nitric acid in its production, although this is lost into the product, not to the environment.
Vehicle exhausts (from catalytic converters)
Platinum jewelry

is not large, at about 100 to 150 tons a year. Total world production of platinum has been estimated at about 1500 tons. Naturally occurring levels of platinum in the environment are low.

Background levels in human blood are of the order of 0.1 to 2.8 μg/l. These concentrations are also reported in workers occupationally exposed to platinum, with the exception of hospital workers exposed to cisplatin where blood levels can range up to ten times this concentration. Platinum can be measured in both urine and blood, although in exposed workers urinary levels have been reported without simultaneously detectable plasma levels.[71]

4.13 Selenium (Se)

Selenium occurs in nature and in biological systems as selenate (Se^{6+}), selenite (Se^{4+}), elemental selenium (Se^0), and selenide (Se^{2-}); deficiency leads to cardiomyopathy in mammals, including humans. In most countries, the daily intake of selenium is about 100 μg.[72] Environmental selenium compounds may originate from metal smelting, coal combustion, or the disposal of waste (see Table 15). The Earth's crust contains an average selenium content of 0.05 to 0.09 ppm. Higher concentrations of selenium are found in volcanic rock (up to 120 ppm), sandstone, uranium deposits, and carbonaceous rocks. River water levels of selenium vary depending on environmental and geologic factors; 0.02 ppm has been reported as a representative estimate. Selenium has also been detected in urban air, presumably from sulfur-containing materials.[73] Selenium in foodstuffs provides a daily source of selenium. Seafood (especially shrimp), meat, milk products, and grains provide the largest amounts in the diet. The concentration of selenium in different foodstuffs varies considerably (0.01 to 1.0 mg/kg), depending on the origin of the food. Some plants accumulate selenium.[74,75]

Determination of selenium in blood or urine is often carried out to establish deficiency rather than excess. In selenium deficiency, blood and serum concentrations are below 40 μg/l.[76] Concentrations in bodily fluids vary with dietary intake and geographically, and their biological significance remains to be fully established.[77] Concentrations in plasma and urine may reflect recent exposures, while chronic exposures are better reflected by levels in red blood cells.

4.14 Thallium (Tl)

Thallium occurs in small amounts in sulfur-containing ores and potassium minerals. Following decline of its use as a rodenticide, thallium has virtually no economic or technical importance (see Table 16). Minor uses include cement additives, deep temperature thermometers, acid-resistant alloys, low-melting glasses, semiconductors, and scintigraphy of

TABLE 15

Human Exposures to Selenium

Industrial/Occupational
Mining, milling, smelting, and refining
Burning of fossil fuel
Electrolytic refining of copper
Rectifiers and burned-out rectifiers
Glass manufacture (flat-glass, pressed or blown glass, and glassware)
Catalyst for oxidation in chemical industry
Electronic usage (photoconductors and photoelectric cells)
Inorganic pigments (in plastics, paints, enamels, inks, rubber, and ceramics)
Color copper and copper alloy
Lubricants
Environmental
Emission from selenium mining, milling, smelting, and refining
Coal combustion
Waste disposal
Insecticide
Soil additive
Domestic
Foodstuffs
Food additives
Veterinary pharmaceuticals
Lubricants
Medical usages (drugs and radioactive diagnostic agent)

TABLE 16

Human Exposures to Thallium

Industrial/Occupational
Extraction from residues from the electrolytic smelting of copper, lead, zinc, and iron ores
Rodenticide (now discontinued)
Accidental, deliberate, and criminal misuse of thallium led to prohibition of its use in most countries in the 1970s and 1980s
Environmental
Some vegetation damage was reported around a cement plant in Germany which was associated with thallium emissions. Elevated thallium levels were seen in soils, plants, and cattle.[78]

the heart and circulatory systems. Some thallium is found in copper and selenium minerals, but generally thallium minerals are rare. Determination of urinary thallium is considered to be preferable to blood as an indicator of exposure but not a great deal is known about this relationship.[79]

4.15 Tin (Sn)

Tin is rather unique in the wide variety of its compounds and applications. Ever since the beginning of the bronze age, the metal and its alloys have been of importance to humans. Occupational exposure occurs during, for example, tin-plating (a major industrial application) and the production of alloys and solders (see Table 17). In general, most of the operations

TABLE 17

Human Exposures to Tin

Industrial/Occupational
Tin extraction from ore
Tin alloy
Tin ore packing (tin dust)
Smelting operation
Tin-plate containers (such as cans) production
Manufacture of collapsible tubes in the pharmaceutical and cosmetic industries
Corrosion resistant coating
Solder
Babbit metal
Brass
Bronze
Catalyst
Modent in dyeing of silk
Deposition of SnO_2 film on glass
Organotin: heat stabilizer (such as PVC stabilizer); transformer oil stabilizer; catalysts; biocides
Environmental
Tin extraction industry
Smelting operation
Agricultural use of organotin (e.g., fungicides, bactericides, and slimicide)
Disposal of tin-containing materials
Garbage dumping
Organotin residues
Domestic
Food packing
Tin-plate containers
Ceramics
Organ pipes
Toothpaste and dental preparation
Ceramic pacifier and pigment
Die casting
Cooking utensils
To line lead pipe for distilled water, beer, carbonated beverages, and some chemicals.

associated with tin extraction and treatment of the tin ore are wet processes. Organic tin compounds also exist and are used as stabilizers in plastics and in some pesticides.[80] Tin is dispersed in very small amounts in silicate rocks containing 2 to 50 ppm, the ore casserite (SnO_2) being of major commercial importance. The Earth's crust contains about 2 to 3 ppm tin. Tin is rarely detected in air, except near industrial emission points. The concentration of tin in soil generally ranges from 2 to 200 mg/kg. In fresh water and in most foodstuffs, the concentration is low and difficult to measure.[81] Trace amounts of tin are present in most natural foods. The normal tin intake is still uncertain. It is well known, however, that consumption of canned foods may increase that daily intake of tin considerably. Estimates for the average daily intake vary between 0.2 mg and 1 mg.[82] Tin has not been widely measured in biological media, and the usefulness of biological monitoring to determine human exposures in relation to health effects has not been developed.[83]

4.16 Uranium (U)

Uranium is mined almost entirely for energy production in fission reactors or for military uses in fission or nuclear bombs. Minor uses include being utilized as a negative contrast in

TABLE 18

Human Exposures to Uranium

Industrial/Occupational
Extraction
Enrichment and transformation into the hexafluoride for the selective enrichment of ^{235}U and ^{239}U
Radioactivity in mine tailings
Recycling of spent nuclear fuel
Environmental
Marginal radiotoxicity of nuclear fallout
Use of fission bombs
Accidental release of radionuclides (such as during the Chernobyl accident)[84]
Burning of coal

electron microscopy and possibly in superconductors (see Table 18). Low levels of uranium are found in igneous rocks such as granite. The metal only exists as unstable radioactive isotopes that undergo a long chain of radioactive decay to end up finally as stable isotopes of lead. Uranium (and indeed all actinides) are hazardous after enrichment. Urinary uranium detection has been used to determine recent exposures to uranium compounds.[85]

4.17 Vanadium (Va)

Vanadium is considered a trace element in some species and presumed to be an essential mineral in humans, although no deficiency symptoms have been reported. The metal is implicated in disruption of coenzyme A activity, Na^+-K^+ ATPase inhibition, and synthesis of cystine and cysteine. The major use of vanadium is in the production of steel (see Table 19). Vanadium steel contains up to 3% vanadium, is strong and heat resistant, and can withstand strain and vibration. The metal readily dissolves in iron, and with a boiling point of 3000°C, there is little vanadium in the fumes produced. Vanadium is also an alloying element in high-strength titanium alloys, usually at a concentration of about 4%. Vanadium pentoxide is used as a catalyst in a variety of reactions, with the most important use being in the oxidation of sulfur dioxide to sulfur trioxide in the production of sulfuric acid. Other compounds of vanadium are used as catalysts in the production of plastics. Vanadium is a typical rare element, fairly widely distributed at low concentrations in the Earth's crust, and is found in the sulfide and oxidized forms in nature. A few ore deposits that exist in Canada, South Africa, and Russia can be mined commercially, but most vanadium is recovered from the mining of other minerals, for example from some ores of iron, aluminum, uranium, magnesium, titanium, lead, and zinc. Vanadium is also found in some crude oils, and the ash from the combustion of such fuels can be a commercial source of vanadium pentoxide.

The background level of vanadium in human urine is below 3 μg/l. Exposed workers (sweeps exposed to vanadium-containing soot) have been reported with urinary levels fluctuating up to 13 μg/l. The background level of vanadium in human blood is below 2.5 μg/l. Exposed metal workers had blood levels ranging up to 55 μg/l. Urinary levels in these workers were also high, and the cystine content of fingernails was lower than normal. A rather wide range of vanadium concentrations has been reported in biological fluids which may be due to problems of analysis rather than being indicative of large differences across individuals or groups. Urinary determinations are preferred, and levels can increase markedly from the beginning to the end of a shift where exposure is occurring. Longer-term exposure is probably better determined by estimating concentrations two days after exposure ceases.[89]

TABLE 19

Human Exposures to Vanadium

Occupational
Recovery from ores
Production of vanadium compounds[86]
Metal and catalyst production
Vanadium pentoxide production
Plastics manufacture
Boiler cleaning[87]
Handling of vanadium-containing wastes (vanadium-rich slags produced in steel production, vanadium-rich sludges produced in titanium alloy production, disposal of spent vanadium catalysts, handling of vanadium-containing ashes, and vanadium contamination of plastics, up to 500 ppm
Environmental
Fly ash from the combustion of fossil fuels[88]
Leaching from soil or strata containing the element
Levels in water near areas of industrial activity or from deposits where industrial wastes containing vanadium have been dumped

4.18 Zinc (Zn)

Zinc is a nutritionally essential metal, and deficiency results in severe health consequences. Excessive levels of zinc are relatively uncommon and require high exposures.[90] Human exposures to zinc are possible in a wide variety of environmental and occupational situations, including soldering, battery manufacture, dentistry, pharmacological manufacture, electroplating, pigment manufacture, and rubber production (see Table 20). The most common zinc-related condition is metal fume fever, caused by inhalation of zinc-containing fumes by welders. Metal fume fever, as the name implies, is associated with other metals such as aluminum, copper, manganese, nickel, and others.[91] Wastage results from all stages of production and processing of zinc, leading to emissions into the atmosphere, water, and solid wastes. Certain amounts of zinc can be recycled, mainly from the melting of alloys. Zinc is ubiquitous in the environment, being present in most foodstuffs, water, and air. Zinc does not accumulate with continued exposure, but body content is modulated by homeostatic mechanisms that act principally on absorption and liver levels. Zinc atmospheric levels are increased over industrial areas. Zinc concentrations of noncontaminated soils ranging from 10 to 300 mg/kg are comparable with those of rocky subsoils; on average, levels of 20 mg/kg are found. Zinc applied to soil is taken up by growing vegetables. Water exposure may be increased by contact with galvanized copper or plastic pipes.[90,92] The average American daily intake is approximately 12 to 15 mg, primarily from food.[93] While determination of zinc in blood and urine has been used to detect work-related exposure, the significance of the levels in regard to any health effects have not been established.

5 SUMMARY

In the technological development of the human race, the availability of metals has been pivotal. The use of bronze took the human race out of the stone age, and the use of iron produced significant advances in agriculture, warfare, and technology. Further, the development of the industrial revolution would not have been possible without new processes which relied on the use of an ever-increasing number of metals.

TABLE 20

Human Exposures to Zinc

Industrial/Occupational
Metallic mining and metal refining
Alloy production
Soldering
Electroplating
Battery manufacture
Dentistry
Pharmacological manufacture
Pigment and rubber production
Environmental
Volcano emission
Mining operations
Waste recycling
Sewage and garbage sludge
Waste dumping ground
Ceramics
Domestic
Artwork (pigments, painting, and sculpture)
Ceramics
Alloy as dental materials
Tobacco smoking
Dermal cream
Foodstuffs
Contaminated tap water[93]

The metals constitute a group of materials that have a significant range of toxicities, from being relatively innocuous to having significant toxicity. The spectrum of health effects includes most, if not all, body organs and tissues. A distinct group of toxic metals has emerged over many years (even centuries), owing to a history of occupational, environmental, domestic, and criminal use. New metals are added to this group as newer technologies use increasing amounts of metals, which in turn produces workplace conditions that cause adverse effects on health.

Some exposures to metals are universal — for example, in mining, extraction, refining, and smelting. These activities can produce exposures to workers that cause injury, disease, and even death. They can also cause environmental contamination which even today is a sad legacy of exploitation, ignorance, and mismanagement.

Other exposures are common to many metals, such as the use of the transition metals in catalyst manufacture and battery production and the use of virtually all metals in the production of alloys or the production of compounds used as inorganic pigments, pesticides, and ceramic additives. A range of metals is used for the manufacture of electronics, printed circuit boards, and semiconductors. Welding is an example of an industrial activities in which workers can be exposed to a range of metals, most with common health consequences.

Other exposures are unique. The use of beryllium in the nuclear industry, the radiotoxicity of uranium enrichment, the use of lithium in the production of tritium, the use of lanthanides for cigarette lighter flints, the use of cisplatin in the treatment of cancer, the use of tin for tin-plate, the use of tungsten in electric light filaments, and the use of zirconium in deodorants are just some examples.

Some other uses include mercurials in millinery, lead arsenate as a pesticide, and thallium as a rodenticide. New technologies such as superconductors have the potential to increase even more the amounts and numbers of metals in commercial use.

The environment has been affected by metals. Natural processes such as volcanism, weathering, and sedimentation have assisted in the distribution of metals over the planet. However, anthropogenic activities have greatly added to this. Because of the long use of metals over the centuries, wasteful extraction technologies, poorly controlled technologies, and inappropriate waste disposal methods have seen metals redistributed into every environmental compartment, often with disastrous consequences. Mine tailings, poorly controlled industrial processes, production of hazardous wastes, badly controlled combustion of fossil fuels, and domestic activities such as automobile use also contribute to the pollution. Further anthropogenic activities such as acid rain and nuclear fallout increase the unpredictability by which environmental metal levels may influence air quality, water quality, biodiversity, and human health. It is hoped that a better understanding of the issues will result in more effective methods of dealing with them.

REFERENCES

1. Leckey, T. D. and Venugopal, B., *Metal Toxicity in Mammals,* Vol. 1, 105, Plenum Press, New York, 1977, p. 7.
2. Winder, C., Toxicity of metals, in *Occupational Toxicology,* Stacey, N. H., Ed., Taylor and Francis, London, 1993, pp. 165–175.
3. Winder, C. and Yeung, P., Health problems in aluminum smelter workers: hazards, exposures and respiratory disease, *Occupat. Health Safety Austr. N.Z.,* 5, 391–402, 1989.
4. Sorenson, J. R. J., Campbell, I. R., Tepper, L. B., and Lingg, R. D., Aluminum in the environment and human health, *Environ. Health Persp.,* 8, 3–95, 1974.
5. Siderman, S. and Manor, D., The dialysis dementia syndrome and aluminum concentration, *Nephron,* 31, 1–10, 1982.
6. Martyn, C. N., The epidemiology of Alzheimer's disease in relation to aluminum, *Ciba Found. Symp.,* 169, 69–86, 1992.
7. Schlatter, C., Biomedical aspects of aluminum, *Medical Lavoro,* 83, 470–474, 1992.
8. Lauwerys, R. R. and Hoet, P., *Industrial Chemical Exposure — Guideline for Biological Monitoring,* 2nd ed., Lewis Publishers, Boca Raton, FL, 1993, p. 16.
9. Elinder, C.-G., Friberg, L., Kjellstroem, T., Nordberg, G., and Oberdoerster, G., *Biological Monitoring of Metals,* IPCS Chemical Safety Monographs, World Health Organization, Geneva, Switzerland, 1994, pp. 41–42.
10. ILO, *Encyclopedia of Occupational Health and Safety,* Vols. I and II, International Labour Office, Geneva, Switzerland, 1983, p. 179.
11. Elinder, C.-G., Friberg, L., Kjellstroem, T., Nordberg, G., and Oberdoerster, G., *Biological Monitoring of Metals,* IPCS Chemical Safety Monographs, World Health Organization, Geneva, Switzerland, 1994, pp. 44–47.
12. Winder, C., Toxicity of metals, in *Occupational Toxicology,* Stacey, N. H., Ed., Taylor and Francis, London, 1993, pp. 165–175.
13. Peterson, P. J., Girling, C. A., Benson, L. M., and Zieve, R., Metalloids, in *Effect of Heavy Metal Pollution on Plants,* Vol. 1, Lepp, N. W., Ed., Applied Science, London, 1981, pp. 213–322.
14. USEPA, EPA's Integrated Risk Information System (IRIS) on Arsenic, Inorganic (CAS 7440-38-2), National Library of Medicine's TOXNET System, November 1, 1994.
15. Goyer, R. A., Toxic effects of metals, in *Casarett and Doull's Toxicology — The Basic Science of Poisons,* 4th ed., Amdur, M. O., Doull, J., and Klaassen, C. D., Eds., Pergamon Press, New York, 1991, pp. 623–680.
16. Leonard, A., Arsenic, in *Metals and Their Compounds in the Environment — Occurrence, Analysis, and Biological Relevance,* Merian, E., Ed., VCH, Weinheim, 1991, pp. 751–774.
17. Hanusch, K., Grossmann, H., Herbst, K.-A., Rose, B., and Wolf, H. U., Arsenic and arsenic compounds, in *Ullmann's Encyclopedia of Industrial Chemistry,* 5th ed., Vol. A3, VCH Verlagsgesellschaft, Basal, 1985, pp. 113–141.
18. NRCC, *Effect of Arsenic in the Canadian Environment,* Report NRCC 15391, National Research Council Canada, Ottawa, 1978.
19. Lauwerys, R. R. and Hoet, P., *Industrial Chemical Exposure — Guideline for Biological Monitoring,* 2nd ed., Lewis Publishers, Boca Raton, FL, 1993, pp. 23–26.
20. IPCS, *Beryllium,* Environmental Health Criteria 106, World Health Organization International Program on Chemical Safety, Geneva, Switzerland, 1990.
21. Goyer, R. A., Toxic effects of metals, in *Casarett and Doull's Toxicology — The Basic Science of Poisons,* 4th ed., Amdur, M. O., Doull, J., and Klaassen, C. D., Eds., Pergamon Press, New York, 1991, pp. 623–680.

22. *The Merck Index,* 10th ed., Merck Co., Rahway, NJ, 1983, p. 166.
23. Griffitts, W. R. and Skilleter, D. N., Beryllium, in *Metals and Their Compounds in the Environment — Occurrence, Analysis, and Biological Relevance,* Merian, E., Ed., VCH, Weinheim, 1991, pp. 775–787.
24. USEPA, *Health Assessment Document for Beryllium,* EPA 600/8-84-026F, U.S. Environmental Protection Agency, U.S. Government Printing Office, Washington, D.C., 1987, pp. 3–4.
25. Brown, K. W., Evans, Jr., G. B., and Frentrup, B. D., Eds., *Hazardous Waste and Land Treatment,* Butterworth Publishers, Boston, MA, 1983, p. 244.
26. USEPA, *Ambient Water Quality Criteria Document: Beryllium,* EPA 440/5-80-024, U.S. Environmental Protection Agency, U.S. Government Printing Office, Washington, D.C., 1980, p. A-1.
27. National Research Council, *Drinking Water and Health,* Vol. 1, National Academy Press, Washington, D.C., 1977, p. 232.
28. Lauwerys, R. R. and Hoet, P., *Industrial Chemical Exposure — Guidelines for Biological Monitoring,* 2nd ed., Lewis Publishers, Boca Raton, FL, 1993, p. 29–31.
29. IPCS, *Chromium,* Environmental Health Criteria 61, World Health Organization International Program on Chemical Safety, Geneva, Switzerland, 1987.
30. Bonde, J. P. and Christensen, J. M., Chromium in biological samples from low-level exposed stainless steel and mild steel welders, *Arch. Environ. Health,* 46, 225–229, 1991.
31. Lauwerys, R. R. and Hoet, P., *Industrial Chemical Exposure — Guidelines for Biological Monitoring,* 2nd ed., Lewis Publishers, Boca Raton, FL, 1993, pp. 43–45.
32. IPCS, *Cadmium,* Environmental Health Criteria 134, World Health Organization International Program on Chemical Safety, Geneva, Switzerland, 1992.
33. Goyer, R. A., Toxic effects of metals, in *Casarett and Doull's Toxicology — The Basic Science of Poisons,* 4th ed., Amdur, M. O., Doull, J., and Klaassen, C. D., Eds., Pergamon Press, New York, 1991, pp. 634–638.
34. Schaller, K. H. and Angere, J., Biological monitoring in the occupational setting — relationship to cadmium exposure, in *Cadmium in the Human Environment: Toxicology and Carcinogenicity,* Nordberg, G. F., Herber, R. F. M., and Alessio, L., International Agency for Research on Cancer, Lyon, France, 1992, pp. 53–64.
35. Elinder, C.-G., Friberg, L., Kjellstroem, T., Nordberg, G., and Oberdoerster, G., *Biological Monitoring of Metals,* IPCS Chemical Safety Monographs, World Health Organization, Geneva, Switzerland, 1994, pp. 48–49.
36. Lauwerys, R. R. and Hoet, P., *Industrial Chemical Exposure — Guidelines for Biological Monitoring,* 2nd ed., Lewis Publishers, Boca Raton, FL, 1993, pp. 32–37.
37. Lauwerys, R. R. and Hoet, P., *Industrial Chemical Exposure — Guidelines for Biological Monitoring,* 2nd ed., Lewis Publishers, Boca Raton, FL, 1993, p. 32.
38. Goyer, R. A., Toxic effects of metals, in *Casarett and Doull's Toxicology — The Basic Science of Poisons,* 4th ed., Amdur, M. O., Doull, J., and Klaassen, C. D., Eds., Pergamon Press, New York, 1991, pp. 623–680.
39. Schrauzer, G. N., Cobalt, in *Metals and Their Compounds in the Environment — Occurrence, Analysis, and Biological Relevance,* Merian, E., Ed., VCH, Weinheim, 1991, pp. 879–892.
40. Elinder, C.-G., Friberg, L., Kjellstroem, T., Nordberg, G., and Oberdoerster, G., *Biological Monitoring of Metals,* IPCS Chemical Safety Monographs, World Health Organization, Geneva, Switzerland, 1994, pp. 52–54.
41. Seiler, H. G., Sigel, H., and Sigel, A., Eds., *Handbook on the Toxicity of Inorganic Compounds,* Marcel Dekker, New York, 1988, p. 252.
42. Lauwerys, R. R. and Hoet, R., *Industrial Chemical Exposure — Guideline for Biological Monitoring,* 2nd ed., Lewis Publishers, Boca Raton, FL, 1993, pp. 48–49.
43. IPCS, *Lead: Biological Effects,* Environmental Health Criteria (Draft), World Health Organization International Program on Chemical Safety, Geneva, Switzerland, 1994.
44. USEPA, *Air Quality Criteria for Lead,* Vols. I–IV, U.S. Environmental Protection Agency, U.S. Government Printing Office, Washington, D.C., 1986.
45. Annest, J. L., Pirkle, J. L., Makuc, D., Nesse, J. W., Bayse, D. D., and Kovar, M. G., Chronological trend in blood levels between 1976 and 1980, *N. Engl. J. Med.,* 308, 1373–1377, 1983.
46. Huynh-Ngoc, L., Whitehead, N. E., and Oregoni, B., Low levels of copper and lead in a highly industrialized river, *J. Toxicol. Environ. Chem.,* 17, 223–236, 1988.
47. CDC, *Preventing Lead Poisoning in Children,* U.S. Centers for Disease Control, U.S. Government Printing Office, Washington, D.C., 1991.
48. Tyrer, S. P., Lithium in the treatment of mania, *J. Affective Dis.,* 8, 251–257, 1985.
49. Carson, B. L., Ellis, II, H. V., and McCann, J. L., *Toxicology and Biological Monitoring of Metals in Humans —Including Feasibility and Need,* Lewis Publishers, Ann Arbor, MI, 1986, pp. 136–139.
50. Elinder, C.-G., Friberg, L., Kjellstroem, T., Nordberg, G., and Oberdoerster, G., *Biological Monitoring of Metals,* IPCS Chemical Safety Monographs, World Health Organization, Geneva, Switzerland, 1994, pp. 44–47.
51. Goyer, R. A., Toxic effects of metals, in *Casarett and Doull's Toxicology — The Basic Science of Poisons,* 4th ed., Amdur, M. O., Doull, J., and Klaassen, C. D., Eds., Pergamon Press, New York, 1991, pp. 623–680.

52. National Research Council, *Drinking Water and Health,* Vol. 1, National Academy Press, Washington, D.C., 1977, p. 268.
53. Lauwerys, R. R. and Hoet, P., *Industrial Chemical Exposure — Guidelines for Biological Monitoring,* 2nd ed., Lewis Publishers, Boca Raton, FL, 1993, pp. 71–73.
54. Elinder, C.-G., Friberg, L., Kjellstroem, T., Nordberg, G., and Oberdoerster, G., *Biological Monitoring of Metals,* IPCS Chemical Safety Monographs, World Health Organization, Geneva, Switzerland, 1994, pp. 59–60.
55. Goyer, R. A., Toxic effects of metals, in *Casarett and Doull's Toxicology — The Basic Science of Poisons,* 4th ed., Amdur, M. O., Doull, J., and Klaassen, C. D., Eds., Pergamon Press, New York, 1991, pp. 623–680.
56. IPCS, *Mercury — Environmental Aspects,* Environmental Health Criteria 86, World Health Organization, International Program on Chemical Safety, Geneva, Switzerland, 1989, pp. 13–15.
57. IPCS, *Inorganic Mercury,* Environmental Health Criteria 118, World Health Organization International Program on Chemical Safety, Geneva, Switzerland, 1991, pp. 15–19.
58. Elinder, C.-G., Friberg, L., Kjellstroem, T., Nordberg, G., and Oberdoerster, G., *Biological Monitoring of Metals,* IPCS Chemical Safety Monographs, World Health Organization, Geneva, Switzerland, 1994, pp. 60–64.
59. Lauwerys, R. R. and Hoet, P., *Industrial Chemical Exposure — Guidelines for Biological Monitoring,* 2nd ed., Lewis Publishers, Boca Raton, FL, 1993, pp. 74–81.
60. IPCS, *Nickel,* Environmental Health Criteria 108, World Health Organization International Program on Chemical Safety, Geneva, Switzerland, 1991.
61. Warner, J. S., Occupational exposure to airborne nickel in producing and using primary nickel products, in *Nickel in the Human Environment, Proceedings of a Joint Symposium,* International Agency for Research on Cancer Scientific Publications, Vol. 52, 1984, pp. 419–437.
62. NAS, *Nickel, Medical and Biological Effects of Environmental Pollutants,* National Academy of Sciences, Washington, D.C., 1975.
63. Di Toro, D. M., Mahony, J. D., Krichgraber, P. R., O'Byrne, A. L., and Pasquale, L. R., Effect of nonreversibility, particle concentration and ionic strength on heavy metal sorption, *Environ. Sci. Technol.,* 20, 55–61, 1986.
64. Smart, G. A. and Sherlock, J. C., Nickel in foods and diet, *Food Additives Contam.,* 4, 61–71, 1987.
65. Bernacki, E. J., Parsons, G. E., Roy, B. R., Mikac-Devic, M., Kennedy, C. D., and Sunderman, F. W., Urine nickel concentrations in nickel exposed workers, *Ann. Clin. Lab. Sci.,* 8, 184–189, 1974.
66. Lauwerys, R. R. and Hoet, P., *Industrial Chemical Exposure — Guidelines for Biological Monitoring,* 2nd ed., Lewis Publishers, Boca Raton, FL, 1993, pp. 82–84.
67. Elinder, C.-G., Friberg, L., Kjellstroem, T., Nordberg, G., and Oberdoerster, G., *Biological Monitoring of Metals,* IPCS Chemical Safety Monographs, World Health Organization, Geneva, Switzerland, 1994, pp. 65–66.
68. IPCS, *Platinum,* Environmental Health Criteria 125, World Health Organization International Program on Chemical Safety, Geneva, Switzerland, 1991.
69. Vennitt, S., Crofton-Sleigh, C., Hunt, J., Speechley, V., and Briggs, K., Monitoring exposure of nursing and pharmacy personnel to cytotoxic drugs: urinary mutation assays and urinary platinum as markers of absorption, *Lancet,* 1, 74–77, 1984.
70. Brooks, S. M., Baker, D. B., Gann, P. H., Jarabek, A. M., Hertzberg, V., Gallagher, J., Biagini, R. E., and Bernstein, I. L., Cold air challenge and platinum skin reactivity in platinum refinery workers, *Chest,* 79, 1401–1407, 1990.
71. Carson, B. L., Ellis, II, H. V., and McCann, J. L., *Toxicology and Biological Monitoring of Metals in Humans —Including Feasibility and Need,* Lewis Publishers, Ann Arbor, MI, 1986, p. 184.
72. Elinder, C.-G., Friberg, L., Kjellstroem, T., Nordberg, G., and Oberdoerster, G., *Biological Monitoring of Metals,* IPCS Chemical Safety Monographs, World Health Organization, Geneva, Switzerland, 1994, pp. 44–47.
73. Fishbein, L., II.25 Selenium, in *Metals and Their Compounds in the Environments — Occurrence, Analysis, and Biological Relevance,* Merian, E., Ed., VCH, Weinheim, 1991, pp. 751–774.
74. Goyer, R. A., Toxic effects of metals, in *Casarett and Doull's Toxicology — The Basic Science of Poisons,* 4th ed., Amdur, M. O., Doull, J., and Klaassen, C. D., Eds., Pergamon Press, New York, 1991, pp. 623–680.
75. IPCS, *Selenium,* Environmental Health Criteria 58, World Health Organization International Program on Chemical Safety, Geneva, Switzerland, 1987, pp. 36–65.
76. Elinder, C.-G., Friberg, L., Kjellstroem, T., Nordberg, G., and Oberdoerster, G., *Biological Monitoring of Metals,* IPCS Chemical Safety Monographs, World Health Organization, Geneva, Switzerland, 1994, p. 67.
77. Lauwerys, R. R. and Hoet, P., *Industrial Chemical Exposure — Guidelines for Biological Monitoring,* 2nd ed., Lewis Publishers, Boca Raton, FL, 1993, p. 87.
78. Schoer, J., Thallium, in *The Handbook of Environmental Chemistry,* Vol. 3, Hutzinger, O., Ed., Springer-Verlag, Berlin, 1984, pp. 143–214.
79. Lauwerys, R. R. and Hoet, P., *Industrial Chemical Exposure — Guideline for Biological Monitoring,* 2nd ed., Lewis Publishers, Boca Raton, FL, 1993, p. 93.

80. ILO, *Encyclopedia of Occupational Health and Safety,* Vols. I and II, International Labour Office, Geneva, Switzerland, 1983, p. 2177.
81. Elinder, C.-G., Fbierg, L., Kjellstroem, T., Nordberg, G., and Oberdoerster, G., *Biological Monitoring of Metals,* IPCS Chemical Safety Monographs, World Health Organization, Geneva, Switzerland, 1994, pp. 68–69.
82. Bulten, E. J. and Meinema, H. A., Tin, in *Metals and Their Compounds in the Environment — Occurrence, Analysis, and Biological Relevance,* Merian, E., Ed., VCH, Weinheim, 1991, pp. 1243–1259.
83. Elinder, C.-G., Friberg, L., Kjellstroem, T., Nordberg, G., and Oberdoerster, G., *Biological Monitoring of Metals,* IPCS Chemical Safety Monographs, World Health Organization, Geneva, Switzerland, 1994, p. 69.
84. Eisenbud, M., *Environmental Radioactivity,* Academic Press, Orlando, FL, 1987.
85. Lauwerys, R. R. and Hoet, P., *Industrial Chemical Exposure — Guidelines for Biological Monitoring,* 2nd ed., Lewis Publishers, Boca Raton, FL, 1993, p. 94.
86. NAS, *Vanadium,* National Academy of Sciences, Washington, D.C., 1974.
87. IPCS, *Vanadium,* Environmental Health Criteria 81, World Health Organization International Program on Chemical Safety, Geneva, Switzerland, 1988.
88. Nriagu, J. O. and Davidson, C. I., in *Toxic Metals in the Atmosphere,* John Wiley & Sons, New York, 1986.
89. Lauwerys, R. R. and Hoet, P., *Industrial Chemical Exposure — Guideline for Biological Monitoring,* 2nd ed., Lewis Publishers, Boca Raton, FL, 1993, p. 95.
90. Goyer, R. A., Toxic effects of metals, in *Casarett and Doull's Toxicology — The Basic Science of Poisons,* 4th ed., Amdur, M. O., Doull, J., and Klaassen, C. D., Eds., Pergamon Press, New York, 1991, pp. 623–680.
91. Winder, C., Toxicity of metals, in *Occupational Toxicology,* Stacey, N. H., Ed., Taylor and Francis, London, 1993, pp. 165–175.
92. Friberg, L., Nordlberg, G. F., Kessler, E., and Vouk, V. B., Eds., *Handbook of the Toxicology of Metals,* 2nd ed., Vols. I and II, Elsevier Science, Amsterdam, 1986.
93. Ohnesorge, F. K. and Wilhelm, M., Zinc, in *Metals and Their Compounds in the Environment — Occurrence, Analysis, and Biological Relevance,* Merian, E., Ed., VCH, Weinheim, 1991, pp. 1309–1342.

CHAPTER 3

METAL METABOLISM AND TOXICITIES

Denis M. Medeiros, Robert Wildman, and Rebecca Liebes

CONTENTS

0-8493-4493-X/97/$0.00+$.50

1 INTRODUCTION

The heavy metals such as lead, cadmium, and mercury have long been known for their potential toxicities; however, there are trace elements known to have toxic effects that are also essential dietary nutrients. Several of these elements have specific deficiency symptoms and can lead to death of the organism. These same essential elements are toxic when an organism has an increased intake, whether by dietary means or other exposure routes. Trace metals such as copper, zinc, iron, and selenium are known for their deficiencies. With respect to selenium, early work focused on its toxicity, especially in the livestock industry where selenosis in specific regions of the world is likely to occur due to the naturally high selenium content of the soil. On the other hand, copper, zinc, and iron have been primarily evaluated in regard to deficiency aspects. Toxicity due to these minerals is known to occur via industrial and environmental exposures. Also, due to the propensity of many individuals to consume nutrient supplements, toxicity symptoms and specific levels known to lead to these symptoms have been of increased interest.[1-3]

Toxic effects of trace elements depend on many factors: chemical form, exposure level, route of intake, and organ and subcellular storage sites within the body that are targeted. The storage site under "normal" levels of intake for the essential trace minerals often determine the type and extent of toxic effects when excess levels of the metals are acquired by the organism. Many of the essential trace elements are components of enzymes, and consequently the potential toxicities may be influenced by disturbances in the activities of these enzymes.

A potential confounding factor in the study of trace metal toxicology includes the influence of other trace elements upon the relative toxicities. For example, increased zinc intake may lead to decreased copper absorption by the gut or other organs. This leads to the question as to whether the observed toxic effects of high metal intake are due to a secondary influence of the lack of another mineral or to a high level of the metal per se. Exposure of organisms to large intakes of some of the macroelements may lead to a decreased uptake of some of the trace metals, and other dietary components may influence their relative toxicities. For instance, excess calcium, fiber, and fat may decrease the bioavailability of the metals for absorption and could theoretically lessen potential toxicities if excess levels of these other nutrients are consumed.

Toxicities of these elements may also be related to an inability to eliminate these trace elements or to regulate their absorption from the diet. Genetic disorders of metabolism may lead to a build-up of some metals in specific organs, such as what happens with liver copper levels in patients afflicted with Wilson's disease.[4] Similarly, iron uptake enhanced by a failure to regulate mucosal absorption is known to result in hemochromatosis. In both of these examples, "normal" levels of these minerals lead to toxicities. The heavy metals such as lead, mercury, and cadmium produce well-defined toxic effects in humans. While these elements

TABLE 1

Copper Metabolism and Toxicity

Metabolic Aspects	Affected Site/Action	Ref.
Digestion and absorption	Stomach and small intestine	6, 8–10
Storage	Liver	6, 11
Elimination	Kidneys	12
Transport	Blood	6, 13
Enzymatic influences	Organ cells	6, 11, 12, 14
Toxicity	Lung	16–18
	Liver/mitochondria	19–23
	Cardiovascular/heart and pulse/	24–27
	blood pressure	28
	Reproductive/teratogen	29–35
	Nervous/neural tube defects	36
	Bone/ossification defects	36
Genetic disorders/animal models		
Menkes' disease		12, 37
Wilson's disease		38, 39
Brindled mouse		40
Bedlington terrier		42, 43
Long-Evans Cinnamon rats		39, 44–46

are not essential to the health and existence of the organism, each of these metals has specific storage sites, upon which the toxic effects they exert is in part dependent. Additionally, the chemical forms of the metal to which the organism is exposed will often determine the toxic effects. This is because the storage sites may be altered by different chemical forms. A good example of this is mercury. Methylmercury and other organic forms are highly toxic to humans, with the nervous system being a primary target due to the enhanced lipid solubility of this chemical form. This is in contrast to the lower toxicity of inorganic mercury, where most of this form is deposited in the kidney and liver. Another aspect to consider is that many of these metals are pro-oxidants, and the production of free radicals may exert their toxic influences. Recent concern with respect to iron supplementation and the potential influence upon heart disease is an illustration of this as a practical consequence. To understand the physiological forces that impact the mineral toxicities, "normal" metabolism in terms of how the body absorbs, utilizes, stores, and eliminates each of the trace metals will be considered for each metal.

2 COPPER

Copper toxicity is usually due to an imbalance between intake into and subsequent excretion from the body.[5] Exposure of humans to copper in the environment can occur via water, food, soil, or air. Many times the chemical form of copper is copper-sulfate. Copper may exist as Cu^{+1} or Cu^{+2}, with very small amounts as Cu^{3+}.[6] Cu^{+} is insoluble and often complexed. Normally it is the +2 form of copper that is found in plant and animal cells, and it is this form that is usually complexed to metalloenzymes. (See Table 1.)

2.1 METABOLISM

Copper is found in foods such as nuts, shellfish, organ meats, and legumes. Grains and grain products, as well as chocolate, have appreciable levels of copper. While these food items are good to excellent sources of copper, the absolute amount of copper absorbed may be influenced by other dietary components. Excess dietary iron may decrease copper absorption; conversely, too much copper may cause an iron deficiency. Zinc excess may also decrease copper absorption. The mechanism by which dietary zinc inhibits the uptake of copper involves the intestinal protein metallothionein (MT), a small-molecular-weight peptide of about 3000. As dietary zinc levels increase, MT synthesis in the mucosal cells of the small intestine increases. This results in a binding and trapping of zinc in the mucosal cell. The binding is not easily dissociated and when mucosal cells are sloughed off in the normal turnover, zinc is lost through the fecal compartment. However, the MT peptide has a greater affinity for copper. Thus, when excess zinc is consumed, the increased MT production will cause copper to be sequestered in the mucosal cells.[6]

Molybdenum excess can produce copper deficiency, in addition to its own inherent toxicity. Ruminants have been reported to have this occur. With respect to humans, vitamin C supplementation results in decreased copper status, and in rats, large doses of vitamin C can lead to copper deficiency.[6] Other dietary components have an influence upon copper status, but not necessarily absorption. Feeding rats either sucrose or fructose, as opposed to glucose or cornstarch, decreases copper status and exacerbates the signs of copper deficiency.[7]

Copper may be absorbed by both the stomach and small intestinal mucosa, with most absorbed by the small intestine.[6,8] Active transport and passive uptake mechanisms are thought to be responsible for the absorption, with MT regulating some of the absorption as illustrated above.[9,10] The average level of copper stored in the body is from 50 to 120 mg.[11] Most of this is found in the liver bound to albumin or transcuprin, and the copper is incorporated into the liver protein ceruloplasmin.[6] Excess copper could bind to MT, thereby acting to detoxify the actions of too much copper, which suggests that the MT role for copper homeostasis is similar to that of other trace metals. Excess copper can also lead to increased kidney levels; however, little copper is excreted via the urine. Copper excreted through the glomerulus may be reabsorbed by the kidney tubules.[12] Copper is usually transported from the liver to other organs in the form of ceruloplasmin, a blood copper containing protein synthesized in the liver.[13]

Most copper is excreted via the bile that is released into the gastrointestinal tract, with minimal copper reabsorbed by intestinal cells. The uptake of copper and elimination through the bile allows copper to be conserved and tightly regulated.[6]

Copper is utilized by most cells as a component of the enzyme cytochrome C oxidase and superoxide dismutase.[6] Copper is also a constituent of many other cuproenzymes, including lysyl oxidase, involved in crosslinking of connective tissue proteins elastin and collagen fibrils; dopamine β hydroxylase, involved in the conversion of dopamine to norepinephrine; and ceruloplasmin.[11,12,14] It is not surprising that some influences of copper toxicity are in part due to an influence upon these enzymes.

2.2 Toxicity

As indicated, copper uptake is regulated at the gut level and excreted through the bile. An increase in tissue copper to toxic levels is caused by an imbalance among these two processes.[5] Normally the first organ deleteriously affected by these increased levels is the liver; however, there appears to be species differences in terms of the relative levels of copper required to induce a toxic response. Sheep, for instance, are animals perhaps the most sensitive to copper toxicity (in terms of dietary copper), with a narrow tolerance range between deficiency and toxicity in comparison to other mammals; hemolytic crisis often occurs in these animals. On the other hand, rats appear to be the most resistant to the toxic effects of copper, being able to withstand on a per weight basis several magnitudes more copper than sheep. At 250 μg Cu per g diet, no toxicity signs have been reported. Pigs and cattle are moderately tolerant to large dosages of copper; however, goats are more similar to sheep in terms of their levels of toxicity.[15]

Serum levels of glutamic oxaloacetic transaminase, liver-specific arginase, glutamate dehydrogenase, and sorbitol dehydrogenase are increased. Lysosomes appear to increase in relation to copper content, and during the subsequent hemolytic crisis, blood levels of methemoglobin and blood urea nitrogen increase. Release of lysosomal enzymes may contribute to the copper toxicity in the liver. Kidney tubule breakdown follows subsequent to this release.[15]

While the majority of toxicity studies appear to focus on diet intake, another potential route of copper intake is through the lung. Such a route is problematic in that the regulatory systems of the gastrointestinal tract are essentially bypassed, leading to the potential for increased toxicity; inhalation of copper dust does irritate the upper airways.[16,17] Copper intake through the lungs may occur via the use of copper-containing pesticide sprays. Romeo-Moreno et al.[18] reported that inhalation of copper sulfate in Wistar rats was similar to injecting rats with copper sulfate in terms of copper concentration among selected organs. In both instances, the copper appeared to bind to metallothionein in both the liver and kidneys.

Hirano et al.[17] reported that acute inhalation of copper oxide in rats resulted in metallothionein induction within 1/2 to 3 days after exposure. The level of MT produced was proportional to the dose. A half-time for elimination was determined to be 37 hr. Indices of lung inflammation, such as the number of macrophages, lactate dehydrogenase, and β-glucuronidase activities of lavage fluid, resulted in a peak within a 1/2 day after exposure and up to 3 days after, depending on the inhalation dose. Followup studies indicated that copper sulfate exposure gave results similar to exposure to copper oxide. Metallothionein probably played a limited role in the handling of the copper because most of the copper was solubilized and rapidly cleared from the lung.

With respect to the effect of copper upon liver toxicity, Sokol et al.[19] suggested that mitochondria lipid peroxidation and decreased cytochrome C oxidase activity could partly explain the toxic effects. Mitochondria respiration was decreased in rats given in excess of 2000 μg Cu per g diet for 8 wk. Decreases in both state 3 respiration and the respiratory control ratio were reported. The authors suggested that peroxidation of mitochondria cardiolipin, which regulates cytochrome C oxidase activity, may be a factor. An inhibition of the oxidoreductase function may have resulted in the decreased state 3 respiration observed.

Copper toxicity has an apparent affect upon liver lysosomes. Copper becomes bound to the MT in the lysosomes, and normally the copper is released from the lysosomes by exocytosis to the bile. The exocytosis of the lysosomal copper is a major excretory pathway for copper elimination. The lysosomes may sequester copper, which may result in lysosomal damage via increased fragility.[20-23] Lysosomal membrane fluidity increases and membrane lipid peroxidation increases, along with an increase in polyunsaturated fatty acids (PUFA) content and a decrease in saturated fatty acid (SFA) content of lysosomal membranes. The increased peroxidation is thought to account for these observations and the altered lysosomal morphology. Copper in the free intracellular form may react with hydrogen peroxide to generate hydroxyl radicals.[23]

Other organ systems may be susceptible to copper toxicity. Cardiovascular disease has been the focus of much research as it relates to copper deficiency;[24-26] however, excess copper also has been shown to cause cardiac dysfunction. Rabbits infused with copper sulfate up to levels of 10 mg/kg body weight had decreased contractile force, heart rate, and pulse pressure, leading eventually to shock at the highest dose.[27] Liu and Medeiros[28] demonstrated that excess copper (100 mg/kg diet) significantly increased systolic blood pressure in Wistar and in the Spontaneously Hypertensive rat.

The effect of copper toxicity as a teratogen has been studied in cattle, pigs, and sheep without apparent effects on offspring.[29] Copper injections in lab animals have been found to be teratogenic.[30,31] Copper-containing intrauterine devices do not demonstrate any teratogenic effects in laboratory animals.[32,33] Wilson's disease patients experience complications during pregnancy.[34,35]

Female rats fed 0.185% copper acetate in drinking water for 7 weeks and then mated demonstrated liver and renal inflammation, and the embryos had moderate growth retardation and differentiation, particularly with regard to the neural tube.[36] A reduced number of ossification centers in the vertebrae, sternum, and fore and hind limb phalanges were reported.[36]

2.3 Genetic Conditions of Copper Toxicity

Two well-known genetic diseases affecting copper metabolism should be addressed. Menkes' kinky-hair disease, characterized and reported in the early 1960s,[12,37] is a problem with copper transport or absorption. Wilson's disease, described in the early part of the 1900s,[38] is characterized by increased liver copper content, leading to severe hepatic damage followed by increased brain copper levels and neurological problems.[39] Menkes' disease, however, results in a pathology resembling copper deficiency as opposed to copper toxicity, as in the case for Wilson's disease. Fortunately, animal models exist to study these genetic conditions. The brindled mouse is a well-researched model for Menkes' disease. Another model, the macular mutant mouse, is also a useful model for the disease. In both of these models, the copper levels are high in the kidney and intestine and correspond to elevations in renal copper metallothionein.[40] Shiraishi et al.[40] demonstrated that the macular mutant mouse was more sensitive to high doses of copper, in terms of toxicity, as compared to the normal mouse. Intraperitoneal injection of copper into 6 to 8 day old macular mice at relatively high doses resulted in nearly 100% mortality within 10 days. The mortality rate for normal mice injected with 28 mg Cu per kg was 38% after 1 day, and 83% for the macular mutant mice. Heterozygote mice had an intermediate mortality rate of 47% 1 day after injection. There were no differences among the three strains of mice with respect to MT synthesis nor MT-1 mRNA levels after copper injection. The different mortalities due to copper toxicity among the three strains was not readily explained by these results.

A gene has been reported to be deleted in subjects with Menkes' disease. The gene apparently codes for an ATPase of the P type, which involves cation transport.[41] Copper toxicity of Wilson's disease has several animal models. The Bedlington terrier has excess liver copper bound to lysosomal metallothionein. In both Wilson's disease patients and in the Bedlington terrier, there may be an inability to degrade MT. There is very little copper content in the bile produced;[42,43] however, Wilson's disease patients have little ceruloplasmin, whereas the terriers appear to have normal levels.[5] Wilson's disease patients appear to accumulate the copper in the liver cytoplasm, in contrast to the lysosomes for the dogs, as indicated previously.[5] As already mentioned, sheep are very sensitive to copper intake levels and toxicity, which could be due to the limited capacity of sheep to synthesize MT. Another more recently studied model is the Long-Evans Cinnamon (LEC) mutant rat, which has a single autosomal recessive mutation.[44] Hepatitis along with jaundice and anemia are clinical outcomes, and the condition is highly lethal.[45,46] Okavasu et al.[39] reported high levels of hepatic copper compared to normal Long-Evans rats, and serum copper levels and ceruloplasmin activity were significantly lower than control Long-Evans rats at 4 months of age. However, kidney levels of copper did not increase until a much later age in the LEC rats (e.g., 12 months of age). Brain and small intestine copper levels did increase at 13 months of age. For further discussion on copper toxicity, the reader is referred to the review of Awing and Metra.[47]

3 ZINC

3.1 Metabolism

Most of the research concerning zinc and human health has centered on deficiency, marginal diet intakes, and the consequences of these aspects. Furthermore, zinc requirements as influenced by age, gender, health status, physical activity, and other dietary components

TABLE 2

Zinc Metabolism and Toxicity

Metabolic Aspects	Affected Site/Action	Ref.
Organ distribution	—	48–50
Endocrine	Insulin, glucagon	10, 54
	Glucocorticoids	51–53
	Growth hormone	55, 56
Genetic toxicity	Small intestine	58, 59
Absorption	Small intestine	60–63
Toxicity	Gastrointestinal	66
	Heart	66
	Blood	66
	Pancreas	67
	Kidney	68
	Immune	71
	HDL cholesterol	72, 73
	Mineral imbalances	76–82
	Nervous	83–87

have been the subject of many investigations. A basic aspect, and perhaps a fundamental problem as well, is the issue that zinc is a co-factor for more than 200 enzymes, many dealing with various aspects of protein synthesis or hormone function of some type. While the studies pertaining to outright zinc deficiency, the consequences of marginal dietary zinc consumption, and health aspects often relate to the effect zinc has upon these enzymes, zinc excess may not have a direct influence upon these target enzymes and in fact could produce secondary problems. This is well documented in the previous discussion on copper and zinc antagonism. Excess diet zinc may be secondary in its ability to lead to a copper deficiency, which has much different symptoms. A challenge, therefore, is to separate out the direct effects of zinc toxicity from the indirect or secondary effects that it may have by perturbing the balance of another essential element. (See Table 2.)

In humans, most of the body's zinc is in the skeletal muscle (about 60%) and one third in the bone. Skin, liver, brain, kidneys, and the heart have small total amounts in this regard.[48] Zinc is intracellular in its primary distribution, and the concentration of zinc in extracellular fluids is low.[48] These distributions are apparently age dependent. Shaw[49] documented that liver zinc in a newborn infant was greater than for an adult man. Widdowson[50] reported that 25% of the zinc in newborns is found in the liver. More zinc is probably also found in the bones of newborns as compared with adults.[49]

Hormonal balance may influence the zinc distribution between extracellular and intracellular fluids. Cousins[10] has reported that insulin, glucagon, and glucocorticoids will likely influence liver zinc levels. For example, Failla and Cousins[51,52] reported earlier that glucocorticoids stimulated zinc uptake in hepatic cultured cells, and Henkin et al.[53] also suggested this for the human liver *in vivo*. Glucagon may stimulate zinc uptake by liver cells.[54] Other hormones for which zinc plays a fundamental role include growth hormones.[55] It is well known that zinc deficiency leads to dwarfism in humans.[56]

There are two known genetic disorders of zinc metabolism. Danbolt and Closs[57] reported a decreased ability of zinc to be absorbed by the small intestine, thereby leading to acrodermatitis enteropathica; zinc supplements can reverse this. The familial disorder, hyperzincaemia,[58] results in elevated serum zinc levels but apparently does not produce any toxicity. Zinc overdoses that result in similar plasma zinc levels have been known to be fatal.[59]

Humans are able to regulate the uptake of zinc such that there is relatively little variation in body zinc in proportion to the variation in dietary zinc. If zinc intake is low, proportionately more dietary zinc is absorbed and vice versa.[60-63] Jackson et al.[63] reported that elevated dietary zinc results in reduced fractional absorption of zinc but an increased rate of gastrointestinal zinc excretion.

There is a point, however, when the body is unable to respond to excess zinc intake. Excess loss of zinc may lead to more zinc deposited in hair.[64] Excess zinc intake also may lead to increased hair zinc levels but other studies dispute this.[65] Prior zinc status and actual diet levels most likely account for such different results. Zinc is usually bound to albumin when it is transported in the plasma;[66] some is bound to amino acids. Zinc uptake from plasma to hepatocytes appears energy dependent.[51] How it gets into other tissues from the plasma remains obscure.

3.2 Toxicity

With respect to human toxicity of zinc, the incidence appears infrequently with some acute toxicities reported.[66] Much of our knowledge of zinc toxicities comes from animal studies and through supplementation with mineral preparations. Using animals that are given high diet zinc results in a large decrease in food intake, perhaps due to unpalatibility. Growth rate and even weight loss follow, but this is due to the decreased food intake. However, excess zinc consumed by animals has been reported to result in rough hair, achromotrichia, emphysema, diarrhea, arthritis-like symptoms, abortive fetuses, stillbirths, etc. A microcytic anemia occurs, but this could be due to a decreased uptake of either copper or iron, both of which, if limited, can lead to this type of anemia. Hemolytic anemia is not uncommon. Renal fibrosis, fatty liver, and liver necrosis have been reported. Hypercholesterolemia sometimes occurs with excess zinc, which may be due to a copper-deficiency. Excess zinc normally would produce symptoms that mimic deficiencies of copper, iron, manganese, and/or calcium. When one or more of these elements are increased in the diet, the signs of zinc toxicity appear markedly reduced.

Acute toxic effects of zinc in humans have been reported; intakes above 15 mg/l in water can produce nausea (see below). As the level of zinc increases, vomiting and diarrhea result.[67] Tachycardia, hemolytic anemia, pancreatitis, renal damage, and death have been reported on occasion.[68] Acute intakes of 4 to 28 g/day have resulted in a variation of these toxic effects.

Subchronic and chronic effects have been investigated, but to a limited extent.[69,70] Impaired immune response has been reported in response to excess diet zinc by Chandra.[71] Other basic studies pertaining to chronic toxicity have focused on the decrease in high-density-lipoprotein (HDL) cholesterol reported by several laboratories and on copper deficiency as a consequence of excess zinc intake. Zinc supplements can lower HDL cholesterol levels in serum, as demonstrated by Hooper et al.[72] Zinc levels as low as 50 mg Zn per day for 12 weeks has resulted in decreased HDL cholesterol levels.[73] This level is often found in over-the-counter mineral supplements. Zinc supplements have been known to produce nausea and vomiting in subjects. Brown et al.[74] documented vomiting and diarrhea when foods and beverages were contaminated by zinc from galvanized containers. When 28 g of zinc sulfate were ingested by a woman, vomiting, tachycardia, hyperglycemia, and eventually death resulted, due to pancreatic hemorrhage and renal damage.[68] Ingestion of 12 g of zinc by an adolescent male resulted in lethargy, lightheadedness, and altered gait.[75]

Copper deficiency is a likely result of pharmacological zinc doses in the range of 100 to 300 mg Zn per day. Low blood copper levels, anemia, and neutropenia result. Lowering zinc levels and/or increasing copper may alleviate this problem.[76,77] The sensitive influence of zinc intake upon copper balance was studied by Sandstead[78] and Festa et al.[79] Less than 20 mg zinc per day may lead to increased fecal copper and thereby decreased copper retention in the human body. 50 mg of zinc as zinc gluconate has been shown to depress erythrocyte Cu, Zn superoxide dismutase (SOD) activity in human males.[80] Low serum iron and copper levels leading to anemia have been reported by several groups with excess zinc intakes (754

TABLE 3
Iron Metabolism and Toxicity

Metabolic Aspects	Affected Site/Action	Ref.
Digestion and absorption	Small intestine	91–98
Transport	Blood	99, 100
Ferritin metabolism	Synthesis	102, 103
Iron overload	HLA-hemochromatosis	100, 104, 105
	Congenital	106, 107
	Genetic	108–111
	Anemia	112, 113
	Blood transfusions	114, 115
	Neonatal iron overload	116–118
Animal models of overload	—	119, 120
Toxicity		
Free radical and liver peroxidation	Liver	121–129, 133, 138–141
	Plasma	127, 130
	Kidney	128, 129
	Spleen	131
	Muscle	128
	Skin	128
	Liver	89
Cancer	Iron overload	132, 142–144
Heart disease	Increased ferritin levels	145, 146

to 450 mg Zn per day for a 2-year period).[76,81] Petrie and Row[82] reported that a dialysis patient exposed to zinc in water that was used for the dialysis suffered severe anemia (3 g hemoglobin per dl).

Some recent attention has been paid to the potential neurotoxicity of zinc.[83] Zinc has been reported to impair the neuroexcitation of the *N*-methyl-D-aspartate receptors and increase the alpha amino-3-hydroxy-5-methyl-4-isoazoleproprionate receptors.[84,85] Attenuation of gamma amino butyric acid (GABA) receptors has been reported.[85] Corticol neurons have been reported to be destroyed by excess zinc intake.[86,87]

4 IRON

With regard to a physiological iron imbalance in humans, it is the deficiency state that receives the most attention. Iron-deficiency anemia is one of the most recognized nutrition-related diseases worldwide;[88] however, the prevalence of hereditary-based iron overload makes it one of the most common metal-related toxicity disorders.[89] (See Table 3.)

Iron is found in food and the human body in the ferrous (Fe^{2+}) and ferric (Fe^{3+}) state. Iron's oxidation-reduction properties make it ideal for participation in complex cellular systems such as mitochondrial electron transport. Likewise, it is probably iron's oxidation-reduction potential properties that result in the pathological alterations associated with overload.

4.1 Metabolism

While there have been reports of certain occupational environments in which iron dust may be inhaled, under most situations iron is introduced into the human body by ingesting iron-containing foods. Since iron does not have a major excretory pathway, human iron balance is regulated at the point of absorption to offset daily losses of approximately 1.0 mg for adult men and 1.5 mg for menstruating women.[98]

The iron content of the Western diet has been estimated at about 7 mg of iron per 1000 kcals.[91] Iron is present in foods in one of two forms: heme iron and nonheme iron. Heme iron is derived from animal sources such as meat, fish, and poultry and is part of hemoglobin, myoglobin, cytochromes, and other heme-containing molecules. Heme iron crosses the luminal membrane of enterocytes intact; once inside the cell, the iron is liberated from the protoporphyrin ring by the action of heme oxygenase.[92] Nonheme iron is derived from both plant and animal sources and includes the iron complexed into ferritin, hemosiderin, and iron-containing enzymes and salts. Nonheme iron is released from food components by digestive secretions; two suggested mechanisms for the movement of ionic iron into enterocytes are (1) a facilitative transport system mediated by a 160,000-M_r glycoprotein,[93] and (2) integrin-involved movement of iron through the plasma membrane to a 56,000-M_r protein in the intracellular compartment.[94,95] Gastric and intestine luminal factors such as gastric acid, ascorbic acid, and components of meat, fish, and poultry appear to increase the efficiency of nonheme iron absorption.[91] Once inside the enterocyte, iron derived from either diet form may be retained by cellular ferritin or transferred across the basolateral membrane to plasma transferrin for systemic distribution.[96]

Iron absorption is directly related to physiological iron need.[91] Absorption may be regulated systemically by the level of plasma transferrin receptors and the rate of erythropoiesis,[97] although some evidence exists for local regulatory factors in mucosal tissues as well.[98] Iron is transported in the plasma to various tissues bound to transferrin, a 79,550-M_r glycoprotein.[99] Transferrin can accommodate two atoms of iron; however, only about 30% of the iron-binding sites are occupied with iron under normal conditions.[100]

The total body content of iron is about 2 to 5 g depending on gender, diet, size, and menstrual status,[101] and it is distributed between either metabolic or structural and transport compartments. Iron is engaged in the metabolic operations in all human cells. Greater than 60% and up to 10% of the iron in the human body can be found in erythrocyte hemoglobin and muscle myoglobin, respectively, while other heme and nonheme enzymes contribute about 2 to 4% of body iron.[101]

Iron transport as transferrin and storage in the form of ferritin and hemosiderin make up the remaining 20 to 30% of body iron.[101] Iron bound to transferrin makes up a very small portion (3 to 4 mg) of this compartment. Apoferritin, a spherical 440,000-M_r protein, contains approximately 24 subunits and can hold approximately 4500 atoms of iron.[102] The primary sites of ferritin synthesis are the liver, spleen, bone marrow, and intestine. Some tissue-derived ferritin can be found in the plasma and is used to gauge body iron stores, as 1 μg of ferritin/l of serum equals 10 mg of iron stores.[103]

The other iron storage protein is hemosiderin, which may be derived from ferritin. The ferritin:hemosiderin ratio in the liver is believed to reflect iron storage, as the ratio increases with decreasing cellular iron content and vice versa.[89,91]

4.2 Iron Overload

Iron toxicity has been reported in both humans and animals. Human leukocyte antigen (HLA)-linked hemochromatosis appears to be one of the most common inborn errors in metabolism among Caucasians of European descent.[104] The hemochromatosis locus is linked to the HLA region on the short arm of chromosome 6 and is an autosomal recessive trait.[100] The prevalence of this genetic abnormality may be as high as 12 in 1000[105] and is characterized by excessive iron absorption, elevated plasma iron concentration and transferrin saturation, and high iron content in liver parenchyma cells.[100] Contrarily, macrophage iron content is relatively low.

Congenital atransferrinemia is an extremely rare disorder characterized by a nearly complete lack of transferrin.[106] This disorder is probably an autosomal-recessive anomaly and is accompanied by hypochromic anemia and iron overload involving the liver, heart, and pancreas and an almost complete lack of iron in bone marrow. This disorder along with a mouse model of hypotransferrin are very suggestive that plasma transferrin is not necessary for the transport of absorbed iron.[107]

More isolated examples of human genetic disposition for iron overload have also been described. One third of the members of a large Melanesian family have been reported to have developed iron overload.[108] Although many characteristics are similar to HLA-linked hemochromatosis, the mode of inheritance appears to be an autosomal-dominant transmission. Another instance of inherited iron overload hereditary was reported in two siblings in a Yemenite Jewish family.[109] This inherited trait is also believed not to be HLA-linked hemochromatosis.

Iron overload has been reported in at least 15 sub-Saharan African countries.[100] The overload is the result of drinking locally brewed beer with a high iron content. The histological alterations to the liver are distinct from alcohol-related insult, and iron accumulates in both hepatic parenchyma cells and macrophages. Necropsy evaluation estimated the incidence of iron overload-induced liver cirrhosis to be greater than 10% in these geographic regions.[110] Further investigation suggested that most likely there is an underlying genetic factor concomitant with a high dietary iron consumption.[111]

The forms of inherited anemia — homozygous β-thalassemia, β-thalassemia/hemoglobin E, and hemoglobin H disease — all result in ineffectual erythropoiesis in bone marrow. Although the mechanisms are unclear, the ineffective erythropoiesis ultimately leads to augmented iron absorption.[112,113] The treatment of these diseases involves multiple blood transfusions which contribute even more iron to these individuals. Initially the overloading of iron results in deposits in the liver; however, with time, iron accumulates in other organs such as the heart and pancreas.

Other forms of anemia associated with ineffective erythropoiesis can increase iron absorption and potentially lead to overload. These anemias include congenital dyserythropoietic anemias, a number of siderblastic anemias, and many anemias associated with poor iron incorporation into hemoglobin.[90]

Iron overload may also be induced clinically by frequent blood transfusions in patients with aplastic anemia, pure red cell anemia, Blackfan-Diamond syndrome, myelodysplasia, and sickle cell disease.[114,115] The iron is derived primarily from erythrocyte hemoglobin, and excessive iron initially accumulates in macrophages and then liver parenchyma cells.[114] Neonatal iron overload has been described as being associated with certain perinatal metabolic disorders such as hypermethionemia[116] and fatal liver disease.[117,118]

Animal models have been developed to study iron overload and its related pathology.[119] Rats fed a diet enriched with 2 to 3% elemental (carbonyl) iron over a period of 2 to 4 months develop hepatic nonheme iron concentrations 50 to 100 times normal. Excessive iron deposition in cardiac and pancreatic tissue is modest, and nonhepatic organ toxicity is not evident.[120]

4.3 Toxicity

The exact mechanisms of toxicity from iron overload are not completely understood; however, many investigators agree that the pathological alterations associated with iron overload are probably the result of increased free radical activity initiated by excessive iron. Under normal situations, iron is almost entirely found bound to proteins; however, unbound iron in the reduced ferrous form is believed to contribute to free radical activity by participating in the Fenton reaction which results in the production of the highly reactive hydroxyl radical ($OH^{\cdot}$):

$$Fe^{2+} + H_2O_2 \rightarrow Fe^{3+} + OH^- + OH^{\cdot}$$

Many investigators have reported the products of lipid peroxidation in various tissue, including liver,[121-129] plasma,[127,130] kidney,[128,129] spleen,[131] muscle,[128] and skin.[128] Britton et al.[89] suggested a paradigm for iron overload associated hepatic tissue pathology. Increased iron absorption results in hepatic iron overload, which ultimately leads to organelle dysfunction and injury; lipocyte collagen synthesis leading to fibrosis; and possibly alterations in hepatic DNA initiating tumor formation.[89]

Although iron deposition during overload deposits in many tissue such as the heart, lung, kidney, and brain, the liver has received the most investigative attention, most likely because of its prominent involvement in iron storage and also because cirrhosis is recognized as one of the most common causes of death with genetic-based hemochromatosis in humans.[132] Rats fed a diet enriched with 2 to 3% carbonyl for 2 to 4 months develop hepatic iron concentrations of 3 to 6000 μg Fe per g liver.[120] The iron preferentially deposits in periportal hepatocytes similar to early HLA-linked hemochromatosis and African iron overload.[120,121] The iron-overloaded rats present direct evidence of mitochondrial and microsomal lipid peroxidation,[121,123] along with an increase in the low-molecular-weight pool of catalytically active iron.[133] Further, at a hepatic iron concentration at which lipid peroxidation is observed, specific mitochondrial membrane-associated activities such as oxidative metabolism and Ca^{2+} sequestering are decreased.[123,133,134] Similarly, microsomal membranes demonstrate decreased cytochrome concentrations, enzyme activities, and Ca^{2+} sequestration.[124,134,135]

Iron overload results in excessive accumulation of iron in hepatocellular lysosomes and appears to increase their fragility.[136,137] This increase in fragility results in the release of hydrolytic enzymes into the cytosol of hepatocytes and initiates cellular damage. Myers et al.[126] reported that experimental iron-overloaded rat hepatocytes presented lysosomes that were more fragile, enlarged, and misshapen.[126] These membranes also demonstrated decreased fluidity and increased lipid peroxidation as determined by malondialdehyde content.

At liver iron concentrations similar to those observed in HLA-linked hemochromatosis (3000 to 6000 μg Fe per g liver), experimental iron-overloaded rat mitochondria show increase lipid peroxidation, as demonstrated *in vivo* by the presence of conjugated dienes in phospholipid extracts.[121-123] These investigative efforts also resulted in the determination of a hepatic iron concentration threshold for the presence of lipid peroxidation in mitochondria (1000 to 1500 μg Fe per g liver) and microsomes (3000 μg Fe per g liver).[122] It has also been reported that mitochondrial malondialdehyde content is also increased several fold in experimental iron-overloaded rats and that this increase in malondialdehyde is likely due not only to increased lipid peroxidation but also to an impairment in malondialdehyde metabolism.[125] Furthermore, at modest increases in hepatic liver iron concentration, there was a significant impediment of mitochondrial electron transport as exemplified by a 70% reduction in cytochrome C oxidase activity and a 48% decrease in cellular oxygen consumption.[138]

Iron overload also results in hepatic fibrosis.[89] The mechanisms of fibrogenesis in this condition are poorly understood, as efforts with experimental iron-overloaded rats and baboons have failed to demonstrate a consistent relationship with prolyl hydrolase activity.[138] Morphological investigation of experimental iron overload has revealed that hepatic fibrosis is recognized at 8 months, and by 1 year periportal fibrosis is pronounced and concomitant to the identification of cirrhosis in some animals.[120] Investigators have found that the hepatic levels of type I procollagen mRNA are augmented[139,140] and that nonparenchymal cells are predominantly involved, most likely activated lipocytes.[141]

Iron overload is also associated with a greater incidence of cancer; humans with HLA-linked hemochromatosis are at about a 200 times greater risk of hepatocellular carcinoma.[132,142] Experimental iron overload rats have presented evidence of an increase in DNA strand breaks with a liver iron concentration of 3130 μg/g tissue but not at lower liver iron concentrations (≈ 600 μg/g).[143] Further, a synergistic carcinogenic effect was reported with the combination of iron in conjunction polychlorinated biphenyls.[144]

Recent concern has been over the reported association of serum ferritin levels with increased myocardial infarction.[145] In a study of over 1900 Finnish males from 40 to 64 years of age, serum ferritin levels greater than 200 μg/l had a 2.2 times greater risk of myocardial infarct compared to males with lower levels. This was after adjustment for other known risk factors such as cigarette smoking, higher systolic blood pressure, lipoprotein cholesterol levels, etc. In fact, those males with a serum low-density-lipoprotein (LDL) cholesterol level greater than 193 mg/100 ml had even a greater risk with the added high serum ferritin levels. The mechanism apparently may be related to the role of iron in free radical generation, as reviewed above. Oxidation of LDL cholesterol is known to result in greater cholesterol uptake by macrophages, which is a key mechanism in foam cell production and subsequent plaque formation.[146]

5 SELENIUM

5.1 METABOLISM

Selenium is efficiently absorbed in the gastrointestinal tract in several organic forms; however, two distinct chemical forms, selenomethionine and selenite, have been traced in humans using stable isotopes. Selenomethionine is a selenium analog of a sulfur-containing amino acid. Selenium and sulfur are exchanged due to their chemical similarities, and selenium is absorbed as selenite. The major site of absorption is the duodenum, although some selenium is absorbed in the ileum and jejunum. Selenium absorption does not occur in the stomach; there is no known regulatory mechanism for selenium absorption.[147] A range of 50 to 100% absorption of these two forms has been demonstrated.[148]

In the form of selenium dioxide, a white crystalline material that melts at 340°C, selenium can be absorbed in the respiratory tract. It is then reduced to selenium metal, which can cause liver damage if prolonged exposure occurs. Symptoms of overexposure to selenium dioxide include a garlic odor of perspiration and breath.[149]

When selenium is absorbed as selenomethionine, it is incorporated into a plasma protein called selenoprotein P. This protein functions to transport and store selenium.[150] As selenite, selenium becomes incorporated into the metalloenzyme glutathione peroxidase. Glutathione peroxidase functions to reduce organic and hydrogen peroxides. This is especially important for phagocytic cells such as leukocytes and macrophages. In these cells, peroxides are the byproducts of the oxidative destruction of foreign matter; therefore, glutathione peroxidase protects these cells from being destroyed as they function.[151]

Another site of glutathione peroxidase activity is at the platelet. Here it acts in an antiaggregative capacity. This metalloenzyme reduces fatty acid peroxide formation, and the ratio of prostacyclin (an antiaggregating factor) to thromboxane (a proaggregant) becomes increased.[152] Through this mechanism, selenium is linked to cardiovascular disease by decreasing platelet aggregation which reduces clots and atherosclerosis.

Of selenium excreted, 50 to 60% is through the urine, the remaining 40 to 50% of selenium being excreted through the feces. Body stores of selenium greatly influence renal clearance of this mineral; hence, the kidneys appear to be the regulatory mechanism for selenium homeostasis.[153] Endogenous selenium is lost through the feces, which was demonstrated by showing that fecal selenium excretion remains the same regardless of dietary intake. At toxic intakes of dietary selenium, volatile selenium compounds such as dimethylselenide are exhaled and can escape through the skin.[153] (See Table 4.)

TABLE 4

Selenium Metabolism and Toxicity

Metabolic Aspects	Affected Site/Action	Ref.
Digestion and absorption	Small intestine	147, 148
Inhalation	Lungs	149
Transport	Blood	150
Storage	Glutathione peroxidase	151, 152
Excretion	Kidneys, colon, lung, skin	153
Toxicity	Symptoms	154–156
	Chemical toxic forms	157, 158
	Cases in China	159–161
	Target organs	162
	Factors affecting	162

5.2 Toxicity

There are three forms of selenium toxicity: acute selenosis, subacute selenosis, and chronic selenosis. Acute selenosis occurs when excess amounts of selenium are ingested over a short length of time. Symptoms of acute selenosis include an unsteady gait, cyanosis of the mucous membranes, and difficulty breathing which can lead to death. Autopsy reports of acute stenosis describe liver congestion, endocarditis, myocarditis, and smooth muscle degeneration in the gastrointestinal tract, gallbladder, and bladder. Long bone erosion was also reported in these cases.[154]

When large doses of selenium are ingested over a long time frame, subacute selenosis is observed. Symptoms of subacute selenosis include neurologic dysfunction such as vision impairment, ataxia, and disorientation; respiratory distress is often seen as well. Subacute selenosis is commonly seen in livestock that graze on selenium-accumulating plants. These seleniferous plants are concentrated in the western U.S. — Montana, Colorado, Wyoming, New Mexico, and Arizona.[155]

Chronic selenosis occurs when moderate doses of selenium are ingested over a considerable length of time. This condition is characterized by skin lesions and dermatitis such as alopecia and hoof necrosis (in livestock), emaciation, chronic fatigue, anorexia, gastroenteritis, liver dysfunction, and spleen enlargement.[156]

The most toxic forms of selenium are noted to be sodium selenite, sodium selenate, selenomethionine, and selenodiglutathione;[157] however, there are wide variations in selenium toxicity with respect to the valence state of the molecule. Recently, a multitude of selenium compounds have been synthesized as chemopreventive/anticarcinogenic substances. These are being tested for their toxic effects.[158]

An area of China has unusually high concentrations in the soil of selenium[159-161] which becomes incorporated into the food supply. Residents were evaluated for clinical and biochemical indications of selenium intoxication. The average daily selenium intake was estimated to be 1.4 mg for adult males and 1.2 mg for adult females. When comparing this group to those whose selenium intakes were 0.07 and 0.06 mg, respectively, for men and women, increased clotting time and reduced serum glutathione were observed. Clinical signs that were observed consisted of garlic odor in the breath and urine, brittle or lost nails, lowered hemoglobin levels, and nervous system problems such as peripheral anesthesia, acroparesthesia, and pain in the extremities.[158]

In livestock, symptoms of selenium toxicity are observed in the nervous system as ataxia, tremors, hypersensitivity, and convulsions. In humans, nervousness, chills, numbness, impaired nerve conduction, and peripheral anesthesia are symptoms. Mottled teeth have been observed in humans with selen[illegible] demonstrated to have steatosis and necrosis associate[illegible] this has not been reported in humans.[158,162]

Kidney problems suc[illegible] [illegible]rosis, and calcinosis have been reported only in animals. T[illegible] by selenium toxicity, resulting in myocarditis in rats and brady[illegible] [illegible]ock have shown respiratory disturbances such as congestion, ed[illegible] and hydrothorax. The skin is affected by selenium toxicity, seen as thic[illegible] [illegible]als; dry, brittle hair; hair loss; red, swollen hands and feet. In animals, crac[illegible] [illegible]noticed, as well as dermatosis and alopecia. Anemia, increased prothrombin time, and decreased hemoglobin are hematological parameters observed in both humans and animals. Decreased immune function has been demonstrated in rats with selenium toxicity. In both animals and humans with selenium toxicity, loose stools, diarrhea, excessive salivation, and dyspepsia have been observed. Deformities of fetal chicks and ducks have been documented in selenium toxicity.[162]

5.3 Interactions Between Selenium and Other Nutritional/Environmental Substances

Vitamin E has been shown in animals to spare selenium and reduce the amount necessary in the diet. Other factors with such a role have also been identified and include: decreased food intake, high protein intake, high levels of vitamin A and vitamin C, and synthetic antioxidants. Conversely, there are substances which are known to antagonize dietary selenium: heavy metals, sulfate, mercaptans, and chlorinated hydrocarbons as well as deficiencies of vitamin E, riboflavin, vitamin B-6, and methionine.[162]

There are substances known to affect selenium toxicity. These may act as methyl donors which synthesize selenium metabolites and are simply excreted. The methyl donors include methionine, betaine, choline, creatinine, and amidinoglycine. Also, the heavy metals mercury, cadmium, lead, silver, and arsenic along with the trace elements copper, zinc, and iron substitute for selenium and reduce the toxic potential of selenium. Furthermore, antioxidants such as vitamin E, diphenyl-p-phenylene diamine (DPPD), and beta-hydroxy-toluene (BHT) help the antioxidant role of glutathione peroxidase and thereby reduce toxic effects of selenium.[162]

6 MERCURY

6.1 Chemical Form as Related to Toxicity and Sources of Exposure

Mercury toxicity is enhanced in the organic chemical form. Organomercurial concentration in food chains was first discovered and reported in Japan in 1959. Shellfish contaminated with mercury were consumed by a small Japanese population in the area of Minimata Bay. This resulted at the time in 46 fatalities along with disorders such as mental depression and tremors in other affected individuals. Minimata disease[163] is the term applied to this chronic alkylmercury poisoning. In the U.S., some instances of mercury poisoning have been reported, such as the well-known case of a New Mexico farmer who accidentally poisoned his family when he fed contaminated waste seed to his hogs, which in turn were eaten by family members.[164]

Mercury has been used for centuries, but it was not until the industrial revolution that its use became extensive.[165] The occurrence of mercurials in the environment can generally be traced to three major sources: (1) alkylmercurials, found in agricultural pesticides; (2)

TABLE 5
Mercury Metabolism and Toxicity

Aspects	Ref.
Historical toxicology	163–164
Chemical forms in the environment	166, 167
Chemical forms and relative toxicities	165, 168
Excretion	163, 168
Organs affected	
Kidney	169
Central nervous system	163, 168, 170, 171
Learning disabilities	172
Endocrine	185–186
Intracellular distribution	175
Cellular influences in selected organs	166, 174–181

arylmercurials, which occur in paints and are used in the manufacture of paper; and (3) inorganic divalent mercury from chlor-alkali plants.[166] Many of these compounds find their way into aquatic systems through run-off and discharge of liquid effluences into surface waters. The U.S. Environmental Protection Agency (USEPA) banned the use of mercurials in pesticides and fungicides in 1972.[167]

The organic mercurials are the most toxic known to humans and other vertebrates. These forms of mercury have a lower water solubility than the inorganic ones but a higher lipid solubility. The alkyl compounds have a greater water solubility and volatility than aryl compounds possessing the same X or side groups.[165,168] Among the methyl and ethyl compounds, the most toxic are the phosphate derivatives, followed by the chloride and cyano forms.[165] (See Table 5.)

6.2 Toxicity

The toxic effects of the mercurials are apparently related to the excretion rates. Phenyl- and alkoxyalkylmercurials are not persistent in vertebrates and have biological half-lives of approximately 3 to 4 days, which is about the same as for inorganic mercury.[168] Alkylmercury compounds are usually more stable and excreted more slowly, showing half-lives of 15 days in the rat and 23 to 27 days in poultry.[168] Excretion rates in rats of aryl and inorganic mercury are similar to one another. Excretion rates of methylmercury are lower than rates observed for aryl and inorganic mercury.[163]

Among higher vertebrates, including humans, inorganic and alkoxyalkyl compounds cause kidney damage, which usually leads to death. Some uptake of inorganic mercury by kidney cells suggests that active transport is involved, since energy is required, but most is by diffusion.[169] Chronic poisoning by elemental mercury is characterized by progressive renal and central nervous system damage. Symptoms include mental depression, irritability, and tremors. Chronic levels of alkylmercury compounds produce different effects. There is a latency period that lasts from one to several weeks, during which no symptoms are apparent. Then, damage to the central nervous system expressed as poor muscular coordination, loss of a sense of positioning and equilibrium, and impaired hearing are observed.[163,168]

Methylmercury compounds appear to be the most toxic of the mercurials, tending to be retained in the body, especially in the brain.[163,168] Sensory malfunctions include paresthesia, astereognosis, and constriction of visual field. Motor manifestations range from impairment of fine coordination to gross ataxia, depending on degree of exposure.[170] Neonatal exposure to mercury vapors in rats results in behavioral changes, such as increased locomotive activity, but decreased rearing of young when tested at 4 months of age.[171] Learning ability is impaired as well. Prenatally exposed rats demonstrated similar findings when tested at similar later ages.[172] Some of the neurotoxicity of mercury may be due to decreased central nervous system (CNS) glutamate uptake by astrocytes and spinal cord, as demonstrated in cell culture experiments.[173] The efficient absorption of methylmercuric and phenylmercuric compounds from food may be due to the lipid solubility of their chloride complexes.[174] Various organo-mercurials are converted to inorganic mercury once in the body, but the precise mechanism appears unclear.

The intracellular distribution of mercurials varies with the type of mercurial present. Lysosomes may be one of the main organelles to concentrate mercuric chloride, followed by the mitochondria and microsomes. The microsomal fraction usually has a greater amount of methylmercury.[175]

The biological properties of the short-chain alkylmercury compounds are related to their ability to cross cell membranes and to their slow conversion to inorganic mercury in the body. The almost complete absorption of these compounds from food and their subsequent rapid passage across the blood-brain and placental barriers accounts for damage to the CNS in both adult and fetal mammals. Low levels of inorganic mercury have been shown to be mutagenic in Chinese hamster ovary cells;[176] these levels were not cytotoxic. The apparent neurotoxicity of methylmercury appears related to the catalyzed hydrolysis of phospholipids composing the cell membranes of neurons. In the liver and kidney, the high affinity of mercury to thiol groups appears to be the most significant chemical property explaining their toxic effects. While they have a high specificity in terms of sulfhydryl groups, these compounds are nonspecific in terms of proteins they target, since almost all proteins contain sulfhydryl groups that are metal reactive. Mercurials are consequently potent but nonspecific inhibitors of enzymes.[166,174,175] The consequence of these properties is apparent inhibition of energy metabolism, formation of cell structural proteins, and a variety of cellular processes.[175] The mercury ion is known to promote oxidation of kidney cells and to disrupt renal mitochondrial function.[177] Increased H_2O_2 production by rat renal mitochondria is an indirect effect of inorganic mercury.[177] Renal mitochondria from rats treated with mercuric chloride (1.5 mg/kg i.p.) have a twofold increase in H_2O_2 but reduced glutathione content. Thiobarbiturate reactive substances were increased by more than two thirds. Depolarization of the inner mitochondrial membrane was reported.[178] Hyperpolarization of cultured renal cell membranes exposed to mercury ions have been reported previously[179] along with an increase in cell membrane potassium selectivity. However, another study supported the concept that mercury thiol complexes in the kidney possess redox activity and promote porphyrinogen oxidation, leading to excess porphyrins in the urine.[180] The affinity of mercury for thiol groups accounts for the accumulation of large amounts of inorganic mercury in the kidneys without much damage on a metabolic basis. However, there has been evidence that some damage may occur to the extent that both renin- and angiotensin-I-converting enzyme activities were reduced[181] and may modify systemic hemodynamics. Metallothionein in the kidney may afford some degree of protection from the toxic effects of mercurials.

Recent studies have contended that mercury may have deleterious influences upon lymphocytes and may be related to renal autoimmune disease in genetically predisposed animals.[182,183] A human study reported increased lymphocyte micronuclei in mercury-exposed chloralkali workers.[184]

The influence of mercury upon endocrine function appears minimal. McGregor and Mason[185] reported in a population exposed to mercury vapors a lack of relation between pituitary and thyroid endocrine function with blood and urinary mercury levels. Moszczynski et al.[186] studied 89 men, 21 to 57 years old, and grouped them into levels according to the length of mercury exposure. No relation among various clinical, hematological, and biochemical measures was noted in these subjects. All appeared clinically healthy.

TABLE 6

Cadmium Metabolism and Toxicity

Metabolic Aspects	Affected Site/Action	Ref.
Digestion and absorption	Small intestine	187
Transport	Blood	189
Storage	Liver	189
Elimination	Kidneys	187, 191
"Itai-itai" disease	Bone	192
Symptoms	—	187, 196, 197
Relation with other trace elements		
	Zinc	192
	Iron	198–200, 202
	Cobalt	201
Toxicity		
	Cardiovascular	
	Hypertensive effects	203–210
	No blood pressure effects	211, 212
	Biphasic blood pressure effects	213, 214
	Hypertensive mechanisms	215, 216
	Epidemiology evidence and hypertension	196, 197, 218, 219, 221, 222
	Reproductive	
	Testicular and prostate tumors	187, 223, 224
	Testicular pathology	225–230
	Sex hormones	231
	Fetal development	187, 232
	Anemia	236–238

7 CADMIUM

7.1 SOURCES OF EXPOSURE

The interest in cadmium toxicity has centered on occupational exposure of workers and exposure of a local population through some type of industrial activity. Exposure to cadmium is common for workers engaged in such occupations as electroplating (coating steel, iron, copper, brass, etc., in order to make it corrosion resistant) and the manufacture of cadmium batteries, plastics, paints, textiles, and phosphate fertilizers. Mining activities can lead to the exposure of workers to cadmium, and the surrounding area could be exposed to large amounts of cadmium through runoff. Accidental exposure of a Japanese population through rice contaminated by runoff from a mine upstream to the paddy is perhaps the best publicized case of cadmium toxicity. (See Table 6.)

Cigarette smoke is a large source of cadmium intake. Between 0.1 and 0.2 μg of cadmium per cigarette has been reported, but the amount inhaled is dependent on the number of puffs and inhalation pressure. Furthermore, more cadmium is found in the particulate phase than in the gaseous phase.[187]

In the nonsmoking population, food is the major source of cadmium intake.[188] Some areas of Japan have the largest intake of foodborne cadmium of any other area studied. In the U.S., the amount of cadmium in diets was fairly constant from 1920 to 1945. From 1945 to 1975, a 20% decline in the average daily cadmium intake was observed. This decline is thought to reflect changes in the dietary patterns of Americans rather than a change in the total amount of available cadmium in the environment.

Water contributes slightly to cadmium intake. Water concentrations below 10 parts per billion (ppb, or μg/l) will contribute little to daily intake. At a concentration of 20 ppb, a daily intake of 20 to 40 μg is expected if water consumption is in the amount of 1 to 2 l per day. Plumbing is a factor that must be considered in assessing cadmium intake, because both metal and plastic pipes contain some cadmium.[187]

Whatever the source, cadmium usually accumulates as the body ages, up to 50 years. At this age, one who has essentially been unexposed to cadmium may have accumulated 20 to 30 mg of cadmium in the body. In human newborns, the total body content of cadmium is less than 1 μg.[187]

7.2 METABOLISM

Experimental animal studies have led to the estimate that less than 10% of an oral dose of cadmium is absorbed. Many studies report an absorption rate of 2%. Inhalation may result in the uptake of more cadmium, with studies reporting 10 to 40% retention of inhaled cadmium.[187] Absorption will occur regardless of the total body burden of cadmium, and little of the absorbed cadmium is excreted through the urine or intestinal tract. For example, less than 2 μg of cadmium per day is excreted in the urine by the average person.[187] An estimated biological half-life for cadmium of between 16 and 33 years has been computed.

Following absorption, cadmium is transported primarily to the liver, where it is bound to metallothionein, a protein composed of sulfur, copper, mercury, zinc, and other metals.[189] The protein has a low molecular weight of approximately 3000. Metallothionein is thought to be composed of three protein subunits that strongly bind cadmium and one protein subunit that loosely binds zinc. Cadmium, therefore, has a greater affinity to metallothionein.[190]

After cadmium is sequestered by the liver, small amounts of metallothionein-bound cadmium appear in the plasma, where it is cleared efficiently by the kidney. Cadmium will accumulate in the renal tubules,[191] where it may interfere with zinc-dependent enzymes (e.g., leucine-aminopeptidase, which is thought to play a role in renal handling of protein). With increased renal cadmium content, less protein will be catabolized or reabsorbed, causing tubular proteinuria. When this occurs, cadmium excretion will increase because less metallothionein will be reabsorbed.[187] When cadmium exposure is moderate, renal concentration rises slowly to about 200 μg of cadmium per gram of tissue. At this concentration, renal tubular injury may occur. Cadmium-exposed workers usually have greater concentrations of β-2-microglobulin in the urine and greater urine and plasma metallothionein concentrations, which suggest renal damage.[191]

Cadmium toxicity is also known to affect calcium metabolism, but the metabolic antagonism between calcium and cadmium may not be a mimicking phenomenon. Excess loss of calcium in the urine through renal injury can lead to calcium mobilization from skeletal stores to maintain serum calcium levels leading to osteomalacia. This condition is often termed "itai-itai disease" and was first discovered in Japan as a result of cadmium toxicity. Translated, it means "ouch-ouch" disease. The victims are often of a decreased height, and deformities and numerous microfractures develop in the skeleton with increasing brittleness due to the loss of calcium and phosphate. Pregnant women are more susceptible to this condition, because they have increased calcium requirements, and an inadequate supply of dietary calcium may further aggravate the problem. Lack of sunlight could result in diminished vitamin D synthesis required to aid in calcium absorption. Cadmium may also have a direct effect on bone,[192] but further studies are needed to confirm this.

Pulmonary damage is known to occur in rats intratracheally treated with cadmium chloride. Polymorphonuclear leukocyte numbers and permeability increased.[193] Pulmonary tumors have been reported to develop in genetically susceptible mice when exposed to cadmium.[194] Chronic exposure to cadmium can cause lung emphysema.[195]

7.3 Clinical Evaluation

How can the cadmium burden of an individual be evaluated? This is difficult to answer. As mentioned previously, the concentration of cadmium in the urine may not indicate exposure, because little of the element is eliminated through the kidney until renal injury has occurred. Blood cadmium levels may be helpful but, again, may represent transient levels. It is not clear whether blood levels reflect short- or long-term accumulation of the element. Hair concentration of cadmium and of other elements has been suggested as a viable alternative, because it may represent a longer term of exposure than blood does and may act as a recording filament. One difficulty arises from the lack of information on normal values. The use of hair to evaluate the exposure of populations to cadmium has been studied;[196,197] however, the use of hair elemental analysis for assessing cadmium burden for an individual has not been established. Until further research has been carried out, the use of hair analysis for clinical evaluation should be viewed with caution.

When chronic cadmium poisoning occurs (e.g., itai-itai disease), several subjective symptoms may become apparent.[187] Back and joint pain, lumbago, disturbance of gait, restriction of spinal movement, decreased height, and pain when pressure is applied to an area are some of these. Roentgenograms will often reveal Milkman's pseudofractures, thinned bone, cortex decalcification, deformation, and fish-bone vertebrae. A urinalysis may reveal proteinuria and glucosuria but a decreased phosphorus/calcium ratio. Analysis of the serum may reveal an increase in alkaline phosphatase and a decrease in serum inorganic phosphate. In many cases, detectable roentgenographic signs of osteomalacia are not observable in the early stages of the disease, but analysis of serum may demonstrate changes.

7.4 Relationship with Other Trace Elements

A discussion of any trace element is not complete unless the interrelationship with other elements is considered, and cadmium is no exception. Zinc deficiency can increase cadmium toxicity, whereas extra zinc may protect against this negative effect to some extent. Selenium may also protect against cadmium toxicity through the formation of strong metal-selenium bonds.[192]

In the absence of copper and iron, 25 ppm of cadmium caused mortality in chicks.[198] When these two elements were added to normal levels, 200 ppm of cadmium were required before an increase in mortality was achieved. The addition of zinc appeared to reverse some of the toxic effects of cadmium, such as growth depression and gizzard abnormalities.

Pregnant mice given cadmium in drinking water exhibited fetal growth retardation and anemia, but iron-supplemented diets prevented these effects.[199] The addition of vitamin C to quail diets lowered the toxic effects of cadmium due to increased iron uptake. Iron-deficient rats given intragastric doses of $^{109}CdCl_2$ exhibited greater cadmium uptake than did animals with normal iron status.[200] Similar results were reported in mice. Cadmium inhibited cobalt uptake in a similar manner.[201] In humans, cadmium absorption is greater for those with lower body iron stores than for those with normal body stores.[202] Thus, it appears that subjects with lower iron and zinc status could have increased cadmium uptake.

7.5 Cardiovascular Toxicity

This area of cadmium research has perhaps generated the most data and interest. Several studies on rats have demonstrated the blood pressure-elevating effects of cadmium administration, either through the drinking water or by intraperitoneal injections.[203-207] Cadmium at a concentration of 5 ppm in drinking water produced systolic hypertension, and a zinc chelate reversed the effect in rats.[203,204] Doses as low as 1 ppm of cadmium have produced hypertension in rats, and concentrations as low as 0.1 ppm of cadmium in drinking water had a pressor effect on rats.[208] Studies with monkeys have produced similar findings.[209,210]

Not all cadmium-feeding trials, however, have produced hypertension in experimental animals.[211] Rats made hypertensive by means of unilateral nephrectomy and given 1% saline as drinking water had lower blood pressure when injected with cadmium as compared to rats not injected with cadmium. In fact, cadmium-treated rats remained normotensive.[212] The results of these studies are in apparent conflict with studies in which blood pressure elevation was found.[203-208]

Some of the inconsistencies as to whether cadmium can raise blood pressure may be explained by the work of Kopp et al.[213] Results from their laboratory revealed that at lower cadmium concentrations in drinking water, comparable to environmental exposure (e.g., 0.01 to 0.5 ppm), blood pressure was elevated in rats. Exposure to cadmium concentrations from 0.5 to 50 ppm lowered rather than raised blood pressure. A series of studies on cardiac function and tissue metabolism documented greater changes at 1 ppm of cadmium than at 5 ppm of cadmium in drinking water, whereas liver metabolism appeared to be unaffected. These data suggest that at lower doses, cadmium accumulates to concentrations that affect cardiovascular tissues without the appearance of overall systemic toxicity. Earlier work by Perry and Erlanger[214] supported the conclusion that larger doses of cadmium are hypotensive or vasodepressive, whereas small doses have a hypertensive effect.

Several ideas on the mechanisms of cadmium-induced hypertension have been suggested. For example, sodium ions accumulate in the kidney due to cadmium deposition, which could influence the excretion and reabsorption of renal sodium, resulting in elevated blood pressure.[215] In another mechanism affecting the kidney, Perry and Erlanger[216] suggested that cadmium increases circulating renin activity leading to elevated blood pressure.

Revis[217] reported that two enzymes, monoamine oxidase and catechol-o-methyl transferase, which are involved in norepinephrine and epinephrine catabolism, were inhibited in the aortic tissue of rats receiving cadmium, either through injection or in their drinking water. The binding of norepinephrine to aortic membranes was stimulated in these animals. Thus, cadmium could exert its hypertensive effect by inhibiting the catabolism of norepinephrine while promoting norepinephrine binding in the aorta.

Another study[211] revealed that sodium chloride intake was greater for rats given 5 ppm of cadmium in their drinking water for 23 to 42 weeks as compared to controls. This result suggests that the control mechanism for fluid intake could be altered in cadmium exposure.

The question remains whether or not cadmium is involved in either hypertension or heart disease in humans. Much of the data are epidemiological. For example, populations living in hard water areas have a lower incidence of heart disease. This is thought to be due to ions, such as calcium, blocking the uptake of cadmium and other deleterious elements or compounds. Borgman et al.[196] demonstrated a positive correlation between cadmium concentration in the hair of adolescents in South Carolina and the incidence of heart disease in their respective home counties. Our laboratory has observed elevated hair cadmium concentrations of adult black hypertensive women as compared to weight- and age-matched black normotensive women.[197] Hair cadmium levels in the babies of hypertensive mothers have been reported to be three times as high as in their hypertensive mothers.[218]

Lead and cadmium may interact and lead to increased risk for heart disease. Voors et al.[219] demonstrated elevated liver cadmium and aortic lead levels in North Carolina among victims dying from cardiovascular disease. Furthermore, Revis et al.[220] demonstrated that lead and cadmium could induce aortic atherosclerosis and hypertension in pigeons.

Not all human studies, however, have suggested a link between cadmium and hypertension. Ostegaard[221] reported in postmortem analysis that hypertensives between the ages of 45 and 65 years had lower renal cadmium concentrations than did normotensives who were accident victims. In France, one study evaluated blood cadmium concentrations in 29 hypertensive men matched to controls for sex, age, and smoking habits.[222] The results indicated no difference in blood cadmium concentrations between the two groups.

In view of the study by Kopp et al.,[213] the experimental approach to evaluating the effects of cadmium on human hypertension must be addressed. Since a biphasic rather than a linear response may be operating, negative data may have to be re-evaluated. It is apparent that studies on humans must be better designed in order to address these problems.

7.6 Reproductive Toxicity

Cadmium is known to have an adverse effect on reproduction. Epidemiological studies on occupational exposure to cadmium have suggested excessive deaths due to prostate cancer. Injections of large amounts of cadmium into experimental rats can cause sarcoma at the injection sites or testicular damage and eventually testicular tumors. Long-term exposures at low cadmium doses, however, have not resulted in testicular or prostate tumors in experimental animals in some studies,[223] but more recent studies suggest findings to the contrary.[224]

Cadmium can concentrate in the testes and prostate during heavy exposure and cause a decrease in testosterone synthesis. Excess exposure may also interfere with a zinc/hormone relationship in the prostate. Evidence suggests that direct action of cadmium on prostate cells is unlikely, and it is also unlikely that low-level exposure to cadmium is a causative factor for prostate cancer.[187]

Mice given a subcutaneous injection of $^{109}CdCl_2$ exhibit effects such as karyolysis on the seminiferous epithelium. Degenerative spermatids with vacuolated nuclei are often observed, but Leydig's cells appear to be unaffected. Some studies have suggested effects on testicular blood vessels.[225-227] Berlinder and Jones-Witters,[228] using electron microscopy, demonstrated that in gerbils cadmium acts on the interstitial capillary bed rather than on the seminiferous tubules. It does appear that scrotal testes are more sensitive to cadmium than crytochid testes.[229,230] There have also been conflicting studies suggesting that estrogen and progesterone offer some protection against the toxic action of cadmium on the male reproductive system.[231]

Studies on the effect of cadmium on the fetus have produced mixed results. Some animal experiments have suggested that the placenta constitutes a barrier against transfer of cadmium when small doses are given. When large doses are given, however, cadmium may destroy the placental barrier and enter the fetus.[187] A study of 102 mothers and their newborns revealed decreased birth weight and an increase in hair cadmium levels of newborns. This was especially evident with cases of placental calcification.[232] The reader is encouraged to refer to Robards and Worsfold[233] for a comprehensive review of cadmium toxicology.

8 LEAD

Lead is a divalent metal and often competes with other divalent ions such as iron, calcium, and zinc with respect to absorption and biochemical physiological processes. It may also substitute for the "normal" roles some of these other ions have, but with deleterious consequences. Overall, lead appears clinically to exert its toxic effects more in some tissues as opposed to others. The nervous, renal, and circulatory system appear to be sites where lead appears to have its greatest toxic impact.[234] (See Table 7.) Furthermore, much like some of the other minerals discussed, lead can cross the placenta and have consequences upon the developing fetus. Age appears to be a strong factor in predicting the relative toxicity of lead. Children are much more sensitive to lead's toxic effects than are adults. Of special concern

TABLE 7

Lead Metabolism and Toxicity

Organ	Pathology	Ref.
Blood	Heme biosynthesis	236–238
Kidney	Vitamin D_3 impairment, protein-lead complexes, gout	238–240
	Nephron	238, 241–243
	Protein and nucleic acids	234, 246–251
Cardiovascular	Hypertension	252–259
	Cardiac conduction	261, 262
	Cardiac calcium influx	234
	Electrocardiograms	264
	Cardiac Na/K ATPase	265
Bone	Dental development	267
	Skeletal development	268
	Osteocalcin and protein synthesis	269, 270
	Cellular activity and homeostasis	271–286
Central nervous system	Memory loss and learning difficulties	287–290, 296–298, 308–310
	Hippocampal and corticol effects	291, 292, 296–298
	Visual effects	293–295
	Brain cellular morphology	299–300
	Functional deficits	293, 301–303
	Pre-synaptic neurotransmitter release	304–307
	Encephalopathy	301, 310, 311
	Cognitive intelligence	312–317
	Dopamine, acetylcholine, and GABA release	318–321
	Second messenger substitution effects, Protein C kinase	322–326
	Edema	327–329
Reproductive	Spontaneous abortions	236, 320
	Sperm morphology and count	331–335
	Pregnancy, gestation period, implantation influences	337–339

has been the influence of even small levels of lead exposure in young children and later effects upon learning processes, which are impaired.[235]

Lead often leads to anemia by several mechanisms: (1) competing for absorption with the ferrous iron form; (2) inhibiting heme synthesis, as detailed below; and (3) altering the relative composition of cell membranes, including red blood cells, that make them more fragile and likely to hemolyze when passing through tiny capillary spaces. Besides the impact that lead has upon the red blood cells, it may also affect white blood cells and impair immune function. Separately from this, lead can bind tightly to antibodies, compromising the ability to ward off infections.[235]

As with the other trace elements discussed, the toxicity of lead depends upon other dietary factors, including the levels of other metals. Low calcium diets may result in greater lead levels of various organs. This could be due to lack of competition for uptake in the small intestine. The amount of lead absorbed is usually greater when the stomach is empty. Iron deficiency, severe or mild, probably has the largest impact on the potential toxicity of lead. Less dietary iron conceivably allows more lead to be absorbed. The added lead to hemopoietic cells will decrease heme synthesis, which has already been compromised by iron deficiency.

While lead may be present in food, it is the industrialization of society that has been the prime factor leading to an increased incidence of lead intoxication. Lead-based paints, pesticides, auto emissions, and other industrialized byproducts contribute to everyday lead exposure. Much of this environmental lead can end up in soils, leading to its accumulation in plants and animals.

8.1 Lead Effects Upon Blood

Anemia is a classic indication of lead toxicity.[236,237] Heme biosynthesis impairment results because lead can inhibit the enzyme delta-aminolevulinic acid dehydratase.[236] This enzyme is involved in synthesis of the porphyrin units. As an overview, porphyrin biosynthesis begins with the condensation of glycine and succinyl CoA, a tri-cyclic acid (TCA) cycle intermediate, to form α-amino-β-ketoadipic acid in the presence of the enzyme aminolevulinic acid synthase and vitamin B_6. The complex undergoes decarboxylation to form delta-aminolevulinic acid. This reaction occurs in the mitochondria. Subsequently, the compound goes to the cytoplasm where two molecules of delta-aminolevulinic acid condense to form porphobilinogen via the dehydratase enzyme. This porphobilinogen synthesis is significantly impaired by lead toxicity. This enzyme is also a zinc-requiring enzyme. The porphyrin heme ring is formed essentially by condensation of four monopyrroles synthesized from the porphobilinogen. Thus impairment of porphobilinogen by lead will decrease heme biosynthesis by this mechanism.

Another mechanism by which lead may cause anemias is through its inhibition of the mitochondrial enzyme ferrochelatase. This enzyme facilitates the transfer of the iron in ferritin into the protoporphyrin ring to produce heme.[238] The protoporphyrin accumulates in the red blood cells of human subjects and leads to intoxication.

8.2 Renal Toxicity

One target organ of lead exposure is the kidney, which given sufficient lead exposure levels over a sufficient length of time may succumb to renal failure. Goyer and Ryne[238] reported that the proximal tubules of the nephron are lead sensitive. Furthermore, it is well known that the active form of vitamin D_3 (1,25-dihydroxycholcalciferol) is produced in the proximal tubules. Decreased calcium absorption by the gut is one result that will affect bone; however, as discussed later, lead has other effects upon bone. Rosen et al.[239] reported that this activation is impaired at blood levels of 25 μg/100 ml blood. Higher blood lead levels can lead to protein-lead complexes in the tubules which appear as dense accumulations.[238] Gout may be a symptom of such toxicity due to increased reabsorption of uric acid.[238] Continued accumulation of lead by the kidneys often leads to an increased accumulation of fibrotic connective tissue.[240] Typical measures of renal failure (e.g., blood urea nitrogen, creatinine) are elevated as a consequence of this lead-induced renal failure.

The mitochondria appear to be altered histologically in the proximal tubules as a result of lead accumulation.[241,242] Vascular lesions and atrophy of various portions of the nephron may appear,[238,241] and inclusion of bodies are commonly reported upon histological examination of the glomeruli.[243] Apparently, other tissue sites in the body have the same appearance of these inclusion bodies upon lead exposure. Osteoclasts and neuroblastoma cells often have these features as a result of lead toxicity.[234,244,245] Subsequent studies have revealed that agents such as cyclohexamide and actinomycin-D impair the formation of these bodies, suggesting that lead leads to protein synthesis processes of these bodies.[246] RNA and DNA levels may also increase.[247,248] Lead-binding proteins are thought to be induced in both cytoplasmic and nuclear portions of tubular cells, as reviewed elsewhere.[234]

Unlike some of the other metals reviewed, lead does not appear to induce metallothionein synthesis in the kidney; however, liver metallothionein appears to increase.[249,250] Goering and Fowler[251] reported that lead will bind to renal metallothionein after its induction by cadmium.

While it is not entirely clear as to why these various proteins are increased in the synthesis, a protective effect may be exerted by the production of these lead-binding proteins by the kidneys. The toxic effects of lead upon the kidney may lead to problems with other organ systems, such as hypertension and deleterious alterations in both circulating hormones and bone metabolism. The influence of other metals (e.g., calcium, copper, zinc, etc.) has an impact upon kidney damage as a result of lead toxicity. Therefore, when discussing the two pathologies below, the toxic influences of lead upon the renal system and the diet of the individual must be considered in the final analysis.

8.3 Cardiovascular Toxicity

The association of lead with hypertension has been linked by some epidemiological studies. Pirkle et al.[252] reported that even low lead levels are associated with elevated blood pressure. Increased lead absorption leading to hypertension was reported in one study.[253] Some of lead's influence upon blood pressure may be related to the associated renal toxicity reviewed previously. Low levels of lead exposure have been reported to result in hypertension when the toxicity signs are absent[254-256] and have been reported by animal studies.[253,257,258] One study with rats suggested that while chronic lead ingestion may lead to increased blood pressure, a concomitant ingestion of sucrose as the major source of carbohydrates potentiates the response.[259] They suggested that, in contrast to other proposals, renal involvement did not appear to be significant due to lack of urine protein. Blood urea nitrogen and serum creatinine did not increase in the rats receiving lead in drinking water. Aviv et al.[260] earlier reported significant renal damage in rats consuming similar lead levels. Thus, it appears that while there may be an association between lead-induced renal damage with hypertension, a clear cause-and-effect relationship has yet to be established.

Rats exposed to lead in drinking water have decreased cardiac conduction[261] and increased sensitivity to arrhythmias produced by catecholamine administration.[262] Lal et al.[263] using rats as a model, reported only minor changes in the heart itself in response to graded lead levels. Calcium influx across papillary and atrial muscle appeared increased when lead was given to rats. Electrocardiogram (ECG) abnormalities were significantly altered in the lead-exposed rats for this same study, which could be due to the influence upon calcium transport; Myerson and Elsenhauer[264] earlier had reported similar findings in humans. Goyer[265] postulated that the inhibition of Na/K ATPase by lead may alter the intracellular concentrations of sodium and calcium. This is thought to elevate plasma renin and cause hypertension. A similar influence of lead upon Na/K ATPase in cardiac myocytes cannot be discounted.

Calcium and lead are known to have antagonistic biochemical roles, as reviewed elsewhere;[234] however, the influence of calcium upon lead-induced increases in blood pressure appears paradoxical. Feeding rats a high level of calcium does not appear to block or dampen the effect of lead hypertension. Furthermore, Bogden et al.[266] reported that such a combination actually led to increased renal tumors and nephrocalcinosis.

8.4 Bone Integrity

Lead has a high affinity for bone, partly due to its antagonistic relationship with calcium. The greatest deposit site for lead in the body is the skeleton, which acts as a reservoir. The influence of lead upon bone must be considered from the perspective of vitamin D_3, the metabolism of which is significantly altered in lead toxicity. Lead is known to influence the various biochemical and physiological events involved with bone remodeling.

Lead may accumulate in bone beginning in fetal life. Young children with growing bones are apparently more susceptible to the toxicity of lead than are adults. Impaired dental development and delayed skeletal maturation have been reported in congenital lead poisoning cases.[267] In children, the NHANES II (second National Health and Nutrition Examination Survey) study documented decreased height and chest circumference with increased blood lead levels for children 7 years and younger.[268] Levels of plasma osteocalcin, a bone protein, are lower in children exposed to toxic levels of lead.[269]

Bone matrix synthesis is impaired with low to high lead exposure. Decreased bone formation rate and radial closure were reported in beagle dogs by Anderson et al.[270] Osteoblastic activity is enhanced, in addition to a decreased bone formation rate.[271] Trabecular bone appears less dense in rabbits exposed to lead. Many of the processes of bone turnover are not only attributed to the direct influences of lead on bone dynamics but also indirect mechanism, as lead may influence hormones targeting calcium and bone homeostasis.[272] The active form of vitamin D_3, 1,25-dihydroxycholcalciferol, appears to be lower from children with elevated blood lead levels.[273,274] These observations have also been reported for rats fed lead acetate in drinking water which resulted in depressed plasma 1,25-dihydroxycholcalciferol levels.[275] Children with elevated blood levels have demonstrated increased parathyroid hormone levels and a concomitant decline in blood ionized calcium.[273] The antagonistic hormone, calcitonin, which inhibits bone resorption, inhibits the hypercalcemia induced by high lead levels.[276-278] As mentioned already, the osteoid proteins are decreased in bone from animals and humans exposed to lead. Type I collagen synthesis is impaired by lead.[271,279-282] Osteocalcin plasma levels are lower from children with toxic lead levels, but treatment with EDTA returns them to a normal range.[269] Lead inclusion bodies have been reported in osteoblasts and osteocytes.[283,284] These bodies appear similar to other organ sites reviewed earlier in this section. Carbonic anhydrase facilitates the acid environment of the osteoclasts, and this activity appears reduced by lead when measured outside of the osteoclasts; however, it is unknown how this observation would be if tested on osteoclasts directly.[285,286] Many studies have been conducted to investigate calcium and lead interactions. While calcium homeostasis is clearly perturbed by lead, as reported above, the mechanisms are not always direct. Plasma membrane calcium channels, mitochondrial calcium pump, and calcium-activated ATPase changes are direct effects of lead. Changes in adenylate cyclase and Na/K ATPase are indirect, and decreased heme levels may be secondary. For an excellent review of these mechanisms, the reader is encouraged to read the review of Pounds et al.[272]

8.5 Central Nervous System Toxicity and Behavior and Learning Consequences

Exposure to low lead levels in growing organisms (e.g., children) lead to behavioral and learning difficulties. The ability of lead to have such consequential effects at relatively low levels gives testimony to its reputation as a potent neurotoxin. Memory loss and learning difficulties have been the subject of numerous reports.[287-290] Both the hippocampus and corticol brain regions appear to be targeted by lead.[291,292] Rodents and children may also exhibit impaired visual function.[293-295] The hippocampus is the region of the brain most often associated with the cellular basis of memory and learning.[296-298] Learning was impaired along with hippocampal long-term potentiation of afferents when rats were exposed to lead in prenatal life or during early postnatal development, with continued exposure until adulthood.[296] If exposure to lead began at day 16 of life, no such impairment due to lead exposure was revealed. Hotzman et al.[299] reported morphology alterations in the glial cells by lead exposure. Swelling of astrocytes and the presence of cytoplasmic electron-dense bodies and intranuclear inclusions are cell responses to lead toxicity. A study with rats in which lead acetate was given to pups at day 7 (via the mother's milk) was continued for up to 90 days. Starting at day 60, hippocampal astroglial alterations were noted, thereby demonstrating that even postnatal exposure to lead could cause brain alterations.[300] While the developing nervous system is sensitive to lead, even when exposure has ceased, the functional effects are likely long term and may lead to a permanent deficit.[293,301,302] Rodent studies in which lead exposure is introduced at the early postnatal stage demonstrated significant decrements in the motor skills and exploratory behavior of rats.[303] Reduced forepaw grasping ability and ambulation and rearing in an open field were reported at lead levels that were similar to some doses reported for children.

The mechanism by which lead may affect brain physiology and biochemistry could be due either to a direct influence upon nerve endings or by influence on neurotransmitter release in some fashion. Low lead concentrations enhance the release of neurotransmitters from presynaptic endings.[304-307]

Depending on the level of lead exposure, children have been reported to have symptoms such as ataxia, convulsions, headache, and learning disabilities and tend to exhibit hyperactive behaviors.[308-310] Encephalopathy has been reported in lead-toxic children. Learning and behavioral deficits in children have been reported.[301,310,311] Blood levels of up to 1.5 μM, which are comparatively low, could result in such dysfunctions in children.

Blood lead levels have demonstrated an inverse relation to the neuropsychological performance of children.[312-316] Children 2 to 4 years of age had lower mental development as blood lead levels increased, even after adjustment for confounding factors.[301,317] In the same studies, children with blood lead levels of 1.45 μM had a decrease of 3.3 points on the Bayley Mental Development Index (3.2%) at 2 years of age, and 7.2 points (6.7%) on the McCarthy General Cognitive Index at 4 years of age, when compared to children of similar ages with blood lead levels of 0.48 μM.[301,317] A followup study on 494 of these children at age 7 revealed an inverse relation between IQ scores and blood lead levels determined antenatally and postnatally. This was significant even after adjustment for socioeconomic factors, maternal IQ, birth weight, birth order, method of infant feeding, gender, parent's education level, etc.

Lead has been shown to stimulate the release of dopamine, acetylcholine, and gamma-aminobutyric acid (GABA).[318-320] Some of these effects may be due to the ability of lead to alter calcium entry into nerve cells or by an increase in the intracellular calcium level.[308] Lead may enter the cells through calcium channels.[321] Calcium may also competitively inhibit lead uptake in nonexcitatory cells such as the adrenal medulla and cannot be discounted as a potential mechanism in nerve cells.[321] Calcium channel blockers likewise may inhibit lead uptake in the same cells.

Lead is known to substitute for calcium as a second messenger and can bind to calmodulin. In fact, calmodulin has a greater affinity for lead than for calcium.[322,323] The calcium-calmodulin complex may activate a kinase referred to as calmodulin protein kinase which is high in nervous tissue and may regulate neurotransmitter release.[324] Synapsin I when phosphorylated is believed to have a role in neurotransmitter release.[325] If lead acts as calcium in the activation of this kinase, this may explain the ability of lead to result in neurotransmitter release.[325]

As indicated previously, lead appears to impair hippocampal voltage potentials. Protein kinase C can regulate this activity. Lead may serve to activate this kinase and consequently inhibit this potential, thereby impairing learning in children.[326]

The encephalopathy induced by lead toxicity is most likely due to a compromise in the blood-brain barrier. Brain edema occurs in the interstitial area and appears due to compromised blood vessel integrity. The brain capillaries and blood vessels have endothelial cells that contain tight junctions and act as a seal or barrier that excludes many plasma proteins and organic molecules and impedes Na and K exchange.[327] Elevated lead levels disrupt these vessels, and plasma proteins such as albumin enter the interstitial spaces, as do some ions. This increases osmotic pressure, and water accumulates in response. The lack of lymphatic structures within the central nervous system means that the fluid flows into the cerebrospinal fluid. This edema causes an increase in intracranial pressure and restricts blood flow to the brain, resulting in ischemia.[328,329] The direct mechanism by which the blood-brain barrier and blood vessels that compose the barrier may be compromised may be due to the astrocytes appearing to be vulnerable to the toxic effects of lead. The astrocytes cover the vascular walls of the brain vessels, and lead can injure these structures, as reviewed earlier.

8.6 Reproductive Toxicity

Lead toxicity is known to influence male and female reproductive organs in both laboratory animals and humans. An increased incidence of spontaneous abortions have been documented in female lead workers and also in the wives of male lead workers.[236,330] Male lead workers with blood lead concentrations of 53 to 75 μg/dl have been reported to have decreased sperm counts as well as altered morphology of the sperm.[331-333] Several animal studies with both laboratory rodents and nonhuman primates support these findings. A long-term study of lead exposure to rats did not reveal any morphological alterations in the testis and epididymis after 9 months of consuming 1% lead acetate in drinking water; however, there was significantly less spermatozoa in all regions of the epididymis as compared to control rats.[334] The spermatozoa also demonstrated reduced oxidative-reductive enzyme activities in the midsection. Other researchers have failed to demonstrate reproductive effects in rats (e.g., structural alterations of gonads, fertility) either by diet (0.3 mg lead acetate in drinking water) or inhalation (5 mg/m^3 lead oxide) after 70 days of exposure.[335] Male Cynomolgus monkeys administered lead acetate in gelatin capsules over different periods of the life cycle (infancy, post-infancy, and lifetime exposure) all demonstrated increased lipid droplets within the secretory cells of the seminal vesicles. The infancy and post-infancy period were the most affected by lead exposure.[336]

In rats, injecting pregnant dams daily with lead acetate did not appear to alter the gestation period or number of pups per litter.[337] Mice, on the other hand, have been observed to have reduced pregnancy and implantation.[338] One study on rats has suggested that inhalation of lead by rats during pregnancy has minimal effects upon the reproductive function of male offspring.[339] No change in litter size, death rate, or malformations were reported in the same study. Despite some differences among rodent studies, it would appear from human studies that lead does appear to affect reproductive performance and fertility, and males appear sensitive to the toxic effects of lead.

9 SUMMARY

The trace minerals discussed all exert toxic effects. The distinguishing characteristics are that copper, zinc, iron, and selenium are essential nutrients for humans, whereas mercury, lead, and cadmium are not. Attention historically has been paid to the latter three elements because of known environmental releases and subsequent deleterious actions upon the environment. Selenium originally was investigated because of its toxic effects in an earlier part of this century, and later its role in normal nutrition and interest in the deficiency aspects began to receive attention. Copper, zinc, and iron toxicities have also been studied but have not had the same level of public awareness as some of these other elements. However, given the propensity for the American adult population to use vitamin and mineral supplements and also some genetic conditions, more attention has been focused on potential toxicities. Clearly, these three trace elements exert toxic effects upon humans. More research information would be of value, especially in determining some of the upper levels of safe intake for these elements.

For all of the above elements reviewed, there appears to be a lack of studies reporting on the reversibility of many of these toxic effects. Such information would be of clinical value and would help evaluate the toxic implications as related to the burden placed on the health-care system for potential treatments, both short and long term.

The above list of elements is by no means exhaustive. Other trace elements are known to have toxic effects upon human health. Iodine, fluoride, and chromium have been studied, but many of the investigations have focused on requirements and deficiency aspects. Iodine-induced hyperthyroidism has been known to occur, as reviewed by Clugston and Hetzel.[340] Chromium is thought to be rather nontoxic to humans, but cases of skin irritations have been reported in some individuals through external contact.[341] Nickel, arsenic, and molybdenum also have received increased interest in terms of their human toxicity. Arsenic has enjoyed the dubious distinction of being a highly toxic element, but in reality it is rather nontoxic in terms of the absolute amounts needed and is less toxic than selenium.[342] For more complete discussions on these and other trace elements as related to human toxicity, the reader is referred to the reviews of Nielsen.[341,342]

REFERENCES

1. Levy, A. S. and Schucker, R. E., Patterns of nutrient intake among dietary supplement users: attitudinal and behavioral correlates, *J. Am. Diet. Assoc.,* 87, 754, 1987.
2. Read, M. H., Medeiros, D. M., Bendel, R. et al., Mineral supplementation practices of adults in seven western states, *Nutr. Res.,* 6, 375–383, 1986.
3. Medeiros, D. M., Bock, M. A., Oritz, M., Raab, C., Read, M., Schutz, H. G., Sheehan, E. T., and Willams, D. K., Vitamin and mineral supplementation practice of adults in seven western states, *J. Am. Diet. Assoc.,* 89, 383–386, 1989.
4. Mason, K. E., A conspectus of research on copper metabolism and requirements of man, *J. Nutr.,* 109, 1979–2066, 1979.
5. Naderbragt, H., Van den lugh, T. S. G. A. M., and Wensvoort, P., Pathobiology of copper toxicity, *Vet. Quart.,* 6, 179–185, 1984.
6. Turnland, J. R., *Copper in Modern Nutrition in Health and Disease,* 8th ed., Shils, M. E., Olson, J. A., and Shike, M., Eds., Lea & Febiger, Philadelphia, 1994, pp. 231–241.
7. Reiser, S., Smith, J. C., Mertz, W., Holbrook, T. T., Schofield, D. J., Powell, A. S., Canfield, W. K., and Canary, J. J., Induction of copper status in human consuming a typical American diet containing either fructose or starch, *Am. J. Clin. Nutr.,* 42, 242–251, 1986.
8. Linder, M. C., *The Biochemistry of Copper,* Plenum Press, New York, 1990.
9. Fischer, P. W. F., Giroux, A., and L'Abbe, M. R., Effects of zinc on mineral copper binding and on the kinetics of copper absorption, *J. Nutr.,* 113, 462–469, 1983.
10. Cousins, R. J., Absorption, transport and hepatic metabolism of copper and zinc: special reference to metallothionein and ceruloplasmin, *Physiol. Rev.,* 65, 238–309, 1985.
11. Davis, G. K. and Mertz, W., Copper, in *Trace Elements in Humans and Animal Nutrition,* Vol. 1, 5th ed., Mertz, W., Ed., Academic Press, San Diego, 1987, pp. 301–364.
12. Danks, D. M., Copper deficiency in humans, *Ann. Rev. Nutr.,* 8, 235–257, 1988.
13. Harris, E. D. and Percival, S. B., Copper transport: insights into a ceruloplasmin-based delivery system, in *Copper Bioavailability and Metabolism,* Kies, C., Ed., Plenum Press, New York, 1990, pp. 95–102.
14. Prohaska, J. R., Biochemical changes in copper deficiency, *J. Nutr. Biochem.,* 1, 452–461, 1990.
15. Abdel-Mageed, A. and Oehme, F. W., A review of the biochemical rules, toxicity and interaction of zinc, copper and iron. II. Copper, *Vet. Hum. Toxicol.,* 32, 230–234, 1970.
16. Cohen, S. R., A review of the health hazard from copper exposure, *J. Occup. Med.,* 16, 621–624, 1974.
17. Hirano, S., Ebihara, H., Sakai, S., Kodama, N., and Suzuki, K. T., Pulmonary clearance and toxicity of intratracheally instilled cupric oxide in rats, *Arch. Toxicol.,* 67, 312–317, 1993.
18. Romeo-Moreno, A., Aguilar, C., Arola, L. I., and Mas, A., Respiratory toxicity of copper, *Environ. Health Perspect.,* 102(Suppl. 3), 339–340, 1994.
19. Sokol, R. J., Devereaux, M. A., O'Brien, K., Khandwala, R. A., and Loehr, J. P., Abnormal hepatic mitochondrial respiration and cytochrome C oxidase activity in rats with long-term copper overload, *Gastroenterology,* 105, 178–187,993.
20. Gross, J. B., Myers, B. M., Lost, L. T., Kuntz, S. M., and LaRusson, N. F., Biliary copper excretion by hepatocyte lysosomes in the rat; major excretory pathway in experimental copper overload, *J. Clin. Invest.,* 83, 30–39, 1989.
21. Mal, I. T. and Weglicki, W. B., Characterization of iron-medicated peroxidative injury in isolated hepatic lysosomes, *J. Clin. Invest.,* 75, 58–63, 1989.
22. LeSage, G. D., Kost, L. J., Barham, S. S., and Russo, N. F., Biliary excretion of iron from hepatocyte lysosomes in the rat: a major excretory pathway in experimental iron overload, *J. Clin. Invest.,* 77, 90–97, 1986.
23. Myers, B. M., Prendergast, F. G., Human, R., Kuntz, S. M., and Larusson, N. F., Alterations in hepatocyte lysosomes in experimental hepatic copper overload in rats, *Gastroenterology,* 105, 1814–1823, 1993.
24. Klevay, L. M., Ischemic heart disease. A major obstacle to becoming old, *Clin. Geriat. Med.,* 3, 361–371, 1987.
25. Klevay, C. M. and Viestenz, K. E., Abnormal electrocardiogram in rats deficient in copper, *Am. J. Physiol.,* 240, H185–H189, 1981.

26. Medeiros, D. M., Davidson, J., and Jenkins, J. E., A unified perspective on copper deficiency and cardiomyopathy, *Proc. Soc. Exp. Biol. Med., 203, 262–273, 1993.*
27. Rhee, H. M. and Dunlap, M., Acute cardiovascular toxic effects of copper in anesthetized rabbits, *Neurotoxicity,* 1, 355–360, 1990.
28. Liu, C. C. R. and Medeiros, D. M., Excess diet copper increase systolic blood pressure in rats, *Biol. Trace Elements Res.,* 9, 15–24, 1986.
29. Hurley, L. S. and Kien, C. L., Teratogenic effect of copper, in *Copper in the Environment,* Part 2, Nriagu, J., Ed., John Wiley & Sons, New York, 1979.
30. Ferm, V. I. T. and Hanolon, D. P., Toxicity of copper salt in hamster embryos, *Dev. Biol. Rep.,* 11, 97–101, 1974.
31. O'Shea, K. S. and Kaufman, M. H., Influence of copper on the early post implantation mouse embryo: one *in vivo* and *in vitro* study, *Arch. Devel. Biol.,* 186, 297–308, 1979.
32. Chang, C. C. and Tatung, H. J., Absence of teratogenicity of intrauterine copper wire in rat, hamster and rabbits, *Contraception,* 7, 413–434, 1973.
33. Barrlow, S. M., Knight, A. F., and House, I., Intrauterine exposure to copper I.U.D.s and prenatal development in the rat, *J. Rep. Fert.,* 62, 123–130, 1981.
34. Marseden, C. D., Wilson's disease, *Q. J. Med.,* 65, 959–966, 1987.
35. Walsh, J. M., Pregnancy in Wilson's disease, *Q. J. Med.,* 46, 73–83, 1977.
36. Haddad, D. S., Al-Alousi, L. A., and Kantarjan, A. H., The effect of copper loading on pregnant rats and their offspring, *Funct. Develop. Morphol.,* 1, 17–32, 1991.
37. Menkes, J. H., Alter, M., Steigleder, G. K., Weakley, D. R., and Sung, J. H., A sex-linked recessive disorder with retardation of growth, peculiar hair, and focal cerebral degeneration, *Pediatrics,* 29, 764–779, 1962.
38. Wilson, S. A. K., Progressive lenticular degeneration of familial nervous disease associated with cirrhosis of the liver, *Brain,* 34, 395–409, 1912.
39. Okavasu, T., Tochimaru, H., Hyuga, T., Tahahashi, T., Takekoshi, Y., Li, Y., Togashiy, Taheichi, N., Kasai, N., and Arashima, J., Inherited copper-toxicity in Long-Evans Cinnamon rats exhibiting spontaneous hepatitis: a model of Wilson's disease, *Pediatr. Res.,* 31, 253–257, 1992.
40. Shiraishi, N., Taguchi, T., and Kinebuch, H., Copper-induced toxicity in macular mutant mouse: an animal model for Menkes' kinky-hair disease, *Toxicol. Appl. Pharmacol.,* 110, 89–96, 1991.
41. Vulpe, C., Levinson, B., Whitney, S., Packman, S., and Gitschier, J., Isolation of a candidate gene for Menkes' disease and evidence that it encodes a copper transporting ATPase, *Nature Genetics,* 3, 7–14, 1994.
42. Frommer, D. J., Defective biliary excretion of copper in Wilson's disease, *Gut,* 15, 125–129, 1974.
43. Su, L. C., Owen, C. A., Zullman, P. E., and Hardy, R. M., A defect in biliary excretion of copper in copper-laden Bedlington terriers, *Am. J. Physiol.,* 243, G231–236, 1982.
44. Yoshida, M. C., Masuda, R., Sasaki, M., Takeichi, N., Kobaysashi, H., Dempo, K., and Mori, M., New mutation causing hereditary hepatitis in the laboratory rat, *J. Hered.,* 78, 361–365, 1987.
45. Scheinberg, I. H. and Sternlieb, I., *Wilson's Disease,* W.B. Saunders, Philadelphia, 1984, pp. 114–125.
46. Sternlieb, I. and Scheinberg, I. H., Prevention of Wilson's disease in asymptomatic patients, *N. Engl. J. Med.,* 278, 353–359, 1968.
47. Awing, D. R. and Metra, R. K., Host defenses against copper toxicity, *Int. Rev. Exp. Pathol.,* 31, 47–83, 1990.
48. Jackson, M. J., Physiology of zinc: general aspects, in *Zinc in Human Biology,* Mills, C. F., Ed., Springer-Verlag, London, 1987, pp. 1–14.
49. Shaw, J. C. L., Trace elements in the foetus and young infant, *Am. J. Dis. Child.,* 13, 1260–1268, 1979.
50. Widdowson, E. M., Chan, H., Harrison, G. E. et al., Accumulation of Cu, Zn, Mn, Cr and Co in the human liver before birth, *Biol. Neonate,* 20, 360–367, 1972.
51. Failla, M. L. and Cousins, R. J., Zinc uptake by isolated rat liver parenchymal cells, *Biochem. Biophys. Acta,* 538, 435–444, 1978.
52. Failla, M. L. and Cousins, R. J., Zinc accumulation and metabolism in primary cultures of the rat liver cells regulation by glucorticords, *Biochem. Biophys. Acta,* 543, 292–304, 1978.
53. Henkin, R. I., Foster, D. M., Aamodt, R. A., and Berman, M., Zinc metabolism in adrenal cortical insufficiency: effects of carbohydrate active steroids, *Metabolism,* 33, 491–501, 1984.
54. Kuipers, P. J. and Cousins, R. J., Zinc accumulation in rat liver parenchymal cells in primary culture and response to glucagon and dexamethasone, *Fed. Proc.,* 43, 1403, 1984.
55. Kirchgessner, M. and Roth, H. P., Influences of zinc depletion and zinc status on serum growth hormone levels in rats, *Biol. Trace Elements Res.,* 7, 263–268, 1985.
56. Prasad, A. S., Clinical and biochemical spectrum of zinc deficiency in human subjects, in *Clinical, Biochemical and Nutritional Aspects of Trace Elements,* Prasad, A. S., Ed., Alan R. Liss, New York, 1982, pp. 3–62.
57. Danbolt, N. and Closs, K., Acrodermatitis enteropathica, *Acta Derm. Venerol.,* 23, 127–129, 1942.
58. Smith, Jr., J. C., Zeller, J. A., Brown, E., and Dandong, S. C., Elevated plasma zinc: a heritable anomaly, *Science,* 193, 496–498, 1976.
59. Brooks, A., Reid, H., and Glazer, G., Acute intravenous zinc poisoning, *Br. Med. J.,* 1, 1390–1391, 1977.
60. Nelder, K. H. and Hambidge, K. M., Zinc therapy of acrodermatitis enteropathica, *N. Engl. J. Med.,* 292, 879–882, 1995.

61. Istfan, N. W., Janghorbani, M., and Young, V. R., Absorption of stable ^{70}Zn in healthy young men in relation to zinc intake, *Am. J. Clin. Nutr.,* 38, 187–194, 1983.
62. Baer, M. T. and King, J. C., Tissue zinc levels and zinc excretion during experimental zinc depletion in young men, *Am. J. Clin. Nutr.,* 39, 556–570, 1984.
63. Jackson, M. J., Jones, D. A., Edwards, R. H. T., Swainbank, I. G., and Cleman, M. L., Zinc homeostasis in man: studies using a new stable isotope dilution technique, *Br. J. Nutr.,* 51, 199–208, 1984.
64. Jackson, M. J., Physiology of zinc: general aspects, in *Zinc in Human Biology,* Mills, C., Ed., Springer-Verlag, London, 1989, pp. 1–14.
65. Medeiros, D. M., Mazhar, A., and Brunett, E. W., Failure of oral zinc supplementation to alter hair zinc levels among healthy human males, *Nutr. Res.,* 7, 1109–1115, 1987.
66. Fox, M. R. S., Zinc excess, in *Zinc in Human Biology,* Mills, C., Ed., Springer-Verlag, London, 1989, pp. 365–369.
67. Brown, M. A., Thom, J. V., Orth, G. L., Cova, P., and Juarez, J., Food poisoning involving zinc contamination, *Arch. Environ. Health,* 8, 657–660, 1964.
68. Cowan, G. A. B., Unusual case of poisoning by zinc sulphate, *Br. Med. J.,* 1, 451–452, 1947.
69. Fosmire, G. J., Zinc toxicity, *Am. J. Clin. Nutr.,* 51, 225–227, 1990.
70. Samman, S. and Roberts, D. C., The effect of zinc supplement on plasma zinc and copper levels and the reported symptoms in healthy volunteers, *Med. J. Aust.,* 146, 246–269, 1987.
71. Chandra, R. K., Excess intake of zinc impairs immune responses, *J. Am. Med. Assoc.,* 252, 1443–1446, 1984.
72. Hooper, P. L., Visconti, L., Garry, P. J., and Johnson, G. E., Zinc lowers high-density lipoprotein-cholesterol levels, *J. Am. Med. Assoc.,* 244, 1960–1961, 1969.
73. Black, M. R., Medeiros, D. M., Brunett, E., and Welke, R., Zinc supplements and serum lipids in young adult white males, *Am. J. Clin. Nutr.,* 47, 970–975, 1988.
74. Brown, M. A., Thom, J. V., Orth, G. L., Cova, P., and Juarez, J., Food poisoning involving zinc contamination, *Arch. Environ. Health,* 8, 657–660, 1964.
75. Murphy, J. V., Intoxication following ingestion of elemental zinc, *J. Am. Med. Assoc.,* 212, 2119–2120, 1970.
76. Prasad, A. S., Brewer, G. J., Schoomacher, E. B., and Rabbani, P., Hypocupremia induced by zinc therapy in adults, *J. Am. Med. Assoc.,* 240, 2166–2168, 1978.
77. Porter, K. G., McMaster, D., Elmes, M. E., and Love, A. H. G., Anemia and low serum-copper during zinc therapy, *Lancet,* II, 774, 1977.
78. Sandstead, H. H., Copper bioavailability and requirements, *Am. J. Clin. Nutr.,* 35, 809–814, 1982.
79. Festa, M. D., Anderson, H. K., Dowdy, R. P., and Ellersiech, M. R., Effect of zinc intake on copper excretion and retention in men, *Am. J. Clin. Nutr.,* 41, 285–292, 1985.
80. Fischer, P. W. F., Grioux, A., and L'Abbe, M. R., Effect of zinc supplementation on copper status in adult man, *Am. J. Clin. Nutr.,* 40, 743–746, 1984.
81. Patterson, W. P., Winkelmann, M., and Perry, M. C., Zinc-induced copper deficiency: megamineral sideroblastic anemia, *Ann. Intern. Med.,* 103, 385–386, 1985.
82. Petrie, J. T. B. and Row, P. G., Dialysis anaemia caused by subacute zinc toxicity, *Lancet,* 1, 1178–1180, 1977.
83. Koh, J. Y. and Choi, D. W., Zinc toxicity on cultured cortical neurons: involvement of *N*-methyl-D-aspartate receptors, *Neuroscience,* 60, 1049–1052, 1994.
84. Peters, S., Koh, J., and Choi, D. W., Zinc selectivity blocks the action at *N*-methyl-D-aspartate on central neurons, *Science,* 236, 589–593, 1987.
85. Westbrook, G. L. and Mayer, M. L., Micromolar concentrations of Zn antagonize NMDA and GABA responses of hippocampal neurons, *Nature,* 328, 640–643, 1987.
86. Choi, D. W., Yukoyama, M., and Koh, J., Zinc neurotoxicity in cortical cell culture, *Neuroscience,* 24, 67–79, 1988.
87. Yokoyama, M., Koh, I., and Choi, D. W., Brief exposure to zinc is toxic to cortical neurons, *Neurosci. Lett.,* 71, 351–355, 1986.
88. Finch, C. A. and Huebers, H., Perspectives in iron metabolism, *N. Engl. J. Med.,* 306, 1520–1528, 1982.
89. Britton, R. S., Ramm, G. A., Olynyk, J., Singh, R., O'Neill, R., and Bacon, B. R., Pathophysiology of iron toxicity, in *Progress in Iron Research,* Hershko, C., Konijn, A. M., and Aisen, P., Eds., Plenum Press, New York, 1994, pp. 239–253.
90. Bothwell, T. H., Charlton, R. W., Cook, J. D., and Finch, C. A., *Iron Metabolism in Man,* Blackwell Scientific, Oxford, 1979.
91. Fairbanks, V. F., Iron in medicine and nutrition, in *Modern Nutrition in Health and Disease,* Vol. 1, 8th ed., Shils, M. E., Olson, J. A., and Shike, M., Eds., Lea & Febiger, Philadelphia, 1994, pp. 185–213.
92. Raffin, S. B., Woo, C. H., Roost, K. T., Price, D. C., and Schmid, R., Intestinal absorption of hemoglobin iron-heme cleavable by mucosal heme oxygenase, *J. Clin. Invest.,* 54, 1344–1352, 1974.
93. Teichmann, R. and Stemmel, W., Iron uptake by human upper small intestine microvillus membrane vesicles, *J. Clin. Invest.,* 86, 2145–2153, 1990.
94. Conrad, M. E. and Umbrieti, J., Iron absorption — the mucin-mobilferrin-integrin pathway. A competitive pathway for metal absorption, *Am. J. Hematol.,* 42, 67–73, 1993.

95. Conrad, M. E., Umbrieti, J., Moore, E. G., and Rooning, C. R., Newly identified iron binding protein in human duodenal mucosa, *Blood,* 79, 224, 1992.
96. Hartman, R. S., Conrad, M. E., and Hartman, R. E., Ferritin-containing bodies in human small intestine epithelium, *Blood,* 22, 397–405, 1963.
97. Huebers, H. A., Beguin, Y., and Pootrakul, P., Intact transferrin receptors in human plasma and their relation to erythropoiesis, *Blood,* 75, 102–107, 1990.
98. Peters, T. J., Raja, K. B., and Simpson, R. J., Mechanisms and regulation of intestinal iron absorption, *Ann. N.Y. Acad. Sci.,* 526, 141–147, 1988.
99. MacGilivray, R. T. A., Mendez, E., and Sinha, B. L., The complete amino acid sequence of human serum transferrin, *Proc. Natl. Acad. Sci. USA,* 79, 2504–2508, 1982.
100. Gordeuk, V. R., McLaren, G. D., and Samowitz, W., Etiologies, consequences, and treatment of iron overload, *Crit. Rev. Clin. Lab. Sci.,* 31, 89–133, 1994.
101. Dall, P. R., Iron, in *Present Knowledge in Nutrition,* 6th ed., International Life Sciences Institute, Nutrition Foundation, Washington, D.C., 1990, pp. 241–250.
102. Munro, H., The ferritin gene: their response to iron status, *Nutr. Rev.,* 51, 65–71, 1993.
103. Cook, J. D. and Skikne, B. S., Serum ferritin: a possible model for the assessment of nutrient stores, *Am. J. Clin. Nutr.,* 35, 1180–1185, 1982.
104. Whittaker, P., Skikne, B. S., and Covell, A. M., Duodenal iron proteins in idiopathic hemochromatosis, *J. Clin. Invest.,* 83, 261–267, 1989.
105. Edwards, C. O., Griffen, L. M., and Kaplan, J., Twenty-four hour variation of transferrin saturation in treated and untreated haemochromatosis homozygotes, *J. Intern. Med.,* 226, 373–379, 1989.
106. Goya, N., Miyazaki, S., and Kordate, S., A family of congenital transferrinemia, *Blood,* 40, 239–245, 1972.
107. Buys, S. S., Martin, C. B., and Eldridge, M., Iron absorption in hypotransferrinic mice, *Blood,* 78, 3288–3290, 1991.
108. Eason, R. J., Adams, P. C., and Aston, C. E., Familial iron overload with possible autosomal dominant inheritance, *Aust. N.Z. J. Med.,* 20, 226–230, 1990.
109. Mevorach, D., Unpublished observations, 4th Int. Conf. Hemochromatosis and Clinical Problems in Iron Metabolism, Jerusalem, Israel, 1993.
110. Gordeuk, V., Heredity and nutritional iron overload, *Ballere's Clin. Heamatol.,* 5, 169–186, 1992.
111. Gordeuk, V., Mukiibi, J., and Hasstedt, S. F., Iron overload in Africa — interaction between a gene and dietary iron content, *N. Engl. J. Med.,* 326, 95–110, 1992.
112. Pootrakul, P., Kitcharoen, K., and Yansukon, P., The effect of erythroid hyperplasia on iron balance, *Blood,* 71, 1124–1129, 1988.
113. Fiorelli, G., Fargion, S., and Piperno, A., Iron metabolism in thalessemia intermedia, *Haematologica,* 75 (Suppl.), 89–95, 1990.
114. Schafer, A. I., Cheron, R. G., and Dluhy, R., Clinical consequences of acquired transfusional iron overload in adults, *N. Engl. J. Med.,* 304, 319, 1981.
115. Cohen, A. and Schwartz, E., Iron chelation therapy in sickle cell anemia, *Am. J. Hematol.,* 7, 69–76, 1979.
116. Perry, T. L., Hardwick, C. B., and Dixon, G. H., Hypermethioninemia: a metabolic disorder associated with cirrhosis, islet cell hyperplasia and renal tubular acidosis, *Pediatrics,* 36, 236–250, 1965.
117. Goldfischer, S., Grotsky, H. W., and Chang, C., Idiopathic neonatal iron storage involving the liver, pancreas, heart, and endocrine and exocrine organs, *Hepatology,* 1, 58–64, 1981.
118. Blisard, K. S. and Bartow, S. A., Neonatal hemochromatosis, *Hum. Pathol.,* 17, 376–383, 1986.
119. Bacon, B. R. and Britton, R. S., The pathology of hepatic iron overload: a free radical related process?, *Hepatology,* 11, 127–137, 1990.
120. Park, C. H., Bacon, B. R., Brittenham, G. M., and Tavill, A. S., Pathology of dietary carbonyl iron overload in rats, *Lab. Invest.,* 57, 555–563, 1987.
121. Bacon, B. R., Tavill, A. S., Brittenham, G. M., Park, C. H., and Recknagel, R. O., Hepatic lipid peroxidation *in vivo* in rats with chronic iron overload, *J. Clin. Invest.,* 71, 429–439, 1983.
122. Bacon, B. R., Brittenham, G. M., Tavill, A. S., McLaren, C. E., Park, C. H., and Recknagel, R. O., Hepatic lipid peroxidation *in vivo* in rats with chronic dietary iron overload is dependent on hepatic iron concentration, *Trans. Assoc. Am. Phys.,* 96, 146–154, 1983.
123. Bacon, B. R., Park, C. H., Brittenham, G. M., O'Neill, R., and Tavill, A. S., Hepatic mitochondrial oxidative metabolism in rats with chronic dietary iron overload, *Hepatology,* 5, 789–797, 1985.
124. Bacon, B. R., Healey, J. F., Brittenham, G. M., Park, C. H., Nunnari, J., Tavill, A. S., and Bonkovsky, H. L., Hepatic microsomal function in rats with chronic dietary iron overload, *Gastroenterology,* 90, 1844–1853, 1986.
125. Britton, R. S., O'Neill, R., and Bacon, B. R., Hepatic mitochondrial malondialdehyde metabolism in rats with chronic iron overload, *Hepatology,* 11, 93–97, 1990.
126. Myers, B. M., Prendergast, F. G., Holman, R., Kuntz, S. M., and LaRusso, N. F., Alterations in the structure, physicochemical properties and pH of hepatocyte lysosomes in experimental iron overload, *J. Clin. Invest.,* 88, 1207–1215, 1991.

127. Houglum, K., Filip, M., Witztum, J. L., and Chojkier, M., Malondialdehyde and 4-hydroxynonenal protein adducts in plasma and liver of rats with iron overload, *J. Clin. Invest.*, 86, 1991–1998, 1990.
128. Goldberg, L., Martin, L. E., and Batchelor, A., Biochemical changes in the tissues of animals injected with iron. 3. Lipid peroxidation, *Biochem. J.*, 83, 291–298, 1962.
129. Hultcrantz, R., Ericson, J. L. E., and Hirth, T., Levels of malondialdehyde production in rat liver following loading and unloading of iron, *Virchows Arch.*, 45, 139–146, 1984.
130. Young, I. S., Trouton, T. G., Tourney, J. J., Callender, M. E., and Trimble, E. R., Antioxidant status in hereditary hemochromotosis (Abstr.), *Free Rad. Res. Commun.*, 16(Suppl. 1), 187, 1992.
131. Heys, A. D. and Dormandy, T. L., Lipid peroxidation in iron-overloaded spleens, *Clin. Sci.*, 60, 295–301, 1981.
132. Niederau, R., Fischer, R., Sonnenberg, A., Stremmel, W., Trampisch, H. J., and Strohmeyer, G., Survival and causes of death in cirrhotics and in noncirrhotic patients with primary hemochomatosis, *N. Engl. J. Med.*, 313, 1256–1262, 1985.
133. Britton, R. S., Ferrali, M., Magiera, C. J., Recknagel, R. O., and Bacon, B. R., Increased prooxidant action of hepatic cytosolic low-molecular-weight iron in experimental iron overload, *Hepatology*, 11, 1038–1043, 1990.
134. Britton, R. S., O'Neill, R., and Bacon, B. R., Chronic dietary iron overload in rats results in impaired calcium sequesteration by hepatic mitochondria and microsomes, *Gastroenterology*, 101, 806–811, 1991.
135. Bonkovsky, H. L., Healy, J. F., Lincoln, B., Bacon, B. R., Bishop, D. F., and Elder, G. H., Hepatic heme synthesis in a new model of experimental hemochromatosis: studies of rats fed finely divided iron, *Hepatology*, 7, 1195–1203, 1987.
136. Peters, T. J. and Seymour, C. A., Acid hydrolase activities and lysosomal integrity in liver biopsies from patients with iron overload, *Clin. Sci. Mol. Med.*, 50, 75–78, 1976.
137. Peters, T. J., O'Connell, M. J., and Ward, R. J., Role of free-radical mediated lipid peroxidation in the pathogenesis of hepatic damage by lysosomal disruption, in *Free Radicals in Liver Injury*, Poli, G., Cheeseman, K. H., Dianzani, M. U., and Slater, T. F., Eds., IRL Press, Oxford, 1985.
138. Bacon, B. R., O'Neill, R., and Britton, R. S., Hepatic mitochondrial energy production in rats with chronic iron overload, *Gastroenterology*, 105(4), 1134–1140, 1994.
139. Pietrangelo, A., Rocchi, E., Schiaffonati, L., Ventura, E., and Cario, G., Liver gene expression during chronic dietary iron overload in rats, *Hepatology*, 11, 798–804, 1990.
140. Pietrangelo, A., Gualdi, R., Geerts, A., DeBleser, P., Casalgrandi, G., and Ventura, E., Enhanced hepatic collagen gene expression in a rodent model of hemochromatosis (abstr.), *Gastroenterology*, 102, A868, 1992.
141. Li, S. C. Y., O'Neill, R., Britton, R. S., Kobayashi, Y., and Bacon, B. R., Lipocytes from rats with chronic iron overload have increased collagen and protein production (abstr.), *Gastroenterology*, 102, A841, 1992.
142. Bradbear, R. A., Bain, C., Siskind, V., Scholfield, F. D., Webb, S., Axelsen, E. M., Halliday, J. W., Bassett, M. L., and Powell, L. W., Cohort study of internal malignancy in genetic hemochromatosis and other chronic nonalcoholic liver diseases, *J. Natl. Cancer Inst.*, 75, 81–84, 1985.
143. Edling, J. E., Britton, R. S., Grisham, M. B., and Bacon, B. R., Increased unwinding of hepatic double-stranded DNA (dsDNA) in rats with chronic dietary iron overload (abstr.), *Gastroenterology*, 98, A585, 1990.
144. Faux, S. P., Francis, J. E., Smith, A. G., and Chipman, J. K., Induction of 8-hydroxydeoxyguanosine in Ah-responsive mouse liver by iron and Aroclor, *Carcinogenesis*, 13, 247–250, 1992.
145. Salonen, J. T., Nyyssonen, K., Kurpela, H., Tuomilehto, J., Seppancn, R., and Salonon, R., High stored iron levels are associated with excess risk of myocardial infarction in eastern Finnish men, *Circulation*, 86, 803–811, 1992.
146. Steinberg, D., Parthasarathy, S., Carew, T. E., Khoo, J. C., and Witztum, J. L., Beyond cholesterol: modification of low density lipoprotein that increase atherogenicity, *N. Engl. J. Med.*, 320, 915–923, 1989.
147. McAdam, P. A., McAdam, P. A., and Lewis, S. A., Absorption of selenite and L-selenomethionine in healthy young men, using a ^{74}Se tracer, *Fed. Proc.*, 44, 1671, 1985.
148. Levander, O. A. and Burk, R. F., Selenium, in *Modern Nutrition in Health and Disease*, 8th ed., Shils, M. E., Olson, J. A., and Shike, M. D., Eds., Lea & Febiger, Philadelphia, PA, 1994.
149. Streilwieser, A. J. and Heathcock, C. H., *Introduction to Organic Chemistry*, 2nd ed., Macmillan, New York, 1981.
150. Motsenbocker, M. A. and Tapper, A. L., A selenocystine-containing selenium transport protein in rat plasma, *Biochim. Biophys. Acta*, 719, 147–153, 1982.
151. Jacob, R. A., Trace elements, in *Fundamentals of Clinical Chemistry*, 3rd ed., Tietz, N. W., Ed., W.B. Saunders, Philadelphia, PA, 1987.
152. Virtamo, J. and Huttunen, J. K., Minerals, trace elements, and cardiovascular disease, *Ann. Clin. Res.*, 20, 102–113, 1988.
153. Robinson, J. R., Robinson, M. F., and Levander, O. A., Urinary excretion of selenium by New Zealand and North American human subjects on different intakes, *Am. J. Clin. Nutr.*, 41, 1023–1031, 1985.
154. Franche, K. W. and Moxon, A. L., A comparison of the minimum fatal doses of selenium, tellurium arsenic and vanadium, *J. Pharmacol. Exp. Ther.*, 58, 454–459, 1936.

155. Rosenfeld, I. and Beath, O. A., *Selenium: Geobotany, Biochemistry Toxicity and Nutrition,* Academic Press, New York, 1964, pp. 198–208.
156. Harr, J. R. and Muth, O. H., Selenium poisoning in domestic animals and its relationship to man, *Clin. Toxicol.,* 5, 175–176, 1972.
157. Combs, Jr., G. F. and Combs, S. B., *The Role of Selenium in Nutrition,* Academic Press, New York, 1986, pp. 463–525.
158. Poirier, K. A., Summary of the derivation of the reference dose for selenium, in *Risk Assessment of Essential Elements,* Mertz, W., Abernathy, C. O., and Olin, S. S., Eds., ILSI Press, Washington, D.C., 1994.
159. Yang, G., Wang, S., Zhou, R., and Sun, S., Endemic selenium intoxication of humans in China, *Am. J. Clin. Nutr.,* 37, 872–889, 1983.
160. Yang, G., Yin, S., Zhou, R. et al., Studies of safe maximal daily dietary selenium intake in a seleniferous area in China. 2. Relation between Se-intake and the manifestations of clinical signs and certain biochemical alterations in blood and urine, *J. Trace Elements Electrolytes Health Dis.,* 3, 123–130, 1989.
161. Yang, G., Zhou, R., Yin, S. et al., Studies of safe maximal daily dietary selenium intake in a seleniferous area in China. 1. Selenium intake and tissue levels of the inhabitant, *J. Trace Elements Electrolytes Health Dis.,* 3, 77–87, 1989.
162. Combs, G. F., Essentiality and toxicity of selenium: a critique of the recommended dietary allowance and the reference dose, in *Risk Assessment of Essential Elements,* Mertz, W., Abernathy, C. O., and Olin, S. S., Eds., ILSI Press, Washington, D.C., 1994.
163. Peakall, D. B. and Lovett, R. J., Mercury: its occurrence and effects in the ecosystem, *Bioscience,* 22, 2–25, 1972.
164. Curley, A., Sedlak, V. A., Girlins, E. F., Hau, R. E., Barthel, W. F., Pierce, P. E., and Likosky, W. H., Organic mercury identified as the cause of poisoning in humans and hogs, *Science,* 172, 65–67, 1971.
165. Trachtenberg, I. M., *Chronic Effects of Mercury on Organisms,* U.S. Dept. of Health, Education, and Welfare, U.S. Government Printing Office, Washington, D.C., 1974.
166. Elder, J. A. and Gaufin, G. R., The toxicity of three mercurials to *Pteronarcys california*, Newport, and some possible physiological effects which influence the toxicities, *Environ. Res.,* 7, 169–175, 1974.
167. ReVelle, C. and ReVelle, P., *Sourcebook in the Environment: The Scientific Perspective,* Houghton Mifflin, Boston, MA, 1974.
168. Walker, C., *Environmental Pollution by Chemicals,* Hutchinson Educational, London, 1971.
169. Endo, T., Sakat, M., and Shaikh, Z. A., Mercury uptake by primary cultures of rat renal cortical epithelial cells. I. Effects of cell density, temperature, and metabolic inhibitors, *Toxicol. Appl. Pharm.,* 132, 36–43, 1995.
170. Weiss, B. and Doherty, B. A., Methylmercury poisoning, *Teratology,* 12, 311–313, 1975.
171. Fredriksson, A., Dahlgren, I., Danielsson, B., Eriksson, P., Dencher, L., and Archer T., Behavioural effects of neonatal metallic mercury exposure in rats, *Toxicology,* 74, 151–160, 1992.
172. Danielsson, B. R., Fredriksson, A., Dahlgren, I., Garlund, A. T., Olsson, L., Dencker, L., and Archer, T., Behavioural effects of prenatal metallic mercury inhalation exposure to rats, *Neurotoxicol. Teratol.,* 15, 391–396, 1993.
173. Brookes, N., The *in vitro* evidence for the role of glutamate in the CNS toxicity of mercury, *Toxicology,* 76, 245–256, 1992.
174. Clarkson, T. W., The biological properties and distribution of mercury, *Biochem. J.,* 130, 61–63, 1972.
175. Ferens, M. C., *A Review of the Physiological Impact of Mercurials,* U.S. Environmental Protection Agency, U.S. Government Printing Office, Washington, D.C., 1974.
176. Ariza, M. F., Holliday, J., and Williams, M. V., Mutagenic effect of mercury (II) in eukaryotic cells, *In Vivo,* 8, 559–563, 1994.
177. Miller, D. M. and Lund, B. O., Reactivity of Hg (II) with superoxide: evidence for the catalytic dismutation of super oxide by Hg (II), *J. Biochem. Toxicol.,* 6, 293–298, 1991.
178. Lund, B. O., Miller, D. M., and Woods, J. S., Studies on HG (II)-induced H_2O_2 formation and oxidative stress *in vivo* and *in vitro* in rat kidney mitochondria, *Biochem. Pharmacol.,* 45, 2017–2024, 1993.
179. Jungwirth, A., Ritter, M., Paulmichl, M., and Lang, F., Activation of cell membrane potassium conductance by mercury in cultural renal epithelioid (MDCK) cells, *J. Cell. Physiol.,* 146, 25–33, 1991.
180. Miller, D. M. and Wood, J. S., Redox activities of mercury-thiol complexes: implication for mercury induced porphyria and toxicity, *Chem. Biol. Interact.,* 88, 23–35, 1993.
181. Carmignani, M., Boscolo, P., Artese, I., Del Rosso, G., Procelli, G., Felaco, M., Volpe, A. R., and Givliano, G., Renal mechanisms in the cardiovascular effects of chronic exposure to organic mercury in rats, *Br. J. Indust. Med.,* 49, 226–232, 1992.
182. Kosuda, L. I., Wayne, A., Mahounou, M., Greiner, D. L., and Bigazzi, P. E., Reduction of the R16.2+ subset of T lymphocytes in brown Norway rats with mercury-induced renal autoimmunity, *Cell Immunol.,* 135, 154–167, 1991.
183. Kosuda, I. L., Hasseinzadeh, H., Greiner, D. L., and Bigazzi, P. B., Role of RT6+ T lymphocytes in mercury-induced renal autoimmunity: experimental manipulations of "susceptible" and "resistant" rats, *J. Toxicol. Environ. Health,* 42, 303–321, 1994.

184. Barregard, I., Hogsted, B., Schultz, A., Karlsson, A., Sallsten, G., and Thiringer, G., Effects of occupational exposure to mercury vapor on lymphocyte micronuclei, *Scand. J. Work Environ. Health*, 17, 263–268, 1991.
185. McGregor, A. J. and Mason, H. J., Occupational mercury vapor exposure and testicular, pituitary and thyroid endocrine function, *Hum. Exp. Toxicol.*, 10, 199–203, 1991.
186. Moszczynski, P., Moszczynski, P., Jr., Slowinski, S., Bem, S., and Bartus, R., Parameters of immunity acute phase reaction in men in relation to exposure duration to mercury vapours, *J. Hyg. Epidemiol. Microbiol. Immunol.*, 35, 351–360, 1991.
187. Friberg, L., *Cadmium in the Environment*, CRC Press, Cleveland, OH, 1971.
188. Travis, C. C. and Etnier, E. L., Dietary intake of cadmium in the United States: 1920–1975, *Environ. Res.*, 27, 1–9, 1982.
189. Fox, M. R. S., Cadmium metabolism — a review of aspects pertinent to evaluating dietary cadmium intake by man, in *Trace Elements in Human Health and Disease*, Vol. 2, Prasad, A. S., Ed., Academic Press, New York, 1976.
190. Shaikh, Z. A. and Lucis, O. J., Cadmium and zinc binding in mammalian liver and kidney, *Arch. Environ. Health.*, 24, 419–425, 1972.
191. Nordberg, G. F., Garvey, J. S., and Chang, C. C., Metallothionein in plasma and urine of cadmium workers, *Environ. Res.*, 28, 179–182, 1982.
192. Bryce-Smith, D., *Nutrition and Killer Diseases*, Noyes Publications, Park Ridge, IL, 1982, pp. 128–138.
193. Gavett, S. H. and Oberdorster, G., Cadmium chloride and cadmium-induced pulmonary injury and recruitment of polymorphonuclear leukocytes, *Exp. Lung Res.*, 20, 517–537, 1994.
194. Waalkes, M. P. and Rehm, S., Chronic toxic and carcinogenic effects of cadmium chloride in male DBA/2Cr and NFS/NCr mice: strain-dependent association with tumors of the hematopoietic system, injection site, liver and lung, *Fund. Appl. Toxicol.*, 23, 21–31, 1994.
195. Leduc, D., Gressier, B., Gosset, P., Lheureux, P., deVuyst, P., Wallaert, B., and Yernault, J. C., Oxidant radical release by alveolar macrophases after cadmium chloride exposure *in vitro*, *J. Appl. Toxicol.*, 14, 381–385, 1994.
196. Borgman, R. F., Lightsey, S. F., and Roberts, W. R., Hair element concentrations and hypertension in South Carolina, *Roy. Soc. Health J.*, 101, 1–2, 1982.
197. Mederiros, D. M. and Pellum, L. K., Elevation of cadmium, lead, and zinc in the hair of adult black female hypertensives, *Bull. Environ. Contam. Toxicol.*, 32, 525–532, 1984.
198. Hill, C. H., Matrone, G., Payne, W. L., and Barber, C. W., *In vivo* interactions of cadmium with copper, zinc, and iron, *J. Nutr.*, 80, 227–235, 1963.
199. Webster, W. S., Cadmium-induced fetal growth retardation in mice and the effects of dietary supplements of zinc, copper, iron and selenium, *J. Nutr.*, 109, 1646–1651, 1979.
200. Valberg, L. S., Sorbie, J., and Hamilton, D. L., Gastrointestinal metabolism of cadmium in experimental iron deficiency, *Am. J. Physiol.*, 231, 462–467, 1976.
201. Hamilton, D. L. and Valberg, L. S., Relationship between cadmium and iron absorption, *Am. J. Physiol.*, 227, 1033–1037, 1974.
202. Flanagan, P. R., McLellan, J. S., Haist, J., Cheriam, G., Chamberlain, M. J., and Valberg, L. S., Increased dietary cadmium absorption in mice and human subjects with iron deficiency, *Gastroenterology*, 74, 841–846, 1978.
203. Schroeder, H. A. and Buckman, J., Cadmium hypertension: its reversal in rats by a zinc chelate, *Arch. Environ. Health*, 14, 693–697, 1967.
204. Schroeder, H. A. and Vinton, W. H., Hypertension in rats induced by small doses of cadmium, *Am. J. Physiol.*, 202, 515–518, 1962.
205. Perry, Jr., H. M., Erlanger, M., Yunice, A., and Perry, E. F., Mechanisms of the acute hypertensive effect of intra-arterial cadmium and mercury in anesthetized rats, *J. Lab. Clin. Med.*, 70, 963–972, 1967.
206. Perry, Jr., H. M. Erlanger, M., and Perry, E. F., Hypertension following chronic, very low dose cadmium feeding, *Proc. Soc. Exp. Biol. Med.*, 156, 173–176, 1977.
207. Perry, Jr., H. M. and Erlanger, M. W., Prevention of cadmium-induced hypertension by selenium, *Fed. Proc.*, 33, 357, 1974.
208. Perry, Jr., H. M. and Erlanger, M. W., Effect of diet on increases in systolic pressure induced in rats by chronic cadmium feeding, *J. Nutr.*, 112, 1983–1989, 1982.
209. Akahori, F., Masaoka, T., Arai, S., Nomiyama, K., Nomiyama, H., Kobayashi, K., Nomura, Y., and Suzuki, T., A nine-year chronic toxicity study of cadmium in monkeys. II. Effects of dietary cadmium in circulatory function, plasma cholesterol and triglyceride, *Vet. Hum. Toxicol.*, 36, 290–294, 1994.
210. Masaoka, T., Akahori, F., Arai, S., Nomiyama, K., Nowiyama, H., Kobayashi, K., Nomura, Y., and Suzuki, T., A nine-year chronic toxicity study of cadmium ingestion in monkeys. I. Effect of dietary cadmium on the general health of monkeys, *Vet. Hum. Toxicol.*, 36, 189–194, 1994.
211. Doyle, J. J., Bernhoft, R. A., and Sandstead, H. H., The effects of low levels of dietary cadmium on blood pressure, ^{24}Na, ^{42}K and water retention in growing rats, *J. Lab. Clin. Med.*, 86, 57–63, 1975.
212. Hall, C. E. and Nasseth, D., Effect of cadmium on salt hypertension in rats, *J. Environ. Pathol. Toxicol.*, 2, 789–797, 1979.

213. Kopp, S. J., Glonek, T., Perry, H. M., Jr., Erlanger, M., and Perry, E. F., Cardiovascular actions of cadmium at environmental exposure levels, *Science,* 217, 837–838, 1982.
214. Perry, Jr., H. M. and Erlanger, M. W., Hypertension and tissue metal levels after intraperitoneal cadmium, mercury and zinc, *Am. J. Physiol.,* 221, 808–811, 1971.
215. Lener, J. and Musil, J., Cadmium influence on the excretion of sodium by kidneys, *Experentia,* 27, 902, 1971.
216. Perry, Jr., H. M. and Erlanger, M. W., Elevated circulating renin activity in rats following doses of cadmium known to induce hypertension, *J. Lab. Clin. Med.,* 82, 399–405, 1973.
217. Revis, N., A possible mechanism for cadmium-induced hypertension in rats, *Life Sci.,* 22, 479–488, 1978.
218. Huel, G., Boudene, C., and Ibrahim, H. A., Cadmium and lead content of maternal and newborn hair: relationship to parity, birth weight and hypertension, *Arch. Environ. Health,* 36, 221–227, 1981.
219. Voors, A. W., Shuman, M. S., and Johnson, W. D., Additive statistical effects of cadmium and lead on heart related disease in a North Carolina autopsy series, *Arch. Environ. Health,* 37, 99–102, 1982.
220. Revis, N. W., Zinsmeister, A. R., and Bull, R., Atherosclerosis and hypertension induction by lead and cadmium ions: an effect prevented by calcium ion, *Proc. Natl. Acad. Sci.,* 78, 6494–6498, 1981.
221. Ostegaard, K., Renal cadmium concentration in relation to smoking habits and blood pressure, *Acta Med. Scand.,* 203, 379–383, 1978.
222. Dally, S., Maury, P., Bordard, D., Eacle, S., and Gaultier, K., Blood cadmium level and hypertension in humans, *Clin. Toxicol.,* 13, 403–409, 1978.
223. Piscator, M., Role of cadmium in carcinogenesis with special reference to cancer of the prostate, *Environ. Health Perspec.,* 40, 107–120, 1981.
224. Waalkes, M. P., Rehm, S., Perantoi, A. O., and Coogan, T. P., Cadmium exposure in rats and tumours of the prostate, *IARC Scientific,* 118, 391–400, 1992.
225. Chiqiconine, A. D., Observations on the early events of cadmium necrosis of the testes, *Anat. Res.,* 149, 23–36, 1974.
226. Fende, P. L. and Nievivenhius, R. J., An electron microscopic study of the effects of cadmium chloride on cryptochid testes of the rat, *Biol. Reprod.,* 16, 298–305, 1977.
227. Gunn, S. A., Gould, T. C., and Anderson, W. A. D., Maintenance of the structure and function of the caudo-epididymidis and contained spermatozoa by testosterone following cadmium-induced testicular necrosis in the rat, *J. Reprod. Fert.,* 21, 443–448, 1970.
228. Berlinder, A. F. and Jones-Witters, D., Early effects of a lethal cadmium dose on gerbil testes, *Biol. Reprod.,* 13, 240–247, 1975.
229. Parizak, J., Effects of cadmium salts on testicular tissue, *Nature,* 117, 1036–1037, 1956.
230. Gunn, S. A., Gould, T. C., and Anderson, W. A., Zinc protection against cadmium injury to rat testes, *Arch. Pathol.,* 71, 274–281, 1961.
231. Shiraishi, N., Barter, R. A., Uno, H., and Waalkes, M. P., Effect of progesterone pretreatment on cadmium toxicity in male Fisher (F344/NCr) and Wistar (WF/NCr) rats, *Environ. Health Perspect.,* 102(Suppl. 3), 272–280, 1994.
232. Frery, N., Nessmann, C., Girard, F., Lafond, J., Moreau, T., Blot, P., Lellouch, J., and Huel, G., Environmental exposure to cadmium and human birthweight, *Toxicology,* 79, 109–118, 1993.
233. Robards, K. and Worsfold, P., Cadmium: toxicology and analysis, *Analyst,* 116, 549–568, 1991.
234. Nolan, C. V. and Shaikh, Z. A., Lead nephrotoxicity and associated disorders: biochemical mechanisms, *Toxicology,* 73, 127–146, 1992.
235. Whitney, E. N., Cataldo, C. B., and Rolfes, S. R., *Understanding Normal and Clinical Nutrition,* 2nd ed., West Publishing, St. Paul, MN, 1987, pp. 463–465.
236. Landrigan, P. J., Current issues in the epidemiology and toxicology of occupational exposure to lead, *Toxicol. Indust. Health,* 7, 9–14, 1991.
237. Baker, Jr., E. L., Landrigan, P. J., Barbour, A. G., Cox, D. H., Folland, D. S., Ligo, R. N., and Throckmorton, J., Occupational lead poisoning in the United States: clinical and biochemical findings related to blood lead levels, *Br. J. Ind. Med.,* 36, 314–322, 1979.
238. Goyer, R. A. and Ryne, B., Pathological effects of lead, *Int. Rev. Exp. Pathol.,* 12, 1–77, 1973.
239. Rosen, J. F., Chessney, R. W., Hamstra, A., DeLuca, H. F., and Mahaffey, K. R., Reduction in 1,25-dihydroxyvitamin D in children with increased lead absorption, *N. Engl. J. Med.,* 302, 1128–1131, 1980.
240. Goyer, R. A., Weinberg, C. R., Victery, W. M., and Miller, C. R., Lead-induced nephrotoxicity: kidney calcium as an indication of tubular injury, Proc. of Third International Conference on Nephrotoxicity, August, 1987, Surrey, England.
241. Goyer, R. A., Lead and the kidney, *Curr. Top. Pathol.,* 55, 147, 1971.
242. Choie, D. D. and Richter, G. W., Effects of lead on the kidney, in *Lead Toxicity,* Singhal, R. L. and Thomas, J. A., Eds., Urban and Schwarzenberg, Baltimore, MD, 1980.
243. Murakami, M., Kawamura, R., Nishii, S., and Katsunuma, H., Early appearance and localization of intranuclear inclusions in the segments of renal proximal tubules of rats following ingestion of lead, *Br. J. Exp. Pathol.,* 64, 144–155, 1983.
244. Van Mullen, P. J. and Stadhouders, A. M., Bone marking and lead intoxication: early pathological changes in osteoclasts, *Virchows. Arch. B,* 15, 345, 1974.

245. Klann, E. and Shelton, K. R., The effect of lead on the metabolism of a nuclear matrix protein which becomes prominent in lead-induced intranuclear inclusion bodies, *J. Biol. Chem.*, 264, 16969–16972, 1989.
246. McLachlin, J. R., Goyer, R. A., and Cherian, M. G., Formation of lead-induced inclusion bodies in primary rat kidney epithelial cell cultures: effect of actinomycin-D and cycloheximide, *Toxicol. Appl. Pharmacol.*, 56, 418–431, 1980.
247. Kuliszewski, M. J. and Nicholls, D. M., Translation of mRNA from rat kidney following acute exposure to lead, *Int. J. Biochem.*, 15, 657–662, 1983.
248. Hitzfeld, B., Planas-Bohne, F., and Taylor, D., The effect of lead on protein and DNA metabolism of normal and lead adapted rat kidney cells in culture, *Biol. Trace Elements Res.*, 21, 87–95, 1989.
249. Ikebuchi, H., Teshima, R., Suzuki, K., Sawada, J. I., Terao, T., and Yamane, Y., An immunological study of a lead thionein-like protein in rat liver, *Biochem. Biophys. Res. Commun.*, 136, 535–541, 1986.
250. Ikebuchi, H., Teshima, R., Suzuki, K., Terao, T., and Yamane, Y., Simultaneous induction of lead-metallothionein-like protein and zinc-thionein in the liver of rats given lead acetate, *Biochem. J.*, 233, 541–546, 1986.
251. Goering, P. L. and Fowler, B. A., Metal constitution of metallothionein influences inhibition of delta amino levulinic acid dehydratase porphobilinogen synthetase by lead, *Biochem. J.*, 245, 339–345, 1987.
252. Pirkle, J. L., Schwartz, J., Landis, R., and Harlan, W. R., The relationship between blood lead levels and blood pressure and its cardiovascular risk implication, *Am. J. Epidemiol.*, 121, 246–258, 1985.
253. Victery, W., Vander, A. J., and Shulak, A. M., Lead, hypertension and the renin-angiotensin system in rats, *J. Lab. Clin. Med.*, 99, 354–362, 1982.
254. Beevers, D. G., Erskine, E., Robertson, M., Golbert, A., Campbell, B. C., and Moore, M. R., Blood lead and hypertension, *Lancet*, 2, 1–3, 1976.
255. Harlan, W. R., Landis, T. R., Schmouder, R. L., Goldstein, N. J., and Harlan, L. C., Blood lead and blood pressure: relationship in adolescent and adult U.S. population, *J. Am. Med. Assoc.*, 253, 530–534, 1985.
256. Orssaud, G., Claude, J., Moreau, T., Lellouch, J., Juguet, B., and Festy, B., Blood lead concentrations and blood pressure, *Br. Med. J.*, 290, 244–248, 1985.
257. Perry, Jr., H. M., Erlanger, M. W., and Perry, F. F., Increase in the blood pressure of rats chronically fed low levels of lead, *Environ. Health Perspect.*, 78, 107–111, 1988.
258. Victery, W., Evidence for effects of chronic lead exposure in blood pressure in experimental animals: an overview, *Environ. Health Perspect.*, 78, 71–76, 1988.
259. Preuss, H. G., Jiang, G., Jones, J. W., MaCarthy, P. O., Andrews, P. M., and Gondal, J. A., Early lead challenge and subsequent hypertension in Sprague-Dawley rats, *J. Am. Coll. Nutr.*, 13, 578–583, 1994.
260. Aviv, A., John, E., Bernstein, J., Goldsmith, D. I., and Spitzer, A., Lead intoxication during development: its late effect on kidney function and blood pressure, *Kidney Int.*, 17, 430–437, 1980.
261. Kopp, S. J., Barany, M., Erlanger, M., Perry, F. F., and Perry, Jr., H. M., The influence of chronic low level cadmium and lead feeding on myocardial contractility related to phosphorylation of cardiac myofibrillar proteins, *Toxicol. Appl. Pharmacol.*, 54, 48–56, 1980.
262. Hejtmancih, M. R. and Williams, B. J., Effect of chronic lead exposure on the direct and indirect components of the cardiac response to norepinephrine, *Toxicol. Appl. Pharmacol.*, 51, 239–245, 1979.
263. Lal, B., Murthy, R. C., Anand, M., Chandra, S. V., Kumar, R., Tripathi, O., and Srimal, R. C., Cardiotoxicity and hypertension in rats after oral lead exposure, *Drug Chem. Toxicol.*, 14, 305–318, 1991.
264. Myerson, R. M. and Elsenhauer, J. H., Atrioventricular condition defects in lead poisoning, *Am. J. Cardiol.*, 11, 409–412, 1963.
265. Goyer, R. A., Mechanism of lead and cadmium nephrotoxicity, *Toxicol. Lett.*, 46, 153–162, 1989.
266. Bogden, J. D., Gertner, S. B., Kemp, F. W., McLeod, R., Bruening, K. S., and Chung, H. R., Dietary lead and calcium: effects on blood pressure and renal neoplasia in Wistar rats, *J. Nutr.*, 121, 718–728, 1991.
267. Pearl, M. and Boxt, L. M., Radiographic findings in congenital lead poisoning, *Radiology*, 136, 83–84, 1980.
268. Schwartz, J., Angle, C. R., Pirkle, J. L., and Pitcher, H., Childhood blood-lead levels and stature, *Pediatrics*, 77, 81–288, 1986.
269. Markowitz, M. E., Gundberg, C. M., and Rosen, J. F., Sequential osteocalcin (Oc) sampling as a biochemical marker of the success of treatment in moderately lead (Pb) poisoned children, *Pediatr. Res.*, 23, 393A, 1988.
270. Anderson, C., Path, M. R. C., and Danylchuk, K. D., The effect of chronic low level lead intoxication on the Haversian remodeling system in dogs, *Lab. Invest.*, 37, 466–469, 1977.
271. Hass, G. M., Landerholm, W., and Hemmens, A., Inhibition of intercellular matrix synthesis during ingestion of organic lead, *Am. J. Pathol.*, 50, 815–819, 1967.
272. Pounds, J. G., Long, G. J., and Rosen, J. F., Cellular and molecular toxicity of lead in bone, *Environ. Health Perspect.*, 91, 17–32, 1991.
273. Rosen, J. F., Chesney, R. W., Hamstra, A., DeLuca, H. F., and Mahaffey, K. R., Reduction in 1,25-dihydroxyvitamin D in children with increased lead absorption, *N. Engl. J. Med.*, 302, 1128–1131, 1980.
274. Mahaffey, K. R., Rosen, J. F., Chesney, R. W., Peeler, J. T., Smith, C. M., and DeLuca, H. F., Association between age, blood lead concentration, and serum 1,25-dihydroxycholcalciferol levels in children, *Am. J. Clin. Nutr.*, 35, 1327–1331, 1982.
275. Smith, C. M., DeLuca, H. F., Tanaka, Y., and Mahaffey, K. R., Effects of lead ingestion on functions of vitamin D and its metabolites, *J. Nutr.*, 111, 1321–1325, 1981.

276. Talmage, R. V. and VanderWiel, C. J., A study of the action of calcitonin by its effect on lead-induced hypercalcemia, *Calcif. Tissue Int.*, 29, 219–224, 1979.
277. Talmage, R. V. and VanderWiel, C. J., Reduction of lead-induced hypercalcemia by calcitonin: comparison between thyroid-intact and thyroidectomized rats, *Calcif. Tissue Int.*, 34, 97–102, 1982.
278. Norimatsu, H. and Talmage, R. V., Influence of calcitonin on the initial uptake of lead and mercury by bone, *Proc. Soc. Exp. Biol. Med.*, 161, 94–98, 1979.
279. Goldberg, R. L., Kaplan, S. R., and Fuller, G. C., Effect of heavy metals on human rheumatoid synovial cell proliferation and collagen synthesis, *Biochem. Pharmacol.*, 15, 2763–2766, 1983.
280. Miyahara, T., Oh-e, Y., Takaine, E., and Kozuka, H., Interaction between cadmium, zinc, copper or lead in relation to the collagen and mineral content of embryonic check bone in tissue culture, *Toxicol. Appl. Pharmacol.*, 676, 41–48, 1983.
281. Radolfo-Sioson, S. A. and Ahrens, F. A., The effects of lead on collagen biosynthesis in neonatal rats, *Res. Commun. Chem. Pathol. Pharmacol.*, 29, 317–318, 1980.
282. Vistica, D. T., Ahrens, F. A., and Ellison, W. R., The effects of lead upon collagen synthesis and proline hydroxylation in the mouse 3T6 fibroblast, *Arch. Biochem. Biophys.*, 179, 15–23, 1977.
283. Hsu, F. S., Krook, L., Shively, J. N., and Duncan, J. R., Lead inclusions in osteoclasts, *Science*, 181, 447–448, 1973.
284. Bonucci, E., Barckhaus, R. H., Silvestrini, G., Ballanti, P., and DiLorenzo, G., Osteoclast changes induced by lead poisoning (saturnism), *Appl. Pathol.*, 1, 241–250, 1983.
285. Calhoun, L. A., Livesey, D. L., Mailer, K., and Addetica, R., Interaction of lead ions with bovine carbonic anhydrase: further studies, *J. Inorg. Biochem.*, 25, 262–275, 1985.
286. Mailer, K., Calhoun, I. A., and Livesey, D. L., The interaction of lead ions with bovine carbonic anhydrase, *Int. J. Peptide Protein Res.*, 19, 233–239, 1982.
287. Altmann, L., Gutowski, M., and Wiegand, H., Effects of maternal lead exposure on functional plasticity in the visual cortex and hippocampus of immature rats, *Brain Res.*, 81, 50–56, 1994.
288. Bornschein, R., Pearson, D., and Reiter, L., Behavioral effects of moderate lead exposure in children and animal models, *CRC Crit. Rev. Toxicol.*, 7, 43–152, 1980.
289. Needleman, H. L., Schell, A., Bellinger, D., Leviton, A., and Allred, E. N., The long-term effects of exposure to low doses of lead in childhood, *N. Engl. J. Med.*, 322, 83–88, 1990.
290. Petit, T. L., Developmental effects of lead: its mechanism in intellectual functioning and neural plasticity, *Neurotoxicology*, 7, 483–496, 1986.
291. Costa, L. G. and Fox, D. A., A selective decrease of cholinergic muscarinic receptors in the visual cortex of adult rats following developmental lead exposure, *Brain Res.*, 276, 259–266, 1983.
292. Petit, T. L., Alfano, D. P., and LeBoutillier, J. C., Early lead exposure and the hippocampus: a review and recent advances, *Neurotoxicology*, 4, 79–94, 1983.
293. Fox, D. A., Lewkowski, J. P., and Cooper, G. P., Acute and chronic effects of neonatal lead exposure on development of the visual evoked response in rats, *Toxicol. Appl. Pharmacol.*, 40, 449–461, 1977.
294. Fox, D. A., Wright, A. A., and Costa, L. G., Visual acuity defects following neonatal lead exposure: cholinergic interactions, *Neurobehav. Toxicol. Teratol.*, 4, 689–693, 1982.
295. Otto, D. A., Robinson, G., Baumann, S., Schroeder, S., Mushak, P., Kleinbaum, D., and Boone, L., Five-year followup study of children with low to moderate lead absorption: electrophysiological evaluation, *Environ. Res.*, 38, 168–186, 1985.
296. Altmann, L., Weinsberg, F., Sveinsson, K., Lilienthal, H., Wiegand, H., and Winneke, G., Impairment of long-term potentiation and learning following chronic lead exposure, *Toxicol. Lett.*, 66, 105–112, 1993.
297. Lynch, G. and Baudry, M., The biochemistry of memory: a new and scientific hypothesis, *Science*, 224, 1057–1063, 1984.
298. Collingridge, G., The role of NMDA receptors in learning and memory, *Nature*, 33, 604–605, 1987.
299. Holtzman, D., DeVries, C., Nguyen, H., Olson, J., and Bensch, K., Maturation of resistance to lead encepholopathy: cellular and subcellular mechanisms, *Neurotoxicology*, 5, 97–124, 1984.
300. Selvin-Testa, A., Lopez-Costa, J. J., Nessi de Avinson, A. C., and Saavedra, J. P., Astroglial alterations in rat hippocampus during chronic lead exposure, *Glia*, 4, 384–392, 1991.
301. McMichael, A. T., Baghurst, P. A., Wigg, N. R., Vimpani, G. V., Roberson, F. F., and Roberts, R. J., Port Pirie cohort study: environmental exposure to lead and children's abilities at the age of four years, *N. Engl. J. Med.*, 319, 468–475, 1988.
302. Palmer, M. R., Bjorklund, H., Freedman, R., Taylor, D. A., Marwaha, J., Olson, L., Seiger, A., and Hoffer, B., Permanent impairment of spontaneous Purkinje cell discharge in cerebellar grafts caused by chronic lead exposure, *Toxicol. Appl. Pharmacol.*, 60, 431–440, 1991.
303. Luthman, J., Oskarsson, A., Olson, L., and Hoffer, B., Postnatal lead exposure affects motor skills and exploratory behavior in rats, *Environ. Res.*, 58, 236–252, 1992.
304. Cooper, G. P., Suszkiw, J. B., and Manalis, R. S., Heavy metals: effects on synaptic transmission, *Neurotoxicology*, 5, 247–266, 1984.
305. Manalis, R. S. and Cooper, G. P., Presynaptic and postsynaptic effects of lead at the frog neuromuscular junction, *Nature*, 243, 354–355, 1973.

306. Atchison, W. D. and Narabushi, T., Mechanism of action of lead on neuromuscular junction, *Neurotoxicology*, 5, 267–282, 1984.
307. Kostial, K. and Vouk, V. B., Lead ions and synaptic transmission in the superior cervial ganglion of the cat, *Br. J. Pharmacol. Chemother.*, 12, 219–222, 1957.
308. Bressler, J. P. and Goldstein, G. W., Mechanism of lead neurotoxicity, *Biochem. Pharmacol.*, 41, 479–484, 1991.
309. Blackman, S. A., The lesions of lead encephalitis in children, *Bull. Johns Hopkins Hosp.*, 61, 1–61, 1937.
310. Needleman, H. L., Schell, A., Bellinger, D., Leviton, A., and Allred, E. N., The long-term effects of exposure to low doses of lead in childhood, *N. Engl. J. Med.*, 322, 83–88, 1990.
311. Bellinger, D., Leviston, A., Waterman, S. C., Needleman, H., and Rabinowitz, M., Longitudinal analyses of prenatal and postnatal lead exposure and early cognitive development, *N. Engl. J. Med.*, 316, 1037–1043, 1987.
312. USEPA, *Air Quality Criteria for Lead*, (EPA-600/8-83/028aF), U.S. Environmental Protection Agency, Environmental Criteria and Assessment Office, Research Triangle Park, NC, 1986.
313. Childhood lead poisoning — United States report of the Congress by the Agency for Toxic Substances and Disease Registry, *J. Am. Med. Assoc.*, 260, 1523–1533, 1988.
314. Lee, W. R. and Moore, M. R., Low level exposure to lead: the evidence for harm accumulates, *Br. Med. J.*, 301, 504–506, 1990.
315. Needleman, H. L. and Gatsonis, C. A., Low level lead exposure and the IQ of children: a meta-analysis of modern studies, *J. Am. Med. Assoc.*, 263, 673–678, 1990.
316. Needleman, H. L. and Bellinger, D., The health effects of low level exposure to lead, *Ann. Rev. Public Health*, 12, 111–140, 1991.
317. Wigg, N. R., Vimpani, G. V., McMichael, A. J., Baghurst, P. A., Robertson, E. F., and Roberts, R. J., Port Pirie cohort study: childhood blood lead and neuropsychological development at age two years, *J. Epidemiol. Commun. Health*, 42, 213–219, 1988.
318. Minnema, D. J., Greenland, R. D., and Michaelson, I. A., Effect of *in vitro* inorganic lead on dopamine release from superfused rat striatal synaptosomes, *Toxicol. Appl. Pharmacol.*, 84, 400–411, 1986.
319. Minnema, D. J., Michaelson, I. A., and Cooper, G. P., Calcium efflux and neurotransmitter release from rat hippocampal synaptosomes exposed to lead, *Toxicol. Appl. Pharmacol.*, 92, 351–357, 1988.
320. Minnema, D. J. and Michaelson, I. A., Differential effects of inorganic lead and δ-aminolevulinic acid *in vitro* on synaptosomal δ-aminobutyric acid release, *Toxicol. Appl. Pharmacol.*, 86, 437–447, 1986.
321. Simons, T. J. B. and Pocock, G., Lead enters bovine adrenal medullary cells through calcium channels, *J. Neurochem.*, 48, 383–389, 1987.
322. Richardt, G., Federolf, G., and Habermann, E., Affinity of heavy metal ions to intracellular Ca^{2+}-binding proteins, *Biochem. Pharmacol.*, 35, 1331–1335, 1986.
323. Habermann, E., Crowell, K., and Janicki, P., Lead and other metals can substitute for Ca^{2+} in calmodulin, *Arch. Toxicol.*, 54, 61–70, 1983.
324. Erondu, N. E. and Kennedy, M. B., Regional distribution of type II Ca^{2+} calmodulin-dependent protein kinase in rat brain, *J. Neurosci.*, 5, 3270–3277, 1985.
325. Hemmings, H. C., Nairn, A. C., McGuinness, T. L., Huganir, R. L., and Greengard, P., Role of protein phosphorylation in neuronal signal transduction, *FASEB J.*, 3, 1583–1592, 1989.
326. Markovacm J. and Goldstein, G. W., Picomolar concentrations of lead stimulate brain protein kinase, *Nature*, 334, 71–73, 1988.
327. Bradbury, M. W. B., The structure and function of the blood brain barrier, *Fed. Proc.*, 43, 186–190, 1984.
328. Clasen, R. A., Hartmann, J. F., Starr, A. J., Coogan, P. S., Pandolfi, S., Laing, I., Becker, R., and Hass, G. M., Electron microscopic and chemical studies of the vascular changes and edema of lead encephalopathy, *Am. J. Pathol.*, 74, 215–240, 1973.
329. Goldstein, G. W., Asbury, A. K., and Diamond, I., Pathogenesis of lead encephalopathy. Uptake of lead and reaction of brain capillaries, *Arch. Neurol.*, 31, 382–389, 1974.
330. Hamilton, A. and Hardy, H. L., *Industrial Toxicology*, Publishing Sciences Group, Acton, MA, 1974.
331. Lancranjan, I., Popescu, H. I., Gavenescu, O., Klepsh, I., and Serbanescu, M., Reproductive ability of workmen occupationally exposed to lead, *Arch. Environ. Health*, 30, 396–401, 1975.
332. Cullen, M. R., Kayne, R. D., and Robbin, J. J., Endocrine and reproductive dysfunction in men associated with occupational inorganic lead intoxication, *Arch. Environ. Health*, 39, 431–440, 1984.
333. Assennato, G., Paci, G., Baser, M. E., Molinini, R., Candela, R. B., Altamura, B. M., and Giogini, R., Sperm count suppression without endocrine dysfunction in lead-exposed men, *Arch. Environ. Health*, 4, 387–390, 1986.
334. Marchlewic, M., Protasowicki, M., Rosewicka, L., Piasecka, M., and Laszczynska, M., Effect of long-term exposure to lead on testis and epididymis in rats, *Fol. Hist. Cytobiol.*, 31, 55–62, 1993.
335. Pinon-Lataillade, G., Thoreux-Manlay, A., Coffigny, H., Manchaux, G., Masse, R., and Soufir, J. C., Effect of ingestion and inhalation of lead on the reproductive system and fertility of adult male rats and their progeny, *Hum. Exp. Toxicol.*, 12, 165–172, 1993.
336. Cullen, C., Singh, A., Dykeman, A., Rice, D., and Foster, W., Chronic lead exposure induces ultrastructural alterations in the monkey seminal vesicle, *J. Submicrosc. Cytol. Pathol.*, 25, 127–135, 1993.

337. Wiebe, J. P., Barr, K. J., and Buckingham, K. D., Lead administration during pregnancy and lactation affects steroidogenesis and hormone receptors in the testes of offspring, *J. Toxicol. Environ. Health,* 10, 653–666, 1982.
338. Jacquet, P., Leonard, A., and Gerber, G., Embryonic death in mouse due to lead exposure, *Experentia,* 31, 1312–1313, 1975.
339. Coffigny, H., Thoreux-Manlay, A., Pinon-Lataillade, G., Monchaux, G., Mass, R., and Soufir, J., Effects of lead poisoning of rats during pregnancy on the reproductive system and fertility of their offspring, *Hum. Exp. Toxicol.,* 13, 241–246, 1994.
340. Clugston, G. A. and Hetzel, B. S., Iodine, in *Modern Nutrition in Health and Disease,* 8th ed., Shils, M. E., Olson, J. A., and Shike, M., Eds., Lea & Febiger, Philadelphia, PA, 1994, pp. 252–263.
341. Nielsen, F. H., Chromium, in *Modern Nutrition in Health and Disease,* 8th ed., Shils, M. E., Olson, J. A., and Shike, M., Eds., Lea & Febiger, Philadelphia, PA, 1994, pp. 264–268.
342. Nielsen, F. H., Ultratrace minerals, in *Modern Nutrition in Health and Disease,* 8th ed., Shils, M. E., Olson, J. A., and Shike, M., Eds., Lea & Febiger, Philadelphia, PA, 1994, 268–286, 1994.

CHAPTER 4

CELLULAR AND MOLECULAR MECHANISMS OF METAL TOXICITIES

M. Duane Enger and Y. James Kang

CONTENTS

0-8493-4493-X/97/$0.00+$.50

1 INTERACTIONS OF CARCINOGENIC TRACE ELEMENTS WITH SIGNAL TRANSDUCTION PATHWAYS

M. Duane Enger

Interactions of carcinogenic trace elements with cells and organisms include effector-type intersections with signal transduction pathways (STPs). An understanding of the relationships between these intersections and cellular, physiological, and genetic responses to metals is necessary to define adequately the extent to which metal-induced alterations in proliferation, apoptosis, differentiation, DNA repair, or metabolism of other xenobiotics contribute to metal-induced carcinogenesis. Such effects may complement direct mutagenic and clastogenic actions of the metal, or in some instances may represent the salient mechanisms by which the metal contributes to tumor formation and progression.

Studies of such interactions benefit from a rapidly developing understanding of the ways by which physiological effectors of STPs induce or inhibit cellular responses important to normal development and function. These studies, however, suffer from the same inability to trace completely such pathways from beginning to end, and from the same potential complexities that attend responses to physiological effectors. These complexities are reflected in differing responses to a given effector as a function of cell type, history, and presence of other effectors. They involve also multiple, intersecting, and redundant STPs and may be pleiotropic, resulting in opposite responses even in very closely related cells.

Importantly, responses to physiological and toxicological effectors may also be opposite in nature as a function of dose. Thus, elucidation of whether and where metals may intersect with STPs is important to understanding and defining dosage effects in metal-induced carcinogenesis. Another factor important to understanding and defining dose dependencies in metal-induced carcinogenesis is the specificity of metal-STP interactions. As will be delineated below, studies of metal-STP interactions are not limited by a lack of defined STP targets — numerous receptors, second messengers, and even transcription factors bind carcinogenic metals at relatively high affinity *in vitro* and have in many instances correspondingly altered form or function. These studies are, however, challenged by the need to define which interactions are important at doses relevant to the carcinogenic process *in vivo*.

A review of metal-STP interactions must therefore emphasize the doses and dose ranges employed in the various studies discussed. In this context, it is also important to seek definition of the doses at which responses change from those that involve catalytic interactions with STPs to those more stoichiometric in nature, involving damage adequate to cause loss of cell integrity and viability. Finally, it is to be expected that metal-STP interactions will become more complex as doses increase, and targets with lower relative affinity bind significant amounts of metal. Whether it is the highest affinity target or one or more of many, lower affinity targets that is of determinative importance in metal-induced carcinogenesis is a vital but obviously difficult question.

These considerations frame discussion of the interactions of carcinogenic trace metals with signal transduction pathways. The metals to be discussed are primarily those established as human carcinogens — chromium, nickel, arsenic, beryllium, and cadmium.[1] Discussion of the interactions of these elements with STPs begins with those most likely to occur at the cell surface (that is, with receptors or ion channels) and proceeds "downstream" to interactions with G proteins, potential second messengers including Ca^{2+} ligands, kinases, and phosphatases, and, finally, transcription factors. The consequences of these interactions are then discussed in the context of their effects on specific gene expression and/or physiological and pathological consequences. In the next chapter, interactions that result more directly in genotoxic effects will be discussed in the context of mechanisms of DNA damage and repair and clastogenic responses to metals.

1.1 Interaction of Carcinogenic Trace Metals with Receptors

1.1.1 Arsenic

Arsenic interacts specifically with steroid receptors to modulate their function *in vivo*. Such specificity of interaction was first suggested by the observation that arsenite inhibits DNA binding by specific receptor-glucocorticoid complexes, those that require the presence of yet another cytosolic factor for activity, but does not inhibit DNA binding by receptor-glucocorticoid complexes already possessing the additional cytosolic factor.[2]

The interaction of arsenite, but not arsenate, with the glucocorticoid receptor blocks steroid binding. Pretreatment of HTC cells with 7 μM arsenite for 30 minutes produces half-maximal inhibition of dexamethasone binding to its receptor.[3] The interaction of arsenite with the receptor appears to be mediated by reaction with vicinal dithiols, as it is reversed by much lower concentrations of dithiothreitol than of mercaptoethanol. Relatively low concentrations of Cd^{2+} and selenite also inhibit binding, but Zn^{2+} at 300 μM is without effect. Arsenite has been reported also to inhibit binding *in vitro* of triiodothyronine to its nuclear receptor, albeit at relatively high (millimolar) concentrations.[4]

Subsequent studies showed that arsenite is very selective in its effects on steroid binding to receptors — 100 μM completely inhibits steroid binding to glucocorticoid receptors but has no effect on androgen, progesterone, estrogen, or mineralocorticoid receptors.[5] The basis

TABLE 1

Arsenical-Receptor Interactions

Effector(s)	Effector Concentration	Receptor(s)	Effect(s)	Ref.
Arsenite	1–25 mM (IC_{50} > 25 mM)	Glucocorticoid	Reduced DNA binding	2
Arsenite (arsenate: NE; arsenite > 10× Cd^{2+}; selenite = arsenite; 300 μM Zn^{2+}: NE)	7 μM (IC_{50})	Glucocorticoid	Steroid binding blocked	3
Cacodylate > arsenite > arsenate	10 μM	Muscarinic	Muscarinic binding (after disulfide reduction) further diminished	9
Arsenite, heat shock	200 μM (43°C)	Glucocorticoid	Nuclear translocation	10
Arsenite	200 μM	Glucocorticoid	Increased gene expression	12
Arsenite	100 μM (2 hr)	Progesterone	Altered heat shock protein binding, increased activity	11

Note: NE = no effect; IC_{50} = concentration for half maximal inhibition.

for this selectivity is the presence in the steroid-binding core of the glucocorticoid receptor of a vicinal dithiol not present in the other steroid receptors.[5,6] The residues comprising this dithiol have been identified as cys-656 and cys-661.[7,8] This selectivity does not obtain when selenite is used.[5]

The susceptibility of muscarinic receptors to sulfhydryl reagents predicts that their ligand binding would also be sensitive to arsenicals. Prior treatment with disulfide-reducing reagents is, however, necessary for an effect of arsenite on muscarinic binding.[9]

In addition to binding specifically to the glucocorticoid receptor to inhibit steroid binding, arsenite (and heat shock) affects its subcellular localization. A shift of unliganded receptor from the cytosol to the nucleus was observed in L929 and WCL2 cells exposed to arsenite or heat shock.[10]

Although arsenite does not bind to the progesterone receptor to affect steroid binding, it affects progesterone activity through its induction of specific heat shock protein (HSP) synthesis. Several of these heat shock proteins (HSP 70 and 90) associate *in vivo* with the progesterone receptor and provide thereby a mechanism for heat shock and arsenite modulation of progesterone activity.[11] That is, heat/arsenite shock results in the association of new and different HSP proteins with the receptor, resulting in enhanced activity.

A third modality for arsenite modulation of glucocorticoid activity is through effects on receptor-mediated gene expression.[12] A relatively high concentration of arsenite (200 μM) was observed to enhance expression in LMCAT2 cells of a reporter plasmid with a dexamethasone-responsive promoter. The effect appears to occur after binding of the receptor to high affinity nuclear sites.[12-14] (See Table 1.)

1.1.2 Chromium

Chromium (Cr^{3+}) plays a special role in the maintenance of glucose tolerance. Deficiency results in moderate to severe glucose intolerance accompanied by increased insulin levels and insulin resistance.[15,16] Supplementation with 150 to 200 μg daily reverses these effects of the deficient state[15,16] in normal subjects as well as in glucose-intolerant humans.[17] Dietary chromium levels may often be inadequate and contribute to adult onset diabetes as well as arteriosclerosis.[18,19] Adequate nicotinic acid appears to be requisite for the chromium effect on glucose tolerance.[20]

TABLE 2

Chromium-Receptor Interactions

Effector	Effector Concentration	Receptor	Effect	Ref.
Cr^{3+} (tripicolinate form most effective)	50 ng/ml	Insulin	Increased insulin binding and response	23

Although the primary effect of Cr^{3+} is to enhance insulin response, its biologically active forms and biochemical targets are yet to be elucidated;[21,22] whether it interacts directly with insulin receptors, acts "downstream" of insulin receptors in their STPs, or affects insulin receptor synthesis is yet to be defined. In any event, these receptor interactions involving Cr^{3+} are reported to be widely beneficial. Interactions of Cr^{5+} or Cr^{6+}, forms with which chromium carcinogenesis is associated, with hormone or growth factor receptors have not been reported. (See Table 2.)

1.1.3 Cadmium

A number of divalent cations interact with the androgen receptor *in vitro* to alter it from one sedimentation form (size) to another.[24] Zn^{2+} at 200 μM stabilizes an 8.6S form of the receptor and reverses the dissociation into a 4.7S species that occurs in low salt and the presence of 200 μM Ca^{2+} or 1 mM Mg^{2+}. This dissociation is reversed also by 10 μM Cd^{2+}.

Chronic ingestion of Cd^{2+} delays clearance of particles and of soluble materials presenting Fc fragments in mice.[25] Correspondingly, *in vitro* studies with macrophages show a Cd^{2+} inhibition of E-IgG and E-IgMC binding to sheep red blood cells.[25] Receptor binding is affected only at the higher concentrations of Cd^{2+} (100 μM).

Zn^{2+} uptake by brush-border-membrane vesicles appears to be carrier mediated and to involve a specific membrane protein receptor.[26] Cadmium, but not cupric, ferric, or ferrous, ions inhibit uptake via this presumed receptor competitively.[26]

Long-term exposure to Cd^{2+} is cytotoxic to a human neuroblastoma cell line grown *in vitro*.[27] The Cd^{2+} cytotoxic response is significantly greater for Cd^{2+} than for lead or aluminum. Nonetheless, Cd^{2+} markedly increases the expression of alpha-bungarotoxin and muscarinic receptors.[27]

The DNA binding domain of the glucocorticoid receptor reversibly binds two Zn^{2+} or Cd^{2+}.[28] As reported above, Cd^{2+} as well as arsenite blocks the binding of steroid to the glucocorticoid receptor.[3]

A basis for possible effects of Cd^{2+} on the insulin receptor is provided by the observation that it contains a Zn^{2+} binding "finger-loop" domain similar to that of many transcription factors.[29] Cd^{2+} also stimulates glucose transport in rat adipocytes.[30] This effect, however, appears to be due to a "post-receptor/kinase mechanism".[30]

A dramatic effect of Cd^{2+} on activation of early steps of several signal transduction pathways occurs in human skin fibroblasts.[31,32] In these cells, Cd^{2+} and other divalent cations elicit a large, transient increase in cytosolic Ca^{2+} in a fashion resembling the response to bradykinin. This increase is accompanied by an increase in Ca^{2+} efflux — 0.1 μM Cd^{2+} suffices to half-maximally stimulate efflux. Cd^{2+} treatment also increases inositol triphosphate fourfold within 15 seconds of treatment. These effects are antagonized by Zn^{2+}.[33] Cd^{2+} appears to bind to an external site ("orphan receptor") to elicit these responses.

Cd^{2+} and Zn^{2+} modulate the responses of cultured mouse hippocampal neurones to excitatory amino acids.[34] Both antagonize responses to *N*-methyl-D-aspartate (NMDA) but potentiate responses to kainate and quisqualate at a concentration of 50 μM, with Zn^{2+} and Cd^{2+} reducing responses to NMDA to 0.19 and 0.39 times control, respectively. The KDs for antagonism are, correspondingly, 13 and 48 μM. These cations act as noncompetitive antagonists that do not interfere with ligand binding but bind presumably to a site on the receptor moiety of the receptor-channel complex to regulate channel activity.

Possible effects of metal ions on the binding of ligands to the opioid receptor of the rat brain were suggested by the observation that Zn^{2+} at physiological levels modulates binding.[35] Accordingly, it was observed that copper, cadmium, and mercury cations inhibit the binding of opioid receptor agonists (Tyr-D-ala-gly-methyl-phe-glyol)-enkephalin and (Tyr-D-ser-gly-phe-leu-thr)-enkephalin.[35] Whereas physiological levels of Zn^{2+} modulate mu but not delta or kappa opioid receptors, the other cations effectively inhibit the binding of both mu and delta receptors at comparable concentrations.

Cu^{2+} (EC_{50} = 200 μM) and Cd^{2+} (EC_{50} = 20 μM) effectively inhibit binding of endothelin to its solubilized receptor, whereas concentrations of the divalent cations of Fe, Ca, Zn, Mg, Ni, and Mn of up to 10 m*M* show little response.[36] The effect of Cd^{2+} on endothelin binding is noncompetitive. Although Cd^{2+} also induces relaxation in endothelin-treated aortic strips, the relative importance of Cd^{2+} interaction with the endothelin receptor in this response is yet to be defined.

Zn^{2+}, Cd^{2+}, Ni^{2+}, and Mn^{2+} inhibit the response of spinal cord neurons to gamma-amino butyric acid (GABA). This inhibition involves direct interaction with the GABA receptor to modulate a Cl^- channel.[37] The nature of the response is such as to suggest an allosteric mechanism of action, involving Cd^{2+} binding to a site distinct from either the Cl^- channel or the GABA binding site.

Estrogen receptors show a zinc dependence for DNA binding. This zinc is contained not in the classical "zinc-finger", but rather in a "zinc-twist" structure.[38] The zinc dependence can be shown by chelation, resulting in a loss of DNA binding and of restored DNA binding following dialysis against 10 μM Zn^{2+}, Cd^{2+}, or Co^{2+} (but not Cu^{2+} or Ni^{2+}). Cu^{2+} and Ni^{2+} can, however, compete for Zn^{2+} during reconstitution with resulting decreases in DNA binding.

A single, interperitoneal administration of cadmium significantly affects glucocorticoid receptor capacity and glucocorticoid receptor DNA binding activity in liver cytosol isolates 24 or 48 hours after treatment.[39] The effect appears due to interaction with thiols, but whether such are in the receptor and/or in receptor-associated proteins is not clear.[39]

Osteoclasts show a Ca^{2+} response mediated by a Ca^{2+} receptor that involves Ca^{2+}-induced increases in cytosolic Ca^{2+}.[40] Remarkably, although treatment of rat osteoclasts with millimolar Ca^{2+} is requisite to the response, micromolar concentrations of Cd^{2+} or Ni^{2+} suffice to provoke a similar initial response. The kinetics of the responses show differences also, with the Ca^{2+} response being more sustained. Further, stepwise increases in Ca^{2+} result in stepwise increases in cytosolic Ca^{2+}, but an initial exposure to Ni^{2+} or Cd^{2+} results in a lack of response to a subsequent increase in concentration of these cations, even when the increment in concentration is, in the case of Cd^{2+}, 100 times. The low effector concentrations of Cd^{2+} and Ni^{2+} needed to provoke the cytosolic Ca^{2+} transients take on additional significance from the fact that these cations are added in the presence of 1.25 m*M* Ca^{2+}. Cd^{2+} and Ni^{2+} are calculated to be factors of 1900 and 85, respectively, greater in potency than Ca^{2+} in provoking the response. These results suggest a perhaps meaningful target for Cd^{2+} toxicity *in vivo* and may provide a mechanistic basis for the known effects of Cd^{2+} on bone metabolism, such as the pronounced osteomalacia and osteoporosis that occur when Cd^{2+} accumulation is sufficiently great that its concentration in kidney cortex exceeds 200 ppm. The osteoclast Ca^{2+} response is suggested as being of possible mechanistic importance in the negative regulation of bone resorption.[40] Should an *in vivo* exposure to Cd^{2+} result in a state of refraction to subsequent increases in Ca^{2+}, dysregulation could occur. Unfortunately, whether Cd^{2+} exposure results in a lack of response in the cultured osteoclasts to subsequent increases in Ca^{2+} was not reported. The observation of a Cd^{2+}-sensitive Ca^{2+} receptor whose activation results in a rapid cytosolic Ca^{2+} increase in osteoclasts also raises the question of whether a similar receptor is involved in Cd^{2+}-induced Ca^{2+} transients in fibroblasts and epithelial cells;[31,32] however, Cd^{2+} reportedly induces the Ca^{2+} response in these cells by interacting with a zinc site on a lectin-binding receptor.[33,41]

The heavy metals mercury, lead, and cadmium inhibit binding of ouabain to its receptor on isolated rat brain microsomes, with concentrations needed for maximum response of 1, 100, and 100 μ*M*, respectively.[42] Cadmium inhibits carbachol-stimulated synthesis of inositol phosphates in rat cochlea.[43] Because this response is insensitive to inhibition by verapamil, it is suggested that the Cd^{2+} effect is not due to blockage of Ca^{2+} channels. Also, Cd^{2+} does not act by displacing ligands from muscarinic sites. The effects of Cd^{2+} and Hg^{2+} are synergistic.

Cd^{2+} treatment markedly affects synthesis of the estrogen receptor in MCF cells.[44] Both estrogen receptor protein and mRNAs are reduced more than 50% by 24-hour exposure to 1 μ*M* Cd^{2+}, a decrease comparable to that produced by one n*M* estradiol. The same treatments result in threefold increases in progesterone receptor (PgR) in these cells. Cd^{2+} and estradiol treatments increase mRNA levels for cathepsin D as well as for PgR. These effects are attributable to changes in transcription of the relevant genes. A functional estrogen receptor is requisite to the observed changes in gene expression induced by Cd^{2+}, suggesting that Cd^{2+} acts through the estrogen receptor. Because Cd^{2+} also stimulates gene activity in an introduced construct carrying an estrogen response element (ERE), it is suggested that "the effects of cadmium are mediated by the estrogen receptor through an ERE".[44] The effects are not mimicked by 25 or 100 μ*M* Zn^{2+}, indicating specificity and that the response is not mediated by a heavy metal response element, which would be responsive to both elements. Finally, these molecular responses are associated with a five- to sixfold increase in cell growth, providing a mechanism by which Cd^{2+} could, by increasing cell proliferation, contribute to carcinogenesis.

Among cations tested, Cd^{2+}, Zn^{2+}, and Cu^{2+} were found to inhibit specific antagonist binding by 5-HT3 receptors in rat cortical membranes.[45] Cations of Co, Ni, Ba, Ca, Mg, or Mn are without effect. The effective cations function apparently by reducing affinity of the receptor for the specific antagonist (GR 65630). (See Table 3.)

1.1.4 Nickel

An analysis of the effects of a variety of divalent cations (and sodium metavanadate) on vascular resistance in perfused rat hearts shows Ni^{2+} to be the most active.[46] In this context, nickel interacts with beta-, but not alpha-, adrenergic receptors to induce vasoconstriction *in vitro*.[47,48] Nickel cations also have the greatest relative activity (among Ni^{2+}, Co^{2+}, Mn^{2+}, and Zn^{2+}) in antagonizing NMDLA (*N*-methyl-D,L-aspartic acid) responses evoked at NMDA receptors in amphibian spinal cords.[49] The response is not attributable to blockage of Ca^{2+} channels.

Relatively significant divalent cation effects have been shown in studies of diazepam binding to benzodiazepine receptors.[50] In the absence of 1 m*M* Ni^{2+}, Co^{2+}, or Zn^{2+}, a single class of binding is shown (KD = 5.3 n*M*). In their presence, however, a novel "super-high-affinity" site is revealed (KD = 1.0 n*M*) that is not detected in their absence. The result is an enhanced binding of ligand. The binding of beta-carboline-3-carboxylate ethyl ester is, however, inhibited by Ni^{2+}, Zn^{2+}, and Co^{2+}.[51] The effect of Ni^{2+} on benzodiazepine receptors was shown to be specific for type 2 receptors when solubilized preparations from cow brain are challenged for binding of flunitrazepam.[52] Studies on mouse brain benzodiazepine receptors suggest that Ni^{2+} acts through a Ca^{2+}-binding site to modulate diazepam binding.[53]

The binding of cis-methyldioxolane to muscarinic acetylcholine receptors in the synaptic membrane fractions of pig caudate nuclei is affected by Ni^{2+} in a biphasic manner.[54] A two- to threefold enhanced binding is seen between 0.1 and 10 m*M*, and inhibition obtains above 10 m*M* Ni^{2+}. One m*M* Co^{2+}, Mn^{2+}, and Zn^{2+} also enhance binding. Of 18 cations tested, only Cd^{2+}, Hg^{2+}, and Cu^{2+} show inhibition at a concentration of 1 m*M*. Nickel-enhanced

TABLE 3
Cadmium-Receptor Interactions

Effector(s)	Effector Concentration	Receptor(s)	Effect(s)	Ref.
Cd^{2+}	10 μ*M*	Androgen	8.6S receptor form stabilized	24
Zn^{2+}	200 μ*M*			
Cd^{2+}	10 μ*M* (IC_{50})	Fc, Complement	Internalization of receptor-particle (erythrocyte Ig [C/MC]) complexes inhibited	25
Cd^{2+}	100 μ*M* (IC_{50})	Zn^{2+}	Zn^{2+} uptake inhibited	26
Cd^{2+}	1.7 μ*M*	Bungarotoxin, muscarinic	Receptor expression increased	27
Cd^{2+}	5 μ*M* (IC_{50})	Bungarotoxin	Ligand binding decreased	28
Pb^{2+}	20 m*M*	Bungarotoxin		
Al^{3+}	16 m*M*	Bungarotoxin		
Cd^{2+}	> 100 μ*M*	Muscarinic	Ligand binding decreased	28
Pb^{2+}	> 20 m*M*	Muscarinic		
Al^{3+}	> 60 m*M*	Muscarinic		
Cd^{2+}	Predicted effect	Insulin	Nuclear translocation and DNA binding of insulin-receptor complex modulated	29
Cd^{2+}	1 m*M*	Postreceptor?	Glucose transport stimulated	30
Zn^{2+}	1 m*M*			
Hg^{2+}	0.1 m*M*			
$Cd^{2+} > Co^{2+} > Ni^{2+} > Fe^{2+} > Mn^{2+}$	0.1 μ*M* (EC_{50})	"Orphan"	Inositol triphosphate formation, Ca^{2+} mobilization, Ca^{2+} efflux stimulated	31
Cd^{2+} (Zn^{2+} 4× Cd^{2+}); KDs for antagonism of NMDA response, Zn^{2+} = 13 μ*M*, Cd^{2+} = 48 μ*M*	50 μ*M*	Excitatory amino acid	Neuronal responses to *N*-methyl-D-aspartate (NMDA); noncompetitively antagonized; responses to kainate and quisqualate potentiated	34
Cd^{2+}	20 μ*M* (IC_{50})	Mu opioid	Agonist ([Tyr-D-ala-gly-methyl-phe-glyol]-enkephalin) binding inhibited	35
Zn^{2+}	37 μ*M*			
Cu^{2+}	23 μ*M*			
Hg^{2+}	9 μ*M*			
Cd^{2+}	58 μ*M* (IC_{50})	Delta opioid	Agonist ([Tyr-D-ala-gly-methyl-phe-glyol]-enkephalin) binding inhibited	
Zn^{2+}	550 μ*M*			
Cu^{2+}	33 μ*M*			
Hg^{2+}	14 μ*M*			
Cd^{2+}	20 μ*M* (IC_{50})	Endothelin	Ligand binding to solubilized receptor decreased	36
Cu^{2+} (Fe^{2+}, Ca^{2+}, Zn^{2+}, Mg^{2+}, Ni^{2+}, Mn^{2+}: NE up to 10 m*M*)	200 μ*M*			
Cd^{2+}	27 μ*M*	Gamma-amino-butyric acid (GABA)	GABA-induced inward current inhibited	37
Zn^{2+}	50 μ*M*			
Ni^{2+}	270 μ*M*			
Mn^{2+} (Ba^{2+}, Ca^{2+}, Mg^{2+}: NE)	> 10 m*M*			
Zn^{2+}	10 μ*M*[a]	Estrogen/estrogen DNA binding domain	DNA binding *in vitro* enhanced	38
Cd^{2+}				
Co^{2+} (Cu^{2+}, Ni^{2+}: NE)				
Cd^{2+}	2 mg/kg (i.p.)	Glucocorticoid	Cytosolic receptor capacity/DNA binding reduced	39
Cd^{2+} [b]	0.5–50 μ*M*	Ca^{2+}	Cytosolic Ca^{2+} increased	40
Ni^{2+} [b]	25–750 μ*M*			
Ca^{2+} [b]	1–15 m*M*			
Cd^{2+}	100 μ*M*	Ouabain	Binding inhibited	42
Pb^{2+}	100 μ*M*			

TABLE 3 (continued)

Cadmium-Receptor Interactions

Effector(s)	Effector Concentration	Receptor(s)	Effect(s)	Ref.
Hg^{2+}	1 μ*M*[c]			
Cd^{2+}	—	M3 cholinoceptor	Carbachol stimulation of inositol triphosphates inhibited	43
Cd^{2+}, estradiol	1 μ*M*	Estrogen	Estrogen mRNA and protein decreased; progesterone mRNA and protein increased; cell growth enhanced	44
Cd^{2+}	0.3 m*M*	5-HT3	Antagonist binding inhibited	45
Zn^{2+}	0.1 m*M*			
Cu^{2+} (Co^{2+}, Ni^{2+}, Ba^{2+}, Ca^{2+}, Mg^{2+}, Mn^{2+},: NE)	0.1 m*M*			

Note: NE = no effect; IC_{50} = concentration for half maximal inhibition; EC_{50} = concentration for half maximal effect.
[a] Concentration during dialysis to reconstitute receptor.
[b] In 1.25 m*M* Ca^{2+}.
[c] Concentration for maximum inhibition.

binding is abrogated by GppNHp, indicating the involvement of a guanyl nucleotide-sensitive site. Kinetic analyses suggest that nickel enhances the number of sites (but not their affinity) available for binding. Correspondingly, 2 m*M* Ni^{2+} enhances muscarinic binding by endogenous neurotransmitter in rat cerebral cortex membranes.[55] Ni^{2+} increases binding by converting low into high affinity agonist binding sites.

Pretreatment of a crude rat synaptic membrane preparation with divalent cations of Ni, Zn, Cu, or Pb results in decreased thyrotropin-releasing hormone (TRH) receptor activity.[56] Nickel is found in solubilized TRH receptors, suggesting a Ni^{2+} binding site.

The analgesic effect of morphine injected into the mouse is potentiated by chloride salts of Mn, Ni, Gd, and La.[57] It is suggested that this effect involves interaction of the metal salts with a specific opiate receptor site.

Nickel enhances also the binding capacity of arginine-vasopressin (AVP) receptors in rat brain membranes.[58] Solubilized receptors bind AVP only in the presence of divalent cations, with Ni^{2+} showing maximal stimulation followed by level of effect by Co^{2+}, Zn^{2+}, and Fe^{2+}. Mg^{2+}, Cu^{2+}, Mn^{2+}, and Ca^{2+} are without effect. Enhanced binding of agonist, but not antagonist, to AVP receptors is seen also with rat microsomal and membrane preparations treated with divalent cations.[59] In these preparations Co^{2+} is most active, followed by Mn^{2+}, Ni^{2+}, and then Mg^{2+}.

T-cell proliferation is enhanced by 100 μ*M* Pb^{2+}, Zn^{2+}, or Ni^{2+} — apparently by several different mechanisms.[60] Whereas Ni^{2+} and Zn^{2+} enhance synthesis and/or secretion of the IL-2 receptor, Pb^{2+} is without significant effect. Correspondingly, mononuclear cells from nickel-sensitive patients are observed to produce more IL-2 and have more IL-2 receptors than those from control subjects.[61]

In addition to its carcinogenic activity, nickel is of significant health effects concern because of its role as the "major cause of metal-induced contact allergy".[62] It is in this context that identification of lymphocyte subsets affected, as indicated by surface markers, and of surface receptors responsive to Ni^{2+} is important. An aid to identification of such subsets is provided by their nickel response *in vitro* — lymphocyte subsets from allergic patients show a proliferative response. CD^{4+} and $CD45RO^{+}$ cells are overrepresented in these responsive populations; $CD^{8+}CD11b^{+}$ and $CD^{4+}CD45R^{+}$ cells are under-represented.[62] Nickel-reacting T cells are found also to use predominately the T-cell receptor alpha-beta heterodimer.[61,62]

Solubilized uterine estrogen receptor proteins show high affinity for Ni^{2+} *in vitro*.[63] The interaction is of such nature as to make Ni^{2+} affinity columns useful in isolation and purification. The high affinity sites map to the DNA-binding domain primarily, with some activity also in the steroid binding domain of the receptor.[63,64] Similarly, the estradiol receptor from calf uteri has binding sites for Ni^{2+} and several related metals.[65] The metal binding modulates hormone binding, receptor transformation, and nuclear binding *in vitro* at micromolar concentrations.

Prolonged ingestion of $NiCl_2$ by rats results in altered insulin binding and response in isolated adipocytes, suggesting nickel effects at both insulin receptor and postreceptor levels.[66] Studies of insulin-dependent substrate phosphorylation and insulin receptor autophosphorylation in permeabilized rat adipocytes show Ni^{2+} to increase both activities at a concentration of 1 m*M* and to inhibit autophosphorylation at 5 m*M*.[67]

Nickel ions show a concentration-dependent inhibition of histamine-induced accumulation of inositol monophosphate in slices of both rat and guinea pig cerebral cortex.[68] The kinetics of inhibition are noncompetitive, consistent with allosteric interaction with the histamine H1 receptor.

Whereas micromolar concentrations of Zn^{2+}, Cu^{2+}, and Fe^{2+} attenuate NMDA-mediated Ca^{2+} uptake and toxicity in rat cerebellar granule cells, Ni^{2+} potentiates activity.[69] At much higher levels, Ni^{2+} is inhibitory also. These findings are significant in light of the role of NMDA-induced Ca^{2+} uptake in neurotoxic responses.[69]

High Ca^{2+} elicits increases in cytosolic Ca^{2+} in osteoclasts, apparently as an initial event in a pathway leading to inhibition of bone resorption. 50 μ*M* to 5 m*M* Ni^{2+} elicits a similar increase in cytosolic Ca^{2+} in the presence of 1.25 m*M* Ca^{2+} and 0.8 m*M* Mg^{2+}.[70] It apparently binds to a single site to stimulate release of Ca^{2+} from internal stores. In the absence of Ca^{2+} and Mg^{2+}, a higher Ni^{2+} sensitivity and a "hooked" Ni^{2+} dose response (with inhibition at higher concentrations) are seen. Addition of Ca^{2+} or Mg^{2+} shifts the dose responses to higher concentrations. In the complete absence of Ca^{2+}, nickel stimulates increases in cytosolic Ca^{2+} at concentrations as low as 5 p*M*.[70]

Chronic ingestion of Co^{2+} or Ni^{2+} in rats reduces the responses of isolated vas deferens to exogenous adenosine triphosphate (ATP), noradrenaline, or l-phenylephrine.[71] The primary targets for the metal-induced modulation of this response are not defined.

Ni^{2+}, as well as Ca^{2+}, Mg^{2+}, Zn^{2+}, and Co^{2+}, induce protein p62 phosphorylation in mouse keratinocytes, apparently by interacting directly at the membrane with a "cationic receptor".[72] Possible downstream STP events are not defined.

Nickel as well as a number of other divalent cations compete with Ca^{2+} for binding to the kinase insert domain of the alpha-platelet-derived growth factor receptor.[73] Binding results in a configurational change in this domain as well as enhanced association with the p85 N-SH2 domain, suggesting a means by which metal cation binding can modulate the first steps of the signal transduction pathway responsive to platelet derived growth factor.

The transition metal ions Ni^{2+}, Co^{2+}, and Mn^{2+} interact with the receptor-agonist complex to modulate salt taste responses by fibers of frog glossopharyngeal nerves.[74] These cations enhance responses to Ca^{2+}, Mg^{2+}, and Na^{+} when present in concentrations from 0.05 to 5.0 m*M*. The order of effectiveness at 1.0 m*M* is $Ni^{2+} > Co^{2+} > Mn^{2+}$. Ni^{2+} ions enhance the response to 0.1 m*M* Ca^{2+}, Mg^{2+}, and Na^{+} when present at 0.02 to 0.1 m*M* but inhibit in the concentration range of 0.5 to 5 m*M*.

The allergic reaction to Ni^{2+} involves responses in dermal and epidermal cells, as well as lymphocytes.[75] In this regard, 0.1 to 20 μg/ml Ni^{2+} is as effective as 10 IU/ml interferon-gamma in inducing the expression of intercellular adhesion molecule-1 (ICAM-1) on the surface of cultured keratinocytes,[75] providing another mechanistic basis for Ni^{2+}-provoked inflammation. Nickel treatment induces the production of interleukin-1 by cultured keratinocytes, also.[75] (See Table 4.)

TABLE 4

Nickel-Receptor Interactions

Effector(s)	Effector Concentration	Receptor(s)	Effect(s)	Ref.
Ni^{2+}	EC_{50} = 0.03 μ*M*	Beta-adreno-receptor	Increased coronary resistance	46, 47, 48
Co^{2+}	EC_{50} = 0.01 μ*M*			
Hg^{2+}	EC_{50} = 0.16 μ*M*	Alpha-receptor		
VO^{3-}	EC_{50} = 0.2 μ*M*			
Cu^{2+}	EC_{50} = 15 μ*M*			
Zn^{2+}	EC_{50} = 50 μ*M*			
Fe^{2+} (NE)				
Cd^{2+} (NE)				
$Ni^{2+} > Co^{2+} > Mg^{2+} > Mn^{2+}$	50–250 μ*M*	*N*-methyl-D-aspartate (NMDA)	Synaptic potential altered, NMDLA antagonized	49
Ni^{2+}, Co^{2+}, Cu^{2+}, Zn^{2+} (Li^{+}, Na^{+}, Mg^{2+}, Ca^{2+}, Cr^{3+}, Mn^{2+}, Fe^{2+}, and Fe^{3+}: NE)	1 m*M*	Benzodiazepine	Diazepam binding (to higher affinity site) enhanced	50
Ni^{2+}, Co^{2+}, Zn^{2+}	1 m*M*	Benzodiazepine	Beta-carboline-3-carboxylate ethyl ester binding inhibited	51
Ni^{2+}, Cu^{2+}, 4–5×, Ca^{2+}, Zn^{2+}, Mn^{2+}, Ba^{2+}, Mg^{2+}	—	Benzodiazepine	Flunitrazepam binding increased	52
Ni^{2+}	0.1–10 m*M*	Muscarinic acetylcholine	Agonist binding enhanced	54
Ni^{2+}	>10 m*M*	Muscarinic acetylcholine	Agonist binding inhibited	54
Co^{2+}, Mn^{2+}, Zn^{2+}	1 m*M*	Muscarinic acetylcholine	Agonist binding enhanced	54
Cd^{2+}, Hg^{2+}, Cu^{2+}	1 m*M*	Muscarinic acetylcholine	Agonist binding inhibited	54
Ni^{2+}	2 m*M*	Muscarinic acetylcholine	Agonist binding increased	55
Ni^{2+}, Cu^{2+}, Pb^{2+}, Zn^{2+}	10 m*M*	TRH	Decreased TRH binding	56
Ni^{2+}	20 μg/mouse	Mu	Morphine binding *in vivo* potentiated	57
Mn^{2+}	30 μg/mouse			
Gd^{3+}	5 μg/mouse			
La^{3+}	10 μg/mouse			
Ni^{2+}, Co^{2+}, Zn^{2+}, Fe^{2+} (Mg^{2+}, Cu^{2+}, Mn^{2+}, Ca^{2+}: NE)	100 μ*M* (EC_{50})	AVP	Binding capacity enhanced	58
$Co^{2+} > Mn^{2+}$; $Ni^{2+} > Ca^{2+}$ = control	1 m*M*	AVP	Agonist binding increased	59
$Mn^{2+} > Co^{2+} > Ca^{2+} > Mg^{2+} > Ni^{2+}$	1 m*M*	AVP	Antagonist binding decreased	59
Ni^{2+}, Zn^{2+}	100 μ*M*	IL-2	Receptor expression increased	60
Ni^{2+} ($NiSO_4$,$6H_2O$)	6.25 μg/ml	IL-2	IL-2 levels increased; IL-2 receptor levels increased	61
$Ni^{2+} > Cu^{2+} \gg Zn^{2+}$	NA	Estrogen	Metals bind receptor	63
Zn^{2+}, Ni^{2+}, Co^{2+}, Cu^{2+} (Fe^{2+}, Cd^{2+}: NE)	μM	Estradiol	Metals bind receptor; hormone binding; nuclear binding of HR complex modulated	64
Ni^{2+}	200 μg/ml (drinking water, 3 generations)	Insulin	Binding *in vitro* (adipocytes) increased	66

TABLE 4 (continued)

Nickel-Receptor Interactions

Effector(s)	Effector Concentration	Receptor(s)	Effect(s)	Ref.
Zn^{2+} Ni^{2+} Co^{2+}	100 *M* 1 m*M* 1–5 m*M*	Insulin	Insulin-dependent substrate phosphorylation as well as insulin receptor autophosphorylation increased	67
Zn^{2+} Ni^{2+}	1 m*M* 5 m*M*	Insulin	Autophosphorylation inhibited	67
Ni^{2+}	250 μ*M* (IC_{50})	Histamine H1	Histamine-induced inositol phosphate (IP_1) accumulation (in brain slices) inhibited	68
$Zn^{2+} > Cu^{2+} > Fe^{2+}$ (Cd^{2+}, Fe^{3+}, Al^{3+}: NE), Ni^{2+}	low μ*M*	*N*-methyl-D-aspartate (NMDA)	Cellular Ca^{2+} uptake attenuated (Ni^{2+} potentiates at μ*M*, inhibits at higher concentration)	69
Ni^{2+}	5 p*M*–50 μ*M* (without Ca^{2+} or Mg^{2+}); 50 μ*M*–5 m*M* (in 1.25 μ*M* Ca^{2+}, 0.8 m*M* Mg^{2+})	Ca^{2+}, Mg^{2+}	Cytosolic Ca^{2+} increased	70
Ni^{2+}, Co^{2+}	Subchronic	ATP, noradrenaline, 1-phenylephrine	Agonist response suppressed	71
Ca^{2+}, Mg^{2+}, Zn^{2+}, Ni^{2+}, Co^{2+}	0.15–0.3 m*M*	Cationic	Protein p62 phosphorylated	72
$Ho^{3+} = Zn^{2+} > Ni^{2+} > Ca^{2+} = Mn^{2+} > Mg^{2+}$, Ba^{2+}	0.5 μ*M*	Platelet derived growth factor receptor kinase insert domain	p85 N-SH2 domain associates with receptor	73
$Ni^{2+} > Co^{2+} > Mn^{2+}$	1 m*M*	Salt taste	Salt taste response enhanced	74
Ni^{2+} (in 0.1 m*M* Ca^{2+})	0.02–0.1 m*M* 0.5–5 m*M*	Salt taste	Ca^{2+} response enhanced	74
Ni^{2+}	0.1–20 μg/ml	Intercellular adhesion molecule-1 (ICAM-1)	Expression and interleukin-1 production increased	75

Note: EC_{50} = concentration for half maximal effect; TRH = thyrotropin-releasing hormone; AVP = arginine-vasopressin; IL-2 = interleukin-2; NA = not applicable; IC_{50} = concentration for half maximal inhibition; ATP = adenosine triphosphate.

1.2 Interaction of Carcinogenic Trace Metals with Ion Channels

Metal ions have the potential to interact with and modulate ion channels in a variety of fashions. Interaction with proteins that regulate channel activity can change channel responses to normal, physiological effectors of channel activity. Metals can interact with channel proteins directly to alter their configuration, resulting in noncompetitive, allosteric modulation of activity. They can also pass through the channel pore, with varying effects on channel availability for passage of physiological ions. These effects may be competitive in nature. An effect on a given type of channel can be highly specific, and metal ion interaction sites of ion channels may be highly selective for particular metal ions.[76] Because of the importance of

Ca^{2+} channel activity in signal transduction, interaction with Ca^{2+} channels provides an important opportunity for metal ion modulation of signal transduction processes. Among the carcinogenic metals, only cadmium and nickel are represented significantly in studies of metal ion modulation of ion channels.

1.2.1 Cadmium

Cadmium appears widely in ion channel literature because of its use as a Ca^{2+} channel blocker. There is evidence that in some instances Cd^{2+} passes directly through the affected Ca^{2+} channel but binds transiently in the pore, resulting in an obstruction of Ca^{2+} passage.[77] In squid neurons, the half-blocking concentration for this effect is 125 μM at 0 mV — the Cd^{2+} channel dwell time (but not entry rate) depends on transmembrane potential.[77] The kinetics of voltage-dependent, reversible blockage of Ca^{2+} channels in squid neurons indicate that Cd^{2+} does not slow channel closing, and channels can be closed when occupied by Cd^{2+}.[77] The model for Cd^{2+} response involves permeation of the Ca^{2+} channel, with transient binding to obstruct Ca^{2+} passage during the time of residence. Cd^{2+} dwell time in the pore is affected by transmembrane potential, being shorter at more negative internal potential.[77] The binding site for Cd^{2+} is predicted to be near the outer end of the Ca^{2+} pore.[77]

There is evidence that Cd^{2+} can block Ca^{2+} channels by in some instances binding to both potential-dependent and -independent sites.[78] In cat sensory neurons, Cd^{2+} blocks through a high-affinity site (dissociation constant of 16 μM at 0 mV) located within the membrane field and through interaction with a low-affinity (dissociation constant of 106 μM), potential-independent site. Cadmium block at the high affinity site is reversed by hyperpolarization.[78] Cadmium channel blockage of Ca^{2+} channels is also voltage dependent in chicken dorsal root ganglion cells,[79] in which Cd^{2+}-blocked channels are cleared by more negative membrane potentials. The block of voltage-operated calcium (VOC) channels includes T-, L-, and N-type channels.[80] Not all voltage-dependent Ca^{2+} channels are blocked, however. Only one of two types of voltage-dependent channels in smooth muscle cells from guinea pig taenia coli is susceptible to cadmium block.[81] In these cells, Cd^{2+} blocks a large conductance channel but is without effect on one of smaller conductance. The specificity for Ca^{2+} channel type is also dependent upon Cd^{2+} concentration. Cd^{2+} at 20 to 50 μM provides greater than 90% block of N and L currents in dorsal root ganglion cells, but less than a 50% block of T currents.[82] 200 μM Cd^{2+} suffices to block N, T, and L channels. In smooth muscle cell single Ca^{2+} channels, external but not internal Cd^{2+} (even at 10 mM) blocks Ca^{2+} transport.[83]

Cadmium ions may activate as well as inactivate Ca^{2+} channels. A positive effect of Cd^{2+} on Ca^{2+} channel activity is seen in neuroblastoma cells, where Cd^{2+} (or omega-conotoxin) exposure results in recruitment of functional Ca^{2+} channels to the plasma membrane.[85]

Relatively few studies have addressed the effects of Cd^{2+} on receptor-operated Ca^{2+} channels (ROC). Analysis of ATP-activated Ca^{2+} channels in smooth muscle indicates that in this instance such ROC are inhibited by neither nifedipine or cadmium.[84]

Cadmium increases active Na^+ transport through bullfrog abdominal skin,[86] but inhibits sodium channel activity in bullfrog cardiac nerves and calf Purkinje fibers.[87,88] It reduces both Ca^{2+} transport and the activity of Ca^{2+}-dependent potassium currents in Helix neurons.[89] Cadmium also shows a dose-dependent reduction of K^+ current in natural killer cells and a concomitant effect on natural killing activity toward target cell lines.[90,91] The K^+ current and killing activity are reduced by other Ca^{2+} channel blockers such as verapamil.

Agents which activate Na^+ influx in brain synaptoneurosomes also induce phosphoinositide breakdown. Cadmium ions (200 μM) block these responses with varying degrees of concordance, depending on the agents used to stimulate Na^+ influx.[92]

Cadmium increases the contractile response and enhances myotonia in the mouse diaphragm.[93] Inhibitor studies suggest these effects are due to Cd^{2+}-induced decreases in membrane Cl^- conductance.

A cadmium-induced increase in conductance occurs in cultured mouse neuroblastoma cells exposed to a low concentration (1 μ*M*) of Cd^{2+}.[94] The type of ion carried by this metal ion-activated (MIA) channel is not defined.

Cadmium also enhances conductance in renal epithelioid cells.[95] In these cells, cadmium increases K^+ conductance at a concentration that is significant relative to possible occupational/environmental acute exposure (the concentration eliciting half-maximal response is 0.2 μ*M*).

Divalent cations of cadmium as well as zinc and mercury block cardiac "fast sodium currents" as studied in Purkinje fibers and ventricular cells.[96] Depression of INa is voltage independent but is modulated by external Na^+. This block by group IIb cations is suggested to be a typical property of cardiac Na^+ channels and characterizes the cardiac as opposed to other types of Na^+ channel.[96]

A variety of divalent cations rapidly and reversibly inhibits carbachol-induced inward currents (Ins, ACh) in guinea pig ilea smooth muscles.[97] The order of effect is $Cd^{2+} \geq Ni^{2+} \gg Co^{2+} \geq Mn^{2+} \gg Mg^{2+}$. It is postulated that the cations inhibit Ins, ACh channels through direct interaction with the channel's proteins.

In addition to the sensitive response seen in renal epithelial cells, Cd^{2+} affects K^+ conductance also in a number of other experimental systems. In cat ventricular myocytes, 0.2 μ*M* Cd^{2+} increases outward K^+ current (IK),[98] reflecting again a quite sensitive and possibly significant (in toxicological terms) response.

In inside-out patches of mouse neuroblastoma cells, Cd^{2+} (and Co^{2+} and Pb^{2+}) causes the opening of small (SK) but not large (BK) conductance Ca^{2+}-activated K^+ channels.[99] An indirect effect of Cd^{2+} on K^+ channel activity is seen in vascular smooth muscle, where low concentrations of Cd^{2+} and Ni^{2+} (100 n*M*) inhibit BK (Ca) channels.[100]

Both inward and outward rectifying K^+ currents are diminished in activated B lymphocytes by exposure to 20 μ*M* Cd^{2+}.[101] It is proposed that this is a mechanism by which Cd^{2+} can elicit toxic/cell cycle effects in B lymphocytes.[101]

Cd^{2+}, Hg^{2+}, and MeHg induce a "short-circuit" current (Isc) in rat mucosal preparations.[102] This response is blocked by inhibitors of Cl^- transport.

Cadmium's effects on Ca^{2+}, Na^+, Cl^-, and K^+ transport are associated with a number of changes in cell and organismal metabolism and function. Because of the important role of Ca^{2+} transport and Ca^{2+} modulation of other ion channels in cell excitation, a number of studies have addressed the possibility of concomitant effects of Cd^{2+} on Ca^{2+} channels and nerve function and viability. Both Cd^{2+} and Pb^{2+} competitively inhibit Ca^{2+}-evoked transmitter release in frog sciatic nerve-frog muscle preparations.[103] Although Cd^{2+} inhibits Ca^{2+}-induced release at low concentration (KD = 1.7 μ*M* for Cd^{2+} binding), prolonged exposure to 100 μ*M* Cd^{2+} produces increased spontaneous release of acetylcholine.[104] Similarly, in mouse neuromuscular junctions Cd^{2+} antagonizes Ca^{2+} to reduce K^+-stimulated increases in miniature end-plate potentials (m.e.p.p.), but in the absence of K^+ it increases m.e.p.p.[105] Exposure to 50 μ*M* Cd^{2+} for 60 min in the presence of 10 m*M* KCl results in a 50% decrease in acetylcholine content of the muscle fibers.[105] Cadmium and other (Ni^{2+}, Mn^{2+}, Co^{2+}, La^{3+}, and Mg^{2+}) Ca^{2+} channel blockers also depress synaptic transmission in guinea pig olfactory cortex slices.[106]

Taken together, studies of Cd^{2+} and other heavy metal effects on nerve transmission indicate preferential effects on presynaptic, Ca^{2+}-dependent and spontaneous neurotransmitter release due to blocking of Ca^{2+} entry in presynaptic terminals.[107]

Because glutamate toxicity is mediated by Ca^{2+} uptake in neuronal cells, Cd^{2+} (10 μ*M*) completely blocks L-glutamate-induced cell lysis in a neuronal cell line.[108] Studies on inhibition of glutamate-induced increases in phosphoinositides by Cd^{2+} and other Ca^{2+} channel blockers in rat brain synaptoneurosomes show that the Ca^{2+} channels activated by glutamate and K^+ are susceptible to blocking by Cd^{2+} but not other "classical" Ca^{2+} antagonists.[109] Similarly, Ca^{2+}-induced contractile responses in rat aortic rings are modified by Cd^{2+}, Co^{2+}, and Ni^{2+} but not by a number of organic Ca^{2+} channel antagonists.[110] Also, whereas the effect of a dihydropyridine Ca^{2+} channel blocker (nitrendipine) on Ca^{2+}-induced catecholamine release from rat adrenals is polarization dependent, the effect of Cd^{2+} is not.[111]

Cd^{2+} at relatively low concentrations may act to diminish neuronal survival. In chick neurons, Cd^{2+} (ED_{50} = 5.8 μM) acts through its block of Ca^{2+} influx to inhibit potassium mediated survival.[112] This same exposure to Cd^{2+}, however, does not block survival mediated by growth factors.

In addition to modulation of Ca^{2+} channel activity by Cd^{2+}, an important aspect of Cd^{2+}-Ca^{2+} channel interactions takes the form of Cd^{2+} uptake through Ca^{2+} channels. A majority of Cd^{2+} uptake in GH4C1 cells occurs through dihydropyridine-sensitive, voltage-dependent Ca^{2+} channels.[113] Consequently, blockage of such channels with nimodipine, nifedipine, verapamil, or diltiazem reduces Cd^{2+} uptake and toxicity, as does an increase in extracellular Ca^{2+} — an increase from 20 μM to 10 mM reduces the LD_{50} for Cd^{2+} fivefold.[113] Conversely, treatment with high potassium ions and BAY K8644 enhances Cd^{2+} uptake and toxicity.

Cadmium uptake in human lung carcinoma A549 cells is reduced also by both verapamil and diamide, suggesting that uptake of Cd^{2+} by Ca^{2+} channels is dependent on sulfhydryl status (GSH/GSSG ratio).[114] The effects of diamide and verapamil on cadmium uptake are not additive when each agent is used at levels providing maximum response. In these cells, approximately one third of cadmium uptake in a 30-min exposure is diamide sensitive. Similarly, verapamil reduces 30-min Cd^{2+} uptake by 31% in primary cultures of rat hepatocytes.[115] Although both Hg^{2+} and Cd^{2+} inhibit Ca^{2+} uptake, calcium channel blockers do not affect Hg^{2+} uptake by the hepatocyte cultures. Treatment with a Ca^{2+} channel agonist (20 nM) increases Cd^{2+} accumulation in these cells by 15%.[115]

Other inorganic ion Ca^{2+} channel blockers (other than Hg^{2+} and besides Cd^{2+}) are, however, taken up in several instances via Ca^{+} channels.[116] These include Mn^{2+} and Co^{2+} but not Ni^{2+}. La^{2+} uptake in the rat melanotrophs studied is K^{+} dependent. It is concluded that Ni^{2+} therefore is among these inorganic ions for use as an agent to block Ca^{2+} channels.[116]

Further studies on Cd^{2+} uptake using the fluorescent dyes Fura 2 and Quin 2 show the importance and specificity of voltage-gated Ca^{2+} channels specifically in Cd^{2+} uptake.[117] That is, agents that modulate L-type Ca^{2+} channel activity similarly affect Cd^{2+} uptake and toxicity in cells possessing these Ca^{2+} channels but do not in cells in which they are absent.[118] (See Table 5.)

1.2.2 Nickel

As indicated above, nickel ions block voltage-activated Ca^{2+} channels but are not taken up through them.[116] As a consequence of this inhibitory effect on Ca^{2+} channel activity, Ni^{2+} has a number of effects on excitatory and nonexcitatory cells. In sino-atrial pacemaker cells, for example, excitation involving Ca^{2+} (and Na^{+}) transport through "slow membrane" channels is inhibited by Ni^{2+}, Co^{2+}, and Mn^{2+}.[119] Two types of Ca^{2+} currents, transient and long lasting, are observed in rabbit sino-atrial node cells. Ni^{2+} (40 μM) and tetramethrin block the former contribution to pacemaker potentials in these cells.[120] Exposure to 40 μM Ni^{2+} also induces bradycardia in these cells. Ni^{2+} affects Ca^{2+} currents in latent pacemaker cells of the cat right atrium.[121] In these cells, Ni^{2+} acts to inhibit ICa and T currents and to increase pacemaker cycle length.

The effects of Ni^{2+} on Ca^{2+} channel activity and the importance of Ca^{2+} currents in nerve cell excitation and neurotransmitter release have resulted in the use of Ni^{2+} as a pharmacological agent in studies of neuronal cell function.[122-134] These studies have focused much more on elucidation of the roles of various Ca^{2+} channel activities in neurotransmission than on the mechanisms and toxicological importance of Ni^{2+} effects.

Studies on the effects of 10 mM Ni^{2+} on miniature endplate currents (m.e.p.c.s) in the frog neuromuscular junction, for example, show complex and somewhat discordant effects on aspects of this response.[122] The high level of Ni^{2+} employed decreases the amplitude of

TABLE 5

Cadmium-Ion Channel Interactions

Effector(s)	Concentration	Ion Channel	Effect(s)	Ref.
Cd^{2+}	125 μ*M* (IC_{50})	Ca^{2+}	Channel blocked	77
Cd^{2+}	KD = 16 μ*M* (at 0 mV)	Ca^{2+}	Channel blocked	78
Cd^{2+}	KD = 106 μ*M*	Ca^{2+}	Channel blocked	78
Cd^{2+}	20 μ*M*	Ca^{2+}	Channel blocked	79
Cd^{2+}	1 m*M*	T-, L-, and N-type, voltage operated, Ca^{2+} channels (VOC)	Channels blocked	80
Cd^{2+}	—	High conductance Ca^{2+}	Channel blocked	81
Cd^{2+}	—	Low conductance Ca^{2+}	Channel not blocked	81
Cd^{2+}	20–50 μ*M* 20–50 μ*M*	N-, L-type Ca^{2+} T-type Ca^{2+}	90% block <50% block	82
Cd^{2+}	200 μ*M*	N-, L-, T-type Ca^{2+}	Channels blocked completely	82
Cd^{2+} (external)	KD = 36 μ*M* (at –20 mV)	Ca^{2+}	Channel blocked	83
Cd^{2+} (internal)	0.1–10 m*M*	Ca^{2+}	Channel not blocked	83
Cd^{2+}, omega-conotoxin	10–150 μ*M*	VOC, Ca^{2+}	Membrane recruitment	85
Cd^{2+}	1 m*M*	Na^+	Na^+ transport increased	86
Cd^{2+}	100–300 μ*M*	Na^+	Activity decreased	87
Cd^{2+}	182 μ*M* (IC_{50})	Na^+	Channel blocked	88
Cd^{2+}	22 μ*M* (IC_{50})	Ca^{2+} and K^+[IK(Ca)]	Transport decreased	89
Cd^{2+}	300 μ*M*	Ca^{2+}	K^+ current; NK-mediated cell killing inhibited	90, 91
Cd^{2+}	200 μ*M* (IC_{50} = 48 μ*M*)	Na^+	Na^+ flux and phosphoinositide breakdown blocked	92
Cd^{2+}	0.1 m*M* (EC_{50})	Cl^-	Myotonia; contractile response potentiated; conductance lowered	93
Cd^{2+} Pb^{2+} Al^{3+}	1 μ*M* 1–200 μ*M* 50 μ*M*	Metal ion-activated (MIA)	Inward current induced	94
Cd^{2+}	0.2 μ*M*	K^+	Conductance increased	95
$Hg^{2+} > Zn^{2+} > Cd^{2+}$	Cd^{2+} IC_{50} = 0.1 m*M* in 140 m*M* Na^+	Na^+ (INa)	Channel blocked	96
Cd^{2+} Ni^{2+} Co^{2+} Mn^{2+} Mg^{2+}	KD = 98 μ*M* KD = 131 μ*M* KD = 700 μ*M* KD = 1 m*M* KD = 10 m*M*	Ins, ACh	Carbachol-induced inward current inhibited	97
Cd^{2+}	0.2 μ*M*	K^+	Outward current increased	98
Cd^{2+}, $Pb^{2+} > Co^{2+}$ (Fe^{2+}, Mg^{2+}: NE)	1–200 μ*M*	Ca^{2+}-activated K^+	Small conductance channel (SK) opened	99
$Pb^{2+} > Co^{2+}$ (Cd^{2+}, Fe^{2+}, Mg^{2+}: NE)	1–100 μ*M*	Ca^{2+}-activated K^+	Large conductance channel (BK) opened	99
Cd^{2+}, Ni^{2+}	0.1 μ*M* (in 1 *M* Ca^{2+})	BK(Ca) K^+	Conductance inhibited	100
Cd^{2+}	20 μ*M*	K^+	Inward and outward rectifying currents reduced	101
Cd^{2+}, Hg^{2+}, MeHg	50 μ*M*	Cl^-	Increased short-circuit current (Isc)	102
Cd^{2+}	KD = 1.7 μ*M*	Ca^{2+}	Ca^{2+}-evoked neurotransmitter release inhibited	103, 104

TABLE 5 (continued)

Cadmium-Ion Channel Interactions

Effector(s)	Concentration	Ion Channel	Effect(s)	Ref.
Cd^{2+}	IC_{50} = 50 μ*M* (in 10 m*M* KCl)	Ca^{2+}	Acetylcholine in neuromuscular junction reduced	105
Cd^{2+} >	27 μ*M* (IC_{50})	Ca^{2+}	Synaptic transmission depressed	106
Ni^{2+} >	600 μ*M* (IC_{50})			
Mn^{2+} >	1.36 m*M* (IC_{50})			
Co^{2+}>	1.52 m*M* (IC_{50})			
La^{3+} >	2–5 m*M* (IC_{50})			
Mg^{2+}	8.3 m*M* (IC_{50})			
Cd^{2+}	10 μ*M*	Ca^{2+}	Ca^{2+}-dependent glutamate toxicity reduced	108
Cd^{2+}	35 μ*M* (IC_{50})	Ca^{2+}	Glutamate- and K^+-stimulated phosphoinositide formation blocked	109
Cd^{2+}	15.5 μ*M* (IC_{50})	Ca^{2+}	Ca^{2+}-induced aortic contraction attenuated	110
Cd^{2+}	5.8 μ*M* (ED_{50})	Ca^{2+}	K^+-mediated neuronal survival blocked	112
Nifedipine	20 n*M*	Ca^{2+}	Cd^{2+} transport blocked	113
Verapamil	4 μ*M*			
Diltiazem	7 μ*M*			
Ca^{2+}	10 m*M*			
Cd^{2+}	4 μ*M* (IC_{50})	Ca^{2+}	Ca^{2+} transport blocked	113
Verapamil	1 μ*M*	Ca^{2+}	Cd^{2+} uptake reduced	114
Diamide	0.5 m*M*			
Diltiazem	50–250 μ*M*	Ca^{2+}	Cd^{2+} (but not Hg^{2+}) uptake reduced	115
Verapamil	50–250 μ*M*			
Nifedipine	25–100 μ*M*			
Nitrendipine	25–100 μ*M*			
Nimodipine	30 μ*M* (LC_{50})	L-type Ca^{2+}	Cd^{2+} uptake reduced	118
BAY K8644	6 μ*M* (LC_{50})	L-type Ca^{2+}	Cd^{2+} uptake enhanced	118

Note: IC_{50} = concentration for 50% maximal inhibition; EC_{50} = concentration for 50% maximal effect; KD = equilibrium constant; ED_{50} = dose for 50% maximal effect; LC_{50} = concentration for 50% lethal effect.

the m.e.p.c.s and single channel current by 64%, shifts the reversal potential for the response, and increases channel lifetime and the time constant for m.e.p.c. decay.[122]

In contrast, 1 m*M* Ni^{2+} increases the amplitude of m.e.p.c.s in the insect neuromuscular junction and potentiates the response to L-glutamate.[123]

Nickel ions (2.5 m*M*) block both pre- and postsynaptic Ca^{2+} channels in the rat hippocampus *in vitro*, even though they respond differentially to several other Ca^{2+} channel blockers and appear to be of different types.[124]

Three types of Ca^{2+} channels (N, L, and T) can be distinguished in chick sensory neurons on the basis of voltage-dependency, ion permeability, and response to inhibitors.[125] Ni^{2+} (100 μ*M*) reduces T- but not N- or L-type currents, while Cd^{2+} (20 to 50 μ*M*) affects N and L currents to a much greater extent than those of the T-type channel.[125]

Currents through voltage-activated Ca^{2+} channels have been studied in rat dorsal ganglion X mouse neuroblastoma hybrid cells using Ba^{2+} as a charge carrier. Two inward currents, a transient and a sustained one, were identified.[126] Cd^{2+} and Ni^{2+} block the transient current at equivalent concentrations, but the threshold for Ni^{2+} block of the sustained current is 10× that for Cd^{2+} (1.0 vs. 0.1 μ*M*).[126]

Chronic treatment with 100 μ*M* Ni^{2+} produces an additional effect on Ca^{2+} channels in neuroblastoma X glioma hybrid cells.[127] That is, both fast as well as slowly inactivating Ca^{2+} channels are reversibly and completely down-regulated. This down-regulation of Ca^{2+} currents is accompanied by a decrease in binding specific for channel proteins, and the reappearance after removal of Ni^{2+} does not occur in the absence of protein synthesis. Thus, continued Ni^{2+} treatment appears to reduce the number of channel proteins present.

A lack of specificity and relative potency for blockage of T-type, voltage-activated calcium channels is indicated by the fact that eight trivalent cations tested block these channels in neural rat and human cells with significantly lower IC_{50} values than Ni^{2+} (see Table 6).[128]

Blockage of Ca^{2+} channels by exposure to pharmacological levels of Ni^{2+} affects a number of responses in excitable as well as nonexcitable cells. Included are effects on the steroidogenic response of adrenal glomerulose cells;[129] inotropic responses to histamine in atrial cells;[130] contractile responses in aortic rings,[131] smooth muscle,[132] and trachea;[133] carbachol-induced release of noradrenaline in neuroblastoma cells;[134] and sodium/calcium exchange in myocytes.[135] Ni^{2+} at pharmacological levels also serves in the short term to protect cells from damage induced by ATP depletion;[136] however, Ni^{2+} is deleterious to the cell during the period of ATP regeneration following reversal of depletion.[136]

An exception to the use of pharmacological levels of Ni^{2+} to elicit cellular or physiological responses via blockage of Ca^{2+} channels is the use of 100 n*M* Ni^{2+} (or Cd^{2+}; see above, also) to modulate significantly the activity of Ca^{2+}-activated K^{+} channels in smooth muscle of bovine mesenteric arteries.[100] It is suggested that this response represents a mechanism by which these divalent cations affect vascular tension.

Intracellular Ni^{2+} markedly potentiates cGMP activation of rod photoreceptor channels expressed in *Xenopus* oocytes.[137] The Ni^{2+} binding site is a histidine residue (H420) in what is believed to be the intracellular mouth of the rod channel. It is proposed that this binding stabilizes the open configuration of the channel.[137] A difference in the effects of external and internal nickel on the activity of cGMP-activated channels has been demonstrated in salamander retinal rods.[138] Whereas blocking kinetics for internal Ni^{2+} are unaffected by the fraction of channels open, blocking by external nickel occurs more rapidly when the channels are open.[138] Low concentrations of cytoplasmic Ni^{2+} potentiate the channel response to cGMP in this system also, and the potentiation is ascribed to an increased rate of cGMP binding.

1.3 G Protein Interactions

Despite the importance of G proteins as signal transducers in many signal transduction pathways leading to alterations in gene expression, few if any studies have explored the interactions of carcinogenic trace metals with G proteins. None have been reported that involve arsenicals, beryllium (except as BeF_3^-),[139] nickel, or chromium ions.

1.3.1 Cadmium

It has been suggested that the Cd^{2+} signal that provokes an increase in cytosolic inositol triphosphate (IP_3) and Ca^{2+} in human fibroblasts is transduced via G protein complexes.[140] The basis for this hypothesis is that Cd^{2+}-induced, transient increases in cytosolic Ca^{2+} are not sensitive to genistein, an inhibitor of tyrosine kinases, but are mimicked by bradykinin, which acts through G proteins to mobilize Ca^{2+}. There is no indication that Cd^{2+} interacts directly with a G protein. Rather, it appears to interact with a membrane receptor that presumably is G protein associated.

It is evident that the involvement of direct interactions of carcinogenic metals with G proteins in metal induced alterations in gene expression or metabolism is an area not explored. Further, there is but one suggestion for G proteins serving as a transducer of signals generated by metal-receptor interactions.

TABLE 6

Nickel-Ion Channel Interactions

Effector(s)	Concentration	Ion Channel	Effect(s)	Ref.
Ni^{2+}, Co^{2+}, Mn^{2+}	1 m*M*	Ca^{2+}	Pacemaker cell excitation inhibited	119
Ni^{2+}	40 μ*M*	Ca^{2+}	Pacemaker transient current blocked	120
Tetramethrin	0.1 μ*M*			
Ni^{2+}	40 μ*M*	ICa, T	Channel blocked, pacemaker cycle length increased	121
Ni^{2+}	10 m*M*	Ca^{2+}	Neuromuscular junction miniature end-plate currents decreased	122
Ni^{2+}	1 m*M*	—	Neuromuscular junction glutamate potential, m.e.p.c.s. increased	123
Ni^{2+}	2.5 m*M*	Ca^{2+}	Pre- and postsynaptic channel blocks	124
Ni^{2+}	100 μ*M*	Ca^{2+} (T-type)	Current reduced strongly	125
Ni^{2+}	100 μ*M*	Ca^{2+} (N-, L-type)	Current not affected	125
Cd^{2+}	20–50 μ*M*	N, L (Ca^{2+})	> 90% block	125
Cd^{2+}	20–50 μ*M*	T (Ca^{2+})	< 50% block	125
Cd^{2+}	200 μ*M*	N, T, L (Ca^{2+})	Blocked	125
Ni^{2+}	1 μ*M*	Ca^{2+} (voltage activated)	Sustained Ba^{2+} current blocked	126
Cd^{2+}	0.1 μ*M*			
Ni^{2+}	IC_{50} = 64 (major), 1.1 (minor) μ*M*	Ca^{2+} (voltage activated)	Transient Ba^{2+} current blocked	126
Cd^{2+}	IC_{50} = 82 (major), 0.3 (minor) μ*M*			
Ni^{2+}	100 μ*M*	Ca^{2+}	Channel down-regulated	127
Ho^{3+}	IC_{50} = 0.1 μ*M*	Ca^{2+}, T-type, voltage gated	Channel blocked	128
Y^{3+}	IC_{50} = 0.12			
Yb^{3+}	IC_{50} = 0.12			
Er^{3+}	IC_{50} = 0.15			
Gd^{3+}	IC_{50} = 0.27			
Nd^{3+}	IC_{50} = 0.43			
Ce^{3+}	IC_{50} = 0.73			
La^{3+}	IC_{50} = 1.0			
Ni^{2+}	IC_{50} = 5.7			
Ni^{2+}	0.2–0.5 m*M*	Ca^{2+}	Inotropic histamine response repressed	130
Ni^{2+}	0.5–1 m*M*	Ca^{2+}	Ca^{2+}-induced aortic contraction attenuated	131
Ni^{2+}	360 μ*M*	Ca^{2+}	Endothelin-induced smooth muscle contraction inhibited	132
Ni^{2+}	1 m*M*	Ca^{2+}	Acetylcholine-induced tracheal contraction blocked	133
Ni^{2+}	2 m*M*	Ca^{2+}	Carbachol- and depolarization-induced release of noradrenaline inhibited	134
Ni^{2+}	5 m*M*	Ca^{2+}	Na/Ca exchange blocked	135
Ni^{2+}	0.5–1 m*M*	Ca^{2+}	Cell damage due to ATP depletion reduced	136
Mg^{2+}	3–10 m*M*			
Ni^{2+}	10 μ*M*	Rod	Rod photoreceptor ion channel activated	137
Ni^{2+}	100 μ*M* (internal)	cGMP activated	Open, closed channels equally blocked	138
Ni^{2+}	40–190 μ*M* (external)	cGMP activated	Open channels preferentially blocked	138
Ni^{2+}	1–10 μ*M*	cGMP activated	cGMP response potentiated	138

Note: IC_{50} = concentration for half maximal inhibition; ATP = adenosine triphosphate.

1.4 Metal-Kinase/Phosphatase Interactions

Because of the ubiquitous role of phosphorylation/dephosphorylation reactions in signal transduction (and regulation of enzyme activity), STP kinases and phosphatases represent important downstream targets for modulatory interactions with signal transduction pathways.

1.4.1 *Arsenic*

Alterations in STP kinase/phosphatase activities are important in terms of effects on gene expression. In this context, it has been reported that both heat shock and arsenite exposure produce HeLa cell lysates with altered kinase activity toward the C-terminal domain peptide (CTD) of RNA polymerase II and toward histone H1.[141] It is suggested that arsenite (stress) activation of CTD kinase might, through hyperphosphorylation, alter the specificity of RNA polymerase and contribute to the onset of the preferential transcription of heat-shock genes.[142] The accumulation of hyperphosphorylated RNA polymerase occurs in the presence of transcriptional inhibition, indicating that synthesis of new proteins is not requisite.[143]

Arsenite as well as heat shock, phorbol esters, or tumor necrosis factor (TNF) induce phosphorylation of HSP27 and increase the level of HSP27 kinase in HeLa cells.[144] This kinase activity purifies with 45- and 54-kDa, mitogen-activated, protein kinase-activated protein (MAPKAP) kinase-2[145] and is induced also by not only heat shock or TNF but also H_2O_2 and growth factors.[145]

Elevation of adenosine monophosphate (AMP) due to ATP depletion results in the activation of a kinase responsible for inactivating enzymes in key metabolic paths.[146] Heat shock and arsenite treatment of rat hepatocytes result in elevation of both AMP and this AMP-activated kinase.[147]

Interaction of arsenicals with protein phosphatases may be a salient effect of arsenic responses *in vivo*. Arsenate-resistant bacterial mutants, for example, are found to be constitutive for alkaline phosphatase synthesis.[148] Arsenate acts as a phosphate analogue at the active site of alkaline phosphatase,[149] and arsenate and arsenite are competitive inhibitors of its hydrolysis of *p*-nitrophenyl phosphate.[150] That such inhibition of phosphatase activity can modify signal transduction is indicated by the fact that arsenite mimics TNF in inducing specific protein phosphorylation in primary human fibroblasts.[151] P32-labeled isoforms of HSP27 are not dephosphorylated during cold chase when cells are exposed either to TNF or to phosphatase inhibitors such as arsenite or okadaic acid,[151] and inhibitors of protein kinases do not suffice to block TNF/arsenite-induced increases in phosphorylated isoforms, indicating that the changes are due primarily to phosphatase inhibition. Arsenite also inhibits dephosphorylation of HSP27 *in vitro*.[151] The residues of HSP27 that are phosphorylated to a greater extent after heat or arsenite treatment of HeLa cells have been identified with serine residues 78 and 82.[151]

As will be discussed below, treatment of cells with arsenite and other stressors such as heat results in increased synthesis of heat shock proteins. There is a corresponding increase in the phosphorylation of heat shock transcription factor-1 following exposure to 200 μM arsenite, resulting in the transactivation of heat shock mRNA synthesis.[152] Whether this increase is due to altered kinase and/or phosphatase activity was not reported. (See Table 7.)

1.4.2 *Beryllium*

Be^{2+} inhibits rat liver cAMP-independent casein kinase 1 *in vitro* and *in vivo*.[153] *In vitro*, the cytoplasmic and nuclear forms are inhibited with KIs of 2.5 and 29 μM, respectively. There is not a corresponding effect on casein kinase 2 or cAMP-dependent protein kinase. Administering rats an LD_{50} dose of Be^{2+} results in an approximately 50% reduction in the activities of kinase 1 isolated from partially hepatectomized livers.[153] Casein kinase 1 of rat liver cytosol contains a 31-kDa protein for which the phosphorylation by casein kinase S is

TABLE 7

Arsenic-Kinase/Phosphatase Interactions

Effector(s)	Concentration	Kinase/Phosphorylase	Effect(s)	Ref.
Arsenite Heat	200–900 μ*M* (1 hr) 45°C (45 min)	Kinase for RNA polymerase II, histone H1	Activity stimulated; polymerase II specificity altered	141–143
Arsenite Heat (NGF, EGF: NE)	500 μ*M* (30 min) 45°C (15 min)	MAPKAP Kinase 2	HSP27 phosphorylation increased; kinase level increased	144–145
Arsenite, heat	—	AMP-activated protein kinase	Activation	146
Arsenate, arsenite	2 eq./dimer (purified enzyme)	Alkaline phosphatase	Inhibition	149
Arsenite	—	Phosphatase	Inhibition	150
Arsenite	200 μ*M*	Kinase/phosphatase	Heat shock, transcription factor-1 phosphorylation increased	152

Note: NGF = nerve growth factor; EGF = epidermal growth factor; NE = no effect; MAPKAP = mitogen-activated, protein kinase-activated protein.

TABLE 8

Beryllium Kinase/Phosphatase Interactions

Effector(s)	Concentration	Kinase/Phosphatases	Effect(s)	Ref.
Be^{2+}	2.5 μ*M* (Ki)	Cytoplasmic casein kinase 1	Activity inhibited	153
Be^{2+}	29 μ*M* (Ki)	Nuclear casein kinase 1	Activity inhibited	153
Be^{2+}	0.1 m*M*	Casein kinase S	Phosphorylation of casein kinase I inhibited	154
Be^{2+}	5 m*M*	Phosphatidate phosphohydrolase	Activity inhibited	158
Be^{2+}	1.1 μ*M* (Ki)	*p*-Nitrophenyl phosphatase	Activity inhibited	160

prevented by beryllium.[154] Fluoride complexes of beryllium inhibit kinases and phosphatases, as well as G proteins.[155-157] BeF_3 interacts as a phosphate analogue, interrupting ATP hydrolysis cycles.[156] Be^{2+} itself also inhibits a variety of phosphatases, phosphohydrolases, and ATPase.[158-162] (See Table 8.)

1.4.3 Cadmium

Cadmium interacts with both kinases and kinase-dependent STPs within a broad range of concentrations and with effects on gene expression, protein synthesis, and cell function. For example, Cd^{2+} inhibits lymphocyte activation and proliferation, with concomitant inhibition of phorbol ester binding to protein kinase C and altered metabolism of phosphatidylinositols.[163] 10 μ*M* Cd^{2+} inhibits the activity and ligand binding of protein kinase C (PKC) *in vitro*[164] and directly inhibits both the activity of purified PKC and PKC-mediated responses in intact macrophages.[165] However, Cd^{2+} may mimic, enhance, or antagonize Ca^{2+} stimulation of protein kinase C or myosin light chain kinase, depending on concentration,[166] and at micromolar concentrations it enhances the binding of phorbol ester to PKC in cells and lysates.[167] Cd^{2+} also potentiates the effects of phorbol esters on nuclear PKC activity in mouse fibroblasts.[168] It stimulates the binding of PKC to a nuclear protein of 105 kDa.[168] It is proposed that in this instance Cd^{2+} binds to a Zn^{2+} binding site in PKC to effect a change in its protein binding site.

Increased phosphorylation of human red cell proteins by Cd^{2+} and Hg^{2+} is sensitive to inhibition of cAMP-dependent protein kinase, indicating a stimulation of this kinase by the cations.[169] Cadmium and trivalent lanthanides mimic the stimulatory effect of Ca^{2+} on muscle phosphorylase kinase.[170] The stimulation by Cd^{2+} is reversed at higher concentrations.

Cd^{2+}, Hg^{2+}, and Pb^{2+} inhibit protein synthesis in reticulocyte lysates within the concentration range of 2.25 to 10 μM.[171] This inhibition is ascribed in part to phosphorylation of eukaryotic initiation factor 2 (eIF2) catalyze by eIF2 alpha kinase.[171,172] The primary site of metal interaction leading to alpha kinase activation is not defined.

Both Cd^{2+} and Hg^{2+} inhibit the activity of Na^+-K^+-ATPase *in vitro*.[173] In the case of Cd^{2+}, this inhibition is ascribed to direct interaction with and inhibition of the phosphatase moiety (K^+-*p*-nitrophenyl phosphatase) of Na^+-K^+-ATPase.

A most sensitive response to Cd^{2+} obtains in its effects on Ca^{2+}-induced curvature reversal in rat sperm.[174] This calcium response is blocked completely by 2×10^{-10} M Cd^{2+}.

Cd^{2+}, Zn^{2+}, and Hg^{2+} stimulate both glucose transport and cAMP phosphodiesterase activity of rat adipocytes.[175] The increased glucose transport is ascribed to transfer of glucose transporters to the plasma membrane. Administration of Cd^{2+} to rats at levels that result in overt toxic responses (hepatotoxicity) causes suppression of DNA synthesis and thymidine kinase activity following partial hepatectomy.[176]

Cadmium at 0.5 to 4 μM induces accumulation of c-*myc* mRNA in cultured fibroblasts.[177-179] The response of this induction to kinase inhibitors of differing specificities indicates that the activity of PKC is requisite, indicating that Cd^{2+} acts upstream of, and through, PKC to induce specific c-*myc* gene expression.[179] Cd^{2+} induces accumulation of c-*fos*, also.[178-180] The induction of c-*fos* by Cd^{2+} (and arsenite and heat shock) is mediated by the serum response element and may involve activation of casein kinase 2.[180]

Calcium calmodulin represents a target for primary intersection of Cd^{2+} with STPs and is immediately "upstream" of some kinases and phosphatases. Cadmium readily replaces Ca^{2+} in calmodulin *in vitro* and *in vivo*.[181-184] Cd^{2+} binding favors complex formation with troponin or phosphodiesterase.[181] In line with the proposal that Cd^{2+} activation/binding of CaM may alter normal Ca^{2+} regulation of CaM activity,[181] Cd^{2+} inhibition of (Ca^{2+} + Mg^{2+}) ATPase appears to be due to Cd^{2+} modulation of CaM activity.[182]

Cd^{2+} interaction with CaM closely mimics that of Ca^{2+}. The site preferences for binding are the same,[183] and the effects of Ca^{2+} and Cd^{2+} on CaM secondary and tertiary structures are identical.[184] Cd^{2+}-induced conformational changes are as extensive as those due to Ca^{2+}.[185]

The consequences of Cd^{2+} replacement for Ca^{2+} in CaM may be stimulatory as well as inhibitory. Maximal stimulation of phosphorylase kinase by CaM is greater with Cd^{2+} than with Ca^{2+} or Tb^{3+}.[170] *In vitro*, Cd^{2+} and CaM effect 80% of the maximal phosphodiesterase activity induced by Ca^{2+} and CaM.[186] *In vivo*, calmodulin inhibitors protect mice from Cd^{2+}-induced testicular damage,[187] and the suggestion is made therefore that calmodulin-dependent pathways are involved in the cytotoxic responses to Cd^{2+}. Cd^{2+}-induced inhibition of ciliary activity and epithelial cell damage in trachea cultures is mimicked by, and is additive with, inhibition of CaM activity.[188]

Although both Cd^{2+} and Ca^{2+} CaM can activate phosphodiesterase *in vitro*, the Cd^{2+} complex has lower activity, and treatment of cells with Cd^{2+} may result in reduced CaM activity with consequent effects on metabolic and signal transduction pathways.[189] An example is Cd^{2+} inhibition of Ca^{2+}-ATPase in monkey brain membranes due to interaction not of Cd^{2+} with the enzyme, but with CaM.[190] Similarly, Cd^{2+} can enhance muscle kinase activity induced by partially Ca^{2+}-replete CaM, but higher concentrations of Cd^{2+} cause inhibition.[191] Another *in vivo* Cd^{2+} response mediated by CaM is an increase in dopamine in the brains of mice administered Cd^{2+} IVT.[192] Chronic exposure to Cd^{2+} results also in reduced CaM, (Ca^{2+} + Mg^{2+}) ATPase, and PDE activity in rat brain.[193]

Troponin C (TnC) is also a potential primary target for Cd^{2+}-induced changes in kinases and phosphatases. Ca^{2+}, Pb^{2+}, and Cd^{2+} activation of skeletal muscle Ca^{2+}-ATPase is mediated by TnC.[194] This activity can, by stimulating myofibrillar precipitation, constitute a potential mechanism for toxicity.[194] (See Table 9.)

1.4.4 *Chromium*

With the exception of studies that employ Cr-ATP as a Mg-ATP analogue to elucidate aspects of ATPase, kinase, and phosphatase structure and function,[195-202] there are no reports of Cr interactions with kinases or phosphatases. Chromium activates calmodulin to a slight degree at micromolar concentrations but to a much greater extent (up to 76% of maximum possible activation) at nanomolar concentrations,[203] comparable to those which might be obtained in exposed workers. It is suggested, therefore, that chromium-calmodulin interactions may be relevant in chromium toxicity. (See Table 10.)

1.4.5 *Nickel*

A nickel-phosphatase interaction suggested to be of possible physiological significance is that of Ni^{2+} with the Ca^{2+} and CaM-dependent phosphoprotein phosphatase calcineurin.[204-213] Its activation by Ni^{2+} is not dependent on calmodulin, but the presence of CaM enhances activation 20-fold.[204] Nickel induces a conformational change in calcineurin.[204,205] The beta subunit of calcineurin "plays a critical role" in Ni^{2+}-stimulated phosphatase activity.[205] Sulfhydryl inactivation analyses indicate that a single cysteine residue in the catalytic subunit of calcineurin is essential for Ni^{2+}-induced conformational change.[209]

Nickel binds calmodulin, also.[214] The physiological and/or pharmacological effects of this binding are not defined, but injection of a single s.c. dose of Ni^{2+} into rats results in a transient increase in membrane-bound calmodulin in the kidney.[215]

Ni^{2+}, Zn^{2+}, and Co^{2+} at pharmacological concentrations stimulate insulin receptor auto- and substrate phosphorylation in permeabilized rat adipocytes.[216] With this exception, N^{2+}-kinase interactions are unexplored. (See Table 11.)

1.5 Specific Gene Expression Induced by Carcinogenic Trace Elements

1.5.1 *Arsenic*

Arsenite induces the synthesis of specific proteins electrophoretically identical to many whose synthesis is induced by agents as diverse as heat shock, copper chelating agents, and the metal ions of copper, cadmium, mercury, and zinc.[217] It has been proposed that these agents share a common sulfhydryl-containing target.

Studies of the response to arsenite by cultured chick fibroblasts showed that the mRNAs for the induced proteins are increased, also, and that the synthesis of the proteins is blocked by actinomycin.[218] The proteins closely resemble those (heat shock proteins) that are induced in *Drosophila* by heat shock, although the heat shock proteins induced by heat and arsenite partition differently between nucleus and cytoplasm.[219] Several of the heat shock proteins also appear *in vivo* in response to stress. Injection of sodium arsenite, or a 3°C increase in temperature, induces accumulation of 74 K HSP mRNA in rabbit kidney, heart, and liver but not brain.[220]

Arsenate as well as arsenite induce heat shock proteins in rat myoblasts, including several not seen after heat shock.[221] On the other hand, heat shock but not arsenite induces increased synthesis of histone H2B in cultured *Drosophila* cells.[222] Similar but not identical patterns of heat shock protein synthesis in response to heat shock or arsenite are also seen in myotube cultures.[223]

TABLE 9

Cd^{2+}-Kinase/Phosphatase Interactions

Effector(s)	Concentration	Kinases/Phosphatases	Effect(s)	Ref.
Cd^{2+}	—	Ca^{2+}-phospholipid-dependent protein kinase (PL-Ca-PK)	Phorbol ester binding reduced, PI metabolism altered	163
Cd^{2+}, Cu^{2+}, Hg^{2+}, Zn^{2+}	10 μM	PKC	Activity and ligand binding inhibited	164
Auranofin	0.1μM	PKC	Activity inhibited	165
Cd^{2+}, Pb^{2+}	100 μM			
Hg^{2+}	1 μM			
Cd^{2+}	—	MLCK, PKC, PL-Ca-PK	Cd^{2+} responses modulated biphasically	166
Zn^{2+}, Cd^{2+}	μM	PKC	Phorbol ester binding enhanced	167
Cd^{2+}	100 pM	Nuclear PKC	Binding of 105-kDa protein enhanced	168
Cd^{2+}, Hg^{2+}	mM	cAMP-dependent protein kinase (PKA)	Phosphorylation of membrane proteins	169
Cd^{2+}	~1–30 μM (stimulate); >100 μM (inhibit)	Phosphorylase kinase	Activity modulated biphasically	170
Tb^{3+}, Gd^{3+}, Pr^{3+}, Ce^{3+}	<1 mM	Phosphorylase kinase	Stimulation	170
Cd^{2+}	IC_{50} = 2.25–10 μM	Eucaryotic initiation factor-2 (eIF-2) alpha kinase	eIF2 phosphorylated, protein synthesis inhibited	171, 172
Hg^{2+}	IC_{50} = 2.25–10 μM			
Pb^{2+}	IC_{50} = 2.25–10 μM			
Cu^{2+}	IC_{50} = 40 μM			
Fe^{3+}	IC_{50} = 250 μM			
Zn^{2+}	IC_{50} = 300 μM			
Cd^{2+}	IC_{50} = 32 μM	Na^{+}-K^{+}	Activity inhibited	173
CH_3HgCl	IC_{50} = 6 μM	ATPase		
Cd^{2+}	2 × 10^{-10} μM (0.1 nM)	cAMP kinase A	Ca^{2+}-induced flagellar curvature blocked	174
Zn^{2+}, Cd^{2+}, Hg^{2+}	1 mM	cAMP	Activity stimulated	175
	0.1 mM	Phosphodiesterase		
Cd^{2+}	2.5 mg/kg (i.p.)	Liver thymidine kinase	Activity reduced	176
Cd^{2+}	0.5–4 μM	PKC	c-*myc* mRNA accumulation	179
Cd^{2+}	300 mM	Casein kinase 2	c-*fos* mRNA accumulation	180
Cd^{2+}	4.5 μM (KD for CaM binding)	Phosphodiesterase	Complex with CaM induced	181
Cd^{2+}	μM	(Ca^{2+} + Mg^{2+}) ATPase	Activity inhibited (via CaM) noncompetitively	182
Cd^{2+} + CaM	100 μM	Phosphodiesterase	Activation to 80% of Ca^{2+} CaM	186
Hg^{2+} > Cd^{2+} > Pb^{2+} > Mn^{2+} > Al^{3+}	3.0, 6.6, 7.9, 9.5, 12.3 μM (IC_{50})	Ca^{2+} ATPase	Activity (CaM dependent) inhibited	190

Note: PKC = protein kinase C; MLCK = myosin light chain kinase; IC_{50} = concentration for half maximal inhibition.

TABLE 10

Chromium-Calmodulin Interactions

Effector	Concentration	Effect	Ref.
Cr	μM	Slight activation	203
Cr	nM	Activation to 76% maximal level	203

TABLE 11

Nickel-Kinase/Phosphatase Interactions

Effector	Concentration/Kinetic Constants	Kinase/Phosphatase	Effect	Ref.
Ni^{2+} Ca^{2+}	K_m = 4.2 μM K_m = 2.2 μM (with histone H1 substrate)	Calcineurin (phosphoprotein-phosphatase)	Activation	204
Ni^{2+} + CaM	K_m = 0.9 μM (phosphotyrosyl muscle myosin substrate)	Calcineurin	Activation	210
Ni^{2+}	95 μmol/kg (s.c.)	Calmodulin (phosphatase activator)	Membrane fraction increased	215
Zn^{2+}	100 μM 1 mM 1–5 mM	Insulin receptor	Substrate and autophosphorylation increased	216

Heat shock and arsenite induce overlapping but distinct patterns of stress protein synthesis in *Xenopus* epithelial cells.[224] For mRNA and protein synthesis of the 68- to 73- and 29- to 31-kDa heat shock proteins the response to heat shock and arsenite is synergistic.

Tumor promoters as well as heat shock or arsenite induce the synthesis of a 32-kDa protein in BALB/c cells.[225] This 32- (or 34-) kDa protein is synthesized prior or parallel to other heat shock proteins in human (or murine) melanoma cells in response to arsenite, metals, or thiol reagents but not to the amino acid analogues L-azetidine-2-carboxylic acid or L-canavanine.[226]

Arsenite, Cd^{2+}, and heat shock induce synthesis of mRNA for heat shock proteins in plants as well as animal cells and tissues.[227] Arsenite and heat shock treatment both induce cytoskeletal alterations in cultured neuroblastoma cells.[228] It is suggested that reorganization of the cytoskeleton during stress requires heat shock protein synthesis, as blocking gene expression prevents reorganization.[228]

Inducers of the heat shock-like stress response, including arsenite, induce the accumulation of c-*fos* mRNA and protein in HeLa cells.[229] The increase appears due to both post-transcriptional stabilization of the mRNA, possibly due to protein synthesis inhibition,[229,230] and to activation of transcription.[236] Inhibition of Na^+/H^+ exchange partially inhibits the response.[230]

Regions 1 and 2 of the immediate early (IE) transcription unit of the human cytomegalovirus, stably incorporated into rat 9G cells, are transcriptionally activated by heat shock or arsenite.[231] Stably integrated genes under control of the HIV long terminal repeat (LTR) are also induced by inhibition of protein synthesis, heat shock, arsenite, and other stress agents.[232] In contrast, heat shock and arsenite inhibit expression of SV40 in infected CV1 cells, apparently at the level of transcription.[233]

Synthesis of the major, 32-kDa stress protein is induced in human fibroblasts by UV radiation and hydrogen peroxide, as well as by arsenite.[234] This induced protein has been identified with heme oxygenase.[234] Induction of this protein may constitute part of a general response to oxidative stress. An increase in heme oxygenase following oxidative stress or arsenite is due to increased transcription.[235]

The *mdr* (multiple drug resistance) gene product plays important roles in drug resistance following chemotherapy in tumor cells and in detoxification of xenobiotics in normal cells. The membrane *p*-glycoprotein coded for by this gene functions by pumping a variety of drugs from the cell by an active transport process. Expression of the human *mdr* gene (*MDR*1) is regulated in cultured rodent and human carcinoma cells by arsenite as well as cadmium chloride and heat shock.[237] Arsenite-induced *MDR*1 gene expression is lost when a 60-bp promoter region containing two heat shock elements is deleted.[238]

Arsenite (As^{3+}) but not arsenate (As^{5+}) induces metallothionein (MT) synthesis by rat liver *in vivo*.[239] As^{3+} also enhances Zn^{2+}-induced MT expression but without causing an increase in MT mRNA.[239]

Post-translational modifications of proteins may also occur following exposure to arsenite or heat shock. These treatments result in changes in the patterns of core histone methylation and acetylation in cultured *Drosophila* cells.[240]

The STPs and transcriptional regulators involved in arsenite-induced specific gene expression are defined only partially. At the DNA level, sequences involved in the heat shock response have been identified.[241] These "heat shock elements" are bound by transcriptionally activating factors. Agents which activate the HSP genes also cause increased lipid perioxidation, leading to the suggestion that the initial and common event in induction of specific gene expression is oxidative damage.[242]

Introduction of a gene construct containing the *Drosophila* heat shock promoter results in inducibility of that gene by either heat shock or arsenite, indicating a common DNA (enhancer-promoter) target for these agents.[243] Nonetheless, the temporal patterns of gene expression following heat shock or arsenite differ,[244] and the transcripts decay at differing rates following induction of 31-kDa proteins in rat cells by heat shock or arsenite.[245] In tetrahymena, although the patterns of protein synthesis induced by heat shock or arsenite are largely (but not completely) similar, the effects on normal protein synthesis are quite different. Arsenite is not as repressive.[246] Differential effects on splicing of mRNA *in vitro* are also seen. Inhibition is seen in extracts following heat shock but not arsenite treatment of HeLa cells.[247]

Upstream events are not well defined for either heat shock- or arsenite-induced specific gene expression. RNA polymerase C-terminal-domain kinase activity is increased in extracts of heat- or arsenite-treated HeLa cells, indicating a potential mechanism for alteration in polymerase II activity;[248-250] however, a cause-effect relationship between an increase in this kinase activity and heat shock protein synthesis has not been shown.

In the instance of arsenite-induced HSP70 synthesis, some aspects of transcriptional control are defined. A negative regulator, CHBF, binds the heat shock element in DNA constitutively. Arsenite, through an undefined path, activates the binding of the heat shock factor (HSF) to the heat shock element and also serves to dissociate CHBF from it.[251] (See Table 12.)

1.5.2 Beryllium

The effects of beryllium on gene expression have been addressed in relatively few studies, despite the fact that it has been known for some time that a low, perhaps toxicologically relevant, concentration of Be^{2+} can affect regulation of gene expression.[252] That is, exposure of cultured hepatoma cells to 1 μM $BeSO_4$ reduces glucocorticoid induction of tyrosine transaminase by approximately half.[252] A significantly higher concentration of beryllium (50 μM) affects DNA synthesis and cell division in cultured cells without effects on synthesis of the division-associated protooncogene c-*myc*.[253] Be^{2+} was observed actually to prevent down-regulation of c-*myc* expression following serum stimulation.[253] (See Table 13.)

TABLE 12

Arsenical-Induced Gene Expression

Effector(s)	Concentration	Gene(s)	Effect(s)	Ref.
Arsenite	20 μM	—	3 to 4 specific proteins synthesized	217
Zn^{2+}	25 μM			
Hg^{2+}	10 μM			
Cu^{2+}	100 μM			
Cd^{2+}	10 μM			
Disulfiram	2 × 10^{-7} μM			
Heat shock (Co^{2+}, Ni^{2+}, Fe^{2+}, Fe^{3+}, Mn^{2+}, Pb^{2+}: NE up to 50 mM)	45°C			
Arsenite, heat shock	50 μM	—	mRNAs for 27-, 35-, 73-, 89-kDa stress-specific proteins increased	218
Arsenite	0.8 mg/kg	Heat shock proteins (HSP)	mRNA for 74-kDa protein in liver, heart, and kidney increased	220
LSD	100 g/kg			
Phorbol esters (OAG)	100 g/ml	HSP	Synthesis of 32-kDa protein increased	225
Heat shock	45°C (10 min)			
Arsenite, $CdCl_2$	100 μM (2 hr)			
Arsenite	6–96 μM	32- (34-) kDa stress protein	Synthesis increased	226
Cu^{2+}, heat shock	1 mM			
A23187 (L-azetidine-2-carboxylate, L-canavanine: NE)	43°C (30–60 min) 10^{-8}–10^{-3} μM			
Zn^{2+}	50–100 μM			
Cd^{2+}	10–100 μM			
OAG (1-oleoyl-2-acetyl glycerol)				
Heat shock	33–35°C	87-, 73-, 70-, 54-, 31-, 30-kDa HSP	mRNA and protein levels increased	224
Arsenite, Zn^{2+}, Cd^{2+}, Cu^{2+}	25–100 μM	73-, 70-kDa HSP	mRNA and protein levels increased	224
Heat shock, arsenite	30°C, 10 μM	68- to 75-, 29- to 31-kDa HSP	mRNA and protein levels increased synergistically	224
Arsenite	75 μM (1 hr)	HSP	Cytoskeletal alterations	228
Heat shock	43°C (30 min)			
Arsenite	80 μM	c-*fos*	mRNA and protein increased	229, 230
Cd^{2+} + Zn^{2+}	20–200 μM			
Heat shock	42–46°C			
Heat shock	42°C (12 min)	Immediate early transcription unit of integrated human cytomegalovirus	Transcriptional activation	231
Arsenite	200 μM (2 hr)			
Arsenite	50 μM (30 min)	Heme oxygenase	mRNA, protein synthesis increased	234, 235
H_2O_2	100 μM (30 min)			
UVA	2 × 10^5 J/m^2			
Cd^{2+}	100 μM (3 hr)			
Iodoacetamide	50 μM (30 min)			
Menadione	500 μM (30 min)			
Arsenite (arsenate: NE)	—	Metallothionein	Expression increased	239
Arsenite, heat shock	—	Histone	Histone modification altered	240
Arsenite vs. heat shock	50 μM, 33°C	HSP	Temporal patterns of HSP synthesis differ	244

Note: NE = no effect.

TABLE 13
Beryllium Effects on Gene Expression

Effector(s)	Concentration	Gene(s)	Effect(s)	Ref.
Be^{2+}	1 μM (IC_{50})	Tyrosine aminotransferase	Induction by glucocorticoids reduced	252
Be^{2+}	50 μM	c-*myc*	Down-regulation (following serum stimulation) reduced	253

Note: IC_{50} = concentration for half maximal inhibition.

1.5.3 Cadmium

In contrast to the paucity of studies on the effects of Be^{2+} on gene expression, there exists a significant body of literature on Cd^{2+} modulation of specific gene expression. In particular, the effects of Cd^{2+} on the expression of metallothioneins, small, cysteine-rich proteins that serve to detoxify intracellular Cd^{2+}, have been analyzed extensively.[254-279] MT synthesis in cultured hamster cells is increased approximately 30-fold within 8 hours following exposure to 2 μM Cd^{2+}, a nonlethal, nontoxic dose for the cells examined.[254] This increase is sensitive to inhibition of RNA synthesis and increases in mRNA levels slightly precede increases in MT synthesis, indicating an effect at the level of transcription. In fact, mRNA levels and thioneine synthesis rates are coordinate during induction, de-induction, and super-induction, indicating that MT synthesis is controlled by the levels of MT mRNA throughout.[254,255] MT gene transcription is enhanced by Zn^{2+}, Cu^{2+}, and Hg^{2+},[256] and by glucocorticoids,[257] as well as by Cd^{2+},[257] *in vivo* and *in vitro.*

The genes for MTs and their associated promoter regions have been sequenced.[257,264] Their expression is blocked in some cultured cells by DNA methylation,[258] but enhanced by DNA amplification that follows growth and/or selection in toxic levels of Cd^{2+}.[259] Amplified genes respond to Cd^{2+} as an inducer of enhanced transcription, but not to glucocorticoids.[260] Gene constructs containing the mouse MT-1 promoter also respond to heavy metals but not to glucocorticoids.[261,262] The sequences necessary for this response are within 90 bp of the transcription start site.[263]

Probe analyses have shown that electrophoretically resolved MTs, such as MT-1, actually represent individual members of closely related gene families.[264] The major hamster MTs are 80% identical in their protein coding sequences, but are only 35% homologous in their 3′ and 5′ untranslated regions,[265] indicating a gene duplication event that occurred at least 45 million years ago. MT-1 and -2 genes show differential responses to cadmium and glucocorticoids.[266] Also, not all metals that induce MT synthesis are detoxified by it,[267] indicating that the genes are not auto-regulated. MT synthesis is also induced by bacterial lipopolysaccharide (LPS) endotoxin *in vivo*.[268,269] DNA sequences involved in the responses to metals (the metal response elements, or MREs) and in basal MT expression have been determined.[270] Multiple MREs are present in MT promoters and act as positive elements in response to Cd^{2+}.[271,272] The transcription factor Sp1 is involved in regulating basal expression.[272] The mouse MT-1 MRE interacts specifically with a 108-kDa nuclear protein.[273] It has been shown that general inhibition of protein kinases C does not block cadmium- or dexamethasone-induced MT synthesis,[274] and that activators of PKC do not induce MT;[275] however, other upstream events regulating the association and activation of MRE-binding transcription factors are largely undefined. Association of the mouse MRE-binding factor, MTF-1, does require Zn^{2+}, suggesting the possibility that Zn^{2+} or Cd^{2+} activates MTF-1 directly.[276-279]

In addition to MTs, Cd^{2+} induces the synthesis of a number of proteins that are induced also by heat shock proteins. The 68-kDa protein whose synthesis is induced in L-132 cells by heat is identical with that induced by Cd^{2+}.[280] The human HSP70 is also induced by heat shock and cadmium.[281] This induction occurs at the level of transcription and requires the presence of DNA sequences between –107 and –68. Located within this region are both a sequence homologous to a heat shock element (HSE) in *Drosophila* and a sequence homologous to the core of the human MT-2 HSE. Although Cd^{2+} induces HSPs, heat shock does not induce MT.[282]

Intracellular expression of an inhibitor specific for cAMP-dependent protein kinase A inhibits basal-, heat-, and cadmium-induced transcription of a human HSP promoter-driven gene.[283] Agents that elevate cAMP levels activate the HSP promoter. Thus, both Cd^{2+} and heat shock can activate HSP70 transcription through a pathway involving protein kinase A. The primary target for Cd^{2+} in its intersection with this pathway is not defined.

Cd^{2+} at doses between 0.125 and 0.25 μ*M* produces a delayed mitogenic response in cultured normal rat kidney fibroblasts (NRK-49F) and modulates epidermal growth factor (EGF)-induced DNA synthesis, suggesting that it might directly or indirectly induce the expression of proliferation-associated protooncogenes such as c-*fos* and c-*myc*.[287] Indeed, Cd^{2+} induces accumulation of mRNAs for c-*jun* and c-*myc* in cultured myoblasts.[177] This accumulation is blocked by actinomycin but stimulated by cycloheximide, indicating primary induction of these mRNAs. Cd^{2+} also induces accumulation of mRNAs specific for genes induced in cultured 3T3 cells by the phorbol ester, tetradecanoyl phorbol acetate (TPA). These include the mRNA for the protooncogene product c-*fos*, as well as for a zinc finger transcription factor.[178] Down-regulation of protein kinase C by prolonged exposure to phorbol esters does not block the response.[178] Cd^{2+} also induces the accumulation of c-*myc* transcripts in NRK-49F cells.[288] This response is accompanied by cellular hypertrophy. This c-*myc* accumulation is blocked by the protein kinase inhibitor H7 but not by the inhibitor HA1004, indicating that protein kinase C mediates the response.[179] In addition to c-*myc*, Cd^{2+} induces the accumulation of mRNAs for the protooncogene products c-*fos* and c-*jun* in NRK-49F cells in a dose-dependent temporal pattern.[179]

As noted above in the section on arsenic, a number of stressors, including Cd^{2+}, induce the synthesis of heme oxygenase in cultured human skin fibroblasts.[234,235] The other agents include H_2O_2, ultraviolet-A radiation, iodoacetamide, and menadione. In mouse hepatoma cells, heme and Cd^{2+} induce accumulation of heme oxygenase and its mRNA by stimulating transcription.[289] Cd^{2+} induces transcription of the heme oxygenase gene in HeLa cells, also.[290] An upstream transcription factor (USF) has been isolated from these cells that binds to a *cis*-acting element 34 bp from the transcription initiation site and stimulates cell free transcription of the heme oxygenase gene.[290] Further studies on gene constructs containing the heme oxygenase promoter and a reporter gene have shown that a region between 4.5 and 4 kb upstream of the transcription initiation site is necessary for Cd^{2+} induction of transcription.[291] Within this region, a ten-base sequence was found that is required for Cd^{2+} response and *in vitro* binding to a presumptive transcription factor. An AP1 binding site is present, also. It has been reported both not to be involved in the Cd^{2+} response[291] and to be essential.[292] Although both MT and heme oxygenase genes are transcriptionally activated by Cd^{2+}, different promoter elements are involved in the responses.[293]

The complete domain of Cd^{2+}-responsive genes has probably not yet been determined. In addition to MT, HSPs, proliferation-associated protooncogenes, and heme oxygenase, Cd^{2+} has been reported to induce the synthesis of several other proteins. As reported above,[237,238] Cd^{2+} as well as heat shock and arsenite activate the multiple drug resistance gene *MDR*1. Cd^{2+} and a variety of other stress agents reduce osteonectin/SPARC gene expression[294] in osteoligament cells but at the level of translation (increased protein but not mRNA). Cd^{2+} also induces interleukin-8 (IL-8) production in peripheral blood mononuclear cells.[296] Because reactive oxygen species appeared also within 10′ of Cd^{2+} addition and the radical scavenger *N*-acetyl-cysteine inhibits the appearance of IL-8 in response to Cd^{2+}, it is suggested that Cd^{2+} induces IL-8 through the production of active oxygen species.[296] Exposure of human breast cancer cells to a relatively low concentration (1 μ*M*) of Cd^{2+} stimulates

TABLE 14

Cadmium Effects on Gene Expression

Effector(s)	Concentration	Gene(s)	Effect(s)	Ref.
Cd^{2+}	2 μ*M*	MT	Synthesis increased 30×	254
Cd^{2+}, Zn^{2+}, Cu^{2+}, Hg^{2+}	0.1–10 mg/kg	MT	mRNA and protein levels increased	256
DNA methylation	NA	MT	Transcription repressed	258
DNA amplification	NA	MT	Transcription induced by Cd^{2+} but not glucocorticoid enhanced	259, 260, 261
Zn^{2+}, Cu^{2+}, Hg^{2+}, Ag^{2+}, Co^{2+}, Ni^{2+}, Bi^{2+}	—	MT	Synthesis induced	267
Cd^{2+}, LPS, dexamethasone	Injection, dose not given	MT	Synthesis induced *in vivo*	269
Cd^{2+}, heat shock	—	P68 HSP	Protein synthesis induced	280
Cd^{2+}, heat shock	1–20 μg/ml, 41–43°C (1–3 hr)	HSP70	Transcription activated	281, 282
Cd^{2+}	15 μ*M*	Rb, N-*myc*	mRNA decreased	284
Cd^{2+}	15 μ*M*	HSP70, HSP	mRNA increased	284
Cd^{2+}	2 mg/kg	HSP70, 90, 110	Protein(s) increased	285
Cd^{2+}	1.5–24 μ*M* (6 hr)	HSP60	mRNA increased	286
Cd^{2+}	0.125–0.25 μ*M*	—	DNA synthesis induced	287
Cd^{2+}	5–10 μ*M*	c-*jun*, c-*myc*	mRNA accumulation	177
Cd^{2+}	0.5–4 μ*M*	c-*myc*	mRNA accumulation, cellular hypertrophy	288
Cd^{2+}	1–50 μ*M*	c-*fos*, zinc-finger transcription factor	mRNA accumulation	178
TPA (phorbol ester)	81 n*M*			
Cd^{2+}	5–10 μ*M*	Heme oxygenase	Transcription increased	234, 235, 289, 290
Heme	5–10 μ*M*			
Cd^{2+}	2 ng/ml (24 hr)	Osteonectin/SPARC	Translation decreased	294
Arsenite	100 μ*M* (1.5 hr)			
Heat shock	45°C (1.5 hr)			
AZC	5 m*M* (12 hr)	Osteonectin/SPARC	Translation increased	294
Cd^{2+}	1–10 pmol/50 μl	*ada*	Transcription inhibited	295
Hg^{2+}	1–10 pmol/50 μl			
Cd^{2+}	1.0–100 *M*	Interleukin-8	Synthesis increased	296
Cd^{2+}	1 μ*M*	Estrogen receptor	mRNA and protein levels decreased one half	297
Cd^{2+} (Zn^{2+}: NE)	1 μ*M*	Progesterone receptor	Receptor and mRNA levels increased 3×	297

Note: MT = metallothionein; NA = not applicable; LPS = lipopolysaccharide; AZC = L-azetidine-2-carboxylic acid; SPARC = secreted protein, acidic, rich in cysteine.

proliferation and has opposing effects on mRNA and protein levels for the estrogen and progesterone receptors.[297] Estrogen mRNA and receptor levels are halved, but progesterone mRNA and receptor levels are increased threefold. (See Table 14.)

1.5.4 Chromium

RNA synthesis is stimulated when DNA or chromatin is treated with Cr^{3+} prior to the addition of RNA polymerase.[298] This effect obtains over a broad range of Cr^{3+} concentrations (1 μ*M* to 1 m*M*). Pre-incubation of RNA polymerase, however, results in inhibition of RNA synthesis.[298] Pretreatment with Cr^{3+} (5 mg/kg, i.p.) similarly enhances RNA synthesis in regenerating rat liver.[299] Nucleolar synthesis is especially enhanced.

TABLE 15

Effects of Chromium on Gene Expression

Effector(s)	Concentration	Gene(s)	Effect(s)	Ref.
Cr^{3+}	1 μM–1 mM	—	DNA-directed RNA synthesis enhanced	298
Cr^{3+}	5 mg/kg (i.p.)	—	Nucleolar RNA synthesis enhanced	299
Cr^{6+} (CrO_4^{2-})	0.2 mmol/kg embryo	5-aminolevulinic acid synthase, cytochrome P-450	Constitutive mRNA levels increased; induction suppressed	300, 301, 302
Cr^{6+}	50–100 μmol/kg embryo	MT	Zn^{2+}-induced expression inhibited	303
Cr^{6+}	30 μM	GRP78 (glucose-regulated protein)	No effect (96% survival)	304
Cr^{6+}	150 μM	GRP78 (glucose-regulated protein)	Induced expression reduced (54% survival)	304
Cr^{6+}	300 μM	GRP78 (glucose-regulated protein)	Induced expression suppressed completely (8% survival)	304
Cr^{6+}	50 μmol/kg embryo	PEPK	Transcription suppressed	305

Note: MT = metallothionein; PEPK = phosphoenolpyruvate carboxykinase.

Cr^{6+} suppresses the induction of 5-aminolevulinic acid synthase and cytochrome P-450 in chick embryo liver but enhances constitutive mRNA levels for these gene products.[300-302] These effects occur at the level of transcription and are correlated first with monoadduct formation and later with DNA-DNA and DNA-protein cross-link formation. Similarly, Cr^{6+} does not affect or increases constitutive MT RNA or protein synthesis in chick embryos, depending on the developmental time when exposed, but it inhibits zinc-induced synthesis of MT mRNA and protein.[303]

Differing effects on the induction of glucose regulated protein (GRP78), general RNA synthesis, and survival are seen in CHO cells treated with varying levels of Cr^{6+}.[304] A concentration (30 μ*M*) allowing 96% survival has no effect on GRP78 production or general RNA synthesis. However, treatment with 150 μ*M* Cr^{6+} results in suppressed GRP78 induction, a 60 to 75% reduction in general RNA synthesis, and 54% survival. 300 μ*M* totally suppresses induction, blocks general RNA synthesis by 80 to 90%, and allows only 8% survival. Again, these effects are accompanied by Cr-adduct and DNA-protein cross-link formation.

Cr^{6+} pretreatment affects both basal and glucocorticoid-induced expression of phosphoenolpyruvate carboxykinase (PEPCK) in chick embryo liver.[305] These effects occur at the level of transcription. (See Table 15.)

1.5.5 Nickel

Exposure of cultured cells to Ni^{2+} results in altered DNA-protein interactions as evidenced by a greater content of tightly bound, nonhistone proteins in isolated chromatin preparations.[306] Despite this early indication of potential for effects on gene expression, few reports indicate such in nonmicrobial systems. Ni^{2+} as well as Co^{2+} and hypoxia stimulate the production of erythropoietin (Epo) in cultured human hepatoma cells.[307] It is suggested that these agents all stimulate via ligand-induced conformational changes in heme, which in turn regulates Epo gene expression.

The patterns of gene expression and of nickel-binding proteins are altered in cultured cells selected for resistance to Ni^{2+}.[308] These changes are accompanied by visible chromosomal alterations. The genes involved in nickel resistance in these cells are undefined. In a separately derived nickel-resistant cell line, changes in cytokeratin gene expression are apparent, but the

TABLE 16
Effects of Nickel on Gene Expression

Effector(s)	Concentration	Gene(s)	Effect(s)	Ref.
Ni^{2+}	0.5–2.5 m*M* (3 hr)	Total chromatin (from Ni^{2+} exposed cells)	DNA-protein interactions altered	306
Ni^{2+} Co^{2+}, hypoxia	—	Epo	mRNA and protein increased	307
Anthralin, iodoacetic acid, BHT, TPA, Tween 60, Ni^{2+}, t-butyl hydroperoxide, benzylperoxide	Equal to those effective in cell transformation	Proliferin	mRNA levels increased	11
Ni^{2+}	500 μ*M*	*LvS*1 (dorsal gene)	Under-expression (in embryo)	312
Ni^{2+}	500 μ*M*	*ExtoV* (ventral gene)	Over-expression (in embryo)	312
Hypoxia ≫ Co^{2+} Ni^{2+}	3% O_2 50 μ*M* 400 μ*M*	Epo	mRNA and protein levels increased	313

Note: BHT = butylated hydroxy toluene; TPA = tetradecanoyl phorbol acetate; Epo = erythropoietin.

significance of these changes in the context of nickel resistance mechanisms and/or in the context of nickel-specific regulation of gene expression is not defined.[309]

In addition to inducing changes in protein binding by DNA, Ni^{2+} may at least theoretically alter gene expression through the induction of conformational changes in DNA or through participation in active oxygen metabolism.[310] These suggested mechanisms have not as yet found experimental verification. Nickel sulfate as well as a number of organic tumor promoters induce the accumulation of proliferin transcripts in cultured fibroblasts when administered at concentrations that promote cell transformation.[311]

The ability of a relatively high concentration of Ni^{2+} to disturb development in sea urchin embryos is associated with changes in specific gene expression.[312] Exposure of embryos to 500 μ*M* Ni^{2+} prevents the development of dorsoventral polarity at midgastrula. This effect is accompanied by underexpression of transcripts of the dorsal gene, *LvS*1, and overexpression of the ventral ectodermal gene, *ExtoV*. Erythropoietin mRNA and protein levels are increased in primary rat hepatocyte cultures exposed to Ni^{2+};[313] however, the concentration of nickel is relatively high, 500 μ*M*, and the effect is less than 7% that observed following treatment with 3% O_2. (See Table 16.)

1.6 Summary and Conclusions

A review of carcinogenic metal ion interactions with signal transduction pathways (STPs) reveals a richness of effects on STP receptors, ion channels, kinases/phosphatases, and gene expression. In almost all instances where such interactions have been explored, however, the metal ion concentrations employed are of pharmacological but not toxicological significance. Further, in those few studies that address effects elicited at possibly relevant concentrations, effects on STPs are not related to consequent alterations in cell proliferation, apoptosis, or fidelity of DNA replication/repair of the kind that can clearly be related causally to the carcinogenic process.

An overall *ad hoc* rather than systematic approach to the study of carcinogenic trace metal/STP interactions is also revealed. While interactions with some signal transduction pathway targets, such as ion channels, have been explored extensively, interactions with several other classes of STP targets remain largely unstudied. Receptors for peptide growth factors such as platelet-derived growth factor (PDGF) and transforming growth factor (TGF), for example, present an array of cell surface chemical groups reactive with carcinogenic metal ions, but this class of interactions has as yet not been addressed. Similarly, interactions with several classes of proteins, such as G proteins, ras, raf, etc., that are involved in transducing signals from receptors to kinases or transcription factors remain largely of undetermined significance. Careful analyses of the metal ion/STP interactions that occur in model systems where low (submicromolar) concentrations of metal ions are found to alter DNA replication or repair, cell growth or proliferation, and apoptosis may perhaps best reveal STP targets that are primary in metal-induced carcinogenesis, but need much more nearly complete development to reveal causal mechanisms.

REFERENCES

1. Boffetta, P., Carcinogenicity of trace elements with reference to evaluations made by the International Agency for Research on Cancer, *Scand. J. Work Environ. Health*, 19(1), 67–70, 1993.
2. Cavanaugh, A. H. and Simons, Jr., S. S., Glucocorticoid receptor binding to calf thymus DNA. 2. Role of a DNA-binding activity factor in receptor heterogeneity and a multistep mechanism of receptor activation, *Biochemistry*, 29(4), 996–1002, 1990.
3. Simons, Jr., S. S., Chakraborti, P. K., and Cavanaugh, A. H., Arsenite and cadmium(II) as probes of glucocorticoid receptor structure and function, *J. Biol. Chem.*, 265(4), 1938–1945, 1990.
4. Takagi, S., Hummel, B. C., and Walfish, P. G., Thionamides and arsenite inhibit specific T3 binding to the hepatic nuclear receptor, *Biochem. Cell Biol.*, 68(3), 616–621, 1990.
5. Lopez, S., Miyashita, Y., and Simons, Jr., S. S., Structurally based, selective interaction of arsenite with steroid receptors, *J. Biol. Chem.*, 265(27), 16039–16042, 1990.
6. Chakraborti, P. K., Hoeck, W., Groner, B., and Simons, Jr., S. S., Localization of the vicinal dithiols involved in steroid binding to the rat glucocorticoid receptor, *Endocrinology*, 127(5), 2530–2539, 1990.
7. Chakraborti, P. K., Garabedian, M. J., Yamamoto, K. R., and Simons, Jr., S. S., Role of cysteines 640, 656, and 661 in steroid binding to rat glucocorticoid receptors, *J. Biol. Chem.*, 267, 11366–11373, 1992.
8. Stancato, L. F., Hutchison, K. A., Chakraborti, P. K., Simons, Jr., S. S., and Pratt, W. B., Differential effects of the reversible thiol-reactive agents arsenite and methyl methane thiosulfonate on steroid binding by the glucocorticoid receptor, *Biochemistry*, 32(14), 3729–3736, 1993.
9. Fonseca, M. I., Lunt, G. G., and Aguilar, J. S., Inhibition of muscarinic cholinergic receptors by disulfide reducing agents and arsenicals, *Biochem. Pharmacol.*, 41(5), 735–742, 1991.
10. Sanchez, E. R., Heat shock induces translocation to the nucleus of the unliganded glucorticoid receptor, *J. Biol. Chem.*, 267, 17–20, 1992.
11. Edwards, D. P., Estes, P. A., Fadok, V. A., Bona, B. J., Onate, S., Nordeen, S. K., and Welch, W. J., Heat shock alters the composition of heteromeric steroid receptor complexes and enhances receptor activity *in vivo*, *Biochemistry*, 31(9), 2482–2491, 1992.
12. Sanchez, E. R., Hu, J. L., Zhong, S., Shen, P., Greene, M. J., and Housley, P. R., Potentiation of glucocorticoid receptor-mediated gene expression by heat and chemical shock, *Mol. Endocrinol.*, 8(4), 408–421, 1994.
13. Cumming, D. V., Ord, M. G., and Stocken, L. A., Protein kinase activities in rat liver nuclei: effects of age and partial hepatectomy, *Biosci. Rep.*, 6(6), 565–571, 1986.
14. Schild, H. O., The effect of metals on the S-S polypeptide receptor in depolarized rat uterus, *Br. J. Pharmacol.*, 36(2), 329–349, 1969.
15. Freund, H., Atamian, S., and Fischer, J. E., Chromium deficiency during total parental nutrition, *J. Am. Med. Assoc.*, 241(5), 496–498, 1979.
16. Riales, R. and Albrink, M. J., Effect of chromium chloride supplementation on glucose tolerance and serum lipids including high-density lipoprotein of adult men, *Am. J. Clin. Nutr.*, 34(12), 2670–2678, 1981.
17. Wallach, S., Clinical and biochemical aspects of chromium deficiency, *J. Am. Coll. Nutr.*, 4(1), 107–120, 1985.
18. Anderson, R. A., Chromium metabolism and its role in disease processes in man, *Clin. Physiol. Biochem.*, 4(1), 31–41, 1986.
19. Dubois, F. and Belleville, F., Chromium: physiologic role and implications in human pathology, *Pathol. Biol.*, 39(8), 801–808, 1991.
20. Urberg, M. and Zemel, M. B., Evidence for synergism between chromium and nicotinic acid in the control of glucose tolerance in elderly humans, *Metabolism*, 36(9), 896–899, 1987.
21. Anderson, R. A., Recent advances in the clinical and biochemical effects of chromium deficiency, *Prog. Clin. Biol. Res.*, 380, 221–234, 1993.
22. Metz, W., Chromium in human nutrition: a review, *J. Nutr.*, 123(4), 626–633, 1993.
23. McCarty, M. F., Homologous physiological effects of phenformin and chromium picolinate, *Med. Hypotheses*, 41(4), 316–324, 1993.

24. Wilson, E. M., Interconversion of androgen receptor forms by divalent cations and 8 S androgen receptor promoting factor. Effects of Zn^{2+}, Cd^{2+}, Ca^{2+} and Mg^{2+}, *J. Biol. Chem.*, 260(15), 8683–8689, 1985.
25. Levy, L., Vredevoe, D. L., and Cook, G., *In vitro* reversibility of cadmium-induced inhibition of phagocytosis, *Environ. Res.*, 41(2), 361–371, 1986.
26. Blakeborough, P. and Salter, D. N., The intestinal transport of zinc studied using brush-border-membrane vesicles from the piglet, *Br. J. Nutr.*, 57(1), 45–55, 1987.
27. Gotti, C., Cabrini, D., Sher, E., and Clementi, F., Effects of long-term *in vitro* exposure to aluminum, cadmium or lead on differentiation and cholinergic receptor expression in a human neuroblastoma cell line, *Cell. Biol. Toxicol.*, 3(4), 431–440, 1987.
28. Freedman, L. P., Luisi, B. F., Korszun, Z. R., Basavappa, R., Sigler, P. B., and Yamamoto, K. R., The function and structure of the metal coordination sites within the glucocorticoid receptor DNA binding domain, *Cell. Biol. Nature,* 334(6182), 543–546, 1988.
29. Pan, T., Freedman, L. P., and Coleman, J. E., Cadmium-^{113}NMR studies of the DNA binding domain of the mammalian glucocorticoid receptor, *Biochemistry,* 29(39), 9218–9225, 1990.
30. Ezaki, O., IIb group metal ions (Zn^{2+}, Cd^{2+}, Hg^{2+}) stimulate glucose transport activity by post-insulin receptor kinase mechanism in rat adipocytes, *J. Biol. Chem.*, 264(27), 16118–16122, 1989.
31. Smith, J. B., Dwyer, S. D., and Smith, L., Cadmium evokes inositol polyphosphate formation and calcium mobilization. Evidence for a cell surface receptor that cadmium stimulates and zinc antagonizes, *J. Biol. Chem.*, 264(13), 7115–7118, 1989.
32. Dwyer, S. D., Zhuang, Y., and Smith, J. B., Calcium mobilization by cadmium or decreasing extra cellular Na^+ or pH in coronary endothelial cells, *Exp. Cell Res.*, 192(1), 22–31, 1991.
33. Smith, L., Pijüan, V., Zhuang, Y., and Smith, J. B., Reversible desensitization of fibroblasts to cadmium receptor stimuli. Evidence that growth in high zinc represses a xenobiotic receptor, *Exp. Cell Res.*, 202(1), 174–182, 1992.
34. Mayer, M. L., Vyklicky, Jr., L., and Westbrook, G. L., Modulation of excitatory amino acid receptors by group IIB metal cations in cultured mouse hippocampal neurones, *J. Physiol. Lond.*, 415, 329–350, 1989.
35. Tejwani, G. A. and Hanissian, S. H., Modulation of mu, delta and kappa opioid receptors in rat brain by metal ions and histidine, *Neuropharmacology,* 29(5), 445–452, 1990.
36. Wada, K., Fujii, Y., Watanabe, H., Satoh, M., and Furuichi, Y., Cadmium directly acts on endothelin receptor and inhibits endothelin binding activity, *FEBS Lett.*, 285(1), 71–74, 1991.
37. Celentano, J. J., Gyenes, M., Gibbs, T. T., and Farb, D. H., Negative modulation of the gamma-aminobutyric acid response by extracellular zinc, *Mol. Pharmacol.*, 40(5), 766–773, 1991.
38. Predki, P. F. and Sarkar, B., Effect of replacement of "zinc finger" zinc on estrogen receptor DNA interactions, *J. Biol. Chem.*, 267(9), 5842–5846, 1992.
39. Dunderski, J., Stanosevic, J., Ristic, B., Trajkovic, D., and Matic, G., *In vivo* effects of cadmium on rat liver glucocorticoid receptor functional properties, *Int. J. Biochem.*, 24(7), 1065–1072, 1992.
40. Shankar, V. S., Bax, C. M., Alam, A. S., Bax, B. E., Huang, C. L., and Zaidi, M., The osteoclast Ca^{2+} receptor is highly sensitive to activation by transition metal cations, *Biochem. Biophys. Res. Commun.*, 187(2), 913–918, 1992.
41. Chen, Y. C. and Smith, J. B., A putative lectin-binding receptor mediates cadmium-evoked calcium release, *Toxicol. Appl. Pharmacol.*, 117(2), 249–256, 1992.
42. Chetty, C. S., Stewart, T. C., Cooper, A., Rajanna, B., and Rajanna, S., *In vitro* interaction of heavy metals with ouabain receptors in rat brain microsomes, *Drug Chem. Toxicol.*, 16(1), 101–110, 1993.
43. Bartolami, S., Planche, M., and Pujol, R., Sulphydryl-modifying reagents alter ototoxin block of muscarinic receptor-linked phosphoinositide turnover in the cochlea, *Eur. J. Neurosci.*, 5(7), 832–838, 1993.
44. Garcia-Morales, P., Saceda, M., Kenney, N., Kim, N., Salomon, D. S., Gottardis, M. M., Solomon, H. B., Sholer, P. F., Jordan, V. C., and Martin, M. B., Effect of cadmium on estrogen receptor levels and estrogen-induced responses in human breast cancer cells, *J. Biol. Chem.*, 269(24), 16896–16901, 1994.
45. Nistio, H., Negishi, Y., Inoue, A., and Nakata, Y., Differential effects of divalent cations on specific ^{3}H-Gr 65630 binding to 5-HT3 receptors in rat cortical membranes, *Neurochem. Int.*, 24(3), 259–265, 1994.
46. Bakos, M. and Rubanyi, G., Effect of Ni^{2+}, Co^{2+}, Zn^{2+}, Fe^{2+}, Cd^{2+} Hg^{2+}, Cu^{2+} and VO^{3-} on coronary vascular resistance in the isolated perfused rat heart, *Acta. Physiol. Acad. Sci. Hung.*, 59(2), 175–180, 1982.
47. Koller, A., Rubanyi, G., Ligeti, L., and Kovach, A. G., Effect of verapamil and phenoxygenzamine on nickel-induced coronary vasoconstriction in the anaesthetized dog, *Acta. Physiol. Acad. Sci. Hung.*, 59(3), 287–290, 1982.
48. Rubanyi, G., Hajdu, K., Pataki, T., and Bakos, M., The role of adrenergic receptors in Ni^{2+}-induced coronary vasoconstriction, *Acta. Physiol. Acad. Sci.*, 59(2), 161–167, 1982.
49. Smith, P. A., The use of low concentrations of divalent cations to demonstrate a role for *N*-methyl-D-aspartate receptors in synaptic transmission in amphibian spinal cord, *Br. J. Pharmacol.*, 77(2), 363–373, 1982.
50. Mizuno, S., Ogawa, N., and Mori, A., Super high affinity binding site for [^{3}H]diazepam in the presence of Co^{2+}, Ni^{2+}, Cu^{2+}, or Zn^{2+}, *Neurochem. Res.*, 7(12), 1487–1493, 1982.
51. Mizuno, S., Ogawa, N., and Mori, A., Differential effects of some transition metal cations on the binding of beta-carboline-3-carboxylate and diazepam, *Neurochem. Res.*, 8(7), 873–880, 1983.

52. Lo, M. M. and Snyder, S. H., Two distinct solubilized benzodiazepine receptors: differential modulation by ions, *J. Neurosci.*, 3(11), 2270–2279, 1983.
53. Hirsch, J. D. and Kochman, R. L., Calcium alters divalent cation modulation of brain benzodiazepine receptors, *Arch. Int. Pharmacodyn. Ther.*, 265(2), 211–218, 1983.
54. Nukada, T., Haga, T., and Ichiyama, A., Muscarinic receptors in porcine caudate nucleus. I. Enhancement by nickel and other cations of [^{3}H]*cis*-methyldioxolane binding to guanyl nucleotide-sensitive sites, *Mol. Pharmacol.*, 24(3), 366–373, 1983.
55. Gurwitz, D., Klogg, Y., and Sokolovsky, M., Recognition of the muscarinic receptor by its endogenous neurotransmitter: binding of [3-]acetylcholine and its modulation by transition metal ions and guanine nucleotides, *Proc. Natl. Acad. Sci.*, 81(12), 3650–3654, 1984.
56. Ogawa, N., Mizuno, S., Kishimoto, T., Mori, A., Kuroda, H., and Ota, Z., Effects of transition metals on TRH-receptor interaction, *Neurosci. Res.*, 1(5), 363–368, 1984.
57. Chichenkov, O. N., Porodenko, N. V., and Zaitsev, S. V., Potentiating action of bi- and trivalent metal salts on the analgesic effect of morphine, *Bull. Eksp. Biol. Med.*, 100(9), 313–315, 1985.
58. Junig, J. T. and Abood, L. G., Solubilization and purification of the Ni-stimulated arginine-vasopressin binding site of rat brain membranes, *Neurochem. Res.*, 12(9), 809–817, 1987.
59. Gopalakrishnan, V., McNeill, J. R., Sulakhe, P. V., and Triggle, C. R., Hepatic vasopressin receptor: differential effects of divalent cations, guanine nucleotides, and *N*-ethylmaleimide on agonist and antagonist interactions with the V1 subtype receptor, *Endocrinology*, 123(2), 922–931, 1988.
60. Warner, G. L. and Lawrence, D. A., The effect of metals on IL-2-related lymphocyte proliferation, *Int. J. Immunopharmacol.*, 10(5), 629–637, 1988.
61. Karttunen, R., Silvennoinen-Kassinen, S., Juutinen, K., Andersson, G., Ekre, H. P., and Karvonen, J., Nickel antigen induces IL-2 secretion and IL-2 receptor expression mainly on CO^{4+} cells, but no measurable gamma interferon secretion in peripheral blood mononuclear cell cultures in delayed type hypersensitivity to nickel, *Clin. Exp. Immunol.*, 74(3), 387–391, 1989.
62. Silvennoinen-Kassinen, S., Ikaheimo, I., Karvonen, J., Kauppinen, M., and Kallioninen, M., Mononuclear cell subsets in the nickel-allergic reaction *in vitro* and *in vivo*, *J. Allergy Clin. Immunol.*, 89(4), 794–800, 1992.
63. Hutchens, T. W. and Li, C. M., Estrogen receptor interaction with immobilized metals: differential molecular recognition of Zn^{2+}, Cu^{2+} and Ni^{2+} and separation of receptor isoforms, *J. Mol. Recog.*, 1(2), 80–92, 1988.
64. Hutchens, T. W., Li, C. M., Sato, Y., and Yip, T. T., Multiple DNA-binding estrogen receptor forms resolved by interaction with immobilized metal ions. Identification of a metal-binding domain, *J. Biol. Chem.*, 264(29), 17206–17212, 1989.
65. Medici, N., Minucci, S., Nigro, V., Abbondanza, C., Armetta, I., Molinari, A. M., and Puca, G. A., Metal binding sites of the estradiol receptor from calf uterus and their possible role in the regulation of receptor function, *Biochemistry*, 28(1), 212–219, 1989.
66. Mayor, P., Cabrera, R., Ribas, B., and Calle, C., Effect of long-term nickel ingestion on insulin binding and antilipolytic response in rat adipocytes, *Biol. Trace Elem. Res.*, 22(1), 63–70, 1989.
67. Mooney, R. A. and Bordwell, K. L., Differential dephosphorylation of the insulin receptor and its 160-kDa substrate (pp 160) in rat adipocytes, *J. Biol. Chem.*, 267(20), 14054–14060, 1992.
68. Arias-Montano, J. A. and Young, J. M., Locus of action of Ni^{2+} on histamine-induced inositol phosphate formation in brain slices and in HeLa cells, *Eur. J. Pharmacol.*, 245(3), 221–228, 1993.
69. Eimeri, S. and Schramm, M., Potentiation of ^{45}Ca uptake and acute toxicity mediated by the *N*-methyl-D-aspartate receptor: the effect of metal binding agents and transition metal ions, *J. Neurochem.*, 61(2), 518–525, 1993.
70. Shankar, V. S., Bax, C. M., Bax, B. E., Alam, A. S., Moonga, B. S., Simon, B., Pazianas, M., Huang, C. L., and Zaidi, M., Activation of the Ca^{2+} "receptor" on the osteoclast by Ni^{2+} elicits cytosolic Ca^{2+} signals: evidence for receptor activation and inactivation, intracellular Ca^{2+} redistribution, and divalent cation modulation, *J. Cell Physiol.*, 155(1), 120–129, 1993.
71. Mutafova-Yambolieva, B., Staneva-Stoytcheva, D., Lasova, L., and Radomirov, R., Effects of cobalt or nickel on the sympathetically mediated contractile responses in rat-isolated vas deferens, *Pharmacology*, 48(2), 100–110, 1994.
72. Filvaroff, E., Calautti, E., Reiss, M., and Dotto, G. P., Functional evidence for an extracellular calcium receptor mechanism triggering tyrosine kinase activation associated with mouse keratinocyte differentiation, *J. Biol. Chem.*, 269(34), 21735–21740, 1994.
73. Mahadevan, D., Thanki, N., Aroca, P., McPhie, P., Yu, J. C., Beeler, J., Santos, E., Wlodawer, A., and Heidaran, M. A., A divalent metal ion binding site in the kinase insert domain of the alpha-platelet-derived growth factor receptor regulates its association with SH2 domains, *Biochemistry*, 34(7), 2095–2106, 1995.
74. Kitada, Y., Enhancing effects of transition metals on the salt taste responses of single fibers of the frog glossopharyngeal nerve: specificity of and similarities among Ca^{2+}, Mg^{2+} and Na^{2+} taste responses, *Chem. Senses*, 19(3), 265–277, 1994.
75. Gueniche, A., Viac, J., Lizard, G., Charveron, M., and Schmitt, D., Effect of nickel on the activation state of normal human keratinocytes through interleukin-1 and intercellular adhesion molecule 1 expression, *Br. J. Dermatol.*, 131(2), 250–256, 1994.

76. Evans, D. H., Weingarten, K. E., and Walton, J. S., The effect of atropine on cadmium- and nickel-induced constriction of vascular smooth muscle of the dogfish shark ventral aorta, *Toxicology,* 62(1), 89–94, 1990.
77. Chow, R. H., Cadmium block of squid calcium currents. Macroscopic data and a kinetic model, *J. Gen. Physiol.,* 98(4), 751–770, 1991.
78. Taylor, W. R., Permeation of barium and cadmium through slowly inactivating calcium channels in cat sensory neurons, *J. Physiol. Lond.,* 407, 433–452, 1988.
79. Swandulla, D. and Armstrong, C. M., Calcium channel block by cadmium in chicken sensory neurons, *Proc. Natl. Acad. Sci.,* 86(5), 1736–1740, 1989.
80. Tranchand-Bunel, D., Blasquez, C., Delbende, C., Jegou, S., and Vaudry, H., Involvement of voltage-operated calcium channels in alpha-melanocyte-stimulating hormone (alpha-MSH) release from perfused rat hypothalamic slices, *Brain Res. Mol. Brain Res.,* 6(1), 21–29, 1989.
81. Yabu, H., Yoshimo, M., Someya, T., and Totsuka, M., Two types of Ca channels in smooth muscle cells isolated from guinea pig taenia coli, *Adv. Exp. Med. Biol.,* 255, 129–134, 1989.
82. Fox, A. P., Nowycky, M. C., and Tsien, R. W., Kinetic and pharmacological properties distinguishing three types of calcium currents in chick sensory neurons, *J. Physiol. Lond.,* 394, 149–172, 1987.
83. Huang, Y., Quayle, J. M., Worley, J. F., Standen, N. B., and Nelson, M. T., External cadmium and internal calcium block of single calcium channels in smooth muscle cells from rabbit mesenteric artery, *Biophys. J.,* 56(5), 1023–1028, 1989.
84. Benham, C. D. and Tsien, R. W., A novel receptor-operated Ca^{2+}-permeable channel activated by ATP in smooth muscle, *Nature,* 328(6127), 275–278, 1987.
85. Passafaro, M., Clementi, F., Pollo, A., Carbone, E., and Sher, E., Omega-conotoxin and Cd^{2+} stimulate the recruitment to the plasmamembrane of an intracellular pool of voltage-operated Ca^{2+} channels, *Neuron,* 12(2), 317–326, 1994.
86. Takada, M. and Hayashi, H., Interaction of cadmium, calcium, and ameloride in the kinetics of active sodium transport through frog skin, *Jpn. J. Physiol.,* 31(3), 285–303, 1981.
87. Bowers, C. W., A cadmium-sensitive, tetrodotoxin-resistant sodium channel in bullfrog autonomic axons, *Brain Res.,* 5, 340(1), 143–147, 1985.
88. DeFrancesco, D., Ferroni, A., Visentin, S., and Zaza, A., Cadmium-induced blockade of the cardiac fast Na channels in calf Purkinje fibres, *Proc. Roy Soc.,* 22, 223(1233), 475–484, 1985.
89. Gola, M. and Ducreux, C., D600 as a direct blocker of Ca-dependent K currents in helix neurons, *Proc. Natl. Acad. Sci.,* 83(2), 451–455, 1985.
90. Schichter, L., Sidell, N., and Hagiwara, S., Potassium channels mediate killing by human natural killer cells, *Proc. Natl. Acad. Sci.,* 83(2), 451–455, 1986.
91. Sidel, N., Schlichter, L. C., Wright, S. C., Hagiwara, S., and Golub, S. H., Potassium channels in human NK cells are involved in discrete stages of the killing process, *J. Immunol.,* 137(5), 1650–1658, 1986.
92. Gusovsky, F., McNeal, E. T., and Daly, J. W., Stimulation of phosphoinositide breakdown in brain synaptoneurosomes by agents that activate sodium influx: antagonism by tetrodotoxin, saxitoxin, and cadmium, *Mol. Pharmacol.,* 32(4), 479–487, 1987.
93. Fu, W. M., Day, S. Y., and Lin-Shiau, S. Y., Studies on cadmium-induced myotonia in the mouse diaphragm, *Naunyn Schmiedebergs Arch. Pharmacol.,* 340(2), 191–195, 1989.
94. Oortgiesen, M., van-Kleef, R. G., and Vijverberg, H. P., Novel type of ion channel activated by Pb^{2+}, Cd^{2+} and Al^{3+} in cultured mouse neuroblastoma cells, *J. Membr. Biol.,* 113(3), 261–268, 1990.
95. Jungwirth, A., Paulmichi, M., and Lang, F., Cadmium enhances potassium conductance in cultured renal epithelioid (MDCK) cells, *Kidney Int.,* 37(6), 1477–1486, 1990.
96. Visentin, S., Zaza, A., Ferroni, A., Tromba, C., and Di Francesco, C., Sodium current block caused by group IIb cations in calf Purkinje fibres and in guinea pig ventricular myocytes, *Plfugers Arch.,* 417(2), 213–222, 1990.
97. Inoue, R., Effect of external Cd^{2+} and other divalent cations on carbachol-activated nonselective cation channels in guinea pig ileum, *J. Physiol.,* 442–447(63), 1990.
98. Follmer, C. H., Lodge, N. J., Cullinan, C. A., and Colatsky, T. J., Modulation of the delayed rectifier, IK, by cadmium in cat ventricular myocytes, *Am. J. Physiol.,* 262, C75–83, 1992.
99. Leinders, T., van-Kleef, R. G., and Vijverberg, H. P., Divalent cations activate small-(SK) and large-conductance (BK) channels in mouse neuroblastoma cells: selective activation of SK channels by cadmium, *Plfugers Arch.,* 422(3), 217–222, 1992.
100. Stockand, J., Sultan, A., Molony, D., DuBose, Jr., T., and Sansom, S., Interactions of cadmium and nickel with K channels of vascular smooth muscle, *Toxicol. Appl. Pharmacol.,* 121(1), 30–35, 1993.
101. McCarthy, D. C., Noelle, R. J., Gallagher, J. D., and McCann, F. B., Effects of cadmium on potassium currents in activated B lymphocytes, *Cell Signal.,* 5(4), 417–424, 1993.
102. Bohme, M., Diener, M., and Rummel, W., Chloride secretion induced by mercury and cadmium: action sites and mechanisms, *Toxicol. Appl. Pharmacol.,* 114(2), 295–301, 1992.
103. Cooper, G. P. and Manalis, R. S., Interactions of lead and cadmium on acetylcholine release at the frog neuromuscular junction, *Toxicol. Appl. Pharmacol.,* 74(3), 411–416, 1984.

104. Cooper, G. P. and Manalis, R. S., Cadmium effects on transmitter release at the frog neuromuscular junction, *Eur. J. Pharmacol.*, 99(4), 251–256, 1984.
105. Nishimura, M., Tsutsui, L., Yagasaki, O., and Yanagiya, I., Transmitter release at the mouse neuromuscular junction stimulated by cadmium ions, *Arch. Int. Pharmacodyn. Ther.*, 271(1), 106–121, 1984.
106. Kuan, Y. F. and Scholfield, C. N., Ca^{++}-channel blockers and the electrophysiology of synaptic transmission of the guinea-pig olfactory cortex, *Eur. J. Pharmacol.*, 130(3), 273–278, 1986.
107. Atchison, W. D., Effects of neurotoxicants on synaptic transmission: lessons learned from electrophysiological studies, *Neurotoxicol. Teratol.*, 10(5), 393–416, 1988.
108. Murphy, T. H., Malouf, A. T., Sastre, A., Schaar, R. L., and Coyle, J. T., Calcium-dependent glutamate toxicity in a neuronal cell line, *Brain Res.*, 444(2), 325–332, 1988.
109. Guiramand, J., Vignes, M., and Recasens, M., A specific transduction mechanism for the glutamate action on phosphoinositide metabolism via the quisqualate metabotropic receptor in rat brain synaptoneurosomes. II. Calcium dependency, cadmium inhibition, *J. Neurochem.*, 57(5), 1501–1509, 1991.
110. Lawson, K. and Cavero, I., Contractile responses to calcium chloride in rat aortic rings bathed in K^+-free solution are resistant to organic calcium antagonists, *Br. J. Pharmacol.*, 96(1), 17–22, 1989.
111. Lopez, M. G., Moro, M. A., Castillo, C. F., Artalejo, C. R., and Garcia, A. G., Variable, voltage-dependent, blocking effects of nitrendipine, verapamil, diltiazem, cinnarizine and cadmium on adrenomedullary secretion, *Br. J. Pharmacol.*, 96(3), 725–731, 1989.
112. Collins, F. and Lile, J. D., The role of dihydropyridine-sensitive voltage-gated calcium channels in potassium-mediated neuronal survival, *Brain Res.*, 502(1), 99–108, 1989.
113. Hinkle, P. M., Kinsella, P. A., and Osterhoudt, K. C., Cadmium uptake and toxicity via voltage-sensitive calcium channels, *J. Biol. Chem.*, 262(34), 16333–16337, 1987.
114. Kang, Y-J., Liu, M-S., and Enger, M. D., Diamide reduces cadmium accumulation by human lung carcinoma A549 cells, *Toxicology*, 62, 53–58, 1990.
115. Blazka, M. E. and Shaikh, Z. A., Differences in cadmium and mercury uptakes by hepatocytes: role of calcium channels, *Toxicol. Appl. Pharmacol.*, 110(2), 355–363, 1991.
116. Shibuya, I. and Douglas, W. W., Calcium channels in rat melanotrophs are permeable to manganese, cobalt, cadmium, and lanthanum, but not to nickel: evidence provided by fluorescence changes in fura-2-loaded cells, *Endocrinology*, 131(4), 1936–1941, 1992.
117. Hinkle, P. M., Shanshala, E. D., and Nelson, E. J., Measurement of intracellular cadmium with fluorescent dyes. Further evidence for the role of calcium channels in cadmium uptake, *J. Biol. Chem.*, 267(35), 25553–25559, 1992 (published erratum appears in *J. Biol. Chem.*, 15-268(8), 6064, 1993).
118. Hinkle, P. M. and Osborne, M. E., Cadmium toxicity in rat phyenochromocytoma cells: studies on the mechanism of uptake, *Toxicol. Appl. Pharmacol.*, 124(1), 91–98, 1994.
119. Kohlhardt, M., Figulla, H. R., and Tripathi, O., The slow membrane channel as the predominant mediator of the excitation process of the sinoatrial pacemaker cell, *Basic Res. Cardiol.*, 71(1), 17–26, 1976.
120. Hagiwara, N., Irisawa, H., and Kameyama, M., Contribution of two types of calcium currents to the pacemaker potentials of rabbit sino-atrial node cells, *J. Physiol. Lond.*, 395, 233–263, 1988.
121. Zhou, Z. and Lipsius, S. L., T-type calcium current in latent pacemaker cells isolated from cat right atrium, *J. Mol. Cell Cardiol.*, 26(9), 1211–1219, 1994.
122. Magleby, K. L. and Weinstock, M. M., Nickel and calcium ions modify the characteristics of the acetylcholine receptor-channel complex at the frog neuromuscular junction, *J. Physiol. Lond.*, 299, 203–218, 1980.
123. Miyamoto, T. and Washio, H., Postsynaptic effects of nickel ions at the insect neuromuscular junction, *Comp. Biochem. Physiol.*, 81(1), 11–17, 1985.
124. Jones, R. S. and Heinemann, U. H., Differential effects of calcium entry blockers on pre- and postsynaptic influx of calcium in the rat hippocampus *in vitro*, *Brain Res.*, 416(2), 257–266, 1987.
125. Fox, A. P., Nowycky, M. C., and Tsien, R. W., Kinetic and pharmacological properties distinguishing three types of calcium currents in chick sensory neurones, *J. Physiol.*, 394, 149–172, 1987.
126. Boland, L. M. and Dingledine, R., Multiple components of both transient and sustained barium currents in a rat dorsal root ganglion, *J. Physiol. Lond.*, 420, 223–245, 1990.
127. Eckert, R., Hescheler, J., Krautwurst, D., Schultz, G., and Trautwein, W., Calcium currents of neuroblastoma x glioma hybrid cells after cultivation with dibutyryl cyclic AMP and nickel, *Pflugers Arch.*, 417(3), 329–335, 1990.
128. Mlinar, B. and Enyeart, J. J., block of current through T-type calcium channels by trivalent metal cations and nickel in neural rat and human cells, *J. Physiol. Lond.*, 469, 639–652, 1993.
129. Schiffrin, E. I., Gutkowska, J., Lis, M., and Genest, J., Relative roles of sodium and calcium ions in the steroidogenic response of isolated rat adrenal glomerulose cells, *Hypertension*, 4(3, Pt. 2), 36–42, 1982.
130. Hattori, Y. and Kanno, M., Effect of Ni^{2+} on the multiphasic positive inotropic responses to histamine mediated by H1-receptors in left atria of guinea pigs, *Naunyn Schmiedebergs Arch. Pharmacol.*, 329(2), 188–194, 1985.
131. Lawson, K. and Cavero, I., Contractile responses to calcium chloride in rat aortic rings bathed in K^+-free solution are resistant to organic calcium antagonists, *Br. J. Pharmacol.*, 96(1), 17–22, 1989.
132. Blackburn, K. and Highsmith, R. F., Nickel inhibits endothelin-induced contractions of vascular smooth muscle, *Am. J. Physiol.*, 258(6, Pt. 1), C1025–C1030, 1990.

133. Cuthbert, N. J., Gardiner, P. J., Nash, K., and Poll, C. T., Roles of Ca^{2+} influx and intracellular Ca^{2+} release in agonist-induced contractions in guinea pig trachea, *Am. J. Physiol.*, 266(6, Pt. 1), L620–L627, 1994.
134. Murphy, N. P., Ball, S. G., and Vaughan, P. F., The effect of calcium antagonists on the release of [^{3}H] noradrenaline in the human neuroblastoma, SH-SY5Y, *Neurosci. Lett.*, 129(2), 229–232, 1991.
135. Levi, A. J., A role for sodium/calcium exchange in the action potential shortening caused by strophanthidin in guinea pig ventricular myocytes, *Cardiovasc. Res.*, 27(3), 471–481, 1993.
136. Kristensen, S. R., Cell damage caused by ATP depletion is reduced by magnesium and nickel in human fibroblasts — a nonspecific calcium antagonism?, *Biochim. Biophys. Acta,* 1091(3), 285–293, 1991.
137. Gordon, S. E. and Zagotta, W. N., A histidine residue associated with the gate of the cyclic nucleotide-activated channels in rod photoreceptors, *Neuron,* 14(1), 177–183, 1995.
138. Karpen, J. W., Brown, R. L., Stryer, L., and Baylor, D. A., Interactions between divalent cations and the gating machinery of cyclic GMP-activated channels in salamander retinal rods, *J. Gen. Physiol.*, 101(1), 1–25, 1993.
139. Bigay, J., Deterre, P., Pfister, C., and Chabre, M., Fluoride complexes of aluminium or berylium act on G-proteins as reversibly bound analogues of the gamma phosphate of GTP, *EMBO J.*, 6(10), 2907–2913, 1967.
140. Lyu, R. M. and Smith, J. B., Genistein inhibits calcium release by platelet-derived growth factor but not bradykinin or cadmium in human fibroblasts, *Cell Biol. Toxicol.*, 9(2), 141–148, 1993.
141. Legagneux, V., Morange, M., and Bensaude, O., Heat-shock and related stress enhance RNA polymerase II C-terminal-domain kinase activity in HeLa cell extracts, *Eur. J. Biochem.*, 193(1), 121–126, 1990.
142. Dubois, M. F., Bensaude, O., and Morange, M., IIa/IIo conversion of RNA polymerase II during heat shock, *C. R. Acad. Sci. III,* 313(3), 165–170, 1991.
143. Dubois, M. F., Bellier, S., Seo, S. J., and Bensaude, O., Phosphorylation of the RNA polymerase II largest subunit during heat shock and inhibition of transcription in HeLa cells, *J. Cell Physiol.*, 158(3), 417–426, 1994.
144. Landry, J., Lambert, H., Zhou, M., Lavoie, J. N., Hickey, E., Weber, L. A., and Anderson, C. W., Human HSP27 is phosphorylated at serines 78 and 82 by heat shock and mitogen-activated kinases that recognize the same amino acid motif as S6 kinase II, *J. Biol. Chem.*, 267(2), 794–803, 1992.
145. Rouse, J., Cohen, P., Trigon, S., Morange, M., Alonso, Llamazares, A., Zamanillo, D., Hunt, T., and Nebreda, A. R., A novel kinase cascade triggered by stress and heat shock that stimulates MAPKAP kinase-2 and phosphorylation of the small heat shock proteins, *Cell,* 78(6), 1027–1037, 1994.
146. Huot, J., Lambert, H., Lavoie, J. N., Guimond, A., Houle, F., and Landry, J., Characterization of 45-kDa/54-kDa HSP27 kinase, a stress-sensitive kinase which may activate the phosphorylation-dependent protective function of mammalian 27-kDa heat-shock protein HSP27, *Eur. J. Biochem.*, 227(1–2), 416–427, 1995.
147. Corton, J. M., Gillespie, J. G., and Hardie, D. G., Role of the AMP-activated protein kinase in the cellular stress response, *Curr. Biol.*, 4(4), 315–324, 1994.
148. Yagil, E. and Be'eri, H., Arsenate-resistant alkaline phosphatase-constitutive mutants of *Escherichia coli*, *Mol. Gen. Genet.*, 154(2), 185–189, 1977.
149. Gettins, P. and Coleman, J. E., ^{113}Cd NMR. Arsenate binding to Cd(II) alkaline phosphate, *J. Biol. Chem.*, 259(8), 4987–4990, 1984.
150. Pappas, P. W. and Leiby, D. A., Competitive, uncompetitive and mixed inhibitors of the alkaline phosphatase activity associated with the isolated brush border membrane of the tapeworm *Hymenolepis diminuta*, *J. Cell Biochem.*, 40(2), 239–248, 1989.
151. Guy, G. R., Cairns, J., Ng, S. B., and Tan, Y. H., Inactivation of a redox-sensitive protein phosphatase during the early events of tumor necrosis factor/interleukin-1 signal transduction, *J. Biol. Chem.*, 268(3), 2141–2148, 1993.
152. Mivechi, N. F., Koong, A. C., Giaccia, A. J., and Hahn, G. M., Analysis of HSF-1 phosphorylation in A549 cells treated with a variety of stresses, *Int. J. Hyperthermia,* 10(3), 371–379, 1994.
153. Cummings, B., Kaser, M. R., Wiggins, G., Ord, M. G., and Stocken, L. A., Beryllium toxicity. The selective inhibition of casein kinase 1, *Biochem. J.*, 208(1), 141–146, 1982.
154. Meggio, F., Agostinis, P., and Pinna, L. A., Casein kinases and their protein substrates in rat liver cytosol: evidence for their participation in multimolecular systems, *Biochim. Biophys. Acta,* 846(2), 248–256, 1985.
155. Garin, J. and Vignais, P. V., Characterization of the inhibition of rabbit muscle adenylate kinase by fluoride and beryllium ions, *Biochemistry,* 32(27), 6821–6827, 1994.
156. Baukrowitz, T., Hwang, T. C., Nairn, A. C., and Gadsby, D. C., Coupling of CFTR Cl-channel gating to an ATP hydrolysis cycle, *Neuron,* 12(3), 473–482, 1994.
157. Sibirnyi, A. A. and Shavlovskii, G. M., Inhibition of alkaline phosphatase I of *Pichia guilliermondii* yeast *in vitro* and *in vivo*, *Ukr. Biokhim. Zh.*, 50(2), 212–217, 1978.
158. Spitzer, H. L. and Johnston, J. M., Characterization of phosphatidase phosphohydrolase activity associated with isolated lamellar bodies, *Biochim. Biophys. Acta,* 531(3), 275–285, 1978.
159. Perevoshchikova, K. A., Prokoph, H., Hering, B., Koen, I. M., and Zbarskii, I. B., Effect of heparin, spermidine and Be^{2+} ions on the phosphatase and RNAse activity of rat liver cell nuclei, *Bull. Eksp. Biol. Med.*, 87(6), 542–544, 1979.

160. Castle, A. G., Ling, A., and Chibber, R., Properties of *p*-nitrophenyl phosphatase activity from porcine neutrophils, *Int. J. Biochem.*, 16(4), 411–416, 1984.
161. Price, D. J. and Joshi, J. G., Ferritin: protection of enzymatic activity against the inhibition by divalent metal ions *in vitro*, *Toxicology*, 31(2), 151–163, 1984.
162. Yora, T. and Sakagishi, Y., Comparative biochemical study of alkaline phosphatase isozymes in fish, amphibians, reptiles, birds and mammals, *Comp. Biochem. Physiol.*, 85(3), 649–658, 1986.
163. Grazia-Cifone, M., Alesse, E., Procopio, A., Paolini, R., Marrone, S., Di-Eugenio, R., Santoni, E., and Santoni, A., Effects of cadmium on lymphocyte activation, *Biochim. Biophys. Acta*, 101(1), 25–32, 1989.
164. Speizer, L. A., Watson, M. J., Kanter, J. R., and Brunton, L. L., Inhibition of phorbol ester binding and protein kinase C activity by heavy metals, *J. Biol. Chem.*, 264(10), 5581–5585, 1989.
165. Lison, D., Raguzzi, F., and Lauwerys, R., Comparison of the effects of auranofin, heavy metals and retinoids on protein kinase C *in vitro* and on a protein kinase C mediated response in macrophages, *Pharmacol. Toxicol.*, 67(3), 239–242, 1990.
166. Mazzei, G. J., Girard, P. R., and Kuo, J. F., Environmental pollutant Cd^{2+} biphasically and differentially regulates myosin light chain kinase and phospholipid/Ca^{2+}-dependent protein kinase, *FEBS Lett.*, 173(1), 124–128, 1984.
167. Forbes, I. J., Zalewski, P. D., Giannakis, C., Petkoff, H. S., and Cowled, P. A., Interaction between protein kinase C and regulatory ligand is enhanced by a chelatable pool of cellular zinc, *Biochim. Biophys. Acta*, 105(2–3), 113–117, 1990.
168. Block, C., Freyermuth, S., Beyersmann, D., and Malviya, A. N., Role of cadmium in activating nuclear protein kinase C and the enzyme binding to nuclear protein, *J. Biol. Chem.*, 267(28), 19824–19828, 1992.
169. Suzuki, K., Ikebuchi, H., and Terao, T., Mercuric and cadmium ions stimulate phosphorylation of band 4.2 protein on human erythrocyte membranes, *J. Biol. Chem.*, 260(7), 4526–4530, 1986.
170. Sotiroudis, T. G., Lanthanide ions and Cd^{2+} are able to substitute for Ca^{2+} in regulating phosphorylase kinase, *Biochem. Int.*, 13(1), 59–64, 1986.
171. Hurst, R., Schatz, J. R., and Matts, R. L., Inhibition of rabbit reticulocyte lysate protein synthesis by heavy metal ions involves the phosphorylation of the alpha-subunit of the eukaryotic initiation factor 2, *J. Biol. Chem.*, 262(33), 15939–15945, 1987.
172. Matts, R. L., Schatz, J. R., Hurst, R., and Kagen, R., Toxic heavy metal ions activate the heme-regulated eukaryotic initiation factor-2 alpha kinase by inhibiting the capacity of hemin-supplemented reticulocyte lysates to reduce disulfide bonds, *J. Biol. Chem.*, 266(19), 12695–12702, 1991.
173. Ahammadsahib, K. I., Ramamurthi, R., and Dusaiah, D., Mechanism of inhibition of rat brain (Na^{+}-K^{+})-stimulated adenosine triphosphatase reaction by cadmium and methyl mercury, *J. Biochem. Toxicol.*, 2, 169–180, 1987.
174. Lindemann, C. B., Gardner, T. K., Westbrook, E., and Kanous, K. S., The calcium-induced curvature reversal of rat sperm is potentiated by cAMP and inhibited by anti-calmodulin, *Cell Motil. Cytoskeleton*, 20(4), 316–324, 1991.
175. Ezaki, O., IIB group metal ions (Zn^{2+}, Cd^{2+}, Hg^{2+}) stimulate glucose transport activity by post-insulin receptor kinase mechanism in rat adipocytes, *J. Biol. Chem.*, 264(27), 16118–16122, 1989.
176. Theocharis, S. E., Margeli, A. P., Ghiconti, K., and Varonos, D., Liver thymidine kinase activity after cadmium-induced hepatotoxicity in rats, *Toxicol. Lett.*, 63(2), 181–190, 1992.
177. Jin, P. and Ringertz, N. R., Cadmium induces transcription of proto-oncogenes c-*jun* and c-*myc* in rat L6 myoblasts, *J. Biol. Chem.*, 265(24), 14061–14064, 1990.
178. Epner, D. E. and Herschman, H. R., Heavy metals induce expression of the TPA-inducible sequence (TIS) genes, *J. Cell Physiol.*, 148(1), 68–74, 1991.
179. Tang, N. and Enger, M. D., Cd^{2+}-induced c-*myc* mRNA accumulation in NRK-49F cells is blocked by the protein kinase inhibitor H7 but not by HA1004, indicating that protein kinase C is a mediator of the response, *Toxicology*, 81(2), 155–164, 1993.
180. van-Delft, S., Coffer, P., Kruijer, W., and van-Wijk, R., c-*fos* induction by stress can be mediated by the SRE, *Biochem. Biophys. Res. Commun.*, 197(2), 542–548, 1993.
181. Suzuki, Y., Chao, S. H., Zysk, J. R., and Cheung, W. Y., Stimulation of calmodulin by cadmium ion, *Arch. Toxicol.*, 57(3), 205–211, 1985.
182. Akerman, K. E., Honkaniemi, J., Scott, I. G., and Andersson, L. C., Interaction of Cd^{2+} with the calmodulin-activated (Ca^{2+} Mg^{2+})-ATPase activity of human erythrocyte ghosts, *Biochim. Biophys Acta*, 845(1), 48–53, 1985.
183. Martin, S. R., Linse, S., Bayley, P. M., and Forsen, S., Kinetics of cadmium and terbium dissociation from calmodulin and its tryptic fragments, *Eur. J. Biochem.*, 161(3), 595–601, 1986.
184. Martin, S. R. and Bayley, P. M., The effects of Ca^{2+} and Cd^{2+} on the secondary and tertiary structure of bovine testis calmodulin. A circular-dichroism study, *Biochem. J.*, 238(2), 485–490, 1986.
185. Akiyama, K., Sutoo, D., and Reid, D. G., A 1H-NMR comparison of calmodulin activation by calcium and by cadmium, *J. Pharmacol.*, 53(3), 393–401, 1990.
186. Buccigross, J. M., O'Donnell, C. L., and Nelson, D. J., A flow-dialysis method for obtaining relative measures of association constants in calmodulin-metal-ion systems, *Biochem. J.*, 235-(3), 677–684, 1986.

187. Niewenhuis, R. J. and Prozialeck, W. C., Calmodulin inhibitors protect against cadmium-induced testicular damage in mice, *Biol. Reprod.*, 37(1), 127–133, 1987.
188. Lag, M. and Helgeland, K., Ion transport and cadmium-induced inhibition of ciliary activity and induction of swelling of epithelial cells in mouse trachea organ culture, *Toxicology,* 47(3), 247–248, 1987.
189. Flik, G., van-de-Winkel, J. G., Part, P., Bonga, S. E., and Lock, R. A., Calmodulin-mediated cadmium inhibition of phosphodiesterase activity *in vitro*, *Arch. Toxicol.,* 59(5), 353–359, 1987.
190. Vig, P. J., Nath, R., and Desaiah, D., Metal inhibition of calmodulin activity in monkey brain, *J. Appl. Toxicol.,* 9(5), 313–316, 1989.
191. Kostrzewska, A. and Sobieszek, A., Diverse actions of cadmium on the smooth muscle myosin phosphorylation system, *FEBS Lett.,* 263(2), 381–384, 1990.
192. Sutoo, D., Akiyama, K., and Imamiya, S., A mechanism of cadmium poisoning: the cross effect of calcium and cadmium in the calmodulin-dependent system, *Arch. Toxicol.,* 64(2), 161–164, 1990.
193. Vig, P. J. and Nath, R., *In vivo* effects of cadmium on calmodulin and calmodulin-regulated enzymes in rat brain, *Biochem. Int.,* 23(5), 927–934, 1991.
194. Chao, S. H., Bu, C. H., and Cheung, W. Y., Activation of troponin C by Cd^{2+} and Pb^{2+}, *Arch. Toxicol.,* 64(6), 490–496, 1990.
195. Hamer, E. and Schoner, W., Modification of the E1ATP binding site of Na^+K^+-ATPase by the chromium complex of adenosine 5-[beta, gamma-methylene]triphosphate blocks the overall reaction but not the partial activities of the E2 conformation, *Eur. J. Biochem.,* 213(2), 743–748, 1993.
196. Smith, G. M. and Mildvan, A. S., Nuclear magnetic resonance studies of the nucleotide binding sites of porcine adenylate kinase, *Biochemistry,* 21(24), 6119–6123, 1982.
197. Meshitsuka, S., Studies of ATP-utilizing enzymes using Cr-ATP and Co-ATP complexes as substrate analogues, *Seikagaku,* 55(5), 320–324, 1993.
198. Pecoraro, V. L., Rawlings, J., and Cleland, W. W., Investigation of substrate specificity of creatine kinase using chromium (III) and cobalt (III) complexes of adenosine 5′-diphosphate, *Biochemistry,* 23(1), 153–158, 1984.
199. Kwok, F. and Churchich, J. E., The interaction of paramagnetic ions chelated to ATP with pyridoxal analogues. Fluorescence studies of pyridoxal kinase, *Eur. J. Biochem.,* 199(1), 157–162, 1991.
200. Gregory, J. D. and Serpersu, E. H., Arrangement of substrates at the active site of yeast phosphoglycerate kinase. Effect of sulfate ion, *J. Biol. Chem.,* 268(6), 3880–3888, 1993.
201. Lu, Z., Shorter, A. L., and Dunaway-Mariano, D., Investigations of kinase substrate specificity with aqua Rh(III) complexes of adenosine 5′-triphosphate, *Biochemistry,* 32(9), 2378–2385, 1993.
202. DiPolo, R. and Beauge, L., Effects of some metal-ATP complexes on Na^+-Ca^{2+} exchange in internally dialyzed squid axons, *J. Physiol.,* 462, 71–86, 1993.
203. MacNeil, S., Dawson, R., Lakey, T., and Morris, B., Activation of calmodulin by the essential trace element chromium, *Cell Calcium,* 8(3), 207–216, 1987.
204. King, M. M. and Huang, C. Y., Activation of calcineurin by nickel ions, *Biochem. Biophys. Res. Commun.,* 114(3), 955–961, 1983.
205. Matsui, H., Pallen, C. J., Adachi, A. M., Wang, J. H., and Lam, P. H., Demonstration of different metal ion-induced calcineurin conformations using a monoclonal antibody, *J. Biol. Chem.,* 260(7), 4174–4179, 1985.
206. Pallen, C. J. and Wang, J. H., Regulation of calcineurin by metal ions. Mechanism of activation by Ni^{2+} and an enhanced response to Ca^{2+}/calmodulin, *J. Biol. Chem.,* 259(10), 6134–6141, 1984.
207. Chan, C. P., Gallis, B., Blumenthal, D. K., Pallen, C. J., Wang, J. H., and Krebs, E. G., Characterization of the phosphotyrosylprotein phosphatase activity of calmodulin-dependent protein phosphatase, *J. Biol. Chem.,* 261(21), 9890–9895, 1986.
208. Li, H. C. and Chan, W. W., Activation of train calcineurin towards proteins containing Thr(P) and Ser(P) by Ca^{2+}, calmodulin, Mg^{2+} and transition metal ions, *Eur. J. Biochem.,* 144(3), 447–452, 1984.
209. King, M. M., Modification of the calmodulin-stimulated phosphatase, calcineurin, by sulfhydryl reagents, *J. Biol. Chem.,* 261(9), 4081–4084, 1986.
210. Chan, C. P., Gallis, B., Blumenthal, D. K., Palen, C. J., Wang, J. H., and Krebs, E. G., Characterization of the phosphotyrosyl protein phosphatase activity of calmodulin-dependent protein phosphatase, *J. Biol. Chem.,* 261(21), 9890–9895, 1986.
211. Mukai, H., Ito, A., Kishima, K., Kuno, T., and Tanaka, C., Calmodulin antagonists differentiate between Ni^{2+}- and Mn^{2+}-stimulated phosphatase activity of calcineurin, *J. Biochem. Tokyo,* 110(3), 402–406, 1991.
212. Yokoyama, N., Ali, Z., and Wang, J. H., Isozyme-specific monoclonal antibodies of CaM-stimulated phosphatase: identification of an enzyme domain involved in metal ion activation, *Arch. Biochem. Biophys.,* 300(2), 615–621, 1993.
213. Yokoyama, N. and Wang, J. H., The role of the autoinhibitory domain in differential metal ion activation of calmodulin-stimulated phosphatase, *FEBS Lett.,* 337(2), 128–130, 1994.
214. Raos, N. and Kasprzak, K. S., Allosteric binding of nickel(II) to calmodulin, *Fundam. Appl. Toxicol.,* 13(4), 816–822, 1989.

215. Raos, N. and Kasprzak, K. S., Effect of nickel(II) acetate on distribution of calmodulin in the rat kidney, *Toxicol. Lett.,* 48(3), 275–282, 1989.
216. Mooney, R. A. and Bordwell, K. L., Differential dephosphorylation of the insulin receptor and its 160-kDa substrate (pp160) in rat adipocytes, *J. Biol. Chem.,* 267(20), 14054–14060, 1992.
217. Levinson, W., Oppermann, H., and Jackson, J., Transition series metals and sulfhydryl reagents induce the synthesis of four proteins in eukaryotic cells, *Biochim. Biophys. Acta,* 606, 170–180, 1980.
218. Johnson, D., Oppermann, H., Jackson, J., and Levinson, W., Induction of four proteins in chick embryo cells by sodium arsenite, *J. Biol. Chem.,* 255, 6975–6978, 1980.
219. Vincent, M. and Tanguay, R. M., Different intracellular distributions of heat-shock and arsenite-induced proteins in *Drosophila* Kc cells, *J. Mol. Biol.,* 162, 365–378, 1982.
220. Brown, I. R. and Rush, S. J., Induction of a "stress" protein in intact mammalian organs after the intravenous administration of sodium arsenite, *Biochem. Biophys. Res. Commun.,* 120, 1, 1984.
221. Kim, Y. J., Shuman, J., Sette, M., and Przybyla, A., Arsenate induces stress proteins in cultured rat myoblasts, *J. Cell Biol.,* 96(2), 393–400, 1983.
222. Tanguay, R. M., Camato, R., Lettre, F., and Vincent, M., Expression of histone genes during heat shock and in arsenite-treated *Drosophila* Kc cells, *Can. J. Biochem. Cell Biol.,* 61(6), 414–420, 1983.
223. Atkinson, B. G., Cunningham, T., Dean, R. L., and Somerville, M., Comparison of the effects of heat shock and metal-ion stress on gene expression in cells undergoing myogenesis, *Can. J. Biochem. Cell Biol.,* 61(6), 404–413, 1983.
224. Heikkila, J. J., Darasch, S. P., Mosser, D. D., and Bols, N. C., Heat and sodium arsenite act synergistically on the induction of heat shock gene expression in *Xenopus laevis* A6 cells, *Biochem. Cell Biol.,* 65(4), 310–316, 1987.
225. Hiwasa, T. and Sakiyama, S., Increase in the synthesis of a M_r 32,000 protein in BALB/c 3T3 cells after treatment with tumor promoters, chemical carcinogens, metal salts, and heat shock, *Cancer Res.,* 46(5), 2474–2481, 1986.
226. Caltabiano, M. M., Koestler, T. P., Poste, G., and Greig, R. G., Induction of 32- and 34-kDa stress proteins by sodium arsenite, heavy metals, and thiol-reactive agents, *J. Biol. Chem.,* 261(28), 13381–13386, 1986.
227. Gurley, W. B., Czarnecka, E., Nagao, R. T., and Key, J. L., Upstream sequences required for efficient expression of a soybean heat shock gene, *Mol. Cell Biol.,* 6(2), 559–565, 1986.
228. van-Bergen-en-Henegouwen, P. M. and Linnemans, A. M., Heat shock gene expression and cytoskeletal alterations in mouse neuroblastoma cells, *Exp. Cell Res.,* 171(2), 367–375, 1987.
229. Andrews, G. K., Harding, M. A., Calvet, J. P., and Adamson, E. D., The heat shock response in HeLa cells is accompanied by elevated expression of the c-*fos* proto-oncogene, *Mol. Cell Biol.,* 7(10), 3462–3468, 1987.
230. Gubits, R. M. and Fairhurst, J. L., c-*fos* mRNA levels are increased by the cellular stressors, heat shock, and sodium arsenite, *Oncogene,* 3(2), 163–168, 1988.
231. Geelen, J. L., Boom, R., Klaver, G. P., Minnaar, R. P., Feltkamp, M. C., van Milligen, F. J., Sol, C. J., and vander-Noordaa, J., Transcriptional activation of the major immediate early transcription unit of human cytomegalovirus by heat shock, arsenite and protein synthesis inhibitors, *J. Gen. Virol.,* 68(Pt11), 2925–2931, 1987.
232. Geelen, J. L., Minnaar, R. P., Boom, R., van-der-Noordaa, J., and Goudsmit, J., Heat-shock induction of the human immunodeficiency virus long terminal repeat, *J. Gen. Virol.,* 69(Pt. 11), 2913–2917, 1988.
233. Angelidis, C. E., Lazaridis, I., and Pagoulatos, G. N., Specific inhibition of simian virus-40 protein synthesis by heat and arsenite treatment, *Eur. J. Biochem.,* 172(1), 27–34, 1988.
234. Keyse, S. M. and Tyrrell, R. M., Heme oxygenase is the major 32-kDa stress protein induced in human skin fibroblasts by UVA radiation, hydrogen peroxide, and sodium arsenite, *Proc. Natl. Acad. Sci.,* 86(1), 99–103, 1989.
235. Keyse, S. M., Applegate, L. A., Tromvoukis, Y., and Tyrrell, R. M., Oxidant stress leads to transcriptional activation of the human heme oxygenase gene in cultured skin fibroblasts, *Mol. Cell Biol.,* 10(9), 4967–4969, 1990.
236. Colotta, F., Polentarutti, N., Staffico, M., Fincato, G., and Mantovani, A., Heat shock induces the transcriptional activation of c-*fos* proto-oncogene, *Biochem. Biophys. Res. Commun.,* 168(3), 1013–1019, 1990.
237. Chin, K. V., Tanaka, S., Darlington, G., Pastan, I., and Gottesman, M. M., Heat shock and arsenite increase expression of the multidrug resistance (*MDR*1) gene in human renal carcinoma cells, *J. Biol. Chem.,* 265(1), 221–226, 1990.
238. Kioka, N., Yamano, Y., Komano, T., and Ueda, K., Heat shock responsive elements in the induction of the multidrug resistance gene (*MDR*1), *FEBS Lett.,* 301(1), 37–40, 1992.
239. Albores, A., Koropatnick, J., Cherian, M. G., and Zelazowski, A. J., Arsenic induces and enhances rat hepatic metallothionein production *in vivo*, *Chem. Biol. Interact.,* 85(2–3), 127–140, 1992.
240. Desrosiers, R. and Tanquay, R. M., Further characterization of the post-translational modifications of core histones in response to heat and arsenite stress in *Drosophila, Biochem. Cell Biol.,* 64(8), 750–757, 1986.
241. Burdon, R. H., Temperature and animal cell protein synthesis, *Symp. Soc. Exp. Biol.,* 41, 113–133, 1987.
242. Burdon, R. H., Gill, V. M., and Rice-Evans, C., Oxidative stress and heat shock protein induction in human cells, *Free Radic. Res. Commun.,* 3(1–5), 129–139, 1987.

243. Asano, M., Nagashima, H., Iwakura, Y., and Kawada, Y., Interferon production under the control of heterologous inducible enhancers and promoters, *Microbiol. Immunol.,* 32(6), 589–596, 1988.
244. Darasch, S., Mosser, D. D., Bols, N. C., and Heikkla, J. J., Heat shock gene expression in *Xenopus laevis* A6 cells in response to heat shock and sodium arsenite treatments, *Biochem. Cell. Biol.,* 66(8), 862–870, 1988.
245. Shuman, J. and Przybyla, A., Expression of the 31-kD stress protein in rat myoblasts and hepatocytes, *DNA,* 7(7), 475–482, 1988.
246. Amaral, M. D., Galego, L., and Rodrigues-Pousada, C., Stress response of *Tetrahymena pyriformis* to arsenite and heat shock: differences and similarities, *Eur. J. Biochem.,* 171(3), 463–470, 1988.
247. Bond, U., Heat shock but not other stress inducers leads to the disruption of a sub-set of snRNPs and inhibition of *in vitro* splicing in HeLa cells, *EMBO J.,* 1:7(12), 4020, 1988.
248. Legagneux, V., Morange, M., and Bensaude, O., Heat-shock and related stress enhance RNA polymerase II C-terminal-domain kinase activity in HeLa extracts, *Eur. J. Biochem.,* 193(1), 121–126, 1990.
249. Dubois, M. F., Bellier, S., Seo, S. J., and Bensaude, O., Phosphorylation of the RNA polymerase II largest subunit during heat shock and inhibition of transcription in HeLa cells, *J. Cell Physiol.,* 158(3), 417–426, 1994.
250. Trigon, S. and Morange, M., Different carboxyl-terminal domain kinase activities are induced by heat-shock and arsenite — characterization of their substrate specificity, separation by mono Q chromatography, and comparison with the mitogen-activated protein kinases, *J. Biol. Chem.,* 270, 22, 13091–13098, 1995.
251. Liu, R. Y., Corry, P. M., and Lee, Y. J., Regulation of chemical stress-induced *HSP70* gene expression in murine L929 cells, *J. Cell Sci.,* 107, 2209–2214, 1994.
252. Perry, S. T., Klukarni, S. B., Lee, K. L., and Kenney, F. T., Selective effect of the metallocarcinogen beryllium on hormonal regulation of gene expression in cultured cells, *Cancer Res.,* 42(2), 473–476, 1982.
253. Skilleter, D. N., Barrass, N. C., and Price, R. J., c-*myc* expression is maintained during the G1 phase cell cycle block produced by beryllium, *Cell Prolif.,* 24(2), 229–237, 1991.
254. Enger, M. D., Rall, L. B., and Hildebrand, C. E., Thioneine gene expression in Cd^{++}-variants of the CHO cell: correlation of thioneine synthesis rates with translatable mRNA levels during induction, deinduction and superinduction, *Nucleic Acids Res.,* 7(1), 271–288, 1979.
255. Enger, M. D., Rall, L. B., Walters, R. A., and Hildebrand, C. E., Regulation of induced thioneine gene expression in cultured mammalian cells: effects of protein synthesis inhibition of translatable thioneine mRNA levels in regulatory variants of the CHO cell, *Biochem. Biophys. Res. Commun.,* 93(2), 343–348, 1980.
256. Durnam, D. M. and Palmiter, R. D., Transcriptional regulation of the mouse metallothionein-1 gene by heavy metals, *J. Biol. Chem.,* 256(11), 5712–5716, 1981.
257. Glanville, N., Durnam, D. M., and Palmiter, R. D., Structure of mouse metallothionein-1 gene and its mRNA, *Nature,* 292(5820), 267–269, 1981.
258. Compere, S. J. and Palmiter, R. D., DNA methylation controls the inducibility of the mouse metallothionein-I gene lymphoid cells, *Cell,* 25(1), 233–240, 1981.
259. Beach, L. R. and Palmiter, R. D., Amplification of the metallothionein-I gene in cadmium-resistant mouse cells, *Proc. Natl. Acad. Sci.,* 78(4), 211–2114, 1981.
260. Mayo, K. E. and Palmiter, R. D. Glucocorticoid regulation of the mouse metallothionein I gene is selectively lost following amplification of the gene, *J. Biol. Chem.,* 25, 257(6), 3061–3067, 1982.
261. Gick, G. G. and McCarty, Sr., K. S., Amplification of the metallothionein-I gene in cadmium- and zinc-resistant Chinese hamster ovary cells, *J. Biol. Chem.,* 257(15), 9049–9053, 1982.
262. Palmiter, R. D., Chen, H. Y., and Brinster, R. L., Differential regulation of metallothionein-thymidine kinase fusion genes in transgenic mice and their offspring, *Cell,* 29(2), 701–710, 1982.
263. Brinster, R. L., Chen, H. Y., Warren, R., Sarthy, A., and Palmiter, R. D., Regulation of metallothionein-thymidine kinase fusion plasmids injected into mouse eggs, *Nature,* 296(5852), 39–42, 1982.
264. Andersen, R. D., Birren, B. W., Ganz, T., Piletz, J. E., and Herschman, H. R., Molecular cloning of the rat metallothionein 1 (MT-1) mRNA sequence, *DNA,* 2(1), 15–22, 1983.
265. Griffith, B. B., Walters, R. A., Enger, M. D., Hildebrand, C. E., and Griffith, J. K., cDNA cloning and nucleotide sequence comparison of Chinese hamster metallothionein I and II mRNAs, *Nucleic Acids Res.,* 11(3), 901–910, 1983.
266. Karin, M. and Richards, R. I., The human metallothionein gene family: structure and expression, *Environ. Health Perspect.,* 54, 111–115, 1984.
267. Durnam, D. M. and Palmiter, R. D., Induction of metallothionein-I mRNA in cultured cells by heavy metals and iodoacetate: evidence for gratuitous inducers, *Mol. Cell Biol.,* 4(3), 484–491, 1984.
268. Seguin, C., Felber, B. K., Carter, A. D., and Hamer, D. H., Competition for cellular factors that activate metallothionein gene transcription, *Nature,* 312(5996), 781–785, 1984.
269. Searle, P. F., Davison, B. L., Stuart, G. W., Wilkie, T. M., Norstedt, G., and Palmiter, R. D., Regulation, linkage, and sequence of mouse metallothionen I and II genes, *Mol. Cell Biol.,* 4(7), 1221–1230, 1984.
270. Varshney, U., Jahroudi, N., Foster, R., and Gedamu, L., Structure, organization, and regulation of human metallothionein IF gene: differential and cell-type-specific expression in response to heavy metals and glucocorticoids, *Mol. Cell Biol.,* 6(1), 25–37, 1986.
271. Karin, M., Haslinger, A., Heguy, A., Dietlin, T., and Cooke, T., Metal-responsive elements act as positive modulators of human metallothionein-IA enhancer activity, *Mol. Cell Biol.,* 7(2), 606–613, 1987.

272. Andersen, R. D., Taplitz, S. J., Wong, S., Bristol, G., Larkin, B., and Herschman, H. R., Metal-dependent binding of a factor *in vivo* to the metal-responsive elements of the metallothionein I gene promoter, *Mol. Cell Biol.,* 7(10), 3574–3581, 1987.
273. Seguin, C. and Prevost, J., Detection of a nuclear protein that interacts with a metal regulatory element of the mouse metallothionein I gene, *Nucleic Acids Res.,* 16(22), 10547–10560, 1988.
274. Zinn, K., Keller, A., Whittemore, L. A., and Maniatis, T., 2-Aminopurine selectively inhibits the induction of beta-interferon, c-*fos*, and c-*myc* gene expression, *Science,* 240(4849), 210–213, 1988.
275. Hanke, T., Tyers, M., and Hanley, C. B., Metallothionein RNA levels in HL-60 cells. Effect of cadmium, differentiation, and protein kinase C activation, *FEBS Lett.,* 24(1–2), 159–163, 1988.
276. Radtke, F., Heuchel, R., Georgiev, O., Hergersberg, M., Gariglio, M., Dembic, Z., and Schaffner, W., Cloned transcription factor MTF-1 activates the mouse metallothionein I promoter, *EMBO J.,* 12(4), 1355–1362, 1993.
277. Palmiter, R. D., Regulation of metallothionein genes by heavy metals appears to be mediated by a zinc-sensitive inhibitor that interacts with a constitutively active transcription factor, MTF-1, *Proc. Natl. Acad. Sci.,* 91(4), 1219–1223, 1994.
278. Heuchel, R., Radtke, F., Georgiev, O., Stark, G., Aguet, M., and Schaffner, W., The transcription factor MTF-1 is essential for basal and heavy metal-induced metallothionein gene expression, *EMBO J.,* 13(12), 2870–2875, 1994.
279. Brugnera, E., Georgiev, O., Radtke, F., Heuchel, R., Baker, E., Sutherland, G. R. and Schaffner, W., Cloning, chromosomal mapping and characterization of the human metal-regulatory transcription factor MTF-1, *Nucleic Acids Res.,* 22(15), 3167–3173, 1994.
280. Cervera, J., Induction of self-tolerance and enhanced stress protein synthesis in L-132 cells by cadmium chloride and by hyperthermia, *Cell Biol. Int. Rep.,* 9(2), 131–141, 1985.
281. Wuk, B. J., Kingston, R. E., and Morimoto, R. I., Human HSP70 promoter contains at least two distinct regulatory domains, *Proc. Natl. Acad. Sci.,* 83(3), 629–633, 1986.
282. Misra, S., Zafarullah, M., Price-Haughey, J., and Gedamu, L., Analysis of stress-induced gene expression in fish cell lines exposed to heavy metals and heat shock, *Biochim. Biophys. Acta,* 1007(3), 325–333, 1989.
283. Choi, H. S., Li, B., Lin, Z., Huang, E., and Liu, A. Y., cAMP and cAMP-dependent protein kinase regulate the human heat shock protein 70 gene promoter activity, *J. Biol. Chem.,* 266(18), 11858–11865, 1991.
284. Murakami, T., Yano, O., Takahashi, H., Akiya, S., and Higashi, K., Effects of cadmium on the gene expression of retinoblastoma (Y79) cells in culture, *Nippon Ganka Gakkai Zasshi,* 96(6), 737–741, 1992.
285. Goering, P. L., Fisher, B. R., and Kish, C. L., Stress protein synthesis induced in rat liver by cadmium precedes hepatotoxicity, *Toxicol. Appl. Pharmacol.,* 122(1), 139–148, 1993.
286. Hiranuma, K., Hirata, K., Abe, T., Hirano, T., Matsuno, K., Hirano, H., Suzuki, K., and Higashi, K., Induction of mitochondrial chaperonin, *HSP60,* by cadmium in human hepatoma cells, *Biochem. Biophys. Res. Commun.,* 194(1), 531–536, 1993.
287. Enger, M. D., Flomerfelt, F. A., Wall, P. L., and Jenkins, P. S., Cadmium produces a delayed mitogenic response and modulates the EGF response in quiescent NRK cells, *Cell Biol. Toxicol.,* 3(4), 407–416, 1987.
288. Tang, N. and Enger, M. D., Cadmium induces hypertrophy accompanied by increased *myc* mRNA accumulation in NRK-49F cells, *Cell Biol. Toxicol.,* 7(4), 401–411, 1991.
289. Alam, J., Shibahara, S., and Smith, A., Transcriptional activation of the heme oxygenase gene by heme and cadmium in mouse hepatoma cells, *J. Biol. Chem.,* 264(11), 6371–6375, 1989.
290. Sato, M., Ishizawa, S., Yoshida, T., and Shibahara, S., Interaction of upstream stimulatory factor with the human heme oxygenase gene promoter, *Eur. J. Biochem.,* 188(2), 231–237, 1990.
291. Takeda, K., Ishizawa, S., Sato, M., Yoshida, T., and Shibahara, S., Identification of a *cis*-acting element that is responsible for cadmium-mediated induction of the human heme oxygenase gene, *J. Biol. Chem.,* 269(36), 22858–22867, 1994.
292. Alam, J., Multiple elements with in the 5′ distal enhancer of the mouse heme oxygenase-1 gene mediate induction by heavy metals, *J. Biol. Chem.,* 269(40), 25049–25056, 1994.
293. Takeda, K., Fujita, H., and Shibahara, S., Differential control of the metal-mediated activation of the human heme oxygenase-1 and metallothionein IIA genes, *Biochem. Biophys. Res. Commun.,* 207(1), 160–167, 1995.
294. Sauk, J. J., Norris, K., Kerr, J. M., Somerman, M. J., and Young, M. F., Diverse forms of stress result in changes in cellular levels of osteonectin/SPARC without altering mRNA levels in osteoligament cells, *Calif. Tissue Int.,* 49(1), 58–62, 1991.
295. Suzuki, M., Takahashi, K., Kawazoe, Y., Sakumi, K., and Sekiguchi, M., Inhibitory effect of cadmium and mercury ions on transcription of the *ada* gene, *Biochem. Biophys. Res. Commun.,* 179(3), 1517–1521, 1991.
296. Horiguchi, H., Mukaida, N., Okamoto, S., Teranishi, H., Kasuya, M., and Matsushima, K., Cadmium induces interleukin-8 production in human peripheral blood mononuclear cells with the concomitant generation of superoxide radicals, *Lymphokine Cytokine Res.,* 12(6), 421–428, 1993.
297. Garcia-Morales, P., Saceda, M., Kenney, N., Kim, N., Satomon, D. S., Gottardis, M. M., Solomon, H. B., Sholler, P. F., Jordan, V. C., and Martin, M. B., Effect of cadmium on estrogen receptor levels and estrogen-induced responses in human breast cancer cells, *J. Biol. Chem.,* 269(24), 16896–16901, 1994.
298. Okada, S., Ohba, H., and Taniyama, M., Alterations in ribonucleic acid synthesis by chromium(III), *J. Inorg. Biochem.,* 15(3), 223–231, 1981.

299. Okada, S., Tsukada, H., and Ohba, H., Enhancement of nucleolar RNA synthesis by chromium(III) in regenerating rat liver, *J. Inorg. Biochem.*, 21(2), 113–124, 1984.
300. Wetterhahn, K. E. and Hamilton, J. W., Molecular basis of hexavalent chromium carcinogenicity: effect on expression, *Sci. Total Environ.*, 86(1–2), 113–129, 1989.
301. Hamilton, J. W. and Wetterhahn, K. E., Differential effects of chromium(VII) on constitutive and inducible gene expression in chick embryo liver *in vivo* and correlation with chromium(VI)-induced DNA damage, *Mol. Carcinog.*, 2(5), 274–286, 1989.
302. Wetterhahn, K. E., Hamilton, J. W., Aiyar, J., Borges, K. M., and Floyd, R., Mechanism of chromium(VI) carcinogenesis. Reactive intermediates and effect on gene expression, *Biol. Trace Elem. Res.*, 21, 405–411, 1989.
303. Alcedo, J. A., Misra, M., Hamilton, J. W., and Wetterhahn, K. E., The genotoxic carcinogen chromium(VI) alters the metal-inducible expression but not the basal expression of the metallothionein gene *in vivo*, *Carcinogenesis*, 15(5), 1089–1092, 1994.
304. Manning, F. C., Xu, J., and Patierno, S. R., Transcriptional inhibition by carcinogenic chromate: relationship to DNA damage, *Mol. Carcinog.*, 6(4), 270–279, 1992.
305. McCaffrey, J., Wolf, C. M., and Hamilton, J. W., Effects of the genotoxic carcinogen chromium(VI) on basal and hormone inducible phosphoenolpyruvate carboxykinase gene expression *in vivo*: correlation with glucocorticoid and developmentally regulated expression, *Mol. Carcinog.*, 19(4), 189–198, 1994.
306. Patierno, S. R. and Costa, M., Effects of nickel(II) on nuclear protein binding to DNA in intact mammalian cells, *Cancer Biochem. Biophys.*, 9(2), 113–126, 1987.
307. Goldberg, M. A., Dunning, S. P., and Bunn, H. F., Regulation of the erythropoietin gene: evidence that the oxygen sensor is a heme protein, *Science*, 242(4884), 1412–1415, 1988.
308. Imbra, R. J., Wang, X. W., and Costa, M., Characterization of a nickel resistant mouse cell line, *Biol. Trace Element. Res.*, 21, 97–103, 1989.
309. Blouin, R., Swierenga, S. H., and Marceau, N., Evidence for post-transcriptional regulation of cytokeratin gene expression in a rat liver epithelial cell line, *Biochem. Cell Biol.*, 70(1), 1–9, 1992.
310. Bieboer, E., Tom, R. T., and Rossetto, F. E., Superoxide dismutase activity and novel reactions with hydrogen peroxide of histidine-containing nickel(I)-oligopeptide complexes and nickel(II)-induced structural changes in synthetic DNA, *Biol. Trace Element. Res.*, 21, 23–33, 1989.
311. Parfett, C. L., Induction of proliferin gene expression by diverse chemical agents that promote morphological transformation in C3H/10T1/2 cultures, *Cancer Lett.*, 30, 64(1), 1–9, 1992.
312. Hardin, J., Coffman, J. A., Black, S. D., and McClay, D. R., Commitment along the dorsoventral axis of the sea urchin embryo is altered in response to $NiCl_2$, *Development*, 116(3), 671–685, 1992.
313. Eckard, K. U., Pugh, C. W., Radcliffe, P. J., and Kurtz, A., Oxygen-dependent expression of the erythropoietin gene in rat hepatocytes *in vitro*, *Pflugers Arch.*, 423(5–6), 356–364, 1993.

2 MECHANISMS OF GENOTOXICITY

M. Duane Enger

2.1 Chromosomal Alterations Induced by Carcinogenic Metals

Carcinogenic trace metals vary considerably in chemical properties. They might, therefore, be expected to show a spectrum of genotoxic mechanisms for cell transformation. Prominent as a possible genotoxic mechanism is direct or indirect induction of DNA damage, involving DNA breaks, deletions, rearrangements, and base changes. Direct damage might be expected to ensue as a result of metal-DNA adduct formation, metal-catalyzed DNA-DNA or DNA-protein cross-linking, or metal interference with replication. Indirect mechanisms could involve the production of active oxygen species as the primary metal reaction.

Alternatively, carcinogenic metals might, through binding to signal transduction intermediates or transcription factors, induce error-prone repair of DNA damage. Metal interaction with repair enzymes could also inhibit this function. Either type of activity would result in a synergistic genotoxic effect when combined with or following exposure to other mutagens.

Carcinogenic metals may be genotoxic also via clastogenic mechanisms not involving DNA damage, but through effects on the mitotic apparatus and process. Aneuploidy would be the expected result in such instances. Also included in the spectrum of carcinogenic metal mechanisms would be protooncogene activation via induction of gene amplification, involving interactions of metal ions with DNA replication intermediate forms and enzyme complexes.

Finally, interaction of metal ions with DNA methylases might result in hypo- or hyper-methylation of DNA, thereby altering gene expression in an epigenetic but heritable (in the somatic cell lineage) fashion. As will be shown in the following review of the literature, there is evidence for metal-induced cell transformation involving all these mechanisms. Clastogenic responses and mechanisms will be reviewed first, followed by discussion of metal-induced DNA damage and repair.

2.1.1 *Arsenic*

The clastogenicity of arsenic has been established both by studies of chromosome abnormalities in humans exposed to carcinogenic levels of arsenicals and by *in vitro* analyses of chromosome alterations in arsenic exposed cultured cells.[1-31] Iatrogenic[4-6] and occupational[7-10] exposure to arsenicals results in increased frequencies of sister chromatid exchanges (SCEs) and/or visible chromosomal alterations in lymphocytes cultured from the exposed individuals. Although the carcinogenic potential of arsenic is not evidenced in animals, there is at least one report of clastogenic effects *in vivo*, involving exposure of laboratory mice to sodium meta-arsenite with resultant increases in micronuclei in somatic cells.[11]

Clastogenic responses are readily demonstrated when human or rodent cells are exposed to arsenicals *in vitro*.[1,3,12-31] The trivalent form is significantly more effective in inducing aberrations in cultured mammalian or human leukocytes or fibroblasts.[12-14] Arsenite is co-clastogenic with a variety of agents, as demonstrated in cultured cells.[15-27] Incubating ultraviolet-exposed Chinese hamster ovary (CHO) cells in arsenite increases synergistically the frequency of chromosome aberrations as well as cytotoxicity.[15] Similarly, CHO and human skin fibroblasts exposed to DNA cross-linking agents plus ultraviolet-A show an enhanced clastogenic response when treated subsequently to arsenite.[16] Post- but not pretreatment of CHO cells with arsenite enhances the clastogenicity and cytotoxicity of methyl methanesulfonate treatment.[17] The clastogenic effects of ethyl methanesulfonate *in vitro* are enhanced by post-treatment with caffeine or 3-aminobenzamide as well as arsenite.[18] The co-clastogenic response *in vitro* occurs when cells are exposed to arsenite in G2 following exposure to S-phase specific clastogens,[19] requires protein synthesis,[20,21] and does not occur if cells are in stationary phase during arsenite exposure.[22] The degree to which arsenite potentiates a clastogenic response to diepoxybutane in normal human lymphocytes varies significantly among individuals.[23] Studies with synchronized cells show that the late G1/early G2 period is when cells are most susceptible to the co-clastogenic effects of ultraviolet and arsenite.[25] It is suggested that the co-clastogenicity of arsenite is due, at least in part, to inhibition of DNA repair.[23,27]

Arsenicals induce cell transformation at subtoxic as well as toxic levels, with arsenite being tenfold more effective than arsenate.[28,29] Arsenic-induced gene mutations are not evident at doses that result in cell transformation. Chromosomal aberrations are induced, however, and with the same dose response as for transformation.[28,29] Cytogenetic effects induced by arsenite include endoreduplication as well as chromosome aberrations.[28-31] (See Table 17.)

2.1.2 *Beryllium*

Despite an early report of increased chromosome aberrations in animals exposed to a toxic dose of beryllium,[32] this aspect of beryllium genotoxicity is largely unexplored with the exception of a study on the separate and combined effects of beryllium and X-rays on cultured CHO cells.[33] This study showed a significant increase in chromosome aberrations at 1.0 but not 0.2 m*M* Be^{2+} (LD_{50} = 1.1 m*M*) and a synergistic response to X-rays and Be^{2+}. (See Table 18.)

2.1.3 Cadmium

Studies on environmentally and occupationally exposed humans, on animals, and on cultured cells have established that cadmium is a clastogen. However, whether a positive response is seen is often more dependent upon conditions than is the case with other clastogens, and numerous citations exist for negative clastogenic responses to cadmium under conditions that provide for a positive response to other agents.

This variation is seen when environmentally and occupationally exposed humans are screened for cytogenetic abnormalities and has been ascribed to variations in co-factors and/or exposure profiles.[34] Chromosomal aberrations (CAs) have been seen in lymphocytes from humans working in cadmium plants where exposure to other metals such as lead also occurs[35-37] and in lymphocytes of humans whose environmental exposure has resulted in overt toxicity.[36,38] However, in one instance where 40 workers were exposed to cadmium salts for over three decades and in another involving 14 workers exposed to cadmium occupationally for 6 to 25 years, no significant increases in aberrations were observed.[39,40] Conversely, in a study of the relationship between urine Cd levels and CAs in an environmentally exposed population, both a significant clastogenic response and a correlation (r = .463) of urine Cd levels with the frequency of CAs were evident.[41] Analysis of a larger cohort (105) of occupationally exposed workers has also indicated that higher levels of cadmium exposure produce an increase in CAs.[42]

Animal studies also produce both positive and negative responses for Cd^{2+}-induced CAs. Treatment of mice with 0.5 to 3.0 mg/kg (i.p.) was found not to result in chromosome rearrangements in spermatocytes,[43] and 4 or 15 mg/kg (i.p.) produced no increase in CAs in murine bone marrow.[44] Chromosome aberrations were, however, detected in the bone marrow of mice kept on a low calcium diet containing 0.5% each Zn^{2+} and Pb^{2+} and 0.06% Cd^{2+}.[45] This regimen produced also 50% lethality at 30 days, the time of CA analysis. A concentration-dependent increase in SCEs and CAs was observed in somatic and germ cells of mice given 0.42 to 6.75 mg/kg Cd^{2+} (i.p.).[46] Increased SCEs, CAs, and nucleolar organizing regions (NORs) are not evident in maternal bone marrow and fetal livers/lungs of mice given CD^{2+} on gestational days 8 to 10 at doses up to 8.4 mg/kg, and only CAs are increased at 11.4 mg/kg.[47] Cd^{2+} also fails to increase erythrocyte micronuclei in mice under conditions producing a positive response to colchicine, hydroquinone, and vinblastine.[48] Cd^{2+} produces germline aneuploidy in *Drosophila*[49] and mice,[50] an increase in numerical aberrations in the testicular tissue of mice 3 days after treatment with a single, carcinogenic dose,[51] and an increase in micronuclei (Mn) and CAs in murine bone marrow.[52]

Cd^{2+} causes chromosomal damage and alterations in cultured cells,[53-68] but again the extent and kinds of damage are very condition dependent. Three-hour treatment of cultured hamster cells with 100 μM Cd^{2+} has a stathmokinetic effect, and chromosome aberrations are evident 12 hours after treatment with 100 μM Cd^{2+} for 1 hr.[54] Lethality is evident when cells are exposed to 10 μM Cd^{2+} for 16 hr, but no aberrations occur when the cells are exposed to 5 μM Cd^{2+} or less for this amount of time. It is thus not clear whether in these studies aberrations occur at nonlethal levels of exposure. Extended exposure of cultured human lymphocytes to concentrations of Cd^{2+} ranging from 5 to 250 μM result in an increase in aberrations, with fragments representing the most common aberration.[55] The toxic Cd^{2+} dose in this instance is reported to be 500 μM. Aberrations are also seen in unstimulated human lymphocytes exposed to 10 to 1000 μM Cd^{2+} for 3 hr.[56]

CHO cells cultured in 15% newborn calf serum show a significant increase in chromosome aberrations 12 hr after exposure to 1 μM Cd^{2+}.[57] However, this concentration also affects cell doubling immediately, and exposure to the threshold concentration of Cd^{2+} for toxicity, 0.2 μM, does not result in increased aberrations.

A marked increase in aberrations occurs in hamster cells cultured for 22 hours following exposure to 1 to 50 μM Cd^{2+} for 2 hr.[58] The ability of Cd^{2+} to induce CAs was found to be

TABLE 17

Arsenic-Induced Chromosomal Alterations

Effector(s)	Concentration	Target(s)	Effect(s)	Ref.
Arsenic	1% (ingestion)	Human lymphocytes *in vivo*	SCEs increased	4
Arsenic, Na_2HAsO_4	Chronic exposure, 1–10 μg/ml	Human lymphocytes *in vivo* and *in vitro*	CAs increased	5
Arsenic	Occupational (chronic) and iatrogenic (300–1200 mg total dose) exposure	Human lymphocytes *in vivo*	CAs increased	6–9
As^{3+}	1–2 μ*M*	Human lymphocytes *in vitro*	SCEs increased	10
As^{3+} ($NaAsO_2$)	0.5–10 μg/kg	Murine lymphocytes *in vivo*	CAs increased	11
$As^{3+} \gg As^{5+}$	0.6–7.2 μ*M* As^{3+} (24 hr)	Cultured cells	Breaks increased	12
As^{3+} (As^{5+}: NE)	Urine level in copper smelter workers	Human lymphocytes *in vitro*	CAs increased	13
$As^{3+} > As^{5+}$	1–10 μ*M* (As^{3+}) (22 hr)	Cultured CHO	CAs increased	14
As^{3+} (post-UV)	10 μ*M*	CHO cells	UV-induced CAs increased synergistically	15, 16
As^{3+} (post-CL)	1–10 μ*M*	CHO cells, skin fibroblasts	CL induced, CAs enhanced	16, 27
As^{3+} (pre-MMS)	10 μ*M*	CHO cells	MMS-induced CAs enhanced	17
As^{3+} (pre-MMS)	10 μ*M*	CHO cells	MMS-induced CAs not altered	17
As^{3+} (post-MMS)	10 μ*M*	CHO cells	MMS-induced CAs enhanced	17
As^{3+} (post-treatment)	—	CHO cells	EMS-induced CAs enhanced	18
3-aminobenzamide (post-treatment)	—	CHO cells	EMS-induced CAs enhanced	18
Caffeine (post-treatment	—	CHO cells		
As^{3+}	10–20 μ*M*	CHO cells	UV or 4-nitroquinoline-1-oxide-induced CAs enhanced	19
As^{3+}	—	CHO cells	EMS- or UV-induced CAs enhanced	20, 21, 25
As^{3+} and cycloheximide	—	CHO cells	EMS- or UV-induced CAs not enhanced	20, 21, 25
As^{3+}	1 μ*M*	Human lymphocytes *in vitro*	SCEs increased	23
As^{3+}	2 μ*M*	Human lymphocytes *in vitro*	SCEs and CAs increased	23
DEB	—	Human lymphocytes *in vitro*	SCEs and CAs increased	23
As^{3+}	1–2 μ*M*	Human lymphocytes *in vitro*	DEB-induced SCEs and CAs increased synergistically	23
As^{3+}	—	Human lymphocytes *in vitro*	X-ray- and UV-induced CAs and SCEs potentiated	24
N-NO-AAF	0.1 m*M*	CHO cells	CAs and SCEs increased	26
As^{3+} (hydroxyurea, cytosine arabinoside: NE)	10 μ*M*	CHO cells	N-NO-AAF-induced CAs and SCEs enhanced	26
As^{3+} 10× As^{5+}	3–10 μ*M* As^{3+}	Syrian hamster embryo cells	CAs with transformation	28, 29
As^{3+}	50–200 μ*M*	G2 phase human skin fibroblasts	Endoreduplication with increased CAs (40–70% cell death)	30

Note: CHO = Chinese hamster ovary; CL = cross-linking; SCE = sister chromatid exchange; CA = chromosome aberrations; NE = no effect; MMS = methyl methane sulfonate; EMS = ethyl methane sulfonate; UV = ultraviolet radiation; DEB = diepoxybutane; N-NO-AAF = *N*-nitroso-2-acetylaminefluorene.

TABLE 18
Beryllium-induced Chromosomal Alterations

Effector(s)	Concentration	Target(s)	Effect(s)	Refs.
Be^{2+}	—	Animal bone marrow cells	CAs increased	32
Be^{2+}	0.2 m*M* (LD_{50} = 1.1 m*M*)	CHO cells	NE	33
Be^{2+}	1.0 m*M*	CHO cells	CAs increased ~ 2×	33
X-ray	—	CHO cells	0.14 CA per cell per Gy	33
Be^{2+} + X-ray	—	CHO cells	CAs increased synergistically	33

Note: CHO = Chinese hamster ovary; NE = no effect; CAs = chromosome aberrations.

comparable to that induced by benzo(a)pyrene at equitoxic doses.[58] It is suggested from these studies that a period of DNA synthesis following Cd^{2+} exposure is requisite for "efficient detection" of CAs induced by this agent.

The effects of various active oxygen scavengers on the induction of CAs by Cd^{2+} suggests an involvement of H_2O_2.[59] That is, cadmium-induced CAs are reduced by catalase, mannitol, and butylated hydroxytoluene, but not by superoxide dismutase or dimethylfuran, indicating that singlet oxygen and superoxide anion are not intermediary in the effect.

Both fibroblastic and epithelioid cell lines transformed to tumorigenicity by Cd^{2+} exposure are hyperdiploid-hypertriploid.[60] Whether the increased ploidy is a cause or consequence of transformation is not defined; however, it has been shown that Cd^{2+} inhibits microtubule formation *in vitro*, as do a number of known aneuploidogens,[61] and induces micronuclei in cultured Chinese hamster cells,[62] as do most tested spindle poisons. On the other hand, Cd^{2+} does not induce chromosome malsegregation in *Aspergillus*, as do a number of known spindle poisons.[63]

Careful examination of the induction of CAs relative to cytotoxic effects shows that doses of Cd^{2+} that reduce colony formation by less than 20% produce at least 15% of cells with aberrations.[64] The time at which CAs are scored is a factor important to efficient detection of CAs. A time 17 to 24 hr after the beginning of a 3-hr exposure to Cd^{2+} appears optimal.[64-66] Cd^{2+} shows a pronounced co-clastogenic effect in mitomycin C-treated CHO K1 cells, but does not show such in excision repair-deficient human XP20SSV cells, suggesting an inhibition of repair as a mechanism for the co-clastogenicity.[68] (See Table 19.)

2.1.4 Chromium

Forms of chromium elicit chromosomal aberrations in animals. Chronic or acute exposure of rats to potassium bichromate results in a significant increase in bone marrow CAs.[69] Fish exposed to sodium dichromate i.m. or in their water also show an increase in CAs.[70] Calcium chromate is positive in the *Drosophila* sex-linked recessive lethal test but does not produce an increase in translocations.[71] The hexavalent form of chromium produces an increase in mitotic recombination in *Drosophila* not seen with Cr^{3+}.[72] It also has a clastogenic effect on bone marrow cells of albino mice 24 hr after exposure.[73]

Human studies also suggest a clastogenic response to Cr^{6+}. Lymphocytes cultured from persons exposed to Cr^{6+} show an increased frequency of CAs.[74,75] Most studies on the clastogenicity of chromium have been performed using cultured cells.[76-94] Cr^{6+} is significantly more toxic to cultured cells than Cr^{3+}, and even at equitoxic doses Cr^{6+} produces a greater number of CAs.[78] Cr^{6+} but not Cr^{3+} is clastogenic in total embryonic hamster cell cultures,[79] and a number of hexavalent Cr compounds are significantly more clastogenic toward cultured human leukocytes than are trivalent forms.[80] A similar pattern of response to Cr^{6+} and Cr^{3+} obtains in cultured mammalian cells.[81,82] Chromosome aberrations are increased tenfold when

TABLE 19

Cadmium-Induced Chromosomal Alterations

Effector(s)	Concentration	Target(s)	Effect(s)	Ref.
Cd + Pb	Occupational exposure	Human lymphocytes *in vivo*	CAs increased	35, 37
Cd	Occupational, environmental exposure	Human lymphocytes *in vivo*	CAs increased	36, 38, 42
Cd^{2+}	Occupational exposure	Human lymphocytes *in vivo*	CAs unchanged	39, 40
Cd^{2+}	0.5, 1.75, 3.0 mg/kg (i.p.)	Murine spermatocytes	CAs unchanged	43
Cd^{2+}	4, 15 mg/kg (i.p.)	Murine bone marrow	CAs unchanged	44
Cd^{2+}	0.06% of diet	Murine bone marrow	CAs unchanged	45
Zn^{2+}	0.5% of diet			
Pb^{2+}	0.5% of diet			
Ca^{2+}	0.03% of diet			
Cd^{2+}	0.42–6.75 mg/kg (i.p.)	Murine somatic and germ cells	CAs, SCEs increased	46
Cd^{2+}	8.4 mg/kg	Murine maternal bone marrow, fetal liver and lung	SCEs, NORs, CAs not increased	47
Cd^{2+}	11.4 mg/kg	Murine maternal bone marrow, fetal liver and lung	SCEs, NORs, CAs unchanged; CAs increased	47
Cd^{2+}	1–10 mg/kg	Murine bone marrow	Mn unchanged	48
Colchicine	0.05–1.0 mg/kg	Murine bone marrow	Mn increased	48
Hydroquinone	30–100 mg/kg			
Vinblastine	0.1–2.0 mg/kg			
Cd^{2+}	20–60 ppm (in food)	*Drosophila* germ cells	Germline aneuploidy increased	49
Acetonitrile	2–50 ppt			
Carbendazim	5–50 ppt			
DMSO	1–50 ppt			
Methylmercury chloride	0.3–100 ppt			
Methoxyethyl acetate	5–42 ppt			
Propionitrile	1.2–40 ppt			
Cd^{2+}	1–6 mg/kg	Murine germ cells	Germline aneuploidy increased	50
Diazepam <	100–150 mg/kg			
Colchicine	1.5–6 mg/kg			
Hydroquinone	80–120 mg/kg			
Vinblastine	0.5–2.0 mg/kg			
Cd^{2+}	1 mg/kg (i.p.) (carcinogenic dose)	Murine tests	CAs increased	51
Cd^{2+}	5–20 mg/kg	Murine bone marrow	Mn and CAs increased	52
Thimerosol	5–25 mg/kg			
Cd^{2+}	—	Human cell cultures	Chromosome damage	53
Cd^{2+}	10–100 μM (16 hr)	Human cell cultures	Stathmokinetic effect; lethality; CAs increased 12 hr postexposure	54
Cd^{2+}	5–250 μM	Stimulated human lymphocytes	CAs increased	55
Cd^{2+}	10–1000 μM	Unstimulated human lymphocytes	CAs increased	56
Cd^{2+}	1.0 μM	CHO	CAs increased	57
Cd^{2+}	1–50 μM (2 hr)	Cultured Chinese hamster cells	CAs increased 22 hr post-treatment	58

TABLE 19 (continued)

Cadmium-Induced Chromosomal Alterations

Effector(s)	Concentration	Target(s)	Effect(s)	Ref.
Cd^{2+}	20 μ*M* (2 hr)	Cultured Chinese hamster cells (V79)	CAs increased	59
Cd^{2+} with catalase, mannitol, or butylated hydroxytoluene	20 μ*M* (2 hr)	Cultured Chinese hamster cells (V79)	CAs increased less (50–100%)	59
Cd^{2+}	NA	Cd^{2+}-transformed fibroblasts and epithelial cells	Hyperdiploidy-hypertriploidy	60
Cd^{2+}	5–50 m*M*	Microtubule preparations	Microtubule assembly *in vitro* inhibited	61
Colchicine	5–50 m*M*			
Acetonitrite	5–50 m*M*			
Cd^{2+}	2–3 μg/ml	Chinese hamster cells	Micronuclei induced	62
Diazepam	20–80 μg/ml			
Thiabendazole	10–25 μg/ml			
Vinblastine	0.3–0.8 ng/ml			
Hydroquinone	1–4.5 μg/ml			
Pyrimethamine	25–50 μg/ml			
Econazole (thimerosol: NE)	6–10 μg/ml			
Chloral hydrate, thiabenzadole, thimerosol, econazole hydroquinone (Cd^{2+}: NE)	—	*Aspergillus midulans*	Chromosome malsegregation induced	63
Cd^{2+} mitomycin C, adriamycin, 2,5-diaminotoluene	Concentrations that reduce CFE ≤ 20%	CHO cells	CAs increased in ≥ 15% cells	64
Cd^{2+}	1–3 μ*M* (3 hr)	CHO cells	CAs increased 12–21 hr after 3-hr exposure	65, 66
Cd^{2+} = 10× Pb^{2+}	15 μ*M* (IC_{50})	CHO cells	SCEs and CAs increased	67
Cd^{2+}	<28 μ*M*	CHO K1	CAs not increased	68
Cd^{2+}	<3.5 μ*M*	CHO K1	CAs induced by mitomycin C increased	68
Cd^{2+}	<3.5 μ*M*	XP2055V (repair-deficient human cells)	CAs induced by 4-nitroquinoline-1-oxide not increased	68

Note: CAs = chromosome aberrations; SCEs = sister chromatid exchanges; NORs = nucleolar organizing regions; Mn = micronuclei; NA = not applicable; CHO = Chinese hamster cells; CFE = colony-forming efficiency; IC_{50} = concentration for half maximal inhibition.

CHO cells are exposed to 1.0 μg/ml Cr^{6+},[83] with increases seen in single chromatid gaps, as well as breaks and interchanges that are proportional to concentration. Studies on the responses of cultured BHK and CHO cells show Cr^{6+} to be 100- to 1000-fold more lethal than Cr^{3+}, and that Cr^{6+} induces SCEs whereas Cr^{3+} does not.[84] Further, CAs are increased 10× by 1.0 μg/ml Cr^{6+} but only 2× by 150 μg/ml Cr^{3+} in these cell lines.[84] Significant increases in SCEs are seen also in human fibroblasts exposed to 10^{-7} to 10^{-6} *M* Cr^{3+} for 48 hr, and CAs are increased at concentrations of 0.8 to 3 μ*M* Cr^{6+} but not Cr^{3+}.[85]

Although Cr^{3+} shows little clastogenic activity per se, contamination of Cr^{3+} compounds such as industrial chromite with Cr^{6+} results in their showing a significant response *in vitro*.[88,89] A possible mechanistic explanation for Cr^{3+} induction of CAs in the instances where such is observed is provided by the observation that CAs induced in stimulated human lymphocytes by this agent are blocked by superoxide dismutase, catalase, and mannitol.[91] These results suggest an indirect action mediated by active oxygen species.

Pretreatment of Chinese hamster V79 cells with vitamin E protects against Cr^{6+}-induced CAs and mutations.[92] Pretreatment with vitamin B_2, however, exacerbates the clastogenicity of Cr^{6+}, possibly through reduction of Cr^{6+} to Cr^{5+}.[93] The clastogenicity of particulate Cr^{6+} (lead chromate) is also ameliorated by pretreatment with Vitamin E.[94] (See Table 20.)

2.1.5 Nickel

Epidemiological evidence linking nickel exposure to chromosome aberrations derives primarily from studies of occupational exposure.[95–102] Nickel refinery workers exposed to nickel-containing dusts and aerosols for an extended period of time show increased chromosome breaks and gaps in peripheral lymphocytes 4 to 15 years after retirement.[96] These workers were exposed to dusts containing Ni_3S_2 and NiO and aerosols containing $NiCl_2$ and $NiSO_4$ at air concentrations that have exceeded 1.0 mg/m^3 and for periods of over 25 years. An increase in chromosome breaks is also seen in lymphocytes from workers using nickel wire for welding steel[98] and in chemical plant workers exposed to NiO.[99] There is also an increase in CAs in the lymphocytes of stainless-steel workers,[100,101] who are exposed to fumes containing both nickel and chromium. Whether the effects of such combined exposure are additive, antagonistic, or synergistic is not defined. Occupational exposure to nickel is associated with an increased risk of lung cancer,[102] but the degree to which nickel-induced chromosomal alterations contributes to such is difficult to assess because, as will be seen below, nickel also induces other types of DNA damage and inhibits repair. Effects on the immune system may also contribute to nickel carcinogenesis.[102]

Nickel-induced animal tumors show increased chromosomal aberrations;[103-106] however, such CAs are associated with most cancers regardless of cause, and studies on fully developed tumors cannot distinguish cause from consequence. In this context, it has been reported that $NiCl_2$ and $Ni(NO_3)_2$ either do not[107] or do[108] induce chromosome aberrations in mice, and that inhalation of nickel refinery waste results in increased CAs in rat alveolar macrophages.[109]

That nickel compounds are indeed clastogenic is well established from studies *in vitro*. A variety of nickel compounds readily induces CAs in cultured cells.[110-112] Particularly telling is the fact that a specific form of chromosomal aberration is emphasized in nickel-exposed cells. That is, heterochromatic regions are affected preferentially, and the long arm of chromosome X is fragmented selectively, especially in cells exposed to particulate forms such as crystalline NiS.[111] This response is attributed to the ability of ingested particulates to deliver a higher concentration of Ni^{2+} to nuclear regions rich in heterochromatin.[111,112] $NiCl_2$ alone produces CAs at a lower level than does crystalline NiS, and selective fragmentation of the long arm of X is not seen.[111,112] When liposome is encapsulated, however, $NiCl_2$ produces fragmentation and decondensation of the long, heterochromatic region of X.[111,112] At levels that do not produce detectable single- or double-strand breaks, $CaCrO_4$, crystalline NiS, and $NiCl_2$ produce an increase in SCEs in cultured CHO.[113] Chinese hamster embryo cells transformed to anchorage independence with NiS or $NiCl_2$ show patterns of CAs distinct from those observed in 3-methylcholanthrene-transformed clones.[114] Associated with a preferential transformation of male cultures by Ni is a high frequency of complete or partial deletions of X in the male and translocations involving X in male and female clones. Cells whose X chromosomes do not have as much X-specific heterochromatin exhibit no preferential transformation of male cells in response to nickel compounds, and transformed male Chinese hamster embryo cells most often show deletion of the long arm of X as a common deletion.[115] Transfer of a normal human or hamster X chromosome to male, nickel-transformed cells carrying an X chromosome deletion results in senescence of immortal lines.[116,117] Methylation of this X during culture diminishes this senescence effect reversibly.[116] Pretreatment of cells

TABLE 20

Chromium-Induced Chromosomal Alterations

Effector(s)	Concentration	Target(s)	Effect(s)	Ref.
$K_2Cr_2O_7$ (acute and chronic)	—	Rat bone marrow	CAs increased	69
$Na_2Cr_2O_7$ (i.m. and in water)	1–5 mg/kg, 30.5 and 24 ppm	Fish	CAs increased	70
$CaCrO_4$	—	*Drosophila*	Sex-linked recessive lethal test positive; translocations not increased	71
Cr^{6+}	1–10 m*M*	*Drosophila*	Mitotic recombination increased	72
Cr^{3+}	5–50 m*M*	*Drosophila*	Mitotic recombination not increased	72
Cr^{6+}	20 mg/kg	Mouse bone marrow *in vivo*	CAs increased	73
Cr^{6+}	—	Lymphocytes from exposed humans	CAs increased	74, 75
$Cr^{6+} > Cr^{3+}$	0.015–0.1 μg/ml 0.04–0.4 μg/ml	Mouse fetal cells (tertiary culture)	CAs increased	78
$K_2Cr_2O_7$ (Cr^{3+}: NE)	0.1–0.5 μg/ml	Hamster embryonic cell culture	CAs increased	79
$K_2Cr_2O_7$	0.5–4 μ*M*	Human lymphocyte cultures	Chromosome breaks increased	80
$K_2CrO_4 \gg$	4–8 μ*M*			
$Cr(CH_3COO)_3$	16–32 μ*M*			
$CR(NO_3)_3$	32 μ*M*			
$CrCl_3$	32 μ*M*			
Cr^{6+}	0.5–0.78 μg/ml 0.3–3 μ*M*	Cultured mammalian cells	CAs increased	81, 82
Cr^{3+}: NE	50–200 μg/ml, 32–1000 μ*M*			
Cr^{6+} ($Na_2Cr_2O_7$ and $K_2Cr_2O_7$)	0.1–1.0 μg/ml	CHO cells	CAs increased up to 10×	83
Cr^{6+}	1.0 μg/ml	BHK and CHO cells	CAs increased 10×	84
Cr^{3+}	150 μg/ml	BHK and CHO cells	CAs increased 2×	84
Cr^{6+} (Cr^{3+}: NE)	0.1–1.0 μ*M*	Human fibroblasts	SCEs increased	85
Cr^{6+} (Cr^{3+}: NE)	0.8–3.0 μ*M*	Human fibroblasts	CAs increased	85
$CaCrO_4 \gg$	0.01–0.02 μg/ml	Cultured human lymphocytes	SCEs increased	86
$CrO_3 >$	0.025–0.1 μg/ml			
$K_2Cr_2O_7$	0.025–0.1 μg/ml			
$CaCrO_4$	<10 μ*M*	Cultured cells	Spindle formation and chromosome segregation aberrant	90
Na_2CrO_4	<100 μ*M*			
$CrCl_3$	2.5 μg/ml (3 days, subtoxic)	PHA-stimulated human lymphocytes	CAs increased	91
$CrCl_3$ plus SOD, catalase, or mannitol	2.5 μg/ml (3 days)	PHA-stimulated human lymphocytes	CAs not induced	91
Na_2CrO_4	2.5–5 μ*M*	Chinese hamster V79 cells	CAs induced	92
Vitamin E	25 μ*M* (24 hr)	Chinese hamster V79 cells	Na_2CrO_4-induced CAs diminished	92
Vitamin B_2	200 μ*M* (24 hr)	Chinese hamster V79 cells	Na_2CrO_4 induced CAs increased	93

Note: CAs = chromosome aberrations; CHO = Chinese hamster ovary, BHK = baby hamster kidney cell line; PHA = phytohemagluttinin; SCEs = sister chromatid exchanges.

with vitamin E reduces NiS- but not $NiCl_2$-induced CAs,[118] suggesting a role for active oxygen in NiS induction of CAs.

It has been proposed that the mechanisms by which nickel causes damage to heterochromatic regions of chromosomes involve both protein oxidation and protein-DNA cross-linking.[119] The cross-links are attributed to Ni acting in a catalytic fashion.[119] Consistent with this suggested mechanism of action is the observation of increased DNA-protein cross-links in workers exposed to welding fumes and in animals treated with chromate.[120] Because pretreatment with deferoxamine obviates nickel-induced DNA damage but not lipid peroxidation, it appears that strand breakage may be due to hydroxyl ions generated by the Fenton reaction.[121]

Exposure of an already immortal but nontumorigenic osteoblast human cell line to 36 μM $NiSO_4$ for 48 to 96 hr results in tumorigenic transformation. Cell lines isolated from tumors formed in nude mice by the transformants show altered karyotypes with an aberrant chromosome 16.[122] (See Table 21.)

2.2 DNA Damage and Repair

2.2.1 *Arsenic*

Arsenic compounds do not by themselves produce mutagenic responses in most test systems. Forms of arsenic are, however, comutagenic, and a number of studies have shown arsenicals to inhibit DNA repair. The effects of sodium arsenite on survival after ultraviolet radiation of *Escherichia coli* strains lacking different repair functions indicates that 0.1 m*M* sodium arsenite preferentially affects a rec-A-dependent function.[123] At sublethal concentrations of sodium arsenite, inhibition of ultraviolet-induced DNA strand breaks in wild type *E. coli* and of post-replication repair in ultraviolet-resistant (uvr) strains also occurs.[124] Arsenate and arsenite also reduce ultraviolet-induced lethality in some strains of *E. coli*.[123,125] This effect is ascribed to arsenical-induced inhibition of DNA synthesis, thus providing additional time for repair.[125] Further, by inhibiting induction of the SOS error-prone repair pathway in *E. coli*, arsenite can effectively function as an antimutagen.[126] Arsenite by itself does not induce trp+ revertants in *E. coli* or ouabain- or thioguanine-resistant mutants in Chinese hamster cells.[126] Arsenic trioxide and pentavalent sodium arsenate at physiological and lethal concentrations inhibit ultraviolet thymine dimer excision and enhance ultraviolet lethality in normal and xeroderma pigmentosum human fibroblasts.[127] An exception to enhanced lethality is seen in XP group A cells. Doses of these arsenicals that partially repress repair also enhance ultraviolet mutagenesis in V79 Chinese hamster cells.[127]

A specific effect of arsenite on DNA ligase activity is seen in Chinese hamster C79 cells.[128] This inhibition is suggested to be the mechanism for arsenite inhibition of excision repair and may thereby contribute to the comutagenicity and carcinogenicity of arsenite.[128]

The nonmutagenicity of arsenite is evident in cultured cells in which multilocus deletions as well as point mutations are detectable;[129] however, the comutagenicity is evident in cells treated with *N*-methyl-*N*-nitrosourea (MNU) and is associated with an increase in DNA strand breaks in cells exposed to arsenite for 3 hr following MNU treatment.[129]

An analysis of the actual DNA base sequence alterations in ultraviolet- and ultraviolet-plus-sodium-arsenite-induced *hprt* mutants in CHO cells indicates a significant increase in ultraviolet-induced transversions and in base changes adjacent to TT and CT sequences in arsenite comutagenesis.[130]

Although arsenicals are not positive in mammalian cell mutagenesis assays, effects on DNA such as the induction of breaks, adducts, protein cross-links, and gene amplification have been reported and related to clastogenic and carcinogenic responses. Arsenite and arsenate both induce methotrexate resistance and dihydrofolate reductase gene amplification

TABLE 21

Nickel-Induced Chromosomal Alterations

Effector(s)	Concentration	Target(s)	Effect(s)	Ref.
Ni_3S_2, NiO dust, $NiCl_2$, $NiSO_4$ aerosols	<1.0 mg/m^3, >25 yr	Human lymphocytes *in vivo*	CAs increased	96
NiO > $NiSO_4$	0.77 mg/m^3, 1.3 mg/m^3	Human lymphocytes *in vivo*	CAs increased	99
Ni_3S_2	10 mg (i.m.)	Rats	Rhabdomyosarcoma with CAs	103
Ni powder	20 mg (i.m.)	Rats	Rhabdomyosarcoma with CAs	104
Crystalline NiS	5 mg (i.m.)	Rats	Rhabdomyosarcoma with CAs	105
Ni_3S_2	20 mg	Rats	Kidney sarcomas and carcinoma with CAs	106
Nickel refinery dust	50 mg/m^3 (5 hr/day, 5 days/wk)	Wistar rat alveolar macrophages	CAs increased	109
$NiCl_2$	6, 8 × 10^{-4} *M*	Mouse mammary carcinoma cells	CAs increased	110
$Ni(OAc)_2$	6, 8 × 10^{-4} *M*			
NiS	4, 6, 8 × 10^{-4} *M*			
$NiCl_2$	1 μ*M*–1 m*M*	Chinese hamster ovary cells	Heterochromatic CAs increased; no specific X effect	111, 112
Crystalline NiS	5–20 μg/ml	Chinese hamster ovary cells	Heterochromatic CAs; fragmentation; X chromosome especially increased	111, 112
$NiCl_2$ (liposome delivery)	100–1000 μ*M*	Chinese hamster ovary cells	Heterochromatic CAs; fragmentation; X chromosome especially increased	111, 112
$CaCrO_4$, crystalline NiS, $NiCl_2$	Concentrations less than those producing DNA breaks	Chinese hamster ovary cells	SCEs increased	113
NiS	10 μg/ml	Chinese hamster embryo cells	Cell transformation with preferential X CAs	114
$NiCl_2$	1 m*M*			
3-methyl cholanthrene	10 μg/ml	Chinese hamster embryo cells	Cell transformation without X CAs	114
Nickel	—	Chinese hamster embryo cells (male)	X deletions in most transformants	115
Vitamin E	25 μ*M*	Chinese hamster cells	NiS-induced CAs increased	118
Vitamin E	25 μ*M*	Chinese hamster cells	$NiCl_2$-induced CAs unchanged	118
Welding fumes	Occupational exposure	Cells from exposed workers	DNA cross-links increased	120
$NiSO_4$	36 μ*M*, 48–96 hr	Human osteoblast cell line HOSTE85	Cell transformation, CAs increased	122
$NiCl_2$	0.56 mmol/kg (s.c.)	Male rats	NE	121
$NiCl_2$	0.75 mmol/kg (s.c.)	Male rats	DNA breakage, lipid peroxidation	121
Deferoxamine	1 g/kg (i.p.)	Male rats	$NiCl_2$-induced DNA breaks ameliorated; lipid peroxidation unchanged	121

Note: CAs = chromosome aberrations; SCEs = sister chromatid exchanges; NE = no effect.

at a high frequency in cultured murine 3T6 cells.[131] DNA amplification is also a sequela of induced arsenite resistance in Leishmania.[132,133]

As_2O_3 produces a low but significant level of DNA protein cross-links in Novikoff hepatoma cells. A very concentration-dependent sodium arsenite induction of cross-links is also seen in human lung fetal fibroblasts.[135] DNA protein cross-links are observed in cells exposed to 1 to 5, but not 10, μ*M* sodium arsenite. Protein-associated DNA breaks are observed also and are maximally induced by 3 μ*M* sodium arsenite. These breaks are suggested to be a possible cause of arsenite-induced CAs.[135] Arsenite at 1 to 10 μ*M* also increases unscheduled DNA synthesis in human fetal lung fibroblasts and enhances unscheduled DNA synthesis (UDS) induced by *N*-methyl-*N'*-nitrosoguanidine (MNNG).[136] Single-strand DNA breaks are induced in human alveolar lung cells in culture by exposure to dimethylarsinic acid but at the very high concentration of 10 m*M*.[137] (See Table 22.)

2.2.2 *Beryllium*

Relatively few studies have addressed the mutagenicity or co-mutagenicity of Be salts. An analysis of the effect of BeCl on the development of 8-azaguanine resistance in Chinese hamster V79 cells shows a low level of mutagenic activity.[138] The background levels of mutation at the *hgprt* locus are increased two- to sixfold by exposure to both beryllium and manganese chloride.[138] $BeCl_2$ also increases the mutation frequency at the *lacI* locus in *E. coli* two- to threefold[139] and is positive in the *Bacillus subtilis* rec assay.[140] The frequency of amber and ochre mutations was found to be induced in *E. coli* to a level 3× background, accompanied by a specific increase in G:C — A:T transitions in the amber mutants.[139] A screen for oncogene and tumor suppressor gene mutations in Be-induced rat lung carcinomas shows an absence of tumor-associated mutations in K-*ras*, *p53*, and c-*raf*-1 genes.[141] Thus, the Be-induced lung cancers in rats do not involve gene dysfunctions commonly associated with human non-small-cell lung cancer.[141] (See Table 23.)

2.2.3 *Cadmium*

Studies using bacterial and mammalian cell test systems show that cadmium induces DNA breaks, has some mutagenic activity, and affects DNA repair. Exposure of cultured Chinese hamster V79 cells to 50 μ*M* $CdCl_2$ results in a marked increase in repairable single-strand (SS) DNA breaks.[142] There is evidence that such metal-induced breakage is less pronounced in human cells, however.[143] Cd^{2+} also induces SS but not double-strands (DS) DNA breaks in CHO cells at nontoxic concentrations.[144] At toxic levels, DS breaks are also induced, possibly due to indirect effects involving activation of enzymes in damaged cells.[144] Oxygen scavengers antagonize Cd^{2+}-induced DNA damage in human diploid fibroblasts.[145] Active oxygen species generated in response to Cd^{2+} exposure thus appear to mediate DNA damage in response to Cd^{2+} in these cells. Low concentrations of Cd^{2+} (0.18 to 1.8 μ*M*) induce unscheduled DNA synthesis in cultured rat hepatocytes.[146] These concentrations range from those that are nontoxic to these cells to marginally toxic levels.

Cadmium chloride induces missense and frameshift mutations in *Salmonella typhimurium* and synergistically increases the mutagenesis of nitrosamines.[147] These effects are evident at a relatively high concentration of 500 μ*M* $CdCl_2$ and are much reduced at 250 μ*M*. The synergistic response suggests that Cd^{2+} may inhibit repair of nitrosamine-induced lesions. This synergism appears to be specific for methylation damage and does not involve effects on SOS repair.[148] In cultured human and simian cells, a relatively low concentration of Cd^{2+} (4 μ*M*) suffices to inhibit repair of ultraviolet lesions.[149] This effect is ameliorated at Zn^{2+} concentrations 5 to 10 times those of Cd^{2+}. (See Table 24.)

TABLE 22
Mutagenic Responses to Arsenicals

Effector(s)	Concentration	Target(s)	Effect(s)	Ref.
Sodium arsenite	>0.1 m*M*	*E. coli*, WP2 (wild-type), WWP2 (uvrA), WP6 (polA)	Post-UV survival decreased	123
Sodium arsenite	>0.1 m*M*	*E. coli*, WP10 (recA)	None	123
Sodium arsenite	>0.1 m*M*	*E. coli*, WP5 (exrA)	Post-UV survival enhanced	123
Sodium arsenite	Nonlethal	*E. coli*, WP2 (wild-type) WP6 (polA)	Post-UV SS breaks reduced	124
Sodium arsenite	Nonlethal	*E. coli*, WP2 (uvrA)	Post-replication repair inhibited	124
Arsenite	1–10 m*M*	*E. coli* B (argF$^-$)	No mutagenic response	125
Arsenate	1–50 m*M*			
Arsenite	1–10 m*M*	*E. coli* H/r 30R (wild-type, Exc$^+$ Rec$^+$)	UV-induced mutations reduced	125
Arsenate	1–50 m*M*			
Arsenite	1–10 m*M*	*E. coli* Hs 30R (uvrA$^-$, Esc$^-$, Rec$^+$)	Post-UV mutation frequency, survival not changed	125
Arsenate	1–50 m*M*			
Arsenite	1–10 m*M*	*E. coli* NG 30 (recA$^-$, Exc$^+$, Rec$^-$)	Post-UV survival increased	125
Arsenate	1–50 m*M*			
Arsenite	0.5–100 μ*M* (toxic threshold, 0.5 μ*M*)	V79 cells	No mutagenic activity	126
Arsenite	0.1–25 m*M* (toxic threshold, 0.4 m*M*)	*E. coli*	Antimutagenic	
Arsenic trioxide As_2O_3 (As^{3+} 10× As^{6+})	1 μg/ml (physiological)	Normal and XP human fibroblasts	Post-UV thymine dimer excision inhibited 30–40%; lethality enhanced	127
Arsenic trioxide As_2O_3 (As^{3+} 10× As^{6+})	5 μg/ml (supralethal)	Normal and XP human fibroblasts	Post-UV thymine dimer excision inhibited 100%; lethality enhanced	127
As^{3+}	0.5 μg/ml	V79 Chinese hamster cells	Co-mutagenic effect (with UV)	127
As^{5+}	5 μg/ml			
Arsenite	10 μ*M* (3 hr)	V79 Chinese hamster cells	DNA ligase II activity reduced	128
Arsenite	10 μ*M*	Normal and G12 Chinese hamster V79 cells	Nonmutagenic	129
Arsenite	5 μ*M* (24 hr), 10 *M* (3 hr), (nontoxic)	Normal and G12 Chinese hamster V79 cells	Co-mutagenic with MNU; DNA breaks increased	129
Arsenite	10 μ*M* (IC_{50} for growth)	Chinese hamster ovary K1 cells	Post-UV transversions increased	130
Arsenite	0.2–6.2 μ*M*	Cultured 3T6 cells	Methotrexate resistance increased, DHFR gene amplified	131
Arsenate	1–32 μ*M*			
$As^{3+} < Ni^{2+} < Cr^{6+}$	1.0 m*M*	Novikoff hepatoma cells	DNA-protein cross-links induced	134
Arsenite	1–5 μ*M* (3 *M* maximal; 10 μ*M*: NE)	Human fetal lung fibroblasts	DNA protein cross-links and DNA breaks	135
Arsenite	1–10 μ*M*	Human fetal lung fibroblasts	UDS	136
Dimethylarsinic acid	10 m*M* (10 hr)	Human alveolar type II cells (L-132)	SS DNA breaks	137

Note: UV = ultraviolet radiation; SS = single-strand; MNU = *N*-methyl-*N*-nitrosourea; DHFR = dihydrofolate reductase; NE = no effect; UDS, unscheduled DNA synthesis; IC_{50} = concentration for half maximal inhibition.

TABLE 23

Mutagenicity of Beryllium Compounds

Effector(s)	Concentration	Target(s)	Effect(s)	Ref.
$BeCl_2$	2.0 m*M*	Chinese hamster V79 cells	Mutations increased 2–6×	138
$MnCl_2$	1.0 m*M*			
$BeCl_2$	50 μ*M*	*E. coli*	Mutations increased 2–3×	139
$MnCl_2$	25 μ*M*			
$K_2Cr_2O_7$	—			
$BeSO_4$	10 m*M*	*B. subtilis*	Rec assay positive	140

TABLE 24

Mutagenicity of Cadmium Compounds

Effector(s)	Concentration	Target(s)	Effect(s)	Ref.
$CdCl_2$	50 μ*M* (2 hr)	Chinese hamster V79 cells	SS breaks	142
$CdSO_4$	30 μ*M* (nontoxic)	CHO cells	SS but not DS breaks	144
$CdSO_4$	300 μ*M* (toxic)	CHO cells	SS and DS breaks	144
$CdCl_2$	1 m*M*	Human diploid fibroblasts	DNA breaks	145
$MgCl_2$	10 m*M*			
$MnCl_2$	10 m*M*			
$K_2Cr_2O_7$	1.0 μ*M*			
$ZnCl_2$	2 m*M*			
Na_2SeO_3	500 μ*M*			
Cd^{2+}	0.18–1.8 μ*M*	Cultured hepatocytes	UDS	146
Cu^{2+}	7.9–78.5 μ*M*			
Cd^{2+}	500 μ*M*	*S. typhimurium*	Mutations	147
Cd^{2+} with MNN	500 μ*M*	*S. typhimurium*	Mutations increased synergistically	147
Cd^{2+}	4 μ*M*	Cultured human and simian cells	UV damage repair inhibited	149

Note: CHO = Chinese hamster ovary; MNN = *N*-methyl-*N'*-nitrosamine; SS = single-strand; DS = double-strand; UDS = unscheduled DNA synthesis; UV = ultraviolet.

2.2.4 *Chromium*

Chromium ions produce DNA damage involving adducts, breaks, and cross-links and inhibit repair of DNA damage induced by other agents such as ultraviolet radiation in bacteria and mammalian cells. Cr^{6+} is usually found to have both much greater toxicity and genotoxicity than Cr^{3+} in tests employing these targets. Tests for genotoxic effects of Cr^{2+} at sublethal levels usually prove negative. In the case of some organic ligands of Cr^{3+}, however, DNA damage and mutations are elicited in microbial test systems.[150] Aromatic amine ligands were found to be especially active in these tests.[150] When the activities of inorganic Cr^{6+} and Cr^{3+} were examined in a seven-test battery involving DNA, bacteria, and cultured cells, however, Cr^{6+} was found to express genotoxic activity in most of the assays employed, but Cr^{3+} was negative in all instances unless a direct interaction with purified DNA was involved.[151] Cr^{3+} also fails to elicit DNA damage in permeabilized mammalian cells under conditions that enhance Cr^{6+} genotoxicity.[152] DNA damage involving breaks and cross-links is induced to a greater extent in repair-deficient than wild-type CHO cells exposed to $CaCrO_4$.[153] Cross-links are repaired in the wild-type cells within 24 hr but persist in the repair-deficient cells, whereas single-strand breaks are rapidly repaired in both. Cytotoxicity in these cells correlates with

DNA damage, raising the question of which is cause or consequence. It is suggested that the DNA damage causes cytotoxicity.[153]

Cr^{6+} but not Cr^{3+} induces in *E. coli* expression of genes for RecA protein, error-prone repair, and inhibition of cell division and thus is active in inducing the SOS system.[154] There is, however, evidence that the co-mutagenicity of azide and Cr^{6+} in *S. typhimurium* is not mediated by effects of Cr^{6+} on the SOS system, as inhibition or enhancement of this system is without effect on the co-mutagenic response.[155]

At levels that significantly affect survival (150 to 300 μ*M*, 54 to 58% survival), Cr^{6+} also has early effects on specific gene induction and on general RNA synthesis, whereas such effects are absent at marginally toxic exposure levels (30 μ*M*, 96% survival).[156] DNA breaks and UDS are seen in cultured hepatocytes at toxic but not subtoxic exposure levels.[157]

An analysis of fractionated chromatin from CHO cells exposed to 150 μ*M* Cr^{6+} for 2 hr shows the nuclear matrix to be enriched in Cr-DNA adducts and to be the site of DNA protein cross-links.[158] This level of exposure results also in internucleosomal DNA breakage characteristic of apoptosis.[159]

Reduction to lower valency forms that may participate in Fenton-like reactions or that react directly with DNA has been implicated in mechanisms by which Cr^{6+} produces adducts, DNA breaks, and cross-links.[160-166] *In vitro*, Cr^{6+} in the presence of ascorbate and H_2O_2 generates OH radicals and ascorbate-derived radicals that cause DNA breaks and, in the instance of OH, 2′-deoxyguanine hydroxylation.[160] Similarly, incubation of ascorbate and Cr^{6+} with DNA results in adducts and breaks[161] and a corresponding production of lower valency forms of Cr. This analysis suggests that Cr^{5+} is responsible for the formation of adducts, and carbon-based radicals for the single-strand DNA breaks.[161,163] Electron paramagnetic resonance spectroscopy (EPR) studies of the reduction products of Cr^{6+} exposed to NADH and NADPH, however, indicate that neither the VI or V valency forms damage DNA directly.[164] The V form does, however, react with H_2O_2 to produce hydroxyl radicals that react with DNA.[164] Correspondingly, Cr^{5+} levels and DNA single-strand breaks are both reduced in CHO cells resistant to H_2O_2 relative to wild-type CHO.[165] A hot spot for mutations induced in human lymphoblasts by Cr^{6+} is common with that for H_2O_2 mutations.[166]

Cr^{6+} induces DNA cross-links in the liver and single-strand breaks and 8-oxo-2-deoxyguanosine formation in the red cells of 14-day chick embryos.[167] Depletion of glutathione reduces 8-oxo-2-dG formation in the chick red cells.[167] Exposure of CHO cells to Cr^{6+} results in stable DNA glutathione and amino acid complexes.[168] These complexes apparently represent noncovalent Cr-GSH or amino acid-DNA ternary complexes.[168] (See Table 25.)

2.2.5 Nickel

Although weakly mutagenic, nickel does induce DNA damage and is strongly co-mutagenic due to its inhibition of DNA repair.[169,170] Repair synthesis but no DNA damage is evident in CHO or Syrian hamster embryo (SHE) cells exposed to crystalline NiS or Ni_3S_2 and to $NiCl_2$, but not to amorphous NiS.[171] A defined mutagenic response is seen in NRK cells infected with a retrovirus that is conditionally defective in RNA splicing.[172] Exposure of these cells to $NiCl_2$ results in revertants with altered RNA splicing patterns, with in one instance a defined mutation involving a 70-base duplication. Nickel mutagenesis is also revealed in a cellular test system with a relatively low background of spontaneous mutations, *hprt*-deficient V79 hamster cells containing one copy of a bacterial *gpt* gene.[173] Exposure of these cells to crystalline NiS results in *gpt*-mutants at a high level (80×) relative to background. Soluble nickel sulfate produces no observable mutations in this system. Similarly, mutations are produced in a Chinese hamster ovary cell line lacking *hgprt* but carrying bacterial *gpt*;[174] however, in this system soluble $NiSO_4$ and $NiOH_2$ as well as insoluble Ni_3S_2 produce mutations. These Ni compounds produce a higher ratio of deletions to point mutations than is observed with ethyl methane sulfonate.

TABLE 25

Mutagenic Responses to Chromium

Effector(s)	Concentration	Target(s)	Effect(s)	Ref.
Aromatic amine Cr^{3+} complexes	100–2000 nmol per plate	*E. coli*	DNA damage	150
Aromatic amine Cr^{3+} complexes	100–2000 nmol per plate	*S. typhimurium*	Mutations	150
Cr^{6+} (Cr^{3+}: NE)	—	Microbiol and mammalian cells	Mutations, cell transformation, SCE	151
Cr^{6+} (Cr^{3+}: NE)	—	Permeabilized BHK cells	DNA damage, SCE	152
$CaCrO_4$	10 μ*M* (4 hr)	CHO cells (wild-type)	SSB, CL	153
$CaCrO_4$	10 μ*M* (4 hr)	Repair-deficient CHO	SSB enhanced	153
$K_2Cr_2O_7 > K_2CrO_4 > CrO_3$ (Cr^{3+}: NE)	0–3 μmol per 3 ml	*E. coli*	SOS genes induced	154
Cr^{6+}	1–7 μ*M* (subtoxic)	*S. typhimurium*	Co-mutagenic (with azide)	155
Cr^{6+}	30–300 μ*M* (2 hr)	CHO cells	SSB, CL DNA adducts	156
Cr^{6+}	30 μ*M*	CHO cells	96% survival	156
Cr^{6+}	150 μ*M*		54% survival	156
Cr^{3+}	>50 μ*M* (20 hr)	Rat hepatocytes	Cytotoxicity	157
Cr^{6+}	2.5 μ*M*	Rat hepatocytes	UDS	157
Cr^{6+}	10–40 μ*M* (1 hr)	Rat hepatocytes	SSB	157
Cr^{6+}	150 μ*M* (2 hr)	CHO cells	Cr-DNA adducts	158
Cr^{6+}	150 μ*M* (2 hr)	Nuclear matrix fraction	Adducts enriched 4× over chromatin; CL	158
Cr^{6+} (Na_2CrO_4)	150 μ*M* (2 hr)	CHO ovary cells	CFE = 54%; DNA breaks	159
Cr^{6+} (Na_2CrO_4)	300 μ*M* (2 hr)	CHO ovary cells	CFE = 8%; DNA breaks	159
Cr^{6+}	400 μ*M*	CHO > CHO(R)	SSB	165
Cr^{6+}	5 μ*M* (5 hr)	Human lymphoblast cells (TK6)	*hprt* mutants	166
Cr^{6+}	100 μmol/kg	14-day chick embryo — liver cells red cells	CL, SSB, 8-oxo-dG	167
Cr^{6+}	<25 μ*M*	CHO cells	DNA-GSH-Cr complexes formed	168

Note: SCE = sister chromatid exchange; CHO = Chinese hamster ovary; SSB = single-strand breaks; CL = cross-linking; UDS = unscheduled DNA synthesis; CFE = colony-forming efficiency; GSH = glutathione.

Epigenetic mechanisms are also observed to account for gene silencing in cells exposed to carcinogenic nickel compounds.[175] Nickel-induced 6-thioguanine resistance in G12 cells spontaneously reverts, as does nickel-induced DNA condensation. Demethylation with 5-azacytidine enhances these reversions. Increased DNA methylation is also seen in the nickel-induced 6-TG resistant cells.

Ni^{2+} is comutagenic with alkylating agents in bacterial test systems through effects on DNA repair.[176] In animal cells, Ni^{2+} is significantly co-mutagenic with ultraviolet and cis-DDP but not methyl methane sulfonate (MMS).[177,178] This effect is ascribed to inhibition of DNA repair following ultraviolet but not MMS.[177,178] That arsenite and nickel differentially affect enzymes involved in DNA repair is shown by their relative effects on repair of DNA strand breaks due to ultraviolet and MMS damage.[179] Nickel inhibits repair of ultraviolet damage only, whereas arsenite inhibits repair of damage due to both agents. The inhibition of ultraviolet repair is specifically associated with an interference with the incision step of pyrimidine dimer excision by Ni^{2+}.[180] This inhibition is partially reversed by Mg^{2+}. Ni^{2+} synergistically increases glutathione (GSH) levels in MMS- but not ultraviolet-exposed cells.[181] Because increased GSH modulates the effects of Ni^{2+}, this observation is offered as a basis for the differential effects of Ni^{2+} on ultraviolet and MMS damage. Nickel and polyamine depletion synergize to inhibit both ultraviolet and X-ray DNA damage.[182] (See Table 26.)

2.3 Summary and Conclusions

All known carcinogenic metals induce clastogenic, mutagenic, and/or co-mutagenic responses *in vivo* and *in vitro* to such extent as to provide a mechanistic explanation for cell transformation. As expected, mechanistic details differ, and quite specific mechanisms are induced by some metal ions but not others, such as the induction of gene amplification by arsenite and DNA methylation by Ni^{2+}. Active oxygen appears to play a role in metal-induced DNA damage by cadmium, chromium, and nickel ions, at least at higher concentrations. Mechanisms for inhibition of repair are also commonly seen but differ in detail. Effects on the mitotic apparatus are suggested for Cd^{2+} but not examined or reported for the other metal ions. Although Be ions appear to be both mutagenic and clastogenic, their effects on repair and the possible role of active oxygen species have not been reported. Because these metal ions also affect many signal transduction pathways in such a fashion as to potentially alter cell proliferation and apoptosis, they may well affect cell transformation in a multifactorial fashion and be involved in different ways at several stages of carcinogenesis.

TABLE 26

Mutagenic Responses to Nickel Compounds

Effector(s)	Concentration	Target(s)	Effect(s)	Ref.
Crystalline NiS	1–5 μg/ml (24 hr)	CHO and SHE cells	UDS	171
Amorphous NiS	10 μg/ml (24 hr)	CHO and SHE cells	NE	171
Crystalline Ni_3S_2	10 μg/ml (24 hr)	CHO and SHE cells	UDS	171
$NiCl_2$	100 μ*M*	CHO and SHE cells	UDS	171
$NiCl_2$	20–120 μ*M*	NRK cells infected with MuSVts 110	Altered RNA splicing	172
Crystalline NiS	0.1–0.6 μg/cm²	G12 cells (*hprt*-V79 with bacterial *gpt*)	Mutations	173
$NiSO_4$	0.05–0.25 m*M*	G12 cells (*hprt*-V79 with bacterial *gpt*)	NE	173
$NiSO_4 > Ni(OH)_2 > Ni_3S_2$	58.7, 1.7, 7.3 μg Ni per ml	AS52 cells	Gene deletions	174
Nickel compounds	—	G12 cells	DNA methylation and compaction	175
Ni^{2+}	3–30 μ*M*	*E. coli*	Co-mutagenic with MMS	176
$NiCl_2$	0.5–2 m*M*	Chinese hamster V79 cells	Co-mutagenic with UV, cis-DPP, but not with MMS	177
$NiCl_2$	0.5–4 m*M*	CHO cells	Repair of UV damage inhibited	178
Ni^{2+}	2 m*M*	CHO cells	Repair of UV damage inhibited	179
Ni^{2+}	2 m*M*	CHO cells	Repair of MMS damage not inhibited	179
Arsenite	40 μ*M*	CHO cells	Repair of UV and MMS damage inhibited	179
Ni^{2+}	500 μ*M* (nontoxic)	CHO cells	Incision repair of UV damage inhibited	180
Ni^{2+}	4 m*M*	CHO cells	GSH content of MMS-treated cells increased synergistically	181
Ni^{2+} with polyamine depletion	25–1000 μ*M* (2 hr)	HeLa cells	UV, X-ray repair decreased synergistically	182

Note: CHO = Chinese hamster ovary; SHE = Syrian hamster embryo; UDS = unscheduled DNA synthesis; NE = no effect; NRK = normal rat kidney; DPP = *cis*-diamminedichloroplatinum (II); MMS = methyl methane sulfonate; UV = ultraviolet; GSH = glutathione.

References

1. Petres, J. and Hundeiker, M., Chromosome pulverization induced *in vitro* in cell cultures by sodium diarsenate, *Arch. Kin. Exp. Dermatol.,* 231(4), 366–370, 1968.
2. Petres, J., Schmid-Ullrich, K., and Wolf, U., Chromosome aberrations in human lymphocytes in cases of chronic arsenic poisoning, *Dtsch. Med. Wochenschr.,* 95(2), 79–80, 1970.
3. Paton, G. R. and Allison, A. C., Chromosome damage in human cell cultures induced by metal salts, *Mutat. Res.,* 16(3), 332–336, 1972.
4. Burgdorf, W., Kurvink, K., and Cervenka, J., Elevated sister chromatid exchange rate in lymphocytes of subjects treated with arsenic, *Hum. Genet.,* 36(1), 69–72, 1977.
5. Petres, J., Baron, D., and Hagedorn, M., Effects of arsenic on cell metabolism and cell proliferation: cytogenetic and biochemical studies, *Environ. Health Perspect.,* 19, 223–227, 1977.
6. Nordenson, I., Salmonsson, S., Brun, E., and Beckman, G., Chromosome aberrations in psoriatic patients treated with arsenic, *Hum. Genet.,* 48(1), 1–6, 1979.
7. Beckman, G., Beckman, L., and Nordenson, I., Chromosome aberrations in workers exposed to arsenic, *Environ. Health Perspect.,* 19, 145–146, 1977.
8. Nordenson, I., Beckman, G., Beckman, L., and Nordstrom, S., Occupational and environmental risks in and around a smelter in northern Sweden. II. Chromosomal aberrations in workers exposed to arsenic, *Hereditas,* 88(1), 47–50, 1978.
9. Nordenson, I. and Beckman, L., Occupational and environmental risks in and around a smelter in northern Sweden. VII. Reanalysis and follow-up of chromosomal aberrations in workers exposed to arsenic, *Hereditas,* 96(2), 175–181, 1982.
10. Zanzoni, F. and Jung, E. G., Arsenic elevates the sister chromatid exchange (SCE) rate in human lymphocytes *in vitro, Arch. Dermatol. Res.,* 267(1), 91–95, 1980.
11. Deknudt, G., Leonard, A., Arany, J., Jenar-Du-Buisson, G., and Delavignette, E., *In vivo* studies in male mice on the mutagenic effects of inorganic arsenic, *Mutagenesis,* 1(1), 33–34, 1986.
12. Nakamuro, K. and Sayato, Y., Comparative studies of chromosomal aberration induced by trivalent and pentavalent arsenic, *Mutat. Res.,* 88(1), 73–80, 1981.
13. Nordenson, I., Sweins, A., and Beckman, L., Chromosome aberrations in cultured human lymphocytes exposed to trivalent and pentavalent arsenic, *Scand. J. Work Environ. Health,* 7(4), 277–281, 1981.
14. Wan, B., Christian, R. T., and Soukup, S. W., Studies of cytogenetic effects of sodium arsenicals on mammalian cells *in vitro, Environ. Mutagen.,* 4(4), 493–498, 1982.
15. Lee, T. C., Huang, R. Y., and Jan, K. Y., Sodium arsenite enhances the cytotoxicity, clastogenicity, and 6-thioguanine-resistant mutagenicity of ultraviolet light in Chinese hamster ovary cells, *Mutat. Res.,* 148(1–2), 83–89, 1985.
16. Lee, T. C., Lee, K. C., Tzeng, Y. J., Huang, R. Y., and Jan, K. Y., Sodium arsenite potentiates the clastogenicity and mutagenicity of DNA cross-linking agents, *Environ. Mutagen.,* 8(1), 119–128, 1986.
17. Lee, T. C., Wang, W-S., Huang, R. Y., Lee, K. C., and Jan, K. Y., Differential effects of pre- and post-treatment of sodium arsenite on the genotoxicity of methyl methanesulfonate in Chinese hamster ovary cells, *Cancer Res.,* 46, 1854–1857, 1986.
18. Jan, K. Y., Huang, R. Y., and Lee, T. C., Different modes of action of sodium arsenite, 3-aminobenzamide, and caffeine on the enhancement of ethyl methanesulfonate clastogenicity, *Cytogenet. Cell Genet.,* 41(4), 202–208, 1986.
19. Lee, T. C., Tzeng, S. F., Chang, W. J., Lin, Y. C., and Jan, K. Y., Post-treatments with sodium arsenite during G2 enhance the frequency of chromosomal aberrations induced by S-dependent clastogens, *Mutat. Res.,* 163(3), 263–269, 1986.
20. Huang, R. Y., Lee, T. C., and Jan, K. Y., Cycloheximide suppresses the enhancing effect of sodium arsenite on the clastogenicity of ethyl-methanesulphonate, *Mutagenesis,* 1(6), 467–470, 1986.
21. Lee, T. C., Kao, S. L., and Yih, L. H., Suppression of sodium arsenite-potentiated cytotoxicity of ultraviolet light by cycloheximide in Chinese hamster ovary cells, *Arch. Toxicol.,* 65(8), 640–645, 1991.
22. Huang, R. Y., Jan, K. Y., and Lee, T. C., Post-treatment with sodium arsenite is coclastogenic in log phase but not in stationary phase, *Hum. Genet.,* 75(2), 159–162, 1987.
23. Wiencke, J. K. and Yager, J. W., Specificity of arsenite in potentiating cytogenetic damage induced by the DNA crosslinking agent diepoxybutane, *Environ. Mol. Mutagen.,* 19(3), 195–200, 1992.
24. Jha, A. N., Noditi, M., Nilsson, R., and Natarajan, A. T., Genotoxic effects of sodium arsenite on human cells, *Mutat. Res.,* 16, 284(2), 215–221, 1992.
25. Huang, H., Huang, C. F., Huang, J. S., Wang, T. C., and Jan, K. Y., The transition from late G1 to early S phase is most vulnerable to the coclastogenic effect of ultraviolet radiation plus arsenite, *Int. J. Radiat. Biol.,* 61(1), 57–62, 1992.
26. Lin, J. K. and Tseng, S. F., Chromosomal aberrations and sister-chromatid exchanges induced by N-nitroso-2-acetylaminofluorene and their modifications by arsenite and selenite in Chinese hamster ovary cells, *Mutat. Res.,* 265(2), 203–210, 1992.

27. Yager, J. W. and Wencke, J. K., Enhancement of chromosomal damage by arsenic: implications for mechanism, *Environ. Health Perspect.*, 101(Suppl. 3), 79–82, 1993.
28. Lee, T. C., Oshimura, M., and Barrett, J. C., Comparison of arsenic-induced cell transformation, cytotoxicity, mutation and cytogenetic effects in Syrian hamster embryo cells in culture, *Carcinogenesis*, 6(10), 1421–1426, 1985.
29. Barrett, J. C., Lamb, P. W., Wang, T. C., and Lee, T. C., Mechanisms of arsenic-induced cell transformation, *Biol. Trace Elem. Res.*, 21, 421–429, 1989.
30. Huang, R. N., Ho, I. C., Yih, L. H., and Lee, T. C., Induction of histone hypophosphorylation and chromosome endoreduplication by sodium arsenite in human fibroblasts, *Metal Ions Biol. Med.*, 123–128, 1994.
31. Huang, R. N., Ho, I. C., Yih, L. H., and Lee, T. C., Sodium arsenite induces chromosome endoreduplication and inhibits protein phosphatase activity in human fibroblasts, *Environ. Mol. Mutagen.*, 25, 3, 1995.
32. Nikiforova, V. and Voronin, S. A., The mutagenic effect of beryllium on animals, *Toxicol. Genet.*, 23(4), 27–30, 1989.
33. Brooks, A. L., Griffith, W. C., Johnson, N. F., Finch, G. L., and Cuddihy, R. G., The induction of chromosome damage in CHO cells by beryllium and radiation given alone and in combination, *Radiat. Res.*, 120(3), 494–507, 1989.
34. Forni, A., Chromosomal effects of cadmium exposure in humans, *IARC Sci. Publ.*, (118), 377–383, 1992.
35. Deknudt, G. and Leonard, A., Cytogenetic investigations on leucocytes of workers from a cadmium plant, *Environ. Physiol. Biochem.*, 5(5), 319–327, 1975.
36. Bui, T. H., Lindsten, J., and Nordberg, G. F., Chromosome analysis of lymphocytes from cadmium workers and itai-itai patients, *Environ. Res.*, 9(2), 187–195, 1975.
37. Bauchinger, M., Schmidt, E., Einbrodt, H. J., and Dresp, J., Chromosome aberrations in lymphocytes after occupational exposure to lead and cadmium, *Mutat. Res.*, 40(1), 57–62, 1976.
38. Shiraishi, Y., Cytogenetic studies in 12 patients with itai-itai disease, *Humangenetik*, 27(1), 31–44, 1975.
39. O'Riordian, M. L., Hughes, E. G., and Evans, H. J., Chromosome studies on blood lymphocytes of men occupationally exposed to cadmium, *Mutat. Res.*, 58(2–3), 305–311, 1978.
40. Fleig, I., Rieth, H., Stocker, W. G., and Thiess, A. M., Chromosome investigations of workers exposed to cadmium in the manufacturing of cadmium stabilizers and pigments, *Ecotoxicol. Environ. Safety*, 7(1), 106–110, 1983.
41. Tang, X. M., Chen, X. Q., Zhang, J. X., and Qin, W. Q., Cytogenetic investigation in lymphocytes of people living in cadmium-polluted areas, *Mutat. Res.*, 241(3), 243–249, 1990.
42. Alessio, L., Apostoli, P., Forni, A., and Toffoletto, F., Biological monitoring of cadmium exposure — an Italian experience, *Environ. Health*, 19(Suppl. 1), 27–33, 1993.
43. Gilliavod, N. and Leonard, A., Mutagenicity tests with cadmium in the mouse, *Toxicology*, 5(1), 43–47, 1975.
44. Vilkima, G. A., Pomerantseva, M. D., and Ramaiia, L. K., Absence of a mutagenic effect of cadmium and zinc salts in mouse somatic and sex cells, *Genetika*, 14(12), 2212–2214, 1978.
45. Deknudt, G. and Gerber, G. B., Chromosomal aberrations in bone-marrow cells of mice given a normal or a calcium-deficient diet supplemented with various heavy metals, *Mutat. Res.*, 68(2), 163–168, 1979.
46. Mukherjee, A., Giri, A. K., Sharma, A., and Talukder, G., Relative efficacy of short-term tests in detecting genotoxic effects of cadmium chloride in mice *in vivo*, *Mutat. Res.*, 206(2), 285–295, 1988.
47. Nayak, B. N., Ray, M., Persaud, T. V., and Nigli, M., Embryotoxicity and *in vivo* changes following maternal exposure to cadmium chloride in mice, *Exp. Pathol.*, 36(2), 75–80, 1989.
48. Adler, D., Kliesch, U., van Hummelen, P., and Kirsch-Volders, M., Mouse micronucleus tests with known and suspect spindle poisons results from two laboratories, *Mutagenesis*, 6(1), 47–53, 1991.
49. Osgood, C., Zimmering, S., and Mason, J. M., Aneuploidy in *Drosophila*. II. Further validation of the FIX and ZESTE genetic test systems employing female *Drosophila melanogaster*, *Mutat. Res.*, 259(1), 147–163, 1991.
50. Miller, B. M. and Adler, I. D., Aneuploidy induction in mouse spermatocytes, *Mutagenesis*, 7(1), 69–76, 1992.
51. Selypes, A., Serenyi-P., Boldog, I., Bokros, F., and Takacs, S., Acute and "long-term" genotoxic effects of $CdCl_2$ on testes of mice, *J. Toxicol. Environ. Health*, 36(4), 401–409, 1992.
52. Marrazzini, A., Betti, C., Bernacchi, F., Barrai, I., and Barale, R., Micronucleus test and metaphase analyses in mice exposed to known and suspected spindle poisons, *Mutagenesis*, 9(6), 515, 1994.
53. Paton, G. R. and Allison, A. C., Chromosome damage in human cell cultures induced by metal salts, *Mutat. Res.*, 16(3), 332–336, 1972.
54. Rohr, G. and Bauchinger, M., Chromosome analyses in cell cultures of the Chinese hamster after application of cadmium sulphate, *Mutat. Res.*, 40(2), 125–130, 1976.
55. Deknudt, G. and Deminatti, M., Chromosome studies in human lymphocytes after *in vitro* exposure to metal salts, *Toxicology*, 10(1), 67–75, 1978.
56. Gasiorek, K. and Bauchinger, M., Chromosome changes in human lymphocytes after separate and combined treatment with divalent salts of lead, cadmium, and zinc, *Environ. Mutagen.*, 3(5), 513–518, 1981.
57. Deaven, L. L. and Campbell, E. W., Factors affecting the induction of chromosomal aberrations by cadmium in Chinese hamster cells, *Cytogenet. Cell Genet.*, 26(2–4), 251–260, 1980.

58. Ochi, T., Mogi, M., Watanabe, M., and Ohsawa, M., Induction of chromosomal aberrations in cultured Chinese hamster cells by short-term treatment with cadmium chloride, *Mutat. Res.,* 137(2–3), 103–109, 1984.
59. Ochi, T. and Ohsawa, M., Participation of active oxygen species in the induction of chromosomal aberrations by cadmium chloride in cultured Chinese hamster cells, *Mutat. Res.,* 143(3), 137–142, 1985.
60. Terracio, L. and Nachtigal, M., Oncogenicity of rat prostate cells transformed *in vitro* with cadmium chloride, *Arch. Toxicol.,* 61(6), 450–456, 1988.
61. Sehgal, A., Osgood, C., and Zimmering, S., Aneuploidy in *Drosophila*. III. Aneuploidogens inhibit *in vitro* assembly of taxol-purified *Drosophila* microtubules, *Environ. Mol. Mutagen.,* 16(4), 217–224, 1990.
62. Antoccia, A., Degrassi, F., Battistoni, A., Cilutti, P., and Tanzarella, C., *In vitro* macronucleus test with kinetochore staining: evaluation of test performance, *Mutagenesis,* 6(4), 319–324, 1991.
63. Crebelli, R., Conti, G., Conti, L., and Carere, A., *In vitro* studies with nine known or suspected spindle poisons: results in tests for chromosome malsegregation *Aspergillus nidulans, Mutagenesis,* 6(2), 131–136, 1991.
64. Armstrong, M. J., Bean, C. L., and Galloway, S. M., A quantitative assessment of the cytoxicity associated with chromosomal aberration detection in Chinese hamster ovary cells, *Mutat. Res.,* 265(1), 45–60, 1992.
65. Bean, C. L., Armstrong, M. J., and Galloway, S. M., Effect of sampling time on chromosome aberration yield for seven chemicals in Chinese hamster ovary cells, *Mutat. Res.,* 265(1), 31–44, 1992.
66. Galloway, S. M., Evaluation of the need for a late harvest time in the assay for chromosome aberrations in Chinese hamster ovary cells, *Mutat. Res.,* 292(1), 3–16, 1993.
67. Lin, R. H., Lee, C. H., Chen, W. K., and Lin-Shiau, S. Y., Studies on cytotoxic and genotoxic effects of cadmium nitrate and lead nitrate in Chinese hamster ovary cells, *Environ. Mol. Mutagen,* 23(2), 143–149, 1994.
68. Yamada, H., Miyahara, T., and Sasaki, Y. F., Inorganic cadmium increases the frequency of chemically induced chromosome aberrations in cultured mammalian cells, *Mutat. Res.,* 302(3), 137–145, 1993.
69. Bigaliev, A. B., Elemesova, M. Sh., and Bigalieva, R. K., Chromosome aberrations in the somatic cells of mammals evoked by chromium compounds, *Tsitol. Genet.,* 10(3), 222–224, 1976.
70. Krishnaja, A. P. and Rege, M. S., Induction of chromosomal aberrations in fish *Boleophthalmus dissumieri* after exposure *in vivo* to mitomycin C and heavy metals mercury, selenium and chromium, *Mutat. Res.,* 102(1), 71–82, 1982.
71. Zimmering, S., Mason, J. M., Valencia, R., and Woodruff, R. C., Chemical mutagenesis testing in *Drosophila*. II. Results of 20 coded compounds tested for the National Toxicology Program, *Environ. Mutagen.,* 7(1), 87–100, 1985.
72. Graf, U., Heo, O. S., and Ramirez, O. O., The genotoxicity of chromium(VI) oxide in the wing spot test of *Drosophila melanogaster* is over 90% due to mitotic recombination, *Mutat. Res.,* 266(2), 197–203, 1992.
73. Sarkar, D., Sharma, A., and Talukder, G., Differential protection of chlorophyllin against clastogenic effects of chromium and chlordane in mouse bone marrow *in vivo*, *Mutat. Res.,* 301(1), 33–38, 1993.
74. Bigaliev, A. B., Chromosome aberrations in a lymphocyte culture from persons in contact with chromium, *Tsitol. Genet.,* 15(6), 63–68, 1981.
75. Sarto, F., Cominato, I., Bianchi, V., and Levis, A. G., Increased incidence of chromosomal aberrations and sister chromatid exchanges in workers exposed to chromic acid (CrO_3) in electroplating factories, *Carcinogenesis,* 3(9), 1011–1016, 1982.
76. Kaneko, T., Chromosome damage in cultured human leukocytes induced by chromium chloride and chromium trioxide, *Sangyo Igaku,* 18(2), 136–137, 1976.
77. Bigaliev, A. B., Elemesova, M. Sh., and Turebaev, M. N., Assessment of the mutagenic activity of chromium compounds, *Gig. Tr. Prof. Zabol.,* 6, 37–40, 1977.
78. Raffetto, G., Parodi, S., Parodi, C., De-Ferrari, M., Troiano, R., and Brambilla, G., Direct interaction with cellular targets as the mechanism for chromium carcinogenesis, *Tumori,* 63(6), 503–512, 1977.
79. Tsuda, H. and Kato, K., Chromosomal aberrations and morphological transformation in hamster embryonic cells treated with potassium dichromate *in vitro*, *Mutat. Res.,* 46(2), 87–94, 1977.
80. Nakamuro, K., Yoshikawa, K., Sayato, Y., and Kurata, H., Comparative studies of chromosomal aberration and mutagenicity of trivalent and hexavalent chromium, *Mutat. Res.,* 58(2–3), 175–181, 1978.
81. Newbold, R. F., Amos, J., and Connell, J. R., The cytotoxic, mutagenic and clastogenic effects of chromium-containing compounds on mammalian cells in culture, *Mutat. Res.,* 67(1), 55–63, 1979.
82. Umeda, M. and Nishimura, M., Inducibility of chromosomal aberrations by metal compounds in cultured mammalian cells, *Mutat. Res.,* 67(3), 221–229, 1979.
83. Majone, F. and Levis, A. G., Chromosomal aberrations and sister-chromatid exchanges in Chinese hamster cells treated *in vitro* with hexavalent chromium compounds, *Mutat. Res.,* 67(3), 231–238, 1979.
84. Levis, A. G. and Majone, F., Cytotoxic and clastogenic effects of soluble chromium compounds on mammalian cell cultures, *Br. J. Cancer,* 40(4), 523–533, 1979.
85. MacRae, W. D., Whiting, R. F., and Stich, H. F., Sister chromatid exchanges induced in cultured mammalian cells by chromate, *Chem. Biol. Interact.,* 26(3), 281–286, 1979.
86. Gomez-Arroyo, S., Altamirano, M., and Villatobos-Pietrini, R., Sister chromatid exchanges induced by some chromium compounds in human lymphocytes *in vitro*, *Mutat. Res.,* 90(4), 425–431, 1981.

87. Stella, M., Montaldi, A., Rossi, R., Rossi, G., and Levis, A. G., Clastogenic effects of chromium on human lymphocytes *in vitro* and *in vivo*, *Mutat. Res.*, 101(2), 151–164, 1982.
88. Levis, A. G. and Majone, F., Cytotoxic and clastogenic effects of soluble and insoluble compounds containing hexavalent and trivalent chromium, *Br. J. Cancer*, 44(2), 219–235, 1981.
89. Venier, P., Montaldi, A., Majone, F., Bianchi, V., and Levis, A. G., Cytotoxic, mutagenic and clastogenic effects of industrial chromium compounds, *Carcinogenesis*, 3(11), 1331–1338, 1982.
90. Nijs, M. and Kirsch-Volders, M., Induction of spindle inhibition and abnormal mitotic figures by Cr(II), Cr(III) and Cr(VI) ions, *Mutagenesis*, 1(4), 247–252, 1986.
91. Friedman, J., Shabtai, F., Levy, L. S., and Djaldetti, M., Chromium chloride induces chromosomal aberrations in human lymphocytes via indirect action, *Mutat. Res.*, 191(3–4), 207–210, 1987.
92. Sugiyama, M., Lin, X. H., and Costa, M., Protective effect of vitamin E against chromosomal aberrations and mutation induced by sodium chromate in Chinese hamster V79 cells, *Mutat. Res.*, 260(1), 19–23, 1991.
93. Sugiyama, M., Tsuzuki, K., Lin, X. H., and Costa, M., Potentiation of sodium chromate(VI)-induced chromosomal aberrations and mutation by vitamin B_2 in Chinese hamster V79 cells, *Mutat. Res.*, 283(3), 211–214, 1992.
94. Wise, J. P., Sr., Stearns, D. M., Wetterhahn, K. E., and Patierno, S. R., Cell-enhanced dissolution of carcinogenic lead chromate particles: the role of individual dissolution products in clastogenesis, *Carcinogenesis*, 15(10), 2249–2254, 1994.
95. Waksvik, H. and Boysen, M., Cytogenetic analyses of lymphocytes from workers in a nickel refinery, *Mutat. Res.*, 103(2), 185–190, 1982.
96. Waksvik, H., Boysen, M., and Hogetveit, A. C., Increased incidence of chromosomal aberrations in peripheral lymphocytes of retired nickel workers, *Carcinogenesis*, 5(11), 1525–1527, 1984.
97. Cai, D. C., Jin, M., Han, L., Wu, S., Xie, Z. Q., and Zheng, X. S., Cytogenetic analysis in workers occupationally exposed to nickel carbonyl, *Mutat. Res.*, 188(2), 149–152, 1987.
98. Elias, Z., Mur, J. M., Pierre, F., Gilgenkrantz, S., Schneider, O., Baruthio, F., Daniere, M. C., and Fontana, J. M., Chromosome aberrations in peripheral blood lymphocytes of welders and characterization of their exposure by biological samples analysis, *J. Occup. Med.*, 31(5), 477–483, 1989.
99. Senft, V., Losan, F., and Tucek, M., Cytogenetic analysis of chromosomal aberrations of peripheral lymphocytes in workers occupationally exposed to nickel, *Mutat. Res.*, 279(3), 171–179, 1992.
100. Jeimert, O., Hansteen, I. L., and Langard, S., Chromosome damage in lymphocytes of stainless steel welders related to past and current exposure to manual metal arc welding fumes, *Mutat. Res.*, 320(3), 223–233, 1994.
101. Knudsen, L. E., Boisen, T., Christensen, J. M., Helnes, J. E., Jensen, G. E., Jensen, J. C., Lundgren, K., Lundsteen, C., Pedersen, B., and Wassermann, B. et al., Biomonitoring of genotoxic exposure among stainless steel welders, *Mutat. Res.*, 279(2), 129–143, 1992.
102. Shen, H. M. and Zhang, Q. F., Risk assessment of nickel carcinogenicity and occupational lung cancer, *Environ. Health Perspect.*, 102(Suppl. 1), 275–282, 1994.
103. Yamashiro, S., Gilma, J. P., Basrur, P. K., and Abandowitz, H. M., Growth and cytogenetic characteristics of nickel sulphide-induced rhabdomyosarcomas in rats, *Acta Pathol. Jpn.*, 28(3), 435–444, 1978.
104. Pot-Deprun, J., Paupon, M. F., Sweeney, F. L., and Chouroulinkiv, I., Growth, metastasis, immunogenicity, and chromosomal content of a nickel-induced rhabdomyosarcoma and subsequent cloned cell lines in rats, *J. Natl. Cancer Inst.*, 71(6), 1241–1245, 1983.
105. Christie, N. T., Tummolo, D. M., Biggart, N. W., and Murphy, E. C., Jr., Chromosomal changes in cell lines from mouse tumors induced by nickel sulfide and methylcholanthrene, *Cell Biol. Toxicol.*, 4(4), 427–445, 1988.
106. Sunderman, F. W., Jr., Hopfer, S. M., Nichols, W. W., Selden, J. R., Allen, H. L., Anderson, C. A., Hill, R., Bradt, C., and Williams, C. J., *Ann. Clin. Lab. Sci.*, 20(1), 60–72, 1990.
107. Deknudt, G. and Leonard, A., Mutagenicity tests with nickel salts in the male mouse, *Toxicology*, 25(4), 289–292, 1982.
108. Dhir, H., Aganwal, K., Sharma, A., and Talukder, G., Modifying role of *Phyllanthus emblica* and ascorbic acid against nickel clastogenicity in mice, *Cancer Lett.*, 59(1), 9–18, 1991.
109. Chorvatovicova, D. and Kovacikova, Z., Inhalation exposure of rats to metal aerosol. II. Study of mutagenic effect on alveolar macrophages, *J. Appl. Toxicol.*, 12(1), 67–68, 1992.
110. Nishimura, M. and Umeda, M., Induction of chromosomal aberrations in cultured mammalian cells by nickel compounds, *Mutat. Res.*, 68(4), 337–349, 1979.
111. Sen, P. and Costa, M., Induction of chromosomal damage in Chinese hamster ovary cells by soluble and particulate nickel compounds: preferential fragmentation of the heterochromatic long arm of the X chromosome by carcinogenic crystalline NiS particles, *Cancer Res.*, 45(5), 2320–2325, 1985.
112. Sen, P. and Costa, M., Pathway of nickel uptake influences its interaction with heterochromatic DNA, *Toxicol. Appl. Pharmacol.*, 84(2), 278–285, 1986.
113. Sen, P. and Costa, M., Incidence and localization of sister chromatid exchanges induced by nickel and chromium compounds, *Carcinogenesis*, 7(9), 1527–1533, 1986.
114. Conway, K. and Costa, M., Nonrandom chromosomal alterations in nickel-transformed Chinese hamster embryo cells, *Cancer Res.*, 49(21), 6032–6038, 1989.

115. Conway, K. and Costa, M., The involvement of heterochromatic damage in nickel-induced transformation, *Biol. Trace Elements Res.,* 21, 437–444, 1989.
116. Klein, C. B., Conway, K., Wang, X. W., Bhamra, R. K., Lin, X. H., Cohen, M. D., Annab, L., Barrett, J. C., and Costa, M., Senescence of nickel-transformed cells by an X chromosome: possible epigenetic control, *Science,* 251(4995), 796–799, 1991.
117. Wang, X. W., Lin, X., Klein, C. B., Bhamra, R. K., Lee, Y. W., and Costa, M., A conserved region in human and Chinese hamster X chromosomes can induce cellular senescence of nickel-transformed Chinese hamster cell lines, *Carcinogenesis,* 13(4), 555–561, 1992.
118. Lin, X. H., Sugiyama, M., and Costa, M., Differences in the effect of Vitamin E on nickel sulfide or nickel chloride-induced chromosomal aberrations in mammalian cells, *Mutat. Res.,* 260(2), 159–164, 1991.
119. Costa, M., Salnikow, K., Cosentino, S., Klein, C. B., Huang, X., and Zhuang, Z., Molecular mechanisms of nickel carcinogenesis, *Environ. Health Perspect.,* 102(Suppl. 3), 127–130, 1994.
120. Costa, M., Molecular targets of nickel and chromium in human and experimental systems, *Scand. J. Work Environ. Health,* 19(Suppl. 1), 71–74, 1993.
121. Stinson, T. J., Jaw, S., Jeffery, E. H., and Plewa, M. J., The relationship between nickel chloride-induced peroxidation and DNA strand breakage in rat liver, *Toxicol. Appl. Pharmacol.,* 117(1), 98–103, 1992.
122. Rani, A. B., Qu, D. Q., Sidhu, M. K., Panagakos, F., Shah, V., Klein, K. M., Brown, N., Pathak, S., and Kumar, S., Transformation of immortal nontumorigenic osteoblast-like human osteosarcoma cells to the tumorigenic phenotype by nickel sulfate, *Carcinogenesis,* 14(5), 947–953, 1993.
123. Rossman, T., Meyn, M. S., and Troll, W., Effects of sodium arsenite on the survival of UV-irradiated *Escherichia coli*: inhibition of a recA-dependent function, *Mutat. Res.,* 30(2), 157–162, 1975.
124. Fong, K., Lee, F., and Bockrath, R., Effects of sodium arsenite on single-strand DNA break formation and post-replication repair in *E. coli* following UV irradiation, *Mutat. Res.,* 70(2), 151–156, 1980.
125. Okada, S., Yamanaka, K., Ohba, H., and Kawazoe, Y., Effect of inorganic arsenics on cytotoxicity and mutagenicity of ultraviolet light on *Escherichia coli* and mechanism involved, *J. Pharm. Dyn.,* 6(7), 496–504, 1983.
126. Rossman, T. G., Stone, D., Molina, M., and Troll, W., Absence of arsenite mutagenicity in *E. coli* and Chinese hamster cells, *Environ. Mutagen.,* 2(3), 371–379, 1980.
127. Okui, T. and Fujiwara, Y., Inhibition of human excision DNA repair by inorganic arsenic and the co-mutagenic effect in V79 Chinese hamster cells, *Mutat. Res.,* 172(1), 69–76, 1986.
128. Li, J. H. and Rossman, T. G., Inhibition of DNA ligase activity by arsenite, a possible mechanism of its comutagenesis, *Mol. Toxicol.,* 2(1), 1–9, 1989.
129. Li, J. H. and Rossman, T. G., Mechanism of comutagenesis of sodium arsenite with *N*-methyl-*N*-nitrosourea, *Biol. Trace Elements Res.,* 21, 373–381, 1989.
130. Yang, J. L., Chen, M. F., Wu, C. W., and Lee, T. C., Post-treatment with sodium arsenite alters the mutational spectrum induced by ultraviolet light irradiation in Chinese hamster ovary cells, *Environ. Mol. Mutagen.,* 20(3), 156–164, 1992.
131. Lee, T. C., Tanaka, N., Lamb, P. W., Gilmer, T. M., and Barrett, J. C., Induction of gene amplification by arsenic, *Science,* 241(4861), 79–81, 1988.
132. Detke, S., Katakura, K., and Chang, K. P., DNA amplification in arsenite-resistant Leishmania, *Exp. Cell Res.,* 180(1), 161–170, 1989.
133. Lee, S. Y., Lee, S. T., and Chang, K. P., Transkinetoplatidy — a novel phenomenon involving bulk alterations of mitochondrion-kinetoplast DNA of a trypanosomatid protozoan, *J. Protozool.,* 39(1), 190–196, 1992.
134. Wedrychowski, A., Schmidt, W. N., and Hnilica, L. S., DNA protein crosslinking by heavy metals in Novikoff hepatoma, *Arch. Biochem. Biophys.,* 251(2), 397–402, 1986.
135. Dong, J. T. and Luo, X. M., Arsenic-induced DNA-strand breaks associated with DNA-protein crosslinks in human fetal lung fibroblasts, *Mutat. Res.,* 302(2), 97–102, 1993.
136. Dong, J. T. and Luo, X. M., Effects of arsenic on DNA damage and repair in human fetal lung fibroblasts, *Mutat. Res.,* 315(1), 11–15, 1994.
137. Tezuka, M., Hanioka, K., Yamanaka, K., and Okada, S., Gene damage induced in human alveolar type II cells by exposure to dimethylarsinic acid, *Biochem. Biophys. Res. Commun.,* 191(3), 1178–1183, 1993.
138. Miyaki, M., Akamatsu, N., Ono, T., and Koyama, H., Mutagenicity of metal cations in cultured cells from Chinese hamster, *Mutat. Res.,* 68(3), 259–263, 1979.
139. Zakour, R. A. and Glickman, B. W., Metal-induced mutagenesis in the *lacI* gene of *Escherichia coli, Mutat. Res.,* 126(1), 9–18, 1984.
140. Kanematsu, N., Hara, M., and Kada, T., Rec assay and mutagenicity studies on metal compounds, *Mutat. Res.,* 77(2), 109–116, 1980.
141. Nickel-Brady, C., Hahn, F. F., Finch, G. L., and Belinsky, S. A., Analysis of K-*ras*, *p53* and c-*raf*-1 mutations in beryllium-induced rat lung tumors, *Carcinogenesis,* 15(2), 257–262, 1994.
142. Ochi, T., Takayanagi, M., and Ohsawa, M., Cadmium-induced DNA single-strand scissions and their repair in cultured Chinese hamster cells, *Toxicol. Lett.,* 18(1–2), 177–183, 1983.
143. Hamilton-Koch, W., Snyder, R. D., and Lavelle, J. M., Metal-induced DNA damage and repair in human diploid fibroblasts and Chinese hamster ovary cells, *Chem. Biol. Interact.,* 59(1), 17–28, 1986.

144. Bradley, M. O., Taylor, V. I., Armstrong, M. J., and Galloway, S. M., Relationships among cytotoxicity, lysosomal breakdown, chromosome aberrations, and DNA double-strand breaks, *Mutat. Res.,* 189(1), 69–79, 1987.
145. Snyder, R. D., Role of active oxygen species in metal-induced DNA strand breakage in human diploid fibroblasts, *Mutat. Res.,* 193(3), 237–246, 1988.
146. Denizeau, F. and Marion, M., Genotoxic effects of heavy metals in rat hepatocytes, *Cell Biol. Toxicol.,* 5(1), 15–25, 1989.
147. Mandel, R. and Ryser, H. J., Mutagenicity of cadmium in *Salmonella typhimurium* and its synergism with two nitrosamines, *Mutat. Res.,* 138(1), 9–16, 1984.
148. Mandel, R. and Ryser, H. J., Mechanisms of synergism in the mutagenicity of cadmium and *N*-methyl-*N*-nitrosourea in *Salmonella typhimurium*: the effect of pH, *Mutat. Res.,* 176(1), 1–10, 1987.
149. Nocentini, S., Inhibition of DNA replication and repair by cadmium in mammalian cells. Protective interaction of zinc, *Nucleic Acids Res.,* 15(10), 4211–4225, 1987.
150. Warren, G., Schultz, P., Bancroft, D., Bennett, K., Abbott, E. H., and Rogers, S., Mutagenicity of a series of hexacoordinate chromium(III) compounds, *Mutat. Res.,* 90(2), 111–118, 1981.
151. Bianchi, V., Celotti, L., Lanfranchi, G., Majone, F., Marin, G., Montaldi, A., Sponza, G., Tanino, G., Venier, P., Zantedeschi, A., and Levis, A. G., Genetic effects of chromium compounds, *Mutat. Res.,* 117(3–4), 279–300, 1983.
152. Bianchi, V., Zantedeschi, A., Montaldi, A., and Majone, F., Trivalent chromium is neither cytotoxic nor mutagenic in permeabilized hamster fibroblasts, *Toxicol. Lett.,* 23(1), 51–59, 1984.
153. Christie, N. T., Cantoni, O., Evans, R. M., Meyn, R. E., and Costa, M., Use of mammalian DNA repair-deficient mutants to assess the effects of toxic metal compounds on DNA, *Biochem. Pharmacol.,* 33(10), 1661–1670, 1984.
154. Llagostera, M., Garrido, S., Guerrero, R., and Barbe, J., Induction of SOS genes of *Escherichia coli* by chromium compounds, *Environ. Mutagen.,* 8(4), 571–577, 1986.
155. LaVelle, J. M., Chromium(VI) comutagenesis: characterization of the interaction of K_2CrO_4 with azide, *Environ. Mutagen.,* 8(5), 717–725, 1986.
156. Manning, F. C., Xu, J., and Patierno, S. R., Transcriptional inhibition by carcinogenic chromate: relationship to DNA damage, *Mol. Carcinog.,* 6(4), 270–279, 1992.
157. Geo, M., Binks, S. P., Chipman, J. K., and Levy, L. S., Hexavalent chromium produces DNA strand breakage but not unscheduled DNA synthesis at sub-cytotoxic concentrations in hepatocytes, *Toxicology,* 77(1–2), 171–180, 1993.
158. Xu, J., Manning, F. C., and Patierno, S. R., Preferential formation and repair of chromium-induced DNA adducts and DNA-protein crosslinks in nuclear matrix DNA, *Carcinogenesis,* 15(7), 1443–1460, 1994.
159. Manning, F. C., Blankenship, L. J., Wise, J. P., Xu, J., Bridgewater, L. C., and Patierno, S. R., Induction of internucleosomal DNA fragmentation by carcinogenic chromate: relationship to DNA damage, genotoxicity, and inhibition of macromolecular synthesis, *Environ. Health Perspect.,* 102, 3, 159–167, 1994.
160. Shi, X., Mao, Y., Knapton, A. D., Ding, M., Rojanasakui, Y., Gannett, P. M., Dalai, N., and Liu, K., Reaction of Cr(VI) with ascorbate and hydrogen peroxide generates hydroxyl radicals and causes DNA damage: role of a Cr(IV)-mediated Fenton-like reaction, *Carcinogenesis,* 15(11), 2475–2478, 1994.
161. Kawanishi, S., Inoue, S., and Yamamoto, K., Active oxygen species in DNA damage induced by carcinogenic metal compounds, *Environ. Health Perspect.,* 102, 3, 17–20, 1994.
162. Stearns, D. M., Kennedy, L. J., Courtney, K. D., Giangrande, P. H., Phieffer, L. S., and Wetterhahn, K. E., Reduction of chromium(VI) by ascorbate leads to chromium-DNA binding and DNA strand breaks *in vitro*, *Biochemistry,* 34(3), 910–919, 1995.
163. Stearns, D. M., Courtney, K. D., Giangrande, P. H., Phieffer, L. S., and Wetterhahn, K. E., Chromium(VI) by reduction by ascorbate: role of reactive intermediates in DNA damage *in vitro*, *Environ. Health Perspect.,* 102, 3, 21–25, 1994.
164. Molyneux, M. J. and Davies, M. J., Direct evidence for hydroxyl radical-induced damage to nucleic acids by chromium(VI)-derived species: implications for chromium carcinogenesis, *Carcinogenesis,* 16(4), 875–882, 1995.
165. Tsuzuki, K., Sugiyama, M., and Haramaki, N., DNA single-strand breaks and cytotoxicity induced by chromate(VI), cadmium(II), and mercury(II) in hydrogen peroxide-resistant cell lines, *Environ. Health Perspect.,* 102, 3, 341–342, 1994.
166. Chen, J. and Thilly, W. G., Use of denaturing-gradient gel electrophoresis to study chromium-induced point mutations in human cells, *Environ. Health Perspect.,* 102, 3, 227–229, 1994.
167. Misra, M., Alcedo, J. A., and Wetterhahn, K. E., Two pathways for chromium(VI)-induced DNA damage in 14 day chick embryos: Cr-DNA binding in liver and 8-oxo-2-deoxyguanosine in red blood cells, *Carcinogenesis,* 15(12), 2911–2917, 1994.
168. Zhitkovic, A., Voitkun, V., and Costa, M., Glutathione and free amino acids form stable complexes with DNA following exposure of intact mammalian cells to chromate, *Carcinogenesis,* 16(4), 907–913, 1995.
169. Shen, H. M. and Zhang, Q. F., Risk assessment of nickel carcinogenicity and occupational lung cancer, *Environ. Health Perspect.,* 102, 1, 275–282, 1994.

170. Hartwig, A., Kruger, I., and Beyersmann, D., Mechanisms in nickel genotoxicity: the significance of interactions with DNA repair, *Toxicol. Lett.*, 72(1–3), 353–358, 1994.
171. Robison, S. H., Cantoni, O., Heck, J. D., and Costa, M., Soluble and insoluble nickel compounds induce DNA repair synthesis in cultured mammalian cells, *Cancer Lett.*, 17(3), 273–279, 1983.
172. Chiocca, S. M., Sterner, D. A., Biggart, N. W., and Murphy, E. C., Jr., Nickel mutagenesis: alteration of the MuSvs110 thermosensitive splicing phenotype by a nickel-induced duplication of the 3′ splice site, *Mol. Carcinog.*, 4(1), 61–71, 1991.
173. Lee, Y. W., Pons, C., Tummolo, D. M., Klein, C. B., Rossman, T. G., and Christie, N. T., Mutagenicity of soluble and insoluble nickel compounds at the *gpt* locus in G12 Chinese transfer cells, *Environ. Mol. Mutagen.*, 21(4), 365–371, 1993.
174. Rossetto, F. E., Turnbull, J. D., and Nieboer, E., Characterization of nickel-induced mutations, *Sci. Total Environ.*, 148(2–3), 201–206, 1994.
175. Lee, Y. W., Klein, C. B., Kargacin, B., Salnikow, K., Kithahara, J., Dowjat, K., Zhitkovich, A., Christie, N. T., and Costa, M., Carcinogenic nickel silences gene expression by chromatin condensation and DNA methylation: a new model for epigenetic carcinogens, *Mol. Cell. Biol.*, 15(5), 2547–2557, 1995.
176. Dubins, J. S. and LaVelle, J. M., Nickel(II) genotoxicity: potentiation of mutagenesis of simple alkylating agents, *Mutat. Res.*, 162(2), 187–199, 1986.
177. Hartwig, A. and Beyersmann, D., Enhancement of UV-induced mutagenesis and sister-chromatid exchanges by nickel ions in V79 cells: evidence for inhibition of DNA repair, *Mutat. Res.*, 217(1), 65–73, 1989.
178. Lee-Chen, S. F., Wang, M. C., Yu, C. T., Wu, D. R., and Jan, K. Y., Nickel chloride inhibits the DNA repair of UV-treated but not methyl methanesulfonate-treated Chinese hamster ovary cells, *Biol. Trace Elements Res.*, 37(1), 39–50, 1993.
179. Lee-Chen, S. F., Yu, C. T., Wu, D. R., and Jan, K. Y., Differential effects of luminal, nickel, and arsenite on the rejoining of ultraviolet light and alkylation-induced DNA breaks, *Environ. Mol. Mutagen.*, 23(2), 116–120, 1994.
180. Hartwig, A., Mullenders, L. H., Schlepegrell, R., Kasten, U., and Beyersmann, D., Nickel(II) interferes with the incision step in nucleotide excision repair in mammalian cells, *Cancer Res.*, 54(15), 4045–4051, 1994.
181. Lynn, S., Yew, F. H., Hwang, J. W., Tseng, M. J., and Jan, K. Y., Glutathione can rescue the inhibitory effects of nickel on DNA ligation and repair synthesis, *Carcinogenesis*, 15(12), 2811–2816, 1994.
182. Synder, R. D., Effects of metal treatment on DNA repair in polyamine-depleted HeLa cells with special reference to nickel, *Environ. Health Perspect.*, 3, 51–55, 1994.

3 MECHANISMS OF CELL DEATH INDUCED BY METALS

Y. James Kang

Cell death (irreversible loss of vital cellular structure and function) is a fundamental phenomenon of biological organisms. Cell death occurs as a physiologic process during organogenesis in embryos and cell turnover in adults, and as a pathologic process in response to various injuries. There are two fundamental types of cell death, namely, apoptosis and necrosis. Apoptosis is a distinct mode of multifocal single-cell death that occurs during the normal development of vertebrates and invertebrates. Apoptosis is considered to be the major process responsible for cell death in various physiological events, such as embryonic tissue remodeling and adult cell turnover and differentiation. This phenomenon has also been observed in various pathologic states, including viral hepatitis, death of tumor cells, graft-vs.-host disease, and other unidentified conditions. Necrosis, on the other hand, occurs as a consequence of an injurious environment. This mode of cell death is characterized by irreversible loss of metabolic functions (e.g., loss of ion gradients and adenosine triphosphate [ATP] depletion) and structural integrity of the plasma membrane. It has been well known that metals at their toxic concentrations cause necrosis in a variety of tissues. Recent studies have also demonstrated that metals also induce apoptosis as a mode of toxicity. This section focuses on pathologic processes of cell death including both apoptosis and necrosis in response to injuries induced by metals or metal-related deleterious conditions.

The mechanisms involved in apoptosis and necrosis differ in many aspects. The same mode of cell death involves different mechanisms in various cell types. A most distinct biochemical process that differs between apoptosis and necrosis is ATP generation and protein synthesis, which are inhibited in necrosis but preserved in apoptosis. There are, however, some shared mechanisms involved in toxicant-induced apoptosis and necrosis, especially the cell death induced by metals. Such mechanisms include alteration in intracellular ion homeostasis and oxidative stress.

3.1 Apoptosis and Necrosis

Several outstanding reviews have described distinct characteristics of apoptosis and necrosis.[1-5] These reviews also discuss the cellular, biochemical, and molecular mechanisms of apoptosis and necrosis. In this section, only those distinct aspects of metals-induced cell death will be discussed.

3.1.1 Apoptosis

Apoptosis induced by metals and metal-related deleterious conditions has only recently been studied. Information about metal-induced apoptosis is therefore very limited. Several interesting phenomena, however, have been observed. Biological essential metals such as copper, iron, and zinc are associated with apoptosis under conditions of their deficiency; overload of these metals often causes necrosis but not apoptosis. Other metals such as cadmium, chromium, and selenium induce both apoptosis and necrosis when biological systems are exposed to these metals. In general, low-dose exposure to these metals most likely causes apoptosis and high doses are associated with necrosis. Exposure dose, therefore, is a major determinant for the occurrence of apoptosis or necrosis. In the case of apoptosis, several common mechanisms have been indicated for the apoptosis induced by essential metal-deficiency and that induced by metal toxicity. For instance, copper deficiency induces apoptosis in rat pancreatic acinar tissue through activation of apoptosis-associated endonucleases.[8] Similarly, exposure to low-dose selenium activates endonucleases such as Ca^{2+} and Mg^{2+}-dependent endonuclease, leading to apoptosis.[10] Alterations in ionic calcium levels may be important contributing factors to endonuclease activation associated with toxic metal-induced apoptosis. Cadmium induces calcium mobilization,[11] which would in turn cause activation of endonucleases. Another common mechanism is oxidative stress. Copper deficiency and toxic metal exposure both generally result in production of reactive-free radicals such as reactive oxygen species, which represents another apoptotic mechanism shared by essential metal deficiency and toxic metal exposure.

Some metals appear to exhibit unique mechanisms for induction of apoptosis. For example, zinc plays an important role in preventing apoptosis. It prevents endonucleosomal fragmentation and subsequent cytolysis;[9] therefore, zinc deficiency causes apoptosis by removal of protective factors. The carcinogen chromium directly interacts with DNA to form a DNA-chromium adduct.[12] It is possible that this represents another mechanism for metal-induced apoptosis. Although the information about metal-induced apoptosis is limited, current studies indicate that metal-induced apoptosis is mediated by several common mechanisms shared by other toxicant-induced apoptosis. In this context, metal-induced apoptosis may serve as a model for studying the mechanisms involved in apoptosis. The content in Table 27 summarizes current studies of metal-induced apoptosis.

3.1.2 Necrosis

Necrosis induced by metals has been well documented. In general, the mechanisms involved in metal-induced necrosis include alteration in ion homeostasis, oxidative injury, mitochondrial dysfunction, and ATP depletion. Inhibition of DNA, RNA, and protein synthesis often occurs during metal toxicosis. This inhibitory effect may not be the primary effect

TABLE 27
Apoptosis

Effector	Concentration	System	Mechanism	Ref.
Iron	Iron deprivation by deferoxamine B mesylate (10^{-6} ~ 10^{-4} *M*)	HL-60 cells	c-*myc* and c-*fos* expression	6
		Leukemic CCRF-CEM cells	—	7
Copper	Copper deficiency by 0.6% triethylenetetramine tetrahydrochloride	F-344 rats; pancreatic acinar tissue	Perturbations in inter- and intracellular milieu, leading to activation of apoptosis-associated genes and endonuclease activities	8
Zinc	Zinc deficiency diet or Zn^{2+} chelators	Mice and cultured cells	Zinc prevents endonucleosomal fragmentation and subsequent cytolysis	9
Selenium	Selenite (20 μ*M*)	Mouse leukemic L1210 cells	Ca^{2+}, Mg^{2+}-dependent, endonuclease-mediated DNA double-strand breakage	10
Cadmium	8–10 μ*M*	T cell, line CEM-C12	Calcium mobilization and/or replacing calcium signaling	11
Chromium	150–300 μ*M*	CHO AA8 cells	DNA-chromium adducts and inhibition of DNA synthesis	12

of metals, with an exception of carcinogenic metals such as chromium. The ultimate event of metal-induced necrosis is plasma membrane damage and cell lysis.

Iron poisoning, both acute and chronic, has been extensively studied. A number of mechanisms have been advanced to account for the toxicity and eventual cell death from this metal. Chronic iron poisoning has been shown to affect the parenchymal tissues of several organs including liver, heart, pancreas, joints, and endocrine glands. Excess deposition of iron in these tissues results in cell injury and functional insufficiency, eventually leading to necrosis. Iron-induced lipid peroxidation has been shown to be involved in this necrosis process.[13] Acute toxicity of iron most often occurs by ingestion of iron compounds. Many mechanisms have been postulated for the acute toxicity of iron,[14] but no one has successfully explained its pathogenesis; however, it has been shown that acute iron poisoning causes necrosis.[14] It has been known for a long time that copper overload also causes necrosis in humans and animals.[15-18] The most popular hypothesis for the mechanism of copper poisoning is production of free radicals. Copper also induces DNA strand breaks in the presence of hydrogen peroxide.[18] Cadmium induces necrosis in multiple organs. The predominant damage of chronic cadmium exposure occurs in the kidney. Studies have shown that the major source of renal cadmium in chronic cadmium exposure is likely derived from hepatic cadmium which is transported in the form of cadmium-metallothionein in blood plasma.[19] The cadmium taken into kidneys in the form of cadmium-metallothionein is eliminated from the kidney by cell death of the proximal tubules and by their resultant defluxion.[20] Mechanisms for the cadmium-induced hepatic necrosis have been studied.[21,23] These include inhibition of protein synthesis and lipid peroxidation. Whether these mechanisms are also involved in cadmium-induced necrosis in the proximal tubules of the kidney is unknown.

Chromium has a wide range of toxic effects on living systems. It causes carcinogenesis, apoptosis, and necrosis. Cellular mechanisms involved in chromium-induced necrosis include alteration of metabolism,[24] production of reactive free radicals,[26] and damage to DNA and mitochondria.[27] Chromium-derived hydroxyl radical also causes damage to nucleic acids, suggesting a role for this chromium-derived species in the carcinogenesis. Mitochondrial damage induced by chromium such as sodium chromate may result from NADH depletion as a result of direct oxidation by chromium(V) as well as decreased production of NADH due to specific enzyme inhibition. The chromium-inhibited enzymes include alpha-ketoglutarate dehydrogenase (IC_{50} = 3 to 5 μM Na_2CrO_4) and beta-hydroxybutyrate and pyruvate dehydrogenase with high concentrations (20–70 μ*M*);[27] therefore, chromium-induced necrosis is mediated by multiple mechanisms. Other metals that cause necrosis by overt inhibitory effects include mercury, nickel, vanadium, lead, thallium, and platinum. A common primary event induced by metals that cause necrosis is alteration of intracellular calcium homeostasis and subsequent lipid peroxidation. Increase in cellular Ca^{2+} is often observed during metal toxicities. Two fundamental actions would each produce increases in cellular Ca^{2+}. First, many metals directly interact with calcium channels and/or calcium-dependent ATPases to alter Ca^{2+} homeostasis. Second, metals cause oxidative damage to plasma membranes, leading to calcium release from intracellular organelles such as mitochondria and endoplasmic reticula. In summary, some key events in metal-induced necrosis include alteration of ion homeostasis, oxidative stress, mitochondrial dysfunction, and eventually cell membrane damage. These events will be discussed in the subsequent sections. (See Table 28.)

3.2 Oxidative Stress

Oxidative stress is defined as a disturbance in the pro-oxidant-anti-oxidant balance in favor of the pro-oxidant. Production of reactive oxygen species during metabolism is associated with normal aerobic life. Under normal physiological conditions, these toxic oxygen species are destroyed by anti-oxidant defense systems. Aerobic life is therefore characterized by a steady formation of pro-oxidants balanced by a similar rate of their consumption by anti-oxidants. Cells encounter oxidative stress under the conditions of (1) production of reactive oxygen species that overwhelm the defense systems; (2) compromised anti-oxidant systems; or (3) the combination of both conditions. Oxidative stress can lead to a diversity of cell damages.

Reactive oxygen species of biological significance include superoxide anion($O_2^{-\cdot}$), perhydroxyl radical ($HO_2^{\cdot}$), hydrogen peroxide (H_2O_2), hydroxyl radical ($HO^{\cdot}$), alkoxy radical ($RO^{\cdot}$), peroxy radical ($ROO^{\cdot}$), organic hydroperoxide (ROOH), and singlet molecular oxygen. Most studies have focused on understanding the metabolism and toxicities of superoxide anion, hydrogen peroxide, and hydroxyl radical. These species would be considered as basic reactive oxygen species. They are continuously produced during aerobic metabolism and destroyed by anti-oxidant systems. Accumulation of these species, however, results in oxidative damage. The anti-oxidants in biological systems include enzymatic and nonenzymatic components. The major enzymatic components include superoxide dismutase (SOD), catalase, and glutathione peroxidase (GSHpx). The nonenzymatic components include ascorbate, alpha-tocopherol, glutathione (GSH), and metallothionein (MT). SOD converts superoxide anion to hydrogen peroxide, which in turn is metabolized by catalase or GSHpx to water and molecular oxygen. The reaction catalyzed by GSHpx requires GSH as a co-factor. In this reaction, GSH is oxidized to GSSG. In the presence of glutathione reductase (GR), GSSG is reduced to GSH. This GSH redox cycle plays a major role in detoxifying cytosolic hydrogen peroxide. A major biological function of tocopherol as an anti-oxidant is its reactivity as a peroxyl radical quencher. Ascorbate interacts with tocopherol to protect it from peroxidation. Both GSH and MT are strong free radical scavengers. Under oxidative stress conditions, especially chronic oxidative stress, some of these anti-oxidants are overexpressed to compensate for the overwhelming production of reactive oxygen species and their damage. On the other hand, these anti-oxidants are compromised under certain toxic conditions, resulting in oxidative injuries.

Energy has been proposed to play a role in the ability of cells to defend against oxidative stress. It is believed that a tissue's ability under an oxidative stress to maintain a normal balance between the reduced and oxidized forms of various compounds is energy dependent. It has been shown that oxidants are more toxic to energy-depleted hypoxic cells than to the corresponding normoxic cells; however, under sufficient energy production, it is the supply of reducing equivalents that determines the ability of a tissue to defend against oxidative

TABLE 28

Necrosis

Effector	Concentration	System	Mechanism	Ref.
Iron	Iron overload by hereditary and secondary hemochromatosis (dietary and medications)	Human parenchymal tissues	Peroxidative injury of organelles such as mitochondria	13, 14
Copper	4–6 mg/kg	Rat	Production of free radicals	15–18
	1 g–4 ounces	Human	Induction of DNA strand breaks and base modification	
	Wilson's disease	Human		
Cadmium	2, 3, 3.9 mg/kg	Rat	Inhibition of protein synthesis, RNA synthesis, ATPase, and oxidative injury	19–23
Chromium	15 mg/kg	Animals	Free radical injury, alteration of metabolism, and mitochondria and DNA damage	24–28
	48 g	Human		
Mercury	6 g mercury vapor	Human	Oxidative stress, alteration of Ca^{2+} homeostasis, and DNA damage	29–33
Nickel	63, 83, 125 μ*M*/kg	Rat	Inhibition of ATPase in target tissue; ↓ RNA polymerase activity and oxidative injury	34–38
Vanadium	Occupational	Human	Inhibition of Na^+-K^+-ATPase, Ca^{2+}-ATPase, and H^+-ATPase, and induction of lipid peroxidation	39–41
	2 mmol/l	Rat		
Lead	Lead and opium pills	Human	↓ Cytochrome P-450, ↓ GSH, ↓ GSH reductase, ↓ GSH peroxidase; damage of cellular aerobic energy metabolism; mitochondria dysfunction	42–45
	100 mmol/kg	Rat		
Zinc	240 mg Zn per kg	Sheep	Displacement of other metals from important intracellular sites	46–48
	706 μg milk	Holstein veal calves		
Arsenic	9 mg/l	Human	Oxidative injury	49–51
	Environmental exposure		Inhibition of sulfhydroxyl enzymes	
Manganese	Occupational exposure, 2 m*M*	Human	Perturb intracellular Ca^{2+} distribution and significant oxidative stress	52–54
Selenium	5–160 mg/kg	Nubian goat	Reaction with thiols and production of free radicals	55, 56
Aluminum	0.225 mmol/week	Rabbit	Free radical reactions	57–59
	5–20 μ*M*/kg	Rat	Inhibition of heme synthesis, alteration of metabolism	
Tin	2 mg/kg	Rabbits	Oxidative stress	60, 61
			Alterations of mitochondrial oxidative phosphorylation and membranes	62–64
Cobalt	250/500 mg/kg	Chicken	Thiol oxidation and free radical injury	62
	0.3–30 mg/m^3	Rat		
Silver	10–40 μ*M*, silver nitrate	Rat	Lipid peroxidation, interaction with other toxins	65
Thallium	10–15 mg/kg	Rat	↑ Lipid peroxidation, ↓ activity of Na^+/K^+-ATPase; inhibition of sulfhydryl enzymes	66–68
Indium	1.3 mg I per kg	Rat	Inhibition of gap junction, microsomal, and heme metabolism; mitochondrial respiratory dysfunction	69–71

TABLE 28 (continued)
Necrosis

Effector	Concentration	System	Mechanism	Ref.
Tellurium	20–30 mg/kg, sodium tellurate	Rat	Nerve fiber demyelination; ↓ acetylcholinesterase, ↓ monoamine oxidase, ↓ GSH, ↓ GST	72–75
Platinum	6 mg/kg cisplatin	Rat	Inhibition of DNA synthesis, glucogenesis, and Na^+/K^+-ATPase activity; GSH depletion and lipid peroxidation	76–78
	50 mg/m^3, cisplatin	Human		

Note: GSH = glutathione; GST = glutathione-*S*-transferase.

injury. Mitochondria are thus the major subcellular organelles that play an important role in oxidative stress. Mitochondria have two significant impacts on the cellular response to oxidative stress. The first is that mitochondria are the major source of cellular reducing equivalents, and the second is that mitochondria oxidize energy-linked substances for energy production. Both energy and reducing equivalents are transferred via the pyridine nucleotides. The pyridine nucleotide NADPH is utilized primarily for reductive biosynthesis, and NADH is used for energy production. There are linkages that allow the interconversion of NADH to NADPH, although the factors controlling these linkages are poorly understood.

Three major factors, therefore, that determine or affect the occurrence of oxidative injury are enhanced production of reactive free radicals, compromised anti-oxidant systems, and depletion of energy. Metals have been shown to increase the production of reactive oxygen species and/or to inhibit anti-oxidant activities, directly or indirectly leading to oxidative injuries.

Those metals that directly increase the production of reactive oxygen species include iron, copper, chromium, vanadium, and manganese. Iron[79,80] and copper[81] are important co-factors in the Fenton reaction to produce reactive oxygen species. Chromium has been shown to facilitate the formation of hydroxyl radical from hydrogen peroxide.[86,87] Vanadium generates hydroxyl radical via a Fenton-like reaction.[92] Manganese acts as a catalyst for free radical generation.[101] It also generates glutathionyl radicals to deplete the glutathione pool, leading to compromised anti-oxidant capacity.[102] Some metals such as cadmium and nickel increase tissue iron levels, thereby increasing reactive oxygen species generation. Others such as cobalt[107] and aluminum[105] promote free radical generation. A large group of metals induce oxidative injuries by inhibiting anti-oxidant activities; the major mechanisms are depletion of cellular glutathione and inhibition of glutathione related enzymes. These metals include cadmium, mercury, nickel, silver, thallium, and platinum.[78,84,88,89,109,112] Catalase and superoxide dismutase are also inhibited by many metals such as cadmium and mercury.[84,85,89] Mercury also decreases tissue levels of ascorbate and tocopherol.[90] An interesting aspect of metal-related oxidative stress is that zinc deficiency induces lipid peroxidation,[97] because zinc acts as membrane stabilizer[97] and anti-oxidant.[98] (See Table 29.)

3.3 Intracellular Ion Homeostasis

Metal-induced alteration in intracellular ion homeostasis is a major intracellular toxic effect. In non-neural cells, disturbances in the ion homeostasis would result in impaired signal transduction, altered cellular metabolism, changes in cell membrane permeability and integrity,

TABLE 29

Oxidative Stress

Effectors	Concentration	System	Mechanism	Ref.
Iron	25 mg, iron dextran (s.c.)	Rats	Catalyst for the Fenton reaction; enhancement of lipid peroxidation, facilitating the formation of OH· radical involved in the generation of $O_2^{\cdot-}$ and H_2O_2	79, 80
Copper	0.2 m*M*, $CuSO_4$	Liposomes and erythrocyte membrane	Cofactor for oxidase and oxygenase; catalyst in the formation of reactive oxygen species and peroxidation of membrane lipids	81
	10^{-9}–10^{-5} *M*, copper salts	*Salmonella*	Accelerated drug oxidation	82
Cadmium	10–100 μ*M*	Rat hepatocytes	Direct action on the peroxidation reaction; increased iron content,	83
	30 μmol/kg	Rat testicular Leydig cells	↓ Glutathione reductase, ↓ catalase	84
	10–100 μ*M*	Rat kidney cells	↓ SOD activities	85
Chromium	25 m*M*, $K_2Cr_2O_7$	*In vitro*	Generation of hydroxyl radical from hydrogen peroxide	86
	1 m*M*, Cr(III)	*In vitro*		87
Mercury	4 mg/kg	Rat kidney	Depletion of glutathione; ↓SOD, ↓ catalase, and ↓ glutathione peroxidase activity	88
	1.5–2.5 mg/kg	Rat kidney		89
	15 μmol/kg	Rat kidney	↓ Vitamin C and vitamin E contents	90
Nickel	50–400 μ*M*/kg, $NiCl_2$	Rat liver	↑ Hydroxyl radical and lipid peroxidation, ↓ glutathione peroxidase, ↑ tissue iron levels	91
Vanadium	0.2–4.8 m*M*	*In vitro*	Generation of hydroxyl radical via a Fenton-like reaction	92
	0.2–1.0 m*M*	Rat	Increased lipid peroxidation and decreased L-ascorbic acid levels	93
	2 mmol/l	Perfused rat liver	Lipid peroxidation	94
Lead	1000–2000 ppm	Chicken	Fatty acid composition changes and increased lipid peroxidation	95
	2% lead in drinking water	Rat brain	Lipid peroxidation	96
Zinc	Zinc-deficient diet	Rat liver	Enhanced lipid peroxidation, with zinc as membrane stabilizer and anti-oxidant	97, 98
Arsenic		Patients with Blackfoot disease	Increased lipid peroxidation	99
	10 mg/kg	Rat	Mixture with 2.6 mg/kg Cd enhanced lipid peroxidation	100
Manganese	10^{-4} *M*	Rat	Catalyst for free radical generation; ↑ lipid peroxidation	101, 102
Selenium		PC-12 cells	Selenium compounds have pro-oxidative activity *in vitro* and/or *in vivo* under toxic conditions	103, 104
Aluminum	12.5–100 μ*M*	*In vitro*	Mediates Fe^{2+}-supported peroxidation	105
Tin	500 μ*M*/kg	Rat	Induces heme oxygenase through oxidative stress	106
Cobalt	10–500 m*M*	Rat	Promotes free radical generation and thiol oxidation	107
	1–1000 μg/kg	Hamster		108
Silver	30–70 μ*M*	Rat hepatocytes	Decreased thiol concentration and glutathione peroxidase activity	109
				110
Thallium	10 mg/kg (p.o.)	Rat	Lipid peroxidation	111
	5 mg/kg	Rat	Deficiency of glutathione	112
Platinum	2 m*M*, cisplatin	Rat renal cortical slices	Decreased glutathione level; enhanced lipid peroxidation	78

Note: SOD = superoxide dismutase.

and disturbance of vital function. In neural cells, the same changes would occur as in non-neural cells; however, the most sensitive targets would be at the synaptic connection and the action of neural transmitters. The intracellular ions whose homeostases are subject to changes induced by metals include calcium, potassium, and magnesium. Chloride homeostasis is also affected by some metals. Because of the ionic nature of metals, a fundamental mechanism for metal-induced disruption of ion homeostasis is the substitution of these metals for the affected intracellular ions. The major target molecules affected by metals include ion-dependent ATPases and ion channels on the cell membrane, and intracellular ion-dependent proteins such as calcium-binding proteins. The action of metals on ion-dependent ATPases generally inhibits the function of these enzymes leading to disturbances in ion homeostasis. Metals also block ion channels. Some metals such as lead, cadmium, cobalt, iron, and magnesium mimic the action of calcium to open calcium-activated potassium channels. The substitution of metals at the ion-binding sites of intracellular ion-binding proteins likely results in increased levels of intracellular free ions. Such an effect has been observed following exposure to cadmium. Cadmium substitutes for calcium at its binding site on various calcium-binding proteins to increase cellular calcium levels. This substitution also alters calcium-binding protein activity and function. Table 30 summarizes the effects of metals on ion homeostasis in both non-neural and neural cells.

3.4 Mitochondrial Dysfunction and ATP Depletion

Mitochondrial dysfunction occurs in most, if not all, toxic insults. Studies on metal-induced mitochondrial dysfunction and its subsequent role in cytotoxicity are limited. Because of the importance of mitochondria in cellular metabolism and function, toxicological studies related to mitochondria would be expected to develop further. Mitochondria are the major source of energy for cellular metabolism and function. They are also a major source of free radicals. On the other hand, mitochondria contain anti-oxidant systems independent of the rest of the cell. The redox pair glutathione (GSH)/glutathione disulfide (GSSG) was considered to be of particular importance to mitochondrial anti-oxidant defenses because these organelles were believed to lack catalase. Recent studies have shown that rat heart mitochondria contain catalase. The levels, however, are relatively low, and GSH is clearly important for mitochondrial defenses against peroxides that catalase does not degrade.

The anti-oxidant systems in mitochondria include superoxide dismutase, glutathione, glutathione peroxidase, and glutathione reductase. Mitochondrial glutathione is derived solely from the cytosol. During oxidative stress GSH is oxidized to become GSSG, which is retained by mitochondria. Mitochondria have their own pools of NADP(H) and NAD(H), which support the reaction catalyzed by glutathione reductase to convert GSSG to GSH. The effects of toxicants on mitochondria have two fundamental endpoints: the first is ATP depletion due to effects of toxic agents on mitochondrial membrane respiration, and the second is oxidative injury due to inhibition of anti-oxidant systems with or without concomitant increase in production of free radicals. Both endpoints have been observed in metal toxicities. As shown in Table 31, metals affect mitochondrial function by directly and indirectly disrupting the membrane respiration function to decrease oxidative phosphorylation and ATP production. The direct effect has been observed in toxicities of metals such as chromium which inhibits enzymes functioning in mitochondrial respiration. Many metals cause alterations in mitochondrial ion homeostasis which in turn result in disrupted mitochondrial respiration, a indirect metal effect. Another important aspect of metal-induced mitochondrial dysfunction is that metals decrease mitochondrial glutathione content. This is the most dominant effect of metals on mitochondrial anti-oxidant systems, leading to oxidative damage. This oxidative damage eventually results in mitochondrial membrane dysfunction and structural alteration. Whether metals directly increase the production of free radicals in mitochondria is not clear.

TABLE 30A

Ion Homeostasis: Non-Neural

Metal	Concentration	System	Effect	Ref.
Ag(I), Hg(II)	10–25 μM	Canine cardiac sarcoplasmic reticulum	Induced Ca^{2+} release; Ag(I): activated Ca^{2+} release, inhibited Ca^{2+}/Mg^{2+}-ATPase	113
Au(III), Ag(I), Cu(II), Hg(II)	0.1 mM	Renal proximal tubule suspension	Rapid net K^+ efflux; Ag(I), Hg(II): stimulated net Ca^{2+} influx	114
Cd(II), Cu(II), Ni(II)	200 μM	Canine cardiac sarcoplasmic reticulum	No effect on Ca^{2+} release	115
Cd(II), Cu(II), V(IV), Zn(II)	IC_{50} = 40–650 μM	Rat hepatic microsomal vesicle	Blocked Ca^{2+} uptake; stimulated rapid Ca^{2+} efflux in preloaded vesicle	116
Cd(II), Pb(II) > Co(II) ≫ Fe(II), Mg(II)	1–100 μM	Human RBC	Opened Ca^{2+}-activated K^+ (CaK) channel via Ca^{2+} regulatory site	117
Fe(II), Co(II) Ni(II)	0.5 mM 2 mM	Mouse cell myotube	Blocked dihydropyridine-sensitive Ca^{2+} channel	118
Cd(II)	0.21 mM	Rabbit kidney cell	Inhibited Na^+/K^+-ATPase by competing with K^+ binding site	119
Cd(II)	50 mM	MDCK cell	↑ K^+ conductance by activating K^+ channel	120
Cd(II)	50 mM	Hepatocyte	↓ Cellular K^+	121
Cu(II)	IC_{50} = 51 μM	Human skeletal muscle cells	Inhibited Na^+/K^+-ATPase, increased cellular Na^+, reversed Na^+/Ca^{2+} exchange, increased cellular Ca^{2+}	122
Cu(I)	1–5 μM	Rabbit renal brush-border membrane vesicle	Stimulated Cl^- efflux by functioning as Cl^-/OH^- exchange ionophore	123
Fe(III)	100 μM	Rat hepatocyte	↑ Cytosolic Ca^{2+}	124
Hg(II) > Pb(II) > Cd(II)	Pb: IC_{50} = 95 nM; Cd: IC_{50} = 950 nM	Rabbit muscle sarcoplasmic reticulum	Inhibited Ca^{2+}-ATPase	125
Hg(II)	1–10 μM	Rabbit renal brush-border membrane vesicle	Stimulated Cl^- efflux by functioning as Cl^-/OH^- exchange ionophore	123
Hg(II)	IC_{50} = 110 μM	Rabbit muscle sarcoplasmic reticulum	Inhibited Ca^{2+}/Mg^{2+}-ATPase by Ca^{2+} substitution	126
Hg(II)	10–50 μM	Liposome	Inhibited Na^+/K^+-ATPase on cytoplasmic side	126
Hg(II)	IC_{50} = 200 nM	Myocyte	Inhibited Na^+/K^+-ATPase via high-affinity Hg(II)-binding site	127
Hg(II)	1 mM	MDCK cell	↑ K^+ conductance by activating K^+ channel	128
Hg, PCMBs (organic)	1 × 10^{-17} mol/cell	RBC	Inhibited Na^+/K^+-ATPase via membrane sulfhydryls	129, 130
Hg, PCMBs (organic)	1 × 10^{-5} mol/cell	RBC	Inhibited passive Na^+ and K^+ permeability	131
MeHg	IC_{50} = 50 μM	Rabbit muscle sarcoplasmic reticulum	Inhibited Ca^{2+}/Mg^{2+}-ATPase by sulfhydryl binding	122
Ni(II)	IC_{50} = 5.65 mM	Rat, human thyroid C cell	Blocked T-type voltage-gated Ca^{2+} channel	132
Pb(II)	0.06–0.08 mM	Human RBC	Stimulated K^+ leakage	133
T1(I)		Human RBC	Competitive inhibition of K^+ transport	134
UO_2	KD = 1 mM	Barnacle muscle fiber	Blocked Ca^{2+} depolarization	135
Zn(II)	KD = 30 μM	Calf myocyte	Blocked cardiac Na^+ channel	136

Note: IC_{50} = concentration for half maximal inhibition; MDCK = Madin-Darby canine kidney (cells); PCMBS = *P*-chloromercuribenzoate; RBC = red blood cells.

TABLE 30B

Ion Homeostasis: Neural

Metal	Concentration	System	Effect	Ref.
Cd(II), Pb(II) > Co(II) ≫ Mg(II), Fe(II)	1–100 μM	Mouse neuroblastoma cell	Opened low-conductance (SK) Ca^{2+}-activated K^+ channel	137, 138
Pb(II) > Co(II) ≫ Cd(II), Mg(II), Fe(II)	1–100 μM	Mouse neuroblastoma cell	Opened high-conductance (BK) Ca^{2+}-activated K^+ channel	137, 138
Pb(II), Cd(II) > Co(II) > Mg(II), Fe(II)	1–100 μM	Human RBC	Opened intermediate-conductance Ca^{2+}-activated K^+ channel	138
Zn(II) > Cd(II) > Mn(II) > Co(II) > Ba(II) > Ni(II)	10–100 μM	Rat mast cell	Blocked Ca^{2+} uptake	139
Al(III)	50–800 μM	Rat brain	Inhibited Ca^{2+}/Mg^{2+}-ATPase; ↓ Ca^{2+} uptake	140
Cd(II)	IC_{50} = 50 μM	Rat brain microsome	Inhibited Na^+/K^+-ATPase	141
MeHg	10–100 μM	Synaptosome fraction	Increased non-Ca^{2+} polyvalent ion efflux; increased Ca^{2+} influx	142
Pb(II)	2.5–50 μM	Rat hippocampal neuron	Selectively inhibited *N*-methyl-D-aspartate (NMDA) receptor ion channel complex	143
Pb(II)	0.01–100 μM	Rat cerebellar and microsomal membrane	Inhibited IP_3-mediated Ca^{2+} release; inhibited microsomal Ca^{2+} reuptake, Ca^{2+} pump, and Na^+/K^+-ATPase	144
Pb(II)	IC_{50} = 2 mM	Rat brain fraction	Inhibited Na^+/K^+-ATPase	145
Tl(I)	>1 mM	Squid axon	Stimulated K^+ leakage	146
Tl(III)	36–360 nM	Brain myelin	Generated pH gradient, then increased Cl^- permeability by functioning as a Cl^-/OH^- exchange ionophore	147
Zn(II)	2–200 μM	Suprachiasmic nucleus neuron	Potentiated voltage-dependent transient K^+ current	148

Note: IC_{50} = concentration for half maximal inhibition; RBC = red blood cells.

The predominant effects of metals on ATP production and anti-oxidant systems in mitochondria provide obstacles for investigating the relationship between exposure of metals and mitochondrial free radical production.

3.5 Plasma Membrane Damage

Metal-induced plasma membrane damage occurs directly through interaction with membrane components such as ion-dependent ATPases and ion channels and indirectly as a consequence of overt cytosolic damage. In general, toxic irreversible plasma membrane damage is the last event of cell injury or cell death. For instance, most toxicants interact with intracellular target molecules to cause changes in cellular metabolism and function, which eventually lead to cell membrane damage and cell lysis. In metal toxicities, the cell membrane may be a primary target. Cadmium, lead, and aluminum can directly interact with the cell membrane to alter its function. The targets on cell membrane with which these metals interact are primarily ion channels and pumps. The interaction between metals and ion channels

TABLE 31

Mitochondrial Dysfunction

Metal	Concentration	System	Effect	Ref.
Cd(II), Zn(II) > Cu(II), Hg(II), Ag(I)	60 μ*M*	Beef myocyte	Mitochondrial K^+ accumulation	149
Al(III)		Rat brain	↓ Mitochondrial membrane conductance	150
Cd(II)	60 μ*M*	Beef myocyte	Decreased oxidative phosphorylation; inhibited massive Ca^{2+} uptake; mitochondrial Mg^{2+} accumulation	151
Cd(II)	5–10 μ*M*	Rat hepatocyte	Potentiated acidic pH-induced respiration inhibition	152
Cd(II)	5–25 μ*M*	Isolated rat hepatocyte mitochondria	Eliminated mitochondrial membrane potential	152
Cd(II)	1 m*M*	Rat hepatocyte	↓ ATP/ADP ratio; ↑ mitochondrial lipid peroxidation	83
Cu(II)		Rat	↑ Hepatic mitochochondrial lipid peroxidation	153
Cr(IV)	10–50 μ*M*	Rat myocyte mitochondria	↓ Mitochondrial O_2 consumption	27
Cr(IV)	3–5 μ*M*	Rat myocyte mitochondria	Inhibited α-ketoglutarate dehydrogenase	27
Cr(IV)	20–70 μ*M*	Rat myocyte mitochondria	Inhibited β-hydroxybutyrate and pyruvate dehydrogenase	27
Cr(V)	30 μ*M*	Rate myocyte mitochondria	Depleted NADH due to direct oxidation	27
Fe(III)	100 μ*M*	Rat hepatocyte	↓ Mitochondrial membrane potential	124
Hg(II)	12–30 nmol/mg protein	Rat renal cortical cell	Dose-dependent mitochondrial GSH depletion; increased mitochondrial H_2O_2; lipid peroxidation under impaired respiratory chain electron transport	154
Hg(II)	10 μ*M*	Kidney	↑ Mitochondrial Ca^{2+} efflux; ↓ mitochondrial membrane potential	155
Hg(II)	1.5 mg/kg	Rat (i.p.)	Depleted mitochondrial GSH; increased mitochondrial H_2O_2; lipid peroxidation	89
Hg(II)	100–500 μ*M*	Fertilized sea urchin egg	Time- and dose-dependent Ca^{2+} influx; uncoupled oxidative phosphorylation; depleted cellular ATP	156
MeHg	10–100 μ*M*	Rat forebrain mitochondria	Blocked ATP-independent Ca^{2+} uptake; stimulated Ca^{2+} efflux in preloaded mitochondria	157
MeHg	10–100 μ*M*	Rat brain synaptosome	Depolarized intrasynaptosomal mitochondria	158
Pb(II)	0.1, 1 m*M*	Cultured astroglia	Dissipated mitochondrial electrochemical gradient	159
Pb(II)	50 m*M*	Hepatocyte	↓ Mitochondrial membrane potential; ↑ mitochondrial Ca^{2+} release	160
Pb(II)	0.06–0.08 m*M*	Human RBC	Depleted cellular ATP	133
Pt(II)	2 m*M*, cisplatin	Rat renal cortical slice mitochondria	Dose-dependent mitochondrial GSH depletion and lipid peroxidation	161
Tl(I)	4–20 m*M*	Rat hepatic mitochondria	Uncoupled oxidative phosphorylation by active Tl^+ accumulation	162
Zn(II)	60 μ*M*	Beef myocyte	Decreased oxidative phosphorylation; inhibited massive Ca^{2+} uptake; mitochondrial Mg^{2+} accumulation	151

Note: RBC = red blood cells; GSH = glutathione; ATP = adenosine triphosphate.

and/or pumps does not necessarily result in structural damage to the cell membrane. Rather, it causes functional alterations. These, in turn, lead to changes in cellular metabolism and function and eventually structural damage to the cell membrane. Therefore, the mode of membrane damage induced by many metals is initial and direct functional damage followed by delayed and indirect structural damage. For other metals such as iron and copper, their primary targets are in the cell. These metals first cause overt cellular changes and then cell membrane damage. The effects of some metals on the plasma membrane and possible mechanisms for these effects are summarized in Table 32.

TABLE 32

Membrane Damage

Metal	Dose	System	Type of Damage	Mechanism	Ref.
Cadmium	5 m*M*	Hepatocytes	Decreased membrane potential	Acidification	152
	9.4 μ*M*	Shark rectal gland	Decreased Na-K-ATPase	Competes with a Mg site	119
	0.21 m*M*	Rabbit kidney cells	Decreased Na-K-ATPase	Competes with a K-binding site	119
		Rat brain synaptosomal plasma membrane	Lipid peroxidation	Free radicals	163
	4 mg/kg	Rat liver	Decreased microsomal membrane fluidity; lipid peroxidation		164
	50 m*M*	MDCK cells	Increased K^+ conduction	Activates K^+ channel	120
	50 m*M*	Hepatocytes	Decreased cellular K content; release of LDH	Acidification	121
	89 m*M*	Hepatocytes	Decreased mitochondria membrane potential; release of LDH		165
	0.25 m*M*	Fish gills	Decreased Ca^{2+} transport	Inhibits Ca^{2+}-transporting enzymes	166
	0.4 mg/kg daily for 30 days	Rat liver and kidney	Lipid peroxidation	Superoxide radical	85
	1 m*M*	Rat hepatocytes	Decreased ATP/ADP ratio; lipid peroxidation in mitochondria	Free radicals	83
Mercury	1.5 mg/kg (i.p., one dose)	Rat kidney	Lipid peroxidation in mitochondria	H_2O_2 formation	89
	5–200 m*M*	Aplysia neurons	Decreased calcium current	Acting on Ca^{2+} channel	167
		Muscles	Decreased transmembrane sodium gradient		168
	200 m*M*	Cerebellar granule cells	Lipid peroxidation		32
	1 m*M*	MDCK Cells	Increased K^+ conductance	Acting on K^+ channel	128
Mercury	10 μ*M in vivo* or 4 mg/kg (i.p.)	Kidney	Increased mitochondrial Ca^{2+} efflux		155
	4 mg/kg (i.p.)		Decreased ψ in mitochondria		
	0.3 m*M* methylmercury	Isolated myocardia	Lipophilicity; decreased Na/K-ATPase	Binding SH^- groups	169
Lead	500–2000 ppm	Chick	Lipid peroxidation	Tissue fat acid alterations	170
	10–100 m*M*	Mouse neuroblastoma cells	Decreased Ca^{2+} channel; increased Ach receptor	Ca^{2+}-Pb^{2+} interaction	171

TABLE 32 (continued)

Membrane Damage

Metal	Dose	System	Type of Damage	Mechanism	Ref.
			Increased intracellular Ca^{2+}-K channel		172
	0.05–1 m*M*	Nervous cells	Increased protein kinase C		173
	IC_{50} = 61 m*M*	Aplysia neurons	Blocked Ca^{2+} channel		167
	15–39 mg	Fish	Increased membrane lability		174
	50 mg/kg	Rat brain	Lipid peroxidation	Free radicals	175
	5000 mg/l	Rats	RBC membrane Na-K-ATPase changes		176
	50 m*M*	Hepatocyte	Decreased mitochondrial membrane potential; mitochondrial Ca^{2+} release; decreased Ca^{2+}-ATPase		160
Iron	Mild iron overload	Rats	Lipid peroxidation in liver and kidney		177
	1.5 g Fe per kg	Guinea pig	Lysosomal membrane fragility in liver and heart; lipid peroxidation		178
	100 μ*M*, $FeCl_3$	Rat hepatocytes	Decreased mitochondrial membrane potential; increased cytosolic Ca^{2+} level	Lipid peroxidation	124
Iron	Sixfold increase	Rat hepatocyte lysosomes	Decreased fluidity of lysosomal membrane; increased lysosomal membrane fragility	Lipid peroxidation	179
		Liver nuclei	Decreased NADPH- and NADH-dependent cytochrome C-reductases		180
		Rat	Increased hepatic mitochondrial malondialdehyde		181
Copper	51 m*M*	Cultured human skeletal muscle cells	Decreased Na^+/K^+-ATPase; increased cytosolic Ca^{2+}		182
		Human red blood cells	Increased rigidity of membrane lipid biolayers; membrane protein changes		183
	0.125%	Hepatocyte lysosomes	Increased lysosomal fragility; decreased lysosomal membrane fluidity; lipid peroxidation		184
		Rat	Liver mitochondrial lipid peroxidation		153
Aluminum	50–800 μ*M*, $AlCl_3$	Rat brain	Decreased Ca^{2+}/Mg^{2+}-ATPase; decreased Ca^{2+} uptake		140
	400 μ*M*, Al^{3+}	Human skin fibroblasts	Lipid peroxidation		185
		Rat brain	Decreased conduction of mitochondrial membrane		150

Note: MDCK = Madin-Darby canine kidney (cells); LDH = lactic dehydrogenase; ATP = adenosine triphosphate; ADP = adenosine diphosphate; IC_{50} = concentration for half maximal inhibition; RBC = red blood cells.

References

1. Corcoran, G. B., Fix, I., Jones, D. P., Treinen-Moslen, M., Nicotera, P., Oberhammer, F. A., and Buttyan, R., Apoptosis: molecular control point in toxicity, *Toxicol. Appl. Pharmacol.*, 128, 169–181, 1994.
2. Buja, L. M., Eigenbrodt, M. L., and Eigenbrodt, E. H., Apoptosis and necrosis: basic types and mechanisms of cell death, *Arch. Pathol. Lab. Med.*, 117, 1208–1214, 1993.
3. Kerr, J. F. R., Winterford, C. M., and Harmon, B. V., Apoptosis: its significance in cancer and cancer therapy, *Cancer*, 73, 2013–2026, 1994.
4. Wolfe, J. T., Ross, D., and Cohen, G. M., A role for metals and free radicals in the induction of apoptosis in thymocytes, *FEBS Lett.*, 352, 58–62, 1994.
5. Rosser, B. G. and Gores, G. J., Liver cell necrosis: cellular mechanisms and clinical implications, *Gastroenterol.*, 108, 252–275, 1995.
6. Fukuchi, K., Tomoyasu, S., Tsuruoka, N., and Gomi, K., Iron deprivation-induced apoptosis in HL-60 cells, *FEBS Lett.*, 350(1), 139–142, 1994.
7. Haq, R. U., Wereley, J. P., and Chitambar, C. R., Induction of apoptosis by iron deprivation in human leukemic CCRF-CEM cells, *Exp. Hematol.*, 23(5), 428–432, 1995.
8. Rao, M. S., Yelandandi, A. V., Subbarao, V., and Reddy, J. K., Role of apoptosis in copper deficiency-induced pancreatic involution in the rat, *Am. J. Pathol.*, 142(6), 1952–1957, 1993.
9. Sunderman, Jr., F. W., The influence of zinc on apoptosis, *Ann. Clin. Lab. Sci.*, 25(2), 134–142, 1995.
10. Lu, J., Kaeck, M., Jiang, C., Wilson, A. C., and Thompson, H. J., Selenite induction of DNA strand breaks and apoptosis in mouse leukemic L1210 cells, *Biochem. Pharmacol.*, 47(9), 1531–1535, 1994.
11. El Azzouri, B., Tsangaris, G. T., Pellegrini, O., Manuel, Y., Benveniste, J., and Thomas, Y., Cadmium induces apoptosis in a human T cell line, *Toxicology*, 88(1–3), 127–139, 1994.
12. Blankenship, L. J., Manning, F. C., Orenstein, J. M., and Patierno, S. R., Apoptosis is the mode of cell death caused by carcinogenic chromium, *Toxicol. Appl. Pharmacol.*, 126(1), 75–83, 1994.
13. Bacon, B. R. and Britton, R. S., Hepatic injury in chronic iron overload. Role of lipid peroxidation, *Chem. Biol. Interact.*, 70(3–4), 183–226, 1989.
14. Whitten, C. F. and Brough, A. J., The pathophysiology of acute iron poisoning, *Clin. Toxicol.*, 4, 585–595, 1971.
15. Toyokuni, S., Okada, S., Hamazaki, S., Fujioka, M., Li, J. L., and Midorikawa, O., Cirrhosis of the liver induced by cupric nitrilotriacetate in Wistar rats. An experimental model of copper toxicosis, *Am. J. Pathol.*, 134(6), 1263–1274, 1989.
16. Chuttani, H. K., Gupta, P. S., Gulati, S., and Gupta, D. N., Acute copper sulfate poisoning, *Am. J. Med.*, 39, 849–854, 1965.
17. Scheinber, I. H. and Sternlieb, I., *Wilson's Disease*, W.B. Saunders, Philadelphia, PA, 1984.
18. Ozawa, T., Ueda, J., and Shimazu, Y., DNA single strand breakage by copper(II) complexes and hydrogen peroxide at physiological conditions, *Biochem. Mol. Biol. Int.*, 31(3), 455–461, 1993.
19. Chan, H. M., Zhu, L. F., Zhong, R., Grant, D., Goyer, R. A., and Cherian, M. G., Nephrotoxicity in rats following liver transplantation from cadmium-exposed rats, *Toxicol. Appl. Pharmacol.*, 123(1), 89–96, 1993.
20. Hayashi, T. and Sudo, J., Interrelations of cadmium contents and histopathological changes in kidneys following single intravenous injection of cadmium-saturated metallothionein II in rats, *J. Toxicol. Sci.*, 19(1), 45–53, 1994.
21. Dudley, R. E., Svoboda, D. J., and Klaassen, C. D., Time course of cadmium-induced ultrastructural changes in rat liver, *Toxicol. Appl. Pharmacol.*, 76(1), 150–160, 1984.
22. White, T. E., Baggs, R. B., and Miller, R. K., Central nervous system lesions in the Wistar rat fetus following direct fetal injections of cadmium, *Teratology*, 42(1), 7–13, 1990.
23. Andersen, H. R. and Andersen, O., Effect of cadmium chloride on hepatic lipid peroxidation in mice, *Pharmacol. Toxicol.*, 63(3), 173–177, 1988.
24. Wedeen, R. P. and Qian, L. F., Chromium-induced kidney disease, *Environ. Health Perspect.*, 92, 71–74, 1991.
25. Van Heerden, P. V., Jenkins, I. R., Woods, W. P., Rossi, E., and Cameron, P. D., Death by tanning — a case of fatal basic chromium sulphate poisoning, *Intensive Care Med.*, 20(2), 145–147, 1994.
26. Molyneux, M. J. and Davies, M. J., Direct evidence for hydroxyl radical-induced damage to nucleic acids by chromium(VI)-derived species: implications for chromium carcinogenesis, *Carcinogenesis*, 16(4), 875–882, 1995.
27. Ryberg, D. and Alexander, J., Mechanism of chromium toxicity in mitochondria, *Chem. Biol. Interact.*, 75(2), 141–151, 1990.
28. Rajaram, R., Nair, B.U., and Ramasami, T., Chromium(III)-induced abnormalities in human lymphocyte cell proliferation: evidence for apoptosis, *Biochem. Biophys. Res. Commun.*, 210(2), 434–440, 1995.
29. Sauder, P., Livardjani, F., Jaeger, A., Kopferschmitt, J., Heimburger, R., Waller, C., Mantz, J. M., and Leroy, M., Acute mercury chloride intoxication. Effects of hemodialysis and plasma exchange on mercury kinetic, *J. Toxicol. Clin. Toxicol.*, 26(3–4), 189–197, 1988.
30. Kanluen, S. and Gottlieb, C. A., A clinical pathologic study of four adult cases of acute mercury inhalation toxicity, *Arch. Pathol. Lab. Med.*, 115(1), 56–60, 1991.

31. Tan, X. X., Tang, C., Castoldi, A. F., Manzo, L., and Costa, L. G., Effects of inorganic and organic mercury on intracellular calcium levels in rat T lymphocytes, *J. Toxicol. Environ. Health,* 38(2), 159–170, 993.
32. Sarafian, T. and Verity, M. A., Oxidative mechanism underlying methyl mercury neurotoxicity, *Int. J. Dev. Neurosci.,* 9(2), 147–153, 1991.
33. Williams, M. V., Winters, T., and Waddell, K. S., *In vivo* effects of mercury(II) on deoxyuridine triphosphate nucleotidohydrolase, DNA polymerase (alpha, beta), and uracil-DNA glycosylase activities in cultured human cells: relationship to DNA damage, DNA repair, and cytotoxicity, *Mol. Pharmacol.,* 31(2), 200–207, 1987.
34. Knight, J. A., Plowman, M. R., Hopfer, S. M., and Sunderman, Jr., F. W., Pathological reactions in lung, liver, thymus, and spleen of rats after subacute parenteral administration of nickel sulfate, *Ann. Clin. Lab. Sci.,* 21(4), 275–283, 1991.
35. Gitlitz, P. H., Sunderman, F. W., Jr., and Goldblatt, P. J., Aminoaciduria and proteinuria in rats after a single intraperitoneal injection of Ni(II), *Toxicol. Appl. Pharmacol.,* 34(3), 430–440, 1975.
36. Sunderman, Jr., F. W., A review of the metabolism and toxicology of nickel, *Ann. Clin. Lab. Sci.,* 7(5), 377–398, 1977.
37. Stinson, T. J., Jaw, S., Jeffery, E. H., and Plewa, M. J., The relationship between nickel chloride-induced peroxidation and DNA strand breakage in rat liver, *Toxicol. Appl. Pharmacol.,* 117(1), 98–103, 1992.
38. Coogan, T. P., Latta, D. M., Snow, E. T., and Costa, M., Toxicity and carcinogenicity of nickel compounds, *Crit. Rev. Toxicol.,* 19(4), 341–384, 1989 (published erratum appears in *Crit. Rev. Toxicol.,* 20(2), 1351, 1989).
39. Levy, B. S., Hoffman, L., and Gottsegen, S., Boilermaker's bronchitis. Respiratory tract irritation associated with vanadium pentoxide exposure during oil-to-coal conversion of a power plant, *J. Occup. Med.,* 26(8), 567–570, 1984.
40. Zaporowska, H. and Wasilewski, W., Haematological effects of vanadium on living organisms, *Comp. Biochem. Physiol. C,* 102(2), 223–231, 1992.
41. Younes, M. and Strubelt, O., Vanadate-induced toxicity towards isolated perfused rat livers: the role of lipid peroxidation, *Toxicology,* 66(1), 63–74, 1991.
42. Beattie, A. D., Briggs, J. D., Canavan, J. S., Doyle, D., Mullin, P. J., and Watson, A. A., Acute lead poisoning: five cases resulting from self-injection of lead and opium, *Q. J. Med.,* 44(174), 275–284, 1975.
43. Abdel-Aal, S. F., Shalaby, S. A., Badawy, A. H., and Sammour, S. A., Effect of lead nitrate administration on liver and kidney structure in rats, *J. Egypt Soc. Parasitol.,* 19(2), 689–699, 1989.
44. Bompart, G. and Orfila, C., Cisplatin nephrotoxicity in lead-pretreated rats: enzymatic and morphological studies, *Toxicol. Lett.,* 50(2–3), 237–247, 1990.
45. Holtzman, D., De Vries, C., Nguyen, H., Olson, J., and Bensch, K., Maturation of resistance to lead encephalopathy: cellular and subcellular mechanisms, *Neurotoxicity,* 5 (3), 97–124, 1984.
46. Smith, B. L. and Embling, P. P., Sequential changes in the development of the pancreatic lesion of zinc toxicosis in sheep, *Vet. Pathol.,* 30(3), 242–247, 1993.
47. Graham, T. W., Homlberg, C. A., Keen, C. L., Thurmond, M. C., and Clegg, M. S., A pathologic and toxicologic evaluation of veal calves fed large amounts of zinc, *Vet. Pathol.,* 25(6), 484–491, 1988.
48. Borovansk'y, J. and Riley, P. A., Cytotoxicity of zinc *in vitro*, *Chem. Biol. Interact.,* 69(2–3), 279–291, 1989.
49. Franzblau, A. and Lilis, R., Acute arsenic intoxication from environmental arsenic exposure, *Arch. Environ. Health,* 44(6), 385–390, 1989.
50. Hindmarsh, J. T. and McCurdy, R. F., Clinical and environmental aspects of arsenic toxicity, *Crit. Rev. Clin. Lab. Sci.,* 23(4), 315–347, 1986.
51. Lin, T. H., Huang, Y. L., and Tseng, W. C., Arsenic and lipid peroxidation in patients with blackfoot disease, *Bull. Environ. Contam. Toxicol.,* 54(4), 488–493, 1995.
52. Nemery, B., Metal toxicity and the respiratory tract, *Eur. Respir. J.,* 3(2), 202–219, 1990.
53. Huang, C. C., Lu, C. S., Chu, N. S., Hochberg, F., Lilienfeld, D., Olanow, W., and Calne, D. B., Progression after chronic manganese exposure, *Neurology,* 43(8), 1479–1483, 1993.
54. Gavin, C. E., Gunter, K. K., and Gunter, T. E., Manganese and calcium efflux kinetics in brain mitochondria. Relevance to manganese toxicity, *Biochem. J.,* 266(2), 329–324, 1990.
55. Ahmed, K. E., Adam, S. E., Idrill, O. F., and Wahbi, A. A., Experimental selenium poisoning in Nubian goats, *Vet. Hum. Toxicol.,* 32(3), 249–251, 1990.
56. Spallholz, J. E., On the nature of selenium toxicity and carcinostatic activity, *Free Radic. Biol. Med.,* 17(1), 45–64, 1994.
57. Wills, M. R., Hewitt, C. D., Sturgill, B. C., Savory, J., and Herman, M. M., Long-term oral or intravenous aluminum administration in rabbits. I. Renal and hepatic changes, *Ann. Clin. Lab. Sci.,* 23(1), 1–16, 1993.
58. Zaman, K., Zaman, A., and Batcabe, J., Hematological effects of aluminum on living organisms, *Comp. Biochem. Physiol. C,* 106(2), 285–293, 1993.
59. Hewitt, C. D., Savory, J., and Wills, M. R., Aspects of aluminum toxicity, *Clin. Lab. Med.,* 10(2), 403–422, 1990.
60. Chmielnicka, J., Zareba, G., Polkowska-Kulesza, E., Najder, M., and Korycka, A., Comparison of tin and lead toxic action on erythropoietic system in blood and bone marrow of rabbits, *Biol. Trace Elem. Res.,* 36(1), 73–87, 1993.

61. Winship, K. A., Toxicity of tin and its compounds, *Adverse Drug React. Acute Poisoning Rev.*, 7(1), 19–38, 1988.
62. Diaz, G. J., Julian, R. J., and Squires, E. J., Lesions in broiler chickens following experimental intoxication with cobalt, *Avian Dis.*, 38(2), 308–316, 1994.
63. Bucher, J. R., Elwell, M. R., Thompson, M. B., Chou, B. J., Renne, R., and Ragan, H. A., Inhalation toxicity studies of cobalt sulfate in F344/N rats and B6C3F1 mice, *Fund. Appl. Toxicol.*, 15(2), 357–372, 1990.
64. Lewis, C P., Demedts, M., and Nemery, B., The role of thiol oxidation in cobalt(II)-induced toxicity in hamster lung, *Biochem. Pharmacol.*, 43(3), 519–525, 1992.
65. Rungby, J., An experimental study on silver in the nervous system and on aspects of its general cellular toxicity, *Dan. Med. Bull.*, 37(5), 442–449, 1990.
66. Herman, M. M. and Bensch, K. G., Light and electron microscopic studies of acute and chronic thallium intoxication in rats, *Toxicol. Appl. Pharmacol.*, 10(2), 199–222, 1967.
67. Mourelle, M., Favari, L., and Amezcua, J. L., Protection against thallium hepatotoxicity by silymarin, *J. Appl. Toxicol.*, 8(5), 351–354, 1988.
68. Woods, J. S., Fowler, B. A., and Eaton, D. L., Studies on the mechanisms of thallium-mediated inhibition of hepatic mixed function oxidase activity. Correlation with inhibition of NADPH-cytochrome C (P-450) reductase, *Biochem. Pharmacol.*, 33(4), 571–576, 1984.
69. Blazka, M. E., Dixon, D., Haskins, E., and Rosenthal, G. J., Pulmonary toxicity to intratracheally administered indium trichloride in Fischer 344 rats, *Fund. Appl. Toxicol.*, 22(2), 231–239, 1994.
70. Guo, X. B., Ohno, Y., Kawanishi, T., Sunouchi, M., and Takanaka, A., Indium inhibits gap junctional communication between rat hepatocytes in primary culture, *Toxicol. Lett.*, 60(1), 99–106, 1992.
71. Fowler, B. A., Kardish, R. M., and Woods, J. S., Alteration of hepatic microsomal structure and function by indium chloride. Ultrastructural, morphometric, and biochemical studies, *Lab. Invest.*, 48(4), 471–478, 1983.
72. Crewenk, E. A., and Cooper, W. C., Toxicology of selenium and tellurium and their compounds, *Arch. Environ. Health*, 3, 189–200, 1961.
73. Browning, E., *Toxicity of Industrial Metals*, 2nd ed., Butterworths, London, 1969.
74. Bouldin, T. W., Samsa, G., Earnhardt, T. S., and Krigman, M. R., Schwann cell vulnerability to demyelination is associated with internodal length in tellurium neuropathy, *J. Neuropathol. Exp. Neurol.*, 47(1), 41–47, 1988.
75. Srivastava, R. C., Srivastava, R., Srivastava, T. N., and Jain, S. P., Effect of organo-tellurium compounds on the enzymatic alterations in rats, *Toxicol. Lett.*, 16(3–4), 311–316, 1983.
76. Choie, D. D., Longnecker, D. S., and Del Campo, A. A., Acute and chronic cisplatin nephropathy in rats, *Lab. Invest.*, 44(5), 397–402, 1981.
77. Lee, T. C., Hook, C. C., and Long, H. J., Severe exfoliative dermatitis associated with hand ischemia during cisplatin therapy, *Mayo. Clin. Proc.*, 69(1), 80–82, 1994.
78. Zhang, J. G., Zhong, L. F., Zhang, M., and Xia, Y. X., Protection effects of procaine on oxidative stress and toxicities of renal cortical slices from rats caused by cisplatin *in vitro*, *Arch. Toxicol.*, 66(5), 354–358, 1992.
79. Alleman, M. A., Koster, J. F., Wilson, J. H., Edixhoven-Bosdijk, A., Slee, R. G., Kroos, M. H., and Von-Eijk, H. G., The involvement of iron and lipid peroxidation in the pathogenesis of HCB induced porphyria, *Biochem. Pharmacol.*, 34(2), 161–166, 1985.
80. Halliwell, B. and Gutteridge, J. M. C., Iron and free radical reactions: two aspects of anti-oxidant protection, *Trends Biochem. Sci.*, 11, 372–375, 1986.
81. Chan, P. C., Peller, O. G., and Kesner, L., Copper(II)-catalyzed lipid peroxidation in liposomes and erythrocyte membranes, *Lipids*, 17(5), 331–337, 1982.
82. Yourtee, D. M., Elkins, L. L., Nalvarte, E. L., and Smith, R. E., Amplification of doxorubicin mutagenicity by cupric ion, *Toxicol. Appl. Pharmacol.*, 116(1), 57–65, 1992.
83. Muller, L., Consequences of cadmium toxicity in rat hepatocytes: mitochondrial dysfunction and lipid peroxidation, *Toxicology*, 40(3), 285–295, 1986.
84. Koizumi, T. and Li, Z. G., Role of oxidative stress in single-dose, cadmium-induced testicular cancer, *J. Toxicol. Environ. Health*, 37(1), 25–36, 1992.
85. Hussain, T., Shukla, G. S., and Chandra, S. V., Effects of cadmium on superoxide dismutase and lipid peroxidation in liver and kidney of growing rats, *in vivo* and *in vitro* studies, *Pharmacol. Toxicol.*, 60(5), 355–358, 1987.
86. Shi, X. G. and Dalal, N. S., On the hydroxyl radical formation in the reaction between hydrogen peroxide and biologically generated chromium(V) species, *Arch. Biochem. Biophys.*, 277(2), 342–350, 1990.
87. Shi, X., Dalas, N. S., and Kasprzak, K. S., Generation of free radicals from hydrogen peroxide and lipid hydroperoxides in the presence of Cr(III), *Arch. Biochem. Biophys.*, 302(1), 294–299, 1993.
88. Gstraunthaler, G., Pfaller, W., and Kotanko, P., Glutathione depletion and *in vitro* lipid peroxidation in mercury or maleate induced acute renal failure, *Biochem. Pharmacol.*, 32(19), 2969–2972, 1983.
89. Lund, B. O., Miller, D. M., and Woods, J. S., Studies on Hg(II)-induced H_2O_2 formation and oxidative stress *in vivo* and *in vitro* in rat kidney mitochondria, *Biochem. Pharmacol.*, 45(10), 2017–2024, 1993.
90. Fukino, H., Hirai, M., Hsueh, Y. M., and Yamane, Y., Effect of zinc pretreatment on mercuric chloride-induced lipid peroxidation in the rat kidney, *Toxicol. Appl. Pharmacol.*, 73(3), 395–401, 1984.

91. Athar, M., Hasan, S. K., and Srivastava, R. C., Evidence for the involvement of hydroxyl radicals in nickel mediated enhancement of lipid peroxidation: implications for nickel carcinogenesis, *Biochem. Biophys. Res. Commun.,* 147(3), 1276–1281, 1987.
92. Shi, X. and Dalal, N. S., Vanadate-mediated hydroxyl radical generation from superoxide radical in the presence of NADH: Haber-Weiss vs. Fenton mechanism, *Arch. Biochem. Biophys.,* 307(2), 336–341, 1993.
93. Zaporowska, H., Wasilewski, W., and Slotwi'nska, M., Effect of chronic vanadium administration in drinking water to rats, *Biometals,* 6(1), 3–10, 1993.
94. Younes, M. and Strubelt, O., Vanadate-induced toxicity towards isolated perfused rat livers: the role of lipid peroxidation, *Toxicology,* 66(1), 63–74, 1991.
95. Lawton, L. J. and Donaldson, W. E., Lead-induced tissue fatty acid alterations and lipid peroxidation, *Biol. Trace Elements Res.,* 28(2), 83–97, 1991.
96. Shafiq-Ur, R., Lead-induced regional lipid peroxidation in brain, *Toxicol. Lett.,* 21(3), 333–337, 1984.
97. Sullivan, J. F., Jetton, M. M., Hahn, H. K., and Burch, R. E., Enhanced lipid peroxidation in liver microsomes of zinc-deficient rats, *Am. J. Clin. Nutr.,* 33(1), 51–56, 1980.
98. Bray, T. M. and Bettger, W. J., The physiological role of zinc as an antioxidant, *Free Radic. Bio. Med.,* 8(3), 281–291, 1990.
99. Lin, T. H., Huang, Y. L., and Tseng, W. C., Arsenic and lipid peroxidation in patients with blackfoot disease, *Bull. Environ. Contam. Toxicol.,* 54(4), 488–493, 1995.
100. Y'anez, L., Carrizales, L., Zanatta, M. T., and Mej'ia, J. J., Arsenic-cadmium interaction in rats: toxic effects in the heart and tissue metal shifts, *Toxicology,* 67(2), 227–234, 1991.
101. Sun, A. Y., Yang, W. L., and Kim, H. D., Free radical and lipid peroxidation in manganese-induced neuronal cell injury, *Ann. N.Y. Acad. Sci.,* 679, 358–363, 1993.
102. Shi, X. L. and Dalal, N. S., The glutathionyl radical formation in the reaction between manganese and glutathione and its neurotoxic implications, *Med. Hypotheses,* 33(2), 83–87, 1990.
103. Tappel, A. L. and Caldwell, K. A., Redox properties of selenium compounds related to biochemical function, in *Selenium in Biomedicine,* Muth, O. H., Oldfield, J. E. et al., Eds., AVI Publishing, Westport, CN, 1967, pp. 345–361.
104. Seko, Y., Saito, Y. et al., Active oxygen generation by the reaction of selenite with reduced glutathione *in vitro,* in *Selenium in Biology and Medicine,* Wendel, A., Ed., Springer-Verlag, Berlin, 1989, pp. 70–73.
105. Oteiza, P. I., A mechanism for the stimulatory effect of aluminum on iron-induced lipid peroxidation, *Arch. Biochem. Biophys.,* 308(2), 374–379, 1994.
106. Neil, T. K., Abraham, N. G., Levere, R. D., and Kappas, A., Differential heme oxygenase induction by stannous and stannic ions in the heart, 57(3), 409–414, 1995.
107. Wang, X., Yokoi, I., Liu, J., and Mori, A., Cobalt(II) and nickel(II) ions as promoters of free radicals *in vivo*: detected directly using electron spin resonance spectrometry in circulating blood in rats, 306(2), 402–406, 1993.
108. Lewis, C. P., Demedts, M., and Nemery, B., Indices of oxidative stress in hamster lung following exposure to cobalt(II) ions: *in vivo* and *in vitro* studies, *Am. J. Respir. Cell Mol. Biol.,* 5(2), 163–169, 1991.
109. Baldi, C., Minoia, C., DiNucci, A., Capodaglio, E., and Manzo, L., Effects of silver in isolated rat hepatocytes, *Toxicol. Lett.,* 41(3), 261–268, 1988.
110. Rungby, J., An experimental study on silver in the nervous system and on aspects of its general cellular toxicity, *Dan. Med. Bull.,* 37(5), 442–449, 1990.
111. Mourelle, M., Favari, L., and Amezcua, J. L., Protection against thallium hepatotoxicity by silymarin, *J. Appl. Toxicol.,* 8(5), 351–354, 1988.
112. Hasan, M. and Ali, S. F., Effects of thallium, nickel, and cobalt administration of the lipid peroxidation in different regions of the rat brain, *Toxicol. Appl. Pharmacol.,* 1981 57(1), 8–13, 1981.
113. Prabhu, S. D. and Salama, G., The heavy metal ions Ag^+ and Hg^{2+} trigger calcium release from cardiac sarcoplasmic reticulum, *Arch. Biochem. Biophys.,* 277(1), 47–55, 1990.
114. Kone, B. C., Brenner, R. M., and Gullans, S. R., Sulfhydryl-reactive heavy metals increase cell membrane K^+ and Ca^{2+} transport in renal proximal tubule, *J. Membr. Biol.,* 113(1), 1–12, 1990.
115. Zhang, G. H., Yamaguchi, M., Kimura, S., Higham, S., and Kraus-Friedmann, N., Effects of heavy metal on rat liver microsomal Ca^{2+}-ATPase and Ca^{2+} sequestering. Relation to SH groups, *J. Biol. Chem.,* 265(4), 2184–2189, 1990.
116. Leinders, T., van Kleef, R. G., and Vijverberg, H. P., Distinct metal ion binding sites on Ca^{2+}-activated K^+ channels in inside-out patches of human erythrocytes, *Biochim. Biophys. Acta.,* 1112(1), 75–82, 1992.
117. Wiengar, B. D., Kelly, R., and Lansman, J. B., Block of current rough single calcium channels by Fe, Co, and Ni. Location of the transition metal binding site in the pore, *J. Gen. Physiol.,* 97(2), 351–367, 1991.
118. Benders, A. A., Li, J., Lock, R. A., Bindels, R. J., Bonga, S. E., and Veerkamp, J. H., Copper toxicity in cultured human skeletal muscle cells: the involvement of Na^+/K^+-ATPase and the Na^+/Ca^{2+}-exchanger, *Pflugers Arch.,* 428(5–6), 461–467, 1994.
119. Kinne-Suffran, E., Hulseweh, M., Pfaff, C., and Kinne, R. K., Inhibition of Na^+/K^+-ATPase by cadmium: different mechanisms in different species, *Toxicol. Appl. Pharmacol.,* 121(1), 22–29, 1993.

120. Jungwirth, A., Paulmichi, M., and Lang, F., Cadmium enhances potassium conductance in cultured renal epithelioid (MDCK) cells, *Kidney Int.*, 37(6), 1477–1486, 1990.
121. Koizumi, T., Yokota, T., and Suzuki, K. T., Mechanism of cadmium-induced cytotoxicity in rat hepatocytes. Cd-induced acidification causes alkalinization accompanied by membrane damage, *Biol. Trace Elements Res.*, 42(1), 31–41, 1994.
122. Karnishki, L. P., Hg^{2+} and Cu^{+} are ionophores, mediating Cl^{-}/OH^{-} exchange in liposomes and rabbit renal brush border membranes, *J. Biol. Chem.*, 267(27), 19218–19225, 1992.
123. Hechtenberg, S. and Beyersmann, D., Inhibition of sarcoplasmic reticulum Ca^{2+}-ATPase activity by cadmium, lead and mercury, *Enzyme*, 45(3), 109–115, 1991.
124. Carini, R., Bellomo, G., Dianzani, M. U., and Albano, E., The operation of Na^{+}/Ca^{2+} exchanger prevents intracellular Ca^{2+} overload and hepatocyte killing following iron-induced lipid peroxidation, *Biochem. Biophys. Res. Commun.*, 208(2), 813–818, 1995.
125. Shamoo, A. E. and Ryan, T. E., Isolation of ionophores from ion-transport systems, *Ann. N.Y. Acad. Sci.*, 264, 83–97, 1975.
126. Anner, B. M. and Moosmayer, M., Mercury inhibits Na-K-ATPase primarily at the cytoplasmic side, *Am. J. Physiol.*, 262(5, Pt. 2), F843–848, 1992.
127. Anner, B. M., Moosmayer, M., and Imesch, E., Mercury blocks Na-K-ATPase by a ligand-dependent and reversible mechanism, *Am. J. Physiol.*, 262(5, Pt. 2), F830–836, 1992.
128. Jungwirth, A., Ritter, M., Paulmichi, M., and Lang, F., Activation of cell membrane potassium conductance by mercury in cultured renal epithelioid (MDCK) cells, *J. Cell. Physiol.*, 146(1), 25–33, 1991.
129. Skou, J. C., Studies on the Na^{+} + K^{+} activated ATP-hydrolyzing enzyme system. The role of SH-groups, *Biochem. Biophys. Res. Commun.*, 10, 79–84, 1963.
130. Weed, R. I., Eber, J., and Rothstein, A., Membrane sulfhydryl groups and ATPase, *Fed. Proc.*, 22, 213, 1963.
131. Sutherland, R. M., Rothstein, A., and Weed, R. I., Erythrocyte membrane sulfhydryl groups and cation permeability, *J. Cell. Physiol.*, 69(2), 185–198, 1967.
132. Milnar, B. and Enyeart, J. J., Block of current through T-type calcium channels by trivalent metal cations and nickel in neural rat and human cells, *J. Physiol. Lond.*, 469, 639–652, 1993.
133. Passow, H. and Schutt, L., Versuche uber den einfuss von komplexbildern auf die kaliumpermabiliatat bleivergifter menschenerythrozyten, *Pflugers Arch. Physiol.*, 262, 193–202, 1956.
134. Cavieres, J. D. and Ellory, J. C., Thallium and the sodium pump in human red cells, *J. Physiol. Lond.*, 243(1), 243–266, 1974.
135. Higiwara, S. and Takashi, K., Surface density of calcium ions and calcium spikes in the barnacle muscle fiber membrane, *J. Gen. Physiol.*, 50, 583–601, 1967.
136. Schild, L. and Moczydlowski, E., Competitive binding interaction between Zn^{2+} and saxitoxin in cardiac Na^{+} channels. Evidence for a sulfhydryl group in the Zn^{2+}/saxitoxin binding site, *Biophys. J.*, 59(3), 523–537, 1991.
137. Leinders, T., van Kleef, R. G., and Vijverberg, H. P., Divalent cations activate small-(SK) and large-conductance (BK) channels in mouse neuroblastoma cells: selective activation of SK channels by cadmium, *Pflugers Arch.*, 422(3), 217–222, 1992.
138. Vijverberg, H. P., Leinders-Zufall, T., and van Kleef, R. G., Differential effects of heavy metal ions on Ca^{2+}-dependent K^{+} channels, *Cell. Mol. Neurobiol.*, 14(6), 841–857, 1994.
139. Hide, M. and Beaven, M. A., Calcium influx in a rat mast cell (RBL-2H3) line. Use of multivalent metal ions to define its characteristics and role in exocytosis, *J. Biol. Chem.*, 266(23), 15221–15229, 1991.
140. Mundy, W. R., Kodavanti, P. R., Dulchinos, V. F., and Tilson, H. A., Aluminum alters calcium transport in plasma membrane and endoplasmic reticulum from rat brain, *J. Biochem. Toxicol.*, 9(1), 17–23, 1994.
141. Chetty, C. S., Cooper, A., McNeil, C., and Rajanna, B., The effects of cadmium *in vitro* on adenosine triphosphate system and protection by thiol reagents in rat brain microsome, *Arch. Environ. Contam. Toxicol.*, 22(4), 456–458, 1992.
142. Denny, M. F., Hare, M. F., and Atchison, W. D., Methylmercury alters intrasynaptosomal concentrations of endogenous polyvalent cations, *Toxicol. Appl. Pharmacol.*, 122(2), 222–232, 1993.
143. Alkondon, M., Costa, A. C., Radhakrishnan, V., Aronstam, R. S., and Albuquerque, E. X., Selective blockade of NMDA-activated channel currents may be implicated in learning deficits caused by lead, *FEBS Lett.*, 261(1), 124–130, 1990.
144. Vig, P. J., Pentyala, S. N., Chetty, C. S., Rajanna, B., and Desaiah, D., Lead alters inositol polyphosphate receptor activities: protection by ATP, *Pharmacol. Toxicol.*, 75(1), 17–22, 1994.
145. Rajanna, B., Chetty, C. S., Stewart, T. C., and Rajanna, S., Effects of lead on pH and temperature-dependent substrate-activation kinetics of ATPase system and its protection by thiol compounds in rat brain, *Biomed. Environ. Sci.*, 4(4), 441–451, 1991.
146. Mullins, L. J. and Moore, R. D., The movement of thallium ions in muscle, *J. Gen. Physiol.*, 43, 759–773, 1960.
147. Diaz, R. S. and Monreal, J., Thallium mediates a rapid chloride/hydroxyl ion exchange through myelin lipid bilayers, *Mol. Pharmacol.*, 46(6), 1210–1216, 1994.
148. Huang, R. C., Peng, Y. W. and Yau, K. W., Zinc modulation of a transient potassium current and histochemical localization of the metal in neurons of the suprachiasmatic nucleus, *Proc. Natl. Acad. Sci. USA*, 90(24), 11806–11810, 1993.

149. Brierley, G. P., Ion transport by heart mitochondria. VII. Activation of the energy-linked accumulation of Mg^{2+} by Zn^{2+} and other catons, *J. Biol. Chem.*, 242(6), 1115–1122, 1967.
150. Mirzabekov, T., Ballarin, C., Nicolini, M., Zatta, P., and Sorgato, M. C., Reconstitution of the native mitochondrial outer membrane in planar bilayers. Comparison with the outer membrane in a patch pipette and effect of aluminum compounds, *J. Membr. Biol.*, 133(2), 129–143, 1993.
151. Brierley, G. P. and Settlemire, C. T., Ion transport by heart mitochondria. IX. Induction of the energy-linked uptake of K^+ by zinc ion, *J. Biol. Chem.*, 242(19), 4324–4328, 1967.
152. Koizumi, T., Yokota, T., Shirakura, H., Tatsumoto, H., and Suzuki, K. T., Potential mechanism of cadmium-induced cytotoxicity in rat hepatocytes: inhibitory action of cadmium on mitochondrial respiratory activity, *Toxicology*, 92(1–3), 115–125, 1994.
153. Sokol, R. J., Devereaux, M., Mierau, G. W., Hambidge, K. M., and Shikes, R. H., Oxidant injury to hepatic mitochondrial lipids in rats with dietary copper overload. Modification by vitamin E deficiency, *Gastroenterology*, 99(4), 1061–1071, 1990.
154. Lund, B. O., Miller, D. M., and Woods, J. S., Mercury-induced H_2O_2 production and lipid peroxidation *in vitro* in rat kidney mitochondria, *Biochem. Pharmacol.*, 42(Suppl.), S181–S187, 1991.
155. Chavez, E., Zazueta, C., Osornio, A., Holguin, J. A., and Miranda, M. E., Protective behavior of captopril on Hg^{2+}-induced toxicity on kidney mitochondria, *In vivo* and *in vitro* experiments, *J. Pharmacol. Exp. Ther.*, 256(1), 385–390, 1991.
156. Walter, P., Allemand, D., De Renzis, G., and Payan, P., Mediating effect of calcium in $HgCl_2$ cytotoxicity in sea urchin egg: role of mitochondria in Ca^{2+}-mediated cell death, *Biochim. Biophys. Acta*, 1012(3), 219–226, 1989.
157. Levesque, P. C. and Atchison, W. D., Disruption of brain mitochondrial calcium sequestration by methyl mercury, *J. Pharmacol. Exp. Ther.*, 256(1), 236–242, 1991.
158. Levesque, P. C., Hare, M. F., and Atchison, W. D., Inhibition of mitochondrial Ca^{2+} release diminishes the effectiveness of methyl mercury to release acetylcholine from synaptosomes, *Toxicol. Appl. Pharmacol.*, 115(1), 11–20, 1992.
159. Legare, M. E., Barhoumi, R., Burghardt, R. C., and Tiffany-Castiglioni, E., Low-level lead exposure in cultured astroglia: identification of cellular targets with vital fluorescent probes, *Neurotoxicology*, 14(2–3), 267–272, 1993.
160. Albano, E., Bellomo, G., Benedetti, A., Carini, R., Fulceri, R., Gamberucci, A., Parola, M., and Comporti, M., Alterations of hepatocyte Ca^{2+} homeostasis by triethylated lead (Et_3Pb^+): are they correlated with cytotoxicity?, *Chem. Biol. Interact.*, 90(1), 59–72, 1994.
161. Zhang, J. G. and Lindup, W. E., Role of mitochondria in cisplatin-induced oxidative damage exhibited by rat renal cortical slices, *Biochem. Pharmacol.*, 45(11), 2215–2222, 1993.
162. Melnick, R. L., Monti, L. G., and Motzkin, S. M., Uncoupling of mitochondria oxidative phosphorylation by thallium, *Biochem. Biophys. Res. Commun.*, 69(1), 68–73, 1976.
163. Fasitsas, C. E., Theocharis, S. E., Zoulas, D., Chrissimou, S., and Deliconstantinos, G., Time-dependent cadmium neurotoxicity in rat brain synaptosomal plasma membranes, *Comp. Biochem. Physiol. C*, 100(1–2), 271–275, 1991.
164. Theocharis, S., Margel, A., Fasitsas, C., Loizideu, M., and Deliconstantinos, G., Acute exposure to cadmium causes time-dependent liver injury in rats, *Comp. Biochem. Physiol. C*, 99(1–2), 127–130, 1991.
165. Martel, J., Marion, M., and Denizeau, F., Effect of cadmium on membrane potential in isolated rat hepatocytes, *Toxicology*, 60(1–2), 161–172, 1990.
166. Verbost, P. M., Flik, G., Lock, R. A., and Wendelaar-Bonga, S. E., Cadmium inhibits plasma membrane calcium transport, *J. Membr. Biol.*, 102(2), 97–104, 1988.
167. Busselberg, D., Evans, M. L., Rahmann, H., and Carpenter, D. O., Effects of inorganic and triethyl lead and inorganic mercury on the voltage activated calcium channel of Aplysia neurons, *Neurotoxicology*, 12(4), 733–744, 1991.
168. Borseth, J. F., Aunaas, T., Einarson, S., Nordtug, T., Olsen, A. J., and Zachariassen, K. E., Pollutant-induced depression of the transmembrane sodium gradient in muscles of mussels, *J. Exp. Biol.*, 169, 1–18, 1992.
169. Halbach, S., Mercury compounds: lipophilicity and toxic effects on isolated myocardial tissue, *Arch. Toxicol.*, 64(4), 315–319, 1990.
170. Lawton, L. J. and Donaldson, W. E., Lead-induced tissue fatty acid alterations and lipid peroxidation, *Biol. Trace Elements Res.*, 29, 83–97, 1991.
171. Oortgiesen, M., Leinders, T., van Kleef, R. G., and Vijverberg, H. P., Differential neurotoxicological effects of lead on voltage-dependent and receptor-operated ion channels, *Neurotoxicology*, 14(2–3), 87–96, 1993.
172. Simons, T. J., Lead-calcium interactions in cellular lead toxicity, *Neurotoxicology*, 14(2–3), 77–85, 1993.
173. Laterra, J., Bressler, J. P., Indurti, R. R., Belloni-Olivi, L., and Goldstein, G. W., Inhibition of astroglia-induced endothelial differentiation by inorganic lead: a role for protein kinase C, *Proc. Natl. Acad. Sci. USA*, 89(22), 10748–10752, 1992.
174. Tabche, L. M., Martinez, C. M., and Sanchez, Hidalgo, E., Comparative study of toxic lead effect on gill and hemoglobin of tilapia fish, *J. Appl. Toxicol.*, 10(3), 193–195, 1990.

175. Sandhir, R., Julka, D., and Gill, K. D., Lipoperoxidative damage on lead exposure in rat brain and its implications on membrane bound enzymes, *Pharmacol. Toxicol.*, 74(2), 66–71, 1994.
176. Khalil-Manesh, F., Tartaglia-Erler, J., and Gonick, H. C., Experimental model of lead nephropathy. IV. Correlation between renal functional changes and hematological indices of lead toxicity, *J. Trace Elements Electrolytes Health Dis.*, 8(1), 13–19, 1994.
177. Galleano, M. and Puntarulo, S., Effect of mild iron overload on liver and kidney lipid peroxidation, *Braz. J. Med. Biol. Res.*, 27(10), 2349–2358, 1994.
178. Adams, E. T. and Schwartz, K. A., Iron-induced myocardial and hepatic lysosomal abnormalities in the guinea pig, *Toxicol. Pathol.*, 21(3), 321–326, 1993.
179. Myers, B. M., Prendergast, F. G., Holman, R., Kuntz, S. M., and LaRusso, N. F., Alterations in the structure, physicochemical properties, and pH of hepatocyte lysosomes in experimental iron overload, *J. Clin. Invest.*, 88(4), 1207–1215, 1991.
180. Galleano, M. and Puntarulo, S., Mild iron overload effect of rat liver nuclei, *Toxicology*, 93(2–3), 125–134, 1994.
181. Britton, R. S., O'Neill, R., and Bacon, B. R., Hepatic mitochondrial malondialdehyde metabolism in rats with chronic iron overload, *Hepatology*, 11(1), 93–97, 1990.
182. Veerkamp, J. H., Copper toxicity in cultured human skeletal muscle cells: the involvement of Na^+/K^+-ATPase and the Na^+/Ca^{2+} exchanger, *Pflugers Arch.*, 428(5–6), 461–467, 1994.
183. Gwozdzinski, K., A spin label study of the action of cupric and mercuric ions on human red blood cells, *Toxicology*, 65(3), 315–323, 1991.
184. Myers, B. M., Prendergast, F. G., Holm, A. R., Kuntz, S. M., and LaRusso, N. F., Alterations in hepatocyte lysosomes in experimental hepatic copper overload in rats, *Gastroenterology*, 105(6), 1814–1823, 1993.
185. Dominguez, M. C., Sole, E., Goni, C., and Ballabriga, A., Effect of aluminum and lead salts on lipid peroxidation and cell survival in human skin fibroblasts, *Biol. Trace Elements Res.*, 47(1–3), 57–67, 1995.

4 ALTERATION OF ANTI-OXIDANT SYSTEMS

Y. James Kang

Anti-oxidant systems play a major role in cellular protection against injuries induced by a diversity of toxicants. Many studies have addressed the effect of metals on anti-oxidant systems. The anti-oxidants in biological systems include enzymatic and nonenzymatic components. The major enzymatic components include superoxide dismutase (SOD), catalase, and glutathione peroxidase (GSHpx). The nonenzymatic components include ascorbate, alpha-tocopherol, glutathione (GSH), and metallothionein (MT). The reaction catalyzed by GSHpx requires GSH as a co-factor. In this reaction, GSH is oxidized to GSSG, which is converted back to GSH in the presence of glutathione reductase (GR), utilizing NADPH as co-factor; therefore, GR is often considered an enzymatic component of anti-oxidants. Recent studies have shown that heat shock proteins (HSPs) also play a role in cytoprotection against toxic insults, and several metals induce HSP synthesis; thus, HSP induction by metals is also considered in this section. The effect of metals on tocopherol and ascorbate will not be discussed in this section, because such studies are very limited.

4.1 Anti-oxidant Enzymes

In general, the effect of metals on anti-oxidant enzyme activities is a biphasic phenomenon. Acute exposure to metals is associated with decreases in the enzyme activities, and chronic exposure displays enhanced enzyme activities. Doses are also important determinants of changes in enzyme activities. Low-dose and long-term exposure likely increases enzyme activities. High-dose exposure often decreases enzyme activities, even under long-term exposure. The mechanisms for these effects of metals on anti-oxidant enzyme activities have not been elucidated. It is possible that some metals directly interact with the enzymes to inhibit their activities. All of the anti-oxidant enzymes are metal dependent: superoxide dismutase is

TABLE 33
Anti-oxidant Enzymes

Metal	Concentration	System	Effect	Ref.
Cd(II) > Zn(II) > Ag(I) > Hg(I) > Pb(II) > Fe(II), Mn(II), As(III)		Cell-free	Inhibited GSHpx activity	1
Ag(I)	30–70 μ*M*	Rat hepatocyte	↓ GSHpx activity	2
Cd(II)	10–100 μ*M*	Rat hepatocyte, renal cell	↓ SOD activity	3
Cd(II)	50 ppm feed Se-deficient (p.o.)	Rat	↓ Cardiac GSHpx, SOD activities	4
Cd(II)	0.4 mg/kg (i.p.)	Rat	↓ Hepatic, renal GR, CAT, SOD activities	3
Cd(II)	0.4 mg/kg (i.p.)	Rat	45 days: ↓ hepatic, renal, testis, some brain region GSHpx, SOD activities	5
Cd(II)	30 μmol/kg (s.c.)	Rat	↓ Testicular GR, CAT activities; ↑ testicular GSHpx activity	6
Cr(II)	1500 mg/kg (3 days), (p.o.)	Rat	↓ Renal, intestinal epithelial GR, GSHpx, CAT, SOD activities	7, 8
Cr(II)	300 mg/kg (30 days), (p.o.)	Rat	↑ Renal, intestinal epithelial GR, GSHpx, CAT, SOD activities	7, 8
Hg(II)	4 mg/kg (s.c.)	Rat	↓ Renal CAT, SOD, GSHpx activities	9
Ni(II)	200 mmol/kg (s.c.)	Rat	↓ Hepatic GSHpx activity	10
Ni(II)	50–400 mmol/kg (s.c.)	Rat	Dose-dependent ↓ hepatic GSHpx activity	10
Pb(II)	1.25, 2.5 mmol/kg	Chick embryo	9 hr: ↓ hepatic GR, ↑ hepatic, cardiac, brain SOD; 72 hr: ↑ hepatic, cardiac, brain CAT, SOD, activities, ↓ hepatic, brain GSHpx activities	11
Pb(II)	100 mg/kg (i.p.)	Mouse	↓ Hepatic GSH	12
PB(II)	57.1 mg/dl (mean blood)	Human (occupational)	↓ CAT, SOD activities in RBC	13
Pb(II)	100 mg/dl	Human RBC	↑ SOD, CAT GSHpx activities	13
Zn(II)	0.5 m*M*	Cell-free	Inhibited GR activity	14

Note: GSH_{px} = glutathione peroxidase; SOD = superoxide dismutase; GR = glutathione reductase; CAT = catalase; GSH = glutathione; RBC = red blood cells.

either Cu,Zn- or Mn-dependent, catalase (CAT) and glutathione reductase are iron dependent, and glutathione peroxidase is Se dependent (but not the iso-enzymes of GSH-*S*-transferase). Toxic metals may substitute for the required metals at their binding sites on the enzymes or interrupt the interactions between enzymes and their co-factors. Another possible mechanism is that metals may alter transcriptional and translational regulation of these enzymes. Long-term treatment with metals has been shown to result in gene amplification and/or up-regulation of anti-oxidant enzymes, leading to enhanced enzyme activities.

Cadmium has a wide range of effects on anti-oxidant enzymes. Treatment *in vivo* with cadmium is often associated with decreases in all anti-oxidant enzyme activities in multiple organs.[3-6] Interestingly, cadmium at 3.5 μmol/kg, i.p. decreased the activity of GHSpx in testis of rat,[5] while 30 μmol/kg, s.c., increased the rat testicular GSHpx activity.[6] Another study to determine the inhibitory effect of several metals on GSHpx in a cell-free system showed that cadmium is the strongest inhibitor of GSHpx.[1] The order of magnitude of GSHpx inhibition by metals is Cd > Zn > Ag > Hg > Pb > Fe, Mn, or As.[1] Effects of other metals on anti-oxidant enzyme activities are summarized in Table 33.

4.2 Effect of Metals on Cellular Glutathione

Glutathione plays a major role in cytoprotection against the toxicities of many metals. It is also an important cellular target for metal toxicities. In general, acute exposure to metals, especially *in vivo* treatment, decreases GSH levels due to formation of metal-GSH complexes and/or consumption by the GSHpx reaction under oxidative stress induced by metals. Alternatively, long-term treatment has been associated with elevation of cellular GSH levels by some metals. This is often observed in cultured cells treated with cadmium. Recent studies have shown that the elevation of GSH levels in human lung carcinoma A549 cells results from up-regulation of γ-glutamylcysteine synthetase (γ-GCS) transcription.[23] The enzyme γ-GCS catalyzes the rate-limiting step in GSH *de novo* synthesis.

In cell-free systems, cadmium, zinc, lead, and mercury bind to GSH to form metal-GSH complexes.[16,17] Copper, cobalt, and nickel each can form a complex with GSSG.[19] The vanadyl cation binds to both GSH and GSSG;[18,20] however, not all of these complexes have been identified in cultured cells and *in vivo*. For instance, it is uncertain whether cadmium forms a complex with GSH *in vivo*. Most *in vivo* studies related to the effect of metals on GSH have focused on the liver and kidney, and to a much lesser extent on the testis and intestines. As shown in Table 34, different doses have varying effects on tissue GSH concentrations, even in the same tissue. In general, an initial and transient effect of metals such as cobalt, nickel, and platinum on hepatic GSH is to decrease its concentration, followed by a prolonged increase. Whether this increase in GSH concentration is due to enhanced synthesis capacity is not clear. Current studies have not attempted to investigate the molecular mechanisms underlying the effect of metals on GSH synthesis and metabolism. Another important aspect of the effect of metals on tissue levels of GSH is target organ specificity. The liver is a major organ for GSH synthesis and export. It would be expected that the effect of metals on GSH concentration in the liver is different from that in the intestine and other organs.

4.3 Induction of Metallothionein

Metallothionein is a low-molecular-weight and cysteine-rich protein. Since its discovery in 1957, studies of MT have often been associated with metals. It has long been believed that MT is a metal-inducible protein, which binds to metals and plays a role in detoxification of metals. Metal ions that induce MT synthesis include Cd, Zn, Cu, Hg, Au, Ag, Co, Cr, Ni, Bi, As, Fe, Pb, Mn, Mo, Pt, Se, Te, and Sn. It has been determined that among these MT inducers, Cd, Zn, Cu, and Hg also bind to it. Others such as Co, Ni, Mn, Cr, and Fe induce MT but do not bind to the newly synthesized protein, which contains Zn as the bound metal. The affinity for the metal-binding follows the order of Hg, Ag > Cu > Cd > Zn. Metals are released from MT upon lowering of pH, and the free metals become bound to MT at neutral or alkaline pH.

Whether induction of MT is a protective mechanism against the toxicities of metals is not clear; however, it has been demonstrated that MT functions in interorgan transport of metals. For example, cadmium induces MT synthesis in the liver and binds to the newly synthesized MT. This Cd-MT complex is then transported via blood to the kidney. The Cd-MT complex is taken up by kidney cells and enters the lysosome. The MT is then degraded to its component amino acids, and the cadmium is released from the lysosomes into cytosol. Accumulation of cadmium in the kidney to concentrations higher than 200 μg/g results in kidney damage. It has also been demonstrated that MT functions in the storage and transport of essential elements such as iron, copper, and zinc, suggesting an important physiological role of MT.

The study of the molecular mechanism for metal-induced MT production is still at its infant stage. Several studies have shown that metals such as arsenic, cadmium, copper, and

TABLE 34

Glutathione

Metal	Concentration	System	Effect	Ref.
Cd(II), Cu(II), Cr(IV)	20, 75, 400 μmol/kg/day (i.p.)	Rat	No effect on hepatic or renal GSH	15
Cd(II), Hg(II), MeHg, Pb(II), Zn(II), V(IV)		Cell-free	GSH-complex formed	16, 17, 18
Cu(II), Co(II), Ni(II), V(IV)		Cell-free	GSSG-complex formed	19, 20
Hg(II), Fe(II), Mn(II), Ni(II), Pb(II), Zn(II)	5, 120, 200, 250, 300, 400 μmol/kg/day (i.p.)	Rat	↑ Renal GSH	15
Co(II), Ni(II), Pt(II)	250 μmol/kg, 125 μmol/kg (s.c.)	Rat	6 hr: transient ↓ hepatic GSH; 24 hr: ↑ GSH above control	21
Ag(I)	30–70 μ*M*	Rat hepatocyte	↓ Cellular GSH	2
Ag(I)	65 μmol/kg/day (i.p.)	Rat	↑ Hepatic GSH; no effect on renal GSH	15
As(III)	2–40 μ*M*	3T3 mouse cell	Biphasic GSH response (dose-dependent): initial ↓, then ↑ due to enhanced synthesis	22
Cd(II)	Step-wise increase	A549 cell	↑ Cellular GSH and γ-GCS mRNA level	23
Cd(II)	0.22 μ*M*	CHO cell	↑ Cellular GSH	24
Cd(II)	10 μ*M*	3T3 mouse cell	↑ Cellular GSH (dose-dependent) due to enhanced synthesis	25
Cd(II)	1.5 mg/kg (s.c.)	Rat	72 hr: ↓ hepatic GSH	26
Cd(II)	30 μmol/kg (s.c.)	Rat	12 hr: ↓ testicular GSH, ↑ testicular GSSG	6
Cd(II)	50 ppm (p.o.)	Rat	↑ Cardiac GSH	4
Cd(II)	2.0 mg/kg (i.p.)	Guinea pig	72 hr: No effect on hepatic or renal GSH	27
Co(II)	100 μmol/kg/day (i.p.)	Rat	↑ Hepatic, renal GSH	15
Cr(IV)	10 m*M*	Erythrocyte	↓ Cellular GSH	28
Cr(IV)	1500 mg/kg (3 days, p.o.)	Rat	Large ↓ intestinal epithelial GSH; no effect on renal GSH	7, 8
Cr(IV)	300 mg/kg (30 days, p.o.)	Rat	Slight ↓ intestinal epithelial GSH	8
Hg(II)	1–100 m*M*	Human oral fibroblast	↓ Cellular GSH	29
Hg(II) > MeHg	25 μ*M*, 6 μ*M*	Rabbit renal prox. tubule cell	↓ Cellular GSH	30
Hg(II)	4 mg/kg (s.c.)	Rat	24 hr: ↓ renal GSH	9
Mn(II) > Cr(IV) > V(IV) > Fe(II)	500, 500, 1500, 2000 m*M*	Rat hepatocyte	↓ Cellular GSH	31
Ni(II)	2 m*M*	3T3 mouse cell	↑ Cellular GSH (dose-dependent)	25
Ni(II)	50–400 mg/kg (s.c.)	Rat	↓ Hepatic GSH	10
Ni(II)	6 mg/kg (i.p.)	Rat	1.5 hr: ↓ hepatic GSH	32
Ni(II)	14.8 mg/kg (s.c.)	Guinea pig	16 hr: ↑ hepatic GSH, ↓ renal GSH	27
Pb(II)	100 mg/dl	Human RBC	24 hr: no effect	13
Pb(II)	0.1, 1 m*M*	Cultured astroglia	7–24 hr: ↓ cellular GSH; 48 hr: normal; 6–9 days: ↑ cellular GSH	33
Pb(II)	1.25, 2.5 μmol/kg	Chick embryo	9 hr: ↓ hepatic, renal, brain GSH	11
Pb(II)	100 mg/kg (i.p.)	Mouse	↓ Hepatic GSH	34
Pb(II)	57.1 mg/dl	Human (occupational)	↓ GSH in RBC	13
Pt(II)	2 m*M* cisplatin	Rat renal cortical slice	↓ Cellular GSH	35

Note: CHO = Chinese hamster ovary; GSH = glutathione; GSSG = glutathione disulfide; GCS = glutamylcysteine synthetase; RBC = red blood cells.

zinc induce MT mRNA synthesis as well as MT protein production. There are multiple MT genes and proteins. In humans, several metal responsive elements (MREs) are present in the promoter regions of all *MT1* and *MT2* genes. The MREs respond to Zn, Cu, Cd, Ni, and Pb. Another important aspect of MT gene regulation is DNA methylation, which has been shown to play a role in the induction and tissue specificity of MT expression. As indicated previously, the mechanisms of metal induction of MT synthesis and the role of MT in cellular responses to metal toxicities are not fully understood. More studies are required to characterize the role of metals in MT production and function. (See Table 35.)

4.4 Induction of Heat Shock Proteins

Heat shock proteins have been shown to be associated with stress responses of living systems. These proteins have also been proposed as markers for toxicities. Most metals studied so far induce a 70,000-Da HSP (HSP70). Some metals preferentially induce different classes of HSPs. Arsenic has been shown to selectively induce HSP72,[63-65] which is also induced by zinc and by cadmium in cultured cells.[73] Under severe toxic conditions, cadmium induces HSP68 production. The mechanisms for metal induction of HSPs is not fully understood.

The pattern of HSP synthesis in relation to cytotoxicity of cadmium has been studied.[66] Sublethal concentrations of cadmium can inhibit protein synthesis and simultaneously increase the synthesis of HSPs. The pattern of HSP induction changes when cadmium exposure conditions become more severe. For instance, HSP68 protein is synthesized under conditions that lead to severe inhibition of protein synthesis by cadmium; however, HSP68 mRNA is synthesized prior to when the severe protein synthesis inhibition has occurred. This study suggests that cadmium-induced HSP68 transcription and translation are under different mechanisms of regulation. Further analysis revealed that cadmium increases the binding of HSF (heat shock factor) to HSE (heat shock element) on the promoter, which may relate to the enhanced transcription of HSPs.

Iron induces HSP70 in cultured neuroblastoma cells.[69] Incubation of these cells with iron induces an elevation in content of HSP70 mRNA, an induction that can be blocked by alpha-tocopherol; iron-mediated cell damage also was observed in association with the HSP production. This study suggests that the production of HSP70 is related to cytoprotective responses to metal toxicities. Generation of free radicals may be involved in the induction of HSP70. Other possible mechanisms related to the differential metal-induced HSP responses remain to be investigated. (See Table 36.)

Acknowledgments

The author (YJK) thanks Yan Chen, Emiko Hatcher, Qiangrong Liang, and Xianhua Yin for the excellect information processing and assistance in completing sections 3 and 4 of this chapter, and KayLynn Bousher and Judith Alexander for manuscript preparation.

TABLE 35
Metallothionein

Effector	Dose	System	Metallothionein (MT) Induction: Protein	mRNA	Mechanism	Ref.
Arsenic (As)	As^{3+}, 85 µmol/kg (s.c.)	Mice	Increase	Increase	Directly or altering post-transcriptional event	36
	As^{3+}, 25–90 µmol/kg (s.c.)	Male SD rats	Increase	Increase		37
	As^{3+}, 5 µ*M*	Hela cells		Increase	Oxidative stress	38
Bismuth (Bi)	Bi^{+}, 1–10 µ*M*	Cultured bovine aortic endothelial cells	Increase			39
	Bi^{3+}, 50 µmol/kg (i.p.)	Female A/J mice	Increase			40
Cadmium (Cd)	Cd^{2+}, 50 µ*M*	Human U373MG astrocytoma cells	Increase	Increase		41
	Cd^{2+}, 2–8 µ*M*	Cultured rat hepatocytes	Increase			42
	Cd^{2+}, 0.3 mg/kg (i.v.)	Male SD rat	Increase			43
	Cd^{2+}, 3 mg/kg	Mice	Increase			51
Chromium (Cr)	Cr^{3+}, 5 mg/kg (i.p.)	Male SD rats	Increase			44
	Cr^{3+}, 150 µmol/kg (i.p.)	Chicks	Increase		Inflammation	45
Cobalt (Co)	Co^{2+}, 150 µmol/kg (i.p.)	Chicks	Increase			45
Copper (Cu)	Cu^{2+}, 100 µ*M*	Hela cells	Increase			46
	Cu^{2+}, 2.5–25 µ*M*	LEC rat liver parenchymal cells		Increase		47
	Cu^{2+}, 1.76 or 3 mg/kg	Male hooded Listar rats	Increase			48
Iron (Fe)	Fe^{2+} and Fe^{3+} 10 mg/kg (i.p.)	Chicks	Increase		Inflammation	45
	Fe^{+}, Fe^{2+}, and Fe^{3+}, 10 mg/kg (i.p.)	Male chicks	Increase		Stress response	49
Lead (Pb)	400 µmol/kg (i.p.)	Young male rats	Increase			50
Manganese (Mn)	Mn^{2+}, 150 µmol/kg (i.p.)	Chicks	Increase			45
	10 mg/kg	Chicks	Increase			51
Mercury (Hg)	Hg^{2+}, 0.3 mg/kg (i.v.)	Male SD rats	Increase			43
	100 n*M*	Rainbow trout hepatocytes	Increase			52
	10 µ*M*	Human lymphocytes	Increase			53
Molybdenum (Mo)	Na_2MoO_4, 1.24 mmol/kg	Rats	Increase		Enhancement of Cd-MT induction	54
Nickel (Ni)	Ni^{2+}, 500–2000 µ*M*	Cultured rat hepatocytes	Increase			42
		Rats	Increase			50
	Ni^{2+}, 150 µmol/kg (i.p.)	Chicks	Increase			45
	200 µ*M*	Human lymphocytes	Increase			53
Platinum (Pt)	Cis-DDP 9 mg/kg	Male BALB/c mice	Increase		Directly or through displacing zinc or copper ions	55
	Cisplatin and K_2PtCl_4	Rabbit	Increase			56
Selenium (Se)	Selenite, 1.5 mg (i.p.)	Male SD rats	Increase			57
	Selenite, 20 µmol/kg	Mice	Increase			58
Zinc (Zn)	Zn^{2+}, 500 µ*M*	Human U377MG astrocytoma cells	Increase	Increase		41
	Zn^{2+}, 100 µ*M*	Hela cells	Increase			59
	1000 µmol/kg (s.c.)	Mice	Increase			60
	Zn^{2+} 10 mg/kg (s.c.)	Male wistar rats	Increase			61
	Zn^{2+}, 5 mg/kg (i.p.)	Male SD rats	Increase			44
	300 mg Zn aspartate (oral)	Human	Increase			62

TABLE 36

Heat Shock Proteins

Effector	Dose	System	Heat Shock Protein (HSP) Induction		Mechanism	Ref.
			Protein	mRNA		
Arsenic (As)	5.25 mg, $NaAsO_2$/kg (i.p.)	Female ND_4 mice	Increase			63
	6 mg, $NaAsO_2$ per kg (i.v.)	Male SD rats	Increase			64
	10–60 μM, As^{3+}	Cultured rat hepatocyte	Increase			65
Cadmium (Cd)	3–100 μM, $CdCl_2$	Reuber H35 hepatoma cell line	Increase	Increase	Increased binding of HSF to HSE on the promoter	66
	2–8 μM, Cd^{2+}	Cultured rat hepatocytes	Increase			65
	1.25/kg/day, $CdCl_2$	SJL/J mice	Increase			67
Copper (Cu)	200 μM, $CuSO_4$	Hela cells	Increase			68, 73
Iron (Fe)	10 μM, $FeSO_4$	Cl300 N2A		Increase	Free radical generation	69
Lead (Pb)	100 mg/kg, Pb^{2+}	*Oniscus asellus*	Increase			70
Nickel (Ni)	500–2000 μM	Cultured hepatocytes	Increase			65
Tin (Sn)	100 μM, $SnCl_2$	Human Hep G_2 and Hep 3B hepatoma cells		Increase		71
Zinc (Zn)	200 μM, Zn^{2+}	Nonthermotolerant HeLa cells	Increase	Increase		72
	100–1500 μM, Zn^{2+}	Cultured rat hepatocytes	Increase			65
	5–100 μM, Zn^{2+}	Rat primary cortical astrocyte culture	Increase			73
	100 μM, Zn^{2+}	Bovine articular chondrocytes		Increase	Protein kinase C → heat shock factor phosphorylation	74

Note: HSF = heat shock factor; HSE = heat shock element.

REFERENCES

1. Splittgerber, A. G. and Tappel, A. L., Inhibition of glutathione peroxidase by cadmium and other metal ions, *Arch. Biochem. Biophys.*, 197(2), 534–542, 1979.
2. Baldi, C., Minoia, C., Di-Nucci, A., Capodaglio, E., and Manzo, L., Effects of silver in isolated rat hepatocytes, *Toxicol. Lett.*, 41(3), 261–268, 1988.
3. Hussain, T., Shukla, G. S., and Chandra, S. V., Effects of cadmium on superoxide dismutase and lipid peroxidation in liver and kidney of growing rats: *in vivo* and *in vitro* studies, *Pharmacol. Toxicol.*, 60(5), 355–358, 1987.
4. Jamall, I. S. and Smith, J. C., Effects of cadmium on glutathione peroxidase, superoxide dismutase and lipid peroxidation in the rat heart: a possible mechanism of cadmium cardiotoxicity, *Toxicol. Appl. Pharmacol.*, 80, 33–42, 1985.
5. Shukla, G. S., Hussain, T., Srivastava, R. S., and Chandra, S. V., Glutathione peroxidase and catalase in liver, kidney, testis and brain regions of rats following cadmium exposure and subsequent withdrawal, *Ind. Health,* 27(2), 59–69, 1989.
6. Koizumi, T. and Li, Z. G., Role of oxidative stress in single-dose, cadmium-induced testicular cancer, *J. Toxicol. Environ. Health,* 37(1), 25–36, 1992.

7. Sengupta, T., Chattopadhyay, D., Ghosh, N., Maulik, G., and Chatterjee, G. C., Impact of chromium on lipid peroxidative processes and subsequent operation of the glutathione cycle in the rat renal system, *Indian J. Biochem. Biophys.*, 29(3), 287–290, 1992.
8. Sengupta, T., Chattopadhyay, D., Ghosh, N., Das, M., and Chatterjee, G. C., Effect of chromium administration on glutathione cycle of rat intestinal epithelial cells, *Indian J. Exp. Biol.*, 28(12), 1132–1135, 1990.
9. Gstraunthaler, G., Pfaller, W., and Kotanko, P., Glutathione depletion and *in vitro* lipid peroxidation in mercury or maleate induced acute renal failure, *Biochem. Pharmacol.*, 32(19), 2969–2972, 1983.
10. Athar, M., Hasan, S. K., and Srivastava, R. C., Evidence for the involvement of hydroxyl radicals in nickel mediated enhancement of lipid peroxidation: implications for nickel carcinogenesis, *Biochem. Biophys. Res. Commun.*, 147(3), 1276–1281, 1987.
11. Somashekaraiah, B. V., Padmaja, K., and Prasad, A. R., Lead-induced lipid peroxidation and antioxidant defense components of developing chick embryos, *Free Rad. Biol. Med.*, 13(2), 107–114, 1992.
12. Nakagawa, K., Modification by phenobarbital of decreased glutathione content and glutathione S-transferase activity in livers of lead-treated mice, *Toxicol. Lett.*, 62(1), 63–71, 1992.
13. Sugawara, E., Nakamura, K., Miyake, T., Fukumura, A., and Seki, Y., Lipid peroxidation and concentration of glutathione in erythrocytes from workers exposed to lead, *Br. J. Ind. Med.*, 48(4), 239–242, 1991.
14. Serrano, A., Rivas, J., and Losadas, M., Purification and properties of glutathione reductase from the cyanobacterium *Anabaena* sp. strain 7119, *J. Bacteriol.*, 158(1), 317–324, 1984.
15. Eaton, D. L., Stacey, N. H., Wong, K. L., and Klaassen, C. D., Dose response effects of various metal ions on rat liver metallothionein, glutathione, heme oxygenase and cytochrome P-450, *Toxicol. Appl. Pharmacol.*, 55(2), 393–402, 1980.
16. Fuhr, B. J. and Rabenstein, D. L., Nuclear magnetic resonance studies of the solution chemistry of metal complexes. IX. The binding of cadmium, zinc, lead, and mercury by glutathione, *J. Am. Chem. Soc.*, 95(21), 6944–6950, 1973.
17. Rabenstein, D. L. and Fairhurst, M. T., Nuclear magnetic resonance studies of the solution chemistry of metal complexes. XI. The binding of methyl mercury by sulfhydryl-containing amino acids and by glutathione, *J. Am. Chem. Soc.*, 97(8), 2086–2092, 1975.
18. Ferrer, E. G., Williams, P. A., and Baran, E. J., A spectrophotometric study of the VO^{2+}-glutathione interactions, *Biol. Trace Elements Res.*, 30(2), 175–183, 1991.
19. Formicka-Kozlowski, G., Kozlowski, H., and Jezowska-Trzebiatowska, B., Nickel(II), cobalt(II), and copper(II) complexes with oxidized glutathione, *Acta Biochim. Pol.*, 26(3), 239–248, 1979.
20. Ferrer, E. G., Williams, P. A., and Baran, E. J., The interaction of the VO^{2+} cation with oxidized glutathione, *J. Inorg. Biochem.*, 50(4), 253–262, 1993.
21. Maines, M. D. and Kappas, A., Regulation of heme pathway enzymes and cellular glutathione content by metals that do not chelate with tetrapyrroles: blockade of metal effects by thiols, *Proc. Natl. Acad. Sci. USA*, 74(5), 1875–1878, 1977.
22. Li, W. and Chou, I. N., Effects of sodium arsenite on the cytoskeleton and cellular glutathione levels in cultured cells, *Toxicol. Appl. Pharmacol.*, 114(1), 132–139, 1992.
23. Hatcher, E. L., Chen, Y., and Kang, Y. J., Cadmium resistance in A549 cells correlates with elevated glutathione content but not antioxidant enzymatic activities, *Free Rad. Biol. Med.*, 19(6), 805–812, 1995.
24. Seagrave, J., Hildebrand, C. E., and Enger, M. D., Effects of cadmium on glutathione metabolism in cadmium sensitive and cadmium resistant Chinese hamster cell lines, *Toxicol.*, 29(1–2), 101–197, 1993.
25. Li, W., Zhao, Y., and Chou, I. N., Alterations in cytoskeletal protein sulfhydryls and cellular glutathione in cultured cells exposed to cadmium and nickel ions, *Toxicology*, 77(1–2), 65–79, 1993.
26. Asokan, P., Effect of partial hepatectomy on hepatic metallothionein and glutathione levels and cadmium distribution in different tissues, *Toxicol. Lett.*, 12(1), 35–40, 1982.
27. Iscan, M., Coban, T., and Eke, B. C., Differential combined effect of cadmium and nickel on hepatic and renal glutathione S-transferases of the guinea pig, *Environ. Health Perspect.*, 102(Suppl. 9), 69–72, 1994.
28. Wiegand, H. J., Ottenwalder, H., and Bolt, H. M., The reduction of chromium(VI) to chromium(III) by glutathione: an intracellular redox pathway in the metabolism of the carcinogen chromate, *Toxicology*, 33(3–4), 341–348, 1984.
29. Liu, Y., Cotgreave, I., Atzori, L., and Grafstrom, R. C., The mechanism of Hg^{2+} toxicity in cultured human oral fibroblasts: the involvement of cellular thiols, *Chem. Biol. Interact.*, 85(1), 69–78, 1992.
30. Aleo, M. D., Taub, M. L., and Kostyniak, P. J., Primary cultures of rat renal proximal tubule cells. III. Comparative cytotoxicity of inorganic and organic mercury, *Toxicol. Appl. Pharmacol.*, 112(2), 310–317, 1992.
31. Liu, J., Kershaw, W. C., and Klaassen, C. D., Protective effects of zinc on cultured rat primary hepatocytes to metals with low affinity for metallothionein, *J. Toxicol. Environ. Health*, 35(1), 51–62, 1992.
32. Herrero, M. C., Alvarez, C., Cartana, J., Blade, C., and Arola, L., Nickel effects on hepatic amino acids, *Res. Commun. Chem. Pathol. Pharmacol.*, 79(2), 243–248, 1993.
33. Legare, M. E., Barhoumi, R., Burghardt, R. C., and Tiffany-Castiglioni, E., Low-level lead exposure in cultured astroglia: identification of cellular targets with vital fluorescent probes, *Neurotoxicology*, 14(2–3), 267–272, 1993.

34. Somashekaraiah, B. V., Padmaja, K., and Prasad, A. R., Lead-induced lipid peroxidation and antioxidant defense components of developing chick embryos, *Free. Read. Biol. Med.*, 13(2), 107–114, 1992.
35. Zhang, J. G., Zhong, L. F., Zhang, M., and Xia, Y. X., Protection effects of procaine on oxidative stress and toxicities of renal cortical slices from rats caused by cisplatin *in vitro*, *Arch. Toxicol.*, 66(5), 354–358, 1992.
36. Kreppel, H., Bauman, J. W., Liu, J., McKim, Jr., J. M., and Klaassen, C. D., Induction of metallothionein by arsenicals in mice, *Fundam. Appl. Toxicol.*, 20(2), 184–189, 993.
37. Albores, A., Koropatnick, J., Cherian, M. G., and Zelazowski, A. J., Arsenic induces and enhances rat hepatic metallothionein production *in vivo*, *Chem. Biol. Interact.*, 85(2–3), 127–140, 1992.
38. Guzzo, A., Karatizios, C., Diorio, C., and DuBow, M. S., Metallothionein-II and ferritin H mRNA levels are increased in arsenite-exposed Hela cells, *Biochem. Biophys. Res. Commun.*, 205(1), 590–595, 1994.
39. Kaji, T., Suzuki, M., Yamamoto, C., Imaki, Y., Mishima, A., Fujiwara, Y., Sakamoto, M., and Kozuka, H., Induction of metallothionein synthesis by bismuth in cultured vascular endothelial cells, *Res. Commun. Mol. Pathol. Pharmacol.*, 86(1), 25–35, 1994.
40. Satoh, M., Kondo, Y., Mita, M., Nakagawa, I., Naganuma, A., and Imura, N., Prevention of carcinogenicity of anticancer drugs by metallothionein induction, *Cancer Res.*, 53(20), 4767–4768, 1993.
41. Sawada, J., Kikuchi, Y., Shibutani, M., Mitsumori, K., Inoue, K., and Kasahara, T., Induction of metallothionein in astrocytes by cytokines and heavy metals, *Biol. Signals*, 3(3), 157–168, 1994.
42. Bauman, J. W., Liu, J., and Klaassen, C. D., Production of metallothionein and heat shock protein in response to metals, *Fundam. Appl. Toxicol.*, 21(1), 15–22, 1993.
43. Chan, H. M., Satoh, M., Zalups, R. K., and Cherian, M. G., Exogenous metallothionein and renal toxicity of cadmium and mercury in rats, *Toxicology*, 76(1), 15–26, 1992.
44. Hanna, P. M., Kadiiska, M. B., Jordan, S. J., and Mason, R. P., Role of metallothionein in Zinc(II) and chromium(II) mediated tolerance to carbon tetrachloride hepatotoxicity: evidence against a trichloromethyl radical-scavenging mechanism, *Chem. Res. Toxicol.*, 6(5), 711–717, 1993.
45. Fleet, J. C., Golemboski, K. A., Dietert, R. R., Andrews, G. K., and McCormick, C. C., Induction of hepatic metallothionein by intraperitoneal metal injection: an associated inflammatory response, *Am. J. Physiol.*, 258(6), G926–G933, 1990.
46. Hatayama, T., Tsukimi, Y., Wakatsuki, T., Kitamura, T., and Imahara, H., Different induction of 70,000-Da heat shock protein and metallothionein in HeLa cells by copper, *J. Biochem.*, 110(5), 726–731, 1991.
47. Kanno, S., Suzuki, J. S., Aoki, Y., and Suzuki, K. T., Selective enhancement of metallothionein mRNA expression by copper in primary cultured liver parenchymal cells of LEC rats, *Res. Commun. Chem. Pathol. Pharmacol.*, 84(2), 153–162, 1994.
48. Bremner, I. and Young, B. W., Isolation of (copper, zinc)-thioneins from the livers of copper-injected rats, *Biochem. J.*, 157(2), 517–520, 1976.
49. McCormick, C. C., The tissue-specific accumulation of hepatic zinc metallothionein following parenteral iron loading, *Proc. Soc. Exp. Biol. Med.*, 176(4), 392–402, 1984.
50. Tandon, S. K., Khandelwal, S., Jain, V. K., and Mathur, N., Influence of dietary iron deficiency on nickel, lead and cadmium intoxication, *Sci. Total Environ.*, 148(2–3), 167–173, 1994.
51. Matsubara, J., Alteration of radiosensitivity in metallothionein-induced mice and a possible role of Zn-Cu-thioneine in GSH-peroxidase system, *EXS*, 52, 603–612, 1987.
52. Gagne, F., Marion, M., and Denizeau, F., Metallothionein induction and metal homeostasis in rainbow trout hepatocytes exposed to mercury, *Toxicol. Lett.*, 51(1), 99–107, 1990.
53. Yamada, H. and Koizumi, S., Metallothionein induction in human peripheral blood lymphocytes by heavy metals, *Chem. Biol. Interact.*, 78(3), 347–354, 1991.
54. Yamane, Y., Fukuchi, M., Li, C. K., and Koizumi, T., Protective effect of sodium molybdate against the acute toxicity of cadmium chloride, *Toxicology*, 60(3), 235–243, 1990.
55. Farnworth, P. G., Hillcoat, B. L., and Roos, I. A., Metallothionein induction in mouse tissues by cis-dichlorodiammineplatinum(II) and its hydrolysis products, *Chem. Biol. Interact.*, 69(4), 319–332, 1989.
56. Zhang, B., Huang, H., and Tang, W., Interaction of *cis*-and *trans*-diamminedichloroplatinum with metallothionein *in vivo*, *J. Inorg. Biochem.*, 58(1), 1–8, 1995.
57. Chen, C. L. and Whanger, P. D., Interaction of selenium and arsenic with metallothionein effect of vitamin B12, *J. Inorg. Biochem.*, 54(4), 267–276, 1994.
58. Iwai, N., Watanabe, C., Suzuki, T., Suzuki, K. T., and Tohyama, C., Metallothionein induction by sodium selenite at two different ambient temperatures in mice, *Arch. Toxicol.*, 62(6), 447–451, 1988.
59. Hatayama, T., Tsukimi, Y., Wakatsuki, T., Kitamura, T., and Imahara, H., Characteristic induction of 70,000-Da heat shock protein and metallothionein by zinc in HeLa cells, *Mol. Chem. Biochem.*, 112(2), 143–153, 1992.
60. Kreppel, H., Liu, J., Liu, Y., Reichl, F. X., and Klaassen, C. D., Zinc-induced arsenite tolerance in mice, *Fundam. Appl. Toxicol.*, 23(1), 32–37, 1994.
61. Du, X. H. and Yang, C. L., Mechanism of gentamicin nephrotoxicity in rats and the protective effect of zinc-induced metallothionein synthesis, *Nephrol. Dial. Transplant.*, 9(4), 135–140, 1994.
62. Mulder, T. P., VanderSluys-Veer, A., Verspaget, H. W., Griffioen, G., Pena, A. S., Janssens, A. R., and Lamers, C. B., Effect of oral zinc supplementation on metallothionein and superoxide dismutase concentrations in patients with inflammatory bowel disease [see comments], *J. Gastroenterol. Hepatol.*, 9(5), 472–477, 1994.

63. Lappas, G. D., Karl, I. E., and Hotchkiss, R. S., Effect of ethanol and sodium arsenite on HSP-72 formation and on survival in a murine endoxin model, *Shock,* 2(1), 34–39, 1994.
64. Ribeiro, S. P., Villar, J., Downey, G. P., Edelson, J. D., and Slutsky, A. S., Sodium arsenite induces heat shock protein-72 kilodalton expression in the lungs and protects rats against sepsis, *Crit. Care Med.,* 22(6), 922–929, 1994.
65. Bauman, J. w., Liu, J., and Klaassen, C. D., Production of metallothionein and heat shock protein in response to metals, *Fundam. Appl. Toxicol.,* 21(1), 15–22, 1993.
66. Ovelgonne, J. H., Souren, J. E., Wiegant, F. A., and Van-Wijk, R., Relationship between cadmium-induced expression of health shock genes, inhibition of protein synthesis and cell death, *Toxicology,* 99(1–2), 19–30, 1995.
67. Weiss, R. A., Madaio, M. P., Tomaszewski, J. E., and Kelly, C. J., T cells reactive to an inducible heat shock protein induced disease in toxin-induced interstitial nephritis, *J. Exp. Med.,* 180(8), 2239–2250, 1994.
68. Hatayama, T., Tsukimi, Y., Wakatsuki, T., Kitamura, T., and Imahara, H., Different induction of 70,000-Da heat shock protein and metallothionein in HeLa cells by copper, *J. Biochem.,* 110(5), 726–732, 1991.
69. Uney, J. B., Anderton, B. H., and Thomas, S. H., Changes in heat shock protein 70 and ubiquitin mRNA levels in C1300 N2A mouse neuroblastoma cells following treatment with iron, *J. Neurochem.,* 60(2), 659–665, 1993.
70. Kohler, H. R., Triebskorn, R., Stocker, W., Kloetzel, P. M., and Alberti, G., The 70-kDa heat shock protein 70 in soil invertebrates: a possible tool for monitoring environmental toxicants, *Arch. Environ. Contam. Toxicol.,* 22(3), 334–338, 1992.
71. Mitani, K., Fujita, H., Fukuda, Y., Kappas, A., and Sassa, S., The role of inorganic metals and metalloporphyrins in the induction of haem oxygenase and heat shock protein 70 in human hepatoma cells, *Biochem. J.,* 290 (Pt. 3), 819–825, 1993.
72. Hatayama, T., Asai, Y., Wakatsuki, T., Kitamura, T., and Imahara, H., Regulation of HSP70 synthesis induced by cupric sulfate and zinc sulfate in thermotolerant Hela cells, *J. Biochem.,* 114(4), 592–597, 1993.
73. Swanson, R. A. and Sharp, F. R., Zinc toxicity and induction of the 72-kDa heat shock protein in primary astrocyte culture, *Glia,* 6(3), 198–205, 1992.
74. Zafarullah, M., Su, S., and Gedamu, L., Basal and inducible expression of metallothionein and heat shock protein 70 genes in bovine articular chondrotyes, *Exp. Cell. Res.,* 208(2), 371–377, 1993.

CHAPTER 5

Tissue Uptake and Subcellular Distribution of Mercury

Edward J. Massaro

CONTENTS

Methylmercury is one of the most toxic forms of mercury to humans.[17,36] Its use as a fungicide and indiscriminate disposal as a byproduct of chemical manufacture has been responsible for human mortality and morbidity of near epidemic proportions in Japan,[75,129] Iraq, Pakistan, Guatemala, Ghana, and the former Soviet Union, and isolated incidents have been reported elsewhere.[3,36,45] Environmental application and overt dumping appear to have subsided and, currently, biotic and abiotic methylation of inorganic mercury to methylmercury in aquatic sediments appears to be the major source of environmental contamination by methylmercury.[18,46,75,82,129,133]

Methylmercury is readily bioaccumulated and biomagnified in food chains/webs.[71,133] It is accumulated by aquatic species from the water and their diet.[61,106,130] Biomagnification in food chains/webs can result in high-dose exposure to top predators as exemplified by the Minamata (Japan) disaster.[75,129] Indeed, substantial concentrations of methylmercury are still found in freshwater and marine fishes and, via consumption, may adversely affect humans, especially the fetus and young child.[107,133] However, as stated by Kevorkian et al.:[56] "[Although] knowledge of mercury levels in fishes is doubtlessly important as a potential source of humans ills ... it is analysis of human tissues which will determine with finality whether environmental mercury pollution is of sufficient magnitude to pose a real human threat, all other biological hosts not withstanding." In addition, knowledge of regional trends in the mercury concentration of human tissues is the ultimate sentinel of localized environmental

0-8493-4493-X/97/$0.00+$.50

contamination of potential human consequence. Unfortunately, information on the mercury content of human tissues is limited.[10,21,35,39,41,49,54,56,59,67,68,76,84,100,104,114,116,138]

If the Kevorkian et al.[56] statement is axiomatic, it would be useful to have sufficient data on human tissue mercury to establish a no-observed-effect level (as a function of chemical compound, if possible) for at least the irreplaceable organs most susceptible to irreversible damage. In addition, since knowledge of long-term, low-level exposure is limited and recent evidence on lead suggests deleterious effects of such exposure,[50,57] alteration of human tissue mercury levels should be monitored as a function of time, especially in areas in which significant environmentally available mercury levels have been detected. Little systematic information is available on long-term trends in human tissue mercury levels and, for the foreseeable future, it does not appear that there will be.

It is observed that the incidence of apparently noninfectious human diseases varies geographically.[133] The factors responsible for the variation have been difficult to ascertain;[16,86,96] however, diet[115] and environmental contamination, both natural and anthropogenic, appear to play an etiological role.[36,37] Unfortunately, with the exception of a limited number of conditions (e.g., arsenic, cadmium, lead, and mercury poisoning), the magnitude of the potential contribution of environmental factors pertaining to the incidence of geographical pathologic phenomena are far from clear.[4,11,36,37,51,72,73,75,135]

In nonacute, metal-induced intoxication, identification of the culpable agent may be confounded by the disruption of tissue elemental homeostasis. Indeed, both theory[47] and experimental evidence[22,60,72,73,113,116] indicate complex multi-element interactions following nonessential (apparently) element-induced homeostatic imbalance (e.g., Pb-Zn-Fe-Cu; Hg-Cd-Zn). Unfortunately, with the exception of blood (for which there is a large clinical database), hair, and nails, little information is available on tissue element concentrations in homeostasis,[30,126-127b,132,138] let alone homeostatic imbalance.[78,124,133] Furthermore, except for blood, hair, and nails, the availability of biopsy specimens from healthy human tissues is limited not only in number and quantity, but also in tissue type. In addition, although blood, hair, and nail mercury levels may be useful indicators of relatively recent environmental exposure to mercury,[133] evidence that the mercury concentration in these tissues accurately reflects conditions of long-term, low-level exposure is limited.[14,132] In any case, such relationships can be established unequivocally only by comparison with the tissue levels themselves, and it is extremely difficult or impossible to obtain healthy human tissues for analysis. Therefore, knowledge of the elemental composition of most human tissues has been obtained, almost exclusively, from postmortem sources deceased as the result of a broad spectrum of conditions (known and unknown). Unfortunately, morbidity and death alter tissue composition.[69] In addition, the elemental composition of tissues may be modified iatrogenically (e.g., dental amalgam quantity and tissue mercury levels[19,28,40,44]), occupationally,[109,136] and/or via self-imposition.[62,85] Also, duration, level, and route of exposure affect the concentration of certain elements in certain tissues. It is well known that renal cortical cadmium increases with age, reaching a maximum between 40 and 50 years, and relatively high concentrations of aluminum and titanium are found in pulmonary tissue as an apparent consequence of their primary portal of entry being the respiratory tract.[31,64,126,127a] Moreover, information on the condition of the donor prior to tissue harvesting and tissue handling prior to analysis often is not comprehensive. Indeed, histological characterization of the material to be analyzed is rarely reported. Almost certainly, perfusion of human donor organs prior to tissue harvesting is not performed. In addition, quantification of tissue element concentrations is impacted by level of tissue hydration. Tissue hydration may be significantly altered by the time interval between death (or biopsy) and analysis, as well as method of preservation (e.g., freezing, formalin fixation, etc.); therefore, the validity of quantification of trace element levels on a wet weight basis may be questionable. In addition, both the method of analysis and the analyst affect the data,[56] a problem amenable to amelioration through employment of certified standards. From the above, it may be concluded that, no matter how precise the elemental analyses, the data obtained can be less than an accurate reflection of the living state. Nevertheless, consistency and care in tissue handling, preparation, and analysis favors precision. High-precision analysis of a large sample size of tissues obtained from long-time resident donors from the same geographical area (*vide infra*) provides the basis for meaningful statistical analysis of tissue element levels and distribution as well as baseline data for comparative purposes (e.g., alterations in levels as a function of time within the geographic area).

With the above caveats in mind, the accumulated evidence indicates that the concentration of elements in human tissues correlates with geographic origin (presumably following sufficient exposure to the local environment to achieve steady-state tissue levels) and, for some elements, with age.[76,91,92,105,127b,132] It has been reported that the concentration ranges of essential metals (e.g., calcium, copper, iron, magnesium, manganese, molybdenum) in human tissues and organs (e.g., aorta, brain, heart, kidney, liver, lung, spleen) are more constant than that of nonessential metals (e.g., aluminum, barium, cadmium, chromium, lead, nickel, silver, tin, titanium, vanadium).[91,105,126-127b] In addition, it has been reported[63] that: (1) essential trace elements have a statistically normal tissue distribution ("maintained by some biological process") while nonessential trace elements have a log-normal distribution (apparently through evasion of homeostatic mechanisms), and (2) the nature of its tissue distribution can be utilized to determine whether an element is essential or nonessential. However, in light of subsequent research establishing the essential nature of elements such as chromium, nickel, and vanadium[83] and the limited safety margin between potentially beneficial and deleterious intake of many essential trace elements,[2,58,74,79,80,90] the utility of these early hypotheses is, at best, limited. In addition, animal studies indicate that relatively short-term exposure to relatively high levels of nonessential elements can affect tissue levels of a spectrum of essential trace elements.

1 TISSUE/ORGAN ACCUMULATION

Mercury occurs in the environment in different chemical and physical forms. Apparently, as a consequence of its ubiquitous distribution in the environment[127] and large thiol association constant ($K_{assoc} = 10^{16}$),[53] mercury has been found in all human tissues examined, even from populations with no known exposure other than background.[29,39,56,67,76,138] As shown in Tables 1 to 6 and Figures 1 and 2, tissue mercury levels varied from region to region within countries (compare Table 1 with Tables 2 and 3 and Table 4 with Table 5) and between countries (compare Tables 1 to 3 with Tables 4 and 5). The variability of the data on kidney and liver mercury levels from studies reported between 1940 and 1974 is summarized in Table 6. It is of interest to note that the data of Glomski et al.[39] and Massaro et al.[67] (Table 1), obtained from the same geographical area, were similar. In addition, the lack of a significant difference between subadult (0 to 14 years) and adult (≥15 years) brain mercury levels in the data of Massaro et al.[67] also has been reported by Ehmann et al.[29] Although their data are highly variable, Kevorkian et al.[56] showed logarithmic declines in tissue mercury levels from the second decade of this century to 1970 (Figures 1 and 2). It was suggested that the declines may be due to the virtual elimination of coal burning as a means of home heating and the consequent pollution of the environment from mercury-containing soot and smoke. However, Kevorkian et al.[56] also reported an age-dependent biphasic pattern of human tissue mercury levels with peaks occurring in childhood and middle age. A similar pattern emerges from the mean tissue mercury level data of Mottet and Bõdy who also reported a much higher kidney burden for some individuals between the ages of 30 and 60.[76]

It has been estimated that, for humans exposed to methylmercury, the blood mercury level above which there is onset of symptomology (i.e., the apparent threshold) is 0.2 μg/g (approximately 0.4 μg Hg per g RBCs).[71] However, Skerfving[110-112] has pointed out uncertainties in the estimation. Furthermore, individuals with RBC total mercury levels >1.0 μg/g,

TABLE 1

Human Tissue/Body Fluid Mercury Levels[a]

Tissue	Number of Samples	Group Range (ppm wet weight)	Mean[g]	SEM[h]
Frontal cortex[b]	100	0.03–1.69	0.336	0.038
Frontal cortex[c]	50	0.04–1.50	0.329	0.043
Cerebellar cortex[b]	100	0.08–2.59	0.358	0.047
Cerebellar cortex[c]	50	0.04–1.65	0.347	0.049
Muscle[b]	100	0.13–0.87	0.276	0.026
Medulla[d]	30	0.04–1.29	0.300	0.090
Pons[d]	30	0.04–1.66	0.341	0.091
Midbrain[d]	30	0.04–1.20	0.336	0.078
Corpus striatum[d]	30	0.03–0.61	0.135	0.038
Thalamus[d]	30	0.04–1.47	0.269	0.098
Thalamus[c]	30	0.04–0.11	0.074	0.006
Callosal white matter[d]	30	0.02–0.15	0.072	0.011
Callosal white matter[e]	30	0.04–0.11	0.070	0.010
Gyri[d]	30	0.02–0.15	0.079	0.016
Gyri[b]	30	0.04–0.11	0.077	0.013
RBC[f]	50	0.003–0.017	0.008	$\geqslant$0.001
Plasma[f] + RBC Washings	50	0.002–0.011	0.007	$\geqslant$0.001
Urine	50	0.003–0.017	0.008	$\geqslant$0.001

[a] Instrumental Neutron Activation Analysis by the method of Pillay et al.[87] Tissue samples (brain and muscle) were obtained from cadavers ranging in age from neonates to 91 years and from apparently healthy volunteers (blood and urine) ranging in age from 16 to 34 years. None of the samples was exposed to fixatives. All individuals had lived within a 60-mile radius of Buffalo, NY, for at least 9 months of the year for 5 years prior to tissue harvesting. Eighteen brain samples (unspecified as to region) from children ranging in age from neonates to 4 months were obtained from the Eleonorenstiftung Universitäts-Kinderklinik, Kinderspital, Zürich, Switzerland, through the courtesy of Dr. Edmond Werder. They had been sealed individually in a polyethylene bag, frozen on dry ice, and shipped by air.

[b] Frontal cortex, cerebellar cortex, and gastrocnemius muscle obtained from each of 100 adult (15–91 years) cadavers.

[c] Frontal cortex and cerebellar cortex obtained from each of 50 subadult (0–14 years) cadavers.

[d] Medulla, pons, midbrain, corpus striatum, thalamus, callosal white matter, and angular and supramarginal gyri from each of 30 adult cadavers selected at random (random numbers table) from the original 100.

[e] Thalamus, callosal white matter, and angular and supramarginal gyri from each of 30 subadult cadavers selected at random from the original 50.

[f] Blood was collected in heparin vacutainer tubes. The RBCs and plasma were separated and the cells washed three times (with 0.013 *M* NaCl). The plasma and combined washings were made 10% in bovine serum albumin (final pH 6.9–7.1) and lyophilized prior to activation analysis. Spiked ($^{203}HgCl_2$ or $Me^{203}HgCl$) samples of plasma and washings (+BSA) exhibited no significant loss of radioactivity upon lyophilization.

[g] Data were not normally distributed and were normalized (logarithmic transformation) prior to statistical analysis. Where appropriate, the data were analyzed by analysis of variance (simple randomized design for unequal Ns), factor analysis, and t-test methods.

[h] SEM = standard error of the mean.

Note: The statistical analysis revealed:

1. No significant differences ($p > 0.05$) in mercury levels among the members of the following groupings:
   a. Adult (15–91 years) and subadult (0–14 years) frontal and cerebellar cortex, medulla, pons, and midbrain and adult thalamus.
   b. Subadult thalamus, adult and subadult callosal white matter, and angular and supramarginal gyri.
   c. The RBCs, plasma, and urine obtained from apparently healthy volunteers. Also, there was no correlation among the mercury levels of the RBCs, plasma, and urine of any single individual.
2. A significantly higher ($p < 0.001$) mercury level in the tissues listed in 1.a above compared to the tissues/body fluids of 1.b and 1.c.
3. A significantly higher ($p < 0.001$) mercury level in the tissues listed in 1.b compared to tissues/body fluids of 1.c.

TABLE 1 (continued)

Human Tissue/Body Fluid Mercury Levels[a]

4. A significantly higher mercury level in the adult gastrocnemius muscle and corpus striatum compared to the tissue/body fluids listed in 1.b ($p < 0.01$) and 1.c. ($p < 0.001$), but a significantly lower mercury level ($p < 0.01$) in muscle and striatum compared to the tissues listed in 1.a.
5. A significantly positive effect of age on thalamic mercury levels (adult > subadult; $p < 0.01$), but, in general, no influence of age, weight, height, sex, race, national origin, occupation, religion, cause of death, or tissue handling parameters on mercury levels.

Source: From Massaro, E. J. et al., *Life Sci.,* 14, 1939–1948, 1974. With permission.

TABLE 2

Overall Means, Ranges, and Standard Errors (SE) by Site of Tissue (μg Hg/g Wet Tissue)[a]

Tissue	Mean	Sample Size	Range	± 2 SE	Standard Deviation	Mean ± 2 SE
Kidney	0.757	95	0.006 → 6.40 = 6.394	± 0.246	1.1979	0.511, 1.003
Liver	0.250	95	0.008 → 1.43 = 1.422	± 0.059	0.2871	0.191, 0.309
Heart	0.102	57	0.012 → 0.704 = 0.692	± 0.036	0.1350	0.066, 0.138
Muscle	0.126	69	0.0 → 0.954 = 0.954	± 0.064	0.2640	0.062, 0.190
Lung	0.251	77	0.0 → 2.763 = 2.763	± 0.104	0.4564	0.147, 0.355
Spleen	0.122	41	0.012 → 1.205 = 1.193	± 0.080	0.2570	0.042, 0.202
Pancreas	0.065	32	0.007 → 0.511 = 0.504	± 0.016	0.0930	0.049, 0.081
Cerebellum	0.132	60	0.006 → 0.965 = 0.959	± 0.050	0.1937	0.082, 0.182
Cerebrum	0.081	61	0.008 → 0.470 = 0.462	± 0.027	0.1050	0.054, 0.108
Cord	0.087	59	0.010 → 0.803 = 0.793	± 0.036	0.1399	0.051, 0.123
Skin	0.193	60	0.004 → 1.275 = 1.271	± 0.041	0.3152	0.152, 0.234

[a] Human autopsy tissues were stored at –20°C in sealed mercury-free plastic Petri dishes for approximately one month prior to analysis by flameless atomic absorption spectroscopy.

Source: From Mottet, N. K. and Bõdy, R. L., *Arch. Environ. Health,* 29, 18–24, 1974. With permission.

TABLE 3

Means, Standard Errors (SE), and Range for Kidney and Liver by Age (μg Hg/g Wet Tissue)[a]

	Sample Size		Means		Range		± 2 SE		Mean ± 2 SE	
Age (yr)	Kidney	Liver	Kidney	Liver	Kidney	Liver	Kidney	Liver	Kidney	Liver
Premature	17	17	0.218	0.176	0.699	0.452	0.093	0.068	0.125, 0.311	0.108, 0.244
0–10	10	10	0.508	0.230	1.916	0.876	0.370	0.173	0.138, 0.878	0.057, 0.403
11–20	6	6	0.175	0.094	0.560	0.205	0.167	0.068	0.008, 0.342	0.026, 0.162
21–30	7	7	0.164	0.141	0.249	0.293	0.071	0.089	0.093, 0.235	0.052, 0.230
31–40	7	7	1.274	0.329	2.958	1.082	0.903	0.307	0.371, 2.177	0.022, 0.636
41–50	9	9	1.469	0.424	6.374	1.149	1.327	0.280	0.142, 2.796	0.144, 0.704
51–60	19	19	0.933	0.222	4.791	0.489	0.631	0.072	0.302, 1.564	0.150, 0.294
61–70	12	12	1.115	0.341	6.271	1.073	1.015	0.221	0.100, 2.130	0.120, 0.562
71–80	2	2	0.681	0.233	1.109	0.207	—	—	—, —	—, —
81–90	5	5	1.215	0.401	1.796	1.333	0.644	0.516	0.571, 1.859	–0.115, 0.917
Totals	94	94								

[a] Human autopsy tissues were stored at –20°C in sealed mercury-free plastic Petri dishes for approximately one month prior to analysis by flameless atomic absorption spectroscopy.

Source: From Mottet, N. K. and Bõdy, R. L., *Arch. Environ. Health,* 29, 18–24, 1974. With permission

TABLE 4

Total Mercury and Methylmercury Concentration in Japanese Human Tissues[a,b]

	Total Mercury						Methylmercury					
Organ or Part	Sex	No.	Average	Range	Mean ± SD	Median	Sex	No.	Average	Range	Mean ± SD	Median
Cerebrum	M	10	0.11	0.039–0.17	0.10 ± 0.042	0.097	M	9	0.022	0.0015–0.069	0.016 ± 0.014	0.012
	F	10	0.099				F	11	0.010			
Cerebellum	M	10	0.11	0.048–0.23	0.10 ± 0.045	0.093	M	9	0.028	0.0015–0.096	0.019 ± 0.020	0.014
	F	11	0.089				F	11	0.012			
Trachea	M	12	0.036	0.015–0.11	0.047 ± 0.029	0.036		c	d	d	c	c
	F	4	0.079									
Lung	M	15	0.081	0.015–0.30	0.080 ± 0.054	0.070	M	5	0.0083	0.0023–0.015	0.0065 ± 0.0034	0.0060
	F	13	0.078				F	6	0.0050			
Heart	M	15	0.054	0.023–0.13	0.069 ± 0.028	0.069	M	7	0.011	0.0030–0.027	0.0092 ± 0.0066	0.0070
	F	14	0.085				F	6	0.0067			
Liver	M	15	0.42	0.16–1.3	0.47 ± 0.25	0.42	M	15	0.058	0.012–0.080	0.044 ± 0.019	0.042
	F	15	0.52				F	15	0.041			
Pancreas	M	15	0.077	0.023–0.29	0.083 ± 0.048	0.077	M	14	0.010	0.0013–0.033	0.010 ± 0.0078	0.0083
	F	15	0.09				F	15	0.010			
Spleen	M	13	0.073	0.021–0.14	0.068 ± 0.028	0.062		c	d	d	c	c
	F	15	0.064									
Kidney	M	15	0.97	0.18–2.6	1.1 ± 0.67	0.98	M	15	0.029	0.010–0.080	0.023 ± 0.015	0.019
	F	15	1.24				F	14	0.018			
Adrenal gland	M	12	0.12	0.03–0.33	0.14 ± 0.073	0.15		c	d	d	c	c
	F	12	0.16									
Small intestine	M	12	0.057	0.024–0.19	0.069 ± 0.037	0.064	M	12	0.016	0.0030–0.069	0.014 ± 0.017	0.0082
	F	13	0.08				F	12	0.012			
Large intestine	M	14	0.078	0.032–0.16	0.083 ± 0.037	0.075	M	9	0.0086	0.0018–0.026	0.0065 ± 0.0061	0.0044
	F	13	0.09				F	11	0.0047			
Testicles	M	14	0.067	0.029–0.12	0.067 ± 0.029	0.070		c	d	d	c	c
Ovary	F	14	0.069	0.028–0.13	0.069 ± 0.029	0.070						

Muscle	M	13	0.056	0.018–0.15	0.060 ± 0.027	0.057	M	6	0.0090	0.0041–0.019	0.0078 ± 0.0043	0.0064
	F	14	0.064				F	6	0.0065			
Skin	M	15	0.051	0.017–0.15	0.059 ± 0.034	0.048		[c]	[d]	[d]	[c]	[c]
	F	12	0.066									
Blood	M	9	0.054	0.016–0.11	0.059 ± 0.026	0.058	M	6	0.012	0.0036–0.026	0.011 ± 0.073	0.0092
	F	10	0.064				F	6	0.010			
Hair	M	14	5.4	1.4–15.0	4.1 ± 2.6	3.4	M	14	3.4	0.63–10.4	2.6 ± 2.1	2.0
	F	15	3.0				F	14	1.8			

Age and Sex of Subjects

Group	Age (years)	Male	Female
—	0–9	—	—
Young	10–19	2	—
	20–29	2	4
Total		**4**	**4**
Middle	30–39	4	6
	40–49	3	1
Total		**7**	**7**
Old	50–59	1	3
	Over 60	3	1
Total		**4**	**4**

[a] Expressed as micrograms per gram wet tissue.

[b] Human autopsy tissues were obtained from subjects that had lived in Hyogo Prefecture (central Japan). Death was caused by traumatic cerebral hemorrhage (4 individuals), fractured skull and brain injury (4 individuals), loss of blood (12 individuals), sleeping-pill intoxication (2 individuals), suffocation (6 individuals), CO intoxication (2 individuals). The tissues were rinsed free of blood with distilled water and stored in polyethylene bags or glass bottles below –10 °C until analyzed by flameless atomic absorption spectroscopy (mercury) and gas chromatography (methylmercury).

[c] Not calculated because there were less than five samples available or there was no mean (testicles and ovary).

[d] Not measured.

Source: From Sumino, K. et al., *Arch. Environ. Health,* 30, 487–494, 1975. With permission.

TABLE 5

Concentration (μg/g wet weight) of Mercury in Japanese Human Tissues[a]

Organ/Tissue	Sex	Mercury				
		$\overline{X}$[b]	Standard Deviation[c]	Median	Range	Number[c]
Muscle	M	0.77	1.8	—[d]	≤4.4	6
	F	—		—	—	4
	Total	0.46	1.4	—	≤4.4	10
Pancreas	M	0.08	0.17	—	≤0.45	7
	F	0.27	0.47	—	≤0.81	3
	Total	0.14	0.28	—	≤0.81	10
Spleen	M	0.71	0.86	0.66	≤2.1	9
	F	0.43	0.72	0.12	≤1.5	4
	Total	0.63	0.80	0.23	≤2.1	13
Lung	M	0.87	0.98	0.32	0.28–2.0	3
	F	0.03	0.04	0.03	≤0.06	2
	Total	0.53	0.83	0.28	≤2.0	5
Aorta	M	0.01	0.02	—	≤0.04	3
Liver	M	0.60	0.70	0.48	≤1.7	5
	F	2.3	2.4	1.0	≤6.0	7
	Total	1.6	2.0	0.78	≤6.0	12
Kidney	M	2.0	2.4	0.97	≤6.6	20
	F	2.4	1.3	2.6	0.98–3.6	3
	Total	2.1	2.3	1.1	≤6.6	23
Cerebrum	M	1.9	2.6	0.71	≤7.4	8
	F	1.2	1.5	0.80	≤3.1	4
	Total	1.9	2.4	0.71	≤7.4	12
Cerebellum	M	3.3	3.3	2.7	≤9.3	6
	F	2.2	0.9	2.3	1.1–3.0	4
	Total	2.9	2.6	2.9	≤9.3	10
Heart	M	0.29	0.36	0.20	≤1.1	10
	F	0.26	0.38	0.12	≤0.81	4
	Total	0.28	0.35	0.20	≤1.1	14

Age and Sex of Individual Subjects

Age Group (yr)	Male Subjects' Age[e]							Female Subjects' Age				
0–9	7	5	3	4 m[f]	15 d[g]			1	7 m	15 m		
10–19	18	17	12	11					18	15	13	
20–29	29	28	27	26	22	21		22	20			
30–39	39	37	35	35	35	31	30	30				
40–49	47	40	43	47				46	41	40		
50–59	58	56	55	51				59	59	56	51	51
60–69	69	65	61					68				
70–79	76	75	72	70	76	70		76				
80–89	85	80						80				
Total	**41 specimens**							**22 specimens**				

[a] Human autopsy tissues were obtained from subjects undergoing forensic medical examination in medical centers in Chiba and Tokyo. Causes of death were pulmonary tuberculosis, subarachnoid hemorrhage, cerebral hemorrhage, cerebral tumor, myocarditis, myocardial infarction, acute heart failure, chronic hypertensional heart failure, *Pseudomonas pycocyanea* intoxication, chronic purulent bronchiectasis and drowning. The tissues were stored in polyethylene bags or glass bottles below −10°C until analyzed by neutron activation analysis and semiconductor gamma-ray spectrometry. For long-term irradiation, the samples were partially thawed and freeze-dried for 24 hours. For short-term irradiation, the frozen samples were ground to powder in a clean agate mortar.

TABLE 5 (continued)

Concentration (μg/g wet weight) of Mercury in Japanese Human Tissues[a]

[b] $\overline{X}$ = mean.
[c] Number = number of specimens.
[d] — indicates below detection limit.
[e] All numerical values represent age in years unless otherwise indicated.
[f] m = month.
[g] d = day.

Source: From Yukawa, M. et al., *Arch. Environ. Health,* 35, 36–44, 1980. With permission.

TABLE 6

Review of the Literature on Organ and Tissue Mercury Concentrations Prior to 1974

Year, Locality	No. of Cases	Method	Average Burden (μg/g wet weight)[a]		Other Organs Examined	Ref.
			Kidney	Liver		
1940, Berlin	6	Micrometric	0.62	.041	27 tissues	114
1949, Indianapolis	92	Diphenylthiocarbazone	1.0	<.2	None	35
1950, Los Angeles	69	Diphenylthiocarbazone	.75	.06	None	10
1954, Los Angles	15	Spectrography	4.1[b]	.74[b]	Spleen	41
1964, Stockholm	12	Neutron activation	—	.02	None	103
1964, New York	11	Spectrography	7.5	—	4 cases of brain, heart, lung, spleen	21
1966, Minamata, Japan	10	Diphenylthiocarbazone	1.19	.57	Brain	59
1967, Glasgow, Scotland	22	Neutron activation	1.89[b]	1.06[b]	3 to 26 tissues from various cases	49
1967, New York	39	Diphenylthiocarbazone	2.75	.30	10 other tissues	54
1971, Buffalo, NY	8	Neutron activation	—	—	Brain only	39
1972, Detroit	59	Atomic absorption	8.8 (highly variable)	3.0 (highly variable)	14 tissues	56
1974, Seattle	113	Atomic absorption	.83	.26	12 tissues	76

[a] Various authors reported results in different units. All have been converted to μg/g wet weight.
[b] Dry weight was converted to wet weight by multiplying kidney mean by 0.21 and liver by 0.29.

Source: From Mottet, N. K. and Bōdy, R. L., *Arch. Environ. Health,* 29, 18–24, 1974. With permission.

resulting from exposure to methylmercury through the consumption of contaminated fish, exhibited no symptoms of methylmercury intoxication.[36] In the apparently asymptomatic population studied by Massaro et al.,[67] RBC total mercury ranged from 0.003 to 0.017 μg/g, while plasma levels ranged from 0.002 to 0.011 μg/g (Table 1). The concentration of mercury in hair at onset of symptoms of methylmercury intoxication has been calculated to be ≥200 μg/g and corresponds to whole blood levels ≤0.7 μHg per g.[71]

The tissue/organ distribution of methylmercury is characteristically different from that of Hg^{+2} (Figure 4).[7-9,116] The mammalian brain appears to be the primary target of methylmercury intoxication.[118] It is relatively impermeable to Hg^{+2}, which accumulates primarily in the kidney.[6,17,36,45] Although the brain takes up methylmercury at a slower rate than other tissues/organs, it also releases mercury at a slower rate than other tissues/organs. It has been reported that, following a single dose of $Hg(NO_3)_2$, the kidney of the rat rapidly (hours)

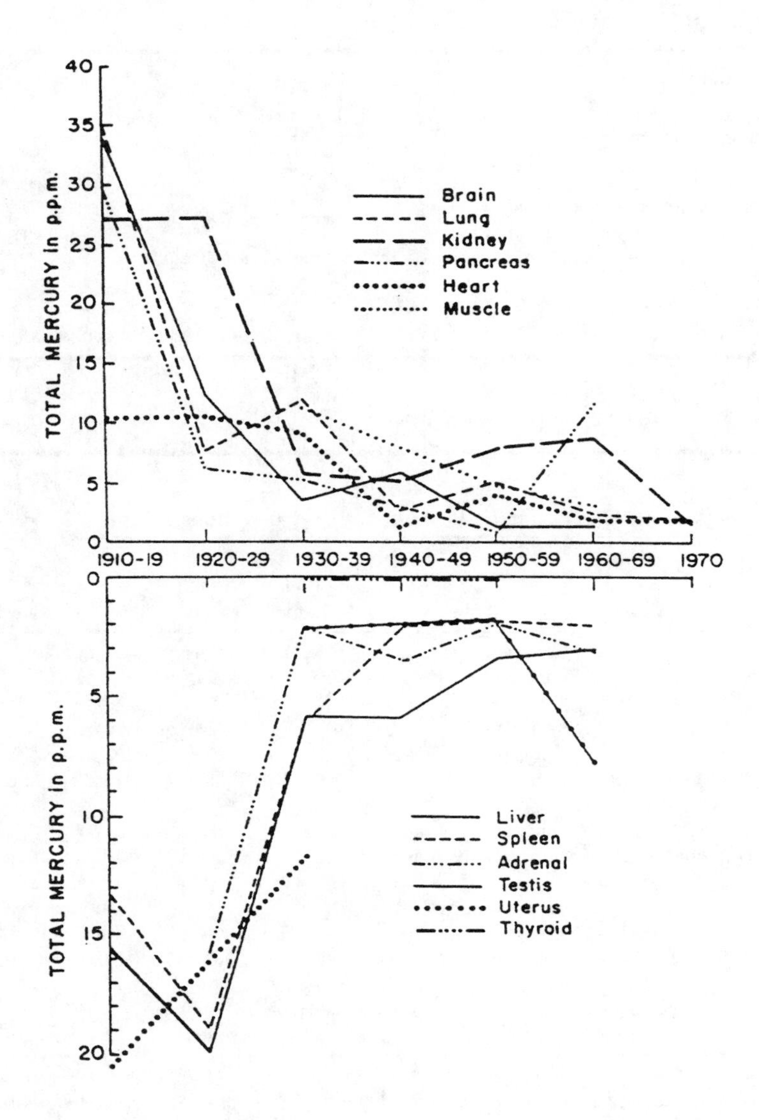

FIGURE 1
Total mercury content of various human tissues averaged for each organ according to decade of autopsy. Tissues were obtained at autopsy from 59 cases (death due to various causes), fixed initially in 10% formalin, and stored in 95% ethyl alcohol. Samples were blotted dry *en bloc* and wrapped in gauze or absorbent paper for several days or weeks prior to analysis. The samples were analyzed by flameless atomic absorption spectrometry. The samples analyzed were devoid of adventitia or supportive tissues and large blood vessels. (From Kevorkian, J. et al., *Am. J. Public Health,* 62, 504–513, 1972. With permission.)

accumulated mercury to a level 300-fold greater than that of the blood, while the brain level increased 10-fold.[120] Because the kidney retains mercury longer than other tissues, its proportion of the mercury body burden increases as a function of time.[102]

Åberg et al.[1] estimated the lowest level of methylmercury in the brain capable of inducing onset of symptoms of intoxication to be 6 μg/g. Twelve victims of the Minimata disaster were reported to have brain concentrations ranging from 2.6 to 24 μg Hg per g.[123] "Normal" brain mercury levels (neutron activation analysis) in the eastern U.S. based on autopsy samples from seven individuals (age range: 33 to 79 years) ranged from 0.02 to 2.0 μg/g wet weight.[39] The range for a full-term stillborn was 0.04 to 0.05 ng Hg per g. Analysis of similar brain areas (Table 1) from a larger sampling (30 to 100 individuals ranging in age from neonates

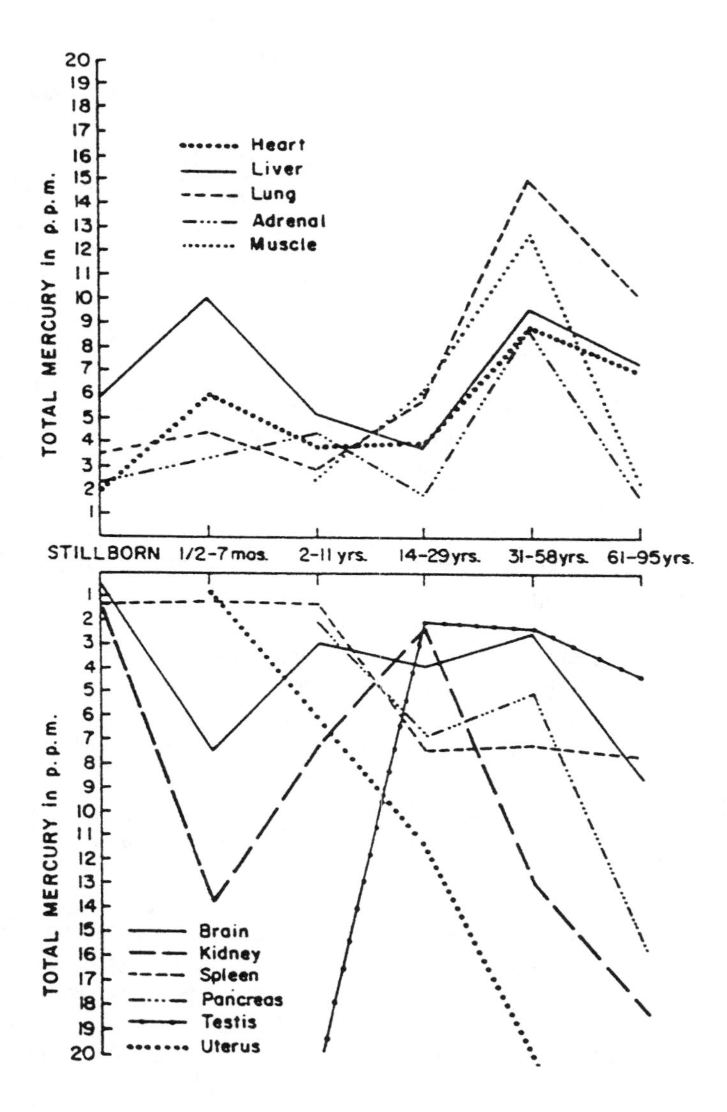

FIGURE 2
Total mercury content in various human tissues averaged for each tissue and organ according to individual age groups. Tissue handling and analysis as in Figure 1. (From Kevorkian, J. et al., *Am. J. Public Health,* 62, 504–513, 1972. With permission.)

to 91 years) from the same study area revealed a range 0.02 to 2.59 μg Hg per g.[67] A similar range has been reported more recently by Ehmann et al.[29]

Methylmercury is a potent developmental toxicant. It crosses the placental barrier and accumulates in the conceptus;[9,15,38,45,55,66,83,87,88,101,117,119] thus, human neonates born to mothers exposed to methylmercury through consumption of contaminated fish or grain were found to have higher RBC mercury levels than their mothers.[117,125] Furthermore, numerous reports indicate developmental toxicity in the absence of apparent maternal effects.[3,33,45,71,129]

Methylmercury crosses the placenta and accumulates in the tissues of the fetal mouse, rat, and macaque (*Macaca mulatta*).[3,7-9,15,33,38,45,66,71,129] In these studies, methylmercury was administered to the pregnant female once or twice during gestation or, in the case of the macaque, directly to the fetus via catherization of the umbilical vein.[93] Analysis of the kinetics of fetal methylmercury uptake and mercury elimination via placental transfer in experimental animals revealed that:

- In the mouse, at a dietary methylmercury content above 0.05 ppm, fetal mercury sequestration increases in direct proportion to the content of methylmercury in the diet.[13] In addition, fetal central nervous system (CNS) mercury levels increased in proportion to maternal dose and were higher than maternal CNS levels.[83]
- In the rat, the net rate of methylmercury transfer from the maternal circulation to fetal tissues during late gestation increases as a function of gestational age at the time of administration.[66]
- In the monkey, the rate of methylmercury transfer from the maternal circulation to the fetal circulation is greater than the reverse transfer rate at equal concentration gradients.[101]

Unfortunately, none of the times selected for administration coincided with the period of organogenesis. However, fetal sensitivity to methylmercury-induced neurological damage appears to be related to enhanced CNS sequestration of the toxicant compared to maternal CNS tissue. Also, the basic tenet of cancer chemotherapeutics, that cycling cells are more sensitive to toxic insult than vegetative cells, probably also applies to the developing embryo/fetus.

To investigate fetal uptake of methylmercury during organogenesis, Olson and Massaro[87,88] exposed pregnant Swiss-Webster CFW mice to a single s.c. dose of methylmercury (5 ppm mercury: 5 mg Hg per kg maternal body weight) in phosphate-buffered saline (PBS: 0.13 M NaCl, 0.01 M NaH_2PO_4-Na_2HPO_4, pH 7.4) on days 7, 8, 9, 10, 11, 12, or 13 of gestation. Fetal mercury accumulation increased with fetal age at time of administration at least up to day 13 of gestation. Methylmercury exposure on days 10 to 13 of gestation produced a fetal mercury concentration higher than that of maternal blood concentrations 7 days after administration on days 10 or 11 of gestation and 5 days after administration on days 12 or 13 of gestation. Fetal mercury concentration peaked 3 days after administration, averaging 4.8 ppm in animals injected on day 13 of gestation. Placental mercury concentration was always equal to or higher than that of maternal blood. An increase in maternal net mercury elimination rate (1.3 and 3.4 μg Hg per day following administration on day 7 or 13 of gestation, respectively) was associated with increased fetal sequestration of mercury as a function of fetal age.

To determine mercury uptake via suckling, offspring of dams administered 5 ppm methylmercury on day 12 of gestation and control dams were cross-fostered. The mercury concentrations of the tissues of methylmercury × methylmercury crosslings were essentially identical to those of the dams at 10 and 21 days postpartum. The mercury concentration of the tissues of methylmercury × PBS crosslings (*in utero* mercury exposure only) were 15 to 30 times greater than that of PBS × methylmercury crosslings (suckling exposure only) indicative of greater mercury transfer *in utero.*

2 SUBCELLULAR DISTRIBUTION

Consistent with their large thiol binding constant, mercurials tend to be present in all subcellular fractions (i.e., nuclear, mitochondrial, lysosomal, soluble) prepared by centrifugation;[5,6,43,65,70,81,89,122,134,137] however, as a function of time, mercury accumulates in the lysosomal/peroxisomal fraction (rat tissues) to a greater extent following exposure to a single dose of $HgCl_2$ compared to methylmercury.[81] Winroth et al.[134] observed that at 3 weeks post-administration of a single subtoxic dose of $Me^{203}HgOH$ (1 to 3 mg/kg body weight), 75% of the radioactivity in squirrel monkey (*Saimiri sciureus*) brain was associated with the particulate fraction obtained by ultracentrifugation. In the soluble fraction, 9% of the total radioactivity was associated with glutathione and the remainder with high-molecular-weight compounds.

Only trace amounts of mercury were found associated with low-molecular-weight compounds (M.W. = 100 to 300) in the kidney of rats administered $HgCl_2$ or ethylmercury.[20,52]

In equilibrium dialysis experiments, Clarkson and Magos[20] reported that over 99% of the kidney mercury of rats administered $HgCl_2$ was present in nondiffusible form. It is generally accepted that $HgCl_2$ perturbs proximal tubule water channel function.[94] Apparently, mercurials inhibit water flow in the proximal tubule by interfering with the function of vasopressin-sensitive water channels.[48] Furthermore, it has been reported that mercurials potentiate the angiotensin II-induced increase in $[Ca^{2+}]_i$ (the second messenger for angiotensin II) in proximal tubule cells and that this may be mediated through phospholipase C activation.[32]

Following exposure to methylmercury, it has been reported that mercury is present almost completely in the protein fraction of rat brain.[43,137] Only a small percentage of the total mercury was found in the lipid and nucleic acid fractions.

Mercurials have been shown to interact with phospholipid monolayers and model membranes.[23,97] Hg^{2+} interacts specifically with phosphatidylserine (PS) and phosphatidylcholine (PC), phospholipids that contain a primary amine in their polar head group.[24-27]

Mercurials also form complexes with purine and pyrimidine bases, nucleosides, nucleotides, and nucleic acids[11,12,37] and are mutagenic.[77,98,99,108]

REFERENCES

1. Åberg, B., Ekman, L., Falk, R., Greitz, U., Persson, G., and Snihs, J.-O., Methylation of methyl mercury (^{203}Hg) compounds in man, excretion and distribution, *Arch. Environ. Health*, 19, 478–484, 1969.
2. Avtsyn, A. P., An insufficiency of essential trace elements and its manifestations in pathology (in Russian), *Arkh. Patol.*, 52, 3–8, 1990.
3. Bakir, F., Damluji, S. F., Amin-Zaki, L., Murtadha, M., Khalidi, A., Ai-Rawi, N. Y., Tikriti, S., Dhahir, H. I., Clarkson, T. W., Smith, J. C., and Doherty, R. A., Methyl mercury poisoning in Iraq, *Science*, 181, 230–241, 1973.
4. Bellinger, D., Leviton, A., Waternaux, C., Needleman, H., and Rabinowitz, M., Low-level lead exposure, social class, and infant development, *Neurotoxicol. Teratol.*, 10, 497–503, 1988.
5. Berlin, M., Blomstrand, C., Grant, C. A., Hamberger, A., and Trofast, J., Tritiated methyl mercury in the brain of squirrel monkeys. Cellular and subcellular distribution, *Arch. Environ. Health*, 30, 591–597, 1975.
6. Berlin, M., Nordberg, G., and Hellberg, J., The uptake and distribution of methylmercury in the brain of *Saimiri sciureus* in relation to behavior and morphological changes, in *Mercurials and Mercaptans*, Miller, M. W. and Clarkson, T. W., Eds., Charles C Thomas, Springfield, IL, 1973, pp. 187–208.
7. Berlin, M. and Ulberg, S., Accumulation and retention of mercury in the mouse. I. An autoradiographic study after a single intravenous injection of mercuric chloride, *Arch. Environ. Health*, 6, 589–601, 1963.
8. Berlin, M. and Ulberg, S., Accumulation and retention of mercury in the mouse. II. An autoradiographic comparison of phenylmercuric acetate with inorganic mercury, *Arch. Environ. Health*, 6, 602–609, 1963.
9. Berlin, M. and Ulberg, S., Accumulation and retention of mercury in the mouse. III. An autoradiographic comparison of methylmercuric dicyandiamide with inorganic mercury, *Arch. Environ. Health*, 6, 610–616, 1963.
10. Butt, E. M. and Simonsen, D. G., Mercury and lead storage in human tissues, *Am. J. Clin. Pathol.*, 20, 716–723, 1950.
11. Caldwell, J., Gajanayake, I., Caldwell, P., and Peiris, T., Sensitization to illness and the risk of death: an explanation for Sri Lanka's approach to good health for all, *Soc. Sci. Med.*, 28, 365–379, 1989.
12. Carrabine, J. A. and Sundaralingman, M., Mercury binding to nucleic acids. Crystal and molecular structures of 2:1 complexes of uracil-mercuric chloride and dihydrouracil-mercuric chloride, *Biochem.*, 10, 292–299, 1971.
13. Carty, A. J. and Malone, S. F., The chemistry of mercury in biological systems, in *The Biogeochemistry of Mercury in the Environment*, Nriagu, J. O., Ed., Elsevier/North Holland, Amsterdam, 433–479, 1979.
14. Chaudhary, K., Ehmann, W. D., Rengan, K., and Markesbery, W. R., Trace element correlations between human brain and fingernails, *J. Trace Microprobe Technol.*, 10, 225–235, 1992.
15. Childs, E. A., Kinetics of transplacental movement of mercury fed in tuna matrix to mice, *Arch. Environ. Health*, 27, 50–53, 1973.
16. Cislaghi, C., Decarli, A., La Vecchia, C., Mezzanotte, G., and Vigotti, M. A., Trends surface models applied to the analysis of geographical variations in cancer mortality, *Rev. Epidemol. Sante Publ.*, 38, 57–69, 1990.
17. Clarkson, T. W., The pharmacology of mercury compounds, *Ann. Rev. Pharmacol.*, 12, 375–406, 1972.
18. Clarkson, T. W., Mercury: major issues in environmental health, *Environ. Health Perspect.*, 100, 31–38, 1993.

19. Clarkson, T. W., Hursch, J. B., and Nylander, M., The prediction of intake of mercury vapor from amalgams, in *Biological Monitoring of Toxic Metals,* Clarkson, T. W., Friberg, L., Nordberg, G. F., and Sager, P. R., Eds., Plenum Press, New York, 1988, pp. 247–264.
20. Clarkson, T. W. and Magos, L., Studies on the binding of mercury in tissue homogenates, *Biochem. J.,* 99, 62–70, 1966.
21. Dal Corvito, L. A., Weinberg, S. B., Giaquinta, P., and Jacobs, M. B., Mercury levels in normal human tissues, *J. Forensic Sci. Soc.,* 9, 501–510, 1964.
22. Davis, G. K., Microelement interactions of zinc, copper, and iron in mammalian species, in *Micronutrient Interactions: Vitamins, Minerals and Hazardous Elements,* Vol. 355, Levander, O. and Cheng, L., Eds., Annals of the New York Academy of Sciences, New York, 1980, pp. 130–139.
23. Delnomdedieu, M. and Allis, J. W., Interaction of inorganic mercury salts with model and red cell membranes: importance of lipid binding sites, *Chem.-Biol. Interact.,* 88, 71–87, 1993.
24. Delnomdedieu, M., Boudou, A., Desmazès, J. P., and Georgescauld, D., Interaction of mercury chloride with the primary amine group of model membranes containing phosphatidylserine and phosphatidylethanolamine, *Biochim. Biophys. Acta,* 986, 191–199, 1989.
25. Delnomdedieu, M., Boudou, A., Desmazès, J. P., Faucon, J. F., and Georgescauld, D., Inorganic mercury-phospholipidic membrane interactions: fundamental role of headgroups bearing a primary amine. A fluorescence polarization study, in *Heavy Metals in the Environment,* Vol. 1, Vernet, J. P., Ed., CEP Consultants, Edinburgh, 1989, 578–581.
26. Delnomdedieu, M., Boudou, A., Georgescauld, D., and Dufourc, E. J., Specific interactions of mercury chloride with membranes and other ligands as revealed by mercury-NMR, *Chem. Biol. Interact.,* 81, 243–269, 1992.
27. Delnomdedieu, M., Georgescauld, D., Boudou, A., and Dufourc, E. J., Mercury-199 NMR: a tool to follow chemical speciation of mercury compounds, *Bull. Magn. Reson.,* 11, 420, 1990.
28. Eggleston, D. W. and Nylander, M., Correlation of dental amalgam with mercury in brain tissue, *J. Prosthet. Dent.,* 58, 704–707, 1987.
29. Ehmann, W. D., Kasarskis, E. J., and Markesbery, W. R., Mercury imbalances in patients with neurodegenerative diseases, in *Mercury Pollution: Integration and Synthesis,* Watras, C. J. and Huckabee, J. W., Eds., Lewis Publishers, Boca Raton, FL, 1994, pp. 651–663.
30. Ehmann, W. D., Markesbery, W. R., Hossain, T. I. M., Aluddin, M., and Goodin, D. T., Trace elements in human brain tissue by INAA, *J. Radioanal. Chem.,* 70, 57–65, 1982.
31. Elinder, C. G., Normal values for cadmium in human tissues, blood, and urine in different countries, in *Cadmium and Health: A Toxicological and Epidemiological Appraisal,* Vol. 1, Freiberg, L., Elinder, C. G., Kjellström, T., and Nordberg, G. F., Eds., CRC Press, Boca Raton, FL, 1985, pp. 81–102.
32. Endou, H. and Jung, K. Y., Effect of mercuric chloride on angiotensin II-induced Ca^{2+} transient in the proximal tubule of rats, in *Advances in Mercury Toxicology,* Suzuki, T., Imura, N., and Clarkson, T. W., Eds., Plenum Press, New York, 1991, pp. 299–314.
33. Engelson, G. and Herner, T., Alkyl mercury poisoning, *Acta Paediat. Scand.,* 41, 289–294, 1952.
34. Fleming, L. E., Watkins, S., Bean, J. A., Stephens, D., and Hubbard, C., *Health Study to Assess the Human Health Effects of Mercury Exposure to Fish Consumed from the Everglades, Final Report,* Florida State Department of Health and Rehabilitative Services, Tallahassee, FL, 1995.
35. Forney, R. B. and Harger, R. N., Mercury content of human tissues from routine autopsy material, *Fed. Proc.,* 8, 292, 1949.
36. Friberg, L. and Vostal, J., Eds., *Mercury in the Environment,* CRC Press, Boca Raton, FL, 1972.
37. Friberg, L., Piscator, M., and Nordberg, G., Eds., *Cadmium in the Environment,* CRC Press, Boca Raton, FL, 1971.
38. Garcia, J. D., Yang, M. N., Wang, J. H. C., and Belo, P. S., Translocation and fluxes of mercury in neonatal and maternal rats treated with methyl mercuric chloride during gestation, *Proc. Soc. Exp. Biol. Med.,* 147, 224–231, 1974.
39. Glomski, C. A., Brody, H., and Pillay, K. K. S., Distribution and concentration of mercury in autopsy specimens of human brain, *Nature,* 232, 200–201, 1971.
40. Goering, P. L., Galloway, W. D., Clarkson, T. W., Lorscheider, F. L., Berlin, M., and Rowland, A. S., Toxicity assessment of mercury vapor from dental amalgams, *Fundam. Appl. Toxicol.,* 19, 319–329, 1992.
41. Griffith, G. C., Butt, E. M., and Walker, J., The inorganic element content of certain human tissues, *Ann. Intern. Med.,* 41, 501–509, 1954.
42. Gruenwedel, D. W. and Davidson, N., Complexing and denaturation of DNA by methylmercuric hydroxide. I. Spectrophotometric studies, *J. Mol. Biol.,* 21, 129–144, 1966.
43. Gruenwedel, D. W., Glaser, J. F., and Cruikshank, M. K., Binding of methylmercury by HeLa S3 suspension-culture cells: intracellular methylmercury levels and their effect on DNA replication and protein synthesis, *Chem. Biol. Interact.,* 36, 259–274, 1981.
44. Hahn, L. J., Kloiber, R., Vimy, M. J., Takahashi, Y., and Lorscheider, F. L., Dental "silver" tooth fillings: a source of mercury exposure revealed by whole-body image scan and tissue analysis, *FASEB J.,* 3, 2641–2646, 1989.

45. *Health Effects of Methylmercury,* MARC Report No. 24, Monitoring and Assessment Research Center, Chelsea College, University of London, 1981.
46. Hecky, R. E., Ramsey, D. J., Bodaly, R. A., and Strange, N. E., Increased methylmercury contamination in fish in newly formed freshwater reservoirs, in *Advances in Mercury Toxicology,* Suzuki, T., Imura, N., and Clarkson, T. W., Eds., Plenum Press, New York, 1991, pp. 33–52.
47. Hill, C. H. and Matrone, G., Chemical parameters in the study of *in vivo* and *in vitro* interaction of transition elements, *Fed. Proc.,* 29, 1474–1481, 1970.
48. Hoch, B. S., Gorfien, P. C., Linzer, D., Fusco, M. J., and Levine, S. D., Mercurial reagents inhibit flow through ADH-induced water channels in toad bladder, *Am. J. Physiol.,* 256, F948–953, 1989.
49. Howie, R. A. and Smith, H., Mercury in human tissue, *J. Forensic Sci. Soc.,* 7, 90–96, 1996.
50. Hu, H., Aro, A., Payton, M., Korrick, S., Sparrow, D., Weiss, S. T., and Rotnitzky, A., The relationship of bone and blood lead in hypertension, *J. Am. Med. Assoc.,* 275, 1171–1176, 1996.
51. Huel, G., Derriennic, F., Ducimetiere, P., and Lazar, P., Water hardness and cardiovascular mortality. Discussion of evidence from geographical pathology, [French], *Rev. Epidemol. Sante Publ.,* 26, 349–359, 1978.
52. Jakubowski, K., Piotrowski, J., and Trojanowska, B., Binding of mercury in the rat: studies using $^{203}HgCl_2$ and gel filtration, *Toxicol. Appl. Pharmacol.,* 16, 743–753, 1970.
53. Jocelyn, P. C., *Biochemistry of the SH Group,* Academic Press, New York, 1972.
54. Joselow, M. M., Goldwater, L. J., and Weinberg, S. B., Absorption and excretion of mercury in man. XI. Mercury of "normal" human tissues, *Arch. Environ. Health,* 15, 64–66, 1967.
55. Kelman, B. J. and Sasser, L. B., Methylmercury movements across the perfused guinea pig placenta in late gestation, *Toxicol. Appl. Pharmacol.,* 39, 119–127, 1977.
56. Kevorkian, J., Cento, D. P., Hyland, J. R., Bagozzi, W. M., and van Hollebebe, E., Mercury content of human tissues during the twentieth century, *Am. J. Public Health,* 62, 504–513, 1972.
57. Kim, R., Rotnitzsky, A., Sparrow, D., Weiss, S. T., Wagner, C., and Hu, H., A longitudinal study of low-level lead exposure and impairment of renal function *J. Am. Med. Assoc.,* 275, 1177–1181, 1996.
58. Kimura, K., Role of essential trace elements in the disturbance of carbohydrate mtabolism (in Japanese), *Nippon Rinsho.,* 54, 79–84, 1996.
59. Kitamura, S., Determination of mercury content in bodies of inhabitants, cats, fishes and shells in Minimata district and in mud of Minimata Bay, in *Minimata Disease,* Study Group of Minimata Disease, Kumamaoto University, Japan, 1968, pp. 257–266.
60. Klevay, L. M., Interaction of copper and zinc in cardiovascular disease, in *Micronutrient Interactions: Vitamins, Minerals and Hazardous Elements,* Vol. 355, Levander, O. and Cheng, L., Eds., Annals of the New York Academy of Sciences, New York, 1980, pp. 140–151.
61. Lange, T. R., Royals, H. E., and Conner, L. L., Mercury accumulation in largemouth bass (*Micropterus salmoides*) in a Florida lake, *Arch. Environ. Contam. Toxicol.,* 27, 466–471, 1994.
62. Lapenna, D., de Gioia, S., Mezzetti, A., Ciofani, G., Consoli, A., Marzio, L., and Cuccurullo, F., *Am. J. Respir. Crit. Care Med.,* 151, 431–435, 1995.
63. Liebscher, K. and Smith, H., Essential and nonessential trace elements, *Arch. Environ. Health,* 17, 881–890, 1968.
64. López-Artíguez, M., Cameán, A., González, G., and Repetto, M., Cadmium concentrations in human renal cortex tissue (necropsies), *Bull. Environ. Contam. Toxicol.,* 54, 841–847, 1995.
65. Madsen, K. M. and Hansen, J. C., Subcellular distribution of mercury in the rat kidney cortex after exposure to mercuric chloride, *Toxicol. Appl. Pharmacol.,* 54, 443–453, 1980.
66. Mansour, M. M., Dyer, N. C., Hoffman, L. H., Davies, J., and Brill, A. B., Placental transfer of mercuric nitrate and methylmercury in the rat, *Am. J. Obstet. Gynecol.,* 119, 557–562, 1974.
67. Massaro, E. J., Yaffe, S. J., and Thomas, Jr., C. C., Mercury levels in human brain, skeletal muscle and body fluids, *Life Sci.,* 14, 1939–1948, 1974.
68. Matsumoto, H., Koya, G., and Takeuchi, T., Fetal Minimata disease: a neuropathological study of two cases on intrauterine intoxication by methyl mercury compound, *J. Neuropath. Exp. Neurol.,* 24, 563–574, 1965.
69. Mayer, D. R., Kosmus, W., Pogglitsch, H., Mayer, D., and Beyer, W., Essential trace elements in humans. Serum arsenic concentrations in hemodialysis in comparison to healthy controls, *Biol. Trace Element Res.,* 37, 27–38, 1993.
70. Mehra, M. and Choi, B. H., Distribution of mercury in subcellular fractions of brain, liver, and kidney after repeated oral administration of ^{203}Hg-labeled methylmercuric chloride in mice, *Exp. Mol. Pathol.,* 35, 435–447, 1981.
71. *Methyl Mercury in Fish,* Report from an Expert Group, *Nord. Hyg. Tidskr.* (Suppl. 4), Stockholm, 1971.
72. Miller, G. D., Massaro, T. F., and Massaro, E. J., Interactions between lead and essential elements: a review, *Neurotoxicology,* 11, 99–120, 1990.
73. Miller, G. D., Massaro, T. F., Koperek, E., and Massaro, E. J., Low-level lead exposure and the time-dependent organ/tissue distribution of essential elements in the neonatal rat, *Biol. Trace. Elements Res.,* 6, 519–529, 1984.
74. Milner, J. A., Trace minerals in the nutrition of children, *J. Pediatr.,* 117, S147–155, 1990.
75. *Minimata Disease,* Study Group of Minimata Disease, Kumamaoto University, Japan, 1968.

76. Mottet, N. K. and Bõdy, R. L., Mercury burden of human autopsy organs and tissues, *Arch. Environ. Health,* 29, 18–24, 1974.
77. Mulvihill, J., Congenital and genetic disease in domestic animals, *Science,* 176, 132–137, 1972.
78. Muto, H., Shinada, M., Tokuta, K., and Takizawa, Y., Rapid changes in concentrations of essential elements in organs of rats exposed to methylmercury chloride and mercuric chloride as shown by simultaneous multielement analysis, *Br. J. Ind. Med.,* 48, 382–388, 1991.
79. Neve, J., Clinical implications of trace elements in endocrinology, *Biol. Trace Elem. Res.,* 32, 173–185, 1992.
80. Nielsen, F. H., New essential trace elements for the life sciences, *Biol. Trace Elem. Res.,* 26–27, 599–611, 1990.
81. Norseth, T., Studies of intracellular distribution of mercury, in *Chemical Fallout,* Miller, M. W. and Berg, G. G., Eds., Charles C Thomas, Springfield, IL, 1969, pp. 408–419.
82. Nriagu, J. O., Ed., *The Biogeochemistry of Mercury in the Environment,* Elsevier/North Holland, Amsterdam, 1979.
83. Null, D. H., Gartside, P. S., and Wei, E., Methylmercury accumulation in brains of pregnant, nonpregnant, and fetal rats, *Life Sci.,* 12, 65–72, 1973.
84. Okinaka, S., Yoshikawa, M., Mozai, T., Mizuno, Y., Terao, T., Watanabe, H., Ogihara, K., Hirai, S., Yoshino, Y., Inose, T., Anzai, S., and Tsuda, M., Encephalomyelopathy due to an organic mercury compound, *Neurology,* 14, 69–76, 1964.
85. Oldereid, N. B., Thomassen, Y., and Purvis, K., Seminal plasma lead, cadmium and zinc in relation to tobacco consumption, *Int. J. Androl.,* 17, 24–28, 1994.
86. Oliver, M. A., Muir, K. R., Webster, R., Parkes, S. E., Cameron, A. H., Stevens, M. C., and Mann, J. R., A geostatistical approach to the analysis of pattern in rare disease, *J. Public Health Med.,* 14, 280–289, 1992.
87. Olson, F. C. and Massaro, E. J., Effects of methyl mercury on murine fetal amino acid uptake, protein synthesis and palate closure, *Teratology,* 16, 187–194, 1977.
88. Olson, F. C. and Massaro, E. J., Pharmacodynamics of methylmercury in the murine maternal/embryo: fetal unit, *Toxicol. Appl. Pharmacol.,* 39, 263–273, 1977.
89. Omata, S., Sato, M., Sakimura, K., and Sugano, H., Time-dependant accumulation of inorganic mercury in subcellular fractions of kidney, liver and brain of rats exposed to methylmercury, *Arch. Toxicol.,* 44, 231–241, 1980.
90. Oskarsson, A. and Sandström, B., A Nordic project — risk evaluation of essential trace elements: essential versus toxic levels of intake, *Analyst,* 120, 911–912, 1995.
91. Perry, H. M., Tipton, I. H., Schroeder, H. A., and Cook, M. J., Variability in the metal content of human organs, *J. Lab. Clin. Med.,* 60, 245–253, 1962.
92. Perry, H. M., Tipton, I. H., Schroeder, H. A., Steiner, R. L., and Cook, M. J., Variation in the concentration of cadmium in human kidney as a function of age and geographic origin, *J. Chron. Dis.,* 14, 259–271, 1961.
93. Pillay, K. K. S., Thomas, C. A., Sondel, J. A., and Hyche, C. M., Determination of mercury in biological and environmental samples by neutron activation analysis, in *Mercury Poisoning,* I, MSS Information Corp., New York, 1973, pp. 22–38.
94. Pratz, J., Ripoche, P., and Corman, B., Evidence for proteic water pathways in the luminal membrane of kidney proximal tubule, *Biochim. Biophys. Acta,* 856, 259–266, 1986.
95. Ptashekas, R. S. and Ptasheka, Iu. R., Problem of the relationship of probable and real ecologic pathology (in Russian), *Arkh. Patol.,* 54, 5–9, 1992.
96. Ptashekas, R. S. and Ptashekas, Iu. R., Problem of the relationship of probable and real ecologic pathology, *Arkh. Patol.,* 54, 5–9, 1992.
97. Rabenstein, D. L. and Isab, A. A., A protein nuclear magnetic resonance study of the interaction of mercury with intact human erythrocytes, *Biochim. Biophys. Acta,* 721, 374–384, 1982.
98. Ramel, C., Genetic effects of organic mercury compounds. I. Cytological investigations on allium roots, *Hereditas,* 61, 208–230, 1969.
99. Ramel, C., Methylmercury as a mitosis disturbing agent, *J. Jpn. Med. Assoc.,* 61, 1072–1077, 1969.
100. Report of an International Committee on Maximum Allowable Concentrations of Mercury Compounds, *Arch. Environ. Health,* 19, 891–905, 1969.
101. Reynolds, W. A. and Pitkin, R. M., Transplacental passage of methylmercury and its uptake by primate fetal tissues, *Proc. Soc. Exp. Biol. Med.,* 148, 523–526, 1975.
102. Rothstein, A. and Hayes, A. D., The metabolism of mercury in the rat studied by isotope techniques, *J. Pharmacol. Exp. Ther.,* 130, 166–176, 1960.
103. Samsahl, K. and Brune, D., Simultaneous determination of 30 trace elements in cancerous and non-cancerous human tissue samples by neutron activation analysis, *J. Appl. Radiat. Isot.,* 16, 273–281, 1965.
104. Schroeder, H. A. and Nason, A. P., Interactions of trace metals in mouse and rat tissues: zinc, chromium, copper, and manganese with 13 other elements, *J. Nutr.,* 106, 198–203, 1976.
105. Schroeder, H. A. and Balassa, J. J., Abnormal metals in man: cadmium, *J. Chron. Dis.,* 14, 236–258, 1961.
106. Schuhmacher, M., Batiste, J., Bosque, M. A., Domingo, J. L., and Corbella, J., Mercury concentrations in marine species from the coastal area of Tarragona Province, Spain. Dietary intake of mercury through fish and seafood consumption, *Sci. Total Environ.,* 156, 269–273, 1994.

107. Sever, L. E., Looking for the causes of neural tube defects: where does the environment fit in?, *Environ. Health Perspect.*, 103, 165–171, 1995.
108. Simpson, R. B., Association constants of methylmercuric and mercuric ions with nucleosides, *J. Am. Chem. Soc.*, 86, 2059–2065, 1966.
109. Skerfving, S., Exposure to mercury in the population, in *Advances in Mercury Toxicology*, Suzuki, T., Imura, N., and Clarkson, T. W., Eds., Plenum Press, New York, 1991, pp. 411–425.
110. Skerfving, S., *Toxicity of Methylmercury with Special Reference to Exposure via Fish*, Government Publishing House, Stockholm, 1972.
111. Skerfving, S., "Normal" concentrations of mercury in human tissue and urine, in *Mercury in the Environment*, Friberg, L. and Vostal, J., Eds., CRC Press, Boca Raton, FL, 1972, pp. 109–139.
112. Skerfving, S., Organic mercury compounds — relation between exposure and effects, in *Mercury in the Environment*, Friberg, L. and Vostal, J., Eds., CRC Press, Boca Raton, FL, 1972, pp. 141–168.
113. Southon, S., Wright, A. J. A., and Fairweather-Tait, S. J., Effects of dietary iron, calcium, and folic acid supplementation on apparent ^{65}Zn absorption and zinc status in pregnant rats, *Br. J. Nutr.*, 62, 415–423, 1989.
114. Stock, A., Der quecksilbergehalt des menschlichen organismus, *Biochem. Z.*, 304, 73–80, 1940.
115. Strikumar, T. S., Kallgard, A., Lindeberg, S., Ockerman, P. A., and Akesson, B., Trace element concentrations in hair of subjects from two South Pacific Islands, Atafu (Tokelau) and Kitava (Papua, New Guinea), *J. Trace Elem. Electrolytes Health Dis.*, 8, 21–26, 1994.
116. Sumino, K., Hayakawa, K. Shibata, T., and Kitamura, S., Heavy metals in normal Japanese tissues, *Arch. Environ. Health*, 30, 487–494, 1975.
117. Suzuki, T., Matsumoto, N., and Miyama, T., Neurological symptoms and mercury concentration in the brain of mice fed with methylmercury salt, *Industr. Health*, 9, 51–58, 1971.
118. Suzuki, T., Neurological symptoms from concentration of mercury in the brain, in *Chemical Fallout*, Miller, M. W. and Berg, G. G., Eds., Charles C Thomas, Springfield, IL, 245–257, 1969.
119. Suzuki, T., Matsumoto, N., Miyama, T., and Katsunuma, H., Placental transfer of mercuric chloride, phenyl mercury acetate and methyl mercury acetate in mice, *Industr. Health*, 5, 149–155, 1967.
120. Swensson, A. and Ulfvarson, U., Distribution and excretion of mercury compounds in rats over a long period after a single injection, *Acta Pharmacol.*, 26, 273–283, 1968.
121. Swensson, Å., Investigations on the toxicity of some organic mercury compounds which are used as seed disinfectants, *Acta Med. Scand.*, 143, 365–384, 1952.
122. Syversen, T. L. M., Distribution of mercury in enzymatically characterized subcellular fractions from the developing rat brain after injections of methylmercuric chloride and diethylmercury, *Biochem. Pharmac.*, 23, 2999–3007, 1976.
123. Takeuchi, T., Pathology of Minimata disease, in *Minimata Disease*, Study Group of Minimata Disease, Kumamoto University, Japan, 1968, pp. 141–228.
124. Tandon, L., Kasarkis, E. J., and Ehmann, W. D., INAA for interelement correlations in rats after mercuric chloride exposure, *J. Radioanal. Nuc. Chem.*, 161, 39–49, 1992.
125. Tejning, S., The mercury contents in blood corpuscles and in blood plasma in mothers and their new-born children, Report 70-05-20, Dept. of Occupational Medicine, University Hospital, S-211 85 Lund, 1970, (stencils).
126. Tipton, I. H., Cook, M. J., Perry, H. M., and Schroeder, H. A., Trace elements in human tissue. III. Subjects from Africa, the Near East, and Europe, *Health Phys.*, 11, 403–451, 1965.
127a. Tipton, I. H., Steiner, R. L., Boye, G. A., Cook, M. J., Perry, H. M., and Schroeder, H. A., Trace elements in human tissue. I. Methods, *Health Phys.*, 9, 89–101, 1963.
127b. Tipton, I. H. and Cook, M. J., Trace elements in human tissue. II. Adult subjects from the United States, *Health Phys.*, 9, 103–145, 1963.
128. *Toxicological Profile for Mercury*, Agency for Toxic Substances and Disease Registry, U.S. Public Health Service, Washington, D.C., 1989.
129. Tsubaki, T. and Irukayama, K., Eds., *Minimata Disease*, Elsevier Scientific, New York, 1977.
130. USEPA, *Proceedings of a National Forum on Mercury in Fish*, New Orleans, LA, September 27–29, 1994, Environmental Protection Agency, U.S. Government Printing Office, Washington, D.C., 1995.
131. USEPA, *U.S. Cancer Mortality Rates and Trends, 1950–1979*. Vol. IV. *Maps*, EPA/600/1-83/015e, U.S. Environmental Protection Agency, U.S. Government Printing Office, Washington, D.C., 1987.
132. Vance, D. E., Ehmann, W. D., and Markesbery, W. R., Trace element content in fingernails and hair of a nonindustrialized U.S. control population, *Biol. Trace Element Res.*, 17, 109–121, 1988.
133. Watras, C. J. and Huckabee, J. W., Eds., *Mercury Pollution: Integration and Synthesis*, Lewis Publishers, Boca Raton, FL, 1994.
134. Winroth, G., Carlstedt, I., Karlsson, H., and Berlin, M., Methyl mercury binding substances from the brain of experimentally exposed squirrel monkeys (*Saimiri sciureus*), *Acta Pharmacol. Toxicol.*, 49, 168–173, 1981.
135. Wu, M. M., Kuo, T. L., Hwang, Y. H., and Chen, C. J., Dose-response relationship between arsenic concentration in well water and mortality from cancers and vascular disease, *Am. J. Epidemiol.*, 130, 1123–1132, 1989.

136. Yamamura, Y., Yoshida, M., and Yamamura, S., Blood and urine mercury levels as indicators of exposure to mercury vapor, in *Advances in Mercury Toxicology*, Suzuki, T., Imura, N., and Clarkson, T. W., Eds., Plenum Press, New York, 1991, pp. 427–437.
137. Yoshino, M., Mozai, T., and Nakao, K., Distribution of mercury in the brain and its subcellular units in experimental organic mercury poisonings, *J. Neurochem.*, 13, 397–406, 1966.
138. Yukawa, M., Susuki-Yasumoto, M., Amano, K., and Terai, M., Distribution of trace elements in the human body determined by neutron activation analysis, *Arch. Environ. Health*, 35, 36–44, 1980.

PART 2

Nutrition and Toxicology

Gregory D. Miller, Editor

CHAPTER 6

USING FOOD CONSUMPTION DATA TO DETERMINE EXPOSURE TO TOXINS

J. S. Douglass, K. H. Fleming, L. M. Barraj, J. T. Heimbach

CONTENTS

0-8493-4493-X/97/$0.00+$.50

1 INTRODUCTION

Potential for human exposure can be assessed for a variety of food toxins, including those resulting from use of pesticides, additives, or animal drugs; materials migrating from packaging; environmental contaminants; microbial toxins; mycotoxins; naturally occurring toxins; and nutrients present at abnormally high concentrations in foods. The objective of dietary exposure assessment generally is to determine whether daily population exposure to the chemical or chemicals in question falls within the range of intake thought to be safe. In the U.S., the estimated daily intake (EDI) for pesticide residues is compared with the reference dose (RfD) set by the U.S. Environmental Protection Agency (USEPA); for additives, the EDI is compared with the acceptable daily intake (ADI) set by the U.S. Food and Drug Administration (USFDA).

Many methods and tools are available for determining whether population intakes of food toxins fall within safe levels, but the relative accuracy and reliability of these methods have been the subjects of much research and debate. Some methods, such as the Budget Method developed in Denmark for use with additives,[1] have been designed as "screening" methods. The Budget Method is an inexpensive means of predicting additive intake because it relies on estimates of physiological requirements for total food and liquid rather than on food consumption survey data. The Budget Method yields a worst-case intake estimate, the theoretical maximum daily intake (TMDI).

Slightly more complex methods for determining whether population intakes of food toxicants fall within safe ranges involve the application of maximum allowed additive levels or maximum residue limits (MRLs) for pesticides[2] to amounts of foods consumed in "model" diets created to be representative of national, regional, or international diets. Screening methods and model diet methods for assessing toxin exposure sacrifice accuracy of estimate for speed, simplicity, and known overestimation of exposure. Because intake is overestimated, results indicating toxin exposure less than the ADI are interpreted as indicating that exposure is safe, and it can be assumed that there is no need to expend resources to apply more sophisticated techniques in search of greater accuracy.

Food toxin exposure estimates are most accurate and reliable when based on methods combining toxicology, analytical chemistry, and food consumption survey data (Figure 1). These methods are based on the following equation:

$$\text{Exposure} = \text{toxin concentration in food} \times \text{consumption of food}$$

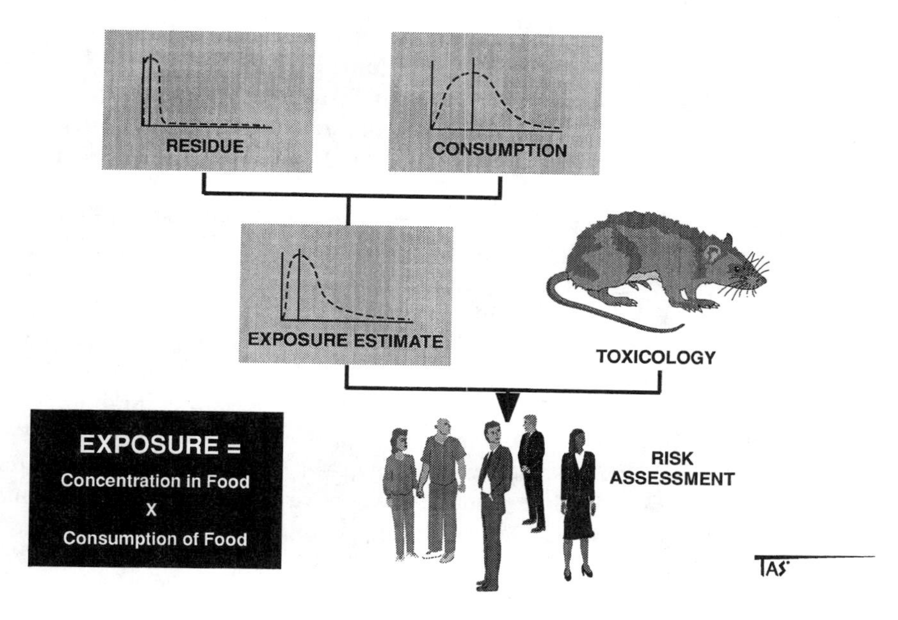

FIGURE 1
Food toxin exposure estimation.

In this chapter, we show how the specific method or model to be used in assessing exposure to a particular substance is selected based on the characteristics of the substance, on the quality and quantity of data available on the concentration of the substance in foods, and on the quality and quantity of data available on consumption of these foods.

2 NEED FOR INFORMATION ON TOXIN CHARACTERISTICS

Before estimating intake of a potentially harmful substance in food, it is necessary to define or characterize the substance in terms of attributes such as structure, volatility, solubility, and biological activity. Questions of concern related to substance properties include: What are the metabolic pathways in plant and animal systems? Does the substance break down during storage? During processing? Are breakdown products of toxicological significance?

Potential toxic effects from dietary intake of the substance in question must be carefully considered in planning an exposure assessment. Factors of interest include dose-response relationships, the length of exposure required to produce an adverse effect, potentially sensitive populations, and variability and uncertainty factors. For substances with acute toxic effects, it is important to estimate intake by people who actually eat the food, rather than intake averaged over a whole population whether or not they eat the food (per capita intake). For substances with chronic toxic effects, on the other hand, per capita intake is calculated and usually annualized to estimate long-term exposure to the substance.

3 NEED FOR INFORMATION ON ORIGIN OF TOXINS IN FOODS

The origin of a particular toxic substance in food is important to consider when characterizing the distribution of the substance in the food supply. Naturally occurring nutrients and toxins are present *inherently* in a particular plant or animal. Concentrations of such substances may vary due to differences in cultivar, season, or region of growth, but the substances generally would be expected to occur at some level in all raw plants or uncooked meat of the same type (an exception would be a nutrient or other natural toxin bred into plants by genetic engineering). The extent to which naturally occurring toxins are present in processed forms of the relevant foods depends on volatility and breakdown characteristics of the substances.

Toxic substances may be present in or on food *accidentally* due to environmental factors such as air pollution, water pollution, accidental contamination, or microbial contamination. The presence and concentration of such substances in foods would be expected to vary significantly depending on geographic location or other factors.

Substances present in food may be there *incidentally*, added indirectly to foods as a result of package migration or a processing step. Toxic substances incorporated incidentally are more likely to be present in processed foods than in raw agricultural commodities.

Toxic substances may be added *intentionally* to crops, livestock feeds, or foods. Pesticides and direct food additives fall in this category, although it should be noted that few food additives are toxic, even at several orders of magnitude above estimated intakes. Animal-based food toxins may differ in type and character due to differences in animal feeds, metabolism, and growing conditions.

4 SOURCES OF INFORMATION ON TOXIN CONCENTRATIONS IN FOODS

Exposure to potentially toxic food additives and pesticides may be estimated using the maximum levels permitted in foods; however, estimates based on such levels must be considered "worst-case". Exposure estimates based on actual toxin concentration levels in foods provide a more realistic picture.

The best method for obtaining existing food concentration data on naturally occurring toxins is to search the scientific literature. The most comprehensive sources of data on food levels of other types of toxins are manufacturers and government monitoring or surveillance programs.

4.1 Manufacturers' Test Data

As part of the package of information submitted to the USEPA for a tolerance (maximum residue limit) for a particular pesticide in a particular crop, the manufacturer must provide field trial data documenting the extent to which the pesticide and pesticide byproducts remain on the crop after harvest. Because the purpose is to determine the maximum residue concentration level resulting from legal use of the product, field trials are conducted under extreme conditions of pesticide use, i.e., at the maximum application rate, maximum application frequency, and minimum pre-harvest interval.

When the only source of pesticide residue concentration data available for use in exposure assessment is pesticide manufacturers' field trials, correction factors often are applied to adjust for regional and seasonal differences in use to yield "anticipated residue" concentration data. Unfortunately, although many states require documentation of pesticide use, there is no central source of pesticide use data.

During food processing, pesticide residues are generally reduced. In a study of pesticides in processed foods, 81.2% of 85,000 raw and finished products had no detectable residues, and 93% of 20,310 processed food products had no detectable residues.[3] Commercial washing appears to be the major cause for decrease in residues during processing.[4] It has been theorized that processes promoting hydrolysis contribute to degradation of residues, further decreasing total content.[3] Cooking may reduce the residues in foods but increase contents of harmful metabolites.[5]

In some cases, residues concentrate during processing. This is most likely to occur when processing results in the loss of moisture, leaving a smaller mass of product containing the same total amount of residue. If the residue concentrates, the manufacturer must petition the USEPA for a higher tolerance for processed forms of the food.

Pesticide manufacturers must document the extent to which pesticides proposed for use on crops destined for use as animal feeds are incorporated into muscle meat, organ meat, milk, and eggs. Animal feeding studies are required for all pesticides used on animal feeds. Animals are dosed with the pesticide for 30 days, then sacrificed; edible animal parts are analyzed for residue content.

The USFDA's testing requirements for proposed new direct additive uses are similar to those required by the USEPA for new pesticide uses. Food additive petitions and GRAS (generally recognized as safe) petitions usually include proposed maximum concentration levels for the substance in test foods; however, as with pesticides, maximum allowable levels may exaggerate the levels actually found in foods.

4.2 Monitoring and Surveillance Programs

Data obtained from food chemical monitoring or surveillance programs are generally more indicative of chemical levels actually found in foods than are manufacturers' test data. Most countries conduct various types of monitoring or surveillance for toxic chemicals in commodities or in foods. Monitoring and surveillance studies are conducted to assess compliance with state, federal, or international regulations that govern which pesticides may be used on which crops and the maximum allowable concentration in those crops.

At the U.S. federal level, the U.S. Department of Agriculture (USDA) monitors residue levels in domestic and imported meat and poultry products, and the USFDA monitors residue levels in all other foods. California, Florida, and a number of other states have pesticide monitoring programs; a national database, FOODCONTAM, incorporates data from monitoring programs in ten states.[6]

Depending on the specific U.S. monitoring program, foods or commodities may be sampled at the point of entry to the country, at the farm gate, at the food processing plant, or at the retail level (Figure 2). Since it is logistically and economically impossible to analyze each and every shipment of every food and commodity, studies are often conducted on target samples suspected to be out of compliance. Data on such samples cannot be considered representative of the food supply, but often these are the only data available for residue levels of specific pesticides on certain crops. It should be noted, however, that the majority of USFDA and FOODCONTAM samples have not contained detectable residues.[6,7]

4.3 Market Basket Surveys

Market basket surveys are conducted in the U.S. and other countries to obtain food chemical concentration data that may be used in exposure assessment. A core group of foods representative of national dietary patterns is obtained and analyzed to determine the concentrations of the substances of interest. Generally, samples of food are purchased at retail outlets in different regions of the country and prepared as for consumption.

Since 1961, the USFDA has conducted a yearly market basket survey, the Total Diet Study (TDS). Although not statistically based, the TDS does yield data useful in exposure assessment. Samples of 265 foods chosen to represent the U.S. food supply are collected four times each year from three cities in each of four U.S. regions. Samples of individual foods

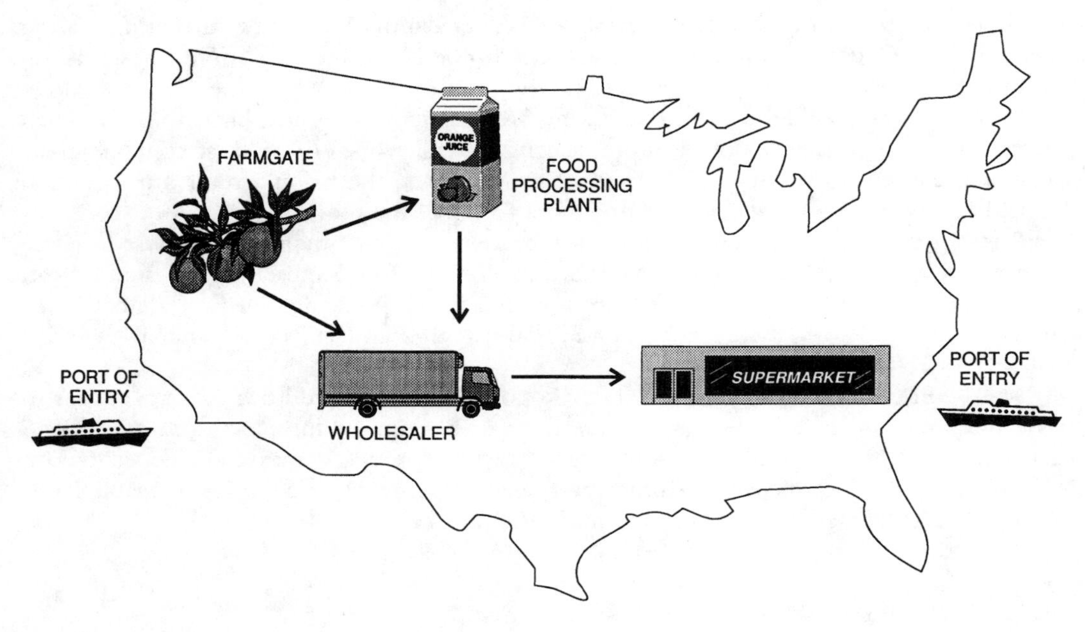

FIGURE 2
Focal points for food monitoring.

from these three cities are composited for analysis. All composited samples are shipped to the USFDA laboratory in Kansas City, MO, for analysis.

The USFDA uses TDS results mainly for identifying trends in concentrations of pesticide residues, contaminants, and nutrients in the food supply and for identifying trends in population exposures to these substances.[8] Because the TDS uses only a few hundred foods to represent thousands of foods, it is not appropriate to make inferences from the amounts of a contaminant in the sampled foods to the amounts of that chemical in the food supply; however, concentration data on the foods sampled can be used as reference points in exposure assessment.

A number of other countries conduct total diet studies, but the rationale and methods for conducting total diet studies vary from country to country. While the U.S. TDS is based on analysis of individual food items, studies performed elsewhere are based on analysis of food composites. Total diet studies in some countries are performed using a "duplicate portion" approach, in which all foods representing the national diet are processed into a single composite for analysis.[9]

Smaller market basket surveys are also undertaken. For example, the National EBDC (ethylene-bis-dithiocarbamate) Food Survey, designed and conducted by TAS, Inc., collected samples of 30 foods from over 300 statistically selected U.S. retail grocery stores to obtain actual residue values for the EBDC class of pesticides close to the point at which food is consumed.

5 SELECTION OF APPROPRIATE FOOD CONSUMPTION DATA

5.1 Considerations in Selecting Food Consumption Data

In addition to basic information on chemical characteristics and chemical concentrations in food, information on consumption of foods containing a particular chemical is required for realistic assessment of exposure. Thus, we must first answer the question, "Who is eating which foods and how much are they eating?".

Food consumption survey data are collected for a number of purposes. Commodity monitoring data are used to track agricultural production and imports; food consumption data on individuals or groups of individuals are collected to characterize expenditures, dietary intake, or nutritional status of those individuals; food consumption data may also be collected to estimate population intakes of specific foods or commodities, especially foods with constituents hypothesized to be related to specific health or disease states. For example, data may be collected on intake of specific foods thought to be protective against or associated with certain types of cancer by individuals at risk of developing those types of cancer.

Data collected with one purpose in mind may, nevertheless, be useful for answering questions outside the scope of the original study. Most food consumption data are not collected for evaluation of exposure to food contaminants but may be used for that purpose when limitations of the data, particularly the population studied and the methodology used, are taken into account.

There are a number of specific factors to consider when selecting food consumption data to estimate exposure:

- When (how long ago) were the data collected?
- Are current patterns of consumption of the foods of interest similar enough to patterns existing when these data were obtained?
- For which population were the food consumption data collected?
- For which country or countries?
- For which geographical regions?
- Were all population groups included?
- Were data collected during all seasons?
- How much detail about specific foods was included?
- Was the quantity of each food estimated?

It should be noted that food consumption data appropriate for use in estimating exposure to additives which have the potential to change drastically the nature of the food supply may be difficult to identify. If, for example, industry use of an additive will result in creation of new foods that are very low in calories or fat or have other characteristics attractive to consumers, what data should be used to estimate exposure to the additive? We must consider a range of factors in determining the food consumption data to use, including which segments of the population would probably consume the products and which foods will no longer be consumed because the new products are available. Postmarket surveillance — food consumption data collected after the additive is introduced into the food supply — can be used to assess the accuracy of the premarket estimates.

Information about the level of detail in food consumption data available from a particular survey is an important determinant of the potential for accuracy in estimating exposure to a specific toxic chemical. For example, if a toxin is found in beef liver, consumption data on beef liver should be used to assess toxin exposure. If the only available consumption data are on "beef", estimates of chemical consumption from beef liver will be less accurate. The level of detail available in data from food consumption surveys varies depending in part on the survey methodology.

5.2 Types of Food Consumption Surveys

Food consumption survey methods have been categorized using a number of different schemes. When evaluating food consumption survey data for use in exposure assessment, we use four main categories: food supply surveys (market disappearance), household or community inventory, household or individual food use, and individual food intake surveys.

5.2.1 *Food Supply Surveys*

Food supply surveys, also called food balance sheets or disappearance data, describe a country's food supply during a specified time period. Food balance sheets published by the Food and Agriculture Organization (FAO) of the United Nations describe the food supply in countries on all continents. European food balance sheets are also prepared by the Organization for Economic Cooperation and Development (OECD) and the Statistical Office of the European Communities (EUROSTAT). Food supply data in the U.S. are developed by the USDA's Economic Research Service.

Food supply data are derived from economic surveys using the following equation:

$$\text{Food availability} = (\text{food production} + \text{imports} + \text{beginning inventory}) - (\text{exports} + \text{ending inventory} + \text{nonfood uses})$$

Nonfood uses include animal feed, pet food, seed, and industrial use.

Availability is determined at different points for different foods or commodities. Mean per capita availability of a food or commodity is calculated by dividing total availability of the food by the country's total population.

These surveys provide data on food availability or disappearance rather than actual food consumption, but may be used to indirectly estimate amounts of foods consumed by the country's population. Food supply data may be useful for setting priorities, analyzing trends, developing policy, and formulating food programs. For some countries, food supply data are the only accessible data representing the country's food consumption. Because similar methods are used around the world, these data may be used to make comparisons internationally and also may be useful in some epidemiological studies.[10]

The value of these data in assessment of exposure to pesticide chemicals in foods is limited for several reasons. First, waste at the household and individual levels usually is not considered; therefore, intake estimates based on food supply data are higher than estimates based on actual food consumption survey data, with the magnitude of the error depending on the quantity of waste produced. Perhaps more importantly for exposure assessment, users of foods cannot be distinguished from nonusers. Therefore, individual variations in intake cannot be assessed nor can exposure of potentially sensitive subpopulations be estimated. Finally, food availability is usually reported in terms of raw agricultural commodities. Processed forms of foods are not considered, nor is there any way to distinguish use of foods as ingredients.

5.2.2 *Household Studies*

The design and methodology of food consumption surveys conducted at the household level overlap, to a certain extent, with surveys conducted at the community level and, in the other direction, of the individual. Household surveys generally can be categorized as (1) household or community inventories, or (2) household or individual food use.

5.2.2.1 *Household or Community Inventories*

Inventories are accounts of what foods are available in the household or, in some cases, in the community. What foods enter the household (or community)? Were they purchased, grown, or obtained some other way? What foods are used up by the household? Were they used by household members, guests, tenants? Were they fed to animals?

Inventories vary in precision with which data are collected. Questionnaires may or may not ask about forms of the food (canned, frozen, fresh), source (grown, purchased, provided through food program), cost, preparation, etc. Quantities of foods may be inventoried as purchased, grown (with inedible parts included or removed), cooked, or raw.

5.2.2.2 *Household or Individual Food Use*

Food use studies, usually conducted at the household or family level, are often used to provide economic data for policy development and planning for feeding programs. Survey methods used include food accounts, inventories, records, and list recalls.[11,12] These methods account for all foods used in the home during the survey period. This includes foods used from what was on hand in the household at the beginning of the survey period and foods brought into the home during the survey period.

Household food-use data may be used in the gross estimation of food or food constituent intakes by a population. The USDA has conducted household food consumption surveys approximately every 10 years since 1935, using the results to develop family food plans such as the Thrifty Food Plan, on which Food Stamp Program allotments and other food policies are based. Interviews are conducted with the "household food manager" to assess household characteristics, shopping practices, participation in food assistance programs, income, and expenditures. An inventory is taken of foods used from the household food supply for the previous week. For each food, the amount used, form, source, unit of purchase, and price are noted.

Although household food-use data have been used for a variety of purposes, including exposure assessment, serious limitations associated with data from these surveys should be noted. Food waste often is not accounted for. Food purchased and consumed outside the household may or may not be considered. Users of a food within a household cannot be distinguished, and individual variation cannot be determined. Intakes by subpopulations based on age, sex, health status, and other variables for individuals can only be estimated based on standard proportions or equivalents for age/sex categories.

5.2.3 *Individual Intake Studies*

Individual intake studies provide data on food consumption by specific individuals. Methods for assessing food intakes of individuals may be retrospective (such as 24-hour or other short-term recalls, food frequencies, and diet histories), prospective (food diaries, food records, or duplicate portions), or a combination thereof. The most commonly used studies are those using the recall or record method and the food frequency.

5.2.3.1 *Recall Method (24-Hour Recall)*

Recalls are used to collect information on foods consumed in the past. The unit of observation is the individual or the household. The subject is asked to recall what foods and beverages he/she or the household consumed during a specific period, usually the preceding 24 hours. Since this method depends on memory, foods are quantified retrospectively, often with the aid of pictures, household measures, or two- or three-dimensional food models.

Advantages of this method are the ease of administration and the time necessary for completion (about 20 minutes or less for a 24-hour recall).[13,14] Recalls have been used successfully with individuals as young as six years, and interviewer-administered recalls are usually the method employed for populations with limited literacy or for those for whom English is not the native language. When individuals are not available for an interview or are unable to be interviewed due to age, infirmity, or temporary absence from the household, surrogate respondents are often used.[15]

The main disadvantage of the recall method is the potential for error due to faulty memory of respondents. Items that were consumed may be forgotten. Or, the respondent may recall items consumed that actually were not consumed during the time investigated. To aid recall memories, the interviewer may "probe" for certain foods or beverages that are frequently forgotten, but this probing has also been shown to introduce potential bias by encouraging reporting of items not actually consumed.

5.2.3.2 Food Record/Diary Method

Records are used to collect information about current food intake. The subject is asked to keep a record of foods and beverages as they are consumed during a specific period. Quantities of foods and beverages consumed are entered in the record usually after weighing, measuring, or recording package sizes. Occasionally, photographs or other recording devices are employed.

5.2.3.3 Food Frequency Questionnaire Method

The food frequency questionnaire (FFQ), or checklist, determines the frequency of consumption of a limited number of foods, usually less than 100. This retrospective method estimates whether and how often foods are usually eaten by an individual during a particular period of time. Subjects indicate how many times a day, week, or month they usually consume each food. Semiquantitative FFQs also estimate amounts consumed by allowing subjects to indicate whether their usual portion size is small, medium, or large compared to a stated "medium" portion. The size of the medium portion is usually standardized for various age/sex groups according to mean intakes of large populations.

Since FFQs collect data about intake of relatively few foods, a carefully tailored FFQ might provide useful data for assessment of chronic exposure if the chemical in question is concentrated in only a few foods and if the food frequency instrument has been designed to target those foods. Depending on the number of foods and whether or not an attempt is made to quantify intake, the FFQ can be administered in 10 to 15 minutes. Information from FFQs cannot be used to estimate intake of acutely toxic chemicals, since data are collected on single food items or types, not on food combinations eaten at the same time.

In general, data from 24-hour and other short-term recalls and from food diaries, which collect detailed information on the kinds and quantities of foods consumed, are the most accurate and flexible data to use in assessment of exposure to food toxins. Data from these surveys can be used to estimate either acute or chronic exposure; averages and distributions can be calculated; and exposure estimates can be calculated for subpopulations based on age, sex, ethnic background, socioeconomic status, and other demographic variables, provided that such information is collected for each individual.

Each of the methods discussed above under Individual Intake Studies may also be applied at the family or household levels. This application is most appropriate when household members eat from a common pot.

In general, data from large surveys using recall/record methods provide the most accuracy in assessment of exposure to food toxins when the chemical under investigation is found in many foods and when most of these foods are consumed on a regular basis. It is more difficult to capture intake of infrequently consumed foods using short-term recalls or records. If an infrequently consumed food, such as liver, is a potent source of the chemical under investigation, exposure estimates for that chemical may be low. In this case, a well-designed FFQ that targets the specific foods of concern should be considered instead. Two methods that are often used to study food consumption but are rarely used to estimate exposure to toxins are the diet history and the duplicate plate methods.

5.2.3.4 Dietary History Method

The diet history is used to obtain information from the respondent about the usual pattern of eating over an extended period of time.[16,17] It is used primarily in epidemiological research. The diet history is a more in-depth and time-consuming procedure than the recall, record, and FFQ methods. A recall or FFQ may be included as a diet history component.

5.2.3.5 Duplicate Plate Method

Subjects in duplicate plate surveys deposit in collection containers an identical portion of all foods and beverages consumed during a specified time.[18] These duplicate portions are kept in cold storage and are then shipped to a laboratory for analysis.

Although duplicate plate methods have the potential to provide the most accurate food intake estimations, there are serious limitations to using of these data in exposure assessment. Duplicate plate surveys have high respondent burden and great potential for respondent bias. There may be reluctance on the part of respondents to "waste" food, particularly more expensive food when preparing portions for the collection container. Selection of foods eaten on collection days may not be representative of habitual intake. The types of individuals willing to commit to the effort required probably also are not typical of the general population. Most importantly, since duplicate plate studies are very expensive to conduct, the number of subjects in these studies usually is too low to be useful for exposure assessment. These methods are seldom used outside the clinical setting.

For each type of food consumption survey, a number of methods of administration are possible. Populations that are literate and fairly highly motivated may be surveyed by a self-administered questionnaire or food record or diary; however, the researcher is less able to probe for adequate descriptions, measurable serving sizes, and forgotten items. Interviewer-administered methods, conducted in person or by telephone, involve the additional expense of an interviewer but may produce more complete and accurate descriptions of foods and beverages consumed.[19] Questions and responses may be recorded on paper, computer, or audio- or videotape.[20,21,22,23]

5.3 U.S. National Surveys of Food Intake

Two large food consumption surveys are conducted by the federal government: the Nationwide Food Consumption Survey (NFCS), conducted by the USDA beginning in 1935, and the National Health and Nutrition Examination Survey (NHANES), undertaken by the U.S. Department of Health and Human Services beginning in 1971. Both surveys employ multi-stage area probability sampling procedures to obtain a sample representative of the noninstitutionalized population of the U.S. Specific years in which these surveys have been conducted are listed in the Appendix.

Although both surveys contain a dietary intake component, they are quite dissimilar in purpose and in elements other than dietary intake. The purpose of the NFCS is to "measure the food and nutrient content of the diet and the money value of food used by U.S. households and the food and nutrient intakes at home and away from home of individuals."[24] The purpose of the NHANES is to "develop information on the total prevalence of a disease condition or a physical state, to provide descriptive or normative information, and to provide information on the interrelationships of health and nutrition variables within the population groups."[24] These two major surveys will be summarized separately.

5.3.1 *Nationwide Food Consumption Survey*

The NFCS has two major components — the Household component, which obtains information on food used by the household over a 7-day period, and the Individual component, which collects data on foods and beverages consumed by individuals in the household.

The series of Continuing Surveys of Food Intakes by Individuals (CSFIIs) does not include the lengthy Household Food Use component but does collect detailed data on household demographics. For all of the surveys in the NFCS/CSFII series conducted by the U.S. Department of Agriculture, data are collected for both components by a trained interviewer.

Each household in the NFCS is contacted about one week before the interview to schedule a date for the interview and to ask the member of the household most familiar with food shopping and meal planning (the "household food manager") to keep receipts and records that would help with the household food use component of the survey. The household food use survey interview is conducted with the household food manager and includes questions about the composition and characteristics of the household and its members, shopping practices, participation in food assistance programs, income, and expenditures. An inventory is then made of foods used from the home food supplies during the previous week — the amount used, the form (including fresh, frozen, canned, dried), the source, unit of purchase (if bought), and price.

Individual intake data for both the NFCSs and the CSFIIs are collected by interview with each household member eligible for survey participation. Prior to 1994, each individual was asked to recall all foods and beverages and quantities consumed during the previous day (24-hour recall) and then to keep a 2-day record of intake for the day of the interview and the following day. For the 1994 to 1996 CSFII, 24-hr recalls are conducted on two consecutive days. Serving sizes are estimated using rulers; household measures such as teaspoons, tablespoons, and cups; and a food instruction booklet, which explains to the respondent how to quantify each type of food. In addition, questions are asked about eating patterns, dietary supplements, activity level, and other areas specific to that particular survey. The household food manager or other responsible adult answers for respondents less than 12 years of age or for individuals unable to answer for themselves.

The NFCS and CSFII surveys provide data that are quite appropriate and useful for assessment of exposure to food toxins for a variety of reasons:

- Foods coded are numerous and very specific.
- Foods are coded in a hierarchical structure, allowing easy aggregation and disaggregation of data in estimating exposure.
- Data on multiple days of intake allow estimation of acute or chronic exposure. (Calculation of usual exposure from data collected on consecutive days is complicated by the nonindependence of intakes over these days. For this reason, the most recent cycle of the CSFII collected data on 2 nonconsecutive days.)
- Distributions of intake may be calculated.
- Data on age, sex, race, region, season, pregnancy, and nursing status allow exposure estimation for a wide variety of subpopulations.
- Body-weight data allow exposure estimations per unit body weight.

5.3.2 National Health and Nutrition Examination Survey

The NHANES, conducted by the National Center for Health Statistics, U.S. Department of Health and Human Services, includes five major components: a household questionnaire, a medical history questionnaire, a dietary questionnaire, a physical examination, and clinical tests. Parts of the survey are conducted in the home, the rest in specially designed mobile examination centers (MECs), where the dietary and clinical procedures are conducted.

The dietary component consists of a recall of dietary intake for the preceding 24 hours administered by a dietitian in the MEC, a food frequency questionnaire for the previous 1 to 3 months (depending on the specific NHANES survey), and questions about special diets, medications, and nutritional supplements. Serving sizes are estimated using food models and geometric shapes. The major limitation of NHANES data is that only one 24-hour recall is obtained for each person, so chronic exposure estimates cannot be made with any accuracy.

5.4 Food Consumption Survey Data: Summary

In general, data from large national surveys using a 24-hour or other short-term recall or record provide the most accuracy in assessment of exposure to food toxins. Data from these surveys can be used to estimate either acute or chronic exposure at the national level. Averages and distributions can be calculated for subpopulations based on age, sex, ethnic background, socioeconomic status, and other demographic variables.

5.5 Validity, Reliability, and Sources of Error

5.5.1 *Validity*

Food consumption data used in assessing exposure to food toxins should be both valid (accurate) and reliable (precise). Validity is the ability of an instrument to measure what it is intended to measure. For food consumption research, investigators are usually interested in knowing what a person's usual intake is or has been; however, it is not possible to know a person's true "usual" food intake. Therefore, the validity of a measurement instrument is usually assessed by comparing results using that instrument with those of another instrument.

Numerous studies attempting to validate one survey method relative to another have been reported.[25-42] For example, the FFQ, a more recent survey method, has been validated by comparing results of dietary intake with repeated multiple day food records, which has served as an estimate of usual intake.[13,43,44] Results of the validity studies indicated that FFQs could provide useful information about individual nutrient intakes; however, these studies generally have shown better correlations between methods for groups than for individual survey participants. The validity of estimates of food toxin intakes is less certain.

Validation of the 24-hour recall, diaries, records, and frequencies has also been reported after comparing estimates of dietary intake obtained using one of these survey instruments to the subjects' actual intakes. Actual intakes may be based on surreptitious observation in a cafeteria, congregate meal cite, or other facility.[45-50]

Survey methodology has also been validated by use of biological markers associated with dietary intake. Possible sources of biological markers include urine, feces, blood, and tissue samples, but the most easily accessible and therefore most commonly used is urine. Nitrogen content of urine has been used to verify protein intake. If protein intake calculated from the reported food intake is in agreement with protein intake calculated from nitrogen excretion, it is assumed that intake of other nutrients is valid.[51]

5.5.2 *Reliability*

Reliability, or reproducibility, is the ability of a method to produce the same or similar estimate on two or more different occasions,[52,53] whether or not the estimate is accurate; however, a method cannot achieve an accurate answer every time unless it gives the same answer every time. The reliability of food consumption survey data for estimating "usual" intake of a population depends somewhat on the number of days of dietary intake data collected for each individual in the population. The number of days of food consumption data required for reliable estimation of population intakes is related to each subject's day-to-day variation in diet (intra-individual variation) and the degree to which subjects differ from each other in their diets (interindividual variation).[54,55] When intra-individual variation is small relative to interindividual variation, population intakes can be reliably estimated with consumption data from a smaller number of days than should be obtained if both types of variation are large. Intake of toxins and other contaminants can be reliably estimated with fewer days of data when those toxins are present in many foods that are commonly consumed.

In assessing food intake, it is generally accepted that mean intake of a population may be reasonably estimated using a one-day recall or diary if the number of subjects is sufficiently large; however, the percentage of the population estimated to be at risk of toxic effects from a chemical will be higher when food intake is assessed using a 1-day recall than with a multiday record or dietary history. This is because extreme levels of intake (e.g., 90th or 95th percentiles) are invariably higher for a single day than they are for multiple days. In addition, large intra-individual variation associated with one-day surveys may limit the power to detect differences between different population groups.[56-58]

5.5.3 *Sources of Error*

Error in individual food consumption surveys may be due to chance or to measurement factors. Data variability due to chance may be related to the survey sample; any sample randomly drawn from a population will differ from any other sample, with the degree of difference depending upon the size of the sample and the homogeneity of the population from which it was drawn. Error due to chance also arises from data collection at different times of the day, on different days of the week, or at different seasons of the year.

Measurement error may be introduced by the survey instrument, the interviewer, or the respondent. The instrument may bias results if questions are not clear, if probes "lead the subject" to give a desired answer, if questions are culture specific, or if the questions do not follow a logical sequence. For self-administered questionnaires, responses will be influenced by the readability level, the use of abbreviations or unfamiliar jargon, clarity of instructions, and amount of space provided for answers. Interviewer bias may be introduced if interviewers make the respondent uncomfortable, are judgmental, or do not use a standard method and/or standard probes.

Respondents may introduce bias if they omit reporting foods they actually ate because they are reluctant to report certain foods or beverages (alcoholic beverages are a good example) or if they are forgetful. Alternatively, they may report the food but understate the quantity consumed. Foods consumed away from home, particularly on occasions when the focus of attention is on the event rather than on the food, are especially difficult for people to remember. Quantities may be underestimated for similar reasons. Foods and beverages that were not consumed may be reported as consumed because of faulty memories, desire to impress the interviewer, or confusion with similar foods.

Measurement errors also include errors in coding. Coding errors may be due to unclear handwritten records or to erroneous data entry. It is important when using food consumption data from surveys that have already been conducted to be aware of the potential for error when making decisions based on those data. When designing food consumption research, the potential for error should be minimized by standardizing and testing all instruments for validity and reliability.

6 SELECTION OF APPROPRIATE EXPOSURE ANALYSIS MODEL

A number of models and tools are available for combining toxin concentration data and food consumption data to estimate exposure of a population to toxins in foods. The model most appropriate for use in assessing exposure to a particular chemical depends in large part on the characteristics of the chemical, the quantity and quality of data available on concentrations in foods, and the quantity and quality of data available on consumption of these foods.

The method used also may depend on whether the population in question is considered to be particularly vulnerable to the toxin. Concern over children's exposure to toxins, especially to pesticides, prompted the U.S. Congress to request the National Academy of Sciences, National Research Council (NAS-NRC), to conduct a thorough review of scientific and policy issues concerning pesticides in the diets of infants and children. The NAS-NRC committee formed to conduct this review recommended that methods be developed for evaluating distribution of exposure in vulnerable populations, for assessing exposure from all foods combined rather than from individual commodities, and for assessing combined exposure from pesticides with similar toxic effects.[59] Several sophisticated new methods for assessing exposure have been developed in response to these recommendations.

6.1 Point Exposure

A point estimate of exposure to a specific chemical by a particular population is a broad estimate generated using one number to represent concentration of the chemical in each food and one number to represent intake of these foods by that population. In estimating chronic exposure, the arithmetic mean of residue concentrations is most commonly used; however, if the distribution of pesticide concentrations is known to be skewed, use of the median (or 50th percentile) concentration is more appropriate.[60] Exposure to acutely toxic chemicals frequently is assessed using the 95th percentile residue concentration in order to produce a "worst-case" estimate.

6.2 Simple Distribution

A simple distribution of exposure can be calculated by either of two procedures. A single number chosen to represent concentration of the toxin in each of the foods of interest may be applied to a distribution of intake levels for each food. Or, a single number chosen to represent food intakes can be applied to a distribution of toxin concentration values. The USEPA currently uses a simple distribution model when this is appropriate to assess exposure to pesticide residues.[61]

6.3 Joint Distribution

Joint distribution analysis relies on sophisticated calculations to allow combination of representative data using a distribution of toxin concentrations and a distribution of food intakes. In joint distribution analysis, best-case and worst-case scenarios of intake are shown on the same table or graph and therefore can be reviewed simultaneously.

Several methods for calculating joint distribution have been proposed. The method used most frequently is based on an approach known as Monte Carlo analysis in which a population is characterized based on data on a random sample from that population. In calculating joint distribution, the Monte Carlo approach assumes that both the residue and consumption distributions belong to a parametric family (e.g., normal or lognormal). Calculations are performed as follows:

1. Random samples are generated from the theoretical distributions.
2. Each value from the sample of consumption values is multiplied by each value from the sample of residue values.

Steps 1 and 2 are repeated thousands of times, and the resulting distributions are merged together to produce an estimate of the exposure distribution.

6.4 Exposure to Multiple Chemicals

Estimates of combined exposure involve considerations not required for exposure to one chemical. Residues of chemicals cannot be simply summed for use in exposure assessment; the chemicals may have different toxic effects and/or different potencies. Toxicity must first be standardized by applying a toxicity equivalency factor (TEF) to convert to a common activity level.

For example, to assess combined exposure to azinphos-methyl and chlorpyrifos (two cholinesterase inhibitors) from apples, celery, and grapes, we must determine the relative potency of the chemicals.

One unit of azinphos-methyl is equivalent to 0.433 units of chlorpyrifos (Figure 3); therefore, the 13 ppb of chlorpyrifos are standardized to 5.63 ppb. The standardized residues can then be summed to produce a toxic equivalency quotient.

Using multiple exposure software and data from the USDA Pesticide Data Program,[62] one can calculate standardized residue distributions of the two chemicals. In this example, the distribution of residues for chemical A (azinphos-methyl) and for chemical B (chlorpyrifos),

Commodity:	Apple
Chemical:	Azinphos-methyl Chlorpyrifos
Common Effect:	ChE Inhibition

Chemical	Detected Levels (ppb)	RfD (mg/kg BW)	TEF	Standardized Residues (ppb)
Azinphos-methyl	24	0.0013	1	24
Chlorpyrifos	13	0.0030	0.433	5.63
TEQ	--	--	--	29.63

FIGURE 3
Illustration of the total equivalency approach (USDA-PDP, 1992).

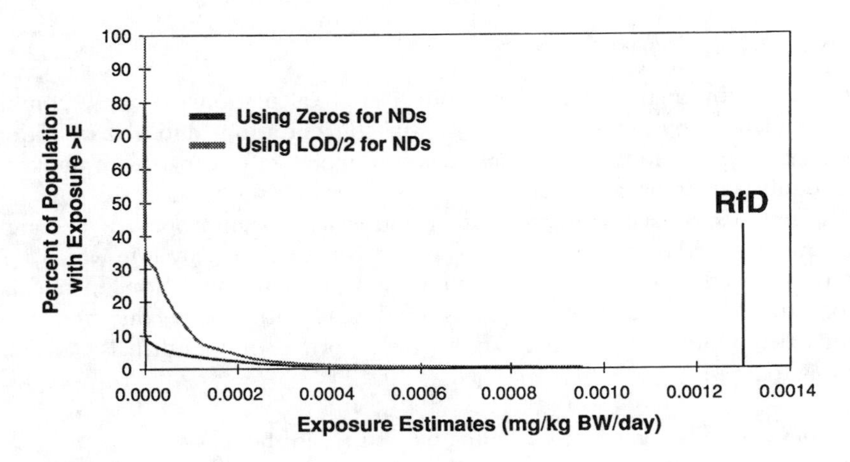

FIGURE 4
Estimates of the exposure distribution for women 19–44 years. Residues data: USDA-PDP, 1992; consumption data: NFCS, 1987–1988; foods: apples, celery, and grapes; chemicals: azinphos-methyl, chlorpyrifos.

to which the TEF has been applied to make residues comparable to those of chemical A, are combined to produce a standardized residue distribution based on the two chemicals together. The resulting standardized residue distributions can then be combined with the appropriate food consumption data to estimate dietary exposure of the two chemicals combined.

Figure 4 shows dietary exposure calculated for combined azinphos-methyl and chlorpyrifos on three foods: apples, celery, and grapes using JDA for women 19 to 44 years old. The food consumption data used are from the 1987 to 88 NFCS.[63] Exposure was estimated in mg/kg body weight. Two different methods of expressing exposure were used when no residues could be detected (ND), either using zero or, more conservatively, using one-half the limit of detection (LOD/2). Using either, the estimate of exposure was well below the RfD.

Finally, regardless of the exposure analysis model used, the uncertainty associated with exposure estimates should be evaluated and presented. Uncertainty can be characterized quantitatively (i.e., what thought processes were used to select or reject specific data) or quantitatively (i.e., ranges of exposure).[64] Uncertainty may result from missing or incomplete data, measurement error, sampling error, use of surrogate data, gaps in scientific theory used to make predictions, and how well the theory or model represents the situation being assessed. Analysis of uncertainty provides decision-makers with information concerning potential variability in exposure estimates and the effect of data gaps on exposure estimates.

6.5 Commodity Contribution

The methods described above may be used to assess toxin exposure from one food or from multiple foods. When exposure from multiple foods is assessed, the relative contribution of consumption of each food to total toxin exposure may be of interest. Commodity contribution analyses provide information useful in planning regulatory strategies for new pesticide tolerances and/or in prioritizing pesticide use reduction efforts.

7 APPLICATIONS

The chief objective in using food consumption data to assess dietary exposure to food toxins generally is to provide accurate, reliable data for determining whether daily population exposure to the chemical or chemicals in question falls within the range of intake thought to be safe. These assessments may be performed for any of a wide variety of purposes, including research, regulatory support, and public interest. Exposure assessments are required components of information packages submitted by pesticide manufacturers to the USEPA for new pesticide tolerances; such assessments are also required components of direct and indirect additive petitions and GRAS affirmation petitions.

No matter what the purpose for the exposure assessment, the food consumption data to be used should be selected with the nuances and limitations of the data in mind. In this chapter, we have presented the factors affecting the use of food consumption in assessing exposure to food toxins. The specific method or model for which the food consumption data are to be used in assessing exposure to a particular chemical must be selected based on the characteristics of the chemical, on the quality and quantity of data available, on concentration of the chemical in foods, and on the quality and quantity of data available regarding consumption of these foods.

REFERENCES

1. Hansen, S. C., Conditions for use of food additives based on a budget for an Acceptable Daily Intake, *J. Food Protection*, 42, 429, 1979.
2. Rees, N. and Tennant, D., Estimation of food chemical intake, in *Nutritional Technology*, Raven Press, New York, 1994.
3. Chin, H. B., The effect of processing on residues in foods: the food processing industry's residue database, in *Pesticide Residues and Food Safety: A Harvest of Viewpoints*, Tweedy, B. G., Dishburger, H. J., Ballantine, L. G., and McCarthy, J., Eds., American Chemical Society, Washington, D.C., 1991, p. 175.
4. Elkins, E. R., Effect of commercial processing on pesticide residues in selected fruits and vegetables, *J. Assoc. Off. Anal. Chem.*, 72, 533, 1989.
5. Tomerlin, J. R. and Engler, R., Estimation of dietary exposure to pesticides using the Dietary Risk Evaluation System, in *Pesticide Residues and Food Safety: A Harvest of Viewpoints*, Tweedy, B. G., Dishburger, H. J., Ballantine, L. G., and McCarthy, J., Eds., American Chemical Society, Washington, D.C., 1991, p. 192.
6. Minyard, J. P., Jr., and Roberts, W. E., FOODCONTAM: A state data resource on toxic chemicals in foods, in *Pesticide Residues and Food Safety: A Harvest of Viewpoints*, Tweedy, B. G., Dishburger, H. J., Ballantine, L. G., and McCarthy, J., Eds., American Chemical Society, Washington, D.C., 1991, p. 151.
7. U.S. Food and Drug Administration Pesticide Program, *Residues in Foods 1990*, U.S. Food and Drug Administration, U.S. Government Printing Office, Washington, D.C., 1991.
8. Pennington, J. A. T., The 1990 revision of the Total Diet Study, *J. Nutr. Educ.*, 244, 173, 1992.

9. World Health Organization, *Principles for the Safety Assessment of Food Additives and Contaminants in Food*, Environmental Health Criteria Document No. 70, World Health Organization, Geneva, Switzerland, 1987.
10. Sasaki, S. and Kestelloot, H., Value of Food and Agriculture Organization data on food-balance sheets as a data source for dietary fat intake in epidemiologic studies, *Am. J. Clin. Nutr.*, 56, 716, 1992.
11. Pao, E. M., Sykes, K. E., and Cypel, Y. S., *USDA Methodological Research for Large-Scale Dietary Intake Surveys, 1975–88*, Home Economics Research Report No. 49, U.S. Department of Agriculture, Human Nutrition Information Service, U.S. Government Printing Office, Washington, D.C., 1989.
12. Lee, R. D. and Nieman, D. C., *Nutritional Assessment*, W. C. Brown, Dubuque, IA, 1993.
13. Block, G., Human dietary assessment: methods and issues, *Prev. Med.*, 18, 653, 1989.
14. Dwyer, J. T., Assessment of dietary intake, in *Modern Nutrition in Health and Disease*, 7th ed., Shils, M. E. and Young, V. R., Eds., Lea & Febiger, Philadelphia, PA, 1988.
15. Samet, J. M., Surrogate measures of dietary intake, *Am. J. Clin. Nutr.*, 50, 1139, 1989.
16. Burke, B. S., The dietary history as a tool in research, *J. Am. Diet. Assoc.*, 23, 1041, 1947.
17. Hankin, J. H., Development of a diet history questionnaire for studies of older persons, *Am. J. Clin. Nutr.*, 50, 1121, 1989.
18. Kim, W. W., Mertz, W., Judd, J. T., Marshall, M. W., Kelsay, J. L., and Prather, E. S., Effect of making duplicate food collections on nutrient intakes calculated from diet records, *Am. J. Clin. Nutr.*, 40, 1333, 1984.
19. Shucker, R. E., Alternative approaches to classic food consumption measurement methods: telephone interviewing and market data bases, *Am. J. Clin. Nutr.*, 35, 1306, 1982.
20. Brown, J. E., Tharp, T. M., Dahlberg-Luby, E. M., Snowdon, D. A., Ostwald, S. K., Buzzard, I. M., Rysavy, D. M., and Wieser, M. A., Videotape dietary assessment: validity, reliability, and comparison of results with 24-hour dietary recalls from elderly women in a retirement home, *J. Am. Diet. Assoc.*, 90, 1675, 1990.
21. Elwood, P. C. and Bird, G., A photographic method of diet evaluation, *Hum. Nutr. Appl. Nutr.*, 37A, 474, 1983.
22. Kolasa, K. M. and Miller, M. G., New developments in nutrition education using computer technology, *J. Nutr. Educ.*, 28, 7, 1996.
23. Stockley, L., Chapman, R. I., Holley, M. L., Jones, F. A., Prescott, E. H. A., and Broadhurst, A. J., Description of a food recording electronic device for use in dietary surveys, *Hum. Nutr. Appl. Nutr.*, 40A, 13, 1986.
24. National Research Council, *Nutrient Adequacy Assessment Using Food Consumption Surveys*, Subcommittee on Criteria for Dietary Evaluation, Coordinating Committee on Evaluation of Food Consumption Surveys, Food and Nutrition Board, Commission on Life Science, National Academy Press, Washington, D.C., 1989.
25. Bingham, S., Wiggins, H. S., Englyst, H., Seppanen, R., Helms, P., Strand, R., Burton, R., Jorgensen, I. M., Poulsen, L., Paerregaard, A., Bjerrum, L., and James, W. P., Methods and validity of dietary assessments in four Scandinavian populations, *Nutr. Cancer*, 4, 23, 1982.
26. Blake, E. C. and Durnin, J. V. G. A., Dietary values from a 24-hour recall compared to a 7-day survey on elderly people, *Proc. Nutr. Soc.*, 22, I, 1963.
27. Block, G., A review of validations of dietary assessment methods, *Am. J. Epidemiol.*, 115, 492, 1982.
28. Bransby, E. R., Daubney, C. G., and King, J., Comparison of results obtained by different methods of individual dietary survey, *Br. J. Nutr.*, 2, 89, 1948.
29. Fanelli, M. T. and Stevenhagen, K. J., Consistency of energy and nutrient intakes of older adults: 24-hour recall vs. 1-day food record, *J. Am. Diet. Assoc.*, 86, 664, 1986.
30. Grewal, T., Gopaldas, T., Gadre, V. J., Shrivastava, S. N., Pranjpe, B. M., Chatterjee, B. N., and Srinivasan, N., A comparison of weighment and questionnaire dietary survey methods for rural preschool children, *Ind. J. Nutr. Diet.*, 11, 224, 1974.
31. Hussain, M. A., Abdullah, M., Huda, N., and Ahmad, K., Studies on dietary survey methodology — a comparison between recall and weighing method in Bangladesh, *Bangladesh Med. Res. Council Bull.*, 6, 53, 1980.
32. Karvetti, R. L. and Knuts, L. R., Validity of the 24-hour recall, *J. Am. Diet. Assoc.*, 85, 1437, 1985.
33. Lubbe, A. M., A survey of the nutritional status of white school children in Pretoria: description and comparative study of two dietary survey techniques, *S. A. Med. Journal Suppl.*, 616, 1968.
34. Mahalko, J. P., Johnson, L. K., Gallagher, S. K., and Milne, D. B., Comparison of dietary histories and seven-day food records in a nutritional assessment of older adults, *Am. J. Clin. Nutr.*, 42, 542, 1985.
35. Meredith, A., Matthews, A., Zickefoose, M., Weagley, E., Wayave, M., and Brown, E. G., How well do school children recall what they have eaten?, *J. Am. Diet. Assoc.*, 27, 749, 1951.
36. Morgan, K. J., Johnson, S. R., Rizek, R. L., Reese, R., and Stampley, G. L., Collection of food intake data: an evaluation of methods, *J. Am. Diet. Assoc.*, 87, 888, 1987.
37. Morrison, S. D., Russell, F. C., and Stevenson, J., Estimating food intake by questioning and weighing: a one-day survey of eight subjects, *Proc. Nutr. Soc.*, 7, v, 1949.
38. Nettleton, P., Day, K. C., and Nelson, M., Dietary survey methods. 2. A comparison of nutrient intakes within families assessed by household measures and the semi-weighed method, *J. Hum. Nutr.*, 34, 349, 1980.
39. Rasanen, L., Nutrition survey of Finnish rural children. VI. Methodological study comparing the 24-hour recall and the dietary history interview, *Am. J. Clin. Nutr.*, 32, 2560, 1979.
40. Russell-Briefel, R., Caggiula, A. W., and Kuller, L. H., A comparison of three dietary methods for estimating vitamin A intake, *Am. J. Epidemiol.*, 122, 628, 1985.

41. Willett, W. C., Sampson, L., Stampfer, M. J., Rosner, B., Bain, C., Witschi, J., Hennekens, C. H., and Speizer, F. E., Reproducibility and validity of a semi-quantitative food frequency questionnaire, *Am. J. Epidemiol.*, 122, 51, 1985.
42. Young, C. M., Hagan, G. C., Tucker, R. E., and Foster, W. D., A comparison of dietary study methods. II. Dietary history vs. seven-day record vs. 24-hour recall, *J. Am. Diet. Assoc.*, 28, 218, 1952.
43. Pietinen, P., Hartman, A. M., Haapa, E., Rasanen, L., Haapakoski, J., Palmgren, J., Albanes, D., Virtamo, J., and Huttenen, J. K., Reproducibility and validity of dietary assessment instruments. II. A qualitative food frequency questionnaire, *Am. J. Epidemiol.*, 128, 667, 1988.
44. Willett, W. C., Sampson, L., Browne, M. L., Stampfer, M. J., Rosner, B., Hennekens, C. H., and Speizer, F. E., The use of a self-administered questionnaire to assess diet four years in the past, *Am. J. Epidemiol.*, 127, 188, 1988.
45. Baranowski, T., Dworking, R., Henske, J. C., Clearman, D. R., Dunn, J. K., Nader, P. R., and Hooks, P. C., The accuracy of children's self-reports of diet: Family Health Project, *J. Am. Diet. Assoc.*, 86, 1380, 1986.
46. Gersovitz, M., Madden, J. P., and Smiciklas-Wright, H., Validity of the 24-hour dietary recall and seven-day record for group comparisons, *J. Am. Diet. Assoc.*, 73, 48, 1978.
47. Greger, J. L. and Etnyre, G. M., Validity of 24-hour recalls by adolescent females, *Am. J. Public Health*, 68, 70, 1978.
48. Madden, J. P., Goodman, S. J., and Guthrie, H. A., Validity of the 24-hr. recall, *J. Am. Diet. Assoc.*, 68, 143, 1976.
49. Samuelson, G., An epidemiological study of child health and nutrition in a northern Swedish county. 2. Methodological study of the recall technique, *Nutr. Metab.*, 12, 321, 1970.
50. Stunkard, A. J. and Waxman, M., Accuracy of self-reports of food intake, *J. Am. Diet. Assoc.*, 79, 547, 1981.
51. Bingham, S. and Cummings, J. H., Urine nitrogen as an independent validatory measure of dietary intake: a study of nitrogen balance in individuals consuming their normal diet, *Am. J. Clin. Nutr.*, 42, 1276, 1985.
52. Block, G. and Hartman, A. M., Issues in reproducibility and validity of dietary studies, *Am. J. Clin. Nutr.*, 50, 1133, 1989.
53. Pietinen, P., Hartman, A. M., Haapa, E., Rasanen, L., Haapakoski, J., Palmgren, J., Albanes, D., Virtamo, J., and Huttenen, J. K., Reproducibility and validity of dietary assessment instruments. I. A self-administered food use questionnaire with a portion size booklet, *Am. J. Epidemiol.*, 128, 655, 1988.
54. Basiotis, P. P., Welsh, S. O., Cronin, J., Kelsay, J. L., and Mertz, W., Number of days of food intake records required to estimate individual and group nutrient intakes with defined confidence, *J. Nutr.*, 117, 1638, 1987.
55. Nelson, M., Black, A. E., Morris, J. A., and Cole, T. J., Between- and within-subject variation in nutrient intake from infancy to old age: estimating the number of days required to rank dietary intakes with desired precision, *Am. J. Clin. Nutr.*, 50, 155, 1989.
56. Liu, K., Stamler, J., Dyer, A., McKeever, J., and McKeever, P., Statistical methods to assess and minimise the role of intra-individual variability in obscuring the relationship between dietary lipids and serum cholesterol, *J. Chron. Dis.*, 31, 399, 1978.
57. Beaton, G. H., Milner, J., Corey, P., McGuire, V., Cousins, M., Stewart, E., de Ramos, E., Hewitt, D., Grambsch, P. V., Kassim, N., and Little, J. A., Sources of variance in 24-hour dietary recall data: implications for nutrition study design and interpretation, *Am. J. Clin. Nutr.*, 32, 2456, 1979.
58. van Staveren, W. A., de Boer, J. O., and Burema, J., Validity and reproducibility of a dietary history method estimating the usual food intake during one month, *Am. J. Clin. Nutr.*, 42, 554, 1985.
59. National Research Council, *Pesticides in the Diets of Infants and Children,* Committee on Pesticides in the Diets of Infants and Children, Board on Agriculture and Board on Environmental Studies and Toxicology, Commission on Life Science, National Academy Press, Washington, D.C., 1993.
60. Mosteller, F. and Tukey, J. W., *Data Analysis and Regression,* Addison-Wesley, Reading, MA, 1977.
61. Saunders, D. S. and Petersen, B. J., *Introduction to the Tolerance Assessment Systems,* U.S. Environmental Protection Agency, U.S. Government Printing Office, Washington, D.C.
62. USDA, Dataset, U.S. Department of Agriculture, Agricultural Marketing Service, Pesticide Data Program, Washington, D.C., 1993.
63. USDA, Human Nutrition Information Service 1987-88 Nationwide Food Consumption Survey Data Tape, U.S. Department of Agriculture, Washington, D.C., 1988.
64. U.S. Environmental Protection Agency, Guidelines for exposure assessment notice, *Fed. Reg.*, 57, 11888, 1992.

APPENDIX. NATIONAL FOOD CONSUMPTION SURVEYS CONDUCTED IN THE U.S.

U.S. Department of Agriculture — Nationwide Food Consumption Surveys

1965	Individual food intakes collected in Spring quarter; 19,015 indivicuals.
1977–1978	Individual food intakes collected in all four quarters; 36,142 individuals.
1977–1978	Low Income Survey: 12,042 individuals (Food Stamp Program participants and nonparticipants are identified).
1977	Puerto Rico: 7828 individuals, July–December.
1978	Alaska: Winter quarter (January–March), 2305 individuals.
1978	Hawaii: Winter quarter (January–March), 3024 individuals.
1979–1980	Low Income Survey: differs from 1977–1978 Low Income Survey because the 1979 Food Stamp Act changed the purchase requirements and elibility standards for Food Stamp Program participation; 9093 individuals (Food Stamp Participants and nonparticipants are included).
1985	Continuing Survey of Food Intakes by Individuals (CSFII): Women 19–50 years and their children 1–5 years. There were 1503 women and 548 children in the sample for the first dietary recall. This was a longitudinal survey, and data are available for many women and children over a period of 1 year. Four nonconsecutive days of data are available for 1088 women and 371 of their children.
1985	CSFII: Men 19–50 years; 1 day; 1134 individuals; Summer (July–September).
1985	CSFII: Low income; 1 day; women 19–50 years and their children 1–5 years; 1 day; 2120 women and 1314 children; Spring (April–June).
1986	CSFII: Women 19–50 years and their children 1–5 years; 1510 women and 547 children in the sample for the first dietary recall. This was a longitudinal survey and data are available for many women and children over a period of 1 year. Four nonconsecutive days of data are available for 1164 women and 375 of their children; (April–March).
1986	CSFII: Low income; 1 day; women 19–50 years and their children 1–5 years; 1329 women and 816 children; Spring (April–June).
1987–1988	All year; 10,172 individuals.
1988	Bridging Study: 697 women 20–49 years. Experimental design with methodogy bridging 1977 and 1987 surveys.
1989–1991	CSFII: all year; 14,490 individuals.
1994–1996	CSFII: all year.

U.S. Department of Health and Human Services National Health and Nutrition Examination Surveys (NHANES) — Specific Surveys

1971–1975 NHANES I: 10,126 individuals; 1–74 years.

1976–1980 NHANES II: 9983 individuals; 6 months to 74 years.

1982–1984 HHANES: (Hispanic Hanes).

Mexican–American: 8554 interviewed

Cuban–American: 1766 interviewed

Puerto Rican: 3369 interviewed

1988–1994 NHANES III: proposed sample size of 40,000 over the 6 years conducted in two national cycles: (1) Cycle I, 1988–1991; (2) Cycle II, 1991–1994. Followup Supplemental Nutrition Survey to be conducted, with examinees 50 years and older to obtain a second 24-hour recall by telephone in Cycle I (1988–1991).

CHAPTER 7

Nutrition, Pharmacology, and Toxicology: A Dialectic

Daniel J. Raiten

CONTENTS

0-8493-4493-X/97/$0.00+$.50

DEDICATION

This chapter is dedicated to the late Dr. Daphne Roe, a true pioneer whose indomitable spirit and intellect helped to light the path of knowledge for all who follow in her footsteps.

1 INTRODUCTION

The body's ability to process foreign substances obtained through the diet and the environment is dependent on an intricate metabolic system that is inextricably dependent on dietary factors and, ultimately, the nutritional status of the individual. A teleological explanation of the synergism between diet and the detoxification of foreign substances has been offered that is based on the evolutionary change to a complex diet paradoxically rich in essential nutrients yet also containing botanical sources of potentially toxic chemicals.[1] The need to seek out these sources of essential nutrition was linked to the need to develop mechanisms for detoxifying the accompanying toxins. These mechanisms, in turn, became dependent on many of the same essential nutrients, creating an interdependence between nutrition and detoxification. In the modern day human environment, food, air, and water continue to be sources of exposure to potentially harmful toxicants, and this synergism has attained an even more critical role in human health and disease.

In addition to the potential toxicants in the environment, humans must also cope with the exposure to pharmacological substances which, in much the same way as the early botanicals, are being used on a trial-and-error basis to improve the human condition. As was the case with those early exposures to potentially poisonous plants, the exposure to modern medicines can also have a healthful or hurtful outcome. The response, either therapeutic or toxicologic, to any foreign substance, whether a contaminant or a drug, is contingent on numerous factors including stage of development, genetics, general health, and nutritional status.

Understanding the importance of nutrition is a critical component in the appreciation of how the body processes and, in the case of drugs, utilizes these compounds. The examination of the importance of nutrition in the body's response to exposure to foreign substances (referred to as "nutritional toxicology"[2]) continues to evolve as a critical area for study and concern. While many of the mechanisms of detoxification of environmental contaminants and drug utilization and elimination are the same, particular attention will be paid in this chapter to issues related to the role of nutrition in clinical pharmacology.

Clinical pharmacology, or the use of substances to ameliorate specific symptomatology, is a major issue in the chronic care of numerous segments of the U.S. population. The concern about iatrogenic disease is constant and a core element in the often contentious doctor/patient relationship. As will be shown, the relationship between nutrition and pharmacology can be positive in terms of beneficial clinical response to treatments or negative in terms of the potential for iatrogenic or treatment-related detrimental outcomes.

A particular emphasis of this chapter will be the dichotomous relationship between nutrition and toxicology/pharmacology, the impact of diet on xenobiotic processing, and the impact of exposures to xenobiotics, particularly drugs, on nutritional status. The first section of the chapter covers basic concepts about drug metabolism and the effects of nutrition and its processes on drug delivery, metabolism, and therapeutics. A discussion of the importance of relevant developmental changes in physiology to the handling of xenobiotics is included. Finally, a review of issues related to drug-nutrient interactions, including coverage of specific examples of drug-nutrient interactions, is included.

2 OVERVIEW OF DRUG METABOLISM

Nutrition has been defined as the sum total of the processes involved in the ingestion and utilization of nutrients that are subsequently involved in growth, repair, and maintenance of the activities of the body.[3] The processes of nutrition, i.e., ingestion, absorption, metabolism, functional utilization (or distribution), and elimination, are the same processes involved in drug utilization[4] and detoxification of other xenobiotic substances (foreign substances including chemicals or other toxicants) in the environment.[5] A synergism exists between these processes and the ability of both drug and nutrients to perform their respective biological roles; consequently, if a nutrient interferes with the processing of a drug, there will be a change in the effect of the drug. Drugs that affect the processes of nutrition will impact on the nutritional status of the individual which can, in a cyclical manner, have an impact on the health of the individual, as well as on the metabolism and utilization of the subsequent exposures to the drug.

The clinical significance of any drug-nutrient interaction must be viewed in light of the observations by Pinto[6] that the impact on either clinical efficacy or any other functional nutritional or pharmacological parameter of such interactions is contingent on two factors. Significant changes in both drug absorption and bioavailability must occur and the drug in question must have a relatively narrow therapeutic range.[6] Without these two factors such an interaction would have little meaningful impact on drug efficacy. Because drugs have their greatest nutritional impact on micronutrients, the same relationship would presumably hold for assessing the precipitation of a functionally significant drug-nutrient interaction.

As with nutrients, utilization of orally administered drugs depends on the processes of ingestion, digestion, and absorption and, as will be seen, is affected by similar factors. The process of xenobiotic metabolism has two phases: Phase I involves oxidation-reduction reactions resulting in activation, deactivation, or preparation for Phase II, or conjugation, reactions.[5] The Phase II conjugation step results in the attachment of sulfate, glucuronide, or other substances which yields a more polar and water-soluble substance that can subsequently be eliminated.[7] While Phase I reactions may occur in other tissues (e.g., lungs, kidneys, gastrointestinal [GI] tract) and involve other enzyme systems, the majority of these reactions occur in the liver and consist of a series of biotransformations by the microsomal mixed function oxidase system (MFO), culminating in either the activation or deactivation of a given xenobiotic substance.

This microsomal system utilizes the synergism between three primary components, P-450, NADPH-P-450 reductase, and phospholipid (phosphatidylcholine or lecithin).[8] The phospholipid component provides stability for these membrane-bound microsomal enzymes. The dependence on stable membranes introduces the potential for damage due to lipid peroxidation, which implies a role for antioxidants in the protection of the integrity of the MFO system. While research has established their impact on the P-450 enzyme systems, studies of the anti-oxidant properties of vitamin C, vitamin E, selenium (particularly as the latter two are related to glutathione peroxidase activity), and others as they relate to the functional integrity of the MFO systems have been limited. The importance of anti-oxidants in lipid peroxidation has been recently reviewed.[9]

The human P-450-dependent MFO system consists of 31 isoenzyme components that catalyze numerous reactions including epoxidations, *N*-dealkylation, *O*-dealkylations, *S*-oxidations, and hydroxylations of aliphatic and aromatic substances.[10,11] The mechanisms of P-450 reactions have been recently reviewed.[11] It should be noted that these Phase I transformations can be characterized as both toxification (converting otherwise harmless substances such as acetaminophen into potentially toxic substances) and detoxification (deactivation of biologically active substances).[5]

TABLE 1

Mechanisms by which Nutrients and Drugs Can Influence Each Other

Process	Mechanism
Ingestion	Both drugs and disease can cause changes in appetite and nutrient intake; resultant malnutrition can impact on drug efficacy.
Absorption	Drugs and foods can have mechanical effect, via binding or adsorption, that can influence the absorptive processes, resulting in ↑ or ↓ drug and nutrient absorption. Some drugs can affect GI motility, thereby ↑ or ↓ absorption of nutrients. Chemical factors, in particular pH of the stomach contents and the influence of foods therein, can affect the subsequent absorption of drugs.
Transluminal transport	The ability of drugs and nutrients to be transported can depend on such factors as lipid solubility and competition for amino acid transport systems.
Metabolism	The effectiveness of the mixed function oxidase (MFO) and conjugase systems in the liver and elsewhere for converting drugs and nutrients into their active and, ultimately, excretory forms is dependent on the availability of specific nutrient cofactors. In addition, certain drugs can increase the activity of the MFO systems required to convert nutrient precursors into their active forms. Nonnutritive components in foods can induce MFO activity, thereby affecting drug metabolism.
Distribution	The utilization of both drugs and nutrients depends on body composition, the availability and functional integrity of transport proteins, receptor integrity, and intracellular metabolic machinery.
Elimination	Drugs and nutrients can synergistically and competitively interact to cause increased or decreased excretion. Systemic factors such pH and physiological state (e.g., sweating) can dictate whether a drug or nutrient is excreted or resorbed.

Table 1 is a listing of general mechanisms by which nutrients and drugs can affect each other. It is within this context of the processes of nutrition and drug utilization that evidence of drug-nutrient or nutrient-drug interactions will be presented. It must also be stressed that, within the context of clinical pharmacology, awareness should be directed towards the potential for these interactions to enhance or diminish drug efficacy or to increase potential toxicity of a particular drug.

3 ROLE OF NUTRITION IN METABOLISM OF XENOBIOTICS

3.1 Influence of Dietary Factors on Xenobiotic Absorption and Delivery

Table 2 is a listing of some of the numerous examples of the impact of food and specific dietary components on the processes associated with the effective utilization of some pharmacotherapeutic agents. Both the timing and composition of meals can affect drug absorption.[5] In addition, the general nutritional status of any individual will also affect the many components involved in the absorptive process. Marginal nutrient status can interfere with the delivery and metabolism of drugs.[12] Other factors that may affect drug absorption include meal composition (i.e., high fat vs. low fat), the presence or absence of food in the stomach, changes in acidity or alkalinity, or high fiber content which may adsorb therapeutic agents as well as decrease intestinal transit time, thereby decreasing absorption.[5,6,13]

The morphological integrity of the GI tract depends on adequate nutrition. Changes in GI morphology, such as those seen in celiacs disease or HIV infection, affect absorption of both drugs and nutrients. Similarly, malnutrition will influence not only GI morphology but other aspects of the absorptive process, including carrier mechanisms involved in membrane transport and ultimate passage into the circulation.

TABLE 2

Examples of Nutrient Impact on Drugs

Process	Nutrient(s)	Drug or Drug Class	Effect/Comment
Absorption	Calcium (dairy foods)	Tetracycline	↓ Absorption due to formation of insoluble salts.
	Food; food that ↑ urine pH	Aspirin (salicylates)	↓ Absorption, ↓ circulating concentrations; the opposite effect will occur with foods that ↓ urine pH.
	Sodium (salty foods)	Lithium	↓ Absorption and ↑ excretion with increased sodium intake. ↑ Absorption after sodium depletion; ↑ risk of toxicity associated with ↑ reabsorption during sodium depletion.
	Amino acids	L-dopa	↓ Absorption due to competition for intestinal and blood-brain barrier transport system.
	High-fat diet	Theophylline	↑ Absorption, ↑ peak serum levels.
	Food, dairy foods	Isotretinoin	↑ Absorption.
	Foods with pH > 5.5	Drugs with enteric coating	May dissolve enteric coating, thereby resulting in destruction of the drug.
	Food, fiber (bran)	Digoxin	↓ Absorption.
Metabolism	Vitamin B_6	L-dopa	Vitamin B_6 ↑ decarboxylation of L-dopa to dopamine; because dopamine cannot cross the blood barrier, this ↓ therapeutic efficacy.
	Protein	Theophylline	↑ Metabolism and ↓ therapeutic effect observed with high-protein diets.
	Charcoal-broiled meats		↑ Metabolism associated with the induction of liver metabolism by byproducts of the charcoal-broiling process.
	Green vegetables (broccoli, cabbage, Brussels sprouts)	Acetaminophen	↑ Metabolism associated with induction of liver metabolism.
Utilization/ therapeutic effect/toxicity	Caffeine	Neuroleptics	↓ Therapeutic response may be due to ↓ absorption or ↑ liver metabolism.
	Vitamin K-rich foods, e.g., broccoli	Warfarin	↓ Anticoagulant activity.
	Dopamine-rich foods, e.g., fava beans	MAOI	Counteracts therapeutic efficacy of the MAOI by ↑ levels of catecholamines.
	Licorice	Digoxin	Licorice can cause hypokalemia, resulting in potential toxicity of digoxin.
	Caffeine, cola drinks	Theophylline	Potential ↑ in adverse side effects.
	Caffeine, chocolate	MAOI	May precipitate hypertensive reactions.

Note: MAOI = monamine oxide inhibitors.

Dietary substances which bind drugs in the GI tract will prevent absorption and increase excretion. High-fat intake may increase the absorption of fat-soluble drugs, while a low-fat meal may have the opposite effect, as fat and fat-soluble substances must undergo a series of specific steps to allow them to enter the bloodstream. Additionally, free fatty acids and concentrations of the various (i.e., low-density, LDL; high-density, HDL; very-low-density, VLDL) lipoproteins influence the binding and subsequent transport of lipophilic drugs.[6] Factors such as dietary fiber that increase the contractions or motility of the GI tract decrease absorption by speeding up the passage and elimination of GI contents. The interaction between dairy foods and antibiotics is another example of a mechanical effect of a nutrient

TABLE 3
Examples of the Impact of Specific Nutrients on MFO Metabolism

Nutrient	Effect on MFO Metabolism	Potential Mechanism(s)
Protein	Deficiency: can ↓ rate of metabolism; excess: can ↑ rate of metabolism.	↓ Protein synthesis; ↓ in synthesis of other elements, such as hormones, involved in enzyme induction.
Lipids	Deficiency (or diet high in saturated fatty acids): ↓; excess (or diet high in polyunsaturated fatty acids): ↑ activity and inducibility of MFO enzymes.	↓ Activity of MFO possibly connected to the requirement for polyunsaturated fatty acid in the β-position of phosphatidylcholine (lecithin) which is an essential component of the MFO system.
Carbohydrates	Excess: ↓	Secondary effect due to ↓ protein or possibly inhibition of P-450 via ↓ in supporting enzyme components.
Vitamin C	Deficiency: ↓; excess: ↓ MFO activity	Alterations in activities of P-450 and P-450 reductase.
Vitamin B_6	Deficiency: ↓	↓ Synthesis of heme; possible impairment of protein synthesis.
Thiamine (vitamin B_1)	Deficiency: ↑ activity of cytochrome P-450; excess: ↓ (both reductase and P-450)	↑ Activity of specific P-450 isozymes and perhaps other enzymes in deficiency by an unknown mechanism. Effect of excess may be due to ↓ substrate binding.
Riboflavin (vitamin B_2)	Deficiency: ↓ or ↑, depending on the severity.	↓ Reductase activity but ↓ P-450 activity such that metabolism of some drugs will be ↑ while others may be ↓.
Vitamin E	Deficiency: ↓	Because activities of P-450 and reductase are unaffected, it may be due to maintenance of the lecithin component.
Iron	Deficiency: ↓ and ↓; excess: ↓ in microsomal lipid peroxidation	Differential effects on various components of the MFO system. ↑ Lipid peroxidation could lead to damage to the integrity of the system.

Note: MFO = mixed function oxidase.

on drug availability. Because calcium interferes with the absorption of tetracycline via the formation of insoluble calcium compounds, a decrease in the therapeutic response to tetracycline and loss of calcium can be expected to occur when it is given with a glass of milk.

3.2 The Role of Specific Nutrients in Xenobiotic Metabolism

Beyond absorption, the distribution and metabolic fate of drugs and other xenobiotics ultimately depends on nutritional factors. Table 3 provides a list of the impacts of some specific nutrients on drug metabolism. As previously mentioned, xenobiotic metabolism often consists of two phases: the oxidation-reduction reactions of Phase I and the conjugation steps associated with Phase II. Of the two phases, the Phase I components tend to be more susceptible to nutritional effects. Yang et al.[1] described several potential sites for dietary effects on the MFO systems. Nutrients may interfere with (1) the genetic transcription and translation steps for synthesis of the enzymes, (2) the degradation of P-450 mRNA and protein, or (3) MFO enzymes, resulting in either increased or decreased enzymatic activity. The latter effect is demonstrative of the delicate balance and exquisite responsiveness of the body to fluctuations in nutrient availability.

For a given micronutrient, whether vitamin or mineral, there may be a paradoxical effect from an excess or deficiency (Table 3). For example, riboflavin (vitamin B_2) deficiency may result in either an increase or a decrease in the activity of particular components of the MFO system, and these effects may be contingent upon the degree of fluctuation. A mild riboflavin deficiency may result in a decrease in the activity of NADPH reductase, along with an increase in the activity of several other enzymes; a severe deficiency will result in decreased activity of several other MFO enzymes.[1]

The therapeutic efficacy of several classes of drugs involves the ability to act as an antimetabolite or antagonist for specific nutrients. For example, isoniazid and hydralazine are both vitamin B_6 antagonists. Excess intake of vitamin B_6, either through food or with supplements, can affect the efficacy of these drugs. Another example of this type of interaction is between the anticoagulant, warfarin, and vitamin K. The coumarin anticoagulants function by interfering with vitamin K metabolism; consequently, an increase in the intake of green leafy vegetables or use of vitamin K supplements would counteract the anticoagulant effects.

Riboflavin metabolism has been linked to the effectiveness of several drugs. In particular, the effectiveness of the antibiotic Adriamycin® (used as an antineoplastic agent) and antimalarial drugs (e.g., quinacrine) has been tied to their activities in reducing riboflavin availability.[14,15] The clinical trade-off for therapeutic efficacy in these cases may be increased risk for riboflavin deficiency, particularly in the form of the active coenzymatic forms, flavin adenine dinucleotide (FAD) and (FMN). The consequence of this iatrogenic deficiency may be increased toxicity of these drugs to specific tissues such as heart and skeletal muscle.[15] Interestingly, as previously mentioned, riboflavin is a key component of the MFO systems. The question of how this treatment-induced deficiency of riboflavin impacts on subsequent exposures to other concurrent treatments has not been explored.

As co-factors in many of the oxidation-reduction reactions of the MFO system, niacin, riboflavin, pantothenic acid, iron, and copper are required. For the maintenance of membrane integrity and critical supporting components of the MFO systems, protein, lipid, calcium, zinc, magnesium, and vitamins A, E, C, and B_6 are also required.[5] Other nutrients, such as vitamin B_6 (through its action in heme synthesis), provide peripheral support for the MFO systems through their role in biosynthesis of MFO components. Evidence of changes in drug metabolism consequent to changes in the availability of any of these nutrients may be found in numerous studies examining these relationships.[16-20]

3.3 Influence of Dietary Substances on Xenobiotic Metabolism

An extension of the paradoxical relationships between the essential nutrients and potential toxic substances found in food is the interaction between some of these same food-bound chemicals and drugs. Factors other than essential nutrients (i.e., vitamins and minerals) have been identified in foods that influence the activity of the drug modifying conjugase and MFO systems. For example, there have been reports of substances found in cruciferous vegetables (e.g., broccoli, Brussels sprouts, and cabbage) and others found in charcoal-broiled meats that increase (via enzyme induction) the activity of the MFO systems.[21,22]

Caffeine can increase the risk of adverse effects and potential toxicity of theophylline, which is in the same family of chemicals. In addition, caffeine consumption has been linked to decreased therapeutic response to psycho-active medications such as neuroleptics. This may result from an induction of liver metabolism and/or the formation of insoluble precipitates.[23,24] As caffeine consumption in the form of coffee, tea, and soft drinks is a common phenomenon, these contingencies may have considerable clinical importance.

4 DEVELOPMENTAL CHANGES IN XENOBIOTIC METABOLISM

4.1 Age-Related Changes in Physiology and Metabolism

Generically, it is clear that nutritional factors have an intimate role in drug metabolism; however, direct evidence is lacking to address the question of whether (or not) age-dependent physiological changes in infants, children, and the elderly influence the role of nutritional status in pharmacokinetics. The role of nutrition in these systems during these critical periods of development has not been studied. Similarly, the impact of nutrition on the decreased rates of drug metabolism in the elderly has not been systematically addressed.[13]

Evidence exists to support the importance of developmental differences, particularly in newborn and infant animals, to the response to pharmacotherapeutics. Several studies have reported adverse effects to neurological parameters in fetal and newborn animals exposed *in utero* to various classes of drugs including antidepressants and antihypertensives.[25,26] Other studies have reported differences in neurodevelopmental parameters (i.e., neurotransmitter concentrations, receptor number and function) in malnourished infant animals exposed to drugs neonatally.[27] It is clear that early exposure to some drugs can have deleterious outcomes in humans as illustrated by a recent report of developmental problems in children exposed *in utero* to a potent anticonvulsant.[28] It is less clear whether these changes in neurophysiology are consequent to a toxic effect of the unadulterated drug, a developmental aspect of drug metabolism, or a synergism between these two factors.

A direct link has not been established between the changes in requirements for specific nutrients that occur during the life-cycle and drug delivery and/or metabolism. Evidence is well established for developmental changes (i.e., increased activity during the neonatal period) in both stage I (oxidative processing associated with the MFO systems) and stage II (conjugation).[29,30] Sonawane et al.[31] observed that the adult progeny of rats fed diets differing in lipid content had changes in microsomal MFO-driven drug metabolism. Sonawane and Catz[32] further reviewed the research on the developmental aspects of the drug-nutrient relationship and concluded that there may be an increased susceptibility for children to the effects of malnutrition on drug metabolism.

The stage of development has a profound impact on an individual's ability to handle exposure to drugs or other xenobiotics. Table 4 summarizes available information with regard to the developmental changes associated with the processes of drug and nutrient utilization in infants, children, and the elderly and is based on data presented in several excellent reviews.[13,33-36] It should be noted that the factors itemized in Table 4 do not reflect the influence of genetics or environmental factors such as smoking, alcohol consumption, and diet.

4.2 Age-Related Dietary Factors

In addition to the normal changes that occur throughout the life cycle, including a diminution of taste and smell acuity contributing to decreased food intake in the elderly,[37] a substantial body of literature attests to the high risk of specific nutrient deficiencies that may occur at various points in the aging process. Growth spurts throughout infancy, childhood, and adolescence put significant stress on the body and its function, as well as creating the potential for increased requirements for specific nutrients such as protein, energy, calcium, and iron. Changes associated with aging, such as those outlined in Table 4, also create the potential for nutritional inadequacies.

Data from Phase I of the third National Health and Nutrition Examination Survey (NHANES III) provide evidence of significant age- and sex-dependent differences in several essential nutrients including iron, calcium, carotenoids, and vitamin C.[38] Previous studies identified additional nutrients that are particularly susceptible to deficiency as a result of inadequate intake during infancy (the potential for individual nutrient deficiencies is dependent on method of feeding, i.e., breast milk vs. formula), childhood (calcium, iron, zinc, vitamin C, and vitamin A), adolescence (vitamin A, vitamin B_6, riboflavin, iron, calcium, and zinc) and the elderly (multinutrient deficiencies and nutrient imbalances due to decreased intake and physiological changes).[39] Again, the risk of nutritional problems within age groups is also influenced by gender and demographics (socio-economic status, ethnicity, use of alcohol, etc.), which must be factored into any evaluation of the diet/drug relationship.

TABLE 4

Age-Dependent Changes in the Processes of Nutrition and Drug Utilization

Process	Factors	Infants (<2 years)	Children (2–18 years)	Elderly (>65 years)
Absorption	Gastric pH	↓	↑	↑ Gastric pH
	GI motility	↑ (Particularly in premature infants)	↑ GI motility patterns in younger children	↓ GI motility patterns
	Other	↓ Gastric absorptive capacity, ↓ pancreatic enzyme function and bile acid secretion		↓ GI absorptive surface, ↓ blood to GI tract
Metabolism				Factors include ↓ liver mass, liver blood flow, and hepatic enzyme activity and inducibility
Distribution	% body fat	↑	↑ Through early childhood; ↓ in females through age 17. At about 13 years, males have approx. twice the % body fat of females.	↑ Along with ↓ lean body mass
	% body water	↓ (Estimated at from 75–85% compared with 55% in adults)	↓ From infancy through about 12 years, at which time adult levels are achieved	↓
	Plasma proteins	↓ Plasma albumin, total protein concentrations, and α-1 glycoproteins	Little change in plasma protein concentrations in healthy children	↓ Serum albumin
	Other	Acid-base disturbance, competition for binding sites by endogenous substances, and differences in the nature of transport proteins		↓ Cardiac output, ↓ cerebral blood flow, and changes in protein binding
Elimination	GFR[a]	↓ (≈ 20% of adult value in newborn and ↑ to adult levels at about 2 years). Renal function is not fully developed at birth; infants do not have diurnal rhythm in renal function; they do have lower pH which will affect drug excretion.	GFR remains steady through to adulthood	↓ (↓ Number of glomeruli), ↓ in renal blood flow and renal tubular secretory function. Other factors related to hepatobiliary excretion pathways include ↓ liver size and blood flow.

[a] GFR = glomerular filtration rate.

5 SPECIFIC ISSUES IN DRUG-NUTRIENT INTERACTIONS

The other side of the nutrition-pharmacology dichotomy is the effect that drugs have on both the processes of nutrition and the availability and viability of specific nutrients. Table 5 is a list of some examples of the impact of specific drugs on the various processes of nutrition.

Malnutrition may be a consequence of changes in any of the processes of nutrition and may result in either over- or undernutrition. Overnutrition leading to obesity may be the result of decreased activity, psychosocial influences, genetics, metabolic changes secondary to an illness, overeating, or drug use. Studies have demonstrated that obesity can be associated with changes in drug disposition, particularly hepatic drug clearance and subsequent renal clearance.[40,41] In addition to pharmacokinetic parameters, obesity can influence drug efficacy in other ways. For example, neuroleptic and antipsychotic drugs can generally be characterized as lipophilic or fat-soluble, and, consequently, their absorption can be facilitated by the presence of fat in the diet. Furthermore, their lipophilic nature allows them to be deposited in adipose or fat tissue. This deposition becomes clinically significant both during drug withdrawal when storage depots of the drug can be mobilized and during periods of weight loss.[42]

Undernutrition, associated with deficiencies of one or more nutrients has the following five primary causes: (1) inadequate ingestion, (2) inadequate absorption, (3) inadequate utilization (metabolism), (4) increased excretion, and (5) increased requirement.[43] As will be demonstrated, drugs can influence any of these stages. Clinically, the major problem with detecting a potential drug-nutrient interaction, aside from a sensitivity to the possibility of such a potential side effect, is that any manifestation of a drug-induced nutrient deficiency may be subtle and/or easily confused with symptoms of the underlying condition for which the drug was prescribed.

The stages of a nutritional deficiency have been described as progressing from such generalized signs as weight loss or lethargy to specific biochemical and anatomical lesions and eventually death.[44] A nutritional deficiency might appear as a detectable change in circulating metabolites or vitamins or may be masked as an unrelated manifestation of the illness. For example, someone receiving a psycho-active medication might appear listless, restless, anxious, agitated, drowsy, depressed, apathetic, or lethargic.[45] Several nutrient deficiencies have been shown to have similar symptoms. Symptoms associated with deficiencies of various vitamins include depression, apathy, anxiety, irritability (thiamine, riboflavin, niacin, pyridoxine, biotin, vitamin B_{12}, folic acid), and personality disorder (riboflavin and vitamin C).[46] It is conceivable that some of the side effects associated with psychopharmacological interventions are nutritionally mediated. The key to a differential diagnosis is a sensitivity to the possibility that the symptoms or signs demonstrated in a given patient might in fact be a secondary effect of the treatment.

5.1 DRUGS AND DIETARY INTAKE

Numerous mechanisms exist by which drugs might affect dietary intake. Table 6 is a list of some examples of drugs that affect food intake categorized by three primary mechanisms: appetite regulation, alterations in taste, and alterations in smell. The ability of these mechanisms to influence dietary intake involves a complex interplay between neurotransmitters and endogenous and exogenous substances. The existing theories about how drugs may influence any of the factors associated with appetite involve modulations in both central and peripheral nervous system function and have recently been reviewed.[37,47] In addition to the prime mechanisms of appetite, taste, and smell, many drugs can have an indirect effect on intake through changes such as depression, lethargy, musculoskeletal problems (such as rigidity), dry mouth, gastrointestinal disturbances, nausea, and vomiting.[48]

TABLE 5

Examples of Drug-Nutrient Interactions

Process	Drug or Drug Class	Nutrient(s)	Effect(s)/Comment
Absorption	Antibiotics (tetracycline)	Calcium and other divalent cations	↓ Chelation and formation of insoluble complexes
	Mineral oil	Fat-soluble vitamins, particularly β-carotene	↓ Sequestration of fat-soluble vitamins and decreased transit time
	Laxatives	Calcium and potassium	↓; ↓ transit time
	Cholestyramine	Fat-soluble vitamins	↓; used to ↓ cholesterol absorption; drug-induced ↓ in bile acid activity causes ↓ fat digestion
	Neomycin	Fat	↓; inhibits pancreatic lipase activity, ↓ fat digestion and absorption
	Methotrexate	Folate; possibly other nutrients	↓; GI mucosal injury
	Salicylate (aspirin)	Glucose, vitamin B_{12}, vitamin C	↓; causes bleeding and damage to GI tract
	Antacids	Folate, iron, phosphate; riboflavin	↓ Absorption due to ↑ pH; binding
	Penicillamine	Vitamin B_6, zinc, copper	↓; Chelation
	Oral contraceptives	Copper	↑; may result in ↑ blood levels
Metabolism	Amitriptyline, chlorpromazine	Riboflavin	May interfere with conversion to FAD/FMN, the coenzymatic forms
	Methotrexate	Folate	Inhibition of dihydrofolate reductase, key enzyme in folate metabolism
	Isoniazid	Vitamin B_6	Deficiency caused by formation of a complex and interference with metabolism and absorption
	Alcohol	Vitamin A; vitamin D	Decreases conversion to active retinal; changes renal activation of vitamin D
	Theophylline	Vitamin B_6	Inhibition of pyridoxal kinase results in ↓ in circulating levels of the coenzyme pyridoxal-5′-phosphate (PLP)
	Phenytoin	Vitamin D	Interferes with activation of vitamin D
Transport and excretion	Alcohol	Vitamin B_6	Acetaldehyde displaces the vitamin from its binding protein
	Boric acid	Riboflavin	Displacement from carrier protein resulting in ↑ free riboflavin and ↑ excretion.
	Aspirin	Folic acid	↓ Binding to plasma protein
	Chelaters (penicillamine, EDTA)	Trace minerals (copper, zinc)	Chelation therapy for heavy metal toxicity (e.g., lead) can result in chelation of other essential minerals and ↑ excretion

TABLE 6

Examples of Drugs that Affect Food Intake

Mechanism	Drug	Use
Taste	Azidothymidine (AZT)	Antiviral drug used in HIV infection
	Acyclovir	Antiviral used in treatment of herpes simplex
	Diltiazem (Cardizem®)	Antihypertensive (calcium channel blocker)
	Amitriptyline (Elavil®)	Antidepressant
	Fluoxetine (Prozac®)	Antidepressant
	Amphetamine	Sympathomimetic; drug of abuse
	Captopril	Antihypertensive (ACE inhibitor)
	Allopurinol	Anti-inflammatory
	Dexamethasone	Anti-inflammatory
	Gold	Anti-inflammatory; antirheumatic
	Ampicillin	Antimicrobial
	Tetracyclines	Antimicrobial
	Lithium	Bipolar disorder
	Phenytoin	Anticonvulsant
Smell	Opiates (codeine, morphine)	Analgesia; drugs of abuse
	Nifedipine	Antihypertensive (calcium-channel blocker)
	Amitriptyline	Antidepressant
	Amphetamine	Sympathomimetic; drug of abuse
	Streptomycin	Antimicrobial
Appetite	Methylphenidate (Ritalin®)	↓; amphetamine-like drug used for treatment of hyperactivity in children
	Fenfluramine	Appetite suppression
	Ganciclovir	↓; antiviral for treatment of cytomegalovirus in HIV patients, and herpes viruses
	Fluoxetine (Prozac®)	↓; antidepressant
	Lithium	↓; treatment of bipolar disorder
	Alprazolam (Xanax®)	↓; antianxiolytic (anxiety disorders)
	Cyclosporine	↑; an immunomodulator used in HIV infection
	Vinblastine	↓; anticancer
	Cyproheptadine (Periactin®)	↓; antihistamine
	Diazepam (Valium®)	↓; anti-anxiolytic
	Chlorpromazine	↓; antipsychotic

5.2 Drugs and Nutrient Absorption and Transport

Drugs can affect nutrient absorption in several ways: (1) exerting changes in intestinal motility, thereby changing intestinal transit time; (2) changing the GI morphology or mucosal surface, resulting in changes in the absorptive area; (3) affecting intestinal or transluminal transport; or (4) altering the chemical environment in the GI tract either through changes in pH or through chemical or physical reactions, such a sequestration or binding. Examples of each of these effects are shown in Table 5.

The transport of nutrients often relies on the same proteins and carrier mechanisms as described for drugs. In some cases, drugs can compete with nutrients for binding sites on transport proteins. When this happens, as in the cases of aspirin and folate[49] or alcohol and vitamin B_6,[50] the nutrients are more susceptible to metabolic breakdown and excretion, resulting in deficiency.

5.3 Drugs and Nutrient Metabolism: Case Studies of Commonly Used Drugs

Chronic drug therapy may lead to impairment of liver function resulting in an alteration in the ability of this organ to produce enzymatically the coenzymatic forms of many vitamins which are essential for carrying out normal metabolic functions peripherally as well as in the central nervous system (CNS). Drugs may also directly interfere with the conversion of dietary forms of essential vitamins into the biologically active coenzymatic forms. Several case studies of particular classes of commonly prescribed drugs and their impact on specific nutrients will help to demonstrate the potential importance of this relationship.

5.3.1 Psycho-active Drugs

Neuroleptic drugs, also referred to as antipsychotic drugs or tranquilizers, constitute one of the major classes of drugs prescribed for the amelioration of behavioral anomalies in both children and adults who presumably have not responded to alternative, drug-free therapies. Neuroleptic drugs are the most commonly prescribed pharmacological treatment for children with severe developmental disabilities such as autism or pervasive developmental disability.[51] Depending on the dose and duration, neuroleptics have been associated with metabolic and physiological side effects that may interfere with the processes of nutrition. These side effects include abdominal pain and distention and vomiting. A more chronic problem is constipation caused by the anticholinergic effect of neuroleptics.[42] This side effect is often treated with laxatives, e.g., mineral oil or other cathartic drugs, thereby creating another potentially dangerous nutritionally related problem. Laxatives, such as mineral oil, can bind nutrients or prevent their absorption, thereby creating the possibility of nutrient deficiencies.[13] In addition, the therapeutic effect of these drugs is decreased transit time in the GI tract, which can cause decreased nutrient absorption.

Perhaps the most common nutrition-related side effect of neuroleptic therapy is obesity. Martinez et al.[52] reported significant increases in the incidence of obesity in both male and female psychiatric patients receiving psychotropic medications compared with drug-free patients. Many of the metabolic effects that have been associated with chronic neuroleptic therapy have been related to abnormalities in the endocrine status of the individual. Neuroleptics have been shown to stimulate prolactin secretion,[53] which has been implicated in producing a hyperglycemic-diabetogenic effect and enhancing fat deposition. Others have suggested that weight gain associated with chronic neuroleptic therapy may be caused by the combination of fluid retention, fat redistribution, and altered glucose sensitivity.[54]

Awad[55] suggested several other possible drug-induced mechanisms that might lead to obesity. Improved mental status, mouth dryness associated with increased consumption of soft drinks, and decreased activity are the most commonly cited possibilities. Another explanation might be that neuroleptic-induced changes in central neurochemistry are associated with appetite control. Aside from the direct impact of obesity, Mallick[56] pointed out some of the deleterious consequences of weight control and diets on growth, development, and mental functioning of children and adolescents. Furthermore, as behaviorally impaired adults are at risk for the same chronic disease as the general population, obesity will ultimately affect the length and quality of life of these individuals.

The hepatotoxic effects of neuroleptics constitute one of the more serious side effects of prolonged treatments and are often accompanied by jaundice and elevated serum alkaline phosphatase levels. Little information is available regarding the indirect effects of neuroleptics on the metabolic integrity of the liver in terms of its ability to form the necessary coenzymatic forms of many of the water-soluble or B vitamins; however, it is worth noting that recent evidence supports the role of elevated alkaline phosphatase levels in the production of vitamin B_6 deficiency.[57]

Perhaps the most well-known and commonly acknowledged drug-nutrient interaction involves the class of antidepressants known as monoamine oxide inhibitors (MAOI). As is the case with similar types of drugs (e.g., the antiparkinsonian drug, carbidopa, which contains an enzyme inhibitor for aromatic amino acid decarboxylase), these drugs act via a nonspecific inhibition of a systemic enzyme — in this case, monoamine oxidase. This enzyme is responsible for the de-activation of the so-called biogenic amines (dopamine, norepinephrine, serotonin, tyramine, and histamine). All of these substances are neuroactive. Some act primarily on nervous system activity (dopamine, norepinephrine), while others have systemic effects on functions such as blood pressure regulation (tyramine, serotonin). In addition to their use as antidepressants, MAOIs are also used as antihypertensive medications and antineoplastic and antimicrobial agents.[58]

By blocking the activity of the oxidase, MAOIs used as antidepressants are presumed to increase the brain concentrations of serotonin, thereby ameliorating symptomatology associated with depression; however, because the oxidase is ubiquitous, its inhibition also causes increased concentrations of the other monoamines. The nutrient-related concern often associated with this action is the tyramine content of certain foods such as wine, smoked meats, and aged cheeses. Tyramine is a vasoactive amine that causes vasoconstriction and, potentially, elevations in blood pressure; therefore, individuals taking these drugs are cautioned to abstain from eating these foods in order to avoid the precipitation of a hypertensive episode.[39,58] The use of MAOIs is a situation where a class of drugs, through its therapeutic effect, changes a metabolic function of the body and thereby increases the potential toxicity of a commonly found substance in foods.

A more direct and insidious effect of both neuroleptic drugs and certain antidepressants has been demonstrated by Pinto et al.[59] Prompted by the similar structures of riboflavin and the drugs chlorpromazine, imipramine, and amitriptyline, these investigators sought to determine any effects these drugs may have on the conversion of riboflavin to its active coenzyme form, flavin adenine dinucleotide (FAD). The results indicated that all three drugs inhibited the activation of riboflavin in liver, cerebrum, and cerebellum. After periods of 3 to 7 weeks at doses comparable on a weight basis to those used clinically, chlorpromazine treatment led to a deficiency even in animals fed 30 times their recommended daily allowance of vitamin B_2.[59] Of the three, chlorpromazine was shown to have an accelerating effect on the development of riboflavin deficiency that was enhanced by an increase in urinary excretion of the vitamin.[60] These authors concluded that their results "raise the possibility that drug-induced nutritional deficiency may be an unrecognized and undesirable result of antipsychotic drug therapy, particularly when treatment is prolonged." More recently these same investigators reported an increase in the urinary excretion of riboflavin in male human volunteers given chlorpromazine, indicating that the animal findings translate to humans.[6]

The studies of the interaction between riboflavin and the antimalarial, antineoplastic, and psycho-active drugs conducted by Pinto et al.[14,15,59,60] reflect the importance of "drug mimicry"[6] as a factor in both drug efficacy and, potentially, toxicity. In the case of the antimalarial and antineoplastic drugs, the beneficial effect comes from the ability to prevent conversion of riboflavin to its active coenzyme form, FAD; however, in the case of the psycho-active drugs, it is not clear whether this effect enhances or ultimately detracts from the therapeutic efficacy of these drugs. The more likely possibility is the latter as the therapeutic effects of these drugs are all centered around interference with the neurotransmitter metabolism (either serotonin or the catecholamines, dopamine and norepinephrine) in the CNS.

Recall that the toxicity of the antimalarial and chemotherapeutic drugs has been shown to be increased in specific tissues such as the heart and skeletal muscle and related to the drug-induced riboflavin deficiency.[15] An additional concern with regard to the use of the psycho-active drugs is that in addition to the role of riboflavin as an essential vitamin involved in numerous aspects of intermediary metabolism, riboflavin is also a key component (as the co-factors, FAD and FMN) in the MFO system. Consequently, a definable deficiency of riboflavin, particularly of the potential magnitude described in these studies,[59,60] would have significant consequences to the patient receiving chronic therapy with these drugs. The clinical significance of the relationship between riboflavin and the psycho-active drugs remains to be elucidated.

In addition to psycho-active drugs prescribed for the treatment of affective and behavioral disorders, the use of "recreational" drugs or drugs of abuse can also affect the nutritional health of the user. Alcohol and the opiate derivatives, morphine, cocaine, and heroin, can all have profound effects on nutritional status, particularly metabolic changes, that are independent of the life-style circumstances (e.g., unemployment, homelessness, or underlying emotional problems) of the user.[61] In the case of alcohol, some of the more severe outcomes of chronic consumption, such as Wernicke-Korsakoff syndrome disease, are associated with deficiencies of essential nutrients such as thiamine.[62] Alcohol can influence nutritional status through a variety of ways including changes in absorption, metabolism, and transport of essential nutrients.[63]

5.3.2 *Anticonvulsants*

Another major class of drugs that has been associated with nutritional problems is the anticonvulsants. Paramount among these effects is the disturbance in normal vitamin D metabolism and function. Normal functioning of vitamin D requires hepatic conversion of the dietary or endogenous provitamin to the 25-hydroxy-D3 form. This in turn is converted to the biologically active 1,25-hydroxy-D3 form in the kidney.[39] This active form of vitamin D plays a pivotal role in calcium absorption and bone homeostasis. Reduced serum calcium and elevated alkaline phosphatase was reported in individuals maintained on chronic anticonvulsant therapy.[64] Other studies have demonstrated that anticonvulsant therapy results in a reduction in circulating forms of vitamin D in epileptic patients maintained on single and combined regimens.[65,66] Several reports have shown this impairment to culminate eventually in osteomalacia[67] and rickets[66] in adults and young individuals, respectively. Although numerous factors have been implicated in the development of these clinical manifestations (i.e., diet, ambulation, and exposure to sunlight), anticonvulsants appear to be the major contributing factor responsible for the observed manifestations.

Chronic anticonvulsant therapy has also been associated with impaired red blood cell synthesis or hemopoiesis. In particular, anticonvulsant drugs have been shown to precipitate megaloblastic anemia, characterized by abnormal red cell maturation and often associated with low serum folic acid.[68] Morphological changes in the epithelial lining of the GI tract, lip, tongue, and inner surface of the mouth have also been observed in folate deficiency associated with anticonvulsant therapy.[69] Since folic acid plays a major role in DNA synthesis, it is not surprising that epithelial tissues with such a high rate of turnover and normal processes involved in red blood cell synthesis are affected earlier and to a greater extent than other tissues as a result of a drug-induced folate deficiency.

It is important to recognize, however, that, prior to the appearance of severe clinical manifestations, a deficiency of folic acid has been associated with various neurological and behavioral symptoms, including mild polyneuropathies, fatigue, hypotonia, depression, and impaired intellectual performance.[70] Other investigators[71] have suggested the presence of thiamin deficiency in folate-deficient patients, while others[72] reported microcytic hypochromic anemia (characteristic of pyridoxine deficiency and possibly resulting from elevations in alkaline phosphatase activity) in a young adolescent patient treated for petit mal seizures. Thus, it is unclear to what extent the observed changes in neurological functions and behavioral symptoms can be attributed solely to a deficiency of folic acid as opposed to a multinutrient deficit. Additional studies have implicated anticonvulsant therapies in deficiencies of biotin,[73] zinc, and vitamin E.[74] These findings raise the possibility of multiple deficiencies, especially in individuals receiving a marginal diet.

5.3.3 *Theophylline*

Since 1987, a group of investigators has reported on the impact of theophylline therapy, the primary treatment modality for asthmatics and respiratory problems in premature infants, on vitamin B_6 metabolism.[75-78] In particular, these studies reported a reduction in plasma concentrations of the active co-enzymatic form of vitamin B_6, pyridoxal-5′-phosphate (PLP), but not pyridoxal (PL), levels consequent to theophylline therapy in normal and asthmatic subjects. The mechanism of this effect is not clear at this time, but may be due to a competitive inhibition of pyridoxal kinase. In addition to the reported decrease in plasma and erythrocyte PLP concentrations, Ubbink et al.[76,77] also observed significant changes in other indirect or functional indices of vitamin B_6, including increased levels of urinary xanthurenic acid in response to tryptophan loading and reduced activities of both of the vitamin B_6-dependent aminotransferases, aspartate aminotransferase (AST) and alanine aminotransferase (ALT).

The evidence provided by the studies by Ubbink and colleagues,[75-77] indicates that chronic theophylline therapy, maintained at therapeutic levels, causes an inhibition of a key enzyme in the conversion of PL to PLP, thereby creating an apparent deficiency of vitamin B_6 as reflected by decreased levels of PLP. All of the biochemical changes in vitamin B_6 associated with theophylline therapy reportedly have been reversed with pyridoxine (PN) supplementation.[76-78] The fact that supplementation with PN overcomes changes in such parameters as tryptophan metabolism and aminotransferase activity may indicate that theophylline not only results in a biochemical deficiency, (i.e., reduced PLP levels), but may also have functional consequences.

Weir et al.[78] noted that this drug-nutrient interaction may be responsible in part for the toxicity associated with chronic theophylline therapy. An additional concern is raised by a series of reports that have implicated theophylline in both behavioral and cognitive problems in asthmatic children.[79-81] Other neurological signs observed with theophylline administration — increased excitability by electroencephalograms (EEG) and reductions in rapid eye movement (REM) sleep — are similar to side effects observed in vitamin B_6 deficiency.[82] Because functional deficiency of vitamin B_6 has been associated with similar effects, the potential adverse effects of the vitamin B_6/theophylline relationship warrants further attention, especially in children.

6 CONCLUSIONS

We must recognize that clinical pharmacology has other consequences besides the amelioration of specific symptoms. The potential functional impact of drug- or diet-related nutrient deficiencies on clinical outcomes has received minimal attention. Moreover, the efficacy of a given treatment modality is inextricably linked to the nutritional status of the patient. Future studies should address these issues, while clinicians and caregivers must be vigilant in their recognition and monitoring of those clinical parameters that may indicate a potential problem.

The dynamic and dialectical relationship between nutrition and xenobiotics is often beneficial as well as potentially harmful. As human exposure to a more complex array of food and chemicals increases, it is vital that our understanding of the intricate interaction between these substances continues to develop. Environmental and evolutionary pressure will result in unforeseen changes in an already enigmatic system. In order to respond to the health problems that such changes engender, the basic and clinical study of these changes must be expanded.

ACKNOWLEDGMENT

The author wishes to acknowledge the kind support and editorial assistance of Ms. Carol Rilley, Administrator, FASEB/LSRO.

REFERENCES

1. Yang, C. S., Brady, J. F., and Hong, J-Y., Dietary effects on cytochromes P450, xenobiotic metabolism and toxicity, *FASEB J.*, 6, 737–744, 1992.
2. Hathcock, J. N., Nutritional toxicology: definition and scope, in *Nutritional Toxicology*, Vol. 1, Hathcock, J. N., Ed., Academic Press, New York, 1982, pp. 1–15.
3. Raiten, D. J., Nutrition and HIV infection, *Nutr. Clin. Pract.*, 6(Suppl), 1S-94S, 1991.
4. Hayes, J. R. and Borzelleca, J. F., Nutrient interaction with drugs and other xenobiotics, *J. Am. Diet. Assoc.*, 85, 335–339, 1985
5. Hathcock, J. N., Metabolic mechanisms of drug-nutrient interactions, *Fed. Proc.*, 44, 124–129, 1985.
6. Pinto, J. T., The pharmacokinetic and pharmacodynamic interactions of food and drugs, *Top. Clin. Nutr.*, 6, 14–33, 1991.
7. Guengerich, F. P., Effects of nutritive factors on metabolic processes involving bioactivation and detoxication of chemicals, *Ann. Rev. Nutr.*, 4, 207–231, 1984.
8. Kaschnitz, R. M. and Coon, M. J., Drug and fatty acid hydroxylation by solubilized human liver microsomal cytochrome P450: phospholipid requirement, *Biochem. Pharmacol.*, 24, 295–297, 1975.
9. Halliwell, B., Free radicals and antioxidants: a personal view, *Nutr. Rev.*, 52, 253–265, 1994.
10. Guengerich, F. P., Characterization of human cytochrome P450 enxymes, *FASEB J.*, 6, 745–748, 1992.
11. Hollenberg, P. F., Mechanisms of cytochrome P450 and peroxidase-catalyzed xenobiotic metabolism, *FASEB J.*, 6, 686–694, 1992.
12. Campbell, T. C., Nutrition and drug-metabolizing enzymes, *Clin. Pharm. Therapeut.*, 22, 699–706, 1977.
13. Roe, D. A., Drugs and nutrition in the elderly, in *Geriatric Nutrition*, Roe, D. A., Ed., Prentice Hall, Englewood Cliffs, NJ, 1992, pp. 182–207.
14. Dutta, P., Raiczyk, G. B., and Pinto, J., Inhibition of riboflavin metabolism in cardiac and skeletal muscles of rats by quinacrine and tetracycline, *J. Clin. Biochem. Nutr.*, 4, 203–208, 1988.
15. Pinto, J., Raiczyk, G. B., Huang, Y. P., Rivlin, R. S., New approaches to the possible prevention of side effects of chemotherapy by nutrition, *Cancer*, 58, 1911–1914, 1986.
16. Becking, G. C., Hepatic drug metabolism in iron-, magnesium-, and potassium-deficient rats, *Fed. Proc.*, 35, 2480–2485, 1976.
17. Bidlack, W. R. and Smith, C. H., The effect of nutritional factors on hepatic drug and toxicant metabolism, *J. Am. Diet. Assoc.*, 84, 892–898, 1984.
18. Bidlack, W. R., Brown, R. C., and Mohan, C., Nutritional parameters that alter hepatic drug metabolisim, conjugation, and toxicity, *Fed. Proc.*, 45, 142–148, 1986.
19. Campbell, T. C. and Hayes, J. R., The effect of quantity and quality of dietary protein on drug metabolism, *Fed. Proc.*, 35, 2470–2474, 1976.
20. Varma, D. R., Protein deficiency and drug interactions, *Drug Dev. Res.*, 1, 183–198, 1981.
21. Anderson, K. E., Influences of diet and nutrition on clinical pharmacokinetics, *Clin. Pharmacol.*, 14, 325–346, 1988.
22. Anderson, K. E., Pantuck, E. J., Conney, A. H., and Kappas, A., Nutrient regulation of chemical metabolism in humans, *Fed. Proc.*, 44, 130–133, 1985.
23. Lasswell, W. L., Weber, S. S., and Wilkins, J. M., *In vitro* interaction of neuroleptics and tricylic antidepressants with coffee, tea, and gallotannic acid, *J. Pharmaceut. Sci.*, 73, 1056–1058, 1984.
24. Mikkelsen, E. J., Caffeine and schizophrenia, *J. Clin. Psych.*, 39, 732–735, 1978.
25. Mirmiran, M., Brenner, E., Van der Gugten, J., and Swaab, D. F., Neurochemical and electophysiological disturbances mediate developmental behavioral alterations produced by medicines, *Neurobehav. Toxicol. Teratol.*, 7, 677–683, 1985.
26. Mirmiran, M. and Swaab, D. F., Influence of drugs on brain neurotransmitters and behavioral states during development, *Dev. Pharmacol. Ther.*, 10, 377–384, 1987.
27. Goodlett, C. R., Valentino, M. L., Resnick, O., and Morgane, P. J., Altered development of responsiveness to clonidine in severely malnourished rats, *Pharmacol. Biochem. Behav.*, 23, 567–572, 1985.
28. Scolnik, D., Nulman, I., Rovet, J., Gladstone, D., Czuchta, D., Gardner, H. A., Gladstone, R., Ashby, P., Weksberg, R., Einarson, T., and Koren G., Neurodevelopment of children exposed *in utero* to phenytoin and carbamazepine monotherapy, *J. Am. Med. Assoc.*, 271, 767–770, 1994.
29. Dutton, G. J., Developmental spects of drug conjugation, with special reference to glucuronidation, *Ann. Rev. Pharmacol. Toxicol.*, 18, 17–35, 1978.
30. Neims, A. H., Warner, M., Loughnan, P. M., and Aranda, J. V., Developemental aspects of the hepatic cytochrome P450 monoxygenase system, *Ann. Rev. Pharmacol. Toxicol.*, 16, 427–445, 1976.
31. Sonawane, B. R., Coates, P. M., Yaffe, S. J., and Koldovsky, O., Influence of perinatal nutrition on hepatic drug metabolism in the adult rat, *Dev. Pharmacol. Ther.*, 6, 323–332, 1983.
32. Sonawane, B. R. and Catz, C., Nutritional status and drug metabolism during development, in *Drug Metabolism in the Immature Human*, Soyka, L. F. and Redmondm, G. P., Eds., Raven Press, New York, 1981, pp. 87–99.

33. Crom, W. R., Pharmacokinetics in the child, *Environ. Health Perspect.*, 102(Suppl. 11), 103–106, 1994.
34. Mayerson, M., Pharmacokinetics in the elderly, *Environ. Health Perspect.*, 102(Suppl. 11), 119–124, 1994.
35. Milsap, R. L. and Jusko, W. J., Pharmacokinetics in the infant, *Environ. Health Perspect.*, 102(Suppl. 11), 107–110, 1994.
36. Roberts, J. and Tumer, N., Age and diet effects on drug action, *Pharmac. Ther.*, 37, 111–149, 1988.
37. Schiffman, S., Changes in taste and smell: drug interactions and food preferences, *Nutr. Rev.*, 52, S11–S14, 1994.
38. Alaimo, K., McDowell, M. A., Briefel, R. R., Bischof, A. M., Caughman, C. R., Loria, C. M., and Johnson, C. L., Dietary intake of vitamins, minerals, and fiber of persons ages 2 months and over in the United States: Third National Health and Nutrition Survey, Phase I, 1988–91, *NCHS Adv. Data Vital Health Stat.*, No. 258, 1991.
39. Mahan, L. K. and Arlin, M., *Krause's Food Nutrition and Diet Therapy*, 8th ed., W.B. Saunders, Philadelphia, PA, 1992.
40. Abernathy, D. R. and Greenblatt, D. J., Pharmacokinetics of drugs in obesity, *Clin. Pharmacokin.*, 7, 108–124, 1982.
41. Abernathy, D. R., Greenblatt, D. J., Divoll, M., Harmatz, J. S., and Shader, R. I., Alterations in drug distribution and clearance due to obesity, *J. Pharmacol. Exp. Ther.*, 217, 681–685, 1981.
42. Seeman, M. V., Pharmacological features and effects of neuroleptics, *Can. Med. Assoc. J.*, 125, 821–826, 1981.
43. Herbert, V., The five possible causes of all nutrient deficiency: illustrated by deficiencies of vitamin B-12 and folic acid, *Am. J. Clin. Nutr.*, 26, 77–88, 1973.
44. Brin, M., Drugs and environmental chemicals in relation to vitamin needs, in *Nutrition and Drug Interrelations*, Hathcock, J. N. and Coon, J., Eds., Academic Press, New York, 1978.
45. Wasserman, A. L., Principles of psychiatric care of children and adolescents with medical illnesses, in *Psychiatric Disorders in Children and Adolescents*, Garfinkel, B. D., Carlson, G. A., and Weller, E. B., Eds., W.B. Saunders, Philadelphia, PA, 1990, pp. 486–502.
46. Raiten, D. J., Medical basis for nutrition and behavior, in *Psychiatric Disorders in Children and Adolescents*, Garfinkel, B. D., Carlson, G. A., and Weller, E. B., Eds., W.B. Saunders Co., Philadelphia, PA, 1990, pp. 410–427.
47. Sullivan, A. C. and Gruen, R. K., Mechanisms of appetite modulation by drugs, *Fed. Proc.*, 44, 139–144, 1985.
48. Raiten, D. J., Nutritional correlates of HIV infection, *Eur. J. Gastroenterol. Hepatol.*, 4, 428–442, 1992.
49. Lawrence, V. A., Loewenstein, J. E., and Eichner, E. R., Aspirin and folate binding: *in vivo* and *in vitro* studies of serum binding and urinary excretion of endogenous folate, *J. Lab. Clin. Med.*, 103, 944–947, 1984.
50. Lumeng, L., The role of acetaldehyde in mediating the deleterious effects of ethanol on pyridoxal 5′ phosphate, *J. Clin. Invest.*, 62, 286–293, 1977.
51. Weiner, J. M., Psychopharmacology in childhood disorders, *Psychiat. Clin. N. Am.*, 7, 831–843, 1984.
52. Martinez, J. A., Valasco, J. J., and Urbistondo, D., Effects of pharmacological therapy on anthropometric and biochemical status of male and female institutionalized psychiatric patients, *J. Am. Coll. Nutr.*, 13, 192–197, 1994.
53. Hays, S. E. and Rubin, R. T., Differential prolactin responses to haloperidol and TRH in normal adult men, *Psychoneuronendocrinol.*, 6, 45–52, 1981.
54. Simpson, G. M., Pi, E. H., and Sramek, J. J., Adverse effects of antipsychotic agents, *Drugs*, 21, 138–151, 1981.
55. Awad, A. G., Diet and drug interactions in the treatment of mental illness — a review, *Can. J. Psychiat.*, 29, 609–613, 1984.
56. Mallick, M. J., Health hazards of obesity and weight control in children: a review of the literature, *Am. J. Pub. Health.*, 73, 78–82, 1983.
57. Reynolds, R. D., Biochemical methods for status assessment, in *Vitamin B_6 in Pregnancy, Lactation, and Infancy*, Raiten, D. J., Ed., CRC Press, Boca Raton, FL, 1995, pp. 41–59.
58. McCabe, B. J., Dietary tyramine and other pressor amines in MAOI regimens: a review, *J. Am. Diet. Assoc.*, 86, 1059–1064, 1986.
59. Pinto, J., Huang, Y. P., and Rivlin, R. S., Inhibition of riboflavin metabolism in rat tissues by chlorpromazine, imipramine, and amitriptyline, *J. Clin. Invest.*, 67, 1500–1506, 1981.
60. Pelliccione, N., Pinto, J., Huang, Y. P., and Rivlin, R. S., Accelerated development of riboflavin deficiency by treatement with chloropromazine, *Biochem. Pharmacol.*, 32, 2949–2953, 1983.
61. Watson, R. R. and Mohs, M. E., Effects of morphine, cocaine, and heroin on nutrition, *Prog. Clin. Biol. Res.*, 325, 413–418, 1990; Mezey, E., Metabolic effects of alcohol, *Fed. Proc.*, 44, 134–138, 1985.
62. Blass, J. P. and Gibson, G. E., Abnormality of a thiamine-requiring enzyme in patients with Wernicke-Korsakoff syndrome, *N. Eng. J. Med.*, 297, 1367–1370, 1977.
63. Mezey, E., Metabolic effects of alcohol, *Fed. Proc.*, 44, 134–138, 1985.
64. Richens, A. and Rowe, D. J. F., Disturbance of calcium metabolism by anticonvulsant drugs, *Br. Med. J.*, 4, 73–76, 1970.

65. Christensen, C. K., Lund, B. I., Lund, B. J., Sorensen, O. H., Nielsen, H. E., and Mosekilde, L., Reduced 1,25-dihydroxy vitamin D and 24,25-dihydroxy vitamin D in epileptic patients receiving chronic combined anticonvulsant therapy, *Metab. Bone Dis. Relat. Res.,* 3, 17–22, 1981.
66. Morijiri, Y. and Sato, T., Factors causing rickets in institutionalized handicapped children on anticonvulsant therapy, *Arch. Dis. Child.,* 56, 446–450, 1981.
67. Tolman, K. G., Jubiz, W., Sannella, J. J., Madsen, J. A., Belsey, R. E., Goldsmith, R. S., and Freston, J. W., Osteomalacia associated with anticonvulsant drug therapy in mentally retarded children, *Pediatrics,* 56, 45–50, 1975.
68. Waxman, S., Corcino, J., and Herbert, V., Drug, toxins, and dietary amino acids affecting vitamin B_{12} or folic acid absorption or utilization, *Am. J. Med.,* 48, 599–608, 1970.
69. Mallek, H. M. and Nakamoto, T., Dilantin and folic acid status, *J. Periodontol.,* 52, 255–259, 1981.
70. Botez, M. I., Botez, T., Leveille, J., Bielmann, P., and Cadotte, M., Neuropsychological correlates of folic acid deficiency: facts and hypotheses, in *Folic Acid in Neurology, Psychiatry, and Internal Medicine,* Botez, M. I. and Reynolds, E. H., Eds., Raven Press, New York, 1979, pp. 435–461.
71. Thomson, A. D., Baker, H., and Leevy, C., Folate-induced malabsorption of thiamin, *Gastroenterology,* 60, 756, 1971.
72. John, G., Transient osteosclerosis asssociated with sodium valproate, *Develop. Med. Child. Neurol.,* 23, 234–236, 1981.
73. Krause, K. H., Berlit, P., and Bonjour, J. P., Impaired biotin status in anticonvulsant therapy, *Ann. Neurol.,* 12, 485–486, 1982.
74. Higashi, A., Ikeda, T., Matsukura, M., and Matsuda, I., Serum zinc and vitamin E in handicapped children treated with anticonvulsants, *Develop. Pharmacol. Therapeut.,* 5, 109–113, 1982.
75. Delport, R., Ubbink, J. B., Serfontein, W. J., Becker, P. J., and Walters, L., Vitamin B_6 nutritional status in asthma: the effect of theophylline therapy on plasma pyridoxal-5′-phosphate and pyridoxal levels, *Intern. J. Vit. Nutr. Res.,* 58, 67–72, 1988.
76. Ubbink, J. B., Delport, R., Becker, P. J., and Bissbort, S., Evidence of a theophylline-induced vitamin B_6 deficiency caused by noncompetitive inhibition of pyridoxal kinase, *J. Lab. Clin. Med.,* 113, 15–22, 1989.
77. Ubbink, J. B., Delport, R., Bissbort, S., Vermaak, W. J. H., and Becker, P. J., Relationship between vitamin B_6 status and elevated pyridoxal kinase levels induced by theophylline therapy in humans, *J. Nutr.,* 120, 1352–1359, 1990.
78. Weir, M. R., Keniston, R. C., Enriquez, J. I., and McNamee, G. A., Depression of vitamin B_6 levels due to theophylline, *Ann. Allergy,* 65, 59–62, 1990.
79. Furukawa, C. T., DuHamel, T. R., Weimer, L., Shapiro, G. G., Pierson, W. E., and Bierman, C. W., Cognitive and behavioral finding in children taking theophylline, *J. Allergy Clin. Immunol.,* 81, 83–88, 1988.
80. Furukawa, C. T., Shapiro, G. G., DuHamel, T., Weimer, L., Pierson, W. E., Bierman, C. W., Learning and behavior problems associated with theophylline therapy, *Lancet,* 1, 621, 1984.
81. Rachelefsky, G. S., Wo, J., Adelson, J., Mickey, M. R., Spector, S. L., Katz, R. M., Siegel, S. C., and Rohr, A. S., Behavior abnormalities and poor school performance due to oral theophylline usage, *Pediatrics,* 78, 1133–1138, 1986.
82. Ubbink, J. B., Vermaak, W. J. H., Delport, R., and Serfontein, W. J., The relationship between vitamin B_6 metabolism, asthma, and theophylline therapy, *Ann. N.Y. Acad. Sci.,* 585, 285–290, 1990.

CHAPTER 8

TOXIC AGENTS, CHEMOSENSORY FUNCTION, AND DIET

Krystyna M. Rankin and Richard D. Mattes

CONTENTS

1 INTRODUCTION

Alterations in gustatory and olfactory functions may result from injection, ingestion, or topical application of pharmaceutical agents and toxic compounds. Animal studies have yielded insights into several possible mechanisms by which these substances may affect odor[1] and

0-8493-4493-X/97/$0.00+$.50

taste[2] perception, although for most compounds the mode of action has not been identified. In part, this is due to the fact that knowledge of chemosensory transduction mechanisms and coding processes in the periphery and central nervous system (CNS) are not fully understood.

This chapter will (1) provide a brief overview of the anatomy and physiology of the gustatory and olfactory systems, (2) describe clinical manifestations of chemosensory disorders, (3) outline potential mechanisms of drug actions on the senses of taste and smell, (4) compile tables with information about specific drug/toxin effects on the chemical senses, and (5) highlight potential dietary implications and effects on quality of life that may result from chemosensory alterations.

2 TASTE

2.1 Overview of the Anatomy and Physiology

Taste sensation is transduced by receptor cells that occur predominantly in specialized end organs, taste buds, within the lingual epithelium. They have also been identified on the hard and soft palate, esophagus, pharynx, and larynx. Taste buds are located in fungiform papillae (on the tip and sides of the anterior 2/3 of the tongue), foliate papillae (along the lateral margins of the base of the tongue), and circumvallate papillae (arranged in an inverted "V" at the back of the tongue). Taste cells have a rapid turnover rate of about 10 days[3] which renders them highly susceptible to metabolic toxins.

Three cranial nerves, the chorda tympani branch of the facial (VII), vagus (X), and the glossopharyngeal (IX) innervate the taste buds and carry gustatory information to the central nervous system. The lingual branch of the chorda tympani nerve subserves the fungiform papillae, and the glossopharyngeal nerve innervates the foliate and the circumvallate papillae. Taste buds on the extreme posterior tongue, esophagus, and epiglottis are innervated by the superior laryngeal branch of the vagus nerve. The three nerves mediating taste project to the nucleus of the solitary tract (NST) of the medulla.[4] Projections from the NST to other regions of the CNS have not been fully described in humans, although pathways to the thalamus[5] and primary gustatory cortex[6] have been identified.

Different transduction mechanisms have been proposed for each of the four basic taste qualities: salty, sour, sweet, and bitter. Umami, the taste of monosodium glutamate, has been proposed as a fifth basic taste, but this has not been confirmed. The transduction of salt, specifically Na^+ salts, primarily involves passage of sodium ions into the taste cell through amiloride-blockable, apical Na^+ channels, resulting in depolarization of the cell. Alternative mechanisms also subserve salt taste but have not been characterized.

Sour taste is probably a function of the hydrogen ion concentration.[7] Transduction involves acid blockage of apical K^+ channels with a resultant accumulation of positive charges within the receptor cell and consequent depolarization of the membrane.[8]

The transduction of sweetness appears to involve specific membrane receptors. It is hypothesized that a taste molecule binds to a sweet receptor coupled to a G-protein that activates the second messenger, cyclic adenosine monophosphate (cAMP), causing closure of K^+ channels. This leads to taste cell depolarization.[8] Several lines of evidence suggest there is more than one type of sweet receptor.[9-11]

Bitter transduction is also believed to involve specific receptors.[12] Several transduction mechanisms are possible.[13] Some bitter compounds have been found to interfere with potassium conductance: for example, quinine blocks K+ currents in taste cells isolated from mudpuppy,[14] and denatonium chloride blocks K+ current in taste cells isolated from mice.[15] Another mechanism uses receptor/G-protein-coupled responses to accelerate production of the second messenger IP_3 (inositol-1,4,5,-trisphosphate). This second messenger releases calcium ions from organelles within the taste cell.[12]

The responsivity of taste cells to sapid stimuli varies in different regions of the oral cavity; however, subjective reports and electrophysiological recordings following stimulation of single taste cells with varying taste qualities demonstrate that individual cells and regions respond to multiple taste qualities.[16] Consequently, damage to a specific region of the tongue or a specific gustatory nerve does not result in the loss of responsivity to a specific taste quality.

2.2 Clinical Manifestations of Gustatory Disorders

The most common form of taste dysfunction is *hypogeusia,* which refers to partial loss of taste sensitivity to all or specific taste qualities. Complete loss of taste, *ageusia,* is quite rare because the sense of taste is organized to maintain constancy of perceived sensation even when large receptive areas are lost. Fewer than 1% of the patients who suffer from chemosensory disorders have ageusia.[17,18] *Dysgeusia* refers to the presence of a distorted, often unpleasant, taste when a normally neutral or pleasant taste is present or a sensation occurring in the absence of any taste stimuli (gustatory hallucination or phantom taste). Other, less common taste disorders include *hypergeusia,* an increased taste sensitivity to some or all tastants, and *agnosia,* the inability to verbally classify, contrast, or identify a tastant.

Overall, taste dysfunctions are relatively rare. When they do occur, they are frequently caused by pharmacological agents, particularly antirheumatic and antiproliferative medications. Among the specific disorders attributable to these compounds, the most common is hypogeusia, which may progress to ageusia. Dysgeusia is often superimposed on a sensory diminution, but can occur in isolation.

Drugs that have been implicated in taste abnormalities are listed in Table 1. Whenever available, the type of dysfunction, frequency of occurrence, and references are also listed. Available evidence does not implicate exposure to toxic environmental agents[17] or use of tobacco as contributors to significant taste impairment.[19,20]

Taste abnormalities due to other etiologies are extremely rare, but the etiologies may include viral or bacterial infection[17] or damage to neural structures resulting from head trauma that mediate taste.[21] Frequently, complaints of taste abnormalities reflect decreased olfactory sensitivity.[17,22] Indeed, findings from patients evaluated at taste and smell clinics, with primary complaints of chemosensory dysfunctions attributable to etiologies other than pharmacological drugs, indicate that over 96% of demonstrable cases of gustatory deficits are accompanied by olfactory dysfunction. The mistaken attribution to taste is due to the fact that much of the sensation of food flavors derives from retronasal olfactory stimulation by substances in the oral cavity.

2.3 Sites for Disruption of Gustatory Functions

Compounds capable of inducing changes in taste sensation are heterogeneous, thus suggesting that more than one mechanism is involved. Some of the hypothesized mechanisms include the effect of drugs on (1) salivary flow and composition, (2) cell turnover, (3) interference with sensory transmission (e.g., affecting membrane conductance or the activity of second messengers), and (4) processing of gustatory input in the central nervous system.

Changes in salivary flow may modify gustatory function because saliva serves as a vehicle carrying taste stimuli to taste receptors.[23] Changes in salivary composition may alter taste sensitivity because taste receptors adapt to certain levels of salivary components, such as sodium,[24] and respond only to stimuli differing from levels in saliva by a given amount. In addition, chemical interactions between taste compounds and salivary constituents may modify

TABLE 1

Drugs that Affect the Taste System

Drug Name	Effect on Taste	Frequency (%, except as indicated)	Ref.
Acetazolamide (with carbonated drinks)	Altered	1 case	79
Acetylsalicylic acid	Bitter hypergeusia	—	80
Albuterol	Unusual dysgeusia	1–3	81
Allopurinol	Metallic dysgeusia	—	69
Amethocain	Hypogeusia for sweet, hypergeusia for bitter	—	82
Amiloride and its analogs	Hypergeusia for salty	—	83, 84
	Suppression of sourness in:	—	
	Na^+ salts		84a
	Li^+ salts		84b,c
Amiodarone	Bitter dysgeusia	1–3	81
Ampicillin (long term use)	Hypogeusia	Uncommon	85
Amphetamines	Hypogeusia for sweet, hypergeusia for bitter	—	82
	Unpleasant dysgeusia	1–3	81
Amphotericin B	Hypogeusia, ageusia	—	69
Amylocain (at lower concentrations)	Ageusia for salty	—	82
Antacids	Chalky dysgeusia	< 3	81
Antidepressants (tricyclic)	Unpleasant dysgeusia	< 3	81
Appetite suppressants	Unpleasant dysgeusia	> 3	81
Auranofin	Ageusia, metallic dysgeusia	1–3	81
Aurothioglucose	Metallic dysgeusia	< 3	81
Azathioprine	Dysgeusia, ageusia for sweet and salty	—	69
Azelastine	Altered	2–26	86
Baclofen	Hypogeusia for sweet and bitter, ageusia	—	69
Bamifylline HCL	Bitter dysgeusia	Frequent	69
Beclomethasone	Unpleasant dysgeusia	—	81
Benoxaprofen	Altered	0.6	87
Benzocaine	Hypergeusia for sour	—	82
Biguanide	Metallic dysgeusia	3	69
Bitolterol	Unpleasant dysgeusia	1–3	81
Bleomycin	Altered	10	88
Bleomycin and cisplatin	Altered	37	25
Bleomycin + actinomycin D + vindesine + dacarbazine (DTIC) combination	Hypergeusia (low concentrations), decreased ability to differentiate between qualities (high concentrations)	—	28
Bronchodilators, andrenergic	Unpleasant dysgeusia	1–3	81
Bupivacaine	Metallic dysgeusia	8	89
Calcium supplements	Metallic dysgeusia	—	81
Captopril	Ageusia	13	90
	Foul and metallic dysgeusia	40	91, 92
	Ageusia	15	93
	Salty and bitter dysgeusia	16	93
	Ageusia, dysgeusia	2-4	94
	Ageusia, dysgeusia	19	95
	Ageusia	6	96
Carbamazepine	Hypogeusia	1 case	97
	Hypogeusia, ageusia	—	69
Carbimazole	Hypogeusia	1 case	98
	Metallic dysgeusia, burning	—	99
Carmustine	Metallic dysgeusia	2	100
Cefamandole	Unpleasant dysgeusia	1 case	101
Chlofibrate	Hypergeusia	—	102

TABLE 1 (continued)

Drugs that Affect the Taste System

Drug Name	Effect on Taste	Frequency (%, except as indicated)	Ref.
Chlormezanone	Ageusia	—	69
Chlorhexidine digluconate mouth rinses	Metallic dysgeusia	< 3	81
	Hypogeusia for salty	—	103
Chlorodiazepoxide and amitriptyline	Unpleasant dysgeusia	< 3	81
Chlorpheniramine maleate	Altered	—	42
Chromolyn	Unpleasant dysgeusia	< 3	81
Clarithromycin	Altered dysgeusia	1–3	81
Clotazimine	Altered	> 3	81
Cocaine	Hypogeusia for sweet and bitter	—	82, 104
Copper supplements	Metallic dysgeusia	—	81
Cyclobenzaprine	Unpleasant dysgeusia	> 3	81
Cyclophosfate	Bitter dysgeusia	29	105
Dexamethasone	Decreased thresholds	—	106
	Unpleasant dysgeusia	—	81
Dextro-amphetamine sulfate (Dexedrine®)	Decreased threshold for bitter	—	107
Diamox (with carbonated drinks)	Altered	100	108
Diazoxide	Altered	—	42
Dichlorphenamide	Metallic dysgeusia	< 3	81
Diclofenac	Bitter dysgeusia	> 1	81
Diltiazem	Ageusia	1 case	109
Dimethyl sulfoxide	Garlicky dysgeusia	1–3	81
Dipyridamole	Dysgeusia	3 cases	110, 111
Disulfiram	Metallic and garlicky dysgeusia	> 3	81
Doxorubicin + methotrexate combination	Ageusia	—	112
Doxycycline	Dysgeusia	1 case	113
Enalapril	Ageusia, dysgeusia	<1	94, 114
Ergonovine	Unpleasant dysgeusia	1–3	81
Ethacrynic acid	Metallic dysgeusia	9	115
Ethambutol HCL	Metallic dysgeusia	—	69
Etidronate	Ageusia	52	116
	Metallic dysgeusia	1–3	81
Eucain	Hypogeusia for bitter and sweet	—	82
Fenoprofen	Bitter dysgeusia	> 1	81
Flecainide	Altered	5	117
Floctafenine	Bitter dysgeusia	1–3	81
Flunisolide inhaler	Bitter dysgeusia	21	118
	Unpleasant dysgeusia	< 3	81
Flurbiprofen	Bitter dysgeusia	> 1	81
Flunitrazepam	Bitter dysgeusia	2.4	119
Fluoxetine	Altered	1–3	81
Flurazepam	Bitter dysgeusia	25	119
Gallium nitrate	Metallic dysgeusia	0.5	120
Germine monoacetate	Unpleasant dysgeusia	Common	121
Glipizide	Altered	0.9	122
Gold	Metallic dysgeusia	—	69
Griseofulvin	Bitter dysgeusia, hypogeusia	2 cases	123, 124
Hexetidine	Altered	—	125
Histamine	Metallic dysgeusia	< 3	81
Hydrochlorothiazde	Altered	1–3	81
Hydrocortisone	Increased thresholds	—	106
Ibuprofen	Bitter dysgeusia	—	81
Idoxuridine	Garlic dysgeusia	6	126
Indomethacin	Bitter dysgeusia	—	81
Interferons (alpha)	Metallic dysgeusia	1–3	81

TABLE 1 (continued)

Drugs that Affect the Taste System

Drug Name	Effect on Taste	Frequency (%, except as indicated)	Ref.
Iohexol	Metallic dysgeusia	> 3	81
Ipratropium	Metallic and unpleasant dysgeusia	> 3	81
Iron sorbitex (at high doses)	Bad and metallic dysgeusia	Frequent	127
Iron supplements	Metallic dysgeusia	< 3	81
Ketoprofen	Bitter dysgeusia	> 1	81
Labetalol	Altered	1–3	81
Levamisole	Altered	17	128
	Metallic dysgeusia	< 3	81
Levodopa	Unpleasant and bitter dysgeusia,	4.1	129
	Ageusia	0.4	129
Levothyroxine	Dysgeusia, burning mouth	—	17
Lidocaine	Bitter dysgeusia	—	130
Lignocain	Hypogeusia for salty and sweet	—	82
Lincomycin	Hypergeusia	—	102
	Altered	—	69
Lithium carbonate	Metallic	5	131
	Unpleasant dysgeusia	1 case; 1	132, 133
Magaldrate	Chalky dysgeusia	< 3	81
Meclofenamate	Bitter dysgeusia	> 1	81
Meclorehamine	Metallic dysgeusia	1—3	81
Mefanamic acid	Bitter dysgeusia	—	81
5-Mercaptopyridoxal	Hypogeusia	—	134
Mesna	Unpleasant dysgeusia	1–3	81
Metaproterenol	Unpleasant dysgeusia	1–3	81
Metformin	Metallic dysgeusia	—	135, 136
Methadone (chronic use)	Bitter hypogeusia,	42	137
	Salty hypogeusia	33	137
Methazolamide	Metallic dysgeusia	< 3	81
Methimazole	Hypogeusia	1 case	98
	Ageusia	3 cases	138
Methotrexate (MTX)	Ageusia	1 case	139
	Sour dysgeusia	3	140
	Metallic dysgeusia	0.5–3	141
Methotrexate + adriamycin combined	Dysgeusia	3 cases	112
Methylthiouracil	Ageusia	1 case	142
	Ageusia	21 cases	143
Metronidazole	Unpleasant dysgeusia	92	144
	Hypogeusia	—	105
	Bitter and metallic dysgeusia	—	69, 145
Monooctanoin	Metallic dysgeusia	> 1	81
Naproxen	Bitter dysgeusia	—	81
Nedocromil sodium	Dysgeusia	Major side effect	146
Niclosamide	Unpleasant dysgeusia	> 1	81
Nifedipine	Dysgeusia, ageusia	2 cases	147
	Ageusia	9	93
Niridazole	Bitter dysgeusia	1.5	148
Nitrazepam	Bitter dysgeusia	6	119
Nitroglycerin patch	Ageusia	1 case	149
Nylidrin	Metallic dysgeusia	—	81
Oxyfedrine HCL (Ildamen)	Ageusia	—	150
	Hypogeusia, ageusia	—	69, 151
D-penicillamine	Hypogeusia	25–33	152, 153
	Altered	14	154
	Altered	19	155
	Ageusia	9	156
	Hypogeusia for sweet and salt	3	157

TABLE 1 (continued)
Drugs that Affect the Taste System

Drug Name	Effect on Taste	Frequency (%, except as indicated)	Ref.
Pentamidine	Metallic and bitter dysgeusia	1–3	81
Perphenazine and amitriptyline	Unpleasant dysgeusia	< 3	81
Phenformin and derivatives	Sour dysgeusia	1 case	158
	Unpleasant dysgeusia	Infrequent	159
	Altered	6	122
Phenindione	Burning mouth	1 case	160
	Ageusia	—	69
Phenylbutazone	Hypogeusia, ageusia	—	69
	Bitter dysgeusia	—	81
Phenytoin	Hypogeusia	—	42
Pirbuterol	Unpleasant dysgeusia	—	81
Piroxicam	Bitter dysgeusia	—	81
Psilocybin	Hypergeusia for sodium saccharinate	—	161
Procaine HCL (Novocain®)	Altered	—	162
Procaterol	Unpleasant and unusual dysgeusia	—	81
Propafenone	Metallic dysgeusia	9	163
Propatenone	Metallic and bitter dysgeusia	< 3	81
Propylthiouracil	Ageusia	—	69, 164
Protirelin	Unpleasant dysgeusia	< 3	81
Quinapril	Dysgeusia, ageusia	0.3	94
Quinidine	Unpleasant dysgeusia	< 3	81
Salazosulfapyridin	Hypogeusia, ageusia	—	69
Salicylates (dose-related)	Hypogeusia for sweet, hypergeusia for bitter	—	165
Selegiline	Altered	> 3	81
Selenium supplements	Metallic dysgeusia	—	81
Sodium fluoride	Altered	—	166
Sodium lauryl sulfate	Hypogeusia, ageusia	—	69, 167
Spironolactone	Confuse sweet/salty, ageusia	1 case	168
Sulindac	Bitter dysgeusia	> 1	81
Technetium Tc 99m sestamibi	Metallic and bitter dysgeusia	< 3	81
Technetium Tc 99m teboroxime	Metallic dysgeusia	> 3	81
Terbinafine	Ageusia	1 case	169
Terbutaline	Unpleasant dysgeusia	1–3	81
Teriparatide	Metallic dysgeusia	> 1	81
Tetracyclines	Metallic dysgeusia	1 case	170
Thiamazole	Ageusia	1 case	98
	Metallic dysgeusia, burning	—	99
Thiopentone	Onion and garlic dysgeusia	Common	171
5-Thiopyridoxine	Ageusia	8	172
Thiouracil	Ageusia	—	69
Tiaprofenic acid	Bitter dysgeusia	—	81
Tiopronin	Altered	1–3	81
Tolemetin	Bitter dysgeusia	—	81
Trazodone	Unpleasant dysgeusia	< 3	81
Triamcinolone	Unpleasant dysgeusia	—	81
Triazolam	Metallic dysgeusia	—	173
Trifluoperazine	Hypogeusia for sodium saccharinate	—	161
Vancomycin	Unpleasant dysgeusia	< 3	81
Vinblastine + interferon combination	Ageusia	Common	26
Vitamin D (and analogs)	Metallic dysgeusia	—	69
Vitamin K	Unusual dysgeusia	1—3	81
Zopiclone	Bitter dysgeusia	—	174
	Bitter dysgeusia	5 cases	175
	Metallic dysgeusia	—	176
	Bitter dysgeusia	24	119

the availability, conformation, or ionic state of the taste compounds and thereby alter their receptor binding characteristics or ability to pass through appropriate receptor cell channels.

One mode of action of chemotherapeutic agents involves interference with mitotic activity, thereby destroying proliferating cells. Since taste cells are rapidly dividing, they are especially vulnerable to damage during chemotherapy. Taste changes have been reported following treatment with mitotic inhibitors such as methotrexate and bleomycin,[25] and complete loss of taste has been reported after injection of vinblastine.[26] Animal studies have also shown changes in preference for sucrose and saccharin solutions following injections of vinblastine sulfate.[27] Interestingly, cases of hypergeusia for sweet, sour, salty, and bitter substances were recorded after combination chemotherapy using bleomycin, actinomycin D, vindesine, and dacarbazine.[28] An explanation for this finding is not apparent.

Therapeutic radiation to the head and neck region has been shown to decrease the number of taste buds[29] and taste acuity in humans;[30] however, this form of treatment also damages salivary tissue, and noted shifts in taste may be related to altered salivary function, as well. For more information on taste deficits following head and neck irradiation see References 30 to 33.

Drugs may also affect taste responsiveness by interfering with various stages of sensory transmission. The antimicrobial agent amphotericin B, for example, has been shown to alter the permeability of the taste cell membrane. Anesthetics, such as benzocaine, are known to disrupt lipid bilayers of cell membranes, thus altering their permeability and electrical properties. Amiloride, hydrocortisone, and phenytoin modify sodium fluxes through channels in cells.[34] Although there is no direct supporting evidence, drugs theoretically can also modify the activity of second messengers involved in gustatory reception.

3 OLFACTION

3.1 Overview of the Anatomy and Physiology

The olfactory system is anatomically and morphologically distinct from taste. The perception of smell occurs when volatile chemicals interact with receptor proteins located on the dendrites of primary olfactory neurons. The sensory cells that respond to odors are located in a relatively small area in the upper nasal passages called the olfactory epithelium. Olfactory cells are bipolar neurons whose axons form the olfactory nerve (first cranial nerve) and carry information through the cribriform plate directly to the olfactory bulb. Olfactory receptor cells have a rapid turnover rate of about 30 days.[35] They undergo a continual process of replacement, which includes a degeneration of old, as well as formation of new, connections within the CNS.[36]

The sequence of odor transduction events is initiated when odorant molecules interact with receptors and activate one of two second messenger systems, cAMP (adensoine-3′5′-monophosphate) or $I_{ns}P_3$ (inositol-1,4,5-trisphosphate)[37,38] through a process involving G proteins. This results in a depolarization of the receptor cell membrane and generation of action potentials in the olfactory nerve.[38-40] Recent evidence indicates that there may be hundreds of odorant-specific receptors.[41]

3.2 Clinical Manifestations of Olfactory Disorders

The most common manifestation of olfactory dysfunction is *anosmia*, a total loss of the sense of smell. Anosmics comprise about 50% of patients with chemosensory disorders reporting to taste and smell centers.[17] Two other common abnormalities include *hyposmia*, which refers to decreased odor sensitivity, and *dysosmia* (or parosmia), distorted sensation or

TABLE 2

Drugs that Affect the Olfactory System

Drug Name	Effect on Smell	Frequency	Ref.
Acetylcholine and acetyl-betamethychlin	Decreased threshold	—	177
Albuterol	Altered	1–3	81
Amylocaine	Hyposmia, anosmia	—	82
Beta-blockers	Anosmia	1 case	178
Bitolterol	Altered	—	81
Bronchodilators, andrenergic	Altered	—	81
Captopril	Anosmia	2 cases	91
Cocain	Increased threshold	—	82, 104
Codeine	Hyposmia	—	179
Dexamphetamine sulfate	Increased threshold	—	180
Diltiazem	Dysosmia	—	109
Doxycycline	Anosmia, parosmia	3 cases	113
Metaproterenol	Altered	—	81
Methimazole	Hyposmia	2 cases	98, 138
Methylthiouracil	Hyposmia	1 case	142
Morphine	Hyposmia	—	179
Nifedipine	Dysosmia	1 case	147
Phenmetrazine theoclate	Increased threshold	—	180
Pirbuterol	Altered	1–3	81
Procaterol	Altered	—	81
Propylthiouracil	Anosmia	—	164
Streptomycin	Parosmia	—	181
Terbutaline	Altered	—	81
Tetracain	Increased threshold	—	104
Thiamazole	Dysosmia	—	99
Tiopronin	Altered	1–3	81
Tyrothricin	Anosmia, parosmia	8 cases	182
	Parosmia	—	181

the perception of odor in the absence of an exogenous stimulus (phantosmia). Less common dysfunctions are *hyperosmia*, an increased sensitivity to some or all odors, and *agnosia*, an inability to verbally classify, contrast, or identify an odorant.

The most common etiologies of olfactory losses include nasal and/or sinus disease, upper respiratory infection, obstruction in the nasal passages, head trauma with damage to axons of the olfactory receptor cells,[22] and normal aging. A very small number of olfactory dysfunctions can also be related to intake of medications,[34,42] and those are usually accompanied by taste impairment. Table 2 lists the drugs that have been implicated in olfactory dysfunction. Whenever available, type of dysfunction, complication frequency, and references are also provided.

Exposure to environmental toxins has also been shown to affect olfactory function.[17,22,43] A comprehensive review of the literature on the effects of such compounds on olfaction in humans has been compiled by Amoore[43] and reproduced in Tables 3 to 8. Toxic airborne chemicals may exact a greater toll on the olfactory system compared to taste because the olfactory neurons exposed to the environment (and toxins) are the same neurons that comprise the olfactory tract and synapse directly with the olfactory bulb in the CNS.[44] Thus, the olfactory system is not well protected from toxic environmental agents.

3.3 Sites Involved in Altered Olfactory Functioning

Olfactory function may be affected by a variety of mechanisms ranging from the regulation of nasal airflow to the alteration of receptive or transductive mechanisms and modulation of

TABLE 3

Metallurgical Processes Considered Responsible, on Chronic Exposure, for Permanent Hyposmia in Man[a]

Substance	Exposure Time (years)	Olfactometric Method	Incidence of Hyposmia Frequency (%)	Rating[b] (steps)	Assessment
Chromium	1.5	Elsberg	17	–0.7	Below average
Chromium-plating	4	Recognition	Cohort	—	Low normal
Lead	8	Elsberg	33	–0.8[c]	Below average
Lead (severe intoxication)	—	Elsberg	Cohort	–1.2	Below average
Lead (severe intoxication)	—	Elsberg	Cohort	–1.3	Below average
Magnet production[d]	10	Recognition	10 cases	—	Hyposmia
Mercury (chronic intoxication)	29	Elsberg	55	–3.6	Low normal
Nickel-plating	4	Recognition	Cohort	—	Low normal
Nickel refining (electrolytic)	>5	Recognition	33	—	Anosmia
Silver-plating	6	Recognition	Cohort	—	Below average
Steel production	4	Elsberg	6	<–3.3	Low normal
Zinc production	>5	Elsberg	3	<–2.7	Low normal

[a] All original references are in Amoore, 1986 (Reference 43).
[b] Values reflect the olfactory deficit, measured according to pyridine binary dilution step scale.
[c] Just one worker was affected, and he had a unilateral hyposmia.
[d] Iron, aluminum, nickel, cobalt, and chromium powders.

Source: From Amoore, J. E., in *Toxicology of the Nasal Passages,* Barrow, C. S., Ed., Hemisphere Publishing, Washington, D.C., 1986, pp. 155–190. With permission.

activity in the afferent olfactory pathways. It is unclear whether airborne contaminants adversely affect olfaction by damaging the receptor structures or by causing damage to more central structures, such as the olfactory bulb or the olfactory cortex, via the systemic circulation. Animal toxicological studies indicate that exposure to a number of airborne chemicals can result in marked damage to the olfactory neuroepithelium.[45] As in gustatory disorders, drugs or other toxic agents can influence receptor function by affecting the normal turnover of the olfactory receptor cells. Olfactory receptors must be replaced continuously to maintain normal olfactory function.[46] Several reports indicate that drugs that block or interfere with cell development produce anosmia, presumably by disrupting the normal turnover of olfactory receptor neurons.[42] Antidepressive agents, such as amitriptyline, have been shown to interfere with the development of chick and rat olfactory receptor cell neurites *in vitro*.[47] Other studies have noted that human olfactory perception is impaired by treatment with antiproliferative medications that presumably block formation of receptor neurons.[42]

Olfactory transduction involves G proteins coupled with a second messenger system (cAMP).[37,48] Drugs may alter olfactory function by interfering with the activity of second messengers in the chemosensory cells, although at present, there is no direct evidence for this type of activity. Indirect evidence from clinical studies shows that patients who are G_s-protein deficient have an impaired ability to smell, while those with normal G_s-protein activity have normal function.[49]

Animal studies indicate that certain drugs can affect the activity of olfactory receptor cells by modulating neurotransmitter release. The olfactory epithelium has receptors for diazepam, adrenergic and cholinergic ligands, and L-carnosine (a putative neurotransmitter).[50] Exogenous application of neuroactive substances, such as acetylcholine or substance P, increases olfactory receptor cell activity.[51] On the other hand, treating rats with concanavalin A (a lectin) differentially inhibits olfactory responses to a variety of odorants.[52] It is possible that concanavalin A disrupts the activity of at least several odorant receptors by binding to a sugar-specific site on one or more cell surface protein.

TABLE 4

Metallic Compounds Considered Responsible, on Chronic Exposure, for Permanent Hyposmia in Man[a]

Substance	Exposure Conc. (mg/m³)	Exposure Time (years)	Olfactometric Method	Incidence of Hyposmia Frequency (%)	Incidence of Hyposmia Rating[b] (steps)	Assessment
Cadmium compounds	—	—	Elsberg	13	—	Anosmia
Cadmium compounds[c]	1.3	7	Elsberg	66	—	Hyposmia[d]
Cadmium oxide	—	3	Recognition	1 case	—	Anosmia
Cadmium oxide	9[e]	20	Symptom	44	—	Anosmia
Nickel hydroxide	80					
Cadmium oxide	0.5[f]	15	Proetz	27	< –7.6	Hyposmia
Nickel hydroxide	0.01					
Chromate salts	—	18	T&T[g]	27	< –18	Anosmia
Zinc chromate	10	10	Roseburg	30	~ –7	Hyposmia[g]

[a] All original references are in Amoore, 1986 (Reference 43).
[b] Values reflect the olfactory deficit, measured according to pyridine binary dilution step scale.
[c] Oxide, sulfate, carbonate, nitrate, sulfide, selenide, stearate.
[d] The hyposmia was ameliorated by giving caffeine.
[e] Before dust control.
[f] After installation of industrial hygiene equipment.
[g] T&T (Toyota and Takagi) method of measurement.
[h] A followup study on 11 of the same workers 4 years later (after reducing the chrome-dust exposure in the factory) showed no recovery of olfactory performance.

Source: From Amoore, J. E., in *Toxicology of the Nasal Passages,* Barrow, C. S., Ed., Hemisphere Publishing, Washington, D.C., 1986, pp. 155–190. With permission.

TABLE 5

Dusts Considered Responsible, on Chronic Exposure, for Permanent Hyposmia in Man[a]

Substance	Exposure Time (years)	Olfactometric Method	Incidence of Hyposmia Frequency (%)	Incidence of Hyposmia Rating[b] (steps)	Assessment
Cement	—	Proetz	2	—	Hyposmia
Chemicals	6	Proetz	8	<–6.6	Hyposmia
Hardwoods	—	Symptom	5	—	Anosmia
Hardwoods	—	Symptom	—	—	Anosmia
Lime	—	Proetz	6	—	Hyposmia[c]
Printing	—	Naus	24	–3.0	Low normal
Silicosis (first stage)[d]	—	Elsberg	Cohort	–1.2	Below average

[a] All original references are in Amoore, 1986 (Reference 43).
[b] Values reflect the olfactory deficit, measured according to pyridine binary dilution step scale.
[c] Damage to the olfactory epithelium, sensory cells, and bulbar fibers was observed in rats exposed 2 months in the dustiest locations in the factory.
[d] First-, second-, and third-stage silicosis cohorts all showed about the same olfactory deficit.

Source: From Amoore, J. E., in *Toxicology of the Nasal Passages,* Barrow, C. S., Ed., Hemisphere Publishing, Washington, D.C., 1986, pp. 155–190. With permission.

Some of the primary neurotransmitters that have been proposed to exist in the olfactory bulb include acetylcholine, dopamine, DL-homocysteate, met-enkephalin, gamma-amino-butyric acid, glutamate, aspartate, 5-hydroxytryptamine, luteinizing-hormone-releasing hormone,

TABLE 6

Nonmetallic Inorganic Compounds Considered Responsible, on Chronic Exposure, for Permanent Hyposmia in Man[a]

Substance	Exposure		Olfactometric Method	Incidence of Hyposmia		
	Conc. (pm)	Time (years)		Frequency (%)	Rating (steps)[b]	Assessment
Carbon disulfide	—	15	Elsberg	22	–3.4	Low normal
Carbon disulfide (average intoxication)	—	—	Elsberg	Cohort	–2.0[c]	Low normal
Carbon disulfide (intoxication)	62	20	Elsberg	14	—	Hyposmia
Carbon monoxide	—	13	Recognition	1 case	—	Anosmia
Carbon monoxide (intoxication)[d]	—	>100	Elsberg	6	—	Anosmia
Chlorine	—	3	Proetz	70	<–6.6	Hyposmia
Hydrazine	—	—	Elsberg	Cohort	–0.8	Below average
Nitrogen dioxide (NO_x)	3	5	Elsberg	Cohort	–0.7	Below average
Ammonia	30					
Nitrogen dioxide (NO_x) Sulfur dioxide (SO_x)	—	8	Proetz	60	<–6.6	Hyposmia
Sulfur dioxide	90	4	Symptom	14	—	Hyposmia
Sulfur dioxide	155	20	Elsberg	Cohort	–3.4	Low normal
Sulfur dioxide	80	32	T&T[e]	Cohort	–4.4	Hyposmia
Ammonia[f] Sulfur dioxide (SO_x) Nitrogen dioxide (NO_x) Fluorides (HF?)	—	>5	Elsberg	Cohort	–1.2	Below average

[a] All original references are in Amoore, 1986 (Reference 43).
[b] Values reflect the olfactory deficit, measured according to pyridine binary dilution step scale.
[c] The hyposmia could be partially or completely reversed, for 1 or 2 hours by injections of caffeine or eserine.
[d] Includes both acute and chronic intoxications.
[e] T&T (Toyota and Takagi) method of measurement.
[f] Workers exposed to ammonia alone showed no significant olfactory deficit.

Source: From Amoore, J. E., in *Toxicology of the Nasal Passages,* Barrow, C. S., Ed., Hemisphere Publishing, Washington, D.C., 1986, pp. 155–190. With permission.

norepinephrine, somatostatin, and substance P.[53] The significance of the various neurotransmitters to olfactory function is not clear.

The most extensively studied substance in the olfactory bulb is norepinephrine. Indirect evidence suggests that norepinephrine affects the responses of olfactory bulb neurons to odor stimulation. Early studies demonstrated that stimulation of the midbrain with ascending norepinephrine fibers suppresses olfactory bulb activity[54] and modulates responses evoked by olfactory stimulation.[55] Doty and Ferguson-Segall[56] report that odor detection in rats is affected by treatment with d-amphetamine, a drug that mimics the action of norepinephrine. Detection performance is enhanced following a low-dose injection and is depressed by moderate doses of amphetamine. Diminished norepinephrine activity is associated with impairment of olfactory discrimination in Korsakoff's disease.[57]

Other disorders characterized by a significant decrease in olfactory function include Parkinson's and Alzheimer's disease.[58,59] These patients show depressed levels of dopamine in the mesolimbic brain regions and a decreased number of dopamine D-1 and D-2 binding sites.[60,61] Recent pharmacological studies show that systematically administered D-1 and D-2 dopamine receptor agonists enhance and depress, respectively, the odor-detection performances of normal rats.[62,63] These results suggest that D-1 and D-2 receptor agonists may modulate odor detection performance and related behavior.

TABLE 7

Organic Compounds Considered Responsible, on Chronic Exposure, for Permanent Hyposmia in Man[a]

Substance	Exposure		Olfactometric Method	Incidence of Hyposmia		
	Conc. (ppm)	Time (years)		Frequency (%)	Rating[b] (steps)	Assessment
Acetone	—	8	Recognition	1 case	—	Hyposmia
Acetophenone	—	6	Elsberg	12	<–2.3	Low normal
Benzene	—	10	Elsberg	Cohort	–0.8	Below average
Benzine	400	5	Proetz	37	–6	Hyposmia
Benzine						
Ethylacetate Butylacetate	—	60	Elsberg	30	<–3.1	Low normal
Chloromethanes[c]	—	—	Elsberg	44	–2.7	Low normal
Menthol	—	—	Recognition	Cohort	—	Hyposmia
Menthol	—	45	Symptom	1 case	—	Hyposmia
Pentachlorophenol	—	—	Symptom	1 case	—	Anosmia[d]
Trichloroethylene	380	10	Naus	Cohort	–3.1[e]	Low normal
Trichlorethylene (intermittent abuse)	Saturated vapor	9	Symptom	1 case	—	Anosmia

[a] All original references are in Amoore, 1986 (Reference 43).
[b] Values reflect the olfactory deficit, measured according to pyridine binary dilution step scale.
[c] CH_3Cl, CH_2Cl_2, $CHCl_3$, CCl_4.
[d] Tested at beginning of shift, i.e., a permanent hyposmia. When tested at the end of the shift, additional –2.1 steps of temporary hyposmia were demonstrated.
[e] May have been due to upper respiratory tract infection, not occupational exposure.

Source: From Amoore, J. E., in *Toxicology of the Nasal Passages,* Barrow, C. S., Ed., Hemisphere Publishing, Washington, D.C., 1986, pp. 155–190. With permission.

Changes in endocrine status also influence olfactory sensitivity, as evidenced by shifts over the menstrual cycle. It is possible that exogenously administered hormones may also influence olfactory functioning. Administration of estrogen increases and progesterone decreases olfactory detection performance in ovariectomized rats.[64]

Exposure to a number of environmental and industrial agents has been implicated in chronic and acute impairment of olfaction.[17,43,65,66] More recently, a study of 731 workers exposed to acrylates and methacrylates revealed a dose-response relationship between olfactory dysfunction and cumulative exposure scores.[45] Nonsmoking employees exposed to acrylates were over six times as likely to show olfactory dysfunction compared to nonsmoking, non-exposed employees. Cigarette smoke has also been shown to impair olfactory performance. A recent study of 638 adults found that smoking was negatively associated with odor identification ability in a dose-related manner in both current and previous cigarette smokers.[67]

4 RECOVERY FROM CHEMOSENSORY DISORDERS

The prognosis for patients with chemosensory disorders varies depending on the etiology of their problem. Patients with olfactory deficits caused by nasal sinus disease generally experience little improvement.[17] The prognosis for post-traumatic anosmia is also poor.[21] Recovery from post-traumatic ageusia, on the other hand, is much more likely, with different taste modalities recovering at different rates.[68] Drug-induced taste and/or olfactory changes usually resolve after the offending medication is discontinued. Generally, there is a direct association between the duration of the pharmacological insult and recovery time.[69] The same holds for responses to some long-term environmental toxin exposures.[43,45]

TABLE 8

Manufacturing Processes Considered Responsible, on Chronic Exposure, for Permanent Hyposmia in Man[a]

Substance	Exposure Time (years)	Olfactometric Method	Incidence of Hyposmia Frequency (%)	Rating[b] (steps)	Assessment
Acids (organic and inorganic)	7	Proetz	5	<–6.6	Hyposmia
Asphalt (oxidized)	6	Elsberg	18	<–2.3	Low normal
Cutting oils (machining)	—	Elsberg	Cohort	–0.8	Below average
Fragrances	4	Recognition	50	—	Below average
Paint (lead)	—	Naus	56	–2.9	Below average
Paprika	14	Elsberg	4	—	Hyposmia
"Pavinol" (sewing)[c]	—	Elsberg	55	<–2.3	Low normal
Spices	11	Naus	Cohort	–1.0[d]	Below average
Tobacco	12	Elsberg	1	—	Hyposmia
Varnishes	10	Recognition	78	—	Low normal
Varnishes	5	Elsberg	1	—	Hyposmia
Wastewater (refinery)	7	Elsberg	18	<–2.3	Low normal

[a] All original references are in Amoore, 1986 (Reference 43).
[b] Values reflect the olfactory deficit, measured according to pyridine binary dilution step scale.
[c] A synthetic leather; the material contains a slightly volatile plasticizer, dibutyl phthalate, which may be responsible for the hyposmic effect.
[d] Tested at the beginning of the shift, i.e., a permanent hyposmia. When tested at the end of the shift, additional –1.6 steps of temporary hyposmia were demonstrated.

Source: From Amoore, J. E., in *Toxicology of the Nasal Passages,* Barrow, C. S., Ed., Hemisphere Publishing, Washington, D.C., 1986, pp. 155–190. With permission.

5 USE OF PHARMACOLOGICAL DRUGS IN THE TREATMENT OF CHEMOSENSORY DEFICITS

Currently, no treatment is available for most chemosensory disorders; however, recent evidence indicates that several medical interventions hold promise. Topical or systemic corticosteroid administration has provided improvement in olfactory function in anosmic and hyposmic patients, particularly in those with nasal or paranasal sinus disease.[70-72] The mechanism is not clear but is probably due to reduced swelling and nasal obstruction.[70]

Tricyclic antidepressant medication has been used effectively to treat burning mouth syndrome (BMS).[73,74] BMS is characterized by intra-oral pain and occurs primarily in postmenopausal women.[75] Zinc ($ZnSO_4$) has been used as a treatment for hypogeusia,[76] as well as for a variety of other taste and smell dysfunctions;[77] however, there is little evidence that zinc therapy is an effective treatment for any chemosensory dysfunction[78] unless zinc deficiency is the etiology.

6 QUALITY OF LIFE AND DIETARY IMPLICATIONS OF CHEMOSENSORY ABNORMALITIES

Chemosensory abnormalities attributable to medications hold limited long-term health or nutritional implications because the effects are typically transient. When a taste or smell abnormality becomes troublesome to a patient, the medication is usually withdrawn and, in most cases, normalcy is restored. Chemosensory dysfunction resulting from exposure to toxic chemicals may have more long-lasting and serious consequences.

No systematic studies of drug- or toxin-induced chemosensory abnormalities have been reported. Data from taste and smell clinics show that quality of life as well as nutritional status may be compromised in individuals who suffer from chemosensory disorders. Theoretically, the loss of smell could pose a threat to personal safety since anosmic individuals would not detect and avoid potentially dangerous volatile environmental toxins; however, many such agents are also irritants and are detectable through the trigeminal system, which is rarely compromised. Similarly, a loss of smell or taste could present an increased risk for ingestion of spoiled food, but visual cues and minimal attention to storage conditions and timing can minimize this risk. Quality of life could be compromised due, for example, to anxiety associated with the inability to monitor body odors or reduced appreciation of the flavors of foods. Nutritional risk may also be heightened. Approximately 80% of patients reporting to taste and smell clinics complain of a loss of enjoyment of foods.[18] Of patients with a primary chemosensory complaint, 15 to 20% alter their food intake to a degree that body weight changes by more than 10% relative to their predisorder weight.[18] Compensatory dietary responses to alterations of taste or smell may involve increased consumption, most commonly in an attempt to achieve a missed level of sensory stimulation or to mask unpleasant chemosensory sensations. Alternatively, intake may decline due, often, to disinterest or frustration with unpalatable food or because food may be a source of unpleasant sensations. Responses are highly idiosyncratic and may be extreme in a given individual.

7 SUMMARY

1. Drug or toxin exposure may alter gustatory or olfactory function by modifying access of taste or smell stimuli to appropriate receptor sites or channels, influencing interactions between stimuli and binding sites, altering the turnover and/or electrical properties of receptor cells, or disturbing neural transmission and processing. The mode of action for most agents has not been elucidated.
2. A relatively small number of individuals suffer from drug-induced chemosensory dysfunctions. In most cases, drugs affect the sense of taste and only rarely the sense of smell. Generally, chemosensory disturbances caused by drugs do not pose a serious health risk since the effects are reversible.
3. Environmental pollutants generally produce disorders of olfaction more than taste. Hyposmia, the most common olfactory complaint resulting from environmental toxin exposure, may be reversible.
4. Among individuals with long-standing chemosensory disturbances, most develop compensatory responses that minimize the impact of the sensory abnormality on diet, safety, and quality of life; however, for selected individuals, chemosensory complications pose a serious health risk and may compromise the quality of life.

REFERENCES

1. Mair, R. G. and Harrison, L. M., Influence of drugs on smell function, in *The Human Sense of Smell*, Laing, D. G., Doty, R. L., and Breipohl, W., Eds. Springer-Verlag, New York, 1991, pp. 335–359.
2. Catalanotto, F. A., Frank, M. E., and Contreras, R. J., Animal models of taste alteration, in *Clinical Measurement of Taste and Smell*, Meiselman H. L. and Rivlin, R. S., Eds., Macmillan, New York, 1986, pp. 429–442.
3. Beidler, L. M. and Smallman, R. L., Renewal of cells within taste buds, *J. Cell Biol.*, 27, 263–272, 1965.

4. Beckstead, R. M. and Norgren, R., An autoradiographic examination of the central distribution of the trigeminal, facial, glosspharyngeal, and vagal nerves in the monkey, *J. Comp. Neurol.,* 184, 455–472, 1979.
5. Beckstead, R. M., Morse, J. R., and Norgren, R., The nucleus of the solitary tract in the monkey: projections to the thalamus and the brain stem nuclei, *J. Comp. Neurol.,* 190, 259–282, 1980.
6. Pritchard, T. C., Hamilton, R. B., Morse, J. R., and Norgren, R., Projections of thalamic gustatory and lingual areas in the monkey *Macaca fascicularis, J. Comp. Neurol.,* 244(2), 213–228, 1986.
7. Beidler, L. M., Anion influences on taste receptor response, in *Olfaction and Taste: II,* Hayashi, T., Ed., Pergamon Press, New York, 1967, pp. 509–534.
8. Shirley, S. G. and Persaud, K. C., The biochemistry of vertebrate olfaction and taste, *Semin. Neurosci.* 2, 59, 1991.
9. Pfaffmann, C., Specificity of the sweet receptors of the squirrel monkey, *Chem. Senses Flavor,* 1, 61–67, 1974.
10. Schiffman, S. S., Cahn, H., and Lindley, M. G., Multiple receptor sites mediate sweetness: evidence from cross adaptation, *Pharmacol. Biochem. Behav.,* 15, 377–388, 1981.
11. Schiffman, S. S., Diaz, C., and Beeker, T. G., Caffeine intensifies taste of certain sweeteners: role of adenosine receptor, *Pharmacol. Biochem. Behav.,* 24(3), 429–432, 1986.
12. Akabas, M. H., Dodd, J., and Al-Awqati, Q., A bitter substance induces a rise in intracellural calcium in a subpupulation of rat taste cell, *Science,* 242(488), 1047–1050, 1988.
13. Spielman, A. I., Huque, T., Whitney, G., and Brand, J. G., The diversity of bitter taste signal transduction mechanism, in *Sensory Transduction,* Corey, D. P. and Roper, S. D., Eds., Rockefeller University Press, New York, 1992, pp. 307–324.
14. Kinnamon, S. C. and Roper, S. D., Membrane properties of isolated mudpuppy taste cells, *J. Gen. Physiol.,* 91(3), 351–371, 1988.
15. Spielman, A. I., Mody, I., Brand, J. G., Whitney, G., MacDonald, J. F., and Salter, M. W., A method for isolating and patch-clamping single mammalian taste receptor cells, *Brain Res.,* 503(2), 326–329, 1989.
16. Nilsson, B., Taste acuity of the human pallet, *Acta Odontol. Scand.,* 35, 51–62, 1977.
17. Deems, D. A., Doty, R. L., Settle, R. G., Moore-Gillon V., Shaman, P., Mester A. F., Kimmelman, C. P., Brightman, V. J., and Snow, J. B., Smell and taste disorders: a study of 750 patients from the University of Pennsylvania Taste and Smell Center, *Arch. Otolaryngol. Head Neck Surg.,* 117, 519–527, 1991.
18. Mattes, R. D. and Cowart, B. J., Dietary assessment of patients with chemosensory disorders, *J. Am. Diet. Assoc.,* 94, 50–56, 1994.
19. Mela, D. J., Gustatory function and dietary habits in usesrs and nonusers of smokeless tobacco, *Am. J. Clin. Nutr.,* 49(3), 482–489, 1989.
20. Redington K., Taste differences between cigarette smokers and non-smokers, *Pharmacol. Biochem. Behav.,* 21, 203–206, 1984.
21. Costanzo, R. M. and Becker, D. P., Smell and taste disorders in head injury and neurosurgery patients, in *Clinical Measurement of Taste and Smell,* Meiselman H. L. and Rivlin, R. S., Eds., Macmillan, New York, 1986, pp. 565–578.
22. Goodspeed, R. B., Gent, J. F., and Catalanotto, F. A., Chemosensory dysfunction: clinical evaluation results from a taste and smell clinic, *Postgrad. Med.,* 81(1), 251–260, 1987.
23. Nanda, R. and Catalanotto, F. A., Long-term effects of surgical desalivation upon taste acuity, fluid intake, and taste buds in the rat, *J. Dent. Res.,* 60, 69–76, 1981.
24. Morino, T. and Langford, H. G., Salivary sodium correlates with salt recognition thresholds, *Physiol. Behav.,* 21, 45–48, 1978.
25. Lockhart, P. B. and Clark, J. R., Oral complications following neoadjuvant chemotherapy in patients with head and neck cancer, *Nat. Cancer Inst. Monogr.,* 9, 99–101, 1990.
26. Kellokumpu-Lehtinen, P., Nordman, F., and Torvanen, A., Combined interferon and vinblastine treatment of advanced melanoma: evaluation of the treatment results and the effects of the treatment on immunological functions, *Cancer Immunol. Immunother.,* 28, 213–217, 1989.
27. Beidler, L. M. and Smith, J. C., Effects of radiation therapy and drugs on cell turnover and taste, in *Smell and Taste in Health and Disease,* Getchell, T. V., Doty, R. L., Bartoshuk, L. M., and Snow, Jr., J. B., Eds., Raven Press, New York, 1991, pp. 753–763.
28. Mulder, N. H., Smit, J. M., Kreumer, W. M., and Bouman, J., Effects of chemotherapy on taste sensation in patients with disseminated malignant melanoma, *Oncology,* 40, 36–38, 1983.
29. Conger, A. D. and Wells, M. A., Radiation and aging effect of taste structure and function, *Rad. Res.,* 37, 31–49, 1969.
30. Conger, A. D., Loss and recovery of taste acuity in patients irradiated to the oral cavity, *Rad. Res.,* 53, 338–347, 1973.
31. Kalmus, H. and Farnsworth, D., Impairment and recovery of taste following irradiation of the oropharynx, *J. Laryngol. Otol.,* 73, 180–182, 1959.
32. Kalmus, H. and Hubbard, S. J., *The Chemical Senses in Health and Disease,* Charles C Thomas, Springfield, IL, 1960.
33. Mossman, K. L., Gustatory tissue injury in man: radiation dose response relationship and mechanisms of taste loss, *Br. J. Cancer Suppl.,* 53, 9–11, 1986.

34. Schiffman, S. S., Drugs influencing taste and smell perception, in *Smell and Taste in Health and Disease,* Getchell, T. V., Doty, R. L., Bartoshuk, L. M., and Snow, Jr., J. B., Eds., Raven Press, New York, 1991, pp. 845–850.
35. Moulton, D. G., Dynamics of cell populations in the olfactory epithelium, *Ann. N.Y. Acad. Sci.,* 237, 52–61, 1974.
36. Costanzo, R. M. and Graziadei, P. P. C., Development and plasticity of the olfactory system, in *Neurobiology of Taste and Smell,* Finger, T. E. and Silver, W. L., Eds., John Wiley & Sons, New York, 1987.
37. Snyder, S. H., Sklar, P. B., and Pevsner, J., Molecular mechanisms of olfaction, *J. Biol. Chem.,* 263(28), 13971–13974, 1988.
38. Anholt, R. R. H., Molecular neurobiology of olfaction, *CRC Crit. Rev. Neurobiol.,* 7, 1–22, 1993.
39. Firestein, S., Electrical signals in olfactory transduction, *Curr. Opin. Neurobiol.,* 2, 444–448, 1992.
40. Getchell, T. V. and Getchell, M. L., Peripheral mechanisms of olfaction: biochemistry and neurophysiology, in *Neurobiology of Taste and Smell,* Finger, T. E. and Silver, W. L., Eds., John Wiley & Sons, New York, 1987, pp. 91–124.
41. Lancet, D. and Ben-Arie, N., Olfactory receptors, *Curr. Biol.,* 3, 668–476, 1993.
42. Schiffman, S. S., Taste and smell in disease, *N. Engl. J. Med.,* 308, 1275–1279, 1983.
43. Amoore, J. E., Effects of chemical exposure on olfaction in humans, in *Toxicology of the Nasal Passages,* Barrow, C. S., Ed., Hemisphere Publishing, Washington, D.C., 1986, pp. 155–190.
44. Cometto-Muniz, J. E. and Cain, W. S., Influence of airborne contaminants on olfaction and the common chemical sense, in *Smell and Taste in Health and Disease,* Getchell, T. V., Doty, R. L., Bartoshuk, L. M., and Snow, Jr., J. B., Eds., Raven Press, New York, 1991, pp. 765–786.
45. Schwartz, B. S., Doty, R. L., Monroe, C., Frye, R., and Baker, S., Olfactory function in chemical workers exposed to acrylate and methacrylate vapors, *Am. J. Public Health,* 79(5), 613–618, 1989.
46. Gesteland, R. C., Speculations on receptor cells as analyzers and filters, *Experientia,* 42(3), 287–291, 1986.
47. Farbman, A. I., Gonzales, F., and Chuah, M. I., The effect of amitriptyline on growth of olfactory and cerebral neurons *in vitro, Brain Res.,* 457(2), 281–286, 1988.
48. Kurihara, K. and Koyama, N., High activity of adenylate cyclase in olfactory and gustatory organs, *Biochem. Biophys. Res. Commun.,* 48, 30–34, 1972.
49. Ikeda, K., Sakurada, T., Sasaki, Y., Takasaka, T., and Furukawa, Y., Clinical investigation of olfactory and auditory function in type I pseudohypoparathyroidism: participation of adenylate cyclase system, *J. Laryngol. Otol.,* 102(12), 1111–1114, 1988.
50. Getchell, T. V., Functional properties of vertebrate olfactory receptor neurons, *Physiol. Rev.,* 68(3), 772–818, 1986.
51. Bouvet, J. F., Delaleu J. C., and Holley, A., The activity of olfactory receptor cells is affected by acetylcholine and Substance P, *Neurosci. Res.,* 5(3), 214–223, 1988.
52. Shirley, S. G., Polak, E. H., Mather, R. A., and Dodd, G. H., The effect of concanavalin A on the rat electro-olfactogram. Differential inhibition of odorant response, *Biochem. J.,* 245(1), 175–184, 1987.
53. Shepherd, G. M. and Greer, C. A., The olfactory bulb, in *Synaptic Organization of the Brain,* Shepherd, G. M., Ed., Oxford University Press, New York, 1990, pp. 133–169.
54. Yamamoto, C. and Iwama, K., Arousal reaction of the olfactory bulb, *Jpn. J. Physiol.,* 11, 335–345, 1961.
55. Mancia, M., von Baumgarten, R., and Green, J. D., Response patterns of olfactory bulb neurons, *Arch. Ital. Biol.,* 100, 449–462, 1962.
56. Doty, R. L. and Ferguson-Segall, M., Odor detection performance of rats following d-amphetamine treatment: a signal detection analysis, *Psychopharmacology,* 93(1), 87–93, 1987.
57. Mair, R. G. and McEntee, W. J., The amnesic syndrome: a model for the study of central olfactory mechanisms, in *Clinical Measurement of Taste and Smell,* Meiselman H. L. and Rivlin, R. S., Eds., Macmillan, New York, 1986, pp. 550–555.
58. Doty, R. L., Reyes, P. F., and Gregor, T., Presence of both odor identification and detection deficits in Alzheimer's disease, *Brain Res. Bull.,* 18(5), 597–600, 1987.
59. Doty, R. L., Deems, D., and Stellar, S., Olfactory dysfunction in Parkinson's disease: a general deficit unrelated to neurologic signs, disease stage, or disease duration, *Neurology,* 38, 1237–1244, 1988.
60. Rinne J. O., Sako, E., Paljarvi, L., Molsa, P. K., and Rinne, U. K., Brain dopamine D-1 receptors in senile dementia, *J. Neurol. Sci.,* 73(2), 219–230, 1986.
61. Rinne J. O., Sako, E., Paljarvi, L., Molsa, P. K., and Rinne, U. K., Brain dopamine D-2 receptors in senile dementia, *J. Neurol. Transm.,* 65(1), 51–62, 1986.
62. Doty, R. L. and Risser, J., Influence of the D-2 dopamine receptor agonist quinpirole on the odor detection performance of rats before and after spiperone administration, *Psychopharmacology,* 98(3), 310–315, 1989.
63. Doty, R. L., Li, C., Pfeiffer, C., and Risser, J., Enhancement of odor detection performance by the dopamine D-1 agonist SKF 38393, in press.
64. Phillips, D. and Vallowe, H., Cyclic fluctuations in odor detection by female rats and the temporal influences of exogenous steroids on ovariectomized rats, *Proc. Penn. Acad. Sci.,* 49, 160–164, 1975.
65. Halpern, B. P., Environmental factors affecting chemoreceptors: an overview, *Environ. Health Prosp.,* 44, 101–105, 1982.

66. Schiffman, S. S. and Nagle, H. T., Effects of environmental pollutants on taste and smell, *Otolaryngol. Head Neck Surg.*, 106, 693–700, 1992.
67. Frye, R. E., Schwartz, B. S., and Doty, R. L., Dose-related effects of cigarette smoking on olfactory function, *J. Am. Med. Assoc.*, 263(9), 1233–1236, 1990.
68. Sumner, D., Post-traumatic ageusia, *Brain*, 90, 187–202, 1967.
69. Rollin, H., Drug-related gustatory disorders, *Ann. Otol. Rhinol. Laryngol.*, 87, 37–42, 1978.
70. Goodspeed, R. B., Gent., F., Catalanotto, F. A., Cain, W. S., and Zagraniski, R. T., Corticosteroids in olfactory dysfunction, in *Clinical Measurement of Taste and Smell*, Meiselman H. L. and Rivlin, R. S., Eds., Macmillan, New York, 1986, pp. 514–518.
71. Mott, A. E., Topical corticosteriod therapy for nasal poliposis, in *Smell and Taste in Health and Disease*, Getchell, T. V., Doty, R. L., Bartoshuk, L. M., and Snow, Jr., J. B., Eds., Raven Press, New York, 1991, pp. 553–572.
72. Eichel, B. S., Improvement of olfaction following pansinus surgery, *Ear Nose Throat J.*, 73(4), 248–250, 1994.
73. Harris, M., Psychogenic aspects of facial pain, *Br. Dent. J.*, 136, 199–202, 1974.
74. Singer, E., Pain control in dentistry: management of chronic orofacial pain, *Compend. Cont. Ed.*, 114, 116–118, 1987.
75. Lamey, P. J. and Lewi, M. A., Oral medicine in practice: burning mouth syndrome, *Br. Dent. J.*, 167(6), 197–200, 1989.
76. Schechter, P J., Friedewald. W. T., and Bronzert, D. A., Idiopathic hypogeusia: a description of the syndrome and a single blind study with zinc sulfate, *Int. Rev. Neurobiol. (Suppl.)*, 1, 125–140, 1972.
77. Estrem, S. A. and Renner, G., Disorders of taste and smell, *Otolaryngol. Clin. N. Am.*, 20(1), 133–147, 1987.
78. Price, S., The role of zinc in taste and smell, in *Clinical Measurement of Taste and Smell*, Meiselman H. L. and Rivlin, R. S., Eds., Macmillan, New York, 1986, pp. 443–445.
79. Graber, M. and Kellener, S., Side effects of acetazolamide: the champagne blues, *Am. J. Med.*, 84, 979–980, 1988.
80. Bourliere, F., Cendron, H., and Rapaport, A., Action de l'acide acetylsalicylique sur la sensibilite au gout amer chez l'homme, *Rev. Fr. Etudes Clin. Biol.*, 4, 380–382, 1959.
81. USP DI, Drug information for the health care professional, United States Pharmacopeial Convention, Inc., Vol. 1, 1993.
82. Grossman, S. P., Drug effects on taste, olfaction, and food intake, in *A Symposium on Drugs and Sensory Functions*, Herxheimer, A., Ed., Little, Brown & Co., Boston, 1968, pp. 101–130.
83. Mattes, R. D., Christensen, C. M., and Engelman, K., Effects of hydrochlorothiazide and amiloride on salt taste and excretion (intake), *Am. J. Hypertens.*, 3, 436–443, 1990.
84. Schiffman, S. S., Lockhead, E., and Maes, F. W., Amiloride reduces the taste intensity of Na^+ and Li^+ salts and sweeteners, *Proc. Acad. Sci. U.S.A.*, 80, 6136–6140, 1983.
84a. Ossebaard, C. A. and Smith, D. V., Effect of amiloride on the taste of NaCl, Na-gluconate, and KCl in humans: implications for Na^+ receptor mechanisms, *Chem. Senses*, 20(1), 37–46, 1995.
84b. Ossebaard, C. A. and Smith, D. V., Amiloride suppresses the sourness of NaCl and LiCl, *Physiol. Behav.*, 60, 1317–1322, 1996.
84c. Ossebaard, C. A., Polet, I. A., and Smith, D. V., Amiloride effects on taste quality: comparison of single and multiple response category procedures, *Chem. Senses*, in press.
85. Jaffe, I. A., Ampicillin rashes, *Lancet*, 1, 245, 1970.
86. McTavish, D. and Sorkin, E. M., Azelastine. A review of its pharmacodymamic and pharmacokinetic properties and therapeutic potential, *Drugs*, 38(5), 778–800, 1989.
87. Halsey, J. P. and Cardoe, H., Benoxaprofen: side effect profile in 300 patients, *Br. Med. J.*, 284, 1365–1368, 1982.
88. Soni, N. K. and Chatterji, P., Gustotoxicity of bleomycin, *J. Otol. Rhinol. Laryngol. Rel. Spec.*, 47, 101–104, 1985.
89. Tuominen, M., Haasio, J., Hekali, R., and Rosenberg, P. H., Continuous interscalene brachial plexus block: clinical efficacy, technical problems and bupivacaine plasma concentrations, *Acta Anaesthesiol. Scand.*, 33(1), 84–88, 1989.
90. Atkinson, A. B., Brown, J. J., Lever, A. F., and Robertson, J. I. S., Combined treatment of severe intractable hypertension with captopril and diuretic, *Lancet*, 2, 105–108, 1980.
91. Boyd, I., Captopril-induced taste distubance, *Lancet*, 342, 304, 1993.
92. Vlasses, P. H. and Ferguson, R. K., Temporary ageusia related to captopril, *Lancet*, 2, 526, 1979.
93. Coulter, D. M., Eye pain with nifedipine and disturbance of taste with captopril: a mutually controlled study showing a method of postmarketing surveillance, *Br. Med. J.*, 296, 1084–1088, 1988.
94. Materson, B. J., Adverse effects of angiotension-converting enzyme inhibitors in antihypertensive therapy with focus on quinapril, *Am. J. Cardiol.*, 69(10), 46C-53C, 1992.
95. McNeil, J. J., Anderson, A., Christophidis, N., Jarrott, B., and Louis, W. J., Taste loss associated with oral captopril treatment, *Br. Med. J.*, 2, 1555–1556, 1979.
96. Vidt, D. G., Bravo, E. L., and Foud, F. M., Captopril, *New Eng. J. Med.*, 306(4), 214–219, 1982.

97. Halbreich, U., Tegretol dependency and diversion of the sense of taste, *Isr. Ann. Psychiatry,* 12, 328–332, 1974.
98. Erikssen, J., Seegaard, E., and Naess, K., Side-effects of thiocarbamides, *Lancet,* 1, 231–232, 1975.
99. Neundörfer, B., Disturbances of smell and taste during treatment with tiamazole and carbimazole, *Nervenarzt,* 58, 61–62, 1987.
100. Reyes, E. S., Talley, R. W., O'Bryan, R. M., and Gastesi, R. A., Clinical evaluation of 1, 3-bis(2-chloroethyl)-1-nitrosourea (BCNU: NSC-409962) with fluoxymesterone (NSC-12165) in the treatment of solid tumors, *Cancer Chemother. Rep.,* 57, 225–230, 1973.
101. Hodgson, T. G., Bad taste from cefamandole, *Drug Intell. Clin. Pharm.,* 15, 136, 1981.
102. Henkin, R. I., Griseofulvin and dysgeusia: implications?, *Ann. Int. Med.,* 74, 795–796, 1971.
103. Lang, N. P., Catalanottao, F. A, Knopfli, R. U., and Antczak, A. A., (1988) Quality-specific taste impairment following the application of chlorhexidine digluconate mouthrinses, *J. Clin. Periodontol.,* 15, 43–48, 1988.
104. Zilstorff, K., Sense of smell alterations by cocaine and tetracaine, *Arch. Otolaryngol.,* 82, 53–55, 1965.
105. Fetting, J. H., Wilcox, P. M., Sheidler, V. R., Enterline, J. P., Donehower, R. C., and Grochow, L. B., Tastes associated with parenteral chemotherapy for breast cancer, *Cancer Treat. Rep.,* 69, 1249–1251, 1985.
106. Fehm-Wolfsdorf, G., Scheible, E., Zenz, H. Born, J., and Fehm, H. L., Taste thresholds in man are differentially influenced by hydrocortisone and dexamethasone, *Psychoneuroendocrinology,* 14(6), 433–440, 1989.
107. Mata, R., Effect of dextro-amphetamine on bitter taste threshold, *J. Neuropsychiatry,* 4, 315–320, 1963.
108. Wistrand, P. J., The use of carbonic anhydrase inhibitors in ophthalmology and clinical medicine, *Ann. N.Y. Acad. Sci.,* 429, 609–619, 1984.
109. Berman, L. J., Dysosmia, dysgeusia, and diltiazem, *Ann. Intern. Med.,* 102, 717, 1985.
110. Goy, J. J., Finci, L., and Sigwart, U., Dysgeusia after high dose dipyridamole treatment, *Arzeneimittelforschung,* 35, 854, 1985.
111. Willoughby, D. M., Drug-induced abnormalities of taste sensation, *Adverse Drug Reaction Bull.,* 100, 368–371, 1983.
112. Guthrie, D. and Way, S., Treatment of advanced carcinoma of the cervix with adriamycin and methotrexate combined, *Obstet. Gynecol.,* 44, 586–589, 1974.
113. Bleasel, A. F., McLeod, J. G., and Lane-Brown, M., Anosmia after doxycycline use, *Med. J. Austr.,* 152, 440, 1990.
114. McFate-Smith, W., Davies, R. O., Gabriel, M. A., Kramsch, D. M., Moncloa, F., Rush, J. E., and Walker, J. F., Tolerance and safety of enalapril, *Br. J. Clin. Pharmacol.,* 18(Suppl. 2), 249S-255S, 1984.
115. Gifford, R. W., Ethacrynic acid alone and in combination with methyldopa in management of mild hypertension: a report of 23 patients, *Int. Z. Klin. Pharmakol. Ther. Toxikol.,* 3, 255–260, 1970.
116. Jones, P. B., McCloskey, E. V., and Kanis, J. A., Transient taste-loss during treatment with etidronate, *Lancet,* 2, 637, 1987.
117. Neuss, H., Long-term use of flecainide in patients with supraventricular tachycardia, *Drugs,* 4(Suppl. 29), 21–25, 1985.
118. Dry, J., Sors, C., Gervais, P., van Straaten, L., Perrin-Fayolle, M., and Paramelle, B., A comparison of flunisolide inhaler and beclomethasone dipropionate inhaler in bronchial asthma, *J. Int. Med. Res.,* 13(5), 289–293, 1985.
119. Wickstrøm, E. and Giercksky, K.-E., Comparative study of zopiclone, a novel hypnotic and three benzodiazepines, *Eur. J. Clin. Pharmacol.,* 17, 93–99, 1980.
120. Bedikian, A. Y., Valdivieso, M., Bodey, G. P., Burgess, M. A., Benjamin, R. S., Hall, S., and Freireich, E. J., Phase I clinical studies with gallium nitrate, *Cancer Treat. Rep.,* 62(2), 1449–1453, 1978.
121. Cherington, M., Guanidine and germine in Eaton-Lambert syndrome, *Neurology,* 26, 944–146, 1976.
122. Lahon, H. F. J. and Mann, R. D., Glipizide: results of a multicentre clinical trial, *J. Intern. Med. Res.,* 1, 608–615, 1973.
123. Fogan, L., Griseofulvian and dysgeusia: implications?, *Ann. Intern. Med.,* 74, 795–796, 1971.
124. Lance, F., Griseofulvin and dysgeusia. Implications, *Ann. Intern. Med.,* 74(5), 795, 1971.
125. Plath, P. and Otten, E., Untersuchungen uber die Wirksamkeit von Hexetidine bei akuten Erkrankungen des Rachens und der Mundhohle sowie nach Tonsillektomie, *Therapiewocke,* 19, 1565–1566, 1969.
126. Simpson, J. R., Idoxuridine in the treatment of herpes zoste, *Practitioner,* 215, 226–229, 1975.
127. McCurdy, P. R., Parenteral iron therapy. II. A new iron-sorbitol citric acid complex for intra-muscular injection, *Ann. Int. Med.,* 61, 1053–1064, 1964.
128. Runge, L. A., Pinals, R. S., Lourie, S. H., and Tomar, R. H., Treatment of rheumatoid arthritis with levamisole: a controlled trial, *Arthritis. Rheum.,* 20, 1445–1448, 1977.
129. Siegfried, J. and Zumstein, H., Changes in taste under L-dopa therapy, *Z. Neurol.,* 200, 345–348, 1971.
130. Yamada, Y. and Tomita, H., Influences on taste in the area of chorda tympani nerve after transtympanic injection of local anesthetic (4% lidocaine), *Auris. Nasus. Larynx,* 16(Suppl. 1), S41–46, 1989.
131. Bressler, B., An unusual side-effect of lithium, *Psychosomatics,* 21, 688–689, 1980.
132. Duffield, J. E., Side effects of lithium carbonate, *Br. Med. J.,* 1, 491, 1973.
133. Vestergaard, P., Poulstrup, I., and Schou, M., Prospective studies on a lithium cohort. 3. Tremor, weight gain, diarrhea, psychological complaints, *Acta Psychiatr. Scand.,* 78(4), 434–441, 1988.

134. Henkin, R. I. and Bradley, D. F., Regulation of taste acuity by thiols and metal ions, *Proc. Nat. Acad. Sci. USA,* 62, 30–37, 1969.
135. Hermann, L. S., Metformin: a review of its pharmacological properties and therapeutic use, *Diabetes Metab.,* 5, 233–245, 1979.
136. Watkins, P. J., Treatment, *Br. Med. J.,* 284, 1853–1855, 1982.
137. Tallman, J., Willenbring, M., Carlson, G., Boosalis, M., Krahn, D., Levine, A. S., and Morley, J. E., Effects of chronic methadone use in humans on taste and dietary preference, *Fed. Proc.,* 43, 1058, 1984.
138. Hallman, B. L. and Hurst, J. W., Loss of taste as toxic effect of methimazole (Tapazole®) therapy: report of three cases, *J. Am. Med. Assoc.,* 152, 322, 1953.
139. Duhra, P. and Foulds, I. S., Methotrexate-induced impairment of taste acuity, *Clin. Exp. Dermatol.,* 13(2), 126–127, 1988.
140. Rees, R. B., Bennett, J. H., Maibach, H. I., and Arnold, H. L., Methotrexate for psoriasis, *Arch. Dermatol.,* 95, 2–11, 1967.
141. Roenigk, H. H., Fowler-Bregfeld, W., and Curtis, G. H., Methotrexate for psoriasis in weekly oral dose, *Arch. Dermatol.,* 99(1), 86–93, 1969.
142. Leys, D., Hyperthyoidism treated with methyl thiouracil, *Lancet,* 1, 461–464, 1945.
143. Schneeberg, N. G., Loss of sense of taste due to mehtylthiouracil therapy, *J. Am. Med. Assoc.,* 149, 1091–1093, 1952.
144. Brandt, L. J., Bernstein, L. H., Boley, S. J., and Frank, M. S., Metronidazole therapy for perineal Crohn's disease: a follow-up study, *Gastroenterology,* 83(1), 383–387, 1982.
145. Strassman, H. D., Adams, B., and Pearson, A. W., Metronidazole effect on social drinkers, *Quart. J. Stud. Alcohol.,* 31(1), 394–398, 1970.
146. Holgate, S. T., Clinical evaluation of nedocromil sodium in asthma, *Eur. J. Respir. Dis. (Suppl.),* 147, 149–159, 1986.
147. Levinson, J. L. and Kennedy, K., Dysosmia, dysgeusia, and nifedipine, *Ann. Intern. Med.,* 102, 135–136, 1985.
148. Pratra, A., Clinical evaluation of niridazole in *Schistosoma mansoni* infections, *Ann. N.Y. Acad. Sci.,* 160, 660–669, 1969.
149. Ewing, R. C., Janda, S. M., and Henann, N. E., Ageusia associated with transdermal nitroglycerin, *Clin. Pharm.,* 8(2), 146–147, 1989.
150. Rabe, F., Loss of taste and dyskinesia as drug side-effects, *Med. Welt.,* 16, 711–713, 1970.
151. Whittington, J. and Raftery, E. B., A controlled comparison of oxyfedrine, isosorbide dinitrate and placebo in the treatment of patients suffering attacks of angina pectoris, *Br. J. Clin. Pharmacol.,* 10, 211–215, 1980.
152. Henkin, I. R., Keiser, H. R., Jaffe, I. A., Sternlieb, I., and Scheinberg, I. H., Decreased taste sensitivity after D-penicillamine reversed by copper administration, *Lancet,* 1268–1271, 1967.
153. Keiser, H. R., Henkin, R. I., Bartter, F. C., and Sjoerdsma, A., Loss of taste during therapy with penicillamine, *J. Am. Assoc.,* 203, 381–183, 1968.
154. Kean, W. F., Anastassiades, T. P., Dwosh, I. L., Ford, P. M., Kelly, W. G., and Dok, C. M., Efficacy and toxicity of D-penicillamine for rheumatoid disease in the elderly, *J. Am. Geriatr. Soc.,* 94–98, 1982.
155. Manthorpe, R., Horbov, S., Sylvest, J., and Vinterberg, H., Auranofin vs. penicillamine in rheumatoid arthritis; one-year results from a prospective clinical investigation, *Scand. J. Rheumatol.,* 15(1), 13–22, 1986.
156. Shiokawa, Y., Horiuchi, Y., Honma, M., Kageyama, T., Okada, T., and Azuma, T. (1977) Clinical evaluation of D-penicillamine by multicentric double-blind comparative study in chronic rheumatoid arthritis, *Arthritis Rheum.,* 20(8), 1464–1471, 1977.
157. Sternlieb, I. and Scheinberg, I. H., Penicillamine therapy for hepatolenticular degeneration, *J. Am. Med. Assoc.,* 189, 748–754, 1964.
158. Ferguson, A. W., de la Harpe, P. L., and Farquhar, J. W., Dimethyldiguanide in the treatment of diabetic children, *Lancet,* 1, 1367–1369, 1961.
159. Schwartz, M. J., Mirsky, S., and Schaefer, L. E., The effects of phenformin hydrochloride on serum cholesterol and triglyceride levels in the stable adult diabetic, *Metabolism,* 15, 808–822, 1966.
160. Scott, P. J., Glossitis with complete loss of taste sensation during Dindevan treatment: report of a case, *N. Z. Med. J.,* 59, 296, 1960.
161. Fischer, R., Griffin, F., Archer, R. C., Zinsmeister, and S. C., Jastram, P. S., Weber ratio in gustatory chemoreception: an indicator of systemic (drug) reactivity, *Nature,* 207, 1049–1053, 1965.
162. von Skramlik, E., The fundamental substrates of taste, in *Olfaction and Taste,* Zotterman, Y., Ed., Pergamon Press, Oxford, 1963, pp. 125–132.
163. Chow, M. S., Lebsack, C., and Hilleman, D., Propafenone: a new antiarrhythmic agent, *Clin. Pharm.,* 7(12), 869–877, 1988 (published erratum appears in *Clin. Pharm.,* 8(7), 473, 1989).
164. Grossman, S. P., Loss of taste and smell due to propylthiouracil therapy, *N.Y. J. Med.,* 53, 1236, 1953.
165. Niewind, A. and Krondl, M., The effect of salicylates on taste responses of elderly persons with osteoarthritis: implications for food selection, *Fed. Proc.,* 44(2), 937, 1985.
166. Thumfart, W., Plattig, K. H., and Schlict, N., Smell and taste thresholds in older people, *Z. Gerontol.,* 13, 158–188, 1980.

167. DeSimone, J. A., Heck, G. L., and Bartoshuk, L. M., Surface active taste modifiers: a comparison of the physical and psychophysical properties of gymnemic acid and sodium lauryl sulfate, *Chem. Senses,* 5, 317–330, 1980.
168. Clee, M. D. and Burrow, L., Taste and smell in disease, *N. Engl. J. Med.,* 309(17), 1062, 1983.
169. Juhlin, L., Loss of taste and terbinafine, *Lancet,* 339, 1483, 1992.
170. Magnasco, L. D. and Magnasco, A. J., Metallic taste associated with tetracycline therapy, *Clin. Pharm.,* 4, 455–456, 1985.
171. Body, S. C., A taste of allicin?, *Anaesth. Intens. Care,* 14, 94, 1986.
172. Huskisson, E. C., Jaffe, I. A., Scott, J., and Dieppe, P. A., 5-Thiopyridoxine in rheumatoid arthritis: clinical and experimental studies, *Arthritis Rheum.,* 23, 106–110, 1980.
173. Lee, A. and Lader, M., Tolerance and rebound during and after short-term administration of quazepam, triazolam and placebo to healthy human volunteers, *Int. Clin. Psychopharmacol.,* 3(1), 31–47, 1988.
174. Fontaine, R., Beaudry, P., Le Morvan, P., Beauclair, L., and Chouinard, G., Zopiclone and triazolam in insomnia associated with generalized anxiety disorder: a placebo-controlled evaluation of efficacy and daytime anxiety, *Int. Clin. Psychopharmacol.,* 5(3), 173–183, 1990.
175. Goa, K. L. and Heel, R. C., Zopiclone. A review of its pharmacodynamic and pharmacokinetic properties and therapeutic efficacy as an hypnotic, *Drugs,* 32, 48–65, 1986.
176. Lader, M. and Frcka, G., Subjective effects during administration and on discontinuation of zopiclone and temazepam in normal subjects, *Pharmacopsychiatry,* 20(2), 67–71, 1987.
177. Skouby, A. P. and Zilstroff-Pedersen, K., The influence of acetylcholine-like substances, menthol and strychnine, on olfactory receptors in man, *Acta. Physiol. Scand.,* 32, 252–158, 1954.
178. Duran, M. V., Anosmie recurrente sous beta-bloqueurs, *Le Presse Medicale,* 14(40), 2064, 1985.
179. Macht, D. I. and Macht, M. B., Comparison of effect of cobra venom and opiate on olfactory sense, *Am. J. Physiol.,* 129, P411–412, 1940.
180. Turner, P., Some observations on centrally-acting drugs in man, *Proc. Roy. Soc. Med.,* 58, 913–914, 1965.
181. Zilstorff, K. and Herbild, O., Parosmia, *Acta Otolaryngol. (Suppl.),* 360, 40–41, 1979.
182. Seydell, E. M. and McKnight, W. P., Disturbances of olfaction resulting from intranasal use of tyrothricin: a clinical report of seven cases, *Arch Otolaryngol.,* 47, 465–470, 1948.

CHAPTER 9

Essential and Nonessential Mineral Interactions

Gregory D. Miller and Susan M. Groziak

CONTENTS

0-8493-4493-X/97/$0.00+$.50

Six major minerals and 16 trace elements exist (see Table 1).[1,2] The physical and chemical properties of minerals are governed by the quantum number of each electron which in turn determines the energy level (shell), shape of the electron cloud, directional orientation of the electron cloud, and direction in which the electron moves about its axis (see Table 2).[3] The chemical properties of each mineral determine its ability to interact with other minerals by influencing its bioavailability as well as capability to interact directly or indirectly with other minerals. Small molecules with a low ionic charge (such as sodium and potassium) cross the lipid bilayers of cell membranes readily and are almost completely absorbed.[3] Ions with a relatively high ionic charge are more likely to form stable complexes than ions with a lesser charge and consequently are often less well absorbed than ions with a low ionic charge;[3] however, this rule is not absolute. Whereas some minerals which form complexes are poorly absorbed (e.g., calcium), others (e.g., magnesium) may be well absorbed (see Table 3).

TABLE 1

Major Minerals and Trace Elements

Major Minerals	Trace Elements
Sodium, potassium, magnesium, calcium, phosphorus, and chloride	Arsenic, boron, cobalt, chromium, copper, fluoride, iron, iodine, manganese, molybdenum, nickel, selenium, silicon, tin, vanadium, and zinc

TABLE 2

Examples of Minerals with Low and High Ionic Charges

Low Ionic Charge	High Ionic Charge
Sodium, potassium, and fluoride	Chromium, cobalt, and iron

TABLE 3

Absorbability of Mineral Complexes

Form Poorly Absorbed Complexes	Form Well Absorbed Complexes
Calcium, chromium, iron, zinc, copper, manganese, nickel, silicon, tin, and vanadium	Magnesium, arsenic, phosphorus, and molybdenum

TABLE 4

Direct and Indirect Mineral Interactions[a]

Direct	Indirect
Calcium and lead	Copper and iron
Calcium and magnesium	Copper and zinc
Lead and copper	
Lead and manganese	
Lead and zinc	
Lead and iron	
Iron and manganese	

[a] This is not an exhaustive list and is intended only to provide examples of mineral interactions.

Mineral interactions occur between essential and nonessential minerals as well as between essential minerals. The two major types of mineral interactions that occur are direct and indirect (see Table 4). In a direct mineral interaction, two minerals may compete for the same transport site or ligand. In an indirect mineral interaction, a mineral influences the metabolism of another mineral; for example, copper (a component of ceruloplasmin) influences the mobilization of iron from the liver.[4]

Mineral interactions may also be positive (synergistic, where one mineral enhances the bioavailability of another) or negative (antagonistic, where one mineral decreases the bioavailability of another).[5] Consequently, a mineral may either enhance or decrease the toxicity potential of another mineral. This interaction may occur in food, at a transport ligand, or at (or within) the intestinal cell, blood, lymph, or tissues.[6-9]

Researchers have identified four main types of mineral-mineral interactions in food. These interactions include:[3]

1. Displacement of a mineral from a complex with another mineral to form a soluble or insoluble complex
2. Addition of a second or third mineral to a soluble mineral-ligand complex causing precipitation
3. Addition of a mineral causing a mineral-ligand complex to bind to another substrate (ligand) and form a poly-mineral-poly-ligand complex
4. Formation of a poly-mineral-ligand complex, which changes the susceptibility of the mineral-ligand bonds to cleavage by digestive enzymes

In general, minerals with similar orbitals and coordination numbers (and consequently similar physical and chemical properties) antagonize each other directly by competing for the same binding site.[9-11] Copper, zinc, and cadmium have similar orbitals, configurations, and coordination numbers and interact directly.[4] Iron and manganese also interact directly because they have similar chemical parameters;[4] however, similar chemical parameters do not explain

TABLE 5

Minerals that Interact with Calcium

Mineral	
Lead	Manganese
Phosphorus	Fluoride
Magnesium	Cadmium
Iron	Aluminum
Zinc	Tin
Copper	

all observed interactions between minerals. For example, iron and zinc differ in their electronic configuration but interact in an antagonistic manner.

The effects of an excess of one mineral on another may appear quickly or become evident only after an imbalance persists for a long time.[4] Growing individuals or individuals with low mineral stores tend to be more sensitive to developing mineral imbalances.[12] Consequently, many animal studies conducted on mineral interactions utilize growing animals.

1 CALCIUM

Calcium is an essential mineral that plays an important role in the development and maintenance of bones and teeth and additional important roles in nerve conduction, muscle contraction, and blood clotting. Numerous factors influence the intestinal absorption of calcium. These factors include age, hormone status, vitamin D status, and the bioavailability of calcium in food.[13] The Recommended Dietary Allowances (RDAs) are based on the assumption that adults absorb 30 to 40% of dietary calcium and adolescents absorb 40% of calcium consumed;[14] however, recent research indicates that teenage girls and young women absorb lower relative amounts of calcium than the absorption levels assumed by the current RDAs.[15] A wide variety of surveys indicate that children, girls, and women in the U.S. consume less than their RDA for calcium.[16-20] A recent consensus development panel convened by the National Institutes of Health concluded that, to optimize bone health, children, adolescents, and adults need to consume higher levels of calcium than those set by the current calcium RDAs.[21]

Calcium is absorbed by both saturable (cellular) and nonsaturable pathways. Calcium binds to several different proteins at the brush border of the intestine including vitamin D-dependent calbindin, a vitamin D-dependent integral membrane protein, and calmodulin.[22] The transport of calcium across the basolateral membrane of intestinal cells involves a sodium-calcium exchange[23-26] and a high affinity calcium- and magnesium-activated ATPase.[27-29] In food, calcium often occurs primarily in bound forms which may be poorly absorbed. For example, humans on average absorb only 5% of the calcium in spinach compared to an absorption rate of roughly 28% from milk.[30]

Dietary calcium interacts with numerous minerals (see Table 5). The mechanisms by which calcium interacts with other minerals are extremely varied (see Table 6).

1.1 CALCIUM AND LEAD

Research studies conducted on animals as well as humans indicate that calcium and lead interact in a negative (antagonistic) manner. A calcium deficiency raises lead toxicity, and an adequate calcium intake decreases lead toxicity. The exact mechanisms by which calcium

TABLE 6

Proposed Mechanisms for Mineral Interactions with Calcium

Mineral	Proposed Mechanisms	Ref.
Lead	Calcium and lead compete physically for absorption	43
	Calcium promotes urinary lead excretion	37
	Calcium increases release of lead from bones	38
	Lead impairs vitamin D-dependent absorption of calcium	81
	Lead increases urinary calcium excretion	73
	Lead blocks the entrance of calcium into nerve terminals	74
	Lead blocks the calcium efflux from cells potentially by replacing calcium in the calcium-sodium ATP pump	75
Phosphorus	Phosphorus decreases calcium absorption by forming insoluble calcium triphosphate	22, 87
	Phosphorus increases fecal calcium loss by increasing calcium content of digestive secretions	88, 89
	Phosphorus decreases urinary calcium by enhancing calcium reabsorption	90, 91, 92
	Phosphorus increases parathyroid hormone excretion which in turn reduces serum calcium	22
	Calcium increases renal phosphate reabsorption and phosphate mobilization from bone and soft tissue	19, 99, 100
Magnesium	Calcium decreases magnesium absorption by competing for a common transport system	107, 108, 109
Iron	Calcium decreases iron absorption	125, 126, 127
Zinc	Calcium reduces solubility of calcium-phytate-zinc complexes	4
Copper	Calcium reduces copper absorption by raising the pH in the intestine, causing the precipitation of copper hydroxide	134
Manganese	Calcium decreases manganese absorption	135
Fluoride	Calcium decreases intestinal absorption of fluoride by forming insoluble salts	4, 137
Cadmium	Calcium decreases cadmium absorption	142, 143
	Cadmium decreases calcium absorption potentially by inhibiting vitamin D-induced intestinal calcium transport	145
	Cadmium increases urinary calcium excretion	145
Aluminum	Aluminum may decrease calcium absorption by affecting parathyroid hormone and phosphorus metabolism	124, 147, 148, 149

interacts with lead are not all known. Laboratory animal research indicates that calcium decreases lead absorption,[31-36] promotes urinary excretion of lead,[37] increases the release of lead from bones in culture media,[38] and reverses lead's inhibition of acetylcholine release from ganglia.[39] Some research indicates that calcium intake rather than calcium status decreases lead absorption.[31,40] Milk, a good source of calcium, has also been reported to inhibit the accumulation of lead in the intestine of laboratory animals[41] and reduces the short-term retention of ingested lead in man.[42] The U.S. Centers for Disease Control recommends adequate intakes of calcium for the prevention of childhood lead poisoning.[43]

Researchers propose that calcium may inhibit lead absorption via physical competition between calcium and lead for common binding sites on intestinal binding proteins for absorption.[31-33,44] Although some research studies support a common intestinal pathway for calcium and lead absorption,[33] others do not;[45] however, there are at least two mechanisms by which lead moves across the intestinal wall.[32]

It has been demonstrated that increasing calcium intake can decrease tissue lead concentrations in laboratory animals fed lead;[40,46-51] however, study results have been mixed.[52] The studies which indicate that dietary calcium influences lead tissue status link low calcium intakes with increased lead retention in several body tissues, including blood, bone, kidney, brain, and liver.[49-51,53] Based on animal studies, researchers estimate that a low calcium diet can increase susceptibility to lead toxicity by as much as 20-fold.[50]

Human studies conducted on the effect of dietary calcium on lead confirm the findings of animal research with regard to both lead absorption and blood lead levels. In human infants, lead absorption decreases as dietary calcium increases.[54] Higher calcium intakes decrease lead absorption in human adults.[42,44,55,56] Numerous studies conducted on children,[57-59] adults[42,60-67] and pregnant women[60] link higher calcium intakes with lower blood lead levels. A study of Swiss men and women also linked increased dairy product intake with decreased blood lead levels.[68]

Dietary lead, in turn, interacts with calcium in a negative manner. Animal research indicates that lead decreases calcium absorption,[69-71] decreases serum calcium levels,[47,72] increases urinary calcium excretion,[73] blocks the entrance of calcium into nerve terminals,[74] blocks the calcium efflux from cells (potentially by replacing calcium in the calcium-sodium ATP pump),[75] impairs calcium uptake by calcium channels,[76] inhibits mitochondrial uptake of calcium in the heart[77] and brain,[78] displaces calcium in mitochondria, and interferes with calcium messenger systems.[79,80]

Some researchers propose that lead decreases calcium absorption by competing with calcium for binding sites on calcium-binding and receptor proteins such as calmodulin and protein kinase C;[69-71] however, the level of calcium intake appears to affect the influence of lead on calcium absorption. One study found that, whereas dietary lead inhibited calcium absorption, intestinal calbindin D, and alkaline phosphatase synthesis in laboratory animals fed a low calcium diet, exposure to lead actually increased these parameters in animals fed a normal calcium diet.[81] Based on these findings, the researchers speculate that the primary effect of lead occurs at, or prior to, intestinal protein synthesis and most likely involves the cholecalciferol endocrine system rather than any direct interactions between lead and calcium at the intestinal level.[81] Similar to animal studies, the limited research that has been conducted on the effect of dietary lead on calcium status in humans indicates that inorganic lead exposure decreases serum calcium levels.[57,82]

1.2 Calcium and Phosphorus

About 60% of dietary phosphate is absorbed.[22] Phosphorus is more easily absorbed than calcium because there is little physiological control over its absorption.[83] Transport of phosphate across the intestinal cell is driven by a sodium-dependent active transport system[84] and diffusion.[85]

More research has been conducted on the effects of phosphorus on calcium status than on the effects of calcium on phosphorus status. Phosphorus interacts with calcium in both a positive and negative manner. Although research indicates that phosphorus decreases calcium absorption and increases fecal calcium excretion, phosphorus also decreases urinary calcium. High amounts of dietary phosphorus decrease calcium absorption in laboratory animals.[86] It appears that phosphate reduces calcium absorption by interacting with calcium to form a poorly absorbed insoluble complex known as calcium triphosphate.[22,87]

Human research indicates that phosphorus increases the calcium content of digestive secretions and consequently increases endogenous intestinal calcium loss.[88,89] Phosphorus also decreases urinary calcium in humans.[90-92] Phosphate can also stimulate parathyroid hormone excretion which in turn reduces serum calcium.[22]

Calcium reabsorption in the kidneys parallels water reabsorption and involves solvent drag and passive diffusion.[22] Both animal and human studies indicate that increasing phosphorus intake reduces urinary calcium and consequently increases calcium retention.[93-95] There are indications that phosphate decreases urinary calcium by increasing calcium reabsorption in the distal portion of the nephron[94] or through extrarenal mechanisms;[95] however, the net effect of phosphorus on calcium status appears to be neutral. In fact, some studies indicate that increasing phosphorus intake 2.5-fold does not alter calcium balance in adult men, regardless of calcium intake.[96,97]

Little research has been conducted on the effect of calcium on phosphorus. Increasing dietary calcium in pregnant laboratory animals decreases total body phosphorus.[98] Human research indicates that increasing serum calcium increases plasma phosphate levels (presumably by increasing renal phosphate reabsorption and phosphate mobilization from bone and soft tissue).[99,100]

1.3 Calcium and Magnesium

Magnesium and calcium are similar in chemical nature and might be expected to compete for the same ligands.[5] Numerous animal studies demonstrate that a high calcium diet decreases magnesium absorption.[101-105] Magnesium is absorbed from the entire intestine, whereas calcium is absorbed primarily from the duodenum.[106]

Calcium is transported out of the basolateral membrane of intestinal cells, in part, via a high affinity calcium- and magnesium-activated ATPase; however, this is not the mechanism by which researchers propose that magnesium affects calcium status. Researchers theorize that magnesium and calcium compete for a common transport system in the intestine.[107-109] Research indicates that vitamin D influences calcium and magnesium intestinal transport differently and that calcium and magnesium may be also absorbed by different mechanisms.[110,111]

Increasing calcium intake decreases magnesium concentrations in bone[40] and depresses kidney,[40,52] liver, and testis magnesium levels in laboratory animals.[52] Although some studies conducted in humans indicate that high calcium intakes exert a negative effect on magnesium absorption,[112-114] other studies report that a high calcium intake does not alter magnesium absorption and balance in humans beings.[115-123] One reason proposed for the apparent lack of a reported effect of calcium on magnesium absorption in humans compared to animals is the fact that animal studies employ considerably higher magnitudes of calcium in the diet than human studies.[5] Limited research exists on the effect of magnesium on calcium status. One human study reports that magnesium increases fecal calcium excretion in humans.[124]

1.4 Calcium and Iron

Iron occurs in the diet primarily in bound forms that are often poorly absorbed. Increasing calcium intake decreases bone,[124] kidney, liver, and testis iron levels in laboratory animals.[52] Dietary calcium has also been reported to decrease iron absorption in humans;[125-127] however, this inhibitory effect appears to be dose related up to 150 to 300 mg of calcium and is not evident when calcium-rich dairy products are added to the diet of free-living adults.[137]

1.5 Calcium and Zinc

Zinc occurs in the diet primarily in bound form. Researchers propose that calcium does not interact directly with zinc, but instead multiple interactions occur between zinc, calcium, and phytate.[5] Excess dietary calcium reduces the solubility of calcium-phytate-zinc complexes, thereby decreasing zinc bioavailability.[4]

Animal research indicates that high calcium intakes enhance symptoms (e.g., parakeratosis) of a zinc deficiency.[128] Research also indicates that increasing calcium intake decreases bone zinc in laboratory animals.[40] Contrary to animal studies, human studies indicate that dietary calcium has a relatively minor negative effect on zinc status.[125,129] One study found that adding 500 mg of calcium to the diet did not decrease zinc absorption in humans.[125] Another study found that increasing calcium intake to 2000 mg per day had a slight, although not significant, negative effect on zinc absorption and balance.[123,130] Limited research indicates that zinc, in turn, may interact with calcium in a negative manner. A recent human study reports that high intakes of zinc inhibit calcium absorption.[131,132]

1.6 Calcium and Copper

Relatively little research has been conducted on the interactions between calcium and copper. The limited research that exists indicates that calcium decreases copper absorption.[133] Researchers propose that calcium reduces copper absorption by raising the pH in the intestine, causing the precipitation of copper hydroxide.[133] Increasing calcium intake has been reported to decrease bone and kidney copper levels in laboratory animals.[40,52] In human, simultaneous high intakes of calcium and phosphorus reportedly lower copper retention in the body.[130] In turn, a laboratory animal study reports that copper increases fecal calcium and increases the loss of calcium from bone.[134]

1.7 Calcium and Manganese

Similar to calcium and copper, little research has been conducted on the interactions between calcium and manganese. Calcium has been reported to inhibit manganese absorption in laboratory animals.[135] Calcium may decrease manganese retention in humans, but this effect is influenced by other dietary factors.[136]

1.8 Calcium and Fluoride

Calcium and fluoride interact in a negative manner. Animal research indicates that calcium decreases intestinal absorption of fluoride, presumably through the formation of insoluble salts in the intestine.[4,137] High-calcium diets also decrease bone retention of fluoride in laboratory animals.[138,139] At physiological intakes, it appears that calcium exerts little effect on fluoride metabolism in humans.[4] Large doses of calcium slightly decrease fluoride absorption in humans.[129,140] Conversely, fluoride has been reported to have little effect on calcium metabolism in humans.[141]

1.9 Calcium and Cadmium

Calcium helps protect against cadmium toxicity. Calcium has been reported to decrease the accumulation of cadmium in tissues of laboratory animals including the liver and kidneys.[142-144] The mechanism by which calcium exerts this effect is unknown but researchers propose that calcium may interfere with cadmium absorption.[142,143] Cadmium, in turn interacts with calcium in a negative manner. Animal studies report that cadmium both decreases calcium absorption and increases urinary calcium excretion.[145] Researchers propose that cadmium may decrease calcium absorption by inhibiting vitamin D-induced intestinal calcium transport.[145] Cadmium also inhibits calcium deposition in bone of mice.[134] Conversely, high cadmium diets fed throughout pregnancy increase liver and kidney calcium in laboratory rats.[144] Similar to animal studies, the limited studies conducted on humans exposed to cadmium pollution report hypercalciuria.[146]

1.10 Calcium and Aluminum

Calcium and aluminum interact in an antagonistic manner. Most of the studies conducted on the interaction between calcium and aluminum have focused on the negative effect of

TABLE 7
Minerals Demonstrated to Interact with Lead

Mineral
Calcium
Phosphorus
Iron
Zinc
Copper

aluminum on calcium status. Researchers speculate that aluminum may decrease calcium absorption by affecting parathyroid hormone and phosphorus metabolism.[124,147-149]

Some studies indicate that laboratory animals absorb calcium less efficiently when fed high levels of aluminum.[150] Other studies do not confirm this finding.[151-154] Human research has indicated that dietary aluminum increases fecal calcium excretion.[124]

1.11 Calcium and Tin

Very little data exist on the interaction between calcium and tin. The limited data that does exist indicate that increasing tin intake decreases the calcium content of bone and increases kidney calcium levels.[155-158]

1.12 Calcium and Selenium

Increasing selenium intake in humans has no effects on calcium excretion or retention.[159]

2 LEAD

For centuries, lead has been known to be a toxic element for humans. Research links lead ingestion with hypertension, hyperactivity, learning disabilities, aggressive behavior, colic, constipation, weakness, sleep disturbances, and anemia,[53,160-170] in addition to behavioral disorders.[171] Lead exposure is a significant public health concern.[172-175] Roughly 4 to 5 million preschool children and 400,000 pregnant women in the U.S. are affected by lead toxicity.[176] Major sources of ingested lead include airborne dust, soil, lead-based paint, drinking water, and food or water contaminated with lead.[177-179] Lead interacts with a number of other minerals (see Table 7). The majority of interactions are antagonistic in nature (see Table 8).

2.1 Lead and Calcium

See Section 1.1, Calcium and Lead.

2.2 Lead and Phosphorus

Similar to calcium, a phosphorus deficiency has been noted to increase susceptibility to lead toxicity in laboratory animals.[180] Low phosphorus intakes increase tissue retention of lead.[180,181] Human research links ingesting a large amount of phosphorus with lower lead

TABLE 8
Proposed Mechanisms for Mineral Interactions with Lead

Mineral	Proposed Mechanisms	Ref.
Calcium	Calcium and lead compete physically for absorption	44
	Calcium promotes urinary lead excretion	37
	Calcium increases release of lead from bones	38
	Lead impairs vitamin D-dependent absorption of calcium	81
	Lead increases urinary calcium excretion	73
	Lead blocks the entrance of calcium into nerve terminals	74
	Lead blocks the calcium efflux from cells potentially by replacing calcium in the calcium-sodium ATP pump	75
Phosphorus	Phosphorus decreases lead absorption	182
Iron	Lead competes with iron for ferritin binding sites	183
	Iron deficiency enhances lead transport by the iron transport systems of the intestine	185, 186
Zinc	Lead and zinc compete for uptake on the same metallothionein-like transport protein	196
	Lead increases zinc excretion	197
	Lead increases urinary zinc excretion in laboratory animals	197, 198
	Lead decreases the activity of the zinc-dependent enzyme, γ-aminolevulinic acid dehydratase	199
Copper	Lead increases urinary copper excretion	73

absorption.[182] These findings indicate that, similar to calcium, phosphorus may help protect against lead toxicity.

2.3 Lead and Iron

Numerous research studies indicate that lead and iron have an antagonistic relationship. Some researchers propose that lead reduces iron absorption by competing with iron for ferritin binding sites.[183] Iron appears to be preferentially absorbed over lead by intestinal cells;[184-186] consequently, researchers propose that an iron deficiency enhances lead transport by the iron transport systems of the intestine.[185,186] However, not all research supports the theory that lead and iron share a similar transport system. One study, in particular, indicates that iron binds mainly to ferritin and a mucousal transferrin, whereas lead binds to a protein with an intermediate molecular weight.[187]

The majority of animal research conducted on the effects of iron on lead links low iron intakes and a low iron status with increased lead absorption.[184,188,189] Animal research also indicates that increasing iron intake can decrease lead absorption;[170,189] however, not all studies confirm this effect.[190] Low iron intakes and iron deficiency have been reported to increase tissue deposition of lead in laboratory animals.[185,191,192] Researchers project that tissue lead content may increase sixfold when body iron stores are reduced.[44] Conversely, high iron intakes are reported to decrease lead concentrations in kidney and femur of laboratory animals fed lead.[188] Research results indicate that lead competes with iron for ferritin binding sites[183] and increases urinary iron excretion in laboratory animals.[73] Iron does not appear to affect lead excretion in animals.[181,184]

Studies on the effect of iron status on lead absorption in humans have yielded different results, but, for the most part, these studies parallel the findings of animal research.[44] Iron deficiency is often associated with elevated blood lead levels in children.[193,194] Conversely, both higher serum ferritin levels[195] and the use of iron supplements have been linked with lower blood lead levels in pregnant women.[60] Last, similar to the animal research, human research indicates that iron intake has no effect on lead excretion.[193]

TABLE 9
Minerals Demonstrated to Interact with Phosphorus

Mineral
Calcium
Lead
Magnesium
Iron
Fluoride
Aluminum

2.4 Lead and Zinc

Lead and zinc interact in an antagonistic manner. Research indicates that lead and zinc compete for uptake in the intestine, perhaps on the same metallothionein-like transport protein.[196] Lead increases urinary zinc excretion in laboratory animals.[197,198] Lead has also been reported to decrease markedly the activity of the zinc-dependent enzyme, γ-aminolevulinic acid dehydratase.[199] Researchers propose that lead may compete with zinc to bind to this enzyme.[169]

A high lead intake decreases zinc levels in plasma, liver, bone,[200] brain,[198,201] and kidney in laboratory animals.[53] Conversely, a high zinc intake decreases tissue lead levels in laboratory animals.[196,202] A high zinc intake also protects against a lead-induced decrease in activity of γ-aminolevulinic acid dehydratase.[203,204] In turn, a zinc deficiency enhances lead absorption[205] and deposition of lead in tissues,[196] including the bone, spleen,[200] and nervous system tissue.[206] The effect of lead on blood or tissue lead levels in the presence of a zinc deficiency has not been determined in humans due to the ethical issue of conducting such a study.[48] Human studies have reported no effect of zinc supplementation on blood lead levels in adults who had received moderate exposure to lead.[207]

2.5 Lead and Copper

Limited data exists on the interactions between lead and copper. Researchers propose that lead interferes with copper utilization.[208-210] Lead has been reported to increase urinary copper excretion in laboratory animals[73] and decrease brain levels of copper in suckling laboratory rats.[201] In addition, a diet high in lead decreases tissue copper levels and the activity of copper-dependent enzymes in laboratory animals.[211]

The effect of copper on lead is not well elucidated. One study reports that a diet low in copper increases the absorption, tissue levels, and toxic effects of lead in laboratory animals;[212] however, other research links a low copper diet with decreased tissue lead levels in laboratory animals.[213] In addition, dietary copper has been reported to increase, rather than decrease, the severity of lead poisoning in laboratory animals.[213]

3 PHOSPHORUS

Phosphorus is an essential component of bone and plays a critical role in metabolic pathways. As noted earlier, about 60% of dietary phosphate is absorbed.[22] Phosphorus is typically absorbed as free phosphate.[214] (See Tables 9 and 10.)

TABLE 10
Proposed Mechanisms for Mineral Interactions with Phosphorus

Mineral	Proposed Mechanisms	Ref.
Calcium	Phosphorus decreases calcium absorption by forming insoluble calcium triphosphate	22, 87
	Phosphorus increases fecal calcium loss by increasing calcium content of digestive secretions	88, 89
	Phosphorus decreases urinary calcium by enhancing calcium reabsorption	90, 91, 92
	Phosphorus increases parathyroid hormone excretion which in turn reduces serum calcium	22
	Calcium increases renal phosphate reabsorption and phosphate mobilization from bone and soft tissue	99, 100
Lead	Phosphorus decreases lead absorption	180
Magnesium	Phosphorus decreases magnesium absorption	101, 104, 105, 215, 216, 217
	Phosphorus decreases urinary magnesium	218
Iron	Phosphorus may decrease iron absorption when calcium intake is low	219, 220, 221, 222
Fluoride	Phosphorus may decrease the absorption of fluoride by form insoluble salts	4
Aluminum	Aluminum decreases phosphorus absorption	152, 223, 224, 225

3.1 Phosphorus and Calcium

See Section 1.2, Calcium and Phosphorus.

3.2 Phosphorus and Lead

See Section 2.1, Lead and Calcium.

3.3 Phosphorus and Magnesium

Data from the limited research conducted on phosphorus and magnesium interactions indicate that these two minerals behave antagonistically. The results of animal research demonstrate that a high phosphorus intake decreases magnesium absorption and enhances the symptoms of a magnesium deficiency.[101,104,105,215-217] Human studies provide evidence that dietary phosphorus decreases urinary magnesium but demonstrates no overall effect on magnesium balance.[141,218]

3.4 Phosphorus and Iron

The limited data available on the interactions between phosphorus and iron indicate that phosphorus may negatively impact iron metabolism under certain conditions. Human studies report that dietary phosphate can decrease iron absorption and utilization;[219,220] however, high phosphorus intakes are more likely to exert a significant negative effect on iron absorption or retention when calcium intakes are low.[221,222]

3.5 Phosphorus and Fluoride

At high doses, phosphorus interacts with fluoride. Research indicates that phosphorus and fluoride form insoluble salts.[4,137] Large doses of phosphorus decrease intestinal absorption

TABLE 11

Minerals Demonstrated to Interact with Magnesium

Mineral
Calcium
Phosphorus
Manganese
Fluoride

TABLE 12

Proposed Mechanisms for Mineral Interactions with Magnesium

Mineral	Proposed Mechanisms	Ref.
Calcium	Calcium decreases magnesium absorption by competing for a common transport system	107, 108, 109
Phosphorus	Phosphorus decreases magnesium absorption	101, 104, 105, 215, 217
Zinc	Zinc increases fecal excretion of magnesium	8
	Phosphorus decreases urinary magnesium	218
Manganese	Magnesium inhibits manganese absorption	135
Fluoride	Magnesium decreases fluoride absorption by forming insoluble salts	4

of fluoride in animals.[137] At physiological intakes in humans, however, phosphorus appears to exert little effect on fluoride metabolism.[4] Similarly, human studies report little effect of fluoride on phosphorus metabolism.[141]

3.6 Phosphorus and Aluminum

The majority of research examining the interaction between phosphorus and aluminum has focused on the adverse effect of aluminum on phosphorus homeostasis. Animal studies have demonstrated that large doses of aluminum decrease phosphorus absorption and tissue phosphorus levels.[151,223-225] Research conducted in humans notes that the phosphorus depletion induced by dietary aluminum may result in bone pain and fractures.[147,226,227]

4 MAGNESIUM

Magnesium is an essential nutrient for glycolysis, membrane transport, and transmission of the genetic code.[228] Magnesium is also a component of over 300 enzymes.[229] Some researchers propose that magnesium absorption is usually around 20 to 30% in humans, but can increase to 70% when dietary intake and status are low.[230] Other researchers estimate that human adults absorb about 50% of dietary magnesium.[231] Magnesium absorption can be enhanced by vitamin D.[232] (See Tables 11 and 12.)

4.1 Magnesium and Calcium

See Section 1.3, Calcium and Magnesium.

4.2 Magnesium and Phosphorus

See Section 3.3, Phosphorus and Magnesium.

4.3 Magnesium and Zinc

Based on similar orbitals, configurations, and coordination numbers, magnesium and zinc would be expected to directly interact with one another;[9] however, little research has been conducted on the interactions between magnesium and zinc. Large doses of zinc have been reported to increase fecal excretion of magnesium in laboratory animals.[8] Human data indicate that dietary magnesium does not reduce zinc absorption.[8]

4.4 Magnesium and Manganese

Data on the interactions between magnesium and manganese are limited; however, the results of animal research indicate that the relationship between these two minerals is antagonistic in nature. Magnesium has been reported to inhibit manganese absorption in laboratory animals.[135]

4.5 Magnesium and Fluoride

Research indicates that magnesium and fluoride interact to form insoluble salts.[4] In large doses, magnesium decreases intestinal absorption of fluoride in laboratory animals.[137] At physiological intakes, magnesium exerts little effect on fluoride metabolism in humans.[4,116] Fluoride, in turn, exerts no significant effect on magnesium metabolism in humans.[116]

4.6 Magnesium and Nickel

Limited research exists on the interactions between magnesium and nickel. One study indicates that magnesium inhibits nickel-induced carcinogenesis in the kidneys of laboratory rats.[233]

5 IRON

Iron is an essential element involved in innumerable biochemical reactions.[170,171] Iron is a component of heme compounds, cytochromes that function in the electron transport chain, and other metalloproteins.[234,235] Iron absorption is influenced by a number of factors, including body iron stores[236] and the amount and chemical nature of the iron ingested.[237] Many subpopulations in the U.S., including women and girls, fail to consume their Recommended Dietary Allowances for iron.[20] Iron deficiency remains one of the most common nutritional deficiencies among children.[238-242] (See Tables 13 and 14.)

5.1 Iron and Calcium

See Section 1.4, Calcium and Iron.

5.2 Iron and Lead

See Section 2.3, Lead and Iron.

TABLE 13
Minerals Demonstrated to Interact with Iron

Mineral	
Calcium	Cadmium
Lead	Mercury
Phosphorus	Nickel
Zinc	Cobalt
Copper	Iodine
Manganese	

TABLE 14
Proposed Mechanisms for Mineral Interactions with Iron

Mineral	Proposed Mechanisms	Ref.
Calcium	Calcium decreases iron absorption	125, 126, 221
Lead	Lead competes with iron for ferritin binding sites	183
	Iron deficiency enhances lead absorption by increasing lead transport by the iron transport systems of the intestine	186, 280
Phosphorus	Phosphorus may decrease iron absorption when calcium intake is low	219, 220, 221, 222
Zinc	Iron decreases zinc absorption	245, 246
	Iron decreases zinc retention	248
	Zinc inhibits absorption of iron added to food	252
	Zinc decreases the incorporation of iron into ferritin and other storage proteins	258
Copper	Copper is an essential component of ceruloplasmin which mobilizes iron from the liver	5
	Iron and copper may compete for absorption	263
Manganese	Manganese and iron compete for absorption	269
Cadmium	Iron decreases cadmium absorption	48, 274
	Cadmium decreases ferritin tissue levels	276, 277
Nickel	Iron decreases nickel absorption	279
Cobalt	Iron and cobalt compete for absorption	268, 279
Iodine	Iron decreases iodine utilization	208, 209

5.3 Iron and Phosphorus

See Section 3.4, Phosphorus and Iron.

5.4 Iron and Zinc

Although iron and zinc differ in their electronic configuration, they interact in an antagonistic manner.[5] Research indicates that dietary iron decreases both zinc absorption and retention. Iron depletion has been noted to increase zinc absorption in laboratory animals.[243,244] Human research confirms that iron supplementation decreases zinc absorption[245,246] and indicates that ferrous iron inhibits zinc absorption more than ferric iron.[247] Both animal[248] and human research reports that large doses of iron inhibit zinc retention.[245,249] This interaction appears to be strongest for nonheme[8] iron and free zinc;[4,249] however, at physiological doses, iron may have only a minimal effect on zinc status. A number of human studies report that physiological doses of iron have no effect on serum zinc levels.[250,251]

Zinc, in turn, adversely affects iron status more by impairing iron utilization than by affecting iron absorption.[5] Large doses of zinc inhibit absorption of added[252] but not intrinsic[253] iron present in food. High levels of dietary zinc decrease tissue iron levels in both laboratory animals[7,254-256] and humans.[257] Researchers propose that zinc decreases tissue iron levels by decreasing the incorporation of iron into ferritin and other storage proteins[258] and by increasing fecal iron losses.[259]

5.5 Iron and Copper

Although copper is essential for iron absorption and metabolism, copper may actually decrease iron absorption.[5] Copper is a component of ceruloplasmin, an enzyme required to mobilize iron from the liver. Consequently, some researchers contend that a copper deficiency can lead to an iron deficiency;[5] however, animal research links a copper deficiency with increased tissue iron levels.[260-262] Iron and copper may compete for absorption.[263] High iron intakes have been reported to decrease liver copper stores[264,265] and copper status in laboratory animals;[264,266,267] however, iron supplements do not alter serum copper levels in humans.[246]

5.6 Iron and Manganese

Because of their similar orbitals, configurations, and coordination numbers, iron and manganese share a common absorption pathway and mutually inhibit absorption.[268] Iron inhibits the intestinal uptake of manganese in laboratory animals.[269] Conversely, low iron intake increases the concentration of manganese in the liver and intestine of laboratory animals.[270] Human research supports the findings of animal research. Supplemental iron has been observed to decrease both manganese absorption and retention in humans.[271] Animal research indicates that manganese, in turn, inhibits iron absorption but does not appear to alter the transfer of iron to other tissues.[269] Although some human research suggests that manganese decreases iron absorption in humans,[272] other research conducted in humans reports that manganese supplementation does not alter iron status.[273]

5.7 Iron and Cadmium

Researchers have speculated that iron decreases cadmium absorption.[48,274] Both animal and human research links high blood ferritin levels with decreased cadmium absorption.[275] Iron deficiency increases the gastrointestinal absorption of cadmium; however, the mechanism for this effect is, as of yet, unknown.[44] Cadmium in turn, decreases tissue iron levels in laboratory animals.[144,276,277]

5.8 Iron and Mercury

Limited data exist on the interactions between iron and mercury. Research results indicate that methylmercury, but not inorganic mercury, decreases tissue iron content in laboratory animals.[278]

5.9 Iron and Nickel

Data from animal research link an iron deficiency to increased nickel absorption in laboratory animals.[279]

TABLE 15

Minerals Demonstrated to Interact with Zinc

Mineral	
Calcium	Manganese
Lead	Cadmium
Magnesium	Aluminum
Iron	Selenium
Copper	Tin

5.10 Iron and Cobalt

Available data on iron and cobalt indicate that iron and cobalt share a common absorption pathway and mutually inhibit absorption.[268,279]

5.11 Iron and Iodine

Iron has been observed to negatively affect iodine status. Research studies report that iron decreases iodine utilization.[208,209]

6 ZINC

Zinc acts as a co-factor or structural component of more than 200 enzymes[281] and numerous nonenzymatic proteins.[282] On average, 10 to 40% of zinc consumed is retained in humans.[283-285] Symptoms of a zinc deficiency include growth retardation, alopecia, parakeratosis, esophageal lesions, impaired reproductive performance, birth defects, impaired wound healing, developmental bone disorders, hypogeusia, irritability, lethargy, depression and other behavioral disorders.[235] Zinc interacts with numerous minerals (see Tables 15 and 16).

6.1 Zinc and Calcium

See Section 1.5, Calcium and Zinc.

6.2 Zinc and Lead

See Section 2.4, Lead and Zinc.

6.3 Zinc and Magnesium

See Section 4.3, Magnesium and Zinc.

6.4 Zinc and Iron

See Section 5.4, Iron and Zinc.

TABLE 16

Proposed Mechanisms for Mineral Interactions with Zinc

Mineral	Proposed Mechanisms	Ref.
Calcium	Calcium reduces solubility of calcium-phytate-zinc complexes	4
Lead	Lead and zinc compete for uptake on the same metallothionein-like transport protein	196
	Lead increases urinary zinc excretion	197, 198
	Lead decreases the activity of the zinc-dependent enzyme, δ-aminolevulinic acid dehydratase	199
	Lead may compete with zinc to bind to this enzyme	197
Magnesium	Zinc increases fecal excretion of magnesium	8
Iron	Iron decreases zinc absorption	245, 246
	Iron decreases zinc retention	248
	Zinc inhibits absorption of iron added to food	252
	Zinc decreases the incorporation of iron into ferritin and other storage proteins	258
Copper	Zinc increases the level of intestinal metallothionein, which binds copper more strongly than zinc and allows little copper to cross into the body	4, 286, 288, 289
	Zinc competes with copper for binding sites on metallothionein in intestinal cells	263, 286, 287
	Copper may reduce zinc absorption	304
Manganese	Zinc and manganese compete directly for absorption	244
Cadmium	Zinc may increase liver and kidney cadmium tissue levels	276, 309, 310
	Cadmium decreases the absorption of zinc	308
	Cadmium competes with zinc for binding sites on metallothionen thereby decreasing the storage of zinc and transfer of zinc to a fetus	43
Aluminum	Aluminum may decrease zinc absorption	129
Tin	Tin may decrease zinc absorption	313, 314, 315
	Tin may increase fecal zinc excretion	316, 317

6.5 Zinc and Copper

Because of their similar orbitals, configurations, and coordination numbers, zinc and copper interact directly.[4] Based on early research results, researchers propose that zinc may interfere with copper absorption by competing for binding sites on metallothionein in the intestinal mucousal cells.[263,286,287]

Recent research indicates that zinc also interacts indirectly with copper to decrease copper absorption. Zinc interacts with a nuclear protein that governs the synthesis of metallothionein in intestinal mucousal cells. High levels of zinc increase the level of intestinal metallothionein, which in turn binds to copper more strongly than zinc, creating a copper-metallothionein complex that is poorly absorbed.[4,286,288,289]

Animal research has demonstrated that high levels of zinc induce symptoms of a copper deficiency, such as anemia and decreased cytochrome C oxidase activity.[290-292] Excess zinc consumption in laboratory animals decreases the amount of copper absorbed and utilized.[5] High intakes of zinc induce copper deficiency in laboratory animals.[290,291,293-296]

Conversely, animal studies report that zinc deficiency increases plasma copper levels.[297,298] Human research confirms that zinc inhibits the intestinal absorption of copper.[299] High zinc intakes have been reported to induce signs of copper deficiency in humans.[300] High doses of zinc in humans lower serum copper/zinc ratios[301] and increase erythrocyte copper levels,[287] especially when copper intakes are low.[302] Human studies also link low zinc status with elevated blood copper levels.[303]

Although zinc has a strong negative influence on copper bioavailability, copper appears to exert only a minimal effect on zinc bioavailability.[5] Some animal studies indicate that a high copper intake reduces zinc absorption[304] and increases the teratogenic properties of a zinc deficiency.[296] Like animal studies, human research indicates that high copper intakes can inhibit zinc absorption,[8] but this effect is relatively minor compared to the effect of zinc on copper.[304] One study found that a high copper intake did not alter urinary zinc in humans.[305]

6.6 Zinc and Manganese

Zinc and manganese have been observed to compete directly with each other for absorption. Manganese decreases zinc absorption and zinc decreases manganese absorption in laboratory animals.[244] Human research reports no effect of zinc intake on fecal excretion of manganese.[259]

6.7 Zinc and Cadmium

Because zinc and cadmium have similar electronic structures, it has been proposed that these two minerals may compete for absorption.[4,5] Animal research indicates that zinc protects against cadmium toxicity;[306-308] however, zinc has been shown to increase rather than decrease cadmium tissue levels, especially in the liver and kidneys.[275,309,310] In turn, cadmium decreases the absorption of zinc.[308] Cadmium competes with zinc for binding sites on metallothionen, which plays an important role in the storage and transfer of zinc during development.[43]

6.8 Zinc and Aluminum

The reported effects of aluminum on zinc metabolism have been inconsistent.[150,151,153,311] Some,[153] but not all, studies report that aluminum adversely affects zinc metabolism in laboratory animals.[50,151,311] A human study has provided evidence that aluminum may decrease zinc absorption.[129]

6.9 Zinc and Selenium

Limited data are available on the interactions between zinc and selenium. An animal study indicates that high zinc intakes may induce signs of selenium deficiency.[312]

6.10 Zinc and Tin

Data provide evidence that tin decreases zinc absorption and increases zinc excretion.[313-317] Dietary tin depresses zinc levels in bone and soft tissues of laboratory rats.[313-315] A number of studies conducted in humans report that dietary tin increases fecal zinc levels;[316,317] however, other studies do not confirm this effect.[318]

7 COPPER

Copper is an essential element in numerous metabolic processes, including the mitochondrial electron transport chain and iron absorption and mobilization.[319] Between 40 and 60% of dietary copper is absorbed in man.[47] Symptoms of a copper deficiency are species specific and include anemia, neutropenia, skeletal abnormalities, depigmentation, impaired reproductive performance, loss of blood vessel integrity, and disruption of neurological functions.[235, 319] Copper deficiency is uncommon in humans but has been observed in parentally fed infants and malnourished children.[263,320-322] Copper exists in the elemental state and as cuprous (CuI)

TABLE 17

Minerals Demonstrated to Interact with Copper

Mineral	
Calcium	Selenium
Lead	Tin
Iron	Silver
Zinc	Mercury
Cadmium	

TABLE 18

Proposed Mechanisms for Mineral Interactions with Copper

Mineral	Proposed Mechanisms	Ref.
Calcium	Calcium reduces copper absorption by raising the pH in the intestine, causing the precipitation of copper hydroxide	133
Lead	Lead increases urinary copper excretion	73
Iron	Copper is an essential component of ceruloplasmin which mobilizes iron from the liver	5
	Iron and copper may compete for absorption	263
Zinc	Zinc increases the level of intestinal metallothionein, which binds copper more strongly than zinc and allows little copper to cross into the body	4, 286, 288, 289
	Zinc competes with copper for binding sites on metallothionein in intestinal cells	263, 286, 287
	Copper may reduce zinc absorption	304
Cadmium	Cadmium interferes with copper utilization, possibly by decreasing copper absorption	208, 209, 210
Silver	Silver may interfere with copper utilization by decreasing copper absorption	208, 209, 210
Mercury	Mercury may decrease copper absorption	325
	Mercury may interfere with copper utilization	208, 209, 210

and cupric (CuII) ions.[5] Because CuI is a d10 ion, copper would be expected to interact with zinc and cadmium.[5] (See Tables 17 and 18.)

7.1 Copper and Calcium

See Section 1.6, Calcium and Copper.

7.2 Copper and Lead

See Section 2.5, Lead and Copper.

7.3 Copper and Iron

See Section 5.5, Iron and Copper.

7.4 Copper and Zinc

See Section 6.5, Zinc and Copper.

7.5 Copper and Cadmium

Copper and cadmium have similar orbitals, configurations, and coordination numbers and so would be expected to interact directly.[4] Research conducted on the interactions between these two nutrients has focused on the negative effect of cadmium on copper metabolism. It appears that cadmium interferes with copper utilization, possibly by decreasing copper absorption.[208-210] Animal research has shown that high cadmium diets administered during pregnancy decrease plasma and tissue copper levels in both the maternal and fetal organisms.[323]

7.6 Copper and Selenium

The limited research that exists on copper and selenium indicates that high copper intakes induce signs of selenium deficiency in laboratory animals;[312] however, animal research indicates that increasing selenium intake does not alter copper metabolism.[315]

7.7 Copper and Molybdenum

It has been suggested that molybdenum may reduce the bioavailability of copper by forming insoluble complexes.[4] Research conducted on animals reports that molybdenum supplementation does not alter copper metabolism.[324]

7.8 Copper and Tin

It has been observed in animal studies that increasing tin intake depresses copper levels in plasma and soft tissues.[313,315]

7.9 Copper and Silver

Silver and copper have similar electronic structures and so would be expected to interact directly.[5] There is some evidence to indicate that silver may interfere with copper utilization by decreasing copper absorption.[208-210]

7.10 Copper and Mercury

Data have demonstrated that mercury may both decrease copper absorption[52] and interfere with utilization.[208-210] Observations from animal studies show that dietary mercury decreases blood and liver levels of copper and may increase the deposition of copper in the kidney.[325]

8 MANGANESE

Manganese is an essential element for growth and bone development. It is involved in activation of a number of enzymes involved in proteoglycan synthesis.[319, 326] Manganese has been noted to interact with several other minerals (see Tables 19 and 20)

8.1 Manganese and Calcium

See Section 1.7, Calcium and Manganese.

TABLE 19

Minerals Demonstrated to Interact with Manganese

Mineral	
Calcium	Zinc
Magnesium	Cobalt
Iron	Iodine

TABLE 20

Proposed Mechanisms for Mineral Interactions with Manganese

Mineral	Proposed Mechanism	Ref.
Calcium	Calcium decreases manganese absorption	135
Magnesium	Magnesium inhibits manganese absorption	135
Iron	Manganese and iron compete for absorption	268
Zinc	Zinc and manganese compete directly for absorption	244
Cobalt	Manganese and cobalt compete for absorption	268
Iodine	Manganese may adversely affect iodine utilization	208, 209

8.2 Manganese and Magnesium

See Section 4.4, Magnesium and Manganese.

8.3 Manganese and Iron

See Section 5.6, Iron and Manganese.

8.4 Manganese and Zinc

See Section 6.6, Zinc and Manganese.

8.5 Manganese and Cobalt

Manganese and cobalt share a common absorption pathway and mutually inhibit absorption.[210]

8.6 Manganese and Iodine

Researchers have proposed that manganese adversely affects iodine utilization.[208,209]

9 FLUORIDE

Fluoride (as fluorine) is absorbed very quickly from the intestine[327] and is incorporated into bones and teeth. The most well-established benefit of fluoride is its protective effect

TABLE 21
Minerals Demonstrated to Interact with Fluoride

Mineral
Calcium
Phosphorus
Magnesium
Aluminum

TABLE 22
Proposed Mechanisms for Mineral Interactions with Fluoride

Mineral	Proposed Mechanisms	Ref.
Calcium	Calcium decreases intestinal absorption of fluoride by forming insoluble salts	4, 137
Phosphorus	Phosphorus may decrease the absorption of fluoride by form insoluble salts	4
Magnesium	Magnesium decreases fluoride absorption by forming insoluble salts	4
Aluminum	Aluminum decreases fluoride absorption	330, 331

against dental caries.[328] Although fluoride is ubiquitous in nature and found in all foods, the fluoridation of drinking water is the major factor influencing the dietary intake of fluoride by man.[329] Fluoride interacts with a number of minerals (see Tables 21 and 22).

9.1 Fluoride and Calcium

See Section 1.8, Calcium and Fluoride.

9.2 Fluoride and Phosphorus

See Section 3.5, Phosphorus and Fluoride.

9.3 Fluoride and Magnesium

See Section 4.5, Magnesium and Fluoride.

9.4 Fluoride and Aluminum

Animal studies have indicated that aluminum depresses fluoride absorption.[330,331] Aluminum has also been observed to decrease fluoride balance in man.[141]

10 CADMIUM

Cadmium is a toxic mineral that accumulates in the liver and kidney.[48] Research indicates that cadmium has a long biological half-life of approximately 17 to 30 years in man.[48] Cadmium interacts with numerous minerals (see Tables 23 and 24).

TABLE 23

Minerals Demonstrated to Interact with Cadmium

Mineral
Calcium
Iron
Zinc

TABLE 24

Proposed Mechanisms for Mineral Interactions with Cadmium

Mineral	Proposed Mechanisms	
Calcium	Calcium decreases cadmium absorption	142, 143
	Cadmium decreases calcium absorption potentially by inhibiting vitamin D-induced intestinal calcium transport	145
	Cadmium increases urinary calcium excretion	145
Iron	Iron decreases cadmium absorption	48, 274
	Cadmium decreases ferritin	276, 277
Zinc	Zinc may increase liver and kidney cadmium tissue levels	276, 309, 310
	Cadmium decreases the absorption of zinc	308

10.1 Cadmium and Calcium

See Section 1.9, Calcium and Cadmium.

10.2 Cadmium and Iron

See Section 5.7, Iron and Cadmium.

10.3 Cadmium and Zinc

See Section 6.7, Zinc and Cadmium.

10.4 Cadmium and Mercury

Animal research has demonstrated that cadmium protects laboratory animals against the nephrotoxic effects of inorganic mercury.[332] It has also been shown that cadmium decreases the mercury content of some tissues and increases mercury content in others. Cadmium has been observed to decrease the mercury content of kidney protein but increase the accumulation of mercury in liver metallothionein in laboratory animals.[333]

11 Aluminum

Most foods contain some aluminum naturally and a few foods, such as tea and herbs, contain high levels of aluminum.[334,335] In animals, aluminum tends to accumulate in

TABLE 25

Minerals Demonstrated to Interact with Aluminum

Mineral
Calcium
Phosphorus
Zinc
Fluoride

TABLE 26

Proposed Mechanisms for Mineral Interactions with Aluminum

Mineral	Proposed Mechanisms	Ref.
Calcium	Aluminum may decrease calcium absorption by affecting parathyroid hormone and phosphorus metabolism	124, 147, 148, 149
Phosphorus	Aluminum decreases phosphorus absorption	151, 223, 225, 229
Zinc	Aluminum may decrease zinc absorption	129
Fluoride	Aluminum decreases fluoride absorption	330, 331

bones.[153,154,223,336,337] Clinical symptoms of aluminum toxicity include bone pain, increased fracture rate, and increased resistance to vitamin D therapy.[338] The primary effect of aluminum on bone does not appear to involve vitamin D metabolism.[339, 340] Aluminum interacts with a number of other minerals (see Tables 25 and 26).

11.1 Aluminum and Calcium

See Section 1.10, Calcium and Aluminum.

11.2 Aluminum and Phosphorus

See Section 3.6, Phosphorus and Aluminum.

11.3 Aluminum and Zinc

See Section 6.8, Zinc and Aluminum.

11.4 Aluminum and Fluoride

See Section 9.4, Fluoride and Aluminum.

12 SELENIUM

Selenium occurs in the diet primarily in the organic complex selenomethionine. Selenium plays an important role in the activation of the enzyme glutathione peroxidase.[70] Organ meats and seafoods are good dietary sources of selenium.[342] (See Tables 27 and 28.)

TABLE 27

Minerals Demonstrated to Interact with Selenium

Mineral
Zinc
Copper
Tin
Mercury

TABLE 28

Proposed Mechanisms for Mineral Interactions with Selenium

Mineral	Proposed Mechanisms	Ref.
Mercury	Selenium may form a biologically inert complex with mercury	348
Tin	Tin decreases selenium absorption	9, 348

12.1 Selenium and Zinc

See Section 6.9, Zinc and Selenium.

12.2 Selenium and Copper

See Section 7.6, Copper and Selenium.

12.3 Selenium and Mercury

Research indicates that selenium protects against mercury toxicity.[343-347] The exact mechanism is unknown, but researchers speculate that selenium may decrease mercury toxicity either by forming a biologically inert mercury-selenium complex[48,348] or by preventing damage from free radicals generated by mercury toxicity to cell membranes.[349]

12.4 Selenium and Tin

Little is known about the interactions between tin and selenium. The absorption and retention of tin by humans is low.[159,350-352] High intakes of tin depress the absorption of selenium from the intestine.[9] Similarly, human studies indicate that increasing tin intake decreases selenium absorption.[353]

13 SUMMARY

Essential minerals may play a important role in decreasing the risk of toxicity from both essential and nonessential minerals. The level of human exposure to potentially toxic minerals and the ability of essential minerals to reduce this toxicity should be considered in generating dietary recommendations for essential minerals. Health professionals are becoming increasingly aware of the fact that good mineral nutrition is a "balancing act".[354] To benefit consumers, dietary recommendations to optimize mineral nutrition should emphasize a balanced intake of foods from all five food groups and the avoidance of excessive intakes of individual mineral supplements.

REFERENCES

1. Hazell, T., Minerals in foods: dietary sources, chemical forms, interactions, bioavailability, *Wld. Rev. Nut. Diet.,* 46, 1, 1985.
2. Brown, M., Ed., *Present Knowledge in Nutrition,* International Life Sciences Institute, Nutrition Foundation, Washington, D.C., 1990.
3. Clydesdale, F., Mineral interactions in foods, in *Nutrient Interactions,* Bodwell, C. and Erdman, J., Eds., Marcel Dekker, New York, 1988, p. 74.
4. Couzy, F., Keen, C., Gershwin, M., and Mareschi, J., Nutritional implications of the interactions between minerals, *Prog. Food Nutr. Sci.,* 17, 65, 1993.
5. O'Dell, B., Mineral interactions relevant to nutrient requirements, *J. Nutr.* 119, 1932, 1989.
6. Mills, C., Dietary interactions involving the trace minerals, *Ann. Rev. Nutr.,* 5, 173, 1985.
7. Kirchgessner, M., Schwartz, F., and Schnegg, A., Interactions of essential metals in human physiology, in *Clinical, Biochemical and Nutritional Aspects of Trace Elements,* Prasad, A. S., Ed., Alan R. Liss, New York, 1982, pp. 477–512.
8. Solomons, N., Competitive mineral-mineral interaction in the intestine. Implications for zinc absorption in humans, in *Nutritional Bioavailability of Zinc,* ACS Symposium series 210, Inglett, G. E., Ed., American Chemical Society, Washington, D.C., 1983, pp. 247–271.
9. Hill, C. and Matrone, G., Chemical parameters in the study of *in vivo* and *in vitro* interactions of transition elements, *Fed. Am. Soc. Exp. Biol.,* 29, 1474, 1970.
10. Hill, C., Mineral interrelationships, in *Trace Elements in Human Health and Disease,* Prasad, A. and Oberleas, D., Eds., Academic Press, New York, 1976, p. 281.
11. Matrone, G., Chemical parameters in trace-element antagonisms, in *Trace Element Metabolism in Man and Animals,* Vol. 2, Hoekstra, W., Suttie, J., Ganther, H., and Mertz, W., Eds., Baltimore University, 1974, pp. 91–103.
12. Lonnerdal, B. and Keen, C., Trace element absorption in infants: potential and limitations, in *Reproductive and Developmental Toxicity of Metals,* Clarkson, T., Nordberg, G., and Sanger, P. Eds., Plenum Press, New York, 1983, p. 759.
13. National Dairy Council, *Calcium: A Summary of Current Research for the Health Professional,* Washington, D.C., 1989.
14. NRC, *Recommended Dietary Allowances,* 10th ed., Food and Nutrition Board, Commission on Life Sciences, National Research Council, National Academy Press, Washington, D.C., 1989.
15. Weaver, C., Martin, B., Plawecki, K., Peacock, M., Wood, O., Smith, D., and Wastney, M., Differences in calcium metabolism between adolescent and adult females, *Am. J. Clin. Nutr.,* 61, 577, 1995.
16. Life Sciences Research Office, Federation of American Societies for Experimental Biology, *Nutrition Monitoring in the United States — An Update Report on Nutrition Monitoring,* prepared for the U.S. Department of Agriculture and the U.S. Department of Health and Human Services, DHHS Publ. No. (PHS) 89–1255, Public Health Service, U.S. Government Printing Office, Washington, D.C., September 1989.
17. Pennington, J. and Young, B., Total Diet Study nutritional elements, 1982–1989, *J. Am. Diet. Assoc.,* 91, 179, 1991.
18. USDA, *Food and Nutrient Intakes by Individuals in the United States: 1 Day, 1987–88,* Nationwide Food Consumption Survey 1987–88, NFCS Rep. No.87-1-1, U.S. Department of Agriculture, Human Nutrition Information Service, Washington, D.C., 1993.
19. Carroll, M., Abraham, S., and Dresser, C., *Dietary Intake Source Data. United States 1976–1980,* Vital and Health Statistics, Series 11-No. 231 DHHS Pub. No. (PHS) 03-1681, Public Health Service, U.S. Government Printing Office, Washington, D.C., March 1993
20. Alaimo, K., McDowell, M., Briefel, R., Bischof, A., Caughman, C., Loria, C., and Johnson, C., *Advance Data. Dietary Intake of Vitamins, Minerals, and Fiber of Persons Ages 2 Months and Over in the United States,* Third National Health and Nutrition Examination Survey, Phase 1, 1988–91, Centers for Disease Control and Prevention, U.S. Government Printing Office, Washington, D.C., 1994, p. 258.
21. NIH Consensus Conference, Optimal calcium intake: NIH consensus development panel on optimal calcium intake, *J. Am. Med. Assoc.,* 272, 24, 1994.
22. Yanagawa, N. and Lee, D., Renal handling of calcium and phosphorus, in *Disorders of Bone and Mineral Metabolism,* Coe, F. and Favus, M., Eds., Raven Press, New York, 1992, p. 3.

23. Fhijsen, W., DeJong, M., and Van, O., Kinetic properties of Na/Ca exchange in basolateral plasma membranes in rat small intestine, *Biochim. Biophys. Acta.*, 689, 85, 1983.
24. Murer, H. and Hildmann, B., Transcellular transport of calcium and inorganic phosphate in the small intestinal epithelium, *Am. J. Physiol.*, 240, G409, 1982.
25. Hildmann, B., Schmidt, A., and Murer, H., Ca^{++}-transport across basallateral plasma membranes from rat small intestinal epithelial cells, *J. Membr. Biol.*, 65, 55, 1982.
26. Martin, D. and DeLuca, H., Influence of sodium on calcium transport by the rat small intestine, *Am. J. Physiol.*, 216, 1351, 1969.
27. DeJonge, H., Ghijsen, W., and van Os, C., Phosphorylated intermediates of Ca^{2+}-ATPase activity in basolateral plasma membranes of rat duodenum, *Biochim. Biophys. Acta*, 689, 327, 1982.
28. Ghijsen, W., DeJonge, M., and van Os, C., Ca-stimulated ATP-ase in brush border and basolateral membranes of rat duodenum with high affinity sites for Ca ions, *Nature*, 279, 802, 1979.
29. Russell, R., Monod, A., Bonjour, J., and Fleisch, H., Relation between alkaline phosphatase and Ca^{2+}-ATPase in calcium transport, *Nature*, 240, 126, 1972.
30. Heaney, R., Weaver, C., and Recker, R., Calcium absorbability from spinach, *Am. J. Clin. Nutr.*, 47, 707, 1988.
31. Barton, J. C., Conrad, M. E., Harrison, L., Nuby. S., Effects of calcium on the absorption and retention of lead, *J. Lab. Clin. Med.*, 91, 366, 1978.
32. Meredith, P. A., Moore, M. R., and Goldberg, A., The effect of calcium on lead absorption in rats, *Biochem. J.*, 166, 531, 1977.
33. Gruden, N., Static, M., and Buben, M., Influence of lead on calcium and stronium transfer through the duodenal wall in rats, *Environ. Res.*, 8, 203, 1974.
34. Mahaffey, K., Haseman, J., and Goyer, R., Experimental enhancement of lead toxicity by low dietary calcium, *J. Lab. Clin. Med.*, 83, 92, 1973.
35. Morrison, J., Quarterman, J. N., and Humphries, W.R. The effect of dietary calcium and phosphate on lead poisoning in lambs, *J. Comp. Pathol.*, 87, 417, 1977.
36. Quarterman, J., Morrison, J. N., and Humphries, W. R., The influence of high dietary calcium and phosphorus on lead uptake and release, *Environ. Res.*, 17, 60, 1978.
37. Sukhanov, B. P., Korolev, A. A., Marninchuck, A. N., and Merzliakova, N. M., Experimental study of the protective role of calcium in lead poisoning, *Gig. Sanit.*, 12, 47, 1990.
38. Rosen, J. and Wexler, E., Studies of lead transport in bone organ culture, *Biochem. Pharmacol.*, 26, 650, 1977.
39. Kostial, K. and Vouk, V., Lead ions and synaptic transmission in the superior cervical ganglion of the cat, *Br. J. Pharmacol.*, 12, 219, 1957.
40. Bogden, J., Gertner, S., Christakos, S., Kemp, F.W., Yang, Z., Katz, S.R., and Chu, C., Dietary calcium modifies concentrations of lead and other metals and renal calbindin in rats, *Can. J. Nutr.*, 122, 1351, 1992.
41. Henning, S. L. and Cooper, L. C., Intestinal accumulation of lead salts and milk lead by suckling rats, *Proc. Soc. Exp. Biol. Med.*, 187, 110, 1988.
42. Blake, K. and Mann, M. Effect of calcium and phosphorus on the gastrointestinal absorption of 203 Pb in man, *Environ. Res.*, 30, 188, 1983.
43. U.S. Department of Health and Human Services, *Preventing Lead Poisoning in Young Children: A Statement by the Centers for Disease Control,* U.S. Department of Health and Human Services, Public Health Service, Centers for Disease Control, Atlanta, GA, 1991.
44. Committee on Measuring Lead in Critical Populations, Board on Environmental Studies and Toxicology, Commission on Life Sciences, *Measuring Lead Exposure in Infants, Children, and Other Sensitive Populations,* National Academy Press, Washington, D.C., 1993.
45. Gruden, N. and Buben, M., Influence of lead on calcium metabolism, *Bull. Environ. Contam. Toxicol.*, 18, 303, 1977.
46. Chisolm, J. and O'Hara, D., *Lead Absorption in Children: Management, Clinical, and Environmental Aspects,* Urban & Schwarzenberg, Baltimore, MD, 1982.
47. Mahaffey-Six, K.M. and Goyer, R.A., Experimental enhancement of lead toxicity by low dietary calcium, *J. Lab. Clin. Med.*, 76, 933, 1970.
48. Goyer, R., Nutrition and metal toxicity, *Am. J. Clin. Nutr.*, 61(Suppl.), 646S, 1995.
49. Lederer, L. G. and Bing, F. C., Effect of calcium and phosphorus on retention of lead by growing organisms, *J. Am. Med. Assoc.*, 114, 2457, 1940.
50. Mahaffey, K. R., Goyer, R., and Haseman, J. K., Dose response to lead. Ingestion in rats fed low dietary calcium, *J. Lab. Clin. Med.*, 82, 92, 1973.
51. Shields, J. B. and Mitchell, H. H., The effect of calcium and phosphorus on the metabolism of lead, *J. Nutr.*, 21, 541, 1994.
52. Bogden, J. D., Gertner, S. B., Kemp, F. W., McLeod, R., Bruening, K. S., and Chung, H. R., Dietary lead and calcium: effects on blood pressure and renal neoplasia, *J. Nutr.*, 121, 718, 1991.
53. Miller, G., Massaro, T., Koperek, E., and Massaro, E., Low-level exposure and the time-dependent organ tissue distribution of essential elements in the neonatal rat, *Biol. Trace Element Res.*, 6, 519, 1984.
54. Ziegler, E., Edwards, B. R., Jensen, R. L., Mahaffey, K. R., and Fomon, S. J., Absorption and retention of lead by infants, *Pediatr. Res.*, 12, 29, 1978.

55. Heard, M. J. and Chamberlain, A., Effect of minerals and food on uptake of lead from the gastrointestinal tract in humans, *Human Toxicol.,* 1, 411, 1982.
56. Blake, K. C. H. and Mann, M., Effect of calcium and phosphorus on the absorption of 203 lead in man, *Environ. Res.,* 30, 188–194, 1983.
57. Sorrell, M., Rosen, J. F., and Roginsky, M., Interactions of lead, calcium, vitamin D, and nutrition in lead-burdened children, *Arch. Environ. Health,* 32, 160, 1977.
58. Mahaffey, K., Gartside, P., and Glueck, C., Blood lead levels and dietary calcium intake in 1 to 11 year old children: the Second National Health and Nutrition Examination Survey, 1976 to 1980, *Pediatrics,* 78, 257, 1986.
59. Johnson, N. E. and Tenuta, K., Diets and lead blood levels of children who practice pica, *Environ. Res.,* 18, 369, 1979.
60. Baghurst, P. A., Michael, A. J., Vimpani, G. V., Robertson, E. F., Clar, P. D., and Wigg, N. R., Determinants of blood lead concentrations of pregnant women living in Port Pirie and surrounding areas, *Med. J. Aust.,* 146, 69, 1987.
61. Goyer, R. A., Calcium and lead interactions: some new insights, *J. Lab. Clin. Med.,* 91, 363, 1978.
62. Harlan, W. R., Landis, J. R., Schmouder, R. L. et al., Blood lead and blood pressure: Relationship in the adolescent and adult U.S. population, *J. Am. Med. Assoc.,* 253, 530, 1985.
63. Louekari, K., Uusitalo, U., and Pietinen, P., Variation and modifying factors of the exposure to lead and cadmium based on an epidemiological study, *Sci. Total Environ.,* 84, 1, 1989.
64. Ong, C. N. and Lee, W. R., Interaction of calcium and lead in human erythrocytes, *Br. J. Ind. Med.,* 37, 70, 1980.
65. Sartor, F. and Rondia, D., L'evaluation epidemiologique des normes biologiques pour l'exposition du plomb environnant, *Arch. Belg.,* 46, 17, 1988.
66. Silbergeld, E. K., Schwartz, J., and Mahaffey, K., Lead and osteoporosis: mobilization of lead from bone in postmenopausal women, *Environ. Res.,* 47, 79, 1988.
67. Kostial, K., Dekanic, D., Telisman, S., Blanusa, M., Duvancic, S., Prpic-Majic, D., and Pongracic, J., Dietary calcium and blood lead levels in women, *Trace Element,* 81, 181, 1991.
68. Berode, M., Wietlisbach, V., Rickenbach, M., and Guillemin, M. P., Lifestyle and environmental factors as determinants of blood lead levels in a Swiss population. *Environ. Res.,* 55, 1 1991.
69. Fullmer, C., Intestinal interactions of lead and calcium, *Neurotoxicology,* 13, 799, 1992.
70. Habermann, E., Crowell, K., and Janicki, P., Lead and other metals can substitute for Ca^{2+} in calmodulin, *Arch. Toxicol.,* 54, 61, 1983.
71. Chai, S. and Webb, R., Effects of lead on vascular reactivity, *Environ. Health Perpect.,* 78, 85, 1988.
72. Hsu, F. S., Krool, L., Pon, W. G., and Duncan, J. R., Interactions of dietary calcium with toxic levels of lead and zinc in pigs, *J. Nutr.,* 105, 112, 1975.
73. Victery, W., Miller, C. R., Goyer, R. A., Essential trace metal excretion from rats with lead exposure and during chelation therapy, *J. Lab. Clin. Med.,* 107, 129, 1986.
74. Simons, T., Cellular interactions between lead and calcium, *Br. Med. Bull.,* 42, 431, 1986.
75. Simons, T., The role of anion transport in the passive movement of lead across the human red cell membrane, *J. Physiol.,* 387, 287, 1986.
76. Simons, T. and Pocock, G., Lead enters adrenal medullary cells through calcium channels, *J. Neurochem.,* 48, 383, 1987.
77. Parr, D. and Harris, E., The effect of lead on the calcium-handling capacity of rat heart mitochondria, *Biochem. J.,* 158, 289, 1976.
78. Goldstein, G., Lead encephalopathy: the significance of lead inhibition of calcium uptake by brain mitochondria, *Brain Res.,* 136, 185, 1977.
79. Pounds, J., Long, G., and Rosen, J., Cellular and molecular toxicity of lead in bone, *Environ. Health Perspect.,* 91, 17, 1991.
80. Bressler, J. and Goldstein, G., Mechanisms of lead neurotoxicity, *Biochem. Pharmacol.,* 41, 479, 1991.
81. Fullmer, C. S. and Rosen, J. F., Effect of dietary calcium and lead status on intestinal calcium absorption, *Environ. Res.,* 51, 91, 1990.
82. Rosen, J. F., Chesney, R. W., Hamstra, A., Deluca, H. F., and Mahaffey, K. R., Reduction in 1,25-dihydroxy vitamin D in children with increased lead absorption, *N. Engl. J. Med.,* 302, 1128, 1980.
83. Hegsted, D., Present knowledge of calcium, phosphorus and magnesium, *Nutr. Rev.,* 26, 1968, 65.
84. Danisi, G. and Straub, R., Unidirectional influx of phosphate across the mucosal membrane of rabbit small intestine, *Pflugers Arch.,* 385, 117, 1980.
85. Clark, I., Importance of dietary $Ca:PO_4$ ratios on skeletal Ca, Mg, and PO_4 metabolism, *Am. J. Physiol.,* 217, E451, 1979.
86. Hurwitz, S. and Bar, A., Calcium and phosphorus interrelationships in the intestine of the fowl, *J. Nutr.,* 101, 1971, 677.
87. Costa, D. and Motta, S., Minerals, in *Newer Methods of Nutritional Biochemistry,* Albanese, A., Ed., Academic Press, London, 1963.

88. Spencer, H., Kramer, L., Rubio, N., and Osis, D., The effect of phosphorus on endogenous fecal calcium excretion in man, *Am. J. Clin. Nutr.* 43, 844, 1986.
89. Heaney, R., Protein intake and the calcium economy, *J. Am. Diet. Assoc.*, 93(11), 1259, 1993.
90. Albright, F., Bauer, W., and Cockrill, J., Studies in parathryroid physiology. III. The effect of phosphate ingestion in clinical hyperparathyroidism, *J. Clin. Invest.*, 11, 411, 1932.
91. Hully, S., Goldsmith, R., and Ingbar, S., Effect of renal arterial and systemic infusion of phosphate on urinary calcium excretion, *Am. J. Physiol.*, 217, 1570, 1969.
92. Coburn, J., Hartenbower, D., and Massry, S., Modification of calciuretic effect of extracellular volume expansion by phosphate infusion, *Am. J. Physiol.*, 220, 377, 1971.
93. Hegsted, M., Shuette, S., Zemel, M., and Linkswiler, H., Urinary calcium and calcium balance in young men as affected by level of protein and phosphorus intake, *J. Nutr.*, 111, 553, 1981.
94. Wong, N., Quamme, G., Sutton, R., O'Callaghan, T., and Dirks, J., Effect of phosphate infusion on renal phosphate and calcium transport, *Renal Physiol. Basel.*, 8, 30, 1985.
95. Lee, D., Brautbar, N., and Kleeman, C., Disorders of phosphorus metabolism, in *Disorders of Mineral Metabolism,* Vol. III, Bronner, F. and Coburn, J., Eds., Academic Press, New York, 1981, p. 293.
96. Spencer, H., Menczel, J., Lewin, I., and Samachson, J., Effect of high phosphorus intake on calcium and phosphorus metabolism in man, *J. Nutr.*, 86, 125, 1965.
97. Spencer, H., Kramer, L., Osis, D., and Norris, C., Effect of phosphorus on the absorption of calcium and on the calcium balance in man, *J. Nutr.*, 108, 447, 1978.
98. Shackelford, M., Collin, R., Black, R., Ames, M., Dolan, S., Sheikh, N., Chi, R., and O'Donnell, M., Mineral interactions in rats fed AIN-76A diets with excess calcium, *Food Chem. Toxic.*, 32(3), 255, 1994.
99. Hiatt, H. and Thompson, D., Some effects of intravenously administered calcium on inorganic phosphate metabolism, *J. Clin. Invest.*, 36, 573, 1957.
100. Eisenberg, E., Effects of serum calcium level and parathyroid extracts on phosphate and calcium excretion in hypoparathyroid patients, *J. Clin. Invest.*, 44, 942, 1965.
101. Toothhill, J., The effect of certain dietary factors on the apparent absorption of magnesium by the rat, *Br. J. Nutr.*, 17, 125, 1963.
102. Brink, E., Beynen, A., Dekker, P., van Beresteijn, E., and van der Meer, R. Interaction of calcium and phosphate decreases ileal magnesium solubility and apparent magnesium absorption in rats, *J. Nutr.*, 122, 580, 1992.
103. Tufts, E. and Greenberg, D. The biochemistry of magnesium deficiency. II. The minimum magnesium requirement for growth, gestation and lactation and the effect of the dietary calcium thereon, *J. Biol. Chem.*, 122, 715, 1937–1938.
104. O'Dell, B. and Morris, E., Relationship of excess calcium and phosphorus to magnesium requirement and toxicity in guinea pigs, *J. Nutr.*, 81, 175, 1963.
105. Bunce, G., Chiemchaisri, Y., and Phillips, P., The mineral requirement of the dog. IV. Effect of certain dietary and physiologic factors upon magnesium deficiency syndrome, *J. Nutr.*, 76, 23, 1962.
106. Spencer, H. and Osis, D., Studies of magnesium metabolism in man: original data and a review, *Metab. Res.*, 7, 271, 1988.
107. Alcock, N. and MacIntyre, I., Interrelation of calcium and magnesium absorption, *Clin. Sci.*, 22, 185, 1969.
108. Aldor, T. and Moore, E., Magnesium absorption by everted sacs of rat intestine and colon, *Gastroenterology*, 59, 745, 1970.
109. Scott, D., Factors influencing the secretion and absorption of calcium and magnesium in the small intestine of the sheep, *Q. J. Exp. Physiol.*, 59, 312, 1964.
110. Karbach, U., Cellular-mediated and diffusive magnesium transport across the descending colon of the rat, *Gastroenterology*, 96, 1989, 1282.
111. Karbach, U. and Ewe, K., Calcium and magnesium transport and influence of 1,25-dihydroxyvitamin D, *Digestion*, 37, 35, 1987.
112. Kim, Y. and Linkswiller, H., Effect of level of calcium and of phosphorus intake on calcium, phosphorus, and magnesium metabolism in young adult males, *Fed. Proc.*, 39, 895, 1980.
113. Clarkson, E., Warren, R., McDonald, S., and de Wardener, H., The effect of a high intake of calcium on magnesium metabolism in normal subjects and patients with chronic renal failure, *Clin. Sci.*, 32, 11, 1967.
114. MacIntyre, I., Hanna, S., Boothe, C., and Read, A., Intracellular magnesium deficiency in man, *Clin. Sci.*, 20, 297, 1961.
115. Spencer, H., Calcium and magnesium balances in man, *Clin. Chem.*, 25, 1043, 1979.
116. Spencer, H., Kramer, L., Wiatrowki, E., and Osis, D., Magnesium-fluoride interrelationships in man. II. Effect of magnesium on fluoride metabolism, *Am. J. Physiol.*, 234, E343, 1978.
117. Greger, J., Smith, S., and Snedeker, S., Effect of dietary calcium and phosphorus levels on the utilization of calcium, phosphorus, magnesium, manganese, and selenium in adult males, *Nutr. Res.*, 1, 315, 1981.
118. Lewis, N., Marcus, M., Behling, A., and Greger, J. Calcium supplements and milk: effects on acid-base balance and on retention of calcium, magnesium, and phosphorus, *Am. J. Clin. Nutr.*, 49, 527, 1989.
119. Leichsenring, J., Norris, L., and Lamison, S., Magnesium metabolism in college women. Observations on the effect of calcium and phosphorus intake levels, *J. Nutr.*, 45, 477, 1951.

120. Leverton, R., Leichsenring, J., Linkswiler, H., and Meyer, F. Magnesium intakes of young women receiving controlled intakes, *J. Nutr.,* 74,33, 1961.
121. Spencer, H., Lesniak, M., Kramer, L., Coffey, J., and Osis, D. Studies on magnesium metabolism in man, in *Magnesium in Health and Disease,* Cantin, M. and Seelig, M., Eds., Spectrum Press, New York, 1979, p. 911.
122. Spencer, H., Schwartz, R., Norris, C., and Osis, D., Magnesium 28 studies and magnesium balances in man, *J. Am. Coll. Nutr.,* 4, 316, 1985.
123. Spencer, H. Mineral and mineral interactions in human beings, *J. Am. Diet. Assoc.,* 86(7), 864, 1986.
124. Spencer, H., Kramer, L., Norris, C., and Osis, D., Effect of small doses of aluminum-containing antacids on calcium and phosphorus metabolism, *Am. J. Clin. Nutr.,* 36, 32, 1982.
125. Dawson-Hughes, B., Seligson, F., and Hugues, V., Effects of calcium carbonate and hydroxyapatite on zinc and iron retention in post menopausal women, *Am. J. Clin. Nutr.,* 53, 112, 1991.
126. Cook, J., Dassenko, S., and Whittaker, P., Calcium supplementation: effect on iron absorption, *Am. J. Clin. Nutr.,* 53, 106, 1991.
127. Galan, P., Cherouvier, F., Preziosi, P. et al., Effects of the increasing consumption of dairy products upon iron absorption, *Eur. J. Clin. Nutr.,* 45, 553, 1991.
128. Tucker, H. F., and Salmon, W. D., Parakeratosis or zinc deficiency in the pig, *Proc. Soc. Exp. Biol. Med.,* 8, 613, 1955.
129. Grekas, D., Alivanis, P., Balaskas, E., Dombros, N., and Tourkantonis, A., Effect of aluminum hydroxide and calcium on zinc tolerance test in uremic patients, *Trace Element Med.,* 5, 172, 1988.
130. Spencer, H., Kramer, L., Norris, C., and Osis, D., Effect of calcium and phosphorus on zinc metabolism in man, *Am. J. Clin. Nutr.,* 40, 1213, 1984.
131. Spencer, H., Kramer, L., Norris, C., and Osis, D., Inhibitory effect of zinc on the intestinal absorption of calcium, *Clin. Res.,* 33, 72A, 1992.
132. Spencer, H., Norris, C., and Osis, D., Further studies on the effect of zinc on intestinal absorption of calcium in man, *J. Am. Coll. Nutr.,* 11, 561, 1992.
133. Tompsett, S., Factors influencing the absorption of iron and copper from the alimentary tract, *Biochem. J.,* 34, 961, 1940.
134. Wang, C. and Bhattacharyya, M. H., Effect of cadmium on bone calcium and 45 Ca in nonpregnant mice on a calcium-deficient diet: evidence of direct effect of cadmium on bone, *Toxicol. Appl. Pharmacol.,* 120, 228, 1993.
135. Van Barveveld, A. and Van den Hamer, C., The influence of calcium and magnesium on manganese transport and utilization in mice, *Biol. Trace Elem. Res.,* 6, 489, 1984.
136. McDermott, S. and Kies, C., Manganese usage in humans as affected by use of calcium supplements, in *Nutritional Bioavailability of Manganese,* Kies, C., Ed., ACS Symposium Series 354, American Chemical Society, Washington, D.C., 1987, p. 146.
137. Cerklewski, F., Influence of dietary magnesium on fluoride bioavailability in the rat, *J. Nutr.,* 117, 496, 1987.
138. Forsyth, D., Pond, W., and Krook, L., Dietary calcium and fluoride interactions in swine: *in utero* and neonatal effects, *J. Nutr.,* 102, 1637, 1972.
139. Forsyth, D., Pond, W., Wasserman, R., and Krook, L., Dietary calcium and fluoride interactions in swine: effects on physical and chemical bone characteristics, calcium binding protein and histology of adults, *J. Nutr.,* 102, 1623, 1972.
140. Briacon, D., D'Aranda, P., Quillet, P., Duplan, B., Chapuy, M., Arlot, M., and Neunier, P., Comparative study of fluoride bioavailability following the administration of sodium fluoride alone or in combination with different calcium salts, *J. Bone Min. Res.,* 5, S71, 1990.
141. Spencer, H., Kramer, L., Osis, D., Wiatrowski, E., Norris, C., and Lender, M., Effect of calcium, phosphorus, magnesium and aluminum on fluoride metabolism in man, in *Micronutrient Interactions: Vitamins, Minerals and Hazardous Elements,* Levander, O. and Cheng, L., Ed., Vol. 355, Annals of the New York Academy of Sciences, New York, 1980, p. 181.
142. Petering, H., The effect of cadmium and lead on copper and zinc metabolism, in *Trace Element Metabolism in Animals,* Hoekstra, W., Suttie, J., Ganther, H., and Mertz, W. Eds., University Park Press, Baltimore, MD, 1974, p. 119.
143. Powell, G., Miller, J., Morton, J., and Clifton, C., Influence of dietary cadmium level and supplemental zinc on cadmium toxicity in the bovine, *J. Nutr.,* 84, 205, 1964.
144. Pond, W. and Walker, E., Effect of dietary Ca and Cd level of pregnant rats on reproduction and on dam and progeny tissue mineral concentrations, *Proc. Soc. Exp. Biol. Med.,* 148, 665, 1975.
145. Ando, M., Shimizu, M., Sayato, Y., Tanimura, A., and Tobe, M., The inhibition of vitamin D-stimulated intestinal calcium transport in rats after continuous oral administration of cadmium, *Toxicol. Appl. Pharmacol.,* 61, 297, 1981.
146. Buchet, J., Lauwerys, R., Roels, H., Bernard, A., Bruaux, P., Claeys, F., Ducoffre, G., de Plaen, P., Staessen, J., and Amery, A., Renal effects of cadmium body burden of the general population, *Lancet,* 336, 699, 1990.
147. Insogna, K., Bordley, D., Caro, J., and Lockwood, D., Osteomalacia and weakness from excessive antacid ingestion, *J. Am. Med. Assoc.,* 251, 1938, 1980.

148. Cannata, J., Junork, B., Briggs, J., Fell, G., and Beastall, G., Effect of acute aluminum overload on calcium and parathyroid hormone metabolism, *Lancet,* 1, 501, 1983.
149. Cournot-Witmer, G., Zingraff, J., Planchot, J., Escaig, F., Lefevre, R., Boumati, P., Bourdeau, A., Garabedian, M., Galle, P., Bourdon, R., Drueke, T., and Balsan, S. Aluminum localization in bone from hemodialyzed patients: relationship to matrix mineralization, *Kidney Int.,* 20, 375, 1981.
150. Valdivia, R., Ammerman, C., Henry, P., Reaster, J., and Wilcox, C. Effect of dietary aluminum and phosphorus on performance, phosphorus utilization and tissue mineral composition in sheep, *J. Animal Sci.,* 55, 402, 1982.
151. Valdivia, R., Ammerman, C., Wilcox, C., and Henry, P., Effect of dietary aluminum on animal performance and tissue mineral levels in growing steers, *J. Animal Sci.,* 47, 1351, 1978.
152. Greger, J. and Baier, M. Effect of dietary aluminum on mineral metabolism of adult males, *Am. J. Clin. Nutr.,* 38, 411, 1983.
153. Greger, J., Bula, E., and Gum, E., Mineral metabolism of rats fed moderate levels of various aluminum compounds for short periods of time, *J. Nutr.,* 115, 1708, 1985.
154. Greger, J., Gum, E., and Bula, E., Mineral metabolism of rats fed various levels of aluminum hydroxide, *Biol. Trace Elem. Res.,* 9, 67, 1986.
155. Yamaguchi, M., Saito, R., and Okada, S., Dose-effect of inorganic tin on biochemical indices in rats, *Toxicology,* 16, 267, 1980.
156. Yamaguchi, M., Saito, R., and Okada, S., Inorganic tin in the diet affects the femur in rats, *Toxicol. Lett.,* 9, 207, 1981.
157. Yamamoto, T., Yamaguchi, M., and Sato, H., Accumulation of calcium in kidney and decrease of calcium in serum of rats treated with tin chloride, *J. Toxicol. Environ. Health,* 1, 749, 1976.
158. Johnson, M. A. and Greger, J. L., Tin, copper, iron and calcium metabolism of rats fed various dietary levels of inorganic tin and zinc, *J. Nutr.,* 114, 1843, 1985.
159. Johnson, M. A. and Greger, J. L., Effects of dietary tin on tin and calcium metabolism in adult males, *Am. J. Clin. Nutr.,* 35, 655, 1982.
160. De La Burde, B. and Choate, M. S., Early asymptomatic lead exposure and development at school age, *J. Pediatr.,* 87, 639, 1975.
161. David, O., Association between lower level lead concentrations and hyperactivity, *Environ. Health Perspect.,* 7, 17, 1974.
162. Baloh, R., Strum, R., Green, B., and Gleser, G., Neuropsychological effects of chronic asymptomatic increased lead absorption. A controlled study, *Arch. Neurol.,* 32, 326, 1975.
163. Lilis, R., Gavrilscu, N., Nestorescu, B., Dumitriu, C., and Roventa, A., Nephropathy in chronic lead poisoning, *Br. J. Ind. Med.,* 25, 196, 1968.
164. Seppalainen, A. M., Tola, S., Hernberg, S., and Kock, B., Subclinical neuropathy at "safe" levels of lead exposure, *Arch. Environ. Health,* 30, 180, 1975.
165. Vitale, L. F., Joselow, M. M., Wedeen, R. P., and Pawlow, M., Blood lead — an inadequate measure of occupational exposure, *J. Occup. Med.,* 17, 155, 1975.
166. Hanninen, H., Hernberg, S., Mantere, P., Vesnato, R., and Jalkanen, M., Psychological performance of subjects with low-exposure to lead, *J. Occup. Med.,* 20, 683, 1978.
167. Baker, E. L., Landrigan, P. J., Barbour, A. G., Cox, D. H., Folland, D. S., Ligo, R. N., and Throckmorton, J., Occupational Pb poisoning in the United States — clinical and biochemical findings related to blood lead levels, *Br. J. Ind. Med.,* 36, 314, 1979.
168. Mantere, P., Hanninen, H., and Hernberg, S., Subclinical neurotoxic lead effects: two-year follow-up studies with psychological test methods, *Neurobehav. Toxicol. Teratol.,* 4, 725, 1982.
169. Boey, K. W. and Jayaratnam, J., A discriminant analysis of neuropsychological effect of low lead exposure, *Toxicology,* 49, 309, 1988.
170. Miller, G. D., Massaro, T. F., and Massaro, E. J., Interactions between lead and essential elements: a review, *Neurotoxicology,* 11, 120, 1990.
171. Laughlin, N. K., Bushnell, P. J., and Bowman, R. E., Lead exposure and diet: differential effects on social development in the rhesus monkey, *Neurotoxicol. Teratol.,* 13, 429, 1991.
172. APHA, Reducing health risks related to lead exposure, in *Nation's Health,* American Public Health Association, Washington, D.C., 1989, p. 15.
173. Lave, L. B., Lead as a public health problem: is it overestimated?, *J. Health Econom.,* 8, 247, 1989.
174. Needleman, H. L., Schell, A., Bellinger, D., Leviton, A., and Allred, E. N., The long-term effects of exposure to low doses of lead in childhood: an 11-year follow-up report, *N. Engl. J. Med.,* 332, 83, 1990.
175. Silbergeld, E. K., Lead in the environment: coming to grips with multisource risks and multifactorial endpoints, *Risk Anal.,* 9, 137, 1989.
176. Committee on Environmental Health, American Academy of Pediatrics, Lead poisoning: from screening to primary prevention, *Pediatrics,* 92, 176, 1993.
177. Caplun, E., Petit, D., and Picciotto, E., Le plomb dans l'essence, *Recherche,* 15, 270, 1984.
178. Mahaffey, K. R., Annest, J. L., Roberts, J., and Murphy, R. S., National estimates of blood lead levels: United States, 1976–1980. Association with selected demographic and socioeconomic factors, *N. Engl. J. Med.,* 307, 573, 1982.

179. Ratcliffe, J. M., *Lead in Man and the Environment,* Halsted, New York, 1981.
180. Quarterman, J. and Morrison, J., The effects of dietary calcium and phosphorus on the retention and excretion of lead in rats, *Br. J. Nutr.,* 34, 351, 1975.
181. Barltrop, D. and Khoo, H., The influence of dietary minerals and fat on the absorption of lead, *Sci. Total Environ.,* 6, 265, 1976.
182. Heard, M. J., Chamberlain, A. C., and Sherlock, J. C., Uptake of lead by humans and effect of minerals and food, *Sci. Total Environ.,* 30, 245, 1983.
183. Kochen, J. and Greener, Y., Interaction of ferritin with lead and cadmium, *Pediatr. Res.,* 9, 323, 1975.
184. Barton, J. C., Conrad, M. E., Nuby, S., and Harrison, L., Effects of iron on the absorption and retention of Pb, *J. Lab. Clin. Med.,* 91, 536, 1978.
185. Hamilton, D. L., Interrelationships of lead and iron retention in iron-deficiency mice, *Toxicol. Appl. Pharmacol.,* 46, 651, 1978.
186. Miller, G. D., Massaro, T. F., Grandlund, R. W., and Massaro, E. J., Tissue distribution of lead in neonatal rats exposed to multiple doses of lead acetate, *J. Toxicol. Environ. Health,* 11, 121, 1983.
187. Robertson, I. K. and Worwood, M., Lead and iron absorption from rat small intestine: the effect of dietary Fe deficiency, *Br. J. Nutr.,* 40, 253, 1978.
188. Mahaffey-Six, K. and Goyer, R., The influence of iron deficiency on tissue content and toxicity of ingested lead in the rat, *J. Lab. Clin. Med.,* 79, 128, 1972.
189. Hallberg, L., Search for nutritional confounding factors in the relationship between iron deficiency and brain function, *Am. J. Clin. Nutr.,* 50, 598, 1989.
190. Mahaffey, K., Factors modifying susceptibility to lead toxicity, in *Dietary and Environmental Lead: Human Health Effects,* Mahaffey, K., Ed., Elsevier Science, New York, 1985, p. 373.
191. Mahaffey, K., Stone, C., Banks, T., and Reed, G., Reduction in tissue storage of lead in the rat by feeding diets with elevated iron concentration, in *Trace Element Metabolism in Man and Animals,* Vol. 3, Kirchgessner, M., Ed., Arbeitskreis fur Tierenahrungsforschung Weihenstephen, Freising-Weihenstephen, West Germany, 1978, p. 584.
192. Ragan, H. A., Effects of iron deficiency on the absorption and distribution of lead and cadmium in rats, *J. Lab. Clin. Med.,* 90, 70, 1977.
193. Lin-Fu, J. S., Vulnerability of children to lead exposure and toxicity, *N. Engl. J. Med.,* 289, 1229, 1973.
194. Szold, P. D., Plumbism and iron deficiency, *N. Engl. J. Med.,* 290, 520, 1974.
195. Graziano, J. H., Popovac, D., Factor-Litvak, P., Shrout, P., Kline, J., Murphy, M. J., Zhao, Y. H., Mehmeti, A., Ahmedi, X., Rajovic, B. et al. Determinants of elevated blood lead during pregnancy in a population surrounding a lead smelter in Kosovo, Yugoslavia, *Environ. Health Perspect.,* 89, 95, 1990.
196. Cerklewski, F. L. and Forbes, R. M., Influence of dietary zinc on lead toxicity in the fat, *J. Nutr.,* 106, 689, 1976.
197. Victery, W., Thomas, D., Shoeps, P., and Vander A., Lead increases urinary zinc excretion in rats, *Biol. Trace Elem. Res.,* 4, 211, 1982.
198. Victery, W., Miller, C. R., Zho, S., and Goyer, R. A. Effect of different levels and periods of lead exposure on tissue levels and excretion of lead, zinc, and calcium in the rat, *Fund. Appl. Toxicol.,* 8, 506, 1987.
199. Nakao, K., Wada, O., and Yano, Y., γ-aminolevulinic acid dehydratase activity in erythrocytes for the evaluation of lead poisoning, *Clin. Chem. Acta,* 19, 319, 1968.
200. El-Gazzar, R. M., Finelli, V. N., Boinano, J., and Petering, H. G., Influence of dietary zinc on lead toxicity in rats, *Toxicol. Lett.,* 1, 227, 1978.
201. Michaelson, I. A. and Sauerhoff, M. W., The effect of chronically ingested inorganic lead on brain levels of iron, zinc, and manganese of 25 day old rat, *Life Sci.,* 13, 417, 1973.
202. Cerlewski, F. L., Postabsorptive effect of increased zinc on toxicity and removal of tissue lead in rats, *J. Nutr.,* 114, 550, 1984.
203. Meredith, P., Moore, M., and Goldberg, A., The effects of aluminum, lead and zinc on delta-aminolevulinic acid dehydratase, *Biochem. Soc. Trans.,* 2, 1243, 1974.
204. Finelli, V., Klauder, D., Karaffa, M., and Petering, H., Interaction of zinc and lead on delta-aminolevulinate dehydratase, *Biochem. Biophys. Res. Commun.,* 65, 303, 1975.
205. Petering, H., Some observations on the interaction zinc, copper and iron metabolism in lead and cadmium toxicity, *Environ. Health Perspect.,* 25, 141, 1978.
206. Ashraf, M. H. and Fosmire, G. J., Effects of marginal zinc deficiency on subclinical lead toxicity in the rat neonate, *J. Nutr.,* 115, 334, 1985.
207. Lauwerys, R., Roels, H., Bucht, J., Bernard, A., Verhaeven, L., and Konings, J., The influence of orally-administered vitamin C or zinc on the absorption of and the biological response to lead, *J. Occup. Health,* 25, 668, 1983.
208. Davies, N., Anti-nutrient factors affecting mineral utilization, *Proc. Nutr. Soc.,* 38, 121, 1979.
209. Davies, N., Recent studies of antagonistic interactions in the aetiology of trace element deficiency and excess, *Proc. Nutr. Soc.,* 33, 293, 1974.
210. Van Campen, D., Absorption of copper from the gastrointestinal tract, in *Intestinal Absorption of Metal Ions, Trace Elements and Radionuclides,* Skoryna, W.-E., Ed., Pergamon Press, Oxford, 1971.

211. Mylroie, A. A., Boseman, A., and Kyle, J., Metabolic interactions between lead and copper in rats ingesting lead acetate, *Biol. Trace Elem. Res.,* 9, 221, 1986.
212. Klaunder D. S. and Petering, H. G., Protective value of dietary copper and iron against some toxic effects of lead in rats, *Environ. Health Perspect.,* 12, 77, 1975.
213. Cerklewski, F. L. and Forbes, R. M., Influence of dietary copper on lead toxicity in the young rat, *J. Nutr.,* 107, 143, 1977.
214. Avioli, L., Calcium and phosphorus, in *Modern Nutrition in Health and Disease,* 7th ed., Shils, M. and Young Y., Eds., Lea & Febiger, Philadelphia, PA, 1988, p. 142.
215. Morris, E. and O'Dell, B., Magnesium deficiency in the guinea pig. Mineral composition of tissues and distribution of acid-soluble phosphorus, *J. Nutr.,* 75, 77, 1961.
216. O'Dell, B., Morris, E., Pickett, E., and Hogan, A., Diet composition and mineral balance in guinea pigs, *J. Nutr.,* 63, 65, 1957.
217. O'Dell, B., Morris, E., and Regan, W., Magnesium requirement of pigs and rats. Effects of calcium and phosphorus and symptoms of Mg deficiency, *J. Nutr.,* 70, 103, 1960.
218. Spencer, H., Kramer, L., Gatza, C., Norris, C., and Coffey, J., Magnesium-phosphorus interrelations in man, in *Trace Substances in Environmental Health.* XIII. *A Symposium,* Hemphill, D., Ed., University of Missouri, Columbia, 1979, p. 401.
219. Hegsted, D., Finch, C., and Kinney, T., The influence of diet on iron absorption. The interrelationship of iron and phosphorus, *J. Exp. Med.,* 90, 147, 1949.
220. Peters, T., Apt, L., and Ross, J., Effect of phosphates upon iron absorption studies in normal human subjects and in an experimental model using dialysis, *Gastroenterology,* 61, 315, 1971.
221. Monsen, E. and Cook, J., Food iron absorption in human subjects. IV. The effects of calcium and phosphate salts on the absorption of nonheme iron, *Am. J. Clin. Nutr.,* 29, 1142, 1976.
222. Snedeker, S., Smith, S., and Greger, J., Effect of dietary calcium and phosphorus levels on the utilization of iron, copper, and zinc by adult males, *J. Nutr.,* 112, 136, 1982.
223. Ondreicka, R., Kortus, J., and Ginter, E., Aluminum, its absorption, distribution and effects on phosphorus metabolism, in *Intestinal Absorption of Metal Ions, Trace Elements, and Radionuclides,* Skoryna, S. and Waldron-Edward, D., Eds., Permagon Press, Oxford, 1971, p. 293.
224. Clarkson, E., Luck, V., Hynson, W., Bailey, R., Eastwood, J., Woodhead, J., Clements, V., O'Riordan, J., and deWardener, H., The effect of aluminum hydroxide on calcium, phosphorus and aluminum balances, the serum parathyroid hormone concentration and the aluminum content of bone in patients with chronic renal failure, *Clin. Sci.,* 43, 519, 1972.
225. Cam, J., Liuck, V., Eastwood, J., and deWardener, H., The effect of aluminum hydroxide orally on calcium, phosphorus and aluminum metabolism in normal subjects, *Clin. Sci. Mol. Med.,* 51, 407, 1976.
226. Lotz, M., Zisman, E., and Barter, F., Evidence for a phosphorus-depletion syndrome in man, *N. Engl. J. Med.,* 278, 409, 1968.
227. Dent, C. and Winter, C., Osteomalacia due to phosphate depletion from excessive aluminum hydroxide ingestion, *Br. Med. J.,* 1, 551, 1974.
228. Wester, P., Magnesium, *Am. J. Clin. Nutr.,* 45, 1305, 1987.
229. Garfinkel, L. and Garfinkel, D., Magnesium regulation of the glycolytic pathway and the enzymes involved, *Magnesium,* 4, 60, 1985.
230. Seelig, M., The requirement of magnesium by the normal adults, *Am. J. Clin. Nutr.,* 14, 1964, 342.
231. Lonnerdal, B., Magnesium nutrition of infants, *Magnesium Res.,* 8, 99, 1995.
232. Nicar, M. and Pak, C., Oral magnesium load test for the assessment of intestinal magnesium absorption, *Mineral Electrolyte Metab.,* 8, 44, 1982.
233. Kasprzak, K., Diwan, B. A., and Rice, J. M., Iron accelerates while magnesium inhibits nickel-induced carcinogenesis in the rat kidney, *Toxicology,* 90, 129, 1994.
234. Hallberg, L., Iron, in *Present Knowledge in Nutrition,* 5th ed., The Nutrition Foundation, Washington, D.C., 1984, p. 459.
235. Pike, R. L. and Brown, M. L., *Nutrition: An Integrated Approach,* 3rd ed., Wiley, New York, 1984, p. 167.
236. Bothwell, T., Charlton, R., Cook, J., and Finch, C., *Iron Metabolism in Man,* Blackwell, Oxford, 1979; Cook, J., Lipschitz, D., Miles, L., and Finch, C., Serum ferritin as a measure of iron stores in normal subjects, *Am. J. Clin. Nutr.,* 27, 681, 1974.
237. Layrisse, M., Martinez-Torres, C., and Roche, M., Effect of interaction of various foods on iron absorption, *Am. J. Clin. Nutr.,* 21, 1175, 1968.
238. Nelson, W. E., *Textbook of Pediatrics,* W.B. Saunders, Philadelphia, PA, 1969.
239. Owen, G., Lubin, A. H., Garry, P. J., Preschool children in the United States: who has iron deficiency?, *J. Pediatr.,* 79, 563, 1971.
240. Fomon, S., *Infant Nutrition.,* W.B. Saunders, Philadelphia, PA, 1974.
241. Yip, R., Johnson, C., and Dallman, P. R., Age-related changes in laboratory values used in the diagnosis of anemia and iron deficiency, *Am. J. Clin. Nutr.,* 39, 427, 1984.
242. Expert Scientific Working Group, Summary of a report on assessment of the iron nutritional status of the United States population, *Am. J. Clin. Nutr.,* 42, 1318, 1984.

243. Pollack, S., George, J., Reba, R., Kaufman, R., and Crosby, W., The absorption of nonferrous metals in iron deficiency, *J. Clin. Invest.*, 44, 1470, 1965
244. Flanagan, P., Haist, J., and Valberg, L., Comparative effects of iron deficiency induced by bleeding and a low-iron diet on the intestinal absorptive interactions of iron, cobalt, manganese, zinc, lead and cadmium, *J. Nutr.*, 110, 1754, 1980.
245. Solomons, N. and Jacob, R., Studies on the bioavailability of zinc in humans: effects of heme and nonheme iron on the absorption of zinc, *Am. J. Clin. Nutr.*, 34, 475, 1981
246. Meadows, N., Grainger, S., Ruse, W., Keeling, P., and Thompson, R., Oral iron and the bioavailability of zinc, *Br. Med. J.*, 287, 1013, 1983.
247. Solomons, N., Pineda, O., Viteri, F., and Sandstead, H., Studies on the bioavailability of zinc in humans: mechanism of the intestinal interaction of nonheme iron and zinc, *J. Nutr.*, 113, 337, 1983.
248. Lonnerdal, B., Keen, C., Hendrickx, A., Golub, M., and Gershwin, M., Influence of dietary zinc and iron on zinc retention in pregnant rhesus monkeys and their infants, *Obstet. Gynecol.*, 75, 369, 1990.
249. Sandstrom, B., Davidsson, L., Cederblad, A., and Lonnerdal, B., Oral iron, dietary ligands and zinc absorption, *J. Nutr.*, 115, 411, 1985.
250. Yip, R., Reeves, J., Lonnerdal, B., and Keen, C., Does iron supplementation compromise zinc nutrition in healthy infants?, *Am. J. Clin. Nutr.*, 42, 683, 1985.
251. Ballot, D., Mac Phail, A., Bothwell, T., Gillooly, M., and Mayet, F., Fortification of curry powder with NaFe (III) EDTA in an iron-deficient population: report of a controlled iron-fortification trial, *Am. J. Clin. Nutr.*, 49, 162, 1989.
252. Crofton, R., Gvozdanovic, D., Gvozdanovic, S., Khin, C., Brunt, P., Mowat, N., and Aggett, P., Inorganic zinc and the intestinal absorption of ferrous iron, *Am. J. Clin. Nutr.*, 50, 141, 1989.
253. Rossander-Hulten, L., Brune, M., Sandstrom, B., Lonnerdal, B., and Hallberg, L., Competitive inhibition of iron absorption by manganese and zinc in humans, *Am. J. Clin. Nutr.*, 54, 152, 1991.
254. O'Neill-Cutting, M., Bomfor, A., and Munro, H., Effect of excess dietary zinc on tissue storage of iron in rats, *J. Nutr.*, 111, 1981, 1969.
255. Roger, J., Lonnerdal, B., Hurley, L., and Keen, C., Iron and zinc concentrations and Fe retention in developing fetuses of zinc deficient rats, *J. Nutr.*, 117, 1875, 1987.
256. Keen, C., Golub, M., Gershwin, M., Lonnerdal, B., and Hurley, L., Long-term marginal zinc deprivation in rhesus monkeys. III. Use of liver biopsy in the assessment of zinc status, *Am. J. Clin. Nutr.*, 47, 1041, 1988
257. Yadrick, M., Kenney, M., and Winterfelt, E., Iron, copper and zinc status response to supplementation with zinc or zinc and iron in adult females, *Am. J. Clin. Nutr.*, 49, 145, 1989.
258. Cox, D. and Harris, D., Reduction of liver xanthine oxidase activity and iron storage proteins in rats fed excess zinc, *J. Nutr.*, 78, 415, 1962.
259. Greger, J., Zaikis, S., Abernathy, R., Bennett, O., and Huffman, J., Zinc, copper, nitrogen, iron and manganese balance in adolescent females fed two levels of zinc, *J. Nutr.*, 108, 1449, 1978.
260. Lee, C., Nacht, S., Lukens, J., and Cartwright, G., Iron metabolism in copper-deficient swine, *J. Clin. Invest.*, 47, 2058, 1968.
261. Marston, H., Allen, S., and Swaby, S., Iron metabolism in copper-deficient rats, *Br. J. Nutr.*, 25, 15, 1971.
262. Evans, J. and Abraham, P., Anemia, iron storage and ceruloplasmin in copper nutrition in the growing rat, *J. Nutr.*, 103, 196, 1973.
263. Underwood, E., *Trace Elements in Human and Animal Nutrition,* Academic Press, London, 1977.
264. Hedges, J. and Kornegay, E., Interrelationships of dietary copper and iron as measured by blood parameters, tissue stores and feedlot performance of swine, *J. Animal Sci.*, 37, 1147, 1973.
265. Smith, C. and Bidlack, W., Interrelationships of dietary ascorbic acid and iron on the tissue distribution of ascorbic acid, iron and copper in female guinea pigs, *J. Nutr.*, 110, 1398, 1980.
266. O'Dell, B., Bioavailability of and interactions among trace elements, in *Trace Elements in Nutrition of Children,* Chandra, R., Ed., Raven Press, New York, 1985, p. 41.
267. Smith, C. and Bidlack, W., Interrelationships of dietary ascorbic acid and iron on the tissue distribution of ascorbic acid, iron and copper in female guinea pigs, *J. Nutr.*, 110, 1398, 1980.
268. Forth, W. and Rummel, W., Absorption of iron and chemically related metals *in vivo* and *in vitro*: specificity of the iron binding system in the mucosa of the jejunum, in *Intestinal Absorption of Metal Ions, Trace Elements and Radionuclides,* Skoryna, S. and Waldron-Edward, D., Eds., Pergamon Press, Oxford, 1971.
269. Thomson, A. and Valberg, L., Intestinal uptake of iron, cobalt and manganese in the iron-deficient rat, *Am. J. Physiol.*, 223(6), 1327, 1972.
270. Keen, C., Fransson, G., and Lonnerdal, B., Supplementation of milk with iron bound to lactoferrin using weanling mice. II. Effects on tissue manganese, zinc and copper, *J. Ped. Gastroenterol. Nutr.*, 3, 256, 1984.
271. Kies, C., Aldrich, K., Johnson, J., Creps, C., Kowalski, C., and Wang, R., Manganese availability for humans: effect of selected dietary factors, in *Nutritional Bioavailability of Manganese,* Kies, C., Ed., ACS Symposium Series 354, American Chemical Society, Washington, D.C., 1987, p. 136.
272. Keen, C., Zidenberg-Cherr, S., and Lonnerdal, B., Dietary manganese toxicity and deficiency: effects on cellular manganese metabolism, in *Nutritional Bioavailability of Manganese,* Kies, C., Ed., ACS Symposium Series 354, American Chemical Society, Washington, D.C., 1987, p. 21.

273. Davis, C. and Greger, J., Longitudinal changes of manganese-dependent superoxide dismutase and other indexes of manganese and iron status in women, *Am. J. Clin. Nutr.,* 55, 747, 1992.
274. Valberg, L., Sorbie, J., and Hamilton, D., Gastrointestinal metabolism of cadmium in experimental iron deficiency, *Am. J. Physiol.,* 231, 462, 1976.
275. Flanagan, P., McLellan, J., Haist, J., Cherian, M., Chamberlain, M., and Valberg, L., Increased dietary cadmium absorption in mice and human subjects with iron deficiency, *Gastroenterology,* 74, 841, 1978.
276. Stonard, M. and Webb, M., Influence of dietary cadmium on the distribution of the essential metals copper, zinc, and iron in tissues of the rat, *Chem. Biol. Interact.,* 15, 349, 1976.
277. Whanger, P., Effect of dietary cadmium on intracellular distribution of hepatic iron in rats, *Res. Commun. Chem. Pathol. Pharmacol.,* 5, 733, 1973.
278. Bogden, J., Kemp, F., Troiano, R., Jortner, B., Timpone, C., and Giuliani, D., Effect of mercuric chloride and methylmercury chloride exposure on tissue concentrations of six essential minerals, *Environ. Res.,* 21, 350, 1980.
279. Valberg, L. and Flanagan, P., Intestinal absorption of iron and chemically related metals, in *Biological Aspects of Metals and Metal-Related Diseases,* Sarkar, B., Ed., Raven Press, New York, 1983, p. 41.
280. Hamilton, D. L., Interrelationships of lead and iron retention in iron-deficiency mice, *Toxicol. Appl. Pharmacol.,* 46, 651, 1978.
281. Hambridge, K. M., Casey, C. E., Krebs, N. F., Zinc, in *Trace Elements in Human and Animal Nutrition,* Vol. 2, 5th ed., Mertz, W., Ed., Academic Press, New York, 1986, p. 137.
282. Mills, C. F., The biological significance of zinc for man: problems and prospects, in *Zinc in Human Biology,* Mill, C. F., Ed., Springer-Verlag, Great Britain, 1989, p. 371.
283. Sandstead, H., Zinc nutrition in the United States, *Am. J. Clin. Nutr.,* p. 26, 1251, 1973.
284. Solomons, N., Biological availability of zinc in humans, *Am. J. Clin. Nutr.,* 35, 1048, 1982.
285. White, H. and Gynne, T., Utilization of inorganic elements by young women eating iron fortified foods, *J. Am. Diet. Assoc.,* 59, 27, 1971.
286. Hall, A., Young, B., and Bremner, I., Intestinal metallothionein and the mutual antagonism between copper and zinc in the rat, *J. Inorg. Biochem.,* 11, 57, 1979.
287. Fischer, P., Giroux, A., and L'Abbe, M., The effect of dietary zinc on intestinal copper absorption, *Am. J. Clin. Nutr.,* 34, 1670, 1981.
288. Miller, C., Dietary interactions involving the trace elements, *Ann. Rev. Nutr.,* 5, 173, 1985.
289. Cousins, R., Metal elements and gene expression, *Annu. Rev. Nutr.,* 14, 449, 1994.
290. Magee, A. and Matrone, G., Studies on the growth, copper metabolism and iron metabolism of rats fed high levels of zinc, *J. Nutr.,* 72, 233, 1960.
291. Smith, S. and Larson, E., Zinc toxicity in rats. Antagonistic effect of copper and liver, *J. Biol. Chem.,* 163, 29, 1946.
292. Van Reen, R., Effects of excessive dietary zinc in the rat and the interrelationship with copper, *Arch. Biochem. Biophys.,* 46, 337, 1953.
293. Klevy, L., Pond, W., and Medeiros, D., Decreased high density lipoprotein cholesterol and apoprotein A-1 in plasma and ultra structural pathology in cardiac muscle of young pigs fed a diet high in zinc, *Nutr. Res.,* 14, 1227, 1994.
294. Ogiso, T., Moriyama, K., Sasaki, S., Ishimura, Y., Minato, A., Inhibitory effect of high dietary zinc on copper absorption in rats, *Chem. Pharm. Bull.,* 22, 55, 1974.
295. Oestreicher, P. and Cousin, R., Copper and zinc absorption in the rat: mechanism of mutual antagonism, *J. Nutr.,* 115, 159, 1985.
296. Reinstein, N., Lonnerdal, B., Keen, C., and Hurley, L., Zinc-copper interactions in the pregnant rat. Fetal outcome and maternal and fetal zinc, copper and iron, *J. Nutr.,* 114, 1266, 1984.
297. Murthy, L., Klevay, L., and Petering, H., Interrelationships of zinc and copper nutriture in the rat, *J. Nutr.,* 104, 1458, 1974.
298. O'Dell, B., Reeves, P., and Morgan, R., Interrelationships of tissue copper and zinc concentrations in rats nutritionally deficient in one or the other of these elements, in *Trace Substances in Environmental Health,* Vol. X, Hemphill, D., Ed., University of Missouri, Columbia, 1976, p. 411.
299. Sandstead, H., Requirements and toxicity of essential trace elements, illustrated by zinc and copper, *Am. J. Clin. Nutr.,* 61(Suppl.), 621S, 1995.
300. Prasad, A., Brewer, C., Schoomaker, E., and Rabbani, P., Hypocupremia induced zinc therapy in adults, *J. Am. Med. Assoc.,* 240, 2166, 1978.
301. Abdulla, M., Copper levels after oral zinc, *Lancet,* 1, 616, 1979.
302. Dureke, T., Gairard, A., Gueguen, L., Hercberg, S., and Mareschi, J., Average mineral and trace element content in daily adjusted menus (DAM) of French adults, *J. Trace Elem. Electrolytes Health Dis.,* 2, 79, 1988.
303. Halsted, J. and Smith, J., Plasma zinc in health and disease, *Lancet,* 1, 322, 1970.
304. Van Campen, D., Copper interference with the intestinal absorption of zinc-65 by the rat, *J. Nutr.,* 97, 104, 1969.
305. Pratt, W., Omdahl, J., and Sorenson, J., Lack of effects of copper gluconate supplementation, *Am. J. Clin. Nutr.,* 42, 681, 1985.

306. Parizek, J., The destructive effect of cadmium ion on testicular tissue and its prevention by zinc, *J. Endocrinol.*, 15, 56, 1957.
307. Ferm, V. and Carpenter, S., Teratogenic effect of cadmium and its inhibition by zinc, *Nature,* 216, 1123, 1967.
308. Whanger, P., Factors affecting the metabolism of nonessential metals in food, *Nutr. Toxicol.,* 1, 163, 1982.
309. Banis, R., Pond, W., Walker, E., and O'Connor, J., Dietary cadmium, iron, and zinc interactions in the growing rat, *Proc. Soc. Exp. Biol. Med.,* 130, 802, 1969.
310. Deagen, J., Oh, S., and Whanger, P., Biological functions of metallothionein. VI. Interaction of cadmium and zinc, *Biol. Trace Element Res.,* 2, 65, 1980.
311. Rosa, I., Henry, P., and Ammerman, C., Interrelationship of dietary phosphorus, aluminum and iron on performance and tissue mineral composition in lambs, *J. Anim. Sci.,* 55, 1231, 1982.
312. Jensen, L., Precipitation of a selenium deficiency by high dietary levels of copper and zinc, *Proc. Soc. Exp. Biol. Med.,* 149, 113, 1975.
313. Greger, J. L. and Johnson, M. A., Effect of dietary tin on zinc, copper, and iron utilization by rats, *Food Cosmet. Toxicol.,* 19, 163, 1981.
314. Johnson, M. A. and Greger, J. L., Absorption, distribution and endogenous excretion of zinc by rats fed various dietary levels of inorganic tin and zinc, *J. Nutr.,* 114, 1843, 1984.
315. Johnson, M. and Greger, J., Tin, copper, iron and calcium metabolism of rats fed various dietary levels of inorganic tin and zinc, *J. Nutr.,* 115, 615, 1985.
316. Johnson, M. A., Baier, M. J., and Greger, J. L., Effects of dietary tin on zinc, copper, iron, manganese and magnesium metabolism of adult males, *Am. J. Clin. Nutr.,* 35, 665, 1982.
317. Valberg, L. S., Flanagan, P. R., and Chamberlain, M. J., Effects of iron, tin, and copper on zinc absorption in humans, *Am. J. Clin. Nutr.,* 40, 536, 1984.
318. Solomons, N. W., Marchini, J. S., Duarte-Favor, R. M., Vannuchi, H., and Dutra de Oliveira, J. E., Studies on the bioavailability of zinc in humans: intestinal interaction of tin and zinc, *Am. J. Clin. Nutr.,* 37, 566, 1983.
319. O'Dell, B. L., Copper, in *Present Knowledge in Nutrition,* 6th ed., Brown, M., Ed., The Nutrition Foundation, Washington, D.C., 1990, p. 261.
320. Mason, K., A conspectus of research on copper metabolism and requirements of man, *J. Nutr.,* 109, 1979, 1979.
321. Graham, G. G. and Cordano, A., Copper depletion and deficiency in malnourished infants, *Johns Hopkins Med. J.,* 124, 139, 1969.
322. Sivasubramanian, K. N. and Henkin, R. I., Behavioral and dermatologic changes and low serum zinc and copper concentrations in two premature infants after parenteral alimentation, *J. Pediatr.,* 93, 847, 1978.
323. Bremmer, I., Cadmium toxicity — nutritional influences and the role of metallothionein, *World Rev. Nutr. Diet.,* 32, 165, 1978.
324. Yang, M. and Yang, S., Effect of molybdenum supplementation on hepatic trace elements and enzymes of female rats, *J. Nutr.,* 119, 221, 1989.
325. Van Campen, D., Effects of zinc, cadmium, silver, and mercury on the absorption and distribution of copper-64 in rats, *J. Nutr.,* 88, 125, 1966.
326. Hurley, L. and Keen, C., Manganese, in *Trace Elements in Human and Animal Nutrition,* Vol. 1., Mertz, W., Ed., Academic Press, Orlando, FL, 1987.
327. Hodge, H. and Smith, F., *Fluoride Chemistry,* Academic Press, New York, 1965.
328. Hodge, H. and Smith, F., Minerals: fluorine and dental caries, in *Dietary Chemicals vs. Dental Caries,* Gould, R., Ed., Advances in Chemistry Series No. 94, American Chemical Society, Washington, D.C., 1970.
329. Ophaug, R., Fluoride, in *Present Knowledge in Nutrition,* 6th ed., Brown, M., Ed., The Nutrition Foundation, Washington, D.C., 1990, p. 274.
330. Said, A., Slagsvold, P., Bergh, H., Laksesvela, B. High fluoride water to wether sheep maintained in pens, *Nord. Vet. Med.,* 29, 172, 1977.
331. Spencer, H., Kramer, L., Norris, C., and Wiatrowski, E., Effect of aluminum hydroxide on fluoride metabolism, *Clin. Pharmacol. Ter.,* 28, 529, 1980.
332. Webb, M. and Magos, L., Cadmium-thionein and the protection by cadmium against the nephrotoxicity of mercury, *Chem. Biol. Interact.,* 14, 357, 1976.
333. Oh, S., Whanger, P., and Deagen, J., Tissue metallothionein: dietary interaction of cadmium and zinc with copper, mercury, and silver, *J. Toxi. Environ. Health,* 7, 547, 1981.
334. Sorenson, J. R. J., Campbell, I. R., Tepper, L. B., and Lingg, R. D., Aluminum in the environment and human health, *Environ. Health Perspect,* 8, 3, 1974.
335. Schlettwein-Gsell, D. and Mommsen-Straub, S., Spurenelemente in Lebensmitteln. XII. Aluminium, *Int. J. Vitam. Res.,* 43, 251, 1973.
336. Slanina, P., Falkjborn, Y., Frech, W., and Cedergren, A., Aluminum concentrations in the brain and bone of rats fed citric acid, aluminum citrate or aluminum hydroxide, *Food Chem. Toxicol.,* 22, 391, 1984.
337. Slanina, P., Frech, W., Bernhardson, A., Cedergren, A., and Mattson, P., Influence of dietary factors on aluminum absorption and retention in the brain and bone of rats, *Acta Pharmacol. Toxicol.,* 56, 331, 1985.
338. King, S., Savory, J., and Wills, M., The clinical biochemistry of aluminum, *CRC Crit. Rev. Clin. Lab. Sci.,* 14, 1, 1981.

339. Drueke, T., Dialysis osteomalacia and aluminum intoxication, *Nephronology,* 26, 207, 1980.
340. Chan, Y., Alfrey, A., Posen, S., Lissner, D., Hills, E., Dunstan, C., and Evans, R., Effect of aluminum on normal and uremic rats: tissue distribution, vitamin D metabolites and quantitative bone histology, *Calcif. Tissue Int.,* 35, 344, 1983.
341. Hoekstra, W., Biochemical function of selenium and its relation to vitamin E., *Fed. Proc.,* 4, 1670, 1975.
342. Levander, O. and Burk, R., Selenium, in *Present Knowledge in Nutrition,* 6th ed., Brown, M., Ed., The Nutrition Foundation, Washington, D.C., 1990, p. 268.
343. Parizek, J., and Ostadalova, I., The protective effect of small amounts of selenite in sublimate intoxication, *Experientia,* 23, 142, 1967.
344. Parizek, J., Ostadalova, I., Kalouskova, J., Babicky, A., and Benes, J., The detoxifying effects of selenium interrelations between compounds of selenium and certain metals, in *Newer Trace Elements in Nutrition,* Mertz, W. and Cornatzer, W., Eds., Marcel Dekker, New York, 1971, p. 85.
345. Ganther, H. and Sunde, M., Effect of tuna fish and selenium on the toxicity of methyl mercury: a progress report, *J. Food Sci.,* 39, 1, 1974.
346. Ganther, H., Goudie, C., Sunde, M., Kopecky, M., Wagner, P., Sang-Hwan, O., and Hoekstra, W., Selenium: relation to decreased toxicity of methyl mercury added to diets containing tuna, *Science,* 173, 112, 1972.
347. Parizek, J., Kalouskova, A., Babicky, A., Benes, J., and Pavlik, L., Interactions of selenium with mercury, cadmium and other toxic metals, in *Trace Element Metabolism in Animals,* Vol. 2, Hoekstra, W., Suttie, J., Ganther, H., and Mertz, W., Eds., University Park Press, Baltimore, MD, 1974, p. 119.
348. Carmichael, N. and Fowler, B., Effects of separate and combined chronic mercuric chloride and sodium selenate administration in rats: histological, ultrastructural, and X-ray microanalytical studies of liver and kidney *J. Environ. Pathol. Toxicol.,* 3, 399, 1980.
349. Ganther, H., Modification of methyl mercury toxicity and metabolism by vitamin E and selenium, *Environ. Health Perspect.,* 25, 71, 1978.
350. Calloway, D. H. and McMullen, J. J., Fecal excretion of iron and tin by men fed soured canned foods, *Am. J. Clin. Nutr.,* 18, 1, 1966.
351. Tipton, I. H., Steward, P. L., and Dickson, J., Patterns of elemental excretion in long-term balance studies, *Health Phys.,* 16, 455, 1969.
352. Greger, J. L. and Lane, H. W., The toxicology of dietary tin, aluminum, and selenium, *Nutr. Toxicol.,* 2, 223, 1987.
353. Greger, J. L., Smith, S. A., Johnson, M. A., and Baier, M. J., Effects of dietary tin and aluminum on selenium utilization by adult males, *Biol. Trace Elem. Res.,* 4, 269, 1982.
354. Cohen, A., The minerals in your diet: an amazing balancing act, *Environ. Nutr.,* 18, 1, 1995.

CHAPTER 10

Naturally Occurring Orally Active Dietary Carcinogens

Joseph A. Scimeca

CONTENTS

0-8493-4493-X/97/$0.00+$.50

1 INTRODUCTION

In an attempt to provide the reader with handy information and key references in a comprehensible and concise manner on such a broad topic, it is necessary to provide some definitions and assumptions that were used to narrow the subject area covered. To begin with, the title of the chapter contains some terms that are open to interpretation and hence should be defined. "Natural" in the context of this chapter will include all compounds that are found in the food supply, exclusive of synthetic compounds that are either intentionally added or incidentally found in food. This would exclude many synthetic direct and indirect food additives, pesticides, animal drug residues, and most of the so-called environmentally persistent chemicals, such as DDT. Hence, "naturally occurring" includes all nonsynthetic substances that occur through nonanthropogenic activity (i.e., plant- or fungi-derived contaminants) *and* those that are formed endogenously in the food as a result of natural chemical procesess caused by some type of human activity (i.e., cooking). "Orally active" is used in the context of carcinogenesis and excludes the consideration of all compounds that are bereft of evidence of carcinogenic activity via the oral route of administration. Hence, there are many compounds which may be found in food that have carcinogenic activity upon some parenteral route of administration, such as inhalation (e.g., acetaldehyde), but either lack carcinogenic activity upon oral administration or have not been so tested. Since carcinogenic metabolic pathways are so heavily dependent on the route of administration, oral exposure is the only relevant route for food. Given that cancer formation is a multifactorial, multistage process, defining the term "carcinogen" is not so straightforward. For the purposes of this chapter, a rather simple definition was employed. A "carcinogen" is any solitary compound (i.e., not a mixture or physicochemically unidentified substances), exposure to which is capable of increasing the incidence of either benign or malignant neoplasm. Excluded from consideration are all the "co-carcinogens", "promoters", and other amplifiers of the carcinogenic process. For example, certain macro-components of the diet such as alcohol and certain fats are excluded, as well as naturally occurring trace compounds that can occur in the diet, such as the diterpene esters from certain plants. Additionally, compounds that are known to exert their carcinogenic action via a secondary mechanism (i.e., nongenotoxic carcinogens), such as hormone disruptors (e.g., phytoestrogens and thioureas), are also excluded from consideration. In general, effects of nongenotoxic carcinogens occur at high doses, above some threshold level that exceeds human exposure. Finally, somewhat arbitrarily, certain inorganic ions and compounds that are thought to produce cancer in humans (e.g., arsenic) are excluded from consideration.

The general format for each section in the chapter loosely adheres to the following: (1) chemical identification of the specific compound, which may include its structure; (2) dietary natural source(s) of the compound, mostly limited to the U.S.; (3) available information on source level(s) of the compound; (4) evidence of carcinogenicity in laboratory animals, including species and primary carcinogenic target organ identification; (5) mention of mutagenicity evidence, primarily from the Ames' *Salmonella* assay due to its universal reproducibility;

(6) mention of epidemiological carcinogenicity evidence; and (7) the International Agency for Research into Cancer (IARC) carcinogen classification, if available. Each section ends with a brief statement on the relevance of the experimental data to human carcinogenic risk.

2 ALKENYLBENZENES

Safrole is one of the many allylic and propenylic benzene derivatives which occur naturally in many spices, herbs, and vegetables.[1] Additional characteristics of this class of compounds are ring methoxy and/or ring methylenedioxy substitutions.[1] Most of these compounds are found in the essential oils of plants (i.e., the oils obtained from plant materials by steam distillation, solvent extraction, or physical expression).[2,3] Many of these compounds (including synthetic versions) are used as flavor additives and fragrances. A good review of the physical and chemical properties of safrole and related compounds can be found by Woo et al. (1988).[4]

About 30 alkenylbenzenes and related compounds have been found naturally.[5] Eleven have been assessed for carcinogenicity, with six testing positive[6-13] (see Table 1 for chemical name, structure, natural sources with associated references, and references for carcinogenicity studies). The earliest evidence of tumor induction by safrole came in 1961 from three independent investigations, which revealed hepatic tumors (or preneoplastic changes) in rats receiving high doses of safrole in the diet (0.5 to 1%).[14-16] This discovery contributed to the banning of its use (along with isosafrole and the synthetic compound, dihydrosafrole) as a food additive in the U.S.,[17] which included its use as a flavoring agent for root beer (at levels up to 20 mg/l).[18]

Safrole is a major component (from 80 to 93%, depending on the plant variety) of sassafras oil, and is also present in lesser amounts (usually <1 to 10% of the oil) in sweet basil, nutmeg, mace, star anise, ginger, black pepper, and cinnamon leaf.[5] Although banned as a food additive, sassafras bark is still being used as a herbal tea or folk medicine. According to Segleman et al.,[19] a single, commercially bought herbal tea bag contained the equivalent of 200 mg safrole. β-Asarone was once used as a bitter flavor in liqueurs and vermouth at levels up to 10 to 30 ppm,[20] prior to banning by the Food and Drug Administration (FDA) in 1967. Its food use is still permitted in some countries,[21] and calamus drugs containing β-asarone are currently being used in Europe. Estragole, the major constituent of the essential oils of tarragon and basil, has been used in gourmet vinegars[22] and in certain foods such as candy, chewing gum, and ice cream at levels ranging from 2 to 50 ppm.[5] Sesamol is a minor component of sesame seed oil, being found at 4.3 to 45 ppm, depending on the processing method.[23]

Systematic structure-activity investigations of the carcinogenicity of safrole and related compounds were conducted by the Millers and coworkers (see review by Miller and Miller, 1983).[24] They found mice were more susceptible than rats, and that the liver was the carcinogenic target organ for safrole, isosafrole, and estragole;[6,7,9,11] however, the carcinogenic potency of safrole was relatively low, with liver adenomas usually produced in rats and mice fed relatively high dietary levels (0.5 to 1%). In 1976, the IARC determined that safrole and isosafrole were hepatocarcinogenic in rodents.[25] β-Asarone was unusual in that dietary administration produced leiomyosarcomas of the small intestine of the rat,[12] and sesamol induced rodent forestomach tumors.[26] Mutagenicity of safrole and related compounds was determined to be of low activity, low potency, poor consistency, and with little relationship to their carcinogenic activity.[4]

The evidence linking human exposure to naturally occurring alkenylbenzene compounds and potential carcinogenic risk is rather weak. Suggestions that exposure to these compounds may be associated with esophageal cancer are unsupported. Since these compounds occur in the food supply at low part-per-million levels, and in light of their relatively weak mutagenic

TABLE 1
Orally Active Carcinogenic Alkenylbenzenes

Common Name	Chemical Name	Structure	Ref. (Carcinogenicity)	Natural Sources	Ref. (Sources)
Safrole	1-Allyl-3,4-methylenedioxybenzene	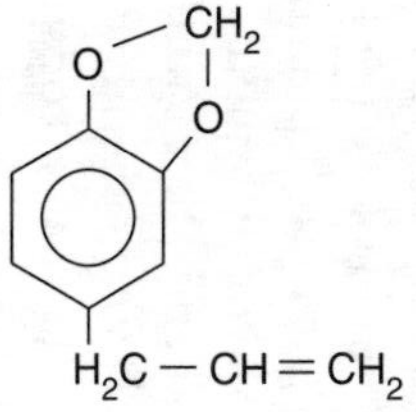 O, CH_2, O; $H_2C-CH=CH_2$	6–9	Sassafras, sweet basil, cinnamon, black pepper, ginger, mace, nutmeg	5, 25
Isosafrole	3,4-Methylenedioxy-1-propenylbenzene	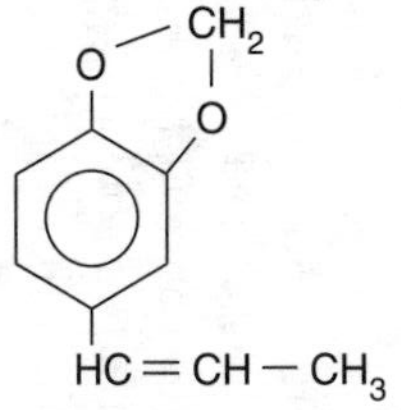 O, CH_2, O; $HC=CH-CH_3$	10, 11	Rarely found in essential oils of some spices; generally similar distribution as safrole	5, 25, 27
Estragole	1-Allyl-4-methoxybenzene	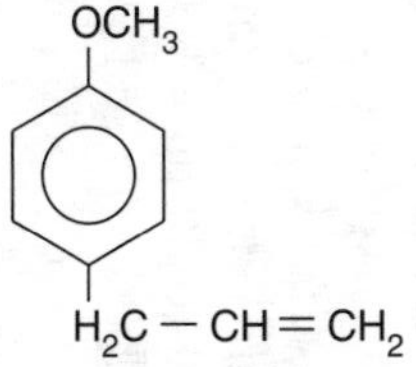 OCH_3; $H_2C-CH=CH_2$	9	Tarragon, sweet basil, anise	5

Methyl eugenol	1-Allyl-3,4-dimethoxybenzene	OCH_3, OCH_3, $H_2C-CH=CH_2$	9	Sweet bay, cloves, lemon grass, black pepper	9, 28
Sesamol	1-Hydroxy-3,4-methylenedioxybenzene	O, CH_2, O, OH	13, 26	Sesame oil	29
β-Asarone	*cis*-1-Propenyl-2,4,5-trimethoxybenzene	OCH_3, OCH_3, CH_3O, $C(H)=C(H)-CH_3$	12, 30	Oil of calamus	12, 30

and carcinogenic activity, it is likely that they make a negligible contribution in the etiology of human cancers.

3 ALLYL ISOTHIOCYANATE

Naturally occurring allyl isothiocyanate (AITC) is found in a large number of commonly consumed vegetables. Most are found in the *Brassica* genus (which is in the Cruciferae family), which includes broccoli, Brussels sprouts, cabbage, cauliflower, collards, kale, mustard, rutabaga, and turnips.[31] AITC is also found in garlic, onion, and horseradish.[31] Normally, AITC and other isothiocyanates are found in the seeds, roots, and leaves of plants as glycosides, which must be enzymatically hydrolyzed to release their active form.[31] For example, the natural form of AITC in mustard seeds is the glucosinolate sinigrin. During processing, cooking, or even maceration, in the presence of water and the enzyme myrosinase, AITC is released[32] and constitutes more than 90% of the resulting volatile mustard seed oil.[33] Food use of synthetically prepared AITC and naturally occurring volatile oil of mustard is quite extensive as flavoring agents at low concentrations (usually not greater than 88 ppm) in pickled products, condiments, and spice flavors.[33] At least 50 chemically different glucosinolates have been identified in the nearly 300 plants in the Cruciferae family.[34]

AITC (food grade, greater than 93% purity) was tested for carcinogenicity by gastric intubation in mice and rats under the National Cancer Institute/National Toxicology Program (NCI/NTP) Bioassay Program.[35] No evidence of tumors was seen in mice, but an increased incidence of urinary bladder tumors (a rare rodent neoplasm) was observed in male rats only. Equivocal evidence of mammary tumors was seen in female rats. Evidence of genotoxicity in short-term tests has been evaluated by the IARC and was found to be limited.[36] Based on the available *in vivo* and *in vitro* data, the NTP and IARC both concluded that there was limited evidence of the carcinogenicity of AITC in experimental animals.[36,37] In the absence of epidemiological studies, confirmatory animal carcinogenicity studies, or mechanistic data, it would be premature to assess the carcinogenicity of dietary AITC in humans based on the limited *in vitro* and *in vivo* animal data available.

4 BRACKEN FERN TOXINS

Bracken fern (BF) (*Pteridium aquilinum*) is sporadically distributed throughout the world, with its consumption, which occurs in many different forms due to varied preservation methods,[38] limited mostly to Japan, New Zealand, Canada, and northeastern U.S.[39] Despite evidence of carcinogenicity for nearly 40 years, consumption remains high. Attempts have been made to reduce its carcinogenicity by boiling, but data indicate this is only partially effective in reducing its carcinogenicity; however, complete reduction has been shown with intensive alkaline treatment or thorough washing in the preparation of the flour.[39,40] Indirect consumption occurs via milk and dairy products in some countries, with experimental evidence demonstrating the transference of the carcinogenic factor(s) into cow's milk.[41,42]

The earliest evidence for the carcinogenicity of BF came via reports of urinary bladder neoplasia in cattle fed bracken fern for extended time periods.[43] Subsequent controlled experiments with laboratory animals[44-47] and cows[48,49] fed bracken fern confirmed these findings. Initially, the small intestine (specifically, the ileum) and the urinary bladder were identified as the target organs, although additional sites of tumorigenicity have since been demonstrated (see Table 2).[50-55] Attempts to isolate and characterize the carcinogenic agent(s) in bracken fern focused on three compounds: quercetin, shikimic acid, and ptaquiloside (Table 3). Upon dietary administration to rodents, each compound produced tumors in the

TABLE 2

Carcinogenic Target Organs for Bracken Fern (BF), Ptaquiloside (PT), Quercetin (QT), and Shikimic Acid (SA)

Species	Small Intestine	Cecum	Large Intestine	Urinary Bladder	Mammary Gland	Lung	Blood Cells	Liver/Bile Duct	Various Sites	Refs.
Rat	BF, PT, QT	BF		BF, PT, QT	BF, PT					44–47, 60–61, 56–58
Mouse	BF					BF	BF		SA[b]	50–52, 59
Hamster	BF	BF								51
Guinea pig	BF			BF						51
Quail	BF[a]	BF[a]	BF[a]							51
Toad	BF							BF		53
Cattle				BF						43, 45, 48, 49
Sheep	BF			BF						51, 55

[a] Hot ethanol extract of BF.

[b] Six glandular stomach tumors, four leukemias, one lung tumor.

intestine and urinary bladder that were grossly and histomorphologically similar to those in BF-fed rats;[56-61] however, the evidence for quercetin and shikimic acid is controversial since subsequent studies failed to confirm ileal and urinary bladder tumors in other rat strains or other species (see Section 6 for more on quercetin).[62-67] On the other hand, the evidence for ptaquiloside as the principal carcinogenic of BF is strong but not conclusive, since confirmatory studies await to be done.

Detailed examination of BF isolates for mutagenicity using the Ames' *Salmonella* assay and mammalian cells did not reveal any significant activity due to shikimic acid.[68] In contrast, Van der Hoeven et al.[69] determined that ptaquiloside was a potent *Salmonella* mutagen, accounting for over half of the mutagenic activity of methanol extracts of BF. Quercetin is also a very potent bacterial mutagen.[70] Epidemiological evidence indicating that the high incidence of stomach cancer in Japan may be partially due to BF is confounded by poor exposure data and the presence of other environmental influences.[71] The IARC has evaluated BF and determined the carcinogenicity evidence to be inadequate in humans and sufficient in animals (classified as a Group 2B carcinogen).[72] Although the putative carcinogenic principle of BF has been identified as ptaquiloside, its possible role in the etiology of human cancer awaits further elucidation.

5 CINNAMIC ACID DERIVATIVES

Cinnamic acid derivatives are phenolic acids that usually occur in nature in the conjugated or esterified form, commonly with quinic acid or sugars. For example, chlorogenic acid, which is particularly abundant in coffee beans (up to 3.8%)[73] is an ester of caffeic and quinic acid and is commonly found in several isomeric and derivatized forms (Figure 1). Coumarin shares the same three carbon side-chain as cinnamic acid, but with the chain formed into a oxygen heterocycle, or lactone (Figure 1). This section will discuss the available evidence of carcinogenicity for caffeic acid and coumarin.

Caffeic acid, chlorogenic acid, and closely related compounds are present in a variety of vegetables, fruits, and seasonings.[74] Ester conjugates may be hydrolyzed upon ingestion, yielding a variable absorption of caffeic acid. Caffeic acid (free and conjugated) is abundant

TABLE 3
Possible Carcinogenic Agents in Bracken Fern (BF)

Compound	Structure	Chemical Name	Level in BF (mg/kg)[a]
Ptaquiloside	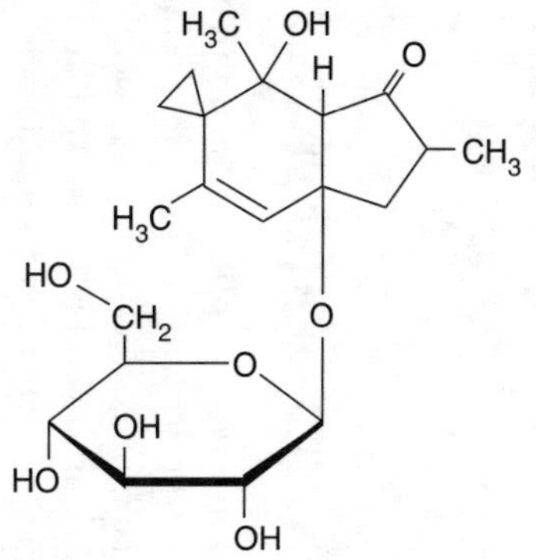	7′a-(β-D-Glucopyranosyloxy)-1′,3′a,4′,7′a-tetrahydro-4′-hydroxy-2′,4′,6′-trimethyspiro-[cyclopropane-1,5′-(5*H*)-inden]-3′-(2′*H*)-one	210–2400
Quercetin	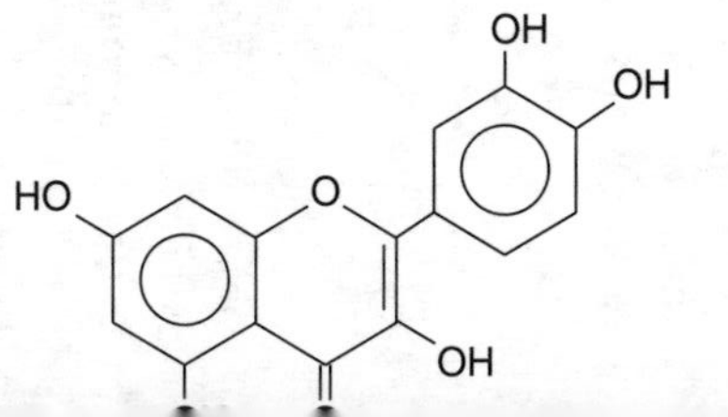	2-(3,4-Dihydroxyphenyl)-3,5,7-trihydroxy-4*H*-1-benzopyran-4-one	570

Shikimic Acid	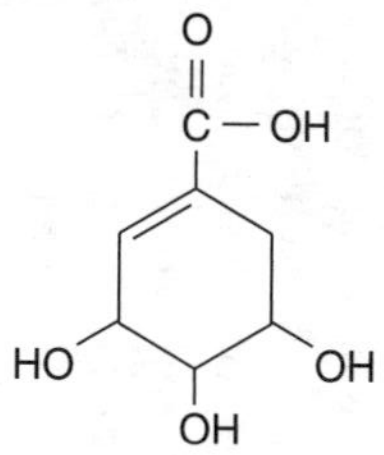 HO, OH, OH; C(=O)—OH (structure)	3,4,5-Trihydroxy-1-cyclohexene-1-carboxylic acid	1440

[a] Dry weight

Source: Adapted from IARC, *IARC Monographs on the Evaluation of the Carcinogenic Risk of Chemicals to Man,* Vol. 40, *Bracken Fern Toxins and Some of Its Constituents,* International Agency for Research on Cancer, Lyon, France, 1986, pp. 45–65.

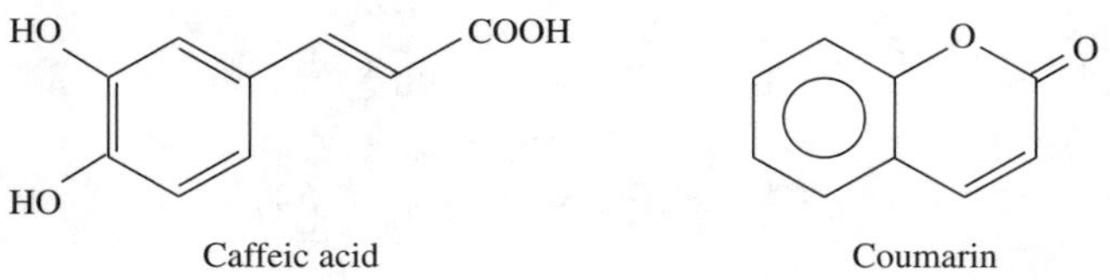

Caffeic acid Coumarin

FIGURE 1
Cinnamic acid derivatives: caffeic acid and coumarin.

(50 to 200 ppm) in apple, carrot, celery, cherry, eggplant, endive, grapes, lettuce, pear, plum, potato (peel), and in a number of herbs (e.g., thyme, basil, aniseed, caraway, rosemary, tarragon, marjoram, savory, sage, dill, and absinthe) at concentrations greater than 1000 ppm.[75,76] Beverages, such as coffee, apple juice, and wine are other significant sources.[75]

Caffeic acid, or 3,4-dihydroxycinnamic acid, is a natural anti-oxidant, with antitumorigenic activity in a number of animal models;[77,78] however, Hagiwara et al.[79] found that the chronic feeding of 2% caffeic acid to rats and mice produced a significant increase in tumors of the forestomach, kidney, and lung (the latter in male mice only and within historical range). Earlier work by the same laboratory using the same model and protocol found only forestomach tumors in rats and mice.[80] The authors concluded that caffeic acid is a weak, nongenotoxic carcinogen that appears to exert a tumorigenic response via secondary mechanisms, and the human carcinogenic risk may be negligible. *In vitro* studies determined that caffeic acid was not mutagenic in the Ames' *Salmonella* assay but did show some weak genotoxic activity in nonbacterial assays.[81-83] There is no epidemiological data available. Despite these limited and tenuous findings of carcinogenicity, the IARC determined that there was sufficient evidence in laboratory animals for the carcinogenicity of caffeic acid, and the overall evaluation was that caffeic acid was "possibly carcinogenic to humans" (i.e., Group 2B).[75] Until further evidence of carcinogenicity becomes available, concerns about the possibility of mechanisms and tumor manifestations unique to the rodent should prevent the consideration of caffeic as a human carcinogen.

Coumarin (2*H*-1-benzopyron-2-one) is present in a variety of plants (e.g., tonka beans, sweet clover, and woodruff) and essential oils (e.g., lavender).[84] Although banned in the U.S. in 1954, coumarin is still used in Europe as a food additive in some alcoholic beverages and wines.[85] Several carcinogenicity studies in rats,[86,87] hamsters,[88] and baboons[89] have found contradictory results. Coumarin is either inactive or weakly active in the Ames' *Salmonella* mutagenic assay.[90] In 1975, the IARC examined the available carcinogenicity evidence for coumarin and concluded that it was carcinogenic in rats;[91] however, in 1990, the NTP felt that the evidence was less than satisfactory and examined the carcinogenicity of coumarin by gavage in mice and rats.[92] Results in F334/N rats indicated that there was some evidence of carcinogenic activity in males and equivocal evidence in females, based on an increased incidences of renal tumors. In B6C3F1 mice, results indicated that there was some evidence of carcinogenicity in males based on lung tumors. Clear evidence of carcinogenicity was evident in female mice based on increased incidences of lung and liver tumors. Nonetheless, given the less than consistent results in the various carcinogenicity assays coupled with concerns about species differences in coumarin metabolism[93] and the lack of epidemiological data, the role that coumarin may have in the etiology of human cancer is uncertain.

6 FLAVONOIDS

Flavonoids, and their glycosides, are a broad group of related polyphenolic compounds that occur widely in the plant kingdom. These hydrophilic substances have an important influence on the flavor and taste of many foods. They possess the fundamental chemical structure of the parent compound, flavone (2-phenylbenzopyrone). Hydroxyl, methoxy, and saccharide moieties are attached at various locations. D-glucose is the most common sugar, but D-galactose, L-rhamnose, L-arabinose, D-xylose, and D-apiose, as well as some uronic acids, can also be found. The flavonoids can be subdivided as flavones, flavonols, flavanones, isoflavones, and catechins. About 2000 individual members of the flavonoid class have been described,[94] with most of the compounds occurring in the form of various glycosides. For example, quercetin, the most common flavonol, has over 70 glycosidic combinations that have been fully characterized, with many more partially identified. The flavonol glycosides, especially those of quercetin (e.g., rutin, isoquercitrin, and quercitrin), and kaempferol (e.g., astragalin and tiliroside), are found at significant concentrations in the edible portion of the majority of plant foods, e.g., fruits, berries, leaf and root vegetables, cereal grains, tea, coffee, and cocoa.[95] For a good review of the levels of quercetin (see Table 3 for structure) and kaempferol in various edible plants see Reference 96. A rough estimate of the total daily intake of flavonoids in the average American diet is about 1 g, with daily intake of quercetin, and its related forms, of about 25 mg per person.[97]

It was this large and widespread consumption of flavonoids that generated considerable interest when in 1977 it was reported that quercetin and kaempferol were mutagenic in the Ames' *Salmonella* assay.[98-100] Nearly 75 aglycones of various flavonoids have been tested in the Ames' *Salmonella* assay, with about one third positive (for a general review of the mutagenicity of flavonoids, see Brown[101] and Nagao et al.[102]). Despite considerable investigation into the structural requirements for mutagenicity in many of these compounds,[103] only a handful have been adequately tested for carcinogenicity. Quercetin and rutin (which is quercetin conjugated to rutinose) have been the focus of the majority of carcinogenicity studies, and aside from three studies by Pamukçu and colleagues,[104-106] all results have been negative. Interestingly, the initial investigations by Pamukçu, Erturk, and coworkers found intestinal and urinary bladder tumors in Norwegian rats,[104] but their later studies found that dietary quercetin significantly increased liver tumors in F344[105,106] and Sprague-Dawley rats.[106] Reasons for this discrepancy might be related to differences in the rat strains used and the dietary levels of quercetin; however, other long-term rat, mouse, and hamster feeding studies of quercetin (and rutin) have not confirmed this carcinogenic effect (see Table 4).

Despite the preponderance of negative studies of carcinogenicity for quercetin, the NTP undertook in 1992 a 2-year carcinogenicity feeding study in rats.[107] Evidence from this study found a dose-responsive increase in renal tumors in male rats only. Based on this, along with positive results from genotoxicity studies, the NTP determined there was some evidence of carcinogenicity. Similarly, the IARC concluded in 1983 there was limited evidence for the carcinogenicity of quercetin in animals, but without epidemiological data no determination of the carcinogenicity of quercetin to humans could be made.[96]

In contrast to the multiple investigations into the carcinogenicity of quercetin, the investigation of kaempferol has been limited to one study, which showed no effect.[108] It is mutagenic in the Ames' *Salmonella* assay, with other limited evidence of genotoxicity.[101] Overall, the IARC determined there was inadequate evidence to determine the carcinogenicity of kaempferol.[109] Epidemiological studies of the relationship between flavonoid intake and human cancer incidence have not been conducted, and hence the role that quercetin or kaempferol may have in the etiology of human cancer is uncertain.

7 FUROCOUMARINS

Furocoumarins encompass a group of secondary plant metabolites which surprisingly have a chemical relationship to aflatoxins by virtue of their common use of the coumarin ring structure. Furocoumarins can be divided into two classes, the larger of which are the linear furocoumarins that are derived from the most common member, psoralen. The second, smaller

TABLE 4

Oral Carcinogenesis Studies of Quercetin, Rutin, Quercitrin, and Kaempferol

Compound Tested	Dietary Level	Species	Result[a]	Ref.
Rutin	1%	Rat	Neg.	110
Quercetin	0.25–1%	Rat	Neg.	111
Quercitrin	0.25–1%	Rat	Neg.	111
Quercetin	0.1%	Rat	Intestinal and bladder tumors	104
Quercetin	2%	Mouse	Neg.	112
Quercetin	1, 5, 10%	Rat	Neg.	113
Rutin	5, 10%	Rat	Neg.	113
Quercetin	5%	Mouse	Neg.[b]	114
Quercetin	1, 4, 10%	Hamster	Neg.	115
Rutin	10%	Hamster	Neg.	115
Quercetin	1, 2%	Rat	Hepatomas and bile duct tumors	105
Quercetin	5%	Rat	Neg.	116
Quercetin	0.1%	Rat	Neg.	108
Kaempferol	0.04%	Rat	Neg.	108
Quercetin	1, 2%	Rat	Hepatic tumors and biliary adenomas	106
Quercetin	0.5%	Rat	Hepatomas	106
Rutin	2%	Rat	Hepatomas	106
Quercetin	1.25, 5%	Rat	Neg.	117
Quercetin	2%	Mouse	Neg.	118
Rutin	4%	Mouse	Neg.	118

[a] Neg. = negative; no difference from control group.
[b] Only lung tissue was examined

group is called angular furocoumarins because the furan ring is attached at an angle to the coumarin ring. Angelicin, which is the unsubstituted ring analog, defines this group. The physical and chemical properties of a number of furocoumarins can be found in Reference 119. Of the nearly 30 different psoralen derivatives that have been isolated from natural sources, 10 have tested for carcinogenicity. Of these, only the three that have been adequately tested by the oral route will be discussed.

Occurrence of furocoumarins is common in a number of plant species, especially those in the Rutaceae and Umbelliferae families. Human exposure to furocoumarins is predominately from limes (upwards of 97% due to lemon-lime flavored beverages), with a secondary source being celery. Parsley, parsnip, carrots, and other citrus fruits contribute a small amount to the diet. Estimated total dietary furocoumarin exposure ranges from 10 to 35 μg/kg body weight (b.w.), depending on a person's age, race, and gender (see Wagstaff[120] for a good review of dietary exposure).

Psoralen is found at relatively low concentrations (1 to 10 ppm) in celery, parsley, and parsnip.[120] 5-Methoxypsoralen (5-MOP, or bergapten) is found in the oil of bergamot (a Mediterranean citrus fruit) and is used as a flavoring agents in foods and as a fragrance in perfumes and other consumer products.[121] Other naturally occurring sources of 5-MOP are from the same Umbelliferae plants as psoralen, but in some cases at 2 to 3 times its concentrations.[120] Dietary intake of 8-methoxypsoralen (8-MOP, or methoxsalen) occurs through the ingestion of parsley and parsnip, with concentrations in parsnip root reported to be 26 to 29 ppm,[122] or up to 1100 ppm.[123] (See Figure 2 for chemical structures.)

Since the psoralens are phototoxic agents, carcinogenicity studies have generally been limited to topical application in combination with ultraviolet radiation.[124-126] Results of these studies clearly indicate that under these conditions psoralen and 5-MOP produced skin tumors. Mutagenicity and genotoxicity studies were also positive, but in most cases light activation was required.[119] Based on the evidence, the IARC has determined that there was

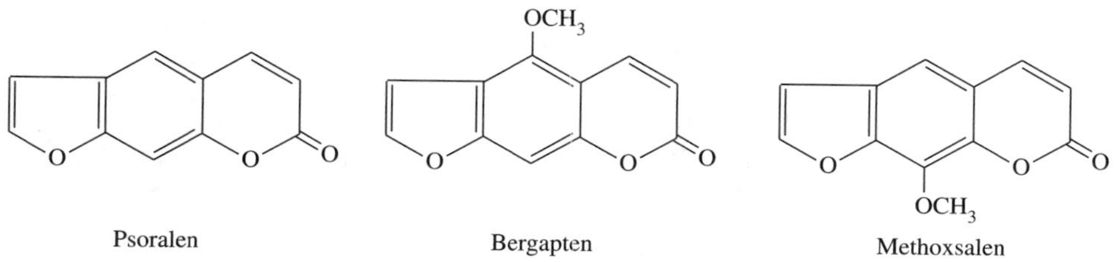

FIGURE 2
Some furocoumarins: psoralen, bergapten, and methoxsalen.

sufficient evidence of carcinogenicity in laboratory animals in combination with ultraviolet or solar-stimulated radiation, but there was inadequate evidence of their carcinogenicity in humans. In addition, the IARC deemed the evidence was inadequate to evaluate their systemic carcinogenicity in animals or humans. The IARC has also determined that there was insufficient evidence of carcinogenicity for other psoralens and angelicins.[119]

Three years after the IARC evaluation, the NTP completed a rat gavage study of 8-MOP (without light activation) and found renal and other tumors in male rats.[127] In addition, 8-MOP was mutagenic without light activation (but with metabolic activation) in the Ames' *Salmonella* assay and produced genotoxicity in other tests.[126] The NTP concluded that there was clear evidence of carcinogenicity for 8-MOP; therefore, based on this evidence, structure-activity relationship data, and the significant human dietary exposure to furocoumarins, there is valid concern about the human carcinogenic risk posed by 8-MOP and possibly other psoralens, under certain conditions.

8 HETEROCYCLIC AMINES

The seminal work in 1977 by Sugimura and colleagues[128,129] led to the realization that the burnt and browned material produced by the cooking of meat was highly mutagenic as measured by the Ames' *Salmonella* assay. The reader is referred to Eisenbrand and Tang[130] and Chen et al.[131] for good overall literature reviews. Further studies by Sugimura and others[132-136] showed that the pyrolysis products of certain amino acids and proteins, as well as the charred surface of meat, contained strongly mutagenic substances, which were eventually isolated and chemically defined.[137-141] These pyrolytic mutagens were found to be formed most efficiently under extreme cooking conditions involving high temperatures (300°F or greater), and, except for charred surfaces, they are normally not found in cooked foods in significant quantities. All these compounds are heterocyclic amines (HAs) (except Lys-P-1, which is a heterocyclic imine), and most can be further subdivided into pyridoindoles and pyridoimidazoles. They require metabolic activation to exert mutagenic activity, and some have demonstrated carcinogenic activity (see Table 5). Additional investigations using commercial beef extract, cooked ground beef, and broiled sardines revealed the presence of a new class of HAs which were formed under more realistic cooking conditions[142-151] (see Table 6). It became apparent that these HAs, which were further subdivided into three subgroups (i.e., imidazoquinolines, imidazoquinoxalines, and imidazopyridines),[148] contribute most of the mutagenic activity of cooked meat. Formation has been hypothesized to occur as result of a browning, or Mallard reaction, and involves creatinine or creatine, free amino acids, and monosaccharides.[152-156]

Heterocyclic amines have been found to be carcinogenic in various organs in mice, rats, and nonhuman primates in long-term feeding studies. Most of the research was done in Japan and generally involved dietary administration with one or two doses, usually at the MTD (maximum tolerated dose) and some large fraction of the MTD.[130,131] The amino acid

TABLE 5

Carcinogenic Heterocyclic Amines Formed by High Temperature Cooking

Structure	Scientific Name	Abbreviation	1° Carcinogenicity Target Site(s)[a]	Food Sources[b]	Carcinogenicity Ref.
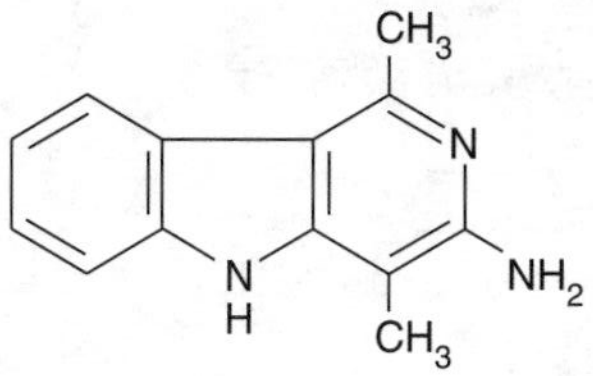	3-Amino-1,4-dimethyl-5*H*-pyrido[4,3-*b*]indole	Trp-P-1	M — liver; R — liver	Broiled, grilled sardines; broiled chicken; broiled, fried beef	191, 192
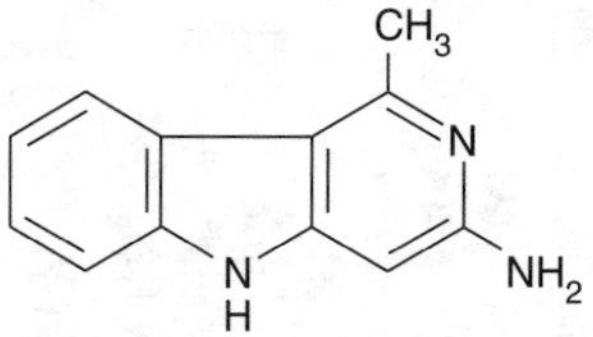	3-Amino-1-methyl-5*H*-pyrido[4,3-*b*]indole	Trp-P-2	M — liver; R — liver, clitoral gland, urinary bladder, mammary gland, hematopoietic system	Broiled, grilled sardines; broiled fried beef; boiled beef extract; broiled chicken; broiled mutton	193, 194
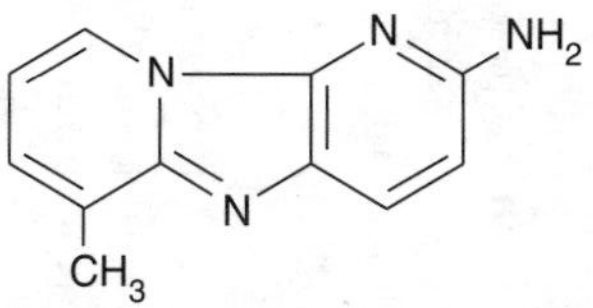	2-Amino-6-methyldipyrido[1,2-α:3,'2'-*d*]imidazole	Glu-P-1	M — liver, blood vessels; R — liver, small and large intestine, brain, clitoral gland, Zymbal gland	Broiled fish	195, 196

Structure	Name	Abbreviation	Target organs	Food sources	Ref.
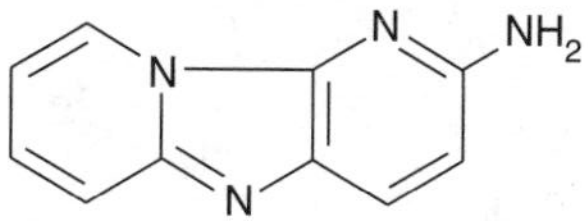	2-Aminodipyrido[1,2-α:3′,2′-*d*]-imidazole	Glu-P-2	M — liver, blood vessels; R — liver, small and large intestine, brain, clitoral gland, Zymbal gland	Broiled, grilled fish	195, 196
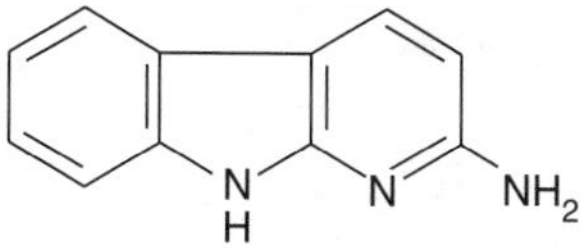	2-Amino-9*H*-pyrido[2,3-*b*]indole	AαC	M — liver, blood vessels	Grilled beef; broiled chicken; broiled mutton; baked, broiled, barbecued salmon; fried fish	195
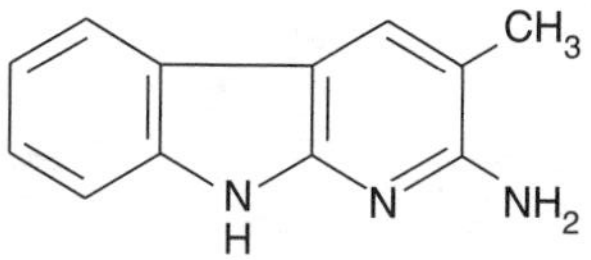	2-Amino-3-methyl-9*H*-pyrido[2,3-*b*]indole	MeAαC	M — liver, blood vessels	Grilled beef; broiled chicken; broiled mutton	195

[a] M = mouse; R = rat.
[b] Representative food sources.

TABLE 6

Carcinogenic Heterocyclic Amines Formed By Moderate Temperature Cooking

Structure	Scientific Name	Abbrev.	1° Carcinogenicity Target Site(s)[a]	Food Sources[b]	Carcinogenicity Ref.
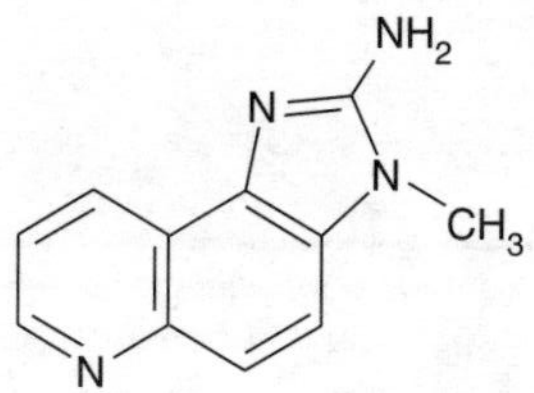	2-Amino-3-methyl-imidazo[4,5-*f*]quinoline	IQ	M — liver, forestomach, lung; R — liver, small and large intestine, mammary gland, Zymbal gland, clitoral gland, skin; P — liver	Broiled, fried beef; boiled beef extract; fried pork; broiled sardines; broiled salmon; fried fish; fried egg	161, 197–199
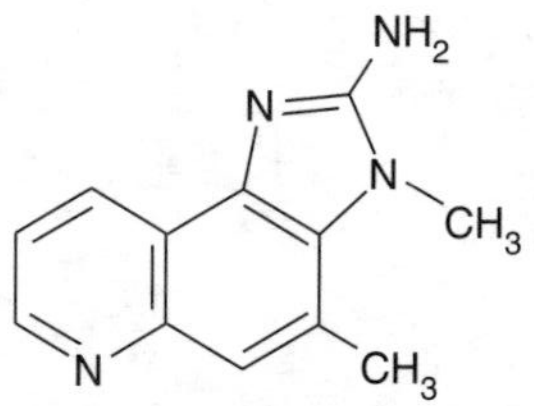	2-Amino-3,4-dimethyl-imidazo[4,5-*f*]quinoline	MeIQ	M — liver, forestomach; R — Zymbal gland, oral cavity, colon, skin, mammary gland	Fried beef; boiled beef extract; fried pork; broiled sardine; broiled salmon	200, 201
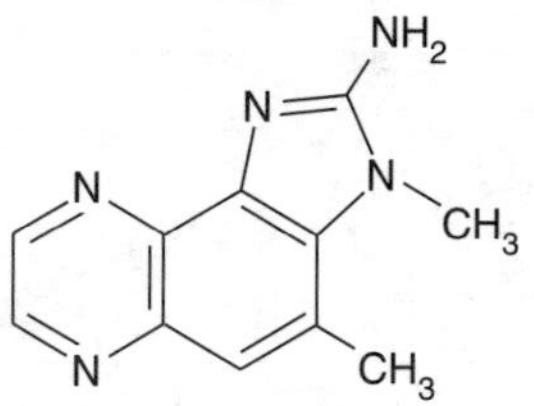	2-Amino-3,4-dimethyl-imidazo[4,5-*f*]quinoxaline	4-MeIQx	M — liver, lung, hematopoietic system	Broiled, fried beef; broiled, barbecued chicken; broiled mutton; boiled beef extract; fried pork; bacon; smoked, dried tuna; fried fish; baked, broiled, barbecued salmon	202, 203

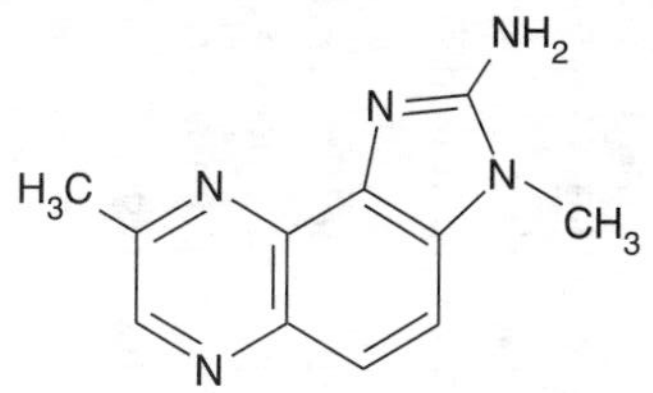	2-Amino-3,8-dimethyl-imidazo[4,5-*f*]quinoxaline	8-MeIQx	R — liver, Zymbal gland, skin	Broiled, fried beef; broiled, barbecued chicken; broiled mutton; boiled beef extract; fried pork; bacon; smoked, dried tuna; fried fish; baked, broiled, barbecued salmon	202, 203
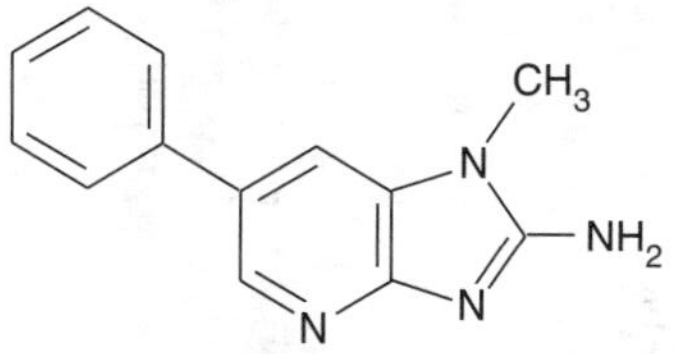	2-Amino-1-methyl-6-phenylimidazo[4,5-*b*]-pyridine	PhIP	M — lymph nodes, lung, spleen; R — colon, mammary gland	Fried beef; broiled chicken; broiled mutton; fried, barbecued pork; bacon; baked, broiled, barbecued salmon; fried fish	204–206

[a] M = mouse; R = rat, P = primate.
[b] Representative food sources.

pyrolysates (Trp-P-1, Trp-P-2, Glu-P-1, Glu-P-2, MeAαC, and AαC) have primarily produced tumors of the liver and blood vessels in mice and the liver and intestines in rats (see Table 5). The murine carcinogenicity target organ for the low-temperature-forming HAs (i.e., IQ, MeIQ, MeIQx, and PhIP) was mainly the liver, but other sites included the forestomach, lung, and hematopoietic system (see Table 6). Rats developed tumors in the liver, intestines, mammary gland, skin, and other organs. It appeared from these studies that females were more susceptible than males and that PhIP was uniquely nonhepatocarcinogenic in both rodent species (see reviews by Sugimura and Sato,[157] Ohgaki et al.,[158] Wakabayashi et al.,[159] and Munro et al.[160]). IQ was tested by gavage in macaque monkeys and produced hepatocellular carcinoma at doses much less than the MTD.[161] Based on tumor potency estimates (i.e., TD_{50} values), the carcinogenic potency of some HAs appears to be comparable to other carcinogens such as *N*-nitrosodimethylamine and dibenzo[*a,h*]anthracene. In evaluating the carcinogenicity of IQ, MeIQ, MeIQx, and PhIP, the IARC determined that there was inadequate evidence in humans but sufficient evidence in experimental animals.[162] The overall evaluation by IARC classified these compounds as being "probably carcinogenic to humans" (Group 2A).

The importance of cooking method, time, and temperature in the formation of HAs has been well demonstrated.[133,163-168] In general, frying or flame-broiling meats produced the greatest mutagenic activity and heterocyclic amine formation.[169-171] Deep frying, roasting, and baking produced a lesser response, and stewing, steaming, poaching, and microwave cooking showed little mutagenic activity.[172] Pariza and coworkers[135] studied the mutagenic activity of hamburgers fried at 143, 191, and 210°C and found activity remained low at all cooking times (4 to 20 min) at the lowest temperature; however, frying at 191 or 210°C for up to 10 min resulted in a considerably greater response. Other researchers have also confirmed the importance of cooking temperature and have generally found a sharp rise in the rate of mutagen formation from 140 to 180°C.[173] In general, frying or broiling of meats results in a 10- to 50-fold increase in the mutagenic activity as opposed to other methods of cooking.[171] For a summary of HA ranges in various cooked meats, see Table 7.

In 1982, Bjeldanes and colleagues[169,170] conducted an extensive survey of mutagen formation in the cooking of various sources of protein in the U.S. diet and found significant amounts in fried ground beef, broiled beef steak, ham, pork chops, bacon, and baked and broiled chicken. Other sources of protein, such as milk, cheese, tofu, and organ meats, produced insignificant amounts of mutagens upon cooking. Analysis of canned food indicated beef- and seafood-containing products exhibited mutagenic activity,[174] while crackers, corn flakes, rice cereal, bread crust, and toast exhibited low activity.[135] Cooking method has only a slight effect on the mutagenic activity for eggs, vegetables, and predominately carbohydrate-based foods.[171]

It should be noted that HAs are remarkably potent frameshift bacterial mutagens, particularly the imadazoquinolines and imadazoquinoxalines.[175] Data indicate a very large variation in HA potency (over 15,000-fold).[157,176] MeIQ is one of the most potent compounds ever tested in the Ames' *Salmonella* assay, with a potency of over 100 times that found for aflatoxin B_1.[177] However, results from carcinogenicity assays indicate that bacterial mutagenic potency does not correlate well with carcinogenic potency.[176] This would be expected based on differences in metabolic processes and toxicokinetics parameters found under *in vivo* conditions[160] and is perhaps an idiosyncratic susceptibility of the Ames' *Salmonella* assay for certain HAs. HAs also have other genotoxic effects that are comprehensively reviewed in Eisenbrand and Tang.[130]

Based on the totality of evidence, it is clear that certain HAs (IQ, MeIQ, MeIQx, and PhIP) are animal carcinogens and should be considered as presumptive human carcinogens, too. It should be realized that normal consumption levels of HAs are minute in comparison to their TD_{50} values (i.e., dose at which 50% of the animals develop tumors). Hence, it is

TABLE 7

Range of Heterocyclic Amine (HA) Concentrations in Cooked Meats and Fish[a]

Food	Cooking Method	Temp. (°C)	Time (min)	HA Range (Median) in ng/g Cooked Food					Ref.
				IQ	8-MeIQx	DiMeIQx	PhIP	AαC	
Ground beef	Fried	150–300	2–12	0–1.8 (0.1)	0 –10.8 (0.8)	0–9.35 (0.25)	0–21.8 (1.2)	—	146, 148, 159, 178, 179, 180, 181, 183, 184, 186, 188
Beef steak	Broiled or fried	190–225	3–6.5	—	0.5–8.3 (5.1)	0.1–2 (1.3)	0.6–48.5 (23.5)	1.2–8.9 (3.2)	159, 184, 187, 190
Ground pork	Fried	180–250	5	0.01–0.04	0.4–1.4	0.24–0.6	1.7–4.5	—	182, 189
Salmon	Fried	200	3–12	—	1.4–4.7 (3.7)	—	1.7–17 (14)	0–9 (4.6)	185
Salmon	Baked	200	20–40	—	0–4.6 (3.1)	—	0–18 (5.9)	0	185
Salmon	Broiled or barbecued	270	4–12	—	0	—	2–73 (69)	2.8–109 (73)	185

[a] Heterocyclic amine range concentrations are included contingent only on two or more independent determinations (most ranges include four determinations and some as many as 20).

likely that only an unusual diet that involved the large consumption of fried and/or broiled meat might pose a significant human carcinogenic risk.

9 HYDRAZINES AND THEIR DERIVATIVES

Hydrazines have long been known to organic chemists due to their high reactivity, but it was not until 1951 that the first example of a naturally occurring nitrogen-nitrogen double bond was reported. Since then, more than 40 compounds have been reported in bacteria, fungi, and higher plants.[207] The most common human dietary exposure to these compounds probably occurs through the ingestion of mushrooms and cycad seeds.

In the early 1960s, agaritine (β-N-[γ-L(+)glutamyl]4-hydroxy-methylphenylhydrazine) was identified and characterized as a naturally occurring hydrazine obtained from the commonly eaten cultivated mushroom, *Agaricus bisporus*, along with other *Agaricus* spp.[208,209] Subsequent experiments indicated that *p*-aminobenzoic acid and glutamic acid may be the precursors of agaritine.[210]

The carcinogenic potential of hydrazines was elegantly demonstrated in a series of experiments by Toth using nearly forty substituted hydrazine analogs. These studies clearly demonstrated the tumorigenic potential of these compounds in mice, rats, and hamsters, with neoplasia developing in intestines, brain, lungs, blood vessels, liver, mammary gland, and kidneys, among other sites. Carcinogenicity of agaritine was found to be negative in mice;[211] however, three breakdown compounds (4-hydroxymethylphenylhydrazine; N^2-[γ-L(+)-glutamyl]-4-carboxyphenylhydrazine; and 4-[hydroxymethyl] benzenediazonium ion) were found positive.[212-215] Another closely related nitrogen-nitrogen bond-containing compound (*p*-hydrazinobenzoic acid) was also shown to be carcinogenic.[216] It should be noted that these studies generally administered the metabolites as various salts, often by gavage or via the drinking water. Finally, chronic feeding of uncooked *A. bisporus* to rats produced tumors in several sites.[217] These results can be explained by the instability of agaritine and its likely near complete destruction through storage and cooking.[218] Unfortunately, this does not hold true for its metabolites, which on average are reduced only 25% by baking.[219] Analysis of *A. bisporus* revealed the presence of agaritine at 360 to 700 ppm,[209,220] while the breakdown products were found at much lower levels (0.6 to 42 ppm).[221-223]

A weak mutagenic effect of agaritine in the Ames' *Salmonella* assay was reported.[224] No epidemiological data is available. Based on the totality of evidence, the IARC determined there was insufficient evidence of carcinogenicity for agaritine, but there was limited evidence for the two metabolites of agaritine (or more precisely, their derivatives).[225] Lacking dose-response tumorigenicity studies and epidemiological studies of any sort, it is extremely difficult to assess the role that *A. bisporus* (or its active ingredients) may have in the etiology of human cancer.

Another example of a hydrazine-containing mushroom comes from the false morel (*Gyromitra esculenta*), which is widely eaten in northern Europe. From this mushroom, gyromitrin (acetaldehyde-*N*-formyl-*N*-methyl-hydrazone) was isolated, and, similar to agaritine, the carcinogenicity of its metabolites was demonstrated in a series of experiments by Toth.[226-228] However, unlike agaritine, gyromitrin administered in 52 weekly intragastric doses was found be tumorigenic in mice.[229] The gyromitrin content of dried false morel is between 0.05 and 0.3%, and *N*-methyl-*N*-formylhydrazine, a metabolite, is present at a concentration of 0.06%.[230] Gyromitrin, but not its metabolites, is destroyed by cooking or drying.[231] Gyromotrin was not mutagenic in bacteria, but a metabolite gave positive results.[232] No epidemiological data is available. Based on the totality of the evidence, the IARC determined that there was sufficient evidence for the carcinogenicity of gyromotrin in experimental animals.[233]

The Cycad family is composed of the surviving members of an ancient line of palm-like plants found throughout the tropical and subtropical regions of the world. Of the nine genera,

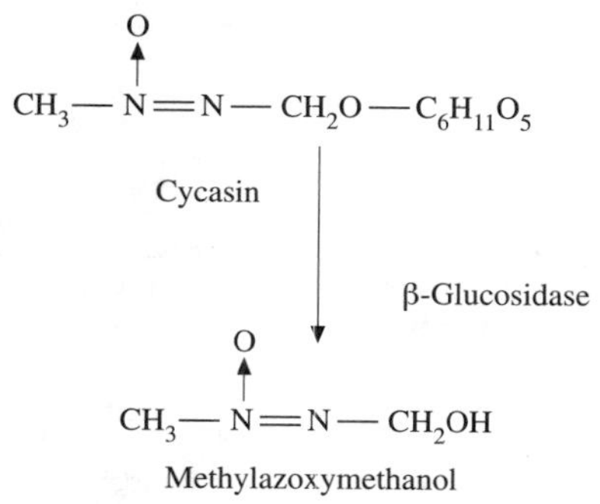

$$CH_3-\overset{\overset{\displaystyle O}{\uparrow}}{N}=N-CH_2O-C_6H_{11}O_5$$

Cycasin

$\downarrow$ β-Glucosidase

$$CH_3-\overset{\overset{\displaystyle O}{\uparrow}}{N}=N-CH_2OH$$

Methylazoxymethanol

FIGURE 3
Chemical structure of cycasin and its aglycone, methylazoxymethanol.

Cycas are the mostly widely distributed, with a region that includes the south Pacific centered around Indonesia and also Indochina and the west coasts of Africa and Australia.[234] The important species are *Cycas revoluta* and *C. circinalis*, with cycad meal used as a source of food in the form of a starch and sometimes in the bean paste, miso; however, in order to avoid toxicity, its preparation requires great care (for a review of the utilization of cycads, see Whiting[235]). Toxicity is caused by either or both of the glycosides, cycasin or macrozamin.[236] These two compounds differ only in carbohydrate moiety, with cycasin containing glucose and macrozamin containing primeverose. Both contain the active component, methyl-azoxymethanol (MAM) as the aglycone (for a good review of the chemistry and biological effects of cycasin and related compounds see Zedeck[237]). Cycasin is a major constituent (ranging from 0.5 to 3.6%) of the seeds, husks, trunk, and leaves of the cycad plant.[238]

Interestingly, cycasin is toxic and carcinogenic only when given orally and after passage through the gastrointestinal tract (thus cleaving the glucosidic moiety; see Figure 3), whereas MAM is active irrespective of the route of administration and in germ-free rats.[239-243] This finding prompted the investigation of dialkylhydrazines, azoalkanes, and azoxyalkanes, which led to the realization that these compounds are potent and selective colon carcinogens.[244] Use of these compounds in experimental models has proven to be of great benefit in investigations of colon carcinogenesis.

Laqueur and his colleagues were the first to reveal that rats fed crude cycad meal developed tumors of the liver and kidney and, rarely, the intestine and lung.[245,246] Subsequent work identified the carcinogen as cycasin and ultimately led to the conclusion that MAM is the proximate carcinogen.[246,247] Carcinogenicity of cycasin has been demonstrated in fish, guinea pigs, hamsters, mice, monkeys, rabbits, and rats (see Table 8). Primary sites of tumor initiation included the kidney and intestine for the rat and the liver for mice and hamsters. Effects observed are dependent on species, sex, age, route of administration, and dosing regimen. MAM was mutagenic in the Ames' *Salmonella* assay, but cycasin was inactive.[258] Cycasin, and MAM, are clearly potent animal carcinogens based on tumor development in single-dose experiments; however, without epidemiological data and sound exposure data, the role that cycasin and MAM may have in the etiology of human cancers is uncertain.

10 MYCOTOXINS

Although there are many hundreds of naturally occurring, mold-produced entities given the moniker "mycotoxin", there are only slightly more than a dozen that have sufficient evidence of carcinogenicity to warrant discussion. These include the aflatoxins (B_1, B_2, G_1, G_2, M_1, and aflatoxicol), sterigmatocystin, ochratoxin A, fumonisin B_1, fusarin C, T-2 toxin,

TABLE 8

Carcinogenicity of Cycad, Cycasin, and MAM[a]

Species	Target Organ	Administration(s)	Ref.
Fish	Liver	Cycad meal in feed, MAM acetate in tank water	248, 249
Guinea pig	Liver, bile duct	Cycad meal in diet or MAM or MAM acetate orally	250
Hamster	Bile duct, liver	Cycasin by gastric intubation, MAM by gastric intubation	250, 251, 252
Mice	Liver	Cycasin by gastric intubation	253
Monkey	Liver, bile duct, kidney, colon	Cycasin and/or MAM acetate orally	254
Rabbit	Liver	Cycad extract by gastric intubation	255
Rat	Intestine (especially colon), kidney, liver	Dietary cycad meal or cycasin, cycasin by single dose gastric intubation, cycasin in drinking water	246, 247, 256, 257

[a] Methylazoxymethanol.

zearalenone, luteoskyrin, cyclochlorotine, rugulosin, and citrinin. For a good recent general review of mycotoxins, see Pohland.[259]

10.1 *Aspergillus* Toxins

A tremendous research effort has gone into the study of a particular group of mycotoxins produced by *Aspergillus* spp., such that more is known about the occurrence and toxicity of this group of toxins than any other natural substance in the food supply. This group of mycotoxins, called aflatoxins, is a collection of more than a dozen closely related compounds that are produced by the mold species, *A. flavus* and *A. parasiticus.* For a good overall review of aflatoxin chemistry, biology, and occurrence in the food supply, see Busby and Wogan.[260] Although some contamination happens prior to crop harvesting, the principle outgrowth of these molds occurs under poor storage conditions. Factors that affect preharvest fungal infection are stress related and include drought and insect damage. Mycotoxin production during storage is related to environmental conditions such as humidity and temperature (see review by Sauer and Tuite[261]).

In the U.S., aflatoxin (AF) contamination is often found in farm products destined for animal feed, such as peanuts, corn, cottonseed, and occasionally in some grains such as rye, sorghum, and wheat.[259,262,263] Contamination of agricultural products intended for human consumption include peanuts, corn, and to a much lesser extent certain tree nuts and, in rare cases, dairy and meat products from animals fed highly contaminated feeds.[259,263,264] There is a strong geographical bias in the U.S., with corn from the southeast having a higher level of aflatoxin contamination (41% incidence) than corn from the midwest (2.5%).[265] Some caution is needed in evaluating evidence of contamination, since laboratory data on mold growth and AF production in certain commodities does not necessary hold true under natural conditions in the field. Evidence seems to indicate that significant human exposure in the U.S. is limited to corn and peanuts.[266] For recent data on aflatoxin presence in corn and peanut in the U.S., see Tables 9 and 10. Worldwide considerations indicate a wider and more extensive occurrence of AF contamination of foodstuff (upwards of 97% of the corn in some severely affected countries)[267] and consequently greater implications for human health effects.

Aflatoxins toxicity was first realized in 1960 when moldy peanuts were identified as the cause of a disease, subsequently called aflatoxicosis, which killed turkeys, ducks, and pheasants.[268,269] In the decades that followed, aflatoxin isolation, purification, characterization, and assessment of biological activity has resulted in literally thousands of publications. Aflatoxins constitute a unique group of heterocylic, oxygen-containing compounds that possess a bis-difurano ring

TABLE 9

Aflatoxin Presence in Corn Intended for Human Consumption in the U.S.[a]

Year	Total No. Tested	Ave. Conc. (ppb)	Incidence of Detection (%)[c]
Raw Shelled Corn[b]			
1991	219	32.2	26
1992	239	15.6	17
1993	248	29.2	11
1994	236	15.3	16
Milled Corn[b]			
1991	188	16.3	10
1992	158	17.6	14
1993	133	58.0	5
1994	168	12.6	5

[a] "Aflatoxin" refers to the sum of aflatoxins B_1, B_2, G_1, and G_2.
[b] Data from U.S. FDA Compliance Program, kindly supplied by Dr. G. E. Wood.
[c] Level of detection is 1 ppb.

system. Four major aflatoxins (B_1, B_2, G_1, G_2) have been isolated and well characterized, with most measurements in foods generally expressed as the sum of these, or alternatively as aflatoxin B_1 (AFB_1), since it is by far the most prevalent congener.[270] Aflatoxins B_2 and G_2 are dihydro derivatives of B_1 and G_1, respectively. Aflatoxin M_1 (AFM_1) and aflatoxin M_2 (AFM_2), hydroxilated derivatives of B_1 and B_2, respectively, are principally found in milk from cows fed contaminated fodder.[264] Results from controlled feeding studies indicated that 0.4 to 4% of AFB_1 ingested by cows as contaminated feed was converted to and excreted as AFM_1 in the milk.[271,272] Surveys conducted in the 1960s, 1970s, and 1980s of the occurrence of AFM_1 in milk from countries from around the world indicate a significant presence, but at levels usually less than 0.05 ppb.[273] Another important metabolite, aflatoxicol, is mutagenic (about 66% of the potency of AFB_1)[274] and possesses approximately 50 to 100% of the carcinogenicity of AFB_1 in trout.[275,276] Of all the aflatoxin metabolites found in food, it is generally believed that only these three (AFB_1, AFM_1, and aflatoxicol) have a significant role in the etiology of human carcinogenesis.

Food processing methods that involve thermal treatment have been shown to have a variable and largely incomplete detoxification effect. Substantial reductions have been achieved for oil- and dry-roasting of peanuts under laboratory conditions that simulate commercial practices.[277] Alkaline treatment of corn (e.g., in the preparation of masa used for tortillas)[278] and the brewing process[279] are other food processes that reduce aflatoxin levels. In contrast, significant detoxification of AFM_1 during the pasteurization, storage and processing of milk has not been found without destroying the milk.[264,280] Under normal cooking conditions, aflatoxins are not readily degraded.[281] See Scott[266] for a good reference on the effect of food processing on aflatoxins and other mycotoxins.

In addition to being the most prevalent congener, AFB_1 is by far the most biologically active (see IARC[273]). Aside from its extremely potent hepatotoxic effects, AFB_1 is one of the most potent hepatocarcinogens ever found, inducing liver tumors in a broad array of laboratory

TABLE 10

Aflatoxin Presence in Peanut Intended for Human Consumption in the U.S.[a]

Year	Total No. Lots Tested	Ave. Conc. (ppb)	Incidence of Detection (%)[d]
Raw Shelled Peanuts[b]			
1987	37,889	4.1	
1988	40,225	1.6	
1989	41,311	2.8	
1990	33,269	15.0	
1991	45,343	2.7	
1992	35,501	1.5	
Peanut Butter[c]			
1991	116	5	34
1992	82	3	22
1993	64	4	3
1994	77	5	30

[a] "Aflatoxin" refers to the sum of aflatoxins B_1, B_2, G_1, and G_2.
[b] Data from National Peanut Council; kindly supplied by Mr. R. Henning.
[c] Data from U.S. FDA Compliance Program; kindly supplied by Dr. G. E. Wood.
[d] Level of detection is 1 ppb.

animals, including rodents (mice, rats, hamsters, and tree shrews), fish (rainbow trout, sockeye salmon, and guppy), birds (ducks), carnivores (marmosets and ferrets), and subhuman primates (rhesus, cynomolgus, African green, and squirrel) by several routes of administration.[260,282,283] In most species, tumors were largely found in the liver, but kidney, lung, and colon neoplasia have also been found. Laboratory experiments have examined a wide variety of parameters that influence aflatoxin carcinogenesis, such as: dose response; route of administration; sex, age, and strain of test animal; diet; hormonal status, liver injury, and enzyme induction; and concurrent administration of other carcinogens and pharmacologically active substances.[260] AFB_1 is one of the most potent mutagens ever tested in the Ames' *Salmonella* assay and has produced positive results in all *in vitro* and *in vivo* genotoxicity tests.[260,273]

AFM_1, which is the hydroxylated derivative of AFB_1, has been isolated and purified from fungal cultures in sufficient quantities to be used in carcinogenicity rodent feeding studies.[284] Results from these studies revealed the hepatocarcinogenicity of AFM_1, to be similar to AFB_1, but with only 2 to 10% the potency. Corroborating evidence indicated genotoxicity in the Ames' *Salmonella* assay (with 1.6 to 3.0% the potency of AFB_1)[274,285] and in mammalian liver cells,[286,287] as well as carcinogencitiy in the trout (about 30% the potency of AFB_1).[288-290]

Epidemiological studies in Uganda,[291] Thailand,[292] Kenya,[293] and Mozambique[294] revealed a strong positive correlation between ingested aflatoxins and incidence of liver cancer, thus providing further evidence of carcinogenicity. There is some concern that many of these studies may be limited due to confounding factors such as hepatitis B virus (HBV) infection. Data from China by Yeh and coworkers[295] suggested that HBV infection may be a necessary co-factor. Based on the totality of scientific evidence, the IARC evaluated AFB_1 and concluded that it is a Group 1 human carcinogen. There was inadequate evidence of human carcinogenicity for AFM_1.[272]

Aflatoxins are not the only carcinogenic toxins produced by *Aspergillus*. Sterigmatocystin (ST) is produced by several *Aspergillus* spp., especially *A. versicolor*, as well as by *Penicillium luteum* and unidentified species of *Bipolaris, Chaetomiun,* and *Emericella*.[296] ST has been found to contaminate cereals, country hams, salami, green coffee beans, cheese, and wheat.[297] Resembling aflatoxin, ST is chemically characterized by a bis-dihydrofuran ring fused to a substituted anthraquinone. It is mutagenic and clastogenic and has between one tenth and one hundredth the hepatocarcinogenic potency of aflatoxin B_1 (see review by Terao[298]). Retaining the key structural elements of aflatoxin which allow epoxide formation and hence DNA binding, it would be reasonable to expect a similar mechanism of action. Analysis of more than 500 samples from 1974 to 1975 by the FDA failed to detect ST contamination;[299] however, occurrence in Japan and elsewhere is greater. Unfortunately, there are no epidemiological studies which have attempted to link ST exposure with any carcinogenic outcome. Hence, the role of ST exposure in human cancer remains speculative at this time.

Ochratoxin A (OT-A) and other related isocoumarin derivatives were first isolated from *Aspergillus ochraceus*,[300] and later from other *Aspergillus* spp.[301] and one species of the genus *Penicillium,* namely *P. verrucosum*.[302] Chemically, OT-A is a dihydroisocoumarin linked to L-beta-phenylalanine.[303,304] Worldwide contamination of food with OT-A includes cereals (wheat, rye, barley, and oats), corn, sorghum, groundnuts, green coffee beans, swine products, and animals feeds, with the latter containing the highest levels.[305,306] With the exception of northern Europe, surveys have not revealed any significant contamination of human foods.[307] A potent nephrotoxin, OT-A has demonstrated carcinogenicity via long-term feeding studies that have resulted in tumors in mice liver and kidneys[308-310] and rat kidneys.[311] It should be noted that the mouse studies suffered from one or more of the following: poor survivability, use of only one sex, high dose in excess of the MTD, presence of impurities with the OT-A (including benzene), and significant increases in tumor incidence still within historical control range (see review by Kuiper-Goodman and Scott[305]). Aside from a single positive study using a modified *Salmonella*-microsome assay,[312] OT-A has tested nonmutagenic in several bacterial and mammalian gene mutation assays (see reviews by Kuiper-Goodman and Scott[305] and Dirheimer and Creppy[313]). The IARC concluded that there was sufficient evidence of carcinogenicity in experimental animals for OT-A.[306] Based on the totality of the evidence, including the investigations[314,315] that have showed an association between ochratoxin and human urinary tract cancers, the IARC concluded that OT-A was a "possible human carcinogen" (Group 2B).[306] Further studies are needed to more fully address the possible genotoxicity of OT-A and its endocrine disruptive effects, prior to reaching any definite conclusions on the human carcinogenic risk.

10.2 *Fusarium* Toxins

The fungus *Fusarium moniliforme* is a ubiquitous contaminant of corn and sorghum, with lesser occurrence for wheat and barley, and is most prevalent in warm, dry years in the presence of insect damage.[316] Several mycotoxins are produced by *F. moniliforme* or a related *Fusarium* species, of which two classes contain putative carcinogens. Fumonsin B_1 is the major toxin of at least seven other fumonisins (B_2, B_3, B_4, A_1, A_2, C, and D) which are known to be produced by *F. moniliforme*-contaminated corn and culture material.[317] However, only fumonisins B_1, B_2, and B_3 have been detected in milled corn intended for human consumption;[316] of these three, only fumonisin B_1 has been implicated as a carcinogen.[318,319] Fumonsins appear to be poor mutagens[320,321] but good cancer promoters.[322] Human exposure to *F. moniliforme*-contaminated maize has been associated with elevated rates of esophageal cancer in S. Africa, allegedly due to fumonisin intake.[323,324]

Fusarins are the another important class of *Fusarium* mycotoxins, and there are at least five produced by *F. moniliforme*. Within this class, only fusarin A, C, and F are found in unprocessed contaminated corn, with only fusarin C considered as a presumptive carcinogen.[325] Measurements of fusarin C in food are limited as a result of its poor stability.

Evidence for the carcinogenicity of fumonisin B_1 and fusarin C have been evaluated by the IARC, and their conclusion was that there was sufficient evidence to classify "toxins derived from *Fusarium moniliforme*" as possible human carcinogens (the exact chemical identity was unspecified).[318] It is important to note that the IARC found the evidence for the carcinogenicity of the specific toxins fumonin B_1 and fusarin C to be limited, and for fumonisin B_2 the evidence was found inadequate.

Fumonisin B_1 content of corn and corn-based products in the U.S. was determined to range up to 330 ppm, with the greatest levels being reported for animal feed.[318,326] Analysis of corn and corn-based human foods indicate fumonisin B_1 levels in the range of 200 to 800 ppb, with processed foods generally having the lowest amounts.[327,328] In general, food processing methods do not destroy fumonisins;[329] however, calcium-hydroxide treatment (nixtamalization) of corn used to make tortilla flour reduces fumonsin levels through hydrolysis, resulting in uncertain effects on its subsequent carcinogenicity.[330] Although studies of fumonisin residues in meat, milk, and eggs are limited, evidence seems to indicate a low or nonexistent occurrence.[331,332] Quantification of fusarins in corn and corn-based products has been hampered by the instability of these compounds.

T-2 toxin, which rarely occurs in cereals (e.g., wheat and maize) as a toxic metabolite of several *Fusarium* spp. but principally *F. sporotrichioides*,[333] has been found worldwide.[334] T-2 toxin (3α-hydroxy-4β,15-diacetoxy-8α-[3-methylbutyryloxy]-12,13-epoxy-trichothec-9-ene) belongs to a large class of tetracyclic sesquiterpenoid compounds known as trichothecenes, with their carcinogenicity presumably due to an epoxy group and/or an olefinic bond.[335,336] Cancer feeding studies in mice revealed tumors in the liver, lung, and forestomach;[337,338] however, no evidence of neoplasia was found in treated trout.[339] Studies in rats are inadequate for evaluation.[340,341] Evidence of genotoxicity, which for the most part has been limited and not compelling, and has been reviewed by Haschek.[342] The IARC determined there was "limited evidence" of carcinogenicity in experimental animals for T-2 toxin, without any evidence in humans.[343] Although the chemistry of T-2 toxin is suggestive of carcinogenic activity, the limited evidence in animal studies and the absence of any human data make drawing any definite conclusions premature.

Three other well-known *Fusarium* mycotoxins are zearalenone (ZEN), deoxynivalenol (DON), and nivalenol (NIV). Post-harvest contamination of cereals with these toxins is known to occur worldwide as the result of several *Fusarium* spp., with surveys indicating a significant level and frequency of contamination of wheat, barley, oat, rye, corn, and rice.[344] ZEN, chemically described as 6-(10-hydroxy-6-oxo-*trans*-1-undecenyl)-β-resorcylic acid lactone, was first isolated from fungal cultures in 1962.[345] The major ZEN-producing *Fusarium* spp. are *F. graminearum* and *F. culmorum*, with their natural occurrence in agricultural commodities being the subject of many studies.[261,346-348] Surveys done in the U.S. in the 1960s and 1970s indicate appreciable levels of ZEN in corn and wheat (including corn products intended for food use), with variable effects of food processing.[349] Carcinogenicity studies in rats showed no neoplastic effects; however, studies in mice produced liver and pituitary tumors.[350,351] The IARC classified ZEN with "limited evidence" of carcinogenicity since positive findings were exclusive to only one species.[352] Results in the Ames' *Salmonella* assay were conclusively negative.[349] Overall, the evidence of carcinogenicity for ZEN is weak (with the absence of epidemiological data), and until the ZEN-induced tumorigenesis can be confirmed in another species, it would be premature to make a definite conclusion about its possible carcinogenic activity in humans. Concerning the mycotoxins DON and NIV, the IARC has evaluated the carcinogenicity evidence and deemed them to be of no special concern, although co-occurrence with aflatoxins may lead to an interaction that produces increased carcinogenicity.[352]

10.3 *Penicillium* Toxins

Lesser known carcinogenic mycotoxins are: (−)luteoskyrin, cyclochlorotine, (+)rugulosin, and citrinin. The first two were initially isolated from the growth of *Penicillium islandicum* on rice and were later determined to be hepatocarcinogenic in mice in several studies.[353,354] (+)Rugulosin, first isolated from *P. rugulosum*, is very similar to (−)luteoskyrin, lacking only two hydroxy groups, but is much less potent in producing murine liver tumors.[353,355] Citrinin, isolated from *P. citrinum* (and other *Penicillum* and *Aspergillus* spp.), has produced liver tumors in long-term feeding studies with rats[356] but was found negative in two other carcinogenicity studies,[357,358] as well as several independent Ames' *Salmonella* assays.[359] See the review by Enomoto and Saito[360] for a fuller description of the fungal producers of these compounds. Occurrence of all these compounds in the U.S. food supply is scarce. Based on the limited available data, including the lack of epidemiological studies, any definitive judgment on the role these compounds may play in the etiology of human cancer is untenable.

11 *N*-NITROSO COMPOUNDS

The current interest in the carcinogenic activity of *N*-nitroso compounds (NOCs) stems from the original reports by Magee and Barnes[361,362] in which they demonstrated that the chronic dietary administration of *N*-nitrosodimethylamine (NDMA) produced liver and kidney tumors in rats. Following this report, numerous carcinogenicity studies on a variety of NOCs have been reported (for an in-depth review, see Preussman and Stewart[363]). To date, of the approximately 300 NOCs that have been evaluated for carcinogenicity, about 90% were positive in one or more of 44 different laboratory animal species, including nonhuman primates.[363] The most widely tested NOC, *N*-nitrosodiethylamine (NDEA) has been shown to be carcinogenic in 40 species.[364] No species tested to date have been found to be resistant to their carcinogenic action.

NOCs can be divided into two chemical classes: nitrosamines and nitrosamides (and related compounds). Nitrosamines are *N*-nitroso derivatives of secondary amines, whereas nitrosamides are *N*-nitroso derivatives of substituted ureas, amides, amino acids, and other nitrogen-containing compounds.[365] There is a more important distinction than the chemical classification between these two classes. Nitrosamines must be activated by enzyme systems and can exert their carcinogenic effects at remote sites in the body. Conversely, nitrosamides are direct-acting carcinogens, do not require enzyme activation, and can cause tumor development at their site of application. In practice, NOCs are broadly divided into two groups based on analytical methods, namely, volatile nitrosamines (VNAs) and nonvolatile nitrosamines (NVNAs). The four VNAs most commonly found in food are *N*-nitrosodimethylamine (NDMA), *N*-nitrosopyrrolidine (NPYR), *N*-nitrosopiperidine (NPIP), and *N*-nitrosomorpholine (NMOR).[366] Some NVNAs commonly found in food, or formed *in vivo* from foods due to intragastric nitrosation, are *N*-nitrosothiazolidine (NTHZ), *N*-nitrosothiazolidine-4-carboxylic acid (NTCA), 2-hydroxmethyl-*N*-nitrosothiazolidine-4-carboxylic acid (HMNTCA), *N*-nitrosodibenzylamine (NDBzA), and *N*-nitroso-*N*-methylurea (NMU).[367]

The occurrence of NOCs (principally VNAs) in foods and beverages has led to a worldwide sampling effort, especially for those products preserved with nitrate and/or nitrite (the reader is referred to the following excellent reviews: Gray,[368] Pensabene and Fiddler,[369]

Sen,[370,371] and Hotchkiss[372]). Data on the occurrence of nonvolatile nitrosamines in foodstuff are sparse due to the lack, until recently, of adequate and reliable analytical methods. In regard to nitrosamides, the instability of these compounds makes it unlikely that significant amounts would accumulate and persist in the food supply. NOCs can occur in foods and beverages in three different ways: (1) formation as a result of the use of curing agents, such as nitrate and nitrite, (2) formation during processing, and (3) via migration from secondary sources such as packaging or other materials. Most Western-style foods have been analyzed for volatile nitrosamines, with results indicating that only bacon and beer consistently showing levels greater than 1 ppb. Nitrosamines are formed in bacon (by a process known as *N*-nitrosation) as an indirect result of the addition of nitrite. A detailed study in 1986 of nitrosamine formation and occurrence in bacon found total nitrosamine (based on the analysis of four nitrosamines) content in 39 samples of cooked bacon to be, on average, about 30 ppb.[373] These levels in bacon are approximately tenfold less than the initial determinations made in the early 1970s[374] and are a consequence of food processing changes, principally via a reduction in added nitrite. In addition, humans are exposed to other dietary sources of nitrite and nitrate (which can be enzymatically converted to nitrite) through the consumption of vegetables and via drinking water contaminated with nitrogen-containing fertilizer.

Among those foods that undergo processing conditions that favor nitrosamine formation, those that are directly dried (e.g., malt used in beer manufacture) are the most susceptible. In response to these findings in the late 1970s, the malting and brewing industries altered their processes to reduce the nitrosamine content in beer. An extensive survey of imported and domestic beer in 1990[375] revealed generally very low levels (less than 0.1 ppb), which were approximately 20-fold less than determinations made in the prior decade. Similar efforts were undertaken with other dried foods, such that a recent survey of 57 samples of dried milk averaged less than 1 ppb.[376] Indirect contributions via migration into foods have been reduced to negligible levels.[377] There is fairly widespread occurrence of NOCs in fish and seafood, particularly those that are salted, dried, and smoked. The levels of NOCs detected in such products vary widely by country and preservation method (see reviews by Hotchkiss[372] and Walker[378]).

There are several lines of evidence that nitrosamines and nitrosamides are endogenously formed in humans, in large part due to the ingestion of nitrate (principally from vegetables) and nitrite (from cured meat, vegetables, and cereals). Indeed, preliminary data seem to indicate that the *in vivo* formation of nitrosamines may actually represent the largest source of exposure to NOCs;[379] however, other contradictory evidence[380] suggests that much further research is needed to gain a clear understanding of the quantitative impact of endogenous nitrosation under the realistic conditions found in humans.

It is important to note that human exposure to other nondietary sources of NOCs occurs through products such as cosmetics, vulcanized rubber products (e.g., nursing nipples, tires, gloves, etc.), new automobile interiors, and other consumer products. In fact, exposure to NOCs from foods represents a much smaller potential exposure than from some of these consumer products.[381]

NOCs are a unique group of carcinogens in that they can induce tumor formation in so many organs of so many animal species. Target organ selectivity depends strongly on structure activity relationships and the particular animal species, but other factors such as nutritional status, age, sex, route of administration, dosing schedule, and the presence of modifying agents also play a role. The large interspecies differences in target organs affected by a particular NOC make predictions for humans problematic (for a review of the structure-activity relationships, see Lijinsky[364]). Various dose-response studies have been carried out that show carcinogenic effects even at fairly low doses. Extensive studies conducted by Druckery et al.[382] and Peto et al.[383] have provided relatively reliable data on tumorigenesis for use by regulators in estimating carcinogenic risk to humans. Many NOCs are potent mutagens.[363] In general, there is a reasonably good qualitative, but not quantitative, correlation between NOC mutagenicity in the modified Ames' *Salmonella* assay and the carcinogenicity results.[384] Combined with data of carcinogenicity in single-dose, transplacental, and co-administration studies of NOCs, there is sufficient evidence from *in vivo* and *in vitro* studies to support the conclusion that the human exposure to nitrosamines could have a causal relationship in the development of certain cancers.

Numerous epidemiological investigators have tried to establish a relationship between NOC exposure and the development of cancer. Many studies have used the ingestion of NOC precursors (e.g., nitrite, nitrate, etc.) as a surrogate exposure estimate. Many have focused on gastric cancer because it is the primary site of nitrosation. Two independent reviews by the U.S. Assembly of Life Sciences[385] and by the World Health Organization (WHO)[386] concluded that the epidemiological studies have failed to provide convincing evidence of a link to human cancer.

The clear evidence of carcinogenicity from well-conducted animal experiments should provide a reasonable basis to conclude that NOCs have some level of carcinogenic risk to humans. The magnitude of this risk is related to the exposure to the causative agent. Unfortunately, the state of the science is imperfect in providing accurate determinations of the human carcinogenic risk based on animal studies, and useful epidemiological studies incorporating sound NOC exposure data have been lacking.

12 POLYCYCLIC AROMATIC HYDROCARBONS

Polycyclic aromatic hydrocarbons (PAHs) are relatively simple organic compounds composed of two or more fused aromatic rings, which may or may not have substituted groups.[387] Although hundreds of compounds have been identified as belonging to this class, about 20 compounds have been found to occur in food, of which half have been determined to be carcinogenic in laboratory animals by skin application and/or by injection; however, carcinogenicity experiments using oral administration have identified only three, specifically benzo[*a*]pyrene (BaP), benz[*a*]anthracene (BaA), and dibenz[*a,h*]anthracene (DBahA)[388] (Table 11). Two other compounds shown to be carcinogenic by the oral route (3-methylcholanthrene and 7,12-dimethylbenz[*a*]anthracene)[389] are not normally found in food.[388]

PAHs are ubiquitous environmental pollutants, having been found in air, water, soil, and food.[390] Although this class of compounds is composed of many members, the majority of the toxicological research has focused on BaP, one of the most prevalent of the carcinogenic PAHs. Analytical determinations of PAHs in food are often expressed as either BaP or total PAHs, the latter of which includes many noncarcinogenic compounds. Hence, the BaP content is a more valuable expression of the carcinogenic potential of a particular PAH-containing food, although it may account only for up to 10% of its noncarcinogenic toxicity.[391] Since many of its members are relatively potent carcinogens and due to their extensive, and in some cases heavy, occurrence in a variety of foods, PAHs constitute one of the more important naturally occurring classes of carcinogens. Although anthropogenic contributions of PAHs to the environment should not be discounted, natural processes play a significant role in the presence of PAHs in the diet. In fresh vegetables, PAHs are thought to be end-products of one or more biosynthetic pathways. The sharp increase in the PAH content of decaying vegetables has been used as evidence that they result from catabolic processes.[392] This accumulation of PAHs in decaying vegetables accounts to a large degree for the PAHs in soil and water and, in turn, back into fresh vegetables.[392] This cycle makes vegetables and their oils, cereals, and fruits significant contributors of PAHs to the diet (Table 12).

The other major source of PAHs in the diet is a result of the formation and deposition of PAHs on foods via thermal processes, such as grilling, roasting, and smoking. Formation of PAHs at temperatures less than 400°C is limited, while the amount formed increases linearly in the range of 400 to 1000°C.[393] Hence, "endogenous" formation on the surface

TABLE 11
Orally Active Carcinogenic Polycyclic Aromatic Hydrocarbons (PAHs) Found in Food

Structure	Name	Abbreviation	Mutagenicity[a]	Carcinogenicity[a]	Carcinogenicity Target Organs	Ref.
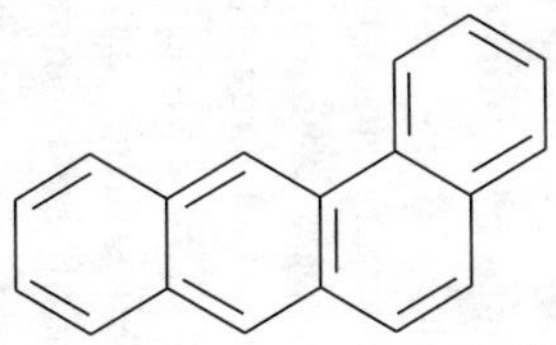	Benz[*a*]anthracene	BaA	+	++	Liver, lung	411
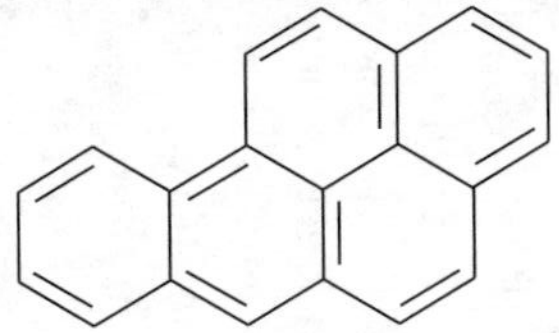	Benzo[*a*]pyrene	BaP	+++	++	Forestomach, mammary gland, lung, hematopoietic system	402–405
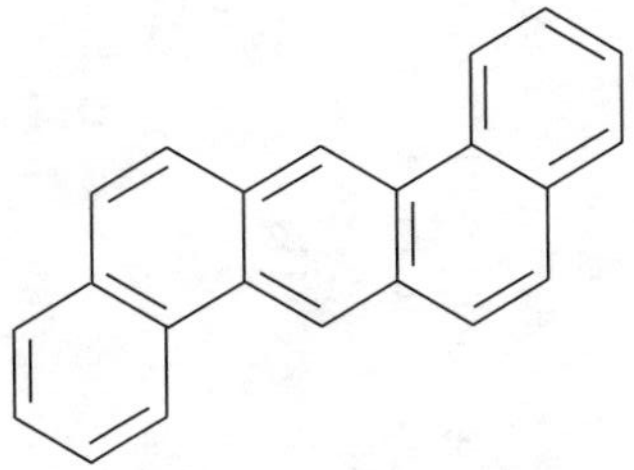	Dibenz[*a,h*]anthracene	DBahA	+	+	Lung, mammary gland, forestomach	412

[a] Relative to PAHs; +, ++, +++ indicate increasing activity.

TABLE 12

Benzo[*a*]pyrene Content in Some Foods

Food[a]	Level (ppb)[b]	Year	Remarks	Ref.
Vegetables				
Cereals	0.2–4.1	1984	—	413
Kale	12.6–48.1; 0.6–4.5[c]	1988; 1984	Surface contamination likely	400; 413
Salad	2.8–5.3	1984	Surface contamination likely	413
Spinach	7.4; 0.09–0.5[c]	1988; 1984	Surface contamination likely	400; 413
Tomatoes	0.2	1984	—	413
Fats and Oils				
Corn oil	ND; 0.7	1966; 1979	Retail products	414; 415
Olive oil	ND; 0.5	1966; 1979	Retail products	414; 415
Peanut oil	ND; 0.6	1966; 1979	Retail products	414; 415
Shortening	ND	1979	Retail products	415
Vegetable oil	ND	1979	Retail products	415
Meats				
Bacon, smoked	ND, 0.42; 0.59	1968; 1970	10 nondetects	397; 416
Beef patty, grilled	18.8–24.1	1979	6 determinations	396
Bologna	ND	1979	—	415
Chicken, grilled	3.7	1967	—	417
Chicken, smoked	TR, 0.5, 0.7; ND	1968; 1979	—	397; 415
Frankfurters, smoked	ND, TR, 0.8; ND	1968; 1979	9 nondetects (beef and pork)	397; 415
Ham, smoked	ND, TR, 0.5–1.5	1968	—	397
Lamb patty, grilled	8.8–12.3	1979	6 determinations	396
Pork chop, grilled	7.9	1967	—	417
Pork patty grilled	25.8–31.6	1979	6 determinations	396
Ribs, barbecued	10.5	1965	—	418
Steak, grilled	5.8, 8.0; 11.1, 50.4	1965; 1967	—	417; 418
Turkey, smoked	TR	1968	—	397
Turkey patty, grilled	ND	1979	6 determinations	396
Misc.				
Cheese, smoked	ND	1968	—	397
Coffee, whole, roasted	ND	1968	—	397
Liquid smoke	ND	1968	—	397

[a] Vegetable analysis from samples collected in the Netherlands; all other samples from U.S.
[b] ND = not detected; TR = trace.
[c] Level of detection ≥ 0.03 ppb.

of food occurs only under certain limited conditions. The conduction of heat in frying and radiation in the electrical broiling of meats do not result in sufficient temperatures under normal cooking conditions to generate significant amounts of PAH. Only when meat was placed in direct contact with open flames did significant amounts (6 to 212 μg BaP/kg) of PAHs form.[394] PAH occurrence in food can also be the result of deposition as a consequence of the fuel combustion. In general, this is nearly insignificant since the amount yielded by charcoal briquettes commonly used in grilling is only zero to 1.0 μg BaP/kg.[394] Somewhat higher levels can result from the embers of a log fire (1 to 25 μg BaP/kg). Lastly, and perhaps most importantly, the occurrence of PAHs can occur when fat drips on the heat source, gets pyrolized into the air, and is subsequently deposited on the food.[395] This was clearly illustrated in an experiment by Doremire and coworkers,[396] in which ground-beef patties containing from 15 to 40% added fat were charcoal-grilled under identical conditions. Analysis of cooked patties indicated BaP levels from 16 to 121 μg/kg, which were directly related to the fat content.

Smoking is a traditional method used to preserve meats and fish, although currently its main purpose is to give certain foods desirable organoleptic traits. Exogenous contamination can occur through absorption of PAHs during the smoking process. Analysis of curing smoke has identified more than 300 compounds, including many PAHs. In general, low-molecular-weight PAHs (e.g., phenanthrene, anthracene, and pyrene) are more frequently found on smoked foods, at total PAH levels of 10 μg/kg. Higher-molecular-weight compounds, including the carcinogens BaP, BaA, and DBahA, are found at much lower concentrations in smoked foods (total PAHs < 1 μg/kg).[388]

The processes used for some distilled spirits can result in the contamination of PAHs, although analysis of bourbon, whisky, and Scotch has not found any[397] or has found very low levels (from 0.03 to 0.08 ppb).[398] Roasting of coffee does not increase the BaP content.[399]

Estimates of dietary intake in the Netherlands, based on duplicate analysis of 50 24-hour diets, revealed total PAH intake of from 1.1 to 22.5 μg per capita.[400] Results also indicated that about 30% of the total ingested PAHs are carcinogens. Another survey of BaA, BaP, and DBahA individual intakes ranged from nondetectable to about 0.5 μg per capita per day.[391] The EPA estimated a daily BaP intake from food of 50 ng.[401] Total exposure to BaP, as well other PAHs, can be substantially elevated depending on environment, habits, and occupation. Overall, consumption of charcoal-broiled and smoked meat and fish does represent a significant exposure medium, but certain atmospheric conditions in urban environments and other nondietary exposures can also serve as substantial sources of BaP and other PAHs.[401]

Among all the PAHs, BaP has received the most biological testing; however, few studies have adequately examined the carcinogenic effect of orally administered BaP. Available studies (which for the most part are from the 1960s) have several shortcomings, such as the less than life-time exposure, small number of test animals, inappropriate controls, and tumors generally limited to the site of application and to an organ that has no structural analogue in humans (i.e., the rodent forestomach). Incidence data from perhaps the best study indicated BaP to be a potent and robust forestomach carcinogen in the mouse;[402] however, other studies by the same researchers also revealed systemic carcinogenicity (lung adenomas and leukemia) in mouse species prone to these particular spontaneous neoplasia.[403,404] Likewise, Huggins and Yang[405] found an increase in spontaneous mammary cancer in rats by a single intragastric dose of BaP (however, poor reporting of this study makes it inadequate for evaluation). Other studies have repeatedly found induction of forestomach cancer by oral administration of BaP.[406-410] Repeated gavage administration of BaA produced murine liver and lung tumors and, to a slight extent, forestomach tumors.[411] Dietary administration of DBahA to mice produced carcinoma of the lung, mammary gland, and forestomach.[412] Mutagenicity results are somewhat varied, with strong positive activity in nearly every test conducted for BaP but mixed and generally weak responses for DBahA. BaA has tested mutagenic in the Ames' *Salmonella* assay, and *in vivo* tests of genotoxicity are also generally but not uniformly positive (see EPA[401] document for summary of genotoxicity testing). Epidemiological data on the carcinogenic effect of PAHs are available from both occupational and community air pollution studies; however, neither addresses oral exposure. Given this lack of human data on oral exposure, one can only speculate on the role that PAHs, individually or in mixtures, may have on human gastrointestinal or other site tumors. Nonetheless, the animal carcinogenicity evidence for BaP, BaA, and DBahA is sufficiently strong enough to warrant serious consideration of their potential risk to human health depending on their intake level.

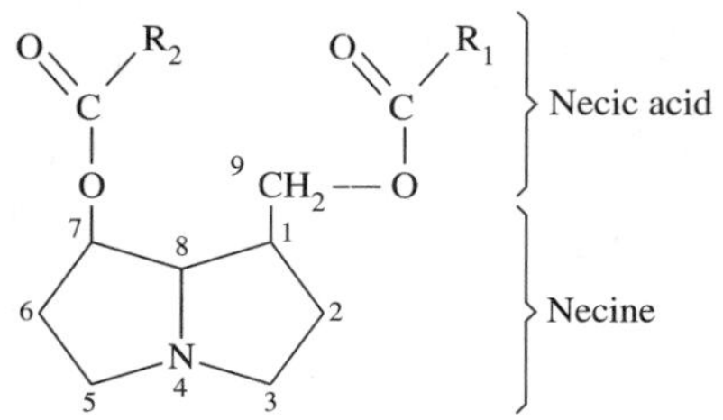

Necic acid

Necine

FIGURE 4
General structure of pyrrolizidine alkaloid. (From Furuya, T. et al., in *Naturally Occurring Carcinogens of Plant Origin*, Hirono, I., Ed., Kodansha, Ltd., Tokyo, 1987, pp. 25–51. With permission.)

13 PYRROLIZIDINE ALKALOIDS

About 250 pyrrolizidine alkaloids (PAs) have been isolated from more than a dozen unrelated plant families (principally Compositae, Boraginaceae, and Leguminosae), encompassing more than 60 genera.[419,420] Although many of these plants have limited or nonexistent consumption by humans, plants such as coltsfoot (*Tussilago farfara*), comfrey (*Symphytum officinale*), and petasites (*Petasites japonicus*) have found fairly widespread use as herbal remedies, dietary supplements, or even as foods.[420,421] Recent analysis of a number of commercial comfrey products (*Symphytum* spp.) obtained at health food stores in the Washington, D.C., area revealed PA content generally under 10 ppm, although bulk root and leaf levels were at several hundred ppm.[422] As an example, a cup (250 ml) of comfrey tea prepared from a root infusion contained from 8.5 to 26 mg of PAs, depending on the preparation.[423] A more detailed review of PA content in samples of *Symphytum* spp. can be found in Mattocks.[419]

The fundamental structure of PAs is composed of two parts, necine and necic acid, combined by an ester link (Figure 4).[420] Necine consists of a basic structure of various hydroxylated congeners of pyrrolizidine, which can be further subdivided into four groups: (1) a trachelanthamidine group of monohydroxylated derivatives, (2) a retronecine group of dihydroxylated derivatives, (3) a rosmarinecine group of trihydroxylated derivatives, and (4) an otonecine group. In most alkaloids, the various necines combine with the various 5 to 10 carbon branched-chain necic acids to form an ester structure. These combinations can also be divided into four groups: (1) nonesters, (2) monoesters, (3) acyclic diesters, and (4) macrocyclic diesters (see Mattocks[419] for a comprehensive review of the chemistry).

PAs were among the first natural products to be proven as animal hepatocarcinogens. Despite this, the total number of PAs tested for carcinogenicity is quite small. Aside from the limited availability of purified material, the pharmacological activity of these compounds has hampered efforts to select the appropriate dosing regimens that would permit sufficient animals to survive long enough for tumor development. Eight PAs have received carcinogenicity testing that yielded evidence of treatment-related tumor development: clivorine,[424] isatidine,[425] lasiocarpine,[426-428] monocrotaline,[429-431] petasitenine,[432] retrosine,[425] senkirkine,[433] and symphytine[433] (see Table 13). Except for senkirkine and symphytine, the remaining compounds were the subject of at least one study employing some type of oral administration (diet, drinking water, or gastric intubation). All were found to be exclusive liver carcinogens except for lasiocarpine, which produced additional tumors of the skin, lung, ileum, and hematopoietic system in rats (see Mattocks[419] for a detailed review of the carcinogenicity data). It should be noted that although the results of these studies are highly suggestive of carcinogenicity, limitations in animal number, survivability, variable dosing regimens (including parenteral administrations), possible mycotoxin contamination of feed, and, in general, poor reporting make drawing definitive conclusions difficult. Bacterial mutagenic activity[434] has not been consistently found for the PAs tested so far (see Woo et al.[435] for a good

TABLE 13

Carcinogenic Pyrrolizidine Alkaloids

Common or Trivial Name	Chemical Structure	Route of Administration	Carcinogenic Target Organ	Example Source	Carcinogenicity Ref.
Clivorine		Drinking water[b]	Liver[a]	*Ligularia dentata*	438
Isatidine		Drinking water[c]	Liver[a]	*Senecio jacobaea* (common ragwort)	439

Lasiocarpine	H C=C CH$_3$ CH$_3$ CO O H CH$_2$OOC HO OH H OCH$_3$ CH$_3$ N	Parenteral[d]	Liver, skin, lung, ileum	*Heliotropium europeum*	440
		Dietary[e]	Liver		441
		Dietary[f]	Liver, hematopoietic system		442
Monocrotaline	H$_3$C CH$_3$ H C—C H$_3$C—C HO OH C O O=C O CH$_2$ O N	Gastric intubation[g]	Liver	*Crotalaria refusa*	443
		Parenteral[h]	Liver, lung, and other sites		444
		Parenteral[i]	Pancreas		445
Petasitenine	CH$_3$ H CH$_3$ H O C C—C—C C H$_3$C H H OH C=O O=C O O O CH$_2$ N CH$_3$	Drinking water[j]	Liver[a]	*Petasites japonicus*	446

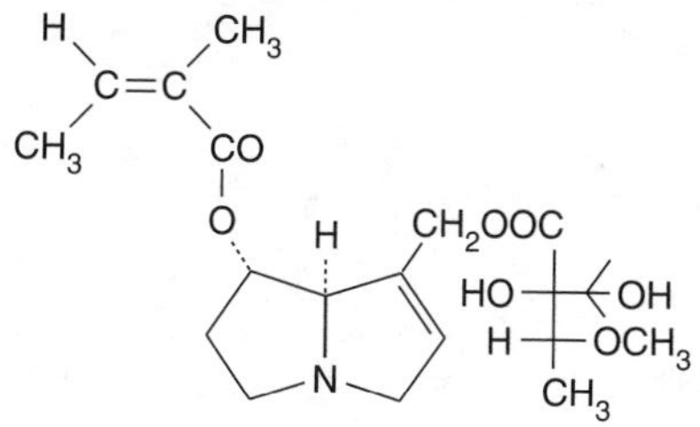

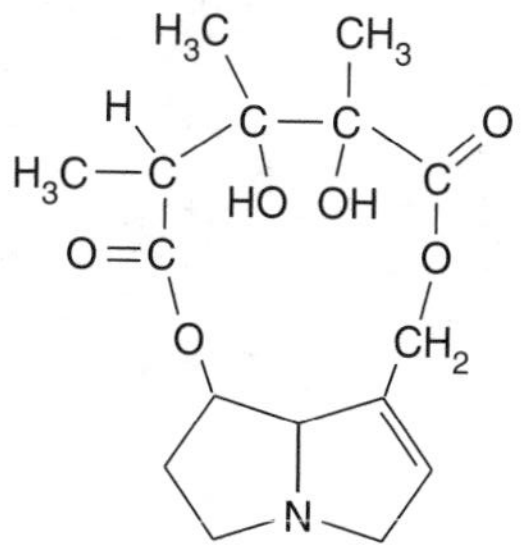

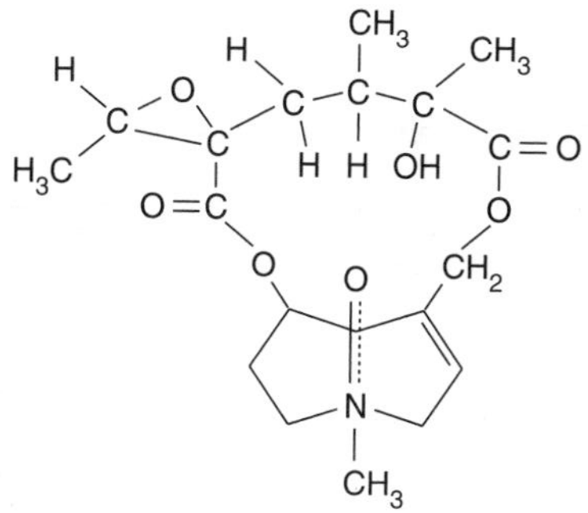

TABLE 13 (continued)

Carcinogenic Pyrrolizidine Alkaloids

Common or Trivial Name	Chemical Structure	Route of Administration	Carcinogenic Target Organ	Example Source	Carcinogenicity Ref.
Retrorsine	CH_3 H CH_2 C OH C C C H_3C CO CO CH_2OH O H CH_2O N	Drinking water[c]	Liver[a]	*Senecio jacobaea* (common ragwort)	439
Senkirkine	CH_3 H H CH_3 C C C C C H_3C H H OH C=O O=C O O CH_2 N CH_3	Parenteral[k]	Liver	*Tussilago farfara* (coltsfoot)	447

Symphytine	CH_3 CH_3 C=C H CO O H CH_2OOC HO HO H N CH_3	Parenteral[k]	Liver	*Symphytum officinale* (comfrey)	447

[a] Study limited by small number of animals in treated and control groups.
[b] 0.005% solution for 340 days (experiment terminated at 480 days).
[c] 0.03 mg/ml solution 3 days per week until death (10–23 months).
[d] Intraperitoneal injections of 7.8 mg/kg b.w. twice per week for 4 weeks and then once per week for 52 weeks (majority of treated animals killed between 60 and 76 weeks).
[e] 50 ppm in diet for 55 weeks (majority of treated animals killed between 48 and 59 weeks).
[f] 7, 15, or 30 ppm in diet until study termination (104 weeks).
[g] Weekly gastric intubations of 25 mg/kg b.w. for 4 weeks and then 8 mg/kg b.w. for 38 weeks.
[h] Subcutaneous injections of 5 mg/kg b.w. biweekly for 12 months (experiment terminated at 24 months).
[i] Single subcutaneous injection of 40 mg/kg b.w. (experiment terminated at 500 days).
[j] 0.01% solution until study termination at 16 months.
[k] Intraperitoneal injections of either 22 mg/kg b.w. (senkirkine) or 13 mg/kg b.w. (symphytine) twice per week for 4 weeks and then once per week for 52 weeks (experiment terminated at 650 days).

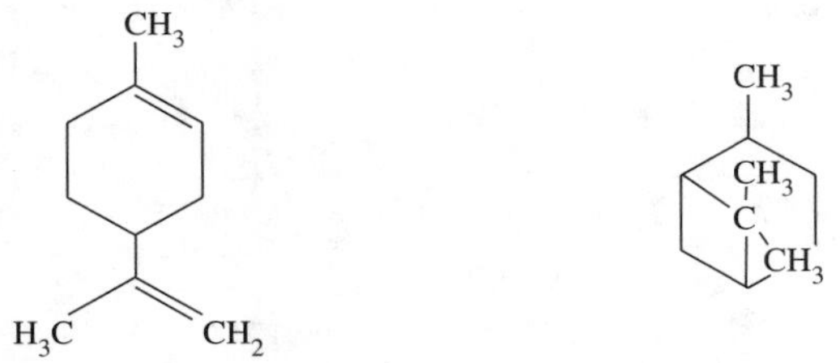

CH_3 H_3C CH_2
Limonene

CH_3 CH_3 C CH_3
Pinene

FIGURE 5
Some terpenes: limonene and pinene.

summary), and no clear structure activity pattern has emerged.[420] The IARC[436,437] has reviewed the available data for each of the above-mentioned seven PAs (clivorine was not evaluated), and determined that in each case there was limited evidence of carcinogenicity in experimental animals, but without epidemiological studies no evaluation of the carcinogenicity to humans could be made. Clearly, these compounds are weak rodent hepatocarcinogens, with evidence that suggests that their carcinogenicity is due to the formation of ultimate carcinogens in the liver; however, to reiterate the IARC evaluation, the possible human carcinogenic hazard posed by the PAs is uncertain due to limited long-term animal studies (which have been exclusive to the rat) and the absolute lack of epidemiological data incorporating sound exposure estimates.

14 TERPENES

Terpenes constitute a class of compounds that includes side chain substitution of cyclohexene with a methyl, isopropenyl, ethenyl, or other group. Terpenes and their derivatives owe their metabolic origin to the five carbon molecule isoprene. Terpenes are the chief constituents of essential oils, which are the volatile oils obtained from plant materials through steam distillation, solvent extraction, or physical expression. Limonene (with the possible exception of α-pinene) is the most frequently naturally occurring monoterpene (or diisoprine) (for structure, see Figure 5), and certainly is the most well-studied. It is a major constituent (in some cases composing over 90%) of the terpenoid fraction of a wide variety of plant materials, especially the oils of citrus fruit peel, such as oranges and lemons.[448] It occurs naturally in both the *d* and *l* forms, with *d*-limonene predominate in the citrus peel oils of orange, grapefruit, and lemon.[449] Commercial technical-grade sources are limited to *d*-limonene[450] and have been used widely as flavor and fragrance additives in food, beverage, perfume, soap, etc., for 50 years. The consumption of *d*-limonene has been estimated to be in the range of 0.2 to 2 mg/kg b.w. per day.[451]

Results of carcinogenicity testing conducted under the auspices of the NTP were negative in mice, while renal neoplasm were found in the male but not female rat.[452] Extensive studies of the mechanism of the carcinogenic action of *d*-limonene revealed a characteristic nephrotoxicity, a key aspect of which is the dose-related accumulation of a protein ($\alpha_{2\mu}$-globulin) which binds *d*-limonene and its metabolite. This phenomena is largely limited to the male rat and is generally believed to have no relevance to humans. Mutagenic activity of *d*-limonene was negative in the Ames' *Salmonella* assay and in other tests of genotoxicity.[448] The IARC concluded that there was limited evidence of carcinogenicity in animals, and that *d*-limonene was not classifiable as to its carcinogenicity to humans (i.e., Group 3).[448] Unless other evidence surfaces, *d*-limonene should not be regarded as a human carcinogen.

15 URETHANE

Urethane, also known as ethyl carbamate, is a naturally occurring contaminant found in alcoholic beverages and other fermented products such as brandy, wine, beer, ale, sake, distilled spirits, liqueurs, soy sauce, bread, yogurt, and olives. Synthetic urethane is used as a chemical intermediate in the manufacture of pesticides, fumigants, and cosmetics. Chemically, it is the ethyl ester of carbamic acid. It should not be confused with the high-molecular-weight polyurethanes used as foams, elastomers, and coatings which are sometimes called urethanes. Such products are not made from, nor do they degrade to, urethane.

Increased levels of urethane in alcoholic beverages have been found to be the result of the use of urea as a yeast nutrient[453] and the use of the food preservative, diethylpyrocarbonate.[454] The latter was banned by the FDA in the early 1970s, and according to the Bureau of Alcohol, Tobacco, and Firearms, urea is no longer used by U.S. producers of alcoholic beverages;[455] however, urethane is a natural byproduct of the fermentation process and can be detected in fermented beverages and foods at generally low but measurable levels.

Ough[456] investigated the occurrence of urethane in fermented foods and reported detectable levels in soy sauce, yogurt, bread, and olives. Concentrations in most samples were between 2 and 5 ppb, or just above the detection limit of 2 ppb. Later analysis by the FDA confirmed these findings, except for soy sauce which had a range of urethane of 0 to 84 ppb (with a mean of 7 ppb). Substantially higher levels have been found for various categories of alcoholic beverages, domestic and imported. Highest levels of urethane were found for fruit brandies, whiskies, liqueurs, and distilled spirits (e.g., vodka, gin, tequila, and rum). Generally, these average urethane levels were below 200 ppb, but a few samples were as high as 2000 to 12,000 ppb. Wines generally had urethane levels under 100 ppb, and most malt beverages had undetectable amounts.

The carcinogenicity of urethane has been suspected since the initial observations by Nettleship et al.[457] Since this observation, extensive carcinogenicity testing in several species by several routes has been carried out by many laboratories. Urethane has been shown to be carcinogenic in rats, mice, hamsters, and monkeys when administered orally, by parenteral injection, and by dermal application. It has produced multiple tumors at both local and systemic sites. These tumors were in the lungs, liver, forestomach, hematopoietic system, mammary gland, and skin, among other organs. Table 14 has a summary of the studies employing the oral route of exposure. Results from mutagenicity studies are either negative or inconclusive, for both bacterial and mammalian cells (for an extensive review of the genotoxicity of urethane see Bateman[458] and Allen et al.[459]). No studies on exposed humans are available. Based on the totality of evidence, the IARC[460] concluded that there is sufficient evidence for the carcinogenicity of urethane in animals, and that it is "possibly carcinogenic to humans" (i.e., Group 2B). The EPA classifies urethane as a "probable human carcinogen"(i.e., a B2 carcinogen). Given the preponderance of evidence, there is sufficient justification to consider urethane as a carcinogenic risk to humans.

16 CONCLUSION

The food supply in the U.S. is the most wholesome, abundant, varied, low cost, and *safest* in the history of mankind. That is not to say that the food supply is devoid of all harmful compounds. It is a fundamental toxicological tenet that all compounds are toxic to some degree, and it is the amount ingested that will determine whether a dietary compound is harmful. The public concern with the development of cancer as a major food hazard may be a bit misguided, but it has some basis. It is known from the elegant epidemiological analysis conducted by Doll and Peto in 1981 that 20 to 50% of all human cancer is associated

TABLE 14

Oral Carcinogenicity Studies of Urethane

Species	Dose[a]	Dosing Duration	Carcinogenic Target Organ(s)	Ref.
Rat	0.15% diet	15 months	Lung	461
Mouse	15 mg/wk i.g.	45 wk	No tumors without co-administered croton oil	462
Mouse	60–900 mg i.g. (total dose)	3–45 wk	Forestomach	463
Mouse[b]	15 mg i.g., or 0.1% diet	1–10 wk, or 6 months	Lung	464
Hamster	0.2% d.w.	55–76 wk	Skin, forestomach	465
Hamster	0.2–0.4% d.w.	42 wk	Skin, forestomach, lung, mammary gland, liver	466
Mouse	0.4% d.w.	42 wk	Hematopoietic system	467
Mouse	0.1–0.3% d.w.	13–31 wk	Lung, hematopoietic system, mammary gland	468
Rat	0.1% d.w.	14–24 wk	Zymbal gland	469
Mouse	2.8 or 5.5 mg i.g.	3 times per wk for 5 wk	Hematopoietic system, liver, lung	470
Mouse	0.4% d.w.	5–30 days	Hematopoietic system, lung, liver, Harderian gland	471
Mouse	0.4% d.w.	10–20 days	Hematopoietic system, mammary gland	472
Mouse	0.4% d.w.	15–20 days	Thymus, lung, Harderian gland, mammary gland	473
Rat	0.2% diet	Lifetime (ave. = 50 wk)	Not statistically significant[c]	474
Rat	0.1% d.w.	Lifetime (ave. = 54 wk)	Cannot be determined due to no control group	475
Mouse	0.1% d.w.	10 wk	Tumor incidence not reported	476
Hamster	0.1% d.w.	Lifetime	Skin, forestomach, cecum, liver, adrenal, gall bladder	477
Mouse	158 mg/kg b.w. i.g. then 600 ppm diet	69–74 wk	Lung, liver, Harderian gland, hematopoietic system	478
Mouse	0.01% d.w.	Lifetime[d]	Lung, bone, hematopoietic system, urinary bladder, mammary gland	479
Rat	0–12.5 mg/kg b.w. in d.w.	Lifetime	Mammary gland	480
Mouse	0–12.5 mg/kg b.w. in d.w.	Lifetime	Lung	480
Monkey[b]	250 mg/kg b.w. i.g.	5 days/wk for ≤ 5 yr	Liver, small intestine	481
Mouse	0–42 mg/kg b.w. i.g.	3 times per wk for 8 wk	Lung	482

[a] i.g. = intragastric; d.w. = drinking water; b.w. = body weight.
[b] Poor reporting of details limits evaluation of this study.
[c] Small group size.
[d] Six-generation study.

with dietary factors. Keeping this in mind, along with a substantial scientific database which includes actual morbidity and mortality data, the FDA has developed a list of food safety priorities. Topping the list are foodborne disease (e.g., pathogenic microbiological contamination) and nutritional imbalances, including overnutrition. Much less of a significant cause of human illness are environmental contaminants and naturally occurring toxicants, including carcinogens. This low-level occurrence of natural "carcinogenic" compounds present in the food supply for thousands of years does not pose a new safety risk. What is relatively new is that we can now crudely determine or estimate this low level of risk, but the fact remains that humans have lived with essentially the same food supply for many centuries without significant harm. Our body's metabolic and other defense systems are well suited to detoxify the low-level exposure to natural carcinogens. Furthermore, the identification and evaluation of nearly all carcinogens have relied on experimental investigation using laboratory animals. This model has several limitations; chief among these is the use of "maximum tolerated dose" (i.e., the MTD). Use of the MTD ensures that even weak carcinogens are not missed; however, its use has the potential to exceed the test organism's natural defense systems and other underlying biological processes and perhaps to produce tumors that are not relevant under conditions normal to humans. Finally, it should not be forgotten that the food supply contains many anticarcinogenic substances as demonstrated by data from hundreds of experimental laboratory studies. Further evidence of the efficacy of these anticarcinogenic substances can be found through the many epidemiological studies showing the clear benefit of consuming a diet high in fruits and vegetables in the prevention of cancer, despite the presence of potentially carcinogenic compounds.

A perusal of this chapter reveals several classes of compounds that have sufficiently potent carcinogenic activity and occur at significantly high enough levels in various foods to warrant attention. These compounds are for the most part limited to mycotoxins, nitrosamines, and compounds produced by cooking or other thermal processes. All of these compounds either receive scrutiny and control by governmental regulatory bodies or are subject to a large number of safeguards, controls, and tests utilized by the food industry. A second tier of compounds includes those that are plant derived but pose uncertainty in relating the animal carcinogenicity evidence to humans. Apart from the uncertainty in species differences in metabolism and sensitivity, the levels of exposures used in the animal studies are many orders of magnitude greater than the levels ingested by humans. In addition, the near total absence of epidemiological studies makes it extremely difficult to assess the human carcinogenic risk. The remaining group of compounds belong to foods that have limited geographical ingestion, typically are tested at unrealistically high levels, and possess evidence of carcinogenicity that may be weak and limited to a few isolated studies. Clearly, the difficulties in determining the human carcinogenic risk posed by the compounds belonging to the latter two groups (and to some extent for the first group of compounds, too) will not be resolved until experimental models are developed to assess carcinogenicity at low exposure levels and greater emphasis is placed on understanding the metabolic pathways of carcinogens and how they may differ among species. Finally, assessing the carcinogenic risk of a particular compound would be greatly enhanced by the inclusion of well-designed epidemiological studies incorporating sound exposure data.

ACKNOWLEDGMENTS

The author is indebted to Dr. Kreutler, Dr. Skrypec, and Ms. White for their helpful advice. The author also thanks Ms. Fahrenbach and Ms. Pacay for their professional assistance, and the expert secretarial staff in the word processing department, especially Ms. Green. Finally, the author thanks Diana, Maggie, and Danny for their patience and understanding in permitting him the many hours away from family time needed in the preparation of this review.

REFERENCES

1. Enomoto, M., Safrole, in *Naturally Occurring Carcinogens of Plant Origin,* Hirono, I., Ed., Kodansha/Elsevier, Tokyo/Amsterdam, 1987, pp. 139–159.
2. Guenther, E., *The Essential Oils,* Vols. 1, 3–6, D. Van Nostrand, New York, 1948–1952.
3. Guenther, E. and Althausen, D., *The Essential Oils,* Vol. 2, D. Van Nostrand, New York, 1949.
4. Woo, Y., Lai, D. Y., Arcos, J. C., and Argus, M. F., *Chemical Induction of Cancer. Structural Bases and Biological Mechanisms,* Vol. IIIC, *Natural, Metal, Fiber, and Macromolecular Carcinogens,* Academic Press, San Diego, CA, 1988.
5. Furia, T. E. and Bellanca, N., Eds., *Fenaroli's Handbook of Flavor Ingredients,* CRC Press, Boca Raton, FL, 1971, p. 610.
6. Borchert, P., Miller, J. A., Miller, E. C., and Shires, T. K., 1-Hydroxysafrole, a proximate carcinogenic metablite of safrole in the rat and mouse, *Cancer Res.,* 33, 590–600, 1973.
7. Wislocki, P. G., Miller, E. C., Miller, J. A., McCoy, E. C., and Rosenkranz, H. S., Carcinogenic and mutagenic activities of safrole, 1-hydroxysafrole and some known or possible metabolites, *Cancer Res.,* 37, 1883–1891, 1977.
8. Boberg, E. W., Miller, E. C., Miller, J. A., Poland, A., and Liem, A., Strong evidence from studies with brachymorphic mice and pentachlorophenol that 1-sulfooxysafrole is the major ultimate electrophilic and carcinogenic metabolite of 1-hydroxysafrole in mouse liver, *Cancer Res.,* 43, 5163–5173, 1983.
9. Miller, E. C., Swanson, A. B., Phillips, D. H., Fletcher, T. L., Liem, A., and Miller, J. A., Structure-activity studies of the carcinogenicities in the mouse and rat of some naturally occurring and synthetic alkenylbenzene derivatives related to safrole and estragole, *Cancer Res.,* 43, 1124–1134, 1983.
10. Hagen, E. C., Jenner, P. M., Jones, W. I., Fitzhugh, O .G., Long, E. L., Brouwer, J. G., and Webb, W. K., Toxic properties of compounds related safrole, *Toxicol. Appl. Pharmacol.,* 7, 18–24, 1965.
11. Innes, J. R. M., Ulland, B. M., Valerio, M. G., Petrucelli, L., Fishbein, L., Hart, E. R., Pallota, A. J., Bates, R. R., Falk, H. L., Gart, L .L., Klein, M., Mitchell, I., and Peter, J., Bioassay of pesticides and industrial chemicals for tumorigenicity in mice: a preliminary note, *J. Natl. Cancer Inst.,* 42, 1101–1114, 1969.
12. Gross, M. A., Jones, W. I., Cook, E. L., and Boone, C. C., Carcinogenicity of oil of calamus, *Proc. Am. Assoc. Cancer Res.,* 8, 24, 1967.
13. Ambrose, A. M., Cox, Jr., A. J., and De Eds, F., Toxicological studies on sesamol, *J. Agric. Food Chem.,* 6, 600–604, 1958.
14. Long, E. L., Hansen, W. H., and Nelson, A. A., Liver tumors produced in rats by feeding safrole, *Fed. Proc.,* 20, 287, 1961.
15. Homburger, F., Kelley, T., Jr., Friedler, G. I., and Russfield, A. B., Toxic and possible carcinogenic effects of 4-allyl-1,2-methylene dioxybenzene (safrole) in rats on deficient diets, *Med. Exp.,* 4, 1–11, 1961.
16. Abbott, D. D., Packman, E. W., Wagner, B. M., and Harrison, J. W. E., Chronic oral toxicity of oil of sassafras and safrole, *Pharmacologist,* 3, 62, 1961.
17. *Fed. Reg.,* 25, 12412, 1960.
18. Wilson, J. B., Determination of safrole and methylsalicylate in soft drinks, *J. Assoc. Off. Agric. Chem.,* 42, 696–698, 1959.
19. Segelman, A. B., Segelman, F. P., Karliner, J., and Sofia, R. D., Sassafras and herb tea. Potential health hazards, *J. Am. Med. Assoc.,* 236, 477, 1976.
20. Miller, J. A., Naturally occurring substances that can induce tumors, in *Toxicants Occurring Naturally in Foods,* 2nd ed., National Academy of Sciences, Washington, D.C., 1973, p. 530.
21. Hall, R. L., Toxicants occurring naturally in spices and flavors, in *Toxicants Occurring Naturally in Foods,* 2nd ed., National Academy of Sciences, Washington, D.C., 1973, p. 448.
22. Hilker, D. M., Carcinogens occurring naturally in food, *Nutr. Cancer,* 2, 217–223, 1981.
23. Fukuda, Y., Nagata, M., Osawa, T., and Namiki, M., Chemical aspects of the antioxidative activity of roasted seed oil, and the effect of using the oil for frying, *Agric. Biol. Chem.,* 50, 857–862, 1986.
24. Miller, J. A., and Miller, E. C., The metabolic activation and nucleic acid adducts of naturally occurring carcinogens: recent results with ethyl carbamate and the spice flavors safrole and estragole, *Br. J. Cancer,* 48, 1–15, 1983.
25. IARC, *Some Naturally Occurring Substances. IARC Monograph on the Evaluation of Carcinogenic Risks of Chemicals to Man,* Vol. 10, *Safrole, Isosafrole, and Dihydrosafrole,* International Agency for Research on Cancer, Lyon, France, 1976, pp. 231–244.
26. Hirose, M., Fukushima, S., Shirai, T., Hasegawa, R., Kato., T., Tanaka, H., Asakawa, E., and Ito, N., Stomach carcinogenicity of caffeic acid, sesamol, and catechol in rats and mice, *Jpn. J. Cancer Res.,* 81, 207–212, 1990.
27. Opdyke, D. L. J., Monographs on fragrance raw materials, *Food Cosmet. Toxicol.,* 14, 307–338, 1976.
28. Russell, G. F., and Jennnings, W. G., Constituents of black pepper. Some oxygenated compounds, *J. Agr. Food Chem.,* 17, 1107–1112, 1969.
29. Budowski, P., and Markley, K. S., The chemical and physiological properties of sesame oil, *Chem. Rev.,* 48, 125–151, 1951.

30. Taylor, J. M., Jones, W. I., Hagan, E. C., Gross, M. A., Davis, D. A., and Cook, E. L., Toxicity of oil of calamus (Jammu variety), *Tox. Appl. Pharm.*, 10, 405, 1967.
31. Clark, G. S., Allyl isothiocyanate, *Perfum. Flavorist,* 17, 107–109, 1992.
32. Food and Drug Research Laboratories, *GRAS (Generally Recognized as Safe) Food Ingredients — Oil of Mustard and Allyl Isothiocyanate, Final Report 1920–1972,* Report No. FDABF-GRAS-015, PB 221 215, Prepared for the Food and Drug Administration, Springfield, VA, National Technical Information Service.
33. Life Sciences Research Office, *Evaluation of the Health Aspects of Mustard and Oil of Mustard as Food Ingredients,* SCOGS-16, Life Sciences Research Office, Bethesda, MD, 1975.
34. VanEtten, C. H., Daxenbichler, M. E., and Wolff, I. A., Natural glucosinolates (thioglucosides) in foods and feeds, *J. Agric. Food Chem.*, 17, 483–491, 1969.
35. Dunnick, J. K., Prejean, J. D., Haseman, J., Thompson, R. B., Giles, H. D., and McConnell, E. E., Carcinogenesis bioassay of allyl isothiocyanate, *Fund. Appl. Toxicol.*, 2, 114–120, 1982.
36. IARC, *IARC Monographs on the Evaluation of the Carcinogenic Risk of Chemicals to Humans,* Vol. 36, *Allyl Isothiocyanate,* International Agency for Research on Cancer, Lyon, France, 1985, pp. 55–58.
37. National Toxicology Program, *Carcinogenesis Bioassay of Allyl Isothiocyanate (CAS No. 57-06-7) in F344/N Rats and B6C3F$_1$ Mice (Gavage Study),* NTP No. 81-36, NIH Publication No. 83-1790, U.S. Department of Health and Human Services, Washington, D.C., 1982.
38. Hodge, W. H., Fern foods of Japan and the problem of toxicity, *Am. Fern. J.,* 63, 77–80, 1973.
39. Hirono, I. and Yamada, K., Bracken fern, in *Naturally Occurring Carcinogens of Plant Origin*, Hirono, I., Ed., Kodansha/Elsevier, Tokyo/Amsterdam, 1987, pp. 87–120.
40. Hirono, I., Shibuya, C., Shimizu, M., and Fushimi, K., Carcinogenic activity of processed bracken used as human food, *J. Natl. Cancer Inst.,* 48, 1245–1250, 1972.
41. Evans, I. A., Widdop, B., Jones, R. S., Barber, G. D., Leach, H., Jones, H., Jones, D. L., and Mainwaring-Burton, R., The possible human hazard of the naturally occurring bracken carcinogen, *Biochem. H.*, 124, 28–29, 1971.
42. Evans, I. A., Jones, R. S., and Mainwaring-Burton, R., Passage of bracken fern toxicity into milk, *Nature* (*Lond.*), 237, 107–108, 1972.
43. Rosenberger, G. and Heeschen, W., Adlerfarn (*Pteris aquilina*) — die ursache des sog. Stallrotes der rinder (haematuria vesicalis bovis chronica), *Dtsch. Tierarztl. Wschr.*, 67, 201–208, 1960.
44. Evans, I. A., and Mason, J., Carcinogenic acitivity of bracken, *Nature,* 208, 913–914, 1965.
45. Price, J. M., and Pamukçu, A. M., The induction of neoplasms of the urinary bladder of the cow and the small intestine of the rat by feeding bracken fern (*Pteris aquilina*), *Cancer Res.*, 28, 2247–2251, 1968.
46. Pamukçu, A. M., and Price, J. M., Induction of intestinal and urinary bladder cancer in rats by feeding bracken fern (*Pteris aquilina*), *J. Natl. Cancer Inst.*, 43, 275–281, 1969.
47. Hirono, I., Shibuya, C., Fushimi, K., and Haga, M., Studies on the carcinogenic properties of bracken, *Pteridium aquilinum, J. Natl. Cancer Inst.*, 45, 179–188, 1970.
48. Pamukçu, A. M., Goksoy, S. K., and Price, J. M., Urinary bladder neoplasms induced by feeding bracken fern (*Pteris aquilina*) to cows, *Cancer Res.*, 27, 917–924, 1967.
49. Pamukçu, A. M., Price, J. M., and Bryan, G. T., Naturally occurring and bracken-induced bovine urinary bladder tumors, *Vet. Pathol.*, 13, 110–122, 1976.
50. Hirono, I., Sasaoko, I. Shibuya, C., Shimizu, M., Fushimi, K., Mori, H., Kato, K., and Haga, M., Natural carcinogenic products of plant origin, *Gann Monogr. Cancer Res.,* 17, 205–207, 1975.
51. Evans, I. A., The radiomimetic nature of bracken toxin, *Cancer Res.,* 28, 2252–2261, 1968.
52. Pamukçu, A. M., Ertürk, E., Price, J. M., and Bryan, G. T., Lymphatic leukemia and pulmonary tumors in female Swiss mice fed bracken fern (*Pteris aquilina*), *Cancer Res.*, 32, 1442–1445, 1972.
53. El-Mofty, M. M., Sadek, I. A., and Bayoumi, S., Improvement in detecting the carcinogenicity of bracken fern using an Egyptian toad, *Oncology,* 37, 424–425, 1980.
54. Dodd, D. C., Adenocarcinoma of the sheep, *N. Z. Vet. J.,* 8, 110–112, 1960.
55. McCrea, C. T. and Head, K. W., II, Experimental production of tumors, *Br. Vet. J.*, 137, 21–30, 1981.
56. Pamukçu, A. M., Yalciner, S., Hatcher, J. F., and Bryan, G. T., Quercetin, a rat intestinal and bladder carcinogen present in bracken fern (*Pteridium aquilinum*), *Cancer Res.*, 40, 3468–3472, 1980.
57. Ertürk, E., Nunoya, T., Hatcher, J. F., Pamukçu, A. M., and Bryan, G. T., Comparison of bracken fern and quercetin carcinogenicity in rats, *Am. Assoc. Cancer Res.*, 24, 53, 1983.
58. Ertürk, E., Hatcher, J. F., Nunoya, T., Pamukçu, A. M., and Bryan, G. T., Hepatic tumors in Sprague-Dawley (SD) and Fischer 344 (F) female rats chronically exposed to quercetin or its glycoside rutin (R), *Am. Assoc. Cancer Res.,* 25, 95, 1984.
59. Evans, I. A., and Osman, M. A., Carcinogenicity of bracken and shikimic acid, *Nature* (*Lond.*), 250, 348–349, 1974.
60. Hirono, I., Aiso, S., Yamaji, T., Mori., H., Yamada, K., Niwa, H., Ojika, M., Wakamatsu, K., Kigoshi, H., Niijama, K., and Uosaki, Y., Carcinogenicity in rats of ptaquiloside isolated from bracken fern, *Gann,* 75, 833–836, 1984.

61. Hirono, I. Ogino, H., Fujimoto, M., Yamada, K. Yoshida, Y., Ikagawa, M., and Okumura, M., Induction of tumors in ACI rats given a diet containing ptaquiloside, a bracken carcinogen, *J. Natl. Cancer Inst.*, 79, 1143–1149, 1987.
62. Hirono, I., Fushimi, K., and Matsubara, N., Carcinogenicity test of shikimic acid in rats, *Toxicol. Lett.*, 1, 9–10, 1977.
63. Saito, D., Shirai, A., Matsushima, T., Sugimura, T., and Hirono, I., Test of carcinogenicity of quercetin, a widely distributed mutagen in food, *Teratogen., Carcinogen., Mutagen.*, 1, 213–221, 1980.
64. Hosaka, S. and Hirono, I., Carcinogenicity test of quercetin by pulmonary-adenoma bioassay in strain A mice, *Gann*, 72, 327–328, 1981.
65. Hirono, I., Ueno, I., Hosaka, S., Takanashi, H., Matsushima, T., Sugimura, T., and Natori, S., Carcinogenicity examination of quercetin and rutin in ACI rats, *Cancer Lett.*, 13, 15–21, 1981.
66. Takanashi, H., Aiso, S., Hirono, I., Matsushima, T., and Sugimura, T., Carcinogenicity test of quercetin and kaempferol in rats by oral administration, *J. Food Safety*, 5, 55–60, 1983.
67. Morino, K., Matsukura, N., Kawachi, T., Ohgaki, H., Sugimura, T., and Hirono, I., Carcinogenicity test of quercetin and rutin in golden hamsters by oral administration, *Carcinogenesis*, 3, 93–97, 1982.
68. Yoshihira, K., Fukuoka, M., Kuroyanagi, M., Natori, S., Umeda, M., Morohoshi, T., Enomoto, M., and Saito, M., Chemical and toxicological studies on bracken fern, *Pteridium aquilinum* var. *latiusculum*. I. Introduction, extraction and fractionation of constituents, and toxicological tests, *Chem. Pharm. Bull.*, 26, 2346–2364, 1978.
69. Van der Hoeven, J. C. M., Lagerweij, W. J., Posthumus, M. A., van Veldhuizen, A., and Holterman, H. A. J., Aquilide A, a new mutagenic compound isolated from bracken fern (*Pteridium aquilinum* (L.) Kuhn), *Carcinogenesis*, 4, 1587–1590, 1983.
70. MacGregor, J. T., Mutagenicity studies of flavonoids *in vitro* and *in vivo*, *Toxicology Appl. Pharmacol.*, 48, A47, 1979.
71. Hirono, I., Natural carcinogenic products of plant origin, *CRC Crit. Rev. Toxicol.*, 8, 235–277, 1981.
72. IARC, *IARC Monographs on the Evaluation of Carcinogenic Risk of Chemicals to Man*, Vol. 40, *Bracken Fern and Some of Its Constituents*, International Agency for Research on Cancer, Lyon, France, 1986, pp. 45–65.
73. Nehlig, A. and Debry, G., Potential genotoxic, mutagenic and antimutagenic effects of coffee: a review, *Muta. Res.*, 317, 145–162, 1994.
74. Sondheimer, E., On the distribution of caffeic acid and the chlorogenic acid isomers in plants, *Arch. Biochem. Biophys.*, 74, 131–138, 1958.
75. IARC, *IARC Monographs on the Evaluation of Carcinogenic Risks to Humans*, Vol. 56, *Some Naturally Occurring Substances: Food Items and Constituents, Heterocyclic Aromatic Amines and Mycotoxins*, International Agency for Research on Cancer, Lyon, France, 1993, pp. 115–134.
76. Ames, B. N., Profet, M., and Gold, L. S., Dietary carcinogens and mutagens from plants, in *Mutagens in Food: Detection and Prevention*, Hayatsu, H., Ed., CRC Press, Boca Raton, FL, 1991, pp. 29–50.
77. Wattenberg, L. W., Coccia, J. B., and Lam, L. K., Inhibitory effects of phenolic compounds on benzo[*a*]pyrene-induced neoplasia, *Cancer Res.*, 40, 2820–2823, 1980.
78. Huang, M. T., Smart, R. C., Wong, C. Q., and Conney, A. H., Inhibitory effect of curcumin, chlorogenic acid, caffeic acid, and feruic acid on tumor promotion in mouse skin by 12-*O*-tetradecanoylphorbol-13-acetate, *Cancer Res.*, 48, 5941–5946, 1988.
79. Hagiwara, A., Hirose, M., Takahashi, S., Ogawa, K., Shirai, T., and Ito, N., Forestomach and kidney carcinogenicity of caffeic acid in F344 rats and C57BL/6N × C3H/HeN F_1 mice, *Cancer Res.*, 51, 5655–5660, 1991.
80. Hirose, M., Fukushima, S., Shirai, T., Hasegawa, R., Kato, T., Tanaka, H., Asakawa, E., and Ito, N., Stomach carcinogenicity of caffeic acid, sesamol and catechol in rats and mice, *Jpn. J. Cancer Res.*, 81, 207–212, 1990.
81. Stich, H. F., Rosin, M. P., Wu, C. H., and Powrie, W. D., A comparative genotoxicity study of chlorogenic acid (3-*O*-caffeoylquinic acid), *Mutat. Res.*, 90, 201–212, 1981.
82. Hanham, A. F., Dunn, B. P., and Stich, H. F., Clastogenic activity of caffeic acid and its relationship to hydrogen peroxide generated during autooxidation, *Mutat., Res.*, 116, 333–339, 1983.
83. Yamada, K., Shirahata, S., Murakami, H., Nishiyama, K., Shinohara, K., and Omura, H., DNA breakage by phenyl compounds, *Agric. Biol. Chem.*, 49, 1423–1428, 1985.
84. Woo, Y., Lai, D. Y., Arcos, J. C., and Argus, M. F., *Chemical Induction of Cancer, Structural Bases and Biological Mechanisms*, Vol. IIIC, *Natural, Metal, Fiber, and Macromolecular Carcinogens*, Academic Press, San Diego, CA, 1988.
85. Windholz, M., Ed., *The Merck Index*, 10th ed., Merck and Co., Rahway, NJ, 1983, p. 367.
86. Hagan E. C., Hansen, W. H., Fitzhugh, O. G., Jenner, P. M., Jones, W. I., Taylor, J. M., Long, E. L., Nelson, A. A., and Brouwer, J. B., Food flavourings and compounds of related structure. II. Subacute and chronic toxicity, *Food Cosmet. Toxicol.*, 5, 141–157, 1967.
87. Griepentrog, F., Pathologisch-anatomische befunde zur karzinogenen wirkung von cumarin im tierversuch, *Toxicology*, 1, 93–102, 1973.
88. Ueno, I. and Hirono, I., Non-carcinogenic response to coumarin in syrian golden hamsters, *Food Cosmet. Toxicol.*, 19, 353–355, 1981.

89. Evans, J. G., Gaunt, I. F., and Lake, B. G., Two-year toxicity study on coumarin in the baboon, *Food Cosmet. Toxicol.,* 17, 187–193, 1979.
90. Haworth, S., Lawlor, T., Mortelmans, K., Speck, W., and Zeiger, E., *Salmonella* mutagenicity test results for 250 chemicals, *Environ. Mutagen. (Suppl.)*, 1, 3–142, 1983.
91. IARC, *Some Naturally Occurring Substances*, Vol. 10, *Coumarin*, International Agency for Research on Cancer, Lyon, France, 1976, pp. 113–119.
92. NTP, *Toxicology and Carcinogenesis Studies of Coumarin (CAS No. 91-64-5) in F344/N Rats and B6C3F$_1$ Mice (Gavage Study),* NTP Technical Report No. 422, NIH Publication No. 93-3153, National Toxicology Program, Research Triangle Park, NC, 1993.
93. Cohen, A. J., Critical review of the toxicology of coumarin with special reference to interspecies differences in metabolism and hepatotoxic response and their significance to man, *Food Chem. Toxicol.*, 17, 277–289, 1979.
94. Harborne, J. B., Mabry, T. J., and Mabry, H., *The Flavonoids,* Chapman & Hall, London, 1975.
95. Herrmann, K., Flavonols and flavones in food plants: a review, *J. Food Technol.*, 11, 433–448, 1976.
96. IARC, *IARC Monographs on the Evaluation of the Carcinogenic Risk of Chemicals to Humans,* Vol. 31, *Quercetin,* International Agency for Research on Cancer, Lyon, France, 1983, pp. 213–229.
97. Kuhnau, J., The flavonoids. A class of semi-essential food components: their role in human nutrition, *World Rev. Nutr. Diet.,* 24, 117–191, 1976.
98. Bjeldanes, L. F. and Chang, G. W., Mutagenic activity of quercetin and related compounds, *Science,* 197, 577–578, 1977.
99. Hardigree, A. A. and Epler, J. L., *Mutagenicity of Plant Flavonols in Microbial Systems, Abstr. 8th Ann. Mtg. Environmental Mutagen Society,* Colorado Springs, CO, 1977, p. 48.
100. Sugimura, T., Nagao, M., Matsushima, T., Yahagi, T., Seino, Y., Shirai, A., Sawamura, M., Natori, S., Yoshihira, K., Fukuoka, M., and Kuroyanagi, M., Mutagenicity of flavone derivatives, *Proc. Jpn. Acad., Ser. B,* 53, 194–197, 1977.
101. Brown, J. P., A review of the genetic effects of naturally occurring flavonoids, anthraquinones and related compounds, *Mutat. Res.,* 75 243–277, 1980.
102. Nagao, M., Morita, N., Yahagi, T., Shimizu, M., Kuroyanagi, M., Fukuoka, M., Yoshihira, K., Natori, S., Fujino, T., and Sugimura, T., Mutagenicities of 61 flavonoids and 11 related compounds, *Environ. Mutagen,* 3, 401–419, 1981.
103. MacGregor, J. T. and Jurd, L., Mutagenicity of plant flavonoids: structural requirements for mutagenic activity in *Salmonella typhimurium, Mutat. Res.,* 54, 297–309, 1978.
104. Pamukçu, A. M., Yalciner, S., Hatcher, J. F., and Bryan, G. T., Quercetin, a rat intestinal and bladder carcinogen present in bracken fern (*Pteridium aquilinum*), *Cancer Res.,* 40, 3468–3472, 1980.
105. Ertürk, E., Nunoya, T., Hatcher, J. F., Pamukçu, A. M., and Bryan, G. T., Comparison of bracken fern and quercetin carcinogenicity in rats, *Proc. Am. Assoc. Cancer Res.*, 24, 53, 1983.
106. Ertürk, E., Hatcher, J. F., Nunoya, T., Pamukçu, A. M., and Bryan, G. T., Hepatic tumors in Sprague-Dawley (SD) and Fischer 344 (F) female rats chronically exposed to quercetin or its glycoside rutin (R), *Proc. Am. Assoc. Cancer Res.*, 25, 95, 1984.
107. NTP, *Toxicology and Carcinogenesis Studies of Quercetin (CAS No. 117-39-5) in F344/N Rats (Feed Studies),* NTP Technical Report No. 409, National Toxicology Program, Research Triangle Park, NC, 1992.
108. Takanashi, H., Aiso, S., and Hirono, I., Carcinogenicity test of quercetin and kaempferol in rats by oral administration, *J. Food Safety,* 5, 55–60, 1983.
109. IARC, *IARC Monographs on the Evaluation of the Carcinogenic Risk of Chemicals to Humans,* Vol. 31, *Kaempferol,* International Agency for Research on Cancer, Lyon, France, 1983, pp. 171–178.
110. Wilson, R. H., Mortarotti, T. C., and Doxtrader, E. K., Toxicity studies on rutin, *Proc. Soc. Exp. Bio. Med.*, 64, 324–327, 1947.
111. Ambrose, A. M., Robbins, D. J., and De Eds, F., Comparative toxicities of quercetin and quercitrin, *J. Am. Pharm. Assoc.*, 41, 119–122, 1952.
112. Saito, D., Shirai, A., Matsushima, T., Sugimura, T., and Hirono, I., Test of carcinogenicity of quercetin, a widely distributed mutagen in food, *Teratog. Carcinog. Mutagen.*, 1, 213–221, 1980.
113. Hirono, I., Ueno, I., Hosaka, S., Takanashi, H., Matsushima, T., Sugimura, T., and Natori, S., Carcinogenicity examination of quercetin and rutin in ACI rats, *Cancer Lett.*, 13, 15–21, 1981.
114. Hosaka, S., and Hirono, I., Carcinogenicity test of quercetin by pulmonary-adenoma bioassay in strain A mice, *Jpn. J. Cancer Res.*, 72, 327–328, 1981.
115. Marino, K., Matsukura, N., Kawachi, T., Ohgaki, H., Sugimura, T., and Hirono, I., Carcinogenicity test of quercetin and rutin in golden hamsters by oral administration, *Carcinogenesis,* 3, 93–97, 1982.
116. Hirose, M., Fukushima, S., Sakata, T., Inui, M., and Ito, N., Effect of quercetin on two-stage carcinogenesis of the rat urinary bladder, *Cancer Lett.,* 21, 23–27, 1983.
117. Ito, N., Hagiwara, A., Tamano, S., Kagawa, M., Shibata, M-A., Kurata, Y., and Fukushima, S., Lack of cacinogenicity of quercetin in F344/DuCrj rats, *Jpn. J. Cancer Res.*, 80, 317–325, 1989.
118. Deschner, E. E., Ruperto, J., Wong, G., and Newmark, H. L. Quercetin and rutin as inhibitors of azoxymethanol-induced colonic neoplasia, *Carcinogenesis,* 12, 1193–1196, 1991.

119. IARC, *IARC Monographs on the Evaluation of the Carcinogenic Risk of Chemicals to Humans. Some Naturally Occurring and Synthetic Food Components, Furocoumarins and Ultraviolet Radiation*, Vol. 40, *Furocoumarins*, International Agency for Research on Cancer, Lyon, France, 1986, pp. 291–371.
120. Wagstaff, D. J., Dietary exposure to furocoumarins, *Regul. Toxicol. Pharmacol.*, 14, 261–272, 1991.
121. Furia, T. E., and Bellanca, N., Eds., *Fenaroli's Handbook of Flavor Ingredients*, CRC Press, Boca Raton, FL, 1971, p. 610.
122. Ivie, G. W., Holt, D. L., and Ivey, M. C., Natural toxicants in human foods: psoralens in raw and cooked parsnip root, *Science*, 213, 909–910, 1981.
123. Ceska, O., Chaudhary, S., Warrington, P., Poulton, G., and Ashwood-Smith, M., Naturally occurring crystals of photocarcinogenic furocoumarins on the surface of parsnip roots sold as food, *Experientia*, 42, 1302–1304, 1986.
124. Young, A. R., Magnus, I. A., Davies, A. C., and Smith, N. P., A comparison of the phototumorigenic potential of 8-MOP and 5-MOP in hairless albino mice exposed to solar simulated radiation, *Br. J. Dermatol.*, 108, 507–518, 1983.
125. Zajdela, E. and Bisagni, E., 5-Methoxypsoralen, the melanogenic additive in sun-tan preparations, is tumorigenic in mice exposed to 365 nm u.v. radiation, *Carcinogenesis*, 2, 121–127, 1981.
126. Cartwright, L. E., and Walter, J. F., Psoralen-containing sunscreen is tumorigenic in hairless mice, *J. Am. Acad. Dermatol.*, 8, 830–836, 1983.
127. Dunnick, J. K., *NTP Technical Report on the Toxicology and Carcinogenesis Studies of 8-Methoxypsoralen (CAS No. 298-81-7) in F344/N Rats (Gavage Studies)*, Publication No. 89-2814, National Institutes of Health, Bethesda, MD, 1989.
128. Sugimura, T., Nagao, M., Kawachi, T., Honds, M., Yahagi, T., Seino, Y., Sato, S., Matsukura, N., Matsushima, T., Shirai, A., Sawamura, M., and Matsumoto, H., Mutagen-carcinogens in food, with special reference to highly mutagenic pyrolytic products in broiled foods, in *Origins of Human Cancer*, Hiatt, H. H., Watson, J. D., and Winston, J. A., Eds., Cold Spring Harbor Laboratory, Cold Spring Harbor, New York, 1977, pp. 1561–1576.
129. Nagao, M., Honda, M., Seino, Y., Yahagi, T., Kawachi, T., and Sugimura, T., Mutagenicity of protein pyrolysates, *Cancer Lett.*, 2, 335–339, 1977.
130. Eisenbrand, G. and Tang, W., Food-borne heterocyclic amines. Chemistry, formation, occurrence, and biological activities. A literature review, *Toxicology*, 84, 1–82, 1993.
131. Chen, C., Pearson, A. M., and Gray, J. I., Meat mutagens, in *Advances in Food Nutrition Research*, Vol 34, Academic Press, New York, 1990, pp. 387–449.
132. Matsumoto, T., Yoshida, D., Mizusaki, S. and Okamoto, H., Mutagenic activity of amino acid pyrolysates in *Salmonella typhimurium* TA98, *Mutat. Res.*, 48, 279–283, 1977.
133. Commoner, B, Vitayathil, A. J., Dolara, P., Nair, S., Madyastha, P., and Cuca, G. C., Formation of mutagens in beef and beef extract during cooking, *Science*, 201, 913–914, 1978.
134. Kosuge, T., Tsuji, Wakabayashi, K., Okamoto, T., Shudo, K., Iitaka, Y., Itai, A., Sugimura, T., Kawachi, T., Nagao, M., Yahagi, T., and Seino, Y., Isolation and structure studies of mutagenic principles in amino acid pyrolysates, *Chem. Pharm. Bull.*, 26, 611–619, 1978.
135. Pariza, M. W., Ashoor, S. H., Chu, F. S., and Lund, D. B., Effects of temperature and time on mutagen formation in panfried hamburger, *Cancer Lett.*, 7, 63–64, 1979.
136. Sugimura, T., Kawachi, T., Nagao, M., Yahagi, T., Seino, Y., Okamoto, T., Shudo, K., Kosuge, T., Tsuji, K., Wakabayashi, K., Iitaki, Y., and Itai, A., Mutagenic principle(s) in tryptophan and phenylalanine pyroysis products, *Proc. Jpn. Acad.*, 53, 58–61, 1977.
137. Sugimura, T., Kawachi, T., Nagao, M., and Yahagi, T., Mutagens in food as causes of cancer, in *Nutrition and Cancer. Etiology and Treatment*, Newell, G. R. and Ellison, N. M., Eds., Raven Press, New York, 1981, pp. 59–71.
138. Sugimura, T., Nagao, M., and Wakabayashi, F., Mutagenic heterocyclic amines in cooked foods, in *Environmental Carcinogens Selected Methods of Analysis*, Egan, H., Fishbein, L., Castegnaro, M., O'Neill I. K., and Bartsh, H., Eds., International Agency for Research on Cancer, Lyon, France, 1981, pp. 251–267.
139. Yamamoto, T., Tsuji, J., Kosuge, T., Okamoto, T., Shjudo, K., Takeda, K., Iitaka, Y., Yamaguchi, K., Seino, Y., Yahagi, T., Nagao, M., and Sugimura, T., Isolation and structure determination of mutagenic substances in L-glutamic acid pyrolysate, *Proc. Jpn. Acad.*, 54, 248–250, 1978.
140. Wakabayashi, K., Tsuji, K., Kosuge, T., Takeda, K., Yamaguchi, K., Shudo, K., Iitaka, Y., Okamoto, Yahagi, T., Nagao, M., and Sugimura, T., Isolation and structure determination of a mutagenic substance in L-lysine pyrolysate, *Proc. Jpn. Acad.*, 54, 569–571, 1978.
141. Yokota, M., Narita, K., Kosuge, T., Wakabayashi, K., Nagao, M., Sugimura, T., Yamaguchi, K., Shudo, K., Iitaka, Y., and Okamoto, T., A potent mutagen isolated from a pyrolysate of L-ornithine, *Chem. Pharm. Bull.*, 29, 1473–1475, 1981.
142. Kasai, H., Nishimura, S., Wakabayashi, K., Nagao, M., and Sugimura, T., Chemical synthesis of 2-amino-3-methylimidazol[4,5-f]quinoline (IQ), a potent mutagen isolated from broiled fish, *Proc. Jpn. Acad.*, 56, 382–384, 1980.

143. Kasai, H., Yamaizumi, Z., and Nishimura, S., A potent mutagen in broiled fish. Part 1. 2-Amino-3-methyl-3H-imidazo[4,5-f]quinoline, *J. Chem. Soc. Perkin Trans.*, 1, 2290–2303, 1980.
144. Kasai, H., Shiomi, T., Sugimura, T., and Nishimura, S., Synthesis of 2-amino-3,8-dimethylimidazo[4,5-f]quinoxaline(MeIQx), a potent mutagen isolated from fried beef, *Chem. Lett.*, 675–678, 1981.
145. Yamaizumi, A., Shiomi, T., Kasai, H., Wakabayashi, K., Nagao, M., Sugimura, T., and Nishimura, S., Quantitative analysis of a novel potent mutagen, 2-amino-3-methylimidazo[4,5-f]quinoline, present in broiled food by GC/MS, *Keonshu-Iyo Masu Kenkuyakai,* 5, 245–248, 1980.
146. Hargraves, W. A. and Pariza, M. W., Purification and mass spectral characterization of bacterial mutagens from commercial beef extract, *Cancer Res,* 43, 1467–1472, 1983.
147. Turesky, R. J., Wishnok, J. S., Tannenbaum, S. R., Pfund, R. A., and Buchi, G. H., Qualitative and quantitative characterization of mutagens in commerical beef extract, *Carcinogenesis,* 4, 863–866, 1983.
148. Felton, J. S., Knize, M. G., Shen, N. H., Andresen, B. D., Bjeldanes, L. F., and Hath, F. T., Identification of the mutagens in cooked beef, *Environ. Health Perspect.*, 67, 17–24, 1986.
149. Becher, G., Knize, M. G., Nes, I. F., and Felton, J. S., Isolation and identification of mutagens from a fried Norwegian meat product, *Carcinogenesis,* 9, 247–253, 1988.
150. Negishi, C., Wakabayashi, K., Tsuda, M., Sato, T., Sugimura, T., Saito, H., Maeda, M., and Jäegerstad, M., Formation of 2-amino-3,7,8-trimethyl-imidazo[4,5-*f*]quinoxaline, a new mutagen, by heating a mixture of creatinine, glucose and glycine, *Mutat. Res. Lett.*, 140, 55–59, 1984.
151. Negishi, C., Wakabayashi, K., Yamaizumi, Z., Saito, H., Sato, S., Sugimura, T., and Jäegerstad, M., Identification of 4,8-DiMeIQx, a new mutagen, *Mutat. Res.*, 147, 267–273, 1985.
152. Shibamoto, T., Nishimura, O., and Mihara, S., Mutagenicity of products obtained from a maltol-ammonia browning model system, *J Agric. Food Chem.*, 29, 643–646, 1981.
153. Jäegerstad, M., Reuterswaerd, A. L., Olsson, R., Grivas, S., Nyhammar, T., Olsson, K., and Dahlqvist, A., Creatin(in)e and Maillard reaction products as precursors of mutagenic compounds: effects of various amino acids, *Food Chem.*, 12, 255–264, 1983.
154. Jäegerstad, M., Grivas, S., Olsson, K., Reuterswaerd, A. L., Negishi, C., and Sato, S., Formation of food mutagens via Maillard reactions, *Prog. Clin. Biol. Res.*, 206, 155–167, 1986.
155. Jäegerstad, M., Reuterswaerd, A. L., Grivas, S., Olsson, K., Negishi, C., and Sato, S., Effect of meat composition and cooking conditions on the formation of mutagenic imidazoquinoxalines (MeIQx and its methyl derivatives), *Proc. Int. Symp. Princess Takamatsu Cancer Fes. Fund.*, 16, 87–96, 1986.
156. Jäegerstad, M., Skog, K., Grivas, S., and Olsson, K., Mutagens from model systems, in *Mutagens and Carcinogens in the Diet,* Pariza, M. W. Aeshbacher, H.-U., Felton, J. S., and Sato, S., Eds., Wiley–Liss, New York, 1990, pp. 71–88.
157. Sugimura, T. and Sato, S., Mutagens-carcinogens in foods, *Cancer Res.*, 43, 2415s-2421s, 1983.
158. Ohgaki, H., Takayama, S., and Sugimura, T., Carcinogenicities of heterocyclic amines in cooked food, *Mutation Res.*, 259, 399–410, 1991.
159. Wakabayashi, K., Ushiyama, H., Takahasi, M., Nukaya, H., Kim, S. B., Hirose, M., Ochai, M., Sugimura, T., and Nagao, M., Exposure to heterocyclic amines, *Environ. Health Perspect.*, 99, 129–133, 1993.
160. Munro, I. C., Kennepohl, E., Erickson, R. E., Portoghese, P. S., Wagner, B. M., Easterday, O. D., and Manley, C. H., Safety assessment of ingested heterocyclic amines: initial report, *Reg. Toxicol. Pharmacol.*, 17, S1–S109, 1993.
161. Adamson, R. H., Takayam, S., Sugimura, T., and Thorgeirsson, U. P., Induction of hepatocellular carcinoma in nonhuman primates by the food mutagen 2-amino-3-methylimidazo[4,5-*f*]quinoline, *Environ. Health Perspect.,* 102, 190–193, 1994.
162. IARC, *IARC Monographs on the Evaluation of Carcinogenic Risk to Humans,* Vol. 56, *Some Naturally Occurring Substances: Food Items and Constituents, Heterocyclic Aromatic Amines and Mycotoxins,* International Agency for Research on Cancer, Lyon, France, 1992, pp. 165–242.
163. Spingarn, N. E. and Weisburger, J. H., Formation of mutagens in cooked foods. I. Beef, *Cancer Lett.*, 7, 259–264, 1979.
164. Bjeldanes, L. F., Morris, M. M., Timourian, H., and Hatch, F. T., Effects of meat composition and cooking conditions on mutagen formation in fried ground beef, *J. Agric. Food Chem.*, 31, 18–21, 1983.
165. Hatch, F. T., Felton, J. S., Stuermer, D. H., and Bjeldanes, L. F., Identification of mutagens from the cooking of food, in *Chemical Mutagens: Principles and Methods for their Detection*, Vol. 9, de Serres, F. J., Ed., Plenum Press, New York, 1984, pp. 111–164.
166. Knize, M. G., Andresen, B. D., Healy, S. K., Shen, N. H., Lewis, P. R., Bjeldanes, L. F., Hatch, F. T., and Felton, J. S., Effect of temperature, patty thickness and fat content on the production of mutagens in fried ground beef, *Food Chem. Toxicol.*, 28, 1035–1040, 1985.
167. Laser-Reutersward, A., Skog, K., and Jagerstad, M., Effects of creatine and creatinine content on the mutagenic activity of meat extracts, bouillons and gravies from different sources, *Food Chem. Toxicol.*, 25, 747–754, 1987.
168. Laser-Reutersward, A., Skog, K., and Jagerstad, M., Carcinogenicity of pan-fried bovine tissue in relation to their content of creatine, creatinine, monosaccharides and free amino acids, *Food Chem. Toxicol.*, 25, 755–762, 1987.

169. Bjeldanes, L. F., Morris, M. M., Felton, J. S., Healy, S., Stuermer, D., Berry, P., Timourian, H., and Hatch, F. T., Mutagens from the cooking of food. II. Survey by the Ames' *Salmonella* test of mutagen formation in the major protein-rich foods of the American diet, *Food Chem. Toxicol.*, 20, 357–363, 1982.
170. Bjeldanes, L. F., Morris, M. M., Felton, J. S., Healy, S., Stuermer, D., Berry, P., Timourian, H., and Hatch, F. T., Mutagens from the cooking of food. III. Survey by Ames' *Salmonella* test of mutagen formation in secondary sources of cooked dietary protein, *Food Chem. Toxicol.*, 20, 365–369, 1982.
171. Doolittle, D. J., Rahn, C. A., Reed, B. A., and Lee, C. K., The effect of cooking methods on the mutagenicity of food and on urinary mutagenicity of humans following consumption (Abstract P14), presented at Mutagens and Carcinogens in the Diet, a satellite symposium of the 5th Int. Conf. Environmental Mutagens, Madison, WI, July 5–8, 1989.
172. Barrington, P. J., Baker, R. S. U., Truswell, A. S., Bonin, A. M., Ryan, A. J., and Paulin, A. P., Mutagenicity of basic fractions derived from lamb and beef cooked by common household methods, *Food Chem. Toxicol.*, 28, 141–146, 1990.
173. Dolara, P., Commoner, B., Vithayathil, A. J., Cuca, C. G., Tuley, E., Madyastha, P., Nair, P., and Kriebel, D., The effect of temperature on the formation of mutagens in heated beef stock and cooked ground beef, *Mutat. Res.*, 60, 231–237, 1979.
174. Krone, C. A. and Iwaoka, W. T., Mutagen formation during the cooking of fish, *Cancer Lett.*, 14, 93–99, 1981.
175. Felton, J. S. and Knize, M. G., Heterocyclic-amine mutagens/carcinogens in foods, in *Chemical Carcinogenesis and Mutagenesis. I. Handbook of Experimental Pharmacology*, Cooper, C. S. and Gover, P. L., Eds., Springer-Verlag, Berlin, 1990, pp. 471–502.
176. Sugimura, T., Wakabayashi, K., Nagao, M., and Ohgaki, H., Heterocyclic amines in cooked food, in *Food Toxicology. A Perspective on the Relative Risks*, Scanlan, R. A., Ed., Marcel Dekker, New York, 1989, pp. 31–55.
177. Sugimura, T. and Wakabayashi, K., Mutagens and carcinogens in food, in *Mutagens and Carcinogens in the Diet*, Pariza, M. W., Aeshbacher, H.-U., Felton, J. S., and Sato, S., Eds., Wiley-Liss, New York, 1990, pp. 1–18.
178. Barnes, W. S., Maher, J. C., and Weisburger, J. H., High pressure liquid chromatographic method for the analysis of 2-amino-3-methylimidazo[4,5-f]quinoline, a mutagen formed from the cooking of food, *J. Agric. Food Chem.*, 31, 883–886, 1983.
179. Felton, J. S., Knize, M. G., Wood, C., Wuebbles, B. J., Healy, S. K., Steurmer, D. H., Bjeldanes, L. F., Kimble, B. J., and Hatch, F. T., Isolation and characterization of new mutagens from fried ground beef, *Carcinogenesis*, 5, 95–102, 1984.
180. Murray, S., Gooderham, N. J., Boobis, A. R., and Davies, D. S., Measurement of MeIQx and DiMeIQx in fried beef by capillary column gas chromatography electron capture negative ion chemical ionisation mass spectrometry, *Carcinogenesis*, 9, 321–325, 1988.
181. Gross, G. A., Philippossian, G., and Aeschbacher, H. U., An efficient and convenient method for the purification of mutagenic heterocyclic amines in heated meat products, *Carcinogenesis*, 10, 1175–1182, 1989.
182. Vahl, M., Gry, J., and Nielsen, P. A., Mutagens in fried pork and the influence of the frying temperature, Proc. XVII Annual Meeting of the European Environmental Mutagen Society (EEMS), Zurich, 1987, p. 99.
183. Felton, J. S., Knize, M. G., Shen, N. H., Lewis, P. R., Andresen, B. D., Happe, J., and Hatch, F. T., The isolation and identification of a new mutagen from fried ground beef: 2-amino-1-methyl-6-phenylimidazo[4,5-b] pyridine (PhIP), *Carcinogenesis*, 7, 1081–1086, 1986.
184. Hayatsu, H., Arimoto, S., and Wakabayashi, K., Methods for separation and detection of heterocyclic amines, in *Mutagens in Food, Detection and Prevention*, Hayatsu, H., Ed., CRC Press, Boca Raton, FL, 1991, pp. 101–112.
185. Gross, G. A. and Grüter, A., Quantitation of mutagenic/carcinogenic heterocyclic amines in food products, *J. Chromatogr.*, 592, 271–278, 1992.
186. Knize, M. G., Dolbare, F. A., Carroll, K. L., and Felton, J. S., Effect of cooking time and temperature on the heterocyclic amine content of fried-beef patties, *Food Chem. Toxicol.*, 32, 595–603, 1995.
187. Gross, G. A., Simple methods for quantifying mutagenic heterocyclic aromatic amines in food products, *Carcinogenesis*, 11, 1597–1603, 1990.
188. Lynch, A. M., Knize, M. G., Boobis, A. R., Gooderham, N. J., Davies, D. S., and Murray, S., Intra- and interindividual variability in systemic exposure in humans to 2-amino-3,8-dimethylimidazo[4,5-f]quinoxaline and 2-amino-1-methyl-6-phenylimidazo[4,5-b] pyridine, carcinogens present in cooked beef, *Cancer Res.*, 52, 6216–6223, 1992.
189. Dragsted, L. O., Exposure and carcinogenicity of heterocyclic amines, Proc. of the Toxicology Forum Annual Meeting, Copenhagen, Denmark, 1992, pp. 141–148.
190. Murray, S., Lynch, A. M., Knize, M. G., and Gooderham, N. J., Quantification of the carcinogens MeIQx, DiMeIQx, and PhIP in food using a combined assay based on capillary column gas chromatography negative ion mass spectrometry, *J. Chrom.*, 616, 211–219, 1993.
191. Matsukura, N., Kawachi, T., Morino, K., Ohgaki, H., Sugimura, T., and Takayama, S., Carcinogenicity in mice of mutagenic compounds from a tryptophan pyrolyzate, *Science*, 213, 346–347, 1981.
192. Takayama, S., Nakatsuru, Y., Ohgaki, H., Sato, S., and Sugimura, T., Carcinogenicity in rats of a mutagenic compound, 3-amino-1,4-dimethy-5H-pyrido[4,5-*b*]indole, from tryptophan pyrolysate, *Gann*, 76, 815–817, 1985.

193. Hosaka, S., Matsushima, T., Hirono, I., and Sugimura, T., Carcinogenic acitivity of 3-amino-1-methyl-5H-pyrido[4,3-*b*]indole (Trp-P-2), a pyrolysis product of tryptophan, *Cancer Lett.*, 13, 23–28, 1981.
194. Takahashi, M., Toyoda, K., Aze, Y., Furuta, K., Mitsumori, K., and Hayashi, Y., The rat urinary bladder as a new target of heterocyclic amine carcinogenicity: tumor induction by 3-amino-1-methyl-5H-pyrido[4,3-*b*]indole acetate, *Jpn. J. Cancer Res.*, 84, 852–858, 1993.
195. Ohgaki, H., Matsukura, N., Morino, K., Kawachi, T., Sugimura, T. and Takayama, S., Carcinogenicity in mice of mutagenic compounds from glutamic acid and soybean globulin pyrolysates, *Carcinogenesis (Lond.),* 5, 815–819, 1984.
196. Takayama, S., Masuda, M., Mogami, M., Ohgaki, H., Sato, S., and Sugimura, T., Induction of cancers in the intestine, liver and various other organs of rats by feeding mutagens from glutamic acid pyrolysate, *Gann,* 75, 207–213 1984.
197. Ohgaki, H., Kusama, K., Matsukura, N., Morino, K., Hasegawa, H., Sato, S., Takayama, S., and Sugimura, T., Carcinogenicity in mice of a mutagenic compound, 2-amino-3-methylimidazo[4,5-*f*]quinoline, from broiled sardine, cooked beef and beef extract, *Carcinogenesis (Lond.),* 5, 921–924, 1984.
198. Takayama, S., Nakatsuru, Y., Masuda, M., Ohgaki, H., Sato, S., and Sugimura, T., Demonstration of carcinogenicity in F344 rats of 2-amino-3-methylimidazo[4,5-*f*]quinoline from broiled sardine, fried beef and beef extract, *Gann.,* 75, 467–470, 1984.
199. Tanaka, T., Barnes, W. S., Williams, G. M., and Weisburger, J. H., Multipotential carcinogenicity of the fried food mutagen 2-amino-3-methylimidazo[4,5-*f*]quinoline in rats, *Jpn. J. Cancer Res.*, 76, 570–576, 1985.
200. Ohgaki, H., Hasegawa, H., Suenaga, M., Kato, T., Sato, S., Takayama, S., and Sugimura, T., Induction of hepatocellular carcinoma and highly metastatic squamous cell carcinomas in the forestomach of mice by feeding 2-amino-3,4-dimethylimidazo[4,5-*f*]quinoline, *Carcinogenesis (Lond.),* 7, 1889–1893, 1986.
201. Kato, T., Migita, H., Ohgaki, H., Sato, S., Takayama, S., and Sugimura, T., Induction of tumors in the Zymbal gland, oral cavity, colon, skin and mammary gland of F344 rats by a mutagenic compound, 2-amino-3,4-dimethylimidazo[4,5-*f*]quinoline, *Carcinogenesis (Lond.),* 10, 601–603, 1989.
202. Ohgaki, H., Hasegawa, H., Suenaga, M., Sato, S., Takayama, S., and Sugimura, T., Carcinogenicity in rats of a mutagenic compound, 2-amino-3,8-dimethylimidazo[4,5-*f*]quinoxaline (MeIQx) from cooked foods, *Carcinogenesis (Lond.),* 8, 665–668, 1987.
203. Kato, T., Ohgaki, H., Hasegawa, H., Sato, S., Takayama, S., and Sugimura, T., Carcinogenicity in rats of a mutagenic compound, 2-amino-3,8-dimethylimidazo[4,5-*f*]quinoxaline, *Carcinogenesis (Lond.),* 9, 71–73, 1988.
204. Esumi, H., Ohgaki, H., Kohzen, E., Takayama, S., and Sugimura, T., Inducation of lymphoma in CDF_1 mice by the food mutagen, 2-amino-1-methy-6-phenylimidazo[4,5-*b*]pyridine, *Jpn. J. Cancer Res.,* 80, 1176–1178, 1989.
205. Ito, N., Hasegawa, H., Sano, M., Tamano, S., Esumi, H., Takayama, S., and Sugimura, T., A new colon and mammary carcinogen in cooked food, 2-amino-1-methyl-6-phenylimidazo[4,5-*b*]pyridine, *Carcinogenesis (Lond.),* 12, 1503–1506, 1991.
206. Hasegawa, R., Sano, M., Tamano, S., Imaido, K., Shirai, T., Nagao, M., Sugimura, T., and Ito, N., Dose-dependence of 2-amino-1-methyl-6-phenylimidazo[4,5-*b*]-pyridine, *Carcinogenesis,* 14, 2553–2557, 1993.
207. LaRue, T. A., Naturally occurring compounds containing a nitrogen-nitrogen bond, *Lloydia,* 40, 307–321, 1977.
208. Levenberg, B., Structure and enzymatic cleavage of agaritine, a phenylhydrazide of L-glutamic acid isolated from *Agaricaceae, J. Am. Chem. Soc.,* 83, 503–504, 1961.
209. Levenberg, B., Isolation and structure of agaritine, a gamma-glutamyl-substituted aryl-hydrazine derivative from *Agaricaceae, J. Biol. Chem.*, 239, 2267–2273, 1964.
210. Shutte, H. R., Liebisch, H. W., Mersch, O., and Senf, L., Untersuchungen zur biosynthese des agaritins in *Agaricus bisporus, Anales de Quimica,* 68, 899–903, 1972.
211. Toth, B., Raha, C. R., Wallcave, L., and Nagel, D., Attempted tumor induction with agaritine in mice, *Anticancer Res.*, 1, 255–258, 1981.
212. Toth, B., Nagel, D., Patil, K., Erickson, J., and Antonson, K., Tumor induction with the *N*-acetyl derivative of 4-hydroxymethylphenylhydrazine a metabolite of agaritine of *Agaricus bisporus, Cancer Res.*, 38, 177–180, 1978.
213. Toth, B., Patil, K., and Jae, H. S., Carcinogenesis of 4-(hydroxymethyl) benzenediazonium ion (tetrafluoroborate) of *Agaricus bisporus, Cancer Res.*, 41, 2444–2449, 1981.
214. Toth, B., Carcinogenesis by N^2-[γ-L(+)-glutamyl]-4-carboxyphenylhydrazine of *Agaricus bisporus* in mice, *Anticancer Res.*, 6, 917–920, 1986.
215. Toth, B., Cancer induction by sulfate form of 4-(hydroxymethyl) benzenediazonium ion of *Agaricus bisporus, In Vivo,* 1, 39–42, 1987.
216. McManus, B. M., Toth, B., and Patil, K., Aortic rupture and aortic smooth muscle tumors in mice: induction by *p*-hydrazinobenzoic acid hydrochloride of the cultivated mushroom *Agaricus bisporus, Lab. Invest.,* 57, 78–85, 1987.
217. Toth, B. and Erickson, J., Cancer induction in mice by feeding of the fresh uncooked cultivated mushroom of commerce *Agaricus bisporus, Cancer Res.,* 46, 4007–4011, 1986.

218. Liu, J. W., Beelman, R. B., Lineback, D. R., and Speroni, J. J., Agaritine content of fresh and processed mushrooms (*Agaricus bisporus* [Lange] Imbach), *J. Food Sci.*, 47, 1542–1548, 1982.
219. Gannett, P. M. and Toth, B., Heat sensitivity of chemicals of the *Agaricus bisporus* (AB) mushroom, *Proc. Am Assoc. Cancer Res.*, 32, 116, 1991.
220. Ross, A. E., Nagel, D., and Toth, B., Occurrence, stability, and decomposition of β-N[γ-L(+)-glutamyl]-4-hydroxymethylphenylhydrazine (agaritine) from the mushroom *Agaricus bisporus, Food Chem. Toxicol.*, 20, 903–907, 1982.
221. Ross, A. E., Nagel, D., and Toth, B., Evidence for the occurrence and formation of diazonium ions in the *Agaricus bisporus* mushroom and its extracts, *J. Agric. Food Chem.*, 30, 521–525, 1982.
222. Chauhan, Y., Nagel, D., Issenberg, P., and Toth, B., Identification of *p*-hydrazinobenzoic acid in the commercial mushroom *Agaricus bisporus, J. Agric. Food Chem.*, 32, 1067–1069, 1984.
223. Chauhan, Y., Nagel, D., Gross, M., Cerny, R., and Toth B., Isolation and N^2-[γ-L(+)-glutamyl]-4-carboxyphenylhydrazine in the cultivated mushroom *Agaricus bisporus, J. Agric. Food Chem.*, 33, 817–820, 1985.
224. De Flora, S., Study of 106 organic and inorganic compounds in the *Salmonella*/microsome test, *Carcinogenesis,* 2, 283–298, 1981.
225. IARC, *IARC Monographs on the Evaluation of the Carcinogenic Risk of Chemicals to Humans. Some Food Additives, Feed Additives, and Naturally Occurring Substances,* Vol. 31, *Agaritine,* International Agency for Research on Cancer, Lyon, France, 1983, pp. 65–69.
226. Toth, B., Hydrazine, methylhydrazine and methylhydrazine sulfate carcinogenesis in Swiss mice. Failure of ammonium hydroxide to interfere in the development of tumors, *Int. J. Cancer,* 9, 109–118, 1972.
227. Toth, B. and Shimizu, H., Methylhydrazine tumorigenesis in Syrian golden hamsters and the morphology of malignant histiocytomas, *Cancer Res.,* 33, 2744–2753, 1973.
228. Toth, B., Hepatocarcinogenesis by hydrazine mycotoxins of edible mushrooms, *J. Toxicol. Environ. Health,* 5, 193–202, 1979.
229. Toth, B., Smith, J. W., and Patil, K. D., Cancer induction in mice with acetaldehyde methylformylhydrazone of the false morel mushroom, *J. Natl. Cancer Inst.*, 67, 881–887, 1981.
230. Schmidlin-Meszaros, J., Gyromitrin in dried false morels (*Gyromitra esculenta* sicc.), *Mitt. Gebiete Lebensmittel. Hyg.*, 65, 453–465, 1974.
231. Pyysalo, H., Test for gyromitrin, a poisonous compound in false morel *Gyromitra esculenta, Z. Lebensmittel. Untersuch.-Forsch.*, 160, 325–330, 1976.
232. von Wright, A., Niskanen, A., and Pyysalo, H., The toxicities and mutagenic properties of ethylidene gyromitrin and *N*-methylhydrazine with *Escherichia coli* as test organism, *Mutat. Res.*, 56, 105-110, 1977.
233. IARC, *IARC Mongraphs on the Evaluation of the Carcinogenic Risk of Chemicals to Humans. Some Food Additives, Feed Additives, and Naturally Occurring Substances.* Vol. 31, *Gyromitrin,* International Agency for Research on Cancer, Lyon, France, 1983, pp. 163–169.
234. Fosberg, F. R., Resume of the cycadaceae, *Fed Proc.*, 23, 1340–1342, 1964.
235. Whiting, M. G., Toxicity of cycads, *Econ. Bot.*, 17, 271–302, 1963.
236. Laqueur, G. L., Oncogenicity of cycads and its implications, in *Environmental Cancer,* Kraybill, H. F. and Mehlman, M. A., Eds., Hemisphere, Washington, D.C., 1977, pp. 231–241.
237. Zedeck, M. S., Hydrazine derivatives, azo and azoxy compounds, and methyazoxymethanol and cycasin, in *Chemical Carcinogens,* Searle, C. S., Ed., ACS Monograph 182, American Chemical Society, Washington, D.C., 1984, pp. 915–944.
238. Palekur, R. S. and Dastur, D. K., Cycasin content of *Cycas circinalis, Nature,* 206, 1363–1365, 1965.
239. Kobayashi, A. and Matsumoto, H., Studies on methylazoxymethanol, the aglycone of cycasin, *Arch. Biochem. Biophys.*, 110, 373–380, 1965.
240. Laqueur, G. L. and Matsumoto, H., Neoplasms in female Fischer rats following intraperitoneal injections of methylazoxymethanol, *J. Natl. Cancer Inst.*, 37, 217–232, 1966.
241. Spatz, M., Laqueur, G. L., and Holmes, J. M., Carcinogenic effects of methylazoxymethanol (MAM) in hamsters, *Proc. Am. Assoc. Cancer Res.*, 10, 86, 1969.
242. Narisawa, T. and Nakano, H., Carcinoma of the large intestine of rats induced by rectal infusion of methylazoxymethanol, *Gann,* 64, 93–95, 1973.
243. Fushima, K., Tumor induction in rats given consecutive injections of methylazoxymethanol acetate, *Acta Schol. Med. Univ. Gifu,* 22, 729–750, 1974.
244. Zedeck, M. S. and Sternberg, S. S., A model system for studies of colon carcinogenesis: tumor induction by a single injection of methylazoxymethanol acetate, *J. Natl. Cancer Inst.,* 53, 1419–1421, 1974.
245. Laqueur, G. L., Mickelsen, O., Whiting, M. G., and Kurland, L. T., Carcinogenic properties of nuts from *Cycas circinalis* L. indigenous to Guam, *J. Natl. Cancer Inst.,* 31, 919–951, 1963.
246. Laqueur, G. L., Carcinogenic effects of cycad meal and cycasin, methylazoxymethanol glycoside, in rats and effects of cycasin in germfree rats, *Fed. Proc.,* 23, 1386–1387, 1964.
247. Laqueur, G., The induction of intestinal neoplasms in rats with the glycoside cycasin and its aglycone, *Virchows Arch. Path. Anat.*, 340, 151–163, 1965.
248. Stanton, M. F., Hepatic neoplasms of aquarium fish exposed to *Cycas circinalis, Fed. Proc.*, 25, 661, 1966.

249. Aoki, K. and Matsudaira, H., Induction of hepatic tumors in a teleost (*Oryzias latipes*) after treatment with methylazoxymethonal acetate, *J. Natl. Cancer Inst.*, 59, 1747–1749, 1977.
250. Laqueur, G. L. and Spatz, M., Oncogenicity of cycasin and methylazoxymethanol, *Gann Monogr. Cancer Res.*, 17, 189–204, 1975.
251. Hirono, I., Hayashi, K., Mori, H., and Miwa, T., Carcinogenic effects of cycasin in Syrian golden hamsters and the transplantability of induced tumors, *Cancer Res.*, 31, 283–287, 1971.
252. Hirono, I., Carcinogenicity and neurotoxicity of cycasin with special reference to species differences, *Fed. Proc.*, 31, 1493–1497, 1972.
253. Hirono, I., Shibuya, C., and Fushimi, K., Tumor induction in C57BL/6 mice by a single administration of cycasin, *Cancer Res.,* 29, 1658–1662, 1969.
254. Sieber, S. M. Correa, P., Dalgard, D. W., McIntire, K. R., and Adamson, R. H., Carcinogenicity and hepatotoxicity of cycasin and its aglycone methylazoxymethanol acetate in nonhuman primates, *J. Nat. Cancer Inst.,* 65, 177–189, 1980.
255. Watanabe, K., Iwashita, H., Muta, K., Hamada, Y., and Hamada, K., Hepatic tumors of rabbits induced by cycad extract, *Gann,* 66, 335–339, 1975.
256. Hirono, I., Laqueur, G. L., and Spatz, M., Tumor induction in Fischer and Osborne-Mendel rats by a single administration of cycasin, *J. Natl. Cancer Inst.,* 40, 1003–1010, 1968.
257. Fukunishi, R., Terashi, S., Watanabe, K., and Kawaji, K., High yield of hepatic tumors in rats by cycasin, *Gann,* 63, 575–578, 1972.
258. Smith, D. W. E., Mutagenicity of cycasin aglycone (methylazoxymethonal), a naturally occurring carcinogen, *Science,* 152, 1273–1274, 1966.
259. Pohland, A. E., Mycotoxins in review, *Food Additives and Contaminants,* 10, 17–28, 1993.
260. Wogan, G. N., Mycotoxins and other naturally occurring carcinogens, in *Environmental Cancer,* Vol. 3, *Advances in Modern Toxicology,* Krybill, H. F. and Mehlman, M. A., Eds., John Wiley & Sons, New York, 1977, pp. 263–290.
261. Sauer, D. B. and Tuite, J. F., Conditions that affect growth of *Aspergillus flavus* and production of aflatoxin in stored maize, in *Aflatoxin in Maize,* Zuber, M. S., Lillehoj, E. B., and Renfro, B. L., Eds., CIMMYT, Mexico, 1987, pp. 41–50.
262. Scott, P. M., Mycotoxins in feeds and ingredients and their origin, *J. Food Protect,* 41, 385–398, 1978.
263. Stoloff, L., Mycotoxins as potential environmental carcinogens, in *Carcinogens and Mutagens in the Environment,* Vol. I, Stich, H. F., Ed., CRC Press, Boca Raton, FL, 1982, pp. 97–120.
264. Appelbaum, R. S., Brackett, R. E., Wiseman, D. W., and Marth, E. H., Aflatoxin: toxicity to dairy cattle and occurrence in milk and milk products — a review, *J. Food Prot.*, 45, 752–777, 1982.
265. Bullerman, L. B., Significance of mycotoxins to food safety and human health, *J. Food Prot.*, 42, 65–86, 1979.
266. Scott, P. M., Effects of food processing on mycotoxins, *J. Food Prot.,* 47, 489–499, 1984.
267. Stoloff, L., Report on mycotoxins, *J. Assoc. Off. Anal. Chem.*, 60, 348–353, 1977.
268. Siller, W. G. and Ostler, D. C., The histopathology of an heterohepatic syndrome in turkey poults, *Vet. Rec.*, 73, 134–138, 1961.
269. Wannop, C. C., The histopathology of turkey "X" disease in Great Britain, *Avian Dis.*, 5, 371–381, 1961.
270. Shotwell, O. L., Goulden, M. L., and Hesseltine, C. W., Aflatoxin M1. Occurrence in stored and freshly harvested corn, *J. Agric. Food Chem.,* 24, 683–684, 1976.
271. Masri, M. S., Garcia, V. C., and Page, J. R., The aflatoxin M content of milk from cows fed known amounts of aflatoxin, *Vet. Rec.,* 84, 146–147, 1969.
272. Kiermeier, F., Uber die Aflatoxin-M-Ausscheidung in Kuhmilch in Abhangigkeit von der aufgenommenen Aflatoxin-B1 Menge, *Milchwissenschaft,* 28, 683–685, 1973.
273. IARC, *Some Naturally Occurring Substances: Food Items and Constituents, Heterocyclic Aromatic Amines and Mycotoxins,* Monograph 56, International Agency for Research on Cancer, Lyon, France, 1993.
274. Coulombe, R. A., Shelton, D. W., Sinnhuber, R. O., and Nixon, J. E., Comparative mutagenicity of aflatoxins using *Salmonella*/trout hepatic enzyme activation systems, *Carcinogenesis,* 3, 1261–1264, 1982.
275. Schoenhard, G. L., Hendricks, J. D., Nixon, J. E., Lee, D. J., Wales, J. H., Sinnhuber, R. O., and Pawlowski, N. E., Aflatoxicol-induced hepatocellular carcinoma in rainbow trout (*Salmon garidneri*) and the synergistic effects of cyclopropenoid fatty acids, *Cancer Res.*, 41, 1011–1014, 1981.
276. Hendricks, J. D., Wales, J. H., Sinnhuber, R. O., Nixon, J. E., Loveland, P. M., and Scanlan, R. A., Rainbow trout (*Salmon gairdneri*) embryos: a sensitive animal model for experimental carcinogenesis, *Fed. Proc.*, 39, 3222–3229.
277. Lee, L. S., Cucullu, A. F., Franz, Jr., A. O., and Pons, W. A., Destruction of aflatoxins in peanuts during dry and oil roasting, *J. Agric. Food Chem.,* 17, 451–453, 1969.
278. Ullosa-Sosa, M. and Schroeder, H. W., Note on aflatoxin decomposition in the process of making tortillas from corn, *Cereal Chem.,* 46, 397–400, 1969.
279. Chu, F. S. C., Chang, C. C., Ashoor, S. H., and Prentice, N., Stability of aflatoxin B1 and ochratoxin A in brewing, *Appl. Microbiol.,* 29, 313–316, 1975.
280. Stoloff, L., Aflatoxin M in perspective, *J. Food Protect.*, 43, 226–230, 1980.
281. Goldblatt, L. A., Ed., *Aflatoxin. Scientific Background, Control and Implications,* Academic Press, New York, 1969.

282. Wogan, G. N., Mycotoxins and other naturally occurring carcinogens, in *Environmental Cancer,* Vol. 3, *Advances in Modern Toxicology,* Krybill, H. F., and Mehlman, M. A., Eds., John Wiley & Sons, New York, 1977, pp. 263–290.
283. Hendricks, J. D., Carcinogenicity of aflatoxins in nonmammalian organisms, in *The Toxicology of Aflatoxins: Human Health, Veterinary, and Agricultural Significance,* Eaton, D. L. and Groopman, J. D., Eds., Academic Press, San Diego, CA, 1994, pp. 101–136.
284. Wogan, G. N. and Palianlunga, S., Carcinogenicity of synthetic aflatoxin M1 in rats, *Food Cosmet. Toxicol.*, 12, 381–384, 1974.
285. Wong, J. J. and Hsieh, D. P. H., Mutagenicity of aflatoxins related to their metabolism and carcinogenic potential, *Proc. Natl. Acad. Sci. USA,* 73, 2241–2244, 1976.
286. Lutz, W. K., Jaggi, W., Luthy, J., Sagelsdorff, P., and Schlatter, C., *In vivo* covalent binding of aflatoxin B1 and aflatoxin M1 to liver DNA of rat, mouse, and pig, *Chem. Biol. Interactions,* 32, 249–256, 1980.
287. Green, C. E., Rice, D. W., Hsieh, D. P. H., and Byard, J. L., The comparative metabolism and toxic potency of aflatoxin B1 and aflatoxin M1 in primary cultures of adult rat hepatocytes, *Food Chem. Toxicol.,* 20, 53–60, 1982.
288. Sinnhuber, R. O., Lee, D. J., Wales, J. H., and Landers, M. K., Aflatoxin M1, a potent liver carcinogen for rainbow trout, *Fed. Proc.*, 29, 68, 1970.
289. Sinnhuber, R. O., Lee, D. J., Wales, J. H., Landers, M. K., and Keyl, A. C., Hepatic carcinogenesis of aflatoxin M1 in rainbow trout (*Salmon gaidneri*) and its enhancement by cyclopropene fatty acids, *J. Natl. Cancer Inst.,* 53, 1285–1288, 1974.
290. Canton, J. H., Kroes, R., van Logten, M. J., van Schothorst, M., Stavenutter, J. F. C., and Verhuledonk, C. A. H., The carcinogenicity of aflatoxin M1 in rainbow trout, *Food Cosmet. Toxicol.*, 13, 441–443, 1975.
291. Alpert, M. E., Hutt, M. S., Wogan, G. N., and Davidson, C. S., Association between aflatoxin content of food and hepatoma frequency in Uganda, *Cancer,* 28, 253–260, 1971.
292. Schank, R. C., Siddhichai, P., Subhamani, B., Bhamarapravati, N., Gordon, E., and Wogan, G. N., Dietary aflatoxins and human liver cancer. V. Duration of primary liver cancer and prevalence of hepatomegaly in Thailand, *Food Cosmet. Toxicol.,* 10, 181–191, 1972.
293. Peers, F. G. and Linsell, C. A., Dietary aflatoxins and liver cancer — a population based study in Kenya, *Br. J. Cancer,* 27, 473–484, 1973.
294. Van Rensburg, S. J., Van Der Watt, J. J., Purchase, I. F. H., Coutinho, L. P., and Markham, R., Primary liver cancer rate and aflatoxin intake in a high cancer area, *S. Afr. Med. J.*, 48, 2508a-2508d, 1974.
295. Yeh, F. S., Yu, M. C., Mo, C. C., Luo, S., Tong, M. J., and Henderson, B. E., Hepatitis B virus, aflatoxins and hepatocellular carcinoma in Southern Guangxi China, *Cancer Res.*, 49, 2506–2509, 1989.
296. Heathcote, J. G., Aflatoxins and related toxins, in *Mycotoxins — Production, Isolation, Separation, and Purification,* Betina, V., Ed., Elsevier Science, Amsterdam, 1984, pp. 89–130.
297. Scott, P. M., van Walbeek, W., Kennedy, B., and Anyeti, D., Mycotoxins (ochratoxin A, citrinin and sterigmatocystin) and toxigenic fungi in grains and other agricultural products, *J. Agric. Food Chem.*, 20, 1103–1107, 1972.
298. Terao, K., Sterigmatocystin: a masked potent carcinogenic mycotoxin, *J. Toxicol. Toxin Rev.*, 2, 77–110, 1983.
299. Stoloff, L., Occurrence of mycotoxins in foods and feeds, in *Mycotoxins and Other Fungi Related Food Problems,* Rodricks, J. V., Ed., Advances in Chemistry Series, No. 149, American Chemical Society, Washington, D.C., 1976.
300. ver der Merwe, K. J., Steyn, P. S., and Fourie, L., Mycotoxins. Part II. The constitution of ochratoxins A, B, and C, metabolites of *Aspergillus ochraceus* Wilh., *J. Chem. Soc.*, 7083–7088, 1965.
301. Ciegler, A., Fennell, D. I., Mintzlaff, H.-J., and Leistner, L., Ochratoxin synthesis by *Penicillium* species, *Naturwissenschaften,* 59, 365–366, 1972.
302. Frisvad, J. C. and Filtenborg, O., Terverticillate penicilla: chemotaxonomy and myctoxin production, *Mycologia,* 81, 837–861, 1989.
303. Steyn, P. S. and Holzapfel, C. W., The synthesis of ochratoxins A and B, metabolites of *Aspergillus ochraceus* Wilh., *Tetrahedron,* 23, 4449–4461, 1967.
304. Roberts, J. C. and Woollven, P., Studies in mycological chemistry. Part XXIV. Synthesis of ochratoxin A, a metabolite of *Aspergillus ochraceus* Wilh., *J. Chem. Soc.*, (C), 278–281, 1970.
305. Kuiper-Goodman, T. and Scott, P. M., Risk assessment of the mycotoxin ochratoxin A, *Biomed. Environ. Sci.,* 2, 179–248, 1989.
306. IARC, *IARC Monographs on the Evaluation of Carcinogenic Risk to Humans,* Vol. 56, *Some Naturally Occurring Substances: Food Items and Constituents, Heterocyclic Aromatic Amines and Mycotoxins,* Vol. 56, *Ochratoxin A,* International Agency for Research on Cancer, Lyon, France, 1989, pp. 488–521.
307. Pohland, A. E., Mycotoxins in review, *Food Add. Contam.*, 10, 17–28, 1993.
308. Kanisawa, M. and Suzuki, S., Induction of renal and hepatic tumors in mice by ochratoxin A, a mycotoxin, *Gann,* 69, 599–600, 1978.
309. Kanisawa, M., Synergistic effect of citrinin and hepatorenal carcinogenesis of OA in mice, in *Toxigenic Fungi — Their Toxins and Health Hazard,* Kurata, H. and Ueno, Y., Eds., Kodansha, Tokyo/Elsevier, Amsterdam, 1984, pp. 245–254.
310. Bendele, A. M., Carlton, W. W., Krogh, P., and Lillehoj, E. B., Ochratoxin A carcinogenesis in the (C57BL/6J X C3H)F1 mouse, *J. Natl. Cancer Inst.*, 75, 733–742, 1985.

311. NTP, *Toxicology and Carcinogenesis Studies of Ochratoxin A (CAS No. 303-47-9) in F344/N Rats (Gavage Studies)*, NTP TR 358, DHHS Publication No. (NIH) 89-2813, National Toxicology Program, Research Triangle Park, NC, 1989.
312. Hennig, A., Fink-Gremmels, J., and Leistner, L., Mutagenicity and effects of ochratoxin A on the frequency of sister chromatid exchange after metabolic activation, in *Mycotoxins, Endemic Nephropathy and Urinary Tract Tumours*, IARC Scientific Publication No. 115, Castegnaro, M., Plestina, R., Dirheimer, G., Chernozemsky, I. N., and Bartsch, H., Eds., International Agency for Research on Cancer, Lyon, France, 1991, pp. 255–260.
313. Dirheimer, G. and Creppy, E. E., Mechanism of action of ochratoxin A, in *Mycotoxins, Endemic Nephropathy and Urinary Tract Tumours*, IARC Scientific Publication No. 115, Castegnaro, M., Plestina, R., Dirheimer, G., Chernozemsky, I. N., and Bartsch, H., Eds., International Agency for Research on Cancer, Lyon, France, 1991, pp. 171–186.
314. Petkova-Bocharova, T. and Castegnaro, M., Ochratoxin A contamination of cereals in an area of high incidence of Balken endemic nephropathy in Bulgaria, *Food Add. Contam.*, 2, 267–270, 1985.
315. Tanchev, Y. and Dorossiev, D., The first clinical description of Balkan endemic nephropathy (1956) and its validity 35 years later, in *Mycotoxins, Endemic Nephropathy and Urinary Tract Tumours*, IARC Scientific Publication No. 115, Castegnaro, M., Plestina, R., Dirheimer, G., Chernozemsky, I. N., and Bartsch, H., Eds., International Agency for Research on Cancer, Lyon, France, 1991, pp. 21–28.
316. Miller, J. D., Epidemiology of *Fusarium* ear diseases of cereals, in *Mycotoxins in Grain. Compounds Other than Aflatoxin*, Miller, J. D. and Trenholm, H. L., Eds., Eagan Press, St. Paul, MN, 1994, pp. 19–36.
317. Nelson, P. E., Desjardins, A. E., and Plattner, R. D., Fumonsins, mycotoxins produced by *Fusarium* species: biology, chemistry, and significance, *A. Rev. Phytopath.*, 31, 233–252, 1993.
318. Gelderbom, W. C. A., Kriek, N. P. J., Marasas, W. F. O., and Thiel, P. G., Toxicity and carcinogenicity of the *Fusarium* moniliforme metabolite, B1, in rats, *Carcinogenesis*, 12, 1247–1251, 1991.
319. IARC, *IARC Monographs on the Evaluation of Carcinogenic Risk to Humans, Some Naturally Occurring Substances: Food Items and Constituents, Heterocyclic Aromatic Amines and Mycotoxins*, Vol. 56, *Toxins Derived from* Fusarium moniliforme: *Fumonisins B1 and B2 and Fusarin C*, International Agency for Research on Cancer, Lyon, France, 1993, pp. 445–466.
320. Gelderblom, W. C. A. and Snyman, S. D., Mutagenicity of potentially carcinogenic mycotoxins produced by *Fusarium moniliforme*, *Mycotoxin Res.*, 7, 46–52, 1991.
321. Park, D. L., Rua, Jr., S. M., Mirocha, C. J., Abd-Alla, E. A. M., and Weng, C. Y., Mutagenic potentials of fumonisin contaminated corn following ammonia decontamination procedure, *Mycopathologia*, 117, 105–108, 1992.
322. Gelderblom, W. C. A., Semple, E., Marasas, W. F. O., and Farber, E., The cancer-initiating potential of the fumonisin B mycotoxins, *Carcinogenesis*, 13, 433–437, 1992.
323. Marasas, W. F. O., Jaskiewicz, K., Venter, F. S., and van Schalkwyk, D. J., *Fusarium moniliforme* contamination of maize in oesaphageal cancer areas in Transkei, *S. Afr. Med. J.*, 74, 110–114, 1988.
324. Thiel, P. G., Marasas, W. F. O., Sydenham, E. W., Shephard, G. S., and Gelderblom, W. C. A., The implications of naturally occurring levels of fumonisins in corn for human and animal health, *Mycopathologia*, 117, 3–9, 1992.
325. Li, M.-X., Jian, Y.-Z., Han, N.-J., Fan, W.-G., Ma, J.-L., and Bjeldanes, L. E., Fusarin C induced esophageal and forestomach carcinoma in mice and rats, *Chin. J. Oncol.*, 14, 27–29, 1992.
326. Bullerman, L. B. and Tsai, W. J., Incidence and levels of *Fusarium moniliforme, Fusarium proliferatum*, and fumonisins in corn and corn-based foods and feeds, *J. Food Protect.*, 57, 541–546, 1994.
327. Sydenham, E. W., Shephard, G. S., Thiel, P. G., Marasas, W. F. O., and Stochenstrom, S., Fumonisin contamination of commercial corn-based human foodstuffs, *J. Agric. Food Chem.*, 38, 1900–1903, 1991.
328. Stack, M. E. and Eppley, R. M., Liquid chromatographic determination of fumonisins B1 and B2 in corn and corn products, *J. Ass. Off. Anal. Chem. Int.*, 75, 834–837, 1992.
329. Scott, P. M. and Lawrence, G. A., Stability and problems of recovery of fumonisins added to corn-based foods, *J. Assoc. Off. Anal. Chem. Int.*, 77, 541–545, 1994.
330. Hendrich, S., Miller, K. A., Wilson, T. M., and Murphy, P. A., Toxicity of *Fusarium proliferatum*-fermented nixtamalized corn-based diets fed to rats: effect of nutritional status, *J. Agric. Food Chem.*, 41, 1649–1654, 1993.
331. Prelusky, D. B., Residues in food products of animal origin, in *Mycotoxins in Grain*, Miller, J. D. and Trenholm, H. L., Eds., Eagan Press, St. Paul, MN, 1994, pp. 405–420.
332. Scott, P. M., Delgado, T., Prelusky, D. B., Trenholm, H. L., and Miller, J. D., Determination of fumonisins in milk, *J. Environ. Sci. Health*, 29, 989–998, 1994.
333. Marasas, W. F. O., Nelson, P. E., and Toussoun, T. A., *Toxigenic Fusarium Species: Identity and Mycotoxicology*, Pennsylvania State University Press, University Park, PA, 1984.
334. Scott, P. M., The natural occurrence of trichothecenes, in *Trichothecene Mycotoxicosis: Pathophysiologic Effects*, Vol. 1., Beasley, V. R., Ed., CRC Press, Boca Raton, FL, 1989, pp. 1–26.
335. Bamburg, J. R., Riggs, N. V., and Strong, F. M., The structures of toxins from two strains of *Fusarium tricinctum*, *Tetrahedron*, 24, 3329–3336, 1968.
336. Bamburg, J. R. and Strong, F. M., 12,13-Epoxytrichothecenes, in *Microbiological Toxins*, Vol. 7, *Algal and Fungal Toxins*, Kadis, S., Ciegler, A., and Ajl, S. J., Eds., Academic Press, New York, 1971, pp. 207–292.

337. Schiefer, H. B., Rousseaux, C. G., Handcock, D. S., and Blakley, B. R., Effects of low-level long-term oral exposure to T-2 toxin in CD-1 mice, *Food Chem. Toxicol.*, 25, 593–601, 1987.
338. Yang, S. and Xia, Q. J., Papilloma of forestomach induced by *Fusarium* T-2 toxin in mice, *Chin. J. Oncol.*, 10, 339–341, 1988.
339. Marasas, W. F. O., Bamburg, J. R., Smalley, E. B., Strong, F. M., Ragland, W. L., and Degurse, P. E., Toxic effects in trout, rats and mice of T-2 toxin produced by the fungus *Fusarium tricinctum* (Cd.), *Toxicol. Appl. Pharm.*, 15, 471–482, 1969.
340. Schoental, R., Joffe, A. Z., and Yagen, B., Cardiovascular lesions and various tumors found in rats given T-2 toxin, a trichothecene metabolite of *Fusarium, Cancer Res.*, 39, 2179–2189, 1979.
341. Li, M.-X., Zhu, G.-F., Cheng, S.-J., Jian, Y.-Z., and Fan, W.-G., Mutagenicity and carcinogenicity of T-2 toxin, a trichothecene produced by *Fusarium* fungi, *Chin. J. Oncol.*, 10, 326–329, 1988.
342. Haschek, W. M., Mutagenicity and carcinogenicity of T-2 toxin, in *Trichothecene Mycotoxicosis: Pathophysiologic Effects*, Vol. 1, Beasley, V. R., Ed., CRC Press, Boca Raton, FL, 1989, pp. 63–72.
343. IARC, *IARC Monographs on the Evaluation of Carcinogenic Risk to Humans: Some Naturally Occurring Substances: Food Items and Constituents, Heterocyclic Aromatic Amines and Mycotoxins*, Vol. 56, *T-2 Toxin*, International Agency for Research on Cancer, Lyon, France, 1993, pp. 467–488.
344. Tanaka, T., Hasegawa, A., Yamamoto, S., Lee, U.-S., Sugiura, Y., and Ueno, Y., Worldwide contamination of cereals by the *Fusarium* mycotoxins nivalenol, deoxynivalenol, and zearelenone. 1. Survey of 19 countries, *J. Agric. Food Chem.*, 36, 979–983, 1988.
345. Stob, M., Baldwin, R. S., Tuite, J., Andrews, F. N., and Gillette, K. G., Isolation of an anabolic uterotrophic compound from corn infected with *Gibberella zeae, Nature (Lond.)*, 196, 1318, 1962.
346. Bennett, G. A. and Shotwell, O. L., Zearalenone in cereal grains, *J. Am. Oil Chem. Soc.*, 56, 812–819, 1979.
347. Shotwell, O. L., Assay methods for zearalenone and its natural occurrence, in *Mycotoxins in Human and Animal Health*, Rodericks, J. V., Hesseltine, C. W., and Mehlman, M. A., Eds., Pathotox, Park Forest South, IL, 1977, pp. 403–413.
348. Senti, F. R., Global perspective on mycotoxins, in *Perspective on Mycotoxins*, Food and Agricultural Organization of the United Nations, Rome, 1979, pp. 15–120.
349. Kuiper-Goodman, T., Scott, P. M., and Watanabe, H., Risk assessment of the mycotoxin zearalenone, *Reg. Toxicol. Pharm.*, 7, 253–306, 1987.
350. NTP, *Carcinogenesis Bioassay of Zearalenone in F344/N rats and B6C3F1 Mice*, Tech. Rep. Series No. 235, National Toxicology Program, Department of Health and Human Services, Research Triangle Park, NC, 1982.
351. Becci, P. J., Voss, K. A., Hess, F. G., Gallo, M. A., Parent, R. A., and Stevens, K. R., Long-term carcinogenicity and toxicity study of zearalenone in the rat, *J. Appl. Toxicol.*, 2, 247–254, 1982.
352. IARC, *IARC Monographs on the Evaluation of the Carcinogenic Risk of Chemicals to Humans: Some Food Additives, Feed Additives and Naturally Occurring Substances*, Vol. 31, International Agency for Research on Cancer, Lyon, France, 1983, pp. 279–291.
353. Uraguchi, K., Saito, M., Noguchi, K., Takahashi, M., Enomoto, M., and Tatsuno, T., Chronic toxicity and carcinogenicity in mice of the purified mycotoxins, luteoskyrin and cyclochlorotine, *Food Cosmet. Toxicol.*, 10, 193–207, 1972.
354. Enomoto, M. and Ueno, I., *Penicillium islandicum* (toxic yellowed rice)-luteoskyrin-islanditoxin-cyclochlorotine, in *Mycotoxins*, Purchase, I. F. H., Ed., Elsevier Scientific, Amsterdam, 1974, pp. 314–325.
355. Tazima, Y., Naturally occurring mutagens of biological origin: a review, *Mutation Res.*, 26, 225–234, 1974.
356. Arai, M. and Hibino, T., Tumorigenicity of citrinin in male F344 rats, *Cancer Lett.*, 17, 281–287, 1983.
357. Kanisawa, M., Synergistic effect of citrinin on hepatorenal carcinogenesis of ochratoxin A in mice, in *Toxigenic Fungi — Their Toxins and Health Hazard*, Kurata, H. and Ueno, Y., Eds., Elsevier, Amsterdam, 1984, pp. 245–254.
358. Shinohara, D. K., Arai, M., Hirao, K., Sugihara, S., Nakanishi, K., Tsonoda, H., and Ito, N., Combination effect of citrinin and other chemicals on rat kidney tumorigenesis, *Gann*, 67, 147–155, 1976.
359. Frank, H. K., Citrinin, *Z. Ernahrungswiss*, 31, 164–177, 1992.
360. Enomoto, M. and Saito, M., Carcinogens produced by fungi, *Ann. Rev. Microbiol.*, 26, 279–312, 1972.
361. Magee, P. N. and Barnes, J. M., The production of malignant primary hepatic tumours in the rat by feeding dimethylnitrosamine, *Br. J. Cancer*, 10, 114–122, 1956.
362. Magee, P. N. and Barnes, J. M., The experimental production of tumours in the rat by dimethylnitrosamine (*N*-nitrosodimethylamine), *Acta Unio Int. Contra Cancrum*, 15, 187–190, 1959.
363. Preussman, R. and Stewart, B. W., *N*-Nitroso carcinogens, in *Chemical Carcinogens*, 2nd ed., Searle, C. E., Ed., ACS Monograph 182, Vol. 2, American Chemical Society, Washington, D.C., pp. 643–828, 1984.
364. Lijinsky, W., Structure-activity relations in carcinogenesis by *N*-nitroso compounds, *Canc. Metas. Rev.*, 6, 301–356, 1987.
365. Mirvish, S. S., Formation of *N*-nitroso compounds: chemistry, kinetics, and *in vivo* occurrence, *Tox. Appl. Pharm.*, 31, 325–351, 1975.
366. Sen, N. P., Analysis and occurrence of *N*-nitroso compounds in foods, in *N-Nitroso Compounds: Biology and Chemistry*, Bhide, S. V. and Rao, K. V. K., Eds., Omega Scientific, New Delhi, 1989, pp. 3–18.

367. Sen, N. P. and Kubacki, S. J., Review of methodologies for the determination of nonvolatile *N*-nitroso compounds in foods, *Food Add. Contamin.*, 4, 357–383, 1987.
368. Gray, J. I., Formation of *N*-nitroso compounds in foods, in *N-Nitroso Compounds in Foods*, Scanlan, R. A. and Tannenbaum, S. R., Eds., ACS Symposium Series 174, American Chemical Society, Washington, D.C., 1981, pp 165–180.
369. Pensabene, J. W. and Fiddler, W., *N*-Nitrosothiazolidine in cured meat products, *J. Food Sci.*, 48, 1870–1874, 1983.
370. Sen, N. P., Formation and occurrence of nitrosamines in food, in *Diet, Nutrition, and Cancer: A Critical Evaluation,* Vol. II, *Micronutrients, Nonnutritive Dietary Factors, and Cancer,* Reddy, B. S. and Cohen, L. A., Eds., CRC Press, Boca Raton, FL, 1986, pp. 135–160.
371. Sen, N. P., Analysis and occurrence of *N*-nitroso compounds in foods, in *N-Nitroso Compounds: Biology and Chemistry,* Bhide, S. V. and Rao, K. V. K., Eds., Omega Scientific, New Delhi, 1990, pp. 3–18.
372. Hotchkiss, J. H., A review of current literature on *N*-nitroso compounds in foods, *Adv. Food Res.*, 31, 53–115, 1987.
373. Vecchio, A. J., Hotchkiss, J. H., and Bisogni, C. A., *N*-nitrosamine ingestion from consumer-cooked bacon, *J. Food Sci.,* 51(3), 754–756, 1986.
374. Havery, D. C., Fazio, T., and Howard, J. W., Trends in levels of *N*-nitrosopyrrolidine in fried bacon, *J. Assoc. Off. Anal. Chem.*, 61, 1379–1382, 1978.
375. Scanlan, R. A., Barbour, J. F., and Chappel, C. I., A survey of *N*-nitrosodimethylamine in U.S. and Canadian beers, *J. Agric. Food Chem.,* 38, 442–443, 1990.
376. Havery, D. C., Hotchkiss, J. H., and Fazio, T., A rapid method for the determination of volatile *N*-nitrosamines in non-fat dried milk, *J. Dairy Sci.,* 65, 182–185, 1982.
377. Sen, N. P., Migration and formation of *N*-nitrosamines from food contact materials, in *Food and Packaging Interactions,* ACS Symposium Series 365, Hotchkiss, J. H., Ed., American Chemical Society, Washington, D.C., pp., 1988, 146–158.
378. Walker, R., Nitrates, nitrites, and *N*-nitroso compounds: a review of the occurrence in food and diet and the toxicological implications, *Food Add. Contam.*, 7, 717–768, 1990.
379. Wagner, D. A. and Tannenbaum, S. R., *In vivo* formation of *N*-nitroso compounds, *Food Technol.,* 39(1), 89–90, 1985.
380. Licht, W. R. and Deen, W. M., Theoretical model for predicting rates of nitrosamine and nitrosamide formation in the human stomach, *Carcinogenesis,* 9, 2227–2237, 1988.
381. Hotchkiss, J. H. and Cassens, R. G., Nitrate, nitrite, and nitroso compounds in foods, *Food Technol.*, 41, 127–136, 1987.
382. Druckery, H., Preussmann, R., Ivankovic, S., and Schmähl, D., Organotrope carcinogene wirkungen bei 65 verschiedenen *N*-nitroso-verbindungen an BD-ratten., *Z. Krebsforsch,* 69, 103–201, 1967.
383. Peto, R., Gray, R., Brantom, P., and Grasso, P., *Nitrosamine Carcinogenesis in 5120 Rodents: Chronic Administration of Sixteen Different Concentrations of NDEA, NDMA, NPYR and NPIP in the Water of 4440 Inbred Rats, With Parallel Studies on NDEA Alone of the Effect of Age of Starting (3, 6, or 20 Weeks) and of Species (Rats, Mice, or Hamsters),* IARC Scientific Publication No. 57, International Agency for Research on Cancer, Lyon, France, 1984, pp. 627–665.
384. Lijinsky, W., Structural relations and dose response studies in nitrosamine carcinogenesis, *Adv. Modern Environ. Toxicol.,* 10, 215–241, 1987.
385. Assembly of Life Sciences, *The Health Effects of Nitrate, Nitrite, and N-Nitroso Compounds,* part 1 of a two-part study by the Committee on Nitrite and Alternative Curing Agents in Food, National Research Council, Washington, D.C., 1981.
386. WHO, *Health Hazards from Nitrates in Drinking Water,* Report of a World Health Organization meeting, Copenhagen, 5–9 March, 1984.
387. Guillen, M. D., Polycyclic aromatic compounds: extraction and determination in food, *Food Add. Contam.,* 11, 669–684, 1994.
388. Larsen, J. C. and Poulsen, E., Mutagens and carcinogens in heat-processed food, in *Toxicological Aspects of Food,* Miller, K., Ed., 1987, pp. 205–252.
389. Huggins, C., Grand, L. C., and Brillantes, F. P., Mammary cancer induced by as single feeding of polynuclear hydrocarbons, and its suppresions, *Nature,* 189, 204–207, 1961.
390. Tilgner, D. J., Food in a carcinogenic environment, *Food Manuf.*, 45, 47–50, 1970.
391. Adrian, J., Billaud, C., and Rabache, M., Part of technological processes in the occurrence of benzo(*a*)pyrene in foods, *World Rev. Nutr. Diet,* 44, 155–184, 1984.
392. Grimmer, G. and Hildebrandt, A., Content of polycyclic hydrocarbons in different vegetables. II. Hydrocarbons in the human surroundings, *Dt. Lebensmitt Rdsch.*, 61, 237–239, 1965.
393. Toth, L. and Potthast., K., Chemical aspects of the smoking of meat and meat products, *Adv. Food Res.*, 29, 87–158, 1984.
394. Larsson, B. K., Sahlberg, G. P., Eriksson, A. T., and Busk, L. A., Polycyclic aromatic hydrocarbons in grilled food, *J. Agric. Food Chem.,* 31, 867–873, 1983.

395. Fretheim, K., Polycyclic aromatic hydrocarbons in grilled meat products: a review, *Food Chem.*, 10, 129–139, 1983.
396. Doremire, M. E., Harmon, G. E., and Pratt, D. E., 3,4-Benzopyrene in charcoal grilled meats, *J. Food Sci.*, 44, 622–623, 1979.
397. Malanoski, A. J., Greenfield, E. L., and Barnes, C. J., Worthington, J. M., and Joe, Jr., F. L., Survey of PAH in smoked foods, *J. Assoc. Off. Anal. Chem.*, 51, 114–121, 1968.
398. Masuda, Y., Mori, K., Hirohata, T., and Kuratsune, M., Carcinogenesis in the esophagus. III. PAH and phenols in whiskey, *Gann,* 57, 549–557, 1966.
399. Van der Stegen, G. H. D. and Van Overbruggen, G. J. J., *A Study on Coffee Roasting and 3,4-Benzopyrene,* 10e Colloque Internationale sur le Cafe, Salvador-Bahia, 1982, Association Internationale du Cafe, Paris, 1983, pp. 347–354.
400. Vaessen, H. A. M. G., Jekel, A. A., and Wilbers, A. A. M. M., Dietary intake of polycyclic aromatic hydrocarbons, *Toxicol. Environ. Chem.,* 16, 281–294, 1988.
401. EPA, *An Exposure and Risk Assessment for Benzo[a]pyrene and Other Polycyclic Aromatic Hydrocarbons,* Vol IV, EPA 440/4-85-020-V4, U.S. Environmental Protection Agency, Office of Water Regulations and Standards, Washington, D.C., 1985.
402. Neal, R. and Rigdon, R. H., Gastric tumors in mice fed benzo(a)pyrene: a quantitative study, *Tex. Rep. Biol. Med.*, 25, 553–557, 1967.
403. Rigdon, R. H. and Neal, J., Gastric carcinomas and pulmonary adenomas in mice fed benzo(a)pyrene, *Tex. Rep. Biol. Med.*, 24, 195–207, 1966.
404. Rigdon, R. H., Neal, J., and Mack, J., Leukemia in mice fed benzo(*a*)pyrene, *Tex. Rep. Biol. Med.*, 25, 422–431, 1967.
405. Huggins, C. and Yang, N. C., Induction and extinction of mammary cancer, *Science,* 137, 257–262, 1962.
406. Pierce, W. E. H., Tumor-production by lime oil in the mouse forestomach, *Nature (Lond.),* 189, 497–500, 1961.
407. Frankfurt, O. S., Mitotic cycle and cell differentiation in squamous cell carcinomas, *Int. J. Cancer,* 2, 304–310, 1967.
408. Field, W. E. H. and Roe, F. J. C., Tumor promotion in the forestomach epithelium of mice by oral administration of citrus oil, *J. Natl. Cancer Inst.*, 35, 771–787, 1965.
409. Roe, F. J. C., Levy, L. S., and Carter, R. L., Feeding studies on sodium cyclamate, saccharin, and sucrose of carcinogenic and tumor promoting activity, *Food Cosmet. Toxicol.,* 8, 135, 1970.
410. Wattenburg, L. W. and Leong, J. L., Inhibition of the carcinogenic action of benzo[a]pyrene by flavones, *Cancer Res.*, 30, 1922–1929, 1970.
411. Klein, M., Susceptibility of strain B6AF1/J hybrid infant mice to tumorigenesis with 1,2-benzanthracene, deoxycholic acid, and 3-methylcholanthrene, *Cancer Res.*, 23, 1701–1707, 1963.
412. IARC, *IARC Monographs on the Evaluation of Carcinogenic Risk of the Chemical to Man: Certain Polycyclic Aromatic Hydrocarbons and Heterocyclic Compounds*, Vol. 3, International Agency for Research on Cancer, Lyon, France, 1972.
413. Vaessen, H. A. M. G., Schuller, P. L., Jekel, A. A., and Wilbers, A. A. M. M., Polycyclic aromatic hydrocarbons in selected foods, analysis and occurrence, *Toxicol. Environ. Chem.,* 7, 297–324, 1984.
414. Howard, J. W., White, R. H., Fry, B. E., and Turicchi, E., Extraction and estimation of PAH in smoked foods. II. Benzo(*a*)pyrene, *J. Assoc. Off. Anal. Chem.,* 49, 611–617, 1966.
415. Joe, F. L., Roseboro, E. L., and Fazio, T., Survey of some market basket commodities for polynuclear aromatic hydrocarbon content, *J. Assoc. Off. Anal. Chem.,* 62, 615–629, 1979.
416. Rhee, K. and Bratzler, L., Benzo(*a*)pyrene in smoked meat products, *J. Food Sci.*, 35, 146–149, 1970.
417. Lijinsky, W. and Ross, A., Production of carcinogenic polynuclear hydrocarbons in the cooking of food, *Food Cosmet. Toxicol.,* 5, 343–347, 1967.
418. Lijinsky, W. and Shubik, P., The detection of PAH in liquid smoke and some foods, *Toxicol. Appl. Pharmacol.*, 7, 337–343, 1965.
419. Mattocks, A. R., *Chemistry and Toxicology of Pyrrolizidine Alkaloids*, Academic Press, London, 1986.
420. Furuya, T., Asada, Y., and Mori, H., Pyrrolizidine alkaloids, in *Naturally Occurring Carcinogens of Plant Origin*, Hirono, I., Ed., Kodansha, Ltd., Tokyo, 1987, pp. 25–51.
421. Huxtable, R. J., Human health implications of pyrrolizidine alkaloids and the herbs containing them, in *Toxicants of Plant Origin,* Vol I, *Alkaloids*, Cheek, P. R., Ed., CRC Press, Boca Raton, FL, 1989, pp. 31–86.
422. Betz, J. M., Eppley, R. M., Taylor, W. C., and Andrzejewski, D., Determination of pyrrolizidine alkaloids in commercial comfrey products (*Symphytum* sp.), *J. Pharmaceut. Sci.,* 83(5), 649–653, 1994.
423. Roitman, J. N., Comfrey and liver damage, *Lancet,* April, 944, 1981.
424. Kuhara, K., Takanaski, H., Hirono, I., Furuya, T., and Asada, Y., Carcinogenic activity of clivorine, a pyrrolizidine alkaloid isolated from *Ligularia dentata*, *Cancer Lett.*, 10, 117–122, 1980.
425. Schoental, R. and Head, M. A., Progression of liver lesions produced in rats by temporary treatment with pyrrolizidene (senecio), alkaloids, and the effects of betaine and high casein diets, *Br. J. Cancer,* 11, 535–544, 1957.
426. Svoboda, D. J. and Reddy, J. K., Malignant tumour in rats given lasiocarpine, *Cancer Res.,* 32, 908–913, 1972.
427. Rao, M. S. and Reddy J. K., Malignant neoplasms in rats fed laisocarpine, *Br. J. Cancer,* 37, 289–293, 1978.
428. NCI, *Bioassay of Lasiocarpine for Possible Carcinogenicity,* NCI Carcinogenesis Tech. Rep. Ser. No. 39, DHEW Publication No. (NIH) 78-839, National Cancer Institute, Betheseda, MD, 1978.

429. Newberne, P. M. and Rogers, A. E., Nutrition, monocrotaline and aflatoxin B_1 in liver carcinogenesis, *Plant Foods for Man*, 1, 23–31, 1973.
430. Shumaker, R. C., Robertson, K. A., Hsu, I. C., and Allen, J. R., Neoplastic transformation in tissues of rats exposed to monocrotaline or dehydroretronecine, *J. Natl. Cancer Inst.*, 56, 787–790, 1976.
431. Hayashi, Y., Shinada, M., and Katayama, H., Experimental insuloma in rats after a single administration of monocrotaline, *Toxicol. Lett.*, 1, 41–44, 1977.
432. Hirono, I., Mori, H., Yamada, K., Hirata, Y., Haga, M., Tatematsu, H., and Kanie, S., Carcinogenic activity of petasitenine, a new pyrrolizidine alkaloid isolated from *Petasites japonicus* Maxim, *J. Natl. Cancer Inst.*, 58, 1155–1157, 1977.
433. Hirono, I., Haga, M., Fujii, M., Matsuura, S., Matsubara, N., Nakayama, M., Furuya, T., Hikichi, M., Takanashi, H., Uchida, E., Hosaka, S., and Ueno, I., Induction of hepatic tumors in rats by senkirkine and symphytine, *J. Natl., Cancer Inst.*, 63, 469–472, 1979.
434. Yamanaka, H., Nagao, M., Sugimura, T., Furuya, T., Shirai, A., and Matsushima, T., Mutagenicity of pyrrolizidine alkaloids in the *Salmonella*/mammalian-microsome test, *Mutat. Res.*, 68, 211–216, 1979.
435. Woo, Y., Lai, D. Y., Arcos, J. C., and Argus, M. F., *Chemical Induction of Cancer. Structural Bases and Biological Mechanisms*, Academic Press, San Diego, CA, 1988, pp. 212–266.
436. IARC, *IARC Monographs on the Evaluation of Carcinogenic Risk of Chemicals to Man: Some Naturally Occurring Substances*, Vol. 10, International Agency for Research on Cancer, Lyon, France, 1976.
437. IARC, *IARC Monographs on the Evaluation of Carcinogenic Risk of Chemicals to Humans*, Vol. 31, International Agency for Research on Cancer, Lyon, France, 1983.
438. Kuhara, K., Takanaski, H., Hirono, I., Furuya, T., and Asada, Y., Carcinogenic activity of clivorine, a pyrrolizidine alkaloid isolated from *Ligularia dentata*, *Cancer Lett.*, 10, 117–122, 1980.
439. Schoental, R. and Head, M. A., Progression of liver lesions produced in rats by temporary treatment with pyrrolizidene (senecio), alkaloids, and the effects of betaine and high casein diets, *Br. J. Cancer*, 11, 535–544, 1957.
440. Svoboda, D. J. and Reddy, J. K., Malignant tumour in rats given lasiocarpine, *Cancer Res.*, 32, 908–913, 1972.
441. Rao, M. S. and Reddy J. K., Malignant neoplasms in rats fed laisocarpine, *Br. J. Cancer*, 37, 289–293, 1978.
442. NCI, *Bioassay of Lasiocarpine for Possible Calcinogenicity*, NCI Carcinogenesis Tech. Rep. Ser. No. 39, DHEW Publication No. (NIH) 78-839, National Cancer Institute, Betheseda, MD, 1978.
443. Newberne, P. M. and Rogers, A. E., Nutrition, monocrotaline and aflatoxin B_1 in liver carcinogenesis, *Plant Foods for Man*, 1, 23–31, 1973.
444. Shumaker, R. C., Robertson, K. A., Hsu, I. C., and Allen, J. R., Neoplastic transformation in tissues of rats exposed to monocrotaline or dehydroretronecine, *J. Natl. Cancer Inst.*, 56, 787–790, 1976.
445. Hayashi, Y., Shinada, M., and Katayama, H., Experimental insuloma in rats after a single administraiton of monocrotaline, *Toxicol. Lett.*, 1, 41–44, 1977.
446. Hirono, I., Mori, H., Yamada, K., Hirata, Y., Haga, M., Tatematsu, H., and Kanie, S., Carcinogenic activity of petasitenine, a new pyrrolizidine alkaloid isolated from *Petasites japonicus* Maxim, *J. Natl. Cancer Inst.*, 58, 1155–1157, 1977.
447. Hirono, I., Haga, M., Fujii, M., Matsuura, S., Matsubara, N., Nakayama, M., Furuya, T., Hikichi, M., Takanashi, H., Uchida, E., Hosaka, S., and Ueno, I., Induction of hepatic tumors in rats by senkirkine and symphytine, *J. Natl. Cancer Inst.*, 63, 469–472, 1979.
448. IARC, *Some Naturally Occurring Substances: Food Items and Consituents, Heterocyclic Aromatic Amines and Mycotoxins*, Monograph 56, International Agency for Research on Cancer, Lyon, France, 1993, pp. 135–162.
449. Verghese, J., The chemistry of limonene and of its derivaties. Part I, *Perfum. Essent. Oil. Rec.*, 59, 439–454, 1968.
450. *Product Data Sheet: d-Limonene*, Florida Chemical Co., Lake Alfred, FL, 1991.
451. FEMA, *d-Limonene Monograph*, Flavor and Extract Manufacturers' Assoc., Washington, D.C., 1991, pp. 1–4.
452. NTP, *Toxicology and Carcinogenesis Studies of d-Limonene (CAS No. 5989-27-5) in F344/N Rats and B6C3F1 Mice (Gavage Studies)*, NTP TR 347, NIH Publication No. 90-2802, National Toxicology Program, Research Triangle Park, NC, 1990.
453. Conacher, H. B. S. and Page, B. D., *Ethyl Carbamate in Alcoholic Beverages: A Canadian Case History*, Proc. Euro Food Toxicology, II, Oct. 15–18, 1986, Zurich, Switzerland, 1986, pp. 237–242.
454. *Fed. Reg.*, 37, 15426, 1972.
455. Coker, Jr., C. E., *United States Policy on Urethane in Alcholic Beverages*, text of presentation before an international symposium of experts on ethyl carbamate, sponsored by the Office of International de la Vigne et du Vin, Paris, France, May 3, 1988.
456. Ough, C. S., Ethylcarbamate in fermented beverages and foods. I. Naturally occurring ethylcarbamate, *J. Agric. Food Chem.*, 24, 323–327, 1976.
457. Nettleship, A., Henshaw, P. S., and Meyer, H. L., Induction of pulmonary tumors in mice with ethyl carbamate (urethane), *J. Nat. Cancer Inst.*, 4, 309–319, 1943.
458. Bateman, A. J., The mutagenic action of urethane, *Mutat. Res.*, 39, 75–96, 1976.
459. Allen, J. W., Sharief, Y., and Langenbach, R. J., An overview of ethyl carbamate (urethane) and its genotoxic activity, in *Genotoxic Effects of Airborne Agents*, Tice, R. R., Costa, D. L., and Schaich, K. M., Eds., Plenum Press, New York, 1982, pp. 443–460.

460. IARC, *IARC Monographs on the Evaluation of the Carcinogenic Risk of Chemicals to Man*, Vol. 7, *Some Anti-Thyroid and Related Substances, Nitrofurans and Industrial Chemicals*, International Agency for Research on Cancer, Lyon, France, 1974, pp. 111–140.
461. Jaffe, W. G., Carcinogenic action of ethyl urethane on rats, *Cancer Res*, 7, 107–111, 1947.
462. Haran-Ghera, N. and Berenblum, I, The induction of the initiating phase of skin carcinogenesis in the mouse by oral administration of urethane (ethyl carbamate), *Br. J. Cancer*, 10, 57–60, 1956.
463. Berenblum, I. and Haran-Ghera, N., A quantitative study of the systemic initiating action of urethane (ethyl carbamate) in mouse skin carcinogenesis, *Br. J. Cancer*, 11, 77–84, 1957.
464. van Esch, G. J., van Genderen, H., and Vink, H. H., The production of skin tumors in mice by oral treatment with urethane, isopropyl-*N*-phenyl carbamate or isopropyl-*N*-chlorophenyl carbamate in combination with skin painting with croton oil and Tween 60, *Br. J. Cancer*, 12, 355–362, 1958.
465. Pietra, G. and Shubik, P., Induction of melanotic tumors in the Syrian Golden Hamster after administration of ethyl carbamate, *J. Natl. Cancer Inst.*, 25, 627–630, 1960.
466. Toth, B., Tomatis, L., and Shubik, P., Multipotential carcinogenesis with urethan in the Syrian Golden Hamster, *Cancer Res.*, 21, 1537–1541, 1961.
467. Toth, B., Della Porta, G., and Shubik, P., The occurrence of malignant lymphomas in urethan-treated Swiss mice, *Br. J. Cancer*, 15, 322–326, 1961.
468. Tannenbaum, A. and Maltoni, C., Neoplastic response of various tissues to the administration of urethane, *Cancer Res.*, 22, 1105–1112, 1962.
469. Tannenbaum, A., Vesselinovitch, S. D., Maltoni, C., and Mitchell, D. S., Multipotential carcinogenicity of urethane in the Sprague-Dawley rat, *Cancer Res.*, 22, 1362–1371, 1962.
470. Klein, M., Induction of lymphocytic neoplasms, hepatomas and other tumors after oral administration of urethan to infant mice, *J. Natl Cancer Inst.*, 29, 1035–1046, 1962.
471. Della Porta, G., Capitano, J., Montipo, W., and Parmil, L., Studio sull'azione cancerogena dell'uretano nel topo (a study of the carcinogenic action of urethane in mice), *Tumor*, 49, 413–428, 1963.
472. Della Porta, G., Capitano, J., and Strambio De Castillia, P., Studies on leukemogenesis in urethane-treated mice, *Acta Un Int Cancr*, 19, 783–785, 1963.
473. Della Porta, G., Capitano, J., Parmi, L., and Colnaghi, M. I., Cancerogenesi da uretano in topi neonati, lattanti e adulti de ceppi C57B1, C3H, BC3F, C3Hf e SWR (urethane carcinogenesis in newborn, suckling and adult mice of C57B1, C3H, $BC3F_1$, C3Hf and SWR strains), *Tumori*, 53, 81–102, 1967.
474. Newberne, P. M., Hunt, C. E., and Wogan, G. N., Neoplasms in the rat associated with administration of urethan and aflatoxin, *Exp. Mol. Pathol.*, 6, 285–299, 1967.
475. Adenis, L., Demaille, A., and Driessens, J., Pouvoir cancérigène de l'uréthane chez le rat Sprague, *Cancer Res. Soc. Biol.*, 162, 458–461, 1968.
476. Brooks, R. E., Pulmonary adenoma of strain A mice: an electron microscopic study, *J. Natl. Cancer Inst.*, 41, 719–742, 1968.
477. Toth, B. and Boreisha, I., Tumorigenesis with isonicotinic acid hydrazide and urethan in the Syrian golden hamsters, *Eur. J. Cancer*, 5, 164–171, 1969.
478. Innes, J. R. M., Ulland, B. M., Valerio, M. G., Petrucelli, L., Fishbein, L., Hart, E. R., Pallotta, A. J., Bates, R. R., Falk, H. L., Gart, J. J., Klein, M., Mitchell, I., and Peters, J., Bioassay of pesticides and industrial chemicals for tumorigenicity in mice: a preliminary note, *J. Natl. Cancer Inst.*, 42, 1101–1114, 1969.
479. Tomatis, L., Turusov, V., Day, N., and Charles, R. T., The effect of long-term exposure to DDT on CF-1 mice, *Int. J. Cancer*, 10, 489–506, 1972.
480. Schmahl, D., Port, R., and Wahrendorf, J., A dose-response study on urethane carcinogenesis in rats and mice, *Int. J. Cancer*, 19, 77–80, 1977.
481. Adamson, R. H. and Siber, S. M., Chemical carcinogenesis studies in nonhuman primates, in *Organ and Species Specificity in Chemical Carcinogenesis*, Langenback, R. and Nesnow, S., Eds., Plenum Press, New York, 1982, pp. 129–156.
482. Bull, R. J., Robinson, M., Laurie, R. D., Stoner, G. D., Grisinger, E., Meier, J. R., and Stober, J. A., Carcinogenic effects of acrylamide in Sencar and A/J mice, *Cancer Res.*, 44, 107–111, 1984.

PART 3

INHALATION TOXICOLOGY

Daniel L. Morgan, Editor

CHAPTER 11

STRUCTURE AND FUNCTION OF THE RESPIRATORY TRACT

Kent E. Pinkerton, Charles G. Plopper, Dallas M. Hyde, Jack R. Harkema, Walter S. Tyler, Kevin T. Morgan, Judith A. St. George, . Michael Kay and Andrew Mariassy

CONTENTS

1 INTRODUCTION

The respiratory system serves as an important interface between the organism and the environment. The internal surfaces of the human lungs cover an area approximately 25 times greater than the external surfaces of the body. From the nose to the alveolus, the respiratory tract facilitates a number of functions including gas exchange, conditioning of the air, removal of inhaled particles, metabolism of xenobiotic compounds, and protection against noxious and injurious agents. The cellular composition and unique anatomical makeup of the nasopharynx, tracheobronchial tree, and parenchyma are optimally designed to bring about each of these functions. The purpose of this section is to discuss the structure and function of the lower respiratory tract. The macroscopic anatomy of the nasopharynx and conducting airways and the cells populating these structures will be reviewed, along with the tissues and cells forming the gas exchange portions of the lung parenchyma.

0-8493-4493-X/97/$0.00+$.50

TABLE 1

Interspecies Comparison of Nasal Cavity Characteristics

	Sprague-Dawley Rat	Guinea Pig	Beagle Dog	Rhesus Monkey	Human
Body weight	250 g	600 g	10 kg	7 kg	~70 kg
Naris cross-section	0.7 mm²	2.5 mm²	16.7 mm²	22.9 mm²	140 mm²
Bend in naris	40°	40°	30°	30°	
Length	23 cm	3.4 cm	10 cm	5.3 cm	7–8 cm
Greatest vertical diameter	9.6 mm	12.8 mm	23 mm	27 mm	40–45 mm
Surface area (both sides of nasal cavity)	10.4 cm²	27.4 cm²	220.7 cm²	61.6 cm²	181 cm²
Volume (both sides)	.4 cm³	.9 cm³	20 cm³	8 cm³	16–19 cm³ (does not include sinuses)
Bend in nasopharynx	15°	30°	30°	80°	~90°
Turbinate complexity	Complex scroll	Complex scroll	Very complex membranous	Simple scroll	Simple scroll

Source: From Proctor, D. F. and Chang, J. C. F., in *Nasal Tumors in Animals and Man,* Vol. III, *Experimental Nasal Carcinogenesis,* Reznik, G. and Stinson, S. F., Eds., CRC Press, Boca Raton, FL, 1983, pp. 1–26. With permission.

2 OVERVIEW

The respiratory tract begins at the nasal cavity and ends at the alveolus. The respiratory system consists of three general anatomical locations: the nasal pharynx, conducting airways, and gas exchange portions of the lungs. The general functions of the lungs are to provide for the exchange of gases, filtering of blood, and storage of blood from the right side of the heart before entering into systemic circulation. Also, metabolic functions which the lungs perform include the processing of proteins and lipids, activation or inactivation of hormones, and processes involved in the metabolism of xenobiotic compounds which may reach the lungs via the airways or through the vasculature. Another important function of the respiratory system is that of defense to provide a mechanism of clearance for the lung and also to mount an immune response for other constituents that may enter into the body via the respiratory system. (see Table 1.)

The upper respiratory tract consists of the pharynx, larynx, and cervical trachea. These structures serve to conduct the air to the lower respiratory tract. They also serve an important function in conditioning the air that is inhaled by warming and humidifying the air prior to reaching the lower respiratory tract. The nasal cavity, pharynx, and larynx also serve to clear particulates from the air as they pass through these structures as well as removing soluble gases. Many toxic substances that may be inhaled that are highly reactive are actually quite effectively removed from the air through absorption in the upper respiratory tract structures. The functions of olfaction are an important aspect of the anatomical make-up of the respiratory tract as well as the formation in that function that is served by the larynx.

The trachea is immediately distal to the larynx as a hollow cylindrical tube approximately 12 cm in length in the adult human male. The trachea is composed of four basic elements: (1) a pseudostratified columnar epithelium lining the inner surfaces of the tracheal lumen, (2) submucosal glands opening onto the tracheal surfaces via epithelial-lined ducts, (3) C-shaped cartilaginous rings that support and maintain the patency of the tracheal lumen, and (4) a connective tissue wall composed of collagen, elastin, smooth muscle, blood vessels, lymphatic channels, and nerves intermeshed in a network of noncellular matrix.

TABLE 2

Terminology of Lung Lobes

Humans	Animals	
NA	NAV	Old
Left Lung		
Superior	Cranial[a]	
	Cranial part	Apical
	Caudal part	Cardiac
Inferior	Caudal[a]	Diaphragmatic
Right Lung		
Superior	Cranial[b]	Apical
Middle	Middle	Cardiac
Inferior	Caudal	Diaphragmatic
	Accessory	Intermediate

Abbreviations: NA = *Nomina Anatomica* terms (1983); NAV = *Nomina Anatomica Veterinaria* terms (1983); Old = old terminology found in many references.

[a] The left lungs of small laboratory mammals (mouse, rat, hamster, gerbil) are not divided, but those of larger ones (guinea pig, rabbit) are.

[b] The right cranial lobe of ruminates is divided into cranial and caudal parts.

The structural features of the trachea extend several generations down the airways entering into the lobes of the lungs. These airways are known as bronchi. The right and left bronchi arising from the trachea of the lungs are primary or mainstem bronchi. The airways distal to the left and right mainstem bronchi which establish the regional branching patterns within each lobe of the lungs are known as tertiary or segmental bronchi. In the human lung, a total of 10 segmental bronchi are found in the lobes of the right lung and 10 in the lobes of the left lung. The right lung has an upper (superior), middle, and lower (inferior) lobe, while the left lung has only an upper and lower lobe. Each segmental bronchus supplies a unique portion of the lungs. (See Tables 2 to 4.)

Only a few generations beyond the tertiary bronchi (generations four to six from the trachea) the airways decrease in cross-sectional diameter and begin to lose some of the features of the trachea and upper bronchi. Cartilaginous rings in the trachea and primary bronchi give way to smaller plates and islands of cartilage in the smaller bronchi. These cartilaginous plates decrease as a function of increasing airway generation. Submucosal glands also decrease in size and frequency. Smooth muscle within the walls of each new airway generation increase as cartilaginous plates become less frequent. This shift in cartilage and smooth muscle down the airways is reflected in the relative rigidity of the upper bronchi vs. the more compliant nature of the bronchioles. The network of blood vessel and nerves continues within the walls of each airway. Lymph nodes and patchy masses of lymphoid tissue within the connective tissues of the trachea are also evident within the walls of bronchi entering into each lobe of the lungs. Cartilaginous plates and submucosal glands may continue for six to eight generations beyond the tracheal bifurcation.

TABLE 3[43]

Yeh Model of Human Airways

Generation (Trachea = 1)	Number of Airways	Diameter (mm)	Length (mm)	θ[a] (°)	ψ[b] (°)	S (cm²)
1	1	20.1	100	0	0	3.17
2	2	15.6	43.6	33	20	3.82
3	4	11.3	17.8	34	31	4.01
4	8	8.27	9.65	22	43	4.30
5	16	6.51	9.95	20	39	5.33
6	32	5.74	10.1	18	39	8.28
7	64	4.35	8.90	19	40	9.51
8	128	3.73	9.62	22	36	13.99
9	256	3.22	8.67	28	39	20.85
10	512	2.57	6.67	22	45	26.56
11	1024	1.98	5.56	33	43	31.56
12	2048	1.56	4.46	34	45	39.14
13	4096	1.18	3.59	37	45	44.79
14	8192	0.92	2.75	39	60	54.46
15	16,384	0.73	2.12	39	60	68.57
16	32,768	0.60	1.68	51	60	92.65

[a] θ = branching angle, angle between path of daughter and that of parent.
[b] ψ = orientation to gravity with 90° corresponding to horizontal.

TABLE 4[42]

Yeh Model of Airways of the Rat

Generation (Trachea = 1)	Number of Airways	Diameter (mm)	Length (mm)	θ[a] (°)	ψ[b] (°)	S (cm²)
1	1	3.40	26.80	0	86	0.091
2	2	2.90	7.15	15	90	0.132
3	3	2.63	4.00	43	86	0.163
4	5	2.03	1.76	36	71	0.162
5	8	1.63	2.08	32	59	0.170
6	14	1.34	1.17	22	58	0.197
7	23	1.23	1.14	16	61	0.273
8	38	1.12	1.30	17	58	0.374
9	65	0.95	0.99	20	55	0.461
10	109	0.87	0.91	15	58	0.648
11	184	0.78	0.96	16	61	0.879
12	309	0.70	0.73	17	56	1.189
13	521	0.58	0.75	17	58	1.377
14	877	0.49	0.60	22	58	1.654
15	1477	0.36	0.55	24	57	1.503
16	2487	0.20	0.35	44	58	0.781

[a] θ = branching angle, angle between path of daughter and that of parent.
[b] ψ = orientation to gravity with 90° corresponding to horizontal.

As more distal airways lose submucosal glands and cartilaginous plates, the internal diameter of the airways also continues to decrease. These airways without submucosal glands or cartilaginous plates are known as bronchioles. In general, bronchioles form 9 to 12 generations of the bronchial tree. The noncartilaginous bronchioles in humans are typically

TABLE 5

Number and Dimensions of Mammalian Terminal Bronchioles

Species	Number	Diameter (mm)	Length (mm)	Ref.
Human	65,000	0.6	1.65	54
	33,000	0.6	1.68	55
	27,000	0.5–0.66	1.18–1.25	43
Dog	18,000–36,000	0.2	0.5	56
	21,000	0.2–0.27	—	57
Ferret	1700	—	—	58
Rat	2500	0.2	0.35	42
Hamster	1200	0.2–0.3	0.4	59

1 to 2 mm in diameter and end with the first airway containing alveolar outpocketings within their walls (Table 5). Airways containing alveolar outpocketings are respiratory bronchioles and form the first structures along the airway path that facilitate gas exchange. Respiratory bronchioles extend for one to four generations and provide a gradual interface between airways involved expressly in air conduction to airways involved in gas exchange.

By volume, 90% of the lungs is composed of a delicate, lacy tissue engaged in gas exchange. These tissues form the walls of approximately 300 million alveoli packaged within the confines of the human lungs. In the adult, this surface covers a surface equivalent to the size of a tennis court or 50 m^2. Respiratory bronchioles are the first structures along the bronchial tree where alveoli are encountered as outpocketings within the walls of the airway. The alveolarization of the walls increase with each succeeding respiratory bronchiole generation until the entire wall is composed of alveolar outpocketing with the absence of a bronchiolar wall. These passages form the alveolar ducts, a series of channels upon which the bulk of the alveoli open (Table 6). Several generations of alveolar ducts may occur before reaching the terminal ends of the airway path as alveolar sacs. It is relevant to note that alveolar sacs or terminations of alveolar ducts are found throughout the lungs in addition to subpleural locations. The lobular arrangement of the lungs is a consequence of collections of gas exchange units. Lung lobules in the human lung may occasionally be circumscribed by a sheath of connective tissue, which may not always be present. Such an arrangement may play an essential role in the confinement of infectious agents or disease processes to localized regions of the lungs.

The pleura covers the outer surfaces of the lungs and consists of a continuous serosal surface of mesothelial cells, a single cell layer thick and covering a fibroelastic connective tissue layer with its own vascular supply and lymphatic plexus. The mesothelium and fibroelastic connective tissue reflects onto every surface of the lungs as the visceral pleura, as well as onto the mediastinal space, diaphragm, and costal surfaces of the thorax as the parietal pleura. This architectural arrangement creates a closed space in the thorax as the parietal pleura. This architectural arrangement creates a closed space in the thorax to create a vacuum relative to the external atmospheric pressure, with the movement of the rib cage and diaphragm altering the volume of the thorax and the relative pressure within the pleural space of the thorax. Such a change (referred to as the transpulmonary pressure) facilitates the passive movement of air into the lungs. Changes in the volume of the thoracic space permit the flow of air in and out of the lungs in the act of breathing as a spontaneous and natural process.

3 COMPOSITION OF THE CONDUCTING AIRWAYS

The arrangement of epithelial cells in the conducting airways varies from pseudostratified columnar to simple cuboidal. All epithelial cells in the conducting airways rest on the basal

TABLE 6

Comparative Lung Biology: Morphologic Features of Pleura, Interlobular, and Segmental Septa and Distal Airways

	Human	Macaque Monkey	Dog, Cat	Ferret	Mouse, Rat, Gerbil, Hamster, Guinea Pig, Rabbit	Horse, Sheep	Ox, Pig
Pleura	Thick	Thin	Thin	Thin	Thin	Thick	Thick
Interlobular and segmental connective tissue	Extensive; interlobular partially surrounds many lobules	Little	Little, if any	Little	Little, if any	Extensive;[a] interlobular partially surrounds many lobules	Extensive; interlobular surrounds many lobules completely
Nonrespiratory bronchiole (nonalveolarized)	Several generations	Fewer generations, commonly only one	Fewer generations	Several generations	Several generations	Several generations	Several generations
	TB ends in respiratory bronchioles	TB ends in respiratory bronchioles	TB ends in respiratory bronchioles	TB ends in respiratory bronchioles	TB ends in alveolar ducts or very short respiratory bronchioles	TB ends in alveolar ducts or very short respiratory bronchioles	TB ends in alveolar ducts or very short respiratory bronchioles
Respiratory bronchiole (alveolarized)	Several generations	Several generations	Several generations	Several generations	Absent or a single short generation	Absent or a single short generation	Absent or a single short generation

Note: TB = terminal nonrespiratory bronchiole.

[a] The interlobular connective tissue of the sheep appears extensive, and lobules appear completely separated in gross preparations but not in LM, SEM, or HRCT.

TABLE 7

Height of the Airway Epithelium in Selected Mammalian Species

	Trachea		Bronchi						
	Generation 0		Generation 1–2		Generation 4–6		Generation 7–11		
	No.	Mean ± SD	No.	Mean ± SD	No.	Mean ± SD	No.	Mean ± SD	Ref.
Rat	3	24 ± 1	20	13	6	13	—	—	12, 21, 30
Rabbit	—	21 ± 1	4	22 ± 3	6	9 ± 2	5	9 ± 1	31
Sheep	5	59 ± 7	6	40 ± 3	8	32 ± 4	6	30 ± 5	6, 15
Rhesus monkey	5	28 + 5	5	19 ± 1	5	17 ± 1	5	13 ± 1	32
Human	3	43 ± 2	5	30	—	—	—	—	24, 30, 53

Source: Adapted from Mariassy, A. T., in *Treatise on Pulmonary Toxicology, Comparative Biology of the Normal Lung*, Vol. 1, Parent, R. A., Ed., CRC Press, Boca Raton, FL, 1992, pp. 63–76. With permission.

lamina. In the trachea and first few generations of bronchi, a proportion of the cells do not reach the surface of the epithelium, and the nuclear position for other cells relative to the basal lamina varies to give the epithelial cells a "layered" look, and thus the designation of pseudostratified columnar epithelium. In smaller airways, the layered appearance of cells disappears with the decrease in basal cell number. The epithelial cells typically are of a simple columnar shape with a single layer of cells. Farther down the airways, the height of the epithelium lessens, the number of ciliated cells decreases, and the epithelium becomes a simple cuboidal layer of cells. The average height of the epithelium lining specific airway generations is given in Table 7 for the rat, rabbit, sheep, monkey, and human.

Various epithelial cell types form the lining surface of the airways (Table 8).[20] Some epithelial cells are found in all airway generations, while others are limited to certain generations of bronchi or bronchioles. The four most common epithelial cell types of the upper airways are ciliated, mucous, serous, and basal (Table 9). Brush cells are neuroendocrine cells are more rarely found along the airways, but tend to be present in most airway generations. Nonciliated bronchiolar epithelial cells (commonly known as Clara cells) are most commonly found in the lower airways of most species (Table 10). The relative population densities of each cell type are shown in Tables 11 to 13 for the trachea, mainstem, lobar, and segmental bronchi, respectively.

Ciliated cells are the most abundant cell type along the majority of the pulmonary airways. Microscopically, these cells are identified by the numerous cilia covering their apical surfaces and extending into the airway lumen. Cilia contain the "9 + 2" arrangement of microtubules, which provide the process for the beating motion of individual cilia to facilitate the movement of fluids up the airways. Synchronization of ciliary beat is critical to the unidirectional movement of materials within the fluid lining layer of the airways. Although ciliated cells are key to the movement of fluids and viscous materials, it may also be true that the surface area formed by cilia may facilitate other functions. Without doubt, the presence of cilia creates an exponential increase in surface area for the presentation and processing of inhaled compounds. The walls of the airways are an important reservoir for immunocompetent cells found as intermittent accumulations of lymphoid cell aggregates (the BALT system) along the airway tree. The processing of inhaled materials as a ready immune defense is likely to involve the interaction of these cells with the airway epithelium including ciliated cells.

Unlike ciliated cells, secretory cells are more heterogeneous in composition, with a proportion of cells containing mucin droplets of various compositions and other cells containing serous components within their secretory granules. Cells containing mucins are referred to as mucous or goblet cells, while cells containing less viscous carbohydrate-deficient granules are serous cells or small granule cells. Each of these cells expel their granules to form the periciliary and mucin lining layers of the airways (Table 14)-. It is assumed that these

TABLE 8

Abundance and Percentage (%) of Epithelial Cell Types in Transitional and Respiratory Epithelium of Bonnet Monkey Anterior Nasal Septum (Mean ± Standard Error of the Mean)

			Percent (%)						
Epithelium	No.	Total No. of Nuclei[a]	Basal Cells	Small Mucous Granule Cells	Nonciliated Cells without Secretory Granules	Nonciliated Cells with Few Secretory Granules	Goblet Cells	Ciliated Cells	Cells with Intracytoplasmic Lumina
Transitional	4	436 ± 26	35.5 ± 1.5	22.2[b] ± 4.0	31.5 ± 5.0	7.6 ± 2.2	3.2[b] ± 0.6	—	—
Respiratory	6	358 ± 32	39.5 ± 1.7	4.9 ± 1.2	—	—	19.1 ± 2.0	35.2 ± 2.1	1.3 ± 0.9

[a] Number of nuclei per mm of basal lamina.
[b] Significantly different ($p < 0.05$) from respiratory epithelium.
Source: From Harkema, J. R. et al., *Am. J. Anat.*, 180, 266–279, 1987. With permission.

TABLE 9

Comparison of Species Differences in Cell Composition of the Tracheal Epithelium

		Cells/mm	Epithelial Cells (%)									
Species	No.	Basal Lamina (Mean ± SD)	Basal	Ciliated	Mucous (Goblet)	Serous	Other	Clara	Brush	Not Identified	Ref.
Mouse	6	215	10	39	<1	0	1	49	<1	—	22
Rat	5	168 ± 12	21	32	2	42	1	0	>1	1	21
Hamster	3	151 ± 11	6	48	0	0	0	41	<1	5	14, 30
Guinea pig	5	307 ± 5	34	32	5	0	29	0	<0	—	24
Rabbit	—	211 ± 30	28	43	1	0	0	18	Rare	9	31
Cat	5	273 ± 15	37	36	20	0	5	0	<1	1	24
Sheep	5	414 ± 33	29	31	5	0	36	0	<1	0	15
Pig	5	303 ± 17	31	43	3	0	23	0	<1	—	24
Horse	3	307 ± 23	31	46	5	0	18	0	—	—	24
Cow	5	323 ± 24	31	42	4	0	23	0	<1	—	24
Monkey	5	181 + 51	42	33	17	0	4	0	<1	4	32
Human	3	303 + 20	33	49	9	0	9	0	—	—	24, 53

Note: Cells classified as "Other" are typically seromucous or nonciliated epithelial cells.
Source: Adapted from Mariassay, A. T., in *Treatise on Pulmonary Toxicology, Comparative Biology of the Normal Lung,* Vol. 1, Parent, R. A., Ed., CRC Press, Boca Raton, FL, 1992, pp. 63–76. With permission.

TABLE 10

Comparison of Species Differences in Cellular Composition of Centriacinar Bronchiolar Epithelium

	Cell Types in Terminal Bronchiole				
Species	Clara	Ciliated	Goblet	Basal	Ref.
Mouse	>50%	<50%	–	–	22, 23
Rat	>50%	<50%	–	–	3, 39
Hamster	>50%	<50%	–	–	30, 61
Guinea pig	>50%	<50%	–	–	61
Rabbit	>50%	<50%	–	–	9, 61
Cat	>95%	<5%	–	–	10, 27
Dog	>95%	<5%	–	–	27
Sheep	>60%	<40%	–	–	15, 27
Pig	+	+	–	–	1
Cow	>50%	<50%	–	–	16, 27
Horse	>50%	>50%	–	–	27
Macaque monkey	–	±50%	±20%	±10%	32, 36
Human	–	+	+	+	5, 11, 61

Note: – = not present; + = present in variable amounts.

secretions are sufficient to provide a complete blanket of fluids covering virtually all the cells of the airways. The various compositions of this lining layer are essential to permitting movement of the mucous blanket by the beating rhythm of the cilia. The mixture of these secretory cells is critical in maintaining the proper balance of airway secretions.

Basal cells are generally considered to be the primary stem cell for the repair and regeneration of new epithelium in the airways, although it is now evident that secretory cells

TABLE 11

Population Densities of the Bronchial Surface Epithelium in Selected Mammalian Species (Mainstem, Primary Bronchi)

		Cells/mm	Epithelial Cells (%)								
		Basal Lamina			Mucous					Not	
Species	No.	(Mean ± SD)	Basal	Ciliated	(Goblet)	Serous	Other	Clara	Brush	Identified	Ref.
Rat	20	126	27	35	<1	21	16	0	—	—	12
Rabbit	4	194 ± 17	27	43	1	0	7	22	<1	—	31
Sheep	6	285 ± 24	18	48	4	0	30	0	<1	0	15
Rhesus monkey	5	175 ± 46	32	44	15	0	5	0	—	3	32
Human	5	—	6 ± 1	56 ± 10	26 ± 5	Scarce	31 ± 2	0	—	—	7

Note: Cells classified as "Other" are typically seromucous or nonciliated epithelial cells.
Source: Adapted from Mariassy, A. T., in *Treatise on Pulmonary Toxicology, Comparative Biology of the Normal Lung,* Vol. 1, Parent, R. A., Ed., CRC Press, Boca Raton, FL, 1992, pp. 63–76. With permission.

TABLE 12

Population Densities of the Bronchial Surface Epithelium in Selected Mammalian Species (Lobar Bronchi, Generations 2 to 6)

		Cells/mm	Epithelial Cells (%)								
		Basal Lamina			Mucous					Not	
Species	No.	(Mean ± SD)	Basal	Ciliated	(Goblet)	Serous	Other	Clara	Brush	Identified	Ref.
Rat	6	116	14	53	<1	20	12	0	Rare	0	12
Rabbit	6	114 ± 12	2	49	0	0	7	41	<1	—	31
Sheep	8	284 ± 23	19	39	12	0	30	0	<1	0	15
Rhesus monkey	5	184 ± 49	32	47	15	0	5	0	—	2	32
Human	—	—	—	—	—	—	—	—	—	—	—

Note: Cells classified as "Other" are typically seromucous or nonciliated epithelial cells.
Source: Adapted from Mariassy, A. T., in *Treatise on Pulmonary Toxicology, Comparative Biology of the Normal Lung,* Vol. 1, Parent, R. A., Ed., CRC Press, Boca Raton, FL, 1992, pp. 63–76. With permission.

TABLE 13

Population Densities of the Bronchial Surface Epithelium in Selected Mammalian Species (Segmental Bronchi, Generations 7 to 11)

		Cells/mm	Epithelial Cells (%)								
		Basal Lamina			Mucous					Not	
Species	No.	(Mean ± SD)	Basal	Ciliated	(Goblet)	Serous	Other	Clara	Brush	Identified	Ref.
Rat	—	—	—	—	—	—	—	—	—	—	—
Rabbit	5	147 ± 18	0	49	0	0	4	47	—	0	31
Sheep	8	223 ± 17	18	43	8	0	31	0	<1	0	15
Rhesus monkey	5	158 ± 15	29	49	14	0	3	0	—	2	32
Human	—	—	—	—	—	—	—	—	—	—	—

Note: Cells classified as "Other" are typically seromucous or nonciliated epithelial cells.
Source: Adapted from Mariassy, A. T., in *Treatise on Pulmonary Toxicology, Comparative Biology of the Normal Lung,* Vol. 1, Parent, R. A., Ed., CRC Press, Boca Raton, FL, 1992, pp. 63–76. With permission.

TABLE 14

Comparison of Estimated Quantities (nl/mm² of Basal Lamina) of Intraepithelial Mucosubstances in Anterior Nasal Airway and Nasopharynx of Rat and Monkey

	Anterior Nasal Cavity						
	Septum		Maxilloturbinate		Nasopharynx		
	Nonciliated	Ciliated	Nonciliated	Ciliated	Septum	Lateral Wall	Ref.
Fischer-344 rat[a]	NP	5.8 ± 0.4	0.04 ± 0.01	NP	NP	2.5 ± 0.12	40
Bonnet monkey[b]	1.1 ± 0.8	3.0 ± 0.4	1.9 ± 0.2	6.3 ± 0.5	10.6 ± 0.8	8.5 ± 1.2	41

Note: NP = Tissue not present at that location.

may also serve as progenitor cells of the airway epithelium. Only ciliated cells are unable to divide, being terminally differentiated cells. Basal cells may also serve as anchoring cells for the entire epithelial layer of upper airways.[6] Each cell type is identified by its shape, ultrastructural characteristics, and location relative to the basal lamina upon which it resides.

Two less common epithelial cell types of the trachea and bronchi are the brush cell and small granule or neuroendocrine cell. Brush cells have large, stubby microvilli of unknown function. It has been speculated that these cells may be involved in some neurological or sensory process since they have been found to be in close association with nerve fibers. The function of neuroendocrine cells has also not been clearly defined. Like the brush cells, they have been found to be in close association with nerve fibers and typically appear as a small cluster of cells as a portion of the airway epithelium. Due to the paucity of these cell types in the airways, their role and significance in airway function have not been definitively established.

Epithelial cells of the trachea and bronchi also extend into submucosal glands within the walls of these airways. The nature of this epithelium changes markedly to almost exclusively secretory cells. Submucosal glands are predominantly serous and mucous cells. A review of the relative composition of these glands suggests that they can be mucous, serous, or mixed. The presence of these glands in the upper airways plays a significant role in the composition and thickness of the lining layer. Morphologic and immunocytochemical studies also indicate distinct differences between mucous and serous cells. The cytoplasm of mucous cells are packed with large coalescing electron lucent granules, while serous cells contain smaller, discrete electron granules. Both cell types may contribute to the uniquely different portions of the mucous lining layer to form the highly viscous gel phase and the underlying watery sol phase of the lining blanket. Without these unique characteristics of the airway lining layer, movement of mucin through the action of ciliary beating would be impossible.

An important epithelial cell of the bronchioles is the Clara cell. Although not present in all airway generations, the Clara cell plays significant roles for a variety of functions, including serving as the progenitor cell of the lower airway epithelium, as a secretory cell adding to the fluid lining layer of the bronchioles, and as a site for the metabolism of inhaled toxicants. The Clara cell is a nonciliated cell with a dome-shaped apical cytoplasmic process protruding into the airway lumen. Abundant amounts of smooth endoplasmic reticulum typically fill the apical portions of the cell, while rough endoplasmic reticulum is found in the lateral and basal portions of the cell. Electron dense granules found throughout the cell cytoplasm are thought to represent secretory granules, but the contents of these granules are not well defined. Clara cells are a primary site for cytochrome P-450-mediated metabolism. Immunocytochemical studies have demonstrated that these cells in the airways contain high levels of cytochrome P-450 mono-oxygenases. The relative percentages of Clara cells found in the lower airways of the lungs for different species are given in Tables 15 and 16.

TABLE 15

Comparison of Numerical Density and Percentage of Clara Cells in the Bronchiolar Epithelial Population of Adults

Species	Bronchiolar Epithelium Density (no. per mm²)[a]	Clara Cells Density (no. per mm²)[a]	Clara Cells Cells (%)	Ref.
Mouse	9759 ± 1700	8730 ± 1966	89.5	62
Hamster	14,238 ± 2794	8248 ± 2106	57.9	62
Rat (Sprague-Dawley)	18,813 ± 2722	14,028 ± 2918	82.2	62
Rat (Fischer 344)	17,070 ± 791	4336 ± 201	25.4	3, 39
Rabbit	15,073 ± 706	9261 ± 434	61.44	9
Cat	19,532 ± 383	19,532 ± 383	100	10
Bonnet monkey	9565 ± 304	8800 ± 280	92	63

[a] Mean ± 1 standard deviation.

TABLE 16

Comparison of Relative Proportions (Percentages) of Cellular Components in Clara Cells[a]

Species	Nucleus[b]	Agranular Endoplasmic Reticulum[b]	Secretory Granules[a]	Cytoplasmic Glycogen[b]	Mitochondria[b]	Large Mitochondria[c]	Lateral Cytoplasmic Extensions[c]
Mouse	21.8 ± 6.5	54.8 ± 7.5	+	0	34.7 ± 6.4	+	+
Hamster	25.2 ± 6.1	79.3 ± 7.6	+	0	10.7 ± 4.4	–	+
Rat	28.5 ± 10.4	66.2 ± 9.4	+	0.1 ± 0.4	16.3 ± 6.0	+	+
Guinea pig	28.6 ± 8.9	58.3 ± 9.0	+	0 ± 8.5	25.1	+	+
Rabbit	23.8 ± 8.8	61.6 ± 5.4	+	7.0 ± 5.4	19.1 ± 7.6	+	+
Dog	23.4 ± 11.4	24.7 ± 8.6	+	57.1 ± 13.7	8.0 ± 6.7	–	+
Cat	26.7 ± 9.2	10.7 ± 6.7	–	61.3 ± 10.1	19.5 ± 9.6	+	+
Macaque monkey	28.6 ± 4.4	5.2 ± 3.3	+	0	14.1 ± 2.8	–	+
Sheep	26.6 ± 10.3	64.6 ± 18.3	+	6.8 ± 12.1	13.8 ± 8.8	–	+
Pig	?	?	+	?	?	+	+
Horse	8.6 ± 4.1	70.6 ± 4.5	+	0	10.6 ± 4.4	–	+
Cow	27.7 ± 10.1	21.9 ± 10.1	+	62.3 ± 11.5	12.0 ± 7.5	–	+
Human	41.9 ± 10.0	3.1 ± 3.5	+	4.6 ± 5.8	15.3 ± 6.6	–	+

[a] As percent of cytoplasmic volume.
[b] Mean ± SD.
[c] +, as present; –, not present.
Source: Compiled from References 63 and 64.

4 COMPOSITION OF THE PULMONARY ACINUS AND ALVEOLI

Branching of the tracheobronchial tree increases the cumulative surface area in an exponential fashion. From the last conducting airway, also known as the terminal bronchiole, the branching of the respiratory bronchioles and alveolar ducts beyond these airways increases the surface area at an even greater rate. Collectively, the respiratory bronchioles and alveolar ducts arising from a single terminal bronchiole form the pulmonary acinus, the basic unit of ventilation in the lungs. The extent of the interdigitation between the conducting airways

TABLE 17[35]

Comparison of Species Differences in Microenvironment of Bronchiolar Cells: Centriacinar Organization and Tracheobronchial Distribution

Species	Transitional or Respiratory Bronchiole: Extensive	Transitional or Respiratory Bronchiole: Minimal	Ref.
Mouse		+	
Rat		+	
Hamster		+	
Guinea pig		+	
Rabbit		+	
Ferret	+		60
Cat	+		
Dog	+		
Sheep	+	+	8
Pig		+	
Cow		+	16
Horse		+	
Macaque monkey	+		36
Human	+		

and gas-exchange regions of the lungs formed by the transitional or respiratory bronchioles differs significantly among species (Table 17). In humans, monkeys, cats, dogs, and ferrets, respiratory bronchioles are well developed and may extend for up to three to five generations. In contrast, respiratory bronchioles are poorly developed in the lungs of rodents, pigs, cows, and horses. The transition between the bronchiolar airways and the alveolar ducts in these species is abrupt, with little overlap between the two structures.

The transition zone represented by the respiratory bronchiole is an airway in which the bronchiolar epithelium is interdigitated with alveolar outpockets. Whether this transition zone is short or extensive, it represents the primary orifice through which gases must pass to reach the alveolar portions of the lungs. In the monkey, the bronchiolar epithelium is typically polarized in relation to the position of the pulmonary arteriole.[36] A pseudostratified population of bronchiolar epithelial cells, including ciliated cells, lines the side adjacent to the pulmonary arteriole. Away from the pulmonary arteriole, the aveolar outpocketings of the respiratory bronchiole are surrounded by simple cuboidal bronchiolar epithelial cells, while aveolar epithelial cells line the alveoli.

The surfaces of the alveoli are covered by three epithelial cell types, referred to as alveolar type I, type II, and type III. Alveolar type I cells cover approximately 95% of the gas-exchange surface with thin, squamous-like cytoplasmic extensions over a large surface area (10,000 μg^3/cell). The attenuated cytoplasmic extensions of type I cells contribute to the thin air-to-blood tissue barrier critical for the exchange of gases between the airspaces of the alveoli and the capillary bed within the walls of the alveolar septa (Table 18). Alveolar type II cells are cuboidal in shape and cover approximately 3 to 4% of the alveolar surfaces. These cells contain numerous secretory granules which, when released from the cells, form the surfactant lining film to significantly reduce alveolar surface tension. These cells are also the progenitor cells for new type II cells and can undergo transition to form new type I cells. Alveolar type III cells are a rare cell type, cuboidal in shape and having short, broad apical villi containing filamentous bundles arising from the cytoplasmic portions of the cell. The purpose or function of these cells is unknown, although it has been hypothesized that they may serve as sensory receptors to inhaled gases or substances. In rats, these cells have been found to be present in the region of the terminal bronchiole and proximal alveolar regions.[2]

Although alveolar type I cells have thin, attenuated cytoplasm in contrast to the cuboidal alveolar type II cells, type I cells are two to four times larger in volume and cover 20 to 40

TABLE 18

Comparative Anatomy of the Lung Parenchyma and Air-Blood Tissue Barrier[a]

Species	No.	Body Weight (g)	Lung Volume	Alveolar Surface Area (Both Lungs) (cm²)	Capillary Surface Area (Both Lungs) (cm²)	T_h Tissue (μm)	Ref.
Shrew (*Suncus etrascus*)	4	2.6 ± 0.2	0.10 ± 0.01	170 ± 10	130 ± 15	0.27 ± 0.02	75
White mouse (*Mus musculus*)	5	23 ± 2	0.74 ± 0.07	680 ± 85	590 ± 60	.032 ± 0.01	76
Waltzing mouse (*Mus wagneri*)	5	13 ± 1	0.58 ± 0.06	630 ± 40	540 ± 30	0.26 ± 0.002	76
Syrian golden hamster (*Mesocricetus auratus*)	4	118 ± 7	2.81 ± 0.24	2760 ± 250	2410 ± 190	0.39 ± 0.10	77
White rat (*Ratus rattus*)	8	140 ± 7	6.34 ± 0.25	3880 ± 190	4070 ± 200	0.37 ± 0.02	78
White rat (Sprague-Dawley)	6	360 ± 4	10.82 ± 0.38	4865 ± 380	4270 ± 385	0.40 ± 0.02	79
White rat (Fischer 344)							
Male (5 months)	4	289 ± 13	8.60 ± 0.31	3915 ± 390	3830 ± 395	0.38 ± 0.03	25
Female (5 months)	4	182 ± 5	7.48 ± 0.10	3420 ± 125	3260 ± 185	0.34 ± 0.01	25
Male (26 months)	4	391 ± 11	12.67 ± 0.74	4630 ± 440	4490 ± 485	0.37 ± 0.01	25
Female (26 months)	4	298 ± 7	9.39 ± 0.40	4020 ± 25	3570 ± 165	0.37 ± 0.01	25
Guinea pig (*Cavia porcellus*)	15	429 ± 11	13.04 ± 3.03	9100 ± 280	7400 ± 230	0.42 ± 0.01	80
Rabbit (*Oryctolagus cuniculus*)	6	3560	79.2	58,600 ± 12,400	47,000 ± 8800	0.50 ± 0.04	81
Dwarf mongoose (*Helogale pervula*)	3	52,800 ± 9800	30.6 ± 5.6	16,100 ± 2600	14,600 ± 3400	0.39 ± 0.02	81
Genet cat (*Genetta tigrina*)	2	137,200 ± 4300	99.0 ± 12.2	56,300 ± 6400	42,300 ± 1600	0.51 ± 0.02	81
Dog (*Canis familiaris*)	3	5400	284.2	182,000 ± 135,000	141,000 ± 111,000	0.43 ± 0.02	81
Dog (*Canis familiaris*)	8	11,200 ± 400	736 ± 25	407,000 ± 39,000	329,000 ± 16,000	0.46 ± 0.01	82
Dog (*Canis familiaris*)	4	16,000 ± 3,000	1322 ± 64	510,000 ± 10,000	570,000 ± 20,000	0.45 ± 0.01	4
Dog (*Canis familiaris*)	6	22,800 ± 600	1501 ± 74	897,000 ± 69,000	718,000 ± 69,000	0.48 ± 0.01	82
Dog (*Canis familiaris*)	5	46,100	2888	1,769,000 ± 456,000	1,319,000 ± 375,000	0.53 ± 0.08	81
Camel (*Camelus dromedarus*)	2	231,700 ± 2700	15,900 ± 1400	4,305,000 ± 584,000	2,726,000 ± 292,000	0.60 ± 0.06	81
Giraffe (*Giraffa camelopardalis*)	1	383,000	21,000	6,361,000	5,516,000	0.60	81
Suni (*Nesotragus moschatus*)	2	3300 ± 300	209.4 ± 0.6	96,900 ± 5500	81,300 ± 13,000	0.56 ± 0.09	81
Dik-dik (*Madoqua kirkii*)	2	4200 ± 100	313.4 ± 1.2	146,000 ± 700	130,000 ± 6550	0.43 ± 0.02	81
Wildebeest (*Connochaetes tauriras*)	1	102,000	7678	3,908,000	2,813,000	0.37	81
Waterbuck (*Kobus defassa*)	2	109,800 ± 16,300	7835 ± 1550	3,829,000 ± 950,000	3,378,000 ± 460,000	0.46 ± 0.04	81
African goat (*Capra hircus*)	2	20,900 ± 1000	1370 ± 15	449,000 ± 12,000	439,000 ± 12,000	0.54 ± 0.03	81
African sheep (*Ovis aries*)	2	21,800 ± 200	17,055 ± 435	671,000 ± 71,000	645,000 ± 139,000	0.53 ± 0.05	81
Zebu cattle (*Bos indicus*)	4	192,500 ± 24,000	10,145 ± 1960	3,850,000 ± 420,000	3,795,000 ± 392,000	0.50 ± 0.04	81
Swiss cow (*Bos taurus*)	1	700,000	22,450	12,830,000	11,380,000	0.51	81
Horse (*Equis caballus*)	2	510,000 ± 0	37,650 ± 1050	24,560,000 ± 124,000	16,630,000 ± 1,080,000	0.60 ± 0.02	83
Monkey (*Macaca irus*)	6	3710	184.2	133,000 ± 12,700	116,000 ± 15,400	0.50 ± 0.03	81
Baboon (*Papio papio*)	5	29,000 ± 3000	2393 ± 100	496,000 ± 77,000	386,000 ± 95,000	0.67 ± 0.06	79
Man (*Homo sapiens*)	8	74,000 ± 4000	4341 ± 285	1,430,000 ± 120,000	1,260,000 ± 120,000	0.62 ± 0.04	84

[a] All values are mean ± SEM.

TABLE 19

Characteristics of Cells from the Alveolar Region of Normal Mammalian Lungs[a]

	Fischer 344 Rat[b]	Sprague-Dawley Rat	Dog	Baboon	Human[c]
Total no. cells per lung, 10^9	$0.67 \pm 0.02_a$	$0.89 \pm 0.04_a$	$114 \pm 13_b$	$99 \pm 9_b$	230 ± 25
Total lung cells (%)					
Alveolar type I	$8.1 \pm 0.3_a$	$8.9 \pm 0.9_a$	$12.5 \pm 107_b$	$11.8 \pm 0.6_b$	$8.3 \pm 0.6_a$
Alveolar type II	$12.1 \pm 0.7_a$	$14.2 \pm 0.7_{a,b}$	$11.8 \pm 0.6_b$	7.7 ± 1.0	$15.9 \pm 0.8_b$
Endothelial	$51.1 \pm 1.7_a$	$42.2 \pm 1.1_a$	$45.7 \pm 0.8_a$	36.3 ± 2.4	30.2 ± 2.4
Interstitial	$24.4 \pm 0.7_a$	$27.7 \pm 1.8_a$	$26.6 \pm 0.7_a$	41.8 ± 2.7	36.1 ± 1.0
Macrophage	$4.3 \pm 1.0_a$	$3.0 \pm 0.3_a$	$3.4 \pm 0.6_a$	$2.3 \pm 0.7_a$	9.4 ± 2.2
Alveolar surface covered (%)					
Alveolar type I	$96.4 \pm 0.5_a$	$96.2 \pm 0.5_a$	$97.3 \pm 0.4_a$	$96.0 \pm 0.6_a$	92.9 ± 1.0
Alveolar type II	$3.6 \pm 0.3_a$	$3.8 \pm 0.5_a$	$2.7 \pm 1.0_a$	$4.0 \pm 0.6_a$	$7.1 \pm 1.0_a$
Average cell volume (μm^3)					
Alveolar type I	$1530 \pm 121_a$	$2042 \pm 374_a$	$1196 \pm 88_a$	$1224 \pm 136_a$	1764 ± 155
Alveolar type II	$455 \pm 108_a$	$433 \pm 80_a$	$428 \pm 37_a$	$539 \pm 184_a$	889 ± 101
Endothelial	$275 \pm 25_a$	$387 \pm 30_a$	$343 \pm 19_a$	$365 \pm 61_a$	632 ± 64
Interstitial	$427 \pm 55_a$	$331 \pm 67_{a,b}$	$440 \pm 47_a$	$227 \pm 30_b$	637 ± 26
Macrophage	$639 \pm 131_a$	$1058 \pm 257_a$	$654 \pm 116_a$	$1059 \pm 287_a$	2492 ± 167
Average cell surface area (μm^2)					
Alveolar type I	$7287 \pm 755_b$	$5320 \pm 694_{a,b}$	$3794 \pm 487_a$	$4004 \pm 383_a$	$5098 \pm 659_a$
Alveolar type II	$185 \pm 56_{a,b}$	$123 \pm 20_a$	$107 \pm 15_a$	$285 \pm 85_b$	$183 \pm 14_{a,b}$
Endothelial	$1121 + 95_a$	$1105 + 72_a$	$1137 + 127_a$	$1040 + 209_a$	$1353 + 67_a$

[a] All data are mean ± SEM. For comparisons between species, letter subscripts indicate those values that are not different from other values having the same letter subscript.
[b] Data from Pinkerton, K. E. et al., *Am. J. Anat.*, 164, 155–174, 1982.
[c] Data from Crapo, J. D. et al., *Am. Rev. Respir. Dis.*, 128, 542–546, 1983.

times greater surface area per cell than do type I cells. The uniformity of epithelial cell size between species (Table 19) is remarkable, considering the marked differences in alveolar size and number among species.[33] The rat alveolus has a surface area of approximately 14,000 μm^2 and is covered by an average of two type I cells and three type II cells. In contrast, the human alveolus has a surface area of 200,000 μm^2 with an average of 40 type I cells and 67 type II cells.[33] These findings suggest that the size and shape of alveolar type I and II cells are independent of alveolar size.

The air-to-blood tissue barrier consists of three tissue compartments: the alveolar epithelium composed of type I and type II cells, the underlying interstitium, and the squamous-like endothelial cells forming the walls of the capillary bed coursing through the alveolar septa. The design of these tissue compartments is ideal for the direct transport of gases across exceeding thin epithelial, interstitial, and endothelial compartments averaging a total of less than 1 µm to reach the capillary lumen filled with plasma and red blood cells.

The architectural integrity of the alveolar wall is maintained in large measure by the interstitium. Interstitial cells represent approximately 25 to 40% of the total parenchymal cell population. The resident cells of the interstitium form a heterogenous population of cells predominantly made up of fibroblasts and pericytes. The fibroblasts are the primary cell involved in the synthesis and laying down of collagen and elastin in the interstitial space. The pericytes are intimately associated with the endothelial cells of the capillaries and blood vessels and may play a role in the regulation of local blood flow by altering the shape of the vascular lumen through contraction or relaxation.[13] Interspersed among these cells in lesser numbers are interstitial macrophages, monocytes, and lymphocytes. The differential of these cells in the interstitium of rats and humans is given in Table 20; the similarity between two species of such vastly different size is striking.

TABLE 20
Differential Count of Parenchymal Interstitial Cells of the Lungs

	Fischer 344 Rat[b] (12)	Human[c] (8)
Total interstitial cell number per lung[a]	$142 \pm 12 \times 10^6$	$84 \pm 10 \times 10^9$
Percent distribution of cell types		
Fibroblast	58%	51%
Pericyte	21%	17%
Monocyte-macrophage	16%	12%
Unknown	3%	7%

[a] Data are mean ± SEM.
[b] Data from Pinkerton, K. E. et al., *Am. J. Anat.*, 164, 155–174, 1982.
[c] Data from Pinkerton, K. E. et al., *Am. Rev. Respir. Dis.*, 137, 344, 1988.

5 RELATIONSHIP BETWEEN LUNG STRUCTURE AND FUNCTION

Billions of cells forming the human airways and alveoli are all involved in the highly diverse functions of the lungs. The location of each cell type is due in large measure to the structural arrangement of the tissues at each level of the lungs from the trachea down to the alveolus. The heterogeneity of epithelial cell types along the tracheobronchial tree and alveoli is actually maintained in a balance under normal conditions. Perhaps one of the most important consequences of disease and injury in the lungs is upsetting the normal balance of these cells along the airways or in the alveoli.

It is relevant to note that on a per-surface basis, the density of cells lining the airways is in striking contrast to the sparse number of epithelial cells lining the alveoli. Such differences are critical to the proper function of each anatomical component of the lungs. Columnar to cuboidal cells lining the airways facilitate the movement, rather than uptake, of air and trap foreign particles within their secretions lining the airway surfaces, while the cuboidal to squamous epithelium of the alveoli provides a highly optimal environment for the efficient exchange of gases from the air space to the blood through a thin epithelial barrier bathed in a fluid that minimizes the alveolar surface tension to lower the risk of collapse during normal respiration.

The total number of cells found within the alveolar portions of the lungs still constitutes 75% of the total cell population of the lungs and constitutes a surface of far greater dimensions than is present in the conducting airways.[34] The diversity of cell types found within the respiratory tract allows several important functions, including warming and humidifying the air, providing for gas exchange, defending or providing immunity against inhaled pathogens, and metabolizing air and blood-borne compounds. A less appreciated function is serving as a reservoir or storage site for blood to the left side of the heart prior to movement to the systemic circulation.

Ciliated cells are found along the entire length of the pulmonary airways. These cells are key to the movement of the mucous blanket up the airways. Although the central function of cilia is the movement of fluids and viscous materials, it is also possible that the tremendous surface area created by the presence of cilia may function in other capacities; however, little information is available to provide insights into possible other functions. The airways are an important reservoir for immunocompetent cells found within the walls of the airways as accumulations of lymphoid cell aggregates (the BALT system). The processing of inhaled materials as a ready immune defense is likely to involve the interaction of these cells with the airway epithelium. A unique cell type possessing cilia-like structures, and referred to as a brush cell, may be involved in the processing and recognition of inhaled substances and may facilitate neurological or endocrinological responses.

In situ preservation of the mucous lining layer in the airways for microscopic examination has proven to be rather difficult; however, some studies suggest that upper airways, including those with submucosal glands, have the greatest amounts of mucin present in this lining layer. The precise physiology and biochemical makeup of airway mucin has been hampered by the fact that several cell types from a variety of locations contribute to this lining layer. These include secretory cells within the submucosal glands, cells from the airways under investigation, and cells from more distal airways in which their secretions have been carried via the ciliary beating to more proximal locations in the bronchial airway tree.

Warming and humidifying of air begins in the nasal cavity and continues in the trachea and conducting airways of the lungs. Epithelial cells serving a secretory function in the conducting airways produce the lubricants and viscous materials to protect the airways by coating the airway surfaces. Different cell types involved in secretion are essential in order to form a lining layer with the proper composition to allow for the impaction and trapping of inhaled materials but at the same time having the consistency to allow for the free movement of this blanket of materials along the airways and the eventual movement out of the lungs of these trapped materials (Table 21). The vascular bed within the walls of the airways serves in part to warm the air as it moves to the gas exchange portions of the lungs. The variety of cell types involved in secretion provides the unique composition of the lining layer of the airways. Carbohydrate-rich glycoproteins form the sticky, viscous portions of the airway mucin, while the watery secretions allows for the free movement of cilia within the sol phase. The combination of these two layers allows for the rather efficient continuous movement of mucin up the airways through the synchronized beating of cilia in an upward direction towards the nasopharynx. (See Tables 22 to 23.)

The primary function of the respiratory system is gas exchange. This function is made possible by bringing the vascular supply of the body into close proximity to the external air. An impressive surface area forming the air-tissue interface is generated by the extensive branching of the airways leading to alveolar duct channels with an equally impressive branching pattern within the confines of the thoracic space. The walls of the alveolar ducts are formed by the outpocketings of alveoli. The configuration of the alveolar surfaces with the pulmonary capillary bed brings two surfaces into close proximity over an immense area to facilitate the exchange of gases from the air to the blood and vice versa through an extremely thin but structurally sound barrier.

An important anatomical consideration in the response of the lungs to inhaled gases and pollutants is acinar size. Mercer and Crapo have demonstrated acinar size to be different among species (Table 24).[18] These differences among species may be as much as 200-fold. In addition, the heterogeneity of ventilatory unit size within a single species is evident with a threefold difference in the range of ventilatory unit volume.[19] Differences in acinar size among species, in addition to differences in the extent of the transitional zone or respiratory bronchioles of the pulmonary acinus among species, are likely to play a critical role in the degree of injury caused by the inhalation of highly reactive gases or particles. Exposure to similar concentrations of gases or particles may result in very different degrees of cellular injury, inflammation, and tissue remodeling within the proximal portions of the lungs, due to the differences in the magnitude of gases that must pass through this orifice to ventilate the acinus; therefore, differences in the arrangement of the acinus must be carefully considered when attempting to extrapolate the effects of inhaled toxicants to the human.

TABLE 21

Distribution of Bronchial Arteries in Mammals

	Lung Type I (Thick Pleura)			Lung Type II (Thin Pleura)			Variant Type II (Thin Pleura)			Lung Type III (Thick Pleura)	
	Cow	Lamb	Pig	Dog	Cat	Rhesus Monkey	Rabbit	Rat	Guinea Pig	Horse	Human
Pleura	+	+	+	–	–	–	–	–	–	+	+
Septa	+	+	+	–	–	–	–	–	–	+	+
Bronchi	+	+	+	+	+	+	+	+	+	+	+
Terminal bronchioles	+	+	+	+	+	+	–	+	+	+	+
Alveoli	–	–	–	–	–	–	–	–	–	+	+
Hilar nodes	+	+	+	+	+	+	+	+	+	+	+
Vasa vasorum	+	+	+	+	+	+	+	+	+	+	+
BPA	+	+	+	–	–	–	+	+	+	+	+

Abbreviations: +, distribution to structure; –, not distributed to structure; BPA, bronchial artery-pulmonary artery anastomoses.
Source: From McLaughlin, R. F., *Am. Rev. Resp. Dis.*, 128, 557, 1983. With permission.

TABLE 22

Carbohydrate Content of Tracheal Epithelium

Species	Cell Type	Abundance	Carbohydrate Content			Ref.
			PAS	AB	HID	
Hamster	Clara	+++	+	–	–	44
	Mucous	+	+	+	–	
Rat	Serous	+++	+	–	–	45
	Mucous	+	+	+	–	
Mouse	Mucous	+	+	+	–	45
Rabbit	Mucous	+	+	+	+	46
	Clara	+++	±	–	–	
Canine	Mucous	++	+	+	+	45
Cat	Mucous	++	+	+	+ and –	48
	Serous	+	ND	ND	ND	
Pig	Mucous	+	+	+ and –	+ and –	49
Sheep	Mucous	++	+	+	+ and –	50
Rhesus	Mucous	++	+	+	+	51
Human	Mucous	+++	+	+	+ and –	52

Note: PAS = periodic acid Schiff, which reacts with vicinal hydroxyl groups; AB = alcian blue, which at pH 2.6 reacts with acidic glycoconjugates; HID = high iron diamine, which reacts with acidic glycoconjugates containing sulfate esters; ND = not determined.

TABLE 23

Carbohydrate Content of Tracheal Submucosal Glands

Species	Abundance	Secretory Cell	Carbohydrate Content			Ref.
			PAS	AB	HID	
Hamster	±	Mucous	+	+	–	44
Rat	+	Serous	+	–	–	45
		Mucous	+	+	+ and –	
Mouse	±	Serous	+	–	–	45
		Mucous	+	+	+ and –	
Rabbit	±	Mucous	+	+	+	46
Canine	++	Serous	+	–	–	47
		Mucous	+	+	+	
Cat	++++	Serous	+	+	+ and –	48
		Mucous	+	+	+	
Pig	++	Serous	+	–	–	49
		Mucous	+	+	+ and –	
Sheep	++	Serous	+	–	–	50
		Mucous	+	+	+	
Rhesus	++	Serous	+	–	–	51
		Mucous	+	+	+	
Human	+++	Serous	+	–	–	52
		Mucous	+	+	+ and –	

TABLE 24
Acinar Morphometery

Species	Fixation[a]	No. per Lung	V (mm^3)	Acinar Size $(mm)^2$	No. Alveoli per Acinus	Alveolar Duct Generations	Ref.
Human			1.33–30.9		15,000		65
		27,992			10,714	6	65,67
	75% TLC	23,000	160.8	7.04 (L)	14,000–20,000	9	68,69
		80,000	15.6			2–5	70
				5.1 (L)	7100	8–12	71
	TLC	26,000–32,000	187.0	8.8 (L)	10,344	9	8
	FRC	43,000	51.0	6.0 (D)	8000	9	
Rabbit		17,900	2.54				72
	55% TLC	18,000	3.46	1.95 (L)		6	74
Guinea pig		5100	1.25				72
	FRC	4097	1.09	1.56 (D)	6890	9–12	
Rat		2500	1.3				72
		2487	5.06				42
	FRC	2020	1.9	1.5 (D)	5243	10–12	85
	70% TLC	5993	1.46	1.5 (L)		6	74

Note: TLC = total lung capacity; FRC = functional residual capacity; V = volume; D = diameter; L = length.
[a] Volume of lung at fixation.

REFERENCES

1. Baskerville, A. and Olson, I. A., The ultrastructure of the bronchiolar epithelium of the pig (abstr.), *Proc. Anat. Soc. G. Br. Ire.*, November, 15–16, 1969.
2. Chang, L.-Y., Mercer, R. R., and Crapo, J. D., Differential distribution of brush cells in the rat lung, *Anat. Rec.*, 216, 49–54, 1986.
3. Chang, L.-Y., Mercer, R. R., Stockstill, B. L., Miller, F. J., Graham, J. A., Ospital, J. J., and Crapo, J. D., Effects of low levels of NO_2 on terminal bronchiolar cells and its relative toxicity compared to O_3, *Toxicol. Appl. Pharmacol.*, 96, 451–464, 1988.
4. Crapo, J. D., Young, S. L., Pinkerton, K. E., Barry, B. E., and Crapo, R. O., Morphometric characteristics of cells in alveolar region of mammalian lungs, *Am. Rev. Respir. Dis.*, 128, S42–S46, 1983.
5. Cutz, E. and Conen, P. E., Ultrastructure and cytochemistry of Clara cells, *Am. J. Pathol.*, 62, 127–134, 1971.
6. Evans, M. J., Cox, R. D., Shami, S. G., Wilson, B., and Plopper, C. G., The role of basal cells in attachment of columnar cells to the basal lamina of the trachea, *Am. J. Respir. Cell. Mol. Biol.*, 1, 463, 1989.
7. Gilljam, H. A., Motakefi, A.-M., Robertson, B., and Strandvik, B., Ultrastructure of the bronchial epithelium in adult patients with cystic fibrosis, *Eur. J. Respir. Dis.*, 71, 187, 1987.
8. Haefeli-Bleuer, B. and Weibel, E. R., Morphometry of the human pulmonary acinus, *Anat. Rec.*, 220, 401–414, 1988.
9. Hyde, D. M., Plopper, C. G., Kass, P. H., and Alley, J. L., Estimation of cell numbers and volumes of bronchiolar epithelium during rabbit lung maturation, *Am. J. Anat.*, 167, 359–370, 1983.
10. Hyde, D. M., Plopper, C. G., Weir, A. J., Murnane, R. D., Warren, D. L., and Last, J. A., Peribronchiolar fibrosis in lungs of cats chronically exposed to diesel exhaust, *Lab. Invest.*, 52(2), 195–206, 1985.
11. Jarkovska, D., Ultrastructure of the epithelium of the respiratory bronchioles in man, *Folia Morphol.*, 18, 352–358, 1970.
12. Jeffery, P. K. and Reid, L., New observations of rat airway epithelium: a quantitative and electron microscopic study, *J. Anat.*, 120, 295, 1975.
13. Kapanci, Y., Assimacopoulos, A., Irle, C., Zwahlen, A., and Gabbiani, G., Contractile interstitial cells in pulmonary alveolar septa: a possible regulator of ventilation/perfusion ratio? Ultrastructural immunofluorescence and *in vitro* studies, *J. Cell. Biol.*, 60, 375–392, 1974.
14. Kennedy, A. R., Desrosiers, A., Terzaghi, M., and Little, J. B., Morphometric and histological analysis of the lungs of Syrian golden hamsters, *J. Anat.*, 125, 527, 1978.

15. Mariassy, A. T. and Plopper, C. G., Tracheobronchial epithelium of the sheep. I. Quantitative light microscopic study of epithelial cell abundance, and distribution, *Anat. Rec.,* 205, 263–275, 1983.
16. Mariassy, A. T., Plopper, C. G., and Dungworth, D. L., Characteristics of bovine lung as observed by scanning electron microscopy, *Anat. Rec.,* 183, 13–26, 1975.
17. Mariassy, A. T., Epithelial cells of trachea and bronchi, in *Treatise on Pulmonary Toxicology, Comparative Biology of the Normal Lung,* Vol. 1, Parent, R. A., Ed., CRC Press, Boca Raton, FL, 1992, pp. 63–76.
18. Mercer, R. R. and Crapo, J. D., Structure of the gas exchange region of the lungs determined by three-dimensional reconstruction, in *Toxicology of the Lung,* Raven Press, New York, 1988, pp. 43–70.
19. Mercer, R. R., Anjilvel, S., Miller, F. J., and Crapo, J. D., Inhomogeneity of ventilatory unit volume and its effects on reactive gas uptake, *J. Appl. Physiol.,* 70, 2193–2205, 1991.
20. Mercer, R. R., Russell, M. L., Roggli, V. L., and Crapo, J. D., Cell number and distribution in human and rat airways, *Am. J. Respir. Cell. Mol. Biol.,* 10, 613–624, 1994.
21. Nikula, K. J., Wilson, D. W., Giri, S. N., Plopper, C. G., and Dungworth, D. L., The response of the rat tracheal epithelium to ozone exposure, *Am. J. Pathol.,* 131, 373, 1988.
22. Pack, R. J., Al-Ugaily, L. H., Morris, G., and Widdicombe, J. G., The distribution and structure of cells in the tracheal epithelium of the mouse, *Cell Tissue Res.,* 208, 65–84, 1980.
23. Pack, R. J., Al-Ugaily, L. H., and Morris, G., The cells of the tracheobronchial epithelium of the mouse: a quantitative light and electron microscopic study, *J. Anat.,* 132, 71–84, 1981.
24. Pavelka, M., Ronge, H. R., and Stockinger, G., Vergleichende Untersuchungen am Trachealepithel Verschiedener Sauger, *Acta Anatomica,* 94, 262, 1976.
25. Pinkerton, K. E., Barry, B. E., O'Neil, J. J., Raub, J. A., Pratt, P. C., and Crapo, J. D., Morphologic changes in the lung during the lifespan of Fischer 344 rats, *Am. J. Anat.,* 164, 155–174, 1982.
26. Pinkerton, K. E., Mercer, R. R., and Crapo, J. D., Interstitial characteristics of the normal human lung, *Am. Rev. Respir. Dis.,* 137, 344, 1988.
27. Plopper, C. G., Mariassy, A. T., and Hill, L. H., Ultrastructure of the nonciliated bronchiolar epithelial (Clara) cell of mammalian lung. II. A comparison of horse, steer, sheep, dog and cat, *Exp. Lung. Res.,* 1, 155, 1980.
28. Plopper, C. G., Mariassay, A. T., and Hill, L. H., Ultrastructure of the nonciliated bronchiolar epithelial (Clara) cell of mammalian lung. III. A study of man with comparison of 15 mammalian species, *Exp. Lung Res.,* 1, 171, 1980.
29. Plopper, C. G., Mariassy, A. T., and Lollini, L. O., Structure as revealed by airway dissection: a comparison of mammalian lungs, *Am. Rev. Respir. Dis.,* 128, S4, 1983.
30. Plopper, C. G., Mariassy, A. T., Wilson, D. W., Alley, J. L., Nishio, S. J., and Nettesheim, P., Comparison of nonciliated tracheal epithelial cells in six mammalian species: ultrastructure and population densities, *Exp. Lung Res.,* 5, 281, 1983.
31. Plopper, C. G., Halsebo, J. E., Berger, W. J., Sonstegard, K. S., and Nettesheim, P., Distribution of nonciliated bronchiolar epithelial (Clara) cells in intra- and extrapulmonary airways of the rabbit, *Exp. Lung Res.,* 4, 79, 1983.
32. Plopper, C. G., Heidsiek, J. G., Weir, A. J., St. George, J. A., and Hyde, D. M., Tracheobronchial epithelium in the adult Rhesus monkey: a quantitative histochemical and ultrastructural study, *Am. J. Anat.,* 18, 31, 1989.
33. Stone, K. C., Mercer, R. R., Freeman, B. A., Chang, L. Y., and Crapo, J. D., Distribution of lung cell numbers and volumes between alveolar and nonalveolar tissue, *Am. Rev. Respir. Dis.,* 146, 454–456, 1992.
34. Stone, K. C., Mercer, R. R., Gehr, P., Stockstill, B., and Crapo, J. D., Allometric relationships of cell numbers and size in the mammalian lung, *Am. J. Respir. Cell. Mol. Biol.,* 6, 235–243, 1992.
35. Tyler, W. S., Comparative subgross anatomy of lungs: pleuras, interlobular septa, and distal airways, *Am. Rev. Respir. Dis.,* 128, S32–S36, 1983.
36. Tyler, N. K. and Plopper, C. G., Morphology of the distal conducting airways in rhesus monkey lungs, *Anat. Rec.,* 211, 295–303, 1985.
37. Weibel, E. R. and Taylor, C. R., Design and structure of the human lung, in *Pulmonary Diseases and Disorders,* Vol. 1, 2nd ed., Fishman, A. P., Ed., McGraw-Hill, New York, 1988.
38. Widdicombe, J. G. and Pack, R. J., The Clara cell, *Eur. J. Respir. Dis.,* 63, 202–220, 1982.
39. Young, S. L., Fram, E. K., and Randell, S. H., Quantitative three-dimensional reconstruction and carbohydrate cytochemistry of rat nonciliated bronchiolar (Clara) cells, *Am. Rev. Respir. Dis.,* 133, 899–907, 1986.
40. Harkema, J. R., Hotchkiss, J. R., and Henderson, R. F., Effects of 0.12 and 0.80 ppm ozone on rat nasal and nasopharyngeal epithelial mucosubstances, *Toxicol. Pathol.,* 17, 525, 1989.
41. Harkema, J. R., Plopper, C. G., Hyde, D. M., St. George, J. A., and Dungworth, D. L., Effects of an ambient level of ozone on primate nasal and epithelial mucosubstances: quantitative histochemistry, *Am. J. Pathol.,* 127, 90, 1987.
42. Yeh, H. C., Schum, F. M., and Duggan, M. T., Anatomic models of the tracheobronchial and pulmonary regions of the rat, *Anat. Res.,* 195, 482–492, 1979.
43. Yeh, H. C. and Schum, G. M., Models of human lung airways and their application to inhaled particle deposition, *Bull. Math. Biol.,* 42, 461–480, 1980.
44. Emura, M. and Mohr, U., Morphological studies on the development of tracheal epithelium in the Syrian golden hamster, *Z. Versuchstiesk Bd.,* 17, 14, 1975.

45. McCarthy, C. and Reid, L. M., Acid mucopolysaccharides in the bronchial tree in the mouse and rat (sialomucins and sulphate), *Q. J. Exp. Physiol.*, 49, 81, 1964.
46. Plopper, C. G., St. George, J. A., Nishio, S. J., Etchison, J. R., and Nettesheim, P., Carbohydrate cytochemistry of tracheobronchial airway epithelium of the rabbit, *J. Histochem. Cytochem.*, 32, 209, 1984.
47. Spicer, S. S., Chakrin, L. W., Wardell, J. R., and Kendrick, W., Histochemistry of mucosubstances in the canine and human respiratory tract, *Lab. Invest.*, 25, 483, 1971.
48. Jeffery, P. K., Structure and function of mucus-secreting cells of cat and goose airway epithelium, in *Respiratory Tract Mucus*, Elsevier-North Holland, New York, 1977, p. 5–19.
49. Jones, R. A., Baskerville, A., and Reid, L. M., Histochemical identification of glycoproteins in pig bronchial epithelium: (a) normal and (b) hypertrophied from enzootic pneumonia, *J. Pathol.*, 116, 1, 1975.
50. Mariassy, A. T., St. George, J. A., Nishio, S. J., and Plopper, C. G., Tracheobronchial epithelium of the sheep. III. Carbohydrate histochemical and cytochemical characterization of secretory epithelial cells, *Anat. Rec.*, 221, 540, 1988.
51. St. George, J. A., Nishio, S. J., and Plopper, C. G., Carbohydrate cytochemistry of the rhesus monkey tracheal epithelium, *Anat. Rec.*, 210, 293, 1984.
52. Lamb, D. and Reid, L. M., Histochemical types of acidic glycoprotein produced by mucous cells of the tracheobronchial glands in man, *J. Pathol.*, 98, 213, 1969.
53. Korhonnen, L. K., Holopainen, E., and Paavolainen, M., Some histochemical characteristics of tracheobronchial tree and pulmonary neoplasms, *Acta Histochem.*, 32, 57–73, 1969.
54. Weibel, E. R., *Morphometry of the Human Lung*, Academic Press, New York, 1963.
55. Horsfield, K., Dart, G., Olson, D. E., Filley, F. G., and Cumming, G., Models of the human bronchial tree, *J. Appl. Physiol.*, 31, 207–217, 1971.
56. Horsfield, K. and Cumming, G., Morphology of the bronchial tree in the dog, *Respir. Physiol.*, 26, 173–182, 1976.
57. Ross, B. B., Influence of bronchial tree structure on ventilation in the dog's lung as inferred from measurements of a plastic cast, *J. Appl. Physiol.*, 10, 1–14, 1957.
58. McBride, J., Postpneumonectomy airway growth in the ferret, *J. Appl. Physiol.*, 58, 1010–1014, 1985.
59. Raabe, O. G., Yeh, H. C., Schum, G. M., and Phalen, R. M., *Trancheobronchial Geometry: Human, Dog, Rat, Hamster*, Lovelace Foundation, Albuquerque, NM, 1976.
60. Hyde, D. M., Samuelson, D. A., Blakeney, W. H., and Kosch, P. C., A correlative light microscopy, transmission and scanning electron microscopy study of the ferret lung, *Scan. Electron Microsc.*, 3, 891–898, 1979.
61. Plopper, C. G., Moriassy, A. T., and Hill, L. H., Ultrastructure of the nonciliated bronchiolar epithelial (Clara) cell of mammalian lung. I. A comparison of rabbit, guinea pig, rat, hamster, and mouse, *Exp. Lung Res.*, 1, 139–154, 1980.
62. Plopper, C. G., Macklin, J., and Nishio, S. J., Relationship of cytochrome P450 activity to Clara cell toxicity. III. Morphometric comparison of changes in the epithelial populations of terminal bronchioles and lobal bronchi in mice, hamsters, and rats after parental administration of naphthalene, *Lab. Invest.*, 67(5), 553–565, 1992.
63. Moffatt, R. K., Hyde, D. M., Plopper, C. G., Tyler, W. S., and Putney, L. F., Ozone-induced adaptive and reactive cellular changes in respiratory bronchioles of Bonnet monkeys, *Exp. Lung Res.*, 12, 57–74, 1987.
64. Hyde, D. M. and Plopper, C. G., Cell and organellar volumes of nonciliated bronchiolar cells of rabbits, cats, and monkeys, *Anat. Rec.*, 208, 77A, 1984.
65. Pump, K. K., The morphology of the finer branches of the bronchial tree of the human lung, *Dis. Chest*, 46, 379, 1964.
66. Horsfield, K. and Cumming, G., Morphology of the bronchial tree in man, *J. Appl. Physiol.*, 229(2), 529–536, 1975.
67. Parker, H., Horsfield, K., and Cumming, G., Morphology of distal airways in the human lung, *J. Appl. Physiol.*, 31, 386–391, 1971.
68. Hansen, J. E. and Ampaya, E. P., Human air space shapes, sizes, areas, and volumes, *J. Appl. Physiol.*, 38, 990–995, 1975.
69. Hansen, J. E., Ampaya, E. P., Bryant, G. H., and Navin, J. J., Branching pattern of airways and air spaces of a single human terminal bronchiole, *J. Appl. Physiol.*, 38(6), 983–989, 1975.
70. Boyden, E. A., The structure of the pulmonary acinus in a child of six years and eight months, *Am. J. Anat.*, 132, 275–300, 1972.
71. Shreider, J. P. and Raabe, O. G., Structure of the human respiratory acinus, *Am. J. Anat.*, 162, 221–232, 1981.
72. Kliment, V., Similarity and dimensional analysis, evaluation of aerosol deposition in the lungs of laboratory animals and man, *Folia Morphologica*, 21(1), 59–64, 1973.
73. Pump, K. K., Morphology of the acinus of the human lung, *Dis. Chest*, 56(2), 126–134, 1969.
74. Rodriguez, M. Bur, S., Favre, A., and Weibel, C. R., Pulmonary acinus: geometry and morphometry of the peripheral airway system in rat and rabbit, *Am. J. Anat.*, 180(2), 143–155, 1987.
75. Gehr, P., Sehovic, S., Burri, P. H., Claassen, H., and Weibel, E. R., The lung of shrews: morphometric estimation of diffusion capacity, *Respir. Physiol.*, 40, 33–47, 1980.

76. Geelhaar, A. and Weibel, E. R., Morphometric estimation of pulmonary diffusion capacity. III. The effect of increased oxygen consumption in Japanese waltzing mice, *Respir. Physiol.*, 11, 354–366, 1971.
77. Gehr, P., unpublished data.
78. Burri, P. H. and Weibel, E. R., Morphometric estimation of pulmonary diffusion capacity. II. Effect of PO_2 on the growing rat lung to hypoxia and hyperoxia, *Respir. Physiol.*, 11, 247–264, 1971.
79. Crapo, J. D., Berry, B. E., Foscue, H. A., and Shelburne, J., Structural and biochemical changes in rat lungs occurring during exposures to lethal and adaptive doses of oxygen, *Am. Rev. Respir. Dis.*, 122, 123–145, 1980.
80. Forrest, J. B. and Weibel, E. R., Morphometric estimation of pulmonary diffusion capacity. VII. The normal guinea pig lung, *Respir. Physiol.*, 24, 191–202, 1975.
81. Gehr, P., Mwangi, D. K., Ammann, A., Maloiy, G. M. O., Taylor, C. R., and Weibel, E. R., Design of the mammalian respiratory system. V. Scaling morphometric pulmonary diffusing capacity to body mass: wild and domestic mammals, *Respir. Physiol.*, 44, 61–86, 1981.
82. Siegwart, B., Gehr, P., Gil, J., and Weibel, E. R., Morphometric estimation of pulmonary diffusion capacity. IV. The normal dog, *Respir. Physiol.*, 13, 141–159, 1971.
83. Gehr, P. and Erni, H., Morphometric estimation of pulmonary diffusion capacity in the horse lung, *Respir. Physiol.*, 41, 199–210, 1980.
84. Gehr, P., Bachofen, M., and Weibel, E. R., The normal human lung: ultrastructure and morphometric estimation of diffusion capacity, *Respir. Physiol.*, 32, 121–140, 1978.
85. Mercer, R. R. and Crapo, J. D., Three-dimensional reconstruction of the acinus of the rat, *J. Appl. Physiol.*, 63, 785–797, 1987.

CHAPTER 12

DISPOSITION OF INHALED TOXICANTS

Richard B. Schlesinger, A. Ben-Jebria, Alan R. Dahl, M. B. Snipes, and J. Ultman

CONTENTS

0-8493-4493-X/97/$0.00+$.50

1 INTRODUCTION

The development of biological responses following the inhalation of chemical toxicants is dependent upon the disposition of these materials within the respiratory system. This, in turn, depends upon their patterns of deposition (the sites within which they initially come into contact with airway epithelial surfaces and the amounts removed from the inhaled air at these sites), any metabolism, and their clearance (the rates and routes by which dissolved gases or deposited particles are physically removed from the respiratory tract). For chemicals that exert their action upon surface contact, the initial deposition is a predictor of toxic response. In many other cases, however, it is the net result of deposition and clearance (namely, retention, or the amount of material remaining in the respiratory tract at specific times after exposure) that influences toxicity. This chapter provides information on the deposition of inhaled particles and gases, the metabolism of deposited materials by the respiratory tract, clearance mechanisms and rates, and ultimate retention patterns.

2 DEPOSITION OF INHALED MATERIALS

2.1 GASES AND VAPORS

This section summarizes the deposition of xenobiotic gases and vapors inhaled into the respiratory tract. A *vapor* is a material that is below its critical temperature, whereas a *gas* is a material that is above its critical temperature. Vapors can, therefore, be present in the liquid phase over a full range of compositions from 0 to 100 mol%, while gases are only capable of limited solubilities, usually a fraction of 1.0 mol%. Aside from this subtle difference, the mechanisms by which gases and vapors are transported are identical. Both are subject to the constraints of thermodynamic equilibrium and to the dynamics of diffusion and chemical reaction. In this section, the term "gas" is, therefore, used to refer to both gases and vapors.

To provide consistency, the term "deposition" is used to describe migration of an inhaled material to the surface of an airway or airspace. Unlike particles, however, gases consist of individual molecules that do not ordinarily accumulate at the surface itself. It is more correct to think of gases absorbing through an interface rather than depositing onto an interface.

The information summarized in this section is restricted to *in vivo* studies on intact animals and humans. Although some measurements have been made on isolated tissues (e.g., Postlethwait and Bidani[189] and Ben-Jebria et al.[20-22]), these *in vitro* preparations generally lacked blood perfusion and mucociliary activity that are necessary to ensure physiologically relevant dose data.

There are two basic methods for evaluating the *in vivo* uptake of a xenobiotic gas — directly measuring the amount that resides in the tissue following exposure (e.g., Santrock et al.[208]) or inferring this amount by monitoring changes in gas phase concentration and flow during the exposure (e.g., Yokoyama and Frank[268]). The direct method requires that a stable tracer molecule be available and that a tissue specimen be taken. The inferential method is frequently more practical, because it requires only that airway gas be sampled and analyzed by a standard method, such as gas chromatography, mass spectrometry, or chemiluminescence. Data obtained by the inferential method can be normalized to obtain uptake efficiency, and this dosimetric has been selected for tabulation. Those studies for which results are not available or which cannot be recomputed in terms of uptake efficiency have been excluded.

Deposition data for various gases are outlined in Table 1. The gas is classified as insoluble when it has limited physical solubility in water; otherwise, it is considered to be soluble. A gas that is known to react with biochemical substrates in the lung, either spontaneously or

Guide to Tables 1A to 1D

Gas or vapor classification
- IS = insoluble
- S = soluble
- NR = nonreactive
- R = reactive

Study population
- People — human:number of subjects; m = male, f = female
- Laboratory Animal — species:number of subjects; m = male, f = female

Exposure conditions
- Un = unencumbered, exposure concentration in ppm
- Fx = breathing fixture, exposure concentration in ppm

Access route (for air)
- Or = oral
- Na = nasal
- On = oronasal
- Tr = tracheal

Breathing pattern
- In = steady unidirectional inspiratory flow in ml/sec
- Ti = tidal breathing; tidal volume in ml, average flow in ml/sec

Respiratory site: uptake efficiecy
- Ua_I = upper airways during inspiration
- Ua_E = upper airways during expiration
- Ua_T = upper airways during complete breath
- Lu = lung during complete breath
- Rs = total respiratory system during a tidal breath

with the aid of an enzyme, is designated as reactive; if such a reaction is not known, the gas is categorized as nonreactive. Gases, such as sulfur dioxide, which are sparingly soluble but rapidly combine with water to form charged ions, are classified as nonsoluble, but reactive. The study population is designated as either human or laboratory animal. Measurements on people have been made only on healthy subjects without regard to ethnic background. Thus, gender is the only subclassification of the study group.

Environmental exposure conditions generally encompass the concentration-time pattern of the gas as well as air temperature and humidity. Since exposure durations have only been 2 hours or less, with air temperatures and humidities maintained within the thermal comfort zone, the most distinctive exposure parameter has been environmental concentration. Exposures have most frequently occurred at a constant inhaled concentration. For those few studies that employed an inhaled gas bolus (denoted by *), the peak inhaled concentration is tabulated as the exposure parameter. Also of concern in detailing exposure is the method of gas delivery. When a subject's breathing is encumbered by the use of a mouthpiece, mask, or some other breathing fixture, the dosimetry data may be different than during unencumbered breathing, such as within a head hood or full-body chamber.

The route of air access is another important element of exposure. The nose, with its relatively large absorption surface and intense perfusion, is thought to be a better gas absorber than the mouth. Whereas many animals breathe exclusively through their noses, people only breathe nasally at low levels of physical activity. With progressively increasing levels of exercise, breathing switches from nasal to oronasal and eventually to oral. In animal experiments, it is possible to surgically bypass the nose and mouth so that inspired air enters directly into the trachea.

In dosimetry measurements made during continuous exposure, gas concentration and flow are almost always measured at the airway opening (i.e., the nose or mouth) and are frequently monitored at an internal airway sampling site. The position of the internal site constitutes an operational boundary between the upper airways (Ua) and the lungs (Lu). The exact anatomical make of Ua and of Lu, therefore, depends on the precise location of

TABLE 1A
Ozone, O_3 [IS, R][a]

Study Population	Exposure Conditions[a]	Access Route	Breathing Pattern	Respiratory Site: Uptake Efficiency				Ref.
				Ua_I	Ua_T	Lu	Rs	
Human: 9 m	Fx: 4.0*	Or	Ti: 500, 250	—	50	—	88	108
Human: 9 m	Fx: 3.0*	Or	Ti: 500, 150	—	65	—	91	107
			Ti: 500, 250	—	55	—	87	
			Ti: 500, 500	—	27	—	82	
			Ti: 500, 750	—	15	—	78	
			Ti: 500, 1000	—	10	—	76	
Human: 9 m	Fx: 0.4–3.4*	Or	Ti: 500, 250	—	50	—	88	110
Human: 9 m	Fx: 2.0*	Or	Ti: 500, 250	—	53	—	88	
		Na	Ti: 500, 250	—	78	—	94	
Human: 10 m	Fx: 2.0*	Or	Ti: 500, 250	—	51	—	89	39
Human: 10 f	Fx: 2.0*	Or	Ti: 500, 250	—	63	—	91	39
Human: 18 m	Un: 0.1–0.4	Or	Ti: 800, 500	40	—	91	—	86
		Na	Ti: 800, 500	36	—	91	—	
		On	Ti: 800, 500	43	—	91	—	
		Or, Na, On	Ti: 800, 350	41	—	93	—	
		Or, Na, On	Ti: 800, 640	38	—	89	—	
Human: 20 m	Un: 0.4	Or	Ti:1500, 1400	39	—	65	—	85
Human: 9 m, 1 f	Un: 0.4	Or	Ti: 800, 300	27	—	96	91	84
Human: 10 m	Fx: 0.3	Or	Ti: 700, 500	—	—	—	77	
		Na	Ti: 700, 500	—	—	—	73	
Dog: 11 m and f	Fx: 0.3–0.8	Or	In: 100	34–27	—	—	—	268
			In: 670	12–10	—	—	—	
		Na	In: 100	72–59	—	—	—	
			In: 670	37–27	—	—	—	
		Tr	Ti: 170, 110	—	—	87–84	—	
			Ti: 170, 170	—	—	83–80	—	
Rabbit: 19 m, 14 f	Un: 0.1–2.0	Na	In: 17	50	—	—	—	147
Guinea pig: 58 m	Un: 0.1–2.0	Na	In: 2.7	50	—	—	—	147
	Un: 2.0–3.0	Na	In: 2.7	41	—	—	—	
Rat: 30 m	Fx: 0.3–1.0	Na	Ti: 2.7, 13	—	—	—	41	261
Rat: 41 m	Fx: 0.3–0.6	Na	Ti: 2.7, 12	—	—	—	45	260
Guinea pig: 6 m	Fx: 0.3	Na	Ti: 2.4, 6.3	—	—	—	53	260

[a] See Guide to Tables 1A to 1D for key to abbreviations.
[b] Studies employing an inhaled gas bolus are indicated with an asterisk (*).

the sampling site. In animals, gas is usually sampled at a tracheostomy site, so that Ua and Lu are comparable between different studies. In humans, a narrow-gauge conduit extending from the nares has been used to access an internal gas sampling site. In this case, there is considerable variation in the anatomic components that constitute Ua and Lu, depending on where the tip of the conduit is positioned (e.g., Gerrity et al.[86] and Gerrity et al.[85]).

Employing such a dual sampling point technique, uptake efficiency in the upper airways, the lungs, or the entire respiratory system (Rs) can be defined by the general equation:

$$\text{Uptake efficiency} = (1 - M_{out}/M_{in}) \times 100$$

where M_{out} and M_{in} are the amounts of the gas obtained by integrating the product of concentration and flow at appropriate measurement points (Figure 1). Depending on whether gas concentrations are monitored during inhalation, exhalation, or throughout a tidal breath, four calculations of uptake are possible:

TABLE 1B

Other Gases [IS, R][a]

Gas		Study Population	Exposure Conditions	Access Route	Breathing Pattern[b]	Respiratory Site: Uptake Efficiency Ua_I	Lu	Rs	Ref.
Nitrogen dioxide	NO_2	Human: 10	Fx: 0.3	Or	Ti: —, 167	—	—	73	17
					Ti: —, 500	—	—	88	
Nitrogen dioxide	NO_2	Dog: 5	Fx: 0.5–1.0	Na	In: 50	90	—	—	251
Nitrogen dioxide	NO_2	Dog: 6	Un: 1.0	Or	Ti: —, 33	63	96	—	116
					Ti: —, 67	57	90	—	
					Ti: —, 133	52	88	—	
					Ti: —, 200	51	86	—	
				Na	Ti: —, 33	90	96	—	
					Ti: —, 67	77	90	—	
					Ti: —, 133	65	88	—	
					Ti: —, 200	59	86	—	
Nitric oxide	NO	Dog: 5	Fx: 1.5–2.0	Na	In: 50	73	—	—	251
Sulfur dioxide	SO_2	Dog: 5	Fx: 0.5	Na	In: 50	100	—	—	251
Sulfur dioxide	SO_2	Dog: 12	Fx: 1.1–141	On	Ti: —, —	—	4.5	—	12
Sulfur dioxide	SO_2	Rabbit: 4	Un: 0.05	On	Ti: —, —	40	80	—	
		Rabbit: 10	Un: 700		Ti: —, 17	95	98	—	
Sulfur dioxide	SO_2	Human: 7 m	Fx: 16.0	Na	Ti: 850, 142	100	—	—	225
Sulfur dioxide	SO_2	Dog: 7	Fx: 1.0–10	Na	Ti: —, 58	100	—	—	78
			Fx: 1.0		Ti: —, 580	97	—	—	
				Or	Ti: —, 58	99	—	—	
			Fx: 10			97	—	—	
					Ti: —, 580	34	—	—	

[a] See Guide to Tables 1A to 1D for key to abbreviations.
[b] — denotes no data available.

1. The uptake efficiency of the upper airways during an inspiratory-directed steady flow or the inhalation portion of a tidal breath (Ua_I) is computed by utilizing M_{in} at the airway opening and M_{out} at the internal airway site.
2. The uptake efficiency of the upper airways during an expiratory-directed steady flow or the exhalation portion of a tidal breath (Ua_E) is computed by employing M_{in} at the trachea and M_{out} at the airway opening.
3. The uptake efficiency of the lungs during a tidal breath (Lu) is computed by using M_{in} at the trachea monitored during inhalation and M_{out} at the trachea monitored during exhalation.
4. The uptake efficiency of the entire respiratory tract during a tidal breath (Rs) is computed by applying M_{in} at the airway opening monitored during inhalation and M_{out} at the airway opening monitored during exhalation.

There is generally no endogenous production of exposure gas, so that net deposition during a complete breath is positive in both the Lu and Rs regions. Similarly, net deposition in the upper airways is positive, except when Ua_E is determined during the expiratory portion of a tidal breath. In that case, the gas entering the upper airways from the lungs, typically containing a low concentration of exposure gas, mixes with higher concentration gas remaining in the upper airways from the previous inhalation. This elevates the gas concentration reaching the airway opening relative to the entering concentration, causing Ua_E to be negative. For example, in quietly breathing people, Ua_E measured at an internal sampling site just distal to the larynx was –200% for ozone.[84] In other words, the amount of ozone leaving at the mouth during exhalation was three times the amount entering at the proximal trachea.

TABLE 1C

Aldehyde Vapors [S, R][a]

Vapor		Study Population	Exposure Conditions	Access Route	Breathing Pattern	Respiratory Site: Uptake Efficiency				Ref.
						Ua_I	Ua_T	Lu	Rs	
Acetaldehyde	CH_3CHO	Dog: 4 m and f	Fx: 224–448	On	Ti: 100–240, 12–280	50–55	54–68	—	44–61	67
				Tr		—	—	47–59	—	
Acetaldehyde	CH_3CHO	Rat: 5 m	Fx: 1.0–10	Na	In: 50	93	—	—	—	155
			Fx: 100			63	—	—	—	
			Fx: 1000			55	—	—	—	
			Fx: 1.0		In: 100	88	—	—	—	
			Fx: 10			59	—	—	—	
			Fx: 100–1000			37	—	—	—	
			Fx: 1.0		In: 200	65	—	—	—	
			Fx: 10			38	—	—	—	
			Fx: 100–1000			25	—	—	—	
			Fx: 1.0		In: 300	47	—	—	—	
			Fx: 10			31	—	—	—	
			Fx: 100–1000			18	—	—	—	
			Fx: 1.0		Ti: 1, 100	—	71	—	—	
			Fx: 10			—	45	—	—	
			Fx: 100			—	40	—	—	
			Fx: 1000			—	26	—	—	
Acrolein	CH_2CHCHO	Dog: 4 m and f	Fx: 180–440	On	Ti: 100–200, 12–67	78–84	67–80	—	80–85	68
				Tr		—	—	63–70	—	
Formaldehyde	CHCHO	Dog: 4 m and f	Fx: 166–415	On	Ti: 100–200, 12–67	94–97	100	—	100	68
				Tr		—	—	95–100	—	
Propionaldehyde	CH_3CH_2CHO	Dog: 4 m and f	Fx: 126–420	On	Ti: 100–200, 12–67	54–66	64–80	—	70–80	68
				Tr		—	—	65–92	—	

[a] See Guide to Tables 1A to 1D for key to abbreviations.

TABLE 1D

Other Vapors[a]

Xenobiotic Vapor		Study Population	Exposure Conditions	Access Route	Breathing Pattern	Respiratory Site: Uptake Efficiency				Ref.
						Ua_I	Ua_E	Lu	Rs	
Acetone [S,NR]	CH_3COCH_3	Dog: 5	Fx: —	Na	In: 22	52	—	—	—	1
					In: 222	9	—	—	—	
Acetone [S,NR]	CH_3COCH_3	Rat: 4 m	Fx: 342–760	Na	In: 52	45	—	—	—	157
					In: 100	34	—	—	—	
					In: 302	12	—	—	—	
		Guinea pig: 4 m			In: 100	20	—	—	—	
					In: 200	13	—	—	—	
					In: 576	7	—	—	—	
Acetone [S,NR]	CH_3COCH_3	Mouse: 19 m	Fx: 632–8432	Na	In: 21	25	—	—	—	158
		Mouse: 12 m			In: 33	20	—	—	—	
		Mouse: 18 m			In: 70	14	—	—	—	
Acetone [S,NR]	CH_3COCH_3	Hamster: 3 m	Fx: 342–760	Na	In: 35	22	—	—	—	159
					In: 70	15	—	—	—	
					In: 200	10	—	—	—	
Ethanol [S,NR]	CH_3CH_2OH	Rat: 4 m	Fx: 441–771	Na	In: 52	84	—	—	—	157
					In: 100	59	—	—	—	
					In: 302	32	—	—	—	
		Guinea pig: 4 m			In: 100	62	—	—	—	
					In: 200	54	—	—	—	
					In: 576	28	—	—	—	
Ethanol [S,NR]	CH_3CH_2OH	Dog: 3 m and f	Fx: 500	Na	Ti: 300, 70	88	–4.3	0.3	84	59
Ethyl acetate [S,R]	$CH_3CO_2CH_2CH_3$	Rat: 8–12 m	Fx: 180–2879	Na	In: 50	34	—	—	—	159
					In: 100	20	—	—	—	
					In: 300	9	—	—	—	
Ethyl acetate [S,R]	$CH_3CO_2CH_2CH_3$	Hamster: 8–12 m	Fx: 180–2879	Na	In: 35	71	—	—	—	159
					In: 70	47	—	—	—	
					In: 200	21	—	—	—	
2,4-Dimethyl pentane [IS,S]	$(CH_3)_2CHCH_2CH(CH_3)_2$	Dog: 3–5 m and f	Fx: 500	Na	Ti: 242, 48	–0.5	0.03	0.2	–0.3	55
1,3-Dioxolane [S]	$C_3H_6O_2$	Dog: 3 m and f			Ti: 303, 68	58	–0.7	0.3	58	55
Propyl ether	$CH_3OC_2H_6$	Dog: 3 m and f			Ti: 286, 58	7	–0.2	0.5	7.3	55
Butanone [S]	$CH_3CH_2COCH_3$	Dog: 3 m			Ti: 271, 60	50	–0.3	0.3	50	55

[a] See Guide to Tables 1A to 1D for key to abbreviations.

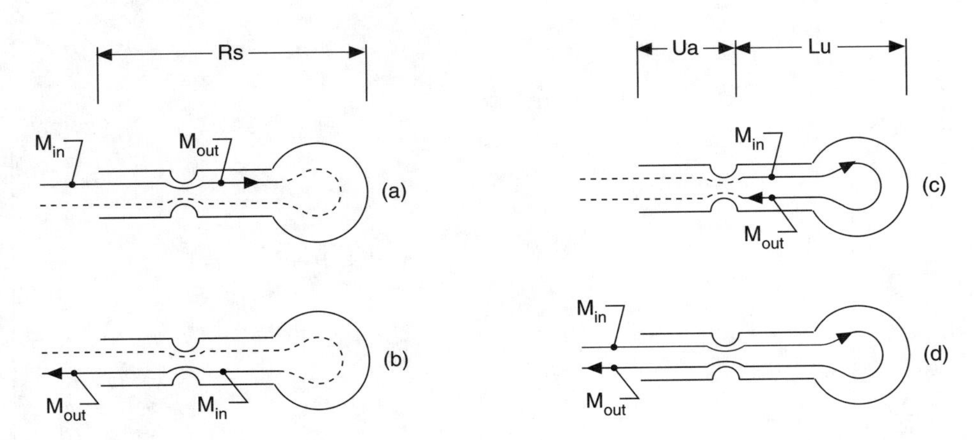

FIGURE 1
Strategies for measurement of uptake efficiencies during continuous exposure to xenobiotic gases.

Deposition data in ventilated dogs indicates that Ua_E is also negative for some soluble vapors.[55] Because ozone is so reactive with tissue, its negative Ua_E is probably due to residual gas present in the airway lumen. On the other hand, the negative Ua_E values associated with soluble vapors are probably due to inhaled gas that dissolves in tissue and then diffuses back into the airway lumen during exhalation.

The inhalation of gas boluses instead of a continuous stream of exposure gas allows the estimation of regional uptake without the need for internal gas sampling. Following a bolus input, M_{in} and M_{out} are obtained by integrating the inhaled and the exhaled concentration data, respectively. A series of such measurements made at different bolus penetrations yields a longitudinal distribution of uptake efficiency. The uptake efficiency at a penetration of 50 ml beyond the airway opening is associated with gas transported to the larynx during inhalation and then returned to the airway opening during exhalation. This upper airway efficiency for a complete breath, Ua_T, must be less than either of the unidirectional upper airway efficiencies, Ua_I and Ua_E, obtained with the more conventional dual sampling point technique. To estimate Rs from bolus measurements, the entire distribution of uptake efficiency must be integrated over a penetration range corresponding to the tidal volume of interest; this additional computation is made prior to tabulating the bolus uptake data.

2.2 Particles

There are five significant mechanisms by which particles may deposit in the respiratory tract. These are depicted schematically in Figure 2.

Impaction is the inertial deposition of a particle onto an airway surface. It occurs when the momentum of the particle prevents it from changing course in an area where there is a rapid change in the direction of bulk airflow. Impaction is the main mechanism by which particles having ≥0.5-μm diameters deposit in the upper respiratory tract and at or near tracheobronchial tree branching points. The probability of impaction increases with increasing air velocity, rate of breathing, particle size, and density.

Sedimentation is deposition due to gravity. When the gravitational force on an airborne particle is balanced by the total of forces due to air buoyancy and air resistance, the particle will fall out of the air stream at a constant rate, known as the terminal settling velocity. The probability of sedimentation is proportional to the particle's residence time within the airway, particle size, and density and decreases with increasing breathing rate. Sedimentation is an important deposition mechanism for particles with ≥0.5-μm diameters which penetrate to airways where air velocity is relatively low, e.g., mid- to small-sized bronchi and bronchioles.

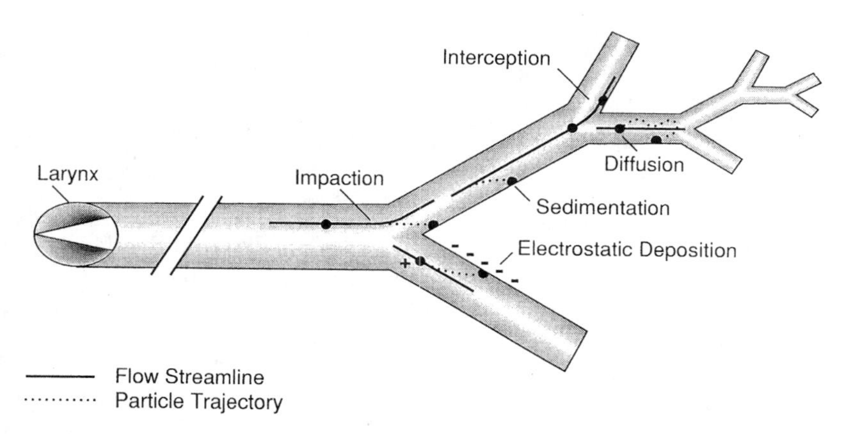

FIGURE 2
Schematic diagram of particle deposition mechanisms. (Adapted from Schlesinger, R. B., in *Concepts in Inhalation Toxicology,* 2nd ed., McClellan, R. O. and Henderson, R. F., Eds., Taylor & Francis, Washington, D.C., 1995.)

Submicrometer-sized particles, especially ultrafines, which are those having diameters <0.1 μm, acquire a random motion due to bombardment by surrounding air molecules; this motion may then result in particle contact with the airway wall. The displacement sustained by the particle is a function of a parameter known as the diffusion coefficient, which is inversely related to particle size, specifically cross-sectional area, but is independent of particle density. The probability of deposition by diffusion increases with increasing particle residence time within the airway, and diffusion is a major deposition mechanism where bulk flow is low or absent (for example, in bronchioles and the pulmonary region, or alveolated airways). However, extremely small ultrafine particles can show significant deposition in the upper respiratory tract, trachea, and larger bronchi; this likely occurs by turbulent diffusion.

Some freshly generated particles can be electrically charged and may exhibit enhanced deposition over that expected from size alone. This can be due to image charges induced on the surface of the airway by these particles and/or to space-charge effects, whereby repulsion of particles containing like charges results in increased migration towards the airway wall. The effect of charge on deposition is inversely proportional to particle size and airflow rate. Since most ambient particles become neutralized naturally due to the presence of air ions, electrostatic deposition is generally a minor contributor to overall particle collection by the respiratory tract; it may, however, be important in some laboratory studies.

Interception is a significant deposition mechanism for fibrous particles, which are those having length-to-diameter ratios > 3:1. While fibers are also subject to all of the same deposition mechanisms as are more spherical or compact particles, they have the additional possibility of deposition when an edge contacts, or intercepts, an airway wall. The probability of interception increases as airway diameter decreases, but it can also be fairly significant in both the upper respiratory tract and upper tracheobronchial tree. While interception probability increases with increasing fiber length, the aerodynamic behavior of a fiber and impaction/sedimentation probability are more influenced by fiber diameter.

From the discussion above, it should be evident that a major particle characteristic that influences deposition is size. But it is important that this be expressed in the proper manner. The deposition probability for particles with geometric diameters ≥0.5 μm is governed largely by their equivalent aerodynamic diameter (D_{ae}), while that for smaller ones is by actual physical diameter; therefore, it follows that aerodynamic diameter is the most appropriate size parameter for describing particles subject to deposition by sedimentation and impaction but not diffusion. Since particles are generally inhaled not singly, but as constituents of aerosols, the mass median aerodynamic diameter (MMAD) is an appropriate parameter to use for those aerosols in which most particles have actual diameters ≥0.5 μm, while the median size of aerosols containing particles with diameters less than this should be expressed in terms of a diffusion diameter, such as thermodynamic equivalent diameter or by using actual geometric size.

The distribution of particle sizes within an aerosol, which is generally characterized as either monodisperse ($\sigma_g = 1$) or polydisperse ($\sigma_g > 1$), is also important in terms of ultimate deposition pattern. If the σ_g of a polydisperse aerosol is < 2, the total amount of deposition within the respiratory tract will probably not differ substantially from that for a monodisperse aerosol having the same median size.[65] However, size distribution is critical in determining the spatial pattern of deposition, since the latter depends upon the sequential removal of particles within each region of the respiratory tract which, in turn, depends upon the actual particle sizes present within the aerosol.

A particle characteristic that may dynamically alter its size after inhalation is hygroscopicity. Hygroscopic particles will grow substantially while they are still airborne within the respiratory tract and will deposit according to their hydrated, rather than their initial dry, size.

Various techniques have been used to measure particle deposition in the respiratory tract of humans and experimental animals. Unfortunately, the use of different experimental methods and assumptions, especially in assessment of regional deposition, has resulted in large variations in reported values, even within the same species.

Total respiratory tract deposition has often been determined by comparing the concentration of test particles administered in inhaled air with that in collected exhaled air, the difference representing the total amount deposited. If assumptions are made about mixing and dead space, estimates of regional deposition can be derived from measurements of particle concentration in different volume fractions of the expired air, but such assumptions cannot be validated.

Specific particle characteristics may be used to measure deposition. Most commonly, radioactively tagged tracer particles are used with various types of detector systems. Total deposition is estimated by monitoring the thoracic and head regions immediately after exposure, whereas regional deposition is usually defined functionally on the basis of subsequent clearance. For example, it is often assumed that any particles remaining in the thorax 24 to 48 hours after exposure are in the pulmonary (alveolar) region, and particles that deposited in the tracheobronchial region were cleared from the lungs prior to this time. This is a reasonable assumption for healthy subjects but may not be for subjects with disease states where clearance is slower, and its use could result in an overestimation of pulmonary deposition and an underestimation of tracheobronchial deposition.

Deposition in the upper respiratory tract is inferred from measurements on the head immediately after exposure. Since this region clears rapidly to the stomach, even the first measurement may not accurately reflect actual deposition; accordingly, some investigators include an initial measurement of material in the gastrointestinal tract in their reported value for upper respiratory deposition. The upper respiratory tract, however, as defined in various studies, may include any or all of the following anatomic regions: nasopharynx, oropharynx, larynx, or upper trachea.

Another technique for deposition analysis in experimental animals is chemical and/or radiological assay of tissues or whole organs removed by dissection after exposure. Obtaining accurate deposition values requires immediate sacrifice and the assumption of no particle translocation (except to the gastrointestinal tract) prior to or during dissection.

It is difficult to compare systematically the deposition patterns obtained from reported studies in one species, and it is even harder to do this between species because of variations in experimental protocols, measurement techniques, definitions of specific respiratory tract regions, and so on. For example, tests with humans are generally conducted under protocols that standardize tidal volume and breathing frequency (although the standardization parameters often differ in different laboratories), whereas those using experimental animals involve a wider variation in respiratory exposure conditions (for example, spontaneous breathing vs. controlled breathing as well as various degrees of sedation). Much of the variability in the reported data for individual species is due to the lack of normalization for specific respiratory parameters during exposure.

In addition, experimental inhalation studies use different exposure techniques, such as nasal mask, oral mask, oral tube, or tracheal intubation. Regional deposition fractions are affected by the exposure route and delivery technique used. Even the specific size of the delivery device can affect inspired airflow rates, which influence the extent of deposition in the upper respiratory tract and the degree of particle penetration into the lungs.

Compilations of experimentally determined deposition values in humans and those experimental animals commonly used in inhalation toxicology studies are shown in Figures 3 and 4, respectively. Not all deposition studies reported in the literature were included in this survey. Only studies where regional deposition values as a fraction of the amount of particles inhaled were provided, or could be derived, were included. Most studies describe regional fractions as a percentage of total deposition rather than in terms of amount of material inhaled and were, therefore, excluded. In addition, only studies using nonhygroscopic, nonviable, nonfibrous aerosols and reporting an aerodynamic or diffusion-related diameter were included. Most studies with humans used monodisperse aerosols, whereas many of those with experimental animals used polydisperse aerosols. Since some of these latter may have consisted of particles of widely different sizes, it is often difficult to evaluate deposition based upon the median size alone; however, it is necessary to include some of these studies, since a substantial amount of the existing database is derived using such aerosols. Finally, although the aerosols in some studies were not charge neutralized, data using these tracer aerosols were included. The presence of electrical charges could account for some of the variability between different studies using the same species and similar size particles.

In regard to Figures 3 and 4, it should be mentioned that in evaluating interspecies deposition, expressing deposition merely as a percentage of the total inhaled (i.e., deposition efficiency) may not be adequate information for relating results between species. For example, total respiratory tract deposition for the same-sized particle can be quite similar in humans and many experimental animals; however, different species exposed to identical particles at the same exposure concentration will not receive the same initial mass deposition. If the total amount of deposition is divided by body (or lung) weight, smaller animals would receive greater initial particle burdens per unit weight per unit exposure time than would larger ones. For example, the initial deposition of 1-μm particles in the rat will be 5 to 10 times that of humans, and for the dog will be 3 times that of humans, if deposition is calculated on a per unit lung or body weight basis.[184] This is discussed further in Section 5.

Figure 3, parts a to d, presents experimentally determined values for spherical particle deposition within the human respiratory tract as a function of the median size of the inhaled aerosol. All values are expressed as deposition efficiency, the percentage deposition of the total amount inhaled.

The effects of hygroscopicity upon deposition deserves mention. If Figure 3a is examined, it is evident that hygroscopic particles inhaled at diameters of 0.1 to 0.5 μm would tend to show a decrease in total deposition if they grow to < 0.5 μm and will show a deposition increase only if their final hydrated diameter is > 1 μm. On the other hand, since particles > 5 μm may grow only minimally in one respiratory cycle, they may not show an increase in deposition at all inhaled sizes compared to nonhygroscopic material.[71] Hygroscopic particles inhaled at 0.2 to 0.5 μm may show substantial changes in their deposition probability, particularly in the tracheobronchial and pulmonary regions.

The deposition of ultrafine particles is of great interest in inhalation toxicology, since these particles present a large surface area for potential adsorption of other toxicants for delivery to the respiratory tract. There are a few studies in humans using ultrafine aerosols in the diameter range of 0.1 to 0.01 μm, less for smaller sizes; the latter is partly due to the

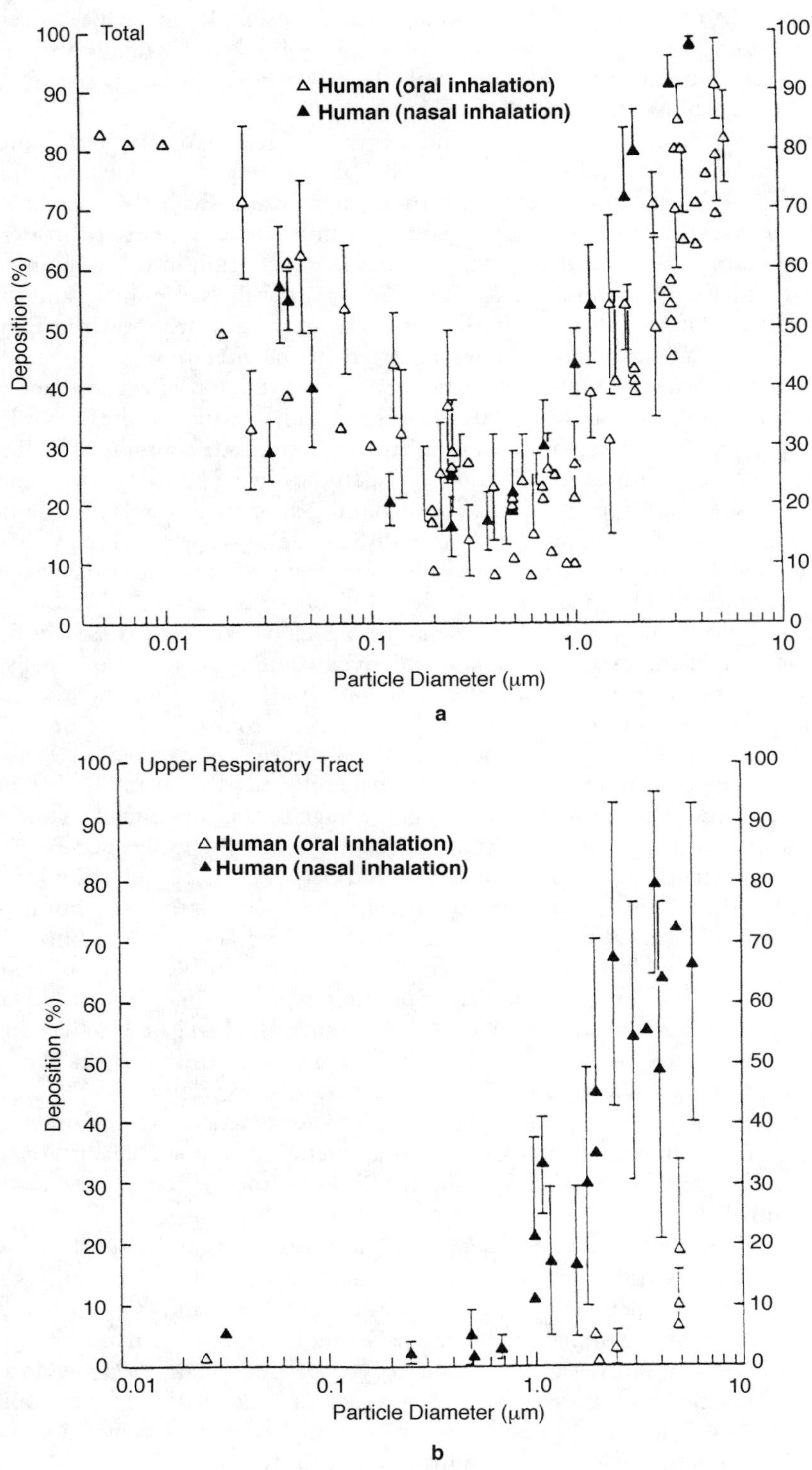

FIGURE 3
Particle deposition in the human respiratory tract. Deposition efficiency, i.e., the percentage deposition of the amount inhaled, is plotted as a function of particle size for: **(a)** total respiratory tract, **(b)** upper respiratory tract, **(c)** tracheobronchial tree, and **(d)** pulmonary region. Particle diameters are aerodynamic for those ≥0.5 μm and diffusion equivalent for those <0.5 μm. (Adapted from Schlesinger, R. B., in *Concepts in Inhalation Toxicology,* 2nd ed., McClellan, R. O. and Henderson, R. F., Eds., Taylor & Francis, Washington, D.C., 1995.)

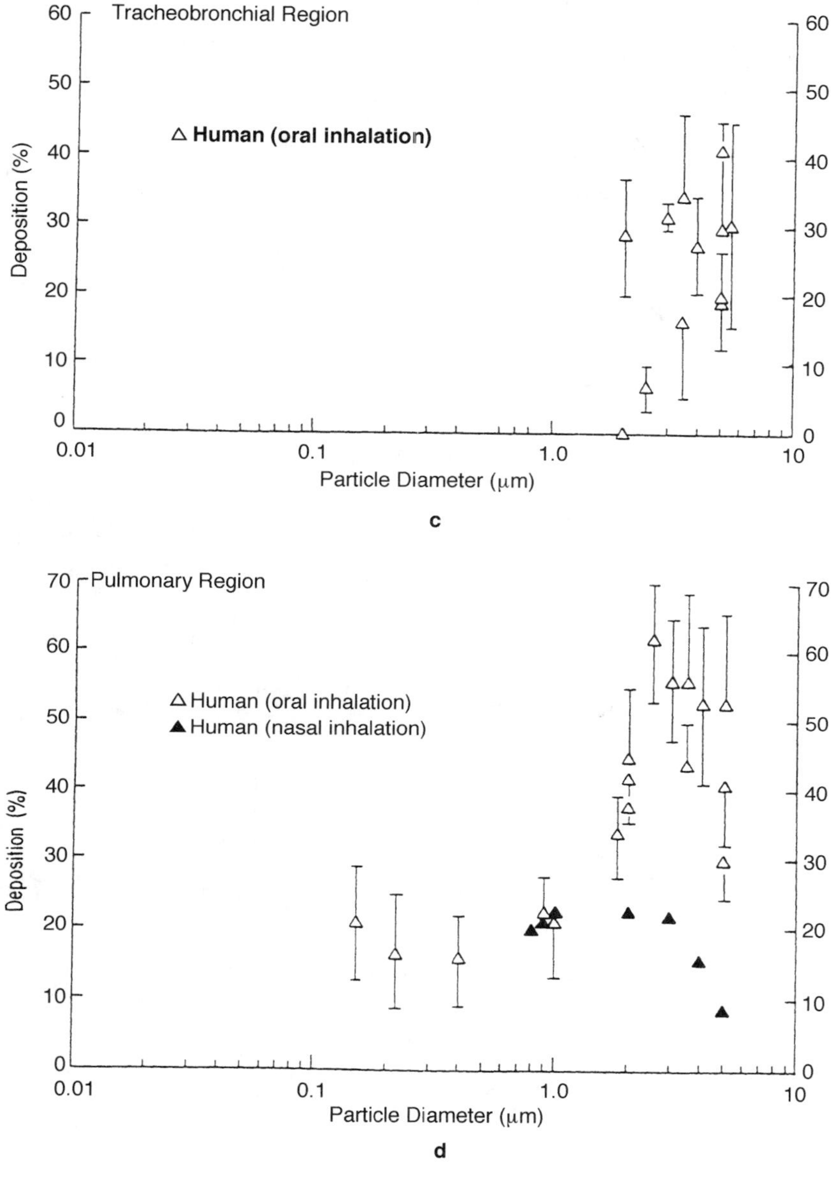

FIGURE 3 (continued)

technical difficulties in producing high quality monodisperse aerosols within this size range in sufficient quantity to allow evaluation of deposition. From Figure 3a, it can be seen that total respiratory tract deposition increases as particle size decreases below 0.2 μm.

The regional deposition of ultrafine particles in humans has been examined using only mathematical and physical models.[43-45,171,238] These indicate that as particle size decreases below 0.2 μm, deposition within the upper respiratory tract and tracheobronchial tree increases substantially, while deposition within the pulmonary region is progressively reduced. Deposition efficiency in the nasal passages can be quite high, reaching over 80% for particles below about 0.002 μm, from a low of about 2% for particles in the 0.1- to 0.2-μm size range. Similar to larger particles, the deposition efficiency for ultrafine particles within the upper respiratory tract with oral breathing is somewhat less than with nasal breathing; oral deposition is likely to be 70 to 90% of nasal deposition for comparable inspiratory flow rates.

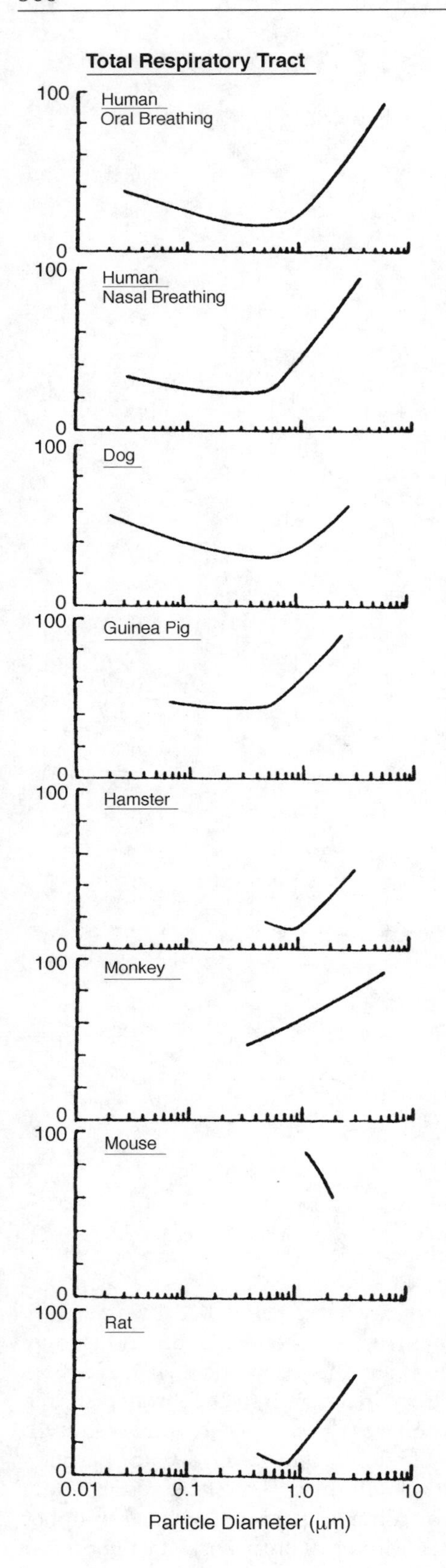

FIGURE 4
Particle deposition efficiencies for experimental animals often used in inhalation toxicological protocols plotted as a function of particle size for: **(a)** total respiratory tract, **(b)** upper respiratory tract, **(c)** tracheobronchial tree, and **(d)** pulmonary region. Each curve represents an eye fit through mean values (or centers of ranges) of the data compiled by Schlesinger. Similar curves for humans are shown for comparison. Particle diameters are aerodynamic for those ≥0.5 μm and diffusion equivalent for those <0.5 μm. (Adapted from Schlesinger, R. B., in *Concepts in Inhalation Toxicology,* 2nd ed., McClellan, R. O. and Henderson, R. F., Eds., Taylor & Francis, Washington, D.C., 1995.)

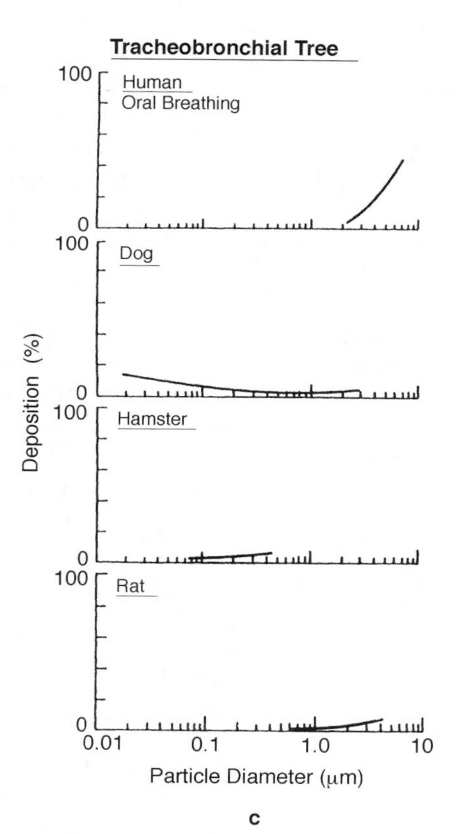

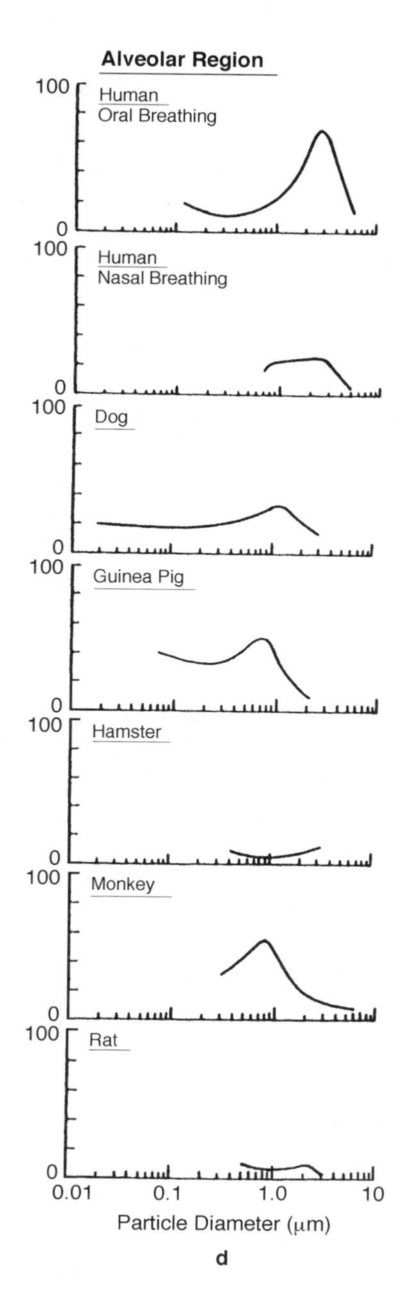

FIGURE 4 (continued)

The deposition efficiencies presented in Figure 3 are for spherical or compact particles. Due to the potential toxicity of fibrous particle shapes, experimental fiber deposition data in humans is not available. However, studies in animals and the use of mathematical and physical models provide some general indication of deposition patterns.[4,5,100,150,236] Long fibers (> 10 μm) tend to show enhanced deposition in the tracheobronchial tree and reduced deposition in the pulmonary region, compared to shorter fibers. But fibers which are very long (e.g., > 50 μm) and thin (e.g., < 0.5 μm) can reach distal conducting airways, and significant amounts of such particles can deposit in the pulmonary region. The deposition of fibers is much more complex than that for spherical particles. For example, fiber shape is important, since straight fibers penetrate more distally than do curly ones.

Particle deposition may not occur in a homogeneous manner along airway surfaces. Specific patterns of enhanced local deposition are important in determining dose, which depends on the surface density of deposition. Nonuniformity implies that the initial dose delivered to specific sites may be greater than that occurring if a uniform density of surface deposit is assumed. This is important for inhaled particles that affect tissues on contact, e.g., irritants, and may be a factor in the site selectivity of certain diseases, e.g., bronchogenic carcinoma.[212]

In the upper respiratory tract, enhanced deposition occurs at areas characterized by constrictions, directional changes, and high air velocities, e.g., the larynx, oropharyngeal bend, and nasal turbinates.[237,239] Likewise, the deposition of aerosols in the tracheobronchial tree is not homogeneous. In humans, air turbulence produced by the larynx results in enhanced localized deposition in the upper trachea and larger bronchi, while deposition is also greatly enhanced at bronchial bifurcations, especially along the carinal ridges, relative to the tubular airway segments.[211] This occurs for spherical particles > 0.5-μm diameter due to impaction, and for fibers due to both impaction and interception;[4] however, enhanced deposition at bifurcations is also seen with submicrometer particles having diameters down to about 0.1 μm[48] due to turbulent diffusion.

The data on localized deposition patterns for the pulmonary region indicate that fibers show nonuniform deposition in distal airways of animals, preferentially depositing on bifurcations of alveolar ducts near the bronchio-alveolar junction.[34,255] While this has yet to be demonstrated in human lungs, the presence of early fiber-related lesions in similar regions suggests that it may occur in these as well.[35]

A concern in inhalation toxicology involves differences in deposition between children and adults. A number of attempts have been made to estimate the influence upon deposition of anatomical and ventilatory changes during postnatal growth in humans.[51,104,185] They indicate that the relative effectiveness of the major deposition mechanisms differs at various times during growth and that this, in turn, may alter regional deposition patterns. Taking into account anatomical differences and the greater ventilation per unit body weight in children, the deposition fractions for some particle sizes, especially those > 1 μm, within certain regions of the growing respiratory tract could be quite different from (and sometimes well above) those found based upon studies with adults; such differences would become even more significant when deposition is expressed on a per-unit-surface-area basis. Since there are also regional differences in clearance rates, this infers that the dose to specific lung compartments from some inhaled particles may vary with age from newborn to adult. Anatomical changes with aging postmaturity may also affect deposition for particles > 1 μm, increasing pulmonary region deposition in older adults compared to younger ones.[185] On the other hand, the deposition of ultrafine particles may not show dramatic differences between children and adults, nor with aging.[185,238]

Any differences in deposition between children and adults may be influenced by activity levels, due to the manner by which breathing pattern changes. For example, increased ventilation with increasing activity in children occurs to a greater extent by increased respiratory frequency, while adults show greater increases in tidal volume. Since increased frequency is associated with decreased deposition of particles > 1 μm in diameter, the greater total respiratory tract deposition with increasing activity levels seen in adults is not seen in young children, and the latter may actually show somewhat of a decline.[19]

3 METABOLISM OF DEPOSITED MATERIALS

The respiratory tract is capable of metabolizing a host of chemicals that deposit on airway surfaces, and this can occur in all segments of the tract, from the nasal passages to the pulmonary region. The result can be either activation or deactivation of a chemical agent which, in turn, affects doses to the respiratory tract or to extrapulmonary sites following translocation. There can be major differences in the rates of metabolism at different sites within one species, as well as between species; this latter may explain certain interspecies differences in site-specific toxicity. Induction or inhibition of respiratory tract enzymes will affect metabolites of deposited toxicants and the ultimate toxicity of an inhaled chemical.

Tables 2 to 4 summarize the regional distribution of xenobiotic metabolizing enzymes in the respiratory tract. These enzymes consist largely of cytochrome P-450-dependent monooxygenases (Phase I enzymes), which play a central role in the oxidative metabolism of many chemicals. Other enzymes which play a role in xenobiotic metabolism include hydrolases, esterases, dehydrogenases, and transferases.

4 CLEARANCE OF DEPOSITED PARTICLES

4.1 Clearance Pathways

Particles that deposit upon airway surfaces may be cleared from the respiratory tract completely or may be translocated to other sites within this system by various regionally distinct processes. These clearance mechanisms, which are listed in Table 5, can be categorized as either absorptive (i.e, dissolution) or nonabsorptive (i.e., transport of intact particles) and may occur simultaneously or with temporal variations. It should be mentioned that particle solubility in terms of clearance refers to solubility *in vivo* within respiratory tract fluids or cells. Thus, an "insoluble" particle is considered to be one for which the rate of clearance by dissolution is insignificant compared to its rate of clearance as an intact particle. For the most part, all deposited particles are subject to clearance by the same mechanisms, with their ultimate fate being a function of deposition site, physicochemical properties (including any toxicity), and sometimes deposited mass or number concentration. Clearance routes from the various regions of the respiratory tract are schematically outlined in Figures 5 and 6.

The clearance of insoluble particles deposited in the nonolfactory portion of the nasal passages occurs via mucociliary transport, and the general flow of mucus is backwards, i.e., towards the nasopharynx. The epithelium of the most anterior portion of the nasal passages, however, is not ciliated, and mucus flow just distal to this is forward, clearing deposited particles to a site (vestibular region) where removal is by sneezing (a reflex response), wiping, or blowing (mechanisms known as extrinsic clearance).

Soluble material deposited on the nasal epithelium will be accessible to underlying cells if it can diffuse to them through the mucus prior to removal via mucociliary transport. Dissolved substances may be subsequently translocated into the bloodstream following movement within intercellular pathways between epithelial cell tight junctions or by active or passive transcellular transport mechanisms. The nasal passages have a rich vasculature, and uptake into the blood from this region may occur rapidly.

Clearance of insoluble particles deposited in the oral passages is by swallowing into the gastrointestinal tract. Soluble particles are likely rapidly absorbed after deposition.[237]

Insoluble particles deposited within the tracheobronchial tree are cleared primarily by mucociliary transport, with the net movement of fluid towards the oropharynx, followed by swallowing. Some insoluble particles may traverse the epithelium by endocytotic processes, entering the peribronchial region.[141,224] Clearance may also occur following phagocytosis by airway macrophages located on or beneath the mucus lining throughout the bronchial tree which then move cephalad on the mucociliary blanket or via macrophages which enter the airway lumen from the bronchial or bronchiolar mucosa.[200] As in the nasal passages, soluble particles may be absorbed through the mucus layer of the tracheobronchial airways and into the blood, via intercellular pathways between epithelial cell tight junctions or by active or passive transcellular transport mechanisms.

TABLE 2

Some Xenobiotic Metabolizing Enzymes in the Nasal Cavity

Enzyme	Test Reaction or Other Method for Detection	Test System	Notes	Ref.
15-Lipoxygenase	Arachidonic acid metabolism	Human nasal cells	Nasal epithelial cells more active than bronchial cells	103
P-450	Diethylnitrosamine deethylase	Human nasal respiratory tissue microsomes	Nasal activity per nmol P-450 10–25 times that of liver	138
P-450	Five dealkylases	Human nasal respiratory tissue microsomes	HMPA and aminopyrine best substrates; ethoxycoumarin and ethoxyresorufin next best substrates; pentoxyresorufin poor substrate	87
Epoxide hydrolase	Safrole oxide hydrolase	Human nasal respiratory tissue homogenate	Activity higher than that in rats	87
Glutathione S-transferase	1-Chloro-2,4-dinitro-benzene conjugation	Human nasal respiratory tissue homogenate	Activity higher than that in rats	87
DT-diaphorase	Dichlorophenol-indophenol metabolism	Human nasal respiratory tissue homogenate	Activity much less than that in rats	87
UDP-glucuronyl transferase	1-naphthol conjugation	Human nasal respiratory tissue homogenate	Absent in humans; present in rats	87
NADPH-cytochrome C-reductase	Cytochrome C reduction	Human nasal respiratory tissue homogenate	Activity about 25% that of rat nasal mucosa	87
Rhodanese	Metabolism of cyanide to thiocyanate	Human nasal respiratory tissue homogenate	Activity in nonsmokers twofold higher than that in smokers	134
P-450PB-B; NADPH-cytochrome P-450 reductase	Immunohistochemistry	Male Holtzman rat olfactory and respiratory tissues, Bowman's glands, and seromucous glands	P-450PB-B is homologous with IIB1; Bowman's glands and apex of olfactory epithelial cells contained high concentrations of reductase	253
P-450c	Immunocytochemistry	Male Alp/Apk rat olfactory epithelium; Bowman's glands	Homologous with P-450IA1; not induced by phenobarbital, clofibrate, or β-naphthoflavone	76
P-450βNF-B P-450PB-B P-450PCN-E	Immunohistochemistry	Male Holtzman rat respiratory and olfactory epithelium; Bowman's glands and seromucus glands	P-450βNF-B (homologous with IA1) present in olfactory tissue at a higher level than in respiratory tissue; consistent with this, aryl hydrocarbon hydroxylase activity; PB-B (IIB1) intensely stained both tissues; PCN-E (IIIA) stained less intensely, but about equally in both tissues; apical portions of epithelial cells and subepithelial glands stained relatively intensely	16 15 15
P-450d	Induction of encoding mRNA	S-D rat olfactory tissue microsomes	P-450d (IA2) but not P-450c (IA1) was induced to detectable levels	89
P-450olf1	cDNA library probe and sequence analysis	S-D rat olfactory tissue microsomes	Termed IIG1; rabbit form may be P-450NMb; olfactory tissue specific	173, 174

P-450olf2	Immunoblots	Wistar rat olfactory tissue	Homologous with IIA family	124
P-450IIE1	Immunohistochemistry	Male F344 rat	Glands in lamina propia heavily labeled; olfactory sustentacular cells apically labeled; ciliated cells of respiratory epithelium and nonsecretory cells of transitional epithelium labeled; luminal surface of olfactory epithelium in vomeral nasal organ labeled	106
Aromatase and 5a-reductase	Testosterone metabolism to estradiol and dihydrotestosterone	Measled S-D rat olfactory epithelium	Castration decreased estradiol production; activity restored by testosterone replacement	140
FAD-containing mono-oxygenase	Dimethylamine and N,N-dimethylaniline metabolism	Male F344 rat olfactory mucosa microsomes	Dimethylaniline apparently metabolized only by FAD-MO; dimethylamine metabolized by P-450, as well	145
Aldehyde dehydrogenase; formaldehyde dehydrogenase	Formaldehyde and acetaldehyde dehydrogenation; histochemistry	Male F344 rat respiratory and olfactory mucosa	Multiple forms; formaldehyde dehydrogenase most abundant in olfactory mucosa; epithelial cell cytoplasm and olfactory sensory cell nuclei; Bowman's and seromucous glands weakly positive; acetaldehyde dehydrogenase most abundant in respiratory mucosa; present in Bowman's glands and olfactory basal cells; absent from sensory cells and sustentacular cells	27, 42, 113
Carboxylesterase	Ester hydrolysis; histochemistry	F344 rat nasal tissue	k_m values ranged from 1–35 m*M*; V_{max} from 0.03 to 0.06; present in all nasal cells except olfactory neurons	26, 58, 197, 230
Carbonic anhydrase	Histochemistry	Rat olfactory tissue	Present in receptor cells; absent in sustentacular cells	37
Epoxide hydrolase; UDP-glucuronyl transferase; glutathione S-transferase forms B, C, and E	Styrene oxide; 7-hydroxycoumarin; styrene oxide; immunohistochemistry	Male F344 or Holtzman rat nasal tissue homogenates	Epoxide hydrolase probably form A; GSH-T B, C, and E probably forms 5,5, 1,1, 3,3, respectively; GSH-T form C — which metabolizes ΔE^5-androstene-3, 17-dione — was at highest levels	15, 16, 30
Rhodanese	Cyanide metabolism to thiocyanate; immunohistochemistry	F344 rat nasal mucosa	Highest in apical portion of olfactory epithelium; absent from receptor cells; negligible in Bowman's glands; present in respiratory epithelium also	60, 133
P-4503a, P-450 form 2, P-450 form 4, P-450 form 5	Immunochemistry; immunoblot; enzyme assays	Male New Zealand white rabbit nasal mucosa	Homologues IIE; Forms 2, 4, 5, by homology, are also termed IIB1, 1A2, and IVB1, respectively; forms 2 and 5 occur in both respiratory and olfactory tissue; form 4 found in olfactory tissue only; form 6 (homologous with IA1) absent in all tissues	64, 203
FAD-containing mono-oxygenase	Immunoblot	Male New Zealand white rabbit olfactory and respiratory mucos	Approximately equal amounts in respiratory and olfactory tissues	203

TABLE 2 (continued)

Some Xenobiotic Metabolizing Enzymes in the Nasal Cavity

Enzyme	Test Reaction or Other Method for Detection	Test System	Notes	Ref.
P-450 form NMa, P-450 form NMb, P-450 form 2, P-450 form 3a, P-450 form 3b, P-450 form 4, P-450 form 6	Immunochemistry; testosterone metabolism; HMPA and phenacetin metabolism	Male New Zealand white rabbit nasal respiratory and olfactory epithelium	Only form 2 (IIB1) found in respiratory tissue; NMa very active for HMPA and phenacetin dealkylation; only 3% of liver P-450 is NMa; NMb (IIG1) occurs only in olfactory tissue; forms 3a and 4 are homologous with IIE1 and IA2, respectively; forms 3b and 6 (homologous to IIC3 and IA1, respectively) absent in nasal tissue	61–63, 262
Carboxylesterase	Ester hydrolysis	Male and female New Zealand white rabbit nasal mucosa	Less activity than for mice, rats, or dogs. Activities in both mucosae similar to that in liver	58, 230
P-450	*p*-Nitroanisole demethylase; aniline hydroxylase; carbon monoxide difference spectra	CD1 mouse nasal tissue microsomes	Compared to dog, rabbit, guinea pig, rat, and Syrian hamster, aniline better substrate in mouse than in any other species except Syrian hamsters	98
P-450; NADPH-cytochrome C-reductase	7-Ethoxycoumarin deethylase; carbon monoxide difference spectrum; cytochrome C-reductase	MFI mouse olfactory epithelium	Very high nasal activities relative to those in liver; activities higher in males than in females	198
Carboxylesterases	Ethylene glycol monomethylether acetate	Male and female $B6C3F_1Cr1Br$ mouse nasal mucosa homogenates	Mouse activity greater than that of rats	230
Carboxylesterases	*p*-Nitrophenyl butyrate hydrolysis; histochemistry	$B6C3F_1$ mouse; both sexes; respiratory and olfactory tissue	Also examined rats; mouse and rat activity (V/k_m) similar; olfactory activity fivefold that of respiratory; V_{max} olfactory 0.5 μmol/min/mg protein, k_m 20–25 m*M*	26
P-450	*p*-Nitroanisole, O-demethylase, and aniline hydroxylase; carbon monoxide difference spectra	Syrian hamster nasal tissue microsomes	Very high activities relative to dog, rabbit, guinea pig, rat, and mouse; values (in nmol P-450 per mg protein): olfactory — 0.36, respiratory — 0.13, trachea — 0.26, liver — 1.10	98, 137
Arylhydrocarbon oxidases/hydroxylase	Benzo(*a*)pyrene metabolism *in vivo*; olfactory and respiratory tissue microsomal activity	Syrian hamster nasal cavity	Also examined *in vitro* metabolism in 5 non-nasal tissues; olfactory tissue highest (4000 pmol/g tissue per hr); metabolites *in vivo* included tetrols, diols, quinones, oxides, and phenols	57

P-450; NADPH cytochrome C-reductase; cytochrome b5	7-Ethoxycoumarin and ethoxyresorufin dealkylase; hexabarbitone oxidase; aniline hydroxylase; cytochrome C-reductase; difference spectra	Syrian hamster olfactory tissue	Very high activities compared to those in rats and female mice; olfactory activities (but not P-450 content) higher than in liver	198, 199
Carboxylesterases	Ester hydrolysis of acetate esters and lactones; ethyl acetate uptake *in vivo*	Syrian hamster olfactory nasal cavity	Hamster nasal activity much higher than rat or rabbit nasal activity with amyl acetate, but not with β-butyrolactone; ester uptake in hamster nose sensitive to enzyme inhibition (63–90%)	58, 159
Alcohol dehydrogenase	Propanol metabolism *in vivo*	Syrian hamster tissue homogenates	k_m is 0.1 m*M*; V_{max} is 4 nmol/mg protein per min, 36 nmol/nose per min	156

Source: Adapted from Dahl, A. R. and Hadley, W. M., *Toxicology,* 21, 345, 1991.

TABLE 3

Summary of P-450 Isozymes Reported in the Rat and Rabbit Nasal Cavities

	Alternate Name[a]		Nasal Tissue	
Isozyme[b]	Rat	Rabbit	Rat	Rabbit
IA1	βNF-B	Form 6	Respiratory and olfactory	Absent
IA2	ISF-G,d	Form 4	Olfactory	Olfactory
IIA	olf2	—	Olfactory	Not reported
IIB1	PB-B,b	Form 2	Respiratory and olfactory	Respiratory and olfactory
IIC3	—	3b	Not reported	Absent
IIE1	j	3a,P-450ALC	Respiratory and olfactory	Olfactory
IIG1	olf1	P-450LM3c	Respiratory and olfactory	Not reported
IIIA	PCN-E			
IVB1	Form 5	Form 5	Not reported	Respiratory and olfactory

[a] See References 62, 64, 96, 172.
[b] Recommended nomenclature.[172]
Source: Adapted from Dahl, A. R. and Hadley, W. M., *Toxicology,* 21, 345, 1991.

TABLE 4

Some P-450 Isozymes Reported in Lungs of Various Species

	Isozyme	Comments	Ref.
Mouse	1A1	Induced in type II cells	75
	1A2	Induced in type II cells and endothelial cells	75
	2B1	Constitutive in type II cells and Clara cells	75
	2B2	Constitutive in type II cells and Clara cells	75
	4B1	Rabbit form activates ipomeanol and 2-aminofluorene	46
	"mN"	In Clara cells, metabolizes naphthalene	46
	"m50b"	In Clara cells, major naphthalene-metabolizing enzyme	170
Rat	1A1	Induced in bronchial epithelium, Clara cells, and type II cells	14
	2A3	Absent in rat liver	115
	2B1	Constitutive at highest levels in Clara cells	14
	2E1	Induced by hyperoxygen	248
	3A1/2	Induced in bronchial epithelium, Clara cells, and type II cells	14
	4B1	Absent in rat liver	176
	"FI"	Constitutive; induced by O_3	242
	"FII"	Cross reacts with rat anti 2B1; constitutive; induced by O_3	
Rabbit	1A1	Highly inducible; occurs in endothelial cells without reductase	180
	2B1	With 4B1, accounts for 80% of uninduced P-450 in lung	66
	2E1, 2E2	2E1, 5% that of liver; 2E2, 2.5% that of liver	188
	4A4	P-450 prostaglandin ω-hydroxylase; occurs only in pregnant rabbits or after induction by progesterone	166
	4B1	Activates 2-aminofluorene and ipomeanol	54
Hamster	"MC"	Possibly IA1; highly inducible	81
	2B	Along with reductase, absent from mesothelium in adults	233
	4B	Present in mesothelium in adult	233
Human	1A1	Inducibility related to smoking and lung cancer	3
	2E1	Racial differences in 2E1 polymorphisms	177
	2F1	Ethoxycoumarin and pentoxyresorufin and 3-methylindole are substrates	176, 246
	4B1	Unlike rabbit form, does not activate 2-aminofluorene	111

Source: Adapted from Dahl, A. R. and Lewis, J. L., *Ann. Rev. Pharmacol. Toxicol.,* 32, 383, 1993.

TABLE 5

Overview of Respiratory Tract Particle Clearance and Translocation Mechanisms

Extrathoracic Region

Mucociliary transport
Sneezing
Nose wiping and blowing
Dissolution (for "soluble" particles) and absorption into blood

Tracheobronchial Region

Mucociliary transport
Endocytosis by macrophages/epithelial cells
Coughing
Dissolution (for "soluble" particles) and absorption into blood

Alveolar Region

Macrophages, epithelial cells
Interstitial pathways
Dissolution for "soluble" and "insoluble" particles (intra- and extracellular)

Source: Adapted from Schlesinger, R. B., in *Concepts in Inhalation Toxicology,* McClellan, R. O. and Henderson, R. F., Eds., Taylor & Francis, Washington, D.C., 1995, pp. 191–224.

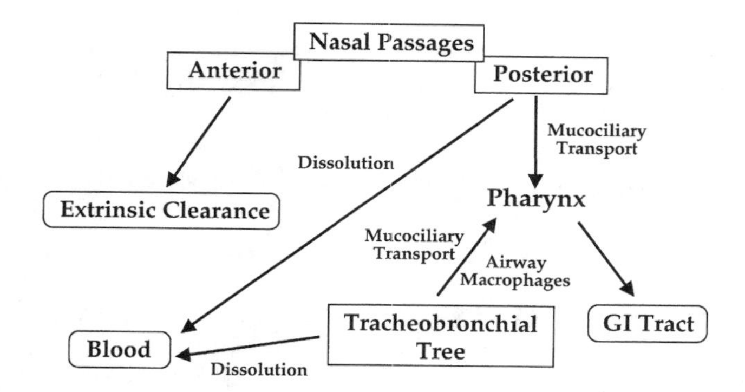

FIGURE 5
Major clearance pathways from the upper respiratory tract and tracheobronchial tree.

Another method of clearance from the tracheobronchial region, under some circumstances, is cough, which can be triggered by receptors located in the area from the trachea through the first few bronchial branching levels. While cough is generally a reaction to some inhaled stimulus, in some cases, especially respiratory disease, it can also serve to clear the upper bronchial airways of deposited substances by dislodging mucus from the airway surface.

Clearance from the pulmonary region occurs via a number of mechanisms and pathways. Particle removal by macrophages comprises the main nonabsorptive clearance process in this region. Alveolar macrophages reside on the epithelium, where they phagocytize and transport deposited material that they contact by random motion, or more likely via directed migration under the influence of local chemotactic factors. Contact may be facilitated as some deposited particles are translocated to sites where macrophages congregate.[182,213]

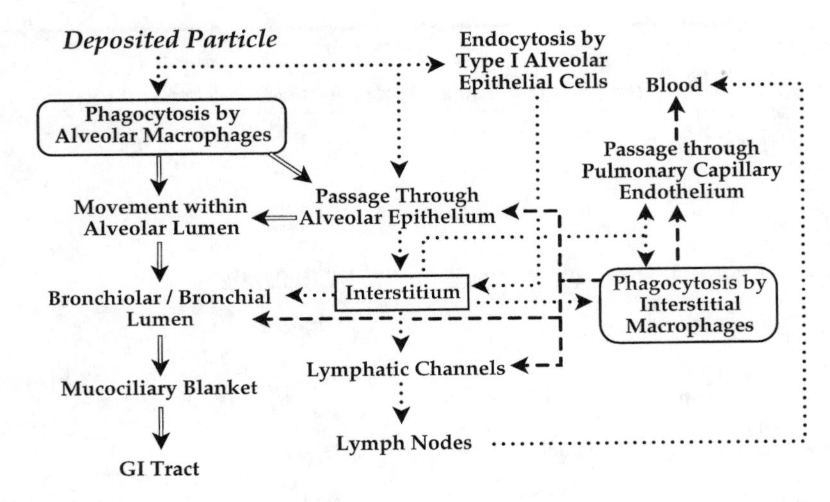

FIGURE 6
Diagram of known and suspected mechanical clearance pathways for insoluble particles depositing in the pulmonary region (dissolution is not included). (Adapted from Schlesinger, R. B., in *Concepts in Inhalation Toxicology,* 2nd ed., McClellan, R. O. and Henderson, R. F., Eds., Taylor & Francis, Washington, D.C., 1995.)

Particle-laden macrophages may be cleared from the pulmonary region along a number of pathways (Figure 6). One route is cephalad transport via the mucociliary system after the cells reach the distal terminus of the mucus blanket; however, the manner by which macrophages actually attain this is not certain. The possibilities are chance encounter; passive movement along the alveolar surface due to surface tension gradients between the alveoli and conducting airways; directed locomotion along a gradient produced by chemotactic factors released by macrophages ingesting deposited material; or passage through the alveolar epithelium and the interstitium, perhaps through aggregates of lymphoid tissue located at bronchioalveolar junctions.[38,49,94,102,114,224]

Some of the cells that follow interstitial clearance pathways likely are resident interstitial macrophages which have ingested particles that were transported through the alveolar epithelium, probably via endocytosis by Type I pneumocytes.[32,36] Particle-laden interstitial macrophages can also migrate across the alveolar epithelium, becoming part of the alveolar macrophage cell population.

Macrophages that are not cleared via the bronchial tree may actively migrate within the interstitium to a nearby lymphatic channel or, along with uningested particles, be carried in the flow of interstitial fluid towards and into the lymphatic system.[102] Passive entry into lymphatic vessels is fairly simple, since the vessels have loosely connected endothelial cells with wide intercellular junctions.[123] Lymphatic endothelium may also actively engulf particles from the surrounding interstitium.[128] Particles within the lymphatic system may be translocated to tracheobronchial lymph nodes, which often become reservoirs of retained material. Particles penetrating the nodes and subsequently reaching the postnodal lymphatic circulation may enter the blood.

Two general categories of lymphatic tissues are recognized in the lungs of mammals. One category is comprised of well-organized, encapsulated lymph nodes. The other category is lymphoid tissue integrally associated with the pulmonary interstitium. The latter type of lymphoid tissue is defined as bronchus-associated lymphoid tissue (BALT). Particles may be brought into the pulmonary interstitium and returned to bronchi through BALT,[38] while BALT may also be a key factor in the processes or mechanisms by which liquids and phagocytic cells pass into and out of the pulmonary interstitium.[94,95] In essence, BALT may represent the terminal component of an excretory pathway for particles or effete phagocytic cells.

Macrophages in the pulmonary interstitium or lymphatic vessels may transport foreign material to BALT, located at the junctions of respiratory and terminal bronchioles, and pass through the BALT to enter the lumen of the airways and thereby gain access to the mucociliary transport system. There appears to be considerable variability among species with regard to the presence of BALT, the type of BALT, and the function of BALT.[169,216] To further complicate the issue, the amount of BALT present in the lung and its function may be influenced by environmental conditions. It appears probable that BALT could represent an important type of lung tissue with functions relevant to species differences in retention and physical clearance of materials deposited in the pulmonary region.

Uningested particles or macrophages in the interstitium may traverse the alveolar-capillary endothelium, directly entering the blood;[105,194] however, endocytosis by endothelial cells followed by exocytosis into the vessel lumen seems to be restricted to particles < 0.1 μm in diameter and may increase with increasing lung burden.[129,179] Once in the systemic circulation, transmigrated macrophages, as well as uningested particles, can travel to extrapulmonary organs. Some mammalian species have pulmonary intravascular macrophages, which can remove particles from circulating blood and may play some role in the clearance of material deposited in the alveoli.[256]

Uningested particles and macrophages within the interstitium may travel to perivenous, peribronchiolar, or subpleural sites, where they become trapped, increasing particle burden. The migration and grouping of particles and macrophages within the lungs can lead to the redistribution of initially diffuse deposits into focal aggregates. Some particles can be found in the pleural space, often within macrophages which have migrated across the visceral pleura.[99,214] Resident pleural macrophages do occur, but their role in clearance, if any, is not certain.

Clearance by the absorptive mechanism involves dissolution in the alveolar surface fluid, followed by transport through the epithelium and into the interstitium, and diffusion into the lymph or blood. Some soluble particles translocated to and trapped in interstitial sites may be absorbed there. Although the factors affecting the dissolution of deposited particles are poorly understood, it is influenced by the surface-to-volume ratio of the particle and other surface properties.[146,162] Thus, materials generally considered to be relatively insoluble may have high dissolution rates and short dissolution half-times if the particle size is small.

Some deposited particles may undergo dissolution in the acidic milieu of the phagolysosomes after ingestion by macrophages, and such intracellular dissolution may be the initial step in translocation from the lungs for such particles.[139] Following dissolution, the material can be absorbed into the blood. Dissolved particles may then leave the lungs at rates which are more rapid than would be expected based upon their normal dissolution rate in lung fluid. Because of this, the clearance rate of such a material can vary with the form in which it is inhaled. Finally, some particles can bind to epithelial cell membranes or macromolecules or other cell components, delaying translocation from the lungs.

4.2 Regional Clearance Kinetics

4.2.1 *Nasal Airways*

Quantitative measurements of clearance rates from the posterior nasal airways of humans have been reported.[80,135] Complete clearance of the posterior nasal airways requires about 3 hours. Because clearance of the posterior nasal airways is dependent upon mucus clearance rates, mucus flow velocities can provide reasonable estimates for rates of nasal clearance from the posterior nasal airways. Table 6 summarizes mucus velocities measured in humans after instillation or inhalation of poorly soluble test particles. Most of these measurements focused on the posterior nasal airways, and results present a broad range of mucus velocities, indicative

TABLE 6

Measured Mucus Velocities in the Posterior Nasal Airways of Normal Humans

Test Material	Number of Subjects	Age Range in Years	Mucus Velocity (mm/min)		Ref.
			Mean	Range	
Colored powder	~100[a]	NA[b]	5.0	3–8	250
Colored powder	92[c]	12–80	4.9	0–12	69
	67[d]	14–78	3.5	0–8.6	69
^{131}I-labeled and unlabeled dye	36[d]	19–52	10.6	4–100	13
^{99m}Tc-labeled resin beads	23[e]	19–55	7.8	0–15	193
^{99m}Tc-labeled resin beads	58[a]	21–26	8.4	2–24	2
^{99m}Tc-labeled polystyrene latex	12[a]	31–69	6.8	2–19	24
^{99m}Tc-labeled resin or saccharine	181[a]	18–46	5.3	0.5–20	191

[a] Smoking status not defined in referenced publication.
[b] NA = information not available in referenced publication.
[c] Nonsmokers.
[d] Smokers.
[e] Smokers and nonsmokers combined.

of individual variability or variability among locations measured in the nasal airways. No dependence of mucus velocity on smoking status, age, or gender has been described. Since the posterior nasal airway is about 6 cm long, an average time to clear this area is about 10 minutes.[80]

A limited number of *in vivo* nasal clearance studies have been conducted in laboratory animals, namely rats and dogs.[154,258] Mucus clearance velocities were similar to results presented in Table 6 for humans, but mucus clearance was more rapid in the respiratory region than in the olfactory region of the nasal airways.

4.2.2 Tracheobronchial Airways

Clearance of poorly soluble particles from the trachea is summarized in Table 7. Clearance rates were variable among these studies, and it is not certain if the differences resulted from intersubject variability or differences in methodology used to make the measurements. Mucus velocities in the trachea are sufficiently fast to transport mucus from the distal to the proximal end of the human trachea within a few minutes in nonsmoking, healthy humans.

Figure 7 presents a comparison of tracheal mucus velocities for several mammalian species. Tracheal mucus velocity varies among species as a function of body weight to the 0.39 power. Correlations with tracheal length, tracheal diameter, and tracheal surface area have been performed, but the best correlation was with tracheal surface area. Thus, reasonable scaling among mammalian species is possible and, like humans, other mammalian species appear to have rapid tracheal mucus clearance velocities which would allow transport of materials from the distal to the proximal end of the trachea to be completed within a few minutes.

Age may be a factor in tracheal mucus clearance rate; slower tracheal clearance rates were noted in older humans and dogs.[91,257] The relationship between age and tracheal mucus velocities is shown in Figure 8;[263] however, the change in velocity is only about a factor of 2 and should not have a significant impact on dosimetry of the trachea, because the trachea clears rapidly in both young and aged subjects.

Direct measurements of mucus velocities in the nasal airways and trachea have provided useful information about clearance rates. Only one study reported clearance velocities for human main bronchi,[77] but no direct measurements have been reported for smaller airways.

TABLE 7

Measured Mucus Clearance Velocities in Tracheas of Healthy, Nonsmoking Adult Humans

Measurement Technique	No. of Subjects	Mean Age in Years (Range)	Clearance Velocity, (mm/min, mean ± SE)	Ref.
External counting of radiolabeled particles using a collimated NaI(T1) detector	5	NA[a] (25–65)	14.0 ± NA	161
Fiberoptic bronchoscopy, Teflon discs	16	27 (20–44)	21.5 ± 0.3	207
Fiberoptic bronchoscopy, Teflon discs	10	27 (18–45)	22.9 ± 2.0	204
Fiberoptic bronchoscopy, Teflon discs	20	27 (18–44)	20.1 ± 1.4	265
Gamma camera imaging of inhaled ^{99m}Tc-albumin microspheres	42	28 (20–43)	3.6 ± 0.4	266
Fluoroscopy of instilled radiopaque Teflon discs	7	25 (21–30)	11.4 ± 1.4	79
Gamma camera imaging of inhaled ^{99m}Tc-albumin microspheres	7	26 (19–31)	4.2 ± 0.6	264
Fluoroscopy of instilled radiopaque Teflon discs	10	23 (19–28)	10.1 ± 1.1	91
	7	63 (56–70)	5.8 ± 1.0	91
Gamma camera imaging of instilled ^{99m}Tc-albumin microspheres	6	29 (NA)	15.5 ± 0.7	47
Fiberoptic bronchoscopy, Teflon discs	7	48 (26–70)	6.5 ± 0.7	205
Gamma camera imaging of inhaled ^{99m}Tc-ferric oxide	7	24 (NA)	5.5 ± 0.4	77
External counting of ^{99m}Tc-labeled Fe_2O_3 particles using collimated NaI(T1) detectors	22	33 (20–64)	5.1 ± 0.6	267
Fiberoptic bronchoscopy, polyethylene discs	7	41 (35–55)	18.5 ± 2.3	249
External counting of ^{99m}Tc-labeled Fe_2O_3 particles using collimated NaI(T1) detectors	10	28 (18–45)	4.3 ± 0.6	132
External counting of ^{99m}Tc-labeled Fe_2O_3 particles using a collimated Phoswich detector	11	NA	4.8 ± 0.5	83

[a] NA = information not available in referenced publication.

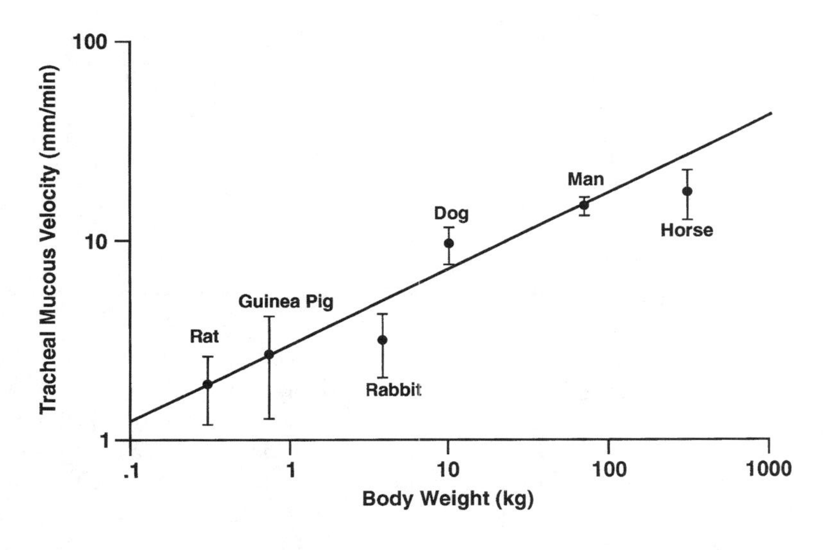

FIGURE 7
Tracheal mucus velocities (TMV) vs. body weight (BW) for six mammalian species. The same techniques were used to measure tracheal mucus velocities for all six species. The function $TMV = 3.0(BW)^{0.39}$ defines the relation between TMV and BW, with a correlation coefficient of 0.94. (From Wolff, R. K., in *Comparative Biology of the Normal Lung,* Vol. I, Parent, R. A., Ed., CRC Press, Boca Raton, FL, 1992, pp. 659–680. With permission.)

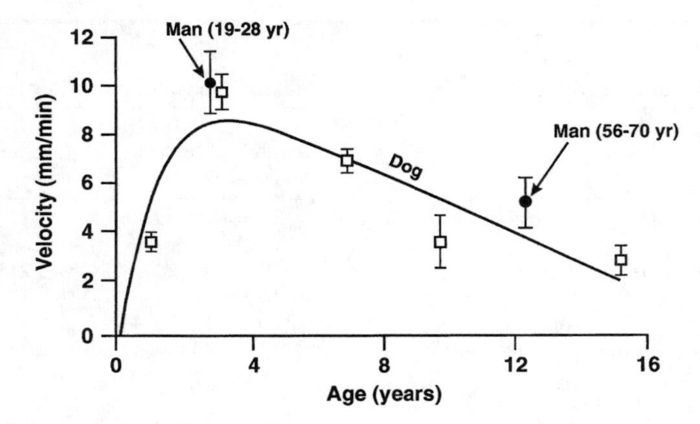

FIGURE 8
Tracheal mucus velocity vs. age for Beagle dogs and humans. Values are mean ± SE. The fitted function is expressed by $V(t) = 11[1-\exp(-0.9t)] - 0.6t$. The ages for humans were transformed to be equivalent to dogs. (From Wolff, R. K., in *Comparative Biology of the Normal Lung,* Vol. I, Parent, R. A., Ed., CRC Press, Boca Raton, FL, 1992, pp. 659–680. With permission.)

TABLE 8
Calculated Mucus Velocities for Tracheobronchial Airways of Humans

	Effective Mucus Clearance Velocities (mm/min)		
Airway Generation	Harley and Pasternack[101]	Lee et al.[131]	Cuddihy and Yeh[52]
1 (trachea)	15.0	5.5	5.5
2	7.9	4.1	3.5
3	2.5	3.0	2.0
4	2.5	2.2	1.1
5	0.9	1.4	0.9
6	0.9	0.88	0.9
7	0.9	0.55	0.7
8	0.24	0.34	0.6
9	0.24	0.21	0.4
10	0.23	0.13	0.3
11	0.01	0.074	0.2
12	0.01	0.044	0.1
13	0.01	0.025	0.05
14	0.01	0.015	0.02
15	0.01	0.0082	0.007
16 (terminal bronchioles)	0.01	0.0046	0.001

Alternatively, mucus clearance velocities have been estimated for individual bronchial airways using information on respiratory tract dimensions, predicted particle deposition patterns in the tracheobronchial region, and thoracic clearance measurements. The results of three sets of model calculations are presented in Table 8. All three models predict a considerable slowing of mucus velocities between the trachea and the terminal bronchioles; mucus clearance from the terminal bronchioles is calculated to be 3 to 4 orders of magnitude slower than in the trachea. Thus, as long as 2 days may be required for clearance of that portion of tracheobronchial deposition that occurs in the terminal bronchioles.[52] Calculated mucus clearance velocities of 3.5 to 8.0 mm/min for the main bronchi are two to three times higher than the measured mucus clearance velocity of 2.4 mm/min.[77]

An *in vitro* study conducted in dogs demonstrated decreasing velocities in conducting airways from the trachea to small bronchi.[6] Measured mucus velocities for the trachea were comparable to those presented in Table 7, and the slower velocities in the bronchi were somewhat proportional to the diameters of the bronchi, with the slowest velocities measured in the more distal airways. These results confirm that mucociliary velocities become slower in the smaller airways; comparable *in vitro* results are not available for humans. As was noted above for nasal airways, mucociliary velocities in the tracheobronchial region also diminish with age in healthy, nonsmoking humans;[192] however, the change is only about a factor of 2 for the age range 25 to 65 yr and would not have a major impact on clearance of inhaled materials that deposit in this region.

4.2.3 Pulmonary Region

Many measurements of pulmonary clearance have been conducted on humans and a variety of laboratory animal species. In some cases, at least two laboratory animal species were exposed to the same aerosolized material, so direct comparisons among species are possible. Few human inhalation exposures to the same materials as used for the animal studies have been performed, so only a limited number of direct comparisons are possible between laboratory animals and humans.

Table 9 summarizes selected results for pulmonary clearance of inhaled materials after single inhalation exposures to small masses of poorly soluble particles (studies of less than about 3 months' duration were not included). The variability in these results was caused by several factors. In many cases, the reported results did not allow division of the pulmonary burden between short- and long-term clearance. Also, for most studies, dissolution and absorption of the exposure materials either were not known or were not reported. The broad range of particle sizes would have influenced deposition patterns and dissolution-absorption rates but probably not the physical clearance of particles from the pulmonary region.

The information shown in Table 9 was used to approximate biological clearance rates for particles inhaled by the species listed in Table 10. In addition, approximations are included for the fractions of pulmonary burdens initially deposited in the pulmonary region that were subjected to short- or long-term clearance. These trends clearly will not apply to all types of inhaled particles. For example, in some cases, deposition and clearance may be influenced by the physicochemical and/or biological characteristics of the inhaled material; however, the generalizations that led to Table 10 allow comparisons for the consequences of chronic inhalation exposures among these animal species and humans that might not otherwise be possible.

Physical clearance patterns for pulmonary burdens of particles are similar for guinea pigs, monkeys, dogs, and humans. For these species, about 20 to 30% of the initial burden of particles clears with a half-time on the order of 1 month, the balance clearing with a half-time of several hundred days. Mice, hamsters, and rats clear about 90% of the deposited particles with a half-time of about 1 month, the remaining 10% having a half-time greater than 100 days. The relative division of the pulmonary burden between short- and long-term clearance represents a significant difference between most rodents and larger mammals and has considerable impact on long-term patterns for retention of material acutely inhaled, as well as for accumulation patterns for materials inhaled in repeated exposures (discussed below).

Some information is available concerning rates of clearance from the pulmonary region along specific pathways. Rates of macrophage-mediated clearance via the mucociliary pathway are species dependent, with small mammalian species generally exhibiting faster clearance than that exhibited by larger species, including humans.

Mechanisms associated with lymphatic transport of particles appear to be common to all mammalian species, and all types of particles appear be cleared from the pulmonary region to local lymphatics. The rates of transport and amounts of particles cleared to the thoracic

TABLE 9

Comparative Pulmonary Clearance Parameters for Poorly Soluble Particles Inhaled by Laboratory Animals and Humans

Species	Aerosol Matrix	Particle Size		Pulmonary Burden[a]				Study Duration (days)	Ref.
		μm	Measure	P_1	T_1	P_2	T_2		
Mouse	FAP	0.7	AMAD	0.93	34	0.07	146	850	219
	FAP	1.5	AMAD	0.93	35	0.07	171	850	219
	FAP	2.8	AMAD	0.93	36	0.07	201	850	219
	Pu oxide	0.38	CMD	0.88	28	0.12	230	490	9
	Pu oxide	0.2	CMD	0.86	20	0.14	460	525	9
Hamster	FAP	1.2	CMD	0.73	50	0.27	220	463	8
Rat	Diesel soot	0.12	MMAD	0.37	6	0.63	80	330	130
	FAP	1.2	CMD	0.62	20	0.38	180	492	8
	FAP	0.7	AMAD	0.91	34	0.09	173	850	219
	FAP	1.5	AMAD	0.91	35	0.09	210	850	219
	FAP	2.8	AMAD	0.91	36	0.09	258	850	219
	FAP	1.2	AMAD	0.83	33	0.17	310	365	72
	FAP	1.4	AMAD	0.76	26	0.24	210	180	73
	Fibers	1.2–2.3	AMAD	—	—	1.00	46–76	101–171	150
	Latex	3.0	CMD	0.39	18	0.61	63	190	221
	Pu oxide	<1.0	CMD	0.20	20	0.80	180	350	122
	Pu oxide	2.5	AMAD	0.75	30	0.25	250	800	206
	U_3O_8	~1–2	CMD	0.67	20	0.33	500	768	82
	Co_3O_4	2.69	MMAD	0.70	19	0.30	125	180	117
Guinea pig	FAP	2.0	AMAD	0.22	29	0.78	385	1100	220
	Diesel soot	0.12	MMAD	—	—	1.00	>2000	432	130
	Latex	3.0	CMD	—	—	1.00	83	190	221
Dog	Coal dust	2.4	MMAD	—	—	1.00	1000	160	88
	Coal dust	1.9	MMAD	—	—	1.00	~700	301–392	160
	Ce oxide	0.09–1.4	MMD	—	—	1.00	>570	140	234
	FAP	2.1–2.3	AMAD	0.09	13	0.91	440	181	25
	FAP	0.7	AMAD	0.15	20	0.85	257	850	219
	FAP	1.5	AMAD	0.15	21	0.85	341	850	219
	FAP	2.8	AMAD	0.15	21	0.85	485	850	219
	FAP	2.01	MMAD	0.05	—	0.95	910	1000	118
	Nb oxide	1.6–2.5	AMAD	—	—	1.00	>300	128	53
	Pu oxide	1–5	CMD	—	—	1.00	1500	280	11
	Pu oxide	4.3	MMD	—	—	1.00	300	300	10
	Pu oxide	1.1–4.9	MMAD	—	~1	—	400	468	161
	Pu oxide	0.1–0.65	CMD	0.10	200	0.90	1000	~4000	181
	Pu oxide	0.72	AMAD	0.10	3.9	0.90	680	730	97
	Pu oxide	1.4	AMAD	0.32	87	0.68	1400	730	97
	Pu oxide	2.8	AMAD	0.22	32	0.78	1800	730	97
	Pu oxide	4.3	MMD	0.50	20	0.50	1600	270	9
	Tantalum	4.0	AMAD	0.40	1.9	0.60	860	155	23
	U_3O_8	0.3	CMD	0.47	4.5	0.53	120	127	74
	Zr oxide	2.0	AMAD	—	—	1.0	340	128	254
Monkey	Pu oxide	2.06	CMAD	—	—	1.0	500–900	200	178
	Pu oxide	1.6	AMAD	—	—	1.0	770–1100	990	120
Human	FAP	1	CMD	0.14	40	0.86	350	372–533	7
	FAP	4	CMD	0.27	50	0.73	670	372–533	7
	Latex	3.6	CMD	0.27	30	0.73	296	~480	28
	Latex	5	CMD	0.42	0.5	0.58	150–300	160	31
	Pu oxide	0.3	MMD	—	—	1.00	240	300	109
	Graphite and PuO_2	6	AMAD	—	—	1.00	240–290	566	196
	Pu oxide	<4–5	CMD	—	—	1.00	1000	427	175
	Th oxide	<4–5	CMD	—	—	1.00	300–400	427	175
	Teflon	4.1	CMD	0.30	4.5–45	0.70	200–2500	300	187
	Zr oxide	2.0	AMAD	—	—	1.00	224	261	254

TABLE 9 (continued)

Comparative Pulmonary Clearance Parameters for Poorly Soluble Particles Inhaled by Laboratory Animals and Humans

Note: FAP = fused aluminosilicate particles; AMAD = activity median aerodynamic diameter; MMAD = mass median aerodynamic diameter; CMD = count median diameter; MMD = mass median diameter. Some aerosols were monodisperse, but most were polydisperse, with geometric standard deviations in the range of 1.5 to 4. Clearance half-times are approximations for biological clearance, the net result of dissolution-absorption and physical clearance processes. In some examples, the original data were subjected to a computer curve-fit procedure to derive the values for P_1 and T_1 presented in this table.

[a] Pulmonary burden = $P_1 \cdot e^{(-\ln 2)t/T_1} + P_2 \cdot e^{(-\ln 2)t/T_2}$, where P_1 and P_2 are fractions of pulmonary burden in fast and slow clearing components constrained to total 1.00, T_1 and T_2 equal clearance half-times for P_1 and P_2 in days, and t equals days after exposure.

TABLE 10

Average Pulmonary Clearance Parameters for Poorly Soluble Particles Inhaled by Selected Laboratory Animal Species and Humans

	Pulmonary Clearance Parameters[a]			
Species	P_1	T_1	P_2	T_2
Mouse	0.9	30	0.1	240
Rat, hamster	0.9	25	0.1	210
Guinea pig	0.2	29	0.8	570
Monkey, dog, human	0.3	30	0.7	700

[a] Pulmonary burden (fraction of initial deposition) = $P_1 \cdot e^{(-\ln 2)t/T_1} + P_2 \cdot e^{(-\ln 2)t/T_2}$, where:

P_1 and P_2 = fractions of pulmonary burden in fast and slow-clearing components
T_1 and T_2 = clearance half-times (days) for P_1 and P_2
t = time in days after an acute inhalation exposure

lymph nodes, however, depend to a significant extent on factors that include species, physical/chemical properties of the particles, occurrence of biological reactions to the particles, and the amounts of particles deposited in the pulmonary region.

Rates for particle translocation from the pulmonary region to lymph nodes appear to vary considerably among species. Rats and mice have particle translocation rates that are quite different from those of guinea pigs, dogs, and possibly humans. Translocation begins soon after an acute inhalation exposure, but after a few days the transport of particles from the pulmonary region to the lymph nodes appears to be negligible in mice and rats[219] and continues at a constant rate in guinea pigs and dogs.[219,222] While no experimental information is available about the rates of translocation in humans, data for amounts of particles accumulated in the lungs of humans exposed repeatedly to dusty environments[49,50,144,229] suggest that poorly soluble particles accumulate in lymph nodes of humans at rates that may be comparable to those observed for guinea pigs, dogs, and monkeys.

Physical movement of particles from the pulmonary region to the thoracic lymph nodes affords the opportunity to transport particles out of the lung, but the result often is sequestering, or trapping, the particles in what is generally perceived to be a dead-end compartment.

Because the lymph nodes represent traps for particles cleared from the lung, particles can accumulate to high concentrations. Translocation of particles from the pulmonary region to the nodes results in concentrations of particles in the nodes that can be more than two orders of magnitude higher than concentrations in the lung.[244,245]

Exposure conditions can influence the translocation of material to the lymph nodes. Low exposure concentrations of insoluble particles favor a primary removal pathway for particle-laden macrophages via the mucociliary transport system, at least in rats; however, with increasing exposure levels, there is an increase in the translocation of free particles to the pulmonary lymphatic system.[18,70,168,232,252] Translocation from the pulmonary region to lymph nodes is independent of the type of particle or fiber, unless the constituents are cytotoxic. All inhalation exposures to respirable particles or fibers result in the potential for interstitialization of some portion of the deposited dust. The larger the amount of deposited dust, or the larger the numbers of dust particles deposited in the lung, the greater the potential for the dust to penetrate into the interstitium and be available for transport to the nodes. The amount of dust that penetrates into the pulmonary interstitium and is subsequently transported to lymph nodes increases as the pulmonary burden increases, but the relationships between amounts of material in the lung and transport rates have not been fully defined and may depend on the chemical and physical characteristics of the specific particles.

5 RETENTION OF PARTICLES

5.1 Influence of Particle Size

A broad range of particle sizes may be inhaled and deposited in the respiratory tract, so it is relevant to know whether or not particle size is an important factor in physical clearance processes. Any relationship between particle size and physical clearance processes has not been clearly demonstrated. Macrophages can phagocytize a broad spectrum of particle sizes and appear to function normally until they have accumulated amounts of particles that cause them to become immobile. The amount of phagocytized material that can immobilize a macrophage presumably is the same whether the material is comprised of one particle or fiber, or several; however, an important consideration with toxic materials is the type and amount of the toxic material phagocytized by the macrophage. Macrophages could be damaged to the extent that their functional abilities become impaired long before they are immobilized, because of the mass of phagocytized material.[90,148] Fibers have physical attributes that make mass loading a secondary issue. Long fibers cannot be completely phagocytized by macrophages, whereas short fibers can; therefore, the biokinetics of fibers are more complex than for particles and are markedly influenced by fiber size.

The effect of particle size on physical clearance from the pulmonary region has been the subject of several studies. Physical clearance of spherical particles in the size range of about 0.3- to 0.4-μm geometric diameter was not size selective.[41,119,149,161,167,219] Results using larger particles yielded different results; the larger particles may have been phagocytized by macrophages but were not efficiently transported within the lung.[218,220,240,241] Additionally, physical transport from the pulmonary region to lymph nodes decreased as the particle size increased, dropping dramatically for particle sizes larger than about 5 μm.[218,220,241]

In contrast to physical clearance of regularly shaped particles from the pulmonary region, physical clearance of fibers has been demonstrated to be size dependent.[127,152,153,201,202,217,247] Fibers less than about 5 μm were preferentially cleared, as compared with fibers longer than about 20 μm. Studies evaluating fibers that have been removed from the lung in broncho-alveolar lavage fluid support the concept of preferential clearance of short fibers from the deep lung.[112,243]

Dissolution-absorption of fibers in rat lungs, which was determined on the basis of changes in the size distributions of the fibers over time, was dependent on both fiber size and composition. Morgan et al.[151] attributed a dependency of dissolution on fiber length to differences in pH encountered by the fibers. The shorter fibers retained within macrophages were presumed to be exposed to a lower pH than nonphagocytized fibers in extracellular fluid. These results indicate that physical and chemical attributes of the fibers were important factors in processes that dissolve or etch them. Additionally, most fibers found in lymph nodes were less than 10 μm long and present in macrophages as single fibers.[125,126] This was a clear demonstration that biological action *in vivo* reduced the more labile types of fibers to sizes which had biokinetics resembling moderately soluble particles and showed that the subunits of fibers could be physically translocated to lymph nodes.

5.2 Retention in Conducting Airways

Most particles that deposit in the upper respiratory tract (head airways) clear rapidly in all mammalian species, but a small percentage of particles may be retained for long time periods.[219,221,235,259] In mice, rats, and dogs exposed by inhalation to monodisperse or polydisperse ^{134}Cs-labeled fused aluminosilicate particles, 0.001 to 1% of the initial internally deposited burden of particles was retained in the head airways and was removed only by dissolution-absorption.[219] Retained particles were in close proximity to the basement membrane of nasal airway epithelium. In another study,[221] 3-, 9-, and 15-μm latex microspheres were inhaled by rats and guinea pigs. About 1 and 0.1% of all the sizes of microspheres were retained in the head airways of the rats and guinea pigs, respectively. For rats, the 9- and 15-μm microspheres cleared with half-times of 23 days; for guinea pigs, the same microspheres cleared with half-times of about 9 days. The 3-μm microspheres were cleared from the head airways of the rats and guinea pigs with biological half-times of 173 and 346 days, respectively. The smaller particles are apparently more likely to penetrate the epithelium and reach long-term retention sites. In dogs, 3-μm polystyrene latex particles instilled onto the epithelium of the maxillary and ethmoid turbinates were retained at both sites after 30 days to the extent of 0.1% of the amount initially deposited.[259] Particles were retained in the epithelial submucosa of both regions.

It is generally concluded that most inhaled particles that deposit in the tracheobronchial region clear within hours or days; however, small portions (generally < 5%) of particles that deposit in, or are cleared through, the tracheobronchial region are retained with half-times on the order of weeks or months.[92,93,183,228] These particles were incorporated into the airway epithelium. Rapid phagocytosis of particles deposited on airway surfaces after inhalation or by other means of exposure suggests that macrophages could be involved in particle transport into airway epithelium.

Stahlhofen et al.[226,227] conducted inhalation studies with humans to assess directly the deposition and retention of poorly soluble particles that deposit in the tracheobronchial region. Human subjects inhaled small volumes of aerosols using procedures that theoretically allowed deposition to occur at specific depths in the tracheobronchial region but not in the pulmonary region. Results of those studies suggested that as much as 50% of the particles that deposited in the tracheobronchial region clear slowly, presumably because they become incorporated into the airway epithelium. Smaldone et al.[215] reported the results from gamma camera imaging analyses of aerosol retention in normal and diseased human subjects, and also suggested that particles deposited on central airways of the human lung do not completely clear within 24 hours. There have also been a few reports indicating that poorly soluble particles associated with cigarette smoke are retained in the epithelium of the tracheobronchial tree of humans.[48,136,195] The cumulative results of these studies strongly suggest that a portion of particles that deposit on the conducting airways can be retained for long periods of time, or indefinitely.

TABLE 11

Physical Clearance Parameters[a] for Modeling Pulmonary Clearance of Particles Inhaled by Humans and Selected Animal Species

Species	Clearance via Mucociliary Transport Pathway	Clearance to Thoracic Lymph Nodes
Mouse[b]	$0.023\ \exp^{-0.008t} + 0.0013$	$0.0007\ \exp^{-0.5t}$
Rat,[b] hamster[c]	$0.028\ \exp^{-0.01t} + 0.0018$	$0.0007\ \exp^{-0.5t}$
Guinea pig[b]	$0.007\ \exp^{-0.03t} + 0.0004$	0.00004
Monkey,[d] dog,[b] human[d]	$0.008\ \exp^{-0.022t} + 0.0001$	0.0002

[a] Fraction of existing pulmonary burden physically cleared per day.
[b] Adapted from Snipes.[222]
[c] Clearance parameters assumed to be the same as for rats.
[d] Clearance parameters assumed to be the same as for dogs.

5.3 Model Projections for Retention of Particles in the Respiratory Tract

5.3.1 Conducting Airways

Insufficient data are available to model adequately the retention of particles deposited in the conducting airways of any mammalian species. It is probable that some portion of particles that deposit in the head airways and tracheobronchial region during an inhalation exposure is retained for long times and represents significant dosimetry concerns. Additionally, some of the particles that are cleared from the pulmonary region via the mucociliary transport pathway may become trapped in the tracheobronchial epithelium during their transit through the airways.

5.3.2 Pulmonary Region and Thoracic Lymph Nodes

Model projections are possible for the pulmonary region and thoracic lymph nodes using information relevant to deposition, retention, and clearance of inhaled particles. Table 11 summarizes physical pulmonary clearance parameters for humans and six laboratory animal species. Pulmonary clearance curves produced using the parameters in Table 11 agree with curves produced using the parameters in Table 10. An advantage to using the parameters in Table 11 is that they separate physical clearance from the pulmonary region into its two components: clearance to the gastrointestinal tract and clearance to lymph nodes. To model the pulmonary biokinetics of a specific type of particle, the physical clearance parameters in Table 11 were integrated with a dissolution-absorption parameter to derive rates for effective clearance from the pulmonary region and for particles translocated to the lymph nodes. Available information suggests that the dissolution-absorption rate for a specific type of particle is the same if the particle is in the pulmonary region or in a lymph node; therefore, the same values for dissolution-absorption parameters were used for lungs and lymph nodes in this modeling effort.

Modeling was done using three distinct dissolution-absorption half-times, indicative of particles that are relatively soluble (10-day half-time), moderately soluble (100-day half-time), and poorly soluble (1000-day half-time). Results of the modeling effort are presented for humans in Figure 9, and for rats in Figure 10. Model projections for mice and hamsters would be very similar or identical to those for rats; model projections for guinea pigs, monkeys, and dogs would be the same as for humans, with the exception that lymph node burdens

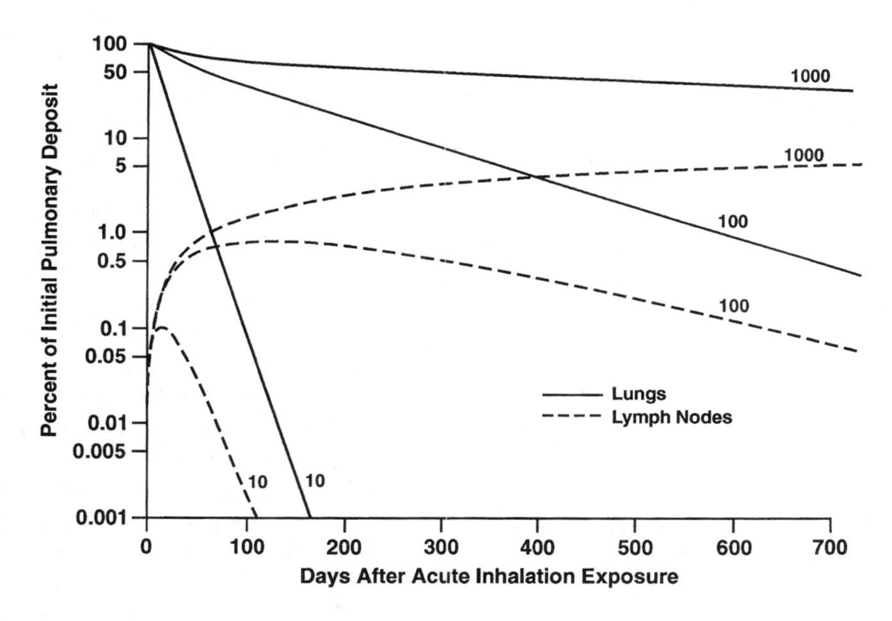

FIGURE 9
Predicted pulmonary and thoracic lymph node burdens of particles in humans as a function of time after acutely inhaling particles having dissolution-absorption half-times of 10, 100, or 1000 days.

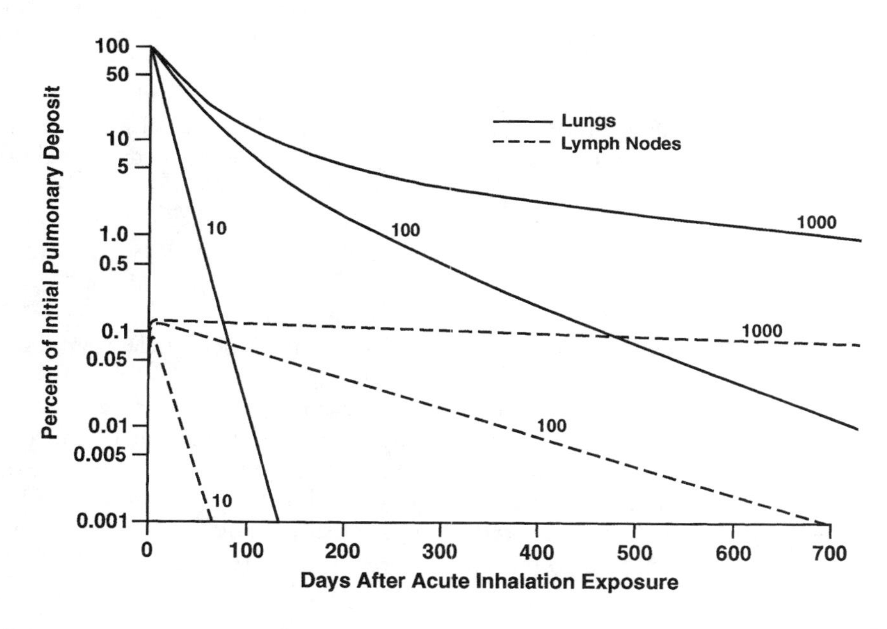

FIGURE 10
Predicted pulmonary and thoracic lymph node burdens of particles in rats as a function of time after acutely inhaling particles having dissolution-absorption half-times of 10, 100, or 1000 days.

would be lower for guinea pigs. The model projections in Figures 9 and 10 clearly demonstrate the importance of dissolution-absorption rates on retention of particles deposited acutely in the pulmonary region. Additionally, the model projections demonstrate significant species differences in the amounts of particles cleared from the pulmonary region to the lymph nodes.

5.4 Particle Overloading

When small amounts of poorly soluble materials are deposited in the lungs, macrophage-mediated pulmonary clearance removes the particles at a rate sufficient to prevent substantial accumulations of particles to occur. Even repeated or chronic exposures at low exposure rates do not overwhelm the functional abilities of macrophages; however, with exposures to large amounts of respirable particles, macrophage-mediated pulmonary clearance is overwhelmed and altered, to the extent that the pulmonary burden of particles increases progressively. The lungs are in a condition termed "lung overload".

The concept of altered pulmonary clearance has been discussed to various degrees by numerous investigators over the past several decades. For example, an early suggestion was that a mass of particles in the rat lung greater than 1.5 mg would significantly alter or block pulmonary clearance mechanisms.[121] While it was recognized that large pulmonary burdens of dust could alter pulmonary clearance, Bolton et al.[29] were the first to develop a clearance overload hypothesis, which was discussed in detail by Morrow.[163-165] Morrow reviewed early and contemporary work related to lung overload and concluded that the amount of dust associated with overloading of rat lungs appears to be about 1 mg particles per gram of lung.

Lung tumors have been noted in several studies in which rats chronically inhaled large amounts of poorly soluble types of respirable particles;[142,143] however, with the exception of quartz, which is a toxic dust, rats that developed lung tumors had accumulated pulmonary burdens of several milligrams of particles per gram of lung or more during chronic exposures equivalent to at least 121 $mg{\cdot}hr/m^3{\cdot}wk$. As an example, this level of exposure could be achieved by chronically exposing rats 8 hr/day, 5 days/wk to about 3 mg particles per m^3. It is important to note that altered pulmonary clearance has been reported for other mammalian species, but the complete spectrum of nonspecific pulmonary responses associated with lung overload may be restricted to rats.

The relevance of lung overload to other laboratory animal species and humans is unclear. Mauderly[143] summarized information available from eight studies of coal miners in the U.S. and Europe that included a total of 1225 subjects. Substantial lung burdens of coal dust were measured after the coal miners were autopsied, and the overall average specific pulmonary burden was between about 7 and 14 mg dust per gram lung. Specific pulmonary burdens of coal dust in these human subjects were sufficiently high to produce lung tumors if the human pulmonary response to accumulations of inhaled particles parallels the well-documented responses reported for rats.[143]

The exposure conditions or accumulated pulmonary burdens of particles that cause altered clearance and lung overload have not yet been conclusively determined. One of two exposure scenarios appears to trigger lung overload. The first scenario involves repeated or chronic exposures to poorly soluble materials until a critical pulmonary burden of the material is reached. This scenario relates to the hypothesis that normal pulmonary clearance exists until a critical burden of material has accumulated; this critical burden then triggers the biological changes in the lung that are associated with lung overload.

Alternatively, the amount of poorly soluble material deposited daily in the lungs may be the dominant factor in lung overload. Muhle et al.[168] proposed that overloading was primarily a function of the rate of pulmonary deposition. Pulmonary clearance was retarded by exposure concentrations of 3 mg/m^3 or higher of respirable particles. The lowest pulmonary burdens associated with reduced rates of pulmonary clearance were 0.8 mg of fly ash or 0.93 mg of glass fibers. In summarizing results from several chronic inhalation studies with rats in which pulmonary clearance had been evaluated during exposures to a variety of poorly soluble aerosols for long periods of time, it was concluded that chronic exposures of rats to a weekly average of 10 mg/m^3 for 1 yr would result in lung overload.[190] This could be the result of a continuous exposure to 0.2 mg particles per m^3 or to 0.8 mg/m^3, 8 hr/day, 5 days/wk.

Yu et al.[269] described their concept of "critical deposition rate", which was based on mathematical analyses of pulmonary clearance rates from several chronic inhalation studies.

They described critical deposition rate as the "… rate above which the overload condition will be present if the exposure time is sufficient." The higher the inhaled concentration of particles, the shorter the time required to reach the overload condition.

The concept of critical deposition rate can also be described independent from pulmonary burden. An alternative definition of critical deposition rate is the deposition rate above which macrophage-mediated pulmonary clearance is overwhelmed. This concept of lung overload can apply to early times after initiation of repeated or chronic inhalation exposures to poorly soluble respirable materials. With repeated or chronic inhalation of relatively small amounts of material, macrophages are able to phagocytize most of the particles within hours of the exposure; some of the particles become incorporated into fixed tissue constituents of the lung, and some particles are cleared via the mucociliary transport pathway. Under these conditions, pulmonary clearance can occur at a normal rate, even though the lung is accumulating a sequestered burden of poorly soluble particles. If the daily deposition rate of particles increases, numbers of macrophages in the lung become elevated as a result of recruitment.[33] At some exposure rate (perhaps the critical deposition rate), the capacity of macrophages is overwhelmed. Additionally, elevated numbers of neutrophils are present in the lung when exposure rates are high, which may be one consequence of overwhelming the functional abilities of macrophages. This may represent the exposure-response scenario that triggers lung overload, which potentially could occur very soon after initiating a chronic inhalation study if the exposure concentration of respirable particles is sufficiently high.

Altered pulmonary clearance, inflammation, and lung tumors are notable exposure-related biological effects that are seen in rats exposed chronically to high concentrations of poorly soluble particles. These biological responses have been noted specifically in studies with rats exposed to diesel soot, carbon black, and titanium dioxide (TiO_2), and it is probable that the same or similar responses result from repeated or chronic exposures to large amounts of other kinds of respirable particles, as well as fibers. The phenomenon of particle overloading in the lungs of other mammalian species continues to be evaluated, and the importance of this phenomenon for human exposures to poorly soluble particles must be determined to provide a firm basis for using information from rodent studies that resulted in pulmonary overload in developing risk evaluations for humans.

5.5 Normalizing Exposures in Different Species to Equalize Retention

An important issue in inhalation toxicology is the relationship between an inhalation exposure and the resulting pulmonary burden of exposure material achieved in the human lung vs. the lungs of other laboratory animal species. It is generally assumed that the magnitude of the pulmonary burden of particles produced during an inhalation exposure is an important determinant of biological responses to the inhaled particles. Therefore, understanding the basis for differences in pulmonary burdens among species resulting from well-defined inhalation exposures will provide a better understanding of pulmonary burdens that would result from exposures of various mammalian species to the same particles. Alternatively, the exposure conditions could be tailored for each species to produce desired pulmonary burdens of particles.

The broad spectrum of mammals used in inhalation toxicology research have body weights ranging upwards from a few grams to hundreds of kilograms; these mammals also exhibit a broad range of respiratory parameters. Table 12 lists body weights, lung weights, and respiratory minute volumes for humans and selected laboratory animal species. Important variables for inhalation toxicology are lung weight and ventilation parameters, which dictate the amounts of inhaled materials potentially deposited in the lung, as well as the specific pulmonary burden that will result from the inhalation exposure. The inverse relationship between body size and metabolic rate is demonstrated by the values for the respiratory minute volume per gram of lung (column 7 of Table 12). For example, liters of air inhaled per minute per gram

TABLE 12

Body Weight, Lung Weight, and Respiratory Minute Volume for Selected Laboratory Animals[a] and Humans[b]

Species	Body Weight (kg)	Lung Weight (g)	Lung Weight % Body Weight	Minute Volume (l/min)	Minute Volume per kg Body Weight (l/min/kg)	Minute Volume per g Lung (l/min·g lung)
Mouse	0.03	0.2	0.67	0.03	1.0	0.15
Hamster	0.12	0.8	0.67	0.11	0.92	0.14
Rat	0.23	1.3	0.57	0.16	0.70	0.12
Guinea pig	0.70	4.0	0.57	0.26	0.37	0.065
Monkey	2.45	22	0.90	0.70	0.29	0.032
Dog	10	110	1.1	3.4	0.34	0.031
Adult man, resting	70	1000	1.4	7.5	0.11	0.0075
Adult man, light activity	70	1000	1.4	20	0.29	0.020

[a] Approximations for body weight, respiratory minute volume, and lung weight for laboratory animals were adapted from Phalen[186] and the cumulative published and unpublished data from the Inhalation Toxicology Research Institute. These approximations are assumed reasonable for animals in exposure scenarios but are likely to depend on levels of activity, sedation, anesthesia, or confinement.

Source: Adapted from Schlesinger, R. B., *J. Toxicol. Environ. Health,* 15, 197–214, 1985; Snyder, W. S. et al., *International Commission on Radiological Protection, No. 23: Report of the Task Group on Reference Man,* Pergammon Press, Oxford, 1975.

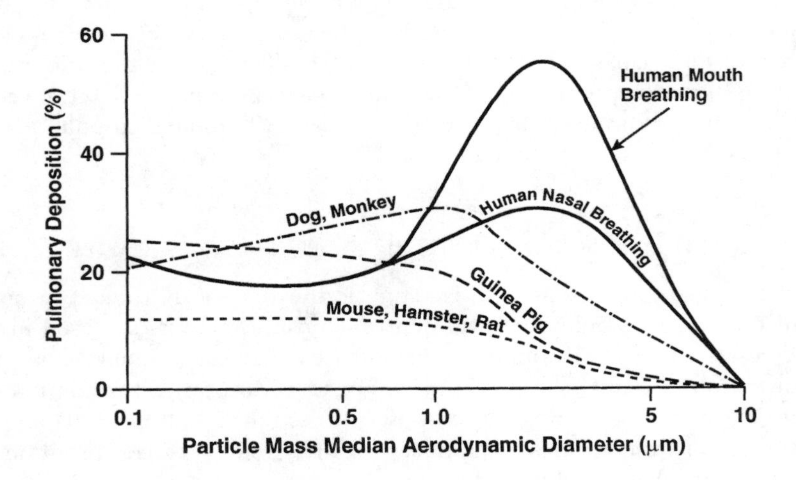

FIGURE 11
Pulmonary deposition of particles inhaled by the mouse, hamster, rat, guinea pig, monkey, dog, and human as a function of particle aerodynamic diameter.

of lung are about 20 times higher for resting mice than for resting humans, which is an important factor to consider relative to potential amounts of particles deposited in the respiratory tract per unit of time during inhalation exposures, as discussed earlier.

Another important factor is pulmonary deposition, defined here as the fraction of inhaled material that deposits in the pulmonary region. Pulmonary deposition fractions as a function of particle aerodynamic diameter are presented in Figure 11, and representative values are presented in Table 13. The range of particle sizes presented in this figure and table encompasses most respirable sizes of particles that might be encountered in the workplace and general environment.

TABLE 13

Fraction of Inhaled Particles Deposited in the Pulmonary Region as a Function of Particle Aerodynamic Diameter for Humans and Selected Animal Species

	Pulmonary Deposition Fraction as a Function of Particle Aerodynamic Diameter (μm)							
Species	0.1	0.5	1	2	3	5	8	12
Mouse, hamster, rat	0.12	0.12	0.10	0.07	0.04	0.02	0.01	0
Guinea pig	0.25	0.23	0.20	0.12	0.06	0.03	0.01	0
Monkey, dog	0.21	0.24	0.30	0.24	0.18	0.08	0.03	0.01
Human, nose-breathing	0.23	0.18	0.23	0.30	0.30	0.18	0.06	0.01
Human, mouth-breathing	0.23	0.18	0.30	0.52	0.55	0.25	0.07	0.01

TABLE 14

Predicted Rates of Pulmonary Deposition for Particles Inhaled by Selected Animal Species and Humans with an Aerosol Concentration of 1 mg Particles per m^3

	Calculated Pulmonary Deposition Rates (ng/min·g lung) for Particle Aerodynamic Diameter (μm)							
Species	0.1	0.5	1	2	3	5	8	12
Mouse	18.0	18.0	15.0	11.0	6.0	3.0	1.5	0
Hamster	17.0	17.0	14.0	9.8	5.6	2.8	1.4	0
Rat	14.0	14.0	12.0	8.4	4.8	2.4	1.2	0
Guinea pig	16.0	15.0	13.0	7.8	3.9	2.0	0.7	0
Monkey	6.7	7.7	9.6	7.7	5.8	2.6	1.1	0.3
Dog	6.5	7.4	9.3	7.4	5.6	2.5	1.1	0.3
Human, nose-breathing[a]	4.6	3.6	4.6	6.0	6.0	3.6	1.2	0.2
Human, mouth-breathing[a]	4.6	3.6	6.0	10.0	11.0	5.0	1.4	0.2

[a] Light activity, minute volume 20 l/min.

The data in Tables 12 and 13 were used to produce the predicted values for rate of pulmonary deposition as a function of particle size shown in Table 14. The aerosol concentration used for these calculations was arbitrarily designated 1 mg/m^3, or 1 μg/l. A sample calculation for a mouse exposed to this aerosol, having 0.1 μm particles, is

$$(0.15\ \text{l/min·g lung}) \times (0.12\ \text{deposition}) \times (1\ \mu\text{g/l}) = 18\ \text{ng/min} \times \text{g lung}$$

Table 14 clearly demonstrates the different rates of pulmonary deposition among these seven mammalian species. The combined effects of minute volume per gram lung and species-dependent pulmonary deposition mean that the rates of particle deposition in the pulmonary region per gram of lung (e.g., ng particles per min·g lung) can range from 0 to 18 ng/min·g lung for particles in the size range of 0.1 to 12 μm and an aerosol concentration of 1 mg/m^3.

The rates for particle deposition in humans for nose-breathing and mouth-breathing were averaged for each particle size, and the result was assumed to be representative for human inhalation exposures to that specific particle size. The average values for pulmonary deposition rates for the human were next divided by the rates for the other six mammals for the same particle size. The results are presented in Figure 12. The interpretation of this figure is that

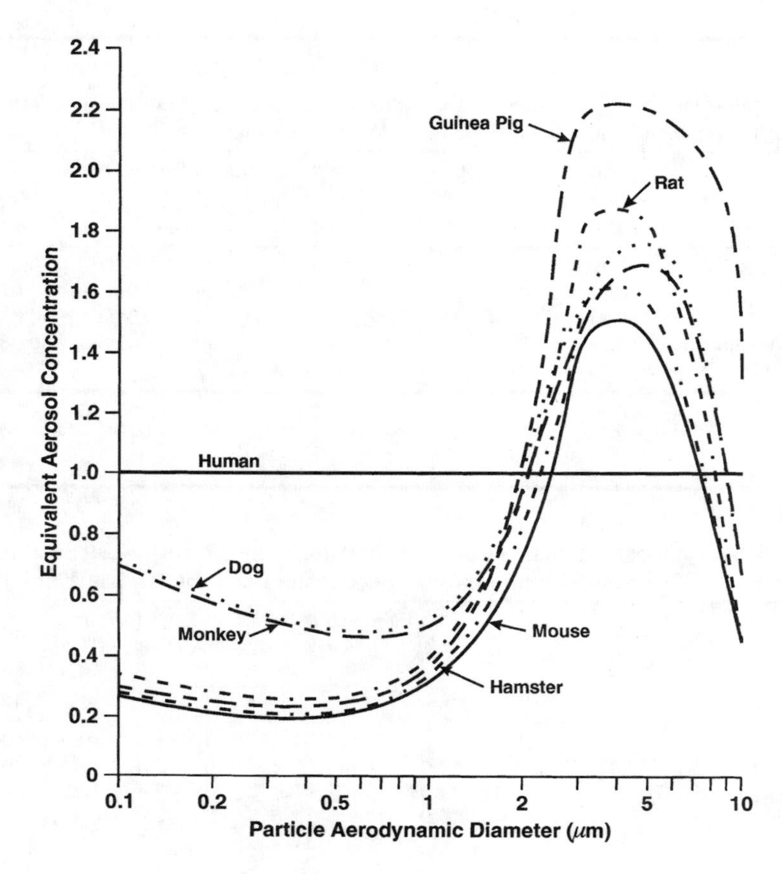

FIGURE 12
Equivalent aerosol concentrations as a function of particle aerodynamic diameter for the mouse, hamster, rat, guinea pig, monkey, and dog. This indicates the aerosol concentration that would yield the same pulmonary deposition rate (amount of particles per min·g lung) relative to the deposition rate for a standard human exposed by inhalation to the same aerosol.

a defined aerosol would produce a pulmonary deposition rate (e.g., ng particles per min·g lung) in the human lung that would depend on particle aerodynamic diameter. The same pulmonary deposition rate could be achieved in the other six mammalian species by adjusting the aerosol concentration. The aerosol concentration that would produce the same rate of pulmonary deposition as in the human is defined as the "equivalent aerosol concentration" and depends strongly on particle aerodynamic diameter. For example, to achieve the same lung-weight-normalized rate of pulmonary deposition of 0.5-μm particles in the dog as in the human, the dog would be exposed to 0.5 times the aerosol concentration of concern for human exposures. The aerosol concentration would have to be about 1.7 times higher for dogs than for humans if the exposure atmosphere contained 5 μm particles. This figure summarizes the relationship between aerosol concentrations and particle sizes that allows adjusting exposure conditions to achieve comparable pulmonary burdens of particles for defined acute exposure conditions or to help understand the extent to which they may be different. Note that no particle size yields the same particle deposition rate for all seven species shown, but particles having about 2-μm aerodynamic diameter would yield the most similar deposition rates among these species.

Predictable deposition, retention, and clearance patterns are possible for acute inhalation exposures of laboratory animal species and humans. Repeated exposures also occur for humans

TABLE 15

Summary of Common and Specific Exposure Parameters Used for Predicting Pulmonary and Thoracic Lymph Node Burdens of Particles in Mice, Rats, Hamsters, Guinea Pigs, Monkeys, Dogs, and Humans

Common Parameters	
Exposure atmosphere	1.0 mg particles/m^3
Particle aerodynamic diameter	0.1, 0.5, 1, 2, or 5 μm
Particle dissolution-absorption half-time	10, 100, or 1000 days
Chronic inhalation exposure pattern	8 hr/day, 5 days/wk
Duration of chronic exposure	4, 13, or 104 wk

	μg Particles Deposited in Pulmonary Region (per day·g lung) for Particle Aerodynamic Diameter (μm)				
Species	0.1	0.5	1	2	5
Mouse	8.64	8.64	7.20	5.28	1.44
Hamster	8.16	8.16	6.72	4.70	1.34
Rat	6.72	6.72	5.76	4.03	1.15
Guinea pig	7.68	7.20	6.24	3.74	0.96
Monkey	3.22	3.70	4.61	3.70	1.25
Dog	3.12	3.55	4.46	3.55	1.20
Human[a]	2.21	1.73	2.55	3.84	2.06

[a] Light activity, breathing 50% through the nose and 50% through the mouth.

and are used routinely in laboratory animals to study the inhalation toxicology of a broad spectrum of hazardous and potentially hazardous particles. The predicted biokinetics of particles acutely inhaled can be readily extrapolated to repeated exposures; however, the predictions become increasingly questionable as exposure conditions deviate away from those used for acute inhalation exposures. As indicated above, pulmonary clearance rates can be altered during exposures of several hours per day and several days per week for extended periods of time. Unfortunately, there has been no clear demonstration of inhalation exposure conditions that trigger the physiological responses in the lung that are associated with altered pulmonary clearance. The following predictions for repeated inhalation exposures are, therefore, intended to be comparative, rather than absolute, and were made using the assumption that physical clearance parameters for the pulmonary region are the same for acute and repeated inhalation exposures.

Table 15 summarizes common and specific parameters used for predicting pulmonary burdens for exposures of humans and six laboratory animal species to an aerosol containing 1.0 mg particles per m^3. Five particle aerodynamic diameters were selected, ranging from 0.1 to 5 μm, to represent particle sizes typical in the environment or used experimentally. Daily pulmonary deposition was expressed in units of μg particles per gram lung to normalize deposition rates among the species. The normalized daily deposition rates were considered appropriate for making direct comparisons among the species in terms of the specific pulmonary burdens that result from repeated exposures and can be extrapolated to amounts of particles in the total lung, using lung weight data available in Table 12. Particle dissolution-absorption rates were also varied; half-times of 10, 100, and 1000 days were used to simulate particles that are relatively soluble, moderately soluble, and poorly soluble, respectively. The pulmonary clearance parameters used for predicting the results of repeated exposures were the same as the ones used for predicting the consequences of acute inhalation exposures and are provided in Table 11. Predictions also considered exposure duration, which was fixed at

TABLE 16

Pulmonary Burdens of Particles for Inhalation Exposures 8 hr/day, 5 days/wk to 1.0 mg Particles per m^3, with Particle Dissolution-Absorption Half-Time of 10 days[a]

Species	Weeks of Exposure	μg of Particles per g Lung for Particle Aerodynamic Diameter (μm) 0.1	0.5	1	2	5
Mouse	4	64	64	53	39	11
	13	70	70	58	42	12
Hamster	4	58	58	48	33	9.6
	13	63	63	52	36	10
Rat	4	48	48	41	29	8.2
	13	52	52	44	31	8.8
Guinea pig	4	66	62	54	32	8.3
	13	76	71	61	37	9.5
Monkey	4	27	32	39	32	11
	13	31	36	45	36	12
Dog	4	27	30	38	30	10
	13	30	34	43	35	12
Human[b]	4	19	15	22	33	18
	13	21	17	25	37	20

[a] Equilibrium pulmonary burdens of particles are achieved in about 50 days for all seven species; therefore, values for 104 wk are predicted to be the same as predicted for 13 wk.

[b] Light activity, breathing 50% through the nose and 50% through the mouth.

8 hr/day, 5 days/wk, with endpoints of 4, 13, and 104 weeks to simulate typical study times used in contemporary inhalation toxicology programs. Tables 16 to 21 present the predictions for specific pulmonary burdens and lymph node burdens of particles for these simulated repeated exposures. The results demonstrate the importance of respiratory, deposition, and physical clearance parameters, as well as the dissolution-absorption characteristics of the inhaled particles. This combination of factors results in significant species differences in pulmonary accumulation patterns of inhaled particles during the course of 4-, 13-, and 104-wk repeated exposures which must be considered in experiments designed to achieve equivalent pulmonary burdens or in evaluating the results of inhalation exposures of different mammalian species to the same aerosolized test materials.

Another comparison of the results of the projections for pulmonary burdens of particles is presented in Tables 22 to 24. These tables show predicted relative air concentrations of particles that, when inhaled by these six laboratory animal species and humans, produce the same specific pulmonary burdens of the particles. A substantial range of relative exposure concentrations is required to produce the same specific pulmonary burdens in these mammalian species, and the relative exposure concentrations are different for different exposure times.

Some of these tables and figures were intentionally normalized to pulmonary deposition rates or specific pulmonary burdens of particles predicted for humans; however, the data in the tables and figures can also be normalized to an alternative species — rats, for example. To further this example, assume an inhalation study is being designed to include rats and mice. The information in this section can be used to define exposure concentrations that will achieve the same specific pulmonary burdens of particles in the rats and mice after 4, 13, or 104 wk. Alternatively, the extent of differences in the specific pulmonary burdens of particles after repeated exposures can be calculated using this information.

TABLE 17

Pulmonary Burdens of Particles for Inhalation Exposures 8 hr/day, 5 days/wk to 1.0 mg Particles per m^3, with Particle Dissolution-Absorption Half-Time of 100 days[a]

Species	Weeks of Exposure	μg of Particles per g Lung for Particle Aerodynamic Diameter (μm)				
		0.1	0.5	1	2	5
Mouse	4	119	119	99	73	20
	13	198	198	165	121	33
	104	245	245	206	150	41
Hamster	4	106	106	87	61	17
	13	167	167	137	96	27
	104	204	204	168	117	33
Rat	4	87	87	75	52	15
	13	137	137	118	82	24
	104	168	168	144	101	29
Guinea pig	4	130	122	105	63	16
	13	311	292	253	152	39
	104	612	574	497	298	76
Monkey	4	54	62	77	62	21
	13	125	144	179	144	49
	104	240	276	344	276	93
Dog	4	52	59	74	59	20
	13	121	138	173	138	47
	104	233	265	333	265	90
Human[b]	4	37	29	43	64	34
	13	86	67	99	149	80
	104	165	129	191	287	154

[a] Equilibrium pulmonary burdens of particles are achieved in about 6 months for mice, hamsters, and rats and after about 1 year for guinea pigs, monkeys, dogs, and humans.
[b] Light activity, breathing 50% through the nose and 50% through the mouth.

TABLE 18

Pulmonary Burdens of Particles for Inhalation Exposures 8 hr/day, 5 days/wk to 1.0 mg Particles per m^3 with Particle Dissolution-Absorption Half-Time of 1000 days[a]

Species	Weeks of Exposure	μg of Particles per g Lung for Particle Aerodynamic Diameter (μm)				
		0.1	0.5	1	2	5
Mouse	4	128	128	107	78	21
	13	243	243	202	148	40
	104	413	413	344	253	69
Hamster	4	114	114	94	66	19
	13	203	203	167	117	33
	104	333	333	274	192	55
Rat	4	94	94	81	56	16
	13	167	167	143	100	29
	104	274	274	235	164	47
Guinea pig	4	141	132	114	69	18
	13	406	381	330	198	51
	104	2190	2050	1780	1070	274
Monkey	4	58	67	83	67	23
	13	163	187	233	187	63
	104	858	985	1230	985	333
Dog	4	56	64	81	64	22
	13	158	179	225	179	61
	104	831	946	1190	946	320
Human[b]	4	40	31	46	69	37
	13	112	87	129	194	104
	104	589	461	679	1020	549

[a] Equilibrium pulmonary burdens of particles are achieved in about 18 months for mice, hamsters, and rats but are not achieved for guinea pigs, monkeys, dogs, and humans during 2 years of exposure.

[b] Light activity, breathing 50% through nose and 50% through mouth.

TABLE 19

Thoracic Lymph Node (TLN) Burdens of Particles for Inhalation Exposures 8 hr/day, 5 days/wk to 1.0 mg Particles per m^3 with Particle Dissolution-Absorption Half-Time of 10 days[a]

Species	Weeks of Exposure	μg of Particles in TLNs per g Lung for Particle Aerodynamic Diameter (μm)				
		0.1	0.5	1	2	5
Mouse	4	0.09	0.09	0.08	0.06	0.01
	13	0.11	0.11	0.09	0.07	0.02
Hamster	4	0.09	0.09	0.07	0.05	0.01
	13	0.10	0.10	0.08	0.06	0.02
Rat	4	0.07	0.07	0.06	0.04	0.01
	13	0.08	0.08	0.07	0.05	0.02
Guinea pig	4	0.02	0.02	0.02	0.01	0.01
	13	0.04	0.04	0.03	0.02	0.02
Monkey	4	0.05	0.06	0.07	0.06	0.02
	13	0.09	0.10	0.12	0.10	0.03
Dog	4	0.05	0.06	0.07	0.06	0.02
	13	0.08	0.10	0.12	0.10	0.03
Human[b]	4	0.04	0.03	0.04	0.06	0.03
	13	0.06	0.05	0.07	0.10	0.06

[a] Equilibrium TLN burdens of particles are achieved in about 50 days for all seven species; therefore, values for 104 wk are predicted to be the same as predicted for 13 wk.

[b] Light activity, breathing 50% through the nose and 50% through the mouth.

TABLE 20

Thoracic Lymph Node (TLN) Burdens of Particles for Inhalation Exposures 8 hr/day, 5 days/wk to 1.0 mg Particles per m^3 with Particle Dissolution-Absorption Half-Time of 100 days

Species	Weeks of Exposure	μg of Particles in TLNs per g Lung for Particle Aerodynamic Diameter (μm)				
		0.1	0.5	1	2	5
Mouse	4	0.19	0.19	0.19	0.12	0.03
	13	0.54	0.54	0.45	0.33	0.09
	104	1.16	1.16	0.97	0.71	0.19
Hamster	4	0.18	0.18	0.15	0.10	0.03
	13	0.51	0.51	0.42	0.29	0.08
	104	1.09	1.09	0.90	0.63	0.18
Rat	4	0.15	0.15	0.13	0.09	0.03
	13	0.42	0.42	0.36	0.25	0.07
	104	0.90	0.90	0.77	0.54	0.15
Guinea pig	4	0.07	0.06	0.06	0.03	0.01
	13	0.55	0.51	0.44	0.27	0.07
	104	3.45	3.24	2.81	1.68	0.43
Monkey	4	0.14	0.17	0.21	0.17	0.06
	13	1.11	1.28	1.59	1.28	0.43
	104	6.79	7.81	9.73	7.81	2.64
Dog	4	0.14	0.16	0.20	0.16	0.05
	13	1.08	1.22	1.54	1.22	0.41
	104	6.58	7.49	9.41	7.49	2.53
Human[a]	4	0.10	0.08	0.11	0.17	0.09
	13	0.76	0.60	0.88	1.32	0.71
	104	4.66	3.65	5.38	8.10	4.35

[a] Light activity, breathing 50% through the nose and 50% through the mouth.

TABLE 21

Thoracic Lymph Node (TLN) Burdens of Particles for Inhalation Exposures 8 hr/day, 5 days/wk to 1.0 mg Particles per m^3, with Particle Dissolution-Absorption Half-Time of 1000 days

Species	Weeks of Exposure	μg of Particles in TLNs per g Lung for Particle Aerodynamic Diameter (μm)				
		0.1	0.5	1	2	5
Mouse	4	0.21	0.21	0.18	0.13	0.04
	13	0.72	0.72	0.60	0.44	0.12
	104	4.71	4.71	3.92	2.88	0.78
Hamster	4	0.20	0.20	0.16	0.11	0.03
	13	0.67	0.67	0.55	0.39	0.11
	104	4.40	4.40	3.63	2.54	0.72
Rat	4	0.16	0.16	0.14	0.10	0.03
	13	0.55	0.55	0.48	0.33	0.09
	104	3.63	3.63	3.11	2.17	0.62
Guinea pig	4	0.08	0.07	0.06	0.04	0.01
	13	0.79	0.74	0.64	0.38	0.10
	104	31.0	29.1	25.2	15.1	3.87
Monkey	4	0.16	0.18	0.23	0.18	0.06
	13	1.60	1.84	2.29	1.84	0.62
	104	60.4	69.4	86.5	69.4	23.5
Dog	4	0.16	0.18	0.22	0.18	0.06
	13	1.55	1.77	2.22	1.77	0.60
	104	58.6	66.6	83.7	66.6	22.5
Human[a]	4	0.11	0.09	0.13	0.19	0.10
	13	1.10	0.86	1.27	1.91	1.02
	104	41.5	32.5	47.9	72.1	38.7

[a] Light activity, breathing 50% through the nose and 50% through the mouth.

TABLE 22

Relative Air Concentrations Predicted to Result in the Same Specific Pulmonary Burden (μg particles per g lung) in Laboratory Animals as in Humans[a] for Inhalation Exposures 8 hr/day, 5 days/wk, with Particle Dissolution-Absorption Half-Time of 10 days[b]

Species	Weeks of Exposure	Relative Aerosol Concentration for Particle Aerodynamic Diameter (μm)				
		0.1	0.5	1	2	5
Mouse	4	0.29	0.23	0.41	0.84	1.64
	13	0.31	0.24	0.43	0.88	1.72
Hamster	4	0.32	0.25	0.45	0.98	1.84
	13	0.34	0.27	0.48	1.03	1.94
Rat	4	0.39	0.31	0.53	1.14	2.14
	13	0.42	0.33	0.56	1.20	2.26
Guinea pig	4	0.29	0.24	0.40	1.02	2.13
	13	0.28	0.24	0.40	1.01	2.12
Monkey	4	0.69	0.47	0.55	1.04	1.65
	13	0.69	0.47	0.55	1.04	1.65
Dog	4	0.71	0.49	0.57	1.08	1.72
	13	0.71	0.49	0.57	1.08	1.72

[a] Light activity, breathing 50% through the nose and 50% through the mouth.
[b] Equilibrium pulmonary burdens of particles are achieved in about 50 days for all seven species.

TABLE 23

Relative Air Concentrations Predicted to Result in the Same Specific Pulmonary Burden (μg particles per g lung) in Laboratory Animals as in Humans[a] for Inhalation Exposures 8 hr/day, 5 days/wk, with Particle Dissolution-Absorption Half-Time of 100 days[b]

Species	Weeks of Exposure	Relative Aerosol Concentration for Particle Aerodynamic Diameter (μm)				
		0.1	0.5	1	2	5
Mouse	4	0.31	0.24	0.43	0.88	1.74
	13	0.43	0.34	0.60	1.24	2.43
	104	0.67	0.53	0.93	1.91	3.76
Hamster	4	0.35	0.27	0.49	1.05	1.97
	13	0.51	0.40	0.72	1.55	2.92
	104	0.81	0.63	1.14	2.45	4.60
Rat	4	0.42	0.33	0.57	1.22	2.30
	13	0.63	0.49	0.84	1.81	3.41
	104	0.98	0.77	1.33	2.85	5.36
Guinea pig	4	0.28	0.24	0.40	1.01	2.12
	13	0.28	0.23	0.39	0.99	2.06
	104	0.27	0.23	0.38	0.96	2.01
Monkey	4	0.69	0.47	0.55	1.04	1.65
	13	0.69	0.47	0.55	1.04	1.65
	104	0.69	0.47	0.55	1.04	1.65
Dog	4	0.71	0.49	0.57	1.08	1.72
	13	0.71	0.49	0.57	1.08	1.72
	104	0.71	0.49	0.57	1.08	1.72

[a] Light activity, breathing 50% through the nose and 50% through the mouth.

[b] Equilibrium pulmonary burdens of particles are achieved in about 6 months for mice, hamsters, and rats and after about 1 year for guinea pigs, monkeys, dogs, and humans.

TABLE 24

Relative Air Concentrations Predicted to Result in the Same Specific Pulmonary Burden (μg particles per g lung) in Laboratory Animals as in Humans[a] for Inhalation Exposures 8 hr/day, 5 days/wk, with Particle Dissolution-Absorption Half-Time of 1000 days[b]

Species	Weeks of Exposure	Relative Aerosol Concentration for Particle Aerodynamic Diameter (μm)				
		0.1	0.5	1	2	5
Mouse	4	0.31	0.24	0.43	0.89	1.75
	13	0.46	0.36	0.64	1.31	2.57
	104	1.42	1.11	1.97	4.05	7.96
Hamster	4	0.35	0.27	0.49	1.06	1.99
	13	0.55	0.43	0.77	1.66	3.12
	104	1.77	1.39	2.48	5.34	10.04
Rat	4	0.43	0.33	0.57	1.23	2.32
	13	0.67	0.52	0.90	1.94	3.64
	104	2.15	1.68	2.89	6.23	11.70
Guinea pig	4	0.28	0.24	0.40	1.01	2.12
	13	0.27	0.23	0.39	0.98	2.05
	104	0.27	0.22	0.38	0.96	2.00
Monkey	4	0.69	0.47	0.55	1.04	1.65
	13	0.69	0.47	0.55	1.04	1.65
	104	0.69	0.47	0.55	1.04	1.65
Dog	4	0.71	0.49	0.57	1.08	1.72
	13	0.71	0.49	0.57	1.08	1.72
	104	0.71	0.49	0.57	1.08	1.72

[a] Light activity, breathing 50% through the nose and 50% through the mouth.
[b] Equilibrium pulmonary burdens of particles are achieved in about 18 months for mice, hamsters, and rats but are not achieved for guinea pigs, monkeys, dogs, and humans during 2 years of exposure.

REFERENCES

1. Aharonson, E. F., Menkes, H., Gurtner, G., Swift, D. L., and Proctor, D. F., Effect of respiratory airflow rate on removal of soluble vapors by the nose, *J. Appl. Physiol.*, 37, 654–657, 1974.
2. Andersen, I., Lundqvist, G. R., and Proctor, D. F., Human nasal mucosal function in a controlled climate, *Arch. Environ. Health*, 23, 408–420, 1971.
3. Antila, S., Hietanen, E., Vainio, H., Camus, A.-M., Gelboin, H. V. et al., Smoking and peripheral type of cancer are related to high levels of pulmonary cytochrome P-450IA in lung cancer patients, *Int. J. Cancer*, 47, 681, 1991.
4. Asgharian, B. and Yu, C. P., Deposition of fibers in the rat lung, *J. Aerosol Sci.*, 20, 355–366, 1989.
5. Asgharian, B. and Yu, C. P., Deposition of inhaled fibrous particles in the human lung, *J. Aerosol. Med.*, 1, 37–50, 1988.
6. Asmundsson, T. and Kilburn, K. H., Mucociliary clearance rates at various levels in dog lungs, *Am. Rev. Respir. Dis.*, 102, 388–397, 1970.
7. Bailey, M. R., Fry, F. A., and James, A. C., Long-term retention of particles in the human respiratory tract, *J. Aerosol Sci.*, 16, 295–305, 1985.
8. Bailey, M. R., Hodgson, A., and Smith, H., Respiratory tract retention of relatively insoluble particles in rodents, *J. Aerosol. Sci.*, 16, 279–293, 1985.
9. Bair, W. J. and McClanahan, B. J., Plutonium inhalation studies, *Arch. Environ. Health*, 2, 48–55, 1961.
10. Bair, W. J., Willard, D. H., Herring, J. P., and George, L. A., III, Retention, translocation and excretion of inhaled $^{239}PuO_2$, *Health Phys.*, 8, 639–649, 1962.
11. Bair, W. J., Deposition, retention, translocation and excretion of radioactive particles, in *Inhaled Particles and Vapours*, Davies, C. N., Ed., Pergamon Press, Oxford, 1961, pp. 192–208.

12. Balchum, O. J., Dybicki, J., and Meneely, G. R., Absorption and distribution of $S^{35}O_2$ inhaled through the nose and mouth by dogs, *J. Appl. Physiol.*, 197, 1317–1321, 1959.
13. Bang, B. G., Mukherjee, A. L., and Bang, F. B., Human nasal mucus flow rates, *Johns Hopkins Med. J.*, 121, 38–48, 1967.
14. Baron, J. and Voight, J. M., Localization, distribution and induction of xenobiotic-metabolizing enzymes and aryl hydrocarbon hydroxylase activity within lung, *Pharmacol. Ther.*, 47, 419, 1990.
15. Baron, J., Burke, J. P., Guengerich, F. P., Jakoby, W. B., and Voight, J. M., Sites for xenobiotic activation and detoxication within the respiratory tract: implications for chemically induced toxicity, *Toxicol. Appl. Pharmacol.*, 93, 493, 1988.
16. Baron, J., Voight, J. M., Whitter, T. B., Kawabata, T. T., Knapp, S. A., Guengerich, F. P., and Jakoby, W. B., Identification of intratissue sites for xenobiotic activation and detoxication, *Adv. Exp. Med. Biol.*, 197, 119, 1986.
17. Bauer, M. A., Utell, M. J., Morrow, P. E., Speers, D. M., and Gibb, F. R., 0.3 ppm nitrogen dioxide inhalation potentiates exercise-induced bronchospasm in asthmatics, *Am. Rev. Respir. Dis.*, 129, A151, 1984.
18. Beattie, J. and Knox, J. F., Studies of mineral content and particle size distribution in the lungs of asbestos textile workers, in *Inhaled Particles and Vapours,* Davies, C. N., Ed., Pergamon Press, Oxford, 1961, pp. 419–433.
19. Becquemin, M. H., Yu, C. P., Roy, M., and Bouchikhi, A., Total deposition of inhaled particles related to age: comparison with age-dependent model calculations, *Radiat. Prot. Dosim.*, 38, 23–28, 1991.
20. Ben-Jebria, A., Crozet, Y. S., Eskew, M. L., and Rudeen, B. L., Acrolein-induced smooth muscle hyperresponsiveness and eicosanoid release in excised ferret trachea, *Toxicol. Appl. Pharmacol.*, 135, 35–44, 1995.
21. Ben-Jebria, A., Full, A. P., DeMaria, D. D., Ball, B. A., and Ultman, J. S., Dynamics of sulfur dioxide absorption in excised porcine tracheae, *Environ. Res.*, 53, 119–134, 1990.
22. Ben-Jebria, A., Hu, S.-C., Kitzmiller, E. L., and Ultman, J. S., Ozone absorption into excised porcine and sheep tracheae by a bolus-response method, *Environ. Res.*, 56, 144–157, 1991.
23. Bianco, A., Gibb, F. R., Kilpper, R. W., Landman, S., and Morrow, P. E., Studies of tantalum dust in the lungs, *Radiology,* 112, 549–556, 1974.
24. Black, A., Evans, J. C., Hadfield, E. H., Macbeth, R. G., Morgan, A., and Walsh, M., Impairment of nasal mucociliary clearance in woodworkers in the furniture industry, *Br. J. Ind. Med.*, 31, 10–17, 1974.
25. Boecker, B. B. and McClellan, R. O., The effects of solubility on the bioassay for inhaled radionuclides, in *Diagnosis and Treatment of Deposited Radionuclides,* Kornberg, H. A. and Norwood, W. D., Eds., Excerpta Medica Foundation, New York, 1968, pp. 234–242.
26. Bogdanffy, M. S., Randall, H. W., and Morgan, K. T., Biochemical quantitation and histochemical localization of carboxylesterase in the nasal passages of the Fischer-344 rat and $B6C3F_1$ mouse, *Toxicol. Appl. Pharmacol.*, 88, 183, 1987.
27. Bogdanffy, M. S., Randall, H. W., and Morgan, K. T., Histochemical localization of aldehyde dehydrogenase in the respiratory tract of the Fischer-344 rats, *Toxicol. Appl. Pharmacol.*, 82, 560, 1986.
28. Bohning, D. E., Atkins, H. L., and Cohn, S. H., Long-term particle clearance in man: normal and impaired, *Ann. Occup. Hyg.*, 26, 259–271, 1982.
29. Bolton, R. E., Vincent, J. H., Jones, A. D., Addison, J., and Beckett, S. T., An overload hypothesis for pulmonary clearance or UICC amosite fibres inhaled by rats, *Br. J. Ind. Med.*, 40, 264–272, 1983.
30. Bond, J. A., Some biotransformation enzymes responsible for polycyclic aromatic hydrocarbon metabolism in rat nasal turbinates: effects on enzyme activities of *in vitro* modifiers and intraperitoneal and inhalation exposure of rats to inducing agents, *Cancer Res.*, 43, 4805, 1983.
31. Booker, D. V., Chamberlain, A. C., Rundo, J., Muir, D. C. F., and Thomson, M. L., Elimination of 5-μm particles from the human lung, *Nature,* 214, 30–33, 1967.
32. Bowden, D. H. and Adamson, I. Y. R., Pathways of cellular efflux and particulate clearance after carbon instillation to the lung, *J. Pathol.*, 143, 117–125, 1984.
33. Brain, J. D., The effects of increased particles on the number of alveolar macrophages, in *Inhaled Particles III,* Vol. 1, Walton, W. H., Ed., Unwin Brothers, Surrey, UK, 1971, pp. 209–223.
34. Brody, A. R. and Roe, M. W., Deposition pattern of inorganic particles at the alveolar level in the lungs of rats and mice, *Am. Rev. Respir. Dis.*, 128, 724–729, 1983.
35. Brody, A. R. and Yu, C. P., Particle deposition at the alveolar duct bifurcations, in *Extrapolation of Dosimetric Relationships for Inhaled Particles and Gases,* Crapo, J. D., Miller, F. J., Smolko, E. D., Graham, J. A., and Wallace Hayes, A., Eds., Academic Press, San Diego, CA, 1989, pp. 91–99.
36. Brody, A. R., Hill, L. H., Adkins, Jr., B., and O'Connor, R. W., Chryostile asbestos inhalation in rats: deposition pattern and reaction of alveolar epithelium and pulmonary macrophages, *Am. Rev. Respir. Dis.*, 123, 670–679, 1981.
37. Brown, D., Garcia-Segura, L.-M., and Orci, L., Carbonic anhydrase is present in olfactory receptor cells, *Histochemistry,* 80, 307, 1984.
38. Brundelet, P. J., Experimental study of the dust-clearance mechanism of the lung. I. Histological study in rats of the intra-pulmonary bronchial route of elimination, *Acta Pathol. Microbiol. Scand.*, 1(Suppl. 175), 1–141, 1965.

39. Bush, M. L., Ben-Jebria, A., Asplund, P. T., Miles, K., and Ultman, J. S., Longitudinal distribution of ozone absorption in the lung: gender differences and intersubject variability, *J. Appl. Physiol.*, 81, 1651–1657, 1996.
40. Carlberg, J. R., Crable, J. V., Limtiaca, L. P., Norris, H. B., Holtz, J. L., Mauer, P., and Wolowicz, F. R., Total dust, coal, free silica, and trace metal concentrations in bituminous coal miner's lungs, *Am. Ind. Hyg. Assoc. J.*, 32, 432–440, 1971.
41. Cartwright, J. and Skidmore, J. W., The size distribution of dust retained in the lungs of rats and dust collected by size-selecting samplers, *Ann. Occup. Hyg.*, 7, 151–167, 1964.
42. Casanova-Schmitz, M., David, R. M., and Heck, H. D., Oxidation of formaldehyde and acetaldehyde by NAD^+-dependent dehydrogenases in rat nasal mucosal homogenates, *Biochem. Pharmacol.*, 33, 1137, 1984.
43. Cheng, Y.-S., Su, Y.-F., and Yeh, H. C., Deposition of thoron progeny in human head airways, *Aerosol Sci. Technol.*, 18, 359–375, 1993.
44. Cheng, Y.-S., Yeh, H.-C., and Swift, D. L., Aerosol deposition in human nasal airway for particles 1 nm to 20 μm: a model study, *Radiat. Prot. Dosim.*, 38, 41–47, 1991.
45. Cheng, Y. S., Yamada, Y., Yeh, H. C., and Swift, D. L., Diffusional deposition of ultrafine aerosols in a human nasal cast, *J. Aerosol Sci.*, 19, 741–751, 1988.
46. Chichester, C. H., Philpot, R. M., Weir, A. J., Buckpitt, A. R., and Plopper, C. G., Characterization of the cytochrome P-450 monooxygenase system in nonciliated bronchiolar epithelial (Clara) cells isolated from mouse lung, *Am. J. Respir. Cell Mol. Biol.*, 4, 179, 1991.
47. Chopra, S. K., Taplin, G. V., Elam, D., Carson, S. A., and Golde, D., Measurement of tracheal mucociliary transport velocity in humans — smokers versus non-smokers, *Am. Rev. Respir. Dis.*, 119(Suppl. 4), 205, 1979 (abstract).
48. Cohen, B. S., Harley, N. H., Schlesinger, R. B., and Lippmann, M., Nonuniform particle deposition on tracheobronchial airways: implications for lung dosimetry, *Ann. Occup. Hyg.*, 32, 1045–1053, 1988.
49. Corry, D., Kulkarni, P., and Lipscomb, M. F., The migration of bronchoalveolar macrophages into hilar lymph nodes, *Am. J. Pathol.*, 115, 321–328, 1984.
50. Cottier, H., Meister, F., Zimmermann, A., Kraft, R., Burkhardt, A., Gehr, P., and Poretti, G., Accumulation of anthracotic particles along lymphatics of the human lung: relevance to "hot spot" formation after inhalation of poorly soluble radionuclides, *Radiat. Environ. Biophys.*, 26, 275–282, 1987.
51. Crawford, D. J., Identifying critical human subpopulations by age groups: radioactivity and the lung, *Phys. Med. Biol.*, 27, 539–552, 1982.
52. Cuddihy, R. G. and Yeh, H. C., Respiratory tract clearance of particles and substances dissociated from particles, in *Inhalation Toxicology: The Design and Interpretation of Inhalation Studies and Their Use in Risk Assessment*, Dungworth, D., Kimmerle, G., Lewkowski, J., McClellan, R., and Stöber, W., Springer-Verlag, New York, 1988, pp. 169–193.
53. Cuddihy, R. G., Deposition and retention of inhaled niobium in Beagle dogs, *Health Phys.*, 34, 167–176, 1978.
54. Czerwinski, M., McLemore, T. E., Philpot, R. M., Nhamburo, P. T., Korzekwa, K. et al., Metabolic activation of 4-ipeomeanol by complimentary DNA-expressed human cytochromes P-450: evidence for species-specific metabolism, *Cancer Res.*, 51, 4636, 1991.
55. Dahl, A. R. and Hadley, W. M., Nasal cavity enzymes involved in xenobiotic metabolism: effects on the toxicity of inhalants, *Toxicology*, 21, 345, 1991.
56. Dahl, A. R. and Lewis, J. L., Respiratory tract uptake of inhalants and metabolism of xenobiotics, *Ann. Rev. Pharmacol. Toxicol.*, 32, 383, 1993.
57. Dahl, A. R., Coslett, D. S., Bond, J. A., and Hesseltine, G. R., Metabolism of benzo(*a*)pyrene on the nasal mucosa of Syrian hamsters: comparison to metabolism by other extrahepatic tissues and possible role of nasally produced metabolites in carcinogenesis, *J. Nat. Cancer Inst.*, 75, 135, 1985.
58. Dahl, A. R., Miller, S. C., and Petridou-Fischer, J., Carboxylesterases in the respiratory tracts of rabbits, rats and Syrian hamsters, *Toxicol. Lett.*, 35, 129, 1987.
59. Dahl, A. R., Snipes, M. B., and Gerde, P., Sites for uptake of inhaled vapors in Beagle dogs, *Toxicol. Appl. Pharmacol.*, 109, 263–275, 1991.
60. Dahl, A. R., The cyanide-metabolizing enzyme rhodanese in rat nasal respiratory and olfactory mucosa, *Toxicol. Lett.*, 45, 199, 1989.
61. Ding, X. and Coon, M. J., Cytochrome P-450-dependent formation of ethylene from *N*-nitrosoethylamines, *Drug Metab. Dispos.*, 16, 265, 1988.
62. Ding, X. and Coon, M. J., Immunochemical characterization of multiple forms of cytochrome P-450 in rabbit nasal microsomes and evidence for tissue-specific expression of P-450s NMa and NMb, *Mol. Pharmacol.*, 37, 489, 1990.
63. Ding, X. and Coon, M. J., Purification and characterization of two unique forms of cytochrome P-450 from rabbit nasal microsomes, *Biochemistry*, 27, 8330, 1989.
64. Ding, X. X., Koop, D. R., Crump, B. L., and Coon, M. J., Immunochemical identification of cytochrome P-450, isozyme 3a (P-450ALC) in rabbit nasal and kidney microsomes and evidence for differential induction by alcohol, *Mol. Pharmacol.*, 30, 370, 1986.
65. Diu, C. K. and Yu, C. P., Respiratory tract deposition of polydisperse aerosols in humans, *Am. Ind. Hyg. Assoc. J.*, 44, 62–65, 1983.

66. Domin, B. A., Devereux, R. R., and Philpot, R. M., The cytochrome P-450 monooxygenase system of the rabbit lung: enzyme components, activities, and induction in the nonciliated bronchiolar epithelial (Clara) cell, alveolar type II, and alveolar macrophage, *Mol. Pharmacol.,* 30, 296, 1986.
67. Egle, J. L., Retention of inhaled acetaldehyde in the dog, *Arch. Environ. Health,* 24, 354–357, 1972.
68. Egle, J. L., Retention of inhaled formaldehyde, propionaldehyde, and acrolein in the dog, *Arch. Environ. Health,* 25, 119–124, 1972.
69. Ewert, G., On the mucus flow rate in the human nose, *Acta Otolaryngol. (Suppl.),* 200, 1–62, 1965.
70. Ferin, J., Effect of particle content of lung on clearance pathways, in *Pulmonary Macrophage and Epithelial Cells,* Sanders, C. L., Schneider, R. P., Dagle, G. E., and Ragan, H. A., Eds., Energy Research and Development Association Technical Information Center, Springfield, VA, 1977, pp. 414–423.
71. Ferron, G. A., Kreyling, W. G., and Haider, B., Influence of the growth of salt aerosol particles on the deposition in the lung, *Ann. Occup. Hyg.,* 32(Suppl. 1), 947–955, 1988.
72. Finch, G. L., Haley, P. J., Hoover, M. D., Snipes, M. B., and Cuddihy, R. G., Responses of rat lungs to low lung burdens of inhaled beryllium metal, *Inhal. Toxicol.,* 6, 205–224, 1994.
73. Finch, G. L., Nikula, K. J., Chen, B. T., Barr, E. B., Chang, I.-Y., and Hobbs, C. H., Effect of chronic cigarette smoke exposure on lung clearance of tracer particles inhaled by rats, *Fundam. Appl. Toxicol.,* 24, 76–85, 1995.
74. Fish, B. R., Inhalation of uranium aerosols by mouse, rat, dog and man, in *Inhaled Particles and Vapours,* Davies, C. N., Ed., Pergamon Press, Oxford, 1961, pp. 151–166.
75. Forkert, P. G., Vessey, M. L., Park, S. S., Gelboin, H. V., and Cole, S. P. C., Cytochromes P-450 in murine lung: an immunohistochemical study with monoclonal antibodies, *Drug Metab.,* 17, 551, 1989.
76. Foster, J. R., Elcombe, C. R., Boobis, A. R., Davies, D. S., Sesardic, A., McQuade, J., Robson, R. T., Haward, C., and Lock, E. A., Immunocytochemical localization of cytochrome P-450 in hepatic and extra-hepatic tissues of the rat with a monoclonal antibody against cytochrome P-450, *Biochem. Pharmacol.,* 35, 4543, 1986.
77. Foster, W. M., Langenback, E., and Bergofsky, E. H., Measurement of tracheal and bronchial mucus velocities in man: relation to lung clearance, *J. Appl. Physiol.,* 48, 965–971, 1980.
78. Frank, N. R., Yoder, R. F., Brain, J. D., and Yokoyama, E., SO_2 (^{35}S-labeled) absorption by the nose and mouth under conditions of varying concentration and flow, *Arch. Environ. Health,* 18, 315–322, 1969.
79. Friedman, M., Stott, F. D., Poole, D. O., Dougherty, R., Chapman, G. A., Watson, H., and Sackner, M. A., A new roentgenographic method for estimating mucous velocity in airways, *Am. Rev. Respir. Dis.,* 115, 67–72, 1977.
80. Fry, F. A. and Black, A., Regional deposition and clearance of particles in the human nose, *J. Aerosol Sci.,* 4, 113–124, 1973.
81. Fujii, H., Sagami, I., Ohmachi, T., Kikuchi, H., and Watanabe, M., Tissue difference in expression of cytochrome P-450 between liver and lung of Syrian golden hamsters treated with 3-methylcholanthrene, *Adv. Exp. Med. Biol.,* 283, 413, 1991.
82. Galibin, G. P. and Parfenov, Y. D., Inhalation study on metabolism of insoluble uranium compounds, in *Inhaled Particles III,* Vol. 1, Walton, W. H., Ed., Unwin Brothers, Surrey, U.K., 1971, pp. 201–208.
83. Gerrard, C. S., Levandowski, R. A., Gerrity, T. R., Yeates, D. B., and Klein, E., The effects of acute respiratory virus infection upon tracheal mucous transport, *Arch. Environ. Health,* 40, 322–325, 1985.
84. Gerrity, T. R., Biscardi, F., Strong, A., Garlington, A. R., and Bromberg, P. A., Bronchoscopic determination of ozone uptake in humans, *J. Appl. Physiol.,* 79, 852–860, 1995.
85. Gerrity, T. R., McDonnell, W. F., and House, D. E., The relationship between delivered dose and functional responses in humans, *Toxicol. Appl. Pharmacol.,* 124, 275–283, 1994.
86. Gerrity, T. R., Weaver, R. A., Berntsen, J., House, D. E., and O'Neal, J., Extrathoracic and intrathoracic removal of O_3 in tidal-breathing humans, *J. Appl. Physiol.,* 65, 393–340, 1988.
87. Gervasi, P. G., Longo, V., Ursino, F., and Panattoni, G., Drug metabolizing enzymes in respiratory mucosa of humans: comparison with rats, in *Cytochrome P-450: Biochemistry and Biophysics,* Schuster, I., Ed., Taylor & Francis, New York, 1989, p. 97.
88. Gibb, F. R., Beiter, H. B., and Morrow, P. E., *Studies of Coal Dust Retention in the Lungs Utilizing Neutron-Activated Coal,* UR-3490-679, Energy Research and Development Association Technical Information Center, Springfield, VA, 1975.
89. Gillner, M., Brittebo, E. B., Brandt, I., Söderkvist, P., Appelgren, L. E. and Appelgren, J.-Å, Uptake and specific binding of 2,3,7,8-tetrachlorodibenzo-*p*-dioxin in the olfactory mucosa of mice and rats, *Cancer Res.,* 47, 4150, 1987.
90. Goodglick, L. A. and Kane, A. B., Role of reactive oxygen metabolites in crocidolite asbestos toxicity to mouse macrophages, *Cancer Res.,* 46, 5558–5566, 1986.
91. Goodman, R. M., Yergin, B. M., Landa, J. F., Golinvaux, M. H., and Sackner, M. A., Relationship of smoking history and pulmonary function tests to tracheal mucous velocity in nonsmokers, young smokers, ex-smokers, and patients with chronic bronchitis, *Am. Rev. Respir. Dis.,* 117, 205–214, 1978.
92. Gore, D. J. and Patrick, G., A quantitative study of the penetration of insoluble particles into the tissue of the conducting airways, *Ann. Occup. Hyg.,* 26, 149–161, 1982.

93. Gore, D. J. and Thorne, M. C., The distribution and clearance of inhaled uranium dioxide particles in the respiratory tract of the rat, in *Inhaled Particles IV,* Part 1, Walton, W. H. and McGovern, B., Eds., Pergamon Press, Oxford, 1977, pp. 275–283.
94. Green, G. M., Alveolobronchiolar transport mechanisms, *Arch. Intern. Med.,* 131, 109–114, 1973.
95. Green, G. M., Alveolobronchiolar transport: observations and hypothesis of a pathway, *Chest,* 59(Suppl.), 1S, 1971.
96. Guengerich, F. P., Enzymology of rat liver cytochromes P-450, in *Mammalian Cytochromes P-450,* Vol. 1, Guengerich, F. P., Ed., CRC Press, Boca Raton, FL, 1987, p. 1.
97. Guilmette, R. A., Diel, J. H., Muggenburg, B. A., Mewhinney, J. A., Boecker, B. B., and McClellan, R. O., Biokinetics of inhaled $^{239}PuO_2$ in the Beagle dog: effect of aerosol particle size, *Int. J. Radiat. Biol.,* 45, 563–581, 1984.
98. Hadley, W. M. and Dahl, A. R., Cytochrome P-450-dependent monooxygenase activity in rat nasal epithelial membranes, *Toxicol. Lett.,* 10, 417, 1982.
99. Hagerstrand, I. and Siefert, B., Asbestos bodies and pleural plaques in human lungs at autopsy, *Acta. Pathol. Microbiol. Scand.,* A81, 457–460, 1973.
100. Hammad, Y., Diem, J., Craighead, J., and Weill, H., Deposition of inhaled man-made mineral fibres in the lungs of rats, *Ann. Occup. Hyg.,* 26, 179–187, 1982.
101. Harley, N. H. and Pasternack, B. S., Experimental absorption measurements applied to lung dose from radon daughters, *Health Phys.,* 23, 771–782, 1972.
102. Harmsen, A. G., Muggenburg, B. A., Snipes, M. B., and Bice, D. E., The role of macrophages in particle translocation from lungs to lymph nodes, *Science,* 230, 1277–1280, 1985.
103. Henke, D., Danilowicz, R. M., Curtis, J. F., Boucher, R. C., and Eling, T. E., Metabolism or arachidonic acid by human nasal and bronchial epithelial cells, *Arch. Biochem. Biophys.,* 267, 426, 1988.
104. Hoffman, W., Mathematical model for the postnatal growth of the human lung, *Respir. Physiol.,* 49, 115–129, 1982.
105. Holt, P. F., Transport of inhaled dust to extrapulmonary sites, *J. Pathol.,* 133, 123–129, 1981.
106. Hotchkiss, J. A., Personal communication, 1991.
107. Hu, S.-C., Ben-Jebria, A., and Ultman, J. S., Longitudinal distribution of ozone absorption in the lung: effects of respiratory flow, *J. Appl. Physiol.,* 77, 574–583, 1994.
108. Hu, S.-C., Ben-Jebria, A., and Ultman, J. S., Longitudinal distribution of ozone absorption in the lung: quiet respiration in healthy subjects, *J. Appl. Physiol.,* 73, 1655–1661, 1992.
109. Johnson, L. J., Dean, P. N., and Ide, H. M., *In vivo* determination of the late-phase lung clearance of ^{239}Pu following accidental exposure, *Health Phys.,* 22, 410–412, 1972.
110. Kabel, J. R., Ben-Jebria, A., and Ultman, J. S., Longitudinal distribution of ozone absorption in the lung: comparison of nasal and oral quiet breathing, *J. Appl. Physiol.,* 77, 2584–2592, 1994.
111. Kato, S., Shields, P. G., Caporaso, N. E., Hoover, R. N., Trump, B. F. et al., Cytochrome P-450IIE1 genetic polymorphisms, racial variation, and lung cancer risk, *Cancer Res.,* 52, 6712, 1992.
112. Kauffer, E., Vigneron, J. C., Hesbert, A., and Lemonnier, M., A study of the length and diameter of fibres, in lung and in broncho-alveolar lavage fluid, following exposure of rats to crysotile asbestos, *Ann. Occup. Hyg.,* 31, 233–240, 1987.
113. Keller, D. A., Heck, H. D. A., Randall, H. W., and Morgan, K. T., Histochemical localization of aldehyde dehydrogenase in the respiratory tract of the Fischer-344 rats, *Toxicol. Appl. Pharmacol.,* 85, 560, 1986.
114. Killburn, K. H., A hypothesis for pulmonary clearance and its implications, *Am. Rev. Respir. Dis.,* 98, 449–463, 1968.
115. Kimuar, S., Kozak, C. A., and Gonzalez, F. J., Identification of a novel P-450 expressed in rat lung: cDNA cloning and sequence, chromosome mapping and induction by 3-methylcholanthrene, *Biochemistry,* 28, 3798, 1989.
116. Kleinman, M. T. and Mautz, W. J., Upper airway scrubbing at rest and exercise, in *Susceptibility to Inhaled Pollutants,* Utell, M. J. and Frank, R., Eds., American Society for Testing and Materials, Philadelphia, PA, 1989, pp. 100–110.
117. Kreyling, W. G., Cox, C., Ferron, G. A., and Oberdörster, G., Lung clearance in Long-Evans rats after inhalation of porous, monodisperse cobalt oxide particles, *Exp. Lung Res.,* 19, 445–467, 1993.
118. Kreyling, W. G., Schumann, G., Ortmaier, A., Ferron, G. A., and Karg, E., Particle transport from the lower respiratory tract, *J. Aerosol Med.,* 1, 351–370, 1988.
119. Kreyling, W. G., Interspecies comparison of lung clearance of "insoluble" particles, *J. Aerosol Med.,* 3(Suppl. 1), S93–S110, 1990.
120. LaBauce, R. J., Brooks, A. L., Mauderly, J. L., Hah, F. F., Redman, H. C., Macken, C., Slauson, D. O., Mewhinney, J. A., and McClellan, R. O., Cytogenetic and other biological effects of $^{239}PuO_2$ inhaled by the Rhesus monkey, *Radiat. Res.,* 82, 310–335, 1980.
121. LaBelle, C. W. and Brieger, H., Patterns and mechanisms in the elimination of dust from the lung, in *Inhaled Particles and Vapours,* Davies, C. N., Ed., Pergamon Press, Oxford, 1961, pp. 356–368.
122. Langham, W. H., Determination of internally deposited radioactive isotopes from excretion analyses, *Am. Ind. Hyg. Assoc. Quart.,* 17, 305–318, 1956.

123. Lauweryns, J. M. and Baert, J. H., The role of the pulmonary lymphatics in the defenses of the distal lung: morphological and experimental studies of the transport mechanisms of intratracheally instilled particles, *Ann. N.Y. Acad. Sci.*, 221, 244–275, 1974.
124. Lazard, D., Zupko, K., Poria, Y., Nef, P., Lasarovits, J., Horn, S., Khen, M., and Lancet, D., Odorant signal termination by olfactory UDP glucuronosyl transferase, *Nature,* 349, 790–793, 1991.
125. Le Bouffant, L., Daniel, H., Henin, J. P., Martin, J. C., Normand, C., Tichoux, G., and Trolard, F., Experimental study on long-term effects of inhaled MMMF on the lungs of rats, *Ann. Occup. Hyg.*, 31, 765–790, 1987.
126. Le Bouffant, L., Henin, J. P., Martin, J. C., Normand, C., Tichoux, G., and Trolard, F., Distribution of inhaled MMMF in the rat lung, in *Biological Effects of Man-Made Mineral Fibres,* Vol. 2, World Health Organization, Copenhagen, 1984, pp. 143–168.
127. Leadbetter, M. R. and Corn, M., Particle size distribution of rat lung residues after exposures to fiberglass dust clouds, *Am. Ind. Hyg. Assoc. J.*, 35, 511–522, 1972.
128. Leak, L. V., Lymphatic removal of fluids and particles in the mammalian lung, *Environ. Health Perspect.*, 35, 55–76, 1980.
129. Lee, K. P., Trochimowicz, H. J., and Reinhardt, C. F., Transmigration of titanium dioxide (TiO_2) in rats after inhalation exposure, *Exp. Mol. Pathol.*, 42, 331–343, 1989.
130. Lee, P. S., Chan, T. L., and Hering, W. E., Long-term clearance of inhaled diesel exhaust particles in rodents, *J. Toxicol. Environ. Health,* 12, 801–813, 1983.
131. Lee, P. S., Gerrity, T. R., Hass, F. J., and Lourenco, R. V., A model for tracheobronchial clearance of inhaled particles in man and a comparison with data, *IEEE Trans. Biomed. Eng.*, BME-26(11), 624–629, 1979.
132. Leikauf, G., Yeates, D. B., Wales, K. A., Spektor, D., Albert, R. E., and Lippmann, M., Effects of sulfuric acid aerosol on respiratory mechanics and mucociliary particle clearance in healthy nonsmoking adults, *Am. Ind. Hyg. Assoc. J.*, 42, 273–281, 1981.
133. Lewis, J. L., Rhoades, C. E., Bice, D. E., Harkema, J. R., Hotchkiss, J. A., Sylvester, D., and Dahl, A. R., Localization of the cyanide metabolizing enzymes, rhodanese, within the olfactory epithelium, *Chem. Senses,* 15, 1990.
134. Lewis, J. L., Rhoades, C. E., Gervasi, P. G., Griffith, W. C., and Dahl, A. R., The cyanide-metabolizing enzyme rhodanese in human nasal respiratory mucosa, *Toxicol. Appl. Pharmacol.*, 108, 114, 1991.
135. Lippmann, M., Deposition and clearance of inhaled particles in the human nose, *Ann. Otol. Rhinol. Laryngol.*, 79, 519–528, 1970.
136. Little, J. B., Radford, E. P., Jr., McCombs, H. L., and Hunt, V. R., Distribution of polonium^{210} in pulmonary tissues of cigarette smokers, *N. Engl. J. Med.*, 273, 1343–1351, 1965.
137. Löftberg, B. and Tjälve, H., The disposition and metabolism of *N*-nitrosodiethylamine in adult, infant and foetal tissue of the Syrian golden hamster, *Acta Pharmacol. Toxicol.*, 54, 104, 1984.
138. Longo, V., Pacifici, G. M., Panattoni, G., Ursino, F., and Gervasi, P. G., Metabolism of diethylnitrosamine by microsomes of human respiratory nasal mucosa and liver, *Biochem. Pharmacol.*, 38, 1867, 1989.
139. Lundborg, M., Eklund, A., Lind, B., and Camner, P., Dissolution of metals by human and rabbit alveolar macrophages, *Br. J. Ind. Med.*, 42, 642–645, 1985.
140. Lupo, D., Lodi, L., Canonaco, M., Valenti, A., and Dessi-Fulgheri, F., Testosterone metabolism in the olfactory epithelium of intact and castrated male rats, *Neurosci. Lett.*, 69, 259, 1986.
141. Masse, R., Ducousso, R., Nolibe, D., Lafuma, J., and Chretien, J., Passage transbronchique des particules metalliques, *Rev. Fr. Mal. Respir.*, 1, 123–129, 1974.
142. Mauderly, J. L., Cheng, Y. S., and Snipes, M. B., Particle overload in toxicological studies: friend or foe?, *J. Aerosol. Med.*, 3(Suppl. 1), S169–S187, 1990.
143. Mauderly, J. L., Contribution of inhalation bioassays to the assessment of human health risks from solid airborne particles, in *Toxic and Carcinogenic Effects of Solid Particles in the Respiratory Tract,* Mohr, U., Dungworth, D. L., Mauderly, J. L., and Oberdörster, G., Eds., International Life Sciences Institute Press, Washington, D.C., 1994, pp. 355–365.
144. McInroy, J. F., Stewart, M. W., and Moss, W. D., Studies of plutonium in human tracheobronchial lymph nodes, in *Radiation and the Lymphatic System,* Ballou, J. E., Ed., Energy Research and Development Association Technical Information Center, Washington, D.C., 1976, pp. 54–58.
145. McNulty, M. J., Casanova-Schmitz, M., and Heck, H. D., Metabolism dimethylamine in the nasal mucosa of the Fischer 344 rat, *Drug Metab. Dispos.*, 11, 421, 1983.
146. Mercer, T. T., On the role of particle size in the dissolution of lung burdens, *Health Phys.*, 13, 1211–1221, 1967.
147. Miller, F. J., McNeal, C. A., Kirtz, J. M., Gardner, D. E., Coffin, D. L., and Menzel, D. B., Nasopharyngeal removal of ozone in rabbits and guinea pigs, *Toxicology,* 14, 273–281, 1979.
148. Misra, V., Rahman, Q., and Viswanathan, P. N., Biochemical changes in guinea pig lungs due to amosite asbestos, *Environ. Res.*, 16, 55–61, 1978.
149. Morgan, A., Black, A., Moores, S. R., and Lambert, B. E., Translocation of ^{239}Pu in mice following inhalation of sized $^{239}PuO_2$, *Health Phys.*, 50, 535–539, 1986.

150. Morgan, A., Evans, J. C., and Holmes, A., Deposition and clearance of inhaled fibrous minerals in the rat, studies using radioactive tracer techniques, in *Inhaled Particles IV,* Part 1, Walton, W. H. and McGovern, B., Eds., Pergamon Press, Oxford, 1977, pp. 259–274.
151. Morgan, A., Holmes, A., and Davison, W., Clearance of sized glass fibres from the rat lung and their solubility *in vivo, Ann. Occup. Hyg.,* 25, 317–331, 1982.
152. Morgan, A., Talbot, R. J., and Holmes, A., Significance of fibre length in the clearance of asbestos fibers from the lung, *Br. J. Ind. Med.,* 35, 146–153, 1978.
153. Morgan, A., Effect of length on the clearance of fibres from the lung and on body formation, in *Biological Effects of Mineral Fibres,* Vol. 1, Wagner, J. C. and Davis, W., Eds., IARC Scientific Publ. No. 30, International Agency for Research on Cancer, Geneva, 1980, pp. 329–335.
154. Morgan, K. T., Patterson, D. L., and Gross, E. A., Responses of the nasal mucociliary apparatus to airborne irritants, in *Toxicology of the Nasal Passages,* Barrow, C. S., Ed., Hemisphere Publishing, Washington, D.C., 1986, pp. 123–133.
155. Morris, J. B. and Blanchard, K. T., Upper respiratory tract deposition of inspired acetaldehyde, *Toxicol. Appl. Pharmacol.,* 114, 140–146, 1992.
156. Morris, J. B. and Cavanagh, D. G., Metabolism and deposition of propanol and acetone vapors in the upper respiratory tract of the hamster, *Fundam. Appl. Toxicol.,* 9, 34, 1987.
157. Morris, J. B., Clay, R. J., and Cavanagh, D. G., Species differences in upper respiratory tract deposition of acetone and ethanol vapors, *Fundam. Appl. Toxicol.,* 7, 671–680, 1986.
158. Morris, J. B., Deposition of acetone vapor in the upper respiratory tract of the B6C3F$_1$ mouse, *Toxicol. Lett.,* 56, 187–196, 1991.
159. Morris, J. B., First-pass metabolism of inspired ethyl acetate in the upper respiratory tracts of the F344 rat and Syrian hamster, *Toxicol. Appl. Pharmacol.,* 102, 331–345, 1990.
160. Morrow, P. E. and Yuile, C. L., The disposition of coal dusts in the lungs and tracheobronchial lymph nodes of dogs, *Fundam. Appl. Toxicol.,* 2, 300–305, 1982.
161. Morrow, P. E., Gibb, F. R., and Gazioglu, K., The clearance of dust from the lower respiratory tract of man. An experimental study, in *Inhaled Particles and Vapours II,* Davies, C. N., Ed., Pergamon Press, Oxford, 1967, pp. 351–369.
162. Morrow, P. E., Alveolar clearance of aerosols, *Arch. Intern. Med.,* 131, 101–108, 1973.
163. Morrow, P. E., Dust overloading of the lungs: update and appraisal, *Toxicol. Appl. Pharmacol.,* 113, 1–12, 1992.
164. Morrow, P. E., Possible mechanisms to explain dust overloading of the lungs, *Fundam. Appl. Toxicol.,* 10, 369–384, 1988.
165. Morrow, P. E., The setting of particulate exposure levels for chronic inhalation toxicity studies, *J. Am. Coll. Toxicol.,* 6, 533–544, 1986.
166. Muerhoff, A. S., Williams, D. E., and Masters, B. S. S., Purification and properties of pregnancy-inducible rabbit lung cytochrome P-450 prostaglandin β-hydroxylase, *Meth. Enzymol.,* 187, 253, 1990.
167. Muhle, H. and Bellmann, B., Pulmonary clearance of inhaled particles in dependence of particle size, *J. Aerosol. Sci.,* 17, 346–349, 1986.
168. Muhle, H., Bellmann, B., and Heinrich, U., Overloading of lung clearance during chronic exposure of experimental animals to particles, *Ann. Occup. Hyg.,* 32(Suppl. 1), 141–147, 1988.
169. Murray, M. J. and Driscoll, K. E., Immunology of the respiratory system, in *Comparative Biology of the Normal Lung,* Vol. 1, Parent, R. A., Ed., CRC Press, Boca Raton, FL, 1992, pp. 725–746.
170. Nagata, K., Martin, B. M., Gillette, J. R., and Sasame, H. A., Isozymes of cytochrome P-450 that metabolize naphthalene in liver and lung of untreated mice, *Drug Metab. Dispos.,* 18, 557, 1990.
171. NCR, *Comparative Dosimetry of Radon in Mines and Homes,* National Academy Press, Washington, D.C., 1991.
172. Nebert, D. W., Nelson, D. R., Adesnik, M., Coon, M. J., Estabrook, R. W., Gonzalez, F. J., Guengerich, F. P., Gunsalus, I. C., Johnson, E. F., Kemper, B., Levin, W., Phillips, I. R., Stato, R., and Waterman, M. R., The P-450 superfamily: updated listing of all genes and recommended nomenclature for the chromosomal loci, *DNA,* 8, 1–13, 1989.
173. Nef, P., Heldman, J., Lazard, D., Margalit, T., Jaye, M., Hanukoglu, I., and Lancet, D., Olfactory-specific cytochrome P-450: cDNA cloning of a novel neuroepithelial enzyme possibly involved in chemoreception, *J. Biol. Chem.,* 264, 6780, 1989.
174. Nef, P., Larabee, T. M., Kagimoto, K., and Meyer, U. A., Olfactory-specific cytochrome P-450 (P-450olf1; IIG1): gene structure and developmental regulation, *J. Biol. Chem.,* 265, 2903, 1990.
175. Newton, D., A case of accidental inhalation of protactinium-231 and actinium-227, *Health Phys.,* 15, 11–17, 1968.
176. Nhamburo, P. T., Gonzalez, F. J., McBride, O. W., Gelboin, H. V., and Kimura, S., Identification of new P-450 expressed in human lung: complete cDNA sequence, cDNA-directed expression, and chromosome mapping, *Biochemistry,* 28, 8060, 1989.

177. Nhamburo, P. T., Kimuar, S., McBride, O. W., Kozak, C. A., Gelboin, H. V., and Gonzalez, F. J., The human CYP2F subfamily: identification of a cDNA coding for a new cytochrome P-450 expressed in lung, cDNA-directed expression and chromosome mapping, *Biochemistry,* 29, 5491, 1990.
178. Nolibe, D., Metivier, H., Masse, R., and LaFuma, J., Therapeutic effect of pulmonary lavage *in vivo* after inhalation of insoluble radioactive particles, in *Inhaled Particles IV,* Part 2, Walton, W. H. and McGovern, B., Eds., Pergamon Press, Oxford, 1977, pp. 597–613.
179. Oberdörster, G., Lung clearance of inhaled insoluble and soluble particles, *J. Aerosol Med.,* 1, 289–330, 1988.
180. Overby, L. H., Nishio, S., Weir, A., Carver, G. T., Plopper, C. G. et al., Distribution of cytochrome P-450 1A1 and NADPH-cytochrome P-450 reductase in lungs of rabbits treated with 2,3,7,8-tetrachlorodibenzo-*p*-dioxin: ultrastructural immunolocalization and *in situ* hybridization, *Mol. Pharmacol.,* 41, 1039, 1992.
181. Park, J. F., Bair, W. J., and Busch, R. H., Progress in beagle dog studies with transuranium elements at Battelle-Northwest, *Health Phys.,* 22, 803–810, 1972.
182. Parra, S. C., Burnette, R., Price, H. P., and Takaro, T., Zonal distribution of alveolar macrophages, type II pneumocytes, and alveolar septal connective tissue gaps in adult human lungs, *Am. Rev. Respir. Dis.,* 113, 908–912, 1986.
183. Patrick, G. and Stirling, C., The retention of particles in large airways of the respiratory tract, *Proc. R. Soc. Lond.,* (Series B), 198, 455–462, 1977.
184. Phalen, R., Kenoyer, J., and Davis, J., Deposition and clearance of inhaled particles: comparison of mammalian species, in *Proceedings of the Annual Conference on Environmental Toxicology,* Vol. 7, AMRL-TR-76-125, National Technical Information Service, Springfield, VA, 1977, pp. 159–170.
185. Phalen, R. F., Oldham, M. J., and Schum, G. M., Growth and ageing of the bronchial tree: implications for particle deposition calculations, *Radiat. Prot. Dosim.,* 38, 15–21, 1991.
186. Phalen, R. F., *Inhalation Studies: Foundations and Techniques,* CRC Press, Boca Raton, FL, 1984.
187. Philipson, K., Falk, R., and Camner, P., Long-term lung clearance in humans studies with Teflon particles labeled with chromium-51, *Exp. Lung Res.,* 9, 31–42, 1985.
188. Porter, T. D., Khani, S. C., and Coon, M. J., Induction and tissue-specific expression of rabbit cytochrome P-450IIE1 and IIE2 genes, *Mol. Pharmacol.,* 36, 61, 1989.
189. Postlethwait, E. M. and Bidani, A., Reactive uptake governs the pulmonary air space removal of inhaled nitrogen dioxide, *J. Appl. Physiol.,* 68, 594–603, 1990.
190. Pritchard, J. N., Dust overloading — a case for lowering the TLV of nuisance dusts?, *J. Aerosol. Sci.,* 20, 1341–1344, 1989.
191. Proctor, D. F., Andersen, I., and Lundqvist, G., Nasal mucociliary function in humans, in *Respiratory Defense Mechanisms, Part I,* Brain, J. D., Proctor, D. F., and Reid, L. M., Eds., Marcel Dekker, New York, 1977, pp. 427–452.
192. Puchelle, E., Zahm, J.-M., and Bertrand, A., Influence of age on mucociliary transport, *Scand. J. Respir. Dis.,* 60, 307–313, 1979.
193. Quinlan, M. F., Salman, S. D., Swift, D. L., Wagner, Jr., H. N., and Proctor, D. F., Measurement of mucociliary function in man, *Am. Rev. Respir. Dis.,* 99, 13–23, 1969.
194. Raabe, O. G., Deposition and clearance of inhaled aerosols, in *Mechanisms in Respiratory Toxicology,* Witschi, H. and Nettesheim, P., Eds., CRC Press, Boca Raton, FL, 1982, pp. 27–76.
195. Radford, E. P. and Martell, E. A., Polonium-210:lead-210 ratios as an index of residence times of insoluble particles from cigarette smoke in bronchial epithelium, in *Inhaled Particles IV, Part 2,* Walton, W. H. and McGovern, B., Eds., Pergamon Press, Oxford, 1977, pp. 567–581.
196. Ramsden, D., Bains, M. E. D., and Fraser, D. C., *In vivo* and bioassay results from two contrasting cases of plutonium-239 inhalation, *Health Phys.,* 19, 9–17, 1970.
197. Randall, H. W., Bogdanffy, M. S., and Morgan, K. T., Enzyme histochemistry of the rat nasal mucosa embedded in cold glycol methacrylate, *Am. J. Anat.,* 179, 10, 1987.
198. Reed, C. J., Lock, E. A., and DeMatteis, F., NADPH: Cytochrome P-450 reductase in olfactory epithelium. Relevance to cytochrome P-450-dependent reactions, *Biochem. J.,* 240, 585, 1986.
199. Reed, C. J., Lock, E. A., and DeMatteis, F., Olfactory cytochrome P-450. Studies with suicide substrates of the haemoprotein, *Biochem. J.,* 253, 569, 1988.
200. Robertson, B., Basic morphology of the pulmonary defense system, *Eur. J. Respir. Dis.,* 61(Suppl. 107), 21–40, 1980.
201. Roggli, V. L. and Brody, A. R., Changes in numbers and dimensions of chrysotile asbestos fibers in lungs of rats following short-term exposure, *Exp. Lung Res.,* 7, 133–147, 1984.
202. Roggli, V. L., George, M. H., and Brody, A. R., Clearance and dimensional changes of crocidolite asbestos fibers isolated from lungs of rats following short-term exposure, *Environ. Res.,* 42, 94–105, 1987.
203. Sabourin, P. J., Tynes, R. E., Philpot, R. M., Winquist, S., and Dahl, A. R., Distribution of microsomal monooxygenases in the rabbit respiratory tract, *Drug Metab. Dispos.,* 16, 557, 1988.
204. Sackner, M. A., Landa, J., Hirsch, J., and Zapata, A., Pulmonary effects of oxygen breathing — a 6-hour study in normal men, *Ann. Intern. Med.,* 82, 40–43, 1975.
205. Sackner, M. A., Yergin, B. M., Brito, M., and Januszkiewicz, A., Effect of adrenergic agonists on tracheal mucous velocity, *Bull. Eur. Physiopath. Resp.,* 15, 505–511, 1979.

206. Sanders, C. L., Dagle, G. E., Cannon, W. C., Craig, D. K., Powers, G. J., and Meier, D. M., Inhalation carcinogenesis of high-fired $^{239}PuO_2$ in rats, *Radiat. Res.*, 68, 349–360, 1976.
207. Santa Cruz, R., Landa, J., Hirsch, J., and Sackner, M. A., Tracheal mucous velocity in normal man and patients with obstructive lung disease: effects of terbutaline, *Am. Rev. Respir. Dis.*, 109, 458–463, 1974.
208. Santrock, J., Hatch, G. E., Slade, G. E., and Hayes, J. M., Incorporation and disappearance of oxygen 18 in lung tissue from mice allowed to breathe 1.0 ppm $^{18}O_3$, *Toxicol. Appl. Pharmacol.*, 98, 75–80, 1989.
209. Schlesinger, R. B., Deposition and clearance of inhaled particles, in *Concepts in Inhalation Toxicology*, 2nd ed., McClellan, R. O. and Henderson, R. F., Eds., Taylor & Francis, Washington, D.C., 1995, pp. 191–224.
210. Schlesinger, R. B., Comparative deposition of inhaled aerosols in experimental animals and humans: a review, *J. Toxicol. Environ. Health*, 15, 197–214, 1985.
211. Schlesinger, R. B., Gurman, J. L., and Lippmann, M., Particle deposition within bronchial airways: comparisons using constant and cyclic inspiratory flow, *Ann. Occup. Hyg.*, 26, 47–64, 1982.
212. Schlesinger, R. B. and Lippmann, M., Selective particle deposition and bronchogenic carcinoma, *Environ. Res.*, 15, 424–431, 1978.
213. Schurch, S., Gehr, I. M., Hof, V., Geiser, M., and Green, F., Surfactant displaces particles toward the epithelium in airways and alveoli, *Respir. Physiol.*, 80, 17–32, 1990.
214. Sebastien, P., Fondimare, A., Bignon, J., Monchaux, G., Desbordes, J., and Bonnand, G., Topographic distribution of asbestos fibers in human lung in relation to occupational and nonoccupational exposure, in *Inhaled Particles IV, Part 2*, Walton, W. H., Ed., Pergamon Press, Oxford, 1977, pp. 435–446.
215. Smaldone, G. C., Perry, R. J., Bennett, W. D., Messina, M. S., Zwang, J., and Ilowite, J., Interpretation of "24-hour lung retention" in studies of mucociliary clearance, *J. Aerosol Med.*, 1, 11–20, 1988.
216. Sminia, T., van der Brugge-Gamelkoorn, G. J., and Jeurissen, S. H. M., Structure and function of bronchus-associated lymphoid tissue (BALT), *CRC Crit. Rev. Immunol.*, 9, 119–150, 1989.
217. Smith, D. M., Ortiz, L. W., Archuleta, R. F., and Johnson, N. F., Long-term health effects in hamsters and rats exposed chronically to man-made vitreous fibres, *Ann. Occup. Hyg.*, 31, 731–754, 1987.
218. Snipes, M. B. and Clem, M. F., Retention of microspheres in the rat lung after intratracheal instillation, *Environ. Res.*, 24, 33–41, 1981.
219. Snipes, M. B., Boecker, B. B., and McClellan, R. O., Retention of monodisperse or polydisperse aluminosilicate particles inhaled by dogs, rats, and mice, *Toxicol. Appl. Pharmacol.*, 69, 345–362, 1983.
220. Snipes, M. B., Chavez, G. T., and Muggenburg, B. A., Disposition of 3-, 7-, and 13-μm microspheres instilled into lungs of dogs, *Environ. Res.*, 33, 333–342, 1984.
221. Snipes, M. B., Olson, T. R., and Yeh, H. C., Deposition and retention patterns for 3-, 9-, and 15-μm latex microspheres inhaled by rats and guinea pigs, *Exp. Lung Res.*, 14, 37–50, 1988.
222. Snipes, M. B., Long-term retention and clearance of particles inhaled by mammalian species, *CRC Crit. Rev. Toxicol.*, 20, 175–211, 1989.
223. Snyder, W. S., Cook, M. J., Karhausen, L. R., Naset, E. S., Howells, G. P., and Tipton, I. H., *International Commission on Radiological Protection No. 23: Report of the Task Group on Reference Man*, Pergamon Press, Oxford, 1975.
224. Sorokin, S. P. and Brain, J. D., Pathways of clearance in mouse lungs exposed to iron oxide aerosols, *Anat. Rec.*, 181, 581–626, 1975.
225. Speizer, F. E. and Frank, N. R., The uptake and release of SO_2 by the human nose, *Arch. Environ. Health*, 12, 725–728, 1966.
226. Stahlhofen, W., Gebhart, J., Heyder, J., Philipson, K., and Camner, P., Intercomparison of regional deposition of aerosol particles in the human respiratory tract and their long-term elimination, *Exp. Lung Res.*, 2, 131–139, 1981.
227. Stahlhofen, W., Gebhart, J., Rudolf, G., and Scheuch, G., Measurement of lung clearance with pulses of radioactively-labeled aerosols, *J. Aerosol Sci.*, 17, 333–336, 1986.
228. Stirling, C. and Patrick, G., The localization of particles retained in the trachea of the rat, *J. Pathol.*, 131, 309–320, 1980.
229. Stöber, W., Einbrodt, H. J., and Klosterkötter, W., Quantitative studies of dust retention in animal and human lungs after chronic inhalation, in *Inhaled Particles and Vapours II*, Davies, C. N., Ed., Pergamon Press, Oxford, 1967, pp. 409–418.
230. Stott, W. T. and McKenna, M. J., Hydrolysis of several glycol ether acetates and acrylate esters by nasal mucosal carboxylesterase *in vitro*, *Fundam. Appl. Toxicol.*, 5, 399, 1985.
231. Strandberg, L. G., SO_2 absorption in the respiratory tract. Study on the absorption in rabbit, its dependence on concentration and breathing pulse, *Arch. Environ. Health*, 9, 160–166, 1964.
232. Strom, K. A., Chan, T. L., and Johnson, J. T., Pulmonary retention of inhaled submicron particles in rats: diesel exhaust exposures and lung retention model, *Ann. Occup. Hyg.*, 32(Suppl. 1), 645–657, 1988.
233. Strum, J. M., Ito, T., Philpot, R. M., DeSanti, A. M., and McDowell, E. M., The immunocytochemical detection of cytochrome P-450 monooxygenase in the lungs of fetal, neonatal, and adult hamsters, *Am. J. Respir. Cell. Mol. Biol.*, 2, 493, 1990.
234. Stuart, B. O., Casey, H. W., and Bair, W. J., Acute and chronic effects of inhaled $^{144}CeO_2$ in dogs, *Health Phys.*, 10, 1203–1209, 1964.

235. Stuart, B. O., Promethium oxide inhalation studies, in *Pacific Northwest Laboratory Annual Report for 1965 in the Biological Sciences,* BNWL-280m, Battelle-Northwest, Richland, WA, 1966.
236. Sussman, R. G., Cohen, B. S., and Lippmann, M., Asbestos fiber deposition in a human tracheobronchial cast. I., *Experimental. Inhal. Toxicol.,* 3, 145–160, 1991.
237. Swift, D. L. and Proctor, D. F., A dosimetric model for particles in the respiratory tract above the trachea, *Ann. Occup. Hyg.,* 32(Suppl. 1), 1035–1044, 1988.
238. Swift, D. L., Montassier, N., Hopke, P. K., Karpen-Hayes, K., Cheng, Y. S., Su, Y. F., Yeh, H. C., and Strong, J. C., Inspiratory deposition of ultrafine particles in human nasal replicate cast, *J. Aerosol Sci.,* 23, 65–72, 1992.
239. Swift, D. L., Aerosol deposition and clearance in the human upper airways, *Ann. Biomed. Eng.,* 9, 593–604, 1981.
240. Takahashi, S., Asaho, S., Kubota, Y., Sato, H., and Matsuoka, O., Distribution of ^{198}Au and ^{133}Ba in thoracic and cervical lymph nodes of the rat following intratracheal instillation of ^{198}Au-colloid and $^{133}BaSO_4$, *J. Radiat. Res.,* 28, 227–231, 1987.
241. Takahashi, S., Kubota, Y., and Hatsuno, H., Effect of size on the movement of latex particles in the respiratory tract following local administration, *Inhal. Toxicol.,* 4, 113–123, 1992.
242. Takahashi, Y. and Miura, T., Responses of cytochrome P-450 isozymes of rat lung to *in vivo* exposure to ozone, *Toxicol. Lett.,* 54, 327, 1990.
243. Teschler, H., Friedrichs, K. H., Hoheisel, G. B, Wick, G., Soltner, U., Thompson, A. B., Konietzko, N., and Costabel, U., Asbestos fibers in bronchoalveolar lavage and lung tissue of former asbestos workers, *Am. J. Respir. Crit. Care Med.,* 149, 641–645, 1994.
244. Thomas, R. G., Tracheobronchial lymph node involvement following inhalation of alpha emitters, in *Radiobiology of Plutonium,* Stover, B. J. and Jee, W. S. S., Eds., JW Press, Salt Lake City, UT, 1972, pp. 231–241.
245. Thomas, R. G., Transport of relatively insoluble materials from lung to lymph nodes, *Health Phys.,* 14, 111–117, 1968.
246. Thornton-Manning, J. R., Ruangyuttikarn, W., Gonzalez, F. J., and Yost, G. S., Metabolic activation of the pneumotoxin, 3-methylindole, by vaccinia-expressed cytochrome P-450s, *Biochem. Biophys. Res. Commun.,* 181, 100, 1991.
247. Timbrell, V. and Skidmore, J. W., The effect of shape on particle penetration and retention in animal lungs, in *Inhaled Particles III,* Vol 1, Walton, W. H., Ed., Unwin Brother, Surrey, U.K., 1971, pp. 49–57.
248. Tindberg, N. and Ingelman-Sundberg, M., Cytochrome P-450 and oxygentoxicity. Oxygen-dependent induction of ethanol-inducible cytochrome P-450(2E1) in rat liver and lung, *Biochemistry,* 28, 4499, 1989.
249. Toomes, H., Vogt-Moykopf, I., Heller, W. D., and Ostertag, H., Measurement of mucociliary clearance in smokers and nonsmokers using a bronchoscopic video-technical method, *Lung,* 159, 27–34, 1981.
250. van Ree, J. H. L. and van Dishoeck, H. A. E., Some investigations on nasal ciliary action, *Pract. Oto-Rhino-Laryngol.,* 24, 383–390, 1962.
251. Vaughan, T. R., Jennelle, L. F., and Lewis, T. R., Long-term exposure to low levels of air pollutants, *Arch. Environ. Health,* 19, 45–50, 1969.
252. Vincent, J. H., Jones, A. D., Johnston, A. M., McMillan, C., Bolton, R. E., and Cowie, H., Accumulation of inhaled mineral dust in the lung and associated lymph nodes: implications for exposure and dose in occupational lung disease, *Ann. Occup. Hyg.,* 31, 375–393, 1987.
253. Voight, J. M., Guenegerich, F. P., and Baron, J., Localization of a cytochrome P-450 isozyme (cytochrome P-450 PB-B) and NADPH-cytochrome P-450 reductase in rat nasal mucosa, *Cancer Lett.,* 27, 241, 1985.
254. Waligora, S. J., Jr., Pulmonary retention of zirconium oxide (^{95}Nb) in man and Beagle dogs, *Health Phys.,* 20, 89–91, 1971.
255. Warheit, D. B. and Hartsky, M. A., Species comparisons of alveolar deposition patterns of inhaled particles, *Exp. Lung Res.,* 16, 83–99, 1990.
256. Warner, A. E. and Brain, J. D., The cell biology and pathogenic role of pulmonary intravascular macrophages, *Am. J. Physiol.,* 258, L1–L12, 1990.
257. Wahley, S. L., Muggenburg, B. A., Seiler, F. A., and Wolff, R. K., Effect of aging on tracheal mucociliary clearance in Beagle dogs, *J. Appl. Physiol.,* 62, 1331–1334, 1987.
258. Wahley, S. L., Wolff, R. K., and Muggenburg, B. A., Clearance of nasal mucus in nonanesthetized and anesthetized dogs, *Am. J. Vet. Res.,* 48, 204–206, 1987.
259. Whaley, S. L., Wolff, R. K., Muggenburg, B. A., and Snipes, M. B., Mucociliary clearance and particle retention in the maxillary and ethmoid turbinate regions of beagle dogs, *J. Toxicol. Environ. Health,* 19, 569–580, 1986.
260. Wiester, M. J., Tepper, J. S., King, M. E., Meanache, M. G., and Costa, D. L., Comparative study of ozone uptake in three strains of rats and in the guinea pig, *Toxicol. Appl. Pharmacol.,* 96, 140–146, 1988.
261. Weister, M. J., Williams, T. B., King, M. E., Menache, M. G., and Miller, F. J., Ozone uptake in awake Sprague-Dawley rats, *Toxicol. Appl. Pharmacol.,* 89, 429–437, 1987.
262. Williams, D. E., Ding, X., and Coon, M. J., Rabbit nasal cytochrome P-450 NMa has high activity as a nicotine oxidase, *Biochem. Biophys. Res. Commun.,* 166, 945, 1990.
263. Wolff, R. K., Mucociliary function, in *Comparative Biology of the Normal Lung,* Vol. I, Parent, R. A., Ed., CRC Press, Boca Raton, FL, 1992, pp. 659–680.

264. Wong, J. W., Keens, T. G., Wannamaker, E. M., Crozier, D. N., Levison, H., and Aspin, N., Effects of gravity on tracheal mucus transport rates in normal subjects and in patients with cystic fibrosis, *Pediatrics,* 60, 146–152, 1977.
265. Wood, R. E., Wanner, A., Hirsch, J., and Farrell, P. M., Tracheal mucociliary transport in patients with cystic fibrosis and its stimulation by terbutaline, *Am. Rev. Respir. Dis.,* 111, 733–738, 1975.
266. Yeates, D. B., Aspin, N., Levison, H., Jones, M. T., and Bryan, A. C., Mucociliary tracheal transport rates in man, *J. Appl. Physiol.,* 39, 487–495, 1975.
267. Yeates, D. B., Pitt, B. R., Spektor, D. M., Karron, G. A., and Albert, R. E., Coordination of mucociliary transport in human trachea and intrapulmonary airways, *J. Appl. Physiol.,* 51, 1057–1064, 1981.
268. Yokoyama, E. and Frank, R., Respiratory uptake of ozone in dogs, *Arch. Environ. Health,* 25, 132–138, 1972.
269. Yu, C. P., Chen, Y. K., and Morrow, P. E., An analysis of macrophage mobility kinetics at dust overloading of the lungs, *Fundam. Appl. Toxicol.,* 13, 452–459, 1989.

CHAPTER 13

Respiratory Responses to Inhaled Toxicants

Jane Q. Koenig and Daniel L. Luchtel

CONTENTS

0-8493-4493-X/97/$0.00+$.50

1 INTRODUCTION

Although there have been some successes in decreasing chronic diseases in the U.S. during the last two decades, respiratory disease rates have increased. One of every 5 persons in the U.S. has a form of chronic pulmonary disease. Symptoms include reductions in the maximal amount of air expelled, chest tightness, wheezing, and possibly cancer. One of the most notable instances of this increase is seen with asthma. In most of the industrialized world, both asthma morbidity and mortality have increased 30 to 40% over the last 10 years.[24,50] Also due to circumstances not understood at this time, mortality from asthma among black Americans is significantly higher than among whites.[24] Another indicator of the serious nature of respiratory disease is the fact that occupational lung disease is the most common occupationally associated disease. Upper airway diseases, primarily nasal disease, also have strong occupational associations. This emphasizes the importance of an understanding of inhaled toxicants which cause and/or aggravate respiratory disease.

2 SOURCES OF RESPIRATORY HEALTH EFFECTS DATA

The four primary sources of data on respiratory effects of inhaled irritant gases and particles are animal toxicology, *in vitro* systems using either animal or human cells, controlled human laboratory studies, and epidemiology. Each field of endeavor has its obvious advantages and disadvantages, as described in Table 1.

TABLE 1

Advantages and Disadvantages of the Three Fields of Health Effects Research

Advantages	Disadvantages
Animal Toxicology and *In Vitro* Systems	
Advantages	**Disadvantages**
Morphological examinations	Extrapolation to humans (or from cells to the intact human)
Biochemical assays	Wide interspecies differences in cell types, lung morphology, enzyme systems
Can use higher doses for mechanistic purposes	Differences in breathing patterns, age span, and response to inhaled materials
Chronic controlled studies	Cells removed from normal physiological interactions
Combination of pollutants	
Life-time exposures	
Human Controlled Exposure Studies	
Advantages	**Disadvantages**
Known subject characteristics	Short-term exposures
Controlled pollutant concentrations	Usually noninvasive measurements
Known activity levels	Small sample size
Exact measurements of effects	
Epidemiology	
Advantages	**Disadvantages**
Study of chronic effects	Inadequate air monitoring data
Realistic exposures	Poor characterization of the actual composition of pollutants over time
Real-time lung function and respiratory symptom measurements	Inadequate health endpoints based on recall of symptoms or diaries with poor quality assurance
	Relative infrequency of some chronic respiratory diseases
	Multifactorial nature of these chronic diseases
	Difficulty in estimating biologically effective dose
	Confounding variables
	Tobacco smoke exposure
	Occupational exposure
	Variation in diets
	Respiratory infections
	Genetic and racial variation

2.1 Animal Toxicology

The selection of species for study is crucial in determining the outcome of a study. Following is a list of several mammalian species commonly used in inhalation studies.

2.1.1 Rodent

Advantages: Widely used in modern inhalation studies; large background literature on structure, physiology, and biochemistry; adaptable to exercise. *Disadvantages:* Pulmonary function measurements on awake animals are difficult; animals do not have well-developed respiratory bronchioles; nasal pharyngeal anatomy not similar to human; tendency toward confounding spontaneous respiratory infections (especially rats). *Other:* Guinea pig with unusually abundant bronchial smooth muscle is a good model for bronchoconstriction; short lifespans (can be advantage or disadvantage).

2.1.2 Dog

Advantages: Convenient size for many physiological measurements; cooperative temperament; normal physiology; morphology well understood; pulmonary aerosol deposition similar to human; useful as a model for tracheal mucous movement. *Disadvantages:* Requires spacious housing; care, feeding is expensive; favorite subject of animal rights groups; subgross lung and nasal anatomy unlike that of humans. *Others:* Has unusually large collateral ventilation.

2.1.3 Cat

Advantage: Convenient intermediate size. *Disadvantage:* Little used in air pollution studies; favorite subject of animal rights groups. *Other:* Unusually large, abundant mucous cells and mucous glands.

2.1.4 Rabbit

Advantage: Convenient intermediate size; well-characterized physiology. *Disadvantage:* Little used in air pollution research.

2.1.5 Ferret

Advantages: Well-developed respiratory bronchioles and lung anatomy similar to humans; well-known husbandry requirements; runs well on treadmill; laboratory-quality purebred animals less expensive compared to dogs. *Disadvantages:* Somewhat disagreeable temperament and scent; not used extensively in the past for inhalation toxicology. *Other:* Very long trachea in relation to lung size; large lung volume per unit body weight.

2.1.6 Horse

Advantages: Subgross lung structure similar to human; large size useful for many procedures; chronic lung diseases similar to those of humans; many pulmonary function characteristics similar to humans; husbandry and care well understood. *Disadvantages:* Requires large housing area; expensive relative to other species.

2.1.7 Bovine (Sheep)

Advantages: Pulmonary function measurement techniques available; mucous cell and gland density similar to those of human; passive nature in laboratory permitting use of a

TABLE 2

Criteria for Selecting an Animal Model

1. Similarity to humans with respect to any or all of the following: anatomy, physiology (specifically pulmonary), and susceptibility to injury or disease
2. Lifespan appropriate to the particular study
3. Possession of unique qualities, such as great sensitivity to certain classes of toxins, or anatomical or physiological properties leading to increased accuracy or precision in empirical measurements
4. Existence of a large or appropriate database on that animal model
5. Existence of proven procedures or apparatus pertaining to the model
6. Simplicity that lends to ease in mathematically modeling the animal system or its responses to airborne pollutants

Source: From Phalen, R. F., *Inhalation Studies: Foundation and Techniques,* CRC Press, Boca Raton, FL, 1984, p. 211. With permission.

TABLE 3

Parameters to Fulfill in Experimental Design

1. One or more species that handle the test material similarly to humans should be used; metabolism, absorption, excretion, storage, and other physiological factors should be considered.
2. Several dose levels should be used, unless a large, multiple maximum, realistic human exposure dose is found to be nontoxic.
3. The biologically insignificant (regarding risk) dose should be found and a safety factor above this dose applied for human protection.
4. Randomization of dosed and control animals must be done to ensure the validity of statistical tests.
5. The route(s) of administration to animals should match those anticipated for humans.

Source: From Phalen, R. F., *Inhalation Studies: Foundation and Techniques,* CRC Press, Boca Raton, FL, 1984, p. 216. With permission.

variety of procedures on unanesthetized individuals. *Disadvantages:* Requires large housing space.

2.1.8 Nonhuman Primate

Advantages: Nasal anatomy similar to human; convenient size; pulmonary function techniques available; commonly used in inhalation studies. *Disadvantages:* Husbandry practices not well standardized; expense and lack of general availability; transmission of serious diseases to handlers; subgross pulmonary anatomy differs from human; reputation for being less than cooperative.

Table 2 lists some criteria to be considered in selection of the appropriate animal model for a specific study. Table 3 describes some parameters to fulfill in constructing an acceptable experimental design.

One of the greatest concerns with the use of animal toxicologic data in predicting human responses is the problem of extrapolation from nonhuman species to humans. Because cell types, enzyme systems, life span, and genetic diversity are all quite different among animals and humans, extrapolation must be made with the utmost care. Following are some factors which reduce the uncertainty.

TABLE 4
Biochemical Markers of Pulmonary Effects of Inhaled Toxins

Source of Marker	Marker	Indications
BALF[a]	Altered cytology	Various types of inflammatory responses
	Extracellular cytoplasmic enzymes	Cytoxicity
	Elevated serum proteins	Damaged alveolar/capillary barrier
	Extracellular lysosomal enzymes	Activated or lysed macrophage
	Elevated *a*-glutamyl-transpeptidase activity	Damage to Clara cells
	Elevated alkaline phosphatase activity	Damage to type II cells or transudation of serum proteins
	Elevated histamine	Allergic response
BALF[a] cells	Increased release of cytokine (TNF, IL-1, Fn) and oxygen radicals	Fibrotic processes
Blood	Elevated lysozyme activity	Chronic bronchitis
	Elevated PIIINP	Fibrotic processes
Urine	Elevated hydroxyproline, hydroxylysine	Connective tissue breakdown (not lung specific)
	Elevated desmosine, isodesmosine	Emphysematous process
	Increased bacterial mutagenicity	Exposure to high levels of mutagens or their precursors
Lung tissue	Elevated mRNAs for extracellular matrix proteins	Fibrotic process
	Elevated mRNA for TNF	Fibrotic process
Altered DNA	Activated K-*ras* oncogene	Cell transformation
	Mutation of *p53* gene	Loss of tumor suppressor activity
	Overexpression of $p185^{neu}$	Decreased survival of patients with adenocarcinomas
	Hypermethylation of DNA	Decrease in gene transcription
	Hypomethylation of DNA	Increase in gene transcription
	Restriction length fragment polymorphisms	Loss of heterozygosity, susceptibility marker

Note: The biological markers listed here have not all been validated. Some are listed as promising areas for research.
[a] BALF = bronchial alveolar lavage fluid.
Source: From Henderson, R. F. and Belinsky, S. A., in *Toxicology of the Lungs,* Gardiner, D. E. et al., Eds., Raven Press, New York, 1993, chpt. 9. With permission.

1. Study of several species such that a generalizable mammalian response pattern can be identified and the human response can be extrapolated on the basis of body size, lifespan, etc.
2. Good understanding of the biochemical or physiologic mechanisms of the experimental animal along with the relevant animal and human similarities and differences in normal biochemistry or physiology
3. Choice of a toxin which belongs to a chemical or physical class in which other members have been studied and animal and human responses have been directly compared previously

Biological markers of exposure, effects, or susceptibility offer promise of early detection of pollutant-induced respiratory disease caused by toxic agents.[114] Table 4 lists some potential sources of biomarkers, while Table 5 lists specific biomarkers by chemical agents. Biomarkers are useful in four fields of health effects research.

2.2 *In Vitro* Systems

Three major thrusts are driving the use of *in vitro* systems for the study of adverse effects of inhaled irritants. One is the animal rights movement, which has caused researchers to reconsider the use of animals in toxicological studies and has created an impetus for using

TABLE 5

Chemical-Specific Biological Markers of Respiratory Tract Exposures

Toxin	Marker
Irritant gases	
O_3	^{18}O
NO_x	NO-heme
SO_2	$S-SO_3$
CO	Carboxyhemoglobin
VOC	Parent compound in exhaled air, urinary metabolites, protein adducts, DNA adducts
Particles	Electron probe analysis, MPG

Source: From Henderson, R. F. and Belinsky, S. A., in *Toxicology of the Lung,* Gardiner, D. E. et al., Eds., Raven Press, New York, 1993, chpt. 9. With permission.

TABLE 6

In Vitro Systems for Study of the Respiratory System Cultures

Cell culture	
Parenchymal	Type II cell, type I cell, fibroblast, mesothelial cell, macrophage
Airway	Ciliated, mucous, dedifferentiated
Vascular	Endothelial cell (from large and small blood vessels)
Cell lines	BEAS 2B, A-549
Explant culture	Tracheal/bronchial ring, parenchymal
Implant culture	Trachael
Organ culture	Parenchymal slices
Other	
Nasal or tracheobronchial lavage	
Inflammatory cells (macrophages, neutrophils, lymphocytes)	
Isolated, perfused, ventilated lung	
Improvements in cell culture	
Methods of dissociating cells from tissue fragments	
Isolating single cell types to initiate cultures	
Types of media and addition of specific growth factors and vitamin supplements	
Culture substrates such as porous membranes for the growth of cells	
Methods of gene transfection in eukaryotic cells in culture	

nonanimal alternatives; the use of animal cells or human cell lines is an attractive alternative. Second, cell culture techniques are now sufficiently advanced such that cells can be grown in culture so they differentiate in culture and behave much the same as in the body. The third important advance in *in vitro* system studies is the increasing use of cells derived from human tissues. Both human nasal and bronchial epithelial cells are now grown successfully in culture. Also, several immortalized cell lines are available such as the BEAS 2B and the A549 cells. A summary of some parameters to consider with *in vitro* research is given in Table 6.

2.3 Controlled Human Laboratory Studies

A complete description of this field can be found in Chapter 14. Table 7 summarizes the factors which must be considered and controlled in such studies. Exposure systems deserve some additional discussion. Mouthpiece exposure using blow-by atmosphere delivery has many advantages. The major advantages are that only a small amount of test atmosphere must be generated and, with use of a pneumotachograph, minute ventilation can be recorded

TABLE 7

Methods for Assessing the Health Effects of Air Pollutants in Human Subjects

Subject Selection
Healthy subjects, such as high performance athletes
Susceptible (at risk) subjects, such as asthmatics, chronic obstructive pulmonary disease (COPD) patients, smokers, elderly, children
Mode of Exposure
Whole-body chamber
Unencumbered breathing
Expensive
Blow-by system
Easy to control
Continuous measure of minute ventilation ($\dot{V}_E$)
Mouthpiece or mask exposure (unrealistic, artificial breathing)
Head dome
Unencumbered breathing
Inexpensive
With care, can measure $\dot{V}_E$
Assessment of Effects
Pulmonary function tests (PFTs)
Mechanisms of airway narrowing
Biomarker tests
Inflammation
Lung injury
Bronchial reactivity tests
Bronchoalveolar lavage
Nasal lavage
Symptom ratings
Protocol Considerations
Single pollutants vs. combined
Rest vs. exercise
Duration
Timing of PFT recording sessions
Delayed effects

continuously during exposure, allowing more precise calculation of inhaled dose. However, one obvious disadvantage is the artificial nature of breathing. Also, the differential scrubbing effect of nasal vs. oral or oronasal breathing has been targeted as a consideration in extrapolating results to actual ambient exposures.[66,68] Brain and Sweeney[19] suggest that the loss of filtering capacity resulting from the switch to oral breathing may contribute to the phenomenon of exercise-induced bronchoconstriction. Niinimaa and co-workers[119] estimated that 70 to 85% of healthy adults breathe oronasally during moderate exercise.

Whole-body chambers remove the disadvantage of the artificial breathing pattern; however, chambers have some disadvantages of their own, the major disadvantage being the initial and ongoing expense of the accessory instrumentation required to monitor both the generation and distribution of the test atmosphere. Also, chambers are poorly suited to the study of unstable compounds or contaminants which in very low concentration may be diluted by subject rebreathing.[12] Finally, chamber exposures do not allow easy measurement of minute ventilation. The facemask has been used both in studies of nasal vs. oronasal inhalation with SO_2[81] and in determining the regional deposition of inhaled nitric and sulfuric acid fogs;[18] however, little research has been done on the delivery of contaminants to human subjects in a system allowing oronasal breathing other than whole-body chambers.[121]

TABLE 8

Sources of Variation in Measurements of Lung Function

Variation	Source of Variation
Within individual	
Technical	Within and between instruments
	Within and between observers in administration and reading of tests
	Curve selection
	Temperature
Biologic	Comprehension and/or cooperation of the subject
	Circadian, weekly, and seasonal effects
	Endocrine, other
Between individuals	
	All of the above sources
Subject	Size, sex, age, respiratory muscularity
	Race and other genetic characteristics
	Past and present health
	Habits (e.g., smoking, physical activity)
Environmental	Residence (income, ambient pollution, etc.)
	Indoor pollution (smoking, gas stoves, etc.)
	Occupational exposures
Between populations	All the above sources
	Selection into or out of the target or study population

Source: From Bates, B. V., *Respiratory Function in Disease,* 3rd ed., W.B. Saunders, Philadelphia, PA, 1989. With permission.

The development of a method that minimizes operation costs without sacrificing the validity of data would greatly expand the arena of air pollution research. One such system is a lightweight head dome which can be suspended over the shoulders and sealed around the neck. The prototype head dome chamber was developed by Dr. Stephen Bowes and coworkers.[18] The chamber consists of a plastic cylinder of approximately 40 l which rests on the shoulders, counterweighted by an overhead support allowing relatively free movements of the subject. The method allows a relatively inexpensive alternative to whole body chambers, thereby expanding the scope of inhalation studies in research settings. This system has been used successfully in controlled exposures of subjects to sulfuric acid and ozone.[17,79,105]

Thus, each of the methods has advantages or disadvantages related to the degree of precision in assessing dose, loss of contaminant to the subject or system, ease of operation, or cost of initial setup and routine maintenance costs. Particularly in the case of studies of human health effects, the need for optimizing the attainment of research objectives for each exposure session becomes of great importance. For instance, in the study of acidic compounds, neutralization of the acid by oral or body ammonia is a major concern.[57,90]

Although correctly measured pulmonary function tests have an intrasubject variation of only ±5%,[48] host factors, technician competence, and physical conditions do affect the values. Table 8 lists some sources of variation in measurement of lung function.

An accurate exposure assessment is necessary to calculate the dose of pollutant that may reach the target tissue. The important distinction between the concentration of an irritant measured in ambient or workplace air and the actual dose delivered to target tissue is laid out in a recent review.[116] (Pulmonary disposition of inhaled irritants is the topic of Chapter 12).

As mentioned earlier, for several years it has been hoped that biomarkers could be identified that would document exposure, effects, and susceptibility to inhaled irritants.[114] This search has been very active in the controlled human studies of air pollutants. With respect to biomarkers of exposure, investigators are searching for some exogenous agent or its metabolite. These biomarkers have been difficult to demonstrate in air pollution studies. The prototypic biomarker of exposure for air pollutants is carboxyhemoglobin. Some urinary biomarkers are available for volatile organic carbons (VOC). Recently, Morgan and co-workers[110] have explored exhaled breath analysis as a source of exposure biomarkers. With respect to biomarkers of effects, even measurements as noninvasive as pulmonary function tests (PFTs) have been categorized as biomarkers. In studies of the effects of inhaled gases and aerosols on the respiratory system, likely biomarkers of effect are the presence of inflammatory mediators retrieved from the lung or nasal passages after breathing the pollutant. The list of potential inflammatory mediators is long (see Table 9 for selected candidates).

As shown in Table 9 pre-existing disease is a biomarker for increased susceptibility for adverse respiratory effects from inhaled irritants. Some characteristics and definitions of human lung disease are given in Table 10.

2.4 EPIDEMIOLOGY

Epidemiology is the primary discipline that allows investigators to explore the long-term effects of air pollution on public health. As pointed out by Hill,[68] epidemiology cannot prove causation; however, "causal" relationships between exposure and effect are usually plausibly inferred by a number of aspects of a study. These include the:

- Strength of the association
- Consistency of data
- Specificity of results
- Temporality of observations
- Demonstration of a biologic gradient
- Plausibility and coherence of results

The emerging field of molecular epidemiology seeks biologic markers such as covalent adducts formed by the binding to toxicants and macromolecules at or near the site of toxic action.[114] Such adducts could prove to be valuable markers of biologically effective doses in the case of DNA adducts or of exposure in the case of hemoglobin adducts.

3 HEALTH EFFECTS OF IRRITANT POLLUTANTS

The remainder of this chapter contains a description of the toxicological effects of common airborne irritants. The irritants (also referred to as air pollutants) are divided into four categories:

1. Common outdoor air pollutants regulated by the U.S. Environmental Protection Agency (USEPA) as "criteria" pollutants: National Ambient Air Quality Standards (NAAQS) are set for these pollutants and given in Table 11.
2. Common indoor air pollutants (see Table 12).
3. Air contaminants regulated by Maximum Achievable Control Technology, 1989 "Air Toxics": This category of pollutants was regulated under National Emission Standards for Hazardous Air Pollutants (NESHAPS) prior to the passage of the 1990 Clean Air Act (CAA) amendments.
4. Occupational irritants (see Tables 42 and 43): The Occupational Safety and Health Act (OSHA) authorized the agency to set standards for workplace exposures. Additionally, threshold limit values (TLVs) are set by the American Conference of Governmental Industrial Hygienists. Both 8-hour time-weighted average (TWA) concentrations and short-term exposure limits (STELs) are set.

TABLE 9

Biomarkers of Air Pollutant-Induced Effects in Human Subjects

Biomarker of Exposure

- Deposited material
  - Blood, urine, teeth, hair
  - Respiratory tract fluids
    - Sputum
    - Saliva
  - Exhaled air

Biomarkers of Effects

- PFTs (measures of pulmonary mechanics)
  - Forced expiratory volume in 1 sec (FEV_1)
  - Forced vital capacity (FVC)
  - Airway resistance (R_{aw})
  - Total respiratory resistance (R_T)
- Measures of airway reactivity
  - Methacholine or histamine challenge
  - Cold air
  - Hypo-osmolar saline
  - SO_2
- Deposition and clearance
- Nasal lavage
- Bronchoalveolar lavage
- Injury to blood-air barrier
- Changes in airway permeability
- Potential inflammatory mediators (selected examples)
  - Histamine
  - Leukotrienes, e.g., LTE_4
  - Prostaglandins, e.g., PGD_2 or $PGDF_{2a}$
  - Platelet activating factor (PAF)
  - Interleukins
  - Cells: neutrophils, mast cells, macrophages
  - Receptors: beta adrenergic, muscarinic
- Markers of injury
  - Lactose dehydrogenase (LDH)
  - Glutathione
  - Uric acid

Biomarkers of Susceptibility

- Asthma
  - Elevated IgE
  - Bronchial hyperresponsiveness, determined by a standardized methacholine challenge test
  - Exercise-induced bronchospasm
  - Intradermal tests for inhalant allergens
- Chronic obstructive pulmonary disease (COPD)
- Cystic fibrosis
- Idiopathic pulmonary fibrosis
- Age
- Ethnicity
- Genetic susceptibilities
- Pre-existing disease

3.1 Outdoor Criteria Pollutants

Two aspects of the outdoor air pollutant standards are important: concentration selected and averaging time. The averaging time reflects the fate and duration of the pollutant in the atmosphere. Currently the USEPA is considering whether to set a short-term (5-min) standard

TABLE 10

Summary of Characteristics of Human Lung Diseases

Inflammation: Acute or chronic response to cell/tissue injury, accompanied by influx of inflammatory cells (macrophages, neutrophils, lymphocytes, eosinophils) and mediators

Bronchitis/bronchiolitis: Inflammation of airways, enlargement of mucous glands, and increased mucous cells; accompanied by airway narrowing by mucous plugs and air blockage, edema, inflammation, and fibrosis

Edema: Leakage of fluid from vascular compartment to interstitial and air compartments

Bronchial asthma: Heightened reactivity of bronchial tree to stimuli leading to bronchial constriction and inflammation; wheezing, cough, dyspnea, and occasionally tenacious sputum may be present

Emphysema: Destruction and permanent loss of septal walls of alveoli, alveolar ducts, and respiratory bronchioles

Allergic alveolitis (hypersensitivity pneumonitis): Tissue reaction to inhaled antigenic organic aerosols; leads to infiltration of lung by inflammatory cells and can result in edema, fibrosis, and emphysema

Fibrosis: Irreversible thickening of alveolar septa/airways by formation of fibrotic (scar-like) tissue; may be associated with edema, presence of inflammatory cell response

TABLE 11

U.S. National Ambient Air Quality Standards

Pollutant	Primary Standard	Averaging Time
Sulfur dioxide (SO_2)	.14 ppm (365 μg/m³)	24 hr; may be exceeded once per year
	.03 ppm (80 μg/m³)	1 year
Particulate matter, ≤10 μm (PM_{10})	150 μg/m³	24 hr; may be exceeded once per year
	50 μg/m³	1 year
Carbon monoxide (CO)	35 ppm (40 μg/m³)	1 hr; may be exceeded once per year
	9 ppm (10 μg/m³)	8 hr; may be exceeded once per year
Ozone (O_3)	.12 ppm (240 μg/m³)	1 hr; may be exceeded once per year
Nitrogen dioxide (NO_2)	.053 ppm (100 μg/m³)	1 year
Airborne lead	1.5 μg/m³	Quarterly

TABLE 12

Sources for Indoor Air Pollution

Pollutant	Source of Pollutant
Environmental tobacco smoke (ETS)	Cigarette, cigar, pipe smoking
Nitrogen dioxide (NO_2)	Gas cooking stoves
Carbon monoxide (CO)	Home-heating devices, including fireplaces and wood-burning stoves
Fine particulate matter	ETS, wood stoves
Formaldehyde	Particle board cabinets, etc.; foam insulation (especially mobile homes)
Organic compounds	Paints, solvents, glues, hobby materials
Pesticides	Flea bombs, tracked-in garden pesticides

for SO_2. Also, the USEPA is reviewing the NAAQS for ozone and particulate matter (PM_{10}). As of July 1994, the new ozone criteria document was in draft form and undergoing external review. The new particulate matter criteria document is being drafted. In both cases, there is support in the scientific community for lowering the concentration and changing the averaging time. With regard to the particulate matter standard, there is discussion of changing the regulated particle size. A $PM_{2.5}$ standard has been mentioned.

TABLE 13

Ozone Formation in the Troposphere

Sources of NMOC or NMHC
Combustion
Chemical processing
Gasoline pumping
Oil refining
Dry cleaning
Sources of NO_x
Primarily combustion, both mobile and stationary
Other photochemical oxidants
Peroxyacetyl nitrate (PAN)
Nitrous acid (HONO)
Nitric acid (HNO_3)

Note: NMOC = nonmethane organic compounds; NMHC = nonmethane hydrocarbons.

3.1.1 Ozone (O_3)

The process of ozone formation in the troposphere is shown below as a reaction between nonmethane organic compounds (NMOC) or nonmethane hydrocarbons (NMHC) and nitrogen oxides in the presence of ultraviolet light.

$$NMOC + NO_x + h\nu \rightarrow O_3 + \text{other pollutants}$$

Ozone is a common source of occupational exposure. Table 13 gives a list of sources of ozone precursors and other photochemical oxidants. A list of sites, uses, and occupational sources of ozone exposure is given in Table 14. Standards for exceedances of the NAAQS and occupational standards are given in Table 15.

3.1.1.1 Animal Toxicology

Mechanisms of ozone toxicity include: (1) oxidation of polyunsaturated fatty acids, especially in membranes; (2) formation of free radicals;[108,125] (3) formation of secondary toxic compounds through oxidation of lipids; and (4) oxidation of sulfhydryl compounds. Animal studies (acute exposures to less than 1 ppm of ozone) have documented the following adverse respiratory effects summarized in the O_3 criteria document currently under review:[40]

- Inflammatory changes throughout the airways, but particularly bronchioles and proximal alveoli
- Injury to alveolar macrophages with secondary effect of reduced resistance to infection
- Impaired muco-ciliary clearance
- Impaired lung function
- Genotoxicity in a variety of assay systems, but results inconsistent
- Development of murine pulmonary adenomas and other hyperplastic nodules in the lungs of nonhuman primates

Chapter 6 of the USEPA[40] document contains many tables useful to an understanding of the toxic effects of ozone inhalation. Lung inflammation and permeability changes have been seen at concentrations as low as 0.1 ppm for short-term exposures. For instance,

TABLE 14
List of Sites, Uses, and Occurrences of Ozone

Environmental
Stratosphere (up to 10 ppm from ultraviolet effect on oxygen)
Troposphere (photochemical smog, electrical storms, stratospheric mixing)
Occupational
Oxidizing agent in chemical manufacturing
Peroxide manufacturing
Disinfectant (drinking water, food in cold storage rooms, sewage treatment)
Deodorizing agent (air, sewer gas, feathers)
Industrial waste treatment
Bleaching agent (paper pulp, oils, textiles, waxes, flour, starch, sugar)
Aging of liquor and wood
Contamination of high-altitude aircraft cabins
Mercury vapor lamps
Photocopy and FAX machines, laser printers
Electric arc welding
High voltage electrical equipment
Linear accelerators
X-ray generators
Indoor ultraviolet sources
Electrostatic air cleaners

TABLE 15
Exposure Limits and Guidelines for Ozone

Environmental	
National Ambient Air Quality Standards	0.12 ppm (1-hr avg.)
Recommended episode criteria (smog alert levels)	
Stage 1 (alert)	0.20 ppm (1-hr avg.)
Stage 2 (warning)	0.40 ppm
Stage 3 (emergency)	0.50 ppm
Emergency exposure limit (NAS)	1 ppm (1-hr avg.)
Occupational	
Threshold limit value (ACGIH)	0.10 ppm (8-hr TWA)
Permissible exposure limit (OSHA)	0.10 ppm (8-hr TWA)
Short-term exposure limit (OSHA)	0.30 ppm (15-min avg.)
Immediately dangerous to life and health (NIOSH)	10 ppm (30-min avg.)

Note: NAS = National Academy of Scientists; ACGIH = American Conference of Governmental Industrial Hygienists; OSHA = Occupational Safety and Health Administration; NIOSH = National Institute for Occupational Safety and Health; TWA = time-weighted average.

increased numbers of alveolar macrophages were seen in rabbits and rats. Increased levels of prostaglandins have been seen in rabbits, cows, and dogs. Increased polymorphonuclear lymphocytes (PMNs) have been seen in most species studied at and below the present NAAQS value of .12 ppm. Immunological effects on host defense mechanisms have been reported in rats and mice at higher concentrations of ozone, mainly above 0.5 ppm. There also are numerous studies documenting pathological and morphological changes in airways in rats,

TABLE 16

Health Effects of Ozone — Controlled Human Laboratory Studies

Decreased lung volume at < .120 ppm	
Pain on deep inspiration	
Influx of inflammatory cells and mediators into nose and lungs using bronchioalveolar or nasal lavage[a]	
Increased permeability of lung membranes	
Attenuation of effects on PFTs with repeated exposures (exposure to O_3 on 5 consecutive days)	
Day 1	Significant effect
Day 2	Greater effect
Day 3	Same as Day 2
Day 4	Lesser effect, but still significant
Day 5	No effect
Days 6–9	No exposure
Day 10	Exposure to same concentration as above, significant decrements in PFTs
Increased susceptibility in asthmatics[b]	
Altered regional particle deposition patterns[c]	

Note: Some of these changes are summarized in Table 17

[a] See Koren et al.,[85] Graham et al.,[54] and Bascom et al.[8]

[b] See Koenig et al.[84]

[c] See Foster.[44]

mice, and nonhuman primates. In general, chronic exposure to ozone at concentrations below 1 ppm causes a smaller, stiffer lung. Some morphological effects such as increased centriacinar lesions and thickening of alveolar septa are seen at durations less than 2 weeks. Ozone exposure also causes changes in lung lipids, proteins, and anti-oxidants.

3.1.1.2 Controlled Laboratory Studies

Ozone has been studied in controlled laboratory settings more extensively than any criteria pollutant. For up-to-date information on many aspects of ozone health effects, the reader is referred again to the new criteria document for ozone and other photochemical oxidants being considered by the USEPA.[40] A summary of some of the primary effects of ozone inhalation seen in controlled human studies is given in Table 16.

A recent review of toxic air pollutants summarized a number of respiratory effects seen after ozone exposure in controlled laboratory settings (Table 17).[100] From this table, it is important to note that significant decrements in pulmonary function are seen after exposures to concentrations considerably lower (80 ppb) than the present NAAQS for ozone (120 ppb).

As mentioned, airway inflammation is one of the hallmarks of human exposure to ozone. Figure 1 shows the relative increase after ozone compared to air exposure in various markers of inflammation. The data are from bronchoalveolar lavage (BAL) fluid obtained from healthy subjects exposed to 0.4 ppm ozone for 2 hr during intermittent exercise.[85] Similar changes have been seen after concentrations of ozone as low as 0.08 ppm.[30] There is also one report that exposure to ozone alters regional function and particle dosimetry in the human lung.[44] Studies using healthy adult subjects have shown wide intersubject variability in pulmonary response to inhaled ozone (see, for instance, McDonnell et al.[106]). Approximately 10% of study populations show exaggerated pulmonary responses. Most attempts to identify predictors for apparent susceptibility have failed, with the exception of a recent study which found that younger age (15 vs. 30 years) was a risk factor for sensitivity to inhaled ozone.[107] Ethnic variations in response to common air pollutants are virtually unstudied, although Seal et al.[143] did find greater pulmonary function decrements in black adult males than in comparable white subjects. This same study found no difference between males and females in their

TABLE 17
Human Responses to Single O_3 Exposures

Response	Subjects	Exposure Conditions
5–10% mean decrement in FEV_1	Healthy young men	180 ppb with intermittent heavy exercise for 2 hr, O_3 in purified air
		100 ppb with moderate exercise for 6.6 hr, O_3 in purified air
		100 ppb with very heavy exercise for 0.5 hr, O_3 in ambient air
	Healthy children	100 ppb, normal summer camp program, O_3 in ambient air
Increased cough	Healthy young men	120 ppb with intermittent heavy exercise for 2 hr, O_3 in purified air
	Healthy young men	80 ppb moderate exercise for 6.6 hr, O_3 in purified air
	Healthy young men and women	120–130 ppb heavy exercise for 16–28 min, O_3 in purified air
Reduced athletic performance	Healthy young men	180 ppb with exercise at V_E of 54 l/min for 30 min, 120 l/min for 30 min, O_3 in purified air
	Healthy young men and women	120–130 ppb with exercise at V_E of 30–120 l/min for 16–28 min, O_3 in purified air
	Healthy young men	80 ppb with moderate exercise for 6.6 hr, O_3 in purified air
	Young adult men with allergic rhinitis	180 ppb with heavy exercise for 2 hr, O_3 in purified air
Increased airway permeability	Healthy young men	400 ppb with intermittent heavy exercise for 2 hr, O_3 in purified air
Increased airway inflammation	Healthy young men	80 ppb with moderate exercise for 6.6 hr, O_3 in purified air
Accelerated tracheobronchial particle clearance	Healthy young men	200 ppb with intermittent light exercise for 2 hr, O_2 in purified air

Note: FEV_1 = forced expiratory flow at 1 sec.
Source: Lippmann, M., in *Environmental Toxicants — Human Exposures and Their Health Effects,* Lippmann, M., Ed., Van Nostrand-Reinhold, New York, 1992, chpt. 16. With permission.

responses to ozone,[143] although another research group has reported gender differences with females being more susceptible.[109]

There is some controversy as to whether subjects with asthma are more susceptible to adverse respiratory effects from ozone exposures than subjects with no signs or symptoms of bronchial hyperresponsiveness. A recent review of the literature[84] reveals at least five studies indicating that asthmatics show an increased sensitivity to ozone. Silverman[146] found consistent decreases in maximal expiratory flow rate in a subset of asthmatic subjects exposed to 0.25 ppm ozone for 2 hr at rest. Kreit and co-workers[88] reported greater pulmonary function changes (FEV_1, FEV_1/FVC ratio, and $FEV_{25\text{-}75}$) in asthmatic subjects than in healthy subjects after 2-hr exposures to 0.40 ppm ozone during intermittent exercise. (Peak concentrations of ozone in Los Angeles are now in the range of 0.30 to 0.35 ppm.) Hackney and co-workers[56] reported that more subjects who showed sensitivity to 0.18 ppm ozone exposure and bronchial hyperresponsiveness had allergies than did those not sensitive to ozone. Aris and co-workers[3] found an increased frequency of bronchial hyperresponsiveness in ozone responders recruited for a controlled laboratory study.

Ozone has been shown to affect the upper airways as well as the lung. Since ozone is relatively water insoluble, it was assumed that most of the ozone in inhaled air was transported

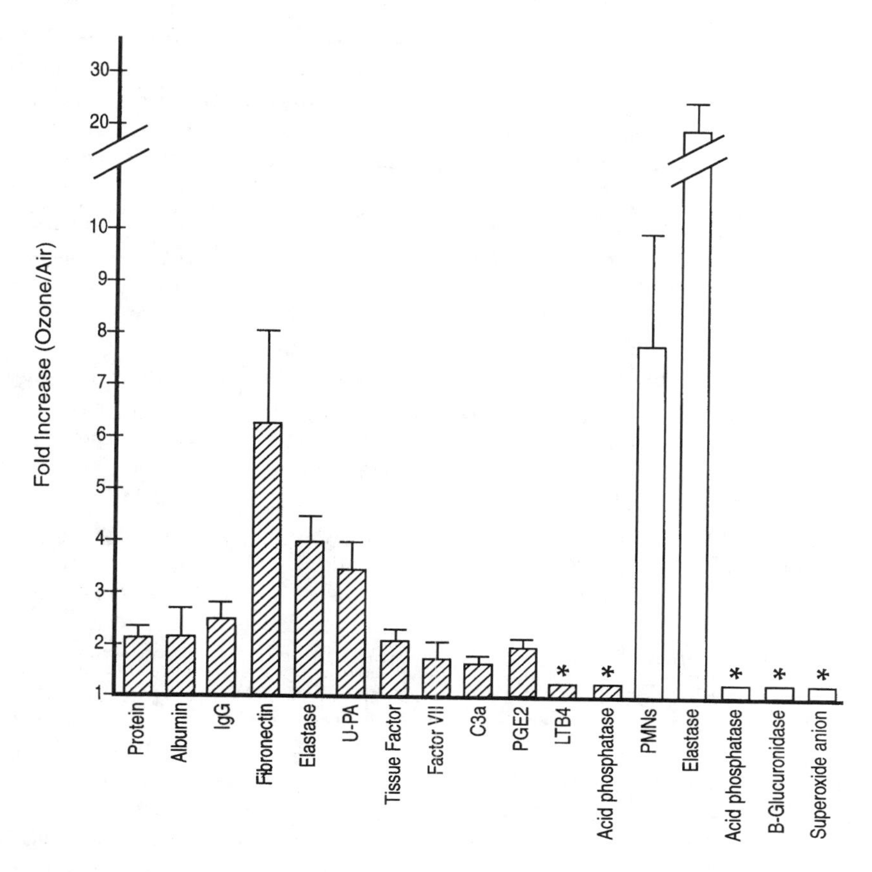

FIGURE 1
Cellular and biochemical changes in BAL from subjects exposed to ozone or air (fold increase). (From Koren, H. S. et al., *Am. Rev. Respir. Dis.*, 139, 407–415, 1989. With permission.)

deep into the lung, however, Gerrity et al.,[51] in a study of extrathoracic and intrathoracic removal of O_3 in tidal-breathing humans, showed that approximately 40% of inhaled ozone is removed in the nasal passages. Subsequently inflammatory responses to inhaled ozone in the nose have been identified by assessment of nasal lavage fluid cells.[8,55,105] One study found similar signs of inflammation in nasal and bronchial lavage fluid after ozone exposure,[55] indicating that the less invasive nasal lavage technique may be suggestive of lower airway effects. A recent study compared ozone exposures (0.12 and 0.24 ppm) in asthmatic and nonasthmatic young adult subjects and found increased inflammatory cells in nasal lavage fluid only among asthmatic subjects.[105]

3.1.1.3 Epidemiological Studies of Ozone

In a data set of 154 children studied over a 2-month period, Kinney and co-workers[76] fit child-specific linear regressions that related short-term pulmonary function changes to air pollution and temperature. Even though ozone concentrations did not exceed 0.078 ppm, this study found significant transient decreases in FVC, FEV_1, and measures of maximal flow which may have long-term significance. Both fine particulate matter and sulfate were measured, as well; however, these pollutants were not strongly correlated with lung function changes. One large positive outlier was identified who happened to be a child with asthma, although the study found no significant difference between the children with asthma and the other children. Presence of asthma in the children was determined by questionnaire; therefore, no detailed information about their asthmatic status is available. In another field study, Lioy and co-workers[98] studied 39 children at a summer camp where ozone was monitored continuously and spirometry was performed daily for 1 month. For each child, a linear relationship was found between peak ozone concentration for a given day and spirometric parameters for that day. There was a general tendency for decreased pulmonary function with increasing ozone concentration. For the 22 girls, the decrement in peak flow was highly significant ($p < 0.01$). (Data for the boys were not discussed.) Peak ozone concentrations during a 5-day episode were 0.143, 0.186, 0.165, 0.135, and 0.112 ppm. The significant decreases in peak flow persisted for at least 3 days. No information was given on allergic status of these subjects. A more recent study by Spektor and co-workers[152] also found significant decreases in peak flow and FEV_1 in children at summer camp, although the ozone concentration during the study period did not exceed 0.12 ppm. The latest study from this group suggests that the significant decrements in lung function seen in the camp studies at relatively low ozone concentrations are due to the persistence of effects from the prior day's exposures.[153]

Some epidemiological evidence implicating outdoor air pollution as a factor in aggravation of asthma was reported by Whittemore and Korn,[167] who showed that daily asthma attacks in asthmatics residing in Los Angeles were strongly associated with daily levels of photochemical oxidants and suspended particulate matter. Bates and Sizto[9] studied admissions in Southern Ontario, Canada, hospitals serving a population of 7 million and observed increased rates of admissions for asthmatics in the summer that correlated with both ozone and suspended sulfates. These authors suggest that a general "acid summer haze" may be responsible for air pollutant-induced respiratory effects. The strongest associations were between asthma visits and ozone and NO_2. A later analysis of the same data set but including more recent years confirmed these results.[21] A recent study in New Jersey found a significant association between visits to emergency departments for asthma during 1988 and 1989 and daily ozone concentrations.[26] During the study period, ozone levels exceeded the 0.12 ppm national standard 34 days in 1988 and 9 days in 1989. Hospital admissions in Buffalo, Albany, and New York City during 1988 and 1989 were associated with summer haze pollution (H^+, sulfate, and ozone).[157] A study of emergency department visits in Atlanta found that the average number of visits for asthma or reactive airway disease was 37% higher on the days after ozone levels equaled or exceeded 0.11 ppm.[166]

One report provides evidence that seasonal ozone exposure may cause bronchial hyperresponsiveness.[170] This recent study evaluated the respiratory health, allergic sensitization, and immune response in 218 Austrian children from a high ozone community (Group A) and compared them with 281 children from a community with lower ozone levels (Group B).[171] Bronchial hyperresponsiveness, measured with a methacholine challenge test, occurred more frequently in the Group A children. Comparison of serum IgE levels showed no difference between the two groups; however, there was a significantly lower helper/suppressor T-cell ratio in Group A children. One limitation of this study is the lack of personal exposure assessment.

Epidemiologic data also support a causative relationship between ozone and upper airway disease. Recent research in Mexico City reported histopathologic changes of the nasal mucosa in southwest metropolitan Mexico City inhabitants after six months' residence in that community with high ozone concentration.[22]

In summary, pulmonary function decrements have been documented in healthy subjects after exposure to concentrations of ozone as low as 0.08 ppm when the exposure lasted over 6 hours.[65] Increased incidence of emergency department visits for asthma were seen at concentrations both above and below the present standard of 0.12 ppm.[26] Individuals who live in areas with higher ozone concentrations appear to show that accelerated lung function declines with age.[156] The relationship between chronic ozone exposure and increased mortality is unclear.

3.1.2 *Nitrogen Dioxide (NO_2)*

Some chemical properties of NO_2 are

- Combines with water to form nitric (HNO_3) and nitrous (HNO_2) acids
- Direct oxidation of lecithin and unsaturated fatty acids, which constitute major elements of cell membranes
- Peroxidation of cell membranes causes increased membrane permeability, leakage of essential electrolytes and enzymes, inhibition of cellular metabolism, and swelling and disintegration of intracellular organelles including mitochondria, lysozomes, and endoplasmic reticulum
- Oxidizes a variety of biologic molecules, generating free radicals that can, in addition to damaging cell membranes, undermine the structural and functional integrity of enzymes and other proteins (particularly those containing thiol groups), nucleic acids, and other biomolecules including elastin and collagen, the structural proteins of lung

Nitrogen oxides are air pollutants derived from the combustion of fossil fuels or wood. Occupational exposure is common, as shown in Table 18. Exposure limits for nitric oxide and nitrogen dioxide are given in Table 19.

3.1.2.1 *Animal Toxicology*

- Extent and severity of structural effects are dose related, affecting various anatomic levels.
- Cell types usually sustaining the greatest damage with exposures to low concentrations (2 ppm) are type I alveolar and ciliated cells in the bronchiolar epithelium, principally at the junction of the terminal airways and proximal alveoli.
- Histological injury may include inflammation of bronchioles and alveoli and increased capillary permeability, predisposing to edema.
- Exposure to higher concentrations (30 ppm) produces transient acute bronchiolitis and alveolitis, which subsequently evolves into a patchy centriacinar emphysema with remodeling of alveoli and small airways and mild intersitial fibrosis.
- Impaired muco-ciliary clearance of particulate matter occurs.
- Impaired resistance to infection occurs.
- Impaired lung function occurs.
- Nitrogen dioxide generates free radical reactions and is genotoxic *in vitro* and may therefore be potentially carcinogenic or cocarcinogenic, although direct evidence is lacking.

A summary of some of the respiratory effects seen in animals exposed to NO_2 is given in Table 20. Effects include morphological, physiological, and biochemical changes.

3.1.2.2 *Controlled Human Laboratory Studies*

Are people with asthma particularly sensitive to inhaled NO_2? In comparison to SO_2 and O_3, it is obvious that after short-term exposure SO_2 causes the most dramatic effects. As documented above, there is an increased susceptibility to ozone in subjects with asthma. A paucity of studies have been designed to test the relative susceptibility to short-term controlled exposures to NO_2 in subjects with asthma. Most studies measuring only pulmonary function changes have not documented effects;[78] however, there is one report of enhanced decrements in pulmonary function after a 4 hr-exposure to 0.3 ppm NO_2 in elderly subjects with chronic obstructive pulmonary disease (COPD) compared to healthy elderly subjects.[111]

Some investigators have shown that exposure to near-ambient concentration of NO_2 causes an increase in bronchial reactivity as measured by histamine, methacholine, or carbachol challenges; however, other investigators have failed to show this effect. A recent meta analysis

TABLE 18

Exposures and Sites of Occurrence of Nitrogen Oxides

Occupational
Combustion of fossil fuels (e.g., automobile garages, ice resurfacing machines in skating rinks, other internal combustion engines, boilers)
Electric arc fixation of nitrogen
Decomposition of aqueous nitrous acid
High temperature oxidation of ammonia
Nitric acid production and transportation
Manufacture of lacquers and dyes
Other chemical manufacturing uses (nitrating agent, oxidizing agent, catalyst, inhibitor of acrylate polymerization)
Manufacture or use of explosives
Missile fuel oxidizer
Agriculture (silo filling)
Mining (diesel exhaust, shot-firing at coal seams)
Arc welding
Firefighting (exposure to smoke from plastics, shoe polish, nitrocellulose film, or fossil fuels)
Nonoccupational
Gas- and oil-fired household appliances
Kerosene heaters
Motor vehicle exhaust
Cigarette smoke
Ice skating rinks
Industrial boilers

TABLE 19

Nitrogen Oxides Exposure Limits

Nitrogen Oxide	Limit
Nitric oxide	
Threshold limit value (ACGIH 8-hr TWA)	25 ppm
Permissible exposure limit (OSHA 8-hr TWA)	25 ppm
Immediately dangerous to life and health (NIOSH 30-min avg.)	100 ppm
Nitrogen dioxide	
Threshold limit value (ACGIH 8-hr TWA)	3 ppm
Permissible exposure limit (OSHA ceiling, 15-min)	1 ppm
Immediately dangerous to life and health (NIOSH 30-min)	50 ppm
National Ambient Air Quality Standard (USEPA annual average)	0.053 ppm

Note: ACGIH = American Conference of Governmental Industrial Hygienists; TWA = time-weighted average; OSHA = Occupational Safety and Health Administration; NIOSH = National Institute for Occupational Safety and Health; USEPA = U.S. Environmental Protection Agency.

concluded that subjects with asthma do show enhanced airway hyperresponsiveness after exposure to NO_2.[42] One study has shown that there is a significant influx of inflammatory cells into BAL fluid following exposure to 4 ppm NO_2.[135]

Since increased bacterial and viral infectivity is a feature of the animal toxicology data with NO_2 exposures, a series of controlled exposure studies has investigated the relationship between infectivity and NO_2 in controlled human studies:

TABLE 20

Respiratory Effects of NO_2 in Animal Studies

Measurement	Concentration	Species	Effect	Ref.
Respiratory tract absorption	41 ppm	Dog/Rabbit	42.1% uptake	23, 39
		Rats	25–28% uptake	
	4–41 ppm	Rat	36% uptake	
		Dogs	100% at high vent	
Mortality	Acute, 720 ppm	Several	Death	39
	Chronic, 710 ppm	Several	Death	
Respiratory effects	1.0	Rabbits	No effect	138
Mucociliary effects	2–10 ppm	Rat	Loss of cilia	5, 129
Effects on alveolar macrophages (AMs)	4.0–25 ppm		Morphological increase in numbers	64
	2.3–10 ppm	Mouse	Loss of phagocytic activity	138
	0.3–1.0	Rabbit		
Immunologic effects	At concentrations as low as 0.5 ppm	Various species	Suppression of T-cell populations	39
Increased infectivity	0.5	Mouse	Increased mortality	38
	12 months	Mouse	From *K. pneumonia*	62
	1.0–2.5 ppm	Mouse	Impaired bacterial activity *S. aurens*	71
	1.0–3.0 ppm	Mouse	Streptococcus	69, 145
Biochemical Effects				
Pulmonary function	0.2–2.0 ppm	Rat, mouse	Decrements in lung volumes and capacities	39
Morphology	0.5–2.0 ppm	Rat	Signs of inflammation	
	>7.0 ppm	Dog	Edema	37

TABLE 21

Nitrogen Dioxide Exposure *In Vivo* and Human Alveolar Macrophage Inactivation of Influenza Virus *In Vitro*

In Vivo	
NO_2	0.60 ppm for 3 hr, N = 9
	0.05 ppm with three 15-min peaks of 2.0 ppm
Measurements	PFTs
	BAL: 3.5 hr postexposure

In Vitro
AMs retrieved from BAL fluid exposed to influenza virus by incubation
Endpoint measured was the ability of the AMs to inhibit infection (by counting viral plaques); also assayed for interleukin-1 as a test for the viability of the AMs
AMs retrieved after NO_2 exposure tended to inactivate virus less effectively than after air exposure in 4/9 subjects

Note: Subjects were healthy, nonsmokers of both genders. PFTs = pulmonary function tests; BAL = bronchoalveolar lavage; AMs = alveolar macrophages.

Source: From Frampton, M. W. et al., *Environ. Res.*, 48, 179–192, 1989. With permission.

TABLE 22
Uses and Sites of Occurrence of Sulfur Oxides

Sulfur Dioxide
Chemical manufacturing (sulfuric acid, sulfites, thiosulfates, hydrosultites, sulfonation of oils, and other uses)
Recovery of volatile materials
Reducing agent and anti-oxidant
Paper manufacturing
Metal and ore refining
Portland cement manufacture
Disinfectant
Food preservative (to inhibit bacterial growth, browning, and enzyme-catalyzed reactions)
Derived From
Combustion of sulfur-containing fuels (particularly in power plants and oil refineries)
Purification (and comparison) of SO_2 gas from smelting
Roasting pyrites

1. NO_2 appeared to inhibit the activity of macrophages[45] (see Table 21).
2. NO_2 did not significantly increase viral shedding as compared to air exposure in a controlled laboratory study, although after NO_2 exposure 91% of the subjects became infected compared to 71% after air exposure.[52]
3. Pinkston and co-workers[122] exposed human alveolar macrophages *in vitro* to NO_2 at relatively high concentrations of 5, 10, and 15 ppm for 3 hr. This study showed no effects on viability or the spontaneous release of neutrophil chemotactic factor.
4. Rubenstein et al.[130] exposed volunteers to 0.6 ppm NO_2 for 2 hr during intermittent exercise. Pollutant-induced effects were measured using standard pulmonary function tests (PFTs). These investigators saw no change in PFTs and no reported symptoms. They also looked at circulating and BAL lymphocytes and saw no differences between baseline and NO_2 exposures.

3.1.2.3 Epidemiological Studies of NO_2

Most of the concern about chronic exposure to NO_2 centers around children who live in a home with a gas-cooking stove. A recent meta analysis of studies investigating the effects of chronic NO_2 exposure on respiratory disease in children concluded that there was a small but significant association between exposure and signs of respiratory distress.[59] Three studies published since the meta analysis support that conclusion by finding an increased risk for respiratory disease at concentrations of NO_2 at or above 15 ppb;[46,70,117] however, another recent study of the association between respiratory disease and NO_2 exposure in children found no relationship in a large number of infants in Albuquerque.[133]

Another population at risk for adverse effects from NO_2 exposure is individuals involved in ice skating. Brauer and co-workers[20] reported that the median NO_2 level inside ice-skating rinks in 70 northeastern rinks was 180 ppb, which is more than 10 times the median outdoor concentration. Soparkar et al.[149] have reported toxic respiratory effects associated with similar exposures.

In summary, increased airway reactivity has been seen after short-term (1-hr) exposures to 0.3 ppm NO_2 in subjects with asthma.[43] Increased signs of respiratory disease in children have been seen at annual concentrations of 15 ppb and above.[70,117]

3.1.3 Sulfur Dioxide (SO_2)

Sulfur dioxide is a common air pollutant in industrialized countries. The uses and sources of SO_2 air pollution are listed in Table 22. Occupational and community exposure standards are given in Table 23.

TABLE 23

Sulfur Dioxide Exposure Limits

Exposure	Limit
Threshold limit value (ACGIH, 8-hr TWA, 15-min ceiling)	2 ppm; 5 ppm
Permissible exposure limit (OSHA, 8-hr TWA)	2 ppm
NIOSH recommendation	0.5 ppm
Immediately dangerous to life or health (NIOSH, 30-min avg.)	100 ppm
National Ambient Air Quality Standard (USEPA)	
24-hr avg.	0.14 ppm
Annual avg.	0.03 ppm
Washington state (1-hr avg.)	0.25 ppm

Mechanisms of Sulfur Oxides Toxicity

Following absorption, the gas reacts with water, forming a weak acid solution that contains sulfurous (H_2SO_3) and sulfuric (H_2SO_4) acids and sulfite (SO_3^{2-}), bisulfate (HSO_3^-), and hydrogen (H^+) ions; the relative contribution of these ions to the toxicity produced its uncertain

Bisulfate ion reacts with many biological molecules via nucleophilic substitution; sulfur-containing free radicals may also produce cellular damage

Efficiently absorbed in the upper respiratory tract

Deep lung penetration and toxicity are enhanced by oxidation and adsorption to submicron particles

Note: ACGIH = American Conference of Governmental Industrial Hygienists; TWA = time-weighted average; OSHA = Occupational Safety and Health Administration; NIOSH = National Institute for Occupational Safety and Health; USEPA = U.S. Environmental Protection Agency.

3.1.3.1 *Animal Toxicology*

- Short-term exposures of 5 ppm cause airway narrowing — the effect in the nasal passages is due to swelling of the mucous lining and excessive secretions, while that of the lower airways is due primarily to smooth muscle contraction (bronchoconstriction).
- Long-term exposures to 0.1–5 ppm have produced few changes in airway caliber, distensibility, or histological appearance of the lung.
- Desquamation of cilia, impairment of ciliary beat, and induction of a more acid mucin occur.[132]
- SO_2 does not appear to be a carcinogen but may be a promoter of the carcinogenic effects of polycyclic aromatic hydrocarbons (PAHs).

3.1.3.2 *Controlled Laboratory Studies*

Sulfur dioxide exposure causes dramatic respiratory effects in subjects with asthma. These effects include decreases in FVC and FEV_1 consistent with bronchoconstriction and increased nasal resistance consistent with nasal congestion. Subjects with asthma respond to SO_2 at concentrations as low as 0.25 for exposures at durations as short as 2.5 min. Responses have been seen in adolescent subjects at 0.1 ppm SO_2 when preceded by ozone exposure.[77] Healthy subjects often do not show pulmonary function changes after exposure to 5.0 ppm SO_2. Cold air has been shown to potentiate SO_2 effects in subjects with asthma.[14,96] Even though the NAAQS averaging times are annual and 24 hr, some states have set shorter averaging times for their SO_2 ambient air standards. For instance, Washington state has both a 1-hr standard of 0.25 and a 5-min standard of 1.0 ppm. The results of some representative studies of SO_2 exposure are summarized in Table 24.

The reader is referred to two reviews of the bronchoconstrictor effects of short-term exposure to SO_2 in subjects with asthma.[66,83] Upper airway effects have been reported after SO_2 exposure which is not surprising since almost 99% of inhaled SO_2 is removed in the nasal

TABLE 24
Effects of Inhaled SO_2 in Controlled Human Exposure Studies

Study	Ref.
Bronchoconstriction	
Asthmatic adolescents	81, 82
1.0 and 0.5 ppm	
Mouthpiece (oral inhalation)	
PFTs	
–23% in FEV_1	
+67% in R_T	
–44% in max. flow	
Adult males (methacholine responders)	67
1 ppm, chamber	
0.5, 1, 2, and 5 min; exercise	
$(s)R_{aw}$: 9.4% increase after 0.5 min; 29.7% after 5 min	
Adult males (methacholine responders)	128
0, 0.25, 0.5, and 1.0 ppm; chamber	
Exposure time = 60 min	
Three 10-min exercise periods at 0.25, and 50 min	
$(s)R_{aw}$	
Twofold and threefold increase in $(s)R_{aw}$ after 0.5 and 1.0 ppm, respectively	
Inflammation	
Healthy males	
8 ppm; chamber	
Exposure time = 20 min during exercise	136
BAL 4, 8, 24, and 72 hr postexposure	
Significant increase in both mast cells and lymphocytes at 4 and 24 hr postexposure	

Note: R_T = total respiratory resistance; PFTs = pulmonary function tests; BAL = bronchoalveolar lavage; $(s)R_{aw}$ = (specific) airway resistance.

passages during quiet breathing.[150] Inhalation of SO_2 at 1 ppm during moderate exercise increased the work of nasal breathing in adolescent subjects with asthma.[81]

3.1.3.3 Epidemiology

Emission sources and meteorologic conditions are such that elevated concentrations of SO_2 and particulate matter tend to be very strongly correlated. As described below, there is a strong body of literature showing associations between adverse respiratory effects and elevated concentrations of particulate matter. Most studies have found it impossible to statistically determine independent effects of SO_2. The SO_2-particulate matter relationship is discussed in a paper by Ware and co-workers,[164] who found adverse effects of outdoor SO_2 emissions in children. Also, a study examining respiratory effects of chronic SO_2 exposure in healthy adults found a positive correlation between concentrations of SO_2 and persistent cough and production of phlegm.[25] In summary, adverse effects in subjects with asthma have been seen at concentrations of SO_2 of 0.25 ppm for as little as 2.5 min[66] and at .10 ppm when subjects were previously exposed to 0.12 ppm ozone for 45 min.[77]

3.1.4 Particulate Matter

Particulate matter (PM) in the outdoor ambient air is both a primary and secondary pollutant. This category of air pollution is regulated by the USEPA as the respirable fraction of particulate matter defined as particles less than or equal to 10 μm in diameter and designated

TABLE 25

Effects of Chronic Exposure of Rabbits, 250 μg/m³ H_2SO_4[a]

Endpoint	Exposure (4 months)	Exposure (8 months)	Exposure (12 months)	Postexposure (3 months)
Mucociliary clearance	–	–	–	– –
Airway sensitivity	+	++	++	NM
Air diameter	–	–	– –	NC
Secretory cell number	+	++	++	+
Secretory cell pH	–	–	–	–

Note: (–) decreased; (+) increased; NM, not measured; NC, no change.

[a] 1 hr/day, 5 days/wk.

Source: From Gearhart, J. M. and Schlesinger, R. B., *Environ. Health Persp.*, 79, 127–136, 1989. With permission.

PM_{10}. Common primary sources of PM_{10} are combustion, industrial processes, and woodstoves. Secondary forms of aerosols regulated as PM_{10} are acid aerosols, both sulfate aerosols which predominate along the eastern coast of the U.S. and nitrate aerosols which predominate in California.

3.1.4.1 Animal Toxicology

Animal toxicological studies of PM_{10} are not common due to the technical difficulty of recreating in a laboratory setting the complex mixture of particles and adsorbed organic material seen in the PM_{10} category. For one example of particle toxicity, the reader is referred to the discussion of the effects of diesel exhaust exposure in Section 3.4.1. Another example of particulate matter toxicity in rodents used community exposures. Saldiva et al.[131] attempted to assess the adverse effects of urban levels of air pollution by placing rats in cages in the center of Sao Paulo for 6 months. These animals showed increased respiratory pathology and mortality compared to control animals housed in a more rural area. No air pollution values were reported. Ziegler and associates[169] compared the effects of exposure to a resuspension of road dust from street sweepings and ammonium nitrate particles on pathology of alveolar macrophages (AMs) in Fischer 344 rats. AMs retrieved after road dust exposure showed a 25% decrease in Fc receptors compared to air or nitrate exposures, indicating a blunting of the immune response. A number of animal studies have investigated the toxic effects of acid particles, mainly sulfuric acid. A summary of results from studies by Gearhart and Schlesinger[49] is provided in Table 25. A series of studies by Amdur and associates[1] has shown that SO_2 combined with metallic aerosols have greater toxicity than SO_2 alone. Comparative toxicities of SO_2 and H_2SO_4 are summarized in Table 26. Schlesinger[139] has reviewed the interactions of inhaled toxicants and mucociliary clearance.

3.1.4.2 Controlled Human Studies

The same difficulty in generation of a controlled atmosphere representative of PM_{10} has resulted in no studies of human volunteers. Some insight into the acute effects of short-term exposures to fine particles can be gleaned from controlled studies of acid aerosols, both sulfates and nitrates. For instance, Hanley et al.[57] studied 22 adolescents with asthma during intermittent exercise for 40 to 45 min on a treadmill. The test atmospheres were air and 70 or 136 μg/m³ H_2SO_4. Significant decrements in FEV_1 (p = 0.0016) and FVC (p = 0.039) were seen after H_2SO_4 exposure compared with air exposure; an average decrease of 37 ml for each μmol/m³ [H^+] was obtained.

TABLE 26

Comparative Toxicity of SO_2 and H_2SO_4 Acute Studies

	S (µg/m³)	
	SO_2	H_2SO_4
Guinea pigs: 1 hr, 10% increase in airway resistance	206	33
Donkeys: 30 min; 1 hr altered bronchial clearance	284,000	66
Normal subjects: 7 min; 1 hr altered bronchial clearance	16,640	33
Normal subjects: 10 min; 5% decrease in tidal volume	768	40
Adolescent asthmatics: 40 min; increase in airway resistance	650	33

Source: From Amdur, M. O., in *Casarett and Doull's Toxicology: The Basic Science of Poisons,* 4th ed., Pergamon Press, New York, 1991. With permission.

Other studies have not shown such significant effects.[4] Exposure of subjects with asthma to 100 and 1000 µg/m³ H_2SO_4 has been associated with decreases in mucociliary clearance.[151] Utell[158] has reviewed the effects of acid aerosols on human respiratory responses. Neutralization of H_2SO_4 by oral ammonia needs to be taken into account in these studies.[57,90]

3.1.4.3 *Epidemiology*

Currently there is great concern about the health effects of PM_{10} due to a recent series of articles documenting an association between increased mortality from respiratory and cardiac causes and elevated concentrations of PM_{10}. The results of these studies are summarized in Tables 27 and 28 from a recent review.[32] In general, these studies have found a 1% increase in mortality with each 10-µg/m³ increase in PM_{10}. Another recent review of the associations between PM_{10} and mortality was published by Schwartz.[141] An assessment of the categories of individuals dying on high air pollution days found that the relative risk of dying was 1.08, whereas the risk was higher for those with COPD (1.25) or pneumonia (1.13).[142] The mortality rate associated with fine particles and sulfates was higher (1.26) than the rate with PM_{10} in a study of individuals over a 14- to 16-yr interval.[31] Assessments of epidemiological evidence for aggravation and promotion of COPD by acid air pollution show a weak association, but more data are needed to make a better interpretation.[34] Outdoor exposure to particulate matter also has been associated with declines in lung function in children,[29,33,36] increases in medication use by patients with asthma,[124] increases in emergency room visits for asthma,[140] and increases in absenteeism in elementary school children.[127]

In summary, present ambient concentrations of PM_{10} in many cities have been associated with increased mortality and morbidity from respiratory and cardiac causes. Remaining questions include identifying the constituent of PM_{10} responsible for the mortality and the individuals who are dying.

3.2 Indoor Air Pollutants

A review of indoor air pollution was published recently.[134] Since most individuals spend much more time indoors than in the outdoors, indoor air pollutant exposures account for much of the total exposure assessment of an individual. Common indoor air pollutants are given in Table 12.

TABLE 27

Acute Effects of Particulate Matter on Hospital Usage

Measure of Hospital Usage	Location and Period	Particulate Matter	Change in Hospital Usage for Each 10 μg/m^3 Increase in PM_{10} (%)
Hospital admissions			
Asthma	New York City; Buffalo, NY	Daily mean SO_4	1.9 (0.4, 3.4)
			2.1 (–0.6, 5.0)
	Toronto, Canada; summer 1986–1988	Daily mean $PM_{2.5}$	2.1 (–0.8, 5.1)
	Combined		1.9
All respiratory	New York City; Buffalo, NY	Daily Mean SO_4	1.0 (0.2, 1.8)
			2.2 (0.6, 3.8)
	Toronto, Canada; summer 1986–1988	Daily mean $PM_{2.5}$	3.4 (0.4, 6.4)
	Southern Ontario; summer 1983–1988	Daily mean SO_4	0.8 (0.4, 1.1)
	Combined		0.8
Emergency department visits			
Asthma (<65 yr)	Seattle, WA; 1989–1990	Daily mean PM_{10}	3.4 (0.9, 6.0)
Respiratory disease	Steubenville, OH	Daily mean TSP	0.5 (0.0, 1.0)
Chronic obstructive pulmonary disease (COPD)	Barcelona, Spain; winter 1985–1989	British smoke	2.3 (1.4, 3.2)
	Combined		1.0

Note: TSP = total suspended particulate; PM_{10} = particulate matter ≤ 10 μm in diameter; $PM_{2.5}$ = particulate matter ≤ 2.5 μm in diamater.

Source: From Dockery, D. W. and Pope, III, C. A., *Ann. Rev. Public Health,* 15, 107–132, 1994. With permission.

3.2.1 Environmental Tobacco Smoke

Environmental tobacco smoke (ETS) is a known respiratory irritant, although the causative constituent(s) are not known.[115] Table 29 lists a variety of studies showing the effects of passive smoking on respiratory symptoms in children and adolescents. More recent data indicate that prenatal exposure and exposure during infancy are risk factors for the development of childhood asthma.[112,113] Table 30 summarizes the effects of ETS on pulmonary function tests in children and adolescents.[115] Table 31 gives data on ETS exposures in subjects with asthma.

Recently, ETS has been listed as a human carcinogen by the USEPA.[41] That report concluded that up to 3000 deaths from lung cancer per year in the U.S. could be attributed to exposure to ETS. A summary from that report of estimated lung cancer mortality attributed to ETS is given in Tables 32 and 33.

3.2.2 Formaldehyde and Other Aldehydes

Formaldehyde is one of the top 10 organic chemical feedstocks in the U.S. and as such is a common source of human exposure in the home, in outdoor air, and in occupational settings. Concern for formaldehyde exposure in the home comes from its use in numerous building and decorating materials. In outdoor air, formaldehyde is a secondary pollutant formed by photochemistry from automotive emissions. The health effects of formaldehyde have been reviewed recently by Bardana and Montanaro[7] and by Leikauf.[94] Tables 34 to 36 from Leikauf summarize some effects in animals and humans. Formaldehyde causes nasal adenoma in rats[75] and has been implicated in nasal cancer in humans.[159,160] Table 34 gives a

TABLE 28

Combined Effect Estimates of Daily Mean Particulate Matter Pollution

Effect	Change in Health Indicator per Each 10 μg/m³ Increase in PM_{10} (%)
Increase in daily mortality	
Total deaths	1.0
Respiratory deaths	3.4
Cardiovascular deaths	1.4
Increase in hospital usage (all respiratory)	
Admissions	0.8
Emergency department visits	1.0
Exacerbation of Asthma	
Asthmatic attacks	3.0
Bronchodilator	2.9
Emergency department visits[a]	3.4
Hospital admissions	1.9
Increase in respiratory systems reports	
Lower respiratory	3.0
Upper respiratory	0.7
Cough	1.2
Decrease in lung function	
Forced expired volume	0.15
Peak expiratory flow	0.08

[a] One study only.

Source: From Dockery, D. W. and Pope, III, C. A., *Ann. Rev. Public Health,* 15, 107–132, 1994. With permission.

representative sample of indoor sources of formaldehyde.[94] Acrolein, another aldehyde commonly involved in human exposure, is considered to be at least 10 times as toxic as formaldehyde.[168]

3.2.3 VOCs in Indoor Air

Many commonly encountered VOCs are solvents. Inhalation of solvents has been associated with various human health endpoints; benzene exposure is associated with leukemia.[53] Toluene diissocyanate is known to induce suggestions of mutagenesis, such as chromosome aberration and base-pair substitutions.[102] Toluene diisocyanate has been associated with causation and persistence of occupational asthma.[73,120,123] An array of respiratory effects has been reported after solvent exposure (see Table 37). In real-life situations, mixtures of solvents are often found rather than single compounds.

Several indices of exposure to solvents have been identified. In fact, toluene and perchloroethylene (PCE) have been found in blood, and methyl ethyl ketone (MEK) has been found in urine of exposed workers.[72] Methylhippuric acid, a toluene metabolite, has been identified in urine.[148] As mentioned elsewhere, solvents can be analyzed in exhaled breath up to 16 hours after exposure.[110]

Several attempts have been made to study VOC exposures in a controlled laboratory setting. Controlled exposures to a mixture of 22 common VOCs have been conducted by Harving et al.[58] and Koren et al.[86] Both research groups found adverse respiratory effects after the VOC exposure. Harving et al.[58] found FEV_1 decrements, and Koren and Devlin[86]

TABLE 29

Effects of Passive Smoking on Respiratory Symptoms: Selected Cross-Sectional Studies Involving Children/Adolescents

Source of Subjects	Subjects	Exposure Assessment	Findings
Aylesbury, UK; seven public schools; 1971	1328 boys, 270 girls; ages 6–14 years	Self-administered questionnaire from parents	Close association of child cough and parent winter morning phlegm Prevalence of cough: 15.6% no smokers; 22.2% both parents smoke
Tucson, AZ; stratified cluster random sample of households; 1972–1973	1655 households; Anglo-white; 1252 children < 16, 2516 children > 15	Self-administered NHLBI questionnaire from children > 15; otherwise, from parents	Prevalence of cough in young children: 7.8% no smokers, 10.4% smokers ($p < 0.05$); significance gone when parental symptoms considered
Survey of towns in Connecticut and South Carolina	816 children in 376 families; 607 children < 16, 109 children > 15	Respiratory symptom questionnaire administered by interviewer	No effect of parental smoking on children's cough or wheeze Prevalence of wheeze in young children related to parental wheeze ($p < 0.01$)
Derbyshire, U.K.; 48 secondary schools; 1974	2847 boys, 2988 girls; 12 years old	Questionnaire self-administered by child	Prevalence of cough: 16% no smokers, 19% one smoker, 23.5% two smokers ($p < 0.01$)
East Boston, MA; random sample in schools 1975–1977	444 children; ages 5–9 years	NHLBI questionnaire administered by interviewer; if age < 10, parent answered	No increase in respiratory illness with parental smoking
1979	650 children; ages 5–9 years		Persistent wheeze: 1% no smokers, 6.8% one smoker, 11.8% two smokers ($p < 0.02$)
Three towns in Arizona; survey of schools; 1978–1979	558 children; ages 8–10 years	Self-administered by parents	Child's wheeze, ($p < 0.05$), sputum ($p < 0.05$), and cough ($p < 0.01$) related to parental smoking
Pennsylvania; survey of schools	4071 children; ages 5–14	Self-administered by parents	Trend with number of smoking parents not significant for any symptoms
Six U.S. cities; different regions survey of schools; 1974–1979	10,106 children; ages 6–13 years	Self-administered by parents	20–35% increased risk of all respiratory illness and symptoms with maternal smoking

Note: NHLBI = National Heart, Lung, and Blood Institute.
Source: From National Research Council (NRC), Hornig, G., Ed., *Environmental Tobacco Smoke: Measuring Exposures and Assessing Health Effects,* National Academy Press, Washington, D.C., 1986. With permission.

found increased neutrophils in nasal lavage fluid. Data from a study by Koren et al.[87] are shown in Table 38.

One epidemiologic study of the effects of ambient volatile organic compounds found statistical differences between levels of respiratory and irritant symptoms in young children enrolled in schools in a valley with high air pollution compared with similar children enrolled in schools outside the valley.[165] The VOCs included emissions from chemical manufacturing plants.

Also, statistically significant increases in respiratory cancers (trachea, bronchus, and lung) were seen in a cohort of workers employed in shoe manufacturing where toluene and other solvent exposure is common.[162] Mortality in 5365 workers in the dry-cleaning industry increased slightly for esophagus standard mortality rate (SMR) = 2.1 and larynx SMR = 1.5.[16]

TABLE 30

Effect of Passive Smoking on Pulmonary Function: Selected Cross-Sectional Studies Involving Children/Adolescents

Source of Subjects	Subjects	Exposure Assessment	Findings
Tucson, AZ; stratified cluster random sample of households; 1972–1973	1655 households; Anglo-white; 1252 children < 16, 2516 children > 15	Self-administered NHLBI questionnaire from children > 15; otherwise, from parents	No relationship of FEV_1 with parental smoking when household aggregation of body mass taken into account
Survey of towns in Connecticut and South Carolina	816 children in 376 families; 607 children < 16, 209 children > 15	Respiratory symptom questionnaire administered by interviewer	$MEF_{50\%}$ lower in younger children with maternal smoking ($p < 0.05$); FEV_1, and PEF not significant
East Boston, MA; random sample in schools 1975–1977, 1979	444 children; ages 5–9 years	NHLBI questionnaire administered by interviewer; if age < 10, parent answered	Lower z-scores for $FEF_{25-75\%}$ in children with smoking parents
CHESS study; seven cities; survey of schools; 1970–1973	16,689 children; ages 5–13 years	Self-administered by parent (usually mother)	$FEV_{0.75}$ dose-response relationship with mother's smoking
Three towns in Arizona; survey of schools; 1978–1979	558 children; ages 8–10 years	Self-administered by parents	No effect of parental smoking on any parameters; cough ($p < 0.01$) related to parental smoking
Six U.S. cities; different regions; survey of schools; 1974–1979	10,106 children; ages 6–13 years	Self-administered by parents	FEV_1 significantly negative; FVC positive relation to maternal smoking
Shanghai, PRC; survey of two schools; 1984	571 children; ages 8–16 years	Self-administered questionnaire by parents	Paternal lifetime smoking related to z-scores of FEV_1, MMEF, and $FEF_{62.6-87.5\%}$
Los Angeles County, CA, survey of four areas in city; 1973	971 nonsmoking nonasthmatic children; ages 7–17 years	Modified NHLBI questionnaire administered by interviewer	Inconsistent effect of maternal smoking in younger boys and older girls

Note: NHLBI = National Heart, Lung, and Blood Institute; MEF = maximal expiratory flow; PEF = peak expiratory flow; FEV_1 = forced expiratory flow at 1 sec; $FEF_{25-75\%}$ = mean of forced expiratory flow at 25% and 75% vital capacity; FVC = forced vital capacity; MMEF = midmaximal expiratory flow.

Source: From National Research Council (NRC), Hornig, G., Ed., *Environmental Tobacco Smoke: Measuring Exposures and Assessing Health Effects,* National Academy Press, Washington, D.C., 1986. With permission.

3.2.4 Wood Smoke

Wood smoke is both an indoor and an outdoor air pollutant. Mutagenicity testing has shown that extracts from wood smoke are mutagenic in the Ames test.[95] Source apportionment of the particulate matter in Seattle showed that more than 80% of that air pollutant is produced by wood-burning devices during the nighttime hours.[91] Furthermore, indoor-outdoor ratios of fine particles are high in residential communities in homes without cigarette smokers. Three studies document such infiltration. Quackenboss and colleagues[126] have reported an indoor/outdoor ratio of 0.63 for PM_{10} in homes of nonsmokers in Tucson, AZ, and Dockery and Spengler[35] reported a similar ratio of 0.55 for total suspended particulate matter in Steubenville, OH. Sexton et al.[144] characterized the indoor/outdoor ratio in residences using woodstoves and found indoor-outdoor ratios ranging from 0.50 to 0.70. The health effects of wood-smoke exposure have been reviewed recently.[91] That review found that increased symptoms of respiratory distress, increased cases of lower respiratory infection, and decrements in lung function are associated with wood-smoke exposure of children (Table 39). One study found an average 1.7-ml decrement in FVC for each $\mu g/m^3$ increase in $PM_{2.5}$.[79]

TABLE 31

Experimental Studies of Acute Environmental Tobacco Smoke (ETS) Exposure for Asthmatic Patients

Population	Exposure	Findings
14 patients from the Gage Research Institute (9 male, 5 female); mean age 37 years	Room: 14.6 m^3 Time: 2 hr Cigarettes: 7 CO: 24 ppm	Changes in pulmonary function slight; slight decrease in total lung capacity (helium mixing, $p < 0.02$)
10 patients from St. Louis Univ. Hospital Allergy Clinic; ages 16–39 years; 10 controls, ages 24–53 years	Room: 30 m^3 Time: 1 hr Cigarettes: NG CO: 15–20 ppm	Linear decrease in pulmonary function over time in patients; FEV_1 decreased 21.4%; $FEF_{25-75\%}$, 19.2%; FVC, 20%; no change in controls
6 patients (4 males, 2 females); mean age 25.5 years	Details not given	Significant decrease in 3/6 subjects; PC_{20} FEV_1 significantly decreased with histamine
9 patients with near normal lung function; ages 19–30 years	Room: 4.25 m^3 Time: 1 hr Cigarettes: NG CO: 40–50 ppm	No change in expiratory flow rates; small decrease in bronchial reactivity; PD_{20} FEV_1 increased from 0.25 to 0.79 with methacholine

Note: NG = not given; FEV_1 = forced expiratory flow at 1 sec; $FEF_{25-75\%}$ = mean of forced expiratory flow at 25% and 75% vital capacity; FVC = forced vital capacity; PC_{20} = provocative concentration causing a 20% reduction in FEV_1; PD_{20} = provocative dose causing a 20% reduction in FEV_1.
Source: From National Research Council (NRC), Hornig, G., Ed., *Environmental Tobacco Smoke: Measuring Exposures and Assessing Health Effects,* National Academy Press, Washington, D.C., 1986. With permission.

3.2.5 *Nitrogen Dioxide (NO_2)*

NO_2 also is an indoor and outdoor air pollutant. Indoor exposure occurs from combustion sources such as gas cooking stoves within the home. The USEPA[39] reports levels as high as 1843 μg/m^3 (nearly 1000 ppb) for a 3-min average in the proximity of gas stoves. The population in the U.S. potentially exposed to indoor concentrations of NO_2 far exceeds the population potentially exposed outdoors, where concentrations rarely exceed 300 ppb. A recent meta analysis of studies investigating the relationship between adverse respiratory effects and NO_2 exposure in children found a significant increased risk (see Table 40).[59]

In summary, residents indoors are exposed to significant concentrations of ETS, formaldehyde, VOCs, wood smoke, and NO_2. All these compounds are associated with adverse respiratory effects depending on the concentrations and the durations of exposure.

3.3 Hazardous Air Pollutants

These pollutants, found in ambient air usually around point sources, are often referred to as "air toxics". In the 1990 amendments of the Clean Air Act, 189 air toxics were identified and the USEPA was charged with regulating them by 1996 (see Table 41). As mentioned earlier, the USEPA does not plan to set individual standards for these "air toxics"; rather, regulation is technology based instead of health based and sources are required to reach maximum achievable control technology (MACT). A recent article describes what we know and do not know about the 189 "air toxics".[74] Table 42 lists hazardous air pollutants measured in ambient air in the Airborne Toxics Element and Organic Substances (ATEOS) study conducted in New Jersey from 1981 to 1982. There are very few data on the respiratory effects of these compounds, and in many cases the respiratory system is not the target organ of concern. The VOCs on the list are discussed under either Section 3.2 or 3.4.

TABLE 32

Summary of Epidemiologic Studies of Risk-Based Exposure Assessed by Spouse Smoking Habits, When Available, or Smoking by the Household Cohabitants

Location	Sex	Lung Cancers in "Exposed" Group		O – E	Var. of	Risk[a]	95% Confidence Limits	
		Obs.	Exp.		(O – E)		Lower	Upper
Case-Control Studies								
Hong Kong	F	34	37.7	−3.7	13.01	0.75	0.44	1.30
Greece	F	38	29.3	8.7	11.70	2.13	1.18	3.78
U.S.	F	14	10.6	3.4	4.75	2.03	0.83	5.03
	M	2	1.2	0.8	0.98	2.29	0.31	16.50
U.S.	F	13	13.7	−0.7	3.06	0.79	0.26	2.43
	M	5	5.0	0.0	1.52	1.00	0.20	4.90
U.S.	F	33	34.1	−1.1	4.78	0.80	0.32	1.99
	M	5	6.6	−1.6	2.37	0.50	0.14	1.79
U.S.	F	92	89.5	2.5	22.33	1.12	0.74	1.69
Sweden	F	33	29.6	3.4	13.88	1.28	0.75	2.16
Japan	F	73	67.4	5.6	14.19	1.48	0.88	2.50
	M	3	1.8	1.2	1.38	2.45	0.46	13.06
Hong Kong	F	51	45.3	5.7	13.19	1.54	0.90	2.64
England	F	22	21.9	0.1	4.71	1.03	0.41	2.47
	M	8	7.3	0.7	2.56	1.30	0.38	4.42
Overall for case-control studies		426	401.0	25.0	114.40	1.24	1.04	1.50
Cohort, Prospective Studies								
U.S.	F	88	81.8	6.2	30.82	1.18[b]	0.90	1.54
Scotland	F	6	6.0	0.0	1.58	1.00[b]	0.20	4.91
	M	4	2.3	1.7	1.40	3.25[b]	0.60	17.65
Japan	F	146	129.5	16.5	34.83	1.63	1.25	2.11
	M	7	3.3	3.7	3.02	2.25	1.04	4.85
Overall for prospective studies		251	222.9	28.1	71.65	1.44	1.20	1.72
Overall for all studies		692	637.7	53.1	186.0	1.34	1.18	1.53

Note: Obs. = observed; Exp. = expected.

[a] Risk is given as calculated odds ratios for case-control studies and published relative risks for cohort, prospective studies.

[b] Ratio of age-standardized mortality rates.

Source: From EPA, *Respiratory Health Effects of Passive Smoking: Lung Cancer and Other Disorders,* EPA/600/6-90/006 F, U.S. Environmental Protection Agency, U.S. Government Printing Office, Washington, D.C., 1992.

3.4 Occupational Air Pollutants

It has been determined that a wide variety of agents in the workplace are associated with occupational lung disease. Occupational lung disease is the most common occupational disorder in the U.S. Figure 2 shows the incidence of occupational asthma and other respiratory disorders plotted at one large occupational medicine clinic.[27] Major industrial air pollutants are given in Table 43. The epidemiology of occupational asthma was reviewed recently by Beckett.[11] Estimates of the incidence of asthma in exposed populations range from 0.2/1000

TABLE 33

Estimated Female Lung Cancer Mortality by Attributable Sources for U.S., 1985, Using the Pooled Relative Risk Estimate from 11 U.S. Studies[a]

		Lung Cancer Mortality[b]					
Smoking Status[c]	Exposed to Spousal ETS	(1) Number at Risk (in millions)	(2) Non-Tobacco-Smoke-Related Causes[d]	(3) Background ETS	(4) Spousal ETS	(5) Ever-Smoking	Total
NS	No	12.92	1220 (3.2)	410 (1.1)			
NS	Yes	19.38	1830 (4.8)	620 (1.6)	470 (1.2)		
ES	—	25.69	2420 (6.4)			31,030[e] (81.7)	
Total	—	58.00	5470 (14.4)	1030 (2.7)	470 (1.2)	31,030 (81.7)	38,000

Note: ETS = environmental tobacco smoke.

[a] Percentage of grand total (38,000) in parentheses in columns (2), (3), (4), and (5).

[b] The nonblank entries in the table are the product of an individual's attributable risk cancer from non-tobacco-smoke-related causes (38,000/58,000,000), the number at risk in column (1), and column-specific multiples.

[c] NS = never-smokers; ES = ever-smokers.

[d] Background sources in the absence of tobacco smoke (i.e., in a zero-ETS environment).

[e] This figure attributes all lung cancer in ever-smokers above the background non-tobacco-smoke-related rate to ever-smoking.

Source: From EPA, *Respiratory Health Effects of Passive Smoking: Lung Cancer and Other Disorders,* EPA/600/6-90/006 F, U.S. Environmental Protection Agency, U.S. Government Printing Office, Washington, D.C., 1992.

TABLE 34

Indoor Sources of Formaldehyde Exposure

Sources	Concentration
Cigarette smoke	40 ppm in 40-ml puff
Dose per pack for smoke	0.38 mg/pack
Environmental tobacco smoke	0.25 ppm
Clothing made with synthetic fibers	
Men's polyester-cotton blend	2.7 μg/g per day
Women's dress	3.7 μg/g per day
Furnishings	
Particleboard[a]	0.4–8.1 μg/g per day
Plywood	1.5–5.3 μg/g per day
Paneling	0.9–21.0 μg/g per day
Draperies	0.8–3.0 μg/g per day
Carpet/upholstery fabric	≤ 0.1 ppm

[a] Made with urea-formaldehyde resin.

Source: From Leikauf, G. D., in *Environmental Toxicants: Human Exposure and Their Health Effects,* Lippmann, M., Ed., Van Nostrand-Reinhold, New York, 1992, chpt. 10. With permission.

TABLE 35

Symptomatic Responses in Humans to Formaldehyde

Formaldehyde Concentration (ppm)	Response
	Eye irritation (blinking rate, lacrimation, conjunctivitis)
0.01	Detectable by some
0.30	Slight but tolerable response
0.50	Intermediate response
0.80	Severe response
1.7–2.0	Marked eye blinking
	Upper respiratory tract irritation (nasal secretion or dryness, throat irritation)
0.03	Minimal or no effect
0.25–1.39	Moderate irritation
1.7–2.1	Significant throat irritation
3.1	Severe to intolerable irritation
	Odor threshold
0.05	Odor threshold
0.17	Detected by 50% exposed
1.5	Detected by all subjects

Source: From Leikauf, G. D., in *Environmental Toxicants — Human Exposure and Their Health Effects,* Lippmann, M., Ed., Van Nostrand-Reinhold, New York, 1992, chpt. 10. With permission.

to as high as 10% in some studies of toluene diisocyanate. One fatal case has been reported.[42] The prognosis of occupational asthma indicates that many but not all cases resolve upon removal from exposure. A wide variety of industrial air pollutants is associated with occupational asthma (see Table 44).

3.4.1 Occupational Lung Cancer

Estimates of the influence of occupational agents in etiology of lung cancer range from 4 to 10% in one report.[28] Table 45 lists established causes of cancer from workplace exposure. Table 46 lists other occupational agents suspected to cause human cancer. Some common occupational air pollutant categories will be discussed individually.

3.4.2 Diesel Exhaust

It can be noted that diesel exhaust contains both particulate phase and gas phase pollutants and should be considered a complex mixture of pollutants similar to wood smoke and ETS. Numerous animal studies (at very high concentrations) have seen increased lung tumors. The data are stronger for lung tumors in rats than in mice. Also, five studies investigating tumor growth in hamsters exposed to diesel exhaust have shown negative results. Epidemiologic studies of cancer risk in human populations exposed to diesel exhaust suggest that long-term exposure is associated with a 20 to 50% increased risk for lung cancer.[104]

3.4.3 Metals

Many metals are associated with adverse human health effects. Following is a brief summary of some of these effects (see Table 48). Lead effects are well known; however, there are no known respiratory effects. The health effects of certain trace elements have been reviewed recently.[15] For many of these, ingestion is a greater threat than inhalation.

TABLE 36

Respiratory Function of Humans Exposed to Formaldehyde

Formaldehyde Concentration (ppm)	Length of Exposure (min)	Route of Exposure	Findings[a] (no. of subjects)
Healthy Individual (Nonsmoking) in Clinical Studies			
0.3–1.6	300	Oronasal (chamber)	No change in FVC, $FEV_{1.0}$, $FEF_{25-75\%}$, or R_{aw} (n = 16)
0.5–3.0	180	Oronasal (chamber, with or without exercise)	No change in FVC, $FEV_{1.0}$, $FEF_{25-75\%}$, or (s)G_{aw} (n = 9–10)
2.0	40	Oronasal (chamber)	No change in FVC, $FEV_{1.0}$, or MMEF (n = 15)
3.0	60	Oronasal (chamber, with exercise)	2.5–3.8% change in FVC and $FEV_{1.0}$ (n = 22)
3.0	180	Oronasal (chamber, with exercise)	2.7% change in FVC and $FEV_{1.0}$, in 60 min; no change in FVC and $FEV_{1.0}$, at 180 min
7.5	2	Oral (mouth-piece)	No change in $FEV_{1.0}$, or R_{aw}
Individual with Asthma in Clinical Studies			
2.0	40	Oronasal (chamber, with exercise)	No change in FVC, $FEV_{1.0}$, or MMEF (n = 12)
3.0	10	Oral (mouth-piece with exercise)	No change in (s)R_{aw}
3.0	60	Oronasal (chamber)	No change in FVC, $FEV_{1.0}$, $FEF_{25-75\%}$, or (s)G_{aw}
Individual with Occupational Exposure			
0.3–0.6	8 hr work	Oronasal (at work)	Small (200 ml) change in FVC and $FEV_{1.0}$ (n = 38)
2.0	40	Oronasal (chamber, with exercise)	No change in FVC, $FEV_{1.0}$, or MMEF (n = 15)
0.1–3.0	20	Oronasal (face mask)	12 ± 2% decrease in $FEV_{1.0}$, as compared to 8 ± 3% in control; little or no change in MMEF and other measures
1.8	30	Oronasal (chamber)	13% decrease in $FEV_{1.0}$ (n = 4)
2.0	20	Oronasal (chamber)	Decrease in R_{aw} in 12 of 230 (5%) persons
3.2	30	Oronasal (chamber)	12% decrease in $FEV_{1.0}$ (n = 8); 3 with decrease > 25% in $FEV_{1.0}$

[a] Abbreviations: FVC = forced vital capacity; $FEV_{1.0}$ = forced expiratory flow at 1.0 sec; $FEF_{25-75\%}$ = mean of forced expiratory flow at 25% and 75% vital capacity; (s)R_{aw} = (specific) airway resistance; MMEF = midmaximal expiratory flow; (s)G_{aw} = (specific) airway conductance.

Source: From Leikauf, G. D., in *Environmental Toxicants — Human Exposure and Their Health Effects,* Lippmann, M., Ed., Van Nostrand-Reinhold, New York, 1992, chpt. 10. With permission.

TABLE 37

Respiratory Effects of Solvent Exposure

Chemical	Effect	Ref.
Toluene	Increase annual average loss in PFTs	120
Toluene	Increase % variation in PEFR	92
Toluene	Asthma	6, 73
Toluene	Methacholine responsiveness not related to TDI exposure	161
Toluene	Increased respiratory symptoms	2

Note: PFTs = pulmonary function tests; PEFR = peak expiratory flow rate; TDI = toluene diisocyanate.

TABLE 38

Changes in PMN Counts in the BAL Fluid of Subjects Exposed to Air and VOC[a]

Subject	Air Exposure				VOC Exposure			
	Pre 1	Pre 4	Post 0	Post 18	Pre 1	Pre 4	Post 0	Post 18
1	0.1	0.4	0.5	0.08	0.01	0.6	11.8	438.3
2	0.6	3.6	1.1	1.9	2.8	2.6	3.6	80.7
3	0.02	0.01	0.03	0.01	1.8	0.08	0.01	3.6
4	64.1	30.3	27.0	10.2	68.5	49.6	129.8	267.5
5	73.1	78.2	64.9	17.8	173.1	89.9	35.2	98.8
6	0.01	0.04	0.03	1.2	0.1	0.06	0.03	0.01
7	63.9	26.5	7.0	17.2	746.1	120.3	64.8	544.9
8	14.3	1.1	0.8	7.9	0.06	0.2	1.8	0.2
9	5.9	10.0	21.4	39.4	1.4	3.5	9.1	34.0
10	4.8	61.4	41.5	2.7	81.3	0.3	3.8	2.6
11	35.8	0.8	0.06	1.1	3.3	1.7	110.8	28.7
12	58.0	36.6	40.4	70.5	63.4	0.01	50.6	66.7
13	0.4	0.08	0.1	11.0	1.0	0.06	2.4	13.5
14	0.02	2.1	0.1	1.5	1.8	1.4	4.8	9.6
Average	22.9	17.9	14.1	13.0	88.0	19.3	30.6	113.5
Standard error	8.1	7.1	5.9	5.5	54.9	10.8	12.0	48.9

Note: PMN = polymorphonuclear leukocyte (neutrophil); BAL = bronchial alveolar lavage; VOCs = volatile organic compounds.

[a] Values are expressed as total PMNs × 10^4 recovered from each lavage and are derived by multiplying the PMNs/ml times the number of milliliters of BAL fluid recovered.

Source: From Koren, H. S. and Devlin, R. B., *Ann. N.Y. Acad. Sci.*, 641, 215–224, 1992. With permission.

TABLE 39

Summary of Studies of Respiratory Effects of Exposure to Wood Smoke

Age of Subjects	No. of Subjects	Endpoints	Results Measured
1–7 years	34 with stoves, 34 without	Symptoms	More symptoms in children with stoves ($p < 0.001$)
5–11 years[a]	258 with stoves, 141 without	Symptoms	Risk ratio = 1.1, showing no significant effect
1 year and older	455 high smoke,[b] 368 low smoke	Symptoms, disease prevalence	No significant effects; trend in children ages 1–5 years
8–11 years	296 healthy,[c] 30 asthmatic	Spirometry	Significant association between fine particles and lung function in asthmatics in an area heavily impacted by wood smoke ($p = 0.05$)
1–5½ years	59	Symptoms	Significant correlation between woodstove use and wheeze and cough frequency ($p = 0.01$)
< 24 months	58 pairs	Respiratory disease	Woodstove significant risk factor for lower respiratory infection
8–11 years	410[c]	Spirometry	Significant decrease in PFTs with elevated wood smoke
8–11 years	495	Spirometry	Significant relation of function decrease with increasing TSP
Mean age = 46 years	182	Symptoms	Significant association

Note: PFTs = pulmonary function tests; TSP = total suspended particulate.

[a] Described as kindergarten through grade 6.

[b] Geographical areas chosen as having high or low wood-smoke pollution.

[c] Grades 3 through 6.

TABLE 40

Summary of the Results of the Effects of Nitrogen Dioxide Exposure on Respiratory Disease in Children

Location (Year)	NO_2 Exposure Measure Used in Analysis	Age (years)	Sample size	Odds Ratio for Respiratory Disease	95% Confidence Limits
28 areas of England and Scotland (1973)	Gas stove vs. electric stove	6–11	5658	1.31	1.16 to 1.48
28 areas of England and Scotland (1977)	Gas stove vs. electric stove	5–10	4827	1.24	1.09 to 1.42
Middlesborough, England (1978)	NO_2 measured with Palmes tubes	6–7	103	1.53	1.04 to 2.24
Middlesborough, England (1980)	NO_2 measured with Palmes tubes	5–6	188	1.11	0.83 to 1.49
London (1975 to 1978)	Gas stove vs. electric stove	<1	390	0.63	0.36 to 1.10
Six U.S. cities (1974–1979)	Gas stove vs. electric stove	6–10	8240	1.08	0.96 to 1.37
Six U.S. cities (1983–1986)	NO_2 measured with Palmes tubes	7–11	1286	1.47	1.17 to 1.86
Tayside region, Scotland (1980)	Gas stove vs. electric stove	<1	1565	1.14	0.86 to 1.50
Iowa City, IA	Gas stove vs. electric stove	6–12	1138	1.10	0.79 to 1.53
Netherlands, (1986)	NO_2 measured with Palmes tubes	6–12	775	0.94	0.66 to 1.33
Columbus, OH (1978)	Gas stove vs. electric stove	< 12	553	1.10	0.74 to 1.54

Source: From Hassalblad, V. et al., *J. Air Waste Manage. Assoc.*, 42, 662–671, 1992. With permission.

3.4.4 *Abestiform and Asbestos Fibers*

3.4.4.1 *Physical Characteristics*

- *Asbestiform fibers:* Naturally occurring crystalline minerals with a fibrous habit (form, structure). Such minerals may also occur in a nonfibrous form. There are two classes of asbestiform fibers: serpentine and amphibole.
- *Asbestos fibers:* Commercial term for asbestiform minerals of economic and industrial importance. The following six varieties are usually regarded as asbestos, although only the first three are of significant commercial importance.
  - Serpentine: Chrysotile
  - Amphiboles: Crocidolite
Amosite
Anthophylite
Tremolite
Actinolite
- *Serpentine asbestos:* Wavy, serpentine morphology; distinctive crystalline structure and shape compared to amphiboles.
- *Amphibole asbestos:* Fibers appear straight and rigid with parallel sides; all have similar crystalline structure; distinguished readily only on basis of variation in chemical composition

Physical characteristics regarding asbestos fibers are important in determining their toxicity; following are three estimates. Stanton hypothesis:[154,155] Long, thin durable fibers, regardless of chemical composition, are carcinogenic. Highly carcinogenic fibers of a length > 8 μm and diameter ≤0.25 mm; moderately carcinogenic fibers are of a length > 4 μm and diameter of ≤1.5 mm. Walton,[163] after an extensive review of the literature, concluded that carcinogenic fibers have the following dimensions:

TABLE 41

A List of 189 Air Toxics Cited in the Clean Air Act Amendments of 1990

Compound	CAS No.	Compound	CAS No.
Acetaldehyde	75-07-0	Diethanolamine	111-42-2
Acetamide	60-35-5	Diethyl sulfate	64-67-5
Acetonitrile	75-05-8	3,3-Dimethoxybenzidine	119-90-4
Acetophenone	98-86-2	Dimethylamino-azobenzene	60-11-7
2-Acetylaminofluorene	53-96-3	*N,N*-dimethylaniline	121-69-7
Acrolein	107-02-8	3,3-dimethyl benzidine	119-93-7
Acrylamide	79-06-1	Dimethylcarbamoyl chloride	79-44-7
Acrylic acid	79-10-7	Dimethyl formamide	68-12-2
Acrylonitrile	107-13-1	1,1-Dimethyl hydrazine	57-14-7
Allyl chloride	107-05-1	Dimethyl phthalate	131-11-3
4-Aminobiphenyl	92-67-1	Dimethyl sulfate	77-78-1
Aniline	62-53-3	4,6-Dinitro-*o*-cresol and salts	534-52-1
o-Anisidine	90-04-0	2,4-Dinitrophenol	51-28-5
Asbestos	1332-21-4	2,4-Dinitrotoluene	121-14-2
Benzene	71-43-2	1,4-Dioxane	123-91-1
Benzidine	92-87-5	1,2-Diphenylhydrazine	122-66-7
Benzotrichloride	98-07-7	Epichlorohydrin	106-89-8
Benzyl chloride	100-44-7	1,2-Epoxybutane	106-88-7
Biphenyl	92-52-4	Ethyl acrylate	140-88-5
bis-(2-ethylhexyl)-phthalate	117-81-7	Ethyl benzene	100-41-4
bis-(chloromethyl)-ether	542-88-1	Ethyl carbamate	51-79-6
Bromoform	75-25-2	Ethyl chloride	75-00-3
1,3-Butadiene	106-99-0	Ethylene dibromide	106-93-4
Calcium cyanamide	156-62-7	Ethylene dichloride	107-06-2
Caprolactam	105-60-2	Ethylene glycol	107-21-1
Captan	133-06-2	Ethylene imine	151-56-4
Carbaryl	63-25-2	Ethylene oxide	75-21-8
Carbon disulfide	75-15-0	Ethylene thiourea	96-45-7
Carbon tetrachloride	56-23-5	Ethylidene dichloride	75-34-3
Carbonyl sulfide	463-58-1	Formaldehyde	50-00-0
Catechol	120-80-9	Heptachlor	76-44-8
Chloramben	133-90-4	Hexachlorobenzene	118-74-1
Chlordane	57-74-9	Hexachlorobutadiene	87-68-3
Chlorine	7782-50-5	Hexachlorocyclopentadiene	77-47-4
Chloroacetic acid	79-11-8	Hexachloroethane	67-72-1
2-Chloroacetophenone	532-27-4	Hexamethylene-1,6-diisocyanate	822-06-0
Chlorobenzene	108-90-7	Hexamethylphosphoramide	680-31-9
Chlorobenzilate	510-15-6	Hexane	110-54-3
Chloroform	67-66-3	Hydrazine	302-01-2
Chloroemthyl methyl ether	107-30-2	Hydrochloric acid	7647-01-0
Chloroprene	126-99-8	Hydrogen fluoride	7664-39-3
Cresols/cresylic acid	1319-77-3	Hydroquinone	123-31-9
o-Cresol	95-48-7	Isophorone	78-59-1
m-Cresol	108-39-4	Lindane (all isomers)	58-89-9
p-Cresol	106-44-5	Maleic anhydride	108-31-6
Cumene	98-82-8	Methanol	67-56-1
2,4-D salts and esters	94-75-7	Methoxychlor	72-43-5
DDE	3547-04-4	Methyl bromide	74-83-9
Diazomethane	334-88-3	Methyl chloride	74-87-3
Dibenzofurans	132-64-9	Methyl chloroform	71-55-6
1,2-Dibromo-3-chloropropane	96-12-8	Methyl ethyl ketone	78-93-3
Dibutylphthalate	84-74-2	Methyl hydrazine	60-34-4
1,4-Dichlorobenzene (*p*)	106-46-7	Methyl iodide	74-88-4
3,3-Dichlorobenzidine	91-94-1	Methyl isobutyl ketone	108-10-1
Dichloroethyl ether	111-44-4	Methyl isocyanate	624-83-9
1,3-Dichloropropene	542-75-6	Methyl methacrylate	80-62-6

TABLE 41 (continued)

A List of 189 Air Toxics Cited in the Clean Air Act Amendments of 1990

Compound	CAS No.	Compound	CAS No.
Dichlorvos	62-73-7	Methyl-t-butyl ether	1634-04-4
4,4-Methylene-*bis*-(2-chloraniline)	101-14-4	2,4-Toluenediamine	95-80-7
Methylene chloride	75-09-2	2,4-Toluenediisocyanate	584-84-9
Methylene diphenyl diisocyanate	101-68-8	*o*-Toluidine	95-53-4
4,4-Methylenedianiline	101-77-9	Toxaphene	8001-35-2
Naphthalene	91-20-3	1,2,4-Trichlorobenzene	120-821
Nitrobenzene	98-95-3	1,1,2-Trichloroethane	79-00-5
4-Nitrobiphenyl	92-93-3	Trichloroethylene	79-01-6
4-Nitrophenol	100-02-7	2,4,5-Trichlorophenol	95-95-4
2-Nitropropane	79-46-9	2,4,6-Trichlorophenol	88-06-2
N-nitroso-*N*-methylurea	684-93-5	Triethylamine	121-44-8
N-nitrosodimethylamine	62-75-9	Trifluralin	1582-09-8
N-nitrosomorpholine	59-89-2	2,2,4-Trimethylpentane	540-84-1
Parathion	56-38-2	Vinyl acetate	108-05-4
Pentachloronitrobenzene	82-68-8	Vinyl bromide	593-60-2
Pentachlorophenol	87-86-5	Vinyl chloride	75-01-4
Phenol	108-95-2	Vinylidene chloride	75-35-4
p-Phenylenediamine	106-50-3	Xylenes (isomers and mixtures)	1330-20-7
Phosgene	75-44-5	*o*-Xylene	95-47-6
Phosphine	7803-51-2	*m*-Xylene	108-38-3
Phosphorous	7723-14-0	*p*-Xylene	106-42-3
Phthalic anhydride	85-44-9	Antimony compounds	—
Polychlorinated biphenyls	1336-36-3	Arsenic compounds	—
1,3-Propane sultone	1120-71-4	Beryllium compounds	—
beta-Propiolactone	57-57-8	Cadmium compounds	—
Propionaldehyde	123-38-6	Chromium compounds	—
Propoxur	114-26-1	Cobalt compounds	—
Propylene dichloride	78-87-5	Coke oven emissions	—
Propylene oxide	75-56-9	Cyanide compounds	—
1,2-Propylenimine	75-55-8	Glycol ethers	—
Quinoline	91-22-5	Lead compounds	—
Quinone	106-51-4	Manganese compounds	—
Styrene	100-42-5	Mercury compounds	—
Styrene oxide	96-09-3	Mineral fibers	—
2,3,7,8-Tetrachlorodibenzo-*p*-dioxin	1746-01-6	Nickel compounds	—
1,1,2,2-Tetrachloroethane	79-34-5	Polycylic organic matter	—
Tetrachloroethylene	127-18-4	Radionuclides (including radon)	—
Titanium tetrachloride	7550-45-0	Selenium compounds	—
Toluene	108-88-3		

Length: >5–10 μm, up to 100 μm
Diameter: <1.5–2 μm, with no minimum specified
Aspect ratio: >5:1 to 10:1

Lippmann,[99] after review of literature for human exposures and animal experiments, concluded that asbestos fibers of varying dimensions cause the three major diseases due to asbestos inhalation. Asbestosis results from fibers of a length of > 2 μm and diameter of 0.15 to 2.0 μm. Lung cancer results from fibers of a length > 10 μm and diameter of 0.3 to 0.8 μm. Mesothelioma can result from fibers of a length of > 5 μm and diameter of < 0.1 μm.

Asbestosis is a common occupational lung disease. Millions of workers were exposed to asbestos in construction and ship building in the 1980s. Due to the long latency in the development of asbestosis, asbestos-induced lung cancer, and mesothelium, many cases are

TABLE 42

Summary of Data Obtained During Airborne Toxic Element and Organic Substance Project

Variable	Description, if Applicable (units)
Inhalable particulate matter (IPM)	Particles of aerodynamic-size diameter, ≤15 μm (μg/m³)
Fine particulate matter	Particles of aerodynamic-size diameter, ≤2.5 μm (μg/m³)
Elements	(μg/m³)
Lead	
Manganese	
Copper	
Vanadium	
Cadmium	
Zinc	
Iron	
Nickel	
Sulfate	Total sulfate concentration in IPM (μg/m³)
Organic fractions	
Cyclohexane	Cyclohexane-soluble or extractable organic fraction of IPM (μg/m³)
Dichloromethane	Dichloromethane-soluble or extractable organic fraction of IPM (μg/m³)
Acetone	Acetone-soluble or extractable organic fraction of IPM (total μg/m³)
Alkylating agents	Alkylating activity of acetone-soluble fraction (μg/m³)
Carbon monoxide	Maximum 1-hr daily CO concentration (parts per thousand)
Sulfur dioxide	Average daily sulfur dioxide concentration (ppm)
Ozone	Maximum 1-hr daily ozone concentration (ppm)
Extractable particulate organic fractions	
Polycyclic aromatic hydrocarbons (PAHs)	All PAHS measured were contained in the cyclohexane-soluble fraction (ng/m³)
Cyclopenta(*c,d*)pyrene	
Benzo(*a*)anthracene	
Benzo(*k*)fluoranthene	
Benzo(*a*)pyrene	
1,2,5,6-Dibenzanthracene	
Benzo(*g,h,i*)perylene	
Indeno(*1,2,3-c,d*)pyrene	
1,2,4,5-Dibenzpyrene	
Coronene	
Dibenzo(*a,c*)anthracene	
Chrysene	
Anthracene	
Benzo(*e*)pyrene	
Benzo(*j*)fluoranthene	
Perylene	
Benzo(*b*)fluoranthene	
Volatile organic compounds (VOCs)	All VOCs were measured during the first two studies (ppb)
Vinyl chloride	
Vinylidene chloride	
Acrylonitrile	
Methylene chloride	
Chloroform	
1,2-Dichloroethane	
Benzene	
Carbon tetrachloride	
Trichloroethylene	
Dioxane	
Toluene	
1,2-Dibromoethane	
Tetrachloroethylene	
Chlorobenzene	
Ethylbenzene	

TABLE 42 (continued)

Summary of Data Obtained During Airborne Toxic Element and Organic Substance Project

Variable	Description, if Applicable (units)
m-Xylene	
p-Xylene	
Styrene	
o-Xylene	
1,1,2,2-Tetra-chloroethane	
o-Chlorotoluene	
p-Chlorotoluene	
p-Dichlorobenzene	
o-Dichlorobenzene	
Nitrobenzene	

Source: From Lioy, P. J. et al., in *Toxic Air Pollution,* Lioy, P. J. and Daisey, J. M., Eds., Lewis Publishers, Chelsea, MI, 1987, pp. 3–43. With permission.

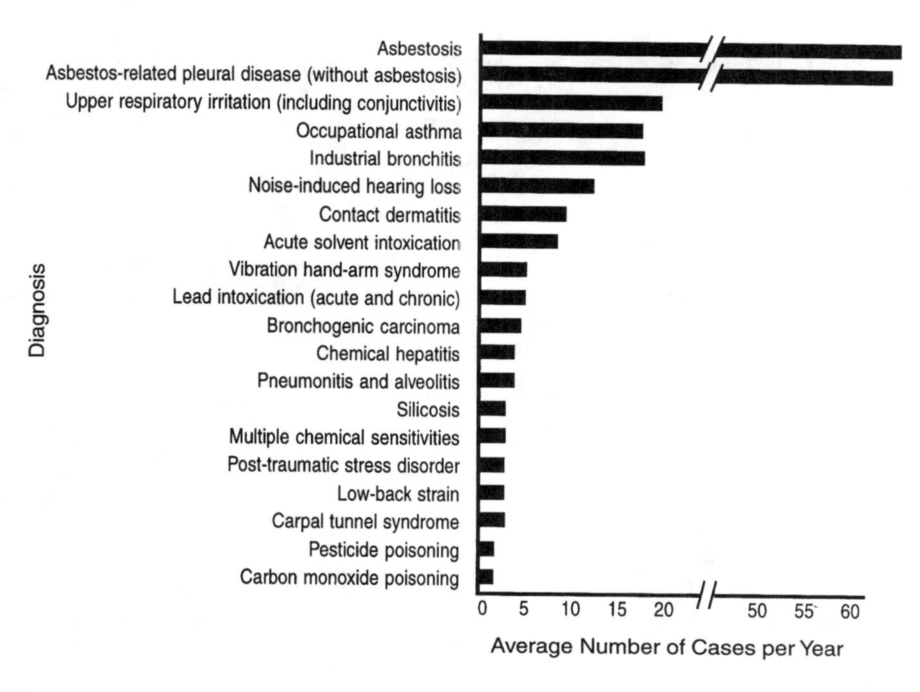

FIGURE 2
The 20 most common occupational diseases diagnosed at the Yale Referral Clinics between 1979 and 1987. (From Cullen, M. R. et al., *N. Engl. J. Med.,* 322, 594–601, 1990. With permission.)

still being diagnosed. Common types of inhalable fibers are given in Table 49. The detection abilities for different types of asbestos fibers are shown in Table 50. Table 51 gives some average concentrations of asbestos fibers found in various sites. Many individuals in the U.S. have the potential of being exposed to asbestos fibers. Some of the categores of these individuals are listed in Table 52. The use and monitoring of asbestos fibers are regulated by the USEPA as a form of hazardous air pollutant. Regulatory guidelines and standards for asbestos are shown in Table 53.

TABLE 43

Major Industrial Air Pollutants

Acrylonitrile	Freon-13
Arsenic	Hexachlorocyclopentadiene
Asbestos	Manganese
Benzene	Mercury
Beryllium	Methyl chloroform
1,3-Butadiene	Methylene chloride
Cadmium	Nickel
Carbon tetrachloride	Perchloroethylene
Chlorobenzenes	Polycyclic organic matter
Chloroform	Radionuclides
Chromium	Toluene
Coke-oven emissions	Trichloroethylene
Dioxin	Vinylidene
Epichlorohydrin	Vinyl chloride
Ethylene dichloride	Metals
Ethylene oxide	

TABLE 44

Materials Causally Linked to Asthma in the Workplace

Vegetable	Animal	Plastic or Chemical	Metal	Pharmaceutical
Grain dust	Danders	Acid anhydrides	Stainless steel	Penicillins
Flour	Insects	Epoxy resins	Galvanized steel	Cephalosporins
Fig plants	Silkworm larva	Diisocyanates	Aluminum fluoride	Piperazine
Wood dust	Shellfish	Persulfates	Vanadium	Psyllium
Seafood	Pig or chicken excreta	*para*-Phenylenediamine	Cobalt	Methyldopa
Green coffee beans	Fish feed	Phthalic anhydride	Tungsten carbide (cobalt)	Spiramycin
Fungal spores	Animal enzymes	Dimethyl ethanolamine	Platinum salts	Tetracycline
Tragacanth		Azobisformamide	Nickel	Amprolium
Castor bean		Azodicarbonamide	Chromium	Cimetidine
Tea		Formaldehyde		Isoniazid
Tobacco		Ethylenediamine		Phenylglycine
Flax		Acrylates		
Hemp		Henna		
Cotton				
Hops				
Bacterial enzymes				
Colophony				

Source: From Cullen, M. R. et al., *New Engl. J. Med.*, d322, 594–601, 1990. With permission.

3.4.4.2 Animal Toxicology

Asbestos exposure has been studied carefully in animal exposures, although inhalation exposures are difficult to perform without danger to researchers. Other routes of exposure have been utilized, as summarized in Table 54. Also, asbestos toxicity has been studied using *in vitro* exposure systems. These studies have identified several biomarkers of asbestos injury (see Table 55).

3.4.4.3 Human Health Effects

One estimate of mortality (carcinogenicity of asbestos) from occupational exposure to asbestos is based on exposures from 1940 to 1979.[118] In that paper, 27,500,000 individuals

TABLE 45

Established Causes of Cancer from Workplace Exposure

Agent	Industries and Trades with Proved Excess Cancers	Primary Affected Site
para-Aminodiphenyl	Chemical manufacturing	Urinary bladder
Asbestos	Construction, asbestos mining and milling, production of friction products and cement	Pleura, peritoneum, bronchus
Arsenic	Copper mining and smelting	Skin, bronchus, liver
Alkylating agents (mechlorethamine hydrochloride and *bis*[chloromethyl] ether)	Chemical manufacturing	Bronchus
	Chemical and rubber manufacturing, petroleum refining	Bone marrow
Benzidine, beta-naphthylamine, and derived dyes	Dye and textile production	Urinary bladder
Chromium and chromates	Tanning, pigment making	Nasal sinus, bronchus
Ionizing radiation, gamma rays	Nuclear, health care	Skin, thyroid, bronchus, bone marrow
Radon	Uranium and hematite mining	Bronchus
Radium	Watch painting	Bone
Nickel	Nickel refining	Nasal sinus, bronchus
Polynuclear aromatic hydrocarbons (from coke, coal tar, shale, mineral oils, and creosote)	Steel making, roofing, chimney cleaning	Skin, scrotum, bronchus
Vinyl chloride monomer	Chemical manufacturing	Liver
Wood dust	Cabinet making, carpentry	Nasal sinus

Source: From Cullen, M. R. et al., *New Engl. J. Med.,* 322, 675–683, 1990. With permission.

TABLE 46

Widely Used Suspected Human Carcinogens in the Workplace

Agent	Industries and Trades	Suspected Human Sites
Beryllium	Beryllium processing, aircraft manufacturing, electronics, secondary smelting	Bronchus
Cadmium	Smelting, battery making, welding	Bronchus
Ethylene oxide	Hospitals; production of hospital supplies	Bone marrow
Formaldehyde	Plastic, textile, and chemical production; health care	Nasal sinus, bronchus
Synthetic mineral fibers (e.g., fibrous glass)	Manufacturing, insulation	Bronchus
Polychlorinated biphenyls	Electrical-equipment production and maintenance	Liver
Organochlorine, pesticides (e.g., chlordane, dieldrin)	Pesticide manufacture and application, agriculture	Bone marrow
Silica	Casting, mining, refracting	Bronchus

Source: From Cullen, M. R. et al., *New Engl. J. Med.,* 322, 675–683, 1990. With permission.

had potential exposures at work; 18,800,000 had exposures in excess of 2 to 3 fiber-yr/cm^3. The current annual asbestos-related cancer deaths are estimated to be 8200. This number is expected to rise to about 9700 deaths annually by the year 2000. Thereafter, the annual numbers will decrease but remain substantial for 3 decades.

TABLE 47

Abbreviated List of Classes of Compounds in Diesel Exhaust

Compound class
Particulate phase
Elemental carbon
Heterocyclics, hydrocarbons (C_{14}–C_{35}), and polycyclic aromatic hydrocarbons and derivatives
Acids
Alcohols
Aldehydes
Anhydrides
Esters
Ketones
Nitriles
Quinones
Sulfonates
Halogenated and nitrated compounds
Inorganic sulfates and nitrates
Metals
Gas and vapor phases
Acrolein
Ammonia
Carbon dioxide
Carbon monoxide
Benzene
1,3-Butadiene
Formaldehyde
Formic acid
Heterocyclics, hydrocarbons (C_1–C_{18}), and derivatives (as listed above)
Hydrogen cyanide
Hydrogen sulfide
Methane
Methanol
Nitric and nitrous acids
Nitrogen oxides
Sulfur dioxide
Toluene
Water

Source: From Mauderly, J. L., in *Environmental Toxicants — Human Exposures and Their Health Effects,* Lippmann, M., Ed., Van Nostrand-Reinhold, New York, 1992, chpt. 5. With permission.

The exact mechanism responsible for the carcinogenicity of asbestos fibers is not known. Potential causes are

- Mutagenic activity
- Clastogenic activity, possible mechanism being interference of ingested fibers with microtubules of mitotic apparatus
- Altered growth factor production/responsiveness
- Membrane-associated affects
  - Increase in activity of the plasma membrane enzyme Na^+-K^+-dependent ATPase
  - Associated decrease in endogenous protein phosphorylation suggesting that asbestos-exposed cells may be stimulated to proliferate through a membrane-triggered response
  - Lipid peroxidation and subsequent production of reactive oxygen and superoxide radicals that may subsequently damage cellular DNA

TABLE 48

Health Effects of Trace Metals Found in the Workplace

Aluminum: Common source of exposures is smelters; both pulmonary fibrosis and bronchial asthma have been observed in Al-exposed workers.
Beryllium: Ubiquitous sources; dermatological, pulmonary, and systemic toxicity of beryllium has been recognized for years. Inhalation can cause a "chemical pneumonia". Chronic beryllium disease is associated with dyspnea at rest. IARC has rated this element as an animal and human carcinogen.
Cadmium: One source of occupational exposure is electroplating. Respiratory effects associated with exposure are dyspnea, coughing, and tightness in the chest and possible subsequent pulmonary edema.
Chromium: Hexavalent Cr(VI) is the most toxic. This compound is corrosive and can cause chronic ulcerations and perforation of the nasal septum and also nasal carcinomas.
Mercury: Exposure can come from natural degassing of the Earth's crust, from combustion of fossil fuels, and from the dental profession. Mercury exposure can cause erosive bronchitis or bronchiolitis with interstitial pneumonitis.
Nickel: Primary exposure is in refining. Main acute effect is usually dermatitis; however, chronic exposure has been associated with respiratory cancer. A fivefold increase in incidence of lung cancer and a 150-fold increase in incidence of nasal cancer have been reported.

[a] IARC = International Association for Research on Cancer; see Reference 89.

TABLE 49

Common Types of Inhalable Fibers

Asbestos fibers
Manmade mineral fibers (MMMFs): glass, mineral wool, ceramic fibers
Other naturally occurring fibers: attapulgite (palygorscite), erionite (fibrous zeolite), nemalite (fibrous brucite-magnesium hydroxide), dawsonite (dihydroxyaluminum carbonate sodium), wollastonite (monocalcium silicate), halloysite (hydrated aluminum silicate), ferroactinolite fibers
Other synthetic fibers
  Inorganic: aluminum oxide, potassium octatitanate (Fybex), silicon carbide, calcium sodium metaphosphate fibers
  Organic: aramid, carbon, polyolefin fibers
Nonfibrous silicate minerals: talc, vermiculite (both may contain various concentrations of asbestos or asbestiform fibers)

Controversies remaining in the understanding of the toxicity of asbestos include:

- Amphibole hypothesis
- Measurement of airborne asbestos
- Environmental exposures (to asbestos in buildings)
- Regulatory policy

The industrial hygiene definition of asbestos (i.e., through light microscopic resolution): fibers > 5 μm in length, > 0.25 μm in diameter, with aspect (length/width) ratio > 3:1. Percentage of airborne asbestos fibers that meet the above definition, as determined by transmission electron microscopy (TEM), are listed in Table 50.

3.4.5 Manmade Mineral Fibers

The generic term for fibrous inorganic substances made primarily from glass, slag, rock, or clay is manmade mineral fibers (MMMFs), also referred to as manmade vitreous fibers (MMVFs). The various types of MMMFs have different chemistries with a range of durabilities in biological systems as described below.

TABLE 50

Airborne Asbestos Fibers

Detection of Airborne Asbestos Fibers

Fiber	Percent (Range) (%)
Chrysotile	9 (1–50)
Crocidolite	4 (1–18)
Amosite	25 (8–43)

Unit Size of Asbestos Fibers as Determined by Transmission Electron Microscopy

Fiber	Range (μm)
Chrysotile	0.02–0.05
Crocidolite	0.04–0.15
Amosite	0.06–0.35

Source: From Berman, D. W. and Chatfield, E. J., *Interim Superfund Method for the Determination of Asbestos in Ambient Air.* Part 2. *Technical Background Document,* EPA/68/01-7290, U.S. Environmental Protection Agency, U.S. Government Printing Office, Washington, D.C., 1989.

TABLE 51

Average Indoor and Outdoor Concentrations of Airborne Asbestos Fibers Longer Than 5 μm as Determined by Transmission Electron Microscopy[a]

Sites	Number of Buildings	Fibers (per m^3)[b]
All buildings	198	300
School	48	500
Residence	96	200
Public and commercial	54	200
Urban outdoor air	—	100
Rural outdoor air	—	10

[a] Summarized from Reference 60.
[b] Humans at rest breathe an average of 10 to 12 m^3 of air per day.

TABLE 52

Categories of Building Occupants[60]

- Bystanders or nonoccupationally exposed building occupants, e.g., office workers, visitors, students, and teachers
- Housekeeping or custodial employees, who may disturb materials in the course of routine cleaning and service functions
- Maintenance or skilled workers, who may disturb asbestos-containing material (ACM) in the course of making repairs or installing new equipment or during minor renovation activity
- Abatement workers or others involved in the removal or renovation of structures with ACM
- Firefighters and other emergency personnel who may be present during or after the fabric of the building has been extensively damaged by fire, wind, water, or earthquake

TABLE 53

Regulatory Guideline and Standards for Asbestos

Group	Year	Limit
ACGIH	1946	5×10^6 particles per ft^3
ACGIH	1968[a]	12 fibers per ml or 2×10^6 particles per ft^3[c]
ACGIH	1970,[a] 1974[b]	5 fibers per ml
OSHA	1972	5 fibers per ml
OSHA	1976	2 fibers per ml
NIOSH	1976	0.1 fibers per ml
ACGIH	1978,[a] 1980[b]	0.2 fibers per ml for crocidolite
		0.5 fibers per ml for amosite
		2.0 fibers per ml for chrysotile and other forms
OSHA	1986	0.2 fibers per ml
OSHA	1994	0.1 fibers per ml

Note: ACGIH = American Conference of Governmental Industrial Hygienists; OSHA = Occupational Safety and Health Administration; NIOSH = National Institute of Occupational Safety and Health.

[a] Notice of intent.

[b] Adopted as threshold limit value (TLV).

[c] All fiber limits based on phase-contrast optical determination at 400–450× magnification.

[d] Proposed.

TABLE 54

Routes of Exposure to Asbestos Fibers in Experimental Studies

Inhalation
Intratracheal instillation
Intrapleural inoculation
Intrapleural implantation
Intraperitoneal injection
Subcutaneous injection
Gavage
Feeding

TABLE 55

In Vitro Responses of Alveolar Macrophages to Toxicant Exposure: Reactive Oxygen Species, Cytokines, and Eicosanoids

Test Material	Exposure Conditions	Species Tested	Culture Conditions (± Serum)	Key Observation
Chrysotile, crocidolite	0.4 mg/ml per 2×10^6 cells	Hamster	+	Both chrysotile and crocidolite asbestos stimulated O_2^-
Crocidolite, erionite, code 100 glass fibers, sepiolite, riebeckite, mordenite, glass beads	2.5–25 mg/cm²	Hamster, rat	+	Rat cells were more responsive than hamster for O_2^- release Fibrous materials stimulated significantly more O_2^- release than nonfibrous particles

TABLE 55 (continued)

In Vitro Responses of Alveolar Macrophages to Toxicant Exposure: Reactive Oxygen Species, Cytokines, and Eicosanoids

Test Material	Exposure Conditions	Species Tested	Culture Conditions (± Serum)	Key Observation
Chrysotile, crocidolite, silica (Min-u-sil), latex beads	25 mg/ml per 10^6 cells	Guinea pig	+	Chrysotile but not the other particles stimulated significant increases in O_2^- Pretreatment of chysotile with IgG significantly enhanced the O_2^- response
Crocidolite, silica (Min-u-sil), titanium dioxide (Anatase), aluminum oxide (Valumina)	10–1000 mg/ml per 10^6 cells	Rat	–	Crocidolite and silica but not titanium dioxide or aluminum oxide stimulate TNF and LTB_4 release IL-1 release was not stimulated by mineral dust
Chrysotile, silica latex beads	1–100 mg/ml per 10^5 cells	Rat	+	Chrysotile and silica but not latex beads stimulated release of TNF and LTB_4 Treatment with lipoxygenase inhibitors decreases the TNF release
Silica (a-quartz), silica (Min-u-sil), crocidolite, titanium dioxide	10–250 mg/ml per 5 × 10^5 cells	Rat	+	Silica and asbestos but not titanium dioxide stimulated IL-1 release
Silica, crocidolite, chrysotile	10–500 mg/ml per 5 × 10^6 cells	Rat	+	All dusts stimulated release of IL-1 and fibroblast growth factor activity The fibroblast growth factor activity was partially attributed to IL-1
Chrysotile	100 mg/ml	Guinea pig	+	Chrysotile stimulated release of fibroblast growth factor activity
Chrysotile, carbonyl iron	15 mg/cm^2, 1 mg/cm^2	Rat	–	Chrysotile and carbonyl iron both stimulate release of a platelet-derived growth factor-like molecule
Chrysotile, carbonyl iron beads	16 mg/cm^2, 0.6 mg/cm^2	Rat	–	Chrysotile and carbonyl iron beads stimulated release at archidonic acid metabolites Both cyclo-oxygenase and lipoxygenase metabolites were detected in a profile similar to that seen with calcium ionophone A23187
Carbonyl iron beads	0.6 mg/cm^2	Rat	–	Carbonyl iron bead-induced eicosanoid release was stimulated by the binding to the cell membrane Membrane sialic acid residues were critical to the activation
Chrysotile	10, 100, 1000 mg/ml per 10^6 cells	Guinea pig	–	Chrysotile activated phospholipase A activity and increased release of PGE_2
Cotton bract extract (ether)	Extract from 100, 200, 300, 400 mg bract per 10^6 cells	Rabbit	—	Cotton bract extract stimulated release of PGF_{2a}
Ozone	0.12– >1.0 ppm	Rabbit	—	Ozone stimulated release of PGE_2 and PGF_{2a} but not LTB_4

Note: LTB_4 = leukotriene B_4.

Source: From Leikauf, G. D. and Driscoll, K., in *Toxicology of the Lung,* Gardiner, D. E. et al., Eds., Raven Press, New York, 1993, chpt. 12. With permission.

TABLE 56

Research Studies of Fiber Carcinogenicity

Cell Culture Studies
Length-diameter characteristics important: long (>10 μm), thin (<1 μm) fibers more cytotoxic compared to coarse (>5-μm diameter) fibers
Chemical composition influences toxicity
Induces neoplastic transformation
Causes chromosomal alterations in eukaryotic cells but not mutagenic in bacterial assays
Intrapleural/Intraperitoneal Injection Studies
Virtually all durable fibrous materials of long, thin dimensions (regardless of their physical or chemical composition) induce mesothelioma
Intratracheal Injection Studies
Some evidence for increases in fibrosis and lung tumors
Inhalation Exposure Studies
Glass fibers — virtually all studies show lack of tumor induction
Mineral wool — all studies show that mineral wool is not tumorigenic
Refractory ceramic fibers — some evidence for induction of fibrosis, lung cancer, and mesothelioma

- *Continuous filament (textile glass fiber):* Produced by being drawn or extruded from holes in a container, rather than being spun or blown. Used to reinforce other materials, especially plastics. Fiber diameter: 6 to 15 (or, in another study, 9 to 25) μm.
- *Glass wool:* Produced by spinning or blowing molten glass. Fiber size selected according to intended use and weight and volume considerations. Used in thermal and acoustical insulation. Fiber diameter: 2 to 9 (or, 1 to 6) μm.
- *Specialty glass wool fiber (fine fiber, glass microfiber):* Specialty uses in filtration, aerospace insulation. Constitutes <1% of total fibrous glass production. Fiber diameter: 0.1 to 3 μm diameter.
- *Mineral wool:* Produced by melting raw materials and centrifuging, drawing, or blowing the molten material into desired fibrous form. Used primarily for thermal and acoustical insulation and fire protection applications. Fiber diameters: 2 to 9 (or, 6 to 8) μm. Types of mineral wool include:
  - Slag wool: Produced from byproducts of metal smelting; constitutes majority of the mineral wool produced in U.S.
  - Rock wool: produced by melting of igneous rock containing high levels of calcium and magnesium; constitutes majority of the mineral wool produced in Europe.
- *Ceramic (aluminum silicate) fibers:* Produced by spinning molten mixtures of blends of alumina and silica, with other refractory oxides added. Used to control heat flow in high temperature areas, e.g., furnace wall liner. Fiber diameters: 1.2 to 3.5 (or, 1 to 12) mm. Types of ceramic fibers include:
  - Kaolin clay-based
  - Blends of alumina, silica, and refractory metal oxides (e.g., chromous and zirconia oxides) or, less commonly, from nonoxide materials, such as silicon carbide or silicon nitride

3.4.5.1 *Animal and* In Vitro *Carcinogenicity of MMMFs*

Various types of nonhuman studies carried out to explore the carcinogenicity of MMMFs are listed in Table 56.

TABLE 57
IARC Classification of MMMFs

MMMF	Group[a]	Human Evidence	Animal Evidence
Glass wool	2B	Inadequate	Limited
Continuous filament	3	Inadequate	Inadequate
Slag wool	2B	Limited	Inadequate
Rock wool	2B	Limited	Limited
RCF	2B	Inadequate	Limited

Note: IARC = International Association for Research on Cancer; MMMFs = manmade mineral fibers; RCF = refractory ceramic fibers.

[a] IARC categories for classifying an agent as a potential carcinogen: *Group 1* — sufficient evidence of human carcinogenicity; *Group 2A* — probably carcinogenic to humans; *Group 2B* — possibly carcinogenic to humans; *Group 3* — not classifiable as to human carcinogenicity; *Group 4* — probably not carcinogenic to humans.

3.4.5.2 Human Health Effects

Both nonmalignant mortality and lung cancer have been seen in studies of exposure to MMMFs. The International Agency for Research on Cancer (IARC) classifications for MMMFs are given in Table 57. As discussed below, a U.S. study found increases in mortality in comparison with local and national mortality patterns.[103] On the other hand, European studies came to a different conclusion.[137,147] Those studies found an overall mortality excess, mainly due to deaths from violent causes, with the excess concentrated among short-term employees. These studies investigating human effects of MMMFs were conducted on large groups of workers in MMMF manufacturing. These studies looked at workers exposed to mineral wool, glass wool, and glass fibers. Marsh et al.[103] conducted a mortality study of almost 17,0000 U.S. workers, many with long-term exposures of up to 40 years, at 17 U.S. glass and mineral wool manufacturing plants (includes 14,800 glass fibers workers and 1846 mineral wool workers). Saracci et al.[137] and Simonato et al.[147] reported a mortality study of almost 25,000 workers (2836 deaths) employed in 13 European factories (includes 11,852 glass fiber production workers and 10,115 mineral wool production workers).

In the U.S. study, there was a statistically significant increase in mortality among workers exposed to mineral wool. Features not consistent with a causal relationship included the fact that only the European study showed a relationship with time from first exposure; neither study showed a relationship with the duration of employment or a relationship with cumulative exposure. No excess mortality was seen in glass wool workers when compared with local rates, but there were statistically significant increases when compared with national rates, a positive relationship with time from first exposure that was not statistically significant, no relationship with duration of employment, and no relationship with cumulative exposure. Both studies showed no increase in lung cancer mortality in workers exposed to glass fibers. No evidence indicates that pleural or peritoneal mesotheliomas are associated with occupational exposure to MMMF.

REFERENCES

1. Amdur, M. O., Air pollutants, in *Casarett and Doull's Toxicology: The Basic Science of Poisons,* 4th ed., Amdur, M. O., Doull, J., Klaassen, C. D., Eds., Pergammon Press, New York, 1991.
2. Angerer, P., Marstaller, H., Barhemann-Hoffmeister, A., Rommel, G., Hoppe, P., and Kessel, R., Alterations in lung function due to mixtures of organic solvents used in floor laying, *Int. Arch. Occup. Environ. Health,* 63, 43–50, 1991.

3. Aris, R., Christian, D., Hearne, P. Q., Kerr, K., Finkbeiner, W. E., and Balmes, J. R., Ozone-induced airway inflammation in human subjects as determined by airway lavage and biopsy, *Am. Rev. Respir. Dis.*, 148, 1363–1372, 1993.
4. Avol, E. L., Linn, W. S., Shamoo, D. A., and Anderson, K. R., Respiratory responses of young asthmatic volunteers in controlled exposures to sulfuric acid aerosol, *Am. Rev. Respir. Dis.*, 142, 343–348, 1990.
5. Azoulzy, E., Soler, P., and Blayo, M. C., The absence of lung damage in rats after chronic exposure to 2 ppm nitrogen dioxide, *Bull. Eur. Physiolpathol. Respir.*, 14, 311–325, 1978.
6. Banks, D. E., Rando, R. J., and Barkman, Jr., H. W., Persistence of toluene diisocyanate-induced asthma despite negligible workplace exposures, *Chest*, 97, 121–125, 1990.
7. Bardana, Jr., E. J. and Montanaro, A., Formaldehyde: an analysis of its respiratory, cutaneous, and immunologic effects, *Ann. Allergy*, 66, 441–452, 1991.
8. Bascom, R., Naclerio, R. M., Fitzgerald, T. K., Kagey-Sobotka, A., and Proud, D., Effect of ozone inhalation on the response to nasal challenge with antigen of allergic subjects, *Am. Rev. Respir. Dis.*, 142, 594–601, 1990.
9. Bates, D. V. and Sizto, R., Air pollution and hospsital admissions in southern Ontario: the acid summer haze effect, *Environ. Res.*, 43, 317–331, 1987.
10. Bates, D. V., *Respiratory Function in Disease*, 3rd ed., W.B. Saunders, Philadelphia, PA, 1989.
11. Beckett, W. S., The epidemiology of occupational asthma, *Eur. Respir. J.*, 7, 161–164, 1994.
12. Bell, K. A., Avol, R. L., Bailey, R. M., Kleinman, M. T., Landis, D. A., and Heiser, S. L., Design, operation and dynamics of aerosol exposure facilities for human subjects. Generation of aerosols and facilities for exposure experiments, Willeke, K., Ed., Ann Arbor Science, Ann Arbor, MI, 1980, pp. 475–491.
13. Berman, D. W. and Chatfield, E. J., *Interim Superfund Method for the Determination of Asbestos in Ambient Air.* Part 2. *Technical Background Document*, EPA-68/01-7290, U.S. Environmental Protection Agency, U.S. Government Printing Office, Washington, D.C., 1989.
14. Bethel, R. A., Sheppard, D., and Epstein, J., Interaction of sulfur dioxide and dry cold air in causing bronchoconstriction in asthmatic subjects, *J. Appl. Physiol.*, 57, 491–523, 1984.
15. Bharma, R. K. and Costa, M., Trace elements — aluminum, arsenic, cadmium, mercury and nickel, in *Environmental Toxicants — Human Exposure and Their Health Effects*, Lippmann, M., Ed., Van Nostrand-Reinhold, New York, 1992, chpt. 19.
16. Blair, A., Stewart, P. A., Tolbert, P. E., Grauman, D., Moran, F. X., Vaught, J., and Rayner, J., Cancer and other causes of death among a cohort of cleaners, *Br. J. Ind. Med.*, 47, 162–168, 1990.
17. Bowes, III, S. M., Francis, M., Laube, B. L., and Frank, R., Acute exposure to acid fog: influence of breathing pattern on effective dose, *Am. Ind. Hyg. Assoc. J.*, 56, 143–150, 1995.
18. Bowes, III, S. M., Frank, R., and Swift, D. L., The head dome: a simplified method for human exposures to inhaled air pollutants, *Am. Ind. Hyg. Assoc. J.*, 79, 151–157, 1989.
19. Brain, J. D. and Sweeney, T. D., Effects of ventilatory patterns and pre-existing disease on deposition of inhaled particles, in *Extrapolation of Dosimetric Relationships for Inhaled Particles and Gases*, Crapo, J. D., Smolko, E. D., Miller, F. J., Graham, J. A., and Hayes, A. W., Eds., Academic Press, San Diego, CA, 1989, pp. 167–186.
20. Brauer, M. and Spengler, J. D., Nitrogen dioxide exposures inside ice skating rinks, *Am. J. Public Health*, 84, 429–433, 1994.
21. Burnett, R. T., Dales, R. E., Raizenne, M. E., Krewski, D., Summers, P. W., Roberts, G. R., Raad-Young, M., Dann, T., and Brook, J., Effects of low ambient ozone and sulfates on the frequency of respiratory admissions to Ontario hospitals, *Environ. Res.*, 65, 172–194, 1994.
22. Calderon-Garciduenas, L., Osormo-Velazquez, A. et al., Histopathologic changes of the nasal mucosa in southwest metropolitan Mexico City inhabitants, *Am. J. Pathol.*, 140, 225–232, 1992.
23. Cavanagh, D. G. and Morris, J. B., Mucus protection and airway peroxidation following nitrogen dioxide exposure in the rat, *J. Toxicol. Environ. Health*, 22, 313–328, 1987.
24. Centers for Disease Control, Asthma — United States, 1980–1987, *Morbid. Mortal Wkly. Rep.*, 139, 493–497, 1990.
25. Chapman, R. S., Calafiore, D. C., and Hasselblad, V., Prevalence of persistent cough and phlegm in young adults in relation to long-term ambient sulfur oxide exposure, *Am. Rev. Respir. Dis.*, 132, 261–267, 1985.
26. Cody, R. P., Weisel, C. P., Birnbaum, G., and Lioy, P. J., The effect of ozone associated with summertime photochemical smog on the frequency of asthma visits to hospital emergency departments, *Environ. Res.*, 58, 184–194, 1992.
27. Cullen, M. R., Cherniack, M. G., and Rosenstock, L., Occupational medicine, part 1, *N. Eng. J. Med.*, 322, 594–601, 1990.
28. Cullen, M. R., Cherniack, M. G., and Rosenstock, L., Occupational medicine, part 2, *N. Eng. J. Med.*, 322, 675–683, 1990.
29. Dassen, W., Brunekreef, B., Hoek, G. et al., Decline in children's pulmonary function during an air pollution episode, *J. Air Pollut. Contr. Assoc.*, 36, 1223–1227, 1986.
30. Devlin, R. B., McDonnell, W. F., Mann, R., Becker, S., House, D. E., Schreinemachers, D., and Koren, H. S., Exposure of humans to ambient levels of ozone for 6.6 hours causes cellular and biochemical changes in the lung, *Am. J. Respir. Cell Mol. Biol.*, 4, 72–81, 1991.

31. Dockery, D. W., Pope, III, C. A., Xu, X., Spengler, J. D., Ware, J. H., Fay, M. E., Ferris, Jr., B. J., and Speizer, F. E., An association between air pollution and mortality in six U.S. cities, *N. Engl. J. Med.*, 329, 1753–1759, 1993.
32. Dockery, D. W. and Pope, III, C. A., Acute respiratory effects of particulate air pollution, *Ann. Rev. Public Health*, 15, 107–132, 1994.
33. Dockery, D. W., Speizer, F. E., Stram, D. O., Ware, J. H., Spengler, J. D., and Ferris, Jr., B. G., Effects of inhalable particles on respiratory health of children, *Am. Rev. Respir. Dis.*, 139, 587–594, 1989.
34. Dockery, D. W. and Speizer, F. E., Epidemiological evidence for aggravation and promotion of COPD by acid air pollution, *Lung Biol. Health Dis.*, 43, 201–225, 1988.
35. Dockery, D. W. and Spengler, J. D., Indoor-outdoor relationships of respirable sulfates and particles, *Atmos. Environ.*, 15, 335–343, 1982.
36. Dockery, D. W., Ware, J. H., Ferris, Jr., B. G. et al., Change in pulmonary function in children associated with air pollution episodes, *J. Air Pollut. Contr. Assoc.*, 32, 937–942, 1982.
37. Dowell, A. R., Kilburn, K. H., and Pratt, P. C., Short-term exposure to nitrogen dioxide: effects on pulmonary ultrastructure, compliance, and the surfactant system, *Arch. Intern. Med.*, 128, 74–80, 1971.
38. Ehrlich, R., Interaction between environmental pollutants and respiratory infections, *Environ. Health Perspect.*, 35, 89–100, 1980.
39. EPA, *Air Quality Criteria for Oxides of Nitrogen*, EPA/600/8-91/049cA, U.S. Environmental Protection Agency, U.S. Government Printing Office, Washington, D.C., 1991.
40. EPA, *Air Quality Criteria for Ozone and Related Photochemical Oxidants,* EPA/600/AP-93/004, U.S. Environmental Protection Agency, U.S. Government Printing Office, Washington, D.C., 1993.
41. EPA, *Respiratory Health Effects of Passive Smoking. Lung Cancer and Other Disorders,* EPA/600/6-90/006F, U.S. Environmental Protection Agency, U.S. Government Printing Office, Washington, D.C., 1992.
42. Fabbri, L. M., Danieli, D., Cresciioli, S., Bevilacqua, P., Meli, S., Saetta, M., and Mapp, C. E., Fatal asthma in a subject sensitized to toluene diisocyanate, *Am. Rev. Respir. Dis.*, 137, 1494–1498, 1988.
43. Folinsbee, L. J., Does nitrogen dioxide exposure increase airways responsiveness?, *Toxicol. Ind. Health,* 8, 273–283, 1992.
44. Foster, W. M., Silver, J. A., and Groth, M. L., Exposure to ozone alters regional function and particle dosimetry in the human lung, *J. Appl. Physiol.*, 75, 1938–1945, 1993.
45. Frampton, M. W., Smeglin, A. M., Roberts, Jr., N. J., Finkelstein, J. N., Morrow, P. E., and Utell, M. J., Nitrogen dioxide exposure *in vivo* and human alveolar macrophage inactivation of influenza virus *in vitro, Environ. Res.*, 48, 179–192, 1989.
46. Frischer, T., Studnicka, M., Beer, E., and Neumann, M., The effects of ambient NO_2 on lung function in primary schoolchildren, *Environ. Res.*, 62, 179–188, 1993.
47. Fuchs, J., Wullenweber, U., Hengstler, J. G., Bienfait, H. G., Hiktk, G., and Oesch, F., Genotoxic risk for humans due to work place exposure to ethylene oxide: remarkable individual differences in susceptibility, *Arch. Toxicol.*, 68, 343–348, 1994.
48. Gardner, R. M., Hankinson, J. L., Clausen, J. L., Crapo, R. O., Johnson, R. L., and Epler, G. R., Standardizatrion of spirometry — 1987 update, *Am. Rev. Respir. Dis.*, 136, 1285–1298, 1987.
49. Gearhart, J. M. and Schlesinger, R. B., Sulfuric acid-induced changes in the physiology and structure of the tracheobronchial airways, *Environ. Health Persp.*, 79, 127–136, 1989.
50. Gergen, P. J. and Weiss, K. B., The increasing problem of asthma in the United States, *Am. Rev. Respir. Dis.*, 146, 823–824, 1992.
51. Gerrity, T. R., Weaver, R. A., Berntsen, J., House, D. E., and O'Neil, J. J., Extrathoracic and intrathoracic removal of O_3 in tidal-breathing humans, *J. Appl. Physiol.*, 65, 393–400, 1988.
52. Goings, S. A., Kulle, T. J., Bascom, R., and Sauder, L. R., Effects of nitrogen dioxide exposure on susceptibility to influenza A virus infection in healthy adults, *Am. Rev. Respir. Dis.*, 139, 1075–1081, 1989.
53. Goldstein, B. D., Benzene, in *Environmental Toxicants: Human Exposures and Their Health Effects,* Lippmann, M., Ed., Van Nostrand-Reinhold, New York, 1992, chpt. 3.
54. Graham, D., Henderson, F., and House, D., Neutrophil influx measured in nasal lavages of humans exposed to ozone, *Arch. Environ. Health,* 43, 228–233, 1988.
55. Graham, D. E. and Koren, H. S., Biomarkers of inflammation in ozone-exposed humans: comparison of the nasal and bronchoalveolar lavage, *Am. Rev. Respir. Dis.*, 142, 152–156, 1990.
56. Hackney, J. D., Linn, W. S., Shamoo, D. A., and Avol, E. L., Responses of selected reactive and nonreactive volunteers to ozone exposure in high and low pollution seasons, in *Atmospheric Ozone Research and Its Policy Implications,* Schneider, T., Ed., Elsevier, Amsterdam, 1989, pp. 311–318.
57. Hanley, Q. S., Koenig, J. Q., Larson, T. V., Anderson, T. L., van Belle, G., Rebolledo, V., Covert, D. S., and Pierson, W. E., Response of young asthmatics to inhaled sulfuric acid, *Am. Rev. Respir. Dis.*, 145, 326–331, 1992.
58. Harving, H., Dahl, R., and Molhave, L., Lung function and bronchial reactivity in asthmatics during exposure to volatile organic compounds, *Am. Rev. Resp. Dis.*, 143, 751–754, 1991.
59. Hasselblad, V., Eddy, D. M., and Kotchmar, D. J., Synthesis of environmental evidence, nitrogen dioxide epidemiology studies, *J. Air Waste Manage. Assoc.*, 42, 662–671, 1992.

60. HEI-AR, *Asbestos in Public and Commercial Buildings: A Literature Review and Synthesis of Current Knowledge,* Health Effects Institute-Asbestos Research, Cambridge, MA, 1991.
61. Henderson, R. F. and Belinsky, S. A., Biological markers of respiratory tract exposure, *Toxicology of the Lung,* Gardiner, D. E., Crapo, J. D., and McClellan, R. O., Eds., Raven Press, New York, 1993, chpt. 9.
62. Henry, M. C., Findlay, J., Spangler, J., and Ehrlich, R., Chronic toxicity of NO_2 in squirrel monkeys. III. Effect on resistance to bacterial and viral infection, *Arch. Environ. Health,* 20, 566–570, 1970.
63. Hill, A. B., The environment and disease: association or causation?, *Proc. R. Soc. Med.,* 58, 295–300, 1965.
64. Hooftman, R. N., Kuper, C. F., and Appelman, L. M., Comparative sensitivity of histo-pathology and specific lung parameters in the detection of lung injury, *J. Appl. Toxicol.,* 8, 59–65, 1988.
65. Horstman, D. H. and Folinsbee, L. J., Ozone concentration and pulmonary response relationships for 6.6 hour exposures with five hours of moderate exercise to 0.08, 0.10, and 0.12 ppm, *Am. Rev. Respir. Dis.,* 142, 1158–1163, 1990.
66. Horstman, D. H. and Folinsbee, L. J., Sulfur dioxide-induced bronchoconstriction in asthmatics exposed for short durations under controlled conditions: a selected review, in *Susceptibility to Inhaled Pollutants,* Utell, M. J. and Frank, R., Eds., American Society for Testing and Materials, Philadelphia, PA, 1989.
67. Horstman, D. H. et al., The relationship between exposure duration and SO_2-induced bronchoconstriction in asthmatic subjects, *Am. Ind. Hyg. Assoc. J.,* 49, 38–47, 1988.
68. Hynes, B., Silverman, F., Cole, P., and Corey, P., Effects of ozone exposure: a comparison between oral and nasal breathing, *Arch. Environ. Health,* 43, 357–359, 1988.
69. Illing, J. W., Miller, F. J., and Garner, D. E., Decreased resistance to infection in exercised mice exposed to NO_2 and O_3, *J. Toxicol. Environ. Health,* 6, 843–851, 1980.
70. Infante-Rivard, C., Childhood asthma and indoor environmental risk factors, *Am. J. Epidemiol.,* 137, 834–844, 1993.
71. Jakob, G. J., Modulation of pulmonary defense mechanisms by acute exposures to nitrogen dioxide, *Environ. Res.,* 42, 215–228, 1987.
72. Jang, J. Y., Kand, S. K., and Chung, H. K., Biological exposure indices of organic solvents for Korean workers, *Int. Arch. Occup. Environ. Health,* 65, S219–S222, 1993.
73. Karol, M. H., Tollerud, D. J., Campbell, T. P., Fabbri, L., Maestrelli, P., Saetta, M., and Mapp, C. E., Predictive value of airways hyperresponsiveness and circulating IgE for identifying types of responses to toluene diisocyanate inhalation challenge, *Am. J. Repir. Crit. Care Med.,* 143, 611–615, 1994.
74. Kelly, T. J., Mukund, R., Spicer, C. W., and Pollack, A. J., Concentrations and transformations of hazardous air pollutants, *Environ. Sci. Technol.,* 28, 378A–387A, 1994.
75. Kerns, W. D., Pavkov, K. L., Donofrio, D. J., Gralla, E. J., and Swenberg, J. A., Carcinogenicity of formaldehyde in rats and mice after long-term inhalation exposure, *Cancer Res.,* 43, 4382–4392, 1983.
76. Kinney, P. L., Ware, J. H., Spengler, J. D., Dockery, D. W., Speizer, F. E., and Ferris, Jr., B. G., Short-term pulmonary function change in association with ozone levels, *Am. Rev. Respir. Dis.,* 139, 56–61, 1989.
77. Koenig, J. Q., Covert, D. S., Hanley, Q., van Belle, G., and Pierson, W. E., Prior exposure to ozone potentiates subsequent response to sulfur dioxide in adolescent asthmatic subjects, *Am. Rev. Respir. Dis.,* 141, 377–380, 1990.
78. Koenig, J. Q., Covert, D. S., Marshall, S. G., van Belle, G., and Pierson, W. E., The effects of ozone and nitrogen dioxide on pulmonary function in healthy and asthmatic adolescents, *Am. Rev. Respir. Dis.,* 136, 1152–1157, 1987.
79. Koenig, J. Q., Dumler, K., Rebolledo, V., Williams, P. V., and Pierson, W. E., Respiratory effects of inhaled sulfuric acid on senior asthmatics and nonasthmatics, *Arch. Environ. Health,* 48, 171–175, 1993.
80. Koenig, J. Q., Larson, T. V., Hanley, Q. S., Rebolledo, V., Dumler, K., Checkoway, H., Wang, S.-Z., Lin, D., and Pierson, W. E., Pulmonary function changes in children associated with fine particle matter, *Environ. Res.,* 63, 26–38, 1993.
81. Koenig, J. Q., Morgan, M. S., Horike, M., and Pierson, W. E., The effects of sulfur oxides on nasal and lung function in adolescents with extrinsic asthma, *J. Allergy Clin. Immunol.,* 76, 813–818, 1985.
82. Koenig, J. Q., Pierson, W. E., Horike, M., and Frank, R., Effects of SO_2 plus NaCl aerosol combined with moderate exercise on pulmonary function in asthmatic adolescents, *Environ. Res.,* 25, 340–348, 1981.
83. Koenig, J. Q. and Pierson, W. E., Pulmonary effects of inhaled sulfur dioxide in atopic adolescent subjects: a review, in *Inhalation Toxicology of Air Pollution: Clinical Research Considerations,* Frank, R., O'Neil, J. J., Utell, M. J., Hackney, J. D., Van Ryzin, J., and Brubaker, P. E., Eds., American Society for Testing and Materials, Philadelphia, PA, 1985, pp. 85–91.
84. Koenig, J. Q., The effect of ozone on respiratory responses in subjects with asthma, *Environ. Health Perspect.,* 103(Suppl 2), 103–105, 1995.
85. Koren, H. S., Devlin, R. B., Graham, D. E. et al., Ozone-induced inflammation in the lower airways of human subjects, *Am. Rev. Respir. Dis.,* 139, 407–415, 1989.
86. Koren, H. S. and Devlin, R. B., Human upper respiratory tract response to inhaled pollutants with emphasis on nasal lavage, *Ann. N.Y. Acad. Sci.,* 641, 215–224, 1992.
87. Koren, H. S., Graham, D. E., and Devlin, R. B., Exposure of humans to a volatile organic mixture. III. Inflammatory response, *Arch. Environ. Health,* 47, 39–44, 1992.

88. Kreit, J. W., Gross, K. B., Moore, T. B., Lorenzen, T. J., D'Arcy, J., and Eschenbacher, W. L., Ozone-induced changes in pulmonary function and bronchial responsiveness in asthmatics, *J. Appl. Physiol.,* 66, 217–222, 1989.
89. Lang, L., Beryllium: a chronic problem, *Environ. Health Perspect.,* 102, 526–531, 1994.
90. Larson, T. V., Hanley, Q. S., Koenig, J. Q., and Bernstein, O., Calculation of acid aerosol dose, in *Advances in Controlled Clinical Inhalation Studies,* Mohr, U., Ed., Springer-Verlag, Berlin, 1993.
91. Larson, T. V. and Koenig, J. Q., Wood smoke: emissions and noncancer respiratory effects, *Ann. Rev. Public Health,* 15, 133–156, 1994.
92. Lee, H. S. and Phoon, W. H., Diurnal variation in peak expiratory flow rate among workers exposed to toluene diisocyanate in the polyurethane foam manufacturing industry, *Br. J. Ind. Med.,* 49, 423–427, 1992.
93. Leikauf, G. and Driscoll, K., Cellular approaches in respiratory tract toxicology, in *Toxicology of the Lung,* Gardiner, D. E., Crapo, J. D., and McClellan, R. O., Eds., Raven Press, New York, 1993, chpt. 12.
94. Leikauf, G. D., Formaldehyde and other aldehydes, in *Environmental Toxicants—Human Exposures and Their Health Effects,* Lippmann, M., Ed., Van Nostrand-Reinhold, New York, 1992, chpt. 10.
95. Lewis, C. W., Baumgardner, R. E., Claxton, L. D., Lewtas, J., and Stevens, R. K., The contribution of wood smoke and motor vehicle emissions to ambient aerosol mutagenicity, *Environ. Sci. Technol.,* 22, 968–971, 1988.
96. Linn, W. S., Shamoo, D. A., Venet, T. G., Bailey, R. M., Weightman, L. H., and Hackney, J. D., Comparative effects of sulfur dioxide exposures at 5°C and 25°C in exercising asthmatics, *Am. Rev. Respir. Dis.,* 129, 234–239, 1984.
97. Lioy, P. J., Daisey, J. M., Morandi, M. T., Harkov, R. D., Greenberg, A., Bozzelli, J., Kebbekus, B., Louis, J., and McGeorge, L. J., The airborne toxic element and organic substances (ATEOS) study design, in *Toxic Air Pollution,* Lioy, P. J. and Daisey, J. M., Eds., Lewis Publishers, Chelsea, MI, 1987, chpt. 1, pp. 3–43.
98. Lioy, P. J., Vollmuth, T. A., and Lippmann, M., Persistence of peak flow decrement in children following ozone exposures exceeding the national ambient air quality standard, *J. Air Pollut. Contr. Assoc.,* 35, 1068–1071, 1985.
99. Lippmann, M., Asbestos exposure indices, *Environ. Res.,* 46, 86–106, 1988.
100. Lippmann, M., Ozone, in *Environmental Toxicants — Human Exposures and Their Health Effects,* Lippmann, M., Ed., Van Nostrand-Reinhold, New York, 1992, chpt. 16.
101. Lundberg, I., Hogstedt, C., Liden, C., and Nise, G., Organic solvents and related compounds, in *Textbook of Clinical, Occupational and Environmental Medicine,* Rosenstock, L. and Cullen, M. R., Eds., W.B. Saunders, Philadelphia, PA, 1994, chpt. 31.
102. Marczynski, B., Czuppon, A. B., Marek, W., and Baur, X., Indication of DNA strand breaks in human white blood cells after *in vitro* exposure to toluene diisocyanate, *Toxicol. Indust. Health,* 8, 157–169, 1992.
103. Marsh, G. M., Enterline, P. E., Stone, R. A., Henderson, V. L., Mortality among a cohort of U.S. man-made mineral fiber workers: 1985 follow-up, *J. Occup. Med.,* 32, 594–604, 1990.
104. Mauderly, J. L., Diesel exhaust, in *Environmental Toxicants — Human Exposures and Their Health Effects,* Lippman, M., Ed., Van Nostrand-Reinhold, New York, 1992, chpt. 5.
105. McBride, D. E., Koenig, J. Q., Luchtel, D. L., Williams, P. V., and Henderson, Jr., W. R., Inflammatory effects of ozone in the upper airways of subjects with asthma, *Am. Rev. Respir. Crit. Care Med.,* 149, 1192–1197, 1994.
106. McDonnell, W. F., Horstman, D. H., Hazucha, M. J., Seal, Jr., E., Haak, E. D., Salaam, S. A., and House, D. E., Pulmonary effects of ozone exposure during exercise: dose-response characteristics, *J. Appl. Physiol.,* 54, 1345–1352, 1983.
107. McDonnell, W. F., Muller, K. E., Bromberg, P. A., and Shy, C. A., Predictors of individual differences in acute response to ozone exposure, *Am. Rev. Respir. Dis.,* 147, 818–825, 1993.
108. Menzel, D. B., The toxicity of air pollution in experimental animals and humans: the role of oxidative stress, *Toxicol. Lett.,* 72, 269–277, 1994.
109. Messineo, T. D. and Adams, W. C., Ozone inhalation effects in females varying widely in lung size: comparison with males, *J. Appl. Physiol.,* 69, 96–103, 1990.
110. Morgan, M. S., Dills, R. L., and Kalman, D. A., Evaluation of stable isotope-labeled probes in the study of solvent pharmacokinetics in human subjects, *Int. Arch. Occup. Environ. Health,* 65, S139–S142, 1993.
111. Morrow, P. E., Utell, M J., Bauer, M. A., Smeglin, A. M., Frampton, M. W., Cox, C., Speers, D. M., and Gibb, F. R., Pulmonary performance of elderly normal subjects and subjects with chronic obstructive pulmonary disease exposed to 0.3 ppm nitrogen dioxide, *Am. Rev. Respir. Dis.,* 145, 291–300, 1992.
112. Murray, A. B. and Morrison, B. J., Passive smoking by asthmatics: its greater effect on boys than on girls and on older than younger children, *Pediatrics,* 84, 451–459, 1989.
113. Murray, A. B. and Morrison, B. J., The effect of cigarette smoke from the mother on bronchial responsiveness and severity of symptoms in children with asthma, *J. Allergy Clin. Immunol.,* 77, 575–581, 1986.
114. National Research Council, Henderson, R., Ed., *Biomarkers in Pulmonary Toxicology,* National Academy Press, Washington, D.C., 1989.
115. National Research Council, Hornig, D., Ed., *Environmental Tobacco Smoke: Measuring Exposures and Assessing Health Effects,* National Academy Press, Washington, D.C., 1986.
116. National Research Council, *Human Exposure Assessment for Airborne Pollutants,* National Academy Press, Washington, D.C., 1991.

117. Neas, L. M., Dockery, D. W., Ware, J. H., Spengler, J. D., Speizer, F. E., and Ferris, Jr., B. G., Association of indoor nitrogen dioxide with respiratory symptoms and pulmonary function in children, *Am. J. Epidemiol.*, 134, 204–219, 1991.
118. Nicholson, W. J., Perkel, G., and Selikoff, I. J., Occupational exposure to asbestos: population at risk and projected mortality: 1980–2030, *Am. J. Ind. Med.*, 8, 259–311, 1982.
119. Niinimaa, V., Cole, P., Mintz, S., and Shephard, R. J., The switching point from nasal to oronasal breathing, *Respir. Physiol.*, 42, 61–71, 1980.
120. Omae, K., Higashi, T., Nakadate, T., Tsuhave, S., Nakaza, M., and Sakurai, H., Four-year follow-up of effects of toluene diisocyanate exposure on the respiratory system in polyurethane foam manufacturing workers. II. Four-year changes in the effects on the respiratory system, *Int. Arch. Occup. Environ. Health*, 63, 565–569, 1992.
121. Phalen, R. F., *Inhalation Studies: Foundation and Techniques*, CRC Press, Boca Raton, FL, 1984.
122. Pinkston, P., Smeglin, A., Roberts, Jr., N. J., Gibb, F. R., Morrow, P. E., and Utell, M. J., Effects of *in vitro* exposure to nitrogen dioxide on human alveolar macrophage release of neutrophil chemotactic factor and interleukin-1, *Environ. Res.*, 47, 48–58, 1988.
123. Pisati, G., Baruffini, A., and Zedda, S., Toluene diisocyanate induced asthma: outcome according to persistence or cessation of exposure, *Br. J. Ind. Med.*, 50, 60–64, 1993.
124. Pope, III, C. A., Dockery, D. W., Spengler, J. D., and Raizenne, M. E., Respiratory health and PM_{10} pollution: a daily time series analysis, *Am. Rev. Respir. Dis.*, 144, 668–674, 1991.
125. Pryor, W. A., Ozone in all its reactive splendor, *J. Lab. Clin. Med.*, 122, 483–486, 1994.
126. Quackenboss, J. J., Lebowitz, M. D., and Crutchfield, C. D., Indoor-outdoor relationships for particulate matter: exposure classification and health effects, *Environ. Int.*, 15, 353–360, 1989.
127. Ransom, M. R. and Pope, III, C. A., Elementary school absences and PM_{10} pollution in Utah Valley, *Environ. Res.*, 58, 204–219, 1992.
128. Roger, L. J., Kehrl, H. R., Hazucha, M., and Horstman, D. H., Bronchoconstriction in asthmatics exposed to sulfur dioxide during repeated exercise, *J. Appl. Physiol.*, 59, 784–791, 1985.
129. Rombout, P. J. A., Dormans, J. A. M., Marra, M., and van Esch, G. J., Influence of exposure regimen on nitrogen dioxide-induced morphological changes in the rat lung, *Environ. Res.*, 41, 466–480, 1986.
130. Rubinstein, I., Reiss, T. F., Bigby, B. G., Stities, D. P., and Boushey, Jr., H. A., Effects of 0.60 ppm nitrogen dioxide on circulating and bronchoalveolar lavage lymphocyte phenotypes in healthy subjects, *Environ. Res.*, 55, 18–30, 1991.
131. Saldiva, P. H. N., King, M., Delmonte, V. L. C., Macchione, M., Parada, M. A. C., Daliberto, M. L., Sakae, R. S., Criado, P. M. P., Silveria, P. L. P., Zin, W. A., and Bohm, G. M., Respiratory alterations due to urban air pollution: an experimental study in rats, *Environ. Res.*, 57, 19–33, 1992.
132. Samet, J. M. and Cheng, P.-W., The role of airway mucus in pulmonary toxicology, *Environ. Health Perspect.*, 102(Suppl. 2), 89–103, 1994.
133. Samet, J. M., Lambert, W. E., Skipper, B. J., Cushing, A. H., Hunt, W. C., Young, S. A., McLaren, L. C., Schwab, M., and Spengler, J. D., Nitrogen dioxide and respiratory illnesses in infants, *Am. Rev. Respir. Dis.*, 148, 1258–1265, 1993.
134. Samet, J. M. and Spengler, J. D., *Indoor Air Pollution: A Health Perspective*, The Johns Hopkins University Press, Baltimore, MD, 1991.
135. Sandstrom, T., Helleday, R., Bjermer, L., and Stjernberg, N., Effects of repeated exposure to 4 ppm nitrogen dioxide on bronchoalveolar lymphocyte subsets and macrophages in healthy men, *Eur. Respir. J.*, 5, 1092–1096, 1992.
136. Sandstrom, T., Stjernberg, N., Andersson, M.-C., Kolomodin-Hedman, B., Lundgren, R., Rosenhall, L., and Angstrom, T., Cell response in bronchoalveolar lavage fluid after exposure to sulfur dioxide: a time-response study, *Am. Rev. Respir. Dis.*, 140, 1828–1831, 1989.
137. Saracci, R., Simonato, L., Acheson, E. D., Andersen, A., Bertazzi, P. A., Claude, J., Charnay, N., Esteve, J., Frentzel-Beyme, R. R., Gardner, M. J. et al., Mortality and incidence of cancer of workers in the man-made vitreous fibers producing industry: an international investigation at 13 European plants, *Br. J. Ind. Med.*, 41, 425–436, 1984.
138. Schlesinger, R. B. and Gearhart, J. M., Intermittent exposures to mixed atmospheres of nitrogen dioxide and sulfuric acid: effect on particle clearance from the respiratory region of rabbit lungs, *J. Toxicol. Environ. Health*, 22, 301–312, 1987.
139. Schlesinger, R. B., The interaction of inhaled toxicants with respiratory tract clearance mechanisms, *CRC Crit. Rev. Toxicol.*, 20, 257–286, 1990.
140. Schwartz, J., Slater, D., Larson, T. V., Pierson, W. E., and Koenig, J. Q., Particulate air pollution and hospital emergency room visits for asthma in Seattle, *Am. Rev. Respir. Dis.*, 147, 826–831, 1993.
141. Schwartz, J., Air pollution and daily mortality: a review and meta analysis, *Environ. Res.*, 64, 36–52, 1994.
142. Schwartz, J., What are people dying of on high air pollution days?, *Environ. Res.*, 64, 26–35, 1994.
143. Seal, Jr., E. McDonnell, W. F., House, D. E., Salaam, S. A., DeWitt, P. J., Butler, S. O., Green, J., and Raggio, L., The pulmonary response of white and black adults to six concentrations of ozone, *Am. Rev. Respir. Dis.*, 147, 804–810, 1993.

144. Sexton, K., Liu, K. S., Trietman, R. D., Spengler, J. D., and Turner, W. A., Characterization of indoor air quality in woodburning residences, *Environ. Int.,* 12, 265–278, 1986.
145. Sherwood, R. L., Lioppert, W. E., Goldstein, E., and Tarkington, B., Effect of ferrous sulfate aerosols and nitrogen dioxide on murine pulmonary defense, *Arch. Environ. Health,* 36, 130–135, 1981.
146. Silverman, F., Asthma and respiratory irritants (ozone), *Environ. Health Perspect.,* 29, 131–136, 1979.
147. Simonato, L., Fletcher, A. C., Cherrie, J. W., Andersen, A., Bertazzi, P., Charnay, N., Claude, J., Dodgson, J., Esteve, J., Frentzel-Beyme, R. et al., The International Agency for Research on Cancer historical cohort study of MMMF production workers in seven European countries: extension of the follow-up, *Ann. Occup. Hyg.,* 31, 603–623, 1987.
148. Skender, L., Karaci, 'cV., and Bosner, B., A selection of biological indicators in occupational exposure to toluene and zylene, *Arh. Hig. Rada. Toksikol.,* 44, 27–33, 1993.
149. Soparkar, G., Mayers, I., Edouard, L., and Hoeppner, V. H., Toxic effects from nitrogen dioxide in ice-skating arenas, *Can. Med. Assoc. J.,* 148, 1181–1182, 1993.
150. Speizer, F. E. and Frank, R., The uptake and release of SO_2 by the human nose, *Arch. Environ. Health,* 12, 725–728, 1966.
151. Spektor, D. M., Leikauf, G. B., Albert, R. E., and Lippmann, M., Effects of submicrometric sulfuric acid aerosols on mucociliary transport and respiratory mechanics in asymptomatic asthmatics, *Environ. Res.,* 37, 174–191, 1985.
152. Spektor, D. M., Lippmann, M., Lioy, P. J., Thurston, G. D., Citak, K., James, D. J., Bock, N., Speizer, F. E., and Hayes, C., Effects of ambient ozone on respiratory function in active, normal children, *Am. Rev. Respir. Dis.,* 137, 313–320, 1988.
153. Spektor, D. M., Thurston, G. D., Mao, J. et al., Effects of single- and multiday ozone exposures on respiratory function in active normal children, *Environ. Res.,* 55, 107–122, 1991.
154. Stanton, M. F., Layard, M., Tegeris, A., Miller, E., May, M., and Kent, E., Carcinogenicity of fibrous glass: pleural response in the rat in relation to fiber dimension, *J. Natl. Cancer Inst.,* 58, 587–603, 1977.
155. Stanton, M. F., Layard, M., Tegeris, A., Miller, E., May, M., Morgan, E., and Smith, A., Relation of particle dimension to carcinogenicity in amphibole asbestos and other fibrous minerals, *J. Natl. Cancer Inst.,* 67, 965–975, 1981.
156. Tashkin, D. P., Detels, R., Simmons, M., Liu, H., Coulson, A. H., Sayre, J., and Rokaw, S., The UCLA population studies of chronic obstructive respiratory disease. XI. Impact of air pollution and smoking on annual change in forced expiratory volume in one second, *Am. Rev. Respir. Crit. Care Med.,* 149, 1209–1217, 1994.
157. Thurston, G. D., Ito, K., Kinney, P. L., and Lippmann, M., A multi-year study of air pollution and respiratory hospital admissions in three New York state metropolitan areas: results for 1988 and 1989 summers, *J. Exp. Anal. Environ. Epidemiol.,* 2, 429–450, 1992.
158. Utell, M. J., Effects of inhaled acid aerosols on lung mechanics: an analysis of human exposure studies, *Environ. Health Perspect.,* 63, 39–44, 1985.
159. Vaughan, T. L., Strader, C., Davis, S., and Daling, J. R., Formaldehyde and cancers of the pharynx, sinus and nasal cavity. I. Occupational exposures, *Int. J. Cancer,* 38, 677–683, 1986.
160. Vaughan, T. L., Strader, C., Davis, S., and Daling, J. R., Formaldehyde and cancers of the pharynx, sinus and nasal cavity. I. Residential exposures, *Int. J. Cancer,* 38, 685–688, 1986.
161. Vogelmeier, C., Baur, X., and Fruhmann, G., Isocyanate-induced asthma: results of inhalation tests with TDI, MDI and methacholine, *Int. Arch. Occup. Environ. Health,* 63, 9–13, 1991.
162. Walker, J. T., Bloom, T. F., Stern, F. B., Okun, A. G., Fingerhut, M. A., and Halperin, W. E., Mortality of workers employed in shoe manufacturing, *Scand. J. Work Environ. Health,* 19, 89–95, 1993.
163. Walton, W. H., The nature, hazards and assessment of occupational exposure to airborne asbestos dust: a review, *Ann. Occup. Hyg.,* 25, 117–247, 1982.
164. Ware, J. H., Ferris, B. G., Dockery, D. W., Spengler, J. D., Stram, D. O., and Speizer, F. E., Effects of ambient sulfur oxides and suspended particles on respiratory health of preadolescent children, *Am. Rev. Respir. Dis.,* 133, 834–842, 1986.
165. Ware, J. H., Spengler, J. D., Neas, L. M., Samet, J. M., Wagner, G. R., Coultas, D., Ozkaynak, H., and Schwab, M., Respiratory and irritant health effects of ambient volatile organic compounds. The Kanawha County Health Study, *Am. J. Epidemiol.,* 137, 1287–1301, 1993.
166. White, M. C., Etzel, R. A., Wilcox, W. D., and Lloyd, C., Exacerbations of childhood asthma and ozone pollution in Atlanta, *Environ. Res.,* 65, 56–68, 1994.
167. Whittemore, A. S. and Korn, E. L., Asthma and air pollution in the Los Angeles area, *Am. J. Public Health,* 70, 687–696, 1980.
168. WHO, *Acrolein Environmental Criteria,* World Health Organization, Geneva, 1992.
169. Ziegler, B., Bhalla, D. K., Rasmussen, R. E., Kleinman, M. T., and Menzel, D. B., Inhalation of resuspended road dust, but not ammonium nitrate, decreases the expression of the pulmonary macrophage Fc receptor, *Toxicol. Lett.,* 71, 197–208, 1994.
170. Zwick, H., Popp, W., Wagner, C., Reiser, K., Schmoger, J., Bock, A., Herkner, K., and Radunsky, K., Effects of ozone on the respiratory health, allergic sensitization and cellular immune system in children, *Am. Rev. Respir. Dis.,* 144, 1075–1079, 1991.

CHAPTER 14

Methods in Human Inhalation Toxicology

Lawrence J. Folinsbee, Chong S. Kim, Howard R. Kehrl, James D. Prah, and Robert B. Devlin

CONTENTS

1 INTRODUCTION

The human lung is constantly exposed to a variety of gases and particles throughout life. Many potentially toxic materials are present in the ambient environment and in the workplace. In addition, different particulate and gaseous materials may be used therapeutically to treat pulmonary or other diseases. The evaluation of responses to these inhaled materials is of importance in establishing regulations that govern ambient air quality and workplace air quality. Controlled human inhalation exposure studies are one of the means available to provide this information. Animal toxicology, *in vitro* cellular responses, and human epidemiological studies are also used in this regard.

The use of humans as the ideal animal model for human responses avoids the uncertainties of extrapolation of physiological and biochemical responses from animal models to humans, uncertainties regarding the comparative exposure dosimetry between animals and man, and uncertainties of exposure assessment that are inherent in epidemiological observations. However, unlike many animal models, humans are heterogeneous with regard to physiological responses, and the variability of many response measures is substantial. Nevertheless, potentially sensitive subgroups within the population can be studied, providing a better estimate of the responses of the most sensitive persons.

In practice, the number of subjects that can be studied is usually limited to relatively small numbers compared to epidemiological studies. Furthermore, ethical considerations dictate that the concentrations of toxic materials must not produce more than a fully reversible acute response. Neither chronic effects nor nonreversible effects can be studied. Controlled exposure studies offer the opportunity to study single or multiple pollutants in relatively "pure" form, without the potential interference of other materials, known or unknown. One of the drawbacks of this approach is that true simulation of the "real" environment is impractical, if not impossible. Unlike animal studies, control of subject behavior before and after exposure is problematic. It is important that participants avoid concurrent or intercurrent exposures to other inhaled materials such as tobacco or other smoke, dust, fumes, irritant gases, and medications that may interfere with responses.

The concentrations of particulate or gaseous material used in controlled human exposure studies should, as much as possible, reflect meaningful exposure situations. In addition to concentration, the duration and activity level during exposure should be tailored to be representative of naturally occurring exposure situations. The temperature and humidity during the exposure can influence the outcome measures as well as the behavior of the gases and particles in the exposure chamber. Outcome variables such as lung function, symptoms, bronchoalveolar lavage samples, and so forth need to be measured at the appropriate time during or following exposure. Some responses may peak early and show subsequent attenuation, while others may be very slow to develop and may not reach a peak for several hours after exposure. For comparability of studies it is helpful if some common reproducible tests (e.g., standard tests of lung function) are included which are similar to those being used by other investigators.

For regulatory purposes it is usually necessary to establish a dose-response relationship for a toxic substance. In many cases, such studies involve a repeated-measures design with the subject serving as her own control. Because many functional responses vary with time of day (e.g., body temperature, pulmonary function) it is important to take this into consideration in the study design. If the typical or worst-case exposure scenario is likely to include work or exercise, this should also be factored into the exposure. Methodology for human exposure studies has been reviewed previously by a number of investigators.[1-4]

2 INHALATION EXPOSURE TECHNIQUES

Controlled human exposures to inhaled toxic gases and vapors have been conducted over the past four decades. In the past 20 years, several specialized human exposure facilities have been developed. This section deals with some of the factors that must be considered in the controlled exposure of humans to inhaled toxicants. Exposure systems run the gamut from small-volume, personalized mouthpiece or facemask configurations to large, elaborate chambers requiring regular maintenance and operation staff. The capabilities, advantages, and disadvantages of each are discussed in Table 1. Control of the chamber air supply, temperature, and humidity is not discussed here, not because this is unimportant, but because the operating characteristics and controls are likely to be specific to each system; an extensive variety of methods are available and discussed in detail elsewhere.[5,7-9] Generally, the supply air should be treated to remove gaseous and particulate contaminants. This is typically accomplished by filtration — usually a series of filters including a high efficiency (>99.9% for submicron particles) particulate air (HEPA) filters — and adsorption or chemisorption of contaminants (charcoal, permanganate/alumina, etc.) from the incoming air. Other methods include oxidizing trace contaminants using a heated catalytic bed (see Reference 6 for a more comprehensive discussion). In large chamber systems, replacement of air filtration and scrubbing systems can be costly and time consuming. In order to control the humidity, it is necessary to remove water vapor from the inlet air which is commonly done by cooling the air to remove the water by condensation. In small volume systems, solid dessicants such as silica gel or calcium sulfate can be used. Establishment of the desired humidity can then be accomplished by adding an appropriate amount of water vapor in the form of steam back into the supply air that has already been heated or cooled to the desired temperature.

Exposure to toxic gases, vapors, and particles is accomplished by adding these materials to the cleaned and conditioned supply air. It is important that the supply system or the chamber design provides adequate mixing of the added contaminants with the supply air. The gases may be available as mixtures or pure gases and can be metered from compressed-gas cylinders; some materials may have to be vaporized from liquids (or solids) by heating or evaporation. Other methods include diffusion from a permeation tube and generation of the gas as a byproduct of a chemical reaction. An example of this is the generation of ozone using a high-voltage corona discharge across a tube through which oxygen is flowing. Ozone can also be generated with ultraviolet (UV) irradiation of oxygen (although ozone can be generated from air, this procedure is not recommended). When compressed-gas mixtures are used, it is important to ascertain their stability and to ensure that the mixtures within the cylinders remain mixed. Occasional rolling of the cylinders can prevent stratification of the mixtures. Some gases are highly corrosive (e.g., NO_2), and the valves and controllers must be inspected and replaced as necessary. The flow controllers and flow measurement devices obviously should be selected based on the type of exposure facility and the range of inlet flows necessary. In Table 2, methods of monitoring, generating, and analyzing calibration for various toxic[10] and background gases and vapors are summarized.

2.1 SAFETY

Safety of human subjects is of utmost importance in a human inhalation toxicology laboratory. Redundant systems and procedures should be in place to minimize the risk of exposing a subject to concentrations in excess of those that have been approved for a particular experimental design. Examples of such safety procedures would include the use of a fixed flow limitation device (such as a critical orifice) downstream from the flow controller that

TABLE 1

Exposure Chamber Facilities

Exposure System Type	Environmental Controls	Air Flow or Turnover Requirements	Pollutant Generation Requirements	Monitoring Requirements	Capabilities	Limitations	Safety Precautions	Example References
Personal Exposure Systems								
Mouthpiece	Temperature and humidity controls for inspired air; chemical scrubbing and particle filtration of supply air; unidirectional flow-by system or a breathing reservoir can be used; separate conditions for the inspired air and ambient air; torso can be cool when inspired air is warm and humid	30–300 l/min minimum, depending on activity level; breathing valve requires less than a flow-by system	Low concentrations of bottled gas or small generators; pollutants completely mixed upstream of breathing zone	Specific analyzers for inspired air immediately upstream from breathing zone	Large-scale environmental controls not needed; inexpensive to build, operate, and maintain	Gas or particle loss in the supply line; reduced mobility and subject discomfort, difficulty in swallowing, dry mouth; forces unusual breathing mode; may alter breathing pattern	Standard subject safety monitoring: EKG, visual monitoring, symptom evaluation; expired air should be exhausted safely	7, 9, 15
Face mask					See above; allows oronasal breathing or separate oral and nasal breathing	Less discomfort than with the mouthpiece, more normal breathing mode; must be properly sealed		3, 8
Head-dome canopy		Usually higher air flow than above: 100-1000 l/min	Larger supply or generation requirements than mouthpiece or face mask	Monitor inspired air and air in the breathing zone	Permits normal breathing mode but less mobility than does the mouthpiece or face mask; better for aerosol delivery	Must ensure adequate mixing and air delivery in breathing zone; cumbersome system		1

Chamber Exposure Systems								
Single-user, body box	Temperature and humidity of single pass or recirculated air; filtered and scrubbed supply air	200 1/min minimum for 5-min turnover in 1-m^3 system	Similar to personal exposure systems; see above	Monitoring of supply air and breathing zone	Permits normal breathing in a small-volume system using expensive or difficult to generate components, e.g., some particles, $^{18}O_2$	Rapid perturbations of temperature and humidity due to small volume; difficulty in exercising; uncomfortable for long exposures	Rapid exit capability needed; standard safety monitoring	6
Small movable chambers, < 20 m^3	See above; filters and scrubbers must be able to handle larger volume; large-capacity heating and cooling equipment	4000 1/min or 4 m^3/min for 5-min turnover time	Computer control with mass flow controllers desirable; larger generators	Monitoring of chamber air; need to assure adequate mixing by design	Permits normal breathing in chamber that will hold more than one person; access to exercise equipment and test equipment in chamber; relatively inexpensive and easy to maintain	Mixing problems may occur without an accessory mixing chamber; absence of air lock may also prevent easy access without disrupting environmental conditions	Standard precautions; exhaust air may have to be scrubbed or filtered	12, 17
Large permanent chambers, 20–70 m^3	Expensive large-capacity heating and cooling equipment; removal of ambient moisture may be necessary; large-capacity scrubbers and filters	15 m^3/min for 2- to 3-min turnover time; slight negative pressure desirable; Rochester-style chambers best for uniform mixing	Premixing chamber, airlock, and built-in toilet facilities desirable; high-capacity generators, highly concentrated sources of pollutants; computer control of generation and monitoring system essential	Redundant monitors for chamber air needed for backup during prolonged exposures; comprehensive monitoring system including gases, particles, temperature, and humidity	Precision control of environment for prolonged periods; extended exposure possible; most "natural" environment	Cost; many such facilities require extended downtime for maintenance calibration and quality assurance; not practical for costly or difficult-to-generate components, e.g., high levels of particles	Dedicated medical monitoring desirable	2, 4, 5, 10, 11, 13, 14, 16

TABLE 1 (continued)

Exposure Chamber Facilities

[1] Bowes, S. M., Frank, R., and Swift, D. L., The head dome: a simplified method for human exposures to inhaled air pollutants, *Am. Ind. Hyg. Assoc. J.*, 51, 257–260, 1990.

[2] Strong, A. A., *Description of the CLEANS Human Exposure System*, EPA-600/1-78-064, U.S. Environmental Protection Agency, Research Triangle Park, NC, 1978.

[3] Amdur, M. O., Silverman, L., and Drinker, P., Inhalation of sulfuric acid mist by human subjects, *Arch. Ind. Hyg. Occup. Med.*, 6, 305–313, 1952.

[4] Utell, M. J., Morrow, P. E., Hyde, R. W., and Schrek, R. M., Exposure chamber for studies of pollutant gases and aerosols in human subjects: design considerations, *J. Aerosol Sci.*, 15, 219–221, 1984.

[5] Kerr, H. D., Diurnal variation of respiratory function independent of air quality, *Arch. Environ. Health*, 26, 144–152, 1973.

[6] Kagawa, J., Respiratory effects of 2-hr exposure to 1.0 ppm nitric oxide in normal subjects, *Environ. Res.*, 27, 485–490, 1982.

[7] Koenig, J. Q., Pierson, W. E., and Frank, R., Acute effects of inhaled SO_2 plus NACl droplet aerosol on pulmonary function in asthmatic adolescents, *Environ. Res.*, 22, 145–153, 1980.

[8] Kirkpatrick, M. B., Sheppard, D., Nadel, J. A., and Boushey, H. A., Effect of the oronasal breathing route on sulfur dioxide-induced bronchoconstriction in exercising subjects, *Am. Rev. Respir. Dis.*, 125, 627–631, 1982.

[9] Jorres, R. and Magnussen, H., Airway response of asthmatics after a 30-min exposure at resting ventilation to 0.25 ppm NO_2 or 0.5 ppm SO_2, *Eur. Respir. J.*, 3, 132–137, 1990.

[10] Morrow, P. E. and Utell, M. J., Technology and methodology of clinical exposures to aerosols, in *Aerosols*, Lee, S. D., Schneider, T., Grant, L. D., and Verkerk, P. J., Eds., Lewis Publishers, Chelsea, MI, 1986, pp. 661–669.

[11] Willeke, K., Ed., *Generation of Aerosols and Facilities for Exposure Experiments*, Ann Arbor-Science, Ann Arbor, MI, 1980.

[12] Avol, E. L., Wightman, L. H., Linn, W. S., and Hackney, J. D., A movable environmental chamber for controlled clinical studies of air pollution exposure, *J. Air Poll. Control Assoc.*, 29, 743–745, 1979.

[13] Utell, M. J., Frampton, M. W., and Morrow, P. E., Quantitative clinical studies with defined exposure atmospheres, in *Toxicology of the Lung*, 2nd ed., Gardner, D. E., Crapo, J. D., and McClellan, R. O., Eds., Raven Press, New York, 1993, pp. 283–309.

[14] Folinsbee, L. J., Human clinical inhalation exposures: experimental design, methodology, and physiological responses, in *Toxicology of the Lung*, Gardner, D. E., Crapo, J. D., and Massaro, E. J., Eds., Raven Press, New York, 1988, pp. 175–199.

[15] DeLucia, A. J. and Adams, W. C., Effects of O_3 inhalation during exercise on pulmonary function and blood biochemistry, *J. Appl. Physiol.*, 43, 75–81, 1977.

[16] Hackney, J. D., Linn, W. S., Buckley, R. D., Pedersen, E. E., Karuza, S. K., Law, D. C., and Fischer, A., Experimental studies on human health effects of air pollutants, *Arch. Environ. Health*, 30, 373–378, 1975.

[17] Bates, D. V., Bell, G., Burnham, C., Hazucha, M. J., Mantha, J., Pengelly, L. D., and Silverman, F., Problems in studies of human exposure to air pollutants, *Can. Med. Assoc. J.*, 103, 833–837, 1970.

TABLE 2

Gas Phase Pollutant Analysis and Generation

Analyzer Type	Measurement Principle	Generation Method	Calibration
Ozone	Chemiluminescent reaction of ethylene and ozone is detected at 300–600 nm by photometry; ultraviolet photometer based on absorption by ozone at 254 nm	Ultraviolet irradiation of oxygen; corona discharge in oxygen	Certified standard ozone photometer, colorimetric method
Sulfur oxide (SO_2)	Fluorescence of SO_2 molecules irradiated by ultraviolet (190–230 nm); flame photometric detector (394 nm) for S in H_2 flame (not specific for SO_2)	Bottled source or pure liquid	Calibrated bottles; permeation tube
Nitrogen oxide (NO, NO_2)	NO_x is converted to NO and measured by chemiluminescent reaction with O_3 (~400 nm); all oxidized nitrogen is measured as NO; not specific for NO_2	Bottled source in N_2 or pure N_2O_4	NO calibration cylinder; NO_2 permeation tube; gas-phase titration
Carbon monoxide	Nondispersive infrared analyzer; electrochemical oxidation; infrared photometer with gas-filter correlation	Bottled source (up to 100%)	Calibrated bottles
Peroxyacylnitrates	Electron capture vapor phase gas chromatography; infrared absorption	Bottled source; degrades rapidly	Bottled source
Volatile organics	Flame ionization gas chromatography; gas chromatograph/mass spectrometer; photo-ionization detector; infrared absorption	Bottled source; vaporization of liquids for large amounts	Calibrated bottles
Nitric acid	NO_x analyzer with nylon prefilter to absorb HNO_3; nitric acid determined by difference from total NO_x	Vaporization of liquid	NO (or HNO_3) permeation tube
Hydrogen sulfide	Electrical conductivity; gas chromatography with S-specific flame photometry; as for SO_2 after oxidation	Bottled gas in N_2	Calibrated bottles; permeation tube
NH_3	Infrared absorption; NO chemiluminescence after oxidation of NH_3 to NO	Bottled source	Permeation tube
Chlorine	Colorimetric chemical method; electrochemical detector	Bottled source	Permeation tube
CO_2	Infrared photometer, mass spectrometer	Background gas	Calibrated bottles
Oxygen	Paramagnetic analyzer; fuel cell; polarographic oxygen analyzer; mass spectrometer	Background gas	Calibrated bottles
Temperature	Thermistor; thermocouple; thermometer	Heating/cooling	Stirred bath and NIST thermometer
Humidity	Dewpoint sensor; psychrometer; solid-state lithium chloride sensor	Dryer/humidifier	NIST gravimetric hygrometer; salt solutions with known P_{H2O}

Note: NIST = National Institute of Standards and Technology.

Source: Data from Nader, J. S., Lauderdale, J. F., and McCammon, C. S., Direct reading instruments for analyzing airborne gases and vapors, in *Air Sampling Instruments for Evaluation of Atmospheric Contaminants,* 7th ed., Hering, S. V., Ed., American Conference of Government and Industrial Hygienists, Cincinnati, OH, 1989, pp. 507–581; Santee, W. R. and Gonzalez, R. R., Characteristics of the thermal environmental, in Pandolf, K. B., Sawka, M. N., and Gonzalez, R. R., *Human Performance Physiology and Environmental Medicine at Terrestrial Extremes,* Cooper Publishing, Carmel, IN, 1986, pp. 1–43.

would limit the maximum input of a toxic gas into an exposure system and prevent accidental overexposure should the flow controller fail. Accidental overexposure could also occur if the ventilation system failed, although this would have somewhat less catastrophic effects in a large-volume system since ventilation is required to deliver the toxic gas. An emergency chamber air-evacuation system could be used to rapidly replace the chamber atmosphere with ambient air should dangerous concentrations of gases or aerosols accumulate. In a small-volume system, such as the mouthpiece exposure system, temporary failure or shutdown of the ventilation system could lead to a marked increase in the concentration of added gases if provision has not been made to simultaneously shut off the supply of added gas. In such a case, restarting the bulk flow would then result in the delivery of a bolus of high concentration of the added gas. A mechanism should be in place that automatically shuts off the entry of pollutant or toxic gas in the event of ventilation system failure. It is important that every control system have a complementary human component trained to act to maximize subject safety (see also Section 4 regarding subject safety).

2.2 Gases and Vapors

Maintenance of the concentration or concentration profile is an important factor in the design of inhalation exposure studies, from the aspects of both subject safety and quality assurance. The two variables that can be most easily regulated are the system airflow and the rate of addition of the toxic gas; however, other factors such as changes in atmospheric pressure, temperature, or humidity or loss of the toxic gas by chemical reaction or on surfaces (e.g., chamber walls, subjects, duct work, etc.) can also affect the control of concentration. Surface losses may be changed as a result of "aging" of the surface, and off-gassing could even be a problem when going from a high to a low concentration. Uniformity of concentration throughout an exposure system will depend on the air-turnover rate and the mixing efficiency. The use, throughout an exposure system, of well-conditioned surfaces that are minimally reactive and minimally absorptive will help to maintain stability of gas mixtures. Greatest attention should be given to maintenance of the toxic gas concentration at the desired level within the human breathing zone. Chamber concentrations should be monitored close to the subject's breathing zone without the possibility of the subject interfering with the quality of the sample. In direct (mouthpiece, facemask) systems, the concentration just upstream of the subject should be controlled. Potential losses in the sample line must be taken into account in calibration of the monitoring instruments or during quality assurance checks of the control system integrity. When continuous monitoring equipment is available for the toxicant being used, continuous feedback control can be used to maintain concentrations. This can be accomplished with manual controls or by using a computer-based system. The response time of the system must be factored in when continuous controls are used. Systems with a slow air-turnover rate will have a slow response time, and thus there is an increased risk of overshooting or undershooting the desired concentration. Turnover rate will also affect the accumulation of gases such as carbon dioxide and ammonia, which are excreted by humans.

2.3 Aerosols and Particulates

Various methods of generating aerosol or particulate atmospheres are summarized in Table 3. Jet nebulizers or atomizers are commonly used for generating aerosols because of the ease of operation and maintenance. Design and operating characteristics of these nebulizers vary widely;[11,13] however, the principle is to atomize liquid solutions or suspensions with compressed air or, if necessary, inert gases such as nitrogen. The operating pressure is usually in the range of 20 to 60 psig and the jet flow rate is less than 10 l/min. An appropriate

TABLE 3

Aerosol Generation Methods for Human Exposure

Exposure Method	Generation Method	Mode of Operation	Output Range	Ref.
Large chamber	Jet nebulizers; jet nebulizers with single-stage liquid impactor	Aqueous solutions or suspensions are atomized with dry compressed air at 20–60 psi. A number of nebulizers (i.e., 100) are grouped together and all or part of them are operated in order to produce a desired output. A single-stage liquid impactor may be used at the outlet of each nebulizer to narrow the size distribution of output aerosols.	Mass median diameter of the initial solution droplets is usually < 5 μm with σ_g = 1.4–2.5. The final dry particles are substantially smaller (< 1.0 μm) than the initial droplets. Output concentration is very high (> 10^7 particles per cm^3), and a substantial dilution is needed to meet the exposure conditions.	1–5
Personal exposure: face mask, head dome, or mouthpiece	Jet nebulizers (same as above)	One nebulizer is usually sufficient.	Same as above.	6–12
	Monodisperse aerosol generator:	These are sophisticated devices which produce uniform size particles by various principles. The standard operating condition of each generator is usually adequate for personal exposure.		
	spinning top generator		1–15 μm dia., < 200 particles per cm^3	
	vibrating orifice generator		1–40 μm dia., < 300 particles per cm^3	
	condensation generator		0.1–8 μm dia., < 10^7 particles per cm^3	
	differential mobility generator		0.0005–0.1 μm dia., < 10^4 particles per cm^3	
	Dry powder dispersion: jet blast disperser, turntable disperser, rotating brush disperser, Wright dust feeder, fluidized bed generator	A bulk of dry powder materials is dispersed into aerosols by various fluid dynamic principles, and large aggregates and poorly dispersed particles are eliminated by inertial separators such as impactors and cyclones.	Output concentrations from the dust dispersers vary widely from a few mg/m^3 to >10 g/m^3, depending on powder materials. Size distribution of dispersed aerosols also varies, depending on condition of powder materials.	

1 Raabe, O. G., Generation of aerosols of fine particles, in *Fine Particles,* Liu, B. Y. H. et al., Eds., Academic Press, New York, 1975, pp. 57–110.
2 Whitby, K. T., Lundgren, D. A., and Peterson, C. M., Homogeneous aerosol generators, *Int. J. Air Water Poll.,* 9, 263–277, 1965.
3 May, K. R., The Collison nebulizer: description, performance, and application, *J. Aerosol Sci.,* 4, 235–243, 1973.
4 Peters, T. M., Chien, H. M., Lundgren, D. A., and Bernsten, J., Submicrometer aerosol generator development for the U.S. Environmental Protection Agency's human exposure laboratory, *Aerosol Sci. Technol.,* 20, 51–61, 1994.
5 Strong, A. A., *Description of the CLEANS Human Exposure System,* EPA-600/1-78-064, U.S. Environmental Protection Agency, U.S. Government Printing Office, Washington, D.C., 1978.
6 Berglund, R. N. and Liu, B. Y. H., Generation of monodisperse aerosol standards, *Environ. Sci. Technol.,* 7, 147–153, 1973.
7 Mitchell, M. P., The production of aerosols from aqueous solutions using the spinning top generator, *J. Aerosol Sci.,* 3, 347–363, 1991.
8 Horton, K. D., Miller, R. D., and Mitchell, J. P., Characterization of a condensation-type monodisperse aerosol generator (MAGE), *J. Aerosol Sci.,* 1, 34–35, 1984.
9 Sinclair, D. and LaMer, V. K., Light scattering as a measure of particle size in aerosols, *Chem. Rev.,* 44, 245–267, 1949.
10 Wright, B. M., A new dust-feed mechanism, *J. Sci. Inst.,* 27, 12–15, 1950.
11 Marple, V. A., Liu, B. Y. H., and Rubow, K. L., A dust generator for laboratory use, *Am. Ind. Hyg. Assoc. J.,* 39, 26–32, 1978.
12 Guichard, J. C., Aerosol generation using fluidized beds, in *Fine Particles,* Liu, B. Y. H. et al., Eds., Academic Press, New York, 1975, pp. 173–193.

supply of compressed air can be easily achieved using an air compressor with moderate capacity. Bottled gases may also be used to operate the nebulizers for a short period. The compressed air must be dried and cleaned, usually by passing it through a silica gel column and HEPA filter. The nebulizers require very little maintenance except for an occasional cleaning of the solution chamber and jet orifice; however, a careful examination of the jet orifice for wear and corrosion is necessary if the aerosolized materials are reactive or abrasive. The mean diameter of the initial droplets discharged from the nebulizer is smaller than a few micrometers, but because dilute solutions are usually used, the size of the residual particles (once the volatile components of the droplets have been vaporized) is much smaller than 1 μm. It should be noted that these nebulizers may not be suitable for generating supra-micron particle aerosols. The size distribution of nebulizer aerosols is broad, ranging from 1.8 to 3.0 in geometric standard deviation. In order to generate aerosols with a narrow size distribution, a single stage impactor is employed at the outlet of the nebulizer. Size distributions with geometric standard deviations as small as 1.4 can be achieved by this method.[12,13] The output concentration of aerosols from the nebulizer is very high, and a large dilution is required in order to bring the concentration down to the levels that can be used in controlled human exposures. One nebulizer is usually sufficient to generate the aerosols needed for personal exposure; however, for chamber exposure studies that require a very large volumetric flow rate, a large number of nebulizers (often > 100) may be needed to supply the necessary quantity of aerosol. Operating flexibility can be achieved by installing nebulizers in several groups (or banks) such that each group consists of different numbers of nebulizers (and thus different aerosol generating capacity) and can run independently. A wide range of exposure conditions can thus be accommodated by operating nebulizers from different groups or in various combinations of more than one group, as in the human exposure facilities of the U.S. Environmental Protection Agency (USEPA).[14] There are many aerosol generation methods that permit generation of aerosols with precise particle sizes and concentrations, as shown in Table 3. These methods are complex and require specific operational procedures. Monodisperse aerosols from such generators are commonly used to test aerosol equipment or to investigate the physical and chemical characteristics of aerosols and their behavior. These methods could be used for personal exposure studies in association with specific particle sizes; however, use of these generators is not practical for large exposure chambers because of their flow output.

Aerosol monitoring can be achieved by a variety of methods, as summarized in Table 4. There are methods to measure particle size distribution, number or mass concentration, or both particle size and concentration. Because operating principles and measurement ranges differ among different methods, a selection of the correct method is essential. Many instruments employ methods for detecting and sizing individual particles (i.e., single particle counter), whereas other instruments measure collective properties of particles. In the former, a substantial dilution of aerosols is necessary to minimize the coincidence error in the measurement. Size distribution is the hallmark of particle measurement. Once a complete size distributions has been determined, many other particle parameters can be obtained by mathematical conversion from the size distribution. For example, mass or number concentrations can be obtained by integrating the size distribution function; however, the measured size distributions are not always accurate and precise and also may not be expressed by compact mathematical forms. The errors in size distribution measurement expand to a great extent during conversion; therefore, it is prudent to use techniques that measure the parameters of interest directly. Direct measurement of mass or number concentration is commonly practiced by light scattering *in situ* or by active collection methods. In using these procedures, prior knowledge of size distributions is necessary to ensure that the size of particles is within the measurable range. Most of the instruments employ the method of continuous monitoring *in situ* and provide the results rapidly. A short-term fluctuation or an accidental change of exposure conditions can thus be monitored; however, methods utilizing the collection of particles on filters or other substrates and subsequent analyses of collected samples (i.e., active sampling method) are also used. These methods are preferred if the primary concern is the particle mass concentration or parameters based on particle mass. The samples may also be used for compositional or microscopic analyses of particles.

2.4 Varying and Monitoring Exposure

Physiological responses to inhaled gases and particles will be governed by a number of factors including the duration of the exposure, the concentration of the material, and the deposition of the material within the lung. Table 5 summarizes various methods of varying exposure doses with an emphasis on changes in subject ventilation, usually by exercise.

In addition to overall respiratory tract exposures systems, a number of techniques can be used to target exposure to specific regions of the respiratory tract. These include nose-only exposure methods, targeted aerosol or gas bolus techniques, and localized spray or instillation methods (Table 6). Methods for monitoring particle deposition and the uptake of gases by the respiratory tract are described in Table 7.

In most cases, toxic gases are present in the ambient atmosphere in combination with other gases and particles. Efforts to simulate these toxic mixtures must take into account the possible interactions of the various components. Phalen[6] suggests some general principles to consider when mixed gas-gas or gas-particle atmospheres are generated. Because mixing the toxic materials at high concentrations may result in unwanted reactions, each component should be diluted with air separately. The monitoring scheme should be prepared with the potential interactions in mind so that these can be followed and quantified if they occur. Finally, control systems should be designed to avoid differences in temperature and humidity among the various air streams which could influence reaction rates or particle growth when the air streams are mixed together.

3 CHARACTERIZING RESPONSES

3.1 Upper Airway

The upper airway is the first portion of the respiratory tract that comes in contact with any inhaled gas or aerosol. This section deals primarily with responses of the nasal airway. Since humans are preferential nasal breathers, the nasal airway is more suited to filtration, humidification, and reaction to inhaled substances.[15]

Physiological responses of the nasal airway can be characterized by measurement of nasal resistance using rhinometry. In addition, geometric characteristics of the airway can be determined using acoustic reflectance methods (Table 8). The effects of inhaled materials on the nasal mucosa can be assessed by biopsy and nasal lavage techniques (Table 9). In some cases, these responses may be useful as surrogates for responses in the lower airways.

Responses of the nasal airway may be challenged in a manner similar to airway inhalation challenge (see Section 3.5). Both nonspecific and specific (allergen) challenges may be performed. The principal differences in approach are the use of nasal response measures discussed here and minor modifications to enable the delivery of challenge gas or aerosol to the nasal airway.

3.1.1 Olfactory Testing

Chemical exposure of the nose alone can affect olfactory, nasal, and pulmonary function. Many inhaled pollutants are absorbed by the nasal passages.[15a,b] The sense of smell and the

TABLE 4

Methods for Aerosol Particle Measurement

Measurement	Methods	Principles	Measurement Range	Operating Range	Ref.
Size distribution	Optical particle counter	Detection of single particles by light scattering: the larger the particle size, the greater the light scattering	0.30–15 μm dia.	< 100 particles per cm^3	1,2
	Laser aerosol spectrometer	Detection of single particles by high-density laser scattering	0.05–10 μm dia.	< 300 particles per cm^3	3,4
	Differential mobility particle sizer	Deflection of singly charged particles in a concentric condenser: greater deflection with smaller particles	0.005–0.1 μm dia.	< 10^8 particles per cm^3	7,8
	Time-of-flight aerodynamic particle sizer	Sizing of individual particles by their velocity: the larger the particle size, the slower the velocity	0.50–30 μm dia.	< 200 particles per cm^3	5,6
	Diffusion battery particle sizer	Loss of particles by Brownian diffusion in laminar tube flow	0.005–0.2 μm dia.	< 10^7 particles per cm^3	9,10
	Cascade impactor	Classification of particle size by sequential multistage inertial impaction	0.10–15 μm dia.	< 10^8 particles per cm^3	11,12
	Cascade quartz crystal microbalance	Sequential collection of particles by inertial impaction and automatic weighing	0.05–25 μm dia.	0.01–100 mg/m^3	13
Total mass concentration	Beta attenuation mass monitor	Attenuation of radiation through particle loading collected on a filter strip	0.02–20 mg/m^3	0.01–20 μm dia.	14,15
	Tapered-element oscillating microbalance	Changes in oscillating frequency of a tapered rod with particle loading collected on a filter mounted at the tip of the rod	5 $\mu m/cm^3$–5 g/m^3	< 10 μm dia.	16
	Quartz crystal aerosol mass monitor	Changes in oscillating frequency of a quartz crystal disc as a function of particle mass collected on the disc	0.01–10 mg/m^3	0.01–10 μm dia.	17,18
	Aerosol photometer	Light scattering from a cloud of aerosol particles	0.001–100 mg/m^3	0.3–10 μm dia.	19
	Filtration	Gravimetric analysis	> 10 μg		
Total number concentration	Condensation nuclei counter	Growth of nuclei particles with vapor condensation and subsequent detection by light scattering	< 10^7 particles per cm^3	0.003–3.0 μm dia.	20,21
	Electrometer	Electric current generated by singly charged particles collected on a Faraday cup	< 10^7 particles per cm^3	0.005–0.1 μm dia.	20,22
	Aerosol photometer	Light scattering from a cloud of monodisperse aerosols; for a given particle size, the intensity of light scattering is proportional to number concentration of particles	< 10^7 particles per cm^3	0.3–10 μm dia.	23
	Optical particle counter	Detection of individual particles by light scattering and accumulation of the number of particles counted over a period time	< 300 particles per cm^3	0.3–10 μm dia.	1,2

1. Willeke, K. and Liu, B. Y. H., Single particle optical counter: principles and application, in *Fine Particles,* Liu, B. Y. H. et al., Eds., Academic Press, New York, 1976, pp. 698–729.
2. Liu, B. Y. H., Berglund, R. N., and Agarwal, J. K., Experimental studies of optical particle counters, *Atmos. Environ.,* 8, 717–732, 1974.
3. Gebhart, J., Heyder, J., Roth, C., and Stahlhofen, W., Optical aerosol size spectroscopy below and above the wavelength of light, in *Fine Particles,* Liu, B. Y. H. et al., Eds, Academic Press, New York, 1976, pp.793–815.
4. Knollenberg, R. G. and Luehr, R., Open cavity laser active scattering particle spectrometry, in *Fine Particles,* Liu, B. Y. H. et al., Eds, Academic Press, New York, 1976, pp.669–696.
5. Agarwal, J. K., Remiarz, R. J., Quant, F. R., and Sem, G. J., Development of an aerodynamic particle size analyzer, *J. Aerosol. Sci.,* 13, 222–223, 1982.
6. Heibrink, W. A., Baron, P. A., and Willeke, K., Coincidence in time-of-flight aerosol spectrometers: phantom particle creation, *Aerosol Sci. Technol.,* 14, 112–126, 1991.
7. Liu, B. Y. H. and Pui, D. Y. H., On the performance of the electrical aerosol analyzer, *J. Aerosol. Sci.,* 6, 249–264, 1975.
8. Knutson, E. O. and Whitby, K. T., Aerosol classification by electric mobility: apparatus, theory, and applications, *J. Aerosol. Sci.,* 6, 443–451, 1975.
9. Sinclair, D., Countess, R. J., Liu, B. Y. H., and Pui, D. Y. H., Experimental verification of diffusion battery theory, *J. Aerosol. Sci.,* 11, 549–556, 1980.
10. Cheng, Y. S., Keating, J. A., and Kanapilly, G. M., Theory and calibration of a screen type diffusion battery, *J. Aerosol. Sci.,* 8, 339–347, 1977.
11. Newton, G. J., Raabe, O. G., and Mokler, B. V., Cascade impactor design and performance, *J. Aerosol. Sci.,* 8, 339–347, 1977.
12. Willeke, K., Performance of the slotted impactor, *Am. Ind. Hyg. Assoc. J.,* 36, 683–691, 1975.
13. Fairchild, C. I. and Wheat, L. D., Calibration and evaluation of a real-time cascade impactor, *Am. Ind. Hyg. Assoc. J.,* 45, 205–211, 1984.
14. Lilienfeld, P., Design and operation of dust measuring instrumentation based on the beta-radiation method, *Staub,* 35, 458–465, 1975.
15. Marple, V. A. and Rubow, K., An evaluation of the GCA respirable dust monitor 101-1, *Am. Ind. Hyg. Assoc. J.,* 39, 17–25, 1978.
16. Patashnick, H. and Rupprecht, G., A new real-time aerosol mass monitoring instrument: the TEOM, *Proc. Advances in Particulate Sampling and Measurement,* EPA-600/9-80-004, Daytona Beach, FL, 1979.
17. Daley, P. S. and Lundgren, D. A., The performance of piezoelectric crystal sensors used to determine aerosol mass concentration, *Am. Ind. Hyg. Assoc. J.,* 36, 518–532, 1975.
18. Sem, G. J., Tsurubayashi, K., and Homma, K., Performance of the piezoelectric microbalance respirable aerosol sensor, *Am. Ind. Hyg. Assoc. J.,* 38, 580–588, 1977.
19. Armbruster, L., Breuer, H., Gebhart, J., and Neulinger, G., Photometric determination of respirable dust concentration without elutriation of coarse particles, *Particle Charact.,* 1, 96–101, 1984.
20. Liu, B. Y. H. and Kim, C. S., On the counting efficiency of condensation nuclei counter, *Atm. Environ.,* 11, 1097–1100, 1977.
21. Stolzenburg, M. R. and McMurry, P. H., An ultrafine aerosol condensation nucleus counter, *Aerosol Sci. Technol.,* 15, 107–111, 1991.
22. Liu, B. Y. H. and Pui, D. Y. H., A submicron aerosol standard and the primary absolute calibration of the condensation nuclei counter, *J. Cell Interface Sci.,* 47, 155–171, 1974.
23. Gebhart, J., Heigwer, G., Heyder, J., Roth, C., and Stahlhofen, W., The use of light scattering photometry in aerosol medicine, *J. Aerosol Med.,* 1, 889–1112, 1988.

TABLE 5

Methods of Varying Exposure

Variable	Ways in Which Exposure Can Be Varied	Ref.
Concentration (C)	C can be varied at different levels of a constant concentration and is the easiest to control and interpret. Variable C can be ramped up and/or down using computer control, or the exposure pattern of the workplace or ambient environment can be simulated. If the goal is to achieve and maintain a physiological concentration of the toxic substance (e.g., HbCO), then an initial bolus followed by a maintenance concentration may be more suitable.	1, 3
Ventilation (V)	Inspired air can enter via the mouth (oral), nose (nasal), or both (oronasal). V can be varied voluntarily, although it is usually necessary to add CO_2 to the inspired air to maintain eucapnia. Alternatively CO_2 can be used to drive increased ventilation, either by having subjects breathe a constant CO_2 concentration or by adding a constant flow of CO_2 to the inspirate. Exercise is more commonly used as a method to increase ventilation during an exposure. Exercise can be continuous or intermittent, at a constant or changing level, and may induce V ranging from 20 to 100 l/min or more. V in the range of 25–45 l/min can be tolerated for prolonged periods, higher levels for brief periods. Treadmills or bicycle ergometers are commonly used, although many modes of work or activity can be stimulated.	2, 4, 8, 11
Duration (D)	D can be varied widely but is constrained by the method of exposure. D in large chambers is theoretically unlimited — exposures lasting 24 hr for several days are feasible, but extremely short exposures are impractical. D in small chambers is practically limited to 4–8 hr. Mask and head-dome exposures should be limited to a couple of hours, and mouthpiece exposures become uncomfortable beyond an hour or so.	5–8
Combinations	Combinations of two or more gases or gas plus particles can either be given together (simultaneously) or one after the other (sequentially) or in a pattern that simulates natural exposure in the environment or workplace.	9,10

[1] Dahms, T. E., Horvath, S. M., and Gray, D. J., Techniques for accurately producing desired carboxyhemoglobin levels during rest and exercise, *J. Appl. Physiol.*, 38, 366–368, 1975.

[2] Bethel, R. A., Sheppard, D., Epstein, J., Tam, E., Nadel, J. A., and Boushey, H. A., Interaction of sulfur dioxide and dry cold air in causing bronchoconstriction in asthmatic subjects, *J. Appl. Physiol.*, 57, 419–423, 1984.

[3] Hazucha, M. J., Seal, E., and Folinsbee, L. J., Effects of steady-state and variable ozone concentration profiles on pulmonary function, *Am. Rev. Respir. Dis.*, 146, 1487–1493, 1992.

[4] Schelegle, E. S. and Adams, W. C., Reduced exercise time in competitive simulations consequent to low level ozone exposure, *Med. Sci. Sports Exer.*, 18, 408–414, 1986.

[5] Adams, W. C., Savin, W. M., and Christo, A. E., Detection of ozone toxicity during continuous exercise via the effective dose concept, *J. Appl. Physiol.*, 51, 415–422, 1981.

[6] Koenig, J. Q., Pierson, W. E., Horike, M., and Frank, N. R., Effects of SO_2 plus NaCl aerosol combined with moderate exercise on pulmonary function in asthmatic adolescents, *Environ. Res.*, 25, 340–348, 1981.

[7] Niinimaa, V., Cole, P., Mintz, S., and Shephard, R., The switching point from nasal to oronasal breathing, *Respir. Physiol.*, 42, 61–67, 1980.

[8] Folinsbee, L. J., McDonnell, W. F., and Horstman, D. H., Pulmonary function and symptom responses after 6.6 hour exposure to 0.12 ppm ozone with moderate exercise, *JAPCA*, 38, 28–35, 1988.

[9] Hazucha, M. J., Folinsbee, L. J., Seal, E., and Bromberg, P. A., Lung function response of healthy subjects following sequential exposures to NO_2 and O_3, *Am. J. Respir. Crit. Care Med.*, 150, 642–647, 1994.

[10] Horvath, S. M., Folinsbee, L. J., and Bedi, J. F., Combined effect of ozone and sulfuric acid on pulmonary function in man, *Am. Ind. Hyg. Assoc.*, 48, 94–98, 1987.

[11] American College of Sports Medicine, *Guidelines for Graded Exercise Testing and Exercise Prescriptions,* 3rd ed., Lea & Febiger, Philadelphia, PA, 1986.

TABLE 6

Isolated Respiratory Tract Segment Exposure

	Methods and Procedures	Interpretation/Significance	Ref.
Nose-only exposure	Not associated with pulmonary exposures; generally limited to directly applied materials as powdered solids, solutions, soaked filter paper, or sprayed aerosols. Using one naris as an input and evacuating the other, the nose can be exposed to gaseous materials with minimal exposure to the lower airway.	Useful for examining nasal allergic responses, nasal clearance mechanisms, and nasal mucosal uptake of toxic substances. Nose-only exposure to gases and odorants permits evaluation of nasal stimuli independant of lower airway responses	10–12
Targeted respiratory tract region	*Aerosol bolus method*: An aerosol bolus as small as 50 ml is injected during inspiration by rapidly switching an aerosol valve with an automatic timing device (usually computer-controlled). Inspiratory volume can be divided into a number of volumetric compartments, and a bolus can be delivered precisely into each of the volumetric compartments by controlling the onset time and duration of valve opening. Aerosol concentrations inhaled and exhaled, as well as respiratory flow rates, are monitored continuously with a laser aerosol photometer and a peumotachograph, respectively. An end-inspiratory breath hold may be used to enhance aerosol deposition in the target regions.	Noninvasive way of exposing different lung regions. Repeated challenges can be performed in the same individuals; however, the airway site of challenge is approximate and may not be warranted if ventilation is uneven. Deposition can be enhanced at the target sites, but deposition also occurs throughout the lung regions proximal to the target sites.	1–4
	Local spray method: A bronchoscope is positioned in the airways of interest and a microspray nozzle connected to the tip of a catheter is inserted through the inner channel of the bronchoscope. A small volume (5–20 μl) of solution or liquid suspension is sprayed by rapidly pushing the air (1–2 ml) through the catheter with a syringe. Spraying may be performed during breathhold at the end of inspiration. The same procedure can be repeated for different airway sites if multiple sites are to be challenged.	Challenge to small, confined airway sites may be assured. Exposure dosages can be determined precisely; however, the size of the airways to be accessed is limited by the diameter of the bronchoscope (~5 mm). The number of challenges in a single individual is limited for this invasive procedure.	5,7
	Instillation method: The techniques are similar to the local spray method except that the materials are instilled instead of sprayed. A bronchoscope is positioned at a desired site in the airways. An instillate (10–50 ml) is delivered slowly and intermittently via an inner channel of the bronchoscope or a catheter inserted through the bronchoscope channel. The delivery may be made over the period of several breaths during inspiration or during the end-expiratory pause.	Same as above except that because of a bulk quantity, local confinement of an instillate may not be warranted. Repetition of the procedure is limited.	6,8,9

1 Stahlhofen, W., Gebhart, J., Rudolf, G., and Scheuch, G., Measurement of lung clearance with pulses or radioactively-labelled aerosols, *J. Aerosol Sci.*, 3, 333–336, 1986.

2 Scheuch, G. and Stahlhofen, W., Particle deposition of inhaled aerosol boluses in the upper human airways, *J. Aerosol Med.*, 1, 29–36, 1988.

3 Anderson, P. J., Blanchard, J. D., Brain, J. D., Feldman, H. A., McNamara, J. C., and Heyder, J., Effect of cystic fibrosis on inhaled aerosol boluses, *Am. Rev. Respir. Dis.*, 140, 1317–1324, 1989.

4 Kim, C. S., Hu, S. C., DeWitt, P., and Gerrity, T. R., Assessment of regional deposition of inhaled particles in human lungs by serial bolus delivery method, *J. Appl. Physiol.*, 81, 2203–2213, 1996.

5 Hoover, M. D., Harkema, J. R., Muggenburg, B. A., Spoo, J. W., Gerde, P., Staller, H. J., and Hotchkiss, J. A., A microspray nozzle for local administration of liquids or suspensions to lung airways via bronchoscopy, *J. Aerosol Med.*, 6, 67–72, 1993.

TABLE 6 (continued)

Isolated Respiratory Tract Segment Exposure

6. Snipes, M. B., Chavez, G. T., and Muggenburg, B. A., Disposition of 3-, 7-, and 13-μm microspheres instilled into lungs of dogs, *Environ. Res.*, 33, 333–342, 1984.
7. Wolff, R. K., Tillquist, H., Muggenburg, B. A., Harkema, J. R., and Mauderly, J. L., Deposition and clearance of radiolabelled particles from small ciliated airways in beagle dogs, *J. Aerosol Med.*, 2, 261–270, 1989.
8. Metzger, W. J., Zavala, D., Richerson, H. B., Mosely, P., Iwamota, P., Monick, M., Sjoerdsma, K., and Hunninghake, G. W., Local allergen challenge and bronchoaveolar lavage of allergic asthmatic lungs, *Am. Rev. Respir. Dis.*, 135, 433–440, 1987.
9. Smith, D. L. and Deshazo, R. D., Bronchoaveolar lavage in asthma: an update and perspective, *Am. Rev. Respir. Dis.*, 148, 523–532, 1993.
10. Solomon, W. R. and McLean, J. A., Nasal provocative testing, in *Provocative Challenge Procedures: Background and Methodology*, Spector, S. L., Ed., Futura Publishing, Mount Kisco, NY, 1989, pp. 569–625.
11. Andersen, I., Effects of airborne substances on nasal function in human volunteers, in *Toxicology of the Nasal Passages*, Barrow, C. S., Ed., Hemisphere, New York, 1986, pp. 143–154.
12. Prah, J. D. and Benignus, V., Olfactory-evoked responses to odorous stimuli of different intensities, *Chem. Senses*, 17, 417–425, 1992.

TABLE 7

Techniques for Measuring Particle Deposition and Gas Uptake Dose in the Lung

	Description	Interpretation/Significance	Ref.
Particle Deposition Dose			
Total lung deposition	*Aerosol photometric techniques:* Generation of monodisperse nonhygroscopic aerosols in the size range of 0.5–10 μm dia. with a concentration ranging from 10^3–10^6 particles per cm^3, depending on particle size (see Table 5 for aerosol generation method). Inhalation of the aerosol with a prescribed breathing pattern as displayed on the computer monitor. Continuous monitoring of aerosol concentration and flow rate during breathing *in situ* with a light-scattering detection device and a pneumotachograph (or spirometer). A spirometer may be used instead of a pneumotachograph when aerosols are inhaled from a bag-in-box system connected to a spirometer. Determines total number of particles inhaled (N_i) and exhaled (N_e) by integrating the product of aerosol concentration and flow rate over the inspiratory and expiratory time, respectively. Total lung disposition is determined as a fraction of inhaled particles lost in the lung, i.e., $(N_i - N_e)/N_i$.	Rapid and precise measurement of total lung deposition fraction. Particles must be uniform in size, spherical in shape, nonhygroscopic and nonlight-absorbing. Particles < 0.3 μm dia. may not be used because of light-scattering limitation. Requires a high concentration of aerosol, 10^3 – 10^6 particles per cm^3.	1–4
	Filter sampling techniques: Aerosols are inhaled from a continuously flowing aerosol circuit or from an aerosol-holding bag and exhaled through a filter. The amount of inspired aerosol may be estimated by sampling aerosols from the inspiratory limb onto a filter for a fixed volume. Filter samples are analyzed with an appropriate method to determine particle mass. This method may be used if the nature of aerosols does not permit use of a continuous aerosol monitor.	No limitations on particle size, but the sampling procedures are often laborious and prone to error. The method is essential if a continuous monitoring of test aerosols is not possible.	5,6

TABLE 7 (continued)

Techniques for Measuring Particle Deposition and Gas Uptake Dose in the Lung

	Description	Interpretation/Significance	Ref.
Regional lung deposition	*Aerosol bolus techniques:* Generation of monodisperse nonhygroscopic aerosols in the same way as described above. Inhalation of clean air with a prescribed breathing pattern displayed on a computer screen and delivery of aerosol bolus (20- to 100-ml) during inspiration by rapidly switching the aerosol valve using an automatic timing device (usually computer controlled). Inspiratory volume can be divided into a number of volumetric compartments, and a bolus can be delivered precisely into each of the volumetric compartments sequentially. The concentrations of bolus aerosols inhaled and exhaled as well as respiratory flow rates are monitored continuously with a laser aerosol photometric device and a pneumotachograph, respectively. The total number of particles inhaled (N_i) and exhaled (N_e) in each breath is determined as described above, and the recovery of bolus (RC) is obtained as N_e/N_i. From the RC values obtained from the sequential volumetric compartments, particle deposition in each compartment is computed.	Rapid, noninvasive method of measuring regional lung deposition. Requirements for aerosol conditions are the same as for the aerosol photometric method. The lung regions may not be defined rigorously, particularly when ventilation is uneven.	7,8
	Radioaerosol imaging method: Aerosols labeled with a selected radionucleid are inhaled with a prescribed breathing pattern, and a whole lung scan is obtained with a gamma camera. Prior to inhalation of the labeled aerosols, a whole lung scan may be obtained with xenon inhalation to define the lung boundary and lung volume distribution in a planar image which may be used as a reference in analyzing lung scan images obtained after aerosol inhalation. The scanned image may be divided into a number of regions and regional deposition is determined by counting activities in each region.	Unique method of visualizing lung deposition; however, precise quantitation of regional deposition may not be warranted with two-dimensional lung images. Delivery dose of aerosol must be carefully monitored to ensure radiation safety.	9–12
Gas Uptake Dose			
Total lung uptake	*Continuous monitoring method: In situ* continuous monitoring of inhaled and exhaled gas concentrations using a rapid response gas detector (time constant < 150 ms) while a person is breathing a test gas via mouthpiece. Integration of the product of flow rate and concentration over the inspiratory and expiratory time determine the total amount of gas inhaled and exhaled breath by breath.	Rapid, noninvasive method of measuring total lung gas uptake *in situ.* Requires a fast responding gas monitor that may be available for only a few selected gases.	13,14
	Bag collection method: Inhalation of a test gas of known concentration from a bag-in-box connected to a spirometer and exhalation into a gas collection bag using a three-way valve for several breaths. Determination of total respiratory uptake by: inhaled volume × (inhaled gas concentration – exhaled gas concentration). This method may be used if the response of gas detector is not fast enough for breath-by-breath analysis.	Does not require a fast responding gas monitor.	—

TABLE 7 (continued)

Techniques for Measuring Particle Deposition and Gas Uptake Dose in the Lung

	Description	Interpretation/Significance	Ref.
Regional respiratory uptake	*Bolus techniques:* Injection of a small gas bolus (10–50 ml) into an inspiratory stream at varying time points during breathing and continuous monitoring of gas concentration and flow rate with a rapid-response gas monitor and pneumotachograph, respectively. Determination of the amount of gas inhaled and exhaled and a fraction of gas lost in the lung as a function of volumetric lung depth to which the bolus was delivered. The onset time and duration of bolus injection may be controlled by a computer or specially built timer as discussed above.	Rapid, noninvasive measurement of regional gas uptake *in situ;* however, precise quantitation of regions of uptake may not be warranted if ventilation is uneven.	15, 16

1 Heyder, J., Armbruster, L., Gebhart, E., Grein, E., and Stahlhofen, W., Total deposition of aerosol particles in the human respiratory tract for nose and mouth breathing, *J. Aerosol Sci.*, 6, 311–328, 1975.

2 Davies, C. N., Heyder, J., and Ramu, M. C., Breathing of half-micron aerosols. I. Experimental, *J. Appl. Physiol.*, 32, 591–600, 1972.

3 Muir, D. C. F. and Davies, C. N., The deposition of 0.5-μm diameter aerosols in the lungs of man, *Ann. Occup. Hyg.*, 10, 161–174, 1967.

4 Altshuler, B., Yarmus, L., and Palmes, E. D., Aerosol deposition in the human respiratory tract, *Arch. Ind. Health,* 15, 293–303, 1957.

5 Wanner, A., Broadnan, J. M., Perez, J., Henke, K. G., and Kim, C. S., Variability of airway responsiveness to histamine aerosol in normal subjects, *Am. Rev. Respir. Dis.*, 131, 3–7, 1985.

6 Hounam, R. F., Black, A., and Walsh, M., The deposition of aerosol particles in the nasopharyngeal region of the human respiratory tract, in *Inhaled Particles III,* Walton, W. H., Ed., Unwin Brothers, London, 1972, pp. 71–80.

7 Scheuch, G. and Stahlhofen, W., Particle deposition of inhaled aerosol boluses in the upper human airways, *J. Aerosol. Med.*, 1, 29–36, 1988.

8 Kim, C. S., Hu, S. C., DeWitt, P., and Gerrity, T. R., Assessment of regional deposition of inhaled particles in human lungs by serial bolus delivery method, *J. Appl. Physiol.*, 81, 2203–2213, 1996.

9 Agnew, J. E., Pavia, D., and Clarke, S. W., Airway penetration of inhaled radioaerosol: an index to small airways function?, *Eur. J. Respir. Dis.*, 62, 239–255, 1981.

10 Dolovich, M. B., Sanchis, J., Rossman, C., and Newhouse, M. T., Aerosol penetrance: a sensitive index of peripheral airways obstruction, *J. Appl. Physiol.*, 40, 468–471, 1976.

11 Emmet, P. C., Love, R. G., Hannan, W. J., Millar, A. M., and Soutar, C. A., The relationship between the pulmonary distribution of inhaled fine aerosols and tests of small airway function, *Bull. Eur. Physiopathol. Respir.*, 20, 325–332, 1984.

12 Agnew, J. E., Aerosol contribution to the investigation of lung structure and ventilatory function, in *Aerosols and the Lung,* Clarke, S. W. and Pavia, D., Eds., Butterworths, London, 1984, chpt. 5.

13 Gerrity, T. R., Weaver, R. A., Berntsen, J., House, D. E., and O'Neil, J. J., Extrathoracic and intrathoracic removal of O_3 in tidal-breathing humans, *J. Appl. Physiol.*, 65, 393–400, 1988.

14 Gerrity, T. R., McDonnell, W. F., and House, D. E., The relationship between delivered ozone dose and functional responses in humans, *Toxicol. Appl. Pharmacol.*, 124, 275–283, 1994.

15 Hu, S. C., Ben-Jebria, A., and Ultman, J. S., Longitudinal distribution of ozone absorption in the lung: quiet respiration in healthy subjects, *J. Appl. Physiol.*, 73, 1655–1661, 1992.

16 Hu, S. C., Ben-Jebria, A., and Ultman, J. S., Longitudinal distribution of ozone absorption in the lung: effects of respiration flow, *J. Appl. Physiol.*, 77, 574–583, 1994.

trigeminal system can also be affected by acute exposure to ambient ozone.[20] Overall respratory effects of nasal stimulation in animals and humans include apnea,[15] modification of the breathing pattern,[15d] and increased nasal resistance.[15e] In a polluted environment, the sense of smell may be diminished by irritant chemicals acting via the trigeminal nerve, causing inhibition of olfaction.[15f] Clearly, the effect of upper airway stimulation must be taken into

TABLE 8
Tests of Nasal Resistance and Airway Geometry

Test	Apparatus	Method	Significance/Interpretation
Anterior rhinometry	Pneumotachograph, pressure transducer, nasal probe, and face-mask	Subject breathes with mouth closed while face mask is applied; single naris resistance determined from monitoring of pressure in occluded naris and face-mask flow through open naris. Total nasal resistance computed using measures from two nares (inverse of combined conductances).	Provides independent measures of both nares; total nasal resistance is a derived value. Requires instrumenting naris, which can affect resistance. Nasal resistance measures are confounded by nasal cycle. Measures before and after decongestant have relevance to mucosal edema and fixed obstruction.[1-3]
Posterior rhinometry	Pneumotachograph, pressure transducer, mouth probe, and face-mask	Subject breathes with mouth closed while face mask in line with pneumotachograph is applied: total nasal resistance determined from monitoring oropharyngeal pressure from mouth probe and face-mask flow through nose.	Provides measure of total nasal resistance. Difficulties in obtaining pressure recordings due to obstruction of mouth probe by oropharyngeal mucosa and secretions. Measure of nasal resistance confounded by nasal cycle.[1-3]
Acoustic rhinometry	Acoustic signal generator, wave tube, microphone, amplifier, low pass filter, A/D converter, computer, and software	Acoustic signal delivered via nosepiece of wavetube to nostril. Sound is reflected due to changes in cross-sectional area of nasal cavity. Signal from microphone is amplified, digitized, and analyzed to create area-distance function.[4]	Provides a measure of cross-sectional area of the nose as a function of distance from the nostril. Provides information about site of obstruction.

[1] Graamans, K., Rhinometry, *Clin. Orolaryngol.*, 6, 291–297, 1981.
[2] Holmstrom, M., Scadding, G. K., and Lund, V. J., The assessment of nasal obstruction: comparison between rhinomanometry and nasal inspiratory peak flow, *Rhinology*, 28, 191–196, 1990.
[3] Clement, P. A. R., Committee report on standardization of rhinometry, *Rhinology*, 22, 151–155, 1984.
[4] See Table 16.

consideration even when the critical endpoints are pulmonary. Given the ability of the nose to absorb and respond to atmospheric pollutants, rhinometry is a valuable but underutilized endpoint.

Because the olfaction can be impaired or modulated by pollutant exposure, testing of this sense should be considered an important endpoint. The systematic testing of the sense of smell can be achieved with methods that range from simple to complex. Selection of the technique depends upon the information required. Table 10 provides basic information about the range of techniques that are currently used. Sniff and puff bottles are closely-related methods for clinical evaluation and psychophysical experiments which examine threshold, intensity, hedonics, and identification ability. They have a history of use since the 19th century and have continued to be used because of their simplicity, reliability, versatility, and ease of use. There is the potential, though, with puff bottles for stimulating the trigeminal receptors via air flow concomitantly with odor stimulation, thereby confounding the response (e.g., Munich Olfactory Test [MOT])[19] and University of Connecticut Odor Test [UCONN]).[16] "Blast" olfactometry,[18] in which the odorant is forced into the nose under pressure, has also fallen from favor for this reason. The University of Pennsylvania Smell Identification Test

TABLE 9
Sampling Techniques for Nasal Fluids and Tissue

Test	Apparatus	Method	Significance/Interpretation
Nasal brush biopsy	Otologic light source, nasal speculum, small plastic curette, or cytological brush	Surface of inferior turbinate is gently scraped with curette or brush under direct visualization. Cells obtained processed for microscopic (light, electron) examination and PCR for specific mRNA cytokine messages.	Minimally invasive with small risk of epistaxis. Allows identification of changes in morphologic features of nasal epithelial cells and cytokine profiles. Technique is complementary to nasal biopsy. Also used to study nasal cilia.[4]
Nasal mucosal biopsy	Otologic light source, nasal speculum, biopsy forceps	Local anesthesia achieved with application of topical lidocaine (2% with 0.025% epinephrine). 2–3 mm biopsy of inferior turbinate performed using biopsy forceps. Biopsy site cauterized and bleeding controlled with a silver nitrate stick.	Biopsy specimens allow characterization of mucosal inflammation, which requires intact tissue samples for immunohistochemical analysis of resident inflammatory cells and *in situ* hybridization studies for local cytokine production. Such analysis cannot be accomplished using surface epithelial cell sampling (scraping). Likewise, epithelial cells used for PCR analysis of epithelial cell cytokine production are not adequately obtained from biopsy specimens. For examination of the inflammatory and cytokine responses of the upper airway to pollutant exposure, these two techniques are complementary but not interchangeable.[5]
Nasal lavage (NAL)[1,3]	Sterile phosphate-buffered saline, blunt-tipped syringes without needles, specimen cups	Instill 5-ml of warmed (37°C) saline into each nasal cavity. After 10 sec, the saline is forcibly expelled into a sterile plastic specimen cup. Subjects must be trained to put pressure on the palate to hold saline in nasal cavity while the head is tilted back. Cells are immediately pelleted for differential and supernatant frozen for later analysis.	Samples can be analyzed for cells and mediators. May be used in field studies and with children. See Section 3.6 (BAL) for possible analyses of these samples. Changes in NAL cells and cytokines may be markers for changes that are occurring in the lower airway but specifically reflect changes occurring in the nose. The test is inexpensive, not technically demanding, easily tolerated, poses almost no risk to the subject, and can be performed repeatedly without adverse effects. In addition to cells and mediators, this procedure could be modified to evaluate mucus and particle content.
Nasal lavage II[2]	Saline in hand-held nasal spray actuator, specimen cups	Subject sprays saline into nostril and snuffs the liquid after each spray. After each five sprays, fluid is forcibly expelled into a cup. A total of 4 ml is delivered (about 40 sprays). Unilateral or bilateral lavage may be used. An aliquot of fluid is reserved for immediate analyses. Supernatant from centrifuged samples is frozen for later analysis.	This method takes longer but does not require the training to hold the saline in the nasal cavity. The improved yield of cells and mediators in this method allows one to collect samples in the field for later analyses in the laboratory, eliminating the need to set up a microscope and cytocentrifuge in the field. This procedure may sample more anteriorly in the nose than the bulk washout method.

TABLE 9 (continued)

Sampling Techniques for Nasal Fluids and Tissue

1 Frischer, T. M., Kuehr, J., Pullwit, A., Meinert, R., Forster, J., Studnicka, M., and Koren, H., Ambient ozone causes upper airways inflammation in children, *Am. Rev. Respir. Dis.*, 148, 961–964, 1993.
2 Peden, D. B., Setzer, R. W., and Devlin, R. B., Ozone exposure has both a priming effect on allergen-induced responses and an intrinsic inflammatory action in the nasal airways of perenially allergic asthmatics, *Am. J. Respir. Crit. Care Med.*, 151, 1336–1345, 1995.
3 Koren, H. S., Hatch, G. E., and Graham, D. E., Nasal lavage as a tool in assessing acute inflammation in response to inhaled pollutants, *Toxicology*, 60, 15–25, 1990.
4 Rutland, J., Dewar, A., Cox, T., and Cole, P., Nasal brushing for the study of ciliary ultrastructure, *J. Clin. Pathol.*, 35, 357–359, 1982.
5 Meltzer, E. O., Orgel, H. A., and Jalowayski, A. A., Cytology, in *Allergic and Non-allergic Rhinitis: Clinical Aspects*, Mygind, N. and Naclerio, R. M., Eds., Saunders, Philadelphia, 1993, pp. 66–81.

(UPSIT),[17] a scratch-and-sniff test, has been used clinically to evaluate odor identification and has the significant advantage of age-based population norms. Dynamic olfactometers,[21] which have versatile control over the stimulus parameters of concentration, flow rate, stimulus duration, and interstimulus interval, have the disadvantage of being very expensive to build and operate but the advantage of computerized data acquisition and control of critical stimulus parameters. They have been used in clinical, psychophysiological, behavioral, and electrophysiological studies of olfaction in both humans and animals.

3.2 Pulmonary Function

Since inhaled substances have their first and often only contact with the body through the surfaces of the respiratory tract, the consequence is that many of these substances have an impact on lung function. Effects may be mediated by lung receptors, through damage to lung tissue, or by formation of some product by reaction with substances (i.e., lipids, proteins, etc.) within the lung or respiratory tract. In this section, various tests of the functional capability of the lung are summarized.[22] Among the tests are measurements of lung volume and its subdivisions, flow rates and timing of forced inspiratory and expiratory maneuvers, measures of airway resistance, diffusing capacity, gas exchange, distribution of ventilation, breathing pattern, assessment of ventilatory control, and airway responses to various provocative agents.

3.2.1 Lung Volume and Flow Rates

The basic subdivisions of lung volume are shown in Figure 1, and techniques for measuring the various lung volumes are described in Table 11. Several of the subdivisions can be simply obtained from the spirometer tracing or its digitized counterpart. The residual volume is usually determined by measuring the functional residual capacity (FRC; i.e., the volume of air remaining in the lungs at the resting end-expiratory position) and then subtracting the expiratory reserve volume (ERV). FRC is measured either by gas dilution methods (nitrogen washout, helium dilution, etc.) or by body plethysmography.

The American Thoracic Society guidelines[23,24] recommend that all volumes and flows be reported at body temperature and ambient pressure saturated with water vapor (BTPS); however, it has been pointed out by a number of workers[27,28] that BTPS correction of spirogram volumes and flows may lead to overestimation of actual values. These effects depend, to a large extent, on the operating temperature of the spirometers. Expired air

TABLE 10
Olfactory Testing Techniques

Technique	Advantages	Disadvantages	Testing Domain	Ref.
Blast olfactometry	Inexpensive; simple to learn; wide stimulus selection	Odorants blasted intranasally under pressure; potentially confounded with air flow, volume, pressure	Clinical and neurological testing; psychophysics	3
Sniff bottles, puff bottles	Inexpensive; stable concentrations; easy to learn; commonly used	Headspace concentration can vary with temperature; headspace must equilibrate; may be difficult to find odor-free diluent for some chemicals; puff bottles may augment or add a trigeminal component	Basic clinical, psychophysics, e.g., threshold, intensity, hedonics, identification	1, 5
University of Pennsylvania Smell Identification Test (UPSIT)	Norms established for a wide age range and both genders; easy to self-administer; stable odor source; forced choice; scratch and sniff	Expensive; concentration may vary with sniffing technique; fixed range of odor qualities	Clinical testing of ability to identify odors	2
Munich Olfactory Test (MOT)	Simple, easy to learn; inexpensive; tests odor quality; odor intensity; detection threshold; odor recognition	Uses 250 -ml squeeze bottles with handmade Teflon nose pieces; bottles not odor free; air flow may add a trigeminal component	Clinical testing of odor identification threshold, quality discrimination threshold	4
University of Connecticut Odor Test (UCONN)	Simple, easy to learn; tests threshold of butanol and identification of common odorants; tests odor and odor plus trigeminal response	Uses squeeze bottles for threshold, which could confound an odor response with a trigeminal component; squeeze bottles not odorless, identification tests use organics that may degrade over time	Clinical testing of odor identification and threshold	1
Dynamic olfactometers	Versatile control of stimulus characteristics, e.g., odor, flow rate, interstimulus interval, duration, intensity	Expensive; no standard olfactometer established; computer control desirable; high maintenance costs	Clinical; psychophysical; electrophysical; behavioral testing	5,6

1 Cain, W. S., Gent, J., Catalanotto, F. A., and Goodspeed, R. B., Clinical evaluation of olfaction, *Am. J. Otolaryngol.*, 4, 252–256, 1983.
2 Doty, R. L., Shaman, P., and Dann, M., Development of the University of Pennsylvania Smell Identification Test: a standardized microencapsulated test of olfactory function, *Physiol. Behav.*, 32, 489–502, 1984.
3 Elsberg, C. A. and Levy, I., The sense of smell. A new and simple method of quantitative olfactometry, *Bull. Neurol. Inst. N.Y.*, 4, 5–19, 1935.
4 Hudson, R., Laska, M., Berger, T., Heye, B., Schopohl, J., and Danek, A., Olfactory function in patients with hypogonadotropic hypogonadism: an all-or-none phenomenon?, *Chem. Senses*, 19, 57–69, 1994.
5 Prah, J. D. and Benignus, V., Effects of ozone exposure on olfactory sensitivity, *Perceptual Motor Skills*, 48, 317–318, 1979.
6 Benignus, V. and Prah, J. D., A computer-controlled vapor dilution olfactometer, *Behav. Res. Meth. Instr.*, 12, 535–540, 1981.

Source: Adapted from Prah, J. D., Walker, J. C., and Sears, S. B., Modern approaches to clinical and research olfactometry, in *Handbook of Clinical Olfaction and Gustation*, Doty, R. L., Ed., Marcel Dekker, New York, 1994.

temperature is less than body temperature, typically about 34 to 35°C, and body temperature may vary according to time of day, meals, and exercise. Changes in spirometer temperature during the course of a series of measurements can result in small errors when the standard BTPS correction factor is used, particularly for dynamic measurements, such as FEV_1 (forced expired volume in 1 sec) or $FEF_{50\%}$ (mean of forced expiratory flow at 50% vital capacity),

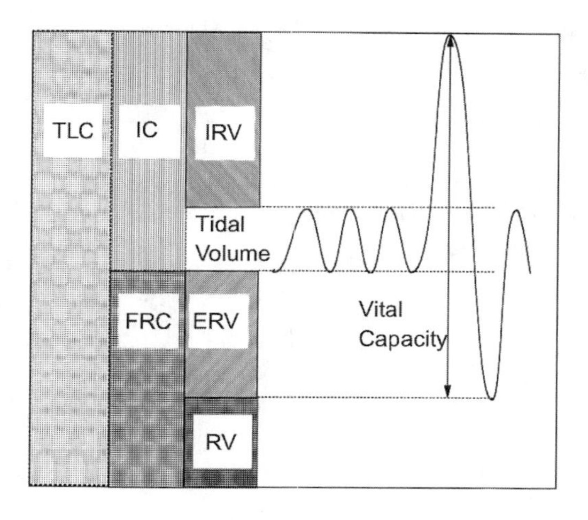

FIGURE 1
Illustration of the major subdivisions of lung volume including total lung capacity (TLC), inspiratory capacity (IC), functional residual capacity (FRC), inspiratory reserve volume (IRV), tidal volume (VT), expiratory reserve volume (ERV), residual volume (RV), and vital capacity (VC). The ordinate is volume and the abscissa is time.

where thermal equilibrium with the spirometer may not be achieved. Spirometer temperature should vary by less than 3°C. The closer the spirometer temperature is to body temperature, the smaller is the error; spirometer temperatures above 23°C introduce a relatively small error. Correction factors for pneumotachographs are not the same as for spirometers, and there is some question whether any correction at all should be applied to measurements obtained from a heated pneumotachograph.[24a] Correction of flow and volume measurements for changes in temperature and water vapor content of expired air depends on the measurement device and calibration method. Table 12 illustrates the correction factors for BTPS under different conditions. Also shown is the STPD (standard temperature and pressure dry) correction factor, since variables such as oxygen uptake and carbon dioxide production are reported at standard temperature and pressure dry.

Dynamic spirometry or forced inspiratory maneuvers produce an additional set of indices related to functional performance of the lung (Table 13). A forced expiratory spirogram is shown in Figure 2, and its corresponding inspiratory and expiratory flow-volume curves are shown in Figure 3. Forced expiratory indices are influenced by changes in posture (standing gives higher values), head flexion (avoid flexion or hyperextension), and effort (maximal effort may give slightly lower FEV_1 than submaximal efforts), as well as the influence of pressure on gas compression and expiratory mechanics.[25] For consistency and comparison of results, it is important to perform identical maneuvers in the same posture and head position, use identical instructions by well-trained technicians, and obtain consistent efforts from the participants. The greatest source of variation within subjects is due to improper test performance. Diurnal variations can be substantial, especially in asthmatics. Normal values and prediction equations for forced vital capacity (FVC) and FEV_1 are shown in Tables 14 and 15. These tests have been widely used in the evaluation of responses to air pollutants such as ozone and sulfur dioxide.

3.2.2 *Airway Resistance*

Various techniques have been used to measure flow resistance across different regions of the respiratory tract. The most commonly used technique is the measurement of airway resistance using a body plethysmograph. Other techniques of estimating airway resistance are summarized in Table 16. In addition to airway and pulmonary resistance, the flow resistance in extrathoracic airways can be measured, including nasal, pharyngeal, and laryngeal resistance.

TABLE 11

Lung Volume Measurements

Test	Apparatus	Method	Significance/Interpretation
Vital capacity (VC or SVC)	Spirometer, pneumotachograph, or other flowmeter and appropriate recording and signal conditioning meeting ATS recommendations	Test subject takes a full complete inspiration and expires fully. The expiration should not be forced and should continue for at least 6 sec and, if possible, until flow has fallen to zero for at least 2 sec. Environmental temperatures should range between 17 and 40°C. Use of noseclips and a standing position are recommended. Measurements should be performed in the same posture.	See prediction equations for normal values based on height and age. VC can be reduced due to inspiratory restriction, edema, fibrosis, COPD, and many other causes. The largest value is usually accepted as best.
Inspiratory capacity (IC)	See above	The IC is calculated as the difference between the resting lung volume and maximum inspiration. Resting lung volume is taken as the average end-expired volume during several breaths, usually prior to the VC maneuver.	—
Expiratory reserve volume (ERV)	See above	The ERV is the difference between the resting end-expired lung volume and the maximum expiratory level.	Errors in this measurement can lead to erroneous values for RV.
Functional residual capacity (FRC)	Spirometer, gas analyzers (He, N_2, or CO_2 and O_2), breathing circuit, etc.	FRC is typically determined by gas dilution methods. Closed-circuit helium dilution determines the remaining lung volume by dilution of the gas in the system and takes about 3 to 5 min. Open-circuit nitrogen washout method determines the original lung volume using the volume of nitrogen washed out of the lung and requires about 5 to 7 min. A rapid equilibration method of N_2 washout has been described and takes only a minute or so to perform. Breath-by-breath analyses of multibreath N_2 washout under strictly controlled conditions of frequency and tidal volume may provide useful information about distribution of ventilation.	In patients with lung disease, extra time usually must be allowed for equilibration. Larger than predicted values may result from hyperinflation due to emphysema or bronchoconstriction. Smaller than predicted values may occur in restrictive disease. Marked variation in FRC with posture — about 25% higher in upright position.
Thoracic gas volume (TGV)	Body plethysmograph system	Subject performs breathing movements against a closed airway and volume of gas in the chest is calculated using Boyle's law, based on the pressure-volume relationships in the lung.	Large volumes of abdominal or thoracic (trapped) gas not in communication with the mouth may cause errors.
Residual volume (RV)	See FRC and TGV	The volume of gas remaining in the lung after a maximal expiration: RV = FRC – ERV. TGV may be substituted for FRC when measured during quite breathing.	Usually limited by chest wall elastic recoil in youth; also limited by small airway closure in adults.
Total lung capacity (TLC)	See FRC and VC	The sum of vital capacity and residual volume.	May be reduced in restrictive disease or by inhaled materials that reduce maximal inhalation or cause pulmonary edema.

Note: See References 1 to 3 in Table 13 for further information.

TABLE 12

Correction Factors for Temperature and Humidity in Spirometry and Ventilation

T_a (°C)	PsH_2O	STPD	BTPS 37	BTPS 38	BTPS 37i
18	15.477	0.919	1.112	1.120	1.124
19	16.477	0.915	1.107	1.115	1.119
20	17.535	0.910	1.102	1.109	1.115
21	18.650	0.906	1.096	1.104	1.110
22	19.827	0.901	1.091	1.099	1.106
23	21.068	0.897	1.085	1.093	1.101
24	22.377	0.892	1.080	1.087	1.096
25	23.756	0.887	1.074	1.082	1.092
26	25.209	0.883	1.069	1.076	1.087
27	26.739	0.878	1.063	1.070	1.082
28	28.349	0.873	1.057	1.064	1.077

Note: T_a = air temperature; P_sH_2O = vapor pressure in air saturated with water vapor at specific temperature; STPD = temperature and pressure correction from saturated ambient air; BTPS 37 = correction for saturated gas in a spirometer at the specified spirometer temperature assuming saturated air at body temperature = 37°C; BTPS 38 = same as BTPS 37 but for a body temperature of 38°C; BTPS 37i = correction factor for inspired air via pneumotachograph calibrated with ambient air (room temperature with relative humidity = 50%) to saturated air in the lung at 37°C.

Methods to estimate changes in airway diameter by acoustic reflection can be used as an alternative to direct imaging methods. Changes in the resistance of airways due to constriction, congestion, or mucus plugging can be caused by inhalation of toxic gases or aerosols. As the first line of defense, the nose may be particularly susceptible to a wide variety of inhaled material, although the physiological responses of the nose have not been widely studied in humans (see also Section 3.1). Methods for measuring compliance of the lungs are also described in Table 16.

3.2.3 *Diffusion Capacity*

The exchange of gases between the environment and the lung and tissues depends on the convective exchange of gases between the lung and the environment, the ability of gases to transfer across the alveolar capillary membrane, and the subsequent transport of the gases by the blood to the tissues. The transfer of gases across the alveolar capillary membrane is assessed by measurement of diffusing capacity or transfer factor. Diffusion of a gas is inversely proportional to gas density and diffusion distance and directly proportional to solubility in lung tissue, surface area for diffusion, the diffusion coefficient, and the partial pressure gradient of the gas across the alveolar capillary membrane. Inhaled toxic materials could alter diffusion by increasing the diffusion distance through interstitial edema and by decreasing diffusion surface area by damaging alveolar epithelial cells. Oxygen transfer across the alveolar capillary membrane can be incomplete during severe hypoxia and during exercise. The measurement of diffusion capacity is typically made with carbon monoxide, and the diffusing capacity for oxygen is calculated from this measurement. The diffusion capacity for oxygen can be subdivided into two components, that due to the membrane diffusing capacity and that portion dependant on pulmonary capillary blood volume. Diffusion capacity is increased several-fold during exercise due to the increase in pulmonary blood flow and the more uniform distribution of blood flow within the lung. Various methods of determining diffusing capacity

TABLE 13

Some Indices of Lung Function Derived from Spirometry

Test	Equipment/Definition/ Typical Values	Method	Significance/Interpretation	Ref.
Forced expiratory maneuver	Spirometer, pneumotachograph, or other flow sensor, usually as part of a computer-based system (see Table 14 for normal values)	The subject, while breathing through a mouthpiece, inspires maximally and then expires forcefully and as rapidly and completely as possible and continues the maneuver for at least 6 sec (in healthy subjects) or until volume no longer increases.	FVC should be similar to VC or SVC and should not deviate by more than 5% from the second-best maneuver. FVC will typically be slightly smaller than SVC.	1–3, 7
Forced expired volume in t sec, FEV_t	FEV_1: amount of air expired in the first second of the forced expiratory maneuver; also FEV_2 and FEV_3 (see Table 14 for normal values)	Maximal inspiration and maintenance of expiratory flow without cough are necessary for optimal results. Extension of the neck and consistent forced effort are also needed. Flow maintenance is more important than extreme effort over the final portion of the maneuver.	Reduced flow occurs with airway obstruction or volume restriction. Reduced FEV_t can also indicate poor effort or improper technique. In obstruction, the ratio of FEV_t/FVC will be low, whereas in restrictive responses, the ratio will be in the normal range. Submaximal effort can result in increased FEV_1. Very reproducible and widely used.	1–3, 7
FEV_t/FVC	Ratio	Calculated ratio: FEV_1/FVC or FEV_3/FVC	Normal adult range 75–85%. Ratio decreases with increasing age and vital capacity; high values may occur with restrictive disease or absence of sustained effort. Ratios below 70% are usually indicative of airway obstruction or loss of elastic recoil. FEV_3/FVC may be useful in detecting changes in small airways.	1–3, 7
Forced expired flow % ($FEF_{XX\%}$; also: VXX%VC)	$FEF_{XX\%}$: forced expired flow at XX% of vital capacity (expired); $\dot{V}25\%VC = FEF_{75\%}$	Flow rates may be expressed over volume ranges (e.g., 200–1200 ml) or relative ranges (e.g., 25–75% or middle half of FVC) or at relative points in the expiratory maneuver (e.g., $FEF_{75\%}$), indicating instantaneous flow at that point.	Flow rates during the early phase of forced expiration are effort dependant, in part. $FEF_{200-1200}$ and PEF will be reduced in obstruction. Poor effort will result in decreased values, although submaximal effort may result in paradoxical increase in flow. Flow measurements are typically more variable than timed volumes.	1–3, 7
Maximal midexpiratory flow: $FEF_{25-75\%}$	Mean flow over the middle half of the vital capacity (for prediction, see Table 15)	See Figure 2. In cases where the TGV may change, the flow can be expressed relative to the original absolute lung volume. (i.e., %TLC)	$FEF_{25-75\%}$ indicates the flow characteristics of medium to small airways and is reduced in cases of airway obstruction. Note that $FEF_{25-75\%}$ can be reduced by a decrease in VC (also TLC). Isovolume measurements (i.e., at the same absolute lung volume) ensure that flow represents primarily obstruction. Value typically derived from curve with the largest sum of FVC + FEV_1.	1–3, 7
Peak expiratory flow rate (PEFR) or FEF_{max}	Maximum flow achieved during expiration; peak flow meter is required; may also be taken from flow-volume curve if system has good frequency response	A brief maximal blast into the peak flowmeter is sufficient. Small portable peak flowmeters can be used to track function on a regular basis at home or in the workplace. Peak flow achieved in the maximal expiratory maneuver is an index of effort.	Highly effort-dependent test. Excess variability in PEFR can be indicative of airway hyperresponsiveness. Peak flows from flow-volume curves may be lower than with peak flowmeters. This test may be useful in tracking responses to inhaled toxic materials over a prolonged period of time.	1–3, 7

Forced inspiratory maneuver	A maximal inspiratory effort beginning at minimum lung volume (RV) and inhaling to a maximal inspiratory position	Can be performed independently but is best done immediately after the MEFV maneuver to ensure minimum volume. Corresponding indices include $FIV_{0.5}$, FIF_{max}, FIVC, $FIF_{50\%}$, etc.	Not widely used clinically. Useful in detecting upper airway obstruction. Useful in identifying responses to inhaled substances that stimulate receptors that impede or impair maximal inspiration.	1,2
Density dependance of MEFV	Alteration in flow-volume relationships when a less dense gas, such as helium (in helium-oxygen mixture), is breathed	After 3–5 full breaths of a 80%/20% helium-oxygen mixture is breathed, an MEFV maneuver is performed. The helium and air curves are overlaid and differences in flow at 50% and 25% of VC (Δ50% and Δ25%) and the crossover point (V of isoflow) are determined.	Turbulent flow dominates larger airways and is thus improved by less dense gas. Large differences in 50% and 25% are expected in healthy lungs. Small or negative differences indicate flow limitation in small airways. A large volume of isoflow also indicates small airway flow limitations. Reproducibility is poor compared to other tests.	1–3
MVV	Maximal voluntary ventilation	Subject breathes in and out as rapidly and forcefully as possible, usually over a period of 12–15 sec. Breathing rate should exceed 75/min. Standardized rates should be used when tests will be repeated. Tests may also be conducted over longer periods, up to 4 min. In such cases, supplemental CO_2 may be necessary.	The test is highly effort dependent. It is a useful test of the overall breathing system and tests the ability to sustain a high level of ventilation, as may be necessary in heavy exercise. MVV can be reduced by airway obstruction, loss of elastic recoil, hyperinflation, or muscle weakness. Tests conducted over sustained periods may be used to assess ventilatory muscle fatigue or conversely ventilatory muscle endurance. Typical maximal exercise is < 70–80% of MVV but may be higher in exceptional athletes for short periods of time.	1–3
Flow-volume (MEFV) curves	The graphical representation of flow (Y) and volume (X) derived from a maximal expiratory maneuver and the derived indices (see above).	Typically displayed as maximal inspiratory (MIFV) and expiratory (MEFV) loops. See Figure 3. Visual evaluation of the overall shape of the curve as well as various parameters illustrated in Figure 3 are useful indicators, as noted above. For compression-free techniques, see Table 16.	Shape and area parameters are useful visual indicators of response. A linear segment from 75% to 0% of VC is normal. A concave downward curvilinear segment in the same range is indicative of obstruction. Increase in the ratio, ½ ($FEF_{50\%}/FEF_{25\%}$), is indicative of this shape change. Rapid termination of the MEFV curve may indicate poor effort or may be indicative of a mechanical limit to expiration in a healthy young subject.	4,5
Partial EFV curves	Similar to above	The test is similar to an MEFV maneuver but is initiated after a normal inspiration. Flows may be compared at the same absolute lung volume in both MEFV and PEFV curves.	Deep inhalation may cause a reduction in airway resistance and thus alter the measured variables in the expiratory maneuver. PEFV curves begun without a deep breath permit evaluation of flow at low lung volumes not influenced by lung inflation. Inhalation of some toxicants may cause a greater effect on PEFV curves than on MEFV curves.	6

[1] Ruppel, G. L., *Manual of Pulmonary Function Testing,* Mosby, St. Louis, 1991.
[2] Cotes, J. E., *Lung Function: Assessment and Application in Medicine,* Blackwell Scientific, Oxford, 1993.
[3] Clausen, J. L., Ed., *Pulmonary Function Testing: Guidelines and Controversies,* Academic Press, New York, 1982.
[4] Hyatt, R. E., Rodarte, J. R., Mead, J., and Wilson, T. A., Changes in lung mechanics: flow-volume relations, in *The Lung in Transition Between Health and Disease,* Macklem, P. T. and Permutt, S., Eds., Marcel Dekker, New York, 1979, pp. 73–112.
[5] Permutt, S. and Menkes, H. A., Spirometry: analysis of forced expiration within the time domain, in *The Lung in Transition Between Health and Disease,* Macklem, P. T. and Permutt, S., Eds., Marcel Dekker, New York, 1979, pp. 113–152.
[6] Fish, J. E., Ankin, M. G., Kelly, J. F., and Peterman, V. I., Regulation of bronchomotor tone by lung inflation in asthmatic and non-asthmatic subjects, *J. Appl. Physiol.,* 50, 1079–1086, 1981.
[7] American Thoracic Society, Lung function testing: selection of reference values and interpretive strategies, *Am. Rev. Respir. Dis.,* 144, 1202–1218, 1991.

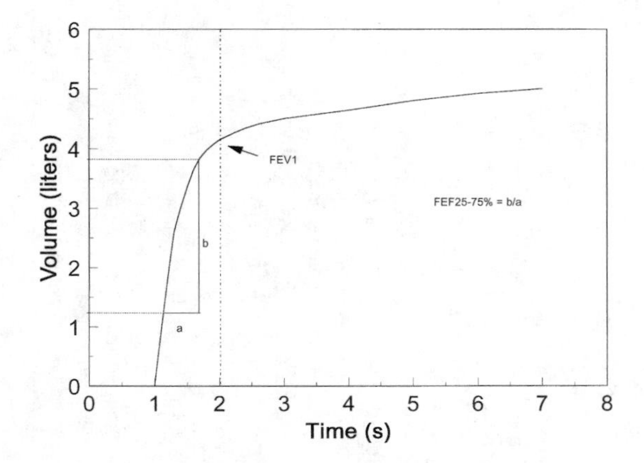

FIGURE 2
Volume-time tracing from a forced expiratory maneuver. The graphical method of determining the forced expired volume in one second (FEV_1) and maximal midexpiratory flow ($FEF_{25-75\%}$) (b/a) is illustrated.

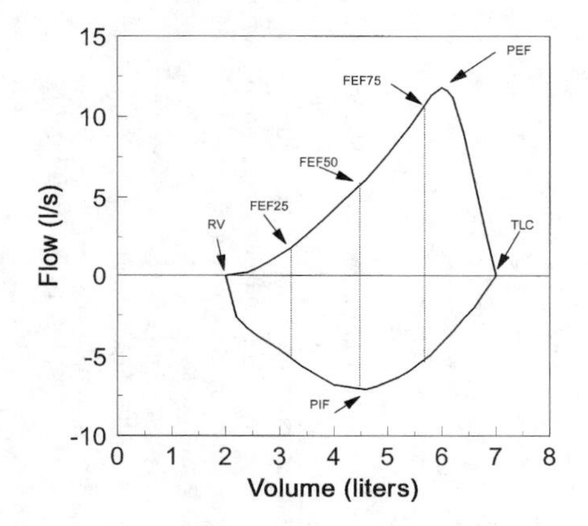

FIGURE 3
Maximal inspiratory and expiratory flow volume loops. Tidal breathing is not illustrated. Peak expiratory flow is achieved early in expiration, whereas peak inspiratory flow is achieved near midinspiration. Some of the various indices that can be derived from the flow-volume curve are illustrated. The vertical dotted lines divide the FVC into fourths.

are briefly summarized in Table 17. Table 18 lists some properties of gases that are relevant to gas exchange.

3.2.4 Gas Exchange

Adequate distribution of inhaled gas within the lung improves gas exchange. Ideally, ventilation needs to be matched with perfusion. Table 19 summarizes some of the methods used to evaluate distribution of gases within the lung. The general principle used in each of these methods is to evaluate the washout of gas from the lung. This may done with a single-breath technique (single-breath nitrogen washout or closing volume) or a multiple-breath technique (multiple-breath nitrogen or xenon washout). The xenon-133 washout method, in combination with gamma camera imaging, can also be used to evaluate distribution of pulmonary blood flow when the xenon is dissolved in saline and administered intravenously. The distribution of ventilation-perfusion ratios can also be estimated using a multiple inert gas elimination technique.[26]

TABLE 14

FVC and FEV_1 in Male and Female African-Americans and Caucasions

	Black Females		Black Males		White Females		White Males	
Height (cm)	FEV_1	FVC	FEV_1	FVC	FEV_1	FVC	FEV_1	FVC
1.500	2.520	2.780	2.635	2.875	2.725	3.130	2.630	3.125
1.525	2.591	2.876	2.745	3.029	2.810	3.237	2.780	3.312
1.550	2.662	2.971	2.856	3.183	2.896	3.344	2.930	3.499
1.575	2.733	3.067	2.966	3.336	2.981	3.451	3.080	3.685
1.600	2.804	3.162	3.076	3.490	3.066	3.558	3.230	3.872
1.625	2.875	3.258	3.186	3.644	3.151	3.665	3.380	4.059
1.650	2.946	3.353	3.297	3.798	3.237	3.772	3.530	4.246
1.675	3.017	3.449	3.407	3.951	3.322	3.879	3.680	4.432
1.700	3.088	3.544	3.517	4.105	3.407	3.986	3.830	4.619
1.725	3.159	3.640	3.627	4.259	3.492	4.093	3.980	4.806
1.750	3.230	3.735	3.738	4.413	3.578	4.200	4.130	4.993
1.775	3.301	3.831	3.848	4.566	3.663	4.307	4.280	5.179
1.800	3.372	3.926	3.958	4.720	3.748	4.414	4.430	5.366
1.825	3.443	4.022	4.068	4.874	3.833	4.521	4.580	5.553
1.850	3.514	4.117	4.179	5.028	3.919	4.628	4.730	5.740
1.875	3.585	4.213	4.289	5.181	4.004	4.735	4.880	5.926
1.900	3.656	4.308	4.399	5.335	4.089	4.842	5.030	6.113
1.925	3.727	4.404	4.509	5.489	4.174	4.949	5.180	6.300
1.950	3.798	4.499	4.620	5.643	4.260	5.056	5.330	6.487

Note: Representative values for FVC (forced vital capacity) and FEV_1 (air expired in 1 sec of forced expiratory volume) for young adult Euro-American and Afro-American ethnic groups aged 18–35 years. For older subjects, see prediction equations in Table 15. Values for Asian-Americans will be intermediate between black and white predicted values.

Source: Adapted from McDonnell, W. F. and Seal, E., Relationship between lung function and physical characteristics in young black and white males and females, *Eur. Respir. J.*, 4, 279–289, 1991.

3.2.5 *Ventilatory Control*

Some of the many reflexes that control ventilation and the breathing pattern are influenced by receptors in the lung. Inhaled toxic materials may stimulate, suppress, or damage these receptors and hence lead to an altered pattern of breathing. Stimulation of these receptors can also alter heart rate and blood pressure. Table 20 summarizes some methods that have been used to evaluate the control of breathing in man. Many irritants induce a pattern of rapid shallow breathing which can be quantified by measurement of the breathing pattern, including volume flow, and duration of various parts of the respiratory cycle. In humans, this is most effectively done during exercise. Table 21 presents data illustrative of the volume, flow, and timing components of the breathing pattern associated with exercise in humans. Changes in ventilation with progressive incremental exercise tests (anaerobic threshold) may also be useful in evaluating responses to substances that could alter oxygen transport. Ventilation may also be stimulated by hypercapnia or hypoxia. Changes in the breathing pattern induced by irritants or other substances may be more easily evaluated in combination with the increased ventilation associated with exercise, hypercapnia, or hypoxia. Measurement of occlusion pressure can provide an estimate of central ventilatory drive independent of changes in ventilation. The separate measurement of nasal and oral ventilation can be useful in dosimetry, particularly nasal dosimetry, of inhaled substances.

TABLE 15

Normal Values for FVC, FEV, and $FEF_{25-75\%}$

Test	Height (cm)	Age (years)	Constant	Gender/Age	Reference Value	Ref.
FVC	0.062	–0.024	–5.38	M/25	4.87	4
	0.0583	–0.025	–4.241	M/25	5.34	6
	0.0600	–0.0214	–4.65	M/25	5.32	1
	0.0567	–0.0206	–4.37	M/70	4.11	3
	0.037	–0.018	–2.171	F/25	3.39	4
	0.0453	–0.024	–2.852	F/25	3.91	6
	0.0491	–0.0216	–3,590	F/25	3.85	1
	0.0365	–0.0330	–0.70	F/70	2.92	3
FEV_1	0.046	–0.028	–3.18	M/25	4.17	4
	0.0362	–0.032	–1.26	M/25	4.28	6
	0.0414	–0.0244	–2.19	M/25	4.45	1
	0.035	–0.018	–1.845	M/60/Jp	3.20	5
	0.0378	–0.0271	–1.73	M/70	2.99	3
	0.028	–0.021	–1.066	F/25	2.96	4
	0.035	–0.026	–1.932	F/25	3.13	6
	0.0342	–0.0255	–1.578	F/25	3.34	1
	0.0281	–0.0325	–0.09	F/70	2.20	3
FVC	HT^2 (1.75 – 0.287 SEX – 0.00135 AGE – 0.0001008 AGE^2)			M/25	5.16	2
				F/25	3.15	
				M/70	3.82	
				F/70	2.53	
FEV_1	HT^2 (1.541 – 0.209 SEX – 0.00406 AGE – 0.0000614 AGE^2)			M/25	4.29	2
				F/25	3.69	
				M/70	2.93	
				F/70	1.97	
$FEF_{25-75\%}$	0.0185	–0.045	+2.513	M/25	4.63	6
	0.0204	–0.038	+2.133	M/25	4.65	1
	0.0236	–0.030	+0.551	F/25	3.64	6
	0.0154	–0.046	+2.683	F/25	4.04	1

Note: All equations take the form: A (Ht cm) – B (Age yr) ± Constant, except for Dockery et al. equations, which include a gender term. Reference values are for a 25- (70)-year-old, 175 cm tall male, or a 25- (70)-year-old, 162.5 cm tall female. Enright equations are for ages 65–85. Marcus equations are for Japanese-American males, 45–68 years old. Morris and Crapo equations are for lifetime nonsmoker Caucasians. Predicted values for Afro-Americans are typically 10–15% lower than for Euro-Americans. Asian-American values are intermediate between Afro-Americans and Euro-Americans.

[1] Crapo, R. O., Morris, A. H., and Gardner, R. M., Reference spirometric values using techniques and equipment that meets ATS recommendations, *Am. Rev. Respir. Dis.*, 123, 659–664, 1981.

[2] Dockery, D. W., Ware, J. H., Ferris, B. G., Glicksberg, D. S., Fay, M. E., Spiro, A., and Speizer, F. E., Distribution of forced expiratory volume in one second and forced vital capacity in healthy white adult never-smokers in six U.S. cities, *Am. Rev. Respir. Dis.*, 131, 511–520, 1985.

[3] Enright, P. L., Kronmal, R. A., Higgins, M., Schenker, M., and Haponik, E. F., Spirometry reference values for women and men 65 to 85 years of age, *Am. Rev. Respir. Dis.*, 147, 125–133, 1993.

[4] Higgins, M. W. and Keller, J. B., Seven measures of ventilatory lung function, *Am. Rev. Respir. Dis.*, 108, 258–272, 1973.

[5] Marcus, E. B., Buist, A. S., Curb, J. D., McClean, C. J., Reed, D. M., Johnson, L. R., and Yano, K., Correlates of FEV_1 and prevalence of pulmonary conditions in Japanese-American men, *Am. Rev. Respir. Dis.*, 138, 1398–1404, 1988.

[6] Morris, J. F., Koski, A., and Johnson, L. C., Spirometric standards for healthy non-smoking adults, *Am. Rev. Respir. Dis.*, 103, 57–67, 1971.

TABLE 16

Tests Used to Assess Resistance and Compliance

Test	Equipment/Definition	Method	Significance/Interpretation	Ref.
Body plethysmography (pressure box)	Rigid body plethysmograph, flow transducer-shutter assembly, pressure transducers (three), venting system, pressure ballast, computer data acquisition system and software	In closed box, subject pants on mouthpiece with shutter open and again with shutter closed. R_{aw} is determined from the ratios of P_{alv}:P_{box} and P_{mo}:P_{box}. Resistance increases with increasing flow; a standard flow rate of 0.5 l/sec is used. R_{aw} also varies with panting rate; a fixed rate between about 70 and 90 bpm should be used. These measurements can also be performed during quiet breathing, although R_{aw} may be higher.	The panting technique minimizes changes in temperature and humidity of respired gas and helps to maintain an open glottis, thus minimizing the effect of upper airway structures (i.e., glottis and larynx) on measured resistance. The panting maneuver may be difficult for some subjects but most can perform it with practice. Resistance can be determined for both inspiratory and expiratory loops, although they are usually equal in healthy humans. R_{aw} measurements may be useful in evaluating responses to airway challenge, although they are probably no better than spirometry and are more cumbersome. Changes in R_{aw} reflect primarily changes in more central airways. Typical range is 0.2 to 2.5 cmH_2O/1/sec.	1–3
Body plethysmography (volume box)	As above, but with an additional flow transducer or a rapid responding spirometer to measure changes in box volume transmurally	Principally used to examine gas compression effects on flow volume curves. Using the box, the true flow-volume curve of the thorax free of gas compression can be evaluated.	The use of this technique has not been exploited in the evaluation of inhaled toxicants. In patients with obstruction, the flow-volume curve measured by a plethysmograph may differ considerably from that measured at the mouth.	1–3
Forced oscillation technique	Oscillating pump or loudspeaker is used to provide a flow oscillation to the lung; a flow transducer and pressure transducer	Subject breathes normally while an oscillation is applied to the mouth or chest wall. The frequency range of the oscillator should be in the range of 4–32 (or higher) Hz. Oscillation amplitude is about 40 ml.	The pressure developed during forced oscillation is a function of the total impedance of the pulmonary system, including resistance, compliance, and inertance. The compliance component decreases with frequency and the inertance component increases with frequency. There is a resonant frequency (~8 Hz) at which compliance and inertance cancel each other and the impedance is almost entirely due to resistance. The tissue resistance can be determined by subtracting the airway resistance determined plethysmographically from the total resistance determined by forced oscillation.	2
Pressure-flow method	Esophageal balloon catheter, pressure transducer, flow transducer, spirometer; shutter for compliance measurements	Alveolar pressure is estimated from esophageal pressure, and resistance is calculated accordingly. Compliance of the lung can also be determined using P_{es} as an estimate of intrapleural pressure at different lung volumes.	Positioning of the esophageal balloon and its volume are of critical importance in this technique. Many subjects have difficulty swallowing the balloon, and the pressure measurement is affected by balloon volume and position within the esophagus. In addition to pulmonary resistance from the pressure-flow relationship and pulmonary compliance from the pressure-volume relationship, maximum static elastic recoil pressure can be measured. Dynamic compliance can also be determined at various frequencies. These techniques are technically demanding, time consuming, and not often used in the evaluation of acute responses to inhaled substances.	2

TABLE 16 (continued)

Tests Used to Assess Resistance and Compliance

Test	Equipment/Definition	Method	Significance/Interpretation	Ref.
Interrupter method	Rotating shutter alternately closes and opens flow at a rate of ~10–20/sec; pressure and flow transducers	Alveolar pressure is determined during brief interruption of flow by a shutter at the mouth. In theory, when flow = 0, $P_m = P_{alv}$.	The resistance obtained is not purely airway resistance but is closer to total lung resistance. When the airway resistance is high (as in obstructive disease) the pressure at the mouth is not equal to alveolar pressure during the interruption of flow; seldom used.	2
Laryngeal resistance	Low-frequency sound source, flow transducer, pressure transducer, two surface microphones	Sound pressure amplitude is sampled above and below the larynx. Changes in pressure differential reflect changes in laryngeal resistance if lower airway resistance is not altered.	This method may be useful to examine the specific location of changes in airway resistance; not widely used.	8,9
Acoustic reflectance technique for cross-sectional area	Data acquisition computer, a device to produce an acoustic pulse, a rigid wave tube, microphone(s), and signal processing (filter, amplifier, A/D converter)	The procedure involves generation of a multifrequency sound pulse while a subject breathes. By analyzing the reflected sound, the cross-sectional area of the airways up to the carina (possibly farther) can be estimated.	This procedure permits examination of dynamic changes in the central airway in response to provocative agents. It has also been used to examine changes in the diameter of the nasal airway. The measurements made with this technique have been validated radiographically.	4,6,7

1 Ruppel, G. L., *Manual of Pulmonary Function Testing,* Mosby, St. Louis, 1991.

2 Cotes, J. E., *Lung Function: Assessment and Application in Medicine,* Blackwell Scientific, Oxford, 1993.

3 Clausen, J. L., Ed., *Pulmonary Function Testing, Guidelines and Controversies,* Academic Press, New York, 1982.

4 Jackson, A. C., Butler, J. P., Miller, E. J., Hoppin, F. G., and Dawson, S. V., Airway geometry by analysis of acoustic pulse response measurements, *J. Appl. Physiol.,* 43, 523–536, 1977.

5 Coates, A. L., Desmond, K. J., Demizio, D., Allen P., and Beaudry, P. H., Sources of error in flow-volume curves, *Chest,* 94, 976–9892, 1988.

6 Louis, B., Glass, G. M., Kresen, B., and Fredberg, J. J., Airway area by acoustic reflection: the two microphone method, *J. Biomech. Eng.,* 115, 278–285, 1993.

7 Hoffstein, V. and Fredberg, J., The acoustic reflection technique for noninvasive assessment of upper airway area, *Eur. Respir. J.,* 4, 601–611, 1991.

8 Sekizawa, K., Shindoh, C., Hida, W., Suzuki, S., Akaizawa, Y., Shimizu, Y., Sasaki, H., and Takishima, T., Noninvasive method for detecting laryngeal narrowing with low frequency sound, *J. Appl. Physiol.,* 55, 591–597, 1983.

9 White, D. P., Lombard, R. M., Cadieux, R. J., and Zwillich, C. W., Pharyngeal resistance in normal humans: influence of gender, age, and obesity, *J. Appl. Physiol.,* 58, 365–371, 1985.

TABLE 17

Tests Used to Assess Diffusion

Test	Equipment/Definition	Method	Significance/Interpretation	Ref.
Steady-state CO diffusion capacity (D_LCO-SS)	CO analyzer, blood gas analyzer (CO_2), CO_2 gas analyzer, O_2 analyzer, volume measurement, valves, etc.	Subject breathes a constant concentration of CO in air (~1000 ppm). Test usually takes 3–6 min. $D_LCO = \dot{V}_{CO}/P_ACO$. The dead space is used to estimate the P_ACO from expired concentrations.	Method is very sensitive to changes in the dead space:tidal volume ratio. Best suited for exercise testing. Also affected by uneven ventilation distribution and CO back pressure. CO back pressure can be measured by an oxygen rebreathing technique. The physiological dead space can be estimated using the Bohr equation and either end-tidal or arterial CO_2.	2
Single-breath CO diffusion capacity (D_LCO-SB)	CO, CO_2, He analyzers, spirometer, bag-in-box, valves etc.; gas mix containing 0.3% CO, tracer gas (e.g., He), 17–21% CO_2, balance N_2	Subjects breathe out to RV, inhale test gas to total lung capacity (TLC), hold breath for 8–10 sec, then expire fully at a steady rate. Measurements made at high and low O_2 concentrations permit determination of membrane diffusion capacity and pulmonary capillary blood volume.	Most widely used method and available on a variety of computer-based systems. The long breathhold is impractical for exercise and in patients with severe dyspnea. Method requires no blood sample and is relatively easy to perform. Better than steady-state method when ventilation is unevenly distributed. Modifications of this procedure are available that require no breathholding.	1–4
Rebreathing CO diffusion capacity (D_LCO-RB)	Gas mixture containing 0.1% CO, He, O_2, and N_2; Ne can also be used as a trace gas (~0.5%)	Subject begins breathing at RV and rebreathes for 35–45 sec at 30/min. Helium is used to calculate dilution of test gas.	This method is not widely used but may be useful for patients with severe maldistribution of ventilation or very small vital capacity. Method gives results that are systematically lower than the SB method.	2
Diffusion capacity for NO and CO (D_LNO/CO)	CO, NO, He, O_2 analyzers; gas mixture containing 0.1–0.2% CO, ~8 ppm NO, 21% O_2, 3–5% He, balance N_2	The method is similar to the SB technique above. The breathhold (BH) time can be as brief as 3 sec. In relatively healthy subjects, BH time of 3 sec and NO concentration of 8 ppm are suggested.	D_LNO is about 5 times higher than D_LCO. D_LNO is believed to be a good estimate of membrane diffusing capacity. Thus, from the combined measurement, D_m and cardiac output (Q_C) can be determined in the same maneuver. The breathhold time must be brief because the rapid uptake of NO may result in an exhaled concentration that is too low to measure. This method may be useful in exercising subjects.	2,5

1 Ruppel, G. L., *Manual of Pulmonary Function Testing,* Mosby, St. Louis, 1991.
2 Cotes, J. E., *Lung Function: Assessment and Application in Medicine,* Blackwell Scientific, Oxford, 1993.
3 Clausen, J. L., Ed., *Pulmonary Function Testing: Guidelines and Controversies,* Academic Press, New York, 1982.
4 ATC Committee, Single breath carbon monoxide diffusion capacity (transfer factor): recommendations for a standard technique, *Am. Rev. Respir. Dis.,* 136, 1299–1307, 1987.
5 Guenard, H., Varene, N., and Vaida, P., Determination of lung capilliary blood volume and membrane diffusing capacity in man by the measurements of NO and CO transfer, *Respir. Physiol.,* 70, 113–120, 1987.

TABLE 18

Some Properties of Gases Used in Respiratory Tests

Substance	Molecular Weight	Solubility (ml/ml H_2O)[a]	Partition Coefficient[b]	Rate of Diffusion[c]
Oxygen	32	.02386 [0.0214][P]	5.0	1.0
C_2H_2	26	0.747 [0.74][B]	—	34.8
A	40	0.026	5.3	0.97
CO_2	44	0.56 [0.515][P]	1.6	20.3
CO	28	0.0184	—	0.83
He	4	0.0085 [0.0098][B]	1.7	1.01
Kr	85[d]	[0.059][B]	9.6	1.15
Nitrogen	28	0.0123 [0.0147][B]	5.2	0.55
NO	30	0.041	—	1.76
N_2O	44	0.47 [0.473][B]	3.2	13.9
Xe	133[d]	0.085	20.0	1.75

Note: Refer to the following references for more information.

Altman, P. L. and Ditmer, D. S., Eds., *Respiration and Circulation, Biological Handbook,* American Society of Experimental Biology, Bethesda, MD, 1971, pp. 16–21.

Butler, J., Measurement of cardiac output using soluble gases, in *Handbook of Physiology,* Vol. II, Fenn, W. O. and Rahn, H., Eds., American Physiology Society, Washington, D.C., 1965, pp. 1489–1504.

Cotes, J. E., *Lung Function: Assessment and Application in Medicine,* Blackwell Scientific, Oxford, 1993, p. 372. (Table modified from Cotes with permission.)

[a] Solubility coefficient in ml gas per ml fluid at 37 to 38°C; P = plasma; B = blood.
[b] Oil/water or fat/blood partition coefficient.
[c] Rate of diffusion relative to oxygen.
[d] Radioactive isotope.

3.3 Particle Clearance and Permeability Measurement Techniques

Particles are cleared from the lung via two major routes: mucociliary and interstitial pathways. Particles in the ciliated airways are usually cleared from the lung within 24 hours, whereas particles deposited beyond the ciliated airways take from months to years to clear out, primarily via interstitial pathways. Mucus transport velocity (MTV) varies along the ciliated airways and is faster in the proximal airways and slower in the periphery; however, because of the large size of the airway and its easy accessibility, MTV is measured almost exclusively in the trachea. Both MTV and particle clearance rate from the lung are frequently measured to assess the effects of inhaled agents. Four methods of measuring MTV are summarized in Table 22. Three of the four methods utilize direct placement of tracer particles in the trachea, either by insufflation or instillation via a bronchoscope. Movement of individual particles are monitored and recorded using an external detector, and time is measured for particle movement over two reference points in order to determine the transport velocity. The measurement method itself is straightforward, but recorded images of particles are often fuzzy, which may cause an error in identifying particles. Also, use of invasive bronchoscopy to introduce tracer particles may be a confounding factor. The radioaerosol bolus method is noninvasive; however, it is often difficult to generate a well-defined aerosol bolus and even more difficult to deposit the aerosol specifically in airways of the desired size range. In all of these methods, MTV can be measured over a short period of time, usually in an hour; therefore, repeated measurements may be used to assess changes over time. Methods for measuring particle clearance from the lung are summarized in Table 23. The principal method involves inhaling insoluble-radiolabelled aerosols and subsequent monitoring of radioactivity

TABLE 19

Tests Used To Assess Distribution of Ventilation

Test	Equipment/Definition	Method	Significance/Interpretation	Ref.
Single-breath nitrogen washout (SBNW)	Oxygen, bag-in-box system, spirometer/pneumotachograph, nitrogen analyzer/mass spectrometer, recorder/computer, flow-rate indicator; may be part of a system	Subject exhales to RV, then inhales 100% oxygen at a steady rate to total lung capacity (TLC), pauses briefly, and expires to RV, flow rate 0.3–0.5 l/sec. Adequate control of flow and maximum and minimum lung volume are important.	Measurements obtained include slope of the alveolar nitrogen plateau slope of phase III), closing volume (CV), closing capacity (CC = CV + RV) CV/VC, CC/TLC. The flatter the slope of phase III and the smaller the CV, the more uniform is ventilation. CV is large in children, is small or undetectable in adolescents and young adults, and increases with age, smoking, and airway obstruction. Typically, CV = 5–6% of VC.	1–3, 5
SB washout with bolus	Similar to SBNW but uses a bolus of a trace gas (He, Ne, Ar, or Xe) instead of oxygen	A small bolus (~300 ml) of trace gas is given at the start of inspiration from RV. Determine CV with analysis of trace gas expired curve.	CV for trace gas is almost always detectable even if SBNW does not detect CV. CV is usually higher with this method.	6
Multiple-breath nitrogen washout (MBNW)	Oxygen, nitrogen analyzer, volume measurement; an automated open-circuit system that allows control of V_T and f_B is desirable	Subject, at FRC, begins breathing 100% oxygen for 4–7 min. Test is complete when the nitrogen level falls below 1.2%. Leak control is essential.	Rapid clearance of nitrogen is an index of ventilatory uniformity. The FRC can also be calculated. Analysis of the rate of clearance of N_2 can be used to derive indices of ventilatory nonhomogeneity.	2–4
Distribution of inhaled radioactive gases	Gamma scintillation camera, ^{133}Xenon or other gas	Subject rebreathes from bag containing ^{133}Xe for 1–2 min (up to 20 min in COPD) until a stable distribution is achieved, and then the washout of the activity is monitored.	Interpretation of overall washout is similar to other tests. Unique feature is the ability to monitor regional or segmental washout of tracer. A related technique of infusing ^{133}Xe dissolved in saline may be used to examine regional differences in distribution of pulmonary blood flow.	2,7

Note: f_B = breathing frequency; V_T = tidal volume; RV = residual volume; FRC = functional residual capacity; COPD = chronic obstructive pulmonary disease.

1. Ruppel, G. L., *Manual of Pulmonary Function Testing,* Mosby, St. Louis, 1991.
2. Cotes, J. E., *Lung Function: Assessment and Application in Medicine,* Blackwell Scientific, Oxford, 1993.
3. Clausen, J. L., Ed., *Pulmonary Function Testing: Guidelines and Controversies,* Academic Press, New York, 1982.
4. Lewis, S. M., Evans, J. W., and Jalowayski, A. A., Continuous distributions of specific ventilation recovered from inert gas washout, *J. Appl. Physiol.,* 34, 416–423, 1978.
5. Buist, S. A. and Ross, B. J., Predicted values for closing volumes using a modified single breath nitrogen test, *Am. Rev. Respir. Dis.,* 107, 744–752, 1973.
6. Travis, D. M., Green, M., and Don, H., Simultaneous comparison of helium and nitrogen expiratory closing volumes, *J. Appl. Physiol.,* 34, 304–308, 1973.
7. Ball, W. C., Stewart, P. B., Newsham, L. G. S., and Bates, D. V., Regional pulmonary function studied with xenon^{133}, *J. Clin. Invest.,* 41, 519–531, 1962.

TABLE 20

Tests Used To Assess Control of Breathing

Test	Equipment/Definition	Method	Significance/Interpretation	Ref.
Breathing pattern	Flow and pressure transducers, breath detection software (alternatively: magnetometers, capacitance vest)	Measurement of tidal volume, breathing frequency, inspiratory duration, expiratory duration. Permits calculation of numerous additional variables. Also measures instantaneous peak inspiratory and expiratory flow.	Many inhaled toxic materials cause changes in breathing pattern, typically to a rapid shallow or tachypneic mode. In humans, this is best assessed during exercise. Overall ventilation may or may not be altered. Changes in resistive or elastic loading also influence breathing pattern.	1–3
Anaerobic threshold	Standard open-circuit or breath-by-breath oxygen uptake system	Determination of the breakpoint from linearity of the $\dot{V}_E/\dot{V}O_2$ or the $\dot{V}O_2/\dot{V}CO_2$ relationship.	The "ventilatory anaerobic threshold" during incremental exercise is thought to represent the onset of metabolic acidosis due to anaerobic metabolism.	4
Ventilatory response to CO_2	CO_2 analyzer, spirometer, or pneumotachograph, CO_2 mixtures	Steady-state method A: measures ventilation while breathing CO_2 at several different concentrations (2–6% CO_2) for 7–10 min. Steady-state method B: measures V_E while breathing air supplemented with a small, steady flow of 100% CO_2. Rebreathing method: subject rebreathes from a bag of about 7 l containing O_2 and about 7% CO_2.	End-tidal (typically) or arterial CO_2 is plotted against ventilation to determine response slope and intercept. The rebreathing method is the most widely used since it can be performed rapidly and gives results similar to steady-state method A. The steady-state methods are preferable during exercise studies. Simple breathholding methods have also been utilized but are highly dependant on subject cooperation and training.	2,5, 9
Ventilatory response to hypoxia	O_2 analyzer, flow and pressure transducers, valves, etc.	Measures ventilation during progressive isocapnic hypoxia. The hyperbolic response $P_AO_2 - V_E$ response curve is evaluated by its shape. A simplified method is to plot S_aO_2 vs. V_E, which is linear.	Interpretation of these response parameters is difficult since they are highly variable. Alternatively, the effect of hypoxia on CO_2 responses can be informative.	5
Occlusion pressure	Pressure transducer, occlusion device	The pressure in the occluded airway is randomly measured during the first 100 msec of inspiration. Subject must not be aware that the occlusion has occurred prior to the onset of inspiration.	The occlusion pressure is an index of ventilatory drive and can be useful in determining changes in effort that are not reflected in changes in ventilation or flow rate. For example, during elastic loading, the ventilatory drive will increase, but the ventilation may decrease.	6,7
Oral/nasal transition or partitioning	Face mask divided into oral and nasal chambers, flow transducers, pressure transducers	Separately measures the oral and nasal components of ventilation.	The specific nasal and oral ventilation may be useful in evaluating the dose of an inhaled substance which enters via mouth or nose. The transition from nasal-only to oral and nasal ventilation may be altered by irritants, nasal congestion/decongestion, etc.	8

1. Ruppel, G. L., *Manual of Pulmonary Function Testing,* Mosby, St. Louis, 1991.
2. Cotes, J. E., *Lung Function: Assessment and Application in Medicine,* Blackwell Scientific, Oxford, 1993.
3. Clausen, J. L., Ed., *Pulmonary Function Testing: Guidelines and Controversies,* Academic Press, New York, 1982.
4. Schneider, D. A., Phillips, S. E., and Stoffolano, S., The simplified V-slope method of detecting the gas exchange threshold, *Med. Sci. Sports Exer.,* 25, 1180–1184, 1993.
5. Rebuck, A. S. and Slutsky, A. S., Measurement of ventilatory responses to hypercapnia and hypoxia, in *Regulation of Breathing,* Hornbein, T. F., Ed., Marcel Dekker, New York, 1981, pp. 745–772.
6. Ward, S. A., Aqleh, K. A., and Poon, C.-S., Breath-to-breath monitoring of inspiratory occlusion pressure in humans, *J. Appl. Physiol.,* 51, 520–523, 1981.
7. Whitelaw, W. A., Derenne, J. P., and Milic-Emili, J., Occlusion pressure as a measure of respiratory center output in conscious man, *Resp. Physiol.,* 23, 181–199, 1975.
8. Ninimaa, V., Cole, P., Mintz, S., and Shephard, R., The switching point from nasal to oronasal breathing, *Respir. Physiol.,* 42, 61–67, 1980.
9. Fenn, W. O. and Craig, A. B., Effect of CO_2 on respiration using a new method of administering CO_2, *J. Appl. Physiol.,* 18, 1023–1024, 1963.

TABLE 21

Breathing Patterns at Rest and During Exercise

Condition	No.	V_E (l/min)	V_T (l)	f_B	T_{TOT}	T_I	T_E	T_I/T_{TOT}	V_T/T_I	O_2	CO_2	Ref.
Rest	75			16–18								1
Rest	10	6.8	0.59	11	5.25	1.74	3.51	0.33	0.34			2
Rest	9	5.0	0.32	15	4.07	1.67	2.40	0.40	0.38			5
37 W	9	15.4	0.75	21	2.87	1.16	1.55	0.46	1.11	0.80	0.68	5
37 W	6F	9.5	0.50	24	2.50	1.10	1.40	0.44	0.47			5
60 W	11	23.2	1.15	20	3.00	1.22	1.78	0.41	0.94			7
90 W	11	28.7	1.32	22	2.80	1.16	1.64	0.41	1.14			7
120 W	11	35.1	1.52	23	2.61	1.12	1.49	0.43	1.36			7
150 W	11	42.0	1.66	25	2.39	1.04	1.35	0.44	1.60			7
180 W	11	48.9	1.84	27	2.27	1.01	1.26	0.45	1.82			7
210 W	11	57.7	2.01	29	2.11	0.94	1.17	0.45	2.14			7
240 W	11	67.2	2.11	32	1.89	0.85	1.04	0.45	2.48			7
270 W	11	82.3	2.12	39	1.55	0.72	0.83	0.46	2.92			7
Max	11	100.5	2.19	46	1.32	0.62	0.70	0.47	3.53			7
45% max	14	41.3	1.53	27	2.33	1.12	1.21	0.48	1.37	1.55	1.36	4
60% max	14	60.4	1.88	32	1.93	0.97	0.96	0.50	1.94	2.26	2.06	4
75% max	14	77.7	2.04	38	1.66	0.86	0.80	0.52	2.38	2.69	2.56	4
100% max	14	134.4	2.30	58	1.03	0.53	0.50	0.51	4.34	3.55	3.94	4
100% max	14T	151.7	2.58	59	1.03	0.53	1.50	0.51	4.87	4.52	5.01	4
100% max	5	136.0	2.77	49	1.27	0.62	0.65	0.49	4.36	3.53	4.23	3
100% max	7T	183.0	2.91	63	0.96	0.50	0.46	0.52	5.82	5.39	5.86	3
100% max	8TF	109.9	2.21	50	1.20	0.49	0.71	0.41	4.52	—	3.77	6

Note: F = female; T = trained; W = watts; max = $\dot{V}O_2$ max; V_E = ventilation in l/min; V_T = tidal volume; f_B = breathing frequency; T_{TOT} = duration of total breath cycle(s); T_I = duration of inspiration(s); T_E = duration of expiration(s); T_I/T_{TOT} = fractional inspiratory duty cycle; V_T/T_I = mean inspiratory flow; O_2 = oxygen uptake; CO_2 = CO_2 output.

[1] Mead, J., Control of respiratory frequency, *J. Appl. Physiol.*, 15, 325–336, 1960.

[2] Bechbache, R. R., Chow, H. H. K., Duffin, J., and Orsini, E. C., The effects of hypercapnia, hypoxia, exercise and anxiety on the pattern of breathing in man, *J. Physiol.*, 293, 285–300, 1979.

[3] Folinsbee, L. J., Wallace, E. S., Bedi, J. F., and Horvath, S. M., Exercise respiratory pattern in elite cyclists and sedentary subjects, *Med. Sci. Sports Exer.*, 15, 503–509, 1983.

[4] Joyner, M. J., Jilka, S. M., Taylor, J. A., Kalis, J. K., Nittolo, J., Hicks, R. W., Lohman, T. G., and Wilmore, H. H., β-Blockade reduces tidal volume during heavy exercise in trained and untrained men, *J. Appl. Physiol.*, 62, 1819–1825, 1987.

[5] Goldstein, S. A., Weissman, C., Askanazi, J., Rothkopf, M., Milic-Emili, J., and Kinney, J. M., Metabolic and ventilatory responses during very low level exercise, *Clin. Sci.*, 73, 417–424, 1987.

[6] Szal, S. E. and Schoene, R. B., Ventilatory response to rowing and cycling in elite oarswomen, *J. Appl. Physiol.*, 67, 264–269, 1989.

[7] Lind, F. G. and Hesser, C. M., Breathing pattern and occlusion pressure during moderate and heavy exercise, *Acta Physiol. Scand.*, 122, 61–69, 1984.

in the lung as a function of time over a period of at least 24 hr or longer. The clearance curve (lung activity vs. time) consists of two phases: fast and slow clearance. The fast phase represents clearance from the ciliated airways, whereas the slow phase represents clearance from the alveolar regions. The fraction of particles cleared from the conducting airways is determined by the *y*-intercept of a horizontal line drawn from the 24-hr clearance value or an extension of the slope of the slow phase (see Figure 4). The same measurement procedure can be used with magnetite aerosol particles. Magnetic particles may pose less of a health risk than radioactive particles, but the magnetic particle detection system is very complex and of very limited accessibility.

Pulmonary permeability is frequently measured to assess airway injury or structural integrity. Although this methodology does not provide the information required for the computation of permeability coefficients, the measurement obtained is considered to reflect

TABLE 22

Techniques for Measuring Mucociliary Clearance from the Lung

	Radio-aerosol Scintigraphy[1-10]	Magnetopneumography[11,12]
Aerosol materials: material names followed by gamma-emitting radionucleids used and a half-life of the nucleids in hours (hr), days (d), minutes (min)	Teflon: ^{99m}Tc (6.04 hr) Sulfur colloid: ^{99m}Tc Fe_2O_3: ^{99m}Tc, ^{198}Au (2.7 d), ^{51}Cr (27.7 d) Human serum albumin: ^{99m}Tc, ^{131}I (8.04 d), ^{125}I (59.4 d) Human albumin microsphere: ^{99m}Tc Polystyrene latex: ^{99m}Tc, ^{82}Br (35.4 hr), ^{131}I, ^{51}Cr Gold colloid: ^{198}Au, $^{54}MnO_2$ (312.2 d), ^{59}Fe (45 d), ^{38}F (112 min)	Ferromagnetic particles: Fe_3O_4 (magnetite)
Aerosol generation	Nebulization of liquid suspension of insoluble radiolabelled particles or generation of monodisperse aerosols with spinning top generator. Particle size ranges from 0.5–5 μm in diameter.	Same as column 2 and direct dispersion of dry powder
Inhalation procedure	Inhalation of the radiolabelled aerosols until the initial lung burden amounts to 30,000–50,000 counts per min. Normal spontaneous breathing or controlled breathing is used to achieve a suitable initial lung deposition.	Inhalation of magnetite aerosol until an initial lung burden of about 0.02 mg
Initial lung scan[8-10]	Equilibrium lung scan with Xe to define the lung boundary and regional lung volume, i.e., central (C) vs. peripheral (P), prior to aerosol inhalation. Whole lung scan with gamma camera is done immediately after aerosol inhalation, as well as calculation of the C/P ratio as an index of lung deposition pattern.	—
Retention measurement	Whole lung activity count with widely collimated scintillation probes or with a gamma camera every 15 min for the first hour followed by an hourly measurement for a period of 6 hr and measurements after 24 and 48 hr; for long-term clearance studies, an hourly activity count is done for the first few hours after inhalation followed by a daily measurement for several days and weekly or monthly measurements over a period of up to one year.	Magnetization of deposited particles (~20 sec) in the lung with a strong (i.e., 660 *g*) magnet followed by 3-point RMF (remnant magnetic field) measurements (2 min each) in a shielded room with a SQUID (superconducting quantum interference device); timing of the RMF measurement is the same as column 2
Data plotting and analysis	Construction of a retention curve: normalized chest retention (% of initial chest count) after correction for physical decay as a function of time. Two-phase analysis of the retention curve: a fast clearance phase for the initial 24 hr followed by a slow clearance phase thereafter. Each phase may be fitted with a single exponential function (see Figure 1).	Same as column 2, except that chest retention is assessed by RMF measurement

TABLE 22 (continued)

Techniques for Measuring Mucociliary Clearance from the Lung

	Radio-aerosol Scintigraphy[1-10]	Magnetopneumography[11,12]
Tracheobronchial clearance	Defined by the fraction of activity cleared during the initial 24 hr or the fraction determined by extrapolating the slow clearance curve to t = 0 (see Figure 1). The rate of tracheobronchial (TB) clearance is determined by the slope of the fast phase curve and is often expressed by $T_{1/2}$, the time required to clear half of the TB deposition.	Same as column 2.
Alveolar clearance	Alveolar deposition fraction at t = 0 is obtained by 1 – TB deposition fraction, defined above. The rate of clearance is then obtained by the slope of the slow phase curve. $T_{1/2}$ of alveolar clearance in normal lungs ranges from 60–300 days.	Same as column 2
Interpretation/significance	Clearance measurements are variable with the initial lung deposition pattern that is difficult to control. Potential errors in TB and alveolar clearance if particles deposited in the TB fail to clear out completely within 24 hr. A long-term retention study may not be warranted because of potential health risks and also detectability of low activity.	Same as column 2; however, a long-term retention study may be permissible because of a minimal health risk of magnetite particles

1 Morrow, P. E., Gibb, F. R., and Gazioglu, K. M., A study of particle clearance from the human lungs, *Am. Rev. Respir. Dis.*, 96, 1209–1221, 1967.

2 Albert, R. E., Lippmann, M., Peterson, H. T., Berger, J., Sanborn, K., and Bohning, D., Bronchial deposition and clearance of aerosols, *Arch. Intern. Med.*, 131, 115–127, 1973.

3 Lippmann, M. and Albert, R. E., The effect of particle size on the regional deposition of inhaled aerosols in the human respiratory tract, *Am. Ind. Hyg. Assoc. J.*, 30, 257–275, 1969.

4 Stahlhofen, W., Gebhart, J., and Heyder, J., Experimental determination of the regional deposition of aerosol particles in the human respiratory tract, *Am. Ind. Hyg. Assoc. J.*, 41, 385–398, 1980.

5 Baily, M. R., Fry, F. A., and James, A. C., Long-term retention of particles in the human respiratory tract, *J. Aerosol Sci.*, 6, 295–305, 1985.

6 Pavia, D., Bateman, J. R. M., and Clarke, S. W., Deposition and clearance of inhaled particles, *Bull. Eur. Physiopathol. Resp.*, 16, 335–366, 1980.

7 Albert, R. E., Lippman, M., Spiegelman, J., Strehlow, C., and Briscoe, W., The clearance of radioactive particles from the lung, in *Inhaled Particles and Vapors*, Vol. II, Davis, C. N., Ed., Pergamon Press, London, 1967, pp. 361–377.

8 Thompson, M. L. and Pavia, D., Particle penetration and clearance in the human lung, *Arch. Environ. Health*, 29, 214–219, 1974.

9 Sanchis, J., Dolovich, M., Chalmers, R., and Newhouse, M., Quantitation of regional aerosol clearance in the normal human lung, *J. Appl. Physiol.*, 33, 757–762, 1972.

10 Smaldone, G. C., Perry, R. J., Bennett, W. D., Messinna, M. S., Zwang, J., and Ilowite, J., Interpretation of 24-hour lung retention in studies of mucociliary clearance, *J. Aerosol Med.*, 1, 11–20, 1988.

11 Cohen, D., Arai, S., and Brain, J. D., Smoking impairs long-term dust clearance from the lung, *Science*, 204, 514, 1979.

12 Cohen, D., Ferromagnetic contamination in the lungs and other organs of the human body, *Science*, 180, 745–748, 1973.

TABLE 23

Techniques for Measuring Mucus Transport Velocity

Cinebronchofiberscopic Techniques[4]	Roentgenographic Techniques[2]	Radioisotopic Technique Utilizing a Bronchoscope[5]	Radioaerosol Boli Technique[6]
Insufflation of Teflon discs (0.68 mm in diameter, 0.13 mm in thickness, and 0.13 mg) onto the tracheal mucosa through the inner channel of the fiberoptic bronchoscope; filming of the movement of the discs via bronchoscope; time measurement for the cephalic movement of the discs (10–20 discs) between two reference points	Insufflation of radio-opaque bismuth trioxide-impregnated Teflon discs (1 mm in diameter, 0.8 mm thick, and 1.8 mg) onto the tracheal mucosa through the inner channel of the fiberoptic bronchoscope; external viewing of the discs via a fluoroscope and videotaping of the movement of the particles; time measurement for particle movement over two reference points	Instillation of a minute suspension of ^{99m}Tc-labelled albumin microspheres (5–7 µm in diameter) on the mucosal surface at the lower end of the trachea via the bronchoscope; taking sequential pictures with a gamma camera at 1- to 2-min intervals until particles reach the larynx; divide the length of the trachea by time to measure the tracheal mucus velocity	Inhalation of radiolabelled aerosols (5–7 µm in diameter) with special breathing maneuvers, i.e., fast and shallow near TLC or end inspiration delivery followed by breathhold to ensure deposition of the particles mainly in the large airways; monitoring and recording of particle boli transport using a gamma camera or a specially designed multidetector probe located over the trachea.
Normal values: 21.5 ± 5.5 mm/min[5]	Normal values: 11.4 ± 3.8 mm/min; values in various subject groups[3]	Normal values: 15.5 ± 0.7 mm/min	Normal values: 4.4 ± 1.3 mm/min
Advantages: Short observation period, direct and precise positioning of tracer particles; *disadvantages:* invasive, limited to tracheal measurement	*Advantages:* same as the first column; *disadvantages:* same as the first column	*Advantages:* same as the first column; *disadvantages:* same as the first column	*Advantage:* non-invasive; *disadvantage:* difficult to achieve well-defined particle boli

1 Chopra, S. K., Taplin, G. V., Elam, D., Carson, S. W., and Golde, D., Measurement of tracheal mucociliary transport velocity in humans: smokers vs. nonsmokers, *Am. Rev. Respir. Dis.*, 119(Suppl.), 205, 1979.

2 Friedman, M., Stott, F. D., Poole, D. O., Dougherty, R., Chapman, G. F., Watson, H., and Sackner, M. A., A new roentgenographic method for estimating mucous velocity in airways, *Am. Rev. Respir. Dis.*, 115, 67–72, 1977.

3 Goodman, R. M., Yergin, B. M., Landa, J. F., Golinvaux, M. H., and Sackner, M. A., Relationships of smoking history and pulmonary function tests to tracheal mucous velocity in nonsmokers, young smokers, ex-smokers, and patients with chronic bronchitis, *Am. Rev. Respir. Dis.*, 117, 205–214, 1978.

4 Sackner, M. A., Rosen, M. J., and Wanner, A., Estimation of tracheal mucous velocity by bronchofiberscopy, *J. Appl. Physiol.*, 34, 495–499, 1973.

5 Santa Cruz, R., Landa, J., Hirsch, J., and Sackner, M. A., Tracheal mucous velocity in normal man and patients with obstructive lung disease: effects of terbutaline, *Am. Rev. Respir. Dis.*, 109, 458–463, 1974.

6 Yeats, D. B., Aspin, N., Levison, H., Jones, M. T., and Bryan, A. C., Mucociliary tracheal transport rates in man, *J. Appl. Physiol.*, 39, 487–495, 1975.

TABLE 24
Techniques for Measuring Pulmonary Permeability

	Technical Description	Ref.
Aerosol preparation	Preparation of ^{99m}Tc-labeled diethylene triamine penta-acetic acid (^{99m}Tc-DTPA) solution by combining sodium pertechnetate (TcO_4) with DTPA in a commercial reaction vial. Test of the binding efficiency by paper chromatography. Nebulization of the solution for aerosol particles less than 2-μm diameter. Aerosols with a larger particle size may be used for bronchial clearance measurements.	1–3
Aerosol inhalation	Inhalation of the aerosol with normal tidal breathing for 1 to 2 min and until initial lung activity of about 20,000 counts per min. A longer inhalation time may result in a loss of the initial clearance measurement.	
Lung scanning	Whole lung scan with a gamma camera every 15–30 sec post inhalation for a period of 20–30 min. A longer scanning period is not warranted because of potential complications with absorbed activities in the circulating blood and extrapulmonary tissue.	
Data analysis	Plot of normalized lung activity (% of the initial lung count) vs. time and fit of the curve to single exponential function. Calculation of clearance rate (%/min) based on the first 15-min curve. Normal pulmonary clearance rate: 0.6 to 1.56 (average ~0.8 to 0.9) %/min.	1
Causes of increased pulmonary permeability	*Disease:* sarcoidosis, adult respiratory distress syndrome (ARDS), hyaline membrane disease, bronchopulmonary dysplasia, pneumocystis carinii pneumonia, pulmonary fibrosis, bone marrow transplant. *Environmental effects:* cigarette smoking; exposure to asbestos, oxygen, and ozone; increase of lung volume from exercise; positive end expiratory pressure.	4–6
Interpretation/ significance	The following factors are the sources of variability of permeability measurement: Radiopharmaceutical preparation and purity Aerosol particle size and deposition site Particle concentration Body position Scanning period length Method of correction for background activity of lung tissue Lung volumes during aerosol inhalation and scanning Method of computation of clearance rate	7–15

1 Smith, R. J., Hyde, R. W., Waldman, D. L., Freund, G. G., Weber, D. A., Utell, M. J., and Morrow, P. E., Effect of pattern of aerosol inhalation on clearance of technetium-99m-labeled diethylenetriamine pentaacetic acid from the lungs of normal humans, *Am. Rev. Respir. Dis.,* 145, 1109–1116, 1992.

2 O'Brodovich, H. and Coats, G., Pulmonary clearance of ^{99m}Tc-DTPA: a non-invasive assessment of epithelial integrity, *Lung,* 165, 1–16, 1987.

3 Chopra, S. K., Taplin, G. V., Tashkin, D. P., and Elam, D., Lung clearance of soluble radioaerosols of different molecular weights in systemic sclerosis, *Thorax,* 34, 63–67, 1979.

4 Jacobs, M. P., Baughmann, R. P., and Hughes, J., Radioaerosol lung clearance in patients with active sarcoidosis, *Am. Rev. Respir. Dis.,* 131, 687–689, 1985.

5 Mason, G. R., Uszler, J. M., Effros, R. M., and Reid, E., Rapidly reversible alterations of pulmonary epithelial permeability induced by smoking, *Chest,* 83, 6–11, 1983.

6 Griffith, D. E., Holden, W. E., Morris, J. F., Min, L. K., and Krishnamurthy, G. T., Effects of common therapeutic concentrations of oxygen on lung clearance of ^{99m}Tc-DTPA and bronchoalveolar lavage albumin concentration, *Am. Rev. Respir. Dis.,* 134, 233–237, 1986.

7 Kehrl, H. R., Vincent, L. M., Kowalsky, R. J., Horstman, D. H., O'Neil, J. J., McCartney, W. H., and Bromberg, P. A., Ozone exposure increases respiratory epithelial permeability in humans, *Am. Rev. Respir. Dis.,* 135, 1124–1128, 1987.

8 Marks, J. D., Luce, J. M., Lzer, N. M., Nago-Sun, W. J., Lipausky, A. J. A., and Murray, J. F., Effect of increase in lung volume of clearance of aerosolized solute from human lungs, *J. Appl. Physiol.,* 59, 1242–1248, 1985.

9 Waldman, D. L., Weber, D. A., Oberdoerster, G. et al., Chemical breakdown of technetium-99m DTPA during nebulization, *J. Nucl. Med.,* 28, 378–382, 1987.

10 Bennett, W. D. and Illowite, J. S., Dual pathway clearance of ^{99m}TC-DTPA from the bronchial mucosa, *Am. Rev. Respir. Dis.,* 139, 1132–1138, 1989.

11 Oberdoerster, G., Utell, M. J., Morrow, P. E., Hyde, R. W., and Weber, D. A., Bronchial and alveolar absorption of inhaled ^{99m}Tc-DTPA, *Am. Rev. Respir. Dis.,* 134, 944–950, 1986.

TABLE 24 (continued)

Techniques for Measuring Pulmonary Permeability

12 Oberdoerster, G., Utell, M. J., Weber, D. A., Ivanovich, M., Hyde, R. W., and Morrow, P. E., Lung clearance of inhaled ^{99m}Tc-DTPA in the dog, *J. Appl. Physiol.*, 57, 589–595, 1984.
13 Groth, S., Mortensen, J., Lange, P., Vest, S., Rossing, N., and Swift, D., Effect of change in particle number on pulmonary clearance of aerosolized ^{99m}Tc-DTPA, *J. Appl. Physiol.*, 66, 2750–2755, 1989.
14 Meignan, M., Rosso, J., and Robert, R., Lung epithelial permeability to aerosolized solutes: relation to position, *J. Appl. Physiol.*, 62, 902–911, 1987.
15 Coats, G. and O'Brodovich, H., Extrapulmonary radioactivity in lung permeability measurements, *J. Nucl. Med.*, 28, 903–906, 1987.

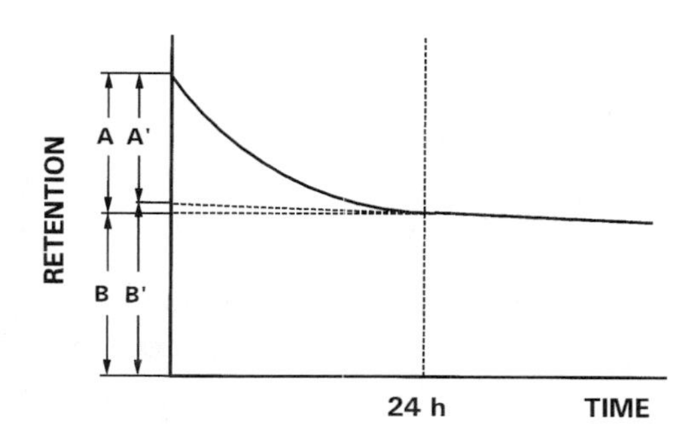

FIGURE 4
Clearance curve for inhaled radioactive aerosol indicating rapidly cleared fraction (A or A') and slowly cleared fraction (B or B').

the permeability of the respiratory epithelium. The method involves inhalation of a radiolabelled probe molecule (^{99m}Tc-labeled DTPA [diethylene triamine penta-acetate] aerosol, <2 μm in diameter) and subsequent monitoring of lung activity with sequential gamma camera imaging for about 30 min (see Table 24). A rapid clearance rate indicates damage to the airway epithelial structure. It is important to use aerosols with a small particle size and also a slow and deep inhalation maneuver to maximize pulmonary deposition. Because of potential complications with absorbed activities in the circulating blood and extrapulmonary tissue, lung activity measured more than 30 min postinhalation may not be accurate.

3.4 Symptoms

Evaluation of subject symptoms is an important aspect of inhalation toxicology. Examination of symptoms either at rest or during or after exercise is important not only because of the clues about response mechanisms that may be deduced, but also because symptom responses are readily understood by lay people and administrators from the public health or environmental sectors. Furthermore, symptom responses can be assigned a monetary value by economists seeking to derive economic risk benefit ratios. However, the collection of symptom data in human inhalation toxicology studies is not standardized and may vary considerably from laboratory to laboratory depending on the technician administering the questionnaire and the extent and type of instruction provided. In general, specific individual symptoms or categories of symptoms are evaluated using some form of severity scale. In addition, ability to perceive changes in loads to the respiratory system may be altered by some

TABLE 25A

Procedures Used To Assess Symptom Responses to Breathing Difficulty or Exercise

Test	Method	Significance/Interpretation	Ref.
Graphic scales	Use a graphic, mechanical, electronic, or computer-generated scale on which levels of symptoms such as breathlessness, cough, chest discomfort, muscle fatigue, etc. can be rated.	This kind of approach lends itself to computerized scoring of symptoms. Subject training is necessary to achieve reproducible results with minimal intersubject variability.	2,3
Category scale	Symptom categories are presented on a list, usually with a numerical value. Descriptors are used which categorize the severity of symptoms.	Often used to describe level of dyspnea or breathing difficulty and level of perceived exertion during exercise experiments. The number of categories on the scale may need to be limited for the subjects to make meaningful use of it.	1,2
Respiratory load detection	Determine the ability to detect changes in volume or respiratory loads (resistance, elastance) using psychophysical procedures such as magnitude estimation or magnitude production.	Ability to detect loads or volume changes may be altered by inhalation of irritant gases. Responses may be suggestive of the type of receptors involved.	2,4, 5,6

1 Borg, G., Psychophysical bases of perceived exertion, *Med. Sci. Sports Exer.*, 14, 377–381, 1982.
2 Adams, L. and Guz, A., Dyspnea on exertion, in *Pulmonary Physiology and Pathophysiology,* Whipp, B. J. and Wasserman, K., Eds., Marcel Dekker, New York, 1991.
3 Hackney, J. D. and Linn, W. S., Collection and analysis of symptom data in clinical air pollution studies, in *Inhalation Toxicology of Air Pollution: Clinical Research Considerations,* Frank, R., O'Neill, J. J., Utell, M. J., Hackney, J. D., Van Ryzin, J., and Brubaker, P. E., Eds., ASTM STP 872, American Society for Testing and Materials, Philadelphia, PA, 1985, pp. 73–82.
4 Guilford, J. P., *Psychometric Methods,* McGraw-Hill, New York, 1954.
5 Folinsbee, L. J., Gliner, J. A., and Horvath, S. M., Perceptual cues used in the reproduction of inspired volume, *Percept. Psychophys.*, 32, 449–453, 1982.
6 Gliner, J. A., Folinsbee, L. J., and Horvath, S. M., Accuracy and precision of matching inspired lung volume, *Percept. Psychophys.,* 29, 511–515, 1981.

TABLE 25B

Borg Category Scale To Describe Symptom Severity

0	Nothing at all
0.5	Just noticeable
1	Very weak
2	Weak
3	Moderate
4	Somewhat strong
5	Strong
6	
7	Very strong
8	
9	
10	Very, very strong (almost maximal)
*	Maximal or intolerable

inhaled materials. Table 25A briefly describes some methods used to evaluate symptoms that have been used in inhalation toxicology studies. Table 25B shows one version of the commonly used Borg scale to describe symptom severity. A list of symptoms that may be experienced during or following exposure to irritant gases is provided in Table 26.

TABLE 26

Symptoms that May Result from Exposure to Irritant Gases

"Bad" air	Nausea
"Burning" sensation in nose or throat	Runny nose
Chest tightness	Sputum production or phlegm
Cough	Strong or pungent odor
Dyspnea	Substernal pain or irritation
Eye irritation	Tearing or lacrimation
Faintness	Throat dryness
Fatigue or tiredness	Throat irritation
Headache	Unable to take deep breath
Nasal irritation	Wheeze
Nasal stuffiness or congestion	

3.5 Airway Responsiveness/Reactivity

Airway hyperresponsiveness is a hallmark of asthma but may also be encountered after acute exposure to toxic inhalants or after chronic occupational exposure to a variety of inhalable gaseous or particulate materials. Nonspecific bronchial hyperresponsiveness refers to the increased airway response to "nonspecific" chemical stimuli, most commonly methacholine or histamine.[31,32] Specific responses to a variety of allergens and occupational chemicals (plicatic acid, toluene diisocyanate, platinum salts, etc.) can also be evaluated; however, specific allergen challenges are not covered in detail in this section, in part because of the space limitations and, in part, because of the potential risk to the subject. The reader is referred to Spector[33] for a detailed review of bronchial challenge procedures and methodology, including allergen challenge. In addition to chemical challenges, exercise, hyperpnea of dry/cold air, or aerosols of hypo- or hypertonic saline can induce bronchoconstriction, primarily in asthmatics. The induction of airway hyperresponsiveness after exposure to an inhaled toxicant can be evaluated in nonasthmatics as well as asthmatics.

For research purposes, the choice of agonist depends upon the question being asked; however, for general evaluation of bronchial responsiveness, histamine and methacholine are the two most commonly used agonists. For asthmatics, the responses are well correlated, and both agonists are equally useful. In nonasthmatics, the high concentrations of histamine necessary to provoke bronchoconstriction also cause systemic side effects, which may include flushing, headache, syncope, and tachycardia. Repeated challenges with either drug in nonasthmatic subjects may lead to decreased responses to the same concentrations. Acetylcholine must be freshly prepared daily and is not commonly used.

3.5.1 Inhalation Challenge

The general procedure for conducting an inhalation challenge is similar for a variety of mediators such as histamine, methacholine, carbachol, adenosine, and a variety of others listed in Table 27. In many cases, a stock solution of the bronchoconstricting agent will have been made up beforehand. It may be desirable to wait until the subject has been cleared to participate before making up the serial dilutions, although with stable inexpensive mediators this is not necessary. Prior to the challenge, the subject should receive a brief physical examination and it should be determined that there are no contraindications to performing the inhalation challenge (see Table 28). If informed consent has not already been obtained, this must be done prior to the challenge. Cardiopulmonary resuscitation equipment should be immediately available and any necessary medication prepared. Injectable bronchodilators (adrenaline, terbutaline) should be prepared in advance (mainly for allergen challenge), and

TABLE 27
Inhalation Challenge

Mediator/Provocative Agent (Pharmacologic)	Chemical Name	Actions	Metabolism	Concentration Range (mg/ml)
Acetylcholine	Acetylcholine	+ Muscarinic receptors, bronchoconstrictor	Rapidly hydrolyzed by acetylcholine esterase (AChE)	0.1 to 32
Carbachol	Carbamyl choline	+ Nicotinic, bronchoconstrictor	Resistant to AchE hydrolysis; stable if refrigerated	0.1 to 10
Methacholine (Provocholine®)	Acetyl methylcholine	+ Muscarinic receptors in airway smooth muscle, bronchoconstrictor	Slowly hydrolyzed by AchE; stable if refrigerated	0.03 to 25
Adenosine	Adenosine or adenosine monophosphate (6-amino-9-D-ribofuranosyl-9-H-purine)	+ Stimulates mast cell mediator release, bronchoconstrictor; peak response: 2–5 min; highly selective, does not cause bronchospasm in nonasthmatics	Rapidly metabolized to inosine or AMP; half-time in blood ~10 sec	0.4 to 1600 (start at 0.03 with continuous nebulization)
Metabisulfite	Sodium metabisulfite	+ Stimulates sensory endings	—	0.6 to 160
Bradykinin	Bradykinin (nonapeptide)	+ Potent bronchoconstrictor in asthmatics, cough, chest tightness, reflex effect via C fibers, neuropeptides; B_2 receptor agonist	Dipeptidyl carboxypeptidase kininase II, very short half-life (~15 sec in plasma)	0.005 to 1.0
Histamine	Aminoethylimidazole	+ H_1 receptor, agonist, bronchoconstriction	Histamine-*N*-methyl transferase	0.125 to 25
Physical	**Conditions**	**Actions**	**Duration of Effect**	**Stimulus Range**
Exercise	Cool-dry environment; heat and humidity blunt the response. Repeated testing causes refractoriness. ECG monitoring for safety. Treadmill or cycle ergometer and ventilation monitoring system desirable. Appropriate safety guidelines for exercise testing should be followed. A noseclip should be used to force mouth breathing.	Moderate to heavy exercise leads to airway narrowing. Mechanism not established; possibly related to airway cooling/heat loss and/or airway drying and changes in osmolarity.	Effect peaks several min after exercise. Spontaneous recovery usually occurs in 30–60 min. Occasionally a severe response may occur. Standardization of inspired air humidity may reduce test variability. The use of cold dry air in combination with exercise is more asthmogenic.	6–8 min of exercise at ~70% of maximum aerobic capacity is an optimal stimulus. Exercise heart rate should be ~170/min (adjusted for age). Running (best), walking, or cycling are preferred exercise modes. A stepwise approach is recommended for naive subjects, especially unfit adults, to allow for warmup and careful observation of cardiovascular responses.

Cold-dry air hyperpnea	Cold, dry air is supplied to the subject to breathe via mouthpiece or face mask. Hypocapnia is prevented by the addition of CO_2 (~5%) to the inspired air.	3–6 min of hyperpnea with cool, dry air leads to airway narrowing possibly due to cooling and/or drying of the airways.	Peak effect occurs after hyperpnea; recovery usually spontaneous.	V_E levels range from 15 to 75 l/min. Duration of hyperventilation ranges from 3–6 min. Some means of providing feedback are necessary for subjects to attain the "target" ventilation.
Hypotonic or hypertonic solutions	Aerosols generated by continuous ultrasonic nebulizer and administered via mouthpiece. Subject wears noseclip. Nebulizer frequency 0.5–4 MHz; higher frequency, smaller aerosol. Use smooth-bore tubing.	Inhalation of distilled water aerosol or hypertonic saline solutions will lead to airway narrowing. Small or no effect in nonasthmatics. Effect elicited through mediator release (e.g., mast cells). Response blocked by cromolyn.	Response peaks about 90 sec after inhalation. Subjects are refractory on repeated testing (refractoriness can be blocked with indomethacin).	Subjects inhale increasing volumes of hypertonic saline (4.5%). Progressive doubling of the duration of breathing hypertonic saline is recommended (e.g., 30 sec and 1, 2, 4, and 8 min). Another method is to increase tonicity of aerosol (e.g., 0.9, 1.8, 2.7, ... 5.4%).
Other	**Conditions**	**Actions**	**Metabolism**	**Concentration Range**
Sulfur dioxide	Hyperventilation of SO_2 through a mouthpiece during exercise or voluntary eucapnic hyperpnea. SO_2 added by rotameter from bottled source; concentration monitored at the mouth.	Causes reflex increase in airway resistance. Blocked by atropine. Peak response occurs in about 5 min.	Removed by circulatory system and excreted via kidneys. Rapidly absorbed by moist surfaces in the nose and mouth.	Primarily used in combination with exercise and only in asthmatics. At V_E of ~30 l/min, SO_2 conc. of 0.1, 0.2, 0.4, 0.8, 1.6 ppm are used.

Note: Data from Cropp, J. A., The exercise bronchoprovocation test: standardization of procedures and evaluation of response, *J. Allergy Clin. Immunol.,* 64, 627–633, 1979; Smith, C. M. and Anderson, S. D., Inhalational challenge using hypertonic saline in asthmatic subjects: a comparison with responses to hyperpnoea, methacholine, and water, *Eur. Respir. J.,* 3, 144–151, 1990; Simonsson, B. G., Skoogh, B. E., Bergh, N. P., Andersson, R., and Svedmyr, N., *In vivo* and *in vitro* effect of bradykinin on bronchial motor tone in normal subjects and patients with airways obstruction, *Respiration,* 30, 378–388, 1973. See also References 29 to 39.

TABLE 28

Safety Issues, Contraindications, and Equipment

Subject Requirements Before Challenge
Asymptomatic
Absence of obstruction
FEV_1 should be > 80% predicted (when tests are necessary for clinical purposes, somewhat lower values may be satisfactory, but they should not be less than 80% of previous "best" value)
No recent viral illness or vaccination (6 weeks)
No recent toxicant/air pollution exposure (48 hr)
No recent allergen exposure (4 weeks)
Resting (no vigorous exercise within past 3 hr)
No concurrent asthma attack
No reaction to challenge with diluent alone
Adequate pulmonary reserve (FEV_1 > 2 l)
Normal blood pressure
Normal 12-lead ECG
Absence of pregnancy
Safety Monitoring and Equipment
ECG monitoring
Pulse oximetry
End tidal PCO_2
Resuscitation equipment (oxygen, drugs, syringes, IV sets, defibrillator, bag-mask hand respirator, laryngoscope, tracheal tubes, sphygmomanometer)
Bronchodilator therapy (nebulizer, metered dose inhaler, or IV injectables)
Medical supervision of test
All personnel certified in CPR
Medication Withholding (Varies with Inhalant)
Cromolyn (disodium cromoglycate or nedocromil sodium) (12 hr)
β-Sympathomimetic agonists (12 hr)
Anti-inflammatories (e.g., indomethacin) (12 hr)
Antihistamines (H_1 antagonists) (48 hr)
Oral corticosteroids
Inhaled corticosteroids
Theophylline (48 hr)
Anticholinergics (12 hr)
Alpha-adrenergic blockers
Protocol Issues
Do not suggest responses to subjects.
Subjects should avoid coffee, cola, chocolate, smoking.
Responses may vary with time of day, phase of menstrual cycle.
Nebulizer exhaust and patient exhalation should be vented or filtered.
Personnel should not be allergic asthmatics (for allergen challenges).
Glassware and nebulizers must be clean.
Nebulizers must be periodically calibrated.
Aerosol size (MMAD) should range from 3–6 μm (jet nebulizers typically produce somewhat smaller aerosols and have a lower mass output (μg/min) than the ultrasonic nebulizers).
Avoid cross-contamination of high- and low-concentration containers and nebulizers.
Avoid diluents with preservatives.
Check pH of diluent.
Subjects should breathe through the mouth with a noseclip in place.

TABLE 28 (continued)

Safety Issues, Contraindications, and Equipment

Inhalation Challenge Apparatus
Jet-type nebulizer (e.g., DeVilbiss 646™)
Mouthpiece and noseclip
Dosimeter (commercially available or custom-built); includes control circuitry or microprocessor, flow sensor or pneumotachograph, adjustable timing circuit, solenoid valve, pressure regulator, compressed medical grade air, volume measure, or flow integrator.
Set of nebulizers with different concentration of inhalant (or set of vials and one nebulizer)
Ultrasonic continuous nebulizer
Facemask and smooth-bore tubing
Set of nebulization containers with different concentrations of inhalant
Exhaust filter and/or an exhaust system (e.g., vacuum snorkel or hood) in well-ventilated room
Chair and equipment stand
Apparatus for Dilution Preparation
Electronic balance (for bulk material)
Pipettes, syringes, or autopipette
Set of dilution tubes
Vortex mixer
Mask, gloves
Saline (no preservatives)

inhaled bronchodilators (isoproterenol, albuterol) should also be available. Preferably, the subject should have been resting for 30 to 60 min prior to the challenge and should not have performed vigorous exercise within the past 3 hr.

Dilutions should be prepared according to a well-defined written procedure. In the case of methacholine (Provocholine®), which is supplied in 100-mg units, the dilution schedule is supplied and can be either followed directly or modified slightly as necessary. In many cases, however, the bronchoconstrictor is supplied as a powder and must be weighed and then made into solution. In such cases, extreme care must be exercised to avoid weighing errors that could lead to overdose. Fresh, preferably factory-sealed, diluent should be used to make each batch of dilutions. Diluent should be discarded after a single use to avoid any possibility of contamination.

Baseline pulmonary function measures (e.g., spirometry, plethysmography) should be performed first to ascertain that function is within acceptable guidelines (for research purposes, a baseline FEV_1 > 80% predicted and an FEV_1/FVC ratio of > 70% are typically used). If these criteria are met, the subject will then inhale the diluent only (typically, isotonic saline). If there is a response (e.g., a >5% decline in FEV_1) to the diluent, the test should be discontinued since the results will be uninterpretable. The subject may be responsive to suggestion or possibly reacting to a preservative or contaminant.

Delivery of the aerosolized solution may be accomplished in a number of ways (see also Section 2). The simplest method is to use a continuously operating nebulizer (ultrasonic) with a known output and to have the subject inhale the aerosol during tidal breathing for a set period of time. Dosimetry is somewhat problematic unless tidal volume and breathing rate are monitored and controlled. The same nebulizer and solution volume (–5 ml) should be used in all tests. Even the type of container used to hold the nebulized solution can cause variation in nebulizer output.

The dosimeter method uses a jet nebulizer that is activated for a prescribed period of time (usually ~0.6 sec but not more than 50% of inspiration time) at a constant pressure (10 to 20 psi or 70 to 140 kPa) during a slow inhalation from resting expiration (i.e., FRC [functional residual capacity]) to full inspiration (i.e., TLC [total lung capacity]). Typically, five consecutive breaths are used to deliver a given concentration. Depending on the protocol, subjects may or may not hold their breath for 2 to 5 sec at the end of inspiration. The activation pressure and air flow (5 to 10 l/min) through the nebulizer should be known and held constant. In some dosimeters, the inspired volume or time at which the nebulizer is activated can be adjusted so that the volume penetration of the aerosol can be influenced (aerosol delivered later in inspiration will be distributed primarily to the conducting airways, whereas aerosol delivered early in inspiration will penetrate to the lung periphery). Thus, the dose of inhaled material can be accurately quantified and distribution of the aerosol can be qualitatively controlled. With an appropriately designed dosimeter, the apparatus is easy to operate and maintain.

A third method of aerosol delivery uses a reservoir system to temporarily store the aerosol in a water-sealed spirometer or a gas bag, allowing precise measurement of aerosol volume delivery. The aerosol must be flushed between breaths in order to keep it "fresh", making this method rather cumbersome. Other hybrid systems (e.g., a continuous nebulizer with control of tidal volume and number of breaths) have also been used.[34]

Nebulizers should be adequately cleaned, dried, and inspected between challenges. Periodically, the output of the nebulizer should be characterized at the pressure and flow settings that will be used in the challenge.[38,39] Particle size (distribution and mass median aerodynamic diameter, MMAD), and total volume output should be determined. Aerosol sizing and counting instrumentation can be used for this purpose (see Section 2.2). An alternative method for characterizing the mass output of the smaller jet nebulizers is to weigh the nebulizer before and after a period of intermittent operation using a representative airflow through the nebulizer; however, this procedure does not replace the need for characterizing aerosol size.

3.5.2 Dosing Schedules

Various dosing schedules have been utilized, including varying solution concentrations, numbers of breaths, breath volume, etc. For each system, the delivered mass of drug or antigen should be determined as accurately as possible based on the calibrated characteristics of the nebulizer or dosimeter. A safe starting concentration will depend upon the system and the type of subjects to be included.[29] In any case, when using a new aerosol generation system, it is best initially to test nonreactive subjects (nonallergic, nonasthmatics) using a conservative dosage schedule (i.e., begin with a much lower concentration than you would expect to produce a response in the most sensitive subjects). Conservative dosing schedules should be used with new subject groups or when testing the influence of unknown factors (e.g., air pollutants) until sufficient experience is gained so that the range of response is established. Dosing schedules can be constructed in a variety of ways, the most common being successive doubling of concentrations. The doubling dose method has the added benefit of ease of preparation of the dilutions. An example of a dosing schedule using methacholine for asthmatic and healthy subjects is shown in Table 29.

For exercise challenge, usually a single level of exercise is used. Standard precautions for exercise testing should be followed.[30,37] Nonasthmatics usually do not respond to exercise testing with bronchoconstriction. In unfit adults or anyone experiencing an exercise test for the first time, the work load should be incremented stepwise until the desired exercise level is achieved. If cold air is used in combination with exercise, a "dose-response relationship" can be determined based on the estimated respiratory heat loss and change in lung function.

Commonly used response measures to evaluate bronchoconstriction caused by airway challenge include forced spirometry and airway resistance. Procedures for these measurements are covered in Sections 3.2.1 and 3.2.2. The evaluation of hyperresponsiveness requires the construction of dose-response curves. A frequently used endpoint, and the simplest to

TABLE 29

Sample Dosing Schedule for Methacholine or Histamine Challenge

Time	Conc. (mg/ml)	No. of Breaths	Breath Units[a] (mg/ml)	Cumulative Breath Units
0:00	Saline	5	0	0
0:05	0.075	5	0.375	.375
0:10	0.15	5	0.75	1.125
0:15	0.30	5	1.50	2.625
0:20	0.62	5	3.1	5.725
0:25	1.25[b]	5	6.25	12.0
0:30	2.50	5	12.5	24.5
0:35	5.00[c]	5	25.0	50
0:40	10.0	5	50.0	100
0:45	20.0	5	100.0	200
0:55	β-Agonist		If needed	

[a] One breath unit (BU) is one breath of aerosol generated from a solution with a concentration of 1 mg/ml.
[b] Nonasthmatics start here (after saline).
[c] Asthmatics stop here or when target response level is achieved, whichever is lower.

determine, is the provocative concentration (PC) of bronchoconstrictor that causes a 20% decline in FEV_1; this is referred to as the PC_{20}. Similarly, the concentration that causes a 100% increase in specific airways resistance may be called the PC_{100}. Since the duration of action of methacholine (30 to 60 min) is such that the dose accumulates during a challenge, the "dose" is also expressed in cumulative breath units or cumulative inhalation units (i.e., the sum of all doses given). Breath units may be defined as one inhalation of a given concentration of agonist, e.g., one breath of aerosol from a 1 mg/ml solution. A true cumulative "delivered" dose requires a precise estimate of the amount of aerosol actually delivered to the trachea and beyond; such determinations may be useful for certain research purposes but are seldom done for standard testing of hyperresponsiveness. Cumulative dose would be given in units of milligrams or micromoles of drug delivered. In practice, the log of agonist concentration or cumulative dose vs. the percent change in the response measure is plotted and the provocative dose (PD) or PC is determined by linear or logarithmic interpolation. Additional measures include the shape of the dose-response curve, slope of the dose-response curve, area under the dose-response curve, and maximal response plateau (*not done in asthmatics*).[36] The *sensitivity* has been defined as the provocative dose and the *reactivity* as the slope of the dose-response curve.[35]

3.6 Airway Sampling Techniques

The feasibility and utility of bronchoscopy for the study of airways and lung disease and injury has been well documented and is being increasingly recognized. For more complete information, the reader is referred to the articles by the National Institutes of Health[43] and Smith and DeShazo.[45] Bronchoscopy research applications include bronchoalveolar lavage (BAL), segmental airway lavages, endobronchial brushings, and mucosal biopsies, as well as evaluation of tracheal mucus transport velocity, regional gas exchange, gas and particle dosimetry, airway physiology, and placement of measurement devices. Analysis of the cell and chemical nature of specimens obtained with bronchoscopy provides an array of information about airway cell populations and their activity, levels of soluble mediators, and markers of tissue damage and repair (see Table 30). The technical aspects of performing the procedure, methods of analysis of the specimens obtained, and reporting of results have not been standardized and the reader is encouraged to review recent literature.[41–44] The following is a brief synopsis of current methodology and procedures.

Before bronchoscopy, subjects should undergo a complete medical evaluation that includes spirometric testing of lung function. Contraindications to bronchoscopy include significant underlying cardiovascular disease, clotting disorder, sensitivity to local anesthesia, and respiratory infection within the previous 4 weeks. Relative contraindications include other chronic illnesses such as renal disease or hypertension. Persons with underlying lung disease, including asthmatics, have been successfully studied with bronchoscopic techniques.[45,47-49] Due to safety concerns, clinical studies are generally limited to subjects with mild to moderate lung function impairment. Considerations in evaluating asthmatics for bronchoscopy include frequency and severity of asthma attacks, specific allergies, past and present medication use, lung function, and degree of nonspecific bronchial reactivity. Current recommendations for asthmatics include (1) no history of status asthmaticus or mechanical ventilation, (2) subject clinically stable for 4 weeks preprocedure, (3) $FEV_1 > 60\%$ of predicted, and (4) exclusion of individuals with extreme bronchial hyperresponsiveness. Unless the use of such drugs would interfere with the results of the study, asthmatics should receive preprocedure medication with atropine and beta-agonist bronchodilator.

The procedure should be performed by an experienced bronchoscopist with facilities and personnel available to manage medical emergencies and cardiopulmonary resuscitation. The cardiac rhythm, blood pressure, and oxygen saturation should be monitored and intravenous access must be secured. Premedication with atropine, beta-agonist bronchodilator, and analgesics/sedatives can be either omitted or administered as required by the research protocol and the comfort and safety of the subject. Upper and lower airway topical anesthesia should be obtained with the lowest possible dose of local anesthetic; no more than 300 mg of lidocaine should be administered. Although administration of intravenous anxiolytic/analgesic medications can be used to facilitate subject comfort, it introduces the additional risk of adverse effects of the medication. Supplemental oxygen should be administered as necessary to maintain oxygen saturation >90% during and after the procedure.

3.6.1 *Bronchoalveolar Lavage*

Bronchoalveolar lavage should be performed prior to forceps or brush biopsy. To reduce cough, the lavage fluid (sterile buffered or unbuffered saline) is prewarmed to 37°C. The bronchoscope is wedged into a 4th or 5th order bronchus; the right middle lobe, lingula, and lower lobes are generally preferred as the return of lavage fluid is greater than from the upper lobes. The lavage fluid typically is instilled in 50- to 60-ml aliquots (although smaller volumes may be used) and aspirated with gentle suction; it is recommended that the total amount of fluid instilled not exceed 400 ml total.

Lavage specimens should be put on ice immediately after recovery from the patient and processed as soon as possible to prevent degradation of soluble components. Anti-proteases should be added at this time if especially labile proteins will be analyzed. If saline is instilled more than one time, or in more than one lobe, samples may be kept separate or pooled to increase recovery. Samples are centrifuged at 1000 *g* for 10 min in a refrigerated centrifuge to pellet cells. Supernatants are immediately aliquoted into small fractions (typically 1 to 10 ml) and frozen at –80°C for later analysis of soluble mediators. A small amount of fluid may be kept on ice for analysis of soluble mediators which are rendered inactive upon freezing. Cells are usually resuspended in a small volume (e.g., 5 ml) of a basal medium, such as RPMI, and counted, and medium is added to give a final concentration of 1×10^6 viable cells per

TABLE 30

Cells and Soluble Mediators Found in Bronchoalveolar Lavage[a]

Test/Assay	Equipment Needed	Method	Significance/Interpretation
Cells and Markers of Activated Cells			
Cell counts/viability	Microscope	Cells are mixed with Trypan blue and placed in a hemocytometer, and the number of viable (Trypan blue negative) cells determined.	Increases or decreases from the normal range can be indicative of certain disease states, infections, or inflammation and lung damage caused by xenobiotics.
Cell differential	Microscope, cytocentrifuge	Cells are pelleted on a glass slide using a cytocentrifuge. Slides are dipped into a modified Wright's stain to stain specific cell types differentially.	Cell differentials can determine the presence of specific cell types such as PMNs, macrophages, eosinophils, lymphocytes, epithelial cells, mast cells, and basophils. The presence or absence of specific cells can be indicative of acute clinical problems, such as infection or inflammation, or chronic problems, such as sarcoidosis or asthma, depending on the cell type.
Markers of activated PMNs: elastase and myeloperoxidase (MPO)	ELISA reader spectrophotometer	MPO levels in bronchoalveolar lavage fluid (BALF) are measured using commercially available immunoassay kits. Elastase activity is measured using specific colorimetric substrates.	MPO and elastase are enzymes that are secreted by activated PMNs. These potent proteases may damage surrounding cells and tissues.
Markers of activated eosinophils: esosinophilic cationic protein (ECP)	ELISA reader	ECP levels in BALF are measured using commercially available immunoassay kits.	ECP is a protein secreted by activated eosinophils which can cause direct epithelial cell damage and mast cell degranulation.
Markers of activated mast cells: histamine and tryptase	ELISA reader	Histamine and tryptase levels in BALF are measured using commercially available immunoassay kits.	Histamine and tryptase are proteins secreted by activated mast cells. Histamine is a principal mediator in allergic reactions and can cause increased vascular permeability and bronchoconstriction.
Markers of activated macrophages: superoxide anion and altered phagocytic capacity	Microscope, spectrophotometer, flow cytometer, centrifuge	For superoxide release, macrophages are mixed with reduced cytochrome C and oxidation of the compound is followed in a spectrophotometer. For phagocytosis, macrophages are mixed with particles. (e.g., latex beads, microorganisms) and phagocytosis is monitored by a microscope or using flow cytometry.	Macrophages are the lung's first line of defense against inhaled microorganisms. Activated macrophages have enhanced phagocytic capability and produce compounds such as superoxide which aid in killing bacteria. In contrast, damaged macrophages may result in susceptibility to infection.
Markers of activated lymphocytes: interleukin-2/IL-2 receptor and HLA-DR	Flow cytometer, centrifuge	BAL cells are incubated with appropriate antibodies, and flow cytometry is used to quantify the expression of these molecules on lymphocytes.	Expression of these markers indicates an immune signal has been delivered to lymphocytes, which can further amplify a specific immune or nonspecific inflammatory response.

TABLE 30 (continued)

Cells and Soluble Mediators Found in Bronchoalveolar Lavage[a]

Test/Assay	Equipment Needed	Method	Significance/Interpretation
Soluble Mediators			
Cell damage: lactate dehydrogenase (LDH)	Spectrophotometer	LDH activity in BALF is measured using commercially available kits.	LDH is an enzyme that leaks from damaged cells. An increase in LDH activity above normal ranges is indicative of damage to pulmonary cells.
Vascular permeability/edema: albumin and total protein	Spectrophotometer	Albumin and total protein levels in BALF are measured colorimetrically with commercially available kits.	Increased levels of albumin or total protein in BALF indicate leakage of vascular components into the airways or alveoli, although total protein may also be increased by products of lung cells.
Cytokines: interleukins, especially IL-1, IL-2, IL-6, IL-8; tumor necrosis factor; and chemokines, especially GRO, RANTES, MIP, MCP	ELISA reader, cultured cells	Cytokines in BALF are measured with commercially available immunoassay kits. Alternatively, bioassays may be used to measure the amount of biologically active cytokines.	Cytokines modulate the inflammatory and immune response of the lung, depending on the cytokine. Not all immunoactive cytokines are biologically active; conversely, bioassays may not measure total cytokines in the lavage fluid if inhibitors are present. Therefore, both immunoassays and bioassays may have to be performed to obtain a more complete profile of cytokine levels.
Growth factors: platelet-derived growth factors (PDGF), transforming growth factors (TGF), colony-simulating factors (GM-CSF), and insulin-like growth factors (IGF)	ELISA reader, cultured cells	Growth factor levels are quantified in BALF using commercially available immunoassay kits or by bioassay.	These peptides are involved in a wide range of activities, including tissue repair, inflammation, elaboration of extracellular matrix proteins, and regulation of cell growth.
Eicosanoids: PGE_2; thromboxane; LTB_4, C_4, D_4; and 5- 12-, 15-HETEs	ELISA reader, HPLC, GC-MS, and scintillation counter	Eicosanoid levels are quantified in BALF using commercially available immunoassay or radioimmune assay kits that measure amounts of specific eicosanoids. Alternatively, HPLC can be used to measure whole classes of eicosanoids with a single run.	Eicosanoids are responsible for a variety of effects, including modulation of the inflammatory and immune response, and pulmonary function changes.
Other lipid mediators/enzymes: phospholipase A2, platelet activating factor (PAF), and lysophospholipids (LP)	Scintillation counter, TLC, GC-MS, platelet, and aggregometer	PAF is measured in BALF with radioimmunoassay or bioassay. PLA2 activity is measured by hydrolysis of radiolabelled phospholipid substrates. LP are derivitized and measured by GC-MS.	PAF is highly pro-inflammatory and causes pulmonary function changes. PLA2 mobilizes arachidonic acid for eicosanoid metabolism. LP are also pro-inflammatory as well as being potent detergents.
Enzyme inhibitors: α-1 antitrypsin (AT) and α-2 macroglobulin (MG)	ELISA reader	AT and MG levels are quantified in BALF using commercially available immunoassay kits.	These inhibitors bind to and inactive elastase and can act to minimize cell damage caused by elastase.

Anti-oxidants: catalase; glutathione; and vitamins A, C, E	Spectrophotometer, HPLC	Vitamins A, C, E and glutathione are measured in BALF using HPLC. Catalase is measured with a colorimetric assay using specific substrates.	These soluble compounds are capable of scavenging free radicals and other oxidants produced either intracellularly in disease states or via external xenobiotics such as ozone.
Coagulation factors: tissue factor (TF), factor VII (FVII)	Fibrometer	TF and FVII activity is measured in clotting assays, using plasma as a substrate.	Extravascular coagulation and fibrin deposition play a role in many disease states and lung injury. TF and FVII are members of the coagulation pathway, and their presence in lavage fluid is suggestive of fibrin deposition.
Fibrinolytic factors: tissue plasminogen activator (tPA)	ELISA reader, spectrophotometer	tPA activity in BAL cells or fluid is measured using commercially available kits.	tPA plays a key role in fibrinolysis, dissolving fibrin deposited during lung injury.
Neuropeptides: substance P	ELISA reader	Substance P levels are quantified in BALF using a commercially available immunoassay kit.	This tachykinin, thought to be released by C fibers in the lung, has many actions including smooth muscle contraction, vascular permeability, and mucus secretion. It may also play a role in neurogenically mediated inflammation.
Fibronectin	ELISA reader	Fibronectin levels are quantified in BALF using a commercially available immunoassay kit.	Fibronectin has been reported as being able to recruit fibroblasts to the lung and acting as a competency factor I stimulating them to divide. It has also been reported to play a role in repair of lesions caused by sloughing of lung epithelial cells.
Complement factors: C3, C5	ELISA reader or scintillation counter	C3 and C5 levels are quantified in BALF using commercially available radioimmunoassay kits.	The complement system functions as a potent mediator of inflammation and may play a role in enhancing lung injury. The presence of specific complement factors such as C3 and C5 indicate the system has been activated.
Mucin	ELISA reader	Mucin present in lavage fluid can be quantified by ELISA using antibodies directed against specific mucin epitopes; however, commercial ELISAs are not yet available to measure mucins.	Mucin is overexpressed in certain disease states and following many kinds of lung injury.

Note: HLA = human lymphocyte antigen; PMN = polymorphonuclear (lymphocyte); MIP = macrophage inflammatory protein; MCP = monocyte chemotactic factor; ELISA = enzyme-linked immunosorbent assay; GMCSF = granulocyte-macrophage colony stimulating factor; PGE_2 = prostaglandin E2; LTB4 = leukotriene B4; 12-HETE = 12-hydroxyeicosatetraenoic acid; GCMS = gas chromatograph mass spectrometer; HPLC = high performance liquid chromatography; PLA2 = phospholipase A2; C3 = complement factor.

[a] Data derived from following references:

Hunninghake, G., Gadek, J. E., Kawamani, O., Ferrans, V. J., and Crystal, R. G., Inflammatory and immune processes in the human lung in health and diseases: evaluation by bronchoalveolar lavage, *Am. J. Pathol.*, 97, 149–191, 1979.

Klech, H. and Pohl, W., Technical recommendations and guidelines for bronchoalveolar lavage: European Working Party Working Report, *Eur. Respir. J.*, 2, 561–585, 1989.

Reynolds, H. Y., Bronchoalveolar lavage, *Am. Rev. Respir. Dis.*, 135, 250–263, 1987.

Stanley, M. W., Henry-Stanley, M. J., and Iber, C., *Bronchoalveolar Lavage: Cytology and Clinical Applications,* Igaku-Shoin, Tokyo, 1991.

Utell, M. J., Frampton, M. W., and Morrow, P. E., Quantitative clinical studies with defined exposure atmospheres, in *Toxicology of the Lung,* Gardiner, D. E., Ed., Raven Press, New York, 1993, pp. 283–308.

Walters, E. H. and Gardner, D. E., Bronchoalveolar lavage as a research tool, *Thorax,* 46, 613–618, 1991.

Walters, E. H. and Gardner, D. E., Bronchoalveolar lavage in inhalation lung toxicity, in *Pathophysiology and Treatment of Inhalation Injuries,* Loke, J., Ed., Marcel Dekker, New York, 1988, pp. 207–237.

ml. Cells may then be used for functional assays or for analysis of cellular components such as enzymes, surface proteins, or specific mRNAs.

3.6.2 Biopsy

Bronchial biopsies should always be performed under direct visualization. Mucosal biopsies are taken from a combination of the main carina and one or more segmental or subsegmental carina, and up to six biopsies have been obtained during a single procedure without sequelae. Information on the safety of more than six biopsies is not available, and exceeding this number is not recommended. In general, the longer the duration of the procedure, the greater the risk to the subject. The biopsy specimens obtained can be placed in a variety of fixatives and/or cell media depending on the assays planned. Cytology brushings for sampling of mucosal epithelial cells are usually taken from the trachea and mainstem bronchi. Multiple passes are performed with the same brush. The performance of several cytology brushings adds minimal risk to the bronchoscopy procedure.

Cells can be removed from the cytology brush by shaking the brush gently in a basal medium such as RPMI. Cells are dissaggregated by incubation in mild protease, which also destroys damaged or dead cells. Cells are then pelleted, washed, and suspended in a small volume of medium. Cells may be assayed directly or cultured. Typically 200,000 to 500,000 cells are obtained from each brush, requiring the use of assays that utilize small numbers of cells such as polymerase chain reaction (PCR), immunohistochemistry, *in situ* hybridization, and comet assay for DNA single-strand breaks.

With respect to multiple procedures, two bronchoscopies that utilized both lavage and bronchial biopsies have been safely performed within a 24-hr period without clinically significant problems. Serial bronchoscopies with lavage and biopsies have been performed successfully in the same subject at 4-wk intervals. Investigative bronchoscopy is an invasive procedure. In performing this procedure, the safety of the subject is of primary importance and should not be compromised for the sake of investigative data. Potential complications during bronchoscopy include layrngospasm, bronchoconstriction, hypoxemia, hypopnea, bleeding, and drug reactions. If significant bronchoconstriction occurs, the procedure is aborted and the subject treated with inhaled bronchodilator. The procedure is also terminated for clinically significant arrhythmias and for oxygen desaturation that is not easily correctable with supplemental oxygen. Bleeding is an unusual complication that generally occurs only with bronchial brushings or biopsy; the blood loss is usually clinically insignificant. A post-bronchoalveolar lavage flu-like syndrome sometimes occurs; this is characterized by malaise, fever, and chest X-ray infiltrate that resolves within 24 hr and is responsive to acetaminophen or nonsteroidal anti-inflammatory drugs. Bacterial pneumonia is a rare complication of bronchoscopic procedures. In asthmatics, studies have shown that although bronchoscopy with lavage often causes short-term decrements in lung function it does not appear to affect airway inflammation or nonspecific bronchial reactivity.

Following the bronchoscopy, the recovery of the subject must be supervised and monitored at regular intervals (for at least 2 hr). Subjects are not discharged until their condition is stable, and the gag reflex and swallowing mechanism have returned. In asthmatics, the absence of airflow obstruction of clinical significance should be documented with forced expiratory spirometry. Subjects should be given the name and telephone number of a physician to contact in the event of untoward symptoms or problems. A phone interview the following day to confirm the lack of postprocedure complications is prudent.

Investigative bronchoscopy has several limitations. The procedure is invasive, which restricts the ability to evaluate the time-course of a dynamic process. In addition, the usual bronchoalveolar lavage mainly samples the alveolar epithelial surfaces and does not necessarily reflect events in the airway mucosa, which may be the tissue of interest. Likewise, bronchial biopsies are limited to larger airways and may not be indicative of findings in peripheral airways. Finally, there is no standard way of normalizing information from either biopsies or lavages.

Other research applications for bronchoscopic techniques often require specialized expertise and/or equipment.[49,50] Isolated airway lavage[40] permits sampling of airway material only; however, this procedure is technically difficult to perform, increases the risk of significant hypoxemia, and samples only a 3-cm segment of the airway using a customized double-balloon catheter. An alternative technique which approximates an airway lavage is to instill small volumes of lavage fluid (about 10 to 20 ml) during the first washing. Intrapulmonary monitoring of gases may be accomplished by placement of a gas sampling line via the bronchoscope. The capability to accurately measure regional gas exchange or gas (e.g., ozone) dosimetry depends on a rapid sample rate and response time of the gas analyzer. In addition, dosimetry also requires a gas control and delivery system. The bronchoscope has also been used to provide a measure of peripheral airflow resistance;[48] 5% CO_2 in air is administered at a known flow rate via one channel of a double-lumen catheter passed through a wedged bronchoscope while pressure is monitored through the second channel. This technique also requires monitoring of esophageal pressures with an esophageal balloon-tipped catheter. The clinical significance of the measurement has not been clearly defined. For further information on bronchoscopic applications in evaluating tracheal mucous velocity, refer to Table 23. Use of the bronchoscope in isolated airway segment exposure is covered in Table 6.

3.6.3 *Other Airway Sampling Techniques*

In addition to direct sampling via bronchoscopy, airway materials may be obtained using sputum samples that are induced by the inhalation of hypertonic saline aerosol. Analyses of expired air for trace gases and volatile compounds can also provide information regarding byproducts of reactions of inhaled gases with airway surfaces and liquids (Table 31), as well as production of endogenous substances.

4 USE OF HUMAN SUBJECTS

The procedural guidelines for the use of human subjects in experimental studies have evolved considerably over the past decade or more. In addition to the important issues of informed consent and institutional review of experimental protocols, a number of other issues must be addressed. These include recruitment procedures, payment of subjects for participation, characterization of the study population, adequate training of subjects in the experimental procedures, guidelines for inclusion or exclusion of subjects, and confidentiality of subject records.

Table 32 presents some considerations for the planning, review, approval, and conduct of a human research study. After developing an experimental plan, it should be reviewed not only for scientific quality but also for safety, experimental design, and quality assurance. The inclusion of noncaucasian subjects, female subjects, and subjects of different age groups is an important consideration to avoid the development of a database that deals solely with young adult white males. Volunteers may or may not be compensated for their time, but in no case should the payment be coercive. A simple question that can be asked is whether the subject would participate in the study if she was only compensated for her time (at an average hourly rate) and received no additional incentive. Caution should be exercised in accepting subjects into a study who are uncooperative, exhibit personality disorders, or are unusually curious about liability for injury. In many cases, there may be competition for the subject pool and care must be exercised that the inducement to participate does not become excessive. Adequate provisions must be made for privacy of subject information, subject safety, and medical treatment of subjects, if this becomes necessary.

TABLE 31

Noninvasive Techniques for Identifying Airway Inflammatory, Cellular, and Metabolic Responses

Test	Apparatus	Method	Significance/Interpretation	Ref.
Induced sputum	High-volume nebulizer, hypertonic saline (3–5%), spirometry system, sputum container, beta agonist inhaler, dithiothreitol solution (10%) (DTT)	Pretreat with agonist to limit bronchospasm. Breathe saline aerosol for up to 30 min using either constant conc. or increasing conc. of saline. Whenever necessary, subject coughs and interrupts procedure to expectorate. If FEV_1 falls > 10% or if symptoms occur, treat with agonist. Samples are treated with DTT solution and processed.	This method is suitable for harvesting sputum from both healthy and asthmatic subjects. Since hypertonic saline is also used as an airway challenge, caution must be exercised with asthmatics. Cellular and biochemical analyses suggest similar findings in sputum samples and in BAL fluid. Method is noninvasive and could be used in field studies; however, the procedure may take up to 30 min to generate mucus plugs.	1,2
Expired air markers of inflammation (nitric oxide)	Chemiluminescent NO analyzer	The concentration of NO in the expired air is monitored. Normal levels are in the 3–8 ppb range. Care must be exercised to exclude gases derived from the digestive tract (e.g., via eructation) and the nose.	NO levels increase with breathholding and hyperventilation. Expired amount (but not conc.) increases with exercise. Levels are higher in asthmatics but lower in smokers. NO may be a marker of airway inflammation, though a large portion may be derived from pulmonary vascular cells.	4
Expired air markers of inflammation (H_2O_2)	Mouthpiece, tubing, cold trap, fluorimeter	A cold trap is used to collect about 3 ml of expired air condensate. Samples are analyzed fluorimetrically.	H_2O_2 is elevated in acute hypoxemic respiratory failure, in pediatric asthmatics, and more so in ill pediatric asthmatics, suggesting it is a marker for inflammation.	3
Expired air collection of nonrespiratory gases	Cryogenic, wet chemical, or absorptive trap, GC-MS, purified air source	Subject breathes purified air, and expired alveolar air is sampled downstream after removal of water vapor. The sample is analyzed by gas chromatography/ mass spectrometry (GC-MS)	Various gases can be detected including hydrogen, carbon disulfide, numerous (~ 400) volatile organic compounds including pentane and ethane as lipid peroxidation markers. Pentane excretion is elevated after oxygen breathing, and high levels are associated with increased mortality in premature infants. This technology may be useful in identifying exposure to exotic chemicals or in identifying chemical reaction products resulting from chemical exposure or biological processes induced by such exposure.	5

1 Fahy, J. V., Liu, J., Wong, H., Boushey, H. A., Cellular and biochemical analysis of induced sputum from asthmatic and healthy subjects, *Am. Rev. Respir. Dis.*, 147, 1126–1131, 1993.

2 Pin, I., Freitag, A. P., O'Bryne, P. M., Girgis-Gabardo, A., Watson, R. M., Dolovich, J., Denburg, J. A., and Hargreave, P. E., Changes in the cellular profile of induced sputum after allergen-induced asthmatic responses, *Am. Rev. Respir. Dis.*, 145, 1265–1269, 1992.

3 Dohlman, A. W., Black, H. R., and Royall, J. A., Expired breath hydrogen peroxide is a marker of acute inflammation in pediatric patients with asthma, *Am. Rev. Respir. Dis.*, 148, 955–960, 1993.

4 Persson, M. G., Wiklund, N. P., and Gustafsson, L. E., Endogenous nitric oxide in single exhalations and the change during exercise, *Am. Rev. Respir. Dis.*, 148, 1210–1214, 1993.

5 Phillips, M. and Greenberg, J., Ion-trap detection of volatile organic compounds in alveolar breath, *Clin. Chem.*, 38, 60–65, 1992.

TABLE 32
Planning, Reviewing, Approving, and Conducting a Human Research Study

Research Plan Development	Final Protocol Development Considerations	Preparation of Consent Form Considerations	Institutional Review Board

Principal and co-investigators
Scientific peer review
Facility and personnel needs
Medical review
Safety review
  Exposure systems
  Test procedures
  Radiation safety
  Carcinogenicity
  Mutagenicity
  Teratogenicity
  Risk/benefit ratio
Statistical review
  Experimental design
  Database needs
Quality assurance review
	Gender equity
	Racial equity
	Confidentiality of data, subject privacy
	Subject recruiting methods
	Arrangements for medical treatment of subjects if necessary
	Methods for training subjects to ensure reliability and reproducibility of data
		Simple, nontechnical language
		Clear descriptions of procedures
		Names of principal and other investigators; noninvestigator physician responsible for subject welfare
		Presentation of all potential risks
		Determine compensation for volunteers, if planned; must not be coercive. Plans for partial payment in event of withdrawal or disqualification of subject
		Provision for subject copy of consent form
		Describe methods for preserving confidentiality
		Indicate number of participants
		Describe: Research purpose, duration, planned procedures, all risks and discomforts no matter how minor or unlikely, plans to minimize risks, potential benefits to society or individuals, arrangements to provide results or information to participants costs to be incurred by subjects, rights to withdraw at any time for any reason, procedures for dealing with injury or illness.
			Review and approval
			  Scientific review
			  Legal review
			  Ethics review
			Approval
			  Perform research study
			  Recruit subjects
			  Administer consent
			  Train subjects
			  Conduct experiments
			  Report adverse events or problems to IRB

TABLE 33
Subject Demographics

	Anthropometrics/Demographics	Clinical
Minimal subject information	Height, weight, age, gender	Existing chronic disease, recent acute disease, current medication and medication sensitivity, smoking history
Optimal	Also include sitting height, % body fat, race, ethnicity, blood pressure	Allergies to common allergens, nonspecific bronchial responsiveness, occupational history, spirometry, plethysmography
Additional	$\dot{V}O_2$ max, IgG, place of birth, place of residence, foreign travel history	EKG, chest X-ray, CBC with differential, lipid profile, exercise stress test, urine pregnancy test, radiation exposure history, substance use/abuse history

The safety of subjects participating in human research studies must be a primary consideration. Several aspects of the exposure and evaluation of responses must be considered. The laboratory environment must be free from objective hazards. Careful attention to the most recent toxicological and carcinogenic evaluations of any chemicals that humans will be exposed to must be performed before, during, and after completion of exposure studies. Special precautions must be taken to ensure that the subjects are never exposed to excessive concentrations of gases or particles. If gases are supplied in bottle form, a certified analysis of the contents should be provided. As an added precaution, the gas should be analyzed again after it is received to ensure that the concentrations are correct and that there are no impurities present. Similar precautions apply to permeation tubes, solutions which are to be nebulized or evaporated, and dry chemical mixtures which are to be suspended in the air. As noted previously, a complete end-to-end check of the monitoring system must be performed periodically to ensure that levels monitored at the monitoring instruments indeed reflect concentrations in the chamber or exposure system. Finally, in addition to any automated monitoring a full-time human should monitor the exposure atmospheres at all times. During exposures, the subjects should be continuously monitored by personnel trained to recognize medical problems and untoward side effects. For exposures involving exercise, a standard safety precaution includes electrocardiogram monitoring.

Many of the procedures involved in monitoring responses may be generally recognized as innocuous. Even though standard pulmonary function tests are unlikely to result in injury, maintaining cleanliness of pulmonary function and other respiratory equipment by employing respiratory filters and sterilized hoses and mouthpieces is necessary to avoid transmission of disease. Some procedures are invasive and involve special safety considerations. See, for example, the discussion of bronchoalveolar lavage and airway challenge with nonspecific chemical stimuli. Blood sampling entails risk to the subjects as well as to the laboratory personnel. The use of radioactive materials will follow federal safety guidelines and is typically dealt with through campus radiation safety committees. The risks and benefits of all procedures must be thoroughly evaluated prior to their use.

Safety of subjects must never be overlooked. Even seemingly minor problems such as performing apparently risk-free tests approved for another study but not for your study could result in temporary suspension or even termination of research. Failure to follow good laboratory practice such as proper labeling of all chemical materials could result in inadvertent injury to a subject. The repercussions of subject safety failures may extend beyond an individual study or an individual laboratory.

Each institutional review board will have its own specific set of guidelines for the development of a statement of informed consent. Some generic guidelines are included in Table 32. Each experimental plan and consent form will be subject to a series of reviews before final approval is given to proceed with the study.

In research reports, the subjects' characteristics are seldom emphasized. Table 33 presents some of the characteristics of subjects that may be considered for inclusion in reports. Adequate description of the subject population is an important adjunct to the evaluation of the results of any study, especially when the subjects represent a unique subpopulation (see Table 34). In selecting the subject population, it is important to develop criteria for acceptance or exclusion of subjects. These criteria may be related to physical abilities and characteristics, health status, previous or current disease, sex, occupation, allergies or sensitivities, and other factors. Table 36 provides a list of subject characteristics that may be useful in evaluating the suitability of subjects for inclusion in a study.

TABLE 34

Characteristics and Potential Sensitivities of Different Subject Groups

Subject Group	Characteristics and Possible Sensitivity Factors	Potential Health Effects of Inhaled Substances
Healthy	Absence of acute or chronic illness, normal airway reactivity, few or no allergies or other sensitivity factors	Normal population response
Healthy with above-normal outdoor activity	Athletes, outdoor workers; increased ventilation during exposure to inhalants in ambient air or the workplace; lower body fat	Normal population responses exacerbated by increased "dose" due to excess ventilation
Subjects with allergies (not asthmatic)	Allergic rhinitis; increased nasal responses; possible increased airway reactivity	Above-normal nasal and airway inflammatory responses
Subjects with hyperreactive airways	Increased nonspecific airway responsiveness; excitable airways	Excess airway narrowing in response to inhaled materials
Subjects with asthma	Allergies, airway hyperreactivity, sensitivity to exercise and cold air inhalation; very excitable airways; possible below-normal lung function status; medication usage; chronic respiratory symptoms	Highly excessive airway narrowing in response to some inhaled materials; excessive inflammatory response, especially when sensitized
Smokers	Status rated by pack-years; increased airway permeability, excess mucus production, impaired clearance, small airway damage, possible diminished airway receptor sensitivity	May raise risk of carcinogenesis in response to other materials; may have smaller acute response to irritants
Patients with chronic obstructive lung disease	Often smokers; may have associated cardiovascular disease; small airway damage, alveolar destruction, excess mucus secretion, chronic productive cough; airway wall thickening, nonuniform ventilation, poor exercise capacity	Healthy parts of the lung may receive disproportionately higher dose of inhalant; limited lung function reserve may exacerbate small effects; increased risk to inhalants with vascular effects
Patients with ischemic vascular disease	Coronary artery disease, peripheral vascular disease; diminished exercise capacity; exercise-related ischemia, arrhythmia	Increased risk with inhalants that cause hypoxemia
Gender-specific groups	Generally smaller lungs in women; cyclical changes in hormone levels	Possible cyclical variation in response
Race-specific groups	Limited information; higher incidence of asthma among African-Americans, for example	Further information needed on how race may be linked with response
Healthy younger subjects	Incomplete lung development, higher incidence of asthma, small stature; possibly immature lung defenses	Possible effects on development, increased airway sensitivity
Healthy older subjects	Age-related decline in lung function; decline in exercise capacity; reduced maximal heart rate; possible reduced airway receptor responses and reduced pulmonary defenses	Well-ventilated parts of the lung may receive greater dose of inhalant; increased risk of infection
Patients with acute upper respiratory infection	Normally healthy subjects with acute infections; epithelial damage, airway inflammation	Increased responsiveness to inhaled materials; exacerbation of cough and other symptoms

TABLE 35
Potential Exclusionary/Inclusionary Criteria

Abnormal hemoglobin variants (sickle cell disease)	Inability to perform test procedures
Age limit (too young or too old)	Lung function not within normal limits
Albinism	Medication use (current/past) that interferes with responses or test procedures
Allergies (number and severity)	Migraine
Anemia	Missing organs
Bronchial hyperresponsiveness	Mouth sores (herpes, gingivitis, apthos ulcers)
Cardiovascular disease/disorder	Nasal obstruction
Orthostatic hypotension, positive family history of arteriosclerotic disease prior to age 50	Orthopedic injury or disability
Positive/negative cardiovascular exam, positive/negative stress test	Past or current occupational or job-related exposures to particles, chemicals, etc.
Raynaud's disease	Physical fitness limitations
Rheumatic fever	Pregnancy
Chronic respiratory disease	Psychological profile
Asthma	Radiation exposure exceeding federal guidelines
Bronchitis	Recent travel to countries with risk of blood-borne pathogen infection
COPD	Risks for blood-borne pathogen (blood transfusion, tattoos)
Other	Malaria
Current/recent exposure to high levels of air pollution	Sensitivity or lack of sensitivity to a specific chemical
Current/recent acute illness (e.g., upper respiratory infection)	Smoking history
Dermatitis (eczema, psoriasis)	Tobacco
Diabetes	Other substances
Dyslipidemia (e.g., elevated serum cholesterol, triglycerides)	Passive smoke exposure
Epilepsy, history of seizures	Substance abuse
G6PD deficiency	Syncope, especially with phlebotomy
Glaucoma	Thoracic cage abnormalities (kyphoscoliosis, pectus excavatum)
Hay fever	Thyroid disorders
Height or weight outside normal limits	Tuberculosis
Hepatitis	Venereal disease
HIV	
Hypertension	

Note: This list is not exhaustive and is in no way intended to supplant professional judgment. Many of the listed criteria include those that otherwise healthy individuals may have. We do not intend to imply that all laboratories should test for all of the above. Clearly, a much more (or less) extensive list of criteria can be developed. One of the aims is to exclude subjects who would put themselves at greater than acceptable risk by participating. In some cases, it may be necessary to limit the study population to homogeneous subject groups such as mild allergic asthmatics or patients with coronary artery disease.

REFERENCES

1. Frank, R., O'Neil, J. J., Utell, M. J., Hackney, J. D., Van Ryzin, J., Brubacker, P. E., Eds., *Inhalation Toxicology of Air Pollution: Clinical Research Considerations*, ASTM STP 872, American Society for Testing and Materials, Philadelphia, PA, 1985.
2. Folinsbee, L. J., Human clinical inhalation exposures: experimental design, methodology, and physiological responses, in *Toxicology of the Lung*, Gardner, D. E., Crapo, J. D., and Massoro, E. J., Eds., Raven Press, New York, 1988, pp. 175–199.
3. Hackney, J. D., Linn, W. S., Buckley, R. D., Pedersen, E. S., Karuza, S. K., Law, D. C., and Fischer, A., Experimental studies on human health effects of air pollutants, *Arch. Environ. Health*, 30, 373–378, 1975.
4. Utell, M. J., Frampton, M. W., and Morrow, P. E., Quantitative clinical studies with defined exposure atmospheres, in *Toxicology of the Lung*, 2nd ed., Gardner, D. E., Crapo, J. D., and McClellan, R. O., Eds., Raven Press, New York, 1993, pp. 382–309.

5. Nelson, G. O., *Controlled Test Atmospheres,* Ann Arbor Science, Ann Arbor, MI, 1971.
6. Phalen, R. F., *Inhalation Studies: Foundations and Techniques,* CRC Press, Boca Raton, FL, 1984.
7. McQuiston, F. C. and Parker, J. D., *Heating, Ventilating, and Air Conditioning,* 4th ed., John Wiley & Sons, New York, 1994.
8. ASHRAE, *Heating, Ventilating, and Air Conditioning Applications,* American Society of Heating, Refrigeration, and Air Conditioning Engineers, Atlanta, GA, 1991.
9. ASHRAE, *Heating, Ventilating, and Air Conditioning Systems and Equipment,* American Society of Heating, Refrigeration, and Air Conditioning Engineers, Atlanta, GA, 1992.
10. Hering, S. V., Ed., *Air Sampling Instruments for Evaluation of Atmospheric Contaminants,* American Conference of Government Industrial Hygienists, Cincinnati, OH, 1989.
11. Raabe, O. G., Generation of aerosols of fine particles, in *Fine Particles,* Liu, B. Y. H. et al., Eds., Academic Press, New York, 1975, pp. 57–110.
12. Whitby, K. T., Lundgren, D. A., and Peterson, C. M., Homogenous aerosol generators, *Int. J. Air Water Pollut.,* 9, 263–277, 1965.
13. Peters, T. M., Chien, H. M., Lundgren, D. A., and Bernsten, J., Submicrometer aerosol generator development for the U.S. Environmental Protection Agency's human exposure laboratory, *Aerosol Sci. Technol.,* 20, 51–61, 1994.
14. Strong, A. A., *Description of the CLEANS Human Exposure System,* EPA-600/1-78-064, U.S. Environmental Protection Agency, U.S. Government Printing Office, Washington, D.C., 1978.
15. Miller, F. et al., Eds., *Nasal Toxicity and Dosimetry of Inhaled Xenobiotics: Implications for Human Health,* Taylor & Francis, Washington, D.C., 1995.

15a. Brain, J. D., The uptake of inhaled gases by the nose, *Ann. Otol. Rhinol. Laryngol.,* 79, 1–11, 1970.

15b. Scott, W. T. and McKenna, M. J., The comparative absorption and excretion of chemical vapors by the upper, lower, and intact respiratory tract of rats, *Fund. Appl. Toxicol.,* 4, 594–602, 1984.

15c. Cometto-Muniz, J. E. and Cain, W., Perception of nasal pungency in smokers and nonsmokers, *Physiol. Behav.,* 29, 727–731, 1982.

15d. Alarie, Y., Irritating properties of airborne materials to the upper respiratory tract, *Arch. Environ. Health,* 13, 433–449, 1966.

15e. Lundblad, L., Lundberg, J. M., Brodin, E., and Anggard, A., Origin and distribution of capsaicin-sensitive substance P-immunoreactive nerves in the rat mucosa, *Acta Otolaryngol.,* 96, 485–493, 1983.

15f. Cain, W. and Murphy, C., Interaction between chemoreceptive modalities of odour and irritation, *Nature,* 284, 255–257, 1980.

16. Cain, W. S., Gent, J., Catalanotto, F. A., and Goodspeed, R. B., Clinical evaluation of olfaction, *Am. J. Otolaryngol.,* 4, 252–256, 1983.
17. Doty, R. L., Shaman, P., and Dann, M., Development of the University of Pennsylvania Smell Identification Test: a standardized microencapsulated test of olfactory function, *Physiol. Behavior.,* 32, 489–502, 1984.
18. Elsberg, C. A. and Levy, I., The sense of smell. A new and simple method of quantitative olfactometry, *Bull. Neurol. Inst. N.Y.,* 4, 5–19, 1935.
19. Hudson, R., Laska, M., Berger, T., Heye, B., Schopohl, J., and Danek, A., Olfactory function in patients with hypogonadotropic hypogonadism: an all-or-none phenomenon?, *Chemical Senses,* 19, 57–69, 1994.
20. Prah, J. D. and Benignus, V., Effects of ozone exposure on olfactory sensitivity, *Perceptual Motor Skills,* 48, 317–318, 1979.
21. Prah, J. D., Walker, J. C., and Sears, S. B., Modern approaches to clinical and research olfactometry, in *Handbook of Clinical Olfaction and Gustation,* Doty, R. L., Ed., Marcel Dekker, New York, 1994.
22. Ferris, B. G., Epidemiology Standardization Project. III. Recommended standardized procedure for pulmonary function testing, *Am. Rev. Respir. Dis.,* 118(Suppl. 2), 55–88, 1978.
23. American Thoracic Society, ATS Statement: snowbird workshop on standardization of spirometry, *Am. Rev. Respir. Dis.,* 119, 831–838, 1979.
24. American Thoracic Society, Standardization of spirometry — 1987 update, *Am. Rev. Respir. Dis.,* 136, 1285–1298, 1987.

24a. Dawson, A., Pneumotachography, in *Pulmonary Function Testing: Guidelines and Controversies,* Clausen, J. L. and Zarins, L. P., Eds., Academic Press, New York, 1982.

25. Krowka, M. J., Enright, P. L., Rodarte, J. R., and Hyatt, R. E., Effect of effort on measurement of forced expiratory volume in one second, *Am. Rev. Respir. Dis.,* 136, 829–833, 1987.
26. Wagner, P. D., Saltzman, H. A., and West, J. B., Measurements of continuous distributions of ventiliation-perfusion ratios: theory, *J. Appl. Physiol.,* 36, 588–599, 1974.
27. Pincock, A. C. and Miller, M. R., The effect of temperature on recording spirograms, *Am. Rev. Respir. Dis.,* 128, 894–898, 1983.
28. Hankinson, J. L. and Viola, J. O., Dynamic correction factors for spirometric data, *J. Appl. Physiol.,* 55, 1354–1360, 1983.
29. Eiser, N. M., Kerrebijn, K. F., and Quanjer, P. H., Guidelines for standardization of bronchial challenges with (nonspecific) bronchoconstricting agents, *Bull. Eur. Pathophysiol. Resp.,* 19, 495–514, 1983.

30. Godfrey, S., Bronchial challenge by exercise or hyperventilation, in *Provocative Challenge Procedures: Background and Methodology,* Spector, S. L., Ed., Futura Publishing, Mount Kisco, NY, 1989, pp. 365–394.
31. Subcommittee on Bronchial Inhalation Challenges, Guidelines for bronchial inhalation challenges with pharmacologic and antigenic agents, *ATS News,* Spring, 11–19, 1980.
32. Cockcroft, D. W., Killian, D. N., Mellon, J. J. A., and Hargreave, F. E., Bronchial reactivity to inhaled histamine: a method and clinical survey, *Clin. Allergy,* 7, 235–243, 1977.
33. Spector, S. L., Ed., *Provocative Challenge Procedures: Background and Methodology,* Futura Publishing, Mount Kisco, NY, 1989.
34. Strong, A. A., Hazucha, M. J., Lundgren, D. A., and Cerini, E. R., An aerosol generator system for inhalation delivery of pharmacologic agents, *Med. Instrument.,* 87, 189–194, 1987.
35. Orehek, J., Gayrard, P., Smith, A. P., Grimaud, C., Charpin, J., Airway response to carbachol in normal and asthmatic subjects. Distinction between bronchial sensitivity and reactivity, *Am. Rev. Respir. Dis.,* 115, 937, 1977.
36. Hopp, R. J., Weiss, S. J., Nair, N. M., Bewtra, A. K., and Townley, R. G., Interpretation of the results of methacholine inhalation challenge tests, *J. Allergy Clin. Immunol.,* 80, 821–830, 1987.
37. American College of Sports Medicine, *Guidelines for Graded Exercise Testing and Exercise Prescription,* 3rd ed., Lea & Febiger, Philadelphia, PA, 1986.
38. Merkus, P. J., Essen-Zandvliet, E., van Parleviet, E., Borsboom, G., Sterk, P. J., Kerrebijn, K. F., and Quanjer, Ph. H., Changes of nebulizer output over the years, *Eur. Respir. J.,* 5, 488–491, 1992.
39. Hollie, M. C., Malone, R. A., Skufca, R. M., and Nelson, H. S., Extreme variability in aerosol output of the DeVilbiss 646 jet nebulizer, *Chest,* 100, 1339–1344, 1991.
40. Gravelyn, T. R., Pam, P. M., and Eschenbacher, W. L., Mediator release in an isolated airway segment in subjects with asthma, *Am. Rev. Respir. Dis.,* 137, 641–646, 1988.
41. Hunninghake, G., Gadek, J. E., Kawamani, O., Ferrans, V. J., and Crystal, R. G., Inflammatory and immune processes in the human lung in health and diseases: evaluation by bronchoalveolar lavage, *Am. J. Pathol.,* 97, 149–191, 1979.
42. Klech, H. and Pohl W., Technical recommendations and guidelines for bronchoalveolar lavage: European Working Party Working Report, *Eur. Respir. J.,* 2, 561–585, 1989.
43. National Institutes of Health, Workshop summary and guidelines: investigative use of bronchoscopy, lavage, and bronchial biopsies in asthma and other airway diseases, *J. Allergy Clin. Immunol.,* 88, 808–814, 1991.
44. Stanley, M. W., Henry-Stanley, M. J., and Iber, C., *Bronchoalveolar Lavage: Cytology and Clinical Applications,* Igaku-Shoin, New York, 1991.
45. Smith, D. L. and DeShazo, R. D., Bronchoalveolar lavage in asthma, *Am. Rev. Respir. Dis.,* 148, 523–532, 1993.
46. Utell, M. J., Frampton, M. W., and Morrow, P. E., Quantitative clinical studies with defined exposure atmospheres, in *Toxicology of the Lung,* Gardiner, D. E. et al., Eds., Raven Press, New York, 1988, pp. 283–308.
47. Van Vyve, T., Chanez, P., Bousquet, J., Lacoste, J.-Y., Michel, F.-B., and Goddard, P., Safety of bronchoalveolar lavage and bronchial biopsies in patients with asthma of variable severity, *Am. Rev. Respir. Dis.,* 146, 116–121, 1992.
48. Wagner, E. M., Liu, M. C., Weinmann, G. G., Permutt, S., and Blecker, E. R., Peripheral resistance in normal and asthmatic subjects, *Am. Rev. Respir. Dis.,* 141, 584–588, 1990.
49. Walters, E. H. and Gardiner, P. V., Bronchoalveolar lavage as a research tool, *Thorax,* 46, 613–618, 1991.
50. Young, K. R. and Reynolds, H. Y., Bronchoalveolar lavage in inhalation lung toxicity, in *Pathophysiology and Treatment of Inhalation Injuries,* Loke, J., Ed., Marcel Dekker, New York, 1988, pp. 207–237.

Disclaimer

This chapter has been reviewed in accordance with U.S. Environmental Protection Agency policy and approved for publication. It does not necessarily reflect the views of the agency, and no official endorsement should be inferred. Mention of trade names or commercial products does not constitute endorsement or recommendation for use.

PART 4

IMMUNOTOXICOLOGY

Kathleen E. Rodgers, Editor

CHAPTER 15

INTRODUCTION

Kathleen E. Rodgers

CONTENTS

Immunotoxicology is the study of injury to the immune system that can result from occupational, inadvertent, or therapeutic exposure to a variety of compounds. The field has two broad research areas: one involves studies of the suppression of immunity; the other examines the effects of enhanced or excessive immune responses as occurs in allergy or autoimmunity. Suppression of the immune system can result in an increased incidence of infectious disease or neoplasia. When the immune system generates an immune response to a xenobiotic, adverse effects, such as respiratory tract allergies or contact dermatitis, can develop.

The field of immunotoxicology has grown considerably since the early 1970s. The discipline developed through studies that combined knowledge of immunology and toxicology to determine the effect of xenobiotics on the functioning of the immune system. The field now includes initial identification of suspected immunotoxic chemicals; the development of sensitive, quantitative animal assays to assess any chemically induced immunologic effects; and determinations of the mechanisms by which xenobiotics can alter immune function. The methodologies developed are discussed in the first section of this subsection, and data resulting from utilization of these methods are outlined in subsequent sections.

In a few instances, the immune system has been shown to be the target organ for a chemical. For these chemicals, a significant effect has been shown on the immune response in animal models at exposures that did not alter other commonly tested measures of toxicity, such as body weight, blood chemistries, or organ systems previously identified as the target of the compound.

The immune system can be affected by a variety of conditions, substances, and agents. Those most commonly encountered are radiation, immunosuppressive drugs, infectious

0-8493-4493-X/97/$0.00+$.50

agents, neoplasia, smoking, and pregnancy. There are a number of hereditary syndromes of immune deficiency. Hereditary and environmental factors, such as stress and smoking, must be evaluated carefully in the assessment of suspected chemical-induced dysfunction. The putative causes of immune modulation, particularly in human populations, are diverse. Evaluation of data from immunotoxicity studies in humans must, at least, include considerations of emotional distress as a contributing factor.

1 IMMUNE HYPERSENSITIVITY

Hypersensitivity disorders are by far the most widely recognized manifestations of immunotoxicity. These disorders can result from exposure to environmental contaminants or occupational exposure to chemicals and have been widely reported and documented. Despite the relatively high morbidity and long history of these disorders (particularly in the workplace), much of the pathophysiology has only recently been established, and many questions remain to be answered. Progress in addressing this important health problem has been impaired by the lack of development of appropriate animal models. Consequently, clinical research has been depended upon to a large extent to establish mechanisms and effective modes of treatment.

Several features distinguish hypersensitivity from immunosuppression. First, hypersensitivity disorders are more readily defined and studied; therefore, there is a great deal more evidence for their occurrence. Second, they usually entail a specific antibody, receptor, cell population, or target tissue. Third, because of this specificity, diagnostic and epidemiologic approaches tend to be problem oriented. Lastly, biologic markers of susceptibility, exposure, and effect are more often defined by suspected mechanisms.

Although clinical signs of hypersensitivity generally provide adequate markers of effect, linkage to the particular chemical has presented a challenge. The natural history, epidemiology, and clinicopathologic aspects are pertinent to the consideration of biologic markers of these disorders.

2 AUTOIMMUNE DISEASES

Autoimmune diseases are those in which an individual's own immune system attacks one or more tissues or organs, resulting in a functional impairment and sometimes permanent tissue damage. Autoimmune diseases result from the loss of immune tolerance to self-antigens, and an immune response to one or more relevant tissue antigens can be demonstrated. The autoimmune response can produce disease directly by means of circulating antibody and indirectly through the formation of immune complexes, or as a consequence of cell-mediated immunity. In most cases, more than one pathogenetic mechanism manifests itself in the generation of autoimmune disease. The etiology of autoimmune disease is complex and generally involves a genetic susceptibility to development of the disease.

3 IMMUNOSUPPRESSION

There is an increasing awareness and concern within the scientific and public communities that chemical pollutants can suppress immune processes and thus cause increased development of neoplastic and infectious diseases. Adverse effects on humans treated with immunosuppressive drugs, numerous studies employing experimental animals, and, to a lesser extent, isolated cases of altered immune function in humans inadvertently or occupationally exposed to xenobiotic substances support these concerns. There is no definitive evidence, as yet, that persons who live near contaminated sites or chemical-manufacturing plants have been immunologically compromised to the extent that they are at increased risk of disease. Nevertheless, there is reason to believe that chemical-induced damage to the immune system might be associated with pathologic conditions, some of which could become detectable only after a long latency. Likewise, exposure to immunotoxic xenobiotics may present an additional risk to individuals with immune systems that are already fragile, for example, because of primary immunodeficiency, infancy, or old age. Most of the experimental data on the effects of xenobiotics on immune function have been generated in animal models, and the tables in the sections to follow reflect this; however, as in other area of toxicology, it is difficult to extrapolate change in a given area of immune function in experimental animals to the incidence of clinical or pathologic effects in humans.

CHAPTER 16

IMMUNOTOXICOLOGY METHODS

Robert V. House

CONTENTS

0-8493-4493-X/97/$0.00+$.50

1 INTRODUCTION

Many excellent review articles and, indeed, several textbooks have been written on the subject of methodologies used in immunotoxicity assessment, suggesting that nothing more can be said concerning the subject; however, this chapter will differ somewhat from those preceding it. The toxicological and immunological bases for the various assays are described, along with suggestions on how the results might be interpreted. Also, the assay techniques themselves are described and appropriately referenced, along with suggestions for appropriate positive and negative controls. In addition, potential confounding factors and methodological problems are discussed. This chapter differs, however, in the presentation of a number of immunological techniques which have not, until now, been extensively utilized as tools for investigative immunotoxicology. In addition, suggestions on how these techniques may be incorporated into actual study protocol are discussed.

The chapter covers immunotoxicology assessment in laboratory rodents and humans only. Although certain other species are utilized in selected assays such as hypersensitivity (e.g., guinea pigs), the intent of this chapter is to provide a unified series of immune function assays; this has necessitated discussion of the mouse and human only. Many of the immune functions tests described for the mouse are readily adaptable to the rat; likewise, many human assays may be employed with nonhuman primates. Most important, it is hoped that these descriptions will motivate interested investigators to adapt, validate, and incorporate these various techniques into both routine and mechanistic immunotoxicology investigations.

2 IMMUNOMODULATION IN LABORATORY ANIMALS

2.1 Assessment of Immunopathology

2.1.1 *Body/Organ Weights and Cellularities*

Quantification of body and organ weights forms an integral part of any toxicological study, providing an initial assessment of overall animal health status as well as potential pathology. Descriptions of the tier approach to immunotoxicity evaluation incorporated measurement of body weight, as well as weights of spleen, thymus, kidney, and liver in an initial screen.[74,75] Spleen and thymus weights are included due to their obvious role in immune reactions (B-cell function and T-cell maturation and differentiation, respectively). Liver and kidney determinations are included because of the presence of immune system components (Kupffer cells, mesangial cells), as well as their importance in overall toxicity assessment.[123] Organ weights are routinely expressed as absolute weights as well as percent of body weight; other ratios (organ/organ) may also be employed for comparative purposes. In general, gross observations such as organ weight alone have not been found to be good indicators of immunotoxicity.[75] In comparison, relative ratios such as spleen/body weight or thymus/body weight exhibit a greater concordance and provide more predictive value.[75] In addition to weights, lymphoid organ cellularities are generally included in most experimental designs. As with weights, cellularities may be expressed as actual number of cells per organ or as relative values. A recent review provides an excellent overview of various considerations in assessment of changes in organs weights as they relate to immunotoxicology.[10]

2.1.2 *Histopathology*

The preponderance of assays employed for assessment of immunomodulation is based on observation of functional changes, in particular as they relate to host defense; however, structural changes resulting from xenobiotic exposure are well known. Evaluation of these morphologic changes, in the context of functional alterations, may assist in understanding the mechanism of action of immunotoxicants. Pathology of the immune tissues may take a variety of forms. For example, proliferation of lymphoid cells or associated tissue may proceed normally or in a hyperplastic or neoplastic manner. Cell death may proceed via normal apoptosis (discussed in the next section), or by degeneration and necrosis following toxic insult. In addition, secondary changes may result from inflammatory processes. It follows that proper evaluation of pathological changes requires an intimate knowledge of the structure of both primary (e.g., bone marrow) and secondary (e.g., spleen) immune tissues.

The dynamic nature of the immune system renders it susceptible to toxic damage. The thymus and bone marrow, each with a high rate of cellular proliferation and differentiation, appear to be particularly sensitive to toxic insults and may serve as early indicators of immunotoxicity.[122] Because of the regenerative nature of the immune system, toxicity to these elements may be relatively transitory. Thus, an important consideration in immunopathology is the elapsed time between toxicant exposure and evaluation. A number of other intrinsic factors may significantly affect the morphology of immune system tissues, including stress, the action of steroid hormones, antigenic load, nutritional status, and age;[123] these factors must be taken into account when interpreting the results of histopathological assays.

Comprehensive descriptions of methodology for histological assays in immunotoxicology have been published.[66,121,123] Tier-type immunotoxicology paradigms usually include histological evaluation of "internal" immune system organs such as spleen, thymus, lymph node, and bone marrow; however, the immune system also includes a variety of "external" or secretory immune tissues, generically grouped together as mucosa-associated lymphoid tissue (MALT). This includes elements in the bronchus (BALT), skin (SALT), gut (GALT), and nasal passages (NALT). For a thorough evaluation of immunohistopathology, as many of these tissues as possible should be included. In certain cases, however, it may not be practical to include all of these tissues. For example, SALT includes elements such as the Langerhans cells, which may be difficult to isolate. Thus, it may be necessary to determine the need for such evaluation on a case-by-case basis.

A recent review[123] has surveyed the use of histopathology for immunotoxicology assessment. In general, the authors found histopathology to be less sensitive in detecting immunotoxic substances than some functional assays. This may have been due to the relatively lower doses utilized, as higher exposure levels employed for toxicity tests resulted in detection of toxicity by histology. Still, detection of structural changes provides mechanistic insight into other observed changes in immune competence. In addition, histopathological changes detected in the course of routine toxicology screens may readily be evaluated for immune functional changes on an as-needed basis.[14] Thus, for the foreseeable future, histopathology assessment is expected to remain an important component of overall immunotoxicity assessment.

2.1.3 Assessment of Apoptosis

Apoptosis is one of two forms of cell death currently recognized, the other form being necrosis. In necrosis, cells lose membrane integrity, leading to catastrophic loss of intracellular contents. By way of contrast, apoptosis involves membrane blebbing or zeiosis, cytoplasmal condensation, and activation of endogenous endonucleases which cleave cellular DNA into discrete fragments. It is the most common form of cell death in eukaryotes and is important in pathology, normal growth, and differentiation.[8] In particular, apoptosis is important in normal immune function,[22] where it is involved in thymic selection and the function of natural killer (NK) and cytotoxic T-lymphocyte (CTL) cells. In addition, the study of apoptosis is receiving increasing interest in the field of toxicology.[24] Although apoptosis has numerous potential applications in both toxicology and immunology, its use in the field of immunotoxicology has remained largely unexplored. The demonstrated role of apoptosis in cellular regulation suggests its potential application in immunotoxicology, especially in conjunction with more traditional methods of histopathology assessment.

A variety of techniques has been described for the quantitation of apoptosis, including morphological analysis, analysis of DNA degradation, flow cytometric determination, 3′-OH end labeling, and analysis of endonucleases.[128] These various techniques are all rather specialized and are best suited for dedicated mechanistic evaluation. Recently, commercial enzyme-linked immunosorbent assays (ELISAs) have become available which are based upon detection of histone-associated DNA fragments resulting from internucleosomal degradation of DNA or, alternatively, on the detection of Bcl-2. ELISAs offer several advantages in immunotoxicology, including rapid results, simplicity, and cost-effectiveness. For circumstances where a rapid screening type of application is required, these kits may represent one alternative for quantifying apoptosis. At present, none of these kits has been validated for immunotoxicology assessment.

2.1.4 Surface Marker Analysis

The combination of highly specific monoclonal antibodies and flow cytometry has provided a powerful tool for assessing changes in the intrinsic cell populations of the immune system. The immune system responds to changing host resistance requirements by controlling the intricate traffic of lymphocytes and other immune cells throughout the body.[36] By assessing the relative populations of these cells, including T cells, B cells, NK cells, and monocytes/macrophages, a generalized view of the immune status of the animal is obtained. Although these various populations may be quantified microscopically, flow cytometry greatly accelerates the process and allows for more accurate and detailed analysis such as T-cell subpopulations, etc.[26]

Quantitation of splenic T- and B-cell numbers was originally classified as a Tier II or mechanistic-type assay under the original National Toxicology Program (NTP) classification scheme.[74] When this parameter was evaluated for its predictive potential in combination with other tests, changes in surface markers exhibited a high degree of correlation with immunotoxicity.[75] Quantitation of total T or B cells alone was not as informative as evaluation of CD4:CD8 ratios. Even in the absence of studies to correlate various new markers with generalized immunotoxicity, results with the various surface markers may still be valuable. For example, alterations in NK-cell activity could be compared with expression of NK cell-related surface antigens; this would allow delineation of altered function vs. altered cell numbers.

Various important points must be considered when interpreting the results of surface marker analysis, especially in the context of immunotoxicity assessment. Surface marker analysis is usually performed on cells from either blood or spleen. However, any alterations detected in one tissue may not reflect pathology in other tissues; for example, it is conceivable that xenobiotic exposure may alter cellular traffic, resulting in alterations in immune function. In addition, marker analysis often does not take into account the heterogeneity of lymphocytes (i.e., different ages, different natural histories). Thus, alterations in selected populations, although of critical importance functionally, may not be reflected in gross observations of whole populations. Finally, and perhaps most importantly, functional alterations are often difficult to infer from alterations in cellular ratios.[144] These observations highlight some of the issues and uncertainties that must be addressed in interpretation of such data.

2.2 Assessment of Cell-Mediated Immunity

2.2.1 Lymphoproliferation

2.2.1.1 Background

Proliferation of lymphocytes following *in vitro* exposure to mitogenic stimuli is an important methodology; activation of lymphocytes followed by proliferation indicates a certain

TABLE 1
Stimulants Commonly Used in Lymphoproliferative Assays

Mitogen	Suggested Concentrations	Notes
B lymphocytes		
Staphylococcus aureus (SAE)	1:5000 dilution	Human mitogen
E. coli lipopolysaccharide (LPS)	10–100 μg/ml	Human and rodent mitogen, also others
Salmonella typhimurium mitogen (STM)	10–50 μg/ml	Rat mitogen
Anti-IgM/rIL-4	—	—
T lymphocytes		
Phytohemagglutinin (PHA)	1–10 μg/ml	Works for most species tested
Concanavalin A (ConA)	1–5 μg/ml	Works for most species tested
Tetanus toxoid	—	Human mitogen
Anti-CD3 monoclonal antibody	10–100 μg/ml	Species-specific; works via the antigen receptor

degree of functionality. This assay has enjoyed frequent use in immunotoxicology studies because of its ease of performance and relative reproducibility and provides a good system for comparative studies.[68,133]

2.2.1.2 Methodology

Lymphoproliferation assays in their current form were first described by Anderson et al.[3] Proliferation assays for T lymphocytes have conventionally employed plant lectins such as phytohemagglutinin (PHA) or concanavalin A (ConA), whereas B lymphocytes are stimulated with bacteria (e.g., *Staphylococcus aureus*) or their constituent components (e.g., lipopolysaccharide, LPS) or plant lectins (e.g., pokeweed mitogen, PWM).[54] More recently, monoclonal antibodies specific for the CD3 antigen of the T-cell antigen receptor have been employed for T-cell activation. The binding of this antibody to its cognate antigen results in cellular activation and subsequent proliferation. While this is not strictly analogous to antigen-specific cellular activation, it is arguably more physiologically relevant than mitogenic lectins. Lymphoproliferation assays (Table 1) are routinely performed in a microculture format in which the cells (e.g., splenic lymphocytes, lymph node cells) are cultured in 96-well microculture plates, allowing for ready adaptation to semi- or full automation. Replicate wells of these plates also receive aliquots of the appropriate stimuli. In general practice at least three concentrations of antigen, antigen analog, or mitogen are also added to the wells. Separate wells contain culture medium only, and serve as negative controls. The plates are incubated at 37°C for 3 to 4 days to allow a strong proliferative response to develop. The degree of relative proliferation is subsequently measured.

Another variation of the lymphoproliferation assay is the mixed leukocyte response (MLR), also known as the mixed lymphocyte culture (MLC). This assay measures T-cell proliferation in response to alloantigens (the clinical correlate of this reaction is the transplantation reaction). The MLR response has been demonstrated to be a good indicator of generalized immunocompetence[74] and is often included in the Tier I approach to immunotoxicology testing. The technique is based on the work of Bach and Voynow[6] using one-way mixed lymphocyte cultures. In essence, cells from the treated animal (i.e., responder cells) are cultured in replicate wells of microculture plates. These cells are stimulated with an optimum number of genetically dissimilar cells (i.e., stimulators) which have been metabolically inactivated, usually by irradiation or treatment with a DNA intercalating agent such as mitomycin C. Following incubation for several days, the proliferative response of the responder cells is measured.

TABLE 2

Colorimetric Indicator Systems for Assessing Cellular Proliferation

Reagent System	Nature of Reaction	Ref.
MTT/PMS[a]	Tetrazolium salt/coupling agent metabolized by mitochondria to an insoluble formazan	90
XTT/PMS[b]	Same as MTT/PMS except for aqueous-soluble end product	112
MTS/PMS[c]	Same as XTT/PMS but more stable in solution	42

[a] MTT: 3-(4,5-dimethylthiazol-2-yl)-5-(3-carboxymethoxyphenyl)-2-(4-sulfophenyl)-2H-tetrazolium; PMS: phenazine methosulfate.
[b] XTT: sodium (2,3-bis(2-methoxy-4-nitro-5-sulfophenyl)-2H-tetrazolium.
[c] MTS: 3-(4,5-dimethylthiazol-2-yl)-2,5-diphenyltetrazolium bromide.

2.2.1.3 Colorimetric Methodology for Assessment of Proliferation

The methods for assessing cellular proliferation have undergone some modification in recent years, and these changes merit a brief discussion. Assessment of proliferation has traditionally been performed by adding tritiated thymidine to the cell cultures, usually 18 to 24 hr prior to assay termination (i.e., harvest). Proliferating cells incorporate the radiolabel into DNA in proportion to the degree of cellular replication. At harvest, the cells are harvested on glass fiber filters, precipitated with trichloroacetic acid, and dried. The filters containing precipitated radiolabel are subsequently quantified by scintillation spectrometry. Although this technique has the advantage of measuring cellular proliferation as a function of DNA replication, it has numerous disadvantages. Of prime concern is the generation of significant quantities of radioactive and toxic waste streams. In addition, this technique is both time- and labor-intensive. More recently, colorimetric microculture methods have been developed for assessing cellular proliferation. One of the most widely recognized of these is the MTT assay.[90] MTT is a tetrazolium salt which is cleaved by the mitochondria of viable cells to a colored formazan end product. There are several advantages to this assay, including: (1) a quick turnaround, with the endpoint determination occurring 2 to 4 hr following reagent addition to the cultures; (2) a microculture format with all steps performed directly in the microplate wells; (3) rapid readout: minutes in a microplate reader vs. hours or days in a scintillation counter; and (4) no radioactive waste stream and only minimal amounts of toxic waste products.

As originally described by Mosmann, the MTT system yielded a good correlation between results obtained between colorimetric and radiometric assays. Subsequent work by Wemme et al.[143] suggested that, although MTT gave good results, it was not a replacement for thymidine incorporation. The MTT assay does have one major flaw, namely that the formazan end product is aqueous insoluble. Although various improvements have been suggested,[95] the results are often questionable, and necessary manipulations and reagents may alter the results, canceling any advantages gained. A marked improvement in the colorimetric assay came with the description of a new tetrazolium compound, XTT,[112] which produces a soluble formazan, obviating the solubilization step. Unfortunately, XTT lacks stability in culture due to rapid recrystallization and must be added immediately following preparation. The most recent refinement in colorimetric assays is the reagent MTS,[20a] which, unlike XTT, is stable in culture.[42] (See Table 2.)

2.2.2 Cytotoxic T-Lymphocyte Function

2.2.2.1 Background

Cytotoxic T lymphocytes (CTL) are a population of CD8-positive lymphocytes characterized by specific cytotoxicity for target cells. Similar to natural killer cells described below,

TABLE 3

Potential Mechanisms of Altered CTL Function

Modulation of CTL precursor frequency or function
Specific or nonspecific cytotoxicity to CD4 or CD8 cells
Modulation of T-helper cell function
Inhibition of activation by CD4 or CD8 cells
Alteration in the interaction of CD4 and CD8 cells
Inhibition of CTL effector function

CTL are able to directly lyse their targets; however, unlike NK cells, CTL lyse their targets in an MHC-restricted manner[91] and require a sensitization period in which precursor cells differentiate into effector cells.[12] CTL are able to destroy a variety of targets, including tumor cells and virally infected cells.[81] The exact mechanism of target cell destruction is the source of some contention, although it appears that virally infected and tumor cells experience enforced cell death, whereas other cells experience induced cell death.[100] In some cases, CTL may represent the major form of immunity to infection; thus, measurement of their function may contribute to the evaluation of total host resistance to certain forms of disease. CTL are also thought to be responsible for the *in vivo* elimination of neoplastic cells, and alterations in CTL function are significantly correlated with *in vivo* susceptibility to *Listeria monocytogenes* and transplantable tumors.[76]

Cytotoxic T lymphocytes are derived from a precursor population (CTL_p), the majority of which are specific for H-2K or H-2D determinants on stimulator/target cells, although limited reactivity to H-2I has also been described.[91] Following stimulation with the appropriate antigen, maturation and differentiation into effector cells proceeds in the absence of DNA synthesis.[78] The induction of CTL from CTL_p involves specific interactions of several cell types and cytokines. The result of this complexity is that assessment of CTL function may reveal deficits not only in the effector phase of the immune response, but also functional abnormalities in cellular activation and regulatory pathways. (See Table 3.)

2.2.2.2 Methodology

Methods for the induction of cytotoxic cells were first reported by Wunderlich and Canty[146a] and later modified by Brunner et al.[19] The method described below is a simplified version utilized in our laboratory which yields consistent control values, limited assay-to-assay variability, low background cytotoxicity, and demonstrated sensitivity to modulation by xenobiotics.[49] An important benefit of this *in vitro* induction procedure is the ability to expose cells *in vitro*, which makes it practical to evaluate the effect of test agents too toxic, expensive, or rare to test *in vivo*. In addition, it facilitates greater control over the timing of exposure (and thus the target cells affected) as well as the delivered dose of a test agent.

Methodology for the *in vitro* induction of murine CTL has recently been reviewed.[48] Murine splenocytes (responder cells) are co-cultured for 5 days with mitomycin C-inactivated P815 mastocytoma cells (stimulator cells). The use of living stimulator cells in this instance provides a strong inductive signal for the development of CTL. This *in vitro* induction period offers an opportunity to assess the ability of xenobiotics to directly affect cell-mediated immunity, as chemicals may be added directly to the culture flasks. Following this induction, the CTL (effector cells) are washed to remove cellular debris. These cells are adjusted to a known concentration, and several dilutions are prepared from this preparation. These effector cell dilutions are subsequently added to replicate wells of 96-well, round-bottom microculture plates. Fresh P815 target cells are radiolabeled with ^{51}Cr, and free radiolabel is washed away. (Alternative endpoints have been reported by Ohmori et al.[98] and Hussain et al.[50]). The target cells are adjusted to a known concentration and added to the plates containing the effector cells. Following a 4-hr incubation, the cell-free supernatant fluids are collected, and radiolabel release (or an alternative endpoint) is quantified. Percent specific cytotoxicity is calculated as percent of total releasable counts, corrected for spontaneous release. In general, the final effector:target cell ratios are 25:1, 12.5:1, and 6.25:1, although theoretically any series of ratios could be used. The use of multiple cell ratios allows for the calculation of lytic units. Whereas many immunotoxicology investigators report cytotoxicity of CTL and NK in terms of percent lysis, this method may not be particularly useful when comparing cytotoxicity at different effector:target ratios. The lytic unit facilitates expression of cytotoxic activity independent of these ratios.[20]

The procedures described above have been used extensively in many laboratories to evaluate the effect of various test agents on cell-mediated immunity, as well as serving as the basis of more detailed mechanistic immunotoxicology studies. These assays have been optimized to provide highly reproducible results with an optimal efficiency of time and materials.[48]

2.2.3 *Quantitation of Cytokines*

2.2.3.1 *Background*

Cytokines are peptides mediating a wide range of functions, most conspicuously the regulation of immunity and inflammation.[2] Although produced by, and acting on, a wide variety of cell types, cytokines are generally considered to be primarily immune system mediators. They are produced in the greatest proportion by $CD3^+/CD4^+$ T-helper lymphocytes. Furthermore, T-helper cells exhibit a function diversity in their cytokine production. For example, T-helper-1 (TH1) cells exclusively secrete interleukin-2 (IL-2) and interferon-gamma (IFN-γ), whereas TH2 cells exclusively secrete IL-4, IL-5, and IL-10. Both populations produce IL-3, granulocyte-macrophage colony-stimulating factor (GM-CSF) and tumor necrosis factor (TNF). A third population, TH0, appears to be an early precursor of TH1 and TH2 and demonstrates an intermediate pattern of cytokine production.[125] This diversity is of limited importance when cytokine production is used as an immunotoxicology screening tool; however, this pattern becomes much more important when designing and performing mechanistic studies.

Given their central role in the induction and maintenance of specific immunity, assessment of cytokines provides a valuable tool for both screening and mechanistic evaluations in immunotoxicology. A number of mechanisms may be envisioned in which exposure of cells to xenobiotics may result in modulation of cytokine activity. For example, cytokine-producing cells such as T-helper cells may be destroyed or inactivated by the immunotoxic agent, or cell surface-active compounds might modulate cytokine release from the cell or interfere with the binding of cytokine with its receptor. A more subtle effect might be a specific modulation of cytokine production in the absence of other functional effects. Finally, immunotoxic agents may induce cells to produce inhibitory molecules which inactivate the immune response. Immunomodulation may be mediated by predominantly stimulatory cytokines (e.g., IFN-γ) or by cytokines the function of which appears to be primarily suppressive. Suppressive cytokines may be either nonspecific or specific. An example of a specific inhibitory cytokine is IL-10, which is able to inhibit the production, but not the action, of other cytokines. Specific cytokine inhibitors have also been described, such as interleukin-1 receptor antagonist, IL-1ra,[69] which blocks the action of IL-1α and IL-1β in a dose-responsive manner.[3a] Other examples of inhibitors include cytokine scavenger or chaperone molecules (e.g., α_2-macroglobulin), soluble receptors (e.g., TNF-BP), or circulating anti-cytokine antibodies. It is likely that other specific inhibitors await characterization.

2.2.3.2 *Methodology*

There are currently three major types of cytokine assays used, namely bioassays, immunoassays, and molecular probes. Each of these assay types exhibits advantages and disadvantages, and no one type of assay is best suited for all applications. For example, bioassays tend to be more

TABLE 4
Advantages and Disadvantages of Cytokine Assays

Assay Type	Advantages	Disadvantages
Bioassays	Detects only functional molecules Sensitive (pg/ml or less)	May lack specificity Potential variability in indicator cell responses Requires cell culture Time-consuming
Immunoassays	Rapid Monospecific No cell culture required	May not detect functional molecules Sensitivity may be limited
Molecular probes	Highly specific Can detect production at single-cell level Detects alterations early	Can be expensive May not detect functional molecules

sensitive and detect only functional cytokine molecules, immunoassays such as ELISAs are highly specific and convenient, and molecular methods such as polymerase chain reaction (PCR) are able to detect genetic messages, making them the most time-sensitive.[21,28,47,140,145]

The type of assay chosen is subjective and will depend on the capabilities of the laboratory, as well as the type of information to be gained. For example, the multiplicity of cytokine action makes bioassays slightly impractical, since most cell lines will respond to several cytokines, thus presenting a major difficulty when evaluating the complex mix of cytokines produced *in vivo.* Molecular probes, although highly specific and able to detect alterations in genetic messages, are time consuming and expensive; moreover, these assays do not necessarily detect functional alterations in protein expression. For most situations, cytokine immunoassays represent the most practical choice at present. These assays are highly specific (when based on monoclonal antibodies), sensitive for all but the most demanding needs, and convenient. ELISA-format kits are currently available for almost all described cytokines, obviating the need to develop and validate such methodology. A second important consideration in assessing alterations in cytokine production following xenobiotic exposure is the source of the cytokines. For example, lymphocytes or macrophages may be exposed to xenobiotics in an *in vitro* system, allowing for precise control of test material exposure, number and type of cells, etc. Quantitation of cytokines in culture supernatant fluids often simplifies the process, since the cytokines produced are usually not lost as quickly due to metabolism, receptor-mediated uptake, and other mechanisms. In addition, circulating cytokine levels are difficult to measure accurately or reproducibly, and baseline levels have not yet been determined for many of these factors. Conversely, *in vitro* models are not necessarily accurate models of the *in vivo* condition. For example, cytokines are designed to work primarily at a local level and thus the concentrations produced *in vitro* are probably artificially high. A more detailed consideration of these factors is provided in the subsequent section on human immunotoxicology assessment. (See Table 4.)

2.3 Assessment of Humoral Immunity

2.3.1 *Anti-SRBC Antibody-Forming Cell Assay*

2.3.1.1 *Background*

Humoral immunity refers broadly to immune reactions involving the production and action of antigen-specific antibodies. A number of different assays is available for measuring humoral immunity; some of these techniques include hemagglutination assays, the quantification of total antibody levels (e.g., circulating IgM and IgG levels), *in vitro* antibody production (polyclonal or antigen-specific), and analysis of cell size increase and MHC expression. However, the assay which has received greatest acceptance in immunotoxicology is the anti-SRBC antibody-forming cell (AFC) assay, also called the plaque-forming cell (PFC) assay. Due to its wide acceptance and the considerable amount of immunotoxicology data already obtained with this assay, only the AFC assay (and its variant, the anti-SRBC [sheep red blood cell] ELISA) will be discussed in this section.

The AFC assay is particularly useful for immunotoxicology measurement because the induction of antibody-forming cells requires the precise interaction of a number of cell types and soluble factors (cytokines). In brief, an antigen (i.e., SRBC) is phagocytized, degraded, and presented along with MHC antigen on the surface of an antigen-presenting cell. The antigen is recognized by T-helper cells and B cells of the appropriate specificity, leading to differentiation of the B cells into antibody-producing plasma cells. Although the AFC assay is generally thought of as a measure of humoral immune function, it may in fact be sensitive to toxicity directed against a variety of cell types and immunological processes. The AFC assay is routinely included in screening-type (Tier I) assays as a measure of humoral immunity[74] and has been validated with an extensive panel of chemicals. In studies on the predictive value of immune function assays for alterations in host resistance, the AFC assay has a high correlation with host resistance.[75,77]

2.3.1.2 *Methodology*

The basic methodological format for the IgM AFC assay was first described by Jerne and Nordin[56] and later modified by Cunningham and Szenberg.[27] In brief, animals are injected intravenously with washed SRBCs. The animals are euthanized 4 to 5 days later, and single-cell suspensions of the spleens are prepared in culture medium. An aliquot of these cells is mixed with fresh SRBC and complement (generally guinea pig complement). This mixture is then placed into glass chambers comprising glass slides separated by spacers. This results in the formation of a thin lawn of cells composed primarily of SRBC. The chambers are sealed to prevent evaporation and are then incubated at 37°C for approximately 1 hr. During incubation, responsive plasma cells secrete anti-SRBC antibodies, which attach to neighboring SRBC. In the presence of complement, localized areas of hemolysis are produced ("plaques"), which are then enumerated visually. Results are generally expressed as AFC per 10^6 cells or as AFC per spleen.

For quantitation of IgG anti-SRBC antibodies, a modification of the above technique is used.[127] The animals are rested for an additional 2 days following immunization with SRBC, and an optimum concentration of rabbit anti-mouse IgG antibody is added to the cell mixture. When assessing IgG, it is necessary to concomitantly measure IgM response and then subsequently correct the final count for IgM plaques. For practicality, in studies involving *in vivo* dosing with test material, two cohorts of animals are used, with one cohort receiving immunization with SRBC prior to the second cohort. The first cohort is then used for IgG quantitation, and the second for IgM quantitation.

A further modification which may be useful in mechanistic-type studies is the evaluation of the so-called T-independent antibody response. In this approach, antibody production by B cells is induced by agents that cross-link Ig receptors, rather than acting in an antigen-specific manner. This technique[108] involves the immunization of animals with a trinitrophenylated compound such as TNP-LPS (trinitophenylated lipopolysaccharide) or TNF-Ficoll. The standard AFC assay is then performed, except that SRBC haptenated with a trinitrophenyl group are used as the target. LPS is a "type 1" antigen which has intrinsic polyclonal activating properties for all B cells, including immature populations. Ficoll, a "type 2" antigen, has no polyclonal activating property and therefore stimulates only mature B-cell populations.[118]

2.3.2 *Anti-SRBC ELISA*

2.3.2.1 *Background*

In spite of its demonstrated sensitivity and reproducibility, the standard AFC assay has several disadvantages. Primarily, it is a labor-intensive technique requiring some specialized skills (e.g., intravenous injection of antigen, visualization of plaques). In some situations, this may limit the utility of this assay for screening purposes. In addition, the AFC is a terminal-phase assay, requiring that animals be euthanized. This is disadvantageous for studies in which a time course examination of changes is required. To address this situation, a number of investigators have utilized an ELISA technique for measuring specific antibodies to SRBC. This technique lends itself to semi-automation, potentially enhancing its utility in screening applications, as well as providing a quantitative endpoint measurement less susceptible to readout error. In addition, it provides the option of time course experiments, since blood samples may be collected at multiple time points from the same animal. In spite of these various advantages, it must be stressed that the results obtained using the AFC and anti-SRBC ELISA are not completely interchangeable. For example, the AFC is organ specific, measuring only antibody production by spleen cells; in contrast, the ELISA measures systemic antibody production. As pointed out by Temple et al.,[134] this difference may be significant when a xenobiotic differentially affects a specific organ, since antibody production in the bone marrow peaks later than splenic antibody production.

2.3.2.2 *Methodology*

Temple et al.[134,135] have reported the most detailed methodology development of an anti-SRBC ELISA technique, as well as a comparison of this technique with the standard AFC assay using cyclophosphamide as a positive control. In brief, animals are immunized with SRBC, and on Day 4 or 5 blood is collected via retro-orbital puncture. Serum is isolated and diluted, and these dilutions are added to ELISA plates previously coated with SRBC membrane antigens. Following incubation to allow binding, excess serum is washed away. Next, horseradish peroxidase-conjugated, anti-mouse antibody is added, and the plates are incubated to allow binding. Excess antibody is washed away, and peroxidase substrate is added. The substrate is cleaved by the peroxidase to a colored product, and the colorimetric reaction is read with a spectrophotometer. In most ELISAs, experimental values are subsequently read from a reference curve run concomitantly with the unknown samples; however, due to the current lack of commercial standardized reagents, data analysis becomes more subjective. For example, Temple et al.[135] recommend the construction of monoclonal anti-SRBC IgM antibodies to serve as a reference preparation. Unfortunately, this is an involved technique and may be impractical for some investigators. In addition, SRBC antigen for coating the plates must be prepared by each individual and may lack consistency between batches. Another practical consideration for this assay is the use of SRBC as an antigen. Although SRBC in a potent immunogen which elicits a specific B-cell response, there may be significant variability in response to different lots of SRBC (this is true for the AFC, also). Other antigens such as ovalbumin (OVA) may be used for immunization, with antibody production subsequently measured by ELISA. However, OVA is a relatively weak immunogen that elicits a nonspecific B-cell response, which may represent a drawback. None of these disadvantages is insurmountable to the interested investigator, but at present they hamper a widespread acceptance of the anti-SRBC ELISA. Although the technique has not yet been fully validated with a series of chemicals with known activity on humoral immunity, it appears to hold potential for future immunotoxicology studies.

2.4 Assessment of Natural (Innate) Immunity

2.4.1 *Natural Killer Cell Function*

2.4.1.1 *Background*

Natural killer (NK) cells are a population of non-B, non-T lymphocytes which exhibit cytotoxicity toward a variety of target cells, including certain tumor cells. NK cells are morphologically distinct, being larger than other lymphocytes and possessing numerous granules; this phenotype leads to their designation as large granular lymphocytes or LGL.[72] Unlike CTL, NK cells are able to lyse their targets in an MHC-unrestricted manner and do not require previous contact with the target. They therefore represent an important mediator of nonspecific host defense. NK cells produce and respond to a variety of cytokines, most importantly IL-2 and IFN-γ, both of which have been demonstrated to enhance basal NK-cell function.[93] NK cells appear to be intimately involved in resistance to infectious organisms[124] and cancer.[102] This range of protective functions suggests that NK cells are important for immune homeostasis and thus is an important parameter in immunotoxicity assessment. The NK cell assay has been used extensively in immunotoxicology assessment for a number of years and has proven to be reproducible and reliable.[74-76] Alteration of NK-cell function has been shown to correlate well with changes in resistance to experimental neoplasia such as PYB6 sarcoma and B16F10 melanoma.[74]

2.4.1.2 *Methodology*

Assessment of murine NK-cell activity is based on the work of Reynolds and Herberman.[106] Murine splenocytes are generally used, although other tissue sources (e.g., lungs) may be used also. The cells are adjusted to a known concentration, and several dilutions are prepared from this preparation. These effector cell dilutions are subsequently added to replicate wells of 96-well, round-bottom microculture plates. Fresh YAC-1 tumor target cells are radiolabeled with ^{51}Cr, and free radiolabel is washed away. The target cells are adjusted to a known concentration and added to the plates containing the effector cells. Following a 4-hr incubation, the cell-free supernatant fluids are collected, and radiolabel release is quantitated. Nouri et al.[96] have described a colorimetric modification of this assay using MTT as the indicator rather than radiolabel release. This variation of the assay has not been utilized extensively but may be of interest to investigators wishing to limit or eliminate use of radioisotopes. If this avenue is pursued, a more effective alternative may be the use of MTS rather than MTT, as described in the previous section on lymphoproliferation. Regardless of the indicator system used, percent specific cytotoxicity is calculated as percent of total releasable counts (or absorbance), corrected for spontaneous release. The final effector:target cell ratios used are generally 100:1, 33:1, and 11:1, although other dilution schemes may be used. As with the CTL assay, the use of multiple cell ratios allows for the calculation of lytic units.

It is important to note that exposure of animals to xenobiotics may result in either enhancement or suppression of NK-cell activity. Although a decrease in NK-cell function is certainly a biological disadvantage, it does not follow that enhanced NK-cell function is necessarily advantageous. Certain compounds, such as synthetic nucleic acids, are known to enhance NK-cell function; therefore, in designing immunotoxicology studies, it is often useful to include negative and positive controls. An effective *in vivo* positive control for suppression of NK-cell activity is afforded by injecting animals with anti-asialo GM_1 antibody as described by Barlozzari et al.[7] One to four days prior to assay, animals are injected i.v. with titrated antibody, which results in essentially complete abrogation of NK-cell activity. A useful positive control for *in vivo* augmentation of NK-cell activity is the synthetic polynucleotide polyinosinic:polycytidylic acid (poly(I)·poly(C)), a potent interferon inducer. Animals are injected intraperitoneally with 100 μg per animal of poly(I)·poly(C) approximately 24 hr prior to assay, which results in significant enhancement of NK-cell function.[30]

2.4.2 *Macrophage Function*

2.4.2.1 *Background*

The term "macrophage" encompasses an eclectic group of cells, including monocytes, tissue macrophages, Kupffer cells, and alveolar macrophages, as well as related cells such as mesangial cells, Langerhans cells, microglia, and osteoclasts.[71] To date, immunotoxicology studies have tended to concentrate primarily on peritoneal macrophages, as this population in rodents is relatively easy to isolate in large numbers. Assessment of macrophage function was previously included in Tier II-type assessment of nonspecific host resistance,[74] although it is not consistently performed in these studies. Often overlooked, however, is the fact that many specific immune function assays also evaluate macrophage function, if only indirectly. Macrophages or macrophage-like cells serve as antigen-presenting cells, as well as producing a raft of immunomodulatory factors necessary for immune function acquisition.[92] Thus, for example, lymphoproliferation, CTL induction, and production of antibodies all require the cooperation of competent macrophages.

2.4.2.2 *Methodology*

Due to the multiplicity of macrophage actions, as well as the diversity of cell types covered by the term, an extensive range of assays is available for functional assessment. Arguably, no particular assay could be characterized as "best". Therefore, whenever possible, several functions should be measured when determining immunotoxicity. The precise methodology for these assays is often involved and presumes a working knowledge of macrophage biology; for this reason, the techniques will not be elaborated here. Some suggestions for assays which have previously been used for immunotoxicity assessment include tumor cell cytolysis,[34] phagocytosis,[94] and respiratory burst.[111] In addition, molecular immunology assays are now available that hold promise for use in mechanistic immunotoxicology studies.[43]

2.4.3 *Neutrophil Function*

2.4.3.1 *Background*

Neutrophils, also known as polymorphonuclear (PMN) cells, represent the "first line" of host defense. They are the predominant phagocytic cells in the circulation and are responsible for initial defense against bacterial infection.[70] This vital role would suggest that neutrophils would represent an important measurement for immunotoxicity assessment, especially since neutropenia or altered neutrophil function may result in a disposition to infection. Moreover, neutrophils have also been implicated in pathological inflammation, again suggesting that cellular dysregulation (as might result from exposure to xenobiotics) may result in altered immune function;[132] however, in spite of the recognized importance of this cell for host defense, to date assessment of neutrophil function paradoxically has not been extensively employed in immunotoxicology assessment. Likewise, no reports have been published on the sensitivity or predictive utility of neutrophil function for immunotoxicology. It is anticipated that such assays will, for the foreseeable future, remain the purview of mechanistic immunology/immunotoxicology.

2.4.3.2 *Methodology*

An extensive panel of functional assays is available for measuring neutrophil function; however, as mentioned above, such assays are not routinely employed for immunotoxicology studies. Three relatively simple assays which have been employed for human neutrophils and may be modified for murine studies include respiratory burst,[103] tumor cytostasis,[87] and inducible adherence.[97]

2.4.4 *Acute-Phase Proteins and Complement*

2.4.4.1 *Background*

The acute-phase response is not, strictly speaking, a specific immune response. Rather, the release of various proteins associated with the acute-phase response (e.g., C-reactive protein, serum amyloid A, C-1 inhibitor) and complement activation generally occurs in response to tissue damage or infection and represents an important nonspecific host defense mechanism. The acute-phase response is initiated and controlled in large part by ancillary immune mechanisms such as macrophage activation and cytokine production;[11,67] it is therefore conceivable that xenobiotic-related perturbations in the immune system may be mediated at least partially through the acute-phase response.

2.4.4.2 *Methodology*

Although the acute-phase response and complement activation are important for overall host defense, to date assessment of these functions has not been extensively utilized in immunotoxicology studies. This is probably not due to technical difficulties, as many of these factors may now be easily quantified with commercially available ELISA kits. One potential disadvantage in their evaluation is the multiplicity of factors secreted during these responses; it may be impractical or undesirable to screen an extensive panel of factors in a blind manner. However, in cases where immunotoxicant exposure is demonstrated to alter host resistance or to alter inflammatory processes, inclusion of acute-phase protein or complement measurements may be a valuable adjunct to mechanistic immunotoxicology.

2.5 Assessment of Hypersensitivity

The continuum of immune responsiveness ranges from immunosuppression to immunostimulation. In large part, the preceding sections of this chapter have described approaches traditionally employed to determine the ability of drugs or other xenobiotics to suppress the immune response; however, increasing interest is being given to the opposite extreme of immunotoxicity, i.e., immunostimulation. Whereas modest amounts of immune stimulation may actually be beneficial in maintaining host resistance, enhanced activity may lead to pathology as well. A prime example of this is hypersensitivity (allergy), in which the immune system responds inappropriately to otherwise innocuous substances. In the context of immunotoxicology, two types of hypersensitivity response are receiving great interest, these being contact hypersensitivity and respiratory sensitivity.[59]

2.5.1 *Contact Hypersensitivity*

2.5.1.1 *Background*

The guinea pig traditionally has been the animal model of choice for assessment of contact sensitizing potential of test materials. The validated assays available include the Magnusson and Kligman assay, the Buehler assay, the guinea pig maximization test, and the adjuvant/patch test, among others.[63,83] An extensive body of literature has resulted from the use of these assays, and they have generally proven to be reliable as predictors of human contact sensitization. Unfortunately, most of these assays share a number of problems: they can be expensive and time-consuming; they routinely employ a subjective, rather than an objective, endpoint; and they are sensitive to colored test materials. Recent efforts have been made to develop and validate murine models of contact sensitization which use objective endpoint determinations that are more easily compared between laboratories. At present, two assay systems are being employed to evaluate the contact sensitizing potential of xenobiotics: the Mouse Ear Swelling Test (MEST) and the Murine Local Lymph Node Assay. Each of these tests is discussed in detail below.

2.5.1.2 Mouse Ear Swelling Test

Currently, most approved animal models for assessing contact sensitization potential of test materials measure cutaneous edema/erythema, usually in guinea pigs. These models, although providing good correlation with results of human testing, are cumbersome and expensive and rely on a subjective endpoint. The use of mice for research on delayed-type (contact) hypersensitivity was pioneered by the work of Asherton and Ptak,[4] as well as others. The MEST was subsequently developed as a potential alternative to the more traditionally utilized guinea pig models of dermal sensitization, most notably by Gad et al.[37] In the classical MEST, BALB/c mice are shaved over the abdominal region, and a single sensitizing dose of test material is applied (induction). Five days later the mice are challenged with the same material by application to the pinnae (elicitation). Ear thickness is evaluated with a micrometer at various time points prior to and following challenge. This test has been validated and widely employed for assessment of dermal sensitization. The principal disadvantage of the MEST is its potential for variability associated with measuring ear thickness. Subtle differences in technique between laboratories and even between technicians make inter-laboratory comparisons difficult. A radioisotopic modification of the technique has been reported, although this technique was not found to be reliable.[25] Another modification was reported by Garrigue and colleagues,[40] who reported enhanced detection of weak sensitizers by applying the potential sensitizer three times rather than once as reported by Gad et al.[37] More recently, Thorne et al.[138,139] reported "refinements" to the assay (which they refer to as the Mouse Ear Swelling Assay, or MESA) for increasing its detection of weak sensitizers. The current status of the MEST and its use in safety testing have recently been reviewed.[38]

In addition to testing for the sensitization potential of test materials, a modification of the MEST has also been used extensively to assess the general effectiveness of cell-mediated immunity. Assessment of DTH was included in the original NTP protocol as a Tier II assay.[74] This assay utilized a radioisotopic endpoint to measure cellular influx into skin challenged with keyhole limpet hemocyanin. This method was found to be less sensitive than either MLR or CTL in detecting immune dysregulation following chemical exposure. In addition, the use of radioisotopes is not always desirable. A more recent development is an assay utilizing skin thickness as an endpoint. In this assay, mice are sensitized to the potent sensitizer oxazolone on the flank or abdomen. Several days later, the animals are challenged with oxazolone on the pinna, and reaction is measured as alterations in skin thickness using a micrometer.[4,13a] Animals may be treated with test material at any point in this procedure, but in most cases the assay would be scheduled so that the animals would be exposed to test material during the sensitization phase. This assay provides a relatively simple assessment of cell-mediated immune function, in particular immune function related to induction of a T-helper-1 type of response. Alterations in this response exhibit correlation with changes in a number of host resistance parameters.[76]

2.5.1.3 Murine Local Lymph Node Assay

More recently, a newer model for determining sensitization potential — namely, the murine local lymph node assay (LLNA) — has been developed. The LLNA is simple in concept. During induction of contact hypersensitivity, allergens bind to Langerhans cells in the epidermis, and these cells migrate via the efferent lymphatics to the regional node (i.e., the auricular lymph node). In the lymph node, these cells then initiate antigen-specific T-lymphocyte proliferation. Quantitation of the degree of proliferation is then correlated with the potential of the test material to induce sensitization. Although very similar to the MEST,

TABLE 5

Examples of Compounds Tested for Contact Sensitizing Potential Using the Murine Local Lymph Node Assay

Compounds Tested	Ref.
Photoallergens	119
Metal salts	51
Rubber chemicals	52
Petrochemicals	32

the LLNA differs in two main points. First, the LLNA endpoint measurement is taken during the induction, rather than the challenge, portion of the hypersensitivity response. Second, the LLNA is based on a more quantitative measurement: radioisotope incorporation (see below). The LLNA offers a number of diverse advantages over other animal models employed to assess contact sensitization, including reduced time for assay completion, utilization of a quantitative endpoint, reduced animal use, substantial reductions in labor and physical resources, and insensitivity to colored test materials.

The LLNA, as first described in detail by Kimber and Weisenberger,[61] relied on measurement of cellular proliferation *in vitro*. The assay was subsequently modified to eliminate the need for cell culture[61a] and has remained essentially unchanged to the present. As currently performed, CBA mice are exposed via both pinnae to test materials in a suitable vehicle, usually a 4:1 mixture of acetone and olive oil, for 3 consecutive days. Following a 2-day rest, the animals are then injected i.v. with tritiated thymidine, and 5 hours later the auricular lymph nodes are excised and pooled for each group. The nodes are disaggregated and treated with trichloroacetic acid, and the precipitated radiolabel is quantified by β-scintillation spectrometry. A stimulation index is calculated by comparing test results with the appropriate control. Traditionally, a stimulation index of 3 or greater has been utilized as the criterion for classifying a compound as a contact sensitizer.

Since its initial description, the LLNA has been the subject of numerous validation studies using standardized protocols.[9,62] The results of these studies have revealed that the assay is reproducible between laboratories. More important, with few exceptions, predictions of strong and moderate contact sensitizing potential based on results of the LLNA are in close agreement with the results of studies using guinea pigs. More recent studies have modified the standard LLNA in an effort to examine its utility in risk assessment.[41] In this modification, mice are treated with test material daily for 4 consecutive days, and lymph node proliferation is assessed on day 5. In addition, proliferation of cells from individual nodes is quantitated, rather than using pooled populations. This modification allows for statistical evaluation. Using this technique, test materials were evaluated for contact sensitization potential with good correlation to existing data.

Recent validation studies have compared several variations of the assay[63,64,73] and have revealed that the assay can accommodate minor procedural alterations without significantly affecting test performance. It is hoped that these validation studies will aid in greater acceptance of the LLNA for safety testing. (See Table 5.)

2.5.2 *Respiratory Hypersensitivity*

2.5.2.1 *Background*

Occupational exposure to a number of substances has been associated with the development of asthma and other allergic reactions, with an incidence of between 5 and 50%.[58] Although the immunologic reactions and treatment of respiratory allergies have been well characterized, less is understood concerning the mechanisms involved. The ability to screen for potential sensitizers is obviously an important immunotoxicological concern, requiring sensitive and validated methodology.

2.5.2.2 *Methodology*

Validated methodologies have been described using the guinea pig as a model animal; the reader is directed to detailed descriptions of these procedures to assess potential respiratory sensitizers.[18,116] Recent efforts have focused on developing a mouse model for this purpose. The methodology is based on the observations that contact sensitizers and respiratory sensitizers appear to induce differentially cytokine production by T-helper cells. Specifically, contact sensitizers induce a T-helper-1 type response with production of IL-2 and interferon-gamma, whereas respiratory sensitizers tend to induce a T-helper-2 type of response with production of IL-4 and IL-10.[60] Concurrent with this differential cytokine production, it was observed that exposure to respiratory sensitizers also resulted in increased serum IgE concentrations, as well as the presence of antigen-specific IgE molecules.

Whereas evaluation of cytokine production following exposure to potential contact or respiratory sensitizers is not technically demanding, it is labor intensive and may be impractical for screening purposes. To address these issues, a method to assess the IgE response in mice has been described.[29] In this assay, animals are treated epicutaneously with test material. Following a period of sensitization, the animals are bled, and the concentration of total IgE in the serum is quantified. This test is currently being validated at a number of laboratories, using a variety of chemicals. Although this assay displays promise, its predictability and future role in risk assessment are unknown at present.

2.6 Models of Autoimmunity

2.6.1 *Background*

As described above for hypersensitivity, autoimmunity represents an extreme in the continuum of immune responsiveness. In the case of autoimmunity, the hyperresponsiveness of the immune system is thought to become directed toward self antigens, a condition that is not easily reversible. In humans, autoimmunity is known to be a sequela of exposure to various drugs[85] and chemicals;[13] however, to date, there have been few well-documented observations of autoimmunity induced specifically by chemical exposure in laboratory rodents. Often, induction of autoimmunity requires a long-term exposure to the inducing material, and the limited life-span of rodents may preclude observation of this reaction. A more efficient approach, therefore, may be the use of animals which are predisposed toward autoimmune disease. In this case, animals are treated with a control substance or test materials prior to the anticipated appearance of disease. The time to the onset of disease or change in severity or duration may then provide information on a test material's immunomodulatory potential. At present, several models are available. Two of these are discussed below.

2.6.2 *Methodology*

At least two rodent models of spontaneous autoimmune disease have been well-characterized to date; namely, the non-obese diabetic (NOD) mouse and the New Zealand Black (NZB) mouse. The NOD mouse develops autoimmune type I insulin-dependent diabetes spontaneously.[55] As with humans, this disease is under polygenic control and is subject to environmental factors such as microbial flora and diet. NZB mice spontaneously develop a condition similar to human systemic lupus erythematosus, with the most consistent feature being the development of immune complex-mediated glomerulonephritis.[137] In addition to these two models of spontaneously arising autoimmunity, several other induced models are

TABLE 6

Rodent Models of Autoimmunity

Model	Condition Modeled
Organ-Specific	
Non-obese diabetic (NOD) mice	Insulin-dependent diabetes
Buffalo rat (spontaneous)	Thyroiditis
CFA[a] + brain (induced)	Allergic encephalomyelitis
Systemic	
NZB[b] × W mice (spontaneous)	SLE[c]
CFA + anti-DNA (induced)	SLE

[a] CFA: complete Freund's adjuvant.
[b] NZB: New Zealand Black mouse.
[c] SLE: systemic lupus erythematosus.

available.[115] In addition, molecular immunology techniques may facilitate detailed mechanistic studies,[15,33] although such assays still require extensive validation before their routine use in immunotoxicology assessment. (See Table 6.)

2.7 Use of Immunological Variants

2.7.1 *Background*

Immunologically normal rodents (both mice and rats) have proven to be reliable models for assessment of immune function and immunotoxicity; however, this normalcy may work to the disadvantage of the model as well. For example, humans sometimes exhibit conditions of suboptimal immunocompetence. It is reasonable to assume that such deficits could render these individuals more susceptible to potential immunotoxicity by xenobiotics. Perhaps of lesser importance in the context of immunotoxicology, but still of scientific utility, is the ability to examine selectively the effect of limited or comprehensive modification of the immune phenotype. In these instances, the use of immunological variants may be considered. These immunological variants may be broadly classified as either natural or engineered.

2.7.2 *Natural Mutations*

Natural immunological mutants are sometimes referred to as "experiments of nature". A considerable number of naturally occurring mutants has been described as affecting the immune response; these are reviewed by Schultz.[120] Some of the more well-known mutants include the athymic (nude) mouse (*nu*), the beige mouse (*bg*), and the X-linked immunodeficient mouse (*xid*). The nude mouse is deficient in functional T cells due to a lack of expression of class II MHC antigens on thymic stroma and the resultant defect in T-cell maturation.[120] Nude mice are particularly useful in experimental protocols examining the role of T cells in immunotoxicology, especially those protocols including host resistance assays. The beige mouse is most useful in studies examining defects in NK-cell function. The X-linked immunodeficient mouse exhibits subnormal B-cell function due to its lack of certain mature B-cell populations. Although the latter two models could be readily adapted to excellent immunotoxicology models, they have as yet been underutilized in this respect.

Two additional natural variants with potential for immunotoxicology studies are mice with more extensive immune deficiencies, namely the triple-deficient (*bg/nu/xid*) mouse and the severe combined immunodeficient (*scid*) mouse. Triple-deficient mice combine the T-cell and NK-cell deficiencies of the nude and beige mice with the B-cell deficiency of the *xid*

TABLE 7
Selected Immunological Mutations in Mice

Designation	Nature of Defect	Phenotype
Nude (*nu*)	Lack class II MHC on thymic stroma	Athymic; lack functional T cells; enhanced NK
Beige (*bg*)	Altered lysosomal biogenesis	Decreased NK and CTL function
X-linked immunodeficiency (*xid*)	Lack subset of mature B cells	Abnormal B-cell responses
Triple-deficient (*nu/bg/xid*)	Combination of above	Multiple defects
Severe combined immunodeficiency (*scid*)	Defective heavy chain recombinase	Lack functional T and B cells

mouse.[57,120] *scid* mice possess a mutation in chromosome 16 which disrupts the recombinase responsible for heavy chain gene rearrangement. This defect leads to a lack of antigen-specific receptors and thus a lack of functional T and B cells.[16] Both of these models are of special interest due to the demonstration that they may be engrafted with human lymphoid cells, resulting in a chimeric animal possessing a functional — albeit incomplete — human immune system. *scid* mice subjected to this modification are known as *scid-hu* mice, and they are finding increasing utility for a variety of immunological investigations.[89] This leads to the intriguing possibility of using these animals as models for human immunotoxicity. Heretofore, any evidence for human immunotoxicity has been, by necessity, retrospective. Exposure of human cells to test materials in *in vitro* culture may provide some suggestive evidence of toxicity, but questions of metabolism and secondary interactions are not directly addressable by such a system. Initial steps in the development of the *scid-hu* mouse have been made by Pollock et al.[101] Although this model demonstrated some promise, high inter-animal variability limited the interpretation of the results. Given refinements in the methodology, this system may hold considerable promise in the future of immunotoxicology.[126] (See Table 7.)

2.7.2.1 Engineered Mutations

Rather than rely on evolution to produce useful genetic mutants, the sophistication of modern molecular biology has led to the advent of custom-engineered immunological mutants. Two overlapping techniques of particular importance are homologous recombination ("knockout") and transgenic animals. These techniques allow for the selective ablation or inclusion of highly specific genetic sequences, resulting in well-defined immune abnormalities. Knockout technology, although technically demanding, is well established and within the technical reach of most laboratories.[39] Reduced to its primary components, homologous recombination involves engineering a targeting vector containing a selectable marker delineated by genetic sequences homologous to the gene of interest. The package is transfected into an embryonic stem cell line, where the original gene is replaced by the new sequence. The modified stem cells are implanted into a blastocyst and contribute to the developing embryo (including the germ cell line). Genetic chimeras are subsequently selected and crossbred to obtain homozygous animals. A variety of immunological mutants has been created in this way.[142]

2.7.3 Potential Problems

Although the utility of natural and engineered immunological mutants is limited only by the imagination of the investigator, their use is accompanied by a variety of potential drawbacks. Of immediate practical concern is the propensity of these animals to become infected with opportunistic pathogens, necessitating germ-free maintenance. This first factor contributes to the second factor, namely the considerable expense involved in either creating or purchasing these animals. Thus, for the foreseeable future, these animals probably will not be used in screening-type assays, but will be limited primarily to mechanistic immunotoxicology. Finally, it must be recognized that, as in all of immunology, absolutes are elusive. For example, approximately 15% of *scid* mice express detectable immunoglobulin, a condition termed "leaky".[16] In addition, some techniques do not appear to be consistent and reproducible, as exemplified by the work of Pollock et al.[101] Nonetheless, immunological variants exhibit exceptional promise as models.

2.8 Molecular Immunotoxicology

2.8.1 Background

The field of molecular biology has contributed immeasurably to the advancement of immunology. It is reasonable to extrapolate that the discipline of immunotoxicology can also benefit from the use of the powerful tools supplied by this rapidly expanding field.[84] Presented below are two representative approaches which are already contributing to our mechanistic understanding of immunotoxicity.

2.8.2 Polymerase Chain Reaction

Polymerase chain reaction (PCR) is a powerful molecular biology tool with an increasing presence in immunotoxicology. PCR is a DNA amplification method with three defined stages. In the first step, DNA is denatured to two strands using heat. In the second step, the temperature is reduced to allow single-strand oligonucleotide primers to anneal to their complementary sequences on the DNA strands. In the final step, DNA polymerase is added to allow synthesis of the full-length DNA from the primer sequence.[104] PCR enables one to expand a genetic sequence many-fold and facilitates genetic analysis of changes produced by experimental treatment, such as exposure to xenobiotics.

2.8.3 In Situ *Hybridization*

Another important molecular technique is *in situ* hybridization. This technique is very similar to the well-established method of immunocytohistochemistry, except that nucleic acids (RNA or DNA) serve as the substrate.[86] Tissues or cells are treated to preserve the essential morphology, and the cells are treated to allow entry of the probe. Cloned probes (RNA, DNA, or synthetic oligonucleotides), labeled either with radioisotopic tags or nonisotopically (e.g., biotin, 2-acetylaminofluorene), are added to the tissue/cells, and hybridization is allowed to take place. Detection of hybridization is subsequently performed by methods appropriate for the type of probe label. *In situ* hybridization allows detection of genetic message expression *in situ*, which makes it a highly sensitive and selective tool for immunotoxicology studies, especially for mechanistic immunotoxicology.

3 IMMUNOMODULATION IN HUMANS

3.1 Introduction

Although animal models of immunotoxicity have been extensively utilized, the increasing emphasis on human risk assessment requires that those potential health hazards, including immunomodulation, be identified whenever possible. In addition, extrapolation of alterations in immune function observed in laboratory animals to human health is associated with various uncertainties.[126] For example, experimental animals are often inbred; this lessens inter-animal variability, simplifying statistical evaluation of observations. Humans, on the other hand, are highly outbred, with a consequent high degree of inter-individual variability in immune response. In addition, laboratory studies in rodents are highly controlled for environment, diet, and health status, whereas any human study must take into account extreme variability in all of these parameters.

Immune function assays for rodents have been fairly well validated using a number of chemicals, but most human immunotoxicity assays have not. Whereas most human assays are technically similar to their rodent counterparts, ethical considerations preclude certain procedures, such as immunization with hazardous antigens or experimental challenge with infectious organisms. In addition, biological material as a source of effector cells is somewhat limited in humans, consisting of peripheral blood leukocytes and, in exceptional cases, biopsy material. Given these limitations, two options emerge: (1) retrospective assessment, in which the immune function of exposed individuals is compared to some reference, and (2) *in vitro* studies, in which cells from "normal" individuals are exposed *in vitro* to chemicals or drugs of interest. Many of the assays described in the following sections may be utilized for either type of study.

3.2 Assessment of Immunopathology

Except under extraordinary circumstances, most measures of immunopathology (e.g., lymphoid organ weights, histopathology) in humans are impractical or serverely restricted due to ethical considerations. Two viable options in such cases are routine complete blood counts (CBC) and differentials[114] and the assessment of potential alterations in cell surface markers. CBC and differential analysis can provide a simple and inexpensive measurement of potential alterations in the total number of bone marrow-derived immune cells, which may aid in detection of myelotoxicity. In addition, the relative proportions of the various populations may be determined, giving some indication of selectivity by an immunotoxicant. For more detailed analysis of cellular subsets, flow cytometry can provide a more sensitive quantitation of subset ratios (e.g., CD4/CD8). Another area of increasing interest that may be important in human immunotoxicology is the study of T-cell memory.[79] Study of this phenomenon may be facilitated by analysis of surface markers, especially CD45RA and CD45RO. As previously mentioned for rodents, one of the most important obstacles to extensive use of lymphocyte markers in human immunotoxicology at present is the lack of inclusive reference ranges. This situation is improving, however, with the publication of numerous studies on normal reference ranges in humans.[65,99,105]

3.3 Assessment of Cell-Mediated Immunity

3.3.1 *Lymphoproliferation*

3.3.1.1 *Background*

Lymphoproliferation is considered to be a general measure of immune responsiveness. Once lymphoid cells (either T or B cells) are activated by agents such as mitogens, antigens, phorbol esters, superantigens, or cytokines, the cells are induced to proliferate in anticipation of mounting a specific immune response. As with many of the assays described in this chapter, lymphoproliferative assays may be utilized in essentially two ways. In the first, the overall immune status of an individual may be estimated based on responsiveness to either specific or nonspecific stimuli. The common difficulty with this approach is the establishment of a normal or baseline response; for example, if the assay is performed retrospectively following xenobiotic exposure, comparative analysis may be difficult. The second use of this assay would be to determine the effect of direct, or *in vitro*, exposure to the chemical or drug on human cells. This approach may represent a natural progression from evaluating effects in *in vitro* and *in vivo* animal studies to examining the effects on human cells.

3.3.1.2 Methodology

Quantitation of human lymphoproliferation is essentially identical to the procedure described above for laboratory animals. In addition to phytolectins such as ConA and PHA, recall antigens such as tetanus toxoid, diphtheria toxoid, and tuberculin purified protein derivative (PPD) may be used, since many individuals have experienced prior exposure to these antigens. In addition, other antigens may be used if the individual has a known history of exposure. Another example is anti-CD3 monoclonal antibody (either soluble or substrate-bound), which may be used to stimulate T cells polyclonally via the antigen receptor. Finally, the response to allogeneic lymphocytes (the MLR reaction) may be employed by stimulating the cells with inactivated lymphoid cells from an allogeneic (or possibly xenogeneic) donor, or with cloned lymphoid cells from cell culture. As with the laboratory animal proliferation assays described previously, the human response may also be measured colorimetrically.

3.3.2 Cytokine Production

3.3.2.1 Background

Because of their ubiquitous role in regulation of immunity, inflammation, and natural host defense, measurement of circulating cytokines may arguably provide one of the most basic measurements of human immune responsiveness and disease states.[21,145] Cytokines appear to serve primarily as local mediators of immunity and inflammation; thus, detection of elevated levels of particular cytokines in the circulation may be evidence of immunopathology. Detection of alterations in circulating cytokine levels is straightforward (see below); however, given the local-acting nature of these molecules, detection in the circulation may also suggest substantial immunomodulation. If so, then less pronounced immunomodulation may not be detected due to the limited range of most assays. Newer molecular techniques allow for detection of cytokine alterations at far lower levels[28] but may be impractical for screening studies. Another important consideration is that, analogous to the murine system, the phenotypic dichotomy of T-helper lymphocytes appears to hold true for the human system as well.[113] This may be of greater concern in mechanistic vs. screening type studies for human immunotoxicology, particularly when exposure to xenobiotics appears to be associated with increased susceptibility to infection.[88]

3.3.2.2 Methodology

In general, the comments previously made in the section on rodent immunotoxicology are applicable to human cytokine assessment. To reiterate, cytokines may be quantitated by bioassay, immunoassay, or by a variety of molecular biology techniques. Bioassays, although sensitive to low cytokine concentrations and responsive only to functional molecules, are notoriously nonspecific. This is especially problematic when analyzing complex mixtures, such as serum or cell culture supernatants. Except for exceptional circumstances, they are probably the least desirable choice. Conversely, some immunoassays approach the sensitivity level of bioassays but are generally highly specific for individual molecules. From a practical aspect, ELISAs are more accessible, since most commercially available ELISA kits, as well as reagents for the construction of ELISAs, are specific for human cytokines. In addition, the microwell format of ELISAs makes them ideally suited to screening studies or large-scale mechanistic studies. Molecular biology techniques for cytokine evaluation are probably best employed only in detailed mechanistic studies.

Although ELISA technology is well understood by most investigators, several technical points which may be of special importance to immunotoxicity assessment must be considered.

The first point is the method of sample collection. If cytokines are generated in an *in vitro* system, sample collection is relatively straightforward; however, when collecting biological fluids such as serum or plasma, the method of collection has been demonstrated to affect subsequent cytokine measurement.[107,136] Second, certain endogenous molecules may interfere with measurement. For example, α_2 macroglobulin, thought to serve as a cytokine "chaperone" molecule systemically, may result in an underestimation of cytokine levels when present in experimental samples.[53] Third, and related to the consideration just described, is the presence of natural inhibitors such as IL-1RA or endogenous anti-cytokine antibodies.[117] Some sensitive and specific ELISAs may still underestimate cytokine levels. For example, ELISAs designed to detect 17-kDa IL-1β (the mature form) have been found to be relatively insensitive to 35-kDa IL-1β (pro-IL-1β); pro-IL-1β is the precursor to mature IL-1β and is a significant component of total IL-1β production.[44] Another important consideration when assessing cytokine production is the current lack of a comprehensive database of "normal" values. In addition, the complex redundancy and multiplicity of action by most cytokines also contribute to potential difficulties in proper interpretation of results. Finally, in humans we have the unique situation of cytokines being the "xenobiotic", since these molecules are increasingly being utilized as potential therapeutics.[23] It should be apparent, therefore, that although cytokine assessment holds great promise for assessing human immunotoxicity, much work remains to be done before its routine implementation.

3.3.3 Cytotoxic T-Lymphocyte Function

3.3.3.1 Background

While cytokine quantitation can provide information concerning the regulatory competence of T cells, measures are also required to evaluate the effector function of cell-mediated immunity. CTL function serves this purpose well, since the induction of antigen-specific cytolytic function requires the interaction of several types and an intricate cascade of secreted cytokines.[5,12] Although CTL are actively involved in rejection of allografts,[82] this may have more clinical significance than true host-resistance importance. Indeed, as previously mentioned, CTL may serve primarily as a virus control mechanism, with only a secondary role in controlling allografts or neoplasia.[81] Measuring CTL function therefore serves two roles. First, detection of alterations in overall cytolytic gives a measure of cell-mediated immunocompetence. Second, the multiplicity of potential targets allows for the possibility of sophisticated mechanistic immunotoxicology studies.

3.3.3.2 Methodology

The standard antigen-specific CTL assay described above for mice is sometimes impractical for routine human studies, primarily due to the quantity of effector cells required and secondarily to the choice of an appropriate target cell. One good choice is assessment of antigen-mediated cytolytic activity against viral proteins, as described by Shearer et al.[129] in what is sometimes referred to as "redirected" CTL. Redirected CTL is based on the ability of CTL to lyse target cells in the presence of an anti-CD3 specific monoclonal antibody. The actual lysis of the target cells may be antigen-mediated or antigen-independent, since the lytic mechanism appears to be triggered by bypassing the antigen receptor. In one version of the assay, radiolabeled, EBV-transformed B lymphoblastoid cells are sensitized with viral antigens (or alternatively synthetic viral proteins). Effector cells are then co-cultured with the target cells in the presence of anti-CD3 monoclonal antibody. Following incubation, the released radiolabel is quantified as previously described for the murine CTL assay. A somewhat simpler option which we have utilized with success in our laboratory is an assay described by Hoffman et al.,[45] in which human lymphocytes are incubated in microculture plates with radiolabeled murine hybridomas secreting an anti-human CD3 monoclonal antibody; this approach obviates the need for a separate target cell. Following an optimum co-culture period, the culture supernatants containing released radiolabel are collected, and cytotoxicity is determined as a percentage of the total releasable counts corrected for spontaneous release. The obvious utility of this assay is its ability to effectively bypass the requirement for specific antigen binding, thus detecting CTL of various specificities.

The most obvious problem, as with most of the human assays described here, is the uncertainly over what constitutes a "normal" or baseline response. A more practical use of this technique may be in mechanistic immunotoxicology, in which cells from naive humans are exposed to test material *in vitro*, followed by assessment of CTL activity. Whereas this approach is removed from *in vivo* conditions, it may still provide useful information.

3.3.4 Delayed-Type Hypersensitivity

Evaluation of delayed-type hypersensitivity (DTH) is often used as one of the first measures of cell-mediated immunity in humans, since it is a simple and inexpensive procedure. Many humans develop sensitivity to certain agents such as fungi (e.g., *Candida albicans*) or bacteria (e.g., streptococci).[1,114] Given the inherent variability in responsiveness to recall antigens, a test panel of such antigens has been developed.[35] These preparations are typically administered intradermally, and the response is measured approximately 48 hr later. The degree of reaction (e.g., erythema, induration) may be used as a measure of cell-mediated immunity to a recall antigen. Unfortunately, this testing approach has never been standardized for immunotoxicology assessment, and its potential remains uncertain.[114]

3.4 Assessment of Humoral Immunity

3.4.1 Background

Production of antigen-specific antibodies by mature plasma cells represents an important immune effector function. As a major arm of the immune response in humans, it is critical to ascertain potential toxic effects of xenobiotics on humoral immunity. As with most other measures of human immunity, antibody levels (actual concentrations as well as relative proportion of isotypes) exhibit considerable variability, particularly with age and general physiological constitution.[114] It is, therefore, extremely important that appropriate controls are utilized in any evaluation of humoral immunity.

3.4.2 Methodology

A relatively simple method for examining B-cell function is the determination of total immunoglobulin levels. This is readily performed using ELISA technology, which allows not only for determination of total levels, but also relative levels of various immunoglobulin subclasses. Although simple and quick, this method is antigen nonspecific, detecting polyclonal antibody production. An alternative is the proactive quantitation of specific antibody production. In certain cases it is possible to evaluate *in vivo* antigen-specific antibody production. This approach depends on measuring pre- and postimmunization titers to protein antigens such as diphtheria toxoid and provides good evidence of B-cell dysfunction.[1] A more practical approach may be an *in vitro* antibody production system such as one described by Yarchoan et al.[147] In this system, normal human leukocytes are incubated with antigen (e.g., influenza virus) for several days. The culture supernatants are subsequently collected and assayed for the presence of specific antibody by ELISA. An additional advantage of this assay is its ready adaptability to *in vitro* studies in which the naive cells are exposed *in vitro* to test materials. A less desirable modification of this assay is quantitation of polyclonal antibody production following *in vitro* stimulation with a polyclonal activator, such as a mitogen.[141]

3.5 Assessment of Natural Killer Cell Function

3.5.1 Background

As with the murine system, human NK-cell function represents an important nonspecific host defense mechanism. Well-defined alterations in human health and disease have been associated with alterations in NK cell number and activity.[109,146] Although there is a paucity of data on specific alteration of NK cell number or function with xenobiotic exposure, it is reasonable to assume that this parameter could be affected by immunotoxic compounds. The relative ease of performance and the limited volume of clinical sample necessary renders the NK assay an excellent inclusion in any evaluation of human immunocompetence.

3.5.2 Methodology

The methodology for assessing human NK-cell function is essentially identical to that for the mouse, with the exception of the target cell; in almost all cases, the K562 cell line is used as the target cell for human NK cell assays. Nouri et al.[96] have recently described the use of an epithelial cell line (Fen cells) as an alternative to the use of K562. These authors claim a significant increase in sensitivity using this cell line; at present, this line has not been used extensively, and its application in human immunotoxicity assessment remains to be validated. Another modification of the assay which may be of some utility in the context of immunotoxicology has been described by Mariani et al.,[80] who describe a standardized microcytotoxicity assay that uses tenfold fewer cells than standard assays. This assay would be of particular usefulness when the amount of a clinical sample is limited.

For obvious reasons, the use of *in vivo* positive controls for NK-cell activity is precluded in humans; however, viable *in vitro* alternatives are available. The Daudi cell line is relatively insensitive to NK cell-mediated lysis; for this reason, it is often used as a negative control. For example, lymphokine-activated killer (LAK) cells, which exhibit significantly enhanced cytolytic activity, will often destroy Daudi cells in the absence of prior exposure, whereas NK cells will not. Conversely, a positive control may be provided *in vitro* by adding exogenous cytokines, primarily IL-2 or IFN, which have been demonstrated to enhance NK-cell activity.[93]

3.6 Assessment of Autoimmunity

3.6.1 Background

Autoimmune disease affects approximately 5 to 7% of the population and is a significant cause of chronic disease.[131] The condition has been described as a failure of self-tolerance because the immune response becomes directed against self antigens. Although the basic biology of the condition is gradually being elucidated,[148] much remains to be learned. The development of autoimmunity has been associated with exposure to certain xenobiotics, in particular drugs and certain chemicals.[85] The process of establishing the autoimmune basis of human disease is involved, consisting of demonstration of autoantibodies (the easiest step), following by identification of the specific antigen, a much more difficult process. Injection of animals with an equivalent antigen, following by reproduction of the clinical disease, supports this basis.[114]

3.6.2 Methodology

Methodologies for assessing autoimmunity have been reviewed.[31] Although many autoimmune diseases are diagnosed on the basis of clinical presentation (e.g., diabetes), *in vitro* tests are available for some parameters. These include such tests as determination of anti-nuclear antibodies including antibodies to double-stranded DNA (e.g., SLE), anti-globulin antibodies, or rheumatoid factors (e.g., rheumatoid arthritis). Using *in vitro* techniques, it may also be demonstrated in cases for which the drug or chemical has rendered self proteins immunogenic. Finally, molecular immunology techniques such as Western blots[85] have shown some promise in elucidating the causes of these diseases. At present, however, a great deal of additional work remains to be done in this area, and no method has been completely validated as yet.

4 CONCLUSIONS

This chapter has attempted to provide an overview of experimental methodologies available for assessing the effects of experimental or inadvertent exposure to xenobiotics on the immune response in humans and laboratory rodents. The majority of immunotoxicology studies performed to date have utilized laboratory animals, and thus a large database of normal (baseline) values is available to assess accurately the toxic effects of chemical or drug exposure. The significant homology of the rodent and human immune systems has historically allowed us to make decisions on the immunotoxic potential of test materials based on the results of these experimental models. However, the expanding sophistication of risk assessment strategies dictates that the effects of xenobiotics on humans are assessed in a more direct manner. As pointed out in this chapter, a variety of assays is available for assessing human immune function; however, only a limited number of these assays has been employed in the context of immunotoxicology assessment. Consequently, the database of human responses, especially responses to known or suspected immunotoxicants, is far less complete than that for experimental animals. Therefore, it is anticipated that the development, validation, and utilization of human immunotoxicology methods will be increasingly important in the future.

REFERENCES

1. Adelman, D. C., Functional assessment of mononuclear leukocytes, *Immunol. Allergy Clin. N. Am.*, 14, 241–263, 1994.
2. Aggarwal, B. B. and Puri, R. K., Common and uncommon features of cytokines and cytokine receptors: an overview, in *Human Cytokines: Their Role in Disease and Therapy*, Aggarwal, B. B. and Puri, R. J., Eds., Blackwell Science, Cambridge, MA, 1995, pp. 3–24.
3. Anderson, J., Moller, G., and Sjoberg, O., Selective induction of DNA synthesis in T and B lymphocytes, *Cell. Immunol.*, 4, 381–393, 1972.

3a. Arend, W. P., Welgus, H. G., Thompson, R. C., and Eisenberg, S. P., Biological properties of recombinant human monocyte-derived interleukin-1 receptor antagonist, *J. Clin. Invest.*, 85, 1694–1697, 1990.

4. Asherton, G. and Ptak, G., Contact and delayed hypersensitivity in the mouse. I. Active sensitization and passive transfer, *Immunology*, 15, 405–416, 1968.
5. Apasov, S., Redegeld, F., and Sitkovsky, M., Cell-mediated cytotoxicity: contact and secreted factors, *Curr. Opin. Immunol.*, 5, 404–410, 1993.
6. Bach, F. H. and Voynow, N. K., One-way stimulation in mixed leukocyte cultures, *Science*, 153, 545–547, 1966.
7. Barlozzari, T., Leonhardt, J., Wiltrout, R. H., Herberman, R. B., and Reynolds, C. W., Direct evidence for the role of LGL in the inhibition of experimental tumor metastases, *J. Immunol.*, 134, 2783–2789, 1985.
8. Barr, P. J. and Tomei, L. D., Apoptosis and its role in human disease, *Bio/Technology*, 12, 487–493, 1994.
9. Basketter, D. A., Scholes, E. W., Kimber, I., Botham, P. A., Hilton, J., Miller, K., Robbins, M. C., Harrison, P. T. C., and Waite, S. J., Interlaboratory evaluation of the local lymph node assay with 25 chemicals and comparison with guinea pig test data, *Toxicol. Meth.*, 1, 30–43, 1991.
10. Basketter, D. A., Bremmer, J. N., Buckley, P., Kammuller, M. E., Kawabata, T., Kimber, I., Loveless, S. E., Magda, S., Stringer, D. A., and Vohr, H.-W., Pathology considerations for, and subsequent risk assessment of, chemicals identified as immunosuppressive in routine toxicology, *Food Chem. Toxicol.*, 33, 239–243, 1995.
11. Baumann, H. and Gauldie, J., The acute phase response, *Immunol. Today*, 15, 74–80, 1994.
12. Berke, G., The cytolytic T lymphocyte and its mechanism of action, *Immunol. Lett.*, 20, 169–178, 1989.
13. Bigazzi, P. E., Autoimmunity induced by chemicals, *Clin. Toxicol.*, 26, 125–156, 1988.

13a. Blaylock, B. L., Kouchi, Y., Comment, C. E., Pollock, P. L., and Luster, M. I., Topical application of T-2 toxin inhibits the contact hypersensitivity response in BALB/c mice, *J. Immunol.*, 150, 5135-5143, 1993.

14. Bloom, J. C., Thiem, P. A., and Morgan, D. G., The role of conventional pathology and toxicology in evaluating the immunotoxic potential of xenobiotics, *Toxicol. Pathol.*, 15, 283–293, 1987.

15. Boitard, C., Sempé, P., Villà, M. C., Becourt, C., Richard, M. F., Timsit, J., and Bach, J. F., Monoclonal antibodies: probes for studying experimental autoimmunity in animals, *Res. Immunol.*, 142, 495–503, 1991.

16. Bosma, G. C., Fried, M. R., Custer, R. P., Carroll, A., Gibson, D. M., and Bosma, M. J., Evidence of functional lymphocytes in some (leaky) *scid* mice, *J. Exp. Med.*, 167, 1016–1033, 1988.

17. Bosma, M. J. and Carroll, A. M., The *scid* mouse mutant: definition, characterization and potential uses, *Annu. Rev. Immunol.*, 9, 323–350, 1991.

18. Briatico-Vangosa, G., Braun, C. L. J., Cookman, G., Hofmann, T., Kimber, I., Loveless, S. E., Morrow, T., Pauluhn, J., Sorensen, T., and Niessen, H. J., Respiratory allergy: hazard identification and risk assessment, *Fund. Appl. Toxicol.*, 23, 145–158, 1994.

19. Brunner, K. T., Engers, H. D., and Cerottini, J. C., The 51Cr-release assay as used for the quantitative measurement of cell-mediated cytolysis *in vitro*, in In Vitro *Methods of Cell-Mediated and Tumor Immunity*, Bloom, B. R. and David, J. R., Eds., Academic Press, New York, 1976, pp. 423–428.

20. Bryant, J., Day, R., Whiteside, T. L., and Herberman, R. B., Calculation of lytic units for the expression of cell-mediated cytotoxicity, *J. Immunol. Meth.*, 146, 91–103, 1992.

20a. Buttke, T. M., McCubrey, J. A., and Owen, T. C., Use of an aqueous soluble tetrazolium/formazan assay to measure viability and proliferation of lymphokine-dependent cell lines, *J. Immunol. Meth.*, 157, 233–240, 1993.

21. Cannon, J. G., Nerad, J. L., Poutsiaka, D. D., and Dinarello, C. A., Measuring circulating cytokines, *J. Appl. Physiol.*, 75, 1897–1902, 1993.

22. Cohen, J. J., Duke, R. C., Fadok, V. A., and Sellins, K. S., Apoptosis and programmed cell death in immunity, *Annu. Rev. Immunol.*, 10,267–293, 1992.

23. Cohen, R. B., Siegel, J. P., Puri, R. K., and Pluznik, D. H., The immunotoxicology of cytokines, in *Clinical Immunotoxicology*, Newcombe, D. S., Rose, N. R., and Bloom, J. C., Eds., Raven Press, New York, 1992, pp. 93–108.

24. Corcoran, G. B., Fix, L., Jones, D. P., Moslen, M. T., Nicotera, P., Oberhammer, F. A., and Buttyan, R., Apoptosis: molecular control point in toxicity, *Toxicol. Appl. Pharmacol.*, 128, 169–181, 1994.

25. Cornacoff, J. B., House, R. V., and Dean, J. H., A comparison of a radioisotopic incorporation method and the mouse ear swelling test (MEST) for contact sensitivity to weak sensitizers, *Fund. Appl. Toxicol.*, 10, 40–44, 1987.

26. Cornacoff, J. B., Graham, C. S., and LaBrie, T. K., Phenotypic identification of peripheral blood mononuclear leukocytes by flow cytometry as an adjunct to immunotoxicity evaluation, in *Methods in Immunotoxicology*, Vol. 1, Burleson, G. R., Dean, J. H., and Munson, A. E., Eds., Wiley-Liss, New York, 1995, pp. 211–226.

27. Cunningham, A. J., Further improvements in the plaque technique for detecting single antibody-forming cells, *Immunology*, 14, 599–601, 1968.

28. Dallman, M. J., Montgomery, R. A., Larsen, C. P., Wanders, A., and Wells, A. F., Cytokine gene expression: analysis using Northern blotting, polymerase chain reaction and *in situ* hybridization, *Immunol. Rev.*, 119, 163–179, 1991.

29. Dearman, R. J., Basketter, D. A., and Kimber, I., Variable effects of chemical allergens on serum IgE concentrations in mice. Preliminary evaluation of a novel approach to the identification of respiratory sensitizers, *J. Appl. Toxicol.*, 12, 317–323, 1992.

30. Djeu, J. Y., Heinbaugh, J. A., Holden, H. T., and Herberman, R. B., Augmentation of mouse natural killer cell activity by interferon and interferon inducers, *J. Immunol.*, 122, 175–188, 1979.

31. Druet, P., Diagnosis of autoimmune disease, *J. Immunol. Meth.*, 150, 177–184, 1992.

32. Edwards, D. A., Soranno, T. M., Amoruso, M. A., House, R. V., Tummey, A. C., Trimmer, G. W., Thomas, P. T., and Ribeiro, P. L., Screening petrochemicals for contact hypersensitivity potential: a comparison of the murine local lymph node assay with guinea pig and human test data, *Fund. Appl. Toxicol.*, 23, 179–187, 1994.

33. Feldmann, M., Brennan, M., Chantry, D., Haworth, C., Turner, M., Katsikis, P., Londei, M., Abney, E., Buchan, G., Barrett, K., Corcoran, A., Kissonerghis, M., Zheng, R., Grubeck-Loebenstein, B., Barkley, D., Chu, C. Q., Field, M., and Maini, R. N., Cytokine assays: role in evaluation of the pathogenesis of autoimmunity, *Immunol. Rev.*, 119, 105–123, 1991.

34. Ferrari, M., Fornasiero, M. C., and Isetta, A. M., MTT colorimetric assay for testing macrophage cytotoxic activity *in vitro*, *J. Immunol. Meth.*, 131, 165–172, 1990.

35. Frazer, I. H., Collins, E. J., Fox, J. S., Jones, B., Oliphant, R. C., and Mackay, I. R., Assessment of delayed-type hypersensitivity in man: a comparison of the "multitest" and conventional intradermal injection of six antigens, *Clin. Immunol. Immunopathol.*, 35, 182–190, 1985.

36. Freitas, A. A., Rocha, B., and Coutinho, A. A., Lymphocyte population kinetics in the mouse, *Immunol. Rev.*, 91, 5–37, 1986.

37. Gad, S. C., Dunn, B. J., Dobbs, D. W., Reilly, C., and Walsh, R. D., Development and validation of an alternative dermal sensitization test: the mouse ear swelling test (MEST), *Toxicol. Appl. Pharmacol.,* 84, 93–114, 1986.
38. Gad, S. C., The mouse ear swelling test (MEST) in the 1990s, *Toxicology,* 93, 33–46, 1994.
39. Galli-Taliadoros, L. A., Sedgwick, J. D., Wood, S. A., and Körner, H., Gene knock-out technology: a methodological overview for the interested novice, *J. Immunol. Meth.,* 181, 1–15, 1995.
40. Garrigue, J.-L., Nicolas, J.-F., Fraginals, R., Benezra, C., Bour, H., and Schmitt, D., Optimization of the mouse ear swelling test for *in vivo* and *in vitro* studies of weak contact sensitizers, *Cont. Dermatitis,* 30, 231–237, 1994.
41. Gerberick, G. F., House, R. V., Fletcher, R., and Ryan, C. A., Examination of the local lymph node assay for use in contact sensitization risk assessment, *Fund. Appl. Toxicol.,* 19, 438–445, 1992.
42. Goodwin, C. J., Holt, S. J., Downes, S., and Marshall, N. J., Microculture tetrazolium assays: a comparison between two new tetrazolium salts, XTT and MTS, *J. Immunol. Meth.,* 179, 95–103, 1995.
43. Gordon, S., Clarke, S., Greaves, D., and Doyle, A., Molecular immunobiology of macrophages: recent progress, *Curr. Opin. Immunol.,* 7, 24–33, 1995.
44. Herzyk, D. J., Berger, A. E., Allen, J. N., and Wewers, M. D., Sandwich ELISA formats designed to detect 17-kDa IL-1β significantly underestimate 37-kDa IL-1β, *J. Immunol. Meth.,* 148, 243–254, 1992.
45. Hoffman, R. W., Bluestone, J. A., Leo, O., and Shaw, S., Lysis of anti-T3-bearing murine hybridoma cells by human allospecific cytotoxic T cell clones and inhibition of that lysis by anti-T3 and anti-LFA-1 antibodies, *J. Immunol.,* 135, 5–8, 1985.
46. Holobaugh, P. A. and McChesney, D. C., Effect of anticoagulants and heat on the detection of tumor necrosis factor in murine blood, *J. Immunol. Meth.,* 135, 95–99, 1990.
47. House, R. V., Cytokine bioassays for assessment of immunomodulation: applications, procedures and practical considerations, in *Methods in Immunotoxicology,* Vol. 1, Burleson, G. R., Dean, J. H., and Munson, A. E., Eds., Wiley-Liss, New York, 1995, pp. 251–276.
48. House, R. V. and Thomas, P. T., *In vitro* induction of cytotoxic T-lymphocytes, in *Methods in Immunotoxicology,* Vol. 1, Burleson, G. R., Dean, J. H., and Munson, A. E., Eds., Wiley-Liss, New York, 1995, pp. 159–171.
49. House, R. V., Lauer, L. D., Murray, M. J., and Dean, J. H., Suppression of T-helper cell function in mice following exposure to the carcinogen 7,12-dimethylbenz[a]anthracene and its restoration by interleukin-2, *Int. J. Immunopharmacol.,* 9, 89–97, 1987.
50. Hussain, R. F., Nouri, A. M. E., and Oliver, R. T. D., A new approach for measurement of cytotoxicity using colorimetric assay, *J. Immunol. Meth.,* 160, 89–96, 1993.
51. Ikarashi, Y., Tsuchiya, T., and Nakamura, A., Detection of contact sensitivity of metal salts using the murine local lymph node assay, *Toxicol. Lett.,* 62, 53–61, 1992.
52. Ikarashi, Y., Tsuchiya, T., and Nakamura, A., Evaluation of contact sensitivity of rubber chemicals using the murine local lymph node assay, *Cont. Dermatitis,* 28, 77–80, 1993.
53. James, K., Milne, I., Cunningham, A., and Elliott, S.-F., The effect of α_2 macroglobulin in commercial cytokine assays, *J. Immunol. Meth.,* 168, 33–37, 1994.
54. Janossy, G. and Greaves, M. F., Lymphocyte activation. I. Response of T and B lymphocytes to phytomitogens, *Clin. Exp. Immunol.,* 9, 483–498, 1971.
55. Jaramillo, A., Gill, B. M., and Delovitch, T. L., Insulin-dependent diabetes mellitus in the non-obese diabetic mouse: a disease mediated by T cell anergy?, *Life Sci.,* 55, 1163–1177, 1994.
56. Jerne, N. K. and Nordin, A. A., Plaque formation in agar by single antibody-producing cells, *Science,* 140, 405, 1963.
57. Kamal-Reid, S. and Dick, J. E., Engraftment of immune-deficient mice with human hematopoietic stem cells, *Science,* 242, 1706–1709, 1988.
58. Karol, M. H., Occupational asthma and allergic reactions to inhaled compounds, in *Principles and Practices of Immunotoxicology,* Miller, K., Turk, J. L., and Nicklin, S., Eds., Blackwell Scientific, Oxford, 1992, pp. 228–241.
59. Kimber, I. and Dearman, R. J., Assessment of contact and respiratory sensitivity in mice, in *Immunotoxicology and Immunopharmacology,* 2nd ed., Dean, J. H., Luster, M. I., Munson, A. E., and Kimber, I., Eds., Raven Press, New York, 1994, pp. 721–732.
60. Kimber, I. and Dearman, R. J., Immune responses to contact and respiratory allergens, in *Immunotoxicology and Immunopharmacology,* 2nd ed., Dean, J. H., Luster, M. I., Munson, A. E., and Kimber, I., Eds., Raven Press, New York, 1994, pp. 663–679.
61. Kimber, I. and Weisenberger, C., A murine local lymph node assay for the identification of contact allergens. Assay development and results of an initial validation study, *Arch. Toxicol.,* 63, 274–282, 1989.
61a. Kimber, I., Hilton, J., and Weisenberger, C. The murine local lymph node assay for identification of contact allergens: a preliminary evaluation of *in situ* measurement of lymphocyte proliferation, *Cont. Dermatitis,* 21, 215–220, 1989.

62. Kimber, I., Hilton, J., Botham, P. A., Basketter, D. A., Scholes, E. W., Miller, K., Robbins, M. C., Harrison, P. T. C., Gray, T. J. B., and Waite, S. J., The murine local lymph node assay: results of an inter-laboratory trial, *Toxicol. Lett.*, 55, 203–213, 1991.
63. Kimber, I., Dearman, R. J., Scholes, E. W., and Basketter, D. A., The local lymph node assay: developments and applications, *Toxicology*, 93, 13–31, 1994.
64. Kimber, I., Hilton, J., Dearman, R. J., Gerberick, G. F., Ryan, C. A., Basketter, D. A., Scholes, E. W., Ladics, G. S., Loveless, S. E., House, R. V., and Guy, A., An international evaluation of the murine local lymph node assay and comparison of modified procedures, *Toxicology*, 103, 63–73, 1995.
65. Kotylo, P. K., Sample, R. B., Redmond, N. L., and Hibner, G. C., Reference ranges for lymphocyte subsets, *Arch. Pathol. Lab. Med.*, 115, 181–184, 1991.
66. Kuper, C. F., Schuurman, H.-J., and Vos, J. G., Pathology in immunotoxicology, in *Methods in Immunotoxicology*, Vol. 1, Burleson, G. R., Dean, J. H., and Munson, A. E., Eds., Wiley-Liss, New York, 1995, pp. 397–436.
67. Kushner, I., Regulation of the acute phase response by cytokines, *Perspect. Biol. Med.*, 36, 611–622, 1993.
68. Lang, D. S., Meier, K. L., and Luster, M. I., Comparative effects of immunotoxic chemicals on *in vitro* proliferative responses of human and rodent lymphocytes, *Fund. Appl. Toxicol.*, 21, 535–545, 1993.
69. Larrick, J. W., Native interleukin 1 inhibitors, *Immunol. Today*, 10, 61–66, 1989.
70. Lehrer, R. I., Ganz, T., Selsted, M. E., Babior, B. M., and Curnutte, J. T., Neutrophils and host defense, *Ann. Intern. Med.*, 109, 127–142, 1988.
71. Lewis, J. G., Introduction to the mononuclear phagocyte system, in *Methods in Immunotoxicology*, Vol. 2., Burleson, G. R., Dean, J. H., and Munson, A. E., Eds., Wiley-Liss, New York, 1995, pp. 3–13.
72. Lotzova, E., Definition and functions of natural killer cells, *Nat. Immun.*, 12, 169–176, 1993.
73. Loveless, S. E., Ladics, G. S., Gerberick, G. F., Ryan, C. A., Basketter, D. A., Scholes, E. W., House, R. V., Dearman, R. J., and Kimber, I., Further evaluation of the local lymph node assay in the final phase of an international collaborative trial, *Toxicology*, 108, 141–152, 1996.
74. Luster, M. I., Munson, A. E., Thomas, P. T., Holsapple, M. P., Fenters, J. D., White, K. L., Lauer, L. D., Germolec, D. R., Rosenthal, G. J., and Dean, J. H., Development of a testing battery to assess chemical-induced immunotoxicity: National Toxicology Program's guidelines for immunotoxicity evaluation in mice, *Fund. Appl. Toxicol.*, 10, 2–19, 1988.
75. Luster, M. I., Portier, C., Pait, D. G., White, K. L., Gennings, C., Munson, A. E., and Rosenthal, G. J., Risk assessment in immunotoxicology. I. Sensitivity and predictability of immune tests, *Fund. Appl. Toxicol.*, 18, 200–210, 1992.
76. Luster, M. I., Portier, C., Pait, D. G., Rosenthal, G. J., Germolec, D. R., Corsini, E., Blaylock, B. L., Pollock, P., Kouchi, Y., Craig, W., White, K. L., Munson, A. E., and Comment, C. E., Risk assessment in immunotoxicology. II. Relationships between immune and host resistance tests, *Fund. Appl. Toxicol.*, 21, 71–82, 1993.
77. Luster, M. I., Portier, C., Pait, D. G., and Germolec, D. R., Use of animal studies in risk assessment for immunotoxicology, *Toxicology*, 92, 229–243, 1994.
78. MacDonald, H. and Lees, R. K., Primary generation of cytolytic T lymphocytes in the absence of DNA synthesis, *J. Exp. Med.*, 150, 196–201, 1979.
79. MacKay, C. R., Immunological memory, *Adv. Immunol.*, 53, 217–265, 1993.
80. Mariani, E., Monaco, M. C. G., Sgobbi, S., de Zwart, J. F., Mariani, A. R., and Facchini, A., Standardization of a micro-cytotoxicity assay for human natural killer cell lytic activity, *J. Immunol. Meth.*, 172, 173–178, 1994.
81. Martz, E. and Howell, D. M., CTL: virus control cells first and cytolytic cells second?, *Immunol. Today*, 10, 79–86, 1989.
82. Mason, D., The roles of T cell subpopulations in allograft rejection, *Transplant. Proc.*, 20, 239–242, 1988.
83. Maurer, T., Arthur, A., and Bentley, P., Guinea-pig contact sensitization assays, *Toxicology*, 93, 47–54, 1994.
84. Meredith, C., Molecular immunotoxicology, in *Principles and Practices of Immunotoxicology*, Miller, K., Turk, J. L., and Nicklin, S., Eds., Blackwell Scientific, Oxford, 1992, pp. 344–356.
85. Merk, H. F., Gleichmann, E., and Gleichmann, H., Adverse immunological effects of drugs and other chemicals and methods to detect them, in *Principles and Practices of Immunotoxicology*, Miller, K., Turk, J. L., and Nicklin, S., Eds., Blackwell Scientific, Oxford, 1994, pp. 86–103.
86. Mitchell, B. S., Dhami, D., and Schumacher, U., *In situ* hybridization: a review of methodologies and applications in the biomedical sciences, *Med. Lab. Sci.*, 49, 107–118, 1992.
87. Miyake, Y., Ajitsu, S., Yamashita, T., and Sendo, F., Enhancement by recombinant interferon-γ of spontaneous tumor cytostasis by human neutrophils, *Mol. Biother.*, 1, 37–42, 1988.
88. Modlin, R. L. and Nutman, T. B., Type 2 cytokines and negative immune regulation in human infections, *Curr. Opinion Immunol.*, 5, 511–517, 1993.
89. Mosier, D. E., Immunodeficient mice xenografted with human lymphoid cells: new models for *in vivo* studies of human immunobiology and infectious diseases, *J. Clin. Immunol.*, 10, 185–191, 1990.
90. Mosmann, T., Rapid colorimetric assay for cellular growth and survival: application to proliferation and cytotoxicity assays, *J. Immunol. Meth.*, 65, 55–63, 1983.
91. Nabholz, M. and MacDonald, H. R., Cytolytic T lymphocytes, *Ann. Rev. Immunol.*, 1, 273–306, 1983.
92. Nathan, C. F., Secretory products of macrophages, *J. Clin. Invest.*, 79, 319–326, 1987.

93. Nauem, B. and Espevik, T., Immunoregulatory effects of cytokines on natural killer cells, *Scand. J. Immunol.*, 40, 128–134, 1994.
94. Neldon, D. L., Lange, R. W., Rosenthal, G. J., Comment, C. E., and Burleson, G. R., Macrophage nonspecific phagocytosis assays, in *Methods in Immunotoxicology*, Vol. 2, Burleson, G. R., Dean, J. H., and Munson, A. E., Eds., Wiley-Liss, New York, 1995, pp. 39–57.
95. Nikš, M. and Otto, M., Towards an optimized MTT assay, *J. Immunol. Meth.*, 130, 149–151, 1990.
96. Nouri, A. M. E., Mansouri, M., Hussain, R. F., Dos Santos, A. V. L., and Oliver, R. T. D., Super-sensitive epithelial cell line and colorimetric assay to replace the conventional K562 target and chromium release assay for assessment of non-MHC-restricted cytotoxicity, *J. Immunol. Meth.*, 180, 63–68, 1995.
97. Oez, S., Welte, K., Platzer, E., and Kalden, J. R., A simple assay for quantifying the inducible adherence of neutrophils, *Immunobiology*, 180, 308–315, 1990.
98. Ohmori, H., Takai, T., Tanigawa, T., and Honma, Y., Establishment of an enzyme release assay for cytotoxic T lymphocyte activity, *J. Immunol. Meth.*, 147, 119–124, 1992.
99. Peters, R. E. and Al-Ismail, S., Immunophenotyping or normal lymphocytes, *Clin. Lab. Hematol.*, 16, 21–32, 1994.
100. Podack, E. R., Functional significance of two cytolytic pathways of cytotoxic T lymphocytes, *J. Leukocyte Biol.*, 57, 548–552, 1995.
101. Pollock, P. L., Germolec, D. R., Comment, C. E., Rosenthal, G. J., and Luster, M. I., Development of human lymphocyte-engrafted SCID mice as a model for immunotoxicity assessment, *Fund. Appl. Toxicol.*, 22, 130–138, 1994.
102. Pross, H. F. and Lotzova, E., Role of natural killer cells in cancer, *Nat. Immun.*, 12, 279–292, 1993.
103. Pruett, S. B. and Loftis, A. Y., Characteristics of MTT as an indicator of viability and respiratory burst activity of human neutrophils, *Int. Arch. Allergy Appl. Immunol.*, 92, 189–192, 1990.
104. Rapley, R., Theophilus, B. D. M., Bevan, I. S., and Walker, M. R., Fundamentals of the polymerase chain reaction: a future in clinical diagnosis?, *Med. Lab. Sci.*, 49, 119–128, 1992.
105. Reichert, T., DeBruyère, M., Deneys, V., Tötterman, T., Lydyard, P., Yuksel, F., Chapel, H., Jewell, D., Van Hove, L., Linden, J., and Buchner, L., Lymphocyte subset reference ranges in adult Caucasians, *Clin. Immunol. Immunopathol.*, 60, 190–208, 1991.
106. Reynolds, C. W. and Herberman, R. B., *In vitro* augmentation of rat natural killer (NK) cell activity, *J. Immunol.*, 126, 1581–1585, 1981.
107. Riches, P., Gooding, R., Millar, B. C., and Rowbottom, A. W., Influence of collection and separation of blood samples on plasma IL-1, IL-6 and TNF-α concentrations, *J. Immunol. Meth.*, 153, 125–131, 1992.
108. Rittenberg, M. B. and Pratt, K. L., Anti-trinitrophenyl (TNP) plaque assay. Primary response of BALB/c mice to soluble and particulate immunogen, *Proc. Soc. Exp. Biol. Med.*, 132, 575–581, 1969.
109. Robertson, M. J. and Ritz, J., Biology and clinical relevance of human natural killer cells, *Blood*, 76, 2421–2438, 1990.
110. Roder, J. and Duwe, A., The beige mutation in the mouse selectively impairs natural killer cell function, *Nature*, 278, 451–453, 1979.
111. Rodgers, K., Measurement of the respiratory burst of leukocytes for immunotoxicologic analysis, in *Methods in Immunotoxicology*, Vol. 2, Burleson, G. R., Dean, J. H., and Munson, A. E., Eds., Wiley-Liss, New York, 1995, pp. 67–77.
112. Roehm, N. W., Rodgers, G. H., Hatfield, S. M., and Glasebrook, A. L., An improved colorimetric assay for cell proliferation and viability utilizing the tetrazolium salt XTT, *J. Immunol. Meth.*, 142, 257–265, 1991.
113. Romagnani, S., Biology of human T_H1 and T_H2 cells, *J. Clin. Immunol.*, 15, 121–129, 1995.
114. Rose, N. R. and Margolick, J. B., The immunological assessment of immunotoxic effects in man, in *Clinical Immunotoxicology*, Newcombe, D. S., Rose, N. R., and Bloom, J. C., Eds., Raven Press, New York, 1992, pp. 9–25.
115. Rose, N. R. and Bhatia, S., Autoimmunity: animal models of human autoimmune disease, in *Methods in Immunotoxicology*, Vol. 2, Burleson, G. R., Dean, J. H., and Munson, A. E., Eds., Wiley-Liss, New York, 1995, pp. 427–445.
116. Sarlo, K. and Clark, E. D., Evaluating chemicals as respiratory allergens: using the tier approach for risk assessment, in *Methods in Immunotoxicology*, Vol. 2, Burleson, G. R., Dean, J. H., and Munson, A. E., Eds., Wiley-Liss, New York, 1995, pp. 411–426.
117. Satoh, H., Chizzonite, R., Ostrowski, C., Ni-Wu, G., Kim, H., Fayer, B., Mae, N., Nadeau, R., and Liberato, D. J., Characterization of anti-IL-1α autoantibodies in the sera from healthy humans, *Immunopharmacology*, 27, 107–118, 1994.
118. Scher, I., B-lymphocyte ontogeny, *Crit. Rev. Immunol.*, 1, 287–320, 1981.
119. Scholes, E. W., Basketter, D. A., Lovell, W. W., Sarll, A. E., and Pendlington, R. U., The identification of photoallergic potential in the local lymph node assay, *Photodermatol. Photoimmunol. Photomed.*, 8, 249–254, 1992.
120. Schultz, L. D., Immunological mutants of the mouse, *Am. J. Anat.*, 191, 303–311, 1991.
121. Schuurman, H.-J., de Weger, R. A., van Loveren, H., Krajnc-Franken, M. A. M., and Vos, J. G., Histopathological approaches, in *Principles and Practices of Immunotoxicology*, Miller, K., Turk, J. L., and Nicklin, S., Eds., Blackwell Scientific, Oxford, 1992, pp. 279–303.

122. Schuurman, H.-J., van Loveren, H., Rozing, J., and Vos, J. G., Chemicals trophic for the thymus: risk for immunodeficiency and autoimmunity, *Int. J. Immunopharmacol.,* 14, 369, 1992.
123. Schuurman, H.-J., Kuper, C. F., and Vos, J. G., Histopathology of the immune system as a tool to assess immunotoxicity, *Toxicology,* 86, 187–212, 1994.
124. Scott, P. and Trinchieri, G., The role of natural killer cells in host-parasite interactions, *Curr. Opinion Immunol.,* 7, 34–40, 1995.
125. Seder, R. A. and Paul, W. E., Acquisition of lymphokine-producing phenotype by CD4+ T cells, *Annu. Rev. Immunol.,* 12, 635–673, 1994.
126. Selgrade, M. K., Cooper, K. D., Devlin, R. B., van Loveren, H., Biagini, R. E., and Luster, M. I., Immunotoxicity — bridging the gap between animal research and human health effects, *Fund. Appl. Toxicol.,* 24, 13–21, 1995.
127. Sell, S., Park, A. B., and Nordin, A. A., Immunoglobulin classes of antibody-forming cells in mice. I. Localized hemolysis-in-agar plaque-forming cells belonging to five immunoglobulin classes, *J. Immunol.,* 104, 483, 1970.
128. Sgonc, R. and Wick, G., Methods for the detection of apoptosis, *Int. Arch Allergy Immunol.,* 105, 327–332, 1994.
129. Shearer, G. M., Salahuddin, S. Z., Markham, P. D., Joseph, L. J., Payne, S. J., Kriebel, P., Bernstein, D. C., Biddison, W. E., Sarngadharan, M. G., and Gallo, R. C., Prospective study of cytotoxic T lymphocyte responses to influenza and antibodies to human T lymphotropic virus-III in homosexual men, *J. Clin. Invest.,* 76, 1699–1704, 1985.
130. Shultz, L. D., Immunological mutants of the mouse, *Am. J. Anat.,* 191, 303–311, 1991.
131. Sinha, A. A., Lopez, M. T., and McDevitt, H. O., Autoimmune diseases: the failure of self tolerance, *Science,* 248, 1380–1388, 1990.
132. Smith, J. A., Neutrophils, host defense, and inflammation: a double-edged sword, *J. Leukocyte Biol.,* 56, 672–686, 1994.
133. Snyder, C. A. and Valle, C. D., Lymphocyte proliferation assays as potential biomarkers for toxicant exposures, *J. Toxicol. Environ. Health,* 34, 127–139, 1991.
134. Temple, L., Kawabata, T. T., Munson, A. E., and White, K. L., Comparison of ELISA and plaque-forming cell assays for measuring the humoral immune response to SRBC in rats and mice treated with benzo[a]pyrene or cyclophosphamide, *Fund. Appl. Toxicol.,* 21, 412–419, 1993.
135. Temple, L, Butterworth, L., Kawabata, T. T., Munson, A. E., and White, Jr., K. L., ELISA to measure SRBC specific serum IgM: method and data evaluation, in *Methods in Immunotoxicology,* Vol. 1, Burleson, G. R., Dean, J. H., and Munson, A. E., Eds., Wiley-Liss, New York, 1995, pp. 137–157.
136. Thavasu, P. W., Longhurst, S., Joel, S. P., Slevin, M. L., and Balkwill, F. R., Measuring cytokine levels in blood: importance of anticoagulants, processing and storage conditions, *J. Immunol. Meth.,* 153, 115–124, 1992.
137. Theofilopoulos, A. N. and Dixon, F. J., Murine models of systemic lupus erythematosus, *Adv. Immunol.,* 37, 269–389, 1985.
138. Thorne, P. S., Hawk, C., Kaliszewski, S. D., and Guiney, P. D., The noninvasive mouse ear swelling assay. I. Refinements for detecting weak contact sensitizers, *Fund. Appl. Toxicol.,* 17, 790–806, 1991.
139. Thorne, P. S., Hawk, C., Kaliszewski, S. D., and Guiney, P. D., The noninvasive mouse ear swelling assay. II. Testing the contact sensitizing potency of fragrances, *Fund. Appl. Toxicol.,* 17, 807–820, 1991.
140. Thorpe, R., Wadhwa, M., Bird, C. R., and Mire-Sluis, A. R., Detection and measurement of cytokines, *Blood Rev.,* 5, 133–148, 1992.
141. Tosato, G., Magrath, I. T., Koski, I. R., Dooley, N. J., and Blaese, R. M., B cell differentiation and immunoregulatory T cell function in human cord blood lymphocytes, *J. Clin. Invest.,* 66, 383–390, 1980.
142. Viney, J. L., Transgenic and knockout models for studying diseases of the immune system, *Curr. Opin. Gen. Devel.,* 4, 461–465, 1994.
143. Wemme, H., Pfeifer, S., Heck, R., and Müller-Quernheim, J., Measurement of lymphocyte proliferation: critical analysis of radioactive and photometric methods, *Immunobiology,* 185, 78–89, 1992.
144. Westermann, J. and Pabst, R., Lymphocyte subsets in the blood: a diagnostic window on the lymphoid system?, *Immunol. Today,* 11, 406–410, 1990.
145. Whiteside, T. L., Cytokine measurements and interpretation of cytokine assays in human disease, *J. Clin. Immunol.,* 14, 327–339, 1994.
146. Whiteside, T. L. and Herberman, R. B., Role of human natural killer cells in health and disease, *Clin. Diag. Lab. Immunol.,* 1, 125–133, 1994.
146a. Wunderlich, J. R. and Canty, T. G., Cell mediated immunity induced *in vitro, Nature* (*Lond.*), 288, 62, 1970.
147. Yarchoan, R., Murphy, B. R., Strober, W., Schneider, H. S., and Nelson, D. L., Specific anti-influenza virus antibody production *in vitro* by human peripheral blood mononuclear cells, *J. Immunol.,* 127, 2588–2593, 1981.
148. Zouali, M., Kalsi, J., and Isenberg, D., Autoimmune diseases — at the molecular level, *Immunol. Today,* 14, 473–476, 1993.

CHAPTER 17

Immunotoxicology of Therapeutic Proteins

Elizabeth E. Sikorski and Helen G. Haggerty

CONTENTS

1 INTRODUCTION

The remarkable advances in molecular and cell biology occurring over the past four decades have opened up many areas from which to draw new therapeutic strategies. One rapidly growing area involves the use of therapeutic proteins. This chapter will focus on the immunotoxicology of therapeutic proteins that are being developed based on their potential to modulate immunological and inflammatory responses and provide therapeutic benefit in cancer as well as inflammatory, autoimmune, and infectious diseases. The three classes of agents that will be reviewed in this chapter include (1) recombinant cytokine proteins, including cytokine fusion proteins, soluble cytokine receptors, and cytokine antagonists; (2) monoclonal antibodies; and (3) immunoconjugates/immunotoxins.

0-8493-4493-X/97/$0.00+$.50

The objective of this chapter is to review the immunotoxicology of protein therapeutics which includes hypersensitivity as well as effects on both specific and nonspecific immune responses. In general, this review will include any efficacious or toxicological changes in immune parameters listed in a journal article. This was done to stress that the immunomodulatory effects of cyokines must be considered for an individual patient within the context of a particular study. As an example, augmentation of the immune response by IL-2 may be beneficial in a cancer patient but may produce deleterious side effects if the patient has a concomitant autoimmune or inflammatory disease that may also be augmented by this cytokine.

Most information included in this chapter has been obtained from published human clinical studies; however, with certain proteins on which there is limited human data, studies in nonhuman primates have been included as well. Also included are reports of other toxicities that may not appear related to the immune system. These were included as additional information for it is known that stress and toxicity in other organ systems can produce secondary effects on the immune system.

A close evaluation of the studies listed will indicate that immunological baseline parameters of patients can differ. These differences could be due to a number of reasons, including the clinical condition of the patient and/or the therapy (including surgery) the patient is receiving, both of which have been shown to alter immune parameters. Baxevanis et al.[14] have demonstrated *in vitro* alterations in various cellular immune parameters following surgery. In addition, deficient autologous mixed lymphocyte reaction (AMLR) and natural killer (NK) cell function have been observed in patients with several diseases, including cancer and autoimmunity.[5,12,15,16]

The toxicological effects of protein therapeutics, including effects on the immune system, are in many cases related to the unique features of these agents. Each of the three classes of therapeutic proteins will be discussed briefly below to point out these features.

2 RECOMBINANT CYTOKINE PROTEINS

The introduction of recombinant gene technology has made many cytokines available in large quantities and has opened up the possibility of large-scale clinical testing for their therapeutic potential in a variety of malignant, infectious, or inflammatory diseases. The recombinant cytokine proteins generally act as the endogenous proteins (cytokines, soluble cytokine receptors, or cytokine antagonists) which have specific surface receptors on immune, inflammatory, or cancer cells. Cytokine therapy may be used to augment the activities of the immune system to destroy tumors or infectious agents. Cytokines may also be used to suppress responses in inflammatory, hypersensitivity, or autoimmune disease. To illustrate this point, cytokines have been used in many ways in cancer therapy. Cytokines may (1) promote nonspecific antitumor activity via macrophages, eosinophils, or neutrophils; (2) promote antigen-specific responses against tumor antigens; or (3) have direct cytotoxic or antiproliferative effects on the tumor. In cancer therapy, cytokines may also be useful in reversing the myelosuppressive effect of chemotherapy by their ability to induce growth and differentiation of a variety of immune cells and/or their progenitors. In this chapter, studies on interleukins, interferons, tumor necrosis factor, and colony stimulating factors (growth factors) are reviewed in Table 1 for their toxicities and effects on immune parameters when used as therapeutic agents.

Therapeutically, a number of other recombinant proteins such as soluble cytokine receptors, cytokine antagonists, and cytokine fusion proteins are being tested for their ability to mimic or modulate cytokine activity. Soluble cytokine receptors and cytokine antagonists are endogenous molecules that regulate cytokine activity *in vivo*. In this chapter, the soluble receptor for TNF-α and the IL-1 receptor antagonist are reviewed as examples of proteins

TABLE 1

Recombinant Therapeutic Proteins: Cytokines, Soluble Cytokine Receptors, Cytokine Antagonists, and Cytokine Fusion Proteins

Cytokine	Route/Dose/ Schedule	Clinical Condition or Species	Parameter	Comments	Ref.
rhL-1α	i.v. 1, 3, or 10 μg/m²/d continuous i.v. for 4 d	Ovarian cancer	Increase in PB monocytes Increase in monocyte superoxide production Increase in monocyte secretion of IL-1α and β No change in monocyte secretion of TNF-α	Up to ~3-fold increase Following phorbol ester stimulation Increase in mRNA levels seen as well; LPS stimulation did not alter basal secretion Basal or LPS-stimulated	100
rhIL-1α	i.v. Low dose 0.1–1.0 μg/m²/d continuous infusion High dose 3–10 μg/m²/d	Ovarian cancer	Increase in chemotactic response of granulocytes to FMLP and zymosan-activated serum Slight effects on granulocyte H_2O_2 production Suppression of PMA-stimulated granulocyte H_2O_2 production No change in chemotactic response of granulocytes to FMLP and zymosan-activated serum Dramatic effects on basal granulocyte H_2O_2 Suppression of PMA-stimulated granulocyte H_2O_2 production	Improvement to control levels Normalized duration of H_2O_2 production Rates of H_2O_2 production suppressed	22
IL-1α	i.v. 0.03 μg/kg i.v. over a 30-min interval daily for 5 d	Advanced malignancy	Dramatic increase in plasma TNFsrp55 levels Increase in IL-1ra levels in serum	Peaked within 1 hr; declined by 24 hr Peaked within 2 hr; declined by 24 hr	168

TABLE 1 (continued)

Recombinant Therapeutic Proteins: Cytokines, Soluble Cytokine Receptors, Cytokine Antagonists, and Cytokine Fusion Proteins

Cytokine	Route/Dose/ Schedule	Clinical Condition or Species	Parameter	Comments	Ref.
rhIL-1α	i.v. 0.1–10 μg/kg/d daily, 6-hr infusions for 1–2 wk	Autologous transplantation in cancer patients	Fever duration	5 d at low dose and 10–11 d at the higher doses; fever occurred between 2 and 6 hr	182
			Hypotension	Seen in 3/7 patients at 0.1 μg with an increase to 12/13 patients at 3.0 μg; dose-limiting at 10 μg	
			Decrease in days to ANC 100	Decrease from 18 d at 0.1 μg to 11 d at 1.0 and 3.0 μg	
			Decrease in days to ANC 500	Decrease from 24 to 12 d at dose of 3.0 μg	
			Change in days to platelet independence	Increase at 1.0 μg from 38 to 63 d with a decrease at 3.0 μg to 29 d	
			Change in days to RBC independence	Increase at 1.0 μg from 31 to 45 d with a decrease at 3.0 μg to 23 d	
			No change in bone marrow cellularity		
			Increase in BM granulocytic precursors	Increase in CFU-GM/10^5 from 2.5 to 14.2 at 3.0 μg	
			Increased plasma G-CSF concentrations	Dose dependent	
			No change in plasma INF-γ or TNF-α		
			Increase in plasma IL-6	Increase at all doses with 341 pg IL-6 per ml at 3.0 μg	
			Shortened median hospital stay	Median stay of 25–37 d	
rhIL-1β	i.v. 0.01–0.05 μg/kg/d 30-min infusion once daily for 5 d	Leukemia patients undergoing bone marrow transplantation	Moderate toxicity observed in all patients		123
			Fever and chills	All patients developed within 30 min	
			Hypotension	Seen in 14/17 (82%) of patients within 5–8 hr	
			Decrease in days to achieve a neutrophil count >500 ml	Decrease from 34 to 25 d	
			No change in clinical chemistries		
			Reduced incidence of infection		
			Marginal increase in survival	Between days 0 to 28 after infusion Increased from 20 to 30%	
rhIL-1ra (IL-1 receptor antagonist)	i.v. 1–10 mg/kg Continuous infusion for 3 hr	Healthy humans	No toxicities reported by subjects		65
			No change in plasma levels of IL-1β, IL-2, IL-6, IL-8, TNF-α, or GM-CSF		
			No changes in PB mononuclear cell phenotypes		
			Decrease in IL-6 production by LPS-stimulated PBMC	Seen at all doses after 3 hr	
			No changes in T-cell proliferation		
			No detectable anti-IL-1ra antibodies	Evaluated between 2 and 6 weeks	

rhIL-1ra (IL-1 receptor antagonist)	i.v. 100-mg loading dose followed by 72-hr infusion of 1–2 mg/kg/hr	Sepsis syndrome	Increase in survival time	Observed at highest dose	49
rhIL-1ra (IL-1 receptor antagonist)	i.v. 400–3200 mg/d continuous 24-hr infusion for 7 d	Graft vs. host disease (GVHD)	Reduction of acute GVHD by at least one grade	Seen in 10/16 (63%) of patients	6
			Decrease in IL-1β message in PB mononuclear cells	Seen in responders only	
			Decrease in GVHD-induced increase in TNF-α message in PB monocytes	Responders had TNF levels below baseline vs. the 16-fold increase in nonresponders	
			Elevation of transaminases in serum	2/16 (12%) of patients; reversible	
rhIL-1ra (IL-1 receptor antagonist)	s.c. 1–6 mg/kg Once daily for 28 d	Rheumatoid arthritis	Decrease in mean number of tender joints	Seen at 7 d; decrease from 24 to 10	98
			Decrease in serum C-reactive protein	Decreased from 2.9 to 1.9; 12 patients with decreases ≥ 50%	
			Decrease in ESR	Decreased from 48 to 31; 5 patients with decreases ≥ 50%	
			Skin reactions at injection site		
			No evidence of IL-1ra antibodies		
rIL-2	i.v. 3.3–40 μg/m²/d Continuous for 90 d	Advanced cancer	Selective increase in number of PB NK cells	Progressive, preferential increase in CD56^{hi} NK cells at 10, 30, and 40 μg/m²/d; dose dependent	25
			No changes in NK cell CD16 expression		
			CD56⁺ cells lacked CD3, CD4, and CD5 expression	Some CD56⁺ cells expressed CD16 and others did not	
			CD56⁺ cells expressed CD2 and some expressed CD8		
			Increase in PB NK cytotoxicity against K562	Gradual increase over time	
			Increase in PB NK cytotoxicity against NK-resistant target	Increase seen at high dose	
			No change in the numbers of T and B cells in PB	No lymphocytopenia or lymphocytosis seen with this low-dose therapy	
			Increase in IL-2R p75 density on NK cells but not on T cells	Effects on NK cells appear to be selective	
			Proliferation of CD56⁺ T cells not altered by IL-2		

TABLE 1 (continued)

Recombinant Therapeutic Proteins: Cytokines, Soluble Cytokine Receptors, Cytokine Antagonists, and Cytokine Fusion Proteins

Cytokine	Route/Dose/ Schedule	Clinical Condition or Species	Parameter	Comments	Ref.
IL-2	i.v. Low dose 250 U/kg/hr continuous infusion for 4 d	Advanced cancers	Decrease in PMN migration to FMLP	Seen in 2/2 patients after 3 d of therapy	79
			No change in PMN phagocytosis of *Staphylococcus aureus*		
			Slight decrease in PMN-mediated ADCC	No change at 2 hr after initial infusion but decreased by >50% after 3 d of treatment in 26% of patients	
	High dose 10^5 U/kg every 8 hr for 3 d		Decrease in PMN migration to FMLP	Seen in 4/4 patients	
			Decrease in superoxide production after PMA stimulation	Seen in 67% of patients after 3 d of therapy	
			No change in PMN phagocytosis of *S. aureus*		
			Decrease in PMN FcγRII and III expression	Seen in 100 and 80% of patients, respectively, after 3 d	
			No change in PMN markers My4, My7, ICAM-1, LFA-1, 3, or Mac1		
rhIL-2	i.v. Low dose 3 mU/m²/d continuous infusion for intermittent periods of 4–5 d	Cancer (patients undergoing leukophoresis)	Increase in catheter-related bacteremia	3/17 patients	137
	High dose 100,000 U/kg as 15-min infusion every 8 hr for two cycles of 5 d		Increase in catheter-related bacteremia	3/8 patients	
IL-2	i.v. 10,000–50,000 U/kg every 8 hr by i.v. drip	Cancer and immune-mediated diseases	Anorexia, fever, chills	Seen in 100% of patients	28
			Weight gain and hypotension	Seen in 82% of patients	
			Liver function abnormalities	Seen in 82% of patients	
			Grade III to IV granulocytopenia	Seen in 27% of patients	
			Eosinophilia (>4.0 cells/nl)	Seen in 73% of patients	
			Leucopenia	Seen in 18% of patients	

rhIL-2	i.v. 1 and 3 × 10^6 U/m^2/d continuous infusion 4 d/wk for 4 wk	Carcinomas or melanomas	Fever	Seen in all patients at high dose	160
			Rigors	Seen in 6/8 patients at high dose only	
			Fatigue		
			Anemia		
			Weight gain	Due to fluid retention; occurred in all patients	
			Hypotension	Seen at both dose levels	
			GI toxicity	Nausea and vomitng	
			Cutaneous reactions	5/8 pateints; diffuse erythematous rash at high dose only	
			Decrease in hemoglobin	Seen in all patients	
			Neutropenia	Seen in 3/11 patients; developed within first 28 d	
			Grade II and IV granulocytopenia		
			Eosinophilia in serum	Seen in all patients	
			Changes in lymphocyte counts	Initial decrease within 24 hr of start of cycle with rebound 24 hr after IL-2 stopped; rebound higher at high dose	
			Decrease in serum T and B lymphocytes		
			Increase in lymphocyte lysis of Daudi targets by PBL	100-fold increase in cytotoxic potential	
			Increase in liver function tests	Seen in all patients; reversible; not clinically significant	
			No renal or hepatic dysfunction noted by chemistries		
			Partial antitumor response in one patient		
rIL-2	i.v. 1.4 × 10^6 U/m^2 bolus injections 3×/d for 5–6 consecutive d	Malignant melanoma	Increase in lysis of K562 cells by PBL		54
			Increase in lysis of Daudi cells by PBL	~3-fold increase; indicative of increase in LAK cell activity	
			Increase in lysis of allogeneic melanoma cells		
rIL-2	i.v. 3 × 10^6 U/m^2/d continuous 120-hr infusion once or twice per month for up to 5 cycles	Advanced cancer	Splenic enlargement	5/9 patients had enlargement from 24 to 65% of the initial length	134
			Increase in serum eosinophils	15/16 patients	
			Rebound lymphocytosis	11/16 patients	
rIL-2	i.v. 18 × 10^6 U/m^2/d continuous 5-d infusion	Colorectal carcinoma	Increase in body weight	Indicative of vascular leak syndrome; median time to peak was 96 hr; mean gain 4.6 kg	34
			Increase in serum TNF-α	Median time to peak level was 72 hr; max level was 151 pg/ml	
			Increase in serum C-reactive protein	Median time to peak level was 72 hr; evidence of acute-phase response	
			Circulating levels of complement components C3 and C4	Fell progressively during 5 d of dosing; reversible within 10 d	
			Increase in serum IL-6	Peaked within 12–48 hr from start; median peak was 42 pg/ml	
			Decrease in serum albumin		

TABLE 1 (continued)

Recombinant Therapeutic Proteins: Cytokines, Soluble Cytokine Receptors, Cytokine Antagonists, and Cytokine Fusion Proteins

Cytokine	Route/Dose/ Schedule	Clinical Condition or Species	Parameter	Comments	Ref.
rIL-2	i.v. 160–600 μg/m²/d continuous i.v. for one day	Leukemia and myelomas	PBL had increased expression of IL-4 and IL-6 mRNA	Occurred during infusion	63
			Increase in serum IL-6 with no change in serum IL-4		
			Abrogated IgM or IgG primary antibody response to KLH		
			Decrease in response to recall antigen	Seen in 1/6 treated with IL-2 vs. ~90% of controls	
			Modest decrease in IgG and IgA serum levels during IL-2 infusion	Return to normal within 48 hr after stopping IL-2	
			Modest decrease in PB lymphocytes of all phenotypes	Occurred during infusion	
			No effect on PWM-induced IgG and IgM secretion from PBL		
			Decrease in percentage of $CD45^-UCHL1^+$ (memory) cells	Consistent decrease seen during IL-2 infusion	
IL-2 ± LAK cells	i.v. Low dose 1–1.25 mg/m²/d continuous i.v. for 5–6 d for two cycles	Metastatic cancer	Increased incidence of bacterial sepsis	Seen in 3/7 patients receiving LAK cells; 1/11 not receiving LAK cells	90
			Increase in basal superoxide production by PMN	Increase after second cycle of treatment only	
			Increase in PMA stimulated superoxide by PMN	Increase seen at end of second cycle	
			No change in neutrophil degranulation in response to FMLP		
			No change in phagocytosis of *S. aureus*		
			No change in ability to kill *S. aureus*		
			No change in random PMN migrattion		
			Decrease in PMN chemotaxis in response to FMLP	Seen near end of both exposure cycles	
			Decrease in PMN chemotaxis to zymosan-activated serum	Seen at end of first cycle of treatment	

rhIL-2 + LAK cells	i.v. 4.5–6 × 10^6 U/m^2 bolus every 8 hr on days 1–5 for several cycles	Advanced cancer	Initial decrease in circulating BFU-E and CFU-GM	Seen from 0–5 d; 8 and 15% of pretreatment levels	170
			Rebound increase in circulating BFU-E and CFU-GM	4- to 7-fold above baseline at days 5–10; peaked at days 8–10 when IL-2 discontinued	
			Initial decrease in PB lymphocytes with rebound	Seen with CD3$^+$, CD4$^+$, CD8$^+$, and CD56$^+$ lymphocytes; rebound seen at 5–10 d	
			No significant changes in serum levels of IL-3, GM-CSF, or IL-1β	One patient had increase in GM-CSF on days 5 and 8	
			Increase in plasma IL-6	Seen in all patients; levels peaked at day 5 and decreased by days 8–10	
			Increase in plasma G-CSF	Marked increase at days 3 and 5; decreased on day 8	
			Slight decrease in IL-4 levels	75% of patients had decrease on days 3 and 5 but increased by day 10	
			Slight increase in PB neutrophils		
			Marked increase in eosinophils	Increase on days 15–18 of treatment	
			Decrease in platelets	Lowest at day 5	
rhIL-2 + LAK cells	i.v. 18 × 10^6 U/m^2/d continuous infusion for 5 d; several cycles	Metastatic renal cell carcinoma	Thyroid hypofunction developed		181
			Low serum levels of T4, T3 and FTI	Seen in 7/15 (47%) patients; detected between 60 to 120 d; authors suggest autoimmune mechanism may be responsible	
			High serum levels of TSH		
			Elevated levels of antithyroid antibodies and antimicrosomal antibodies	Seen in patients with thyroid hypofunctions	
rIL-2 ± LAK cells	i.v. or s.c. Continuous infusion	Progressive metastatic cancer	i.v. route developed anti-rIL-2 antibodies	Seen in 45/85 patients; 4% were neutralizing antibodies	147
			s.c. route developed anti-IL-2 antibodies	Seen in 58/120 patients; 10% were neutralizing antibodies; affected native IL-2, as well	
			Both groups developed anti-rIL-2 antibodies with similar frequencies and titers that did not affect clinical outcome		
rhIL-2	s.c. Escalating doses 1.8 × 10^6 U/m^2/d every 12 hr; 6 d/wk for 4 wk; in week 5, maintained at 14.4 × 10^6 once/wk	Progressive metastatic cancer	Increase in soluble IL-2 receptor	Increase seen by 2 wk and still elevated by 8 wk; mean increase of 700 pM	73
			Increase in serum IFN-γ levels		
			No change in serum TNF-α levels		
	Pulsed doses 4.8 × 10^6 U/m^2/d 0–3 doses per day for up to 44 d		Increase in soluble IL-2 receptor	Increase seen by 2 wk and still elevated by 8 wk; mean increase of 580 pM	
			Increase in serum IFN-γ levels		
			No change in serum TNF-α levels		

TABLE 1 (continued)

Recombinant Therapeutic Proteins: Cytokines, Soluble Cytokine Receptors, Cytokine Antagonists, and Cytokine Fusion Proteins

Cytokine	Route/Dose/ Schedule	Clinical Condition or Species	Parameter	Comments	Ref.
rhIL-2	s.c. 4.8 million U/m^2 1–3 doses/day, 5 d/week, up to 44 d	Cancer	No evidence of vascular leak syndrome	Generally characterized by weight gain and hypotension	151
			Grade I and II anemia	Seen in 4/20 and 1/20 patients, respectively	
			Grade I and II granulocytopenia	Both seen in 1/20 cancer patients	
			Grade I thrombocytopenia	Seen in 1/20 patients	
			Significant leucocytosis (mean peak 16.2 leukocytes/nl)		
			Rebound lymphocytosis	Occurred in 90% of patients after second or third week of therapy	
			Eosinophilia — >1, >5, >10 cells/nl	Seen in 100, 47, and 20% of patients, respectively	
rIL-2	s.c. 9 × 10^6 U, twice daily for 3 d	Major surgery in cancer patients	Increase in serum IL-6	Started to rise by 48 hr; peaked at 96 hr; remained elevated for at least 10 more days	33
			Increase in serum C-reactive protein	Started to rise by 48 hr; peaked at 96 hr; remained elevated for at least 7 more days	
			Increase in soluble IL-2 receptors in serum	Increase seen at 48 hr; peaked by 96 hr; still elevated by 21 d	
			No changes in serum IL-1β		
			Increase in serum TNF-α	Increase in 4/18 patients with range of 32–47 pg/ml	
			No change in serum albumin levels		
			Significant increase in the percentage of CD3$^+$ lymphocytes expressing CD25 and HLA-DR	Peaked at 3–4 d following start	
			In vitro neutrophil phagocytosis augmented	Prevented surgery-induced suppression of phagocytosis	
rIL-2	Injection in perimastoid region 100,000 U/d for 7–10 d	Carcinoma	Increase in anti-tumor cytotoxicity in cervical lymph nodes	Seen in 12/12 patients; increase seen in cytotoxicity of K562 and Daudi cells; mainly due to CD16$^+$ lymphocyte fraction	138
			Induction of a T-cell population in lymph nodes with ability to lyse autologous tumor cells		
			No change in spontaneous proliferation of cervical lymph nodes		
			No changes in lymph node morphology		
			Increase in IL-2 receptors and Leu19$^+$ cells in the cervical lymph nodes		

rhIL-2	i.p. injection 1000-U bolus injection once daily for 14 d	Lung cancer	Increase in eosinophils in pleural fluid	Peaked on days 3–5; decreased by 1/3 at 28 d	121
			Marked lymphocytosis in pleural fluid	Increase measured on day 14; still elevated 14 d later	
			Moderate increase in PB eosinophils that was present at 14 d following last injection	Peaked on days 9 to 19; still elevated at day 28	
			Increase in BM eosinophil cell lineage	Measured on day 14 of treatment; 3/6 patients; decreased by 28 d	
			No change in neutrophil and monocyte CSF activity in pleural fluids		
			Increase in eosinophil CSF activity in pleural fluid	Appeared on day 3 and persisted 5 d after cessation of treatment	
			No eosinophil progenitors in pleural cavity or blood		
			Increased eosinophil chemotactic activity in pleural fluid	4-fold higher than pretreatment activity	
rhIL-2 and pegylated rhIL-2	i.d. injection 10 μg daily for 4 wk	HIV positive	Increase in keratinocyte expression of MHC Class II and IP-10		165
			No change in WBC count, CD4 or CD8 numbers or percentages, or CD4:CD8 ratio during 30 d		
			No significant improvement in DTH response to *Candida albicans* or tetanus toxoid		
			Increase in NK cell activity of PBMC to K562	Seen after 4 wk of rIL-2	
			No significant changes in LAK cell activity		
			No significant toxic reactions	Fever, flu, grade II thrombocytopenia	
			Injection site showed induration and erythema	Increased in size within 3–4 d to 1000 mm²; characterized by mononuclear cell infiltrate with lymphocytes predominating over monocytes	
	Pegylated derivative 9 μg twice weekly for for 4 wk		Increase in keratinocyte expression of MHC Class II and IP-10		
			No change in WBC count, CD4 or CD8 numbers or percentages, or CD4:CD8 ratio during 30 d	Greater than response with rhL-2	
			Improved DTH responses to *Candida albicans* and tetanus toxoid but not PPD		
			No significant changes in total $CD3^+$, $CD14^+$ monocytes, $HLA\text{-}DR^+CD3^+$ cells, or number of $CD25^+$ cells		
			Increased NK cell activity of PBMC to K562	Seen after 4 wk of PEG IL-2	
			No significant changes in LAK cell activity		
			Increase in PB proliferative response to PHA	3-fold increase seen after 2 wk	
IL-2	i.v. Continuous 5 d infusion	AML patients	Analysis of peripheral blood	Analyzed on days 0, 2, and 5	57
			Reversible decrease in the CD4/CD8 ratio		
			Moderate increase in $CD56^+$ and $CD16^+$ NK cells	10–90% increase	
			Consistent increase in $CD3^+/CD8^+/CD56^+$ cells	10–300% increase on day 2; return to baseline by day 5	

TABLE 1 (continued)

Recombinant Therapeutic Proteins: Cytokines, Soluble Cytokine Receptors, Cytokine Antagonists, and Cytokine Fusion Proteins

Cytokine	Route/Dose/ Schedule	Clinical Condition or Species	Parameter	Comments	Ref.
IL-3	i.v. Continuous infusion 1–30 μg/kg/d for 14 — 93 d	Rhesus monkey	Development of anti-IL-3 antibodies with all three dosing schedules; some antibodes were neutralizing antibodies	Dose is important as all animals at the two highest doses in the three dosing schedules developed antibodies; IL-3 given three times a day s.c. tended to induce earlier antibody formation than a single s.c. administration; IL-3 given for a prolonged period as a dose of 1 μg/kg/d by continuous i.v. did not elicit antibody formation	175
	s.c. 3–100 μg/kg/d Daily injection for 14–30 d				
	s.c. 3–100 μg/kg/d Divided into 2–3 injections/day for 14–30 d				
rhIL-3	s.c. 11–100 μg/kg/d Three times daily for 14 d	Rhesus monkey	Increase in WBC count in PB	Peaked during second week with 2- to 3-fold increase	109
			Increase in basophils in PB	Dose dependent; 50- to 110-fold increase	
			Increase in eosinophils in PB	Dose dependent; 2- to 10-fold	
			No significant changes in neutrophil or lymphocyte counts		
			Increase in total histamine levels in blood	Steady increase during treatment up to 400- to 700-fold; may be due to IL-3 effects on basophils	
			All monkeys developed anti-rhIL-3 antibody	IgG isotypes detectable by day 10	
rhIL-3	s.c. 100 μg/kg/d Twice daily for 7d	Baboon	No adverse reactions		139
			No changes in serum chemistries or liver function		
			Increase in WBC count	Progressive increase during week 2; still above baseline at 28 d	
			No change in platelet counts		

rhIL-3	i.v. 125–500 μg/m² Single bolus	Cancer	Increase in serum IL-6 levels	Dose-dependent; peaked at 1 hr (60–90 pg/ml); returned to baseline at 8 hr	103
			Increase in serum C-reactive protein	Dose-dependent increase measured at 24 hr	
	s.c. 125–500 μg/m²/d Single bolus for 13 d		Increase in serum IL-6 levels	Dose-dependent increase seen on day 6 (32–70 pg/ml); still elevated on day 14	
			Increase in serum IgM but not IgA or IgG levels	Measured on day 14	
rhIL-3	s.c. 60 and 125 μg/m² 3×/wk for 12 wk	Myelodysplastic syndromes	Increase in leukocytes	Seen in 4/7 patients	154
			Increase in lymphocytes	Seen in 3/7 patients	
			Increase in monocytes	Seen in 3/7 patients	
			Increase in platelet counts in serum	Seen in 2/7 patients	
			Increase in reticulocytes	Seen in 3/7 patients	
			No change in soluble IL-2 receptors		
	250 and 500 μg/m² Daily for 15 d		Increase in leukocytes	Seen in all patients	
			Increase in lymphocytes	Seen in 2/8 patients	
			Increase in monocytes	Seen in 4/8 patients	
			Increase in platelet counts in serum	Seen in 2/8 patients	
			Increase in reticulocytes	Seen in 1/8 patients	
			Elevated soluble IL-2 receptors	May be an indication that IL-3 activates lymphocytes	
rhIL-3	i.v. 30 to 1000 μg/m² 4-hr infusion for 28 d	Bone marrow failure	Response in all cell lineages in BM	Type and degree of response varied between individuals	96
			Minimal side effects — low-grade fever and mild headaches		
rhIL-3	s.c. 30–500 μg/m² Single bolus daily for 15 d	Cancer patients with normal bone marrow function	Increase in PB platelet counts	Dose-dependent increase in 15/18 patients, from 1.3- to 1.9-fold increase	55
			Increase in PB leukocyte counts	Dose dependent with 2.8-fold increase at highest dose	
			Increase in PB neutrophils	Dose dependent with 2.7-fold increase at highest dose	
			Increase in PB lymphocytes	Both Th and Ts cells increased 1.9- to 2.4-fold	
			Increase in PB eosinophils	Occurred at all dose levels	
			Increase in PB basophils	Increased up to 12-fold	
			Delayed increase in reticulocyte counts	Seen on day 10 in 14/18 patients; not dose dependent	
			Increase in serum IgM levels	Seen in 15/18 patients	
			No changes in serum IgG, IgA, and IgE		
			Increase in bone marrow cellularity		
			Increase in numbers of megakaryocytes and eosinophils in BM		

TABLE 1 (continued)

Recombinant Therapeutic Proteins: Cytokines, Soluble Cytokine Receptors, Cytokine Antagonists, and Cytokine Fusion Proteins

Cytokine	Route/Dose/ Schedule	Clinical Condition or Species	Parameter	Comments	Ref.
		Cancer patients with bone marrow failure	Increase in PB platelet counts	Dose dependent; 1.5- to 14.3-fold increase seen in 5/8 patients	
			Increase in PB leucocyte counts	Seen in all patients; more delayed than those with BM function	
			Increase in PB neutrophils	Median time to peak was 19 d	
			Increase in PB eosinophils, basophils, and monocytes	Peaked at median time of 13–16 d	
			Increase in reticulocyte counts	Max at median of 8 d and rose 1.2- to 16.6-fold	
			Increase in bone marrow cellularity	Seen in 6/8 patients	
			Increase in the percentage of immature myeloid cells and megakaryoctyes in BM	Seen in 4/8	
			Increase in eosinophils in BM		
			Mild side effects in both groups of patients	Fever, headache, and flushing	
rhIL-3	s.c. 250 and 500 μg/m^2 Bolus injection daily for 15 d	Myelodysplastic syndrome	Increase in total leukocytes	Determined at end of 15-d treatment	55,56
			Increase in PB granulocytes	Not dose dependent; median time to maximal counts (1.1- to 5.3-fold increase) 15 d	
			Increased PB eosinophils	Median time to maximal counts 15 d	
			Increase in PB basophils	2.2- to >66-fold increase at median of 15 d in 8/9 (89%) of patients	
			Increase in PB lymphocytes	Similar changes in Th and Ts cells	
			Elevation of serum IgM and IgA but not IgG		
			Transient increase in circulating blast cells	Seen in 2/9 of patients with median time to peak of 15.5 d	
			Modest changes in platelets		
			Increase in bone marrow cellularity	Seen in 7/9 (79%) of patients	
			Increase in BM eosinophils	Seen in all patients	
			Increase in BM basophils	4- to 6-fold increase in 4/9 patients	
			Increase in megakaryocytes	Seen in 4/9 patients	
			Mild toxicity	Fever, headache	
			Local erythema at injection site		

rhIL-3+ carboplatin	s.c. 60–500 μg/m²/d Once daily for 2 d	Ovarian cancer	No change in serum neutrophil counts		141
			No change in serum basophil counts		
			Increase in serum eosinophil counts		
			Mild to moderate toxic effects at doses below 250 μg	Fever, myalgia, rigors, facial flushing, headache	
			Skin reactions	Pruritis and rash	
			Dose limiting effects at 500 μg/m²	Flu-like symptoms and headaches	
			No benefit in terms of myelosuppression		
rhIL-4	s.c. 0, 1, 5, 25 μg/kg/d for 30 d	Cynomolgus monkeys	Increase in prothrombin time	Seen at high dose on days 8, 22, and 28	62
			Decrease in plasma fibrinogen	Seen at high dose on day 22	
			Change in platelet counts	Decrease at high dose and increase at low dose	
			Significant increase in total leukocytes	Seen on days 22 and 28 at high dose	
			Significant increase in neutrophil counts	Seen on days 22 and 28 at high dose	
			Increase in lymphocyte counts	Seen at 5 and 25 μg/kg on days 22 and 28	
			Appearance of lymphoblasts, plasma cells, and large lymphocytes in serum	May be indicative of antigenic or mitogenic stimulation	
			Significant increase in eosinophils	Seen at high dose on day 22	
			Significant decreases in total serum protein	At high dose on day 8 and lower doses on days 22 and 28	
			Decrease in serum albumin	Seen at high dose at days 8, 22, and 28 and at 5 μg on days 22 and 28	
rhIL-4	i.v. 0.25–2.0 μg/kg/d Continuous infusion for several days	Cancer patients	Moderate toxicity at all levels	Fever/chills, arthralgia, fluid retention, periorbital edema; not dose limiting	23
			No changes in WBC, neutrophil, or platelet numbers		
			No increase in monocytes		
			Enhanced IL-β and TNF-α production *in vitro* by PBMC	Seen in 2/16 and 1/16 patients, respectively, on day 4 of therapy	
rhIL-4	i.v. and s.c. 40–400 μg/m²/d Single i.v. bolus on day 1 followed by continuous infusion on days 4 and 5 followed by s.c. for 2 wk	Advanced cancer	Mild toxicity		59
			Phenotypic analysis of PB cells showed no consistent changes or trends		
			Decrease in PBMC K562 and Daudi cytotoxicity	Lowest cytotoxicity seen on day 22	
			No change in activity of cytotoxic precursor cells		

TABLE 1 (continued)

Recombinant Therapeutic Proteins: Cytokines, Soluble Cytokine Receptors, Cytokine Antagonists, and Cytokine Fusion Proteins

Cytokine	Route/Dose/ Schedule	Clinical Condition or Species	Parameter	Comments	Ref.
rhIL-6	s.c. 15 μg/kg/d Once daily for 3 wk	Rhesus monkeys on chemotherapy	Increase in circulating platelet counts	Lessens chemotherapy-induced thrombocytopenia; first apparent on day 13	184
			Shortens chemotherapy-induced PB neutropenia	9 vs. 21 d	
			Increase in development of anemia		
			No change in circulating reticulocyte counts		
			Reversed chemotherapy-induced hypocellularity in BM	Returned to normal or slightly hypercellular; evidence of trilineage hematopoeisis with presence of many megakaryocytes	
			Enhanced recovery of progenitor cells in bone marrow	Progenitors comprised of multilineage colonies, erythroid burst forming cells, and megakaryocyte colonies	
			Decrease in food intake and weight loss due to chemotherapy	Weight reduced by 16% during third and fourth week	
			Gastrointestinal toxicity		
			Decrease in serum albumin		

rhIL-6	s.c. 15 μg/kg/d Once daily for 28 d	Middle-aged female rhesus monkeys	Lethargy		163
			Decrease in body weight	Average 10.9% loss	
			Decrease in serum hemoglobin	Seen as early as 1 d into treatment; return to normal by day 42	
			Increase in RBC count	Peaked at day 7; returned to normal by day 42	
			Increase in WBC count	Peaked at day 7; returned to normal by day 14	
			Increase in platelet counts	All had ~2-fold increase between 7–21 d	
			No change in bone marrow cellularity		
			No change in megakaryocytes in bone marrow		
			Decrease in PBMC NK activity	Significant decrease (~50% of pretreatement activity) in first 2 wk; gradual recovery	
			Decrease in proliferative response to the mitogens ConA, PHA, and PWM	Significant decrease (to < 10% of pretreatment value) on day 7 with return to baseline on day 14	
			Significant decrease in serum albumin levels		
			Significant increase in alkaline phosphatase levels		
			Increase in total protein		
			Decreases in cholesterol, glucose, BUN, SGPT		
			Lethargy		
		Old female rhesus monkeys	Decrease in body weight	Average 10.9% loss	
			Decrease in serum hemoglobin	Seen as early as 1 d into treatment; return to normal by day 42	
			Increase in RBC count	Peaked at day 7; returned to normal by day 42	
			Increase in WBC count	Peaked at day 7; returned to normal by day 14	
			Increase in platelet counts	All had ~2-fold increase between 7–21 d	
			No change in bone marrow cellularity		
			No change in megakaryocytes in bone marrow		
			Increase in serum alkaline phosphatase levels		
			Decrease in PBMC NK activity	Decrease only seen at day 28 and was ~60% of pretreatement level	
			Decrease in proliferative response to the mitogens ConA, PHA, and PWM	Significant decrease to 60% of pretreatment level on day 7 with return to baseline on day 14	
			Significant decrease in serum albumin levels		
			Significant increase in alkaline phosphatase levels		
			Decrease in total protein	Differed from response in middle age monkeys	
			Decreases in cholesterol, glucose, BUN, SGPT		

TABLE 1 (continued)

Recombinant Therapeutic Proteins: Cytokines, Soluble Cytokine Receptors, Cytokine Antagonists, and Cytokine Fusion Proteins

Cytokine	Route/Dose/ Schedule	Clinical Condition or Species	Parameter	Comments	Ref.
rhIL-6	s.c. 5–80 μg/kg/d Twice daily for 14 d	Cynomolgus monkey	Increase in platelet counts	Dose dependent; began on day 7 and continued through injections; at dose >20 μg the increase was biphasic with peaks at days 7 and 16; single peak for 5 and 10 μg occurred at 10 and 14 d; at all doses increase was ≥ 2-fold	9
			Moderate increase in neutrophils	Seen at higher doses	
			No changes in basophils and eosinphils		
			Myeloid hyperplasia in BM		
			Increase in size of megakaryocytes in BM	Suggestive of megakaryocyte maturation	
			Increase in C-reactive protein	Normalized within 1 wk after cessation of therapy	
			No changes in renal or liver function tests		
rhIL-6	s.c. 25–500 μg/kg/d Daily injections for 9 wk or 1000 μg/kg for 4 wk	Marmoset	No clinical toxic effects		142
			No changes at injection site		
			Preferential increase in serum platelets	~2-fold increase seen at all doses which appeared at 2 wk and lasted to ~6 wk	
			Modest increase in circulating WBC	Dose dependent; occurred within 1–2 wk	
			No change in bone marrow cellularity or megakaryocyte count		
			Increased size of megakaryocytes in BM	Increases seen when measured at weeks 1 and 4	
			Increase in circulating soluble IL-2 receptors	Seen when measured at weeks 1 and 4	
			Elevated serum immunoglobulin levels		
			Increased expression of CD25, MHC Class II antigen, and NK markers	Seen in spleen and lymph nodes at 9 wk	
			Increase in follicle size in cervical lymph nodes	Normal architecture not changed	
			Increase in follicle size in spleen		
			No alterations in histopathology of thymus		
			Decrease in albumin and increase in γ-globulin	Indicative of an acute-phase response	
			No indications of liver damage	Via serum chemistries or microscopic analysis	
			No alterations in renal function		
			Anti-IL-6 antibodies found	Appeared after 2 wk and increased up to 9 wk; neutralizing antibodies were produced which limited length of studies	

rhIL-6	s.c. 5 or 1000 μg/kg Daily for 3 weeks	Marmoset	Well tolerated at all doses and time courses		91
			Did not induce fever at any doses		
			No change in body weight		
			Decrease in circulating RBCs and hemoglobin	Dose dependent; RBC down at 1 wk; hemoglobin down at 1–4 wk	
	25, 100, or 500 μg/kg Daily for 3 months		Progressive increase in platelet counts	Increase seen at 2 wk (except 100 μg/kg) and lowest for all doses at 4 wk; effect lost at 7 wk	
			Significant increase in WBC counts	Seen at all doses at 1 and 2 wk; effect decreased at 4 wk except at high dose	
			Significant increase in neutrophils first week only	Seen at all doses	
			Increase in basophils at weeks 2, 4, and 7	Seen at all doses	
			No change in monocyte counts		
			Decrease in AST activity	Dose dependent; seen after 2 wks of treatment	
			Increase in total plasma protein	Dose dependent	
			Anti-rhIL-6 Ab appeared in week 2 and increased up to 9 weeks	Dose dependent; antibodies may account for return of responses to baseline values	
rhIL-6	i.v. and s.c. 0.5–20 μg/kd/d Continuous infusion for 1 d followed by daily s.c. for 6 d	Breast or lung cancer	Fever and flu-like symptoms	Seen in 3/7 patients at 0.5–1.0 μg and all patients at higher doses; fever tended to be higher and occur earlier (3–4 hr) at higher doses	174
			Local erythema at injection site	Seen in all patients; resolved within 48 hr	
			Increase in platelet counts	Initial decrease at day 3 with subsequent dose-dependent increase peaking days 8–15; significant at doses >1.0 μg	
			Slight increase in leukocyte counts	Seen at doses >2.5 μg and occurred on days 3 and 5; due to increase in neutrophils on day 3, monocytes on days 3 and 8, and lymphocytes on day 15	
			Increase in T, NK, and $CD25^+$ cells in PB	Measured on day 8; seen at doses 1.0–2.5 μg	
			No change in serum IgM or IgG levels		
			Slight increase in serum IgA levels	Reached maximum on day 15	
			No effect on basophils or eosinophils in PB		
			Rapid decrease in hemoglobin levels	Dose dependent	
			No change in BFU-E and CFU-GM		
			Increase in serum AST and ALT	Maximum values at day 8; reversible	
			Increase in antinuclear antibodies	Seen in only one patient	
			Increase in C-reactive protein	Rapid dose-dependent increase that quickly decreased after cessation of therapy	

TABLE 1 (continued)

Recombinant Therapeutic Proteins: Cytokines, Soluble Cytokine Receptors, Cytokine Antagonists, and Cytokine Fusion Proteins

Cytokine	Route/Dose/ Schedule	Clinical Condition or Species	Parameter	Comments	Ref.
rhIL-6	s.c. 0.5–20 μg/kg/d Daily for 1 wk; 2 wk of observation and 4 wk of daily treatment	Colon and pancreatic cancer	—	Blood samples taken at days 8, 22, 36, 50, and 78; changes measured at day 8	148
			Significant reduction in NK cell activity to K562 target cells	Seen at doses >5 μg; marked rebound (>2-fold) seen at highest dose following cessation of treatment	
			Reduction in LAK activity as measured by Daudi	Seen at doses >5 μg	
			Reduction in IL-2-induced LAK activity in PBMC	Seen at doses >5 μg; rebound observed	
			Reduction in PBMC proliferative response to IL-2	Seen at doses >2.5 μg	
			No change in PBMC response to PHA, PWM, and SAC at any doses		
			No change in lymphocyte counts at any dose		
			No change in PBMC percentage of CD4, CD8, or HLA-DR positive cells at any dose		
			Increase in serum gG, IgA, and IgM levels at all doses		
			Increase in serum IgE levels at all doses		
			Increase in C-reactive protein at all doses	Reached values above 8 mg/dl at doses 1 and 2.5 μg and above 20 mg/dl at doses above 2.5 μg	
			Antinuclear factor antibodies found	Increased in one patient at 2.5 μg from a titer of 100 to 300; developed in one patient at 20 μg	

rIL-6	s.c. 3 μg/kg/d Once daily for 7 d, 7-d rest, 7-d dose	Advanced malignancies	Fever, chills, and minor fatigue No antitumor response		180
	10 μg/kg/d		Fever, chills, and minor fatigue Significant increase in C-reactive protein Significant increase in fibrinogen Significant increase in platelet counts Decreases in albumin and hemoglobin		
	30 μg/kg/d		Fever, chills, and minor fatigue		
			Grade III major organ toxicity	Hepatotoxicity and cardiac arrhythmias; dose-limiting toxicity in 2/5 patients	
			Significant increase in C-reactive protein Significant increase in fibrinogen Significant increase in platelet counts Decreases in albumin and hemoglobin Increase in IL-2 receptors No antitumor response seen at any dose		
rhIL-11	s.c. 10–75 μg/kg/d for 14 d	Breast cancer (women undergoing chemotherapy)	Fatigue, myalgia, arthralgia — not dose limiting	Seen in all patients at high dose; reversible	61
			Weight gain of 3–5% with extremity edema	Seen at doses > 10 μg though greatest at 75 μg	
			Transient decrease in platelets followed by a subsequent increase over baseline for all doses	Mean decrease of 14%; increases from 76–180% peaking between days 14 and 19	
			No changes in WBC count or differentials		
			Decrease in PB hemoglobin concentration	20% decrease; not dose dependent; reversible	
			Increase in acute phase proteins: C-reactive protein, fibrinogen, and haptoglobin	Increased at all doses; returned to baseline after cessation of treatment	
rhIL-12	i.v. or s.c. 1 μg/kg/d bolus daily for 5 d	Cynomolgus monkeys	Fever	Elevated temperature on days 4 and 6	20
			Decrease in WBC count, primarily due to decrease in lymphocytes and monocytes	Occurred rapidly and continued through treatment; peak decrease on day 6	
			Decrease in lymphocyte counts	Nadir on day 2 and remained below baseline until day 9	
			Decrease in monocyte counts	Mean nadir on day 6 and day 2 for i.v. and s.c., respectively	
			Increase in neutrophil counts	Seen on day 2 and return to or below baseline by day 6	
			Decrease in number of $CD4^+$ and $CD8^+$ PBMC	Nadir on day 2 and still decreased on day 6	
			Decrease in $CD20^+$ PBMC	Decreases seen on days 2, 4, and 6	
			Decrease in circulating platelets	Nadir on day 4; 46 and 33% decrease for i.v. and s.c., respectively; by day 9 platelets began to increase and rebounded above baseline, peaking on day 13 and 11 for i.v. and s.c., respectively	
			Decrease in hematocrit	Mean decrease ~15% below baseline on day 6	

TABLE 1 (continued)

Recombinant Therapeutic Proteins: Cytokines, Soluble Cytokine Receptors, Cytokine Antagonists, and Cytokine Fusion Proteins

Cytokine	Route/Dose/ Schedule	Clinical Condition or Species	Parameter	Comments	Ref.
rhIL-12	s.c.	Squirrel monkeys	No effects at injection site		145
	0.1–50 μg/kg/d once daily for 14 d		Fever	Slight elevation at 10 and 50 μg/kg	
			Decrease in hemoglobin	Seen in males at 50 μg and females at 10 and 50 μg	
			Decrease in erythrocyte counts	Seen in males at 50 μg and females at 1–50 μg	
			Decrease in platelet counts	Only seen in females at 1–50 μg	
			Increase in leukocytes	Increase in males, though not significant; significantly increased in females at 1–50 μg	
			Decrease in total protein	Seen in both males and females at 50 μg	
			Decrease in serum albumin, phosphorus, and calcium	Seen only in females at high doses	
			Generalized lymph node enlargement	Seen in all monkeys; dose related	
			Splenomegaly	Seen in all monkeys; dose related	
			Increase in spleen, liver, kidney, and lung weights	Dose dependent; generally seen at lower dose in females; due to combination of cell infiltration, hypertrophy, and edema	
			Hyperplasia in spleen	Infiltration of red pulp by various cell types	
			Hyperplasia in the lymph nodes	Hyperplasia seen in T, B, and histiocyte areas	
			Atrophy of thymus	Dose dependent	
			Increase in cellularity of BM	Due to trilineage hyperplastic response	
			Mononuclear cell infiltration in septa of lungs	Dose dependent	
			Changes in liver histopathology	Dose dependent; Kupffer cell hypertrophy and hyperplasia; mononuclear cells in sinusoids	
			Development of anti-rhIL-12 antibodies	Seen on day 15; detected in 10- and 50-μg groups; did not correlate with dose but with plasma levels	
			Enhanced LAK activity	Dose dependent; 2.7- to 6.4-fold	
			Enhanced lectin-dependent cytotoxicity	Dose dependent; 3- to 9.7-fold	
			No increase in serum TNF, IL-2, or IFN-γ		
rINF-α	s.c.	Cancer	No evidence of vascular leak syndrome	s.c. administration did not lead to VLS	151
	$3–6 \times 10^6$ U/m^2		Decrease in total serum protein	Seen in 5% of patients	
	Three times/wk for 6 wk		Grade I and II anemia	Seen in 20% of patients	
			Grade I thrombocytopenia	Seen in 15% of patients	
			Grade I leucocytopenia		
			Grade I granulocytopenia		
			Decrease in eosinphil counts		
			Decrease in PB lymphocytes		

rIFN-α + immuno-modulator P40	i.m. 500 μg Six injections at 1-month intervals	AIDS	Mild to severe pain at injection site	Seen in 13/18 patients; lasting 1–3 d	66
			Fever	Seen in 12/17 patients; lasting 1–3 d	
			No alterations in liver or kidney function		
			Elevation of WBC, particularly lymphocytes		
			Increase in CD4 cell counts	Increase seen over 9-month period	
			No significant change in CD8 cell counts		
			No differences in cellular viremia titers		
			CTL activity to allogenic cells remained stable up to one year		
			Elevated response to recall antigen		
			No change in PBMC proliferative response to allogenic cell stimulation		
			Decrease in AIDs-induced increase in circulating IFN-α levels	IFN-α levels generally increased in placebo group	
			No change in serum titers of β2 microglobulin		
			Patients developed DTH response to IFN-α		
			Most patients produced antibodies to IFN-α and HIV antigens	Seen in most patients	
			No additional clinical manifestations		
Natural and rhINF-β	i.v. Natural and recombinant human INF-β 6×10^6 U Single dose	Healthy humans	Similar effects seen with both forms of INF-β		102
			Mild toxicity	Fever, chills, headache; transient	
			Increase in heart rate		
			Decrease in leukocyte counts	Seen in 4/12 patients	
			Granulocyte toxicity	Seen in 3/12 patients; grade I	
			Increase in serum in β2-microglobulin	Above baseline at 24–96 hr post-injection	
			Increase in serum neopterin values	Increased at 24–72 hr post-injection	
			Intracellular rise of 2′,5′-adenylate in PBMC	Increased at 10–48 hr post-injection	
			Intracellular rise in Hu-Mx synthesis	Increased from 10–96 hr post-injection	
			No change in serum levels of IL-1α or IL-1β		
rIFN-γ	i.v. 28 d	Cynomolgus monkey	Increase in body temperature		166
			Decrease in hematopoietic function		
			Decrease in leukocyte counts		
			Thrombocytopenia		
			Decrease in bone marrow cellularity		
			Thymic atrophy		
			Mild to moderate hypertrophy of the liver RES		
			Lymphoid depletion in lymph nodes and splenic follicles		
			Some monkeys developed high titers of neutralizing antibodies during the 28-d study		

TABLE 1 (continued)

Recombinant Therapeutic Proteins: Cytokines, Soluble Cytokine Receptors, Cytokine Antagonists, and Cytokine Fusion Proteins

Cytokine	Route/Dose/ Schedule	Clinical Condition or Species	Parameter	Comments	Ref.
rhIFN-γ	i.v. 0.001 to 1.0 mg/m²/d for approx. 18 d	AIDS	Marked increase in nonopportunistic bacterial infections	Seen in 17/52 patients compared to 0/22 in controls	117
			No change in number of opportunistic infections		
rINF-γ	i.v. 0.01–0.25 mg/m²/d 6-hr infusion daily for 10 d	Cancer	Increase in PB monocyte-mediated cytotoxicity	20–66% increases seen on day 4	89
	i.v. 0.01 or 0.025 mg/m²/d Continuous infusion for 4 weeks		Increase in PB monocyte-mediated antitumor cytotoxicity	2/10 patients; one at each dose	
	i.m. 0.25–1.0 mg/m²/d Daily injections for 42 d		Fever and severe fatigue	Seen in 6/6 patients at the high dose	
			Increase in PB monocyte-mediated antitumor cytotoxicity	Seen at 0.25 and 0.5 mg but not 1.0 mg	
rhIFN-γ	s.c. 0.01–0.5 mg/m² Once daily or every other day for five to six doses	Chronic granulomatous disease	No symptoms or changes in blood chemistry or cell counts		153
			Enhanced neutrophil killing of *S. aureus*	Response persisted and was still evident > 2 wk after cessation of treatment	
			Enhanced monocyte killing of *S. aureus*	Effects persisted for up to 1 week following cessation of treatment	

rhIFN-γ	s.c. 50 μg/m² 3×/wk for up to 1 yr	Chronic granulomatous disease	Reduction in number of patients with serious infections	22% developed infection with IFN-γ compared to 46% in controls; maintained during year of study	53
			Reduction in time to serious infection	77% receiving INF-γ were infection free for 12 months as compared with only 30% in controls	
			Decrease in time of hospitalization for infection	1493 vs. 497 d	
			No significant changes in superoxide production by neutrophils and monocytes		
			No effect on hematologic or biochemical indexes		
			No effects on levels of rheumatoid factor or antinuclear antibodies		
			Antibodies against IFN-γ not detected		
			Mild toxicity	Fevers, chills, headache, erythma at injection site	
rIFN-γ	s.c. 0.2 mg Once daily for 6 months	Lung cancer	Elevated HLA-DR expression on PB monocytes	Increases seen at 4, 8, and 12 weeks	132
			Elevated Fc-γ-receptor on monocytes	Consistently increased reaching 5-fold at week 12	
			No change in the percentage of CD3⁺ cells		
rIFN-γ	s.c. 0.1 mg/m²/d Once daily for 2 d	Chronic granulomatous disease	Increased production of superoxide by PB granulocytes	Occurred in all patients; sustained response that began 2–3 d after treatment and reached a peak after 2–3 wk	48
			Increased production of superoxide by PB monocytes	Peaked after 1 to 2 wk; persisted up to 5 wk	
			Augmentation of *S. aureus* killing by granulocytes	Measured 14 d after treatment	
			Increase in cellular content of phagocyte cytochrome b		
rIFN-γ	i.d. 6 injections over 9 d followed by 100–200 μg per month for 5–10 months	Lepromatous leprosy	Erythema nodosum leprosum (ENL)	Induced in 60% of the patients within 6–7 months compared with 15% of controls	144
			Reduction in number of bacilli in tissues	Seen over first 6 months	
			Induration and erythema at injection site		
			Increase in monocyte respiratory burst	2.5- to 5.1-fold increase in response to agonists	
			Patients with ENL had increased release of TNF-α in plasma	Patients who developed ENL had higher release of TNF-α from monocytes and high levels in serum	
			Mild systemic symptoms	Fatigue, fever, headache, myalgia, and nausea	
			No significant changes in total leukocyte counts		
			No change in percent T cells or monocytes		
			No change in MHC Class II expression on PBMC		
			Increase in spontaneous and PMA-stimulated respiratory burst of PBMC monocytes		
			PBMC response to mitogens not altered		

TABLE 1 (continued)

Recombinant Therapeutic Proteins: Cytokines, Soluble Cytokine Receptors, Cytokine Antagonists, and Cytokine Fusion Proteins

Cytokine	Route/Dose/Schedule	Clinical Condition or Species	Parameter	Comments	Ref.
rIFN-γ	i.d.	Lepromatous leprosy	Local effects in skin		122
	1, 10 μg		Induration and erythema at injection site	Developed by 24–48 hr; peaked at 72 hr	
	Three times daily for 6 d into lesions		T-cell and monocyte infiltration		
			Keratinocyte proliferation		
			Decrease in the number of epidermal LC		
			Elevation of HLA-DR expression in epidermis and dermis		
			Decrease in acid-fast bacilli in lesions		
			Systemic effects		
			Restored H_2O_2 secretion of PB monocytes	Response greater at the high dose	
			No additional toxic effects		
			No antibodies to IFN-γ found at 28 d		
rINF-γ	i.d.	Lepromatous leprosy	Induration at injection site	Size of induration peaked at 2nd or 3rd injection and was smaller on subsequent injections	83
	10 μg				
	Twice daily for 3d		Fever was the only side effect	3/50 patients	
			Progressive infiltration of T cells and monocytes	Seen at injection sites	
			Increase in the percentage of dermis infiltrated	4-fold over baseline, remained elevated for 2 wk	
			Infiltrate had increased number of $CD4^+$ cells	In lepromatous patients, $CD8^+$ cells predominate	
			Thickening of epidermis over injection site		
			Increase in keratinocytes expression of MHC Class II and IP-10		
			No emigration of Langerhans cells into dermis		
			Reduction of bacilli in lesions	Reduction seen by 3 wk ranging from 5- to 1000-fold	
			Persistence of T cells and epithelioid cells at 6 months		

rIFN-γ	Inhalation 250, 500, or 1000 μg daily for 3 d	Nonsmoking normal humans	Increased IP-10 message expression in alveolar macrophages	Seen at 1 hr after administration; undetectable after 24 hr	80
			No IP-10 message detected in blood monocytes		
			No decrements in respiratory function		
			No clinical signs or symptoms		
	s.c. 250 μg Single administration		No change in cellular components of lavage fluid		
			IP-10 mRNA transcripts present in blood monocytes	Seen 1 hr after administration; faintly visible at 4 hr	
			Alveolar macrophages did not express message for IP-10		
			Spectrum of adverse effects ranging from fever, nausea, malaise, local pain, induration, and erythema	All patients experienced fever	
rhM-CSF	s.c. 0.1–25.6 mg/m^2 1×/d on days 1–5 and 8–12 Cycle repeated every 28 d	Metastatic tumors	Mild toxicity	Fever, chills, fatigue, arthralgias at all doses	23,24
			Dose-limiting toxicities at highest dose	Thrombocytopenia (2/42) and iritis (1/42)	
			Decrease in platelets	Dose dependent	
			Increase in monocyte counts	Seen on days 5–15	
			Increase in CD45$^+$/CD14$^+$ monocytes in PB	Dose dependent	
			No change in tumoricidal activity of PB monocytes	Activity against SK-MEL 28 measured	
			Increase in LPS-stimulated IL-1 and TNF-α secretion		
			No change in numbers of lymphocyte subsets in PB		
			NK and LAK activity not enhanced		
rGM-CSF	i.v. 0.5–8 μg/kg/d Single 60-min infusion followed by continuous infusion for 14 d	AIDS	Increase in circulating neutrophils	Dose dependent (2-fold at 0.5 μg to 40-fold at 8 μg)	13
			Primed neutrophils for enhanced oxidative metabolism and *E. coli* killing		
			Reversed neutrophil defects in phagocytosis of opsonized bacteria and intracellular killing of *S. aureus*	Occurred during infusion	
rhGM-CSF	i.v. 1.3–20 × 10^3 U/kg/d Single 60-min infusion followed by continuous i.v. for 14 d	AIDS	Increase in circulating leukocytes	Dose dependent; initial increase rapid (6 hr after bolus) at doses > 2.6 × 10^3 U; peak at 2 wk; return to baseline 3–9 d after therapy	67
			Significant increase in circulating neutrophils	Dose dependent; initial increase rapid (6 hr after bolus) at doses > 2.6 × 10^3 U; peak at 2 wk; return to baseline 2 wk after therapy	
			Significant increase in circulating eosinophils	Dose dependent; initial increase rapid (6 hr after bolus) at 1.3–2.6 × 10^3 U; all doses peaked at 2 wks and returned to baseline 2 wks after therapy	
			Increase in circulating monocytes	Dose dependent; initial increase rapid (6 hr after bolus) at doses > 2.6 × 10^3 U; all doses peaked at 2 wk; returned to baseline 2 wk after therapy	

TABLE 1 (continued)

Recombinant Therapeutic Proteins: Cytokines, Soluble Cytokine Receptors, Cytokine Antagonists, and Cytokine Fusion Proteins

Cytokine	Route/Dose/ Schedule	Clinical Condition or Species	Parameter	Comments	Ref.
			No change in absolute lymphocyte counts		
			No change in RBC counts		
			No change in platelet counts		
			Increase in BM cellularity	4/14 patients	
			Increase in BM eosinophils	8/14 patients; 1–2 patients in each dose group	
			No change in CD4:CD8 ratio		
			No change in viremia		
			Mild side effects: back pain, myalgia, nausea	Not dose dependent; seen in 1–2 patients in each group	
			Fever	11/16 patients	
rhGM-CSF	i.v.	Breast cancer or melanoma	Accelerated myeloid recovery		19
	2–32 μg/kg/d		Increase in leukocyte and granulocyte counts	Dose dependent; rapid rise that remained high until infusion ended	
	Continuous i.v. for 14 d		Increase in appearance of one or more myeloid cell clusters in BM	Seen at day 10	
			Increase in BM cellularity	Began at day 10 and peaked at day 15 by over 20%	
			Lowered morbidity and mortality		
			Decreased bacteremia	In 16 vs. 35% of control patients	
			Decreased elevation in creatinine and bilirubin		
			Mild side effects: rash, myalgia, edema	Not dose dependent; seen in all patients	
			Severe toxicity	Seen at 32 μg/kg; edema, weight gain, hypotension, vascular leakage	
rh-GM-CSF	i.v.	Cancer	Dose-limiting side effects at high dose	Capillary leak syndrome, edema, hypotension	97
	2–32 μg/kg/d		Increase in BM CFU-GM progenitor cells	Dose dependent; seen on days 6 to 16	
	Continuous infusion for 14 d		Increase in BM CFU-GEMM progenitor cells	Seen on day 21	
			No change in BFU-E		
			Increase in PB CFU-GM progenitor cells	Early and sustained increase	
			Increase in PB WBC		
			Increase in PB neutrophils		

rhGM-CSF	i.v. 30 to 240 μg/m²/d 2-hr infusion daily for 14 d	Bone marrow transplant in cancer patients	Increase in neutrophil count > 500	Occurred more rapidly at doses ≥ 60 μg; seen within 2 wk	124
			Hematopoietic reconstitution	Seen in 75% at doses ≤ 30 μg and 100% at doses ≥ 60 μg	
			Mild to moderate toxicity	Pain, fever; fewer days of fever at doses ≥ 60 μg	
			No hepatic, pulmonary, cardiac, or renal toxicity		
			Shortened hospital stay		
rhGM-CSF	i.v. 15–480 μg/m² 1- or 4-hr infusion daily for 7 d or as a 12-hr infusion for 14 d	Aplastic anemia	Modest increase in granulocytes	Median time to peak was 4 d; seen at most doses	7
			Modest increase in monocytes	Median time to peak was 5 d; seen at most doses	
			Modest increase in RBC	Median time to peak was 13 d; seen at higher doses	
			No change in eosinophils and immature myloid cells and myeloblasts in PB		
			Minimal toxicity	Low back discomfort, anorexia, myalgias, and fever	
		Myelodysplastic syndrome	Moderate increase in granulocytes	Median time to peak was 9 d; largest responses at high doses	
			Moderate increase in monocytes	Median time to peak was 6 d; largest responses at high doses	
			Moderate increase in RBC	Median time to peak was 5 d; largest responses at high doses	
			No change in eosinophils, immature myloid cells, or myeloblasts in PB		
			Minimal toxicity	Low back discomfort, anorexia, myalgias, and fever	
			No development of anti-GM-CSF antibodies		
			No hypersensitivity or skin test reactivity		
rhGM-CSF	i.v. 200–250 μg/m²/d Continuous infusion for varying times	Malignant lymphoma	No changes in hemoglobin concentration		76
			No changes in platelet or leukocyte counts		
			Antitumor responses	Complete or partial response in 5/8 patients	
			Enhanced recovery of neutrophils	Seen 7–11 d after start of therapy	
			Increase in total WBC, lymphocytes, eosinophils, and monocytes	All peaked 7–11 d after start of therapy	
			Increase in soluble IL-2R; may indicate GM-CSF activates lymphocytes	Rapid dramatic increase (2.5- to 13.2-fold) in all patients that peaked at 15 d	
			Increase in soluble CD8	Gradual and moderate increase in all patients	
			Significant increase in percentage of $CD25^+$ and $CD4^+$ cells in lymphocyte fraction	Majority of these cells were also $CD3^+$	

TABLE 1 (continued)

Recombinant Therapeutic Proteins: Cytokines, Soluble Cytokine Receptors, Cytokine Antagonists, and Cytokine Fusion Proteins

Cytokine	Route/Dose/ Schedule	Clinical Condition or Species	Parameter	Comments	Ref.
rhGM-CSF	i.v. 60–500 μg/cm²/d Continuous infusions for 2 wk, rest 2 wks, and repeat treatment for up to 4 cycles	Aplastic anemia	Increased WBC, primarily PMNs, eosinophils, and monocytes	Seen in all patients with maximum from 1.6- to 10-fold	172
			Increase in number of neutrophilic granulocytes	Primarily responsible for increase in WBC (1.5- to 20-fold)	
			Increase in eosinophil counts in PB	Markedly increased 12- to >70-fold in all patients	
			Increase in monocyte counts in PB	Increased 2- to 32-fold	
			No change in ratio of Th to Ts		
			Increase in bone marrow cellularity	Seen in 7 of 9 patients	
			Increase in myeloid elements and eosinophils in BM with no change in mature neutrophils or monocytes		
			No change in PB neutrophil locomotion		
			No increase in PB PMN FMLP or zymosan-stimulated chemotaxis		
			Increased serum levels of IL-2 and erythropoietin	In 8/9 and 9/9 of patients, respectively	
			Mild to moderate side effects	Constitutional symptoms, bone pain, and GI disturbances	
rhGM-CSF	s.c. 200 μg/m² × 4 wk followed by 400 μg/m² × 4 wk i.v. 400 μg/m² × 4 wks followed by 800 μg/m² × 4 wk	Aplastic anemia	Both i.v. and s.c. routes produced the same results		75
			Increase in neutrophil counts	Increase was >2-fold over pretreatment values from days 2–6 in 7/8 patients; at end of 8-wk treatment, increased from 8- to 327-fold	
			No significant change in lymphocyte counts and subsets		
			Decrease in CD56⁺ NK cells		
			No change in serum IgG or IgA		
			Increase in serum IgM	Increase seen in 5/8 patients at 4 wk and 6/8 during 8 wk	
			No change in serum IL-6 and IFN-γ		
rhGM-CSF	s.c. 0.5–3 μg/kg/d Daily for 6 wk	Chronic hepatitis B	Reduction in serum hepatitis B DNA levels	Reduction seen during second week to about 50% at all doses and remained decreased until completion of 6-wk therapy	107
			Increase in 2-5A synthetase activity in mononuclear cells		
			Increase in production of basal TNF-α	Seen by week 2; increased at all three dose levels	
			Increase in production of basal IL-1β	Seen by week 4; increased at all three dose levels	
			Increased serum levels of sIL-2 receptor, soluble CD4, and β2 microglobulin	Dose dependent; levels of sIL-2R and β2 microglobulin remained increased during therapy	

rhGM-CSF	s.c. 30–250 μg/m²/d Daily for 14 d; cycle repeated every 28 d	Solid malignancies	Local erythema at injection site		26
			Fever, chills, and malaise		
			Increased monocyte cytotoxicity against HT29 cells	Seen in all patients; increased on days 8, 15, 22, and 29	
			Increase in IFN-γ-induced monocyte cytotoxicity	Increase measured on day 15	
			No change in serum TNF-α or IL-1β		
			No significant changes in Leu-M3 or HLA-DR expression on PB monocytes		
rhGM-CSF	s.c. 400 μg Daily for 7 d	Non-Hodgkin's lymphoma	Reduction in infection	Reduced from 16/30 to 28/69	58
			Increase in probability of remaining infection free	Increased from 48 to 70%	
			Reduction in periods of neutropenia		
			Reduction in days with fever	From 4 to 2.1 d	
			Reduction in days of hospitalization for infection	From 8 to 3.5 d	
			No difference in survival rates		
			Injection site reactions and rash	Allergic reactions	
			No increase in antibodies to rhGM-CSF		
rhGM-CSF	s.c. 400 μg for 7 d	Non-Hodgkin's lymphoma	Reduced the length and nadir of neutropenia		45
			Reduced the length of fever episodes		
			Reduced frequency of all severe infections		
			Reduced the hospitalization and antibiotic use		
			Only side effects were cutaneous reactions		
			Did not increase overall survival		
rhTNF-α	i.v. 0.1 and 10 μg/kg/d Given as divided doses 2× daily for 10 d	African green monkey	Increased severity of simian varicella virus infection	Dose dependent; more extensive viremia, higher and earlier SGOT, more severe histological lesions, and greater mortality	158
rhTNF	i.v. 12.5–175 μg/m²/d Continuous infusion via hepatic arterial catheter for 5 d	Liver metastasis	Hypophosphatemia	Seen in most patients	108
			Dose-limiting toxicities	Hypophosphatemia, cardiovascular, and CNS effects at doses ≥ 125 μg/m²/d	
			Tumor regression and minor antitumor responses	Seen in 14 and 21% of patients, respectively	

TABLE 1 (continued)

Recombinant Therapeutic Proteins: Cytokines, Soluble Cytokine Receptors, Cytokine Antagonists, and Cytokine Fusion Proteins

Cytokine	Route/Dose/ Schedule	Clinical Condition or Species	Parameter	Comments	Ref.
rhTNF	i.v.	Progressive neoplastic disease	Decrease in PB NK-cell activity	Measured at end of 24-hr infusion to K562 target cells	88
	0.04 mg/m²/d		Decrease in percentage of CD16⁺ cells in serum	Measured at end of 24-hr infusion	
	Continuous infusion for 24 hr		Reduced TNF and IL-1 production by PB mononuclear cells	Measured at end of 24-hr infusion	
			Decreased proliferation of PB mononuclear cells to PHA and Con A	Proliferation in response to Con A altered to greater extent	
			No alterations in differential blood cell counts		
rhTNF	Intra-tumor injection	Malignant disease	Rigors and fever	18/21 patients experienced in first hour	129
	Single administration		Nausea and vomiting	Seen in 11/21 patients	
	87–522 μg/m²		Dose-dependent hypotension	Developed in 43% of patients at 12 to 24 hr	
			Elevations in SGOT and SGPT		
			Fluid retention and body weight gain	Seen in all patients at the two highest doses	
			Grade I leucopenia	Seen in 14% of patients	
			Transient lymphocytopenia	Seen in 67% of patients	
			No change in IgG, IgM, IgA, or IgE levels		
			No allergic reactions, positive skin tests, or antibody development		
			Development of local inflammation in tumor	Developed 8–24 hr after injection, progressing in some cases to hemorrhagic necrosis at 2–5 d	
			Tumor effects	Seen in 52% of patients with one (5%) complete regression and four (19%) partial regressions	
rTNF-α	i.v.	Renal cell carcinoma	Dose-limiting toxicities: hypotension, rigors, hallucinations	Seen in 15% of patients	156
	150 μg/m²/d		Granulocytopenia	2/22 patients; Grade I and III	
	30-min infusion for 5 d every other wk		Non-dose-limiting side effects: allergy, fever, hypotension, rigors, nausea, headache		
			Low antitumor response rate	2/22 patients	

rTNF	i.v. 100–500 μg/m² 30-min infusions twice daily for 5 d every other wk for 8 wk	Advanced colorectal carcinoma	No complete or partial antitumor responses		85
			Common side effects: rigors, pain and hyper- and hypotension, chills		
			Increase in bilirubin, alkaline phosphatase, and SGOT levels	Liver function tests; increased in 29, 43, and 64% of patients, respectively	
			Increase in plasma cortisol	Measured 3 hr after rTNF	
			Increase in C-reactive protein	Measured 24 hr after rTNF; acute-phase reactant	
rTNF	Alternating i.v. and i.m. 1–10 μg/m² Maximum of 4 sequential dose escalations/wk over a 4-wk period	Disseminated cancer	Fever and chills	Seen in most patients but generally resolved within 24 hr	18
			Injection site soreness		
			No significant hypotension or weight loss		
			No consistent hemodynamic changes		
			Decrease in platelet counts	Not clinically important	
			Significant decrease in the absolute lymphocyte count	Seen 4 hr after i.v. or i.m. administration; normal by 24 hr	
			Increase in granulocyte counts		
			No circulating antibodies to TNF developed by 4 wk		
rhTNF-α	ip 40–350 μg/m² 1×/wk for up to 2 months	Malignant ascites	Partial to complete resolution of ascites	22/29 of patients	135
			No response to tumors		
			Side effects not dose limiting: fever, chill, nausea, malaise	No correlation between side effects and dose	
			No changes in leukocyte and platelet counts		
			Local erythema and pain at site of injection	11% of patients	
			No change in total number of monocytes in ascites fluid		
			Increase in total number of granulocytes in ascites fluid	24 hr following injections	
			Decrease in total number of granulocytes in ascites fluid	24 hr following injections	
Recombinant human dimeric TNF receptor	1 mg/kg/d for 10 d	Renal allograft in cynomolgus monkeys	Increase in allograft survival	From 8.5 to 23 d	42
			Delay in onset of rejection	Mean time of onset was 11.7 vs. 4.4 d in controls	
			No adverse side effects observed		
PEG-BP30 (pegylated construct of p55 TNF receptor)	i.v. 0.2 or 5.0 mg/kg Single i.v. bolus	Baboons with septic shock	Improved survival in *E. coli* model of bacteremic shock	83% survived the 14-d observation period compared to 33% of the controls	46
			Decrease in plasma TNF-α levels	Decreased to nondetectable levels by 4 hr	
			Decrease in plasma levels of IL-1β, Il-6, and IL-8	Dramatic dose-related decreases seen as early as 1 hr following administration	
			Rapid restoration of *E. coli*-induced panleukopenia	Seen by 4–8 hr	

TABLE 1 (continued)

Recombinant Therapeutic Proteins: Cytokines, Soluble Cytokine Receptors, Cytokine Antagonists, and Cytokine Fusion Proteins

Cytokine	Route/Dose/ Schedule	Clinical Condition or Species	Parameter	Comments	Ref.
PIXY321 (GM-CSF/IL-3 fusion protein)	s.c. 25–1000 μg/m²/d Twice daily for 14 d Two cycles	Cancer patients with cumulative myelo- suppression	Prechemotherapy		171
			Modest increase in WBC count	Gradual rise with peak effect after cessation of treatment; no clear dose response	
			Modest increase in platelets	Dose-related rise occurring during treatment	
			Modest increase in neutrophils		
			Modest increase in BM cellularity		
			Significant increase in proliferation of BM precursors, and committed and multipotential progenitor cells		
			Mobilizes committed and multipotential progenitor cells in the PB		
			Postchemotherapy		
			Neutropenia reduced	Seen at 500–1000 μg/m²	
			Increase in platelets	Dose-dependent; peak at 750 μg/m²/d	
			Reduction in multilineage myelosuppression	Seen over multiple cycles; dose dependent	

being tested for their ability to modulate the activity of the cytokines TNF-α and IL-1 in disease, respectively. Recombinant cytokines in the form of fusion proteins are also being tested. In this review, one such protein (PIXY321) is included as an example. In PIXY321, the active domains from two cytokines, GM-CSF and IL-3, are coupled by a flexible amino acid linker. The rationale for creating such a molecule is based on the different but complementary hematological effects produced by these cytokines.[171] In general, GM-CSF induces a rapid increase in white blood cell (WBC) count which returns to pretreatment levels shortly after cessation of cytokine treatment, and alters the functional activity of mature myeloid cells. IL-3, on the other hand, produces a gradual increase in WBC count and, unlike GM-CSF, the hematopoietic response to IL-3 is sustained for some time following discontinuation of the cytokine.

2.1 Inherent Features

There are several inherent features of recombinant cytokine proteins that can influence the toxicities and effects that these agents can have on immune parameters in humans. Most are related to the fact that these proteins mimic endogenous immunomodulatory molecules. The four features that will be briefly discussed below are (1) dose-response relationships, (2) pleiotropism of cytokines, (3) the complex interrelation between individual cytokines (cascades, synergy, anergy, pleiotropism, and redundancy), and (4) the mechanisms that regulate production, activity, or removal of cytokines.

In the evaluation of classical drugs, clinical trials are based on the concept of a direct relationship between dose and efficacy. In addition, there is generally a dose-response relationship seen with toxicological responses including side effects on the immune system. In contrast, immunomodulatory drugs may not always show a clear relationship between dose administered and biological effect (reviewed by Talmadge and Dean[164]). The biological effects may be schedule dependent, therefore occurring at a reproducible optimal dose and treatment schedule. One example of such a dose-response relationship is seen with administration of rhIFN-γ.[106,188] Regarding immunological parameters, toxicity can also be dose and schedule dependent, as has been shown for IL-2.[21,167] It should be noted that many immune parameters are measured in peripheral blood. Therefore, any changes in the patient produced by the cytokine which results in a shift of immune cells or activity out of the peripheral blood into other body compartments might provide confounding information and generate information in which a dose-response relationship is not readily apparent.

Evaluation of recombinant cytokine proteins has challenged clinical research by the extraordinary complexity and multifunctionality of their biological activities. Many of the cytokines listed have pleiotropic activities and are produced by and act on a variety of cell types.[23] Cytokines not only participate in the regulation of the immune response and intervene in the orchestration of the complex processes of hematopoiesis, wound healing, and inflammation, but also are involved in the normal physiology of nonimmunological and nonhematopoietic organ systems. For most cytokines, receptors exist on a broad variety of cell types, and a broad variety of responses may be observed in different settings. The fact that cytokines participate in many cellular functions and diseases thus provides many opportunities for cytokine therapy; however, the multifunctionality can also be a problem with cytokine therapy because (1) a number of organ systems in the body can be affected by systemic application and (2) the greater the number of biological effects, the greater the number of potential side effects. This is evidenced by the many side effects seen with several cytokines at high doses following systemic administration.

In addition to the pleiotropism of cytokines, there are many complex interactions between cytokines, including synergism and antagonism.[95] Cytokines can trigger cascades of other cytokines with similar pleiotropic activities; therefore, toxicities and effects on immune parameters can result not only from direct effects of the cytokine but from indirect effects due to actions of other cytokines, as well. With IL-2, studies in animal models have demonstrated the central role of an intact immune system in mediating many toxicities of IL-2 in neuroendocrine, cardiovascular, and other systems and the involvement of other cytokines, the production of which is induced by IL-2.[52,155] Many aspects of the complex and overlapping interactions of cytokines have not been fully elucidated. For this reason, *in vivo* effects in humans may not always be predicted from *in vitro* or animal studies. Such predictions can also be complicated by the fact that the clinical condition of the patient may affect cytokine interactions.

Cytokines generally maintain the specificity of their actions on immunological or inflammatory responses partially through their paracrine function.[95] Thus, many of the cytokines are potent inflammatory or cell growth factors when delivered at high concentrations to the responding effector cells. In addition, cytokines have a very short half-life due to their catabolism and and are tightly regulated by a variety of mechanisms. Systemic levels of most cytokines are usually undetectable or barely measurable in the plasma of normal individuals and, if present, are orders of magnitude below the kd for their receptor. These levels, therefore, do not produce systemic effects; however, it is known that under certain conditions such as infection, cancer, or inflammation, cytokine concentration in the circulation can increase with resulting activity at distant tissues. In some of these conditions, cytokines are beneficial to the host, representing an integral component to the immunological defense network (acute phase response). In other cases, however, cytokines can actually cause the most striking manifestations of disease if production is unregulated or action prolonged. For example, the cardiovascular collapse in sepsis syndrome is secondary to massive production of IL-1 and TNF-α stimulated by bacterial lipopolysaccharide;[37,176] therefore, one potential problem that generally leads to effects on immune parameters is the high dose of recombinant cytokines required. These pharmacologic doses have frequently been used to attain biologically significant levels in the tissues. The administration of these agents by the systemic route further contributes to the exaggerated toxicity often reported for molecules many believed would be innocuous because they mimic endogenous proteins.

2.2 Toxicity

As can be seen in Table 1, recombinant cytokines produce a variety of toxicological effects and alterations in immune parameters. Most of these represent exaggerated physiology and pharmacology that at times can have toxic manifestations (reviewed by Cohen et al.[29] and Talmadge and Dean[164]). This is due to the fact that recombinant cytokine proteins mimic endogenous immunomodulatory molecules and when given at pharmacologic doses can reach peak plasma levels that may impact the immune system as well as other organ systems in the body. Due to the fact that toxicities can be the result of exaggerated physiology, toxicity studies must be performed in species that are pharmacologically responsive to the cytokine. Therefore, evaluation of cytokines such as the interferons, which are highly species specific, must be done in an appropriate animal model to adequately predict clinical toxicity. In addition to toxicities that may be expected as a result of exaggerated physiology, unexpected toxicities may be seen. This is illustrated by clinical studies on GM-CSF, where several of the toxic effects, some of them potentially life threatening, did not occur in animal studies even at very high doses of the drug or with routes of administration that closely approximated the routes and schedules used in human trials.[38,39]

Fever and chills are common side effects of recombinant protein therapy. Fever is generally accompanied by other generalized symptoms such as fatigue, myalgias, and arthralgias. There have been reports of fever produced by IL-2 with indications that it may be due to increased serum levels of TNF-α.[113] The mechanism of fever production by other recombinant cytokine proteins has not been well studied but may, in some cases, also involve release of secondary cytokines that are pyrogenic.

Dermatological effects can also be a side effect of recombinant cytokine therapy.[187] This can include local erythema at sites of injection, along with macular rashes, pruritus, and occasional generalized maculopapular eruptions.

One potential dose-limiting toxicity that has been seen with cytokines is vascular leak syndrome (VLS). This has been one of the major toxicities of IL-2 administration and is characterized by an increase in generalized vascular permeability with resulting edema in the interstitium of many organs.[155] It is manifested by an increase in body weight, peripheral edema, ascites, and pleural effusion. Accumulation of excess fluid in the lungs can be a critical outcome in some patients. The mechanism by which IL-2 produces VLS is unknown.[155] VLS is believed to be mediated by competent immune cells, as nude or irradiated mice do not develop VLS,[140] and administration of lymphokine-activated killer cells contributes to IL-2-induced syndrome in mice.[47] The LAK cells may be directly toxic to vascular endothelial cells or may release cytokines which directly or indirectly cause increased vascular permeability. In one study, depletion of lymphocytes of the LAK phenotype abrogated VLS in mice.[128] Vascular leak syndrome has been reported as a side effect of administration of other cytokines such as GM-CSF[8,19,97] and IL-1.[157]

The potential immunogenicity of recombinant proteins is generally evaluated in clinical evaluation of new protein therapeutics. It has been demonstrated that the appearance of antibodies to a recombinant human protein in monkeys can result in a hypersensitivity response or dramatically alter its pharmacokinetics in plasma, resulting in accelerated clearance and reduced blood levels.[32] Fortunately, recombinant human proteins are generally more immunogenic in monkeys than in man. Studies with IL-2 have demonstrated that the recombinant human protein appears to be biologically indistinguishable from the native cytokine in humans, though it has been shown to induce minor antibody formation which generally does not appear to alter functional properties.[183] A case report discussed a patient with metastatic carcinoma under long-term subcutaneous IL-2 therapy who developed antibodies to IL-2. Some of the antibodies were shown to be neutralizing antibodies accompanied by inhibition of immune responsiveness to the therapy.[189] Though some clinical studies have reported the appearance of antibodies to a recombinant cytokine as well as altered pharmacokinetics due to such a response, antibody formation, in general, has not appeared to be a major problem. In addition, dose-limiting pathology or immune complex-mediated disease have not been reported with the appearance of antibodies in human clinical studies. No studies were found demonstrating an IgE-mediated immediate hypersensitivity (type I) response resulting in anaphylaxis following cytokine administration.

When a humoral response is generated by a recombinant cytokine protein, the dose and route of exposure can be important variables in immunogenicity. Studies by Konrad et al.[93] with rhIFN-α demonstrated an 11% incidence of antibody in patients receiving intravenous administration, of which 4% had neutralizing antibody, while subcutaneous exposure resulted in a 31% incidence of antibody, of which 70% were neutralizing. By the intranasal route, IFN-α was weakly immunogenic. It is interesting to note that autoantibodies to cytokines exist in humans,[17] and some of these antibodies have been found to be potent and specific regulators of cytokines; however, the biological role of autoantibodies in regard to cytokines is not understood.

Though cytokine therapy has been shown to be a useful addition to the therapeutic armamentarium in some diseases,[10,94] the toxicities seen with these agents has, to a certain extent, limited progress in this area. Current research efforts are aimed at reducing treatment-related toxicity of cytokines while maintaining efficacy. Changing the dose route or schedule of adminstration may, in some cases, increase efficacy as well as decrease toxicity. In addition, there are several new approaches being investigated to increase local concentrations of cytokines or minimize toxicity, including (1) local injection of cytokines directly into the tumor,[77,185] (2) production of cytokine analogs with greater specificity for receptors on target cells,[74] (3) polyethylene glycol modification of cytokines to increase half-life and decrease toxicity,[84,111] and (4) gene therapy via modification of tumor cells with cytokine genes to enhance host immunity by targeting the therapy to a particular anatomical site.[127,149,177]

3 MONOCLONAL ANTIBODIES

During the past 15 years, a number of clinical trials have been conducted with monoclonal antibodies to elucidate their therapeutic effectiveness and safety in patients with cancer, autoimmune disorders, and inflammatory disease. Generally, infusions of monoclonal antibodies have been well tolerated, and the toxicities observed have been primarily immunological reactions.[35] Some of these studies and the toxicities associated with their treatment are presented in Table 2. Due to the foreign nature of murine monoclonal antibodies and their large size, compared with conventional drugs, strong humoral immune responses are often induced when administered to humans. Significant human anti-monoclonal antibody titers have been shown to alter dramatically the pharmacokinetics of the antibody in the serum by forming immune complexes with the monoclonal antibody, thereby increasing its clearance by the reticuloendothelial system. If the antibodies are directed to the binding domains of the therapeutic monoclonal (anti-idiotypic), they can inhibit the therapeutic's ability to bind to its target, further abrogating its therapeutic efficacy. The antibody response can also diminish some of the toxicity observed with these compounds when given repeatedly, since their overall exposure to the drug will be reduced following subsequent doses. This is a concern in preclinical safety studies in animals because a potential repeat-dose toxicity may be missed due to the generation of an antibody response to the drug. Initially, the decrease in exposure due to the humoral response can be overcome in the clinic by increasing the dose, but generally the therapy has to be discontinued if this response is significant.

The presence of antibodies directed to a therapeutic monoclonal antibody can also induce immunologically mediated toxicities, such as acute hypersensitivity (type I hypersensitivity) or immune complex-mediated reactions (type III hypersensitivity). High titers of IgM and/or IgG to a monoclonal antibody in the presence of the therapeutic can result in the formation of immune complexes. These immune complexes are formed in the circulation and may deposit in vessel walls virtually anywhere in the body and can activate complement, generating C3a and C5a fragments that can cause the release of vasoactive amines from mast cells and basophils, producing an immune complex-mediated anaphylaxis. This acute immune complex syndrome[110] is often characterized by fever, rigor, dypsnea, arthralgia, hypotension, and headaches and is usually observed at the start of the infusion following the generation of an IgM or IgG antibody response to the protein therapeutic via a previous infusion. If these complexes are formed in great excess, such that they cannot be effectively cleared by the reticuloendothelial system, they will circulate in the blood and become deposited in various tissues of the body, inducing local inflammatory reactions in the kidney, joints, and skin. This can result in fever, urticaria, changes in renal function, and joint pain (arthralgia) and swelling. These symptoms usually resolve within a week when the immune complexes are eventually cleared and the antibody titer increases.[35,36,110]

Often, when a monoclonal antibody is first infused into a patient, an immune response occurs within a matter of hours and is characterized by fever, chills, sweats, rigor, prostration, and, sometimes, dypsnea and hypotension. The "first-infusion response" appears to be dose related, as a reduction in dose or infusion rate can decrease the severity of the symptoms. If this response is observed following subsequent infusions, the symptoms are generally much less severe. This response appears to correlate with the presence of circulating target antigen or target cells and their disappearance from the circulation following administration of the monoclonal antibody. The mechanism of this response is thought to be due to either the lysis of the target cells or mass activation of immune cells via the target antigen by the monoclonal antibody which leads to the secondary release of various cytokines and acute

TABLE 2

Monoclonal Antibodies

Monoclonal Antibodies	Route/Dose/ Schedule	Clinical Condition or Species	Parameter	Comments	Ref.
Murine monoclonal anti-idiotype IgG antibodies	i.v. infusion Intrapatient dose escalation 400–3183 mg total dose	B-cell malignancies		Toxicities occurred in patients who had circulating tumor cells, serum idiotype >1 μg/ml, or an anti-mouse antibody response; patients without target protein or cells at time of infusion could tolerate infusion at rates as high as 200 mg/hr without side effects	110
			Chills	Seen in 3/6 patients with serum id > 1 g/ml and 4/5 with presence of human anti-mouse response	
			Fever	Seen in 3/6 patients with serum id > 1 g/ml and 4/5 with presence of human anti-mouse response	
			Vomiting	Seen in 1/6 patients with serum id > 1 g/ml and 1/5 with presence of human anti-mouse response	
			Transient dyspnea	Seen in 2/6 patients with serum id > 1 g/ml and 0/5 with presence of human anti-mouse response	
			Rash	Seen in 1/6 patients with serum id > 1 g/ml and 1/5 with presence of human anti-mouse response	
			Diarrhea	Seen in 0/6 patients with serum id > 1 g/ml and 3/5 with presence of human anti-mouse response	
			Facial palsy	Seen in 0/6 patients with serum id > 1 g/ml and 1/5 with presence of human anti-mouse response	
			Hypotension	Seen in 0/6 patients with serum id > 1 g/ml and 1/5 with presence of human anti-mouse response	
			Azotemia	Seen in 0/6 patients with serum id > 1 g/ml and 1/5 with presence of human anti-mouse response	
			Neutropenia	Seen in 0/6 patients with serum id > 1 g/ml and 1/5 with presence of human anti-mouse response	
			Thrombocytopenia	Seen in 3/6 patients with serum id > 1 g/ml and 1/5 with presence of human anti-mouse response	
			Transaminase elevation	Seen in 1/6 patients with serum id > 1 g/ml and 2/5 with presence of human anti-mouse response	

TABLE 2 (continued)
Monoclonal Antibodies

Monoclonal Antibodies	Route/Dose/ Schedule	Clinical Condition or Species	Parameter	Comments	Ref.
			Headache		
			Nausea		
			Myalgia		
			Low serum complement levels (C3, C4, CH50)	Seen in 3 patients during episodes of clinical toxicity	
			Antibody response to drug	Seen in 5 of 11 patients between days 10 and 24 of therapy; greatest response seen in patient without prior immunosuppressive therapy; response increased clearance of drug and decreased efficacy of subsequent doses	
Anti-IL-6 (BE-8; murine IgG1)	i.v. infusion 10–40 mg/d for 21 d	HIV lymphomas	Thrombocytopenia	Consistent but moderate	44
			Neutropenia	Occasional and moderate	
			Headache	Seen in 1 of 11 patients during 1 hr after administration	
			Antibody response to the drug	Seen in 2 of 11 immunocompromised patients	
CO17-1A (murine Ig)	i.v. infusion 400 mg weekly	Metastatic gastrointestinal cancer	Anaphylactic hypersensitivity reaction: dyspnea, tachycardia, hypotension	Seen in 2 of 20 patients after third dose; reversible after treatment was stopped and medicated	104
			Flushing and tachycardia	Seen in 1 of 20 patients middle second infusion; resolved by slowing infusion rate	
			Nausea and vomiting, or diarrhea with or without cramps	Beginning within 1 hr of infusion and lasted < 24 hr; may relate to Ab binding to normal gastrointestinal mucosa	
			Antibody response to the drug	Seen in 17 of 20 patients within 29 d of first exposure with 11 of 20 having detectable response by day 8; no effect on PK	
Anti-IL-2 receptor (2A3; murine IgG_1)	i.v. infusion 0.1–1.0 mg/kg/d for 7 d	Graft-vs.-host disease following allogeneic marrow transplantation	Fever	Seen in 3 of 11 patients	4
			Respiratory distress	Seen in 3 of 11 patients	
			Hypertension	Seen in 5 of 11 patients	
			Hypotension	Seen in 1 of 11 patients	
			Chills	Seen in 1 of 11 patients	
				The above side effects were seen in 11 of 72 (14%) infusions	
			Antibody response to drug	4 of 8 patients developed IgM and 1 developed IgG antibodies to murine Ig	

Anti-ICAM-1 (CD54; R6.5, murine IgG_{2a})	i.v. 0.01–2 mg/kg/d for 10–12 d	Cynomolgus monkeys with renal allografts	No bone marrow dysfunction		30
			No hepatic dysfunction		
			No vascular lesions		
			No effect on percent or absolute number of T-cell subsets in peripheral blood	R6.5 had no cytolytic action *in vivo*	
			Antibody response to drug	Response detected as early as the 11th day of treatment and persisted as late as 1 month after treatment	
Anti-T4 (CD4; murine IgG_{2a})	i.v. infusion 0.2 mg/kg/d 5×	Multiple sclerosis	Abolished PWM-induced Ig synthesis without lysis of $CD4^+$ T-cell subpopulation	For up to 2 wk after infusion	70
Anti-T11 (CD2; murine IgG_{2b})			Antibody response to drug	Similar to that seen with anti-T11	
			Blocked T-cell activation via the CD2 alternative pathway		
			Antibody response to drug	Significant anti-idiotypic mouse Ig seen in 2 of 3 patients; not observed until days 20–30	
Anti-GD2 (14G2a; murine IgG_{2a})	Continuous i.v. infusion for up to 5 d at 10, 20, or 40 mg/m²/d	Neuroecto-dermal tumors	Severe generalized pain	Seen in 18 of 18 patients; occurring gradually after 1–2 d of treatment and intensified over 5-day interval	119
			Hyponatremia	Seen in 14 of 18 patients	
			Fever (Grade IV)	Seen in 12 of 18 patients	
			Chills	Seen in 4 of 18 patients	
			Parasthesias (Grade IV)	Seen in 10 of 18 patients	
			Rash	Seen in 9 of 18 patients	
			Pruritus	Seen in 5 of 18 patients	
			Fatigue (Grade III)	Seen in 9 of 18 patients	
			Motor weakness	Seen in 3 of 18 patients	
			Anorexia	Seen in 2 of 18 patients	
			Confusion	Seen in 3 of 18 patients	
			Dyspnea with hypoxia (Grade III)	Seen in 3 of 18 patients	
			Hypotension (Grade III)	Seen in 3 of 18 patients; dose limiting; occurring 2–4 weeks after treatment	
			Hypertension	Seen in 4 of 18 patients	
			Dizziness	Seen in 2 of 18 patients	
			Nausea	Seen in 5 of 18 patients	
			Vomiting	Seen in 3 of 18 patients	
			Diarrhea (Grade III)	Seen in 3 of 18 patients	
			Antibody response to drug	Seen in 16 of 18 patients by day 11; high anti-idiotypic response	
			Allergic reaction/symptoms (Grade III–IV)	Seen in 2 of 18 patients	

TABLE 2 (continued)

Monoclonal Antibodies

Monoclonal Antibodies	Route/Dose/ Schedule	Clinical Condition or Species	Parameter	Comments	Ref.
Anti-GD2 (14G2a; murine IgG_{2a})	i.v. 100–400 mg/m^2	Neuroblastoma	Severe abdominal and joint pain	Seen in 9 of 9 patients; required morphine	72
			Transient hypertension	Seen in 4 of 9 patients	
			Bradycardia	Seen in 3 of 9 patients	
			Pruritus	Seen in 9 of 9 patients	
			Urticaria	Seen in 9 of 9 patients	
			Fever	Seen in 4 of 9 patients	
			Serum sickness	Seen in 3 of 9 patients	
			Paralysis of visual accommodation	Seen in 1 of 9 patients; present after 8 months	
			Transient nephrotic syndrome	Seen in 1 of 9 patients	
			Exanthema	Seen in 1 of 9 patients	
			Quincke edema	Seen in 1 of 9 patients	
			Antibody response to drug	Seen in 9 of 9 patients 4–10 d after start of therapy	
			Continuous decrease in serum complement C4, initial decrease in serum complement C3c, and initial increase in serum complement C3a	Seen in 8 of 8 patients	
			Circulating immune complexes	Detected in 3 of 9 patients pre- and post-therapy	
OKT3 (anti-CD3; murine Ig)	i.v. Bolus 5 mg/d; 14×	Renal allograft recipients	Profound depletion of mature peripheral T cells	Seen within 1 hr after injection and lasting as long as serum levels were maintained	27
			Antibody response to drug	Seen in 6 of 6 patients; rapid, strong, and neutralizing despite strong immunosuppression induced by OKT3; resulted in increased clearance of serum OKT3 and an abrupt reappearance of circulating OKT3$^+$ cells; onset dramatically delayed in patients on additional immunosuppressive therapy	
OKT3 (anti-CD3; murine Ig)	i.v. Infusion 1–2 mg/d for 10 d	Renal allograft recipients	Chills and fever		30
			Profound depletion of mature peripheral T cells; antibody response to drug not assessed	Seen in 1 of 2 patients after first infusion only, suggesting lympholysis	

OKT3 (anti-CD3; murine Ig)	i.v. 1–5 mg/d for 10–20 d	Renal allograft recipients	Antibody response to drug; no adverse allergic reactions	Seen in 75% of 20 patients; IgM anti-OKT3 seen in 65% of patients; IgG anti-OKT3 in 50%; peak response 20–33 d after last dose; 60% of patients made an anti-idiotypic response and 45% made a non-idiotypic response; anti-id antibodies neutralized *in vivo* effectiveness in 1 patient; response arising despite immunosuppression	81
Anti-Leu-1 (CD5; murine IgG_{2a})	i.v. Infusion Intrapatient dose escalation; 17 doses of 1, 10, and 20 mg given twice weekly over 10 weeks; total dose 164 mg	Cutaneous T-cell lymphoma	Transient decrease in creatinine clearance	Only seen after first dose in 1 of 1 patients	114
			Transient decrease in WBC after each dose; T cells declined progressively over extended treatment course	Return to pretreatment levels within 24–48 hr; decrease in WBC due to granulocytes, mononuclear cells, and T cells; granulocytes and mononuclear cells return to pretreatment levels within 24–48 hr	
			No antibody response to drug	Patient may have been immunocompromised	
Ab 89 (murine Ig)	i.v. Infusion Intrapatient dose escalation; 3 doses of 25, 75, and 150 given daily and 1500 mg given 1× a month later	Lymphoma	Transient decreases in creatinine clearance following 75 mg and higher antibody infusion	Presumably due to antigen-antibody complexes deposited in the kidney; return to normal within 24 hr	120
			Transient decreases in WBC following 75 mg and higher antibody infusion		
			Lymph node and liver tenderness following 75 mg and higher antibody infusion	Possibly due to reticuloendothelial clearance of antibody-coated tumor cells	
TNF- Mab (murine IgG_1)	i.v. Infusion 7.5 or 15 mg/kg, 1×	Sepsis syndrome	No hypersensitivity or immediate allergic reactions		1
			Serum sickness-like reactions	Seen in 2.5% of patients; resolved with treatment by 28 d after infusion	
			Antibody response to drug	Seen in 86% of patients 28 d after infusion	
ABL364 (murine Ig)	i.v. Infusion 50 or 100 mg/d on days 1, 3, 5, 8, 10, 12	Small-cell lung cancer	Nausea/vomiting (Grade I–III)	Seen in 13 of 19 patients; dose-limiting	162
			Abdominal cramps (Grade (I, III)	Seen in 2 of 19 patients	
			Erythema, itching (Grade I–III)	Seen in 4 of 19 patients	
			Dyspnea bronchospasm (Grade I–III)	Seen in 4 of 19 patients	
			Capillary leak (Grade I–II)	Seen in 3 of 19 patients	
			Fever (Grade III)	Seen in 1 of 19 patients	
			Antibody response to drug	Seen in 10 of 19 patients by day 18; 7 of 8 patients by day 60	

TABLE 2 (continued)

Monoclonal Antibodies

Monoclonal Antibodies	Route/Dose/ Schedule	Clinical Condition or Species	Parameter	Comments	Ref.
Anti-CD4 (cM-T412; chimeric IgG_1)	i.v. Infusion 10–700 mg 1×, 3× every other day, or 7× daily	Rheumatoid arthritis	Transient fever	Seen in 76% of patients either during the infusion or within 24 hr postinfusion	115, 116
			Myalgia	Seen in 36% of patients	
			Chills	Seen in 28% of patients	
			Headache	Seen in 28% of patients	
			Nausea	Seen in 16% of patients	
			Asymptomatic hypotension	Seen in 24% of patients	
			Decline in all circulating lymphoid cells within 1 hr	$CD8^+$ and $CD20^+$ returning to normal within 72 hrs	
			Sustained decreased circulating $CD4^+$ T cells	Still decreased at 18 and 30 months post-treatment	
			Transient elevation IL-6 following infusion		
			No significant anti-drug response	Transiently low levels in 2 of 25 patients	
			No changes in total Ig, CH50, C3, or C4		
			Transient depression of mitogen and antigen response *in vitro*	Immediate post-treatment response significantly depressed	
Anti-CD4 (cM-T412; chimeric IgG_1)	i.v. Infusion 10, 50, or 100 mg for 7 d	Rheumatoid arthritis	Tiredness	Seen in 9 of 32 patients after first infusion	173
			Headache	Seen in 9 of 32 patients after first infusion	
			Nausea	Seen in 9 of 32 patients after first infusion	
			Increase in body temperature	Seen in 8 of 32 patients after first infusion	
			Chills	Seen in 6 of 32 patients after first infusion	
			Increase in serum levels of IL-6 accompanied by fever and chills	Seen in 7 of 32 patients after first infusion; none of 15 patients without side effects had an increase in serum IL-6	
			Transient decrease in blood pressure	Seen in 3 of 32 patients after first infusion	
			No alterations in kidney or liver function	When the infusion rate was diminished, the above side effects were diminished	
			No major infections or neoplasms	Seen in 12–18 months followup	
			Sustained decrease in $CD4^+$ T cells	Strong decrease in within 1 hr after first infusion; recovering to 45% after 7 d of treatment and remaining depressed; 12 months still at 60% of baseline	
			Decrease in circulating monocytes	Decreased to 60–70% of baseline immediately following infusion; returned to baseline by day 7	
			No change in % of circulating $CD8^+$ T cells, B cells, NK cells, granulocytes, erythrocytes, or thrombocytes		

Anti-CD4 (Leu 3a; chimeric IgG_1)	i.v. Infusion 10, 20, 40, or 80 mg/dose 2× wk for 3 wk	Mycosis fungoides	Opportunistic infections	Not clear if drug-related	92
			Antibody response to drug	Low titer response in 3 of 7 patients directed toward mouse variable and human constant regions; no effect on PK; population immunosuppressed with treatment	
			No significant depletion of $CD4^+$ cells		
			Marked suppressed MLR reactivity		
			T-cell proliferative responses not suppressed		
			Did not induce tolerance to co-injected antigen		
Chimeric L6 (IgG_1)	i.v. Infusion 350–700 mg/m² 1× or 4 wkly doses of 350 mg/m²	Non-small-cell lung, colon, and breast cancer	Chills, fever, nausea	During and immediately following infusion lasting 24–48 hr	60
			Decrease platelet and absolute leukocyte counts	Immediately following treatment, returning to pretreatment levels by day 7	
			Rapid decrease in serum C3 and C4	Within 24 hrs following treatment, returning to normal within 7–14 d	
			Antibody response to drug	Seen in 4 of 18 patients, directed to murine variable regions and lower titers than against murine L6	
Chimeric anti-GD2 (ch14.18; human IgG_1)	i.v. Infusion 15 or 45 mg/dose 1×; 50 mg/dose 2×	Metastatic melanoma	Mild nausea during infusion	Seen in 2 of 13 patients	143
			Mild-severe infusion-related abdominal/pelvic pain syndrome (Grade I, III)	Seen in 7 of 13 patients; occurred during infusion subsided within an hour following the end of infusion; required morphine	
			Weak-modest antibody response to drug	Seen in 8 of 13 patients directed at the murine variable region	
Chimeric anti-CD20 (IDEC-C2B8; human IgG_1)	i.v. Infusion 10, 50, 100, 250, or 500 mg/m² 1×	B cell lymphomas	Low-grade fever (Grade I, II)	Seen in 13 of 15 patients during infusion	105
			Nausea (Grade I, II)	Seen in 4 of 15 patients during infusion	
			Chills	Seen in 2 of 15 patients during infusion	
			Headache	Seen in 2 of 15 patients during infusion	
			Rigors (Grade I, II)	Seen in 3 of 15 patients during infusion	
			Myalgia	Seen in 1 of 15 patients during infusion	
			Orthostatic hypertension	Seen in 1 of 15 patients during infusion	
			Bronchospasm (Grade II)	Seen in 1 of 15 patients during infusion	
			Thrombocytopenia (Grade II)	Seen in 1 of 15 patients	
			Transient decreases in serum complement (C3)	Seen in 2 of 15 patients at 24 hr posttreatment	
			Serum IgG, IgM, and IgA were unchanged		
			Neutrophils and T cells were unchanged		
			No antibody response to drug		
			B cells depleted from peripheral blood	Dose-dependent; depleted at 24–72 hr and generally persisting for 1 to > 3 months. One patient followed for 7 mo showed return of B cells	

TABLE 2 (continued)

Monoclonal Antibodies

Monoclonal Antibodies	Route/Dose/Schedule	Clinical Condition or Species	Parameter	Comments	Ref.
Chimeric anti-CD20 (chimeric-2B8; human IgG_1)	i.v. Infusion 0.01 mg/kg for 4 consecutive d	Macaque cynomolgus monkeys	Depleted >50% of peripheral blood B cells; no toxicity	Recovery started at 2 weeks after treatment and required 60 to >90 d for recovery	136
	i.v. Infusion 0.1–1.6 mg/kg for 4 consecutive d		Depleted >95% of peripheral blood B cells; no toxicity	By day 15 with little recovery within the first 30 d	
	i.v. Infusion 0.4 mg/kg for 4 consecutive d or 6.4 mg/kg 1×		34–78% depletion of lymph node B cells; no toxicity		
	i.v. 16.8 mg/kg once wkly for 4 wk		69–87% depletion of lymph node B cells; no toxicity		
CAMPATH-1H (Cdw52; humanized IgG_1)	i.v. Infusion 1–20 mg daily for up to 43 d	Non-Hodgkin lymphoma	No antibody response to drug	Drug itself immunosuppressive	
			No myelosuppression	Following infusion; seen at all doses	
			Fever and malaise	Starting at 5 weeks in 1 of 2 patients	
			Severe rigors	Possibly caused by tumor lysis	
			Hypotension	Developed on 42 dose in 1 of 2 patients ending treatment	
			Urticarial rash		
CAMPATH-1H (Cdw52; humanized IgG_1)	i.v. Infusion 80 mg 3× week for up to 9 months	Non-Hodgkin lymphoma	Anaphylaxis	Seen in 1 patient; was able to continue treating with lower dose	130
			Neutropenia		
			Opportunistic infections	Seen in 1 patient	
			Prolonged bone marrow hypoplasia	Seen in 2 patients	
			Rigors hypotension	Seen in a number of patients; not always associated with first dose and did not recur on challenge dose	

CAMPATH-1H (Cdw52; humanized IgG_1)	i.v. Infusion 4 mg/d 5× followed by 8 mg/d 5× Second course 40 mg 5×	Rheumatoid arthritis	Lymphopenia	Evident within an hr after start of first infusion; remaining suppressed for several months	78
			Fever	Seen with first dose in all patients; milder after second dose	
			Rigors	Seen with first dose in all patients; milder after second dose	
			Nausea	Seen with first dose in all patients; milder after second dose	
			Hypotension	Seen with first dose in 1 of 8 patients	
			Antibody response to drug	No response after first dose; response seen in 3 of 4 patients 6–10 d after retreatment; anti-idiotypic	
Humanized Mik1 (anti-IL-2 receptor - chain)	i.v. Bolus 1 mg/kg every other day for up to 44 d	Primate cardiac allograft	Antibody response to drug	A very low titer was seen in 1 of 6 monkeys	169
MAb C23 (human anti-cytomegalo-virus)	i.v. Infusion 5–80 mg 1× or 60, 20, and 20 mg at 3–1 week intervals	Healthy humans	No antibody response to drug	Seen in 0 of 22 immunocompetent patients	11
SDZ MSL-109 (human IgG_1 anti-cytomegalo-virus)	i.v. Infusion 50, 250, and 500 g/kg at 3 week intervals for 6 months	Bone marrow transplant recipients	Minimal fever	Seen in 1 of 15 patients	40
			Blood pressure alterations	Seen in 5 of 15 patients during or within 30 min after infusion	
			Transient tachycardia	Seen in 3 of 15 patients	
			Transient increase in respiratory rate	Seen in 2 of 15 patients	
			No antibody response to drug	Seen in 0 of 15 patients	
HA-1A (anti-endotoxin; human IgM)	i.v. Infusion 100 mg 1×	Gram-negative bacteremia and septic shock	Transient episode of localized hives	Seen in 1 of 262 patients; near site of infusion 10–15 min after infusion, resolved without therapy	186
			Facial flushing and mild hypotension	Seen in 1 of 262 patients; near end of infusion	
			No antibody response to drug	Seen in 0 of 116 patients	

phase reactants, such as IL-1, IL-2, and IL-6.[2,173] When these circulating antigens or target cells are absent, high doses of monoclonal antibodies can generally be given rapidly without any side effects,[110] further supporting the role of these circulating antigens in elicition of this response. When murine monoclonal antibodies were first administered in clinical trials, there was a great concern that an allergic reaction may be manifested by the development of an IgE antibody response that could result in a severe anaphylactoid reaction. For the most part, this has been relatively rare, only occurring in a small percentage of the population. What has been observed in some patients are episodes of mild, readily reversible reactions consisting of rash, hives, flushing, bronchospasms, and hypotension that may or may not have been IgE-mediated.[35,36,87]

Finally, many of the monoclonal therapeutics that are being developed are directed to tumor-associated antigens which are expressed at a high density on tumor cells; however, these antigens can also be expressed on normal tissues, although generally at a lower density. The antigen-specific binding of monoclonal antibodies to non-neoplastic tissue can lead to unwanted toxicity of that tissue that at times can be dose limiting. Gastrointestinal toxicity characterized by nausea, vomiting, and/or diarrhea has been observed with both ABL356[162] and KS1/4.[43,150] Both of these antibodies are directed to the tumor-associated antigens that are also expressed on normal cells of the gastrointestinal tract. In addition, 14G2, which reacts with GD2 ganglioside expressed on neuroblastomas and melanomas, has also been shown to bind to brain tissue, resulting in dose-limiting neural toxicity.[72,118,119]

To reduce the major limitations due to the generation of anti-murine monoclonal antibody response in repeat-dose clinical trials, genetically engineered chimeric and humanized monoclonal antibodies are now being synthesized. This is accomplished either by replacing the murine constant region with a human constant region or by grafting murine complementary determinate regions onto a human antibody framework, resulting in antibodies that are less immunogenic and have a slower onset, respectively.[60,78] No human antibody responses to human antibodies have been observed to date.[11,40,186] It should be noted that the immunogenicity of an antibody is also dependent on the status of the immune system, such that immunogenicity is generally greatly reduced in a patient who is immunocompromised, either due to a disease state or immunosuppressive therapy.

4 IMMUNOCONJUGATES AND IMMUNOTOXINS

A number of trials have been or are being conducted with monoclonal antibodies and growth factors which are either conjugated to a chemotherapeutic agent, isotope, or natural toxin or genetically fused to a natural toxin. These immunoconjugates and immunotoxins allow for the selective delivery of these cytotoxic moieties to a tumor or diseased cell, decreasing the amount that needs to be administered and, thus, decreasing the toxicity to the host. Several clinical trials that have employed immunoconjugates and immunotoxins and the toxicities associated with their use are presented in Table 3.

Antibodies conjugated to chemotherapeutic agents result in toxicities similar to those seen with the unconjugated antibody, except that toxicities to normal tissues expressing the targeted antigen are enhanced,[43] while the toxicities induced by the unconjugated drug are generally much less severe. The major dose-limiting toxicity associated with radioimmunoconjugate therapy has been myelosuppression, resulting in thrombocytopenia and leukopenia. At times, autologous bone marrow infusions have been given to overcome the myelosuppression.[131] In addition, thyroid dysfunction has been observed with the use of ^{131}I. All other toxicities are those associated with the antibody, such as fever, nausea, and pruritis.

Immunotoxins made with ricin, *Pseudomonas* exotoxin A (PE), and diphtheria toxin and their mutant or chemically modified derivatives have produced a variety of toxicities in clinical trials. These toxicities generally result from the toxin binding to nontargeted cells at high

TABLE 3

Immunoconjugates and Immunotoxins

Immunoconjugates and Fusion Proteins	Route/Dose/ Schedule	Clinical Condition or Species	Parameter	Comments	Ref.
^{131}I-CC49 (murine Ig)	i.v. infusion 75 mCi /m^2 (20 mg MAb)	Colorectal cancer	Nausea (Grade I)	Seen in 2 of 15 patients	118
			Arthralgia (Grade II)	Seen in 3 of 15 patients	
			Transient fever (Grade II)	Seen in 1 of 15 patients	
			Transient chills (Grade I)	Seen in 2 of 15 patients	
			Transient hypertension (Grade I)	Seen in 1 of 15 patients	
			Transient hypotension (Grade I)	Seen in 1 of 15 patients	
			Dyspnea and chest pain (Grade III)	Seen in 1 of 15 patients immediately after infusion persisting for 1 hr until relieved by medication	
			Reversible lymphopenia (Grade III-IV)	Seen in more than one third of patients; nadirs occurring during weeks 4 and 5 and returning to normal by weeks 7 and 8 after treatment	
			Reversible thrombocytopenia (Grade II-IV)	Seen in 7 of 15 patients	
			Reversible granulocytopenia	Seen in 6 of 15 patients	
			Antibody response to drug	Significant response seen in 12 of 13 patients 6–8 weeks post-treatment; 2 of 3 patients retreated had increased clearance and reduced uptake in tumor and enhanced uptake in thymus due to elevated antibody response	
^{131}I-anti-B1 (anti-CD20; murine IgG_{2a})	i.v. infusion (30 min) Pretreatment with unlabeled Ab of 0, 135, or 685 mg followed by 25, 35, or 45 cGy of labeled Ab	B-cell lymphoma	Leukopenia, reversible (Grade I)	Seen in most patients 4–7 wk after therapy	82
			Thrombocytopenia, reversible (Grade I)	Seen in most patients 4–7 wk after therapy	
			Mild rigor and fever	Seen in 1 patient during infusion of radioimmuno-therapeutic dose	
			Decrease in CD20$^+$ B cells	Seen within 24 hr following tracer infusion, returning to baseline 1–3 months following therapy	
			No increase in infection rate		
			No change in serum Ig levels		
			Antibody response to drug	Anti-Ig response seen in 2 patients 53 and 81 d after first infusion	

TABLE 3 (continued)

Immunoconjugates and Immunotoxins

Immunoconjugates and Fusion Proteins	Route/Dose/Schedule	Clinical Condition or Species	Parameter	Comments	Ref.
^{131}I-labeled murine anti-CD20 (B1 and 1F5; IgG_{2a}) and anti-CD37 (MB-1; IgG_1)	i.v. infusion 234–777 mCi (58–1168 mg) Autologous bone marrow support	B-cell lymphoma	Myelosuppression	After all therapeutic infusions	131
			Thrombocytopenia	Occurring 3–4 wk after low doses and 10–14 d after high doses	
			Opportunistic infections	All resolved with antibiotics	
			Mild nausea	Seen in 79% of patients	
			Fever	Seen in 74% of patients	
			Elevated thyrotropin	Seen in 42% of patients	
			Mild alopecia	Seen in 21% of patients	
			Hyperbilirubin	Seen in 37% of patients	
			Transient serum alanine aminotransferase	Seen in 42% of patients	
			Marked asthenia, nausea, diarrhea, and anorexia	Seen in patients at high doses (≥27.25 Gy)	
			Parotitis and ileus	Single occurrences	
			Hemorrhagic pneumonia and cardiomyopathy	Seen in 1 patient 2 months after treatment (27.25 Gy)	
			Severe postural hypotension	Seen in 1 patient (30.75 Gy)	
			Antibody response to drug	Anti-Ig response in 14 of 43 patients (33%), 2–76 wk after exposure	
^{131}I-chimeric B72.3 (human IgG_4)	i.v. 18, 27, or 36 mCi/m^2 1×; second course given 8 wk later	Metastatic colorectal cancer	No acute toxicity related to antibody administration		87, 112
			Bone marrow suppression	Correlated with whole-body radiation dose estimates	
			Thrombocytopenia	Seen in 2 of 3 patients (Grade I-II) at mid dose and seen in 6 of 6 patients (Grade I, III, IV) at high dose	
			Leukopenia	Seen in 2 of 3 patients (Grade I) at mid dose and seen in 5 of 6 patients (Grade I-III) at high dose	
			Antibody response to drug	Seen in 7 of 12 patients following first infusion with response primarily directed to the murine V region; upon second infusion: 2 patients with high antibody response to initial infusion had an anamnestic response altering PK upon retreatment, 1 patient with no response to initial infusion had no response on second and no effect on PK, and 1 patient with mild response to initial infusion had a modest shortening of PK	

KS1/4 (murine IgG_{2a})	i.v. infusion 1–1000 mg/m² 661 mg prior to HSR	Non-small cell lung carcinoma	Acute immune complex-mediated reaction	48 hr after 5th dose (66A1 mg total) of KS1/4 alone; resolved following discontinuation of dosing and aspirin; seen in 1 patient that had high human anti-mouse levels	43
			Fever		
			Muscle ache		
			Acute polyarthritis and synovitis		
			Antibody response to drug	Seen in 5 of 6 patients; no effect on clearance	
			Fever, rigor/chills (mild to moderate)	Seen in 1 of 6 patients	
			Anorexia	Seen in 1 of 6 patients	
			Nausea/vomiting	Seen in 2 of 6 patients	
			Diarrhea	Seen in 1 of 6 patients	
			Anemia (mild)	Seen in 4 of 6 patients	
			Transient elevation in liver transaminase	Seen in 3 of 6 patients	
KS1/4-metho-trexate	i.v. infusion 1–1000 mg/m² 661 mg prior to HSR		Antibody response to drug	Seen in 5 of 5 patients; no effect on clearance	
			Fever, rigor/chills (mild to moderate)	Seen in 4 of 5 patients	
			Anorexia	Seen in 1 of 5 patients	
			Nausea/vomiting	Seen in 3 of 5 patients	
			Diarrhea	Seen in 2 of 5 patients	
			Anemia (mild)	Seen in 3 of 5 patients	
			Transient elevation in liver transaminase	Seen in 1 of 5 patients	
KS1/4-desacetyl-vinblastine (murine IgG_{2a})	i.v. infusion 40–250 mg/m², 1×	Adenocarcinoma	Gastrointestinal toxicity: acute onset of abdominal pain, nausea, emesis, diarrhea	Seen in 8 of 13 patients; possibly due to gastrointestinal targeting of mAb (intense binding by drug); complement deposition on GI epithelium	150
			Antibody response to drug	Seen in 10 of 13 patients; first noted 5–10 d after infusion; peak response 3–4 wk after infusion	
	63 mg/m² every 2–3 d for up to 9 doses		Gastrointestinal toxicity (as above)	Seen in 7 of 9 patients; dose-limiting; possibly due to gastrointestinal targeting of mAb	
			Leukocytosis	Seen in multiple patients	
			Fever	Seen in multiple patients	
			Increase in serum amylase and lipase	Seen in multiple patients	
			Antibody response to drug	Seen in 6 of 9 patients	

TABLE 3 (continued)

Immunoconjugates and Immunotoxins

Immunoconjugates and Fusion Proteins	Route/Dose/Schedule	Clinical Condition or Species	Parameter	Comments	Ref.
XOMAZYME-Mel (murine IgG_{2a} conjugated to ricin-A chain)	i.v. infusion 0.01–0.05 mg/kg/d Daily × 5 d 0.75–1.0 mg/kg/d Daily × 4 d	Melanoma	Malaise/fatigue	Seen in 15 of 22 patients	161
			Fever	Seen in 14 of 22 patients; several hr after infusion	
			Tachycardia	Seen in 14 of 22 patients	
			Decreased appetite	Seen in 12 of 22 patients	
			Nausea	Seen in 6 of 22 patients	
			Weight gain (>5%); often associated with edema	Seen in 6 of 22 patients	
			Myalgia	Seen in 5 of 22 patients	
			Flush	Seen in 2 of 22 patients	
			Death	Seen in 1 of 22 patients; relationship to therapy not determined	
			Pruritus	Seen in 1 of 22 patients	
			Albumin decrease associated with fluid shifts	Seen in 20 of 22 patients; dose limiting	
			Total serum protein decrease	Seen in 20 of 22 patients; dose limiting	
			Low voltage on EKG	Seen in 16 of 22 patients	
			Fibrinogen increase	Seen in 12 of 13 patients	
			Erythrocyte sedimentation increase	Seen in 8 of 11 patients	
			Leukocytosis	Seen in 8 of 22 patients	
			Calcium decrease	Seen in 7 of 22 patients	
			SGOT increase	Seen in 6 of 22 patients	
			LDH increase	Seen in 2 of 22 patients	
			Thrombocytopenia	Seen in 2 of 22 patients	
			PTT prolongation	Seen in 2 of 22 patients	
			C reactive protein increase	Seen in 2 of 2 patients	
			Eosinophilia	Seen in 1 of 22 patients	
			Creatinine increase	Seen in 1 of 22 patients	
			Hemoconcentration	Seen in 1 of 22 patients	
			Metabolic acidosis	Seen in 1 of 22 patients	
XMMCO-791-RTA (murine IgG_{2a} conjugated to ricin A chain)	i.v. infusion 0.05–0.2 mg/kg/d Daily × 5 d	Colorectal cancer	Strong antibody response to drug	Seen in all 7 patients directed to murine Ig and toxin; part of response anti-idiotypic (neutralizing)	41

260F9MAb-rRA (murine IgG_1 conjugated to recombinant ricin-A chain; not glycosylated)	Continuous i.v. infusion 50 or 500 μg/kg/d Daily × 6–8 d	Breast carcinomas	Fever	Seen in 5 of 5 patients	64
			Anorexia	Seen in 5 of 5 patients	
			Malaise	Seen in 5 of 5 patients	
			Weight gain	Seen in 5 of 5 patients; maximal gain on day 5	
			Edema	Seen in 5 of 5 patients; resolved in 3 patients when therapy stopped, other 2 hospitalized with pulmonary and peripheral edema	
			Arthalgias/myalgia	Seen in 5 of 5 patients	
			Dyspnea on exertion	Seen in 3 of 5 patients	
			Finger arthritis, rash, and 7% eosinophilia	Seen in 1 of 5 patients; drug reaction resolved when therapy stopped	
			Hypotension	Seen in 1 of 5 patients	
			Transient nausea and diarrhea	Seen in 1 of 5 patients	
			Hypoalbuminemia	Seen in 4 of 5 patients	
			Minor elevations serum glucose	Seen in 5 of 5 patients	
			Minor elevations liver transaminase	Seen in 3 of 5 patients	
			Minor elevations serum triglycerides	Seen in 5 of 5 patients	
			Debilitating plexopathies and neuropathies	Seen in 3 of 5 patients; most severe 2–3 months after treatment	
			Antibody response to drug	Anti-rRa response seen in 4 of 5 patients; anti-Ig response seen in 3 of 5 patients; no response seen in 1 patient with sepsis who was heavily pretreated	
Anti-B4-br (murine IgG_1 anti-CD19 conjugated to blocked ricin-A chain)	i.v. bolus 1 hr 1–60 μg/kg/d Daily × 5 d 50 μg/kg/d = MTD	Refractory B-cell malignancies	Transient elevated serum hepatic transaminase (Grade III)	Seen in most patients; dose limiting; began 24–48 hr after start of infusions, generally resolving within 7 d after final infusion	69
			Transient thrombocytopenia	Seen in 12 of 25 patients	
			Transient leukopenia	Seen in 1 of 25 patients	
			Fever	Seen in 5 of 25 patients	
			Fatigue, malaise, and nausea	Only seen at MTD (50 μg/kg/d)	
			Hypoalbuminemia	Seen in 24 of 25 patients	
			Pleural effusion and weight gain	Seen in 1 of 25 patients	
			Antibody response to drug	9 of 25 patients; developed anti-Ig and anti-ricin-A response 23–41 d after therapy	
			No allergic manifestations	Including: anaphylaxis, rash, or immune complex formation	

TABLE 3 (continued)

Immunoconjugates and Immunotoxins

Immunoconjugates and Fusion Proteins	Route/Dose/ Schedule	Clinical Condition or Species	Parameter	Comments	Ref.
Anti-B4-br (murine IgG_1 anti-CD19 conjugated to blocked ricin-A chain)	Continuous i.v. infusion 10–70 μg/kg/d Daily × 7 d	Refractory B cell neoplasms	Fever	Seen in 28 of 34 patients	68
			Headache	Seen in 18 of 34 patients	
			Myalgia	Seen in 8 of 34 patients	
			Gastrointestinal		
			Nausea (Grade II)	Seen in 24 of 34 patients	
			Vomiting (Grade II)	Seen in 11 of 34 patients	
			Increased hepatic transaminase (Grade IV)	Seen in 34 of 34 patients; began within 24–48 hr of starting infusion, returning to baseline 7–14 d after infusion	
			Vascular leak syndrome		
			Hypoalbuminemia	Seen in 19 of 34 patients	
			Edema	Seen in 15 of 34 patients	
			Weight gain (≥3 kg)	Seen in 10 of 34 patients	
			Dyspnea (Grade I)	Seen in 5 of 34 patients	
			Pleural effusions	Seen in 3 of 34 patients	
			Thrombocytopenia (Grade IV)	Seen in 21 of 34 patients; returned to normal 3 wk after completion of therapy; due to peripheral platelet destruction	
			Antibody response to drug	24 of 34 patients developed anti-Ig and/or anti-ricin A response; 5 developed anti-Ig alone, and 6 developed anti-ricin-A alone	

IgG-RFB4-dgA (murine IgG_1 anti-CD22 conjugated to deglycosylated ricin-A chain)	i.v. infusion given over 4 hr at 48-hr intervals 2.5–13.9 mg/m² = max. single dose 4.7–142 mg/m² = total dose	B-cell lymphoma, non-Hodgkin's lymphoma	Vascular leak syndrome	VLS was observed in all patients but all symptoms were not observed in all patients; dose limiting	3
			Decreased sera albumin	Seen in 24 of 26 (92%) patients	
			Edema: peripheral	Seen in 14 of 26 (54%) patients	
			pulmonary	Seen in 2 of 26 (8%) patients	
			pleural/pericardial	Seen in 10 of 26 (38%) patients	
			Weight gain	Seen in 24 of 26 (92%) patients	
			Tachycardia	Seen in 18 of 26 (69%) patients	
			Hypotension	See in 6 of 26 (23%) patients	
			Decreased EKG voltage	Seen in 13 of 26 (50%) patients	
			Oliguria	Seen in 12 of 26 (46%) patients	
			Death	Two patients died; VLS exacerbated by pulmonary disease	
			Decreased granulocytes	Seen in 1 patient	
			Decreased platelets	Seen in 1 patient	
			Antibody response to drug	Seen in 9 of 24 patients; 1 only anti-Ig, 2 only anti-ricin-A, and 6 made both	
			Myalgia (Grade I–IV)	Seen in 58% of patients; dose limiting in one	
			Fever (Grade I–II)	Seen in 46% of patients	
			Nausea (Grade I–II)	Seen in 46% of patients	
			Respiratory (Grade I–IV)	Seen in 38% of patients	
			Cardiovascular (Grade I–II)	Seen in 15% of patients	
			Aphasia (Grade I, III)	Seen in 15% of patients	

TABLE 3 (continued)

Immunoconjugates and Immunotoxins

Immunoconjugates and Fusion Proteins	Route/Dose/ Schedule	Clinical Condition or Species	Parameter	Comments	Ref.
IgG-RFB4-dgA (murine IgG_1 anti-CD22 conjugated to deglycosylated ricin-A chain)	i.v. continuous infusion 9.6, 19.2, and 28.8 mg/m^2/192 hr Multiple courses	B-cell lymphoma	Malaise/fever (Grade I–II)	Seen in 25 of 32 courses	146
			Nausea/vomiting (Grade I)	Seen in 5 of 32 courses	
			Vascular leak syndrome	One patient died of complications related to VLS	
			Hypoalbuminemia (Grade I–II)	Seen in 30 of 32 courses	
			Weight gain (Grade I, III–IIV)	Seen in 23 of 32 courses	
			Pulmonary edema (Grade IV–V)	Seen in 3 of 32 courses	
			Hypotension (Grade I, III, V)	Seen in 5 of 32 courses	
			Decreased creatinine clearance (Grade I–II)	Seen in 9 of 32 courses	
			Decreased urine volume (Grade I)	Seen in 11 of 32 courses	
			Neuromuscular		
			Confusion (Grade I)	Seen in 1 of 32 courses	
			Dizziness (Grade I)	Seen in 1 of 32 courses	
			Aphasia (Grade III)	Seen in 1 of 32 courses	
			Myalgia (Grade I)	Seen in 4 of 32 courses	
			Joint discomfort (Grade I)	Seen in 4 of 32 courses	
			Decrease in leukocytes (Grade I–II)	Seen in 3 of 32 courses	
			Decrease in hemaglobin (Grade I)	Seen in 4 of 32 courses	
			Decrease in platelets (Grade I)	Seen in 6 of 32 courses	
			Decrease in $[Na^+]$ (Grade I)	Seen in 1 of 32 courses	
			Decrease in $[K^+]$ (Grade I)	Seen in 1 of 32 courses	
			Decrease in $[Ca^{2+}]$ (Grade I)	Seen in 1 of 32 courses	
			Increase in bilirubin (Grade I)	Seen in 2 of 32 courses	
			Increase in transaminases (Grade I)	Seen in 1 of 32 courses	
			Rash (Grade I)	Seen in 9 of 32 courses	
			Antibody response to drug	Seen in 25 of 32 courses; 12 of 16 (75%) patients developed an anti-Ig or anti-ricin-A response	

Fab′ -RFB4-dgA (murine IgG_1 Fab′ anti-CD22 conjugated to deglycosylated ricin-A chain)	i.v. infusion every 48 hr 25–120 mg/m² Cumulative dose 2–6 courses	B-cell lymphoma	Vascular leak syndrome	Seen in all patients	178
			Hypoalbuminemia	Seen in 15 of 15 patients	
			Edema: peripheral	Seen in 7 of 15 patients	
			pulmonary	Seen in 2 of 15 patients; dose limiting	
			Weight gain	Seen in 15 of 15 patients due to edema	
			Tachycardia	Seen in 6 of 15 patients	
			Decrease EKG voltage	Seen in 6 of 15 patients	
			Death	One death; complicated by pulmonary edema	
			Fever (mild to moderate)	Seen in 10 of 15 patients	
			Aphasia	Seen in 2 of 15 patients	
			Rhabdomyolysis	Seen in 1 of 15 patients; resulting in reversible kidney failure	
			Myalgia (mild to severe)	Seen in 7 of 15 patients	
			Antibody response to drug	3 of 14 patients had a response to ricin-A chain only; 1 patient had a significant response to murine-Ig and ricin-A chain	
CD4-PE40 (soluble CD4-*Pseudomonas* exotoxin derivative fusion protein)	i.v. infusion 40, 80, or 160 μg/m² 3–7× over 10 d	HIV	Increased hepatic transaminase	Dose-dependent and reversible; seen in 1 of 3 patients at high dose	133
			Fever	Reversible with antipyretics and fluids	
			Chills	Reversible with antipyretics and fluids	
			Hypotension	Reversible with antipyretics and fluids	
			Antibody response to drug	Measurable response after 3 weeks; increased clearance of drug	
Anti-CD5 Plus™ (murine IgG_1 conjugated to ricin-A chain)	i.v. infusion 0.20 or 0.33 mg/kg Daily for 5 d	Rheumatoid arthritis	Transient depletion of PB $CD3^+$ T cells	During treatment days 2 and 5, returning to baseline on days 15 to 29	50
			Transient depletion of PB $CD5^+$ B cells		
			No significant effect on monocytes		
			Proliferative response to antigenic, allogeneic, and mitogenic stimuli *in vitro* were depressed and normalized with recovery of T-cell number		
			No significant effect on B-cell *in vitro* function		
OVB3-PE (murine IgG_{2b} conjugated to *Pseudomonas* exotoxin)	i.p. 1–10 μg/kg days 1,4; or 5 μg/kg on days 1,4 or 1,4,7 or 1,3,5,7	Ovarian cancer	Toxic encephalopathy (Grade III–IV) death	Dose-limiting 5 μg/kg (×3) and 5 μg/kg (×2); seen in 3 patients and causing one death	126
			Transient elevation of liver enzymes (Grade I–II)	Seen in 35% of patients	
			Abdominal pain (Grade I–II)	Seen in 83% of patients	
			Fever (Grade I–II)	Seen in 22% of patients	
			Nausea and vomiting (Grade I–II)	Seen in 52% of patients	
			Antibody response to drug	Anti-PE response seen in 100% of patients by day 14; anti-murine Ig response seen in 92% of patients assessed by day 28	

TABLE 3 (continued)

Immunoconjugates and Immunotoxins

Immunoconjugates and Fusion Proteins	Route/Dose/ Schedule	Clinical Condition or Species	Parameter	Comments	Ref.
B3-PE40 (murine IgG_1; B3-tumor reactive mAb conjugated to *Pseudomonas* exotoxin derivative)	i.v. bolus 10, 30, 50 and 100 µg/kg on days 1, 3, and 5	Cynomolgus monkeys	*10–50 µg/kg/d*		125
			Transient loss of appetite and mild weight loss	Seen during first week	
			Elevation of SGOT, SGPT, AP, and LDH	3- to 15-fold over pretreatment levels; peaking day 5 and returning to normal by end of 2 wk	
			Bilirubin and serum amylase normal		
			No significant changes in other blood chemistries		
			No significant changes in cell blood count		
			100 µg/kg/d		
			Death on day 5	Seen in 2 of 2 monkeys	
			Massive liver failure	Pale liver with diffuse hepatic necrosis	
			Elevated liver enzymes	More than 100-fold over pretreatment levels	
			Elevated serum bilirubin	Ranged from 4–5 mg/dl on day 5	
DAB_{486} IL-2 (IL-2-diphtheria toxin derivative fusion protein)	i.v. bolus 0.0007–0.2 mg/kg/d for up to 10 days	Hematologic malignancies	Elevation of hepatice transaminase (Grade II–IV)	Reversible; seen in 6 of 18 patients	101
			Elevated creatinine (Grade II)	Seen in 1 of 18 patients	
			Proteinuria (Grade II)	Seen in 3 of 18 patients	
			Rash	Seen in 1 of 18 patients	
			Nausea	Seen in 1 of 18 patients	
			Chest tightness (Grade II)	Seen in 2 of 18 patients	
			Fever (Grade II–III)	Seen in 5 of 18 patients	
			Antibody response to drug	Seen in 50% of patients	
			No significant effect on immune function	PB phenotyping and lymphocyte blastogenic response	
			No significant changes in complement or serum lg		

DAB_{486} IL-2 (IL-2-diphtheria toxin derivative fusion protein)	i.v. fusion 0.2 mg/kg/d 5× every 21 d	Mycosis fungoides	Increased hepatic transaminases (Grade I–III)	Seen in 75% of courses and occurred in 13 of 14 patients after first course; incidence and magnitude decreased with subsequent courses	51
			Chills and fever	Seen in 83% of courses	
			Hypotension (Grade I–II)	Seen in 28% of courses	
			Nausea and vomiting	Seen in 20% of courses	
			Reversible bronchospasm without hypotension	Seen in 1 patient	
			Hypoalbuminemia	Seen in 58% of courses	
			Mild transient proteinuria	Seen in 4 patients	
			Reversible minor increases in creatinine	Seen in 2 patients	
			Azotemia accompanied by hemolysis and thrombocytopenia	Seen in 1 patient	
			No myelosuppression		
			Antibody response to drug	Variable pre-therapy anti-drug response with 7 of 13 having neutralizing anti-diphtheria toxin antibodies; 11 of 13 patients had an increase in titer	
			Acute hypersensitivity reaction	Seen in 1 patient on day 4 who had a high pre-treatment titer of anti-diphtheria toxin	

concentrations or binding to the targeted antigen expressed on normal tissues. The two major dose-limiting toxicities seen with these compounds have been vascular leak syndrome (VLS) and liver toxicity. Peripheral edema, hypoalbuminemia, increased hematocrit, weight gain, and hyptension characterize VLS. If it becomes severe, it can lead to pulmonary edema and effusions that can be life-threatening. Other symptoms that have been observed and may be related to VLS include: tachycardia that could be induced by extravascular hypovolemia, decreases in EKG voltage that could be a consequence of edema in the chest wall and/or pericardium, anorexia and nausea that could be related to gastric mucosal edema, and transient aphasia that could be caused by cerebral edema.[179] Although the cause has not been determined, immunotoxin-related VLS is believed to be due to the release of cytokines and the direct damage to vascular endothelial cells at high serum concentrations. For ricin, where VLS is most often observed, evidence suggests that it is due to the nontargeted binding of the ricin A-chain to endothelial cells and subsequent killing of the cells and damage to the vessels.[159] The liver toxicity that is most often observed with the pseudomonas endotoxin A may very likely be due to the nonspecific uptake and internalization of proteins by hepatocytes. These toxicities are most severe with the first generation native toxins. To increase specificity and reduce the toxicity to cells that do not express the targeted antigen, these toxins have been genetically or chemically modified so that the cell-binding portion of the toxin is no longer accessible or has been deleted. Immunotoxins are also very immunogenic, with immune responses being generated to both the toxin and the antibody. While the antibody portion of the immunotoxin may be humanized to reduce immunogenicity, the toxin cannot be so easily modified.

5 SUMMARY

In the tables in this chapter, the literature is reviewed regarding effects of recombinant cytokine proteins (Table 1), monoclonal antibodies (Table 2), and immunoconjugates/immunotoxins (Table 3) on immune parameters. As is evident in this review, each class of protein therapeutics is associated with toxicities at certain doses and dose schedules. It has been noted that there are three common toxic effects seen with each class of protein therapeutics which involve either specific or nonspecific immune functions. One is based on the fact that many of these agents are recombinant proteins or monoclonal antibodies and, as such, may be immunogenic, resulting in (1) side effects in the patients (skin reactions, arthralgia, fever, and malaise) and/or (2) a decrease in the efficacy of the agent. This is generally a more important concern with monoclonal antibodies or immunoconjugates/immunotoxins that are composed of antibodies from a species other than humans. However, it should be noted that a therapeutic protein can be immunogenic due to more subtle differences such as being denatured or aggregated. With recombinant human cytokines, the immunogenicity in humans has not generally resulted in dose-limiting side effects or changes in the efficacy of the agent. The second common toxicity associated with administration of therapeutic proteins is fever and chills and the accompanying symptoms of myalgia and arthralgia. The mechanisms involved have not been clearly elucidated and may differ between the the various classes of proteins, with some responses resulting from immune reactions and others resulting from increased cytokine release. A third toxicity commonly seen with some members of each class of therapeutic protein is vascular leak syndrome. It is generally a serious toxicological response associated with immunotoxins and IL-2, though it has been seen with other members of each class of therapeutic proteins. Again, the actual mechanism may differ between the classes of therapeutic proteins but is generally thought to be the result of treatment-related cytokine or inflammatory mediator release that affects vascular endothelium.

In this review, studies have been chosen which illustrate the effects of protein therapeutics on the immune system. These studies were also chosen to illustrate that toxicity can be dependent on route of exposure, dose, and dose schedules. IL-2 is an example of a cytokine where high-dose, intravenous therapy has resulted in severe side effects that are dose limiting and nonefficacious. To overcome these limitations, studies have been conducted with regional or subcutaneous administration to reduce toxicity as well as improve efficacy. Other investigators have also evaluated low-dose or intermittent rather than high-dose, prolonged therapy to decrease side effects.

REFERENCES

1. Abraham, E., Wunderink, R., Silverman, H., Perl, T. M., Nasraway, S., Levy, H. et al., Efficacy and safety of monoclonal antibody to human tumor necrosis factor α in patients with sepsis syndrome, *JAMA,* 273, 934–941, 1995.
2. Alegre, M-L., Lenschow, D. J., and Bluestone, J. A., Immunomodulation of transplant rejection using monoclonal antibodies and soluble receptors, *Dig. Dis. Sci.*, 40, 58–64, 1995.
3. Amlot, P. L., Stone, M. J., Cunningham, D., Fay, J., Newman, J., Collins, R., May, R., McCarthy, M., Richardson, J., Ghetie, V., Ramilo, O., Thorpe, P. E., Uhr, J. W., and Vitteta, E. S., A phase I study of an anti-CD22 deglycosylated ricin A chain immunotoxin in the treatment of B-cell lymphomas resistant to conventional therapy, *Blood,* 9, 2624–2633, 1993.
4. Anasetti, C., Martini, P. J., Hansen, J. A. Appelbaum, F. R., Beatty, P. G., Doney, K. et al., A phase I-II study evaluating the murine anti-IL-1 receptor antibody 2A3 for treatment of acute graft-versus-host disease, *Transplantation,* 50, 49–54, 1990.
5. Anastasopoulos, E., Reclos, G. J., Baxevanis, C. N., Gritzapis, A. D., Tsilivakos, V., Panagiotopoulos, N. et al., Monocyte disorders associated with T cell defects in patients with solid tumors, *Anticancer Res.,* 12, 489–494, 1992.
6. Antin, J. H., Weinstein, H. J., Guinan, E. C., McCarthy, P., Bierer, B. E., Gilliland, D. G. et al., Recombinant human interleukin-1 receptor antagonist in the treatment of steroid-resistant graft-versus-host disease, *Blood*, 84, 1342–1348, 1994.
7. Antin, J. H., Smith, B. R., Holmes, W., and Rosenthal, D. S., Phase I/II study of recombinant human granulocyte-macrophage colony-stimulating factor in aplastic anemia and myelodysplastic syndrome, *Blood*, 72, 705–713, 1988.
8. Antman, K. S., Griffin, J. D., Elias, A. et al., Effect of recombinant human granulocyte macrophage colony-stimulating factor on chemotherapy-induced myelosuppression, *New Engl. J. Med.*, 319, 593–598, 1988.
9. Asano, S., Okano, A., Ozawa, K., Nakahata, T., Ishibashi, T., Koike, K. et al., *In vivo* effects of recombinant human interleukin-6 in primates: stimulated production of platelets, *Blood*, 75, 1602–1605, 1990.
10. Aulitzky, W. E., Schuler, M., Peschel, C. and Huber, C., Interleukins — clinical pharmacology and therapeutic use, *Drugs*, 48, 667–677, 1994.
11. Azuma, J., Kurimoto, T., Tsuji, S., Mochizuki, N., Fujinaga, S., Matsumoto, Y., and Masuho, Y., Phase I study on human monoclonal antibody against cytomegalovirus: pharmacokinetics and immunogenicity, *J. Immunother.*, 10, 278–285, 1991.
12. Balch, C. M., Tilden, A. B., Dougherty, P. A., Cloud, G. A., and Abo, T., Depressed levels of granular lymphocytes with natural killer (NK) cell function in 247 cancer patients, *Ann. Surg.*, 198, 192–199, 1983.
13. Baldwin, G. C., Gasson, J. C., Quan, S. G., Fleischmann, J., Weisbart, R., Oette, D. et al., Granulocyte-macrophage colony-stimulating factor enhances neutrophil function in acquired immunodeficiency syndrome patients, *Proc. Natl. Acad. Sci. USA*, 85, 2763–2766, 1988.
14. Baxevanis, C. N., Papilas, K., Dedoussis, G. V. Z., Pavlis, T., and Papamichail, M., Abnormal cytokine serum levels correlate with impaired cellular immune responses after surgery, *Clin. Immunol. Immunopathol.*, 71, 82–88, 1994.
15. Baxevanis, C. N., Reclos, G. J., Gritzapis, A. D., Dedoussis, G. V. Z., Arsenis, P. Katsiyiannis, A. et al., Comparison of immune parameters in patients with one or two primary malignant neoplasms, *Nat. Immun.*, 12, 41–49, 1993.
16. Baxevanis, C. N., Reclos, G. J., Sfagos, C., Doufexis, E., Papagerogiou, C., and Papamichail, M., Multiple sclerosis. I. Monocyte stimulatory defect in mixed lymphocyte reaction associated with clinical disease activity, *Clin. Exp. Immunol.*, 67, 362–371, 1987.
17. Bendtzen, K., Cytokines and autoantibodies to cytokines, *Stem Cells*, 13, 206–222, 1995.
18. Blick, M., Sherwin, S. A., Rosenblum, M., and Gutterman, J., Phase I study of recombinant tumor necrosis factor in cancer patients, *Canc. Res.*, 47, 2986–2989, 1987.

19. Brandt, S. J., Peters, W. P., Atwater, S. K., Kurtzberg, J., Borowitz, M. J., Jones, R. B. et al., Effect of recombinant human granulocyte-macrophage colony-stimulating factor on hematopoietic reconstitution after high-dose chemotherapy and autologous bone marrow transplantation, *New Engl. J. Med.*, 318, 869–876, 1988.
20. Bree, A. G., Schlerman, F. J., Kaviani, M. D., Hastings, R. C., Hitz, S. L., and Goldman, S. J., Multiple effects on peripheral hematology following administration of recombinant human interleukin 12 to nonhuman primates, *Biochem. Biophys. Res. Comm.*, 204, 1150–1157, 1994.
21. Bruton, J. K. and Koeller, J. M., Recombinant interleukin-2, *Pharmacotherapy*, 14, 635–656, 1994.
22. Buescher, E. S., McIlheran, S. M., Banks, S. M., Kudelka, A. P., Kavanagh, J. J., and Vadhan-Raj, S., The effects of interleukin-1 therapy on peripheral blood granulocyte function in humans, *Canc. Immunol. Immunother.*, 37, 26–30, 1993.
23. Bukowski, R. M., Olencki, T., McLain, D., and Finke, J. H., Pleiotropic effects of cytokines: clinical and preclinical studies, *Stem Cells*, 12, 129–141, 1994.
24. Bukowski, R. M., Budd, G. T., Gibbons, J. A., Bauer, R. J., Child, A., and Antal, J., Phase I trial of subcutaneous recombinant macrophage of colony-stimulating factor: clinical and immunomodulatory effects, *J. Clin. Oncol.*, 12, 97–106, 1994.
25. Caligiuri, M. A., Murray, C., Robertson, M. J., Wang, E., Cochran, K., Cameron, C. et al., Selective modulation of human natural killer cells *in vivo* after prolonged infusion of low dose recombinant interleukin 2, *J. Clin. Invest.*, 91, 123–132, 1993.
26. Chachoua, A., Oratz, R., Hoogmoed, R., Caron, D., Peace, D., Liebes, L. et al., Monocyte activation following systemic administration of granulocyte-macrophage colony-stimulating factor, *J. Immunother.*, 15, 217–224, 1994.
27. Chatenoud, L., Baudrihaye, M. F., Chkoff, N., Kreis, H., Goldstein, G., and Bach, J.-F., Restriction of the human *in vivo* immune response against the mouse monoclonal antibody OKT3, *J. Immunol.*, 137, 830–838, 1986.
28. Chien, C. H. and Hsieh, K. H., Interleukin-2 immunotherapy in children, *Pediatrics*, 86, 937–943, 1990.
29. Cohen, R. B., Siegel, J. P., Puri, R. K., and Pluznik, D. H., The immunotoxicology of cytokines, in *Clinical Immunotoxicology*, Newcombe, D. S., Rose, N. R., and Bloom, J. C., Eds., Raven Press, New York, 1992, pp. 93–108.
30. Cosimi, A. B., Colvin, R. B., Burton, A. R. C., Rubin, R. H., Goldstein, G., Kung, P. C. et al., Use of monoclonal antibodies to T-cell subsets for immunologic monitoring and treatment in recipients of renal allografts, *New Engl. J. Med.*, 305, 308–314, 1981.
31. Cosimi, A. B., Conti, D., Delmonico, F. L., Preffer, F. I., Wee, S.-L., Rothlein, R. et al., *In vivo* effects of monoclonal antibody to ICAM-1 (CD54) in nonhuman primates with renal allografts, *J. Immunol.*, 144, 4604–4612, 1990.
32. Dean, J. H., Cornacoff, J. B., Barbolt, T. A., Gossett, K. A., and LaBrie, T., Pre-clinical toxicity of IL-4: a model for studying protein therapeutics, *Int. J. Immunopharmac.*, 14, 391–397, 1992.
33. Deehan, D. J., Heys, S. D., Simpson, W., Broom, J., McMillan, D. N., and Eremin, O., Modulation of the cytokine and acute-phase response to major surgery by recombinant interleukin-2, *Br. J. Surg.*, 82, 86–90, 1995.
34. Deehan, D. J., Heys, S. D., Simpson, W., Herriot, R., Broom, J., and Eremin, O., Correlation of serum cytokine and acute phase reactant levels with alterations in weight and serum albumin in patients receiving immunotherapy with recombinant IL-2, *Clin. Exp. Immunol.*, 95, 366–372, 1994.
35. Dillman, R. O., Antibodies as cytotoxic therapy, *J. Clin. Oncol.*, 12, 1497–1515, 1994.
36. Dillman, R. O., Beauregard, J. C., Halpern, S. E., and Clutter, M., Toxicities and side effects associated with intravenous infusions of murine monoclonal antibodies, *J. Biol. Resp. Mod.*, 5, 73–84, 1986.
37. Dinarello, C. A., Interleukin-1 and tumor necrosis factor: effector cytokines in autoimmune diseases, *Semin. Immunol.*, 4, 133–145, 1992.
38. Donahue, R. E., Seehra, J., Metzger, M. et al., Human IL-3 and GM-CSF act synergistically in stimulating hematopoiesis in primates, *Science*, 241, 1820–1923, 1988.
39. Donahue, R. E., Wang, E. A., Stone, D. K. et al., Stimulation of hematopoiesis in primates by continuous infusion of recombinant GM-CSF, *Nature Lond.*, 321, 872–875, 1986.
40. Drobyski, W. R., Gottlieb, M., Carrigan, D., Ostberg, L., Grebenau, M., Schran, H. et al., Phase I study of safety and pharmacokinetics of a human anticytomegalovirus monoclonal antibody in allogeneic bone marrow transplant recipients, *Transplantation*, 51, 1190–1196, 1991.
41. Durrant, L. G., Byers, V. S., Scannon, P. J., Rodvien, R., Grant, K., Robins, A. et al., Humoral immune response to XMMCO-791-RTA immunotoxin in colorectal cancer patients, *Clin. Exp. lmmunol.*, 75, 258–264, 1989.
42. Eason, J. D., Wee, S. L., Kawai, T., Hong, H. Z., Powelson, J., Widmer, M. et al., Recombinant human dimeric tumor necrosis factor receptor (TNFR:Fc) as an immunosuppressive agent in renal allograft recipients, *Transplant. Proc.*, 27, 554, 1995.

43. Elias, D. J., Hirschowitz, L., Kline, L. E., Kroener, J. F., Dillman, R. O., Walker, L. E. et al., Phase I clinical comparative study of monoclonal antibody KS1/4 and KS1/4-methotrexate immunoconjugate in patients with non-small-cell lung cancer, *Canc. Res.*, 50, 4154–4159, 1990.
44. Emilie, D., Wijdenes, J., Gisselbrecht, C., Jarrousse, B., Billaud, E., Blay, J.-Y. et al., Administration of an anti-interleukin-6 monoclonal antibody to patients with acquired immunodeficiency syndrome and lymphomas: effect on lymphoma growth and on B clinical symptoms, *Blood*, 84, 2472–2479, 1994.
45. Engelhard, M., Gerhartz, H., Brittinger, G., Engert, A., Fuchs, R., Geiseler, B. et al., Cytokine efficiency in the treatment of high-grade malignant non-Hodgkin's lymphomas: results of a randomized double-blind placebo-controlled study with intensified COP-BLAM ± rhGM-CSF, *Ann. Oncol.*, 5(Suppl. 2), 123–125, 1994.
46. Espat, N. J., Cendan, J. C., Beierle, E. A., Auffenberg, T. A., Rosenberg, J., Russell, D. et al., PEG-BP-30 monotherapy attenuates the cytokine-mediated inflammatory cascade in baboon *Escherichia coli* septic shock, *J. Surg. Res.*, 59, 153–158, 1995.
47. Ettinghausen, S. E., Puri, R. K., Rosenberg, S. A. et al., Increased vascular permeability in organs mediated by the systemic administration of lymphokine-activated killer cells and recombinant interleukin-2 in mice, *J. Natl. Canc. Inst.*, 80, 177–188, 1988.
48. Ezekowitz, R. A. B., Dinauer, M. C., Jaffe, H. S., Orkin, S. H., and Newburger, P. E., Partial correction of the phagocyte defect in patients with x-linked chronic granulomatous disease by subcutaneous interferon gamma, *New Engl. J. Med.*, 319, 146–151, 1988.
49. Fisher, C. J., Dhainaut, J. A., Opal, S. M., Pribble, J. P., Balk, R. A., Slotman, G. J. et al., Recombinant human interleukin 1 receptor antagonist in the treatment of patients with sepsis syndrome, *JAMA*, 271(23), 1836–1843, 1994.
50. Fishwild, D. M. and Strand, V., Administration of an anti-CD5 immunoconjugate to patients with rheumatoid arthritis: effect of peripheral blood mononuclear cells and *in vitro* immune function, *J. Rheumatol.*, 21, 596–604, 1994.
51. Foss, F. M., Borkowski, T. A., Gilliom, M., Stetler-Stevenson, M., Jaffe, E. S., Figg, W. D. et al., Chimeric fusion protein toxin DAB486IL-2 in advanced mycosis fungoides and the Sezary syndrome: correlation of activity and interleukin-2 receptor expression in a Phase II study, *Blood*, 84, 1765–1774, 1994.
52. Fraker, D. L., Langstein, H. N., and Norton, J. A., Passive immunization against tumor necrosis factor partially abrogates interleukin-2 toxicity, *J. Exp. Med.*, 170, 1015–1020, 1989.
53. Gallin, J. I., Malech, H. L., Weening, R. S., Curnette, J. T., Quie, P. G., Jaffe, H. S. et al., A controlled trial of interferon gamma to prevent infection in chronic granulomatous disease, *New Engl. J. Med.*, 324, 509–516, 1991.
54. Gambacorti-Passerini, C., Rivoltini, L., Radrizzani, M., Belli, F., Sciorelli, G., Ravagnani, F. et al., Differences between *in vivo* and *in vitro* activation of cancer patient lymphocytes by recombinant interleukin 2: possible role for lymphokine-activated killer cell infusion in the *in vivo*-induced activation, *Canc. Res.*, 49, 5230–5234, 1989.
55. Ganser, A., Lindemann, A., Seipelt, G., Ottmann, O. G., Herrmann, F., Eder, M. et al., Effects of recombinant human interleukin-3 in patients with normal hematopoiesis and in patients with bone marrow failure, *Blood*, 76, 666–676, 1990.
56. Ganser, A., Seipelt, G., Lindemann, A., Ottmann, O. G., Falk, S., Eder, M. et al., Effects of recombinant human interleukin-3 in patients with myelodysplastic syndromes, *Blood*, 76, 455–462, 1990.
57. Garritsen, H. S. P., Constantin, C., Kolkmeyer, A., de Grooth, B. G., Greve, J., Hiddemann, W. et al., A transient but consistent increase in $CD3^{+}CD8^{+}CD57^{+}$ lymphocytes under IL-2 therapy in AML, *Proc. Am. Assoc. Canc. Res.*, 33, 323, 1992.
58. Gerhartz, H. H., Engelhard, M., Meusers, P., Brittinger, G., Wilmanns, W., Schlimok, G. et al., Randomized, double-blind, placebo-controlled, phase III study of recombinant human granulocyte-macrophage colony-stimulating factor as adjunct to induction treatment of high-grade malignant non-Hodgkin's lymphomas, *Blood*, 82, 2329–2339, 1993.
59. Ghosh, A. K., Smith, N. K., Prendiville, J., Thatcher, N., Crowther, D. and Stern, P. L., A phase I study of recombinant human interleukin-4 administered by the intravenous and subcutaneous route in patients with advanced cancer: immunological studies, *Eur. Cytokine Netw.*, 4, 205–211, 1993.
60. Goodman, G. E., Hellstrom, I., Yelton, D. E., Murray, J. L., O'Hara, S., Meaker, E. et al., Phase I trial of chimeric (human-mouse) monoclonal antibody L6 in patients with non-small-cell lung, colon, and breast cancer, *Canc. Immunol. Immunother.*, 36, 267–273, 1993.
61. Gordon, M. S., Sledge, Jr., G. W., Battiato, L., Breeden, E., Cooper, R., McCaskill-Stevens, W. J. et al., The *in vivo* effects of subcutaneously (sc) administered recombinant human interleukin-11(Neumega™ rhIL-11 growth factor; rhIL-11) in women with breast cancer (bc), *Blood*, 82(Suppl.), 1796, 1993.
62. Gossett, K. A., Barbolt, T. A., Cornacoff, J. B., Zelinger, D. J., and Dean, J. H., Clinical pathologic alterations associated with subcutaneous administration of recombinant human interleukin-4 to cynomolgus monkeys, *Toxicol. Pathol.*, 21, 46–53, 1993.
63. Gottlieb, D. J., Prentice, H. G., Heslop, H. E., Bello, C., and Brenner, M. K., IL-2 infusion abrogates humoral immune responses in humans, *Clin. Exp. Immunol.*, 87, 493–498, 1992.

64. Gould, B. J., Borowitz, M. J., Groves, E. S., Carter, P. W., Anthony, D., Weiner, L. M., and Frankel, A. E., Phase I study of an anti-cancer immunotoxin by continuous infusion: report of a targeted toxic effect not predicted by animal studies, *J. Natl. Canc. Inst.*, 81, 775–781, 1989.
65. Granowitz, E. V., Porat, R., Mier, J. W., Pribble, J. P., Stiles, D. M., Bloedow, D. C. et al., Pharmacokinetics, safety and immunomodulatory effects of human recombinant interleukin-1 receptor antagonist in healthy humans, *Cytokine*, 4, 353–360, 1992.
66. Gringeri, A., Santagostino, E., Mannucci, P. M., Tradati, F., Cultraro, D., Buzzi, A. et al., A randomized, placebo-controlled, blind anti-AIDS clinical trial: safety and immunogenicity of a specific anti-IFNα immunization, *J. Acquired Immune Deficiency Syndr.*, 7, 978–988, 1994.
67. Groopman, J. E., Mitsuyasu, R. T., DeLeo, M. J., Oette, D. H., and Golde, D. W., Effect of recombinant human granulocyte-macrophage colony-stimulating factor on myelopoiesis in the acquired immunodeficiency syndrome, *New Engl. J. Med.*, 317, 593–598, 1987.
68. Grossbard, M. L., Lambert. J. M., Goldmacher, V. S., Spector, N. L., Kinsella, J., Eliseo, L. et al., Anti-B4 blocked ricin: a phase I trial of 7-day continuous infusions in patients with B-cell neoplasms, *J. Clin. Oncol.*, 11, 726–737, 1993.
69. Grossbard, M. L., Freedman, A. S., Ritz, J., Coral, F., Goldmacher, V. S., Eliseo, L. et al., Serotherapy of B-cell neoplasms with anti-B4–blocked ricin: a phase I trial of daily bolus infusion, *Blood*, 79, 576–585, 1992.
70. Hafler, D. A., Ritz, J., Schlossman, S. F., and Weiner, H. L., Anti-CD4 and anti-CD2 monoclonal antibody infusions in subjects with multiple sclerosis: immunosuppressive effects and human anti-mouse response, *J. lmmunol.*, 141, 131–138, 1988.
71. Hale, G., Clark, M. R., Marcus, R., Winter, G., Dyer, M. J. S., Phillips, J. M., Reichmann, L., and Waldmann, H., Remission induction in non-Hodgkin's lymphoma with reshaped human monoclonal antibody CAMPATH-1 H, *Lancet*, 2, 1394–1399, 1988.
72. Handgretinger, R., Baader, P., Dopfer, R., Klingebiel, T., Reuland, P., Treuner, J. et al., A phase I study of neuroblastoma with the anti-ganglioside GD2 antibody 14.G2a, *Canc. Immunol. Immunother.*, 35, 199–204, 1992.
73. Hänninen, E. L., Körfer, A., Hadam, M., Schneekloth, C., Dallmann, I., Menzel, T. et al., Biological monitoring of low-dose interleukin 2 in humans: soluble interleukin 2 receptors, cytokines, and cell surface phenotypes, *Canc. Res.*, 50, 6312–6316, 1991.
74. Heaton, K. M., Ju, G., Morris, D. K., Delisio, K., Bailon, P., and Grimm, E. A., Characterizaion of lymphokine activated killing by human peripheral blood mononuclear cells stimulated with interleukin (IL-2) analogs specific for the intermediate affinity IL-2 receptor, *Cell. Immunol.*, 147, 167–179, 1993.
75. Hibi, S., Yoshihara, T., Nakajima, F., Misu, H., and Imashuku, S., Effect of recombinant human granulocyte-colony stimulation factor (rhG-CSF) on immune system in pediatric patients with aplastic anemia, *Pediatr. Hematol. Oncol.*, 11, 319–323, 1994.
76. Ho, A. D., Haas, R., Wulf, G., Knauf, W., Ehrhardt, R., Heilig, B. et al., Activation of lymphocytes induced by recombinant human granulocyte-macrophage colony-stimulating factor in patients with malignant lymphoma, *Blood*, 75, 203–212, 1990.
77. Huland, E and Huland, H., Local continuous high dose IL-2: a new therapeutic model for the treatment of advanced bladder cancer, *Canc. Res.*, 49, 5469, 1989.
78. Isaacs, J. D., Wafts, R. A., Hazleman, B. L., Hale, G., Keogan, M. T., Cobbold, S. P., and Waldmann, H., Humanized monoclonal antibody therapy for rheumatoid arthritis, *Lancet*, 340, 748–752, 1992.
79. Jablons, D., Bolton, E., Mertins, S., Rubin, M., Pizzo, P., Rosenberg, S. A., and Lotze, M. T., IL-2-based immunotherapy alters circulating neutrophil Fc receptor expression and chemotaxis, *J. Immunol.*, 144, 3630–3636, 1990.
80. Jaffe, H. A., Buhl, R., Mastrangeli, A., Holroyd, K. J., Saltini, C., Czerski, D. et al., Local activation of mononuclear phagocytes by delivery of an aerosol of recombinant interferon-γ to the human lung, *J. Clin. Invest.*, 88, 297–302, 1991.
81. Jaffers, G. J., Fuller, T. C., Cosimi, A. B., Russel, P. S., Winn, H. J., and Colvin, R. B., Monoclonal antibody therapy: anti-idiotypic and non-anti-idiotypic antibodies to OKT3 arising despite intense immunosuppression, *Transplantation*, 41, 572–578, 1986.
82. Kaminski, M. S., Zasadny, K. R., Francis, I. R., Milik, A. W., Ross, C. W., Scott, S. D. et al. Radioimmunotherapy of B-cell lymphoma with [^{131}I] anti-B1 (anti-CD20) antibody, *New Engl. J. Med.*, 329, 459–465, 1993.
83. Kaplan, G., Mathur, N. K., Job, C. K., Nath, I., and Cohn, Z. A., Effect of multiple interferon-γ injections on the disposal of *Mycobacterium leprae*, *Proc. Natl. Acad. Sci. USA*, 86, 8073–8077, 1989.
84. Katre, N. V., Knauf, M. J., and Laird, W. J., Chemical modification of recombinant IL-2 by polyetheylene glycol increases potency in the murine Meth A sarcoma model, *Proc. Natl. Acad. Sci. USA*, 84, 1487, 1987.
85. Kemeny, N., Childs, B., Larchian, W., Rosado, K., and Kelsen, D., A phase II trial of recombinant tumor necrosis factor in patients with advanced colorectal carcinoma, *Cancer*, 66, 659–663, 1990.
86. Khazaeli, M. B., Conry, R. M., and LoBuglio, A. F., Human immune response to monoclonal antibodies, *J. Immunother.*, 15, 42–52, 1994.

87. Khazaeli, M. B., Saleh, M. N., Liu, T. P., Meredith, R. F., Wheeler, R. H., Baker, T. S. et al., Pharmacokinetics and immune response of ^{131}I-chimeric mouse/human B72.3 (human γ4) monoclonal antibody in humans, *Canc. Res.*, 51, 5461–5466, 1991.
88. Kist, A., Ho., A. D., Rath, U., Wiedenmann, B., Bauer, A., Schlick, E. et al., Decrease of natural killer cell activity and monokine production in peripheral blood of patients treated with recombinant tumor necrosis factor, *Blood*, 72, 344–348, 1988.
89. Kleinerman, E. S., Kurzrock, R., Wyatt, D., Quesada, J. R., Gutterman, J. U., and Fidler, I. J., Activation or suppression of the tumoricidal properties of monocytes from cancer patients following treatment with human recombinant γ-interferon, *Canc. Res.*, 46, 5401–5405, 1986.
90. Klempner, M. S., Noring, R., Mier, J. W., and Atkins, M. B., An acquired chemotactic defect in neutrophils from patients receiving interleukin-2 immunotherapy, *New Engl. J. Med.*, 322, 959–965, 1990.
91. Klug, S., Neubert, R., Stahlmann, R., Thiel, R., Ryffel, B., Car, B. D. et al., Effects of recombinant human interleukin 6 (rhIL-6) in marmosets (*Callithrix jacchus*), *Arch. Toxicol.*, 68, 619–631, 1994.
92. Knox, S. J., Levy, R., Hodgkinson, S., Bell, R., Brown, S., Wood, G. S. et al., Observations on the effect of chimeric anti-CD4 monoclonal antibody in patients with mycoses fungoides, *Blood*, 77, 20–30, 1991.
93. Konrad, M. W., Childs, A. L., Merigan, T. C. and Borden, E. C., Assessment of antigenic response in humans to a recombinant mutant interferon beta, *J. Clin. Immunol.*, 7, 365–375, 1987.
94. Kopp, W. C., Holmlund, J. T. and Urba, W. J., Immunological monitoring and clinical trials of biological response modifiers, in *Cancer Chemotherapy and Biological Response Modifiers Annual 15*, Pinedo, H. M., Longo, D. L., and Chabner, B. A., Eds., Elsevier Science, New York, 1994, pp. 226–286.
95. Kroemer, G., de Alboran, I. M., Gonzalo, J. A., and Martinez-A., C., Immunoregulation by cytokines, *CRC Crit. Rev. Immunol.*, 13, 163–191, 1993.
96. Kurzrock, R., Talpaz, M., Estrov, Z., Rosenblum, G., and Gutterman, J. U., Phase I study of recombinant human interleukin-3 in patients with bone marrow failure, *J. Clin. Oncol.*, 9, 1241–1250, 1991.
97. Laughlin, M. J., Kirkpatrick, G., Sabiston, N., Peter, W., and Kurtzberg, J., Hematopoietic recovery following high dose combined alkylating-agent chemotherapy and autologous bone marrow support in patients in phase-I clinical trials of colony-stimulating factors: G-CSF, GM-CSF, IL-1, IL-2, M-CSF, *Ann. Hematol.*, 67, 267–276, 1993.
98. Lebsack, M. E., Paul, C. C., Bloedow, D. C., Burch, F. X., Sack, M. A., Chase, W. et al., Subcutaneous IL-1 receptor antagonist in patients with rheumatoid arthritis, (abstract), *Arthr. Rheum. J.*, 34, 545, 1991.
99. Ledermann, J. A., Begent, R. H. J., Bagshawe, K. D. et al., Repeated antitumour therapy in man with suppression of the host response by cyclosporin A, *Br. J. Canc.*, 8, 654–657, 1988.
100. Lee, A. M., Vadhan-Raj, S., Hamilton, R. F., Scheule, R. K., and Holian, A., The *in vivo* effects of rhIL-1α therapy on human monocyte activity, *J. Leukocyte Biol.*, 54, 314–321, 1993.
101. LeMaistre, C. R., Meneghetti, C., Rosenblum, M., Reuben, J., Parker, K., Shaw, J. et al., Phase I trial of an interleukin-2 (IL-2) fusion toxin (DAB_{486}IL-2) in hematologic malignancies expressing the IL-2 receptor, *Blood*, 79, 2547–2554, 1992.
102. Liberati, A. M., Garofani, P., De Angelis, V., Di Clemente, F., Horisberger, M., Cecchini, M. et al., Double-blind randomized phase I study on the clinical tolerance and pharmacodynamics of natural and recombinant interferon-β given intraveously, *J. Interferon Res.*, 14, 61–69, 1994.
103. Lindemann, A., Ganser, A., Hoelzer, D., Mertelsmann, R., and Herrmann, F., *In vivo* administration of recombinant human interleukin-3 elicits an acute phase response involving endogenous synthesis of interleukin-6, *Eur. Cytokine Net.*, 2, 173–176, 1991.
104. LoBuglio, A. F., Saleh, M., Peterson, L., Wheeler, R., Carrano, R., Huster, W., and Khazaeli, M. B., Phase I clinical trial of C017-1A monoclonal antibody, *Hybridoma*, 5, S117–S123, 1986.
105. Maloney, D. G., Liles, T. M., Czerwinski, D. K., Waldichuk, C., Rosenberg, J., Grillo-Lopez, A., and Levy, R., Phase I clinical trial using escalating single-dose infusion of chimeric anti-CD20 monoclonal antibody (IDEC-C2B8) in patients with recurrent B-cell lymphoma, *Blood*, 84, 2457–2466, 1994.
106. Maluish, A. E., Urba, W. J., Longo, D. L. et al., The determination of an immunologically active dose of interferon gamma in patients with melanoma, *J. Clin. Oncol.*, 6, 434–445, 1988.
107. Martín, J., Quiroga, J. A., Bosch, O., and Carreño, V., Changes in cytokine production during therapy with granulocyte-macrophage colony-stimulating factor in patients with chronic hepatits B, *Hepatology*, 20, 1156–1161, 1994.
108. Mavligit, G. M., Zukiwski, A. A., Charnsangavej, C., Carrasco, C. H., Wallace, S., and Gutterman, J. U., Hepatic arterial infusion of recombinant human tumor necrosis factor in patients with liver metastases, *Cancer*, 69, 557–561, 1992.
109. Mayer, P., Valent, P., Schmidt, G., Liehl, E., and Bettelheim, P., The *in vivo* effects of recombinant human interleukin-3: demonstration of basophil differentiation factor, histamine-producing activity, and priming of GM-CSF-responsive progenitors in nonhuman primates, *Blood*, 74, 613–621, 1989.
110. Meeker, T. C., Lowder, J., Maloney, D. G., Miller, R. A., Thielemans, K., Warnke, R., and Levy, R., A clinical trial of anti-idiotype therapy for B cell malignancy, *Blood*, 65, 1349–1363, 1985.

111. Menzel, T., Schomberg, A., Korfer, A., Hadam, M., Meffert, M., Dallmann, I. et al., Clinical and preclinical evaluation of recombinant PEG-IL-2 in human, *Canc. Biother.*, 8, 199–212, 1993.
112. Meredith, R. F., Khazaeli, M. B., Ploft, W. E., Saleh, M. N., Liu, T., Allen, L. F. et al., Phase I trial of iodine-131-chimeric B72.3 (human IgG4) in metastatic colorectal cancer, *J. Nucl. Med.*, 33, 23–29, 1992.
113. Mier, J. W., Vachino, G., van der Meer, J. W. et al., Induction of circulating tumor necrosis factor (TNF-alpha) as the mechanism for the febrile response to interleukin-2 (IL-2) in cancer patients, *J. Clin. Immunol.*, 8, 426–436, 1988.
114. Miller, R. A. and Levy, R. Response of cutaneous T cell lymphoma to therapy with hybridoma monoclonal antibody, *Lancet*, 2, 226–230, 1981.
115. Moreland, L. W., Pratt, P. W., Bucy, R. P., Jackson, B. S., Feldman, J. W., and Koopman, W. J., Treatment of refractory rheumatoid arthritis with a chimeric anti-CD4 monoclonal antibody, *Arthr. Rheum.*, 37, 834–838, 1994.
116. Moreland, L. W., Bucy, R. P., Tilden, A., Pratt, P. W., LoBuglio, A. F., Khazaeli, M. et al., Use of a chimeric monoclonal antibody anti-CD4 antibody in patients with refractory rheumatoid arthritis, *Arthr. Rheum.*, 36, 307–318, 1993.
117. Murphy, P. M., Lane, H. C., Gallin, J. I., and Fauci, A. S., Marked disparity in incidence of bacterial infections in patients with the acquired immunodeficiency syndrome receiving interleukin-2 or interferon-γ, *Ann. Intern. Med.*, 108, 36–41, 1988.
118. Murray, J. L., Macey, D. J., Kasi, L. P., Rieger, P., Cunningham, J., Bhadkamkar, V. et al., Phase II radioimmunotherapy trial with ^{131}I-CC49 in colorectal cancer, *Cancer*, 73, 1057–1066, 1994.
119. Murray, J. L., Cunningham, J. E., Brewer, H., Mujoo, K., Zukiwski, A. A., Podoloff, D. A. et al., Phase I trial of murine monoclonal antibody 14G2a administration by prolonged intravenous infusion in patients with neuroectodermal tumors, *J. Clin. Oncol.*, 12, 184–193, 1994.
120. Nadler, L. M., Stashenko, P., Hardy, R., Kaplan, W. D., Bufton, L. N., Kufe, D. W. et al., Serotherapy of a patient with a monoclonal antibody directed against a human lymphoma-associated antigen, *Canc. Res.*, 40, 3147–3155, 1980.
121. Nakamura, Y., Ozaki, T., Yanagawa, H., Yasuoka, S., and Ogura, T., Eosinophil colony-stimulating factor induced by administration of interleukin-2 into the pleural cavity of patients with malignant pleurisy, *Am. J. Respir. Cell. Molec. Biol.*, 3, 291–300, 1990.
122. Nathan, C. F., Kaplan, G., Levis, W. R., Nusrat, A., Witmer, M. D., Sherwin, S. A. et al., Local and systemic effects of intradermal recombinant interferon-γ in patients with lepromatous leprosy, *New Engl. J. Med.*, 315, 6–14, 1986.
123. Nemunaitis, J., Appelbaum, F. R., Lilleby, K., Buhles, W. C., Rosenfeld, C., Zeigler, Z. R. et al., Phase I study of recombinant interleukin-1β in patients undergoing autologous bone marrow transplant for acute myelogenous leukemia, *Blood*, 83, 3473–3479, 1994.
124. Nemunaitis, J., Singer, J. W., Buckner, C. D., Hill, R., Storb, R., Thomas, E. D. et al., Use of recombinant human granulocyte-macrophage colony-stimulating factor in autologous marrow transplantation for lymphoid malignancies, *Blood*, 72, 834–836, 1988.
125. Pai, L. H., Batra, J. K., FitzGerald, D. J., Willingham, M. C., and Pastan, I., Antitumor effects of B3-PE and B3-LysPE40 in a nude mouse model of human breast cancer and the evaluation of B3-PE toxicity in monkeys, *Canc. Res.*, 52, 3189–3193, 1992.
126. Pai, L. H., Bookman, M. A., Ozols, R. F., Young, R. C., Smith, II, J. W., Longo, D. L. et al., Clinical evaluation of intraperitoneal *Pseudomonas* exotoxin immunoconjugate OVB3-PE in patients with ovarian cancer, *J. Clin. Oncol.*, 9, 2095–2103, 1991.
127. Pardoll, D. M., Paracrine cytokine adjuvants in cancer immunotherapy, *Ann. Rev. Immunol.*, 13, 399–415, 1995.
128. Peace, D. J. and Cheever, M. A., Toxicity and therapeutic efficacy of high dose interleukin 2. *In vivo* infusion of antibody to NK-1.1 attenuates toxicity without compromising efficacy against murine leukemia, *J. Exp. Med.*, 169, 161–173, 1989.
129. Pfreundschuh, M. G., Steinmetz, H. T., Tüschen, R., Schenk, V., Diehl, V., and Schaadt, M., Phase I study of intratumoral application of recombinant human tumor necrosis factor, *Eur. J. Cancer Clin. Oncol.*, 25, 379–388, 1989.
130. Poynton, C., Mort, D., and Maughan, T. S., Adverse reactions to Campath-1 H monoclonal antibody, *Lancet*, 341, 1037, 1993.
131. Press, O. W., Eary, J. F., Appelbaum, F. R., Martin, P. J., Badger, C. C., Nelp, W. B. et al., Radiolabeled antibody therapy of B-cell lymphoma with autologous bone marrow support, *New Engl. J. Med.*, 329, 1219–1224, 1993.
132. Pujol, J., Gibney, D. J., Su, J. Q., Maksymiuk, A. W., and Jett, J. R., Immune response induced in small-cell lung cancer by maintenance therapy with interferon γ, *J. Nat. Canc. Inst.*, 85, 1844–1849, 1993.
133. Ramachandran, R. V., Katzenstein, D. A., Wood, R., Batts, D. H., and Merigan, T. C., Failure of short-term CD4-PE40 infusions to reduce virus load in humans immunodeficiency virus-infected persons, *J. Infect. Dis.*, 170, 1009–1013, 1994.

134. Ratcliffe, M. A., Roditi, G., and Adamson, D. J. A., Interleukin-2 and splenic enlargement, *J. Nat. Canc. Inst.*, 84, 810–811, 1992.
135. Räth, U., Kaufmann, M., Schmid, H., Hofmann, J., Wiedenmann, B., Kist, A. et al., Effect of intraperitoneal recombinant human tumour necrosis factor alpha on malignant ascites, *Eur. J. Canc.*, 27, 121–125, 1991.
136. Reff, M. E., Carner, K., Chambers, K. S., Chinn, P.C., Leonard, J. E., Raab, R., Newman, R. A., Hanna, N., and Anderson, D. R., Depletion of B cells *in vivo* by a chimeric mouse human monoclonal antibody to CD20, *Blood*, 83, 435–445, 1994.
137. Richards, J. M., Gilewski, T. A., and Vogelzang, N. J., Association of interleukin-2 therapy with staphylococcal bacteremia, *Cancer*, 67, 1570–1575, 1991.
138. Rivoltini, L., Gambacorti-Passerini, C., Squadrelli-Saraceno, M., Grosso, M. I., Cantù, G., Molinari, R. et al., *In vivo* interleukin 2-induced activation of lymphokine-activated killer cells and tumor cytotoxic T-cells in cervical lymph nodes of patients with head and neck tumors, *Canc. Res.*, 50, 5551–5557, 1990.
139. Rosen, B. S., Levine, E. A., Egrie, J. C., Sehgal, L. R., Greenberg, R., Rosen, A. L. et al., Effects of recombinant human erythropoietin and interleukin-3 on erythropoietic recovery from acute anemia, *Exp. Hematol.*, 21, 1487–1491, 1993.
140. Rosenstein, M., Ettinghausen, S. E., and Rosenberg, S. A., Extravasation of intravascular fluid mediated by the systemic administration of recombinant IL-2, *J. Immunol.*, 137, 1735–1742, 1986.
141. Rusthoven, J. J., Eisenhauer, E., Mazurka, J., Hirte, H., O'Connell, G., Muldal, A. et al., Phase I clinical trial of recombinant human interleukin-3 combined with carboplatin in the treatment of patients with recurrent ovarian carcinoma, *J. Nat. Canc. Inst.*, 85, 823–825, 1993.
142. Ryffel, B. and Weber, M., Preclinical safety studies with recombinant human interleukin 6 (rhIL-6) in the primate *Callithrix jacchus* (marmoset): comparison with studies in rats, *J. Appl. Toxicol.*, 15, 19–26, 1995.
143. Saleh, M. N., Khazaeli, M., Wheeler, R. H., Allen, L., Tilden, A. B., Grizzle, W. et al., Phase I trial of the chimeric anti-GD2 monoclonal antibody ch14.18 in patients with malignant melanoma, *Hum. Antibod. Hybridomas*, 3, 19–24, 1991.
144. Sampaio, E. P., Moreira, A. L., Sarno, E. N., Malta, A. M., and Kaplan, G., Prolonged treatment with recombinant interferon-γ induces erythema nodosum leprosum in lepromatous leprosy patients, *J. Exp. Med.*, 175, 1729–1737, 1992.
145. Sarmiento, U. M., Riley, J. H., Knaack, P. A., Lipman, J. M., Becker, J.M., Gately, M. K. et al., Biologic effects of recombinant human interleukin-12 in squirrel monkeys (*Sciureus saimiri*), *Lab. Invest.*, 71, 862–873, 1994.
146. Sausville, E. A., Headlee, D., Stetler-Stevenson, M., Jaffe, E. S., Solomon, D., Figg, W. D. et al., Continuous infusion of the anti-CD22 immunotoxin IgG-RFB4-SMPT-dgA in patients with B-cell lymphoma: a phase I study, *Blood*, 85, 3457–3465, 1995.
147. Scharenberg, J. G. M., Stam, A. G. M., von Blomberg, B. M. E., Evers, M. P. J., Roest, G. J., Palmer, P. A. et al., The development of anti-interleukin-2 (IL-2) antibodies in patients treated with recombinant IL-2 does not interfere with clinical responsiveness, *Proc. Am. Assoc. Canc. Res.*, 34, 464, 1993.
148. Scheid, C., Young, R., McDermott, R., Fitzsimmons, L., Scarffe, J. H., and Stern, P. L., Immune function of patients receiving recombinant human interleukin-6 (IL-6) in a phase I clinical study: induction of C-reactive protein and IgE and inhibition of natural killer and lymphokine-activated killer cell activity, *Canc. Immunol. Immunother.*, 38, 119–126, 1994.
149. Schmidt-Wolf, G. D. and Schmidt-Wolf, I, G. H., Cytokines and gene therapy, *Immunol. Today*, 16, 173–175, 1995.
150. Schnek, D., Butler, F., Dugan, W., Littrel, D., Dorrbecker, S., Peterson, B. et al. Phase I study with a murine monoclonal antibody vinca conjugate (KS1/4-DAVLB) in patients with adenocarcinoma, *Antib. Immunoconjugates Radiopharm.*, 2, 93–100, 1989.
151. Schomburg, A., Kirchner, H., Lopez-Hänninen, E., Menzel, T., Rudolph, P., Körfer, A. et al., Hepatic and serologic toxicity of systemic interleukin-2 and/or interferon-α, *Am. J. Clin. Oncol.*, 17, 199–209, 1994.
152. Schroff, W., Foon, K. A., Beatty, S. M., Oldham, R. K., and Morgan, Jr., A. C., Human anti-murine immunoglobulin responses in patients receiving monoclonal antibody therapy, *Canc. Res.*, 45, 879–885, 1985.
153. Sechler, J. M. G., Malech, H. L., White, C. J., and Gallin, J. I., Recombinant human interferon-γ reconstitutes defective phagocyte function in patients with chronic granulomatous disease of childhood, *Proc. Natl. Acad. Sci. USA*, 85, 4874–4878, 1988.
154. Seipelt, G., Ganser, A., Duranceyk, H., Maurer, A., Ottmann, O. G., and Hoelzer, D., Induction of soluble IL-2 receptor in patients with myelodysplastic syndromes undergoing high-dose interleukin-3 treatment, *Ann. Hematol.*, 68, 167–170, 1994.
155. Siegel, J. P. and Puri, R. K., Interleukin-2 toxicity, *J. Clin. Oncol.*, 9, 694–704, 1991.
156. Skillings, J., Wierzbicki, R., Eisenhauer, E., Venner, P., Letendre, F., Stewart, D. et al., A phase II study of recombinant tumor necrosis factor in renal cell carcinoma: a study of the National Cancer Institute of Canada clinical trials group, *J. Immunother.*, 11, 67–70, 1992.
157. Smith, J. W., Urba, W. J., Curti, B. D., Elwood, L. J., Steis, R. G., Janik, J. E. et al., The toxic and hematologic effects of interleukin-1 alpha administered in a phase I trial to patients with advanced malignancies, *J. Clin. Oncol.*, 10, 1141, 1992.

158. Soike, K. F., Czarniecki, C. W., Baskin, G., Blanchard, J., and Liggitt, D., Enhancement of simian *Varicella* virus infection in African green monkeys by recombinant human tumor necrosis factor alpha, *J. Infect. Dis.*, 159, 331–335, 1989.
159. Soler-Rodriguez, A.-M., Ghetie, M.-A., Oppenheimer-Marks, N., Uhr, J. W., and Vitetta, E. S., Ricin A-chain and ricin A-chain immunotoxins rapidly damage human endothelial cells: implications for vascular leak syndrome 206, 227–234, 1993.
160. Sondel, P. M., Kohler, P. C., Hank, J. A., Moore, K. H., Rosenthal, N. S., Sosman, J. A. et al., Clinical and immunological effects of recombinant interleukin 2 given by repetitive weekly cycles to patients with cancer, *Canc. Res.*, 48, 2561–2567, 1988.
161. Spitler, L. E., del Rio, M., Khentigan, A., Wedel, N. I., Brophy, N. A., Miller, L. L. et al., Therapy of patients with malignant melanoma using a monoclonal anti-melanoma antibody-ricin A chain immunotoxin, *Canc. Res.*, 47, 1717–1723, 1987.
162. Stahel, R. A., Lacroix, H., Sculier, J. P., Morant, R., Richner, J., Janzek, E. et al., Phase I/II study of monoclonal antibody against Lewis Y hapten in relapsed small-cell lung cancer, *Ann. Oncol.*, 3, 319–320, 1992.
163. Sun, W. H., Binkley, N., Bidwell, D. W., and Ershler, W. B., The influence of recombinant human interleukin-6 on blood and immune parameters in middle-aged and old rhesus monkeys, *Lymphokine Cytokine Res.*, 12, 449–455, 1993.
164. Talmadge, J. E. and Dean, J. H., Immunopharmacology of recombinant cytokines, in *Immunotoxicology and Immunopharmacology*, 2nd ed., Dean, J. H., Luster, M. I., Munson, A. E., and Kimber, I., Eds., Raven Press, New York, 1994, pp. 227–247.
165. Teppler, H., Kaplan, G., Smith, K., Cameron, P., Montana, A., Meyn, P. et al., Efficacy of low doses of the polyethylene glycol derivative of interleukin-2 in modulating the immune response of patients with human immunodeficiency virus type 1 infection, *J. Infect. Dis.*, 167, 291–298, 1993.
166. Terrell, T. G. and Green J. D., Comparative pathology of recombinant murine interferon-gamma in mice and recombinant human interferon-gamma in cynomolgus monkeys, *Int. Rev. Exp. Pathol.*, 34(Pt. B), 73–101, 1993.
167. Thompson, J. A., Lee, D. J., Lindgren, C. G., Benz, L. A., Collins, C., Levitt, D., and Fefer, A., Influence of dose and duration of infusion of interleukin-2 on toxicity and immunomodulation, *J. Clin. Oncol.*, 6, 669–678, 1988.
168. Tilg, H., Trehu, E., Shapiro, L., Pape, D., Atkins, M. B., Dinarello, C. A. et al., Induction of circulating soluble tumour necrosis factor receptor and interleukin 1 receptor antagonist following interleukin 1α infusion in humans, *Cytokine*, 6, 215–219, 1994.
169. Tinubu, S. S., Hakimi, J., Kondas, J. A., Bailon, P., Famillefti, P. C., Spence, C. et al., Humanized antibody directed to the IL-2 receptor β-chain prolongs primate cardiac allograft survival, *J. Immunol.*, 153, 4330–4338, 1994.
170. Tritarelli, E., Rocca, E., Testa, U., Boccoli, G., Camagna, A., Calabresi, F. et al., Adoptive immunotherapy with high-dose interleukin-2: kinetics of circulating progenitors correlate with interleukin-6, granulocyte colony-stimulating factor level, *Blood*, 77, 741–749, 1991.
171. Vadhan-Raj, S., PIXY321 (GM-CSF/IL-3 fusion protein): biology and early clinical development, *Stem Cells*, 12, 253–261, 1994.
172. Vadhan-Raj, S., Buescher, S., Broxmeyer, H. E., LeMaistre, A., Lepe-Zuniga, J. L., Ventura, G. et al., Stimulation of myelopoiesis in patients with aplastic anemia by recombinant human granulocyte-macrophage colony-stimulating factor, *New Engl. J. Med.*, 319, 1628–1634, 1988.
173. van der Lubbe, P. A., Reiter, C., Breedveld, F. C., Kroger, K., Schattenkirchner, M., Sanders, M. E., and Riethmüller, G., Chimeric CD4 monoclonal antibody cM-T412 as a therapeutic approach to rheumatoid arthritis, *Arthr. Rheum.*, 36, 1375–1379, 1993.
174. van Gameren, M. M., Willemse, P. H. B., Mulder, N. H., Limburg, P. C., Groen, H. J. M., Vellenga, E. et al., Effects of recombinant human interleukin-6 in cancer patients: a phase I-II study, *Blood*, 84, 1434–1441, 1994.
175. van Gils, F. C. J. M., Westerman, Y., Visser, T. P., Burger, H., van Leen, R. W., and Wagemaker, G., Neutralizing antibodies during treatment of homologous nonglycosylated IL-3 in rhesus monkeys, *Leukemia*, 8, 648–651, 1994.
176. Vasalli, P., The pathophysiology of tumor necrosis factors, *Annu. Rev. Immunol.*, 10, 411–452, 1992.
177. Vieweg, J. and Gilboa, E., Considerations for the use of cytokine-secreting tumor cells preparations for cancer treatment, *Canc. Invest.*, 13, 193–210, 1995.
178. Vitetta, E. S. and Thorpe, P. E., Immunotoxins containing ricin or its A-chain, *Sem. Cell Biol.*, 2, 47–58, 1991.
179. Vitteta, E. S., Stone, M., Amlot, P., Fay, J., May, R., Till, M. et al., Phase I immunotoxin trial in patients with B-cell lymphoma, *Canc. Res.*, 51, 4052–4058, 1991.
180. Weber, J., Yang, J. C., Topalian, S. L., Parkinson, D. R., Schwartzentruber, D. S., Ettinghausen, S. D. et al., Phase I trial of subcutaneous interleukin-6 in patients with advanced malignancies, *Medline Express*, 11, 499–506, 1993.
181. Weijl, N. I., Van Der Harst, D., Brand, A., Kooy, Y., Van Luxemburg, S., Schroder, J. et al., Hypothyroidism during immunotherapy with interleukin-2 is associated with antithyroid antibodies and response to treatment, *J. Clin. Oncol.*, 11, 1376–1383, 1993.

182. Weisdorf, D., Katsanis, E., Verfaillie, C., Ramsay, N. K. C., Haake, R., Garrison, L. et al., Interleukin-1α administered after autologous transplantation: a phase I/II clinical trial, *Blood*, 84, 2044–2049, 1994.
183. Whittington, R. and Faulds, D., Interleukin-2: a review of its pharmacological properties and therapeutic use in patients with cancer, *Drugs*, 46, 446–514, 1993.
184. Winton, E. F., Srinivasiah, J., Kim, B. K., Hillyer, C. D., Strobert, E. A., Orkin, J. L. et al., Effect of recombinant human interleukin-6 (rhIL-6) and rhIL-3 on hematopoietic regeneration as demonstrated in a nonhuman primate chemotherapy model, *Blood*, 84, 65–73, 1994.
185. Yasumoto, L., Miyazaki, K., Nagashima, A., Induction of LAK cells by intrapleural instillations of recombinant IL-2 in patients with malignant pleurisy due to lung cancer, *Canc. Res.*, 47, 2184, 1987.
186. Ziegler, E. J., Fisher, C. J., Sprung, C. I., Straube, R. C., Sadoff, J. C., Foulke, G. E. et al., Treatment of gram-negative bacteremia and septic shock with HA-1A human monoclonal antibody against endotoxin, *New Engl. J. Med.*, 324, 429–436, 1991.
187. Asnis, L. A. and Gaspari, A., Cutaneous reactions to recombinant cytokine therapy, *J. Am. Acad. Dermatol.*, 33, 393–410, 1995.
188. Talmadge, J. E., Black, P. L., Tribble, H., Pennington, R., Bowersox, O., Schneider, M., and Phillips, H., Preclinical approaches to the treatment of metastatic disease: therapeutic properties of rHTNF, rM IFN-gamma, and rH IL-2, *Drugs Exper. Clin. Res.*, 13, 327–337, 1987.
189. Kirchner, H., Körfer, A., Evers, P., Szamel, M. M., Knüver-Hopf, J., Mohr, H., Franks, C. R., Pohl, U., Resch, K., Hadam, M., Poliwoda, H., and Atzpodien, J., The development of neutralizing antibodies in a patient receiving subcutaneous recombinant and natural interleukin-2, *Cancer*, 67, 1862–1864, 1990.

CHAPTER 18

Effects of Drugs on Immune System Parameters

Wen-Jun Wu, Edmond J. Carson, Stephanie D. Collier, Deborah Keil, Paul A. Weiss, and Stephen B. Pruett

CONTENTS

1 INTRODUCTION

The immunotoxic effects of therapeutic drugs and drugs of abuse have been known or suspected for many years. Immunosuppression, hypersensitivity, or autoimmunity may occur in at least some individuals in response to some drugs. Comparison of the immunotoxic effects of drugs in humans and in experimental animals offers a unique opportunity that is not available for most other classes of immunotoxicants to validate animal models. The dosage and duration of exposure of persons to drugs can usually be ascertained more easily and more reliably than exposure to other classes of immunotoxicants. Thus, it should be possible (with appropriate scaling to account for differences in body surface area and/or metabolic rate) to compare the effects of particular dosages of drugs on particular immunological parameters in humans and in experimental animals. There are several examples in this section of drugs that produce similar immunotoxic effects in humans and in experimental animals, and quantitative (dose-response) comparisons would be of interest in these cases; however, it should be emphasized that numerous studies in the immunotoxicology literature demonstrate that the immunotoxic effects of a particular agent are often remarkably species- or even strain-specific. Thus, reports of effects in only one species or case reports of effects in just a few persons may not represent typical effects for other species or for most persons.

0-8493-4493-X/97/$0.00+$.50

Hypersensitivity (allergy) is probably the most commonly reported immunotoxic effect of drugs. Induction of true, persistent autoimmunity apparently occurs rarely, if ever, but induction of autoimmune-like syndromes that persist during exposure to the inducing drug has been reported for a number of drugs (for a more complete listing of drugs for which such effects have been reported, the reader is referred to *Toxicology of the Immune System,* by R. Burrell, D. K. Flaherty, and L. J. Sauers, Van Nostrand-Reinhold, New York, 1992, pp. 79–136).

Immunosuppression is the desired function of some drugs given to prevent allograft rejection or to treat allergies. These drugs are covered extensively in numerous texts and review articles; therefore, there is no attempt in this section to cite immunosuppressive effects of drugs that are given purposely to suppress undesired immunological responses. However, some of these immunosuppressive drugs paradoxically cause immunological hypersensitivity responses (allergic responses) in some persons and some experimental animals. Such effects are noted in this section. Immunosuppressive effects are cited for drugs not designed or commonly used as immunosuppressants. The citations focus on relatively recent work, but it should be noted that there is a large literature over a long period of time regarding the immunosuppressive effects of cancer chemotherapeutic agents, and this work is not extensively cited here.

Because of the strong associations between alcohol and other drugs of abuse and increased incidence of infection, there is an unusually rich literature on the effects of drugs of abuse on the immune system. In this area, in particular, the references cited here are not exhaustive, but simply provide examples of most of the major effects that have been reported for a particular drug.

Immunotoxicity is an important side effect of many drugs, and numerous examples are cited here indicating effects in humans or animal models or both. Many of these effects are familiar to clinicians and researchers, but a few may be unexpected. For example, a number of antibiotics and antiviral agents suppress various immune functions, possibly indicating a reason in addition to the generation of resistant strains of microorganisms for restraint in the clinical use of antibiotics.

ACKNOWLEDGMENTS

The authors wish to thank Amber Oakes and Jason Mussellwhite for their invaluable contributions to this work.

TABLE 1

Effects of Drugs on Immune System Parameters[a]

Compound Dose	Route (Regimen)	Strain (Sex)	Immunological Parameter	Effect[b]	Ref.
Anesthetic Drugs					
Halothane	Single exposure to anesthesia	Balb/c[c]	Resistance to tumors	$\downarrow$	145
			Ab response	$\downarrow$	
Ketamine					
8 mg/kg	i.m. (24 hr)	Human	Cellular immune activity	$\leftrightarrow$	200
Antibiotics, Antifungal, and Antiviral Drugs					
Amikacin					
5, 10, 15 mg/kg	s.c. (2×/d, 5 or 7 d)	Balb/c (M)[c]	DTH	$\downarrow$	150
			IgM and IgG response to SRBC	$\downarrow$	
20 µg/ml	*In vitro*	Human	Anti-SRBC Ab response	$\leftrightarrow$	187
Cefaclor					
50 mg/kg	Oral (1 dose)	Balb/c (M)[c]	IgM response	$\uparrow$	55
			IgG response	$\leftrightarrow$	
Cefotaxime					
20, 40, 60 mg/kg	s.c. (2×/d, 5 or 7 d)	Balb/c (M)[c]	DTH	$\downarrow$	150
			IgM, IgG response	$\downarrow$	
4 g	i.v. (8 hr)	Human	MLR		37
Cephalexin					
0.5 ml (10 or 20 mg/ml)	i.d. (6 hr)	Hartley (M)[d]	Delayed-type hypersensitivity reaction		124
Chloramphenicol					
4 g	i.v. (8 hr)	Human	MLR	$\downarrow$	37
~1.5 mg/mouse	(i.p., 1×/wk, for 3 wk)	CBA/J (40 d old)[c]	2° response to SRBC or tetanus toxoid	$\downarrow$	190
			T-cell response to PPD or Coxsackie virus	$\downarrow$	
Clindamycin					
10, 20, 35 mg/kg	s.c. (2×/d, 5 or 7 d)	Balb/c (M)[c]	Proliferation to ConA and LPS	$\leftrightarrow$	150
			DTH	$\leftrightarrow$	
			IgM and IgG response	$\leftrightarrow$	
90 mg/kg, 3 doses/d	i.m. (5 d)	New Zealand[e]	PMN phagocytosis	$\downarrow$	163
			PMN peritoneal adhesiveness	$\downarrow$	

TABLE 1 (continued)

Effects of Drugs on Immune System Parameters[a]

Compound Dose	Route (Regimen)	Strain (Sex)	Immunological Parameter	Effect[b]	Ref.
Colistin					
~2 mg/mouse	(i.p., 1×/wk, 3 wk)	CBA/J (40 d old)[c]	2° response to SRBC or tetanus toxoid	↓	190
			T-cell response to PPD or Coxsackie virus	↓	
Cytosine arabinoside					
1 μg/ml	*In vitro*	HMC-1 cells	Viability and ^{3}H-thymidine uptake	↓	22
0.1 μg/ml	*In vitro*	HMC-1 cells	Viability and ^{3}H-thymidine uptake	↓	
0.01 μg/ml	*In vitro*	HMC-1 cells	Viability and ^{3}H-thymidine uptake	↓	
20–40 mg/kg/d	s.c. (4 d)	C3H/HeJ (M)[c]	Rejection of allograft	↓	63
20–40 mg/kg/d	s.c. (1×, d 1–3, 2×, d 4)	C3H/HeJ (M)[c]	IgM anti-SRBC response	↓	
20–40 mg/kg/d	s.c. (1×, d 1–3, 2×, d 4)	C3H/HeJ (M)[c]	IgG anti-SRBC response	↓	
Demecocycline					
0.20–0.25 μ*M*	*In vitro* (25 m)	RBL-2H3 cells	IgE and A23187 mediated calcium influx	↔	184
0.20–0.25 μ*M*	*In vitro* (30 m)	RBL-2H3 cells	Arachidonic acid and histamine release	↓	
2′,3′-Dideoxyadenosine					
87.5–350 mg/kg	Gavage (daily, 22 d)	B6C3F1 (F)[c]	IgM response to SRBC	↓	24
2′,3′-Dideoxycytidine	Gavage	B6C3F1 (F)[c]	IgM response to SRBC	↓	109
3′-Azido-3′-deoxythymidine (AZT)	Gavage	B6C3F1[c]	Humoral, cell-mediated, and innate parameters	↔	109
Erythromycin					
50 mg/kg	Oral (a single dose)	Balb/c (M)[c]	IgM response	↔	55
~2 mg/mouse	i.p. (1×/wk, 3 wk)	CBA/J (40 d old)[c]	2° response to SRBC or tetanus toxoid	↓	190
			T-cell response to PPD or Coxsackie virus	↓	
Ethambutol					
3 or 12 μg/ml	*In vitro*	Human	PGE_2 production by PMN	↔	203
Gentamicin					
260 g	i.v. (8 hr)	Human	MLR	↓	37
4 μg/ml	*In vitro*	Human	Ab response to SRBC	↓	187
Gramicidin					
0.2–10.1 ng/ml		Human	Lymphocyte blastogenesis	↔	71
Isoniazid					
1.25 and 5 μg/ml	*In vitro*	Human	PGE_2 production by PMN	↔	203
Metronidazole					
1 g	i.v. (8-hr exposure)	Human	MLR	↓	37
500 mg	Oral (1 hr)	Human (F)	Chills, suprapubic pain, fever erythema, maculopapular rash		95

Mezlocillin					
85, 225, 285 mg/kg	s.c. (2×/d, 5 or 7 d)	Balb/c (M)[c]	Prolifieration to ConA and LPS	↓	150
			DTH	↓	
			IgM and IgE response	↓	
Netilimicin					
8 mg/kg, 3 doses/d	i.m. (5 d)	New Zealand[e]	Blood PMN adhesiveness	↓	
			PMN superoxide production	↓	
Penicillin					
33 IU/ml	*In vitro*	Human	Anti-SRBC Ab response	↔	187
~5 mg/mouse	i.p. (1×/wk, 3 wk)	CBA/J (40 days old)[c]	2° response to SRBC or tetanus toxoid	↔	190
			T-cell response to PPD or Coxsackie virus	↔	
Piperacillin					
65, 130, and 265 mg/kg	s.c. (2×/d, 7 or 5 d)	Balb/c (M)[c]	Proliferation to ConA and LPS	↓	150
			DTH	↓	
			IgM and IgG response	↓	
Pyrazinamde					
25 or 100 μg/ml	*In vitro*	Human	PGE_2 production by PMN	↔	203
Quinacrine					
5–500 μM	*In vitro* (2 hr)	Human	Free radical release by MØ	↓	168
Ribavirin					
25 or 250 mg/kg	i.v. (25 mg/kg/d or 250 mg/kg 2×/wk, 2.5 wk)	NZB/W (F)[c]	Number of IgM-producing cells	↔	93
			Number of IgG-producing cells	↓	
Rifampicin					
25 or 100 μg/ml	*In vitro*	Human	PGE_2 production by PMN	↑	203
	In vitro	Human	PHA-induced blastogenesis	↓	
Rifampin					
7 μg/ml	*In vitro*	Human	Anti-SRBC Ab response	↔	187
Rifamycin					
2 g	i.v. (8 hr)	Human	MLR	↓	37
Streptomycin					
33 μg/ml	*In vitro*	Human	Ab response to SRBC	↓	187
~0.3 mg/mouse	i.p. (1×/wk, 3 wk)	CBA/J (40 d old)[c]	2° response to SRBC or tetanus toxoid	↔	190
			T-cell response to PPD or Coxsackie virus	↔	
1.25 or 5 μg/ml	*In vitro*	Human	PGE_2 production by PMN	↔	203
1.25 or 5 μg/ml	*In vitro*	Human	PHA response	↔	

TABLE 1 (continued)

Effects of Drugs on Immune System Parameters[a]

Compound Dose	Route (Regimen)	Strain (Sex)	Immunological Parameter	Effect[b]	Ref.
TCI (1-thiocarbanmoyl-2-imidazolidinone) (metabolite of Niridazole)					
10^{-9} g/kg	Oral (2 d)	C57BL/6J (M)[c]	DTH reaction to DNFB at 5 d	↓	53
		AKR (M)[c]	DTH reaction to DNFB at 5 d	↑	
10^{-9} g/kg	Oral (2 d)	C57BL/6J (M)[c]	DTH to DNFB at 5 d	↔	53
		AKR (M)[c]	DTH to DNFB at 10 d	↑	
10^{-12}, 10^{-12}–10^{-3}, and 10^{-9}–10^{-1} g/l	*In vitro* (3 d)	C57BL/6J (M)[c] or	ConA-stimulated blastogenesis	↓	53
		AKR (M)[c]	ConA-stimulated blastogenesis	↓	
10^{-13}–10^{-9} g/kg	Oral (24 hr)	C57BL/6J (M)[c] or	DTH to DNFB	↓	53
		AKR (M)[c]	DTH to DNFB	↓	
Tetracycline					
50 mg/kg	Oral (a single dose)	Balb/c (M)[c]	IgM response	↔	55
~0.1 mg/mouse	i.p. (1×/wk, 3 wk)	CBA/J (40 d old)[c]	2° response to SRBC or tetanus toxoid	↔	190
			T-cell response to PPD or Coxsackie virus	↔	
~2 mg/mouse	i.p. (1×/wk, 3 wk)	CBA/J (40 d old)[c]	2° response to SRBC or tetanus toxoid	↔	
			T-cell response to PPD or Coxsackie virus	↔	
Anti-Cancer Drugs					
Adriamycin					
0.4 μg/ml	*In vitro* (84 hr)	Human (M and F)	PHA-induced INF-γ production by PBMC	↓	27
Bactobolin					
0.005–5.0 mg/kg	i.p. (at time of immunization)	CDF_1 (F)[c]	DTH and Ab production in spleen	↔	79
0.08 mg/kg/d	i.p. (10 d)	CDF_1 (F)[c]	Survival of L-1210 tumor inoculation	↑	
0.025–2.5 mg/kg/d	i.p. (4 d)	CDF_1 (F)[c]	DTH and Ab production to SRBC	↔	
5 mg/kg/d	i.p. (1, 2, or 3 d)	CDF_1 (F)[c]	Ab production to SRBC	↓	
1,3 bis(2-chloroethyl)-1-nitrosourea (BCNU)					
20 mg/kg	i.p.	C57BL/6 (F)[c]	Survival after tumor challenge	↑	125
20 mg/kg	i.p. (2 d)	C57BL/6 (F)[c]	T-cell response to ConA or PHA	↓	
20 mg/kg	i.p. (10 d)	C57BL/6 (F)[c]	$CD4^+$ cells	↔	
Bleomycin					
0.1–200 μg/ml	*In vitro* (3 d)	Human	PHA-induced lymphocyte blastogenesis	↓	180

Busulfan					
0.5 mg every 2 wk	i.p. (2 months)	CAF_1 (M)[c]	Mitogenic response to PHA, ConA, or PWM	↓	15
			MLR	↔	
			NK cell activity	↓	
			IgM response to SRBC	↔	
10 mg/kg	Oral (15 d)	ACI (M)[f]	Spleen cell blastogenisis to BCG cell wall	↔	40
Carboplatin					
0.01 mg	i.c.	Human (M and F)	Wheals and erythemas		196
CDDP (cis-diamine-dichloro-platinum II)					
5 mg/kg	i.p. (4–7 d)	BDF_1[c]	Ab production to SRBC	↑	4
			DTH to oxazolone	↔	
2 mg/kg	i.p. (every other d, 4 doses)	Mouse (M and F)	PFC to SIII	↑	8
			PFC to SRBC	↑	
			PFC to TNP-KLH	↔	
			PFC to TNP-KLH	↑	
			Thymus, spleen, and lymph node weight	↔	
12 mg/kg	i.p.	Mouse (M and F)	PFC to SIII	↓	8
			PFC to SRBC	↓	
			PFC to TNP-KLH	↓	
			PFC to TNP-KLH (after pretreatment with SIII)	↔	
			Spleen weight	↓	
2–2.25 mg/kg	i.v.	Human (M)	Allergic response		90
Chlorambucil	Oral	Human	Allergic response		179
Chlorozotocin					
20 mg/kg	i.p.	C57BL/6 (F)[c]	Survival after tumor challenge	↑	125
20 mg/kg	i.p. (2 d)	C57BL/6 (F)[c]	T-cell response to ConA or PHA	↓	125
20 mg/kg	i.p. (5 d)	C57BL/6 (F)[c]	DTH to LSA or EL-4 tumors	↑	125
20 mg/kg	i.p. (10 d)	C57BL/6 (F)[c]	$CD4^+$ cells	↔	125
Colchicine					
1.5–2.0 μ*M*	(25 m)	RBL-2H3 cells	IgE- and A23187-mediated calcium influx	↔	184
1.5–2.0 μ*M*	(30 m)	RBL-2H3 cells	Arachidonic acid and histamine release	↓	184
Cyclophosphamide					
300 mg/m^2	i.v. bolus	Human (F and M)	DTH using KLH	↑	13
			KLH Ab response	↑	
100 mg/kg	i.p. (3-d exposure)	Lewis[f]	IgE response to tree allergen P	↑	7
150 mg/kg	i.p. (4-d exposure)	C57BL/6Cr[c]	DTH response	↑	62
40 mg/kg	i.v. (15 d)	ACI (M)[f]	Spleen cell blastogenesis response to BCG cell wall	↔	40
Cytochalasin					
0.35–0.50 μ*M*	*In vitro* (25 min)	RBL-2H3 cells	IgE- and A23187-mediated calcium influx	↑	184
0.35–0.50 μ*M*	*In vitro* (30 min)	RBL-2H3 cells	Arachidonic acid and histamine release	↑	

TABLE 1 (continued)

Effects of Drugs on Immune System Parameters[a]

Compound Dose	Route (Regimen)	Strain (Sex)	Immunological Parameter	Effect[b]	Ref.
Deoxycoformycin					
0.3 mg/kg/d	Catheterization through the tail into the abdominal cavity (5 d)	C57BL/6J (M)[c]	Mitogen-induced lymphocyte proliferation	↓	166
			Spleen, lymph node, and thymus hypoplasia	↑	
			Leukocyte number	↓	
			Differentiated lymphocyte number	↓	
4–40 mg/m^2	i.v. infusion	Human (M and F)	Hypersensitivity	↑	131
4–40 mg/m^2	i.v. infusion	Human (M and F)	Eosinophils	↑	
2–4 mg/m^2	i.v. infusion	Human	T-cell, B-cell, and CD4 lymphocyte number	↓	96
2–4 mg/m^2	i.v. infusion	Human	NK cell activity	↔	
5-Deoxy-5-fluoridine					
40–400 mg/kg	i.p. (dosed on d 1, 2, 3, and assay on d 7)	C57BL/6J (F)[c]	Pyron-induced inflamation	↔	31
40–400 mg/kg	i.p. (dosed on d 1, 2, 3, and assay on d 9)	C57BL/6J (F)[c]	Pyron-mediated anti-tumor activity	↔	31
40–400 mg/kg	i.p. (dosed on d 1, 2, 3, and assay on d 7)	C57BL/6J (F)[c]	Tumorcidal activity induced by pyron and fluoropyrimidine	↑	31
40–400 mg/kg	i.p. (dosed on d 1, 2, 3, and assay on d 9)	C57BL/6J (F)[c]	Leukopenia and monocyclopenia in pyron-treated mice	↔	31
Doxorubicin hydrochloride					
6 μg/ml	*In vitro*	HMC-1 cells	Viability and ^{3}H-thymidine uptake	↓	
0.6 μg/ml	*In vitro*	HMC-1 cells	Viability and ^{3}H-thymidine uptake	↓	
0.06 μg/ml	*In vitro*	HMC-1 cells	Viability and ^{3}H-thymidine uptake	↓	
Etoposide					
100 mg/m^2	i.v. (2 m)	Human (M)	Acute dyspnea, chest tightness, cutaneous flushing, moderate hypertension		133
100 mg/m^2/d	i.v. (3 d)	Human (M)	Dyspnea, facial flushing, weakness, fainting		88
5-Fluorouracil					
5–37.5 mg/kg	i.p. (dosed on d 1, 2, 3, and assay on d 7)	C57BL/6J (F)[c]	Pyron-induced inflammation	↓	31
5–37.5 mg/kg	i.p. (dosed on d 1, 2, 3, and assay on d 9)	C57BL/6J (F)[c]	Pyron-mediated anti-tumor activity	↔	31
25 and 37.5 mg/kg	i.p. (dosed on d 1, 2, 3, and assay on d 7)	C57BL/6J (F)[c]	Tumorcidal activity treated with pyron and fluoropyrimidine	↑	31
			Leukopenia and monocyclopenia in pyron-treated mice	↔	

Drug/dose	Route/duration	Species/strain	Parameter	Effect	Ref.
15–60 mg/kg	s.c. daily, 6 d	C57BL/6 and DBA/2 (M)[c]	Ab response to SRBC	↓	118
	s.c. daily, 6×/wk, 3–4 wk		TNP-Ficoll- or TNP-LPS-induced PFC	↑	
			Ab response to SRBC and MLC	↔	
15–60 mg/kg			P815 tumor cell-induced CTL	↓	118
40 μg/ml	*In vitro* (84 hr)	Human (M and F)	INF-γ production	↓	27
Megestrol acetate					
2.5 mg every other day	Oral (21–25 d)	Dunkin-Hartley (M and F)[d]	Skin or corneal DTH to ovalbumin	↔	18
			Ab response to ovalbumin	↔	
6-Mercaptopurine					
3.2 μg/ml	*In vitro* (84 hr)	Human (M and F)	INF-γ production	↓	27
Methotrexate					
16 μg/ml	*In vitro* (84 hr)	Human (M and F)	IFN-γ production	↓	27
0.5 mg/kg	i.p. (14 d)	CBA/CaJ[c]	Th cell-mediated B-cell response	↓	48
			Ts cell activity when cultured with mitoxanthrone	↔	
			Ts cell activity	↓	
			Th cell activity with macrophages present	↔	
			Ts cell activity with macrophages present	↓	
Mitomycin C					
2.75 mg/kg	i.p. or i.v.	Sprague-Dawley (M)[f]	C3b receptor binding	↓	17
			Lysosomal enzyme activity in Kupffer cells	↓	
			Phagocytosis by Kupffer cells	↓	
10 μg/ml and 1 μg/ml	Dermal	Human	Vesicular palmoplantar eczema		56
40 mg	(1 month)	Human (M and F)	Eosinophilic cystitis		54
1 mg/kg	i.v. (15 d)	ACI (M)[c]	Spleen cell blastogenic response to BCG cell wall	↔	40
Mitoxanthrone					
0.5 mg/kg daily	i.p. (14 d)	CBA/CaJ and C57BL/6[c]	PFC response to SRBC and ovalbamin	↓	47
			PFC response to TNP-Ficoll and TNP-LPS	↓	
			Splenic B-cell number	↓	
			LPS induced B-cell mitogenisis	↓	
Streptozotocin					
20–200 mg/kg	i.p.	C57BL/6 (F)[c]	Survival after tumor challenge	↓	125
20–200 mg/kg	i.p. (24 or 72 hr)	C57BL/6 (F)[c]	Tumoricidal activity	↔	
0.01–20 m*M*	*In vitro* (48 hr)	C57BL/6 (F)[c]	Tumoricidal activity	↔	
20, 100, and 200 mg/kg	i.p. (2 d)	C57BL/6 (F)[c]	T-cell response to ConA or PHA	↔	
200 mg/kg	i.p. (5 d)	C57BL/6 (F)[c]	DTH to LSA or EL-4 tumors	↓	
20, 100, or 200 mg/kg	i.p. (10 d)	C57BL/6 (F)[c]	$CD4^+$ cell number	↔	

TABLE 1 (continued)

Effects of Drugs on Immune System Parameters[a]

Compound Dose	Route (Regimen)	Strain (Sex)	Immunological Parameter	Effect[b]	Ref.
Tamoxifen					
500–1000 μg/d	s.c.	Fisher[f]	Ab responses	↓	126
6.5 μm	*In vitro* (30 months)	Sprague-Dawley (M)[f]	Histamine and serotonin secretion	↓	188
6.0 μm	*In vitro* (immediate)		Calcium release	↓	
4–6 μm	*In vitro* (30 months)		Serotonin secretion	↓	
500–1000 μg/d	s.c. (14 d)	Fisher (F)[f]	DCNB-induced contact dermatitis	↓	126
500–1000 μg/d	s.c. (10 d)	Fisher (F)[f]	Ab response to SRBC	↓	
Taxol					
5–250 mg/m^2	Infusion (1–6 hr)	Human	Erythema, pruritus, dyspnea, bronchospasm, hypotension		198
15–30 μ*M*	*In vitro* (25 m)	RBL-2H3	IgE- and A23187-mediated calcium influx	↓	184
15–30 μ*M*	*In vitro* (30 m)	RBL-2H3	Arachidonic acid and histamine release	↓	
Teniposide					
165 mg/m^2	i.v. (given on the fourth day of induction therapy)	Human	Degranulation of basophils	↑	130
Cancer therapy regimen		Human (M and F)	Hypersensitivity reactions (occurred in 6.5% of patients)		132
Vinblastine sulfate					
0.05 μg/ml	*In vitro*	HMC-1 cells	Viability and ^{3}H-thymidine uptake	↓	22
0.005 μg/ml	*In vitro*	HMC-1 cells	Viability and ^{3}H-thymidine uptake	↓	
0.0005 μg/ml	*In vitro*	HMC-1 cells	Viability and ^{3}H-thymidine uptake	↓	
0.01–0.05 μ*M*	*In vitro* (25 months)	RBC-2H3 cells	IgE- and A23187-mediated calcium influx	↔	184
0.01–0.05 μ*M*	*In vitro* (30 months)	RBC-2H3 cells	Arachidonic acid and histamine release	↓	
Vincristine					
0.08 μg/ml	*In vitro* (84 hr)	Human (M and F)	PHA-induced INF-γ production by PBMC	↓	27
Anti-Graft Rejection Drugs					
FK506					
0.4–0.04%	Topical	Pig	Inflammation and skin hypersensitivity	↓	115
Anti-Inflammatory/Analgesic Drugs					
Acetaminophen					
1000 mg	Oral (every 3 hr, 9 hr-exposure)	Human (M and F) (aspirin sensitive)	Allergic response		160

Amcinonide					
0.1%	Topical (48/96 hr)	Human (M and F)	Allergic patch test		149
0.01–1% ointment	Dermal	Human (M and F)	Contact allergy, erythema		97
Aspirin					
4 g	p.o. (5 d)	Human (M and F)	Lymphocyte proliferation	↔	43
			T and B cell %	↔	
			Allergic skin test	↔	
0.5 *M*	*In vitro* (60 months)	Human	Complement activity	↓	66
1 g	Oral (10 and 20 months)	Human (M and F)	Hemolytic complement activity	↓	191
30–650 mg	Oral (1 d)	Human (aspirin sensitive)	Allergic response		107
25 mg/kg/d	i.d.	Lewis (M)[f]	Inflammatory response	↓	142
50 and 150 mg/kg	Oral (5 d)	C57BL/6J (F)[c]	SRBC-induced DTH	↓	157
50 μg/ml	*In vitro* (15 months)	Human (M and F)	Plasma prostaglandin levels	↑	199
1 and 5 μ*M*	*In vitro* (15 months)	Human (aspirin sensitive)	Production of prostaglandin by leukocytes	↓	60
100 mg	Oral (3 and 6 hr)	Human (M and F) (aspirin sensitive)	Urinary leukotriene concentration	↑	30
30, 60, and 650 mg	Oral (30, 60, 90, 120 and 180 min)	Human (M and F)	Nasal secretion of leukotrienes and histamine	↑	46
325 mg	Oral (24 hr)	Human	Production of IL-2 and INF-γ by PBLs	↑	77
300 and 600 μ*M*	*In vitro* (72 hr)		Production of IL-2 and INF-γ by PBLs	↑	
6.20 and 25 mg/ml	Inhalation (3 and 6 hr)	Human (aspirin sensitive)	Uninary leukotriene concentration	↑	29
Auranofin					
0.9–2 μg/ml	*In vitro* (15 months)	Human	Degranulation of basophils	↓	170
10^{-5} and 10^{-4} *M*	*In vitro* (15 months)	Hartley[d]	Tracheal tissue response to histamine	↓	110
3 mg	Oral (2×/d, 20 wk)	Human	Leukocyte histamine release	↓	14
Aurothioglucose					
1.5–5 mg/kg	s.c. (daily, 8 d)	CFLP (M)[c]	DTH response	↔	103
			Hemagglutinating Ab response	↔	
2, 5, 20, 25, 50 μg/ml	*In vitro* (7 d)	Human (M)	Lymphocyte transformation	↑	192
Aurothiomalate					
1.5–5 mg/kg	s.c. (given daily for 8 d)	CFLP (M)[c]	DTH response	↔	103
			Hemagglutinating Ab response	↔	
Betamethasone valerate					
10% in petrolatum	Dermal	Human	Contact dermatitis and conjunctivitis	3	
Bucillamine					
10 and 30 μg/ml	*In vitro*	Human	T-cell response to mitogen	↓	67
Budesonide					
0.05, 0.005, and 0.0005% ointment	Topical (2 d)	Human (M and F)	Allergic patch test response		129
0.025% ointment	Topical (2 d)	Human	Contact dermatitis		185

TABLE 1 (continued)

Effects of Drugs on Immune System Parameters[a]

Compound Dose	Route (Regimen)	Strain (Sex)	Immunological Parameter	Effect[b]	Ref.
Cetirizine					
15 mg twice daily	Oral (8 d)	Human (M and F)	Recruitment of inflammatory cells (eosinophils)	↓	147
10^{-2}, 10^{-1}, and 1 μg/ml	*In vitro* (2 hr)	Human (M and F)	FMLP- and PAF-induced eosinophil chemotaxis	↓	41
	In vitro (2 hr)		IgE-dependent parasite cytotoxicity	↓	
10^{-2}, 10^{-1}, and 1 μg/ml	*In vitro* (2 hr)	Human (M and F)	FMLP- and PAF-induced eosinophil chemotaxis	↓	101
10^{-2}, 10^{-1}, and 1 μg/ml	*In vitro* (2 hr)	Human (M and F)	Eosinophil LDH	↔	
10^{-2}, 10^{-1}, and 1 μg/ml	*In vitro* (2 hr)	Human (M and F)	Eosinophil EPO	↔	
Chlobetasone butyrate					
20% in petrolatum	Topical	Human	Eczema		28
Chlorpheniramine					
24 mg/d and 48 mg/d	p.o. (2 or 22 d)	Human (M and F)	Prick skin test to histamine and morphine or antigen	↓	174
Clobetasol-17-propionate					
0.05%, 0.5% ointment	Topical	Human	Hypersensitivity reaction		148
Desonide					
0.1%	Topical (48/96 hr)	Human (M and F)	Allergic patch test reaction		149
Dexamethasone					
0.1%	Topical (48/96 hr)	Human (M)	Allergic patch test recation		149
Epinastine					
20 and 50 mg/kg	p.o. (3 d)	Mice and rats (M)	Types I and II allergic reactions	↓	86
Diclofenac					
2.5%	Topical	Human	Dermatitis		155
Fluocinolone acetonide					
0.025% ointment	Topical	Human	Allergic patch test reaction		148
Flurbiprofen					
1.5 mg/kg		Sprague-Dawley[f]	Delayed-type hypersensitivity	↔	139
Gold					
210–1320 mg	p.o.	Human (M and F)	Lymphocyte transformation	↑	35
			IgE level	↑	
Gold chloride					
(Trimethylphosphine) 1.5–5 mg/kg daily	s.c. (8 d)	CFLP (M)[c]	DTH response	↔	103
			Hemagglutinating Ab response	↔	
Gold salt					
80–4050 mg cumulative dose	(1–30 months)	Human (M and F)	Dermatitis, stomatitis, proteinuria		61

Gold sodium thiomalate					
50 mg/wk	i.m. (1×/wk, 3 months)	Hartley (F)[d]	Airway responsiveness to histamine	↓	202
	i.m.	Human (M andF)	Skin rashes	↑	36
			Eosinophilia	↑	
			IgE	↑	
Gold sodium thiosulfate					
0.5%	Topical	Human	Dermatitis		16
Halcinonide					
0.1% cream	Topical	Human	Allergic patch test reaction		148
Hydrocortisone					
10 and 25% cream	Topical	Human (M and F)	Allergic response (patch test)		2
10% cream	Topical	Human (M and F)	Allergic response (patch test)		
15% cream	Topical	Human (M and F)	Allergic response (patch test)		
1 or 10% cream	Topical	Human	Allergic response (patch test)		148
100 mg	i.v. (30 months)	Human (M and F)	Anaphylactoid reaction		34
10% cream	Topical	Human (M and F)	Allergic response (patch test)		2
15% cream	Topical	Human (M and F)	Allergic response (patch test)		
Ibuprofen					
400 mg, 3 times/d	p.o.	Human (F)	Allergic reaction		9
1, 5, and 10 mg/kg/d	i.p. (14 d)	BDF_1 (F)[c]	Sensitization to DNFB	↑	65
400 mg for adult	Oral (4×/wk, 4 wk)	Human (M and F)	T-cell subsets	↔	85
6 mg/kg for children			Production of interferon	↔	
20 mg/kg	i.v. (5 d)	C3H/HEN and C57BL/6 (M)[c]	DTH response	↑	162
Indomethacin					
1.5 mg/d	p.o. (22 d)	ACI (M)[f]	Tumor growth	↔	40
1 × 10^{-4} *M*	*In vitro* (20 hr)	ACI (M)[f]	Splenic blastogenesis	↔	
0.5 mg/kg	i.p. (14 d)	BDF_1 (F)[c]	Sensitization to DNFB	↑	65
3 mg/kg		Sprague-Dawley[f]	PMN migration	↓	139
			Delayed-type hypersensitivity	↔	
Ketoprofen					
2.5%	Topical (2 d)	Human (M and F)	Allergic response (patch and photopatch tests)		121
Methylprednisolone sodium succinate					
37.5 μg/ml	*In vitro*	HMC-1 cells	Viability and ^{3}H thymidine uptake	↓	22
3.75 μg/ml	*In vitro*	HMC-1 cells	Viability and ^{3}H thymidine uptake	↓	
0.375 μg/ml	*In vitro*	HMC-1 cells	Viability and ^{3}H thymidine uptake	↓	
Naproxen					
250 mg	Oral (6 hr)	Human (M)	Inflammatory interstitial infiltration of lymphocytes and eosinophils		123

TABLE 1 (continued)

Effects of Drugs on Immune System Parameters[a]

Compound Dose	Route (Regimen)	Strain (Sex)	Immunological Parameter	Effect[b]	Ref.
Penicillamine					
0.15 *M* aq.	Topical	Human (M)	Allergic patch test		135
20 mg/ml	i.v.	Human	Immune complexes in arthritis patients	↓	119
Phenacetin					
0.535%	Oral (6–15 wk)	Fischer (M)[f]	Ab response	↔	84
			PHA blastogenic response	↔	
Phenazone					
5%	Oral	Human	Erythema multiforme		98
Phenylbutazone					
50 mg/kg	Oral in rats and mice, s.c. in guinea pigs	Swiss (M)[c]	Pleural exudate	↓	171
		Sprague-Dawley (M)[f]	Leukocytes	↓	
		Dunkin Hartley (M)[d]	PMN	↓	
			Macrophages	↓	
			DTH to PPD	↓	
Prednisolone					
10% cream	Topical	Human (M and F)	Allergic response (patch test)		2
Pyrazinobutazone					
30 mg BID	Oral	Human	Agranulocytosis, liver injury		111
Sulfasalazine					
25, 100, 200 mg/kg/d	Oral (40 d)	Lewis[f]	Experimentally induced autoimmune encephalomyelitis	↑	32
Triamcinolone acetonide					
0.1%	Topical (48/96 hr)	Human (F)	Allergic response (patch test)		149
Cardiovascular Drugs					
Captopril					
1.2 mg/ml	In drinking water (36 wk)	MRL lpr/lpr (F)[c]	Survival of spontaneous autoimmune disease	↑	68
Clonidine					
0.1 mg/d	Transdermal (1 wk)	Guinea pig (F)	Contact hypersensitivity reaction	↓	59
Enalapril					
10 mg/12 hr	Oral (48 hr)	Humans (M and F)	Wheal and flare reactions to allergens	↔	165

Hydralazine					
~400 mg/d	Oral	Human	Systemic Lupus Erythematosis syndrome		140
Procainamide					
Typical therapeutic doses	Oral	Humans	Systemic Lupus Erythematosis syndrome		92
Reserpine					
5 mg/kg	i.p. (18 hr)	Balb/c (F)[c]	Contact sensitivity	↓	116
Verapamil hydrochloride					
8% w/v	Topical (4×/d, 6 d)	Human (M)	DTH to tuberculin	↓	112
Drugs of Abuse					
Cocaine					
See regimen for dosages	i.p. (5 mg/kg/d in wk 1, 20 mg/kg/d in wk 2, 35 mg/kg/d in wk 3, 40 mg/kg/d from wk 4 to wk 11)	C57BL/6 (F)[c]	Intestinal lamina propria IgA^+ cells	↓	105
			Intestinal lamina propria $CD8^+$ cells	↑	
15–60 mg/kg/d	p.o. (5 d)	ICR-SCH (M)[c]	DTH to DNFB	↓	194
			Ab response to SRBC	↓	
5–40 mg/kg/d	i.p. (up to 11 wk)	C57BL/6 (F)[c]	Body weight	↓	104
			$CD8^+$ splenocytes	↓	
			Splenocyte IFN-γ production	↓	
			TNF-α production	↑	
			sIL-2R	↓	
1 mg/kg/d or single dose of 1–5 mg/kg	i.p. (7 d)	Balb/c (F)[c]	NK cell activity	↓	42
			CTL activity	↓	
			CTL generation *in vivo*	↓	
			Resistance to infection	↓	
5–40 mg/kg/d	i.p. (11 wk)	C57BL/6[c]	Percent of $CD4^+$ cells in spleen	↓	195
			Percent of $CD8^+$ cells in spleen	↓	
			Percent of B cells in spleen	↑	
Cocaine and ethanol					
15 mg/kg cocaine and 5 g/kg ethanol	p.o. (5 d)	ICR-SCH (M)[c]	DTH response	↔	194
			Ab response	↓	
Ethanol					
5.5 g/kg	i.p. (1 dose)	Sprague-Dawley (M)[f]	TNF levels	↓	128
6–7 g/kg	Gavage (1 dose)	B6C3F1 (M and F)[c]	Humoral response to SRBC	↓	64
6–7 g/kg	Gavage (1 dose)	B6C3F1 (M and F)[c]	SRBC-induced IL-1, IL-2, and IL-4 mRNA (spleen)	↓	
5–7 g/kg	Gavage (1 dose)	B6C3F1 (F)[c]	Basal and induced NK cell activity	↓	201
3 g/kg	Gavage (1 dose, assay at 0.5–4 hr)	Sprague-Dawley (M)[f]	Neutrophil chemotaxis	↓	89

TABLE 1 (continued)

Effects of Drugs on Immune System Parameters[a]

Compound Dose	Route (Regimen)	Strain (Sex)	Immunological Parameter	Effect[b]	Ref.
Ethanol (continued)					
60 g/d	p.o.	Human (M)	DTH response to 7 antigens	↓	181
0.75 l of 50% ethanol/d	p.o.	Human	Antibody response to KLH	↓	58
0.75 l of 50% ethanol/d	p.o.	Human	DTH response to KLH	↓	
35.5% of calories	p.o. (6 wk)	Lewis	Anti-SRBC antibody response	↓	146
35.5% of calories	p.o. (6 wk)	Fischer 344[f]	Anti-SRBC antibody response	↔	
8–10 g/kg/d	Gavage (4×/d)	Lewis[f]	Spleen and lymph node weight	↓	137
8–10 g/kg/d	Gavage (4×/d)	Lewis[f]	Ratio of lymphocytes to neutrophils in blood	↓	
8–10 g/kg/d	Gavage (4×/d)	Lewis[f]	DTH response to *Borrelia burgdorferi*	↓	
36% of calories	p.o. (14 d)	B6C3F1 (F)[c]	Humoral response to SRBC	↔	73
36% of calories	p.o. (35 d)	B6C3F1 (F)[c]	Humoral response to SRBC	↓	
37% of calories	p.o. (8 d)	C57BL/6 (M)[c]	Spleen and thymus cellularity	↓	83
37% of calories	p.o. (8 d)	C57BL/6 (M)[c]	Humoral response to SRBC	↓	
37% of calories	p.o. (8 d)	C57BL/6 (M)[c]	Humoral response to T-independent antigen	↔	
37% of calories	p.o. (7–8 d)	C57BL/6 (M)[c]	Clearance of *Listeria monocytogenes*	↓	152
40% of calories	p.o. (in water)(1–2 wk)	C57BL/6 (F)[c]	Basal and IL-2-stimulated NK cell activity in spleen	↓	113
7% ethanol diet for 1 wk, followed by 3-wk abstinence and 1 wk of 5% ethanol diet	Oral	C57BL/6[c]	Survival after retrovirus challenge	↓	195
Fentanyl					
0.3 mg/kg	s.c. (3 hr)	Rat (M)	Splenic NK cell activity	↓	10
Morphine					
10 μmol/l	*In vitro*	Swiss (M and F)[c]	Histamine release from sensitized cells	↔	114
		Wistar (M and F)[f]	Histamine release from sensitized cells	↔	
30 mg/kg/d	s.c. (2×/d, 3 wk)	Balb/c (F)[c]	Mitogen-induced lymphocyte proliferation	↓	72
			NK cell function	↓	
			PFC response	↓	
30 mg/kg	s.c. (3- and 6-hr exposure)	F344 (M and F)[f]	Splenic NK cell activity	↓	161
50 mg/kg twice daily	(18 d)	Rat	DTH	↓	138
75 mg pellet/mouse	s.c. implant	C57BL/6J (M)[c]	Ab response to SRBC *in vitro*	↓	44
75 mg pellet/mouse	s.c. implant (12–24 hr exposure)	B6C3F1 (F)[c]	Ab response to SRBC *in vitro*	↓	143
75 mg pellet/mouse	s.c. implant (12–48 hr exposure)	B6C3F1 (F)[c]	Splenic NK cell activity	↓	50
75 mg pellet/mouse	s.c. implant (2–3 d)	C3H/HEN (M)[c]	Thymus and spleen weight	↓	21
			Mitogen responsiveness	↓	

75 mg pellet/mouse	s.c. implant	Mice	DTH response and GVH response	$\downarrow$	19
75 mg pellet/mouse	s.c. implant	Mice	Thymus and spleen weight	$\downarrow$	20
			Adrenal gland weight	$\uparrow$	
75 mg pellet/mouse	s.c. implant	C57Bl-6J[c]	ConA response of splenocytes at 24 and 48 hr	$\downarrow$	158
1×10^{-5} to 1×10^{-7} *M*	In vitro	Balb/c[c]	Ab response to SRBC *in vitro*	$\downarrow$	173
1×10^{-5}, 10^{-6}, 10^{-7} *M*	*In vitro* (48 hr)	Balb/c (M)[c]	Mitogenic response to SEB, PMA, PHA	$\downarrow$	45
75 mg pellet/mouse	s.c. implant (12–72 hr)	Rodents	Spleen and thymus weight	$\downarrow$	19
			Body weight	$\downarrow$	
75 mg pellet/mouse	s.c. implant (12–72 hr)	Rodents	Mitogen-induced lymphocyte proliferation	$\downarrow$	19
			Lymphocyte blastogenesis to ConA and LPS	$\downarrow$	
			MLR	$\downarrow$	
			Ab response to SRBC	$\downarrow$	
			Ab response to TNP-ovalbumin	$\downarrow$	
			B-cell number	$\downarrow$	
			Tumoricidal activity of BCG-elicited macrophages	$\downarrow$	
8, 25, or 75 mg pellet	s.c. implant	B6C3F1 (F)[c]	Blood leukocyte number	$\downarrow$	102
			B- and T-cell number	$\downarrow$	
			Spleen and thymus weight	$\downarrow$	
			NK cell activity	$\downarrow$	
			Phagocytic activity of Kupffer cells	$\downarrow$	
			Resistance to bacterial infection	$\uparrow$	
25 mg/kg	s.c. (2 hr)	C57BL/6J (F)[c]	Splenic NK cell activity	$\downarrow$	25
10 mg/kg	s.c. (2 hr)	Sprague-Dawley (M)[f]	Mitogen-induced lymphocyte activation	$\downarrow$	69
5 mg	i.v. (2 doses in 15 months)	Human	Flaring from i.v. site, burning		33
Norcocaine (metabolite of cocaine)					
10 μ*M* or greater	*In vitro*	B6C3F1[c]	Ab response to SRBC	$\downarrow$	82
			Response to ConA or LPS	$\downarrow$	
Phencyclidine					
0.01–100 μ*M*	*In vitro*	B6C3F1 (F)[c]	IL-2 production	$\downarrow$	178
			IL-4 production	$\downarrow$	
			Induction of cytotoxic T lymphocytes	$\downarrow$	
			B lymphocyte response to anti-IgM + IL-4	$\downarrow$	
			IL-2-augmented NK cell activity	$\downarrow$	
			IL-6 production by macrophages	$\leftrightarrow$	
Phenylcyclohexane (a pyrolysis product of phencyclidine)					
123.6 μmol/kg/d	i.p. (14 d)	CD-1 (M and F)[c]	Humoral and cellular immunity	$\leftrightarrow$	74
317 μmol/kg/d	i.p. (14 d)	CD-1 (M)[c]	Spleen weight	$\downarrow$	
634.5 μmol/kg/d	i.p. (14 d)	CD-1 (M and F)[c]	Thymus weight	$\downarrow$	
634.5 μmol/kg/d			Ab to SRBC	$\downarrow$	
634.5 μmol/kg/d	i.p. (14 d)	CD-1 (F)[c]	Cell-mediated immunity	$\downarrow$	

TABLE 1 (continued)

Effects of Drugs on Immune System Parameters[a]

Compound Dose	Route (Regimen)	Strain (Sex)	Immunological Parameter	Effect[b]	Ref.
Δ^9-Tetrahydrocannabinol					
10 mg/kg/d	p.o. (5 d)	ICR-SCH (M)[c]	DTH	↔	194
			PFC response	↓	
30 mg/kg/d	p.o. (5 d)	ICR-SCH (M)[c]	DTH	↓	194
			PFC response	↔	
22 μ*M*	*In vitro*	B6C3F1 (F)[c]	Response of splenocytes to PMA and ionomycin	↓	87
			Ab response to SRBC *in vitro*	↓	
50–200 mg/kg	Oral	B6C3F1 (F)[c]	Ab response to SRBC or DNP-Ficoll	↓	154
Nervous System Drugs					
Alprazolam					
0.02–1.0 mg/kg	i.p. (2 and 24 hr)	BIO.BR/SgSnJ (M)[c]	NK cell activity, mixed leukocyte reactivity, and mitogen-induced lymphocyte proliferation	↑	52
5–10 mg/kg	i.p. (2 and 24 hr)	BIO.BR/SgSnJ (M)[c]	NK cell activity, mixed leukocyte reactivity, and mitogen-induced lymphocyte proliferation	↔	52
Buspirone (with auditory stress)					
1 mg/kg/d	i.p. (15 d before stress, and after 4, 8, 12, 16, and 20 d of stress)	Balb/c (F)[c]	Stress-induced decrease in thymus and spleen cellularity	↓	51
			Stress-induced decrease in T-cell population	↓	
			Stress-induced decrease in lymphocyte blastogenesis	↓	
Carbamazepine					
5, 10, 15 mg/kg/d	Oral (7 d)	Balb/c (M)[c]	Humoral and cellular immune responses	↓	5
Chlorpromazine					
1 μ*M*	*In vitro* (24 hr)	C3H/HEN (M)[c]	Mitogen-induced lymphocyte proliferation	↓	151
1.3–1.5 × 10^{-5} *M*	*In vitro* (12 or 24 hr)	C57BL/6J (M)[c]	Ab response to SRBC	↓	78
	In vitro (72 or 96 hr)		Ab response to SRBC	↑	
5 × 10^{-7}–5 × 10^{-5} *M*	*In vitro* (12 or 24 hr)	C57BL/6J (M)[c]	Ab response to SRBC	↓	
100 μg	i.p. (2- to 3-hr exposure)	Balb/c (M)[c]	Production of T-cell-derived lymphokines (IL-2IL-4, TNF, IFN-γ, and GM-CSF)	↓	172
			Macrophage-dependent IL-10 production	↑	
0.02 ml of 0.125%	i.d. (0 hr)	Human (F)	Allergic photopatch response		76
0.0625%	Topical (24 hr)	Human (F)	Allergic photopatch response		

7.5 mg/kg/d	i.p. (3 d)	Balb/c (M)[c]	Picryl chloride-induced DTH	↓	151
15 mg/kg/d	i.p. (3 d)	B6C3F1 (M)[c]	Graft vs. host response	↓	
12.5 and 15 mg/kg/d	i.p. (3 d)	CBA/J (M)[c]	Lymphocyte proliferation	↓	
3–4 μ*M*	*In vitro* (25 min)	RBC-2H3 cells	IgE and A23187-mediated calcium influx	↓	184
3–4 μ*M*	*In vitro* (30 min)	RBC-2H3 cells	Arachidonic acid and histamine release	↓	
1 × 10^{-4} *M*	*In vitro* (30 min)	Human (M and F)	Erythrocyte Ab rosettes	↔	183
			ADCC	↓	
Clomipramine					
200 mg/d	Oral	Human	Photo and contact allergy	↓	127
Diazepam					
1.25 mg/kg/d	(6 d) (pregnant females)	Long-Evans[f]	Neonate cellular immune responses	↓	156
8 mg/kg	i.p.	Swiss (F)[c]	Ab response	↓	38
			DTH	↓	
1.1 mg/kg/d	Oral (21 d)	C57BL/6 (M)[c]	DTH	↔	39
			Ab response	↑	
2.2 mg/kg/d	Oral (21 d)	C57BL/6 (M)[c]	DTH	↓	39
			Ab response	↔	
5–10 mg/kg/d	Oral (3 d)	Mice (F)[c]	Recovery of stress and cyclophosphamide-induced suppression of antibody response; Ag-specific T helper activity	↑	134
Dopamine					
0.05–5.0 μg	i.d. or i.v. (15 months)	Human (M and F)	Cutaneous response histamine	↓	26
Haloperidol					
10 μ*M*	*In vitro* (24 hr)	C3H/HEN (M)[c]	Mitogen-induced lymphocyte proliferation	↓	151
Histamine					
1 × 10^{-3} *M*	Human		Mitogen-induced lymphocyte proliferation	↔	193
Lidocaine					
1 × 10^{-2} *M*	*In vitro* (30 months)		Erythrocyte Ab rosettes	↔	183
			ADCC	↓	
0.1%	Dermal	Human (F)	Contact allergy		49
Lithium					
3 mEq/kg twice/d	i.p. (30 d)	Wistar (F)[f]	Arthus reaction, delayed skin reaction, short-term IgM and IgG effect	↓	81
Mepivacaine	s.c.	Human (F)	Type II allergic reaction		94
Meprobamate					
40 mg/kg/d	Oral (21 d)	C57BL/6 (M)[c]	DTH response	↔	38
			Ab response	↔	
80 mg/kg/d	Oral (21 d)	C57BL/6 (M)[c]	DTH response	↔	38
			Ab response	↔	

TABLE 1 (continued)

Effects of Drugs on Immune System Parameters[a]

Compound Dose	Route (Regimen)	Strain (Sex)	Immunological Parameter	Effect[b]	Ref.
Naloxone					
1 mg/kg	i.p.	Mice	Anaphylaxis	↑	114
10 mg/kg			Anaphylaxis	↓	
5 mg/kg (2×/d for 5d)			IgE Ab titer and Ab response to SRBC	↓	
Pergolide mesylate					
400 μg/d	s.c. (3 wk)	Fischer (F)[f]	DCNB-induced contact dermititis	↓	126
			Primary Ab response to SRBC	↓	
Phenytoin					
0.5, 1.0 mg	s.c.	C3H (F)[c]	T-dependent immunoblastic lymphadenopathy	↑	57
	s.c.	C3H (F)[c]	B-cell activity	↑	
	(Anti-epileptic regimen)	Human (M andF)	Ab response to *S. typhi* vaccine	↓	167
	(Anti-epileptic regimen)	Human (M andF)	DTH response	↓	
	(Anti-epileptic regimen)	Human (M andF)	Mitogenic response to PHA	↓	
100, 50, or 25 mg/kg/d	Oral (7 d)	Balb/c (M)[c]	Ab response to SRBC	↓	6
300 mg/m^2	Oral (1 dose, 10–58 hr)	Human (M and F)	PHA-induced blastogenesis	↔	177
			Serum immunoglobulin levels	↔	
			NK and K cell activity	↓	
Sodium valproate					
50 and 150 mg/kg/48 hr	i.p. (3 and 7 weeks)	Mouse	Ab response	↑	144
			Spleen weight	↑	
150 mg/kg/48 hr	i.p. (1 or 3 weeks)		DTH response	↔	
Trifluoperazin					
1 μ*M*	*In vitro* (24 hr)	C3H/HEN (M)[c]	Mitogen-induced lymphocyte proliferation	↓	151
10 and 30 mg/kg	i.p. (3 d)	Balb/C (M)[c]	Picryl chloride induced DTH	↓	151
30 mg/kg	i.p. (3 d)	CBA/J (M)[c]	Lymphocyte proliferation	↓	
30 mg/kg	i.p. (3 d)	B6C3F1 (M)[c]	GVHR	↓	
1.5–4.0 μ*M*	*In vitro* (25 months)	RBL-2H3 cells	IgE- and A23187-mediated calcium influx	↓	184
1.5–4.0 μ*M*	*In vitro* (30 months)	RBL-2H3 cells	Arachidonic acid and histamine release	↓	
Steroid Hormones and Related Drugs					
Danazol					
10^{-5} *M*	*In vitro* (72 hr)	Human (M and F)	PHA-, PWM-, and ConA-induced blastogenesis	↓	70
			MLR	↓	
10^{-6} *M*	*In vitro* (72 hr)	Human (M and F)	PHA- and ConA-induced blastogenesis	↓	
			MLR	↓	

Diethylstilbestrol (DES)					
0.2, 2.0, and 8.0 mg/kg		Mouse (F)	Antibody response	↑	108
0.2, 1.0, and 4.0 mg/kg/d	s.c. (14 d)	B6C3F1 (F)[c]	DTH to KLH	↓	75
			Inflammation to carrageenin	↓	
			Lymphocyte proliferation to ConA and PHA	↓	
			Lymphocyte response to LPS	↓	
			MLR	↓	
8 or 0.2 μ/g	B6C3F1[c]		Resistance to *Trichinella spiralis*	↓	106
40 μg/g	B6C3F1[c]		DTH	↓	
Diethylstilbestrol diphosphate					
500 μg/ml	*In vitro* (72 hr)	Human PBLs (M)	PHA-stimulated ^{3}H-thymidine uptake	↓	1
Dihydrotestosterone					
10^{-9}–10^{-14} *M*		Human (M and F)	Degranulation of eosinophils	↔	91
Estradiol					
10^{-9} *M*	*In vitro* (24 hr)	Human (F)	IFNβ2/IL-6 from IL-1α-induced stromal cells	↓	169
10^{-9}–10^{-4} *M*	*In vitro* (2, 3, 6, 12 hr)	Human (M and F)	Eosinophil degranulation	↔	91
10^{-9}–10^{-7} *M*	*In vitro* (2, 3, 6, 12 hr)	Human (M and F)	Eosinophil degranulation	↔	
1–10 μg/ml	*In vitro* (5-d exposure)	Balb/c and C57BL/6 (M)[c]	Cell-mediated lympholysis	↓	136
10–1000 pg/ml	*In vitro* (24 hr)	Sprague-Dawley (M)[f]	PHA-, ConA-, and PWM-induced blastogenesis	↔	122
10–1000 pg/ml	*In vitro* (96 hr)	Sprague-Dawley (M)[f]	MLR	↔	
75 and 750 mg/kg daily	i.p. (8 d)	Sprague-Dawley (M)[f]	PHA-, ConA-, and PWM-induced blastogenesis	↑	
75 and 750 mg/kg daily	i.p. (4 d)	Sprague-Dawley (M)[f]	MLR	↓	
75 and 750 mg/kg daily	i.p. (3 d)	Sprague-Dawley (M)[f]	MLR	↑	
0.012 mg	s.c. (2×/wk, 3 wk)	C57BL/10$^-$ (M and F)[c]	CD4$^+$ T-cell number	↓	12
17-β-estradiol					
1 μ*M*	*In vitro* (90 months)	Sprague-Dawley (M)[f]	Histamine and serotonin secretion	↑	188
			IgE secretion by mast cells	↔	
10^{-6}–20^{-5} *M*	*In vitro* (90 months)	Sprague-Dawley (M)[f]	Histamine and serotonin secretion	↑	188
10^{-8}–1.0^{-4} *M*		Sprague-Dawley (M)[f]	Histamine and serotonin secretion	↑	
Mestranol					
0.200–0.002 μg/3 ml	*In vitro* (6 d)	Human cells (F)	Lymphoblast response	↑	153
Oxondrolone					
1.1 mg/kg daily	s.c. (5 d)	Sprague-Dawley (M)[f]	Response to i.d. PHA	↓	117
1.1 mg/kg daily	s.c. (10 d)	Sprague-Dawley (M)[f]	Response to i.d. PHA	↑	
Progesterone					
2×10^{-4}	*In vitro* (30 months)	Sprague-Dawley (M)[f]	Serotonin secretion	↑	189
	In vitro (30 months)		Histamine seretion	↔	
2×10^{-4}	*In vitro* (120 months)	Sprague-Dawley (M)[f]	Serotonin secretion	↑	
	In vitro (120 months)		Histamine seretion	↔	
4×10^{-4}	*In vitro* (0 months)	Sprague-Dawley (M)[f]	Intracellular calcium levels	↑	

TABLE 1 (continued)

Effects of Drugs on Immune System Parameters[a]

Compound Dose	Route (Regimen)	Strain (Sex)	Immunological Parameter	Effect[b]	Ref.
4×10^{-4}	*In vitro* (90 months)	Sprague-Dawley (M)[f]	Serotonin secretion	↑	
	In vitro (90 months)		Histamine seretion	↔	
1–10 μg/ml	*In vitro* (5 d)	Balb/c and C57BL/6 (M)[c]	Cell-mediated lympholysis	↓	136
10^{-5} *M*	*In vitro*	Human (M and F)	PBMC ^{3}H-thymidine incorporation	↓	186
1 ng/ml	s.c. (0 d)	Holstein cows (F)	PHA-induced lymphocyte proliferation	↓	164
5 ng/ml	s.c. (12–36 d)	Holstein cows (F)	PHA-induced lymphocyte proliferation	↓	
RU 486 (mifepristone)					
10^{-5} *M*	*In vitro* (72 hr)	Human (M and F)	PBMC ^{3}H-thymidine uptake	↓	186
10 mg/kg/d	(7–14 d)	Human (M)	Total leukocytes	↔	99
			Lymphocyte subsets	↔	
			Lymphocyte proliferation or cytotoxicity	↔	
			Generalized exanthem		
Stanozolol					
1.1 mg/kg daily	s.c. (5 d)	Sprague-Dawley (M)[f]	Response to i.d. PHA	↓	117
1.1 mg/kg daily	s.c. (10 d)	Sprague-Dawley (M)[f]	Response to i.d. PHA	↑	
Testosterone					
10^{-4} *M*	*In vitro* (90 months)	Sprague-Dawley (M)[f]	Histamine and serotonin secretion	↓	188
0.9 mg	s.c. (2×/wk, 3 wk)	C57BL/10⁻ (M and F)[c]	CD8⁺ T-cell number	↑	12
			Nucleated spleen cell number	↓	
250 μg/d	s.c. (10 wk)	NZB (F)[c]	Female PFC to SRBC	↓	120
20 mg	s.c. pellet implant	Balb/c, DBA/2 (M and F)[c]	Female thymus and spleen weight	↓	120
	Balb/c × DBA/2 F1[c]		PFC to SRBC in males	↔	
		NZB (M and F)[c]			
1–10 μg/ml	*In vitro* (5 d)	Balb/c[c]	Cell-mediated lympholysis	↔	136
		C57BL/6 (M)[c]			
1.1 mg/kg/d	s.c. (5 d)	Sprague-Dawley (M)[f]	Response to i.d. PHA	↓	117
1.1 mg/kg/d	s.c. (10 d)	Sprague-Dawley (M)[f]	Response to i.d. PHA	↓	
15 mg capsule	Implant (3 d after castration)	B/W (M)[c]	Bone marrow activity	↑	197
		Balb/c (M)[c]	Thymus atrophy	↑	
		Balb/c nude (M)[c]	Splenic suppressor T-cell activity	↑	
			T-cell response to PHA	↑	
Testosterone propionate					
1.1 mg/kg/d	s.c. (5 d)	Sprague-Dawley (M)[f]	Response to i.d. PHA	↓	117
1.1 mg/kg/d	s.c. (10 d)	Sprague-Dawley (M)[f]	Response to i.d. PHA	↓	

Testrolactone					
1.1 mg/kg/d	s.c. (5 d)	Sprague-Dawley (M)[f]	Response to i.d. PHA	↓	117
1.1 mg/kg/d	s.c. (10 d)	Sprague-Dawley (M)[f]	Response to i.d. PHA	↑	
Miscellaneous Drugs					
Azimexon					
2 courses of 3 × 600 mg	Oral	Human (M and F) (myeloma patients)	NK cell activity	↑	141
Benzylparaben					
	Topical	Human	Allergic responses		182
Cigarette smoking					
>20 cigarettes/d	Inhalation	Human (M and F)	IL-4 production	↑	23
			PBMC IFN-γ production	↔	
			IgE levels in serum	↔	
<20cigarettes/d	Inhalation	Human (M and F)	IL-4 production	↑	23
			PBMC IFN-γ production	↔	
			IgE levels in serum	↔	
Cimetidine hydrochloride					
100 mg/kg twice daily	i.p. (0, 1, and 2 d; 7, 8, and 9 d after sensitization)	Balb/c[c]	Histamine release and mast cell morphology	↔	11
			Ag-presenting cells in epidermis	↔	
100 mg/kg twice daily			Allergic contact hypersensitivity	↑	11
Forphenicinol 5-2-(3-hydroxy-4-hydroxymethylphenyl) glycine					
10 μg/d	p.o. (5 d)	CDF_1 (F)[c]	Survival from mitomycin C-induced leucopenia	↑	80
Levan					
0.1 mg	s.c.	C57BL/6[c]	B16 melanoma tumor resistance	↓	100
10 mg			B16 melanoma tumor resistance	↑	
MESNA					
(2-mercaptoethane sulphonate)	i.v or oral	Human	Persistant rash and decreased platelet number		159
Neurotropin (and stress)					
25 and 50 mg/kg	i.m. (30 months before stress)	C3H/H (F)[c]	Body weight	↔	176
		AKR (F)[c]	Stress-induced decrease in phagocytic activity	↓	
		C3H/H (F)[c]	Stress-induced decrease in T-cell cytotoxic activity	↓	
		C3H/H (F)[c]	Stress-induced alterations in T-cell subsets	↓	
Tetrandrine					
1, 10, 20 μg/ml	*In vitro* (25 months)	Sprague-Dawley (M)[f]	Degranulation of mast cells	↓	175
W-7					

TABLE 1 (continued)

Effects of Drugs on Immune System Parameters[a]

Compound Dose	Route (Regimen)	Strain (Sex)	Immunological Parameter	Effect[b]	Ref.
10–15 μ*M*	*In vitro* (25 months)	RBL-2H3 cells	IgE- and A23187-mediated calcium influx	↓	184
10–15 μ*M*	*In vitro* (30 months)	RBL-2H3 cells	Arachidonic acid and histamine release	↓	

[a] The effects of antigraft rejection drugs and other drugs designed or commonly used as immunosuppressants have been extensively reivewed elsewhere, and they are covered in most pharmacology texts. Thus, this list does not include drugs designed or commonly used as immunosuppressants, unless those drugs exhibit an unexpected immunological effect (e.g., hypersensitivity).

[b] The effects indicated are as follows: ↓ indicates a decrease in the measured parameter; ↑ indicates an increase in the measured parameter; ↔ indicates no significant change in the measured parameter. If the compound causes a hypersensitivity response, this is not indicated by an arrow but is simply listed as the immunological parameter.

[c] Indicates a mouse strain.

[d] Indicates a guinea pig strain.

[e] Indicates a rabbit strain.

[f] Indicates a rat strain.

REFERENCES

1. Ablin, R. J., Bruns, G. R., Guinan, P., Sadoughi, N., and Bush, I. M., Immunosuppressive effect of estrogen on thymic dependent lymphocytic blastogenesis, *Urolog. Res.*, 2, 69–71, 1974.
2. Alani, M. D. and Alani, S. D., Allergic contact dermatitis to corticosteroids, *Ann. Allergy,* 30, 181–185, 1972.
3. Alani, S. D. and Alani, M. D., Allergic contact dermatitis and conjunctivitis to corticosteroids, *Contact Dermat.*, 2, 301–304, 1976.
4. Andrade-Mena, C. E., Orbach-Arbouys, S., and Mathe, G., Enhancement of immune responses in tumor-bearing mice after administration of cis-diamino-dichloro-platinum (II), *Int. Arch. Allergy Appl. Immunol.*, 76, 341–343, 1985.
5. Andrade-Mena, C. E., Sardo-Olmedo, J. A. J., and Ramirez-Lizardo, E. J., Effects of carbamazepine on murine humoral and cellular immune responses, *Epilepsia*, 35, 205–208, 1994.
6. Andrade-Mena, C. E., Sardo-Olmedo, J. A. J., and Ramirez-Lizardo, E. J., Effects of phenytoin administration on murine immune function, *J. Neuroimmunol.*, 50, 3–7, 1994.
7. Anfosso, F., Leyris, R., Soler, M., and Charpin, J., Pollen-allergen IgE response in the inbred rat. Regulation of allergen P-IgE antibodies in high and low responders by suppressive factors, *Int. Arch. Allergy Appl. Immunol.*, 63, 44–51, 1980.
8. Bagasra, O., Currao, L., DeSouza, L. R., Oosterhuis, J. W., and Damjanov, I., Immune response of mice exposed to cis-diamminedichloroplatinum, *Canc. Immunol., Immunother.*, 19, 142–147, 1985.
9. Bar-Sela, S., Levo, Y., Zeevi, D., Slavin, S., and Eliakim, M., A lupus-like syndrome due to ibuprofen hypersensitivity, *J. Rheumatol.*, 7, 379–380, 1980.
10. Beilin, B., Shavit, Y., Cohn, S., and Kedar, E., Narcotic-induced suppression of natural killer cell activity in ventilated and nonventilated rats, *Clin. Immunol. Immunopathol.*, 64, 173–176, 1992.
11. Belsito, D. V., Kerdel, F. A., Potozkin, J., and Soter, N. A., Cimetidine-induced augmentation of allergic contact hypersensitivity reactions in mice, *J. Invest. Dermatol.*, 94, 441–445, 1990.
12. Benten, W. P., Wunderlich, F., Herrmann, R., and Kuhn-Velten, W. N., Testosterone-induced compared with oestradiol-induced immunosuppression against *Plasmodium chabaudi* malaria, *J. Endocrinol.*, 139, 487–494, 1993.
13. Berd, D., Maguire, H. C., and Mastrangelo, M. J., Potention of human cell-mediated and humoral immunity by low-dose cyclophosphamide, *Canc. Res.*, 44, 5439–5443, 1984.
14. Bernstein, D. I., Bernstein, I. L., Bodenheimer, S. S., and Piertrusko, R. G., An open study of auranofin in the treatment of steroid-dependent asthma, *J. Allergy Clin. Immunol.*, 81, 6–16, 1988.
15. Bhoopalam, N., Price, K. S., Norgello, H., and Fried, W., Busulfan-induced suppression of natural killer cell activity, *Exp. Hematol.*, 13, 1127–1132, 1985.
16. Björkner, B., Bruze, M., and Moller, H., High frequency of contact allergy to gold sodium thiosulfate. An indication of gold allergy?, *Contact Dermat.*, 30, 144–151, 1994.
17. Bodenheimer, H., Charland, C., and Leith, J., Alternation of rat Kupffer cell function following mitomycin-c administration, *J. Leuk. Biol.*, 43, 265–270, 1988.
18. Browning, M. J., Herbert, W. J., and White, R. G., Feline miliary eczema: megestrol acetate does not suppress immune responses, *Res. Vet. Sci.*, 35, 245–246, 1983.
19. Bryant, H. U., Bernton, E. W., and Holaday, J. W., Immunomodulatory effects of chronic morphine treatment: pharmacologic and mechanistic studies, *NIDA Res. Mon.*, 96, 131–149, 1990.
20. Bryant, H. U., Bernton, E. W., Kenner, J. R., and Holaday, J. W., Role of the adrenal cortical activation in the immunosuppressive effects of chronic morphine treatment, *Endocrinology,* 128(6), 3253–3258, 1991.
21. Bryant, H. U., Yoburn, B. C., Inturrisi, C. E., Bernton, E. W., and Holaday, J. W., Morphine-induced immunomodulation is not related to serum morphine concentrations, *Eur. J. Pharmacol.*, 149, 165–169, 1988.
22. Butterfield, J. H., and Weiler, D. A., *In vitro* sensitivity of immature human mast cells to chemotherapeutic agents, *Int. Arch. Allergy Appl. Immunol.*, 89, 297–300, 1989.
23. Byron, K. A., Varigos, G. A., and Wootton, A. M., IL-4 production is increased in cigarette smokers, *Clin. Exp. Immunol.*, 95, 333–336, 1994.
24. Cao, W., Sikorski, E. E., Fuchs, B. A., Stern, M. A., Luster, M. I., and Munson, A. E., The B lymphocyte is the immune cell target for 2′,3′-dideoxyadenosine, *Toxicol. Appl. Pharmacol.*, 105, 492–502, 1990.
25. Carr, D. J. J., Gebhardt, B. M., and Paul, D., Alpha adrenergic and mu-2 opioid receptors are involved in morphine-induced suppression of splenocyte natural killer activity, *J. Pharmacol. Exp. Ther.*, 264, 1179–1186, 1993.
26. Casale, T. B., Bowman, S., and Kaliner, M., Induction of human cutaneous mast cell degranulation by opiates and endogenous opioid peptides: evidence for opiate and nonopiate receptor participation, *J. Allergy Clin. Immunol.*, 73, 775–781, 1994.
27. Cesario, T. C., Slater, L. M., Kaplan, H. S., Gupta, S., and Gorse, G. J., Effect of antineoplastic agents on gamma-interferon production in human peripheral blood mononuclear cells, *Canc. Res.*, 44, 4962–4966, 1984.

28. Chalmers, R. J. G., Beck, M. H., and Muston, H. L., Simultaneous hypersensitivity to clobetasone butyrate and clobetasol propionate, *Contact Dermat.*, 9, 317–318, 1983.
29. Christie, P. E., Tagari, P., Ford-Hutchinson, A. W., Black, C., Markendorf, A., Schmitz-Schumann, M., and Lee, T. H., Urinary leukotriene E4 after lysine-aspirin inhalation in asthmatic subjects, *Am. Rev. Resp. Dis.*, 146, 1531–1534, 1992.
30. Christie, P. E., Tagari, P., Ford-Hutchinson, A. W., Charlesson, S., Chee, P., Arm, J. P., and Lee, T. H., Urinary leukotriene E4 concentrations increase after aspirin challenge in aspirin-sensitive asthmatic subjects, *Am. Rev. Resp. Dis.*, 143, 1025–1029, 1991.
31. Connolly, K. M., Diasio, R. B., Armstrong, R. D., and Kaplan, A. M., Decreased immunosuppression associated with antitumor activity of 5-deoxy-5-fluorouridine compared to 5-fluorouracil and 5-fluorouridine, *Canc. Res.*, 43, 2529–2535, 1983.
32. Correale, J., Olsson, T., Bjork, J., Smedegard, G., Hojeberg, B., and Link, H., Sulfasalazine aggravates experimental autoimmune encephalomyelitis and causes an increase in the number of autoreactive T cells, *J. Neuroimmunol.*, 34, 109–120, 1991.
33. Cromwell, T. A. and Zsigmond, E. K., Hypersensitivity to intravenous morphine sulfate, *Plastic Reconstr. Surg.*, 54, 224–227, 1974.
34. Dajani, B. M., Sliman, N. A., Shubair, K. S., and Hamzeh, Y. S., Bronchospasm caused by intravenous hydrocortisone sodium succinate (Solu-Cortef) in aspirin-sensitive asthmatics, *J. Allergy Clin. Immunol.*, 68, 201–204, 1981.
35. Davis, P., Ezeoke, A., Munro, J., Hobbs, J. R., and Hughes, G. R. V., Immunological studies on the mechanism of gold hypersensitivity reactions, *Br. Med. J.*, 3, 676–678, 1973.
36. Davis, P. and Hughes, G. R. V., Significance of *Eosinophilia* during gold therapy, *Arthr. Rheum.*, 17, 964–968, 1974.
37. De Simone, C., Pugnaloni, L., Cilli, A., Forastieri, E. M. A., Bernardini, B., Delogu, G., and Sorice, F., Pharmacokinetic assessment of immunosuppressive activity of antibiotics in human plasma by a modification of the mixed lymphocyte reaction, *Crit. Care Med.*, 12, 483–485, 1984.
38. Descotes, G., Mazue, G., and Richez, P., Drug immunotoxicological approaches with some selected medical products: cyclophosphamide, methylprednisolone, betamethasone, cefoxitine, minor tranquilizers, *Toxicol. Lett.*, 13, 129–137, 1982.
39. Descotes, J., Tedone, R., and Evreux, J. C., Suppression of humoral and cellular immunity in normal mice by diazepam, *Immunol. Lett.*, 5, 41–43, 1982.
40. DeSilva, M. A., Wepsic, H. T., Mizushima, Y., Nikcevich, D. A., and Larson, C. H., Modification of *in vitro* and *in vivo* BCG cell wall-induced immunosuppression by treatment with chemotherapeutic agents or indomethacin, *J. Natl. Canc. Inst.*, 74, 917–921, 1985.
41. DeVos, C., Joseph, M., Leprevost, C., Vorng, H., Tomassini, M., Capron, M., and Capron, A., Inhibition of human eosinophil chemotaxis and of the IgE-dependent stimulation of human blood platelets by cetirizine, *Int. Arch. Allergy Appl. Immunol.*, 88, 212–215, 1989.
42. Di Francesco, P., Pica, F., Croce, C., Favalli, C., Tubaro, E., and Garaci, E., Effect of acute of daily cocaine administration on cellular immune response and virus infection in mice, *Nat. Immun. Cell Growth Regul.*, 9, 397–405, 1990.
43. Duncan, M. W., Person, D. A., Rich, R. R., and Sharp, J. T., Aspirin and delayed type hypersensitivity, *Arthr. Rheum.*, 20, 1174–1178, 1977.
44. Eisenstein, T. K., Bussiere, J. L., Rogers, T. J., and Adler, M. W., Immunosuppressive effects of morphine on immune responses in mice, *Drugs Abuse, Immunity, AIDS*, 335, 41–52, 1993.
45. Eisenstein, T. K., Taub, D. D., Adler, M. W., and Rogers, T. J., The effect of morphine and DAGO on the proliferative response of murine splenocytes, *Adv. Exp. Biol. Med.*, 228, 203–209, 1991.
46. Ferreri, N. R., Howland, W. C., Stevenson, D. D., and Spiegelberg, H. L., Release of leukotrienes, prostaglandins, and histamine into nasal secretions of aspirin-sensitive asthmatics during reaction to aspirin, *Am. Rev. Resp. Dis.*, 137, 847–854, 1988.
47. Fidler, J. M., DeJoy, S. Q., and Gibbons, Jr., J. J., Selective immunomodulation by the antineoplastic agent mitozantrone. I. Suppression of B lymphocyte function, *J. Immunol.*, 137, 727–732, 1986.
48. Fidler, J. M., DeJoy, S. Q., Smith, F. R., and Gibbons, Jr., J. J., Selective immunomodulation by the antineoplastic agent mitoxantrone. II. Nonspecific adherent suppressor cells derived from mitoxantrone-treated mice, *J. Immunol.*, 136, 2747–2754, 1986.
49. Fregert, S., Tegner, E., and Thelin, I., Contact allergy to lidocaine, *Contact Dermat.*, 5, 185–188, 1979.
50. Freier, D. O. and Fuchs, B. A., A mechanism of action for morphine-induced immunosuppression: corticosterone mediates morphine-induced suppression of natural killer cell activity, *J. Pharmacol. Exp. Ther.*, 270, 1127–1133, 1994.
51. Freire-Garabal, M., Belmonte, A., and Suarez-Quintanilla, J., Effects of Buspirone on the immunosuppressive response to stress in mice, *Arch. Int. Pharmacodyn.*, 314, 160–168, 1991.
52. Fride, E., Skolnick, P., and Arora, P. K., Immunoenhancing effects of alprazolam in mice, *Life Sci.*, 47, 2409–2420, 1990.

53. Gautman, S. C., Scissors, D. L., and Webster, J. L. T., Further observations on the effects of 1-thiocarbamoyl-2-imidazolidinone (TCI) on cell-mediated immunity, *Immunopharmacology*, 4, 201–212, 1982.
54. Gelabert-Mas, A., Arango, O., Rosales, A., Coronado, J., and Moreno, A., Eosinophilic cystitis and allergy to mitomycin-c, *Acta Urologica Belgica*, 58, 65–72, 1990.
55. Gillissen, G., Influence of Cefaclor on immune response parameters, *Drug Res.*, 34, 1535–1540, 1984.
56. Giorgini, S., Martinelli, C., and Sertoli, A., Delayed-type sensitivity reaction to mitomycin, *Contact Dermat.*, 24, 378–379, 1991.
57. Gleichmann, H. I. K., T lymphocytes mediate lymphadenopathy and autoimmunization in graft-versus-host reactions and hypersensitivity to diphenylhydantoin, *Agressologie*, 23, 53–55, 1982.
58. Gluckman, S. J., Dvorak, V. C., and MacGregor, R. R., Host defenses during prolonged alcohol consumption in a controlled environment, *Arch. Int. Med.*, 137, 1539–1543, 1977.
59. Goeptar, A. R., De Groot, J., Lang, M., Van Tol, R. G. L., and Scheper, R. J., Suppressive effects of transdermal clonidine administration on contact hypersensitivity reactions in guinea pigs, *Int. J. Immunopharmacol.*, 10, 277–282, 1988.
60. Goetzl, E. J., Valacer, D. J., Payan, D. G., and Wong, M. Y., Abnormal responses to aspirin of leukocyte oxygenation of arachidonic acid in adults with aspirin intolerance, *J. Allergy Clin. Immunol.*, 77, 693–698, 1986.
61. Goldermann, R., Schuppe, H.-C., Gleichmann, E., Kind, P., Merk, H., Rau, R., and Goerz, G., Adverse immune reactions to gold in rheumatoid arthritis: lack of skin reactivity, *Acta Dermatol. Venereol.*, 73, 220–222, 1993.
62. Goto, M., Mitsuoka, A., Sugiyama, M., and Kitano, M., Enhancement of delayed hypersensitivity reaction with varieties of anti-cancer drugs. A common biological phenomenon, *J. Exp. Med.*, 154, 204–209, 1981.
63. Griswold, D. E., Heppner, G. H., and Calabresi, P., Selective suppression of humoral and cellular immunity with cytosine arabinoside, *Canc. Res.*, 32, 298–301, 1972.
64. Han, Y.-C. and Pruett, S. B., Mechanisms of ethanol-induced suppression of a primary antibody response in a mouse model for binge drinking, *J. Pharmacol. Exp. Ther.*, 275, 950–957, 1995.
65. Hansbrough, J., Peterson, V., Zapata-Sirvent, R., and Claman, H. N., Postburn immunosuppression in an animal model. II. Restoration of cell-mediated immunity by immunomodulating drugs, *Surgery*, 95, 290–296, 1984.
66. Hansch, G. M., Voigtlander, V., and Rother, U., Effect of aspirin on the complement system *in vitro*, *Int. Arch. Allergy Appl. Immunol.*, 61, 150–158, 1980.
67. Hashimoto, K. and Lipsky, P. E., Immunosuppression by the disease modifying antirheumatic drug bucillamine: inhibition of human T lymphocyte function by bucillamine and its metabolites, *J. Rheumatol.*, 20, 953–957, 1993.
68. Herlitz, H., Tarkowski, A., Svalander, C., Volkmann, R., and Westberg, G., Beneficial effect of captopril on systemic lupus erythematosis-like disease in MRL lpr/lpr mice, *Int. Arch. Allergy Appl. Immunol.*, 85, 272–277, 1988.
69. Hernandez, M. C., Flores, L. R., and Bayer, B. M., Immunosuppression by morphine is mediated by central pathways, *J. Pharmacol. Exp. Ther.*, 267, 1336–1341, 1993.
70. Hill, J. A., Barbieri, R. L., and Anderson, D. J., Immunosuppressive effects of danazol *in vitro*, *Fertility Sterility*, 48, 414–418, 1987.
71. Hirano, T., Oka, K., and Tamaki, T., Gramicidin as a potential immunosuppressant for organ transplantation: suppression of human lymphocyte blastogenesis *in vitro* and prolongation of heart allograft survival in the rat, *J. Pharmacol. Exp. Ther.*, 273, 223–229, 1995.
72. Ho, W. K. K., Cheung, K. W., Leung, K. N., and Wen, H. L., Suppression of immunological functions in morphine addicted mice, *NIDA Res. Monogr.*, 75, 599–602, 1986.
73. Holsapple, M. P., Eads, M., Stevens, W. D., Wood, S. C., Kaminski, N. E., Morris, D. L., Poklis, A., Kaminski, E. J., and Jordan, S. D., Immunosuppression in adult female B6C3F1 mice by chronic exposure to ethanol in a liquid diet, *Immunopharmacology*, 26, 31–51, 1993.
74. Holsapple, M. P., Munson, A. E., Freeman, A. S., and Martin, B. R., Pharmacological activity and toxicity of phencyclidine (PCP) and phenylcyclohexane (PC), a pyrolysis product, *Life Sci.*, 31, 803–813, 1982.
75. Holsapple, M. P., Munson, A. E., Munson, J. A., and Bick, P. H., Suppression of cell-mediated immunocompetence after subchronic exposure to diethylstilbestrol in female B6C3F1 mice, *J. Pharmacol. Exp. Ther.*, 227, 130–138, 1983.
76. Horio, T., Chlorpromazine photoallergy, *Arch. Dermatol.*, 111, 1469–1471, 1975.
77. Hsia, J., Sarin, N., Oliver, J. H., and Goldstein, A. L., Aspirin and thymosin increase interleukin-2 and interferon-gamma production by human peripheral blood lymphocytes, *Immunopharmacology*, 17, 167–173, 1989.
78. Ichimura, K., Effect of Chlorpromazine on the *in vitro* immune response, *Folia Biologica*, 24, 162–172, 1978.
79. Ishizuka, M., Fukasawa, S., Masuda, T., Sato, J., and Kanbayashi, N., Antitumor effect of bactobolin and its influence on mouse immune system and hematopoietic cells, *J. Antibiotics*, 33, 1054–1062, 1980.
80. Ishizuka, M., Ishizeki, S., Masuda, T., Momose, A., Aoyagi, T., Takeuchi, T., and Umezawa, H., Studies on effects of forphenicinol on immune responses, *J. Antibiotics*, 35, 1042–1048, 1982.

81. Jankovic, B. D., Popeskovic, L., and Isakovic, K., Cation-induced immunosuppression: the effect of lithium on arthrus reactivity, delayed hypersensitivity and antibody production in the rat, *Adv. Exp. Med. Biol.*, 114, 339–344, 1979.
82. Jeong, T. C., Matulka, R. A., Jordan, S. D., Yang, K.-H., and Holsapple, M. P., Role of metabolism in cocaine-induced immunosuppression in splenocyte cultures from B6C3F1 female mice, *Immunopharmacology*, 29, 37–46, 1995.
83. Jerrells, T. R., Smith, W., and Eckardt, M. J., Murine model of ethanol-induced immunosuppression, *Alcohol. Clin. Exp. Res.*, 14, 546–550, 1990.
84. Johansson, S., Cohen, S. M., Yang, J. P. S., Arai, M., and Friedell, G. H., The influence of N-[4-(5-nitro-2-furyl)-2-thiazolyl]formamide and phenacetin on the immune status in male Fischer rats, *Invest. Urol.*, 15, 308–311, 1978.
85. Jorizzo, J. L., Goldblum, R. G., Daniels, J. C., Ichikawa, Y., Langford, M. P., and Fagan, K. M., Evaluation of immune-enhancing effects of ibuprofen in an immunodeficiency model, *Int. J. Dermal.*, 24, 183–187, 1985.
86. Kamei, C., Izushi, K., Adachi, Y., Shimazawa, M., and Tasaka, K., Inhibitory effect of epinastine on the type II-IV allergic reactions in mice, rats, and guinea pigs, *Drug Res.*, 41, 1150–1153, 1991.
87. Kaminski, N. E., Koh, W. S., Kang, K. H., Lee, M., and Kessler, F. K., Suppression of the humoral immune response by cannabinoids is partially mediated through inhibition of adenylate cyclase by a pertussis toxin-sensitive G-protein coupled mechanism, *Biochem. Pharmacol.*, 48, 1899–1908, 1994.
88. Kasperek, C. and Black, C. D., Two cases of suspected immunologic-based hypersensitivity reactions to etoposide therapy, *Ann. Pharmacother.*, 26, 1227–1230, 1992.
89. Kawakami, M., Meyer, A. A., Johnson, M. C., and Rezvani, A. H., Immunologic consequences of acute ethanol ingestion in rats, *J. Surg. Res.*, 47, 412–417, 1989.
90. Khan, A., Hill, J. M., Grater, W., Loeb, E., MacLellan, A., and Hill, N., Atopic hypersensitivity to *cis*-dichlorodiammineplatinum(II) and other platinum complexes, *Canc. Res.*, 35, 2766–2770, 1975.
91. Kita, H., Abu-Ghanzaleh, R., Sanderson, C. J., and Gleich, G. J., Effect of steroids on immunoglobulin-induced eosinophil degranulation, *J. Allergy Appl. Immunol.*, 87, 70–77, 1991.
92. Klajman, A., Farkus, R., Gold, F. F., and Ben-Efraim, S., Procainamide-induced antibodies to nucleoprotein, denatured and native DNA in human subjects, *Clin. Immunol. Immunopathol.*, 3, 525–530, 1975.
93. Klassen, L. W., Williams, G. W., Reinertsen, J. L., Gerber, N. L., and Steinberg, A. D., Ribavirin treatment in murine autoimmune disease, *Arthr. Rheum.*, 22, 145–154, 1979.
94. Klein, C. E. and Gall, H., Type IV allergy to amide-type local anesthetics, *Contact Dermat.*, 25, 45–48, 1991.
95. Knowles, S., Choudhury, T., and Shear, N. H., Metronidazole hypersensitivity, *Ann. Pharmacother.*, 28, 325–326, 1994.
96. Kraut, E. H., Neff, J. C., Bouroncle, B. A., Gochnour, D., and Grever, M. R., Immunosuppressive effects of pentostatin, *J. Clin. Oncol.*, 8, 848–855, 1990.
97. Kubo, Y., Nonaka, S., and Yoshida, H., Contact allergy to amcinonide, *Contact Dermat.*, 15, 109–111, 1986.
98. Landwehr, A. J., Delayed-type allergy to phenazone in a patient with erythema multiforme, *Contact Dermat.*, 8, 283–289, 1982.
99. Laue, L., Lotze, M. T., Chrousos, G. P., Barnes, K., Loriaux, D. L., and Fleisher, T. A., Effect of chronic treatment with the glucocorticoid antagonist RU 486 in man: toxicity, immunological, and hormonal aspects, *J. Clin. Endocrinol. Metabol.*, 71, 1474–1480, 1990.
100. Leibovici, J., Kopel, S., Siegal, A., and Gal-Mor, O., Effect of tumor inhibitory and stimulatory doses of levan, alone and in combination with cyclophosphamide, on spleen and lymph nodes, *Int. J. Immunopharmacol.*, 8(4), 391–403, 1986.
101. Leprevost, C., Capron, M., DeVos, C., Tomassini, M., and Capron, A., Inhibition of eosinophil chemotaxis by a new antiallergic compound (cetirizine), *Int. Arch. Allergy Appl. Immunol.*, 87, 9–13, 1988.
102. LeVier, D. G., McCay, J. A., Stern, M. L., Harris, L. S., Page, D., Brown, R. D., Musgrove, D. L., Butterworth, L. F., White, K. L., and Munson, A. E., Immunotoxicological profile of morphine sulfate in B6C3F1 female mice, *Fundam. Appl. Toxicol.*, 22, 525–542, 1994.
103. Lewis, A. J., Cottney, J., White, D. D., Fox, P. K., McNeillie, A., Dunlop, J., Smith, W. E., and Brown, D. H., Action of gold salts in some inflammatory and immunological models, *Agents Actions*, 10, 63–77, 1980.
104. Lopez, M. C., Chen, G.-J., Huang, D. S., Wang, Y., and Watson, R. R., Modification of spleen cell subsets by chronic cocaine administration and murine retroviruses infection in normal and protein-malnourished mice, *Int. J. Immunopharmacol.*, 14, 1153–1163, 1992.
105. Lopez, M. C. and Watson, R. R., Effect of cocaine and murine AIDS on lamina propria T and B cells in normal mice, *Life Sci.*, 54, PL147–151, 1994.
106. Luebke, R. W., Luster, M. I., Dean, J. H., and Hayes, H. T., Altered host resistance to *Trichinella spiralis* infection following subchronic exposure to diethylstilbestrol, *Int. J. Immunopharmacol.*, 6(6), 609–617, 1984.
107. Lumry, W. R., Curd, J. G., Zeiger, R. S., Pleskow, W. W., and Stevenson, D. D., Aspirin-sensitive rhinosinusitis: the clinical syndrome and effects of aspirin administration, *J. Allergy Clin. Immunol.*, 71, 580–587, 1983.

108. Luster, M. I., Boorman, G. A., Dean, J. H., Luebke, R. W., and Lawson, L. D., The effect of adult exposure to diethylstilbestrol in the mouse: alterations in immunological functions, *J. Reticuloendothelial Soc.*, 28, 561–569, 1980.
109. Luster, M. I., Rosenthal, G. J., Cao, W., Thompson, M. B., Munson, A. E., Prejean, J. D., Shopp, G., Fuchs, B. A., Germolac, D. R., and Tomaszewski, J. I., Experimental studies of the hematologic and immune system toxicity of nucleoside derivatives used against HIV infection, *Int. J. Immunopharmacol.*, 13, 99–107, 1991.
110. Malo, P. E., Wasserman, M., Parris, D., and Pfeiffer, D., Inhibition by auranofin of pharmacologic and antigen-induced contractions of the isolated guinea pig trachea, *J. Allergy Clin. Immunol.*, 77, 371–376, 1986.
111. Maria, V. A., da Silva, P., and Victorino, R. M. M., Agranulocytosis and liver damage associated with pyrazinobutazone with evidence for an immunological mechanism, *J. Rheumatol.*, 16, 1484–1485, 1989.
112. McFadden, J., Bacon, K., and Camp, R., Topically applied verapamil hydrochloride inhibits tuberculin-induced delayed-type hypersensitivity reactions in human skin, *J. Invest. Dermatol.*, 99, 784–786, 1992.
113. Meadows, G. G., Blank, S. E., and Duncan, D. D., Influence of ethanol consumption on natural killer cell activity in mice, *Alcohol. Clin. Exp. Res.*, 13, 476–479, 1989.
114. Mediratta, P. K., Das, N., Gupta, V. S., and Sen, P., Modulation of humoral immune responses by endogenous opioids, *J. Allergy Clin. Immunol.*, 81, 27–32, 1988.
115. Meingassner, J. G. and Stutz, A., Immunosuppressive macrolides of the type FK 506: a novel class of topical agents for treatment of skin diseases, *J. Invest. Dermatol.*, 98, 851–855, 1992.
116. Mekori, Y. A., Weitzman, G. L., and Galli, S. J., Reevaluation of reserpine-induced suppression of contact sensitivity. Evidence that reserpine interferes with T lymphocyte function independently of an effect on mast cells, *J. Exp. Med.*, 162, 1935–1953, 1985.
117. Mendenhall, C. L., Grossman, L. J., Roselle, G. A., Hertelendy, Z., Ghosn, S. J., Lamping, K., and Martin, K., Anabolic steroid effects on immune function: differences between analogues, *J. Steroid Biochem. Mol. Biol.*, 37, 71–76, 1990.
118. Merluzzi, V. J., Last-Barney, K., Susskind, B. M., and Faanes, R. B., Recovery of humoral and cellular immunity by soluble mediators after 5-fluorouracil-induced immunosuppression, *Clin. Exp. Immunol.*, 50, 318–326, 1982.
119. Mohammed, I., Barraclough, D., Holborow, E. J., and Ansell, B. M., Effect of penicillamine therapy on circulating immune complexes in rheumatoid arthritis, *Ann. Rheum. Dis.*, 35, 458–462, 1975.
120. Morton, J. I., Weyant, D. A., Siegel, B. V., and Golding, B., Androgen sensitivity and autoimmune disease. I. Influence of sex and testosterone on the humoral immune response of autoimmune and non-autoimmune mouse strains to sheep erythrocytes, *Immunology*, 44, 661–669, 1981.
121. Mozzanica, N. and Pigatto, P. D., Contact and photocontact allergy to ketoprofen: clinical and experimental study, *Contact Dermat.*, 23, 336–340, 1990.
122. Myers, M. J., Butler, L. D., and Petersen, B. H., Estradiol-induced alteration in the immune system. II. Suppression of cellular immunity in the rats is not the result of direct estrogenic action, *Immunopharmacology*, 11, 47–55, 1986.
123. Nader, D. A. and Schillaci, R. F., Pulmonary infiltrates with eosinophilia due to maproxen, *Chest*, 83, 280–282, 1983.
124. Nagakura, N., Kobayashi, T., Shimizu, T., Masuzawa, T., and Yanagihara, Y., Immunological properties of cephalexin-induced delayed type hypersensitivity reaction in guinea pigs, *Chem. Pharmacol. Bull.*, 38, 3410–3413, 1990.
125. Nagarkatti, M., Toney, D. M., and Nagarkatti, P. S., Immunomodulation by various nitrosoureas and its effect on the survival of the murine host bearing a syngeneic tumor, *Canc. Res.*, 49, 6587–6592, 1989.
126. Nagy, E. and Berczi, I., Immunomodulation by tamoxifen and pergolide, *Immunopharmacology*, 12, 145–153, 1986.
127. Nair, M. P. N. and Schwartz, S. A., Immunoregulation of human natural killer cells (NK) by corticosteroids: inhibitory effect of culture supernatants, *J. Allergy Clin. Immunol.*, 82, 1089–1097, 1988.
128. Nelson, S., Bagby, G. J., Bainton, B. G., and Summer, W. G., The effects of acute and chronic alcoholism on tumor necrosis factor and the inflammatory response, *J. Infect. Dis.*, 160, 422–429, 1989.
129. Noda, H., Matsunaga, K., Noda, T., Abe, M., Ohtani, T., Shimizu, Y., Asahi, K., Iida, Y., Takigami, N., Masutani, M., and Ueda, H., Contact sensitivity and cross-reactivity of budesonide, *Contact Dermat.*, 28, 212–215, 1993.
130. Nolte, H., Carstensen, H., and Hertz, H., VM-26 (Teniposide)-induced hypersensitivity and degranulation of basophils in children, *Am. J. Ped. Hematol./Oncol.*, 10, 308–312, 1988.
131. O'Dwyer, P. J., King, S. A., Eisenhauer, E., Grem, J. L., and Hoth, D. F., Hypersensitivity reactions to deoxycoformycin, *Canc. Chemother. Pharmacol.*, 23, 173–175, 1989.
132. O'Dwyer, P. J., King, S. A., Fortner, C. L., and Leyland-Jones, B., Hypersensitivity reactions to teniposide (VM-26): an analysis, *J. Clin. Oncol.*, 4, 1262–1269, 1986.
133. Ogle, K. M. and Kennedy, B. J., Hypersensitivity reactions to etoposide, *Am. J. Clin. Oncol.*, 11, 663–665, 1988.

134. Okimura, T. and Nagata, I., Effect of benzodiazepine derivatives. I. augmentation of T cell-dependent antibody response by diazepam in mouse spleen cells, *J. Immunopharmacol.*, 8, 327–346, 1986.
135. Oleaga, J. M., Aguirre, A., Gonzalez, M., and Diaz-Perez, J. L., Topical provocation of fixed drug eruption due to sulfamethoxazole, *Contact Dermat.*, 29, 155–167, 1993.
136. Pavia, C., Siiteri, P. K., Perlman, J. D., and Stites, D. P., Suppression of murine allogeneic cell interactions by sex hormones, *J. Repro. Immunol.*, 1, 33–38, 1979.
137. Pavia, C. S., Bittker, S., and Cooper, D., Immune response to Lyme spirochete *Borrelia burgdorferi* affected by ethanol consumption, *Immunopharmacol.*, 22, 165–174, 1991.
138. Pellis, N. R., Harper, C., and Dafny, N., Suppression of the induction of delayed hypersensitivity in rats by repetitive morphine treatments, *Exp. Neurol.*, 93, 92–97, 1986.
139. Perianin, A., Roch-Arveiller, M., Giroud, J.-P., and Hakim, J., *In vivo* effects of indomethacin and flurbiprofen on the locomotion of neutrophils elicited by immune and non-immune inflammation in the rat, *Eur. J. Pharmacol.*, 106, 327–333, 1985.
140. Perry, H. M., Late toxicity to hydralazine resembling systemic lupus erythematosis or rheumatoid arthritis, *Am. J. Med.*, 54, 58–72, 1973.
141. Peter, H. H., Dziuba-Traber, H., and Boerner, D., The effect of BM 12.531 (Azimexon) on natural killer cell activity in myeloma patients, *Eur. J. Canc. Oncol.*, 20, 353–359, 1984.
142. Phadke, K., Carroll, J., and Nanda, S., Effects of various anti-inflammatory drugs on type II collagen-induced arthritis in rats, *Clin. Exp. Immunol.*, 47, 579–86, 1982.
143. Pruett, S. B., Han, Y. C., and Fuchs, B. A., Morphine suppresses primary humoral immune responses by a predominantly indirect mechanism, *J. Pharmacol. Exp. Ther.*, 262, 923–928, 1992.
144. Queiroz, M. L. S. and Mullen, P. W., Effects of sodium valproate on the immune response, *Int. J. Immunopharmacol.*, 14(7), 1133–1137, 1992.
145. Radosevic-Stasic, B., Cuk, M., Mrakovcic-Sutic, I., Baraclatas, V., Muhvic, D., Lucin, P., Petkovic, M., and Rukavina, D., Immunosuppressive properties of halothane anesthesia and/or surgical stress in experimental conditions, *Int. J. Neurosci.*, 51, 235–236, 1990.
146. Razani-Boroujerdi, S., Savage, S., and Sopori, M., Alcohol-induced changes in the immune response: immunological effects of chronic ethanol intake are genetically regulated, *Toxicol. Appl. Pharmacol.*, 127, 37–43, 1994.
147. Rédier, H., Chaney, P., De Vos, C., Rifai, N., Clauzel, A. M., Michel, F. B., and Godard, P., Inhibitory effect of cetirizine on the bronchial eosinophil recruitment induced by allergen inhalation challenge in allergic patients with asthma, *J. Allergy Clin. Immunol.*, 90, 215–224, 1992.
148. Reitamo, S., Lauerma, A. I., Stubb, S., Kayhko, K., Visa, K., and Forstrom, L., Delayed hypersensitivity to topical corticosteroids, *J. Am. Acad. Dermatol.*, 14, 582–589, 1986.
149. Rivara, G., Tomb, R. R., and Foussereau, J., Allergic contact dermatitis from topical corticosteroids, *Contact Dermat.*, 21, 83–91, 1989.
150. Roszkowski, W., Ko, H. L., Roszkowski, K., Jeljaszewicz, J., and Pulverer, G., Antibiotics and immunomodulation: effects of cefotaxime, amikacin, mezlocillin, piperacillin, and clindamycin, *Med. Microbiol. Immunol.*, 173, 279–289, 1985.
151. Roudebush, R. E., Berry, P. L., Layman, N. K., Butler, L. D., and Bryant, H. U., Dissociation of immunosuppression by chlorpromazine and trifluoperazine from pharmacologic activities as dopamine antagonists, *Int. J. Immunopharmacol.*, 13, 961–968, 1991.
152. Saad, A. J., Domiati-Saad, R., and Jerrells, T. R., Ethanol ingestion increases susceptibility of mice to *Listeria monocytogenes*, *Alcohol Clin. Exp. Res.*, 17, 75–85, 1993.
153. Savel, H., Madison, J. F., and Meeker, I. C., Cutaneous eruptions and *in vitro* lymphocyte hypersensitivity, *Arch. Derm.*, 101, 187–190, 1970.
154. Schatz, A. R., Koh, W. S., and Kaminski, N. E., Delta 9-tetrahydrocannabinol selectively inhibits T-cell dependent humoral immune responses through direct inhibition of accessory T-cell function, *Immunopharmacology*, 26, 129–137, 1993.
155. Schiavino, D., Papa, G., Nucera, E., Schinco, G., Fais, G., Pirrotta, L. R., and Patriarca, G., Delayed allergy to diclofenac, *Contact Dermat.*, 26, 357–358, 1992.
156. Schlumpf, M., Parmar, R., Ramseier, H. R., and Lichtensteiger, W., Prenatal benzodiazepine immunosuppression: possible involvement of peripheral benzodiazepine site, *Dev. Pharmacol. Ther.*, 15, 178–185, 1990.
157. Schrier, D. J., Effects of antiarthritic compounds on sheep erythrocyte-induced delayed-type hypersensitivity, *J. Immunopharmacol.*, 7, 313–323, 1985.
158. Sei, Y., McIntyre, T., Fride, E., Yoshimoto, K., Skolnick, P., and Arora, P. K., Inhibition of calcium mobilization is an early event in opiate-induced immunosuppression, *FASEB J.*, 5, 2194–2199, 1991.
159. Seidel, A., Andrassy, K., Ritz, E., Kässer, U., and Lemmel, E.-M., Allergic reactions to MESNA, *Lancet*, 338, 381, 1991.
160. Settipane, R. A. and Stevenson, D. D., Cross sensitivity with acetaminophen in aspirin-sensitive subjects with asthma, *J. Allergy Clin. Immunol.*, 84, 26–33, 1989.

161. Shavit, Y., Martin, F. C., Yirmiya, R., Ben-Eliyahu, S., Terman, G. W., Weiner, H., Gale, R. P., and Liebeskind, J. C., Effects of a single administration of morphine or footshock stress on natural killer cell cytotoxicity, *Brain Behav. Immun.*, 1, 318–328, 1987.
162. Shelbey, J. and Hisatake, G., Effect of ibuprofen and interleukin 2 on transfusion-induced suppression of cell-mediated immunity, *Arch. Surg.*, 123, 1397–1399, 1988.
163. Sheng, F. C., Freischlag, J., Backstrom, B., Kelly, D., and Busuttil, R. W., The effects of *in vivo* antibiotics on neutrophil (PMN) activity in rabbits with peritonitis, *J. Surg. Res.*, 43, 239–245, 1987.
164. Sherblom, K. P., Smagula, R. M., Moody, C. E., and Anderson, G. W., Immunosuppression, sialic acid, and sialyltransferase of bovine serum as a function of progesterone concentration, *J. Repro. Fert.*, 74, 509–517, 1985.
165. Snyman, J. R. and Sommers, D. K., Effect of enalapril on allergen-induced cutaneous hypersensitivity reaction, *Br. J. Clin. Pharmacol.*, 32, 713–716, 1991.
166. Sordillo, E. M., Ikehara, S., Good, R. A., and Trotta, P. P., Immunosuppression by 2′ deoxycoformycin: studies on the mode of administration, *Cell. Immunol.*, 63, 259–271, 1981.
167. Sorrell, T. C., Forbes, I. J., Burness, F. R., and Rischbieth, R. H. C., Depression of immunological function in patients treated with phenytoin sodium (sodium diphenylhydantoin), *Lancet*, 2, 1233–1235, 1971.
168. Struhar, D., Kivity, S., and Topilsky, M., Short communication quinacrine inhibits oxygen radicals release from human alveolar macrophages, *Int. J. Immunopharmacol.*, 14, 275–277, 1992.
169. Tabibzadeh, S. S., Santhanam, U., Sehgal, P. B., and May, L. T., Cytokine-induced production of IFN-beta 2/IL-6 by freshly explanted human endometrial stromal cells. Modulation by estradiol-17 beta, *J. Immunol.*, 142, 3134–3139, 1989.
170. Takaishi, T., Morita, Y., Kudo, K., and Miyamoto, T., Auranofin, an oral chrysotherapeutic agent, inhibits histamine release from human basophils, *J. Allergy Clin. Immunol.*, 74, 296–301, 1984.
171. Tarayre, J. P. and Lauressergues, H., Comparison of the effects of a nonsteroidal anti-inflammatory agent, an immunosuppressive, a corticosteroid and a immunomodulator on various immunological and non-immunological inflammatory experimental models, *Arch. Int. Pharmacodyn.*, 242, 159–176, 1979.
172. Tarazona, R., Gonzalez-Garcia, A., Zamzami, N., Marchetti, P., Frechin, N., Gonzalo, J. A., Ruiz-Gayo, M., van-Rooijen, N., Martinez, C., and Kroemer, G., Chloropromazine amplifies macrophage-dependent IL-10 production *in vivo*, *J. Immunol.*, 154, 861–870, 1995.
173. Taub, D. D., Eisenstein, T. K., Geller, E. B., Adler, M. W., and Rogers, T. J., Immunomodulatory activity of mu- and kappa-selective opioid agonists, *Proc. Natl. Acad. Sci. USA*, 88, 360–364, 1991.
174. Taylor, R. J., Long, W. F., and Nelson, H. S., Development of subsensitivity to chlorpheniramine, *J. Allergy Clin. Immunol.*, 76, 103–107, 1985.
175. Teh, B. S., Seow, W. K., Chalmers, A. H., Playford, S., Ioannoni, B., and Thong, Y. H., Inhibition of histamine release from rat mast cells by the plant alkaloid tetrandrine, *Int. Arch. Allergy Appl. Immunol.*, 86, 220–224, 1988.
176. Teshima, H., Sogawa, H., Kihara, H., Kubo, C., Mori, K., and Nakagawa, T., Prevention of immunosuppression in stressed mice by neurotropin (NSP), *Life Sci.*, 47, 869–876, 1990.
177. Thatcher, N., Wan, H. H., Swindell, R., Wilkinson, P. M., and Crowther, D., Effects of diphenylhydantoin on killer cell activity and other immunological functions, *Int. J. Immunopharmacol.*, 4, 167–174, 1982.
178. Thomas, P. T., House, R. V., and Bhargava, H. N., Phencyclidine exposure alters *in vitro* cellular immune response parameters associated with host defense, *Life Sci.*, 53, 1417–1427, 1993.
179. Thompson-Moya, L., Martin, T., Heuft, H.-G., Neubauer, A., and Herrman, R., Case report: allergic reaction with immune hemolytic anemia resulting from chlorambucil, *Am. J. Hematol.*, 32, 230–231, 1989.
180. Tisman, G., Herbert, V., Go, L. T., and Brenner, L., Marked immunosuppression with minimal myelosuppression by bleomycin *in vitro*, *Blood*, 41, 721–726, 1973.
181. Tonnesen, H., Petersen, K. R., Hojgaard, L., Stockholm, K. H., Nielsen, H. J., Knigge, U., and Kehlet, H., Postoperative morbidity among symptom-free alcohol misusers, *Lancet*, 340, 334–337, 1992.
182. Tosti, A., Fanti, P. A., and Pileri, S., Dermal contact dermatitis from benzylparaben, *Contact Dermat.*, 21, 49–63, 1989.
183. Tsokos, G., Mandyla, H., Xanthou, M., and Papamichail, M., Chlorpromazine and lidocaine inhibit antibody-dependent cell-mediated cytotoxicity but not erythrocyte antibody rosette formation, *Int. Arch. Allergy Appl. Immunol.*, 61, 344–346, 1980.
184. Urata, C. and Siraganian, R. P., Pharmacologic modulation of the IgE or Ca^{2+} ionophore A23187 mediated Ca^{2+} influx, phospholipase activation, and histamine release in rat basophilic leukemia cells, *Int. Arch. Allergy Appl. Immunol.*, 78, 92–100, 1985.
185. Van Hecke, E. and Temmerman, L., Contact allergy to the corticosteroid budesonide, *Contact Dermat.*, 6, 509, 1980.
186. Van Voorhis, B. J., Anderson, D. J., and Hill, J. A., The effects of RU 486 on immune function and steroid-induced immunosuppression *in vitro*, *J. Clin. Endocrinol. Metabol.*, 69, 1195–1199, 1989.
187. Villa, M. L., Rappocciolo, G., Piazza, P., and Clerici, E., The interference of antibiotics with antigen-specific antibody responses in man, *Int. J. Immunopharmacol.*, 8, 805–809, 1986.

188. Vliagoftis, H., Dimitriadou, V., Boucher, W., Rozniecki, J. J., Correia, I., Raam, S., and Theoharides, T. C., Estradiol augments while tamoxifen inhibits rat mast cell secretion, *Int. Arch. Allergy Immunol.*, 98, 398–409, 1992.
189. Vliagoftis, H., Dimitriadou, V., and Theoharides, T. C., Progesterone triggers selective mast cell secretion of 5-hydroxytryptamine, *Int. Arch. Allergy Appl. Immunol.*, 93, 113–119, 1990.
190. Voiculescu, C., Stanciu, L., Voiculescu, M., Rogoz, S., and Dumitriu, I., Experimental study of antibiotic-induced immunosuppression in mice. I. Humoral and cell-mediated immune responsiveness related to *in vivo* antibiotic treatment, *Comp. Immun. Microbiol. Infect. Dis.*, 6, 291–299, 1983.
191. Voigtlander, V., Hansch, G. M., and Rother, U., Effect of aspirin on complement *in vivo*, *Int. Arch. Allergy Appl. Immunol.*, 61, 145–149, 1980.
192. Walzer, R., Feinstein, R., Shapiro, L., and Einbinder, J., Severe hypersensitivity reaction to gold, *Arch. Dermatol.*, 106, 231–234, 1972.
193. Wang, S. R. and Zweiman, B., Inhibitory effects of cortocosteroids and histamine on human lymphocytes, *J. Allergy Clin. Immunol.*, 67, 39–44, 1981.
194. Watson, E. S., Murphy, J. C., El Sohly, H. N., El Sohly, M. A., and Turner, C. E., Effects of the administration of coca alkaloids on the primary immune response of mice: interaction with Δ^9-tetrahydrocannabinol and ethanol, *Toxicol. Appl. Pharmacol.*, 71, 1–13, 1983.
195. Watson, R. R., Prabhala, R. H., Abril, E., and Smith, T. L., Changes in lymphocyte subsets and macrophage functions from high, short-term dietary ethanol in C57BL/6 mice, *Life Sci.*, 43, 865–870, 1988.
196. Weidmann, B., Mulleneisen, N., Bojko, P., and Niederle, N., Hypersensitivity reactions to carboplatin. Report of two patients, review of the literature, and discussion of diagnostic procedures and management, *Cancer*, 73, 2218–2222, 1994.
197. Weinstein, Y. and Berkovich, Z., Testosterone effect on bone marrow, thymus, and suppressor T cells in the (NZB X NZW) F1 mice: its relevance to autoimmunity, *J. Immunol.*, 126, 998–1002, 1981.
198. Weiss, R. B., Donehower, R. C., Wiernik, P. H., Ohnuma, T., Gralla, R. J., Trump, D. L., Baker, J. R., Van Echo, D. A., Von Hoff, D. D., and Leyland-Jones, B., Hypersensitivity reactions from taxol, *J. Clin. Oncol.*, 8, 1263–1268, 1990.
199. Williams, W. R., Pawlowicz, A., and Davies, B. H., Aspirin-sensitive asthma: significance of the cyclooxygenase-inhibiting and protein-binding properties of analgesic drugs, *Int. Arch. Allergy Appl. Immunol.*, 95, 303–308, 1991.
200. Wilson, R. D., Priano, L. L., Traber, D. L., Sakai, H., Daniels, J. C., and Ritzmann, S. E., An investigation of possible immunosuppression from Ketamine and 100 percent oxygen in normal children, *Anesthesia Analgesia*, 50, 464–465, 1971.
201. Wu, W.-J., Wolcott, R. M., and Pruett, S. B., Ethanol decreases the number and activity of splenic natural killer cells in a mouse model for binge drinking, *J. Pharmacol. Exp. Ther.*, 271, 722–729, 1994.
202. Yamauchi, N., Suko, M., Morita, Y., Suzuki, S., Ito, K., and Miyamoto, T., Decreased airway responsiveness to histamine in gold salt-treated guinea pigs, *J. Allergy Clin. Immunol.*, 74, 802–807, 1984.
203. Zeis, B. M., Effects of anti-tuberculosis drugs on the production of prostaglandin E_2 and on mononuclear leucocyte transformation, *Chemotherapy*, 33, 204–210, 1987.

CHAPTER 19

Metal Immunotoxicology

Judith T. Zelikoff and Mitchell D. Cohen

CONTENTS

1 INTRODUCTION

Heavy metals are ubiquitous in the biosphere, where they occur as part of the natural background of chemicals to which human beings are constantly being exposed; however, industrial uses of metals and other industrial and domestic processes have introduced substantial amounts of potentially toxic heavy metals into the atmosphere and into aquatic and terrestrial environments. The toxicological effects associated with high-dose exposure to many metals including arsenic, chromium, and mercury are well known. Adverse effects resulting from long-term subclinical levels of exposure, comparable to environmental conditions, however, are less clear. Within the last few decades, experimental data have shown that low-level exposure to certain metals induces subtle changes within a host, including altered immunological competence. Environmental stressors such as metals may act directly to kill the exposed organisms or indirectly to exacerbate disease states by lowering resistance and allowing the invasion of infectious pathogens or the progression of nascent tumors.

Although effects of metals are dependent upon such variables as host species and exposure parameters, including route, dose, and duration, the conclusion reached in most immunotoxicological studies is that heavy metals act to suppress immunocompetence. The most consistent finding in experimental and epidemiological studies evaluating the effects of metals on immune functions is a decreased host resistance to infectious agents. Immunotoxicity of metals may occur via direct effects on a specific component of the immune system or,

0-8493-4493-X/97/$0.00+$.50

alternatively, via inhibition of immunoregulation, which can result in immunosuppression, hypersensitivity reactions, or autoimmune disorders. It has been postulated that metal toxicity may, at least in part, be due to autoimmunity, since an autoimmune disorder exists for all the major target organs affected by heavy metals.

The purpose of this chapter is not to list exhaustively all of the different metals and to review each of the immunotoxicology studies performed, but rather to provide the reader with a general overview of the immunotoxicity associated with environmentally and/or occupationally relevant metals, so as to understand how individual metals act to bring about alterations in human host defense. These metals include aluminum, arsenic, beryllium, chromium, gold, mercury, inorganic and organic tin, and vanadium. Immunotoxicological studies selected for discussion demonstrated the immunosuppressive/immunoenhancing effects of individual metals on a variety of immunological responses. Although the focus of this book is on human health effects, most of the information gained in metal immunotoxicology has been generated using nonhuman animal models. Thus, data from animal studies have been included in this chapter for comparative and extrapolation purposes. For further information on the immunotoxicity of metals not listed in this section, the reader is referred to a chapter by Zelikoff and Cohen on the "Immunotoxicology of Inorganic Metal Compounds" in *Experimental Immunotoxicology* (R. J. Smialowicz and M. P. Holsapple, Eds., CRC Press, 1996).

2 LIST OF COMMONLY USED ABBREVIATIONS

Ab = antibody

ADCC = antibody-dependent cell cytotoxicity

Ag = antigen

A/PR8 = influenza virus

Anti CD3ε = T-cell receptor surface molecule stimulation

ATP/GTP = adenosine triphosphate/guanidine triphosphate

2B4 = mouse hybridoma cell line

β-Glu = β-glucuronidase

β-NAG = β-*N*-acetyl–β-D-glucosaminidase

B. abortus = *Brucella abortus*

BAL = bronchoalveolar lavage

C3 = complement component (3)

C3b = complement component (3b)

C3bR = complement component 3b receptor

cAMP = cyclic adenosine monophosphate

CD23 = antibody receptor for IgE

CD25 = interleukin-2 receptor

CD3 = T-lymphocyte marker

$CD4^+$ = T-helper cell subpopulation

CD69 = Activator inducer molecule (AIM) marker on T and B cells

$CD8^+$ = T-suppressor cell subpopulation

ConA = Conconavalin A (T-cell mitogen)

CTL = cytotoxic T lymphocyte

DTH = delayed-type hypersensitivity

E. coli = *Escherichia coli*
EAC = erythrocyte-antibody-complement complex
F_cR = antibody F_c receptor
$F_{c\gamma 2a}R$ = IgG_{2a} F_c receptor
$F_{c\gamma 2b}R$ = IgG_{2b} F_c receptor
fMLP = formyl(f)-methionine-leucine-phenylalanine
F4/80 = murine macrophage marker
G6PDH = glucose-6-phosphate dehydrogenase
GSH = glutathione
GSHPX/GSHRX = glutathione peroxidase/glutathione reductase activity ratio
GSSG = oxidized glutathione
H_2O_2 = hydrogen peroxide
HMS = hexose monophosphate shunt
HSV-2 = Herpes simplex virus type 2
Ia = mixed histocompatability antigens on the antigen-presenting cell
ICAM = adhesion molecule
ID = intradermal injection
IFN = interferon
IFN-γ = interferon-gamma
Ig = immunoglobulin (G, M, D, A, E)
IL-1 = interleukin-1
IL-2 = interleukin-2
IL-2R = interleukin-2 receptor
IM = intramuscular injection
IP = intraperitoneal
IT = intratracheal instillation
J774 = macrophage-like cell line
K. pneumoniae = *Klebsiella pneumoniae*
KLH = keyhole limpet hemocyanin
L. monocytogenes = *Listeria monocytogenes*
L929 = mouse fibroblast cells
LDH = lactate dehydrogenase enzyme
LFA-1 = natural killer cell surface adhesion molecule
LGL = large granular lymphocyte
LLC-MK2 = monkey kidney cells
LPS = lipopolysaccharide (B-cell mitogen)
LTB_4 = leukotriene B_4
MAC-1, 2, 3 = macrophage and NK cell surface adhesion molecules (β integrins)
MHC = mixed histocompatability complex
MIF = macrophage inhibitory factor
MLR = mixed lymphocyte response
MNC = mononuclear cell

MØ = macrophage
ND = not determined
NE = no effect
NBT = nitroblue tetrazolium
NK = natural killer cell
$\cdot O_2^-$ = superoxide anion
OZ = opsonized zymosan
P. yoelii = *Plasmodium yoelii*
PAM = pulmonary alveolar macrophage
PBL = peripheral blood lymphocyte
PBM = peripheral blood monocyte/macrophage
PEC = peritoneal exudate cell
PFC = plaque-forming cell
PGE_2 = prostaglandin E_2
PHA = phytohemagglutinin (T-cell mitogen)
PMA = phorbol myristate acetate
PMN = polymorphonuclear leukocyte
PRV = pseudorabies virus
PWM = pokeweed mitogen (B- and T-cell mitogen)
RBC = red blood cells
ROI = reactive oxygen intermediates
RT6:$RT6^+$ = T-lymphocyte differentiation surface marker
S. pneumoniae = *Streptococcus pneumoniae*
SC = subcutaneous injection
sIg = surface immunoglobulin
SRBC = sheep red blood cells
T. spiralis = *Trichinella spiralis*
TCR = T-cell receptor
T_h = T-helper cell ($CD4^+$)
TNF-α = tumor necrosis factor-alpha
TNP-LPS = trinitrophenyl-lipopolysaccharide
T_s = T-suppressor cell ($CD8^+$)
U937 = human monocyte cell line
VSV = vesicular stomatitis virus
WBC = white blood cell

3 TABLES

TABLE 1

Effects of Aluminum Compounds on Immune System Parameters

Compound	Concentration (Exposure Protocol)	Host	Parameter Affected	Range of Effects	Dose-Response	Ref.
In Vivo *Exposures*						
Aluminum chlorhydrate	0.1, 1, or 10 mg; IT (injection every 3 d for 10 d)	Rabbit (New Zealand)	PAM phagocytosis	NE/↑	No	1
			Lavagable lung MNC numbers	NE/↑	No	
			DTH reaction	NE		
Aluminum hydroxide	5 mg/kg; IP (single injection)	Mouse (NMRI)	MNC mitotic index	↑		2
	20 mg/kg; IP (single injection)	Rat (THOM)	MNC mitotic index	↑		
	40 mg/kg; IP (single injection)	Hamster (Chinese)	MNC mitotic index	↑		
Aluminum hydroxychloride	212 μg/l; inhalation (4 hr/d, 5 d)	Rabbit (New Zealand)	Lung PAM/PMN numbers	↑		3
	164 μg/l; inhalation (4 hr/d, 3 d)	Hamster (Syrian Golden)	Lung PAM/PMN numbers	↑		
Aluminum lactate	1 mg/g diet (6 months)	Mouse (Swiss-Webster)	Splenic index	↑		4
			ConA-induced splenocyte IL-2, IFN-γ, and TNF-α production	↓		
			Splenic $CD4^+$ cell numbers	↓		
			Splenic $CD8^+$ cell numbers	NE		
	0.5 or 1 mg/g diet (6 wk)	Mouse (Swiss-Webster)	*L. monocytogenes* resistance	↓	Yes	5
	1, 5, or 10 mg/kg; SC (every 2 d, for 20 d)	Mouse (Swiss-Webster)	*L. monocytogenes* resistance	↓	No	5
	100 mg; inhalation (monthly; 6 months)	Sheep	Total lung immune cell numbers	NE		6
			Total lung PAM, PMN, and lymphocyte numbers	NE		
			Lung Ig levels	NE		
Aluminum metal powder	2.5 mg; IP (single injection)	Mouse (CD1)	PEM numbers	↑		7
			Peritoneal PMN and lymphocyte numbers	NE		

TABLE 1 (continued)
Effects of Aluminum Compounds on Immune System Parameters

Compound	Concentration (Exposure Protocol)	Host	Parameter Affected	Range of Effects	Dose-Response	Ref.
Aluminum oxide (alumina)	2.5 mg; IP (single injection)	Mouse (CD1)	PEM phagocytosis (metal particle)	↓		7
			PEM numbers	↑		
			Peritoneal PMN and lymphocyte numbers	NE		
In Vitro *Exposures*						
Aluminum chlorhydrate	100 μg/ml (Al presensitized)	Rabbits	PAM MIF production	↓		1
			PAM random migration	↓		
	10 mg/ml	Guinea pig	Serum C3 activation	↑		8
	0.1–1 mg/ml	Guinea pig	PEM LDH release	↑	Yes	9
Aluminum chloride	2–4 m*M*	Human	Whole blood culture			10
			Cell cycle time	↑	Yes	
			Mitogen-induced blastogenesis	↓	Yes	
			Mitotic index	↑/↓		
	76 μ*M*–4.8 m*M*	Human	Thymocyte proliferation	NE/↓	Yes	11
			PBL proliferation	NE/↓	Yes	
	2.5–250 μ*M*	Human	PBL mitogenic responsiveness	NE		12
	1–100 μ*M*	Rat	PAM-stimulated O_2 consumption	↓	Yes	13
			PAM ROI production	↓	Yes	
			PAM viability	↓	Yes	
Aluminum citrate	2.5–250 μ*M*	Human	PBL mitogenic responsiveness	↓	Yes	12
Aluminum fluoride	10–20 μ*M*	Mouse	2B4 cell intracellular Ca^{2+} levels	↑	No	14
			TCR phosphorylation	↑	No	
Aluminum hydroxychloride	1–1000 μg/ml (Al presensitized)	Rabbit	PAM MIF production	↑	Yes	15
	3, 10, or 30 μg/ml	Rabbit	PAM enzyme release (β-Glu, β-NAG, acid phosphatase)	↓	Yes	16
			Lymphocyte viability	↓	Yes	
Aluminum hydroxide						
	0.1–1 mg/ml	Guinea pig	PEM LDH release	↑	Yes	9
	15 mg/ml	Guinea pig	Serum C3 activation	↑		8

Aluminum nitrate	1–100 μM	Human	PBL mitogen responsiveness	NE		17
	100 μM		PBM spontaneous TNF-α production	↓		
			PBM spontaneous IL-1-α/β production	NE		
			PHA-induced PBL IL-2 production	NE		
			PHA-induced PBM/PBL IL-1α/β, TNF-α, and IFN-γ production	↑		
Aluminum oxide (alumina)	0.1–0.5 mg/dish	Hamster	PAM viability	NE		18
	0.1–0.5 mg/dish	Rat	PAM viability	NE		18
	1–20 mg/dish	Rabbit	PAM H_2O_2 production	↑	Yes	19
	1–20 mg/dish	Human	PBM H_2O_2 production	↑	Yes	19
			PMN H_2O_2 production	NE		
			PMN $\cdot O_2^-$ production	↓	Yes	
Aluminum sulphate	200 μM	Mouse	Ig cap formation	↓		20

TABLE 2

Effects of Arsenic Compounds on Immune System Parameters

Compound	Concentration (Exposure Protocol)	Host	Parameter Affected	Range of Effects	Dose-Response	Ref.
In Vivo *Exposures*						
Arsenic trioxide	125–1000 μg/m³ (single inhalation, 3 hr)	Mouse (CD-1)	Mortality due to *S. pneumoniae*	↑	Yes	21
			Bactericidal activity vs. inhaled *K. pneumoniae*	↓	No	
	125–1000 μg/m³ (3 hr/d, 5 d/wk, 20 d)	Mouse (CD-1)	Mortality due to *S. pneumoniae*	↑	No	
			Bactericidal activity vs. inhaled *K. pneumoniae*	↓	No	
	5 mg/kg; SC (virus co-injected)	Mouse (Balb/c)	Mortality due to PRV infection	↑	No	22
Sodium arsenite	3, 5, or 6.25 mg/kg; SC (injections 3×/wk, 8 wk)	Mouse (Swiss-Webster)	Mortality due to PRV infection	↑	No	23
	1 or 2 *M*; drinking water	Mouse (CD-1)	Mortality due to PRV infection	↑	No	
	10 ppm; drinking water	Mouse (C3H)	Tumor incidence/tumor growth	↓/↑	No	24
	1 mg/ml; IT	Rat (Sprague-Dawley)	BAL PGE_2/TNF-α levels	NE/NE		25
			PAM induced TNF-α, $\cdot O_2^-$ and PGE_2 production	↓/NE/NE	No	
	1 or 2 mg/kg; IP (prenatal injection)	Mouse (B6C3F1)	Splenic T-cell proliferation	↑	No	26
			Splenic MLR	↑	No	
			Peritoneal MØ phagocytic activity	↑	No	
Sodium arsenate	2.5, 25, or 100 ppm; drinking water (10–12 wk)	Mouse (C57BL/6)	Tumor latency/tumor incidence			
			From injected sarcoma cells	↑/↓	No	27
			From injected virus	NE		
			Cell-mediated cytotoxicity *in vivo*	↑	No	
			Cell-mediated cytotoxicity *in vitro*	↓/↑		
	1 mg/ml; IT (assays 24 hr later)	Rat (Sprague-Dawley)	BAL PGE_2/TNF-α levels	NE		
			PAM-induced TNF-α, $\cdot O_2^-$, and PGE_2 production	↓/↑/NE	No	25

Gallium arsenide	50, 100, or 200 mg/kg; IT (single instillation; assay 24 hr later)	Mouse (B6C3F1)	Primary IgM Ab response to SRBC	↓	No	28
			Splenic cell SRBC Ag presentation and processing	↓	No	
			MØ IA⁺ percentage/expression	NE/↓	No	
			Splenic cell phagocytosis	NE		
			Splenic cell cytochrome C/KLH Ag presentation	NE/NE		
			Splenic cell IL-1 production	NE		
	50, 100, or 200 mg/kg; IT (single instillation; assay 14 d later)	Mouse (B6C3F1)	Spleen cellularity	↑	Yes	29
			Percentage Thy 1.2/Ig⁺/F4/80	↓/↓/↑	Yes	
			IgM, IgG PFC response to SRBC	↓	No	
			Serum complement (C3)	↑	No	
			Splenic T-cell response to ConA and PHA	NE		
			Splenic B-cell response to LPS	↑	No	
			Splenic MLR	↑	Yes	
			DTH response to KLH	↑	Yes	
			NK-cell activity	↑	No	
			PEC number	↓	Yes	
			Percentage of PEC monocytes	↑	No	
			Percentage of PEC lymphocyes	↓	No	
			Covaspheres PEC phagocytosis	↓	No	
			Chicken erythrocyte PEC phagocytosis	↑	No	
			Host resistance to *P. yoelii* and *S. pneumoniae*	NE		
			Host resistance to *L. monocytogenes*	↑	Yes	
			Host resistance to B16F10 cells	↓	No	
	200 mg/kg; IT (single instillation; assay 24 d later)	Mouse	Splenocyte proliferation	↓	No	30
			CD8⁺/CD4⁺ cell numbers	NE/↓		
			LPS-induced B-cell proliferation	NE		
			ConA/PHA-induced proliferation	↓	No	
			Anti-CD3ε/L-2-induced T-cell proliferation	↓	No	
			CD25 expression	↓	No	
			Leukocyte function antigen-1	↓	No	
			Expression of intercellular adhesion molecule-1	↓	No	
Arsenic-associated fly ash	Inhalation (concentration unspecified)	Human	Blood serum Ig (G, M, A) levels	NE		31
			Blood α-1-antitrypsin/α-2-macroglobulin levels	NE		
			Blood transferrin/orosomucoid/ceruloplasm levels	↑		
Arsenic	Drinking water (III — 2% and V — 98%)	Human	Lymphocyte proliferation	↓	No	32
	Drinking water (concentration unspecified)	Human	MNC PHA mitogenic responses (cells from cancer patients)	↑		33

TABLE 2 (continued)

Effects of Arsenic Compounds on Immune System Parameters

Compound	Concentration (Exposure Protocol)	Host	Parameter Affected	Range of Effects	Dose-Response	Ref.
Arsine	0.5, 2.5, or 5 ppm inhalation (6 hr/d, 14 d)	Mice (B6C3F1)	Percentages of splenic lymphocytes	↓	No	34
			Splenic T-cell percentages	↓	Yes	
			Splenic B-cell percentages	↓	No	
			PEM numbers and cytotoxic activity	NE		
			NK-cell activity	↓	Yes	
			Cytotoxic T-lymphocyte activity	↓	Yes	
			T- and B-cell proliferation	NE		
			Host resistance to PYB6 tumor cells, B16F10 melanoma, or influenza virus	NE		
			Host resistance to *L. monocytogenes* and *P. yoelii*	↓	No	
	0.5, 2.5, or 5 ppm inhalation (6 hr/d, 14 d)	Mice (B6C3F1)	RBC numbers, hematocrit, and hemoglobin levels	↓	No	35
			White blood cell counts	↑	No	
			Marrow-associated erythroid precursor numbers	↓	No	
			Splenic erythropoesis	↑	No	
			Bone marrow cellularity	NE		
			Granulocyte/macrophage precursor numbers	↓	No	
In Vitro *Exposures*						
Sodium Arsenite	25 μ*M*	Mouse	PFC response	↓	Yes	36
	1–100 n*M*	Human	PBL proliferation	↓	Yes	37
	0.001 μ*M*	Human	PBL PHA-proliferation	↓	No	
Sodium Arsenite	10–500 ng/ml	Mouse	PFC response	↑/↓	Yes	38
			Viable B- and T-cell numbers	↓	No	
	2–10 μ*M*	Bovine	PBL PHA-induced proliferation	↑/↓	Yes	39
	2–20 μ*M*	Human	PBL PHA-induced proliferation	↑/↓	Yes	
	0.1–400 μg/ml	Rat	PAM-induced $\cdot O_2^-$ production	↓	Yes	25
			PAM-induced TNF-α production	↓	Yes	
			PAM-induced PGE_2 production	NE		
	1–100 μ*M*	Mouse	VSV-induced IFN production	↓	Yes	23
		Rabbit	VSV-induced IFN production	↓	No	
	8 n*M*–1 m*M*	Human	PHA-stimulated lymphocyte DNA synthesis	↑/↓	Yes	40

Sodium arsenate	1–100 nM	Human	PBL proliferation	↓	Yes	37
	0.001 μM	Human	PBL PHA proliferation	↓	No	
	20–152 μM	Bovine	PBL PHA-induced proliferation	↑/↓	Yes	39
	10–600 μM	Human	PBL PHA-induced proliferation	↑/↓	Yes	
	1–400 μg/ml	Rat	PAM induced $\cdot O_2^-$ production	↓	Yes	25
			PAM induced TNF-α production	↓	Yes	
			PAM induced PGE_2 production	↓	No	
	1 mM	Mouse	VSV-induced IFN production	↓	No	23
	0.1 μM–10 mM	Human	PHA-stimulated lymphocyte DNA synthesis	↑/↓	Yes	40
Gallium arsenide	25 μM	Mouse	PFC response	↓	Yes	36
Arsenic trioxide	8 nM–10 μM	Human	PHA-stimulated lymphocyte DNA synthesis	↑/↓	Yes	40

TABLE 3

Effects of Beryllium Compounds on Immune System Parameters

Compound	Concentration (Exposure Protocol)	Host	Parameter Affected	Range of Effects	Dose-Response	Ref.
In Vivo *Exposures*						
Beryllium sulfate	0.100 mg/ml; IV (single dose)	Mouse (CD-1)	Spleen red pulp size/RBC numbers	↓		41
			Splenic white pulp T-cell zone size	↓		
	3 mg/kg; IP (single dose)	Mouse (C57B16 × DBA/2)	PEM ROI production	↑		42
			Splenic index	↑		
			PEM PMA-induced ROI production	NE		
			Splenocyte mitogen responsiveness	↓		
			Splenocyte $BeSO_4$-induced proliferation	↓/NE		
	1–5 μg; IT (single dose; Be presensitized 3 times with 5–50 μg $BeSO_4$; IT)	Mouse (A/J)	Lung immune cell numbers (total)	↑	Yes	43
			Lung monocyte/PAM levels	↑	Yes	
			MAC-1, -2, -3 PAM surface marker expression	↑	Yes	
			Lung lymphocyte levels	↑	Yes	
			Splenic lymphocyte proliferation ($BeSO_4$ antigen)	↑	Yes	
	13 mg/m^3; inhalation (single 1-hr exposure)	Mouse (BALB/c)	Lung PAM levels	↑		44
			PAM membrane damage	↑		
	0.1–10 μg; IP (single dose)	Mouse (multiple strains)	PEM Ia$^+$ expression	↑	Yes	45
	50 mg; ear painting	Guinea pig (Hartley)	DTH reaction ($BeSO_4$ Ag)	↑		46
			Lymph node PMN levels	↑		
	2.6 mg; IP (1 injection/wk, 2 wk)	Guinea pig (Hartley)	DTH reaction (vs. Be salts as Ag)	↑		47
			PEM MIF production ($BeSO_4$ or BeF_2 Ags)	↑		
			PEM lymphocytotoxin production	↑		

	10 μg; ID toe-pad (2 injections/wk, 6 wk)	Guinea pig (Hartley)	DTH reaction (Be salts Ags)	↑		47
			PEM MIF production ($BeSO_4$ or BeF_2 Ags)	↑		
			PEM lymphocytotoxin production	↑		
	5, 15, or 30 μg/m³ (inhalation for 60 d after) 0.2 mg/l water; 36 d	Guinea pig (Hartley)			No	48
			DTH reaction ($BeSO_4$ Ag)	NE		
			DTH reaction (Be-citrate Ag)	NE		
			DTH reaction (Be-albuminate Ag)	NE		
			DTH reaction (Be-aurintricarboxylate Ag)	NE		
	5 μg/injection; ID (12 doses over 6 wk)		DTH reaction ($BeSO_4$ Ag)	↑		
			DTH reaction (Be-albuminate Ag)	↑		
			DTH reaction (Be-aurintricarboxylate Ag)	NE		
			DTH reaction (Be-citrate Ag)	NE/↓		
	100 μg; ID toe-pad (2 doses/wk, 6 wk)	Rabbit (New Zealand)	DTH reaction ($BeSO_4$ Ag)	↑		15
			PAM MIF production	↑		
			Popliteal lymph node lymphocyte mitogen responsiveness	↓		
	10 mg; ID toe-pad (single dose)	Rabbit (New Zealand)	Popliteal lymph node lymphocyte mitogen responsiveness	↓		
	13 mg/m³; inhalation (single 1-hr exposure)	Rat (F344)	PAM numbers	↑		44
			PAM membrane damage	↑		
Beryllium sulphosalicylate	500 μM-treated lymphocytes (4 × 10⁷ injected IP)	Guinea pig (Hartley)	DTH reaction (BeF_2 Ag)	↑		49
Beryllium metal	800 mg/m³; inhalation (single 50 min exposure)	Rat (F344)	Lung PAM and PMN numbers	↑		50
			PAM morphological changes	↑		
			BAL protein and LDH levels	↑		
Beryllium hydroxide	50–150 mg; SC (single dose)	Sheep	Regional lymph node lymphoblast proliferation	↑		51
Beryllium oxide	2 mg/ml; IT (single dose)	Guinea pig (Hartley)	Lung PAM and PMN numbers	↑		46
			DTH reaction ($BeSO_4$ Ag)	↑		
	4, 20, or 100 μg; IT (single dose)	Mouse (A/J)	Lung lymphocyte numbers	↑	No	43
			Splenic lymphocyte proliferation ($BeSO_4$ Ag)	↑	Yes	
Coal fly ash (with Be)	0.3–8 μg Be/m³ (inhalation; 5–10 yr)	Human	Anti-organ autoantibody levels	↑	ND	52
			Serum IgA, IgM, and IgG levels	↑	ND	

TABLE 3 (continued)

Effects of Beryllium Compounds on Immune System Parameters

Compound	Concentration (Exposure Protocol)	Host	Parameter Affected	Range of Effects	Dose-Response	Ref.
In Vitro *Exposures*						
Beryllium sulfate	1–50 μ*M*	Mouse	Splenic lymphocyte viability	↓	Yes	53
			Splenic lymphocyte proliferation ($BeSO_4$ Ag)	↑/↓		
	0.1 μg/ml	Mouse	PEM IL-1α release	↑		54
	1 ng–10 μg	Mouse	PEM viability	↓	Yes	45
			PEM IL-1α release	NE		
			PEM Ia expression	NE		
	10 n*M*–10 m*M*	Mouse	Splenic lymphocyte proliferation ($BeSO_4$ Ag)	↑	Yes	55
			Splenic B-lymphocyte proliferation ($BeSO_4$ Ag)	↑	Yes	
			Thymocyte proliferation ($BeSO_4$Ag)	NE		
	1–200 μ*M*	Mouse	B-lymphocyte Ig cap formation	↓	Yes	20
	0.01–100 μ*M*	Mouse	Splenic lymphocyte proliferation ($BeSO_4$ Ag)	NE		43
	1 μ*M*–10 m*M*	Dog	PAM viability	↓	Yes	56
	5 μg	Guinea pig	PEM MIF production	NE		47
	0.07–7 μ*M*	Human	PBL $BeSO_4$-induced maturation	NE		57
	0.1–100 μg/ml	Human	PBL $BeSO_4$-induced proliferation	NE		58
			Lung lymphocyte $BeSO_4$-induced proliferation	NE		
	1–1000 μg/ml ($BeSO_4$-presensitized host)	Rabbit	PAM membrane integrity	↓	Yes	15
			PAM viability	↓	Yes	
			PAM protein synthesis	↓	Yes	
	3, 10, or 30 μg/ml	Rabbit	PAM enzyme release (β-Glu, β-NAG, acid phosphatase)	↑	Yes	59
			PAM viability	↓	Yes	

Beryllium oxide	1 μ*M*–10 m*M*	Dog	PAM viability	↓	Yes	56
	30–1500 μg/ml	Rabbit	PAM enzyme release (β-Glu, β-NAG, acid phosphatase)	↑	Yes	59
			PAM viability	↓	Yes	
Coal fly ash (with Be)	1–1000 μg Be/ml	Rat	PAM viability	↓	Yes	60
			PAM phagocytic index	↓	Yes	
Beryllium sulphosalicylate	1–50 μ*M*	Mouse	Lymphocyte viability	↓	Yes	53
			Lymphocyte mitogen responsiveness	↑/↓		
Beryllium chloride	0.01–2.5 μg/ml	Rat	PAM uptake of Be	↑	Yes	61
	14–1400 μ*M*	Rat	Splenic B-cell-dependent immune hemolysis	↓	Yes	62

TABLE 4

Effects of Chromium Compounds on Immune System Parameters

Compound	Concentration (Exposure Protocol)	Host	Parameter Affected	Range of Effects	Dose-Response	Ref.
In Vivo *Exposures*						
Calcium chromate	2 μg; IT (single dose)	Rat (Long-Evans)	PAM OZ-induced chemiluminescence	NE		63
			PAM O_2 consumption	NE		
Chromic chloride	1.4–10.9 μg/g; IM (single injection)	Mouse (Swiss albino)	Splenic α-SRBC plaque formation	NE		64
Chromic nitrate	0.6 mg/m³; inhalation (6 hr/d/5 d, 4-6 wk)	Rabbit (New Zealand)	PAM numbers/size	↑		65, 66
			PAM morphological changes (lamellar inclusions, enlarged Golgi)	↑		
			PAM ROI production	↓		
			PAM phagocytic activity	NE		
Chromium phosphate	1 μg/Ci (single IV dose)	Rat (albino)	Serum PMN numbers	↑		67
			Serum MNC numbers	↓		
			α-SRBC synthesis/titers	↓		
			Whole body carbon clearance	↓		
			Skin graft survival	NE		
Chromium trioxide	2 μg; IT (single dose)	Rat (Long-Evans)	PAM OZ-induced chemiluminescence	NE		63
			PAM O_2 consumption	NE		
Sodium chromate	0.9 mg/m³; inhalation (6 hr/d/5 d; 4-6 wk)	Rabbit (New Zealand)	PAM numbers/size	NE		65, 66
			PAM ROI production	NE		
			PAM phagocytic activity	NE		
			PAM morphological changes (lamellar inclusions, enlarged Golgi)	↑		
	13.3 mg/kg; SC (every 2 d, 30 d)	Rat (Sprague-Dawley)	α-*E. coli* T-1 phage Ab formation	↓		68
Sodium dichromate	25 or 50 μg/m³ (22 hr/d, 28 d)	Rat (Wistar)	Serum Ig levels	NE		69
			Splenic α-SRBC plaque formation	NE		
			PAM numbers (% of BAL)	NE		
			Lymphocyte and PMN numbers	↑	Yes	
			PAM phagocytic activity	↑	Yes	

	25, 50, or 200 μg/m³ (22 hr/d; 90 d)	Rat (Wistar)	Serum Ig levels	↑/↓	Yes	69
			Splenic α-SRBC plaque formation	↑/↓	Yes	
			PAM phagocytic activity	↑/↓	Yes	
			Lymphocyte and PMN numbers	↑/↓	Yes	
			PAM numbers (% of BAL)	↓	Yes	
	1–30 mg/dose (per os) (weekly; 3 wk)	Guinea pig (Duncan-Hartley)	Chromium skin tolerance	↑	Yes	70
	10 mg/dose (per os) (1–6 doses/wk, 3 wk)		Chromium skin tolerance	↑		
In Vitro *Exposures*						
Calcium chromate	2 μg/dish	Rat	PAM OZ-induced chemiluminescence	↓		63
			PAM O_2 consumption	↓		
Chromic chloride	0.125–1.0 m*M*	Rat	Thymocyte ATP/GTP production	NE		71
			Thymocyte O_2 consumption	NE		
	1–500 μ*M*	Rat	PAM stimulated O_2 consumption	↓	Yes	72
			PAM ROI production	↓	Yes	
			PAM viability	↓	Yes	
	3.1 m*M*	Rabbit	PAM phagocytosis	↓		73
	10 n*M*–1.0 m*M*	Human	PBL mitogen responsiveness	NE		74
Chromium (metallic)	0.05–0.5 mg/ml	Monkey	LLC-MK2 kidney cell IFN-α/β production	↓	Yes	75
Chromium trioxide	2 μg/dish	Rat	PAM OZ-induced chemiluminescence	↓		63
			PAM O_2 consumption	↓		
Potassium chromate	20 μ*M*	Rat	PAM viability	↓		76
			PAM LDH release	↑		
Potassium dichromate	0.125–1.0 m*M*	Rat	Thymocyte ATP/GTP production	↓	Yes	71
			Thymocyte O_2 consumption	↓	Yes	
	10 n*M*–0.1 m*M*	Human	PBL mitogen responsiveness	↑/↓		74
Sodium chromate	2.5–40 μ*M*	Mouse	PEM random migration	↓	Yes	77
			PEM phagocytic activity	↓	Yes	
			PEM IFN-α/β production	↓	Yes	
	Unspecified	Mouse	L-929 fibroblast IFN-α/β production	↓		78
	0.001–100 μg/ml	Cow	PAM viability/phagocytosis	↓	Yes	79
Sodium dichromate	1 n*M*–1 m*M*	Human	T-cell mitogen responsiveness	NE/↓	Yes	80
			T-cell IL-2 production	↓	Yes	

TABLE 5

Effects of Gold Compounds on Immune System Parameters

Compound	Concentration (Exposure Protocol)	Host	Parameter Affected	Range of Effects	Dose-Response	Ref.
In Vivo *Exposures*						
Triethylphosphine gold (Auranofin)	1.5–5 mg/kg; SC (daily for 8 d)	Mouse (CFLP)	α-SRBC Ab levels	NE		81
			DTH reaction	NE		
	2.5 mg/kg/d (4-d feedings)	Mouse (Balb/c)	PEM Ag presentaition	NE		82
			PEM IL-1 production	NE		
			DTH reaction	NE		
			Splenic lymphocyte mitogen responsiveness	NE		
			Splenic lymphocyte IL-2 production	NE		
			Splenic T_s lymphocyte induction and functionality	NE		
	20 mg/kg (per os)	Rat (Lewis)	Carageenan/kaolin edema	↓		83
			Passive cutaneous anaphylaxis	↓		
			IgG hemagglutinin	↓		
			ADCC Ab levels	↓		
			Splenic hemolysin PFC numbers	↓		
	3 mg/day (per os) (6 months)	Human	PBL mitogen responsiveness	↓		84
			Blood IgG/IgM levels	↓/NE		
			PBM phagocytic activity	↓/↑		
	6 mg/d (per os) (45 wk)	Human	PBL mitogen responsiveness	↓		85
			Lymphocyte blebbing	↑		
Gold sodium thiomalate (GSTM, Myochrysine)	1.5–5 mg/kg; SC (daily for 8 d)	Mouse (CFLP)	α-SRBC Ab levels	NE		81
			DTH reaction	NE		
	22.5 mg/kg; IM (weekly; 12 wk)	Mouse (C57Bl/6; DBA/2)	Antinuclear, antinucleolar serum autoantibody levels	↑		86
	0.1–5.0 μM (single footpad injection)		Popliteal lymph node size	NE		86
	5 mg/kg (per os)	Rat (Lewis)	Carageenan/kaolin edema	↓/NE		83
			Passive cutaneous anaphylaxis	↓		
			IgG hemagglutinin levels	NE		
			ADCC Ab levels	NE/↑		
			Splenic hemolysin PFC numbers	NE		
	50–70 mg/wk; IM (1 yr)	Human	PBL mitogen responsiveness	↓/↑		85

Gold sodium thioglucose (GSTG)	1.5–5 mg/kg; SC (daily for 8 d)	Mouse (CFLP)	α-SRBC Ab levels	NE		81
			DTH reaction	NE		
Aurothiopropanol-sulphonate (ATPS)	20 mg/kg; SC (3×/wk, 8 wk)	Rat (B.N.; Lewis)	B-cell polyclonal activation	↑		87
			B-cell mitogen responsiveness	↑		
			Blood Ig levels/IgE levels	↑		
			Autoantibody formation	↑		
Gold (III) chlorides ($AuCl_3$, $HAuCl_4$)	0.25–1.0 μ*M* (single footpad injection)	Mouse (C57Bl/6; DBA/2)	Popliteal lymph node size	↑	Yes	86
Gold (I) chlorides (AuCl)	0.3–0.5 μ*M* (single footpad injection)	Mouse (C57Bl/6; DBA/2)	Popliteal lymph node size	NE		86
In Vitro *Exposures*						
Triethylphosphine gold (Auranofin)	5–100 μ*M*	Mouse	PEM ROI production	↓	Yes	88
	0.25–2.5 μ*M*	Mouse	PEM Ag presentation	↓	Yes	82
			PEM IL-1 production	↓	Yes	
	0.5–2 μ*M*		Splenic lymphocyte mitogen responsiveness	↓	Yes	
	0.125–0.5 μ*M*		Splenic lymphocyte IL-2 production	↓	Yes	
	2 μ*M*		Splenic T_s lymphocyte induction and functionality	NE		
	1–100 μg/assay	Mouse	PEM F_cR and C3bR expression	↓	Yes	89
			PEM chemotaxis	↓		
	3 n*M*–1 μ*M*	Mouse	Thymocyte IL-1 responsiveness	↓	Yes	90
			Thymocyte IL-1 viability	↓	Yes	
	1 μ*M*–10 m*M*	Mouse	PEM enzyme activity (β-NAG, β-Gluc, LDH, cathepsins B and D)	↓	Yes	81
	1–10 μg/ml	Human	PBM IL-1 production	↓	Yes	91
	0.05–0.80 μg/ml	Human	PBL mitogen responsiveness	↓	Yes	92
			PBL IL-2 responsiveness	↓	Yes	
			PBL IFN-γ production	↓	Yes	
			PBL PGE_2 production	↓	Yes	
	0.25–2.5 μg/ml	Human	PBM ADCC	↓	Yes	93
			PBL ADCC	↓	Yes	
			NK-cell activity	↓/↑		
			DTH reaction	NE		
	0.1–10 μg/chamber	Human	PEM chemotaxis	↓	Yes	94
			PBM F_cR and C3bR expression	↓	Yes	
			PBM ROI production	↓	Yes	
			PBM intracellular cAMP	↑	Yes	

TABLE 5 (continued)

Effects of Gold Compounds on Immune System Parameters

Compound	Concentration (Exposure Protocol)	Host	Parameter Affected	Range of Effects	Dose-Response	Ref.
	0.4–10 μ*M*	Human	PBM ADCC	↑	Yes	95
			PBM-target (ADCC) contacts	↑		
	0.18–1.36 μg/ml	Human	PBM lysosomal enzyme release	↓	Yes	96
	1 μg/ml	Human	MØ (synovial) ROI production	↓	Yes	97
	1.0–2.5 μg/ml	Human	IFN-γ-induced co-mitogenesis	↓	Yes	98
			PBM LPS-induced IL-1α/β production	↓	Yes	
	0.1–2 μg/ml	Human	PBM ROI production	↓/↑		99
	0.1–1 μg/ml	Human	T-cell mitogen responsiveness	↓	Yes	100
			T-cell Ab-mitogen-induced proliferation	↓	Yes	
			T-cell IL-2 production	↓	Yes	
Gold sodium thiomalate (GSTM, Myochrysine)	1 μ*M*–10 m*M*	Mouse	PEM cathepsins B and D activities	↓	Yes	81
			PEM β-Gluc, β-NAG, LDH activities	NE		
	3 n*M*–166 μ*M*	Mouse	Thymocyte IL-1 responsiveness	↓	Yes	90
			Thymocyte IL-1 viability	↓	Yes	
	1–100 μg/ml	Human	PBM ADCC	NE		93
			PBL ADCC	NE		
	0.1–10 μg/chamber	Human	PBM chemotaxis	↓	Yes	94
			PBM F_cR and C3bR expression	↓	Yes	
			PBM NBT reduction	↓	Yes	
			PBM intracellular cAMP	↑	Yes	
	1–40 μg/ml	Human	PBM lysosomal enzyme release	↓	Yes	96, 101
	25–50 μg/ml	Human	MØ (synovial) ROI production	NE		97
	2–20 μg/ml	Human	PBM IL-1 production	↓	Yes	91
	3.2–200 μg/ml	Human	PBL mitogen responsiveness	↓	Yes	92
			PBL IL-2 responsiveness	↓	Yes	
			PBL IFN-γ production	↓	Yes	
	1–100 μg/ml	Human	PBM ROI production	NE		99
	0.1–1 μg/ml	Human	T-cell mitogen/Ab-induced proliferation	↓	Yes	100
			T-cell IL-2 production	↓	Yes	
	2.0–20 μg/ml	Human	IFN-γ-induced co-mitogenesis	↓	Yes	98
			PBM LPS-induced IL-1α/β production	↓	Yes	
	1–100 μg/ml	Human	PBL mitogen responsiveness	↓	Yes	102

TABLE 5 (continued)

Effects of Gold Compounds on Immune System Parameters

Compound	Concentration (Exposure Protocol)	Host	Parameter Affected	Range of Effects	Dose-Response	Ref.
Gold sodium thioglucose (GSTG)	1 μ*M*–1 m*M*	Mouse	PEM ROI production	↓	Yes	88
	50–800 μg/ml	Human	PBL mitogen responsiveness	↓	Yes	92
			PBL IL-2 responsiveness	↓	Yes	
			PBL IFN-γ production	↓	Yes	
	3 n*M*–166 μ*M*	Mouse	Thymocyte IL-1 responsiveness	↓	Yes	90
			Thymocyte IL-1 viability	↓	Yes	
Gold (III) chlorides ($AuCl_3$, $HAuCl_4$)	10–50 μg/ml	Human	PBL mitogen responsiveness	↓	Yes	102

TABLE 6

Effects of Mercury Compounds on Immune System Parameters

Compound	Concentration (Exposure Protocol)	Host	Parameter Affected	Range of Effects	Dose-Response	Ref.
In Vivo *Exposures*						
Mercuric chloride	25–100 μg/ml (water, 4 wk)	Mouse (CD-1)	PEM phagocytic activity	↑	No	103
			Bone marrow lymphoid/total cell ratios	NE		
			Bone marrow B-cell maturation	NE		
			Bone marrow B-cell maturation surface marker expression	NE		
			Splenic B-cell sIg expression	NE		
			Splenic T-cell marker (Thy 1.2, Lyt 2, L3T4) expression	↓/NE	No	
	3, 15, or 75 μg/ml (water, 7 wk)	Mouse (B6C3F1)	Thymic index	↓	Yes	104
			Splenic index	↓	Yes	
			Splenic α-SRBC PFC numbers	↓	No	
			Splenic T-lymphocyte mitogen responsiveness	↓	No	
			Splenic T-lymphocyte MLR	↓	No	
			Splenic B-lymphocyte mitogen responsiveness	↑	No	
			Total blood WBC numbers	↑/↓		
			Total blood RBC numbers	NE		
			Splenic α-LPS PFC numbers	NE		
			Serum Ig levels	NE		
	5 or 20 μg; IP (single dose)	Mouse (Balb/c)	HSV-2 clearance (hepatic)	↓	Yes	105
	5 mg/l (water, 2 wk)	Mouse (SJL/N)	Splenic T-lymphocyte mitogen responsiveness	↑		106
			Splenic B-lymphocyte mitogen responsiveness	↑		
		Mouse (DBA)	Lymphocyte (splenic) mitogen responsiveness	NE		
	0.625–5 mg/l (water, 10 wk)	Mouse (SJL/N)	Antinucleolar Ab formation	↑		107
	1.6 mg/kg; SC (every 3 d; 6 wk)	Mouse (multiple strains)	Antinucleolar Ab formation	↑		108

0.5–2 mg/kg; SC (dose increased each day over 6-d period)	Mouse (multiple strains)	Antinuclear Ab formation	↑	109
1 mg/kg; SC (3/wk, 4 wk)	Mouse			
	(WHT/Ht)	Antinucleolar Ab formation	↑	110, 111
	(ICR)	Antinucleolar Ab formation	↑	
1 mg/kg; SC (3/wk, 4 wk)	Mouse			
	(C57Bl/6)	Antinucleolar Ab formation	NE	110, 111
	(CBA, C3H)	Antinucleolar Ab formation	NE	
	(Balb/c)	Antinucleolar Ab formation	NE	
	(DBA)	Antinucleolar Ab formation	NE	
1 mg/kg; SC (3/wk, 2 wk)	Rat			
	(BN)	Serum α-lamanin Ab levels	↑	112
		Spleen cell numbers (total)	↑	
		Lymph node cell numbers (total)	↑	
		Splenic T-lymphocyte ratio ($RT6^-:RT6^+$)	NE	
		Lymph node T-lymphocyte ratio ($RT6^-:RT6^+$)	NE	
	(Lewis)	Serum α-lamanin Ab levels	NE	
		Spleen cell numbers (total)	↑	
		Lymph node cell numbers (total)	↑	
		Splenic T-lymphocyte ratio ($RT6^-:RT6^+$)	↑	
		Lymph node T-lymphocyte ratio ($RT6^-:RT6^+$)	↑	
0.5 mg/kg; SC (3×/wk, 1–2 months)	Rat (BN)	Serum IgE levels	↓	113
2 mg/kg; SC (3×/wk, 1–2 months)	Rat (Lewis)	Serum IgE levels	NE	113
1 mg/kg; IP (1 or 2 injections within 48-hr period)	Rat	Splenic IFN-γ-producing cell numbers		
		Brown Norway rat	↓	114
		Lewis rat	NE	
300 μg/ml (water, 11 wk)	Chicken (Hubbard)	α-SRBC Ab formation		115
		Primary response	↓	
		Secondary response	↓	
		Serum IgM, IgG levels	↓	
		α-*B. abortus* Ab formation		
		Primary response	↓	
		Secondary response	↑	

TABLE 6 (continued)

Effects of Mercury Compounds on Immune System Parameters

Compound	Concentration (Exposure Protocol)	Host	Parameter Affected	Range of Effects	Dose-Response	Ref.
	3 or 12 mg/kg; IM (5 daily injections)	Chicken (Hubbard)	α-SRBC Ab formation			115
			Primary response	NE		
			Secondary response	NE		
			Serum IgM, IgG levels	NE		
	2 mg/kg; IM (3/wk, 2 wk)	Rabbit (New Zealand)	Serum α-basement membrane autoantibody levels	↑		116
Methylmercury chloride	10 mg/kg; IP (single dose)	Mouse (Balb/c)	Splenic lymphocyte microtubule integrity/assembly	↓	Yes	117
			Splenic lymphocyte mitogen responsiveness	↓	Yes	
	0.4 mg/ml; SC (7 daily injections)	Mouse (B10 and CRJ)	Thymic index	↓		118
			Splenic follicle size	↓		
			Splenic α-SRBC PFC numbers	↓		
			Splenic lymphocyte mitogen responsiveness	↓		
			Splenic index	↑		
			Thymic lymphocyte mitogen responsiveness	↑		
	3.9 μg/g diet (12 wk)	Mouse (Balb/c)	Thymic index	↓		119
			Thymic total cell numbers	↓		
			NK-cell activity (blood/splenic)	↓		
			Lymphocyte (splenic and thymic) mitogen responsiveness	↑		
			Total blood RBC numbers	↑		
			Total blood WBC numbers	NE		
			Splenic index	NE		
	1, 5, or 10 μg/g diet (10 wk)	Mouse (CBA/J)	α-SRBC Ab formation			
			Primary response	↓	Yes	120
			Secondary response	↑	No	
	1, 5, or 10 μg/g diet (10 wk)	Mouse (CBA/J)	α-SRBC Ab formation			
			Primary response	↓	Yes	121
			Secondary response	NE/↓		
	1, 10, or 20 μg/g diet (14 wk)	Rabbit (New Zealand)	α-A/PR8 virus Ab formation			
			Primary response	↓	Yes	122
			Secondary response	NE/↓		
	3.9 μg/g diet (dams fed 11 wk)	Rat pups (Sprague-Dawley)	Placentally exposed only			123
			Thymic total cell number	NE		
			Thymic index	NE		

			Splenic index	NE		
			Splenic total cell numbers	NE		
			NK-cell activity (splenic)	NE		
			Lymphocyte (splenic and thymic) mitogen responsiveness	NE		
			Total blood WBC numbers	↑		
	3.9 μg/g diet (Dams fed 11 wk)	Rat pups (Sprague-Dawley)	Milk-exposed only			123
			Total blood WBC numbers	NE		
			Thymic total cell numbers	NE		
			Thymic index	NE		
			Splenic total cell numbers	NE		
			NK-cell activity (splenic)	NE		
			Splenic index	↓		
			Splenic lymphocyte mitogen responsiveness	↓		
			Thymic lymphocyte mitogen responsiveness	↑		
			Milk- and placentally exposed			
			Total blood WBC numbers	NE		
			Thymic total cell numbers	NE		
			Thymic index	NE		
			Splenic index	NE		
			Splenic total cell numbers	NE		
			NK-cell activity (splenic)	↓		
			Splenic lymphocyte mitogen responsiveness	↓		
			Thymic lymphocyte mitogen responsiveness	↑		
In Vitro *Exposures*						
Mercuric chloride	1.25–40 μ*M*	Mouse	PEM viability	↓	Yes	124
			PEM random migration	↓	Yes	
			PEM phagocytosis	↓	Yes	
	0.01–2.5 μ*M*	Mouse	PEM IFN-α production	↓	Yes	124
	1.25–80 μ*M*	Mouse	PEM viability	↓	Yes	125
			PEM random migration	↓	Yes	
			PEM phagocytosis	↓	Yes	
			PEM morphological changes	↑	Yes	
	2–32 μ*M*	Mouse	PEM viability	↓	Yes	126
			PEM random migration	↓	Yes	
			PEM phagocytosis	NE/↓	Yes	
	1–10 μ*M*	Mouse	PEM ROI production (+ PMA)	↓	Yes	127
			PEM LDH release	↑	Yes	
			PEM plasminogen activator levels	↑/↓	Yes	

TABLE 6 (continued)

Effects of Mercury Compounds on Immune System Parameters

Compound	Concentration (Exposure Protocol)	Host	Parameter Affected	Range of Effects	Dose-Response	Ref.
	0.001–1 μM	Mouse	Splenic B-lymphocyte viability	NE		128
			B-lymphocyte surface marker expression (± LPS) (ICAM, LFA1, MHC Class I, Class II)	NE		
			B-lymphocyte RNA synthesis	↓	Yes	
			B-lymphocyte DNA synthesis	↓	Yes	
			B-lymphocyte cell cycle progression (LPS-stimulated)	↓	Yes	
			B-lymphocyte Ig isotype expression (IgM, IgG_1, IgG_{2a}, IgG_{2b}, IgG_3)	↓	Yes	
	40 pM–40 μM	Mouse	PEM viability	↓	Yes	105
			PEM ROI production (+ PMA)	↓	Yes	
			PEM ROI production (+ HSV-2)	↓	Yes	
			PEM HSV-2 virus-induced cytokine (TNF-α, IFN-α) production	↓	Yes	
	0.001–10 μM	Mouse	Splenic lymphocyte mitogen responsiveness	↓	Yes	129
			Splenic α-TNP-SRBC PFC numbers	↓	Yes	
			Splenic MLR	↓	Yes	
	0.01–10 μM	Mouse	LPS toxicity against PEM	NE/↓	No	130
	1.4–2.7 μg/ml	Mouse	PEM phagocytic index	NE		131
			PEM phagocytic capacity	NE		
			PEM intracellular killing	NE		
			PEM glass adherence capacity	NE		
	10–20 μM	Rabbit	PAM ROI production	↓	Yes	132
	1–100 μM	Rabbit	PAM LTB_4 production	↑	Yes	133
			PAM-free arachidonic acid levels	↑	Yes	
	1–100 μM	Rat	PAM-stimulated O_2 consumption	↓	Yes	72, 134
			PAM ROI production	↓	Yes	
			PAM viability	↓	Yes	
	1 fM–1 μM	Rat (BN)	PEM ROI production	↓	Yes	135
			PEM phagocytic activity	NE/↓	Yes	
			Peritoneal PMN ROI production	NE/↓	Yes	
		Rat (Lewis)	Peritoneal PMN ROI production	↑	Yes	
			PEM ROI production	↑	Yes	

		Peritoneal cell ROI production			
		Nonelicited, preincubated before Hg^{2+} introduction	↑	Yes	
		Without preincubation	↓	Yes	
		PEM phagocytic activity	NE/↓	Yes	
0.01–1 μ*M*	Rat	T-lymphocyte (splenic) Ca^{2+} levels			
		Extracellular Ca^{2+} present	↑	Yes	136
		Extracellular Ca^{2+} absent	↓	No	
0.01–100 μ*M*	Rat	Splenic IFN-γ-producing cells	↓	Yes	114
	(BN)	Splenocyte viability	↓	Yes	
0.1–50 μ*M*	Human	PMN chemiluminescence	↓	Yes	137
		PMN phagocytic activity	↓	Yes	
		PMN lysozyme release	↑		
		PMN β-glucuronidase release	NE		
		PMN viability	NE		
1–5 μ*M*	Human	PMN ROI production	↑	No	138
		fMLP-induced PMN ROI production	↑	No	
10–100 μ*M*	Human	U937 MØ tissue factor activity	↑	Yes	139
		Cytosolic Ca^{2+} levels	↑	Yes	
0.1 n*M*–0.1 μ*M*	Human	PBL IgE PFC formation (–PWM)	NE		140
		PBL IgA, IgG, IgM PFC formation			
		(– PWM)	NE		
		(+ PWM)	NE		
		PBL IgE PFC formation (+ PWM)	↑	Yes	
1–50 ng/ml	Human	PMN-stimulated O_2 consumption	↓	Yes	141
25–250 ng/ml	Human	PMN chemiluminescence	↓	Yes	141
		PMN ROI production (+ fMLP/OZ)	↓	Yes	
		PMN viability	NE/↓	Yes	
1–80 μ*M*	Human	PMN viability	↓	Yes	142
		PMN ROI production	↓	Yes	
		PMN chemotaxis	↓	Yes	
50–1250 ng/ml	Human	PBL (B-cell) mitogen responsiveness	↓	Yes	143
		PBL (B-cell) GSH and GSSG content	↓	Yes	
		PBL (T-cell) GSH and GSSG content	↓	Yes	
		PBL (T-cell) GSH/GSSG ratio	↑	Yes	
		PBL (B-cell) GSH/GSSG ratio	NE		
		PBL (B-cell) GSHPX/GSHRX	NE		
		PBL (T-cell) GSHPX/GSHRX	NE		
50–625 ng/ml		PBM GSH content	↓	Yes	
		PBM GSH/GSSG ratio	↓	Yes	
		PBM GSSG content	NE		
		PBM GSHPX/GSHRX	NE		

TABLE 6 (continued)

Effects of Mercury Compounds on Immune System Parameters

Compound	Concentration (Exposure Protocol)	Host	Parameter Affected	Range of Effects	Dose-Response	Ref.
	5–1000 ng/ml	Human	PBL (B-cell) mitogen responsiveness	↓	Yes	144
			PBL (B-cell) IgG, IgM production	↓	Yes	
	0.5–10 μg/ml	Human	PBL (B-cell) viability	↓	Yes	144
			PBL (B-cell) membrane integrity	↓	Yes	
			PBL (B-cell) mitogen-induced CD23 and transferrin receptor expression	↓	Yes	
			PBL (B-cell) mitogen-induced CD69 expression	NE		
			PBL (B-cell) energy metabolism	NE		
			PBL (B-cell) Ca^{2+} levels	↑	Yes	
			PBL (B-cell) nuclear changes	↑	Yes	
	10–320 ng/ml	Human	PBL (T-cell) IL-2 production	↓	Yes	145
	0.5–10 μg/ml	Human	PBL (T-cell) mitogen-induced CD69, IL-2R, and transferrin receptor expression	↓	Yes	145
	5–1000 ng/ml	Human	PBL (T-cell) mitogen responsiveness	↓	Yes	145
	1–20 μg/ml	Human	PBL (T-cell) viability	↓	Yes	146
			PBL (T-cell) membrane integrity	↓	Yes	
	1–20 μg/ml	Human	PBL (T-cell) energy metabolism	NE		146
			PBL (T-cell) Ca^{2+} levels	↑	Yes	
			PBL (T-cell) nuclear changes	↑	Yes	
			PBM viability/membrane integrity	↓	Yes	
			PBM energy metabolism	↓	Yes	
			PBM Ca^{2+} levels/nuclear changes	↑	Yes	
	3.3 n*M*–1.4 μ*M*	Guinea pig	Lymphocyte (splenic, thymic, and lymph node) proliferation	↑	No	147
	11 n*M*–33 μ*M*	Guinea pig	Lymphocyte (splenic, thymic, and lymph node) proliferation	↑/↓	Yes	148
Methylmercury chloride	1–10 μ*M*	Mouse	Splenic lymphocyte microtubule integrity/assembly	↓	Yes	117
			Splenic lymphocyte mitogen responsiveness	↓	Yes	
	1.25–20 μ*M*	Mouse	PEM viability	↓	Yes	124
			PEM random migration	↓	Yes	
			PEM phagocytosis	↓	Yes	
	0.01–10 μ*M*	Mouse	PEM IFN-α production	↓	Yes	124

	1 nM–1 μM	Mouse	Splenic T-lymphocyte mitogen responsiveness	↑	No	106
		(SJL/N)	Splenic B-lymphocyte mitogen responsiveness	↑	Yes	
	0.1–100 nM	(DBA)	Splenic T-lymphocyte mitogen responsiveness	NE		
			Splenic B-lymphocyte mitogen responsiveness	↓	Yes	
	0.001–10 μM	Mouse	Splenic lymphocyte mitogen responsiveness	↓	Yes	129
			Splenic MLR	↓	Yes	
			Splenic α-TNP-SRBC PFC levels	NE/↓	Yes	
	0.02–2 μM	Rat	Splenic T-lymphocyte (Ca²⁺ levels			
			Extracellular Ca²⁺ present	↑	Yes	136
			Extracellular Ca²⁺ absent	↑	Yes	
	1–80 μM	Human	PMN viability	↓	Yes	142
			PMN ROI production	↓	Yes	
			PMN chemotaxis	↓	Yes	
	5–100 ng/ml	Human	PBL (B-cell) mitogen responsiveness	↓	Yes	144
			PBL (B-cell) IgG, IgM production	↓	Yes	
	0.5–10 μg/ml	Human	PBL (B-cell) viability	↓	Yes	144
			PBL (B-cell) energy metabolism	↓	Yes	
			PBL (B-cell) mitogen-induced CD69 expression	NE		
			PBL (B-cell) mitogen-induced CD23 and transferrin receptor expression	↓	Yes	
			PBL (B-cell) Ca²⁺ levels	↑	Yes	
			PBL (B-cell) nuclear changes	↑	Yes	
	0.5–32 ng/ml	Human	PBL (T-cell) IL-2 production	↓	Yes	145
	0.5–10 μg/ml	Human	PBL (T-cell) mitogen-induced CD69, IL-2R, transferrin receptor expression	↓	Yes	145
	5–100 ng/ml	Human	PBL (T-cell) mitogen responsiveness	↓	Yes	145
	0.1–2 μg/ml	Human	PBL (T-cell) viability	↓	Yes	146
			PBL (T-cell) membrane integrity	↓	Yes	
			PBL (T-cell) energy metabolism	↓	Yes	
			PBL (T-cell) Ca²⁺ levels	↑	Yes	
			PBL (T-cell) nuclear changes	↑	Yes	
			PBM viability	↓	Yes	
			PBM membrane integrity	↓	Yes	
			PBM energy metabolism	↓	Yes	
			PBM Ca²⁺ levels	↑	Yes	
			PBM nuclear changes	↑	Yes	
Phenyl mercuric acetate	10–100 μM	Human	U937 MØ tissue factor activity	↑	Yes	139

TABLE 7
Effects of Tin Compounds on Immune System Parameters

Compound	Concentration (Exposure Protocol)	Host	Parameter Affected	Range of Effect	Dose-Response	Ref.
In Vivo *Exposures*						
Tributyltin oxide (TBTO)	0.5, 2, 5, or 50 mg/kg diet (28-d feeding)	Rat (Sprague-Dawley)	Thymic index	↓	No	149
			α-SRBC PFC numbers	↓	No	
			L. monocytogenes clearance	↓	No	
			DTH reaction	NE		
	2.5–15 mg/kg/d (10 d/gavage)	Rat (Fischer)	α-TNP-LPS PFC numbers	NE		150
			DTH reaction	NE		
			$CD8^+$ cell numbers	↓	Yes	
			α-SRBC PFC numbers	↑	Yes	
	0.5, 5, or 50 mg/kg diet (4–6 months and 15–17 months)	Rat (Wistar)	Ag-induced IgM, IgG levels	NE		151
			T-cell mitogen responsiveness	NE		
			B-cell mitogen responsiveness	NE		
			Splenic index	NE		
			Thymic index	↓	No	
			Ag-induced IgE levels	↓	No	
			T-cell/B-cell ratio	↓	No	
			L. monocytogenes clearance	↓	No	
			T. spiralis clearance	↓	No	
			NK-cell activity (spleen)	↓	No	
	20 or 80 mg/kg diet (6-wk feeding)	Rat (Wistar)	LPS lethality	NE	No	152
			Ag-induced IgM, IgG levels	NE		
			T-cell mitogen responsiveness	NE		
			DTH reaction	↓	No	
			Ag-induced IgE levels	↓	No	
			T-cell numbers (splenic T_h, T_s)	↓	No	
			L. monocytogenes clearance	↓	No	
			T. spiralis clearance	↓	No	
			NK-cell activity (spleen)	↓	No	
			B-cell mitogen responsiveness	↑	No	
	20 or 80 mg/kg diet (6-wk feeding)	Rat (Wistar)	NK-cell activity (lung)	↓	Yes	153

	5, 20, 80, or 320 mg/kg diet	Rat	Thymic index/T-cell numbers	↓	Yes	154
	(6-wk feeding)	(Wistar)	Serum IgG/IgM levels	↓/↑	Yes	
Dioctyltin dichloride	20, 100, or 500 mg/kg	Mouse	Thymic index	↓	No	155
(DOTC)	(weekly gavage, 8 wk)	(Balb/c)	α-SRBC (self) Ab levels	↓	No	
			α-SRBC (rat) Ab levels	NE		
			DTH reaction	NE		
	75 or 100 mg/kg	Rat	Thymic index	↓	No	156
	(gavage)	(PVG)	Thymocyte IL-2 production	↓	No	
	50 or 150 ppm/diet	Rat	α-TNP-LPS PFC numbers	NE		157
	(6-wk feeding)	(Wistar)	PBM particle clearance	NE		
			DTH reaction	↓	Yes	
			Allograft rejection time	↑	Yes	
			α-SRBC PFC numbers	↓	Yes	
			α-LPS Ab levels	↑	No	
	50 or 150 ppm/diet	Rat	Thymic index	↓	Yes	158
	(6-wk feeding)	(Wistar)	Nucleated thymocyte numbers	↓	Yes	
	75 ppm/diet	Rat	Thymic index	↓	No	159
	(8-wk feeding)	(PVG)	Splenic lymphoctye MLR	↓	No	
			T-cell mitogen-responsiveness	↓	No	
			T-cell numbers (circulating T_h)	↓	No	
			α-Rat SRBC Ab levels	NE		
			T-cell numbers (circulating T_s)	NE		
			NK-cell activity (splenic)	NE		
Trioctyltin trichloride	50 or 150 ppm/diet	Rat	Thymic index	NE		160
	(2- or 4-wk feeding)	(Wistar)	Splenic index	NE		
Dibutyltin dichloride	5–35 mg/kg	Rat	Thymic index	↓	Yes	161
(DBTC)	(single gavage)	(Wistar)	Thymic proliferating lymphoblasts	↓		
	50 or 150 ppm/diet	Rat	α-TNP-LPS PFC numbers	NE		157
	(6-wk feeding)	(Wistar)	α-LPS Ab levels	NE		
			α-SRBC PFC numbers	↓	Yes	
			Allograft rejection time	↑	Yes	
Monobutyltin trichloride	10–180 mg/kg	Rat	Thymic index	NE		161
(MBTC)	(single gavage)	(Wistar)	Thymic proliferating lymphoblasts	NE		
Tributyltin trichloride	50 or 150 ppm/diet	Rat	Thymic index	↓	Yes	160
(TBTC)	(2- or 4-wk feeding)	(Wistar)	Splenic index	↓	Yes	
	5–60 mg/kg	Rat	Thymic index	↓	Yes	161
	(single gavage)	(Wistar)	Thymic proliferating lymphoblasts	↓		
	10 or 100 ppm/diet	Mouse	NK-cell activity	↓	No	162
	(7-d feeding)	(C3H/HEN)	LGL numbers (splenic)	↓	No	
			NK:target cell conjugates	↓	No	

TABLE 7 (continued)

Effects of Tin Compounds on Immune System Parameters

Compound	Concentration (Exposure Protocol)	Host	Parameter Affected	Range of Effect	Dose-Response	Ref.
Tripropyltin trichloride	15, 50, or 150 ppm/diet	Rat	Thymic index	↓	Yes	160
	(2- or 4-wk feeding)	(Wistar)	Splenic index	↓	Yes	
Triphenyltin trichloride	15, 50, or 150 ppm/diet	Rat	Thymic index	↓	Yes	160
	(2- or 4-wk feeding)	(Wistar)	Splenic index	↓	Yes	
Trihexyltin trichloride	50 or 150 ppm/diet	Rat	Thymic index	NE		160
	(2- or 4-wk feeding)	(Wistar)	Splenic index	NE		
Triphenyltin hydroxide	25 mg/kg diet	Rat	Thymic index	↓		163
	(3- to 4-wk feeding)	(Wistar)	Splenic index	↓		
			DTH reaction	↓		
			B-cell mitogen responsiveness	↓		
			T-cell mitogen responsiveness	NE		
			Ag-induced Ig generation	NE		
			Allograft rejection time	NE		
			L. monocytogenes clearance	NE		
			L. monocytogenes mortality	↑		
Stannous chloride	167 μmol/kg; IP	Mouse	Splenic index	↑		164
	(3 daily injections)	(ddY)	Splenic PFC (1°) (direct/indirect)	↓		
			Splenic PFC (2°) (direct/indirect)	NE		
	200-μg; IP	Mouse	Splenic α-SRBC rosetting	↑		165
	(single dose)	(C57Bl/6J)	DTH reaction	NE		
Stannic chloride	200-μg; IP	Mouse	Splenic α-SRBC rosetting	↓		165
	(single dose)	(C57Bl/6J)	Splenic PFC (1°) (direct)	↓		
			Splenic PFC (1°) (indirect)	↓		
			DTH reaction	NE		
Methyltin trichloride	200-μg; IP	Mouse	Splenic PFC (1°) (direct)	↓		165
	(single dose)	(C57Bl/6J)	Splenic PFC (1°) (indirect)	↓		
			DTH reaction	NE		
Metallic tin powder	20% (w/v); IP	Rat	Ig-secreting cell numbers	↑		166, 167
	(single dose)	(Lewis)	α-SRBC PFC numbers	↑		

TABLE 8

Effects of Vanadium Compounds On Immune System Parameters

Compound	Concentration Exposure Protocol	Host	Parameter Affected	Range of Effect	Dose-Response	Ref.
In Vivo *Exposures*						
Ammonium metavanadate	2.5 or 10 mg/kg; IP	Mouse	*Listeria* clearance (spleen/liver)	↓	Yes	168
	(every 3 d, 6 wk)	(B6C3F1)	PEM *Listeria* uptake/killing	↓	Yes	
	2.5 or 10 mg/kg; IP	Mouse	PEM HMS/GSH redox (G6PDH,	↓	Yes	169
	(every 3 d, 6 wk)	(B6C3F1)	GSHPX/GSHRX) enzyme activity			
			PEM ROI production	↓	Yes	
			PEM GSSG levels	↑	Yes	
	2.5 or 10 mg/kg; IP	Mouse	PEM acid phosphatase activity	↓	Yes	170
	(every 3 d, 6 wk)	(B6C3F1)	PEM lysosomal enzyme activity	NE		
	2.5 or 10 mg/kg; IP	Mouse	PEM Fcγ2a expression	↓	Yes	171
	(every 3 d, 6 wk)	(B6C3F1)	PEM Fcγ2a-mediated binding	NE		
			PEM Fcγ2b expression	↓	Yes	
			PEM Fcγ2b-mediated binding	↓	Yes	
	2.5 or 10 mg/kg; IP	Mouse	PEM LPS-inducible Ca^{2+} influx	NE		172
	(every 3 d, 6 wk)	(B6C3F1)	PEM resting Ca^{2+} levels	↑	Yes	
	2.5, 5, or 10 mg/kg; IP	Mouse	*Listeria* resistance	↓	Yes	173
	(every 3 d; 3, 6, or 9 wk)	(B6C3F1)	Thymic index	↓	Yes	
			PEM phagocytic activity	↓	Yes	
			Splenic lymphocyte EAC-rosette formation	↑/↓	Yes	
			E. coli endotoxin resistance	↑	Yes	
			Splenic megakaryoctyes	↑	Yes	
	15.5 mg/kg; SC	Mouse	Lymphoid necrosis	↑		174
	(single injection)	(Balb/c)				
	20 mg/kg; SC	Mouse	Lymphoid necrosis	↑		175
	(single injection)	(Balb/c)	Hepatic/splenic indices	↑		
	2 mg/m^3 inhalation	Rat	Lung PMN/PAM recovery	↑		176
	(8 hr/d, 4 d)	(Fischer)	PAM ROI production	↑		
			PAM IFN-γ-primed ROI production	↓		
			PAM inducible TNF-α production	↓		
			PAM IFN-γ-inducible Ia expression	↓		

TABLE 8 (continued)

Effects of Vanadium Compounds On Immune System Parameters

Compound	Concentration Exposure Protocol	Host	Parameter Affected	Range of Effect	Dose-Response	Ref.
	0.3–0.9 mg/kg; SC (every day, 16 d)	Rat (Sprague-Dawley)	Blood monocyte numbers	↑	Yes	177
			Blood lymphocyte numbers	NE		
			Blood total PMN numbers	NE/↓		
			Blood eosiniphil numbers	NE/↓		
			Total WBC numbers	↑/↓/↑		
	0.15 mg/l water (4 wk)	Rat (Wistar)	Blood PMN/lymphocyte numbers	↑		178
			PMN phagocytic activity	↓		
Sodium orthovanadate	1, 10, or 50 mg/l water (1–13 wk)	Mouse (Swiss-Webster)	Splenic Ab-producing cell numbers	↓	Yes	179
			α-SRBC DTH reactions	NE		
			Serum Ig levels	NE		
			Splenic lymphocyte mitogen responsiveness	NE		
Sodium vanadate	1–4 mg/kg; IP (daily for 2 wk)	Rat (Sprague-Dawley)	PAM HMS/GSH redox (G6PDH, GSHPX/GSHRX) enzyme activity	NE/↑		180
Vanadium pentoxide	1–4 mg/kg; IP (daily for 2 wk)	Rat (Sprague-Dawley)	PAM HMS/GSH redox (G6PDH, GSHPX, GSHRX) enzyme activity	↑	No	180
	0.5 or 5 mg/m^3 (inhalation)	Monkey (Cynomolgus)	Lung PMN recovery	↑	No	181, 182
			Lung PAM/lymphocyte recovery	NE		
In Vitro *Exposures*						
Ammonium metavanadate	60–70 μ*M*	Rabbit	PAM phagocytic activity	NE		183
	1 fM–100 μ*M*	Mouse	WEHI-3 TNF-α production	↓	Yes	184
			WEHI-3 IL-1α production	NE/↓	Yes	
			WEHI-3 spontaneous PGE_2 production	NE/↑	Yes	
	1 n*M*–100 μ*M*	Mouse	WEHI ROI production	↑	No	185
			WEHI resting Ca^{2+} levels	↑	No	
			WEHI IFN-γ-inducible Ca^{2+} influx	↓	No	
			WEHI IFN-γ binding	↓		
			WEHI IFN-γR expression	↓		
			WEHI IFN-γ-primed ROI production	↓		
			WEHI IFN-γ-inducible Ia expression	↓		
	3.6–10.8 μg	Mouse	J774 MØ resting Ca^{2+} levels	↑	Yes	172
			J774 MØ LPS-induced Ca^{2+} influx	NE		

Sodium orthovanadate	100 nM–1 mM	Mouse	Splenic lymphocyte mitogen responsiveness	↓	Yes	179
	1–250 μM	Rat	PMN morphological changes	↑	Yes	186
			PMN cell spreading	↓		
	25–100 μM	Human	PBL (T-cell) mitogen responsiveness	↑	Yes	187
	25–300 μM	Human	PBL (T-cell) IFN-γ production	↑	Yes	187
	100–200 μM	Human	T-cell (Jurkat) IL-2 production	↑	Yes	188
	100–200 μM	Human	T-cell (Jurkat) IL-2R expression	↑		188
			T-cell (Jurkat) Ca^{2+} mobilization	↑		
Sodium vanadate	0.1–500 μM	Rabbit	PAM ROI production	↓	Yes	189
			PAM membrane integrity	↓	Yes	
	10–15 μM	Rabbit	PAM ROI production	↓	Yes	132
Vanadium pentoxide	1–100 μg/ml	Rabbit	PAM acid phosphatase activity	↓	Yes	190
			PAM phagocytic activity	↓	Yes	
			PAM lysosomal enzyme activity	NE		
	5–9 μg/ml	Mouse	PAM phagocytic activity	↓	Yes	183
			PAM adherence	NE		
	2 μM–2 mM	Mouse	PEM ROI production	NE		128
			PEM plasminogen activator levels	NE		
	1 nM–100 μM	Mouse	WEHI ROI production	↑	No	185
			WEHI resting Ca^{2+} levels	↑	No	
			WEHI IFN-γ-inducible Ca^{2+} influx	↓	No	
			WEHI IFN-γ-primed ROI production	↓		

REFERENCES

1. Stankus, R. P., Schuyler, M. R., D'Amato R. A., and Salvaggio, J. E., Bronchopulmonary cellular response to aluminum and zirconium salts, *Infect. Immun.*, 20, 847–852, 1978.
2. Nashed, N., Preparation of peritoneal cell metaphases of rats, mice, and Chinese hamsters after mitogenic stimulation with magnesium sulfate and/or aluminum hydroxide, *Mutat. Res.*, 30, 407–416, 1975.
3. Drew, R. T., Gupta, B. N., Bend, J. R., and Hook, G. E., Inhalation studies with a glycol complex of aluminum-chloride-hydroxide, *Arch. Environ. Health*, 28, 321–326, 1974.
4. Golub, M. S., Takeuchi, P. T., Gershwin, M. E., and Yoshida, S. H., Influence of dietary aluminum on cytokine production by mitogen-stimulated spleen cells from Swiss-Webster mice, *Immunopharmacol. Immunotoxicol.*, 15, 605–619, 1993.
5. Yoshida, S., Gershwin, M. E., Keen, C. L., Donald, J. M., and Golub, M. S., The influence of aluminum on resistance to *Listeria monocytogenes* in Swiss-Webster mice, *Int. Arch. Allergy Appl. Immunol.*, 89, 404–409, 1989.
6. Dubois, F., Begin, R., Cantin, A., Masse, S., Martel, M., Bilodeau, G., Dufresne, A., Perreault, G., and Sebastien, P., Aluminum inhalation reduces silicosis in a sheep model, *Am. Rev. Respir. Dis.*, 137, 1172–1179, 1988.
7. Pizzoferrato, A., Vespucci, A., Ciapetti, G., Stea, S., and Tarabusi, C., The effect of injection of powdered biomaterials on mouse peritoneal cell populations, *J. Biomed. Mater. Res.*, 21, 419–428, 1987.
8. Ramanathan, V. D., Badenoch-Jones, P., and Turk, J. L., Complement activation by aluminum and zirconium compounds, *Immunology*, 37, 881–888, 1979.
9. Badenoch-Jones, P., Turk, J. L., and Parker, D., The effects of some aluminum and zirconium compounds on guinea pig peritoneal macrophages and skin fibroblasts in culture, *J. Pathol.*, 124, 51–62, 1978.
10. Yao, X., Jenkins, E. C., and Wisniewski, H. M., Effect of aluminum chloride on mitogenesis, mitosis, and cell cycle in human short-term whole blood cultures: lower concentrations enhance mitosis, *J. Cell. Biochem.*, 54, 473–477, 1994.
11. Norklind, K., Further studies on the ability of different metal salts to influence the DNA synthesis of human lymphoid cells, *Int. Arch. Allergy Appl. Immunol.*, 79, 83–85, 1986.
12. Castranova, V., Bowman, L., Miles, P. R., and Reasor, M. J., Toxicity of metal ions to alveolar macrophages, *Am. J. Indust. Med.*, 1, 349–357, 1980.
13. McGregor, S. J., Naves, M. L., Birly, A. K., Russell, N. H., Halls, D., Junor, B. J., and Brock, J. H., Interaction of aluminum and gallium with human lymphocytes: the role of transferrin, *Biochim. Biophys. Acta*, 1095, 196–200, 1991.
14. O'Shea, J. J., Urdahl, K. B., Luong, H. T., Chused, T. M., Samelson, L. E., and Klausner, R. D., Aluminum fluoride induces phosphatidylinositol turnover, elevation of cytoplasmic free calcium, and phosphorylation of the T-cell antigen receptor in murine T-cells, *J. Immunol.*, 139, 3463–3469, 1987.
15. Kang, K., Bice, D., Hoffmann, E., D'Amato, R., and Salvaggio, J., Experimental studies of sensitization to beryllium, zirconium, and aluminum compounds in the rabbit, *J. Allergy Clin. Immunol.*, 59, 425–436, 1977.
16. Kang, K. and Salvaggio, J., Effects of zirconium and aluminum salts on the alveolar macrophage, *Tohoku J. Exp. Med.*, 126, 317–324, 1978.
17. Theocharis, S., Margeli, A., and Panayiotidis, P., Effects of various metals on DNA synthesis and lymphokine production by human peripheral blood lymphocytes *in vitro*, *Comp. Biochem. Physiol.*, 99, 131–133, 1991.
18. Warshawsky, D., Reilman, R., Cheu, J., Radike, M., and Rice, C., Influence of particle dose on the cytotoxicity of hamster and rat pulmonary alveolar macrophage *in vitro*, *J. Toxicol. Environ. Health*, 42, 407–421, 1994.
19. Gusev, V. A., Danilovskaja, Y. V., Vatolkina, O. Y., Lomonosova, O. S., and Velichkivsky, B. T., Effect of quartz and alumina dust on generation of superoxide radicals and hydrogen peroxide by alveolar macrophages, granulocytes, and monocytes, *Br. J. Indust. Med.*, 50, 732–735, 1993.
20. Morita, K., Inoue, S., Murai, Y., Watanabe, K., and Shima, S., Effect of the metals Be, Fe, Cu, and Al, on the mobility of immunoglobulin receptors, *Experientia*, 38, 1227–1228, 1982.
21. Aranyi, C., Bradof, J. N., and O'Shea, W. J., Effects of arsenic trioxide inhalation exposure on pulmonary antibacterial defenses in mice, *J. Toxicol. Environ. Health*, 15, 163–172, 1985.
22. Gainer, J. H. and Pry, T. W., Effects of arsenicals on viral infections in mice, *Am. J.Vet. Res.*, 33, 2299–2302, 1972.
23. Gainer, J. H., Effects of arsenicals on interferon formation and action, *Am. J. Vet. Res.*, 33, 2579–2586, 1972.
24. Schrauzer, G. N. and Ishmael, D., Effects of selenium and of arsenic on the genesis of spontaneous mammary tumors in inbred C3H mice, *Ann. Clin. Lab. Sci.*, 4, 441–447, 1974.
25. Lantz, R. C., Parliman, G., Chen, G., and Carter, D. E., Effect of arsenic exposure on alveolar macrophage function. I. Effect of soluble As (III) and As (V), *Environ. Res.*, 67, 183–195, 1994.
26. Zelikoff, J. T., Reynolds, C., Bowser, D., Valle, C., and Snyder, C. A., Modulation of tumor surveillance mechanisms by *in utero* exposure to arsenic, *Proc. Eastern Reg. Symp. Mech. Immunotox.*, 4, 12, 1989.

27. Kerkvliet, N. I., Steppan, L. B., Koller, L. D., and Exon, J. H., Immunotoxicology studies of sodium arsenate-effects of exposure on tumor growth and cell-mediated tumor immunity, *Environ. Pathol. Toxicol.*, 4, 65–79, 1980.
28. Sikorski, E. E., Burns, L. A., McCoy, K. L., Stern, M., and Munson, A. E., Suppression of splenic accessory cell function in mice exposed to gallium arsenide, *Toxicol. Appl. Pharmacol.*, 110, 143–156, 1991.
29. Sikorski, E. E., McCay, J. A., White, K. L., Bradley, S. G., and Munson, A. E., Immunotoxicity of the semiconductor gallium arsenide in female B6C3F1 mice, *Fund. Appl. Toxicol.*, 13, 843–858, 1989.
30. Burns, L. A. and Munson, A. E., Gallium arsenide selectively inhibits T cell proliferation and alters expression of CD 15 (IL-2R/p55), *J. Pharmacol. Exper. Ther.*, 265, 178–186, 1993.
31. Bencko, V., Wagner, V., Wagnerova, M., and Batora, J., Immunological profiles in workers of a power plant burning coal rich in arsenic content, *J. Hyg. Epid. Micro. Immunol.*, 32, 137–146, 1988.
32. Ostrosky-Wegman, P., Gonsebatt, M. E., Montero, R., Vega, L., Barba, H., Espinosa, J., Palao, A., Cortinas, C., Garcia-Vargas, G., del Razo, L. M., and Cebrian, M., Lymphocyte proliferation kinetics and genotoxic findings in a pilot study on individuals chronically exposed to arsenic in Mexico, *Mutat. Res.*, 250, 477–482, 1991.
33. Yu, H. S., Chang, K. L., Wang, C. M., and Yu, C. L., Alterations of mitogenic responses of monocuclear cells by arsenic in arsenical skin cancers, *J. Dermatol.*, 19, 710–714, 1992.
34. Rosenthal, G. J., Fort, M. M., Germolec, D. R., Ackermann, M. F., Lamm, K. R., Blair, P. C., Fowler, B. A., Luster, M. I., and Thomas, P. T., Effect of subchronic arsine inhalation on immune function and host resistance, *Inhal. Toxicol.*, 1, 113–127, 1989.
35. Hong, H. L., Fowler, B. A., and Boorman, G. A., Hematopoietic effects in mice exposed to arsine gas, *Toxicol. Appl. Pharmacol.*, 97, 173–182, 1989.
36. Burns, L. A., Sikorski, E. E., Saady, J. J., and Munson, A. E., Evidence for arsenic as the immunosuppressive component of gallium arsenide, *Toxicol. Appl. Pharmacol.*, 110, 157–169, 1991.
37. Gonsebatt, M. E., Vega, L., Herrera, L. A., Montero, R., Rojas, E., Cebrian, M. E., Ostrosky-Wegman, P., Inorganic arsenic effects on human lymphocyte stimulation and proliferation, *Mutat. Res.*, 283, 91–95, 1992.
38. Yoshida, T., Shimamura, T., and Shigeta, S., Enhancement of the immune response *in vitro* by arsenic, *Int. J. Immunopharmacol.*, 9, 411–415, 1987.
39. McCabe, M., Maguire, D., and Nowak, M., The effects of arsenic compounds on human and bovine lymphocyte mitogenesis *in vitro*, *Environ. Res.*, 31, 323–331, 1983.
40. Meng, Z., Effects of arsenic on DNA synthesis in human lymphocytes, *Arch. Environ. Contam. Toxicol.*, 25, 525–528, 1993.
41. Moatamed, F., Karnovsky, M. J., and Unanue, E. R., Early cellular responses to mitogens and adjuvants in the mouse spleen, *Lab. Invest.*, 32, 303–312, 1975.
42. Pfeifer, S., Bartlett, R., Strausz, J., Haller, S., and Muller-Quernheim, J., Beryllium-induced disturbances of the murine immune system reflect some phenomena observed in sarcoidosis, *Int. Arch. Allergy Immunol.*, 104, 332–339, 1994.
43. Huang, H., Meyer, K. C., Kubai, L., and Auerbach, R., An immune model of beryllium-induced pulmonary granulomata in mice, *Lab. Invest.*, 67, 138–146, 1992.
44. Sendelbach, L. E., Witschi, H. P., and Tryka, A. F., Acute pulmonary toxicity of beryllium sulfate inhalation in rats and mice: cell kinetics and histopathology, *Toxicol. Appl. Pharmacol.*, 85, 248–256, 1986.
45. Behbehani, K., Beller, D. I., and Unanue, E. R., The effects of beryllium and other adjuvants on Ia expression by macrophages, *J. Immunol.*, 134, 2047–2049, 1985.
46. Chiappino, G., Cirla, A., and Vigliani, E. C., Delayed-type hypersensitivity reactions to beryllium compounds, *Arch. Pathol.*, 87, 131–140, 1969.
47. Marx, J. J. and Burrell, R., Delayed hypersensitivity to beryllium compounds, *J. Immunol.*, 111, 590–598, 1973.
48. Krivanek, N. and Reeves, A. L., The effect of chemical forms of beryllium on the production of the immunologic response, *Am. Ind. Hyg. Assoc. J.*, 33, 45–52, 1972.
49. Jones, J. M. and Amos, H. E., Antigen formation in metal contact sensitivity, *Nature*, 256, 499–500, 1975.
50. Hall, J. G., Studies on the adjuvant action of beryllium. I. Effects on individual lymph nodes, *Immunology*, 53, 105–113, 1984.
51. Haley, P. J., Finch, G. L., Hoover, M. D., and Cuddihy, R. G., The acute toxicity of inhaled beryllium metal in rats, *Fund. Appl. Toxicol.*, 15, 767–778, 1990.
52. Bencko, V., Vasilieva, E. V., and Symon, K., Immunological aspects of exposure to emissions from burning coal of high beryllium content, *Environ. Res.*, 22, 439–449, 1980.
53. Price, R. J. and Skilleter, D. N., Stimulatory and cytotoxic effects of beryllium on proliferation of mouse spleen lymphocytes *in vitro*, *Arch. Toxicol.*, 56, 207–211, 1985.
54. Unanue, E. R., Kiely, J. M., and Calderon, J., The modulation of lymphocyte functions by molecules secreted by macrophages. II. Conditions leading to increased secretion, *J. Exp. Med.*, 144, 155–166, 1976.
55. Newman, L. S. and Campbell, P. A., Mitogenic effect of beryllium sulfate on mouse B-lymphocytes but not T-lymphocytes *in vitro*, *Int. Arch. Allergy Appl. Immunol.*, 84, 223–227, 1987.

56. Finch, G. L., Verburg, R. J., Mewhinney, J. A., Eidson, A. F., and Hoover, M. D., The effect of beryllium compound solubility on *in vitro* canine alveolar macrophage cytotoxicity, *Toxicol. Lett.,* 41, 97–105, 1988.
57. Deodhar, S. D., Barna, B., and Van Ordstrand, H. S., A study of the immunologic aspects of chronic berylliosis, *Chest,* 63, 309–313, 1973.
58. Saltini, C., Winestock, K., Kirby, M., Pinkston, P., and Crystal, R. G., Maintenance of alveolitis in patients with chronic beryllium disease by beryllium-specific helper T-cells, *N. Engl. J. Med.,* 320, 1103–1109, 1989.
59. Kang, K., Bice, D., D'Amato, R., Ziskind, M., and Salvaggio, J., Effects of asbestos and beryllium on release of alveolar macrophage enzymes, *Arch. Environ. Health,* 34, 133–140, 1979.
60. Finch, G. L., Lowther, W. T., Hoover, M. D., and Brooks, A. L., Effects of beryllium metal particles on the viability and function of cultured rat alveolar macrophages, *J. Toxicol. Environ. Health,* 34, 103–114, 1991.
61. Hart, B. A. and Pittman, D. G., The uptake of beryllium by the alveolar macrophage, *J. Reticuloendothel. Soc.,* 27, 49–58, 1980.
62. Seko, Y., Koyama, T., Ichiki, A., Sugamata, M., and Miura, T., The relative toxicity of metal salts to immune hemolysis in a mixture of antibody-secreting spleen cells, sheep red blood cells, and complement, *Res. Commun. Chem. Pathol. Pharmacol.,* 36, 205–213, 1982.
63. Galvin, J. B. and Oberg, S. G., Toxicity of hexavalent chromium to the alveolar macrophage *in vivo* and *in vitro, Environ. Res.,* 33, 7–16, 1984.
64. Graham, J. A., Miller, F. J., Daniels, M. J., Payne, E. A., and Gardner, D. E., Influence of cadmium, nickel, and chromium on primary immunity in mice, *Environ. Res.,* 16, 77–87, 1978.
65. Johansson, A., Robertson, B., Curstedt, T., and Camner, P., Rabbit lung after inhalation of hexa- and trivalent chromium, *Environ. Res.,* 41, 110–119, 1986.
66. Johansson, A., Wiernik, A., Jarstrand, C., and Camner, P., Rabbit alveolar macrophages after inhalation of hexa- and trivalent chromium, *Environ. Res.,* 39, 372–385, 1986.
67. Kinnaert, P., Desmul, A., Tagnon, A., and Toussaint, C., Effect of ^{32}P chromium phosphate on immune reactions and phagocytic activity in rats, *Transplantation,* 9, 457–462, 1970.
68. Figoni, R. A. and Treagan, L., Inhibitory effect of nickel and chromium upon antibody response of rats to immunization with T-1 phage, *Res. Commun. Chem. Pathol. Pharmacol.,* 11, 335–338, 1975.
69. Glaser, U., Hochrainer, D., Kloppel, H., and Kuhnen, H., Low level chromium(VI) inhalation effects on alveolar macrophages and immune functions in Wistar rats, *Arch. Toxicol.,* 57, 250–256, 1985.
70. van Hoogstraten, I. M., Boden, D., von Blomberg, B. M., Kraal, G., and Scheper, R. J., Persistent immune tolerance to nickel and chromium by oral administration prior to cutaneous sensitization, *J. Invest. Dermatol.,* 99, 608–616, 1992.
71. Lazzarini, A., Luciani, S., Beltrame, M., and Arslan, P., Effects of chromium(VI) and chromium(III) on energy charge and oxygen consumption in rat thymocytes, *Chem. Biol. Interact.,* 53, 273–281, 1985.
72. Castranova, V., Bowman, L., Miles, P. R., and Reasor, M. J., Toxicity of metal ions to alveolar macrophages, *Am. J. Indust. Med.,* 1, 349–357, 1980.
73. Graham, J. A., Gardner, D. E., Waters, M. D., and Coffin, D. L., Effect of trace metals on phagocytosis by alveolar macrophages, *Infect. Immun.,* 11, 1278–1283, 1975.
74. Borella, P., Manni, S., and Giardino, A., Cadmium, nickel, chromium, and lead accumulate in human lymphocytes and interfere with PHA-induced proliferation, *J. Trace Elem. Electrolytes Health Dis.,* 4, 87–95, 1990.
75. Hahon, N. and Booth, J. A., Effect of chromium and manganese particles on the interferon system, *J. Interferon Res.,* 4, 17–27, 1984.
76. Pasanen, J. T., Gustafsson, T. E., Kalliomaki, P. L., Tosavainen, A., and Jarvisalo, J. O., Cytotoxic effects of four types of welding fumes on macrophages *in vitro*: a comparative study, *J. Toxicol. Environ. Health,* 18, 143–152, 1986.
77. Christensen, M. M., Enrst, E., and Ellermann-Ericksen, S., Cytotoxic effects of hexavalent chromium in cultured murine macrophages, *Arch. Toxicol.,* 66, 347–353, 1992.
78. Pribyl, D. and Treagan, L., A comparison of the effect of metal carcinogens chromium, cadmium, and nickel on the interferon system, *Acta Virol.,* 21, 507, 1977.
79. Hooftman, R. N., Arkesteyn, C. W., and Roza, P., Cytotoxicity of some types of welding fume particles to bovine alveolar macrophages, *Am. Occup. Hyg.,* 32, 95–102, 1988.
80. Kucharz, E. J. and Sierakowski, S. J., Immunotoxicity of chromium compounds: effect of sodium dichromate on the T-cell activation *in vitro, Arch. Hig. Rada. Toksikol.,* 38, 239–243, 1987.
81. Lewis, A. J., Cottney, J., White, D. D., Fox, P. K., McNeillie, A., Dunlop, J., Smith, W. E., and Brown, D. H., Action of gold salts in some inflammatory and immunological models, *Agents Action,* 10, 63–77, 1980.
82. Griswold, D. E., Lee, J. C., Poste, G., and Hanna, N., Modulation of macrophage-lymphocyte interactions by the antiarthritic gold compound, auranofin, *J. Rheumatol.,* 12, 490–497, 1985.
83. Walz, D. T., DiMartino, M. J., and Griswold, D. E., Comparative pharmacology and biological effects of different gold compounds, *J. Rheumatol.,* 9(Suppl. 8), 54–60, 1982.
84. Coughlan, R. J., Richter, M. B., and Panayi, G. S., Changes in mononuclear cell function in patients with rheumatoid arthritis following treatment with auranofin, *Clin. Rhematol.,* 3(Suppl. 1), 25–32, 1984.

85. Lorber, A., Jackson, W. H., and Simon, T. M., Effects of chrysotherapy on cell-mediated immune response, *J. Rheumatol.,* 9(Suppl. 8), 37–45, 1982.
86. Schuhmann, D., Kubicka-Muranyi, M., Mirtschewa, J., Gunther, J., Kind, P., and Gleichmann, E., Adverse immune reactions to gold. I. Chronic treatment with an Au(I) drug sensitizes mouse T-cells not to Au(I), but to Au(III), and induces autoantibody formation, *J. Immunol.,* 145, 2132–2139, 1990.
87. Tournade, H., Guery, J.C., Pasquier, R., Nochy, D., Hinglais, N., Guilbert, B., Druet, P., and Pelletier, L., Experimental gold-induced autoimmunity, *Nephrol. Dial. Transplant.,* 6, 621–630, 1991.
88. Sung, C. P., Mirabelli, C. K., and Badger, A. M., Effect of gold compounds on phorbol myristic acetate (PMA) activated superoxide production by mouse peritoneal macrophages, *J. Rheumatol.,* 11, 153–157, 1984.
89. Herman-Hernandes, C. E., Scheinberg, M. A., Benson, M. D., and Finkelstein, A. E., The effect of gold salts on casein-induced macrophage activation and serum amyloid protein SAA levels, *Clin. Exper. Rheumatol.,* 4, 347–350, 1986.
90. Haynes, D. R., Garrett, I. R., Whitehouse, M. W., and Vernon-Roberts, B., Do gold drugs inhibit interleukin-1? Evidence from an *in vitro* lymphocyte activating factor assay, *J. Rheumatol.,* 15, 775–778, 1988.
91. Remvig, L., Enk, C., and Bligaard, N., Effect of auranofin and sodium aurothiomalate on interleukin-1 production from human monocytes *in vitro, Scand. J. Rheumatol.,* 17, 255–262, 1988.
92. Barrett, M. L. and Lewis, G. P., Unique properties of auranofin as a potential anti-rheumatic drug, *Agents Action,* 19, 109–115, 1986.
93. Russell, A.S., Davis, P., and Miller, C., The effect of a new antirheumatic drug, triethylphosphine gold (auranofin), on *in vitro* lymphocyte and monocyte cytotoxicity, *J. Rheumatol.,* 9, 30–35, 1982.
94. Scheinberg, M. A., Santos, L. M., and Finkelstein, A. E., The effect of auranofin and sodium aurothiomalate on peripheral blood monocyte, *J. Rheumatol.,* 9, 366–369, 1982.
95. DiMartino, M. J. and Walz, D. T., Effect of auranofin on antibody-dependent cellular cytotoxicity (ADCC) mediated by rat peripheral blood polymorphonuclear leukocytes and mononuciear cells, *Inflammation,* 4, 279–288, 1980.
96. Finkelstein, A. E., Roisman, F. R., and Walz, D. T., Effect of auranofin, a new antiarthritic agent, on immune complex-induced release of lysosomal enzymes from human leukocytes, *Inflammation,* 2, 143–150, 1977.
97. Harth, M., McCain, G. A., and Orange, J. F., The effects of auranofin and gold sodium aurothiomalate on the chemiluminescent response of stimulated synovial tissue cells from patients with rheumatoid arthritis, *J. Rheumatol.,* 12, 881–884, 1985.
98. Harth, M., McCain, G. A., and Cousin, K., The modulation of interleukin-1 production by interferon gamma, and the inhibitory effects of gold compounds, *Immunopharmacology,* 20, 125–134, 1990.
99. Davis, P. and Johnston, C., Effects of gold compunds on function of phagocytic cells. Comparative inhibition of activated polymorphonuclear leukocytes and monocytes from rheumatoid arthritis and control subjects, *Inflammation,* 10, 311–320, 1986.
100. Hashimoto, K., Whitehurst, C. E., Matsubara, T., Hirohata, K., and Lipsky, P. E., Immunomodulatory effects of therapeutic gold compounds, *J. Clin. Invest.,* 89, 1839–1848, 1992.
101. Finkelstein, A. E., Ladizesky, M., Borinsky, R., Kohn, E., and Ginsburg, I., Antiarthritic synergism of combined oral and parenteral chrysotherapy. II. Increased inhibition of activated leukocyte oxygen burst by combined gold action, *Inflammation,* 12, 383–390, 1988.
102. Lipsky, P. E. and Ziff, M., Inhibition of antigen- and mitogen-induced human lymphocyte proliferation by gold compounds, *J. Clin. Invest.,* 59, 455–466, 1977.
103. Brunet, S., Guertin, F., Flipo, D., Fournier, M., and Krzystyniak, K., Cytometric profiles of bone marrow and spleen lymphoid cells after mercury exposure in mice, *Int. J. Immunopharmacol.,* 15, 811–819, 1993.
104. Dieter, M. P., Luster, M. I., Boorman, G. A., Jameson, C. W., Dean, J. H., and Cox, J. W., Immunological and biochemical responses in mice treated with mercuric chloride, *Toxicol. Appl. Pharmacol.,* 68, 218–228, 1983.
105. Ellermann-Eriksen, S., Christensen, M. M., and Mogensen, S. C., Effect of mercuric chloride on macrophage-mediated mechanisms against infection with herpes simplex virus type 2, *Toxicology,* 93, 269–287, 1994.
106. Hultman, P. and Enestrom, S., Dose-response studies in murine mercury-induced auto-immunity and immune-complex disease, *Toxicol. Appl. Pharmacol.,* 13, 199–208, 1992.
107. Hultman, P. and Johansson, U., Strain differences in the effect of mercury on murine cell-mediated immune reactions, *Food Chem. Toxicol.,* 29, 633–638, 1991.
108. Hultman, P., Bell, L. J., Enestrom, S., and Pollard, K. M., Murine susceptibility to mercury. II. Autoantibody profiles and renal immune deposits in hybrid, backcross, and H-2d congenic mice, *Clin. Immunol. Immunopathol.,* 68, 9–20, 1993.
109. Robinson, C. J., Balazs, T., and Egorov, I. K., Mercuric chloride-, gold sodium thiomalate-, and D-penicillamine-induced antinuclear antibodies in mice, *Toxicol. Appl. Pharmacol.,* 86, 159–169, 1986.
110. Saegusa, J., Kubota, H., and Kiuchi, Y., Antinucleolar autoantibody induced in mice by mercuric chloride. A genetic study, *Indust. Health,* 29, 167–170, 1991.
111. Saegusa, J., Yamamoto, S., Iwai, H., and Ueda, K., Antinucleolar autoantibody induced in mice by mercuric chloride, *Indust. Health,* 28, 21–30, 1990.

112. Kosuda, L. L., Greiner, D. L., and Bigazzi, P. E., Mercury-induced renal autoimmunity: changes in RT6+ T-lymphocytes of susceptible and resistant rats, *Environ. Health Perspect.*, 101, 178–185, 1993.
113. Prouvost-Danon, A., Abadie, A., Sapin, C., Bazin, H., and Druet, P., Induction of IgE synthesis and potentiation of anti-ovalbumin IgE antibody response by $HgCl_2$ in the rat, *J. Immunol.*, 126, 699–702, 1981.
114. van der Meide, P. H., de Labie, M. C., Botman, C. A., van Bennekom, W. P., Olsson, T., Aten, J., and Weening, J. J., Mercuric chloride down-regulates T-cell interferon-γ production in Brown Norway but not in Lewis rats: role of glutathione, *Eur. J. Immunol.*, 23, 675–681, 1993.
115. Bridger, M. A. and Thaxton, J. P., Humoral immunity in the chicken as affected by mercury, *Arch. Environ. Contam. Toxicol.*, 12, 45–49, 1983.
116. Roman-Franco, A. A., Turiello, M., Albini, B., Ossi, E., Milgrom, F., and Andres, G. A., Anti-basement membrane antibodies and antigen-antibody complexes in rabbits injected with mercuric chloride, *Clin. Immunol. Immunopathol.*, 9, 464–481, 1978.
117. Brown, D. L., Reuhl, K. R., Bormann, S., and Little, J. E., Effects of methyl mercury on the microtubule system of mouse lymphocytes, *Toxicol. Appl. Pharmacol.*, 94, 66–75, 1988.
118. Hirokawa, K. and Hayashi, Y., Acute methyl mercury intoxication in mice. Effect on the immune system, *Acta Pathol. Jpn.*, 30, 23–32, 1980.
119. Ilback, N. G., Effects of methyl mercury exposure on spleen and blood natural killer (NK) cell activity in the mouse, *Toxicology*, 67, 117–124, 1991.
120. Koller, L. D., Exon, J. H., and Brauner, J. A., Methylmercury: decreased antibody formation in mice, *Proc. Soc. Exp. Biol. Med.*, 155, 602–604 , 1977.
121. Koller, L. D., Isaacson-Kerkvliet, N., Exon, J. H., Brauner, J. A., and Patton, N. M., Synergism of methylmercury and selenium producing enhanced antibody formation in mice, *Arch. Environ. Health*, 34, 248–252, 1979.
122. Koller, L. D., Exon, J. H., and Arbogast, B., Methylmercury: effect on serum enzymes and humoral antibody, *J. Toxicol. Environ. Health*, 2, 1115–1123, 1977.
123. Ilback, N. G., Sundberg, J., and Oskarsson, A., Methyl mercury exposure via placenta and milk impairs natural killer (NK) cell function in newborn rats, *Toxicol. Lett.*, 58, 149–158, 1991.
124. Christensen, M. M., Ellermann-Eriksen, S., Rungby, J., and Mogensen, S. C., Comparison of the interaction of methyl mercury and mercuric chloride with murine macrophages, *Arch. Toxicol.*, 67, 205–211, 1993.
125. Christensen, M. M., Ellermann-Eriksen, S., Rungby, J., Mogensen, S. C., and Danscher, G., Histochemical and functional evaluation of mercuric chloride toxicity in cultured macrophages, *Prog. Histochem. Cytochem.*, 23, 306–315, 1993.
126. Christensen, M., Mogensen, S. C., and Rungby, J., Toxicity and ultrastructural localization of mercuric chloride in cultured murine macrophages, *Arch. Toxicol.*, 62, 440–446, 1988.
127. Lison, D., Dubois, P., and Lauwerys, R., *In vitro* effect of mercury and vanadium on superoxide anion production and plasminogen activator activity of mouse peritoneal macrophages, *Toxicol. Lett.*, 40, 29–36, 1988.
128. Daum, J. R., Shepherd, D. M., and Noelle, R. J., Immunotoxicology of cadmium and mercury on B-lymphocytes. I. Effects on lymphocyte function, *Int. J. Immunopharmacol.*, 15, 383–394, 1993.
129. Nakatsuru, S., Oohashi, J., Nozaki, H., Nakada, S., and Imura, N., Effect of mercurials on lymphocyte functions *in vitro*, *Toxicology*, 36, 297–305, 1985.
130. Patierno, S. R., Costa, M., Lewis, V. M., and Peavy, D. L., Inhibition of LPS toxicity for macrophages by metallothionein-inducing agents, *J. Immunol.*, 130, 1924–1929, 1983.
131. Tam, P. E. and Hindsill, R. D., Evaluation of immunomodulatory chemicals: alteration of macrophage function *in vitro*, *Toxicol Appl. Pharmacol.*, 76, 183–194, 1984.
132. Geertz, R., Gulyas, H., and Gercken, G., Cytotoxicity of dust constituents towards alveolar macrophages: interactions of heavy metal compounds, *Toxicology*, 86, 13–27, 1994.
133. Kudo, N. and Waku, K., Mercuric chloride induces the production of leukotriene B_4 by rabbit alveolar macrophages, *Arch. Toxicol.*, 68, 179–186, 1994.
134. Castranova, V., Bowman, L., Reasor, M. J., and Miles, P. R., Effects of heavy metal ions on selected oxidative metabolic processes in rat alveolar macrophages, *Toxicol. Appl. Pharmacol.*, 53, 14–23, 1980.
135. Contrino, J., Kosuda, L. L., Marucha, P., Kreutzer, D. L., and Bigazzi, P. E., The *in vitro* effects of mercury on peritoneal leukocytes (PMN and macrophages) from inbred Brown Norway and Lewis rats, *Int. J. Immunopharmacol.*, 14, 1051–1059, 1992.
136. Tan, X., Tang, C., Castoldi, A. F., Manzo, L., and Costa, L. G., Effects of inorganic and organic mercury on intracellular calcium levels in rat T-lymphocytes, *J. Toxicol. Environ. Health*, 38, 159–170, 1993.
137. Baginski, B., Effect of mercuric chloride on microbicidal activities of human polymorphonuclear leukocytes, *Toxicology*, 50, 247–256, 1988.
138. Jansson, G. and Harms-Ringdahl, M., Stimulating effects of mercuric and silver ions on the superoxide anion production in human polymorphonuclear leukocytes, *Free Rad. Res. Commun.*, 18, 87–98, 1993.
139. Kaneko, H., Kakkar, V. V., and Scully, M. F., Mercury compounds induce a rapid increase in procoagulant activity of monocyte-like U937 cells, *Br. J. Haematol.*, 87, 87–93, 1994.

140. Kimata, H., Shinomiya, K., and Mikawa, H., Selective enhancement of human IgE production *in vitro* by synergy of pokeweed mitogen and mercuric chloride, *Clin. Exp. Immunol.,* 53, 183–191, 1983.
141. Malamud, D., Dietrich, S. A., and Shapiro, I. M., Low levels of mercury inhibit the respiratory burst in human polymorphonuclear leukocytes, *Biochem. Biophys. Res. Commun.,* 128, 1145–1151, 1985.
142. Obel, N., Hansen, B., Christensen, M. M., Nielsen, S. L., and Rungby, J., Methyl mercury, mercuric chloride, and silver lactate decrease superoxide anion formation and chemotaxis in human polymorphonuclear leucocytes, *Human Exp. Toxicol.,* 12, 361–364, 1993.
143. Shenker, B. J., Berthold, P., Decker, S., Mayro, J., Rooney, C., Vitale, L., and Shapiro, I. M., Immunotoxic effects of mercuric compounds on human lymphocytes and monocytes. II. Alterations in cell viability, *Immunopharmacol. Immunotoxicol.,* 15, 555–577, 1993.
144. Shenker, B. J., Berthold, P., Rooney, C., Vitale, L., DeBolt, K., and Shapiro, I. M., Immunotoxic effects of mercuric compounds on human lymphocytes and monocytes. III. Alterations in B-cell function and viability, *Immunopharmacol. Immunotoxicol.,* 15, 87–112, 1993.
145. Shenker, B. J., Mayro, J. S., Rooney, C., Vitale, L., and Shapiro, I. M., Immunotoxic effects of mercuric compounds on human lymphocytes and monocytes. IV. Alterations in cellular glutathione content, *Immunopharmacol. Immunotoxicol.,* 15, 273–290, 1993.
146. Shenker, B. J., Rooney, C., Vitale, L., and Shapiro, I. M., Immunotoxic effects of mercuric compounds on human lymphocytes and monocytes. I. Suppression of T-cell activation, *Immunopharmacol. Immunotoxicol.,* 14, 539–553, 1992.
147. Nordlind, K., Effect of metal allergens on the DNA synthesis of unsensitized guinea pig lymphoid cells cultured *in vitro, Int. Arch. Allergy Appl. Immunol.,* 69, 12–17, 1982.
148. Nordlind, K., Stimulating effect of mercuric chloride and nickel sulfate on DNA synthesis of thymocytes and peripheral lymphoid cells from newborn guinea pigs, *Int. Arch. Allergy Appl. Immunol.,* 72, 177–179, 1983.
149. Verdier, F., Virat, M., Schweinfurth, H., and Descotes, J., Immunotoxicity of bis(tri-n-butyltin)oxide in the rat, *J. Toxicol. Environ. Health,* 32, 307–317, 1991.
150. Smialowicz, R. J., Riddle, M. M., Rogers, R. R., Luebke, R. W., Copeland, C. B., and Ernst, G. G., Immune alterations in rats following subacute exposure to tributyltin oxide, *Toxicology,* 64, 169–178, 1990.
151. Vos, J. G., De Klerk, A., Kranjc, E. I., van Loveren, H., and Rozing, J., Immunotoxicity of bis(tri-*n*-butyltin)oxide in the rat: effects on thymus-dependent immunity and on nonspecific resistance following long-term exposure in young vs. aged rats, *Toxicol. Appl. Pharmacol.,* 105, 144–155, 1990.
152. Vos, J. G., De Klerk, A., Kranjc, E. I., Kruizinga, W., van Ommen, B., and Rozing, J., Toxicity of bis(tri-*n*-butyltin)oxide in the rat. II. Suppression of thymus-dependent immune responses and of parameters of nonspecific resistance after short-term exposure, *Toxicol. Appl. Pharmacol.,* 75, 387–408, 1984.
153. van Loveren, H., Krajnc, E. I., Rombout, P. J., Blommaert, F. A., and Vos, J. G., Effects of ozone, hexachlorobenzene, and bis(tri-*n*-butyltin)oxide on natural killer activity in the rat lung, *Toxicol Appl. Pharmacol.,* 102, 21–33, 1990.
154. Krajnc, E. I., Wester, P. W., Loeber, J. G., van Leeuwen, F. X., Vos, J. G., Vaessen, H. A, and van der Heijden, C. A., Toxicity of bis(tri-*n*-butyltin)oxide in the rat. I. Short-term effects on general parameters and on the endocrine and lymphoid systems, *Toxicol. Appl. Pharmacol.,* 75, 363–386, 1984.
155. Miller, K., Maisey, J., and Nicklin, S., Effect of orally-administered dioctyltin dichloride on murine immunocompetence, *Environ. Res.,* 398, 434–441, 1986.
156. Volsen, S. G., Barrass, N., Scott, M. P., and Miller, K., Cellular and molecular effects of di-*n*-octyltin dichloride on the rat thymus, *Int. J. Immunopharmacol.,* 11, 703–715, 1989.
157. Seinen, W., Vos, J. G., van Krieken, R., Penninks, A., Brands, R., and Hooykaas, H., Toxicity of organotin compounds. III. Suppression of thymus-dependent immunity in rats by di-*n*-butyltindichloride and di-*n*-octyltindichloride, *Toxicol. Appl. Pharmacol.,* 42, 213–224, 1977.
158. Seinen, W. and Willems, M. I., Toxicity of organotin compounds. I. Atrophy of thymus and thymus-dependent lymphoid tissue in rats fed di-*n*-octyltindichloride, *Toxicol. Appl. Pharmacol.,* 35, 63–75, 1976.
159. Miller, K. and Scott, M. P., Immunological consequences of dioctyltin dichloride (DOTC)-induced thymic injury, *Toxicol. Appl. Pharmacol.,* 78, 395–403, 1985.
160. Snoeij, N. J., van Iersel, A. A., Penninks, A. H., and Seinen, W., Toxicity of triorganotin compounds: comparative *in vivo* studies with a series of trialkyltin compounds and triphenyltin chloride in male rats, *Toxicol. Appl. Pharmacol.,* 81, 274–286, 1985.
161. Snoeij, N. J., Penninks, A. H., and Seinen, W., Dibutyltin and tributyltin compounds induce thymus atrophy in rats due to a selective action on thymic lymphoblasts, *Int. J. Immunopharmacol.,* 10, 891–899, 1988.
162. Ghoneum, M., Hussein, A. E., Gill, G., and Alfred, L. J., Suppression of murine natural killer cell activity by tributyltin: *In vivo* and *in vitro* assessment, *Environ. Res.,* 52, 178–186, 1990.
163. Vos, J. G., van Logten, M. J., Kreeftenberg, J. G., and Kruizinga, W., Effect of triphenyltin hydroxide on the immune system of the rat, *Toxicology,* 29, 325–336, 1984.
164. Hayashi, O., Chiba, M., and Kikuchi, M., The effects of stannous chloride on the humoral immune response of mice, *Toxicol. Lett.,* 21, 279–285, 1984.
165. Dimitrov, N. V., Meyer, C., Nahhas, F., Miller, C., and Averill, B. A., Effect of tin on immune responses of mice, *Clin. Immunol. Immunopathol.,* 20, 39–48, 1981.

166. Levine, S. and Sowinski, R., Tin salts prevent the plasma cell response to metallic tin in Lewis rats, *Toxicol. Appl. Pharmacol.,* 68, 110–115, 1983.
167. Levine, S., Saad, A., and Rappaport, I., The effects of metallic tin on background plaque-forming cells, immunoglobulins, and the immune response, *Immunol. Invest.,* 16, 201–212, 1987.
168. Cohen, M. D., Chen, C. M., and Wei, C. I., Decreased resistance to *Listeria monocytogenes* in mice following vanadate exposure: effects upon the function of macrophages, *Int. J. Immunopharmacol.,* 11, 285–292, 1989.
169. Cohen, M. D. and Wei, C. I., Effects of ammonium metavandate treatment upon macrophage glutathione redox cycle activity, superoxide production, and intracellular glutathione status, *J. Leukocyte Biol.,* 44, 122–129, 1988.
170. Vaddi, K. and Wei, C. I., Effect of ammonium metavanadate on mouse peritoneal macrophage lysosomal enzymes, *J. Toxicol. Environ. Health,* 33, 65–78, 1991.
171. Vaddi, K. and Wei, C. I., Modulation of F_c receptor expression and function in mouse peritoneal macrophages by ammonium metavanadate, *Int. J. Immunopharmacol.,* 13, 1167–1176, 1991.
172. Vaddi, K. and Wei, C. I., Modulation of macrophage activation by ammonium metavanadate, *J. Toxicology Environ. Health,* in press.
173. Cohen, M. D., Wei, C. I., Tan, H., and Kao, K. J.,, Effect of ammonium metavanadate on the murine immune response, *J. Toxicol. Environ. Health,* 19, 279–298, 1986.
174. Al Bayati, M. A., Culbertson, M. R., Schreider, J. P., Rosenblatt, L. S., and Raabe, O. G., The lymphotoxic action of vanadate, *J. Environ. Pathol. Toxicol. Oncol.,* 11, 19–27, 1992.
175. Wei, C. I., Al Bayati, M. A., Culbertson, M. R., Rosenblatt, L. S., and Hansen, L. D., Acute toxicity of ammonium metavanadate in mice, *J. Toxicol. Environ. Health,* 10, 673–687, 1982.
176. Cohen, M. D., Yang, Z., Zelikoff, J. T., and Schlesinger, R. B., Pulmonary immunotoxicity of inhaled ammonium metavanadate in Fischer 344 rats, *Fund. Appl. Toxicol.,* 33, 254–263, 1996.
177. Al-Bayati, M. A., Giri, S. N., and Rabe, O. G., Time and dose-response study of the effects of vanadate in rats: changes in blood cells, serum enzymes, protein, cholesterol, glucose, calcium, and inorganic phosphate, *J. Environ. Pathol. Toxicol. Oncol.,* 9, 435–455, 1990.
178. Zaporowska, H. and Wasilewski, W., Haematological results of vanadium intoxication in Wistar rats, *Comp. Biochem. Physiol.,* 101C, 57–61, 1992.
179. Sharma, R. P., Bourcier, D. R., Brinkerhoff, C. R., and Christensen, S. A., Effects of vanadium on immunologic functions, *Am. J. Indust. Med.,* 2, 91–99, 1981.
180. Kacew, S., Parulekar, M. R., and Merali, Z., Effects of parenteral vanadium administration on pulmonary metabolism of rats, *Toxicol. Lett.,* 11, 119–124, 1982.
181. Knecht, E. A., Moorman, W. J., Clark, J. C., Hull, R. D., Biagini, R. E., Lynch, D. W., Boyle, T. J., and Simon, S. D., Pulmonary reactivity to vanadium pentoxide following subchronic inhalation exposure in a non-human primate animal model, *J. Appl. Toxicol.,* 12, 427–434, 1992.
182. Knecht, E. A., Moorman, W. J., Clark, J. C., Lynch, D. W., and Lewis, T. R., Pulmonary effects of acute vanadium pentoxide inhalation in monkeys, *Am. Rev. Respir. Dis.,* 132, 1181–1185, 1985.
183. Fisher, G. L., McNeill, K. L., Whaley, C. B., and Fong, J., Attachment and phagocytosis studies with murine pulmonary alveolar macrophages, *J. Reticuloendothel. Soc.,* 24, 243–252, 1978.
184. Cohen, M. D., Parsons, E., Schlesinger, R. B., and Zelikoff, J. T., Immunotoxicity of *in vitro* vanadium exposures: effects on interleukin-1, tumor necrosis factor-α, and prostaglandin E_2 production by WEHI-3 macrophages, *Int. J. Immunopharmacol.,* 15, 477–486, 1993.
185. Cohen, M. D., McManus, T. P., Yang, Z., Qu, Q., Schlesinger, R.B., and Zelikoff, J.T., Vanadium affects macrophage-interferon-γ binding and -inducible responses, *Toxicol. Appl. Pharmacol.,* 138, 110–120, 1996.
186. Bennett, P. A., Dixon, R. J., and Kellie, S., The phosphotyrosine phosphatase inhibitor vanadyl hydroperoxide induces morphological alterations, cytoskeletal rearrangements, and increased adhesiveness in rat neutrophil leucocytes, *J. Cell Sci.,* 106, 891–901, 1993.
187. Evans, G. A., Garcia, G. G., Erwin, R., Howard, O. M., and Farrar, W. L., Pervanadate simulates the effects of interleukin-2 (IL-2) in human T-cells and provides evidence for the activation of two distinct tyrosine kinase pathways by IL-2, *J. Biol. Chem.,* 269, 23407–23412, 1994.
188. Imbert, V., Peyron, J. F., Far, D. F., Mari, B., Auberger, P., and Rossi, B., Induction of tyrosine phosphorylation and T-cell activation by vanadate peroxide, an inhibitor of protein tyrosine phosphatases, *Biochem. J.,* 297, 163–173, 1994.
189. Labedzka, M., Gulyas, H., Schmidt, N., and Gercken, G., Toxicity of metallic ions and oxides to rabbit alveolar macrophages, *Environ. Res.,* 48, 255–274, 1989.
190. Waters, M. D., Gardner, D. E., and Coffin, D. L., Cytotoxic effects of vanadium on rabbit alveolar macrophages *in vitro, Toxicol. Appl. Pharmacol.,* 28, 253–263, 1974.

CHAPTER 20

EFFECTS OF ORGANIC SOLVENTS AND PESTICIDES ON IMMUNE SYSTEM PARAMETERS

John B. Barnett

CONTENTS

1 EFFECT OF ORGANIC SOLVENTS ON IMMUNE SYSTEM PARAMETERS

A more restrictive definition of immunotoxicity is utilized in this section than in the following section on pesticides. Although immunotoxicity is defined as any organic solvent-induced perturbation to the immune response, immune cells, or hemopoiesis, not included in this section are the many citations regarding the genotoxicity of the compounds. There is an enormous number of published studies on virtually all industrial organic compounds focusing strictly on this topic (for example, Reference 1); therefore, it was concluded that the readers would be well served by these sources and to repeat them here would be unnecessarily redundant.

A solvent is defined as "...a liquid substance capable of dissolving or dispersing one or more other substances," as opposed to a solute which is "a dissolved substance". Thus,

0-8493-4493-X/97/$0.00+$.50

applying this definition, any organic compound capable of dissolving a substance should be included in these tables; however, in practice, it was often difficult to determine if a given organic compound should be classified as a solvent. Some guidance in this determination was provided by the *Merck Index*,[2] the *Industrial Solvents Handbook*,[3] or the author(s) of the citation.

There are several organic compounds that are obviously missing from these tables. One group of compounds missing includes the substances of abuse, even though one of their uses is as a solvent (e.g., ethanol). These substances are covered in other chapters in this book. The other obvious omission is dimethyl sulfoxide (DMSO). DMSO has been extensively used (≈8000 references listed in Medline™), primarily as the solvent used to administer a variety of substances to *in vitro* cell cultures. Most studies included DMSO as the vehicle control, but very few studied the toxic effects of DMSO itself. Thus, a listing of DMSO toxicity is beyond the scope of this chapter. Few *in vitro* studies have been carried out using solvents; therefore, Table 3 is fairly brief. This is likely due to the difficulty in administering an organic solvent, which in most instances is water insoluble, to cultures that must grow in an aqueous environment.

2 EFFECT OF PESTICIDES ON IMMUNE SYSTEM PARAMETERS

A very broad definition of "immunotoxicity" was used for this section so that the greatest breadth of information could be included. Immunotoxicity is defined here as any pesticide-induced perturbation to the immune response, immune cells, or hemopoiesis. Thus, perturbations are listed that would, by a more restricted definition, be classified as genotoxicity (chromosomal aberrations) when affecting an immune cell (e.g., lymphocytes) or hematotoxicity (changes in hemopoiesis) when affecting leukocyte production.

The reference used primarily to define the scope of pesticides was the Pesticide Dictionary in the *Farm Chemicals Handbook '95*.[87] This reference book provides a very complete listing of current and previously used pesticides as well as their characteristics as a pesticide. Another useful reference for defining the basic characteristics is the *Basic Guide to Pesticides*.[88] Although not nearly as comprehensive as the previous reference, it contains a useful Index of Pesticide Names that allowed cross-referencing a pesticide to a more common one.

With a few exceptions, all references found regardless of the species tested are covered. The first of these exceptions is human exposures, where the pesticide was not definitively identified. In some epidemiological studies, the effects of pesticides exposure on immune function, without specifically identifying the particular pesticide, were reported. These data were not included in the tables because it is not possible to assign these effects to any particular pesticide. Another omission is the epidemiological studies where many different types of pesticides were manufactured at the plant during the time of exposure (for example, Reference 89); however, many pesticides are sold as mixtures (e.g., monochlorobenzene), and where information was available on these pesticides, it was included.

Many pesticides, at least in their early production, contained significant quantities of impurities which later proved to be very toxic. These include 2,4,5-T, contaminated with tetrachlorodibenzo-*p*-dioxin (for which a vast amount of material on its effects has been published); malathion, contaminated with O,O,S-trimethyl phosphorothioate;[90] lindane, contaminated with the β-isomer of hexachlorocyclohexane;[91] and pentachlorophenol, which was contaminated with furans, etc.[92] No studies involving only the pesticide impurity are listed; however, studies on the technical grade product (specifically identified as "technical" grade) are included, in addition to the studies using the pure form of the pesticide.

The polychlorinated biphenyls (PCB) are considered a particularly pesky environmental contaminant, and a large number of studies have reported on a variety of these compounds; however, Aroclor, which is also a mixture of PCBs, was manufactured and sold as an insecticide until 1971.[87] Therefore, several studies on the various compositions of Aroclor are reported. Excluding the deliberate omissions, stated above, every attempt was made to be as inclusive as possible; however, there are undoubtedly some references that were missed. I apologize for these inadvertent omissions.

3 ABBREVIATIONS USED

↔	no effect
↑	increased
↓	decreased
ADH	adherent
AP	acid phosphatase
ATP	adenosine triphosphate
BFU-E	burst forming unit-erythropoiesis
BM	bone marrow
CFU-E	colony forming unit-erythropoiesis
CFU-GM	colony forming unit-granulocyte/macrophage
CFU-M	colony forming unit-macrophage
chemilum	chemilumescence
CMV	cytomegalovirus
ConA	conconavalin A
CTL	cytotoxic lymphocyte
d	day(s)
derm	dermal
diff'n	differentiation
DTH	delayed-type hypersensitivity
EMCV	encephomyocarditis virus
exp'r	exposure
exp'd	exposed
fMLP	formyl-methionine-leucine-phenylalanine
G.I.	gastrointestinal
GvH	graft vs. host
HA	hemagglutination
HEV	high endothelial venule
HQ	hydroquinone
i.m.	intramuscular
i.p.	intraperitoneal
i.v.	intravenous
Ig	immunoglobulin
imm'z	immunized
KlebsM	membrane preparation of *Klebsiella pneumoniae*
KLH	keyhole limpet hemocyanin
LAK	lymphokine-activated killer (cell)
LN	lymph node
LPS	lipopolysaccharide
m	month(s)
mac.	macrophage
MHV	mouse hepatitis virus
min	minute
mitoblastogenesis	mitogen-induced blastogenesis
MLR	mixed lymphocyte reaction

MSV	Maloney sarcoma virus
NBT	nitroblue-tetrazolium
NK	natural killer
NOEL	no observable effect level
OA	ovalbumin
occup	occupational or environmental
p.n.	prenatal
p.o.	per os
PB	peripheral blood
PBMC	peripheral blood mononuclear cell
pcd	postcoital day
PE	peritoneal exudate
PEC	peritoneal exudate cell
PFC	plaque forming cells
PHA	phytohemagglutinin
PMN	polymorphonuclear (cell)
ppb	parts per billion
ppm	parts per million
pulm	pulmonary
PWM	polk weed mitogen
RBC	red blood cells
resp	respiratory
s.c.	subcutaneous
SCE	sister-chromatid exchange
spln	spleen
SRBC	sheep erythrocytes
TDI	toluene diisocyanate
TPA	12-*O*-tetradecanoyl phorbol-13-acetate
w	week(s)
WBH	white blood cells
y	year(s)

4 TABLES

TABLE 1

Effect of Organic Solvents on Immune System Parameters — *In Vivo* Exposures

Compound (Chemical Name)	Dose (NOEL)	Route	Species/Strain	Sex	Parameter	Effect	Miscellaneous	Ref.
Aniline hydrochloride *(benzenamine-HCl)*	0–100 mg/kg/d	p.o.	Fischer 344	F	Spln weight	↓	13-d exp'r	4
		p.o.	Fischer 344	F	Methemoglobinemia	↑	13-d exp'r	4
		p.o.	Fischer 344	F	Hemopoeitic activity	↑	13-d exp'r	4
		p.o.	Fischer 344	F	RBC numbers	↓	13-d exp'r	4
Benzene	5 ppm	p.n.	Swiss-Webster	M/F	CFU-E hemopoiesis	↔	Exp'd 6–15 pcd; assay @ 2 d of age; liver	5
		p.n.	Swiss-Webster	M	CFU-E hemopoiesis	↔	Exp'd 6–15 pcd; assay @ 6 wk of age; spln and BM	5
	0–10 ppm	p.n.	Swiss-Webster	M	BFU-E hemopoiesis	↑	Exp'd 6–15 pcd; 16 pcd assay	5
		p.n.	Swiss-Webster	M/F	CFU-E hemopoiesis	↑	Exp'd 6–15 pcd; 16 pcd assay	5
		p.n.	Swiss-Webster	F	CFU-GM hemopoiesis	↔	Exp'd 6–15 pcd; @2 d of age	5
	0–10 (5) ppm[a]	p.n.	Swiss-Webster	F	BFU-E hemopoiesis	↑	Exp'd 6–15 pcd; 16 pcd assay	5
	10 ppm	Resp	C57Bl/6	M	BM B-cell numbers	↔	6-d exp'r	6
		Resp	C57Bl/6	M	BM B-cell mitogenesis	↓	6-d exp'r	6
		p.n.	Swiss-Webster	M/F	CFU-E hemopoiesis	↑↓	Exp'd 6–15 pcd; assay @ 2 d of age; liver	5
		p.n.	Swiss-Webster	M	CFU-E hemopoiesis	↑	Exp'd 6–15 pcd; assay @ 6 wk of age; spleen	5
		p.n.	Swiss-Webster	M	CFU-GM hemopoiesis	↓	Exp'd 6–15 pcd; assay @ 2 d of age	5
	10 (31) ppm	Resp	C57Bl/6	M	Spln T-cell numbers	↔	6-d exp'r	6
		Resp	C57Bl/6	M	Spln T-cell mitogenesis	↓	6-d exp'r	6
	10 + 10 ppm	p.n. + resp	Swiss-Webster	M/F	CFU-GM hemopoiesis; BM	↔	Exp'd 6–15 pcd and 10 wk of age for 2 wk	5
		p.n. + resp	Swiss-Webster	M/F	CFU-GM hemopoiesis; spln	↓	Exp'd 6–15 pcd and 10 wk of age for 2 wk	5
		p.n. + resp	Swiss-Webster	F	CFU-E hemopoiesis; BM	↔	Exp'd 6–15 pcd and 10 wk of age for 2 wk	5
		p.n. + resp	Swiss-Webster	F	CFU-GM hemopoiesis; spln	↔	Exp'd 6–15 pcd and 10 wk of age for 2 wk	5
		p.n. + resp	Swiss-Webster	M	CFU-E hemopoiesis; BM	↓	Exp'd 6–15 pcd and 10 wk of age for 2 wk	5
		p.n. + resp	Swiss-Webster	M	CFU-GM hemopoiesis; spln	↔	Exp'd 6–15 pcd and 10 wk of age for 2 wk	5
	0–20 ppm	p.n.	Swiss-Webster	M/F	BFU-E hemopoiesis	↔	Exp'd 6–15 pcd; @ 2 d of age; liver	5
		p.n.	Swiss-Webster	M/F	BFU-E hemopoiesis	↔	Exp'd 6–15 pcd; @ 6 wk of age; spln and BM	5
		p.n.	Swiss-Webster	F	CFU-E hemopoiesis	↔	Exp'd 6–15 pcd; @ 6 wk of age; spln and BM	5
		p.n.	Swiss-Webster	M/F	CFU-GM hemopoiesis	↔	Exp'd 6–15 pcd; @ 6 wk of age; spln and BM	5
		p.n.	Swiss-Webster	M/F	CFU-E hemopoiesis	↓	Exp'd 6–15 pcd; @ 6 wk of age; BM	5
		p.n.	Swiss-Webster	M/F	CFU-GM hemopoiesis	↔	Exp'd 6–15 pcd; 16 pcd assay; liver	5
	20 ppm	p.n.	Swiss-Webster	M	CFU-E hemopoiesis	↔	Exp'd 6–15 pcd; @ 6 wk of age; spln and BM	5
		p.n.	Swiss-Webster	M	CFU-E hemopoiesis	↑	Exp'd 6–15 pcd; @ 2 d of age; liver	5
		p.n.	Swiss-Webster	M/F	CFU-E hemopoiesis	↓	Exp'd 6–15 pcd; 16 pcd assay	5

TABLE 1 (continued)

Effect of Organic Solvents on Immune System Parameters — *In Vivo* Exposures

Compound (Chemical Name)	Dose (NOEL)	Route	Species/Strain	Sex	Parameter	Effect	Miscellaneous	Ref.
		p.n.	Swiss-Webster	F	CFU-E hemopoiesis	↔	Exp'd 6–15 pcd; @ 2 d of age; liver	5
		p.n.	Swiss-Webster	M/F	CFU-GM hemopoiesis	↑	Exp'd 6–15 pcd; @ 2 d of age	5
		p.n.	Swiss-Webster	M/F	BFU-E hemopoiesis	↔	Exp'd 6–15 pcd; 16 pcd assay	5
	0–100 ppm	Resp	C57Bl/6	M	Spln cell phenotypes	↔	10-d exp'r	7
		Resp	C57Bl/6	M	MLR	↓	20-d exp'r	7
		Resp	C57Bl/6	M	PBY6 tumor incidence	↑	10-d exp'r; @ 10^4 cell inoculum	7
		Resp	C57Bl/6	M	Suppressor cell activity	↔	20-d exp'r	7
		Resp	C57Bl/6	M	CTL activity	↔	20-d exp'r; decreased @ 100:1 E:T	7
	0–200 ppm	Resp	Balb/c	M	PB T-cell number	↓	14-d exp'r	8
		Resp	Balb/c	M	Spln:body weight	↓	7-d exp'r	8
		Resp	Balb/c	M	PB T-cell number	↓	7-d exp'r	8
		Resp	Balb/c	M	T-dependent IgG antibody	↓	14-d exp'r; anti-SRBC per spleen	8
		Resp	Balb/c	M	T-dependent IgM antibody	↓	14-d exp'r; anti-SRBC per spleen	8
		Resp	Balb/c	M	PB B-cell number	↓	14-d exp'r	8
		Resp	Balb/c	M	Suppressor cell activity	↔	14-d exp'r	8
		Resp	Balb/c	M	Thymus:body weight	↓	14-d exp'r	8
		Resp	Balb/c	M	Spln:body weight	↓	14-d exp'r	8
		Resp	Balb/c	M	PB WBC	↓	14-d exp'r	8
	0–200 (50) ppm	Resp	Balb/c	M	CHR	↓	14-d exp'r; anti-picryl chloride	8
		Resp	Balb/c	M	Thymus:body weight	↓	7 d exp'r	8
		Resp	Balb/c	M	PB B-cell number	↓	7-d exp'r	8
		Resp	Balb/c	M	PB WBC	↓	7-d exp'r	8
	0–300 (10) ppm	Resp	C57Bl/6	M	Spln B cells	↓	@ d 1, 4, 7; exp'r for 5 d pre-infection	9
		Resp	C57Bl/6	M	Spln B cells	↓	@ d 1, 4, 7; exp'r for 5 d pre- and 7 d postinfection	9
		Resp	C57Bl/6	M	Spln Thy1.2 + cell numbers	↓	@ d 1, 4, 7; exp'r for 5 d pre- and 7 d postinfection	9
		Resp	C57Bl/6	M	Spln lymphocyte numbers	↓	@ d 1, 4, 7; exp'r for 5 d pre- and 7 d postinfection	9
		Resp	C57Bl/6	M	Spln *Listeria* counts	↑	Increased @ d4; exp'r for 5 d pre- and 7 d postinfection	9
		Resp	C57Bl/6	M	Spln Thy1.2 + cell numbers	↓	@ d 1, 4, 7; exp'r for 5 d pre-infection	9
		Resp	C57Bl/6	M	Spln lymphocyte numbers	↓	@ d 1, 4, 7; exp'r for 5 d pre-infection	9

	Resp	C57Bl/6	M	Spln cell numbers	↓	@ d 1, 4, 7; exp'r for 5 d pre-infection	9
	Resp	C57Bl/6	M	Spln cell numbers	↓	Increased @ d 4; exp'r for 5 d pre- and 7 d postinfection	9
0–300	Resp	C57Bl/6	M	Spln esterase + cell numbers	↓	@ d 1, 4, 7; exp'r for 5 d pre-infection	9
(100) ppm	Resp	C57Bl/6	M	Spln esterase + cell numbers	↓	@ d 1, 4, 7; exp'r for 5 d pre- and 7 d postinfection	9
	Resp	C57Bl/6	M	Spln *Listeria* counts	↑	Increased @ d 4; exp'r for 5 d pre-infection	9
300 ppm	Resp	C57Bl/6	M	Thymic cellularity	↑	6-, 30-, 115-d exp'r — 15× increase	10
	Resp	C57Bl/6	M	BM cell number	↑	6-, 30-, 115-d exp'r — 3× increase	10
	Resp	C57Bl/6	M	Spln T-cell mitoblastogenesis	↓	6-, 30-, 115-d exp'r	10
0.25 mg/kg/d	s.c.	Rabbit		PB B-cell numbers	↓	10-d exp'r	11
	s.c.	Rabbit		PB T-cell numbers	↔	10-d exp'r	11
0.5 ml/kg/d	s.c.	Fischer 344	M	PB lymphocyte number	↓	Aroclor 1254-induced	12
100 mg/kg	i.p.	B6C3F1	M	BM adherence cell number	↔	Exp'r 2×/d for 4 d	13
	i.p.	B6C3F1	M	BM cell number	↔	Exp'r 2×/d for 4 d	13
	i.p.	B6C3F1	M	CFU-GM	↓	Exp'r 2×/d for 4 d; exp'd BM cell + norm ADH cell	13
	i.p.	DBA/2J	M	CFU-GM	↔	Exp'r 2×/d for 4 d; exp'd ADH cell + norm BM cell	13
	i.p.	DBA/2J	M	BM adherence cell number	↔	Exp'r 2×/d for 4 d	13
	i.p.	DBA/2J	M	BM cell number	↓	Exp'r 2×/d for 4 d	13
	i.p.	B6C3F1	M	CFU-GM	↓	Exp'r 2×/d for 4 d; exp'd ADH cell + norm BM cell	13
	i.p.	DBA/2J	M	CFU-GM	↑	Exp'r 2×/d for 4 d; exp'd BM cell + norm ADH cell	13
	i.p.	B6C3F1	M	BM CFU-GM	↔	Exp'r 2×/d for 4 d	14
	i.p.	B6C3F1	M	BM stroma GM-CSF	↓	Exp'r 2×/d for 4 d	14
	i.p.	B6C3F1	M	BM PGE2 production	↑	Exp'r 2×/d for 4 d	14
	i.p.	B6C3F1	M	BM B-cell numbers	↓		15
	i.p.	B6C3F1	M	BM pre-B-cell numbers	↑		15
	i.p.	B6C3F1	M	BM culture B-cell numbers	↓	Cultures from treated mice	15
	i.p.	B6C3F1	M	BM culture pre-B-cell numbers	↑	Cultures from treated mice	15
	p.n.	B6C3F1	M/F	Fetal liver pre-B-cell numbers	↓	Exp'r 2×/d 12.5–19.5 dpc	15
	p.n.	B6C3F1	M/F	Fetal liver B-cell numbers	↓	Exp'r 2×/d 12.5–19.5 dpc	15
	i.p.	B6C3F1	F	BM B-cell numbers	↓	Maternal; exp'r 2×/d 12.5–19.5 dpc	15
	i.p.	B6C3F1	F	BM pre-B-cell numbers	↓	Maternal; exp'r 2×/d 12.5–19.5 dpc	15

TABLE 1 (continued)

Effect of Organic Solvents on Immune System Parameters — *In Vivo* Exposures

Compound (Chemical Name)	Dose (NOEL)	Route	Species/Strain	Sex	Parameter	Effect	Miscellaneous	Ref.
		i.p.	B6C3F1	F	BM culture pre-B-cell numbers	↓	Maternal; exp'r 2×/d 12.5–19.5 dpc	15
		p.n.	B6C3F1	M/F	Fetal liver culture pre-B-cell numbers	↓	Exp'r 2×/d 12.5–19.5 dpc	15
		p.n.	B6C3F1	M/F	Fetal liver culture B-cell numbers	↓	Exp'r 2×/d 12.5–19.5 dpc	15
		p.n.	B6C3F1	M/F	Fetal liver ADH stroma colonies	↓	Exp'r 2×/d 12.5–19.5 dpc	15
		p.n.	B6C3F1	M/F	Liver pre-B cells	↑	d8 neonate; exp'r 2×/d 12.5–19.5 dpc	15
		p.n.	B6C3F1	M/F	Neonatal LPS blastogenesis	↓	d8 neonate; exp'r 2×/d 12.5–19.5 dpc	15
	440 mg/kg	i.p.	C57Bl/6	M	LPS blastogenesis	↔	w/Aroclor 1254 induction	16
		i.p.	C57Bl/6	M	ConA blastogenesis	↔	w/Aroclor 1254 induction	16
		i.p.	C57Bl/6	M	T-dependent antibody	↓	Anti-SRBC PFC; w/Aroclor 1254 induction	16
	0–660 mg/kg	i.p.	C57Bl/6	M	T-dependent antibody	↓	Anti-SRBC PFC	16
	0–660 (44) mg/kg	i.p.	C57Bl/6	M	ConA blastogenesis	↓		16
	0–660 (132) mg/kg	i.p.	C57Bl/6	M	LPS blastogenesis	↓		16
	660 mg/kg	i.p.	Balb/c	M	BM leuk. TNF-α production	↓	3-d exp'r; LPS-induced	17
		i.p.	Balb/c	M	BM leuk. IL-1 (peak time)	↑	3-d exp'r; LPS-induced	17
		i.p.	Balb/c	M	BM leuk. IL-6 production	↔	3-d exp'r; LPS-induced	17
		i.p.	Balb/c	M	TNF-α production	↑	CSF-1 + LPS-induced	17
		i.p.	Balb/c	M	BM granulocyte chemotaxis	↑	3-d exp'r; TPA-induced	18
		i.p.	Balb/c	M	BM mac. H_2O_2 production	↑	3-d exp'r; TPA-induced	18
		i.p.	Balb/c	M	BM mac. chemotaxis	↑	3-d exp'r; TPA-induced	18
		i.p.	Balb/c	M	Number mature BM macs.	↑	3-d exp'r	18
		i.p.	Balb/c	M	BM cell phenotypes	↔	3-d exp'r	18
	0–790 (31) mg/l	p.o.	CD-1	M	IL-2 production	↓	28-d exp'r	19
	800 mg/kg	i.p.	Balb/c	M	BM stroma pre-IL1α->IL-1 conversion	↓	2-d exp'r; LPS-induced	20
		s.c.	C57Bl/6	M	Mac. tumoricidal activity	↓	P815-target; IFN-gamma + LPS-induced	21
		s.c.	C57Bl/6	M	Mac. H_2O_2 release	↑	TPA-induced	21
		s.c.	C57Bl/6	M	Mac. Ia induction	↔	IFN-gamma-induced	21

		s.c.	C57Bl/6	M	Fc-mediated mac. phagocytosis	↓	SRBC	21
		i.p.	Balb/c	F	BM cell nitrate production	↑		22
		i.p.	Balb/c	F	iNOS mRNA	↑		22
		i.p.	Balb/c	F	BM cell GM-CSF-induced proliferation	↑		22
	Occup	Resp	Human		PB T-cell numbers	↓	55- to 122-m exposure	23
		Resp	Human		T-cell function	↔	55- to 122-m exposure	23
		Resp	Human		NBT reduction — PMN	↑	Short-term occupational exposure	24
		Resp	Human		NBT reduction — PMN	↓	Long-term occupational exposure	24
Benzenetriol (benzene metabolite)	0–880 mg/kg	s.c.	Fischer 344	M	Thymocyte numbers	↓	1×/d for 7 d	25
Benzene and β-naphthylamine (mixture)	Occup	Resp	Human	M	PB $CD16^+$ percentage	↑		26
		Resp	Human	M	PB $CD4^+$ cell number	↓		26
Benzidine (*[1,1′-biphenyl]-4,4-diamine)*	Occup	Resp	Human	M	PB NK cell activity	↓	Per cell basis	27
		Resp	Human	M	PB Leu $11a^+$ cell number	↑		27
Butyl chloride (1-chlorobutane)	0–120 mg/kg	p.o.	Fischer 344	M/F	Spln lymphocytes	↓		28
	0–250 mg/kg	p.o.	B6C3F1	M/F	Spln lymphocytes	↓		28
Carbon tetrachloride (tetrachloromethane)	0.25 mg/kg	i.m.	AB mouse	M	Thymus cell number	↓		29
		i.m.	AB mouse	M	Thymus weight	↔		29
	0.15 ml 2:3 CCl_4:oil	p.o.	CFLP mouse	M	Macro infiltration to liver	↑		30
	0.1 ml 1:3 CCl_4:oil	s.c.	Balb/c		T-dependent antibody	↑	Anti-SRBC PFC; exp′r 5 hr before immunization	31
		s.c.	Balb/c		T-independent antibody	↓	Anti-TNPandLPS PFC; exp′r 5 hr before immunization	31
		s.c.	Balb/c Nu/Nu		T-independent antibody	↑	Anti-Fl-LPS PFC; exp′r 5 hr before immunization	31
		s.c.	Balb/c		T-dependent antibody	↔	Anti-SRBC PFC; exp′r 5 hr before immunization	31
		s.c.	Balb/c		Macro Ag-presentation	↑	Spln macro; exp′r 5 hr before immunization	31
	0.3 ml/kg	i.p.	Rat	M/F	Indirect agglutination	↑	Anti-mitochondrial Ag	32
	0.7 ml/kg	i.p.	Wistar	M	Liver phagocytosis	↑		33
		i.p.	Wistar	M	Liver ADH cell number	↑	Due to migrational influx	33
		i.p.	Wistar	M	PB WBC	↔		33

TABLE 1 (continued)

Effect of Organic Solvents on Immune System Parameters — *In Vivo* Exposures

Compound (Chemical Name)	Dose (NOEL)	Route	Species/Strain	Sex	Parameter	Effect	Miscellaneous	Ref.
		i.p.	Wistar	M	PB lymphocyte numbers	↓		33
		i.p.	Wistar	M	PB monocyte numbers	↔		33
		i.p.	Wistar	M	T-dependent antibody	↑	Anti-SRBC titers	33
	0.5 ml	s.c.	Rabbit		*M. avium* resistance	↓	2×/wk for 23 exp′r	34
		s.c.	Rabbit		*M. avium* tissue counts	↑	2×/wk for 23 exp′r	34
	0.5 ml/g	s.c.	Rat	M	Kupffer cell TNF-α production	↑	Peak production 6 hr post-exp′r	35
	0.5 ml/2 ml/kg	s.c.	Balb/c	M	Suppressor cells	↑	Inhibition of recipient DTH by adoptive CCl_4-spln cells	36
		s.c.	Balb/c	M	Adherent suppressor cells	↑	Inhibition recipient DTH by adoptive CCl_4-spln adherent cells	36
		s.c.	Balb/c	M	PGE_2 production	↑		36
		s.c.	Balb/c	M	Superoxide production	↑		36
	2 ml/kg	p.o.	Sprague-Dawley	M	Phagocytic index	↑	Collodial carbon clearance	37
Catechol (benzene metabolite)	75 mg/kg	i.v.	C57Bl/6	M	Spln cell number	↓		38
		i.p.	C57Bl/6	M	Spln cell number	↓		38
		i.p.	C57Bl/6	M	BM cell number	↔		38
		i.v.	C57Bl/6	M	BM cell number	↔		38
		i.v.	C57Bl/6	M	T-independent antibody	↓	Anti-LPS	38
		i.p.	C57Bl/6	M	T-independent antibody	↓	Anti-LPS	38
		i.p.	C57Bl/6	M	BM LPS-blastogenesis	↔		38
		i.v.	C57Bl/6	M	BM LPS-blastogenesis	↓		38
		i.v.	C57Bl/6	M	BM DxS-blastogenesis	↑		38
		i.p.	C57Bl/6	M	BM DxS-blastogenesis	↔		38
		i.p.	C57Bl/6	M	BM DxS + LPS-blastogenesis	↔		38
		i.v.	C57Bl/6	M	BM DxS + LPS-blastogenesis	↔		38
Chloroaniline-HCL (*p*-chloroaniline-HCL)	0–80 mg/kg	p.o.	Fischer 344	M/F	BM hemopoiesis	↑	13-wk exp′r	39
		p.o.	Fischer 344	M/F	Spln hemopoiesis	↑	13-wk exp′r	39
		p.o.	Fischer 344	M/F	Liver hemopoiesis	↑	13-wk exp′r	39
		p.o.	Fischer 344	M/F	Spln weight	↑	13-wk exp′r	39
	0–120 mg/kg	p.o.	B6C3F1	M/F	BM hemopoiesis	↔	13-wk exp′r	39
		p.o.	B6C3F1	M/F	Spln weight	↑	13-wk exp′r	39
		p.o.	B6C3F1	M/F	Liver hemopoiesis	↑	13-wk exp′r	39
		p.o.	B6C3F1	M/F	Spln hemopoiesis	↑	13-wk exp′r	39

Chlorobenzene (monochlorobenzene)	0–250 mg/kg	p.o.	B6C3F1	M	Spln weight	↔	13-wk exp'r	40
	0–500 mg/kg	p.o.	B6C3F1	F	Spln weight	↔	13-wk exp'r	40
	0–750 mg/kg	p.o.	Fischer	M	Spln weight	↓	13-wk exp'r	40
	0–750 (500) mg/kg	p.o.	Fischer	F	Spln weight	↓	13-wk exp'r	40
Dichloroethylene (*trans*-1,2-dichloroethylene)	0–2.0 mg/kg	p.o.	CD-1	M	T-dependent antibody	↓	Anti-SRBC PFC; 90-d exp'r	41
		p.o.	CD-1	M	Spln weight	↔	90-d exp'r	41
		p.o.	CD-1	F	T-dependent antibody	↓	Only @ 1.0 mg/kg; anti-SRBC PFC; 90-d exp'r	41
		p.o.	CD-1	F	Spln weight	↓	Only @ 1.0 mg/kg; 90-d exp'r	41
		p.o.	CD-1	M	T-dependent antibody	↔	Anti-SRBC HA; 90-d exp'r	41
		p.o.	CD-1	F	T-dependent antibody	↔	Anti-SRBC HA; 90-d exp'r	41
		p.o.	CD-1	M	ConA blastogenesis	↔	90-d exp'r	41
		p.o.	CD-1	F	ConA blastogenesis	↔	90-d exp'r	41
		p.o.	CD-1	M	LPS blastogenesis	↔	90-d exp'r	41
		p.o.	CD-1	F	LPS blastogenesis	↑	90-d exp'r; only @ 1.0 mg/kg	41
		p.o.	CD-1	M	DTH	↔	Anti-SRBC; 90-d exp'r	41
		p.o.	CD-1	F	DTH	↑	Anti-SRBC; 90-d exp'r; only @ 1.0 mg/kg	41
		p.o.	CD-1	M	Popliteal LN proliferation	↔	90-d exp'r	41
		p.o.	CD-1	F	Popliteal LN proliferation	↔	90-d exp'r	41
		p.o.	CD-1	M	PEC phagocytic index	↔	90-d exp'r	41
		p.o.	CD-1	F	PEC phagocytic index	↔	90-d exp'r	41
	0–210 mg/kg	p.o.	CD-1	M	T-dependent antibody	↔	Anti-SRBC PFC; 14-d exp'r	41
		p.o.	CD-1	M	Spln weight	↔	14-d exp'r	41
		p.o.	CD-1	M	DTH	↔	Anti-SRBC; 14-d exp'r	41
Dimethylaniline (*N,N*-dimethylbenzenamine)	31.25 mg/kg	p.o.	B6C3F1	M/F	Spln weight	↑	13-wk exp'r	42
	62.5 mg/kg	p.o.	Fischer 344	M/F	BM hyperplasia	↑	13-wk exp'r	42
		p.o.	Fischer 344	M/F	Spln extramedullary hemopoiesis	↑	13-wk exp'r	42
		p.o.	Fischer 344	M/F	Spln weight	↓	13-wk exp'r	42
Dimethyl sulfoxide (sulfinylbismethane)	1.5 ml — 30%	i.p.	Syrian hamster	M	Lung PMN infiltration	↓	formyl-norleucyl-leucyl-phenylalanine-induced	43
		i.p.	Syrian hamster	M	Lung leakage	↓	formyl-norleucyl-leucyl-phenylalanine-induced	43
		i.p.	Syrian hamster	M	PMN chemotaxis	↓	Using plasma from DMSO-treated animals	43
		i.p.	Syrian hamster	M	PMN superoxide prdn	↔	Using plasma from DMSO-treated animals	43
		i.p.	Syrian hamster	M	PMN adherence	↔	Endothelial cell adherence	43
	0.05,0.025, 0.05 mg/kg/d	i.p.	SW mouse	F	PB WBC	↔	Doses for d 0–7, 8–14, 14–36, respectively	44

TABLE 1 (continued)

Effect of Organic Solvents on Immune System Parameters — *In Vivo* Exposures

Compound (Chemical Name)	Dose (NOEL)	Route	Species/Strain	Sex	Parameter	Effect	Miscellaneous	Ref.
		i.p.	SW mouse	F	T-dependent antibody	↔	Anti-SRBC HA titers; doses for d 0–7, 8–14, 14–36, respectively	44
		i.p.	SW mouse	F	Spln weight	↔	Doses for d 0–7, 8–14, 14–36, respectively	44
Dinitrochlorotrifluoro-toluene (3,5-dinitro-4-chloro-α,α,α-trifluorotoluene)	150 mg/kg	p.o.	Sprague-Dawley	M	Leukocytosis	↑	PB and spln	45
		p.o.	Sprague-Dawley	M	Spln weight	↑		45
	300 mg/kg	p.o.	Sprague-Dawley	M	BM leukocytosis	↑		45
Ethylbenzene	Occup	Resp	Human		WBC	↔		46
Formaldehyde	15 ppm	Resp	B6C3F1	F	Thymus weight	↔	15-d exp'r	47
		Resp	B6C3F1	F	Spln weight	↔	15-d exp'r	47
		Resp	B6C3F1	F	*Listeria* resistance	↑	15-d exp'r	47
		Resp	B6C3F1	F	B16F10 tumor resistance	↓	15-d exp'r	47
		Resp	B6C3F1	F	PBY6 tumor resistance	↓	15-d exp'r	47
		Resp	B6C3F1	F	DTH	↔	Anti-KLH; 15-d exp'r	47
		Resp	B6C3F1	F	NK-cell activity	↔	15-d exp'r	47
		Resp	B6C3F1	F	Mitoblastogenesis	↔	PHA and LPS; 15-d exp'r	47
		Resp	B6C3F1	F	Spln cell phenotypes	↔	15-d exp'r	47
		Resp	B6C3F1	F	CFU-S	↔	15-d exp'r	47
		Resp	B6C3F1	F	CFU-GM	↔	15-d exp'r	47
Gasoline (mixture of C_4 to C_{12} hydrocarbons)	Occup	Resp	Human		PB PMN numbers	↑		48
		Resp	Human		Total WBC	↑		48
		Resp	Human		PB eosinophil numbers	↑		48
Hydroquinone ([8α,9R]-10,11-dihydro-6′-methoxycinchonan-9-ol)	100 mg/kg	i.p.	C57Bl/6	M	Spln-cell number	↓		38
		i.v.	C57Bl/6	M	Spln-cell number	↓		38
		i.v.	C57Bl/6	M	BM-cell number	↔		38
		i.p.	C57Bl/6	M	BM-cell number	↓		38
		i.p.	C57Bl/6	M	T-independent antibody	↓	Anti-LPS	38
		i.v.	C57Bl/6	M	T-independent antibody	↓	Anti-LPS	38
		i.p.	C57Bl/6	M	BM LPS-blastogenesis	↓		38
		i.v.	C57Bl/6	M	BM LPS-blastogenesis	↓		38
		i.v.	C57Bl/6	M	BM DxS-blastogenesis	↑		38
		i.p.	C57Bl/6	M	BM DxS-blastogenesis	↔		38
		i.p.	C57Bl/6	M	BM DxS + LPS-blastogenesis	↓		38
		i.v.	C57Bl/6	M	BM DxS + LPS-blastogenesis	↓		38

Hydroquinone + phenol (mixture)	50 mg/kg ea.	i.p.	Balb/c	M	BM mac. H_2O_2 production	↔	3-d exp'r; TPA-induced	18
		i.p.	Balb/c	M	BM mac. chemotaxis	↑	3-d exp'r; TPA-induced	18
		i.p.	Balb/c	M	BM granulocyte chemotaxis	↑	3-d exp'r; TPA-induced	18
Isophorone diisocyanate	Occup	Resp	Human	M	Asthma induction	Δ	One subject	48
Methyl cyanate	0–3 ppm	Resp	B6C3F1	M	NK-cell activity	↔	4-d exp'r	49
		Resp	B6C3F1	M	MLR	↓	4-d exp'r; recovered by 120 d	49
		Resp	B6C3F1	M	T-dependent antibody	↔	4-d exp'r; anti-SRBC PFC	49
		Resp	B6C3F1	M	Mitoblastogenesis	↔	4-d exp'r; ConA, LPS, PHA	49
		Resp	B6C3F1	M	Spln and thymus pathology	↔	4-d exp'r	49
		Resp	B6C3F1	M	PB WBC	↔	4-d exp'r	49
		Resp	B6C3F1	M	B16F10 tumor incidence	↔	4-d exp'r	49
		Resp	B6C3F1	M	Influenza resistance	↔	4-d exp'r	49
		Resp	B6C3F1	M	*Plasmodium* resistance	↔	4-d exp'r	49
		Resp	B6C3F1	M	*Listeria* resistance	↔	4-d exp'r	49
Methyl isocyanate (isocyanatomethane)	0–3 ppm	Resp	B6C3F1	F	*Listeria* resistance	↔		50
		Resp	B6C3F1	F	B16F10 tumor resistance	↔		50
		Resp	B6C3F1	F	T-dependent antibody	↔		50
		Resp	B6C3F1	F	Influenza resistance	↔	Increasing susceptibility trend	50
		Resp	B6C3F1	F	Plasmodium resistance	↔		50
		Resp	B6C3F1	F	Mitoblastogenesis	↔	Decreasing trend; LPS, PHA, ConA	50
		Resp	B6C3F1	F	NK-cell function	↔		50
	0–3 (1) ppm	Resp	B6C3F1	F	MLR	↓		50
Nitroaniline (*p*-nitroaniline)	0–9 mg/kg/d	p.o.	Sprague-Dawley	F	Spln weight	↔	2-yr exp'r	51
	0–9 (0.25) mg/kg/d	p.o.	Sprague-Dawley	M	Spln weight	↑	2-yr exp'r	51
	0–90 mg/m³	Resp	Sprague-Dawley	M/F	Spln hemopoiesis	↑		52
	0–90 (10) mg/m³	Resp	Sprague-Dawley	M/F	Spln weight	↓		52
		Resp	Sprague-Dawley	M/F	Erythropoiesis	↔		52
Nitrochlorobenzene (*p*-nitrochlorobenzene)	0–45 mg/m³	Resp	Sprague-Dawley	M/F	Spln hemopoiesis	↑		52
	0–45 (5) mg/m³	Resp	Sprague-Dawley	M/F	Erythropoiesis	↓		52
	0–45 (15) mg/m³	Resp	Sprague-Dawley	M/F	Spln weight	↑		52
Phenol (hydroxybenzene)	100 mg/kg	i.p.	DBA/2J	M	BM adherent cell number	↔	Exp'r 2x/d for 4 d	13
		i.p.	B6C3F1	M	BM adherent cell number	↔	Exp'r 2x/d for 4 d	13
		i.p.	DBA/2J	M	CFU-GM	↔	Exp'r 2x/d for 4 d; exp'd ADH cell + normal BM cell	13
		i.p.	B6C3F1	M	CFU-GM	↔	Exp'r 2x/d for 4 d; exp'd ADH cell + normal BM cell	13
		i.p.	DBA/2J	M	CFU-GM	↓	Exp'r 2x/d for 4 d; exp'd BM cell + normal ADH cell	13

TABLE 1 (continued)

Effect of Organic Solvents on Immune System Parameters — *In Vivo* Exposures

Compound (Chemical Name)	Dose (NOEL)	Route	Species/Strain	Sex	Parameter	Effect	Miscellaneous	Ref.
		i.p.	B6C3F1	M	CFU-GM	↓	Exp'r 2×/d for 4 d; exp'd BM cell + normal ADH cell	13
Shell Sol TD	0–1444 ppm	Resp	Rat	M	Spln weight	↑	@ high doses only	53
Styrene (ethenylbenzene)	1/20–1/5 × LD_{50}	p.o.	Swiss	M	Thymus weight	↔		54
		p.o.	Swiss	M	T-dependent antibody	↓	Anti-SRBC PFC	54
		p.o.	Swiss	M	Spln weight	↓		54
		p.o.	Swiss	M	Spln-cell number	↓		54
		p.o.	Swiss	M	Thymus-cell number	↓		54
		p.o.	Swiss	M	Lymph node-cell number	↔		54
		p.o.	Swiss	M	BM-cell number	↔		54
		p.o.	Swiss	M	PEC-SRBC adherence	↓		54
		p.o.	Swiss	M	DTH reaction	↑	Anti-SRBC	54
		p.o.	Swiss	M	PEC phagocytosis	↓		54
		p.o.	Swiss	M	T-dependent antibody	↓	Anti-SRBC serum HA	54
		p.o.	Swiss	M	Regional lymph node weight	↔		54
	1/20–1/5 × LD_{50} (0.02)	p.o.	Swiss	M	Peripheral lymph node weight	↓		54
		p.o.	Swiss	M	Mitoblastogenesis	↑	PHA and LPS	54
		p.o.	Swiss	M	PEC NBT clearance	↑		54
Toluene (methyl benzene)	0–405 mg/l	p.o.	CD-1	M	Mitoblastogenesis	↓	4-wk exp'r	55
		p.o.	CD-1	M	Thymus weight	↓	4-wk exp'r	55
		p.o.	CD-1	M	Liver weight	↑	4-wk exp'r	55
		p.o.	CD-1	M	Serum anti-SRBC antibody	↔	4-wk exp'r	55
	0–405 (80) mg/l		CD-1	M	IL-2 production	↓	28-d exp'r	19
		p.o.	CD-1	M	IL-2 production	↓	4-wk exp'r	55
		p.o.	CD-1	M	Alloantigen blastogenesis		4-wk exp'r	55
		p.o.	CD-1	M	Spln anti-SRBC PFC	↓	4-wk exp'r	55
	Occup	Resp	Human		T-cell function	↔	55–122-m exposure	23
		Resp	Human		PB T-cell numbers	↓	55–122-m exposure	23
Toluene diisocyanate (2,4-diisocyanatotoluene)	0, 1, 2 ppm	Resp	Guinea pig	M	Eosinophils in tracheal mucosa	↓		56
		Resp	Guinea pig	M	Extravascular PMN in trachea	↑		56

	0.02–1.0 ppm (3 hr/d) (0.2 ppm)	Resp	Guinea pig	F	Anti-TDI antibody	↑		57
	10% solution	Resp	Guinea pig	F	Contact hypersensitivity reaction	↑		
	Occup	Resp	Human	M/F	Eosinophil cell number	↑	TDI-asthma positive	59
		Resp	Human	M/F	Mast cell number	↑	TDI-asthma positive	59
		Resp	Human		Plasma histamine	↔	TDI-asthma positive	60
		Resp	Human		Lymphocyte cAMP dose response	↓	TDI-asthma positive	60
		Resp	Human		Split products of complement	↔	TDI-asthma positive	60
		Resp	Human		Complement levels	↔	TDI-asthma positive	60
		Resp	Human		Bronchospasm	↑	TDI-asthma positive	60
		Resp	Human		Eosinophil numbers in bronchial mucosal	↑	TDI-asthma positive	61
		Resp	Human		Asthma	↑		62
		Resp	Human		IL-2 receptor positive $CD25^+$ cells	↑	TDI-asthma positive; bronchial biopsy	63
		Resp	Human		T-cell phenotypes	↔	TDI-asthma positive; bronchial biopsy	63
		Resp	Human		PMN number	↔	TDI-asthma positive; bronchial biopsy	63
		Resp	Human		CD68 macrophages	↔	TDI-asthma positive; bronchial biopsy	63
		Resp	Human		TDI-specific contact hyper.	↔		64
		Resp	Human		PB eosinophil numbers	↔		64
		Resp	Human		PB Ig-concentrations	↔	Including IgE	64
		Resp	Human		Anti-TDI IgE	↔		64
		Resp	Human		Lung provocation challenge	↑	11/14 workers	64
		Resp	Human		TDI-specific lymph transformation	↔		64
Trichloroethylene	Occup	p.o.	Human		PB $CD3^+$ numbers	↑	Smaller amounts of other solvents present	65
		p.o.	Human		PB $CD4^+$ numbers	↑	Smaller amounts of other solvents present	65
		p.o.	Human		PB $CD8^+$ numbers	↑	Smaller amounts of other solvents present	65
		p.o.	Human		PB CD4:8 ratio	↓	Smaller amounts of other solvents present	65
Trifluoromethylaniline (4-trifluoromethylaniline)	0.125 m*M*/kg		Rat		Mitoblastogenesis	↓		66
Xylene (dimethyl benzene)	Occup	Resp	Human		T-cell function	↔	55- to 122-m exposure	23
		Resp	Human		PB T-cell numbers	↓	55- to 122-m exposure	23
		Resp	Human		Lymphocyte acid phos. redistributed	Δ	AP moved lysosome to cytoplasm	67

[a] NOEL concentration shown in parentheses is in same units as original range.

TABLE 2

Effect of Organic Solvents on Immune System Parameters — No Observable Effect Level (NOEL) *In Vivo* Exposures

Compound (Chemical Name)	Maximum NOEL Dose	Route	Species/Strain	Sex	Parameter	Ref.
Benzene	5 ppm	p.n.	Swiss-Webster	F	Erythropoiesis	5
	10 ppm	p.n.	Swiss-Webster	F	CFU-GM hemopoiesis	5
		Resp	C57Bl/6	M	Spln lymphocyte numbers	9
		Resp	C57Bl/6	M	Spln *Listeria* counts	9
		Resp	C57Bl/6	M	BM B-cell numbers	6
	10 + 10 ppm	p.n. + resp	Swiss-Webster	M/F	CFU-GM, CFU-E hemopoiesis; BM	5
		p.n. + resp	Swiss-Webster	M/F	CFU-GM hemopoiesis; spln	5
	20 ppm	p.n.	Swiss-Webster	M/F	Erythropoiesis	5
		p.n.	Swiss-Webster	M/F	CFU-GM hemopoiesis	5
	31 ppm	Resp	C57Bl/6	M	Spln T-cell numbers and mitogenesis	6
	50 ppm	Resp	Balb/c	M	CHR	8
		Resp	Balb/c	M	Thymus:body weight	8
		Resp	Balb/c	M	PB B-cell number and WBC	8
		p.n.	Swiss-Webster	M/F	CFU-GM hemopoiesis	5
	100 ppm	Resp	C57Bl/6	M	Spln-cell phenotypes	7
		Resp	C57Bl/6	M	Suppressor cell activity	7
		Resp	C57Bl/6	M	CTL activity	7
		Resp	C57Bl/6	M	Spln esterase + cell numbers	9
		Resp	C57Bl/6	M	Spln *Listeria* counts	9
	200 ppm	Resp	Balb/c	M	Suppressor cell activity	8
	0.25 mg/kg/day	s.c.	Rabbit		PB T-cell numbers	11
	44 mg/kg	i.p.	C57Bl/6	M	ConA blastogenesis	16
	100 mg/kg	i.p.	B6C3F1	M	BM total and adherence cell numbers	13
		i.p.	DBA/2J	M	BM adherence cell number	13
		i.p.	B6C3F1	M	BM CFU-GM	14
	132 mg/kg	i.p.	C57Bl/6	M	LPS blastogenesis	16
	440 mg/kg	i.p.	C57Bl/6	M	LPS and ConA blastogenesis	16
	660 mg/kg	i.p.	Balb/c	M	BM leuk. IL-6 production and cell phenotypes	17
	800 mg/kg	s.c.	C57Bl/6	M	Mac. Ia induction	21
	31 mg/l	p.o.	CD-1	M	IL-2 production	19
	Occup	Resp	Human		T-cell function	23
Carbon tetrachloride (tetrachloromethane)	0.25 mg/kg	i.m.	AB mouse	M	Thymus weight	29
	0.1 ml 1:3 CCl_4:oil	s.c.	Balb/c		T-dependent antibody	31
	0.7 ml/kg	i.p.	Wistar	M	PB WBC and monocyte numbers	33

Catechol (benzene metabolite)	75 mg/kg	i.p.	C57Bl/6	M	BM cell number	38
		i.p.	C57Bl/6	M	BM LPS- and DxS-blastogenesis	38
		i.p.	C57Bl/6	M	BM DxS + LPS-blastogenesis	38
Chloroaniline-HCL (*p*-chloroaniline-HCL)	0–120 mg/kg	p.o.	B6C3F1	M/F	BM hemopoiesis	39
Chlorobenzene (monochlorobenzene)	0–250 mg/kg	p.o.	B6C3F1	M	Spln weight	40
	0–500 mg/kg	p.o.	B6C3F1	F	Spln weight	40
Dichloroethylene (*trans*-1,2-dichloroethylene)	2.0 mg/kg	p.o.	CD-1	M	Spln weight	41
		p.o.	CD-1	M	T-dependent antibody	41
		p.o.	CD-1	F	T-dependent antibody	41
		p.o.	CD-1	M/tF	ConA blastogenesis	41
		p.o.	CD-1	M	LPS blastogenesis	41
		p.o.	CD-1	M	DTH	41
		p.o.	CD-1	M/F	Popliteal LN proliferation	41
		p.o.	CD-1	M/F	PEC phagocytic index	41
	210 mg/kg	p.o.	CD-1	M	T-dependent antibody	41
		p.o.	CD-1	M	Spln weight	41
		p.o.	CD-1	M	DTH	41
Dimethyl sulfoxide (sulfinylbismethane)	1.5 ml — 30%	i.p.	Syrian hamster	M	PMN superoxide production and adherence	43
	0.05, 0.025, 0.05 mg/kg/d	i.p.	Swiss-Webster	F	PB WBC	44
		i.p.	Swiss-Webster	F	T-dependent antibody	44
		i.p.	Swiss-Webster	F	Spln weight	44
Ethylbenzene	Occup	Resp	Human		WBC	46
Formaldehyde	15 ppm	Resp	B6C3F1	F	Thymus and spln weight	47
		Resp	B6C3F1	F	T-cell and NK-cell function	47
		Resp	B6C3F1	F	Mitoblastogenesis	47
		Resp	B6C3F1	F	Spln-cell phenotypes	47
		Resp	B6C3F1	F	CFU-S	47
		Resp	B6C3F1	F	CFU-GM	47
Hydroquinone ([8α,9R]-10,11-dihydro-6′methoxycinchonan-9-ol)	100 mg/kg	i.v.	C57Bl/6	M	BM cell number and DxS blastogenesis	38
Hydroquinone + phenol (mixture)	50 mg/kg ea.	i.p.	Balb/c	M	BM mac. H_2O_2 production	18
Methyl cyanate	3 ppm	Resp	B6C3F1	M	NK-cell activity	49
		Resp	B6C3F1	M	T-dependent antibody	49
		Resp	B6C3F1	M	Mitoblastogenesis	49
		Resp	B6C3F1	M	Spln and thymus pathology	49
		Resp	B6C3F1	M	PB WBC	49
		Resp	B6C3F1	M	B16F10 tumor incidence	49
		Resp	B6C3F1	M	Influenza, *Plasmodium*, and *Listeria* resistance	49

TABLE 2 (continued)

Effect of Organic Solvents on Immune System Parameters — No Observable Effect Level (NOEL) *In Vivo* Exposures

Compound (Chemical Name)	Maximum NOEL Dose	Route	Species/Strain	Sex	Parameter	Ref.
Methyl isocyanate (isocyanatomethane)	3 ppm	Resp	B6C3F1	F	Influenza, *Plasmodium*, and *Listeria* resistance	50
		Resp	B6C3F1	F	B16F10 tumor resistance	50
		Resp	B6C3F1	F	T-cell and NK-cell function	50
		Resp	B6C3F1	F	Mitoblastogenesis	50
	1 ppm	Resp	B6C3F1	F	MLR	50
Nitroaniline (*p*-nitroaniline)	0.25 mg/kg/d	p.o.	Spague-Dawley	M	Spln weight	51
	10 mg/m^3	Resp	Spague-Dawley	M/F	Spln weight	52
		Resp	Spague-Dawley	M/F	Erythropoiesis	52
Nitrochlorobenzene (*p*-nitrochlorobenzene)	0-45 (15) mg/m^3	Resp	Spague-Dawley	M/F	Spln weight	52
	0-45 (5) mg/m^3	Resp	Spague-Dawley	M/F	Erythropoiesis	52
Phenol (hydroxybenzene)	100 mg/kg	i.p.	DBA/2J and B6C3F1	M	BM adherent cell number	13
		i.p.	DBA/2J and B6C3F1	M	CFU-GM	13
Styrene (ethenylbenzene)	1/20 × LD_{50}	p.o.	Swiss	M	Thymus and LN weight	54
		p.o.	Swiss	M	LN and BM cell number	54
Toluene (methyl benzene)	405 mg/l	p.o.	CD-1	M	Serum anti-SRBC antibody	55
	Occup	Resp	Human		T-cell function	23
Toluene diisocyanate (2,4-diisocyanatotoluene)	Occup	Resp	Human		Plasma histamine	60
		Resp	Human		Split products of complement	60
		Resp	Human		Complement levels	60
		Resp	Human		T-cell phenotypes	63
		Resp	Human		PMN number	63
		Resp	Human		CD68 macrophages	63
		Resp	Human		TDI-specific contact hyper.	64
		Resp	Human		PB eosinophil numbers	64
		Resp	Human		PB Ig concentrations	64
		Resp	Human		Anti-TDI IgE	64
		Resp	Human		TDI-specific lymph. transformation	64
Xylene (dimethyl benzene)	Occup	Resp	Human		T-cell function	23

TABLE 3

Effect of Organic Solvents on Immune System Parameters — *In Vitro* Exposures

Compound (Chemical Name)	Dose (NOEL)	Species	Cell Type	Parameter	Effect	Miscellaneous	Ref.
Alkane (nC6-)	0–10 (5) m*M*	Rabbit (New Zealand)	Pulm macrophage	Lysosomal enzyme release	↑		68
	0–10 (10) m*M*	Sprague-Dawley	Pulm macrophage	Lysosomal enzyme release	↑		68
Alkane (nC7-)	0–10 (1) m*M*	Rabbit (New Zealand)	Pulm macrophage	Lysosomal enzyme release	↑		68
	0–10 (2) m*M*	Sprague-Dawley	Pulm macrophage	Lysosomal enzyme release	↑		68
Alkane (nC8-)	0–10 (.5) m*M*	Rabbit (New Zealand)	Pulm macrophage	Lysosomal enzyme release	↑		68
	0–10 (1) m*M*	Sprague-Dawley	Pulm macrophage	Lysosomal enzyme release	↑		68
Alkane (nC9-)	0–10 (0.2) m*M*	Rabbit (New Zealand)	Pulm macrophage	Cell respiration	↓		68
		Sprague-Dawley	Pulm macrophage	Cell respiration	↓		68
	0–10 (2.5) m*M*	Rabbit (New Zealand)	Pulm macrophage	Lysosomal enzyme release	↑		68
	0–10 (5) m*M*	Sprague-Dawley	Pulm macrophage	Lysosomal enzyme release	↑		68
Alkane (nC10-)	0–10 (0.2) m*M*	Rabbit (New Zealand)	Pulm macrophage	Cell respiration	↓		68
		Sprague-Dawley	Pulm macrophage	Cell respiration	↓		68
	0–10 (10) m*M*	Sprague-Dawley	Pulm macrophage	Lysosomal enzyme release	↑		68
	0–10 (5) m*M*	Rabbit (New Zealand)	Pulm macrophage	Lysosomal enzyme release	↑		68
Allylbenzene (styrene metabolite)	0–500 (100) μ*M*	C57Bl/6	Splenocyte	CTL activity	↓		69
	500 μ*M*	C57Bl/6	Splenocyte	ConA blastogenesis	↔		69
		C57Bl/6	Splenocyte	LPS blastogenesis	↑		69
Benzene	0–100 × 10^{-5} *M*	Human	BM	BFU-E produced by adherent stroma	↓		70
	10 m*M* (LD_{50})	Rat	Pulm macrophage	Toxicity	↑		68
	5 m*M* (LD_{50})	Rabbit	Pulm macrophage	Toxicity	↑		68
	5 × 10^{-3} *M*	Human	BM	BFU-E numbers	↓		70
Benzenetriol (1,2,4-benzenetriol [benzene metabolite])	0–100 (25) μ*M*	C57Bl/6	PE macrophage	Fc-mediated phagocytosis	↓	IgG-SRBC	71
	0–30 × 10^{-6} *M*	Mouse	Spln	ConA blastogenesis	↓		12
	0–100 × 10^{-7} (10^{-6}) *M*	Fischer 344	Spln lymphocyte	PHA blastogenesis	↑		72
	0–100 (20) μ*M*	Fischer 344	Spln lymphocyte	Tubulin polymerization	↓		73
	10^{-9}–10^{-6} *M*	Fischer 344	BM cells	PHA blastogenesis	↓	>PHA conc. required to achieve equal stimulation	73
Benzoquinone *(1,4-benzoquinone [benzene metabolite])*	0–100 μ*M*	C57Bl/6	PE macrophage	Fc-mediated phagocytosis	↓	IgG-SRBC	71
	0–25 (12.5) μ*M*	C57Bl/6	PE macrophage	Cellular F-actin content	↓		71
	0–50 μ*M*	C57Bl/6	PE macrophage	CR-mediated phagocytosis	↓		71
	5 μ*M*	C57Bl/6	PE macrophage	Fc-mediated IgG-SRBC binding	↔		71

TABLE 3 (continued)

Effect of Organic Solvents on Immune System Parameters — *In Vitro* Exposures

Compound (Chemical Name)	Dose (NOEL)	Species	Cell Type	Parameter	Effect	Miscellaneous	Ref.
	0–30 × 10^{-6} *M*	Mouse	Spln	ConA blastogenesis	↓		12
	0–100 × 10^{-7} *M*	Fischer 344	Spln lymphocyte	PHA blastogenesis	↓		73
	0–100 × 10^{-7} (6) *M*	Fischer 344	Spln lymphocyte	PHA agglutination	↓		73
	5–80 (5) μ*M*	Fischer 344	Spln lymphocyte	Tubulin polymerization	↓		73
Butanol	0–60 m*M*	Guinea pig	PB PMN	Superoxide release	↓		74
(1-butyl alcohol)	0.25–0.50%	Human	PB PMN	TNF-α-induced adhesion	↓		75
		Human	PB PMN	TNF-α-induced CD11b expression	↓		75
		Human	PB PMN	TNF-α-induced membrane fluidity	↓		75
		Human	PB PMN	TNF-α-induced migration inhibition′	↓		75
	55 m*M*	Guinea pig	PB PMN	Protein kinase activity	↓		74
	70 m*M* (ID_{50})	CFW mice	PEC	Subtilisin-induced spreading	↓		76
Butanol (2-butyl alcohol)	0–60 m*M*	Guinea pig	PB PMN	Superoxide release	↔		74
Butanol	0–60 m*M*	Guinea pig	PB PMN	Superoxide release	↔		74
(t-butyl alcohol)	17–60 m*M*	Mouse	CXBGABMCT-1	Arachidonic acid metabolism	↔	LTC_4, 6-*trans*-LTB_4, LTB_4	77
Caffeic acid phenethyl ester	10 m*M*	Human	PMN	5-Lipoxygenase	↓		78
		Human	PMN	Reactive oxygen production	↓		78
		Human	PMN	Xanthine/xanthine oxidase	↓		78
Carbon tetrachloride	0.05–1 (0.1) m*M*	Wistar	PEC	TxB_2 formation	↑		79
(tetrachloromethane)	0.05–1 (0.2) m*M*	Wistar	PEC	β-glucoronidase release	↑		79
Catechol	0–100 μ*M*	C57Bl/6	PE macrophage	Fc-mediated phagocytosis	↔	IgG-SRBC	71
(1,2-benzenediol	0–30 × 10^{-6} *M*	Mouse	Spln	ConA blastogenesis	↓		12
[benzene metabolite])	10^{-10}–10^{-2} *M*	C57Bl/6	BM hemopoietic cells	CFU-GM numbers	↔		80
	10^{-6}–10^{-2} (10^{-4}) *M*	C57Bl/6	BM hemopoietic cells	CFU-M numbers	↑		80
	10^{-5}–10^{-3} *M*	Fischer 344	Spln lymphocyte	PHA blastogenesis	↓		72
	10^{-7} *M*	Fischer 344	Spln lymphocyte	PHA blastogenesis	↑		72
	0–100 × 10^{-7} *M*	Fischer 344	Spln lymphocyte	PHA blastogenesis	↔		73
	0–100 × 10^{-7} *M*	Fischer 344	Spln lymphocyte	PHA agglutination	↔		73
Decanol (1-decyl alcohol)	0.2 m*M* (ID_{50})	CFW mice	PEC	Subtilisin-induced spreading	↓		76
Ethylbenzene	500 μ*M*	C57Bl/6	Splenocyte	ConA blastogenesis	↑		69
(styrene metabolite)		C57Bl/6	Splenocyte	CTL activity	↔		69
		C57Bl/6	Splenocyte	LPS blastogenesis	↔		69

Hexanol (1-hexyl alcohol)	4.6 m*M*	CFW mice	PEC	Subtilisin-induced spreading	↓		76
Hydroquinone ([8α,9R]-10,11-dihydro-6′–methoxycinchonan-9-ol [benzene metabolite])	0–10 μ*M*	B6C3F1	BM stroma	B-cell generation	↓		81
		B6C3F1	BM stroma	Growth factor production	↓		81
		B6C3F1	BM macrophage	IL-1 production	↓		81
		B6C3F1	BM stroma	IL-4 production	↓		81
	0–100 (25) μ*M*	C57Bl/6	PE macrophage	Fc-mediated phagocytosis	↓	IgG-SRBC	71
	0–100 × 10^{-5} *M*	Human	BM	BFU-E produced by adherent stroma	↓		70
	0–10^{-3} *M*	B6C3F1	BM stroma	PGE_2 levels	↔		14
	0–10^{-3} (10^{-6}) *M*	B6C3F1	BM stroma	CFU-GM induction	↓		14
	0–10^{-4} (10^{-7}) *M*	B6C3F1	BM macrophage	CFU-GM induction	↓		82
		B6C3F1	BM macrophage	IL-1 production	↓		82
	0–10^{-5} *M*	B6C3F1	BM stroma	BL-CFC numbers	↓		83
		B6C3F1	BM stroma	BM adherent cell numbers	↓		83
		B6C3F1	BM stroma	sIgM + cell numbers	↓		83
		B6C3F1	BM stroma	Small pre-B-cell numbers	↑		83
	0–30 × 10^{-6} *M*	Mouse	Spln	ConA blastogenesis	↓		12
	0.01–10 μ*M*	Human	HL-60	HL-60 → granulocyte differentiation	↔	DMSO-induced	84
		Human	HL-60	HL-60 → granulocyte differentiation	↔	IL-1β-induced	84
		Human	HL-60	HL-60 → macrophage differentiation	↓	TPA-induced	84
		Human	HL-60	HL-60 → macrophage differentiation	↓	1,25-$(OH)_2D_3$-induced	84
		Human	HL-60	HL-60 → macrophage differentiation	↔	Retinoic acid-induced	84
	0.5–5 μ*M*/l	DBA/2	P388D1	pre-IL-1α → IL-1 conversion	↓	LPS-induced	20
		C57BL/6	BM stromal macrophage	pre-IL-1α → IL-1 conversion	↓	LPS-induced	20
	10^{-12}–10^{-6} *M*	C57BL/6	BM hemopoietic cells	CFU-GM numbers	↓	Lineage-depleted cells	80
	10^{-5}–10^{-3} *M*	C57BL/6	BM hemopoietic cells	CFU-GM numbers	↓		80
	10^{-9}–10^{-5} *M*	C57BL/6	BM hemopoietic cells	CFU-GM numbers	↑		80
	5 × 10^{-5} *M*	Human	BM	BFU-E numbers	↓		70
	10^{-10}–10^{-6} *M*	C57BL/6	BM hemopoietic cells	HQ-induced CFU-GM numbers	Δ	Prevents HQ-induced increase	80
	10^{-10}–10^{-4} *M*	C57BL/6	BM hemopoietic cells	CFU-GM numbers	Δ	Optimal conc. required for CFU-GM production decreased	80
	>1.3–10^{-6} *M*	Fischer 344	Spln lymphocyte	PHA blastogenesis	↓		72
	<1.3 × 10^{-6} *M*	Fischer 344	Spln lymphocyte	PHA blastogenesis	↑		72

TABLE 3 (continued)

Effect of Organic Solvents on Immune System Parameters — *In Vitro* Exposures

Compound (Chemical Name)	Dose (NOEL)	Species	Cell Type	Parameter	Effect	Miscellaneous	Ref.
	10^{-7}–10^{-3} (10^{-5}) *M*	Fischer 344	Spln lymphocyte	Intracellular ATP	↓		72
	>1.3–10^{-6} *M*	Fischer 344	Spln lymphocyte	PHA-induced agglutination	↓		72
	20–100 × 10^{-7} *M*	Fischer 344	Spln lymphocyte	PHA blastogenesis	↓		73
	4–10 × 10^{-7} *M*	Fischer 344	Spln lymphocyte	PHA blastogenesis	↑		73
	20–100 × 10^{-7} *M*	Fischer 344	Spln lymphocyte	PHA agglutination	↓		73
	4–10 × 10^{-7} *M*	Fischer 344	Spln lymphocyte	PHA agglutination	↔		73
	20–100 × 10^{-7} *M*	Fischer 344	Spln lymphocyte	Tubulin polymerization	↓		73
Methanol (methyl alcohol)	0–992 μ*M*	Rat	PE PMN	Arachidonic acid metabolism	↓	Endogenous	85
	190 m*M*	Human	PB PMN	6-*trans*-LTB_4 production	↔		77
		Human	PB PMN	LTB_4 production	↓		77
	220 m*M*	Mouse	CXBGABMCT-1	Arachidonic acid metabolism	↑	LTC_4, 6-*trans*-LTB_4, LTB_4	77
Muconaldehyde (*trans.trans*-muconaldehyde [benzene metabolite])	10^{-10}–10^{-6} *M*	C57BL/6	BM hemopoietic cells	CFU-GM numbers	↔		80
		C57BL/6	BM hemopoietic cells	CFU-M numbers	↔		80
Octanol (1-octyl alcohol)	1.4 m*M* (ID_{50})	CFW mice	PEC	Subtilisin-induced spreading	↓		76
Phenol (hydroxybenzene)	0–100 × 10^{-5} *M*	Human	BM	BFU-E produced by adherent stroma	↓		70
	10^{-6}–10^{-2} (10^{-4}) *M*	C57BL/6	BM hemopoietic cells	CFU-M numbers	↑		80
	10^{-9}–10^{-3} M	C57BL/6	BM hemopoietic cells	CFU-GM numbers	↔		80
	5 × 10^{-4} *M*	Human	BM	BFU-E numbers	↓		70
	10^{-4}–10^{-3} *M*	Fischer 344	Spln lymphocyte	PHA blastogenesis	↑		72
Propanol (*n*-propyl alcohol)	84 m*M*	Mouse	CXBGABMCT-1	Arachidonic acid metabolism	↓	LTC_4, 6-*trans*-LTB_4, LTB_4	77
Styrene (ethenylbenzene)	250–1000 μ*M*	C57Bl/6	Splenocyte	CTL activity	↔		69
	250–1000 (400) μ*M*	C57Bl/6	Splenocyte	NK-cell activity	↓		69
Styrene glycol (1-phenyl-1,2-ethanediol [styrene metabolite])	370–1800 μ*M*	C57Bl/6	Splenocyte	CTL activity	↔		69
		C57Bl/6	Splenocyte	NK-cell activity	↔		69

Styrene oxide	200–800 μ*M*	C57Bl/6	Splenocyte	CTL activity	↔		69
(styrene metabolite)	200–800 (400) μ*M*	C57Bl/6	Splenocyte	NK-cell activity	↓		69
	50–500 (100) μ*M*	C57Bl/6	Splenocyte	Mitoblastogenesis	↓	ConA and LPS	69
Toluene	0–100 × 10^{-5} *M*	Human	BM	BFU-E produced by adherent stroma	↓		70
(methylbenzene)	0.15%	Human	PB lymphocytes	Membrane permeability	↑	1-min exp'r @ 4°C	86
	10 m*M* (LD_{50})	Rat	Pulm macrophage	Toxicity	↑		68
	5 m*M* (LD_{50})	Rabbit	Pulm macrophage	Toxicity	↑		68
	5 × 10^{-4} *M*	Human	BM	BFU-E numbers	↓		70
	500 μ*M*	C57Bl/6	Splenocyte	ConA blastogenesis	↑		69
		C57Bl/6	Splenocyte	CTL activity	↔		69
		C57Bl/6	Splenocyte	LPS blastogenesis	↑		69

TABLE 4

Effect of Pesticides on Immune System Parameters — *In Vivo* Exposures

Compound (Chemical Name)	Dose (NOEL)	Route	Species/Strain	Sex	Parameter	Effect	Miscellaneous	Ref.
2,4-D (dichlorophenoxyacetic acid)	0–200 mg/kg	p.n.	CD-1	F	T-dependent antibody	↔	d 11 exp'r; anti-SRBC PFC @ 6 wk	93
		p.n.	CD-1	F	Mitoblastogenesis	↔	d 11 exp'r; ConA and LPS stimulated index @ 6 wk	93
	1–100 μg	i.p.	Balb/c	F	2,4-D-specific IgE antibody	↑		94
		Derm	Balb/c	F	2,4-D-specific DTH	↔		94
		Derm	Balb/c	F	2,4-D-specific IgE antibody	↔		94
	1/8–1/32 (1/16) LD_{50}	Derm	CD-1	M	BM micronuclei	↔		95
	100 mg/kg	p.o.	Chinese hamster	M	BM SCE	↔	2-wk exp'r	96
		p.o.	Wistar	M	Lymphocyte SCE	↔	2-wk exp'r	96
	17.5–70 mg/kg	i.p.	Wistar	M	BM chromosome aberrations	↑		97
Aldicarb (2-methyl-2-[methylthio] propionaldehyde O-[methylcarbamoyl]oxime)	0.1–10 ppb	p.o.	C57Bl/6	F	Spln T-cell numbers	↓	28-d exp'r	98
		p.o.	C57Bl/6	F	T-dependent antibody	↑	Anti-SRBC PFC, 28-d exp'r	98
		p.o.	C57Bl/6	F	Macrophage phagocytosis	↔	28-d exp'r	98
		p.o.	C57Bl/6	F	Mitoblastogenesis	↑	28-d exp'r	98
		p.o.	C57Bl/6	F	Spln B-cell numbers	↓	28-d exp'r	98
		p.o.	C57Bl/6	F	Spln T-cell percentages	↓	28-d exp'r	98
		p.o.	C57Bl/6	F	Spln T-cell numbers	↔	90-d exp'r	98
		p.o.	C57Bl/6	F	T-dependent antibody	↔	Anti-SRBC PFC, 90-d exp'r	98
		p.o.	C57Bl/6	F	MLR	↔	28-d exp'r	98
		p.o.	C57Bl/6	F	Spln B-cell numbers	↔	90-d exp'r	98
		p.o.	C57Bl/6	F	Phagocytosis	↔	90-d exp'r	98
		p.o.	C57Bl/6	F	Mitoblastogenesis	↔	90-d exp'r	98
		p.o.	C57Bl/6	F	MLR	↔	90-d exp'r	98
		p.o.	C57Bl/6	F	Spln T-cell percentages	↔	90-d exp'r	98
	0.1–1000 ppb	p.o.	B6C3F1	F	T-cell mitoblastogenesis	↔		99
		p.o.	B6C3F1	F	Virus resistance	↔		99
		p.o.	B6C3F1	F	T-dependent antibody	↔	Serum levels	99
		p.o.	B6C3F1	F	T-dependent antibody	↔	Anti-SRBC PFC	99
		p.o.	B6C3F1	F	MLR	↔		99
		p.o.	B6C3F1	F	Leukocyte counts	↔		99
		p.o.	B6C3F1	F	B-cell mitoblastogenesis	↔		99
		i.p.	C3H	F	NK cell tumor cytotoxicity	↔		100

		i.p.	C3H	F	Macrophage tumor cytotoxicity	↓	7-d exp'r	100
		p.o.	Swiss-Webster	F	Virus resistance	↔		99
	0.1–1000 ppb	p.n.	Swiss-Webster	F	T-dependent antibody	↔	Serum levels	99
		p.o.	Swiss-Webster	F	B-cell mitoblastogenesis	↔		99
		p.o.	Swiss-Webster	F	T-cell mitoblastogenesis	↔		99
		p.o.	Swiss-Webster	F	Leukocyte counts	↔		99
		p.o.	Swiss-Webster	F	MLR	↔		99
		p.o.	Swiss-Webster	F	T-dependent antibody	↔	Anti-SRBC PFC	99
	0.806 mg/l	In tank	Rosy barb fish		PB PMN numbers	↑	4-wk exp'r	101
		In tank	Rosy barb fish		PB lymphocyte numbers	↑	4-wk exp'r	101
		In tank	Rosy barb fish		PB monocyte numbers	↓	4-wk exp'r	101
	0.3–48.2 μg/d	p.o.	Human	F	Mitoblastogenesis	↔		102
		p.o.	Human	F	PB CD4:CD8	↓		102
		p.o.	Human	F	PB $CD8^+$ cells	↑		102
		p.o.	Human	F	T-dependent antibody	↔		102
		p.o.	Human	F	Total Ig levels	↔		102
		p.o.	Human	F	MLR	↔		102, 103
		p.o.	Human	F	Lympho-proliferative response	↔	Tetanus	102, 103
	1.2–5.3 μg/d	p.o.	Human	F	PB lymphocyte percentage	↑		104
		p.o.	Human	F	PB total B-cell numbers	↔		103, 104
		p.o.	Human	F	PB $CD4^+$ cell numbers	↔		103, 104
		p.o.	Human	F	$CD2^+$ cell numbers	↑		104
	Occup	p.o.	Human	F	PB $CD8^+$ cells	↑		103
		p.o.	Human	F	T-dependent antibody	↔	Anti-tetanus IgM and IgG	103
		p.o.	Human	F	Total Ig levels	↔		103
		p.o.	Human	F	PB CD4:CD8	↓		103
		p.o.	Human	F	Lympho-proliferative response	↑	*Candida*-specific	103
		p.o.	Human	F	PB lymphocyte percentage	↔		103
		p.o.	Human	F	$CD2^+$ cell numbers	↑		103
		p.o.	Human	F	$CD8^+$ cell numbers	↑		104
Ametryn (2-ethylamino-4-isopropylamino-6-methylthio-*s*-triazine)	87 mg/kg	p.o.	Balb/c	F	T-dependent antibody	↔	Anti-SRBC PFC; 8- or 28-d exp'r	105
	87–870 mg/kg	p.o.	Balb/c	F	T-dependent antibody	↓	Anti-SRBC PFC; exp'd –5, 0, +2 d immunization	105

TABLE 4 (continued)

Effect of Pesticides on Immune System Parameters — *In Vivo* Exposures

Compound (Chemical Name)	Dose (NOEL)	Route	Species/Strain	Sex	Parameter	Effect	Miscellaneous	Ref.
Aminocarb (4-[dimethylamino]-*m*-tolyl methyl carbamate)	1/8–1/32 LD_{50}	Derm	CD-1	M	BM micronuclei	↔		95
	6.6 mg/kg	p.o.	C57Bl/6	F	MHV3-induced MLR inhibition	↔	2 doses given	106
		p.o.	C57Bl/6	F	MLR	↔	2 doses given	106
	9.9 mg/kg	i.p.	C57Bl/6	F	MHV3-induced MLR inhibition	↔		106
		i.p.	C57Bl/6×A/J	F	MHV3 susceptibility	↔		106
	10 mg/kg	i.p.	C57Bl/6	F	MLR	↔	Cells from MHV3-infected mice	106
		i.p.	C57Bl/6	F	MHV3-induced cytotoxicity	↑		107
		i.p.	C57Bl/6	F	MHV3-induced superoxide	↔		107
Aroclor-1232 (mixture of chlorinated biphenyls)	10–2000 mg/kg	i.p.	Wistar	M	Thymus size	↔	14 d after exp'r	108
Aroclor-1242 (mixture of chlorinated biphenyls)	167 ppm	p.o.	Balb/c	M	Endotoxin sensitivity	↑	3- or 6-wk exp'r	109
		p.o.	Balb/c	M	Malarial resistance	↓		109
		p.o.	Balb/c	M	T-dependent antibody (secondary)	↓	Primary immunization 2 wk before 6-wk exp'r	110
		p.o.	Balb/c	M	Serum IgG1, IgA, IgM conc.	↔	After SRBC immunization; 6-wk exp'r	110
		p.o.	Balb/c	M	Serum IgA conc.	↓	Primary immunization 2 wk before 6-wk exp'r	110
		p.o.	Balb/c	M	IgA, IgG1, IgM conc.	↔	After SRBC immunization; 6-wk exp'r	110
		p.o.	Balb/c	M	T-dependent antibody (secondary)	↓	6-wk exp'r	110
		p.o.	Balb/c	M	Spln:body weight	↔	6-wk exp'r	110
		p.o.	Balb/c	M	Thymus:body weight	↔	6-wk exp'r	110
	167 ppm	p.o.	Balb/c	M	Ig conc.	↓	After SRBC imunization; 6-wk exp'r	110
		p.o.	Balb/c	M	T-dependent antibody (primary)	↓	Anti-SRBC PFC	110
	1.25–5 g/kg	p.o.	Osborne-Mendel	M	BM micronuclei	↔		111
	10–2000 mg/kg	i.p.	Wistar	M	Thymus size	↔	14 d after exp'r	108
	1000 mg/kg	i.p.	Balb/c	M	Spln lymphocyte numbers	↓		112
		i.p.	Balb/c	M	Spln weight	↑		112

		i.p.	Balb/c	M	T-cell numbers	↑	GvH induction capacity	112
Aroclor-1248 (mixture of chlorinated biphenyls)	10–2000 mg/kg	i.p.	Wistar	M	Thymus size	↔	14 d after exp'r	108
Aroclor-1254 (mixture of chlorinated biphenyls)	0–100 (1) ppm	p.o.	Trout		Spln white pulp	↓	330 d exp'r	113
	0–170 ppm	p.o.	Rabbit (New Zealand)	M	DTH reaction	↔	Tuberculin	114
		p.o.	Rabbit (New Zealand)	M	Lymph node plasma cell number	↓		114
		p.o.	Rabbit (New Zealand)	M	Thymus size	↓		114
		p.o.	Rabbit (New Zealand)	M	T-dependent antibody	↔	Anti-SRBC	114
		p.o.	Rabbit (New Zealand)	M	Spln germinal center number	↔		114
	0–350 ppm	p.o.	Fischer	M	Spln weight	↓	10-d exp'r	115
	0–80 mg/kg	p.o.	Rhesus	F	T-dependent IgM antibody	↓	Anti-SRBC; 55-wk exp'r	116
		p.o.	Rhesus	F	T-dependent IgG antibody	↓	Anti-SRBC; 55-wk exp'r	116
		p.o.	Rhesus	F	T-dependent antibody	↔	Anti-pneumococcal; 55-wk exp'r	116
		p.o.	Rhesus	F	Mitoblastogenesis	↔	ConA, PHA, PWM; 55-wk exp'r	116
		p.o.	Rhesus	F	MLR	↔	55-wk exp'r	116
		p.o.	Rhesus	F	PB monocyte chemilum	↔	55-wk exp'r	116
		p.o.	Rhesus	F	PB monocyte chemilum (peak)	↓	55-wk exp'r	116
		p.o.	Rhesus	F	PB lymphocyte phenotypes	↔	55-wk exp'r	116
		p.o.	Rhesus	F	Thymosin β-4	↔	55-wk exp'r	117
		p.o.	Rhesus	F	Thymosin β-1	↑	55-wk exp'r	117
		p.o.	Rhesus	F	TNF production	↔	55-wk exp'r	117
		p.o.	Rhesus	F	Complement activity	↑	55-wk exp'r	117
		p.o.	Rhesus	F	NK-cell function	↑	55-wk exp'r	117
	40 mg/kg	p.o.	Rhesus	F	Interferon levels	↓	55-wk exp'r	117
	54 mg/kg	i.p.	Salmon		T-independent antibody	↓	TNP-KLH antigen	118
	1.25–5 g/kg	p.o.	Osborne-Mendel	M	BM micronuclei	↔		111
	10–2000 mg/kg	i.p.	Wistar	M	Thymus size	↔	14 d after exp'r	108
	20 and 80 mg/kg	p.o.	Rhesus	F	Interferon levels	↑	55-m exp'r	117
	280 mg/kg	p.o.	Rhesus	F	Lymph node germinal centers	↓	27- to 28-m exp'r	119
		p.o.	Rhesus	F	BM cellularity	↓	27- to 28-m exp'r	119
		p.o.	Rhesus	F	Spln germinal centers	↓	27- to 28-m exp'r	119

TABLE 4 (continued)

Effect of Pesticides on Immune System Parameters — *In Vivo* Exposures

Compound (Chemical Name)	Dose (NOEL)	Route	Species/Strain	Sex	Parameter	Effect	Miscellaneous	Ref.
	Occup		Human		PB PMN Fc receptor+	↓	Blood conc. >15 ppb	120
			Human		PB PMN complement receptor+	↓	Blood conc. >15 ppb	120
Aroclor-1260 (mixture of chlorinated biphenyls)	10–2000 mg/kg	i.p.	Wistar	M	Thymus size	↔	14 d after exp'r	108
Atrazine (6-chloro-*N*2-ethyl-*N*4-isopropyl-1,3,5-triazine-2,4-diamine)	27.3–875 mg/kg	p.o.	C57Bl/6	F	Spln-cell number	↔	40-d exp'r	121
		p.o.	C57Bl/6	F	Spln T-cell numbers	↔	40-d exp'r	121
		p.o.	C57Bl/6	F	T-dependent antibody	↓	Anti-SRBC; 40-d exp'r	121
		p.o.	C57Bl/6	F	MLR	↓	Early in treatment only; 40-d exp'r	121
		p.o.	C57Bl/6	F	Mitoblastogenesis	↔	40-d exp'r	121
		p.o.	C57Bl/6	F	Spln weight	↔	40-d exp'r	121
		p.o.	C57Bl/6	F	Spln IL-2 production	↔	40-d exp'r	121
		p.o.	C57Bl/6	F	Macrophage phagocytosis	↔	40-d exp'r	121
Azinphos-methyl (guthion) (O,O-dimethyl *S*-[(4-oxo-1,2,3-benzotriazin-3-(4H)-methyl]phosphorodithioate)	3.1 mg/kg	i.p.	C57Bl/6	F	MHV3-induced cytotoxicity	↑		107
		i.p.	C57Bl/6	F	MHV3-induced superoxide	↔		107
	0.5 LD_{50}	i.p.	C57Bl/6×A/J	F	MHV3 susceptibility	↑		106
Benomyl (methyl-1-[butylcarbamoyl]benzimidazol-2-ylcarbamate)	250–1000 mg/kg	i.p.	Wistar	M	BM chromosome aberrations	↔		97
Bentazon TP (3-isopropyl-1H-2,1,3-benzothiadiazin-4[3H]-one 2,2-dioxide)	97.5, 195 mg/kg	p.o.	Sheep		Leukocyte migration index	↓	Daily for 3 m	122
		p.o.	Sheep		Phagocytic activity	↔	Daily for 3 m	122
		p.o.	Sheep		Phagocytic activity	↔	Daily for 3 m	122
Buminafos (O,O-dibutyl-[1-butylaminocyclohexyl]-phosphonate)	175–2000 mg/kg	p.o.	NMRI mouse	M	BM micronuclei	↔	Single dose	123
Carbaryl (1-naphthyl methylcarbamate)	0–150 ppm	p.o.	Rabbit (New Zealand)	M	DTH reaction	↔	Tuberculin	114
		p.o.	Rabbit (New Zealand)	M	Lymph node plasma cell number	↔		114
		p.o.	Rabbit (New Zealand)	M	T-dependent antibody	↔	Anti-SRBC	114

		p.o.	Rabbit (New Zealand)	M	Spln germinal center number	↔		114
	0–150 (20) ppm	p.o.	Rabbit (New Zealand)	M	Thymus size	↓		114
	150 ppm	p.o.	Balb/c	M/F	T-dependent antibody	↔	Anti-SRBC, i.p. immunization	124
		p.o.	Balb/c	M/F	IgG2a, IgA, IgG3, IgM antibody	↑	Anti-SRBC, oral immunization	124
		p.o.	Balb/c	M/F	IgG1, IgG2b antibody	↑	Anti-SRBC, oral immunization	124
		p.o.	Balb/c	M/F	*Giardia* resistance	↔	30-d exp'r	124
	10–50 (10) mg/kg	p.o.	CD rat	M	WBC	↓	Not dose dependent	125
	15.3 mg/kg	p.o.	Balb/c	F	T-dependent antibody	↔	Anti-SRBC PFC; 8- or 28-d exp'r	105
	15.3–153 mg/kg	p.o.	Balb/c	F	T-dependent antibody	↓	SRBC PFC; exp'd + 2-d immunization	105
	3.75–30 mg/kg	i.v.	Sprague-Dawley	M	Phagocytic capacity	↓	Collodial carbon uptake	126
	36–335 mg/kg	Resp	CD rat	M	Thymus weight	↓	2-wk exp'r @ 6 hr/d	125
		Resp	CD rat	M	Spln-cell number	↓	2-wk exp'r @ 6 hr/d	125
		Resp	CD rat	M	T-dependent antibody	↓	Anti-SRBC serum Ab; 2-wk exp'r @ 6 hr/d	125
		Resp	CD rat	M	T-dependent antibody	↓	Anti-SRBC PFC; 2-wk exp'r @ 6 hr/d	125
	40.3 mg/kg	i.p.	C57Bl/6	F	MHV3-induced cytotoxicity	↑		107
		i.p.	C57Bl/6	F	MHV3-induced superoxide	↔		107
	100–1000 mg/kg	Derm	CD rat	M	Thymus weight	↔		125
		Derm	CD rat	M	Spln-cell number	↔		125
		Derm	CD rat	M	T-dependent antibody	↔	Anti-SRBC PFC; 2-wk exp'r	125
		Derm	CD rat	M	T-dependent antibody	↔	Anti-SRBC serum Ab; 2-wk exp'r @ 6 hr/d	125
Carbofuran (2,3-dihydro-2,2-dimethyl-7-benzofuranyl methyl carbamate)	0–20 ppm	p.o.	Rabbit (New Zealand)	M	DTH reaction	↔	Tuberculin	114
		p.o.	Rabbit (New Zealand)	M	Lymph node plasma cell number	↓		114
		p.o.	Rabbit (New Zealand)	M	Thymus size	↔		114
		p.o.	Rabbit (New Zealand)	M	Spln germinal center number	↓		114
		p.o.	Rabbit (New Zealand)	M	T-dependent antibody	↔	Anti-SRBC	114

TABLE 4 (continued)

Effect of Pesticides on Immune System Parameters — *In Vivo* Exposures

Compound (Chemical Name)	Dose (NOEL)	Route	Species/Strain	Sex	Parameter	Effect	Miscellaneous	Ref.
	0.01, 0.5 mg/kg	p.n.	F2 hybrid mice	M/F	Serum Ig levels	↔		127
	0.25 mg/kg	i.p.	Swiss	M	PB lymphocyte numbers	↓		128
		i.p.	Swiss	M	PB basophil numbers	↑		128
		i.p.	Swiss	M	PB PMN numbers	↑		128
		i.p.	Swiss	M	Spln cell numbers	↑		128
	0.6 mg/kg	i.p.	C57Bl/6	F	MHV3-induced cytotoxicity	↑		107
		i.p.	C57Bl/6	F	MHV3-induced superoxide	↔		107
Carboxin (5,6-dihydro-2-methyl-*N*-phenyl-1,4-oxanthiin-3-carboxamide)	95.5–382 mg/kg	i.p.	Wistar	M	BM chromosome aberrations	↑		97
Chlordane (octachloro-4,7-methanoindene)	0.1–8.0 mg/kg	p.o.	B6C3F1	F	T-dependent antibody	↔	Anti-SRBC-PFC; 14-d exp'r	129
		p.o.	B6C3F1	F	Mitoblastogenesis	↔	PHA, LPS; 14-d exp'r	129
		p.o.	B6C3F1	F	DTH reaction	↔	KLH antigen; 14-d exp'r	129
		p.o.	B6C3F1	F	Spln:body weight	↔	14-d exp'r	129
		p.o.	B6C3F1	F	Thymus:body weight	↔	14-d exp'r	129
		p.o.	B6C3F1	F	WBC	↔	14-d exp'r	129
		p.o.	B6C3F1	F	PB leukocyte differential	↔	14-d exp'r	129
	0.1–8.0 (4.0) mg/kg	p.o.	B6C3F1	F	ConA mitoblastogenesis	↑	14-d exp'r	129
		p.o.	B6C3F1	F	MLR	↑	14-d exp'r	129
	0.16–8.0 mg/kg	p.n.	Balb/c	M/F	T-dependent antibody	↔	SRBC PFC	130
	0.16–8.0 (<0.16) mg/kg	p.n.	Balb/c	M/F	Virus resistance	↑	Influenza	131
	0.16–8.0 (<0.16) mg/kg	p.n.	Balb/c	F	Contact hypersensitivity response	↓	Oxazalone	130
	0.16–8.0 (0.16) mg/kg	p.n.	Balb/c	M	Contact hypersensitivity response	↓	Oxazalone	130
	0.16–8.0 (2.0) mg/kg	p.n.	Balb/c	M/F	Primary T-dependent antibody	↑	Anti-influenza	131
	0.16–8.0 (4.0) mg/kg	p.n.	Balb/c	M/F	Secondary T-dependent antibody	↑	Anti-influenza	131
	4.0–16.0 mg/kg	p.n.	Balb/c	F	PHA mitoblastogenesis	↔	@ 30 d of age	132
		p.n.	Balb/c	M/F	Contact hypersensitivity response	↔	Oxazalone @ 30 d of age	132
	4.0–16.0 mg/kg	p.n.	Balb/c	M	ConA mitoblastogenesis	↑	4.0 m/k only @ 30 d of age	132
		p.n.	Balb/c	M	PHA mitoblastogenesis	↔	@ 30 d of age	132
		p.n.	Balb/c	M	LPS mitoblastogenesis	↔	@ 30 d of age	132
		p.n.	Balb/c	F	LPS mitoblastogenesis	↔	@ 30 d of age	132
	4.0–16.0 (<4.0) mg/kg	p.n.	Balb/c	M	MLR	↓	@ 30 d of age	132
		p.n.	Balb/c	F	MLR	↔	@ 30 d of age	132
		p.n.	Balb/c	M/F	Contact hypersensitivity response	↓	Oxazalone @ 100 d of age	132

		p.n.	Balb/c	F	ConA mitoblastogenesis	↑	@ 30 d of age	132
	4.0–8.0 mg/kg	p.n.	Balb/c	M/F	CTL activity	↔	@ 100 d of age	133
		p.n.	Balb/c	M/F	NK-cell function	↔	@ 100 d of age	133
		p.n.	Balb/c	F	NK-cell function	↔	@ 200 d of age	133
	4.0–8.0 (4.0) mg/kg	p.n.	Balb/c	M	NK-cell function	↓	@ 200 d of age	133
	8.0 mg/kg	p.n.	Balb/c	M/F	Macrophage tumor cytotoxicity	Δ	P815 target; time to peak delay	134
		p.n.	Balb/c	M/F	Macrophage H_2O_2 production	Δ	Time to peak delay	134
		p.n.	Balb/c	M/F	Macrophage NO production	↑		134
		p.n.	Balb/c	M/F	Macrophage IP_3 production	Δ	Time to peak delay	134
		p.n.	Balb/c	M/F	Transferrin receptor binding	↑	Increase in resident macs only	135
		p.n.	Balb/c	M/F	5′ nucleotidase activity	↓	Decrease in resident macs only	135
		p.n.	Balb/c	M/F	Fetal liver myelopoiesis	↓	Gender not determined	136
		p.n.	Balb/c	M	BM myelopoiesis	↔	@ 42 d of age	136
		p.n.	Balb/c	F	BM myelopoiesis	↓	@ 42 d of age	136
	8.0–16.0 mg/kg	p.n.	Balb/c	M/F	Antigen-specific T-cell blastogenesis	↔	Influenza	137
		p.n.	Balb/c	M/F	DTH reaction	↓	Anti-influenza	137
		p.n.	Balb/c	M/F	Influenza replication	↔		137
	1/8–1/32 (1/16) LD_{50}	Derm	CD-1	M	BM micronuclei	↑		95
	Occup	Resp	Human	M/F	Circulating CD1 thymocytes	↓	Mean percentage positive lymphocytes	138
		Resp	Human	M/F	CD45RA/T4	↑	Mean percentage positive lymphocytes	138
		Resp	Human	M/F	PB mitoblastogenesis	↑	ConA, PHA, PWM	138
		Resp	Human	M/F	MLR	↑		138
		Resp	Human	M/F	PB NK-cell activity	↔		138
		Resp	Human	M/F	PB cell phenotypes	↔	Other than CD1, CD45R/T4	138
		Resp	Human	M/F	PB NK-cell activity	↓		138
Chlordecone (1,1a,3,3a,4,5,5a,5b,6-decachloro-octachloro-1,3,4-metheno-2H-cyclobutano(cd) pentalen-2-one)	0–100 ppm	p.o.	Sprague-Dawley	M	T-dependent antibody	↑	Spln PFC	139
		p.o.	Sprague-Dawley	M	Spln weight	↑		139
	0.625–10 mg/kg	p.o.	Fischer	M	PHA mitoblastogenesis	↔	10-d exp′r	140
		p.o.	Fischer	M	PWM mitoblastogenesis	↔	10-d exp′r	140
		p.o.	Fischer	M	WBC	↔	10-d exp′r	140
		p.o.	Fischer	M	PB lymphocyte numbers	↔	10-d exp′r	140
	0.625–10 (5) mg/kg	p.o.	Fischer	M	PB monocyte numbers	↔	10-d exp′r	140
		p.o.	Fischer	M	STM mitoblastogenesis	↑	10-d exp′r; *S. typhimurium* mitogen	140

TABLE 4 (continued)

Effect of Pesticides on Immune System Parameters — *In Vivo* Exposures

Compound (Chemical Name)	Dose (NOEL)	Route	Species/Strain	Sex	Parameter	Effect	Miscellaneous	Ref.
		p.o.	Fischer	M	NK-cell function	↓	10-d exp'r	140
	0.625–10 (5) mg/kg	p.o.	Fischer	M	Thymus:body weight	↓	10-d exp'r	140
		p.o.	Fischer	M	PB PMN numbers	↓	10-d exp'r	140
		p.o.	Fischer	M	Spln:body weight	↓	10-d exp'r	140
		p.o.	Fischer	M	ConA mitoblastogenesis	↓	10-d exp'r	140
Chlordimeform (N'-[4-chloro-o-tolyl]-*N,N*-dimethylformamidine)	14.8 mg/kg	p.o.	Balb/c	F	T-dependent antibody	↔	Anti-SRBC PFC; 8- or 28-d exp'r	105
	14.8–148 mg/kg	p.o.	Balb/c	F	T-dependent antibody	↓	SRBC PFC; exp'd 0, +2 d immunization	105
	0–75 (10) mg/kg	i.p.	Sprague-Dawley		Spln:body weight	↓		141
	0–75 mg/kg	i.p.	Sprague-Dawley		Spln cell number	↔		141
		i.p.	Sprague-Dawley		Mitoblastogenesis	↔	ConA and LPS	141
		i.p.	Sprague-Dawley		Natural cytotoxicity	↓	WEHI 164 target	141
	0–75 (1) mg/kg	i.p.	Sprague-Dawley		NK-cell activity	↓	YAC-1 target	141
Chlordimeform metabolite (4-chloro-*o*-toluidine)	0–100 (50) mg/kg	i.p.	Sprague-Dawley		Spln:body weight	↓		141
	0–100 mg/kg	i.p.	Sprague-Dawley		Spln-cell number	↔		141
		i.p.	Sprague-Dawley		Mitoblastogenesis	↔	ConA and LPS	141
		i.p.	Sprague-Dawley		Natural cytotoxicity	↔	WEHI 164 target	141
		i.p.	Sprague-Dawley		NK-cell activity	↔	YAC-1 target	141
Chlorphenvinphos (O,O-diethyl *O*-[2-chloro-1-(2,4-dichlorophenyl)vinyl] phosphate)	140 mg/kg	p.o.	Japanese quail		PB leukocyte numbers	↑		142
Cypermethrin ([±]-α-cyano-3-phenoxybenzyl [±]-*cis-trans*-3-[2,2-dichlorovinyl]-2,2-dimethylcyclopropane-carboxylate)	0–40 (20) mg/kg	p.o.	Rat	M	DTH reaction	↓	@ d 61	143
		p.o.	Rat	M	Leukocyte count	↓	@ d 90	143
		p.o.	Rat	M	T-dependent antibody	↔	Serum antibody @ d 90	143
		p.o.	Rat	M	Spln weight	↓		143
		p.o.	Rat	M	Adrenal weight	↑		143
	0–180 mg/kg	i.p.	Swiss	M/F	BM micronuclei	↔	Single high dose or 3× low dose	144
	1/2–1/40 LD_{50}	p.o.	Rabbit	M	T-dependent antibody	↓	Anti-*S. typhimurium*	145
		p.o.	Rabbit	M	Cell-mediated immunity	↓		145
		p.o.	Rat	M	T-dependent antibody	↓	Anti-SRBC and anti-OVA	145

	360 mg/kg	Derm	Swiss	M/F	BM micronuclei	↑	4 treatments	144
	900 ppm	p.o.	Swiss	M/F	BM micronuclei	↑	7- and 14-d treatment	144
DBDCV (di-*n*-butyl dichlorovos)	1–4 mg/kg	p.o.	Chicken	F	Lymphocytic neurotoxic esterase activity	↓		146
DDT (dichloro diphenyl trichloroethane)	0–100 (20) ppm	p.o.	Hissar mouse	M	T-dependent antibody (primary)	↓	Anti-SRBC PFC @ 8–12 wk	148
	0–100 (20) ppm	p.o.	Hissar mouse	M	T-dependent antibody (secondary)	↓	Anti-SRBC PFC @ 8–12 wk	148
		p.o.	Hissar mouse	M	T-independent antibody	↓	3- to 12-wk exp'r	149
		p.o.	Wistar	M	Thymus weight	↔	After 18-wk exp'r	150
		p.o.	Wistar	M	Spln weight	↓	After 22-wk exp'r	150
		p.o.	Wistar	M	Spln weight	↓	After 18-wk exp'r	150
	0–100 (50) ppm	p.o.	Hissar mouse	M	T-dependent antibody (secondary)	↓	Anti-SRBC serum titer at 3–12 wk	148
		p.o.	Hissar mouse	M	T-dependent antibody (primary)	↓	Anti-SRBC serum titer @ 12 wk	148
		p.o.	Hissar mouse	M	T-independent antibody	↓	LPS antigen	149
		p.o.	Wistar	M	T-dependent antibody	↓	Anti-tetanus toxoid @ 18–22 wk	150
	0–150 ppm	p.o.	Rabbit (New Zealand)	M	Thymus size	↔		114
		p.o.	Rabbit (New Zealand)	M	T-dependent antibody	↔	Anti-SRBC	114
		p.o.	Rabbit (New Zealand)	M	Spln germinal center number	↓		114
		p.o.	Rabbit (New Zealand)	M	Lymph node plasma cell number	↓		114
	0–150 (45) ppm	p.o.	Rabbit (New Zealand)	M	DTH reaction	↓	Tuberculin	114
	100 ppm	p.o.	NH Chicken		T-dependent IgG antibody	↓	Anti-SRBC serum antibody	151
		p.o.	NH Chicken		T-dependent IgM antibody	↑	Anti-SRBC serum antibody	151
	20, 200 ppm	p.o.	Albino guinea pig	F	T-dependent antibody	↔	Anti-diphtheria toxoid; 31-d exp'r	152
		p.o.	Albino guinea pig	F	Mesenteric mast cell numbers	↓	31-d exp'r	152
		p.o.	Albino guinea pig	F	Anaphylaxis severity	↓	31-d exp'r	152
	20–25 mg/kg	i.p.	Albino guinea pig	F	Lung histamine levels	↓		153
		i.p.	Albino guinea pig	F	Lung mast cell numbers	↓		153

TABLE 4 (continued)

Effect of Pesticides on Immune System Parameters — *In Vivo* Exposures

Compound (Chemical Name)	Dose (NOEL)	Route	Species/Strain	Sex	Parameter	Effect	Miscellaneous	Ref.
	200 ppm	p.o.	Rat	M	Spln weight	↓		154
		p.o.	Rat		Spln weight	↓	35-d exp'r	154
		p.o.	Rat		T-dependent antibody	↓	Anti-ovalbumin	154
		p.o.	Rat	M	Liver weight	↑		154
		p.o.	Rat	M	T-dependent antibody	↓	Anti-OA; 35-d exp'r	154
	25–250 ppm	i.p.	Balb/c		BM chromosome aberrations	↑		155
	0.031 mg/kg	p.o.	Swiss	F	T-independent antibody	↑	Anti-LPS PFC; 8–24 wk	156
		p.o.	Swiss	F	Mitoblastogenesis	↑	LPS	156
	0.031–3.16 mg/kg	p.o.	Swiss	F	T-dependent antibody	↑	Anti-SRBC PFC; 8–24 wk	156
	0.31–3.16 mg/kg	p.n. + p.o.	Swiss	M/F	Mitoblastogenesis	↓	Pre- and postnatal exp'r; LPS	156
		p.n. + p.o.	Swiss	M/F	T-independent antibody	↓	Pre- and postnatal exp'r; anti-LPS PFC	156
		p.n.	Swiss	M/F	T-independent antibody	↓	Anti-LPS PFC	156
		p.n.	Swiss	M/F	Mitoblastogenesis	↓	LPS	156
		p.n.	Swiss	M/F	T-dependent antibody	↓	Anti-SRBC PFC	156
		p.o.	Swiss	F	Mitoblastogenesis	↓	LPS; 20–24 wk	156
		p.o.	Swiss	F	Mitoblastogenesis	↑	LPS; to 16 wk	156
		p.o.	Swiss	F	T-independent antibody	↑	Anti-LPS PFC; @ 8–16 wk	156
		p.o.	Swiss	F	T-independent antibody	↓	Anti-LPS PFC; 20–24 wk	156
		p.o.	Swiss	F	T-dependent antibody	↓	Anti-SRBC PFC; 2–24 wk	156
		p.o.	Swiss	F	T-dependent antibody	↑	Anti-SRBC PFC; 8–16 wk	156
		p.n. + p.o.	Swiss	M/F	T-dependent antibody	↓	Pre- and postnatal exp'r; anti-SRBC PFC	156
	1/8–1/32 LD_{50}	Derm	CD-1	M	BM micronuclei	↔		95
	10–20 mg/kg	i.p.	Albino guinea pig	F	T-dependent antibody	↔	Anti-diphtheria toxoid	157
		i.p.	Albino guinea pig	F	Anaphylaxis severity	↓	Diphtheria toxoid sensitization	157
		i.p.	Albino guinea pig	F	Spln weight	↔		157
		i.p.	Albino guinea pig	F	Diphtheria toxin neutralization	↔	In immunized animals	157

		i.p.	Albino guinea pig	F	Anaphylaxis induction	↓	Diphtheria toxoid sensitization	157
	30 mg/kg	p.o.	Balb/c	F	T-dependent antibody	↔	Anti-SRBC PFC; 8- or 28-d exp'r	105
	30–300 mg/kg	p.o.	Balb/c	F	T-dependent antibody	↓	Anti-SRBC PFC; exp'd 2 d post-immunization	105
	Occup		Human		PB lymphocyte chromosomal aberrations	↑		158
Demeton (*O,O*-diethyl *O*-[2-(ethylthio)ethyl] phosphorthioate + *O,O*-diethyl S-[2-(ethylthio)ethyl] phosphorthioate)	1/10–1 LD_{50}	i.p.	Syrian hamster	F	BM chromosomal aberrations	↑		159
Di-butyltin (di-*n*-butyltin dichloride)	0–150 ppm	p.o.	Wistar	F	T-dependent antibody	↔	Anti-SRBC (per 10^6); 4-wk exp'r	160
		p.n. + p.o.	Wistar	M/F	T-dependent antibody	↓	Anti-SRBC PFC @ 39 d of age	160
		p.n. + p.o.	Wistar	M/F	T-dependent antibody	↓	Serum Ab @ 39 d of age	160
		p.n. + p.o.	Wistar	M/F	Spln cell number	↓	@ 39 d of age	160
		p.o.	Wistar	F	Spln cell number	↓	4-wk exp'r	160
		p.o.	Wistar	F	T-dependent antibody	↓	Anti-SRBC (whole spleen); 4-wk exp'r	160
	0–150 (50) ppm	p.o.	Wistar	F	T-dependent antibody	↓	Serum antibody; 4-wk exp'r	160
		p.o.	Wistar	M	Allograft rejection time	↑	6-wk exp'r	160
		p.o.	Wistar	F	T-dependent antibody	↓	Serum antibody @ 39d of age	160
		p.o.	Wistar	M	T-independent antibody	↔	38 d exp'r	160
	0–3 mg/kg	p.o.	Wistar	M/F	Allograft rejection time	↑	9-wk exp'r from 3 d of age	160
Di-octyltin (DOTC) (di-*n*-octyltin dichloride)	0–100 ppm	p.o.	Wistar	M	DTH	↓	Tuberculin; 6-wk exp'r	160
		p.o.	Wistar	M	Spln lymphocytes	↓	Histologically; 6-wk exp'r	160
		p.o.	Wistar	M	Thymic lymphocytes	↓	Histologically; 6-wk exp'r	160
	0–150 (50) ppm	p.o.	Wistar	M	Allograft rejection time	↑	6-wk exp'r	160
		p.o.	Wistar	M	Phagocytosis	↔	Carbon clearance; 7-wk exp'r	160
	50 ppm	p.o.	Hartley guinea pig	F	DTH	↔	Tuberculin	160
	0–120 (30) mg/kg	i.m.	AB mouse	M	Spln weight	↓		161
		i.m.	AB mouse	M	Thymus weight	↓		161
	30 mg/kg	i.m.	AB mouse	M	Thymus cell number	↓		161
	120 mg/kg	i.m.	AB mouse	M	T-dependent antibody	↓	Anti-SRBC HA titers	161
Diazinon (*O,O*-diethyl *O*-[6-methyl-2-(1-methylethyl)-4-pyrimidinyl] phosphorothioate)	0.18–9.0 mg/kg	p.n.	F2 hybrid mice	M/F	Serum IgG1 levels	↔	Up to 800 d of age	127
		p.n.	F2 hybrid mice	M/F	IgG2b and IgM serum Ig conc.	↔		127

TABLE 4 (continued)

Effect of Pesticides on Immune System Parameters — *In Vivo* Exposures

Compound (Chemical Name)	Dose (NOEL)	Route	Species/Strain	Sex	Parameter	Effect	Miscellaneous	Ref.
Dibromochloropropane (DBCP) (1,2-dibromo-3-chloropropane)	0–300 (30) mg/kg	s.c.	Sprague-Dawley	M	Spln lymphocyte numbers	↓	Histologically	162
		s.c.	Sprague-Dawley	M	Lymph node lymphocyte numbers	↓	Histologically	162
	0–300 (30) mg/kg	s.c.	Sprague-Dawley	M	BM myelopoiesis	↓		162
		s.c.	Sprague-Dawley	M	Thymus lymphocytic necrosis	↑		162
		s.c.	Sprague-Dawley	M	Thymus:body weight	↓		162
		s.c.	Sprague-Dawley	M	Spln:body weight	↓		162
Dichlorvos (DDVP) (2,2-dichlorovinyl dimethyl phosphate)	1.5–24.5 mg/l	In tank	Carp		T-dependent antibody	↔	Anti-*Yersinia ruckeri*	163
	1/100–1/50 LD_{50}	p.o.	Wistar	M	BM cell chromosomal aberrations	↑	6-wk exp'r	164
	1/10–1 LD_{50}	i.p.	Syrian hamster	F	BM chromosomal aberrations	↑		159
	1/100–1/50 LD_{50}	p.o.	Wistar	M	BM chromosome aberrations	↑	6-wk exp'r	165
	1/40–1/10 LD_{50}	p.o.	Rabbit	M	T-dependent antibody	↓	Anti-S. typhi; 5×/wk for 6 w	166
		p.o.	Rabbit	M	DTH reaction	↓	Anti-tuberculin; 5×/wk for 6 wk	166
		p.o.	Rabbit	M	DTH reaction	↓	Anti-tuberculin; 5×/wk for 6 wk	166
	1/8–1/32 LD_{50}	Derm	CD-1	M	BM micronuclei	↔		95
Dieldrin ([1R,4S,4aS,5R,6R,7S,8S,8Ar]-1,2,3,10,10-hexachloro-1,4,4a,5,6,7,8,8a-octahydro-6,7-epoxy-1,4:5,8-dimethanonaphthalene)	1–5 ppm	p.o.	Balb/c	M	Leishmanial resistance	↓	*Leishmania topic*	167
		p.o.	Balb/c	M	Malarial resistance	↓	*Plasmodium berghei*	167
		p.o.	Mouse	M	Macrophage antigen processing	↓		167
	0–36 mg/kg	i.p.	C57Bl/6	F	Macrophage phagocytosis	↓		168
		i.p.	C57Bl/6	F	Macrophage antigen presentation	↓		168
		i.p.	C57Bl/6	F	Macrophage antigen processing	↓		168
		i.p.	C57Bl/6	F	Superoxide generation	↑	Zymosan and PMA stimulation	107
		i.p.	C57Bl/6	F	YAC-MHV3-induced PEC number	↔		107
		i.p.	C57Bl/6	F	YAC-MHV3-induced superoxide	↔		107
	0–36 (9) mg/kg	i.p.	C57Bl/6	F	Macrophage phagocytosis	↓	*S. typhimurium*	107
		i.p.	C57Bl/6	F	Antigen processing	↓	Avidin	107
	1–50 mg/kg	i.p.	STS mouse		BM chromosomal abnormalies	↑		169
	16.6 mg/kg	p.o.	C57Bl/6	F	MLR	↓	Two doses given	106
	24.6 mg/kg	i.p.	C57Bl/6	F	MHV3-induced MLR inhibition	↓		106
	30 mg/kg	i.p.	C57Bl/6×A/J	F	MHV33 susceptibility	↑		106

		i.p.	C57Bl6×A/J	F	MLR	↓	Cells from MHV3-infected mice	106
	36 mg/kg	i.p.	A/J	M	MLR	↓		170
		i.p.	A/J	M	Graft-vs.-host disease	↓	Donor treated	170
		i.p.	A/J	F	MLR	↓	Only @ d 7 of 24-d treatment	171
		i.p.	C57Bl/6	F	MHV3-induced cytotoxicity	↑		107
		i.p.	C57Bl/6	F	MHV3-induced superoxide	↑		107
		i.p.	A/J	F	Mitoblastogenesis	↔		171
	0.6 LD_{50}	i.p.	C57Bl/6	M	T-dependent antibody	↓	Anti-SRBC	172
		i.p.	C57Bl/6	M	T-independent antibody	↓	Anti-LPS	172
Dimethoate (*O,O*-dimethyl-*S*-methylcarbamoylmethyl phosphorodithioate)	1/10–1 LD_{50}	i.p.	Syrian hamster	F	BM chromosomal aberrations	↑	Marginal increase	159
	1/100–1/50 LD_{50}	p.o.	Wistar	M	BM chromosome aberrations	↑	6-wk exp'r	165
		p.o.	Wistar	M	BM cell chromosomal aberrations	↑	6-wk exp'r	164
	5–15 mg/kg	p.o.	Wistar		BM hyperplasia	↑		173
	75 mg/kg	i.p.	AB mouse		PB PMN numbers	↑		174
		i.p.	AB mouse		PB lymphocyte number	↓		174
		i.p.	AB mouse		T-dependent antibody	↓	Anti-SRBC serum antibody	174
		i.p.	AB mouse		Thymus weight	↓		174
		i.p.	AB mouse		Spln weight	↓		174
		i.p.	Wistar		Thymus weight	↓		174
		i.p.	Wistar		PB PMN numbers	↑		174
		i.p.	Wistar		T-dependent antibody	↓	Anti-SRBC serum antibody	174
		i.p.	Wistar		Spln weight	↓		174
		i.p.	Wistar		PB lymphocyte number	↓		174
Dinitro-*o*-cresol (DNOC) (2-methyl-4,6-dinitrophenol)	10 mg	i.p.	CFLP mouse	M/F	BM micronuclei	↑	F1; male only exposed	175
		i.p.	CFLP mouse	M/F	BM micronuclei	↑	F4; male of each generation exposed	175
		i.p.	CFLP mouse	M/F	BM micronuclei	↑	F2; no further exp'r	175
		s.c.	CFLP mouse	M	BM micronuclei	↑	@ 1 yr post-exp'r	175
	20 mg	i.p.	CFLP mouse	M	BM micronuclei	↑	@ 1 yr post-exp'r	175
Dursban (*O,O*-diethyl-*O*-[3,5,6-trichloro-2-pyridyl]phosphorothioate)	5–30 mg/kg	i.p.	Swiss	M/F	BM micronuclei	↑		176
	80, 240 ppm	p.o.	Swiss	M/F	BM micronuclei	↑		176
	99 mg/kg	Derm	Swiss	M/F	BM micronuclei	↔	2×/wk for 2 wk	176
Ekatin (thiometon) (*S*-2-[ethylthio] ethyl *O,O*-dimethyl phosphorodithioate)	3.75–15.0 mg/kg	i.p.	Wistar	M	BM chromosome aberations	↓		177
	25–200 mg/kg	i.p.	Wistar	M	BM micronuclei	↓		178
	50 mg/kg	p.o.	Pharaoh quail	M/F	PB basophil numbers	↓	Gender differences noted	177
		p.o.	Pharaoh quail	M/F	PB WBC	↑	Gender differences noted	177
		p.o.	Pharaoh quail	M/F	PB lymphocyte numbers	↓	Gender differences noted	177
		p.o.	Pharaoh quail	M/F	PB monocyte numbers	↓	Gender differences noted	177
		p.o.	Pharaoh quail	M/F	PB eosinophil numbers	↓	Gender differences noted	177
		p.o.	Pharaoh quail	M/F	PB PMN numbers	↑	Gender differences noted	177

TABLE 4 (continued)

Effect of Pesticides on Immune System Parameters — *In Vivo* Exposures

Compound (Chemical Name)	Dose (NOEL)	Route	Species/Strain	Sex	Parameter	Effect	Miscellaneous	Ref.
Endosulfan (6,7,8,9,10,10-hexachloro-1,5,5a,6,9,9 a-hexahydro-6,9-methano-2,4,3-benzodioxanthiepin 3-oxide)	6.3 mg/kg	i.p.	C57Bl/6	F	MHV3-induced cytotoxicity	↔		107
	6.3 mg/kg	i.p.	C57Bl/6	F	MHV3-induced superoxide	↔		107
	0–100 ppm	p.o.	Leghorn	M	T-dependent antibody	↓	90-d exp'r; anti-BSA	179
	1/10–1 LD_{50}	i.p.	Syrian hamster	F	BM chromosomal aberrations	↑		159
Fenitrothion (*O,O*-dimethyl *O*-[4-nitro-*m*-totyl]-phosphorothioate)	15–60 mg/kg	i.p.	Wistar	M	BM chromosome aberations	↓		177
	75–330 mg/kg	i.p.	Wistar	M	BM micronucleir	↓		178
Fenvalerate ([RS]-α-cyano-3-phenoxybenzyl [RS]-2-[4-chlorophenyl]-3-methylbutyrate)	10–1250 (10) ppm	p.o.	B6C3F1	M/F	Microgranulomas	↑	Spln and lymph node	180
	100–200 (100) mg/kg	i.p.	Swiss	M/F	BM micronuclei	↑		181
Hexachlorobenzene (1,2,3,4,5,6-hexachlorobenzene)	5 ppm	p.o.	Balb/c	M	Endotoxin sensitivity	↑	3-wk exp'r	182
	5–100 ppm	p.o.	Balb/c	M	Malarial resistance	↔	*Plasmodium berghei*	167
		p.o.	Balb/c	M	Leishmanial resistance	↓	*Leishmania topic*	167
	167 ppm	p.o.	Balb/c	M	Malarial resistance	↓	*P. berghei*; 3- or 6-wk exp'r	182
		p.o.	Balb/c	M	Endotoxin sensitivity	↑	3- or 6-wk exp'r	182
		p.o.	Balb/c	M	T-dependent antibody (primary)	↓	Anti-SRBC PFC; 6-wk exp'r	183
		p.o.	Balb/c	M	Spln cell numbers	↔	6-wk exp'r	183
		p.o.	Balb/c	M	T-dependent antibody (secondary)	↓	Anti-SRBC PFC; 6-wk exp'r	183
		p.o.	Balb/c	M	Serum IgG1, IgA, IgM conc.	↓	After SRBC immunization; 6-wk exp'r	183
		p.o.	Balb/c	M	Endotoxin sensitivity	↑		109
		p.o.	Balb/c	M	Malarial resistance	↓		109
		p.o.	Balb/c	M	Thymus:body weight	↔	Anti-SRBC PFC; 6-wk exp'r	110
		p.o.	Balb/c	M	Ig conc.	↓	After SRBC imm'z; 6-wk exp'r	110
		p.o.	Balb/c	M	T-dependent antibody (primary)	↓	Anti-SRBC PFC; 6-wk exp'r	110
		p.o.	Balb/c	M	T-dependent antibody (secondary)	↓	Anti-SRBC PFC; 6-wk exp'r	110
		p.o.	Balb/c	M	T-dependent antibody (secondary)	↓	Primary immunization 2 wk before 6-wk exp'r	110
		p.o.	Balb/c	M	T-dependent antibody (secondary)	↓	Anti-SRBC PFC; 6-wk exp'r	110

	p.o.	Balb/c	M	Serum IgA conc.	↓	Primary immunization 2 wk before 6-wk exp'r	110
	p.o.	Balb/c	M	Serum IgG1, IgM conc.	↔	After SRBC immunization; 6-wk exp'r	110
	p.o.	Balb/c	M	Spln:body weight	↔	After SRBC immunization; 6-wk exp'r	110
	p.o.	Balb/c	M	Serum IgA conc.	↓	After SRBC immunization; 6-wk exp'r	110
	p.o.	Balb/c	M	MHV1 resistance	↓	6-wk exp'r	184
	p.o.	Nude	M	MHV1 resistance	↓	6-wk exp'r	184
	p.o.	Balb/c	M/F	*Pneumonia* virus resistance	↔	6-wk exp'r	184
	p.o.	Balb/c	M/F	Mouse CMV resistance	↔	6-wk exp'r	184
	p.o.	Nude	M	Mouse CMV resistance	↔	6-wk exp'r	184
	p.o.	C57Bl/6	M	Spln:body weight	↑	13–24 wk of 41-wk exp'r	185
	p.o.	C57Bl/6	M	Thymus:body weight	↓	@ 41 wk; 41-wk exp'r	185
	p.o.	C57Bl/6	M	GvHR	↓	B6D2F1 receptor; @ wk 37	185
	p.o.	C57Bl/6	M	MLR	↓	19- to 24-wk exp'r	185
	p.o.	C57Bl/6	M	PHA mitoblastogenesis	↓	@ wk 24 of 41-wk exp'r	185
167 ppm	p.o.	C57Bl/6	M	LPS mitoblastogenesis	↓	@ wk 24 of 41-wk exp'r	185
250 ppm	p.o.	Sprague-Dawley	M	AgAb complex binding	↓	Macrophage binding	186
0–5.0 mg/kg	p.n.	Balb/c	M/F	T-cell numbers	↑		187
	p.n.	Balb/c	M/F	B-cell numbers	↓		187
	p.n.	Balb/c	M/F	T-dependent antibody	↔	SRBC PFC	187
	p.n.	Balb/c	M/F	Mitoblastogenesis	↔	ConA, PHA, LPS	187
0–5.0 (<0.5) mg/kg	p.n.	Balb/c	M/F	Contact hypersensitivity response	↓	oxazalone	187
0–5.0 (0.5) mg/kg	p.n.	Balb/c	M/F	MLR	↓		187
0–2000 mg/kg	p.o.	Wistar	M	PB basophil numbers	↑	3-wk exp'r	188
	p.o.	Wistar	M	PB monocyte numbers	↑	3-wk exp'r	188
	p.o.	Wistar	M	Total serum IgG level	↑	3-wk exp'r	188
	p.o.	Wistar	M	PB PMN numbers	↑	3-wk exp'r	188
	p.o.	Wistar	M	Lymph node HEV	↑	3-wk exp'r	188
	p.o.	Wistar	M	Spln white pulp size	↑	3-wk exp'r	188
	p.o.	Wistar	M	Total serum IgM level	↑	3-wk exp'r	188
50–150 mg/kg	p.n. + p.o.	Wistar	F	Allograft rejection	↔	Pre- and postnatal	189
	p.n. + p.o.	Wistar	F	*Listeria* resistance	↓	Pre- and postnatal	189
	p.n. + p.o.	Wistar	F	Anti-*Trichinella* IgG antibody	↑	Pre- and postnatal; *Trichinella spiralis*	189
	p.n. + p.o.	Wistar	F	Anti-*Trichinella* IgM antibody	↔	Pre- and postnatal; *Trichinella spiralis*	189

TABLE 4 (continued)

Effect of Pesticides on Immune System Parameters — *In Vivo* Exposures

Compound (Chemical Name)	Dose (NOEL)	Route	Species/Strain	Sex	Parameter	Effect	Miscellaneous	Ref.
		p.n. + p.o.	Wistar	F	T-independent antibody	↔	Pre- and postnatal; LPS antigen	189
		p.n. + p.o.	Wistar	F	Listerial clearance	↔	Pre- and postnatal	189
		p.n. + p.o.	Wistar	F	Carbon clearance	↔	Pre- and postnatal	189
		p.n. + p.o.	Wistar	F	Thymic mitoblastogenesis	↔	Pre- and postnatal; PHA, ConA, PWM	189
		p.n. + p.o.	Wistar	F	Spln mitoblastogenesis	↔	Pre- and postnatal; PHA, ConA, PWM, LPS	189
		p.n. + p.o.	Wistar	F	T-independent IgE antibody	↔	Pre- and postnatal	189
		p.n. + p.o.	Wistar	F	*Trichinella* resistance	↓	Pre- and postnatal; *Trichinella spiralis*	189
		p.n. + p.o.	Wistar	F	Lymph node HEV	↑	Pre- and postnatal	189
	50–150 (50) mg/kg	p.n. + p.o.	Wistar	F	Adrenal weight	↑	Pre- and postnatal	189
		p.n. + p.o.	Wistar	F	Spln HEV	↑	Pre- and postnatal	189
		p.n. + p.o.	Wistar	F	Liver weight	↑	Pre- and postnatal	189
	1000 mg/kg	p.o.	Rat		Allograft rejection	↔	3-wk exp'r	188
		p.o.	Rat		DTH reaction	↔	3-wk exp'r — tuberculin antigen	188
		p.o.	Rat		Macrophage phagocytosis	↔	3-wk exp'r	188
		p.o.	Wistar	M	Endotoxin susceptibility	↑	3-wk exp'r — marginal increase	188
		p.o.	Wistar	M	T-independent antibody	↔	3-wk exp'r — LPS antigen	188
		p.o.	Wistar	M	T-dependent antibody	↑	3-wk exp'r — tetanus toxoid antigen	188
		p.o.	Wistar	M	*Listeria* resistance	↔	3-wk exp'r	188
		p.o.	Wistar	M	Thymic mitoblastogenesis	↔	3-wk exp'r — PHA, ConA, PWM	188
		p.o.	Wistar	M	Thymic mitoblastogenesis	↑	3-wk exp'r — LPS	188
		p.o.	Wistar	M	PB mitoblastogenesis	↑	3-wk exp'r — ConA	188
	150–450 mg/kg food	p.o.	Wistar	M	Lung NK-cell activity	↓	6-wk exp'r	190
	35 mg/m^3	Resp	Sprague-Dawley		Pulmonary bactericidal activity	↓		191
		Resp	Sprague-Dawley		PHA mitoblastogenesis	↔	In mesenteric lymph nodes	191

	35 mg/m³	Resp	Sprague-Dawley		Alveolar macrophage phagocytosis	↑		191
		Resp	Sprague-Dawley		STM-LPS mitoblastogenesis	↑	In mesenteric lymph nodes	191
		Resp	Sprague-Dawley		PHA mitoblastogenesis	↓	In mesenteric lymph nodes after 16-wk exp'r	191
		Resp	Sprague-Dawley		PHA mitoblastogenesis	↔	In lung associated lymph nodes	191
		Resp	Sprague-Dawley		PHA mitoblastogenesis	↑	In lung associated lymph nodes after 16-wk exp'r	191
	1–1000 (10) mg/dog	p.o.	Beagle	M/F	PB PMN numbers	↑		192
		p.o.	Beagle	M/F	G.I. lymphoid hyperplasia	↑		192
Leptofos (*O*-[4-bromo-2,5-dichlorophenyl] *O*-methyl phenylphosphonothioate)	0–500 ppm	p.o.	Swiss	M	T-dependent antibody	↔	Anti-SRBC PFC; 12-wk exp'r	193
	25–150 mg/kg	p.o.	Chicken	F	Lymphocytic neurotoxic esterase activity	↓	24 hr after exp'r	146
Lindane (γ-1,2,3,4,5,6-hexachloro-cyclohexane)	0–100 (25) ppm	p.o.	Leghorn	M	T-dependent antibody	↓	90-d exp'r; anti-BSA	179
	150 ppm	p.o.	Balb/c	M/F	IgG2a, IgG1, IgA, IgG3, IgM antibody	↔	Anti-SRBC, oral immunization	124
		p.o.	Balb/c	M/F	IgG2b antibody	↑	Anti-SRBC, oral immunization	124
		p.o.	Balb/c	M/F	T-dependent antibody	↔	Anti-SRBC, i.p. immunization	124
		p.o.	Balb/c	M/F	*Giardia* resistance	↓	30-d exp'r	124
	0.012 mg/kg	p.o.	Swiss	F	LPS mitoblastogenesis	↑	@ 8 wk; 24-wk exp'r	194
	0.012–1.2 mg/kg	p.o.	Swiss	F	ConA mitoblastogenesis	↔	@ 12 wk; 24-wk exp'r	194
		p.o.	Swiss	F	LPS mitoblastogenesis	↓	@ 20–24 wk; 24-wk exp'r	194
		p.o.	Swiss	F	T-dependent antibody	↑	Anti-SRBC PFC @ 4 and 8 wk; 24-wk exp'r	194
		p.o.	Swiss	F	LPS mitoblastogenesis	↑	@ 4 wk; 24-wk exp'r	194
		p.o.	Swiss	F	T-independent antibody	↓	Anti-LPS PFC @ 8–24 wk; 24-wk exp'r	194
		p.o.	Swiss	F	T-dependent antibody	↓	Anti-SRBC PFC @ 12–24 wk; 24-wk exp'r	194
		p.o.	Swiss	F	DTH reaction	↑	@ 4 and 8 wk; 24-wk exp'r	194
		p.o.	Swiss	F	DTH reaction	↔	@ 12 and 16 wk; 24-wk exp'r	194
		p.o.	Swiss	F	ConA mitoblastogenesis	↑	@ 4 and 8 wk; 24-wk exp'r	194
		p.o.	Swiss	F	DTH reaction	↓	@ 20 and 24 wk; 24-wk exp'r	194
		p.o.	Swiss	F	ConA mitoblastogenesis	↓	@ 16–24 wk; 24-wk exp'r	194
		p.o.	Swiss	F	ConA mitoblastogenesis	↓	W/verapamil or trifluoroperazine	195
		p.o.	Swiss	F	Calcium influx	↓		195

TABLE 4 (continued)

Effect of Pesticides on Immune System Parameters — *In Vivo* Exposures

Compound (Chemical Name)	Dose (NOEL)	Route	Species/Strain	Sex	Parameter	Effect	Miscellaneous	Ref.
		p.o.	Swiss	F	LPS mitoblastogenesis	↓	@ 12–16 wk; 24-wk exp'r	194
		p.o.	Swiss	F	T-independent antibody	↑	Anti-LPS PFC @ 4 wk; 24-wk exp'r	194
	0.12 mg/kg	p.o.	Swiss	F	LPS mitoblastogenesis	↔	@ 8 wk; 24-wk exp'r	194
	1 mg/kg	p.o.	Trout		Mitoblastogenesis	↔		196
		p.o.	Trout		Phagocytosis by pronephric cells	↓	PMA-induced	196
		p.o.	Trout		PB B-cell numbers	↔		196
	1.2 mg/kg	p.o.	Swiss	F	LPS mitoblastogenesis	↓	@ 8 wk; 24-wk exp'r	194
	1.5–12 mg/kg	p.o.	Rabbit		T-dependent antibody	↓	Anti-*S. typhi* @ 5 wk — 5×/wk for 5 wk	197
	5 mg/kg	p.o.	Wild birds		PB lymphocyte numbers	↓	Sparrow, weaver, parakeet, myna, pigeon, duck	198
		p.o.	Wild birds		Spln-cell numbers	↓	Sparrow, weaver, parakeet, myna, pigeon, duck	198
		p.o.	Wild birds		WBC	↓	Sparrow, weaver, parakeet, myna, pigeon, duck	198
	7 mg/kg	p.o.	Rabbit	M	PB PMN phagocytosis	↓	4-wk exp'r	199
	10 mg/kg	p.n.	Swiss	M/F	DTH reaction	↑	Anti-SRBC; 10 d of age	200
		p.n.	Swiss	M/F	T-dependent antibody	↑	Anti-SRBC PFC; 10 d of age	200
		p.n.	Swiss	M/F	Mitoblastogenesis	↑	ConA and LPS; 10 d of age	200
		p.n.	Swiss	M/F	Spln weight	↑	10 d of age	200
		p.n.	Swiss	M/F	Spln weight	↑	10 d of age	200
	0–100 mg/kg	p.n.	Swiss	M/F	Spln weight	↔	10 d of age	200
		i.p.	Trout		T-dependent antibody	↓	*Yersinia ruckeri* antigen	201
		i.p.	Trout		Mitoblastogenesis B cell	↓		196
		i.p.	Trout		Mitoblastogenesis T cell	↔		196
		i.p.	Trout		Head kidney B-cell numbers	↔		196
	0–1000 mg/kg	p.o.	Carp		T-dependent antibody	↔	*Yersina ruckeri* antigen	202
		p.o.	Carp		Spln weight	↔		202
	100 mg/kg	p.n.	Swiss	M/F	DTH reaction	↓	Anti-SRBC; 10 d of age	200
		p.n.	Swiss	M/F	Mitoblastogenesis	↔	ConA and LPS; 10 d of age	200
		p.n.	Swiss	M/F	T-dependent antibody	↔	Anti-SRBC PFC; 10 d of age	200
		p.n.	Swiss	M/F	Spln weight	↔	10 d of age	200

	150 mg/kg	p.o.	Balb/c		IgG2b T-dependent antibody	↑	Anti-SRBC	124
		p.o.	Balb/c		*G. muris* infection	↑		124
	50–100 (50) mg/kg	i.p.	Chicken		BM micronuclei	↑		203
		p.o.	Chicken		BM micronuclei	↑		203
	0.1 LD_{50}	p.o.	Rabbit		Altered lymphocyte ultrastructure			204
		p.o.	Rabbit		PMN phagocytic activity	↓		204
Malathion (*O,O*-dimethyl phosphorodithioate of diethyl mercaptosuccinate)	0–1600 ppm	p.o.	Leghorn	M	T-dependent antibody	↔	90-d exp'r; anti-BSA	179
	0.5 LD_{50}	i.p.	C57Bl/6×A/J	F	MHV3 susceptibility	↓		106
	1–100 μg	Derm	Balb/c	M/F	Malathion-specific DTH	↔		205
		i.p.	Balb/c	F	IgE production	↑	Malathion-KLH antigen	205
		i.p.	Balb/c	F	Malathion-specific IgE	↑	MMA-KLH sensitization; 3 exposures	205
		Derm	Balb/c	F	Malathion-specific IgE	↑		205
	17.8–177.6 μg/ml	Derm	Balb/c	F	IgE production	↔	Malathion antigen	205
		Derm	Balb/c	F	DTH reaction	↔	Against malathion	205
	0–460 (230) mg/kg	i.p.	Balb/c		BM chromosomal aberrations	↑		206
	0–600 mg/kg	p.o.	C57Bl/6	F	β-hexosaminidase	↑		207
		p.o.	C57Bl/6	F	Mast-cell degranulation	↑		207
		p.o.	C57Bl/6	F	PEC H_2O_2 production	↑		207
	0–900 mg/kg	p.o.	C57Bl/6	F	PEC phagocytosis	↑		207
	1–2 mg/kg	i.p.	W'Furth	M	Malathion-specific IgE	↑		208
		i.p.	Wistar	M	Malathion-specific IgE	↑		208
	5–100 mg/kg	p.o.	Rabbit		T-dependent antibody	↓	Anti-*S. typhi* @ 6 wk — 5×/wk for 6 wk	197
	115–460 mg/kg	i.p.	Balb/c		BM micronuclei	↑		206
	143 mg/kg	p.o.	C57Bl/6	F	Splenocyte number	↔	14-d exp'r	209
	143 mg/kg	p.o.	C57Bl/6	F	Thymocyte number	↓	14-d exp'r	209
		p.o.	C57Bl/6	F	Thymus weight	↔	14-d exp'r	209
		p.o.	C57Bl/6	F	T-dependent antibody	↔	Anti-SRBC; 14-d exp'r	209
		p.o.	C57Bl/6	F	Mitoblastogenesis	↔	ConA and LPS; 14-d exp'r	209
	143–715 mg/kg	p.o.	C57Bl/6	F	CTL activity	↔	Various exp'r regimens	209
	200 mg/kg	i.p.	C57Bl/6	F	MHV3-induced cytotoxicity	↔		107
		i.p.	C57Bl/6	F	MHV3-induced superoxide	↔		107
	715 mg/kg	p.o.	C57Bl/6	F	Splenocyte number	↔	1 exp'r	209
		p.o.	C57Bl/6	F	Thymocyte number	↔	1 exp'r	209
		p.o.	C57Bl/6	F	Thymus weight	↔	1 exp'r	209
		p.o.	C57Bl/6	F	T-dependent antibody	↔	Anti-SRBC; d 1 or 3 post-exp'r	209
		p.o.	C57Bl/6	F	T-dependent antibody	↑	Anti-SRBC; d 5 post-exp'r	209
		p.o.	C57Bl/6	F	Mitoblastogenesis	↑	ConA and LPS; d 1, 3, or 5 post-exp'r	209

TABLE 4 (continued)

Effect of Pesticides on Immune System Parameters — *In Vivo* Exposures

Compound (Chemical Name)	Dose (NOEL)	Route	Species/Strain	Sex	Parameter	Effect	Miscellaneous	Ref.
		p.o.	C57Bl/6	F	ConA and LPS mitoblastogenesis	↑	Effect on adherence cells	210
	715 and 900 mg/kg	p.o.	C57Bl/6	F	PEC H_2O_2 production	↑		210
	1/10–1 LD_{50}	i.p.	Syrian hamster	F	BM chromosomal aberrations	↑	Marginal increase	159
	10% solution	Derm	Human	M	Malathion-specific hypersensitivity	↑	Skin test	211
Mancozeb (coordination product of zinc ion, manganese ethylene bisdithiocarbamate)	Occup		Human	M/F	PB lymphocyte chromosomal aberrations	↑		212
	Occup		Human	M	PB lymphocyte SCE	↑		212
	Occup		Human	F	PB lymphocyte SCE	↔		212
MCPA ([4-chloro-2-methyl]phenoxyacetic acid)	100 mg/kg	p.o.	Hamster		BM SCE	↔	2-wk exp'r	96
		p.o.	Wistar	M	Lymphocyte SCE	↔	2-wk exp'r	96
Methomyl (*S*-methyl *N*-[(methylcarbamoyl)-oxy]thioacetimidate)	8.3 mg/kg	i.p.	C57Bl/6	F	MHV3-induced cytotoxicity	↔		107
		i.p.	C57Bl/6	F	MHV3-induced superoxide	↔		107
Methyl parathion (*O,O*-dimethyl *O*-[4-dintrophenyl] phosphorothioate)	0–23 ppm	p.o.	Rabbit (New Zealand)	M	Lymph node plasma cell number	↓		114
		p.o.	Rabbit (New Zealand)	M	DTH reaction	↔	Tuberculin	114
		p.o.	Rabbit (New Zealand)	M	T-dependent antibody	↔	Anti-SRBC	114
	0–23 (2.6) ppm	p.o.	Rabbit (New Zealand)	M	Thymus size	↔		114
		p.o.	Rabbit (New Zealand)	M	Spln germinal center number	↓		114
	0.5–2.0 mg/kg	i.p.	Wistar	M	BM chromosome aberations	↓		177
	1.0–4.0 mg/kg	i.p.	Wistar	M	BM micronuclei	↓		178
	9.375–75 mg/kg	p.o.	Swiss	F	BM micronuclei	↑		213
	10 mg/kg	i.p.	Q mouse	M	BM chromosomal aberrations	↔		214
	1/100–1/50 LD_{50}	p.o.	Wistar	M	BM chromosome aberrations	↔	6-wk exp'r	165
		p.o.	Wistar	M	BM cell chromosomal aberrations	↔	6-wk exp'r	164
	Occup		Human		PB lymphocyte chromosomal aberrations	↑		158

Mirex (dodecachlorooctahydro-1,3,4-metheno-2*H*-cyclobuta[cd]pentalene)	100 ppm	p.o.	Chicken		T-dependent IgM antibody	↔	Anti-SRBC serum antibody	151
		p.o.	Chicken		T-dependent IgG antibody	↔	Anti-SRBC serum antibody	151
Monochlorobenzene (pendimethalin + imazaquin mix)		Resp	Human		PB lymphocyte DNA damage	↑		215
Monocrotophos (dimethyl [E]-1-methyl-2-[methylcarbamoyl]vinyl phosphate)	0.5–2 mg/kg	i.p.	Wistar	M	BM chromosome aberrations	↔		97
	8.0 mg/kg	i.p.	Swiss	M	PB lymphocyte numbers	↓		128
		i.p.	Swiss	M	PB PMN numbers	↑		128
		i.p.	Swiss	M	PB basophil numbers	↑		128
Paraoxon (*O,O*-diethyl *O-p*-nitrophenyl phosphate)	0.3 mg/kg	i.p.	Q mouse	M	BM chromosomal aberrations	↔		214
	0.75 mg/kg	p.o.	Chicken	F	Lymphocytic neurotoxic esterase activity	↔	24-hr after exp'r	146
	10 mg/kg	i.p.	Q mouse	M	BM chromosomal aberrations	↔		214
Parathion (O,O-diethyl O-[4-nitrophenyl] phosphorothioate)	2.2 mg/kg	p.o.	Balb/c	F	T-dependent antibody	↓	Anti-SRBC PFC; 8- or 28-d exp'r	105
	2.2–22.3 mg/kg	p.o.	Balb/c	F	T-dependent antibody	↓	Anti-SRBC PFC; exp'd 2 d postimmunization	105
	10 mg/kg	i.p.	Q mouse	M	BM chromosomal aberrations	↔		214
Pentachlorophenol (analytical) (2,3,4,5,6-pentachlorophenol)	0–60 ppm	p.o.	C57Bl/6		T-dependent antibody	↔	Anti-SRBC PFC	92
	647/491 ppm	p.o.	Holstein	F	Serum Ig levels	↔		216
		p.o.	Holstein	F	ConA blastogenesis	↔	150-d exp'r	216
		p.o.	Holstein	F	PWM blastogenesis	↔	150-d exp'r	216
		p.o.	Holstein	F	T-dependent antibody	↔	Anti-SRBC titer	216
		p.o.	Holstein	F	MLR	↔		216
		p.o.	Holstein	F	Thymus weight	↓		216
Pentachlorophenol (technical) (2,3,4,5,6-pentachlorophenol)	0–60 ppm	p.o.	C57Bl/6		T-dependent antibody	↓	Anti-SRBC PFC	92
	0–250 (10) ppm	p.o.	DBA/2		T-dependent antibody	↓	Anti-SRBC PFC, 6-wk exp'r	92
	0–250 (100) ppm	p.o.	C57Bl/6	F	Thymus:body weight	↓	8-wk exp'r	217
	0–250 ppm	p.o.	C57Bl/6	F	T-independent antibody	↓	Anti-DNP serum antibody	217
		p.o.	C57Bl/6	F	Spln:body weight	↔	8-wk exp'r	217
		p.o.	C57Bl/6	F	T-independent antibody	↓	Anti-DNP PFC	217
		p.o.	Swiss-Webster	F	T-dependent antibody	↓	Anti-SRBC; secondary response	217
		p.o.	Swiss-Webster	F	T-dependent antibody	↓	Anti-SRBC; @ non-peak, primary response	217
	0–500 ppm	p.o.	C57Bl/6	F	Spln:body weight	↔	8-wk exp'r	218
		p.o.	C57Bl/6	F	Thymus:body weight	↔	8-wk exp'r	218

TABLE 4 (continued)

Effect of Pesticides on Immune System Parameters — *In Vivo* Exposures

Compound (Chemical Name)	Dose (NOEL)	Route	Species/Strain	Sex	Parameter	Effect	Miscellaneous	Ref.
		p.o.	C57Bl/6	M	Sarcoma susceptibility	↑	10- to 12-wk exp'r	219
		p.o.	C57Bl/6	M	MSV-tumor susceptibility	↑	10- to 12-wk exp'r	219
		p.o.	C57Bl/6	M	EMCV susceptibility	↔	10- to 12-wk exp'r	219
		p.o.	Swiss-Webster	F	Thymus:body weight	↔	8-wk exp'r	217
		p.o.	Swiss-Webster	F	Spln:body weight	↔	8-wk exp'r	217
	0–500 (50) ppm	p.o.	C57Bl/6	M	Macrophage phagocytosis	↑	10- to 12-wk exp'r	219
		p.o.	C57Bl/6	M	CTL activity	↓	110- to 12-wk exp'r	219
	100–250 ppm	p.o.	C57Bl/6	F	MLR	↓	8-wk exp'r	218
	250 ppm	p.o.	C57Bl/6	F	NK-cell activity	↔	8-wk exp'r	218
		p.o.	C57Bl/6	F	Macrophage phagocytic activity	↔	8-wk exp'r	218
		p.o.	C57Bl/6	F	CTL activity	↔	8-wk exp'r	218
		p.o.	C57Bl/6	F	Mitoblastogenesis	↔	PHA, ConA, LPS; 8-wk exp'r	218
		p.o.	C57Bl/6	F	BM cellularity	↔	8-wk exp'r	218
	647/491 ppm	p.o.	Holstein	F	ConA blastogenesis	↑	By 150-d exp'r	216
		p.o.	Holstein	F	PWM blastogenesis	↑	By 90-d exp'r	216
	647/491 ppm	p.o.	Holstein	F	T-dependent antibody	↔	Anti-SRBC titer	216
		p.o.	Holstein	F	MLR	↔		216
		p.o.	Holstein	F	Thymus weight	↔		216
		p.o.	Holstein	F	Serum IgG1 levels	↔		216
		p.o.	Holstein	F	Serum IgG2 levels	↓		216
		p.o.	Holstein	F	Serum IgM levels	↓		216
		p.o.	Holstein	F	Serum IgA levels	↔		216
	Occup	Resp	Human	M/F	MLR	↓		220
		Resp	Human	M/F	PB Mitoblastogenesis	↓		220
		Resp	Human	M/F	PB CD25, CD26, CD10	↑	Mean percentage positive lymphocytes	220
		Resp	Human	F	PB NK-cell activity	↑		220
		Resp	Human	M	PB NK-cell activity	↔		220
		Resp	Human	M/F	PB CD29	↓	Mean percentage positive lymphocytes	220
Phorate (*O,O*-diethyl *S*-[(ethylthio)methyl]-phosphorodithioate)	0.075–30.0 mg/kg	i.p.	Wistar	M	BM chromosome aberations	↓		177
	0.25–0.75 mg/kg	i.p.	Wistar	M	BM micronuclei	↓		178

Phosalone (*S*-[(6-chloro-2-oxo-3-(2*H*)-benzoxazolyl)methyl]*O,O*-diethyl phosphorodithioate)	50 ppm	p.o.	Balb/c	M/F	*Giardia* resistance	↔	30-d exp'r	124
		p.o.	Balb/c	M/F	T-dependent antibody	↔	Anti-SRBC, oral immunization	124
		p.o.	Balb/c	M/F	T-dependent antibody	↔	Anti-SRBC, i.p. immunization	124
Phosphamidon (2-chloro-2-diethylcarbamoyl-1-methyl-vinyl)	0–3 mg/kg	i.p.	Wistar	M	BM chromosome aberrations	↑		97
	0–4 mg/kg	i.p.	Wistar	M	BM chromosome aberrations	↑		221
		i.p.	Wistar	M	BM chromosome aberrations	↑		221
Phosphine (aluminum phosphide)	Occup		Human		PB lymph. chromosomal aberrations	↑		222
Piperazine (hexahydropyrazine)	500 mg/ml	p.o.	Balb/c		T-dependent antibody	↔	Influenza-specific	223
		p.o.	Balb/c		CTL activity	↔	Allospecific	223
		p.o.	Balb/c		Helper T cells	↔	Influenza-specific	223
		p.o.	Balb/c		Memory T cells	↔	Influenza-specific	223
		p.o.	Balb/c		CTL activity	↔	Influenza-specific	223
Propanil (*N*-[3,4-dichlorophenyl]propanamide)	0-200 (<50) mg/kg	i.p.	C57Bl/6	F	NK-cell function	↓		224
	0-200 (50) mg/kg	i.p.	C57Bl/6	F	T-independent antibody	↓	Anti-DNP-Ficoll PFC	224
	0-400 (100) mg/kg	i.p.	C57Bl/6	F	Spln-cell number	↑		225
		i.p.	C57Bl/6	F	Spln weight	↑		225
		i.p.	C57Bl/6	F	Thymus weight	↓		225
	0-400 (200) mg/kg	i.p.	C57Bl/6	F	ConA blastogenesis	↓		225
		i.p.	C57Bl/6	F	Contact hypersensitivity response	↓	Oxazolone	225
		i.p.	C57Bl/6	F	MLR	↓		225
		i.p.	C57Bl/6	F	LPS mitoblastogenesis	↓		225
	0-400 (50) mg/kg	i.p.	C57Bl/6	F	T-dependent antibody	↓	SRBC PFC	225
	100-200 (<100) mg/kg	i.p.	C57Bl/6	F	PEC IL-6 production	↓	*Ex vivo* production — 7 d post-exp'r	226
		i.p.	C57Bl/6	F	PEC TNF-α production	↓	*Ex vivo* production — 7 d post-exp'r	226
		i.p.	C57Bl/6	F	IL-6 production	↓	*Ex vivo* production — 7 d post-exp'r	226
	100-200 (100) mg/kg	i.p.	C57Bl/6	F	Thymic $CD3^+CD4^-CD8^+$ cell number	↓	7 d after single dose	227
		i.p.	C57Bl/6	F	Thymic $CD3^+CD4^+CD8^+$ cell number	↓	7 d after single dose	227
		i.p.	C57Bl/6	F	Thymic $CD3^+CD4^+CD8^-$ cell number	↓	7 d after single dose	227
	100-200 (150) mg/kg	i.p.	C57Bl/6	F	Thymic $CD3^-$ percentage	↔	7 d after single dose	227
		i.p.	C57Bl/6	F	Spln $CD4^+$ percentage	↔	7 d after single dose	227

TABLE 4 (continued)

Effect of Pesticides on Immune System Parameters — *In Vivo* Exposures

Compound (Chemical Name)	Dose (NOEL)	Route	Species/Strain	Sex	Parameter	Effect	Miscellaneous	Ref.
		i.p.	C57Bl/6	F	MLN $CD4^+$ percentage	↔	7 d after single dose	227
		i.p.	C57Bl/6	F	Spln $CD8^+$ percentage	↔	7 d after single dose	227
		i.p.	C57Bl/6	F	Thymic $CD3^+CD4^+CD8^-$ percentage	↓	7 d after single dose	227
		i.p.	C57Bl/6	F	Thymic $CD3^+CD4\text{-}CD8^+$ percentage	↓	7 d after single dose	227
		i.p.	C57Bl/6	F	Thymic $CD3^+CD4^+CD8^+$ percentage	↓	7 d after single dose	227
		i.p.	C57Bl/6	F	MLN $CD8^+$ percentage	↔	7 d after single dose	227
		i.p.	C57Bl/6	F	IL-2 production	↓	*Ex vivo* production — 7 d post-exp'r	228
		i.p.	C57Bl/6	F	Interferon-γ production	↔	*Ex vivo* production — 7 d post-exp'r	228
	200 mg/kg	i.p.	C57Bl/6	F	CFU-E	↔		229
		i.p.	C57Bl/6	F	CFU-M, CFU-G	↔		229
		i.p.	C57Bl/6	F	BFU-E	↓		229
		i.p.	C57Bl/6	F	BM cellularity	↓		229
		i.p.	C57Bl/6	F	CFU-S	↓		229
		i.p.	C57Bl/6	F	Macrophage tumor cytotoxicity	↑		230
		i.p.	C57Bl/6	F	Interferon-γ production	↑		230
		i.p.	C57Bl/6	F	TNF-α production	↑	WEHI-164 cytotoxicity	230
		i.p.	C57Bl/6	F	TNF-α production	↑	L-929 cytotoxicity	230
	200 (<50) mg/kg	i.p.	C57Bl/6	F	CFU-GM, CFU-IL3	↓		229
Propoxur (Unden) (2-[1-methylethyoxy]phenyl methylcarbamate)	0.5-5.0 (0.5) mg/kg	p.o.	C57Bl/6	M/F	GvH response	↓	Anti-SRBC; 30-d exp'r	231
		p.o.	C57Bl/6	M/F	T cell numbers	↓	10-d exp'r	231
	0.5-5.0 mg/kg	p.o.	C57Bl/6	M/F	IL-1 production	↔	10-d exp'r	231
	0.5-5.0 mg/kg	p.o.	C57Bl/6	M/F	B-cell numbers	↑	@ 0.5 mg/kg only; 10-d exp'r	231
		p.o.	C57Bl/6	M/F	T-dependent antibody	↓	Anti-SRBC serum Ab; 30-d exp'r	231
		p.o.	C57Bl/6	M	CFU-S	↓	10-d exp'r	231
	0.5-5.0 (2) mg/kg	p.o.	C57Bl/6	M/F	T-dependent antibody	↓	Anti-SRBC; 30-d exp'r	231
Pyraline	150 ppm	p.o.	Balb/c	M/F	*Giardia* resistance	↔	30-d exp'r	124
		p.o.	Balb/c	M/F	T-dependent antibody	↔	Anti-SRBC, i.p. immunization	124

		p.o.	Balb/c	M/F	T-dependent antibody	↔	Anti-SRBC, oral immunization	124
Pyridathion (*O*-ethyl-*O*-isopropyl-*O*-[5-methoxy-1-methyl-6-oxo-1*H*-pyridasine-4H]-thiophosphate)	0.07-0.35 mg/kg	p.o.	Wistar	M	BM micronuclei	↔	3-m exp'r	232
	0.6-24 mg/kg	p.o.	Swiss	F	BM micronuclei	↔		232
	1.0-5.0 mg/kg	p.o.	Swiss	F	BM micronuclei	↔	5 doses	232
	6.0 mg/kg	i.p.	Swiss	F	BM micronuclei	↔		232
Rotenone (1,2,12,12a-tetrahydro-2-α-isopropyl-8,9-dimethoxy[1]-benzopyrano[2,4-b]-furo[2,3-h][1]benzopyrano-6[6aH]-one)	225 ppm	p.o.	Swiss	M/F	BM micronuclei	↑	14-d treatment	144
	2, 3 mg/kg	i.p.	Swiss	M/F	BM micronuclei	↑	1×, 2×, or multi treatments	144
	135 mg/kg	i.d.	Swiss	M/F	BM micronuclei	↑	4 treatments	144
Tributyltin (bis[tri-*n*-butyltin]oxide)	0-80 mg/kg food	p.o.	Wistar	M	Lung NK-cell activity	↓	6-wk exp'r	190
Trichlorfon (dimethyl [2,2,2-trichloro-1-hydroxyethyl]phosphonate)	5 mg/kg	p.o.	Japanese quail		PB WBC	↑		233
	6-50 (10) mg/kg	i.p.	CFLP mouse	M	BM chromosomal aberrations	↑		234
	50 mg/kg	p.o.	Japanese quail		PB leukocyte numbers	↑		142
	100 mg/kg	p.o.	Rabbit		PB lymphocyte numbers	↓		235
		p.o.	Rabbit		PB monocyte numbers	↑		235
		p.o.	Rabbit		PB PMN numbers	↑		235
		p.o.	Rabbit		PB basophil numbers	↔		235
		p.o.	Rabbit		PB eosinophil numbers	↓		235
	15 mg/kg	p.o.	Wistar		BM myeloproliferation	↑		173
		p.o.	Wistar		BM hyperplasia	↑		173
	175 mg/kg	p.o.	Balb/c		CTL activity	↔	Influenza-specific	223
		p.o.	Balb/c		Helper T cells	↔	Influenza-specific	223
		p.o.	Balb/c		T-dependent antibody	↔	Influenza-specific	223
		p.o.	Balb/c		Memory T cells	↔	Influenza-specific	223
		p.o.	Balb/c		CTL activity	↔	Allospecific	223
	1/10-1 LD_{50}	i.p.	Syrian hamster	F	BM chromosomal aberrations	↔		159
Trifluralin (α,α,α-trifluoro-2,6-dinitro-*N*,*N*-dipropyl-*p*-toluidine)	0-2 mg/g	p.o.	CFLP mouse	M	PB chromosomal aberrations	↑		234

TABLE 5

Effect of Pesticides on Immune System Parameters — No Observable Effect Level (NOEL) *In Vivo* Exposures

Compound	Maximum NOEL Dose	Route	Species/Strain	Sex	Parameter	Ref.
2,4-D	100 μg	Derm	Balb/c	F	2,4-D-specific DTH and IgE antibody	94
	100 mg/kg	p.o.	Chinese hamster	M	BM SCE	96
		p.o.	Wistar	M	Lymphocyte SCE	96
	200 mg/kg	p.n.	CD-1	F	T-dependent antibody and mitoblastogenesis	93
	1/132 LD_{50}	Derm	CD-1	M	BM micronuclei	95
Aldicarb	10 ppb	p.o.	C57Bl/6	F	B-cell, T-cell, and macrophage function	98
	1000 ppb	p.o.	B6C3F1	F	B-cell, T-cell, and macrophage function	99
		i.p.	C3H	F	NK-cell function	100
		p.o.	Swiss-Webster	F	B cell, T cell, and macrophage function	99
		p.n.	Swiss-Webster	F	T-dependent antibody	99
	48.2 μg/d	p.o.	Human	F	Total Ig levels/T-dependent Ab/mitogen	102
Ametryn	87 mg/kg	p.o.	Balb/c	F	T-dependent antibody	105
Aminocarb	6.6 mg/kg	p.o.	C57Bl/6	F	MLR	106
	9.9 mg/kg	i.p.	C57Bl/6 and C57Bl/6×A/J	F	MHV3-induced MLR inhibition	106
	1/8 LD_{50}	Derm	CD-1	M	BM micronuclei	95
Aroclor-1241	2000 mg/kg	i.p.	Wistar	M	Thymus size	108
Aroclor-1242	167 ppm	p.o.	Balb/c	M	Spln and thymus weight/Ig conc.	110
	2000 mg/kg	i.p.	Wistar	M	Thymus size	108
	5000 mg/kg	p.o.	Osborne-Mendel	M	BM micronuclei	111
Aroclor-1248	170 ppm	p.o.	Rabbit (New Zealand)	M	Spln germinal centers and DTH Rxtn	114
	2000 mg/kg	i.p.	Wistar	M	Thymus size	108
Aroclor-1254	1 ppm	p.o.	Trout		Spln white pulp	113
	170 ppm	p.o.	Rabbit (New Zealand)	M	T-dependent antibody	114
	80 mg/kg	p.o.	Rhesus	F	B-cell, T-cell, and macrophage function	116
	2000 mg/kg	i.p.	Wistar	M	Thymus size	108
	5000 mg/kg	p.o.	Osborne-Mendel	M	BM micronuclei	111
Aroclor-1260	2000 mg/kg	i.p.	Wistar	M	Thymus size	108
Atrazine	875 mg/kg	p.o.	C57Bl/6	F	B-cell, T-cell, and macrophage function	121
Azinphos-methyl (guthion)	3.1 mg/kg	i.p.	C57Bl/6	F	MHV3-induced superoxide	107
Benomyl	1000 mg/kg	i.p.	Wistar	M	BM chromosome aberrations	97
Bentazon TP	195 mg/kg	p.o.	Sheep		Phagocytic activity	122
Buminafos	2000 mg/kg	p.o.	NMRI mouse	M	BM micronuclei	123
Carbaryl	20 ppm	p.o.	Rabbit (New Zealand)	M	Thymus size	114
	150 ppm	p.o.	Rabbit (New Zealand)	M	B- and T-cell function	114

		p.o.	Balb/c	M/F	*Giardia* resistance and T-cell function	124
	10 mg/kg	p.o.	CD rat	M	WBC	125
	15.3 mg/kg	p.o.	Balb/c	F	T-dependent antibody	105
	65.3 mg/kg	i.p.	C57Bl/6	F	MHV3-induced superoxide	107
	1000 mg/kg	Derm	CD rat	M	B- and T-cell function and numbers	125
Carbofuran	20 ppm	p.o.	Rabbit (New Zealand)	M	B- and T-cell function	114
	0.5 mg/kg	p.n.	F2 hybrid mouse	M/F	Serum Ig levels	127
	0.6 mg/kg	i.p.	C57Bl/6	F	MHV3-induced superoxide	107
Chlordane	≤0.16 mg/kg	p.n.	Balb/c	M/F	Virus resistance and DTH function	131,132
	<4.0 mg/kg	p.n.	Balb/c	F	T-cell function	134
	2.0 mg/kg	p.n.	Balb/c	M/F	Primary T-dependent antibody	135
	4.0 mg/kg	p.o.	B6C3F1	M/F	B-, T-, and NK-cell function	129, 130, 133
		p.n.	Balb/c	M	BM myelopoiesis	136
	16.0 mg/kg	p.n.	Balb/c	F	B- and T-cell function	132
	1/16 LD_{50}	Derm	CD-1	M	BM micronuclei	95
Chlordecone	5 mg/kg	p.o.	Fischer	M	B- and T-cell and macrophage function	140
Chlordimeform	10 mg/kg	i.p.	Sprague-Dawley		Spln:body weight	141
	14.8 mg/kg	p.o.	Balb/c	F	T-dependent antibody	105
	75 mg/kg	i.p.	Sprague-Dawley		Mitoblastogenesis	141
	100 mg/kg	i.p.	Sprague-Dawley		NK-cell activity	141
Cypermethrin	20 mg/kg	p.o.	Rat	M	B- and T-cell function	143
	180 mg/kg	i.p.	Swiss	M/F	BM micronuclei	144
DDT	20 ppm	p.o.	Hissar mouse	M	B- and T-cell and function	148,149
		p.o.	Wistar	M	Spln weight	150
	40 ppm	p.o.	Wistar	M	Thymus weight	150
	45 ppm	p.o.	Rabbit (New Zealand)	M	DTH reaction	114
	50 ppm	p.o.	Wistar	M	T-dependent antibody	150
	150 ppm	p.o.	Rabbit (New Zealand)	M	T-dependent antibody and thymus size	114
	20 mg/kg	i.p.	Albino guinea pig	F	B-cell function	157
	200 ppm	p.o.	Albino guinea pig	F	T-dependent antibody	152
	30 mg/kg	p.o.	Balb/c	F	T-dependent antibody	105
	1/8 LD_{50}	Derm	CD-1	M	BM micronuclei	95
Di-butyltin (DBTC)	50 ppm	p.o.	Wistar	M	B- and T-cell and macrophage function	160
	150 ppm	p.o.	Wistar	F	T-dependent antibody	160
	30 mg/kg	i.m.	AB mouse	M	Spln and thymus weight	161
Diazinon	9.0 mg/kg	p.n.	F2 hybrid mouse	M/F	Serum Ig levels	127
Dibromochloropropane (DBCP)	30 mg/kg	s.c.	Sprague-Dawley	M	Thymus lymphocytic necrosis	162
		s.c.	Sprague-Dawley	M	LN and spln lymphocyte numbers	162

TABLE 5 (continued)

Effect of Pesticides on Immune System Parameters — No Observable Effect Level (NOEL) *In Vivo* Exposures

Compound	Maximum NOEL Dose	Route	Species/Strain	Sex	Parameter	Ref.
		s.c.	Sprague-Dawley	M	Spleen: and thymus:body weight	162
		s.c.	Sprague-Dawley	M	BM myelopoiesis	162
Dichlorvos (DDVP)	30 ppm		Carp		T-dependent antibody	163
	1/8–1/41 LD_{50}	Derm	CD-1	M	BM micronuclei	95
	9 mg/kg	i.p.	C57Bl/6	F	Macrophage function	107
Dieldrin	36 mg/kg	i.p.	A/J	F	Mitoblastogenesis	171
Dinitrochlorobenzene	<37 mg/kg	i.p.	C57Bl/6	F	T-independent antibody	224
Dursban	99 mg/kg	Derm	Swiss	M/F	BM micronuclei	106
Endosulfan	6.3 mg/kg	i.p.	C57Bl/6	F	MHV3-induced superoxide and cytotoxicity	107
Fenvalerate	10 ppm	p.o.	B6C3F1	M/F	Microgranulomas	180
	100 mg/kg	i.p.	Swiss	M/F	BM micronuclei	181
Hexachlorobenzene	100 ppm	p.o.	Balb/c	M	Malarial resistance	167
	167 ppm	p.o.	Balb/c	M	Spln cell numbers	183
		p.o.	Balb/c	M	Thymus and spln:body weight	110
		p.o.	Balb/c	M	Serum Ig conc.	110
		p.o.	Nude and Balb/c	M	CMV resistance	184
		p.o.	Balb/c	M/F	*Pneumonia* virus resistance	184
	<0.5 mg/kg	p.n.	Balb/c	M/F	Contact hypersensitivity response	187
	0.5 mg/kg	p.n.	Balb/c	M/F	MLR	187
	5 mg/kg	p.n.	Balb/c	M/F	Mitoblastogenesis and T-dependent Ab	187
	50 mg/kg	p.n. + p.o.	Wistar	F	Liver and adrenal weight	189
		p.n. + p.o.	Wistar	F	Spln HEV	189
	150 mg/kg	p.n. + p.o.	Wistar	F	B-cell, T-cell, and macrophage function	189
	1000 mg/kg	p.o.	Rat		T-cell and macrophage function	188
		p.o.	Wistar	M	B-cell, T-cell, and macrophage function T-independent antibody	188
	35 mg/m³	Resp	Sprague-Dawley		PHA mitoblastogenesis	191
	10 mg/d	p.o.	Beagle	M/F	PB PMN numbers	192
		p.o.	Beagle	M/F	G.I. lymphoid hyperplasia	192
Leptofos	500 ppm	p.o.	Swiss	M	T-dependent antibody	193
Lindane	25 ppm	p.o.	Leghorn	M	T-dependent antibody	179
	150 ppm	p.o.	Balb/c	M/F	T-dependent antibody	124
	0.012 mg/kg	p.o.	Swiss	F	B- and T-cell function	194
	0.12 mg/kg	p.o.	Swiss	F	LPS mitoblastogenesis	194

	1 mg/kg	p.o.	Trout		Mitoblastogenesis and PB B-cell numbers	196
	50 mg/kg	p.o.	Chicken		BM micronuclei	203
	100 mg/kg	i.p.	Trout		Head kidney B-cell numbers	196
		i.p.	Trout		Mitoblastogenesis T cell	196
		p.n.	Swiss	M/F	B- and T-cell function	200
	1000 mg/kg	p.o.	Carp		Spln weight	202
		p.o.	Carp		T-dependent antibody	202
Malathion	1600 ppm	p.o.	Leghorn	M	T-dependent antibody	179
	100 μg	Derm	Balb/c	M/F	Malathion-specific DTH	205
	177.6 μg/ml	Derm	Balb/c	F	DTH reaction	205
		Derm	Balb/c	F	IgE production	205
	143 mg/kg	p.o.	C57Bl/6	F	B- and T-cell function	209
	200 mg/kg	i.p.	C57Bl/6	F	MHV3-induced superoxide and cytotoxicity	107
	230 mg/kg	i.p.	Balb/c	M	BM chromosomal aberrations	206
	715 mg/kg	p.o.	C57Bl/6	F	B- and T-cell function	209
MCPA	100 mg/kg	p.o.	Hamster		BM SCE	96
		p.o.	Wistar	M	Lymphocyte SCE	96
Methomyl	8.3 mg/kg	i.p.	C57Bl/6	F	MHV3-induced superoxide and cytotoxicity	107
Methyl parathion	2.6 ppm	p.o.	Rabbit (New Zealand)	M	B- and T-cell function	114
	10 mg/kg	i.p.	Q mouse	M	BM chromosomal aberrations	
	1/50 LD_{50}	p.o.	Wistar	M	BM chromosomal aberrations	164,165
		p.o.	Wistar	M	BM chromosomal aberrations	
Mirex	100 ppm	p.o.	Chicken		T-dependent IgM antibody	151
		p.o.	Chicken		T-dependent IgG antibody	151
Monocrotophos	0.5–2 mg/kg	i.p.	Wistar	M	BM chromosomal aberrations	97
Paraoxon	0.3 mg/kg	i.p.	Q mouse	M	BM chromosomal aberrations	
	0.75 mg/kg	p.o.	Chicken	F	Lymphocytic neurotoxic esterase activity	146
	10 mg/kg	i.p.	Q mouse	M	BM chromosomal aberrations	
Pentachlorophenol (analytical)	0–60	p.o.	C57Bl/6		T-dependent antibody	92
	647/491 ppm	p.o.	Holstein	F	B- and T-cell function	216
Pentachlorophenol (technical)	0–250 ppm	p.o.	DBA/2		T-dependent antibody	92
	50 ppm	p.o.	C57Bl/6	M	T-cell and macrophage function	219
	100 ppm	p.o.	C57Bl/6	F	Thymus:body weight	217
	250 ppm	p.o.	C57Bl/6	F	T-cell and macrophage function	218
	500 ppm	p.o.	C57Bl/6 and Swiss-Webster	F	Spln: and thymus:body weight	217,218
		p.o.	C57Bl/6	M	EMCV susceptibility	219
	647/491 ppm	p.o.	Holstein	F	Serum Ig levels	216
		p.o.	Holstein	F	B- and T-cell function	216
Phosalone	50 ppm	p.o.	Balb/c	M/F	T-dependent antibody	124
		p.o.	Balb/c	M/F	*Giardia* resistance	124

TABLE 5 (continued)

Effect of Pesticides on Immune System Parameters — No Observable Effect Level (NOEL) *In Vivo* Exposures

Compound	Maximum NOEL Dose	Route	Species/Strain	Sex	Parameter	Ref.
Piperazine	500 mg/ml	p.o.	Balb/c		B- and T-cell function	223
Propanil	<50 mg/kg	i.p.	C57Bl/6	F	NK-cell function; CFU-GM	224,229
	50 mg/kg	i.p.	C57Bl/6	F	B- and T-cell function	224
	<100 mg/kg	i.p.	C57Bl/6	F	PEC and spln cytokine production	226
	100 mg/kg	i.p.	C57Bl/6	F	Spln and thymus weight and number	225
	100 mg/kg	i.p.	C57Bl/6	F	Thymic cell phenotype numbers	227
	150 mg/kg	i.p.	C57Bl/6	F	Thymic cell phenotype percentage	227
		i.p.	C57Bl/6	F	Spln cytokine production	228
		i.p.	C57Bl/6	F	MLN $CD4^+$ and $CD8^+$ percentage	227
		i.p.	C57Bl/6	F	Spln $CD4^+$ and $CD8^+$ percentage	227
	200 mg/kg	i.p.	C57Bl/6	F	BM myelopoiesis	229
		i.p.	C57Bl/6	F	DTH and MLR response	225
		i.p.	C57Bl/6	F	LPS mitoblastogenesis	225
Propoxur (Unden)	0.5 mg/kg	p.o.	C57Bl/6	M/F	IL-1, GvH response, T-cell numbers	231
	2 mg/kg	p.o.	C57Bl/6	M/F	T-dependent antibody	231
Pyraline	150 ppm	p.o.	Balb/c	M/F	T-dependent antibody	124
		p.o.	Balb/c	M/F	*Giardia* resistance	124
Pyridathion	0.35 mg/kg	p.o.	Wistar	M	BM micronuclei	232
	5.0 mg/kg	p.o.	Swiss	F	BM micronuclei	232
	6.0 mg/kg	i.p.	Swiss	F	BM micronuclei	232
	24 mg/kg	p.o.	Swiss	F	BM micronuclei	232
Trichlorfon	100 mg/kg	p.o.	Rabbit		PB basophil numbers	235
		i.p.	CFLP mouse	M	BM	234
	175 mg/kg	p.o.	Balb/c		B- and T-cell function	223
	1/10–1 LD_{50}	i.p.	Syrian hamster	F	BM chromosomal aberrations	159

TABLE 6

Effect of Pesticides on Immune System Parameters — *In Vitro* Exposures

Compound (Chemical Name)	Dose (NOEL)	Species	Cell Type	Parameter	Effect	Miscellaneous	Ref.
2,4-D (2,4-dichlorophenoxyacetic acid)	0–60 (0.2) μg/ml	Human	PB lymphocyte	SCE	↑		236
	0–60 (50) μg/ml	Human	PB lymphocyte	Chromosomal aberrations	↑		236
	0–250 μg/ml	Human	PB lymphocyte	SCE	↑		237
	0.125–1.250 m*M*	Human	PB lymphocyte	Chromosomal aberrations	↑	Technical grade; without metabolic activation	238
	0.125–0.350 m*M*	Human	PB lymphocyte	Chromosomal aberrations	↔	"Pure" grade; without metabolic activation	238
2,4,5-T (2,4,5-trichlorophenoxyacetic acid)	0–20 (2) μg/ml	Human	Bone marrow	CFU-GM	↓	Only @ d 7 incubation	239
	0–200 μg/ml	Rat	Bone marrow	CFU-GM	↓		239
Alachlor (2-chloro-2′,6′-diethyl-N-[methoxymethyl]-acetanilide)	0.01–1.0 μ*M*	Human	PBMC	LAK function	↔		240
		Human	PBMC	NK-cell activity	↔		240
		Human	PBMC	CTL proliferation	↔		240
		Human	PBMC	Polyclonal IgG and IgM production	↔	PWM stimulation	240
		Human	PBMC	Ag-induced blastogenesis	↔		240
		Human	PBMC	Mitoblastogenesis	↔		240
	0.01–1.0 μg/ml	Human	PB lymphocyte	Cytogenetic damage	↑		240
	0.01–1.0 μg/ml	Human	PB lymphocyte	Chromosomal aberrations	↑		240
	5–1000 μ*M*/ml	Human	PB lymphocyte	SCE	↑	No effect with S9 activation	241
Aldicarb (2-methyl-2-[methylthio] propionaldehyde O-[methylcarbamoyl]oxime)	0.5–50 μ*M*	Mouse	CTLL2	IL-2-dependent proliferation	↓		242
	0–250 μg/ml	Human	PB lymphocyte	SCE	↑	± Metabolic activation	243
	0–300 (50) μg/ml	Human	PB lymphocyte	SCE	↑	Nitroso-metabolite	244
	0–350 μg/ml	Human	PB lymphocyte	SCE	↑	± Metabolic activation	245
Aldicarb metabolite (2-methyl-2-[methylthio] propionaldehyde *O*-[methylcarbamoyl]oxime)	0–300 (50) μg/ml	Human	PB lymphocyte	SCE	↔		244
Aroclor-1242 (mixture of chlorinated biphenyls)	1.0–28.3 (7) ppm	Balb/c	PEC	RNA synthesis	↓	2- to 8-hr incubation	196
	0–10 (1) μg/ml	Sprague-Dawley	PE PMN	IP_3 generation	↑		246
		Sprague-Dawley	PE PMN	Superoxide generation	↑		246
Aroclor 1254 (mixture of chlorinated biphenyls)	0–2.5 ppm	Human	PB PMN	Fc^+ cells	↓		120
		Human	PB PMN	CR^+ cells	↓		120
		Human	PB monocyte	Fc^+ cells	↓		120
		Human	PB monocyte	CR^+ cells	↓		120
	0–10 μ*M*	Human	PB PMN	β-glucuronidase activity	↓		247
	0–100 μ*M*	Human	PB monocyte	Mitochrondial respiration	↓		248

TABLE 6 (continued)

Effect of Pesticides on Immune System Parameters — *In Vitro* Exposures

Compound (Chemical Name)	Dose (NOEL)	Species	Cell Type	Parameter	Effect	Miscellaneous	Ref.
		Human	PB PMN	Mitochrondial respiration	↔		248
		Human	PB lymphocyte	Mitochrondial respiration	↓		248
		Human	PB lymphocyte	Mitochrondial respiration	↓		248
		Human	PB lymphocyte	Intracellular ATP concentration	↓		248
		Human	PB lymphocyte	Lactate production	↔		248
	0.45–0.50 m*M*	Human	Whole PB	Chemilum	↓	Zymosan or TPA activity	249
		Human	Whole PB	Chemilum	↑	Unactivated	249
	0.4, 20 mg/ml	Rat	Splenocyte	IL-2 production	↓	24-hr exp'r	250
		Rat	Splenocyte	NK-cell activiity	↓	24-hr exp'r	250
		Rat	Splenocyte	NK-cell activity	↓		251
Atrazine (6-chloro-N2-ethyl-N4-isopropyl-1,3,5-triazine-2,4-diamine)	0.01–1.0 μg/ml	Human	PB lymphocyte	Chromosomal aberrations	↑		240
		Human	PB lymphocyte	Cytogenetic damage	↑		240
	0–500 μ*M*/ml	Human	PB lymphocyte	SCE	↔		241
Azinphos (O,O-dimethyl S-[(4-oxo-1,2,3-benzotriazin-3(4H)-yl)-methyl]phosphorodithioate)	0–20 μg/ml	Human	Bone marrow	CFU-GM	↓	@ >10-d incubation	239
Benomyl (methyl-1-[butylcarbamoyl]-benzimidazol-2-ylcarbamate)	0–25 μg/ml	Sprague-Dawley	Hepatocyte	Micronucleus frequency	↑		252
	0.25–4.0 μg/ml	Human	PB lymphocyte	Aneuploidy	↑		253
	0.5–2.0 μg/ml	Human	PB lymphocyte	Cell division	↓		253
	0–6 μg/ml	Human	PB lymphocyte	SCE	↔		254
	0.5 mg/ml	Human	PB lymphocyte	Chromosomal aberrations	↔		255
Bromacil (5-bromo-3-sec-butyl-6-methyluracil)	0–20 (0.2) μg/ml	Human	BM	CFU-GM	↓		239
	0–200 (20) μg/ml	Rat	BM	CFU-GM	↓	@ >7-d incubation	239
Captan (*cis-N*-[(trichloromethyl)thio]-4-cyclohexene-1,2-dicarboximide)	3–300 mg/m	Human	PB lymphocyte	SCE	↑		256
Carbaryl (1-naphthyl methylcarbamate)	2.5–12.5 (2.5) μ*M*	Swiss	Macrophage	Prostaglandin release	↓		257
		Swiss	Macrophage	Free arachidonic acid production	↔		257
		Swiss	Macrophage	Superoxide generation	↓		257
	0.5–50 μ*M*	Mouse	CTLL2	IL-2-dependent proliferation	↓		242
		Human	PB NK cell	Cytotoxicity	↓	K562 cell targets	258
		Human	PB NK cell	IL-2-dependent proliferation	↓	Added to large granular lymphocytes at culture	259

	32 μ*M*	Swiss	PE macrophage	Arachidonic acid metabolism	↓	Zymosan activation; >15 min post-activation	260
		Swiss	PE macrophage	5-hydroxyeicosatetraenoic production	↓	Zymosan activation; 2–15 min post-activation	260
		Swiss	PE macrophage	Prostaglandin production	↑	Zymosan activation; 2–15 min post-activation	260
Carbaryl metabolite (1-naphthol)	0.5–50 (5) μ*M*	Mouse	CTLL2	IL-2-dependent proliferation	↓		242
	0.5–50 (5) μ*M*	Human	PB NK cell	IL-2-dependent proliferation	↓	Added to large granular lymphocytes at culture	259
Carbofuran (2,3-dihydro-2,2-dimethyl-7-benzofuranyl methyl carbamate)	0.5–50 (5) μ*M*	Mouse	CTLL2	IL-2-dependent proliferation	↓		242
Chlordane (octachloro-4,7-methanoindene)	0–100 μ*M*	B6C3F1	Bone marrow	Proliferation	↓	3H incorporation ± serum	261
	0–100 (0.1) μ*M*	B6C3F1	Splenocyte	MLR	↓	γ-Isomer	261
	0–100 (0.2) μ*M*	B6C3F1	Splenocyte	T-dependent antibody	↓	*In vitro* anti-SRBC; γ-isomer	261
	0–100 (1.0) μ*M*	B6C3F1	Splenocyte	Mitoblastogenesis	↓	LPS and ConA; γ-isomer	261
	0–20 μ*M*	Guinea pig	PB PMN	Intracellular Ca^{2+}	↑	*cis*-Isomer	262
	0–20 (2.5) μ*M*	Guinea pig	PB PMN	Membrane potential	Δ	*cis*-Isomer > *trans*-isomer	262
		Guinea pig	PB PMN	Superoxide generation	↑	*cis*-Isomer > *trans*-isomer	262
	20 μ*M*	Guinea pig	PB PMN	PKC activity	↔	*cis*-Isomer and *trans*-isomer	262
		Mouse	L-5178Y lymphoma	Forward mutation assay			263
	4 μg/ml	Mouse	L-5178Y lymphoma	Cell cycling	↓	Second round synchronous division	264
		Mouse	L-5178Y lymphoma	DNA synthesis	↔	G2 phase possible site of action	264
Cypermethrin ([±]-α-cyano-3-phenoxybenzyl [±]-*cis-trans*-3-[2,2-dichlorovinyl]-2,2-dimethylcyclopropanecarboxylate)	0–50 μg/ml	Human	PB lymphocyte	SCE	↔		265
		Human	PB lymphocyte	SCE	↔		265
		Human	PB lymphocyte	Proliferation	↓		265
	0.25–400 μg/ml	Mouse	Splenocyte	Chromosomal aberrations	↑		266
		Mouse	Splenocyte	SCE	↑		266
	2–50 μg/ml	Human	PB lymphocyte	Chromosomal aberrations	↔		265
		Human	PB lymphocyte	Proliferation index	↓		265
DDT (dichloro diphenyl trichloroethane)	1.0 μ*M*	Human	PB lymphocyte	Mitochrondial respiration	↓		248
		Human	PB lymphocyte	Intracellular ATP concentration	↓		248
		Rabbit	PB lymphocyte	Mitoblastogenesis	↓		267
	0.06–0.2 μg/ml	Human	PB lymphocyte	Chromosomal aberrations	↔		268
	1–15 μg/ml	Human	PB lymphocyte	Chromosomal aberrations	↑		268
	20 μg/ml	Albino guinea pig	Mast cell	Numbers	↓		153

TABLE 6 (continued)

Effect of Pesticides on Immune System Parameters — *In Vitro* Exposures

Compound (Chemical Name)	Dose (NOEL)	Species	Cell Type	Parameter	Effect	Miscellaneous	Ref.
Deltamethrin ([S]-α-cyano-m-phenoxybenzyl [1R,3R]-3-[2,2-dibromovinyl]-2,2 dimethylcyclopropane carboxylate)	0–60 μg/ml	Human	PB lymphocyte	SCE	↔		254
Diazinon (*O,O*-diethyl *O*-[6-methyl-2-(1-methylethyl)-4-pyrimidinyl]phosphorothioate)		Mouse	L-5178Y lymphoma	Forward mutation assay			263
Dichlorvos (DDVP) (2,2-dichlorovinyl dimethyl phosphate)	0.5–50 μ*M*	Mouse	CTLL2	IL-2-dependent proliferation	↓		242
	0–500 μg/ml	Human	PB lymphocyte	Chromosomal aberrations	↑		269
	220 μg/ml	Human	PB monocyte	Esterase activity	↓	50% inhibition	270
	0–65.3 (10.9) μg/ml	Carp	Leukocyte	Phagocytosis	↓	Spleen and pronephros leukocytes; *Y. ruckeri*	163
	1.5–24.5 (6.1) μg/ml	Carp	Lymphocyte	PHA-mitoblastogenesis	↓	Spleen and pronephros lymphocytes	163
	0–70 μ*M*	Human	PB lymphocyte	Lysosome content	↓		271
		Human	PB lymphocyte	Nucleoli content	↓		271
Dieldrin ([1R,4S,4aS,5R,6R,7S,8S,8Ar]-1,2,3,10,10-hexachloro-1,4,4a,5,6,7,8,8a-octahydro-6,7-epoxy-1,4:5,8-dimethanonaphthalene)	0.1–10.0 ppm	Balb/c	PE macrophage	RNA synthesis	↓		272
	0.1–10 (0.1) ppm	Balb/c	PEC	RNA synthesis	↓	2- to 8-hr incubation	196
	0–100 (10) μ*M*	Sprague-Dawley	PE PMN	Superoxide production	↑		273
		Sprague-Dawley	PE PMN	LDH release	↑		273
Dimethoate (*O,O*-dimethyl *S*-methylcarbamoylmethyl phosphorodithioate)	0–100 μg/ml	Human	PB lymphocyte	SCE	↑		254
DNOC (2-methyl-4,6-dinitrophenol)							239
		Human	Bone marrow	CFU-GM			239
		Rat	Bone marrow	CFU-GM			256
Fenaminosulf (sodium [4-dimethylamino]phenyldiazene sulfate)	2.5–250 mg/ml	Human	PB lymphocyte	SCE	↑		265
Fenvalerate ([RS]-α-cyano-3-phenoxybenzyl [RS]-2-[4-chlorophenyl]-3-methylbutyrate)	0–50 μg/ml	Human	PB lymphocyte	Chromosomal aberrations	↑		265
		Human	PB lymphocyte	SCE	↑		265
		Human	PB lymphocyte	Proliferation	↓		265
	2–50 μg/ml	Human	PB lymphocyte	Proliferation index	↓		265
		Human	PB lymphocyte	Chromosomal aberrations	↓		274
	10–50 μg/ml	Human	PB lymphocyte	Micronucleus frequency	↑		239
Fosetyl-aluminum (aluminum tris [*O*-ethyl phosphonate])	0–20 μg/ml	Human	Bone marrow	CFU-GM	↓		239

Glyphosate (isopropylamine salt of *N*-[phosphonomethyl]glycine)	0.25–25 mg/m^3	Human	PB lymphocyte	SCE	↑		256
Heptachlor (1,4,5,6,7,8,8-heptachloro-3a,4,7,7a-tetrahydro-4,7-methanoindene)		Mouse	L-5178Y lymphoma	Forward mutation assay			263
	0–20 μ*M*	Guinea pig	PB PMN	Intracellular Ca^{2+}	↑		262
		Guinea pig	PB PMN	Superoxide generation	↑		262
		Guinea pig	PB PMN	Membrane potential	Δ		262
	20 μ*M*	Guinea pig	PB PMN	PKC activity	↔		262
Heptchlor-epoxide (1,4,5,6,7,8,8-heptachloro-3a,4,7,7a-tetrahydro-4,7-methanoindene)	0–20 μ*M*	Guinea pig	PB PMN	Superoxide generation	↑		262
	20 μ*M*	Guinea pig	PB PMN	PKC activity	↔		262
	80 μ*M*	Guinea pig	PB PMN	Membrane potential	Δ		262
	2.5–20 μ*M*	Guinea pig	PB PMN	Intracellular Ca^{2+}	↑		262
Lindane (γ-1,2,3,4,5,6-hexachloro-cyclohexane)	16 μ*M*	Swiss	Macrophage	Leukotriene production	↑		275
	0–20 μ*M*	Swiss	Macrophage	Chemiluminescence response	↑		275
	16 μ*M*	Swiss	Macrophage	Prostaglandin production	↑		275
	200 μ*M*/l	Human	PB PMN	Ca^{2+} release	↑		276
	4 × 10^{-4} μ*M*/l	Human	PB PMN	Ca^{2+} release	↑		277
		Human	PB PMN	Lysosomal release	↑		277
	4 × 10^{-4} μ*M*/l	Human	PB PMN	Cytoplast stimulation	↑		277
	0–8 × 10^{-4} μ*M*/l	Human	PB PMN	Superoxide generation	↑		277
		Human	PB PMN	Lindane chemotactic activity	↔		277
	0–250 m*M*/l	Human	PB PMN	Superoxide production	↑		276
	0–250 μ*M*	MEI mouse	Macrophage	Leukotriene production	↑		278
		MEI mouse	Macrophage	Prostaglandin production	↔		278
		MEI mouse	Macrophage	Phospholipid hydrolysis	↑		278
	0.1 m*M*	Human	PB lymphocyte	Mitoblastogenesis	↓	PHA	279
	1.0 m*M*	Mouse	PE macrophage	Protein synthesis	↓		280
		Mouse	PE macrophage	RNA synthesis	↓		280
		Mouse	PE macrophage	Acid phosphatase	↓		280
		Mouse	PE macrophage	Pinocytosis	↓		280
	0–200 (20) μg/ml	Rat	Bone marrow	CFU-GM	↓		239
	0.20 μg/ml	Human	Bone marrow	CFU-GM	↓	≥2.0 μg/ml (lethal)	239
	0–200 μ*M*	Human	PB PMN	Chemotaxis	↔	Lindane-induced	281
		Human	PB PMN	fMLP-induced chemotaxis	↓		281
	0–200 (60) μ*M*	Human	PB PMN	Cap formation	↓	f-Actin and ConA-induced	281
	0–200 (20) μ*M*	Human	PB PMN	Intracellular Ca^{2+} conc.	↑		281
	125 μ*M*	Guinea pig	Macrophage	Phosphoinositol activity	↑		282

TABLE 6 (continued)

Effect of Pesticides on Immune System Parameters — *In Vitro* Exposures

Compound (Chemical Name)	Dose (NOEL)	Species	Cell Type	Parameter	Effect	Miscellaneous	Ref.
	0–20 (0.2) μg/ml	Rat	BM	CFU-GM	↓		283
	0–20 μg/ml	Human	BM	CFU-GM	↓		283
Malathion (pure) (*O,O*-dimethyl phosphorodithioate of diethyl mercaptosuccinate)	0–75 μg/ml	C57Bl/6	Splenocyte	ConA and LPS mitoblastogenesis	↓	Metabolite not effective	210
	0–250 μg/ml	C57Bl/6	PEC	H_2O_2 production	↑	Metabolite not effective	210
		Human	PBMC	H_2O_2 production	↑	Metabolite not effective	210
		Rat	RBL-1	β-hexosamidase levels	↑		207
	0–500 μg/ml	Human	PBMC	ConA and LPS mitoblastogenesis	↓	Metabolite not effective	210
	1–125 μg/ml	C57Bl/6	Splenocyte	CTL activity	↔	1-hr exp'r post *in vitro* sensitization	284
	1–125 (25) μg/ml	C57Bl/6	Splenocyte	CTL activity	↓	1-hr exp'r pre *in vitro* sensitization	284
	10–100 (10) μg/ml	Human	PB lymphocyte	DNA synthesis	↓	PHA stimulated; ^{3}H-Thy incorporation	285
		Human	PB lymphocyte	RNA synthesis	↓	PHA stimulated; ^{3}H-Uridine incorporation	285
	10–70 μg/ml	Human	PB lymphocyte	RNA content	↓	PHA stimulated	286
		Human	PB lymphocyte	DNA content	↓	PHA stimulated	286
		Human	PB lymphocyte	RNA polymerase activity	↓		287
	10–70 (10) μg/ml	Human	PB lymphocyte	RNA synthesis	↓		287
	50–400 μg/ml	Human	B411-4	Chromosomal aberrations	↔		288
		Human	B411-4	Cell proliferation	↓		288
Malathion (technical) (*O,O*-dimethyl phosphorodithioate of diethyl mercaptosuccinate)	500 μg/ml	Human	PB monocyte	Esterase activity	↔	50% inhibition	270
MCPA ([4-chloro-2-methyl]phenoxyacetic acid)		Rat	BM	CFU-GM			239
		Human	BM	CFU-GM			239
Methiocarb (3,5-dimethyl-4-[methylthio]phenyl methylcarbamate)	0.5–50 (5) μ*M*	Mouse	CTLL2	IL-2-dependent proliferation	↓		242
Methyl parathion (*O,O*-dimethyl *O*-[4-dintrophenyl]phosphorothioate)		Rat	BM	CFU-GM			239
		Human	BM	CFU-GM			239
	0–40 μg/ml	Human	Jeff	SCE	↑		289
		Human	B35M	Cell cycling	↓		289

		Human	B35M	SCE	↑		289
		Human	Jeff	Cell cycling	↓		289
Mevinphos (2-carbomethoxy-1 methylvinyl dimethyl phosphate)		Human	BM	CFU-GM			239
		Rat	BM	CFU-GM			239
	0.5–50 (0.5) μ*M*	Mouse	CTLL2	IL-2-dependent proliferation	↓		242
Monocrotophos (dimethyl [E]-1-methyl-2-[methylcarbamoyl]vinyl phosphate)	0–1 nM	Human	PB lymphocyte	Chromosomal aberrations	↑		290
	0.5–50 μ*M*	Mouse	CTLL2	IL-2-dependent proliferation	↔		242
Naled (1,2-dibromo-2,2-dichloroethyl dimethylphosphate)	1500 mg/l	Human	PB monocyte	Esterase activity	↓	50% inhibition	270
Nicotine ([S]-3-[1-methyl-2-pyrrolidyl]pyridine)	500 mg/l	Human	PB monocyte	Esterase activity	↔	50% inhibition	270
Omethoate (*O,O*-diemthyl *S*-[2-(methylamino)-2-oxoethyl] phosphorothioate)	0–120 μg/ml	Human	PB lymphocyte	SCE	↑		254
Oxychlordane (chlordane metabolite)	0–100 (1.0) μ*M*	B6C3F1	Splenocyte	T-dependent antibody	↓	*In vitro* anti-SRBC	261
Paraoxon (*O,O*-diethyl *O-p*-nitrophenyl phosphate)	0–250	Rat	RBL-1	β-hexosamidase levels	↑		207
	0.5–50 (0.5) μ*M*	Mouse	CTLL2	IL-2-dependent proliferation	↓		242
Parathion (*O,O*-diethyl *O*-[4-nitrophenyl] phosphorothioate)	0–250	Rat	RBL-1	β-hexosamidase levels	↑		207
	3.7 mg/l	Human	PB monocyte	Esterase activity	↓	50% inhibition	270
Pendimethaline (*N*-[1-ethylpropyl]-3,4-dimethyl-2,6-dinitrobenzenamine)	0–0.1 m*M*/l	Human	PB lymphocyte	SCE	↔		241
Pentachlorophenol (analytical) (2,3,4,5,6-pentachlorophenol)	0–200 μ*M*	Human	PB lymphocyte	Ig secretion	↓	PWM and KlebsM activation	291
		Human	PB lymphocyte	T-cell mitoblastogenesis	↓	@ suboptimal conc.	291
		Human	PB lymphocyte	IL-2 production	↓	PHA activation	291
		Human	PB lymphocyte	T-cell mitoblastogenesis	↓	@ suboptimal conc.	291
		Human	PB lymphocyte	IL-2 production	↓	PHA activation	291
		Human	PB lymphocyte	Ig secretion	↓	PWM and KlebsM activation	291
Phosphamidon (2-chloro-2-diethylcarbamoyl-1-methyl-vinyl dimethyl phosphate)		Human	PB leukocyte				221
		Mouse					221
Phosphine (aluminum phosphide)	0.4.5 μg/l	Human	PB lymphocyte	Chromosomal aberrations	↑		222
Pirimiphos-methyl (*O*-[2-diethylamino-6-methylpryimidin-4-yl] *O,O*-dimethyl phosphorothioate)	0–50 μg/ml	Sprague-Dawley	Hepatocyte	Micronucleus frequency	↔		252
Propoxur (2-[1-methylethoxy]phenyl methylcarbamate)	15 μg/ml	Human	PB monocyte	Esterase activity	↓	50% inhibition	270
	0–200 μg/ml	Human	PB lymphocyte	Micronucleus frequency	↑		292
	0–200 (50) μg/ml	Human	PB lymphocyte	SCE	↑		292
Propoxur — nitroso derivative	0–200 (50) μg/ml	Human	PB lymphocyte	Micronucleus frequency	↑		292
		Human	PB lymphocyte	SCE	↑		292
Pyrethrum (esters of pyrethrolone, pyrethroic acid, cinerolone, chrysanthemic acid, and jasmoline)	500 mg/l	Human	PB monocyte	Esterase activity	↔	50% inhibition	270

TABLE 6 (continued)

Effect of Pesticides on Immune System Parameters — *In Vitro* Exposures

Compound (Chemical Name)	Dose (NOEL)	Species	Cell Type	Parameter	Effect	Miscellaneous	Ref.
Pyridathion (*O*-ethyl-*O*-isopropyl-*O*-[5-methoxy-1-methyl-6-oxo-1*H*-pyridasine-4*H*]-thiophosphate)	9.2 m*M*	Human	PB lymphocyte	Chromosomal aberrations	↔		232
Simazine (2-chloro-4,6-*bis*[thylamino]-*s*-triazine)	0–50 μ*M*/ml	Human	PB lymphocyte	SCE	↔		241
Tetrachlorovinphos ([Z]-2-chloro-1-[2,4,5-trichlrophenyl]vinyl dimethylphosphate)	2.6 μg/l	Human	PB monocyte	Esterase activity	↓	50% inhibition	270
Trichlorfon (dimethyl [2,2,2-trichloro-1-hydroxyethyl]phosphonate)	0.001–10,000 (125) mg/l	Carp	Leukocyte	Phagocytosis	↓	Spleen and pronephros leukocytes; *Y. ruckeri*	163
	0.01–1000 (1) mg/l	Carp	Lymphocyte	PHA mitoblastogenesis	↓	Spleen and pronephros lymphocytes	163
		Carp	Pronephros	Cell proliferation	↓		163
		Carp	Pronephros	Myeloid cell respiratory burst	↓		163
	30–120 (60) mg/kg	Mouse	BM	SCE	↑		293
	10–60 (10) μg/ml	Human	PB lymphocyte	SCE	↑	Not dose responsive	293

REFERENCES

1. McGregor, Brown, Howgate, McBride, Riach, and Caspary, Responses of the L5178Y mouse lymphoma cell forward mutation assay. V. 27 coded chemicals, *Environ. Mol. Mutagen.*, 17, 196, 1991.
2. Budavari, O'Neil, Smith, and Heckelman, Eds., *The Merck Index*, 11th ed., Merck and Co., Rahway, NJ, 1989.
3. Mellan, I., *Industrial Solvents Handbook*, Noyes Data Corp., Park Ridge, NJ, 1977.
4. Price, Tyl, Marks, Paschke, Ledoux, and Reel, Teratologic and postnatal evaluation of aniline hydrochloride in the Fischer 344 rat, *Toxicol. Appl. Pharmacol.*, 77, 465, 1985.
5. Keller and Snyder, Mice exposed *in utero* to low concentrations of benzene exhibit enduring changes in their colony forming hematopoietic cells, *Toxicology*, 42, 171, 1986.
6. Rozen, Snyder, and Albert, Depressions in B- and T-lymphocyte mitogen-induced blastogenesis in mice exposed to low concentrations of benzene, *Toxicol. Lett.*, 20, 343, 1984.
7. Rosenthal and Snyder, Inhaled benzene reduces aspects of cell-mediated tumor surveillance in mice, *Toxicol. Appl. Pharmacol.*, 88, 35, 1987.
8. Aoyama, Effects of benzene inhalation lymphocyte subpopulations and immune response in mice, *Toxicol. Appl. Pharmacol.*, 85, 92, 1986.
9. Rosenthal and Snyder, Modulation of the immune response to *Listeria* monocytogenes by benzene inhalation, *Toxicol. Appl. Pharmacol.*, 80, 502, 1985.
10. Rozen and Snyder, Protracted exposure of C57BL/6 mice to 300 ppm benzene depresses B- and T-lymphocyte numbers and mitogen responses. Evidence for thymic and bone marrow proliferation in response to the exposures, *Toxicology*, 37, 13, 1985.
11. Irons and Moore, Effect of short term benzene administration on circulating lymphocyte subpopulations in the rabbit: evidence of a selective b-lymphocyte sensitivity, *Res. Commun. Chem. Pathol. Pharmacol.*, 27, 147, 1980.
12. Irons, Greenlee, Wierda, and Bus, Relationship between benzene metabolism and toxicity: a proposed mechanism for the formation of reactive intermediates from polyphenol metabolites, *Adv. Exp. Med. Biol.*, 136(Pt. A), 229, 1981.
13. Gaido and Wierda, Modulation of stromal cell function in DBA/2J and B6C3F1 mice exposed to benzene or phenol, *Toxicol. Appl. Pharmacol.* 81, 469, 1985.
14. Gaido and Wierda, Suppression of bone marrow stromal cell function by benzene and hydroquinone is ameliorated by indomethacin, *Toxicol. Appl. Pharmacol.* 89, 378, 1987.
15. Wierda, King, Luebke, Reasor, and Smialowicz, Perinatal immunotoxicity of benzene toward mouse B-cell development. *J. Am. Coll. Toxicol,* 8, 981, 1989.
16. Wierda, Irons, and Greenlee, Immunotoxicity in C57BL/6 mice exposed to benzene and Aroclor 1254, *Toxicol. Appl. Pharmacol.* 60, 410, 1981.
17. MacEachern and Laskin, Increased production of tumor necrosis factor-α by bone marrow leukocytes following benzene treatment of mice, *Toxicol. Appl. Pharmacol.*, 113, 260, 1992.
18. MacEachern, Snyder, and Laskin, Alterations in the morphology and functional activity of bone marrow phagocytes following benzene treatment of mice, *Toxicol. Appl. Pharmacol.* 117, 147, 1992.
19. Hsieh, Sharma, and Parker, Hypothalamic-pituitary-adrenocortical axis activity and immune function after oral exposure to benzene and toluene, *Immunopharmacol.* 21, 23, 1991.
20. Renz and Kalf, Role for interleukin-1 (IL-1) in benzene-induced hematotoxicity: inhibition of conversion of pre-IL-1α to mature cytokine in murine macrophages by hydroquinone and prevention of benzene-induced hematotoxicity in mice by IL-1α, *Blood,* 78, 938, 1991.
21. Klan, Adams, and Lewis, Effects of exposure to benzene *in vivo* on the murine mononuclear phagocyte system, *Toxicol. Appl. Pharmacol.* 103, 198, 1990.
22. Punjabi, Laskin, Hwang, MacEachern, and Laskin, Enhanced production of nitric oxide by bone marrow cells and increased sensitivity to macrophage colony-stimulating factor (CSF) and granulocyte-macrophage CSF after benzene treatment of mice, *Blood,* 83, 3255, 1994.
23. Moszczynsky and Lisiewicz, Occupational exposure to benzene, toluene and xylene and the T lymphocyte functions, *Haematologia. (Budap.),* 17, 449, 1984.
24. Moszczynski and Lisiewicz, The nitroblue tetrazolium reduction test in workers exposed to benzene and its homologues, *Med. Interne.,* 21, 217, 1983.
25. Pfeifer and Irons, Effect of benzene metabolites on phytohemagglutinin-stimulated lymphopoiesis in rat bone marrow, *J. Reticuloendothel. Soc.,* 31, 155, 1982.
26. Araki, Tanigawa, Ishizu, and Minato, Decrease of CD4-positive T lymphocytes in workers exposed to benzidine and beta-naphthylamine, *Arch. Environ. Health,* 48, 205, 1993.
27. Tanigawa, Araki, Ishizu, Morita, Okazaki, and Minato, Natural killer cell activity in workers exposed to benzidine and beta- naphthylamine, *Br. J. Ind. Med.,* 47, 338, 1990.
28. National Toxicology Program, Toxicology and carcinogenesis studies of *n*-butyl chloride (CAS NO. 109-69-3) in F344/N rats and B6C3F1 mice (gavage studies), *Natl. Toxicol. Progr. Tech. Rep. Ser.,* 1, 1986.

29. Hennighausen and Lange, Toxic effects of di-*n*-octyltin dichloride on the thymus in mice, *Arch. Toxicol. Suppl.*, 2, 315, 1979.
30. Thompson, Jack, and Patrick, The possible role of macrophages in transient hepatic fibrogenesis induced by acute carbon tetrachloride injury, *J. Pathol.*, 130, 65, 1980.
31. Babb, Billing, Gans, and Yamaguchi, Acute hepatotoxin exposure effects lymphoid and accessory cell types in inbred mice, *Proc. Soc. Exp. Biol. Med.*, 174, 392, 1983.
32. Weir and Suckling, Immunocyto-adherence by spleen cells and tissue antigens in normal and carbon tetrachloride-treated rats, *Clin. Exp. Immunol.*, 3, 837, 1968.
33. Lloyd and Triger, Studies on hepatic uptake of antigen. III. Studies of liver macrophage function in normal rats and following carbon tetrachloride administration, *Immunology*, 29, 253, 1975.
34. Singh, Mathur, Gupta, Srivastava, and Gupta, Increased susceptibility of rabbits to intravenous challenge with *Mycobacterium avium* after mild hepatitis produced by carbon tetrachloride, *J. Med. Microbiol.*, 13, 319, 1980.
35. Armendariz-Borunda, Seyer, Postlethwaite, and Kang, Kupffer cells from carbon tetrachloride-injured rat livers produce chemotactic factors for fibroblasts and monocytes: the role of tumor necrosis factor-alpha, *Hepatology*, 14, 895, 1991.
36. Tajima, Nishimura, and Ito, Suppression of delayed-type hypersensitivity mediated by macrophage-like cells in mice with experimental liver injury, *Immunology*, 54, 57, 1985.
37. Paumgartner, Longueville, and Leevy, Phagocytic activity in experimental liver injury, *Exp. Mol. Pathol.*, 9, 161, 1968.
38. Wierda and Irons, Hydroquinone and catechol reduce the frequency of progenitor B lymphocytes in mouse spleen and bone marrow, *Immunopharmacology*, 4, 41, 1982.
39. Chhabra, Thompson, Elwell, and Gerken, Toxicity of *p*-chloroaniline in rats and mice, *Food Chem. Toxicol.*, 28, 717, 1990.
40. Kluwe, Dill, Persing, and Peters, Toxic responses to acute, subchronic, and chronic oral administrations of monochlorobenzene to rodents, *J. Toxicol. Environ. Health*, 15, 745, 1985.
41. Shopp, Jr., Sanders, White, Jr., and Munson, Humoral and cell-mediated immune status of mice exposed to trans-1,2-dichloroethylene, *Drug Chem. Toxicol.*, 8, 393, 1985.
42. Abdo, Jokinen, and Hiles, Subchronic (13-week) toxicity studies of N,N-dimethylaniline administered to Fischer 344 rats and B6C3F1 mice, *J. Toxicol. Environ. Health*, 29, 77, 1990.
43. Leff, Oppegard, McCarty, Wilke, Shanley, Day, Ahmed, Patton, and Repine, Dimethyl sulfoxide decreases lung neutrophil sequestration and lung leak, *J. Lab. Clin. Med.*, 120, 282, 1992.
44. Caren, Oven, and Mandel, Dimethyl sulfoxide: lack of suppression of the humoral immune response in mice, *Toxicology Lett.*, 26, 193, 1985.
45. Guastadisegni, Mantovani, Ricciardi, Stazi, Maffi, and Salvati, Hematotoxic effects in the rat of a toluene dinitro derivative after short-term exposure, *Ecotoxicol. Environ. Safety*, 17, 21, 1989.
46. Bardodej and Cirek, Long-term study on workers occupationally exposed to ethylbenzene, *J. Hyg. Epidemiol. Microbiol. Immunol.*, 32, 1, 1988.
47. Dean, Lauer, House, Murray, Stillman, Irons, Steinhagen, Phelps, and Adams, Studies of immune function and host resistance in B6C3F1 mice exposed to formaldehyde, *Toxicol. Appl. Pharmacol.*, 72, 519, 1984.
48. Clarke and Aldons, Isophorone diisocyanate induced respiratory disease (IPDI), *Aust. N. Z. J. Med.*, 11, 290, 1981.
49. Luster, Tucker, Germolec, Silver, Thomas, Vore, and Bucher, Immunotoxicity studies in mice exposed to methyl isocyanate, *Toxicol Appl. Pharmacol.*, 86, 140, 1986.
50. Tucker, Bucher, Germolec, Silver, Vore, and Luster, Immunological studies on mice exposed subacutely to methyl isocyanate, *Environ. Health Perspect.*, 72, 139, 1987.
51. Nair, Auletta, Schroeder, and Johannsen, Chronic toxicity, oncogenic potential, and reproductive toxicity of *p*-nitroaniline in rats, *Fundam. Appl. Toxicol.*, 15, 607, 1990.
52. Nair, Johannsen, Levinskas, and Terrill, Subchronic inhalation toxicity of *p*-nitroaniline and *p*-nitrochlorobenzene in rats, *Fundam. Appl. Toxicol.*, 6, 618, 1986.
53. Mullin, Ader, Daughtrey, Frost, and Greenwood, Toxicology update isoparaffinic hydrocarbons: a summary of physical properties, toxicity studies and human exposure data, *J. Appl. Toxicol.*, 10, 135, 1990.
54. Dogra, Khanna, Srivastava, Shukla, and Shanker, Styrene-induced immunomodulation in mice, *Int. J. Immunopharmacol.*, 11, 577, 1989.
55. Hsieh, Sharma, and Parker, Immunotoxicological evaluation of toluene exposure via drinking water in mice, *Environ. Res.*, 49, 93, 1989.
56. Gordon, Sheppard, McDonald, Distefano, and Scypinski, Airway hyperresponsiveness and inflammation induced by toluene diisocyanate in guinea pigs, *Am. Rev. Respir. Dis.*, 132, 1106, 1985.
57. Huang, Aoyama, and Ueda, Experimental study on respiratory sensitivity to inhaled toluene diisocyanate, *Arch. Toxicol.*, 67, 373, 1993.
58. Sugawara, Okamoto, Sawahata, and Tanaka, Skin reactivity in guinea pigs sensitized with 2,4-toluene diisocyanate, *Int. Arch. Allergy Immunol.*, 100, 190, 1993.
59. Di Stefano, Saetta, Maestrelli, Milani, Pivirotto, Mapp, and Fabbri, Mast cells in the airway mucosa and rapid development of occupational asthma induced by toluene diisocyanate, *Am. Rev. Respir. Dis.*, 147, 1005, 1993.

60. Butcher, Karr, O'Neil, Wilson, Dharmarajan, Salvaggio, and Weill, Inhalation challenge and pharmacologic studies of toluene diisocyanate (TDI)-sensitive workers, *J. Allergy Clin. Immunol.*, 64, 146, 1979.
61. Saetta, Di Stefano, Maestrelli, de Margo, Milani, Pivirotti, Mapp, and Fabbri, Airway mucosal inflammation in occupational asthma induced by toluene diisocyanate, *Am. Rev. Resp. Dis.*, 145, 160, 1992.
62. Mapp, Boschetto, Dal Vecchio, Maestrelli, and Fabbri, Occupational asthma due to isocyanates, *Eur. Resp. J.*, 1, 273, 1988.
63. Bentley, Maestrelli, Saetta, Fabbri, Robinson, Bradley, Jeffery, Durham, and Kay, Activated T-lymphocytes and eosinophils in the bronchial mucosa in isocyanate-induced asthma, *J. Allergy Clin. Immunol.*, 89, 821, 1992.
64. Butcher, Salvaggio, Weill, and Ziskind, Toluene diisocyanate (TDI) pulmonary disease: immunologic and inhalation challenge studies, *J. Allergy Clin. Immunol.*, 58, 89, 1976.
65. Byers, Levin, Ozonoff, and Baldwin, Association between clinical symptoms and lymphocyte abnormalities in a population with chronic domestic exposure to industrial solvent-contaminated domestic water supply and a high incidence of leukaemia, *Canc. Immunol. Immunother.*, 27, 77, 1988.
66. Seifert, Mostecka, and Kolar, Trifluoromethylanilines — their effect on DNA synthesis and proliferative activity in parenchymal organs of rats, *Toxicology,* 83, 49, 1993.
67. Moszczynski, The effect of working environment contaminated with organic solvents on the activity of acid phosphatase in lymphocytes, *Med. Interne.*, 21, 37, 1983.
68. Suleiman, Petroleum hydrocarbon toxicity *in vitro*: effect of *n*-alkanes, benzene and toluene on pulmonary alveolar macrophages and lysosomal enzymes of the lung, *Arch. Toxicol.*, 59, 402, 1987.
69. Grayson and Gill, Effect of *in vitro* exposure to styrene, styrene oxide, and other structurally related compounds on murine cell-mediated immunity, *Immunopharmacology,* 11, 165, 1986.
70. Brown, Lutton, Nelson, Abraham, and Levere, Microenvironmental cytokines and expression of erythroid heme metabolic enzymes, *Blood Cells,* 13, 123, 1987.
71. Manning, Adams, and Lewis, Effects of benzene metabolites on receptor-mediated phagocytosis and cytoskeletal integrity in mouse peritoneal macrophages, *Toxicol. Appli. Pharmacol.*, 126, 214, 1994.
72. Pfeifer and Irons, Inhibition of lectin-stimulated lymphocyte agglutination and mitogenesis by hydroquinone: reactivity with intracellular sulfhydryl groups, *Exp. Mol. Pathol.*, 35, 189, 1981.
73. Irons, Neptun, and Pfeifer, Inhibition of lymphocyte transformation and microtubule assembly by quinone metabolites of benzene: evidence for a common mechanism, *J. Reticuloendothel. Soc.*, 30, 359, 1981.
74. Ding and Badwey, Wortmannin and 1-butanol block activation of a novel family of protein kinases in neutrophils, *FEBS Lett.*, 348, 149, 1994.
75. Salyer, Bohnsack, Knape, Shigeoka, Ashwood, and Hill, Mechanisms of tumor necrosis factor-alpha alteration of PMN adhesion and migration, *Am. J. Pathol.*, 136, 831, 1990.
76. Rabinovitch and Destefano, Macrophage spreading *in vitro*. III. The effect of metabolic inhibitors, anesthetics and other drugs on spreading induced by subtilisin, *Exp. Cell Res.*, 88, 153, 1974.
77. Westcott and Murphy, Effect of alcohols on arachidonic acid metabolism in murine mastocytoma cells and human polymorphonuclear leukocytes, *Biochim. Biophys. Acta,* 833, 262, 1985.
78. Sud'ina, Mirzoeva, Pushkareva, Korshunova, Sumbatyan, and Varfolomeev, Caffeic acid phenethyl ester as a lipoxygenase inhibitor with antioxidant properties, *FEBS Lett.*, 329, 21, 1993.
79. Lynch, Blackwell, and Moncada, Carbon tetrachloride-induced eicosanoid synthesis and enzyme release from rat peritoneal leucocytes, *Biochem. Pharmacol.*, 34, 1515, 1985.
80. Irons, Stillman, Colagiovanni, and Henry, Synergistic action of the benzene metabolite hydroquinone on myelopoietic stimulating activity of granulocyte/macrophage colony-stimulating factor *in vitro*, *Proc. Natl. Acad. Sci. USA,* 89, 3691, 1992.
81. King, Landreth, and Weirda, Bone marrow stromal cell regulation of B-lymphopoiesis. II. Mechanisms of hydroquinone inhibition of pre-B cell maturation, *J. Pharmacol. Exp. Ther.*, 250, 582, 1989.
82. Thomas, Reasor, and Wierda, Macrophage regulation of myelopoiesis is altered by exposure to the benzene metabolite hydroquinone, *Toxicol. Appl. Pharmacol.*, 97, 440, 1989.
83. King, Landreth, and Weirda, Hydroquinone inhibits bone marrow pre-B cell maturation *in vitro*, *Mol. Pharmacol.*, 32, 807, 1987.
84. Oliveira and Kalf, Induced differentiation of HL-60 promyelocytic leukemia cells to monocyte/macrophages is inhibited by hydroquinone, a hematotoxic metabolite of benzene, *Blood,* 79, 627, 1992.
85. Dawson, McGee, Smith, and Brooks, Methanol is a potent inhibitor of the metabolism of endogenous arachidonic acid by elicited rat peritoneal leukocytes, *Prostaglandins,* 39, 395, 1990.
86. Burrone, Polypeptide synthesis in toluene-treated lymphocytes, *Eur. J. Biochem.*, 86, 439, 1978.
87. Meister and Sine, Eds., *Farm Chemicals Handbook '95*, Biesterfeld, New York, 1995.
88. Briggs, *Basic Guide to Pesticides: Their Characteristics and Hazards,* Taylor & Francis, Washington, D.C., 1992.
89. Hermanowicz, Nawarska, Borys, and Maslankiewicz, The neutrophil function and infectious diseases in workers occupationally exposed to organochloride insecticides, *Int. Arch. Occup. Environ. Health,* 50, 329, 1982.
90. Rodgers, Imamura, and Devens, Effects of subchronic treatment with O,O,S-trimethyl phosphorothioate on cellular and humoral immune response systems, *Toxicol. Appl. Pharmacol.*, 81, 310, 1985.

91. Cornacoff, Lauer, House, Tucker, Thurmond, Vos, Working, and Dean, Evaluation of the immunotoxicity of β-hexachlorocyclohexane (β-HCH), *Fund. Appl. Toxicol.,* 11, 293, 1988.
92. Kerkvliet, Brauner, and Matlock, Humoral immunotoxicity of polychlorinated diphenyl ethers, phenoxyphenols, dioxins and furans present as contaminants of technical grade pentachlorophenol, *Toxicology,* 36, 307, 1985.
93. Blakley and Blakley, The effect of prenatal exposure to the *n*-butylester of 2,4-dichlorophenoxyacetic acid (2,4-D) on the immune response in mice, *Teratology,* 33, 15, 1986.
94. Cushman and Street, Allergic hypersensitivity to the herbicide 2,4-D in BALB/c mice, *J. Toxicol. Environ. Health,* 10(4-5), 729, 1982.
95. Schop, Hardy, and Goldberg, Comparison of the activity of topically applied pesticides and the herbicide 2,4-D in two short-term *in vivo* assays of genotoxicity in the mouse, *Fundam. Appl. Toxicol.,* 15, 666, 1990.
96. Linnainmaa, Induction of sister chromatid exchanges by the peroxisome proliferators 2,4-D, MCPA, and clofibrate *in vivo* and *in vitro*, *Carcinogenesis,* 5, 703, 1984.
97. Adhikari and Grover, Genotoxic effects of some systemic pesticides: *in vivo* chromosomal aberrations in bone marrow cells in rats, *Environ. Mol. Mutagen.,* 12, 235, 1988.
98. Hajoui, Flipo, Mansour, Fournier, and Krzystyniak, Immunotoxicity of subchronic vs. chronic exposure to aldicarb in mice, *Int. J. Immunopharmacol.,* 14, 1203, 1992.
99. Thomas, Ratajczak, Eisenberg, Furedi-Machacek, Ketels, and Barbera, Evaluation of host resistance and immunity in mice exposed to the carbamate pesticide aldicarb, *Fundam. Appl. Toxicol.,* 9, 82, 1987.
100. Selvan, Dean, Misra, Nagarkatti, and Nagarkatti, Aldicarb suppresses macrophage but not natural killer (NK) cell-mediated cytotoxicity of tumor cells, *Bull. Environ. Contam. Toxicol.,* 43, 676, 1989.
101. Gill, Pande, and Tewari, Hemopathological changes associated with experimental aldicarb poisoning in fish (*Puntius conchonius* Hamilton), *Bull. Environ. Contam. Toxicol.,* 47, 628, 1991.
102. Hong, Effects of environmental toxins on lymphocyte function: studies in rhesus and man, *Ann. Allergy,* 66, 474, 1991.
103. Fiore, Anderson, Hong, Golubjatnikov, Seiser, Nordstrom, Hanrahan, and Belluck, Chronic exposure to aldicarb-contaminated groundwater and human immune function, *Environ. Res.,* 41, 633, 1986.
104. Mirkin, Anderson, Hanrahan, Hong, Golubjatnikov, and Belluck, Changes in T-lymphocyte distribution associated with ingestion of aldicarb-contaminated drinking water: a follow-up study, *Environ. Res.,* 51, 35, 1990.
105. Wiltrout, Ercegovich, and Ceglowski, Humoral immunity in mice following oral administration of selected pesticides, *Bull. Environ. Contam. Toxicol.,* 20, 423, 1978.
106. Fournier, Chevalier, Nadeau, Trottier, and Krzystyniak, Virus-pesticide interactions with murine cellular immunity after sublethal exposure to dieldrin and aminocarb, *J. Toxicol. Environ. Health,* 25, 103, 1988.
107. Krzystyniak, Trottier, Jolicoeur, and Fournier, Macrophage functional activities vs. cellular parameters upon sublethal pesticide exposure, *Mol. Toxicol.,* 1, 247, 1987.
108. Harris, Zacharewski, and Safe, Comparative potencies of Aroclors 1232, 1242, 1248, 1254, and 1260 in male Wistar rats — assessment of the toxic equivalency factor (TEF) approach for polychlorinated biphenyls (PCBs), *Fundam. Appl. Toxicol.,* 20, 456, 1993.
109. Loose, Silkworth, Pittman, Benitz, and Mueller, Impaired host resistance to endotoxin and malaria in polychlorinated biphenyl and hexachlorobenzene treated mice, *Infect. Immun.,* 20, 30, 1978.
110. Loose, Pittman, Benitz, and Silkworth, Polychlorinated biphenyl and hexachlorobenzene induced humoral immunosuppression, *J. Reticuloendothel. Soc.,* 22, 253, 1977.
111. Green, Carr, Palmer, and Oswald, Lack of cytogenetic effects in bone marrow and spermatagonial cells in rats treated with polychlorinated biphenyls (Aroclors 1242 and 1254), *Bull. Environ. Contam. Toxicol.,* 13, 14, 1975.
112. Carter and Clancy, Jr., Acutely administered polychlorinated biphenyls (PCBs) decrease splenic cellularity but increase its ability to cause graft-versus-host reactions in BALB/c mice, *Immunopharmacology,* 2, 341, 1980.
113. Nestel and Budd, Chronic oral exposure of rainbow trout (*Salmon gairdneri*) to a polychlorinated biphenyl (Aroclor 1254): pathological effects, *Can. J. Comp. Med.,* 39, 208, 1975.
114. Street and Sharma, Alternation of induced cellular and humoral immune responses by pesticides and chemicals of environmental concern: quantitative studies of immunosuppression by DDT, Aroclor 1254, carbaryl, carbofuran, and methylparathion, *Toxicol. Appl. Pharmacol.,* 32, 587, 1975.
115. Carter and Mercer, Pair-feeding study of PCB (Aroclor 1254) toxicity in rats, *Bull. Environ. Contam. Toxicol.,* 31, 686, 1983.
116. Tryphonas, Luster, Schiffman, Dawson, Hodgen, Germolec, Hayward, Bryce, Loo, Mandy et al., Effect of chronic exposure of PCB (Aroclor 1254) on specific and nonspecific immune parameters in the rhesus (*Macaca mulatta*) monkey, *Fundam. Appl. Toxicol.,* 16, 773, 1991.
117. Tryphonas, Luster, White, Jr., Naylor, Erdos, Burleson, Germolec, Hodgen, Hayward, and Arnold, Effects of PCB (Aroclor 1254) on non-specific immune parameters in rhesus (*Macaca mulatta*) monkeys, *Int. J. Immunopharmacol.,* 13, 639, 1991.

118. Arkoosh, Clemons, Myers, and Casillas, Suppression of B-cell mediated immunity in juvenile chinook salmon (*Oncorhynchus tshawytscha*) after exposure to either a polycyclic aromatic hydrocarbon or to polychlorinated biphenyls, *Immunopharmacol. Immunotoxicol.,* 16, 293, 1994.
119. Tryphonas, Arnold, Zawidzka, Mes, Charbonneau, and Wong, A pilot study in adult rhesus monkeys (*M. mulatta*) treated with Aroclor 1254 for two years, *Toxicol. Pathol.,* 14, 1, 1986.
120. Chang, Hsieh, Lee, and Tung, Immunologic evaluation of patients with polychlorinated biphenyl poisoning: determination of phagocyte Fc and complement receptors, *Environ. Res.,* 28, 329, 1982.
121. Fournier, Friborg, Girard, Mansour, and Krzystyniak, Limited immunotoxic potential of technical formulation of the herbicide atrazine (AAtrex) in mice, *Toxicol. Lett.,* 60, 263, 1992.
122. Mikula, Pistl, and Kacmar, The immune response of sheep to subclinical chronic exposure to the herbicide Bentazon TP, *Vet. Hum. Toxicol.,* 34, 507, 1992.
123. Nehez, Fischer, Scheufler, Schmidt, Sipos, and Desi, Investigations of the acute toxic, cytogenetic, and embryotoxic activity of buminafos, *Ecotoxicol. Environ. Safety,* 24, 13, 1992.
124. Andre, Gillon, Andre, Lafont, and Jourdan, Pesticide-containing diets augment anti-sheep red blood cell nonreaginic antibody responses in mice but may prolong murine infection with *Giardia muris, Environ. Res.,* 32, 145, 1983.
125. Ladics, Smith, Heaps, and Loveless, Evaluation of the humoral immune response of CD rats following a 2-week exposure to the pesticide carbaryl by the oral, dermal, or inhalation routes, *J. Toxicol. Environ. Health,* 42, 143, 1994.
126. Pipy, Beraud, and Gaillard, Phagocytic activity of the reticuloendothelial system in the rat after administration of an anticholinesterasic pesticide, carbaryl (author's transl.), *Experientia,* 34, 87, 1978.
127. Barnett, Cranmer, Avery, and Hoberman, Immunocompetence over the lifespan of mice exposed *in utero* to carbofuran or diazinon. I. changes in serum immunoglobulin concentrations, *J. Environ. Pathol. Toxicol.,* 4, 53, 1980.
128. Gupta, Bagchi, Bandyopadhyay, Sasmal, Chatterjee, and Dey, Hematological changes produced in mice by Nuvacron or Furadan, *Toxicology,* 25, 255, 1982.
129. Johnson, Holsapple, and Munson, An immunotoxicological evaluation of gamma-chlordane, *Fundam. Appl. Toxicol.,* 6, 317, 1986.
130. Spyker-Cranmer, Barnett, Avery, and Cranmer, Immunoteratology of chlordane: cell-mediated and humoral immune responses in adult mice exposed *in utero*, *Toxicol. Appl. Pharmacol.,* 62, 402, 1982.
131. Menna, Barnett, and Soderberg, Influenza type A virus infection of mice exposed *in utero* to chlordane; survival and antibody studies, *Toxicol. Lett.,* 24, 45, 1985.
132. Barnett, Soderberg, and Menna, The effect of prenatal chlordane exposure on the delayed hypersensitivity response of BALB/c mice, *Toxicol. Lett.,* 25, 173, 1985.
133. Blaylock, Soderberg, Gandy, Menna, Denton, and Barnett, Cytotoxic T-lymphocyte and NK responses in mice treated prenatally with chlordane, *Toxicol. Lett.,* 51, 41, 1990.
134. Theus, Tabor, Soderberg, and Barnett, Macrophage tumoricidal mechanisms are selectively altered by prenatal chlordane exposure, *Agents Actions,* 37, 140, 1992.
135. Theus, Lau, Tabor, Soderberg, and Barnett, *In vivo* prenatal chlordane exposure induces development of endogenous inflammatory macrophagaes, *J. Leuk. Biol.,* 51, 366, 1993.
136. Barnett, Blaylock, Gandy, Menna, Denton, and Soderberg, Alteration of fetal liver colony formation by prenatal chlordane exposure, *Fundam. Appl. Toxicol.,* 15, 820, 1990.
137. Barnett, Holcomb, Menna, and Soderberg, The effect of prenatal chlordane exposure on specific anti-influenza cell-mediated immunity, *Toxicol. Lett.,* 25, 229, 1985.
138. McConnachie and Zahalsky, Immune alterations in humans exposed to the termiticide technical chlordane, *Arch. Environ. Health,* 47, 295, 1992.
139. Chetty, Brown, Walker, and Woods, Effects of chlordecone and malnutrition on immune response in rats, *Life Sci.,* 52, PL175, 1993.
140. Smialowicz, Luebke, Riddle, Rogers, and Rowe, Evaluation of the immunotoxic potential of chlordecone with comparison to cyclophosphamide, *J. Toxicol. Environ. Health,* 15, 561, 1985.
141. Thomas, Craig, and Stacey, Effects of chlordimeform and its metabolite 4-chloro-*o*-toluidine on rat splenic T, B and tumoricidal effector cells, *Immunopharmacology,* 19, 79, 1990.
142. Gromysz-Kalkowska, Szubartowska, and Kaczanowska, Peripheral blood in the Japanese quail (*Coturnix coturnix* japonica) in acute poisoning by different insecticides, *Comp. Biochem. Physiol. C,* 81, 209, 1985.
143. Varshneya, Singh, Sharma, Bahga, and Garg, Immunotoxic responses of cypermethrin, a synthetic pyrethroid insecticide in rats, *Indian J. Physiol. Pharmacol.,* 36, 123, 1992.
144. Amer and Aboul-ela, Cytogenetic effects of pesticides. III. Induction of micronuclei in mouse bone marrow by the insecticides cypermethrin and rotenone, *Mutat. Res.,* 155, 135, 1985.
145. Desi, Dobronyi, and Varga, Immuno-, neuro-, and general toxicologic animal studies on a synthetic pyrethroid: cypermethrin, *Ecotoxicol. Environ. Safety,* 12, 220, 1986.
146. Schwab and Richardson, Lymphocyte and brain neurotoxic esterase: dose and time dependence of inhibition in the hen examined with three organophosphorus esters, *Toxicol. Appl. Pharmacol.,* 83, 1, 1986.
147. Glick, Antibody-mediated immunity in the presence of Mirex and DDT, *Poult. Sci.,* 53, 1476, 1974.

148. Banerjee, Ramachandran, and Hussain, Sub-chronic effect of DDT on humoral immune response in mice, *Bull. Environ. Contam. Toxicol.,* 37, 433, 1986.
149. Banerjee, Sub-chronic effect of DDT on humoral immune response to a thymus-independent antigen (bacterial lipopolysaccharide) in mice, *Bull. Environ. Contam. Toxicol.,* 39, 822, 1987.
150. Banerjee, Effects of sub-chronic DDT exposure on humoral and cell-mediated immune responses in albino rats, *Bull. Environ. Contam. Toxicol.,* 39, 827, 1987.
151. Rao and Glick, Pesticide effects on the immune response and metabolic activity of chicken lymphocytes, *Proc. Soc. Exptl. Biol. Med.,* 154, 27, 1977.
152. Gabliks, Al-zubaidy, and Askari, DDT and immunological responses. 3. Reduced anaphylaxis and mast cell population in rats fed DDT, *Arch. Environ. Health,* 30, 81, 1975.
153. Askari and Gabliks, DDT and immunological responses. II. Altered histamine levels and anaphylactic shock in guinea pigs, *Arch. Environ. Health,* 26, 309, 1973.
154. Wassermann, Wassermann, Gershon, and Zellermayer, Effects of organochlorine insecticides on body defense systems, *Ann. N.Y. Acad. Sci.,* 160, 393, 1969.
155. Larsen and Jalal, DDT induced chromosome mutations in mice — further testing, *Can. J. Genet. Cytol.,* 16, 491, 1974.
156. Rehana and Rao, Effect of DDT on the immune system in Swiss albino mice during adult and perinatal exposure: humoral responses, *Bull. Environ. Contam. Toxicol.,* 48, 535, 1992.
157. Gabliks, Askari, and Yolen, DDT and immunological responses I. Serum antibodies and anaphylactic shock in guinea pig, *Arch. Environ. Health,* 26, 305, 1973.
158. Rabello, Dealmeida, Pigati, Ungaro, Murata, Perira, and Becak, Cytogenetic study on individuals occupationally exposed to DDT, *Mutat. Res.,* 28, 449, 1975.
159. Dzwonkowska and Hubner, Induction of chromosomal aberrations in the Syrian hamster by insecticides tested *in vivo*, *Arch. Toxicol.,* 58, 152, 1986.
160. Seinen, Vos, van Krieken, Penninks, Brands, and Hooykaas, Toxicity of organotin compounds. III. Suppression of thymus-dependent immunity in rats by di-*n*-butyltindichloride and di-*n*-octyltindichloride, *Toxicol. Appl. Pharmacol.,* 42, 213, 1977.
161. Hennighausen and Lange, Toxic effects of di-n-octyltin dichloride on the thymus in mice, *Arch. Toxicol. Suppl., 2,* 315, 1979.
162. Saegusa, Radiomimetic toxicity of 1,2-dibromo-3-chloropropane (DBCP), *Ind. Health,* 24, 1, 1986.
163. Dunier, Siwicki, and Demael, Effects of organophosphorus insecticides: effects of trichlorfon and dichlorvos on the immune response of carp (*Cyprinus carpio*). III. *In vitro* effects on lymphocyte proliferation and phagocytosis and *in vivo* effects on humoral response, *Ecotoxicol. Environ. Safety,* 22, 79, 1991.
164. Nehez, Toth, and Desi, The effect of dimethoate, dichlorvos, and parathion-methyl on bone marrow cell chromosomes of rats in subchronic experiments *in vivo*, *Ecotoxicol. Environ. Safety,* 29, 365, 1994.
165. Safe, Polychlorinated biphenyls (PCBs) and polybrominated biphenyls (PBBs): biochemistry, toxicology, and mechanism of action, *CRC Crit. Rev. Toxicol.,* 13, 319, 1984.
166. Desi, Varga, and Farkas, The effect of DDVP, an organophosphorus pesticide on the humoral and cell-mediated immunity of rabbits, *Arch. Toxicol. (Suppl.),* 4, 171, 1980.
167. Loose, Macrophage induction of T-suppressor cells in pesticide-exposed and protozoan-infected mice, *Environ. Health Perspect.,* 43, 89, 1982.
168. Krzystyniak, Flipo, Mansour, and Fournier, Suppression of avidin processing and presentation by mouse macrophages after sublethal exposure to dieldrin, *Immunopharmacology,* 18, 157, 1989.
169. Majumdar, Kopelman, and Schnitman, Dieldrin-induced chromosome damage in mouse bone-marrow and WI-38 human lung cells, *J. Hered.,* 67, 303, 1976.
170. Hugo, Bernier, Krzystyniak, Potworowski, and Fournier, Abrogation of graft-versus-host reaction by diedrin in mice, *Toxicol. Lett.,* 41, 11, 1988.
171. Hugo, Bernier, Krzystyniak, and Fournier, Transient inhibition of mixed lymphocyte reactivity by dieldrin in mice, *Toxicol. Lett.,* 41, 1, 1988.
172. Bernier, Hugo, Krzystyniak, and Fournier, Suppression of humoral immunity in inbred mice by dieldrin, *Toxicol. Lett.,* 35, 231, 1987.
173. Stieglitz, Gibel, Werner, and Stobbe, Experimental study on haematotoxic and leukaemogenic effects of trichlorophone and dimethoate, *Acta Haematol.,* 52, 70, 1974.
174. Tiefenbach and Lange, Studies on the action of dimethoate on the immune system, *Arch. Toxicol. (Suppl.),* 4, 167, 1980.
175. Nehez, Selypes, Mazzag, and Berencsi, Additional data on the mutagenic effect of dinitro-o-cresol-containing herbicides, *Ecotoxicol. Environ. Safety,* 8, 75, 1984.
176. Amer and Fahmy, Cytogenetic effects of pesticides. I. Induction of micronuclei in mouse bone marrow by the insecticide Dursban, *Mutat. Res.,* 101, 247, 1982.
177. Malhi and Grover, Genotoxic effects of some organophosphorus pesticides. II. *In vivo* chromosomal aberration bioassay in bone marrow cells in rat, *Mutat. Res.,* 188, 45, 1987.
178. Grover and Malhi, Genotoxic effects of some organophosphorous pesticides. I. Induction of micronuclei in bone marrow cells in rat, *Mutat. Res.,* 155, 131, 1985.

179. Varshneya, Bahga, and Sharma, Effect of insecticides on humoral immune response in cockerels, *Br. Vet. J.*, 144, 610, 1988.
180. Parker, McCullough, Gellatly, and Johnston, Toxicologic and carcinogenic evaluation of Fenvalerate in the B6C3F1 mouse, *Fundam. Appl. Toxicol.*, 3, 114, 1983.
181. Pati and Bhunya, Cytogenetic effects of Fenvalerate in mammalian *in vivo* test system, *Mutat. Res.*, 222, 149, 1989.
182. Loose, Pittman, Benitz, Silkworth, Mueller, and Coulston, Environmental chemical-induced immune dysfunction, *Ecotoxicol. Environ. Safety*, 2, 173, 1978.
183. Loose, Pittman, Benitz, and Silkworth, Polychlorinated biphenyl and hexachlorobenzene induced humoral immunosuppression. *J. ReticuloEndo Soc.*, 22, 253, 1977.
184. Carthew, Edwards, and Smith, Immunotoxic effects of hexachlorobenzene on the pathogenesis of systemic, pneumonic and hepatic virus infections in the mouse, *Hum. Exp. Toxicol.*, 9, 403, 1990.
185. Silkworth and Loose, Environmental chemical induced modification of cell mediated immune responses, *Adv. Exp. Med. Biol.*, 121A, 499, 1979.
186. Ziprin and Fowler, Rosette-forming ability of alveolar macrophages from rat lung: inhibition by hexachlorobenzene, *Toxicol. Appl. Pharmacol.*, 39, 105, 1977.
187. Barnett, Barfield, Walls, Joyner, Owens, and Soderberg, The effect of hexachlorobenzene on the developing immune response of BALB/c mice, *Toxicol. Lett.*, 39, 263, 1987.
188. Vos, van Logten, Kreeftenberg, and Kruizinga, Hexachlorobenzene-induced stimulation of the humoral immune response in rats, *Ann. N. Y. Acad. Sci.*, 320, 535, 1979.
189. Vos, van Logten, Kreeftenberg, Steerenberg, and Kruizinga, Effect of hexachlorobenzene on the immune system of rats following combined pre- and postnatal exposure, *Drug Chem. Toxicol.*, 2, 61, 1979.
190. Van Loveren, Krajnc, Rombout, Blommaert, and Vos, Effects of ozone, hexachlorobenzene, and bis(tri-*n*-butyltin)oxide on natural killer activity in the rat lung, *Toxicol. Appl. Pharmacol.*, 102, 21, 1990.
191. Sherwood, Thomas, O'Shea, Bradof, Ratajczak, Graham, and Aranyi, Effects of inhaled hexachlorobenzene aerosols on rat pulmonary host defenses, *Toxicol. Ind. Health*, 5, 451, 1989.
192. Gralla, Fleischman, Luthra, Hagopian, Baker, Esber, and Marcus, Toxic effects of hexachlorobenzene after daily administration to beagle dogs for one year, *Toxicol. Appl. Pharmacol.*, 40, 227, 1977.
193. Koller, Exon, and Roan, Immunological surveillance and toxicity in mice exposed to the organophosphate pesticide, Leptophos, *Environ. Res.*, 12, 238, 1976.
194. Meera, Rao, Shanker, and Tripathi, Immunomodulating effects of gamma-HCH (lindane) in mice, *Immunopharmacol. Immunotoxicol.*, 14, 261, 1992.
195. Meera, Tripathi, Kamboj, and Rao, Role of calcium in biphasic immunomodulation by gamma-HCH (lindane) in mice, *Immunopharmacol. Immunotoxicol.*, 15, 113, 1993.
196. Dunier, Siwicki, Scholtens, Dal Molin, Vergnet, and Studnicka, Effects of lindane exposure on rainbow trout (*Oncorhynchus mykiss*) immunity. III. Effect on nonspecific immunity and B lymphocyte functions, *Ecotoxicol. Environ. Safety*, 27, 324, 1994.
197. Desi, Vorga, and Farkas, Studies on the immunosuppressive effect of organochlorine and organophosphoric pesticides in subacute experiments, *J. Hyg. Epidemiol. Microbiol. Immunol.*, 22, 115, 1978.
198. Mandal, Chakraborty, and Lahiri, Hematological changes produced by lindane (gamma-HCH) in six species of birds, *Toxicology*, 40, 103, 1986.
199. Grabarczyk, Kopec Szlezak, Szczepanska, Wozniak, and Podstawka, The effect of gamma-hexachlorocyclohexane (lindane) on blood cells, kidney and liver tissues in rabbits, *Haematologia (Budap.)*, 23, 171, 1990.
200. Das, Paul, Saxena, and Ray, Effect of *in utero* exposure to hexachlorocyclohexane on the developing immune system of mice, *Immunopharmacol. Immunotoxicol.*, 12, 293, 1990.
201. Dunier and Siwicki, Effects of lindane exposure on rainbow trout (*Oncorhynchus mykiss*) immunity. I. Effect of lindane on antibody-secreting cells (ASC) measured by ELISPOT assay, *Ecotoxicol. Environ. Safety*, 27, 1, 1994.
202. Cossarini-Dunier, Monod, Demael, and Lepot, Effect of gamma-hexachlorocyclohexane (lindane) on carp (*Cyprinus carpio*). I. Effect of chronic intoxication on humoral immunity in relation to tissue pollutant levels, *Ecotoxicol. Environ. Safety*, 13, 339, 1987.
203. Bhunya and Jena, Genotoxic potential of the organochlorine insecticide lindane (gamma- BHC): an *in vivo* study in chicks, *Mutat. Res.*, 272, 175, 1992.
204. Kopec-Szlezak, Szczepanska, Grabarczyk, and Podstawka, Late toxic effects of long-term exposure to lindane in peripheral blood cells in rabbits. I. Function impairment and structural disturbances in leucocytes, *Mater. Med. Pol.*, 22, 179, 1990.
205. Cushman and Street, Allergic hypersensitivity to insecticide malathion in BALB/c mice, *Toxicol. Appl. Pharmacol.*, 70, 29, 1983.
206. Dulout, Pastori, and Olivero, Malathion-induced chromosomal aberrations in bone-marrow cells of mice: dose-response relationships, *Mutat. Res.*, 122, 163, 1983.
207. Rodgers and Ellefson, Mechanism of modulation of murine peritoneal cell function and mast cell degranulation by low doses of malathion, *Agents Actions*, 35, 57, 1992.

208. Vijay, Mendoza, and Lavergne, Production of homocytotropic antibodies (IgE) to malathion in the rat, *Toxicol. Appl. Pharmacol.*, 44, 137, 1978.
209. Rodgers, Leung, Ware, Devens, and Imamura, Lack of immunosuppressive effects of acute and subacute administration of malathion, *Pestic. Biochem. Physiol.*, 25, 358, 1986.
210. Rodgers and Ellefson, Modulation of respiratory burst activity and mitogenic response of human peripheral blood mononuclear cells and murine splenocytes and peritoneal cells by malathion, *Fundam. Appl. Toxicol.*, 14, 309, 1990.
211. Milby and Epstein, Allergic contact sensitivity to malathion, *Arch. Environ. Health*, 9, 434, 1964.
212. Jablonicka, Polakova, Karelova, and Vargova, Analysis of chromosome aberrations and sister-chromatid exchanges in peripheral blood lymphocytes of workers with occupational exposure to the mancozeb-containing fungicide Novozir Mn80, *Mutat. Res.*, 224, 143, 1989.
213. Mathew, Rahiman, and Vijayalaxmi, *In vivo* genotoxic effects in mice of Metacid 50, an organophosphorus insecticide, *Mutagenesis*, 5, 147, 1990.
214. Degraeve and Moutschen, Absence of genetic and cytogenetic effects in mice treated by the organophosphorus insecticide parathion, its methyl analogue, and paraoxon, *Toxicology*, 32, 177, 1984.
215. Major, Kemeny, and Tompa, Genotoxic effects of occupational exposure in the peripheral blood lymphocytes of pesticide preparing workers in Hungary, *Acta Med. Hung.*, 49, 79, 1992.
216. McConnell, Moore, Gupta, Rakes, Luster, Goldstein, Haseman, and Parker, The chronic toxicity of technical and analytical pentachlorophenol in cattle, *Toxicol. Appl. Pharmacol.*, 52, 468, 1980.
217. Kerkvliet, Baecher-Steppan, Claycomb, Craig, and Sheggeby, Immunotoxicity of technical pentachlorophenol (PCP-T): depressed humoral immune responses to T-dependent and T-independent antigen stimulation in PCP-T exposed mice, *Fund. App. Toxicol.*, 2, 90, 1982.
218. Kerkvliet, Brauner, and Baecher-Steppan, Effects of dietary technical pentachlorophenol exposure on T cell, macrophage and natural killer cell activity in C57Bl/6 mice, *Int. J. Immunopharmacol.*, 7, 239, 1985.
219. Kerkvliet, Baecher-Steppan, and Schmitz, Immunotoxicity of pentachlorophenol (PCP): increased susceptibility to tumor growth in adult mice fed technical PCP-containing diets, *Toxicol. Appl. Pharmacol.*, 62, 55, 1982.
220. McConnachie and Zahalsky, Immunological consequences of exposure to pentachlorphenol, *Arch. Environ. Health*, 46, 249, 1991.
221. Patankar and Vaidya, Evaluation of genetic toxicity of insecticide phosphamidon using *in vitro* and *in vivo* mammalian test systems, *Indian J. Exp. Biol.*, 18, 1145, 1980.
222. Garry, Griffith, Danzl, Nelson, Whorton, Krueger, and Cervenka, Human genotoxicity: pesticide applicators and phosphine, *Science*, 246, 251, 1989.
223. Reiss, Herrman, and Hopkins, Effect of anthelminthic treatment on the immune response of mice, *Lab. Anim. Sci.*, 37, 773, 1987.
224. Barnett, Gandy, Wilbourn, and Theus, Comparison of the immunotoxicity of propanil and its metabolite, 3,4-dichloroaniline, in C57Bl/6 mice, *Fundam. Appl. Toxicol.*, 18, 628, 1992.
225. Barnett and Gandy, The effect of acute propanil exposure on the immune response of C57Bl/6 mice, *Fund. Appl. Toxicol.*, 12, 757, 1989.
226. Xie, Schafer, and Barnett, *In vitro* exposure of macrophages to propanil alters TNF-α and IL-6 production, *submitted.*
227. Zhao, Schafer, Cuff, Gandy, and Barnett, Changes in primary and secondary lymphoid organ T-cell subpopulations resulting from acute *in vivo* exposure to Propanil, *J. Toxicol. Environ. Health*, 46, 101, 1995.
228. Zhao, Schafer, and Barnett, *In vivo* and *in vitro* exposure to propanil reduces IL-2 and IL-6 but not interferon-gamma production, *submitted.*
229. Blyler, Landreth, Lillis, Schafer, Theus, Gandy, and Barnett, Selective myelotoxicity of Propanil, *Fund. Appl. Toxicol.*, 22, 505, 1994.
230. Theus, Tabor, Gandy, and Barnett, Alteration of macrophage cytoxicity through endogenous interferon and tumor necrosis factor α induction by Propanil, *Toxicol. Appl. Pharmacol.*, 118, 46, 1993.
231. Gieldanowski, Kowalczyk-Bronisz, and Bubak, Studies on affinity of pesticide Unden — 2-isopropoxyphenyl *N*-methylcarbamate to immunological system, *Arch. Immunol. Ther. Exp. (Warsz.)*, 39, 85, 1991.
232. Vargova, Polakova, Podstavkova, Siskova, Dolan, Vlcek, and Miadokova, The mutagenic effect of the new insecticide and acaricide pyridathion, *Mutat. Res.*, 78, 353, 1980.
233. Gromysz-Kalkowska, Szubartowska, Sulikowska, and Trocewicz, Peripheral blood changes of the Japanese quail *Coturnix coturnix* japonica following repeated small doses of trichlorfon, *Bull. Environ. Contam. Toxicol.*, 35, 757, 1985.
234. Nehez, Paldy, Selypes, Korosfalvi, Lorinczi, and Berencsi, The mutagenic effect of trifluralin-containing herbicide on mouse bone marrow *in vivo*, *Ecotoxicol. Environ. Safety*, 3, 454, 1979.
235. Szubartowska and Gromysz-Kalkowska, Differences between breeds of rabbits in their susceptibility to the effect of foschlor, *Comp. Biochem. Physiol. C.*, 85, 33, 1986.
236. Korte and Jalal, 2,4-D induced clastogenicity and elevated rates of sister chromatid exchanges in cultured human lymphocytes, *J. Hered.*, 73, 224, 1982.

237. Turkula and Jalal, Increased rates of sister chromatid exchanges induced by the herbicide 2,4-D, *J. Hered.,* 76, 213, 1985.
238. Mustonen, Kangas, Vuojolahti, and Linnainmaa, Effects of phenoxyacetic acids on the induction of chromosome aberrations *in vitro* and *in vivo, Mutagenesis,* 1, 241, 1986.
239. Parent-Massin and Thouvenot, *In vitro* study of pesticide hematotoxicity in human and rat progenitors, *J. Pharmacol. Toxicol. Meth.,* 30, 203, 1993.
240. Meisner, Belluck, and Roloff, Cytogenetic effects of alachlor and/or atrazine *in vivo* and *in vitro, Environ. Mol. Mutag.,* 19, 77, 1992.
241. Dunkelberg, Fuchs, Hengstler, Klein, Oesch, and Struder, Genotoxic effects of the herbicides alachlor, atrazine, pendimethaline, and simazine in mammalian cells, *Bull. Environ. Contam. Toxicol.,* 52, 498, 1994.
242. Casale, Vennerstrom, Bavari, and Wang, Inhibition of interleukin 2 driven proliferation of mouse CTLL2 cells, by selected carbamate and organophosphate insecticides and congeners of carbaryl, *Immunopharmacol. Immunotoxicol.,* 15, 199, 1993.
243. Cid and Matos, Induction of sister-chromatid exchanges in cultured human lymphocytes by Aldicarb, a carbamate pesticide, *Mutat. Res.,* 138, 175, 1984.
244. Gonzalez Cid, Loria, and Matos, Nitroso-aldicarb induces sister-chromatid exchanges in human lymphocytes *in vitro, Mutat. Res.,* 204, 665, 1988.
245. Gonzalez Cid and Matos, Chromosomal aberrations in cultured human lymphocytes treated with aldicarb, a carbamate pesticide, *Mutat. Res.,* 191, 99, 1987.
246. Tithof, Contreras, and Ganey, Aroclor 1242 stimulates the production of inositol phosphates in polymorphonuclear neutrophils, *Toxicol. Appl. Pharmacol.,* 131, 136, 1995.
247. Lee, The inhibition of PMN beta-glucuronidase release by Aroclor 1254, *Environ. Res.,* 25, 386, 1981.
248. Lee and Park, Biochemical basis of Aroclor 1254 and pesticide toxicity *in vitro.* 1. Effects on intracellular ATP concentration, *Res. Commun. Chem. Pathol. Pharmacol.,* 25, 597, 1979.
249. Gyorkos, Brock, and Sparkes, Chemiluminescence in human whole blood: modulation by the cocarcinogens phorbol diester and polychlorobiphenyls, *Immunopharmacol. Immunotoxicol.,* 10, 417, 1988.
250. Exon, Talcott, and Koller, Effect of lead, polychlorinated biphenyls, and cyclophosphamide on rat natural killer cells, interleukin 2, and antibody synthesis, *Fundam. Appl. Toxicol.,* 5, 158, 1985.
251. Talcott, Koller, and Exon, The effect of lead and polychlorinated biphenyl exposure on rat natural killer cell cytotoxicity, *Int. J. Immunopharmacol.* 7, 255, 1985.
252. Piatti, Marabini, and Chiesara, Increase of micronucleus frequency in cultured rat hepatocytes treated *in vitro* with benomyl and pirimiphos-methyl separately and in mixture, *Mutat. Res. Lett.,* 324, 59, 1994.
253. Georgieva, Vachkova, Tzoneva, and Kappas, Genotoxic activity of benomyl in different test systems, *Environ. Mol. Mutagen,* 16, 32, 1990.
254. Dolara, Salvadori, Capobianco, and Torricelli, Sister-chromatid exchanges in human lymphocytes induced by dimethoate, omethoate, deltamethrin, benomyl and their mixture, *Mutat. Res.,* 283, 113, 1992.
255. Lamb and Lilly, An investigation of some genetic toxicological effects of the fungicide benomyl, *Toxicology,* 17, 83, 1980.
256. Vigfusson and Vyse, The effect of the pesticides, Dexon, Captan and Roundup, on sister-chromatid exchanges in human lymphocytes *in vitro, Mutat. Res.,* 79, 53, 1980.
257. de Maroussem, Pipy, Beraud, Souqual, and Forgue, The effect of carbaryl on the arachidonic acid metabolism and superoxide production by mouse resident peritoneal macrophages challenged by zymosan, *Int. J. Immunopharmacol.,* 8, 155, 1986.
258. Casale, Bavari, Gold, and Vitzthum, Inhibition of interleukin-2-stimulated enhancement of human natural killer (NK) cell activity by carbaryl, an anticholinesterase insecticide, *Toxicol. Lett.,* 63, 299, 1992.
259. Bavari, Casale, Gold, and Vitzthum, Modulation of interleukin-2-driven proliferation of human large granular lymphocytes by carbaryl, an anticholinesterase insecticide, *Fund. Appl. Toxicol.,* 17, 61, 1991.
260. Forgue, Pipy, Beraud, Souqual, and Combis, 1-naphthyl *N*-methyl carbamate effect on intra- and extracellular concentrations of arachidonic acid metabolites, and on the chemiluminescence generation by mouse peritoneal macrophages, *Int. J. Immunopharmacol.,* 12, 155, 1990.
261. Johnson, Kaminski, and Munson, Direct suppression of cultured spleen cell responses by chlordane and the basis for differential effects on *in vivo* and *in vitro* immunocompetence, *J. Toxicol. Environ. Health,* 22, 497, 1987.
262. Suzaki, Inoue, Okimasu, Ogata, and Utsumi, Stimulative effect of chlordane on the various functions of the guinea pig leukocytes, *Toxicol. Appl. Pharmacol.,* 93, 137, 1988.
263. McGregor, Brown, Cattanach, Edwards, McBride, Riach, and Caspary, Responses of the L5178Y tk^+/tk^- mouse lymphoma cell forward mutation assay. III. 72 coded chemicals, *Environ. Mol. Mutagen.,* 12, 85, 1988.
264. Brubaker, Flamm, and Bernheim, Effect of gamma-chlordane on synchronized lymphoma cells and inhibition of cell division, *Nature,* 226, 548, 1970.
265. Puig, Carbonell, Xamena, Creus, and Marcos, Analysis of cytogenetic damage induced in cultured human lymphocytes by the pyrethroid insecticides cypermethrin and fenvalerate, *Mutagenesis,* 4, 72, 1989.
266. Amer, Ibrahim, and el-Sherbeny, Induction of chromosomal aberrations and sister chromatid exchange *in vivo* and *in vitro* by the insecticide cypermethrin, *J. Appl. Toxicol.,* 13, 341, 1993.

267. Kannan and Sharma, Defective lymphocyte transformation by DDT: *in vitro* responsiveness of rabbit peripheral blood lymphocytes to PHA, *Indian. J. Exp. Biol.*, 17, 805, 1979.
268. Lessa, Becak, Nazareth Rabello, Pereira, and Ungaro, Cytogenetic study of DDT on human lymphocytes *in vitro*, *Mutat. Res.*, 40, 131, 1976.
269. Perocco and Fini, Damage by dichlorvos of human lymphocyte DNA, *Tumori.*, 66, 425, 1980.
270. Lee and Waters, Inhibition of monocyte esterase activity by organophosphate insecticides, *Blood*, 50, 947, 1977.
271. Grabarczyk and Kopec-Szlezak, Protective action of vitamin E on the subcellular structures of lymphocytes intoxicated with pesticides *in vitro*, *Mater. Med. Pol.*, 24, 237, 1992.
272. Conradt, Muller, Loose, Klein, Coulston, and Korte, Incorporation of [3H]uridine into RNA under the influence of dieldrin and polychlorinated biphenyls, *Ecotoxicol. Environ. Safety*, 3, 10, 1979.
273. Hewett and Roth, Dieldrin activates rat neutrophils *in vitro*, *Toxicol. Appl. Pharmacol.*, 96, 269, 1988.
274. Surralles, Carbonell, Puig, Xamena, Creus, and Marcos, Induction of mitotic micronuclei by the pyrethroid insecticide fenvalerate in cultured human lymphocytes, *Toxicol. Lett.*, 54, 151, 1990.
275. Forgue, Pinelli, Beraud, Souqual, and Pipy, Chemiluminescence response and arachidonic acid metabolism of macrophages induced by gamma-hexachlorocyclohexane (lindane), *Food Addit. Contam.*, 7(Suppl. 1), S97, 1990.
276. Kuhns, Kaplan, and Basford, Hexachlorocyclohexane, potent stimuli of O_2^- production and calcium release in human polymorphonuclear leukocytes, *Blood*, 68, 535, 1986.
277. English, Schell, Siakotos, and Gabig, Reversible activation of the neutrophil superoxide generating system by hexachlorocyclohexane: correlation with effects on a subcellular superoxide-generating fraction, *J. Immunol.*, 137, 283, 1986.
278. Meade, Harvey, Boot, Turner, Bateman, and Osborne, gamma-Hexachlorocyclohexane stimulation of macrophage phospholipid hydrolysis and leukotriene production, *Biochem. Pharmacol.*, 33, 289, 1984.
279. Roux, Treich, Brun, Desoize, and Fournier, Effect of lindane on human lymphocyte responses to phytohemagglutinin, *Biochem. Pharmacol.*, 28, 2419, 1979.
280. Roux, Puiseux-Dao, Treich, and Fournier, Effect of lindane on mouse peritoneal macrophages, *Toxicology*, 11, 259, 1978.
281. Kaplan, Zdziarski, Kuhns, and Basford, Inhibition of cell movement and associated changes by hexachlorocyclohexanes due to unregulated intracellular calcium increases, *Blood*, 71, 677, 1988.
282. Holian, Marchiarullo, and Stickle, gamma-Hexachlorocyclohexane activation of alveolar macrophage phosphatidylinositol cycle, calcium mobilization of O_2-production, *FEBS Lett.* 176, 151, 1984.
283. Parent-Massin, Thouvenot, Rio, and Riche, Lindane haematotoxicity confirmed by *in vitro* tests on human and rat progenitors, *Hum. Exp. Toxicol.*, 13, 103, 1994.
284. Rodgers, Grayson, Imamura, and Devens, *In vitro* effects of malathion and O,O,S-trimethyl phosphorothioate on cytotoxic T-lymphocyte responses, *Pest. Biochem. Physiol.*, 24, 260, 1985.
285. Czajkowska and Walter, Effect of malathion on nucleic acid synthesis in phytohemagglutinin-stimulated human lymphocytes, *Hum. Genet.*, 56, 189, 1980.
286. Czajkowska and Lipecka, Effect of malathion on the genetic material of human lymphocytes stimulated by phytohemagglutinin (PHA), *Hum. Genet.*, 53, 375, 1980.
287. Wiszkowska, Kulamowicz, Malinowska, and Walter, The effect of malathion on RNA polymerase activity of cell nuclei and transcription products in lymphocyte culture, *Environ. Res.*, 41, 372, 1986.
288. Huang, Effect on growth but not on chromosomes of the mammalian cells after treatment with three organophosphorus insecticides, *Proc. Soc. Exp. Biol. Med.*, 142, 36, 1973.
289. Chen, Hsueh, Sirianni, and Huang, Induction of sister-chromatid exchanges and cell cycle delay in cultured mammalian cells treated with eight organophosphorus pesticides, *Mutat. Res.*, 88, 307, 1981.
290. Vaidya and Patankar, Mutagenic effect of Monocrotophos — an insecticide in mammalian test systems, *Indian J. Med. Res.*, 76, 912, 1982.
291. Lang and Mueller Ruchholtz, Human lymphocyte reactivity after *in vitro* exposure to technical and analytical grade pentachlorophenol, *Toxicology*, 70, 271, 1991.
292. Gonzalez Cid, Loria, and Matos, Genotoxicity of the pesticide propoxur and its nitroso derivative, NO-propoxur, on human lymphocytes *in vitro*, *Mutat. Res.*, 232, 45, 1990.
293. Madrigal-Bujaidar, Cadena, Trujillo-Valdes, and Cassani, Sister-chromatid exchange frequencies induced by metrifonate in mammalian *in vivo* and *in vitro* systems (see comments), *Mutat. Res.*, 300, 135, 1993.

PART 5

Reproductive and Developmental Toxicology

James L. Schardein, Editor

CHAPTER 21

Biology of Reproduction and Methods in Assessing Reproductive and Developmental Toxicity in Humans

Iyabo O. Obasanjo and Claude L. Hughes

CONTENTS

0-8493-4493-X/97/$0.00+$.50

1 PHASES OF NORMAL REPRODUCTION: A CONCEPTUAL CONSTRUCT FOR REPRODUCTIVE AND DEVELOPMENTAL TOXICOLOGY

1.1 Gametogenesis and Gonadogenesis

1.1.1 Gamete Development

In mammals, genetic sex determines chromosomal sex, genetic sex, and sexual phenotype. The presence of the Y chromosome results in production of testicular androgens and anti-Müllerian hormone in the embryo which guide sexual differentiation to the male direction. The female phenotype is the default, and no additional factors are required for development of female gametes and sexual characteristics. Primordial germ cells of both sexes undergo mitotic division in the yolk sac until the time of gonadal differentiation in fetal life. The primordial germ cells appear in the wall of the yolk sac at the third week of fetal life. About 2 weeks later, the gonadal ridge appears, which is the earliest sign of gonadal development. The primordial germ cells progress into the gonadal ridge by ameboid movement at about 6 weeks of fetal life. The gonadal ridge grows and surrounds the primordial germ cells, and together they form three histologic components: the surface epithelium, the primitive cords, and the gonadal blastema, made up of primordial germ cells and mesenchymal cells. After the gonads start to differentiate, the primordial germ cells continue to divide; these are called oogonia in females and spermatogonia in males. The oogonia are transformed to oocytes by the initiation of meiotic division during the third to ninth months of fetal life. The oocytes are arrested in the diplotene stage of the first meiotic prophase until puberty. The spermatogonia do not undergo meiotic division until puberty.

1.1.2 Gonadal and Genital Development

The male gonads start to differentiate about 7 weeks after conception, with organization of the gonadal blastema into testicular cords and stroma. The testicular cords are composed of spermatogonia and primitive Sertoli cells. By the eighth week, the stroma between the testicular cords differentiate into Leydig cells, which start to synthesize testosterone almost immediately. This androgen production stimulates the further differentiation of Leydig cells, leading to increased numbers from the ninth to the eighteenth week. Testosterone stimulates the growth and differentiation of the mesonephric ducts (Wolffian ducts), leading to the development of the epididymis, ductus deferens, and the accessory glands of the male reproductive tract.

The Sertoli cells secrete anti-Müllerian hormone; this inhibition of differentiation of the Müllerian, or paramesonephric, ducts preempts development of the female upper reproductive tract. The Leydig cells undergo involution after about 18 weeks of gestation and are not apparent within a few days after birth. The appearance and involution of Leydig cells are controlled by human chorionic gonadotropin (hCG), but the mechanism is not well understood.

During embryogenesis, testicular cords grow in length with adjacent mesonephric ducts and become canalized to form the rete testis. The testicular end of the mesonephric duct differentiates into the epididymis, while the remaining portion becomes the ductus deferens. The mesonephric ducts adjacent to the primitive testes establish contact with the testicular tubules to form the efferent ducts. The portion of the mesonephric duct immediately distal to the efferent ducts becomes elongated and convoluted to form the epididymis. Seminal vesicles develop from the distal mesopheric duct. The terminal portion of the mesonephric duct between the seminal vesicles and urethra becomes the ejaculatory ducts and the ampullae of the vas deferens. The testicular cords are separated from the surface epithelium by the tunica albuginea; the septa grows from the tunica to divide the testis into wedge-shaped lobules. The testicles start to descend into the inguinal canal during the sixth month of fetal life and enter the scrotal swelling shortly before birth. The prostate gland develops from the urogenital sinus as a secretory tubuloalveolar gland and in the adult surrounds the urethra at the base of the bladder.

The female gonads start to differentiate later than those in males. The gonadal blastema begins to differentiate into medullary cords and stroma in the eighth week of fetal life. The medullary cords degenerate and leave mainly connective tissue. The cortex of the early ovary is made up of oogonia and epithelium. By the eleventh week of fetal life, interstitial connective tissues intersect the epithelium leading to compartments containing single oocytes surrounded by a single layer of epithelial cells. These are the primordial follicles. The connective tissue of the ovaries is referred to as the ovarian stroma, and it determines the size of the ovary. The epithelium of the follicles becomes cuboidal, and stroma-derived cells make up the theca layer around the primary follicles. The ovaries descend into the pelvis by the twelfth week of fetal life. The paramesonephric duct differentiates in the female to form the oviduct (fallopian tubes), uterus, and proximal vagina. The distal vagina develops from the urogenital sinus.

1.2 Mating Behavior

Events occurring around puberty instigate the start of mating behavior. At puberty, males and females become aware of each other as sexual beings. Experience as well as hormones are important in maintenance of copulatory behavior. Castration prior to puberty in male mammals abolishes the capacity for copulatory behavior.[1] In postpubertal monkeys, copulatory behavior continues after castration but decreases in frequency over time.

In human males and females, androgen increases libido. Women treated with androgens frequently complain of excessive libido as a side effect, whereas women treated with anti-androgens complain of reduced libido. Sexual activity may occur when testosterone levels are relatively low, although an interval of testosterone priming may be necessary. In all male primates, androgens increase sexual drive. Male macaques exposed to receptive females increase their production of testosterone, while the stress of social subordination reduces testosterone production.[2] Stress reduces plasma testosterone in males and females by acting through the hypothalamic-gonadal axis. Stress indicative of depressed hypothalamic function affects both male and female behavior, decreasing and even eliminating it in some cases.

Environmental influences on reproductive function and mating behavior may occur through visual and olfactory cues. For example, circadian rhythms influence the timing of the periovulatory surge of luteinizing hormone (LH) and ovulatory rates. The LH surge tends to occur at night and in the morning.[3,4] The best evidence of seasonal variation in human reproduction is from studies in Alaskan Eskimos showing seasonal changes in birth rates corresponding to peaks in June and January.[5] Humans do not reproduce in a rigidly seasonal manner, possibly because they have a reduced ability to respond to environmental cues or because they live in modified environments that attenuate the impact of environmental influences.[6]

Macaques, which are seasonal breeders in temperate environments, will ovulate year-round when housed in conditions of controlled temperature and light;[6,7] however, macaques that are reared indoors and exposed to varying photoperiods do not lose their ability to ovulate year-round. This suggests that humans no longer demonstrate a reproductive response to changing photoperiods because of ontogenic and possibly evolutionary experience. Pheromones also may play important roles in human reproduction; however, the only confirmed pheromone-driven reproduction phenomenon is menstrual cycle synchronicity in females sharing the same living environment.[6,8,9]

Hormonal control of mating behavior is well documented in animals. Estrogen and, to a lesser extent, progesterone act in the central nervous system to control lordosis behavior in rats. In female nonhuman primates, attractivity and proceptivity change with the stage of the menstrual cycle or as a result of sex steroid administration. The effects of hormones on receptivity are unclear. The relationship between steroid hormones and behavior seen in animals is assumed to extend to humans, but the influences of social and other environmental factors affect mating behavior and make characterization of the role of sex steroids exceedingly complex.

Studies of male mating behavior have been conducted in a variety of mammalian species, predominantly in rodents. The male mating repertoire involves precopulatory behavior, mounting, intromission, ejaculation, and postejaculatory behavior. Important aspects of male mating behavior are controlled by the pre-optic anterior hypothalamic area. Rhesus monkeys with bilateral lesions in this area do not attempt to copulate but continue to masturbate and achieve erections, indicating a deficit in mating behavior towards the female but no obvious neuroendocrine physiological compromise.[1]

1.3 Gamete Transport

For fertilization to occur, sperm deposited in the vagina must travel to the fallopian tubes, and the ovum must be released from the ovary and transported within the fallopian tubes. Spermatozoa acquire the capacity for motility in the proximal cauda epididymis. The epididymis also serves as a reservoir for sperm. Other maturational changes in the spermatic plasma membrane that occur in the epididymis involve physical and chemical alterations in the membrane lipids. Capacitation is the physiological change that makes spermatozoa capable of fertilization, and this occurs after deposition in the female tract. Capacitation begins when spermatozoa are in the cervical mucus and continues to sequestration in the isthmus. Capacitation involves changes in intracellular ion concentrations (e.g., K^+, Na^+, Ca^+), increased metabolism, stabilization of disulfide bonds in nuclear proteins, and removal of decapacitation factors from the plasma membrane. Seminal and prostatic fluids are added to the spermatozoa during passage through the male reproductive tract to form semen. Semen is deposited in the anterior vagina during coitus. Vaginal mucus protects the spermatozoa in the acidic vaginal environment from being ingested by macrophages. Spermatic movement through the cervix may continue for days, and most spermatozoa die during this maneuver. Prostaglandins deposited with spermatozoa in the semen stimulate the uterine contractions that facilitate movement of surviving spermatozoa through the uterus. Only a small proportion of deposited spermatozoa reach the site of fertilization in the mid-fallopian tube.

During the time preceding ovulation, the vaginal mucus is thin and watery, permitting passage of spermatozoa more easily than the thicker mucus that occurs after ovulation. Spermatozoan motility is necessary for movement up the female reproductive tract to the oviduct. Immotile spermatozoa is an easily detected cause of infertility in males. Another common cause of male infertility, inability of spermatozoa to penetrate the cervical mucus, can be tested *in vitro*.[10]

At ovulation, the ovum, surrounded by the cumulus oophorus, is released from the ovary and moves down the fallopian tubes. Ovulation occurs as a result of an LH surge, which causes luteinization of the granulosa cells of the follicle destined for ovulation. The Graafian follicle increases in size under the influence of follicle-stimulating hormone (FSH) and LH, and the primary oocyte resumes meiosis. The surface of the ovary bulges, and follicular fluid oozes out. The ovarian surface over the follicle gradually opens, and the oocyte with the surrounding cumulus oophorus is detached from the ovary by the oviduct. There, the fimbriae at the tip of the oviduct sweep the cumulus mass into the tubal lumen and thus move the oocyte to the site of fertilization. Since mating can occur at any time in the cycle in humans, the spermatozoa either may be at the fallopian tubes at the time of ovulation, or the ovum may be at the fallopian tubes awaiting the arrival of the spermatozoa. After fertilization, in the isthmus of the fallopian tubes, the zygote moves to the uterus for implantation. Prostaglandins, catecholamines, estrogen, and progesterone, as well as smooth muscle contractions in the walls of the fallopian tubes, aid in moving the zygote to the site of implantation.

1.4 Fertilization

Fertilization is the fusion of the sperm and ovum. The sperm head binds to the egg plasma membrane (oolemma), and the entire spermatozoa becomes incorporated into the ovum cytoplasm (Figure 1). Only capacitated spermatozoa with intact acrosomes can enter and pass through the cumulus oophorus. The acrosome is a membrane-bound, cap-like structure covering the anterior portion of the sperm nucleus. The acrosomal reaction is the release of materials capable of lysing the glycoprotein coat (zona pellucida) surrounding the ovum. This is necessary for fertilization to take place. Before undergoing the acrosomal reaction, sperm go through a type of hypermotility called hyperactivation. The hypermotility involves vigorous, whiplash beating of the tail and linear dashing movements. Hyperactivity appears to begin in the isthmus of the fallopian tubes and is thought to facilitate movement in the fluid of the oviduct (fallopian tubes) and to provide the power necessary to penetrate the cumulus oophorus and zona pellucida.

Usually the first segment of the sperm to make contact with the oolemma is the inner acrosomal membrane, followed by the postacrosomal region. The plasma membrane of the sperm attaches to microvilli on the oolemma. Sperm-egg fusion can be apparent by the reduced movement of the sperm tail.[11,12] *In vivo,* usually only one spermatozoon fuses with the oolemma. Fertilization is temperature and pH dependent. After fusion with the sperm, the ovum leaves its arrested state (metaphase of second meiosis) and completes its division into a haploid daughter cell. After incorporation into the ovum, the sperm nucleus undergoes decondensation, with breakdown of the sperm nuclear membrane. The haploid ovum nuclei decondenses during the meiosis division to form a pronuclei, and this ovum pronuclei fuses with the sperm pronuclei in a process called syngamy. This fusion of the two pronuclei restores the diploid chromosomal number.

1.5 Blastogenesis

Early cleavage is asynchronous in mammalian embryos, resulting in a spectrum of sizes of embryonic cells (Figure 2). Human embryos *in vitro* divide at intervals of 16 to 24 hours during early cleavage. After several cleavages, the embryo undergoes a process of compaction into a relatively dense mass of cells called the morula. The morula consists of an inner mass of cells and a distinct surrounding outer layer of cells. The next step in differentiation is the formation of a cavity within the mass of cells, resulting in the blastocyst. The cells are further partitioned into inner and outer compartments. Fluid begins to accumulate in the space between the cells. The cells of the inner compartment form the embryo proper.

By the eighth day after fertilization, the outer layer of trophectodermal cells is divided into two layers of cells: an inner layer of mononucleated cells (cytotrophoblast) and an outer layer of larger, multinucleated cells that form the implantation syncytium (syncytiotrophoblast). The inner, mononucleated cells continue to form giant cells that are pushed to the outer layers and form the ectoplacental cone and the extra-embryonic ectoderm. The inner layer of cells that form the embryo proper then aggregate at one pole of the blastocyst and divide to form the bilaminar germ disc, with a layer of small cuboidal cells (hypoblasts) and a layer of high columnar cells (epiblasts).

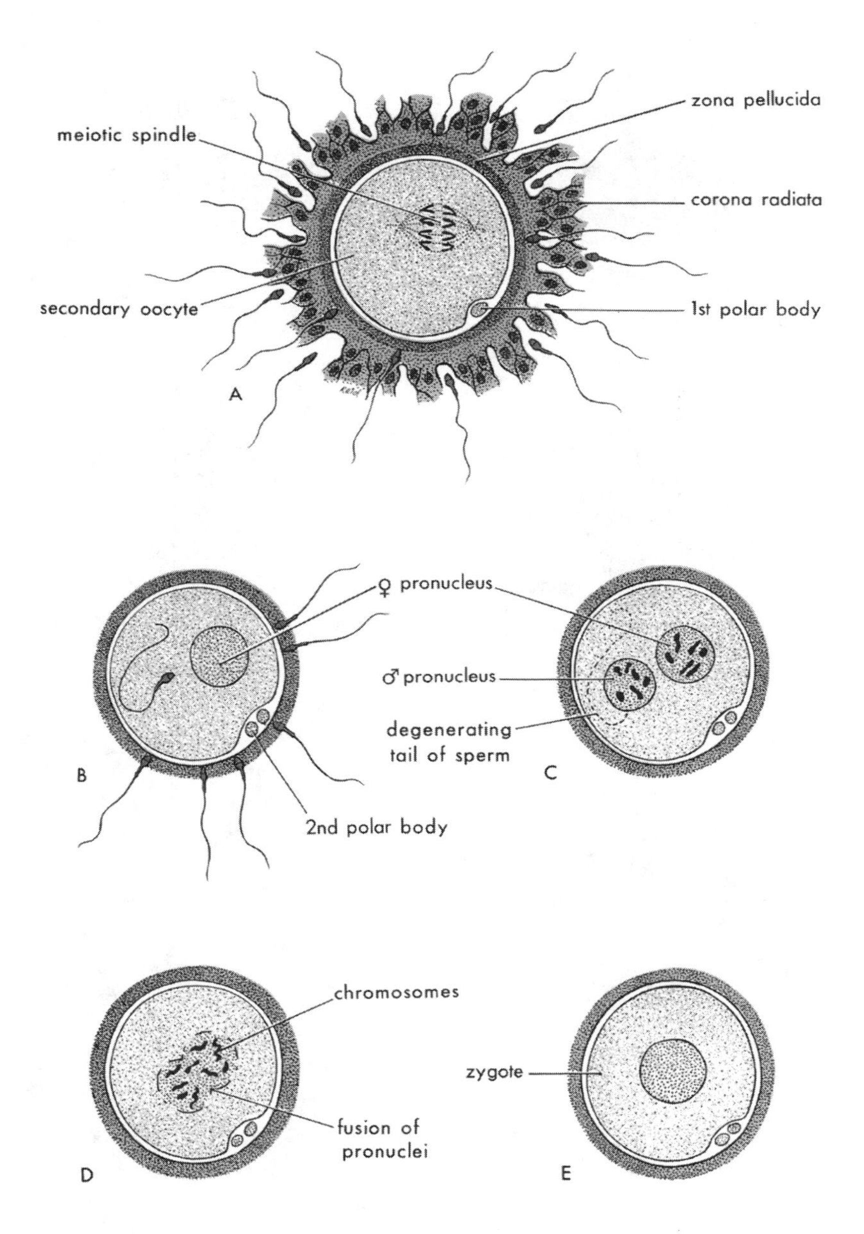

FIGURE 1
Schematic depiction of the process of fertilization, with penetration of the oolemma by a single spermatozoa and completion of the second meiotic division from the metaphase II stage (**A** and **B**). The male and female pronuclei are formed and fuse, resulting in formation of the zygote (**C**, **D**, and **E**). (From Moore, K. L., Ed., *The Developing Human,* W.B. Saunders, Philadelphia, 1988, 30. With permission.)

1.6 Implantation

Placentation is initiated by implantation of the blastocyst within the maternal endometrial wall. This attachment occurs at about the sixth day of embryonic life. Cells of the fetal trophoblast attach to the uterine mucosa by apposition and adhesion. Under the influence of progesterone and estrogen, there is a closure of the uterine lumen, which brings the blastocyst into close contact with the endometrium. Adhesion of the blastocyst trophoblast to

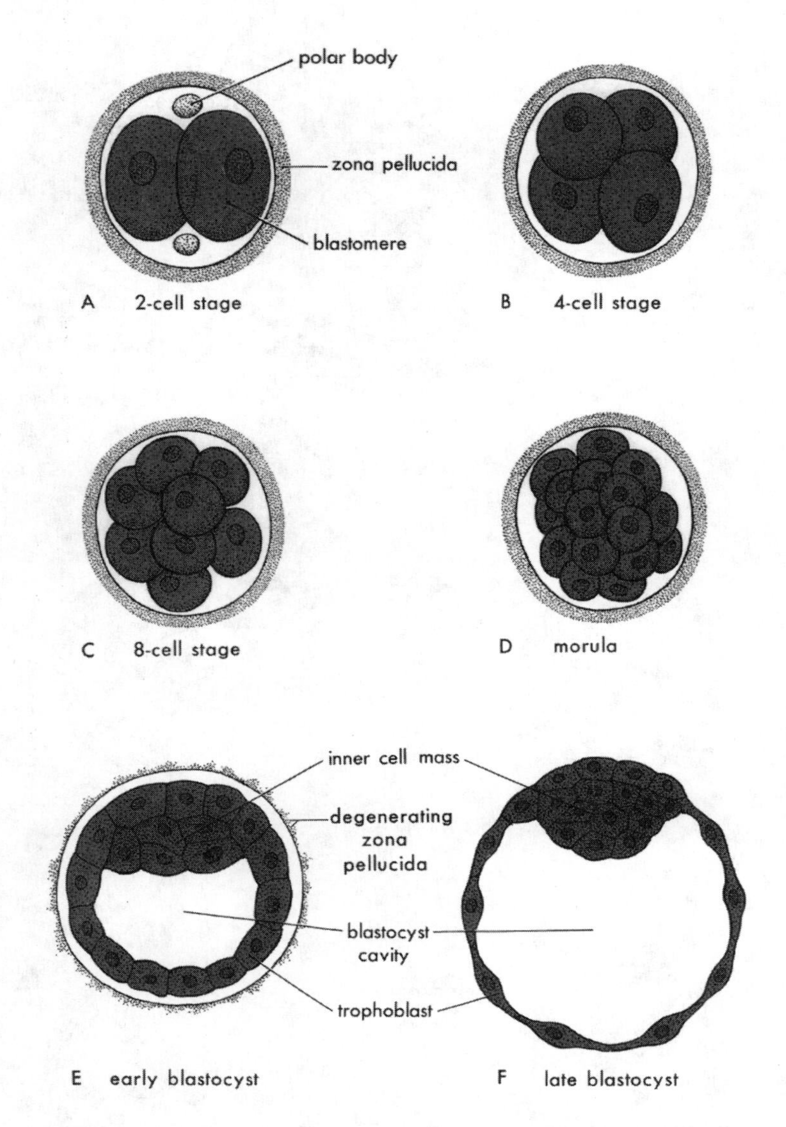

FIGURE 2
Early cleavage (**A** to **C**), morula (**D**), and blastocyst (**E** and **F**) formation. (From Moore, K. L., Ed., *The Developing Human,* W.B. Saunders, Philadelphia, 1988, 32. With permission.)

the uterine epithelium occurs with increasing apposition and involves cell-surface glycoproteins. The uterine epithelium is penetrated by syncytial growths on the trophoblast into the adjacent uterine epithelial cells; subsequently, the trophoblastic membranes share junctional complexes and desmosomes with the uterine epithelial cells. The syncytial trophoblastic processes penetrate the basal lamina of endometrial blood vessels. A group of characteristic changes in the endometrium known as decidualization occurs as a response to implantation. This decidualized endometrium is thickened due to proliferation of endometrial stromal cells, infiltration of inflammatory cells, increased vascular permeability, engorgement of blood vessels, and edema. Sustained exposure of the endometrium to both progesterone and estrogen is necessary for decidualization to occur. Blockage of progesterone action by antiprogestins or by high doses of estrogen, both of which disturb the estrogen/progesterone balance, inhibits decidualization and implantation.

1.7 Embryogenesis

The conceptus is properly called an embryo from the time the bilaminar germ appears during the second week after fertilization to about the eighth week, when most major organ and tissue development has occurred. After the bilaminar germ disc is formed, a small cavity develops among the epiblastic cells and enlarges to become the amniotic cavity. The amnion is the fluid-filled membranous sac immediately surrounding the embryo and fetus. The epiblastic cells are the primordial forms for fetal mesoderm, endoderm, and ectoderm. The yolk sac arises from cells of the cytotrophoblast that form a continuous membrane with the hypoblasts of the bilaminar germ disc and surround the initial blastocyst cavity, now called the exocelomic cavity. The trophoblast gives rise to a layer of cells loosely arranged as the extra-embryonic mesoderm around the amnion. Spaces in the syncytiotrophoblast become filled with a mixture of maternal blood from ruptured, engorged endometrial capillaries and secretions from eroded endometrial glands. The spaces in the syncytiotrophoblast fuse to form networks that are the primordial version of the intervillous spaces of the placenta. Maternal blood seeps in and out of the spaces, beginning the formation of the uteroplacental circulation.

The cytotrophoblast produces extensions called the primary chorionic villi that penetrate the syncytiotrophoblast. The extra-embryonic celom divides the extra-embryonic mesoderm into two layers. The extra-embryonic mesoderm lines the trophoblast and covers the amnion. The extra-embryonic somatic mesoderm and the trophoblast next to it are the chorion, and the extra-embryonic cavity is now the chorionic cavity.

The first indication of cranial differentiation occurs when the hypoblasts at the junction of the amnion and yolk sac form the prochordal plate. The next series of embryonic events includes gastrulation, neurulation, and formation of the major organ systems. The primitive streak appears in the bilaminar disc, and from it mesenchymal cells form the mesodermal layer between the epiblast and hypoblast, giving rise to the trilaminar embryo. The notochord develops from mesenchymal cells and moves to an area of the prochordal plate. Cells from the primitive streak migrate cranially to form the oropharyngeal membrane and the primitive cardiogenic area. The embryonic ectoderm, which is the epiblast of the bilaminar stage, surrounds the developing notochord to form the neural plate. The ectoderm of the neural plate (ultimately to form the brain and spinal cord) grows cranially until it reaches the oropharyngeal membrane. The neural plate invaginates along its central axis and forms the neural tube. The neural tube completely separates from the surface ectoderm, which gives rise to the epidermis of the skin. Some ectodermal cells of the neural tube migrate to each side of the neural tube and form irregular, flattened masses called neural crests. The neural crests give rise to the sensory ganglia of the spinal and cranial nerves.

Because elements that will later form the central nervous system appear earlier than most other systems in the developmental process, disturbances of the neurulation process may result in severe abnormalities of the brain and spinal cord. Other ectodermal thickenings, the otic placode and lens placode, develop into the inner ear and lens, respectively. A series of mesodermal blocks called somites forms around the neural tube, and these somites give rise to the axial skeleton. The intra-embryonic celom appears in the lateral mesoderm and later divides into the pericardial, pleural, and peritoneal cavities. The intermediate mesoderm forms the nephrogenic cord, which later becomes the kidneys.

Prior to the third week, embryonic nutrition occurs by diffusion of maternal blood. At that point, development of the primitive blood and blood vessels begins in the extra-embryonic mesoderm of the yolk sac. These cells differentiate into angioblasts that form cords and clusters that in turn canalize. Cells in the periphery become flattened and form the endothelial cells, whereas the inner cells give rise to the primitive blood cells. By fusion and continuous budding, the extra-embryonic vessels that have contact with maternal circulation establish contact with vessels arising from the embryo proper. The mesenchymal cells surrounding the primitive endothelial cells differentiate to form the muscular and connective tissues of the vessel wall. The primitive heart is formed from mesenchymal cells in the cardiogenic area in a manner similar to the formation of the blood vessels.

The gastrointestinal tract is formed from the endodermal germ layer. The embryo folds cephalo-caudally and laterally to incorporate the endodermal layer into the body cavity. At the cephalic end, the foregut is bounded by the buccopharyngeal membrane. The hindgut terminates at the cloacal membrane.

By the end of the embryonic period in the eighth week, tissues and organ systems have developed and the major features of the external body form have developed. The ectodermal layer has given rise to primordial forms of the central and peripheral nervous systems; the sensory epithelium of the ear, nose, and eye; and the epidermis. The ectodermal layer also gives rise to the mammary and pituitary glands and the enamel of the teeth. In addition to the kidneys, the mesoderm gives rise to cartilage, bone, and connective tissue; all muscles; the heart, blood, and lymph cells; the gonads (ovaries and testes) and the genital ducts; the serous membrane lining the body cavities (pericardial, pleural, and peritoneal); the spleen; and the cortex of the adrenal glands. The endoderm gives rise to the gastrointestinal and respiratory tracts; the tonsils; the thyroid, parathyroid, and thymus glands; the liver; the pancreas; and epithelial linings throughout the body.

Connection of the placenta to the rounded embryonic body is maintained by the umbilical cord. Blood vessels within the allantois become the umbilical vein and arteries. The allantois itself extends from the umbilicus to the urinary bladder, and as the fetus develops it involutes, leaving a thick tube called the urachus. The umbilical cord develops from the connecting stalk of allantois and the yolk sac stalk. The two are pushed together by the developing amniotic cavity as it obliterates the chorionic cavity. The embryonic period is a time of major organ and system development, and any agents that disrupt the development process at this time are potent teratogens.

1.8 Fetal Maturation

The fetal period extends from the ninth week to birth. Many systems already formed in the embryonic period continue to develop further. The fetus undergoes rapid body growth. At the beginning of the fetal period, the head makes up half the length of the fetus. The growth of the rest of the body proceeds rapidly, and relative growth of the head slows. Primary ossification centers appear in the skeleton by 12 weeks, and the limbs undergo further development. The external genitalia of males and females appear similar until the ninth week but are fully different by the twelfth week. Urine starts to form between the ninth and twelfth week. By 16 weeks, the eyes come closer together, in distinction to the lateral orientation of earlier stages. By the seventeenth week, growth slows and by the twentieth week, the fetus is covered in fine, downy lanugo hair. From 21 to 25 weeks, the fetus gains body weight, and the respiratory system rapidly develops. By 26 weeks, the eyes open, hair on the head and body is well developed, and toenails are apparent. The quantity of fat in the body increases, and subcutaneous deposition makes the skin less wrinkled. The site of erythropoiesis now shifts from the spleen to the bone marrow. At 36 weeks, the girth of the torso increases, and there is a further slowing down of the growth process. In this interval, the male fetus grows more rapidly than the female fetus, resulting in a greater weight at birth. From the thirty-eighth week onwards, the full-term male fetus should undergo a full descent of the testes into the scrotum.

1.9 Gestation

1.9.1 Placentation

The normal exchange of substances between the maternal and fetal bloodstreams depends on normal placentation. The fetal portion of the placenta develops from the chorionic sac, while the maternal portion arises from the endometrium.

By the end of the third week, the primary chorionic villi develop. They begin to branch and are surrounded by connective tissue called the secondary chorionic villi. Blood vessels differentiate from spaces within the villi, which become tertiary chorionic villi. Vessels in the chorion become connected to vessels developing in the embryo and to the heart primordia. By the beginning of the embryonic period, oxygen and nutrients in the maternal blood diffuse through the walls of the villi into fetal vessels, while carbon dioxide and waste products diffuse from the fetal capillaries into the vessels of the maternal endometrium. During this period, the villi extend over the entire amniotic cavity. By the end of the embryonic period, maternal blood enters the vessels in the villi from endometrial arteries, while blood is carried away from the villi by endometrial veins. Fetal membranes form a barrier to limit direct contact of fetal and maternal compartments. The fetal portion of the placenta is attached to the maternal part by the cytotrophoblastic shell, which is an extension of the cytotrophoblast through the syncytiotrophoblast. The villi from the cytotrophoblastic shell anchor the placenta to the endometrium. The villi from the syncytiotrophoblast are the main places of exchange between maternal and fetal blood.

By the fifteenth week, the decidua forms septae that project into the villi vessels. The surface of the septae are covered by syncytial cells such that the maternal blood does not directly contact fetal tissue. The formation of septae results in compartmentalization of the placenta into cotyledons. The thin separation between maternal and fetal blood that occurs in the human placenta is called hemochorial placentation.

Apart from its role in maternal-fetal interchange, the placenta is involved in metabolism and endocrine secretion. The placenta synthesizes glycogen, cholesterol, and fatty acids early in gestation. The placental syncytioblast secretes human chorionic gonadotropin (hCG) and human placental lactogen (hPL). hCG maintains the corpus luteum, and hPL gives the fetus priority for utilization of maternal blood glucose. The placenta also produces large quantities of progestins and estrogens.

Substances normally transferred across the placenta include gases, nutrients, certain hormones (e.g., thyroxine), electrolytes, antibodies, and waste products. Many drugs and several infectious agents can also cross the placenta. Maternal steroid hormones generally do not cross the placenta to the fetus. When natural or synthetic steroid hormones do cross the placenta, there is the potential for developmental toxicity.

1.9.2 Maintenance

Maintenance of pregnancy depends on the presence of a well-functioning placenta (see above), intact fetal membranes, and quiescence of the maternal myometrium. The fetal membranes separate the fetus from the mother but also serve as the immediate environment of the fetus without contributing to the fetal body parts. The amniotic fluid is derived from secretion by cells lining the amniotic cavity and fetal urine. Fetal swallowing of amniotic fluid is normal, and the fluid is absorbed into the fetal circulation. Several conditions can be diagnosed before parturition using amniocentesis, and fetal sex can be determined by characterizing the desquamated cells. The yolk sac is involved only in early preplacental transfer of substances between the fetus and mother; otherwise, the yolk sac is important as the site of development of the primordial germ cells.

Hormonal influences appear critical for maintenance of the quiescent state of the maternal myometrium. Certainly progesterone plays a central role, but the applicability of some elegant animal studies to humans remains uncertain.

Preterm labor is of great concern in human obstetrical management but is exceedingly difficult to control. It appears that, excluding structural abnormalities of the uterus and cervix, most preterm labor is microbiologic in origin; thus, local genital tract phagocytic and immune function should be seen as an essential component in the normal maintenance of pregnancy.

1.10 Parturition

During parturition, the fetus, placenta, and fetal membranes are expelled from the mother in a process involving both maternal and fetal factors. While the timing of human parturition is known to occur at 38 ± 1 weeks after ovulation and fertilization, the process by which parturition is normally initiated is not well understood. It is known that oxytocin released from the maternal neurohypophysis evokes uterine contractions needed for the parturition process. The initiation of maternal contractions also may depend on steroids produced from the fetal adrenal cortex. In addition, intrauterine factors that decrease progesterone synthesis in the placenta may be involved. Around the time of parturition, progesterone levels drop and estrogen levels arise, possibly antagonizing the suppressive action of progesterone on the uterus and stimulating the contractile action of oxytocin. Local uterine production of prostaglandins maintains myometrial contractions throughout parturition.

1.11 Lactation

The mammary glands synthesize milk necessary for the nourishment and development of neonates. The mammary gland consists externally of the nipple, where a system of ducts starting from the alveoli open to the skin surface. The alveoli do not fully develop until pregnancy and lactation. Cortisol, insulin, and placental lactogen contribute to alveolar development, but estrogen and progesterone are most important. The cells that line the alveoli synthesize milk, and they are surrounded by a layer of myoepithelial cells that are involved in the ejection of milk.

Until puberty, the mammary glands of males and females are essentially the same. Estrogen stimulation at puberty results in growth of the duct system, and there is increased accumulation of fat cells in females resulting in external development of breasts. Progesterone further stimulates development of the ducts. During pregnancy, estrogen and progesterone cause the mammary gland ductal system to undergo further branching, and the alveolar cells undergo proliferation. In late pregnancy, milk secretion and letdown begin. The hormones involved in maintaining lactation include adrenocorticotropic hormone (ACTH), thyroxine, thyrotropin-releasing hormone, and growth hormone. During lactation, the myoepithelial cells contract in response to suckling-induced oxytocin secretion from the posterior pituitary gland.

1.12 Maturation (Postnatal)

After birth, the human growth processes started at fertilization continue. Organs and tissue systems increase in size to different extents. The neonate has a period of rapid growth following birth, with continuation of the fat deposition that began prior to parturition. This period lasts from about 4 to 6 months. By 1 year, human infants are three times their birth weight, but after this point the rate of growth decreases. A rapid increase in body size is not seen again until puberty. Ultimate adult human size appears to be largely determined by the length of the neonate after the initial growth spurt following birth.

The extended period of human development is related to the general trend in nonhuman primates for delayed development, which produces a larger brain and lower fecundity compared to other species. Sexual dimorphism in human size begins at birth and becomes more pronounced at adolescence. Males have a greater growth rate and a longer growth period due to the delayed closure of epiphysial growth plates. Human body size is thought to be genetically programmed, but a wide range of external and internal factors can modify ultimate adult size.

The adolescent stage begins at puberty, when the capacity to reproduce is acquired. Puberty is determined by several related events, some of which are not well understood. Puberty is hard to pinpoint in primates: it does not occur at a particular time but instead is a series of processes that occur, leading to sexual maturity. Activation of the gonadotropin-releasing hormone (GnRH) pulse generator is the main process leading to puberty.

2 FEMALE REPRODUCTIVE PHYSIOLOGY

2.1 Functional Morphology of the Ovary

The ovary is the source of the female reproductive gamete and hormones that regulate the female reproductive cycle. The adult ovary is made up of three regions: the cortex, containing ovarian follicles; the medulla; and the hilum (the ovarian attachment area). During each menstrual cycle, a primordial, inactive follicle undergoes complete growth to become a Graafian follicle, from which an oocyte is released. Throughout the female reproductive life, maturing follicles at different stages undergo atresia. The medulla consists of the ovarian stroma, which is mainly connective tissue, and the sex steroid-producing interstitial cells.

2.1.1 Folliculogenesis

At birth, the oocytes are enclosed within primary follicles, and before sexual maturity they are arrested in the diplotene phase of the first meiotic division. The follicle — consisting of the oocyte and one thin layer of granulosa cells — begins to grow (Figure 3). The oocyte enlarges, and the granulosa cells become cuboidal and thicken to three or more layers. This is now called the secondary follicle. While the oocyte remains constant in size, the granulosa cells undergo rapid division, and a fluid-filled antrum forms. Such a Graafian or antral follicle is five to six times larger than the secondary follicle. The innermost layer of granulosa cells surrounding the oocyte (the corona radiata) becomes columnar, and these cells form specialized gap junctions with the oocyte plasma membrane that are important in regulation of oocyte development. The oocyte secretes a layer of glycoproteins to form the zona pellucida, which separates the oocyte from the surrounding corona radiata. The connective tissue delineating the follicle from the surrounding ovarian stroma is the theca.

After sexual maturity, folliculogenesis leads to a midcycle LH surge that induces the single follicle destined for ovulation to undergo events preparing it for possible fertilization. First, meiosis is resumed, with completion of the first meiotic division and formation of one daughter cell and a second smaller cell with little cellular content, the polar body. The daughter cell enters into the second meiotic division and arrests in the second metaphase until fertilization. The follicle becomes highly vascularized and a protrusion in the follicular wall (macula pellucida) indicates the location of the rupture, resulting in the release of the oocyte with its surrounding granulosa cells. At ovulation, the oocyte (with its surrounding granulosa cells) is extruded onto the ovarian surface, where the cumulus mass can be retrieved by the fimbriated end of the fallopian tube. After ovulation, the walls of the follicular cavity in the ovary develop into the corpus luteum. If fertilization occurs, the lifespan of the corpus luteum is maintained; otherwise, it disintegrates during the 12 to 15 days of the luteal phase of the menstrual cycle.

In addition to triggering ovulation, the midcycle LH surge induces luteinization of the theca and granulosa cells of the follicle wall. Normal luteal function depends upon both normal folliculogenesis and successful follicular rupture. Vascularization of the follicle wall

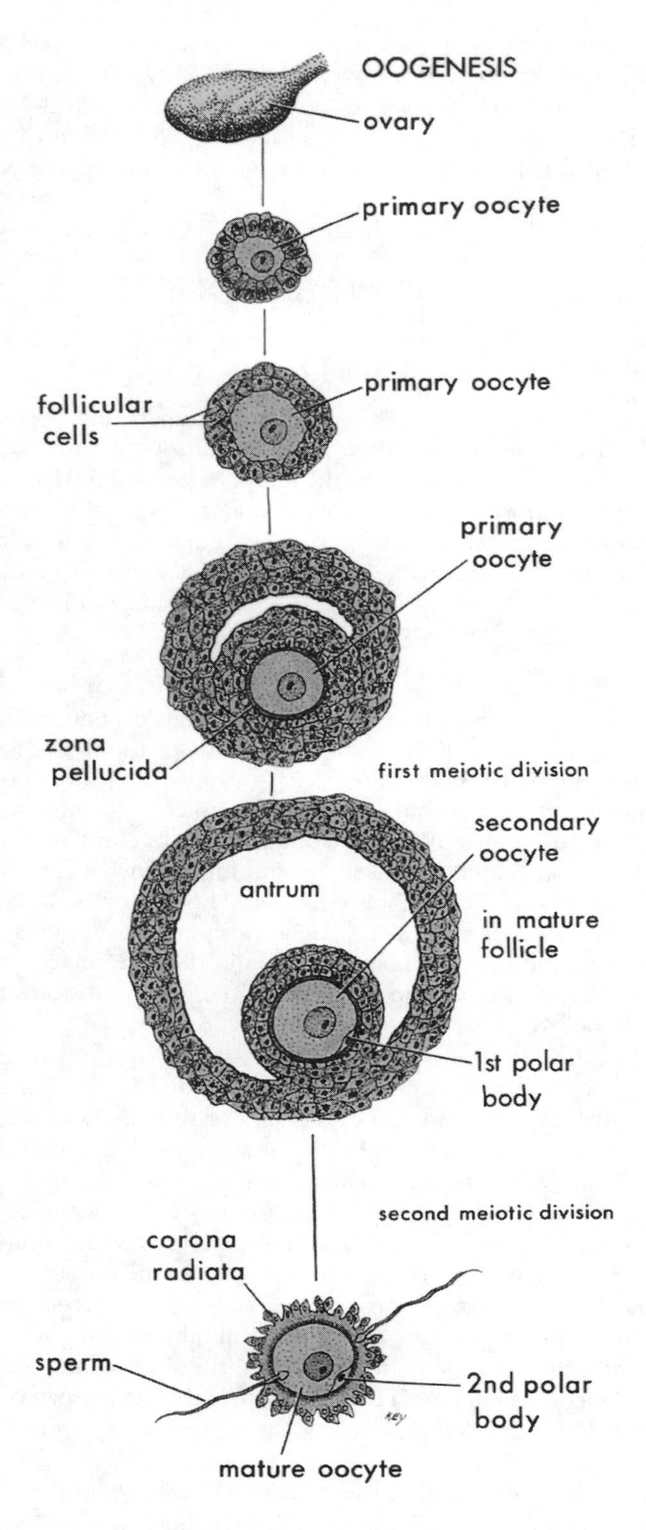

FIGURE 3
Female gamete maturation. (From Moore, K. L., Ed., *The Developing Human,* W.B. Saunders, Philadelphia, 1988, 16. With permission.)

that occurs with ovulation provides the luteinized theca and granulosa cells with low density lipoprotein (LDL) cholesterol from the circulation, which is used to synthesize progesterone.

Throughout postpubertal life, cohorts of follicles continue to be recruited. Some of these will mature to become Graafian follicles, with subsequent ovulation; however, most follicles undergo atresia. Although approximately one million follicles are present in the ovaries at birth, essentially all are depleted by about the age of 50 years, resulting in menopause. Follicular atresia accounts for the vast preponderance of reduction in follicular numbers with aging.

2.1.2 *Intra-Ovarian Signaling*

The principal extra-ovarian regulators controlling intra-ovarian processes are the pituitary gonadotropins, LH and FSH. Within the ovary, reproductive steroid hormones (androgens, estrogens, and progestins) appear to be important *in situ* regulators. The presence of receptors for all three gonadal steroids within the ovary suggests that intra-ovarian signaling occurs via classical steroid hormone-receptor mechanisms.

Apart from being the substrate for estrogen production, androgens exert actions on the granulosa cells. They promote FSH-induced granulosa cell aromatase activity and progestin synthesis.[13,14] Androgens also promote follicular atresia in the absence of gonadotropins and generally act to oppose estrogen action. The thecal cells and the interstitial cells of the ovarian stroma produce androgens under the influence of FSH. Circulating LDL cholesterol is converted to pregnelonone, then progesterone. The 17-alpha hydroxylase/17,20 desmolase activity in the theca-interstitial cells is responsible for converting pregnenolone and progesterone to dehydroepiandrosterone and androstenedione. The androgen produced is mainly androstenedione, which can go directly into the circulation, be converted to testosterone, or be aromatized to estrogens in the granulosa cells by FSH-induced aromatase activity.

The granulosa cells of the ovarian follicles and the corpus luteum are the progesterone-synthesizing cells. The relatively avascular, preovulatory follicle produces little progesterone. Following ovulation, perhaps in association with an increased supply of LDL cholesterol from the circulation, progesterone production by the luteinized granulosa cells drastically increases.

Estrogens promote cellular division in the granulosa cells, prevent follicular atresia, enhance antrum formation, and inhibit ovarian androgen production. Because regulation of all ovarian processes cannot be explained adequately by gonadal steroids alone, it is likely that other intra-ovarian regulators exist. Such putative intra-ovarian regulators have to meet the following criteria: (1) There is evidence that the regulator is produced in the ovaries, (2) receptors for the regulator are present in the ovaries, and (3) there is demonstrated biological action within the ovary. By these criteria, insulin-like growth factor-1 (IGF-1) appears to be an intra-ovarian regulator. It is produced by the ovaries, and IGF-1 receptors are present in the ovaries. IGF-1 acts on both granulosa and theca cells to influence steroidogenesis at both sites.[15,16] Epidermal growth factor (EGF) and transforming growth factor-alpha (TGF-alpha) share the same receptor and may have autocrine and paracrine activities in the ovaries. Epidermal growth factor decreases LH/hCG receptor induction mediated by FSH action.[17] EGF also has mitogenic effects on granulosa cells and may be an important signal for granulosa cell proliferation. TGF-alpha is thought to act in ways similar to EGF. Interleukin-1 is a cytokine produced by macrophages that has been found to suppress the luteinization of granulosa cells in the ovary. Other putative intra-ovarian regulators are transforming growth factor-beta 1, which acts to increase steroidogenesis in granulosa cells[18] and decreases steroidogenesis in theca cells[18] and Leydig cells.[19]

Acidic and basic fibroblast growth factor (bFGF) are related and act through the same receptor. bFGF may act in early embryonic life to instigate differentiation of ectodermal cells to mesoderm. bFGF also may act to induce early granulosa cell growth towards becoming the destined follicle for ovulation before gonadotropin dependence develops.[20] It also may mediate angiogenesis in follicular development.[21]

Other possible ovarian regulators include tumor necrosis factor-alpha, catecholaminergic input, luteinization inhibitor, gonadotropin-binding inhibitors, oocyte maturation inhibitor, and the ovarian renin-angiotensin system. Ovarian renin-angiotensin has been detected in follicular fluid.[22] Receptor sites for angiotensin II, the main active peptide of the renin-angiotensin system, have been found in the ovaries.[22] Ovarian renin-angiotensin also may have an autocrine role in angiogenesis, steroidogenesis, and oocyte maturation. Inhibin is produced by the ovarian follicle and corpus luteum[23] and appears to have some autocrine action on theca and granulosa cells involving steroidogenesis.[24]

Two other putative intra-ovarian regulators are anti-Müllerian hormone and oocyte maturation inhibitor. Anti-Müllerian hormone is produced by the Sertoli cells of the fetal testis and acts to regress the Müllerian duct in male fetuses; thus, it is the first hormonal influence leading towards male gonadal differentiation. Anti-Müllerian hormone also is produced by granulosa cells. By the time the primordial germ cells differentiate to primary follicles in female fetuses, the Müllerian duct is resistant to anti-Müllerian hormone. Its putative role as an intra-ovarian signal is believed to be that of regulating oocyte meiosis arrest, because it is found mostly in the corona radiata in close contact with the oocyte. Oocyte maturation inhibitor is thought to arrest oocytes in the dictyate stage in the first meiotic arrest and in metaphase in the second meiotic arrest.[25]

2.2 Neuroendocrine Regulation of Ovarian Function and Reproductive Cyclicity

2.2.1 *Central Nervous System and Pituitary Processes*

Beginning in fetal life, GnRH release from the hypothalamus influences LH and FSH production by the pituitary gland. The mean peak plasma levels of LH and FSH during gestation are higher in female than in male fetuses. The pulsatile release of LH and FSH secretion is controlled by neurons located in the arcuate nucleus of the hypothalamus which secrete GnRH into the hypothalamic-hypophysial portal system. Among the hypothalamic neurotransmitters important in this function are dopamine and serotonin. Estradiol acts through positive and negative feedback mechanisms to control GnRH release.

2.2.2 *Patterns of Ovarian Response*

Follicular growth, ovulation, and maintenance of pregnancy are controlled by the balanced secretion of FSH and LH. Both hormones act on the ovarian target cells by means of cell surface receptors. In addition, FSH stimulates the aromatization of theca-derived androgens to estrogens by granulosa cells. The main endocrine action of LH is to stimulate progesterone production by granulosa cells. LH stimulation also induces follicular theca cells and stromal interstitial cells to produce androgens, which serve as precursors for FSH-dependent estrogen production by granulosa cells. LH secretion occurs in prepubertal females in a basal fashion with irregular pulsatile occurrences. In cycling females, the ovarian hormonal profile is linked to the menstrual cycle, which is controlled by LH and FSH levels. There is an LH surge before ovulation that is temporally linked to the preceding rise of estradiol. After ovulation, progesterone production accelerates, and serum levels of estrogen increase to a lesser degree until luteolysis, when progesterone and estrogen return to basal levels.

2.2.3 *Effects on Reproductive Tract and Breast*

Pulsatile secretion of gonadotropins starts at puberty, and the subsequent increase in estrogen levels stimulates growth of the rudimentary mammary gland. When progesterone production starts as part of an ovulatory cycle, it stimulates the alveolar buds within the

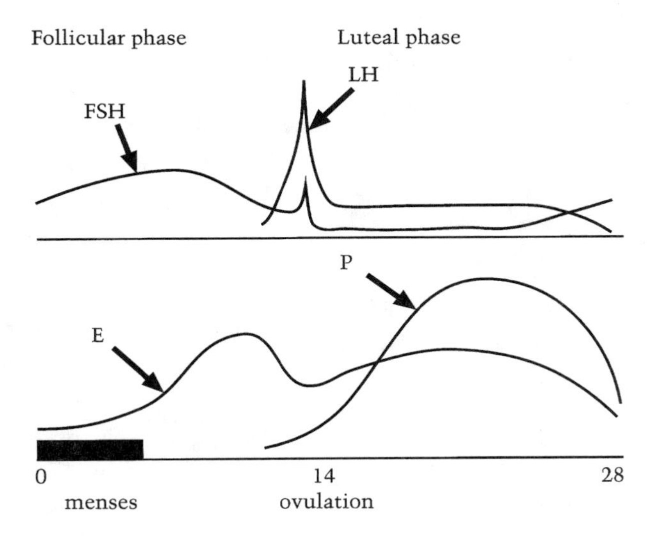

FIGURE 4
Blood levels of FSH, LH, estrogen, and progesterone during the menstrual cycle. Estrogen levels rise before the midcycle LH surge. Ovulation is followed by a rise in progesterone levels. (From Olsen, E. A., Ed., *Disorders of Hair Growth, Diagnosis and Treatment,* McGraw-Hill, New York, 1994, 341. With permission.)

mammary lobules. After puberty, the changing estrogen and progesterone levels in the cycling female affect breast tissue development, with maximal activity in the luteal phase. Since the levels of estrogen and progesterone in the cycling female are under the influence of the hypothalamo-pituitary axis, the neuroendocrine system indirectly influences breast development (Figure 4).

During pregnancy, alveolar cells proliferate under the influence of progesterone, estrogen, and prolactin. Growth hormone, ACTH, and thyroxine are all involved in maintaining lactation, and their production by the pituitary is under the control of peptide hormones from the hypothalamus. The signal for the breast to start producing milk after delivery is probably the rapid decline in serum progesterone and estrogen concentrations that occurs after delivery of the placenta.

3 MALE REPRODUCTIVE PHYSIOLOGY

3.1 Functional Morphology of the Testis

The testes — the site of spermatozoa production and initial transport of spermatozoa for semen production — are paired, encapsulated organs made up of seminiferous tubules and interstitial tissue. They descend from the abdomen and reach the scrotum shortly after birth; ultimately, this process is necessary for spermatogenesis to occur. The embryonic genital ridge is next to the mesonephros. When the mesonephros degenerates, it leaves a band of mesenchyme that extends into the genital ridge. This band of mesenchyme, called the gubernaculum, anchors the fetal testis to the inguinal region. During the third month of gestation, herniation of the fetal celomic cavity occurs through the ventral abdominal cavity and results in formation of the inguinal canal. The descent of the testes from the abdominal cavity through the inguinal canal and into the scrotum starts in the sixth month of fetal life. This process is controlled by anti-Müllerian hormone and androgens. The spermatogonia are found in the basal compartment of the testis, while primary spermatocytes reside in the adluminal compartment (Figure 5).

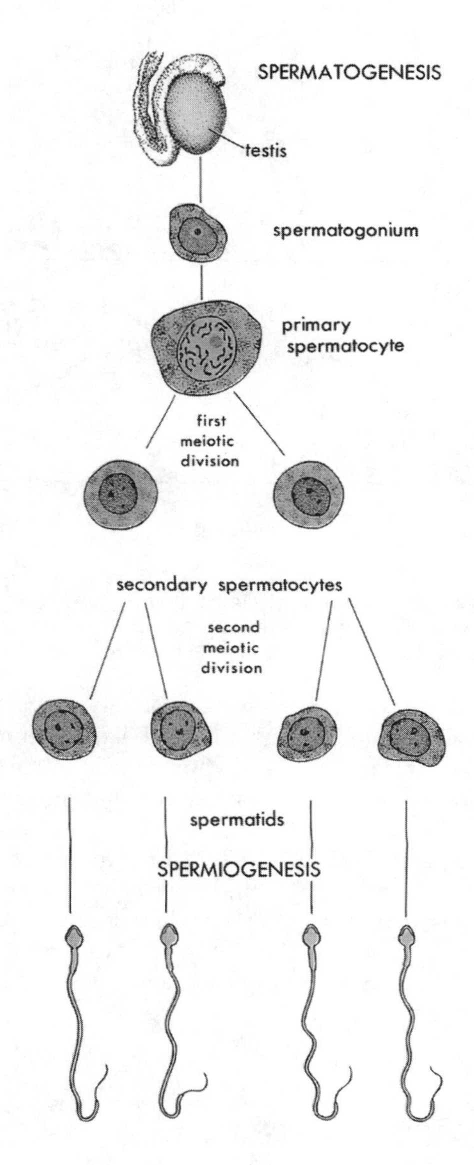

FIGURE 5
Male gamete maturation. (From Moore, K. L., Ed., *The Developing Human,* W.B. Saunders, Philadelphia, 1988, 16. With permission.)

The capsule of the testis is called the tunica albuginea. It is a tri-layered fibrous covering, the inner layers of which are continuous with the interstitial tissue. The tunica albuginea consists of fibroblasts, collagen, and smooth muscle. The capsule maintains pressure on the testis and contracts, both of which are important in spermatic transport.

The seminiferous tubules are about 350 μm in diameter and are arranged in lobules each containing one to four tubules. They are the site of spermatozoa formation. There are muscle cells in the walls of the seminiferous tubules that also aid in spermatic transport. They empty into a network of ducts called the rete testes. The sperm are then transported to the epididymis, which is made up of the caput, corpus, and cauda regions. The interstitial tissue fills the spaces between the seminiferous tubules and contains the blood and lymph vessels and nerves of the testes. It is the site of androgen synthesis via the Leydig cells in the interstitium. The interstitial tissue also contains inflammatory cells.

Sertoli cells and spermatogonia in different stages of maturation are the main components of the seminiferous epithelium. The Sertoli cells are tall, columnar cells with a variety of three-dimensional shapes that accommodate the structural transformations and mobilization of spermatogonia from the basal surface to the surface of the seminiferous epithelium. Tight junctional complexes between Sertoli cells in the basal region of the seminiferous tubules contribute to the blood-testis barrier. The germ cells in the basal region are spermatogonia and early primary spermatocytes. The adluminal area has germ cells in more advanced stages of maturation and lacks the tight junctions between Sertoli cells.

3.1.1 Spermatogenesis

Spermatogenesis is the process by which precursor cells form mature haploid spermatozoa within the seminiferous tubules. Spermatogonia remain quiescent, undergoing mitosis until puberty. At that point, under the influence of testosterone, they undergo differentiation, giving rise to primary spermatocytes that enter the first meiotic division and become secondary spermatocytes. Secondary spermatocytes undergo the second meiotic division and give rise to spermatids, which then differentiate into spermatozoa. Each spermatogonium gives rise to 64 spermatozoa. After puberty, the number of spermatogonia starting the process of becoming spermatozoa daily is about 3 million.[26]

The spermatid is transformed into spermatozoa by a process called spermiogenesis. The nucleus of the spermatid comes to occupy an eccentric position, with an acrosomal cap separating it from the cranial pole of the spermatozoa. The acrosome is formed from the Golgi complex and is closely attached to the spermatic cell membrane. The flagellum is formed from centrioles at the caudal pole of the nucleus. The flagellum portions are divided into the middle, principal, and end pieces. Mitochondria aggregate and form a spiral sheath around the middle piece of the spermatic flagellum. After the flagellum develops and there is reorganization of the cytoplasm and cellular organelles, the nucleus becomes pyknotic. The cytoplasm of the spermatid moves caudally and spermatid cytoplasm is shed in a process termed spermiation. Some cytoplasm remains which contains Golgi complex and endoplasmic reticulum.

Spermatogenesis takes approximately 70 days; the spermatozoa that leave the testes are immature and incapable of fertilization. Hormonal control of spermatogenesis involves LH, FSH, and hCG. FSH acts on the seminiferous tubules and LH acts indirectly by influencing Leydig cell synthesis of testosterone. Testosterone acts on Sertoli cells to activate the process of spermatogenesis.

3.1.2 Intratesticular Signaling

Although the testes also produce androstenedione and dehydroepiandrosterone, the main autocrine hormones in the testes are androgens, the most important of which is testosterone. This testosterone is peripherally converted to the more potent dihydrotestosterone. Testicular androgens produced by Leydig cells act locally to control spermatogenesis and are responsible for early fetal development of the testes. Leydig cells store cholesterol in lipid droplets in their cytoplasm, and these are in turn the source of the cholesterol used to form androgens. To form testosterone, cholesterol is converted to pregnenolone by cleavage of the C_{27} side-chain in the mitochondria under the enzymatic action of cytochrome P-450 enzymes. The other conversions leading to androgen formation take place in the microsomes of the Leydig cells after transfer of pregnenolone to that site. Pregnenolone can be converted to progesterone or 17-alpha hydroxy pregnenolone depending on whether the delta 4 or delta 5 pathway is used. Progesterone is converted to 17-alpha hydroxyprogesterone, then to androstenedione, and then to testosterone. The 17-alpha hydroxy pregnenolone is converted to dehydroepiandrosterone.

Estrogens are produced by Leydig cells and appear to inhibit spermatogenesis. In adult male rats, administration of estrogens suppresses spermatogenesis. Although the effect can be attributed to estrogen-mediated negative feedback via suppression of LH and consequent decrease in testosterone secretion by Leydig cells, some evidence suggests an intratesticular role for estradiol. Specifically, two studies showed that administration of GnRH agonists to rats inhibited spermatogenesis, but this effect was independent of testicular testosterone levels.[27,28] Some of the factors involved in intra-ovarian signaling (see above) also may be involved in intratesticular signaling; for example, IGF-1 may regulate spermatogenesis.[29]

3.2 Neuroendocrine Regulation of Testicular Function

Since LH, FSH, and hCG are involved in spermatogenesis, the control of their pituitary release by the hypothalamus is important in maintaining testicular function. Inhibin produced by Sertoli cells is the major factor in the feedback inhibition of FSH. Testosterone exerts a negative feedback on the GnRH release from the hypothalamus in the regulation of LH secretion. Males release LH from the pituitary in a pulsatile manner, and this induces increases in circulating testosterone. FSH is released steadily; therefore, there may be a FSH release mechanism separate from that of LH in males. Neonatal exposure of the hypothalamus to androgen is necessary for the sexual differentiation of the LH secretory mechanism to respond to either androgen or estrogen.

4 MECHANISMS OF TOXICITY AND METHODS FOR ITS DETECTION AND ASSESSMENT OF RISK

4.1 Overview

4.1.1 Human Studies

For assessment of risk to the human reproductive system, the most appropriate data would be derived from studies conducted in humans. Unfortunately, this is commonly impossible or unethical. Assessments thus derive from occasional clinical reports, epidemiologic studies of human populations, or relevant animal studies. For most agents, data are limited in both type and quantity.

Disorders of reproduction in humans include many outcomes, such as impotence, menstrual disorders, spontaneous abortion, low birth weight, birth defects, premature reproductive senescence, and infertility. Infertility in a couple, which is defined clinically as the failure to conceive after 1 year of unprotected intercourse, is often difficult to assess. If fertility of a couple is decreased, it is often difficult to discern which partner has reduced reproductive capability and whether any environmental factor has been causally involved. Since national surveys about infertility in the U.S. show a high incidence of infertility in the population (an estimated 4.9 million women ages 15 to 44 years; 8.4%),[30] any study that detects an apparent association of decreased fertility with exposure to a particular agent must take careful measures to control for other relevant (clinical) infertility factors.

4.1.2 Animal Studies

In the absence of adequate human data, data from experimental animal studies are used to estimate the risk of reproductive effects in humans. The extrapolation of data from experimental animal studies to humans usually involves five basic assumptions: (1) An agent that produces an adverse reproductive effect in experimental animal studies is assumed to pose a potential reproductive hazard to humans; (2) it is assumed that the most sensitive species is the most appropriate model for humans, unless pharmacokinetic data indicate otherwise; (3) in the absence of specific information to the contrary, it is assumed that a chemical that acts as a reproductive toxicant in one sex may act similarly in the other sex; (4) a threshold is assumed for the dose-response curve for reproductive toxicity (although growing evidence suggests that non-threshold models may be more appropriate); and (5) in both sexes — but especially in the female — the reproductive life cycle may be divided into physiologically distinct phases (fetal, prepubertal, cycling adult, pregnant, lactating, and postreproductive).

4.2 Pubertal Timing and Events

Toxicants that affect feedback mechanisms or gonadotrophs can affect development and timing of puberty. In both sexes, the first signal of puberty is activation of the hypothalamic-pituitary axis, which is initially apparent as nocturnal rises in LH secretion. The development of puberty in males seems to be determined by the exposure of the fetal brain to testosterone, while in females development of the capacity to attain puberty does not occur until sometime in the prepubertal period.

Assessing problems of puberty requires a detailed morphological, physiological, and psychosocial assessment. Since the chronology and onset of puberty vary individually, it is difficult to determine when an actual problem exists. In female rats and mice, the age at vaginal opening is the most commonly measured marker of puberty. This event results from increases in the levels of estradiol in the blood. The ages and weights of females at the first cornified (estrus) vaginal smear, the first diestrus smear, and the onset of vaginal cycles also have been used as endpoints for onset of puberty. In males, preputial separation or appearance of sperm in expressed urine or ejaculates can act as markers of puberty. Body weight at puberty may serve to separate specific delays in puberty from general delays in development.

Puberty can be accelerated or delayed by exogenous agents, and both types of effects may be adverse. For example, an acceleration of vaginal opening may be associated with a delay in the onset of cyclicity, infertility, and with accelerated reproductive aging.[31] Delays in pubertal development in rodents are usually related to delayed maturation or inhibition of function of the hypothalamic-pituitary axis. Adverse effects for males include delay or failure of testis descent, as well as delays in age at preputial separation or appearance of sperm in expressed urine or ejaculates.

4.3 Functional Morphology

4.3.1 Male Organ Weights

Male reproductive organs for which weights may be useful for reproductive risk assessment include the testes, epididymis, pituitary gland, seminal vesicles, and prostate. Organ weight data may be presented as absolute weights or relative weights (i.e., ratios of organ weight to body weight). Changes in testis weights in particular are useful indications of gonadal injury; however, substantial testicular injury can be caused by some agents before measurable changes in weight occur.[32,33] Prostatic and seminal vesicular weights are androgen dependent and may reflect changes in the animal's endocrine status or testicular function. Assessment of pituitary gland weights also can be useful; however, the pituitary contains cell types that are responsible for the regulation of a variety of nonreproductive physiologic functions. Thus, changes in pituitary weight may not necessarily reflect reproductive impairment. If weight changes are observed, gonadotroph-specific histopathologic evaluations may be useful in identifying the affected cell types. This information may then be used to judge whether the observed effect on the pituitary is related to reproductive system function and therefore should be viewed as an adverse reproductive effect.

4.3.2 Histopathologic Evaluations of Male Tissues

Histopathologic evaluations of male reproductive organs from exposed animals are a mainstay in assessing male reproductive toxicity. Such evaluations are a relatively sensitive indicator of damage, may show effects over a short course of exposure, and often provide useful information on the site of toxicity. Many histopathologic evaluations of the testis only detect lesions if the germinal epithelium is severely depleted, degenerating cells are obvious, or sloughed cells are present in the tubule lumen. On the other hand, staging techniques permit qualitative or quantitative assessment of testicular tissues that can identify less obvious lesions, which may accompany exposures to lower doses or less toxic agents.[34,35]

4.3.3 Female Organ Weights and Histopathologic Assessment

4.3.3.1 Ovary

Significant increases or decreases in ovarian weight compared with controls should be considered an indication of female reproductive toxicity. Oocyte and follicle depletion, persistent polycystic ovaries, inhibition of corpus luteum formation, luteal cyst development, reproductive aging, and altered hypothalamic-pituitary function may all be associated with changes in ovarian weight. However, not all adverse histologic alterations in the ovary are associated with changes in ovarian weight.

Histologic evaluation of the ovary includes assessment of follicular, luteal, interstitial, and stromal components.[36,37] Methods are available to quantify the number of follicles and their stages of maturation.[38] These methods are especially useful for quantitative approaches for determining increased follicular atresia.

Adverse ovarian effects of an agent can be detected by increased or decreased ovarian weight, increased incidence of follicular atresia, decreased number of primary follicles, decreased number or lifespan of corpora lutea, or evidence of abnormal folliculogenesis or luteinization, including cystic follicles, luteinized follicles, or failure of ovulation.

4.3.3.2 Uterus

An alteration in uterine weight may be considered an indication of female reproductive organ toxicity. Compounds may decrease uterine weight by inhibiting reproductive cyclicity or may increase uterine weight by having a direct estrogenic effect. Increased uterine weight in prepubertal or ovariectomized rodents is a classical parameter for comparing relative potencies of estrogenic compounds.[39] Additionally, rates of pre- and postimplantation loss can be determined by counting implantation sites and corpora lutea in pregnant or postpartum animals.

Histologic analysis of the uterus may show normal patterns or disturbance of normal cycle-related changes. The uterine endometrium in particular is sensitive to agents that mimic or antagonize estradiol or progesterone.[40]

Effects on the uterus that may be considered adverse include significant dose-related weight alterations, as well as gross anatomic or histologic abnormalities, including infantile or malformed uterus or cervix, endometrial hyperplasia, hypoplasia, or aplasia or decreased number of implantation sites.

4.3.3.3 Oviducts

Typically, the oviducts are not weighed or examined histologically in tests for reproductive toxicity. Hypoplasia of the tubes or altered proliferation of the oviductal epithelium would represent an adverse effect.

4.3.3.4 *Vagina and External Genitalia*

Although vaginal weight changes occur during the normal estrous cycle, cytologic changes of the vaginal epithelium as shown in serial vaginal smears are generally much more valuable in assessing toxicity. The vaginal smear pattern is a useful technique to detect disruption of normal reproductive cyclicity in terms of both neuroendocrine patterns and sexual behavior. Adverse effects that may be seen in the vagina and external genitalia include increases or decreases in weight, altered patterns of vaginal smear cytology, malformed vulva, changes in anogenital distance, vaginal hypoplasia, aplasia or malformations, or altered timing of vaginal opening.

4.3.3.5 *Pituitary*

As in the male, alterations in weight of the pituitary gland may be considered an adverse effect. Pituitary weight normally increases during pregnancy and lactation, and pituitary size may increase with estrogen treatment, especially if accompanied by hyperprolactinemia and constant vaginal estrus. Significant histopathologic damage to the pituitary should be considered an adverse effect, but to be relevant to reproduction the involved cells usually should be the gonadotrophs or lactotrophs.

4.4 Hormone Levels and Patterns

A normal hormonal mileu is important for maintaining reproductive function in males and females. Any substance that disrupts the delicate hormonal balances necessary for normal function can produce adverse effects. Such effects may be due to hormonal agonist or antagonist actions. These effects can be extremely potent, as illustrated by the use of oral contraceptives containing estrogen and progestin analogs to disrupt intentionally normal hormonal patterns and to prevent pregnancy.

Significant alterations in circulating levels of gonadal steroids or gonadotropins may be indicative of central nervous system, pituitary, or gonadal injury. Changes in the levels of reproductive hormones may be helpful in identifying sites or mechanisms of toxicant action.

4.4.1 *Endocrine Evaluations in Males*

Measurement of the reproductive hormones in males is useful in assessing potential reproductive toxicants.[41-43] Those most useful are LH, FSH, and testosterone and occasionally prolactin, inhibin, and sex hormone-binding globulin.

Toxic agents can alter endocrine function by affecting any part of the central nervous system-pituitary-gonadal axis. If a compound affects the central nervous system or pituitary, then serum LH and FSH may be decreased, leading to decreased testosterone concentrations. On the other hand, severe interference with Sertoli cell function or spermatogenesis would be expected to elevate serum FSH levels due to diminished secretion of inhibin.

4.4.2 *Endocrine Evaluations in Females*

Female reproduction is controlled by complex endocrine signaling involving the central nervous system, pituitary, ovaries, and targets of gonadal steroid action, such as the reproductive tract and the mammary glands. Because it is difficult to detect important aspects of female reproduction, such as ovulation and fetal well-being, assessment of endocrine status may be one of the few options available.

Events of the estrous or menstrual cycle are interrelated. Thus, compromised folliculogenesis may include a diminished estradiol rise, leading to a reduced LH surge and impaired establishment of luteal function. The resultant inadequate luteal phase is associated with decreased cycle fecundity, presumably due to a reduced rate of successful implantation or early pregnancy maintenance. Interpretation of endocrine effects is facilitated by multiple samples from well-defined time frames due to cyclic, circadian, and pulsatile patterns of hormone secretion.

4.5 Receptor Numbers and Dynamics

The importance of hormone receptors and receptor mediation of reproductive toxicity is illustrated by the tremendous potency of many hormonal agonists and antagonists as reproductive toxicants. The potency of compounds in this regard depends not only on properties of the compound itself but also on the number and state of receptors in target organs. Reductions in receptor numbers may decrease hormonal action to levels incapable of inducing physiologic action. While quantitation of receptors for reproductive hormones has been used extensively in the classification and study of reproductive hormone-dependent malignancies in humans, signal transduction constituents are not used widely as markers in testing for reproductive toxicity of compounds. Specific reproductive toxicity of an agent can often be explained on the basis of selective actions via one or more receptor sites or pathways.

4.6 Gamete Numbers and Quality

Gamete numbers indicate the current potential for fertilization in the male or near- or long-term potential in the female. Spermatozoa counts (semen analysis) are the common way of assessing the number of viable gametes in males. In females, the only direct option is counting follicles in some portion of ovarian tissue, such as in a biopsy sample.

4.6.1 Spermatozoa

Compared with many other species, human males produce fewer sperm relative to the number of sperm required for fertility.[44,45] Fertility in men appears to decrease detectably at sperm concentrations below 20 million motile sperm per ejaculate. Especially for men who have a low initial count, a modestly toxic exposure may further decrease production of sperm and impair fertility. Since most test animal species produce many more sperm than do humans, negative results in an animal study that is limited to fertility endpoints do not prove that the compound poses no reproductive hazard for men. The parameters important for spermatic evaluations are number, morphologic characteristics, and motility. Similar data can be obtained from human ejaculates, making extrapolation from animals to humans reasonably straightforward.

4.6.2 Oocytes

Abnormal folliculogenesis, failure of ovulation, or subnormal corpus luteum function can disrupt cyclicity, reduce fertility, and (at least in many nonhuman primates) interfere with normal sexual behavior. Additionally, in normal females, all of the follicles (and the resident oocytes) are present at or soon after birth. Most of these follicles undergo atresia, and only a few actually ovulate sometime during the life span. Follicular depletion is irreversible, and acceleration of depletion causes premature ovarian failure in animals or women.

4.7 Ovarian Cyclicity

The pattern of events in the estrous cycle of rodents provides a useful indicator of the normality of reproductive neuroendocrine and ovarian function in the nonpregnant female. In nonpregnant females, repetition of the four stages of the estrous cycle at regular intervals suggests that neuroendocrine control of the cycle and ovarian responses to that control are normal. Daily vaginal smear data can provide useful information on: (1) cycle length, (2) occurrence or persistence of estrus, (3) occurrence or persistence of diestrus, (4) presence of sperm in the vagina as an indication of mating, (5) incidence of spontaneous pseudopregnancy, (6) distinguishing pregnancy from pseudopregnancy (based on the number of days the smear remains leukocytic), (7) fetal death and resorption (detected by the presence of blood in the smear after day 12 of gestation), and (8) acyclicity, indicative of onset of reproductive senescence.[46]

Estrous cycle normality can be monitored in rats and mice by observing cytologic changes in vaginal smears.[47,48] In general, vaginal smear cytology should be examined daily for at least three consecutive estrous cycles prior to treatment, and an equivalent or longer series should be obtained while treatment continues.

Significant evidence that the estrous cycle (or menstrual cycle in primates) has been disrupted by exposure to a compound should be considered an adverse effect. The exposures during development, childhood, or adult life may all be important. Significant, irreversible injury can be identified in some instances by assessment of cycle-associated events in women. For example, if an environmental exposure leads to accelerated oocyte atresia, diminished fertility may be difficult to detect in women who are not attempting to conceive; however, increased follicular depletion may be proceeding and will be detected indirectly through evaluation of the age of menopause. In evaluating the health of aging women, advancement of the age of menopause from exposure to a toxicant certainly must be deemed an adverse health effect.

Numerous diagnostic methods exist to evaluate female infertility in the clinical setting. Some are too cumbersome or invasive to be used for occupational or environmental toxicologic evaluations (e.g., histology of endometrial biopsies and tests of tubal patency), but in particular instances they may be helpful in defining the mechanisms related to female reproductive abnormalities. Clinical observations may often give the initial evidence that some exposure may be linked to female reproductive failure.

In women, reproductive dysfunction can be studied by the evaluation of irregularities of menstrual cycles.[49] The length of the menstrual cycle, particularly the follicular phase, can vary among individuals and may make it difficult to determine significant effects in populations of women.[50,51] Menstrual cyclicity is affected by age, nutritional status, stress, drugs of abuse, and pharmaceuticals, especially those with reproductive hormone activity such as oral contraceptives.

Vaginal cytology in humans is of some use for detecting reproductive cycles but serves better as a bioassay of estrogen action. Changes in cervical mucus viscosity may be prominent in some women and can be used to estimate the occurrence of ovulation and to distinguish the phases of the reproductive cycle. Ovulation can be estimated in some women by the biphasic shift in basal body temperature.

Ovulation and the endocrine status of a woman can be evaluated by the measurement of hormones (estradiol, LH, FSH, and progesterone) in blood, urine, or saliva. Multiple samples provide the most explicit characterization of a cycle; however, a single sample for progesterone determination some 7 to 9 days after the estimated midcycle surge of gonadotropins may provide strong evidence for the presence of a normal corpus luteum and, thus, prior normal folliculogenesis and ovulation.

Significant effects on measures showing an earlier age of onset of reproductive senescence in females should be considered adverse. The principal cause of the age-dependent loss of ovarian cyclicity in humans appears to be the depletion of oocytes.[52] Regardless of age, if acyclicity or irregular cycles are associated with distinctly elevated levels of FSH in postpubertal women, ovarian failure is the clear interpretation.

4.8 Behavioral Indices

Sexual behavior reflects complex neural, endocrine, and reproductive organ interactions and is often difficult to assess reliably in humans because of the emotional and social aspects involved. Changes in sexual behavior can be due to either psychosocial or physiological causes, including a variety of toxic agents, diseases, and pathologic conditions. Data on sexual behavior usually are unavailable from studies of human populations exposed to potentially toxic occupational or environmental agents, nor are such data routinely obtained in animal studies. Nevertheless, since human data are usually lacking, effects of an agent on sexual behavior in animals usually are the only data available. Some evidence does suggest that central nervous system effects can disrupt sexual behavior in both animals and humans.[53,54]

In rats, a relatively simple measure of sexual receptivity (female) and successful mating behavior (male) is the detection of copulatory plugs or sperm-positive vaginal lavages. In the female, a direct measure of sexual receptivity is the lordosis response. The lordosis response reflects sexual receptivity of the female rat that is detectable in normal cycles during the late evening of vaginal proestrus. In the male, more detailed measures of sexual behavior include estimates of latency periods to first mount, mount with intromission, first ejaculation, number of mounts with intromission to ejaculation, and the postejaculatory interval.[55]

4.9 Pregnancy Wastage and Outcomes

4.9.1 *Corpus Luteum*

The corpus luteum arises from the ruptured follicle and serves as the principal source of progesterone required for the maintenance of early pregnancy in the human;[56] therefore, establishment and maintenance of normal corpora lutea are essential to normal reproductive function. However, with the exception of histopathologic evaluations that may establish little more than presence or absence, these structures are not evaluated in routine testing. Additional research is needed to determine the importance of incorporating endpoints that examine direct effects on luteal function in routine toxicologic testing.

4.9.2 *Fertility and Pregnancy Outcomes*

Breeding studies in animals (rodents) are a major source of data on reproductive toxicants. Detrimental effects of an agent on fertility or pregnancy outcome in multi-generational reproduction studies should be considered adverse. Whether effects are on the female reproductive system or directly on the embryo or fetus is often not distinguishable, but the distinction may not be important, because all these effects should be cause for concern. Many rates, indices, and ratios can be calculated from animal breeding studies. Some parameters with relatively comparable endpoints in humans include mating rate (time to pregnancy), pregnancy rate, gestation length, fetal death rate, offspring gender, birth weight, and offspring survival.

Disorders of the male or female reproductive systems also may be manifested as adverse outcomes of pregnancy. For example, it has been estimated that approximately 50% of human conceptuses fail to reach term.[57] Methods that detect pregnancy as early as 9 days after conception have suggested that 35% of postimplantation pregnancies end in embryonic or fetal loss.[58]

Most studies of toxic effects on pregnancy loss have been observational epidemiologic studies. The use of the retrospective approach to assess fetal losses is fraught with biases, since persons who have experienced losses are more likely to report exposures. Prospective studies are much more expensive and time consuming to conduct but provide more persuasive results.

Toxicants that affect embryo and early fetal development resulting in spontaneous abortion are difficult to detect in humans; thus, animal studies are useful in evaluating such abortifacient and teratogenic toxicants. Teratogenic effects depend on the developmental stage in which the conceptus was exposed. Both the blastocyst and the uterus must be ready for implantation, and their synchronous development is critical.[59] Treatments that alter the internal hormonal environment, inhibit protein synthesis, or inhibit mitosis or cell differentiation can block implantation and cause embryo death.

4.10 Neonatal Outcomes

4.10.1 Malformations

Approximately 3% of newborn children have one or more significant congenital malformations at birth, and by the end of the first postnatal year about 3% more are recognized to have serious developmental defects.[60] Of these, it is estimated that 20% are of known genetic transmission, 10% are attributable to known environmental factors, and the remaining 70% result from unknown causes.[61] Also, approximately 7.4% of children have low birth weight (i.e., below 2.5 kg).[62] Offspring survival is dependent on the same factors as postnatal weight, although more severe effects usually are necessary to affect survival.

4.10.2 Paternally Mediated Effects

It is well accepted that exposure of a female to toxic chemicals during gestation or lactation may produce death, structural abnormalities, growth alteration, or postnatal functional deficits in her offspring. Sufficient data for a variety of agents now exist to conclude that exposure of males also can produce deleterious effects in offspring.[63,65] These effects may be the result of direct damage to the sperm or delivery of toxicant (via seminal plasma or the fertilizing sperm) to the female genital tract or even the embryo directly.

4.10.3 Mammary Gland and Lactation

Mammary gland size, milk production and release, and histology can be affected adversely by toxic agents. Reduced growth of young could be caused by reduced milk availability, by ingestion of a toxic agent into the milk, or by other factors unrelated to lactational ability (e.g., poor suckling ability). Because of the tendency for mobilization of lipids from adipose tissue and secretion of those lipids into milk by lactating females, milk may contain persistent lipophilic agents at concentrations equal to or higher than those present in the blood or organs of the mother. Thus, suckling offspring may be exposed to elevated levels of such toxicants.

In animal studies, a simple estimate of milk production may be obtained by measuring litter weights taken after 1 or 2 hours of nursing by milk-deprived pups (6 hours). Cleared and stained whole mounts of the mammary gland can be prepared at necropsy for histologic examination. In addition, the DNA, RNA, and lipid content of the mammary gland and the composition of milk have been measured following toxicant administration as indicators of toxicity to this target organ. Significant reductions in milk production or negative effects on milk quality, whether measured directly or reflected in impaired development of young, should be considered adverse reproductive effects.

4.11 Developmental Assessments of Progeny

4.11.1 Morphological

Chemical or physical agents can affect anatomical development during the prenatal period. Teratological effects may be major or minor and may involve virtually every organ system. Excellent comprehensive summaries of teratogens are available.[66]

4.11.2 Functional

The effects of some reproductive toxicants may not be apparent until long after birth; for example, some neurological damage may not be obvious until learning or normal growth is impaired. In addition, some cases of toxic damage may not be detected until the injured system is challenged. Abnormal neonatal growth could indicate neuroendocrine involvement. Adverse developmental outcomes can result from exposure to toxicants *in utero*, via milk from exposed mothers, or both.

4.12 Epidemiologic Studies

Good epidemiologic studies provide the most relevant information for assessing human risk. There are many different types of epidemiologic studies, and certain features of study design are important to consider. Some of the important factors that enhance a study and thus increase its usefulness for risk assessment have been noted in a number of publications.[43,67-73]

4.12.1 Power of a Study

The power of a study (i.e., its ability to detect a true effect) depends on the size of the study group, the frequency of the outcome in the general population, and the level of excess risk to be identified. In a cohort study, groups are defined by exposure, and their health outcomes examined. In case-referent studies, groups are defined by health status and prior exposures are examined. Study sizes are dependent upon the frequency of exposure within the source population. The confidence one has in the results of a study with negative findings is related directly to the power of the study to detect meaningful differences in the endpoints. Power may be enhanced by combining populations from several studies using a meta-analysis.[74] The combined analysis would increase confidence in the absence of risk for agents with negative findings; however, care must be exercised in the combination of potentially dissimilar study groups. There are statistical methods for assessing study design as part of meta-analysis.

A posteriori determination of the power of an actual study is useful in evaluating negative findings. Negative findings in a study of low power should be given considerably less weight than either a positive study or a negative study with high power. Positive findings from very small studies are open to question because of the instability of the risk estimates and the highly selected nature of the population.

4.12.2 Bias

Bias may result from the way the study group is selected or information collected.[75] Selection bias may occur in the process of identification of subjects for study or when willingness to participate varies with the exposure status or health status of that individual. Information bias may result from misclassification of characteristics of individuals or events identified for study. Recall bias, one type of information bias, may occur when respondents with specific exposures or outcomes recall information differently than those without the exposures or outcomes. Interview bias may result when the interviewer knows *a priori* the category of exposure (for cohort studies) or outcome (for case-referent studies) in which the respondent belongs. Use of highly structured questionnaires and/or "blinding" of the interviewer reduces the likelihood of such bias.

Data from any source may be prone to errors or bias. All types of bias are difficult to assess; however, validation using independent data sources (e.g., vital records or medical charts), or use of biomarkers of exposure or outcome, often increases confidence in the results of the study by showing the degree of bias present. Studies with lower probabilities of bias should carry more weight in risk assessments.[73,76] Differential misclassification may either raise or lower the risk estimate. Nondifferential misclassification will bias the results toward a finding of "no effect".[75]

4.12.3 Confounders and Modifiers of Effects

Known and potential risk factors should be examined to identify those that may be confounders or effect modifiers. A confounder is a variable that is a risk factor for the disease under study and is associated with the exposure under study but is not a consequence of the exposure. A confounder may distort both the magnitude and direction of the measure of association between the exposure of interest and the outcome. An effect modifier is a factor that produces different exposure-response relationships at different levels of that factor. For example, age would be an effect modifier if the risk associated with a given exposure increased with the individual's age. Both confounders and effect modifiers must be controlled in the study design and/or analysis to improve the estimate of the effects of exposure.[75,77] Studies that fail to account for these important factors should be given less weight in risk assessment.

4.12.4 Selection of Study Outcomes

Reproductive outcomes available for epidemiologic examination are limited by a number of factors, including the relative magnitude of the exposure, the size and demographic characteristics of the population, and the ability to observe the reproductive outcome in humans. The most feasible epidemiologic endpoints that reflect reproductive toxicity include: (1) fertility/infertility estimates, (2) reproductive history studies of certain pregnancy outcomes (e.g., spontaneous abortions, birth weight, sex ratio, congenital malformations, post-natal function, and neonatal growth and survival; (3) seminal evaluations, and (4) menstrual history. Improved methods for identifying some outcomes, such as early spontaneous abortion, using ultra-sensitive hCG assays occasionally change the potential outcomes available for study.[60,78]

4.12.5 Reproductive History Studies

Although some human population-based studies may utilize endpoints of reproductive toxicity, such as seminal analyses, reproductive hormone levels, or menstrual cyclicity, assessments of the effects of environmental agents on human fertility and pregnancy outcomes primarily depend on epidemiologic approaches.

4.12.6 Measures of Fertility

Infertility or subfertility may be thought of as a nonevent: a couple is unable to have children within a specific time frame. Therefore, the epidemiologic measurement of reduced fertility typically is indirect and is accomplished by comparing birth rates or time intervals between births or pregnancies. In these evaluations, the couple's joint ability to procreate is estimated. One method, the standardized birth ratio (SBR; also referred to as the standardized fertility ratio), compares the number of births observed to those expected based on the person-years of observation stratified by factors such as time period, age, race, marital status, parity, and contraceptive use.[79-84] Analysis of the time between recognized pregnancies or live births has been suggested as another indirect measure of fertility.[85-88] Although fertility may be affected by alterations in sexual behavior, data associating toxic exposures to behavioral changes in humans that in turn adversely affect fertility on that basis are limited or nonexistent.

4.12.7 Pregnancy Outcomes

Pregnancy outcomes examined in human studies of parental exposures may include spontaneous abortions or fetal loss, congenital malformations, birth weight effects, sex ratio at birth, and possibly postnatal effects (e.g., growth and development, organ or system function, and behavior). A dose-response relationship is usually an important criterion in the assessment of exposure to a potentially toxic agent, but not always.[89] Specifically, with increasing dose, a pregnancy might end in abortion or fetal loss rather than a live birth with malformations, depending on either level or timing of exposure.[89,90]

REFERENCES

1. Eisenberg, L., Physiological and psychological aspects of sexual development and function, in *Basic Reproductive Medicine,* Vol. 1, Hamilton, D. W. and Naftolin, F., Eds., MIT Press, Cambridge, 1981, 118.
2. Rose, R. M., Gordon, T. P., and Bernstein, I. S., Plasma testosterone levels in male rhesus. Influences of sexual and social stimuli, *Science,* 178, 643, 1978.
3. Testart, J., Frydman, R., and Roger, M., Seasonal influence of diurnal rhythms in onset of plasma luteinizing hormone surge in women, *J. Clin. Exp. Metabol.,* 55, 374, 1982.
4. Seibel, M. M., Shine, W., Smith, D. M. and Taynor, M. L., Biological rhythm of the luteinizing hormone surge in women, *Fertil. Steril.,* 37, 709, 1982.
5. Ehrenkranz, J. R. L., Seasonal breeding in humans: birth recordings of the Laborador Eskimos, *Fertil. Steril.,* 40, 485, 1983.
6. Van Vugt, D. A., Influences of the visual and olfactory systems on reproduction, *Semin. Reprod. Endocrinol.,* 8, 1, 1990.
7. Vandenbergh, J. G. and Vessey, S., Seasonal breeding of free ranging rhesus monkeys and related ecological factors, *J. Reprod. Fertil.,* 15, 71, 1968.
8. McClintock, M. K., Menstrual synchrony and suppression, *Nature,* 229, 244, 1971.
9. Quadagno, D. M., Shubeita, H. E., Deck, J., and Francoeur, D., Influence of male social contacts, exercise and all female living conditions on the menstrual cycle, *Psychoneuroendocrinology,* 6, 239, 1981.
10. Eggert-Kruse, W., Leinhos, G., Gerhard, I., Tilgen, W., and Runnebaum, B., Prognostic value of *in vitro* sperm penetration into hormonally standardized human cervical mucus, *Fertil. Steril.,* 51, 317, 1989.
11. Yanagimachi, R., Sperm-egg fusion, *Curr. Top. Membrane Transp.,* 32, 349, 1988.
12. Takano, H., Yanagimachi, R., and Urch, U. A., Evidence that acrosin activity is important for the development of fusibility of mammalian spermatozoa with the oolemma: inhibitor studies using the golden hamster, *Zygote,* 1, 79, 1993.
13. Armstrong, D. T. and Dorrington, J. H., Androgens augment FSH-induced progesterone secretion by cultured rat granulosa cells, *Endocrinology,* 99, 1411, 1976.
14. Lucky, A. W., Schreiber, J. R., Hillier, S. G., Schulman, J. D., and Ross, G. T., Progesterone production by cultured preantral rat granulosa cells; stimulation by androgens, *Endocrinology,* 100, 128, 1977.
15. Adashi, E. Y., Resnick, C. E., Hernandez, E. R., Svoboda, M. E., and Van Wyk, J. J., Potential relevance of insulin-like growth factor 1 to ovarian physiology: from basic to clinical application, *Endocrine Rev.,* 7, 94, 1989.
16. Guthrie, H. D., Grimes, R. W., and Hammond, J. M., Changes in insulin-like growth factor binding protein-1 and -3 in follicular fluid during atresia of follicles grown after ovulation in pigs, *J. Reprod. Fertil.,* 104, 225, 1995.
17. May, J. V. and Schomberg, D. W., The potential relevance of epidermal growth factor and transforming growth factor-alpha to ovarian physiology, *Semin. Reprod. Endocrinol.,* 7, 1, 1989.
18. Knecht, M., Feng, P., and Catt, K. J., Transforming growth factor-beta: autocrine, paracrine, and endocrine effects in ovarian cells, *Semin. Reprod. Endocrinol.,* 7, 20, 1989.
19. Lin, T., Blaisdell, J., and Haskell, J. F., Transforming growth factor-beta inhibits Leydig cell steroidogenesis in primary culture, *Biochem. Biophys. Res. Commun.,* 146, 387, 1987.
20. Gospodarowicz, D., Fibroblast growth factor involvement in early development and ovarian function, *Semin. Reprod. Endocrinol.,* 7, 21, 1989.
21. Koos, R. D., Potential relevance of angiogenic factors to ovarian physiology, *Semin. Reprod. Endocrinol.,* 7, 29, 1989.
22. Lightman, A., Palumno, A., DeCherney, A. H., and Naftolin, F., The ovarian renin-angiotensin system, *Semin. Reprod. Endocrinol.,* 7, 79, 1989.
23. Tsonis, C. G., Hiller, S. G., and Baird, D. T., Production of inhibin bioactivity by human granulosa-lutein cells: stimulation by LH and testosterone *in vitro, J. Endocrinol.,* 112, R11, 1987.

24. Burger, H. G. and Findlay, J. K., Potential relevance of inhibin to ovarian physiology, *Semin. Reprod. Endocrinol.*, 7, 69, 1989.
25. Tsafriri, A. and Adashi, E. Y., Local nonsteroidal regulators of ovarian function, in *The Physiology of Reproduction,* 2nd ed., Vol. 1, Knobil, E. and Neill, J. D., Eds., Raven Press, New York, 1994, 817.
26. Johnson, L., Petty, C. S., and Neaves, W. B., Further quantification of human spermatogenesis: germ cell loss during postprophase of meiosis and its relationship to daily sperm production, *Bio. Reprod.*, 29, 207, 1983.
27. Kerr, J. B. and Sharpe, R. M., Macrophage activation enhances the hCG-induced disruption in the rat, *J. Endocrinol.*, 121, 285, 1989.
28. Labrie, F., Seguin, C., Belanger, A., Pelletier, G., Reeves, J., Kelly, P. A., Lemay, A., and Raynaud, J. P., Antifertility effects of LHRH agonists in the male, *J. Androl.*, 1, 209, 1980.
29. Antich, M., Fabian, E., Sarquella, J., and Bassas, L., Effect of testicular damage induced by cryptorchidism on insulin-like growth factor 1 receptors in rat Sertoli cells, *J. Reprod. Fertil.*, 104, 267, 1995.
30. Mosher, W. D. and Pratt, W. F., *Fecundity and Infertility in the United States, 1965–88,* Report 192, National Center for Health Statistics, Hyattsville, MD, 1990.
31. Gorski, R. A., The neuroendocrinology of reproduction: an overview, *Biol. Reprod.*, 20, 111, 1979.
32. Berndtson, W. E., Methods for quantifying mammalian spermatogenesis: a review, *J. Anim. Sci.*, 44, 818, 1977.
33. Foote, R. H., Schermerhorn, E. C., and Simkin, M. E., Measurement of semen quality, fertility, and reproductive hormones to assess dibromochloropropane (DBCP) effects in live rabbits, *Fund. Appl. Toxicol.*, 6, 628, 1986.
34. Russell, L. D., Ettlin, R., Sinha Hikim, A. P., and Clegg, E. D., *Histological and Histopathological Evaluation of the Testis,* Cache River Press, Clearwater, FL, 1990.
35. Hess, R. A., Quantitative and qualitative characteristics of the stages and transitions in the cycle of the rat seminiferous epithelium: light microscopic observations of perfusion-fixed and plastic-embedded testes, *Biol. Reprod.*, 43, 525, 1990.
36. Kurman, R. and Norris, H. J., Germ cell tumors of the ovary, *Pathol. Annu.*, 13, 291, 1978.
37. Langley, F. A. and Fox, H., Ovarian tumors. Classification, histogenesis, etiology, in *Haines and Taylor's Obstetrical and Gynaecologic Pathology,* Fox, H., Ed., Churchill Livingstone, Edinburgh, 1987, 542.
38. Plowchalk, D. R., Smith, B. J., and Mattison, D. R., Assessment of toxicity to the ovary using follicle quantitation and morphometrics, in *Methods in Toxicology: Female Reproductive Toxicology,* Heindel, J. J. and Chapin, R. E., Eds., Academic Press, San Diego, 1993, 57.
39. Kupfer, D., Critical evaluation of methods for detection and assessment of estrogenic compounds in mammals: strengths and limitations for application to risk assessment, *Reprod. Toxicol.*, 1, 147, 1987.
40. Warren, J. C., Cheatum, S. G., Greenwald, G. S., and Barker, K. L., Cyclic variation of uterine metabolic activity in the golden hamster, *Endocrinology,* 80, 714, 1967.
41. Sever, L. E. and Hessol, N. A., Overall design considerations in male and female occupational reproductive studies, in *Reproduction: The New Frontier in Occupational and Environmental Research,* Lockey, J. E., Lemasters, G. K., and Keye, W. R., Eds., Alan R. Liss, New York, 1984, 15.
42. Heywood, R. and James, R. W., Current laboratory approaches for assessing male reproductive toxicity, in *Reproductive Toxicology,* Dixon, R. L., Ed., Raven Press, New York, 1985, 147.
43. National Research Council, *Biologic Markers in Reproductive Toxicity,* National Academy Press, Washington, D.C., 1989.
44. Amann, R. P., A critical review of methods for evaluation of spermatogenesis from seminal characteristics, *J. Androl.*, 2, 37, 1981.
45. Working, P. K., Male reproductive toxicity: comparison of the human to animal models, *Environ. Health,* 77, 37, 1988.
46. LeFevre, J. and McClintock, M. K., Reproductive senescence in female in rats: a longitudinal study of individual differences in estrous cycles and behavior, *Biol. Reprod.*, 38, 780, 1988.
47. Long, J. A. and Evans, H. M., The oestrous cycle in the rat and its associated phenomena, *Mem. Univ. Calif.*, 6, 1, 1922.
48. Cooper, R. L., Goldman, J. M., and Vandenbergh, J. G., Monitoring of the estrous cycle in the laboratory rodent by vaginal lavage, in *Methods in Toxicology: Female Reproductive Toxicology,* Heindel, J. J. and Chapin, R. E., Eds., Academic Press, San Diego, 1993, 45.
49. Lemasters, G. K., Hagen, A., and Samuels, S., Reproductive outcomes in women exposed to solvents in 36 reinforced plastic companies. 1. Menstrual dysfunction, *J. Occup. Med.*, 27, 490, 1985.
50. Burch, T. K., Macisco, J. J., and Parker, M. P., Some methodologic problems in the analysis of menstrual data, *Int. J. Fertil.*, 12, 69, 1967.
51. Treloar, A. E., Boynton, R. E., Borghild, G. B., and Brown, B. W., Variation in the human menstrual cycle through reproductive life, *Int. J. Fertil.*, 12, 77, 1967.
52. Mattison, D. R., Clinical manifestations of ovarian toxicity, in *Reproductive Toxicology,* Dixon, R. L., Ed., Raven Press, New York, 1985, 109.
53. Rubin, H. B. and Henson, D. E., Effects of drugs on male sexual function, *Adv. Behav. Pharmacol.*, 65, 1979.
54. Waller, D. P., Killinger, J. M., and Zaneveld, L. J. D., Physiology and toxicology of the male reproductive tract, in *Endocrine Toxicology,* Thomas, J. A., Korach, K. S., and McLachlan, J. A., Eds., Raven Press, New York, 1985, 269.

55. Beach, F. A., Animal models for human sexuality, in *Ciba Foundation Symposium No. 62: Sex, Hormones and Behavior,* Elsevier-North Holland, London, 1979, 113.
56. Csapo, A. I. and Pulkkinen, M., Indispensability of the human corpus luteum in the maintenance of early pregnancy: lutectomy evidence, *Obstet. Gynecol. Surv.,* 33, 69, 1978.
57. Hertig, A. T., The overall problem in man, in *Comparative Aspects of Reproductive Failure,* Benirschke, K., Ed., Springer-Verlag, New York, 1967, 11.
58. Wilcox, A. J., Weinburg, C. R., Wehmann, R. E., Armstrong, E. G., Canfield, R. E., and Nisula, B. C., Measuring early pregnancy loss: laboratory and field methods, *Fertil. Steril.,* 44, 366, 1985.
59. Cummings, A. M. and Perreault, S. D., Methoxychlor accelerates embryo transport through the rat reproductive tract, *Toxicol. Appl. Pharmacol.,* 102, 110, 1990.
60. Shepard, T. H., Human teratogenicity, *Adv. Pediatr.,* 33, 225, 1986.
61. Wilson, J. G., Embryotoxicity of drugs in man, in *Handbook of Teratology,* Vol. 1, Wilson, J. G. and Fraser, F. C., Eds., Plenum Press, New York, 1977, 309.
62. Selevan, S. G., Design considerations in pregnancy outcome studies of occupational populations, *Scand. J. Work Environ. Health,* 7, 76, 1981.
63. Hood, R. D., Paternally mediated effects, in *Developmental Toxicity: Risk Assessment and the Future,* Hood, R. D., Ed., Van Nostrand Reinhold, New York, 1989, 77.
64. Nagao, T. and Fujikawa, K., Genotoxic potency in mouse spermatogonial stem cells of triethylenemelamine, mitomycin-C, ethylnitrosourea, procarbazine, and propyl methanesulfonate is measured by F1 congenital defects, *Mutation Res.,* 229, 123, 1990.
65. Davis, D. L., Friedler, G., Mattison, D., and Morris, R., Male-mediated teratogenesis and other reproductive effects: biologic and epidemiologic findings and a plea for clinical research, *Reprod. Toxicol.,* 6, 289, 1992.
66. Schardein, J. L., *Chemically Induced Birth Defects,* Marcel Dekker, New York, 1993.
67. Selevan, S. G., *Evaluation of Data Sources for Occupational Pregnancy Outcome Studies,* University of Cincinnati, University Microfilms, Ann Arbor, MI, 1980.
68. Bloom, A. D., *Guidelines for Reproductive Studies in Exposed Human Populations: Guideline for Studies of Human Populations Exposed to Mutagenic and Reproductive Hazards,* Report of Panel 11, March of Dimes Birth Defects Foundation, White Plains, NY, 1981, 37.
69. Hatch, M. and Kline, J., *Spontaneous Abortion and Exposure to the Herbicide 2,4,5-T: A Pilot Study,* EPA-56016-81006, U.S. Environmental Protection Agency, U.S. Government Printing Office, Washington, D.C., 1981.
70. Wilcox, A. J., Surveillance of pregnancy loss in human populations, *Am. J. Ind. Med.,* 4, 285, 1983.
71. Axelson, O., Epidemiologic methods in the study of spontaneous abortions: sources of data, methods, and sources of error, in *Occupational Hazards and Reproduction,* Hemminki, K., Sorsa, M., and Vaino, H., Eds., Hemisphere, Washington, 1985, 231.
72. Tilley, B. C., Barnes, A. B., Bergstrahl, E., Labarthe, D., Noller, K. L., Colton, T., and Adam, E., A comparison of pregnancy history recall and medical records: implications for retrospective studies, *Am. J. Epidemiol.,* 121, 269, 1985.
73. Kimmel, C. A., Kimmel, G. L., and Frankos, V., Interagency Regulatory Liaison Group workshop on reproductive toxicity risk assessment, *Environ. Health,* 66, 193, 1986.
74. Greenland, S., Quantitative methods in the review of epidemiologic literature, *Epidemiol. Rev.,* 9, 1, 1987.
75. Rothman, K. J., *Modern Epidemiology,* Little Brown, Boston, 1986.
76. Stein, A. and Hatch, M., Biological markers in reproductive epidemiology: prospects and precautions, *Environ. Health,* 74, 67, 1987.
77. Kleinbaum, D. G., Kupper, L. L., and Morgenstern, H., *Epidemiologic Research: Principles and Quantitative Methods,* Lifetime Learning Publications, London, 1982.
78. Sweeney, A. M., Meyer, M. R., Aarons, J. H., Mills, J. L., and LaPorte, R. E., Evaluation of methods for the prospective identification of early fetal losses in environmental epidemiology studies, *Am. J. Epidemiol.,* 127, 843, 1988.
79. Wong, Q., Utidjian, H. M. D., and Karten, V. S., Retrospective evaluation of reproductive performance of workers exposed to ethylene dibromide, *J. Occup. Med.,* 21, 98, 1979.
80. Levine, R. J., Symons, M. J., Balogh, S. A., Arndt, D. M., Kaswandik, N. R., and Gentile, J. W., A method for monitoring the fertility of workers. 1. Method and pilot studies, *J. Occup. Med.,* 22, 781, 1980.
81. Levine, R. J., Symons, M. J., Balogh, S. A., Milby, T. H., and Whorton, M. D., A method for monitoring the fertility of workers. 11. Validation of the method among workers exposed to dibromochloropropane, *J. Occup. Med.,* 23, 183, 1981.
82. Levine, R. J., Blunden, P. B., DalCorso, R. D., Starr, T. B., and Ross, C. E., Superiority of reproductive histories to sperm counts in detecting infertility at a dibromochloropropane manufacturing plant, *J. Occup. Med.,* 25, 591, 1983.
83. Levine, R. J., Methods for detecting occupational causes of male infertility: reproductive history versus semen analysis, *Scand. J. Work Environ. Health,* 9, 371, 1983.
84. Starr, T. B., DalCorso, R. D., and Levine, R. J., Fertility of workers: a comparison of logistic regression and indirect standardization, *Am. J. Epidemiol.,* 123, 490, 1986.

85. Dobbins, J. G., Eifler, C. W., and Buffler, P. A., *The Use of Parity Survivorship Analysis in the Study of Reproductive Outcomes,* Society for Epidemiologic Research Conference, Seattle, WA, 1994.
86. Baird, D. D. and Wilcox, A. J., Cigarette smoking associated with delayed conception, *J. Am. Med. Assoc.,* 253, 2979, 1985.
87. Baird, D. D., Wilcox, A. J., and Weinberg, C. R., Using time to pregnancy to study environmental exposures, *Am. J. Epidemiol.,* 124, 470, 1986.
88. Weinberg, C. R. and Gladen, B. C., The beta-geometric distribution applied to comparative fecundability studies, *Biometrics,* 42, 547, 1986.
89. Wilson, J. G., *Environment and Birth Defects,* Academic Press, New York, 1973.
90. Selevan, S. G. and Lemasters, G. K., The dose-response fallacy in human reproductive studies of toxic exposure, *J. Occup. Med.,* 29, 451, 1987.

CHAPTER 22

Male Reproductive Toxicity

Steven M. Schrader

CONTENTS

1 SITES OF TOXICANT ACTION

Toxicants can attack the male reproductive system at one of several sites or at multiple sites. These sites and the assays associated with their respective functions are discussed individually. This does not necessarily indicate, however, that there exists an absolute one-to-one relationship between a particular measurement and the associated site of action. These sites include the neuroendocrine system, the testes, accessory sex glands, and sexual function.

The establishment of a male reproductive profile for assessing reproductive potential for both individual and population investigations is necessary. The same profile can be used for both types of studies, but there are some basic differences in methodology. The assessment profile illustrated here is that being used by the National Institute for Occupational Safety and Health (NIOSH) to assess populations exposed to potential reproductive toxicants. Differences between assessing the individual vs. the population will be noted. A summary of assessments and specific methodologies follow. If individual data (vs. population comparisons) are to be used, care should be taken to compare the results with the normal range of results of the laboratory conducting the analysis and not published values. If a population-based study is being conducted, a concurrent comparison cohort must be used and the analyses should be blind to exposure status.

1.1 Neuroendocrine System

The endocrine system, in concert with the nervous system, coordinates the function of the various components of the reproductive axis, drawing upon external (e.g., sexual cues, temperature) and internal (e.g., checks and balances between endocrine tissue function, metabolic status) inputs. The reproductive endocrine status of the male is best established by measuring the hormones in the blood, urine, or saliva. The hormones of interest are luteinizing hormone (LH), follicle-stimulating hormone (FSH), and testosterone.

Since the circulating profile of LH is pulsatile, the status of this hormone for the individual, if measured in blood, should be estimated in serial samples. The pooled results of three samples collected at 20-min intervals will provide a reasonable estimate of mean concentration.[1] Alternatively, an integral of the pulsatile LH secretion rate may be obtained by measuring this gonadotropin in urine. If a population is being evaluated, a single blood sample per individual may suffice.[2]

Circulating FSH levels are not as variable as those for LH. This is attributable in part to a longer circulating half-life for FSH compared to LH. Thus, analysis of a single blood sample for an individual will provide a more reliable estimate of FSH than for LH. FSH can also be measured in urine for the sake of convenience.

Approximately 2% of circulating testosterone is free, whereas the remainder is bound to sex hormone binding globulin (SHBG), albumin, and other serum proteins. The free circulating testosterone is the active component and therefore provides a more accurate marker of physiologically available testosterone than does total circulating testosterone under condition when SHBG concentration or binding is altered.[1] Circulating testosterone levels, like those for LH, fluctuate considerably over time. Estimates of free and total testosterone can be determined in single blood samples but are greatly improved by assaying multiple blood samples and pooling the results. Alternatively, a single measurement of a testosterone metabolite in urine (e.g., androsterone, etiocholanolone, or testosterone glucuronide) provides a convenient index of total testosterone.[3] Quantifying testosterone in saliva affords a convenient alternative to blood sampling while providing a measure of the unbound, biologically active component of circulating testosterone.[4] Protein hormones such as the gonadotropins are not exuded into the saliva.

TABLE 1

Endocrine Profile for Assessing Reproductive Toxicant Effects

Hormone	Fluid for Measurement		
	Saliva	Blood	Urine
Luteinizing hormone (LH)		X	X
Follicle-stimulating hormone (FSH)		X	X
Testosterone	X	X	X

TABLE 2

Normal Semen Parameters and Values[a]

Semen Parameter	Normal Value
Semen volume	≥ 2.0 ml
Sperm concentration	≥ 20 million sperm per ml
Total sperm count	≥ 40 million sperm per ejaculate
Percent motility	≥ 50% progressively motile
Morphology	≥ 30% normal sperm
Viability (vital stain)	≥ 75% alive

[a] Based on fertile men as reported by WHO.[184]

If measuring steroid hormone metabolites in urine, consideration must be given to the potential that the exposure being studied may alter the metabolism of excreted metabolites. This is especially pertinent since most metabolites are formed by the liver, a target of many toxicants. Lead, for example, reduced the amount of sulphated steroids that were excreted into the urine.[5]

Blood levels for both gonadotropins become elevated during sleep as the male enters puberty, while testosterone levels maintain this diurnal pattern through adulthood in men.[6] Thus, blood, urine, or saliva samples should be collected at approximately the same time of day to avoid variations due to diurnal secretory patterns.

In summary, Table 1 provides the hormones to be measured. FSH, LH, and testosterone can all be evaluated in a population-based study by assessing the hormone levels in a single blood sample from each man (preferable at about the same time of day). In the study of an individual, three blood samples should be collected 20 min apart, or urinary assessment of testosterone metabolites and gonadotropins should be conducted.

1.2 Testes

Semen analysis provides a useful profile of the function of the male reproductive system. Normal ranges for semen parameters are provided in Table 2 as general information. The various measurements that are routinely used in the assessment of occupational exposure are presented in Table 3. Exact instructions should be provided to each man to ensure that the semen sample is collected by masturbation after a set time of abstinence (usually 2 days) and delivered to the laboratory within 1 hour from the time of ejaculation. The men should be instructed to maintain the semen at room temperature, avoiding any temperature shock to the sperm cells. NIOSH has produced a videotape with these instructions which is shown to all participants of a semen study.[7] At the time of collecting the semen sample, each subject should record the duration of abstinence, time of semen collection, and any information regarding spillage. Providing a label on the jar facilitates the recording of this information.

TABLE 3

Semen Profile for Assessing Reproductive Toxicant Effects

Sperm concentration
Sperm viability
Vital stain
Hypo-osmotic swelling
Sperm motility
Percent motile
Curvilinear velocity
Straight-line velocity
Linearity
Lateral-head amplitude
Sperm size and shape
Morphology
Morphometry
Sperm genetics
DNA stability
Semen parameters
pH
Volume
Marker chemicals from accessory glands
Toxicant or metabolite concentration

Semen analyses can be conducted in two phases. The initial evaluation of the sample should be conducted when the sample arrives at the laboratory (or field site) and should consist of recording the temperature, turbidity, color, liquefaction time, volume, and pH of the semen. Videorecordings for motility assessments, viability estimates, and sperm counts; the preparation of slides; and preservation of seminal plasma should also be conducted at this time. Morphologic and morphometric analyses of sperm on slides and motility and velocity analyses of the sperm recorded on videotapes (either manually or using a computer-assisted sperm analysis system [CASA]) may be conducted at a later time. If CASA is used, several sperm motility variables can be measured (see Table 3). These variables provide useful information on the progression of sperm cells (curvilinear velocity, straight-line velocity, and linearity) as well as the sperm motility pattern. It should be noted that some CASA manufacturers recommend the use of a fluorescent DNA stain to facilitate the isolation of sperm from debris.

Sperm viability may be determined by two methods: Eosin Y stain exclusion[8] and hypo-osmotic swelling (HOS assay).[9] These techniques test for the structural and functional integrity of the cell membrane, respectively.[10] Sperm concentration and motility characteristics should be measured in a chamber at least 10 μm deep in order for the sperm to move freely in all planes.

Measurements of sperm motility and velocity should be conducted using a microscope stage warmed to 37°C. An attempt to record 100 motile sperm per sample is desirable if one is interested in the distribution of velocity measurements, but 50 motile sperm will suffice if means are to be compared. If videotapes are being used to calculate the percent motility, one should avoid "hunting" for motile sperm. All fields examined or searched should be included in the calculations; therefore, recording a certain number of arbitrary fields is advised. Whole semen should be used for measuring sperm motility. If a CASA system is being used for velocity estimates, the number of sperm per field should be reduced to minimize cell collisions. Using a 10- to 20-μm deep chamber, the sperm concentration should be less than 40 million/ml. Diluents (including seminal plasma), however, alter sperm velocity up to a dilution of about 1:1. The current recommendation for CASA of sperm velocity is to dilute *all* samples to one part semen in one part iso-osmotic buffer. If this dilution does not reduce the sperm concentration below 40 million/ml, then an additional dilution in the same buffer should be performed on those concentrated samples. Thus, two recordings should be made: whole semen for percent motility and diluted sperm for sperm velocity.

Sperm morphology should be estimated on air-dried, stained semen smears. During the past 30 years, several schemes have been presented for the assessment of normal and abnormal sperm morphologies. Variation in sperm size and shape are not distinct, but rather a continuum. This provides a challenge within and especially among laboratories to establish a repeatable system for morphological classification.[11-14] With recent advances of computerized image analyses, several methods of sperm morphometry have been introduced.[15-21] These morphometric analysis systems provide objective assessments of individual sperm head size and shape. Comparisons of measurements among different analysis systems should be avoided. Sperm morphometry is now routinely used as part of the assessment of reproductive hazards to the male workers.[22]

Sperm count, sperm morphology, and sperm head morphometry all provide indices of the integrity of spermatogenesis and spermiogenesis. Thus, the number of sperm in the ejaculate is directly correlated with the number of germ cells per gram of testis,[23] while abnormal morphology is probably a result of abnormal spermiogenesis. Azoospermia is probably the most severe observation, as it is often an indication that type-A spermatogonia have been lost and recovery is unlikely. Promising new methods, DNA stability and DNA adducts, may provide information about spermatogenesis at the genetic level.

Some toxicants have been shown to exhibit an effect at the testis/spermatogenesis/spermiogenesis site. Exposure to dibromochloropropane (DBCP) reduced sperm concentration in ejaculates to 46 million cells/ml in exposed workers compared to a median of 79 million cells/ml in unexposed men.[24] Upon removing the workers from the exposure, those with reduced sperm counts experienced a partial recovery, while men who had been azoospermic remained sterile. Testicular biopsy revealed that the target of DBCP was the spermatogonia. This substantiates the severity of the effect when stem cells are the target of toxicants. There were no indications that DBCP exposure of men was associated with adverse pregnancy outcome.[25] Another example of a toxicant targeting the testis was the study of workers exposed to ethylene dibromide (EDB). They had more sperm with tapered heads and fewer sperm per ejaculate than did controls.[26]

Genetic damage is difficult to detect in human sperm. Epidemiological studies of large populations have demonstrated increased frequency of adverse pregnancies in women whose husbands were working in various occupations.[27] Such studies indicate a need for methods to detect genetic damage in human sperm.[28] Some promising methods for assessing sperm genetics are karyotyping of sperm chromosomes,[29-31] the DNA stability assay,[32-34] and the labeling of dipoidy using fluorescent *in situ* hybridization (FISH).[35-37]

1.3 Accessory Sex Glands

Seminal plasma is not essential for fertilization; thus, the artificial insemination of sperm collected from the epididymis results in conception. On the other hand, seminal plasma contributes importantly to the normal coitus-fertilization scenario. Seminal plasma serves as a vehicle for sperm transport, a buffer from the hostile acidic vaginal environment, and an initial energy source for the sperm. Cervical mucus prevents passage of seminal plasma into the uterus. Some constituents of seminal plasma, however, are carried into the uterus to the site of fertilization by adhering to the sperm membrane.

The viability and motility of spermatozoa in seminal plasma is typically a reflection of seminal plasma quality. Alterations in sperm viability, as measured by stain exclusion or by hypo-osmotic swelling, or alterations in sperm motility parameters would suggest an effect on the accessory sex glands.

Biochemical analysis of seminal plasma provides insights into the function of the accessory sex glands. Chemicals that are secreted primarily by each of the glands of this system are typically selected to serve as a marker for each respective gland. For example, the epididymis is represented by glycerylphosphorylcholine (GPC), the seminal vesicles by fructose, and the prostate gland by zinc. Note that this type of analysis provides only gross information on glandular function and little or no information on the other secretory constituents. Measuring semen pH and volume provides additional general information on the nature of seminal plasma.

Seminal plasma may be analyzed for the presence of a toxicant or its metabolite. Heavy metals have been detected in seminal plasma using atomic absorption spectrophotometry,[38] while halogenated hydrocarbons have been measured in seminal fluid by gas chromatography after extraction[38] or protein-limiting filtration.[39]

A toxicant or its metabolite may act directly on accessory sex glands to alter the quality or quantity of their secretions. Alternatively, the toxicant may enter the seminal plasma[40] and thereby affect the sperm and the body of the female partner after intercourse or may be carried to the site of fertilization on the sperm membrane and affect the ova or conceptus.

There are few reports of toxicant effects on the accessory sex glands in humans. Ethylene dibromide (EDB) is one example of a toxicant that exerts post-testicular effects. Short-term exposure to the toxicant reduced sperm velocity and semen volume.[41] Chronic exposure decreased sperm motility and viability, decreased seminal fructose levels, and increased semen pH.[35] An EDB metabolite was present within the semen of some exposed workers.[39] Other potential toxicants that have been detected in semen include: lead, cadmium, hexachlorobenzene, hexachlorocyclohexane, dieldrin, and polychlorinated biphenyls.[38] Cocaine has been shown to bind to the sperm membrane.[42]

There are several sperm assessment methods that assess the function of the sperm.[43] Functional tests may evaluate sperm across more than one of the subjective toxicant site divisions outlined above. The penetration of sperm through cervical mucus (or viscous fluids stimulating cervical mucus),[44-46] the penetration of sperm into a zona-free hamster egg (sperm penetration assay, SPA),[47] and the penetration of sperm through a zona pellucida removed from immature human ova (hemi-zona assay) have been shown to evaluate different sperm functions.[48,49] With the exception of the SPA, these have not been utilized in assessing reproductive toxicants in the field setting. SPA has been used with limited success.[47,50]

1.4 Sexual Function

Human sexual function refers to the integrated activities of the testes and secondary sex glands, the endocrine control systems, and the central nervous system-based behavioral and psychologic components of reproduction (libido). Erection, ejaculation, and orgasm are three distinct, independent, physiological, and psychodynamic events that normally occur concurrently in men. If details regarding functions or mechanisms are desired, several reviews and in-depth reports are available.[51-53]

Assessment of occupational exposure-induced anomalies of sexual function is difficult. The researcher usually must rely on the testimony and recall of the worker regarding his sexual function. This testimony may often be confounded by the bias of the individual to guard his ego or masculine image or to attribute a pre-existing libido problem to exposures at work.

Burris and colleagues[54] reported application of a monitor for assessing erection at home. If these results are substantiated, this may provide a convenient, objective means by which to evaluate erectile function of exposed workers.

The assessment of ejaculate volume may provide information on the integrity of the emission phase of ejaculation. This is, of course, complicated by effects on the accessory sex

TABLE 4

Assessment of Reproductive Function

Assessment Method	Neuroendocrine System	Testes	Accessory Sex Glands	Sexual Function
Follicle-stimulating hormone (FSH)	X			
Luteinizing hormone (LH)	X			
Testosterone	X			
Sperm count		X		
Sperm morphology and morphometry		X		
Sperm chromosomes		X		
DNA stability assay		X		
Sperm viability (vital stain and hypo-osmotic swelling)			X	
Sperm motility (% motile and velocity)		?	X	
Semen biochemistry			X	
Semen pH			X	
Sperm penetration assay		X	X	
Nocturnal penile measurements				X
Personal history				X

glands' secretory capacity. Thus, a semen sample of reduced volume but with a normal ratio of constituents (marker chemicals) supports a diagnosis of an emission phase defect. Table 4 lists the various reproductive measures and the site at which a toxicant might affect this measure.

2 REPRODUCTIVE TOXICANTS

2.1 ENVIRONMENT

Several authors have reported that environmental exposures may have significant effects on the male reproductive system.[55-59] Few human studies, however, have been conducted evaluating environmental exposures.

There are four major indications that environmental exposures are related to deficits in male reproductive function. First, approximately 15% of all couples are infertile,[60,61] with 10 to 20% of unknown etiology. Even when reduced male fertility is identified, the causes of the anomalies are often idiopathic.[62] Second, there are several reports that sperm numbers have declined over the last several decades as environmental exposures have increased.[63-68] Third, several occupational exposures to toxicants have been demonstrated to cause decreases in the reproductive potential of male workers.[27] Fourth, environmental contaminants that are toxic to other physiologic systems have been found in human seminal plasma.[69,70]

Environmental (nonoccupational) effects on male fecundity have been reported in a few studies (Table 5). Environmental toxicant exposures have rarely been studied because of the difficulties of characterizing populations and exposure levels. Occupationally exposed populations tend to have higher exposures which are better defined.

2.2 LIFE STYLE

Life style practices may also have a negative effect on male fecundity (Table 6). Systemic illnesses,[71] infectious diseases,[71] and many prescribed pharmaceuticals[72] have negative effects on male fecundity.

TABLE 5

Environmental Effects on Male Reproduction — Selected References

	Neuroendocrine	Testis	Accessory Sex Glands	Sexual Function
Heat/climate		105, 106, 107, 108, 109		
Stress	110, 111	111, 112, 113	111, 112, 113	110
Clothes		114, 115	114, 115	
Radiation[a]		116, 117, 118		
Pesticides[b]	119	119	69, 70, 119	
Altitude	120	121		

[a] Radiation data were collected from clinical treatment of cancer.
[b] Most of the pesticide data are from occupational exposures.

TABLE 6

Life Style Effects on Male Reproduction — Selected References

	Neuroendocrine	Testis	Accessory Sex Glands	Sexual Function
Alcohol	122, 123	122, 124	122	125, 126
Smoking	127	124, 128	128, 129	
Drug abuse	130	130	130	131
Exercise	132, 133	132 (?)	133 (?)	
Caffeine		134	134, 135	
Sauna		136		

Note: (?) indicates a trend but not a significant difference.

2.3 Occupation

2.3.1 *Initiation of Studies*

Many factors influence the decision to study a group of workers. Studies of DBCP were initiated after informal discussions of infertility problems among wives attending a softball game.[73] A physician noting a cluster of patients from a specific occupation (e.g., professional truck drivers[74]) may initiate a study. The petroleum refinery industry exemplifies a profession in which the workers themselves had concerns regarding their reproductive health.[75] Work-related accidents such as the chemical spill of bromide[76] or the nuclear radiation disaster in Chernobyl[77] have led to studies. Corporations may conduct occupational research to validate previous claims, as with studies on dinitrotoluene (DNT) and toluenediamine (TDA).[78] In Europe, epidemiologic comparisons of fertility and paternal occupation have been made readily as parents' work records are linked to birth records in these countries. The majority of epidemiological and occupational research, however, has been stimulated by data from animal studies indicating that a compound is a reproductive toxicant. Thus, the toxicologist, physician, epidemiologist, and the worker himself, as well as the labor union and the corporation, have been and will continue to be "on the lookout" for potential exposures and study populations. These alerts have triggered several different approaches to defining the nature and extent of the effect. Some of these study designs are described below.

TABLE 7

Population-Based Studies of Paternal Effects

Type of Exposure/Occupation	Association with Exposure[a]	Effect	Ref.
Solvents	–	Spontaneous abortion	137
	+	Pre-term birth	142
Service station	+	Spontaneous abortion	137
Organic solvents	–	Spontaneous abortion	138
Mechanics	+	Spontaneous abortion	139
Food processing	+	Developmental defects	139
Ethylene oxide	+	Spontaneous abortion	140
Petroleum refinery	+	Spontaneous abortion	140
Impregnants of wood	+	Spontaneous abortion	140
Rubber chemicals	+	Spontaneous abortion	140
Metals	+	Child cancer risk	141
Machinists	+	Child cancer risk	141
Smiths	+	Child cancer risk	141
Lead and solvents	+	Pre-term birth	142
Lead	+	Perinatal death	142
	+	Male child morbidity	142

[a] –: no significant association; +: significant association.
Source: Modified from Taskinen, H., *Scand. J. Work Environ. Health,* 16, 297, 1990.

2.3.2 Population-Based Studies

Population-based studies (Table 7) link job title and/or description with reproductive outcome. Because adverse pregnancies are generally rare events, large populations are needed to detect a significant association. Although in many areas of the U.S. birth records do not reveal parental occupation, this type of study is routinely conducted in European countries where the parents' work history is included in medical records. These studies are problematic in that they preclude a control for potentially confounding life-style factors (e.g., alcohol consumption and drug abuse). Another major problem with population-based studies is the possibility of nondifferential misclassification of exposure. For example, pesticide exposure of such heterogeneous groups as crop farmers and a fishery husbandry may be very different, although these two groups are often classified together. Differentiation of exposure is difficult even if the International Standard Industrial Classification System is used. For example, Code 92102 (Refuse Removal Activities) combines cleaning companies, dumps, and incinerators, but cleaning company employees are not subjected to the levels of ash and pyrolysis products to which incinerator operators are commonly exposed. This drawback may make it more difficult to identify correctly a toxic effect on reproduction.

Population-based studies have caused a new surge of interest in male-mediated adverse pregnancy outcomes.[79] Several reports (Table 7) have shown that paternal exposure may affect pregnancy or the health of the offspring. These data have stimulated research into the genetic stability of the sperm cell and the cause/effect relationships of damage to sperm.[80]

2.3.3 Case-Control Studies

Case-control or case-referent studies (Table 8) involve comparing the frequency of toxic exposure of persons who have experienced reproductive dysfunction to those without such a medical history.[81] The relative exposure frequency of each group is determined by interviews

TABLE 8
Case-Control Studies of Paternal Effects

Type of Exposure/Occupation	Association with Exposure[a]	Effect	Ref.
Printing industry	(+)	Cleft lip	144
Paint	(+)	Cleft palate	144
Paint	+	Damage to CNS[b]	145
Solvents	(+)	Damage to CNS[b]	145
Low-level radiation	+	Neural tube defects	146
Organic solvents	+	Spontaneous abortion	147
Aromatic hydrocarbons	+	Spontaneous abortion	147
Dusts	+	Spontaneous abortion	147
Radiation	+	Childhood leukemia	148
Welding	+	Time to conception	149
Agriculture	(+)	Child brain tumor	150
Construction	(+)	Child brain tumor	150
Food/tobacco processing	(+)	Child brain tumor	150
Metal	+	Child brain tumor	150
Lead	(+)	Spontaneous abortion	151
Lead	(+)	Congenital defects	152
Ethylene glycol ether	+	Abnormal spermiogram	153
Metals	+	Cadmium in semen	154

[a] (+): marginally significant association; +: significant association.
[b] CNS = central nervous system.
Source: Modified from Taskinen, H., *Scand. J. Work Environ. Health,* 16, 297, 1990.

and other data sources. Like population-based studies, case-control studies provide the statistics needed to detect a rare event; however, in this type of study individual bias may yield misleading results. A person who has experienced an adverse outcome may report an exposure more readily than one with no history of reproductive dysfunction. As Levin describes,[81] a couple that has recently experienced a stillbirth or congenitally malformed child will be more inclined to search for a previous toxic exposure as the source. Another potential source of bias in these studies exists with the interviewer, who should be "blind" to the reproductive status and the exposure classification of the respondent.

2.3.4 *Standardized Fertility Ratio*

The standardized fertility or birth ratio (SFR) (Table 9) provides a comparison of the number of observed births within a population to the number of expected births. The latter value is obtained using the birth rate of an external population. To monitor a specific workforce group, Wong[82] uses a simple standardization method that provides noninvasive, readily available data. His method uses a couple's reproductive history to measure male fecundity. The number of expected births is obtained from the birth rate for U.S. women of the same age group. There are some problems with this simplified approach. Because the U.S. birth rate as collected by the National Center for Health Statistics does not take into account marital status, birth control, or frequency of intercourse,[83] the potential exists for a positive or negative bias in making such comparisons. Wong's method erroneously assumes all women to be equally libidinous and fecund. Additionally, because only married men are evaluated in SFR studies, there are age-related restrictions in the data as well as a likely underestimate of the number of children born out of wedlock. Some researchers have tried to correct for these variations in the population using different models.[83,84] It appears that the SFR overestimates birth rate in both control and exposed populations, even when a reduction in semen quality

TABLE 9

Standarized Birth Ratio Studies of Male Fertility Effects

Type of Exposure/Occupation	Association with Exposure[a]	Ref.
Ethylene dibromide	(+)	82
DBCP	+	83
TDA and DNT	–	78
Mercury	–	155
Manganese	+	155
Ethylene glycol ether	–	85
Chemical-contaminated sewage	–	156

[a] –: no significant association; (+): marginally significant association; +: significant association.

Source: Modified from Taskinen, H., *Scand. J. Work Environ. Health,* 16, 297, 1990.

exists.[85] Levine[84] suggests SFR would be a reliable indicator in a cohort study in which both control and comparison population are selected on the same basis.

2.3.5 Cohort Studies

Cohort studies (Table 10) evaluate the frequency of adverse reproductive outcome between exposed and unexposed groups. Cross-sectional cohort studies evaluate the groups as they currently exist, whereas historical cohort studies are conducted on previously exposed and unexposed groups. An example of the latter is the 1989 Vietnam experience study in which military veterans were grouped according to whether or not they had served in Vietnam from 1967 to 1972.[86] This study was able to detect subsequent differences between the groups in semen quality and time to pregnancy but revealed little about the reproductive health of the individuals at the time of exposure.

A cohort study may be longitudinal (prospective). In such a study, baseline data are collected, and individuals are studied over time for a specific reproductive outcome. A longitudinal study is exemplified by the study of "summer hire" pesticide applicators.[41] Individuals were evaluated at the beginning of the season before they started working with pesticides and at the conclusion of the spraying season 2 months later. This report illustrates the importance of selecting appropriate variables for a study design. If semen analyses are conducted to predict reproductive outcome, the correct time frame is needed. Since the time for spermatogenesis and delivery of mature sperm to the ejaculate is approximately 72 days, if primary spermatogonia were affected by exposure, this would not be observed in a time frame covering less than 80 to 90 days. Thus, the study of "summer hire" workers could not make valid conclusions regarding the effects of a 2-month exposure to pesticide application on spermatogenesis.

Cohort studies may involve questionnaires, neuroendocrine measurements, and semen analyses or a combination of these. Questionnaire studies are advantageous in that they are the least expensive and least invasive. As with case control studies, however, inherent problems exist, such as possible interviewer bias and the selective memory of respondents who have experienced an adverse outcome. Additionally, many cohort studies lack the number of pregnancies needed to detect differences between groups.[87] Studies evaluating the neuroendocrine system measure male hormones in blood samples. Although blood sampling is a widely accepted medical practice, it is an invasive procedure, and the endocrine profile does not necessarily reflect the status of the male reproductive system.[88] Semen analysis provides

TABLE 10

Cohort Studies of Male Reproductive Effects with Positive Association with Exposure

Type of Study	Type of Exposure	Effect	Ref.
Questionnaire, semen	Lead	Semen quality, sex function	157
Questionnaire, semen, endocrine	DBCP	Semen quality, endocrine	158
Questionnaire, semen	DBCP	Semen quality	159
Questionnaire, semen	Carbaryl (Sevin)	Semen quality	160
Questionnaire, semen	TDA and DNT	Semen quality	161
Questionnaire, records	High voltage	Congenital malformations	162
Questionnaire	DBCP	Sex ratio in offspring (favors ♀)	163
Questionnaire, semen, endocrine	Lead	Semen quality, not endocrine	88
Questionnaire, semen	Ethylene dibromide	Semen quality	164
Questionnaire, semen	Ethylene dibromide	Semen quality	41
Questionnaire, semen, endocrine	Plastic production (styrene and acetone)	Semen quality	165
Questionnaire, semen, endocrine	Ethylene glycol ethers	Semen quality	97
Questionnaire, semen	Ethylene glycol ethers	Semen quality	166
Questionnaire, semen, endocrine	Military (Vietnam)	Semen quality	86
Semen	Welding	Semen effects not reversible	167
Semen	Welding	Semen quality	168
Questionnaire, birth records	Welding	Fertility (mild steel)	169
Semen	Perchloroethylene	Semen quality	170
Questionnaire	Perchloroethylene	Time to pregnancy	171
Endocrine	Lead	Endocrine	172
Questionnaire, endocrine	Mercury vapor	Sex binding globulin	173
Questionnaire, semen	Heat	Semen quality and conception	174
Questionnaire, semen, endocrine	Radar	Semen quality	175
Questionnaire	Lead	Fertility	176
Questionnaire	Spray paint	Birth weight and length	177
Semen	Lead	Semen quality	178
Semen	2,4-Dichlorophenoxy acetic acid (2,4-D)	Semen quality	179
Questionnaire, semen	Carbon disulfide	Sexual function	180

information on spermatogenesis, accessory sex glands, and sperm cell motility.[89] Studies utilizing semen samples require the participants to produce samples by masturbation, and there are concerns regarding participation bias.[83] Additionally, this procedure employs complex scientific equipment and methodologies.[90] Cohort studies that combine all of the above approaches provide the most information and thus the greatest likelihood of detecting adverse reproductive effects. However, such studies are expensive and complex and necessitate a team approach requiring at the very least: (1) an andrologist, (2) an epidemiologist, (3) an industrial hygienist, (4) a physician, and (5) a statistician.[91]

2.3.6 Clinical Studies

Clinical or case studies (Table 11) involve the report by a physician of workers exposed to potentially toxic agents. These reports involve the evaluation of individuals, groups of workers of the same occupation or exposure, or clinical treatment following accidental exposure. While such reports rarely provide a definitive relationship between exposure and male reproductive effects, they can be sentinel reports which initiate further studies.

Some clinical studies provide unique information that would not be observed by using other study methods. The clinical study of a firearms instructor[92] provides possibly the best demonstration of the effect of lead on sperm. The instructor had fathered one son but became infertile as a result of work exposure which elevated his blood lead concentration to 88 μg/dl. During the next 3 years the exposure was decreased, and the man was placed on chelation

TABLE 11

Clinical/Case Studies of Male Reproductive Effects with Positive Association with Exposure

Type of Study	Type of Exposure	Effect	Ref.
Cluster	DBCP	Semen, endocrine	181
Case report	Kepone	Semen	93
Medical monitor	Lead	Semen	182
Case report	Methylene chloride	Semen	183
Case report	Lead	Semen	92
Accident	Bromine vapor	Semen	76
Accident	Radiation	Semen, endocrine	77

therapy. His sperm count increased as his blood levels decreased, and he subsequently fathered another child after the level of lead decreased below 30 μg/dl. Similarly, after men exposed to high levels of kepone in the work environment were treated with cholestyramine to offset the toxic action of kepone, their sperm count and sperm motility increased accordingly.[93]

Work-related accidents may provide case study data on high-level exposures. These in turn may indicate the parameters to be studied at lower exposures.

3 SURVEILLANCE OF MALE REPRODUCTION

A surveillance strategy for evaluating men working with known male reproductive toxicants was proposed and conducted by a team from the University of California;[94] however, this program had many problems and was eventually discontinued.[95] While this first attempt was discouraging, chemicals such as lead and ethylene glycol ethers remain in the U.S. workplace, posing a hazard to the reproductive health of the male worker. A surveillance program is required to monitor those working with these and other occupational toxicants.

3.1 Susceptibility of Subgroups

There is a concern that there may be a subgroup of individuals that may be more susceptible to a toxicant than others in the population. While for male reproductive toxicants in humans this has not been tested, there are some indications that it may in fact happen in some cases. Chapin et al.[96] have shown that different strains of mice respond differently do the same toxicant. Their data indicate that a strain of mice known to be "poor breeders" were more sensitive to the toxicant 2-methoxyethanol than strains of "good breeders". A similar response was noted by Welch et al.,[97] who reported on a study in which workers were exposed to a related compound, 2-ethoxyethanol. In this human study, the exposed workers had a median sperm count similar to the unexposed comparison group; however, when the distribution of sperm counts was evaluated, it appeared that the number of sperm of men with low sperm counts was much lower. This suggested that a man with a lower sperm count suffered a severe decrement compared to others in the population with exposure.

3.2 Declining Sperm Counts

Much has been written about declining male fecundity in recent years.[68,98,99] The reported decline in sperm numbers has received the most attention. Such reports are not necessarily new, as this controversy has been debated before.[63-67] What appears different is that the popular press has sensationalized these reports. While there is evidence that sperm numbers have changed, one must be cautious, remembering that both andrology and laboratory methodology have changed over the last half century. Are the observed changes a reflection of study participants and/or methodology, and are they real?[100-102] Related to this current controversy are reports that this decline is due to male offspring being exposed *in utero* to estrogenic compounds from our environment. While exposure to estrogenic (and other hormonal) compounds *in utero* can have negative effects on sperm production,[103] a direct cause and effect have not been shown between estrogenic compounds and a decline of sperm numbers in the human race. This is an interesting hypothesis that needs much more research.

4 SUMMARY

The male reproductive system is susceptible to toxic insult that results in decreased fecundity and/or an increase in adverse pregnancy. Each year, several hundred new compounds are added to the 70 thousand compounds and 4 million mixtures already in commercial use,[104] suggesting that there is a need to evaluate the effects that a man's workplace and environment has on his fecundity. There are several methods for evaluating male fecundity, often requiring a large research team and complex laboratory assessments. While the problems in conducting such a study may seem unsurmountable, a team approach has allowed this research to be successful.[91]

REFERENCES

1. Sokol, R. Z., Endocrine evaluations in the assessment of male reproductive hazards, *Reprod. Toxicol.*, 2, 217–222, 1988.
2. Schrader, S. M., Turner, T. W., Breitenstein, M. J., and Simon, S. D., Measurement of male reproductive hormones for field studies, *J. Occup. Med.*, 35, 574–576, 1993.
3. Bardin, C. W., Pituitary-testicular axis, in *Reproductive Endocrinology,* Yen, S. S. C. and Jaffe, R. B., Eds., W.B. Saunders, Philadelphia, 1986, pp. 177–199.
4. Raid-Fahmy, D., Read, G. F., Walker, R. F., and Griffiths, K., Steroids in saliva for assessing endocrine function, *Endocrinol. Rev.,* 3, 367–395, 1982.
5. Apostoli, P., Romeo, L., Peroni, E., Ferioli, A., Ferrari, S., Pasini, F., and Aprili, F., Steroid hormone sulphation in lead workers, *Br. J. Ind. Med.*, 46, 204–208, 1989.
6. Plant, T. M., Puberty in primates, in *The Physiology of Reproduction,* Knobil, E. and Neill, J. D., Eds., Raven Press, New York, 1988, pp. 1763–1788.
7. NIOSH, *Collecting a Semen Sample,* National Institute for Occupational Safety and Health, Cincinnati, OH, 1986.
8. Eliasson, R. and Treichl, L., Supravital staining of human spermatozoa, *Fertil. Steril.*, 22, 134–137, 1971.
9. Jeyendran, R. S., Van den Ven, H. H., Perez-Palaez, M., Crabo, B. G., and Zaneveld, L. J. D., Development of an assay to assess the functional integrity of the human sperm membrane and its relationship to other semen characteristics, *J. Reprod. Fertil.*, 70, 219–228, 1984.
10. Schrader, S. M., Platek, S. F., Zaneveld, L. J. D., Perez-Palaez, M., and Jeyendran, R. S., Sperm viability: a comparison of analytical methods, *Andrologia,* 18, 530–538, 1986.
11. MacLeod, J., Semen quality in 1000 men of known fertility and in 800 cases of infertile marriage, *Fertil. Steril.*, 2, 115–139, 1951.
12. Freund, M., Standards for the rating of human sperm morphology, *Int. J. Fertil.*, 11, 97–180, 1966.
13. Fredricson, B., Morphologic evaluation of spermatozoa in different laboratories, *Andrologia,* 11, 57–61, 1979.
14. Hanke, L. J., *Comparison of Laboratories Conducting Sperm Morphology,* Report TA78-28, National Institute for Occupational Safety and Health, Cincinnati, OH, 1981.
15. Schmassmann, A., Mikuz, G., Bartsch, G., and Rohr, H., Quantification of human sperm morphology and motility by means of semi-automatic image analysis systems, *Microscopica Acta,* 82, 163–178, 1979.
16. Katz, D. F., Overstreet, J. W., and Pelprey, R. J., Integrated assessment of the motility, morphology and morphometry of human spermatozoa, *INSERM,* 103, 97–100, 1981.

17. Schrader, S. M., Turner, T. W., Hardin, B. D., Niemeier, R. W., and Burg, J. R., Morphometric analysis of human spermatozoa, *J. Androl.,* 5, 22, 1984.
18. Jagoe, J. R., Washbrook, N. P., and Hudson, E. A., Morphometry of spermatozoa using semiautomatic image analysis, *J. Clin. Pathol.,* 139, 1347–1352, 1986.
19. DeStefano, F., Annest, J. L., Kresnow, M. J., Flock, M. L., and Schrader, S. M., Automated semen analysis in large epidemiologic studies, *J. Androl.,* 8, 24, 1987.
20. Turner, T. W., Schrader, S. M., and Simon, S. D., Sperm head morphometry as measured by three different computer systems, *J. Androl.,* 9, 45, 1988.
21. Moruzzim J. F., Wyrobek, A. J., Mayall, B. H., and Gledhill, B. L., Quantification classification of human sperm morphology by CASA, *Fertil. Steril.,* 50, 142–152, 1988.
22. Schrader, S. M., Ratcliffe, J. M., Turner, T. W., and Hornung, R. W., The use of new field methods of semen analysis in the study of occupational hazards to reproduction: the example of ethylene dibromide, *J. Occup. Med.,* 29, 963–966, 1987.
23. Zukerman, Z., Rodriguez-Rigau, L. J., Weiss, D. B., Chowdhury, A. K., Smith, K. D., and Steinberger, E., Quantitative analysis of the seminiferous epithelium in human testicular biopsies, and the relation of spermatogenesis to sperm density, *Fertil. Steril.,* 30, 448–455, 1978.
24. Whorton, D., Milby, T. H., Krauss, R. M., and Stubbs, H. A., Testicular function in DBCP-exposed pesticide workers, *J. Occup. Med.,* 21, 161–166, 1979.
25. Potashnik, G. and Abeliovich, D., Chromosomal analysis and health status of children conceived to men during or following dibromochloropropane-induced spermatogenic suppression, *Andrologia,* 17, 291–296, 1985.
26. Ratcliffe, J. M., Schrader, S. M., Steenland, K., Clapp, D. E., Turner, T., and Hornung, R. W., Semen quality in papaya workers with long term exposure to ethylene dibromide, *Br. J. Ind. Med.,* 44, 317–326, 1987.
27. Schrader, S. M. and Kanitiz, M. H., Occupational hazards to male reproduction, in *State of the Art Reviews in Occupational Medicine: Reproductive Hazards,* Gold, E., Schenker, M., and Lasley, B., Eds., Hanley & Belfus, Philadelphia, 1994, pp. 405–414.
28. Olshan, A. F. and Mattison, D. R., *Male Mediated Development Toxicity,* Plenum Press, New York, 1994.
29. Martin, R. H., A detailed method for obtaining preparations of human sperm chromosomes, *Cytogenet. Cell Genet.,* 35, 252–256, 1983.
30. Martin, R. H., Detection of genetic damage in human sperm, *Reprod. Toxicol.,* 7(Suppl. 1), 47–52, 1993.
31. Estop, A. M., Marquez, C., Munne, S., Navarro, J., Cieply, K., Vankirk, V., Martorell, M. R., Benet, J., and Templado, C., An analysis of human sperm chromosome breakpoints, *Am. J. Human Genet.,* 56, 452–460, 1995.
32. Evenson, D. P., Flow cytometry of acridine orange-stained sperm is a rapid and practical method for monitoring occupational exposure to genotoxicants, in *Monitoring Occupational Genotoxicity,* Sorsa, M. and Norppa, H., Eds., Alan R. Liss, New York, 1986, pp. 121–132.
33. Evenson, D. P., Jost. L. K., Baer, R. K., Turner, T. W., and Schrader, S. M., Individuality of DNA denaturation patterns in human sperm as measured by the sperm chromatin structure assay, *Reprod. Toxicol.,* 5(2), 115–125, 1991.
34. Spano, M. and Evenson, D. P., Flow cytometric studies in reproductive toxicology, in *New Horizons in Biological Dosimetry,* Gledhill, B. L. and Mauro, F., Eds., Wiley-Liss, New York, 1991, pp. 497–511.
35. Holmes, J. M. and Martin, R. H., Aneuploidy detection in human sperm nuclei usisng fluorescence *in situ* hybridization, *Human Genet.,* 91, 20–24, 1993.
36. Wyrobek, A. J., Robbins, W. A., Mehraein, Y., Pinkel, D., and Weier, H. U., Detection of sex chromosomal aneuploidies X-X, Y-Y, and X-Y in human sperm using two-chromosome fluorescence *in situ* hybridization, *Am. J. Med. Genet.,* 53, 1–7, 1994.
37. Bischoff, F. Z., Nguyen, D. D., Burt, K. J., and Shaffer, L. G., Estimates of aneuploidy using multicolor fluorescence *in situ* hybridization on human sperm, *Cytogenet. Cell Genet.,* 66, 237–243, 1994.
38. Stachel, B., Dougherty, R. C., Lahl, U., Schlosser, M., and Zeschmar, B., Toxic environmental chemicals in human semen: analytical method and case studies, *Andrologia,* 21, 282–291, 1989.
39. Zikarge, A., Cross-Sectional Study of Ethylene Dibromide-Induced Alterations of Seminal Plasma Biochemistry as a Function of Post-Testicular Toxicity with Relationships to Some Indices of Semen Analysis and Endocrine Profile, Dissertation to the University of Texas Health Science Center, Houston, TX, 1986.
40. Mann, T. and Lutwak-Mann, C., Passage of chemicals into human and animal semen: mechanisms and significance, *CRC Crit. Rev. Toxicol.,* 11, 1–14, 1982.
41. Schrader, S. M., Turner, T. W., and Ratcliffe, J. M., The effects of ethylene dibromide on semen quality: a comparison of short term and chronic exposure, *Reprod. Toxicol.,* 2, 191–198, 1988.
42. Yazigi, R. A., Odem, R. R., and Polakoski, K. L., Demonstration of specific binding of cocaine to human spermatozoa, *J. Am. Med. Assoc.,* 266, 1956–1959, 1991.
43. Aitkin, R. J., Development of *in vitro* tests of human sperm function: a diagnostic toll and model system for toxicological analyses, *Toxic. In Vitro,* 4, 560–569, 1990.
44. Katz, D. F., Overstreet, J. W., and Hanson, F. W., A new quantitative test for sperm penetration into cervical mucus, *Fertil. Steril.,* 33, 179–186, 1980.

45. Niederberger, C. S., Lamb, D. J., Glinz, M., Lipshultz, L. I., and Scully, N. F., Tests of sperm function for evaluation of the male — Penetrak-Asterisk and Tru-Trax, *Fertil. Steril.,* 60, 319–323, 1993.
46. Biljan, M. M., Taylor, C. T., Manasse, P. R., Joughin, E. C., Kingsland, C. R., and Lewisjones, D. I., Evaluation of different sperm function tests as screening methods for male fertilization potential — the value of the sperm migration test, *Fertil. Steril.,* 62, 591–598, 1994.
47. Rogers, B. J., Use of SPA in assessing toxic effects on male fertilizing potential, *Reprod. Toxicol.,* 2, 233–240, 1988.
48. Coddington, C. C., Franken, D. R., Burkman, L. J., Oosthuizen, W. T., Kruger, T., and Hodgen, G. D., Functional aspects of human sperm binding to the zona-pellucida using the hemizona assay, *J. Andrology,* 12, 1–8, 1991.
49. Franken, D. R., Acosta, A. A., Kruger, T. F., Lombard, C. J., Ochninger, S., and Hodgen, G. D., The hemizona assay — its role in identifying male factor infertility in assisted reproduction, *Fertil. Steril.,* 59, 1075–1080, 1993.
50. Schrader, S. M., Turner, T. W., and Simon, S. D., Sources of variation of the sperm penetration assay under field study conditions, *Assist. Reprod. Technol. Androl.,* 2(Suppl. 4), 63–74, 1991.
51. deGroat, W. C. and Booth, A. M., Physiology of male sexual function, *Am. Intern. Med.,* 92, 329–331, 1980.
52. Thomas, Jr., A. J., Ejaculatory dysfunction, *Fertil. Steril.,* 39, 445–454, 1983.
53. Krane, R. J., Goldstein, I., and de Tejada, I. S., Impotence, *N. Engl. J. Med.,* 321, 1648–1659, 1989.
54. Burris, A. S., Banks, S. M., and Sherins, R. J., Quantitative assessment of nocturnal penile tumescence and rigidity in normal men using a home monitor, *J. Androl.,* 10, 492–497, 1989.
55. Dixon, R. L., Sherins, R. J., and Lee, I. P., Assessment of environmental factors affecting male fertility, *Environ. Health Perspec.,* 30, 53–68, 1979.
56. Fabro, S., Ed., *Reproductive Toxicology: A Medical Letter on Environmental Hazards to Reproduction,* Reproductive Toxicology Center, Washington, D.C., 1985.
57. Manson, J. and Simons, R., Influence of environmental agents on male reproductive failure, in *Work and the Health of Women,* Hunt, V. R., Ed., CRC Press, Boca Raton, FL, 1979.
58. Sever, L. E. and Hessol, N. A., Toxic effects of occupational and environmental chemicals on the testes, in *Endocrine Toxicology,* Thomas, J. A., Ed., Raven Press, New York, 1985.
59. Thomas, J. A., Reproductive hazards and environmental chemicals: a review, *Toxic Substances J.,* 2, 318–348, 1981.
60. American Fertility Society, What you should know about infertility, *Contemp. Ob.-Gyn.,* 15, 101–105, 1980.
61. Rantala, M. L., Epidemiological and Clinical Studies on the Etiology of Infertility, Dissertation, University of Helsinki, Finland, 1988.
62. Purvis, K and Christiansen, E., Male infertility — current concepts, *Ann. Med.,* 24, 259–272, 1992.
63. Nelson, C. K. M. and Bunge, R. G., Semen analysis: evidence for changing parameters of male fertility potential, *Fertil. Steril.,* 26, 503–507, 1974.
64. Bostofte, E., Serup, J., and Rebbe, H., Has the fertility of Danish men declined through the years in terms of semen quality? A comparison of semen qualities between 1952 and 1972, *Int. J. Fertil.,* 128, 91–95, 1983.
65. Osser, O. Liedholm, P., and Ranstam, J., Depressed semen quality: a study over two decades, *Arch. Androl.,* 12, 113–116, 1984.
66. Menkveld, R., Van Zyl, J. A., Kotze, T. J. W., and Joubert, G., Possible changes in male fertility over 15-year period, *Arch. Androl.,* 17, 143–144, 1986.
67. James, W. H., Secular trends in reported sperm count, *Andrologia,* 12, 381–388, 1980.
68. Carlsen, E., Giwercman, A., Keiding, N., and Skakkebaek, N. E., Evidence for decreasing quality of semen during past 50 years, *Br. Med. J.,* 305, 609–613, 1992.
69. Kaur, S., Effect of environmental pollutants on human semen, *Environ. Contam. Toxicol.,* 40, 102–104, 1988.
70. Stachel, B., Daugherty, R. C., Lahl, U., Schlosser, M., and Zeschmer, B., Toxic environmental chemicals in human semen — analytical method and case studies, *Andrologia,* 21, 282–291, 1989.
71. Gangi, G. R. and Nagler, H. M., Clinical evaluation of the subfertile man, *Infertil. Reprod. Med. Clin. N. Am.,* 3, 299–318, 1992.
72. Drife, J. O., The effects of drugs on sperm, *Drugs,* 33, 610–622, 1987.
73. Whorton, D. and Foliart, D., DBCP: Eleven Years Later, NIOSH Symposium on the Assessment of Reproductive Hazards in the Workplace, Presentation June 16, Cincinnati, OH, 1988.
74. Sas, M. and Szollosi, J., Impaired spermatogenesis as a common finding among professional drivers, *Arch. Androl.,* 3, 57–60, 1979.
75. Rosenberg, M. J., Wyrobek, A. J., Ratcliffe, J., Gordon, L. A., Watchmaker, G., Fox, S. H., Moore, II, D. H., and Hornung, R. W., Sperm as an indicator of reproductive risk among petroleum refinery workers, *Br. J. Ind. Med.,* 42, 123–127, 1985.
76. Potashnik, G., Carel, R., Belmaker, I., and Levine, M., Spermatogenesis and reproductive performance following human accidental exposure to bromine vapor, *Reprod. Toxicol.,* 6, 171–174, 1992.
77. Birioukov, A., Meurer, M., Peter, R. U., Braun-Falco, O., and Plewig, G., Male reproductive system in patients exposed to ionizing irradiation in the Chernobyl accident, *Arch. Androl.,* 3, 99–104, 1993.

78. Hamill, P. V. V., Steinberger, E., Levine, R. J., Rodriguez-Rigau, L. J., Lemeshow, S., and Avrunin, J. S., The epidemiologic assessment of male reproductive hazard from occupational exposure to TDA and DNT, *J. Occup. Med.*, 24, 985–993, 1982.
79. Colie, C. F., Male-mediated teratogenesis, *Reprod. Toxicol.*, 7, 3–9, 1993.
80. Davis, D. L., Friedler, G., Mattison, D., and Morris, R., Male-mediated teratogenesis and other reproductive effects — biologic and epidemiologic findings and a plea for clinical research, *Reprod. Toxicol.*, 6, 289–292, 1992.
81. Levin, S. M., Problems and pitfalls in conducting epidemiological research in the area of reproductive toxicology, *Am. J. Ind. Med.*, 4, 349–364, 1983.
82. Wong, O., Utidijian, H. M. D., and Karten, V. S., Retrospective evaluation of reproductive performance of workers exposed to ethylene dibromide (EDB), *J. Occup. Med.*, 21, 98–102, 1979.
83. Levine, R. J., Blunden, P. B., Dalcorso, D., Starr, L. B., and Ross, C. E., Superiority of reproductive histories to sperm count in detecting infertility at dibromochloropropane manufacturing plant, *J. Occup. Med.*, 25, 591–597, 1983.
84. Levine, R. J., Monitoring fertility to detect toxicity to the male reproductive system, *Reprod. Toxicol.*, 2, 223–227, 1988.
85. Welch, L. S., Plotkin, E., and Schrader, S., Indirect fertility analysis in painters exposed to ethylene glycol ethers: sensitivity and specificity, *Am. J. Ind. Med.*, 20, 229–240, 1991.
86. Destefano, F., Annest, J. L., Kresnow, M.-J., Schrader, S. M., and Katz, D. F., Semen characteristics of Vietnam veterans, *Reprod. Toxicol.*, 3, 165–173, 1989.
87. Schrader, S. M., Ratcliffe, J. M., Turner, T. W., and Hornung, R. W., The use of new field methods of semen analysis in the study of occupational hazards to reproduction: the example of ethylene dibromide, *J. Occup. Med.*, 29, 963–966, 1987.
88. Assennato, G., Paci, C., Baser, M. E., Molinini, R., Candela, G., Altamura, B. M., and Giorgino, R., Sperm count suppression without endocrine dysfunction in lead-exposed men, *Arch. Environ. Health*, 41, 387–390, 1986.
89. Schrader, S. M. and Kesner, J. S., Mechanisms of male reproductive toxicology, in *Occupational and Environmental Reproductive Hazards: A Guide for Clinicians*, Paul, M., Ed., Williams & Wilkins, Baltimore, MD, 1992, pp. 3–17.
90. Schrader, S. M., Chapin, R. E., Clegg, E. D., Davis, R. O., Fourcroy, J. L., Katz, D. F., Rothmann, S. A., Toth, G., Turner, T. W., and Zinaman, M., Laboratory methods for assessing human semen in epidemiologic studies: a consensus report, *Reprod. Toxicol.*, 6, 275–279, 1992.
91. Schrader, S. M., General techniques for assessing male reproductive potential in human field studies, in *Methods in Reproductive Toxicology*, Vol. 3, Chapin, R. and Heindel, J., Eds., Academic Press, San Diego, 1993, pp. 362–371.
92. Fisher-Fischbein, J., Fischbein, A., Melnick, H. D., and Bardin, C. W., Correlation between biochemical indicators of lead exposure and semen quality in a lead-poisoned firearms instructor, *J. Am. Med. Assoc.*, 257, 803–805, 1987.
93. Cohn, W. J., Boylan, J. J., Blanke, R. V., Farris, M. W., Howell, J. R., and Guzelian, P. S., Treatment of chlordecone (kepone) toxicity with cholestyramine; results of a controlled clinical trial, *New Engl. J. Med.*, 298, 243–248, 1975.
94. Schenker, M. B., Samules, S. J., Perkins, C., Lewis, E. L., Katz, D. F., and Overstreet, J. W., Prospective surveillance of semen quality in the workplace, *J. Occup. Med.*, 30, 336–344, 1988.
95. Samuels, S. J., Lessons from a surveillance program of semen quality, *Reprod. Toxicol.*, 2, 229–231, 1988.
96. Chapin, R. E., Morrissey, R. E., Gulati, D. K., Hope, E., Barnes, L. H., Russell, S. A., and Kennedy, S. R., Are mouse strains differentially susceptible to the reproductive toxicity of ethylene glycol monomethyl ether? A study of 3 strains, *Fundam. Appl. Toxicol.*, 21, 8–14, 1993.
97. Welch, L. S., Schrader, S. M., Turner, T. W., and Cullin, M. R., Effects of exposure to ethylene glycol ethers on shipyard painters. I. Male reproduction, *Am. J. Ind. Med.*, 14, 509–526, 1988.
98. van Waeleghem, K., DeClercq, N., Vermeulen, L., Schoonjans, F., and Comhaire, F., Deterioration of sperm quality in young Belgium men during recent decades, *Hum. Reprod.*, 9(Suppl. 4), 73, 1994.
99. Auger, J., Kunstmann, J. M., Czyglik, F., and Jouannet, P., Decline in semen quality among fertile men in Paris during the past 20 years, *New Engl. J. Med.*, 332, 281–285, 1995.
100. Bromwich, P., Cohen, J., Stewart, I., and Walker, A., Decline in sperm counts: an artefact of changed reference range of "normal"?, *Br. Med. J.*, 309, 19–22, 1994.
101. Farrow, S., Falling sperm quality: fact or fiction?, *Br. Med. J.*, 309, 1–2, 1994.
102. Sherins, R. J., Are semen quality and male fertility changing?, *New Engl. J. Med.*, 332, 327, 1995.
103. Schardein, J. L., Hormones and hormonal antagonists, in *Chemically Induced Birth Defects*, Marcel Dekker, New York, 1993, pp. 271–340.
104. USEPA, *Chemical Substance Inventory*, EPA 620-929/0027/1985, U.S. Environmental Protection Agency Office of Toxic Substances, U.S. Government Printing Office, Washington, D.C., 1990.
105. Levine, R., Bordson, B. L., Mathew, R. M., Brown, M. H., Stanley, J. M., and Starr, T. B., Deterioration of semen quality during summer in New Orleans, *Fertil. Steril.*, 49, 900–907, 1988.

106. Levine, R. J., Mathew, R. M., Chenault, C. B., Brown, M. H., Hurtt, M. E., Bentley, K. S., Mohr, K. L., and Working, P. K., Differences in the quality of semen in outdoor workers during summer and winter, *New Engl. J. Med.*, 323, 12–16, 1990.
107. Levine, R. J., Brown, M. H., Bell, M., Shue, F., Greenberg, G. N., and Bordson, B. L., Air-conditioned environments do not prevent deterioration of human seman quality during the summer, *Fertil. Steril.*, 57, 1075–1083, 1992.
108. Ieffendy, W. K., Environmental risk factors in the history of male patients of an infertility clinic, *Andrologia*, 19, 262–265, 1987.
109. Mieusset, R., Association of scrotal hyperthermia with impaired spermatogenesis in infertile men, *Fertil. Steril.*, 48, 1006–1011, 1988.
110. McGrady, A. V., Effects of psychological stress on male reproduction: a review, *Arch. Androl.*, 13, 1–7, 1984.
111. Steeno, O. P. and Pangkahila, A., Occupational influences on male fertility and sexuality, *Andrologia*, 16, 93–101, 1984.
112. Giblin, M. P., Effects of stress and characteristic adaptability on semen quality in healthy men, *Fertil. Steril.*, 49, 127–132, 1988.
113. Harrison, K. L., Stress and semen quality in an *in vitro* fertilization program, *Fertil. Steril.*, 48(4), 633–636, 1987.
114. Sanger, W. G. and Friman, P. C., Fit of underwear and male spermatogenesis — a pilot investigation, *Reprod. Toxicol.*, 4(3), 229–232, 1990.
115. Shafik, A., Ibrahim, I. H., and Elsayed, E. M., Effect of different types of textile fabric on spermatogenesis. I. Electrostatic potentials generated on surface of human scrotum by wearing different types of fabric, *Andrologia*, 24, 145–147, 1992.
116. Mikamo, K., Kamiguchi, Y., and Tateno, H., Spontaneous and *in vitro* radiation-induced chromosome aberrations in human spermatozoa — application of a new method, *Mutat. Environ. B*, 340, 447–456, 1990.
117. Meistrich, M. L., Reduction in sperm levels after testicular irradiation of the mouse: a comparison with man, *Radiat. Res.*, 102, 138–147, 1985.
118. Martin, R. H., Hildebrand, K., Yamamoto, J., Rademaker, A., Barnes, M. et al., An increased frequency of human sperm chromosomal abnormalities after radiotherapy, *Mutat. Res.*, 174, 219–225, 1986.
119. Mattison, D. R., Bogumil, J., Chapin, R., Hatch, M., Hendrix, A., Jarrell, J., Labarbera, A., Schrader, S., and Selevan, S., Reproductive toxicology of pesticides, in *Advances in Modern Environmental Toxicology*, Baker, S. R. and Wilkinson, C. F., Eds., Scientific Press, Princeton, NJ, 1990, pp. 263–348.
120. Beall, C. M., Worthman, C. M., Stallings, J., Strohl, K. P., Brittenham, G. M., and Barragan, M., Salivary testosterone concentration of Aymara men native to 3600 m, *Ann. Human Biol.*, 19(1), 67–78, 1992.
121. Garciahjarles, M. A., Spermatogram and seminal biochemistry of high altitude natives and patients with chronic mountain sickness, *Arch. Biol. Med. Exp.*, 1989.
122. Ucheria, K., Semen analysis in alcohol dependence syndrome, *Andrologia*, 17, 558–563, 1985.
123. Ida, Y., Tsujimaru, S., Nakamaura, K., Shirao, I., Mukasa, H., Egami, H. et al., Effects of acute and repeated alcohol ingestion on hypothalamic-pituitary-gonadal and hypothalamic-pituitary-adrenal functioning in normal males, *Drug Alcohol Dependence*, 31, 57–64, 1992.
124. Savitz, D. A., Schwingl, P. J., and Keels, M. A., Influence of paternal age, smoking, and alcohol consumption on congenital anomalies, *Teratology*, 44(4), 429–440, 1991.
125. Nirenberg, T. D., Liepman, M. R., Begin, A. M., Doolittle, R. H., and Broffman, T. E., The sexual relationship of male alcoholics and their female partners during periods of drinking and abstinence, *J. Stud. Alcohol*, 51(6), 565–568, 1990.
126. Bain, C. L. and Guay, A. T., Reproducibility in monitoring nocturnal penile tumescence and rigidity, *J. Urol.*, 148, 811–814, 1992.
127. Attia, A. M., Cigarette smoking and male reproduction, *Arch. Androl.*, 23, 45–49, 1989.
128. Stillman, R. J., Rosenberg, M. J., and Sachs, B. P., Smoking and reproduction, *Fertil. Steril.*, 46, 545–566, 1986.
129. Holzki, G., Gall, H., and Hermann, J., Cigarette smoking and sperm quality, *Andrologia*, 23, 141–144, 1991.
130. Smith, R. C., Drug abuse and reproduction, *Fertil. Steril.*, 48, 355–373, 1987.
131. Fabro, S., *Drugs and Male Sexual Function. Reproductive Toxicology: A Medical Letter*, Vol. 4, Reproductive Toxicology Center, Washington, D.C., 1985.
132. Ayers, J. W. T., Komesu, Y., Romani, T., and Ansbacher, R., Anthropomorphic, hormonal and psychologic correlates of semen quality in endurance-trained male athletes, *Fertil. Steril.*, 43, 917–921, 1985.
133. Grandi, M. and Celani, M. F., Effects of football on the pituitary-testicular axis (PTA) — differences between professional and non-professional soccer players, *Exp. Clin. Endocrinol.*, 96, 253–259, 1990.
134. Marshburn, P. B. et al., Semen quality and association with coffee drinking, cigarette smoking, and ethanol consumption, *Fertil. Steril.*, 52, 162–165, 1989.
135. Beach, C. A., Bianchine, J. R., and Gerber, N., The excretion of caffeine in the semen of men: pharmacokinetics and comparison in the concentrations in blood and semen, *J. Clin. Pharmacol.*, 24, 120–126, 1984.
136. Brown-Woodman, P. D. C., Post, E. J., Gass, G. C., and White, I. G., The effect of a single sauna exposure on spermatozoa, *Arch. Endrol.*, 12, 9–15, 1984.

137. Lindbohm, M.-L., Hemminki, K., and Kyyronen, P., Parental occupational exposure and spontaneous abortions in Finland, *Am. J. Epidemiol.,* 120, 370–378, 1984.
138. Daniell, W. E. and Vaughan, T. L., Paternal employment in solvent related occupations and adverse pregnancy outcomes, *Br. J. Ind. Med.,* 45, 193–197, 1988.
139. McDonald. A. D., McDonald, J. C., Armstrong, B., Cherry, N. M., Nolin, A. D., and Robert, D., Fathers' occupation and pregnancy outcome, *Br. J. Ind. Med.,* 46, 329–333, 1989.
140. Lindbohm, M. L., Hemminki, K., Bonhomme, M. G., Anttila, A., Rantala, K., Heikkila, P., and Rosenberg, M. J., Effects of paternal occupational exposure on spontaneous abortions, *Am. J. Public Health,* 81, 1029–1033, 1991.
141. Olsen, J. H., Brown, P., Schulgen, G., and Jensen, O. M., Parental employment at time of conception and risk of cancer in offspring, *Eur. J. Canc.,* 27, 958–965, 1991.
142. Kristensen, P., Irgens, L. M., Daltveit, A. K., and Andersen, A., Perinatal outcome among children of men exposed to lead and organic solvents in the printing industry, *Am. J. Epiodemiol.,* 137, 134–144, 1993.
143. Taskinen, H., Effects of parental occupational exposures on spontaneous abortion and congenital malformation, *Scand. J. Work Environ. Health,* 16, 297–314, 1990.
144. Kucera, J., Exposure to fat solvents: a possible cause of sacral agenesis in man, *J. Pediatr.,* 72, 857–859, 1968.
145. Olsen, J., Risk of exposure to teratogens amongst laboratory staff and painters, *Dan. Med. Bull.,* 30, 24–28, 1983.
146. Sever, L. E., Gilbert, E. S., Hessol, N. A., and McIntyre, J. M., A case-control study of congenital malformations and occupational exposure to low-level radiation, *Am. J. Epidemiol.,* 127, 226–242, 1988.
147. Taskinen, H., Antila, A., Lindbohm, M. L., Sallmen, M., and Hemminki, K., Spontaneous abortions and congenital malformations amoung the wives of men occupationally exposed to organic solvents, *Scand. J. Work Environ. Health,* 15, 345–352, 1989.
148. Gardner, M. J., Hall, A. J., Snee, M. P. et al., Methods and basic design of case-control study of leukemia and lymphoma among young people near Sellafield nuclear plant in West Cumbria, *Br. Med. J.,* 300, 429–434, 1990.
149. Bonde, J. P. E., Subfertility in relation to welding — a case referent study among male welders, *Dan. Med. Bull.,* 37, 105–108, 1990.
150. Wilkins, J. R. and Sinks, T., Parental occupation and intracranial neoplasms of childhood: results of a case-control interview study, *Am. J. Epidemiol.,* 132, 275–292, 1990.
151. Lindbohm, M. L., Sallmen, M., Anttila, A., Taskinen, H., and Hemminki, K., Paternal occupational lead exposure and spontaneous abortion, *Scand. J. Work Environ. Health,* 17, 95–103, 1991.
152. Sallmen, M., Lindbohm, M. L., Anttila, A., Taskinen, H., and Hemminki, K., Paternal occupational lead exposure and congenital malformations, *J. Epidemiol. Comm. Health,* 46, 519–522, 1992.
153. Veulemans, H., Steeno, O., Masschelein, R., and Groesneken, D., Exposure to ethylene glycol ethers and spermatogenic disorders in man. A case-control study, *Br. J. Ind. Med.,* 50, 71–78, 1993.
154. Chia, S. E., Ong, C. N., Lee, S. T., and Tsakok, F. H. M., Blood concentration of lead, cadmium, mercury, zinc, and copper and human semen parameters, *Arch. Androl.,* 29, 177–183, 1992.
155. Lauwerys, R., Roels, H., Genet, P., Toussaint, G., Boukaert, A., and DeCoorman, S., Fertility of male workers exposed to mercury vapor or to manganese dust: a questionnaire study, *Am. J. Ind. Med.,* 7, 171–176, 1985.
156. Lemasters, G. K., Zenick, H., Hertzberg, V., Hansen, K., and Clark, S., Fertility of workers chronically exposed to chemically contaminated sewer wastes, *Reprod. Toxicol.,* 5, 31–37, 1991.
157. Lancranjan, I., Popescu, H. I., Gavanescu, O., Klepsch, I., and Serbanescu, M., Reproductive ability of workmen occupationally exposed to lead, *Arch. Environ. Health,* 30, 396–401, 1975.
158. Whorton, M. D., Milby, T. H., Krauss, R. M., and Stubbs, H. A., Testicular function in DBCP-exposed pesticide workers, *J. Occup. Med.,* 21, 161–166, 1979.
159. Milby, T. H. and Whorton, D., Epidemiological assessment of occupationally related, chemically induced sperm count suppression, *J. Occup. Med.,* 22, 77–82, 1980.
160. Wyrobek, A. J., Watchmaker, G., Gordon, L., Wong, K., Moore, II, D., and Whorton, D., Sperm shape abnormalities in carbaryl-exposed employees, *Environ. Health Persp.,* 40, 225–265, 1981.
161. Ahrenholz, S. H. and Meyer, C. R., *Health Hazard Evaluation Determination,* report HE 79-113-728: Olin Chemical Company, Brandernburg, KY, U.S. DHHS Centers for Disease Control, National Institute for Occupational Safety and Health, Cincinnati, OH, 1980.
162. Nordstrom, S., Birke, E., and Gustavsson, L., Reproductive hazards among workers at high-voltage substations, *Bioelectromagnetics,* 4, 91–101, 1983.
163. Goldsmith, J. R. and Potashnik, G., Reproductive outcomes in families of DBCP-exposed men, *Arch. Environ. Med.,* 39, 85–89, 1984.
164. Ratcliffe, J. M., Schrader, S. M., Steenland, K., Clapp, D. E., Turner, T. W., and Hornung, R. W., Semen quality in papaya workers with long-term exposure to ethylene dibromide, *Br. J. Ind. Med.,* 44, 317–326, 1987.
165. Jelnes, J. E., Semen quality in workers producing reinforced plastic, *Reprod. Toxicol.,* 2, 209–212, 1988.
166. Ratcliffe, J. M., Schrader, S. M., Clapp, D. E., Halperin, W. E., Turner, T. W., and Hornung, R. W., Semen quality in workers exposed to 2-ethoxyethanol, *Br. J. Ind. Med.,* 46, 399–406, 1989.
167. Bonde, J. P., Semen quality in welders before and after 3 weeks of non-exposure, *Br. J. Ind. Med.,* 47, 515–518, 1990.

168. Bonde, J. P., Semen quality and sex hormones among mild steel and stainless steel welders — a cross-sectional study, *Br. J. Ind. Med.,* 47, 508–514, 1990.
169. Bonde, J. P., Hansen, K. S., and Levine, R. J., Fertility among Danish male welders, *Scand, J. Work Environ. Health,* 16, 315–322, 1990.
170. Eskenazi, B. E., Wyrobek, A. J., Fenster, L., Katz, D. F., Sadler, M., Lee, J., Hudes, M., and Rempel, D. M., A study of the effect of perchloroethylene exposure on semen quality in dry cleaning workers, *Am. J. Ind. Med.,* 20, 575–591, 1991.
171. Eskenazi, B. E., Fenster, L., Hudes, M., Wyrobek, A. J., Katz, D. F., Gerson, J., and Rempel, D. M., A study of the effect of perchloroethylene exposure on the reproductive outcomes of wives of dry cleaning workers, *Am. J. Ind. Med.,* 20, 593–600, 1991.
172 Ng. T. P., Goh, H. H., Ng, Y. L., Ong, H. Y., and Jeyarathnam, J., Male endocrine functions in workers with moderate exposure to lead, *Am. J. Ind. Med.,* 48, 485–491, 1991.
173. McGregor, A. J. and Mason, H. J., Occupational mercury vapour exposure and testicular, pituitary, and thyroid endocrine function, *Human Exp. Toxicol.,* 10, 199–203, 1991.
174. Figa-Talamamca, I., Dell'Orco, V., Pupi, A., Dondero, F., Gandini, L., Lenzi, A., Lombardo, F., Scavalli, P., and Mancini, G., Fertility and semen quality of workers exposed to high temperature in the ceramics industry, *Reprod. Toxicol.,* 6, 517–523, 1992.
175. Weyandt, T. B., Schrader, S. M., Turner, T. W., and Simon, S. D., Semen analysis of military personnel associated with military duty assignments, *J. Androl.,* 13, 29, 1992.
176. Gennart, J. P., Buchet, J. P., Roels, H., Ghyselen, P., Ceulemans, E., and Lauwerys, R., Fertility of male workers exposed to cadmium, lead, or manganese, *Am. J. Epidemiol.,* 135, 1209–1219, 1992.
177. Hoglund, G. V., Iselius, E. L., and Knave, B. G., Children of male spray painters — weight and length at birth, *Br. J. Ind. Med.,* 49, 249–253, 1992.
178. Lerda, D., Study of sperm characteristics in persons occupationally exposed to lead, *Am. J. Ind. Med.,* 22, 567–571, 1992.
179. Lerda, D. and Rizzi, R., Study of reproductive function in persons occupationally exposed to 2,4-dichlorophenoxyacetic acid (2,4-D), *Mutat. Res.,* 262, 47–50, 1991.
180. Vanhoorne, M., Comhaire, F., and DeBacquer, D., Epidemiological study of the effects of carbon disulfide on male sexuality and repsroduction, *Arch. Environ. Health,* 49, 273–278, 1994.
181. Whorton, D., Krauss, R. M., Marshall, S., and Milby, T. H., Fertility in male pesticide workers, *Lancet,* 2, 1259–1260, 1977.
182. Cullen, M. R., Kaye, R. D., and Robins, J. M., Endocrine and reproductive dysfunction in men associated with occupational inorganic lead intoxication, *Arch. Environ. Health,* 39, 431–440, 1984.
183. Kelly, M., Case reports of individuals with oligospermia and methylene chloride exposures, *Reprod. Toxicol.,* 2, 13–17, 1958.
184. World Health Organization, *WHO Laboratory Manual for the Examination of Human Semen and Sperm-Cervical Mucus Interaction,* Third ed., Cambridge University Press, Oxford, 1992, pp. 43–44.

CHAPTER 23

Female Reproductive Toxicity

Anthony R. Scialli and Christine F. Colie

CONTENTS

1 ADVERSE OUTCOMES AND METHODS FOR THEIR ASCERTAINMENT

1.1 Clinical Settings

The female reproductive system includes organs and organ systems that serve a number of functions related to the creation and maturation of a new organism. Each of the components of the system (Table 1) may be subject to the toxic effects of xenobiotic agents. Note that among the clinical endpoints there is considerable overlap; in fact, these clinical endpoints are crude and nonspecific. Subfertility, for example, is an endpoint associated with abnormal function at any of a number of anatomic sites. In clinical reproductive medicine, the relationship between a patient complaint and a mechanism of dysfunction is often indirect.

In simplistic terms, the purpose of the female reproductive system is to foster the creation of viable offspring. Using this criterion, the appearance of healthy children is arguably the

0-8493-4493-X/97/$0.00+$.50

TABLE 1

Components of the Female Reproductive System

Organ	Functions	Clinical Manifestations of Malfunction
Ovary	Maturation/release of gametes	Infertility
	Production of sex steroids	Menstrual disturbance; hypoestrogenic symptoms; altered libido
Fallopian tube	Transport/nutrition of ovum/embryo	Infertility; ectopic pregnancy
	Transport of sperm	Infertility
Uterus	Implantation of embryo	Infertility; ectopic pregnancy
	Protection of conceptus	Spontaneous abortion; stillbirth
	Expulsion of fetus	Premature or prolonged gestation
Cervix	Production of mucus to store and protect sperm	Infertility
	Barrier to ascending organisms	Intrauterine infection
	Retention of fetus	Premature delivery
Vagina	Coitus	Dyspareunia
	Passage of fetus	
Vulva	Coitus	Dyspareunia
Central nervous system/pituitary	Libido	Sexual dysfunction
	Ovulatory mechanisms	Infertility; menstrual disturbance; hypoestrogenic symptoms

TABLE 2

Human Female (or Couple) Reproductive Parameters

Parameter (Definition)	Typical Values	Ref.
Infertility (proportion of couples not achieving pregnancy in 12 months of unprotected coitus)	0.10 to 0.15	80, 98, 107, 165, 177
Menstrual cycle length (first day of one cycle to first day of the next)	22 to 35 days; mean 29.1 ± 7.46 (SD)	42, 220
Duration of menstrual flow	2 to 8 days	220
Premenstrual symptoms	20 to 90%	88, 112, 119
Disabling premenstrual syndrome	3 to 8%	87, 108, 179, 184
Spontaneous abortion rate (proportion of recognized pregnancies)	0.3	239
Age at menarche (years)	13 ± 1 (mean ± SD)	76, 135, 251
Age at menopause (years)	48 to 55 (median 50)	124

only indicator of normal reproductive function. Although such a standard may be appropriate for experimental of domestic animals, the clinical care of women requires that the definition of normal also include regular menstrual cycles, comfortable sexual activity, and the absence of excessive genital-tract discomfort.

Table 2 lists human female reproductive parameters derived from apparently normal populations in developed countries. Parameters such as infertility show tremendous variations in incidence and are markedly influenced by age, reproductive history, and smoking status.[84,98] The heterogeneity among women within a population is accentuated by the observation that some parameters (age at menarche, for example) change over time and with geographic location. In studies of reproductive function, then, it is important to consider whether groups being compared (for example, women exposed and nonexposed to a putative toxicant) are in fact comparable for the reproductive parameters of interest.

Table 3 presents the main categories of infertility obtained from a large British study.[100] Some populations have a higher proportion of Fallopian tube damage as a cause of infertility,

TABLE 3

Causes of Infertility

Cause	Percentage of Infertile Couples
Male factor	22
Ovulatory problem	18
Tubal damage	14
Endometriosis	6
Coital problem	5
Cervical mucus factor	3
Other/mixed causes	3
Unexplained	28

Source: From Hull, M. G. R. et al., *Br. Med. J.*, 291, 1693, 1985. With permission.

presumably reflecting a higher incidence of sexually transmitted disease. Ovulatory problems increase with age, and older populations will have a higher proportion of couples with this category of infertility. Ovulatory disorders are generally considered not only to include failure of ovum release by the follicle but also inadequate follicular maturation, hormone production, and luteal function, all of which are interrelated. This category of fertility disorder is, in theory, the most likely to be associated with exposure to toxicants.

Population studies, however, differ from fertility studies in individual women. When a couple presents with a concern about fertility, the use of population statistics may not be of any clinical help in achieving a pregnancy. Instead, an orderly evaluation consisting of taking a history, performing a physical examination, and obtaining laboratory tests is used to identify the problem, if one exists, and to plan therapy. Table 4 details a typical protocol for a fertility evaluation and an outline of possible treatments. Of course, the evaluation and therapy are individualized to the particular couple's clinical situation, and the outline in Table 4 is intended only to give a general idea of how an evaluation might be conducted.

1.2 Population Investigations

The investigation of whether a population has experienced a decline in reproductive function, due for example to a toxicant, proceeds along entirely different lines from a fertility work-up for an individual couple. Although population studies can evaluate pregnancy rates or birth rates, which reflect in a general sense all of the parameters in Table 4, more detailed evaluation of reproductive alteration in a population concentrates on ovarian function and, in particular, on ovarian cyclicity.

Disruption of cyclicity can be evaluated using questionnaires or prospectively recorded diaries that provide information on dates of bleeding and clinical signs (such as preovulatory cervical mucus production) evaluable by individual women. Because there is considerable variation in parameters such as cycle length and duration of menstrual flow, rather large excursions from "normal" are necessary in order to suspect a change in reproductive parameters associated with a toxicant. Questionnaires and diaries are supplemented under ideal circumstances with biomarkers of ovarian cycle events. One of the oldest of the methods used to follow the ovarian cycle is based on the increase in basal body temperature associated with the presence of luteal progesterone after ovulation. By measuring her temperature each morning prior to arising, a women often can detect an increase of about 0.5°F a day or so after ovulation. This increased baseline temperature continues until progesterone plasma concentrations fall around the time of menstruation. The presence of this temperature shift

TABLE 4

A Clinical Approach to Infertility

Area of Reproductive Function	Information Obtained on History	Pertinent Physical Findings	Laboratory Tests and Procedures[a]	Therapeutic Options
Ovulatory function	Menstrual regularity Premenstrual symptoms (presence suggests ovulatory status) Menopausal symptoms (e.g., hot flashes) Hirsuitism (may reflect chronic anovulation or adrenal disorder) Galactorrhea (may reflect hyperprolactinemia) Other illnesses (e.g., thyroid, chronic illnesses)	Body habitus (weight extremes associated with anovulation) Hirsuitism Galactorrhea Abdominal or flank masses (e.g., adrenal tumors) Vaginal mucosa, cervical mucus (evidence of estrogenization) Ovarian size	Serum FSH (early follicular levels may be elevated if follicles are inadequately sensitive as may occur with age) Serum progesterone (luteal phase) Basal body temperature recording (presumptive evidence of ovulation) Urinary LH detection (commercial ovulation predictor kits) Sonogram (follicle growth and collapse)	Ovulation induction (clomiphene citrate, gonadotropins) Progesterone supplementation Ovum donation
Tubal factor	Previous pregnancy, including ectopic pregnancy Past pelvic infection Previous abdominal surgery	Pelvic mass or tenderness	Hysterosalpingogram Laparoscopic visualization of fallopian tubes	Salpingoplasty *In vitro* fertilization/embryo transfer
Cervical factor	No specific information is obtained from the history	Cycle-specific appearance of the cervical mucus (e.g., copious and clear in the preovulatory period)	Postcoital test	Estrogen supplementation Intrauterine insemination
Male factor	Previous pregnancies Chronic illnesses (e.g., diabetes mellitus) Medication use Ethanol use Erectile/ejaculatory function	Secondary sexual characteristics Testicular size	Standard semen analysis (ejaculate volume, sperm concentration, motility, morphology) Tests of sperm function (e.g., acrosome reaction, ovum penetration)	Artificial insemination with donor sperm *In vitro* fertilization/embryo transfer Gamete intrafallopian transfer Intracytoplasmic sperm injection
Endometriosis	Dyspareunia Dysmenorrhea	Nodularity, tenderness on pelvis examination	Laparoscopy	Surgical ablation of endometriosis Menstrual suppression (e.g., gonadotropin-releasing hormone analogs)
Unexplained infertility	—	—	—	Ovulation induction/intrauterine insemination *In vitro* fertilization/embryo transfer Gamete intrafallopian transfer

[a] FSH = follicle-stimulating hormone; LH = luteinizing hormone.

Note: Not all tests or procedures will be applicable to all patients with a given problem.

is suggestive of ovulation, although the temperature cannot be used to evaluate follicle or corpus luteum function or the adequacy of hormone levels in plasma.

The use of serial hormone measurements adds additional information to diaries and temperature recording. Sampling plasma every 2 to 3 days would give an informative profile of the cycle with identification of cycles in which follicular or luteal function are abnormal; however, such sampling is impractical for population studies. The use of urinary or salivary hormone determinations is more practical because subjects can collect and store daily samples. It has been shown in a small sample of working women that such a serial collection and storage scheme is achievable.[244] It should be noted, however, that it is common for apparently normal women to have atypical cycles during which ovulation does not occur or occurs followed by low levels of progesterone secretion.[117]

The addition of urinary or salivary hormone measures to a diary of cycle-associated symptoms and observations would permit a profile of the menstrual cycle among women in a population under investigation.[99] Table 5 lists some of the methods of cycle monitoring that have been used or proposed for population reproductive studies of women. The ability of any single test or group of tests to identify the adverse effects of a reproductive toxicant has not been shown, however, and it is not clear how well these tests would screen a population for reproductive impairment. It is more likely that a battery of tests such as those listed in Table 5 would find its greatest use in the testing of a hypothesis about the putative reproductive toxicity of a well-defined exposure when an appropriate comparison group exists.

1.3 Animal Studies in Female Reproductive Toxicology

The number of agents for which human female reproductive toxicity has been defined is small, due in part to the large variation in normal reproductive parameters and the difficulty and expense involved in performing detailed population studies. Most agents for which there is clear evidence of reproductive damage are chemical and physical agents for which nonreproductive toxicity is prominent — for example, cytotoxic cancer chemotherapeutic drugs.

In many instances, human female reproductive toxicity of an agent is suspected based on studies performed in experimental animals. The small experimental animals often used (mouse, rat, hamster) have 4- or 5-day *estrous* cycles rather than month-long *menstrual* cycles: however, the distinction in terminology is based on historic observations of marked cycle-dependency of sexual receptivity (during estrus) and on the endometrial discharge in nonpregnant cycles (during menses) rather than on tremendous differences in reproductive physiology. The neuroendocrinology, steroid biochemistry, and other physiologic events in the females of most species are similar in their susceptibility to disruption by toxicants.

There are differences between laboratory and domestic animals and humans that are worth mentioning, if only to understand better the design and results of some of the experiments that will be listed later in this chapter. Species can be divided based on whether ovulation occurs spontaneously, on a cyclic basis, or only in response to coitus (so called reflex-induced ovulation). Reflex ovulators include the rabbit and cat, among others. These animals have cohorts of maturing follicles in the ovary that are ovulated in response to gonadotropins (chiefly luteinizing hormone, LH) produced in response to coitus or to mechanical stimulation. Spontaneous ovulators, including mice, rats, hamsters, sheep, and primates (including humans), have an orderly and fairly predictable progression of follicular maturation followed by ovulation in response to an LH surge, whether or not coitus has occurred. In some animals (for example, the rodents), there is an absence of a luteal phase after ovulation unless coitus or mechanical stimulation of the genital tract has occurred. This feature gives rise to the possibility of creating a "pseudopregnancy" during which there is prolonged luteal progesterone production and a uterine deciduoma in response to mating with a vasectomized male or simple mechanical stimulation of the cervix.

TABLE 5

Tests of Ovarian Function Potentially Useful in Population Studies

Test	Rationale	Limitations	Ref.
Questionnaire	Women asked questions about menstrual cyclicity may be able to identify changes when comparing their current cycles to a pre-exposure epoch.	Inaccurate recall may be random or biased by the belief that a current exposure is harmful.	26
Diary	Prospective recording of menstrual cycle parameters (bleeding, pain, other symptoms) gives a less biased record of events	Large variability in normal cycle parameters requires large deviations from "normal" in order to raise the suspicion of a possible toxic effect.	26
Basal body temperature	Progesterone production after ovulation causes an increase in body temperature.	Not all ovulatory cycles show a temperature shift; in addition, the shift gives imprecise information about ovulation timing and no information about normalcy.	25, 95, 131, 151
Cervical mucus/vaginal secretion evaluation	Prior to ovulation, mucus is thin and watery with a characteristic crystallization pattern (ferning) on drying; after ovulation, the mucus becomes tacky and cloudy. Electric resistance of vaginal secretions also changes. The electric resistance changes are also applicable to saliva.	Women vary in their ability to detect mucus changes; these alterations reflect the estrogen peak prior to ovulation, not ovulation itself, and can be present in anovulatory cycles. Electric resistance changes are less subjective but require an instrument.	4, 5, 65, 103
Urinary LH	LH appears in high amounts in urine 1 to 2 days prior to ovulation.	Daily urine testing required around the time of anticipated ovulation; correlation with serum LH surge is 80 to 90%.	121, 228
Urinary steroid hormones	Conjugated steroids excreted in urine reflect plasma concentrations during preceding hours; urinary conjugated estrone and estriol peak in concert with the plasma estradiol peak that triggers the preovulatory LH surge. Pregnanediol, a progesterone metabolite, appears after ovulation.	Daily urine collection and storage required; conditions of storage may be important. The urinary steroid hormone measures give information primarily on the presence and timing of ovulation.	4, 18, 45, 109, 116-118, 130, 154, 239, 243
Urinary hCG	Transient rises in urinary hCG, measured by a sensitive assay, reflect early, clinically inapparent pregnancy losses.	Requires daily urine collection and storage.	118, 239
Salivary steroids (principally progesterone)	Salivary hormone levels are correlated with the unbound and, therefore, active fraction of the plasma steroid concentration.	There is disagreement on the predictive value of low salivary progesterone for low plasma progesterone and luteal phase deficiency.	58, 63, 132, 182, 232
Ultrasound imaging	Sonography, particularly using vaginal transducers, can image the ovaries with identification of follicular growth and collapse characteristic of ovulation. The endometrial image also correlates with the phase of the ovarian cycle.	Serial examinations are required, making subject compliance an issue. The instrumentation is expensive, and trained personnel are required.	20, 180, 244

Note: LH = luteinizing hormone; hCG = human chorionic gonadotropin.

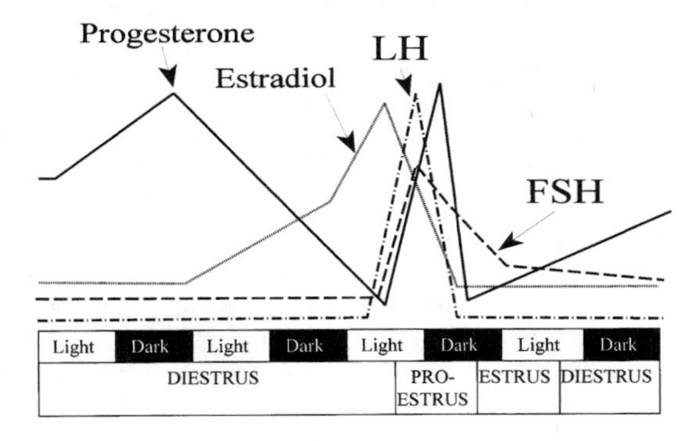

FIGURE 1
The typical rat 4-day estrous cycle. Hormone profiles are drawn in a simplified fashion from data published by Laskey et al.[129] Hormone concentrations are not drawn to relative scale. For comparative purposes, peak LH (luteinizing hormone) is 30 ng/ml, FSH (follicle stimulating hormone) 15 ng/ml, progesterone 45 ng/ml, and estradiol 60 pg/ml.

As indicated above, the estrous cycle is characterized by a period of sexual receptivity, called estrus from the Greek *oistros,* meaning desire. Three phases are generally described (see Figure 1). Diestrus (which is sometimes subdivided into metestrus and diestrus) typically lasts 3 days, during the latter part of which estrogen production begins to increase. Proestrus is the phase during which the LH surge occurs, typically beginning in the late morning or afternoon, followed by ovulation and the production of progesterone. Estrus, the period of maximal sexual receptivity, begins during the night following proestrus. After estrus is a return to diestrus (the portion of diestrus sometimes referred to as metestrus). Estrous cycle stages are evaluated by changes in vaginal epithelial cells collected by lavage. During estrus, most of the cells are cornified. In diestrus, there are leukocytes, cornified cells, and a few nucleated cells. In proestrus, both nucleated and cornified cells are predominant.

Human menstrual cycle disorders often have parallels in conditions found in rodents. For example, anovulatory conditions may manifest as persistent vaginal estrus. Reproductive senescence, either natural or toxicant-induced, is accompanied by anestrous or atrophic vaginal smears.[157] In addition, just as women nearing menopause have menstrual cycles of increasingly variable length, rodents with 4-day cycles will show an increasing number of 5- and 6-day cycles as they age. Such increased variability associated with a treatment would be suggestive of disruption of normal mechanisms of cyclicity.

It is important to remember that cyclicity and reproduction in experimental and domestic animals is often dependent on season, food supply, and laboratory conditions. Experimental animals may be maintained on an artificial schedule of light and dark periods to optimize reproductive potential. In some species, the presence of a male exerts an important influence on cyclicity.

Although reproductive toxicology studies can be conducted using a number of designs tailored to the research question at hand, several protocols have been in common use for studies of a general screening nature. Table 6 presents an overview of some of these study designs, and Table 7 gives the indices of reproductive function that may be calculated from these types of studies. Often a screening study will not define the nature of a reproductive toxic effect or even whether it is the male or the female of the species that is affected by a toxicant; however, these studies can be used to guide the design of more detailed investigations.

Perhaps the most important consideration in designing or evaluating reproductive toxicology studies is the exposure scenario. Exposure considerations include dose and route. The route of administration may be very important in determining whether toxicity will occur,

TABLE 6

Types of Studies Used in Female Reproductive Toxicity Research

Name	Design Features	Endpoints and Limitations
Single-generation test	A breeding pair produces a single litter, which is then evaluated. Males are exposed to the test agent from the age of 5 to 8 wk for a period of 8 to 10 wk before mating. More recent protocols use briefer periods of male exposure prior to mating, in conjunction with testicular histopathology evaluation after mating. Females are exposed for a short period of time prior to mating, and exposure is continued through pregnancy and lactation.	All the indices listed in Table 7 can be derived from the results of single-generation tests. Because of timing of exposure of females, effects produced by exposure of premeiotic or early meiotic stages of germ cell development will not be identified. Single-generation tests do not discriminate among treatment effects due to male reproductive toxicity, female reproductive toxicity, or a combination of both.
Multigeneration	The progeny from a single-generation design are mated to produce a second generation or even a third	Because the progeny (called F_1) of the first treated generation (called F_0 or P) will have had intrauterine exposure, early germ cell stages in the F_1 will have been exposed to the test agent. F_1 fertility parameters will reflect this exposure, and F_2 developmental parameters will reflect toxicity to the offspring from such early germ cell exposure.
Segment III test[a]	Exposure occurs during the latter part of pregnancy and during lactation.	Endpoints include indices of perinatal survival and growth. Although abnormalities produced during late pregnancy may be characterized as developmental toxicity, lactation-associated toxicity may be due either to developmental effects (the compound in milk directly affects the nursling) or adverse effects on the dam's ability to lactate or otherwise care for her pups.
Reproductive assessment by continuous breeding	A cohabiting pair remains together for the production of four or five litters. Treatment begins 1 wk prior to cohabitation. This results in several F_1 litters are often designated by letter (F_{1a}, F_{1b}, etc.). The last F_1 litter may be used for assessment of developmental affects. Any of the F_1 litters may be mated to produce a design analogous to a multigeneration type.	The fertility endpoints may be more sensitive in this design because milder manifestations of fertility impairment may be detected as a longer interval between litters. In addition, there is an opportunity to detect cumulative effects that may only produce effects late in the cohabitation period. This design does not separate male and female effects, although such effects can be investigated by mating members of the pair to untreated animals. There may be a confounding of developmental toxicity (loss of an entire litter due to resorption of abnormal conceptuses) with reproductive toxicity (failure of pregnancy to occur), a limitation common to most reproductive toxicity testing designs that use birth of pups as a criterion for diagnosing pregnancy.
End organ evaluation	Examination of vaginal smears for determination of estrous cycle stage is often used in reproductive studies. Additional evaluation of the female genital tract (for example, organ weight, histology) or neuroendocrine organs (hypothalamus, pituitary) can be performed at necropsy.	Departures from expected 4- or 5-day estrous cycles in rodents may be a sign of reproductive toxicity if these departures are large in magnitude and consistent over the course of the study. Evaluation of end organs requires killing of the animals and is not suited to interim evaluation — for example, in a continuous breeding study — unless the sample size is increased to permit sacrifice of some animals prior to study completion.

[a] The name, Segment III, refers to the three segments of preclinical testing used by the U.S. Food and Drug Administration. Segment I is a single-generation test; Segment II is a teratology test.

TABLE 7
Indices of Female Reproductive Function

Female mating index	$\frac{\text{No. of estrous cycles with copulation}}{\text{No. cycles required for conception}} \times 100$
Female fertility index	$\frac{\text{No. of females presumed pregnant}}{\text{No. females cohabited}} \times 100$
Female fecundity index	$\frac{\text{No. confirmed pregnant}}{\text{No. with copulatory plug/sperm}} \times 100$
Parturition index	$\frac{\text{No. of parturitions}}{\text{No. females confirmed pregnant}} \times 100$
Gestation index	$\frac{\text{No. of females with pups born alive}}{\text{No. confirmed pregnant}} \times 100$
Live litter size	$\frac{\text{No. live offspring}}{\text{No. females with copulatory plug/sperm}} \times 100$
Live birth index	$\frac{\text{Mean pups per litter born alive}}{\text{Mean pups per litter}} \times 100$
Viability index	$\frac{\text{Mean pups per litter alive on day 4}}{\text{Mean pups per litter born alive}} \times 100$
Lactation index[a]	$\frac{\text{Mean pups per litter alive on day 21}}{\text{Mean pups per litter alive on day 4}} \times 100$
Preweaning index	$\frac{\text{Mean pups born per litter} - \text{mean pups per litter weaned}}{\text{Mean pups born per litter}} \times 100$

[a] In many studies, litters are culled to a standard number (e.g., four male and four female pups). In such cases, the "weaning index" is used instead of the lactation index and includes the number of pups kept at day 4 in the denominator.

but it is the issue of dose that has produced the most misunderstanding in the interpretation of experimental animal studies. It is axiomatic that all agents will be toxic at some dose; for example, if a group of animals is so intoxicated as to be moribund, they will not mate, and the fertility index (see Table 7) will be zero. There is little point in performing a reproductive toxicology experiment using doses that cause extreme general toxicity. It is more important to determine if there is unique sensitivity of the reproductive system to a toxicant — that is, to see if reproduction is impaired in the absence of general toxicity or significant toxicity to other organs. It is also useful to estimate from experimental animal studies whether the exposure conditions commonly encountered by humans are likely to be associated with reproductive toxicity.

The quantitative estimation of the exposure conditions, including dose, at which human toxicity may occur is sometimes called risk assessment. Risk assessment uses assumptions about the magnitude of differences in sensitivity among different species and among individuals within a species in order to estimate a dose of a toxicant below which adverse effects are unlikely to occur in humans.[153] When reading the experimental literature regarding reproductive toxicology, it is necessary to consider the dose range used and the presence of general toxicity in evaluating the relevance of a particular study for estimating human risk.

2 THE FEMALE REPRODUCTIVE TOXICITY OF SELECTED AGENTS

Because of the dependence of toxicity on exposure conditions, it is misleading to create lists of "reproductive toxicants". For example, ionizing radiation at the doses used in cancer therapy (thousands of rads) can produce ovarian failure, but at the doses used in diagnostic radiology (millirads), there is no evidence of reproductive effects. If we construct a list of reproductive toxicants, we would not be correct in putting X-ray on the list (because virtually all human X-ray exposure is without adverse reproductive effect), but we also would not be correct in leaving it off (because there are some X-ray exposures that are clearly ovotoxic).

It is more useful to use descriptive designations rather than a listing strategy in order to convey the nature of the known reproductive effects of particular agents. In Table 8, we present descriptive information on a number of agents with female reproductive effects; our listing is accompanied by information, when available, on exposure conditions and specific effects. We have also provided our thoughts about the possible mechanisms of action of some of these agents, although we ask the reader to remember that these thoughts are often little more than hypotheses waiting to be tested. As might be expected, given the amount of attention to experimental animal studies thus far in this chapter, we present experimental data and epidemiologic data on equal footing. In estimating the potential for exposure-related human harm, we need all the help we can get and cannot afford to neglect any part of the data set.

The agents in Table 8 were selected to provide an overview of the range of female reproductive effects that have been associated with medicinal and other chemical exposures. Many of these exposures appear to act through an alteration of hormone production. For example, a number of pharmaceutical agents with dopamine antagonist properties cause an elevation in prolactin plasma concentration with consequent inhibition of gonadotropin secretion and ovarian estrogen production. Of course, the pharmaceutical preparation best known for interfering with female reproductive function is the oral contraceptive, which suppresses gonadotropin production and release and which has other effects on endometrial histology and cervical mucus production. Oral contraceptives are not presented in Table 8; detailed reviews of clinical considerations regarding these products are available elsewhere.[216]

3 ISSUES IN FEMALE REPRODUCTIVE TOXICOLOGY

Although female reproductive physiology is well understood, with increasing elucidation of the molecular mechanisms of normal reproductive function, there is a sense that the evaluation of toxicant-induced alterations has not kept pace. There are several areas to which active inquiry is being directed.

3.1 Test Protocols

The use of experimental animal testing will continue to be relied upon to provide clues about agents that may be capable of disrupting human reproductive function at relevant exposure levels. The protocol selected for evaluation of a particular compound may be the result of economic and regulatory influences as much as scientific deliberation. Whether any single test protocol will be found to be ideal as a screen for female reproductive toxicity potential is unclear but remains a question in which there is much interest.

3.2 Characterization of Exposures

Because the production of toxicity by any compound is a function of the exposure scenario, it is important to characterize exposure in a meaningful way. It is possible, even likely, that acute high-dose exposure to a compound will produce effects different from chronic, low-dose exposure. At present, there is considerable discussion about whether life-long exposure of humans to trace amounts of chemicals with endocrine toxicity in acute tests will have clinically relevant consequences. Treatment of pregnant rats with 1 μg of 2,3,7,8-tetrachlorodibenzo-*p*-dioxin (TCDD) on day 15 of pregnancy produces disruptions of estrous cyclicity in the female offspring,[77] but the utility of this finding with acute high-dose treatment in predicting response to chronic lower doses of TCDD has not been established.

Characterization of exposure has always been a major element of epidemiology research, and occupational exposures can be particularly challenging. Assumptions about exposure are often made from job title, which may provide some information about the agents for which there is exposure but which rarely provides a quantitative estimate of exposure. The use of biologic markers of exposure, such as plasma or urinary levels of putative toxicants or their products, may improve these estimates, but there are few compounds for which such biologic monitoring has been used.

3.3 Distinguishing Reproductive from Developmental Toxicity

Making a distinction between adverse effects on the female reproductive function and adverse effects on the conceptus is considered important in some settings. For example, in clinical medicine, if a couple cannot achieve a diagnosed pregnancy, the evaluation and treatment may be different from that of a couple who can achieve conception but who repeatedly lose the pregnancy to early miscarriage. This distinction, however, may be misleading. A couple with repeated early pregnancy loss may not have any diagnosed pregnancies; that is, the conceptus is lost before a pregnancy test becomes positive, and the presentation may be identical to that of a couple with infertility.

In fact, it is not clear that a distinction between female reproductive and early developmental toxicology is necessary. Reproductive toxicology testing protocols generally involve exposure of experimental animals after mating as well as before mating, which means that prevention of fertilization (a reproductive effect) and resorption of an entire litter (possibly a developmental effect) will give rise to the same endpoint. Similarly, human exposure to therapeutic, occupational, or environmental agents is not discretely confined to periods of ovum maturation, ovulation, and fertilization but is likely to be continued through early gestation.

Perhaps the most compelling reason to distinguish between female reproductive toxicity and developmental toxicity is in the investigation of mechanisms. Elucidation of the manner in which an agent produces abnormal reproductive outcome is an important step in the prevention of human disease, the ultimate goal of research in this field.

TABLE 8

The Female Reproductive Toxicity of Selected Agents

Agent	Species	Proposed Effects	Exposure Scenario	Mechanism	Ref.
Cancer Chemotherapeutic Drugs					
Busulfan	Rat	Destruction of primordial germ cells	10 mg/kg to the dam during pregnancy	DNA alkylation	51, 74, 114, 140, 169, 181, 187, 202, 203, 231
	Human	Ovarian failure, possibly reversible	Cancer chemotherapy, often with other agents	—	
Chlorambucil	Human	Ovarian failure; young age is protective	Cancer chemotherapy, often with other agents	DNA alkylation	39, 68, 153
Cyclophosphamide	Human	Amenorrhea, infertility; older women more likely to be affected than younger women	Cancer chemotherapy or immunosuppression; usual clinical doses	Alkylating agent; presumably interferes with folliculogenesis	15, 35, 128
Doxorubicin	Human	Ovarian failure; did not occur in women under 35 years of age	Combination chemotherapy for breast cancer	Intercalates nucleic acids	197
Etoposide	Human	Ovarian failure	Therapy for gestational trophoblastic disease	Antimitotic	43
5-Fluorouracil (5-FU)	Human	Suppression of menses on combination chemotherapy that included 5-FU; the role of this compound unclear	Clinical treatment of breast cancer	Pyrimidine analog; interferes with DNA synthesis	106
	Rat	Reduced fertile matings	25 mg/kg/day for 3 wk (about twice the human dose on a mg/kg basis)	—	(Package insert)
Mechlorethamine	Human	Ovarian failure; older women more likely to be affected	Combination chemotherapy (MOPP) for Hodgkin's disease	Alkylating agent; interferes with cell replication	198
Melphalan	Human	Impaired ovarian function	Cancer chemotherapy	Alkylation	3, 53, 189, 197
Methotrexate	Human	Amenorrhea	Cancer chemotherapy; other agents co-administered	Folic acid antagonist; inhibition of DNA synthesis may impair folliculogenesis	204
Vinblastine	Human	Ovarian dysfunction	Cancer chemotherapy; other agents co-administered	Antimitotic	197

Vincristine	Human	Infertility	Cancer chemotherapy; other agents co-administered	Antimitotic	168
Other Pharmaceuticals					
Bupropion	Human	Menstrual irregularity	Therapeutic use	Unknown	86
Clomiphene citrate	Mouse, rat	Inhibited implantation	—	Antiestrogenic	24, 48, 54, 199, 205
	Human	Luteal phase defect, early pregnancy loss	Ovulation induction		
Cyclosporine	Rabbit	Ovarian dysfunction	15 mg/kg	Unknown	6
Diclofenac	Rat, rabbit	Inhibition of ovulation (follicle rupture)	*In vitro* and *in vivo* models	Prostaglandin synthesis inhibition	40, 60
Dihydrotestosterone	Rat, monkey	Defeminization of sexually dimorphic regions of brain and subsequently of behavior; anovulation in rat	Prenatal or neonatal treatment	Androgenicity	13, 101, 150, 170, 219
Domperidone	Human	Increased lactation	30 mg/day	Increased prolactin due to dopamine antagonism	96, 173
Fluoxetine	Human	Anorgasmia	Therapy for depression	Unknown	21, 93, 120, 143, 148, 155
Fluphenazine	Human	Amenorrhea, galactorrhea	Therapeutic use	Dopamine antagonism with increased prolactin	(Package insert)
Hydroxyflutamide	Rat	Ovulation inhibition	5 mg every 12 hours for six doses	Inhibition of luteinizing hormone (LH) surge	162
Ibuprofen	Rat, rabbit	Decreased implantation sites	100 mg/kg/day (about 10 times the human dose)	Prostaglandin synthesis inhibitor	85, 160
	Human	Menstrual delay	Clinical use for dysmenorrhea	—	
Labetolol	Human	Altered sexual response	Therapeutic use	β-Adrenergic blockage	183
Leuprolide	Rat, human	Hypogonadism	Therapeutic use	Down-regulation of gonadotropin-releasing hormone receptor; possibly luteolytic in rats	161, 214, 246
	Rat	Spontaneous abortion (not seen in humans)	—	—	
Medroxyprogesterone	Human	Decreased fertility	Therapeutic; serum levels 1 to 3 mg/ml	Gonadotropin suppression; altered cervical mucus; endometrial atrophy	23, 167, 237
Mesoridazine	Human	Menstrual irregularity	Therapeutic use	Dopamine antagonist; increases prolactin	(Package insert)

TABLE 8 (continued)
The Female Reproductive Toxicity of Selected Agents

Agent	Species	Proposed Effects	Exposure Scenario	Mechanism	Ref.
Metoclopromide	Human	Amenorrhea	Clinical use for gastroparesis	Dopamine antagonist; increases prolactin	(Package insert)
Mifepristone	Mouse, rat	Inhibition of ovulation and fertilization, abnormal oviductal transport, impaired early embryo development	Given in humans as a postcoital contraceptive and as an abortifacient	Progesterone antagonist	47, 110, 178, 186, 187, 196, 206, 208, 209, 223, 242, 247
	Monkey, human	Attenuation of LH surge; abortion	—	—	
Miosoprostol	Human	Threatened abortion; menstrual abnormalities	Therapy for peptic disease	Prostaglandin E_1 mimicry	10, 94, 176
Nafarelin	Baboon	Early spontaneous abortion	Comparison to human therapy, but effect not confirmed in humans	Luteolysis	229, 230
	Human	Anovulation	Therapeutic	Downregulation of gonadotropin-releasing hormone receptor	
Naloxone	Rat, hamster, sheep, cow, pig, human	Increased LH, decreased prolactin in some species	Experimental protocol	Blocks endogenous opioids, which provide tonic inhibition of LH secretion	22, 36, 49, 55, 67, 70, 122, 136, 144, 145, 172, 175, 185, 200, 211–213, 222, 235, 236, 248
Neltrexone	Mice, humans	Increased LH	Experimental protocol	Blocks endogenous opioids, which provide tonic inhibition of LH secretion	158, 227
Nitrous oxide	Rat	Resorption	75% nitrous oxide 6 hr/day for 3 days	Both aneuploidy and interference with methionine have been proposed	44, 78, 126, 147, 190, 192
		Impaired fertility	Chronic low dose		
	Human	Spontaneous abortion; reduced fertility	Occupational; without scavenging	—	
Propoxyphene	Rat	Reduced fertility	200 mg/kg/day (about 100 times the human dose; associated with substantial maternal toxicity)	The reduced fertility in this study may well have been due to maternal toxicity and not a specific reproductive effect	59

Sertraline	Rat	Reduced fertility	80 mg/kg/day	Unknown	(Package label)
Spironolactone	Rat and mouse	Delayed sexual maturation; decreased fertility; decreased implantation	100 to 200 mg/kg/day	Anti-aldosterone and anti-androgenic action; perhaps effects on other steroid hormones	50, 156
	Human	Menstrual irregularity	Therapeutic use		
Sulconazole	Rat	Prolongation of estrous cycle	10 mg/kg/day	Effect may be due to general toxicity	123
Sulpiride	Rat, rabbit, baboon, human	Disruption of ovulation cycle; promotion of lactation	Therapy for lactation deficiency	Hyperprolactinemia secondary to dopamine antagonism	9, 11, 12, 37, 97, 133, 134, 164, 249, 250
Thioridazine	Human	Amenorrhea, galactorrhea	Therapeutic use	Dopamine antagonism with increased prolactin	(Package insert)
Tinidazole	Rat	Impaired sexual behavior	300/mg/kg/day (more than 100 times the human dose)	Unknown, possibly general toxicity	33
Verapamil	Human	Galactorrhea (which might be associated with disruption of menses)	Therapeutic use	Hyperprolactinemia	64
Recreational Drugs					
Caffeine	Monkey	Spontaneous abortion	Human equivalent of 5–15 cups of coffee per day	Unknown	72, 73, 102, 166, 238, 241
	Human	Spontaneous abortion	Controversial: some effect identified in some studies at 1 or 2 cups of coffee per day	Unknown	
		Infertility	> 1 cup of coffee per day	Meiotic dysfunction	
Cigarette smoking	Human	Decreased fertility; impaired follicle function; increased spontaneous abortion; earlier menopause	Recreational smoking	Presumed direct toxicity of smoking constituents (e.g., nicotine, cadmium, polycyclic aromatic hydrocarbons) on granulosa cells and trophoblast	19, 30, 105, 146, 225, 240
Δ^9-Tetrahydocannabinol	Rhesus monkey	Spontaneous abortion	2.5 mg/kg/day	Reduced chorionic gonadotropin and progesterone	14

TABLE 8 (continued)

The Female Reproductive Toxicity of Selected Agents

Agent	Species	Proposed Effects	Exposure Scenario	Mechanism	Ref.
Metals					
Copper	Human	Infertility; spontaneous abortion (anecdotal reports)	Wilson's disease	Unknown	31, 174
Lead	Human	Spontaneous abortion	High-level exposures, reported historically; more recent report suggests no such effect with blood lead in the 10- to 20-μg/dl range	Possible direct poisoning of conceptus	79, 188
Other Occupational and Environmental Chemicals					
Aldrin	Mouse, rat	Reduced fertility	12.5 or 25 ppm in the diet	Presumed estrogenicity at high dose	2, 115, 141, 142
	Birds	Abnormal development of sexual organs, infertility	Dipping eggs in the chemical		
Benzo(*a*)pyrene	Mouse	Inhibited follicle growth and ovulation	1 mg/kg	Unknown	149, 217
Carbaryl	Rat, gerbil	Decreased fertility	5000 ppm (rat), 2000 ppm (gerbil) in the diet	Unknown	46
Carbon disulfide	Human	Spontaneous abortion (questionable association); menstrual disorders	Chemical workers	Unknown	1, 90, 92, 226
Cycloheximide	Cow and pig, but not mouse Hamster and rabbit, but not rat	Inhibition of germinal vesicle breakdown Inhibition of follicular steroidogenesis and ovulation	Most of these studies used *in vitro* models of ovulation and germinal vesicle breakdown; the rabbit study used 5 mg/kg 20 hr before ovulation	Protein (receptor) synthesis inhibition	52, 61, 62, 69, 81, 82, 113, 125, 233, 234
Dieldrin	Mouse, rat	Reduced fertility	10 or 12.5 ppm in diet; decreased litter size at 5 ppm; altered gonadotropins at 2 ppm	Presumed estrogenicity at high dose	2, 16, 32, 75
Dimethylbenz(*a*)anthracene	Mouse	Oocyte destruction (direct or transplacental treatment)	3 or 5 mg/kg (whole animal) or 0.1–1.0 μg (intraovarian)	Unknown	201, 207, 221
Fenprostalene	Cow	Synchronization of estrous; abortion	1 or 2 mg/animal	Luteolysis, probably due to prostaglandin-like action	167, 171, 245

Formaldehyde	Human	No increase in spontaneous abortion	Chemical sterilization workers	Not applicable	91, 120
		Menstrual disorders; infertility (anecdotal)	Chemical workers	Unknown	
Hexachlorobenzene	Monkey	Increased menstrual cycle variability; decreased luteal phase progesterone	10 mg/kg/day by mouth	Direct toxicity to the oocyte or granulosa	66, 104
Mercaptobenzothiazole	Rat	Altered estrous cycle; reduced fertility	Unknown	Unknown	7
Methylene chloride	Rat	Negative two-generation study	1500 ppm (inhalation)	Not applicable	159, 218
	Human	Spontaneous abortion	Occupational	Unknown	
Nitrofurazone	Turkey	Reduced LH and egg production	15 mg/kg	Unknown	8, 11
	Mouse, rat	Gonadal hypoplasia	1250 ppm (mouse), 2500 ppm (rat) in diet		
Papaya extract	Rat	Decreased implantation	500 mg/kg/day	Unknown	71
Phenanthroline	Rat, fish	Inhibition of ovulation (follicle rupture)	*In vitro* models	Chelation of metal ions necessary for collagenase function	29, 34
2-Phenoxyethanol	Mouse	Decrease in litters per breeding pair	2.5% in the diet	Selective fetotoxicity	89
Sodium bromide	Rat	Decreased fertility	4800 mg/kg/day chronically	Unknown	224
2,3,7,8-Tetrachlorodibenzo-*p*-dioxin	Rat	Disruption of estrous cycle	1 μg to the dam on day 15 of pregnancy	Presumed estrogenicity or anti-estrogenicity	77
Tetraethylene glycol	Rat	Estrous cycle disruption	≤ 0.02 LD_{50}	Unknown	38
Vinylcyclohexene	Mouse	Oocyte destruction	2.7 mmol/kg reduces small oocyte population by 50%	Unknown; epoxidation appears to be required for oocyte toxicity	83, 215
Zearalenone	Pig, rat	Estrous cycle disruption; reduced fertility; embryo loss	Field studies and studies modeled on grain contaminated with *Fusarium graminearum*	Estrogenic activity	17, 27, 28, 41, 56, 57, 127, 137–139, 193–195

REFERENCES

1. Agadzhanova, A. A., Occupational hygiene of women engaged in the manufacture of rayon fiber, *Gig. Tr. Prof. Zabol.,* 22, 10–13, 1978.
2. Agarwal, S. P. and Ahmad, A., Effects of pesticides on reproduction in mammals, *Pesticides,* 12, 33–38, 1979.
3. Ahmann, D. L., Repeated adjuvant chemotherapy with phenylalanine mustard or 5-fluorouracil, cyclophosphamide and prednisone with or without radiation, *Lancet,* 1, 893–896, 1978.
4. Albertson, B. D. and Zinaman, M. J., The prediction of ovulation and monitoring of the fertile period, *Adv. Contracept.,* 3, 263–290, 1987.
5. Albrecht, B. H., Fernando, R. D., Regas, J., and Betz, G., A new method for predicting and confirming ovulation, *Fertil. Steril.,* 44, 200–205, 1985.
6. Al-Chalabi, H. A., Effect of cyclosporin A on the morphology and function of the ovary and fertility in the rabbit, *Int. J. Fertil.,* 29, 218–223, 1984.
7. Akeksandrov, S. E., Effect of vulcanizing accelerants on embryolethality in rats, *Biull. Eksp. Biol. Med.,* 93, 87–88, 1982.
8. Ali, B. H., Effect of furazolidone and nitrofurazone on egg production, on plasma luteinizing hormone and on prolactin concentrations in turkeys, *Br. Poult. Sci.,* 28, 613–621, 1987.
9. Aliev, M. G. and Kocharli, R. Kh., Effect of sulpiride on the hypothalamic monoaminergic regulation of prolactin formation and milk secretion, *Fiziol. Zh. SSSR.,* 69, 756–760, 1983.
10. Anonymous, Misoprostol, *Med. Lett. Drug Ther.,* 31, 21–22, 1989.
11. Aono, T., Miyake, A., Koike, K., Tasaka, K., and Kurachi, K., Impaired LH-RH release by estrogen in women with sulpiride-induced hyperprolactinemia, *Endocrinol. Jpn.,* 31, 571–557, 1989.
12. Aono, T., Aki, T., Koike, K., and Kurachi, K., Effect of sulpiride on poor puerperal lactation, *Am. J. Obstet. Gynecol.,* 143, 927–932, 1982.
13. Arai, Y., Yamanouchi, K., Mkizukami, S., Yanai, R., Shibata, K., and Nagasawa, H., Induction of anovulatory sterility by neonatal treatment with 5 beta-dihydrotestosterone in female rats, *Acta Endocrinol. (Copenh.),* 96, 439–443, 1981.
14. Asch, R. H. and Smith, C. G., Effects of 9-THC, the principal psychoactive component of marijuana during pregnancy in the rhesus monkey, *J. Reprod. Med.,* 31, 1071–1081, 1986.
15. Ataya, K. and Moghissi, K., Chemotherapy-induced premature ovarian failure: mechanisms and prevention, *Steroids,* 43, 607–626, 1987.
16. Atea, M. M., Zaki, A. A., and Korayem, W. I., Toxic effect of dieldrin on gonadotrophin levels (FSH and LH) in serum of mature female albino rats, *Arch. Exp. Veterinarmed.,* 44, 357–360, 1990.
17. Aucock, H. W., Marasas, W. F., Meyer, C. J., and Chalmers, P., Field outbreaks of hyperestrogenism (vulvovaginitis) in pigs consuming maize infected by *Fusarium graminearum* and contaminated with zearalenone, *J. S. Afr. Vet. Assoc.,* 51, 163–166, 1980.
18. Baird, D. D., Weinberg, C. R., Wilcox, A. J., McConnaughey, D. R., Musey, P. I., and Collins, D. C., Hormonal profiles of natural conception cycles ending in early, unrecognized pregnancy loss, *J. Clin. Endocrinol. Metab.,* 72, 793–800, 1991.
19. Baird, D. D. and Wilcox, A. J., Cigarette smoking associated with delayed conception, *J. Am. Med. Assoc.,* 253, 2979, 1985.
20. Bakos, O., Lundkvist, O., and Bergh, T., Transvaginal sonographic evaluation of endometrial growth and texture in spontaneous ovulatory cycles — a descriptive study, *Hum. Reprod.,* 8, 799–806, 1993.
21. Balogh, S., Hendricks, S. E., and Kang, J., Treatment of fluoxetine-induced anorgasmia with amantadine, *J. Clin. Psychiatry.,* 53, 212–213, 1992.
22. Barb, C. R., Kraeling, R. R., Rampacek, G. B., and Whisnant, C. S., Opioid inhibition of luteinizing hormone secretion in the postpartum lactating sow, *Biol. Reprod.,* 35, 368–371, 1986.
23. Bassol, S., Garza-Flores, J., Cravioto, M. C., and Diaz-Sanchez, V., Ovarian function following a single administration of depot-medroxyprogesterone acetate (DMPA) at different doses, *Fertil. Steril.,* 42, 216–222, 1984.
24. Bateman, B. G., Kolp, L. A., Nunley, W. C., Felder, R., and Burkett, B., Subclinical pregnancy loss in clomiphene citrate-treated women, *Fertil. Steril.,* 57, 25–27, 1992.
25. Bauman, J. E., Basal body temperature: unreliable method of ovulation detection, *Fertil. Steril.,* 36, 729–733, 1981.
26. Bean, J. A., Leeper, J. D., Wallace, R. B., Sherman, B. M., and Jagger, H., Variations in the reporting of menstrual histories, *Am. J. Epidemiol.,* 109, 181–185, 1979.
27. Becci, P. J., Johnson, W. D., Hess, F. G., Gallo, M. A., Parent, R. A., and Taylor, J. M., Combined two-generation reproduction-teratogenesis study of zearalenone in the rat, *J. Appl. Toxicol.,* 2, 201–206, 1982.
28. Belchev, L., Pathomorphological changes in the estrogenic syndrome of swine, *Vet. Med. Nauki,* 16, 33–40, 1970.
29. Berndtson, A. K. and Goetz, F. W., Protease activity in brook trout (*Salvelinus fontinalis*) follicle walls demonstrated by substrate-polyacrylamide gel electrophoresis, *Biol. Reprod.,* 38, 511–516, 1988.

30. Bernstein, L., Pike, M. C., Lobo, R. A. et al., Cigarette smoking in pregnancy results in marked decrease in maternal hCG and oestradiol levels, *Br. J. Obstet. Gynecol.,* 96, 92–96, 1989.
31. Biller, J., Swiontoniowski, M., Brazis, and P. W., Successful pregnancy in Wilson's disease: a case report and review of the literature, *Eur. Neurol.,* 24, 306–309, 1985.
32. Birgo, B. B. and Bellward, G. D., Effects of dietary dieldrin on reproduction in the Swiss-Vancouver (SWV) mouse, *Environ. Physiol. Biochem.,* 5, 440–450, 1975.
33. Boyadzhieva, N., Experimental studies on the influence of bioshik (tinidazole) on some parameters of reproduction, *Eksp. Med. Morfol.,* 29, 53–57, 1990.
34. Brannstrom, M., Woessner, Jr., J. F., Koos, R. D., Sear, C. H., and LeMaire, W. J., Inhibitors of mammalian tissue collagenase and metalloproteinases suppress ovulation in the perfused rate ovary, *Endocrinology,* 122, 1715–1721, 1988.
35. Brincker, H., Mouridsen, H. T., Anderson, K. W., Rose, C., and Dombernowksy, P., Castration induced by cytotoxic chemotherapy, *J. Clin. Oncol.,* 7, 679–680, 1989.
36. Brooks, A. N., Haynes, N. B., Yang, K., and Lamming, G. E., Ovarian steroid involvement in endogenous opioid modulation of LH secretion in seasonally anoestrous mature ewes, *J. Reprod. Fertil.,* 76, 709–715, 1986.
37. Bruguerolle, B., Jadot, G., Vallim M., Bouyard, L., Fabregou-Bergier, P., Perrot, J., and Bouyard, P., Four benzamides' (metoclopramide, sulpiride, sultopride and tiapride) effects on the oestrus cycle of the female rat: a comparative statistical study, *J. Pharmacol.,* 12, 27–36, 1981.
38. Byshovets, T. F., Barilyak, I. R., Korkach, V. I., and Spikovskaya, L. D., Gonadotoxic activity of glycols, *Gig. Sanit.,* 9, 84–85, 1987.
39. Callis, L., Nieto, J., Vila, A., and Rende, J., Chlorambucil treatment in minimal lesion nephrotic syndrome: a reappraisal of its gonadal toxicity, *J. Pediatr.,* 97, 653–656, 1980.
40. Carp, H. J., Fein, A., and Nebel, L., Effect of diclofenac on implantation and embryonic development in the rat, *Eur. J. Obstet. Gynecol. Reprod. Biol.,* 28, 273–277, 1988.
41. Chang, K., Kurtz, H. J., and Mirocha, C. J., Effects of the cyclotoxin zearalenone on swine reproduction, *Am. J. Vet. Res.,* 40, 1260–1267, 1979.
42. Chiazze, Jr., L., Brayer, F. R., Macisco, Jr., J. J., Parker, M. P., and Duffy, B. J., The length and variability of the human menstrual cycle, *J. Am. Med. Assoc.,* 203, 89–92, 1968.
43. Choo, Y. C., Chan, S. Y., Wong, L. C., and Ma, H. K., Ovarian dysfunction in patients with gestational trophoblastic neoplasia treated with short intensive course of etoposide (VP-16-213), *Cancer,* 55, 2348–2352, 1985.
44. Coate, W. B., Kapp, Jr., R. W., and Lewis, T. R., Chronic exposure to low concentrations of halothane-nitrous oxide; reproductive and cytogenetic effects in the rat, *Anesthesiology,* 50, 310–318, 1979.
45. Collins, W. P., Collins, P. O., Kilpatrick, M. J., Manning, P. A., Pike, J., and Tyler, J. P. P., The concentrations of urinary oestrone-3-glucuronide, LH and pregnanediol-3-glucuronide as indices of ovarian function, *Acta Endocrinol.,* 93, 123–128, 1979.
46. Collins, T. F. X., Hansen, W. H., and Keeler, H. V., Effect of carbaryl (Sevin) on reproduction or the rat and gerbil, *Toxicol. Appl. Pharmacol.,* 19, 202–216, 1971.
47. Croxatto, H. B., Salvatierra, A. M., Romero, C., and Spitz, I. M., Late luteal phase administration of RU486 for three successive cycles does not disrupt bleeding patterns or ovulation, *J. Clin. Endocrinol. Metab.,* 65, 1272–1277, 1987.
48. Cummings, A. M., Perreault, S. D., and Harris, S. T., Validation of protocols for assessing early pregnancy failure in the rat: clomiphene citrate, *Fundam. Appl. Toxicol.,* 16, 506–516, 1991.
49. Cuttler, L., Egli, C. A., Styne, D. M., Kaplan, S. L., and Grumbach, M. M., Hormone ontogeny in the ovine fetus. XVIII. The effect of an opioid antagonist on luteinizing hormone secretion, *Endocrinology,* 116, 1997–2002, 1985.
50. de Gasparo, M., Whitebread, S. E., Priswerek, G., Jeunemaitre, X., Corvol, P., and Menard, J., Antialdosterones: incidence and prevention of sexual side effects, *J. Steroid Biochem.,* 32, 223–227, 1989.
51. de Sanctis, V., Galimberti, M., Lucarelli, G., Polchi, P., Ruggiero, L., and Vullo, C., Gonadal function after allogenic bone marrow transplantation for thalassaemia, *Arch. Dis. Child,* 66, 517–520, 1991.
52. Dickmann, Z. and Terranova, P. F., Ovulation blockade through synergism of cycloheximide with assorted anesthetics, *Contraception,* 41, 189–195, 1990.
53. Dobbing, J., Pregnancy and leukaemia, *Lancet,* 1, 11–15, 1977.
54. Dziadek, M., Preovulatory administration of clomiphene citrate to mice causes fetal growth retardation and neural tube defects (exencephaly) by an indirect maternal effect, *Teratology,* 47, 263–273, 1993.
55. Ebling, F. J. and Lincoln, G. A., Endogenous opioids and the control of seasonal LH secretion in Soay rams, *J. Endocrinol.,* 107, 341–353, 1985.
56. Edwards, S., Cantley, T. C., Rottinghaus, G. E., Osweiller, G. D., and Day, B. N., The effects of zearalenone on reproduction in swine. I. The relationship between ingested zearalenone dose and anestrus in non-pregnant, sexually mature gilts, *Theriogenology,* 28, 43–50, 1987.
57. Edwards, S., Cantley, T. C., and Day, B. N., The effects of zearalenone on reproduction in swine. II. The effect on puberty attainment and postweaning rebreeding performance, *Theriogenology,* 28, 51–58, 1987.

58. Ellison, P. T. and Lager, C., Moderate recreational running is associated with lower salivary progesterone profile in women, *Am. J. Obstet. Gynecol.,* 154, 1000–1003, 1986.
59. Emmerson, J. L., Owen, N. V., Koenig, G. R., Markham, J. K., and Anderson, R. C., Reproduction and teratology studies on propoxyphene napsylate, *Toxicol. Appl. Pharmacol.,* 19, 471–479, 1971.
60. Espey, L. L., Comparison of the effect of nonsteroidal and steroidal antiinflammatory agents on prostaglandin production during ovulation in the rabbit, *Prostaglandins,* 26, 71–78, 1983.
61. Espey, L. L., Cycloheximide inhibition of ovulation, prostaglandin biosynthesis and steriodogenesis in rabbit ovarian follicles, *J. Reprod. Fertil.,* 78, 679–683, 1986.
62. Espey, L. L., Norris, C., Forman, J., and Siler-Khodr, T., Effect of indomethacin, cycloheximide, and aminoglutethimide on ovarian steroid and prostanoid levels during ovulation in the gonadotropin-primed immataure rat, *Prostaglandins,* 38, 531–539, 1989.
63. Evans, J. J., Progesterone in saliva does not parallel unbound progesterone in plasma, *Clin. Chem.,* 32, 542–544, 1986.
64. Fearrington, E. L., Rand, Jr., C. H., Rose, J. D., Hyperprolactinemia-galactorrhea induced by verapamiol, *Am. J. Cardiol.,* 51, 1466–1467, 1983.
65. Fernando, R. S., Regas, J., and Betz, G., Prediction of ovulation with the use of oral and vaginal electrical measurements during treatment with clomiphene citrate, *Fertil. Steril.,* 47, 409–415, 1987.
66. Foster, W. G., McMahon, A., Villeneuve, D. C., and Jarrell, J. F., Hexachlorobenzene (HCB) suppresses circulating progesterone concentrations during the luteal phase in the cynomolgus monkey, *J. Appl. Toxicol.,* 12, 13–17, 1992.
67. Fraioli, F., Fabbri, A., Gnessi, L., Moretti, C., Bonifacio, V., Isidori, A., and Dufau, M., Naloxone increases bioactive LH in man: evidence for selective release of early LH pool, *J. Endocrinol. Invest.,* 8, 513–517, 1985.
68. Freckman, H. A., Fry, H. L., Mendez, F. L., and Maurer, E. R., Chloambucil-prednisone therapy for disseminated breast carcinoma, *J. Am. Med. Assoc.,* 189, 23–26, 1964.
69. Fulka, Jr., J. Motlik, J., Fulka, J., and Jilek, F., Effect of cycloheximide on nuclear maturation of pig and mouse oocytes, *J. Reprod. Fertil.,* 77, 281–285, 1986.
70. Gabriel, S. M., Simpkins, J. W., and Kalra, S. P., Modulation of endogenous opioid influences on luteinizing hormone secretion by progesterone and estrogen, *Endocrinology,* 113, 1806–1811, 1983.
71. Garg, S. K. and Garg, G. P., Antifertility screening of plants. VII. Effect of five indigenous plants on early pregnancy in albino rats, *Ind. J. Med. Res.,* 59, 302–306, 1970.
72. Gilbert, S. G. and Rice, D. C., Somatic development of the infant monkey following *in utero* exposure to caffeine, *Fundam. Appl. Toxicol.,* 17, 454–465, 1991.
73. Gilbert, S. G., Rice, D. C., Rufehl, K. R., and Stavric, B., Adverse pregnancy outcome in the monkey (*Macaca fascicularis*) after chronic caffeine exposure, *J. Pharmacol. Exp. Ther.,* 245, 1048–1053, 1988.
74. Giorgiani, G., Bozzola, M., Cisternino, M., Locatelli, F., Gambrana, D., Bonetti, F., Zecca, M., Lorini, R., Sanders, J. E., and the Long-Term Followup Team, Endocrine problems in children after bone marrow transplant for hematologic malignancies, *Bone Marrow Transplant,* 8(Suppl. 1), 2–4, 1991.
75. Good, E. E. and Warte, G. W., Effects of insecticides on reproduction in the laboratory mouse. IV, *Toxicol. Appl. Pharmacol.,* 14, 201–203, 1969.
76. Graber, J. A., Brooks-Gunn, J., and Warren, M. P., The antecedents of menarcheal age: heredity, family environment, and stressful life events, *Child. Dev.,* 66, 346–359, 1995.
77. Gray, Jr., L. E., and Ostby, J. S., *In utero* 2,3,7,8-tetrachlorodibenzo-*p*-dioxin (TCDD) alters reproductive morphology and function in female rat offspring, *Toxicol. Appl. Pharmacol.,* 133, 285–294, 1995.
78. Gray, R. H., Nitrous oxide and fertility, *N. Engl. J. Med.,* 328, 284, 1993.
79. Graziano, J., Popovac, D., Factor-Litvak, P. et al., The Influence of Environmental Lead Exposure on Human Pregnancy Outcome, presented at the Conf. on Lead Research: Implications for Environmental Health, Research Triangle Park, N.C.; cited in Ernhart, C.B., A critical review of low-level prenatal lead exposure in the human. 1. Effects on the fetus and newborn, *Reprod. Toxicol.,* 6, 9–19, 1992.
80. Greenhall, E. and Vessey, M., The prevalence of subfertility: a review of the current confusion and a report of two new studies, *Fertil. Steril.,* 54, 978–983, 1990.
81. Greenwald, G. S. and Limback, D., Effectgs of treatment with cycloheximide at proestrus on subsequent *in vitro* follicular steroidogenesis in the hamster, *Biol. Reprod.,* 30, 1105–1116, 1984.
82. Greenwald, G. S. and Wang, S. C., *In vitro* effects of cycloheximide and luteinizing hormone on the estradiol-to-progesterone shift in follicular steroidogenesis, *Biol. Reprod.,* 40, 729–734, 1989.
83. Grizzle, T. B., George, J. D., Fail, P. A., Seely, J. C., and Heindel, J. J., Reproductive effects of 4-vinylcyclohexene in Swiss mice assessed by a continuous breeding protocol, *Fundam. Appl. Toxicol.,* 122, 122–129, 1994.
84. Hakim, R. B., Gray, R. H., and Zacur, H., Infertility and early pregnancy loss, *Am. J. Obstet. Gynecol.,* 172, 1510–1517, 1995.
85. Halbert, D. R., Menstrual delay and dysfunctional uterine bleeding associated with antiprostaglandin therapy for dysmenorrhea, *J. Repro. Med.,* 28, 592–594, 1983.
86. Halbert, D. R., Menstrual delay and dysfunctional uterine bleeding associated with bupropion treatment, *J. Clin. Psychiatr.,* 52, 15–16, 1991.

87. Hallman, J., The premenstrual syndrome — an equivalent of depression?, *Acta Psychiatr. Scand.*, 73, 403–411, 1986.
88. Hargrove, J. T. and Abraham, G. E., The incidence of premenstrual tension in a gynecologic clinic, *J. Reprod. Med.*, 27, 721–724, 1982.
89. Heindel, J. J., Gulati, D. K., Russell, V. S., Reel, J. R., Lawton, A. D., and Lamb, 4th, J. C., Assessment of ethylene glycol monobutyl and monophenyl ether reproductive toxicity using a continuous breeding protocol in Swiss CD-1 mice, *Fundam. Appl. Toxicol.*, 15, 683–696, 1990.
90. Hemminki, K., Franssila, E., and Vainio, H., Spontaneous abortions among female chemical workers in Finland, *Int. Arch. Occup. Environ. Health*, 46, 93–98, 1980.
91. Hemminki, K., Mutanen, P., Saloniemi, I., Niemi, M. L., and Vainio, H., Spontaneous abortions in hospital staff engaged in sterilizing instruments with chemical agents, *Br. Med. J.*, 285, 1461–1463, 1982.
92. Hemminki, K. and Niemi, M. L., Community study of spontaneous abortions: relation to occupation and air pollution by sulfur dioxide, hydrogen sulfide, and carbon disulfide, *Int. Arch. Occup. Environ. Health*, 51, 55–63, 1982.
93. Herman, J. B., Brotman, A. W., Pollack, M. H., Falk, W. E., Biederman, J., and Rosenbaum, J. F., Fluoxetine-induced sexual dysfunction, *J. Clin. Psychiatr.*, 51, 25–27, 1990.
94. Herting, R. I. and Clay, G. A., Overview of misoprostal clinical experience, *Dig. Dis. Sci.*, 31, 47S–54S, 1986.
95. Hilgers, T. W. and Bailey, A. J., Natural family planning. II. Basal body temperature and estimated time of ovulation, *Obstet. Gynecol.*, 55, 333–339, 1980.
96. Hofmeyr, G. J., Van Iddekinge, B., and Blott, J. A., Domperidone secretion in breast milk and effect on puerperal prolactin levels, *Br. J. Obstet. Gynecol.*, 92, 1414, 1985.
97. Horie, K., Ban, C., Taii, S., Mori, T., and Aso, T., Impaired steroidogenic function of corpora lutea from hyperprolactinemic baboons induced by sulpiride, *Endocrinol. Jpn.*, 33, 211–214, 1986.
98. Howe, G., Westhoff, C., Vessey, M., and Yeates, D., Effects of age, cigarette smoking, and other factors on fertility: findings in a large prospective study, *Br. Med., J.*, 290, 1697–1700, 1985.
99. Hughes, Jr., C. L., Monitoring of ovulation in the assessment of reproductive hazards in the workplace, *Reprod. Toxicol.*, 2, 163–169, 1988.
100. Hull, M. G. R., Glazener, C. M. A., Kelly, N. J., Conway, D. I., Foster, P. A., Hinton, R. A., Coulson, C., Lambert, P. A., Watt, E. M., and Desai, K. M., Population study of causes, treatment, and outcome of infertility, *Br. Med. J.*, 291, 1693, 1985.
101. Iguchi, T. and Takasugi, N., Occurrence of permanent anovulation in mouse ovaries and permanent changes in the vaginal and uterine epithelia following neonatal treatment with large doses of 5 alpha-dihydrotestosterone, *Endocrinol. Jpn.*, 28, 207–213, 1981.
102. Infante-Rivard, C., Fernandez, A., Gauthier, R., David, M., and Rivard, G.-E., Fetal loss associated with caffeine intake before and during pregnancy, *J. Am. Med. Assoc.*, 270, 2940–2943, 1993.
103. Jacobs, M. H., Blaso, L., and Sondeimer, S. J., Ovulation prediction by monitoring salivary and vaginal electrical resistance with the peak ovulation predictor, *Obstet. Gynecol.*, 73, 817–822, 1989.
104. Jarrell, J. F., McMahon, A., Villeneuve, D., Franklin, C., Singh, A., Valli, V. E., and Bartlett, S., Hexachlorobenzene toxicity in the monkey primordial germ cell without induced porphyria. *Reprod. Toxicol.*, 7, 41–47, 1993.
105. Jick, H. and Porter, J., Relation between smoking and age of natural menopause, report from the Boston Collaborative Drug Surveillance Program, Boston University Medical Center, *Lancet*, 1, 1354, 1977.
106. Jochimsen, P. R., Spaight, M. E., and Urdaneta, L. F., Pregnancy during adjuvant chemotherapy for breast cancer, *J. Am. Med. Assoc.*, 245, 1660–1661, 1981.
107. Joffe, M. and Li, Z., Association of time to pregnancy and the outcome of pregnancy, *Fertil. Steril.*, 62, 71–75, 1994.
108. Johnson, S. R., McChesney, C., and Bean, J. A., Epidemiology of premenstrual symptoms in a nonclinical sample. I. Prevalence, natural history and help-seeking behavior, *J. Reprod. Med.*, 33, 340–346, 1988.
109. Jones, G. S., Some newer aspects of the management of inferetility, *J. Am. Med. Assoc.*, 141, 1123–1129, 1949.
110. Juneja, S. C. and Dodson, M. G., *In vitro* effect of RU 486 on sperm-egg interaction in mice, *Am. J. Obstet. Gynecol.*, 163, 216–221, 1990.
111. Kari, F. W., Huff, J. E., Leininger, J., Haseman, J. K., Eustis, S. L., Toxicity and carcinogenicity of nitrofurazone in F344/N rats and B6C3F1 mice, *Food. Chem. Toxicol.*, 276, 129–137, 1989.
112. Kashiwagi, T., McClure, Jr., J. N., Wetzel, R. D., Premenstrual affective syndrome and psychiatric disorder, *Dis. Nerv. Syst.*, 37, 116–119, 1976.
113. Kastrop, P. M., Hulshof, S. C., Bevers, M. M., Destree, O. H., and Kruip, T. A., The effects of alpha-amanitin and cycloheximide on nuclear progression, protein synthesis, and phosphorylation during bovine oocyte maturation *in vitro, Mol. Reprod. Dev.*, 28, 249–254, 1991.
114. Kasuga, F. and Takahashi, M., The endocrine function of rat gonads with reduced number of germ cells following busulphan treatment, *Endocrinol. Jpn.*, 33, 105–115, 1986.
115. Keplinger, M. L., Deichmann, W. B., and Sala, F., Effects of combinations of pesticides on reproduction in mice, *Pest. Symp., Coll. Pap. Inter-Am. Conf. Toxicol. Occup. Med.*, 647, 125–138, 1970.

116. Kesner, J. S., Krieg, Jr., E. F., Knecht, E. A., Wright, D. M., Power analyses and immunoassays for measuring reproductive hormones in urine to assess female reproductive potential in field studies, *Scand. J. Work Environ. Health,* 18(Suppl. 2), 33–36, 1992.
117. Kesner, J. S., Wright, D. M., Schrader, S. M., Chin, N. W., and Krieg, Jr., E. F., Methods of monitoring menstrual function in field studies: efficacy of methods, *Reprod. Toxicol.,* 6, 385–400, 1992.
118. Kesner, J. S., Knecht, E. A., and Krieg, Jr., E. F., Stability of urinary female reproductive hormones stored under various conditions, *Reprod. Toxicol.,* 9, 239–244, 1995.
119. Kessel, N. and Coppen, A., The prevalence of common menstrual symptoms, *Lancet,* 2, 61, 1963.
120. Kline, M. D., Fluoxetine and anorgasmia, *Am. J. Psychiatr.,* 146, 804–805, 1989.
121. Knee, G. R., Feinman, M. A., Strauss, J. F., Blasco, L., and Goodman, D. B. P., Detection of the ovulatory luteinizing hormone (LH) surge with a semiquantitative urinary LH assay, *Fertil. Steril.,* 44, 707–709, 1985.
122. Knight, P. G., Howles, C. M., and Cunningham, F. J., Evidence that opioid peptides and dopamine participate in the suckling-induced release of prolactin in the ewe, *Neuroendocrinology,* 44, 29–35, 1986.
123. Kobayashi, T., Ariyuki, F., Higaki, K., Shibano, T., Kano, M., Kitahara, S., and Nakagawa, H., Reproductive studies in rats and rabbits given sulconazole nitrate (RS 44872), *Oyo Yakuri,* 30, 451–465, 1985.
124. Krailo, M. D. and Pike, M. C., Estimation of the distribution of age at natural menopause from prevalence data, *Am. J. Epidemiol.,* 117, 356, 1983.
125. Kubelka, M., Motik, J., Fulka, Jr., J., Prochazka, R., Rimkevicova, Z., and Fulka, J., Time sequence of germinal vesicle breakdown in pig oocytes after cycloheximide and p-aminobenzamidine block, *Gamete Res.,* 19, 423–431, 1988.
126. Kugel, G., Letelier, C., Atallah, H., and Zive, M., Chronic low level nitrous oxide exposure and infertility, *J. Dent. Res.,* 63, 313, 1989.
127. Kumagai, S. and Shimizu, T., Neonatal exposure to zearalenone causes persistent anovulatory estrus in the rat, *Arch. Toxicol.,* 50, 279–286, 1982.
128. Langevitz, P., Klein, L., Pras, M., and Many, A., The effect of cyclophosphamide pulses on fertility in patients with lupus nephritis, *Am. J. Reprod. Immunol.,* 28, 157–158, 1992.
129. Laskey, J. W., Berman, E., and Ferrell, J. M., The use of cultured ovarian fragments to assess toxicant alterations in steroidogenesis in the Sprague-Dawley rat, *Reprod. Toxicol.,* l9, 141–131, 1995.
130. Lasley, B. L., Mobed, K., and Gold, E. B., The use of urinary hormonal assessments in human studies, *Ann. N. Y. Acad. Sci.,* 709, 299–311, 1994.
131. Leader, A., Wiseman, D., and Taylor, P. J., The prediction of ovulation: a comparison of the basal body temperature graph, cervical mucus score, and real-time pelvic ultrasonography, *Fertil. Steril.,* 43, 385–388, 1985.
132. Lenton, E. A., Gelsthorp, C. H., and Harper, R., Measurement of progesterone in saliva: assessment of the normal fertile range using spontaneous conception cycles, *Clin. Endocrinol.,* 28, 637–646, 1988.
133. Lin, K. C., Kawamura, N., Okamura, H., and Mori, T., Inhibition of ovulation, steroidogenesis and collagenolytic activity in rabbits by sulpiride-induced hyperprolactinaemia, *J. Reprod. Fertil.,* 83, 611–618, 1988.
134. Lin, K. C., Okamura, H., and Mori, T., Inhibition of human chorionic gonadotropin-induced ovulation and steroidogenesis by short-term hyperprolactinemia in female rabbits, *Endocrinol. Jpn.,* 34, 675–683, 1987.
135. Lindgren, G. W., Degerfors, I. L., Frederiksson, A., Loukili, A., Mannerfeldt, R., Nordin, M., Palm, K., Petterson, M., Sundstrand, G. and Sylvan, E., Menarche 1990 in Stockholm schoolgirls, *Acta Paediatr. Scand.,* 80, 953–955, 1991.
136. Lodico, G., Stoppelli, I., Delitala, G., and Maioli, M., Effects of naloxone infusion on basal and breast-stimulation-induced prolactin secretion in puerperal women, *Fertil. Steril.,* 40, 600–603, 1983.
137. Long, G. G., Diekman, M., Tuite, J. F., Shannon, G. M., and Vesonder, R. F., Effect of *Fusarium roseum* corn culture containing zearalenone on early pregnancy in swine, *Am. J. Vet. Res.,* 43, 1599–1603, 1982.
138. Long, G. G. and Diekman, M. A., Characterization of effects of zearalenone in swine during early pregnancy, *Am. J. Vet. Res.,* 47, 184–187, 1986.
139. Long, G. G., Diekman, M., Tuite, J. F., Shannon, G. M., and Vesonder, R. F., Effect of *Fusarium roseum* (*Gibberella zea*) on pregnancy and the estrous cycle in gilts fed moulded corn on days 7–17 post-estrus, *Vet. Res. Commun.,* 6, 199–204, 1983.
140. Lopez-Ibor, B. and Schwartz, A. D., Gonadal failure following busulfan therapy in an adolescent girl, *Am. J. Pediatr. Hematol. Oncol.,* 8, 85–87, 1986.
141. Lutz-Ostertag, Y., Lutz, H., Note préliminaire sur les effets “estrogènes” de l'aldrine sur le tractus uro-genital de l'embryon d'oiseau, *CR Acad. Sci.,* 269, 484–486, 1969.
142. Lutz-Ostertag, Y. and Lutz, H., Sexuality and pesticides, *Ann. Biol.,* 13, 173–185, 1974.
143. Lydiard, R. B., and George, M. S., Fluoxetine-related anorgasmy, *South. Med. J.,* 82, 933–934, 1989.
144. Malven, P. V., Bossut, D. F., and Diekman, M. A., Effects of naloxone and electroacupuncture treatment on plasma concentrations of LH in sheep, *J. Endocrinol.,* 101, 75–80, 1984.
145. Mattioli, M., Conte, F., Galeati, G., and Seren, E., Effect of naloxone on plasma concentrations of prolactin and LH in lactating sows, *J. Reprod. Fertil.,* 76, 167–173, 1986.
146. Mattison, D. R., Shiromizu, K., and Nightingale, M. S., Oocyte destruction by polycyclic aromatic hydrocarbons, *Am. J. Ind. Med.,* 4, 91, 1983.

147. Mazze, R. I., Fujinaga, M., Rice, S. A., Harris, S. B., and Baden, J. M., Reproductive and teratogenic effects of nitrous oxide, halothane, isoflurane, and enflurane in Sprague-Dawley rats, *Anesthesiology,* 62, 226–228, 1986.
148. McCormick, S., Olin, J., and Brotman, A. W., Reversal of fluxetine-induced inorgasmia by cyproheptadine in two patients, *J. Clin. Psychiatr.,* 51, 383–384, 1990.
149. Miller, M. M., Plowchalk, D. R., Weitzman, G. A., London, S. N., and Mattison, D. R., The effect of benzo(a)pyrene on murine ovarian and corpora lutea volumes, *Am. J. Obstet. Gynecol.,* 166, 1535–1541, 1992.
150. Mizukami, S., Yamanouchi, K., Arai, Y., Yanai, R., and Nagasawa, H., Failure of ovulation after neonatal administration of 5-alpha-dihydrotestosterone to female rats, *Endokrinologie,* 79, 1–6, 1982.
151. Moghissi, K. M., Prediction and detection of ovulation, *Fertil. Steril.,* 34, 89–98, 1980.
152. Moore, J. A., Daston, G. P., Faustman, E., Golub, M. S., Hart, W. L., Hughes, C., Jr., Kimmel, C. A., Lamb, J. C., IV, Schwetz, B. A., and Scialli, A. R., An evaluation process for assessing human reproductive and developmental toxicity of agents, *Reprod. Toxicol.,* 9, 61–95, 1995.
153. Morgenfield, M. C., Goldberg, V., Parisier, H., Buynard, J. C., and Bur, G. E., Ovarian lesions due to cytostatic agents during the treatment of Hodgkin's disease, *Surg. Gynecol. Obstet.,* 134, 826–828, 1972.
154. Munro, C. J., Stabenfeldt, G. H., Cragun, J. R., Addiego, L. A., Overstreet, J. W., and Laskey, B. L., Relationship of serum estradiol and progesterone concentrations to the excretion profiles of their major urinary metabolites as measured by enzyme immunoassay and radioimmunoassay, *Clin. Chem.,* 37, 383–844, 1991.
155. Musher, J. S., Anorgasmia with the use of fluoxetine, *Am. J. Psychiatr.,* 147, 948, 1990.
156. Nagi, S. and Virgo, B. B., The effects of spironolactone on reproductive functions in female rats and mice, *Toxicol. Appl. Pharmacol.,* 66, 21–228, 1992.
157. Nelson, J. F., Felico, L. S., Randall, P. K., Sims, C., and Finch, C. E., A longitudinal study of estrous cyclicity in aging C57BL/6J mice. I. Cycle frequency, length and vaginal cytology, *Biol. Reprod.,* 27, 327–339, 1982.
158. Nieder, G. L. and Corder, C. N., Effects of opiate antagonists on early pregnancy and pseudopregnancy in mice, *J. Reprod. Fertil.,* 65, 341–346, 1982.
159. Nitschke, K. D., Eisenbrandt, D. L., Lomax, L. G., and Rao, K. S., Methylene chloride: two-generation inhalation reproductive study in rats, *Fundam. Appl. Toxicol.,* 11, 60–67, 1988.
160. Ono, M., Ogawa, Y., Ogihara, K., Nagase, M., and Asimi, K., Reproductive studies of ibuprofen by rectal administration, *Oyo Yakuri,* 24, 467–473, 539–534, 1982.
161. Ooshima, Y., Yoshida, T., Sugitani, T., Sudo, K., and Ihara, T., Effect of TAP-144-SR on fertility and general reproductive performance of rats, *Yakuri to Chiryo,* 18(Suppl.), S589-607, 1990.
162. Opavsky, M. A., Chandrasekhar, Y., Roe, M., and Armstrong, D. T., Interference with the preovulatory luteinizing hormone surge and blockade of ovulation in immature pregnant mare's serum gonadotropin-primed rats with the anti-androgenic drug, hydroxyflutamide, *Biol. Reprod.,* 36, 636–642, 1987.
163. Ortiz, A., Serum medroxyprogesterone acetate (MPA) concentrations and ovarian function following intramuscular injections of depot-MPA, *J. Clin. Endocrinol. Metab.,* 44, 32–38, 1977.
164. Oseko, F., Morikawa, K., Motohashi, T., and Aso, T., Effects of chronic sulpiride-induced hyperprolactinemia on menstrual cycles of normal women, *Obstet. Gynecol.,* 72, 267–271, 1988.
165. Page, H., Estimate of the prevalence and incidence of infertility in a population: a pilot study, *Fertil. Steril.,* 51, 571–577, 1989.
166. Parazzini, F., Chatenoud, L., and La Vecchia, C., Fetal loss and caffeine intake, *J. Am. Med. Assoc.,* 272, 28, 1994.
167. Patterson, D. J., Corah, L. R., Kiracofe, G. H., Stevenson, J. S., and Brethour, J. R., Conception rate in *Bos taurus* and *Bos indicus* crossbred heifers after postweaning energy manipulation and synchronization of estrus with melengestrol acetate and fenprostalene, *J. Anim. Sci.,* 67, 1138–1147, 1989.
168. Pektasides, D., Rustin, G. J., Newlands, E. S., Begent, R. H., and Bagshawe, K. D., Fertility after chemotherapy for ovarian germ cell tumors, *Br. J. Obstet. Gynaecol.,* 94, 477–479, 1987.
169. Pelloux, M. C., Picon, R., Gangnerau, M. N., and Darmoul, D., Effects of busulfan on ovarian folliculogenesis, steroidogenesis and anti-mullerian activity of rat neonates, *Acta Endocrinol. (Copenh.),* 118, 218–226, 1988.
170. Perakis, A. and Stylianopoulou, F., Effects of a prenatal androgen peak on rat brain sexual differentiation, *J. Endocrinol.,* 108, 281–285, 1986.
171. Peters, A. R., Changes in electrical resistance of the vaginal mucosa in prostaglandin-treated cows, *Vet. Rec.,* 125, 505–507, 1989.
172. Petraglia, F., D'Ambrogio, G., Comitini, G., Facchinetti, F., Volpe, A., and Genazzani, A. R., Impairment of opioid control of luteinizing hormone secretion in menstrual disorders, *Fertil. Steril.,* 43, 534–540, 1985.
173. Petraglia, F., De Leo, V., Sardelli, S., Pieroni, M. L., D'Antona, N., and Genazzani, A. R., Domperidone in defective and insufficient lactation, *Eur. J. Obstet. Gynecol. Reprod. Biol.,* 19, 281–287, 1985.
174. Piussah, C. and Mathieu, M., Teratogenic risk during treatment of Wilson's disease, *J. Genet. Hum.,* 33, 357–362, 1985.
175. Piva, F., Maggi, R., Limonta, P., Motta, M., and Martini, L., Effect of naloxone on luteinizing hormone, follicle-stimulating hormone, and prolactin secretion in the different phases of the estrous cycle, *Endocrinology,* 117, 766–772, 1985.

176. Porro, G. B. and Parente, F., Side effects of anti-ulcer prostaglandins: an overview of the worldwide clinical experience, *Scand. J. Gastroenterol.,* 24(Suppl. 164), 224–231, 1989.
177. Pratt, W. F., Mosher, W. D., Bachrach, C., and Horn, M. C., Infertility — United States, *MMWR,* 34, 197, 1985.
178. Psychoyos, A. and Prapas, I., Inhibition of egg development and implantation in rats after post-coital administration of the progesterone antagonist RU 486, *J. Reprod. Fertil.,* 80, 487–491, 1987.
179. Pullon, S. R. H., Reinken, J. A., and Sparrow, M. J., The prevalence of premenstrual symptoms in Wellington women, *N. Z. Med. J.,* 100, 562–564, 1987.
180. Queenan, J. T., O'Brien, G. D., Bains, L. M., Collins, P. O., Simpson, J., Collins, W. P., and Sampbell, S., Ultrasound scanning of the ovaries to detect ovulation in women, *Fertil. Steril.,* 34, 99–105, 1980.
181. Reddoch, R. B., Pelletier, R. M., Barbe, G. J., and Armstrong, D. T., Lack of ovarian responsiveness to gonadotropic hormones in infantile rats sterilized with busulfan, *Endocrinology,* 119, 879–886, 1986.
182. Riad-Fahmy, D., Read, G. F., Walker, R. F., and Griffiths, K., Steroid in saliva for assessing endocrine function, *Endocr. Rev.,* 3, 367–395, 1982.
183. Riley, A. J. and Riley, E. J., The effect of labetolol and propranolol on the pressor response to sexual arousal in women, *Br. J. Clin. Pharmacol.,* 12, 341–344, 1981.
184. Rivera-Tovar, A. D. and Frank, E., Late luteal phase dysphoric disorder in young women, *Am. J. Psychiatr.,* 147, 1634–1636, 1990.
185. Roberts, A. C., Hastings, M. H., Martensz, N. D., and Herbert, J., Naloxone-induced secretion of LH in the male Syrian hamster: modulation by photoperiod and gonadal steroids, *J. Endocrinol.,* 106, 243–248, 1985.
186. Roblero, L. S., Fernandez, O., and Croxatto, H. B., The effect of RU486 on transport, development and implantation of mouse embryos, *Contraception,* 36, 549–555, 1987.
187. Roblero, L. S. and Croxatto, H. B., Effect of RU486 on development and implantation of rat embryos, *Mol. Reprod. Devel.,* 29, 342–346, 1991.
188. Rom, W. N., Effects of lead on the female and reproduction: a review, *Mt. Sinai J. Med.,* 43, 542–552, 1976.
189. Rose, D. P. and David, P. E., Ovarian function in patients receiving adjuvant chemotherapy for breast cancer, *Lancet,* 1, 1174–1176, 1977.
190. Rowland, A. S., Baird, D. D., Shore, D. L., Weinberg, C. R., Savitz, D. A., and Wilcox, A. J., Nitrous oxide and spontaneous abortion in female dental assistants, *Am. J. Epidemiol.,* 141, 531–537, 1995.
191. Rowland, A. S., Baird, D. D., and Weinberg, C. R., Nitrous oxide and fertility (reply), *N. Engl. J. Med.,* 328, 284, 1993.
192. Rowland, A. S., Baird, D. D., Shore, D. L., Shy, C. M., and Wilcox, A. J., Reduced fertility among women employed as dental assistants exposed to high levels of nitrous oxide, *N. Engl. J. Med.,* 327, 993–997, 1992.
193. Ruddick, J. A., Scott, P. M., and Harwig, J., Teratological evaluation of zearalenone administered orally to the rat, *Bull. Environ. Contam. Toxicol.,* 15, 678–681, 1976.
194. Ruzsas, C., Biro-Gosztonyi, M., Woller, L., and Mess, B., Effect of the fungal toxin (zearalenone) on the reproductive system and fertility of male and female rats, *Acta Biol. Acad. Sci. Hung.,* 30, 335–346, 1979.
195. Ruzsas, C., Mess, B., Biro-Gosztonyi, M., and Woller, L., Effect of pre- and perinatal administration of the fungus F2 toxin on the reproduction of the albino rat, *Dev. Endocrinol.,* 3, 57–60, 1978.
196. Sanchez-Criado, J. E., Bellido, C., Galiot, F., Lopez, F. J., and Gaytan, F., A possible dual mechanism of the anovulatory action of antiprogesterone RU486 in the rat, *Biol. Reprod.,* 42, 877–886, 1990.
197. Schilsky, R. L., Lewis, B. J., Sherins, R. J., and Young, R. C., Gonadal dysfunction in patients receiving chemotherapy for cancer, *Ann. Intern. Med.,* 93, 109–114, 1980.
198. Schilsky, R. L., Sherins, R. J., Hubbards, S. M., Wesley, M. N., Young, R. C., and DeVita, V. T., Long-term follow-up ovarian function in women treated with MOPP chemotherapy for Hodgkin's disease, *Am. J. Med.,* 71, 552–556, 1981.
199. Scialli, A. R., The reproductive toxicity of ovulation induction, *Fertil. Steril.,* 45, 315–323, 1986.
200. Selmanoff, M. and Gregerson, K. A., Suckling-induced prolactin release is suppressed by naloxone and simulated by beta-endorphin, *Neuroendocrinology,* 42, 255–259, 1986.
201. Selvan, R. S. and Rao, A. R., Influence of butylated hydroxyanisole on oocyte depletion induced by 7,12-dimethylbenz[a]anthracene in mice, *Indian J. Exp. Biol.,* 23, 320–322, 1985.
202. Severi, F., Gonadal function in adolescents receiving different conditioning regimens for bone marrow tranplantation, *Bone Marrow Transplant.,* 8(Suppl. 1), 53, 1991.
203. Shalev, O., Rahav, G., and Milwidsky, A., Reversible busulfan-induced ovarian failure, *Eur. J. Obstet. Gynecol. Reprod. Biol.,* 26, 239–242, 1987.
204. Shamberger, R. C., Rosenberg, S. A., Seipp, C. A., and Sherins, R. J., Effects of high-dose methotrexate and vincristine on ovarian and testicular functions in patients undergoing postoperative adjuvant treatment of osteosarcoma, *Canc. Treat. Rep.,* 65, 739–746, 1981.
205. Sharma, R. and Chantler, E., IVR regimens using clomiphene citrate for superovulation reduce embryo estradiol content and implantation in mice, *IRCS Med. Sci.,* 14, 821–822, 1986.
206. Shibata, S., Effect of RU486 on collagenolytic enzyme activities in immature rat ovary, *Nippon Sanka Fujinka Gakkai Zasshi,* 42, 136–142, 1990.

207. Shiromizu, K. and Mattison, D. R., Murine oocyte destruction following intraovarian treatment with 3-methylcholanthrene or 7,12-dimethylbenz(a)anthracene: protection by alpha-naphthoflavone, *Teratogesis. Carcinog. Mutagen.*, 5, 463, 472, 1985.
208. Shoupe, D., Mishell, Jr., D. R., Lahteenmaki, P., Heikinheimo, O., Birgerson, L., Madkour, H., and Spitz, I. M., Effects of the antiprogesterone RU486 in normal women. I. Single-dose administration in the midluteal phase, *Am. J. Obstet. Gynecol.*, 157, 1415–1420, 1987.
209. Shoupe, D., Michell, Jr., D. R., Page, M. A., Madkour, H., Spitz, I. M., and Lobo, R. A., Effects of the antiprogesterone RU486 in normal women. II. Administration in the late follicular phase, *Am. J. Obstet. Gynecol.*, 157, 1421–1426, 1987.
210. Shumilina, A. V., Menstrual and child-bearing functions of female workers occupationally exposed to the effects of formaldehyde, *Gig. Trud. Prof. Zabol.*, 19, 18–21, 1975.
211. Sirinathsinghji, D. J. and Audsley, A. R., Endogenous opioid peptides participate in the modulation of prolactin release in response to cervicovaginal stimulation in the female rat, *Endocrinology*, 117, 549–556, 1985.
212. Sirinathsinghji, D. J. and Martini, L., Effects of bromocriptine and naloxone on plasma levels of prolactin, LH and FSH during suckling in the female rat: responses to gonadotrophin releasing hormone, *J. Endocrinol.*, 100, 175–182, 1984.
213. Sirinathsinghji, D. J., Motta, M., and Martini, L., Induction of precocious puberty in the female rat after chronic naloxone administration during the neonatal period: the opiate "brake" on prepubertal gonadotrophin secretion, *J. Endocrinol.*, 104, 299–307, 1985.
214. Skarin, G., Nillius, S. J., and Wide, L., Failure to induce abortion of early human pregnancy by high doses of superactive LRH agonist, *Contraception*, 26, 457, 1982.
215. Smith, B. J., Mattison, D. R., Sipes, I. G., The role of epoxidation in 4-vinylcyclohexene-induced ovarian toxicity, *Toxicol. Appl. Pharmacol.*, 105, 872–881, 1990.
216. Speroff, L. and Darney, P., *A Clinical Guide for Contraception*, Williams & Wilkins, Baltimore, MD, 1992.
217. Swartz, W. J. and Mattison, D. R., Benzo(a)pyrene inhibits ovulation in C57BL/6N mice, *Anat. Rec.*, 212, 268–276, 1985.
218. Taskinen, H., Lindbolm, M. L., and Hemminki, K., Spontaneous abortion among women working in the pharmaceutical industry, *Br. J. Ind. Med.*, 43, 199–205, 1986.
219. Thorton, J., Coy, R. W., Female-typical sexual behavior of rhesus and defeminization by androgens given prenatally, *Horm. Behav.*, 20, 129–147, 1986.
220. Treloar, A. E., Boynton, R. E., Behn, B. G., and Brown, B. W., Variation of the human menstrual cycle through reproductive life, *Int. J. Fertil.*, 12, 77–125, 1967.
221. Vahakangas, K., Rajaniemi, H., and Pelkonen, O., Ovarian toxicity of cigarette smoke exposure during pregnancy in mice, *Toxicol. Lett.*, 25, 75–80, 1995.
222. van Bergeijk, L., Gooren, I. J., Van Kessel, H., and Sassen, A. M., Effects of naloxone infusion on plasma levels of LH, FSH, and in addition TSH and prolactin in males, before and after oestrogen or anti-oestrogen treatment, *Horm. Metab. Res.*, 18, 611–615, 1986.
223. van der Schoot, P., Bakker, G. H., and Klijn, J. G., Effects of the progesterone antagonist RU486 on ovarian activity in the rat, *Endocrinology*, 121, 1375–1382, 1987.
224. van Leeuwen, F. X., den Tonkelaar, E. M., and Van Logten, M. J., Toxicity of sodium bromide in rats: effects on endocrine system and reproduction, *Food Chem. Toxicol.*, 21, 383–389, 1983.
225. Van Voohis, B. J., Syrop, C. H., Hammitt, D. G., Dunn, M. S., and Synder, G. D., Effects of smoking on ovulation induction for assisted reproductive techniques, *Fertil. Steril.*, 58, 981–985, 1992.
226. Vasileva, I. A., Effect of small concentrations of carbon disulfide and hydrogen sulfide on the menstrual function of women and estrous cycle of experimental animals, *Gig. Sanit.*, 7, 24–7, 1973.
227. Veldhuis, J. D., Rogol, A. D., Perez-Palacios, G., Stumpf, P., Kitchin, J. D., and Dufau, M. L., Endogenous opiates participate in the regulation of pulsatile luteinizing hormone release in an unopposed estrogen milieu: studies in estrogen-replaced, gonadectomized patients with testicular feminization, *J. Clin. Endocrinol. Metab.*, 61, 790–793, 1985.
228. Vermesh, M., Kletzky, O. A., Davajan, V., and Israel, R., Monitoring techniques to predict and detect ovulation, *Fertil. Steril.*, 47, 259–264, 1987.
229. Vickery, B. H., Comparison of the potential utility of LHRH agonists and antagonists for fertility control, *J. Steroid. Biochem.*, 23, 779–791, 1985.
230. Vickery, B. H., McRae, G. I., and Stevens, V. C., Suppression of luteal and placental function in pregnant baboons with agonist analogs of lutenizing hormone-releasing hormone, *Fertil. Steril.*, 36, 664–668, 1981.
231. Viguier-Martinez, M. C., Hochereau-de Reviers, M. T., Barenton, B., and Perreau, C., Effect of prenatal treatment with busulfan on the hypothalamo-pituitary axis, genital tract and testicular histology of prepubertal male rats, *J. Reprod. Fertil.*, 70, 67–73, 1984.
232. Vining, R. F. and McGinley, R. A., Hormones in saliva, *CRC Crit. Rev. Clin. Lab. Sci.*, 23, 95–146, 1986.
233. Wang, S. C. and Greenwald, G. S., Effect of cyclohexamide injected at proestrus on ovarian protein synthesis, peptide and steroid hormone levels, and ovulation in the hamster, *Biol. Reprod.*, 33, 201–211, 1985.

234. Wang, S. C. and Greenwald, G. S., Effect of cyclohexamide during the periovulatory period on ovarian follicular FSH, hCG, and prolactin receptors and on follicular maturation in the hamster, *Proc. Soc. Exp. Biol. Med.,* 185, 55–61, 1987.
235. Whisnant, C. S., Kiser, T. E., Thompson, F. N., and Barb, C. R., Opioid inhibition of luteinizing hormone secretion during the postpartum period in suckled beef cows, *J. Anim. Sci.,* 63, 1445–1448, 1986.
236. Whisnant, C. S., Thompson, F. N., Kiser, T. E., and Barb, C. R., Effect of naloxone on serum luteinizing hormone, cortisol and prolactin concentrations in anestrous beef cows, *J. Anim. Sci.,* 62, 1340–1345, 1986.
237. Wikstrom, A., Green, B., and Johansson, E. D., The plasma concentration of medroxyprogesterone acetate and ovarian function during treatment with medroxyprogesterone acetate in 5 and 10 mg doses, *Acta Obstet. Gynecol. Scand.,* 63, 163–168, 1984.
238. Wilcox, A. J., Weinberg, C. R., and Baird, D. D., Caffeinated beverages and decreased fertility, *Lancet,* 2, 1453–1455, 1988.
239. Wilcox, A. J., Weinberg, C. R., O'Connor, J. F. et al., Incidence of early pregnancy loss, *N. Engl. J. Med.,* 319, 189–194, 1988.
240. Wilcox, A. J., Baird, D. D., and Weinberg, C. R., Do women with childhood exposure to cigarette smoking have increased fecundability?, *Am. J. Epidemiol.,* 129, 1079, 1989.
241. Williams, M. A., Monson, R. R., Goldman, M. B., Mittendorf, R., and Ryan, K. J., Coffee and delayed conception, *Lancet,* 335, 1063, 1990.
242. Wolf, J. P., Hsiu, J. G., Anderson, T. L., Ulmann, A., Baulieu, E. E., and Hodgen, G. D., Noncompetitive antiestrogenic effect of RU486 in blocking the estrogen-stimulated luteinizing hormone surge and the proliferative action of estradiol on endometrium in castrated monkeys, *Fertil. Steril.,* 52, 1055–1060, 1989.
243. World Health Organization Task Force on Methods for the Determination of the Fertile Period, The measurement of urinary steroid glucuronides as indices of the fertile period in women, *J. Steroid. Biochem.,* 17, 695–702, 1982.
244. Wright, D. M., Kesner, J. S., Schrader, S. M., Chin, N. W., Wells, V. E., Krieg, Jr., E. F., Methods of monitoring menstrual function in field studies: attitudes of working women, *Reprod. Toxicol.,* 6, 401–409, 1992.
245. Wright, J. M., Kiracofe, G. H., and Beeman, K. B., Factors associated with shortened estrous cycles after abortion in beef heifers, *J. Anim. Sci.,* 66, 3185–3189, 1988.
246. Yamazaki, I., Serum concentration patterns of an LHRH agonist, gonadotrophins and sex steroids after subcutaneous, vaginal, rectal and nasal administration of the agonist to pregnant rats, *J. Reprod. Fertil.,* 72, 129–136, 1984.
247. Yang, Y. O. and Wu, J. T., RU486 interferes with egg transport and retards the *in vivo* and *in vitro* development of mouse embryos, *Contraception,* 41, 551–556, 1990.
248. Yen, S. S., Quigley, M. E., Reid, R. L., Ropert, J. F., and Cetel, N. S., Neuroendocrinology of opioid peptides and their role in the control of gonadotropin and prolactin secretion, *Am. J. Obstet. Gynecol.,* 152, 485–493, 1985.
249. Ylikorkala, O., Kauppila, A., Kivinen, S., and Viinikka, L., Sulpiride improves inadequate lactation, *Br. Med. J. (Clin. Res. Ed.),* 285, 249–251, 1982.
250. Ylikorkala, O., Kauppila, A., Kivinen, S., and Viinikka, L., Treatment of inadequate lactation with oral sulpiride and buccal oxytocin, *Obstet. Gynecol.,* 63, 57–60, 1984.
251. Zacharias, L., Rand, W. M., and Wurtman, R. J., A prospective study of sexual development and growth in American girls: the statistics of menarche, *Obstet. Gynecol. Surv.,* 31, 325, 1976.

CHAPTER 24

Developmental Toxicity

David A. Beckman, Lynda B. Fawcett, and Robert L. Brent

CONTENTS

0-8493-4493-X/97/$0.00+$.50

TABLE 1

Frequency of Reproductive Risks in the Human

Reproductive Risk	Frequency
Immunologically and clinically diagnosed spontaneous abortions per 10^6 conceptions	350,000
Clinically recognized spontaneous abortions per 10^6 pregnancies	150,000
Genetic diseases per 10^6 births	110,000
Multifactorial or polygenic (genetic-environmental interactions)	90,000
Dominantly inherited disease	10,000
Autosomal and sex-linked genetic disease	1200
Cytogenetic (chromosomal abnormalities)	5000
New mutations	3000
Major congenital malformations per 10^6 births	30,000
Prematurity per 10^6 births	40,000
Fetal growth retardation per 10^6 births	30,000
Stillbirths per 10^6 pregnancies (>20 wk)	20,900
Infertility	15% of couples

1 ADVERSE DEVELOPMENTAL OUTCOMES

Every conception has a risk of ending in abortion or serious congenital anomaly (Tables 1 and 2). Abortion and birth defects have some common etiologies, but in many instances the causes of these two areas of reproductive failure are divergent. In this review, we evaluate the data concerning developmental toxicants that increase the risk of abortion and congenital malformation in the human. The evaluations were made after a review of the available clinical, epidemiological, and experimental data, and the analysis is based on reproducibility, consistency, and biological plausibility.[70] Only key references or reviews are cited which will guide the reader to additional relevant literature.

1.1 Characterization of Adverse Developmental Outcomes

1.1.1 *Spontaneous Abortion*

The definition of spontaneous abortion is based on the stage of embryonic development when viability is not possible outside the uterus. This stage is presently considered to be 20 wk or less of gestation and a fetal weight of less than 500 g, although these criteria are not universally accepted.

The frequency of spontaneous abortion varies with the stage of gestation (Tables 2 and 3). More than 80% of abortions occur in the first trimester. There is a steady decline in the risk of abortion as pregnancy progresses (Table 3); therefore, it is essential that epidemiological studies of the causes of abortion compare control and "exposed" populations with the same mean stage and range of abortion. Two pregnant populations with a 2-wk difference in mean stage of pregnancy will have a different background incidence of abortion. Abortion in human populations includes the following causes (Table 3).

1.1.1.1 *Chromosomal Abnormalities*

The earlier the abortion, the higher the proportion of chromosomal abnormalities.[59,296,322,575] Approximately 53% of spontaneous abortions in the first trimester are due to chromosomal abnormalities, 36% are due to chromosomal abnormalities in the second trimester, and only 5% of stillbirths in the third trimester are due to chromosomal abnormalities. Over 95% of abortuses with chromosomal abnormalities were due to autosomal trisomy,

TABLE 2

Estimated Outcome of 100 Pregnancies vs. Time from Conception

Time from Conception	Percent Survival to Term[a]	Percent Loss During Interval[a]	Last Time for Induction of Selected Malformations[b]
Preimplantation			
0–6 days	25	54.55	—
Postimplantation			
7–13 days	55	24.66	—
14–20 days	73	8.18	—
3–5 wk	79.5	7.56	22–23 days; cyclopia, sirenomelia, microtia 26 days; anencephaly 28 days; meningomyelocele 34 days; transposition of great vessels
6–9 wk	90	6.52	36 days; cleft lip 6 wk; diaphragmatic hernia, rectal atresia, ventricular septal defect, syndactyly 9 wk; cleft palate
10–13 wk	92	4.42	10 wk; omphalocele
14–17 wk	96.26	1.33	12 wk; hypospadias
18–21 wk	97.56	0.85	—
22–25 wk	98.39	0.31	—
26–29 wk	98.69	0.30	—
30–33 wk	98.98	0.30	—
34–37 wk	99.26	0.34	—
38+ wk	99.32	0.68	38+ wk; CNS cell depletion

[a] Data from Kline et al.[391] An estimated 50 to 70% of all human conceptions are lost in the first 30 wk of gestation and 78% are lost before term.[322,575]

[b] Modified from Schardein.[609]

double trisomy, monosomy, triploidy, or tetraploidy.[369,629] Most chromosomal abnormalities are not the cause of repetitive abortion, although in about 4% of couples with two or more spontaneous abortions, a normal-appearing parent could be a carrier for a balanced translocation or may be a mosaic with abnormal cells in the germ cell line. Environmental exposures during pregnancy cannot account for any of these abortions, because most aneuploidies result from meiotic nondisjunction during gametogenesis before conception.

1.1.1.2 Abortions with Normal Chromosomes (Euploidy)

Hertig[322] and may other investigators reported the occurrence of malformed or blighted embryos as a cause of abortion. These embryonic losses may occur later in the first trimester and have been shown to have normal karyotypes in more recent studies.[369] The etiology of these abortions are manifold and include:

1. *Genetic abnormalities:* Dominant mutations (lethals), polygenic genetic abnormalities, or recessive disease may rarely account for repetitive abortion, but in most instances they will occur sporadically. There is an extensive literature concerning lethal mutations in the mouse. A review of gene knockouts and mutations in mice that produce intrauterine death suggests that embryonic death resulted from disrupting basic cellular functions, vascular circulation, hematopoiesis, or nutritional supply from the mother rather than affecting embryonic organ systems.[133]
2. *Maternal diabetes:* Type I (insulin-dependent) diabetes mellitus with poor metabolic control increases the risk of abortion and stillbirths, but there is no increased risk with good metabolic control.[235]

TABLE 3

Etiology of Spontaneous Abortion in the Human

Chromosomal Abnormalities

Chromosomal abnormalities from either the maternal or paternal gonadocytes. The earlier the abortion the higher the proportion of chromosomal abnormalities. Over 95% of abortuses with chromosomal abnormalities were due to autosomal trisomy, double trisomy, monosomy, triploidy, or tetraploidy.

Abortions with Normal Chromosomes (Euploidy)

Some embryonic losses occurring later in the first trimester have been shown to have normal karyotypes.
- Genetic abnormalities
  - Dominant mutations (lethals), polygenic genetic abnormalities, recessive disease from the maternal, paternal, or both parents', gonadocytes; in rare instances, these conditions may account for repetitive abortion, but in most instances they will occur sporadically.
- Maternal disease states
  - Maternal diabetes
  - Maternal hypothyroidism
  - Corpus luteum or placental progesterone deficiency (luteal phase deficiency)
  - Maternal infection which results in fetal infection: *Treponema pallidum, Plasmodium falciparum, Toxoplasma gondii,* herpes simplex virus, parvovirus B19, or cytomegalovirus
  - Severe, debilitating maternal diseases: hepatitis, collagen diseases, untreated hyperthyroidism, severe malnutrition
  - Antiphospholipid antibodies: lupus anticoagulant, anticardiolipin antibodies
  - Maternal-fetal histocompatibility
  - Overmature gametes
  - Mechanical or physical problems: uterine abnormalities, multiple pregnancies, very rarely trauma
  - Cervical incompetence: more likely to result in fetal wastage during second trimester than during first trimester
  - Abnormal placentation: hypoplastic trophoblast, circumvallate implantation
  - Some environmental teratogens (see Table 10)

3. *Maternal hyper- and hypothyroidism:* Abnormal thyroid function is rare in patients with recurrent abortion.
4. *Corpus luteum or placental progesterone deficiency (luteal phase deficiency):* It is controversial whether low hormone levels after implantation result from impending abortion or are the cause of the abortion.
5. *Maternal infection:* Infections of the genital tract could be responsible for abortion, but it is not easy to document causality. The data suggesting that infection with *Chlamydia trachomatis, Borrelia burgdorferi, Mycoplasma hominis, Listeria monocytogenes,* and *Ureaplasma urealyticum* result in abortion are not conclusive. In contrast, maternal disease resulting in fetal infection with *Treponema pallidum, Plasmodium falciparum, Toxoplasma gondii,* herpes simplex virus, parvovirus B19, or cytomegalovirus has the potential to cause stillbirth or spontaneous abortion.[22,38,93,211,298,333, 398,425,472,561,575,625,690,727]
6. *Severe, debilitating maternal diseases:* These include hepatitis, collagen diseases, untreated hyperthyroidism, Wilson's disease, or severe malnutrition.
7. *Antiphospholipid antibodies:* Lupus anticoagulant and anticardiolipin antibodies predispose women to recurrent abortion in both first and second trimesters due to vascular disruption in the placenta.
8. *Maternal-fetal histocompatibility:* It is suggested that embryonic loss increases if the mother and fetus are more histocompatible at the human leukocyte antigen (HLA) locus, resulting in the failure to develop maternal blocking antibodies against paternal antigens.
9. *Overmature gametes:* Either the ovum or sperm could age because insemination occurred a few days prior to ovulation or ovulation occurred prior to insemination. The magnitude of this risk factor as a cause of spontaneous abortion is not known and some investigators are skeptical that this phenomenon is clinically significant.

TABLE 4

Etiology of Human Congenital Malformations Observed During the First Year of Life

Suspected Cause	Percent of Total
Unknown	65–75
Polygenic	
Multifactorial (gene-environment interactions)	
Spontaneous errors of development	
Autosomal and sex-linked inherited genetic disease	
Cytogenetic (chromosomal abnormalities)	
New mutations	
Genetic	15–25
Autosomal and sex-linked inherited genetic disease	
Cytogenetic (chromosomal abnormalities)	
New mutations	
Environmental	10
Maternal conditions: alcoholism, diabetes, endocrinopathies, phenylketonuria, smoking and nicotine, starvation, nutritional deficits	4
Infectious agents: rubella, toxoplasmosis, syphilis, herpes simplex, cytomegalovirus, varicella-zoster, Venezuelan equine encephalitis, parvovirus B19	3
Mechanical problems (deformations): amniotic band constrictions, umbilical cord constraint, disparity in uterine size and uterine contents	1–2
Chemicals: prescription drugs, high-dose ionizing radiation, hyperthermia	<1

Source: Modified from Brent.[65,69,70]

10. *Mechanical or physical problems related to uterine abnormalities, multiple pregnancies, or trauma:* A hostile intrauterine environment can result from submucosal or intramural myomas, adhesions (Asherman's syndrome), multiple embryos, or abnormalities of the uterus (bifid uterus, infantile uterus). Uterine trauma from a direct blow or penetrating injury may rarely be responsible for an abortion, and if this type of injury occurs it would be more likely to result in a stillbirth at midgestation or later.
11. *Cervical incompetence:* More likely to result in second trimester than first trimester abortions.
12. *Abnormal placentation:* Hypoplastic trophoblast and circumvallate implantation increase the risk of fetal loss.
13. *Some environmental teratogens and reproductive toxins:* See Table 10.

1.1.2 Congenital Malformations

The etiology of congenital malformations can be divided into three categories: unknown, genetic, and environmental (Table 4). The etiology of 65 to 75% of human malformations is unknown.[65,302,751] A significant proportion of congenital malformations of unknown etiology is likely to have an important genetic component.[109,469] Malformations with an increased recurrent risk — such as cleft lip and palate, anencephaly, spina bifida, certain congenital heart diseases, pyloric stenosis, hypospadias, inguinal hernia, talipes equinovarus, and congenital dislocation of the hip — fit in the category of multifactorial disease as well as in the category of polygenic inherited disease.[109,216] The multifactorial/threshold hypothesis[216] postulates the modulation of a continuum of genetic characteristics by intrinsic and extrinsic (environmental) factors. Although the modulating factors are not known, they probably include placental blood flow, placental transport, site of implantation, maternal disease states, maternal malnutrition, infections, drugs, chemicals, and spontaneous errors of development.

Spontaneous errors of development may account for some of the malformations that occur without apparent abnormalities of the genome or environmental influence. We postulate that there is some probability for error during embryonic development based on the fact that embryonic development is such a complicated process.[64,216] It is estimated that 75% of all conceptions are lost before term, 50% within the first 3 wk of development.[322,575] The World Health Organization[762] estimated that 15% of all clinically recognizable pregnancies end in a spontaneous abortion, 50 to 60% of which are due to chromosomal abnormalities.[59,629] Finally, 3 to 6% of offspring are malformed, which represents the background risk for human maldevelopment. This means that, as a conservative estimate, 1176 clinically recognized pregnancies will result in approximately 176 miscarriages, and 30 to 60 of the infants will have congenital anomalies in the remaining 1000 live births. The true incidence of pregnancy loss is much higher because undocumented pregnancies are not included in this risk estimate.

Based on his review of the literature, Wilson[751] provided a format of theoretical teratogenic mechanisms: mutation; chromosomal aberrations; mitotic interference; altered nucleic acid synthesis and function; lack of precursors, substrates, or coenzymes for biosynthesis; altered energy sources; enzyme inhibition; osmolar imbalance; alterations in fluid pressures, viscosities, and osmotic pressures; and altered membrane characteristics. Even though an agent can produce one or more of these pathologic processes, exposure to such an agent does not guarantee maldevelopment will occur. Furthermore, it is likely that a drug, chemical, or other agent can have more than one effect on the pregnant woman and the developing conceptus; therefore, the nature of the drug or its biochemical or pharmacologic effects will not in themselves predict a teratogenic effect in the human. In fact, the discovery of human teratogens has come primarily from human epidemiological studies. Animal studies and *in vitro* studies can be very helpful in determining the mechanism of teratogenesis and the pharmacokinetics related to teratogenesis.[71] However, even if one understands the pathologic effects of an agent, one cannot predict the teratogenic risk of an exposure without taking into consideration the developmental stage, the magnitude of the exposure, and the repairability of the embryo.

Viral, bacterial, and parasitic agents may cause maldevelopment in humans, especially rubella, cytomegalovirus, herpes simplex, parvovirus B19, syphilis, toxoplasmosis, varicella-zoster, and Venezuelan equine encephalitis (reviewed by Sever[620]). The lethal or developmental effects of infectious agents are the result of mitotic inhibition, direct cytotoxicity, or necrosis. Repair processes may result in metaplasia, scarring, or calcification, which cause further damage by interfering with histogenesis. Infectious agents appear to be exceptions to some of the principles of teratogenesis, because dose and time of exposure cannot be demonstrated as readily for replicating agents. However, transplacental transmission of an infectious agent does not necessarily result in congenital malformations, growth retardation, or lethality.

Vascular disruption is a rare event associated with intrauterine death and a wide range of structural anomalies, including cerebral infarctions; certain types of visceral and urinary tract malformations; congenital limb amputations of the nonsymmetrical type; and orofacial malformations such as mandibular hypoplasia, cleft palate, and Moebius syndrome, which vary too widely to constitute a recognized syndrome. Some anomalies associated with twin pregnancies can be explained by vascular disruption resulting from placental anastomoses in shared placenta of monozygotic twins, anastomoses in a small percentage of dichorionic placentas in the case of dizygotic twins, or death of one twin resulting in emboli, intravascular coagulation, and altered fetal hemodynamics in the co-twin.[701] Vascular disruption may also result from physical trauma, such as chorionic villous sampling, causing chorion bleeding, and exposure to some developmental toxicants, such as cocaine and misoprostol. Although uterine bleeding during the first trimester may result in fetal anomalies, the malformations associated with vascular disruption can also occur later in gestation. This topic is discussed in greater detail below with specific toxicants.

1.2 Factors that Affect Susceptibility to Developmental Toxicants

A basic tenet of environmentally produced embryo- and fetotoxicity is that the effects of teratogenic or abortagenic milieu have certain characteristics in common and follow certain

TABLE 5

Factors that Influence Susceptibility to Developmental Toxicants

Stage of development
- The developmental period at which an exposure occurs will determine which structures are most susceptible to the adverse effects of the developmental toxicant and to what extent the embryo can repair the damage.

Magnitude of the exposure
- Both the severity and incidence of toxic effects increase with dose. Factors that influence dose-response relationships include:
  - Active metabolites: The concentration of active metabolites may be more pertinent than the dosage of the original chemical.
  - Duration of exposure: A chronic exposure can contribute to an increased risk of fetal anomalies or abortion, whereas an acute exposure may not.
  - Fat-solubility: Fat-soluble substances may produce fetal toxicity for an extended period after the last exposure in a woman because they may have an extended half-life.

Threshold phenomena
- The threshold dose is the dose below which the incidence of death, malformation, growth retardation, or functional deficit is not statistically greater than that of nonexposed subjects.

Pharmacokinetics and metabolism
- The physiologic changes in the pregnant woman and during fetal development and the bioconversion of compounds can significantly influence the developmental toxicity of drugs and chemicals by affecting absorption, body distribution, active metabolites, and excretion.

Maternal disease
- A maternal disease may increase the risk of fetal anomalies or abortion with or without exposure to a developmental toxicant.

Placental transport
- Most drugs and chemicals cross the placenta. The factors that determine the rate and extent to which a drug or chemical crosses the placenta include: molecular weight, lipid solubility, polarity or degree of ionization, plasma protein binding, receptor mediation, placental blood flow, pH gradient between the maternal and fetal serum and tissues, and placental metabolism of the chemical or drug.

Genotype
- The maternal and fetal genotypes may result in differences in cell sensitivity, placental transport, absorption, metabolism, receptor binding, and distribution of an agent and may account for some variations in toxic effects among individual subjects and species.

basic principles. These principles determine the quantitative and qualitative aspects of environmentally produced developmental toxicity (Table 5).

1.2.1 Stage of Development

The induction of developmental toxicity by environmental agents usually results in a spectrum of morphologic anomalies or intrauterine death, which varies in incidence depending on stage of exposure and dose. The developmental period at which an exposure occurs will determine which structures are most susceptible to the deleterious effects of the drug or chemical and to what extent the embryo can repair the damage. The period of sensitivity may be narrow or broad, depending on the environmental agent and the malformation in question. Limb defects produced by thalidomide have a very short period of susceptibility (Table 6), while microephaly produced by radiation has a long period of susceptibility. Our knowledge of the susceptible stage of the embryo to various environmental influences is continually expanding and is vital to evaluating the significance of individual exposures or epidemiological studies.

During the first period of embryonic development, from fertilization through the early postimplantation stage, the embryo is most sensitive to the embryolethal effects of drugs and chemicals. Surviving embryos have malformation rates similar to the controls, not because malformations cannot be produced at this stage but because significant cell loss or chromosome abnormalities at these stages have a high likelihood of killing the embryo. Because of

TABLE 6

Developmental Stage Sensitivity to Thalidomide-Induced Limb Reduction Defects in the Human

Time from Conception for Induction of Defects (days)	Limb Reduction Defects
21–26	Thumb aplasia
23–34	Hip dislocation
24–29	Amelia, upper limbs
24–33	Phocomelia, upper limbs
25–31	Preaxial aplasia, upper limbs
27–31	Amelia, lower limbs
28–33	Preaxial aplasia, lower limbs; phocomelia, lower limbs; femoral hypoplasia; girdle hypoplasia
30–36	Triphalangeal thumb

Source: Modified from Brent, R. L. and Holmes, L. B., *Teratology,* 38, 241–251, 1988.

the omnipotentiality of early embryonic cells, surviving embryos have a much greater ability to have normal developmental potential. Wilson and Brent[753] demonstrated that the all-or-none phenomenon, or marked resistance to teratogens, disappears over a period of a few hours in the rat during early organogenesis and utilizing X-irradiation as the experimental teratogen. The term all-or-none phenomenon has been misinterpreted by some investigators to indicate that malformations cannot be produced at this stage. On the contrary, it is likely that certain drugs, chemicals, or other insults during this stage of development can result in malformed offspring,[237,535] but the nature of embryonic development at this stage will still reflect the basic characteristic of the all-or-none phenomenon, which is a propensity for embryo lethality rather than surviving malformed embryos.

The period of organogenesis (from day 18 through about day 40 postconception in the human) is the period of greatest sensitivity to teratogenic insults and the period when most gross anatomic malformations can be induced. Most environmentally produced major malformations occur before the 36th day of gestation in the human. The exceptions are malformations of the genitourinary system, the palate, or the brain or deformations due to problems of constraint, disruption, or destruction. Severe growth retardation in the whole embryo or fetus may also result in permanent deleterious effects in many organs or tissues.

The fetal period is characterized by histogenesis involving cell growth, differentiation, and migration. Teratogenic agents may decrease the cell population by producing cell death, inhibiting cell division, or interfering with cell differentiation. There is, of course, some overlap in that permanent cell depletion may be produced earlier than the 60th day. Effects such as cell depletion or functional abnormalities, not readily apparent at birth, may give rise to changes in behavior or fertility which are apparent only later in life. The last gestational days on which certain malformations may be induced in the human are presented in Table 2.

1.2.2 Magnitude of the Exposure

The dose-response relationship is extremely important when comparing effects among different species because usage of mg/kg doses are, at most, rough approximations. Dose equivalence among species can only be accomplished by performing pharmacokinetic studies, metabolic studies, and dose-response investigations in the human and the species being studied. Furthermore, the response should be interpreted in a biologically sound manner. One example is that a substance given in large enough amounts to cause maternal toxicity is also likely to have deleterious effects on the embryo such as death, growth retardation, or

TABLE 7

Stochastic and Threshold Dose-Response Relationships of Diseases Produced by Environmental Agents

Relationship	Pathology	Site	Diseases	Risk	Definition
Stochastic phenomena	Damage to a single cell may result in disease	DNA	Cancer, mutation	Some risk exists at all dosages; at low exposures, the risk is below the spontaneous risk	The incidence of the disease increases, but the severity and nature of the disease remain the same
Threshold phenomena	Multicellular injury	High variation in etiology, affecting many cell and organ processes	Malformation, growth retardation, death, chemical toxicity, etc.	No increased risk below the threshold dose	Both the severity and incidence of the disease increase with dose

Source: Modified from Brent, R. L., *Teratology,* 34, 359–360, 1986.

retarded development. Another example is that because the steroid receptors that are necessary for naturally occurring and synthetic progestin action are absent from nonreproductive tissues early in development, the evidence is against the involvement of progesterone or its synthetic analogues in nongenital teratogenesis.[86,329,751,752]

An especially anxiety-provoking concept is that the interaction of two or more drugs or chemicals may potentiate their developmental effects. Although this is an extremely difficult hypothesis to test in the human, it is an especially important consideration because multichemical or multitherapeutic exposures are common. Fraser[217] warns that the actual existence of a threshold phenomenon when nonteratogenic doses of two teratogens are combined could easily be misinterpreted as potentiation or synergism. Potentiation or synergism should be invoked only when exposure to two or more drugs is just below their thresholds for toxicity.

Several considerations affect the interpretation of dose-response relationships:

1. *Active metabolites:* Metabolites may be the proximate teratogen rather than the administered chemical, i.e., the metabolites phosphoramide mustard and acrolein may produce maldevelopment resulting from exposure to cyclophosphamide.[487]
2. *Duration of exposure:* Chronic exposure to a prescribed drug can contribute to an increased teratogenic risk, i.e., anticonvulsant therapy; in contrast, an acute exposure to the same drug may present little or no teratogenic risk.
3. *Fat solubility:* Fat-soluble substances such as polychlorinated biphenyls[480] can produce fetal maldevelopment for an extended period after the last ingestion or exposure in a women because they have an unusually long half-life. Etretinate may present a similar risk,[407] but the data are not as conclusive.

1.2.3 Threshold Phenomena

The threshold dose is the dosage below which the incidence of death, malformation, growth retardation, or functional deficit is not statistically greater than that for controls. The threshold level of exposure is usually from less than one to three orders of magnitude below the teratogenic or embryopathic dose for drugs and chemicals that kill or malform half the embryos. A teratogenic agent therefore has a no-effect dose as compared to mutagens or carcinogens, which have a stochastic dose-response curve. Threshold phenomena are compared to stochastic phenomena in Table 7. The severity and incidence of malformations produced by every exogenous teratogenic agent that has been appropriately tested have exhibited threshold phenomena during organogenesis.[751]

1.2.4 Pharmacokinetics and Metabolism

The physiologic alterations in pregnancy and the bioconversion of compounds can significantly influence the teratogenic effects of drugs and chemicals by affecting absorption, body distribution, active form(s), and excretion of the compound. Physiologic alterations in the mother during pregnancy that affect the pharmacokinetics of drugs include:[347,458,646] (1) decreased gastrointestinal motility and increased intestinal transit time that delay absorption of drugs absorbed in the small intestine due to increased stomach retention and enhanced absorption of slowly absorbed drugs; (2) decreased plasma albumin concentration, which alters the kinetics of compounds normally bound to albumin; (3) increased plasma and extracellular fluid volumes that affect concentration-dependent transfer of compounds; (4) renal elimination, which is generally increased but is influenced by body position during late pregnancy; (5) inhibition of metabolic inactivation in the liver late in pregnancy; and (6) variations in uterine blood flow, although little is known about how this affects transfer across the placenta.

The fetus also undergoes physiologic alterations which affect the pharmacokinetics of drugs:[646] (1) the amount and distribution of fat varies with development and affects the distribution of lipid-soluble drugs; (2) the fetal circulation contains a higher concentration of unbound drug largely because the plasma fetal proteins are lower in concentration than in the adult which may lower drug affinity; (3) the functional development of pharmacologic receptors is likely to proceed at different rates in the various tissues; and (4) drugs excreted by the fetal kidneys may be recycled by swallowing of amniotic fluid.

The role of the placenta in drug pharmacokinetics (reviewed by Juchau and Rettie[367] and Miller[480]) involves: (1) transport (discussed in detail below); (2) the presence of receptor sites for a number of endogenous and xenobiotic compounds (β-adrenergic, glucocorticoid, epidermal growth factor, IgG-Fc, insulin, low-density lipoproteins, opiates, somatomedin, testosterone, transcobalamin II, transferrin, folate, retinoid;[480] and (3) the bioconversion of xenobiotics. Bioconversion of xenobiotics has been shown to be important in the teratogenic activity of several xenobiotics. There is strong evidence that reactive metabolites of cyclophosphamide, 2-acetylaminofluorene, and nitroheterocycles (niridazole) are the proximal teratogens.[306] There is also experimental evidence that suggests that other chemicals undergo conversion to intermediates that have deleterious effects on embryonic development including phenytoin, procarbazine, rifampicin, diethylstilbestrol, some benzhydrylpiperazine antihistamines, adriamycin, testosterone, benzo(*a*)pyrene, methoxyethanol, caffeine, and paraquat.[366,367]

The major site of bioconversion of chemicals *in vivo* is likely to be the maternal liver. Placental P-450-dependent mono-oxygenation of xenobiotics will occur at low rates unless induced by such compounds as those found in tobacco smoke;[367] however, the rodent embryo and yolk sac have been shown to possess functional P-450 oxidative isozymes capable of converting proteratogens to active metabolites during early organogenesis.[764] In addition, P-450-independent bioactivation has been suggested. For example, there is strong evidence that the rat embryo can reductively convert niridazole to an embryotoxic metabolite.[189]

As defined by Juchau,[366] there are several experimental criteria that would suggest that a suspected metabolite is responsible for the *in vivo* teratogenic effects of a chemical or drug: (1) the chemical must be convertible to the intermediate; (2) the intermediate must be found in or have access to the tissue(s) affected; (3) the embryotoxic effect should increase with the concentration of the metabolite; (4) inhibiting the conversion should reduce the embryotoxic effect of the agent; (5) promoting the conversion should increase the embryotoxicity of the agent; (6) inhibiting or promoting the conversion should not alter the target tissues; and (7) inhibition of biochemical inactivation should increase the embryotoxicity of the agent. It is readily apparent why there may exist marked qualitative and quantitative differences in species response to a teratogenic agent.

1.2.5 Maternal Diseases

It may be difficult to determine whether a maternal disease or the treatment for the disease plays a role in the etiology of malformations associated with the treatment for that disease during pregnancy. For example, the genetic and environmental milieu which causes epilepsy may also contribute to the maldevelopment associated with exposure to diphenylhydantoin.[289,623]

1.2.6 Placental Transport

The exchange between the mammalian embryo and the maternal organism is controlled by the placenta, which includes the chorioplacenta, the yolk sac placenta, and the paraplacental chorion. The placenta varies in structure and function among species and for each stage of gestation. As an example, the rodent yolk sac placenta continues to function as an organ of transport for a much greater part of gestation than is true for the human. Thus, differences in placental function and structure may affect our ability to apply teratogenic data developed in one species directly to other species, including the human.[65] As pharmacokinetic techniques and the actual measurement of metabolic products in the embryo become more sophisticated, the appropriateness of utilizing animal data to project human effects may improve.

Historically, a placental barrier was thought to exist which prevented harmful substances from reaching the embryo. It is now clear that there is no "placental barrier" *per se*. The fact is that most drugs and chemicals cross the placenta. It will be a rare substance that will cross the placental barrier in one species and be unable to reach the fetus in another.[68] No such chemical exists except for selected proteins, the actions of which are species specific.

Even before there were chemical techniques to demonstrate the presence of drugs or chemicals in the embryo, there was clear evidence that the drugs had reached the fetus because of clinical manifestations:

1. Anticoagulants such as warfarin can affect the clotting of fetal blood.
2. Many drugs can affect the fetal cardiac rate.
3. Changes in the fetal EEG can be demonstrated to be due to the many drugs that affect the central nervous system.
4. Newborns may exhibit withdrawal symptoms from drugs taken by their mothers, either medications or substances of abuse such as alcohol or opiates.

These observations demonstrate clinically significant placental transport of drugs only in the latter portion of gestation and may not be a means of evaluating embryonic exposure during early organogenesis.

Those factors which determine the ability of a drug or chemical to cross the placenta and reach the embryo include: molecular weight, lipid affinity or solubility, polarity or degree of ionization, protein binding, and receptor mediation. Compounds with low molecular weight, lipid affinity, and nonpolarity and without protein binding properties will cross the placenta with ease and rapidity. As an example, ethyl alcohol is a chemical that reaches the embryo rapidly and in concentrations equal to or greater than the level in the mother.

High-molecular-weight compounds such as heparin, (20,000 Da) do not readily cross the placenta; therefore, heparin is used to replace warfarin-like compounds during pregnancy for the treatment of hypercoagulation conditions. Rose bengal does not cross the placenta. In general, compounds with molecular weights of 1000 or greater do not readily cross the placenta, while 600-Da compounds usually do; most drugs are 250 to 400 Da and cross the placenta.[488]

Besides the particular properties of the drug or chemical, three other conditions affect the quantitative aspect of placental transport: (1) placental blood flow, (2) the pH gradient between the maternal and fetal serum and tissues, and (3) placental metabolism of the chemical or drug. The biotransformation properties of the placenta and/or maternal organism are important because a number of chemicals or drugs are not teratogenic in their original form.

The most important concept with regard to placental transport of teratogens must be re-emphasized. An agent is teratogenic because it affects the embryo directly or indirectly by its ability to produce a toxic effect in the embryo or extraembryonic membranes at exposures attained in the human being, not because it crosses the placenta *per se*.

1.2.7 Genotype

The genetic constitution of an organism is an important factor in the susceptibility of a species to a drug or chemical. More than 30 disorders of increased sensitivity to drug toxicity or effects have been reported in the human due to an inherited trait.[468] The effect of a drug or chemical depends on both the maternal and fetal genotypes and may result in differences in cell sensitivity, placental transport, absorption, metabolism (activation, inactivation, active metabolites), receptor binding, and distribution of an agent. This accounts for some variations in teratogenic effects among species and in individual subjects.

1.3 Potential for Developmental Toxicity in the Human

1.3.1 Evaluation of Data Available for the Human

While chemicals and drugs can be evaluated for fetotoxic potential by utilizing *in vivo* animal studies and *in vitro* systems, it should be recognized that these testing procedures are only one component in the process of evaluating the potential teratogenic risk of drugs and chemicals in the human. The evaluation of the teratogenicity of drugs and chemicals should include, when possible: (1) data obtained from human epidemiological studies, (2) secular trend data in humans, (3) animal developmental toxicity studies, (4) the dose-response relationship for developmental toxicity and the relationship to the human pharmacokinetic equivalent dose in the animal studies, and (5) considerations of biological plausibility (Table 8).[66,70] This method is of greatest value when utilized for the evaluation of chemicals and drugs that have been in use for some time or for evaluating new drugs that have a similar mechanism of action, structure, and pharmacology and a purpose similar to other, extensively studied agents. The ability to establish a casual relationship between an environmental agent and abortigenic effect is more difficult, for the following reasons:

1. Abortion is a very frequent reproductive event; therefore, the incidence can vary considerably within a population. Differences in the abortion incidence between two populations in a single study may be due to chance alone.
2. Most epidemiological studies dealing with abortion ignore the multiple causes of abortion and make no attempt to determine the etiology of the abortions. Since most abortions are due to preconceptual or periconceptual events, it would be necessary to have large increases in a particular etiological category of environmentally induced abortion in order to demonstrate a statistically significant increase in the incidence of spontaneous abortion in an "exposed" population of pregnant women.
3. Confounding factors appear to be more significant in abortion studies than in birth defect studies (cocaine, smoking, alcohol, syphilis, narcotics, caffeine). This further decreases the possibility that the agent being studied has a direct abortigenic effect.

Some investigators and regulatory agencies divide drugs and chemicals into developmentally toxic and nontoxic compounds. In reality, potential developmental toxicity can be evaluated only if one considers, as a minimum, the agent, the dose, the species, and the stage of gestation. Working definitions for developmental toxicity in the human are suggested in Table 9.

TABLE 8

Evaluation of Potential for Developmental Toxicity in the Human

Epidemiological studies
Controlled epidemiologic studies consistently demonstrate an increased incidence of pregnancy loss or a particular spectrum of fetal effects in exposed human populations.
Secular trend data
Secular trends demonstrate a relationship between the incidence of pregnancy loss or a particular fetal effect and the changing exposures in human populations. The percent of the population exposed must be large for this analysis.
Animal developmental toxicity studies
An animal model mimics the human developmental effect at clinically comparable exposures. Since mimicry may occur in only one animal species, if it occurs at all, it would not necessarily be observed during an initial developmental toxicology study. Developmental toxicity studies are indicative of a potential hazard in general rather than the potential for a specific adverse effect on the fetus.
Dose-response relationship
Developmental toxicity in the human increases with dose, and the developmental toxicity in animals occurs at a dose that is pharmacokinetically equivalent to the human dose.
Biological plausibility
The mechanisms of developmental toxicity are understood or the results are biologically plausible.

Source: Modified from Brent.[66,70]

TABLE 9

Definitions of Potential for Developmental Toxicity in the Human

Developmental toxicity
Results from an agent or milieu that has been demonstrated to produce permanent alterations or death in the embryo or fetus following intrauterine exposures that usually occur or are attainable in the human.
Potential developmental toxicity
Results from an agent or milieu that has not been demonstrated to produce permanent alterations or death in the embryo or fetus following intrauterine exposures that usually occur or are attainable in the human, but which can affect the embryo or fetus if the exposure is raised substantially above the usual exposure. Most chemicals and drugs have the potential for interrupting a pregnancy or inducing developmental defects if the exposure is increased sufficiently.
Little or no potential developmental toxicity
Results from an agent or milieu that has not been demonstrated to produce embryo- or fetotoxicity at any attainable dose in the human. In contrast, an environmental agent may be so toxic that it has no developmental toxicity in the human because it kills the mother before or at the same dose that it begins to have adverse effects on the embryo.

Actually, the largest group of drugs are the potential human teratogens and abortifacients because they include all drugs and chemicals that can produce embryotoxic and fetotoxic effects at some exposure. Since these exposures are not utilized or attained in the human, they represent no or minimal risks to the human embryo.

1.3.2 Misconceptions in Evaluating Developmental Toxicity in the Human

Misconceptions of clinicians and scientists have led to confusion regarding the potential effects of even proven teratogens. Examples of erroneous concepts include:

1. If an agent can produce one type of malformation, it can produce any malformation.
2. An agent presents a risk at any dose, once it can be proven to be teratogenic.
3. An agent that is teratogenic is likely to be abortigenic.

These concepts are incorrect. The data clearly indicate that proven teratogens do not have the ability to produce every birth defect. Many teratogens can be identified on the basis of the malformations that are produced. Thus, the concept of the syndrome is probably more appropriate in clinical teratology than any other area of clinical medicine. Some symptoms or signs appear in many teratogenic syndromes, such as growth retardation or mental retardation, and are not very discriminating. On the other hand, rare or specific effects, such as deafness, retinitis, or a pattern of cerebral calcifications, may point to a specific teratogen. It is also true that there is substantial overlap in malformation syndromes which may not always be separable. Environmentally produced birth defects may be confused with genetically determined malformations. Using thalidomide as an example, a patient with bilateral radial aplasia and a ventricular septal defect may have the Holt Oram syndrome or the thalidomide syndrome. It may or may not be possible to make a diagnosis with absolute certainty, even if one has a history of thalidomide ingestion during pregnancy. It is possible, however, to refute the suggestion that thalidomide was responsible for congenital malformations in an individual by the nature of the limb malformation.

The specificity of some environmental teratogens can sometimes point to the mechanism or site of action. For instance, the predominant central nervous system effects of methylmercury are understood when one realizes the propensity for organic mercury to be stored in lipid.

Epidemiologists sometimes use poor judgment when grouping malformations. As an example, limb reduction defects are frequently studied with regard to their association with environmental teratogens. In many of the studies, limb defects that are clearly related to problems of organogenesis are lumped with congenital amputations.[351] Yet it is very unlikely that any agent will be responsible for both types of malformations. It is clear that epidemiological studies could be markedly improved if there was more input from clinical teratologists in planning and performing the studies.

Epidemiological studies concerning spontaneous abortion may contain serious errors: in case-control studies, the populations being studied should be similar with regard to the stage of pregnancy when abortion occurred. This study design diminishes the possibility that the abortion rate will differ on the basis of the selection process and not the environmental agent being studied. Unfortunately, most epidemiological studies dealing with environmentally induced abortion do not attempt to determine the etiology of the abortion.

2 THE TOXICITY OF SELECTED AGENTS

We evaluated the literature concerning chemicals, physical agents, prescription drugs, and therapeutic agents that cause or are suggested to cause congenital malformations in the human. The data included human epidemiological studies, secular trend data in humans where appropriate, and animal developmental toxicity studies. In our analysis, we considered the dose-response relationship of teratogenicity, the relationship to the human pharmacokinetic equivalent dose in the animal studies, and biological plausibility (Table 5).[66,70] Table 10 focuses on these agents, listing their teratogenic and abortigenic potential in the human. Although these agents account for a small percentage of all malformations and abortions, they are important because these exposures may be preventable.

2.1 Alcohol

Adverse effects in offspring from excessive alcohol consumption during pregnancy were recognized more than 200 years ago.[723] It was Jones et al.,[363] however, who defined the fetal alcohol syndrome in children with intrauterine growth retardation, microcephaly, mental

TABLE 10

Developmental Toxicants: Risks of Congenital Malformations and Abortion in the Human

Developmental Toxicant	Reported Effects or Associations and Estimated Risks	Comments[a]
Alcohol	Fetal alcohol syndrome: intrauterine growth retardation, maxillary hypoplasia, reduction in width of palpebral fissures, characteristic but not diagnostic facial features, microcephaly, mental retardation. An increase in spontaneous abortion has been reported, but since mothers who abuse alcohol during pregnancy have multiple other risk factors, it is difficult to determine whether this is a direct effect on the embryo. Consumption of 6 oz of alcohol or more per day constitutes a high risk, but it is likely that detrimental effects can occur at lower exposures.	Quality of available information: good to excellent. Direct cytotoxic effects of ethanol and indirect effects of alcoholism; while a threshold teratogenic dose is likely, it will vary in individuals because of a multiplicity of factors.
Aminopterin, methotrexate	Microcephaly, hydrocephaly, cleft palate, meningomyelocele, intrauterine growth retardation, abnormal cranial ossification, reduction in derivatives of first branchial arch, mental retardation, postnatal growth retardation. Aminopterin can induce abortion within its therapeutic range; it is used for this purpose to eliminate ectopic embryos. Risk from therapeutic doses is unknown but appears to be moderate to high.	Quality of available information: good. Anticancer, antimetabolic agents; folic acid antagonists that inhibit dihydrofolate reductase, resulting in cell death.
Androgens	Masculinization of female embryo: clitoromegaly with or without fusion of labia minora. Nongenital malformations are not a reported risk. Androgen exposures which result in masculinization have little potential for inducing abortion. Based on animal studies, behavioral masculinization of the female human will be rare.	Quality of available information: good. Effects are dose and stage dependent; stimulates growth and differentiation of sex steroid receptor-containing tissue.
Angiotensin-converting enzyme (ACE) inhibitors	The therapeutic use of ACE inhibitors has neither a teratogenic nor an abortigenic effect in the first trimester. Since this group of drugs does not interfere with organogenesis, they can be used in a woman of reproductive age; if the woman becomes pregnant, therapy can be changed during the first trimester without an increase in the risk of teratogenesis. Later in gestation, these drugs can result in fetal and neonatal death, oligohydramnios, pulmonary hypoplasia, neonatal anuria, intrauterine growth retardation, and skull hypoplasia. Risk is dependent on dose and length of exposure.	Quality of available information: good. Antihypertensive agents; adverse fetal effects are related to severe fetal hypotension over a long period of time during the second or third trimester.
Caffeine	Caffeine is teratogenic in rodent species with doses of 150 mg/kg. There are no convincing data that moderate or usual exposures (300 mg/day or less) present a measurable risk in the human for any malformation or group of malformations. On the other hand, excessive caffeine consumption (exceeding 300 mg/day) during pregnancy is associated with growth retardation and embryonic loss.	Quality of available information: fair to good. Behavioral effects have been reported and appear to be transient or temporary; more information is needed concerning the population with higher exposures.
Carbamazepine	Minor craniofacial defects (upslanting palpebral fissures, epicanthal folds, short nose with long philtrum), fingernail hypoplasia, and developmental delay. Teratogenic risk is not known but likely to be significant for minor defects. There are too few data to determine whether carbamazepine presents an increased risk for abortion. Since embryos with multiple malformations are more likely to abort, it would appear that carbamazepine presents little risk because an increase in these types of malformations has not been reported.	Quality of available information: fair to good. Anticonvulsant; little is known concerning mechanism. Epilepsy may itself contribute to an increased risk for fetal anomalies.

Cocaine	Preterm delivery; fetal loss; placental abruption; intrauterine growth retardation; microcephaly; neurobehavioral abnormalities; vascular disruptive phenomena resulting in limb amputation, cerebral infarctions, and certain types of visceral and urinary tract malformations. There are few data to indicate that cocaine increases the risk of first trimester abortion. The low but increased risk of vascular disruptive phenomena due to vascular compromise of the pregnant uterus would more likely result in midgestation abortion or stillbirth. It is possible that higher doses could result in early abortion. Risk for deleterious effects on fetal outcome is significant; risk for major disruptive effects is low, but can occur in the latter portion of the first trimester as well as the second and third trimesters.	Quality of available information: fair to good. Cocaine causes a complex pattern of cardiovascular effects due to its local anesthetic and sympathomimetic activities in the mother. Fetopathology is likely to be due to decreased uterine blood flow and fetal vascular effects. Because of the mechanism of cocaine teratogenicity, a well-defined cocaine syndrome is not likely. Poor nutrition accompanies drug abuse, and multiple drug abuse is common.
Chorionic villous sampling (CVS)	Low but increased risk of orofacial malformations and limb reduction defects of the congenital amputation type as seen in vascular disruption malformations in some series. The risk of abortion following CVS is quite low.	Quality of available information: fair. Excessive bleeding from the chorion is probably related in part to the experience of the operator. Further research is necessary to determine whether CVS is safer for the fetus at certain stages of gestation.
Coumarin derivatives	Nasal hypoplasia; stippling of secondary epiphysis; intrauterine growth retardation; anomalies of eyes, hands, neck; variable CNS anatomical defects (absent corpus callosum, hydrocephalus, asymmetrical brain hypoplasia). Risk from exposure 10 to 25% during 8th to 14th week of gestation. There is also an increased risk of pregnancy loss. There is a risk to the mother and fetus from bleeding at the time of labor and delivery.	Quality of available information: good. Anticoagulant; bleeding is an unlikely explanation for effects produced in the first trimester. Central nervous system defects may occur anytime during second and third trimester and may be related to bleeding.
Cyclophosphamide	Growth retardation, ectrodactyly, syndactyly, cardiovascular anomalies, and other minor anomalies. Teratogenic risk appears to be increased, but the magnitude of the risk is uncertain. Almost all chemotherapeutic agents have the potential for inducing abortion. This risk is dose-related; at the lowest therapeutic doses, the risk is small	Quality of available information: fair. Anticancer, alkylating agent; requires cytochrome P-450 mono-oxydase activation; interacts with DNA, resulting in cell death.
Diethylstilbestrol (DES)	Clear cell adenocarcinoma of the vagina occurs in about 1:1000 to 10,000 females who were exposed *in utero*. Vaginal adenosis occurs in about 75% of females exposed *in utero* before the 9th week of pregnancy. Anomalies of the uterus and cervix may play a role in decreased fertility and an increased incidence of prematurity, although the majority of women exposed to DES *in utero* can conceive and deliver normal babies. *In utero* exposure to DES increased the incidence of genito-urinary lesions and infertility in males. DES can interfere with zygote survival, but it does not interfere with embryonic survival when given in its usual dosage after implantation. Offspring who were exposed to DES *in utero* have an increased risk for delivering prematurely but do not appear to be at increased risk for first trimester abortion.	Quality of available information: fair to good. Synthetic estrogen; stimulates estrogen receptor-containing tissue; may cause misplaced genital tissue which has a greater propensity to develop cancer.
Diphenylhydantoin	Hydantoin syndrome: microcephaly, mental retardation, cleft lip/palate, hypoplastic nails, and distal phalanges; characteristic but not diagnostic facial features. Associations documented only with chronic exposure. Wide variation in reported risk of malformations but appears to be no greater than 10%. The few epidemiological data indicate a small risk of abortion for therapeutic exposures for the treatment of epilepsy. For short-term treatment, i.e., prophylactic therapy for a head injury, there is no appreciable risk.	Quality of available information: fair to good. Anticonvulsant; direct effect on cell membranes, folate, and vitamin K metabolism. Metabolic intermediate (epoxide) has been suggested as the teratogenic agent.

TABLE 10 (continued)

Developmental Toxicants: Risks of Congenital Malformations and Abortion in the Human

Developmental Toxicant	Reported Effects or Associations and Estimated Risks	Comments[a]
Electromagnetic fields (EMF)	The data pertaining to video display terminals indicate that the EMF exposures from these units do not present an increased risk for abortion or congenital malformations. The data on power line and appliance exposures are too varied to draw any conclusions, although the risks appear to be small or nonexistent. Human exposures to video display terminals and power lines are quite low and are unlikely to have reproductive effects.	Quality of available information: fair. Pregnant animals exposed to EMF do not exhibit consistent or reproducible reproductive effects. There are still questions about biologic effects for frequencies and wave forms of magnetic fields that have not been adequately studied.
Infectious agents	The cytotoxic effects and inflammatory responses resulting from fetal infections interfere with organogenesis and/or histogenesis. Fetopathic syndromes are related to the specific tissue localization, pathologic characteristics of the infectious agent, and the duration of the infection in the embryo and fetus. In some instances, the infection may be debilitating to the mother and indirectly result in pregnancy loss.	
Cytomegalovirus	Fetal cytomegalovirus infection presents an increased risk for abortion, but it does not appear that maternal genital infection increases the risk of abortion. Fetal infection occurs in about 20% of maternal infections. Intrauterine growth retardation; risk of brain damage is moderate after fetal infection early in pregnancy; characteristic parenchymal calcification.	Quality of available information: good to excellent. CMV damages organs principally by cellular necrosis.
Herpes simplex virus	Generalized organ infections, microcephaly, hepatitis, eye defects, vesicular rash. Maternal infection can be transmitted *in utero* or perinatally. Herpes simplex 2 is one of the few infections for which it is agreed that the risk of abortion is increased.	Quality of available information: good. Fetal anomalies associated with herpes simplex virus infection appear to be due to disruption rather than malformation.
Human immunodeficiency virus (HIV)	The overall risk of vertical transmission of HIV is 25–40%. Fetal HIV infection and asymptomatic maternal HIV infection are not associated with adverse effect on fetal growth or development. Symptomatic maternal HIV infection, other sexually transmitted diseases, and opportunistic infections may increase the risk of low birth weight of perinatal morbidity.	Quality of available information: fair to good. Prophylactic treatment with zidovudine does not appear to cause permanent adverse effects in the fetus.
Parvovirus B19	Infection can result in erythema infectiosum in children but in the fetus can result in hydrops fetalis and fetal death; congenital anomalies are likely to be very rare. The risk for stillbirth with hydrops has been clearly substantiated. An increased risk for abortion has been suggested but is more difficult to substantiate.	Quality of available information: fair to good. Fetal infection is not common. Infection of red blood cell precursors causes severe anemia.
Rubella virus	Greater than 80% incidence of embryonic infection with exposure in first 12 weeks, 54% at 13 to 14 weeks, 25% at end of second trimester, and 100% at term. Defects include mental retardation, deafness, cardiovascular malformations, cataracts, glaucoma, microphthalmia. Diabetes mellitus or rubella panencephalitis may develop later in life. The abortigenic risk of maternal rubella is uncertain.	Quality of available information: excellent. Rubella has an affinity for specific tissues. Damage is caused by mitotic inhibition, cell death, and interference with histogenesis by repair processes, resulting in calcification and scarring.

Syphilis	Defects in 50% of offspring after early exposure to primary or secondary syphilis and 10% after late exposures. Defects include maculopapular rash, hepatosplenomegaly, deformed nails, osteochondritis at joints of extremities, congenital neurosyphilis, abnormal epiphyses, chorioretinitis. Syphilis can increase the incidence of abortion.	Quality of available information: good. Fetal pathology is associated with maturation of the fetal immune system at about the 20th week.
Toxoplasmosis	Hydrocephaly, microphthalmia, chorioretinitis. Risk is predominantly associated with pregnancies in which the mother acquires toxoplasmosis. Epidemiological studies do not indicate that toxoplasmosis increases the incidence of early abortion, but congenital toxoplasmosis may be responsible for the stillbirth of severely affected fetuses.	Quality of available information: good to excellent. Toxoplasmosis is unlikely to contribute to the risk of repeated abortion.
Varicella-Zoster	Skin and muscle defects; intrauterine growth retardation; limb reduction defects. No measurable increased risk of early teratogenic effects. Incidence of maternal varicella during pregnancy is low, but risk of severe neonatal infection is high if maternal infection occurs in last week of pregnancy. There does not appear to be an increased risk of first trimester abortion.	Quality of available information: fair to good. Virus infection of fetal tissues can cause cellular necrosis.
Venezuelan equine encephalitis	Hydroanencephaly, microphthalmia, central nervous system destructive lesions, luxation of hip. There are not enough data to determine whether infection presents an increased risk of abortion.	Quality of available information: poor to fair. Infection can cause cellular necrosis in fetal tissues but fetal infection is rare.
Lead	There is no indication that serum lead levels below 50 μg% result in congenital malformations in exposed embryos and fetuses, but the developing central nervous system in the fetus and child may be susceptible to lead toxicity, resulting in decreased IQ and behavioral effects. Lead levels above 50 μg% result in anemia, and encephalopathy can have serious effects on CNS development. Lead levels below 50 μg% do not increase the risk of abortion.	Quality of available information: good. While there are human studies indicating small deficiencies in IQ in patients with lead levels above 10 μg%, there could be other explanations for these IQ differences. Furthermore, pathological findings have not been described in the brain at these levels.
Lithium carbonate	Although animal studies have demonstrated a clear teratogenic risk, the effect in humans is uncertain. Early reports indicated an increased incidence of Ebstein's anomaly and other heart and great vessel defects, but as more studies are reported the strength of this association has diminished. Lithium levels within the therapeutic range (<1.2 mg%) do not increase the risk of abortion.	Quality of available information: fair to good. Antidepressant; mechanism has not been defined.
Maternal conditions		
Diabetes	Caudal hypoplasia or caudal regression syndrome; congenital heart disease; anencephaly. Vascular lesions in long-standing diabetics may produce placental dysfunction and result in fetal growth retardation. Documented significant increased risk of abortion. The risk is greatest in untreated diabetics or patients who are poorly controlled; insulin therapy protects the fetus.	Quality of available information: good. The results of *in vitro* studies suggest that diabetic embryopathy has a multifactorial etiology. Adverse fetal effects have not been demonstrated with gestational diabetes.
Endocrinopathy	If condition is compatible with pregnancy, effects are similar to those following administration of high or low doses of the hormone. Hypo- and hyperthyroidism may increase the risk of abortion. Cushing's disease, pituitary tumors, hypothalamic tumors, and androgen-producing tumors do not appear to increase the risk of abortion but may contribute to infertility.	Quality of available information: good. Receptor-mediated exposures to high levels of hormone.

TABLE 10 (continued)

Developmental Toxicants: Risks of Congenital Malformations and Abortion in the Human

Developmental Toxicant	Reported Effects or Associations and Estimated Risks	Comments[a]
Nutritional deprivation	Central nervous system anomalies, intrauterine growth retardation, increased morbidity. In some instances of teratogenesis, abnormal nutrition may be the final common mechanism. Severe malnutrition can contribute to pregnancy loss at any stage of pregnancy.	Quality of available information: fair to good. The high mitotic rate of the fetus in general and the central nervous system in particular is very sensitive to severe alterations in nutrient supply.
Phenylketonuria	Mental retardation, microcephaly, intrauterine growth retardation. Documented significant increased risk of abortion. The risk is greatest in pregnant women who were not treated for their phenylketonuria.	Quality of available information: good. Very high levels of phenylalanine interfere with embryonic cell metabolism.
Mechanical problems	Birth defects such as club feet, limb reduction defects, aplasia cutis, cranial asymmetry, external ear malformations, midline closure defects, cleft palate, and muscle aplasia. Submucosal and intramural myomata, amniotic bands, multiple implantations, bifid uterus, or infantile uterus which contribute to mechanical problems do not result in early abortions.	Quality of available information: good. Physical constraint can result in distortion and a reduction in blood supply and is more frequent in pregnancies with multiple conceptuses, abnormal uterus, amniotic abnormalities, certain placental abnormalities, and oligohydramnios.
Methylmercury	Minamata disease: cerebral palsy, microcephaly, mental retardation, blindness, cerebellar hypoplasia. Does not appear to decrease fertility unless the mother becomes clinically ill from methylmercury poisoning. At low exposures the teratogenic effect predominates, and there are few human data to indicate the risk of abortion.	Quality of available information: good. Organic mercurials accumulate in lipid tissue causing cell death due to inhibition of cellular enzymes, especially sulfhydryl enzymes. Since most cases result from accidental environmental exposure, risk estimation is usually retrospective.
Methylene blue	Hemolytic anemia and jaundice in neonatal period after exposure late in pregnancy. There may be a small risk for intestinal atresia, but this is not yet clear. No indication of increased risk of abortion.	Quality of available information: poor to fair. Used to mark amniotic cavity during amniocentesis.
Misoprostol	Misoprostol is a synthetic prostaglandin analog that has been used by millions of women for illegal abortion. A low incidence of vascular disruptive phenomenon, such as limb reduction defects and Moebius syndrome, has been reported.	Quality of available information: fair. Classical animal teratology studies would not be helpful in discovering these effects, because vascular disruptive effects occur after the period of early organogenesis.
Oxazolidine-2,4-diones (trimethadione, paramethadione)	Fetal trimethadione syndrome: V-shaped eye brows, low-set ears with anteriorly folded helix, high-arched palate, irregular teeth, CNS anomalies, severe developmental delay. Wide variation in reported risk. Characteristic facial features are documented only with chronic exposure. The abortifacient potential has not been adequately studied but appears to be minimal.	Quality of available information: good to excellent. Anticonvulsants; affect cell membrane permeability. Actual mechanism of action has not been determined.

D-penicillamine	Cutis laxa, hyperflexibility of joints. Condition appears to be reversible, and the risk is low. There are no human data on the risk of abortion.	Quality of available information: fair to good. Copper-chelating agent; produces copper deficiency inhibiting collagen synthesis and maturation.
Polychlorinated biphenyls	Cola-colored babies: pigmentation of gums, nails, and groin; hypoplastic deformed nails; intrauterine growth retardation; abnormal skull calcification. Although abortion can be induced at high exposure, most human exposures from environmental contamination are unlikely to increase the risk of abortion.	Quality of available information: good. Environmental contaminants; polychlorinated biphenyls and commonly occurring contaminants are cytotoxic. Body residues in exposed women can affect pigmentation in offspring for up to 4 years after exposure.
Progestins	Masculinization of female embryo exposed to high doses of some testosterone-derived progestins; may interact with progesterone receptors in the liver and brain later in gestation. The dose of progestins present in modern oral contraceptives presents no masculinization or feminization risks. All progestins present no risk for nongenital malformations. Many synthetic progestins and natural progesterone have been used to treat luteal phase deficiency, embryos implanted via IVF, threatened abortion, or bleeding in pregnancy with variable results. Conversely, synthetic progestins that interfere with progesterone function may cause early pregnancy loss; RU-486 is presently used specifically for this purpose.	Quality of available information: good. Stimulate or interfere with sex steroid receptor-containing tissue.
Radioactive isotopes	Tissue- and organ-specific damage is dependent on the radioisotope element and distribution, i.e., ^{131}I administered to a pregnant woman can cause fetal thyroid hypoplasia after the 8th week of development. Radioisotopes used for diagnosis present no risk for inducing abortion because the dose to the embryo and implantation site is too low. There may be unusual circumstances wherein isotopes are introduced into the abdominal cavity in a pregnant woman for the treatment of malignancy. If the resulting dose to the embryo or fetus is substantial, the risk for abortion is increased.	Quality of available information: good to excellent. Higher doses of radioisotopes can produce cell death and mitotic delay. Effect is dependent on dose, distribution, metabolism, and specificity of localization.
Radiation (external irradiation)	Microcephaly, mental retardation, eye anomalies, intrauterine growth retardation, visceral malformations. Teratogenic risk depends on dose and stage of exposure. Exposures from diagnostic procedures present no increased risk of abortion, growth retardation, or malformation. No measurable risk with exposures for 5 rad (5 mGy) or less of X-rays at any stage of pregnancy. In contrast, exposure of the pregnant uterus to therapeutic doses of ionizing radiation significantly increases the risk of aborting the embryo; the fetus is more resistant.	Quality of available information: good to excellent. Diagnostic and therapeutic agents; produces cell death and mitotic delay.
Retinoids, systemic (isotretinoin, etretinate)	Increased risk of CNS, cardio-aortic, ear, and clefting defects. Microtia, anotia, thymic aplasia, and other branchial arch, aortic arch abnormalities, and certain congenital heart malformations. Exposed embryos are at greater risk for abortion. This is plausible since many of the malformations, such as neural tube defects, are associated with an increased risk of abortion.	Quality of available information: fair. Used in treatment of chronic dermatoses. Retinoids can cause direct cytotoxicity and alter programmed cell death; affect many cell types but neural crest cells are particularly sensitive.
Retinoids, topical (tretinoin)	Epidemiological studies, animal studies, and absorption studies in humans do not suggest a teratogenic risk. Regardless of the risks associated with systemically administered retinoids, topical retinoids present little or no risk for intrauterine growth retardation, teratogenesis, or abortion because they are minimally absorbed and only a small percentage of skin is exposed.	Quality of available information: poor. Topical administration of tretinoin in animals in therapeutic doses is not teratogenic, although massive exposures can produce maternal toxicity and reproductive effects. More importantly, topical administration in humans results in nonmeasurable blood levels.

TABLE 10 (continued)

Developmental Toxicants: Risks of Congenital Malformations and Abortion in the Human

Developmental Toxicant	Reported Effects or Associations and Estimated Risks	Comments[a]
Smoking and nicotine	Placental lesions; intrauterine growth retardation; increased postnatal morbidity and mortality. While there have been some studies reporting increases in anatomical malformations, most studies do not report an association. There is no syndrome associated with maternal smoking. Maternal or placental complications can result in fetal death. Exposures to nicotine and tobacco smoke are a significant risk for pregnancy loss in the first and second trimester.	Quality of available information: good to excellent. While tobacco smoke contains many components, nicotine can result in vascular spasm vasculitis which has resulted in a higher incidence of placental pathology.
Sonography (ultrasound)	No confirmed detrimental effects resulting from medical sonography. The levels and types of medical sonography that have been used in the past have no measurable risks. The present clinical use of diagnostic ultrasound presents no increased risk of abortion.	Quality of available information: good to excellent. It appears that if the embryonic temperature never exceeds 39°C, there is no measurable risk.
Streptomycin	Streptomycin and a group of ototoxic drugs can affect the eighth nerve and interfere with hearing; it is a relatively low-risk phenomenon. There are not enough data to estimate the abortigenic potential of streptomycin. Because the deleterious effect of streptomycin is limited to the eighth nerve, it is unlikely to affect the incidence of abortion.	Quality of available information: fair to good. Long-duration maternal therapy during pregnancy is associated with hearing deficiency in offspring.
Tetracycline	Bone staining and tooth staining can occur with therapeutic doses. Persistent high doses can cause hypoplastic tooth enamel. No other congenital malformations are at increased risk. The usual therapeutic doses present no increased risk of abortion to the embryo or fetus.	Quality of available information: good. Antibiotic; effects seen only if exposure is late in the first or during second or third trimester, since tetracyclines have to interact with calcified tissue.
Thalidomide	Limb reduction defects (preaxial preferential effects, phocomelia), facial hemangioma, esophageal or duodenal atresia, anomalies of external ears, eyes, kidneys, and heart, increased incidence of neonatal and infant mortality. The thalidomide syndrome, while characteristic and recognizable, can be mimicked by some genetic diseases. Although there are fewer data pertaining to its abortigenic potential, there appears to be an increased risk of abortion.	Quality of available information: good to excellent. Sedative-hypnotic agent; the etiology of thalidomide teratogenesis has not been definitively determined.
Thyroid: iodides, radioiodine, antithyroid drugs (propylthiouracil), iodine deficiency	Fetal hypothyroidism or goiter with variable neurologic and aural damage. Maternal hypothyroidism is associated with an increase in infertility and abortion. Maternal intake of 12 mg of iodide per day or more increases the risk of fetal goiter.	Quality of available information: good. Fetopathic effect of endemic iodine deficiency occurs early in development. Fetopathic effect of iodides, antithyroid drugs, and radioiodine involves metabolic block, decreased thyroid hormone synthesis, and gland development.
Toluene	Intrauterine growth retardation, craniofacial anomalies, microcephaly. It is likely that high exposures from abuse or intoxication increase the risk of teratogenesis and abortion. Occupational exposures should present no increase in the teratogenic or abortigenic risk. The magnitude of the increased risk for teratogenesis and abortion in abusers is not known because the exposure in abusers is too variable.	Quality of available information: poor to fair. Neurotoxicity is produced in adults who abuse toluene; a similar effect may occur in the fetus.

Valproic acid	Malformations are primarily neural tube defects and facial dysmorphology. The facial characteristics associated with this drug are not diagnostic. Small head size and developmental delay have been reported with high doses. The risk for spina bifida is about 1%, but the risk for facial dysmorphology may be greater. Because therapeutic exposures increase the incidence of neural tube defects, one would expect a slight increase in the incidence of abortion.	Quality of available information: good. Anticonvulsant; little is known about the teratogenic action of valproic acid.
Vitamin A	The same malformations that have been reported with the retinoids have been reported with very high doses of vitamin A (retinol). Exposures below 10,000 I.U. present no risk to the fetus. Vitamin A in its recommended dose presents no increased risk for abortion.	Quality of available information: good. High concentrations of retinoic acid are cytotoxic; it may interact with DNA to delay differentiation and/or inhibit protein synthesis.
Vitamin D	Large doses given in vitamin D prophylaxis are possibly involved in the etiology of supravalvular aortic stenosis, elfin faces, and mental retardation. There are no data on the abortigenic effect of vitamin D.	Quality of available information: poor. Mechanism is likely to involve a disruption of cell calcium regulation with excessive doses.

[a] Quality of available information is modified from TERIS.[682]

retardation, maxillary hypoplasia, flat philtrum, thin upper lip, and reduction in the width of palpebral fissures (cardiac abnormalities were also seen). Many of the children of alcoholic mothers had fetal alcohol syndrome, and all of the affected children evidenced developmental delay.[360,363,671]

A period of greatest susceptibility is not clearly established, but the risk for adverse effects increases with increased consumption,[514] and binge drinking early in pregnancy may be associated with an increased risk of alcohol-related effects.[262,673] The risk of decreased brain growth and differentiation that results from high alcohol consumption is greater during the second and third trimester. Chronic consumption of 6 oz of alcohol per day constitutes a high risk, while the fetal alcohol syndrome is not likely when the mother consumes fewer than two drinks (equivalent to 1 oz of alcohol) per day.[672] Reduction of alcohol consumption or cessation of drinking early in pregnancy will reduce the incidence and severity of alcohol-related effects[15,16,151,209,673] but may not entirely eliminate the risk of some degree of physical or behavioral impairment. The human syndrome is likely to involve the direct effects of alcohol and the indirect effects of genetic susceptibility and poor nutrition. Alcoholism can have maternally deleterious effects on intermediary metabolism and nutrition, especially if alcoholic cirrhosis is present, which can contribute to an adverse milieu for the developing embryo.

Craniofacial anomalies typical of fetal alcohol syndrome are thought to result from excessive cell death in the mesoderm leading to a deficiency in the anteromedial aspect of the embryo which gives rise to the forebrain and midfacial structures.[381,676] Extensive cell death may also result in the increased incidence of urinary tract malformations in children with fetal alcohol syndrome.[224] Several animal models developed to study various aspects of the fetal alcohol syndrome showed that the incidence of congenital defects and the maternal alcohol level were inversely related to maternal alcohol dehydrogenase levels in three mouse strains.[117,423,675,741] Developmental delay in the central nervous system, especially in the mid-brain, may cause abnormal patterns of innervation with resulting detrimental effects on learning and behavior.[775] Acetaldehyde is embryotoxic, but its role in producing teratogenesis is questionable.[51,52,324]

Although alcoholic mothers frequently smoke and consume other drugs, there is little doubt from the human and animal data that alcohol ingestion alone can have a disastrous effect on the developing embryo or fetus. The reported incidence of fetal alcohol syndrome varies widely in different studies but appears to be approximately 6% in offspring of women who drink heavily during pregnancy.[151] Fetal alcohol syndrome may be the most commonly recognized cause of environmentally induced mental deficiency;[121,672] there are at least several hundred children born each year with full fetal alcohol syndrome and probably many more with subtler fetal alcohol effects.

2.2 Aminopterin and Methotrexate

Aminopterin and methotrexate (methylaminopterin) are folic acid antagonists that inhibit dihydrofolate reductase, resulting in cell death during the S-phase of the cell cycle.[634] Four aminopterin-induced therapeutic abortions resulted in malformations (hydrocephalus, cleft palate, meningomyelocele) in the abortuses.[253,683] Thirteen case reports of attempted therapeutic abortions that failed included observations of growth retardation, abnormal cranial ossification, high-arched palate, and reduction in derivatives of the first branchial arch.[476,623,717,722] Similar anomalies were reported in three offspring of women treated with methotrexate during early pregnancy.[161,486,552,647]

Dyban et al.[171] have reported abnormalities of cell proliferation and cytotoxicity employing rat blastomere cultures exposed to different doses of aminopterin. Skalko and Gold[633] demonstrated a threshold effect and a dose-dependent increase in malformations in mice exposed to methotrexate *in utero*. Although malformations were induced in rats at doses exceeding those used in humans,[23] smaller doses than those used in humans have resulted in malformation in rabbits.[252] The risk of adverse effects due to aminopterin and methotrexate in the usual therapeutic range is not known precisely but appears to be moderate to high.[17,591,717,719]

2.3 Androgens

Masculinization of the external genitalia of the female was reported following *in utero* exposure to large doses of testosterone,[276] methyltestosterone, and testosterone enanthate.[332] The masculinization is characterized by clitoromegaly with or without fusion of the labia minora and no indication of nongenital malformations. Affected females experience normal secondary sexual development at puberty.[569]

There are many animal models showing the masculinizing effects of androgens (usually at high doses) in the rat,[267,287] guinea pig,[258,545] pig,[208,563] hamster,[95] rabbit,[365] and monkey.[734] These studies demonstrated the masculinization of the urogenital sinus, its derivatives, and the external genitalia, although there was little effect on the Müllerian ducts and ovarian inversion did not occur. Based on experimental animal studies of altered sexually dimorphic behavior in female guinea pigs, rats, and monkeys,[168,258,259,330,341,545] behavioral masculinization of the female due to prenatal exposure to androgens in the human will be very rare. The available literature indicates that the effects of androgens on the fetus are dependent on the dose and the stage of development during which the exposure occurred.

2.4 Caffeine

Caffeine is a methylated xanthine which acts as a central nervous system stimulant. It is contained in many beverages, including coffee, tea, and colas, as well as chocolate. Caffeine constitutes 1 to 2% of roasted coffee beans, 3.5% of fresh tea leaves, and about 2% of mate leaves.[260,261,655] Caffeine is also present in many over-the-counter medications, such as cold and allergy tablets, analgesics, diuretics, and stimulants, the latter leading to relatively minimal intakes. Caffeine-containing food and beverages are consumed in large quantities by most of the human populations of the world. The per capita consumption of caffeine from all sources is estimated to be about 200 mg/day, or about 3 to 7 mg/kg per day.[26] Consumption of caffeinated beverages during pregnancy is quite common[326] and is estimated to be approximately 144 mg per day.[495]

Current evidence does not appear to implicate the usual exposure of caffeine as a human teratogen; however, associations between maternal coffee drinking during pregnancy and miscarriage or poor fetal growth have been reported in epidemiological studies.[31,87,103,194,451,657,702,725,726,745] In many instances, these associations are largely attributable to confounding effects of maternal cigarette smoking[31] or other factors. Some of these studies have serious methodological limitations.[39] If maternal consumption of caffeine-containing beverages in conventional amounts during pregnancy does have an association with the rate of miscarriage or fetal growth retardation, the effect appears to be relatively small.

In other studies, no association has been found between caffeine consumption during pregnancy and congenital defects.[303,403,404,434,705] For instance, Rosenberg et al.[589] analyzed six selected birth defects in relation to maternal ingestion of more than 8 mg/kg per day of tea, coffee, or cola. The defects were inguinal hernia, cleft lip/cleft palate, cardiac defects, pyloric stenosis, cleft palate (isolated), and neural tube fusion defects. None of the point estimates of relative risk was significantly greater than unity, suggesting that caffeine was not a major teratogen, at least for the defects evaluated.

In animal studies, caffeine is teratogenic only at very high doses. Nishimura and Nakai[520] reported that caffeine injected into mice at a dose of 250 mg/kg at selected times during organogenesis resulted in 43% of the offspring with cleft palate and digital defects. Fetal growth retardation and embryo lethality were observed at maternally toxic doses.[242]

The teratogenic effect in rats and mice has been confirmed by the oral (gavage) route,[43,44,393] by dietary feeding,[221] by addition to the drinking water,[173,534] by fetal injection,[549] and by the subcutaneous route.[222] The teratogenic response varies with the dose, route, and species, but limb hypoplasia and digital defects (ectrodactyly) were a common finding. The rabbit exhibited 9% digital defects following oral administration during the first half of gestation.[44] Teratogenic effects have not been produced in hamsters.[130] An increased frequency of malformations, especially of the limbs and palate, has been observed among the offspring of rats or mice treated with caffeine during pregnancy in doses equivalent to the human consumption of 40 or more cups of coffee/day.[130,504] Fetal death, growth retardation, and skeletal variations are often seen in these experiments after maternal treatment with very high doses of caffeine during pregnancy. In addition, paraxanthine, a metabolite of caffeine, induced increased resorption, cleft palate, and limb defects, similar to the effects of caffeine when given intraperitoneally to rats.[769]

In the majority of animal studies in which lower doses of caffeine were administered (equivalent to 5 to 20 cups of coffee per day), adverse effects on development either were not observed or were mild.[502,523,534] In one study, an increased frequency of cleft palate was observed among the offspring of rats given the human equivalent of 5 to 19 cups of coffee a day during pregnancy.[534] In another study, an increased rate of cardiac defects was observed among the offspring of rats treated during pregnancy with the equivalent of 15 or more cups of coffee per day;[784] however, most investigations do not show an increased frequency of malformations among the offspring of rodents treated during pregnancy with these high doses of caffeine.[129,130,173,375,524] An extensive study showed that the no-observable-effect level for frank teratogenesis in the rat was 40 mg/kg per day on days 0 to 19 postconception and that both the severity and frequency of the effects increase with dose.[130] The overall result of this large study was that embryotoxicity did not occur unless the exposure was significantly above the usual human exposure.

Several groups of investigators believe prenatal caffeine may result in subtle, but lasting, physical and behavioral impairments in rats.[272] Persistent behavioral alterations have been reported to occur among the offspring of rats and mice treated during pregnancy with doses of caffeine equivalent to 10 to 60 cups of coffee a day.[272,273,342,343,644] The developmental toxicity profile of caffeine in animals was reviewed in detail by Nolen.[524]

In primates, teratogenicity due to caffeine consumption has not been observed; however, stillbirths and miscarriages were observed with increased frequency among the offspring of *Macaca fascicularis* female monkeys treated during pregnancy with caffeine in a dose equivalent to 5–7 or 12–17 cups of coffee per day.[243] The cause for the stillbirths was not apparent on necropsy. Behavioral alterations have also been recorded among the offspring of monkeys born to mothers treated with an unspecified dose of caffeine during pregnancy.[570] The relevance of these observations to the risks in infants born to women who drink large amounts of caffeinated beverages during pregnancy is unknown, but many of the whole animal pregnancy studies indicate that if exposures are very high and the administration technique can attain high blood levels, reproductive effects may occur.

Interpretation of the available information pertaining to the animal and human studies regarding the teratogenicity of caffeine leads us to conclude that the usual exposure of caffeine does not present a measurable risk in the human for any one malformation or group of malformations. There is a clear indication that the consumer must ingest a substantial amount of caffeine in order to have an effect on the developing embryo or fetus; total consumption of 300 mg/day may be a safe upper daily limit. Most reviewers and investigators concluded that there is a threshold below which caffeine does not exert a detrimental effect, and the usual human consumption falls in this nontoxic range. The quantity of caffeine consumed in an average cup of coffee, about 1.4 to 2.1 mg/kg,[193] is believed to be below the amount that induces congenital defects in animals. Quantities of caffeine in tea and soft drinks would be even less.

2.5 Carbamazepine

Although epidemiologic and case report studies have not yielded consistent results,[85] exposure to carbamazepine has been associated with minor craniofacial defects; fingernail hypoplasia; developmental delay;[362,518] reduced birthweight, length, and head circumference;[42,325] and neural tube defects.[587] Confounding the issue is the possibility that epilepsy itself may increase the risk for malformations.[363] Animal studies in pregnant mice yielded inconsistent results at similar doses.[430,537] An attempted suicide involving carbamazepine, however, produced blood levels of 27 to 28 μg/ml (the therapeutic range is 8 to 12 μg/ml) during what was estimated to be 3 to 4 wk postconception.[435] The fetus was later determined to have myeloschisis, with carbamazepine being the only known exogenous risk factor. This suggests that carbamazepine has the potential to produce neural tube defects at about two- to threefold the therapeutic level. It appears that the risk for minor defects is significant, but the risk for all teratogenic effects is not known. The risk for abortion is also not known but appears to be small.

2.6 Coumarin Derivatives

Nasal hypoplasia following exposure to several drugs, including warfarin, during pregnancy was reported by DiSaia.[163] Kerber et al.[382] were the first to suggest warfarin as the teratogenic agent. Coumarin anticoagulants have since been associated with nasal hypoplasia, calcific stippling of the secondary epiphysis, and central nervous system abnormalities. Warfarin embryopathy has been described, and an overview of the difficulties in relating a congenital malformation to an environmental cause has been published.[27,29,286,383,542,666,715,716,776] There is an estimated 10% risk for affected infants following exposure during the period from the 8th through the 14th week of pregnancy, although this risk has been reported to be much lower in some series, and other factors besides dose and gestational stage seem to play a role.[286] Low-dose warfarin (5 mg/day or less) throughout pregnancy did not result in any adverse effects in 20 offspring.[135]

Coumarin has been shown to inhibit the formation of carboxyglutamyl residues from glutamyl residues, decreasing the ability of proteins to bind calcium.[663] The inhibition of calcium binding by proteins during embryonic/fetal development, especially during a critical period of ossification, could explain the nasal hypoplasia, stippled calcification, and skeletal abnormalities of warfarin embryopathy.[296] Microscopic bleeding does not seem to be responsible for these problems early in development.[27]

One case report was unique in that the time of exposure to warfarin was between 8 and 12 wk of gestation, and the infant presented Dandy-Walker malformations, eye defects, and agenesis of the corpus callosum.[373] This case report is the clearest evidence for a direct effect of warfarin on the developing central nervous system rather than an effect mediated by hemorrhage, because the exposure was well defined and occurred before the appearance of vitamin K-dependent clotting factors. Further supportive evidence for a direct pathogenic role of warfarin is the report of an infant with an inherited deficiency of multiple vitamin K-dependent coagulation factors whose congenital anomalies were similar to warfarin syndrome without exposure to warfarin.[536] The risk of stillbirths and spontaneous abortions is increased in pregnant women treated with warfarin,[286,604,624,710] but the risk may be less if the exposure is in the last half of pregnancy. The risk of adverse effects due to hemorrhaging increases later in gestation.

No increase in major malformations was reported in mice,[399] rats,[334] or rabbits[275] exposed to warfarin at greater than therapeutic doses. In the rat, however, increased fetal hemorrhage was reported,[334] and maxillonasal hypoplasia occurred after postnatal administration.[335] One difficulty in producing an appropriate animal model appears to be that the prenatal skeletal maturation in the human that is susceptible to the detrimental effects of warfarin occurs postnatally in the rat.

2.7 Cyclophosphamide

Cyclophosphamide is a widely used antineoplastic agent with an apparent increased teratogenic risk, but the magnitude is uncertain.[609] The defects include growth retardation, ectrodactyly, syndactyly, cardiovascular anomalies, and other minor anomalies.[265,614,688] Ten normal pregnancies have been reported after cyclophosphamide exposure.[54]

Experimental animal studies in the rat,[115,632] mouse,[241] and rabbit[220] have shown distinct developmental stage specificity, dose-effect relationships, and a high sensitivity of nervous system and mesenchymal tissues.[487]

The mechanism of cyclophosphamide teratogenesis was reviewed by Mirkes:[487] cytochrome P-450 mono-oxygenases convert cyclophosphamide to 4-hydroxycyclophosphamide, which in turn breaks down to phosphoramide mustard and acrolein. Phosphoramide mustard may produce teratogenic effects by interacting with cellular DNA in an as yet undefined manner, while acrolein acts in a different manner, possibly by affecting sulfhydryl linkages in proteins.[285] Tissue sensitivity to phosphoramide mustard and acrolein is thought to be related to such processes as detoxification and cellular repair.

2.8 Diethylstilbestrol

The first abnormality reported following exposure to diethylstilbestrol (DES) during the first trimester was clitoromegaly in female newborns.[57] Herbst et al.[315,319] and Greenwald et al.[268] later reported an association of vaginal adenocarcinoma in female offspring following first trimester exposures. DES is the only drug with proven transplacental carcinogenic action in the human. Almost all of the cancers occurred after 14 years of age and only in those exposed before the 18th week of gestation.[316,318,698] There is a 75% risk for vaginal adenosis for exposures occurring before the 9th week of pregnancy; the risk of developing adenocarcinoma is about 1:1000 to 1:10,000.[311,474,526,708] While the incidence of vaginal adenosis was related to the amount of DES administered,[526] the incidence of vaginal carcinoma does not appear to be related to the maternal dose.[312]

Although there does not appear to be an adverse effect on menstrual cycle functioning[610] or on the rate of conception,[24] the anatomic abnormalities of the uterus and cervix induced by intrauterine exposure to DES — including T-shaped uterus, transverse fibrous ridges, and uterine hypoplasia — cause the increased incidence of ectopic pregnancies, spontaneous abortions, and premature delivery in pregnancies of women exposed to DES *in utero*.[24,40,152,313,314,433,605,707]

There have been reports that males exposed to DES *in utero* exhibited genital lesions and abnormal spermatozoa.[46,245,246,626] Other studies reported no increase in the risk for the male for genitourinary abnormalities or infertility.[417,708] An association between *in utero* expsosure to DES and testicular cancer in male offspring has been suggested but the data are not conclusive.[239,708] The controversial nature of the effects of DES exposure on the male may be attributable to study design or, more likely, to the fact that dose levels varied greatly according to different regimens: exposures during the first half of pregnancies varied from 1.5 to 150 mg per day, with total doses from 135 mg to 18 g.[317]

Teratogenic and transplacental carcinogenic effects following *in utero* exposure to DES have been demonstrated in several animal species,[448,712] including the rat,[481] mouse,[469,513] hamster,[268] and monkey.[306,309,358,684] A major difficulty in studies of the mechanism of action of DES is the extensive biotransformation that occurs in the adult and fetus.[444,476]

DES is a potent nonsteroidal estrogen and, as in the case of steroidal estrogens, must interact with the receptor proteins present only in estrogen-responsive tissues before exerting its effects by stimulating RNA, protein, and DNA synthesis. The carcinogenic effect of DES is most likely indirect: DES exposure results in the presence of columnar epithelium in the vagina, and this misplaced tissue may have a greater susceptibility to developing adenocarcinoma, much as teratomas and other misplaced tissues are more susceptible to malignant degeneration.

2.9 Diphenylhydantoin

Hanson and Smith[291] characterized the fetal hydantoin syndrome in infants whose mothers were treated for epilepsy with hydantoin anticonvulsants. Chronic exposure to diphenylhydantoin (phenytoin) has been suggested to present a 5 to 10% risk for the full fetal hydantoin syndrome including ocular hypertelorism, flat nasal bridge, and distal digital hypoplasia with nail hypoplasia.[5,190,289,482,653] Cleft lip and palate, congenital heart disease, and microcephaly have been reported, but hypoplasias of the nails and distal phalanges are possibly more common malformations in the exposed fetuses.[29,145,225,327] Although the hydantoin syndrome is observed in 11% of the subjects in some studies, three times as many exhibit mental deficits.[289,290] It should be mentioned that prospective studies demonstrate a much lower frequency of effects, and some do not demonstrate any effect; thus, the overall prospective risk may be much lower for the classically reported effects.

Factors associated with epilepsy may contribute to the etiology of these malformations.[225-228,378,379] Based on the U.S. Collaborative Perinatal Project and a large Finnish registry, the incidence of malformations was 10.5% when the mother was epileptic, 8.3% when the father was epileptic, and 6.4% when neither parent was affected.[623]

Cleft lip and palate, cardiac defects, and skeletal anomalies have been produced in mice, rats, and rabbits,[128,174,201,295,463,593,778] and the malformation rate was dose dependent.[200,201,465] Phenytoin doses that resulted in one or two times the therapeutic blood levels in the monkey were maternally toxic and increased embryonic death but did not increase fetal malformations.[310] Similar exposures in rats and rhesus monkeys produced behavioral changes in offspring.[4] Recent preliminary evidence suggests that exposure to phenytoin alters the expression of several growth and transcription factors in mouse embryos.[503]

The teratogenic action of phenytoin has been postulated to involve the cytochrome P-450 metabolism of phenytoin to produce a reactive epoxide metabolite. The arene oxide would covalently bind to macromolecules and interfere with their function.[50,294,453,654] Further studies have not confirmed this or other hypotheses.

2.10 Infectious Agents

Various viral, bacterial, and parasitic agents are known to cause maldevelopment in humans including rubella, cytomegalovirus, herpes simplex, parvovirus B19, syphilis, toxoplasmosis, varicella-zoster, and Venezuelan equine encephalitis.[620] The lethal or developmental effects of infectious agents are the result of mitotic inhibition, direct cytotoxicity, or necrosis. Repair processes may result in metaplasia, scarring, or calcification, which causes further damage by interfering with histogenesis. Infectious agents appear to be exceptions to some of the principles of teratogenesis because the relevance of dose and time of exposure cannot be demonstrated as readily for replicating teratogenic agents. Transplacental transmission of an infectious agent does not necessarily result in congenital malformations, growth retardation, or lethality.

2.10.1 *Cytomegalovirus*

Cellular necrosis is the principal mechanism by which cytomegalovirus (CMV) damages organs. It appears that 2 to 5% of pregnant women become infected with CMV,[621,772] but the incidence varies considerably with geographical location. Approximately 20% of infected pregnant women have infected fetuses.[288,658,659,772] Reactivation of latent CMV infection is common, as with other herpes viruses, but reinfection is also common due to the antigenic diversity of CMV. Fetal infection may result more commonly from reactivation than from reinfection. Serious clinical manifestations are observed at birth in about 10% of the infected fetuses, but the incidence of CMV-related abnormalities may increase to as high as 25% within the first few years of life.[131,212] Six to twenty of every 10,000 infants born in the U.S. are brain damaged by CMV. Fetal infections result in an incidence of approximately 60% for some eye anomaly, hearing loss, and/or learning disability.[271,659,772]

2.10.2 *Herpes Simplex Virus*

The incidence of fetal herpes virus infections (type 1 or 2) is a very rate event.[22,354] The pattern of anomalies includes intracranial calcifications, microcephaly, microphthalmia, and retinal dysplasia.[22,204,354,608,650] Primary genital infections during pregnancy have been associated with spontaneous abortion, stillbirth, and congenital malformations, while recurrent genital infections may be associated with premature delivery but not usually with fetal abnormalities.[22,91] Disseminated infections adversely affect many organs of the fetus, but fetal death is often due to maternal complications[22] or fetal herpetic hepatitis. Fetal anomalies associated with herpes simplex virus infection appear to be due to disruption rather than malformation.

2.10.3 *Parvovirus B19*

Parvovirus B19 (erythema infectiosum) has been shown to be causal for hydrops fetalis and fetal death.[92,394,472,625] Transplacental infection with parvovirus B19 has been demonstrated.[56] Parvoviruses replicate in cells during a permissive S-phase of DNA mitosis. The infections may cause death in specific groups of cells or cause generalized effects such as hemorrhage, severe anemia, edema, and fetal death.[364] Focal endothelial damage, abnormalities of the eye, and destruction of muscle tissue resulting from intrauterine infection in the human resembles the pathology resulting from parvovirus infection in animals.[299] Fetal infection with parvovirus B19 is not common, and, while there is a minimal increased risk of congenital malformations, there is a 5 to 15% risk of fetal loss.[176,299,494,544,689,732]

2.10.4 *Rubella Virus*

The original rubella syndrome described by Gregg[269] comprised heart disease, deafness, and cataracts, but intrauterine growth retardation, encephalitis, thrombocytopenia, and mental retardation are also observed.[554] The frequency of embryonic rubella infection after maternal rubella with a rash is more than 80% during the first 12 wk of pregnancy, 54% at 13 to 14 wk, 25% at the end of the second trimester, and 100% at term.[481,649,696,697] Adult survivors of intrauterine rubella infection have an increased risk of diabetes mellitus and, more rarely, progressive rubella panencephalitis.[250,475,649] Rubella has an affinity for specific tissues and causes damage by means of several mechanisms: mitotic inhibition, cell death, and interference with histogenesis by repair processes, resulting in calcification and scarring.[507]

2.10.5 *Syphilis*

Congenital syphilis, *Treponema pallidum* infection, occurs in about 0.05% of live births in the U.S. Fetal syphilis infection can cause stillbirth and miscarriage in the second trimester.[333] It is thought that fetal pathology is associated with maturation of the fetal immune system at about the 20th week which permits an intense inflammatory response in the fetus.[38,298] The congenital abnormalities can include deafness, bone and dental deformities, and skin lesions.

2.10.6 *Toxoplasmosis*

Although *Toxoplasmosis gondii* infection can damage the fetus,[211] infants asymptomatic at birth have been shown to progress from mild to severe chorioretinitis and permanent neurologic damage.[750] The risk of transmission increases as pregnancy progresses, but the risk of severe effects, brain calcification, hydrocephaly, and hepatosplenomegaly is greatest for fetuses from mothers infected in the first trimester.[568] It appears to be unlikely that toxoplasmosis contributes to the risk of repeated spontaneous abortion.[387,568,714]

2.10.7 *Varicella-zoster*

Varicella-zoster virus has been reported to cause an unusual pattern of anomalies, including cicatrical lesions, intrauterine growth retardation, limb anomalies, brain defects, and eye abnormalities.[21,240,274] The increased risk of congenital defects is about 1 to 2% for fetuses from women infected during the first trimester.[555] Because the risk is low, it is unusual for a woman who has sought counseling to choose elective abortion. If fetal infection occurs very close to the time of delivery, there is a high risk of severe neonatal infection.

2.10.8 *Venezuelan Equine Encephalitis*

Venezuelan equine encephalitis[437,736] infections can cause localized cell necrosis in fetal tissues and may lead to brain damage, but fetal infection is rare.

2.11 Maternal Disease States

2.11.1 *Diabetes Mellitus*

The introduction of insulin into clinical practice in 1922 made pregnancy possible for many women with diabetes mellitus. Fetal overgrowth quickly was recognized as a complication of the disorder. Pedersen[540] was the first to propose a credible explanation for this fetal overgrowth: maternal hyperglycemia produces fetal hyperglycemia, which in turn stimulates a hypersecretion of insulin by the fetal pancreas. The lipogenic and protein anabolic actions of insulin and glucose result in fetal macrosomia. Fetal growth retardation observed in 5 to 10% of newborns of diabetic mothers is due most probably to the placental vascular lesions in severe diabetics.

The offspring of insulin-dependent diabetic mothers also have an increased incidence of caudal dysplasia or caudal regression syndrome and of malformations such as spina bifida, anencephaly, and transposition of the great vessels.[400,484,540] Epidemiologic and clinical studies show that congenital abnormalities occur in 8 to 12% of offspring of insulin-dependent diabetic women,[484,540] that the abnormalities are induced early in development,[485] and that the incidence is highest when maternal glycemic control is poor early in pregnancy.[266,479,768] Experimental animal investigations concerning the developmental effects of diabetes also show that the uncontrolled diabetic condition plays the primary role in congenital defects, and that insulin therapy can reduce the incidence of malformations in offspring.[181,600] The results of *in vitro* studies suggest that factors other than hypoglycemia or hyperketonemia associated with poorly controlled diabetes make a significant contribution to the embryopathic process.[96] This lends support to the conclusion that diabetic embryopathy has a multifactorial etiology.

2.11.2 Epilepsy

Both epilepsy in pregnant women and its treatment appear to increase the risk of congenital anomalies in offspring. Although all major anti-epileptic drugs (carbamazepine, valproate, phenobarbital, phenytoin, and primidone) have been linked to teratogenic abnormalities, the pregnant woman will continue to receive treatment because the risk of seizures to her health or that of the fetus is greater than the risk of teratogenic effects.

The risk of a congenital anomaly is increased in siblings of children born with defects associated with exposure to an antiepileptic drug.[431,579] It has also been shown that in 25% of the families with a case of a neural tube defect associated with exposure to an anti-epileptic drug, there is a positive family history for neural tube defects.[432] It has not been shown that an epileptic genotype is associated with a predisposition for at least one of the anomalies associated with exposure to antiepileptic drugs, namely neural tube defects. Another possibility is that the decreased serum folate concentration demonstrated in pregnant women treated with antiepileptic drugs may contribute to an increased risk for neural tube defects.[767]

2.11.3 Maternal Endocrinopathies

Endocrinopathies and endocrine tumors may expose the fetus to excessive hormone concentrations. If these conditions do not also interfere with the maintenance of pregnancy, they can be expected to have developmental effects similar to the administration of those synthetic hormones that induce malformations.

2.11.4 Nutritional Perturbation

Investigations of the detrimental effects of famine in war-time Holland on the outcome of pregnancy constitute what is probably the most extensive collection of data concerning intrauterine undernutrition in humans.[662] A famine early in gestation increased prenatal and neonatal mortality, as well as increasing the incidence of central nervous system abnormalities (spina bifida and hydrocephalus) and prematurity. Exposure to famine late in gestation resulted in intrauterine growth retardation. Homeostatic mechanisms in mothers and fetuses appear to minimize variations in fetal growth over a wide range of maternal nutritional states: birth weights were reduced by only about 9% at the peak of the Dutch famine from 1944 to 1945.

Dietary restriction produced central nervous system and skeletal abnormalities[595] or cleft palate[678] in mice. In laboratory animals, embryonic wastage is more commonly induced than congenital malformation.[370]

The role of malnutrition is an important area for investigation because it may be a contributing factor to many teratogenic milieu. Folic acid deficiency is an example; by altering the timing and the duration of folic acid deficiency in experimental animals, the incidence and variety of congenital defects can be altered.[251,370]

A series of investigations provided evidence suggesting that folic acid supplementation could reduce the incidence of recurrence of neural tube defects in the human.[416,639-641] It was later shown convincingly that periconceptional supplementation with folic acid, 4 mg/day, reduces the risk of recurrence of neural tube defects in subsequent siblings of children with neural tube defects.[473] Furthermore, low-dose folic acid supplementation, 0.8 mg/day, was reported to decrease the incidence of neural tube defects in a population not at increased risk for these defects.[144] Although folate supplementation reduces the incidence of neural tube defects, folate supplementation will not prevent all neural tube defects, and folate deficiency has not been proven to be associated with neural tube defects.[588] It is not known whether folic acid supplementation corrects an undefined metabolic defect or a nutritional deficiency.

2.11.5 *Phenylketonuria*

Maternal phenylketonuria[215,338] has been associated with an increased incidence of abnormalities: the offspring of mothers with phenylketonuria have a high frequency of mental retardation, microcephaly, and intrauterine growth retardation.[232,338,418] There is also an increased risk of abortion.

2.12 Mechanical Problems

Birth defects involving the limb, skin, head shape, midline closure, ear, jaw, and muscle may be produced by mechanical problems of constraint. Pathologic states such as oligohydramnios, uterine malformations (infantile uterus, bifid uterus, uterine myoma), velamentous cord insertion, amniotic bands, umbilical cord constrictions, extrinsic pressure, or uterine contractions could result in (1) distortion and hypoplasia of external structures; (2) failure of closure of the lip, palate, neural tube, and/or abdomen; (3) limb amputation or malposition; or (4) limb reduction defects. Not only might there be abnormal pressure externally, but these conditions may also interfere with the vascular supply to structures such as the skin, the limbs, or the ears. There is substantial experimental work in animals to support these hypotheses for the human.[263]

2.13 Methylmercury

There have been several incidences of human exposures to organic mercury as an environmental pollutant or as a fungicide present on seed grain consumed by humans.[293] In Minamata, Japan, the local population was exposed by ingesting fish caught in a bay polluted by methylmercury. In 1953, neurologic abnormalities began to develop in the population, and approximately 38% of the afflicted persons died. About 6% of the children born between 1955 and 1959 were affected: cerebral palsy and associated microcephaly were the common features, with few other congenital defects.[455,643] Synder[643] described the severe damage to the central nervous system in offspring of women who had ingested meat from a pig fed seed grain containing a mercurial fungicide. Seed grain containing a mercury fungicide was responsible for fetal exposures in Iraq, when pregnant women consumed bread inadvertantly prepared using contaminated seed grain.[7,8] Cerebral palsy, as well as motor and mental impairments, were reported.

In vitro studies have shown that methylmercury in human blood interacts with erythrocytes and protein components, primarily lipoproteins. The nature of this association (and the biochemical basis of the biologic effects of organic mercury) was attributed to interactions between methylmercury and –SH or –SS groups of proteins.[343]

In experimental animal studies, methylmercury exposures have been associated with an increased incidence of cleft palate and central nervous system abnormalities.[345,384,499,656] Organic mercury has been reported to concentrate preferentially in the fetal brain and can mimic the central nervous system lesions seen in the human.[390,765] The severity of brain lesions in experimental animals appears to be dose dependent.[97]

2.14 Methylene Blue

Methylene blue may be used to mark the amniotic cavity during amniocentesis but it can also be used in the treatment of methemoglobinemia. There appears to be no increased risk of congenital malformations or abortion from intrauterine exposure to methylene blue. The magnitude of the risk of intestinal atresia associated with exposure to methylene blue during amniocentesis is not known, but if it is causal the risk is likely to be small; however, in rats, intra-amniotic injection of methylene blue increases fetal loss.[546]

In one study, methylene blue was injected into one aminotic cavity of 86 twin pregnancies;[703] jejunal atresia occurred in one of the infants in 17 of these pregnancies. In 15 of these cases it was possible to determine which twin was exposed to methylene blue, and in each case the twin exposed to methylene blue had jejunal atresia. Because of the generally increased incidence of congenital anomalies associated with twin pregnancies and other factors, it is not yet clear whether the association of methylene blue with intestinal atresia is causal or fortuitous.[169,493,516]

Neonatal hemolytic anemia and jaundice have been documented in neonates exposed to methylene blue late in pregnancy.[136,139,619] Hemolytic anemia is a complication of methylene blue therapy in children.

2.15 Misoprostol

Misoprostol is a synthetic prostaglandin E_1 methyl analogue used for the prevention of gastric ulcers induced by nonsteroidal anti-inflammatory drugs. It has known, but not very effective, abortifacient properties.[559] Gonzalez et al.[256] recently reported seven newborns with vascular disruptive phenomena (limb reduction defects, Moebius syndrome) whose mothers used misoprostol early in pregnancy in an attempt to induce abortion. Although there is evidence that misoprostol is used illegally by thousands of pregnant Brazilian women as an abortifacient,[123,134,439] controlled cohort or case-control epidemiological studies of the fetal outcome of failed abortions are not available. Coelho et al.[123] indicate that the World Health Organization estimated that there are 5 million illegal abortions in Brazil each year. Although the data available are not conclusive, the uterine bleeding produced by misoprostol[559] and the type of malformations produced suggest a vascular disruption mechanism for misoprostol-induced teratogenesis.

Most of the animal studies dealing with the teratogenic potential of prostaglandins were negative, although hydrocephaly, anophthalmia, and microphthalmia have been reported at maternally toxic doses (2.0 mg/kg prostaglandin E_1, alprostadil) in pregnant rats.[447] If one is looking for vascular disruption, it will more likely be produced later in gestation; therefore, classical animal teratology experiments will not detect the vascular disruptive effect of drugs or chemicals unless they are exposed beyond the period of early oganogenesis.[73] Furthermore, it has become clear that if an agent produces vascular disruption, it is a rare event and therefore large populations would have to be studied before the effect may be discovered.[203,515]

Previous case reports are also of little assistance. Collins and Mahoney[127] reported an infant with hydrocephalus and attenuated digital phalanges after exposure intravaginally to 15-methyl F_{2alpha} prostaglandin 5 wk after conception. Schuler et al.[611] reported that 29% of women who used misoprostol in Brazil as an abortifacient failed to abort. Seventeen children who failed to abort were observed to have no malformations. Woods et al.[761] reported an infant exposed to oxytocin and prostaglandin E_2 for the purpose of termination to have hydrocephaly and growth retardation. Schonhofer[610] and Fonseca et al.[206] reported five Brazilian infants with defects of the skull and overlying scalp who were exposed to misoprostol *in utero*. These case reports indicate the low risk of misoprostol exposure and the possibility that some of the features reported may or may not be due to misoprostol.[110] It is too early to know the extent of the effects of misoprostol, but it is biologically plausible that they should include all of the features of vascular disruption.

2.16 Ozazolidine-2,4-Diones (Trimethadione, Paramethadione)

Trimethadione and paramethadione are antiepileptic oxazolidine-2,4-diones that distribute uniformly throughout body tissues and exert their effects by means of the action of their metabolites. These drugs affect cell membrane permeability and vitamin K-dependent clotting factors, but their primary mode of action is unknown.

Zackai et al.[774] described the fetal trimethadione syndrome characterized by developmental delay, V-shaped eyebrows, low-set ears with anteriorly folded helix, high arched palate, and irregular teeth. Clinical observations including these and other associated findings, such as cardiovascular, genitourinary, and gastrointestinal anomalies, have been reviewed.[125,192,238,637] The incidence of miscarriage, stillbirth, and infant death was also increased. There are wide variations in reported risk, with estimates as high as 80% for major or minor defects. Because the number of exposures is small, the actual risk could vary considerably from these figures.

Mice exposed to high doses of trimethadione on days 8 to 10 or 11 to 13 of gestation had a high incidence of fetal growth retardation and abnormalities of the viscera and skeleton; especially common were aortic arch and vertebral defects.[90,735] The incidence of malformations was increased in pregnant rats and rhesus monkeys treated with 1 to 10 times the therapeutic dose of trimethadione in the human.[213,711] Behavioral alterations were also observed in the offspring of pregnant rats exposed to 5 times the therapeutic dose of trimethadione.[711]

2.17 D-Penicillamine

D-Penicillamine has been used in the treatment of rheumatoid arthritis and cystinuria. D-Penicillamine is a copper chelator shown to induce cleft palate and skeletal defects in the mouse and rat.[386,446,505,592,661] Copper deficiency appears to be the mechanism for teratogenicity.[346,377,445] Exposure to D-penicillamine can induce a connective tissue defect including generalized cutis laxa, hyperflexibility of the joints, varicosities, and impaired wound healing.[297,429,489,645] The exposure must be long enough to induce a copper deficiency sufficient to inhibit collagen synthesis and maturation; however, the condition appears to be reversible and the risk is low, 5% or less.[177,270,585,591]

2.18 Polychlorinated Biphenyls

"Yusho" is the Japanese term for poisoning by polychlorinated biphenyls (PCBs), first identified in Japan in 1968 and later in Taiwan in 1979.[583] PCBs consumed in contaminated cooking oil by pregnant women resulted in fetal PCB syndrome characterized by pigmented children (cola-colored) with characteristic low birth weight; pigmentation of the gums, nails, and groin; hypoplastic deformed nails; conjunctivitis, with an enlargement of the sebaceous glands of the eyelid; abnormal calcification of the skull; rocker bottom heel; and, possibly, an alteration in calcium metabolism.[583,765]

An important observation is the occurrence of fetal PCB syndrome in subsequent children after the use of the contaminated oil was discontinued: residual contaminants in the bodies of pregnant women resulted in congenital fetal PCB syndrome in 13 children born up to 2 years after the last ingestion of the contaminated oil, and slight brown staining of the skin continued for approximately 4 years.[480] There is evidence that infant learning is impaired,[687] and lower scores in developmental tests have been reported for children exposed to contaminated cooking oil prenatally or by breast feeding.[773] Although reduced birthweight has been reported for newborns of mothers eating contaminated fish from the Great Lakes[191] or exposed to PCBs in the workplace,[680] another study found no association between PCB levels in infants and their birthweight or head circumference.[584]

A complete evaluation of PCB developmental toxicity in the human is difficult because PCB-contaminated cooking oil contained several PCB isomers, as well as toxic polychlorinated quarterphenyls (PCQs) and polychlorinated dibenzofurans (PCDFs). Although dose-response relationships have been demonstrated for the toxic effects, there are no dose-response data for the developmental effects in the human.

Kunita et al.[402] employed nonpregnant rats and monkeys to study the hepatic hypertrophy, immunosuppression, hepatic microsomal enzyme activation, and dermal symptoms resulting from exposures to various combinations of PCBs, PCQs, and PCDFs. Although they concluded that PCDFs are the primary agent causing yusho in rats, there is great variation in the effects of different polychlorinated aromatics in different species.

Occupational exposure to PCBs, which may occur during repair of or accidents involving PCB-containing transformers and capacitors, can result in exposures 10 to 1000 times above those in nonoccupationally exposed persons. Because of the qualitative nature of PCB identification and measurements, it is difficult to estimate the teratogenic risk of occupational exposures to PCBs. It does appear, however, that most acute, accidental occupational exposures present less of a teratogenic risk than do the chronic exposures that resulted in yusho after ingestion of contaminated cooking oil. Transformer oil-contaminated roads, fields, and livestock expose women to doses that are several orders of magnitude lower than those that occurred in Japan.

2.19 Progestins

It is often overlooked that, although various progestins utilized therapeutically as progestational agents act by means of similar receptors, their potential androgenic effects can differ markedly. This point is critical to the evaluation of the virilizing effects of these compounds in the human. It has been shown, for example, that the pharmacokinetic parameters that estimate steroid bioavailability and metabolism show great variability among subjects and between steroids conveniently grouped together, such as "progestins".[210] One must assume that these differences in bioavailability and metabolism reflect differences in the biologic activity of these sterioids in humans.

In contrast to progesterone and 17_{alpha}-hydroxyprogesterone caproate, high doses of some of the synthetic progestins have been reported to cause virilizing effects in humans. Exposure during the first trimester to large doses of 17_{alpha}-ethinyltestosterone has been associated with masculinization of the external genitalia of female fetuses.[746,747] Similar associations result from exposure to large doses of 17_{alpha}-ethinyl-19-nortestosterone (norethadrone)[746] and 17_{alpha}-ethinyl-17-OH-5(10)estren-3-one (Enovid-R).[277] The synthetic progestins, like progesterone, can influence only those tissues with the appropriate steroid receptors. The preparations with androgenic properties may cause abnormalities in the genital development of females only if present in sufficient amounts during critical periods of development.[704,746,747] In 1959, Grumbach et al.[277] pointed out that labioscrotal fusion could be produced with large doses if the fetuses were exposed before the 13th week of pregnancy, whereas clitoromegaly could be produced after this period, illustrating that a specific form of maldevelopment can be induced only when the embryonic tissues are in a susceptible state of development.

The World Health Organization[763] reported that there is a suspicion that combined oral contraceptives or progestogens may be weakly teratogenic but that the magnitude of the relative risk is small. In a large retrospective study, Heinonen et al.[302] reported a positive association between cardiovascular defects and *in utero* exposure to female sex hormones. A re-evaluation of some of the base data by Wiseman and Dodds-Smith,[756] however, did not support the reported association. Another retrospective study conducted by Ferencz et al.[195] did not find a positive association between female sex hormone therapy and congenital heart defects. Although neither study disproved the positive association reported by Heinonen et al.,[302] their findings made the association less likely.

Epidemiologic studies have reported an association between exposures to female sex hormones, oral contraceptives or progestogens, and congenital neural tube defects[230,231] and limb defects.[350,351] Further studies and re-evaluations have not supported either of these associations.[414-416,622] Several reviews have discussed the evidence against the involvement of female sex hormones in nongenital teratogenesis.[86,751,752]

Further support for the absence of a nongenital effect of progestins comes from (1) a negative correlation between sex hormone usage during pregnancy and malformations,[755] (2) no increased incidence in malformations following progesterone therapy to maintain pregnancy,[581] and (3) no increased incidence in malformations following first trimester exposure to progestogens (mostly medroxyprogesterone) administered to pregnant women who had signs of bleeding.[374,771] The U.S. Food and Drug Administration has recently recognized that the evidence does not support an increased risk of limb reduction defects, congenital heart disease, or neural tube defects following exposure to oral contraceptives or progestins.[69]

It is generally accepted that the actions of steroid hormones are mediated by specific steroids,[527] and only those tissues with the specific receptors can be affected by steroid hormones. It has been shown that 17_{alpha}-hydroxyprogesterone caproate[106,616] and medroxyprogesterone[107] do not cause developmental abnormalities in nonreproductive organs of mice. Nonreproductive tissues of the fetal primate are similarly unaffected by exposure to sex steroids because they do not possess estrogen or progesterone receptors during early organogenesis.[329]

2.20 Radioactive Isotopes

The radiation exposure to the embryo from medically administered radioactive isotopes is determined by several factors, including placental exchange, tissue affinity, and nature of the radiation(s) emitted (alpha particle, beta particle, gamma ray). Estimating the absorbed dose and hazard to the fetus is complex because the radioisotope may locate in specific target organs, it may or may not cross the placenta, the distribution of irradiation may not be random, metabolism of the element or compound may be affected by disease or genotype, and the radiation dose rate decreases exponentially with time.

In addition to the administered isotope, background radiation contributes to the total exposure. Background radiation has been estimated to contribute less than 100 mrad (1 mGy) over the course of pregnancy to the dose absorbed by soft tissue, which presents no increased risk of deleterious effects for the embryo.[72] This is an important concept because many of the exposures from nuclear medicine procedures are very low.

Radioactive iodine in the form of ^{131}I is used for uptake studies, radioactive scanning, and cancer therapy. It may be in the form of the inorganic ion or bound to protein. ^{125}I is used to label hormones for *in vivo* and *in vitro* studies. Radioactive iodine is a potential risk to the fetal thyroid, especially once the fetal thyroid begins to concentrate iodide at 10 to 12 weeks of gestation. Inorganic iodides readily cross the placenta and, in time, a substantial amount of bound iodide will be released and become available to the fetus.[348,651] In all likelihood, there is no compound containing radioisotope of iodide which does not expose the fetus to some radioactivity.

Fetal thyroid avidity for iodides is greater than maternal thyroid avidity.[58] Reported fetal effects from therapeutic (ablative) doses of ^{131}I administered to pregnant women include total fetal thyroid destruction. In a retrospective study of fetuses accidentally exposed to ^{131}I during the first or first and second trimesters, 6 neonates out of 178 live births had hypothyroidism, and other anomalies were statistically increased above the general population.[669] Although there are few case reports in the literature, there is a definite risk of thyroid dysfunction in the offspring. The use of radioactive iodine should be avoided during pregnancy unless its utilization is essential for the medical care of the mother.

Inorganic radioactive potassium, sodium, phosphorus, cesium, thallium, selenium, chromium, iron, and strontium cross the placenta readily. Experiments in animals with radioactive phosphorus and strontium indicate that if the dose is large enough, embryonic abnormality and death can result.[627] Only radioactive phosphorus or gold may be utilized therapeutically, e.g., in the treatment of polycythemia or management of malignancies involving peritoneal surfaces. Most new radioactive agents are bound to a complex macromolecule or macroaggregate, cross the placenta in minuscule amounts, and deliver extremely low doses to the embryo.

Estimates of approximate fetal and maternal exposure for standard doses and procedures have been published.[77] In the vast majority of instances, a careful analysis will reveal that the exposure is too low to present a measurable risk to the embryo.

2.21 Radiation (External Irradiation)

The classic effects of radiation are cell death or mitotic delay. These effects are due to direct damage to the cell chromatin and are expressed in the offspring as gross malformations, intrauterine growth retardation, or embryonic death, each having a dose-response relationship and a threshold exposure below which no difference between an exposed and nonexposed control population can be demonstrated.[67] Offspring born to patients receiving radiation therapy for various conditions exhibited growth retardation, eye malformations, and central nervous system defects.[153,254,501] Microcephaly is probably the most common manifestation observed following *in utero* exposure to high levels of radiation in the human.[483] Fetal exposure to radiation at Hiroshima and Nagasaki resulted in microcephaly, growth retardation, and mental retardation,[482,759] but the reported increase in anomalies may not be statistically significant.[533] In a review of radiation teratogenesis, Brent[67] pointed out that no malformation of the limb, viscera, or other tissue has been observed unless the child also exhibits intrauterine growth retardation, microcephaly, and/or eye malformations. The risk of major anatomical malformations is not increased by *in utero* exposure to 5 rads (5 mGy) or less[67] of linear energy transfer (LET) radiation.

Experimental animal models have shown that radiation-induced effects on the developing organism are the result of the direct action of ionizing radiation on the embryo and are not due to a maternal effect.[67] Prior to implantation, the mammalian embryo is minimally sensitive to the teratogenic and growth-retarding effects of radiation and is very sensitive to the lethal effects.[80,596,598] Organogenesis is a stage sensitive to the teratogenic, growth-retarding, and lethal effects, but the embryo has some recuperative capacity.[79,596,597,599] Sensitivity to the teratogenic effects of radiation decreases during the fetal stage, but the fetus may still sustain permanent cell depletion since the fetus has a reduced recuperative capacity.[67] Permanent growth retardation is thus more severe following midgestation radiation.[355,356] Because of its extended periods of organogenesis and histogenesis, the central nervous system retains the greatest sensitivity of all organ systems to the detrimental effects of radiation through the later fetal stages.[78,612] The documented effects of prenatal exposure to ionizing radiation, which leads to histopathological abnormalities of the brain in experimental animals, are cell death and inhibition of cell migration.[197]

2.22 Retinoids, Systemic Administration (Isotretinoin, Etretinate)

Vitamin A congeners, including retinol, retinal, all-*trans*-retinoic acid (tretinoin), and 13-*cis*-retinoic acid (isotretinoin) are all teratogenic in numerous species.[603,681] Both isotretinoin (Accutane, Roche Laboratories), marketed for treating severe acne, and etretinate (Tigason, Sautier Laboratories), marketed for treating psoriasis, contained warnings by the manufacturers against exposure during pregnancy. Unfortunately, exposures occurred. The resulting malformations have been reviewed.[408,585] Human malformations include central nervous system, cardioaortic, microtia, clefting defects, and, more controversially, limb defects.[408,441,574,586] Isotretinoin has a serum half-life of 10 to 12 hr; there is no apparent risk to the fetus if maternal use is terminated before conception.[147] However, the risk for malformations and subnormal results on standard intelligence tests is high for offspring of women who were treated with therapeutic doses of isotretinoin in the first 60 days after conception.[1-3,148]

Etretinate exposure during the first 60 days of pregnancy is associated with a high risk of malformations. Etretinate persists in the body for up to 2 years.[160,572] Whether the lower concentrations of etretinate remaining in the mother's circulation over these extended periods increases the risk of malformations is not known, but if there is an increased risk it is likely to be low.

Experimental evidence suggests that endogenous retinoic acid may act as a natural morphogen. Exogenous retinoids act either directly, resulting in cytotoxicity, or via receptor-mediated pathways to interact with DNA and alter programmed cell death.[6,376,674,677,766] A typical dose-response relationship and developmental stage specificity have been demonstrated in mice.[395,405] Also in mice, the metabolites of isotretinoin, (4-oxo-isotretinoin and tretinoin) are more efficiently transferred across the placenta than isotretinoin and are more potent teratogens.[137,395] It is likely that different specificities of retinoid-binding proteins account for the variations in placental transfer.[531] Although retinoids can influence many types of cells, Lammer[408] emphasized that neuroectodermally derived cells of the rhomboencephalon are particularly sensitive and that the resulting neural crest cell abnormality differs from that resulting in oculoauriculo-vertebral dysplasia or Goldenhar syndrome. The susceptibility of specific cell types to the effects of the retinoids may be determined by the intracellular concentration of cellular retinoic binding protein.[408]

2.23 Retinoids, Topical Administration (Tretinoin)

There have been a number of case reports of congenital malformations occurring in the offspring of mothers who used topical tretinoin during their pregnancy. The U.S. Food and Drug Administration received adverse reaction reports involving approximately 17 infants who were born from mothers who used topical tretinoin during pregnancy, with a higher than expected representation of holoprosencephaly (Rosa, F., personal communication). While there has not been an abundance of epidemiological studies involving topical tretinoin, there are three reports. Using Michigan Medicaid data, the incidence of birth defects in 147 pregnancies exposed to topical tretinoin was compared to the incidence of birth defects in 104,092 nonexposed; the relative risk was 0.8 in the exposed population (Rosa, F., personal communication). A relative risk of 0.7 for birth defects was determined in 215 pregnancies exposed to topical tretinoin as compared with 430 nonexposed mothers in data from Group Health of Puget Sound (Rosa, F., personal communication). De Wals et al.[155] evaluated the association of the occurrence of holoprosencephaly with topical tretinoin exposure. Among 502,189 births there were 31 infants with holoprosencephaly. Eight patients had an abnormal karyotype, and 16 had a normal karyotype. None of the patients with a normal karyotype was exposed to topical tretinoin during pregnancy.

There have been a few animal teratology studies in which topical tretinoin was applied to pregnant animals. Seegmiller et al.[615] reported that very high doses applied via topical administration (100 mg/kg) in pregnant rats resulted in maternal toxicity and malformations. Topical administration of tretinoin (10.5 mg/kg) to hamsters did not increase the incidence of congenital malformations.[749]

Since all teratogens that have been appropriately studied have a no-effect dose, it would be paramount that topical administration of a known teratogen such as tretinoin must be absorbed and produce teratogenic concentrations in the blood. At conventional doses, the blood levels from topical administration are far below the teratogenic dose. It would appear that prudent use of this topical medication presents no risk to the embryo, because there would be no teratogenic exposure. The pharmacokinetics, animal studies, and human studies support this conclusion.

2.24 Smoking and Nicotine

Approximately 30% of all women of childbearing age smoke, and about 25% of all women will continue to smoke after they become pregnant.[553] The evidence in humans indicates that smoking affects the fetus directly in a dose-related manner and probably involves more than one component of smoke.[506] Placental lesions and fetal growth retardation have been consistently reported in epidemiological studies involving pregnant women who smoked cigarettes while pregnant.[114,328,467,668] Fetal death is 20 to 80% higher among women who smoked cigarettes while pregnant;[466,668] however, despite suggestions that smoking during pregnancy may increase the risk of limb anomalies,[143] there is no proven relationship between smoking and specific malformations or malformations in general.[178,410,466,517,685,739,740] One study suggests that infants of women who smoked throughout pregnancy experience an increase in mortality, which continues until at least 5 years of age.[562] Because of the large number of pregnant women who smoke and the documented effects of smoking on the fetus, one can conclude that smoking presents a significant risk to the fetus for growth retardation and abortion.

2.25 Streptomycin

Based upon case reports, there appears to be a small increased risk of sensorineural deafness in offspring of women treated with streptomycin for tuberculosis during pregnancy.[170,642,718] Other congenital anomalies have not been associated with *in utero* exposure to streptomycin in the human[304] or in mice.[180,525] Since auditory nerve damage is a toxic effect of streptomycin in the adult, and animal data show inner ear damage after high exposures *in utero*,[508] it is likely that there is a small increased risk of deafness but not of malformations after *in utero* exposure to streptomycin. A related drug, kanamycin, appears to have minimal risk of causing similar adverse effects.[360,521]

2.26 Tetracycline

The antimicrobial tetracyclines inhibit bacterial protein synthesis by preventing access of aminoacyl transfer RNA to the messenger RNA-ribosome complex.[733] Tetracycline crosses the placenta but is not concentrated by the fetus. Tetracyclines complex with calcium and the organic matrix of newly forming bone without altering the crystalline structure of hydroxyapatite.[126] Although tetracycline has been shown to discolor teeth without affecting the likelihood of developing caries,[19,564] very high doses may depress skeletal bone growth.[126] No congenital malformations of any other organ system have been associated with antenatal tetracycline exposures.[14,303,357] Several cases reports of limb reduction defects in human embryos exposed to tetracycline are not supported by epidemiologic or animal studies. Although one study reported staining of the fetal skeleton and stunting in rats exposed to tetracycline,[45] other experimental animal studies found either no teratogenic effect[344] or ambiguous effects.[199] Therapeutic doses of tetracycline are associated with no or minimal increased risk of congenital malformations, but they are likely to result in some degree of dental staining, which does not appear to have a deleterious effect on the offspring.

2.27 Thalidomide

Lenz and Knapp[422] were the first to associate thalidomide exposure during pregnancy with limb reduction defects and other features of the thalidomide syndrome. Limb defects resulted from exposure limited to a 2-wk period from the 22nd to the 36th days postconception; exposures from the 27th to the 30th days most often affected only the arm, whereas exposures from the 30th to the 33rd days resulted in both leg and arm abnormalities.[83,420,421] Although there was no association of mental retardation, brain malformations, or cleft palate, other abnormalities included facial hemangioma; microtia; esophageal or duodenal atresia; deafness; anomalies of the eyes, kidneys, heart, and external ear; and increased incidence of miscarriages and neonatal mortality.[83,385,392,422,594,638] A high proportion, about 20%, of the fetuses exposed during the critical period were affected. The current use of thalidomide in Brazil for the treatment of leprosy has resulted in more recent cases of embryopathy including at least 29 children born with thalidomide syndrome.[141,255,359]

McCredie[464] proposed that the segmental pattern was determined by the peripheral nerves derived from the neural crest. Stephens and McNulty[665] confirmed that limb development exhibits a segmental pattern; however, recent studies by Strecker and Stephens[670] have refuted the proposed role of peripheral nerve damage in thalidomide-induced embryopathy: a foil barrier was placed lateral to the chick neural tube to block the innervation of the wing field by the brachial plexus. The reduced source of innervation from spinal nerves anterior or posterior to the brachial plexus resulted in muscular atrophy but not in reductions or malformations of the skeleton of the wing.

Lash and Saxen[413] have postulated that thalidomide indirectly exerts its effects on limb chondrogenesis by acting on the kidney primordia. Based on an association between nephric tissue and limb development,[411-413] *in vitro* evidence suggests that thalidomide inhibits an interaction between metanephric tissue and associated mesenchymal tissue necessary for normal limb development. Although the mechanism of teratogenic action for thalidomide is not yet defined, the subject has been critically reviewed by Stephens.[664]

2.28 Thyroid: Iodides, Antithyroid Drugs, Iodine Deficiency

Iodine deficiency is the primary cause of endemic cretinism, but other factors, such as consumption of cassava or millet, may increase the incidence of the disorder.[229,720] Damage to the embryo due to iodine deficiency occurs early in gestation and results in irreversible neurologic and aural damage with variable severity. Goiter in a female of reproductive age due to endemic iodine deficiency is an indication for iodine supplementation prior to conception to prevent harmful developmental effects.

Several drugs used to treat maternal hyperthyroidism (^{131}I and antithyroid drugs) and nonthyroid conditions (especially iodide-containing compounds for bronchitis and asthma) affect thyroid function. *In utero* exposure to these drugs may result in congenitally hypothyroid infants who will not reach their potential for physical or mental development unless treated very early after birth with thyroid hormone.

There are several case reports of congenital goiter due to *in utero* exposures to iodide-containing drugs.[108,449] Maternal intake of as low as 12 mg per day may result in fetal goiter.[106] Iodinated diagnostic X-ray contrast agents used for amniofetography have been reported to affect fetal thyroid function adversely.[582]

Propylthiouracil and methimazole, used to treat thyrotoxicosis, readily cross the placenta.[118] Methimazole has been associated with aplasia cutis,[477,497] but more evidence is needed before a causal association can be demonstrated. Propylthiouracil is safer because the incidence of fetal goiter is low,[99,118] and there have been no observed detrimental effects on cognitive function.[100,461]

2.29 Toluene

Although occupational exposure to toluene has not been associated with congenital malformations in offspring, there are case reports of malformations resulting from the abuse of toluene. The first description of an infant with features similar to fetal alcohol syndrome born to a chronic abuser of toluene appeared in 1979.[691] This case and 22 additional cases have been described in detail.[320,321,538] Thirty-nine percent of the toluene-exposed infants were born prematurely, and 9% died in the perinatal period. In the surviving infants, 52% exhibited growth deficiency, 67% were microcephalic, and 80% exhibited developmental delay. Craniofacial features similar to those in the fetal alcohol syndrome were observed in 89%. An increased incidence of prematurity, perinatal death, growth and developmental delay, and phenotypic features similar to fetal alcohol syndrome were reported in another series of 35 pregnancies of 15 toluene abusers.[10,11,748] Experimental studies in animals report fetal growth retardation, postnatal growth and developmental abnormalities, and a variety of fetal malformations.[146,340,397] Pearson et al.[538] suggest that the clinical and experimental data can be interpreted to imply that alcohol and toluene may have a common mechanism of facial teratogenesis. Toluene appears to have the potential for developmental toxicity in the human but the magnitude of the risk is minimal for usual occupational exposures[18,179] and has not yet been determined for toluene abusers.

2.30 Valproic Acid

Valproic acid (dipropylacetic acid) is used for the treatment of various types of epilepsy. Valproic acid had been identified as a teratogen in animal studies,[94,372,438,510,742] but Dalens et al.[149] were the first to report the association of valproic acid and congenital malformations in the human. Although other reports followed, Robert and colleagues[576-578,580] described the associated malformations, consisting primarily of neural tube defects, and their incidence in detail. The neural tube defect observed is usually spina bifida in the lumbar or sacral region,[431,432] and increased risk appears to be correlated with higher serum levels.[529,530] Other anomalies include postnatal growth retardation, microcephaly, midface hypoplasia, microagnathia, and epicanthal folds.[452] Therapeutic dosages during pregnancy present a teratogenic risk for spina bifida of about 1%,[409] but the risk for facial dysmorphology may be greater.

Valproic acid crosses the human placenta,[157,511] but the fetal serum concentrations are not known. In the Rhesus monkey, the fetus is exposed to approximately one half of the free valproic acid concentration present in the maternal plasma;[307] craniofacial and skeletal anomalies and fetal death are observed.[308,454]

2.31 Vitamin A

Case reports have associated congenital defects in humans with massive vitamin A ingestion during pregnancy.[41,187,496,547,585] Although effects similar to those produced by vitamin A congeners — namely, central nervous system, cardioaortic, microtia, and clefting defects (see discussion of retinoids above) — may be predicted, no pattern of anomalies has emerged. Another 14 infants of women who ingested high doses of vitamin A (25,000 IU or more) during pregnancy had no congenital anomalies, although three additional pregnancies ended in miscarriage.[763]

Historically, vitamin A has played an important role in experimental and clinical teratology: vitamin A deficiency in swine was the first experimental model of teratogenesis in a mammal.[282-284,781] Studies into the effects of excess vitamin A documented dose-dependent teratogenesis — most frequently craniofacial, central nervous system, and skeletal anomalies — and fetal death in rats, mice, and hamsters with doses equivalent to hundreds or thousands times the human recommended daily allowance of 2267 IU of vitamin A as retinol during pregnancy.[371,681]

A review of literature concerning retinoids and birth defects entitled "Recommendations for Vitamin A Use During Pregnancy" was published by the Teratology Society.[681] Supplementation of 8000 IU vitamin A per day should be the maximum during pregnancy, and high dosages (25,000 IU or more) are not recommended. Beta-carotene as a source of vitamin A is likely to be associated with a smaller risk than an equivalent dose of vitamin A as retinol.[681]

2.32 Vitamin D

Excess of vitamin D has been associated with increased incidence of congenital malformations. Huge doses of vitamin D administered for rickets prophylaxis in pregnant women resulted in a marked increased incidence of a syndrome consisting of supravalvular aortic stenosis, elfin faces, and mental retardation in the human.[218,233] Animal studies and additional clinical reports suggest that the teratogenic risk of therapeutic doses of vitamin D is none to minimal.[111,207,219,257]

3 CURRENT ISSUES IN DEVELOPMENTAL TOXICITY

3.1 Recently Recognized Developmental Toxicants

3.1.1 Angiotensin-Converting Enzyme (ACE) Inhibitors

The first ACE inhibitor, captopril, was introduced in 1981 for the treatment of severe refractory hypertension. Since then the number of ACE inhibitors has increased and now includes enalapril, lisinopril, quinapril, peridopril, fosinopril, ramipril, and cilazapril.[460] Their relative effectiveness combined with a paucity of side effects as compared to other antihypertensives have made these drugs extremely popular for the treatment of all types of hypertension as well as congestive heart failure and diabetic nephropathy.[460]

ACE inhibitors are competitive inhibitors of angiotensin-converting enzyme, a carboxypeptidase that forms an integral part of the renin angiotensin system.[460] ACE catalyzes the conversion of angiotensin I to angiotensin II, one of the most potent vasoconstrictors known. ACE, which is the same enzyme as kininase II, also catalyzes the breakdown of bradykinin. A vasodilatory peptide itself, bradykinin stimulates the release of other vasodilatory substances, including prostaglandins and endothelium-derived relaxation factor.[236] Both mechanisms of action contribute to the decrease in blood pressure resulting from ACE inhibition.[236]

Although considered relatively safe for the treatment of hypertension, the use of ACE inhibitors during pregnancy has been associated with adverse fetal outcomes in both humans and experimental animals. A high rate of fetal loss was first reported to occur in rabbits treated with therapeutic doses of captopril in 1980 and was later confirmed in other animal species.[88,89,198] The first case of adverse fetal outcome in humans was reported in 1981.[279] In that report, treatment with captopril began on the 26th week of gestation, oligohydramnios was detected 2 wk later, and a cesarean section was performed the following week. The child was anuric and hypotensive and died on day 7. The kidneys and bladder were morphologically normal but hemorrhagic foci were found in the renal cortex and medulla. Numerous cases of severe and often lethal adverse fetal effects associated with ACE inhibitor use during pregnancy have since been reported.[292,548,556] The most consistent findings have been associated with a disruption of fetal renal function resulting in oligohydramnios and neonatal anuria accompanied by severe hypotension.[292,556] Intrauterine growth retardation, pulmonary hypoplasia, hypocalvaria, persistent patent ductus arteriosis, and renal tubular dysgenesis have also been reported.[28,76,556] Some of these effects may also result from the condition for which ACE inhibitors were prescribed.[292] These effects have been associated with ACE inhibitor treatment only during the second and third trimester. There are no reports of adverse fetal outcome associated with ACE inhibitor use during the first trimester.[76,556] As ACE inhibitors do not appear to affect organogenesis in either humans or in animal studies, it is not a classical teratogen. For this reason, Pryde et al.[556] have proposed the term ACE inhibitor fetopathy to describe the characteristic syndrome that results from ACE inhibitor use during pregnancy.

The majority of adverse fetal effects associated with ACE inhibitor use during pregnancy result from the direct therapeutic action of ACE inhibitors on the fetus. ACE inhibitors readily cross the placenta, where they inhibit fetal ACE activity.[278,556] Animal studies have confirmed that fetal hypotension occurs *in utero* in response to maternal ACE inhibitor treatment.[88] The decreased renal blood flow caused by vasodilation of renal efferent arterioles results in a loss of glomerular filtration pressure, leading to fetal anuria and oligohydramnios.[278,450] This in turn may result in other adverse fetal outcomes, such as pulmonary hypoplasia. Fetal urine production and tubular function do not begin until approximately 9 to 12 weeks of gestation,[278] which probably explains the lack of adverse fetal effects when ACE inhibitor treatment is discontinued in the first trimester.[450] In animal studies, ACE inhibition also decreases uterine blood flow, which may contribute to ACE-inhibitor-associated fetopathy.[88,198] Renal dysplasia, in particular a lack of renal proximal tubule differentiation, has also been noted in some affected fetuses.[28,140,450] This finding is thought to result from decreased renal perfusion and ischemic injury rather than to an effect on development;[28,450] however, a recent *in vitro* studies suggest that angiotensin II inhibits proliferation, induces hypertrophy in cultured proximal tubular cells,[757] and may act as mitogen for medullary cells derived from Henle's loop.[758] Furthermore, angiotensin II may be involved in glomerular growth and maturation.[205] The significance of these findings as they pertain to ACE inhibitor-induced renal defects is not yet known.

ACE inhibitor exposure during pregnancy has also resulted in several cases of hypocalvaria, an ossification defect of the membranous bones of the skull which leaves the fetal brain inadequately protected.[28,76,556] Although the pathogenesis is still unknown, inadequate perfusion of developing bone due to fetal hypotension combined with pressure from uterine muscles due to oligohydramnios may explain this defect.[28,76] It has also been suggested that ACE inhibitors may affect ossification by acting on osteoblast-derived growth factors.[28]

Despite consistent reports of adverse ACE inhibitor fetopathy, there are no controlled studies available to assess the risks associated with the use of ACE inhibitors during pregnancy. Because there is no reported incidence of adverse fetal effects due to ACE inhibitor use during the first trimester of pregnancy, there is no reason not to use ACE inhibitors in women of reproductive age. If the woman becomes pregnant, therapy must be changed to an alternative antihypertensive that is safer for the fetus.

3.1.2 Cocaine

Cocaine (benzoylmethylecgonine) is one of the most commonly used illicit drugs by women of reproductive age. Reported estimates for cocaine use during pregnancy range from 3 to 17%,[635] the highest rates occurring in inner-city populations. Because of its widespread use during pregnancy and the growing cost of caring for cocaine-exposed neonates,[543] there has been increasing concern over the risks to maternal and fetal health that are associated with prenatal cocaine use. Yet, despite numerous clinical studies linking prenatal cocaine use with a variety of adverse maternal and fetal effects, methodological limitations in these studies have made it difficult to establish a causal relationship between these alleged effects and maternal cocaine use. Not only are the timing, frequency, and dose of cocaine use difficult to determine, but adverse effects due to low socio-economic status, poor nutrition, multiple drug use, infections, and a lack of prenatal care are difficult to dissociate from effects due to cocaine use alone.[249,635] As such, the issue of how much risk to the fetus is associated with cocaine use during pregnancy is unresolved. Nonetheless, a growing body of literature supports the concept that cocaine is a developmental toxicant. Adverse effects attributed to prenatal cocaine exposure include a higher incidence of spontaneous abortion, placental abruption, stillbirth, prematurity, low birth weight, growth retardation, decreased head circumference, intracerebral hemorrhage, congenital defects, neurobehavioral abnormalities,[635,770] and a possible association with increased risk of sudden infant death syndrome (SIDS).[247] These effects are reduced, but not eliminated, in mothers receiving appropriate prenatal care.[442] Like other developmental toxicants, outcome is dependent on dose and time of use.

The majority of adverse effects associated with cocaine use during pregnancy appear to be due to high levels of cocaine abuse in later stages of gestation rather than in the first trimester or organogenesis.[361] Moderate usage of cocaine only in the first trimester does not appear to result in adverse fetal outcome and may not pose an increased risk to the fetus;[396] however, both human and animal studies suggest that neurological and urogenital abnormalities can occur due to cocaine usage during this time.[113,249]

When taken systemically, cocaine blocks presynaptic re-uptake of mono-amines, leading to stimulation of the sympathetic nervous system.[573] Cocaine also directly affects the heart.[550] Physiologic effects include vasoconstriction, tachycardia, cardiac dysrhythmias, and hypertension. It is thought that the increased incidence of placental abruption reported in cocaine users results from this latter effect.[361,550] Altered cocaine metabolism during pregnancy may increase the susceptibility of pregnant women and their fetuses to the cardiovascular effects of cocaine. Cocaine, which has a half-life of approximately 40 to 80 min,[573,743] is metabolized primarily by liver and plasma cholinesterases to benzoylecgonine and ecgonine methyl ester, which are readily excreted in the urine.[635] Approximately 20% is also metabolized by an *N*-demethylation pathway to norcocaine, an even more potent vasoconstrictor than cocaine.[301,550] Studies have shown that during pregnancy plasma cholinesterases are reduced,[249] and transformation to norcocaine may be increased, potentially increasing the vasoactive effects of cocaine.[760] Enhanced hypertensive effects due to cocaine have been demonstrated in nonpregnant animals treated with progesterone, suggesting that these changes are hormonally mediated.[760]

Adverse fetal outcomes associated with maternal cocaine use are thought to primarily result from the vasoconstrictive effects of cocaine on both the maternal and fetal vasculature.[361] Vasoconstriction of the uterine arteries, which are normally fully dilated during pregnancy, may compromise fetal growth and development. Studies in animals have confirmed that cocaine reduces uterine artery and placental blood flow, leading to reduced oxygen and nutrient supply to the fetus.[760] In addition, cocaine may directly inhibit the transfer of nutrients such as amino acids by the placenta,[25,156] contributing to growth retardation. Fetal cardiovascular effects resulting from uterine vasoconstriction include hypertension, tachycardia, hypoxia, and an increase in cerebral blood flow.[550,760] Cocaine also crosses the placenta, where it has a direct effect on the fetal vasculature flow.[760] In primates, fetal concentrations are approximately one third that of the mother.[48] The combination of fetal hypertension combined with increased cerebral blood flow may result in intracerebral hemorrhage or infarction, which has been reported to occur in cocaine-exposed fetuses in both human and animal studies.[112,164,167,172,337,730]

Disruption of uterine and fetal vasculature may also lead to a variety of congenital anomalies that have been associated with cocaine abuse. A significant association between cocaine use and an increased incidence of genitourinary tract malformations has been found.[112,116,337,635] Other defects reported include limb reduction defects;[112,337] nonduodenal intestinal atresia;[337] cardiac anomalies; hypospadias; prune belly syndrome, due to urethral obstruction; hydronephrosis;[550] and crossed renal ectopia.[428] Two cases of limb-body wall complex have also been reported.[709] With the exception of genitourinary tract malformations, the sample size in these clinical studies has not been sufficient to determine a statistically significant relationship between cocaine use and these congenital anomalies.[440]

The risk of congenital defects from cocaine use is dependent on the dose and frequency of use. In one study, light to moderate usage with decreased usage after the first trimester did not result in adverse fetal outcome.[571] In other studies, the incidence of anomalies for cocaine-exposed infants was reported to be anywhere from 3 to 6 times the incidence in controls, or higher.[47,112,116,709,770]

In rodents, cocaine is teratogenic during the late organogenic to postorganogenic period.[729] Defects reported in these studies are similar in nature to those reported in human studies and include genitourinary malformations, limb reduction defects, and cerebral hemorrhages.[172,202,729] These defects share similarities in that they have a etiology suggestive of vascular disruption phenomena.[202,337,361,700,729] Decreased uterine/placental blood flow alone or in conjunction with direct effects on fetal vasculature may lead to hemorrhage/edema, followed by necrosis and reabsorbtion of affected tissues,[700,729] resulting in the destruction of already formed structures.[361,700] Temporary clamping of the uterine arteries in the rat results in similar malformations.[73] Recent evidence indicates that reperfusion of tissues following cocaine-induced ischemia may also result in tissue damage due to free radicals generated by the mother[782] and later in gestation by the fetus.[188]

Various neurobehavioral effects have also been described in infants following *in utero* cocaine exposure, including tremors, seizures, irritability, excessive high-pitched crying, poor feeding, sleeping abnormalities, poor regulation of behavior, and abnormal EEG.[541,770] A majority of studies using the Brazelton neonatal assessment scale indicate that neonates exposed to cocaine *in utero* have altered behavioral responses, especially in orientation and habituation, when compared to nonexposed controls,[630,770] although there are conflicting reports. Many of the behavioral patterns observed can be ascribed to the direct toxic/physiologic effects of cocaine on the neonate and disappear as cocaine and its metabolites are eliminated.[166,248] However, impaired organizational ability, orientation, and regulation of behavior have also been observed in neonates exposed to cocaine only during the first trimester, indicating a more direct effect of cocaine on CNS development and maturation.[113,443] Sensory deficits and altered auditory brain-stem response have also been reported but normalized by 3 months of age in the human.[770] Other studies indicate that deficient performance on developmental tests may persist up to 2 years in some exposed groups.[631] One other study demonstrated no difference in IQ scores between the cocaine-exposed and unexposed groups at 3 years of age.[630] Unfortunately, these studies have methodological complications due to confounding variables such as alcohol abuse. Thus, there is no convincing evidence to date that cocaine is a behavioral teratogen in the human.

Studies in rodents, however, suggest that cocaine may produce permanent neurochemical alterations in the brain and may alter behavior and learning.[323,628,651] Studies have also demonstrated that *in utero* exposure to cocaine may lead to lasting changes in cholinergic,[695] dopaminergic,[305] seratonergic,[104,305] and noradrenergic[617] systems. In addition, cocaine reduces DNA synthesis in the brain[9] and inhibits macromolecular synthesis by glial cells *in vitro*.[234] Alterations in brain ornithine decarboxylase activity have also been reported in both fetal rabbits[248] and rats[618] exposed to cocaine. *In utero* cocaine exposure also resulted in delayed auditory brain-stem response in adult[120] and 22-day rat pups,[603] possibly due to delayed myelination. Hypomyelination has been demonstrated in rat pups exposed to cocaine in the fetal period of gestation.[744] These studies and others suggest that cocaine may subtly alter CNS development and maturation, leading to the altered behavior and learning patterns that have been observed. Further studies are necessary to determine if permanent subtle alterations in learning and behavior are indeed apparent in human infants and how long these changes persist. Until carefully controlled followup studies are performed, it is not possible to determine whether cocaine has direct neuroteratogenic properties in the human.

3.2 Potential Developmental Toxicants of Controversial Risk

3.2.1 *Antituberculosis Therapy*

Drugs prescribed for the treatment of tuberculosis include aminoglycosides, ethambutol, isoniazid, rifampin, and ethionamide. The ototoxic effects of streptomycin (discussed above) are the only proven adverse effects of these drugs on the fetus. Other aminoglycosides have not been associated with fetal effects. Neither ethambutol nor rifampin have been associated with an increase in the incidence of growth retardation, premature birth, or malformations.[55,427,642,718]

Early reports did not associate therapeutic exposures to isoniazid with an increased risk of malformations,[642,718] but there is an unconfirmed association with central nervous system dysfunction.[491,706] There was one attempted suicide involving 50 tablets of isoniazid per day during the 12th week that resulted in a stillbirth with arthogyryposis multiplex congenita syndrome.[419] Isoniazid may have a small increased risk for adverse effects on the central nervous system, but there is no apparent increase in risk for congenital malformations or abortions with therapeutic exposures. Only one report associated ethionamide with an increased risk of teratogenic effects;[551] however, this association is tenuous and not supported by other case reports.[780]

As a general observation, the antituberculosis drugs produce adverse effects on the fetus of experimental animals in greater doses than those equivalent to therapeutic exposures and not in all species. Embryotoxicity can be demonstrated for streptomycin in the mouse, kanamycin in the guinea pig, isoniazid in the rat, rifampin in the rat, and ethionamide in the rabbit.[165,508,660]

The only antituberculosis drug with confirmed developmental toxicity is streptomycin, i.e., a small risk of ototoxicity associated with high exposures over a prolonged period. While this does not eliminate the possibility of adverse effects on the fetus following exposure to the other prescribed tuberculostatic medications discussed, therapeutic exposures appear to represent a very small risk of teratogenesis and even less risk of abortion.

3.2.2 *Benzodiazepines*

The benzodiazepines, such as chlordiazepoxide (Librium), diazepam (Valium), xanax, and meprobamate, are widely used as tranquilizers during pregnancy and, therefore, it is not surprising that they have been associated with congenital malformations in some publications.

Chlordiazepoxide was associated with various anomalies after exposure during early pregnancy, but no syndrome was identified.[303,401] Other studies were inconclusive[138] or found no association.[142,300,590] Animal studies reported dose-dependent maternal and developmental toxicity from doses that exceed therapeutic exposures.[102,280,602] Chlordiazepoxide appears to have a minimal risk for congenital anomalies and no increased risk for abortion at therapeutic doses. Higher exposures are likely to increase the risk of adverse effects on the fetus, but the magnitude of the increase is not known.

Some studies reported an association between diazepam and increased incidence of congenital malformations;[60,590] however, a followup study found no associations.[779] The majority of studies of fetal outcome following *in utero* exposure to diazepam are negative.[14,18,142,601,686] Behavior alterations have been reported in infants exposed to benzodiazepines, mostly diazepam,[406] but this observation must be confirmed and the long-term developmental outcome evaluated before this finding can be appropriately interpreted.

Dose-dependent developmental toxicity can be demonstrated in laboratory animals at doses that greatly exceed therapeutic exposures.[244,728] Although third trimester exposure to diazepam can reversibly affect the fetus and neonate,[567] there is minimal increased risk of congenital malformations and no demonstrated increased risk of abortions from therapeutic exposures.

Meprobamate has been weakly associated with a variety of congenital malformations;[357,478,607] other studies found no associations.[32,300,303] Malformations and fetal loss can be induced in experimental animals at doses equivalent to many times the therapeutic range.[331,498,737] Because of inconsistencies, the data are not sufficient to confirm or rule out a small increase risk of malformations due to exposures early in pregnancy.

Benzodiazepines appear to have minimal increased risk of malformations at therapeutic ranges; higher exposures may increase the risk. The risk for abortion is unknown, but given the widespread use of these drugs, it is unlikely that a significant abortigenic effect would have gone unnoticed.

3.2.3 Chorionic Villous Sampling

A National Institute of Child Health and Human Development (NICHD) workshop evaluated reports indicating that there was an increased prevalence of limb reduction defects (LRD) in the offspring of women who had undergone first trimester diagnostic chorionic villous sampling (CVS).[515] While there was not unanimous agreement, there appeared to be a low risk of vascular disruption-type malformations following CVS. Not all of the research groups who studied the malformation rates following CVS were able to corroborate these findings, but the report concluded the following:

1. There was an increase in the prevalence of vascular disruption-type malformations (congenital amputations of the nonsymmetrical type, orofacial malformations such as mandibular hypoplasia, cleft palate, and Moebius syndrome).
2. There appeared to be a sensitive period for the induction of these malformations, probably because the bleeding that accompanies the CVS may result in hypoperfusion of the limbs, face, and other structures sensitive to minimal reductions in organ perfusion. This may indicate that a shift of a few weeks in the timing of CVS might eliminate or markedly reduce the risk to the embryo.
3. The occurrence of these types of malformations is biologically plausible based on the experimental induction of LRD using uterine vascular clamping and uterine trauma[63,66,73,75,81,731] and the description of malformations due to vascular disruption reported in clinical reports.[336]

In the past, threatened abortion and vaginal bleeding were viewed as indications of an abnormal pregnancy. For example, Funderburk et al.[223] found a slight but not significant increase in fetal anomalies delivered by mothers with first and second trimester bleeding. They suggested that "... early gestational vaginal bleeding is one predictor of suboptimal pregnancy outcome," but the authors did not invoke causation. While there has been disagreement about whether pregnancy bleeding increased the risk of congenital malformations,[12,98,186,223,456,457,519,522,532,539,636,648,667,694] only a few investigators thought that the bleeding was responsible for the malformations. Turnbull and Walker[694] stated, "Whether or not bleeding in the early months of pregnancy is a significant cause of fetal deformity remains unsettled." Many researchers who found an increase in malformations in the group of pregnant women who experienced bleeding attributed the bleeding to the fact that the malformation was already present, not that the bleeding was causal. Matsunaga and Shiota[456] examined over 600 embryos that were obtained from pregnancies with first trimester bleeding. The prevalence of LRD was 15 times greater in this group than in over 30,000 embryos from pregnancies without a history of first trimester bleeding. Furthermore, a portion of the 600 embryos were treated with progestational agents, but the frequency of LRD was no different in the treated or untreated groups. Thus, the malformations were present before the administration of the progestagens. The conclusions of the NICHD workshop support the hypothesis that sometimes the bleeding may be causal and not just a marker for the presence of the malformation. Previous investigations and reports support this concept.

In 1960, it was reported that interruption of the vascular supply to the pregnant uterus in the rat resulted in limb reduction defects, jaw hypoplasia, omphalocele, and anencephaly.[73] Furthermore, this procedure resulted in placental or embryonic-site hemorrhage. Although this research technique was a new method of producing and studying congenital malformations, it appeared to be an unlikely cause of human malformations. The originators suggested that uterine vascular clamping had an unlikely clinical application. Ornoy et al.[532] agreed: "A proper experimental model for the study of gestational bleeding on fetal development is lacking. Reduction of oxygen tension or clamping of uterine arteries in pregnant animals are far from being the appropriate comparative experimental model."

Webster et al.[731] produced malformations in an experimental animal model with uterine trauma. It should be noted that all of these techniques and pathologies are low-risk teratogenic insults. For example, the work of Matsunaga and Shiota[456] indicates that although LRD occurred 15 times more frequently in pregnancies with vaginal bleeding than in nonbleeding pregnancies (1/3500), pregnancies with a history of severe bleeding had 199 embryos with normal limbs for each embryo with a LRD. It appears that we have to endorse a concept that was previously not generally accepted; namely, that bleeding from the chorion that deprives the embryo of a portion of its blood supply during certain periods of the first trimester has a low, but real risk of resulting in one or more of the vascular disruptive malformations.

Bleeding from the pregnant genital tract from other sites and at other times may present a risk of a different magnitude and type. Further research is necessary to determine whether CVS is safer for the fetus at other stages of gestation. Since there is an obvious discrepancy in the results obtained by the various research and clinical groups, these differences may be related to the stage of the procedure, the amount of material removed, the experience of the operator, and the technique used. At this time, the consensus is that the procedure presents a real but small risk to the fetus. Since CVS offers advantages to certain patients, it will be important to determine whether a totally safe procedure can be designed.

3.2.4 Electromagnetic Fields

An extensive review of human epidemiological studies, secular trend data, *in vivo* animal studies, and *in vitro* studies has been conducted concerning the reproductive and teratogenic risks of electromagnetic fields (EMF).[74,82,150] The epidemiological studies included populations exposed to video display terminals, power lines, and household appliances. The population that has been studied most frequently is women exposed to video display terminals, but their EMF exposures are extremely low and frequently are at the level of the ambient EMF fields in a house or office. The results of epidemiological studies involving video display terminals are generally negative for the reproductive effects that have been studied. There have been fewer studies concerned with the reproductive risks of power lines, electric substations, and home appliances, but these studies indicate that even these higher exposures to EMF do not generate a measurable increase in reproductive failures in the human population. There are no epidemiologic reproductive studies dealing with exposure to medical resonance imaging and no anecdotal reports of adverse effects to the human fetus. Secular trend analysis of birth defect incidence data indicates that the increasing generation of electric power during this century is not associated with a concomitant rise in the incidence of birth defects. Embryoculture or cell-culture studies are of little assistance in determining the human risk of EMF; however, well-designed, whole-animal teratology studies are appropriate when the epidemiologists and clinical teratologists are uncertain about the environmental risks. While there does not appear to be measurable risk of reproductive failure or birth defects from EMF exposures in humans, investigations are warranted to answer some of the issues that have been raised.

3.2.5 *Human Immunodeficiency Virus*

The incidence of serum antibody to human immunodeficiency virus (HIV) in pregnant women is increasing from the 1991 estimate of 1.5 per 1000 women delivering in the U.S.;[281] the incidence is as high as 31% in pregnant women in some African cities.[61] In a prospective study of pregnant women with HIV infection in Nairobi, Kenya, Braddick et al.[61] examined 177 newborns from HIV-positive mothers and found 3 (2%) stillbirths and 7 (5%) newborns with various congenital malformations; 1% stillbirths and 5% having congenital malformations were observed in 327 control neonates. However, women with symptoms or signs of HIV-related disease had a threefold higher risk of low-birth-weight infants than did asymptomatic HIV-positive women. The low birth weight was not associated with transmission of HIV. Other studies support the conclusion that asymptomatic HIV pregnancies are not associated with an increased risk of congenital malformations, low birth weight, or abortion.[175,185,557] Fetal HIV infection does not appear to affect fetal development or growth.[53] It is likely that sexually transmitted diseases, opportunistic maternal infections, and symptomatic HIV pregnancies may increase the risk of low birth weight and morbidity in noninfected offspring.

There may be wide variations in the drugs prescribed to prevent the *in utero* transmission of HIV, which is estimated to be about 25 to 40%,[162] and opportunistic infections to the fetus. Zidovudine appears to reduce the vertical transmission of HIV infections, and, although it crosses the placenta, it does not appear to cause permanent adverse effects in the fetus.[154]

3.2.6 *Lead*

It is universally accepted that lead represents a potentially toxic agent in mammalian organisms. In the first half of this century, thousands of children who ingested or inhaled lead from various environmental sources developed the symptoms of lead poisoning which include abdominal pain, constipation, disturbed heme synthesis resulting in anemia and basophilic stippling of the red blood cells, free erythrocyte protoporphyrinuria, mild to severe behavioral and cognitive problems, nephropathy, seizures, and acute encephalopathy which can result in death.[119] Through the efforts of governmental agencies, the sources of lead in the environment have been reduced by eliminating lead in paints, cooking and storage utensils, drinking water pipes, glazing compounds, and gasoline. To further reduce lead exposures, more restrictive programs and new concepts have been introduced; however, there is controversy concerning the distinction between "safe exposure" and "toxic exposure". The most recent decision about the risks of lead resulted in a maximum permissible blood level of 10 μg/dl, above which the diagnosis of lead poisoning is made, and a maximum permissible level of 30 μg/dl for pregnant women.[528] It is interesting that the uncertain toxic potential of low exposures to lead has resulted in the situation in which there is no range of exposure between safe and toxic exposures. It has also been hypothesized that there is no threshold for the reproductive effects of lead.[613] There are several reviews of the reproductive toxicity of lead which provide a historical view and an extensive bibliography.[33,105,609]

Mammalian animal studies indicate that lead administration during pregnancy can cause embryonic death, growth retardation, and congenital malformations when very large doses are administered.[196,462,500,777] Some studies that did not demonstrate malformations or growth retardation *per se* reported behavioral effects.[62,380] Growth retardation and fetal death have also been reported in the absence of gross congenital malformations.[49,436] Learning deficits and neurobehavioral effects have been reported following *in utero* lead exposure.[264,566] In other neurobehavioral studies, lead exposure during pregnancy did not result in neurobehavioral effects,[101,679] in spite of the fact that blood lead levels were similar in both the negative and positive studies; namely, in the range of 50 to 55 μg/dl, although many of the positive studies reached blood lead levels that were much higher. Other studies in which pregnant rats were exposed to injected lead or lead in their drinking water did not exhibit reproductive effects, even though the blood lead levels ranged from 35 to 92 μg/dl.[388,481,558] The animal studies clearly demonstrate that if the lead concentration of the blood in the pregnant mother is in the toxic range for the adult, then some type of reproductive effect is expected. It is at low blood levels that reports of developmental toxicity lack consistency.

Similar results occurred in the human studies,[33] namely that while some human epidemiological studies report statistical associations between populations exposed to lead and various reproductive effects, there is a lack of consistency in those exposed populations with blood lead levels below 35 μg/dl. Two studies reported an association of lead exposure and the occurrence of congenital abnormalities.[13,512] The types of malformations reported to be associated with lead exposure in these studies are not consistent, which is what would be predicted if the association was due to chance rather than being causal. Other epidemiological studies have not found an association between lead exposure and the occurrence of congenital malformations.[182-184,470] As Bellinger stated, no one has been able to define a group of malformations that can be causally related to intrauterine lead exposure.[33]

One case report linked a blood lead level of 62 μg/dl in a pregnant mother to the delivery of a child with the VACTERL (a pattern of multiple malformations that includes vertebral anomalies, anal atresia, cardiac defect, tracheoesophageal fistula, renal anomalies, and limb anomalies) complex of malformations.[424] While the mother worked in a glass factory that used lead-containing glass, the blood level reported was very likely in error, since followup blood lead levels were very low.

The most important reproductive effects of lead relate to the production of permanent or temporary neurobehavioral effects. These effects are clearly related to exposures of lead in children that result in blood lead levels > 50 μg/dl. While there are also studies that indicate changes in behavior at lower lead levels,[34,35,159,339,724] there is less consistency in these studies. Furthermore, the behavioral effects reported in some of the studies were transient and no longer present when the children entered school.[20,36,37,132,158,182]

Since lead is not beneficial for humans, it is appropriate to reduce the population exposure to lead to as low a level as is scientifically and economically feasible. While there are studies that indicate a dose-dependent increased incidence of neurobehavioral deficits in infants from exposures that result in blood levels of lead from 0 to 15 μg/dl,[37] and, according to Bellinger,[33] the World Health Organization is suggesting that lead levels in the 0- to 25-μg/dl range will result in a reduction of 1 to 3 IQ points per 10-μg/dl increase in the blood lead level, it is almost impossible to demonstrate a causal relationship for neurobehavioral effects at these low levels unless more sophisticated studies are pursued. It seems reasonable to investigate the potential toxicity of low exposures to lead in experimental animals by correlating neurobehavioral effects with lead concentrations in the brain rather than in the blood. We are confident that this approach will demonstrate that there is a threshold or no-effect exposure for the developmental toxicity of lead.

3.2.7 Lithium Carbonate

Lithium carbonate, widely used for treatment of manic-depressive disorders, was first associated with human congenital malformations in 1970.[426,699] The malformations described include heart and large vessel anomalies, Epstein's anomaly, neural tube defects, talipes, microtia, and thyroid abnormalities.[214,609,721] Lithium does cross the placenta[560] and appears to be a human teratogen at therapeutic dosages. More recent epidemiologic data indicate that it presents a very small risk for malformations.[124,349] The results of a retrospective study suggest that lithium may also increase the risk for premature delivery,[692] but, again, the magnitude of the risk is likely to be small. Transient toxic effects seen in the neonate include congenital goiter, lethargy, hypotonia, and cardiac murmur.[509,754] Because of the value of lithium carbonate for treating manic-depressive psychosis, the risk associated with psychiatric relapse upon removing the drug may be greater than the teratogenic risk. If other drugs are found for the treatment of manic depressives that have no teratogenic potential, they could be substituted during pregnancy.

Lithium can induce abnormal development in several laboratory animals, but the mechanism of the teratogenic action of lithium is not known. The neurotropic activity of lithium suggests that central nervous system malformations may result from cell membrane disturbances which affect neural tube closure.[368]

3.2.8 *Sonography (Ultrasound)*

While it is clear that the levels and types of medical sonography that have been used in the past have no measurable risk,[84,565] it would be inaccurate to label the modality of ultrasound as totally safe regardless of exposure. The most important biologic effect of sonography is hyperthermia, and the two primary factors that determine heat production are intensity and duration of exposure. Thus, the prudent use of sonography means that exposures should be far below those that have been demonstrated to increase the fetal risks in animal studies and are appropriate based on the knowledge obtained about the physical changes that can be produced in humans as the absorbed dose is elevated. Brent et al.[84] evaluated the reproductive risks using five criteria: (1) human epidemiology, (2) secular trend data, (3) animal experiments, (4) dose-response relationships, and (5) biologic plausibility (Table 8). The analysis revealed that human epidemiology studies do not indicate that diagnostic ultrasound presents a measurable risk to the developing embryo or fetus. Animal studies also indicate that diagnostic levels of ultrasound are safe and do not elevate the fetal temperature into the region where deleterious embryonic and fetal effects will occur. The use of diagnostic procedures and the design of sonographic equipment should take into consideration the hyperthermic potential of higher exposures of ultrasound and the hypothetical additional risk of performing sonography on pregnant patients who are febrile. It would appear that if the embryonic temperature never exceeds 39°C, then there is no measurable risk.[84]

3.3 Drugs Used Late in Gestation

3.3.1 *Aspirin*

Aspirin acts principally by inhibiting prostaglandin synthesis by irreversibly acetylating and inactivating fatty acid cyclo-oxygenase. Low-dose aspirin (60 to 150 mg) is used clinically in the prevention or treatment of a variety of conditions that affect fetal health. For instance, aspirin is used for the treatment of pre-eclampsia, a condition associated with vasospasm, increased platelet activation, coagulation, and pregnancy-induced hypertension. Although there is some evidence that aspirin may prevent or treat pre-eclampsia, the effective minimum dose is undetermined; the decreased incidences of pre-eclampsia, hypertension, and fetal death are not always statistically significant.[30]

Another concern is the potential for increased fetal and maternal bleeding associated with low-dose aspirin use near the time of delivery or higher doses used to treat preterm labor. Jankowski et al.[352] reported a study involving 25 women threatened with premature delivery who were given 3.6 g of oral aspirin per day for 4 successive days within 10 days prior to delivery. They and others have found no adverse effect on maternal bleeding at delivery, fetal hemorrhage, or circulatory disorders.

There is some evidence that aspirin combined with dipyridamole may prevent or ameliorate intrauterine growth retardation associated with idiopathic uteroplacental insufficiency.[713] Fetal growth was also improved with aspirin (150 mg per day) alone in fetuses with high, but not extreme, umbilical artery systolic/diastolic ratios.[693] Third trimester exposure to low daily doses of aspirin were not associated with adverse fetal outcome in these studies.[693,713]

High fetal losses are associated with lupus anticoagulant antibodies, anticardiolipin antibodies, and system lupus erythematosus. Surviving fetuses experience an increased incidence of growth retardation, fetal distress, and preterm delivery. Low-dose aspirin (60 to 80 mg per day) in combination with prednisone (20 to 80 mg per day) improves pregnancy outcome and greatly reduces thrombosis.[30]

Case reports suggest 75 to 300 mg of aspirin per day in combination with dipyridamole might reduce the risk of late fetal loss in patients with arterial thromboembolism, thrombotic thrombocytopenia purpura, and idiopathic or essential thrombocythemia.[30]

Although doubts concerning the safety of aspirin during pregnancy have been expressed, aspirin use during the first trimester is not associated with an increased teratogenic risk.[738] Third trimester exposure to 80 mg per day or less is not associated with an early constriction of the ductus arteriosus.[471] Much larger doses of aspirin during the third trimester have been associated with an increased length of gestation, duration of labor, frequency of postmaturity, and blood loss at delivery and justify careful fetal surveillance.[425]

3.3.2 Indomethacin

Oral administration of the prostaglandin synthetase inhibitor, indomethacin (25 mg every 6 hr) is effective in the treatment of polyhydramnios that is either idiopathic or related to maternal diabetes mellitus. The reduction in renal prostaglandin levels achieved using indomethacin would reduce the inhibitory action of the E prostaglandins on the antidiuretic effect of arginine vasopressin and result in decreased urine production by the fetal kidneys. Oligohydramnios, constriction of the ductus arteriosus (prostaglandins are necessary to maintain the patency of the fetal ductus arteriosus), and fetal hydrops are potentially serious side effects of indomethacin and warrant careful fetal surveillance.[490]

Indomethacin may also be used to prevent preterm labor or intraventricular hemorrhage. Its efficacy for preventing intraventricular hemorrhage is controversial. The cord levels achieved after doses used to treat preterm labor (25 mg every 6 hr) are far below those that are likely to provide protection against intraventricular hemorrhage. However, even these lower doses are associated with a 10% incidence of impaired fetal renal function and oligohydramnios and a 50% incidence of ductus arteriosus constriction.[490]

3.3.3 Thioamides (Propylthiouracil)

Thiomides are antithyroid drugs used to treat fetal thyrotoxicosis. Fetal thyrotoxicosis is induced by thyroid-stimulating immunoglobulins produced in euthyroid or hypothyroid women. The adverse fetal effects (e.g., craniosynostosis, intellectual impairment, and increased mortality) are caused by excess fetal thyroid hormone production. The thioamides block thyroid hormone synthesis by inhibiting the oxidation of iodide or iodotyrosyl. Unlike other thioamides, propylthiouracil also inhibits the peripheral deiodination of thyroxine to triiodothyronine. All thioamides are associated with a significant risk of fetal goiter and teratogenesis;[609] however, in the case of propylthiouracil, the fetal goiter can be reduced with intra-amniotic injections of thyroxine[122] which also prevent other abnormalities caused by inhibition of fetal thyroid function.

4 ACKNOWLEDGMENTS

The authors thank Yvonne Edney for her secretarial assistance. This work was supported in part by funds from NIH HD 29902 and Harry Bock Charities.

REFERENCES

1. Adams, J., High incidence of intellectual deficits in 5 year old children exposed to isotretinoin "in utero", *Teratology,* 41, 614, 1990.
2. Adams, J. and Lammer, E. J., Relationship between dysmorphology and neuropsychological function in children exposed to isotretinoin "in utero", in *Functional Neuroteratology of Short Term Exposure to Drugs,* Adams, J. and Lammer, E. J., Eds., Teikyo University Press, Tokyo, 1992.
3. Adams, J., Lammer, E. J., and Holmes, L. B., A syndrome of cognitive dysfunctions following human embryonic exposure to isotretinoin, *Teratology,* 43, 497, 1991.
4. Adams, J., Vorhees, C. V., and Middauagh, L. D., Developmental neurotoxicity of anticonvulsants: human and animal evidence on phenytoin, *Neurotoxicol. Teratol.,* 12, 203–214, 1990.
5. Albengres, E. and Tillement, J. P., Phenytoin in pregnancy: a review of the reported risks, *Biol. Res. Pregnancy Perinatol.,* 4, 71–74, 1983.
6. Alles, A. J. and Sulik, K. K., Retinoic-acid-induced limb-reduction defects: perturbation of zones of programmed cell death as a pathogenetic mechanism, *Teratology,* 40, 163–171, 1989.
7. Amin-Zaki, L., Elhassani, S., Majeed, M. A., Clarkson, T. W., Doherty, R. A., and Greenwood, M., Intrauterine methylmercury poisoning in Iraq, *Pediatrics,* 54, 587–595, 1974.
8. Amin-Zaki, L., Majeed, M. A., Elhassani, S. B., Clarkson, T. W., Greenwood, M. R., and Doherty, R. A., Prenatal methylmercury poisoning. Clinical observations over five years, *Am. J. Dis. Child.,* 133, 172–177, 1979.
9. Anderson-Brown, T., Slotkin, T. A., and Seidler, F. J., Cocaine acutely inhibits DNA synthesis in developing rat brain regions: evidence for direct actions, *Brain Res.,* 537, 197–202, 1990.
10. Arnold, G. and Wilkins-Haug, L., Toluene embryopathy syndrome, *Am. J. Hum. Genet.,* 47, A46, 1990.
11. Arnold, G. L, Kirby, R. S., Langendoerfer, S., and Wilkins-Haug, L., Toluene embryopathy: clinical delineation and developmental follow-up, *Pediatrics,* 93, 216–220, 1994.
12. Asanti, R. and Vesanto, T., Effect of threatened abortion on fetal prognosis, *Acta Obstet. Gynec. Scand.,* 42, 107, 1963.
13. Aschengrau, A., Zierler, S., and Cohen, A., Quality of community drinking water and the occurrence of late adverse pregnancy outcomes, *Arch. Environ. Health.,* 48, 105–113, 1993.
14. Aselton, P., Jick, H., Milunsky, A., Hunter, J. R., and Stergachis, A., First-trimester drug use and congenital disorders, *Obstet. Gynecol.,* 65, 451–455, 1985.
15. Autti-Ramo, I. and Granstrom, M.-L., The effect of intrauterine alcohol exposure of various durations on early cognitive development, *Neuropediatrics,* 22, 203–210, 1991.
16. Autti-Ramo, I. and Granstrom, M.-L., The psychomotor development during the first year of life of infants exposed to intrauterine alcohol of various durations. Fetal alcohol exposure and development, *Neuropediatrics,* 22, 59–64, 1991.
17. Aviles, A., Diaz-Macqueo, J. C., Talavera, A., Guzman, T., and Garcia, E. L., Growth and development of children of mothers treated with chemotherapy during pregnancy: current status of 43 children, *Am. J. Hematol.,* 36, 243–248, 1991.
18. Axelson, G., Lutz, C., and Rylander, R., Exposure to solvents and outcome of pregnancy in university laboratory employees, *Br. J. Ind. Med.,* 41, 305–312, 1984.
19. Baden, E., Environmental pathology of the teeth, in *Thomas' Oral Pathology,* 6th ed., Gorlin, R. J. and Goldman, H. M., Eds., C.V. Mosby, St. Louis, 1970, pp. 189–191.
20. Baghurst, P., McMichael, A., Wigg, N., Vimpani, G., Robertson, E., Roberts, R., and Tong, S.-L., The Port Pirie cohort study, *N. Engl. J. Med.,* 327, 1279–1284, 1992.
21. Bai, P. V. A. and John, J., Congenital skin ulcers following varicella in late pregnancy, *J. Pediatr.,* 94, 65–66, 1979.
22. Baldwin, S. and Whitley, R. J., Teratogen update: intrauterine herpes simplex virus infection, *Teratology,* 39, 1–10, 1989.
23. Baranov, V. S., Characteristics of the teratogenic effect of aminopterin compared with that of other teratogenic agents, *Bull. Exp. Biol. Med. (USSR),* 61, 77–81, 1966.
24. Barnes, A. B., Colton, T., Gundersen, J., Noller, K. L., Tilley, B. C., Strama, T., Toensend, D. E., Hatab, P., and O'Brien, P. C., Fertility and outcome of pregnancy in women exposed *in utero* to diethylstilbestrol, *N. Engl. J. Med.,* 302, 609–613, 1980.
25. Barnwell, S. L. and Sastry, B. V. R., Depression of amino acid uptake by human placental villus by cocaine, morphine and nicotine, *Trophoblast Res.,* 1, 101–202, 1983.
26. Barone, J. J. and Roberts, H., Human consumption of caffeine, in *Caffeine,* Dews, P. B., Ed., Springer-Verlag, New York, 1984, pp. 59–73.
27. Barr, M. and Burdi, A. R., Warfarin-associated embryopathy in a 17-week abortus, *Teratology,* 14, 129–134, 1976.
28. Barr, Jr., M., and Cohen, Jr., M. M., ACE inhibitor fetopathy and hypocalvaria: the kidney-skull connection, *Teratology,* 44, 485–495, 1991.

29. Barr, M., Pozanski, A. K., and Schmickel, R. D., Digital hypoplasia and anticonvulsants during gestation: a teratogenic syndrome, *J. Pediatr.,* 4, 254–256, 1974.
30. Barton, J. R. and Sibai, B. M., Low-dose aspirin to improve perinatal outcome, *Clin. Obstet. Gynecol.,* 34, 251, 1991.
31. Beaulac-Baillargeon, L. and Desrosiers, C., Caffeine-cigarette interaction on fetal growth, *Am. J. Obstet. Gynecol.,* 157, 1236–1240, 1987.
32. Belafsky, H. A., Breslow, S., Hirsch, L. M., Shangold, J. E., and Stahl, M., B., Meprobamate during pregnancy, *Obstet. Gynecol.,* 34, 378–386, 1969.
33. Bellinger, D., Teratogen update: lead, *Teratology,* 50, 367–373, 1994.
34. Bellinger, D., Leviton, A., Waternaux, C., Neddleman, H., and Rabinowitz, M., Longitudinal analyses of prenatal and postnatal lead exposure and early cognitive development, *N. Engl. J. Med.,* 316, 1037–1043, 1987.
35. Bellinger, D. and Neddleman, H., Prenatal and early postnatal exposure to lead. Developmental effects, correlates, and implications, *Int. J. Mental Health,* 14, 78–111, 1985.
36. Bellinger, D., Sloman, J., Leviton, A., Rabinowitz, M., Neddleman, H., and Waternaux, C., Low-level lead exposure and children's cognitive function in the preschool years, *Pediatrics,* 87, 219–227, 1991.
37. Bellinger, D., Stiles, K., and Neddleman, H., Low-level lead exposure, intelligence and academic achievement: a long-term follow-up study, *Pediatrics,* 90, 855–861, 1992.
38. Benirschke, K., Syphilis: the placenta and fetus, *Am. J. Dis. Child.,* 128, 142–143, 1974.
39. Berger, A., Effects of caffeine consumption on pregnancy outcome — a review, *J. Reprod. Med.,* 33, 945–956, 1988.
40. Berger, M. J. and Goldstein, D. P., Impaired reproductive performance in DES-exposed women, *Obstet. Gynecol.,* 55, 25–27, 1980.
41. Bernhardt, I. B. and Dorsey, D. J., Hypervitaminoisis A and congenital renal anomalies in a human infant, *Obstet. Gynecol.,* 43, 750–755, 1974.
42. Bertollini, R., Kallen, B., Mastroiacovo, P., and Robert, E., Anticonvulsant drugs in monotherapy: effect on the fetus, *Eur. J. Epidemiol.,* 3, 164–167, 1987.
43. Bertrand, M., Schwam, E., Frandon, A., Vagne, A., and Alvary, J., Sur un effet teratogene systematique et specifique de la cafeine chez les rongeurs, *C. R. Soc. Biol. (Paris),* 159, 2199–2202, 1966.
44. Bertrand, M., Girod, J., and Rigaud, M. F., Ectrodactylie provoquee par la cafeine chez les rongeurs: role des facteurs specificques et genetiques, *C. R. Soc. Biol. (Paris),* 164, 1488–1489, 1970.
45. Bevelander, G. and Cohlan, S. W., The effect on the rat fetus of transplacentally acquired tetracycline, *Biol. Neonate,* 4, 365–370, 1962.
46. Bibbo, M., Transplacental effects of diethylstilbestrol, in *Perinatal Pathology,* Grundman, E., Ed., Springer-Verlag, New York, 1979, pp. 191–211.
47. Bingol. J., Fuchs, M., Diaz, V., Stone, R. K., and Gromisch, D. S., Teratogenicity of cocaine in humans, *J. Pediatr.,* 10, 93–96, 1987.
48. Binienda, Z., Bailey, J. R., Duhart, H. M., Slikker, Jr., W., and Paule, M. G., Transplacental pharmacokinetics and maternal/fetal plasma concentrations of cocaine in pregnant macaques near term, *Drug Metab. Dispos.,* 21, 364–368, 1993.
49. Bischoff, F., Maxwell, L. C., Evans, R. D., and Nuzum, F. R., Studies on the toxicity of various lead compounds given intravenously, *J. Pharmacol. Exp. Ther.,* 34, 85–109, 1928.
50. Blake, D. A. and Fallinger, C., Embryopathic interaction of phenytoin and trichloropropene oxide in mice, *Teratology,* 13, 17A, 1976.
51. Blakely, P. M., Experimental teratology of ethanol, in *Issues and Reviews in Teratology,* Kalter, H., Ed., Plenum Press, New York, 1988, pp. 237–282.
52. Blakely, P. M. and Scott, Jr., W. J., Determination of the proximate teratogen of the mouse fetal alcohol syndrome. 2. Pharmacokinetics of the placental transfer of ethanol and acetaldehyde, *Toxicol. Appl. Pharmacol.,* 72, 364–371, 1984.
53. Blanche, S., Rouzioux, C., Moscato, M. L. G., Veber, F., Mayhaux, M. J., Jacomet, C., Tricoire, J., Deville, A., Vial, M., Firtion, G., deCrepy, A., Douard, D., Robin, M., Courpotin, C., and Ciraru-Vigneron, N., A prospective study of infants born to women seropositive for human immunodeficiency virus type 1, *N. Engl. J. Med.,* 320, 1643–1648, 1989.
54. Blatt, J., Mulvihill, J. J., Ziegler, J. L., Young, R. C., and Poplack, D. G., Pregnancy outcome following cancer chemotherapy, *Am. J. Med.,* 69, 828–832, 1980.
55. Bobrowitz, I. D., Ethambutol in pregnancy, *Chest,* 66, 20–24, 1974.
56. Bond, P. R., Caul, E. O., Usher, J., Cohen, B. J., Clewley, J. P., and Field, A. M., Intrauterine infection with human parvovirus, *Lancet,* 1, 448–449, 1986.
57. Bongiovanni, A. M., DiGeorge, A. M., and Grumbach, M. M., Masculinization of the female infant associated with estrogenic therapy alone during gestation: four cases, *J. Clin. Endocrinol. Metab.,* 19, 1004–1011, 1959.
58. Book, S. and Goldman, M., Thyroidal adioiodine exposure of the fetus, *Health Phys.,* 29, 874, 1975.
59. Boue, J., Boue, A., and Lazar, P., Retrospective and prospective epidemiological studies of 1,500 karyotyped spontaneous abortions, *Teratology,* 12, 11–26, 1975.

60. Bracken, M. B. and Holford, T. R., Exposure to prescribed drugs in pregnancy and association with congenital malformations, *Obstet. Gynecol.,* 58, 336–344, 1981.
61. Braddick, M. R., Kreiss, J. K., Embree, J. E., Datta, P., Ndinya-Achola, J. O., Pamba, H., Maitha, G., Roberts, P. L., Quinn, T. C., Holmes, K. K., Vercauteren, G., Piot, P., Adler, M. W., and Plummer, F. A., Impact of maternal HIV infection on obstetrical and early neonatal outcome, *AIDS,* 4, 1001–1005, 1990.
62. Brady, K., Herrara, Y. and Zenick, H., Influence of parental lead exposure on subsequent learning ability of offspring, *Pharm. Biochem. Behav.,* 3, 561–565, 1975.
63. Brent, R. L., The indirect effect of irradiation on embryonic development. II. Irradiation of the placenta, *Am. J. Dis. Child.,* 100, 103–108, 1960.
64. Brent, R. L., Drug testing in animals for teratogenic effects: thalidomide in the pregnant rat, *J. Pediatr.,* 64, 762–770, 1964.
65. Brent, R. L., Environmental factors: miscellaneous, in *Prevention of Embryonic, Fetal and Perinatal Disease,* Brent, R. L. and Harris, M. I., Eds., DHEW Pub. No. (NIH) 76-853, Department of Health, Education, and Welfare, Bethesda, MD, 1976, pp. 211–218.
66. Brent, R. L., Method of evaluating alleged human teratogens (editorial), *Teratology,* 17, 83, 1978.
67. Brent, R. L., Radiation teratogenesis, *Teratology,* 21, 281–298, 1980.
68. Brent, R. L., Drugs and pregnancy: are the insert warnings too dire?, *Contemp. Ob-Gyn.,* 20, 42–49, 1982.
69. Brent, R. L., The magnitude of the problem of congenital malformations, in *Prevention of Physical and Mental Congenital Defects.* Part A. *Basic and Medical Science, Education and Future Strategies,* Marois, M., Ed., Alan R. Liss, New York, 1985, pp. 55–68.
70. Brent, R. L., Definition of a teratogen and the relationship of teratogenicity to carcinogenicity (editorial), *Teratology,* 34, 359–360, 1986.
71. Brent, R. L., Predicting teratogenic and reproductive risks in humans from exposure to various environmental agents using *in vitro* techniques and *in vivo* animal studies, *Cong. Anom.,* 28(Suppl.), S41–S55, 1988.
72. Brent, R. L., The effect of embryonic and fetal exposure to X-ray, microwaves, and ultrasound: counseling the pregnant and non-pregnant patient about these risks, *Sem. Oncol.,* 16, 347–369, 1969.
73. Brent, R. L., Relationship between uterine vascular clamping, vascular disruption, and cocaine teratogenicity (editorial), *Teratology,* 41, 757–760, 1990.
74. Brent, R. L., Reproductive effects, in *Health Effects of Low-Frequency Electric and Magnetic Fields,* Davis, J. G., Bennett, W. R., Brady, J. V., Brent, R. L., Gordis, L., Gordon, W. E., Greenhouse, S. W., Reiter, R. J., Stein, G. S., Sussking, C., and Trichopoulos, D., Eds., (ORAU 92/F8), Committee on Interagency Radiation Research and Policy Coordination, Oak Ridge Associated Universities, Washington, D.C., 1992, 1–70.
75. Brent, R. L., What is the relationship between birth defects and pregnancy bleeding? New perspectives provided by the NICHD workshop dealing with the association of chorionic villous sampling and the occurrence of limb reduction defects, *Teratology,* 48, 93–95, 1993.
76. Brent, R. L. and Beckman, D. A., Angiotensin-converting enzyme inhibitors, an embryopathic class of drugs with unique properties: information for clinical teratology counselors, *Teratology,* 43, 543, 1991.
77. Brent, R. L., Beckman, D. A., and Jensh, R. P., The relationship of animal experiments in predicting the effects of intrauterine effects in the human, in *Radiation Risks to the Developing Nervous System,* Kriegel, H., Schmahl, W., Gerber, G. B., and Stieve, F. E., Eds., Gustav Fisher, New York, 1986, pp. 367–397.
78. Brent, R. L., Beckman, D. A., and Jensh, R. P., Relative radiosensitivity of fetal tissues, in *Relative Radiation Sensitivities of Human Organ Systems,* Leet, J. T. and Altman, K. I., Eds., Academic Press, New York, 1987, pp. 239–256.
79. Brent, R. L. and Bolden, B. T., The indirect effect of irradiation on embryonic development. III. The contribution of ovarian irradiation, uterine irradiation, oviduct irradiation, and zygote irradiation to fetal mortality and growth retardation in the rat, *Radiat. Res.,* 30, 759–773, 1967.
80. Brent, R. L. and Bolden, B. T., The indirect effect of irradiation on embryonic development. IV. The lethal effects of maternal irradiation on the first day of gestation in the rat, *Proc. Soc. Exp. Biol. Med.,* 125, 709–712, 1967.
81. Brent, R. L. and Franklin, J. B., Uterine vascular clamping: new procedures for the study of congenital malformations, *Science,* 132, 89–91, 1960.
82. Brent, R. L., Gordon, W. E., Bennett, W. R., and Beckman, D. A., Reproductive and teratologic effects of electromagnetic fields, *Reprod. Toxicol.,* 7, 535–580, 1993.
83. Brent, R. L. and Holmes, L. B., Clinical and basic science lessons from the thalidomide tragedy: what have we learned about the causes of limb defects?, *Teratology,* 38, 241–251, 1988.
84. Brent, R. L., Jensh, R. P., and Beckman, D. A., Medical sonography: reproductive effects and risks, *Teratology,* 44, 123–146, 1991.
85. Briggs, G. G., Freeman, R. K., and Yaffe, S. J., *Drugs in Pregnancy and Lactation,* 3rd ed., Williams & Wilkins, Baltimore, MD, 1990, pp. 502–508.
86. Briggs, M. H. and Briggs, M., Sex hormone exposure during pregnancy and malformations, in *Advances in Steroid Biochemistry and Pharmacology,* Briggs, M. H. and Corbin, A., Eds., Academic Press, London, 1979, pp. 51–89.

87. Brooke, O. G., Anderson, H. R., Bland, J. M., Peacock, J. L., and Stewart, C. M., Effects on birth weight of smoking, alcohol, caffeine, socioeconomic factors and psychosocial stress, *Br. Med. J.*, 298, 795–801, 1989.
88. Broughton-Pipkin, F., Symonds, E. M., and Turner, S. R., The effect of captopril (SQ 14,225) upon mother and fetus in the chronically cannulated ewe and in the pregnant rabbit, *J. Physiol.*, 323, 415–422, 1982.
89. Broughton-Pipkin, F., Turner, S. R., and Symonds, E. M., Possible risk with captopril during pregnancy. Some animal data, *Lancet*, 2, 1256, 1980.
90. Brown, N. A., Schull, G., and Fabro, S., Assessment of the teratogenic potential of trimethadione in the CD-1 mouse, *Toxicol. Appl. Pharmacol.*, 51, 59–71, 1979.
91. Brown, N. A. and Scialli, A. R., Update on caffeine, in *Reproductive Toxicology: A Medical Letter*, Vol. 6, No. 3, Brown, N. A. and Scialli, A. R., Eds., Reproductive Toxicology Center, Washington, D.C., 1987.
92. Brown, T., Anand, A., Ritchie, L. D., Clewley, J. P., and Reid, T. M. S., Intrauterine parvovirus infection associated with hydrops fetalis, *Lancet*, 2, 1033–1034, 1984.
93. Brown, Z. A., Vontver, L. A., Benedetti, J., Critchlow, C. W., Sells, C. J., Berry, S., and Corey, L., Effects on infants of a first episode of genital herpes during pregnancy, *N. Engl. J. Med.*, 317, 1246–1251, 1987.
94. Bruckner, A., Lee, Y. J., O'Shea, K. S., and Henneberry, R. C., Teratogenic effects of valproic acid and diphenylhydantoin on mouse embryos in culture, *Teratology*, 27, 29–42, 1983.
95. Brunner, J. A. and Witschi, E., Testosterone-induced modifications of sexual development in female hamsters, *Am. J. Anat.*, 79, 293–320, 1946.
96. Buchanon, T. A., Denno, K. M., Sipos, G. F., and Sadler, T. W., Diabetic teratogenesis. *In vitro* evidence for a multifactorial etiology with little contribution from glucose *per se*, *Diabetes*, 43, 656–660, 1994.
97. Burbacher, T. M., Rodier, P. M., and Weiss, B., Methylmercury developmental neurotoxicity: a comparison of effects in humans and animals, *Neurotoxicol. Teratol.*, 12, 191–202, 1990.
98. Burge, E. S., The relationship of threatened abortion to fetal anomalies, *Am. J. Obstet. Gynecol.*, 61, 615–621, 1951.
99. Burrow, G. N., Neonatal goiter after maternal propylthiouracil therapy, *J. Clin. Endocrinol.*, 25, 403–408, 1965.
100. Burrow, G. N., Klatski, E. H., and Genel, M., Intellectual development in children whose mothers received propylthiouracil during pregnancy, *Yale J. Biol. Med.*, 41, 151–156, 1978.
101. Bushnell, P. J. and Bowman, R. E., Effects of chronic lead ingestion on social development in infant rhesus monkeys, *Neurobehav. Toxicol.*, 1, 207–219, 1979.
102. Buttar, H. S., Effects of chlordiazepoxide on the pre- and postnatal development of rats, *Toxicology*, 17, 311–321, 1980.
103. Caan, B. J. and Goldhaber, M. K., Caffeinated beverages and low birthweight — a case control study, *Am. J. Public Health*, 79, 1299–1300, 1989.
104. Cabrera, T. M., Yracheta, J. M., Li, Q., Levy, A. D., Van de Kar, L. D., and Battaglia, G., Prenatal cocaine produced deficits in seratonin mediated neuroendocrine responses in adult rat progeny: evidence for long-term functional alterations in brain seratonin pathways, *Synapse*, 15, 158–168, 1993.
105. Cantarow, A. and Trumper, M., *Lead Poisoning*, Williams & Wilkins, Baltimore, MD, 1994.
106. Carbone, J. P. and Brent, R. L., Genital and nongenital teratogenesis of prenatal progestogen therapy: the effects of 17 alpha-hydroxyprogesterone caproate on embryonic and fetal development and endochondral ossification in the C57Bl/6J mouse, *Am. J. Obstet. Gynecol.*, 169, 1292–1298, 1993.
107. Carbone, J. P., Figurska, K., Buck, S., and Brent, R. L., Effect of gestational sex steroid exposure on limb development and endochondral ossification in the pregnant C57Bl/6J mouse. I. Medroxyprogesterone acetate, *Teratology*, 42, 121–130, 1990.
108. Carswell, F., Kerr, M. M., and Hutchinson, J. H., Congenital goiter and hypothyroidism produced by maternal ingestion of iodides, *Lancet*, 1, 1241–1243, 1970.
109. Carter, C. O., Genetics of common single malformations, *Br. Med. Bull.*, 32, 21–26, 1976.
110. Castilla, E. E. and Orioli, I. M., Teratogenicity of misoprostol: data from the Latin-American collaborative study of congenital malformations (ECLAMC), *Am. J. Med. Genet.*, 51, 161–162, 1994.
111. Chan, G. M., Buchino, J. J., Mehlhorn, D., Bove, K. E., Steichen, J. J., and Tsang, R. C., Effect of vitamin D on pregnancy in rabbits and their offspring, *Pediatr. Res.*, 13, 121–126, 1979.
112. Chasnoff, I. J., Chisum, G. M., and Kaplan, W. E., Maternal cocaine use and genitourinary tract malformations, *Teratology*, 37, 201–204, 1988.
113. Chasnoff, I. J. and Griffith, D. R., Cocaine: clinical studies of pregnancy and the newborn, *Ann. N.Y. Acad. Sci.*, 562, 260–266, 1989.
114. Chattingius, S., Does age potentiate the smoking-related risk of fetal growth retardation?, *Early Hum. Dev.*, 20, 203–211, 1989.
115. Chaube, S., Kury, G., and Murphy, M. L., Teratogenic effects of cyclophoshamide (NSC-26271) in the rat, *Cancer Chemother. Rep.*, 51, 363–376, 1967.
116. Chavez, G. F., Mulinare, J., and Cordero, J. F., Maternal cocaine use during early pregnancy as a risk factor for congenital urogenital anomalies, *J. Am. Med. Assoc.*, 262, 795–798, 1989.
117. Chernoff, G. F., The fetal alcohol syndrome in mice: maternal variables, *Teratology*, 22, 71–75, 1980.

118. Cheron, R. G., Kaplan, M. M., Lasen, P. R., Selenkow, H. A., and Crigler, Jr., J. F. Neonatal thyoid function after propylthiouracil therapy for maternal Graves disease, *N. Engl. J. Med.*, 304, 525–528, 1981.
119. Chisolm, Jr., J. J., and Barltrop, D., Recognition and management of children with increased lead absorption, *Arch. Dis. Child.*, 54, 249, 1979.
120. Church, M. W. and Overbeck, G. W., Sensorineural hearing loss as evidenced by the auditory brainstem response following prenatal cocaine exposure in the Long-Evans rat, *Teratology*, 43, 561–570, 1991.
121. Clarren, S. K. and Smith, D. W., The fetal alcohol syndrome, *N. Engl. J. Med.*, 208, 1063–1067, 1978.
122. Clewell, W. P., *In utero* treatment of thyrotoxicosis, in *Fetal Diagnosis and Therapy: Science, Ethics, and the Law*, Evans, M. I., Fletcher, J. C., Ditlen, A. O., and Schulman, J. D., Eds., J.B. Lippencott, Philadelphia, 1984, p. 124.
123. Coelho, H. L. L., Misago, C., Fonseca, W. V. C., Souza, D. S. C., and Araujo, J. M. L., Selling abortifacients over the counter in pharmacies in Fortaleza, Brazil, *Lancet*, 338, 247, 1991.
124. Cohen, L. S., Friedman, J. M., Jefferson, J. W., Johnson, E. M., and Weiner, M. L., A reevaluation of risk of *in utero* exposure to lithium, *J. Am. Med. Assoc.*, 271, 146–150, 1994.
125. Cohen, M. M., Syndromology — an updated conceptual overview. VII. Aspects of teratogenesis, *Int. J. Oral Maxillofac. Surg.*, 19, 26–32, 1990.
126. Cohlan, S. Q., Bevelander, G., and Tiamsic, T., Growth inhibition of prematures receiving tetracycline: clinical and laboratory investigation, *Am. J. Dis. Child.*, 105, 453–461, 1963.
127. Collins, F. S. and Mahoney, M. J., Hydrocephalus and abnormal digits after failed first trimester prostaglandin abortion attempt, *J. Pediatr.*, 102, 620–621, 1983.
128. Collins, M. D., Fradkin, R., and Scott, W. J., Induction of postaxial forelimb ectrodactyly with anticonvulsant agents in A/J mice, *Teratology*, 41, 61–70, 1990.
129. Collins, T., Welsh, J., Black, T., Whitby, K., and O'Donnell, M., Potential reversibility of skeletal effects in rats exposed *in utero* to caffeine, *Food Chem. Toxicol.*, 25, 647–666, 1987.
130. Collins, T. F. X., Welsh, J. J., Black, T. N., and Collins, V. V., A study of the teratogenic potential of caffeine given by oral intubation to rats, *Regul. Toxicol. Pharmacol.*, 1, 355–378, 1981.
131. Conboy, T. J., Pass, R. F., Alfrod, C. A., Myers, G. J., Britt, W. J., McCollister, F. P., Summers, M. N., McFarland, C. E., and Boll, T. J., Early clinical manifestation and intellectual outcome in children with symptomatic congenital cytomegalovirus infection, *J. Pediatr.*, 111, 343–348, 1987.
132. Cooney, G., Bell, A., McBride, W., and Carter, C., Neurobehavioral consequences of prenatal low level exposure to lead, *Dev. Med. Child. Neurol.*, 11, 95–104, 1989.
133. Copp, A. J., Death before birth: clues from gene knockouts and mutations, *TIG*, 11, 87–93, 1995.
134. Costa, S. H. and Vessey, M. P., Misoprostol and illegal abortion in Rio de Janeiro, Brazil, *Lancet*, 341, 1258–1261, 1993.
135. Cotrufo, M., deLuca, T. S. L., Calabro, R., Mastrogiovanni, G., and Lama, D., Coumarin anticoagulation during pregnancy in patients with mechanical valve prostheses, *Eur. J. Cardiothorac. Surg.*, 5, 300–305, 1991.
136. Cowett, R. M., Hakanson, D. O., Kocon, R. W., and Oh, W., Untoward neonatal effect of intraamniotic administration of methylene blue, *Obstet. Gynecol.*, 48(Suppl.), 745–755, 1976.
137. Creech-Kraft, J., Kochhar, D. M., Scott, W. J., and Nau, H., Low teratogenicisty of 13-*cis*-retinoic acid (isotretinoin) in the mouse corresponds to low embryonic concentrations during embryogenesis: comparisons to the all-*trans*-isomer, *Toxicol. Appl. Pharmacol.*, 87, 474–482, 1987.
138. Crombie, D. L., Pinsent, R. J., Fleming, D. M., Rumeau-Rouquette, C., Goujard, J., and Huel, G., Fetal effects of tranquilizers in pregnancy, *N. Engl. J. Med.*, 293, 198–199, 1975.
139. Crooks, J., Haemolytic jaundice in a neonate after intra-amniotic injection of methylene blue, *Arch. Dis. Child.*, 57, 872–886, 1982.
140. Cunniff, C., Jones, K., Phillipson, J., Benirschke, K., Short, S., and Wujek, J., Oligohydramnios sequence and renal tubular malformation associated with maternal enalapril use, *Am. J. Obstet. Gynecol.*, 162, 187–189, 1990.
141. Cutler, J., Thalidomide revisited, *Lancet*, 343, 795–796, 1994.
142. Czeizel, A., Lack of evidence of teratogenicity of benzodiazepine drugs in Hungary, *Reprod. Toxicol.*, 1, 183–188, 1988.
143. Czeizel, A. E., Kodaj, I., and Lenz, W., Smoking during pregnancy and congenital limb deficiency, *Br. Med. J.*, 308, 1473–1476, 1994.
144. Czeizel, A. E. and Dudas, I., Prevention of the first occurrence of neural-tube defects by periconceptional vitamin supplementation, *N. Engl. J. Med.*, 327, 1832–1835, 1992.
145. D'Souzak, S. W., Robertson, I. G., Donnai, D., and Mawer, G., Fetal phenytoin exposure, hypoplastic nails, and jitteriness, *Arch. Dis. Child.*, 65, 320–324, 1990.
146. da Silva, V. A. and Malheiros, L. R., Developmental toxicity of *in utero* exposure to toluene on malnourished and well nourished rats, *Toxicology*, 64, 155–168, 1990.
147. Dai, W. S., Hsu, M., and Itri, L. M., Safety of pregnancy after discontinuation of isotretinoin, *Arch. Dermatol.*, 125, 362–365, 1989.
148. Dai, W. S., LaBraico, J. M., and Stern, R. S., Epidemiology of isotretinoin exposure during pregnancy, *J. Am. Acad. Dermatol.*, 26, 599–606, 1992.

149. Dalens, B., Raynaud, E.-J., and Gauline, J., Teratogenicity of valproic acid, *J. Pediatr.*, 97, 332–333, 1980.
150. Davis, J. G., Bennett, W. R., Brady, J. V., Brent, R. L., Gordis, L., Gordon, W. E., Greenhouse, S. W., Reiter, R. J., Stein, G. S., Sussking, C., and Trichopoulos, D., *Health Effects of Low-Frequency Electric and Magnetic Fields,* (ORAU 92/F8), Committee on Interagency Radiation Research and Policy Coordination, Oak Ridge Associated Universities, Washington, D.C., 1992.
151. Day, N. L. and Richardson, G. A., Prenatal alcohol exposure: a continuum of effects, *Semin. Perinatol.*, 15, 271–279, 1991.
152. deHass, I., Harlow, B. L., Cramer, D. W., and Frigoletto, F. D., Spontaneous preterm birth: a case-control study, *Am. J. Obstet. Gynecol.*, 165, 1290–1296, 1991.
153. Dekaban, A. S., Abnormalities in children exposed to X-radiation during various stages of gestation: tentative timetable of radiation, half-lives of the drug in mother and baby, *J. Pediatr.*, 94, 832–835, 1968.
154. DeSantis, M., Noia, G., Caruso, A., and Mancuso, S., Guidelines for the use of zidovudine in pregnant women with HIV infection, *Drugs,* 50, 43–47, 1995.
155. DeWals, P., Bloch, D., Calabro, A., Calzolari, E., Cornel, M. C., Johnson, Z., Ligutic, I., Nevin, N., Pexieder, T., Stoll, C., Tenconi, R., and Tilmont, P., Association between holoprosencephaly and exposure to topical retinoids: results of the EUROCAT survery, *Neonatal Perinatal Epidemiol.*, 5, 445–447, 1991.
156. Dicke, J. M., Verges, D. K., and Polakoski, K. L., Cocaine inhibits alanine uptake by human placental microvillus membrane visicles, *Am. J. Obstet. Gynecol.*, 169, 515–521, 1993.
157. Dickinson, R. G., Hapland, R. C., Lynn, R. K., Smith, W. B., and Gerber, N., Transmission of valproic acid across the placenta: half-lives of the drug in mother and baby, *J. Pediatr.*, 94, 832–835, 1979.
158. Dietrich, K., Berger, O., and Succop, P., Lead exposure and the motor development status of urban six year old children in the Cincinnati Prospective study, *Pediatrics,* 91, 301–307, 1993.
159. Dietrich, K., Kraft, K., Bornschein, R., Hammond, P., Berger, O., Succop, P., and Bier, M., Low-level fetal lead exposure effect on neurobehavioral development in early infancy, *Pediatrics,* 80, 721–730, 1987.
160. DiGiovanna, J. J., Zech, L. A., Ruddel, M. E., Gantt, G., McClean, S. W., Gross, E. G., and Peck, G. L., Etretinate: persistent serum levels of a potent teratogen, *Clin. Res.*, 32, 579A, 1984.
161. Diniz, E. M. A., Corradini, H. B., Ramos, J. L., and Brock, R., The effects of methotrexate on the developing fetus, *Rev. Hosp. Clin. Fac. Med. Univ. Sao Paulo,* 33, 286–290, 1978.
162. Dinsmoor, M. J., HIV infection and pregnancy, *Clin. Perinatal.*, 21, 85–94, 1994.
163. DiSaia, P. J., Pregnancy and delivery of a patient with a Starr-Edwards mitral valve prosthesis: report of a case, *Obstet. Gynecol.*, 29, 469–472, 1996.
164. Dixon, S. D. and Bejar, R., Brain lesions in cocaine and methamphetamine exposed neonates, *Pediatr. Res.*, 23, 405A, 1988.
165. Dluzniewski, A. and Gastol-Lewinska, L., The search for teratogenic activity of some tuberculostatic drugs, *Diss. Pharm. Pharmacol.*, 23, 383, 392, 1971.
166. Doberczak, T., Shanzer, S., Senie, R., and Kandall, S., Neonatal, neurologic, and electroencephalographic effects of intrauterine cocaine exposure, *J. Pediatr.*, 113, 354–358, 1988.
167. Dogra, V. S., Menon, P. A., Poblete, J., and Smeltzer, J. S., Neurosonographic imaging of small for gestational age neonates exposed and not exposed to cocaine and cytomegalovirus, *J. Clin. Ultrasound,* 22, 93–102, 1994.
168. Dohler, K. D., Hancke, J. L., Srivastava, S. S., Hofman, C., Shryne, J. E., and Gorski, R. R., Participation of estrogens in female sexual differentiation of the brain: neuroanatomical, neuroendocrine and behavioral evidence, *Prog. Brain Res.*, 6, 99–117, 1984.
169. Dolk, H., Methylene blue and atresia or stenosis of ileum and jejunum, *Lancet,* 338, 1021–1022, 1991.
170. Donald, P. R. and Sellars, S. L., Streptomycin ototoxicity in the unborn child, *S. Afr. Med. J.*, 60, 316–318, 1981.
171. Dyban, A. P., Sekirina, G. G., and Golinsky, G. F., The effect of aminopterin on the preimplantation rat embryo cultivated *in vitro, Ontogenz,* 82, 121–127, 1977.
172. El-Bizri, H., Guest, I., and Varma, R., Effects of cocaine on rat embryo development *in vivo* and in cultures, *Pediatr. Res.*, 29, 187–190, 1991.
173. Elmazar, M. M. A., McElhatton, P. R., and Sullivan, F. M., Studies on the teratogenic effects of different oral preparations of caffeine in mice, *Toxicology,* 23, 57–72, 1982.
174. Elshave, J., Cleft palate in the offspring of female mice treated with phenytoin, *Lancet,* 2, 1074, 1969.
175. Embree, J. E., Braddick, M., Datta, P., Murlithi, J., Hoff, C., Kreiss, J. K., Roberts, P. L., Law, B. J., Pamba, H. O., Ndinya-Achola, J. O., and Plummer, F. A., Lack of correlation of maternal human immunodeficiency virus infection with neonatal malformations, *Pediatr. Infect. Dis. J.*, 8, 700–704, 1989.
176. Enders, G. and Biber, M., Parvovirus B19 infections in pregnancy, *Behring Inst. Mitt.*, 85, 74–78, 1990.
177. Endres, W., D-penicillamine in pregnancy — to ban or not to ban, *Klin. Worchenschr.*, 59, 535–538, 1981.
178. Erickson, J. D., Risk factors for birth defects: data from the Atlanta defects case-control study, *Teratology,* 43, 41–51, 1991.
179. Ericson, A., Kallen, B., Zetterstrom, R., Eriksson, M., and Westerholm, P., Delivery outcome of women working in laboratories during pregnancy, *Arch. Environ. Health,* 39, 5–10, 1984.
180. Ericson-Strandvik, B. and Gyllensten, L., The central nervous system of foetal mice after administration of streptomycin, *Acta Pathol. Microbiol. Scand.*, 59, 292–300, 1963.

181. Eriksson, U. J., Dahlstrom, E., Larsson, K. S., and Hellerstrom, C., Increased incidence of congenital malformations in the offspring of diabetic rats and their prevention by maternal insulin therapy, *Diabetes,* 31, 1–6, 1982.
182. Ernhart, C., Morrow-Tlucak, M., Wolf, A., Super, A., and Drotar, D., Low level lead exposure in the prenatal and early preschool periods: intelligence prior to school entry, *Neurotoxicol. Teratol.,* 11, 161–170, 1989.
183. Ernhart, C., Wolf, A., Kennard, M., Erhard, P., Filipovich, H., and Sokol, R., Intrauterine exposure to the low levels of lead: the status of the neonate, *Arch. Environ. Health,* 41, 287–291, 1986.
184. Ernhart, C. B. and Greene, T., Low level lead exposure in the prenatal and early preschool periods: language development, *Arch. Environ. Health,* 45, 342–353, 1990.
185. European Collaborative Study, Mother-to-child transmission of HIV infection, *Lancet,* ii, 1039–1043, 1988.
186. Evans, J. H. and Beicher, N. A., The prognosis of threatened abortion, *Med. J. Aust.,* 2, 165–168, 1970.
187. Fantel, A. G., Shepard, T. H., Newell-Morris, L. L., and Moffett, B. C., Teratogenic effects of retinoic acid in pigtail monkeys (*Macaca nemistrina*), *Teratology,* 15, 65–72, 1977.
188. Fantel, A. G., Barber, C. V., Carda, M. B., Tumbic, R. W., and Mackler, B., Studies on the role of ischemia/reperfusion and superoxide anion radical production in the teratognicity of cocaine, *Teratology,* 46, 293–300, 1992.
189. Fantel, A. G., Person, R. E., and Juchau, M. R., Niridazole metabolism by rat embryos *in vitro, Teratology,* 37, 213–223, 1988.
190. Frederick, J., Epilepsy and pregnancy: a report from Oxford record linkage study, *Br. Med. J.,* 2, 442–448, 1973.
191. Fein, G. G., Jacobson, J. L., Jacobson, S. W., Schwartz, P. M., and Dowler, J. K., Prenatal exposure to polychlorinated biphenyls effects on birth size and gestational age, *J. Pediatr.,* 105, 315–320, 1984.
192. Feldman, G. L., Weaver, D. D., and Lovrien, E. W., The fetal trimethadione syndrome, *Am. J. Dis. Child.,* 131, 1389–1392, 1977.
193. Felts, J. H., Coffee arabica, *N. C. Med. J.,* 42, 281, 1981.
194. Fenster, L., Eskenazi, B., Windham, G. C., and Swan, S. H., Caffeine consumption during pregnancy and fetal growth, *Am. J. Public Health,* 81, 458–461, 1991.
195. Ferencz, C., Matanoski, G. M., Wiolson, P. D., Rubin, J. D., Neill, C. A., and Gutberlet, R., Maternal hormone therapy and congenital heart disease, *Teratology,* 21, 225–239, 1980.
196. Ferm, V. H. and Carpenter, S. J., Developmental malformations resulting from the administration of lead salts, *Exp. Mol. Pathol.,* 7, 208–213, 1967.
197. Ferrer, I., Xumetra, A., and Santamaria, J., Cerebral malformation induced by prenatal X-irradiation: an autoradiographic and Golgi study, *J. Anat.,* 138, 81–93, 1984.
198. Ferris, T. F. and Weir, E. K., Effect of captopril on uterine blood flow and prostaglandin E synthesis in the pregnant rabbit, *J. Clin. Invest.,* 71, 809–815, 1982.
199. Fillippi, B., Antibiotics and congenital malformations: evaluation of the teratogenicity of antibiotics, in *Advances in Teratology,* Woollam, D. H. M., Ed., Academic Press, New York, 1967, pp. 237–256.
200. Finnell, R. H., Phenytoin-induced teratogenesis: a mouse model, *Science,* 211, 483–484, 1981.
201. Finnell, R. H., Abbott, L. C., and Taylor, S. M., The fetal hydantoin syndrome: answers from a mouse model, *Reprod. Toxicol.,* 3, 127–133, 1989.
202. Finnell, R. H., Toloyan, S., van Waes, M., and Kalivas, P. W., Preliminary evidence for cocaine-induced embryopathy in mice, *Teratol. Appl Pharmacol.,* 103, 228–237, 1990.
203. Firth, H. V., Boyd, P. A., Chamberlain, P., MacKenzie, I. Z., Lindenbaum, R. H., and Huson, S. M., Severe limb abnormalities after chorionic villus sampling at 56–66 days gestation, *Lancet,* 337, 762–763, 1991.
204. Florman, A. L., Gershon, A. A., Blackett, P. R., and Nahmias, A. J., Intrauterine infection with herpes simplex virus. Resultant congenital anomalies, *J. Am. Med. Assoc.,* 225, 129–132, 1973.
205. Fogo, A., Yoshida, Y., Yared, A., and Ichikawa, I., Importance of angiogenic action of angiotensin II in the glomerular growth of maturing kidneys, *Kidney Int.,* 38, 1068–1074, 1990.
206. Fonseca, W., Alencar, A. J. C., Mota, F. S. B., and Coelho, H. L. L., Misoprostol and congential malformations, *Lancet,* 336, 56, 1991.
207. Forbes, G. B., Vitamin D in pregnancy and the infantile hypercalcemia syndrome (letter), *Pediatr. Res.,* 13, 1382, 1979.
208. Ford, J. J. and Christenson, R. K., Influences of pre- and postnatal testosterone treatment on defeminization of sexual receptivity in pigs, *Biol. Reprod.,* 36, 581–587, 1987.
209. Forrest, F. and du V. Florey, C., The relation between maternal alcohol consumption and child development: the epidemiological evidence, *J. Public Health Med.,* 13, 247–255, 1991.
210. Fotherby, K., A new look at progestins, *Clin. Obstet. Gynecol.,* 11, 701–722, 1984.
211. Foulon, W., Naessens, A., Maher, T., Mahler, T., DeWaele, M., and DeCatte, L., Prenatal diagnosis of congenital toxoplasmosis, *Obstet. Gynecol.,* 76, 769–772, 1990.
212. Fowler, K. B. and Pass, R. F., Sexually transmitted diseases in mothers of neonates with congenital cytomegalovirus infection, *J. Infect. Dis.,* 164, 259–264, 1991.
213. Fradkin, R., Scott, W. J., and Wilson, J. G., Trimethadione teratogenesis in the rat and rhesus monkey, *Teratology,* 24, 39–40A, 1981.

214. Frankenberg, R. R. and Lipinski, J. F., Congenital malformations, *N. Engl. J. Med.*, 309, 311–312, 1983.
215. Frankenberg, W. K., Duncan, B. R., Coffelt, R. W., Koch, R., Coldwell, J. G., and Son, C. D., Maternal phenylketonuria: implications for growth and development, *J. Pediatr.*, 73, 560–570, 1968.
216. Fraser, F. C., The multifactorial/threshold concept — uses and misuses, *Teratology,* 14, 267–280, 1976.
217. Fraser, F. C., Interactions and multiple causes, in *Handbook of Teratology,* Wilson, J. G. and Fraser, F. C., Eds., Plenum Press, New York, 1977, pp. 445–463.
218. Friedman, W. F., Vitamin D and the supravalvular aortic stenosis syndrome, in *Advances in Teratology,* Woollam, D. H. M., Ed., Academic Press, New York, 1968, pp. 83–96.
219. Friedman, W. F. and Roberts, W. C., Vitamin D and the supravalvular aortic stenosis syndrome. The transplacental effects of vitamin D on the aorta of the rabbit, *Circulation,* 34, 77–86, 1966.
220. Fritz, H. and Hess, R., Effects of cyclophosphamide on embryonic development in the rabbit, *Agents Actions,* 2, 83–86, 1971.
221. Fujii, T. and Nishimura, H., Adverse effects of prolonged administration of caffeine on rat fetus, *Toxicol. Appl. Pharmacol.*, 22, 449–457, 1972.
222. Fujii, T., Sasaki, H., and Nishimura, H., Teratogenicity of caffeine in mice related to its mode of administration, *Jpn. J. Pharmacol.*, 19, 134–138, 1969.
223. Funderburk, S. J., Guthrie, D., and Meldrum, D., Outcome of pregnancies complicated by early vaginal bleeding, *Br. J. Obstet. Gynecol.*, 87, 100–105, 1980.
224. Gage, J. C. and Sulik, K. K., Pathogenesis of ethanol induces hydronephrosis and hydroureter as demonstrated following *in vivo* exposure of mouse embryos, *Teratology,* 44, 299–312, 1991.
225. Gaily, E., Distal phalangeal hypoplasia in children with prenatal phenytoin exposure: results of a controlled anthropometric study, *Am. J. Med. Genet.*, 35, 574–578, 1990.
226. Gaily, E. and Granstrom. M.-L., A transient retardation of early postnatal growth in drug-exposed children of epileptic mothers, *Epilepsy Res.*, 4, 147–155, 1989.
227. Gaily, E., Granstrom, M.-L., Hiilesmass, V., and Bardy, A., Minor anomalies in offspring of epileptic mothers, *J. Pediatr.*, 112, 520–529, 1988.
228. Gaily, E., Kantola-Sorsa, E., and Granstrom, M.-L., Intelligence of children of epileptic mothers, *J. Pediatr.*, 113, 677–684, 1986.
229. Gaitan, E. and Dunn, J. T., Epidemiology of iodine deficiency, *Trends Endocrinol. Metab.*, 3, 170–175, 1992.
230. Gal, I., Risks and benefits of the use of hormonal pregnancy test tablets, *Nature,* 240, 241–242, 1972.
231. Gal, I., Kirman, B., and Stern, J., Hormonal pregnancy tests and congenital malformations, *Nature,* 216, 83, 1967.
232. Gandier, B., Ponte, C., Duquennoy, G., Callens, M., and Ballester, L., Retard de croissance intra-uterin avec microcephalie chez trois enfants nes de mere hyperphenylalaninemique, *Ann. Pediat. (Paris),* 19, 269–276, 1972.
233. Garcia, R. E., Fredman, W. F., Kaback, M. M., and Rowe, R. D., Idiopathic hypercalcemia and supravalvular stenosis: documentation of a new syndrome, *N. Engl. J. Med.*, 271, 117–120, 1964.
234. Garg, U. C., Turndorf, H., and Bansinath, M., Effect of cocaine on macromolecular synthesis and cell proliferation in cultured glial cells, *Neuroscience,* 57, 467–472, 1993.
235. Garner, P., Type I diabetes mellitus and pregnancy, *Lancet,* 346, 157–161, 1995.
236. Gavras, I. and Gavras, H., ACE inhibitors: a decade of clinical experience, *Hosp. Pract. (Off. Ed.),* 28, 117–127, 1993.
237. Generoso, W. M., Rutledge, J. C., Cain, K. T., Hughes, L. A., and Downing, D. J., Mutagen-induced fetal anomalies and death following treatment of females within hours after mating, *Mutat. Res.*, 199, 175–181, 1988.
238. German, J., Kowal, A., and Ehlers, K. H., Trimethadione and human teratogenesis, *Teratology,* 3, 349–362, 1970.
239. Gershman, S. T. and Stolley, P. D., A case-control study of testicular cancer using Connecticut tumour registry data, *Int. J. Epidemiol.*, 17, 738–742, 1988.
240. Gershon, A. A., Chickenpox, measles and mumps, in *Infectious Diseases of the Fetus and Newborn Infants,* Remington, J. S. and Klein, J. O., Eds., W.B. Saunders, Philadelphia, 1990, pp. 395–420.
241. Gibson, J. E. and Becker, B. A., The teratogenicity of cyclophosphamide in mice, *Cancer Res.*, 28, 475–480, 1968.
242. Gilbert, E. F. and Pistey, W. R., Effect on the offspring of repeated caffeine administration to pregnant rats, *J. Reprod. Fertil.*, 34, 495–499, 1973.
243. Gilbert, S. G., Rice, D. C., Reuhl, K. R., and Stavric, B., Adverse pregnancy outcome in the monkey (*Macaca fascicularis*) after chronic caffeine exposure, *J. Pharmacol. Exp. Ther.*, 245(3), 1048–1053, 1988.
244. Gill, T. S., Guram, M. S., and Geber, W. F., Comparative study of the teratogenic effects of chlorodiazepoxide and diazepam in the fetal hamster, *Life Sci.*, 29, 2141–2147, 1981.
245. Gill, W. B., Schumacher, G. F. B., and Bibbo, M., Structural and functional abnormalities in the sex organs of male offspring of mothers treated with diethylstilbestrol (DES), *J. Reprod. Med.*, 16, 147–153, 1976.
246. Gill, W. B., Schumacher, G. F. B., Bibbo, M., Straus, F. H., and Schoenberg, H. W., Association of diethylstilbestrol exposure *in utero* with cryptorchidism, testicular hypoplasia and semen abnormalities, *J. Urol.*, 122, 36–39, 1979.

247. Gingras, J. L., O'Donnell, K. J., and Hume, R. F., Maternal cocaine addiction and fetal behavioral state: a human model for study of sudden infant death syndrome, *Med. Hypoth.*, 33, 227–230, 1990.
248. Gingras, J. L., Weese-Mayer, D. E., Dalley, L. B., and Klemka-Walder, L. M., Prenatal cocaine exposure alters postnatal ornithine decarboxylase activity in rabbit brain, *Biochem. Med. Metab. Biol.*, 50, 284–291, 1993.
249. Gingras, J. L., Weese-Mayer, D. E., Hume, R. F., and O'Donnell, K. J., Cocaine and development: mechanisms of fetal toxicity and neonatal consequences of prenatal cocaine exposure, *Early Hum. Dev.*, 31, 1–24, 1992.
250. Ginsberg-Fellner, F., Witt, M. E., Fedun, B., Taub, F., Dobersen, M. J., McEvoy, R. C., Cooper, L. Z., Notkins, A. L., and Rubinstein, P., Diabetes mellitus and autoimmunity in patients with the congenital rubella syndrome, *Rev. Infect. Dis.*, 7(Suppl. 1), S170–S176, 1985.
251. Giroud, A. and Tuchmann-Duplessis, H., Malformations congenitales role des facteurs exogenes, *Pathol. Biol.*, 10, 119–151, 1962.
252. Goeringer, G. C. and DeSesso, J. M., Developmental toxicity in rabbits of the antifolate aminopterin and its amelioration by leucovorin, *Teratology*, 41, 560–561, 1990.
253. Goetsch, C., An evaluation of amniopterin as an abortifacient, *Am. J. Obstet. Gynecol.*, 83, 1474–1477, 1962.
254. Goldstein, L. and Murphy, D. P., Microcephalic idiocy following radium therapy for uterine cancer during pregnancy, *Am. J. Obstet. Gynecol.*, 18, 189–195, 281–283, 1929.
255. Gollop, T. R., Eigier, A., and Guiduglio-Neto, J., Prenatal diagnosis of thalidomide syndrome, *Prenat. Diagn.*, 7, 295–298, 1987.
256. Gonzalez, C. H., Vargas, F. R., Perez, A. B. A., Kim, C. A., Marques-Dias, M. J., Leone, C. R., Neto, J. C., Llerena, J. C., Jr., and Cabral de Almeida, J. C., Limb deficiency with or without Moebius sequence in seven Brazilian children associated with misoprotol use in the first trimester of pregnancy, *Am. J. Med. Genet.*, 46, 59–64, 1993.
257. Goodenday, L. S. and Gordon, G. S., No risk from vitamin D in pregnancy, *Ann. Intern. Med.*, 75, 807–808, 1971.
258. Goy, R. W., Bercovitch, F. B., and McBrair, M. C., Behavioral masculinization is independent of genital masculinization in prenatally androgenized female rhesus macaques, *Horm. Behav.*, 22, 552–571, 1988.
259. Goy, R. W., Bridson, W. E., and Young, W. C., Period of maximal susceptibility of the prenatal female guinea pig to masculinizing actions of the testosterone propionate, *J. Comp. Physiol. Psychol.*, 57, 166–174, 1964.
260. Graham, H. N., Mate, in *The Methylxanthine Beverages and Foods: Chemistry, Consumption and Health Effects*, Spiller, G. A., Ed., Alan R. Liss, New York, 1984, pp. 179–183.
261. Graham, H. N., Tea: the plant and its manufacture, chemistry and consumption of the beverage, in *The Methylxanthine Beverages and Foods: Chemistry, Consumption and Health Effects*, Spiller, G. A., Ed., Alan R. Liss, New York, 1984, pp. 29–74.
262. Graham, J. M., Jr., The effects of alcohol consumption during pregnancy, in *Prevention of Physical and Mental Congenital Defects*, Part C. *Basic and Medical Science, Education and Future Strategies*, Marois, M., Ed., Alan R. Liss, New York, 1985, pp. 335–339.
263. Graham, J. M., Jr., Causes of limb reduction defects: the contribution of fetal constraints and/or vascular disruption, *Clin. Perinatol.*, 13, 575–591, 1986.
264. Grant, L. D., Kimmel, C. A., West, G. L., Martinez-Vargas, C. M., and Howard, J. L., Chronic low level toxicity in the rat. II. Effects on postnatal physical and behavioral development, *Toxicol. Appl. Pharmacol.*, 56, 42–58, 1980.
265. Greenberg, L. H. and Tanaka, K. R., Congenital anomalies probably induced by cyclophosphamide, *J. Am. Med. Assoc.*, 188, 423–426, 1964.
266. Greene, M. F., Hare, J. W., Cloherty, J. P., Benacerraf, B. R., and Soeldnere, J. S., First-trimester hemoglobin A1 and risk for major malformation and spontaneous abortion in diabetic pregnancy, *Teratology*, 39, 225–231, 1989.
267. Greene, R. R., Burrill, M. W., and Ivy, A. C., Experimental intersexuality: the effect of antenatal androgens on sexual development of female rats, *Am. J. Anat.*, 65, 415–469, 1939.
268. Greenwald, P., Barlow, J. J., Nasca, P. C., andf Burnett, W. S., Vaginal cancer after maternal treatment with synthetic estrogens, *N. Engl. J. Med.*, 285, 390–392, 1971.
269. Gregg, N. M., Congenital cataract following German measles in the mother, *Trans. Ophthalmol. Soc. Aust.*, 3, 35–46, 1941.
270. Gregory, M. C. and Mansell, M. A., Pregnancy and cystinuria, *Lancet*, 2, 1158–1160, 1983.
271. Griffiths, P. D., Baboonian, C., Rutter, D., and Peckham, C., Congenital and maternal cytomegalovirus infections in a London population, *Br. J. Obstet. Gynaecol.*, 98, 135–140, 1991.
272. Grimm, V. E. and Frieder, B., Prenatal caffeine causes long lasting behavioral and neurochemical changes, *Int. J. Neurosci.*, 41, 15–28, 1988.
273. Groisser, D. S., Rosso, P., and Winick, M., Coffee consumption during pregnancy: subsequent behavioral abnormalities of the offspring, *J. Nutr.*, 112, 829–832, 1982.
274. Grose, C. and Itani, O., Pathogenesis of congenital infection with three diverse viruses: varicella-zoster virus, human parvovirus, and human immunodeficiency virus, *Semin. Perinatol.*, 13, 278–293, 1989.
275. Grote, V. W. and Weinmann, I., Examination of the active substances coumarin and rutin in a teratogenic trial with rabbits, *Arzneimittelforsch*, 23, 1319–1320, 1973.

276. Grumbach, M. M. and Conte, F. A., Disorders of sex differntiation, in *Textbook of Endocrinology,* Williams, R. H., Eds., W.B. Saunders, Philadelphia, 1981, pp. 422–514.
277. Grumbach, M. M., Ducharine, J. R., and Moloshok, R. E., On the fetal masculinizing action of certain oral progestins, *J. Clin. Endocrinol. Metab.,* 19, 1369–1380, 1959.
278. Guignard, J. P., Effect of drugs on the immature kidney, *Adv. Nephrol. Neker Hosp.,* 22, 193–211, 1993.
279. Guignard, J. P., Burgener, F., and Calame, A., Persistent anuria in neonate: a side effect of captopril, *Intern. J. Pediatr. Nephrol.,* 2, 133, 1981.
280. Guram, M. S., Gill, T. S., and Geber, W. F., Comparative teratogenicity of chlordiazepoxide, amitriptyline, and combination of the two compounds in the fetal hamster, *Neurotoxicology,* 3, 83–90, 1982.
281. Gwinn, M., Pappaioanou, M., George, J. R., Hannon, W. H., Wasser, S. C., Redus, M. A., Hoff, R., Grady, G. F., Willouhby, A., Novello, A. C., Petersen, L. R., Dondero, T. J., and Curran, J. W., Prevalence of HIV infection in childbearing women in the United States, *J. Am. Med. Assoc.,* 265, 1704–1708, 1991.
282. Hale, F., Pigs born without eyeballs, *J. Hered.,* 24, 105–106, 1933.
283. Hale, F., Relation of vitamin A to anophthalmos in pigs, *Am. J. Ophthalmol.,* 18, 1087–1093, 1935.
284. Hale, F., Relation of maternal vitamin A deficiency to microphthalmia in pigs, *Tex. State J. Med.,* 33, 228–232, 1937.
285. Hales, B. F., Effects of phosphoramide mustard and acrolein, cytotoxic metabolites of cyclophosphamide, on mouse limb development *in vitro, Teratology,* 40, 11–20, 1989.
286. Hall, J. G., Pauli, R. M., and Wilson, R. M., Maternal and fetal sequelae of anticoagulation during pregnancy, *Am. J. Med.,* 68, 122–140, 1980.
287. Hamilton, J. B. and Wolfe, J. M., The effect of male hormone substances upon birth and prenatal development in the rat, *Anat. Rec.,* 70(Suppl. 3), 443–440, 1938.
288. Hanshaw, J. B., Developmental abnormalities associated with congenital cytomegalovirus infection, in *Advances in Teratology,* Vol. 4, Woollam, D. H. M., Ed., Academic Press, New York, 1970, pp. 62–93.
289. Hanson, J. W., Teratogen update: fetal hydantoin effects, *Teratology,* 33, 349–353, 1986.
290. Hanson, J. W., Myrianthopoulos, N. C., Harvey, M. A. S., and Smith, D. W., Risks to the offspring of women treated with hydantoin anticonvulsants, with emphasis on the fetal hydantoin syndrome, *J. Pediatr.,* 89, 662–668, 1976.
291. Hanson, J. W. and Smith, D. W., The fetal hydantoin syndrome, *J. Pediatr.,* 87, 285–290, 1975.
292. Hanssens, M., Keirse, M. J. N. C., Vankelecom, F., and Van Assche, F. A., Fetal and neonatal effects of treatment with angiotensin-converting enzyme inhibitors in pregnancy, *Obstet. Gynecol.,* 79, 128–135, 1991.
293. Harada, M., Congenital Minamata disease: intrauterine methylmercury poisoning, *Teratology,* 18, 285–288, 1979.
294. Harbison, R. D., Proposed mechanism for phenylhydantoin-induced teratogenesis, *Pharmacologist,* 19, 179, 1977.
295. Harbison, R. D. and Becker, B. A., Relation of dosage and time of administration of diphenylhydantoin to its teratogenic effect in mice, *Teratology,* 2, 305–312, 1969.
296. Harlap, S. and Shiono, P. H., Alcohol, smoking and the incidence of spontaneous abortions in the first and second trimester, *Lancet,* 2, 173, 1980.
297. Harpey, J.-P., Jaudon, M.-C., Clavel, J.-P., Galli, A., and Darbois, Y., Cutix laxa and low serum zinc antenatal exposure to penicillamine, *Lancet,* 2, 858, 1983.
298. Harter, C. A. and Benirschke, K., Fetal syphilis in the first trimester, *Am. J. Obstet. Gynecol.,* 124, 705–711, 1976.
299. Hartwig, N. G., Vermeij-Keers, C., Van Elsacker-Niele, A. M. W., and Fleuren, G. J., Embryonic malformations in a case of intrauterine parvovirus B19 infection, *Teratology,* 39, 295–302, 1989.
300. Hartz, S. C., Heinonen, O. P., Shapiro, S., Siskind, V., and Slone, D., Antenatal exposure to meprobamate and chlordiazepoxide in relation to malformations, mental development, and childhood mortality, *N. Engl. J. Med.,* 292, 726–728, 1975.
301. Hawks, R. L., Kopin, I. J., Colburn, R. W., and Thoa, N. B., Norcocaine: a pharmacologically active metabolite of cocaine found in the brain, *Life Sci.,* 15, 2189, 1974.
302. Heinonen, O. P., Sloane, D., Monson, R. R., Hook, E. B., and Shapiro, S., Cardiovascular birth defects and antenatal exposure to female sex hormones, *N. Engl. J. Med.,* 296, 67–70, 1977.
303. Heinonen, O. P., Sloane, D., and Shapiro, S., *Birth Defects and Drugs in Pregnancy,* Publishing Sciences Group, Littleton, MA, 1977.
304. Heinonen, O. P., Sloane, D., and Shapiro, S., *Birth Defects and Drugs in Pregnancy,* Publishing Sciences Group, Littleton, MA, 1977, pp. 516.
305. Henderson, M. G. and McMillen, B., Changes in dopamine, seratonin, and their metabolites in discrete brain areas of rat offspring after *in utero* exposure to cocaine or related drugs, *Teratology,* 48, 421–430, 1993.
306. Hendrickx, A. G., Benirschke, K., Thompson, R. S., Ahern, J. K., Lucas, W. E., and Oi, R. H., The effects of prenatal diethylstilbestrol (DES) exposure on the genitalia of pubertal *Macaca mulatta.* I. Female offspring, *J. Reprod. Med.,* 22, 233–240, 1979.
307. Hendrickx, A. G., Nau, H., Binkerd, P., Rowland, J. M., Rowland, J. R., Cukierski, M. J., and Cukierski, A., Valproic acid developmental toxicity and pharmacokinetics in the rhesus monkey: an interspecies comparison, *Teratology,* 38, 329–345, 1988.

308. Hendrickx, A. G., Nau, H., Binkerd, P., Rowland, J. M., Rowland, J. R., Cukierski, M. J., and Cukierski, M. A., Valproic acid developmental toxicity and pharmacokinetics in the rhesus monkey: an interspecies comparison, *Teratology,* 38, 329–345, 1988.
309. Hendrickx, A. G., Prahalada, S., and Binkerd, P. E., Long-term evaluation of the diethylstilbestrol (DES) syndrome in adult female rhesus monkeys (*Macaca mulatta*), *Reprod. Toxicol.,* 1, 253–261, 1988.
310. Hendrie, T. A., Rowland, J. R., Binkerd, P. E., and Hendrickx, A. G., Developmental toxicity and pharmacokinetics of phenytoin in the rhesus macaque: an interspecies comparison, *Reprod. Toxicol.,* 4, 257–266, 1990.
311. Herbst, A. L. and Anderson, D., Clear cell adenocarcinoma of the vagina and cervix secondary to intrauterine exposure to diethylstilbestrol, *Semin. Surg. Oncol.,* 6, 343–346, 1990.
312. Herbst, A. L., Cole, P., Colton, T., Robboy, S. J., and Scully, R. E., Age incidence and risk of DES-related clear cell adenocarcinoma, *Am. J. Obstet. Gynecol.,* 128, 43–48, 1977.
313. Herbst, A. L., Hubby, M. M., Azizl, F., and Makii, M. M., Reproductive and gynecologic surgical experience in diethylstilbestrol exposed daughter, *Am. J. Obstet. Gynecol.,* 141, 1019, 1981.
314. Herbst, A. L., Hubby, M. M., Blough, R. R., and Azizl, F., A comparison of pregnancy experience in DES-exposed and DES-unexposed daughters, *J. Reprod. Med.,* 24, 62–69, 1980.
315. Herbst, A. L., Kurman, R. J., Scully, R. E., and Poskanzer, D. C., Clear-cell adenocarcinoma of the genital tract in young females, *N. Engl. J. Med.,* 287, 1259–1264, 1972.
316. Herbst, A. L., Poskanzer, D. C., Robboy, S. J., Friedlander, L., and Scully, R. E., Prenatal exposure to stilbestrol: a prospective comparison of exposed female offspring with unexposed controls, *N. Engl. J. Med.,* 292, 334–339, 1975.
317. Herbst, A. L., Robboy, S. J., Scully, R. E., and Poskanzer, D. C., Clear-cell adenocarcinoma of the vagina and cervix in girls: analysis of 170 registry cases, *Am. J. Obstet. Gynecol.,* 119, 713–724, 1981.
318. Herbst, A. L., Scully, R. E., and Robboy, S. J., Effects of maternal DES ingestion on the female genital tract, *Hosp. Pract.,* 10, 51–57, 1975.
319. Herbst, A. L., Ulfelder, H., and Poskanzer, D. C., Adenocarcinoma of the vagina: association of maternal stilbestrol therapy with tumor appearance in young women, *N. Engl. J. Med.,* 284, 878–881, 1971.
320. Hersh, J. H., Toluene embryopathy: two new cases, *J. Med. Genet.,* 26, 333–337, 1989.
321. Hersh, J. H., Podruch, P. E., Rogers, G., and Weisskopf, B., Toluene embryopathy, *J. Pediatr.,* 106, 922–927, 1985.
322. Hertig, A. T., The overall problem in man, in *Comparative Aspects of Reproductive Failure,* Benirschke, K., Ed., Springer-Verlag, Berlin, 1967, pp. 11–41.
323. Heyser, C. J., McKinzie, D. L., Athalie, F., Spear, N. E., and Spear, L. P., Effects of prenatal exposure to cocaine on heart rate and nonassociative learning and retention in infant rats, *Teratology,* 49, 470–478, 1994.
324. Higuchi, Y. and Matsumoto, N., Embryotoxicity of ethanol and acetaldehyde: direct effects of mouse embryo *in vitro, Cong. Anom.,* 24, 9–28, 1984.
325. Hilesmaa, V. K., Teramo, K., Granstrom, M. L., and Bardy, A. H., Fetal head growth retardation associated with maternal antiepileptic drugs, *Lancet,* 2, 165–167, 1981.
326. Hill, R. M., Craig, J. P., Chaney, M. D., Tennyson, L. M., and McCulley, L. B., Utilization of over-the-counter drugs during pregnancy, *Clin. Obstet. Gynecol.,* 20, 281–394, 1977.
327. Hill, R. M., Verland, W. M., Horning, M. G., McCulley, L. B., and Morgan, N. F., Infants exposed *in utero* to antiepileptic drugs, *Am. J. Dis. Child.,* 127, 645–653, 1974.
328. Hjortdal, J. O., Hjortdal, V. E., and Foldspang, A., Tobacco smoking and fetal growth: a review, *Scand. J. Soc. Med.,* 45(Suppl. I-II), 1–22, 1989.
329. Hochner-Celnikier, D., Marandici, A., Iohan, F., and Monder, C., Estrogen and progesterone receptors in the organs of prenatal Cynomolgus monkey and laboratory mouse, *Biol. Reprod.,* 35, 633–640, 1986.
330. Hoepfner, B. A. and Ward, I. L., Prenatal and neonatal androgen exposure interact to affect sexual differentiation in female rats, *Behav. Neurosci.,* 102, 61–65, 1988.
331. Hoffeld, D. R., McNew, J., and Webster, R. L., Effect of tranquilizing drugs during pregnancy on activity of offspring, *Nature,* 218, 357–358, 1968.
332. Hoffman, F., Overzier, C., and Uhde, G., Zur frage der hormonalen erzengung fotaler zwittenbildugen beim menschen, *Geburtshife Frauerheikd,* 15, 1061–1070, 1955.
333. Holder, W. R. and Knox, J. M., Syphilis in pregnancy, *Med. Clin. N. Am.,* 56, 1151–1160, 1972.
334. Howe, A. M. and Webster, W. S., Exposure of the pregnant rat to warfarin and vitamin K1: an animal model of intraventricular hemorrhage in the fetus, *Teratology,* 42, 413–420, 1990.
335. Howse, A. M. and Webster, W. S., The warfarin embryopathy: a rat model showing maxillonasal hypoplasia and other skeletal disturbances, *Teratology,* 46, 379–390, 1992.
336. Hoyme, E. H., Jones, K. L., Van Allen, M. I., Saunders, B. S., and Benirscke, K., The vascular pathogenesis of transverse limb reduction defects, *J. Pediat.,* 101, 839–843, 1982.
337. Hoyme, E. H., Jones, K. L., and Dixon, S. D., Prenatal cocaine exposure and prenatal vascular disruption, *Pediatrics,* 85, 743, 1990.
338. Hsia, D. Y., Phenylketonuria and its variants, in *Progress in Medical Genetics,* Steinberg, A. G. and Bearn, E. G., Eds., Grune and Stratton, New York, 1970, pp. 53–68.

339. Hu, H., Knowledge of diagnosis and reproductive history among survivors of childhood plumbism, *Am. J. Public Health,* 81, 1070–1072, 1991.
340. Hudak, A. and Ungvary, G., Embryotoxic effects of benzene and its methyl derivatives: toluene, xylene, *Toxicology,* 11, 55–63, 1978.
341. Huffman, L. and Hendricks, S. E., Prenatally injected testosterone propionate and sexual behavior of female rats, *Physiol. Behav.,* 26, 773–778, 1981.
342. Hughes, R. N. and Beveridge, I. J., Behavioral effects of prenatal exposure to caffeine in rats, *Life Sci.,* 38, 861–868, 1986.
343. Hughes, R. N. and Beveridge, I. J., Effects of prenatal exposure to chronic caffeine on locomotor and emotional behavior, *Psychobiology,* 15(2), 179–185, 1987.
344. Hurley, L. S. and Tuchmann-Duplessis, H., Influence de la tetracycline sur la developpement pre- et post-natal du rat, *C. R. Acad. Sci.,* 257, 302–304, 1963.
345. Inouye, M., Teratology of heavy metals: mercury and other contaminants, *Congenital Anom.,* 29, 333–344, 1989.
346. Irino, M., Sanada, H., Tashiro, S.-I., Yasuhira, K., and Takeda, T., D-pencillamine toxicity in mice. III. Pathological study of offspring of pencillamine-fed pregnant and lactating mice, *Toxicol. Appl. Pharmacol.,* 65, 273–285, 1982.
347. Jackson, M. J., Drug absorption, in *Drug and Chemical Action in Pregnancy: Pharmacologic and Toxicologic Principles,* Fabro, S. and Scialli, A. R., Eds., Marcel Dekker, New York, 1986, pp. 15–36.
348. Jacobson, A. G. and Brent, R. L., Radioiodine concentration by the fetal mouse thyroid, *Endocrinology,* 65, 408–416, 1959.
349. Jacobson, S. J., Jones, K., Johnson, K., Ceolin, L., Kaur, P., Sahn, D., Donnefeld, A. E., Rieder, M., Santelli, R., Smythe, J., Pastuszak, A., Einarson, T., and Koren, G., Prospective multicentre study of pregnancy outcome after lithium exposure during first trimester, *Lancet,* 339, 530–533, 1992.
350. Janerich, D. T., Dugan, J. M., Standfast, S. J., and Strite, L., Congenital heart disease and prenatal exposure to exogenus sex hormones, *Br. Med. J.,* 1, 1058–1060, 1977.
351. Janerich, D. T., Piper, J. M., and Glebatis, D. M., Oral contraceptives and congenital limb-reduction defects, *N. Engl. J. Med.,* 291, 697–700, 1974.
352. Jankowski, A., Skublicki, S., Wichlinski, L. M., and Szymanski, W., Clinical pharmacokinetic investigations of acetylsalicylic acid in cases of imminent premature delivery, *J. Clin. Hosp. Pharm.,* 10, 361, 1985.
353. Janz, D., Antiepileptic drugs and pregnancy: altered utilization patterns and teratogenesis, *Epilepsia,* 23, S53–S63, 1982.
354. Jeffries, D. J., Intra-uterine and neonatal herpes simplex virus infection, *Scand. J. Infect. Dis. (Suppl.),* 78, 21–26, 1991.
355. Jensh, R. P. and Brent, R. L., Effects of prenatal X-irradiation on postnatal testicular development and function in the Wistar rat: development/teratology/behavior/radiation, *Teratology,* 39, 443–449, 1988.
356. Jensh, R. P. and Brent, R. L., Effects of prenatal X-irradiation on the 14th–18th days of gestation on postnatal growth and development in the rat, *Teratology,* 38, 431–441, 1988.
357. Jick, H., Holmes, L. B., Hunter, J. R., Madsen, S., and Stergachis, A., First trimester drug use and congenital disorders, *J. Am. Med. Assoc.,* 246, 343–346, 1981.
358. Johnson, L. D., Palmer, A. E., King, N. W., and Hertig, A. T., Vaginal adenosis in *Cebus apella* monkeys exposed to DES *in utero, Obstet. Gynecol.,* 57, 629–635, 1981.
359. Jones, G. R. N., Thalidomide: 35 years on and still deforming, *Lancet,* 343, 1041, 1994.
360. Jones, H. G., Intrauterine otoxicity. A case report and review of literature, *J. Natl. Med. Assoc.,* 65, 201–203, 1973.
361. Jones, K. L., Developmental pathogenesis of defects associated with prenatal cocaine exposure: fetal vascular disruption, *Clin. Perinatol.,* 18, 139–146, 1991.
362. Jones, K. L., Lacro, R. V., Johnson, K. A., and Adams, J., Pattern of malformations in the children of women treated with carbamazepine during pregnancy, *New Engl. J. Med.,* 320, 1661–1666, 1989.
363. Jones, K. L., Smith, D. W., Streissguth, A. P., and Myrianthopoulous, N. C., Outcome in offspring of chronic alcoholic women, *Lancet,* 1, 1076–1078, 1974.
364. Jordan, E. K. and Sever, J. L., Fetal damage caused by parvoviral infections, *Reprod. Toxicol.,* 8, 161–189, 1994.
365. Jost, A., Problems of fetal endocrinology: the gonadal and hypophyseal hormones, *Rec. Progr. Horm. Res.,* 8, 379–418, 1953.
366. Juchau, M. R., Bioactivation in chemical teratogenesis, *Ann. Rev. Pharmacol. Toxicol.,* 29, 165–187, 1989.
367. Juchau, M. R. and Rettie, A. E., The metabolic role of the placenta, in *Drug and Chemical Action in Pregnancy: Pharmacologic and Toxicologic Principles,* Fabro, S. and Scialli, A. R., Eds., Marcel Dekker, New York, 1986, pp. 153–169.
368. Jurand, A., Teratogenic activity of lithium carbonate: an experimental update, *Teratology,* 38, 101–111, 1988.
369. Kajii, T., Ferrier, A., Niikawa, N., Takahara, H., Ohama, K., and Avirachan, S., Anatomic and chromosomal anomalies in 639 spontaneous abortions, *Hum. Genet.,* 55, 87, 1980.

370. Kalter, H. and Warkany, J., Experimental production of congenital malformations in mammals by metabolic procedure, *Physiol. Rev.,* 39, 69–115, 1959.
371. Kamm, J. J., Toxicology, carcinogenicity, and teratogenicity of some orally administered retinoids, *J. Am. Acad. Dermatol.,* 6, 652–659, 1982.
372. Kao, J., Brown, N. A., Schmid, B., Goulding, W. H., and Fabro, S., Teratogenicity of valproic acid: *in vivo* and *in vitro* investigations, *Teratogen. Carcinogen. Mutagen.,* 1, 367–376, 1981.
373. Kaplan, L. C., Congenital Dandy Walker malformation associated with first trimester warfarin: a case report and literature review, *Teratology,* 32, 333–337, 1985.
374. Katz, K., Lancet, M., Skornick, J., Chemke, J., Mogilner, B. M., and Klingberg, M., Teratogenicity of progestagens given during the first trimester of pregnancy, *Obstet. Gynecol.,* 65, 775–780, 1985.
375. Kavlock, R. J., Chernoff, N., and Rogers, E. H., The effect of acute maternal toxicity on fetal development in the mouse, *Teratogen. Carcinogen. Mutagens.,* 5, 3–13, 1985.
376. Kay, E. D., Craniofacial dysmorphogenesis following hypervitaminosis A in mice, *Teratology,* 35, 105–117, 1987.
377. Keen, C. L., Mark-Savage, P., Lonnerdal, B., and Hurley, L. S., Teratogenesis and low copper status resulting from D-penicilliamine in rats, *Teratology,* 26, 163–165, 1982.
378. Kelly, T. E., Teratogenicity of anticonvulsant drugs. I. Review of the literature, *Am. J. Med. Genet.,* 19, 413–434, 1984.
379. Keneko, S., Antiepileptic drug therapy and reproductive consequences: functional and morphological effects, *Reprod. Toxicol.,* 5, 179–198, 1991.
380. Kennedy, G., Arnold, D., Keplinger, M. L., and Calandra, J. C., Mutagenic and teratogen studies with lead: acute and tetraethyl lead, *Toxicol. Appl. Pharmacol.,* 19, 370, 1971.
381. Kennedy, L. A., The pathogenesis of brain abnormalities in the fetal alcohol syndrome: an integrating hypothesis, *Teratology,* 29, 363–368, 1984.
382. Kerber, I. J., Warr, O. S., and Richardson, C., Pregnancy in a patient with prosthetic mitral valve, *J. Am. Med. Assoc.,* 203, 223–225, 1968.
383. Khera, K. S., Maternal toxicity of drugs and metabolic disorders — a possible etiologic factor in the intrauterine death and congenital malformation: a critique on human data, *CRC Crit. Rev. Toxicol.,* 17, 345–375, 1987.
384. Khera, K. S. and Nera, E. A., Maternal exposure to methyl mercury and postnatal cerebellar development in mice, *Teratology,* 4, 233, 1971.
385. Kida, M., *Thalidomide Embryopathy in Japan,* Kodnasha, Tokyo, 1987.
386. Kilburn, K. H. and Hess, R. A., Neonatal deaths and pulmonary dysplasia due to D-penicillamine in the rat, *Teratology,* 26, 1–9, 1982.
387. Kimball, A. C., Kean, B. H., and Fuchs, F., The role of toxoplasmosis in abortion, *Am. J. Obstet. Gynecol.,* 111, 219–226, 1971.
388. Kimmel, C. A., Grant, L. D., Sloan, C. S., and Gladen, B. C., Chronic low level toxicity in the rat, *Toxicol. Appl. Pharmacol.,* 56, 28–41, 1980.
389. King, A. G., Threatened and repeated abortion, *Status Ther.,* 1, 104, 1953.
390. King, R. B., Robkin, M. A., and Sheppard, T. H., Distribution of ^{203}Hg in the maternal and fetal rat, *Teratology,* 13, 275–290, 1976.
391. Kline, J., Stein, Z., Susser, M., and Warburton, D., Environmental influences on early reproductive loss in a current New York City study, in *Human Embryonic and Fetal Death,* Porter, I. M and Hook, E. M., Eds., Academic Press, New York, 1980, pp. 225.
392. Knapp, K., Lenz, W., and Nowack, E., Multiple congenital abnormalities, *Lancet,* 2, 725, 1962.
393. Knoche, C. and Konig, J., Zur pranatalen toxizitat von diphenylpyralin-8-chlortheophyllinat unterberucksichtgung von erfahrungen mit thalidomid und caffein, *Arzneimittelforschung,* 14, 415–424, 1964.
394. Knott, P. D., Welply, G. A. C., and Anderson, M. J., Serologically proved intrauterine infection, *Br. Med. J.,* 289, 1660, 1984.
395. Kochhar, D. M. and Penner, J., Developmental effects of isotretinoin and 4-*oxo*-isotretinoin: the role of metabolism in teratogenicity, *Teratology,* 36, 67–75, 1987.
396. Koren, G. and Graham, K., Cocaine in pregnancy: analysis of fetal risk, *Vet. Hum. Toxicol.,* 34, 263–264, 1992.
397. Kostas, J. and Hotchin, J., Behavioral effects of low-level perinatal exposure to toluene in mice, *Neurobehav. Toxicol. Teratol.,* 3, 467–469, 1981.
398. Kriel, R. L., Gates, G. A., Wulff, H., Powell, N., Poland, J. D., and Chin, T. D. Y., Cytomegalovirus isolations associated with pregnancy wastage, *Am. J. Obstet. Gynecol.,* 106, 885–892, 1970.
399. Kronic, J., Phelps, N. E., McCallion, D. J., and Hirsh, J., Effects of sodium warfarin administered during pregnancy in mice, *Am. J. Obstet. Gynecol.,* 118, 819–823, 1974.
400. Kucera, J., Rate and type of congenital anomalies among offspring of diabetic women, *J. Reprod. Med.,* 7, 61–70, 1971.
401. Kullander, S. and Kallen, B., A prospective study of drugs and pregnancy, *Acta Obstet. Gynecol. Scand.,* 55, 25–33, 1976.
402. Kunita, N., Hori, S., Obana, H., Otake, T., Nishimura, H., Kashimoto, T., and Ikegami, N., Biological effect of PCBs and PCDFs present in the oil causing Yusho and Yu-Cheng, *Environ. Health Persp.,* 59, 79–84, 1985.

403. Kurppa, K., Holmberg, P. C., Kuosma, E., and Saxen, L., Coffee consumption during pregnancy, *N. Engl. J. Med.*, 306, 1548, 1982.
404. Kurppa, K., Holmberg, P. C., Kuosma, E., and Saxen, L., Coffee consumption during pregnancy and selected congenital malformations: a nationwide case-control study, *Am. J. Public Health*, 73, 1397–1399, 1983.
405. Kwasigroch, T. E. and Bullen, M., Effects of isotretinoin (13-*cis*-retinoic acid) on the development of mouse limbs *in vivo* and *in vitro*, *Teratology*, 44, 605–616, 1991.
406. Laegreid, L., Hagberg, G., and Lundberg, A., Neurodevelopment in late infancy after prenatal exposure to benzodiazepines — a prospective study, *Neuropediatrics*, 23, 60–67, 1992.
407. Lammer, E. J., Developmental toxicity of synthetic retinoids in humans, *Prog. Clin. Biol. Res.*, 281, 193–202, 1988.
408. Lammer, E. J., Schunior, A., Hayes, A. M., and Holmes, L. B., Isotretinoin dose and teratogenicity, *Lancet*, 2, 503–504, 1988.
409. Lammer, E. J., Sever, L. E., and Oakley, Jr., G. P., Valproic acid, *Teratology*, 35, 465–473, 1987.
410. Landesman-Dwyer, S. and Emanuel, I., Smoking during pregnancy, *Teratology*, 19, 119–126, 1979.
411. Lash, J. W., Studies on the ability of embryonic mesonephros explants to form cartilage, *Develop. Biol.*, 6, 219–232, 1963.
412. Lash, J. W., Normal embryology and teratogenesis, *Am. J. Obstet. Gynecol.*, 90, 1193–1207, 1964.
413. Lash, J. W. and Saxen, L., Human teratogenesis: *in vitro* studies on thalidomide-inhibited chondrogenesis, *Dev. Biol.*, 28, 61–70, 1972.
414. Laurence, K. M., Reply to Gal, *Nature*, 240, 242, 1972.
415. Laurence, K. M., James, N., Miller, M. H., Tennant, G. B., and Campbell, H., Double-blind randomized controlled trial of folate treatment before conception to prevent recurrence of neural tube defects, *Br. Med. J.*, 282, 1509–1511, 1981.
416. Laurence, M., Miller, M. H., Vowles, M., Evans, K., and Carter, C., Hormonal pregnancy tests and neural tube malformations, *Nature*, 233, 495–496, 1971.
417. Leary, F. J., Resseguie, L. J., Kurland, L. T., O'Brien, P. C., Emslander, R. F., and Noller, K. L., Males exposed to diethylstilbestrol, *J. Am. Med. Assoc.*, 252, 2984–2989, 1984.
418. Lenke, R. R. and Levy, H. L., Maternal phenylketonuria and hyperphenylalaninemia, *N. Engl. J. Med.*, 303, 1202–1208, 1980.
419. Lenke, R. R., Turkel, S. B., and Monsen, R., Severe fatal deformities associated with ingestion of excessive isoniazid in early pregnancy, *Acta Obstet. Gynecol. Scand.*, 64, 281–282, 1985.
420. Lenz, W., Thalidomide embryopathy in Germany, 1959–1961, in *Prevention of Physical and Mental Congenital Defects.* Part C. *Basic and Medical Science, Education, and Future Strategies,* Maurois, M., Ed., Alan R. Liss, New York, 1985, pp. 77–83.
421. Lenz, W., A short history of thalidomide embryopathy, *Teratology*, 38, 203–215, 1988.
422. Lenz, W. and Knapp, K., Thalidomide embryopathy, *Arch. Environ. Health*, 5, 100–105, 1962.
423. Leonard, B. E., Alcohol as a social teratogen, *Prog. Brain Res.*, 73, 305–317, 1988.
424. Levine, F. and Muenke, M., VACTERL association with high prenatal lead exposure: similarities to animal models of lead teratogenicity, *Pediatrics*, 87, 390–392, 1991.
425. Lewis, R. B. and Schulman, J. D., Influence of acetylsalicylic acid, an inhibitor of prostaglandin synthesis, on the duration of human gestation and labour, *Lancet*, 2, 1159, 1973.
426. Lewis, W. H. and Suris, O. R., Treatment with lithium carbonate. Results in 35 cases, *Tex. Med.*, 66, 58–63, 1970.
427. Lewit, T., Nebel, L., Terracina, S., and Karman, S., Ethambutol in pregnancy: observations on embryogenesis, *Chest*, 66, 25–26, 1974.
428. Lezcano, L., Antia, D. E., Sahdeve, S., and Jhaveri, M., Crossed renal ectopia associated with maternal alkaloid cocaine abuse: a case report. *J. Perinatol.*, 14, 230–233, 1994.
429. Linares, A., Zarranz, J. J., Rodriguez-Alarcon, J., and Diaz-Perez, J. L., Reversible cutix laxa due to maternal D-penicillamine treatment, *Lancet*, 2, 43, 1979.
430. Lindhout, D., Hoppener, R. J. E., A., and Meinardi, H., Teratogenicity of antiepileptic drug combinations with special emphasis on epoxidation (of carbamazepine), *Epilepsia*, 25, 77–83, 1984.
431. Lindhout, D., Omtzigt, J. G. C., and Cornel, M. C., Spectrum of neural tube defects in 34 infants prenatally exposed to antiepileptic drugs, *Neurology*, 42(Suppl. 5), 111–118, 1992.
432. Lindhout, D., Meinardi, H., Meijer, J. W. A., and Nau, H., Antiepileptic drugs and teratogenesis in two consecutive cohorts: changes in prescription policy paralleled by changes in pattern of malformations, *Neurology*, 42(Suppl. 5), 94–110, 1992.
433. Linn, S., Lieberman, E., Schoenbaum, S. C., Monson, R. R., Stubblefield, P. G., and Ryan, K. J., Adverse outcomes of pregnancy in women exposed to diethylstilbestrol *in utero*, *J. Reprod. Med.*, 33, 3–7, 1988.
434. Linn, S., Schoenbaum, S. C., Monson, R. R., Rosner, B., Stubblefield, P. G., and Ryan, K. J., No association between coffee consumption and adverse outcomes of pregnancy, *N. Engl. J. Med.*, 306, 141–144, 1982.
435. Little, B. B., Santos-Ramos, R., Newell, J. F., and Maberry, M. C., Megadose carbamazepine during the period of neural tube closure, *Obstet. Gynecol.*, 82, 705–708, 1993.

436. Logberg, B., Brun, A., Berlin, M., and Schutz, A., Congenital lead encephalopathy in monkeys, *Acta Neuropathol.*, 77, 120–127, 1988.
437. London, W. T., Levitt, N. H., Kent, S. G., Wong, V. G., and Sever, J. L., Congenital cerebral and ocular malformations induced in rhesus monkeys by Venezuelan equine encephalitis virus, *Teratology*, 16, 285–296, 1977.
438. Loscher, W., Nau, H., Marescaux, C., and Vergnes, M., Comparative evaluation of anticonvulsant and toxic potencies of valproic acid and 2-*en*-valproic acid in different animal models of epilepsy, *Eur. J. Pharmacol.*, 99, 211–218, 1984.
439. Luna-Coelho, H. L., Teixeira, A. C., Santos, A. P., Barros-Forte, E., Macedo-Morais, S., La-Vecchia, C., Tognoni, C., and Herxheimer, A., Misoprostol and illegal abortion in Fortaleza, Brazil, *Lancet*, 341, 1261–1263, 1993.
440. Lutiger, B. K., Graham, K., Einarson, T. R., and Koren, G., Relationship between gestational cocaine use and pregnancy outcome: a meta analysis, *Teratology*, 44, 405–414, 1991.
441. Lynberg, M. C., Khoury, M. J., Lammer, E. J., Waller, K. O., Cordero, J. F., and Erickson, J. D., Sensitivity specificity and positive predictive value of multiple malformations in isotretinoin embryopathy surveillance, *Teratology*, 42, 513–519, 1990.
442. MacGregor, S. N., Keith, L. G., Bachicha, J. A., and Chasnoff, I. J., Cocaine abuse during pregnancy: correlation between prenatal care and perinatal outcome, *Obstet. Gynecol.*, 74, 882–885, 1989.
443. MacGregor, S. N., Keith, L. G., Chasnoff, I. J., Rosner, M. A., Chisum, G. M., Shaw, P., and Minogue, J. P., Cocaine use during pregnancy: adverse perionatal outcome, *Am. J. Obstet. Gynecol.*, 157, 686–690, 1987.
444. Madl, R. and Metzler, M., Oxidative metabolites of diethylstilbestrol in the fetal Syrian golden hamster, *Teratology*, 30, 251–357, 1984.
445. Mark-Savage, P., Keen, C. L., and Hurley, L. S., Reduction by copper supplementation of teratogenic effects of D-penicillamine, *J. Nutr.*, 113, 501–510, 1983.
446. Mark-Savage, P., Keen, C. L., Lonnerdal, B., and Hurley, L. S., Teratogenicity of D-penicillamine in rats, *Teratology*, 23, 50A, 1981.
447. Marks, T. A., Morris, D. F., and Weeks, J. R., Developmental toxicity of alprostadil in rats after subcutaneous administration or intravenous infusion, *Toxicol. Appl. Pharmacol.*, 91, 341–357, 1987.
448. Marselos, M. and Tomatis, L., Diethylstilbestrol. II. Pharmacology, toxicology and carcinogenicity in experimental animals, *Eur. J. Canc.*, 29A, 149–155, 1993.
449. Martin, M. M. and Rento, R. D., Iodide goiter with hypothyroidism in 2 newborn infants, *J. Pediatr.*, 61, 94–99, 1962.
450. Martin, R. A., Jones, K. L., Mendoza, A., Barr, M., and Benirschke, K., Effect of ACE inhibition in the fetal kidney: decreased renal blood flow, *Teratology*, 46, 317–321, 1992.
451. Martin, T. R. and Bracken, M. B., The association between low birth weight and caffeine consumption during pregnancy, *Am. J. Epidemiol.*, 126, 813–821, 1987.
452. Martinez-Frias, M. L., Clinical manifestation of prenatal exposure to valproic acid using case reports and epidemiologic information, *A. J. Med. Genet.*, 57, 277–292, 1990.
453. Martz, F., Failinger, C., and Blake, D. A., Phenytoin teratogenesis: correlation between embryopathic effect and covalent binding of putative arene oxide metabolite in gestational tissue, *J. Pharmacol. Exp. Ther.*, 203, 231–239, 1977.
454. Mast, T. J., Cukierski, M. A., Nau, H., and Hendrickx, A. G., Predicting the human teratogenic potential of the anticonvulsant, Valproic acid, from a nonhuman primate model, *Toxicology*, 39, 111–119, 1986.
455. Matsumoto, H., Koya, G., and Takeuchi, T., Fetal Minamata disease: a neuropathological study of two cases of intrauterine intoxication by a methyl mercury compound, *J. Neuropathol. Exp. Neurol.*, 24, 563–574, 1965.
456. Matsunaga, E. and Shiota, K., Threatened abortion, hormone therapy and malformed embryos, *Teratology*, 20, 69–80, 1979.
457. Matsunaga, E. and Shiota, K., Ectopic pregnancy and myoma uteri: teratogenic effects of maternal characteristics, *Teratology*, 20, 61–69, 1980.
458. Mattison, D. R., Physiologic variations in pharmacokinetics during pregnancy, in *Drug and Chemical Action in Pregnancy: Pharmacologic and Toxicologic Principles*, Fabro, S. and Scialli, A. R., Eds., Marcel Dekker, New York, 1986, pp. 37–102.
459. Mau, G. and Netter, P., Kaffee und alkoholkonsum, risikofaktoren in der schwangerschaft, *Geburtsh Frauenheilkd*, 34, 1018–1022, 1974.
460. Maxwell, S. R. J. and Kendall, M. J., ACE inhibition in the 1990s, *Br. J. Clin. Pract.*, 47, 30–37, 1993.
461. McCarroll, A. M., McCarroll, A. M., Hutchinson, M., McAuley, R., and Montgomery, D. A. D., Long-term assessment of children exposed *in utero* to carbimazole, *Arch. Dis. Child.*, 51, 532, 536, 1976.
462. McClain, R. M. and Becker, B. A., Placental transport and teratogenicity of lead in rats and mice, *Fed. Proc.*, 29, 347, 1972.
463. McClain, R. M. and Langhoff, L., Teratogenicity of diphenylhydantoin in the New Zealand White rabbit, *Teratology*, 21, 371–379, 1980.
464. McCredie, J., Sclerotome subtraction: a radiologic interpretation of reduction deformities of the limbs, in *Birth Defects: Original Article Series*, Bergsman, D. and Lowry, R. B., Eds., Alan R. Liss, New York, 1977, pp. 65–77.

465. McDevitt, J. M., Gautieri, R. F., and Mann, D. E., Comparative toxicity of cortisone and phenytoin in mice, *J. Pharm. Sci.,* 70, 631–634, 1981.
466. McIntosh, I. D., Smoking and pregnancy. II. Offspring risks, *Publ. Health Rev.,* 12, 29–63, 1984.
467. McIntosh, I. D., Smoking and pregnancy. I. Maternal and placental risks, *Publ. Health Rev.,* 12, 1–28, 1984.
468. McKusick, V. A., *Mendalian Inheritance in Man: Catalogs of Autosomal Dominant, Autosomal Recessive, and X-Linked Phentotypes,* 8th ed., Johns Hopkins University Press, Baltimore, MD, 1988.
469. McLaughlin, J. A., Prenatal exposure to diethylstilbestrol in mice: toxicological studies, *J. Toxicol. Environ. Health,* 2, 527–537, 1977.
470. McMichael, A., Vimpani, G., Robertson, E., Baghurst, P., and Clark, P., The Port Pirie cohort study: maternal blood lead and pregnancy outcome, *J. Epidemiol. Commun. Health,* 40, 18–25, 1986.
471. McParland, P., Pearce, J. M., and Chamberlain, G. V. P., Dopler ultrasound and aspirin in recognition and prevenion of pregnancy-induced hypertension, *Lancet,* 335, 1552, 1990.
472. Mead, P. B., Parvovirus B19 infection and pregnancy, *Contemp. Ob. Gyn.,* 34, 56–70, 1989.
473. Medical Research Council, Prevention of neural tube defects: results of the Medical Research Council vitamin study, *Lancet,* 338, 131–137, 1991.
474. Melnick, S., Cole, P., Anderson, D., and Herbst, A., Rates and risks of diethylstilbestrol-related clear-cell adenocarcinoma of the vagina and cervix, *N. Engl. J. Med.,* 316, 514–516, 1987.
475. Menser, M. A. and Reye, R. D. K., The pathology of congenital rubella: a review written by request, *Pathology,* 6, 215–222, 1974.
476. Metzler, M., The metabolism of diethylstilbestrol, *CRC Crit. Rev. Biochem.,* 10, 171–212, 1981.
477. Milham, Jr., S., and Elledge, W., Maternal methimazole and congenital defects in children, *Teratology,* 5, 125, 1972.
478. Milkovich, L. and van den Berg, B. J., Effects of prenatal meprobamate and chlordiazepoxide hydrochloride on human embryonic and fetal development, *N. Engl. J. Med.,* 291, 1268–1271, 1974.
479. Miller, E., Hare, J. W., Coherty, J. P., Dunn, J. P., Gleason, R. E., Soeldnere, J. S., and Kitzmiller, J. L., Elevated maternal hemoglobin A1C in early pregnancy and major congenital anomalies in infants of diabetic mothers, *N. Engl. J. Med.,* 304, 131–1334, 1981.
480. Miller, R. K., Placental transfer and function: the interface for drugs and chemicals in the conceptus, in *Drug and Chemical Action in Pregnancy: Pharmacologic and Toxicologic Principles,* Fabro, S. and Scialli, A. R., Eds., Marcel Dekker, New York, 1986, pp. 123–152.
481. Miller, R. K., Heckman, M. E., and McKenzie, R. C., Diethylstilbestrol: placental transfer, metabolism, covalent binding, and fetal distribution in the Wistar rat, *J. Pharmacol. Exp. Ther.,* 220, 358–365, 1982.
482. Miller, R. W., Delayed radiation effects in atomic bomb survivors, *Science,* 166, 569–574, 1969.
483. Miller, R. W. and Mulvihill, J. J., Small head size after atomic irradiation, *Teratology,* 14, 355–358, 1976.
484. Mills, J. L., Malformations in infants of diabetic mothers, *Teratology,* 25, 385–394, 1982.
485. Mills, J. L., Baker, L., and Goldman, A. S., Malformations in infants of diabetic mothers occur before the seventh gestational week: implications for treatment, *Diabetes,* 28, 292–293, 1979.
486. Milunsky, A., Graef, J. W., and Gaynor, M. F., Methotrexate-induced congenital malformations with a review of the literature, *J. Pediatr.,* 72, 790–795, 1968.
487. Mirkes, P. E., Cyclophosphamide teratogenesis: a review, *Teratogen. Carcinogen. Mutagen.,* 5, 75–88, 1985.
488. Mirkin, B. L., Maternal and fetal distribution of drugs in pregnancy, *Clin. Pharmacol. Ther.,* 14, 643–647, 1973.
489. Mjolnerod, O. K., Rasmussen, K., Dommerud, S. A., and Gjeruldsen, S. T., Congenital connective-tissue defect probably due to D-penicillamine treatment in pregnancy, *Lancet,* 1, 673–675, 1971.
490. Moise, K. J., Indomethacin therapy in the treatment of symptomatic polyhydramnios, *Clin. Obstet. Gynecol.,* 24, 310, 1991.
491. Monnet, P., Kalb, J. C., and Pujol, M., Harmful effects of isoniazid on the fetus and infants, *Lyon Med.,* 218, 431–455, 1967.
492. Monson, R. R., Rosenberg, L., Hartz, S. C., Shapliro, S., Heinonen, O. P., and Sloane, D., Diphenylhydantoin and selected malformations, *N. Engl. J. Med.,* 29, 1049–1052, 1973.
493. Moorman-Voestermans, C. G. M., Heig, H. A., and Vos, A., Jujunal atresia in twins, *J. Pediatr. Surg.,* 25, 638–639, 1990.
494. Moey, A. L., Nikcolini, U., Welch, C. R., Economides, D., Chamberlain, P. F., and Cohen, B. J., Parvovirus B19 infection and transient fetal hydrops, *Lancet,* 337, 496, 1991.
495. Morris, M. B. and Weinstein, L., Caffeine and the fetus — is trouble brewing?, *Am. J. Obstet. Gynecol.,* 140, 607–610, 1981.
496. Mounoud, R. L., Klein, D., and Weber, R., A propos d'un case de syndrome de Goldenhar intoxication aigue a la vitamin A chez la mere pendent la grossesse, *J. Genet. Hum.,* 23, 135–154, 1975.
497. Mujtaba, Q. and Burrow, G. M., Treatment of hyperthyroidism in pregnancy with propylthiouracil and methimazole, *Obstet. Gynecol.,* 46, 282–286, 1975.
498. Murai, N., Effect of maternal medication during pregnancy upon behavioral development of offspring, *Tohoku J. Exp. Med.,* 89, 165–272, 1966.
499. Murakami, U., Organic mercury problem affecting intrauterine life, *Adv. Exper. Biol. Med.,* 27, 301–336, 1972.

500. Murakami, U., Kameyama, Y., and Kato, T., Basic processes seen in disturbances of early development of the central nervous system, *Nagoya J. Med. Sci.*, 17, 74–84, 1954.
501. Murphy, D. P. and Goldstein, L., Micromelia in a child irradiated *in utero*, *Surg. Gynecol. Obstet.*, 50, 79–80, 1930.
502. Murphy, S. J. and Benjamin, C. P., The effects of coffee on mouse development, *Microbios Lett.*, 17, 91–99, 1981.
503. Musselman, A. C., Bennett, G. D., Greer, K. A., Eberwine, J. H., and Finnell, R. H., Preliminary evidence of phenytoin-induced alterations in embryonic gene expression in a mouse model, *Reprod. Toxicol.*, 8, 383–395, 1994.
504. Muther, T. F., Caffeine and reduction of fetal ossification in the rat: fact or artifact?, *Teratology*, 37, 239–247, 1988.
505. Myint, B. A., D-penicillamine-induced cleft palate in mice, *Teratology*, 30, 333–340, 1984.
506. Naeye, R. L., Effects of maternal cigarette smoking on the fetus and placenta, *Br. J. Obstet. Gynecol.*, 85, 732–737, 1979.
507. Naeye, R. L. and Blanc, W. A., Pathogenesis of congenital rubella, *J. Am. Med. Assoc.*, 194, 1277–1283, 1965.
508. Nakamoto, Y., Otani, H., and Tanaka, O., Effects of aminoglycosides administered to pregnant mice on postnatal development of inner ear in their offspring, *Teratology*, 32, 34B, 1985.
509. Nars, P. W. and Girard, J., Lithium carbonate intake during pregnancy leading to large goiter in a premature infant, *Am. J. Dis. Child.*, 131, 924–925, 1977.
510. Nau, H. and Spielmann, H., Embryotoxicity testing of valproic acid, *Lancet*, 1, 763–764, 1983.
511. Nau, H., Zierer, R., Spielmann, H., Neubert, D., and Gansau, C., A new model for embryotoxicity testing: teratogenicity and pharmacokinetics of valproic acid following constant-rate administration in the mouse using human therapeutic drug and metabolite concentrations, *Life Sci.*, 29, 2803–2814, 1981.
512. Needleman, H., Rabinowitz, M., Leviton, A., Linn, S., and Schoenbaum, S., The relationship between prenatal exposure to lead and congenital anomalies, *J. Am. Med. Assoc.*, 251, 2956–2959, 1984.
513. Newbold, R. R., Bullock, B. C., and McLachlan, J. A., Exposure of diethylstilbestrol during pregnancy permanently alters the ovary and oviduct, *Biol. Reprod.*, 28, 735–744, 1983.
514. Newman, N. M. and Correy, J. F., Effects of alcohol in pregnancy, *Med. J. Aust.*, 2, 5–10, 1980.
515. NICHD Workshop, CVS and limb reduction defects, *Teratology*, 48(1), 7–13, 1993.
516. Nicolini, U. and Monni, G., Intestinal obstruction in babies exposed *in utero* to methylene blue, *Lancet*, 336, 1258–1259, 1990.
517. Nieburg, P., Marks, J. S., McLaren, N. M., and Remington, P. L., The fetal tobacco syndrome, *J. Am. Med. Assoc.*, 253, 2998–2999, 1985.
518. Nielsen, M. and Froscher, W., Finger- and toenail hypoplasia after carbamazepine monotherapy in late pregnancy, *Neuropediatrics*, 16, 167–168, 1985.
519. Nishimura, H., Incidence of malformations in abortions, in *Congenital Malformations*, Fraser, F. C. and McKusick, V. A., Eds., Excerpta Medica, Amsterdam, 1970, pp. 275–283.
520. Nishimura, H. and Nakai, K., Congenital malformations in offspring of mice treated with caffeine, *Proc. Soc. Exp. Biol. Med.*, 104, 140–142, 1960.
521. Nishimura, H. and Tanimura, T., *Clinical Aspects of the Teratogenicity of Drugs*, Excerpta Medica, New York, 1976.
522. Nishimura, H., Uwabe, C., and Semba, R., Examination of teratogenicity of progestogens and/or estrogens by observation of the induced abortuses, *Teratology*, 10, 93, 1974.
523. Nolen, G. A., A reproduction/teratology study of decaffeinated coffees, *Toxicologist*, 1, 104, 1981.
524. Nolen, G. A., The developmental toxicology of caffeine, in *Issues and Reviews in Teratology*, Vol. 14, Kalter, H., Ed., Plenum Press, New York, 1989, pp. 305–350.
525. Nomura, T., Kimura, S., Kanzaki, T., Tanaka, H., Shibata, K., Nakajima, H., Isa, Y., Kurokawa, N., Hatanaka, T., and Kinuta, M., Induction of tumours and malformations in mice after prenatal treatment with some antibiotic drugs, *Med. J. Osaka Univ.*, 35, 13–17, 1984.
526. O'Brien, P. C., Noller, K. L., Robboy, S. J., Barnes, A. B., Kaufman, R. H., Tilley, B. C., and Townsend, D. E., Vaginal epithelial changes in young women enrolled in the National Cooperation Diethylstilbestrol Adenosis (DESAD) Project, *Obstet. Gynecol.*, 53, 300–308, 1979.
527. O'Malley, B. W. and Schrader, D. T., The receptors of steroid hormones, *Sci. Am.*, 234, 32–43, 1976.
528. Occupational Safety and Health Administration (OSHA), Occupational exposure to lead: final standard, *Fed. Reg.*, 43, 52952–52960, 1978.
529. Omtzigt, J. G. C., Los, F. J., Grobbee, D. E., Pijpers, L., Jahoda, M. G., Brandenburg, H., Stewart, P. A., Gaillar, H. L., Sachs, E. S., and Wladimiroff, J. W., The risk of spina bifida aperta after first-trimester exposure to valproate in a prenatal cohort, *Neurology*, 42(Suppl. 5), 119–125, 1992.
530. Omtzigt, J. G. C., Nau, H., Los, F. J., Pijpers, L., and Lindhout, D., The disposition of valproate and its metabolites in the late first trimester and early second trimester of pregnancy in maternal serum, urine, and amniotic fluid: effect of dose, co-medication, and the presence of spina bifida, *Eur. J. Clin. Pharmacol.*, 43, 381–388, 1992.

531. Ong, D. E. and Chytil, F., Changes in levels of cellular retinol- and retinoic-acid-binding proteins of liver and lung during perinatal development rat, *Proc. Natl. Acad. Sci. USA,* 73, 3976–3978, 1976.
532. Ornoy, A., Benady, S., Kohen-Raz, R. and Russell, A., Association between maternal bleeding during gestation and congenital anomalies, *Am. J. Obstet. Gynecol.,* 124, 474–478, 1976.
533. Otake, M., Schull, W. J., and Neel, J. V., Congenital malformations, stillbirths and early mortality among the children of atomic bomb survivors: a reanalysis, *Radiat. Res.,* 122, 1–11, 1990.
534. Palm, P. E., Arnold, E. P., Rachwall, P. C., Leyczek, J. C., Teague, K. W., and Kensler, C. J., Evaluation of the teratogenic potential of fresh-brewed coffee and caffeine in the rat, *Toxicol. Appl. Pharmacol.,* 44, 1–16, 1978.
535. Pampfer, S. and Streffer, C., Prenatal death and malformations after irradiation of mouse zygotes with neutrons or X-rays, *Teratology,* 37, 599–607, 1988.
536. Pauli, R. M., Lian, J. B., and Mosher, D. F., Association of congenital deficiency of multiple vitamin K-dependent coagulation factors and the phenotype of the warfarin embryopathy: clues to the mechanism of teratogenicity of coumarin derivatives, *Am. J. Hum. Genet.,* 41, 566–583, 1987.
537. Paulson, R. B., Paulson, G. W., and Jreissaty, S., Phenytoin and carbamazepine in production of cleft palates in mice. Comparison of teratogenic effects, *Arch. Neurol.,* 36, 832–836, 1979.
538. Pearson, M. A., Hoyme, H. E., Seaver, L. H., and Rimsza, M. E., Toulene embryopathy: delineation of the phenotype and comparison with fetal alcohol syndrome, *Pediatrics,* 93, 211–215, 1994.
539. Peckham, C. H., Uterine bleeding during pregnancy. 1. When not followed by immediate termination of pregnancy, *Obstet. Gynecol.,* 35, 937–941, 1970.
540. Pedersen, J., Congenital malformations, in *The Pregnant Diabetic and Her Newborn Infant,* Munksgard, Copenhagen, 1977, pp. 191–196.
541. Peters, H. and Theorell, C. J., Fetal and neonatal effects of maternal cocaine use, *J. Obstetr. Gynecol. Neonatal Nursing,* 20, 121–126, 1990.
542. Pettiflor, J. M. and Benson, R., Congenital malformations associated with the administration of oral anticoagulants during pregnancy, *J. Pediatr.,* 86, 459–462, 1975.
543. Phibbs, C. S., Bateman, D. A., and Schwartz, R. M., The neonatal costs of maternal cocaine use, *J. Am. Med. Assoc.,* 266, 1521–1526, 1991.
544. PHLS, Public Health Laboratory Service working on fifth disease. Prospective study of human parvovirus (B19) infection in pregnancy, *Br. Med. J.,* 300, 1166–1170, 1990.
545. Phoenix, C. H., Goy, R. W., Gerall, A. A., and Young, W. C., Organizing action of prenatally administered testosterone propionate on the tissues mediating mating behavior in the female guinea pig, *Endocrinology,* 65, 369–382, 1959.
546. Piersma, A. H., Verhoef, A., DeLiefde, A., van Nesselrooij, B. P. M., and Garbis-Berkvens, J. M., Embryotoxocity of methylene blue in the rat, *Teratology,* 43, 458–459, 1991.
547. Pilotti, G. and Scorta, A., Hypervitaminosis A during pregnancy and neonatal malformation of the urinary apparatus, *Minerva Gynecol.,* 17, 1103–1108, 1965.
548. Piper, J. M., Ray, W. A., and Rosa, F. W., Pregnancy outcome following exposure to angiotensin converting enzyme inhibitors, *Obstet. Gynecol.,* 80, 429–432, 1992.
549. Pitel, M. and Lerman, S., Further studies on the effects of intrauterine vasoconstrictors on the fetal rat lens, *Am. J. Ophthalmol.,* 58, 464–470, 1964.
550. Plessinger, M. A. and Woods, J. R., Maternal, placental, and fetal pathophysiology of cocaine exposure during pregnancy, *Clin. Obstet. Gynecol.,* 36, 267–278, 1993.
551. Potworowska, M., Sianoz-Ecka, E., and Szufladowicz, R., Treatment with ethionamide in pregnancy, *Gruzlica,* 34, 341–347, 1966.
552. Powell, H. R. and Ekert, H., Methotrexate-induced congenital malformations, *Med. J. Aust.,* 2, 1076–1077, 1971.
553. Prager, K., Malin, H., Speigler, D., Van Natta, P., and Placek, P., Smoking and drinking behavior before and during pregnancy of married mothers of liveborn and stillborn infants, *Publ. Health Rep.,* 99, 117–127, 1984.
554. Preblud, S. R. and Alford, C. A., Rubella, in *Infectious Diseases of the Fetus and Newborn Infant,* Remington, J. S. and Klein, J. O., Eds., W.B. Saunders, Philadelphia, 1990, pp. 196–240.
555. Preblud, S. R., Cochi, S. L., and Orenstein, W. A., Varicella-zoster infection in pregnancy (letter to the editor), *N. Engl. J. Med.,* 315, 1415–1418, 1986.
556. Pryde, P. G., Sedman, A. B., Nugent, C. E., and Barr, B., Angiotensin-converting enzyme inhibitor fetopathy, *J. Am. Soc. Nephrol.,* 3, 1575–1582, 1993.
557. Qazi, Q. H., Sheikh, T. M., Fikrig, S., and Menikoff, H., Lack of evidence for craniofacial dysmorphism in perinatal human immunodeficiency virus infection, *J. Pediatr.,* 112, 7–11, 1988.
558. Rabe, R., French, J. H., Sinha, B., and Fersko, R., Functional consequences of prenatal exposure to lead in immature rats, *Neurotoxicology,* 6, 43–54, 1985.
559. Rabe, R., Basse, H., Thuro, H., Kiesel, L., and Runnebaum, B., Wirkung des PGE1-methylanalogens misoprostol auf den schwangeren uterus im erstentrimester, *Geburtsch Frauenheilk,* 47, 324–331, 1987.
560. Rane, A., Tomson, G., and Bjarke, B., Effects of maternal lithium therapy in a newborn infant, *J. Pediatr.,* 93, 296–297, 1974.

561. Ransome-Kuti, O., Malaria in childhood, *Adv. Pediatr.,* 19, 319–340, 1972.
562. Rantakallio, P., The effect of maternal smoking on birth weight and the subsequent health of the child, *Early Hum. Dev.,* 2, 371–382, 1978.
563. Raynaud, A., Observations dur de development normal des ebauches de la glande mammaire des foetus males et femelles de souris, *Ann. Endorinol.,* 8, 349–359, 1947.
564. Rebich, T., Kumar, J., and Brustman, B., Dental caries and tetracycline-stained dentition in an American-Indian population, *J. Dent. Res.,* 64, 462–464, 1985.
565. Reece, E. A., Assimaklpoulos, E., Zheng, X., Hasgay, Z., and Hobbins, J. C., The safety of obstetric ultrasonography: concern for the fetus, *Obstet. Gynecol. Rev.,* 76, 139–145, 1990.
566. Reiter, L. W., Anderson, G. E., Laskey, J. W., and Cahill, D. F., Developmental and behavioral changes in the rat during chronic exposures to lead, *Environ. Health Persp.,* 12, 119–123, 1975.
567. Rementeria, J. L. and Bhatt, K., Withdrawal symptoms in neonates from intrauterine exposure to diazepam, *J. Pediatr.,* 90, 123–126, 1977.
568. Remington, J. S. and Desmonts, G., Toxoplasmosis, in *Infectious Diseases of the Fetus and Newborn Infant,* Remington, J. S. and Klein, J. O., Eds., W.B. Saunders, Philadelphia, 1990, pp. 89–195.
569. Reschini, E., Giustina, G., D'Alberton, A., and Candiani, G. B., Female pseudohermaphroditism due to maternal androgen administration: 25-year follow-up, *Lancet,* 1, 1226, 1985.
570. Rice, D. C. and Gilbert, S. G., Automated behavioral procedures for infant monkeys, *Neurotoxicol. Teratol.,* 12, 429–439, 1990.
571. Richardson, G. A. and Day, N. L., Maternal and neonatal effects of moderate cocaine use during pregnancy, *Neurotoxicol. Teratol.,* 13, 455–460, 1991.
572. Rinck, G., Gollnick, H., and Organos, C. E., Duration of contraception after etretinate, *Lancet,* 1, 845–846, 1989.
573. Ritchie, J. M. and Greene, N. M., Local anesthetics, *The Pharmacological Basis of Therapeutics,* 7th ed., Tillman, A. G., Goodman, L. S., Rall, W., and Murad, F., Eds., MacMillan Publishing, New York, 1985, p. 309.
574. Rizzo, R., Lammer, E. J., Parano, E., Pavone, L., and Argyle, J. C., Limb reduction defects in humans associated with prenatal isotretinoin exposure, *Teratology,* 44, 599–604, 1991.
575. Robert, C. J. and Lowe, C. R., Where have all the conceptions gone?, *Lancet,* 1, 498–499, 1975.
576. Robert, E., Valproic acid and spina bifida: a preliminary report — France, *M.M.W.R.,* 31, 515–566, 1982.
577. Robert, E., Valproic acid as a human teratogen, *Congen. Anom.,* 28(Suppl.), S71–S80, 1988.
578. Robert, E. and Guibaud, P., Maternal valproic acid and congenital neural tube defects, *Lancet,* 2, 1142, 1982.
579. Robert, E., Lofkvist, E., and Mauguiere, F., Valproate and spina bifida, *Lancet,* 2, 1392, 1984.
580. Robert, E., and Rosa, F., Valproate and birth defects, *Lancet,* 2, 1142, 1983.
581. Rock, J. A., Wentz, A. C., Cole, K. A., Kimball, Jr., A. W., Sacur, H. A., Early, S. A., and Jones, C. S., Fetal malformations following progesterone therapy during pregnancy: a preliminary report, *Fertil. Steril.,* 44, 17–19, 1985.
582. Rodesch, F., Camus, M., Ermans, A. M., Dodion, J., and Delange, F., Adverse effect of amniofetography on fetal thyroid function, *Am. J. Obstet. Gynecol.,* 126, 723–726, 1976.
583. Rogan, W. J., PCBs and cola-colored babies: Japan, 1968, and Taiwan, 1979, *Teratology,* 26, 259–261, 1982.
584. Rogan, W. J., Gladen, B. C., McKiney, J. D., Carreras, N., Hardy, P., Thullen, J., Tinglestad, J., and Tully, M., Neonatal effects of transplacental exposure to PCBs and DDE, *J. Pediatr.,* 109, 335–341, 1986.
585. Rosa, F. W., Teratogen update: Penicillamine, *Teratology,* 33, 127–131, 1986.
586. Rosa, F. W., Isotretinoin dose and teratogenicity, *Teratology,* 39, 341–348, 1989.
587. Rosa, F. W., Spina bifida in infants of women treated with carbamazepine during pregnancy, *N. Engl. J. Med.,* 10, 674–677, 1991.
588. Rose, N. C. and Mennuti, M. T., Periconceptional folate supplementation and neural tube defects, *Clin. Obstet. Gynecol.,* 37, 605–620, 1994.
589. Rosenberg, L., Mitchell, A. A., Shapiro, S., and Slone, D., Selected birth defects in relation to caffeine-containing beverages, *J. Am. Med. Assoc.,* 247, 14219–1432, 1982.
590. Rothman, K. J., Fyler, D. C., Glodblatt, A., and Kreidberg, M. B., Exogenous hormones and other drug exposures of children with congenital heart disease, *Am. J. Epidemiol.,* 109, 433–439, 1979.
591. Roubenoff, R., Hoyt, J., Petri, M., Hochberg, M. C., and Hellmann, D. B., Effects of antiinflammatory and immunosuppressive drugs on pregnancy and fertility, *Semin. Arthr. Rheum.,* 18, 88–110, 1988.
592. Rousseaux, C. G. and MacNabb, L. G., Oral administration of D-penicillamine causes neonatal mortality without morphological defects in CD-1 mice, *J. Appl. Toxicol.,* 12, 35–38, 1992.
593. Rowland, J. F., Binkerd, P. E., and Hendrickx, A. G., Developmental toxicity and pharmacokinetics of oral and intravenous phenytoin in the rat, *Reprod. Toxicol.,* 4, 191–202, 1990.
594. Ruffing, L., Evaluation of thalidomide children, *Birth Defects,* 13, 287–300, 1977.
595. Runner, M. N. and Miller, J. R., Congenital deformity in the mouse as a consequence of fasting, *Anat. Rec.,* 124, 437–438, 1956.
596. Russell, L. B., X-ray indiced developmental abnormalities in the mouse and their use in the analysis of embryological patterns. I. External and gross visceral changes, *J. Exp. Zool.,* 114, 545–602, 1950.

597. Russell, L. B., X-ray induced developmental abnormalities in the mouse and their use in the analysis of embryonical patterns. II. Abnormalities of the vertebral column and thorax, *J. Exp. Zool.*, 131, 329–395, 1956.
598. Russell, L. B. and Russell, W. L., The effects of radiation on the preimplantation stages of the mouse embryo, *Anat. Res.*, 108, 521, 1950.
599. Russell, L. B. and Russell, W. L., An analysis of the changing radiation response of the developing mouse embryo, *J. Cell Comp. Physiol.*, 43, 103–149, 1954.
600. Sadler, T. W., Effects of maternal diabetes on early embryogenesis. I. The teratogenic potential of diabetic serum, *Teratology*, 21, 339–347, 1980.
601. Safra, M. J. and Oakley, G. P., Association between cleft lip with or without cleft palate and prenatal exposure to diazepam, *Lancet*, 2, 478–479, 1975.
602. Saito, H., Kobayashi, H., Takeno, S., and Sakai, T., Fetal toxicity of benzodiazepines in rats, *Res. Commun. Chem. Pathol. Pharmacol.*, 46, 437–447, 1984.
603. Salamy, A., Dark, K., Salfi, M., Shah, S., and Peeke, H. V. S., Perinatal cocaine exposure and functional brain stem development in the rat, *Brain Res.*, 598, 307–310, 1992.
604. Salazar, E., Sajarias, A., Gutierrez, N., and Iturbe, I., The problem of cardiac valve prostheses, anticoagulants and pregnancy, *Circulation*, 70(Suppl. I), I169–I177, 1984.
605. Sandberg, E. C., Riffle, N. L., Higdon, J. V., and Getman, C. E., Pregnancy outcome in women exposed to diethylstilbestrol *in utero*, *Am. J. Obstet. Gynecol.*, 110, 194–205, 1981.
606. Sareli, P., England, J. M., and Berk, M. R., Maternal and fetal sequelae of anticoagulation during pregnancy in patients with mechanical heart valve prosthesis, *Am. J. Cardial.*, 63, 1462–1465, 1989.
607. Saxen, I., Assocation between oral clefts and drugs taken during pregnancy, *Int. J. Epidemiol.*, 4, 37–44, 1975.
608. Schaffer, A. J., *Diseases of the Newborn*, 2nd ed., W.B. Saunders, Philadelphia, 1965, pp. 733–734.
609. Schardein, J. L., *Chemically Induced Birth Defects*, Marcel Dekker, New York, 1993.
610. Schonhofer, P. S., Brazil: misuse of misoprostol as an abortifacient may induce malformations, *Lancet*, 337, 1534, 1991.
611. Schuler, L. S., Ashton, P. W., and Sanseverino, M. T., Teratogenicity of misoprostol, *Lancet*, 339, 437, 1992.
612. Schull, W. J., Norton, S., and Jensh, R. P., Ionizing radiation and the developing brain, *Neurotoxicol. Teratol.*, 12, 249–260, 1990.
613. Schwartz, J., Low-level lead exposure and children's IQ: a meta-analysis and search for a threshold, *Environ. Res.*, 64, 42–55, 1994.
614. Scott, J. R., Fetal growth retardation associated with maternal administration of immunosuppressive drugs, *Am. J. Obstet. Gynecol.*, 128, 668–676, 1977.
615. Seegmiller, R. E., Carter, M. W., Ford, W. H., and White, R. D., Induction of maternal toxicity in the rat by dermal application of retinoic acid and its effect on fetal outcome, *Reprod. Toxicol.*, 4, 277–281, 1990.
616. Seegmiller, R. E., Nelson, B. W., and Johnson, C. K., Evaluation of the teratogenic potential of Delalutin (17α-hydroxyprogesterone caproate) in mice, *Teratology*, 28, 201–208, 1983.
617. Seidler, F. J. and Slotkin, T. A., Fetal cocaine exposure causes persistent noradrenergic hyperactivity in rat brain regions: effects on neurotransmitter turnover and receptors, *J. Pharmacol. Exp. Ther.*, 263, 413–421, 1992.
618. Seidler, F. J. and Slotkin, T. A., Prenatal cocaine and cell development in rat brain regions: effect on ornithine decarboxylase and macromolecules, *Brain Res. Bull.*, 30, 91–99, 1993.
619. Serota, F. T., Bernhaum, J. C., and Schwartz, E., The methylene blue baby, *Lancet*, 2, 1142–1143, 1979.
620. Sever, J. L., Infections in pregnancy: highlights from the collaborative perinatal project, *Teratology*, 25, 227–237, 1982.
621. Sever, J. L., Hueber, R. J., Castellano, G. A., and Bell, J. A., Serological diagnosis "en masse" with multiple antigens, *Am. Rev. Resp. Dis. Suppl.*, 88, 342–359, 1962.
622. Sever, L. E., Hormonal pregnancy tests and spina bifida, *Nature*, 242, 410–411, 1973.
623. Shapiro, S., Slone, D., Hartz, S. C., Rosenberg, L., Siskind, V., Monson, R. R., Mitchell, A. A., and Heinonen, O. P., Anticonvulsants and parental epilepsy in the development of birth defects, *Lancet*, 1, 272–275, 1976.
624. Sheikhazadeh, A., Ghabusi, P., Hamim, S. H., and Tarbiat, S., Congestive heart failure in valvular heart disease in pregnancies with and without valvular prostheses and anticoagulant therapy, *Clin. Cardiol.*, 6, 465–470, 1983.
625. Shmoys, S. and Kaplan, C., Parvovirus and pregnancy, *Clin. Obstet. Gynecol.*, 33, 268–265, 1990.
626. Shy, K. K., Stenchever, M. A., Karp, L. E., Berger, R. E., Williamson, R. A., and Leonard, J., Genital tract examinations and zona-free hamster egg penetration tests from men exposed *in utero* to diethylstilbestrol, *Fertil. Steril.*, 42, 772–778, 1984.
627. Sikov, M. R. and Noonan, T. R., Anomalous development induced in embryonic rat by the maternal administration of radiophosphorous, *Am. J. Anat.*, 103, 137, 1958.
628. Simonik, D. K., Robinson, S. R., and Smotherman, W. P., Cocaine alters behavior in the rat fetus, *Behav. Neurosci.*, 107, 867–875, 1993.
629. Simpson, J. L., Genes, chromosomes and reproductive failure, *Fertil. Steril.*, 33, 116–778, 1980.
630. Singer, L., Arendt, R., and Minnes, S., Neurodevelopmental effects of cocaine, *Clin. Perinatol.*, 20, 245–262, 1993.

631. Singer, L. T., Yamashita, T. S., Hawkins, S., Cairns, D., Baley, J., and Kliegman, R., Increased incidence of intraventricular hemorrhage and developmental delay in cocaine-exposed, very low birth weight infants, *J. Pediatr.,* 124, 765–771, 1994.
632. Singh, S., The teratogenicity of cyclophosphamide (Endoran-Asta) in rats, *Indian J. Med. Res.,* 59, 1128–1135, 1971.
633. Skalko, R. G. and Gold, M. P., Teratogenicity of methotrexate in mice, *Teratology,* 9, 159–164, 1974.
634. Skipper, H. T. and Schabel, Jr., F. M., Quantitative and cytokinetic studies in experimental tumor models, in *Cancer Medicine,* Holland, J. F. and Frei, III, E., Eds., Lea & Febiger, Philadelphia, 1973, pp. 629–650.
635. Slutsker, L., Risk associated with cocaine use during pregnancy, *Obstet. Gynecol.,* 79, 778–789, 1992.
636. Smith, D. W., Teratogenicity of anticonvulsive medications, *Am. J. Dis. Chil.,* 131, 1337–1339, 1977.
637. Smith, E. S. O., Dafoe, C. S., Miller, J. R., and Banister, P., An epidemiological study of congenital reduction deformities of the limbs, *Br. J. Prev. Soc. Med.,* 31, 39–41, 1977.
638. Smithells, R. W., Defects and disabilities of thalidomide children, *Br. Med. J.,* 1, 269–272, 1973.
639. Smithells, R. W., Seller, M. J., Nevin, N. C., Sheppard, S., Harris, R., Read, A. P., Fielding, D. W., Walker, S., Schorah, C. J., and Wild, J., Further experience of vitamin supplementation for prevention of neural tube defect recurrences, *Lancet,* 1, 1027–1031, 1983.
640. Smithells, R. W., Sheppard, S., Schorah, C. J., Seller, M. J., Nevin, N. C., Harris, R., Read, A. P., and Fielding, D. W., Apparent prevention of neural tube defects by periconceptional vitamin supplementation, *Arch. Dis. Child.,* 56, 911, 1981.
641. Smithells, R. W., Sheppard, S., Wild, J., and Schorah, C. J., Prevention of neural tube defect recurrences in Yorkshire: final report, *Lancet,* 2, 498–499, 1989.
642. Snider, D. E., Layde, P. M., Johnson, M. W., and Lyle, M. A., Treatment of tuberculosis during pregnancy, *Am. Rev. Respir. Dis.,* 122, 65–79, 1980.
643. Synder, R. D., Congenital mercury poisoning, *N. Engl. J. Med.,* 284, 1014–1016, 1971.
644. Sobotka, T. J., Spaid, S. L., and Brodie, R. E., Neurobehavioral teratology of caffeine exposure in rats, *Neurotoxicology,* 1, 403–416, 1979.
645. Solomon, L., Abrams, G., Dinner, M., and Berman, L., Neonatal abnormalities associated with D-penicillamine treatment during pregnancy, *N. Engl. J. Med.,* 296, 54–55, 1977.
646. Sonawane, B. R. and Yaffe, S. J., Physiologic disposition of drugs in the fetus and newborn, in *Drug and Chemical Action in Pregnancy: Pharmacologic and Toxicologic Principles,* Fabro, S. and Scialli, A. R., Eds., Marcel Dekker, New York, 1986, pp. 103–121.
647. Sosa Munoz, J. L., Perez-Santana, M. T., Sosa Sanchez, R., and Labardini, J. R., Acute leukemia and pregnancy, *Rev. Invest. Clin. (Mex.),* 35, 55–58, 1983.
648. South, J., The effect of vaginal bleeding in early pregnancy on the infant born after the 28th week of pregnancy, *J. Obstet. Gynec.,* 80, 236–241, 1973.
649. South, M. A. and Sever, J. L., Teratogen update: the congenital rubella syndrome, *Teratology,* 31, 297–307, 1985.
650. South, M. A., Thompkins, W. A. F., Morris, C. R., and Rawls, W. E., Congenital malformation of the central nervous system associated with genital type (type 2) herpes virus, *J. Pediatr.,* 75, 13–18, 1969.
651. Spear, L. P., Kirstein, C. L., Bell, J., Yoottanasumpun, V., Greenbaum, R., O'Shea, J., Hoffman, H., and Spear, N. E., Effects of prenatal cocaine exposure on behavior during the early postnatal period, *Neurotoxicol. Teratol.,* 11, 57–63, 1989.
652. Speert, H., Quimbi, E. H., and Werner, S. C., Radioiodine uptake by the fetal mouse thyroid and resultant effects in later life, *Surg. Gynecol. Obstet.,* 93, 230–242, 1951.
653. Speidel, B. D. and Meadow, S. R., Maternal epilepsy and abnormalities of the fetus and newborn, *Lancet,* 2, 839–843, 1972.
654. Speilberg, S. P., Gordon, G. B., Blake, D. A., Mellits, E. D., and Bross, D. S., Anticonvulsant toxicity *in vitro*: possible role of arene oxides, *J. Pharmacol. Exp. Therap.,* 217, 386–389, 1981.
655. Spiller, G. A., The chemical components of coffee, *Prog. Clin. Biol. Res.,* 158, 47–91, 1984.
656. Spyker, J. M. and Smithberg, M., Effects of methyl mercury on prenatal development in mice, *Teratology,* 5, 181–190, 1972.
657. Srisuphan, W. and Bracken, M. B., Caffeine consumption during pregnancy and association with late spontaneous abortion, *Am. J. Obstet. Gynecol.,* 154, 14–20, 1986.
658. Stagno, S., Cytomegalovirus, in *Infectious Diseases of the Fetus and Newborn Infant,* Remington, J. S. and Klein, J. O., Eds., W.B. Saunders, Philadelphia, 1990, pp. 241–281.
659. Stagno, S., Pass, R. F., Dworsky, M. E., Britt, M. J., and Alford, C. A., Congenital and perinatal cytomegalovirus infections: clinical characteristics and pathogenic factors, *Birth Defects,* 20, 65–85, 1984.
660. Steen, J. S. M. and Stainton-Eldis, D. M., Rifampicin in pregnancy, *Lancet,* 2, 604–605, 1977.
661. Steffek, A. J., Verrusio, A. C., and Watkins, C. A., Cleft palate in rodents after maternal treatment with various lathrogenic agents, *Teratology,* 5, 33–40, 1972.
662. Stein, Z. Z., Susser, M., Saenger, G., and Marollla, F., *Famine and Human Development. The Dutch Hunger Winter of 1944/45,* Oxford University Press, New York, 1975.

663. Stenflo, J. and Suttie, J. W., Vitamin K-dependent formation of gamma-carboxyglutamic acid, *Ann. Rev. Biochem.*, 46, 157–172, 1977.
664. Stephens, T. D., Proposed mechanisms of action in thalidomide embryopathy, *Teratology*, 38, 229–239, 1988.
665. Stephens, T. D. and McNulty, T. R., Evidence for a metameric pattern in the development of the chick humerus, *J. Emb. Exp. Morphol.*, 61, 191–205, 1981.
666. Stevenson, R. E., Burtonk, M., Furlauto, G. J., and H.A., T., Hazards of oral anticoagulants during pregnancy, *J. Am. Med. Assoc.*, 243, 1549–1551, 1980.
667. Stevenson, S. S., Worchester, V., and Rice, R. G., Congenitally malformed infants and associated gestational characteristics, *Pediatrics*, 6, 37, 1950.
668. Stillman, R. J., Rosenberg, M. J., and Sachs, B. P., Smoking and reproduction, *Fertil. Steril.*, 46, 545–566, 1986.
669. Stoffer, S. S. and Hamber, J. I., Inadvertent ^{131}I therapy for hyperthyroidism in the first trimester of pregnancy, *J. Nucl. Med.*, 17, 146–149, 1976.
670. Strecker, T. R. and Stephens, T. D., Peripheral nerves do not play a trophic role in limb skeletal morphogenesis, *Teratology*, 27, 159–167, 1983.
671. Streissguth, A. P., Grant, T. M., Barr, H. M., Brfown, Z. A., Martin, J. C., Mayock, D. E., Ramey, S. L., and Moore, L., Cocaine and the use of alcohol and other drugs during pregnancy, *Am. J. Obstet. Gynecol.*, 164, 1239–1243, 1991.
672. Streissguth, A. P., Landesman-Dwyer, C., Martin, J. C., and Smith, D. W., Teratogenic effects of alcohol in humans and laboratory animals, *Science*, 209, 353–361, 1980.
673. Streissguth, A. P., Sampson, P. D., and Marr, H. M., Neurobehavioral dose-response effects of prenatal alcohol exposure in humans from infancy to adulthood, *Ann. N. Y. Acad. Sci.*, 562, 145–258, 1989.
674. Sulik, K. K. and Dehart, D. B., Retinoic-acid-induced limb malformations resulting from apical ectodermal ridge cell death, *Teratology*, 37, 527–537, 1988.
675. Sulik, K. K. and Johnston, M. C., Acute ethanol administration in an animal model results in craniofacial features characteristic of the fetal alcohol syndrome, *Science*, 214, 936–938, 1982.
676. Sulik, K. K. and Johnston, M. C., Sequence of developmental alterations following acute ethanol exposure in mice: craniofacial features of the fetal alcohol syndrome, *Am. J. Anat.*, 166, 257–269, 1983.
677. Sulik, K. K., Johnston, M. C., and Dehart, D. B., Potentiation of programmed cell death by 13-*cis*-retinoic acid: a common mechanism for early craniofacial and limb malformation?, *Teratology*, 35, 32A, 1987.
678. Szabo, K. and Brent, R. L., Species differences in experimental teratogenesis by tranquilising agents, *Lancet*, 1, 565, 1974.
679. Taylor, D. H., Noland, E. A., Brubaker, C. M., Crofton, K. M., and Bull, R. J., Low level lead exposure produces learning deficits in young rat pups, *Neurobehav. Toxicol. Teratol.*, 4, 311–314, 1982.
680. Taylor, P. R., Lawrence, C. E., Hwang, H.-L., and Paulson, A. S., Polychlorinated biphenyls influence on birthweight and gestation, *Am. J. Publ. Health*, 74, 1153–1154, 1984.
681. Teratology Society, Teratology Society position paper: recommendations for vitamin A use during pregnancy, *Teratology*, 35, 269–275, 1987.
682. TERIS, *Teratogenic Effects of Drugs: A Resource for Clinicians*, Friedman, J. M. and Polifka, J. E., Eds., The Johns Hopkins University Press, Baltimore, MD, 1994.
683. Thiersch, J. B., Therapeutic abortions with a folic acid inhibitor (4-amino P.G.A.), *Am. J. Obstet. Gynecol.*, 63, 1298–1304, 1952.
684. Thompson, R. S., Hess, D. L., Binkerd, P. E., and Hendrickx, A. G., The effects of prenatal diethylstilbestrol exposure on the genitalia of pubertal *Macaca mulatta*. II. Male offspring, *J. Reprod. Med.*, 26, 309–316, 1981.
685. Tikkanen, J. and Heinonen, O. P., Maternal exposure to chemical and physical factors during pregnancy and cardiovascular malformations in the offspring, *Teratology*, 43, 591–600, 1991.
686. Tikkanen, J. and Heinonen, O. P., Risk factors for conal malformations of the heart, *Eur. J. Epidemiol.*, 8, 48–57, 1992.
687. Tilson, H. A., Jacobson, J. L., and Rogan, W. J., Polychlorinated biphenyls and the developing nervous system: cross-species comparisons, *Neurotoxicol. Teratol.*, 12, 239–248, 1990.
688. Toledo, T. M., Harper, R. C., and Moser, R. H., Fetal effects during cyclophosphamide and irradiation therapy, *Ann. Intern. Med.*, 74, 87–91, 1971.
689. Torok, T. J., Human parvovirus B19 infections in pregnancy, *Pediatr. Infect. Dis. J.*, 9, 772–776, 1990.
690. Torpin, R., Malaria in pregnancy, *Am. J. Obstet. Gynecol.*, 41, 882–885, 1941.
691. Toutant, C. and Lippman, S., Fetal solvents syndrome, *Lancet*, 1, 1356, 1979.
692. Troyer, W. A., Pereira, G., Lannon, R. A., Belik, J., and Yoder, M. C., Association of maternal lithium exposure and premature delivery, *J. Perinatol.*, 13, 123–127, 1993.
693. Trudinger, B. J., Cook, C. M., Giles, W. B., Connelly, A. J., and Thompson, R. S., Low-dose aspirin in pregnancy, *Lancet*, 1, 1410, 1989.
694. Turnbull, E. P. N. and Walker, J., The outcome of pregnancy complicated by threatened abortion, *J. Obstetr. Gynecol. Br. Emp.*, 63, 553–559, 1956.
695. Tyrala, E. E., Mathews, S. V., and Rao, G. S., Effect of intrauterine exposure to cocaine on acetylcholinesterase in primary cultures of fetal mouse brain cells, *Neurotoxicol. Teratol.*, 14, 229–233, 1992.

696. Ueda, K., Hisanaga, S., Nishida, Y., and Shepard, T. H., Low birth weight and congenital rubella syndrome: effect of gestational age at time of maternal gestation, *Clin. Pediatr.,* 20, 730–733, 1981.
697. Ueda, K., Nishida, Y., Oshima, K., and Shepard, T. H., Congenital rubella syndrome: correlation of gestational age at time of maternal rubella with type of defect, *J. Pediatr.,* 94, 763–765, 1979.
698. Ulfelder, H., DES — transplacental teratogen and possibly also carcinogen, *Teratology,* 13, 101–104, 1976.
699. Vacaflor, L., Lehmann, H. E., and Ban, T. A., Side effects and teratogenicity of lithium carbonate treatment, *J. Clin. Pharmacol.,* 10, 387–389, 1970.
700. Van Allen, M. I., Structural anomalies resulting from vascular disruption, *Pediatr. Clin. N. Am.,* 39, 255–277, 1992.
701. Van Allen, M. I., Siegel-Bartelt, J., Dixon, J., Zuker, R. M., Clarke, H. M., and Toi, A., Constriction bands and limb reduction defects in two newborns with fetal ultrasound evidence for vascular disruption, *Am. J. Med. Genet.,* 44, 598–604, 1992.
702. van der Berg, B. J., Epidemiological observations of prematurity: effects of tobacco, coffee and alcohol, in *The Epidemiology of Prematurity,* Reed, D. M. and Stanley, F. J., Eds., Urban & Schwarzenberg, Baltimore, MD, 1977, pp. 157–177.
703. van der Pol, J. G., Wolf, H., Boer, K., Treffers, P. E., Leschot, N. J., Hey, H. A., and Vos, A., Jejunal atresia related to the use of methylene blue in genetic amniocentesis in twins, *Br. J. Obstet. Gynecol.,* 99, 141–143, 1992.
704. Van Wyk, J. and Grumbach, M. M., Disorders of sex differentiation, in *Textbook of Endocrinology,* Williams, R. H., Ed., W.B. Saunders, Philadelphia, 1968, pp. 537–612.
705. van't Hoff, W., Caffeine in pregnancy, *Lancet,* 1, 1020, 1982.
706. Varpela, E., On the effect exerted by the first-line tuberculosis medicines on the fetus, *Acta Tuberc. Scand.,* 35, 53–69, 1964.
707. Veridiano, N. P., Delk, I., Rogers, J., and Tancer, M. L., Reproductive performance of DES-exposed female progeny, *Obstet. Gynecol.,* 58, 58–61, 1981.
708. Vessey, M. P., Epidemiological studies of the effects of diethylstilbestrol, *Int. Agency Res. Canc. Sci. Publ.,* 96, 335–348, 1989.
709. Viscarello, R. R., Ferguson, D. D., Nores, J., and Hobbins, J. C., Limb-body wall complex associated with cocaine abuse: further evidence of cocaine's teratogenicity, *Obstet. Gynecol.,* 80, 523–526, 1992.
710. Vitali, E., Donatelli, F., Quaini, E., Groppelli, G., and Pellegrini, A., Pregnancy in patients with mechanical prosthetic heart valves, *J. Cardiovasc. Surg.,* 27, 221–227, 1986.
711. Vorhees, C. V., Behavorial teratogenicity testing as a method of screening for hazards to human health: a methodological approach, *Neurobehav. Toxicol. Teratol.,* 5, 469–4474, 1983.
712. Walker, B. E., Animal models of prenatal exposure to diethylstilbestrol, *Int. Agency Res. Canc. Sci. Publ,,* 96, 349–364, 1989.
713. Wallenburg, H. C. S. and Rotmans, N., Prevention of recurrent idiopathic fetal growth retardation by low-dose aspirin and dipyridamole, *Am. J. Obstet. Gynecol.,* 157, 1230, 1987.
714. Warkany, J., *Congenital Malformation Notes and Comments,* Year Book Medical Publishers, Chicago, 1971, pp. 78–81.
715. Warkany, J., A warfarin embryopathy?, *Am. J. Dis. Child.,* 129, 287–288, 1975.
716. Warkany, J., Warfarin embryopathy?, *Teratology,* 14, 205–209, 1976.
717. Warkany, J., Aminopterin and methotrexate: folic acid deficiency, *Teratology,* 17, 353–358, 1978.
718. Warkany, J., Antituberculosis drugs, *Teratology,* 20, 131–138, 1979.
719. Warkany, J., Teratogenicity of folic acid antagonists, *Canc. Bull.,* 33, 76–77, 1981.
720. Warkany, J., Teratogen update: iodine deficiency, *Teratology,* 31, 309–311, 1985.
721. Warkany, J., Teratogen update: lithium, *Teratology,* 38, 593–596, 1988.
722. Warkany, J., Beautry, P. H., and Horstein, S., Attempted abortion with aminopterin (4-aminopteroylglutamic acid), *Am. J. Dis. Child.,* 97, 274–281, 1959.
723. Warner, R. H. and Rosett, H. L., The effects of drinking on offspring: an historical survey of the American and British literature, *J. Stud. Alcohol,* 36, 1395, 1975.
724. Wasserman, G., Graziano, J., Factor-Litvak, P., Popovac, D., Morina, N., Musabegovic, A., Vrenezi, N., Capuni-Paracka, S., Lekic, V., Preteni-Redjepi, E., Hadzialijevic, S., Slavkovich, V., Kline, J., Shrout, P., and Stein, Z., Consequences of lead exposure and iron supplementation on childhood development at age 4 years, *Neurotoxicol. Teratol.,* 16, 233–240, 1994.
725. Watkinson, B. and Fried, P. A., Maternal caffeine use before, during and after pregnancy and effects upon offspring, *Neurobehav. Toxicol. Teratol.,* 7(1), 9–17, 1985.
726. Weathersbee, P. S., Olsen, L. K., and Lodge, J. R., Caffeine and pregnancy. A retrospective survey, *Postgrad. Med.,* 62, 64–69, 1977.
727. Weber, K., Bratzke, H. J., Neubert, U., Wilske, B., and Durray, P. H., *Borrelia burgdorferi* in a newborn despite oral penicillin for Lyme borreliosis during pregnancy, *Pediatr. Infect. Dis. J.,* 7, 826–289, 1988.
728. Weber, L. W. D., Benzodiazepines in pregnancy — academical debate or teratogenic risk?, *Biol. Res. Pregnancy,* 6, 151–167, 1985.

729. Webster, W. S. and Brown-Woodman, P. D. C., Cocaine as a cause of congenital malformations of vascular origin: experimental evidence in the rat, *Teratology,* 41, 689–697, 1990.
730. Webster, W. S., Brown-Woodman, P. D. C., Lipson, A. H., and Ritchie, H. E., Fetal brain damage in the rat following prenatal exposure to cocaine, *Neurotoxicol. Teratol.,* 13, 621–626, 1991.
731. Webster, W. S., Lipson, A. H., and Brown-Woodman, P. D. C., Uterine trauma and limb defects, *Teratology,* 35, 253–260, 1987.
732. Weiland, H. T., Vermey-Keers, C., Salimans, M. M., Fleuren, G. J., and Verwey, R. A., Parvovirus B19 associated with fetal abnormality, *Lancet,* 1, 682–683, 1987.
733. Weisblum, B. and Davies, J., Antibiotic inhibitors of the bacterial ribosome, *Bacteriol. Rev.,* 32, 493–528, 1968.
734. Wells, L. J. and Van Wagenen, G., Androgen-induced female pseudohermaphroditism in the monkey (*Macaca mulatta*). Anatomy of the reproductive organs, *Carnegie Inst. Contrib. Embryol.,* 35, 93–106, 1954.
735. Wells, P. G., Nagal, M. K., and Greco, G. S., Inhibition of trimethadione and dimethadione teratogenicity by the cyclooxygenase inhibitor acetylsalicylic acid: a unifying hypothesis for the teratologic effects of hydantoin anticonvulsants and structurally related compounds, *Toxicol. Appl. Pharmacol.,* 97, 406–414, 1989.
736. Wenger, F., Venezuelan equine encephalitis, *Teratology,* 16, 359–362, 1977.
737. Werboff, J. and Kesner, R., Learning deficits of offspring after administration of tranquilizing drugs to the mothers, *Nature,* 197, 106º107, 1963.
738. Werler, M. M., Mitchell, A., and Shapiro, S., The relation of aspirin use during the first trimester of pregnancy to congenital cardiac defects, *N. Engl. J. Med.,* 321, 1639, 1989.
739. Werler, M. M., Mitchell, A. A., and Shapiro, S., Demographic, reproductive, medical and environmental factors in relation to gastroschisis, *Teratology,* 45, 353–360, 1992.
740. Werler, M. M., Pober, B. R., and Holmes, L. B., Smoking and pregnancy, *Teratology,* 32, 473–481, 1985.
741. West, J. R. and Goodlett, C. R., Teratogenic effects of alcohol on brain development, *Ann. Med.,* 22, 319–325, 1990.
742. Whittle, B. A., Pre-clinical teratological studies on sodium valproate (Epilim) and other anticonvulsants, in *Clinical and Pharmacological Aspects of Sodium Valproate (Epilim) in the Treatment of Epilepsy,* Legg, N. J., Ed., MCS Consultants, Turnbridge Wells, England, 1976, pp. 105–110.
743. Wiggins, R. C., Pharmacokinetics of cocaine in pregnancy and effects on fetal maturation, *Clin. Pharmacokinet.,* 22, 85–93, 1992.
744. Wiggins, R. C. and Ruiz, B., Development under the influence of cocaine. II. Comparison of the effects of maternal cocaine and associated undernutrition on brain myelin development in offspring, *Metab. Brain Dis.,* 5, 101–109, 1990.
745. Wilcox, A. J., Weinberg, C. R., and Baird, D. D., Risk factors for early pregnancy loss, *Epidemiology,* 1, 382–385, 1990.
746. Wilkins, L., Masculinization due to orally given progestins, *J. Am. Med. Assoc.,* 172, 1028–1032, 1960.
747. Wilkins, L., Jones, H. W., Holman, G. H., and Stempfel, R. S., Masculinization of the female fetus associated with administration of oral and intramuscular progestins during gestation: nonadrenal female pseudohermaphrodism, *J. Clin. Endocrinol. Metab.,* 18, 559–585, 1958.
748. Wilkins-Haug, L. and Gabow, P. A., Toluene abuse during pregnancy: obstetric complications and perinatal outcomes, *Obstet. Gynecol.,* 77, 505–509, 1991.
749. Willhite, C. C., Sharma, R. P., Allen, P. V., and Berry, D. L., Percutaneous retinoid absorption and embryotoxicity, *J. Invest. Dermatol.,* 95, 523–529, 1990.
750. Wilson, C. B., Remington, J. S., Stagno, S., and Reynolds, D. W., Development of adverse sequelae in children born with subclinical toxoplasma infection, *Pediatrics,* 66, 767–774, 1980.
751. Wilson, J. G., *Environment and Birth Defects,* Academic Press, New York, 1973.
752. Wilson, J. G. and Brent, R. L., Are female sex hormones teratogenic?, *Am. J. Obstet. Gynecol.,* 114, 567–580, 1981.
753. Wilson, J. G., Brent, R. L., and Jordan, H. C., Differentiation as a determinant of the reaction of rat embryos to X-irradiation, *Proc. Soc. Exp. Biol. Med.,* 82, 67–70, 1953.
754. Wilson, N., Forfar, J. D., and Godman, M. J., Atrial flutter in the newborn resulting from lithium ingestion, *Arch. Dis. Child.,* 58, 538–539, 1983.
755. Wiseman, R. A., Negative correlation between sex hormone usage and malformations, in *Prevention of Physical and Mental Congenital Defects.* Part C. *Basic and Medical Science, Education and Future Strategies,* Maurois, M., Ed., Alan R. Liss, New York, 1985, pp. 171–175.
756. Wiseman, R. A. and Dodds-Smith, I. C., Cardiovascular birth defects and antenatal exposure to female sex hormones: a reevaluation of some base data, *Teratology,* 30, 359–370, 1984.
757. Wolf, G. and Neilson, E. G., Angiotensin II induces cellular hypertrophy in cultured murine proximal tubular cell, *Am. J. Physiol.,* 259, F768–F777, 1990.
758. Wolf, G., Ziyadeh, F. N., Helmchen, U., Zahner, G., Schroeder, R., and Stahl, R. A. K., ANG II is a mitogen for a murine cell line isolated from medullary thick ascending limb of Henle's loop, *Am. J. Physiol.,* 268, F940–F947, 1995.

759. Wood, J. W., Johnson, K. G., and Omori, Y., *In utero* exposure to the Hiroshima atomic bomb. An evaluation of head size and mental retardation: twenty years later, *Pediatrics,* 39, 385–392, 1967.
760. Woods, J. R. and Plessinger, M. A., Maternal-fetal cardiovascular system: a target of cocaine, *NIDA Res. Monogr.,* 108, 7–27, 1991.
761. Woods, J. R., Plessinger, M. A., and Clark, K. E., Effect of cocaine on uterine blood flow and fetal oxygenation, *J. Am. Med. Assoc.,* 257, 957–961, 1987.
762. World Health Organization, Spontaneous and induced abortion, in *Technical Report Series,* No. 461, World Health Organization, Geneva, 1970.
763. World Health Organization, The effect of female sex hormones on fetal development and infant health, in *Technical Report Series,* No. 657, World Health Organization, Geneva, 1981.
764. Yang, H. Y. L., Namkung, M. J., and Juchau, M. R., Cytochrome P450-dependent biotransformation of a series of phenoxazone ethers in the rat conceptus during early organogenesis: evidence for multiple P450 isozymes, *Mol. Pharmacol.,* 34, 67–74, 1988.
765. Yang, M. G., Krawford, K. S., Garcia, J. D., Wang, J. H. C., and Lei, K. Y., Deposition of mercury in fetal and maternal brain, *Proc. Soc. Biol. Med.,* 141, 1004–1007, 1972.
766. Yasuda, Y., Konishi, H., Kihara, T., and Tanimura, T., Developmental anomalies induced by all-*trans*-retinoic acid in fetal mice. II. Induction of abnormal neuroepithelium, *Teratology,* 35, 355–366, 1987.
767. Yerby, M. S., Risk of pregnancy in women with epilepsy, *Epilepsia,* 33, S23–S27, 1992.
768. Ylinen, K., Aula, P., Sstenman, U.-H., Kesaniemi-Kuokkanen, T., and Teramo, K., Risk of minor and major fetal malformations in diabetes with high haemoglobin A1C values in early pregnancy, *Br. Med. J.,* 289, 345–346, 1984.
769. York, R. G., Randall, J. L., and Scott, Jr., W. J., Reduction of caffeine teratogenicity in mice by inducing maternal drug metabolism with b-Naphthoflavone, *Teratology,* 31, 217–225, 1985.
770. Young, S. L., Vosper, H. J., and Phillips, S. A., Cocaine: its effects on maternal and child health, *Pharmacotherapy,* 12, 2–17, 1992.
771. Yovich, J. L., Turner, S. R., and Draper, R., Medroxyprogesterone acetate therapy in early pregnancy has no apparent fetal effects, *Teratology,* 38, 135–144, 1988.
772. Yow, M. D., Williamson, D. W., Leeds, L. J., Thompson, P., Woodward, R. M., Walmas, B. F., Lester, J. W., Six, H. R., and Griffiths, P. D., Epidemiologic characteristics of cytomegalovirus infection in mothers and their infants, *Am. J. Obstet. Gynecol.,* 158, 1189–1195, 1988.
773. Yu, M., Hsu, C., Gladen, B. C., and Rogan, W. J., *In utero* pcb/pcdf exposure: relation of developmental delay to dysmorphology and dose, *Neurotoxicol. Teratol.,* 13, 195–202, 1991.
774. Zackai, E. H., Melman, W. J., Neiderer, B., and Hanson, J. W., The fetal trimethadione syndrome, *J. Pediatr.,* 87, 280–284, 1975.
775. Zajac, C. S., Bunger, P. C., and Moore, J. C., Neuron development in the superior colliculus of the fetal mouse following maternal and alcohol exposure, *Teratology,* 38, 37–43, 1988.
776. Zakzouk, M. S., The congenital warfarin syndrome, *J. Laryngol. Otol.,* 100, 215–219, 1986.
777. Segarska, Z. and Kikowska, K., Development defects in white rats caused by acute lead poisoning, *Folia Morphol. (Warz.),* 33, 23–28, 1974.
778. Zengel, A. E., Keith, D. A., and Tassinari, M. S., Prenatal exposure to phenyltoin and its effect on postnatal growth and craniofacial proportion in the rat, *J. Craniofac. Genet. Dev. Biol.,* 9, 147–160, 1989.
779. Zierler, S. and Rothman, K. J., Congenital heart disease in relation to maternal use of Bendectin and other drugs in early pregnancy, *N. Engl. J. Med.,* 313, 347–352, 1985.
780. Zierski, M., Effects of ethionamide on the development of the human fetus, *Gruzlica,* 34, 349–352, 1966.
781. Zilva, S. S., Golding, J., Drummond, J. C., and Coward, K. H., The relation of the fat-soluble factor to rickets and growth in pigs, *Biochem. J.,* 15, 427–437, 1921.
782. Zimmerman, E. F., Potturi, R. B., Resnick, E., and Fisher, J. E., Role of oxygen free radicals in cocaine-induced vascular disruption in mice, *Teratology,* 49, 192–201, 1994.
783. Zuber, C., Librizzi, R. J., and Vogt, B. L., Outcomes of pregnancies exposed to high doses of vitamin A, *Teratology,* 35, 42A, 1987.
784. Matsuoka, R., Uno, H., Tanaka, H., Kerr, C. S., Nakazawa, K., and Nadal-Ginard, B., Caffeine induces cardiac and other malformations in the rat, *Am. J. Med. Gen.,* (Suppl. 3), 433–443, 1987.

INDEX

INDEX

A

B

C

D

E

F

G

H

I

J

K

N

O

P

Q

R

S

T

U

V

W

X

Y

Z